D1569425

ELECTRONICS ENGINEER'S REFERENCE BOOK

ELECTRONICS ENGINEER'S REFERENCE BOOK

Edited by

L. W. TURNER

C.Eng., F.I.E.E., F.R.T.S.

With specialist contributors

BUTTERWORTHS

LONDON – BOSTON

Sydney – Wellington – Durban – Toronto

The Butterworth Group

United Kingdom **Butterworth & Co. (Publishers) Ltd**
London: 88 Kingsway, WC2B 6AB

Australia **Butterworths Pty Ltd**
Sydney: 586 Pacific Highway, Chatswood, NSW 2067
Also at Melbourne, Brisbane, Adelaide and Perth

Canada **Butterworth & Co. (Canada) Ltd**
Toronto: 2265 Midland Avenue, Scarborough, Ontario M1P 4S1

New Zealand **Butterworths of New Zealand Ltd**
Wellington: T & W Young Building, 77–85 Customhouse Quay, 1,
CPO Box 472

South Africa **Butterworth & Co. (South Africa) (Pty) Ltd**
Durban: 152–154 Gale Street

USA **Butterworth (Publishers) Inc**
Boston: 19 Cummings Park, Woburn, Mass. 01801

First published 1958 by Heywood Books
Second edition 1959, reprinted 1962
Third edition 1967
Fourth edition 1976 by Newnes-Butterworths
Reprinted 1978, 1981

ISBN 0 408 00168 2

Printed in the United States of America

FOREWORD

by
Sir Harold Bishop, CBE, B.Sc. (Eng.), Hon.: F.I.E.E., F.I.Mech.E., F.I.E.E.E.
Past President, Institution of Electrical Engineers.

The appearance of the fourth edition of this Reference Book indicates not only the value of the previous editions but also the huge and continuing strides in the science and practice of electronic engineering. Whatever profession one may practice the difficulty of keeping up-to-date is so well recognised that it is hardly an exaggeration to say that a nagging concern is ever present in those whose academic studies are ten or more years behind them.

Many professions can boast of a history of maybe hundreds of years and in this period a state of knowledge has been built up by the labours of many practitioners. Not so in electronics. Twenty-five years ago the word was hardly in current use. Indeed the older electrical engineer in the power field and even some of those in telecommunications felt some resentment that this new term in electrical science should be given so much attention.

The astonishing developments from the late forties after the war to the present day have created a new and highly technical atmosphere which has made possible extensions of knowledge and application too numerous to mention but now taken for granted by the public at large. All of which, or at any rate those things which are wisely and humanely used, is of value to civilisation generally in improving our way of life.

But to the technological expert, be he engineer, scientist or teacher or a combination of all three, the ever-expanding state of knowledge as techniques become more complex, means inevitably that specialisation is intensified. The trap of over-specialisation is wide open and for the technologist who seeks broad professional leadership it is wise to avoid it. By the nature of any job this is inherently difficult and a conscious effort is needed to achieve a wide spectrum of learning and experience. This is particularly the case in middle life when the tendency is too often to continue to develop the expertise of early academic training.

And so we are led to the idea of professional retraining. To acquire this by a definite break in the money-making rhythm of middle life is exceedingly difficult and one looks for less drastic means. One way of proved value is by maintaining close contact with appropriate reference books and in the expanding electronics field the *Electronics Engineer's Reference Book* is an excellent example.

PREFACE

Since the earlier editions of this reference book were published, the first in 1958, the second in 1959 and the third in 1967, new techniques have emerged, many new electronic devices have been developed and the application of electronic equipment has rapidly extended over ever widening fields. This phenomenal rate of development, which continues, has been rightly described as startling even in a world which has become accustomed to a rapid rate of technological advancement.

In this situation no reference book can be complete or entirely up-to-date. The aim in publishing this fourth edition has been to present within the scope of a single volume of practical size as much as possible of the latest knowledge and techniques to provide a work of reference which will be of value to the electronics engineer and his fellows in other engineering disciplines wishing to assess effectively and quickly the potentialities of electronic solutions to their problems; and also to the scientist, the student, the educational profession, management personnel and the reader with a general interest in electronics and their applications.

This edition has been entirely rewritten. A bibliography has been included which gives reference to nationally and internationally recognised authoritative literature on the various subjects. The editor is indebted to the 61 writers for their contributions and for their ready cooperation in fitting their work into the format of the new edition. All are specialists in their own fields. Their qualifications and activities are given in the List of Contributors.

The arrangement of the Reference Book has been revised to give a more logical grouping and sequence of its 27 sections which broadly follow the general order of: basics, materials and components, devices, circuits, measurements and applications.

The application of electronics is now so vast, there being few areas in our everyday lives where electronic devices are not used in one form or another, such that it is impracticable in a single-volume reference book of this size to include them all. The main applications, including the very wide field of telecommunications embracing colour television and broadcasting generally, are however described in some detail in the last 13 sections, numbers 15 to 27.

As in the case of the companion volumes, the new editions of the *Civil Engineer's*, the *Electrical Engineer's* and the *Mechanical Engineer's Reference Books*, the Publishers have seized the opportunity provided by the complete revision of this fourth edition of the *Electronics Engineer's Reference Book* to issue the work in a new and modernised format giving a more pleasing and clear appearance making the text more easily assimilable and more direct in reference. The index has been simplified and made more easily usable.

L.W.T.

ACKNOWLEDGEMENTS

The production of this reference book would have been impossible without the good will, help and cooperation of the electronics industry, the users of electronic equipment and members of the educational profession. Bare acknowledgements are very inadequate but the editor wishes to thank the following firms and organisations which so readily made available information and illustrations and permitted members of their specialist staffs to write contributions:

Appleton Laboratory, Science Research Council
British Aircraft Corporation
British Broadcasting Corporation
British Railways Board, Research and Development Division
Cambridge Scientific Instruments Ltd.
Clinical Research Centre
Crypton-Triangle Ltd.
English Electric Valve Company Limited
GEC-Marconi Electronics Limited
Gent and Company Ltd.
Greater London Council, Department of Planning and Transportation
Hammond Organ Company
Heriot-Watt University, Edinburgh
Lucas Electrical Company Limited
Meteorological Office
Mullard Limited
Plessey Company Limited
Polytechnic of North London
Post Office
Radio Corporation of America
Rediffusion Limited
Texas Instruments Incorporated
University of Aston
University of Leeds
University of Sussex
University of Technology, Loughborough

Acknowledgement is made to the Director of the International Radio Consultative Committee (C.C.I.R.) for permission to use information and reproduce diagrams and curves from the C.C.I.R. Documents.

Extracts from British Standard publications are reproduced by permission of the British Standards Institution.

L.W.T.

CONTENTS

CONTRIBUTORS

D. ALDOUS, M.A.E.S., M.B.K.S.T.S.
Founder-member No. 1, B.S.R.A.
Technical Editor, Hi-Fi News & Record Review (Section 23)

E. G. M. ALKIN, M.B.K.S
Manager, Sound and Vision Operations,
BBC Welsh Region (Section 16)

S. W. AMOS, B.SC.(HONS.), C.Eng., M.I.E.E.
Freelance Technical Editor, and Author (Section 13)

W. J. BAKER
Technical Editor (Research),
Marconi Research Laboratories (Section 2)

G. T. C. BALL, B.SC. (ELECTROTECHNOLOGY)
Head of ECM Group,
Travelling Wave Tube Section,
English Electric Valve Company Limited (Section 7)

H. BARBER, B.Sc.(Eng.)
Senior Lecturer,
University of Technology, Loughborough (Section 24)

R. M. BERGSLIEN, Licensed Professional Engineer, B.S.E.E.
Manager,
Product Systems Engineering,
Hammond Organ Company (Section 17)

G. L. BIBBY, B.Sc.(Eng.), C.Eng., M.I.E.E.
Director of the Standards Laboratory,
British Calibration Service,
Lecturer, University of Leeds (Section 14)

B. B. BOWEN, B.Sc., C.Eng., M.I.E.E.
Chief Electronic Designer,
Crypton-Triangle Ltd. (Section 25)

P. A. BRADLEY, B.Sc., M.Sc., C.Eng., M.I.E.E.
Principle Scientific Officer,
Appleton Laboratory, Science Research Council (Section 5)

P. M. CHALMERS, B.Eng., A.M.I.E.E.
Assistant Manager, Travelling Wave Tube Section,
English Electric Valve Company Limited (Section 7)

M. J. H. CHANDLER, C.Eng., M.I.E.E., M.I.E.R.E.
Project Leader, Area Traffic Control and Signals,
Greater London Council (Section 25)

H. W. COLE
Senior Radar Systems Engineer,
Marconi Radar Systems Limited (Section 18)

A. P. O. COLLIS, B.A., C.Eng., M.I.E.E.

Manager,
Power Grid Division,
English Electric Valve Company Limited (Section 7)

B. F. COOPER, B.Sc., C.Eng., M.I.E.E.

Manager,
Millimetre Wave Tubes,
English Electric Valve Company Limited (Section 7)

A. COX, A.C.T.(Birm.), Dip.Tech.(Eng.)

Senior Technical Author (Electronics),
Lucas Electrical Company Limited (Section 25)

D. L. CROOM, Ph.D., C.Eng., M.I.E.E.

Principal Scientific Officer,
Appleton Laboratory,
Science Research Council (Section 20)

B. C. CULLEN

Network Manager,
Control Data Europe Incorporated (Section 15)

G. W. A. DUMMER, M.B.E., C.Eng., F.I.E.E., F.I.E.E.E., F.I.E.R.E.

Electronics Consultant (Section 6)

F. FAULKNER, B.Sc.(Hons.), M.Inst.P., M.I.E.E., C.Eng.

Principal Engineer,
Airadio Division,
Marconi-Elliott Avionic Systems Limited (Section 25)

K. FIRTH, B.Sc., Ph.D., F.Inst.P.

Section Chief,
Lasers,
The Marconi Company Limited (Sections 10, 15, 24, 27)

W. M. FRASER., C.Eng., M.I.E.R.E., Dip.Ed.

Marketing Manager,
Electronic Systems Group,
British Aircraft Corporation (Section 21)

R. P. GABRIEL, B.Sc., C.Eng., F.I.E.E., M.I.E.R.E., F.I.E.E.E.

Chief Engineer,
Rediffusion Limited (Section 15)

E. E. GELBSTEIN C.Eng., M.I.E.E., M.Sc.

Senior Principal Scientific Officer,
Deputy Head of the Signalling Section,
British Railways Board, Research and Development Division (Section 25)

C. L. S. GILFORD, M.Sc., Ph.D., F.Inst.P., C.Eng., M.I.E.E.

Reader in Acoustics,
Department of Building,
University of Aston (Section 14)

R. L. GRIMSDALE, M.Sc., Ph.D., C.Eng., M.I.E.R.E., F.B.C.S.

Professor of Electronics
University of Sussex (Section 24)

J. E. HARRY, B.Sc.(Eng.), Ph.D.
Electricity Council Senior Lecturer in Electroheat, University of Technology, Loughborough (Section 24)

E. W. HEROLD, B.Sc., M.Sc., D.Sc., F.I.E.E.E.
Chairman, U.S. Dept. of Defense Advisory Group on Electron Devices; Chairman, Board of Trustees, Palisades Institute for Research Services (Section 7)

D. J. HILLS
Quality Assurance Engineer, Cambridge Scientific Instruments Ltd. (Section 14)

A. S. HUDSON, M.A., D.M.S., A.I.Ceram.
Technical Manager, Specialised Components Division, Marconi Communication Systems Limited (Section 6)

P. J. W. HYDE
Manager, Demonstration Unit, Cambridge Scientific Instruments Ltd. (Section 14)

A. K. JEFFERIS, B.Sc., C.Eng., M.I.E.E.
Head of Section, Space Communications Systems Division, Telecommunications Development Department, Post Office Telecommunication Headquarters (Section 15)

P. J. KILLINGWORTH, B.Sc.
Technical Manager, Cambridge Scientific Instruments Ltd. (Section 14)

A. LANDMAN, Dipl.Ing., F.I.E.E.
Consultant Engineer (Section 15)

J. A. LANE, D.Sc., C.Eng., F.I.E.E.
Senior Principal Scientific Officer, Appleton Laboratory, Science Research Council (Section 5)

J. B. LETTS, C.Eng., M.I.E.R.E.
Chief Development Engineer, Gent & Company Limited (Section 24)

J. H. LINDARS, C.Eng., M.I.E.E.
Technical Training Manager, Plessey Marine Research Unit, The Plessey Company Limited (Section 25)

E. W. McWHORTER, M.Sc., M.B.A.
Process-Control Instrumentation Engineer, Texas Eastman Company (Section 24)

D. MEYERHOFER, B.Eng., Ph.D.

Member, Technical Staff,
RCA Laboratories,
Radio Corporation of America
(Section 10)

R. C. MILLS

Radiophonic Workshop,
BBC
(Section 17)

H. B. MOORE, Licensed Professional Engineer, B.S.E.E., M.B.A.

Vice President of Operations,
Hammond Organ Company
(Section 17)

C. N. O'LOUGHLIN, B.Sc.

Manager,
Microwave Beam Tube Department,
English Electric Valve Company Limited
(Section 7)

A. G. ORBELL, D.M.S., A.M.B.I.M.

Head of Digital Switching Systems Section,
Telecommunications Systems Strategy Dept.,
Post Office Telecommunications Headquarters
(Section 15)

L. W. OWERS, C.Eng., M.I.E.R.E.

Senior Technical Resident Engineer
Mullard Limited
(Section 22)

W. T. PARKMAN, A.M.I.E.E., B.Sc.(Hons.)

Senior Principal Scientific Officer,
Deputy Head of the Automation Section,
British Railways Board, Research and Development Division
(Section 25)

A. F. PETRIE, B.S.E.E., M.Sc.

Manager,
Advanced Product Development,
Hammond Organ Company
(Section 17)

D. C. PRESSEY, B.Sc.

Senior Technical Officer,
Division of Bioengineering,
Clinical Research Centre
(Section 27)

P. RAINGER, B.Sc.(Eng.), C.Eng., F.I.E.E.

Head,
BBC Engineering Research Department
(Section 15)

C. F. REYNOLDS

Coordinator and Quality Controller,
Gent & Company Limited
(Section 24)

M. J. ROSE, B.Sc.(Eng.)

Print Production Manager,
Central Technical Services,
Mullard Limited
(Sections 6, 7, 8, 9, 10, 11, 12)

P. RYDER, B.Sc.(Hons.), Ph.D.
Principal Scientific Officer,
Clouds Physics Branch,
U.K. Meteorological Office (Section 19)

M. G. SAY, Ph.D., M.Sc., C.Eng., A.C.G.I., D.I.C., F.I.E.E., F.R.S.E.
Professor Emeritus of Electrical Engineering,
Heriot-Watt University, Edinburgh (Sections 3, 26)

M. J. B. SCANLAN, B.Sc., A.R.C.S.
Group Chief,
Applied Physics Group,
Marconi Research Laboratories (Section 6)

H. V. SIMS, C.Eng., F.I.E.E., F.I.E.R.E.
Engineering and Training Consultant (Section 15)

C. R. SPICER, A.M.I.E.E.
Designs Engineer,
BBC Designs Department (Section 15)

K. R. STURLEY, Ph.D., B.Sc., F.I.E.E., F.I.E.E.E.
Telecommunications Consultant (Section 15)

F. P. THOMSON, O.B.E., C.Eng., M.I.E.R.E., A.M.B.I.M.
Communications, Banking and Security Adviser
to various Bodies in the U.K. and abroad.
Associate Consultant, Communications & Equipment
Consultants Ltd. (Section 24)

W. E. TURK, B.Sc., C.Eng., F.I.E.E., F.R.T.S., M.Inst.M.
Manager,
Phototube Marketing,
English Electric Valve Co (Section 7)

L. W. TURNER, C.Eng., F.I.E.E., F.R.T.S.
Formerly Head of Engineering Information Department, BBC
Consultant Engineer (Sections 1, 4, 7)

J. A. van RAALTE, B.Sc., M.Sc., Ph.D., M.Amer.Phys.Soc., M.A.A.A.S., S.M.I.E.E.E.
Head,
Displays and Device Concepts Research,
RCA Laboratories,
Radio Corporation of America (Section 10)

R. C. WHITEHEAD, C.Eng., M.I.E.E.
Senior Lecturer,
Polytechnic of North London (Section 14)

F. J. WEAVER, B.Sc., M.I.E.E.
Head,
Magnetron Research and Development,
The English Electric Valve Company Limited (Section 7)

A. C. YOUNG
Consultant,
Musical Standards,
Hammond Organ Company (Section 17)

1 GENERAL INFORMATION

GENERAL INFORMATION 1

1 GENERAL INFORMATION

TERMINOLOGY

Electronics

Electronics may be defined as that branch of science and technology which deals with the study of the phenomena of *conduction* of electricity in a vacuum, in a gas, and in semiconductors, and with the utilisation of devices based on these phenomena. British Standard 204:1960–11029.

For other terminology and definitions in the field of electronics and associated subjects the reader is referred to the British Standards listed below.

BS 204:1960 *Glossary of terms used in telecommunication (including radio) and electronics.*

Note: This standard is under progressive revision and is gradually being replaced by BS 4727 which is being issued in a number of separate parts.

BS 204:Supplement No. 1:1964 *Terms used in wired broadcast and broadcast relay.*

BS 204:Supplement No. 3:1966 *Colour television terms.*

BS 4727 *Glossary of electrotechnical, power, telecommunications, electronics, lighting and colour terms.*

Part 1 *Terms common to power, telecommunications and electronics.*
- Group 01:1971 Fundamental concepts.
- Group 02:1971 General technological terminology.
- Group 03:1971 Relay terminology.
- Group 04:1971 Measurement terminology.
- Group 05:1972 Semiconductor terminology.
- Group 06:1973 Electronic tube terminology.

Part 3 *Terms particular to telecommunications and electronics.*
- Group 01:1971 General telecommunication and electronics terminology.
- Group 02:1971 Telephony terminology.
- Group 03:1971 Telegraphy, including facsimile terminology.
- Group 04:1971 Broadcasting, radio and television terminology.
- Group 05:1971 Propagation and media terminology.
- Group 06:1971 Radio location and navigation terminology.
- Group 07:1971 Radiocommunication terminology.

Part 4 *Terms particular to lighting and colour.*
- Group 01:1971 Radiation and photometry.
- Group 02:1971 Vision and colour terminology.
- Group 03:1972 Lighting technology terminology.

Terminology for electronics reliability

Some terms recommended in the U.K. by the British Standards Institution (BSI) and internationally by the International Electrotechnical Commission (IEC) are given below. Full details are given in BS 4200 Part 2: 1967.

RELIABILITY

General definition The ability of an item to perform a required function under stated conditions for a stated period of time.
Note: The term reliability is also used as a reliability characteristic denoting a probability of success, or a success ratio.
Observed reliability—of non-repaired items For a stated period of time, the ratio of the number of items which performed their functions satisfactorily at the end of the period to the total number of items in the sample at the beginning of the period.
Observed reliability—of repaired item or items The ratio of the number of occasions on which an item or items performed their functions satisfactorily for a stated period of time to the total number of occasions the item or items were required to perform for the same period.
Assessed reliability The reliability of an item determined by a limiting value or values of the confidence interval associated with a stated confidence level, based on the same data as the observed reliability of nominally identical items.
Extrapolated reliability Extension by a defined extrapolation or interpolation of the observed or assessed reliability for durations and/or conditions different from those applying to the observed or assessed reliability.
Predicted reliability For the stated conditions of use, and taking into account the design of an item, the reliability computed from the observed, assessed or extrapolated reliabilities of its parts.

FAILURES

General definition The termination of the ability of an item to perform a required function.
Failure cause The circumstances during design, manufacture or use which have led to failure.
Failure mode The effect by which a failure is observed. For example, an open or shortcircuit condition, or a gain change.
Failure mechanism The physical, chemical or other process which results in failure.
Misuse failure Failure attributable to the application of stresses beyond the stated capabilities of the item.
Inherent weakness failure Failure attributable to weakness inherent in the item itself when subjected to stresses within the stated capabilities of the item.
Primary failure Failure of an item, not caused either directly or indirectly by the failure of another item.
Secondary failure Failure of an item, caused either directly or indirectly by the failure of another item.
Wear-out failure Failure whose probability of occurrence increases with the passage of time and which occurs as a result of processes which are the characteristic of the population.
Sudden failure Failure that *could not* be anticipated by prior examination or monitoring.
Gradual failure Failure that *could* be anticipated by prior examination or monitoring.
Partial failure Failure resulting from deviations in characteristic(s) beyond specified limits but *not such* as to cause complete lack of the required function.
Complete failure Failure resulting from deviations in characteristic(s) beyond specified limits *such* as to cause complete lack of the required function.
Intermittent failure Failure of an item for a limited period of time, following which the item recovers its ability to perform its required function without being subjected to any external corrective action.
Catastrophic failure Failure which is both sudden and complete.

Degradation failure Failure which is both gradual and partial.

Note: In time such a failure may develop into a complete failure.

Early failure period That possible early period, beginning at a stated time and during which the failure rate decreases rapidly in comparison with that of the subsequent period.

Constant failure rate period That possible period during which the failures occur at an approximately uniform rate.

Wear-out failure period That possible period during which the failure rate increases rapidly in comparison with the preceding period.

Observed failure rate For a stated period in the life of an item, the ratio of the total number of failures in a sample to the cumulative observed time on that sample. The observed failure rate is to be associated with particular, and stated time intervals (or summation of intervals) in the life of the items, and with stated conditions.

Observed mean life The mean value of the lengths of observed times to failure of all items in a sample under stated conditions.

Observed mean time to failure (MTTF) For a stated period in the life of an item, the ratio of the cumulative time for a sample to the total number of failures in the sample during the period, under stated conditions.

Observed mean time between failures (MTBF) For a stated period in the life of an item, the mean value of the length of time between consecutive failures, computed as the ratio of the cumulative observed time to the number of failures under stated conditions.

Terminology for computer systems

Peripheral equipment (or 'Peripherals') The various items of equipment which are connected to a computer to make up any particular system.

Central data processing system The assembly of equipment—including computers and peripherals—which form the electronic 'heart' of the complete scheme.

Computer store (or memory) A vital part of a computer used to retain (or remember) operating data and instructions during computation processes.

Disc file store One particular peripheral device used as an additional electronic memory for retaining data and operating instructions within the system.

Software A recognised name for computer programmes—the information having been prepared by 'programmers' and punched on to paper tape.

Tape reader The device to convert the information on the paper tape into electrical input signals to the computer.

Display 'back-up' The waveform generation and processing equipment connected between the display indicator units and the computer system.

'On-line' operation The mode of operation in which the computing system is directly connected to the plant or process which is to be controlled.

'Off-line' operation The mode of operation in which the computing system provides data to permit control or processing operations to be carried out but is *not* directly connected to the plant or process which is being controlled.

'Real-time' operation The mode of operation where the computing system is required to provide data to control and/or monitor events at the instant they occur.

Interface equipment Equipment interposed between two other equipments (or systems) to ensure compatibility of operation.

Terminology for integrated (micro-) circuits

Digital (integrated) circuit A circuit which operates with digital variables at the input(s) and output(s) and which is characterised by the inter-relationships between the states of the digital variables at the input and output terminals.

Note: The digital variable may be voltage, or current, or impedance, etc.

Binary (digital) circuit A digital circuit in which the digital variable at each input or output terminal may take one of only two states.

Note: The pairs of ranges of values of the digital variable may, however, be different at different terminals.

Combinatorial (digital) circuit A digital circuit in which for each possible combination of the states of the digital variable at the input(s), there is one, and only one, combination of the states of the digital variable at the output(s).

Sequential (digital) circuit A digital circuit in which there exists at least one combination of the states of the digital variable at the input(s) for which there is more than one corresponding combination of the states of the digital variables at the output(s).

Note: These combinations at the outputs are determined by the previous electrical history (including internal memory, delay, etc.).

Monostable (binary digital) circuit A sequential circuit which has one stable state and which requires an appropriate excitation to remain in another defined state during a determined time interval.

Note 1: This time interval may be independent of the duration of the excitation.

Note 2: The stable state has an unlimited duration when no excitation is applied.

Bistable (binary digital) circuit A sequential circuit which has two internal states both of which are stable and which requires an appropriate excitation when in either state to cause a transition to the other state.

Multistable (binary digital) circuit A sequential circuit which has more than two internal states each of which is stable and which requires an appropriate excitation when in any state to cause a transition to another state.

Delay (binary digital) circuit A sequential circuit for which the changes in the states of the digital variable at the output(s) are delayed for a determined time with regard to the respective changes in the states of the digital variable at the input(s).

Note: The delay may or may not be the same ıor changes of the digital variable to each state.

Gate A network having one or more inputs which opens or closes a channel according to the combination of stimuli applied to the input(s).

Positive—AND gate (Negative—OR gate) A binary digital gate whose output is in the High state if, and only if all its inputs are in the High state.

Positive—OR gate (Negative—AND gate) A binary digital gate whose output is in the Low state if, and only if all its inputs are in the Low state.

Positive—NAND gate (Negative—NOR gate) A binary digital gate whose output is in the Low state if, and only if all its inputs are in the High state.

Positive—NOR gate (Negative—NAND gate) A binary digital gate whose output is in the High state if, and only if all its inputs are in the Low state.

Analogue integrated circuit A circuit for which a continuous relationship (either linear or non-linear) exists between its input(s) and output(s).

Terminology for microelectronics

Microelectronics The concept of the construction and use of highly miniaturised electronic circuits.

Micro-circuit A micro-electronic device, having a high equivalent circuit-element and/or component density which is considered as a single unit.

Note: A micro-circuit may be a micro-assembly or an integrated (micro-) circuit.

Integrated (micro-) circuit A micro-circuit in which a number of circuit elements are inseparably associated and electrically interconnected such that for the purposes of specification and testing, commerce and maintenance, it is considered indivisible.

Note 1: For this definition, a circuit element does not include envelope or external connection and is not specified or sold as a separate item.

Note 2: Where no misunderstanding is possible, the term Integrated Micro-circuit may be abbreviated to Integrated Circuit.

Note 3: Further qualifying adjectives may be used to describe the technique used in the manufacture of a specific micro-circuit.

Examples of the use of qualifying adjectives:

Semiconductor monolithic integrated circuit
Semiconductor multi-chip integrated circuit
Thin film integrated circuit
Thick film integrated circuit
Hybrid integrated circuit.

Micro-assembly A micro-circuit in which the various components and/or integrated micro-circuits are constructed separately and can be tested before being assembled and packaged.

Note 1: For this definition it is assumed that a component has external connections and possibly an envelope as well and that it can also be specified and sold as a separate item.

Note 2: Further qualifying adjectives may be used to describe the form of the components and/or the assembly technique used in the construction of the specified micro-assembly.

Examples of the use of qualifying adjectives:

Semiconductor multi-chip micro-assembly
Discrete component micro-assembly.

UNITS

International unit system

The International System of Units (SI) is the modern form of the metric system agreed at an international conference in 1960. It has been adopted by the International Standards Organisation (ISO) and the International Electrotechnical Commission (IEC) and its use is recommended wherever the metric system is applied. It is now being adopted throughout most of the world and is likely to remain the primary world system of units of measurement for a very long time. The indications are that SI Units will supersede the units of existing metric systems and all systems based on Imperial Units.

SI Units and the rules for their application are contained in *ISO Resolution* R1000 (1969) and an informatory document *SI-Le Systeme International d'Unités*, published by the Bureau International des Poids et Mesures (BIPM). An abridged version of the former is given in British Standards Institution (BSI) publication PD 5686 *The use of SI Units* (1969) and BS 3763 *International System (SI) Units*; BSI (1964) incorporates information from the BIPM document.

The adoption of SI presents less of a problem to the electronics engineer and the electrical engineer than to those concerned with other engineering disciplines as all the practical electrical units were long ago incorporated in the metre-kilogram-second (MKS) unit system and these remain unaffected in SI.

The SI was developed from the metric system as a fully coherent set of units for science, technology and engineering. A coherent system has the property that corresponding equations between quantities and between numerical values have exactly the same form, because the relations between units do not involve numerical conversion factors. In constructing a coherent unit system, the starting point is the selection and definition of a minimum set of independent 'base' units. From these, 'derived' units are obtained by forming products or quotients in various combinations, again without numerical factors. Thus the base units of length (metre), time (second) and mass (kilogramme) yield the SI units of velocity (metre/second), force (kilogramme–metre/second-squared) and so on. As a result there is, for any given physical quantity, only one SI unit with no alternatives and with no numerical conversion factors. A single SI unit (joule = kilogramme metre-squared/second-squared) serves for energy of any kind, whether it be kinetic, potential, thermal, electrical, chemical . . ., thus unifying the usage in all branches of science and technology.

The SI has seven base units, and two supplementary units of angle. Certain important derived units have special names and can themselves be employed in combination to form alternative names for further derivations.

Each physical quantity has a quantity-symbol (e.g., *m* for mass) that represents it in equations, and a unit-symbol (e.g., kg for kilogramme) to indicate its SI unit of measure.

BASE UNITS

Definitions of the seven base units have been laid down in the following terms. The quantity-symbol is given in italics, the unit-symbol (and its abbreviation) in roman type.

Length: *l*; metre (m). The length equal to 1 650 763.73 wavelengths in vacuum of the radiation corresponding to the transition between the levels $2p_{10}$ and $5d_5$ of the krypton-86 atom.

Mass: *m*; kilogramme (kg). The mass of the international prototype kilogramme (a block of platinum preserved at the International Bureau of Weights and Measures at Sèvres).

Time: *t*; second (s). The duration of 9 192 631 770 periods of the radiation corresponding to the transition between the two hyperfine levels of the ground state of the caesium-133 atom.

Electric current: *i*; ampere (A). The current which, maintained in two straight parallel conductors of infinite length, of negligible circular cross-section and 1 m apart in vacuum, produces a force equal to 2×10^{-7} newton per metre of length.

Thermodynamic temperature: *T*; kelvin (K). The fraction 1/273.16 of the thermodynamic (absolute) temperature of the triple point of water.

Luminous intensity: *I*; candela (cd). The luminous intensity in the perpendicular direction of a surface of 1/600 000 m^2 of a black body at the temperature of freezing platinum under a pressure of 101 325 newtons per square metre.

Amount of substance: *Q*; mole (mol). The amount of substance of a system which contains as many elementary entities as there are atoms in 0.012 kg of carbon-12. The elementary entity must be specified and may be an atom, a molecule, an ion, an electron, etc., or a specified group of such entities.

SUPPLEMENTARY ANGULAR UNITS

Plane angle: $\alpha, \beta \ldots$; radian (rad). The plane angle between two radii of a circle which cut off on the circumference an arc of length equal to the radius.

Solid angle: Ω; steradian (sr). The solid angle which, having its vertex at the centre of a sphere, cuts off an area of the surface of the sphere equal to a square having sides equal to the radius.

Force: The base SI unit of electric current is in terms of force in newtons (N). A force of 1 N is that which endows unit mass (1 kg) with unit acceleration (1 m/s^2). The newton is thus not only a coherent unit; it is also devoid of any association with gravitational effects.

TEMPERATURE

The base SI unit of thermodynamic temperature is referred to a point of 'absolute zero' at which bodies possess zero thermal energy. For practical convenience, two points on the Kelvin temperature scale, namely 273.15 K and 373.15 K, are used to define the Celsius (or Centigrade) scale (0 °C and 100 °C). Thus in terms of temperature *intervals*, 1 K = 1 °C; but in terms of temperature *levels*, a Celsius temperature θ corresponds to a Kelvin temperature (θ + 273.15) K.

DERIVED UNITS

Nine of the more important SI derived units with their definitions are given

Quantity	*Unit Name*	*Unit Symbol*
Force	newton	N
Energy	joule	J
Power	watt	W
Electric charge	coulomb	C
Electric potential difference and EMF	volt	V
Electric resistance	ohm	Ω
Electric capacitance	farad	F
Electric inductance	henry	H
Magnetic flux	weber	Wb

Newton That force which gives to a mass of 1 kilogram an acceleration of 1 metre per second squared.
Joule The work done when the point of application of 1 newton is displaced a distance of 1 metre in the direction of the force.
Watt The power which gives rise to the production of energy at the rate of 1 joule per second.
Coulomb The quantity of electricity transported in 1 second by a current of 1 ampere.
Volt The difference of electric potential between two points of a conducting wire carrying a constant current of 1 ampere, when the power dissipated between these points is equal to 1 watt.
Ohm The electric resistance between two points of a conductor when a constant difference of potential of 1 volt, applied between these two points, produces in this conductor a current of 1 ampere, this conductor not being the source of any electromotive force.
Farad The capacitance of a capacitor between the plates of which there appears a difference of potential of 1 volt when it is charged by a quantity of electricity equal to 1 coulomb.
Henry The inductance of a closed circuit in which an electromotive force of 1 volt is produced when the electric current in the circuit varies uniformly at a rate of 1 ampere per second.
Weber The magnet flux which, linking a circuit of one turn, produces in it an electromotive force of 1 volt as it is reduced to zero at a uniform rate in 1 second.

Some of the simpler derived units are expressed in terms of the seven basic and two supplementary units directly. Examples are listed in *Table 1.1*.

Table 1.1 DIRECTLY DERIVED UNITS

Quantity	*Unit-name*	*Unit-symbol*
Area	square metre	m^2
Volume	cubic metre	m^3
Mass density	kilogramme per cubic metre	kg/m^3
Linear velocity	metre per second	m/s
Linear acceleration	metre per second square	m/s^2
Angular velocity	radian per second	rad/s
Angular acceleration	radian per second squared	rad/s^2
Force	kilogramme metre per second squared	$kg\ m/s^2$
Magnetic field strength	ampere per metre	A/m
Concentration	mole per cubic metre	mol/m^3
Luminance	candela per square metre	cd/m^2

Units in common use, particularly those for which a statement in base units would be lengthy or complicated, have been given special shortened names. Those that are named from scientists and engineers have an initial capital letter: all others are in small letters.

Table 1.2 NAMED DERIVED UNITS

Quantity	*Unit-name*	*Unit-symbol*	*Derivation*
Force	newton	N	kg m/s^2
Pressure	pascal	Pa	N/m^2
Power	watt	W	J/s
Energy	joule	J	N m, W s
Electric charge	coulomb	C	A s
Electric flux	coulomb	C	A s
Magnetic flux	weber	Wb	V s
Magnetic flux density	tesla	T	Wb/m^2
Electric potential	volt	V	J/C, W/A
Resistance	ohm	Ω	V/A
Conductance	siemens	S	A/V
Capacitance	farad	F	A s/V, C/V
Inductance	henry	H	V s/A, Wb/A
Luminous flux	lumen	lm	cd sr
Illuminance	lux	lx	lm/m^2
Frequency	hertz	Hz	l/s

The named derived units are used to form further derivations. Examples are:

Table 1.3 FURTHER DERIVED UNITS

Quantity	*Unit-name*	*Unit-symbol*
Torque	newton metre	N m
Dynamic viscosity	pascal second	Pa s
Surface tension	newton per metre	N/m
Power density	watt per square metre	W/m^2
Energy density	joule per cubic metre	J/m^3
Heat capacity	joule per kelvin	J/K
Specific heat capacity	joule per kilogramme kelvin	J/(kg K)
Thermal conductivity	watt per metre kelvin	W/(m K)
Electric field strength	volt per metre	V/m
Magnetic field strength	ampere per metre	A/m
Electric flux density	coulomb per square metre	C/m^2
Current density	ampere per square metre	A/m^2
Resistivity	ohm metre	Ω m
Permittivity	farad per metre	F/m
Permeability	henry per metre	H/m

Names of SI units and the corresponding EMU and ESU CGS units are given in *Table 1.4*.

Table 1.4

Quantity	*Symbol*	*SI*	*EMU & ESU*
Length	l	metre (m)	centimetre (cm)
Time	t	second (s)	second
Mass	m	kilogram (kg)	gram (g)
Force	F	newton (N)	dyne (dyn)
Frequency	f, ν	hertz (Hz)	hertz
Energy	E, W.	joule (J)	erg (erg)
Power	P	watt (W)	erg/sec. (erg/s)
Pressure	p	newton/metre^2 (N/m^2)	dyne/centimetre^2 (dyne/cm^2)
Electric charge	Q	coulomb (C)	coulomb (C)
Electric potential	V	volt (V)	volt
Electric current	I	ampere (A)	ampere
Magnetic flux	Φ	weber (Wb)	maxwell (Mx)
Magnetic induction	B	tesla (T)	gauss (G)
Magnetic field strength	H	ampere turn/metre (At/m)	oersted (Oe)
Magneto-motive force	F_m	ampere turn (At)	gilbert (Gb)
Resistance	R	ohm (Ω)	ohm
Inductance	L	henry (H)	henry
Conductance	G	mho (Ω^{-1}) (siemens)	mho
Capacitance	C	farad (F)	farad

GRAVITATIONAL AND ABSOLUTE SYSTEMS

There may be some difficulty in understanding the difference between SI and the Metric Technical System of units which has been used principally in Europe. The main difference is that while mass is expressed in kg in both systems, weight (representing a force) is expressed as kgf, a gravitational unit, in the MKSA system and as N in SI. An absolute unit of force differs from a gravitational unit of force because it induces unit acceleration in a unit mass whereas a gravitational unit imparts gravitational acceleration to a unit mass.

A comparison of the more commonly known systems and SI is shown in *Table 1.5*.

Table 1.5 COMMONLY USED UNITS OF MEASUREMENT

	SI (*absolute*)	*FPS* (*gravitational*)	*FPS* (*absolute*)	*cgs* (*absolute*)	*Metric technical units* (*gravitational*)
Length	metre (m)	ft	ft	cm	metre
Force	newton (N)	lbf	poundal (pdl)	dyne	kgf
Mass	kg	lb or slug	lb	gram	kg
Time	s	sec	sec	sec	sec
Temperature	°C K	°F	°F °R	°C K	°C K

Table 1.5—*continued* COMMONLY USED UNITS OF MEASUREMENT

		SI (absolute)	*FPS (gravitational)*	*FPS (absolute)*	*cgs (absolute)*	*Metric technical units (gravitational)*
Energy	mech.	joule*	ft lbf	ft pdl	dyne cm = erg	kgf m
	heat		Btu	Btu	calorie	k cal.
Power	mech.	watt	hp	hp	erg's	metric hp
	elec.		watt	watt		watt
Electric current		amp	amp	amp	amp	amp
Pressure		N/m^2	lbf/ft^2	pdl/ft^2	$dyne/cm^2$	kgf/cm^2

* 1 joule = 1 newton metre or 1 watt second.

EXPRESSING MAGNITUDES OF SI UNITS

To express magnitudes of a unit, decimal multiples and submultiples are formed using the prefixes shown in *Table 1.6*. This method of expressing magnitudes ensures complete adherence to a decimal system.

Table 1.6 THE INTERNATIONALLY AGREED MULTIPLES AND SUBMULTIPLES

Factor by which the unit is multiplied		*Prefix*	*Symbol*	*Common everyday examples*
One million million (billion)	10^{12}	tera	T	
One thousand million	10^{9}	giga	G	gigahertz (GHz)
One million	10^{6}	mega	M	megawatt (MW)
One thousand	10^{3}	kilo	k	kilometre (km)
One hundred	10^{2}	hecto*	h	
Ten	10^{1}	deca*	da	decagramme (dag)
UNITY	1			
One tenth	10^{-1}	deci*	d	decimetre (dm)
One hundredth	10^{-2}	centi*	c	centimetre (cm)
One thousandth	10^{-3}	milli	m	milligramme (mg)
One millionth	10^{-6}	micro	μ	microsecond (μs)
One thousand millionth	10^{-9}	nano	n	nanosecond (ns)
One million millionth	10^{-12}	pico	p	picofarad (pF)
One thousand million millionth	10^{-15}	femto	f	
One million million millionth	10^{-18}	atto	a	

* To be avoided wherever possible.

AUXILIARY UNITS

Certain auxiliary units may be adopted where they have application in special fields. Some are acceptable on a temporary basis, pending a more widespread adoption of the SI system. *Table 1.7* lists some of these.

Table 1.7 AUXILIARY UNITS

Quantity	*Unit-symbol*	*SI equivalent*
Day	d	86 400 s
Hour	h	3 600 s
Minute (time)	min	60 s
Degree (angle)	°	$\pi/180$ rad
Minute (angle)	′	$\pi/10\ 800$ rad
Second (angle)	″	$\pi/648\ 000$ rad
Are	a	$1\ dam^2 = 10^2\ m^2$
Hectare	ha	$1\ hm^2 = 10^4\ m^2$
Barn	b	$100\ fm^2 = 10^{-28}\ m^2$
Standard atmosphere	atm	101 325 Pa
Bar	bar	$0.1\ MPa = 10^5\ Pa$
Litre	l	$1\ dm^3 = 10^{-3}\ m^3$
Tonne	t	$10^3\ kg = 1\ Mg$
Atomic mass unit	u	$1.660\ 53 \times 10^{-27}$ kg
Angström	Å	$0.1\ nm = 10^{-10}$ m
Electron-volt	eV	$1.602\ 19 \times 10^{-19}$ J
Curie	Ci	$3.7 \times 10^{10}\ s^{-1}$
Röntgen	R	2.58×10^{-4} C/kg

NUCLEAR ENGINEERING

It has been the practice to use special units with their individual names for evaluating and comparing results. These units are usually formed by multiplying a unit from the cgs or SI system by a number which matches a value derived from the result of some natural phenomenon. The adoption of SI both nationally and internationally has created the opportunity to examine the practice of using special units in the nuclear industry, with the object of eliminating as many as possible and using the pure system instead.

As an aid to this, ISO draft Recommendations 838 and 839 have been published, giving a list of quantities with special names, the SI unit and the alternative cgs unit. It is expected that as SI is increasingly adopted and absorbed, those units based on cgs will go out of use. The values of these special units illustrate the fact that a change from them to SI would not be as revolutionary as might be supposed. Examples of these values together with the SI units which replace them are shown in *Table 1.8*.

Table 1.8 NUCLEAR ENGINEERING

Special unit			*SI Replacement*
Name		*Value*	
Angström	(Å)	10^{-10} m	m
Barn	(b)	$10^{-28}\ m^2$	m^2
Curie	(Ci)	$3.7 \times 10^{10}\ s^{-1}$	s^{-1}
Electron-volt	(eV)	$(1.602\ 10 \pm 0.000\ 07) \times 10^{-19}$ J	J
Röntgen	(R)	2.58×10^{-4} C/kg	C/kg

UNIVERSAL CONSTANTS IN SI UNITS

Table 1.9

The digits in parentheses following each quoted value represent the standard deviation error in the final digits of the quoted value as computed on the criterion of internal consistency. The unified scale of atomic weights is used throughout ($^{12}C = 12$). C = coulomb; G = gauss; Hz = hertz; J = joule; N = newton; T = tesla; u = unified nuclidic mass unit; W = watt; Wb = weber. For result multiply the numerical value by the SI unit.

Constant	*Symbol*	*Numerical value*	*SI unit*
Speed of light in vacuum	c	2.997 925(1)	10^8 m/s^{-1}
Gravitational constant	G	6.670(5)*	10^{-11} N m^2/kg^{-2}
Elementary charge	e	1.602 10(2)	10^{-19} C
Avogadro constant	N_A	6.022 52(9)	10^{26} kmol^{-1}
Mass unit	u	1.660 43(2)	10^{-27} kg
Electron rest mass	m_e	9.109 08(13)	10^{-31} kg
		5.485 97(3)	10^{-4} u
Proton rest mass	m_p	1.672 52(3)	10^{-27} kg
		1.007 276 63(8)	u
Neutron rest mass	m_n	1.67 482(3)	10^{-27} kg
		1.008 665 4(4)	u
Faraday constant	F	9.648 70(5)	10^4 C mol^{-1}
Planck constant	h	6.625 59(16)	10^{-34} J s
	$h/2\pi$	1.054 494(25)	10^{-34} J s
Fine-structure constant	α	7.297 20(3)	10^{-3}
	$1/\alpha$	137.038 8(6)	
Charge-to-mass ratio for electron	e/m_e	1.758 796(6)	10^{11} C kg^{-1}
Quantum of magnetic flux	hc/e	4.135 56(4)	10^{-11} Wb
Rydberg constant	R_∞	1.097 373 1(1)	10^7 m^{-1}
Bohr radius	a_0	5.291 67(2)	10^{-11} m
Compton wavelength of electron	$h/m_e c$	2.426 21(2)	10^{-12} m
	$\lambda c/2\pi$	3.861 44(3)	10^{-13} m
Electron radius	$e^2/m_e c^2 = r_e$	2.817 77(4)	10^{-15} m
Thomsen cross section	$8\eta r_e^2/3$	6.651 6(2)	10^{-29} m^2
Compton wavelength of proton	$\lambda c, p$	1.321 398(13)	10^{-15} m
	$\lambda c, p/2\pi$	2.103 07(2)	10^{-16} m
Gyromagnetic ratio of proton	γ	2.675 192(7)	10^8 rad s^{-1} T^{-1}
	$\gamma/2\pi$	4.257 70(1)	10^7 Hz T^{-1}
(Uncorrected for diamagnetism H_2O)	γ'	2.675 123(7)	10^8 rad s^{-1} T^{-1}
	$\gamma'/2\pi$	4.257 59(1)	10^7 Hz T^{-1}
Bohr magneton	μB	9.273 2(2)	10^{-24} J T^{-1}
Nuclear magneton	μN	5.050 50(13)	10^{-27} J T^{-1}
Proton moment	μ_p	1.410 49(4)	10^{-26} J T^{-1}
	$\mu_p/\mu N$	2.792 76(2)	

* The universal gravitational constant is not, and cannot in our present state of knowledge, be expressed in terms of other fundamental constants. The value given here is a direct determination by P. R. Heyl and P. Chrzanowski, *J. Res. Natl. Bur. Std.* (*U.S.*) 29, 1 (1942).

The above values are extracts from *Review of Modern Physics* Vol. 37 No. 4 October 1965 published by the American Institute of Physics.

Table 1.9—*continued*

The digits in parentheses following each quoted value represent the standard deviation error in the final digits of the quoted value as computed on the criterion of internal consistency. The unified scale of atomic weights is used throughout ($^{12}C = 12$). C = coulomb; G = gauss; Hz = hertz; J = joule; N = newton; T = tesla; u = unified nuclidic mass unit; W = watt; Wb = weber. For result multiply the numerical value by the SI unit.

Constant	*Symbol*	*Numerical value*	*SI unit*
(Uncorrected for diamagnetism in H_2O sample)		2.792 68(2)	
Gas constant	R_0	8.314 34(35)	J deg^{-1} mol^{-1}
Boltzmann constant	k	1.380 54(6)	10^{-23} J deg^{-1}
First radiation constant ($2\eta hc^2$)	c_1	3.741 50(9)	10^{-16} W/m^2
Second radiation constant (hc/k)	c_2	1.438 79(6)	10^{-2} m/deg
Stefan-Boltzmann constant	σ	5.669 7(10)	10^{-8} W m^{-2} deg^{-4}

METRIC TO IMPERIAL CONVERSION FACTORS

Table 1.10

SI units	*British units*
SPACE AND TIME	
Length:	
1 μm (micron)	= 39.37 × 10^{-6} in
1 mm	= 0.039 370 1 in
1 cm	= 0.393 701 in
1 m	= 3.280 84 ft
1 m	= 1.093 61 yd
1 km	= 0.621 371 mile
Area:	
1 mm^2	= 1.550 × 10^{-3} in^2
1 cm^2	= 0.155 0 in^2
1 m^2	= 10.763 9 ft^2
1 m^2	= 1.195 99 yd
1 ha	= 2.471 05 acre
Volume:	
1 mm^3	= 61.023 7 × 10^{-6} in^3
1 cm^3	= 61.023 7 × 10^{-3} in^3
1 m^3	= 35.314 7 ft^3
1 m^3	= 1.307 95 yd^3
Capacity:	
10^6 m^3	= 219.969 × 10^6 gal
1 m^3	= 219.969 gal
1 litre (l)	= 0.219 969 gal = 1.759 80 pint

Table 1.10—*continued*

SI units	*British units*
Capacity flow:	
$10^3/m^3/s$	$= 791.9 \times 10^6$ gal/h
1 m^3/s	$= 13.20 \times 10^3$ gal/min
1 litre/s	= 13.20 gal/min
1 m^3/kW h	= 219.969 gal/kW h
1 m^3/s	= 35.314 7 ft^3/s (cusecs)
1 litre/s	$= 0.588\ 58 \times 10^{-3}$ ft^3/min (cfm)
Velocity:	
1 m/s	= 3.280 84 ft/s = 2.236 94 mile
1 km/h	= 0.621 371 mile/h
Acceleration:	
1 m/s^2	= 3.280 84 ft/s^2
MECHANICS	
Mass:	
1 g	= 0.035 274 oz
1 kg	= 2.204 62 lb
1 t	= 0.984 207 ton = 19.684 1 cwt
Mass flow:	
1 kg/s	= 2.204 62 lb/s = 7.936 64 klb/h
Mass density:	
1 kg/m^3	= 0.062 428 lb/ft^3
1 kg/litre	= 10.022 119 lb/gal
Mass per unit length:	
1 kg/m	= 0.671 969 lb/ft = 2.015 91 lb/yd
Mass per unit area:	
1 kg/m^2	= 0.204 816 lb/ft^2
Specific volume:	
1 m^3/kg	= 16.018 5 ft^3/lb
1 litre/tonne	= 0.223 495 gal/ton
Momentum:	
1 kg m/s	= 7.233 01 lbft/s
Angular momentum:	
1 kg m^2/s	= 23.730 4 $lbft^2$/s
Moment of inertia:	
1 kg m^2	= 23.730 4 $lbft^2$
Force:	
1 N	= 0.224 809 lbf
Weight (force) per unit length:	
1 N/m	= 0.068 521 8 lbf/ft = 0.205 566 lbf/yd
Moment of force (or torque):	
1 Nm	= 0.737 562 lbf/ft
Weight (force) per unit area:	
1 N/m^2	= 0.020 885 lbf/ft^2
Pressure:	
1 N/m^2	$= 1.450\ 38 \times 10^{-4}$ lbf/in^2
1 bar	= 14.5038 lbf/in^2
1 bar	= 0.986 923 atmosphere
1 mbar	= 0.401 463 in H_2O
	= 0.029 53 in Hg

Table 1.10—*continued*

SI units	*British units*
Stress:	
1 N/mm^2	= 6.474 90 × 10^{-2} tonf/in^2
1 MN/m^2	= 6.474 90 × 10^{-2} tonf/in^2
1 hbar	= 0.647 490 tonf/in^2
Second moment of area:	
1 cm^4	= 0.024 025 in^4
Section modulus:	
1 m^3	= 61 023.7 in^3
1 cm^3	= 0.061 023 7 in^3
Kinematic viscosity:	
1 m^2/s	= 10.762 75 ft^2/s = 10^6 cSt
1 cSt	= 0.038 75 ft^2/h
Energy, work:	
1 J	= 0.737 562 ft lbf
1 MJ	= 0.372 5 hph
1 MJ	= 0.277 78 kW h
Power:	
1 W	= 0.737 562 ft lbf/s
1 kW	= 1.341 hp = 737.562 ft lbf/s
Fluid mass:	
(Ordinary) 1 kg/s	= 2.204 62 lb/s = 793 6.64 lb/h
(Velocity) 1 kg/m^2 s	= 0.204 815 lb/ft^{2}s
HEAT	
Temperature:	
(Interval) 1 °K	= 9/5 deg R (Rankine)
1 °C	= 9/5 deg F
(Coefficient) 1 °R^{-1}	= 1 deg F^{-1} = 5/9 deg C
1 °C^{-1}	= 5/9 deg F^{-1}
Quantity of heat:	
1 J	= 9.478 17 × 10^{-4} Btu
1 J	= 0.238 846 cal
1 kJ	= 947.817 Btu
1 GJ	= 947.817 × 10^3 Btu
1 kJ	= 526.565 CHU
1 GJ	= 526.565 × 10^3 CHU
1 GJ	= 9.478 17 therm
Heat flow rate:	
1 W(J/s)	= 3.412 14 Btu/h
1 W/m^2	= 0.316 998 Btu/ft^2 h
Thermal conductivity:	
1 W/m °C	= 6.933 47 Btu in/ft^2 h °F
Coefficient and heat transfer:	
1 W/m^2 °C	= 0.176 110 Btu/ft^2 h °F
Heat capacity:	
1 J/°C	= 0.526 57 × 10^{-3} Btu/°R
Specific heat capacity:	
1 J/g °C	= 0.238 846 Btu/lb °F
1 kJ/kg °C	= 0.238 846 Btu/lb °F
Entropy:	
1 J/K	= 0.526 57 × 10^{-3} Btu/°R

Table 1.10—*continued*

SI units	*British units*
Specific Entropy:	
1 J/kg °C	= 0.238 846 × 10^{-3} Btu/lb °F
1 J/kg °K	= 0.238 846 × 10^{-3} Btu/lb °R
Specific energy/Specific latent heat:	
1 J/g	= 0.429 923 Btu/lb
1 J/kg	= 0.429 923 × 10^{-3} Btu/lb
Calorific value:	
1 kJ/kg	= 0.429 923 Btu/lb
1 kJ/kg	= 0.773 861 4 CHU/lb
1 J/m^3	= 0.026 839 2 × 10^{-3} Btu/ft^3
1 kJ/m^3	= 0.026 839 2 Btu/ft^3
1 kg/litre	= 4.308 86 Btu/gal
1 kJ/kg	= 0.009 630 2 therm/ton
ELECTRICITY	
Permeability:	
1 H/m	= $10^7/4\,\Pi\,\mu o$
Magnetic flux density:	
1 tesla	= 10^4 gauss = 1 Wb/m^2
Conductivity:	
1 mho	= 1 reciprocal ohm
1 siemen	= 1 reciprocal ohm
Electric stress:	
1 kV/mm	= 25.4 kV/in
1 kV/m	= 0.025 4 kV/in

SYMBOLS AND ABBREVIATIONS

Table 1.11 QUANTITIES AND UNITS OF PERIODIC AND RELATED PHENOMENA
(Based on ISO recommendation R31)

Symbol	*Quantity*
T	periodic time
$\tau, (T)$	time constant of an exponentially varying quantity
f, ν	frequency
η	rotational frequency
ω	angular frequency
λ	wave length
$\sigma\ (\tilde{\nu})$	wave number
k	circular wave number
$\log e\,(A_1/A_2)$	natural logarithm of the ratio of two amplitudes
$10 \log_{10} (P_1/P_2)$	ten times the common logarithm of the ratio of two powers
δ	damping coefficient
Λ	logarithmic decrement
α	attenuation coefficient
β	phase coefficient
γ	propagation coefficient

Table 1.12 SYMBOLS FOR QUANTITIES AND UNITS OF ELECTRICITY AND MAGNETISM (Based on ISO recommendation R31)

Symbol	*Quantity*
I	electric current
Q	electric charge, quantity of electricity
e	volume density of charge, charge density
σ	surface density of charge
$E, (K)$	electric field strength
V, ϕ	electric potential
$U, (V)$	potential difference, tension
E	electromotive force
D	displacement (rationalised displacement)
D'	non-rationalised displacement
ψ	electric flux, flux of displacement (flux of rationalised displacement)
ψ'	flux of non-rationalised displacement
C	capacitance
ϵ	permittivity
ϵ_0	permittivity of vacuum
ϵ'	non-rationalised permittivity
ϵ'_0	non-rationalised permittivity of vacuum
ϵ_r	relative permittivity
χ_e	electric susceptibility
χ'_e	non-rationalised electric susceptibility
P	electric polarisation
$p, (p_e)$	electric dipole moment
$J, (S)$	current density
$A, (\alpha)$	linear current density
H	magnetic field strength
H'	non-rationalised magnetic field strength
U_m	magnetic potential difference
F, F_m	magnetomotive force
B	magnetic flux density, magnetic induction
Φ	magnetic flux
A	magnetic vector potential
L	self inductance
M, L_{12}	mutual inductance
$k, (x, k)$	coupling coefficient
σ	leakage coefficient
μ	permeability
μ_0	permeability of vacuum
μ'	non-rationalised permeability
μ'_0	non-rationalised permeability of vacuum
μ_r	relative permeability
x, k	magnetic susceptibility
x', k'	non-rationalised magnetic susceptibility
m	electromagnetic moment (magnetic moment)
H_i, M	magnetisation
B_i, J	magnetic polarisation
J'	non-rationalised magnetic polarisation
ω	electromagnetic energy density
S	Poynting vector
c	velocity of propagation of electromagnetic waves in vacuo

Table 1.12—*continued*

Symbol	*Quantity*
R	resistance (to direct current)
G	conductance (to direct current)
e	resistivity
γ, σ	conductivity
R, R_m	reluctance
$A, (P)$	permeance
N	number of turns in winding
m	number of phases
p	number of pairs of poles
ϕ	phase displacement
Z	impedance (complex impedance)
$[Z]$	modulus of impedance (impedance)
X	reactance
R	resistance
Q	quality factor
Y	admittance (complex admittance)
$[Y]$	modulus of admittance (admittance)
B	susceptance
G	conductance
P	active power
$S, (P_s)$	apparent power
$Q, (P_q)$	reactive power

Table 1.13 SYMBOLS FOR QUANTITIES AND UNITS OF ACOUSTICS
(Based on ISO recommendation R31)

Symbol	*Quantity*
T	period, periodic time
f, v	frequency, frequency interval
ω	angular frequency, circular frequency
λ	wavelength
k	circular wave number
ρ	density (mass density)
Ps	static pressure
p	(instantaneous) sound pressure
$\epsilon, (x)$	(instantaneous) sound particle displacement
u, v	(instantaneous) sound particle velocity
a	(instantaneous) sound particle acceleration
q, U	(instantaneous) volume velocity
c	velocity of sound
E	sound energy density
$P, (N, W)$	sound energy flux, sound power
I, J	sound intensity
$Z_s, (W)$	specific acoustic impedance
$Z_a, (Z)$	acoustic impedance
$Z_m, (w)$	mechanical impedance
$L_p, (L_N, L_w)$	sound power level
$L_p, (L)$	sound pressure level
δ	damping coefficient

Table 1.13—*continued*

Symbol	*Quantity*
Λ	logarithmic decrement
α	attenuation coefficient
β	phase coefficient
γ	propagation coefficient
δ	dissipation coefficient
r, τ	reflection coefficient
γ	transmission coefficient
$\alpha, (\alpha_a)$	acoustic absorption coefficient
R	sound reduction index sound transmission loss
A	equivalent absorption area of a surface or object
T	reverberation time
$L_N, (\Lambda)$	loudness level
N	loudness

Table 1.14 SOME TECHNICAL ABBREVIATIONS AND SYMBOLS

Quantity	*Abbreviation*	*Symbol*
Alternating current	a.c.	
Ampere	A or amp	
Amplification factor		μ
Amplitude modulation	a.m.	
Angular velocity		ω
Audio frequency	a.f.	
Automatic frequency control	a.f.c	
Automatic gain control	a.g.c.	
Bandwidth		Δf
Beat frequency oscillator	b.f.o.	
British thermal unit	Btu	
Cathode-ray oscilloscope	c.r.o.	
Cathode-ray tube	c.r.t.	
Centigrade	C	
Centi-	c	
Centimetre	cm	
Square centimetre	cm^2 or sq cm	
Cubic centimetre	cm^3 or cu cm or c.c.	
Centimetre-gramme-second	c.g.s.	
Continuous wave	c.w.	
Coulomb	C	
Deci-	d	
Decibel	dB	
Direct current	d.c.	
Direction finding	d.f.	
Double sideband	d.s.b.	
Efficiency		η
Equivalent isotropic radiated power	e.i.r.p.	
Electromagnetic unit	e.m.u.	
Electromotive force instantaneous value	e.m.f.	E or V, e or v

Table 1.14—*continued*

Quantity	*Abbreviation*	*Symbol*
Electron volt	eV	
Electrostatic unit	e.s.u.	
Fahrenheit	F	
Farad	F	
Frequency	freq.	f
Frequency modulation	f.m.	
Gauss	G	
Giga-	G	
Gramme	g	
Henry	H	
Hertz	Hz	
High frequency	h.f.	
Independent sideband	i.s.b.	
Inductance-capacitance		L-C
Intermediate freq.	i.f.	
Kelvin	K	
Kilo-	k	
Knot	kn	
Length		l
Local oscillator	l.o.	
Logarithm, common		log or $\log_{10}$
Logarithm, natural		ln or $\log_e$
Low frequency	l.f.	
Low tension	l.t.	
Magnetomotive force	m.m.f.	F or M
Mass		m
Medium frequency	m.f.	
Mega-	M	
Metre	m	
Metre-kilogramme-second	m.k.s.	
Micro-	μ	
Micromicro	p	
Micron		μ
Milli-	m	
Modulated continuous wave	m.c.w.	
Nano-	n	
Neper	N	
Noise factor		N
Ohm		Ω
Peak to peak	p–p	
Phase modulation	p.m.	
Pico-	p	
Plan-position indication	PPI	
Potential difference	p.d.	V
Power factor	p.f.	
Pulse repetition frequency	p.r.f.	

Table 1.14—*continued*

Quantity	*Abbreviation*	*Symbol*
Radian	rad	
Radio frequency	r.f.	
Radio telephony	R/T	
Root mean square	r.m.s.	
Short-wave	s.w.	
Single sideband	s.s.b.	
Signal frequency	s.f.	
Standing wave ratio	s.w.r.	
Super-high frequency	s.h.f.	
Susceptance		*B*
Travelling-wave tube	t.w.t.	
Ultra-high frequency	u.h.f.	
Very high frequency	v.h.f.	
Very low frequency	v.l.f.	
Volt	V	
Voltage standing wave ratio	v.s.w.r.	
Watt	W	
Weber	Wb	
Wireless telegraphy	W/T	

MATHEMATICAL SIGNS AND SYMBOLS

Table 1.15 MATHEMATICAL SIGNS AND SYMBOLS FOR USE IN TECHNOLOGY
(Based on ISO recommendation R31)

Sign, symbol	*Quantity*
$=$	equal to
╪ $\neq$	not equal to
$\equiv$	identically equal to
Δ	corresponds to
$\approx$	approximately equal to
$\rightarrow$	approaches
$\simeq$	asymptotically equal to
$\sim$	proportional to
∞	infinity
$<$	smaller than
$>$	larger than
$\leq$	smaller than or equal to
$\geq$ $\geqslant$ $\geqq$	larger than or equal to
$\ll$	much smaller than
$\gg$	much larger than
$+$	plus
$-$	minus
$\cdot$ $\times$	multiplied by

Table 1.15—*continued*

Sign, symbol	*Quantity*
$\frac{a}{b}$ a/b	a divided by b
$\lvert a \rvert$	magnitude of a
a^n	a raised to the power n
$a^{\frac{1}{2}}$ $a^{1/2}$ $\sqrt{a}$ $\sqrt{a}$	square root of a
$a^{1/n}$ $a^{\frac{1}{n}}$ $\sqrt[n]{a}$ $\sqrt[n]{a}$	n'th root of a
$\bar{a}$ $\langle a \rangle$	mean value of a
$p!$	factorial p, $1 \times 2 \times 3 \times \ldots \times p$
$\binom{n}{p}$	binomial coefficient, $\frac{n(n-1)\ldots(n-p+1)}{1 \times 2 \times 3 \times \ldots \times p}$
Σ	sum
Π	product
$f(x)$ $f(x)$	function f (of f) of the variable x
$[f(x)]_a^b$ $f(x)/_a^b$	$f(b) - f(a)$
$\lim_{x \to a} f(x)$; $\lim_{x\to a} f(x)$	the limit to which $f(x)$ tends as x approaches a
Δx	delta x = finite increment of x
δx	delta x = variation of x
$\frac{df}{dx}$; df/dx; $f'(x)$	differential coefficient of $f(x)$ with respect to x
$\frac{d^nf}{dx^n}$; $f^{(n)}(x)$	differential coefficient of order n of $f(x)$
$\frac{\partial f(x, y, \ldots)}{\partial x}$; $\left(\frac{\partial f}{\partial x}\right)_{y \ldots}$	partial differential coefficient of $f(x, y, \ldots)$ with respect to x, when $y, \ldots$ are held constant
df	the total differential of f
$\int f(x)dx$	indefinite integral of $f(x)$ with respect to x
$\int_a^b f(x)dx$; $\int_a^b f(x)dx$	definite integral of $f(x)$ from $x = a$ to $x = b$
e	base of natural logarithms
e^x; $\exp x$	e raised to the power x
$\log_a x$	logarithm to the base a of x
$\ln x$; $\log_e x$	natural logarithm (Napierian logarithm) of x
$\lg x$; $\log x$; $\log_{10} x$	common (Briggsian) logarithm of x
$\mathrm{lb}\, x$; $\log_2 x$	binary logarithm of x
$\sin x$	sine of x
$\cos x$	cosine of x
$\tan x$; $\mathrm{tg}\, x$	tangent of x
$\cot x$; $\mathrm{ctg}\, x$	cotangent of x
$\sec x$	secant of x
$\mathrm{cosec}\, x$	cosecant of x
$\arcsin x$	arc sine of x
$\arccos x$	arc cosine of x
$\arctan x$, $\mathrm{arctg}\, x$	arc tangent of x
$\mathrm{arccot}\, x$, $\mathrm{arcctg}\, x$	arc cotangent of x
$\mathrm{arcsec}\, x$	arc secant of x
$\mathrm{arccosec}\, x$	arc cosecant of x
$\sinh x$	hyperbolic sine of x
$\cosh x$	hyperbolic cosine of x
$\tanh x$	hyperbolic tangent of x
$\coth x$	hyperbolic cotangent of x

Table 1.15—*continued*

Sign, symbol	*Quantity*
sech x	hyperbolic secant of x
cosech x	hyperbolic cosecant of x
arsinh x	inverse hyperbolic sine of x
arcosh x	inverse hyperbolic cosine of x
artanh x	inverse hyperbolic tangent of x
arcoth x	inverse hyperbolic cotangent of x
arsech x	inverse hyperbolic secant of x
arcosech x	inverse hyperbolic cosecant of x
i, j	imaginary unity, $1^2 = -1$
$Re\, z$	real part of z
$Im\, z$	imaginary part of z
$\|z\|$	modulus of z
arg z	argument of z
z^*	conjugate of x, complex conjugate of z
$\tilde{\mathrm{A}}$	transpose of matrix A
A*	complex conjugate matrix of matrix A
A†	Hermitian conjugate matrix of matrix A
Aa	vector
$\|\mathbf{A}\|$, A	magnitude of vector
$\mathbf{A} \cdot \mathbf{B}$	scalar product
$\mathbf{A} \times \mathbf{B}$, $\mathbf{A} \wedge \mathbf{B}$	vector product
∇	differential vector operator
$\nabla\phi$, grad ϕ	gradient of ϕ
$\nabla \cdot A$, div **A**	divergence of **A**
$\nabla \times \mathbf{A}$, $\nabla \wedge \mathbf{A}$, curl **A**, rot **A**	curl of **A**
$\nabla^2\phi$, $\Delta\phi$	Laplacion of ϕ

MATHEMATICAL FORMULAE

Algebraic and Trigonometric Formulæ

$$\sin^2 A + \cos^2 A = \sin A \operatorname{cosec} A = 1 \qquad 1 + \tan^2 A = \sec^2 A.$$

$$\sin A = \frac{\cos A}{\cot A} = \frac{1}{\operatorname{cosec} A} = \sqrt{1 - \cos^2 A}. \qquad 1 + \cot^2 A = \operatorname{cosec}^2 A.$$

$$\cos A = \frac{\sin A}{\tan A} = \frac{1}{\sec A} = \sqrt{1 - \sin^2 A}. \qquad 1 - \sin A = \text{coversin}\, A.$$

$$\text{tangent}\, A = \frac{\sin A}{\cos A} = \frac{1}{\cot A}; \qquad \tan \theta/2 = t; \qquad \sin\theta = \frac{2t}{1+t^2}; \qquad \cos\theta = \frac{1-t^2}{1+t^2}$$

$$\text{cotangent}\, A = \frac{1}{\tan A}, \qquad \text{secant}\, A = \frac{1}{\cos A}, \qquad \text{cosecant}\, A = \frac{1}{\sin A}$$

$$\sin(A \pm B) = \sin A \cos B \pm \cos A \sin B; \qquad \tan(A \pm B) = \frac{\tan A \pm \tan B}{1 \pm \tan A \tan B};$$

$$\cos(A \pm B) = \cos A \cos B \pm \sin A \sin B \qquad \sinh u = (e^u - e^{-u}) \div 2;$$

$$\cosh u = \frac{e^u + e^{-u}}{2}; \qquad \tanh u = \frac{e^u - e^{-u}}{e^u + e^{-u}}. \quad \cot(A \pm B) = \frac{\cot A \cot B \pm 1}{\cot B \pm \cot A};$$

$$\cosh^2 u - \sinh^2 u = 1; \qquad \sin A + \sin B = 2 \sin \tfrac{1}{2}(A + B) \cos \tfrac{1}{2}(A - B);$$

$\sin^2 A - \sin^2 B = \sin(A+B)\sin(A-B)$; $\tan A \pm \tan B = \dfrac{\sin(A \pm B)}{\cos A \cos B}$.

$\sin A - \sin B = 2\cos\frac{1}{2}(A+B)\sin\frac{1}{2}(A-B)$; $e^{i\theta} = \cos\theta + i\sin\theta$;

$\cos A + \cos B = 2\cos\frac{1}{2}(A+B)\cos\frac{1}{2}(A-B)$; $e^{i\theta} = \cos\theta - i\sin\theta$;

$\cos B - \cos A = 2\sin\frac{1}{2}(A+B)\sin\frac{1}{2}(A-B)$. versine $A = 1 - \cos A$ $\cos\theta = \dfrac{e^{i\theta} + e^{-i\theta}}{2}$

$\sin\theta = \dfrac{e^{i\theta} - e^{-i\theta}}{2i}$, $i = \sqrt{-1}$, $e^{in\theta} = \cos n\theta + i\sin n\theta$ $(\cos\theta + i\sin\theta)^n = \cos n\theta + i\sin n\theta$;

$\cot A \pm \cot B = \dfrac{\sin(B \pm A)}{\sin A \sin B}$ $\sin 2A = 2\sin A\cos A$; $\cos 2A = \cos^2 A - \sin^2 A$;

$\cos^2 A - \sin^2 B = \cos(A+B)\cos(A-B)$; $\tan 2A = \dfrac{2\tan A}{1-\tan^2 A}$ $\sin\frac{1}{2}A = \sqrt{\dfrac{1-\cos A}{2}}$,

$\cos\frac{1}{2}A = \pm\sqrt{\dfrac{1+\cos A}{2}}$; $\tan\frac{1}{2}A = \dfrac{\sin A}{1+\cos A}$; $\sin^2 A = \dfrac{1-\cos 2A}{2}$; $\cos^2 A = \dfrac{1+\cos 2A}{2}$;

$\tan^2 A = \dfrac{1-\cos 2A}{1+\cos 2A}$; $\dfrac{\sin A \pm \sin B}{\cos A + \cos B} = \tan\frac{1}{2}(A \pm B)$; $\dfrac{\sin A \pm \sin B}{\cos B - \cos A} = \cot\frac{1}{2}(A \pm B)$.

Angle	*0*	*30°*	*45°*	*60°*	*90°*	*180°*	*270°*	*360°*
Radians	0	$\pi/6$	$\pi/4$	$\pi/3$	$\pi/2$	π	$3\pi/2$	2π
Sine	0	$\frac{1}{2}$	$\frac{1}{2}\sqrt{2}$	$\frac{1}{2}\sqrt{3}$	1	0	-1	0
Cosine	1	$\frac{1}{2}\sqrt{3}$	$\frac{1}{2}\sqrt{2}$	$\frac{1}{2}$	0	-1	0	1
Tangent	0	$\frac{1}{3}\sqrt{3}$	1	$\sqrt{3}$	∞	0	∞	0

APPROXIMATIONS FOR SMALL ANGLES

$\sin\theta \simeq (\theta - \theta^3/6\ldots)$; $\tan\theta \simeq (\theta + \theta^3/3\ldots)$; $\cos\theta \simeq (1 - \theta^2/2\ldots)$ θ in radians; $\sin 14\frac{1}{2}° = \frac{1}{4}$; $\sin 19\frac{1}{2}° = \frac{1}{3}$.

QUADRATIC EQUATION

If $ax^2 + bx + c = 0$, then $x = \dfrac{-b \pm \sqrt{b^2 - 4ac}}{2a}$

TO FIND THE SUM OF ANY NUMBER OF TERMS IN AN ARITHMETICAL PROGRESSION

$T_n = a + (n-1)d$; $S = n(a + l)/2 = n[2a + (n-1)d]/2$.
where a = first term; l = last term; n = number of terms; T_n = nth term; S = sum; d = common difference.

TO FIND THE SUM OF ANY NUMBER OF TERMS IN A GEOMETRICAL PROGRESSION

Let r = common ratio, then $T_n = ar^{n-1}$; $S = \dfrac{a(r^n - 1)}{r-1} = \dfrac{a(1-r^n)}{1-r}$.

COMBINATIONS AND PERMUTATIONS

$${}_nC_r = \frac{n!}{r!\,(n-r)!} = {}_nC_{n-r}.$$

The number of permutations of n things r at a time is ${}_nP_r$.

$${}_nP_n = n(n-1)(n-2) \text{———} 3.2.1 = n!$$
$${}_nP_r = n(n-1)(n-2) \text{———} (n-r+1).$$

BINOMIAL THEOREM

$$(1 \pm x)^n = 1 \pm nx + \frac{n(n-1)}{1.2}x^2 \pm \frac{n(n-1)(n-2)}{1.2.3}x^3 + \ldots$$

MACLAURIN'S THEOREM

$$f(x) = f(o) + xf'(o) \frac{x^2}{1.2}f''(o) + \ldots$$

PROPERTIES OF 'e'

$$e = 1 + 1 + \frac{1}{2!} + \frac{1}{3!} \ldots = 2.718\,28; e^x = 1 + x + \frac{x^2}{2!} + \frac{x}{3!} + \ldots$$
$\log_{10}e = 0.434\,29$, $\log_e 10 = 2.302\,59$, $e^{i\theta} = \cos\theta + i\sin\theta$, $(i^2 = -1)$

$e^u = \cosh u + \sinh u.$

Derivatives and Integrals

y	$\frac{dy}{dx}$	$\int y dx$
x^n	nx^{n-1}	$x^{n+1}/n+1$
$1/x = x^{-1}$		$\log_e x$
$\sin\omega x$	$\omega\cos\omega x$	$-\cos\omega x/\omega$
$\cos\omega x$	$-\omega\sin\omega x$	$\sin\omega x/\omega$
$\tan\omega x$	$\omega\sec^2\omega x$	$-\log\cos\omega x/\omega$
$\tan x$	$\sec^2 x$	$-\log\cos x$ or $\log\sec x$
$\cot x$	$-\text{cosec}^2 x$	$\log\sin x$
$\sec x$	$\tan x\sec x = \sin x/\cos^2 x$	$\log_e(\sec x + \tan x)$
$\text{cosec}\, x$	$-\cot x\,\text{cosec}\, x = -\frac{\cos x}{\sin^2 x}$	$\log_e(\text{cosec}\, x - \cot x)$
$\sin^{-1}\left(\frac{x}{a}\right)$	$\frac{1}{\sqrt{a^2-x^2}}$	$x\sin^{-1}\frac{x}{a} + \sqrt{a^2-x^2}$
$\cos^{-1}\left(\frac{x}{a}\right)$	$\frac{1}{-\sqrt{a^2-x^2}}$	$x\cos^{-1}\frac{x}{a} - \sqrt{a^2-x^2}$
$\left(\frac{x}{a}\right)$	$\frac{a}{a^2+x^2}$	$x\tan^{-1}\frac{x}{a} - a\log_e\sqrt{a^2+z^2}$
e^x	e^x	e^x
e^{ax}	ae^{ax}	e^{ax}/a
$\log_e x$	$1/x$	$x(\log_e x - 1)$
$\log_a x$	$\frac{1}{x}\log_a e$	$x\log_a\frac{x}{e}$

$$\int \frac{dx}{x^2 + a^2} = \frac{1}{a} \tan^{-1} \frac{x}{a}, \qquad \int \frac{x\,dx}{ax^2 + b} = \frac{1}{2a} \log_e (ax^2 + b)$$

$$\int \sqrt{x^2 \pm a^2}\, dx = \tfrac{1}{2}\left[x\sqrt{x^2 \pm a^2} \pm a^2 \log_e \left(x + \sqrt{x^2 \pm a^2}\right)\right]$$

$$\int \sqrt{a^2 - x^2}\, dx = \tfrac{1}{2}\left(x\sqrt{a^2 - x^2} + a^2 \sin^{-1}\left(\frac{x}{a}\right)\right)$$

$$\int \frac{dx}{\sqrt{x^2 \pm a^2}} = \log_e\left(x + \sqrt{x^2 \pm a^2}\right). \qquad \int \frac{dx}{\sqrt{a^2 - x^2}} = \sin^{-1} \frac{x}{a}$$

$$\int x\sqrt{x^2 \pm a^2}\, dx = \tfrac{1}{3}\sqrt{(x^2 \pm a^2)^3}. \qquad \int x\sqrt{a^2 - x^2}\, dx = -\tfrac{1}{3}\sqrt{(a^2 - x^2)^3}$$

$$\int \frac{x\,dx}{\sqrt{a^2 - x^2}} = -\sqrt{a^2 - x^2}. \qquad \int \frac{dx}{\sqrt{2ax - x^2}} = \cos^{-1} \frac{a - x}{a}$$

$$\int \sin^2 x\,dx = \tfrac{1}{2}(x - \sin x \cos x) = \tfrac{1}{2} x - \tfrac{1}{4} \sin 2x. \qquad \int \cos^2 x\,dx = \tfrac{1}{2}(x + \sin x \cos x) = \tfrac{1}{2} x + \tfrac{1}{4} \sin 2x.$$

If $y = f(u)$, $u = \phi(x)$ then $\dfrac{dy}{dx} = \dfrac{dy}{du}\dfrac{du}{dx}$

$\oint$ integral round a curve.

If y is a product, uv, then $\dfrac{dy}{dx} = v\dfrac{du}{dx} + u\dfrac{dv}{dx}$.

$$\oint a \cos \theta\, ds = 0$$

If y is a quotient, $\dfrac{u}{v}$, then $\dfrac{dy}{dx} = \left(v\dfrac{du}{dx} - u\dfrac{dv}{dx}\right)\Big/v^2$.

$$\int u \frac{dv}{dx}\, dx = [uv] - \int v \frac{du}{dx}\, dx.$$

Trigonometric solution of triangles

RIGHT-ANGLED TRIANGLES (RIGHT ANGLE AT C)

$\sin A = \dfrac{a}{c}$. $\cos B = \dfrac{a}{c}$. $b = \sqrt{c^2 - a^2} = \sqrt{(c + a)(c - a)}$

Area $= \dfrac{a}{2}\sqrt{c^2 - a^2} = \dfrac{ab}{2} = \dfrac{a^2 \cot A}{2} = \dfrac{b^2 \tan A}{2}$. $\tan A = \dfrac{a}{b}$.

$B = 90° - A$, $b = a \cot A$, $c = \dfrac{a}{\sin A}$ $c = \sqrt{a^2 + b^2}$. covers $A = \dfrac{b - a}{b}$

$a = c \sin A$, $b = c \cos A$, Area $= \dfrac{c^2 \sin A \cos A}{2}$, covers $A = \dfrac{c - b}{c}$

TRIANGLES

$\sin \tfrac{1}{2} A = \sqrt{\dfrac{(s - b)(s - c)}{bc}}$; $\cos \tfrac{1}{2} A = \dfrac{s(s - a)}{bc}$; $s = \dfrac{a + b + c}{2}$

$\tan \tfrac{1}{2} A = \sqrt{\dfrac{(s - b)(s - c)}{s(s - a)}}$, similar values for Angles B and C

Area $= \sqrt{s(s - a)(s - b)(s - c)} = \tfrac{1}{2} ab \sin C = \dfrac{a^2 \sin B \sin C}{2 \sin A}$

$b = \dfrac{a \sin B}{\sin A}$; $c = \dfrac{a \sin C}{\sin A} = \dfrac{a \sin (A + B)}{\sin A} = \sqrt{a^2 + b^2 - 2ab \cos C}$

$\tan A = \dfrac{a \sin C}{b - a \cos C}$; $\tan \tfrac{1}{2}(A - B) = \dfrac{a - b}{a + b} \cot \tfrac{1}{2} C$

$a^2 = b^2 + c^2 - 2bc \cos A$; $b^2 = a^2 + c^2 - 2ac \cos B$; $c^2 = a^2 + b^2 - 2ab \cos C$

Fourier transforms

Among other applications, these are used for converting from the time domain to the frequency domain.

Basic formulae:

$$\int_{-\infty}^{\infty} U(f) \exp(j2\pi ft)df = u(t) \risingdotseq U(f) = \int_{-\infty}^{\infty} u(t) \exp(-j2\pi ft)dt.$$

Change of sign and complex conjugates:

$$u(-t) \risingdotseq U(-f),\ u^*(t) \risingdotseq U^*(-f).$$

Time and frequency shifts (τ and ϕ constant):

$$u(t-\tau) \risingdotseq U(f) \exp(-j2\pi f\tau),\ \exp(j2\pi\phi t)u(t) \risingdotseq U(f-\phi).$$

Scaling (T constant):

$$u(t/T) \risingdotseq TU(fT).$$

Products and convolutions:

$$u(t)\dagger v(t) \risingdotseq U(f)V(f),\ u(t)v(t) \risingdotseq U(f)\dagger V(f).$$

Differentiation:

$$u'(t) \risingdotseq j2\pi fU(f), -j2\pi tu(t) \risingdotseq U'(f),$$
$$\partial u(t,\alpha)/\partial\alpha \risingdotseq \partial U\text{-}f,\alpha)/\partial\alpha.$$

Integration ($U(0) = 0$, a and b real constants):

$$\int_{-\infty}^{t} u(\tau)d\tau \risingdotseq U(f)/j2\pi f,\ \int_{a}^{b} v(t,\alpha)d\alpha \risingdotseq \int_{a}^{b} V(f,\alpha)d\alpha.$$

Interchange of functions:

$$U(t) \risingdotseq u(-f).$$

Dirac delta functions:

$$\delta(t) \risingdotseq 1,\ \exp(j2\pi f_0 t) \risingdotseq \delta(f-f_0).$$

Rect(t) (unit length, unit amplitude pulse, centred on $t = 0$):

$$\text{rect}(t) \risingdotseq \sin \pi f/\pi f.$$

Gaussian distribution:

$$\exp(-\pi t^2) \risingdotseq \exp(-\pi f^2).$$

Repeated and impulse (delta function) sampled waveforms:

$$\sum_{-\infty}^{\infty} u(t-nT) \risingdotseq (1/T)\ U(f) \sum_{-\infty}^{\infty} \delta(f-n/T);$$
$$u(t)\sum_{-\infty}^{\infty} \delta(t-nT) \risingdotseq (1/T) \sum_{-\infty}^{\infty} U(f-n/T).$$

Parseval's lemma:

$$\int_{-\infty}^{\infty} u(t)v^*(t)dt = \int_{-\infty}^{\infty} U(f)V^*(f)df,\ \int_{-\infty}^{\infty} |u(t)|^2 dt = \int_{-\infty}^{\infty} |U()|^2 d_j.$$

RELATION BETWEEN DECIBELS, CURRENT AND VOLTAGE RATIO, AND POWER RATIO

Table 1.16

$$\text{dB} = 10 \log \frac{P_1}{P_2} = 20 \log \frac{V_1}{V_2} = 20 \log \frac{I_1}{I_2}.$$

dB	I_1/I_2 or V_1/V_2	I_2/I_1 or V_2/V_1	P_1/P_2	P_2/P_1	*dB*	I_1/I_2 or V_1/V_2	I_2/I_1 or V_2/V_1	P_1/P_2	P_2/P_1
0.1	1.012	.989	1.023	.977	15.0	5.62	.178	31·6	.031 6
0.2	1.023	.977	1.047	.955	15.5	5.96	.168	35·5	.028 2
0.3	1.035	.966	1.072	.933	16.0	6.31	.158	39.8	.025 1
0.4	1.047	.955	1.096	.912	16.5	6.68	.150	44.7	.022 4
0.5	1.059	.944	1.122	.891	17.0	7.08	.141	50.1	.020 0
0.6	1.072	.933	1.148	.871	17.5	7.50	.133	56.2	.017 8
0.7	1.084	.923	1.175	.851	18.0	7.94	.126	63.1	.015 8
0.8	1.096	.912	1.202	.832	18.5	8.41	.119	70.8	.014 1
0.9	1.109	.902	1.230	.813	19.0	8.91	.112	79.4	.012 6
1.0	1.122	.891	1.259	.794	19.5	9.44	.106	89.1	.011 2
1.1	1.135	.881	1.288	.776	20.0	10.00	.100 0	100	.010 0
1.2	1.148	.871	1.318	.759	20.5	10.59	.094 4	112	.008 91
1.3	1.162	.861	1.349	.741	21.0	11.22	.089 1	126	.007 94
1.4	1.175	.851	1.380	.724	21.5	11.88	.084 1	141	.007 08
1.5	1.188	.841	1.413	.708	22.0	12.59	.079 4	158	.006 31
1.6	1.202	.832	1.445	.692	22.5	13.34	.075 0	178	.005 62
1.7	1.216	.822	1·479	.676	23.0	14.13	.070 8	200	.005 01
1.8	1.230	.813	1.514	.661	23.5	14.96	.066 8	224	.004 47
1.9	1.245	.804	1.549	.645	24.0	15.85	.063 1	251	.003 98
2.0	1.259	.794	1.585	.631	24.5	16.79	.059 6	282	.003 55
2.5	1.334	.750	1.778	.562	25.0	17.78	.056 2	316	.003 16
3.0	1.413	.708	1.995	.501	25.5	18.84	.053 1	355	.002 82
3.5	1.496	.668	2.24	.447	26.0	19.95	.050 1	398	.002 51
4.0	1.585	.631	2.51	.398	26.5	21.1	.047 3	447	.002 24
4.5	1.679	.596	2.82	.355	27.0	22.4	.044 7	501	.002 00
5.0	1.778	.562	3.16	.316	27.5	23.7	.042 2	562	.001 78
5.5	1.884	.531	3.55	.282	28.0	25.1	.039 8	631	.001 58
6.0	1.995	.501	3.98	.251	28.5	26.6	.037 6	708	.001 41
6.5	2.11	.473	4.47	.224	29.0	28.2	.035 5	794	.001 26
7.0	2.24	.447	5.01	.200	29.5	29.8	.033 5	891	.001 12
7.5	2.37	.422	5.62	.178	30.0	31.6	.031 6	1 000	.001 00
8.0	2.51	.398	6.31	.158	31.0	35.5	.028 2	1 260	7.94×10^{-4}
8.5	2.66	.376	7.08	.141	32.0	39.8	.025 1	1 580	6.31×10^{-4}
9.0	2.82	.355	7.94	.126	33.0	44.7	.022 4	2 000	5.01×10^{-4}
9.5	2.98	.335	8.91	.112	34.0	50.1	.020 0	2 510	3.98×10^{-4}
10.0	3.16	.316	10.00	.100	35.0	56.2	.017 8		3.16×10^{-4}
10.5	3.35	.298	11.2	.089 1	36.0	63.1	.015 8	3 980	2.51×10^{-4}
11.0	3.55	.282	12.6	.079 4	37.0	70.8	.014 1	5 010	2.00×10^{-4}

Table 1.16—*continued*

dB	I_1/I_2 or V_1/V_2	I_2/I_1 or V_2/V_1	P_1/P_2	P_2/P_1	dB	I_1/I_2 or V_1/V_2	I_2/I_1 or V_2/V_1	P_1/P_2	P_2/P_1
11.5	3.76	.266	14.1	.070 8	38.0	79.4	.012 6	6 310	1.58×10^{-4}
12.0	3.98	.251	15.8	.063 1	39.0	89.1	.011 2	7 940	1.26×10^{-4}
12.5	4.22	.237	17.8	.056 2	40.0	100.0	.010 0	1.00×10^{4}	1.00×10^{-4}
13.0	4.47	.224	20.0	.050 1	50.0	316.0	.003 16	1.00×10^{5}	1.00×10^{-5}
13.5	4.73	.211	22.4	.044 7	60.0	1 000.0	.001 00	1.00×10^{6}	1.00×10^{-6}
14.0	5.01	.200	25.1	.039 8	70.0	3 160.0	.000 32	1.00×10^{7}	1.00×10^{-7}
14.5	5.31	.188	28.2	.035 5	80.0	10 000.0	.000 10	1.00×10^{8}	1.00×10^{-8}

METRIC AND DECIMAL EQUIVALENT OF FRACTIONS OF AN INCH

Table 1.17

Inches		mm	Inches		mm
$\frac{1}{64}$	*.016*	0.396	$\frac{33}{64}$	*.516*	13.096
$\frac{1}{32}$	*.031*	0.793	$\frac{17}{32}$	*.531*	13.492
$\frac{3}{64}$	*.047*	1.190	$\frac{35}{64}$	*.547*	13.890
$\frac{1}{16}$	*.063*	1.587	$\frac{9}{16}$	*.563*	14.287
$\frac{5}{64}$	*.078*	1.984	$\frac{37}{64}$	*.578*	14.683
$\frac{3}{32}$	*.094*	2.381	$\frac{19}{32}$	*.594*	15.080
$\frac{7}{64}$	*.109*	2.778	$\frac{39}{64}$	*.609*	15.477
$\frac{1}{8}$	*.125*	3.175	$\frac{5}{8}$	*.625*	15.875
$\frac{9}{64}$	*.141*	3.571	$\frac{41}{64}$	*.641*	16.271
$\frac{5}{32}$	*.156*	3.968	$\frac{21}{32}$	*.656*	16.667
$\frac{11}{64}$	*.172*	4.365	$\frac{43}{64}$	*.672*	17.064
$\frac{3}{16}$	*.188*	4.762	$\frac{11}{16}$	*.688*	17.462
$\frac{13}{64}$	*.203*	5.159	$\frac{45}{64}$	*.703*	17.858
$\frac{7}{32}$	*.219*	5.556	$\frac{23}{32}$	*.719*	18.255
$\frac{15}{64}$	*.234*	5.955	$\frac{47}{64}$	*.734*	18.652
$\frac{1}{4}$	*.25*	6.350	$\frac{3}{4}$	*.75*	19.050
$\frac{17}{64}$	*.266*	6.746	$\frac{49}{64}$	*.766*	19.446
$\frac{9}{32}$	*.281*	7.143	$\frac{25}{32}$	*.781*	19.842
$\frac{19}{64}$	*.297*	7.540	$\frac{51}{64}$	*.797*	20.239
$\frac{5}{16}$	*.313*	7.937	$\frac{13}{16}$	*.813*	20.637
$\frac{21}{64}$	*.328*	8.334	$\frac{53}{64}$	*.828*	21.033
$\frac{11}{32}$	*.344*	8.730	$\frac{27}{32}$	*.844*	21.429
$\frac{23}{64}$	*.359*	9.127	$\frac{55}{64}$	*.859*	21.827
$\frac{3}{8}$	*.375*	9.525	$\frac{7}{8}$	*.875*	22.225
$\frac{25}{64}$	*.391*	9.921	$\frac{57}{64}$	*.891*	22.621
$\frac{13}{32}$	*.406*	10.318	$\frac{29}{32}$	*.906*	23.017
$\frac{27}{64}$	*.422*	10.715	$\frac{59}{64}$	*.922*	23.414
$\frac{7}{16}$	*.438*	11.112	$\frac{15}{16}$	*.938*	23.812
$\frac{29}{64}$	*.453*	11.508	$\frac{61}{64}$	*.953*	24.208
$\frac{15}{32}$	*.469*	11.905	$\frac{31}{32}$	*.969*	24.604
$\frac{31}{64}$	*.484*	12.302	$\frac{63}{64}$	*.984*	25.002
$\frac{1}{2}$	*.5*	12.700			

GREEK ALPHABET AND SYMBOLS

alpha	A	α	angles, coefficients, area
beta	B	β	angles, coefficients
gamma	Γ	γ	specific gravity
delta	Δ	δ	density, increment, finite difference operator
epsilon	E	ϵ	napierian logarithm, linear strain, permittivity, error, small quantity
zeta	Z	ζ	co-ordinates, coefficients, (Cap) impedance
eta	H	η	magnetic field strength, efficiency
theta	Θ	θ	angular displacement, time
iota	I	ι	inertia
kappa	K	κ	bulk modulus, magnetic susceptibility
lambda	Λ	λ	permeance, conductivity, wavelength
mu	M	μ	bending moment, coefficient of friction, permeability
nu	N	ν	kinematic viscosity, frequency, reluctivity
xi	Ξ	ξ	output coefficient
omicron	O	o	
pi	Π	π	circumference ÷ diameter
rho	P	ρ	specific resistance
sigma	Σ	σ	(Cap) summation, radar cross section, standard deviation
tau	T	τ	time constant, pulse length
upsilon	Y	υ	
phi	Φ	ϕ	flux, phase
chi	X	χ	(Cap) Reactance
psi	Ψ	ψ	angles
omega	Ω	ω	angular velocity, ohms

2 HISTORY OF ELECTRONICS

HISTORY OF ELECTRONICS 2

2 HISTORY OF ELECTRONICS

In 1897 Sir J. J. Thomson confirmed the existence of the electron as a negative charge of electricity; the word *electronic* was probably coined about this time and was certainly used by Professor J. A. (later, Sir Ambrose) Fleming in 1902. The expression *electronics*, however, found no great favour in Britain for many years. It began to appear about 1940 as an import from the U.S.A. but did not come into general acceptance until after World War II.

As everything in the known universe is electronic in character, the expression is not a felicitous one to apply to a section of electrical science, in which context it has had to acquire a special meaning. The definition of *electronics* (see Section 1) in practice needs considerable qualification; for instance, a Martian observer, with no background of common usage to draw upon, would be perplexed to find that magnetohydrodynamic (m.h.d.) power generation, radioactivity, fluorescent lamps and neon signs (all of which fulfil the criteria) are not generally regarded as electronic devices. He would also be confused to discover that the overwhelming majority of the components used in electronic engineering—resistors, capacitors, inductors, etc.—are wholly electrical devices.

ELECTRONICS IN THE NINETEENTH CENTURY

Although electronics is very much a twentieth-century activity, the fundamental building blocks, namely photoelectric, thermionic, semiconductor and gas discharge phenomena, were entirely nineteenth-century discoveries. They were, however, during that period, mainly scientific curiosities, having few applications outside of the laboratory, and it has been the role of the present century to develop them into commercial devices, equipments and systems.

Photoelectricity

Probably the earliest electronic device on record (although not recognised as such at the time) was the photovoltaic cell produced by Becquerel in 1839. The discovery of the photoconductive behaviour of selenium occurred in 1861 but did not receive much attention until 1873 when Willoughby-Smith read a paper to the IEE on the subject. The first practical application came in 1878 when Dr. Graham Alexander Bell and Sumner Tainter used a selenium cell in their Photophone (telephony via light waves). Hertz was the first to observe the photoemissive effect (1887) and confirmation was provided by Wiedemann and Ebert. Hallwachs and also Elster and Geitel studied the effect intensively; in 1890 the last-mentioned co-workers produced the prototype of the modern photoemissive tube.

Thermionics

The first step towards the thermionic valve came in 1873 when Guthrie noted that a heated metal ball, when placed near a charged electroscope, discharged it. In 1880 Elster and Geitel enclosed a lamp filament and a metal plate in a vacuum and found that a tiny current flowed from filament to plate across the vacuum. In the early 1880s both

Edison and Professor J. A. Fleming (who was at that time doing research work for Edison) were investigating the premature blackening of the glass envelopes of the Edison lamps. In an experimental version a metal plate having an external connection had been inserted into the lamp; Edison and Fleming both noted that when the plate was given a positive potential with respect to the filament a direct current flowed across the vacuum and through the external circuit; when the polarity was reversed, no current flowed. Neither man investigated the phenomenon at the time, although it was recorded and passed into history as the Edison Effect. The diode which they had unwittingly manufactured remained unrecognised as such for another 20 years.

Gas discharges

From the 1850s onward the behaviour of high voltage electricity through gases was the subject of considerable research, following Plucker's discovery that when such a voltage was connected between two separated electrodes in a near-evacuated glass cylinder a purple glow could be seen on the wall of the tube. Hittorf, Goldstein and Sir William Crookes investigated further—the latter with the familiar Crookes tube with its *Maltese Cross* electrode—and established that the glow was produced by the impact of invisible rays upon the glass and that these rays travelled from the negative electrode to the positive. Crookes believed that these *kathode rays* as they were called were negatively charged particles; this was confirmed by Hallwachs and (in 1897) by Sir. J. J. Thomson. The particles were termed *electrons*.

Research now divided. Tesla and others concentrated on experiments with gases to obtain a maximum overall glow for lighting purposes (the modern fluorescent lamp is one outcome), while some researchers studied the rays themselves, finding that they could cause certain materials to fluoresce. In 1897 Braun developed a much improved cathode-ray tube in which a mica screen coated with phosphors was mounted on the end wall; in 1902 he added deflection coils (Fleming had shown that the beam could be moved by varying the current through such coils) and thus produced the forerunner of the modern oscilloscope. To this, Wehnelt contributed two notable improvements, namely, a cylindrical control grid and also an oxide-coated filament to provide a copious supply of electrons. By so doing the cathode-ray tube was converted into a thermionic device—the first with a practical application.

X-rays

An off-shoot of the work on cathode rays was the discovery of X-rays by Professor Röntgen in 1895. These electromagnetic waves have a wavelength (10^{-1} nm to 10^{-3} nm), much shorter than that of visible light. X-rays are generated by focusing high-velocity electrons emitted from a cathode on to a target (anode) which is commonly of tungsten or rhenium.

The penetrating effect of X-rays and their ability to activate a photographic emulsion have been used in medical diagnosis since the early days; more recently these same properties have been used for non-destructive testing of materials (castings, weldings, etc.). All X-rays have an ionising effect and are biologically destructive; these properties are used in medical therapy and for sterilisation.

With the availability of particle accelerators and radio-active sources of high energy particles, the use of conventional X-ray tubes for therapy has now declined.

Semiconductors

In 1874 Braun discovered the semiconductor effect in some metallic sulphides, notably lead sulphide and iron sulphide. The first application of this effect was made fortuitously by Professor D. E. Hughes in 1878–79. Hughes discovered a means of generating electromagnetic waves (although he knew nothing of Clerk Maxwell's work) and devised various detectors, most of which were re-invented many years after. One of these was an iron needle in contact with a globule of mercury; in this, almost certainly, an oxide film on the mercury provided genuine rectification. Although Hughes anticipated Hertz in the generation of radio waves by five years and others by 10 years in the means of detecting them, he was discouraged from publishing and so his achievements went unrecognised for many years.

In 1835 Munk had discovered that certain metallic powders changed in electrical conductivity when an electric spark occurred near them. Over the years the phenomenon was resurrected at intervals; by S. A. Varley in 1866; by Professor Calzecchi-Onesti in 1884 and by Branly in 1890. No great attention was paid to it until Professor Minchin and Professor (later Sir Oliver) Lodge independently suggested that the effect might be caused by electromagnetic waves emanating from the spark discharge. In 1894, Lodge took a tube of iron filings as used by Branly in his experiments and demonstrated that it could be used as a detector of Hertzian waves. Lodge called the device a coherer because, under the influence of the waves, the filings clung together and had to be tapped before the device could receive another wavetrain. (It was not a rectifier but a relay which shortcircuited on encountering radio-frequency signals.) The coherer became widely used in early wireless telegraphy but it is perhaps of interest to note that when, in 1901, Marconi received the first signals across the Atlantic he used an iron–mercury globule rectifier which had been re-invented in 1899 by Professor Tommasina. This detector was variously known as the *Castelli*, the *Solari* or the *Italian Navy* self-restoring coherer.

ELECTRONICS IN THE TWENTIETH CENTURY

Paradoxically, radio communication, from which electronics sprang, was not at the outset recognisably electronic in character. When, in 1895–96, Marconi developed the first practical system of wireless telegraphy, his apparatus, with the possible exception of the coherer, was basically electro-mechanical. Others soon produced rival systems; some, like Marconi's, were spark-derived, while others, a little later, employed arc or r.f. alternator methods and although both spark and arc fall within the specification of the official definition of electronics, they contained no elements which would be familiar to today's electronics engineers. All three approaches—spark, arc and r.f. alternator—entirely monopolised long-distance radio transmissions until the 1920s.

The birth of thermionic electronics occurred in 1904 when Fleming (then technical consultant to Marconi) found one of the old experimental Edison lamps in a cupboard and had the idea of trying it as a rectifier of radio signals. It worked. Fleming made an improved version which worked even better and then patented it on behalf of the Marconi Company.

Marconi was not unduly impressed. His magnetic detector, developed two years earlier, would do everything the diode could do, and was more robust (in fact, Fleming's valve never superseded it). Then in 1907 Dr. Lee de Forest announced the Audion, a valve with a third electrode (nicknamed the *grid-iron*, later abbreviated to *grid*). This valve, claimed its inventor, could amplify weak signals. The Marconi Company began litigation against de Forest, claiming that the Fleming diode was the master-invention and that the grid was merely an appendage; the initial verdict was for the Marconi Company, but this was overset in another court—and so it went on, with action and counter-action. Not until the 1920s was the issue settled, and then only by compromise; thus, in the early years, valve development was stultified because no one had a clear right to manufacture the triode.

At the outset, the principles of action of the triode were imperfectly understood; it was erratic in performance—and performance at best was poor. Gradually however, improvements were made, particularly in high vacuum techniques, and in 1913 came the great discovery that the triode could be made to oscillate. H. J. Round of the Marconi Company demonstrated valved radio telephony in that year but A. Meissner of Germany was the first to patent the valve oscillator, just a little ahead of several other workers.

World War I gave a tremendous impetus to valve development; by 1919 direction finding equipment and relatively low powered radio telephony transmitters were in considerable use by the armed services, while for reception the crystal detector (first used by Dunwoodie in 1906) was commonly used in conjunction with one or two stages of valve amplification, or superseded altogether by the triode detector. Higher transmitting powers were being achieved by connecting valves in parallel and in 1915 a 300-valve transmitter at Arlington, U.S.A., succeeded in spanning the Atlantic by radio-telephony. Progress was such that by 1919 a British transmitter did this using only three main valves.

Although it was not realised at the time, the end of the war was a turning point in electronics history. Hitherto, virtually the only applications of electronic devices had been in radio communication, but now an era of diversification was about to begin.

Radio communications

In November 1921 a newly-designed 100 kW valved transmitter at the Marconi station at Caernarvon established contact with Australia and in doing so relegated existing spark, arc and r.f. alternator systems to obsolescence. A further revolution came in 1924 with the experimental introduction of the Marconi–Franklin h.f. beam system, which, when fully developed, was to become the backbone of long distance point-to-point radio communication and remained so until the comparatively recent advent of satellite techniques. (One of Franklin's many inventions in connection with the beam system was the concentric feeder, which, as the coaxial cable, is ubiquitous today.)

The success of the h.f. system encouraged research into the even higher frequencies as potential message-carriers and this led in turn to the development of valves which would oscillate at these frequencies. World War II brought a vast acceleration in this and all other branches of electronics research and the knowledge thus gained was put to commercial usage in the post-war period—the klystron and travelling-wave-tube, for example, were key devices in opening world markets in v.h.f., u.h.f. and s.h.f. point-to-point multichannel links. Other new communication methods introduced include the ionospheric and tropospheric scatter systems, but by far the most significant innovation of recent years has been communication via satellite. Various factors combined to make this possible; the inventions of the transistor, the microcircuit and the printed board; the rocket techniques developed by Germany during the war and subsequently improved upon by the U.S.A. and the U.S.S.R. during the height of the 'cold war'—all these made possible the evolution of ultra light-weight compact equipments and provided the capability of putting the satellites into orbit.

On 10 July 1962 the Telstar satellite clearly demonstrated the advantages of the new approach and the subsequent introduction of geostationary satellites emancipated long-distance circuits from the vagaries of ionospheric reflection, provided multichannel working on a round-the-clock basis and have proved themselves to be economically viable.

While the electronics in the satellites are a hybrid of reasonably conventional thermionic (usually travelling wave tube) and solid-state microcircuit techniques, special wideband low-noise amplifiers are necessary at the earth terminals. At first, masers (Microwave Amplification by the Stimulated Emission of Radiation) were used but today the trend is towards parametric amplifiers. The maser concept was introduced by Townes in 1951, while the history of the parametric amplifier dates back to Lord Rayleigh (and even, perhaps, to Faraday). The modern approach to a practical device is due to the work of Suhl and Weiss in the late 1950s but workable *paramps* were first made by Uhler and others in 1958 and considerable development has taken place since. Even at room temperature the parametric amplifier has a low noise performance, but the use of refrigeration gives a noise figure comparable to that of a maser, but with a much greater bandwidth.

Sound broadcasting

In 1916 David Sarnoff of the American Marconi Company (later the Radio Corporation of America) put forward a plan for entertainment broadcasting, but this was shelved. After World War I, various radio companies developed new telephony transmitters, foreseeing a peacetime market for message-carrying by this means, and it became the custom for their engineers to relieve the monotony of reciting lists into a microphone (as a means of range testing) by playing the occasional gramophone record. To everyone's surprise, requests began to come in from amateurs (many of whom had been in signals units during the war), asking for more music. The demand grew, and in this fashion broadcasting began—a classic instance of serendipity.

Broadcasting to the general public began in the U.S.A. in 1920 (station KDKA) but in Britain, despite successful concerts from the Marconi Works at Chelmsford in 1920 (Melba and Melchior were among the artistes) the Post Office vetoed further transmissions on the grounds that they might interfere with legitimate services. Only after a petition by the amateur wireless societies was the embargo lifted and in January 1922 permission was granted to the Marconi Company to broadcast a concert for a halfhour once a week. On 14 February the Marconi station 2MT started broadcasting under these terms. A second permit followed and in May 1922, station 2LO at Marconi House, London, was brought into service. Its success, and that of 2MT, encouraged 23 other radio manufacturers to apply for permits—a circumstance that would clearly lead to chaos. As a solution a consortium of radio manufacturers was formed, named the British Broadcasting Company, which constituted a single broadcasting authority. This organisation took over responsibility for 2LO on 14 November 1922, after which other stations were quickly inaugurated in other parts of the country. As from 31 December 1926 the BBC was reconstituted to become a Corporation and the radio manufacturing companies ceased to be directly responsible for broadcasting in the U.K.

The impact of sound broadcasting was tremendous. Such was the public demand for receivers that two new sub-industries were created almost overnight; one catering for factory-built domestic receivers and the other for components for home construction. Aided by a spate of periodicals, home construction became a major British hobby; at first the simple crystal set was highly popular but was gradually superseded by battery-driven valve receivers. By the 1930s, the employment of mass production techniques in factory receivers had reduced costs to a figure competitive with home construction; circuits were becoming more complex (the supersonic heterodyne or *superhet* principle, invented by Armstrong in 1919 was coming into favour) and these factors, in conjunction with the arrival of the *all-mains* receiver, caused the wane of home construction. By this time, however, the domestic receiver and components industries had stabilised and sound broadcasting had become an integral part of everyday life.

The broadcasting boom gave great impetus to research, particularly in valve technology, and multi-electrode thermionic devices were rapidly introduced in the 1920s and 1930s. Applications other than for broadcasting soon became apparent and a hiving-off process began, which has continued to the present day. The gramophone industry for example, which had been dealt a severe blow by sound broadcasting, derived a new lease of life from electronic sound recording and reproduction and today is a major industry in its own right. Audio amplifier techniques were applied in the film industry, resulting in the *talking picture*. Public address techniques sprang from the same source. Another sub-industry, devoted to test equipment and instrumentation, was emerging; thermionic valve amplifiers were becoming more and more commonplace in general research laboratories and in advanced medical centres. Line communication systems were putting amplifiers to good use; in short, electronic devices were emerging from the confines of radio communication and were becoming ubiquitous tools.

On the transmission side of sound broadcasting, great strides were also being made. One important development was frequency modulation (f.m.). F.M. has its historical roots in 1902 but there was a long period of neglect until Westinghouse investigated it briefly in the 1920s. It is to Armstrong, however, that the credit must go, not only for its development but his years of perseverance to get the technique accepted in the U.S.A. It was not until 1950 that the BBC's first f.m. broadcasting station (at Wrotham Hill, Kent) was completed, after which a network of f.m. stations was provided throughout the U.K.

Television broadcasting

This also owes its inception to sound broadcasting although its concept goes back much earlier even than wireless telegraphy. As far back as 1847 practical systems of transmitting still pictures were in being and the wide attention given to the photoconductive properties of selenium in the 1870s brought renewed interest in trying to devise means of transmitting pictures having movement. Among the experimenters were Carey, Ayrton and Perry, de Paiva, Leblanc, Senlac and Nipkow; none succeeded but all made contributions to the state of the art. Indeed, Nipkow's ideas of 1884 would have worked, had he possessed thermionic amplifiers and a better photocell. The word *television* was coined by Perskyi in 1900.

In 1907 Professor Rosing, a Russian scientist, succeeded in transmitting images of geometrical shapes, but any attempt at movement produced picture break-up. Rosing's transmitter embodied mirror drum scanning (invented by Professor Weiller in 1889) and, at the receiver, electromagnetic scanning and a cathode-ray tube display.

A. A. Campbell Swinton in 1908 and again in 1911 drew up a remarkable blueprint for the future; his proposed system used a photomosaic form of camera picture tube at the transmitting end and a cathode-ray tube display at the receiver. As far as is known the equipment was never built, and the brilliant concept, upon which the television system of today is based, was all but forgotten. It was in 1911 also that it was first suggested (by A. Sinding-Larsen) that radio waves might be used as a carrier for picture impulses.

J. L. Baird in 1926 was the first to demonstrate true television pictures. His apparatus was essentially Nipkow's rotating disc system of 1884, with an improved photocell and the addition of amplifiers. In quick succession Baird also demonstrated Noctovision (televising in darkness), colour and stereoscopic television. The pictures were small, of low definition and had no entertainment value; nevertheless his successes encouraged many others to experiment with mechanical systems. In 1930 the BBC began experimental transmissions using the Baird apparatus.

In the U.S.A. in the 1920s two men began working (quite independently) on all-electronic scanning; Farnsworth on his 'Image Dissector' and Zworykin (a former pupil of Rosing) on his 'Iconoscope'. The practical difficulties in both devices were enormous and even by 1933, when Zworykin was able to demonstrate a 240-line picture, large-scale manufacture of his Iconoscope was still not possible. It was, however, the pattern for the future as it possessed the important feature of energy-storage between scans, which neither the Image Dissector nor the various mechanical systems had. Television owes a great debt to Zworykin, both for his camera tube and for the notable improvements he made to the cathode-ray tube.

In England in the early 1930s Baird in his laboratory had turned his attention to high-definition television, but still relied on mechanical scanning. However, by 1935 he faced strong competition from Marconi-EMI, for a brilliant Electric and Musical Industries team led by Shoenberg had developed the Emitron, a camera tube similar in principle to Zworykin's and this, allied to Marconi expertise in wideband modulation and v.h.f. transmission, provided a complete high-definition system.

In 1935 the Selsdon Committee recommended that the U.K. should have a high-definition service and by 1936 the BBC's first station was built at Alexandra Palace, London. A public trial of both the Baird and Marconi-EMI systems was arranged and the station began official programme service on 12 November 1936, each system taking its turn. By February 1937 the Marconi-EMI system was declared the better and thereafter carried the service exclusively. In September 1939 the Alexandra Palace station closed for the duration of the war; it did not reopen until 7 June 1946, after which the BBC pushed rapidly ahead with plans to provide television coverage throughout the U.K.

On 30 July 1954 the Independent Television Authority (now Independent Broadcasting Authority) was formed and on 22 September 1955 its London station came into service, to be followed by a U.K. network. In 1964 the BBC opened its second network, BBC2, operating a 625-lines at u.h.f. The IBA now also have a u.h.f. network. Colour television came into official service in Britain in the summer of 1967, using the PAL system, a German development of the American NTSC system which had been in public service in the U.S.A. since 1954.

Among the many important technical developments of recent years is the electronic field-store converter, first developed by the BBC in 1967 to permit the exchange of television programmes between countries using systems having different line standards and field frequencies. Another notable advance was the introduction of video recording on tape. The first audio tape recorder (operating with steel wire or ribbon) was exhibited by Poulsen at the Paris Exhibition of 1900 but remained virtually undeveloped for many years. The technique was resurrected in the 1920s. World War II saw further considerable developments, with magnetic oxide impregnated paper and plastic tapes coming into use. After the war—and particularly after the invention of the transistor—audio tape recorders for domestic and commercial purposes came into wide usage. The problem of recording video signals on tape is a much more difficult one and it was not until 1956 that a really satisfactory process was evolved (by the Ampex Corporation in the U.S.A.). Since then the process, with its facility of instant playback, has become general in television programme production, although film is also widely used.

Electronic navigational aids

Professor Elihu Thomson took out the first patent for directional reception of radio waves in 1899, but the basic principles of loop reception were known to Hertz. One of the most successful of the early direction-finding systems was that of Bellini and Tosi; just before World War I, H. J. Round developed a valved form of Bellini-Tosi equipment and this became the war's *secret weapon* (it was responsible for the battle of Jutland and the destruction of Zeppelins). Shipborne and airborne direction finders soon evolved.

Today, ships and aircraft carry a considerable complement of electronic navigational aids in addition to the normal communications equipment. A modern aircraft for example, may carry electronic instrumentation for automatic take-off and landing, in addition to automatic direction finders (radio compasses), radio altimeters, weather radar, doppler navigators and other devices. A military aircraft tends to be packed with electronics for communication, navigation, and offensive and defensive action.

Radar

This term did not come into use in the U.K. until well into the World War II period. Of American origin, it is an acronym (RAdio Detection And Ranging). The former British terms were RDF (Radio Direction Finding), or Radiolocation.

In 1888, Hertz found that radio waves could be reflected by objects in their path. Tesla, in 1900, suggested that this attribute could be used for detecting moving objects, while in 1904 Hülsmeyer patented a primitive form of radar. Marconi, in 1922, suggested means by which ultra-short radio waves could be used to detect ships but did no work on it at that time.

In 1924, Appleton and Barnett used similar techniques to those later employed for c.w. radar for their ionospheric sounding work and in 1925 Breit and Tuve developed a pulsed system for similar research. In Britain the first known proposal for a radar system came from Butement and Pollard in January 1931. Although their equipment produced short-range results the work was abandoned for lack of government support.

In 1935 Watson-Watt (later Sir Robert) suggested that radio waves might be used to detect aircraft at a distance and outlined a means of doing so. Intensive research began and by 1939 Britain possessed a defensive chain of highly secret RDF stations, while gun-laying and airborne equipments were also being developed. Contrary to popular belief, other countries, notably Germany, France and the U.S.A. possessed radar prior to the war but lacked Britain's highly developed control system. In 1941 the development of the resonant cavity magnetron by Boot and Randall gave the Allies a powerful new weapon in high-power centimetric radar. Later in the war the proximity fuse and radar-aligned guns successfully contained the German V1 flying bomb menace. Shipping and aircraft also made extensive use of radar for attack and defence.

In the post-war years highly complex radar installations have come into use as air traffic control aids at major airports. Shipborne radar is commonplace, while meteorological radar now fulfils a valuable role in weather forecasting. Speed-checking equipment, using radar principles, is used by many police forces.

Transistors

The semiconductor oscillator predates the discovery of the thermionic valve oscillator, for in 1909–11 Dr. Eccles in the course of his work on semiconductors produced an oscillating crystal. No significant further progress was made however until the 1920s when Scott-Taggart, Lossev and Podliasky were among those who revived interest in the matter.

The foundations of modern semiconductor techniques were laid by Planck's quantum theory of 1901 and by the papers of others over the years, particularly those of Schrödinger and Heisenberg in 1926–27. Shottky's work on semiconductor rectifier theory (1940) provided a further impetus. Meanwhile, however, Professor Lilienfeld, in 1930, had patented a semiconductor device similar to an insulated gate field-effect transistor. The device did not work, largely because of the inadequacy of the semiconductor material Lilienfeld had at his disposal. It was not until 1947 that Bardeen, Brattain and Shockley of Bell Telephone Laboratories produced the point-contact transistor (the word, a telescoping of *transfer resistor* was coined by Pierce, of Bell Telephones). In 1951 Shockley patented the junction transistor; in 1959 came another vital step when the planar technique was developed.

The subsequent years have seen monumental further improvements in the technology and performance of the transistor, which has now ousted the thermionic valve from most of its former applications in the lower power categories.

Printed circuits and microcircuits

During World War II the need for lighter, more compact, equipments led to research into means of achieving these aims. One approach was to sub-miniaturise components, while retaining their conventional forms; another was to use thin film-techniques (in rudimentary form) to manufacture certain components, while a third was to replace the conventional wiring by printing metallised paths on to insulating boards. Both thin-film and printed board techniques continued in development after the war but the break-through came with the invention of the transistor, for the low voltages and currents associated with it permitted the use of resistors and capacitors of much smaller physical dimensions, suitable for wiring on to printed boards, and enabled such components to be encapsulated into modules.

In the 1950s the cold war and the consequent acceleration of guided missile and space vehicle programmes by the U.S.A. and the U.S.S.R. brought the concept of producing whole circuits in one tiny block. The two main approaches to this were the thin-film techniques and the integrated semiconductor method; the latter had the advantage that not only passive components could be formed but active ones (transistors and diodes) also. At first, the small size and light weight were the prime advantages but it was soon realised that well-made microcircuits were inherently far more reliable than conventional components because of the great reduction in the number of inter-component connections.

When microcircuit techniques became available to commercial electronics, a new type of industry gradually evolved which owes much more to photolitho techniques than to conventional processes of manufacture. A further benefit accrued in that the processes are eminently suitable to mass production. Just as the transistor edged out the thermionic valve, so now is the large-scale integrated circuit approach taking over from discrete transistors and components in many areas.

Computers

The first computer was the *Difference Engine* designed by Charles Babbage in 1822. This was wholly mechanical and was never completed, but it did embody some of the features of the modern computer.

No further progress was made for over 100 years, and it was not until after World War II that the first automatic calculators (employing thermionic valves) appeared; one of these, ENIAC, embodied 18 000 valves (usually some failed every time the equipment was switched on). This computer was built at the Pennsylvania University in 1946–47.

The first computers to work under the control of a program stored in the computer itself were built at Cambridge and Manchester universities in 1949. Commercial development in the U.K. was begun by Ferranti and Elliott Brothers and other companies followed soon after. The first deliveries began in 1954. The innovations of transistors and printed circuits saw the introduction of a second generation of computers; the first British transistor-based machines began to appear in 1960. The third generation, now in production, embodies microcircuits, which provide even greater reliability and higher switching speeds.

Lasers

The laser was a logical extension of the development of the maser (Microwave Amplification by Stimulated Emission of Radiation), the 'L' standing for 'Light'. The theoretical possibility of producing coherent light was stated by Einstein in 1917 but little was done in this direction until Schawlaw and Townes suggested (in 1958) that an optical maser was theoretically possible. In 1960 Dr. Maiman demonstrated the first true laser; this used a ruby crystal and produced a pulsed output. Another laser, using trivalent uranium, was also described in that year by Sorkin and Stevenson.

In 1961 the first gas discharge laser (which provided continuous emission), was produced by Bennett, Herriott and Javan and in the following year several research laboratories in the U.S.A. announced the development of a continuous-wave laser which used gallium arsenide in semiconductor junction form.

All this work was done in the U.S.A. although several countries were soon producing lasers. Unlike the maser, the laser did not immediately find a significant application but was rather the outcome of scientific curiosity. However, applications were soon forthcoming and many others will doubtless follow; among the areas in which the device is already in use, or has potential applications, are communications, holography, radar, aviation, control systems and medical research and surgery.

Conclusion

In the preceding pages an outline has been given of some of the main stages in the development of electronics. Today, electronics is an essential part of the fabric of living, serving the community in such diverse applications as line communications, railway signalling, machine tool control, medical diagnosis, chemical analysis, traffic engineering, process control and, indeed, in almost every facet of modern life, each of which has a history in its own right.

3 GENERAL PHYSICAL BACKGROUND

GENERAL PHYSICAL BACKGROUND

3

3 GENERAL PHYSICAL BACKGROUND

PHYSICAL QUANTITIES

Engineering physics

Engineering technology involves energy associated with physical materials. The object is to employ materials—solid and fluid—to convert, transport or radiate energy. A comprehensive all-embracing theory or process is quite impossible, for energy has many forms, and materials differ profoundly in their physical nature. Many separate technologies have been built up around specific engineering systems, and for materials in a bulk macroscopic form, or in microstructure, or in molecular, atomic and sub-atomic structure. Each technology applies only to its own forms of energy and material structure, i.e., to its specific technological system.

Energy

Like 'force' or 'time' or 'mass', energy is a unifying *concept* invented to explain systematically certain important physical phenomena. Appreciation of the meaning of the concept is largely intuitive, aided by a study of 'energetic' physical systems.

Mechanical force systems possess potential energy, and tend to positions of equilibrium in which the potential energy is a minimum. When the potential energy is reduced, work is done by the system as kinetic energy of motion; when the potential energy is increased, energy is taken in from outside the system, i.e., work is done on the system. Energy is associated with gravitational, chemical, thermal, magnetic and electric systems; it can be stored, transported, radiated and converted from one form to another. In these processes it is conserved, i.e., the total quantity does not change.

Energy almost defeats precise definition, but statements such as the following aid the intuitive grasp of the meaning:

ENERGY is the capacity for doing *work*, using the word 'work' in the widest sense of 'action'.

WORK is the measure of the change of energy *state*.

STATE is the measure of the energy condition of a *system*.

SYSTEM is the ordered arrangement of related physical entities or processes, or of their *model*.

MODEL is the pictorial diagram used to describe the system, or the mathematical statement set up to describe its *behaviour*.

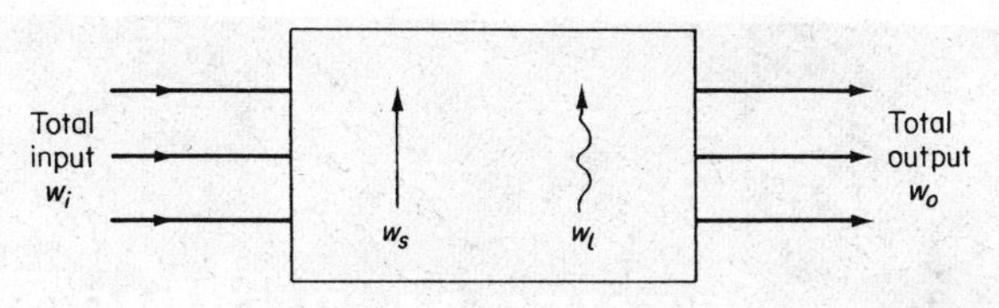

Figure 3.1 Energy balance

BEHAVIOUR is the verbal or mathematical description of the energy processes involved in changes of state. *Storage* occurs if the work done on a physical system is recoverable in the original form. *Conversion* takes place when related changes in state concern different forms of energy and if the action is reversible. *Dissipation* is an irreversible conversion into heat. *Transmission* and *radiation* are forms of energy transport in which there is a finite propagation time.

In a physical system, *Figure 3.1*, there will be identifiable energy inputs w_i and outputs w_o of any form. The system itself may store energy w_s and dissipate energy w_l. Then the conservation principle states that

$$w_i = w_s + w_l + w_o$$

Comparable statements can be made for energy changes Δw, and for energy rates $\Delta w/\Delta t = p$, i.e., the powers:

$$\Delta w_i = \Delta w_s + \Delta w_l + \Delta w_o \quad \text{and} \quad p_i = p_s + p_l + p_o$$

ELECTRICITY

Most of the observed electrical phenomena are explicable in terms of electric *charge* at rest, in motion and in acceleration. Static charges give rise to an *electric field* of force; charges in motion carry an electric field accompanied by a *magnetic field* of force; charges in acceleration develop a further field of *radiation.*

Modern physics has established the existence of elemental charges and their responsibility for observed phenomena. Modern physics is complex: it is customary to explain phenomena of engineering interest at a level adequate for a clear and reliable concept, based on the electrical nature of matter.

Molecules, atoms and electrons

Material substances, whether solid, liquid or gaseous, are conceived as composed of very large numbers of *molecules*. A molecule is the smallest portion of any substance which cannot be further subdivided without losing its characteristic material properties. In all states of matter molecules are in a state of rapid continuous motion. In a *solid* the molecules are relatively closely 'packed' and the molecules, although rapidly moving, maintain a fixed mean position. Attractive forces between molecules account for the tendency of the solid to retain its shape. In a *liquid* the molecules are less closely packed and there is a weaker cohesion between them, so that they can wander about with some freedom within the liquid, which consequently takes up the shape of the vessel in which it is contained. The molecules in a *gas* are still more mobile, and are relatively far apart. The cohesive force is very small, and the gas is enabled freely to contract and expand. The usual effect of heat is to increase the intensity and speed of molecular activity so that 'collisions' between molecules occur more often; the average spaces between the molecules increase, so that the substance attempts to expand, producing internal pressure if the expansion is resisted.

Molecules are capable of further subdivision, but the resulting particles, called *atoms*, no longer have the same properties as the molecules from which they came. An atom is the smallest portion of matter that can enter into chemical combination or be chemically separated, but it cannot generally maintain a separate existence except in the few special cases where a single atom forms a molecule. A molecule may consist of one, two or more (sometimes many more) atoms of various kinds. A substance whose molecules are composed entirely of atoms of the same kind is called an *element*. Where atoms of two or more kinds are present, the molecule is that of a chemical *compound*. At present 102 atoms are recognised, from combinations of which every conceivable substance is made.

As the simplest example, the atom of hydrogen has a mass of 1.63×10^{-27} kg and a molecule (H_2), containing two atoms, has twice this mass. In one gram of hydrogen there are about 3×10^{23} molecules with an order of size between 1 and 0.1 nm.

Electrons, as small particles of negative electricity having apparently almost negligible mass, were discovered by J. J. Thomson, on a basis of much previous work by many investigators, notably Crookes. The discovery brought to light two important facts: (1) that atoms, the units of which all matter is made, are themselves complex structures, and (2) that electricity is atomic in nature. The atoms of all substances are constructed from particles. Those of engineering interest are: *electrons*, *protons* and *neutrons*. Modern physics concerns itself also with *positrons*, *mesons*, *neutrinos* and many more. An *electron* is a minute particle of negative electricity which, when dissociated from the atom (as it can be) indicates a purely electrical, nearly mass-less nature. From whatever atom they are derived, all electrons are similar. The electron charge is $e = 1.6 \times 10^{-19}$ C, so that 1 C $= 6.3 \times 10^{18}$ electron charges. The apparent rest mass of an electron is 1/1850 of that of a hydrogen atom, amounting to $m = 9 \times 10^{-28}$ g. The meaning to be attached to the 'size' of an electron (a figure of the order of 10^{-13} cm) is vague. A *proton* is electrically the opposite of an electron, having an equal charge, but positive. Further, protons are associated with a mass the same as that of the hydrogen nucleus. A *neutron* is a chargeless mass, the same as that of the proton.

Atomic structure

The mass of an atom is almost entirely concentrated in a nucleus of protons and neutrons. The simplest atom, of hydrogen, comprises a nucleus with a single proton, together with one associated electron occupying a region formerly called the K-shell. Helium has a nucleus of two protons and two neutrons, with two electrons in the K-shell. In these cases, as in all normal atoms, the sum of the electron charges is numerically equal to the sum of the protons charges, and the atom is electrically balanced. The neon atom has a nucleus with 10 protons and 10 neutrons, with its 10 electrons in the K- and L-shells.

The *atomic weight* A is the total number of protons and neutrons in the nucleus. If there are Z protons there will be $A - Z$ neutrons: Z is the *atomic number*. The nuclear structure is not known, and the forces that keep the proton together against their mutual repulsion are conjectural.

A nucleus of atomic weight A and atomic number Z has a charge of $+Ze$ and is normally surrounded by Z electrons each of charge $-e$. Thus copper has 29 protons and 35 neutrons ($A = 64$, $Z = 29$) in its nucleus, electrically neutralised by 29 electrons in an enveloping cloud. The atomic numbers of the known elements range from 1 for hydrogen to 102 for nobelium, and from time to time the list is extended. This multiplicity can be simplified: within the natural sequence of elements there can be distinguished groups with similar chemical and physical properties. These are the *halogens* (F 9, Cl 17, Br 35, I 53); the *alkali metals* (Li 3, Na 11, K 19, Rb 37, Cs 55); the *copper* group (Cu 29, Ag 47, Au 79); the *alkaline earths* (Be 4, Mg 12, Ca 20, Sr 38, Ba 56, Ra 88); the *chromium* group (Cr 24, Mo 42, W 74, U 92); and the *rare gases* (He 2, Ne 10, A 18, Kr 36, Xe 54, Rn 86). In the foregoing the brackets contain the chemical symbols of the elements concerned followed by their atomic numbers. The difference between the atomic numbers of two adjacent elements within a group is always 8, 18 or 32. Now these three bear to one another a simple arithmetical relation: $8 = 2 \times 2 \times 2$, $18 = 2 \times 3 \times 3$ and $32 = 2 \times 4 \times 4$. Arrangement of the elements in order in a periodic table beginning with an alkali metal and ending with a rare gas shows a remarkable repetition of basic similarities. The periods are I, 1–2; II, 3–10, III, 11–18; IV, 19–36; V, 37–54, VI, 55–86; VII, 87–?.

An element is often found to be a mixture of atoms with the same chemical property but different atomic weights (*isotopes*). Again, because of the convertibility of mass and energy, the mass of an atom depends on the energy locked up in its compacted nucleus. Thus small divergences are found in the atomic weights which, on simple grounds, would be expected to form integral multiples of the atomic weight of hydrogen. The atomic weight of oxygen is arbitrarily taken as 16.0, so that the mass of the proton is 1.007 6 and that of the hydrogen atom is 1.008 1.

Atoms may be in various energy states. Thus the atoms in the filament of an incandescent lamp may emit light when excited, e.g., by the passage of an electric heating current, but will not do so when the heater current is switched off. Now heat energy is the kinetic energy of the atoms of the heated body. The more vigorous impact of atoms may not always shift the atom as a whole, but may shift an electron from one orbit to another of higher energy level within the atom. This position is not normally stable, and the electron gives up its momentarily-acquired potential energy by falling back into its original level, releasing the energy as a definite amount of light, the *light-quantum* or *photon*.

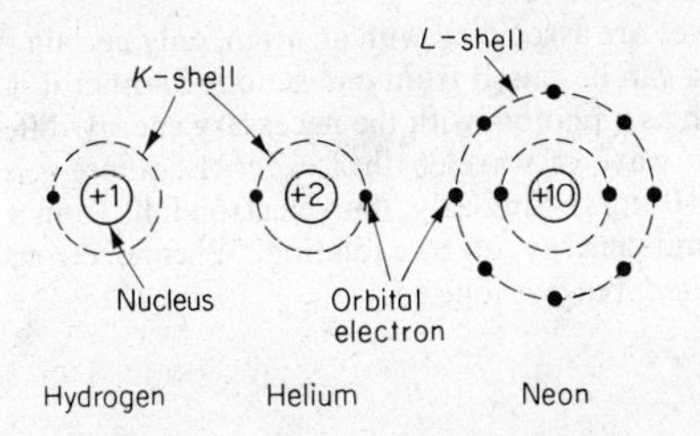

Figure 3.2 *Atomic structure. The nuclei are marked with their positive charges in terms of total electron charge. The term 'orbital' is becoming obsolete*

Electron: *mass* $m = 9 \times 10^{28}$ Y
charge $e = -1.6 \times 10^{-19}$ C
Proton: *mass* $= 1.63 \times 10^{-24}$ g
charge $= +1.6 \times 10^{-19}$ C
Neutron: mass as for proton; no charge

Among the electrons of an atom those of the outside peripheral shell are unique in that, on account of all the electron charges on the shells between them and the nucleus, they are the most loosely bound and most easily *removable*. In a variety of ways it is possible so to excite an atom that one of the outer electrons is torn away, leaving the atom *ionised* or converted for the time into an *ion* with an effective positive charge due to the unbalanced electrical state it has acquired. Ionisation may occur due to impact by other fast-moving particles, by irradiation with rays of suitable wavelength and by the application of intense electric fields.

The three 'structures' of *Figure 3.2* are based on the former 'planetary' concept, now modified in favour of a more complex idea derived from consideration of wave-mechanics. It is still true that, apart from its mass, the chemical and physical properties of an atom are given to it by the arrangement of the electron 'cloud' surrounding the nucleus.

Wave mechanics

The fundamental laws of optics can be explained without regard to the nature of light as an electromagnetic wave phenomenon, and photo-electricity emphasises its nature as a stream or ray of corpuscles. The phenomena of diffraction or interference can only be explained on the wave concept. *Wave mechanics* correlates the two apparently conflicting ideas into a wider concept of 'waves of matter'. Electrons, atoms and even molecules participate in this duality, in that their effects appear sometimes as corpuscular, sometimes as of a wave nature. Streams of electrons behave in a corpuscular fashion in photo-emission, but in certain circumstances show the diffraction effects familiar in wave action. Considerations of particle mechanics led de Broglie to write several theoretic papers (1922–6) on the parallelism between the dynamics of a particle and geometrical optics, and suggested that it was necessary to admit that classical dynamics could not interpret phenomena involving energy quanta. Wave mechanics was established by Schrödinger in 1926 on de Broglie's conceptions.

When electrons interact with matter they exhibit wave properties: in the free state they act like particles. Light has a similar duality, as already noted. The hypothesis of de Broglie is that a particle of mass m and velocity u has wave properties with a wavelength $\lambda = h/mu$, where h is the Planck constant, $h = 6.626 \times 10^{-34}$ J s. The mass m is relativistically affected by the velocity.

When electron waves are associated with an atom, only certain fixed-energy states are possible. The electron can be raised from one state to another if it is provided, by some external stimulus such as a photon, with the necessary energy-difference Δw in the form of an electromagnetic wave of wavelength $\lambda = hc/\Delta w$, where c is the velocity of free-space radiation (3×10^8 m/s). Similarly, if an electron falls from a state of higher to one of lower energy, it emits energy Δw as radiation. When electrons are raised in energy level, the atom is *excited*, but not ionised.

Electrons in atoms

Consider the hydrogen atom. Its single electron is not located at a fixed point, but can be anywhere in a region near the nucleus with some probability. The particular region is a kind of shell or cloud, of radius depending on the electron's energy state.

With a nucleus of atomic number Z, the Z electrons can have several possible configurations. There is a certain radial pattern of electron probability cloud distribution (or shell pattern). Each electron state gives rise to a cloud pattern, characterised by a definite energy level, and described by the series of quantum numbers n, l, m_l and m_s. The number $n(= 1, 2, 3 \ldots)$ is a measure of the energy level; $l(= 0, 1, 2 \ldots)$ is concerned with angular momentum; m_l is a measure of the component of angular momentum in the direction of an applied magnetic field; and m_s arises from the electron spin. It is customary to condense the nomenclature so that electron states corresponding to $l = 0$, 1, 2 and 3 are described by the letters s, p, d and f and a numerical prefix gives the value of n. Thus boron has 2 electrons at level 1 with $l = 0$, two at level 2 with $l = 0$, and one at level 3 with $l = 1$: this information is conveyed by the description $(1s)^2(2s)^2(2p)^1$.

The energy of an atom as a whole can vary according to the electron arrangement. The most stable state is that of minimum energy, and states of higher energy content are *excited*. By Pauli's *exclusion principle* the maximum possible number of electrons in states 1, 2, 3, 4 ... n are 2, 8, 18, 32 ... $2n^2$ respectively. Thus only 2 electrons can occupy the 1s state (or K-shell) and the remainder must, even for the normal minimum-energy condition, occupy other states. Hydrogen and helium, the first two elements, have respectively 1 and 2 electrons in the 1-quantum (K) shell; the next, lithium, has its third electron in the 2-quantum (L) shell. The passage from lithium to neon (*Figure 3.2*) results in the filling up of this shell to its full complement of 8 electrons. During the process, the electrons first enter the 2s subgroup, then fill the 2p subgroup until it has 6 electrons, the maximum allowable by the exclusion principle (see *Table 3.2*).

Very briefly, the effect of the electron-shell filling is as follows. Elements in the same chemical family have the same number of electrons in the subshell that is incompletely filled. The rare gases (He, Ne, A, Kr, Xe) have no uncompleted shells. Alkali metals (e.g., Na) have shells containing a single electron. The alkaline earths have two electrons

Period	*Atomic Number*	*Name*	*Symbol*	*Atomic Weight*
I	1	Hydrogen	H	1.008
	2	Helium	He	4.002
II	3	Lithium	Li	6.94
	4	Beryllium	Be	9.02
	5	Boron	B	10.82
	6	Carbon	C	12.00
	7	Nitrogen	N	14.008
	8	Oxygen	O	16.00
	9	Fluorine	F	19.00
	10	Neon	Ne	20.18
III	11	Sodium	Na	22.99
	12	Magnesium	Mg	24.32
	13	Aluminium	Al	26·97
	14	Silicon	Si	28·06
	15	Phosphorus	P	31.02
	16	Sulphur	S	32.06
	17	Chlorine	Cl	35.46
	18	Argon	A	39.94
IV	19	Potassium	K	39.09
	20	Calcium	Ca	40.08
	21	Scandium	Sc	45.10
	22	Titanium	Ti	47.90
	23	Vanadium	V	50.95
	24	Chromium	Cr	52.01
	25	Manganese	Mn	54.93
	26	Iron	Fe	55.84
	27	Cobalt	Co	58.94
	28	Nickel	Ni	58.69
	29	Copper	Cu	63.57
	30	Zinc	Zn	65.38
	31	Gallium	Ga	69.72
	32	Germanium	Ge	72.60
	33	Arsenic	As	74.91
	34	Selenium	Se	78.96
	35	Bromine	Br	79.91
	36	Krypton	Kr	83.7
V	37	Rubidium	Rb	85.44
	38	Strontium	Sr	87.63
	39	Yttrium	Y	88.92
	40	Zirconium	Zr	91.22
	41	Niobium	Nb	92.91
	42	Molybdenum	Mo	96.0
	43	Technetium	Tc	99
	44	Ruthenium	Ru	101.7
	45	Rhodium	Rh	102.9
	46	Palladium	Pd	106.7
	47	Silver	Ag	107.9
	48	Cadmium	Cd	112.4
	49	Indium	In	114.8
	50	Tin	Sn	118
V	51	Antimony	Sb	121.8
	52	Tellurium	Te	127.6
	53	Iodine	I	126.9
	54	Xenon	Xe	131.3
VI	55	Caesium	Cs	132.9
	56	Barium	Ba	137.4
	57	Lanthanum	La	138.9
	58	Cerium	Ce	140.1
	59	Praseodymium	Pr	140.9
	60	Neodymium	Na	144.3
	61	Promethium	Pm	147
	62	Samarium	Sm	150.4
	63	Europium	Eu	152.0
	64	Gadolinium	Gd	157.3
	65	Terbium	Tb	159.2
	66	Dysprosium	Dy	162.5
	67	Holmium	Ho	163.5
	68	Erbium	Er	167.6
	69	Thulium	Tm	169.4
	70	Ytterbium	Yb	173.0
	71	Lutecium	Lu	175.0
	72	Hafnium	Hf	178.6
	73	Tantalum	Ta	181.4
	74	Tungsten	W	184.0
	75	Rhenium	Re	186.3
	76	Osmium	Os	191.5
	77	Iridium	Ir	193.1
	78	Platinum	Pt	195.2
	79	Gold	Au	197.2
	80	Mercury	Hg	200.6
	81	Thallium	Tl	204.4
	82	Lead	Pb	207.2
	83	Bismuth	Bi	209.0
	84	Polonium	Po	210
	85	Astatine	At	211
	86	Radon	Rn	222
VII	87	Francium	Fr	223
	88	Radium	Ra	226.0
	89	Actinium	Ac	227
	90	Thorium	Th	232.1
	91	Protoactinium	U_{X2}	234
	92	Uranium	U	238.1
	93	Neptunium	Np	239
	94	Plutonium	Pu	242
	95	Americium	Am	243
	96	Curium	Cm	243
	97	Berkelium	Bk	245
	98	Californium	Cf	246
	99	Einsteinium	Es	247
	100	Fermium	Fm	256
	101	Mendelevium	Md	256
	102	Nobelium	No	—

Table 3.2 TYPICAL ATOMIC STRUCTURES

Element and Atomic Number		*Principal and Secondary Quantum Numbers*									
		1*s*	2*s*	2*p*	3*s*	3*p*	3*d*	4*s*	4*p*	4*d*	4*f*
H	1	1									
He	2	2									
Li	3	2	1								
C	6	2	2	2							
N	7	2	2	3							
Ne	10	2	2	6							
Na	11	2	2	6	1						
Al	13	2	2	6	2	1					
Si	14	2	2	6	2	2					
Cl	17	2	2	6	2	5					
A	18	2	2	6	2	6					
K	19	2	2	6	2	6		1			
Mn	25	2	2	6	2	6	5	2			
Fe	26	2	2	6	2	6	6	2			
Co	27	2	2	6	2	6	7	2			
Ni	28	2	2	6	2	6	8	2			
Cu	29	2	2	6	2	6	10	1			
Ge	32	2	2	6	2	6	10	2	2		
Se	34	2	2	6		6	10	2	4		
Kr	36	2	2	6		6	10	2	6		
		1	2	3	4*s*	4*p*	4*d*	4*f*	5*s*	5*p*	
Rb	37	2	8	18	2	6					
Xe	54	2	8	18	2	6	10		2	6	

in uncompleted shells. The good conductors (Ag, Cu, Au) have a single electron in the uppermost quantum state. An irregularity in the ordered sequence of filling (which holds consistently from H to A) begins at potassium (K) and continues to Ni, becoming again regular with Cu, and beginning a new irregularity with Rb.

Energy levels

The electron of a hydrogen atom, normally at level 1, can be raised to level 2 by endowing it with a particular quantity of energy most readily expressed as 10.2 eV. (1 eV = 1 electron-volt = 1.6×10^{-19} J is the energy acquired by a free electron falling through a potential difference of 1 V, which accelerates it and gives it kinetic energy.) The 10.2 V is the *first excitation potential* for the hydrogen atom. If the electron is given an energy of 13.6 eV it is freed from the atom, and 13.6 V is the *ionisation potential.* Other atoms have different potentials in accordance with their atomic arrangement.

Electrons in metals

An approximation to the behaviour of metals assumes that the atoms lose their valency electrons, which are free to wander in the ionic lattice of the material to form what is called an electron gas. The sharp energy-levels of the free atom are broadened into wide bands by the proximity of others. The potential within the metal is assumed to be smoothed out, and there is a sharp rise of potential at the surface that prevents the electrons from escaping: there is a potential-energy step at the surface that the electrons cannót normally overcome: it is of the order of 10 eV. If this is called W, then the energy of an electron wandering within the metal is $-W + \frac{1}{2}mu^2$.

The electrons are regarded as undergoing continual collisions on account of the thermal vibrations of the lattice, and on Fermi-Dirac statistical theory it is justifiable to treat the energy states (which are in accordance with Pauli's principle) as forming an energy-continuum. At very low temperatures the ordinary classical theory would suggest that electron energies spread over an almost zero range, but the exclusion principle makes this impossible and even at absolute zero of temperature the energies form a continuum, and physical properties will depend on how the electrons are distributed over the upper levels of this energy range.

Conductivity

The interaction of free electrons with the thermal vibrations of the ionic lattice (called 'collisions' for brevity) causes them to 'rebound' with a velocity of random direction but small compared with their average velocities as particles of an electron gas. Just as a difference of electric potential causes a drift in the general motion, so a difference of temperature between two parts of a metal carries energy from the hot region to the cold, accounting for thermal conduction and for its association with electrical conductivity. The free-electron theory, however, is inadequate to explain the dependence of conductivity on crystal axes in the metal.

At absolute zero of temperature (zero K $= -273$ °C) the atoms cease to vibrate, and free electrons can pass through the lattice with little hindrance. At temperatures over the range 0.3–10 K (and usually round about 5 K) the resistance of certain metals, e.g., Zn, Al, Sn, Hg and Cu, becomes substantially zero. This phenomenon, known as *superconductivity*, has not been satisfactorily explained.

Superconductivity is destroyed by moderate magnetic fields. It can also be destroyed if the current is large enough to produce at the surface the same critical value of magnetic field. It follows that during the superconductivity phase the current must be almost purely superficial, with a depth of penetration of the order of 10 μm.

Electron emission

A metal may be regarded as a potential 'well' of depth $-V$ relative to its surface, so that an electron in the lowest energy state has (at absolute zero temperature) the energy $W = Ve$ (of the order 10 eV): other electrons occupy levels up to a height ϵ^* (5–8 eV) from the bottom of the 'well'. Before an electron can escape from the surface it must be endowed with an energy not less than $\phi = W - \epsilon^*$, called the *work function*.

Emission occurs by *surface irradiation* (e.g., with light) of frequency ν if the energy quantum $h\nu$ of the radiation is at least equal to ϕ. The threshold of photo-electric emission is therefore with radiation at a frequency not less than $\nu = \phi/h$.

Emission takes place at *high temperatures* if, put simply, the kinetic energy of electron normal to the surface is great enough to jump the potential step W. This leads to an expression for the emission current i in terms of temperature T, a constant A and the thermionic work-function ϕ:

$$i = AT^2 \exp(-\phi/kT)$$

Electron emission is also the result of the application of a *high electric-field intensity* (of the order 1–10 GV/m) to a metal surface; also when the surface is bombarded with electrons or ions of sufficient kinetic energy, giving the effect of *secondary* emission.

Electrons in crystals

When atoms are brought together to form a crystal, their individual sharp and well-defined energy levels merge into energy *bands*. These bands may overlap, or there may be gaps in the energy levels available, depending on the lattice spacing and inter-atomic bonding. Conduction can take place only by electron migration into an empty or partly filled band: filled bands are not available. If an electron acquires a small amount of energy from the externally applied electric field, and can move into an available empty level, it can then contribute to the conduction process.

Insulators

In this case the 'distance' (or energy increase Δw in electron volts) is too large for moderate electric applied fields to endow electrons with sufficient energy, so the material remains an insulator. High temperatures, however, may result in sufficient thermal agitation to permit electrons to 'jump the gap'.

Semiconductors

Intrinsic semiconductors (i.e., materials between the good conductors and the good insulators) have a small spacing of about 1 eV between their permitted bands, which affords a low conductivity, strongly dependent on temperature and of the order of one-millionth that of a conductor.

Impurity semiconductors have their low conductivity provided by the presence of minute quantities of foreign atoms (e.g., 1 in 10^8) or by deformations in the crystal structure. The impurities 'donate' electrons of energy-level that can be raised into a conduction band (*n*-type); or they can attract an electron from a filled band to leave a 'hole', or electron deficiency, the movement of which corresponds to the movement of a positive charge (*p*-type).

Magnetism

Modern magnetic theory is very complex, with ramifications in several branches of physics. Magnetic phenomena are associated with moving charges. Electrons, considered as particles, are assumed to possess an axial spin, which gives them the effect of a minute current-turn or of a small permanent magnet, called a Bohr *magneton*. The gyroscopic effect of electron spin develops a precession when a magnetic field is applied. If the precession effect exceeds the spin effect, the external applied magnetic field produces less magnetisation than it would in free space, and the material of which the electron is a constituent part is *diamagnetic*. If the spin effect exceeds that due to precession, the material is *paramagnetic*. The spin effect may, in certain cases, be very large, and high magnetisations are produced by an external field: such materials are *ferromagnetic*.

An iron atom has, in the $n = 4$ shell (N), electrons that give it conductive properties. The K, L and N shells have equal numbers of electrons possessing opposite spin-directions, so cancelling. But shell M contains 9 electrons spinning in one direction and 5 in the other, leaving 4 net magnetons. Cobalt has 3, and nickel 2. In a solid metal, further cancellation occurs and the average number of unbalanced magnetons is: Fe, 2.2; Co, 1.7; Ni, 0.6.

In an iron crystal the magnetic axes of the atoms are aligned, unless upset by excessive thermal agitation. (At 770 °C for Fe, the Curie point, the directions become random and ferromagnetism is lost.) A single Fe crystal magnetises most easily along a cube edge of the structure. It does not exhibit spontaneous magnetisation like a permanent magnet, however, because a crystal is divided into a large number of *domains* in which the various magnetic directions of the atoms form closed paths. But if a crystal is exposed to an external applied magnetic field, (i) the electron spin axes remain initially unchanged, but those domains having axes in the favourable direction grow at the expense of the others (domain-wall displacement); and (ii) for higher field intensities the spin axes orientate into the direction of the applied field.

If wall movement makes a domain acquire more internal energy, then the movement will relax again when the external field is removed. But if wall-movement results in loss of energy, the movement is non-reversible—i.e., it needs external force to reverse it. This accounts for hysteresis and remanence phenomena.

The closed-circuit self-magnetisation of a domain gives it a mechanical strain. When the magnetisation directions of individual domains are changed by an external field, the strain directions alter too, so that an assembly of domains will tend to lengthen or shorten. Thus readjustments in the crystal lattice occur, with deformations (e.g., 20 parts in 10^6) in one direction. This is the phenomenon of *magnetostriction*.

The practical art of magnetics consists in control of magnetic properties by alloying, heat-treatment and mechanical working to produce variants of crystal structure and consequent magnetic characteristics.

Simplified electrical theories

In the following paragraphs, a discussion of electrical phenomena is given in terms adequate for the purpose of simple explanation.

Consider two charged bodies separated in air, *Figure 3.3*. Work must have been done in a physical sense to produce on one an excess and on the other a deficiency of

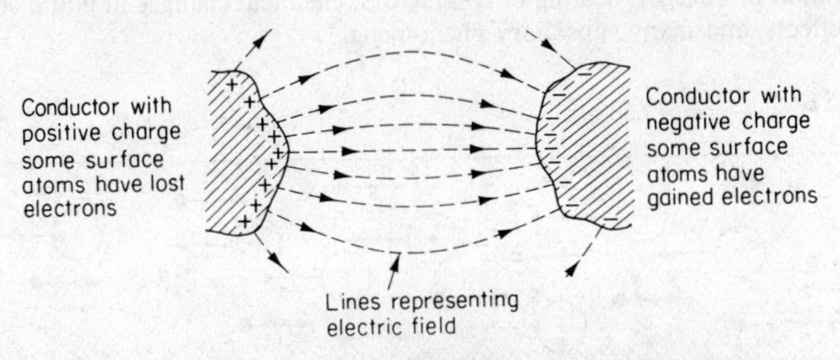

Figure 3.3 Charged conductors and their electric field

electrons, so that the system is a repository of potential energy. (The work done in separating charges is measured by the product of the charges separated and the difference of electrical potential that results.) Observation of the system shows certain effects of interest: (1) there is a difference of electric potential between the bodies depending on the amount of charge and the geometry of the system; (2) there is a mechanical force of attraction between the bodies. These effects are deemed to be manifestations of the *electric field* between the bodies, described as a special state of space and depicted by *lines of force* which express in a pictorial way the strength and direction of the force effects. The lines stretch between positive and negative elements of charge through the medium (in this case, air) which separates the two charged bodies. The electric field is only a concept—for the lines have no real existence—used to calculate various effects produced when charges are separated by any method which results in excess and deficiency states of atoms by electron transfer. Electrons and protons, or electrons and positively ionised atoms, attract each other, and the stability of the atom may be considered due to the balance of these attractions and dynamic forces such as electron spin. Electrons are repelled by electrons and protons by protons, these forces being summarised in the rules, formulated experimentally long before our present knowledge of atomic structure, that 'like charges repel and unlike charges attract one another'.

Conductors and insulators

In substances called *conductors*, the outer-shell electrons can be more or less freely interchanged between atoms. In copper, for example, the molecules are held together comparatively rigidly in the form of a 'lattice'—which gives the piece of copper its permanent shape—through the interstices of which outer electrons from the atoms can be interchanged within the confines of the surface of the piece, producing a random movement of free electrons called an 'electron atmosphere'. Such electrons are responsible for the phenomenon of electrical conductivity.

In other substances called *insulators* all the electrons are more or less firmly bound to their parent atoms so that little or no relative interchange of electron charges is possible. There is no marked line of demarcation between conductors and insulators, but the copper-group metals in the order silver, copper, gold, are outstanding in the series of conductors.

Conduction

Conduction is the name given to the movement of electrons, or ions, or both, giving rise to the phenomena described by the term *electric current*. The effects of a current include a redistribution of charges, heating of conductors, chemical changes in liquid solutions, magnetic effects, and many subsidiary phenomena.

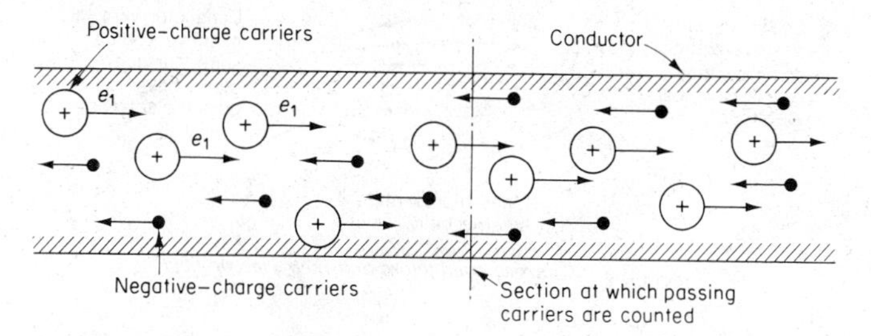

Figure 3.4 Electric current as the result of moving charges

If at some point on a conductor (*Figure 3.4*), n_1 carriers of electric charge (they can be water-drops, ions, dust particles, etc.) each with a positive charge e_1 arrive per second, and n_2 carriers (such as electrons) each with a negative charge e_2 arrive in the opposite direction per second, the total rate of passing of charge is $n_1e_1 + n_2e_2$, which is the charge per second or *current*. A study of conduction concerns the kind of carriers and their behaviour under given conditions. Since an electric field exerts mechanical forces on charges, the application of an electric field (i.e. a potential difference) between two points on a conductor will cause the movement of charges to occur, i.e., a current to flow, so long as the electric field is maintained.

The discontinuous particle nature of current flow is an observable factor. The current carried by a number of electricity carriers will vary slightly from instant to instant with the number of carriers passing a given point in a conductor. Since the electron charge is 1.6×10^{-19} C, and the passage of one coulomb per second (a rate of flow of *one ampere*) corresponds to $10^{19}/1.6 = 6.3 \times 10^{18}$ electron charges per second, it follows that the discontinuity will be observed only when the flow comprises the very rapid movement of a few electrons. This may happen in gaseous conductors, but in metallic conductors the flow is the very slow drift (measurable in mm/s) of an immense number of electrons.

A current may be the result of a two-way movement of positive and negative particles. Conventionally the direction of current flow is taken as the same as that of the positive charges and against that of the negative ones.

Conduction in metallic conductors

Reference has been made above to the 'electron atmosphere' of electrons in random motion within a lattice of comparatively rigid molecular structure in the case of copper, which is typical of the class of good metallic conductors. The random electronic motion, which intensifies with rise in temperature, merges into an average shift of charge of almost (but not quite) zero continuously (*Figure 3.5*). When an electric field is applied

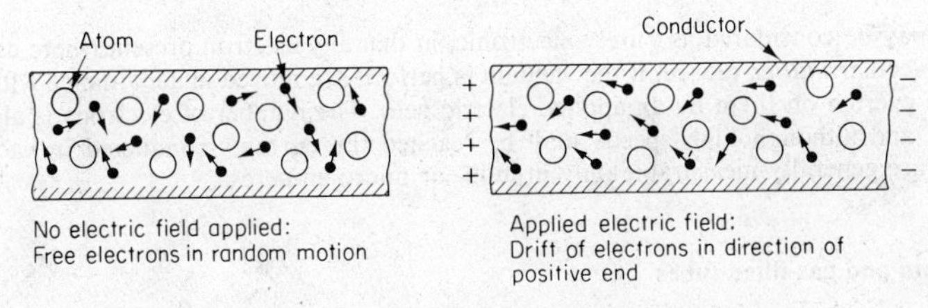

Figure 3.5 *Electronic conduction in metals*

along the length of a conductor (as by maintaining a potential difference across its ends), the electrons have a *drift* towards the positive end superimposed upon their random digressions. The drift is slow, but such great numbers of electrons may be involved that very large currents, entirely due to electron drift, can be produced by this means. In their passage the electrons are impeded by the molecular lattice, the collisions producing heat and the opposition called *resistance*. The conventional direction of current flow is actually opposite to that of the drift of charge, which is exclusively electronic.

Conduction in liquids

Liquids are classified according to whether they are *non-electrolytes* (non-conducting) or *electrolytes* (conducting). In the former the substances in solution break up into electrically balanced groups, whereas in the latter the substances form ions, each a part of a single molecule with either a positive or a negative charge. Thus common salt, NaCl, in a weak aqueous solution breaks up into sodium and chlorine ions. The sodium ion Na^+ is a sodium atom less one electron, the chlorine ion Cl^- is a chlorine atom with one electron more than normal. The ions attach themselves to groups of water molecules. When an electric field is applied the sets of ions move in opposite directions, and since they are much more massive than electrons the conductivity produced is markedly inferior to that in metals. Chemical actions take place in the liquid and at the electrodes when current passes. Faraday's Electrolysis Law states that the mass of an ion deposited at an electrode by electrolyte action is proportional to the quantity of electricity which passes and to the *chemical equivalent* of the ion.

Conduction in gases

Gaseous conduction is strongly affected by the pressure of the gas. At pressures corresponding to a few centimetres of mercury gauge, conduction takes place by the movement of positive and negative ions. Some degree of ionisation is always present due to stray radiations (light, etc.). The electrons produced attach themselves to gas atoms and the sets of positive and negative ions drift in opposite directions. At very low gas pressures the electrons produced by ionisation have a much longer free path before they collide with a molecule, and so have scope to attain high velocities. Their motional energy may be enough to *shock-ionise* neutral atoms, resulting in a great enrichment of the electron stream and an increased current flow. The current may build up to high values if the effect becomes cumulative, and eventually conduction may be effected through a *spark* or *arc*.

Conduction in vacuum

This may be considered as purely electronic, in that any electron present (there can be no molecular *matter* present if the vacuum is perfect) are moved in accordance with the forces exerted on them by an applied electric field. The number of electrons is always small, and although high speeds may be reached the currents conducted in vacuum tubes are generally measurable only in milli- or micro-amperes.

Vacuum and gas-filled tubes

Some of the effects described above are illustrated in *Figure 3.6*. At the bottom is an electrode, the *cathode*, from the surface of which electrons are emitted, generally by heating the cathode material. At the top is a second electrode, the *anode*, and an electric field is established between anode and cathode, which are enclosed in a vessel which contains a low-pressure inert gas. The electric field causes electrons emitted from the cathode to move upwards. In their passage to the anode these electrons will encounter gas molecules. If conditions are suitable, the gas atoms are ionised, becoming in effect positive charges associated with the nuclear mass. Thereafter the current is increased by

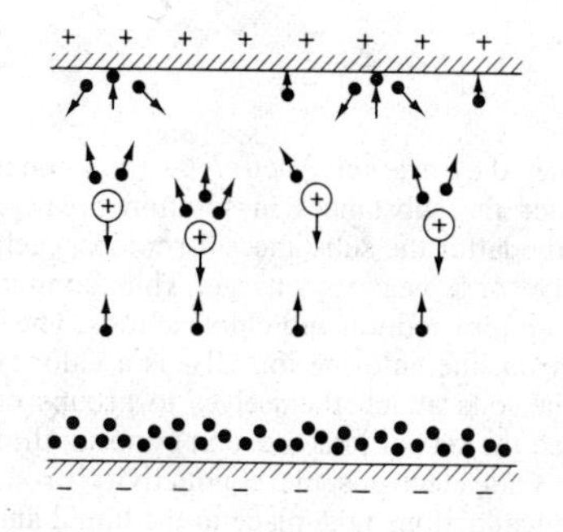

Figure 3.6 Electrical conduction in gases at low pressure

the detached electrons moving upwards and by the positive ions moving more slowly downwards. In certain devices (such as the mercury-arc rectifier) the impact of ions on the cathode surface maintains its emission. The impact of electrons on the anode may be energetic enough to cause the *secondary emission* of electrons from the anode surface. If the gas molecules are excluded and a vacuum established, the conduction becomes purely electronic.

Convection currents

Charges may be moved by mechanical means, on discs, endless belts, water-drops, dust or mist particles. A common example is the electron beam between anode and screen in the cathode-ray oscilloscope. Such a motion of charges, independent of an electric field, is termed a *convection* current.

Displacement and polarisation currents

If an electric field is applied to a perfect insulator, whether solid, liquid or gaseous, the electric field affects the atoms by producing a kind of 'stretching' or 'rotation' which displaces the electrical centres of negative and positive in opposite directions. This polarisation of the dielectric insulating material may be considered as taking place in the manner indicated in *Figure 3.7*. Before the electric field is applied, in (*a*), the atoms of

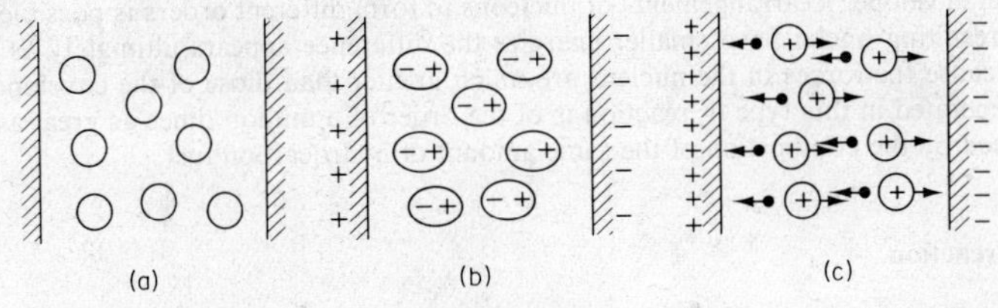

Figure 3.7 Polarisation, displacement and breakdown in a dielectric material (*a*) *no electric field; atoms unstrained* (*b*) *electric field applied; polarisation* (*c*) *intensified electric field; atoms ionised*

the insulator are neutral and unstrained; (*b*) as the potential difference is raised the electric field exerts opposite mechanical forces on the negative and positive charges and the atoms become more and more highly strained. On the left face the atoms will all present their negative charges at the surface: on the right face, their positive charges. These surface polarisations are such as to account for the effect known as *permittivity*. The small displacement of the electric charges is an electron shift, i.e., a *displacement current* flows while the polarisation is being established. *Figure 3.7* (*c*) shows that under conditions of excessive electric field atomic disruption or ionisation may occur, converting the insulator material into a conductor, resulting in *breakdown*.

Radiation

Reference has been made to the emission of energy as light when an electron returns from an outer shell to its normal shell nearer the nucleus of the parent atom. The ability to convert to *radiation* energy of other forms is a fundamental atomic property, and the converse transformation is also possible. Radiation has both particle and wave interpretation, the latter in connection with its transmission through empty space and through transparent media. A basic distinction between various kinds of radiation is that certain types (those in the *electromagnetic wave spectrum*) travel in empty space at the natural constant velocity of 3×10^8 m/s. The others do not, their velocity depending on circumstances.

NUCLEAR PHYSICS

The first hint of nuclear energy resources came in 1905 when Einstein stated the equivalence of energy and mass in the equation $W = mc^2$. This suggests that 1 kg of matter, if transformed, would equate to 100 TJ or 30 000 MWh. In a sense all the familiar energy sources are nuclear (having been stemmed from the sun), but the immediate source in most power plants employs the combustion of a hydrocarbon fuel, a process of atomic rearrangement. The fuel constituents are oxygen and a complex hydrocarbon. The latter originated by aid of solar energy, which is stored in the molecules by virtue of their binding energy. At moderate temperatures the oxygen–hydrogen mixture is stable, but at elevated temperatures a triggering effect takes place. The heated molecules absorb energy and break up: the fragments combine with oxygen to form arrangements of lower intrinsic energy content. The released energy appears as radiation or as increased fragment velocity and the result is the production of heat, which in turn augments the trigger effect. If the heat energy is drained off too slowly the process rapidly becomes explosive.

Nuclear energy is derived from atomic nuclei, the sub-nuclear particles (*nucleons*) in the structure of which rival in order and complexity those of the electrons of the external envelope. Rearrangements of nucleons to form different orders is possible, and if the resulting nuclei have smaller energies the difference appears ultimately as heat. But because the forces in the nucleus are much greater than those of the envelope, the heat generated in this type of reaction is of the order of a million times as great as that produced by the combustion of the same amount of hydrocarbon fuel.

Chain reaction

Protons and neutrons are bound together in the nucleus by forces not yet understood. When energy is supplied to the nucleus one or more of either particle may be expelled. The neutron carries no charge and there are no large repulsive forces on it. It may therefore strike another nucleus, and cause a corresponding ejection which may suffice to maintain a *chain reaction.*

Fission

Fission can take place if the neutron hits a suitable target, such as U235 (a 0.7 per cent isotope of natural uranium), or Pu239. The target nucleus disintegrates into two, possibly three and rarely four parts with the liberation of great quantities of energy (about 150 MeV per nucleus); moreover each fission releases an average of $2\frac{1}{2}$ further neutrons making possible a self-sustaining chain-reaction, provided that the neutrons are not too fast.

Nuclear reactor

Using natural uranium (or slightly enriched uranium), there are 100–140 atoms of U238 to act as a diluent; the fission products are fast neutrons; and surface losses will reduce the average neutron production per fission. But by using a *moderator* to slow down the neutrons they will have an increased probability of causing fission. By increasing the bulk of the material a critical size is reached for which the neutron losses are low and a chain-reaction is just sustained. Thereafter the reactor can be built to any size for which the heat developed can be removed at an adequate rate.

Bred fuel

Neutrons in the reaction described that are captured by non-fissionable U238 atoms are instrumental in initiating an important subsidiary reaction

$$\mathrm{U238} + n \rightarrow \mathrm{U239} \rightarrow \mathrm{Np239} + e \rightarrow \mathrm{Pu239} + e$$

where n stands for the addition of a neutron and e represents an electron charge. Plutonium (Pu) is stable and easily fissionable, so is an important fuel. Complicated processes have been developed to remove the few grammes of Pu from the tons of original U fuel rods.

Another possible fuel is an isotope U233 formed as an end-product from a reaction starting with the relatively abundant material thorium (Th232).

Reactors

The reactions described are those of *thermal reactors*, making use of thermal neutrons obtained by moderating fast ones. If highly enriched fuel is available, with relatively little U238 as a diluent, a moderator is unnecessary and *fast reactors* can be built. By choosing the appropriate proportion of U238, a greater number of Pu atoms can be generated than the U235 atoms disintegrated: such reactors are called *breeder reactors*, and play an increasing part in nuclear power production.

Fission products

The fragments of nuclear disintegration of U235 are usually approximately 2/5 and 3/5 of the whole (i.e., Sr and Xe), but almost all the isotopes of all the elements may occur. The separation of these is becoming a major industry.

PHYSICAL CONSTANTS

The following table lists a number of values of physical constants used in electrophysics and electrotechnology.

Table 3.3 PHYSICAL CONSTANTS

Quantity	*Symbol*	*Numerical value*
Avogadro number	N_A	2.69×10^{25} m^{-3}
Bohr magneton	μ_B	1.165×10^{-29} Wb m
energy	β	0.927×10^{-23} J/T
Boltzmann constant	k	$1.380\,5 \times 10^{-23}$ J/K
Electron: charge	$-e$	$1.602\,1 \times 10^{-19}$ C
rest-mass	m_e	$9.109\,1 \times 10^{-31}$ kg
charge/rest-mass	e/m_e	$1.758\,8 \times 10^{11}$ C/kg
Faraday constant	F	$96\,520 \times 10^3$ C/mol

Table 3.3 PHYSICAL CONSTANTS—*continued*

Quantity	*Symbol*	*Numerical value*
Free space: electric constant	ϵ_0	$8.854\,2 \times 10^{-12}$ F/m
intrinsic impedance	Z_0	376.9 Ω
magnetic constant	μ_0	$4\pi \times 10^{-7}$ H/m
speed of e.m. waves	c_0	$2.997\,9 \times 10^{8}$ m/s
Ideal-gas constant	R	8 314.4 J/(K mol)
Neutron rest-mass	m_n	$1.674\,8 \times 10^{-27}$ kg
Planck constant	h	$6.625\,6 \times 10^{-34}$ J s
	$h/2\pi$	$1.054\,5 \times 10^{-34}$ J s
Proton: charge	$+e$	$1.602\,1 \times 10^{-19}$ C
rest-mass	m_p	$1.672\,6 \times 10^{-27}$ kg
charge/rest-mass	e/m_p	$0.905\,8 \times 10^{8}$ C/kg
Standard acceleration of free fall	g_n	9.806 6 m/s
Stefan-Boltzmann constant	σ	5.67×10^{-8} J(m² s K)
Universal gravitational constant	G	6.67×10^{-11} N m²/kg²

FURTHER READING

LOTHIAN, G. P., *Electrons in Atoms*, Butterworth (1963)
SEYMOUR, J., *Physical Electrons*, Pitmans (1972)

4 GENERAL BACKGROUND TO ELECTROMAGNETIC AND NUCLEAR RADIATION

GENERAL BACKGROUND TO ELECTROMAGNETIC AND NUCLEAR RADIATION 4

4 GENERAL BACKGROUND TO ELECTROMAGNETIC AND NUCLEAR RADIATION

ELECTROMAGNETIC RADIATION

Light and heat were for centuries the only known kinds of radiation. Today it is known that light and heat radiation form only a very small part of an enormous range of radiations extending from the longest radio waves to the shortest gamma-rays and known as the electromagnetic spectrum, *Figure 4.1*. The wavelength of the radiations extends from about 100 kilometres to fractions of micrometres. The visible light radiations are near the centre of the spectrum. All other radiations are invisible to the human eye.

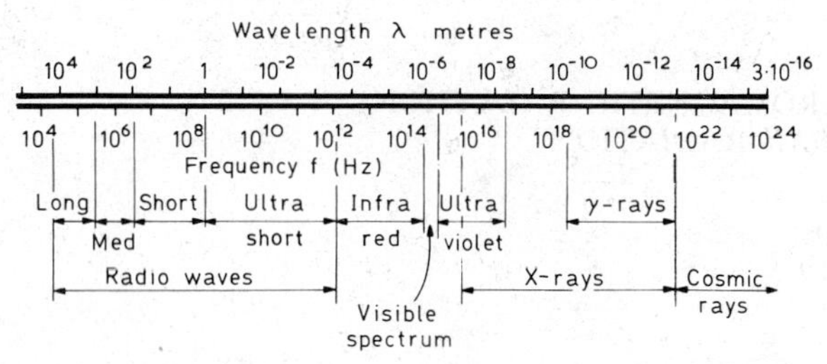

Figure 4.1 Electromagnetic wave spectrum

The researches into electromagnetic radiation can be traced back to 1680, to Newton's theory of the composition of white light. Newton showed that white light is made up from rays of different colours. A prism refracts these rays to varying degrees according to their wavelengths and spreads them out. The result is the visible spectrum of light.

In 1800, William Herschel, during research into the heating effects of the visible spectrum, discovered that the maximum heating was not within the visible spectrum but just beyond the red range. Herschel concluded that in addition to visible rays the sun emits certain invisible ones. These he called infra-red rays.

The next year, the German physicist Ritter made a further discovery. He took a sheet of paper freshly coated with Silver Chloride and placed it on top of a visible spectrum produced from sunlight falling through a prism. After a while he examined the paper in bright light. It was blackened, and it was most blackened just beyond the violet range of the spectrum. These invisible rays Ritter called ultra-violet rays.

The next step was taken in 1805, when Thomas Young demonstrated that light consists of waves, a theory which the Frenchman Fresnel soon proved conclusively. Fresnel showed that the waves vibrated transversely, either in many planes, or in one plane, when the waves were said to be plane-polarised. The plane containing both the direction of propagation and the direction of the electric vibrations is called the plane of polarisation.

In 1831 Faraday showed that when a beam of light was passed through a glass block to which a magnetic field was applied in the same direction as the direction of polarisation the plane of polarisation could be rotated. Moreover, when the magnetic field was increased, the angle of rotation also increased. The close relationship between light, magnetism and electricity was thus demonstrated for the first time.

In 1864 James Clerk Maxwell formulated his theory of electromagnetic waves and laid the foundation of the wave theory of the electromagnetic spectrum as it is known today.

The *Fundamental Maxwell theory* includes two basic laws and the displacement-current hypothesis:

Ampere's law. The summation of the magnetic force H round a closed path is proportional to the total current flowing across the surface bounded by the path:

$$F = \text{line-integral of } H \,.\, \mathrm{d}l = I$$

Displacement current. The symbol I above includes polarisation and displacement currents as well as conduction currents.

Faraday's law. The summation of the electric force E round a closed path is proportional to the rate-of-change of the magnetic flux Φ across the surface bounded by the path:

$$e = \text{line-integral of } E \,.\, \mathrm{d}l = -\mathrm{d}\Phi/\mathrm{d}t$$

The magnetic flux is circuital, and representable by 'closed loops' in a 'magnetic circuit'. The electric flux may be circuital, or it may spring from charges. The total flux leaving or entering a charge Q is Q coulombs.

A metallic circuit is not essential for the development of an e.m.f. in accordance with Faraday's law. The voltage-gradient E exists in the space surrounding a changing magnetic flux. The conductor is needed when the e.m.f. is to produce conduction currents. Again, the existence of a magnetic field does not necessarily imply an associated conduction current: it may be the result of a displacement current.

Maxwell deduced from these laws (based on the work of Faraday) the existence of electromagnetic waves in free space and in material medial. Waves in free space are classified in accordance with their frequency f and their wavelength λ, these being related to the free-space propagation velocity $c \simeq 3 \times 18^8$ m/s by the expression $c = f\lambda$. Radiant energy of wavelength between 0.4 and 0.8 μm (frequencies between 750 and 375 GHz) is appreciated by the eye as *light* of various colours over the visible spectrum between violet (the shorter wavelength) and red (the longer). Waves shorter than the visible are the *ultra-violet*, which may excite visible fluorescence in appropriate materials. *X-rays* are shorter still. At the longer-wave end of the visible spectrum is *infra-red* radiation, felt as *heat.* The range of wavelengths of a few millimetres upward is utilised in *radio* communication.

In 1886 Heinrich Hertz verified Maxwell's theory. At that time Wimshurst machines were used to generate high voltages. A Leyden jar served to store the charge which could be discharged through a spark gap. Hertz connected a copper spiral in series with the Leyden jar; this spiral acted as a radiator of electromagnetic waves. A second spiral was placed a small distance from the first; this also was connected to a Leyden jar and a spark gap. When the wheel of the Wimshurst machine was turned sparks jumped across both gaps. The secondary sparks were caused by electromagnetic waves radiated from the first spiral and received by the second. These waves were what are today called *radio* waves. This experiment was the first of a series by which Hertz established the validity of Maxwell's theory.

In 1895, the German Roentgen found by chance that one of his discharge tubes had a strange effect on a chemical substance which happened to lie nearby: the substance emitted light. It even fluoresced when screened by a thick book. This meant that the tube emitted some kind of radiation. Roentgen called these unknown rays X-rays.

A year later the French physicist Henri Becquerel made a further discovery. He placed a photographic plate, wrapped in black paper, under a compound of uranium. He left it there over night. The plate, when developed, was blackened where the uranium had been. Becquerel had found that there exist minerals which give off invisible rays of some kind.

Later, research by Pierre and Marie Curie showed that many substances had this effect: radio-activity had been discovered. When this radiation was analysed it was found to consist of charged particles, later called alpha- and beta-rays by Rutherford. These particles were readily stopped by thin sheets of paper or metal.

In 1900, Villard discovered another radiation, much more penetrating and able to pass even through a thick steel plate. This component proved to consist of electromagnetic waves which Rutherford called gamma-rays. They were the last additions to the electromagnetic spectrum as is known today.

Waves are classified according to their uses and the methods of their generation, as well as to their frequencies and wavelengths. Radio waves are divided into various bands: the long-, medium- and short-wave bands; and now the v.h.f. and u.h.f. bands, including television and radar which extends to the shortest radio waves, called microwaves. The microwave band overlaps the infra-red band. Actually, *all* wave bands merge imperceptibly into each other; there is never a clear-cut division. Next is the narrow band of visible light. These visible rays are followed by the ultra-violet rays and the X-rays. Again, all these bands merge into each other. Finally, the gamma-rays. They are actually part of the X-ray family and have similar characteristics excepting that of origin.

A convenient unit for the measurement of wavelengths shorter than radio waves is the micrometre (μm) which is 10^{-6} metre. The micrometre is equal to the micron (μ), a term still used but now deprecated. Also deprecated is the term mμ, being 10^{-3} micron. For measurement of still shorter wavelengths, the nanometre (10^{-9} metre) is used. The Ångstrom unit (Å), which is 10^{-10} metre, is commonly used in optical physics. These units of wavelength are compared in *Table 4.1*.

Table 4.1 COMPARISON OF UNITS OF LENGTH

		Å	*nm*	*μm*	*mm*	*cm*	*m*
Å	=	1	10^{-1}	10^{-4}	10^{-7}	10^{-8}	10^{-10}
nm	=	10	1	10^{-3}	10^{-6}	10^{-7}	10^{-9}
μm	=	10^{4}	10^{3}	1	10^{-3}	10^{-4}	10^{-6}
mm	=	10^{7}	10^{6}	10^{3}	1	10^{-1}	10^{-3}
cm	=	10^{8}	10^{7}	10^{4}	10	1	10^{-2}
m	=	10^{10}	10^{9}	10^{6}	10^{3}	10^{2}	1

Electromagnetic waves are generated by moving charges such as free electrons or oscillating atoms. Orbital electrons (see Section 3) radiate when they move from one orbit to another, and only certain orbits are permissible. Oscillating nuclei radiate gamma-rays.

The frequency of an electromagnetic radiation is given by the expression:

$$f = \frac{E}{h}$$

where E is the energy and h is Planck's constant ($h = 6.6 \times 10^{-34}$ joule/s).

The identity of electromagnetic radiations has been established on the following grounds:

(1) The velocity of each *in vacuo* is constant.
(2) They all experience reflection, refraction, dispersion, diffraction, interference and polarisation.
(3) The mode of transmission is by transverse wave action.
(4) All electromagnetic radiation is emitted or absorbed in bursts or packets called quanta (or photons in the case of light).

In connection with (4), Planck established that the energy of each quantum varies directly with the frequency of the radiation (see the above expression).

Modern physics now accepts the concept of the dual nature of electromagnetic radiation, viz. that it has wave-like properties but at the same time it is emitted and absorbed in quanta.

Polarisation. An electromagnetic radiation possesses two fields at right angles to each other as viewed in the direction of the oncoming waves. These are the electric field and the magnetic field. The direction of either of these is known as the polarisation of the field, but the term is more usually related to the electric field, and this is at the same angle as the radiating source. For example, in case of radio waves, a horizontally positioned receiving dipole will not respond efficiently to waves which are vertically polarised.

The same phenomenon occurs with light radiation which is normally unpolarised, i.e., it is vibrating in all transverse planes. A sheet of polaroid allows light to pass through in one plane, due to the molecular structure of the material, and the resulting plane-polarised light is absorbed if a second sheet of polaroid is set at right-angles to the first.

The 'optical window'. It is important to realise that our knowledge of the universe around us depends upon incoming electromagnetic radiation. However, from the entire spectrum of such radiation, only two bands effectively reach the earth's surface:

(1) the visible light spectrum, together with a relatively narrow band of the adjacent ultra-violet and infra-red ranges;
(2) a narrow band of radio waves in the 1 cm to about 10 m band.

Thus the gamma rays, X-rays, most of the ultra-violet and infra-red rays, together with the longer ranges of radio waves fail to reach the earth's surface from outer space. This is mainly due to absorption in the ionosphere and atmosphere. (See Section 5.)

The applications of electromagnetic radiations range over an enormous field. Radio waves are used for telecommunication, sound and television broadcasting, navigation, radar, space exploration, industry, research, etc. Infra-red rays have many applications including security systems, fire detection, dark photography, industry, medical therapy. Ultra-violet rays, of which the sun is the chief source, have wide industrial and medical applications. X-rays and gamma-rays have become the everyday tools of the doctor, the scientist and in industry. And the narrow band of visible rays not only enables the world around us to be seen but together with ultra-violet radiation makes possible the process of photosynthesis by which plants build up and store the compounds of all our food. Thus the laws governing electromagnetic radiation are relevant to life itself. Many of these applications are described in more detail in later sections.

NUCLEAR RADIATION

There are three main types of radiation that can originate in a nucleus: *alpha*, *beta*, and *gamma* radiation.

Alpha radiation

An alpha particle has a charge of two positive units and a mass of four units. It is thus equivalent to a helium nucleus, and is the heaviest of the particles emitted by radioactive isotopes. Alpha (α) particles are emitted mostly by heavy nuclei and can possess only discrete amounts of energy, i.e., they give a line energy spectrum. The probability of collision between particles increases with the size of the particles. Thus the rate of ionisation in a medium traversed by particles emitted from radioactive isotopes, and hence the rate of loss of energy of the particles, also increases with the size of the particles. Consequently the penetrating power of the large alpha particles is relatively poor.

Beta radiation

Beta particles can be considered as very fast electrons. They are thus much smaller than alpha particles and therefore have greater penetrating powers. Beta (β) radiation will be absorbed in about 100 inches of air or half an inch of Perspex.

Unlike α-particles, β-particles emitted in a nuclear process have a continuous energy spectrum, i.e., β-particles can possess any amount of energy up to a maximum determined by the energy equivalent to the change in mass involved in the nuclear reaction. This has been explained by postulating the existence of the *neutrino*, a particle having no charge and negligible mass. According to this theory, the energy is shared between the β-particle and the neutrino in proportions that may vary, thus giving rise to a continuous energy spectrum.

Gamma radiation

Gamma radiation is electromagnetic in nature and has, therefore, no charge or mass. Its wavelength is much shorter than that of light or radio waves, and is similar to that of *X-rays.* The distinction between γ-rays and X-rays is that γ-rays are produced within the nucleus while X-rays are produced by the transition of an electron from an outer to an inner orbit.

γ-radiation has well-defined amounts of energy—that is, it occupies very narrow bands of the energy spectrum—since it results from transitions between energy levels within the nucleus. Characteristic X-rays of all but the very lightest of elements also possess well-defined amounts of energy.

γ-radiation has very great penetrating powers. Significant amounts are able to pass through lead bricks 50 mm thick; γ-photons possessing 1 MeV of energy (see p. 4–9) will lose less than 1% of its energy in traversing half a mile of air.

A rough comparison of the penetrating powers of α-, β-, and γ- radiation is given in *Figure 4.2.*

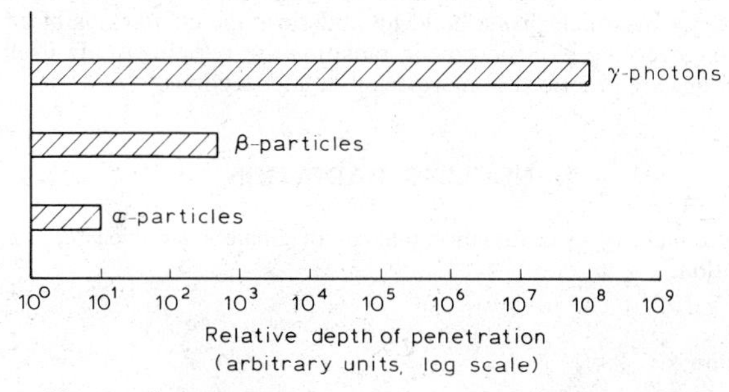

Figure 4.2 Rough comparison of penetrating powers of α-, β-, and γ-radiation

Neutrons

Neutrons were discovered in 1932 as a result of bombarding light elements (for example, beryllium and boron) with α-particles. For laboratory purposes, this is still a convenient method of production, but the most useful and intense source is the nuclear reactor in which the neutrons are produced as a by-product of the fission of fissile materials such as uranium-235.

Free neutrons are unstable, and decay to give a proton and a low-energy β-particle. Neutrons when they are produced may have a wide range of energy, from the several millions of electron volts of *fast neutrons* to the fractions of electron volts of *thermal neutrons*.

Neutrons lose energy by elastic collision. An *elastic collision* is one in which the incident particles rebound—or are scattered—without the nucleus that is struck having been excited or broken up. An *inelastic collision* is one in which the struck nucleus is excited, or broken up, or captures the incoming particle. For neutrons, the loss of energy in an elastic collision is greater with light nuclei; for example, a 1 MeV neutron loses 28% of its energy in collision with a carbon atom, but only 2% in collision with lead.

By successive collisions, the energy of neutrons is reduced to that of the thermal agitation of the nucleus (that is, some 0.025 eV at 20 °C) and the neutrons are then captured.

The consequence of the capture of a neutron may be a new nuclide, which may possibly be radioactive. This is, in fact, the main method of producing radioisotopes. Because they are uncharged, neutrons do not cause direct ionisation, and may travel large distances in materials having a high atomic number. The most efficient materials for shielding against neutron emission are those having light nuclei; as indicated above, these reduce the energy of neutrons much more rapidly than heavier materials. Examples of efficient shielding materials are water, the hydrocarbons and graphite.

FISSION

Fission is the splitting of a heavy nucleus into two approximately equal fragments known as fission products. Fission is accompanied by the emission of several neutrons and the release of energy. It can be spontaneous or caused by the impact of a neutron, a fast charged or a photon:

$$^{235}_{92}u + ^{1}_{0}n \rightarrow ^{93}_{38}\mathrm{Sr} + ^{140}_{54}\mathrm{Xe} + ^{1}_{0}n + ^{1}_{0}n + ^{1}_{0}n$$

Total number of sub-atomic particles is unchanged:

$$235 + 1 = 93 + 140 + 1 + 1 + 1$$

(other combinations of particles are possible)

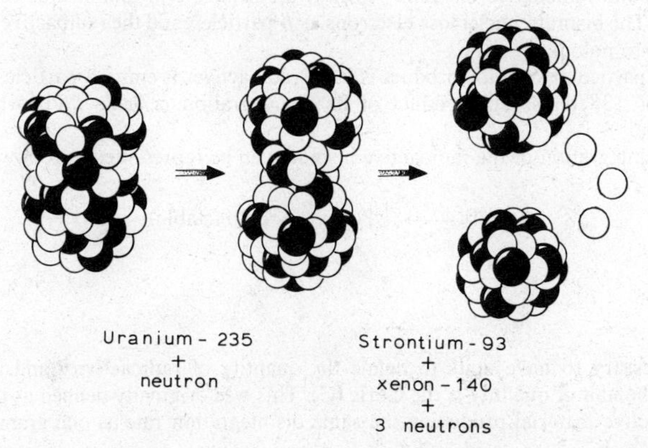

Figure 4.3 Diagramatic representation of the fission of Uranium-235

The number of neutrons is unchanged:

$$92 + 0 = 38 + 54$$

Hence the number of neutrons is also unchanged. *Figure 4.3* is a diagramatical representation of the foregoing.

OTHER MODES OF NUCLEAR DISINTEGRATION

Mention must be made of two other methods of nuclear disintegration:

(1) emission of positively charged electrons or *positrons* (β^+);
(2) *electron capture.*

Positrons interact rapidly with electrons after ejection from the nucleus. The two electrical charges cancel each other, and the energy is released in a form of γ-radiation known as *annihilation radiation.*

In the process of electron capture, the energy of an unstable nucleus is dissipated by the capture into the nucleus of an inner orbital electron. The process is always accompanied by the emission of the characteristic X-rays of the atom produced by electron capture. For example, germanium-71, which decays in this manner, emits gallium X-rays.

Radioactive decay

Radioactive isotopes are giving off energy continuously and if the law of the conserva tion of energy is to be obeyed this radioactive decay cannot go on indefinitely. The nucleus of the radioactive atom undergoes a change when a particle is emitted and forms a new and often non-radioactive product. The rate at which this nuclear reaction takes place decreases with time in such a way that the time necessary to halve the reaction rate is constant for a given isotope and is known as its half-life. The half-life period can be as short as a fraction of a microsecond or as long as ten thousand million years.

Radioactive decay can be illustrated by considering a radioactive form of bismuth, $^{210}_{83}Bi$, which has a half-life of five days. If the number of particles emitted by a sample in one minute is recorded, then after five days, two minutes would be required for the same number to be emitted. After ten days, four minutes would be required, and so on. The amount of the radioactive bismuth, $^{210}_{83}Bi$, in the sample will diminish as the emission proceeds. The bismuth nuclei lose electrons as β-particles, and the radioactive bismuth is converted to polonium, $^{210}_{84}Po$.

In this particular case the product is itself radioactive. It emits α-particles and has a half-life of 138 days. The product of its disintegration is lead, $^{206}_{82}Pb$, which is not radioactive.

The disintegration of the radioactive bismuth can be represented as follows:

$$^{210}_{83}\text{Bi} \xrightarrow[\text{5 days}]{\beta} {}^{210}_{84}\text{Po} \xrightarrow[\text{138 days}]{\alpha} {}^{206}_{82}\text{Pb (stable)}$$

Units

It is necessary to have units to define the quantity of radioactivity and its physical nature. The unit of quantity is the Curie (C). This was originally defined as the quantity of radioactive material producing the same disintegration rate as one gramme of pure radium.

The definition of quantity must be couched in different terms in modern times to include the many artificially produced radioisotopes. The Curie is now defined as the quantity of radioisotope required to produce 3.7×10^{10} disintegrations per second. Quantity measurements made in the laboratory with small sources are often expressed in terms of disintegrations per second (d.p.s.). What is actually recorded by the detector is expressed in counts per second (c.p.s.). The weights of material associated with this activity can vary greatly. For example, 1 Curie of iodine-131 weighs 8 microgrammes, whereas 1 Curie of uranium-238 weighs 2.7 tons.

The unit of energy is the electron volt (eV). This is the kinetic energy acquired by an electron when accelerated through a potential difference of one volt. The electron volt is equivalent to 1.6×10^{-19} joules. With α-, β- and γ-radiation, it is usual to use thousands of electron volts (keV) or millions of electron volts (MeV).

FURTHER READING

FOSTER, K. and ANDERSON, R., *Electro-magnetic Theory Vols. 1 and 2*, Butterworths (1970)
YARWOOD, J., *Atomic and Nuclear Physics*, University Tutorial Press (1973)

5 THE IONOSPHERE AND THE TROPOSPHERE—THEIR INFLUENCE ON RADIO WAVE PROPAGATION

THE IONOSPHERE AND TROPOSPHERE—THEIR INFLUENCE ON RADIO WAVE PROPAGATION 5

5 THE IONOSPHERE AND THE TROPOSPHERE— THEIR INFLUENCE ON RADIO WAVE PROPAGATION

THE IONOSPHERE

The ionosphere is an electrified region of the Earth's atmosphere situated at heights of from about fifty kilometres to several thousand kilometres. It consists of ions and free electrons produced by the ionising influences of solar radiation and of incident energetic solar and cosmic particles. The ionosphere is subject to marked geographic and temporal variations. It has a profound effect on the characteristics of radio waves propagated within or through it. By means of wave refraction, reflection or scattering it permits transmission over paths that would not otherwise be possible, but at the same time it screens some regions that could be illuminated in its absence (see *Figure 5.1*).

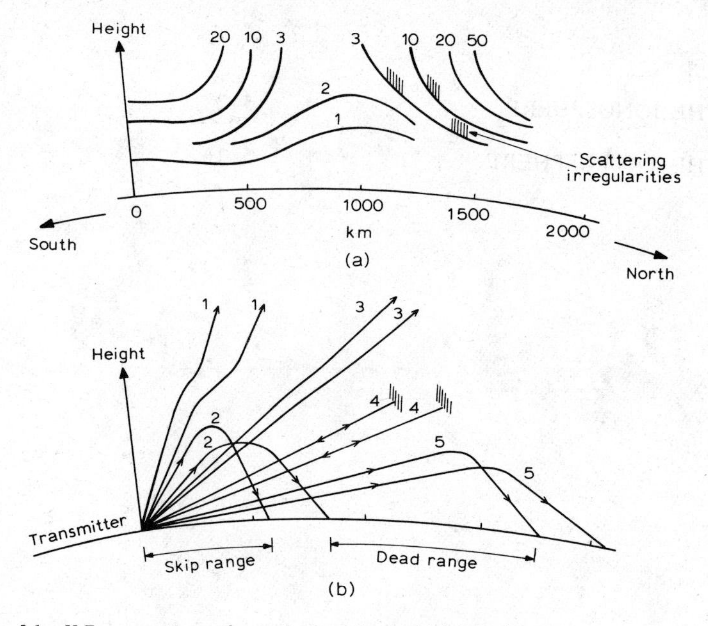

Figure 5.1 *H.F. propagation paths via the ionosphere at high latitudes*
(a) Sample distribution of electron density (arbitrary units) in northern hemisphere high-latitude ionosphere (adapted from Buchau[1]*)*
(b) Raypaths for signals of constant frequency launched with different elevation angles
The ability of the ionosphere to refract, reflect or scatter rays depends on their frequency and elevation angle. Ionospheric refraction is reduced at the higher frequencies and for the higher elevation angles, so that provided the frequency is sufficiently great rays 1 escape whereas rays 2 are reflected back to ground. Rays 3 escape because they traverse the ionosphere at latitudes where the electron density is low (the Muldrew trough[2]*). Irregularities in the F-region are responsible for the direct backscattering of rays 4. The low-elevation rays 5 are reflected to ground because of the increased ionisation at the higher latitudes. Note that for this ionosphere and frequency there are two ground zones which cannot be illuminated.*

The ionosphere is of considerable importance in the engineering of radio communication systems because:

(*a*) It provides the means of establishing various communication paths, calling for system-design criteria based on a knowledge of ionospheric morphology.
(*b*) It requires specific engineering technologies to derive experimental probing facilities to assess its characteristics, both for communication-system planning and management, and for scientific investigations.
(*c*) It permits the remote monitoring by sophisticated techniques of certain distant natural and man-made phenomena occurring on the ground, in the air and in space.

Formation of the ionosphere and its morphology

There is widespread interest in the characteristics of the ionosphere by scientists all over the world. Several excellent general survey books have been published describing the principal known features[3,4] and other more specialised books concerned with aeronomy, and including the ionosphere and magnetosphere, are of great value to the research worker.[5–7] Several journals in the English language are devoted entirely, or to a major extent, to papers describing investigations into the state of the ionosphere and of radio propagation in the ionosphere. See list of further reading.

The formation of the ionosphere is a complicated process involving the ionising influences of solar radiation and solar and cosmic particles on an atmosphere of complex structure. The rates of ion and free-electron production depend on the flux density of the incident radiation or particles, as well as on the ionisation efficiency, which is a function of the ionising wavelength (or particle energies) and the chemical composition of the atmosphere. There are two heights where electron production by the ionisation of molecular nitrogen and atomic and molecular oxygen is a maximum. One occurs at about 100 km and is due to incident X-rays with wavelengths less than about 10 nm and to ultraviolet radiation with wavelengths near 100 nm; the other is at about 170 km and is produced by radiation of wavelengths 20–80 nm.

Countering this production, the free electrons tend to recombine with the positive ions and to attach themselves to neutral molecules to form negative ions. Electrons can also leave a given volume by diffusion or by drifting away under the influences of temperature and pressure gradients, gravitational forces or electric fields set up by the movement of other ionisation. The electron density at a given height is given from the so-called *continuity equation* in terms of the balance between the effects of production and loss.

Night-time electron densities are generally lower than in the daytime because the rates of production are reduced. *Figure 5.2* gives examples of a night-time and a daytime height distribution of electron density. The ionisation is continuous over a wide height range, but there are certain height regions with particular characteristics, and these are known, following E. V. Appleton, by the letters *D*, *E* and *F*. The *E*-region is the most regular ionospheric region, exhibiting a systematic dependence of maximum electron density on solar-zenith angle, leading to predictable diurnal, seasonal and geographical variations. There is also a predictable dependence of its electron density on the changes in solar radiation which accompany the long-term fluctuations in the state of the sun. Maximum *E*-region electron density is approximately proportional to sunspot number, which varies over a cycle of roughly 11 years.

In the daytime the *F*-region splits into two, with the lower part known as the *F1*-region and the upper part as the *F2*-region. This splitting arises because the principal loss mechanism is an ion–atom interchange process followed by dissociative recombination, the former process controlling the loss rates in the *F2*-region and the latter in the *F1*-region. Although maximum production is in the *F1*-region, maximum electron density results in the *F2*-region, where the loss rates are lower. The maximum electron density of the *F1*-region closely follows that of the *E*-region, but there are significant and less predictable changes in its height. The maximum electron density and height of the *F2*-region are subject to large changes which have important consequences to radio wave propagation. Some of these changes are systematic but there are also major day to day variations. It seems likely that the *F2*-region is controlled mainly by ionisation transport to different heights along the lines of force of the Earth's magnetic field under the influence of thermospheric winds at high and middle latitudes [8] and by electric fields at low latitudes.[9] These effects, taken in conjunction with the known variations in atmospheric composition, can largely explain characteristics of the *F2*-region which

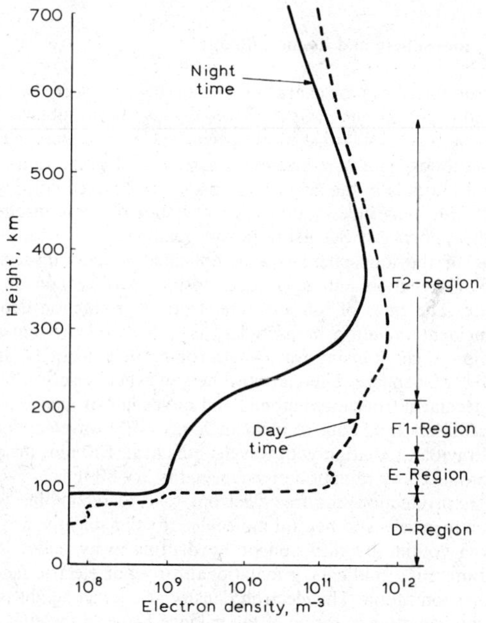

Figure 5.2 Sample night-time and daytime height distributions of electron density at midlatitudes in summer

have in the past been regarded as anomalous by comparison with the *E*-region—namely, diurnal changes in the maximum of electron density in polar regions in the seasons of complete darkness, maximum electron densities at some middle-latitude locations at times displaced a few hours from local noon with greater electron density in the winter than the summer, and at low latitudes longitude variations linked more to the magnetic equator than to the geographic equator, with a minimum of electron density at the magnetic equator and maxima to the north and south where the magnetic dip is about 30°. At all latitudes electron densities in the *F2*-region, like those in the *E*- and *F1*-regions, increase with increase of sunspot number. The electron densities at heights above the maximum of the *F2*-region are controlled mainly by diffusion processes.

The *D*-region shows great variability and fine structure, and is the least well understood part of the ionosphere. The only ionising radiations that can penetrate the upper regions and contribute to its production are hard X-rays with wavelengths less than about 2 nm and Lyman-α radiation at 121.6 nm. Chemical reactions responsible for its formation principally involve nitric oxide and other minor atmospheric constituents.

The *D*-region is mainly responsible for the absorption of radio waves because of the high electron-collision frequencies at such altitudes (see below). While the electron densities in the upper part of the *D*-region appear linked to those in the *E*-region, leading to systematic latitudinal, temporal and solar-cycle variations in absorption, there are also appreciable irregular day to day absorption changes. At middle latitudes anomalously high absorption is experienced on some days in the winter. This is related to warmings of the stratosphere and is probably associated with changes in *D*-region composition. In the lower *D*-region at heights below about 70 km the ionisation is produced principally by energetic cosmic rays, uniformly incident at all times of day. Since the free electrons thereby generated tend to collide and become attached to molecules to form negative ions by night, but are detached by solar radiation in the daytime, the lower *D*-region ionisation, like that in the upper *D*-region, is much greater by day than night. In contrast, however, electron densities in the lower *D*-region, being related to the incidence of cosmic rays, are reduced with increase in the number of sunspots. Additional *D*-region ionisation is produced at high latitudes by incoming particles, directed along the lines of force of the Earth's magnetic field. Energetic electrons, probably originating from the sun, produce characteristic auroral absorption events over a narrow band of latitudes about 10° wide, associated with the visual auroral regions.[10]

From time to time disturbances occur on the sun known as solar flares. These are regions of intense light, accompanied by increases in the solar far ultraviolet and soft X-ray radiation. Solar flares are most common at times of high sunspot number. The excess radiation leads to sudden ionospheric disturbances (SID's), which are rapid and large increases in ionospheric absorption occurring simultaneously over the whole sunlit hemisphere. These persist for from a few minutes to several hours, giving the phenomena of short-wave fadeouts (SWF's), first explained by Dellinger. Accompanied by solar flares are eruptions from the sun of energetic protons and electrons. These travel as a column of plasma, and depending on the position of the flare on the sun's disc and on the trajectory of the Earth, they sometimes impinge on the ionosphere. Then, the protons, which are delayed in transit from fifteen minutes to several hours, produce a major enhancement of the *D*-region ionisation in polar regions that can persist for several days. This gives the phenomenon of polar cap absorption (PCA) with complete suppression of h.f. signals over the whole of both polar regions.[11] Slower particles, with transit times of from 20–40 hours, produce ionospheric storms. These storms, which result principally from movements in ionisation, take the form of depressions in the maximum electron density of the *F2*-region.[12] They can last for several days at a time with effects which are progressively different in detail at different latitudes. Since the sun rotates with a period of about 28 days, sweeping out a column of particles into space when it is disturbed, there is a tendency for ionospheric storms to recur after this time interval.

Additional ionisation is sometimes found in thin layers, 2 km or less thick, embedded in the *E*-region at heights between 90 and 120 km. This has an irregular and patchy structure, a maximum electron density which is much greater than that of the normal *E*-region, and is known as sporadic-*E* (or *Es*) ionisation because of its intermittent occurrence. It consists of patches up to 2 000 km in extent, composed of large numbers of individual irregularities each less than 1 km in size. Sporadic-*E* tends to be opaque to the lower h.f. waves and partially reflecting at the higher frequencies. It results from a number of separate causes and may be classified into different types,[13] each with characteristic occurrence and other statistics. In temperate latitudes sporadic-*E* arises principally from wind shear, close to the magnetic equator it is produced by plasma instabilities and at high latitudes it is mainly due to incident energetic particles. It is most common at low latitudes where it is essentially a daytime phenomenon.

Irregularities also develop in the *D*-region due to turbulence and wind shears and other irregularities are produced in the *F*-region. The *F*-region irregularities can exist simultaneously over a wide range of heights, either below or above the height of maximum electron density and are referred to as spread-*F* irregularities. They are found at all latitudes but are particularly common at low latitudes in the evenings where their occurrence is related to rapid changes in the height of the *F*-region.[14]

Ionospheric effects on radio signals

A radio wave is specified in terms of five parameters: its amplitude, phase, direction of propagation, polarisation and frequency. The principal effects of the ionosphere in modifying these parameters are considered as follows.

REFRACTION

The change in direction of propagation resulting from the traverse of a thin slab of constant ionisation is given approximately by Bouger's law in terms of the refractive index and the angle of incidence. A more exact specification including the effects of the Earth's magnetic field is given by the Haselgrove equation solution.[15] The refractive index is determined from the Appleton-Hartree equations of the magnetoionic theory[16, 17] as a function of the electron density and electron-collision frequency, together with the strength and direction of the Earth's magnetic field, the wave direction and the wave frequency. The dependence on frequency leads to wave dispersion of modulated signals. Since the ionosphere is a doubly-refracting medium it can transmit two waves with different polarisations (see below). The refractive indices appropriate to the two waves differ. Refraction is reduced at the greater wave frequencies, and at v.h.f. and higher frequencies it is given approximately as a function of the ratio of the wave and plasma frequencies, where the plasma frequency is defined in terms of a universal constant and the square root of the electron density.[16] *Table 5.1* lists the magnitude of the refraction and of other propagation parameters for signals at a frequency of 100 MHz which traverse the whole ionosphere.

Table 5.1 EFFECT OF ONE-WAY TRAVERSE OF TYPICAL MID-LATITUDE IONOSPHERE AT 100 MHZ ON SIGNALS WITH ELEVATION ANGLE ABOVE 60 DEGREES[22]

Effect	*Day*	*Night*	*Frequency dependence, f*
Total electron content	5×10^{13} cm^{-2}	5×10^{12} cm^{-2}	
Faraday rotation	15 rotations	1.5 rotations	f^{-2}
Group delay	12.5 μs	1.2 μs	f^{-2}
Change in phase-path length	5.2 km	0.5 km	f^{-2}
Phase change	7 500 radians	750 radians	f^{-2}
Phase stability (peak-to-peak)	±150 radians	±15 radians	f^{-1}
Frequency stability (r.m.s.)	±0.04 Hz	±0.004 Hz	f^{-1}
Absorption (in *D*- and *F*-regions)	0.1 dB	0.01 dB	f^{-2}
Refraction	$\leqslant 1°$	—	f^{-2}

CHANGE IN PHASE-PATH LENGTH

The phase-path length is given approximately as the integral of the refractive index with respect to the ray-path length. Ignoring spatial gradients, the change in phase-path length introduced by passage through the ionosphere to the ground of signals at v.h.f. and higher frequencies from a spacecraft is proportional to the total-electron content. This is the number of electrons in a vertical column of unit cross section.

GROUP DELAY

The group and phase velocities of a wave differ because the ionosphere is a dispersive medium. The ionosphere reduces the group velocity and introduces a group delay which for transionospheric signals at v.h.f. and higher frequencies, like the phase-path change, is proportional to the total-electron content.

POLARISATION

Radio waves that propagate in the ionosphere are called characteristic waves. There are always two characteristic waves known as the ordinary wave and the extraordinary wave; under certain restricted conditions a third wave known as the *Z*-wave can also exist.[16] In general the ordinary and extraordinary waves are elliptically polarised. The polarisation ellipses have the same axial ratio, orientations in space that are related such that under many conditions they are approximately orthogonal, and electric vectors which rotate in opposite directions.[16] The polarisation ellipses are less elongated the greater the wave frequency. Any wave launched into the ionosphere is split into characteristic ordinary and extraordinary wave components of appropriate power. At m.f. and above these components may be regarded as travelling independently through the ionosphere with polarisations which remain related, but continuously change to match the changing ionospheric conditions. The phase paths of the ordinary and extraordinary wave components differ, so that in the case of transionospheric signals when the components have comparable amplitudes, the plane of polarisation of their resultant slowly rotates. This effect is known as Faraday rotation.

ABSORPTION

Absorption arises from inelastic collisions between the free electrons, oscillating under the influence of the incident radio wave, and the neutral and ionised constituents of the atmosphere. The absorption experienced in a thin slab of ionosphere is given by the Appleton-Hartree equations[16] and under many conditions is proportional to the product of electron density and collision frequency, inversely proportional to the refractive index and inversely proportional to the square of the wave frequency. The absorption is referred to as non-deviative or deviative depending on whether it occurs where the refractive index is close to unity. Normal absorption is principally a daytime phenomenon. At frequencies below 5 MHz it is sometimes so great as to completely suppress effective propagation. The absorptions of the ordinary and extraordinary waves differ, and in the range 1.5–10 MHz the extraordinary wave absorption is significantly greater.

AMPLITUDE FADING

If the ionosphere were unchanging the signal amplitude over a fixed path would be constant. In practice, however, fading arises as a consequence of variations in propagation path, brought about by movements or fluctuations in ionisation. The principal causes of fading are:

(*a*) Variations in absorption.
(*b*) Movements of irregularities producing focusing and defocusing.
(*c*) Changes of path length among component signals propagated via multiple paths.
(*d*) Changes of polarisation, such as for example due to Faraday rotation.

These various causes lead to different depths of fading and a range of fading rates. The slowest fades are usually those due to absorption changes which have a period of about 10 minutes. The deepest and most rapid fading occurs from the beating between two signal components of comparable amplitude propagated along different paths. A regularly reflected signal together with a signal scattered from spread-F irregularities can give rise to so-called *flutter* fading, with fading rates of about 10 Hz. A good general survey of fading effects, including a discussion of fading statistics has been produced.[18] On operational communication circuits fading may be combated by space diversity or polarisation-diversity receiving systems and by the simultaneous use of multiple-frequency transmissions (frequency diversity).

FREQUENCY DEVIATIONS

Amplitude fading is accompanied by associated fluctuations in group path and phase path, giving rise to time and frequency-dispersed signals. When either the transmitter or receiver is moving, or there are systematic ionospheric movements, the received signal is also Doppler-frequency shifted. Signals propagated simultaneously via different ionospheric paths are usually received with differing frequency shifts. Frequency shifts for reflections from the regular layers are usually less than 1 Hz, but shifts of up to 20–30 Hz have been reported for scatter-mode signals at low latitudes.[19]

REFLECTION, SCATTERING AND DUCTING

The combined effect of refraction through a number of successive slabs of ionisation can lead to ray reflection. This may take place over a narrow height range as at l.f. or rays may be refracted over an appreciable distance in the ionosphere as at h.f. Weak incoherent scattering of energy occurs from random thermal fluctuations in electron density, and more efficient aspect-sensitive scattering from ionospheric irregularities gives rise to direct backscattered and forward-scatter signals. Ducting of signals to great distances can take place at heights of reduced ionisation between the E- and F-regions, leading in some cases to round-the-world echoes.[20] Ducting can also occur within regions of field-aligned irregularities above the maximum of the F-region.

SCINTILLATION

Ionospheric irregularities act as a phase-changing screen on transionospheric signals from sources such as earth satellites or radiostars. This screen gives rise to diffraction effects with amplitude, phase and angle-of-arrival scintillations.[21]

Communication and monitoring systems relying on ionospheric propagation

Ionospheric propagation is exploited for a wide range of purposes, the choice of system and the operating frequency being largely determined by the type and quantities of data to be transmitted, the path length and its geographical position.

COMMUNICATION SYSTEMS

Radio communication at very low frequencies (v.l.f.) is limited by the available bandwidth, but since ionospheric attenuation is very low, near world-wide coverage can be achieved. Unfortunately the radiation of energy is difficult at such frequencies and complex transmitting aerial systems, coupled with large transmitter powers, are needed to overcome the high received background noise from atmospherics—the electromagnetic radiation produced by lightning discharges. Because of the stability of propagation, v.l.f. systems are used for the transmission of standard time signals and for c.w. navigation systems which rely on direction-finding techniques, or on phase comparisons between spaced transmissions as in the Omega system (10–14 kHz).[23] At low frequencies (l.f.) increased propagation losses limit area coverage, but simpler aerial systems are adequate and lower transmitter powers can be employed because of the reduced atmospheric noise. Low frequency systems are used for communication by on-off keying and frequency-shift keying. Propagation conditions are more stable than at higher frequencies because the ionosphere is less deeply penetrated. Low frequency signals involving ionospheric propagation are also used for communication with submarines below the surface of the sea, with receivers below the ground and with space vehicles not within line-of-sight of the transmitter. Other l.f. systems[24] relying principally on the ground wave, which are sometimes detrimentally influenced by the sky wave at night, include the Decca c.w. navigation system (70–130 kHz), the Loran C pulse navigation system (100 kHz) and long-wave broadcasting.

At medium frequencies (m.f.) daytime absorption is so high as to completely suppress the sky wave. Some use is made of the sky wave at night-time for broadcasting, but generally medium frequencies are employed for ground-wave services. Despite the advent of reliable multichannel satellite and cable systems, high frequencies continue to be used predominantly for broadcasting, fixed, and mobile point-to-point communications, via the ionosphere—there are still tens of thousands of such circuits.

Very high frequency (v.h.f.) communication relying on ionospheric scatter propagation between ground-based terminals is possible. Two-way error-correcting systems with scattering from intermittent meteor trains can be used at frequencies of 30–40 MHz over ranges of 500–1 500 km.[25] Bursts of high-speed data of about 1 s duration with duty cycles of the order of 5% can be achieved, using transmitter powers of about 1 kW. Meteor-burst systems find favour in certain military applications because they are difficult to intercept, since the scattering is usually confined to 5–10 degrees from the great-circle path. Forward-scatter communication systems at frequencies of 30–60 MHz, also operating over ranges of about 1 000 km, rely on coherent scattering from field-aligned irregularities in the *D*-region at a height of about 85 km.[26] They are used principally at low and high latitudes. Signal intensities are somewhat variable, depending on the incidence of irregularities. During magnetically disturbed conditions signals are enhanced at high latitudes, but are little affected at low latitudes. Directional transmitting and receiving aerials with intersecting beams are required. Particular attention has to be paid to avoiding interference from signal components scattered from sporadic-*E* ionisation or irregularities in the *F*-region. Special frequency-modulation techniques involving time division multiplex are used to combat Doppler effects. Systems with 16 channels, automatic error correction and operating at 100 words per minute, now exist.

MONITORING SYSTEMS

High frequency (h.f.) signals propagated obliquely via the ionosphere and scattered at the ground back along the reverse path may be exploited to give information on the characteristics of the scattering region.[27] Increased scatter results from mountains, from cities and from certain sea waves. Signals backscattered from the sea are enhanced when the signal wavelength is twice the component of the sea wavelength along the direction of incidence of the signal, since round-trip signals reflected from successive sea-wave crests then arrive in phase to give coherent addition. The Doppler shift of backscatter returns from sea waves due to the sea motion usually exceeds that imposed on the received signals by the ionosphere, so that Doppler filtering enables the land and sea-scattered signals to be examined separately. This permits studies of distant land–sea boundaries,[28] and since the wavelength of a sea wave is directly related to its velocity, provides a means of synoptic monitoring of distant sea currents. Doppler filtering can also resolve signals reflected or scattered back along ionospheric paths from aircraft, rockets or rocket trails.

Studies of the Doppler shift of stable frequency, h.f., c.w. signals propagated via the ionosphere between ground-based terminals provide important information about infrasonic waves in the *F*-region originating from nuclear explosions,[29] earthquakes,[30] severe thunderstorms[31] and air currents in mountainous regions.

High-altitude nuclear explosions lead to other effects which may be detected by radio means.[32] An immediate wideband electromagnetic pulse is produced which can be monitored throughout the world at v.l.f. and h.f. Also, enhanced *D*-region ionisation, lasting for several days over a wide geographical area, produces an identifiable change in the received phase of long-distance v.l.f. ionospheric signals. Other more localised effects which can be detected include the generation of irregularities in the *F*-region.

Atmospherics may be monitored at v.l.f. out to distances of several thousand kilometres and by recording simultaneously the arrival azimuths at spaced receivers the locations of thunderstorms may be determined and their movements tracked as an aid to meteorological warning services.

Ionospheric probing techniques

There are a wide variety of methods of sounding the state of the ionosphere involving single and multiple-station ground-based equipments, rocket-borne and satellite probes

Figure 5.3 Part of the array of dipoles of the radar antenna at the Jicamarca Radar Observatory, Lima, Peru (U.S. National Bureau of Standards, Cover photo, 1 Feb. 1963, Vol. 139, Science. *Copyright 1963 by the American Association for the Advancement of Science)*

Table 5.2 PRINCIPAL IONOSPHERIC PROBING TECHNIQUES

Height range	*Technique*	*Parameters monitored*	*Site*
	vertical-incidence sounding	up to height of maximum ionisation–electron density	ground
	topside sounding	from height of satellite to height of maximum ionisation–electron density; electron and ion temperatures; ionic composition	satellite
	incoherent scatter	up to few thousand km—electron density; electron temperature; ion temperature; ionic composition; collision frequencies; drifts of ions and electrons	ground
Above 100 km	Faraday rotation and differential Doppler	total-electron content	satellite-ground
	in-situ probes	wide range of parameters	satellite
	c.w. oblique incidence	solar flare effects; irregularities; travelling disturbances; radio aurorae	ground
	pulse oblique incidence	oblique modes by ground backscatter and oblique sounding; meteors; radio aurorae; irregularities and their drifts	ground
	whistlers	out to few Earth radii–electron density; ion temperature; ionic composition	ground or satellite
	vertical-incidence sounding	absorption	ground
	riometer	absorption	ground
	c.w. and pulse oblique incidence	electron density and collision frequency	ground
Below 100 km	wave fields	electron density and collision frequency	rocket-ground
	in-situ probes	electron density and collision frequency; ion density; composition of neutral atmosphere	rocket
	cross-modulation	electron density and collision frequency	ground
	partial reflection	electron density and collision frequency	ground
	lidar	neutral air density; atmospheric aerosols; minor constituents	ground

(see *Table 5.2*). A comprehensive survey of the different techniques has been produced by a Working Group of the International Union of Radio Science (URSI).[33] Some of the techniques involve complex analysis procedures and require elaborate and expensive equipments and aerial systems (*Figure 5.3*); others need only a single radio receiver.

The swept-frequency ground-based *ionosonde* consisting of a co-located transmitter and receiver was developed for the earliest of ionospheric measurements and is still the most widely used probing instrument. The transmitter and receiver frequencies are swept synchronously over the range from about 0.5–1 MHz to perhaps 20 MHz depending on ionospheric conditions, and short pulses typically of duration 100 μs with a repetition rate of 50/s are transmitted. Calibrated film records of the received echoes give the group path and its frequency dependence.

In practice these records require expert interpretation because (*a*) multiple echoes occur, corresponding to more than one traverse between ionosphere and ground (so-called *multiple-hop* modes) or when partially reflecting sporadic-*E* or *F*-region irregularities are present; (*b*) the ordinary and extraordinary waves are sometimes reflected from appreciably different heights, and (*c*) oblique reflections occur when the ionosphere is tilted. Internationally agreed procedures for scaling ionosonde records (ionograms) have been produced.[34]

Since reflection takes place from a height where the sounder frequency is equal to the ionospheric plasma frequency, and since the group path can be related to that height provided the electron densities at all lower heights are known, the data from a full frequency sweep can be used to give the true-height distribution of electron density in the *E*- and *F*-regions up to the height of maximum electron density of the *F2*-region. The conversion of group path to true height requires assumptions regarding missing data below the lowest height from which echoes are received and over regions where the electron density does not increase monotonically with height. The subject of true-height analysis is complex and controversial.[35] Commercially manufactured pulse sounders use transmitter powers of about 1 kW. Sounders with powers of around 100 kW and a lower frequency limit of a few kHz have been operated successfully in areas free from m.f. broadcast interference, to study the night-time *E*-region—this has a maximum plasma frequency of around 0.5 MHz.

Pulse-compression systems[36] and c.w. chirp sounders[37] offer the possibility of improved signal/noise ratios and echo resolution. In the chirp sounder system, originally developed for use at oblique incidence, the transmitter and receiver frequencies are swept synchronously so that the finite echo transit time leads to a frequency modulation of the receiver i.f. signals. These signals are then spectrum-analysed to produce conventional ionograms. Receiver bandwidths of only a few tens of Hertz are needed so that transmitter powers of a few watts are adequate. Other ionosondes have been produced and used operationally which record data digitally on magnetic tape.[34] An ionosonde has been successfully flown in an aircraft to investigate geographical changes in electron density at high latitudes.[38] Over 100 ground-based ionosondes throughout the world make regular soundings each hour of each day; data are published at monthly intervals.[39]

Since 1962 swept-frequency ionosondes have been operated in satellites orbiting the earth at altitudes of around 1 000 km. These are known as *topside sounders* and they give the distributions of electron density from the satellite height down to the peak of the *F2*-region. They also yield other plasma-resonance information, together with data on electron and ion temperatures and ionic composition. Fixed-frequency topside sounders are used to study the spatial characteristics of spread-*F* irregularities and other features with fine structure.

Many different monitoring probes are mounted in satellites orbiting the earth at altitudes above 100 km, to give direct measurements of a range of ionospheric characteristics. These include r.f. impedance, capacitance and upper-hybrid resonance probes for local electron density, modified Langmuir probes for electron temperature, retarding potential analysers and sampling mass spectrometers for ion density, quadrupole and monopole mass spectrometers for ion and neutral-gas analysis and retarding potential analysers for ion temperature measurements.

Ground measurements of satellite beacon signals permit studies of total-electron content, either from the differential Doppler frequency between two harmonically related h.f./v.h.f. signals,[40] or from the Faraday rotation of a single v.h.f. transmission.[41] Beacons on geostationary satellites are valuable for investigations of temporal variations. The scintillation of satellite signals at v.h.f. and u.h.f. gives information on the incidence of ionospheric irregularities, their heights and sizes.[42]

A powerful tool for ionospheric investigations up to heights of several thousand kilometres is the vertical-incidence incoherent-scatter radar. The technique makes use of the very weak scattering from random thermal fluctuations in electron density which exist in a plasma in quasi-equilibrium. Several important parameters of the plasma affect the scattering such that each of these can be determined separately. The power, frequency spectrum and polarisation of the scattered signals are measured and used to give the height distributions of electron density, electron temperature, ionic composition, ion-neutral atmosphere and ion-ion collision frequencies, and the mean plasma-drift velocity. Tristatic receiving systems enable the vertical and horizontal components of the mean plasma drift to be determined. Radars operate at frequencies of 40–1 300 MHz using either pulse or c.w. transmissions. Transmitter peak powers of the order of 1 MW, complex aerial arrays (*Figure 5.3*), and sophisticated data processing procedures are needed. Ground clutter limits the lowest heights that can be investigated to around 100 km.

Electron densities, ion temperatures and ionic composition out to several Earth radii may be studied using naturally occurring whistlers originating in lightning discharges. These are dispersed audio-frequency trains of energy, ducted through the ionosphere and then propagated backwards and forwards along the Earth's magnetic-field lines to conjugate points in the opposite hemisphere. Whistler dispersions may be observed either at the ground or in satellites.[43]

Continuous wave and pulsed signals, transmitted and received at ground-based terminals, may be used in a variety of ways to study irregularities or fluctuations in ionisation. Cross-correlation analyses of the amplitudes on three spaced receivers, of pulsed signals of fixed frequency reflected from the ionosphere at near vertical incidence, give the direction and velocity of the horizontal component of drift.[44] The heights, patch sizes and incidence of *F*-region irregularities responsible for oblique-path forward-scatter propagation at frequencies around 50 MHz may be investigated by means of highly-directional aerials and from signal transit times.[14] Measurements of the Doppler frequency variations of signals from stable c.w. transmitters may be used to study (*a*) ionisation enhancements in the *E*- and *F*-regions associated with solar flares, (*b*) travelling ionospheric disturbances,[45] and (*c*) the frequency-dispersion component of the ionospheric channel-scattering function.[46]

Sporadic-*E* and *F*-region irregularities associated with visual aurorae may be examined by pulsed-radar techniques over a wide range of frequencies from about 6 MHz to 3 000 MHz. They may also be investigated using c.w. bistatic systems in which the transmitter and receiver are separated by several hundred kilometres. Since the irregularities are known to be elongated and aligned along the direction of the Earth's magnetic field and since at the higher frequencies efficient scattering can only occur under restricted conditions, the scattering centres may readily be located. Using low-power v.h.f. beacon transmitters, this technique has proved very popular with radio amateurs. Pulsed meteor radars incorporating Doppler measurements indicate the properties and movements of meteor trains.[25]

Two other oblique-path techniques, giving information on the regular ionospheric regions, are high-frequency ground backscatter sounding and variable-frequency oblique sounding. The former uses a co-located transmitter and receiver, and record interpretation generally involves identifying the skip distance (see *Figure 5.1*) where the signal returns are enhanced because of ray convergence. It is important to use aerials with azimuthal beamwidths of only a few degrees to minimise the ground area illuminated. Long linear aerial arrays with beam slewing, and circularly-disposed banks of log-periodic aerials with monopulsing are used. Oblique-incidence sounders are adaptions of vertical-incidence ionosondes with the transmitter and receiver controlled from stable synchronised sources. Atlases of characteristic records obtained from the two types of sounder under different ionospheric conditions have been produced. Mean models of the ionosphere over the sounding paths may be deduced by matching measured data with ray-tracing results.[47]

So far, no mention has been made of the height region below about 100 km. As already noted, the *D*-region is characterised by a complex structure and high collision frequencies which lead to large daytime absorption of h.f. and m.f. waves. This absorption may be measured using fixed-frequency vertical-incidence pulses,[48] or by monitoring c.w. transmissions at ranges of 200–500 km, where there is no ground-wave component and the dominant signals are reflected from the *E*-region by day and the sporadic-*E* layer by night. There is then little change in the raypaths from day to night so that, assuming night absorption can be neglected, daytime reductions in amplitude are a measure of the prevailing absorption. Multifrequency absorption data give information on the height distributions of electron density.[49] Auroral absorption is often too great to be measured in such ways, but special instruments known as riometers can be used.[50] These operate at a frequency around 30 MHz and record changes in the incident cosmic noise at the ground caused by ionospheric absorption.

D-region electron densities and collision frequencies may be inferred from oblique or vertical-path measurements of signal amplitude, phase, group-path delay and polarisation at frequencies of 10 Hz to 100 kHz, with atmospherics as the signal sources at the lower frequencies. Vertically radiated signals in the frequency range 1.5–6 MHz suffer weak partial reflections from heights of 75–90 km. Measurements of the reflection coefficients of both the ordinary and extraordinary waves, which can be of the order of 10^{-5}, enable electron density and collision-frequency data to be deduced.[51] Pulsed signals with high transmitter powers (50–1 000 kW), highly-directional circularly-polarised aerial systems and very sensitive receivers are needed. As well as *in-situ* probes in rockets, there are a wide range of other schemes for determining electron density and collision frequency, involving the study of wave-fields radiated between the ground and a rocket. These use combinations of frequencies in the v.l.f.–v.h.f. range and include the measurement of differential-Doppler frequency, absorption, differential phase, propagation time and Faraday rotation.

Theory shows that signals propagated via the ionosphere can become cross-modulated by high-power interfering signals which heat the plasma electrons through which the wanted signals pass. This heating causes the electron-collision frequency, and therefore the amplitude of the wanted signal to fluctuate at the modulation frequency of the interfering transmitter. Investigations of this phenomenon (known as the Luxembourg effect after the first identified interfering transmitter) usually employ vertically transmitted and received wanted pulses, modulated by a distant disturbing transmitter radiating synchronised pulses at half the repetition rate. Changes in signal amplitude and phase between successive pulses are measured, and by altering the relative phase of the two transmitters, the height at which the cross-modulation occurs can be varied. Such data enable the height distributions of electron density in the *D*-region to be determined.[52]

Using a Laser radar (Lidar), the intensity of the light backscattered by the atmospheric constituents at heights above 50 km gives the height distributions of neutral-air density and the temporal and spatial statistics of high-altitude atmospheric aerosols. Minor atmospheric constituents may be detected with tunable dye lasers from their atomic and molecular-resonance scattering.

Prediction procedures

Long-term predictions based on monthly median ionospheric data are required for the circuit planning of v.l.f.–h.f. ground-based systems. Estimates of raypath launch and arrival angles are needed for aerial design, and of the relationship between transmitter power and received field strength at a range of frequencies, so that the necessary size of transmitter and its frequency coverage can be determined. Since there are appreciable day to day changes in the electron densities in the *F2*-region, in principle short-term predictions based on ionospheric probing measurements or on correlations with geophysical indices should be of great value for real-time frequency management. In practice, however, aside from the technical problems of devising schemes of adequate accuracy, (*a*) not all systems are frequency agile (e.g. broadcasting); (*b*) effective schemes may require two-way transmissions, and (*c*) only allocated frequencies may be used. An alternative to short-term predictions is real-time channel sounding; certain procedures involve a combination of the two techniques.

LONG-TERM PREDICTIONS

The first requirement of any long-term prediction is a model of the ionosphere. At v.l.f. waves propagate between the Earth and the lower boundary of the ionosphere at heights of 70 km by day and 90 km by night as if in a two-surface waveguide. Very low frequency field-strength predictions are based on a full-wave theory that includes diffraction and surface-wave propagation. For paths beyond 1 000 km range only three or fewer waveguide modes need be considered. A general equation gives field strength as a function of range, frequency, ground-electrical properties and the ionospheric reflection height and reflection coefficients.[53] Unfortunately the reflection coefficients vary in a complex way with electron density and collision frequency, the direction and strength of the Earth's magnetic field, wave frequency and angle of incidence, so that in the absence of accurate *D*-region electron-density data estimates are liable to appreciable error. At l.f. propagation is more conveniently described by wave-hop theory in terms of component waves with different numbers of hops. As at v.l.f. reflection occurs at the base of the ionosphere and the accuracy of the field-strength prediction is largely determined by uncertainties in ionospheric models and reflection coefficients. Medium-frequency signals penetrate the lower ionosphere and are usually reflected from heights of 85–100 km, except over distances of less than 500 km by night when reflection may be from the *F*-region. Large absorptions occur near the height of reflection and daytime signals are very weak. It is now realised that because of the uncertainties in ionospheric models, signal-strength predictions are best based on empirical equations fitted to measured signal-strength data for other oblique paths.

Prediction schemes for h.f. tend to be complicated because they must assess the active modes and elevation angles; these vary markedly with ionospheric conditions and transmitter frequency. Equations are available for the raypaths at oblique incidence through ionospheric models composed of separate layers, each with a parabolic distribution of electron density with height.[54] They are employed in one internationally used

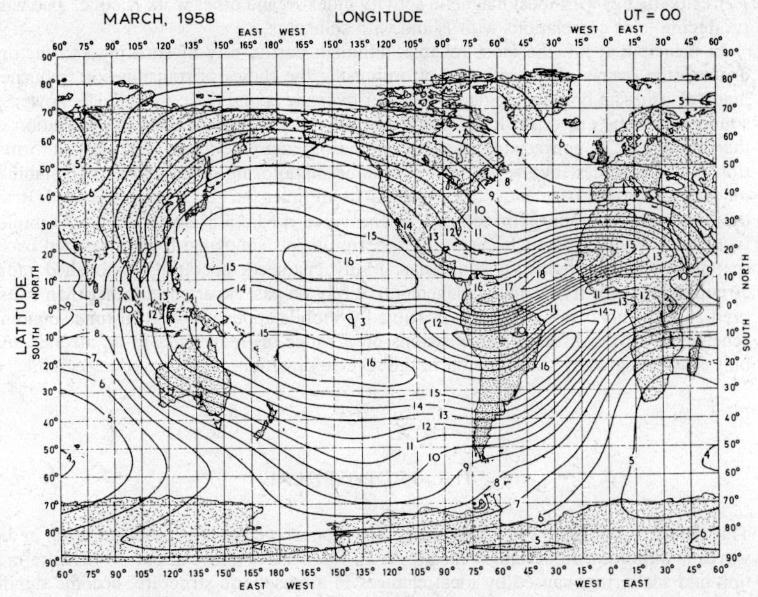

Figure 5.4 Predicted median foF2, MHz for 00h, UT in March 1958 (Reproduced by permission of the Institute for Telecommunication Sciences, Boulder, U.S.A.)

prediction scheme,[55] with the parameters of the parabolas determined from numerical prediction maps of the vertical-incidence ionospheric characteristics, as given by data from the world network of ionosondes,[56] (see *Figure 5.4*). Calculations over a fixed path for a range of frequencies indicate the largest frequency (termed the m.u.f. or maximum usable frequency) that propagates via a given mode. Assuming some statistical law for the day to day variability of the parameters of the model they also give the *availability*, which is the fraction of days that the mode can exist. Received signal strengths are then determined in terms of the transmitter power and a number of transmission loss and gain factors. These include transmitting and receiving-aerial directivity, spatial attenuation, ray convergence gain, absorption, intermediate-path ground-reflection losses and polarisation-coupling losses. Predictions may be further extended by including estimates of the day to day variability in signal intensity. Calculations are prohibitively lengthy without computing aids and a number of computerised prediction schemes have been produced. By means of estimates of background noise intensity, and from the known required signal/noise ratio, the mode reliability may also be determined. This is the fraction of the days that the signals are received with adequate strength. For some systems involving fast data transmission, predictions of the probability of multipath, with two or more modes of specified comparable amplitude with propagation delays differing by less than some defined limit, are also useful and can be made.

SHORT-TERM PREDICTIONS AND REAL-TIME CHANNEL SOUNDING

Some limited success has been achieved in the short-term prediction of the ionospheric characteristics used to give the parameters of the ionospheric models needed for h.f. performance assessment. Schemes are based either on spatial or temporal extrapolation of near real-time data or on correlations with magnetic activity indices. Regression statistics have been produced for the change in the maximum plasma frequency of the *F2*-region (foF2) with local magnetic activity index *K*, and other work is concerned with producing joint correlations with *K* and with solar flux.

In principle at h.f. the most reliable, although costly, way of ensuring satisfactory propagation over a given path and of optimising the choice of transmission frequency involves using an oblique-incidence sounder over the actual path; in practice, however, sounder systems are difficult to deploy operationally, require expert interpretation of their data, lead to appreciable spectrum pollution, and give much redundant information. Some schemes involve low-power channel monitoring of the phase-path stability on each authorised frequency, to ensure that at all times the best available is used. Real-time sounding on one path can aid performance predictions for another. Examples include ray tracing through mean ionospheric models simulated from measured backscatter or oblique-incidence soundings. Many engineers operating established radio circuits prefer, for frequency management, to rely on past experience, rather than to use predictions. This is not so readily possible for mobile applications. Real-time sounding schemes involving ground transmissions on a range of frequencies to an aircraft, but only single-frequency transmission in the reverse direction, have proved successful.[57]

THE TROPOSPHERE

The influence of the lower atmosphere, or troposphere, on the propagation of radio waves is important in several respects. At all frequencies above about 30 MHz refraction and scattering, caused by local changes in atmospheric structure, become significant—especially in propagation beyond the normal horizon. In addition, at frequencies above about 5 GHz, absorption in oxygen and water vapour in the atmosphere is important at certain frequencies corresponding to molecular absorption lines. An understanding of the basic characteristics of these effects is thus essential in the planning of very high frequency communication systems. The main features of tropospheric propagation are summarised from a practical point of view as follows.

Two general problems are: firstly, the influence of the troposphere on the *reliability* of a communication link. Here attention is concentrated on the weak signals which can be received for a large percentage of the time, say 99.99%. Secondly, it is necessary to consider the problem of *interference* caused by abnormal propagation and unusually strong, unwanted signals on the same frequency as the wanted transmission.

In both these aspects of propagation, the radio refractive index of the troposphere plays a dominant role. This parameter depends on the pressure, temperature and humidity of the atmosphere. Its vertical gradient and local fluctuations about the mean value determine the mode of propagation in many important practical situations. Hence the interest in the subject of radio meteorology, which seeks to relate tropospheric structure and radio-wave propagation. In most ground-to-ground systems the height range 0–2 km above the Earth's surface is the important region, but in some aspects of Earth—space transmission, the meteorological structure at greater heights is also significant.

Historical background

Although some experiments on ultra-short-wave techniques were carried out by Hertz and others more than sixty years ago, it was only after about 1930 that any systematic investigation of tropospheric propagation commenced. For a long time it was widely believed that at frequencies above about 30 MHz transmission beyond the geometric horizon would be impossible. However, this view was disputed by Marconi as early as 1932. He demonstrated that, even with relatively low transmitter powers, reception over distances several times the optical range was possible. Nevertheless, theoreticians continued for several years to concentrate on studies of diffraction of ultra-short waves around the Earth's surface. However, their results were found to over-estimate the rate of attenuation beyond the horizon. To correct for this disparity, the effect of refraction was allowed for by assuming a process of diffraction around an Earth with an effective radius of 4/3 times the actual value. In addition, some experimental work began on the effect of irregular terrain and the diffraction caused by buildings and other obstacles.

However, it was only with the development of centimetric radar in the early years of World War II that the limitations of earlier concepts of tropospheric propagation were widely recognised. For several years attention was concentrated on the role of unusually strong refraction in the surface layers, especially over water, and the phenomenon of trapped propagation in a *duct.* It was shown experimentally and theoretically that in this mode the rate of attenuation beyond the horizon was relatively small. Furthermore, for a given height of duct or surface layer having a very large, negative, vertical gradient of refractive index, there was a critical wavelength above which trapping did not occur; a situation analogous to that in waveguide transmission.

Again however it became apparent that further work was required to explain experimental observations. The increasing use of v.h.f., and later u.h.f., for television and radio communication emphasised the need for a more comprehensive approach on beyond-the-horizon transmission. The importance of refractive-index variations at heights of the order of a kilometre began to be recognised and studies of the correlation between the height-variation of refractive index and field strength began in several laboratories.

With the development of more powerful transmitters and antennae of very high gain it proved possible to establish communication well beyond the horizon even in a 'well-mixed' atmosphere with no surface ducts or large irregularities in the height-variation of refractive index. To explain this result, the concept of *tropospheric scatter* was proposed. The trans-horizon field was assumed to be due to incoherent scattering from the random, irregular fluctuations in refractive index produced and maintained by turbulent motion. This procedure has dominated much of the experimental and theoretical work of recent years and it certainly explains some characteristics of troposphere propagation. However, it is inadequate in several respects. It is now known that some degree of stratification in the troposphere is more frequent than was hitherto assumed. The possibility of reflection from a relatively small number of layers or *sheets* of large vertical gradient must be considered, especially at v.h.f. At u.h.f. and s.h.f. strong scattering from a 'patchy' atmosphere, with local regions of large variance in refractive index filling only a fraction of the common volume of the antenna beams, is probably the mechanism which exists for much of the time.

The increasing emphasis on microwaves for terrestrial and space systems has recently focused attention on the effects of precipitation on tropospheric propagation. While absorption in atmospheric gases is not a serious practical problem below 40 GHz, the attenuation in rain and wet snow can impair the performance of links at frequencies of about 10 GHz and above. Moreover, scattering from precipitation may prove to be a significant factor in causing interference between space and terrestrial systems sharing the same frequency. The importance of interference-free sites for Earth stations in satellite links has also stimulated work on the shielding effect of hills and mountains. In addition, the use of large antennae of high gain in space systems requires a knowledge of refraction effects (especially at low angles of elevation), phase variations over the wavefront, and the associated effects of scintillation fading and gain degradation. Particularly at the higher microwave frequencies, thermal noise radiated by absorbing regions of the troposphere (rain, clouds, etc.) may be significant in space communication. Much of the current research is therefore being directed towards a better understanding of the spatial structure of precipitation.

Survey of propagation modes

Figure 5.5 illustrates qualitatively the variation of received power with distance in a homogeneous atmosphere at frequencies above about 30 MHz. For antenna heights of a

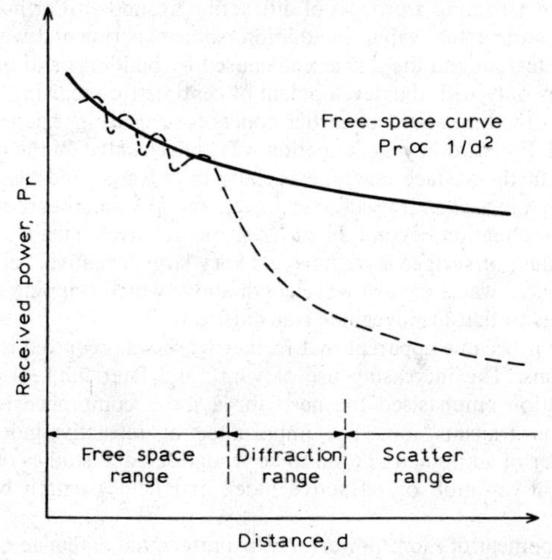

Figure 5.5 Tropospheric attenuation as a function of distance in an homogeneous atmosphere. Direct and ground-reflected rays interfere in the free-space range; obstacle-diffraction effects predominate in the diffraction range; and refractive-index variations are important in the scatter range

wavelength or more, the propagation mode in the free-space range is a *space wave* consisting of a direct and a ground-reflected ray. For small grazing angles the reflected wave has a phase change of nearly 180° at the Earth's surface, but imperfect reflection reduces the amplitude below that of the direct ray. As the path length increases, the signal strength exhibits successive maxima and minima. The most distant maximum will occur where the path difference is $\lambda/2$, where λ is the wavelength.

The range over which the space-wave mode is dominant can be determined geometrically allowing for refraction effects. For this purpose we can assume that the refractive index, n, decreases linearly by about 40 parts in 10^6 (i.e. 40 N units) in the first kilometre. This is the equivalent to increasing the actual radius of the Earth by a factor of 4/3 and drawing the ray paths as straight lines. The horizon distance d, from an antenna at height h above an Earth of effective radius a is

$$d = (2ah)^{\frac{1}{2}} \tag{1}$$

For two antennae 100 m above ground the total range is about 82 km, 15% above the geometric value.

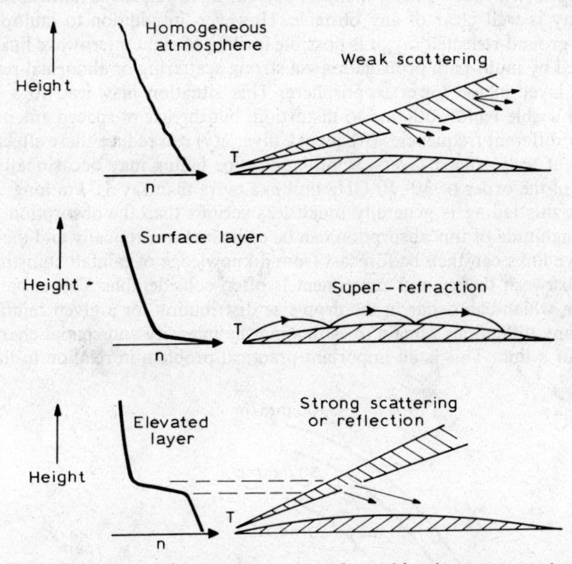

Figure 5.6 Tropospheric propagation modes and height-variation index, n

Beyond the free-space range, diffraction around the Earth's surface and its major irregularities in terrain is the dominant mode, with field strengths decreasing with increasing frequency and being typically of the order of 40 dB below the free space value at 100 km at v.h.f. for practical antenna heights. As the distance increases, the effect of reflection or scattering from the troposphere increases and the rate of attenuation with distance decreases. In an actual inhomogeneous atmosphere the height-variation of n is the dominant factor in the *scatter zone* as illustrated in *Figure 5.6*. However, in practice the situation is rarely as simple as that indicated by these simple models.

Ground-wave terrestrial propagation

When the most distant maximum in *Figure 5.5* occurs at a distance small compared with the optical range, it is often permissible to assume the Earth flat and perfectly reflecting, particularly at the low-frequency end of the v.h.f. range. The space-wave field at E at a distance d is then given by

$$E = (90W^{\frac{1}{2}}h_t h_r)/\lambda d^2 \qquad (2)$$

where W is the power radiated from a $\lambda/2$ dipole, and h_t and h_r are the heights of the transmitting and receiving antennae respectively.

The effects of irregular terrain are complex. There is some evidence that, for short, line-of-sight links, a small degree of surface roughness increases the field strength by eliminating the destructive interference between the direct and ground-reflected rays. Increasing the terrain irregularity then reduces the field strength, particularly at the higher frequencies, as a result of shadowing, scattering and absorption by hills, buildings and trees. However, in the particular case of a single, obstructing ridge visible from both terminals it is sometimes possible to receive field-strengths greater than those over level terrain at the same distance. This is the so-called *obstacle gain.*

In designing microwave radio-relay links for line-of-sight operation it is customary to so locate the terminals that, even with unfavourable conditions in the vertical gradient of refractive index (with *sub-refraction* effects decreasing the effective radius of the Earth), the direct ray is well clear of any obstacle. However, in addition to multipath fading caused by a ground-reflected ray, it is possible for line-of-sight microwave links to suffer fading caused by multi-path propagation via strong scattering or abnormal refraction in an elevated layer in the lower troposphere. This situation may lead to a significant reduction in usable bandwidth and to distortion, but the use of spaced antennae (space diversity) or different frequencies (frequency diversity) can reduce these effects. Even in the absence of well-defined layers, scintillation-type fading may occasionally occur at frequencies of the order of 30–40 GHz on links more than say 10 km long.

However, this fading is generally much less serious than the absorption caused by rain. The magnitude of this absorption can be estimated theoretically and the reliability of microwave links can then be forecast from a knowledge of rainfall statistics. But the divergence between theory and experiment is often considerable. This is partly due to the variation which can occur in the drop-size distribution for a given rainfall rate. In addition, many difficulties remain in estimating the intensity and spatial characteristics of rainfall for a link. This is an important practical problem in relation to the possible

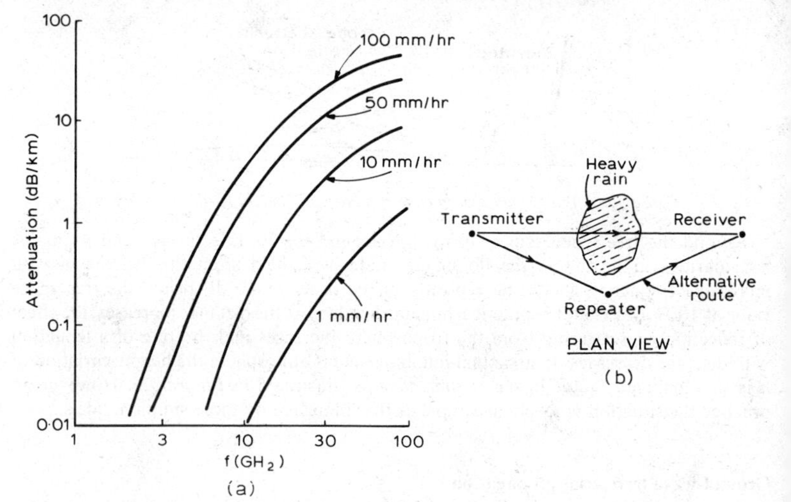

Figure 5.7 (a) Attenuation in rain, and (b) the application of route diversity to minimise effects of fading

use of *route diversity* to minimise the effects of absorption fading. Experimental results show that for very high reliability (i.e. for all rainfall rates less than say 50–100 mm/hr in temperate climates) and terrestrial link operating at frequencies much above 30 GHz must not exceed say 10 km in length. It is possible, however, to design a system with an alternative route so that by switching between the two links the worst effects of localised, very heavy rain can be avoided. The magnitude of attenuation in rain is shown in *Figure 5.7* (*a*) and the principle of route diversity is illustrated in *Figure 5.7* (*b*).

Beyond-the-horizon propagation

Although propagation by surface or elevated layers (see *Figure 5.5*) cannot generally be utilised for practical communication circuits, these features remain important as factors in co-channel interference. Considerable theoretical work, using waveguide mode theory, has been carried out on duct propagation and the results are in qualitative agreement with experiment. Detailed comparisons are difficult because of the lack of knowledge of refractive index structure over the whole path, a factor common to all beyond-the-horizon experiments. Nevertheless, the theoretical predictions of the maximum wavelength trapped in a duct are in general agreement with practical experience. These values are as follows:

$\lambda(max)$ *in cm*	*Duct height in m*
1	5
10	25
100	110

Normal surface ducts are such that complete trapping occurs only at centimetric wavelengths. Partial trapping may occur for the shorter metric wavelengths. Over land the effects of irregular terrain and of thermal convection (at least during the day) tend to inhibit duct formation. For a ray leaving the transmitter horizontally, the vertical gradient of refractive index must be steeper than -157 parts in 10^6 per kilometre.

Even when super-refractive conditions are absent, there remains considerable variability in the characteristics of the received signal. This variability is conveniently expressed in terms of the transmission loss, which is defined as $10 \log (P_T/P_r)$, where P_T and P_r are the transmitted and received powers respectively. In scatter propagation, both slow and rapid variations of field strength are observed. Slow fading is the result of large scale changes in refractive conditions in the atmosphere and the hourly median values below the long-term median are distributed approximately log-normally with a standard deviation which generally lies between 4 and 8 decibels, depending on the climate. The largest variations of transmission loss are often seen on paths for which the receiver is located just beyond the diffraction region, while at extreme ranges the variations are less. The slow fading is not strongly dependent on the radio frequency. The rapid fading has a frequency of a few fades per minute at lower frequencies and a few Hertz at u.h.f. The superposition of a number of variable incoherent components would give a signal whose amplitude was Rayleigh-distributed. This is found to be the case when the distribution is analysed over periods of up to five minutes. If other types of signal form a significant part of that received, there is a modification of this distribution. Sudden, deep and rapid fading has been noted when a frontal disturbance passes over a link. In addition, reflections from aircraft can give pronounced rapid fading.

The long-term median transmission loss relative to the free-space value increases approximately as the first power of the frequency up to about 3 GHz. Also, for most temperate climates, monthly median transmission losses tend to be higher in winter than in summer, but the difference diminishes as the distance increases. In equatorial climates, the annual and diurnal variations are generally small. The prediction of transmission loss, for various frequencies, path lengths, antenna heights, etc., is an important practical problem. An example of the kind of data required is given in *Figure 5.8*.

At frequencies above 10 GHz, the heavy rain occurring for small percentages of the time causes an additional loss due to absorption, but the accompanying scatter from the rain may partly offset the effect of absorption.

Physical basis of tropospheric scatter propagation

Much effort has been devoted to explaining the trans-horizon field in terms of scattering theory based on statistical models of turbulent motion. The essential physical feature of

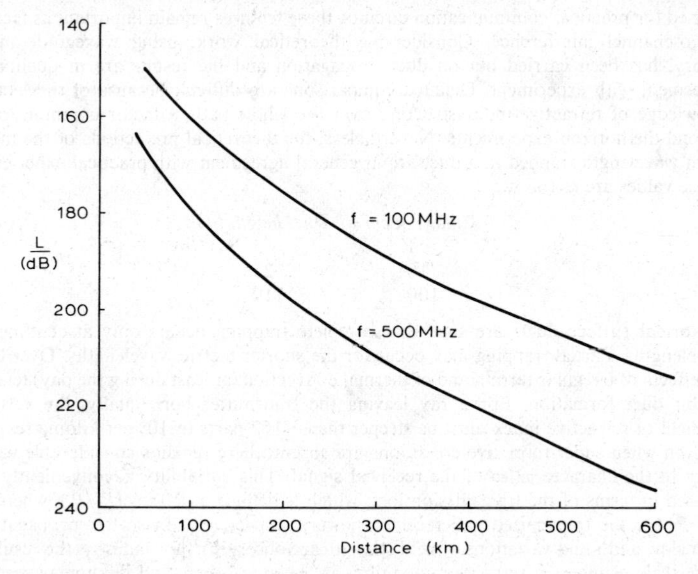

Figure 5.8 Median transmission loss, L, between isotropic antennas in a temperate climate and over an average rolling terrain. The height of the transmitting antenna is 40 m, and the height of the receiving antenna is 10 m

this approach is an atmosphere consisting of irregular *blobs* in random motion which in turn produce fluctuations of refractivity about a stationary mean value. Using this concept, some success has been achieved in explaining the approximate magnitude of the scattered field but several points of difficulty remain. There is now increasing evidence, from refractometer and radar probing of the troposphere, that some stratification of the troposphere is relatively frequent.

By postulating layers of varying thickness, horizontal area and surface roughness, and of varying lifetime it is possible in principle to interpret many of the features of tropospheric propagation. Indeed, some experimental results (e.g. the small distance-dependence of v.h.f. fields at times of anomalous propagation) can be explained by calculating the reflection coefficient of modal layers of constant height and with an idealised height-variation of refractive index such as half-period sinusoidal, exponential, etc. The correlation between field strength and layer height has also been examined and some results can be explained qualitatively in terms of *double-hop* reflection from extended layers. Progress in ray-tracing techniques has also been made. Nevertheless, the problems of calculating the field strength variations on particular links remain formidable, and for many practical purposes statistical and empirical techniques for predicting link performance remain the only solution.

Other problems related to fine structure are space and frequency diversity. On a v.h.f. scatter link with antennas spaced normal to the direction of propagation, the correlation coefficient may well fall to say 0·5 for spacings of 5–30λ in conditions giving fairly rapid fading. Again, however, varying meteorological factors play a dominant role. In frequency diversity, a separation of say 3 or 4 MHz may ensure useful diversity operation in many cases, but occasionally much larger separations are required. The irregular structure of the troposphere is also a cause of gain degradation. This is the decrease in actual antenna gain below the ideal free-space value. Several aspects of the irregular refractive-index structure contribute to this effect and its magnitude depends somewhat on the time interval over which the gain measurement is made. Generally, the decrease is only significant for gains exceeding about 50 dB.

Tropospheric effects in space communications

In space communication, with an Earth station as one terminal, several problems arise due to refraction, absorption and scattering effects, especially at microwave frequencies. For low angles of elevation of the Earth station beam, it is often necessary to evaluate the refraction produced by the troposphere, i.e., to determine the error in observed location of a satellite. The major part of the bending occurs in the first two kilometres above ground and some statistical correlation exists between the magnitude of the effect and the refractive index at the surface. For high-precision navigation systems and very narrow beams it is often necessary to evaluate the variability of refraction effects from measured values of the refractive index as a function of height. A related phenomenon important in tracking systems is the phase distortion in the wave-front due to refractive index fluctuations, a feature closely linked with gain degradation. This phase distortion also affects the stability of frequencies transmitted through the troposphere.

Absorption in precipitation (see *Figure 5.7* (*a*)) has aleady been mentioned in relation to terrestrial systems. Water drops attenuate microwaves both by scattering and by absorption. If the wavelength is appreciably greater than the drop-size then the attenuation is caused almost entirely by absorption. For rigorous calculations of absorption it is necessary to specify a drop-size distribution; but this, in practice, is highly variable and consequently an appreciable scatter about the theoretical value is found in experimental measurements. Moreover, statistical information on the vertical distribution of rain is very limited. This makes prediction of the reliability of space links difficult and emphasises the value of measured data. Some results obtained, using the sun as an extraterrestrial source, are shown in *Figure 5.9*. In addition, scatter from rain (being approximately isotropic) may cause appreciable interference on co-channel terrestrial and space systems even when the beams from the two systems are not directed towards each other on a great-circle path. This aspect of 'scatter propagation' requires further study, particularly in relation to the screening effect of local hills on other modes of propagation.

Because precipitation (and to a smaller extent the atmospheric gases) absorb microwaves, they also radiate thermal-type noise. It is often convenient to specify this in terms of an *equivalent black-body temperature* or simply *noise temperature* for an antenna pointing in a given direction. With radiometers and low-noise receivers it is now possible to measure this tropospheric noise and assess its importance as a factor in limiting the performance of a microwave Earth–space link. For a complete solution, it is necessary to consider not only direct radiation into the main beam but also ground-reflected radiation, and emission from the ground itself, arriving at the receiver via side and back lobes. From the meteorological point of view, radiometer soundings (from the ground, aircraft, balloons or satellites) can provide useful information on tropospheric and stratospheric structure. Absorption in precipitation becomes severe at frequencies

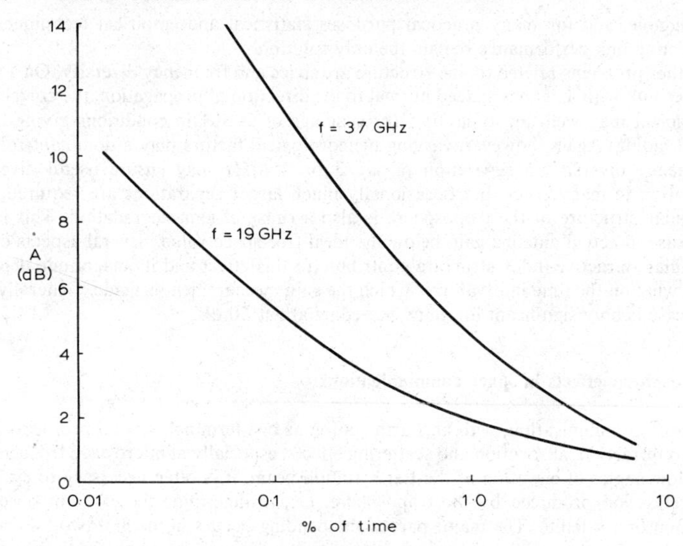

Figure 5.9 Measured probability distribution of attenuation A *on earth-space path at 19 GHz and 37 GHz (S.E. England: Elevation angles 5° to 40°: data from solar-tracking radiometers)*

above about 30 GHz and scintillation effects also increase in importance in the millimetre range. However, for space links in or near the vertical direction the system reliability may be sufficient for practical application even at wavelengths as low as 3–4 mm. Moveover, spaced receivers in a site-diversity system can be used to minimise the effects of heavy rain.

Techniques for studying tropospheric structure

The importance of a knowledge of the structure of the troposphere in studies of propagation is clearly evident in the above sections. The small-scale variations in refractive index and in the intensity of precipitation are two important examples. They form part of the general topic of *tropospheric probing*.

Much useful information on the height-variation of refractive index can be obtained from the radio-sondes carried on free balloons and used in world-wide studies of meteorological structure. However, for many radio applications these devices do not provide sufficient detail. To obtain this detail instruments called refractometers have been developed, mainly for use in aircraft, on captive balloons or on tall masts. They generally make use of a microwave cavity for measuring changes in a resonance frequency, which in turn is related to the refractive index of the enclosed air. Such refractometers are robust, rapid-response instruments which have been widely used as research tools, though they have yet to be developed in a form suitable for widespread, routine use.

High-power, centimetric radar is also a valuable technique. By its use it is possible to detect layers or other regions of strong scatter in the troposphere, and to study their location and structure. Joint radar-refractometer soundings have proved of special interest in confirming that the radar does indeed detect irregularities in clear-air structure. The application of radar in precipitation studies is, of course, a well-known and widely used technique in meteorology; although to obtain the detail and precision necessary for radio applications requires careful refinements in technique.

Optical radar (lidar) and acoustic radar have also been used to probe the troposphere, although the information they provide is only indirectly related to radio refractive index.

The millimetre and sub-millimetre spectrum, as yet not exploited to any significant degree for communications, is nevertheless a fruitful region for tropospheric probing. In particular, the presence of several absorption lines (in water vapour, oxygen and minor constituents such as ozone) makes it possible to study the concentration and spatial distribution of these media. Near the ground, direct transmission experiments are feasible; for example, to study the average water-vapour concentration along a particular path. In addition, it is possible to design radiometers for use on the ground, in an aircraft or in a satellite, which will provide data on the spatial distribution of absorbing atmospheric constituents by measurement of the emission noise they radiate. This topic of *remote probing* is one exciting considerable current interest in both radio and meteorology.

REFERENCES

1. BUCHAU, J., 'Instantaneous versus averaged Ionosphere', *Air Force Surveys in Geophysics No. 241* (*Air Force Systems Command, United States Air Force*), **1** (1972)
2. MULDREW, D. B., 'F-layer Ionisation Troughs deduced from Alouette Data', *J. Geophys. Res.*, **70**, 2635 (1965)
3. DAVIES, K., *Ionospheric Radio Propagation*, Monograph 80, National Bureau of Standards, Washington (1965)
4. RATCLIFFE, J. A., *Sun, Earth and Radio—an Introduction to the Ionosphere and Magnetosphere*, Weidenfeld and Nicolson, London (1970)
5. RISHBETH, H. and GARRIOTT, O. K., *Introduction to Ionospheric Physics*, Academic Press, London (1969)
6. DAVIES, K., *Ionospheric Radio Waves*, Blaisdell, Waltham Mass. (1969)
7. RATCLIFFE, J. A., *An Introduction to the Ionosphere and Magnetosphere*, Cambridge University Press, Cambridge (1972)
8. RISHBETH, H., Thermospheric Winds and the F-region: a Preview, *J. Atmosph. Terr. Phys.*, **34**, 1 (1972)
9. DUNCAN, R. A., 'The Equatorial F-region of the Ionosphere', *J. Atmosph. Terr. Phys.*, **18**, 89 (1960)
10. HARTZ, T. R., 'The General Pattern of Auroral Particle Precipitation and its Implications for high Latitude Communication Systems', *Ionospheric Radio Communications*, ed. K. Folkestad, Plenum, New York, 9 (1968)
11. BAILEY, D. K., 'Polar Cap Absorption', *Planet. Space Sci.*, **12**, 495 (1964)
12. MATSUSHITA, S., 'Geomagnetic Disturbances and Storms', *Physics of Geomagnetic Phenomena*, ed. Matsushita, S., and Campbell, W. H., Academic Press, London, 793 (1967)
13. SMITH, E. K. and MATSUSHITA, S., *'Ionospheric Sporadic-E'*, Macmillan, New York (1962)
14. COHEN, R. and BOWLES, K. L., 'On the Nature of Equatorial Spread-F', *J. Geophys. Res.*, **66**, 1081 (1961)
15. HASELGROVE, J., 'Ray Theory and a New Method for Ray Tracing', *Report on Conference on Physics of Ionosphere*, Phys. Soc. London, 355 (1954)
16. RATCLIFFE, J. A., *'The Magnetoionic Theory'*, Cambridge University Press, Cambridge (1959)
17. BUDDEN, K., *Radio Waves in the Ionosphere*, Cambridge University Press, Cambridge (1961)
18. C.C.I.R. REPORT 266–2, 'Fading of Radio Signals Received via the Ionosphere', *Documents of 12th Plenary Assembly New Delhi*, I.T.U., Geneva (1970)
19. NIELSON, D. L., 'The Importance of Horizontal F-Region Drifts to Transequatorial VHF Propagation', *Scatter Propagation of Radio Waves*, ed. Thrane, E., *AGARD Conference Proceedings No. 37*, NATO, Neuilly-sur-Seine, France (1968)
20. FENWICK, F. B. and VILLARD, O. G., 'A Test of the Importance of Ionosphere Reflections in

Long Distance and Around-the-World High Frequency Propagation', *J. Geophys. Res.*, **68**, 5659 (1963)

21. RATCLIFFE, J. A., 'Some Aspects of Diffraction Theory and their Application to the Ionosphere', *Reports on Progress in Physics*, Phys. Soc., London, **19**, 188 (1956)
22. C.C.I.R. REPORT 263–2, 'Ionospheric Effects upon Earth-space Radio Propagation', *Documents of 12th Plenary Assembly, New Delhi*, I.T.U. Geneva (1970)
23. PIERCE, J. A., 'OMEGA', *I.E.E.E. Trans. Aer. and Elect. Syst.*, **1**, 206 (1965)
24. STRINGER, F. S., 'Hyperbolic Radionavigation Systems', *Wireless World*, **75**, 353 (1969)
25. SUGAR, G. R., 'Radio Propagation by Reflection from Meteor Trails', *Proc. I.E.E.E.*, **52**, 116 (1964)
26. BAILEY, D. K., BATEMAN, R. and KIRBY, R. C., 'Radio Transmission at VHF by Scattering and Other Processes in the Lower Ionosphere', *Proc. I.R.E.*, **43**, 1181 (1955)
27. CROFT, T. A., 'Skywave Backscatter: a Means for Observing Our Environment at Great Distances', *Rev. Geophys. and Space Physics*, **10**, 73 (1972)
28. BLAIR, J. C., MELANSON, L. L. and TVETEN, L. H., 'HF Ionospheric Radar Ground-Scatter Map showing Land-Sea Boundaries by a Spectral Separation Technique', *Electronics Letters*, **5**, 75 (1969)
29. BAKER, D. M. and DAVIES, K., 'Waves in the Ionosphere Produced by Nuclear Explosions', *J. Geophys. Res.*, **73**, 448 (1968)
30. DAVIES, K. and BAKER, D. M., 'Ionospheric Effects Observed Around the Time of the Alaskan Earthquake of March 28 1964', *J. Geophys. Res.*, **70**, 2251 (1965)
31. BAKER, D. M. and DAVIES, K., 'F2-region Acoustic Waves from Severe Weather', *J. Atmosph. Terr. Phys.*, **31**, 1345 (1969)
32. PIERCE, E. T., 'Nuclear Explosion Phenomena and their Bearing on Radio Detection of the Explosions', *Proc. I.E.E.E.*, **53**, 1944 (1965)
33. SMITH, E. K., 'Electromagnetic Probing of the Upper Atmosphere', ed. U.R.S.I. Working Group, *J. Atmosph. Terr. Phys.*, **32**, 457 (1970)
34. PIGGOTT, W. R. and RAWER, K., *U.R.S.I. Handbook of Ionogram Interpretation and Reduction*, 2nd edn., Rep. UAG-23, Dept. of Commerce, Boulder, U.S.A. (1972)
35. BEYNON, W. J. G., 'Special Issue on Analysis of Ionograms for Electron Density Profiles', ed., U.R.S.I. Working Group, *Radio Science*, **2**, 1119 (1967)
36. COLL, D. C. and STOREY, J. R., 'Ionospheric Sounding using Coded Pulse Signals', *Radio Science*, **68D**, 1155 (1964)
37. FENWICK, R. B. and BARRY, G. H., 'Sweep Frequency Oblique Ionospheric Sounding at Medium Frequencies', *I.E.E.E. Trans. Broadcasting*, **12**, 25 (1966)
38. WHALEN, J. A., BUCHAU, J. and WAGNER, R. A., 'Airborne Ionospheric and Optical Measurements of Noontime Aurora', *J. Atmosph. Terr. Phys.*, **33**, 661 (1971)
39. Ionospheric Data—Series FA—published monthly for National Geophysical and Solar-Terrestrial Data Centre, Boulder, U.S.A.
40. GARRIOTT, O. K. and NICHOL, A. W., 'Ionospheric Information Deduced from the Doppler Shifts of Harmonic Frequencies from Earth Satellites', *J. Atmosph. Terr. Phys.*, **22**, 50 (1965)
41. ROSS, W. J., 'Second-Order Effects in High Frequency Transionospheric Propagation', *J. Geophys. Res.*, **70**, 597 (1965)
42. AARONS, J., 'Total-Electron Content and Scintillation Studies of the Ionosphere', ed. *AGARDograph 166*, NATO, Neuilly-sur-Seine, France (1973)
43. HELLIWELL, R. A., *Whistlers and Related Ionospheric Phenomena*, Stanford University Press., Stanford, California (1965)
44. MITRA, S. N., 'A Radio Method of Measuring Winds in the Ionosphere', *Proc. I.E.E.*, **46**, Pt. III, 441 (1949)
45. MUNRO, G. H., 'Travelling disturbances in the ionosphere', *Proc. Roy. Soc.*, **202A**, 208 (1950)
46. BELLO, P. A., 'Some Techniques for Instantaneous Real-Time Measurements of Multipath and Doppler Spread', *I.E.E.E. Trans. Comm. Tech.*, **13**, 285 (1965)
47. CROFT, T. A., 'Special Issue on Ray Tracing', ed. *Radio Science*, **3**, 1 (1968)
48. APPLETON, E. and PIGGOTT, W. R., 'Ionospheric Absorption Measurements during a Sunspot Cycle', *J. Atmosph. Terr. Phys.*, **5**, 141 (1954)

49. BEYNON, W. J. G. and RANGASWAMY, S., 'Model Electron Density Profiles for the Lower Ionosphere', *J. Atmosph. Terr. Phys.*, **31**, 891 (1969)
50. HARGREAVES, J. K., 'Auroral Absorption of H. F. Radio Waves in the Ionosphere—a Preview of Results from the First Decade of Riometry', *Proc. I.E.E.E.*, **57**, 1348 (1969)
51. BELROSE, J. S. and BURKE, M. J., 'Study of the Lower Ionosphere Using Partial Reflections', *J. Geophys. Res.*, **69**, 2799 (1964)
52. FEJER, J. A., 'The Interaction of Pulsed Radio Waves in the Ionosphere', *J. Atmosph. Terr. Phys.*, **7**, 322 (1955)
53. C.C.I.R. REPORT 265–2, 'Sky-Wave Propagation at Frequencies Below About 150 kHz With Particular Emphasis on Ionospheric Effects', *Documents of 12th Plenary Assembly, New Delhi*, I.T.U., Geneva (1970)
54. APPLETON, E. V. and BEYNON, W. J. G., 'The Application of Ionospheric Data to Radio Communications', *Proc. Phys. Soc.*, **52**, 518 (1940); and **59**, 58 (1947)
55. C.C.I.R. REPORT 252–2, 'C.C.I.R. Interim Method for Estimating Sky-Wave Field Strength and Transmission Loss at Frequencies Between the Approximate Limits of 2 and 30 MHz', *Documents of 12th Plenary Assembly, New Delhi*, I.T.U. Geneva (1970)
56. C.C.I.R. REPORT 340–1, 'C.C.I.R. Atlas of Ionospheric Characteristics', *Documents of 12th Plenary Assembly, New Delhi*, I.T.U. Geneva (1970)
57. STEVENS, E. E., 'The CHEC Sounding System', *Ionospheric Radio Communications*, ed. K. Kolkestad, Plenum, New York, 359 (1968)

FURTHER READING

The Ionosphere

Journal of Atmospheric and Terrestial Physics, Pergamon Press, Oxford
Journal of Geophysical Research, American Geophysical Union, Washington
Planetary and Space Science, Pergamon Press, Oxford

Space Research, Proceedings, Plenary Meetings, COSPAR Committee on Space Research, Akademie-Verlag, Berlin
Space Science Reviews, Reidel, Dordrecht, Holland

The Troposphere

BEAN, B. R. and DUTTON, E. J., *Radio Meteorology*, Monograph 92, U.S. Government Printing Office, Washington (1966)
CASTEL, F. DU., *Tropospheric Radiowave Propagation Beyond the Horizon*, Pergamon Press, Oxford (1966)
C.C.I.R. XIITH PLENARY ASSEMBLY, 2, Pt, 1., 'Propagation in Non-Ionised Media', I.T.U. Geneva (1970)
I.E.E. CONFERENCE PUBLICATION, NO. 48, 'Tropospheric Propagation', London (1968)
I.E.E. CONFERENCE PUBLICATION, NO. 98, 'Propagation of Radio Waves at Frequencies above 10 GHz.', London (1973)
SAXTON, J. A., Editor, *Advances in Radio Research*, Academic Press, London (1964)

6 ELECTRONIC MATERIALS AND COMPONENTS

ELECTRONIC MATERIALS AND COMPONENTS 6

6 ELECTRONIC MATERIALS AND COMPONENTS

RESISTIVE MATERIALS AND COMPONENTS

Basic laws of resistivity

Ohm's law states that: The ratio of the potential difference E between the ends of a conductor to the current L flowing in it is a constant, provided the physical conditions of the conductor are unaltered. Ohm defined the resistance, R, of the given conductor as the ratio E/I.

The practical units in which the law is expressed are resistance in ohms, potential difference in volts and current in amperes. Until 1948 the standard of resistance was the International Ohm, the resistance offered to an unvarying electric current by a column of mercury of mass 14.452 1 g, of constant cross-section and 106.300 cm in length, maintained at the temperature of melting ice. Resistors made in this way by the five main national standardising laboratories of the world attain equality to within a few parts in one hundred thousand.

The Absolute Ohm, as from the 1 January 1948, replaced the International Ohm. Onc Absolute Ohm – 0.999 51 International Ohm. The Absolute Ohm has been determined, in England, by rotation of a conductor under specified conditions (Lorenz method).

Electromagnetic theory shows that resistance has the same dimensions as the product of inductance and frequency, and the Lorenz method makes use of this fact by rotating a conductor to cut the entire flux produced by coils of accurately known inductance. The current supplying the inductance coils passes through the standard resistor being calibrated, and the e.m.f. produced by the rotating conductor is arranged, by adjustment of known speed, to balance the IR drop across the resistor. The method is accurate to a few parts in 100 000.

If M is total flux when the current is I amperes, the change in flux linkage per revolution is MI. If the conductor makes n rev/s, the e.m.f. induced in the circuit by rotation $= MIn$.

Potential drop across resistor $= IR$

At balance $MIn = IR$

$$R = Mn$$

M is known, by calculation from the geometry and spacing of the known coils, and n is obtained by accurate frequency measurement.

The resistance R of a given conductor is proportional to its length l and inversely proportional to its area of cross-section a. Thus R varies as l/a and $R = p(l/a)$ where p is a constant of the material known as its *specific resistance* or *resistivity*. When l and a are unity, $R = p$; thus the resistivity of a given material may be defined as the resistance of *unit cube* of the material. In the metric system p may be expressed in *ohms per centimetre cube* or simply *ohm-centimetres*.

It will be seen that resistivity may be expressed in many terms, of which the following are those in general use:

(*a*) Microhms per centimetre cube (microhm/cm)
(*b*) Ohms per square mil per foot (Ω/sq mil foot)
(*c*) Ohms per circular mil per foot (Ωcir mil foot)

In connection with the above terms, there are three points which should be noted:

(i) That a mil is an expression used to denote one thousandth of an inch (0.001 in) and should not be confused with millimetres.
(ii) Microhm/cm is sometimes expressed as microhms/cm^3.
(iii) Resistivity must always be qualified by reference to the temperature at which it is measured, usually 20 °C.

Resistance materials

MASS AND VOLUME RESISTIVITIES

The units of mass resistivity and volume resistivity are interrelated through the density. For copper wires this is stated in the IACS (International Annealed Copper Standard) as 8.89 gram/cm^3 at 20 °C. The volume resistivity, ρ; the mass resistivity, δ; and density, d, are related in the formula

$$\delta = \rho d$$

The IACS in various units of mass and volume resistivity, all at 20 °C, is:

0.153 28	ohm-gram/metre^2
875.20	ohm-pound/mile^2
0.017 241	ohm-millimetre^2/metre
1.724 1	microhm-centimetre
0.678 79	microhm-inch
10.371	ohm-circular mil/foot

TEMPERATURE COEFFICIENT OF RESISTANCE

At a temperature of 20 °C, the coefficient of variation of resistance with temperature of standard annealed copper-measured between two potential points rigidly fixed to the wire—the metal being allowed to expand freely, is given as 0.003 93 = 1/254.45 per degree centigrade. The temperature coefficient of resistance of a copper wire of constant mass and conductivity of 100% at 20 °C is therefore 0.003 93 per degree centigrade.

For other conductivities of copper, the 20 °C temperature of resistance is given by multiplying the decimal number expressing the per cent conductivity by 0.003 93.

The conductivity of copper may be calculated from the coefficient of resistance. The temperature coefficient of resistance, α, for different initial Centigrade temperatures and different conductivities may be calculated from the formula,

$$\alpha t_1 = \frac{1}{\dfrac{1}{\eta(0.003\,93)} + (t_1 - 20)} \qquad (1)$$

in which η = the conductivity expressed decimally
αt_1 = temperature coefficient at t_1

The temperature coefficient of resistivity is generally taken to be 0.006 8 microhm-cm per degree C. Expressed in other values, it is

0.000 597	ohm-gram/metre^2
3.31	ohm-pound/mile^2
0.000 681	microhm-centimetre
0.002 68	microhm-inch
0.040 9	ohm-circular mil/foot

The Fahrenheit equivalent for these constants may be found by dividing them by 1.8. The change of resistivity per degree F is

0.001 49 microhm-inch

RESISTIVITIES OF COMMON METALS AND ALLOYS

For a rough comparison of resistivities *Figure 6.1* shows a chart of the more common materials.

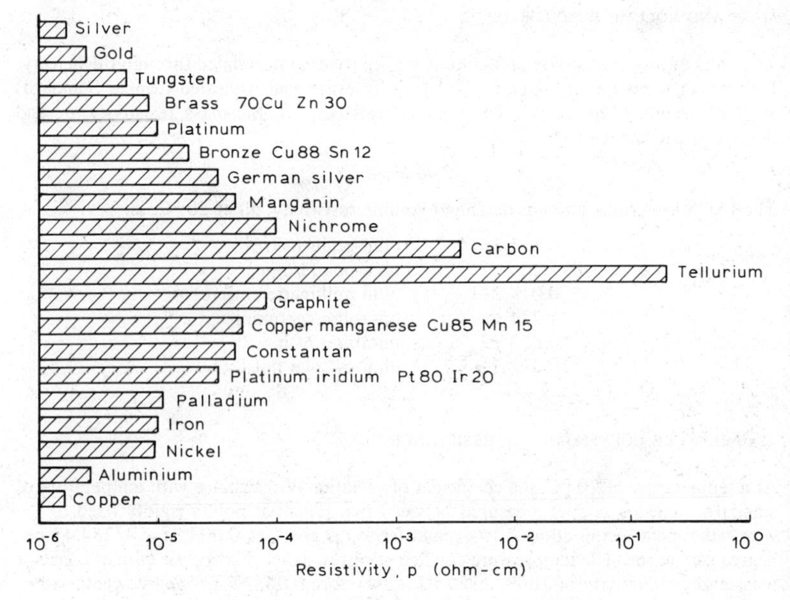

Figure 6.1 Resistivities of common metals and alloys

RESISTIVITIES OF RESISTANCE MATERIALS

A more extensive tabulation of resistivities for most resistive elements, metals and alloys at 20 °C is given in *Table 6.1* (page **6**–5).

RESISTIVITIES OF ELEMENTS, METALS AND ALLOYS

An extensive table of the resistivities of elements, metals and alloys is given in *Table 6.2*. (page **6**–9).

TEMPERATURE COEFFICIENTS OF RESISTANCE FOR METALS AND ALLOYS

A table of temperature coefficient of resistance for many elements, metals and alloys is given in *Table 6.3* (page **6**–18).

Table 6.1 RESISTIVITIES OF RESISTANCE MATERIALS

Material	*Composition*	*Temperature, °C*	*Resistivity, ohm-cm × 10^{-6}*	*Authority*
Carbon		0	3 500	
		500	2 700	
		1 000	2 100	
		2 000	1 100	
		2 500	900	
Constantan	Cn 60, Ni 40	20	49	Bureau of Standards
		−200	42.4	Nicolai
	Cn 60, Ni 40	−150	43.0	Nicolai
		−100	43.5	Nicolai
		−50	43.9	Nicolai
		0	44.1	Nicolai
		+100	44.6	Nicolai
		400	44.8	Nicolai
Copper, commercial:				
annealed		20	1.724 1*	Bureau of Standards
hard drawn		20	1.77	Bureau of Standards
pure, annealed		20	1.692	Wolff, Delinger, 1910
		−258.6	0.014	Nicolai
		−206.6	0.163	Nicolai
		−150	0.567	Nicolai
		−100	0.904	Nicolai
		+100	2.28	Northrup, 1914
		200	2.96	Northrup, 1914
		500	5.08	Northrup, 1914
		1 000	9.42	Northrup, 1914
Copper–manganese	Mn 0.98	0	4.83	Munker, 1912
	Mn 1.49	0	6.66	Munker, 1912
	Mn 4.2	20	17.9	Sebast & Gray, 1916
	Mn 7.4	20	19.7	Sebast & Gray, 1916
	Mn 15	20	50	Klein, 1924
Copper–manganese–iron	Cu 91, Mn 7.1, Fe 1.9	0	20	Blood
	Cu 70.6, Mn 23.2, Fe 6.2	0	77	Blood
Copper–manganese–nickel	Cu 73, Mn 24, Ni 3	0	48	Feussner, Lindeck
Eureka		0	47	Drysdale, 1907
German silver	Cu 60.16, Zn 25.37 Ni 14.03, Fe 0.3 Co and Mn trace	−200	27.9	Dewar, Fleming
		−100	29.3	
		+100	33.1	

Table 6.1—*continued*

Material	*Composition*	*Temperature, °C*	*Resistivity, ohm-cm × 10⁻⁶*	*Authority*
Graphite †		0	800	
		500	830	
		1 000	870	
		2 000	1 000	
		2 500	1 100	
Manganese–copper	Mn 30, Cu 70	0	100	Feussner, Lindeck
Manganin	Cu 84, Mn 12, Ni 4	20	44	Bureau of Standards
		22.5	45	Kimura, Sakamaki
		−200	37.8	Nicolai
		−100	38.5	Nicolai
		−50	38.7	Nicolai
		0	38.8	Nicolai
		100	38.9	Nicolai
		400	38.3	Nicolai
Nickel–chromium	Ni 80, Cr 20	20	110	Bureau of Standards
Nickel		20	7.8	Bureau of Standards
pure		−182.5	1.44	Fleming, 1900
		−78.2	4.31	Fleming, 1900
		0	6.93	Fleming, 1900
		94.9	11.1	Fleming, 1900
		400	60.2	Nicolai, 1907
Nickel–copper–zinc	Ni 12.84, Cu 30.59 Zn 6.57 by volume	0	20.3	Matthiessen
Palladium		20	11	Bureau of Standards
		−183	2.78	Dewar, Fleming
		−78	7.17	Dewar, Fleming
		0	10.21	Dewar, Fleming
		98.5	13.79	Dewar, Fleming
Palladium–copper	Pd 72, Cu 28	20	47	Johansson, Linde
Palladium–gold	Pd 50, Au 50	20	27.5	Sedstrom, Wise
Palladium–silver	Pd 60, Ag 40	20	42	Sedstrom & Svensson
Platinum		20	10	Bureau of Standards
		−203.1	2.44	Dewar, Fleming
		−97.5	6.87	Dewar, Fleming
		0	10.96	Dewar, Fleming
		+100	14.85	Dewar, Fleming
		400	26	Nicolai
		−265	0.10	Nernst
		−253	0.15	Nernst

Table 6.1—*continued*

Material	*Composition*	*Temperature, °C*	*Resistivity, ohm-cm × 10^{-6}*	*Authority*
Platinum		−233	0.54	Nernst
		−153	4.18	Nernst
		−73	7.82	Nernst
		0	11.05	Nernst
		+100	14.1	Pirrani
		200	17.9	Pirrani
		400	25.4	Pirrani
		800	40.3	Pirrani
		1 000	47.0	Pirrani
		1 200	52.7	Pirrani
		1 400	58.0	Pirrani
		1 600	63.0	Pirrani
Platinum–gold	Au 60, Pt 40	20	42.0	Johansson, Linde
	Au 80, Pt 20	20	25.0	Johansson, Linde
Platinum–iridium	Pt 90, Ir 10	0	24	Barnes
	Pt 80, Ir 20	0	31	Barnes
	Pt 65, Ir 35	20	36	Geibel, Carter and Nemilow
Platinum–rhodium	Pt 90, Rh 10	−200	14.40	Dewar, Fleming
		−100	18.05	Dewar, Fleming
		0	21.14	Dewar, Fleming
		+100	24.2	Dewar, Fleming
	Pt 80, Rh 20	20	20	Acken, Nemilow, Voronow and Carter
Platinum–silver	Pt 67, Ag 33	0	24.2	
	Pt 55, Ag 45	20	61	Kurnakow and Nemilow
Platinum–copper	Pt 75, Cu 25	20	92	Sedstron
Rheotan	Cu 53.28, Ni 25.31 Zn 16.80, Fe 4.46 Mn 0.37	0	53	Feussner, Lindeck
Rose metal	Bi 49, Pb 28 Sn 23	0	64	
Silver (electrolytic)	99.98%	18	1.629	Jager, Diesselhorst
		−183	0.390	Dewar, Fleming
		−78	1.021	Dewar, Fleming
		0	1.468	Dewar, Fleming
		+98.15	2.062	Dewar, Fleming
		192.1	2.608	Dewar, Fleming
		−258.6	0.009	Nicolai

Table 6.1—*continued*

Material	*Composition*	*Temperature, °C*	*Resistivity, ohm-cm × 10^{-6}*	*Authority*
Silver (electrolytic)		−200	0.357	Nicolai
		−100	0.916	Nicolai
		0	1.506	Nicolai
		+100	2.15	Northrup
		200	2.80	Northrup
		400	3.46	Northrup
		750	6.65	Northrup
Steel				
aluminium	Al 5, C 0.2	20	65	Portevin, 1909
	Al 15, C 0.9	20	88	Portevin, 1909
chromium	Cr 13, C 0.7	20	60	Portevin, 1909
	Cr 40, C 0.8	20	71	Portevin, 1909
Invar	35% Ni	20	81	Bureau of Standards
manganese		20	70	Bureau of Standards
nickel	Ni 10, C 0.1	20	29	
	Ni 25, C 0.1	20	39	
	Ni 80, C 0.1	20	82	Portevin, 1909
Siemens–Martin		20	18	Bureau of Standards
silicon	Si 2.5%	20	45	
	Si 4%	20	62	
tempered glass-hard			45.7	Stronhal, Barnes
tempered yellow			27	Stronhal, Barnes
tempered blue			20.5	Stronhal, Barnes
tempered soft			15.9	Stronhal, Barnes
titanium	Ti 2.5, C 0.15	20	16	Portevin, 1909
tungsten	W 5, C 0.2	20	20	Portevin, 1909
	W 20, C 0.2	20	24	Portevin, 1909
vanadium	V 5, C 1.1	20	121	Portevin, 1909
Tellurium ‡		19.6	200 000	Matthiessen
Tin-bismuth	Sn 90.5, Bi 9.5	12	16	
	Sn 2, Bi 98	0	244	

* International Annealed Copper Standard.

† N. B. Polycrystalline; the resistivity of a single-crystal in the plane of the hexagonal network is about 60×10^{-6} ohm-cm at 20 °C.

‡ Resistivity is greatly dependent on purity of the specimen.

Table 6.2 TABLE OF RESISTIVITIES OF ELEMENTS, METALS AND ALLOYS

	Resistivity at 20 °C		Conductance, % IACS
	Microhm-cm (*1*)	*Microhm-inch* (*2*)	(*3*)
Elements			
Aluminium (99.996% Al)	2.665	1.05	64.94
Antimony (fully annealed)	(39.8)	15.7	4.3
Arsenic	(37.9)	14.9	4.6
Barium	9.8	3.9	18
Beryllium	(6.66)	2.6	26
Bismuth	(120)	47	1.4
Cadmium (99.9% fully annealed)	(7.42)	2.9	23
Calcium	(3.74)	1.5	46
Carbon (graphite)	1 375 (5)	542 (5)	0.1 (5)
Cerium	78	31	2.2
Cesium	(20.8)	8.2	8.3
Chromium (electro-chromium)	14.1	5.6	12.2
Cobalt (99.91%)	(6.22)	2.5	28
Copper (pure)	1.673 0	0.66	103
Gallium	53.4 (5)	21 (5)	3.2 (5)
Gold	(2.35)	0.93	73
Indium	(9.03)	3.6	19
Iridium	5.3	2.1	33
Iron	9.71	3.8	18
Lead	20.65	8.1	8.3
Magnesium	4.46	1.8	39
Mercury	95.783	38	18
Molybdenum	5.7	2.2	30
Nickel	6.84	2.7	25
Osmium	9.5	3.7	18
Palladium (annealed)	10.8	4.3	16
Platinum	(10.58)	4.2	16
Rhodium	(4.7)	1.9	37
Rubidium	12.5	4.9	14
Silicon	85×10^3	33×10^3	—
Silver (extremely pure, melted and annealed in vacuo)	1.59	0.63	108
Sodium	(4.8)	1.9	36
Strontium	22.76	8.9	7.6
Sulphur (rhombic)	2×10^{23}	7.9×10^{22}	—
Tantalum	(13.52)	5.3	13
Tellurium	2×10^5 (6)	7.9×10^4 (6)	—
Tin	11.5	4.5	15
Titanium	(88)	35	2.0
Tungsten	5.5	2.2	31
Uranium	60 (7)	24 (7)	2.9
Vanadium	26	10	6.6
Zinc (polycrystalline)	5.916	2.3	29
Zirconium	(44.6)	18	3.9

Table 6.2—*continued*

	Resistivity at 20 °C		*Conductance, % IACS*
	Microhm-cm (*1*)	*Microhm-inch* (*2*)	(*3*)
Platinum Alloys			
Platinum, Type A (99.99% Pt)	10.6	4.2	16
Platinum, Type B (99.9% Pt)	10.8	4.2	16
Platinum, Type C (99.5% Pt)	11.4	4.5	15
Platinum, Type D (99.0% Pt)	14.9	5.9	12
Platinum–5% iridium	19	7.5	9.1
Platinum–10% iridium	25	9.8	6.9
Platinum–25% iridium	33	13	5.2
Platinum–5% ruthenium	31.5	12.4	5.5
Platinum–10% ruthenium	43.0	16.9	4.0
Platinum–1% nickel	12.7	5.0	14
Platinum–2% nickel	15.0	5.9	11
Platinum–5% nickel	23.3	9.2	7.4
Platinum–4% tungsten	36.9	14.5	4.7
Iron Alloys			
Carbon steels			
0.06 C, 0.38 Mn (1006 range)	13.0	5.1	13
0.08 C, 0.31 Mn (1010 range)	14.2	5.6	12
0.23 C, 0.635 Mn (1020 range)	16.9	6.6	10
0.415 C, 0.643 Mn (1040 range)	17.1	6.7	10
0.80 C, 0.32 Mn (1078 range)	18.0	7.1	9.6
1.22 C, 0.35 Mn	19.6	7.7	8.8
Alloy steels			
0.23 C, 1.51 Mn, 0.105 Cu	20.8	8.2	8.3
0.325 C, 0.55 Mn, 0.17 Cr, 3.47 Ni	27.1	10.6	6.4
0.33 C, 0.53 Mn, 0.80 Cr, 3.38 Ni	26.8	10.5	6.4
0.325 C, 0.55 Mn, 0.71 Cr, 3.41 Ni	28.0	11.0	6.2
0.34 C, 0.55 Mn, 0.78 Cr, 3.53 Ni, 0.39 Mo	28.9	11.4	6.0
0.315 C, 0.69 Mn, 1.09 Cr, 0.073 Ni	21.0	8.3	8.2
0.35 C, 0.59 Mn, 0.88 Cr, 0.26 Ni	22.3	8.8	7.7
0.485 C, 0.90 Mn, 1.98 Si, 0.637 Cu	42.9	16.9	4.0
High-alloy steels			
1.22 C, 13.0 Mn, 0.22 Si	68.3	26.9	2.5
0.28 C, 0.89 Mn, 28.37 Ni	84.2	33.1	2.0
0.08 C, 0.37 Mn, 19.11 Cr, 8.14 Ni, 0.60 W	71.0	27.9	2.4
0.13 C, 0.25 Mn, 12.95 Cr	50.6	19.9	3.4
0.27 C, 0.28 Mn, 13.69 Cr	52.2	20.5	3.3
0.715 C, 0.25 Mn, 4.26 Cr, 18.45 W, 1.075 V	41.9	16.5	4.1

Table 6.2—*continued*

	Resistivity at 20 °C		Conductance, % IACS
	Microhm-cm (1)	Microhm-inch (2)	(3)
	Iron Alloys (cont.)		
Iron–manganese alloys			
26% Mn	67	26	2.6
30% Mn	68	27	2.5
35% Mn	68.5	27	2.5
40% Mn	69.5	27	2.5
42% Mn	70	28	2.5
45% Mn	71	28	2.4
48% Mn	72	28	2.4
Malleable irons			
ASTM Spec. A47.33, Grade 35018	30	12	5.7
ASTM Spec. A47.33, Grade 32510	32	13	5.4
Nickel steels			
AISI 301	72	28	2.4
AISI 302	72	28	2.4
AISI 302 B	72	28	2.4
AISI 303	72	28	2.4
AISI 304	72	28	2.4
AISI 308	72	28	2.4
AISI 309	72	28	2.4
AISI 310	72	28	2.4
AISI 316	74	29	2.3
AISI 321	72	28	2.4
AISI 347	73	29	2.4
36% Ni (Invar)	81	32	2.1
42% Ni	70	28	2.5
50% Ni	48	19	3.6
4% silicon iron	60	24	2.9
	Other iron Alloys		
Annealed carbon-steel castings			
0.07–0.20% C	13–14	5.1–5.5	13
0.20–0.45% C	14–16	5.5–6.3	11
0.45–1.50% C	16–20	6.3–7.9	9.6
Low-carbon ingot iron	10.7	4.2	16
Bessemer steel	14.0	5.5	12
Cast (carbon) steel	15.0	5.9	11
Malleable iron	32.0	12.6	5.4
Cast iron	100.0	39.4	1.7

Table 6.2—*continued*

Zinc Alloys

	Resistivity at 20 °C		Conductance, % IACS	Temperature coefficient of electrical resistance per °C (4)	
	Microhm-cm (1)	Microhm-inch (2)	(3)	α	Temp, °C
Pure zinc	5.916	2.32	29	0.004 19	0–100
Zamak–3	6.369 4	2.50	27	0.003 774	0–100
Zamak–5	6.535 9	2.57	26	0.003 527	0–100
Zamak–2	6.849 3	2.69	25		
Commercial rolled zinc	6.06–6.10	2.39	28		
Zilloy 40	6.22	2.45	28		
Zilloy 15	6.31	2.48	27		

Lead Alloys

	Resistivity at 20 °C		Conductance, % IACS
	Microhm-cm (1)	Microhm-inch (2)	(3)
Corroding lead (99.73% Pb)*	20.648	8.1	8.3
1% antimonial lead, heat treated, quenched and aged 150 days	22.0	8.7	7.88
Hard lead, heat treated, quenched and aged 150 days	24.0	9.5	7.2
8% antimonial lead, heat treated, quenched and aged 150 days	26.5	10.4	6.5
50–50 soft solder	15.6	6.1	11
Lead-base babbits	28.2–28.7	11.1–11.3	6–6.1

* Temperature coefficient of electrical resistance per °C (α) 0.003 36 from 20 to 40 °C. (4)

Tin Alloys

Pure tin (99.8+) (polycrystalline) †	11.5	4.5	15
Antimonial tin solder (95 Sn–5 Sb)	14.5	5.7	12
Tin–silver solder (95 Sn–5 Ag) ‡	10.4	4.1	17
Eutectic solder (63 Sn–37 Pb)	14.5	5.7	12
Tin foil	12.1	4.8	14
White metal	15.48	6.1	11

† Temperature coefficient of electrical resistance per °C (α) 0.004 47 from 0 to 100 °C. (4)
‡ Temperature coefficient of electrical resistance per °C (α) 0.004 23 from 0 to 100 °C. (4)

Table 6.2—*continued*

Palladium Alloys

	Resistivity at 20 °C		Conductance, % IACS (3)	Temperature coefficient of electrical resistance per °C (4)	
	Microhm-cm (1)	Microhm-inch (2)		α	Temp, °C
Palladium	10.8	4.2	16	0.003 77	0–100
60 Pd–40 Ag	42	17	4.1	0.000 02	20–100
60 Pd–40 Cu (annealed and quenched)	35	14	4.9	0.000 32	20–100
60 Pd–40 Cu (cold worked and heated to 300 °C)	3.5	1.4	49	0.002 24	20–100

Copper Alloys

	Microhm-cm (1)	Microhm-inch (2)	Conductance, % IACS (3)	α	Temp, °C
Copper, pure (spectrographically pure)	1.673 0	0.66	103.06	0.006 8–	20
Copper, oxygen-free, hard (OFHC)					
Copper (annealed)	(1.70)	(0.67–)	101.7		
Copper, electrolytic tough pitch (99.92 Cu–0.04 O)	1.71	0.67	101	0.003 92	20
Copper, deoxidised annealed (99.94 Cu–0.02 P)	2.03	0.80	85		
Bronze, commercial (90 Cu –10 Zn), annealed	3.9	1.5	44	0.001 86	20
Brass, red (85 Cu–15 Zn), annealed	4.7	1.9	37	0.001 6	20

Other Copper Alloys

	Resistivity at 20 °C		Conductance, % IACS
	Microhm-cm (1)	Microhm-inch (2)	(3)
Brass, yellow, annealed	6.4	2.5	27
Bronze, leaded commercial, annealed	4.1	1.6	42
Bronze, manganese, annealed	(7.1)	2.8	24
Brass, aluminium, annealed	(7.5)	3.0	23
Phosphor bronze, 5%, Grade A	9.6	3.8	18
Phosphor bronze, 8%, Grade C	13	5.1	13
Phosphor bronze, 10%, Grade D	16	6.3	11
Phosphor bronze, 1.25%, Grade E	3.6	1.4	48
Cupro-nickel, 30%	37	14.6	4.6
Nickel silver, 18%, Alloy A	29	11.4	6
Nickel silver, 18%, Alloy B	31	12.2	5.5
Silicon bronze, Type A, annealed	(25)	9.8	7

Table 6.2—*continued*

Other Copper Alloys (cont.)

	Resistivity at 20 °C		*Conductance, % IACS*
	Microhm-cm (1)	*Microhm-inch (2)*	*(3)*
Silicon bronze, Type B, annealed	(14)	5.5	12
Aluminium bronze, 5%, annealed	9.8	3.9	17.5
Aluminium bronze, 10%, annealed	13.67	5.4	12.6
Beryllium copper, solution treated, quenched	10	3.9	17
Beryllium copper, solution treated, quenched and precip. hard	6.8–9.8	2.7–3.9	21
Leaded tin bronze	(12)	4.8	14
Leaded tin bearing bronze	(16)	6.2	11
Nickel silver	(34–43)	14–17	4.5
Aluminium bronze	(12)	4.8	14

Nickel Alloys

	Resistivity at 20 °C		*Conductance, % IACS*	*Temperature coefficient of electrical resistance per °C (4)*	
	Microhm-cm (1)	*Microhm-inch (2)*	*(3)*	*α*	*Temp, °C*
Pure nickel (99.95 Ni + Co)	6.84	2.7	25.2	0.006 9	0–100
A nickel (99.4 Ni + Co)	9.5	3.7	18	0.004 74	20–100
D nickel (95 Ni–4.5 Mn)	14	5.5	12		
Z nickel (94 Ni–4.5 Al)	43.3	17.0	4.0		
Cast nickel	21	8.3	8.2		
Monel	48.2	19.0	3.6	0.001 1	20–100
Cast Monel	53.3	21.0	3.2		
K Monel	58.3	22.9	3.0	0.000 18	20–100
S Monel	63.3	24.9	2.7		
Hastelloy A	126.7	49.8	1.4	zero	20–800
Hastelloy B	135	53	1.3		
Hastelloy C	133	52	1.3	zero	20–800
Hastelloy D	113	44	1.5	almost zero	20–800
Inconel	98.1	38.6	1.8	0.000 125	20–500
Inconel X	124 (8)	49	1.4		
Incalloy	(97)	38	1.8		
Nichrome (60 Ni, 24 Fe, 16 Cr)	112	44	1.5	0.000 17	20–100
Nichrome (80 Ni, 20 Cr) IV	107.9	42.4	1.6	0.000 219	20–100
Ni Resist, Type 1	(175) (5)	69 (5)	0.98 (5)		
Ni Resist, Type 2	(175) (5)	69 (5)	0.98 (5)		
35 Ni, 50 Fe, 15 Cr	100	39	1.7	0.000 31	20–500
Constantan (wrought)	49	19	3.5	0.000 025	20–500

Table 6.2—*continued*

	Resistivity at 20 °C		*Con-ductance, % IACS*
	Microhm-cm (*1*)	*Microhm-inch* (*2*)	(*3*)
	Nickel Alloys (Special)		
Temperature compensation alloys			
30% Ni type	80	31	2.2
32.5% Ni type	80	31	2.2
Rectangular hysteresis alloys			
50% Ni type	50	20	3.4
65 Permalloy	25	9.8	6.9
Magnetostrictive alloys, A Nickel	8	3.1	22
Insulated powder alloy 2–81, Permalloy	10^6	3.9×10^5	—
Very-high-permeability alloys			
78.5 Permalloy, 78.5 Ni	16	6.3	11
Modified 79 Permalloy, 79 Ni + 4 Mo	58	23	3.0
Mumetal, 77 Ni + 5 Cu + 1.5 Cr	60	24	2.9
Supermalloy, 79 Ni + 5 Mo	65	26	2.7
High-permeability alloys for higher field strengths			
50% Ni type alloys	45	18	3.8
Monimax	79	31	2.2
Sinimax	89	35	1.9
Constant-permeability alloys 45–25 Perminvar	19	7.5	9.1
7–45–25 Perminvar	80	31	2.2
Conpernik	45	18	3.8
	Miscellaneous Metals		
Alumel	33.3	13	5.2
Brass	4–7	2–3	29
Bronze	13–18	5–7	11
Cast iron	57–114	22–45	3–1.5
Chromel	70–110	28–43	2.5–1.6
Constantan	47–51	18–20	3.5
Ferrite	9.5	3.7	18
Hadfield manganese steel	29–67	11–26	6–2.6
Invar	75	30	2.3
Monel	42.5–45	17–18	3.9
Wood's metal	51.7	20	3.3
	Aluminum Alloys		
99.996% Al	2.654 8	1.04	65
1100–99.0%, Al 1100–0*	2.922	1.15	59

Table 6.2—*continued*

	Resistivity at 20 °C		*Con-ductance, % IACS*
	Microhm-cm (*1*)	*Microhm-inch* (*2*)	(*3*)
Aluminium Alloys (*cont.*)			
1100–H 18†	3.025	1.19	57
3003–O	3.448	1.36	50
3003–H 12	4.105	1.61	42
3003–H 14	4.205	1.65	41
3003–H 18	4.310	1.70	40
2011–T 3 ‡	4.310	1.70	40
2014–O	3.448	1.36	50
2014–T 6	4.310	1.70	40
2017–O	3.831	1.51	45
2017–T 4	5.747	2.26	30
2018–O	3.448	1.36	50
2018–T 61	4.310	1.70	40
2024–O	3.448	1.36	50
2024–T 4	5.747	2.26	30
4032–O	4.310	1.70	40
4032–T 6	4.926	1.94	35
5052–O	4.926	1.94	35
5052–H 38	4.926	1.94	35
5056–O	5.945	2.34	29
5056–H 38	6.386	2.52	27
6061–O	3.831	1.51	45
6061–T 4	4.310	1.70	40
6061–T 6	4.310	1.70	40
7075–T 6	5.747	2.26	30
13 Alloy DC§	4.421	1.74	39
43 Alloy, SC,‖ PM,¶ DC (as cast)	4.660	1.84	37
SC, PM (annealed)	4.105	1.61	42
85 Alloy, DC	6.158	2.43	28
108 Alloy, SC	5.562	2.19	31
Allcast, SC, PM (as cast)	(6.39)	2.52	27
(stress relieved)	(5.75)	2.26	30
Solution heat-treated and aged	(5.75)	2.26	30
Solution heat-treated and stress-relieved	(4.79)	1.89	36
A–108 Alloy, PM	4.660	1.84	37
113 Alloy, SC	5.747	2.26	30
C–113 Alloy, PM	6.386	2.52	27
122 Alloy, SC, condition T–2	4.205	1.66	41
SC, condition T–61	5.225	2.06	33
PM (as cast)	5.071	2.00	34
A–132 Alloy, condition T–551	5.945	2.34	29
142 Alloy, SC, condition T–21	3.918	1.54	44
SC, condition T–571	5.071	2.00	34
SC, condition T–77	4.660	1.84	37
PM, condition T–61	5.388	2.12	32

Table 6.2—*continued*

	Resistivity at 20 °C		*Con-ductance, % IACS*
	Microhm-cm (*1*)	*Microhm-inch* (*2*)	(*3*)
	Aluminium Alloys (cont.)		
195 Alloy, SC, condition T–4	4.926	1.94	35
SC, condition T–62	4.660	1.84	37
B–195 Alloy			
PM, condition T–4	4.926	1.94	35
PM, condition T–6	4.789	1.89	36
214 Alloy, SC	4.926	1.94	35
A–214 Alloy, PM	5.225	2.06	33
218 Alloy, DC	7.184	2.83	24
220 Alloy, SC, condition T–4	8.210	3.24	21
319 Alloy, SC	6.386	2.52	27
PM	6.158	2.42	28
355 Alloy, SC, condition T–51	4.010	1.58	43
SC, condition T–6	4.789	1.89	36
SC, condition T–61	4.660	1.84	37
SC, condition T–7	4.105	1.61	42
PM, condition T–6	4.421	1.74	39
365 Alloy, SC, condition T–51	4.010	1.58	43
SC, condition T–6	4.421	1.74	39
SC, condition T–7	4.310	1.70	40
PM, condition T–6	4.205	1.66	41
Red X–8, SC, PM (as cast)	(6.53)	2.57	26
(stress-relieved)	(5.95)	2.34	29
350 Alloy, DC	4.660	1.84	37
380 Alloy, 4–9 Alloy, DC	6.386	2.52	27
750 Alloy, PM	3.831	1.51	45
40 E Alloy, SC	4.926	1.94	35

* O = annealed † H = strain-hardened ‡ T = heat-treated
§ DC = die-cast ‖ SC = sand-cast ¶ PM = permanent mould

	Magnesium Alloys		
99–80 Mg (polycrystalline)	4.46	1.76	38.6
Mazio AM-C59S, wrought	18.0	7.1	9.7
Mazlo AM-265, Dowmetal H			
As cast	11.5	4.5	15.0
Heat-treated	14.0	5.5	12.3
Heat-treated and aged	12.5	4.9	13.8
Mazlo AM-260, Dowmetal C	14–16.5	5.5–6.5	10.5–12.3
Mazlo AM-263, Dowmetal R	17.0	6.7	10.1
Mazlo AM-403, AM-3S, Dowmetal M	5.0–6.7	2.0–2.6	25.7–34.5
Mazlo AM-C52S, Dowmetal FS-1	9.3	3.7	18.5
Dowmetal JS-1	13.5	5.3	12.8
Mazlo AM-C57S, Dowmetal J-1	14.9	5.9	11.6
Mazlo AM-C58S, Dowmetal O-1	11.8–16.2	4.6–6.4	10.6–14.6
Mazlo AM-65S, Dowmetal D	13.8	5.4	12.5

Table 6.3 TABLE OF TEMPERATURE COEFFICIENTS OF RESISTANCE FOR METALS AND ALLOYS

Material	*Temperature, °C*	*Temperature coefficient of resistance per °C*
*Advance (See Constantan)		
Ni 30, Cr 5, Fe 65	20	+0.000 72
*Alumel		
Ni 94, Mn 2.5, Fe 4.5, Al 2, Si 1	0	0.001 2
Aluminium	18	+0.003 9
	25	0.003 4
	100	0.004
	500	0.005
annealed, highest purity	0–100	0.004 45
Aluminium-bronze		
Cu 97, Al 3		0.001 02
Cu 90, Al 10		0.003 20
Cu 6, Al 94		0.003 80
Antimony	20	0.003 6
*Argentan		
Cu 61.6, Ni 15.8, Zn 22.6	0–160	0.000 387
Arsenic	20	0.004 2
Bismuth	20	0.004
	0–100	0.004 46
Brass	20	0.002
Cu 66, Zn 34	15	0.002
Cu 60, Zn 40	15	0.001
Brightray B	20–500	++0.000 14
Brightray C	20–500	+0.000 079
Brightray F	20–500	+0.000 25
Brightray H	20–500	+0.000 084
Brightray S	20–500	+0.000 061
Bronze		
Cu 88, Sn 12	20	0.000 5
Cadmium	20	0.003 8
drawn	0–100	0.004 24
annealed, pure	0	0.004 2
Carbon		−0.000 5
*Chromax		
Ni 30, Cr 20, Fe 50	20–500	0.000 31
*Chromel A		
Ni 80, Cr 20	20–500	0.000 13
*Chromel C		
Ni 60, Cr 16, Fe 24	20–500	0.000 17
*Chromel D		
Ni 30, Cr 20, Fe 50	20–500	0.000 32
Climax	20	+0.000 7
Cobalt	0	0.003 3
	0–100	0.006 58
Constantan	12	0.000 008

Table 6.3—*continued*

Material	*Temperature, °C*	*Temperature coefficient of resistance per °C*
Constantan (*cont.*)	25	0.000 002
	100	0.000 033
	200	0.000 02
*Copel		
Ni 45, Cu 55	0–100	+0.000 02
Copper, annealed	20	0.003 93
hard drawn	20	0.003 82
	100	0.003 8
	400	0.004 2
	1 000	0.006 2
electrolytic	0	0.004 1
pure, annealed	0–100	0.004 33
Copper–manganese		
Cu 96.5, Mn 3.5		0.000 22
Cu 95, Mn 5		0.000 026
Cu 70, Mn 30		0.000 04
Copper–manganese–iron		
Cu 91, Mn 7.1, Fe 1.9	0	0.000 12
Cu 70.6, Mn 23.2, Fe 6.2	0	0.000 022
Copper–manganese–nickel		
Cu 73, Mn 24, Ni 3	0	−0.000 03
Copper–nickel		
Cu 60, Ni 40	0	±0.000 02
Eureka	0	+0.000 05
*Evanohm		
Cr 20, Al 2.5, Cu 2.5, Ni Bal.	−50 to +100	±0.000 02
Excello	20	0.000 16
*Ferry	20–100	±0.000 02
German-silver		
Ni 18%	20	0.000 04
Cu 60, Zn 25, Ni 15	0	0.000 36
Gold	20	0.003 4
	100	0.003 5
	500	0.003 5
	1 000	0.004 9
Gold–copper–silver		
Au 58.3, Cu 26.5, Ag 15.2	0	0.000 574
Au 66.5, Cu 15.4, Ag 18.1	0	0.000 529
Au 7.4, Cu 78.3, Ag 14.3	0	0.001 830
Gold–silver		
Au 90, Ag 10	0	0.001 2
Au 67, Ag 33	0	0.000 65
Indium	0	0.004 7
Iridium	0–100	0.004 11

Table 6.3—*continued*

Material	*Temperature, °C*	*Temperature coefficient of resistance per °C*
Iron	20	0.005
	0	0.006 2
	25	0.005 2
	100	0.006 8
	500	0.014 7
	1 000	0.005
*Karma		
Ni 80, Cr 20	20	0.000 16
Lead	18	0.004 3
pure	0–100	0.004 22
Lithium	0	0.004 7
	230	0.002 7
*Lohm		
Ni 6, Cu 94	20–100	0.000 71
Magnesium	20	0.004
	0	0.003 8
	25	0.005
	100	0.004 5
	500	0.003 6
	600	0.010
*Mancoloy	200	0.000 5
Manganese–copper		
Mn 30, Cu 70	0	0.000 04
Manganese–nickel		
Mn 2, Ni 98	20–100	0.004 5
Manganin		
Cu 84, Mn 12, Ni 4	12	0.000 006
	25	0.000 000
	100	−0.000 042
	250	−0.000 052
	475	0.000 000
	500	+0.000 11
Mercury	20	0.000 89
	0	0.000 88
*Midohm		
Ni 23, Cu 77	20–100	0.000 18
*Minalpha	40	0.000 000
Molybdenum	25	+0.003 3
	100	0.003 4
	1 000	0.004 8
	0–100	0.004 35
*Monel-metal	20	0.002

Table 6.3—*continued*

Material	*Temperature, °C*	*Temperature coefficient of resistance per °C*
*Nichrome		
Ni 61, Cr 14, Fe 24	20–500	0.000 17
*Nichrome V		
Ni 80, Cr 20	20–500	0.000 13
Nickel	20	0.006
	0	0.006
	25	0.004 3
	100	0.004 3
	500	0.003
	1 000	0.003 7
pure, annealed	0–100	0.006 75
*Ohmax	20–500	0.000 066
Palladium	20	0.003 3
pure	0–100	0.003 77
pure	0	0.003 5
Phosphor–bronze	0	0.004–0.003
Palladium–gold		
Pd 50, Au 50	0–100	0.000 36
Palladium–silver		
Pd 60, Ag 40	0–100	0.000 04
Platinite, nickel steel		
Ni 46–48	0	0.003
Platinum	20	0.003
	0	0.003 7
	0–100	0.003 92
Platinum–iridium		
Pt 90, Ir 10	0	0.001 2
Pt 80, Ir 20	0	0.000 8
Platinum–rhodium		
Pt 90, Rh 10	0	0.001 3
Platinum–silver		
Pt 33, Ag 67	0	0.000 24
Platinum–gold		
Pt 40, Au 60	20	0.000 6
Pt 20, Au 80	20	0.002 5
Platinum–copper		
Pt 75, Cu 25	20	0.000 3
Potassium	0	0.005 5
liquid	100	0.004 2
Rheotan	0	0.000 4

Table 6.3—*continued*

Material	*Temperature, °C*	*Temperature coefficient of resistance per °C*
Rhodium	0–100	0.004 43
*Rose metal		
Bi 48.9, Sn 23.5, Pb 27.6	0	0.002
Rubidium	0	0.006
*Silchrome		
Si, Cr, Fe	20	0.000 025
Silicon bronze	0	0.003 8–0.002 3
Silver	20	0.003 8
	25	0.003
	100	0.003 6
	500	0.004 4
pure, annealed	0–100	0.004 1
Sodium	9	+0.004 4
liquid	120	0.003 3
Steel		
Invar		
Ni 36, C 0.2	0	0.002
piano wire	0	0.003 2
Siemens–Martin	20	0.003
silicon		
Si 4	20	0.000 8
tempered glass-hard	0	0.001 6
tempered blue	0	0.003 3
Tantalum	20	0.003 1
	0–100	0.003 47
Thallium	0	0.004
*Therlo	20	0.000 01
Thorium	20–1 800	0.002 1
Tin	20	0.004 2
Tungsten	18	0.004 5
	500	0.005 7
	1 000	0.008 9
pure, annealed	0–100	0.004 65
*Vacrom		
Ni 80, Cr 20	0–500	0.000 06
*Wood's metal	0	0.002
Zinc	20	0.003 7
	0	0.004
	0–100	0.004 15

* Trade names.

Resistive components—fixed and variable resistors

COLOUR CODES FOR U.K. AND U.S.A. DISCRETE FIXED RESISTORS

Bands or rings of colour are usually placed round the resistor, as shown in *Figure 6.2*. In the U.K. the colours of the first three rings determine the total value of the resistance—the first ring *A* determines the first digit, the second ring *B* determines the

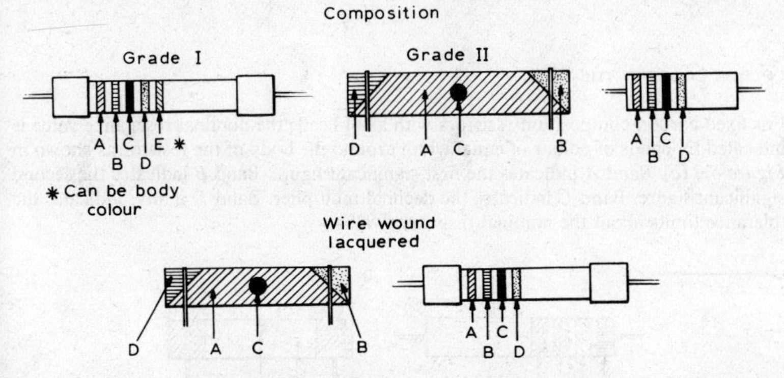

Figure 6.2 British colour coding for fixed resistors

second digit, and the third ring *C* determines the number of noughts. In some resistors the body is colour coded for the first digit, the tip for the second digit and a colour dot on the body indicates the number of noughts. The colour code used universally is given in *Table 6.4*.

Table 6.4 THE BRITISH STANDARD COLOUR CODE

Colour	*A* *1st digit*	*B* *2nd digit*	*C* *Multiplying factor*	*D* *Selection tolerance*	*E** *Stability (Comp. only)*
Brown	1	1	10	±1%	—
Red	2	2	100	±2%	—
Orange	3	3	1 000	—	—
Yellow	4	4	10 000	—	—
Green	5	5	100 000	—	—
Blue	6	6	1 000 000	—	—
Violet	7	7	—	—	—
Grey	8	8	—	—	—
White	9	9	—	—	—
Black	—	0	1	—	—
Gold	—	—	0.1	±5%	—
Silver	—	—	0.01	±10%	—
None	—	—	—	±20%	—
Salmon Pink	—	—	—	—	Grade I (High Stability)

* Can be body colour.

A fourth ring *D*, or end colour, is usually added to denote the tolerance on value, e.g., a gold ring indicates a 5% tolerance, a silver ring a 10% tolerance and the absence of a fourth colour indicates a 20% tolerance. For other tolerance values the normal colour code may be used, e.g. a brown ring would indicate a 1% tolerance. Cracked-carbon resistors and precision wirewound resistors are often required to an accuracy of three digits and then a number of cyphers. The colour coding system is not easily applied to these types of resistor, and so they are not normally colour coded by some manufacturers.

COLOUR CODES IN THE U.S.A.

For fixed carbon composition resistors with axial heads the nominal resistance value is indicated by bands of colour of equal width around the body of the resistor as shown in *Figure 6.3 (a)*. Band *A* indicates the first significant figure. Band *B* indicates the second significant figure. Band *C* indicates the decimal multiplier. Band *D* if any, indicates the tolerance limits about the nominal resistance value.

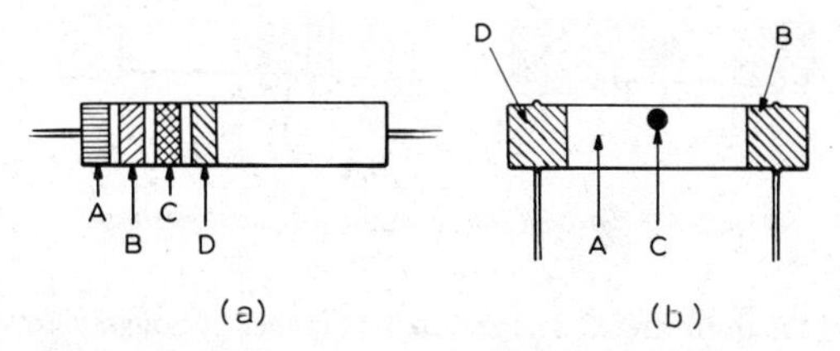

Figure 6.3 *American colour coding for fixed resistors*

Resistors with radial leads may use the same colour code as those with axial leads but alternatively they may be colour coded as in *Figure 6.3 (b)*, where body *A* corresponds to band *A* above, and *B* corresponds to band *B* above, dot *C* (or band *C*) corresponds to band *C* above, and end *D* corresponds to band *D* above.

The colour code is given in *Table 6.5*.

Table 6.5 THE AMERICAN STANDARD COLOUR CODE

Colour	*Significant figure*	*Decimal multiplier*	*Tolerance* (%)
Black	0	1	
Brown	1	10	
Red	2	10^2	
Orange	3	10^3	
Yellow	4	10^4	
Green	5	10^5	
Blue	6	10^6	
Violet	7	10^7	
Grey	8	10^8	
White	9	10^9	
Gold	—	10^{-1}	± 5
Silver	—	10^{-2}	± 10
No colour	—	—	± 20

Body colour: Black indicates uninsulated. Any other colour indicates insulated.

GENERAL CHARACTERISTICS OF DISCRETE FIXED RESISTORS

A summary of electrical characteristics is given in *Table 6.6*.

Fixed resistors are generally placed in one of two categories—high stability or general purpose. High-stability types include the pyrolytic or cracked-carbon resistors, the wirewound resistors and the metal or metal oxide-film resistors, all of which are capable of providing stable resistance to within 1 or 2%. General purpose types are usually of carbon composition and are cheaper and usually smaller. They are not so stable and resistance variations on load may be from 5 to 20%.

In choosing a resistor for a particular application, a knowledge of some of the following characteristics will be needed.

Size. In general, carbon resistors dissipate less power than wirewound resistors of the same resistance value; they are also smaller. The maximum resistance of a wirewound

Table 6.6 SUMMARY OF THE ELECTRICAL CHARACTERISTICS OF FIXED RESISTORS

Resistor type	*Overall stability (after climatic tests)* (%)	*Mfg. accuracy* (%)	*Best selection accuracy* (%)	*Max. noise* ($\mu V/V$)	*Temperature coefficient* (*ppm/°C*)	*Voltage coefficient* (%/*Volt*)	*Max. resistor temp.* (°C)
Moulded carbon composition (Insulated and uninsulated)	25	20	5	2.0 (for low values) to 6.0 (for high values)	±1 200 ±1 200	1 M – 0.025 1 M – 0.05	115
Carbon composition Film type (Insulated)	25	20	5	2.0 (for low values) to 6.0 (for high values)	±1 200	1 M – 0.025 1 M – 0.05	115
Cracked carbon (Insulated and uninsulated)	2	1	0.25	0.03 (for low values) to 0.5 (for high values)	−200 (for low values) to −1 000 (for high values)	0.005	150
Wirewound (General-purpose type)	1	1	0.1	None	±200	None	320
Wirewound (Precision type)	0.01 (if hermetically sealed)	0.01	0.01	None	(Ni/Cr) + 70 (Cu/Ni) + 20	None	70
Metal film (Ni/Cr)	1	1	0.2	Up to 0.3	+220	None	150
Oxide film (Sn/Sb)	2	1	0.5	Up to 0.5	−500 to +500		300

resistor is limited by the length of wire of a given material and the diameter that can be wound upon the available former length. Subminiature cracked-carbon high-stability resistors are made that are comparable in size with the carbon-composition type. Metal-oxide-film resistors are usually larger for high resistance values because of their lower ohms/square.

Power-handling capacity. Composition resistors are commonly available for dissipating up to about 2 W, but rarely over 5 W (except in special resistors of low ohmic values), although sintered types can dissipate high powers. Cracked-carbon resistors are available up to 2 W at normal temperatures. All these ratings have to be reduced when the resistors are used at high ambient temperatures. Small metal-film resistors are made up to 2-W dissipation and small oxide-film resistors, up to 6-W dissipation. Wirewound resistors are invariably used when higher powers are to be dissipated, and some vitreous-enamelled wirewound types will handle powers as great as 300–400 W. It is important to remember that the temperatures reached by the resistors when dissipating these wattages can be very high—of the order of several hundred degrees centigrade. Large oxide-film and metal-film resistors are made to dissipate several hundred watts.

There is a 'critical value' of resistance for each wattage rating given by the equation,

$$R = V^2/W$$

In pulse operation (particularly when the duty cycle is low) only the mean power is effective in raising the internal temperature of a resistor. As the power is supplied in short pulses, very high peak ratings are possible, but the mean power should not exceed the continuous rating wattage. Peak pulse voltages for high-stability (cracked-carbon) resistors should be limited to twice the normal rated d.c. voltage; otherwise the limit set by internal sparking or external corona might be exceeded, whereas for general-purpose carbon-composition resistors, the maximum peak pulse voltage should be no greater that the maximum continuous rating.

Stability. Stability and accuracy are often confused. Stability is the change in resistance under shelf life or working conditions; accuracy is the tolerance to which the value of the resistor is made or selected. For general purposes, the carbon-composition resistor has been used for many years and is therefore known to have an acceptable long-term stability for domestic and many commercial purposes. Changes in resistance under normal working conditions may be of the order of 5%, but in more severe conditions, such as those encountered in military services, changes of up to 25% may occur. It is found that changes caused by high temperature (due either to ambient or self-generated heat) result in a permanent increase in resistance value, whereas exposure to high humidity increases the resistance, but the effect is largely reversible.

The stability of wirewound and cracked-carbon resistors is much higher—of the order of 1 to 2%. Metal-film resistors are comparable in stability to the wirewound types. This stability is dependent mainly on the protection afforded to the resistive element by sealing. Even under the severe conditions encountered in the services, changes are not usually more than 2%. Oxide films have a stability rather better than that of the cracked-carbon film. The highest stability with lowest temperature coefficient is still obtained with wirewound precision resistors, although some of the evaporated metal-film resistors approach this but do not yet equal it.

Accuracy (or tolerance). Carbon-composition resistors are made to approximate target values and then selected to various values after manufacture. Selection tolerances are set up and resistors sorted to ±5%, ±10% and ±20% of the nominal batch value. Carbon-composition resistors cannot be regarded as accurate to better than 5%, because of the lack of precise control in their composition and because of a tendency to drift in value. Pyrolytic or cracked-carbon resistors are usually accurate to 1 or 2% but can be manufactured to about 0.1% if necessary. Wirewound resistors are accurate to 0.25% and can be manufactured to 0.05% or even 0.01% if desired.

If resistors are used to the lowest manufacturing accuracy, the cost will be the minimum. If a greater accuracy is required, it can be provided up to the limits given in *Table 6.7* below. It follows, however, that the resistors will be more expensive, not only by reason of the work involved in selection but by the possibility that the resistors not within the required accuracy may be less readily saleable.

Table 6.7 ACCURACY OF RESISTORS

Type of resistor	*Manufacturing accuracy* %	*Best selection accuracy* %
Carbon composition (solid)	20	5
Carbon composition (film)	20	5
Carbon film (pyrolytic)	1	0.25
Metal film	1	0.1
Oxide film	1	0.25
Wirewound vitreous	5	0.1
Wirewound lacquered	1	0.1
Wirewound precision	0.01	0.01

Maximum operating temperature. Carbon-composition resistors are seriously affected by ambient temperatures over 100 °C, mainly by changes in the structure of the binder used in the resistor mixture. The maximum recommended surface temperature is about 110 °C to 115 °C. This is the total working temperature produced by the power dissipated inside the resistance, the heat from associated valves and components, and the ambient temperature in which the resistor is operating. Cracked-carbon resistors can be operated up to a maximum surface temperature of 150 °C under the same conditions, metal films up to 175 °C, and oxide films up to 200 °C to 250 °C, approximately. Some special metal and metal-oxide-film power resistors can operate at 500 °C to 600 °C when no limiting protective coating is applied.

Wirewound resistors are generally lacquered or vitreous enamelled for protection of the windings. For both types the safe upper limit is set by the protective coating. For lacquered types the maximum recommended temperature is 130 °C (some will work up to 450 °C).

Free circulation of air should be allowed, and the ends of tubular resistors should not be placed flat against the chassis. If the resistors are badly mounted, or if several resistors are placed together, derating is necessary.

Maximum operating voltage. The maximum operating voltage is determined mainly by the physical shape of the resistor and by the resistance value (which determines the maximum current through the resistor and therefore the voltage for a given wattage), that is, the 'critical value' referred to previously.

Commercial ratings at room temperature are some 25% to 50% higher than military ratings, and reference should always be made to the resistor manufacturer for his maximum voltage rating.

Frequency Range. On a.c. carbon-composition resistors (up to about 10 kΩ in value) behave as pure resistors up to frequencies of several MHz. At higher frequencies the self-capacitance of the resistor becomes predominant, and the impedance falls. The inductance of carbon-composition resistors does not usually cause trouble below 100 MHz (except in special cases such as in attenuator resistors). Cracked-carbon resistors specially manufactured with little or no spiral grinding can be operated at frequencies of many hundreds of MHz, but methods of mounting and connection become important at these frequencies. Other film-type resistors such as metal film and metal-oxide film are also suitable for use at high frequencies, and the effect of spiralling of the film is relatively unimportant below 50 MHz.

For wirewound resistors, the inductance of single-layer windings becomes appreciable, and *Ayrton–Perry* or *back-to-back* windings are often used for so-called non-inductive resistors. At high frequencies the capacitive rather than the inductive effect limits the frequency of operation. For example, the reactance of a typical resistor of 6 kΩ with an Ayrton–Perry winding becomes capacitance at 3 MHz.

In all measurements on resistors at high frequencies, the method of mounting the resistor is important. The direct end-to-end capacitance of the resistor and the capacitance of the two leads to the resistor body are included in the total capacitance being measured, and the resistor should therefore be mounted as near as possible as it is to be mounted in use. Ideally the mounting fixtures should be standardised for comparison measurements.

To summarise, for a resistor to be suitable for operation at high frequencies it should meet the following general requirements:

(*a*) Its dimensions should be as small as possible.
(*b*) It should be low in value.
(*c*) It should be of the film type.
(*d*) A long thin resistor has a better frequency characteristic than a short fat one.
(*e*) All connections to the resistor should be made as short as possible.
(*f*) There should be no sudden geometrical discontinuity along its length.

Noise. Carbon-composition resistors generate noise of two types: thermal agitation, or *Johnson* noise, which is common to all resistive impedances, and *current* noise, which is caused by internal changes in the resistor when current is flowing through it. The latter is peculiar to the carbon-composition resistor and other non-metallic films and does not occur in good quality wire-wound resistors. Cracked-carbon resistors generate noise in a similar fashion to the carbon-composition types but at a very much lower level. For low values of resistance (where the film is thick) the noise is difficult to measure. Metal-film and metal-oxide-film resistors generate noise at a very low level indeed.

Measurements have shown that for carbon-composition resistors, current noise increases linearly with current up to about 15 μA. With greater currents the noise curve approximates to a parabola.

Temperature coefficient of resistance. A resistor measured at 70 °C will have a different resistance value from that at 20 °C; the change in value at differing temperatures can be calculated from the temperature coefficient for each class of resistor. Approximate maximum values are given in *Table 6.8*.

The large values for carbon-composition types are partly due to non-cyclic changes, which tend to mask temperature coefficient effects.

Voltage Coefficient. When a voltage is applied across a carbon resistor, there is an immediate change in resistance, usually a decrease. The change, which is not strictly proportional to the voltage, is usually measured at values not less than 100 kΩ. In carbon-composition resistors, the change in resistance value due to the applied voltage is usually within 0.02%/V d.c. With cracked-carbon resistors, particularly the larger sizes, the effect is negligible for low values of resistance, certainly being less than 0.001% on the higher values it can rise to 0.002%, and on very small resistors, where the stress is clearly much greater, the maximum values may approach 0.005%/V. The voltage coefficient is frequently quoted at too high a figure because of the difficulty in separating it from effects due to temperature coefficient. Wirewound resistors do not show this effect, provided they are free from leakage between turns. Metal-film resistors have voltage coefficients from 0.000 1%/V to 0.003%/V depending on wattage, whereas metal-oxide-film resistors approximate from 0.000 1%/V to 0.000 5%/V.

Table 6.8 TEMPERATURE COEFFICIENT OF RESISTANCE OF RESISTORS

Type of material	*Temperature range of measurement* °C	*Temperature coefficient* ppm/°C	%/°C
Carbon composition	+20 to +70	±1 200	±0.12
Cracked carbon	+20 to +70	Ranges from −200 (for low values) to −1 000 (for high values)	−0.02 to −0.1
Wirewound:			
General purpose	+20 to +130	+200	+0.02
Precision	+20 to +130	+20	+0.002
Metal film:			
Gold–platinum	−40 to +150	+250 to +600 *	+0.025 to +0.06
Nickel–chromium	−50 to +150	+150 to 200 *	+0.15 to +0.02
Oxide film	−40 to +300	−500 to +500	−0.05 to +0.05

* Depending upon the composition.

Solderability. There is a change in the value of carbon-composition resistors, and to a smaller extent in cracked-carbon resistors, when they are soldered into equipment. This change, which is due to overheating, can be quite serious in miniature constructions if the connecting leads are short, and permanent changes of up to 25% may be caused. If the soldered joint is made 12 mm away from the resistor, there is usually no excessive overheating.

Shelf life. There is a change in the resistance of most types of resistor during storage. During one year the resistor of a carbon-composition resistor may change by 5%, while a cracked-carbon or wirewound resistor may change by only 0.5%. Metal-film resistors change by as little as 0.1% or less; oxide-film resistors change by less than 0.5%.

Load life (or working life). Resistors are also tested for their change in resistance after 1 000 h at a temperature of 70 °C. Under these conditions the resistance of cracked-carbon resistors may vary from 0.1% (for low values) to 3% (for high values). Wire-wound and oxide-film resistors do not change in value by more than 1%, but carbon-composition resistors may change by as much as 15%.

Thermistors

The name thermistor is an acronym for THERMally-sensitive resISTOR. This name is applied to resistors made from semiconductor materials whose resistance value depends on the temperature of the material. This temperature is determined by both the ambient temperature in which the thermistor is operating and the temperature rise caused by the power dissipation within the thermistor itself. The thermistors first developed had a negative temperature coefficient of resistance, that is the resistance value decreases as the temperature increases. This is in contrast to the behaviour of most metals and therefore of wirewound and metal-film resistors whose resistance increases with an increase of temperature. Subsequently thermistors with a positive temperature coefficient of resistance were developed, with the advantage over normal resistors of having a larger temperature coefficient. Thermistors are used in applications as temperature sensors or as stabilising elements to compensate the effects of temperature changes in a circuit.

THERMISTOR MATERIALS

Negative temperature coefficient (n.t.c.) thermistors are manufactured from the oxides of such materials as iron, chromium, manganese, cobalt and nickel. In the pure state, these oxides have a high resistivity. They can be changed into semiconductor materials, however, by the addition of small quantities of a metal with a different valency. Examples of the materials used in the manufacture of commercially available n.t.c. thermistors are given below.

One group of n.t.c. thermistors is manufactured from ferric oxide Fe_2O_3 in which a small number of trivalent ferric Fe^{3+} ions are replaced by tetravalent titanium Ti^{4+} ions. The titanium ions are compensated by an equal number of bivalent ferrous Fe^{2+} ions to maintain the electrical neutrality of the material. At low temperatures, the extra electrons of the Fe^{2+} ions are situated on iron ions next to the Ti^{4+} ions. At higher temperatures, however, the electrons are loosened from these sites and become free charge carriers. Hence the conductivity of the material increases; that is, the resistance decreases with an increase of temperature. Because electrons act as the charge carriers in the material, this oxide mixture is an *n*-type semiconductor material.

Another group of n.t.c. thermistors is made from nickel oxide NiO or cobalt oxide CoO. A small number of bivalent nickel Ni^{2+} ions or cobalt Co^{2+} ions are replaced by univalent lithium Li^{1+} ions, and these are compensated by an equal number of trivalent nickel Ni^{3+} or cobalt Co^{3+} ions. At low temperatures, the deficiencies of electrons (holes) of the Ni^{3+} or Co^{3+} ions are situated on the Li^{1+} ions, but at higher temperatures the holes are free to move through the crystal as charge carriers. Again there will be a decrease in resistance as the temperature is increased. These oxide mixtures are *p*-type semiconductor materials.

Stabilising oxides may be added to both *n*-type and *p*-type materials to obtain better reproducability and stability of the characteristics of the manufactured thermistor. The choice of material depends on the required resistance value and temperature coefficient for the particular thermistor.

Positive temperature coefficient (p.t.c.) thermistors are made from barium titanate $BaTiO_3$ or a solid solution of barium titanate and strontium titanate, $BaTiO_3$ and $SrTiO_3$. A semiconductor material is formed by substituting ions of higher valency for either the barium or the titanium ions. Trivalent ions such as lanthanum La^{3+} or bismuth Bi^{3+} are used to replace a small number of bivalent barium Ba^{2+} ions, or pentavalent ions such an antimony Sb^{5+} or niobium Nb^{5+} are used to replace a small number of titanium Ti^{4+} ions. When the barium titanate mixture is prepared in the absence of oxygen, an *n*-type semiconductor material with a low negative temperature coefficient is obtained. A positive temperature coefficient over a particular temperature range is obtained by heating the material in an oxygen atmosphere. As the mixture cools, oxygen atoms penetrate along the crystal boundaries. The absorbed oxygen atoms attract electrons from a thin layer of the semiconductor crystals along the boundaries. An electrical potential barrier is formed in this zone, consisting of a negative space charge on both sides of which a positive space charge is formed by the now uncompensated added ions. These barriers cause an increase in resistance with an increase in temperature to give the positive coefficient of resistance for the thermistor.

Another material used for p.t.c. thermistors is silicon. In extrinsic silicon, the conductivity is determined by the density and mobility of the charge carriers. The conductivity σ is given by

$$\sigma = Ne\mu$$

where N is the density of the charge carriers (electrons or holes depending on whether the silicon is *n*-type or *p*-type) and is determined by the doping level; e is the magnitude of the electronic charge; and μ is the mobility of the charge carrier. Thus the variation of conductivity with temperature will be determined by the variation with temperature of the charge carrier density and mobility. The density will increase with temperature because of the thermally generated electron-hole pairs. The mobility of both electrons and holes decreases with increasing temperature because of the increased effectiveness of lattice scattering. At the lower temperatures, the effect of the mobility is greater than that of the thermal generation, and so the conductivity falls with increasing temperature (a positive temperature coefficient is obtained). At higher temperatures, however, the increase in carrier density through thermal generation predominates, and the conductivity increases. The variation of conductivity over a temperature range −50 °C to +250 °C with doping level as parameter has the form shown in *Figure 6.4*. It can be seen that by choosing a suitable doping level, the required positive temperature coefficient can be obtained over various temperature ranges. Because the minimum value of conductivity occurs approximately 20–30 °C higher for *n*-type material than for *p*-type material, thermistors made from extrinsic silicon generally use *n*-type material.

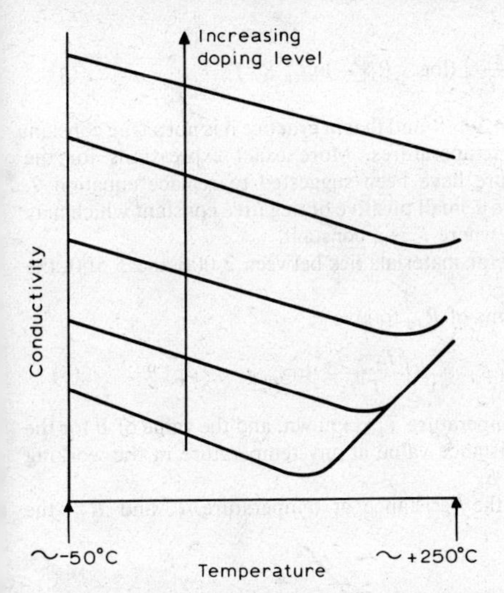

Figure 6.4 Variation of conductivity with temperature for extrinsic silicon with doping level as parameter

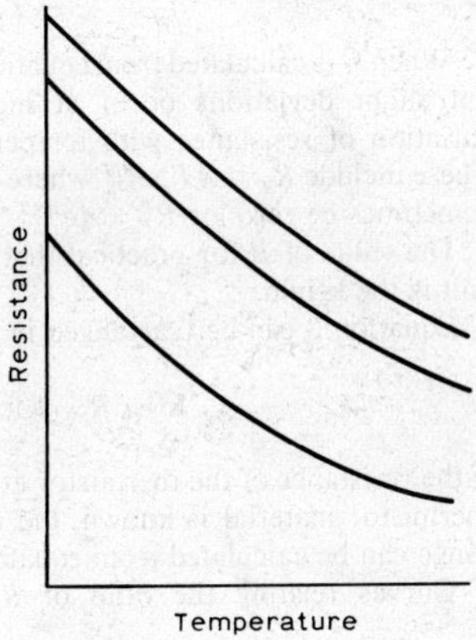

Figure 6.5 Resistance/temperature characteristic for n.t.c. thermistor

It should be remembered that the semiconductor materials from which thermistors are manufactured are polycrystalline and not monocrystalline like those used for transistors and diodes.

ELECTRICAL CHARACTERISTICS OF N.T.C. THERMISTORS

The variation of resistance with temperature for various types of n.t.c. thermistor has the form shown in *Figure 6.5*. The relationship between the resistance and temperature can be expressed as

$$R_T = Ae^{B/T} \tag{2}$$

where R_T is the resistance in ohms at an absolute temperature T in kelvin; e is the base of natural logarithms (2.718); and A and B are constants.

The value of B for a particular thermistor material can be found by measuring the resistance at two values of absolute temperature T_1 and T_2. From equation 2

$$R_{T_1} = Ae^{B/T_1}, \; R_{T_2} = Ae^{B/T_2}.$$

Dividing these two equations gives

$$\frac{R_{T_1}}{R_{T_2}} = e^{(B/T_1 - B/T_2)}.$$

Taking logarithms

$$\log_{10} R_{T_1} - \log_{10} R_{T_2} = \log_{10} e \, . \, B\left(\frac{1}{T_1} - \frac{1}{T_2}\right) \tag{3}$$

from which

$$B = \frac{1}{\log_{10} e} \cdot \frac{\log_{10} R_{T_1} - \log_{10} R_{T_2}}{1/T_1 - 1/T_2}$$

which simplifies to

$$B = 2.303\left(\frac{T_1 T_2}{T_2 - T_1}\right)(\log_{10} R_{T_1} - \log_{10} R_{T_2}) \tag{4}$$

When B is calculated from equation 4, it is found that in practice it is not a true constant but slight deviations occur at high temperatures. More exact expressions for the variation of resistance with temperature have been suggested to replace equation 2. These include $R_T = AT^C e^{B/T}$ where C is a small positive or negative constant which may sometimes be zero, or $R_T = Ae^{B/(T+K)}$, where K is a constant.

The value of B for practical thermistor materials lies between 2 000 and 5 500; the unit is the kelvin.

Equation 3 can be rearranged in terms of R_{T1} to give

$$\log_{10} R_{T_1} = \log_{10} R_{T_2} + B\left(\frac{T_2 - T_1}{T_1 T_2}\right)\log_{10} e \tag{5}$$

If the resistance of the thermistor at temperature T_2 is known, and the value of B for the thermistor material is known, the resistance value at any temperature in the working range can be calculated from equation 5.

Curves relating the ratio of R_T, the resistance at temperature T, and R_{25}, the

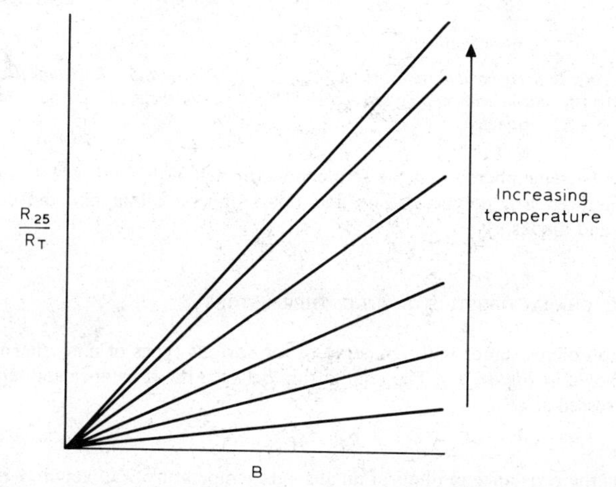

Figure 6.6 Ratio of R_{25}/R_T plotted against B value with temperature as parameter

resistance at 25 °C which is taken as a *standard* value for comparing different types of thermistor, to the value of B with temperature as parameter are sometimes given in published data. From these curves, shown in *Figure 6.6*, and the known value of R_{25} and B for the thermistor type given in the data, the value of R_T can be determined.

A temperature coefficient of resistance can be derived for the thermistor material. This coefficient α is obtained by differentiating equation 2 with respect to temperature:

$$\alpha = \frac{1}{R}\frac{dR}{dT} = -\frac{B}{T^2} \tag{6}$$

At 25 °C, the value of α for practical thermistor materials lies, typically, between 2.5 and 7%. It can be seen from equation 6 that the temperature coefficient varies inversely as the square of the absolute temperature.

The voltage/current characteristic for an n.t.c. thermistor is shown in *Figure 6.7*. The characteristic relates the current through the thermistor with the voltage drop across it after thermal equilibrium has been established in a constant ambient temperature. This static characteristic is plotted on logarithmic scales.

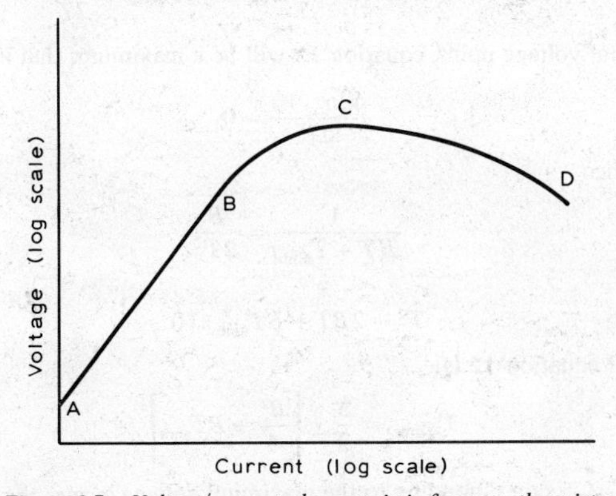

Figure 6.7 *Voltage/current characteristic for n.t.c. thermistor*

Over the low-current part of the characteristic, part $A - B$, the power input to the thermistor is too low to cause any rise in temperature by internal heating. The thermistor therefore acts as a linear resistor. Above point B, however, the current causes sufficient power dissipation within the thermistor to produce a rise in temperature and therefore a fall in resistance. The resistance value is therefore lower than would be expected for a linear resistor. As the current is increased further, the power dissipation within the thermistor causes a progressively larger fall in resistance, so that at some current value the voltage drop across the thermistor reaches a maximum, point C. Above this current value, the fall in resistance caused by the internal heating is large enough to give the thermistor a negative incremental resistance, part $C - D$.

The temperature corresponding to the maximum voltage across the thermistor can be calculated for a particular thermistor material. If it is assumed that the temperature is constant throughout the body of the thermistor, and that the heat transfer is proportional to the temperature difference between the thermistor and its surroundings (which is true for low temperature differences), then equation 2 can be rewritten in terms of natural logarithms as

$$\log_e R_T = \log_e A + \frac{B}{T} \tag{7}$$

At thermal equilibrium, the electrical power input to the thermistor is equal to the heat dissipated; that is

$$VI = D(T - T_{amb}) \tag{8}$$

where V is the voltage drop across the thermistor, I is the current through it, T is the temperature of the thermistor body, T_{amb} is the ambient temperature, and D is the dissipation constant (the power required for unit temperature rise). The dissipation constant may also be represented by the symbol δ.

Since $R_T = V/I$, equation 7 can be rewritten as

$$\log_e V - \log_e I = \log_e A + \frac{B}{T} \tag{9}$$

Taking logarithms of equation 8 gives

$$\log_e V + \log_e I = \log_e D + \log_e (T - T_{amb}) \tag{10}$$

Equations 9 and 10 can be added to give

$$\log_e V = \frac{1}{2} \log_e AD + \frac{1}{2} \log_e (T - T_{amb}) + \frac{B}{2T} \tag{11}$$

At the maximum voltage point, equation 11 will be a maximum; that is

$$\frac{d(\log_e V)}{dT} = 0$$

This occurs when

$$\frac{1}{2(T - T_{amb})} = \frac{B}{2T^2}$$

or

$$T^2 - 2BT + BT_{amb} = 0 \tag{12}$$

The solution of equation 12 is

$$T_{V(max)} = \frac{B}{2} \pm \left[\frac{B^2}{4} - BT_{amb}\right]^{\frac{1}{2}} \tag{13}$$

The value of $T_{V(max)}$ corresponding to the maximum voltage across the thermistor is

$$T_{V(max)} = \frac{B}{2} - \left[\frac{B^2}{4} - BT_{amb}\right]^{\frac{1}{2}}$$

It can be seen from equation 13 that a solution is possible only if $B > 4T_{amb}$. Also, the temperature corresponding to the maximum voltage is determined by the B value only and not by the resistance value. For practical thermistor materials, $T_{V(max)}$ lies between 45 °C and 85 °C.

In many applications, it is necessary to know the time taken for the thermistor to reach equilibrium. Assuming again that the temperature is constant throughout the body of the thermistor, the cooling in a time dt is given by

$$-HdT = D(T - T_{amb})\, dt \tag{14}$$

where H is the thermal capacity of the thermistor in joules/°C. When the thermistor cools from temperature T_1 to T_2, equation 14 gives

$$(T_1 - T_{amb}) = (T_2 - T_{amb})_e - t/\tau$$

where τ is the thermal time-constant of the thermistor and is equal to H/D.

In practice, the temperature is not constant throughout the body of the thermistor, the surface cooling more rapidly than the interior. The time-constant quoted in published data is defined as the time required by the thermistor to change by 63.2% of the total change between the initial and final body temperatures when subjected to an instantaneous temperature change under zero power conditions. Cooling curves showing the rise in resistance with time when the electrical input is removed from the thermistor are often given in published data.

ELECTRICAL CHARACTERISTICS OF P.T.C. THERMISTORS

The variation of resistance with temperature for a p.t.c. thermistor is more complex than that for an n.t.c. thermistor shown in *Figure 6.5*. Because the temperature coefficient of resistance is positive only over part of the temperature range, a typical resistance/temperature characteristic for a p.t.c. thermistor may have the form shown in *Figure 6.8*. Over parts *A–B* (temperature T_1 to T_2) and *C–D* (temperature T_3 to T_4) the

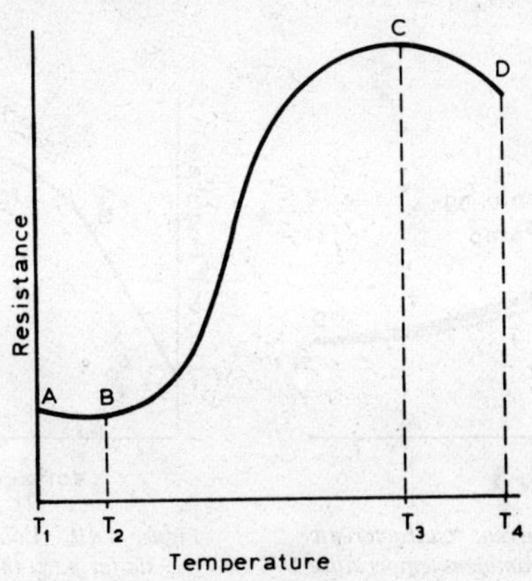

Figure 6.8 Resistance/temperature characteristic for p.t.c. thermistor

coefficient is negative; it is only over part *B–C*, corresponding to the temperature range T_2 to T_3, that the required positive coefficient is obtained. Over this range, the relationship between resistance and temperature can be expressed (approximately) as

$$R_T = A + Ce^{BT} \qquad (15)$$

where R_T is the resistance in ohms at an absolute temperature T in kelvin, e is the base of natural logarithms, A, B and C are constants, and T is restricted in value to $T_2 < T < T_3$.

Equation 15 can be differentiated as before to give the temperature coefficient of resistance α

$$\alpha = \frac{1}{R}\frac{dR}{dT} = \frac{BCe^{BT}}{A + Ce^{BT}}$$

In practice, unfortunately, the resistance/temperature characteristic can seldom be described by such a simple relationship as that of equation 15. Unlike n.t.c. thermistors where equation 2 provides a reasonable approximation to the practical behaviour, any attempt to modify equation 15 to a more accurate form results in complicated expressions. For this reason, graphical methods are often used for design calculations. A quantity called the switch temperature T_s is often quoted in published data. This temperature is the one at which a p.t.c. thermistor begins to have a usable positive temperature coefficient of resistance. T_s is defined as the higher of two temperatures at which the resistance of the thermistor is twice its minimum value.

The voltage/current characteristic for a p.t.c. thermistor is shown in *Figures 6.9* and *6.10*. In both figures the voltage and current axes are interchanged with respect to those of the characteristic for an n.t.c. thermistor shown in *Figure 6.7*. The characteristic shown in *Figure 6.9* is plotted on linear scales, and characteristics for different ambient temperatures are included to show the effect on the thermistor. When the voltage across the thermistor is low, the power dissipation within the thermistor is insufficient to heat it above the ambient temperature. The thermistor behaves as a linear resistor, part *A–B*. When the voltage is increased, the power dissipation causes the thermistor temperature to rise above the switch temperature T_s, point *C*. The resistance of the thermistor rises, and the current falls. Any further increase in voltage results in a progressive fall in current, part *C–D*. The characteristic for a constant ambient temperature plotted on logarithmic scales is shown in *Figure 6.10*.

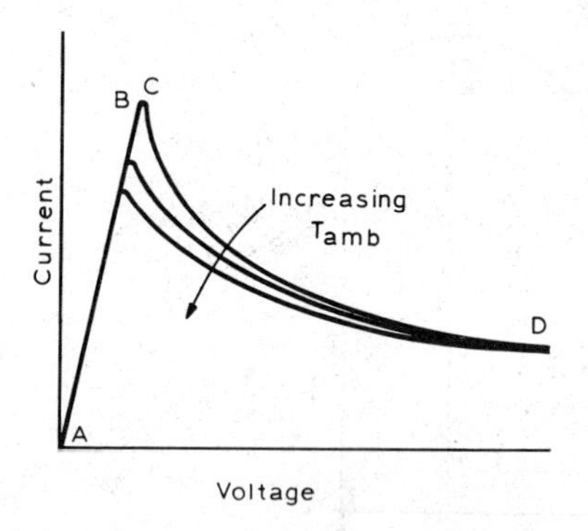

Figure 6.9 Voltage/current characteristic for p.t.c. thermistor with ambient temperature as parameter (*linear scales*)

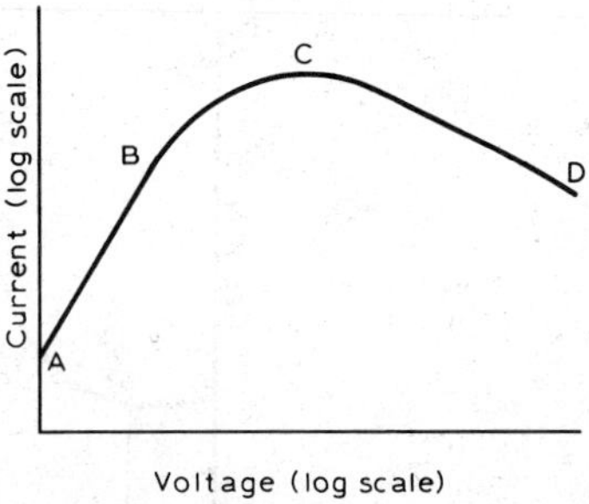

Figure 6.10 Voltage/current characteristic for p.t.c. thermistor (*log scales*)

At higher voltages, p.t.c. thermistors show a voltage dependency, the resistance value being determined by the voltage across the thermistor as well as by the temperature. The behaviour of the thermistor under these conditions can be represented by an equivalent circuit consisting of an ideal p.t.c. thermistor (no voltage dependency) in parallel with an ideal voltage dependent resistor (the voltage and current being related by the expression: $V \propto I^{\beta}$). The voltage/current characteristics of the components of this equivalent circuit are shown in *Figure 6.11* compared with the characteristic of a normal p.t.c. thermistor. The *normal* characteristic is measured at constant ambient temperature under pulse conditions to avoid self-heating of the thermistor. At low voltages, the characteristic of the normal p.t.c. thermistor coincides with the characteristic of the ideal thermistor. At the higher voltages where the voltage dependency becomes effective, the characteristic of the normal thermistor coincides with that of the voltage dependent resistor. The point of intersection of the two 'ideal' characteristics, where the currents through the two components of the equivalent circuit are equal, defines the balance voltage V_b. The value of balance voltage for a specified ambient temperature is given in published data. The voltage dependency of the thermistor β can be calculated from the expression

$$\beta = \frac{\log V_3 - \log V_2}{\log (I_3R - V_3) - \log (I_2R - V_2)}$$

where V_2 is a pulse voltage greater than V_b; V_3 is a pulse voltage greater than V_2; I_2 and I_3 are the currents corresponding to V_2 and V_3 respectively; and R is the initial slope of

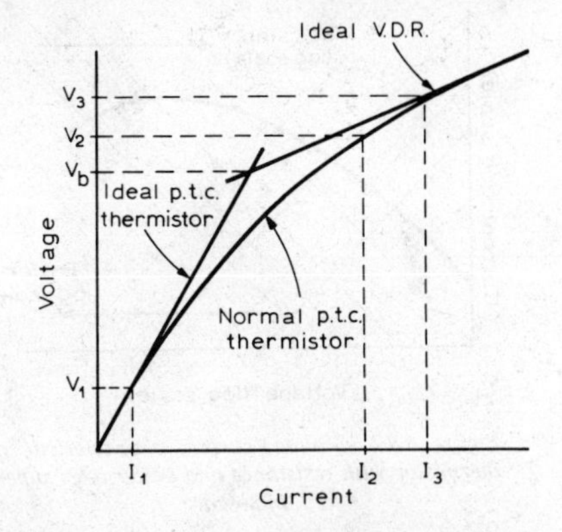

Figure 6.11 Voltage/current characteristic of p.t.c. thermistor compared with ideal *components of equivalent circuit*

the characteristic given by V_1/I_1. As the value of β depends on temperature, when quoted in published data the value is qualified by the relevant temperature.

As with n.t.c. thermistors, a thermal time-constant τ is given in published data for p.t.c. thermistors, defined in the same way as previously described. Cooling curves showing the fall in resistance with time after the electrical input is removed from the thermistor are also given.

CHOICE OF THERMISTOR FOR APPLICATION

When n.t.c. and p.t.c. thermistors are used in circuits, the designer must consider certain requirements before he can select the correct thermistor for his purpose. These requirements include the resistance value, temperature coefficient of resistance and temperature range; the power dissipation required of the thermistor; and the thermal time-constant.

Most of these factors have been discussed in the previous sections describing the characteristics of n.t.c. and p.t.c. thermistors. In the published data a resistance value, usually at 25 °C, is given, supplemented by resistance values at other temperatures, a temperature coefficient of resistance (α or B value), and resistance/temperature curves for various ambient temperatures. Sometimes an operating temperature range or a maximum operating temperature is given. For p.t.c. thermistors, the switch temperature is given. The voltage/current characteristic of the thermistor sometimes has resistance and power axes superimposed, as shown in *Figure 6.12*. The balance voltage V_b and voltage dependency β at a specified temperature are also given for p.t.c. thermistors. A maximum power dissipation may be specified, or a maximum voltage or current. The dissipation factor δ is often given in two forms: for still air assuming cooling by natural convection and radiation, and when mounted on a heatsink to increase the cooling. A thermal time-constant τ is given, supplemented by cooling curves.

From this data, the designer is able to choose a suitable thermistor. For some applications, however, a single thermistor may not meet the requirements, and it is possible to combine a thermistor with a series/parallel combination of linear resistors so that the overall characteristic corresponds to that required. For p.t.c. thermistors, it should be noted that with a series resistance three working points are possible. In *Figure 6.13*, the voltage/current characteristic of a p.t.c. thermistor plotted on linear scales is shown with a resistive loadline superimposed. The resistive loadline intersects the voltage axis at the supply voltage V_s, and the current axis at I_s where I_s is given by V_s/R. Of the three working points given by the intersection of the characteristic and loadline, points P_1 and P_2 are stable, while P_3 is an unstable working point. When the supply voltage is first applied, equilibrium will be established at working point P_1 at which the

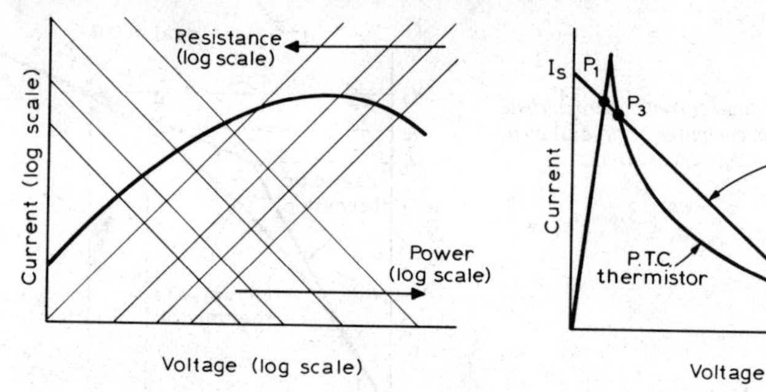

Figure 6.12 Voltage/current characteristic of thermistor with resistance and power axes superimposed

Figure 6.13 Voltage/current characteristic of p.t.c. thermistor with resistive loadline superimposed

load current is relatively high. Working point P_2 can only be reached if the supply voltage increases, the ambient temperature increases, or the resistance in series with the thermistor decreases. In each case, the displacement of the loadline or the change in the thermistor characteristic with temperature must be large enough to allow the peak of the thermistor characteristic to lie under the loadline.

MANUFACTURE OF THERMISTORS

Thermistors are manufactured by sintering the various oxide mixtures previously described (see Thermistor materials). The generally used shapes for unencapsulated thermistors are rods, discs, plates, and beads. Some types are encapsulated to protect the thermistor element, either in plastic or glass.

The oxide mixture used is finely ground and a plastic binder material such as polyvinyl acetate added. Rod thermistors are formed by extrusion through dies of the required diameter, and cutting to length; discs and plates are formed by pressing. The formed element is heated to drive out the binder material, and then sintered at a temperature between 1 000 °C and 1 350 °C. Electrical contacts are generally made with silver, gold, or platinum paste which is cured before the lead-out wires are attached. Typical dimensions of rod-type thermistors are diameters of 1–6 mm, and lengths of 5–50 mm. The diameters of disc-type thermistors can be from 1.5 to 15 mm, with thicknesses between 1 and 6 mm. Plate thermistors have a rectangular outline of a comparable size to disc thermistors. The element of this type of thermistor is generally painted to provide some protection and to colour-code the device.

Bead thermistors are made by placing small masses of a slurry formed from the oxide mixture on two closely-spaced parallel platinum wires. Surface tension draws the masses into a bead shape. The beads are allowed to dry, and then gently heated until strong enough to be sintered. Sintering shrinks the beads firmly on to the platinum wires so that good electrical contact is made. The wires can then be cut to form the lead-out wires for the thermistor, and the element enamelled for protection. A typical diameter for bead thermistors is 1 mm.

Thermistors are encapsulated to provide protection when used in dirty or corrosive atmospheres or liquids. Small glass envelopes are used for bead thermistors. An alternative form of encapsulation uses plastic. This enables the thermistor element to be in good thermal contact with the environment and so have a short response time, and at the same time be electrically insulated.

Typical thermistor shapes and encapsulations are shown in *Figure 6.14* and the circuit symbol for a thermistor is shown in *Figure 6.15*. The $t°$ indicates temperature

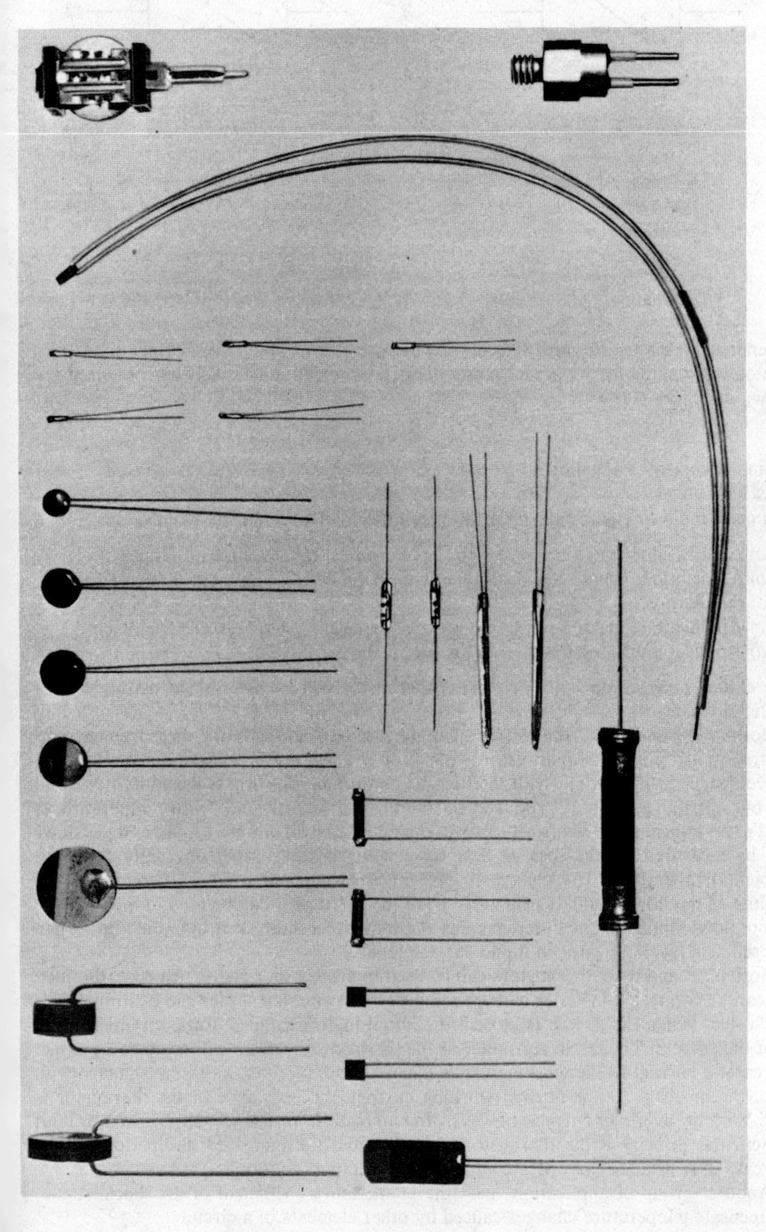

Figure 6.14 Typical thermistors (Courtesy Mullard Limited)

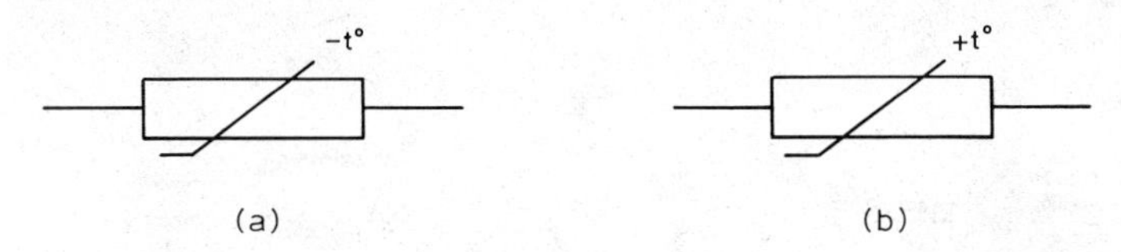

Figure 6.15 *Circuit symbol for (a) n.t.c. thermistor and (b) p.t.c. thermistor*

Figure 6.16 *Old circuit symbols for (a) n.t.c. thermistor (b) p.t.c. thermistor*

dependency, an n.t.c. thermistor being indicated by $-t°$ and a p.t.c. thermistor by $+t°$. The older symbols for n.t.c. and p.t.c. thermistors which may still be encountered are shown in *Figure 6.16*.

APPLICATIONS OF THERMISTORS

It is convenient to classify the applications of thermistors into four groups:

(*a*) Applications where the resistance is determined by the ambient temperature.
(*b*) Applications where the resistance is determined by the power dissipated within the thermistor.
(*c*) Applications making use of the positive or negative temperature coefficient.
(*d*) Applications using the thermal inertia of the thermistor.

The choice between an n.t.c. or a p.t.c. thermistor will be determined mainly by the application.

Both n.t.c. and p.t.c. thermistors can be used in temperature measurement and control applications. The thermistor is operated at a low power level so that self-heating is avoided. A bridge circuit with the thermistor forming one arm is generally used, and the out-of-balance current produced by the change of resistance with temperature is used as an indication of temperature or to operate a control circuit. Liquid and gas flows can be measured by the loss of heat from a thermistor heated internally to above ambient temperature. The change in resistance as a result of the different rates of cooling as the flow conducts heat away from the thermistor can be used to measure the rate of flow. Similarly, the different rates of cooling of a thermistor in liquid and air can be used as a level indicator in liquid storage tanks.

Both n.t.c. and p.t.c. thermistors can be used in protection circuits. An n.t.c. thermistor can be connected in series with a switch so that switching surges can be limited to a safe value. When the switch is closed, the initial high resistance of the thermistor will limit the current. The fall in resistance as the thermistor heats up will produce a gradual increase in current to the working value. Current limit circuits can be constructed with p.t.c. thermistors. At the normal working current, the resistance of the thermistor is low. With an overload current, however, the increase in resistance can be used to limit the current directly if the thermistor can withstand the increased dissipation, or the increase in resistance can operate additional protection circuits.

Another group of applications uses the temperature coefficient of the thermistor to compensate temperature changes caused by other elements in a circuit.

The change in resistance as a thermistor heats up can be used in delay circuits. Connecting an n.t.c. thermistor in series with, for example, a relay can delay the energising of the relay until the resistance of the thermistor has fallen sufficiently to allow the operating current to flow. Similarly the increase in resistance of a p.t.c. thermistor can be used to de-energise a relay after an initial energising.

Voltage-sensitive resistors

These are formed by dry-pressing silicon carbide with a ceramic binder into discs or rods and firing at about 1 200 °C. The ends of the rods, or the sides of the discs, are sprayed with metal (usually brass) to which connections are soldered. They are often known as voltage dependent resistors, the current through the resistor being given by

$$I = KE^n$$

where K = a constant equal to the current in amperes at E = 1 V; and n = a constant dependent on voltage, varying between 3 and 7 for common mixes. It is usually between 4.0 and 5.0.

APPLICATIONS OF VOLTAGE-SENSITIVE RESISTORS

Suppressing voltage surges and quenching contact sparks. The resistor is connected across inductive loads and prevents voltage surges. The space required by the resistor is small and the current normally passing through it is quite small. A similar application is in avoiding sparking of relay contacts although the current permissible is limited to about 0.2 A per unit.

Protection of smoothing capacitors. In the anode circuit of a valve a smoothing capacitor is often connected from the anode to earth to prevent coupling. After switching on it requires some time for the cathode of the valve to reach a high enough temperature to allow the correct anode current to flow and during this time the capacitor is subjected to full input voltage and it must be rated accordingly. A much 'lighter' type of capacitor can be used if it is shunted by a voltage-sensitive resistor which will limit the voltage across the capacitor.

Voltage stabilisation. A resistor is used either directly across a varying power source or in a bridge circuit with linear resistors, but in both cases the loss of energy is somewhat excessive.

Other uses suggested include lightning arrestors, shunting of rectifiers, arc furnaces, thyratrons, armatures, etc.

General characteristics of integrated circuit resistors

There are three types of resistors for integrated circuits:

(*a*) Thick-film resistors.
(*b*) Thin-film resistors.
(*c*) Solid-state resistors.

Of these the most widely used are the thick-film resistors found in most *Hybrid* or *Thick-film* circuits.

THICK-FILM RESISTORS

These are made by screening a resistive paste in a required pattern on to a ceramic base and firing at a high temperature. The resistors so formed are adjusted to the value required by cutting slots either by abrasive material or by laser beams. A thick-film resistor is usually defined as a film thicker than 0.25 mm.

Resistive pastes. The resistive pastes are composed of powdered inorganic solids, such as metals and metal oxides, mixed with a powdered glass binder and suspended in an organic vehicle. Four basic classes of thick-resistor materials are commercially available:

(*a*) A platinum-based system, which fires at about 980 °C.
(*b*) A ruthenium-oxide system, which fires at about 760 °C.
(*c*) A thallium-oxide system, which fires at about 580 °C.
(*d*) Various palladium–silver based systems, which fire at about 760 °C.

Resistor materials are available in the range from 1 Ω to 1 MΩ/square/mil. Inks of the same resistor family can be blended to form intermediate values. By adding semiconductor oxides, the temperature coefficient will be reduced. For special applications, the paste supplier usually modifies the basic resistor properties. Available commercial inks are a compromise of the various properties.

These pastes are deposited on the substrate by the stencil screen process, an outgrowth of the silk screen method of printing. The silk has been replaced by a woven mesh of fine stainless steel wires, which is used to hold the pattern for the resistors to be deposited. The pattern is produced photographically, and the holes in the mesh are blocked by an emulsion wherever the inks are not to be deposited.

Resistor trimming. Abrasive and laser trimming are similar in that the material is removed to increase the resistor value. Some typical patterns are shown in *Figure 6.17*. Abrasive trimming equipment consists of a set of nozzles, any size or shape, positioned

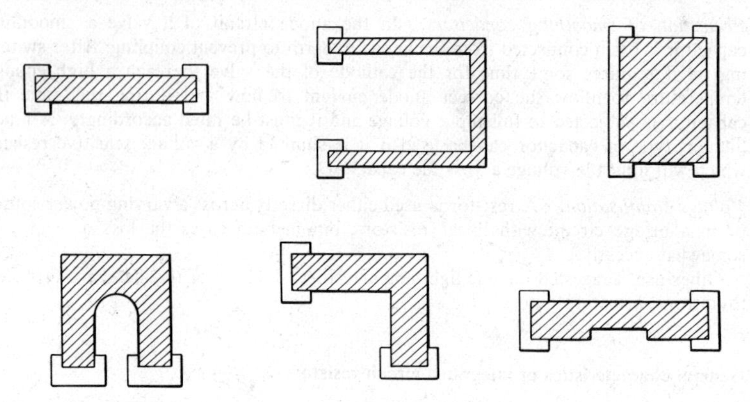

Figure 6.17 Typical film resistor patterns

over the resistor to be trimmed. Aluminium oxide powder, usually 27 micron, is propelled at the resistor and the material is removed (see *Figure 6.18*). The operation is continuously monitored throughout until the proper value is attained, then the abrasive flow is stopped. Tolerances as low as ±0.5% can be realised with abrasive trimming.

The laser process uses a focused laser beam to accomplish the task of trimming. Resistor material in the path of the beam is vaporised, leaving an exposed substrate. Thus, the resistor area that is not required is merely isolated from the conductive path. There are a variety of laser systems; the most common types are YAG (yttrium–aluminum–garnet) and CO_2 (carbon dioxide).

Laser trimming offers some advantages over abrasive techniques because the trimming can be performed on glazed resistors without altering the glaze. In laser trimming,

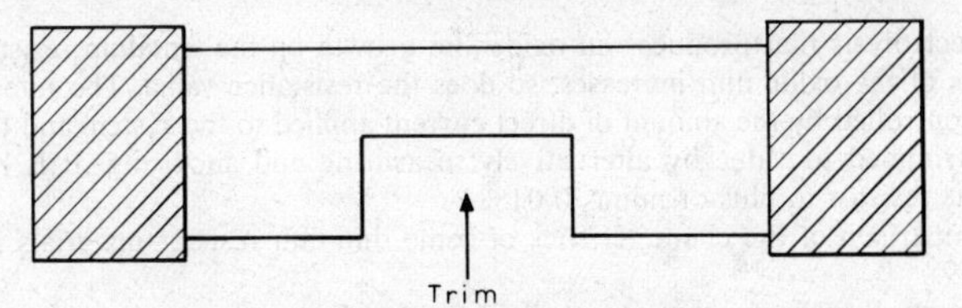

Figure 6.18 Air abrasive trimming of thick-film resistor

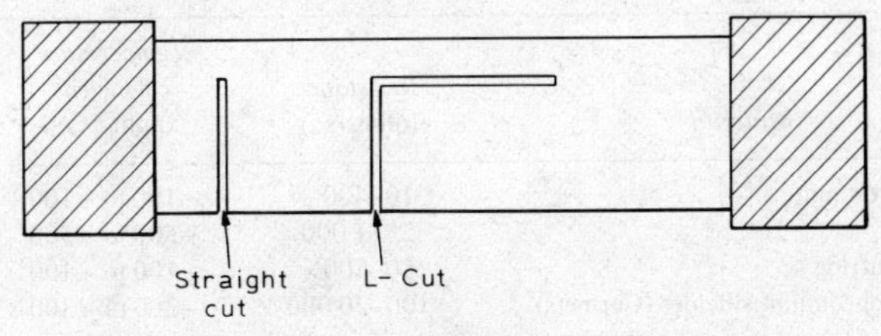

Figure 6.19 Laser trimming of thick-film resistor

narrow trim paths are possible as shown in *Figure 6.19*. Laser-trimmed resistors seem to maintain identical characteristics to the basic resistive material and abrasive units exhibit higher drift and noise parameters. However, the capital investment for a laser system is large and can only be justified for large production requirements.

THIN-FILM RESISTORS

There are two main types of thin-film resistors—evaporated nickel chromium and anodised tantalum. Thin-film resistors are more accurate and stable than thick-film resistors but gas usually more costly to produce.

Evaporated-film resistors. With the evaporated metal alloy or *vacuum deposition* process, the resistive materials are vaporised in a vacuum, then deposited on a substrate through a metal mask or a photographically deposited resist layer. The resulting pattern forms the resistors on a circuit and the film thickness is monitored to the required value while being deposited. The metals evaporated are usually based on chromium–nickel alloys although other alloys can be used. The temperature coefficients of resistance of resistors produced in this method can be very low. Metal–silicon monoxide (Cermet) films can be evaporated by co-evaporation of a metal and silicon monoxide. This is done either by using two independent sources of evaporation (one of chromium and the other of silicon monoxide) or by *Flash* evaporation in which the chromium–silicon monoxide mixture is dropped (as a powder mix) on to a hot plate.

Sputtered-film resistors. In the sputtering system one of the electrodes is made of the material which it is desired to deposit as a thin film. After evacuation an inert gas (such as argon) is fed in which is ionised by a high voltage. The positive ions strike the resistor material, liberating atoms which diffuse and coat the substrate. Masking techniques are again used to provide the resistor pattern. A useful property of sputtering is that by using oxygen gas instead of argon, oxides can be deposited to make oxide-film resistors.

Anodised-film resistors. The metal tantalum can have its surface partially oxidised by a chemical process to form a resistor which is extremely stable and has a low temperature coefficient. A special method of adjusting to value is termed anodising. This process consists of an electrolytic cell, containing electrolyte in which the tantalum resistor (anode) is immersed with an inert cathode. Application of a d.c. voltage to the terminals cause electrolysis that produces an oxide-film growth on the tantalum resistor. As the thickness of the oxide film increases, so does the resistance value. The rate of growth can be controlled by the amount of direct current applied to the system and the resistor can be trimmed to value by alternatively measuring and anodising. It is possible to adjust the resistor to plus or minus 0.01%.

A comparison of the characteristics of some thin-film resistor materials is given in *Table 6.9*.

Table 6.9 CHARACTERISTICS OF SOME THIN-FILM RESISTOR MATERIALS

Material	*Resistance* (ohms/sq)	*Temperature coefficient* (ppm/°C)	*Deposition tolerance* (%)
Nickel–chromium	10–400	−100 to +100	5
Tin oxide	25–1 000	−500 to +500	15
Tantalum nitride	50–500	−100 to +100	10
Tantalum–chromium-silicides (Cermets)	100–20 000	−300 to +300	20

SOLID-STATE RESISTORS

It is possible to fabricate some resistors as part of the diffusion process when making silicon integrated circuits. Thin-film resistors can be deposited directly on silicon and ion implantation can be used to fabricate resistors. In fabricating diffused resistors the resistivity of the silicon material itself is used, but it is subject to some limitations, such as reverse bias operation and high temperature coefficient. Each of the three methods has certain fabrication advantages and performance disadvantages and *Table 6.10* based on a paper by Hans H. Stellrecht and Gary Kelson at the 1973 International Solid-State Circuits Conference gives a comparison of solid-state resistor technologies.

Table 6.10 COMPARISON OF MONOLITHIC-RESISTOR TECHNOLOGIES

Fabrication process	*Nominal sheet resistance* ohms/square	*Matching tolerance*				*Temperature coefficient* ppm/°C
		Deviation (%)		*Mean* (%)		
		10μ	40μ	10μ	40μ	
Diffusion	135	0.44	0.23	−0.1	0.07	+1 500
Thin film	1 000	0.24	0.11	−0.1	0.06	−200
Ion Implantation	1 250	0.34	0.12	−0.04	0.05	+400

General characteristics of variable resistors

TYPES OF VARIABLE RESISTOR

There are two general classes of variable resistor: general purpose and precision. The general-purpose resistors may be sub-divided into wirewound and carbon-composition types. The precision resistors, which are always wirewound, usually follow linear, sine-cosine or other mathematical laws. Linearities as high as 0.01% (for linear) and 0.1% (for sine-cosine and other laws) are obtainable. The general-purpose types usually follow a linear law, but some follow a logarithmic law. They have overall resistance tolerances of 10% for the wirewound types (although much closer tolerances can be obtained) and 20% for the carbon-composition types.

The metal-film and the high-quality molded-track types can also be considered to be in the precision category. Linearities of 0.5% in the molded type and 0.1% in metal-film types (with the aid of trimming) are obtainable.

ELECTRICAL CHARACTERISTICS OF VARIABLE RESISTORS

A summary of the electrical characteristics of most types is given in *Table 6.11*.

RESISTANCE VALUE

For precision-variable resistors the upper limit of resistance value is about 100 kΩ; above this the element size may exceed 150 mm in diameter. General-purpose types are made in values up to 500 kΩ (wirewound) and 5 MΩ (carbon). The lower limit is about 1 Ω for wirewound resistors and about 10 Ω for carbon-composition types.

RESISTANCE LAW

The resistance law is the low relating the change of resistance to the movement of the wiper, and it may be linear, logarithmic, log-log, sine-cosine, secant and the like, depending on the requirement for which the variable resistor is designed.

LINEARITY

There is often confusion between the terms *linearity*, *resolution*, *discrimination* and *accuracy* in discussing variable resistors. An ideal linear variable resistor has a constant resistance change for each equal increment in angular rotation (or linear movement) of the slider. In practice, this relationship is never achieved, and the linearity, or linear accuracy, is the amount by which the actual resistance at any point on the winding varies from the expected straight line of a 'resistance *vs* rotation' graph in a rotary variable resistor or 'resistance *vs* movement' graph in a linear variable resistor. For example, a 1 kΩ variable resistor held to a linearity of $\pm$0.1% would not vary more than 1 Ω on either side of the line of zero error.

The terms *resolution* and *discrimination* are synonymous. Resolution, or discrimination, is the resistance per turn of resistance wire and is thus a function of the number of turns on the variable resistor. For example, a resistor of 100 Ω containing 100 turns of wire has a resolution of 1 Ω. Resolution may be defined more accurately as *resistance resolution*, *voltage resolution* or *angular resolution*. Resistance resolution is the resistance per turn; voltage resolution is the voltage per turn; and angular resolution is the minimum change in slider angle necessary to produce a change in resistance. In general, the resistance resolution is one-half of the linearity. For example, if the linearity (linear accuracy) of a 1 kΩ resistor is to be held to within 0.1%, the resistance resolution should be 0.5% or less, and the winding should have at least 2 000 turns. The word *accuracy* unqualified, has no meaning in defining a variable resistor.

STABILITY

Stability concerns the change of resistance with time, or under severe climatic conditions, as well as the behaviour under normal load conditions. For general-purpose

Table 6.11 SUMMARY OF ELECTRICAL CHARACTERISTICS OF VARIABLE RESISTORS

	Mfg. tolerance (%)	*Selection tolerance* (%)	*Overall stability (after climatic tests)* (%)	*Linearity* (%)	*Resolution* (*in degrees*)	*Life* (*number of sweeps*)
Carbon composition. Coated-track types	±20	±20	±20	±15 (at 50% rotation)	(Stepless)	20 000 minimum. Max. depends on construction
Carbon construction. Moulded solid-track types	±20	±20	±5	±15 (at 50% rotation)	(Stepless)	20 000 minimum. Max. depends on construction
Rotary wirewound general-purpose types	±10	±10	±2	1.0	1.0	20 000 minimum. Max. depends on construction
Wirewound precision linear types	±5	Not applicable	High if sealed	Average 0.5 (can be 0.01)	Average 0.1 (depends on size, wire, etc.)	50 000 minimum. Max. several millions
Wirewound precision toroidal types	±5	Not applicable	High if sealed	Average 0.1 (can be higher)	Average 0.1 (depends on size, wire, etc.)	50 000 minimum. Max. several millions
Wirewound precision helical types	±5	Not applicable	High if sealed	Average 0.25 (can be higher)	0.01 (for 10-turn pot.) (depends on wire, etc.)	50 000 minimum. Max. several millions
Sine–cosine potentiometer. Card-wound types	±5	Not applicable	High if sealed	Average 0.5 (can be higher)	Varies with slider position	50 000 minimum. Max. several millions

carbon-composition variable resistors the stability tolerance is 25% and for general-purpose wirewound resistors it is 2%. The stability of the precision types of wirewound variable resistor is much higher, as these are usually sealed to exclude moisture and dust.

MINIMUM EFFECTIVE RESISTANCE

All variable resistors have some method of *ending off* the resistance element so that the slider goes into a *dead* position at each end, although it may rotate a few degrees more. There is a small jump in resistance, known as the *hop-off* resistance, as the slider touches the element. For general-purpose wirewound types, this should be less than 3% of the nominal resistance, and for carbon composition types it should be less than 5%.

EFFECTIVE ANGLE OF ROTATION

The *dead* positions mentioned in the previous paragraph are known as the *hop-off* angles. For military use the hop-off angle must not exceed 10% of the total angular rotation at either end for general-purpose wirewound resistors and 30% for carbon-composition types. The effective angle of rotation is 360° less the sum of the hop-off angles at the ends and the space allowed for terminations.

LIFE UNDER GIVEN CONDITIONS

Service specifications require that both wirewound and carbon-composition types should withstand 10 000 sweeps at 30 cycles/min with full-load current through the resistance element, totalling 20 000 cycles. After the test the change in resistance should not be more than 2% for the wirewound types and 5% for the composition types. Precision-variable resistors designed for long life—for example, with low brush pressure and carrying little current—have much longer lives—up to two million sweeps and sometimes up to ten million.

PERFORMING UNDER VARIOUS CLIMATIC CONDITIONS

The most frequent causes of failure in variable resistors are corrosion of the metal parts and swelling and distortion of plastic parts such as track moldings, cases and the like, due to moisture penetration. To combat these problems, the variable resistor should embody metal parts made from non-corroding metals, which may be difficult to fabricate, or be sealed in a container with a rotating seal for the spindle. The wattage rating is sometimes lowered slightly because of the sealing, but the life of the component is increased by many times. Some present types are made with solid-molded carbon-composition tracks and bases that resist the effects of humidity.

PERFORMANCE UNDER VIBRATION

In variable resistors difficulties may be experienced due to open circuit or intermittent contact if the slider vibrates off the track or due to change of resistance if the slider moves along the track. In general, the second is much more serious, particularly if the vibration occurs sideways to the potentiometer. Resonant frequencies vary between 100 and 300 Hz for the small 45-mm-diameter wirewound potentiometer and at amplitudes of

5–10 g. The shaft-length and knob-weight also affect the resonant frequency. Reduction in shaft length to 6 mm may raise the resonant frequency to 1 000 Hz or more, and the knob should be as light and as small as possible.

NOISE

Electrical noise in carbon-composition variable resistors is usually due to poor or intermittent contact between the slider and the track. Variations of pressure, or the presence of dust or metal particles, cause changes in contact resistance, resulting in noise. Sealing or at least dust-proofing is necessary to avoid trouble due to dust contact variation. In wirewound variable resistors there are several types of noise contact resistance or constriction resistance noise, loading noise, resolution noise and vibrational noise due to slip-rings (if these are used).

Specifications for fixed and variable resistors

The three basic specifications covering general requirements for all components of assessed quality are:

BS 9000 General requirements for electronic parts of assessed quality
Part 1: 1970. General description and basic rules
Part 2: 1972 Data on generic and detail specifications

BS 9001 1967. Sampling procedures and tables for inspection by attributes for electronic parts of assessed quality

BS 9002 Issue 2: 1971 Qualified parts list for electronic parts of assessed quality (including list of approved firms)

A brief list of specifications covering resistors is given below, but applications in all cases should be made to the BSI (British Standards Institution).

BS 9110 1969. Fixed resistors of assessed quality. Generic data and methods of test

BS 9111 1969. Rules for the preparation of detail specifications for fixed non-wirewound resistors, film type (type 1) of assessed quality

BS 9112 1970. Rules for the preparation of detail specifications for fixed non-wirewound resistors (type 2) of assessed quality. General applications category

BS 9113 1970. Rules for the preparation of detail specifications for fixed wire-wound precision resistors of assessed quality. General applications category

BS 9114 1970. Rules for the preparation of detail specifications for fixed wire-wound resistors (type 2) of assessed quality

BS 9130 1972. Potentiometers of assessed quality: generic data and methods of test

Other BSI specifications of interest are:

BS 3239 1960. Methods of determination of resistivity of metallic electrical conductor materials

BS 3466 1962. Methods of test for resistance per unit length of metallic electrical resistance material

BS 4059 1966. Methods for the determination of mass resistivity and mass conductivity of metallic electrical conductor materials

BS 4119 1967. Method of measurement of current noise generated in fixed resistors

FURTHER READING

CHURCH, H. F., 'The long-term Stability of Fixed Resistors', *I.R.E. Proc. on Compt Parts*, p. 31 (1961)

DUMMER, G. W. A., *Fixed Resistors*, 2nd edn., Pitman, London (1967)

DUMMER, G. W. A., *Variable Resistors*, 2nd edn., Pitman, London (1963)

DUMMER, G. W. A., *Materials for Conductive and Resistive Functions*, Hayden Book Co., New York (1970)

DUMMER, G. W. A. and BURKETT, R. H. W., 'Recent Developments in Fixed and Variable Resistors', *Proc. I.E.E.*, Part B, No. 21, *Supplement* (1962)

KARP, H. R., 'Trimmers Take a Turn for the Better', *Electronics*, p. 79 (17 Jan. 1972)

RAGAN, R., 'Power Rating Calculations for Variable Resistors', *Electronics*, p. 129 (19 July 1973)

WELLAND, C. L., *Resistance and Resistors*, McGraw-Hill, New York (1960)

DIELECTRIC MATERIALS AND COMPONENTS

Characteristics of dielectric materials

GENERAL CHARACTERISTICS

Dielectric materials used for radio and electronic capacitors can be grouped into the following five main classes:

(1) Mica, glass, low-loss ceramic, etc.: used for capacitors from a few pF to a few hundred pF.
(2) High-permittivity ceramic: used for capacitors from a few hundred pF to a few tens of thousands of pF.
(3) Paper and metallised paper: used for capacitors from a few thousand pF up to some μF.
(4) Electrolytic (oxide film): used for capacitors from a few μF to many μF.
(5) Dielectrics such as polystyrene, polythene, polythylene terephthalate, polycarbonate, etc.: range of use from a few hundred pF to many μF.

Many factors affect the dielectric properties of a material when it is used in a capacitor; among them being the permittivity, power factor, leakage current, dielectric absorption, dielectric strength, operating temperature, etc.

SUMMARY OF PROPERTIES OF CAPACITOR DIELECTRICS

A table of the main characteristics of some dielectric materials used in capacitors is given in *Table 6.12*.

PERMITTIVITY (DIELECTRIC CONSTANT)

The permittivity, dielectric constant or specific inductive capacity of any material used as a dielectric is equal to the ratio of the capacitance of a capacitor using the material as a dielectric, to the capacitance of the same capacitor using vacuum as a dielectric. The permittivity of dry air is approximately equal to one. A capacitor with solid or liquid dielectric of higher permittivity (ϵ) than air or vacuum can therefore store ϵ times as

Table 6.12 PROPERTIES OF SOME CAPACITOR DIELECTRIC MATERIALS

	Loss (at room temperature)							
	Limiting frequency of operation		*Power factor (Loss angle* tan δ) at 1 kHz	*Permittivity (over operating frequency range)*	*Dielectric strength (Volts per mil) (Breakdown)*	*Temperature limits* (°C)		
Material	*Approx min*	*Approx max*				*Approx min*	*Approx max*	*Remarks (and some (registered) Trade Names)*
1	2	3	4	5	6	7	8	9
Kraft paper (capacitor tissue)	Poor below 100 Hz (can be d.c.)	1 MHz	0.01 to 0.03	4.5	500 to 1 000 (depends on impregnant)	No limit but impregnant freezes	85 to 100	Capacitor properties depend greatly on impregnant
with mineral oil	—	—	0.003 5	2.23	—	−55	105	Effect of various impregnants
with castor oil	—	—	0.007	4.7	—	−25	65	
with silicone oil	—	—	0.003 5	2.6	—	−60	125	
with polyisobutylene	—	—	0.003	2.2	—	−55	125	
with chloronapthalene	—	—	0.005	5.2	—	−20	55	
with diphenyl	—	—	0.003	4.9	—	−55	85	
Mica (Ruby)	Precision work 100 Hz (can be d.c.)	10 000 MHz	0.000 5	7.0	1 000	No limit	+200	Properties vary according to source of origin
Ceramic (Low-permittivity types) Magnesium silicate	100 Hz*	10 000 MHz	0.001	5.4 to 7.0	200 to 300	No limit	+150 to 200	Steatite, Tempradex, Frequentex, etc.
(Medium-permittivity types) Titania Rutile (TiO_2)	500 Hz*	5 000 MHz	0.001	70 to 90	100 to 150	No limit	+120	Faradex, Condensa, (Has large temperature coefficient, approx −750 ppm/°C) †
(High-permittivity types) Titanate	1 000 Hz*	1 000 MHz	0.01	Approx 1 000 to over 7 000	100	−100	+120	Characteristics dependent on temperature. Sharp max in permittivity at 120 °C.

								high-permittivity materials exist with different peaks
Glass (Soft lead–soda)	200 Hz	10 000 MHz	0.001	6.5 to 6.8	<500	No limit	+200	
Glass (Hard, borosilicate)	100 Hz	10 000 MHz	0.001	4.0	<500	No limit	+200	
Quartz (Fused)	100 Hz	>10 000 MHz	0.000 2	3.8	1 000	No limit	+300	
Plastics								
Polyethylene	d.c.	10 000 MHz	0.000 2	2.3	1 000	Becomes brittle −40	+70	Polythene, Alkathene, Telcothene
Polystyrene film	d.c.	10 000 MHz	0.000 3	2.3	1 000	−40	+65	Distrene, Lustron, Styron, Trolitul
Polycarbonate	d.c.	10 000 MHz	0.000 1	2.8	1 000	−40	+70	
Polymonochlorotrifluoro-ethylene (PCTFE) film	d.c.	10 000 MHz	0.01 to 0.05	2.3 to 2.8	3 000 to 5 000	−195	+180	Kel-F, Hostaflon
Polyethylene terephthalate film	Poor between, say, 0.1 to 10 MHz		0.01 at l.f. 0.001 at h.f.	3.1	4 000 to 5 000	−40	+130	Terylene or Melinex. Could replace paper at l.f.
Paraffin wax	d.c.	10 000 MHz	0.000 1	2.2	1 000	No limit but hardens	60	Capacitor impregnant, coil potting, etc.
Liquid insulants								
Paraffin	d.c.	10 000 MHz	0.001	2.2	1 000	No limit but solidifies	50	Ozokerite (slightly plastic). Used for coil potting
Transformer oil	d.c.	1 MHz	0.001	2.4	1 000	−40	120	
Silicone oil	d.c.	10 000 MHz	0.000 1	2.8	1 000	−50	250	
Nitro-benzene	(see remarks)	10 000 MHz	0.000 5	40	50 to 1 000	Freezes at +3 then low permittivity	150	Note high permittivity. Cannot be used on d.c. because of leakage. Has large negative temperature coefficient. Has specialised use as capacitor dielectric

* Can be d.c. but slow capacitance change possible.

† ppm/°C = parts per million per degree Centigrade.

much energy for equal voltage applied across the capacitor plates. A few typical figures for capacitor dielectrics are:

	Permittivity (ϵ)
Vacuum	1.0
Dry air	1.000 59
Polythene, polystyrene, etc.	2.0 to 3.0
Impregnated paper	4.0 to 6.0
Glass and mica	4.0 to 7.0
Ceramic (magnesium titanate, etc.)	Up to 20
Ceramic (titania)	80 to 100
Ceramic (high-ϵ)* (or High-K)	1 000 upwards

Dielectrics can be classified in two main groups—polar and non-polar materials. Polar materials have a permanent unbalance in the electric charges within the molecular structure. The dipoles within the structure consist of molecules whose ends are oppositely charged. These dipoles therefore tend to align themselves in the presence of an alternating electric field (if the frequency is not too high). The resultant oscillation causes a large loss at certain frequencies and at certain temperatures.

LOSSES IN DIELECTRICS

Losses occur due to current leakage, dielectric absorption, etc., depending on the frequency of operation. For a good non-polar dielectric the curve relating loss with frequency takes the approximate shape given in *Figure 6.20* (*a*). For a polar material the loss-frequency curve may be shown approximately as in *Figure 6.20* (*b*)

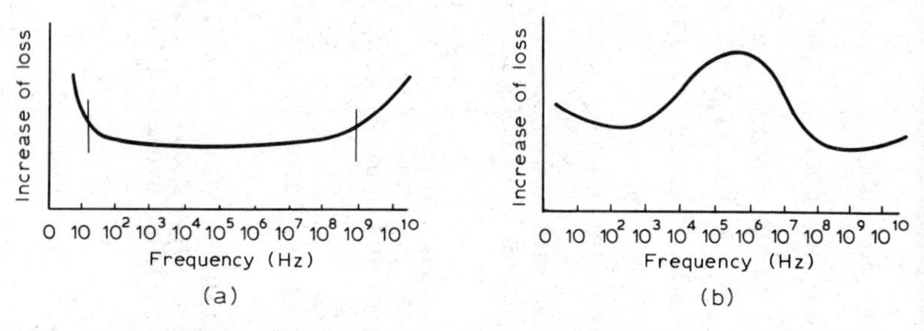

Figure 6.20 Loss/frequency curve for (*a*) *non-polar dielectric* (*b*) *polar dielectric*

The variation of permittivity with frequency is negligible so long as the loss is low. Increased losses occur when the process of alignment cannot be completed, owing to molecular collisions, and in these regions there is a fall in permittivity. Viscous drag in the molecular structure limits the frequency at which full alignment can be carried out. If the applied frequency is comparable with the limiting frequency losses still become high.

Equivalent circuits showing series and parallel loss resistance can be given, but are greatly dependent on the system of measurement at any particular frequency. The important criterion is the ratio:

$$\frac{\text{power wasted per cycle}}{\text{power stored per cycle}}$$

* The high permittivity in high-ϵ ceramic capacitors comes from the fact that the electric charges in the molecular structure of the material are very loosely bound and can move almost freely under the polarising voltage, resulting in high total capacitance.

This is the power factor of the material and for good dielectrics it is independent of frequency.

ABSORPTION

If a capacitor were completely free from dielectric absorption the initial charging or polarisation current when connected to a d.c. supply would be

$$I = (V/R)e^{-t/CR}$$

where I = current flowing after a time t
V = applied voltage
R = capacitor series resistance
C = capacitance
e = base of Napierian logs (2.718)

and the polarisation current would die off asymptotically to zero. If R is small, this takes place in a very short time and the capacitor is completely charged. In all solid-dielectric capacitors it is found that, after a fully-charged capacitor is momentarily discharged and left open-circuited for some time, a new charge accumulates within the capacitor, because some of the original charge has been 'absorbed' by the dielectric. This produces the effect known as dielectric absorption. A time lag is thus introduced in the rate of charging and of discharging the capacitor which reduces the capacitance as the frequency is increased and also causes unwanted time delays in pulse circuits.

LEAKAGE CURRENTS AND TIME CONSTANTS OF CAPACITORS

Losses due to leakage currents when a capacitor is being used on d.c. prevent indefinite storage capacity being realised, and the charge acquired will leak away once the source is removed. The time in which the charge leaks away to $1/e$, or 36.8%, of its initial value is given by RC, where R is the leakage resistance and C is the capacitance. If R is measured in megohms and C in microfarads the time constant is in seconds. This can also be expressed as megohm-microfarads or as ohm-farads. Some typical time-constants for various dielectrics used in capacitors are

Polystyrene	several days
Impregnated paper	several hours
Tantalum-pellet electrolytic capacitors	one or two hours
High-permittivity capacitors (ceramic)	several minutes
Plain-foil electrolytic capacitors	several seconds

It should be borne in mind that below capacitance values of about 0.1 μF, the time constant is generally determined by the structure, leakage paths, etc., of the capacitor assembly itself rather than the dielectric material. Leakage current increases with increase of temperature (roughly exponentially). In good dielectrics at room temperature it is too small to measure, but at higher temperatures the current may become appreciable, even in good dielectrics.

INSULATION RESISTANCE

The insulation resistance of a dielectric material may be measured in terms of surface resistivity in ohms or megohms, or as volume resistivity in ohm-centimetres. A method of measurement is given in BS 771 : 1954.

DIELECTRIC STRENGTH

The ultimate dielectric strength of a material is determined by the voltage at which it breaks down. The stress in kilovolts per inch (or volts per mil), at which this occurs depends on the thickness of the material, the temperature, the frequency and the waveform of the testing voltage, and the method of application, etc., and therefore comparisons between different materials should ideally be made on specimens equal in thickness and under identical conditions of measurement. The ultimate dielectric strength is measured by applying increasing voltage through electrodes to a specimen with recessed surfaces (to ensure that the region of maximum stress shall be as uniform as possible). Preparation of the specimens is important and their previous histories should be known.

The dielectric strength of a material is always reduced when it is operated at high temperatures or if moisture is present. Few materials are completely homogenous and breakdown may take the form of current leak along certain small paths through the material; these become heated and cause rapid deterioration, or flashover along the surface and permanent carbonisation of the surface of organic materials. Inorganic materials such as glass, ceramic and mica are usually resistant to this form of breakdown. The time for which the voltage is applied is important; most dielectrics will withstand a much higher voltage for brief periods. With increasing frequency the dielectric strength is reduced, particularly at radio frequencies, depending on the power factor, etc., of the material.

EFFECT OF FREQUENCY

At very low frequencies, also at very high frequencies, there is an increase of loss which sets a limit to the practical use of a capacitor with any given dielectric. At very low frequencies various forms of leakage in the dielectric material have time to become apparent, such as d.c. leakage currents and long time-constant effects, which have no effect at high frequencies. At very high frequencies some of the processes contributing to dielectric polarisation do not have time to become effective and therefore cause loss. These losses might be simply and approximately represented as in *Figure 6.21*.

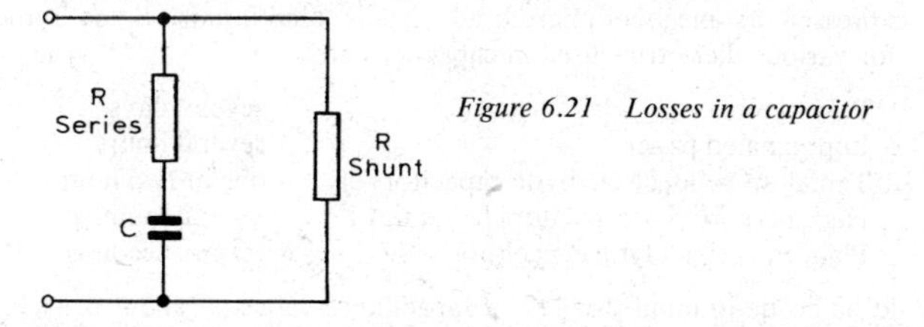

Figure 6.21 Losses in a capacitor

At very low frequencies the circuit is entirely resistive, all the current passing through the shunt resistance (d.c. leakage resistance, etc.). At very high frequencies the current passes through the capacitance *C* but all the volts are dropped across the series resistance and again the circuit is lossy. This series resistance may be due to the resistance of the capacitor leads, the silvering (in the case of silvered mica or ceramic capacitors), contact resistances, etc., in the capacitor assembly itself. These limit the upper frequency independently of the dielectric material used. Similarly, leakage across the case containing the dielectric may limit the lower frequency so that not all the useful range of the dielectric itself may be realised.

The chart in *Figure 6.22* shows the approximate usable frequency ranges for capacitors with various dielectric materials. The construction of the capacitor assembly

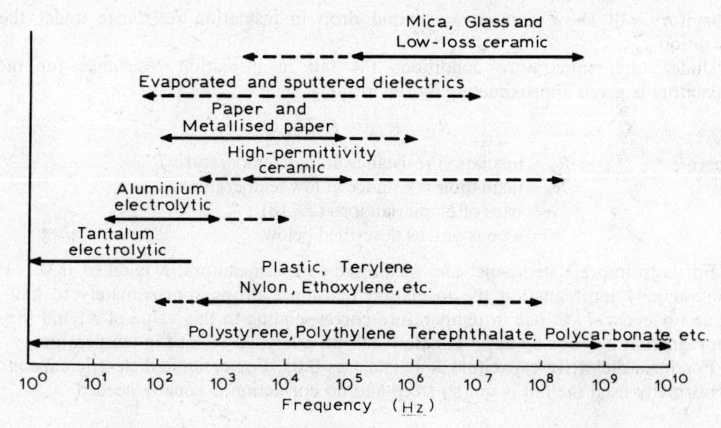

Figure 6.22 Frequency coverage of different classes of capacitor

will affect the frequency coverage to some extent, so that the chart should be regarded as a guide only.

THE IMPEDANCE OF A CAPACITOR

The current in a capacitor when an alternating voltage is applied is given by

$$I = 2\pi fCV \text{ amperes}$$

where C = capacitance in farads
V = voltage in volts
f = frequency in Hertz

and the reactance is given by

$$X_c = -1/2\pi fC \text{ ohms}$$

An ideal capacitor would have enntirely negative reactance but the losses, described previously, due to dielectric, case and leads, preclude this. In addition, inductance is also present in varying amounts and therefore as the frequency is increased the inductive or positive reactance increases, and above a critical frequency the capacitor will behave as an inductor. At the resonant frequency the impedance of the capacitor is controlled by its effective resistance, which in turn is made up of the losses described. Every capacitor will resonate at some given frequency (depending on its construction) and, having inductance and resistance, will exhibit a complex impedance, capacitive in one range of frequencies, resistive in another and inductive in still another.

INSULATION RESISTANCE OF CAPACITORS (OR INSULANCE)

The insulation resistance of the assembled capacitor is important in circuit use. The insulation resistance of any capacitor will be lowered in the presence of high humidity (unless it is sealed), and will be reduced when operated in high ambient temperatures (whether sealed or not).

For perfectly-sealed capacitors used under conditions of high humidity there should be no deterioration, but for imperfectly-sealed capacitors the drop in insulation resistance will be roughly inversely proportional to the effectiveness of the sealing. Unsealed capacitors will show a large and rapid drop in insulation resistance under these conditions.

Under high-temperature conditions the fall in insulation resistance for most capacitors is given approximately by the formula

$$R_T = R_t/e - K(T - t)$$

where R_T = insulation resistance at high temperature T
R_t = insulation resistance at low temperature t
e = base of Napierian logs (2.718)
K = a constant, as described below.

For both impregnated-paper and metallised-paper capacitors, K is taken as 0.1. For mineral jelly impregnation the insulation resistance drops approximately to half its value for every 7 °C rise in temperature corresponding to this value of K, and for oil impregnation it drops by approximately half for every 10 °C rise in temperature.

For mica-dielectric capacitors K is taken as 0.05. For ceramic-dielectric capacitors as normally used the fall is not so steep and no correction is usually needed.

General characteristics of discrete fixed capacitors

Capacitors are generally divided into classes according to their dielectric, e.g., paper, mica, ceramic, etc. It is useful to a designer to know the chief characteristics of these classes of capacitor and the main characteristics are briefly outlined in the following paragraphs. It is important to remember that capacitance is never constant, except under certain fixed conditions. It changes with temperature, frequency and age, and the capacitance value marked on the capacitor strictly applies only at room temperature and at low frequencies. A brief summary of their electrical characteristics is given in *Table 6.13*.

Table 6.13 SUMMARY OF THE ELECTRICAL CHARACTERISTICS OF FIXED CAPACITORS

Capacitor type	*Capacitor stability (after climatic tests) (%)*	*Normal manufacturing tolerance on capacitance (%)*	*Best manufacturing accuracy (%)*	*Permittivity (ϵ)*	*Power factor (at 1 kHz)*	*Temperature coefficient (ppm/°C)*	*Maximum capacitor temperature for long life (°C)*
Impregnated-paper (rect. metal-cased and tubular capacitors)	5	±20	5	approx 5	0.005 to 0.01	+100 to 200	100
Metallised-paper tubular capacitors	5	±25	5	5	0.005 to 0.01	+150 to 200	Normal 85 Special 125
Moulded stacked-mica capacitors	2	±20 or ±10	5	4 to 7	0.001 to 0.005	±200	Up to 120 Depends on casing
Moulded metallised-mica capacitors	1	±10	±2	4 to 7	0.001 to 0.005	±60	Up to 120 Depends on casing
Glass-dielectric capacitors	1	±10	(±1 down to 10 pF)	approx 8	0.001	+150	200

Table 6.13—*continued*

Capacitor type	Capacitor stability (after climatic tests) (%)	Normal manufacturing tolerance on capacitance (%)	Best manufacturing accuracy (%)	Permittivity (ϵ)	Power factor (at 1 kHz)	Temperature coefficient (ppm/°C)	Maximum capacitor temperature for long life (°C)
Glaze-dielectric capacitors	1	±10	1	5 to 10	0.001	+120	150
Ceramic-tubular, normal-ϵ capacitors	1	±10	±1	6 to 15 80 to 90	0.001	+100 −30 −750 (according to mix)	150
Ceramic-tubular, high-ϵ capacitors	20	±20	5 (low values) 1 (high values)	1 500 and 3 000 (may be higher)	0.01 to 0.02 (Varies with temperature, etc.)	−1 500, varies. Non-linear	100
Polystyrene-film, tubular and rectangular, capacitors	>1	±20	5	2.3	0.000 5	−150	60–70
Polyethylene terephthalate (Melinex) capacitors	5	±20	5	2 to 5, depending on frequency	0.01 at 1 kHZ varies with temperature and frequency	Varies with temperature	130
Electrolytic, (normal) capacitors	10	−20 to +100	10	—	0.02 to 0.05	+1 000 to 2 000 approx	70–85
Electrolytic, (tantalum-pellet) capacitors	5	±20	10	—	0.05	+100 to 200	125
Electrolytic, tantalum-foil capacitors	10	±10	5	—	0.05	+500	85
Precision-type air-dielectric capacitors	±0.01	—	0.01	1	0.000 01	+10	20

IMPREGNATED-PAPER CAPACITORS

These are general-purpose paper-dielectric capacitors, made by rolling paper as insulation between metal foils and filling with an impregnant. They are relatively inexpensive, have a high capacitance-to-volume ratio and are capable of working at reasonably high voltages, but their power factor is comparatively high and the selection tolerances are fairly wide. The maximum permissible d.c. working voltage of any impregnated-paper capacitor is dependent on the ambient temperature, and the life of the capacitor is approximately inversely proportional to the fifth power of the operating voltage up to 85 °C. Irrespective of the d.c. working voltage, the maximum a.c. working voltage of a normal impregnated-paper capacitor with solid or semi-solid impregnant is about 300 V r.m.s. at 50 Hz for the tubular type and 600 V r.m.s. for the rectangular type containing two capacitor units in series. If higher a.c. working voltages are required, specially designed capacitors should be used. Smaller types are made for transistor and integrated circuit use.

The insulation resistance of impregnated-paper capacitors is high at room temperature—of the order of 2 000 to 5 000 Ω-F (depending on the paper and impregnant) but falls rapidly as the ambient temperature is increased. Rectangular-cased types of 8 μF capacitance may fall to a few tens of megohms at a temperature of 100 °C, while the tubular types of 0.1 μF capacitance (having a higher initial insulation resistance) may fall to a few hundred megohms at the same temperature. The fall in insulation resistance tends to be inversely proportional to the capacitance from one microfarad upwards, depending upon leakage over the case. The temperature coefficient of impregnated-paper capacitors varies from +100 to +200 parts per million per degree centigrade (ppm/°C). The power factor is about 0.005 to 0.01 at Hz and tends to increase with increase of frequency. The capacitance stability under normal operating conditions is about 0.5% to 5%. The inductance of the tubular types is approximately 0.015 μH per 25 mm length of capacitor (including lead lengths).

METALLISED-PAPER CAPACITORS

These obviate voids between paper insulation and the metal, and were introduced in the late 1940s. In this type of capacitor, one side of the paper is metallised before rolling. The main characteristics are small size and self-healing action under voltage stress. If the paper is punctured the metallising quickly evaporates in the area of the puncture and prevents a shortcircuit. The maximum voltage at which the self-healing action will occur without deterioration of the capacitor properties is termed the test voltage. It is about 1.5 times the working voltage and should never be applied for more than one minute at a time. The maximum voltage which may be applied instantaneously without destroying the capacitor is termed the spark voltage. It is approximately 1.75 times the working voltage and should never be applied for more than a few seconds, as continuous sparking will rapidly destroy the capacitor.

MICA-DIELECTRIC CAPACITORS

The main characteristics of this type of capacitor are low power factor, high voltage operation and excellent long-term stability at room temperature. The stability of the silvered-mica-plate type is about 1% under normal conditions of use and that of the stacked-mica-plate type about 2%. Precision mica capacitors, used as sub-standards, can be adjusted to better than 0.01% for values over 1 μF. They are invariably sealed in cases to prevent moisture, etc., from affecting the stability. Capacitors of this type have remained constant in capacitance within ±0.2 pF on a value of 10 000 pF over a test period of 10 000 h at room temperature. The temperature coefficient is low, between ±100 ppm/°C, but varies according to the source, treatment, etc., of the mica. The silvered-plate capacitor has a better temperature coefficient than the stacked-plate type. Both types, especially the stacked-plate capacitor, show slight non-cyclic capacitance shifts during temperature cycling and, in most of the types available at present, the temperature/capacitance curve is not entirely linear. There is also a wide spread of mean temperature coefficients between different specimens, even of the same batch.

The power factor of mica is approximately 0.000 3 at 1 Hz, but can be as low as 0.000 05 when specially selected and very dry. The permittivity is about 7. The current-carrying capacity of the silvered plate imposes a limit to radio-frequency and pulse loading and the silvered-plate capacitor is therefore less suitable for heavy current work than the stacked-plate type, although the latter is less stable and cannot be made to such a close selection tolerance as the silvered-plate capacitor.

CERAMIC-DIELECTRIC CAPACITORS

Ceramic-dielectric capacitors (ceramic capacitors) are made in three main classes—low permittivity low-loss types, medium-permittivity temperature-compensating types and high-permittivity types.

The low-permittivity low-loss types are generally made of steatite or similar material. Steatite has a permittivity of approximately 8 and other materials may give permittivities between 6 and 15. Their performance at high frequencies, from about 50 Hz upwards, is excellent. The power factor is reasonably low (0.001), approaching that of mica. The temperature coefficient is between +80 and +120 ppm/°C and the capacitors are normally very cyclic in behaviour. The temperature coefficients vary less between different batches than for capacitors of any other dielectric except glass and vacuum. The capacitance stability in normal use is about 1% excluding temperature variations. They operate at comparatively high voltages, 500 V or so (depending on size), over a temperature range from about +150 °C down to extremely low temperatures.

The second class, of medium permittivity (ϵ about 90), are used mainly as temperature-compensating capacitors in tuned circuits and have negative temperature coefficients of the order of −600 to −800 ppm/°C. They are all based on titania or its derivatives. The power factor is again low and may be less than 0.000 3 at radio frequencies. Other temperature coefficients can be obtained by using different mixtures.

The high-permittivity ceramic capacitors provide a very high capacitance in a compact unit. The capacitance and the power factor, however, change widely with temperature, the changes being neither linear nor very cyclic for either property. Capacitors using the $\epsilon = 1\,200$ material, for instance, have a high capacitance peak (Curie point) at about 110 °C, which is two or three times the value at room temperature, with another much smaller one at about −10 °C. The power factor is a minimum around 20 °C to 40 °C and is in general around 2%. High permittivity materials with other permittivities have peaks at other temperatures. In general, the higher the permittivity, the more temperature-sensitive is the capacitor. In addition to changes with temperature, the capacitance is also reduced under d.c. voltage stress, especially at the peak points; at room temperature a reduction in capacitance of 10 to 20% will occur, but up to 50% can be expected at the Curie points. The d.c. working voltage is rather lower than for the low-permittivity ceramic type. The capacitors are subject to hysteresis and accordingly are suitable for working with only very small a.c. voltages. They are used mainly as r.f. bypass capacitors, but can also be used for interstage coupling, provided the capacitance is large enough under all conditions of operation. The properties of high-permittivity capacitors, therefore, vary so much with temperature, voltage stress, etc., that no general electrical characteristics can be given.

GLASS-DIELECTRIC CAPACITORS

These capacitors are formed of very thin glass sheets (approximately 0.000 5 in thick) which are extruded as foil. The sheets are interleaved with aluminium foil and fused together to form a solid block. Their most important characteristics are the high working voltages obtainable and their small size compared with encased mica capacitors.

Glass-dielectric capacitors (glass capacitors) have a positive temperature coefficient of about 150 ppm/°C, and their capacitance stability and Q are remarkably constant. The processes involved in the manufacture of glass can be accurately controlled, ensuring a product of constant quality, whereas mica, which is a natural product, may vary in quality. As the case of a glass capacitor is made of the same material as the dielectric, the Q maintains its value at low capacitances, while the low-inductance direct connections to the plates maintain the Q at high capacitances.

These capacitors are capable of continuous operation at high temperatures and can be operated up to 200 °C. They are also used as high-voltage capacitors in transmitters.

GLAZE- OR VITREOUS-ENAMEL-DIELECTRIC CAPACITORS

Glaze- or vitreous-enamel-dielectric capacitors (glaze- or vitreous-enamel capacitors) are formed by spraying a vitreous lacquer on metal plates which are stacked and fired at a temperature high enough to *vitrify* the glaze. Capacitors made in this way have excellent r.f. characteristics, exceedingly low loss and can be operated at high temperatures, 150 °C to 200 °C. As they are *vitrified* into a monolithic block they are capable of withstanding high humidity conditions and can also operate over a wide temperature range. The total change of capacitance over a temperature range of −55 °C to +200 °C is of the order of 5%. The temperature coefficient is about +120 ppm/°C and the cyclic or retrace characteristics are excellent. As in the glass capacitor the encasing material is the same as the dielectric material and therefore all corona at high voltages is within the dielectric. They are extremely robust and the electrical characteristics cannot normally change unless the capacitor is physically broken.

PLASTIC-DIELECTRIC CAPACITORS

In plastic-dielectric capacitors the dielectric consists of thin films of synthetic polymer material. The chief characteristic of plastic-film capacitors is their very high insulation resistance at room temperature. The main synthetic polymer films used as capacitor dielectrics are.

Polyethylene terephthalate. This is a tough polymer with high tensile strength, free from pinholes and with good insulating properties over a reasonably wide temperature range. This is known under a variety of trade names such as Melinex (ICI), Mylor (Du Pont), Hostaphon (Germany) and Terphane (France). It is commercially available in thin films of 3.5 μm in thickness.

Polycarbonate. This is a polyester of carbonic acid and bisphenols. It combines in good physical properties with a lower loss (or dissipation factor) than polyethylene terephthalate. It has a temperature characteristic nearer to zero and is available in film form down to 2 μm in thickness.

Polystyrene. This is a hydrocarbon material and has a lower permittivity than the previous two dielectrics. It has a better dissipation factor but its tensile strength for winding is much lower and 8 μm is the lowest film thickness available. It is not normally used in metallised capacitors unless heavily derated.

Polypropylene. This is a low-price material and has the lowest dissipation factor of the four films discussed. It is not commercially available in films less than 8 μm thick and it has a lower permittivity and its use is therefore limited.

A comparison of the electrical characteristics of the four film materials at 20 °C is given in *Table 6.14*.

Table 6.14 FILM MATERIAL PROPERTIES

	Polyethylene terephthalate	*Poly-carbonate*	*Poly-styrene*	*Poly-propylene*
Permittivity (at 1 kHz)	3.2	2.8	2.5	2.25
Dissipation factor (at 1 kHz)	0.004	0.001	0.000 2	0.000 5
Dielectric Strength (V/μm)	304	184	200	204
Insulance (ΩF)	1×10^5	3×10^5	1×10^6	1×10^5

ELECTROLYTIC CAPACITORS

The most notable characteristic of these capacitors is the large capacitance obtainable in a given volume, especially if the working voltage is low. Electrolytic capacitors are used for smoothing and bypassing low frequencies, but they can also be used for high-energy-pulse storage applications, such as photoflash and pulsed circuits. The electrical properties change widely under different conditions of use and some indication of these is given below.

Capacitance. There is a slight increase (about 10%) when the temperature is raised from 20 °C to 70 °C; a gradual decrease as the temperature is reduced to −30 °C, and a very rapid decrease at lower temperatures. The capacitance also decreases slightly as the applied frequency is increased from 50 Hz giving a 10% reduction at 10 000 Hz.

Power factor. At 50 Hz and room temperature, the power factor is from 0.02 to 0.05. There is a slight increase at +70 °C and a large increase at −30 °C. A large increase also takes place as the frequency is increased and the power factor becomes about 0.5 at 10 000 Hz.

Leakage Current. This is normally considered instead of insulation resistance, which is very low in this type of capacitor. The leakage current varies directly with temperature, having quite a low value at −30 °C, but at +70 °C it is about ten times the value at room temperature. In addition, the leakage current increases with the applied load, being very high when the load voltage is first applied, but it falls rapidly and after about a minute tends to reach a stable value.

Impedance. There is a gradual increase in impedance as the temperature is reduced, until at −30 °C it is about twice the impedance at room temperature, while at still lower temperatures a much more rapid increase occurs. At temperatures above normal there are only slight variations. The impedance falls rapidly with increase of frequency and at 10 000 Hz is of the order of 2 Ω for a 16 μF capacitor.

The normal type of electrolytic capacitor is made using plain foils of aluminium, but considerably increased capacitance can be obtained by using etched foils or sprayed gauze foils to increase the surface area. Electrolytic capacitors need to be re-formed periodically if they are stored for a considerable time. Reforming is carried out by applying the working voltage through a resistor of approximately 1 000 Ω for one hour.

Tantalum-pellet electrolytic capacitors do not need re-forming and have an expected shelf life of more than ten years. They have the advantage of even greater capacitance in a small volume and the leakage current is extremely small—of the order of a few microamperes, enabling them to be used in circuits such as multivibrators. They have lower voltage ratings, however, and some types are expensive, but they are capable of operating over a temperature range from −55 °C to +125 °C with negligible change in capacitance.

Tantalum-foil electrolytic capacitors are also extremely small in size and have a low leakage current. They can operate at higher voltages than the tantalum-pellet types, but cannot operate over as wide a temperature range. The power factor varies considerably with temperature, also with voltage rating.

Air-dielectric capacitors. Air-dielectric capacitors are used mainly as laboratory standards of capacitance for measurement purposes. With precision construction and use of suitable materials, they can have a permanence of value of 0.01% over a number of years for large capacitance values.

Vacuum and gas-filled capacitors. Vacuum capacitors are used mainly as high-voltage capacitors in airborne radio transmitting equipment and as blocking and decoupling capacitors in large industrial and transmitter equipments. They are made in values up to 500 pF for voltages up to 12 000 V peak. Gas-filled types are used for very high voltages—of the order of 250 000 V. Clean dry nitrogen may be used at pressures up to 150 lb/in.^{2} They are specially designed for each requirement.

General characteristics of integrated circuit capacitors

There are three types of capacitor for possible use in integrated circuits—thin film, thick film and chip capacitors. Of these the chip capacitor is the most widely used in hybrid circuits.

THIN-FILM CAPACITORS

These consist mainly of two types, Silicon monoxide and Tantalum pentoxide, although other materials can be used such as Silicon dioxide and Titanium dioxide. Films can be evaporated, sputtered or anodised. Most thin-film dielectrics are useable up to a frequency of 1 GHz with good stability. Silicon monoxide capacitors change less than 2% after 10 000 hours at 25 °C and less than 3% after 1 000 hours at 85 °C. Although ±5% tolerance in capacitance can be achieved, normal capacitor tolerances are usually ±10% to ±20%. Some properties of thin-film dielectrics due to H. T. Groves and F. Z. Keister are given in *Table 6.15*.

Table 6.15 PROPERTIES OF THIN-FILM DIELECTRICS

Material	*Dielectric constant*	*Dissipation factor*	*Temp. coeff. of capacitance* (ppm/°C)	*Voltage breakdown*	*Dielectric strength* (X 10^6 *volts/cm*)	*Capacitance* (μF/cm^2)	*Thickness range* (nm)
Silicon monoxide (SiO)	5–7	0.01–0.03	150–400	50–100	1–2	0.005–0.02	300–4 000
Silicon dioxide (SiO_2)	4	0.004–0.04	100	50–200	3	0.004–0.02	80–1 000
Tantalum pentoxide (Ta_2O_5)	25–27	0.01–0.5	250–350	50	1–3	0.01–0.5	50–250
Titanium dioxide (TiO_2)	30–100	0.01–1.0	200–800	25–90	0.3–1	0.01–1.0	100–200
Aluminium oxide (Al_2O_3)	8–10	0.20–0.24	200–300	25–120	2–4	0.2–0.25	40–250

THICK-FILM CAPACITORS

There are two types of printed capacitors using fired dielectric and conductive pastes.

(1) Normal parallel plate capacitors where a conductive layer is printed on to the substrate, then a dielectric layer printed and finally another conductive layer on top, giving the required capacitor sandwich.
(2) Planar interdigitated capacitors where only one layer of conductive material is printed in a pattern of interlaced fingers.

Both types are only used for special requirements, as it is easier and cheaper to use attached chip capacitors.

CHIP CAPACITORS

Most widely used for attachment to thick-film circuits are chip capacitors soldered in place. These can be tested beforehand and are relatively cheap. Several types are available, but ceramic dielectrics are widely used because of their high capacitance in a small space, although small solid tantalum and plastic-film types are also made. Using high K materials, thin dielectrics and even thinner metal-film plates produce a range of capacitors useful for most applications. For radio frequency work where small-value capacitor chips (1 to 500 pF, with typical dimensions of 50 × 80 × 50 mils) are usually sufficient, low-loss porcelain dielectrics are used for frequencies well through the X band. At audio frequencies, chips having titanate-ceramic dielectrics with Ks to 50 000 can provide roughly 100 to 50 000 pF; these chips are 30 to 240 mils square and 50 mils thick. For audio-filtering, decoupling and low-frequency bypassing applications, tantalum chip capacitors can be used. They cover a range of values from 0.1 to 100 μF. A 2.2 μF chip measures about 50 × 50 × 100 mils.

Specifications for Fixed Capacitors and Dielectrics

The three basic specifications covering general requirements for all components of assessed quality are:

BS 9000 General requirements for electronic parts of assessed quality
Part 1: 1970. General description and basic rules
Part 2: 1972. Data on generic and detail specifications

BS 9001 1967. Sampling procedures and tables for inspections by attributes for electronic parts of assessed quality

BS 9002 Issue 2: 1971. Qualified parts list for electronic parts of assessed quality (including list of approved firms)

A brief list of specifications covering capacitors is given below, but applications in all cases should be made to the BSI (British Standards Institution).

BS 9070 Fixed capacitors of assessed quality: generic data and methods of test. Sections 1 & 2: 1969 Principles and mandatory procedures. General rules for drafting detail specifications

BS 9073 (Previously section 3) 1970. Tantalum electrolytic capacitors

BS 9074 (Previously section 4) 1971. Polystyrene dielectric capacitors

BS 9075 (Previously section 5) 1971. Ceramic dielectric capacitors

BS 9076 (Previously section 6) 1971. Polycarbonate dielectric capacitors and polyethylene terephthalate dielectric capacitors for d.c. use

BS 9077 (Previously section 7) 1971. Mica dielectric capacitors

BS 9078 (Previously section 8) 1971. Aluminium electrolytic capacitors

Other B.S.I. specifications of interest are:

BS 1650 1971. Capacitors for connection to power-frequency systems
BS 2135 1966. Capacitors for radio interference suppression
BS 2067 1953. Determination of power-factor and permittivity of insulation materials (Hartshorn & Ward method)
BS 4542 1970. Method for the determination of loss tangent and permittivity of electrical insulating materials in sheet form (Lynch method)

FURTHER READING

CAMPBELL, D. S., 'Electrolytic Capacitors', *The Radio & Electronic Engineer*, **Vol. 41**, No. 1, p. 5 (Jan. 1971)

'CAPACITORS—A COMPREHENSIVE EDN REPORT', *Electronic Design News*, U.S.A., p. 139 (May 1966)

CAPACITORS (Supplement), *Electronics Weekly*, p. 14 (21 March 1973)

DUMMER, G. W. A., *Fixed Capacitors*, 2nd edn., Pitman, London (1964) (Contains comprehensive bibliography up to 1962)

DUMMER, G. W. A. and NORDENBERG, H. M., *Fixed and Variable Capacitors*, McGraw-Hill, New York (1960)

GIRLING, D. S., 'Quality Control in Capacitor Production and Testing', *The Radio & Electronic Engineer*, **Vol. 40**, No. 4, p. 173 (Oct. 1970)

'MINIATURE CAPACITOR PROGRESS', *Electron*, p. 15 (29 June 1972)

PAMPLIN, B. F., Capacitor Selection—Facts and Figures', *Electronic Equipment News*, p. 44 (Dec. 1969)

VON HIPPEL, A., *Dielectrics & Waves*, Chapman & Hall, London (1954)

VON HIPPEL, A., *Dielectric Materials & Application*, Chapman & Hall, London (1954)

MATERIALS AND THEIR COMPOUNDS FOR SOLID-STATE DEVICES

Basic construction of materials

In solids there are two categories: crystalline solids and amorphous solids.

CRYSTALLINE SOLIDS

In this construction the atoms are positioned in a regular manner, so that, knowing the position of any one atom, the position of any other atom in the crystal may be determined. This is because in an ideal crystal every lattice point repeats itself periodically in identical surroundings. All semiconductors of interest in electronic devices have a crystalline structure. Crystals can be grouped into seven classic arrangements of their structure according to the relative length of the three axes and the angles between the axes, as follows:

(1) *Triclinic.* Three axes of unequal length not at right angles to each other
(2) *Monoclinic.* Three axes of unequal length, two of which are perpendicular
(3) *Orthorhombic.* Three axes of unequal length, all of which are at right angles to each other
(4) *Rhombohedral.* Three axes of equal length, all of which are at equal but not at right angles to each other

(5) *Hexagonal.* Three coplanar axes of equal length at 120° to each other and a fourth axis of different length at right angle to the others; or two axes of equal length at 120° angle to each other and perpendicular to a third axis of different length
(6) *Tetragonal.* Three mutually perpendicular axes with two of them of equal length
(7) *Cubic.* Three axes of equal length at right angles to each other.

These seven systems are shown diagramatically in *Figure 6.23.*

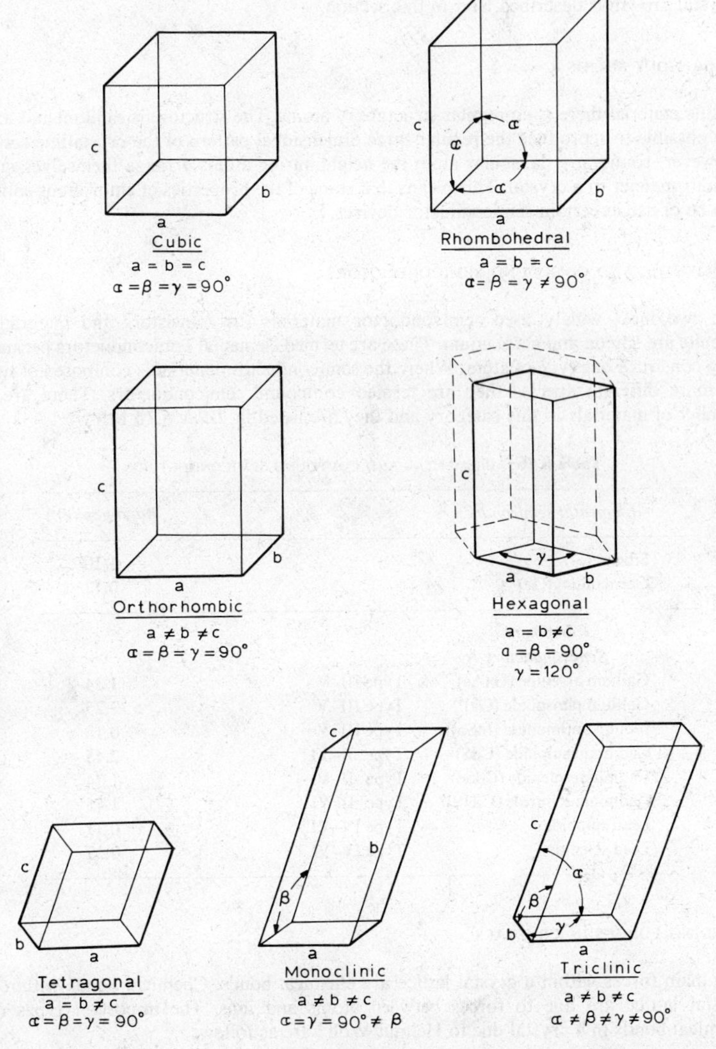

Figure 6.23 The seven crystal systems (arranged according to the lengths of the axes and the angles between them)

POLYCRYSTALLINE SOLIDS

Under natural circumstances, a single crystal structure does not usually occur as a result of the 'chance' way most crystals are formed in nature. The solid organises itself into a large number of small crystallites which are joined together to make up a polycrystalline structure. The boundaries of the crystallites are known as *grain boundaries* and the small crystals as *grains*. Metals such as aluminium and copper are crystalline solids. For use in semiconductor devices, crystalline materials cannot be left to nature, but must be grown with very careful control of impurities and properties. Crystal growth is described later in this section.

AMORPHOUS SOLIDS

In this material there is no regular structure of atoms. The structure is random and it is not possible to reproduce the regular three dimensional pattern of the crystalline solid. However, round any particular atom the neighbouring atoms arrange themselves in a similar manner to a crystal. This means that some of the properties of amorphous solids can be of use in certain semiconductor devices.

ELEMENTAL AND COMPOUND SEMICONDUCTORS

The two most widely used semiconductor materials for transistors and integrated circuits are silicon and germanium. These are termed elemental semiconductors because they consist of one type of atom. Where the semiconductor material is composed of two or more different species they are termed compound semiconductors. There are a number of materials in this category and they are listed in *Table 6.16* below:

Table 6.16 ELEMENTAL AND COMPOUND SEMICONDUCTORS

Elemental semiconductors		*Band gap (eV)*
Silicon (Si)		1.10
Germanium (Ge)		0.72
Compound Semiconductors		
Gallium arsenide (GaAs)	Type III–V	1.34
Gallium phosphide (GaP)	Type III–V	2.25
Indium antimonide (InSb)	Type III–V	0.18
Cadmium sulphide (CdS)	Type II–VI	2.45
Cadmium selenide (CdSe)	Type II–VI	1.74
Cadmium telluride (CdTe)	Type II–VI	1.45
Lead sulphide	Type IV–VI	0.37
Lead selenide	Type IV–VI	0.27

BONDING FORCES IN CRYSTALS

The main forces within a crystal lattice are chemical bonds. Chemical bonds within a crystal lattice are due to forces between atoms and ions. The important types of chemical bonds in a crystal due to Helmut Wolf[1] are as follows:

Ionic bond. If one atom of an interacting pair is electropositive and the other electronegative, so that the first one loses a valence electron to the second, then the attractive force is due to the electrostatic (Coulomb) attraction of two oppositely charged ions and an ionic bond is formed. As a result of this transfer of valence electrons, each ion in an ionic crystal tends to be surrounded by oppositely charged ions while equally charged ions occupy more distant positions. Ionic crystals show high bonding strength due to the strong electrostatic forces between the ions; they normally exhibit high strength and hardness and high melting points. They are brittle because of the directional nature of the bonding forces. Their electrical conductivity is low.

Covalent bond. Covalent bonding between two atoms is due to the sharing of electrons, supplied by one or both of the atoms. The sharing results from the overlap of the bonding orbitals, which lowers the energy of the system. The shared electrons usually serve to fill the outer valence shell of each atom. The covalent bond is very strong, resulting in high hardness and high melting point. In organic molecules (which are covalent in character) intramolecular bonds are strong whereas intermolecular bonds (which are noncovalent) are responsible for their low mechanical strength and low melting point. Bonds which fall between the ionic and covalent types have corresponding characteristics; the degree to which a bond is ionic or covalent is described by the electronegativities of the two atoms, i.e., by their relative abilities to attach an extra electron.

Metallic bond. The bonding in a metal is best described by all the atoms of the crystal taken collectively, with the valence electrons from all the atoms belonging to the crystal as a whole. Many properties can be explained in terms of the free-electron model. The freedom of the valence electrons to move through the metal leads to a high electric conductivity.

Van der Waals bond. Weak van der Waals forces can bind together a solid. These forces arise from the weak attraction between instantaneous electric dipoles due to the motion of the electrons in the atoms or molecules which do not have to have a permanent dipole movement because the fluctuating dipoles average to zero. The binding in such solids is quite weak since these forces are small. Consequently these crystals have low melting points and low cohesive strengths.

Dipole bond. Under certain conditions a hydrogen atom can form a second bond which may be equivalent to the first bond because of resonance between situations where the electron goes to one or the other of the two neighbouring atoms. Dipole bonds are usually quite weak.

IMPERFECTIONS IN CRYSTALS

No crystal is perfect in the sense that no flaws are present and imperfections due to impurity atoms and dislocations play an important part in semiconductor properties.

In any perfect crystal, if an atom is removed from its position in the lattice, a hole that is formed is termed a *vacancy*. If, however, an extra atom is somehow squeezed into the lattice, this extra atom is termed an *interstitial* atom. Both types distort the regular lattice construction.

In the process of crystal growth, impurity atoms are introduced by diffusion into the lattice and these impurities affect the properties of the material for use in transistors and integrated circuits. The mechanical properties of solids are also affected by *dislocations* in which missing atoms lead to strain in the lattice with consequent weak points in the structure.

PROPERTIES OF SILICON, GERMANIUM AND GALLIUM ARSENIDE

These three materials are described in some detail because of their relevance to microelectronics. About 83% of semiconductor devices are made from silicon, 16% from germanium and 1% from other materials.

SILICON

Silicon is a common material found mainly in oxides such as sand, quartz, agate, flint, etc., and in silicates such as granite, asbestos, clay, mica, etc. About 25% of the earth's crust is composed of silicon in oxide or silicate form. As a semiconductor material, it is possible to produce crystalline silicon in extremely pure form and to control the impurities very accurately to produce the required electrical characteristics. Methods of crystal growth are described later in this section.

Crystalline silicon is found as a hard black material with an octahedron structure and as a grey material with tetrahedron structure. It has a high melting point of 1 417 °C. It is relatively inert and is not attacked by acids (except hydrofluoric acid). It has a strong affinity to oxygen. Single crystalline silicon atoms usually have a diamond structure of the cubic crystallographic system. The lattice parameter (atomic spacing) is 0.543 07 nm at 300 K. The number of atoms per cm^3 is 5×10^{22}. The density at 300 K is 2.329 gcm^{-3} and the coefficient of linear thermal expansion at 300 K is $+2.33 \times 10^{-6}$ °C^{-1}. Around 300 K the specific heat is $5.74 + 0.617 \times 10^{-3}T - 1.01 \times 10^{9}T^{-2}$ cal/mole K. At 1 550 K the latent heat of fusion is 1.195×10^4 cal/mole and the latent heat of sublimation is 1.050×10^5 cal/mole.

The intrinsic electrical resistivity at 300 K is 2.3×10^5 Ωcm corresponding to a charge carrier concentration of 3×10^{10} cubic cm (electrons plus holes). The dielectric constant is 11.7.

GERMANIUM

Germanium is of the diamond crystal structure; the lattice constant, i.e., the side of the unit cell, is 0.565 75 nm at 300 K; the number of atoms per cm^3 is 4.42×10^{22} The density is 5.326 7 g/cm^3 at 300 K. The thermal expansion at 300 K is $+5.75 \times 10^{-6}$ K^{-1}; it is not quite isotropic and differences up to 4% are observed; the expansion is greatest in the (100) and least in the (111) direction. The Debye temperature is 406 K; the specific heat, 5.47 cal/mole K at 300 K, varies with temperature as $4.62 + 2.27 \times 10^{-3}$ T. The latent heat of fusion is 8 100 cal/mole and of sublimation 89 000 cal/mole at 1 150 K. At 300 K the intrinsic electrical resistivity is 50 Ωcm corresponding to a charge carrier concentration (electrons plus holes) of 2.4×10^{13} cm^{-3}. The dielectric constant is 16.0.

GALLIUM ARSENIDE

GaAs has the zincblende crystal structure with a lattice constant 0.565 34 nm at 300 K; the number of atoms per cm^3 is 2.21×10^{22}, the density is 5.32 g/cm^3, and the frequency of lattice vibrations 7.2×10^{12} Hz. The thermal expansion at 300 K is $+5.7 \times 10^{-6}$ K^{-1}. The Debye temperature is 344 K, the specific heat is 0.31×10^7 cm^2/sec^2 K. At 300 K the intrinsic electrical resistivity is 7.0×10^7 Ωcm corresponding to a charge carrier concentration (electrons plus holes) of 1.8×10^8 cm^{-3}. The dielectric constant is 10.4.

Materials preparation—crystal growth

High perfection single-crystal material can be produced in bulk or thin-film form. Prior to their preparation the base material must be purified to a high degree. This refining process is normally carried out by zone refining.

ZONE REFINING

This is based on the principle that the maximum impurity concentration of the *molten* material exceeds that of the same material in its solid state. *Figure 6.24* shows an illustration of this process. The rod of polycrystalline semiconductor material placed in

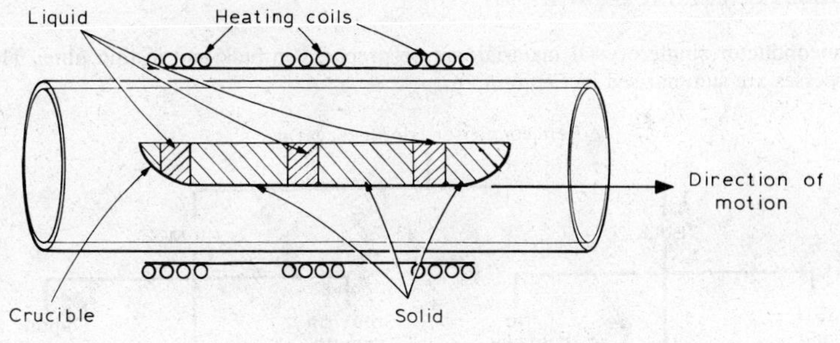

Figure 6.24 Horizontal zone refining

a crucible slowly passes from one end to the other via a heating coil (or coils). The impurities are concentrated in the liquid zones and swept to the end of the crystal which is afterwards cut off and discarded. The process can be repeated until the required purity is obtained. A further method of purification is the floating zone system in which the rod is held vertically as shown in *Figure 6.25*. A single liquid zone is then passed

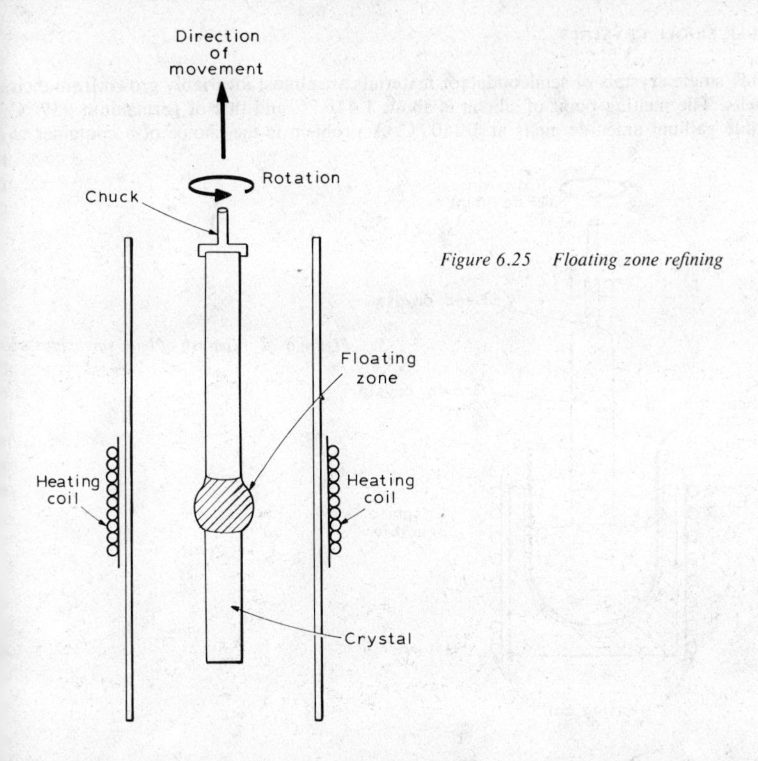

Figure 6.25 Floating zone refining

vertically from one end to the other. This liquid zone is maintained in position by surface tension between the two solid zones. This system avoids any possible contamination from the crucible necessary in the previous process.

METHODS OF CRYSTAL GROWTH

Semiconductor single crystal material can be prepared in bulk and in thin films. The processes are summarised in *Figure 6.26*.

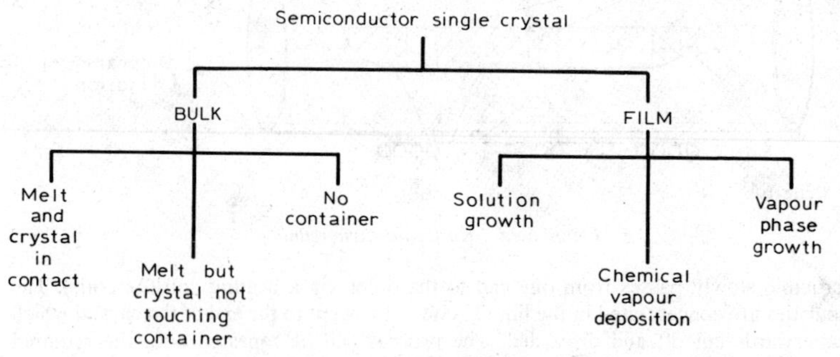

Figure 6.26 Crystal growth processes

BULK SINGLE CRYSTALS

Bulk single crystals of semiconductor materials are almost invariably grown from their melts. The melting point of silicon is about 1 410 °C and that of germanium 937 °C while gallium arsenide melts at 1 240 °C. A problem is the choice of a container to

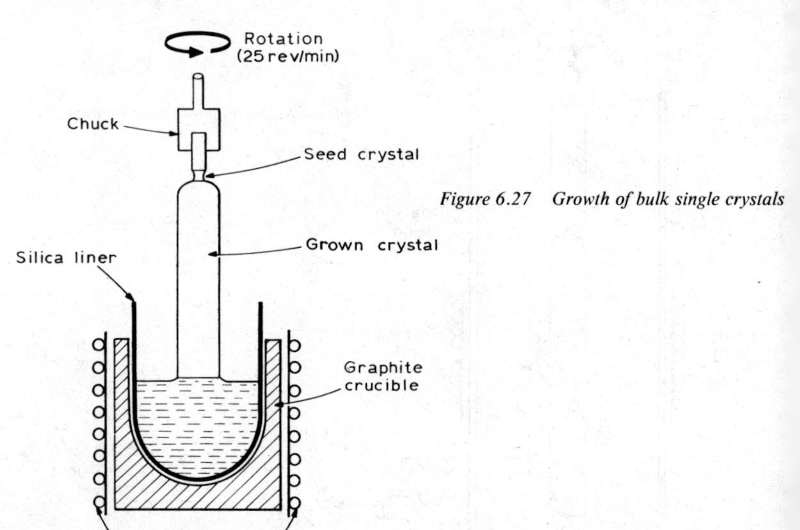

Figure 6.27 Growth of bulk single crystals

contain the melt without contamination of the crystal. Silicon tends to react with almost all container materials, but silica appears to be the best choice. Silicon also requires an inert atmosphere. *Figure 6.27* shows how silicon is melted and pulled to form the required single crystal rod in length of 60 cm or so and diameters up to 5 cm. By means of a heating coil the liquid is maintained just above the melting temperature. A small seed crystal, mounted in a metal chuck, is lowered into the melt and then slowly raised, as the liquid freezes on to the colder seed a single crystal starts to grow with exactly the same structure as the seed crystal. As the crystal grows it is slowly raised and rotated to promote stirring in the melt. The entire apparatus is enclosed in a vessel containing an inert gas or in a vacuum to prevent oxidation or contamination from the atmosphere.

Germanium can be melted without significant contamination in either carbon or silica containers provided it is done in an inert atmosphere or in a vacuum. Gallium arsenide is more difficult, as to prevent decomposition the atmosphere over the melt must contain arsenic at high pressure.

THIN SEMICONDUCTOR FILMS

It is possible to deposit a layer of material known as an epitaxial layer which has the same crystallographic characteristics as the substrate material. Three methods are possible: solution growth, vapour phase growth, and chemical vapour deposition.

SOLUTION OR LIQUID GROWTH

Solution growth is mainly used for gallium arsenide. By lowering a substrate into a crucible containing a solution and then removing as cooling takes place a film is formed having the same atomic structure as the silicon substrate.

An alternative of solution growth is to place the solution and the substrate in a long boat shaped crucible which is then tilted to bring the solution into contact with the substrate for the growth period.

VAPOUR PHASE GROWTH

In this system epitaxial growth is used, i.e., films of Silicon grown on a silicon substrate, Gallium Arsenide on a GaAs substrate, etc. The system can also be used in the preparation of thin films of metals, and insulators in addition to semiconductors. One of the main advantages of epitaxial growth is that the impurity concentrations can be controlled in the films and this is used in the production of integrated circuits to reduce the series resistance of the substrate itself. Epitaxial films can be prepared by evaporation in a vacuum which supplies the atoms directly to the substrate surface, but, more usually, by chemical vapour deposition.

CHEMICAL VAPOUR DEPOSITION

In this system a wafer of silicon single crystal is placed inside a furnace tube and held at approx. 1 250 °C while silicon tetrachloride vapour in a stream of hydrogen gas is passed through the furnace. A chemical reaction takes place and a single crystal film forms on the substrate. The films can be pure or they can be doped *n*-type or *p*-type by bubbling the hydrogen through a weak solution of phosphorus trichloride for *n*-type or boron trichloride for *p*-type doping.

FURTHER READING

BRICE, J. C., 'An Analysis of the Factors Affecting Dislocation Distributions in Pulled Crystals of Gallium Arsenide', *J. Crystal Growth (Netherlands)*, **7**, p. 9, July 1970

BRICE, J. C., and JOYCE, B. A., 'The Technology of Semiconductor Materials Preparation', *The Radio and Electronic Engineer*, **Vol. 43**, No. 1/2, p. 21, Jan./Feb. 1973

BYLANDER, E. G., *Materials for Semiconductor Functions*, Hayden Book Co., New York (1973)

BRICE, J. C., *The Growth of Crystals from Liquids*, North Holland, Amsterdam (1973)

HARPER, C. A., *Handbook of Materials and Processes for Electronics*, McGraw-Hill, New York (1970)

HURLES, D. T. J., 'Mechanisms of Growth of Metal Single Crystals from the Melt', *Progr. Materials Sci.* **10**, No. 2, p. 79 (1962)

LAUDISE, R. A., *The Growth of Single Crystals*, Prentice-Hole, Englewood Cliffs (1970)

NEWMAN, R. C., 'A Review of the Growth and Structure of Thin Films of Germanium and Silicon', *Microelectronics and Reliability*, **3**, 121, Sept. 1964

WOLF, H., *Semiconductors*, John Wiley, New York (1971)

PRINTED CIRCUITS

Printed circuits and printed wiring

Although the description *printed circuits* is widely used as a generic term, *printed wiring* is more correct for the case where the wiring pattern is printed and the components added separately. *Printed circuits* implies the use of certain printed components such as small resistors and capacitors, strain-gauges and microwave components known as *microstrip* or *striplines*. Different forms of printed wiring have existed for many years, but the advent of dip-soldering just after the 1939–45 war made the use of these methods commercially viable.

Subtractive and additive methods

There are two basic systems of printed wiring:

(1) The subtractive method, in which a metal foil (usually of copper, although other metals may be used) is cemented to a plastic base (usually a laminated material). A pattern of the required wiring is printed on to the metal in one of several well-known ways. All the unwanted metal is removed by etching in an acid, leaving the required circuit pattern on the face of the plastic base.

(2) The additive method, in which a piece of plastic (usually a laminate) is punched with the necessary component mounting holes and then the required wiring pattern is built up by plating on to a previously prepared conductive or chemically sensitised layer. Alternatively the wiring pattern may be obtained by hot pressing a metal powder into the plastic, or copper foil may be stamped into the surface with a die, causing the imprint of the wiring to be made.

The first method, known as the *etched-foil* process, gives the better definition and more effort has been devoted to this process. The introduction of microelectronic devices which are small and replaced many larger components also increased the use of printed wiring boards on which the devices are mounted and dip-soldered.

Design and layout factors

As boards may contain single-sided wiring, double-sided wiring or multiple layers o wiring, it is least costly to use single sided or double sided wherever possible. A complex circuit is first broken down into sub-units and an initial layout of point to poin wiring in two dimensions is drawn out. Cross-over wires are reduced to a minimum an it is often convenient to make templates of the components so that 'trial and error methods can be used to finalise the best point to point wiring pattern. Some component can be placed to bridge the wires, but care should be taken with respect to mutua

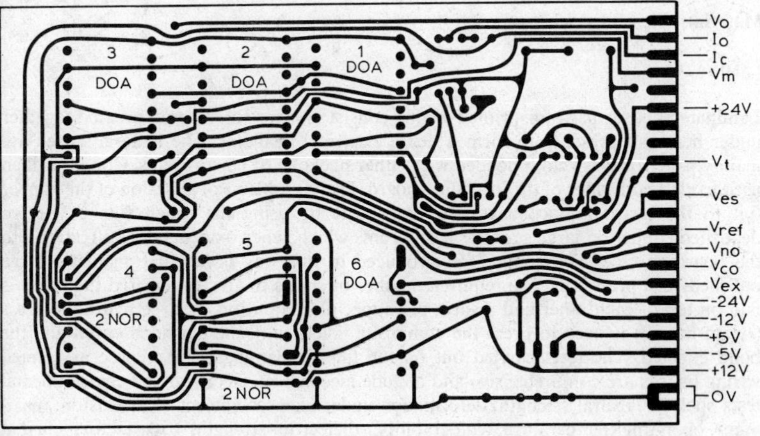

Figure 6.28 Printed wiring board layout

coupling and capacitative effects. The width and spacing of the wires should be sufficient to give high reliability and in general wires of 2.5 mm width spaced not less than 5 mm are usual, although with very miniature constructions tolerances are often higher. Sub-unit boards with contacts on one edge are often plugged into a mother board carrying the main wiring. A typical layout for the printed wiring side of a board is shown in *Figure 6.28* while the component side is shown in *Figure 6.29*. The current carrying capacity of flat wires is quite high and the resistance can be calculated from the expression

$$R = \frac{0.000\,227}{W}$$

where R = resistance in ohms per linear inch
w = width of line in inches.

This calculation is based on a line of 99.5% purity copper 0.002 7 inches thick (or 2 oz copper per square foot in area).

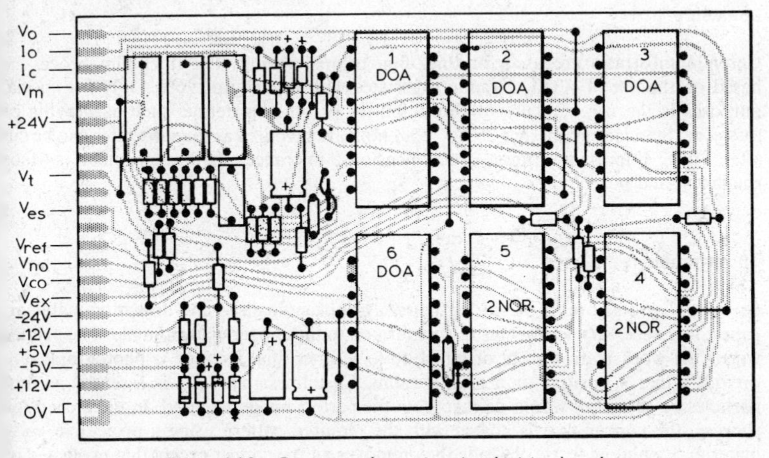

Figure 6.29 Component layout in printed wiring board

Materials for printed wiring

LAMINATES

Laminated boards used for printed wiring consist of layers of material pressed together under heat and pressure to form a dense electrical insulator. The materials used are mainly paper or glass cloth bonded with either phenolic or epoxy resins. Copper is then used to clad both sides of the insulating board. Early problems of adhesion of the copper foil to the board surface have been overcome by using the rough side of electro deposited copper on large stainless steel drums which, when oxided, bonds effectively to the board surface. The copper foil produced in this way is laid flat against a large polished steel press pan. The required number of sheets to give the desired thickness is laid on top of each other and a sheet of copper placed on top. A large hydraulic press (steam heated) is used to press the composite layers until the plastic is cured and the bond effected. The tests carried out on the finished laminates before use as printed wiring boards are comprehensive and include mechanical, electrical and environmental tests such as flexural strength, deformation under load, coefficient of expansion, maximum operating temperature, solderability, dielectric strength, loss factor, current carrying capacity, insulation resistance, etc.

Laminate hole punching and drilling. Before the board can be used for connective wiring, holes have to be punched or drilled to mount the component leads and to carry 'through' connections where required. Paper-base laminates are resilient and there is a tendency to spring-back when holes are pierced, so that the hole is slightly smaller than the punch itself. Glass cloth epoxy laminates have much less spring-back. Boards are usually heated while being punched. Punching is only commercially viable when large quantities of laminates are required. For small quantities or for experimental boards, drilling is usual. Drill point angle and drill material are more important than at first sight, as high-speed steel drills do not last very long when drilling epoxy-glass laminates and solid carbide drills have to be used. Drill chisel edge angle, web thickness, flute and shank diameter, lip clearance and concentricity are all important. If not correct faults such as resin smear, rough bore, bores with fibres inside them and too smooth a bore, will result.

CERAMICS

Ceramic substrates are used for thick-film printed wiring. The thick-film process is based on the use of Al_2O_3 ceramic substrates with a purity of 96%. Only for special purposes is the use of higher or lower purity or other materials such as beryllia or forsterite. Standard sizes, e.g., 25.4 × 25.4 mm, are favoured as the substrates can be cut after firing. Important criteria are dimensions, tolerances, flatness, camber, surface roughness and porosity.

Photographic techniques

Once the basic layout of the wiring pattern has been finalised and drawn on stable ar paper; it is necessary to transfer this pattern on to the copper clad laminate. The printe wiring art work is drawn out or taped up to 8 times final size and is photographed t provide either a positive or a negative film. Cleanliness and care is essential as dus particles or scratches would detract from the perfect image required. In the etched-foi process, the copper face is coated with the required pattern using a protective *resis* material. Etching in ferric chloride then removes all the copper except that protected b the resist and the wiring pattern is therefore left in copper. A second method uses negative of this approach in which the resist is printed on the undesired areas. The uncovered copper is then plated with a metal such as gold, tin, nickel or solder which can at this stage incorporate plated-through holes. The organic resist is then removed and the plated metal operates as a resist to the copper etching process.

Resists

Photosensitive resists are made from organic solutions (mostly from the Kodak Co.) such as KPR (Kodak Photo Resist), KMER (Kodak Metal Etch Resist), KTFR (Kodak Thin Film Resist), etc., which have the property of chemically changing their solubility to solvents when exposed to light of a certain wavelength. The resist is applied as a thin film by dipping, spraying, whirling or roller coating. When exposed to light through the film masking pattern, certain parts are hardened and become insoluble to a developing process. This enables the subsequent etching process to etch away only those parts required. Apart from the photographic masking technique silk screening can be used to place the required resist pattern on to the copper. Silk screening is essentially a stencil process in which the transfer of the resist on to the copper surface is made in the required circuit pattern. The stencil is attached to the surface of a silk or stainless-steel fine-mesh screen stretched on a frame just above the laminate. The resist is then forced through the open areas of the screen by a squeegee wiped across the screen. Hardening and etching then follows in a manner similar to the photographic process. This method is often used on ceramic substrates for *thick-film* or *hybrid* circuits.

Etching processes

To remove the unwanted areas of the wiring pattern, copper is etched away by an etching solution which is normally ferric chloride, although other etchants may be used for plated materials. The etchants are chosen to have the best properties for dissolved copper content, regeneration, useful life, control and economy. Etchants can be ferric chloride, ammonium persulphate, cupric chloride, chromic-sulphuric acids, etc., depending on the process requirements. After thorough cleaning the prepared laminates are passed through the etching solutions in various ways. Immersion etching where the laminates are immersed in the solution until etching is complete. Bubble etching where air is bubbled through the solution to provide continuous fresh etchant at the metal surface. Splash or Paddle etching where paddles throw the solution forcibly against the laminates and spray etching. Spray etching is widely used as it speeds up the etching process and consists of a pump and sprays which continuously etch the copper surface with fresh solution. The process deals with double sided laminates and lends itself to automation.

Plated-through holes

In double-sided laminates it is necessary to connect through holes from one side to the other and this can be done by eyelets, wire leads or rivets. In the ‘plated-through’ hole process, the inside of the holes is made electrically conductive by a chemically applied thin metallic film. The required conductor pattern is printed in reverse on both surfaces of the laminate which has been previously coated with a resist. The laminate is then immersed in an electro-plating bath and copper plated through the holes. Subsequent processes take place as previously described. When ‘plated-through’ holes are used, the joints between the component leads are stronger, as in the soldering process the solder flows up round the component wires in the holes and forms fillets on both sides of the laminate. As with other printed wiring processes, absolute cleanliness is essential and no corrosive or contaminating substance must be present.

Multilayer printed wiring boards

With the increasing complexity of circuits and the wider adoption of integrated circuits, the use of more than one layer of wiring become necessary in order to accommodate the required wiring. Multilayer printed wiring boards consist of a number of layers of thin flexible or laminate systems, stacked one on top of the other with correct registration and bonded together under heat and pressure.

A typical arrangement is shown in *Figure 6.30* which shows plated-through holes and a component mounted on a four-layer board. It will be seen that accuracy of registration is extremely important and also holes must be drilled with appropriate accuracy. Holes of less than 0.5 mm diameter require very careful control. A sharp solid carbide drill of

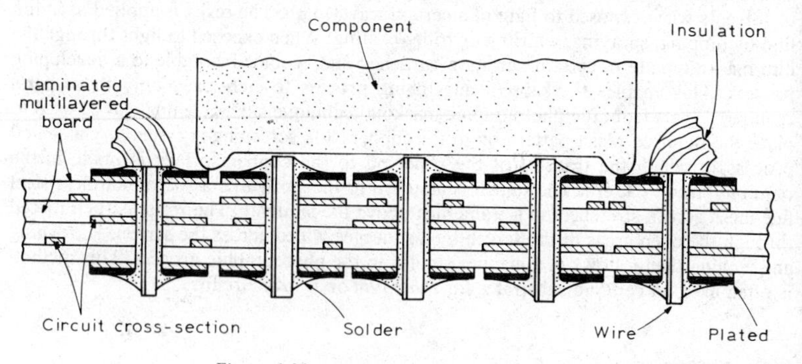

Figure 6.30 Multilayer printed wiring board

0.016 in diameter, rotating at 100 000 r.p.m. with a cutting feed of 0.001 5 in per revolution can produce clean holes, but the drills are weak and those sizes are only used where extremely high packing density is essential. In multilayer boards 'plated-through' holes are usual and are done as previously described. Because of the complexity and high accuracy multilayer boards are expensive and may cost as much as £50 to £100 in certain cases.

Flexible printed wiring

Flexible printed wiring consists of thin metallic foil conductors (usually copper) bonded to a pliable plastic film, with a cover-coat of the same plastic bonded over the copper wiring, leaving only the connector pads exposed. Plastic materials used are polyester, flame retardent polyester, epoxy glass, FEP (fluorinated ethylene propylene), FEP/glass and polyimide. Care must be taken in limiting the heat dissipation of resistors to that which the plastic will stand. Circuits should not be folded into too tight a radius and as much copper left as possible to provide a heat sink for the smaller components.

Soldering systems

HAND SOLDERING

A variety of soldering irons exist for this work ranging from miniature irons of 25 watts or so up to heavy duty irons of 100 or more watts. The original bare copper soldering bits have now been replaced by iron-clad tips. The general rules of using the maximum heat for the minimum time still applies and for temperature sensitive components, heat sinks or heat barriers may be used. Fluxing and cleanliness are important as in all soldering operations.

AUTOMATIC SOLDERING

While hand soldering is still used for small production and for experimental and prototype printed wiring boards, the main advantage of the printed wiring system is that all soldered joints can be made simultaneously by dipping the wiring side of the board through molten solder. After thorough cleaning and usually pre-heating there are five methods of automatic soldering:

(1) Dip Soldering.
(2) Drag Soldering.
(3) Wave Soldering.
(4) Cascade Soldering.
(5) Jet Soldering.

Each has its advantages and disadvantages.

Dip soldering. This is the oldest method of automatic soldering and uses a large soldering pot in which the solder is kept continuously molten. To keep the surface always clean, skimming or coating with an oil is necessary. The pre-fluxed circuit board is dipped at a slow rate into the solder pot and withdrawn from the solder after a specified time. A light tap removes any excess solder (if found necessary).

Drag soldering. In this system the boards are dragged continuously on the surface of the molten solder bath. The board itself floats on top of the solder and the system can be fully mechanised.

Wave soldering. This is the most commonly used system today and is based on the *Flow-soldering* machine developed by Messrs Fry. Instead of dipping the board into the solder, a continuous wave of solder is pumped up from the bath and the board is passed across this solder wave. The width of the wave is equal to the width of the board. The main advantages of this method are that heat distortion of the board is considerably reduced and no skimming operation is necessary, as fresh solder is continuously pumped up.

Cascade soldering. A modification of the wave soldering machine is to have a series of waves pumped up across which the board passes. This gives a multiple soldering effect.

Jet soldering. By using jets of solder it is possible to direct the solder stream in any desired direction and to train it on the area to be wetted. This is normally only used for special applications.

Components for printed circuits

SMALL COMPONENTS AND INTEGRATED CIRCUITS

Small components such as resistors and capacitors with radial or axial wire ends have their wires bent and shaped to fit through the connecting points (usually called *lands*) on the wiring pattern. They are mounted on one side of the board and the connections are all made by dip soldering on the other side. Transistors are similarly mounted while most modern circuits use integrated circuits packaged in multilead packs in which the leads are fitted through the lands. *Beam-lead* packaged integrated circuits are also available in which the connecting leads are brought out as beams which are soldered to the lands. In this case multiple-head soldering irons may be used to solder them into place and also suction de-soldering tools to remove them.

EDGE CONNECTORS

The printed board can be patterned and plated in such a manner as to make the male portion of the contact or contacts. Frequently it is printed only on one side, although commonly the female connectors are shaped, loosely speaking, like a tuning fork and thus contact both sides.

Most connectors ensure alignment of male and female contacts by permitting mechanical float, usually in the female socket, to cover board distortion and 'bowing'. Others ensure sufficient compliance, couple to accurate alignment, to permit mating and unmating without excessive build up of insertion and withdrawal forces.

Printed circuit connectors must have a means of fixing to the printed board, and a polarising feature to provide a maximum number of non-interchangeable connector combinations.

The contacts themselves need to provide at least a thousand matings and unmatings, and still maintain a low-contact resistance of the order of a few milliohms.

PRINTED COILS

Single-turn loops, or multiturn spirals are the only possible forms on flat surfaces and such printed coils are limited to about 50 μH. Dielectric losses in normally used laminates are high and low-loss materials must be used for effective small coils.

ROTARY SWITCHES

It is possible to print the switch contacts as part of the wiring pattern, but it is necessary to plate them with a metal such as rhodium to prevent rapid wear by the wiper arm. Flush-bonding of the device, in which the contacts are embedded flush with the laminate can be done. Special base materials are used such as melamine, epoxy, glass, etc., and switches of this type have been used as high-speed rotary switches in satellite telemetry.

PRINTED MOTORS

Using flush-bonded wiring, a French company has produced rotor coils printed on flat circular plates which form a printed wiring motor.

MICROWAVE PRINTED WIRING

Stripline or *Microstrip* is a method replacing bulky and heavy metal waveguides by smaller and lighter printed circuits. The basis of the printed waveguide is the planar or flat strip. The strip is photo-etched on a supporting dielectric and functions as a conducting guide for the propagation of electromagnetic waves. By variation of the strip pattern, the structure may be adapted to the fabrication of all forms of microwave elements. The technique simplifies waveguide plumbing to the printed circuit and eliminates the high cost of machining conventional waveguide and coaxial systems. Since all models of any one system are reproduced by a photographic process, the accuracy of repetition of mechanical dimensions is very high.

AUTOMATIC ASSEMBLY TECHNIQUES

The extent to which automation can be carried out depends mainly on economics. However the use of a printed wiring board is basic to practically all automatic approaches. In addition a generic grid pattern to govern circuit layout is essential, e.g., a 2.5 mm grid. Components must also be pre-prepared to fit the standard grid insertion holes. Many types of automatic component insertion machines exist. Most use bandoliered resistors, capacitors and small wire ended components, in which the components are strung together for feeding into the machine. The component insertion machine then bends the end wires of each component to the required shape and inserts the ends into the printed circuit board, usually clenching the wiring side over to retain the components in position until soldered by the dip soldering machine. Other components which do not have shapes suitable for these machines must be inserted by hand. Automatic assembly is practical for large production runs of printed wiring boards.

Specifications for printed wiring

Available from, The British Standards Institution.

BS 4025–1966	Specification for the general requirements and methods of test for printed circuits
BS 4597–1970	Specification for the general requirements and methods of test for multilayer printed wiring boards using plated-through holes.

Available from, Her Majesty's Stationery Office, London.

DG 5007–1958	Defence guide for Printed Wiring electrical applications.
DEF 5028–1963	Printed Wiring boards having plated-through holes on plated-up wiring
M O D Interim	*Part I* Test requirements for single- and double-sided printed wiring boards
Defence Standard 59–48	*Part II* Test requirements for multilayer printed wiring boards

Available from, International Electrotechnical Commission.

IEC Pub. 321	Guidance for the design and use of components intended for mounting on boards with printed wiring and printed circuits

FURTHER READING

COOMBS, C. F., Jnr., *Printed Circuits Handbook*, McGraw-Hill, New York (1967)

DUKES, J. M. C., *Printed Circuits—Their Design and Application*, Macdonald, London (1961)

EISLER, P., *Technology of Printed Circuits*, Heywood & Co., London (1959)

FISK, C. J., CASKEY, D. L. and WEST, L. E., 'Taking the Puzzle out of P-C Design', *Electronics*, p. 72 (Sept. 1967)

GREYGOOSE, R. G. and ROBERTSON, F. A., 'Design and Fabrication of Multilayer Printed Circuit Boards', *Solid State Technology*, p. 33 (Nov. 1970)

PROCEEDINGS, *International Electronic Packaging & Production Conferences* (INTERNEPCON) (1973 and annually), Kiver Communications Ltd., Surbiton

RIDER, D. K., 'A Survey of Printed Circuit Processes', *SCP and Solid State Technology*, p. 18 (29 July 1966)

SCHLABACH, T. D. and RIDER, D. K., *Printed & Integrated Circuitry*, McGraw-Hill, New York (1963)

PIEZOELECTRIC MATERIALS AND EFFECTS

Introduction

Piezoelectricity, discovered by the brothers Curie in 1880, takes its name from the Greek piezein, to press, but piezoelectric materials will in general react to any mechanical stress by producing electric charge: in the converse effect, an electric field results in mechanical strain. As a simple introduction, compare small cubes of crystal quartz (a piezoelectric material) and fused quartz (non piezoelectric). If the former is mechanically stressed, in an appropriate way, electric charge will appear on its faces: there will be some elastic change in dimensions, but the amount of this change will depend on the electric boundary conditions, i.e., whether electrodes on the faces of the cube are open—or shortcircuited. Conversely, if a voltage be applied to these electrodes of the unstressed cube, the cube will not only be charged electrically but strained mechanically: as in the first case, the amount of charge stored electrically will depend on whether the cube is clamped or free mechanically. In contrast, the cube of fused quartz behaves purely elastically under mechanical stress, and as a simple dielectric in an electric field: there is no coupling between the two effects.

As in this example, piezoelectric materials are crystalline in structure and anisotropic in many properties (e.g., thermal conductivity, thermal expansion, dielectric constant) while the non piezoelectrics are often amorphous and isotropic. It is also worth noting that the same crystalline structure which results in piezoelectricity results also in optical activity [1] (the linear electro-optic, or Pockels effect is possible with all piezoelectric materials, and only with them) and in pyroelectricity, the generation of electricity by heat (every pyroelectric crystal is piezoelectric, and every ferro-electric crystal, but not every piezoelectric crystal, is pyroelectric. Ferro-electricity, an important sub-class of piezoelectricity, will be further discussed below.)

Piezoelectricity is a much more common phenomenon than is generally realised. Cady [2] reports that over 1 000 piezoelectric crystals have been identified; this is perhaps not so surprising when it is remembered that only 11 out of the 32 crystal classes cannot, because of their symmetrical crystal structure, have any piezoelectric members.

A caution should be given here about notation, both as regards piezoelectricity itself and on crystal structure. Most modern texts follow the IRE 'Standards on Piezoelectric Crystals', dated 1949 and later,[3] but the first edition of Cady [4] (1946) does not, of course, and the second edition [2] (1964) only outlines the standards rather briefly in an appendix. Hueter and Bolt's [5] 'Sonics' use Voigt's notation, dating from 1890, and explains that conversion into other notations can be made as explained in the standards.

On crystal structure, Cady [2] gives a cross reference between the various crystal notations, which will be useful to those not familiar with the subject.

The crystalline basis of piezoelectricity

Piezoelectricity is essentially a phenomenon of the crystalline state and the two subjects are dealt with together in Kittel,[6] Cady,[2] Berlincourt *et al.*,[7] Mason [8] and Nye [9].

Crystals are classified into seven systems, and 32 classes. Of these, eleven have a centre of symmetry, and cannot be piezoelectric: the 21 classes lacking a centre of symmetry all have piezoelectric members. It can therefore be stated as a necessary condition for piezoelectricity that the crystal lacks a centre of symmetry. Only if this is the case will there be a shift of the electrical centre of gravity when the crystal is stressed, and a resultant generation of charge on the crystal faces.

An important sub-class of piezoelectrics is that of the ferro-electrics, which have two or more stable asymmetric states, between which they can be switched by a sufficiently strong electric field. Ferro-electrics have a domain structure, a Curie temperature and show hysteresis; hence their name, by analogy with the ferromagnetics. Cady [2] deals at some length with Rochelle salt, one of the earliest ferro-electrics, and Kittel[6] has a chapter on barium titanate, which has the perovskite structure typical of one family of ferro-electrics.

Another important family (lithium niobate, lithium tantalate, etc.) is of considerable importance for electro-optic purposes as well as (especially lithium niobate) piezoelectricity.

Ferro-electricity can be exploited not only in single crystal material, but also in polycrystalline form, and this gives the commercially important class of piezoelectric ceramics. If the polycrystalline material is *poled*, i.e., subjected to a strong electric field as it is cooled through its Curie point, the domains in the crystals are aligned in the direction of the field. One can in this way have a piezoelectric material whose size, shape and piezoelectric qualities can, within limits, be made to order. These commercially available materials are often solid solutions of lead zirconate and lead titanate with various additives to modify the behaviour. Such a range is marketed under the general name of PZT, with distinguishing numbers, by the Clevite Corporation U.S.A., and under the general name PXE, again with a number, by Mullard in the U.K. A good account of these materials (and of piezoelectric materials in general) is given in Berlincourt *et al.*[8] The Mullard materials are listed in a booklet,[10] together with some application notes.

Finally, it is worth noting that some piezoelectric materials, notably cadmium sulphide, CdS, and zinc oxide, ZnO, can be evaporated or sputtered on to a suitable substrate, on which the deposit is polycrystalline, but oriented.

These materials, 'grown' to a thickness of a half wavelength, can be used as electro-acoustic transducers to launch elastic waves into the substrate at much higher frequencies than can be achieved by the alternative (low frequency) technique of bonding a halfwave thickness of a crystal transducer (quartz, lithium niobate, ceramic) to the substrate.

Piezoelectric constants

In a dielectric medium which is elastic but not piezoelectric (e.g., fused quartz, glass), independent relations exist giving its behaviour in an electric field E, and under a mechanical stress T. These are

$$D = \epsilon E \qquad (1)$$
$$S = sT \qquad (2)$$

where D is the electric displacement, ϵ the dielectric constant, S the strain in the material and s the elastic compliance (the reciprocal of Youngs modulus, if the stress is tension or compression).

In a piezoelectric medium, the strain or the displacement depend linearly on both the stress and the field, and the equations for D and S become

$$D = dT + \epsilon^T E \qquad (3)$$
$$S = s^E T + dE \qquad (4)$$

where d is a piezoelectric constant characteristic of the material, and the superscripts T and E denote that ϵ and s are to be measured at constant stress and constant electric field respectively.

In words, d may be defined as the displacement (charge per unit area) per unit applied stress, the electric field being constant or as the strain per unit applied field, the stress being constant. From the first definition, d is the piezoelectric charge constant.

The piezoelectric voltage constant g can be introduced by an equivalent pair of equations

$$E = -g^T + D/\epsilon^T \qquad (5)$$
$$S = s^D T + gD \qquad (6)$$

By substituting for D from equation 3 into equation 5,

$$d = g\epsilon^T \tag{7}$$

Another very important piezoelectric quantity is the electromechanical coupling coefficient, k, the significance of which is seen as follows. The mechanical energy stored in the piezoelectric medium is, from equation 4

$$U_M = \tfrac{1}{2}ST = \tfrac{1}{2}s^E T^2 + \tfrac{1}{2}dET \tag{8}$$

and the electrical energy stored is, from equation 3

$$U_E = \tfrac{1}{2}DE = \tfrac{1}{2}dTE + \tfrac{1}{2}\epsilon^T E^2.$$

U_M contains a mechanical term ($\tfrac{1}{2}s^E T^2$) and a mixed term ($\tfrac{1}{2}dET$): similarly U_E contains an electrical term ($\tfrac{1}{2}\epsilon^T E^2$) and the same mixed term. By analogy with a transformer, in which the coupling factor is the ratio of the mutual energy to the square root of the product of the primary and secondary energies, k is the ratio of the mixed term to the square root of the products of electrical and mechanical terms, i.e.,

$$k = \frac{\tfrac{1}{2}dET}{\sqrt{\tfrac{1}{4}\epsilon^T s^E E^2 T^2}}$$

or

$$k^2 = \frac{d^2}{\epsilon^T s^E} \tag{9}$$

At low frequency, i.e., below the frequency of any mechanical resonance of the material, k^2 is a measure of how much of the energy supplied in one form (electrical or mechanical) is stored in the other form: it is not an efficiency since the energy applied is all stored in one form or the other.

Piezoelectric notations

For the sake of simplicity, the discussion so far of the piezoelectric constants has implied that there is only one piezoelectric axis, and that the stress or electric field is applied along this axis and the resulting strain or charge is also measured in this direction. This situation is appropriate, for instance, to a cylinder of a ceramic piezoelectric material, poled along its axis, and with the stress or field also along this axis. It is not appropriate to a single crystal piezoelectric, with its anisotropic properties, to which stress is applied in an arbitrary direction. To deal with this situation, a more complicated notation must be set up (IRE Standards, 1949).[3]

Figure 6.31 shows a right-handed orthogonal set of axes X, Y, Z, together with a

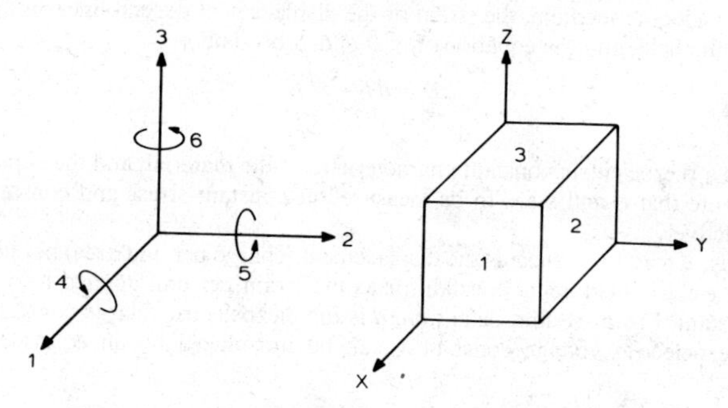

Figure 6.31 Diagram of axes (1, 2, 3), shears (4, 5, 6) and surfaces (1, 2, 3) for piezoelectric notation

small cube with faces *1, 2, 3*, perpendicular to *X, Y* and *Z* respectively. Any stress applied to the cube of material can be resolved into forces *1, 2, 3*, along the axes, or shears *4, 5, 6*, acting about the axes. Any electric field can be resolved into its components along the axes, and any charge resulting from the field or from piezoelectric stress will appear on the faces *1, 2, 3*, as shown. As for stress, there are six ways in which a strain can appear, i.e., tension or compression along the three axes, or shear about the axes.

Hence for the general case, there are six permittivities ϵ_{ij} (*i* and *j*, 1–3), twenty-one elastic compliances s_{ij} (*i* and *j* 1–6) and eighteen piezoelectric constants d_{ij} (*i* 1–3, *j* 1–6). Fortunately, most crystals are symmetrical enough that the number of independent piezoelectric coefficients is greatly reduced: quartz for instance has only two, d_{11} connecting a field along the *X* axis with strain in the same direction, and d_{14}, connecting a field along an axis normal to *Z* and a shear strain in the plane normal to the field.

Berlincourt *et al.*,[8] and Nye[9] give elasto-optic matrices for all 32 of the crystal classes. These are 9 × 9 matrices, symmetrical about the diagonal, so giving the possible 45 elasto-optic constants listed in the previous paragraph. Such matrices are also given in the IRE Standards (1958).[3]

Applications of piezoelectric materials

FREQUENCY STABILISATION

The use of quartz as a frequency standard is still the most important application of piezoelectricity technically and commercially. This use of quartz relies on its high *Q* as a mechanical resonator: Mason[8] shows curves of *Q* against frequency which imply that

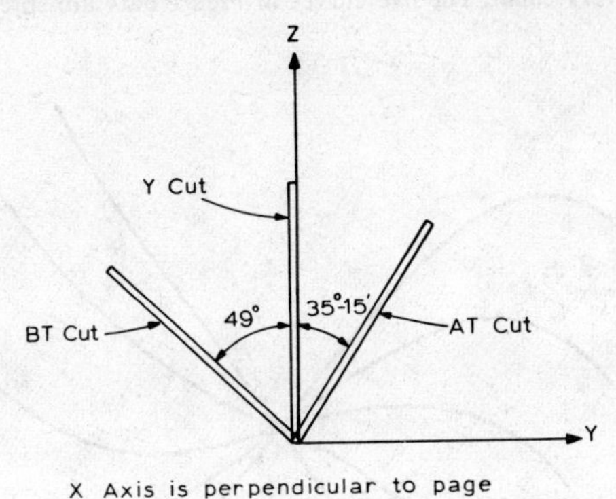

Figure 6.32 Diagram of crystal axes in quartz, illustrating some of the more important crystal cuts

the internal *Q* (i.e., that due to internal frictional losses, and excluding mounting losses and air losses) of an *AT*-cut shear mode crystal can be over 10^7 at 1 MHz. Contributory factors are its high quality as a dielectric, its low dielectric constant, and the relative ease of cutting and polishing. The processes involved in quartz technology are described in some detail by Cady.[2] It is because of the increasing difficulty of mining sufficient high quality natural quartz that the hydrothermal growth of quartz has increased in the last 10 or 15 years. The synthetic quartz can be made with at least as high a *Q* as natural quartz.

The most important cuts of the crystal for frequency standards at frequencies of 1 MHz upwards are the *AT* and *BT* cuts, shown in *Figure 6.32*. A more comprehensive

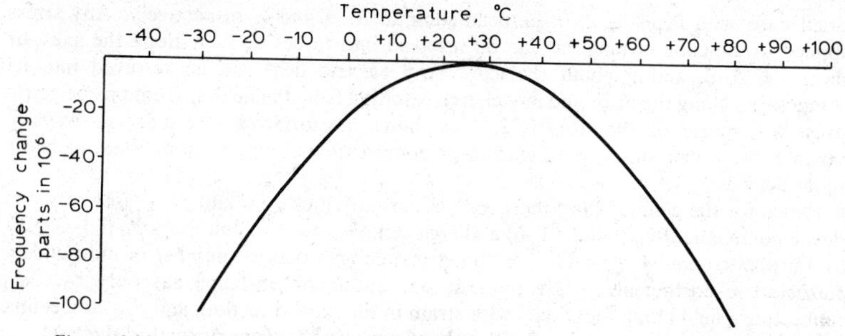

Figure 6.33 The variation of frequency with temperature for a BT cut quartz crystal

diagram, showing the position of a large number of the more important cuts, is in Mason:[8] *AT* and *BT* cuts are *rotated Y-cuts*, i.e., they are obtained by rotating the *Y* plane (perpendicular to *Y*) about the *X* axis, as shown in the diagram. These cuts, which vibrate in a shear mode, are important because they give a zero temperature coefficient of frequency, as shown in *Figures 6.33* (*BT* cut) and *6.34* (*AT* cut). The *BT* temperature coefficient curve is parabolic and is of zero slope at only one temperature: this temperature is a function of the exact angle, and changes by about 1.5 °C per minute of arc. *BT* cuts are preferred to *AT* at frequencies above 10 MHz, since they are more stable against changes in drive level and load capacitance.

The *AT* cut, on the other hand, can be chosen to give two temperatures of zero coefficient, and a wide temperature range (the middle curve of *Figure 6.34*) over which the coefficient is very small. The five curves in *Figure 6.34* are spread over about 20

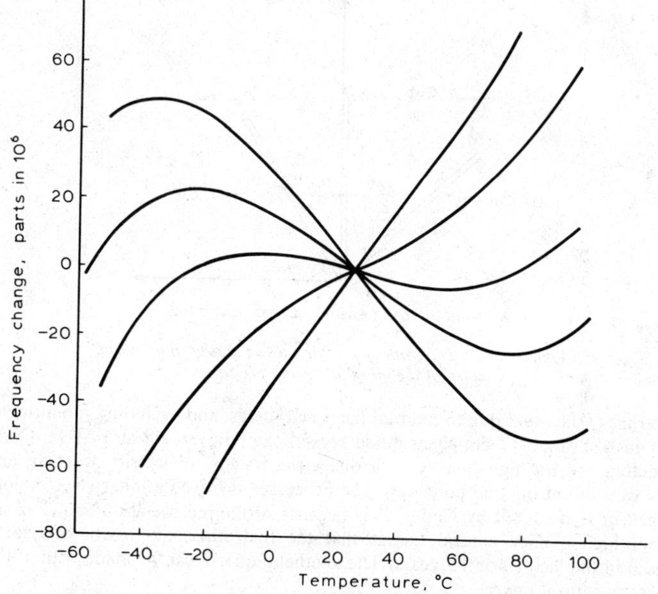

Figure 6.34 Typical family of frequency variation with temperature of AT cut quartz crystals. The curves shown cover an angular range of about 20 minutes of arc

minutes of arc. For best performance as a frequency standard, contoured (i.e., convex on one or both surfaces) *AT* cut plates are used, often operating at a harmonic frequency (e.g., a 5th overtone crystal, abbreviated to 5th *OT*). The crystal, despite its low temperature coefficient, will be in a temperature controlled oven, or, for the highest quality, a double oven. A double oven will reduce temperature fluctuations by a factor of 10^4 or more, i.e., the temperature of the crystal will not vary by more than a few millidegrees.

The frequency of a newly made quartz crystal oscillator will drift quite rapidly for a few months, a process known as ageing.[11] After a few months at constant temperature the ageing rate will be about 1 part in 10^{10} per month or less and the frequency will continue to drift at this, or a very slowly decreasing rate, indefinitely. If conditions are disturbed, e.g., the oscillator or its oven is switched off for a few hours, ageing will recommence at near the initial rate. Hence a quartz frequency standard should never be switched off, nor run at a power output above a few microwatts. The specification of a commercial quartz frequency standard,[12] using a 5th *OT AT* crystal, would include a frequency stability better than 15 parts in 10^{12} at constant voltage and temperature measured over 1 second and an ageing rate of less than 1 part in 10^9 per month after three months. These are of course maximum values, and do not contradict the lower ageing rate quoted above.

Figure 6.35 A micro-miniature quartz crystal oscillator mounted on a TO5 header (*Courtesy The Marconi Company Ltd*)

Because they are not absolute standards (i.e., they require calibration) and because of ageing, quartz crystals have been superseded as frequency/time standards by atomic clocks (rubidium, caesium, hydrogen). However, the short term (1 second) stability of the best quartz crystals is better than that of the atomic standards (except hydrogen) which therefore consist of an atomic clock, used to control and correct the ageing of a good quartz oscillator. It is from the quartz oscillator that the output frequencies are delivered.

Quartz crystal oscillators of high stability are well reviewed in Gerber and Sykes,[13] in an issue of Proceedings IEEE devoted to frequency stability: the issue therefore also reviews atomic frequency standards.

Quartz crystals can be made to any frequency from 1 kHz to 750 kHz, and from 1.5 MHz to 200 MHz, the latter being high harmonic overtone crystals. The frequencies between 750 kHz and 1.5 MHz are difficult to cover as strong secondary resonances occur. There is a tendency to use a higher frequency crystal counted down rather than a very low frequency crystal, which tends to be larger and less stable than is desirable. The units may be packaged in glass envelopes, or in a solder-sealed or a cold-welded metal can: this container may also contain a heater and temperature sensor. *Figure 6.35* shows the interior of a microminiature crystal oscillator (Marconi Type F 3187), providing frequencies in the range 6–25 MHz over the temperature range −55 °C to +90 °C to an accuracy of ±50 parts per million.

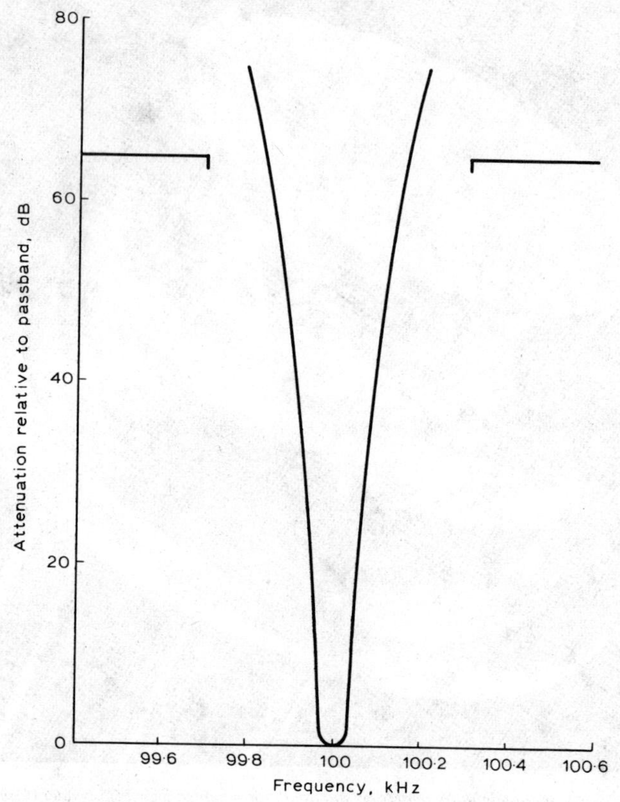

Figure 6.36 Response curve of a very narrow passband crystal filter

CRYSTAL FILTERS

Quartz crystals can also be used, and for the same reasons, in very precise filters, used for defining a passband very accurately, and for rejecting adjacent channel interference. *Figure 6.36* shows a very narrow bandpass filter, with a pass-band of about 50 Hz at a centre frequency of 100 kHz: the 60 dB bandwidth is about 300 Hz. This filter might be used for telegraphy. *Figure 6.37*, on the other hand, shows the pass-band of a filter for a reasonable quality speech circuit. The pass band is 5 kHz wide at 3 dB, and less than 11 kHz at 60 dB. Such filters can be very useful in conditions of severe congestion, or for s.s.b. working.

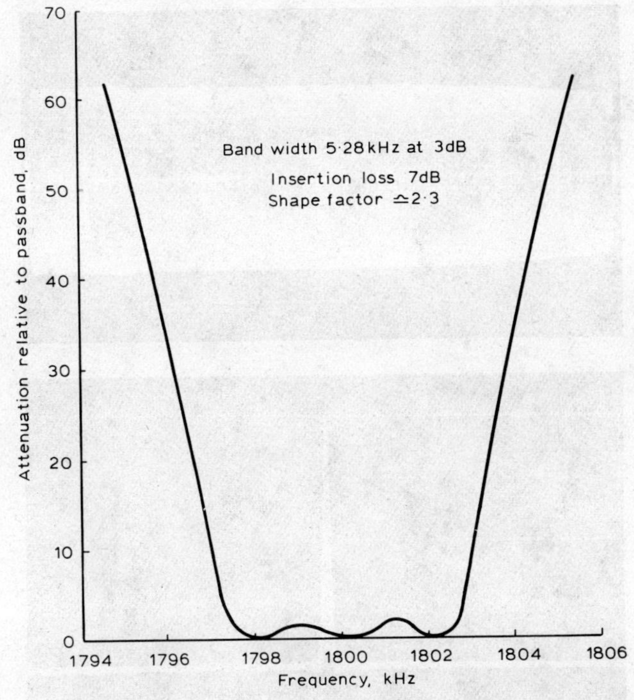

Figure 6.37 Response curve of a multielement narrow passband crystal filter

A more recent development is the monolithic crystal filter, i.e., a single quartz plate on which a number of mechanically-coupled resonators are printed, the whole assembly forming a multiple element filter.

Filters using surface wave techniques will be discussed later, under that heading.

BULK DELAY LINES [13, 14]

Piezoelectric transducers (quartz, lithium niobate, ceramics) are used to launch ultrasonic waves into liquid (water, mercury) or solid (fused quartz, glass, sapphire, spinel) media, in which the velocity of propagation is about 10^5 times slower than that of radio waves in free space. Shear waves are often used, because the velocity is lower and because, with the correct polarisation of the shear axis, the wave is reflected from a boundary as a pure shear wave. The transducers are either ground to thickness (half a wavelength at the required frequency) and then bonded to the medium, or evaporated or sputtered (cadmium sulphide, zinc oxide) to the required thickness. Whereas ground transducers are limited to, say, 100 MHz by their fragility at this frequency, evaporated or sputtered transducers can be deposited to resonate at frequencies up to 10 GHz. Long delays at such frequencies are, however, precluded by high losses in the available media,[15] and by the fact that the transducers can launch only longitudinal waves. Mode conversion from longitudinal to shear mode can be used for intermediate frequencies and delay times.

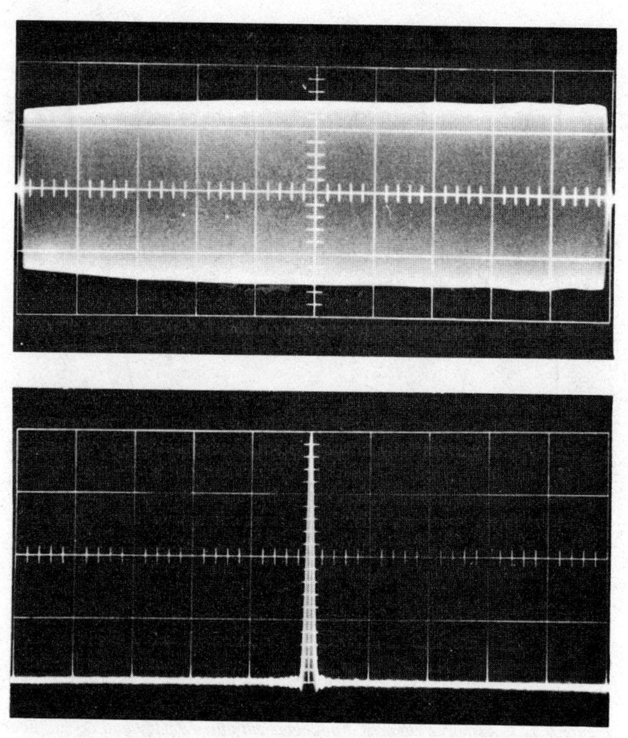

Figure 6.38 Pulse compression from a dispersive ultrasonic delay line. Input (top) is a 5 μs pulse, swept over 25 MHz: the compressed pulse (below) is 52 ns wide, i.e., the compression ratio is about 100 to 1 (Courtesy W. S. Mortley, The Marconi Company Ltd)

Bulk delay lines are used in colour television receivers (~62 μs at the colour subcarrier frequency),[10] in M.T.I. (moving target indication) for the removal of radar clutter from the display, in vertical aperture correction for television cameras, in APEGs (artificial permanent echo generators) for radar performance monitoring, and for a variety of other purposes. Mercury and quartz delay lines were used as data memories in early computers, but are now obsolete in this application. Most bulk wave delay lines are dispersive, but dispersive lines can also be made for pulse compression and other purposes (see *Figure 6.38* and below, under surface waves).

Acoustic waves in bulk media can also be used for the deflection and modulation of light.[16]

ACOUSTIC AMPLIFIERS

Some well-known semiconductor materials (gallium arsenide, GaAs, cadmium sulphide, CdS, etc.) are also piezoelectric. It was discovered by Hutson *et al.*,[17] that if electrons were made to travel by means of a d.c. electric field at the same velocity as an acoustic wave in the material, there could be an interaction between the wave and the electrons, resulting in amplification. Since the acoustic wavelength is small, the gain per unit length of material could be high. These devices do not seem to have fulfilled their early promise, doubtless because of their very low efficiency, which makes it difficult to operate them in a c.w. mode.

More recently, surface wave amplifiers have been made, using a single semiconducting piezoelectric material, or two materials, one piezoelectric (e.g., lithium niobate) the other semiconducting (e.g., silicon), in close proximity.[18] It is too early to say whether these amplifiers will be more successful than the bulk wave versions.

SURFACE ACOUSTIC WAVES

By means of a suitable transducer geometry, a shear or longitudinal wave in a bulk medium may be transferred to the surface of another medium: alternatively, if interdigital transducers (i.e., arrays of fingers one half wave length apart) are laid down on a piezoelectric substrate and driven electrically, a surface wave is again excited.[19] The wave excited in these cases is the Rayleigh wave, travelling as a surface deformation which dies away very rapidly below the surface of the medium. It can be shown fairly simply that the velocity is a little lower than that of a shear wave in the medium, and that the wave is non-dispersive. Moreover, it travels on the surface and so is accessible as required. The velocity, and therefore the wavelength, is about 10^5 times lower than that of the radio wave in free space, giving the possibility, at least, of truly microelectronic circuits at radio frequencies.

This technology, called microsound or SAW (surface acoustic waves), is still in its infancy, but already a considerable literature is accumulating,[20] and a wide variety of devices is being studied, ranging from bandpass filters for the i.f. amplifiers of television receivers to pulse compression, or chirp,[21] filters with time bandwith products of 1 000 or more. The results of chirp, or pulse compression, are shown in *Figure 6.38*, in which a f.m. pulse 5 μs long, swept over 25 MHz, is compressed to a pulse 50 ns long, a compression ratio of 100 (this result was actually achieved with a bulk wave device). Many of the components familiar to the microwave engineer can be realised in SAW:[22] it is important however to realise that the analogy with microwave waveguides can be pushed too far and that in spite of the name microsound, the frequencies involved are in the v.h.f.–u.h.f. region rather than truly microwave.

MISCELLANEOUS PIEZOELECTRIC APPLICATIONS

Besides the major areas dealt with above, piezoelectric materials (usually ceramic materials because of their relatively low cost, ease of fabrication and high coupling coefficient) are used in a wide variety of other applications. These include piezoelectric ('ceramic') pickups for record players[10] (bi-morphs or multimorphs), gas ignition systems, echo sounders, ultrasonic cleaners and soldering irons and ultrasonic drills. (As an alternative to piezoelectric operation, such systems which rely on transmitting high ultrasonic powers, i.e., drills, cleaners and soldering irons, may be driven magnetostrictively.)

Useful accounts of these miscellaneous and high-power applications may be found in books by Crawford[23] and Hueter and Bolt.[5]

Data on piezoelectric materials

Because of its complexity and volume, it is impossible to give here numerical data on even the most important piezoelectric materials: instead, sources of such data are quoted.

Berlincourt *et al.*,[7] give a review of the sources (p. 171) and tables of the more important materials (Tables II–VI inclusive pp. 180–184 covering single crystal materials, and tables VII and VIII (pp. 195 and 202) on ceramic materials). The Mullard Booklet[16] covers the PXE ceramics (pp. 14–15). The American Institute of Physics Handbook[24] lists the piezoelectric strain constants of some 50 or 60 materials (section **9**–97), and the temperature coefficients of a few of these.

REFERENCES

1. KAMINOW, I. P. and TURNER, E. H., 'Electro-optic Light Modulator', *Proc. I.E.E.E.*, **54**, 1374 (1966)
2. CADY, W. G., *Piezoelectricity*, Dover edition, Dover publications (1964)
3. I.R.E. Standards on Piezoelectric Crystals (1949), *Proc. I.R.E.*, **37**, 1378 (1949)
   I.R.E. Standards on Piezoelectric Crystals (1957), *Proc. I.R.E.*, **45**, 354 (1957)
   I.R.E. Standards on Piezoelectric Crystals (1958), *Proc. I.R.E.*, **46**, 765 (1958)
   I.R.E. Standards on Piezoelectric Crystals (1961), *Proc. I.R.E.*, **49**, 1162 (1961)
4. CADY, W. G., *Piezoelectricity*, 1st edn., McGraw-Hill (1946)
5. HUETER, T. F. and BOLT, R. H., *Sonics*, John Wiley (1955)
6. KITTEL, C., *Introduction to Solid State Physics*, Chapman and Hall (1953)
7. BERLINCOURT, D. A., CURRAN, D. R. and JAFFE, H., 'Piezoelectric and Piezomagnetic Materials', in *Physical Acoustics*, Vol. 1, Part A (ed. Mason, W. P.), Academic Press (1964)
8. MASON, W. P., *Piezoelectric Crystals and Their Applications to Ultrasonics*, Van Nostrand (1950)
9. NYE, J. F., *The Physical Properties of Crystals*, Oxford U.P. (1957)
10. VAN RANDERAAT, J., ed., *Piezoelectric ceramics*, Mullard Ltd. (1968)
11. GERBER, E. A. and SYKES, R. A., 'State of the Art—Quartz Crystal Units and Oscillators', *Proc. I.E.E.E.*, **54**, 103 (1966)
12. MARCONI TYPE F.3160, taken from *Catalogue of Quartz Crystal Oscillators, Ovens and Filters*, Marconi Specialised Components Division
13. BROCKELSBY, C. F., PALFREEMAN, J. S. and GIBSON, R. W., *Ultrasonic Delay Lines*, Iliffe Books, Ltd. (1963)
    EVELETH, J. E., 'A Survey of Ultrasonic Delay Lines below 100 MHz', *Proc. I.E.E.E.*, **53**, 1406
14. MAY, J. E., 'Guided Wave Ultrasonic Delay Lines'; and Mason, W. P., 'Multiple Reflection Ultrasonic Delay Lines', in *Physical Acoustics*, Vol. 1, Part A (ed. Mason, W. P.), Academic Press (1964)
15. KING, D. G., 'Ultrasonic Delay Lines for Frequencies above 100 MHz', *Marconi Review*, **34**, 314 (1971)
16. GORDON, E. I., 'A Review of Acoústo Optical Deflection and Modulation Devices', *Proc. I.E.E.E.*, **54**, 1391
17. HUTSON, A. R., MCFEE, J. H. and WHITE, D. L., *Physics Review Letters*, **I**, 237 (1961)
18. LAKIN, K. M. and SHAW, H. J., 'Surface Wave Delay Line Amplifiers', *I.E.E.E. Transactions*, **MTT 17**, 912 (Nov. 1969)
19. STERN, E., 'Microsound Components, Circuits and Applications', *I.E.E.E. Transactions*, **MTT 17**, 912, p. 835 (Nov. 1969)
20. *I.E.E.E. Transactions*, **MTT 17** (Nov. 1969)
    I.E.E.E. Transactions, **MTT 21** (April 1973)

MAINES, J. D. and PAIGE, E. G. S., 'Surface-acoustic-wave components, devices and applications', *I.E.E. Reviews, Proc. I.E.E.*, **120**, No. 10R, 1078 (1973)

21. KLAUDER, J. R. *et al.*, 'The Theory and Design of Chirp Radars', *Bell Sys. Tech. J.*, **39**, 745 (July 1960)
22. MAINES, J. D. and PAIGE, E. G. S., 'Surface-acoustic-wave components, devices and applications', *I.E.E. Reviews, Proc. I.E.E.*, **120**, No. 10R, 1078 (1973)
23. CRAWFORD, A. H., *Ultrasonics for Industry*, Iliffe (1969)
    CRAWFORD, A. H., ed., 'High Power Ultrasonics—International Conference Proceedings', *IPC Press* (1972)
24. *American Institute of Physics Handbook*, 2nd edn., McGraw-Hill (1963)

MAGNETIC MATERIALS—FERRITES

Introduction

In its most general form, the term 'ferrite' may be, and is, used to describe a variety of mixed oxide ceramic materials exhibiting ferromagnetic properties. The term was originally used to describe the several families of binary mixed oxides which are analogues of the naturally occurring mineral *Spinel*—Mg Al_2O_4. These spinel-type materials are of the form M Fe_2O_4, where M may be any one of a number of divalent elements including magnesium, manganese, nickel, copper, cobalt and zinc. More recently the scope of the term 'ferrite' has been extended to include several other important families of mixed oxide compounds, perhaps the most significant of which are the Garnet Ferrites, which take their name from that of the naturally occurring mineral $Mn_3Al_2Si_3O_{12}$, whose crystal structure is adopted by a variety of compounds of general formula $R_3Fe_2Fe_3O_{12}$, known as the Garnet structure, in which R is a trivalent rare earth or yttrium ion. A third family of materials, the hexagonal ferrites, are based upon barium ferrite, Ba $Fe_{12}O_{19}$, and are characterised by a large uniaxial anisotropy rendering them particularly useful as permanent magnet materials.

Ferrites, in common with all chemical compounds, may exist in single crystal or polycrystalline form, although in general it is more difficult to prepare them as single crystals. The difference between the two is that the single crystal comprises material in which the arrangement of ions on the crystal lattice is homogeneous throughout the bulk of the material, a 'single crystal' being perhaps several cubic centimetres in volume. The polycrystalline material, on the other hand, comprises a completely random aggregate of much smaller single crystals, each maybe only a few cubic microns in volume. The properties of single and polycrystalline materials are markedly different, both forms nevertheless finding applications in different spheres of electronics, although those for single crystal material are somewhat limited.

The manufacture of polycrystalline ferrites is based upon conventional ceramic manufacturing processes, with a number of important differences and refinements as described below, whereas that of single crystal material is based upon a number of techniques initially developed for more conventional ceramic crystals. In each class of ferrite wide variations in composition, purity, processing, properties and applications exist, and a number of the more important of these variations are described below.

Crystal structure

SPINEL FERRITES

The generalised molecular formula of a spinel ferrite is $M^2 + Fe_2^3 + O_4$, where M is a divalent metal ion, usually a member of the first transition series. Most frequently the elements Co, Ni, Fe, Mn, Mg and Zn are encountered. The crystal lattice displays cubic symmetry, the unit cell containing eight molecules. The unit cell formula is, therefore $M_8^2 + Fe_{16}^3 + O_{32}$. The metal cations are situated in 24 of the 96 available interstices between the 32 oxygen ions which are arranged on a more or less close packed face-centred cubic lattice. These 96 interstitial sites are divided into 64 tetrahedrally co-ordinated (*a*-) sites and 32 octahedrally coordinated (*b*-) sites. In general, 8 tetrahedral sites and 16 octahedral sites are occupied by the 24 metal ions, but which particular ions occupy *a*- and *b*-sites is a matter for conjecture.

In practice a *normal* spinel is defined as one in which the divalent cations occupy the 8 tetrahedral sites, the trivalent ferric ions occupy the 16 octahedral sites, and the site distribution may then be shown by grouping the cations in the molecular formula

$$[M^{2+}](Fe_2^{3+})O_4 \qquad \text{(normal spinel)}$$

where [ ] and () enclose cations occupying *a*- and *b*-sites respectively.

It is conceivable that, in the unit cell containing 8 M^{2+} and 16 Fe^{3+} ions, the 8 M^{2+} ions may change places with 8 of the 16 Fe^{3+} ions, thus giving rise to the *inverted* spinel structure:

$$[Fe^{3+}](M^{2+}Fe^{3+})O_4 \qquad \text{(inverted spinel)}$$

Most ferrites occur in the inverted spinel form (Mg, Co, Ni) this being a pre-requirement for ferrimagnetism to be observed. Some materials (e.g., $ZnFe_2O_4$), however, take the normal form, and many adopt a structure intermediate between inverted and normal in which the divalent M ions are distributed between the *a*- and *b*-sites. The distribution may be described by invoking a parameter δ, i.e., the fraction of divalent ions occupying tetrahedral (*a*-) sites

$$[M_\delta Fe_{(1-\delta)}](M_{(1-\delta)}Fe_{(1+\delta)})O_4 \qquad \text{(partially inverted spinel)}$$

An example of the partially inverted spinel structure is manganese ferrite $[Mn_{0.8}Fe_{0.2}]$ $(Mn_{0.2}Fe_{1.8})O_4$ ($\delta = 0.8$).

GARNETS

The generalised formula of the garnet ferrites if $R_3Fe_5O_{12}$, where R is a rare earth or yttrium ion, and the crystal structure is similar to the spinel structure in that the symmetry is cubic and there are 8 molecules per unit cell. The unit cell formula is, therefore, $R_{24}Fe_{40}O_{96}$. The structure conforms to space group No. 230-Ia3d, and in contrast to the spinels the cations are arranged on three independent sets of lattice positions, denoted by *c*- (dodecahedral), *a*- (octahedral) and *d*- (tetrahedral) sites respectively. Although the garnet structure is more complicated that the spinel, it does have the advantage that with relatively few but important exceptions, each cation is confined exclusively to one site location, and the structure may be represented as

$$\{R_3\}[Fe_2](Fe_3)O_{12} \qquad \text{(rare earth iron garnet)}$$

where {}, [ ] and () enclose cations situated at dodecahedral (*c*-), octahedral (*a*-) and tetrahedral (*d*-) sites respectively. In practice, only certain ions are able to occupy the *c*-sites as a result of the limited size of the dodecahedral interstices. The rare earths of lower atomic number (La–Pm) are consequently too large and the only ions which can, in fact, form the pure garnet are the rare earths of atomic number 62–71 (elements samarium to lutetium) and the pseudo-rare-earth element yttrium (atomic number 39). Of these pure materials, the most important are yttrium iron garnet (YIG)—$Y_3Fe_5O_{12}$, and gadolinium iron garnet (GdIG)—$Gd_3Fe_5O_{12}$, but the elements samarium, dysprosium, holmium and terbium are also used for certain applications.

HEXAGONAL FERRITES

These compounds are based upon three structures each having hexagonal symmetry, denoted by the symbols M, S and Y:

$$\text{M—Ba}\,Fe_{12}O_{19}$$
$$\text{S—}Me_2Fe_4O_8$$
$$\text{Y—}Ba_2Me_2Fe_{12}O_{22}$$

These three structures form the sub-lattices of hexagonal ferrite materials and may be combined in a number of ways, for example (M + S), (M_2 + S), (M + Y) and (M_2 + Y). Since they are of hexagonal symmetry, they display a usually large anisotropy, and the easy direction of magnetisation may lie along the hexagonal axis, or alternatively in the basal plane perpendicular to the hexagonal axis.

SUBSTITUTION IN FERRITES AND GARNETS

In a given spinel or garnet formulation it is always possible to replace some or all of the metal or iron cations with an alternative cation, in some cases magnetic, and in others not. This process of replacement of one cation by another is known as substitution. In the spinel structure there are two possibilities:

(*a*) Replacement of divalent M′ ions by other divalent ions M″.
(*b*) Replacement of trivalent iron ions by other trivalent ions M″.

$$M'Fe_2O_4 \xrightarrow[xM''']{aM''} M'_{1-a}M''_aFe_{2-x}M'''_xO_4$$

For example, in nickel ferrite (Ni Fe_2O_4), part or all of the nickel content may be replaced by cobalt

$$Ni_{1-a}Co_aFe_2O_4\text{—nickel–cobalt ferrite}$$

or part of the iron may be replaced by a trivalent non-magnetic element such as aluminium, chromium or gallium

$$Ni\,Fe_{2-x}Al_xO_4\text{—nickel–aluminium ferrite}$$

The resultant substituted ferrites are, in fact, equivalent to a solid solution of the two pure end members; thus nickel–cobalt ferrite with the substituent parameter (*a*) equal to 0.5 is equivalent to a solution of one part pure nickel ferrite and one part pure cobalt ferrite

$$2Ni_{0.5}Co_{0.5}Fe_2O_4 = NiFe_2O_4 + CoFe_2O_4$$

and in similar manner nickel aluminium ferrite is a solid solution of pure nickel ferrite and pure nickel aluminate ($NiFe_2O_4$ + Ni Al_2O_4).

The situation is very similar in the garnet structure, which contains three sets of distinct cation lattice sites

$$\underset{c-}{\{Y_3\}}\underset{a-}{[Fe_2]}\underset{d-\ \text{sites}}{(Fe_3)O_{12}} \qquad \text{Pure YIG}$$

It has previously been noted that certain rare earth elements may replace yttrium ions on *c*-sites. Additionally, it is possible to replace either *a*-site Fe or *d*-site Fe or both with certain non-magnetic trivalent elements to give a generalised substituted garnet formulation

$$\{Y_{3-a}R_a\}[Fe_{2-z}M^A_z](Fe_{3-x}M^D_x)O_{12}$$

where M^A and M^D are trivalent elements which display exclusive preference for *a*- and *d*-sites respectively. The most common rare-earth element (R) is gadolinium (Gd), although dysprosium, holmium and terbium are also used to some extent. Examples of M^A and M^D are indium and gallium respectively, indium having recently become of some importance. Finally, it is possible to choose a trivalent non-magnetic element (such as aluminium) which does not display an exclusive preference for one lattice site, but instead distributes itself between *a*- and *d*-sites according to a distribution coefficient f_t (fraction of total substituent occupying tetrahedral (*d*-) sites) which is similar to the coefficient (δ) used above to define the degree of inversion of a spinel, and which is itself a function of total substituent level (y)

$$Y_3Fe_{5-y}Al_yO_{12} = \{Y_3\}[Fe_{2-y(1-f_t)}Al_{(1-f_t)y}](Fe_{3-f_t y}Al_{f_t y})O_{12}$$

The different types of substitution which are possible in either the garnet or spinel structure therefore yield a vast number of different formulations. By choosing the appropriate amounts and types of substituent it is possible to adjust the magnetic and dielectric properties of a basic ferrite or garnet in order to realise almost any desired combination of properties.

'Synthetic' garnets. The considerations employed above have recently been extended to yield a fundamentally different family of magnetic garnet materials. In designing a substituted garnet system, it is necessary to fulfil three basic criteria: (1) the electronic structure of the ion must be suitable for it to occupy the desired lattice site, (2) the size of the ion must be compatible with the size of the lattice site, and (3) overall electronic charge balance must be maintained. In the garnets discussed above, all the cations have been trivalent and the molecule has always been such that there are eight trivalent cations coordinated by twelve divalent anions (oxygen), thus preserving charge balance. It is also possible to preserve charge balance by introducing divalent and pentavalent ions in the ratio 2 : 1 as replacements for trivalent species, and this may be done by substituting calcium for yttrium and vanadium for iron. Also trivalent bismuth may be substituted for any remaining yttrium:

$$\{Y^{3+}_{3-2x}Ca^{2+}_{2x}\}[Fe^{3+}_2](Fe^{3+}_{3-x}V^{5+}_x)O_{12}\text{—YCaVIG}$$
$$\{Bi^{3+}_{3-2x}Ca^{2+}_{2x}\}[Fe_2](Fe_{3-x}V_x)O_{12}\text{—BiCaVIG}$$

In addition to the substitution of calcium, vanadium and bismuth, all the previously mentioned substitutions for conventional garnets may also be made in these 'synthetic' systems, the resulting structures becoming extremely complex.

Important classes of ferrites

The multitude of different ferrite compositions discussed above may be classified very broadly in two groups, the *soft ferrites* which find their application at relatively low frequencies (up to several tens of megahertz) in inductor, transformer and memory devices, and the *microwave ferrites*, which are applied in a quite different way in the frequency range 1 to 40 gigahertz. These latter applications almost always involve the realisation of non-reciprocal networks in which the electromagnetic wave is wholly or partially propagated through the body of the ferrite, the latter being housed in a suitable waveguide, coaxial or stripline transmission line.

Soft ferrites are entirely from the spinel family, the principal types being manganese zinc ferrite, nickel zinc ferrite and magnesium manganese ferrite.

Manganese–zinc : ferroxcube A
Nickel–zinc : ferroxcube B
Magnesium–manganese : ferroxcube D

Microwave Ferrites include both spinel and garnet ferrites, and a much wider range of materials are in common use for microwave applications than is the case with the soft ferrites. Most important among microwave ferrites are the following:

Magnesium–manganese–aluminium spinel
Nickel aluminium (or chromium) spinel
Lithium spinel
Yttrium iron garnet
Yttrium aluminium iron garnet
Yttrium gadolinium iron garnet
Yttrium gadolinium aluminium iron garnet

Properties of Ferrites

The properties of any ferrite are determined by:
(1) Chemical composition including purity of starting materials.
(2) Method of manufacture.

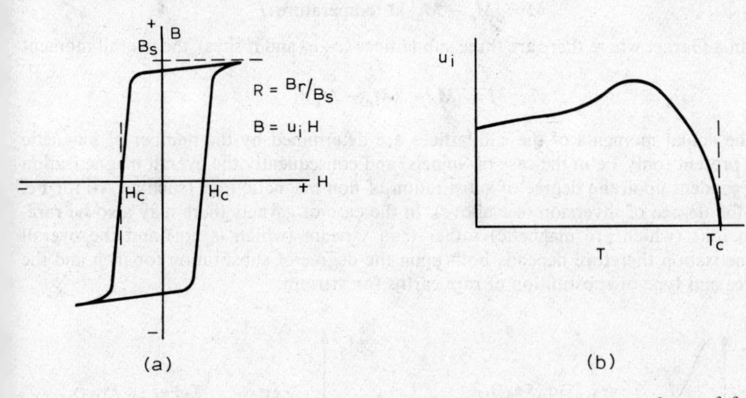

Figure 6.39 *(a) Hysteresis loop of a soft ferrite (b) Permeability* vs. *temperature for a soft ferrite*

Thus it is possible to achieve variations in properties by adjusting chemical composition and maintaining identical manufacturing conditions, by maintaining chemical composition and varying manufacturing conditions, or both.

The following parameters describe the most important properties of ferrites:

Soft ferrites (*Figure 6.39*)

		Range
Initial permeability $\mu_i = B/H$		5–40 000
Loss factor	$\tan \delta/\mu_i$	10^{-6}–10^{-2}
Temperature factor	$\frac{1}{\mu_1^2}\frac{d\mu_i}{dT}$	10^{-6}–10^{-4}/°C
Saturation flux density	B_{sat}	1 000–5 000 Gs
Coercive force	H_c	0.1–1.5 Oe
Curie temperature	T_c	100–500 °C

Microwave ferrites

		Range
Saturation magnetisation	$4\pi M_s$	50–7 000 Gs
Curie temperature	T_c	100–700 °C
Dielectric constant	$\epsilon = \epsilon' - j\epsilon''$	8–17

Dielectric loss tangent	$\tan\delta = \epsilon''/\epsilon'$	10^{-5}–3×10^{-3}
Resonance linewidth	ΔH	1–500 Oe
Gyromagnetic ratio	γ	1.8–2.2
Spin wave linewidth	ΔH_k	0.1–20 Oe
Remanence ratio	$R = B_r/B_{sat}$	0.6–0.99
Coercive force	H_c	0.5–3.0 Oe

SATURATION MAGNETISATION, CURIE TEMPERATURE AND TEMPERATURE FACTOR

The magnetisation at any given temperature is the resultant of the spin magnetic moments of all the magnetic ions on the crystal lattice. Recalling that in Spinels there are two sub-lattices (*a*-sites and *b*-sites) the overall magnetic moment is

$$M = |M_a - M_b| \text{ at temperature } T$$

and in a Garnet where there are three sub-lattices (*c*-, *a*- and *d*-sites), the overall moment is

$$M = |M_c - |(M_d - M_a)||$$

The actual moments of the sub-lattices are determined by the number of magnetic ions present (only Fe in the case of Spinels) and consequently the overall magnetisation is dependent upon the degree of substitution of non-magnetic ions (such as Al) for Fe, and the degree of inversion (see above). In the case of garnets there may also be rare-earth ions (which are magnetic) rather than yttrium (which is not) and the overall magnetisation therefore depends both upon the degree of substitution for iron and the degree and type of substitution of rare earths for yttrium.

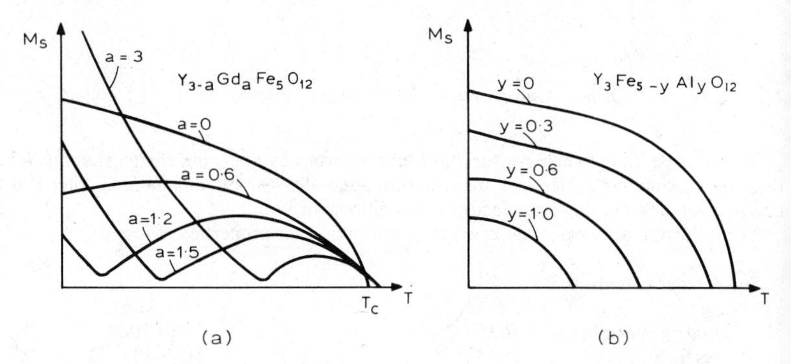

Figure 6.40 Magnetisation vs. temperature for a microwave garnet

The effective magnetic moment of all the ions on the lattice depends upon the exchange forces between the various species of ion and consequently temperature. The overall magnetisation therefore degrades as temperature increases until ultimately the curie temperature (T_c) is reached. In rare-earth containing garnets (such as Y Gd Al IG), the presence of two different magnetic species with different behaviour with respect to temperature can lead to a compensation point at a particular temperature whereat the sub-lattice magnetisations exactly cancel (see *Figure 6.40*). Magnetisation, curie temperature and temperature factor are all crucial parameters in the design of all microwave devices.

FERROMAGNETIC RESONANCE, RESONANCE LINEWIDTH AND GYROMAGNETIC RATIO

In a ferromagnet a spinning electron will precess about the direction of an applied static d.c. field in sympathy with an incident circularly polarised r.f. wave whose sense of polarisation is the same as the sense of precession. Under these circumstances the precession is sustained by energy coupled from the microwave field, and resonance occurs when the microwave and precession frequencies are the same. The resonance frequency (ω) is defined by the strength of the internal d.c. field (H_i) in the ferrite

$$\omega = \gamma H_i \tag{1}$$

where γ = gyromagnetic ratio.

The internal field (H_i) is a function of the externally applied d.c. field (H_a) which is modified by the oppositely directed induced field in the sample. This internal demagnetising field (H_d) is a function of sample shape and magnetisation

$$H_d = N4\pi M_s$$

where N is the shape demagnetising factor.

The shape factor may conveniently be resolved into components N_x, N_y, N_z such as that $\Sigma N_i = 1$ and the internal d.c. field required for resonance in the ferrite is then

$$H_i = \{(H_A - (N_z - N_x)4\pi M_s)(H_A - (N_z - N_y)4\pi M_s)\}^{\frac{1}{2}}$$

Substitution of this expression in equation 1 then gives the Kittel equation for ferromagnetic resonance

$$\omega = \gamma\{(H_A - (N_z - N_x)4\pi M_s)(H_A - (N_z - N_y)4\pi M_s)\}^{\frac{1}{2}} \tag{2}$$

In addition to the demagnetising effect due to sample shape, there is a second source of modification to the internal field which may be included in the resonance equation. This analogous demagnetising field arises from the magnetic anisotropy of the crystal structure, whereby the magnetic behaviour varies with direction in the crystal lattice. Magnetocrystalline anisotropy may be expressed in terms of the energy difference which arises from deviations of the magnetisation M from the direction of easy magnetisation in the crystal. The internal field due to magnetocrystalline anisotropy is

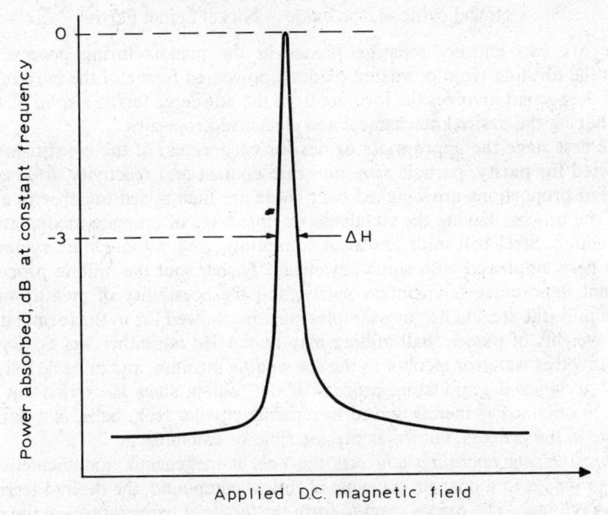

Figure 6.41 Resonance absorption in a microwave garnet

$H_{An} = 2K_1/M_s$ where K_1 is the effective first order anisotropy constant. When this result is inserted into equation 2 the resonance condition becomes

$$\omega = \gamma\{(H_A + H_{An} - (N_z - N_x)4\pi M_s)(H_A + H_{An} - (N_z - N_y)4\pi M_s)\}^{\frac{1}{2}} \qquad (3)$$

When this equation is satisfied, resonance occurs and maximum power is absorbed by the ferrite from the r.f. wave. Non-reciprocal action arises because an r.f. wave whose circular polarisation is of opposite sense to the precession direction will suffer no loss of power since there will be no coupling. If power absorption is plotted as a function of H_A for constant ω the resonance absorption line is obtained, as shown in *Figure 6.41*. The resonance line has a finite breadth ($\triangle H$) which is of considerable significance in device design. This parameter is usually measured at the half power point on the resonance line, giving rise to the '3d*B* linewidth'.

SPIN WAVE LINEWIDTH

Spinwave Linewidth is a measure of a materials' tolerance to high microwave power levels. This parameter is of fundamental importance for high power Radar Systems applications, and a thorough treatment may be found in reference 1.

REMANENCE RATIO AND COERCIVE FORCE

These are the two parameters describing the hysteresis loop (see *Figure 6.39 (a)*), and are of importance only in switching and latching applications.

Manufacture of ferrites

A ferrite, of whatever type, is a 'mixed oxide' chemical compound, and may therefore be prepared from the appropriate physical mixture of component oxides.

Thus $$NiO + Fe_2O_4 \rightarrow Ni\,Fe_2O_4$$
Nickel oxide + Iron oxide → Nickel Spinel Ferrite

There are two entirely separate phases in the manufacturing process; the first involves the forming from powdered oxides a powdered ferrite of the correct composition and the second involves the forming from the powdered ferrite a solid block of high density having the desired mechanical and electrical properties.

In the first stage the appropriate oxides (or carbonates) of the constituent elements are selected for purity, particle size, moisture content and reactivity, following which the desired proportions are weighed out. These are then mixed together in a mill, this stage of the process having the simultaneous objectives of intimate mixing and particle size reduction. Steel ball mills are most commonly used, although air cyclone milling has also been employed with some advantage. Throughout the milling process it is of paramount importance to maintain purity, and the possibility of pick-up of material from the mill (the steel balls, for example) must be allowed for in the formulation of the original weights of oxides. Ball-milling may be carried out either wet or dry. The wet process involves water or alcohol as the suspending medium, and must be followed by a filtering, drying and granulating process. If the milling stage is carried out 'dry', the mixture so obtained is merely sieved to separate powder from balls, and passes to the next stage in the process, known as presintering, or calcining.

The *presintering operation* converts the fine, homogeneous non-magnetic physical mixture of oxides to a coarser magnetic chemical compound, the desired ferrite, by the application of heat. The oxides react to form the ferrite at temperatures in the range 800 to 1 200 °C, and the process is accomplished in a high temperature static or travelling kiln, whereby the mixture dwells for between 1 and 12 hours at the chosen peak temperature. At these elevated temperatures it is possible for some of the oxides to dissociate, thereby losing oxygen, and altering the valence state of the metal cations. This effect will alter drastically the properties of the ferrite, and may be enhanced or suppressed by control of the furnace atmosphere as the presintering cycle proceeds. The atmosphere may be rendered more oxidising by the introduction of pure oxygen, or more reducing by pure nitrogen, depending upon the desired properties of the final ferrite.

The coarse ferrite powder which results from presintering must usually be reground in a similar operation to the initial milling, before proceeding to the second phase of the process whereby the powder is formed and sintered into a solid block of high density.

The *green-forming* stage of a ferrite process is that in which the powdered ferrite is pressed, or extruded, into a part of the desired shape and having a density between 40 and 60% of the final sintered density. A wide variety of techniques are employed to achieve the correct shape—die-pressing, isostatic pressing and extrusion are all commonly used. The powder characteristics of flow, packing density, particle size and moisture content are crucial in obtaining a satisfactory moulding, and to this end a powder preparation stage is often employed. This may simply involve the addition of a binder/lubricant such as polyvinyl alcohol, or may employ more sophisticated techniques of spray or freeze drying. Three methods of green-forming are now discussed.

(1) *Die-pressing* is perhaps the most commonly used green-forming method and has the advantages that it produces a part of closely defined shape, but of somewhat uncontolled density, and it can readily be made automatic.
(2) *Iso-static* pressing is also popular, particularly for parts of complex shape, and has the advantage that, since the part is moulded in a flexible mould immersed in a pressurised liquid, the pressure acts on the powder uniformly in all directions. The part so formed therefore has a uniform density, but its final shape is not so closely defined.
(3) *Extrusion* of a thick ferrite slurry is an extremely economical method forming simple rods, bars and tubes of rather poorly defined mechanical properties.

Following green forming, all that remains is to *sinter* the part to a high density. This is accomplished in a static or travelling hearth kiln at a temperature rather higher by several hundred degrees than that used for pre-sintering. During this process the grains of the ferrite grow, and the air-porosity in the low density green moulding is eliminated, partially or completely, by vacancy diffusion. This is accompanied by shrinkage, during which the part gradually assumes its final dimensions, and it is at this stage also that the final properties of the material are determined. Control of this process to determine final density, grain size distribution, porosity distribution and chemical composition (by atmosphere control) is therefore of critical importance.

Having cooled, the ferrite parts are unloaded from the furnace, and are sampled for test purposes, where a variety of intrinsic and bulk parameters are measured as outlined above. The piece parts may well then be subjected to a machining process to achieve the precise shape and tolerances required by the particular application for which the material is intended.

Applications

SOFT FERRITES

The manganese-zinc and nickel-zinc ferrites find their principle applications in inductors and transformers, and are generally characterised by their high resistivity and the consequent reduction of eddy current losses. Permeability, loss factor, temperature factor, Curie temperature and resistivity all vary from one composition to another within the two families (ferroxcube A and ferroxcube B), but these two classes are distinguished one from the other as follows:

Ferroxcube A—high permeability, low Curie temperature and high temperature factor.
Ferroxcube B—lower peameability, high Curie temperature and low temperature factor.

Grade A is therefore mostly used at lower frequencies up to a maximum of 2 MHz for pot cores, transformers, deflection yokes and toroidal inductors. Grade B is suitable for high frequency applications between 1 and 20 MHz in tuned inductors, transformers and aerial rods.

The magnesium-manganese ferrites (ferroxcube D) are notable for their square hysteresis loop and high remanence ratio. In non-microwave applications they are therefore used extensively in switching and storage applications by utilising the positive and negative remanent states of a toroid to represent binary information of the form 'O' and 'I'.

A comprehensive treatment of the applications of soft ferrites may be found in reference 2.

MICROWAVE FERRITES

The applications of microwave ferrites are nearly all based upon the differential susceptibility (X) which these materials display with respect to the two directions of a microwave signal propagating through the ferrite (*Figure 6.42*).

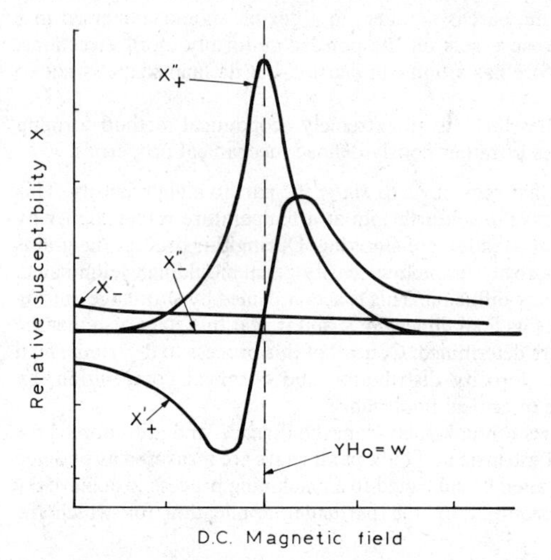

Figure 6.42 Susceptibility of a microwave ferrite

For a given frequency (w) there is a value of applied d.c. field at which resonance will occur. At this field H_{Res} the ferrite material will absorb a microwave signal of one polarisation, while leaving a signal of opposite polarisation almost unattenuated. This effect immediately gives rise to the Resonance Isolator, a device which passes a signal in one direction with low attenuation, while absorbing a signal travelling in the reverse direction.

The differential susceptibility may be used at a level of applied d.c. field other than resonance to realise non-resonant devices including junction circulators, phase-shifters and differential phase shift circulators. These devices are illustrated in *Figure 6.43*.

(*a*)

(*b*)

Figure 6.43 (*a*) *Miniature ferrite isolators and circulators* (*b*) *High power S-band ferrite isolator* (*Courtesy The Marconi Company Ltd*)

Microwave ferrite devices find many applications in radar, communications and microwave power systems. The function of *isolators* is always to protect a signal source or a stage in a system against reflected signals from subsequent stages in the system. *Circulators* are used most often in order to realise some form of signal routing or branching, for example when a common aerial is used both to transmit and receive (*Figure 6.44*), or in filter branching networks.

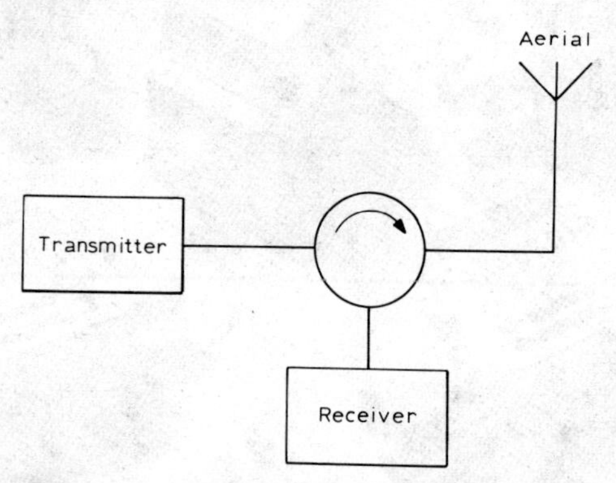

Figure 6.44 Application of a circulator in a common aerial system

Phase shifters, which may be reciprocal or non-reciprocal, are used to introduce variable phase shift into a signal path, and are now finding extensive application in phased-array antennas (reference 3).

A general review and description of microwave ferrite materials may be found in reference 4 and of microwave ferrite components and applications in reference 5.

REFERENCES

1. LAX and BUTTON, *Microwave Ferrites and Ferrimagnetics*, McGraw-Hill (1962)
2. SNELLING, E. C., *Soft Ferrites*, Illiffe (1969)
3. WHICKER, L. R., 'Present Status of ferrite phase shifter technology in U.S.A.', *Proc. of 1973 European Microwave Conference*, I.E.E.

   INCE, W. J., 'Recent Advances in Diode and Ferrite Phaser Technology for Phased-Array Radars', *Microwave J.*, Sept. and Oct. (1972) (in two parts).
4. VON AULOCK, ed., *Handbook of Microwave Ferrite Materials*, Academic Press (1965)
5. HELSZAHN, J., *Principles of Microwave Ferrite Engineering*, Wiley (1969)

7 ELECTRON VALVES AND TUBES

ELECTRON VALVES AND TUBES 7

7 ELECTRON VALVES AND TUBES

SMALL VALVES

The operating principle of all thermionic valves is the controlled flow of electrons in a vacuum from a heated cathode to a positively-charged anode. Intermediate electrodes can be placed between the anode and cathode to control the electron flow. In small valves, these intermediate electrodes are generally in the form of fine-wire meshes, and are called grids. The flow of electrons through the grid can be controlled by the magnitude and polarity of the voltage applied to it.

Development of the thermionic valve

The first thermionic valve was the diode patented by J. A. Fleming in 1904. Fleming's diode arose out of work on the Edison Effect some two decades earlier.

Edison's incandescent lamp consisted of a carbon filament sealed inside an evacuated glass bulb, the filament being heated to incandescence by a d.c. current. After the lamp had been in use for some time, however, it was found that the inside of the bulb became coated with a thin dark layer. Because this layer decreased the amount of light from the bulb, Edison investigated ways of preventing it forming. In one experiment, in 1883, he placed a metal plate close to the filament inside the bulb. He found that if the plate was connected to the positive terminal of the filament through a galvanometer, and the filament was glowing, a current flowed across the vacuum between the filament and the plate. If the plate was connected to the negative terminal of the filament, or if the filament was not glowing, no current flowed. Edison was unable to explain the effect in terms of the current knowledge, and so he reported his observations and continued his work on improving the incandescent lamp.

J. A. Fleming, who at the time was 'electrician' to the Edison Electric Light Company in the U.K., visited Edison in 1884 and was shown this effect. On his return to England, Fleming repeated Edison's experiments and confirmed his results. He also obtained the effect with a platinum filament replacing the carbon filament.

During 1884 and 1885, other workers continued to investigate what had become known as the Edison Effect. Preece (in the U.K.) investigated the voltage/current characteristics of the structure, and Hittorf and Goldstein (in Germany) investigated the effects of different filament materials and working temperatures (that is, thermal emission). About 1894, J. A. Fleming, then Professor of Electrical Engineering at University College London, attempted to explain the conduction of electricity through air at low pressure, but made little progress until J. J. Thompson postulated the existence of the electron in 1897. Papers read by Fleming to the Royal Institution in 1890 and to the Physical Society in 1896 showed that although he was not able to explain the Edison Effect correctly, he was aware that 'the space between the filament and the metal plate is a one-way street for electricity'. From Thompson's work on the electron, O. W. Richardson in 1901 developed a complete explanation of the mechanism of thermal emission. Marconi had demonstrated the feasibility of 'wireless' telegraphy by transmitting across the English Channel in 1899, and in the next few years developed marine wireless telegraphy into a practicable system. This development, however, revealed the need for a more effective detector of wireless signals than the existing devices such as the coherer. In November 1904, Fleming patented an improved structure based on the Edison lamp and plate as a detector for wireless telegraphy. The action of the structure in allowing electrons to flow when the plate was positive, but not allowing them to flow when the plate was negative, was compared by Fleming to the action of a mechanical valve. Since the structure contained two electrodes, it was called a diode valve. A photograph of Fleming's diode is shown in *Figure 7.1*.

The next development of the thermionic valve occurred in the U.S.A. in 1906 when Lee de Forest added a third electrode to the valve. This was the grid, placed between the anode and cathode, which enabled the electron flow between cathode and anode to be

Figure 7.1 Fleming diode, 1904
(Photo: Mullard Limited)

controlled. De Forest called his device the 'Audion'. This triode valve could be used as an amplifier and oscillator as well as a detector, and so opened the way for the future development of electronics.

De Forest continued his development of the triode until 1912. In the meantime, C. D. Child in 1911 made a theoretical analysis of the diode. A further analysis, more related to practical diodes, was made by I. Langmuir in 1913 which generally agreed with that of Child. The result of these analyses was the Child–Langmuir expression relating the anode current and voltage in the diode. Another development in the U.S.A. was made by E. H. Armstrong in 1914 who showed that the presence of small traces of gas in a thermionic valve was not essential for its operation, and Langmuir showed how a higher vacuum could be produced. The resulting 'hard' valves had a superior performance to the previous 'soft' valves.

World War I resulted in the widespread use of 'wireless' for military signalling, with consequent improvements to the valves. By the end of the war, triodes could be produced for operation in transmitters and receivers, and their theory and operation were well understood. The post-war developments were designed towards increasing the amplification and reducing the power consumption, and to increase the frequency of operation.

The maximum frequency at which the triode could be used as an amplifier was limited by the interelectrode capacitance to approximately 4 MHz, the Miller effect. External neutralising circuits, such as the Neutrodyne circuit developed in the U.S.A., were used to balance the interelectrode capacitance. These circuits were not completely satisfactory, particularly when the triode was operating over a wide frequency range. It was suggested by W. Schottky in the U.S.A. in 1919 and by A. W. Hull in the U.K. that an additional grid could be used to shield the control grid from the anode, and so reduce the capacitance. 'Screen grid' valves devised by H. J. Round in the U.K. were introduced in 1926, and were called tetrodes. The tetrode, however, suffered from secondary emission caused by the impact of electrons on the anode after being accelerated by the screen grid. This limited the use of the tetrode. The effect of the secondary emission was overcome by inserting a third grid, the suppressor grid, between the screen grid and the anode. This structure formed the pentode valve, devised by Tellegen and Holst in Holland, and introduced in 1928.

The development of broadcasting for domestic entertainment throughout the 1920s also had important effects on the thermionic valve. The early valves had been manufactured by electric lamp makers, mainly because only they had the plant for producing glass bulbs and for evacuating the assembled valve. After World War I, manufacturers concerned only with valves came into business. The increased demand for valves for use in domestic 'wireless sets' meant increased production and improved methods of manufacture, while the receiver manufacturer demanded improved performances from the valves.

One of the first improvements made to the post-war valves was the reduction of the heater power required. The early valves used filaments made from tungsten wire. These filaments had to operate at high temperatures, approximately 2 200 °C, to obtain the required electron emission. Such a temperature caused mechanical distortion of the filament, and required a large interelectrode spacing to avoid the risk of short-circuits. The large interelectrode spacing is shown in *Figure 7.2*; the valve is a triode from 1920. The working temperature was reduced by the introduction of thoriated tungsten filaments, to approximately 1 600 °C, and further reduced by the oxide-coated filament. The oxides were those of rare earth metals, generally of barium and strontium, and the filament operated at a dull red heat, approximately 700 °C. The oxide-coated filament had been introduced in the U.S.A. in 1918, although the original investigation had been carried out by Wehnelt in Germany in 1904, and it first appeared in valves in the U.K. in 1921. Not only did the lower temperature mean a decrease in the heater power, but the reduced mechanical distortion meant decreased interelectrode spacing and therefore improved performance of the valve. (The slope g_m of a triode is inversely proportional to the grid-to-cathode spacing.) The indirectly-heated cathode, again developed in the U.S.A. and introduced into the U.K. in the mid-1920s, eliminated problems from filament distortion. The filament was now enclosed in a nickel cylinder coated with the emissive oxides so that the electrons were emitted from a rigid surface. The indirectly-heated cathode also enabled the heater supplies to be taken from the a.c. mains supply rather than an accumulator or battery.

By the beginning of the 1930s, the wireless valve had assumed a fairly standard form and method of construction. The electrode structure was mounted vertically between two metal (later mica) discs. The grids were approximately rectangular in cross-section, being wound around two supporting rods the required distance apart. The ends of the rods fitted into insulated holes in the upper and lower discs, but much of the strength of the grid depended on the wire from which it was wound. The grids were surrounded by the cylindrical or rectangular anode which was welded to two supporting rods passing through the lower disc to mount the complete electrode structure on the 'pinch'. The pinch was a flattened glass tube which also contained the connecting wires between the electrodes and the pins in the valve base, and the pumping stem to evacuate the assembled valve. The pinch was fused to the glass bulb, the valve evacuated, and the pumping stem sealed. The valve base was cemented to the bulb, and the connecting wires soldered to the pins.

Figure 7.2 Typical valve of the 1920s. Mullard ORA valve, 1920–22 (Photo: Mullard Limited)

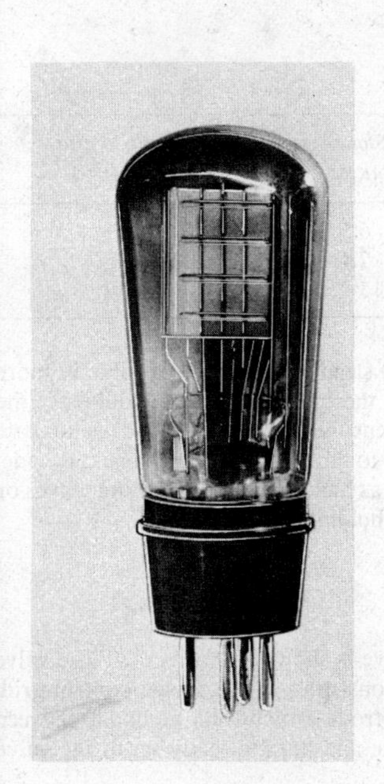

Figure 7.3 Typical wireless valve of the 1930s (Photo: Mullard Limited)

During the 1930s, more complex multigrid valves were developed. The hexode, heptode, and octode were introduced, and combinations of valves such as triode–hexodes and triode–heptodes enclosed in the same bulb. A typical wireless valve of the 1930s is shown in *Figure 7.3*. The use of the valve in telephony and telegraphy also had important results on design. The reliability required of valves in these applications, and the smaller size, led to design improvements that were incorporated into wireless valves. In the mid-1930s, a cylindrical glass envelope was introduced, and the mica discs supporting the electrode structure were extended to make contact with the inside of the glass to give further support. Wireless manufacturers were quick to appreciate the advantages of smaller valves to enable more compact and portable sets to be made. The introduction of television in the U.K. in 1936 also assisted the demand for smaller valves to keep the receiver as compact as possible.

Towards the end of the 1930s, the construction of the valve was changed. The pinch was removed, and the electrode structure was mounted directly on the valve pins which were fitted in a glass baseplate. This was the form of the radio valve used so extensively during World War II.

The post-war years saw a vast increase in the use of electronics in fields other than radio—in fact the radio valve became the electronic valve. And the boom in television in the 1950s as the U.K. service was extended to cover the whole country, and most European countries inaugurated services, led to further improvements to the valve. The operating frequency of valves for domestic use was increased, from the medium-wave broadcasting band to v.h.f. Television receivers became more complex, and more gain was demanded of the valves so that the number of stages in the receiver could be reduced. Thus even closer spacing of cathode and grid was required, and this led to the introduction of techniques like the frame grid. In this, the supporting rods for the grid were supplemented by cross-pieces which formed a rigid frame of the required size. Little support was required of the grid wire itself, and so it could be considerably reduced in diameter. An illustration of the improvements in valve design between pre-war and post-war valves is given in *Table 7.1*.

Table 7.1

Valve type	*Date of introduction*	*Slope* (mA/V)	*Cathode-to-grid spacing* (μm)
EF50	1939	6.5	150
EF80	1950	7.4	110
EF184 (frame grid)	1959	15	50

During the 1960s, the transistor (see Section 8) steadily replaced the valve in more and more applications. By the end of the 1960s, the transistor was undoubtedly the general-purpose electronic device. New equipment is designed around solid-state devices, and production of small valves is expected to fall steadily over the next decade. It is only in the more specialised applications, such as high-power transmitting valves or the microwave valves, that the thermionic valve is holding its own.

Construction of small valve

A cut-away drawing of a typical modern small valve is shown in *Figure 7.4*. The valve shown is a pentode, with an electrode structure consisting of a cathode, control grid, screen grid, suppressor grid, and anode. The electrode structure is mounted between two mica discs, the top mica and the bottom mica, and fitted directly on to the valve pins.

The valve uses an indirectly-heated cathode. The heater is formed by a filament of tungsten wire which is coated with aluminium oxide (alumina) to insulate it electrically from the cathode. The cathode itself is a nickel cylinder coated on the outside with the emissive material, an equal mixture of barium oxide and strontium oxide. The work

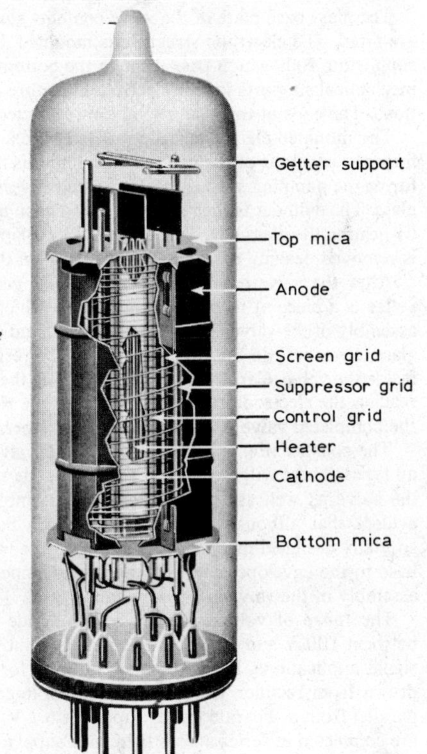

Figure 7.4 Construction of modern pentode (Photo: Mullard Limited)

function ϕ for this oxide-coated cathode is 1.0 eV, and the emissive power efficiency expressed as the electron emission current with heater power is typically 800 mA/W. The operating temperature of the cathode is approximately 700 °C.

The indirectly-heated cathode is used with a.c. heater supplies. It has the advantage that as the cathode is electrically insulated from the heater, the cathode can be maintained at a potential different from that of the heater. The other type of cathode, the directly-heated cathode, has the emissive coating on the heating filament and is used with d.c. heater supplies. This type has the disadvantage that the cathode potential in the valve is always that of the heater supply, and that the potential will vary along the length of the cathode.

The cathode containing the heater is located into a hole in the bottom mica. The control and screen grids, which are roughly rectangular in cross-section, are formed by winding fine wire around two supporting rods the required distance apart. The wire is wound to a small pitch, and is secured at each turn round the supporting rods. The grids are made in long lengths which, after winding, are cut to the length required for the valve. The ends of the supporting rods of the completed grids are inserted into holes in the bottom mica to locate them correctly in relation to the cathode. The suppressor grid is wound to a larger pitch around its supporting rods, and these are located in holes in the bottom mica. Finally the anode, which is a nickel cylinder, is placed over the grids and positioned by lugs which fit into slots in the mica. The electrode assembly is completed by fitting the top mica. Locating holes and slots for the supporting rods and anode lugs are provided, and so the completed electrode structure is held rigidly between the two micas to maintain the correct electrode spacing.

The glass base plate of the valve contains glass-to-metal seals in which the valve pins are fitted. The electrode structure is mounted directly on the valve pins by welding the supporting rods which pass through the bottom mica to the appropriate pins. Thus the mechanical supports for the electrode structure are combined with the electrical connections. The ends of the heater wire are connected to the appropriate pins.

The mounted electrode structure and base are fitted into a glass cylinder which will form the envelope of the valve. The cylinder is fitted with a smaller-diameter tube which forms the pumping stem. A seal is made between the cylinder and the base by fusing the glass. The cylinder is then evacuated by a vacuum pump, and the pumping stem is sealed by heating the glass near the valve until it collapses to form the seal. The rest of the stem is removed, leaving the characteristic 'pip' on the top of the valve shown in *Figure 7.4*.

After the valve has been evacuated, it is *gettered* to remove any residual gas. The getter is a piece of magnesium or barium which is placed on the getter support during assembly of the valve. The getter is ignited and evaporates over the inside surface of the glass, absorbing in the process most of the residual gas atoms to improve the vacuum inside the valve. Care is taken in positioning the getter to ensure that as little as possible reaches the electrode structure to prevent the risk of short-circuits. The pressure inside the completed valve is approximately 10^{-4} torr.

The construction and assembly of small valves as outlined briefly above is typical of all types. There will be differences in electrode structure depending on the application of the valve as well as the type, but the assembly will follow the lines described. It is evident that although some operations can be carried out on batches of valves on specially-designed machines (for example, the vacuum pumping, the sealing of the valve base to the envelope, and the sealing and removal of the pumping stem), much of the assembly of the valve is done by hand.

The range of voltage applied to the anode of a typical signal-processing valve is between 100 V and 250 V. The anode current is between 5 mA and 50 mA for small-signal applications, but as high as 500 mA for output valves. Higher currents can be drawn from rectifier valves. The heater voltage is standardised for valves operated in parallel from a separate heater supply at 6.3 V. In applications where the valve heaters are connected in series across the mains supply, the heater voltage can be higher, say up to 20 V. The heater current is typically between 100 mA and 300 mA.

Diode

The diode is the simplest thermionic valve, containing two electrodes, the anode and the cathode. The circuit symbols for the diode are shown in *Figure 7.5*; (*a*) represents the directly-heated type, while (*b*) represents the type with an indirectly-heated cathode.

When the anode is positive with respect to the cathode, the electrons emitted from the

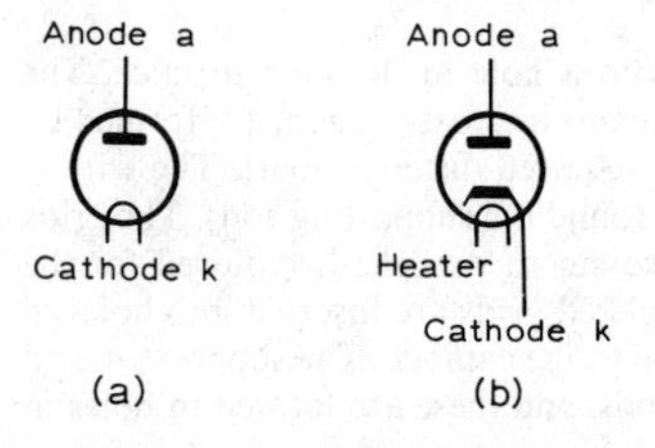

Figure 7.5 Circuit symbol for diode valve: (a) directly-heated cathode (b) indirectly-heated cathode

cathode are attracted to the anode. This electron flow from cathode to anode creates a current through the valve, and so allows a current to flow through a circuit of which the diode is a part. When the anode is negative with respect to the cathode, the emitted electrons are repelled and no current flows. The characteristic of anode current I_a plotted against anode voltage V_a is shown in *Figure 7.6*.

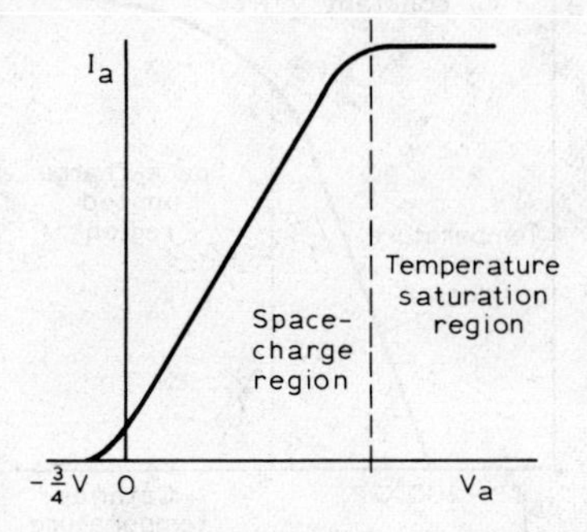

Figure 7.6 Voltage/current characteristic of diode valve

With anode voltages more negative than approximately $-\frac{3}{4}$ V, no current flows. When the anode voltage is highly positive, the anode current remains constant despite any further increases in the anode voltage. A saturation value occurs at which all the emitted electrons are attracted to the anode. In this saturation region, changes of anode current can only occur through changes in the number of electrons emitted, caused by changes in the cathode temperature. The region is therefore called the temperature-saturation region. (The voltage/current characteristic of *Figure 7.6* is plotted for a constant cathode temperature.) The small anode current flowing when the anode voltage is slightly negative occurs because some electrons have sufficient energy when emitted from the cathode to overcome the repulsion of the slightly-negative anode. The part of the characteristic below the temperature-saturation region is called the space-charge region, and in this the relationship between the anode current and voltage is given by

$$I_a \propto V_a^{\frac{3}{2}}$$

This relationship is known as Child's Law or the Child–Langmuir Law. The relationship can be rewritten as

$$I_a = GV_a^{\frac{3}{2}}$$

where G is a constant called the permeance of the valve.

The variation of anode current with cathode temperature for a constant anode voltage is shown in *Figure 7.7*. Thermionic emission occurs when the cathode temperature exceeds approximately 200 °C, and the anode current initially increases linearly with temperature. Above a certain temperature, however, the current does not increase so rapidly and tends towards a saturation value. This saturation is caused by a cloud of electrons forming near the cathode when the rate at which electrons are emitted from the cathode exceeds the rate at which they travel to the anode. This cloud of electrons has a negative charge, and so is called a space charge. The space charge repels electrons emitted from the cathode so that only those emitted with high energies are able to overcome the repulsion. For a constant anode voltage (provided it is not too high), a dynamic equilibrium is reached where the number of electrons leaving the space charge is balanced by the number of electrons leaving the cathode. In other words, the anode current is determined by the space charge rather than the cathode temperature.

The effect of the space charge on the distribution of potential between the cathode and anode is shown in *Figure 7.8*. The broken line represents the potential distribution when no electrons are emitted from the cathode. It follows the expected linear fall of potential with distance from the positive anode to the cathode at earth potential. When electrons are emitted and the space charge has formed, the interelectrode potential is shown by the full line. The minimum in the curve corresponds to the position of the space charge, which forms a virtual cathode.

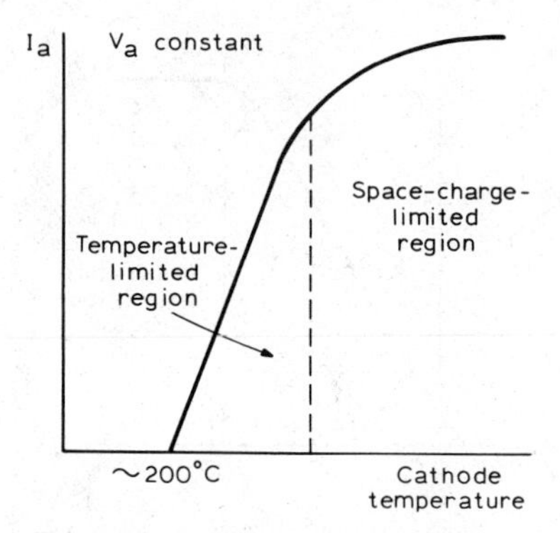

Figure 7.7 *Anode current plotted against cathode temperature for diode valve*

Figure 7.8 *Potential distribution in diode valve: broken line, no electron emission; full line, with electron emission showing effect of space charge*

If the anode voltage is increased so that the diode is operating in the temperature-saturation region, all the electrons are attracted to the anode. The space charge then disappears.

The property of the diode in allowing a current to flow through it in one direction only, enables it to be used in circuits as a rectifier and as a demodulator for amplitude-modulated carriers. In both applications, the thermionic diode has been replaced by the semiconductor diode.

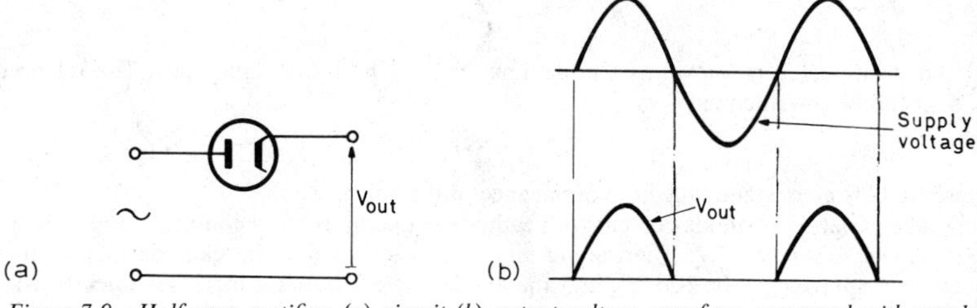

Figure 7.9 *Half-wave rectifier: (a) circuit (b) output voltage waveform compared with supply voltage*

The circuit of a half-wave rectifier is shown in (*a*) of *Figure 7.9* and the output waveform in relation to the supply voltage waveform is shown in (*b*). The diode conducts the positive half-cycle of the supply, and blocks the negative half-cycle. If two diodes are used, with their anodes supplied with voltages 180° out of phase and the cathodes connected together, a full-wave rectifier is obtained. The circuit of a full-wave rectifier is shown in (*a*) of *Figure 7.10* and the voltage waveforms in (*b*). The anodes of the diodes are supplied through a centre-tapped transformer, and conduct on alternate half-cycles. The combined cathode voltages therefore form a series of unidirectional

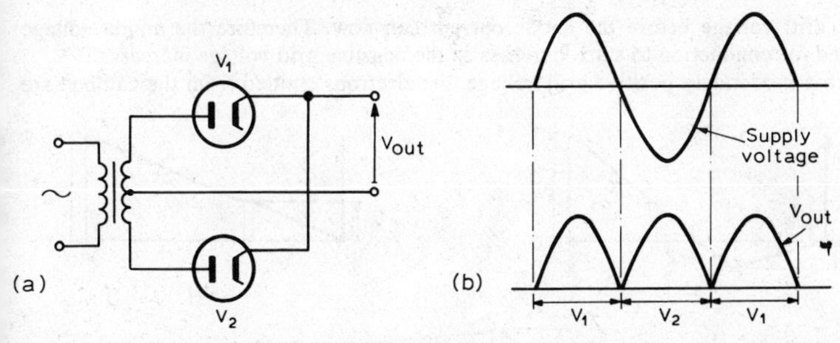

Figure 7.10 Full-wave rectifier: (a) circuit (b) output voltage waveform compared with supply voltage

half-cycles. The pulsating d.c. voltage can be smoothed by a suitable *LC* circuit to provide a constant voltage, as in a practical power supply circuit. The two separate diodes were often combined in one envelope to form a double diode. The circuit symbol for a double diode is shown in *Figure 7.11*.

The published data for a thermionic diode included the characteristic of anode voltage and current. The ratings specified the maximum reverse voltage that could be applied across the diode, called the peak inverse voltage or p.i.v., and the maximum current that could be drawn from the valve.

Triode

The triode contains three electrodes; a grid is placed between the anode and the cathode. The circuit symbol for a triode is shown in *Figure 7.12*.

In the valve, the grid is formed of fine wires placed nearer to the cathode than to the anode. Thus a relatively small voltage on the grid can have a significant effect on the electron flow to the anode, and therefore on the anode current. A schematic cross-section of the triode is shown in *Figure 7.13*.

If no voltage is applied to the grid, the electron flow between the cathode and the positive anode is not affected, and the triode acts as a diode. If the grid is made extremely negative, the electrons from the cathode are repelled by the grid, and no anode current flows. Between these two extremes, the variation of a moderately negative voltage on the grid can be used to control the anode current. The anode current is inversely proportional to the magnitude of the negative grid voltage. This is the normal operating region for the triode.

The distribution of potential between anode and cathode with a moderately negative grid voltage is shown in (*a*) of *Figure 7.14*. Because the negative grid repels electrons emitted from the cathode, the anode voltage must be high enough to neutralise the effect

Figure 7.11 Circuit symbol for double diode

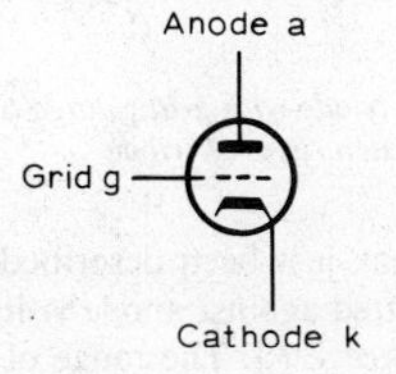

Figure 7.12 Circuit symbol for triode valve

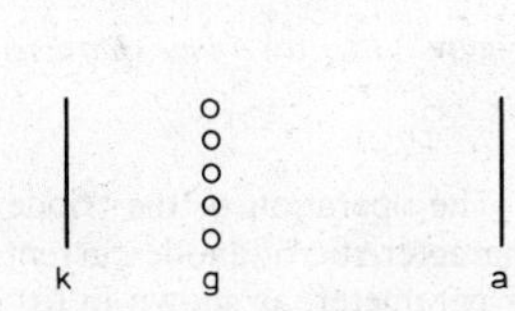

Figure 7.13 Schematic cross-section of triode

of the grid voltage before the anode current can flow. Therefore the anode voltage required for conduction to start increases as the negative grid voltage increases.

With a moderately positive grid voltage, the electrons emitted from the cathode are

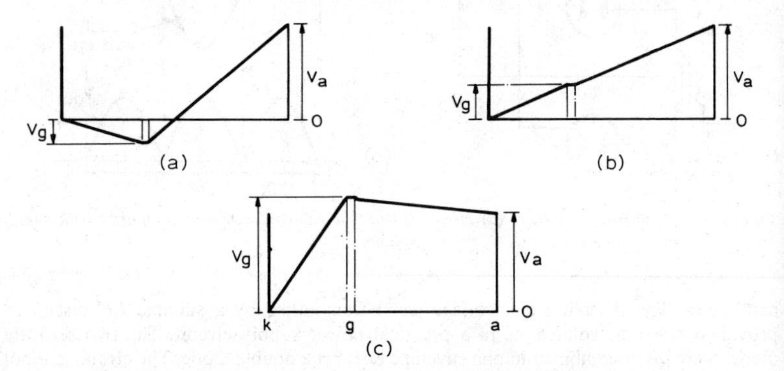

Figure 7.14 *Potential distribution in triode: (a) grid moderately negative (b) grid moderately positive (c) grid extremely positive*

accelerated towards the anode. As the grid voltage adds to the anode voltage, conduction can start even with a slightly negative anode voltage. The distribution of potential between anode and cathode with a moderately positive grid voltage is shown in (*b*). A small grid current will flow when the grid voltage is less than $-\frac{1}{2}$ V.

If the grid voltage is made extremely positive, the grid acts as an anode. Electrons emitted from the cathode flow to the grid. Even those emitted with sufficient energy to pass through the grid can be attracted back to it rather than continue towards the anode. The potential distribution for a highly positive grid voltage is shown in (*c*). A high grid current, higher than the anode current, will flow under these conditions.

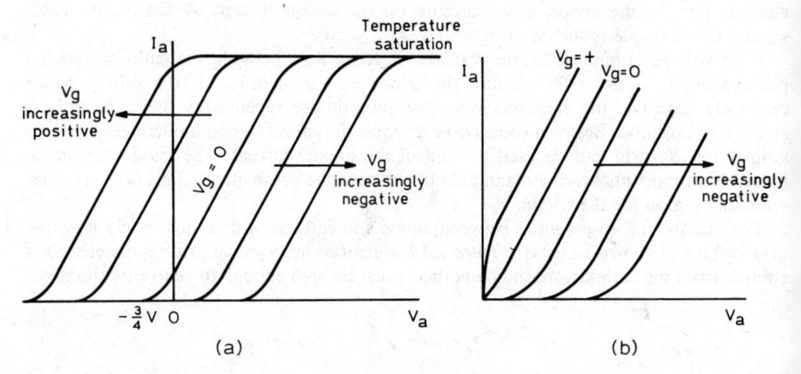

Figure 7.15 *(a) Anode characteristic of triode with grid voltage as parameter; (b) practical anode characteristic of triode*

The operation of the triode that has just been described can be represented by the characteristic of anode current I_a plotted against anode voltage V_a with grid voltage V_g as parameter, as shown in (*a*) of *Figure 7.15*. The range of grid voltage for a practical valve is typically up to -10 V, although higher values up to -20 V may be used with some types of triode. Since the triode is not operated in practice with a negative anode

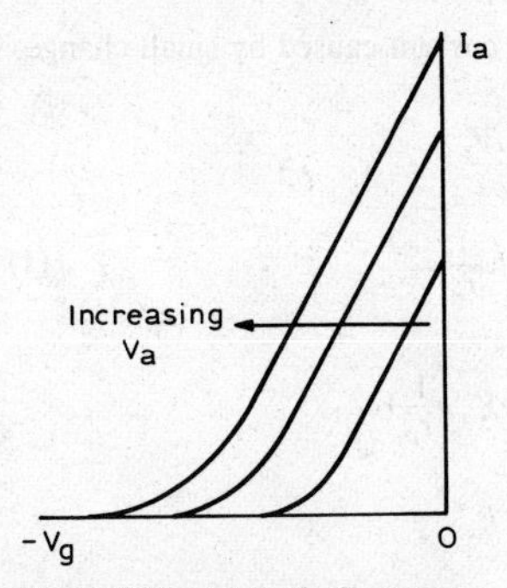

Figure 7.16 *Mutual characteristic of triode with anode voltage as parameter*

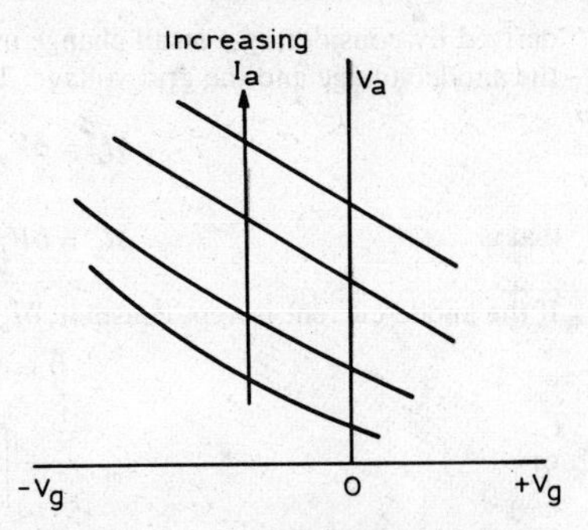

Figure 7.17 *Constant-current characteristic of triode*

voltage or a highly positive grid voltage, the practical anode characteristic is shown in (*b*).

The triode has three principal characteristics. The other two are: the grid or mutual characteristic, the change of anode current with grid voltage, with anode voltage as parameter; and the constant-current characteristic, the change of anode voltage with grid voltage, with anode current as parameter. The mutual characteristic is shown in *Figure 7.16* and the constant-current characteristic in *Figure 7.17*. The mutual characteristic is also known as the transfer characteristic.

The slopes of the three characteristics can be used to obtain the small-signal parameters defining the performance of the triode. As the slopes vary over the length of the characteristic, the value of the parameter should be qualified by the operating point at which it applies.

The slope of the mutual characteristic defines the mutual conductance g_m of the triode. The value of g_m can also be expressed in terms of a small change of anode current with a small change in grid voltage while the anode voltage remains constant. Thus:

$$g_m = \left| \frac{\partial I_a}{\partial V_g} \right|_{V_a}$$

The mutual conductance is also called the *slope* of the triode. The units generally used are mA/V. The value of g_m is determined principally by the cathode-to-grid spacing in the valve; the closer the spacing, the higher the value of g_m.

The slope of the constant-current characteristic defines the amplification factor μ of the triode. Again this can be expressed in terms of small changes in the anode voltage and grid voltage with the anode current remaining constant; that is:

$$\mu = -\left| \frac{\partial V_a}{\partial V_g} \right|_{I_a}$$

Since the value of μ is greater than unity, it can be seen that a small change in the grid voltage will cause a larger change in the anode voltage; in other words, voltage amplification has been obtained. This is the property of the triode that made it important for the development of electronics.

The reciprocal of the slope of the anode characteristic defines the anode slope resistance r_a of the triode. This can be expressed in terms of small changes in anode voltage and anode current while the grid voltage remains constant; that is:

$$r_a = \left| \frac{\partial V_a}{\partial I_a} \right|_{V_g}$$

The small-signal parameters of the triode are interrelated, and the relationship can be derived by considering a small change in the anode current caused by small changes in the anode voltage and the grid voltage. Thus:

$$\delta I_a = \delta V_g \frac{\partial I_a}{\partial V_g} + \delta V_a \frac{\partial I_a}{\partial V_a}$$

that is

$$\delta I_a = \delta V_g \cdot g_m + \delta V_a \cdot \frac{1}{r_a} \qquad (1)$$

If the anode current is kept constant, $\delta I_a = 0$; that is

$$0 = \delta V_g \cdot g_m + \delta V_a \cdot \frac{1}{r_a}$$

or

$$-g_m \cdot r_a = \left| \frac{\delta V_a}{\delta V_g} \right|_{I_a} = -\mu$$

Hence:

$$\mu = g_m \cdot r_a$$

Typical values of the mutual conductance g_m of practical triodes lie between 1 mA/V and 15 mA/V. Typical values of the amplification factor μ lie between 10 and 100, while the anode slope resistance lies between 1 kΩ and 60 kΩ. The variation of the three small-signal parameters with anode current is shown in *Figure 7.18*.

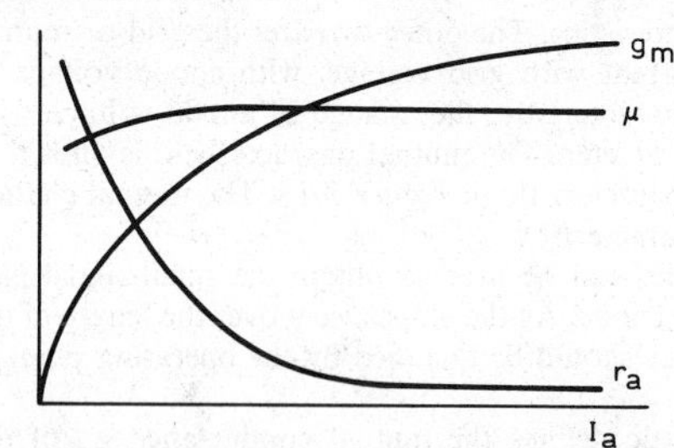

Figure 7.18 Variation of small-signal parameters of triode with anode current

The idealised relationship between the grid current and the grid voltage in a triode is shown in (*a*) of *Figure 7.19*. Electrons from the cathode striking the wires of the grid can cause secondary emission, which will increase as the grid voltage becomes more positive. The grid current will therefore fall as the secondary emission increases, becoming a negative current at some value of grid voltage. However, as the grid voltage is increased further to become comparable with the anode voltage, the triode will start to operate as a diode with the grid acting as the anode. The grid current will therefore increase again. The relationship between grid current and grid voltage taking the secondary emission into account is shown in (*b*). As all valves contain traces of residual gas even after the vacuum pumping and gettering, the ionisation of this gas will produce positive ions which will be attracted to the grid when it is negative. A small grid current, typically a few picoamperes, will flow as shown in (*c*). Thus the practical grid voltage/current characteristic is as shown in (*d*). The ion current has been exaggerated for clarity, and is normally too small to be of consequence unless the triode is connected to a very high grid-to-cathode resistance when the noise voltage developed may be troublesome.

The operating point of the triode on the mutual characteristic is fixed by a constant voltage applied to the grid, the grid bias. The actual voltage on the grid at any instant is thus determined by the sum of the grid bias and the applied signal voltage. Various operating points and ranges can be chosen for the triode, as shown in *Figure 7.20*.

The operating point can be chosen so that the grid signal always remains on the linear part of the characteristic, *Figure 7.20* (*a*). Anode current will flow for the whole cycle. This method of operation is called class *A*. The operating point can also be chosen so that the range of grid signal voltage is on the bottom bend region of the characteristic, (*b*). This is class *AB* operation. If the operating point is at or near the cut-off point, (*c*),

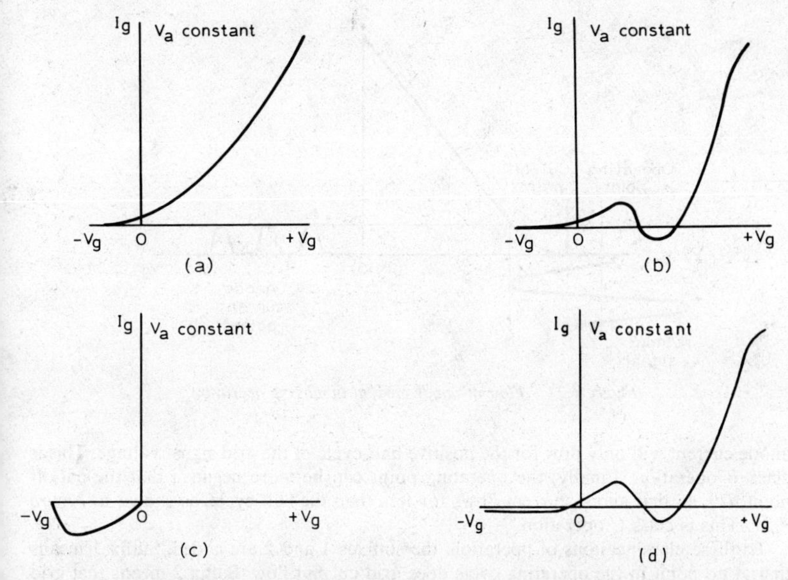

Figure 7.19 Grid voltage/current characteristic of triode: (a) ideal characteristic (b) characteristic including secondary emission (c) ion current (d) practical characteristic

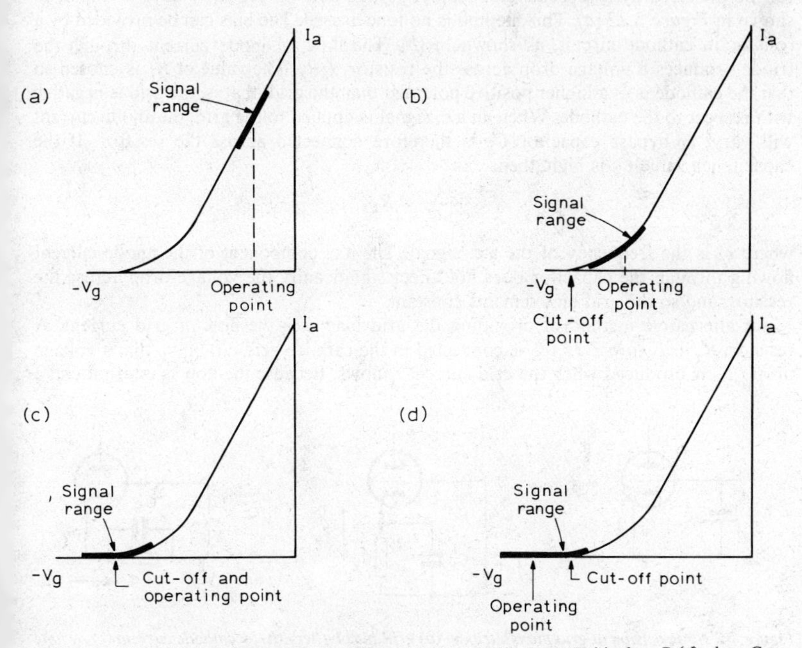

Figure 7.20 Classes of operation for triode: (a) class A (b) class AB (c) class B (d) class C

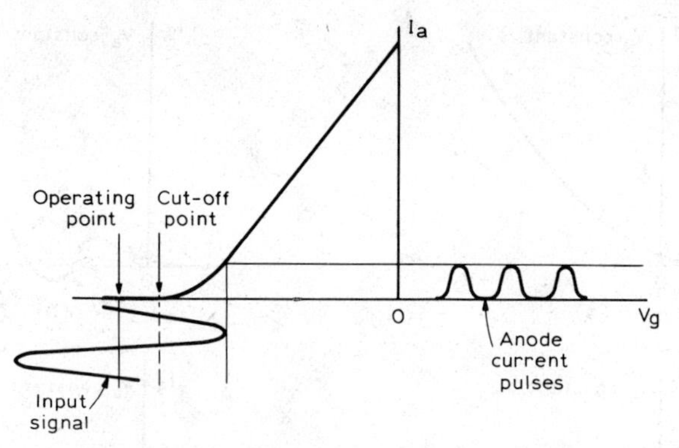

Figure 7.21 *Flow of anode current in class C operation*

anode current will only flow for the positive half-cycle of the grid signal voltage. This is class *B* operation. Finally, the operating point can be more negative that the cut-off point, (*d*), so that anode current flows for less than the half-cycle, as shown in *Figure 7.21*. This is class *C* operation.

To these classifications of operation, the suffixes 1 and 2 are added. Suffix 1 means that at no point in the operating cycle does grid current flow. Suffix 2 means that grid current will flow at some point in the cycle. By convention, if no suffix is given, the suffix 1 is implied.

The grid bias makes the grid always negative with respect to the cathode. The bias can be provided by a separate d.c. supply, as was done in the early days of radio, as shown in *Figure 7.22* (*a*). This method is no longer used. The bias can be provided by a resistor in cathode circuit, as shown in (*b*). The flow of anode current through the triode produces a voltage drop across the resistor, I_aR_k. The value of R_k is chosen so that the cathode is at a higher positive potential than the grid; that is, the grid is negative with respect to the cathode. When an a.c. signal is applied to the grid, the anode current will vary. A bypass capacitor C_k is therefore connected across the resistor. If the capacitance value C_k is high, then:

$$\frac{1}{\omega C_k} \ll R_k$$

where ω is the frequency of the a.c. signal. The a.c. component of the anode current flowing through the capacitor does not affect significantly the voltage drop across the resistor, and so the grid bias remains constant.

An alternative method of providing the grid bias uses the flow of grid current. A resistor, R_g in *Figure 7.22* (*c*), is connected in the cathode-grid circuit so that a voltage drop i_gR_g is produced when the grid current i_g flows. Because the flow is intermittent, a

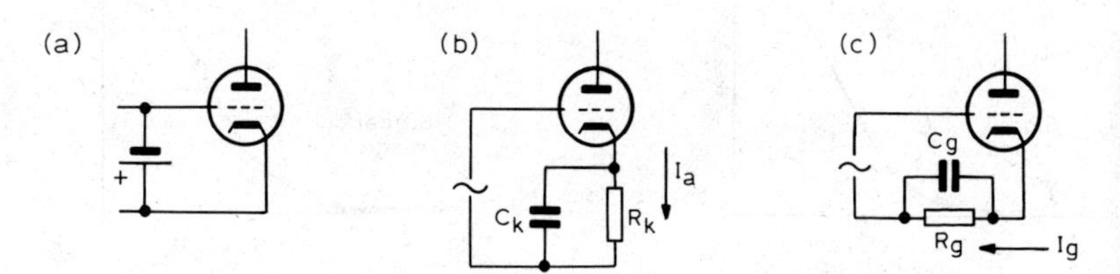

Figure 7.22 *Provision of grid bias in triode:* (*a*) *grid-bias battery* (*b*) *by cathode current* (*c*) *by flow of grid current*

high-value capacitor C_g is connected across the resistor. The time-constant C_gR_g is chosen so that the voltage across the network approximates to a steady value to give the required grid bias.

In a practical circuit, the triode is operated with a load resistor in the anode circuit, as shown in *Figure 7.23*. The changes in grid voltage caused by the signal voltage

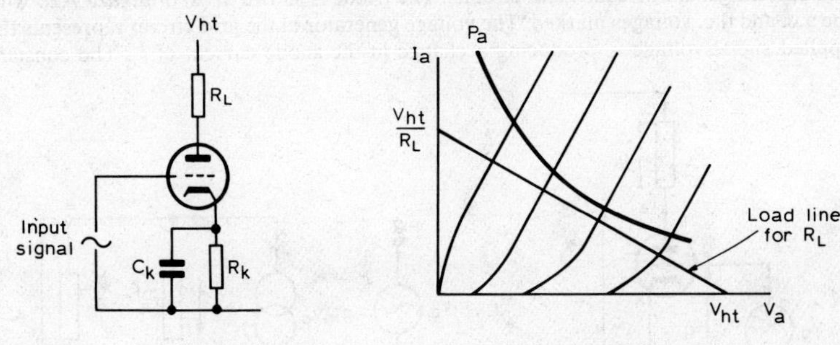

Figure 7.23 Triode amplifier stage

Figure 7.24 Anode characteristic of triode with loadline and anode dissipation curve superimposed

produce changes in the anode current, and these are converted into voltage changes across the load resistor R_L. A loadline representing R_L can be plotted on the anode characteristic, as shown in *Figure 7.24*. This loadline defines the operating point of the triode. Also shown in *Figure 7.24* is the curve of anode dissipation P_a which represents the maximum power that can be extracted from the valve. The operating loadline must lie below the P_a curve.

Various equivalent circuits have been devised for the triode. The simplest is the

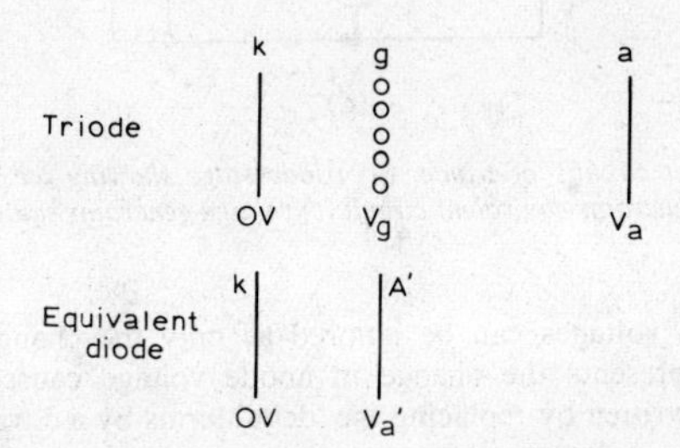

Figure 7.25 Equivalent diode of triode

equivalent diode, shown in *Figure 7.25*. The anode of the equivalent diode A' replaces the grid of the triode, and the voltage V_a' must have the same effect on the grid-to-cathode potential distribution as the anode voltage of the triode had. Thus:

$$V_a' = \frac{V_a}{\mu} + V_g$$

where V_a and V_g are the anode and grid voltages respectively of the triode, and μ is the amplification factor. The total electrode current of the equivalent diode I_a' is

$$I_a' = I_a + I_g$$

where I_a and I_g are the anode and grid currents respectively of the triode. Applying Child's Law:

$$I_a' = G[V_a']^{\frac{3}{2}}$$

or
$$(I_a + I_g) = G\left[\frac{V_a}{\mu} + V_g\right]^{\frac{3}{2}}$$

where G is the permeance of the equivalent diode.

Equivalent circuits that are more useful for circuit analysis are the voltage-generator and current-generator equivalent circuits. The triode is shown in (*a*) of *Figure 7.26* with the a.c. and d.c. voltages marked. The voltage generator in the grid circuit represents the applied signal voltage v_g producing a change in the anode current of i_a. The constant

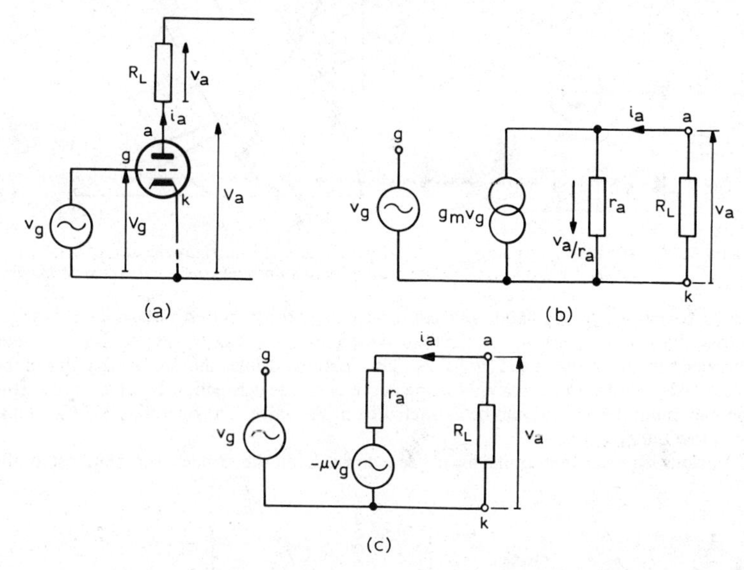

Figure 7.26 Equivalent circuits of triode: (a) triode stage showing a.c. voltages and currents (b) current-generator equivalent circuit (c) voltage-generator equivalent circuit

grid bias and anode voltages can be ignored as only the changes (a.c. values) are considered. If v_a represents the change in anode voltage caused by the signal v_g, equation 1 can be rewritten by replacing the 'delta' terms by a.c. values to give

$$i_a = v_g \cdot g_m + v_a \cdot \frac{1}{r_a} \qquad (2)$$

Equation 2 can be represented by a current generator of magnitude $g_m v_g$ in parallel with a resistance r_a through which a current v_a/r_a is flowing. This equivalent circuit, called the current-generator equivalent circuit, is shown in *Figure 7.26* (*b*).

If equation 2 is multiplied by r_a, then:

$$i_a r_a = v_g \cdot g_m r_a + v_a$$

from which
$$v_a = -\mu v_g + i_a r_a$$

This equation can be represented by a voltage generator of magnitude μv_g in a series with a resistance r_a through which a current i_a is flowing. This is the voltage-generator equivalent circuit shown in *Figure 7.26* (*c*). The negative sign associated with the voltage generator indicates the 180° phase reversal between the grid and anode voltages.

The voltage-generator equivalent circuit can be used to determine the amplification of

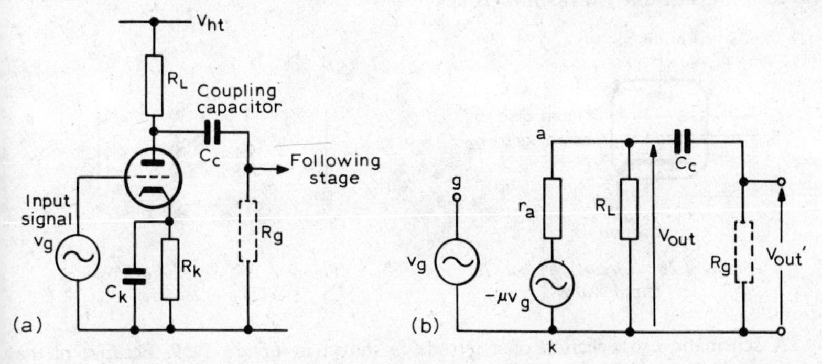

Figure 7.27 Gain of triode amplifier stage: (a) practical triode stage (b) equivalent circuit

the practical triode stage shown in *Figure 7.27* (*a*). The equivalent circuit is shown in (*b*). The output voltage V_{out} is

$$V_{out} = -\mu v_g \left(\frac{R_L}{R_L + r_a}\right)$$

while the input voltage V_{in} is the grid signal voltage v_g. The amplification or gain A is the ratio of V_{out} to V_{in}; that is:

$$A = \frac{V_{out}}{V_{in}} = -\mu \frac{R_L}{R_L + r_a} \qquad (3)$$

This expression is oversimplified as the reactance of the coupling capacitor and the grid resistor of the following valve stage have been neglected. It shows, however, that the gain depends on the value of the amplification factor μ of the triode, and on the relative values of the anode slope resistance of the triode and the load resistor, taking into account the limits imposed by the anode dissipation.

A more exact equation for the gain including the coupling capacitor and grid resistor of the following stage is:

$$A = -\mu \frac{R}{R + r_a} \times \frac{1}{1 - jX_c/P}$$

where X_c is the reactance of the coupling capacitor; R is the resistance of R_L and R_g in parallel; and P is the resistance of R_g in series with the parallel combination of R_L and r_a.

The published data for a triode includes anode characteristic curves (I_a/V_a with V_g as parameter) on which the anode dissipation P_a is plotted, and mutual characteristic curves (I_a/V_g with V_a as parameter). The variations of the three small-signal parameters with anode current are also included. Characteristic values are listed to indicate the performance of the triode at certain operating points. Values are given of such quantities as anode voltage, anode current, grid voltage, and the three small-signal parameters. The interelectrode capacitances are listed, and limiting values are given of anode voltage and dissipation, grid voltage, cathode current, and cathode-to-heater insulation voltage.

Tetrode

The performance of a triode is limited by the interelectrode capacitances. It was for this reason that a second grid, the screen grid, was inserted between the control grid and the anode, and so form the tetrode. The screen grid reduces the grid-to-anode capacitance from about 5 pF (typical for a triode of the time) to 0.01 pF. (The grid-to-anode capacitance of a modern triode is typically 1.5 pF.) The circuit symbol for a tetrode is shown in *Figure 7.28*.

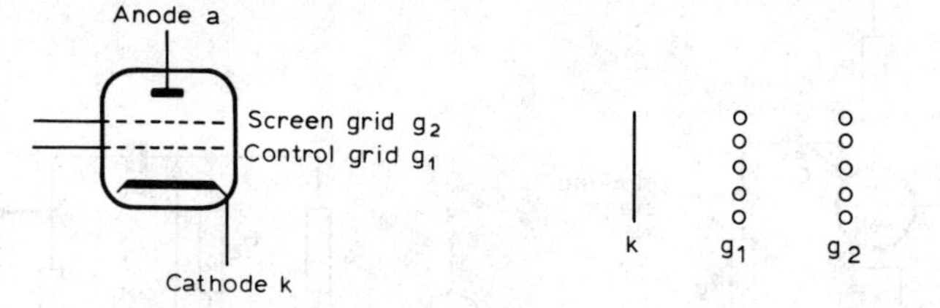

Figure 7.28 Circuit symbol for tetrode valve

Figure 7.29 Schematic cross-section of tetrode

A schematic cross-section of a tetrode is shown in *Figure 7.29.* Because of the screening effect of the two grids, the anode voltage has little effect on the electron emission from the cathode. The control grid is held at a low negative potential, say up to −10 V, which is varied by the applied signal to control the anode current. The screen grid is held at a constant higher positive potential, say +80 V. to accelerate the electrons towards the anode.

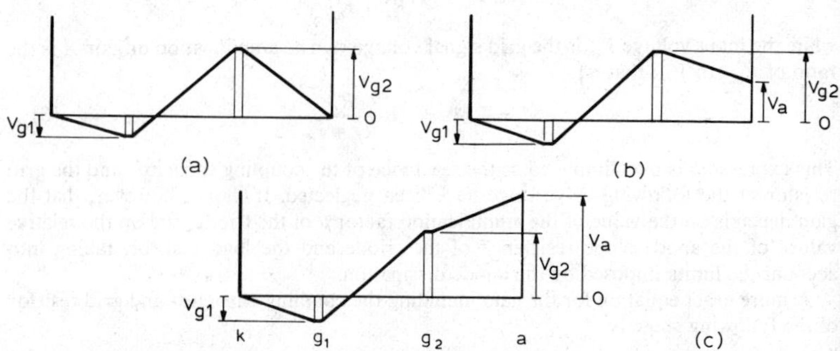

Figure 7.30 Potential distribution in tetrode: (a) anode voltage zero (b) anode voltage less than screen-grid voltage (c) anode voltage greater than screen-grid voltage

The distribution of potential within the tetrode is shown in *Figure 7.30.* If the anode voltage is zero, (*a*), the electrons from the cathode will flow to the screen grid. The anode current I_a will be zero, and a large screen-grid current I_{g2} will flow. If the anode voltage is increased but is still lower than the screen-grid voltage, (*b*), anode current will flow and the screen grid current will be reduced. If the anode voltage is increased further so that it exceeds the screen-grid voltage, (*c*) the anode current will increase further while the screen-grid current will be further reduced. This idealised operation of the tetrode is represented by the voltage/current curves of *Figure 7.31.*

The operation of the tetrode just described is 'ideal' because it neglects the effect of secondary emission. In any valve, when the anode voltage is at its normal operating value, the electrons reaching the anode will have been accelerated to acquire sufficient energy to release secondary electrons on impact. Secondary emission occurs in a triode, but the secondary electrons are repelled by the negative grid and return to the anode. In a tetrode, however, the positive screen grid attracts the electrons to increase the screen-grid current. The secondary emission also causes a fall in anode current. Thus as the anode voltage is increased from zero, initially there will be an increase in anode current and a fall in screen-grid current. When the anode voltage is sufficiently positive to accelerate the electrons enough to cause secondary emission, the anode current will start to fall and the screen-grid current will start to rise again. As the anode voltage is increased to become comparable with the screen-grid voltage, the secondary electrons

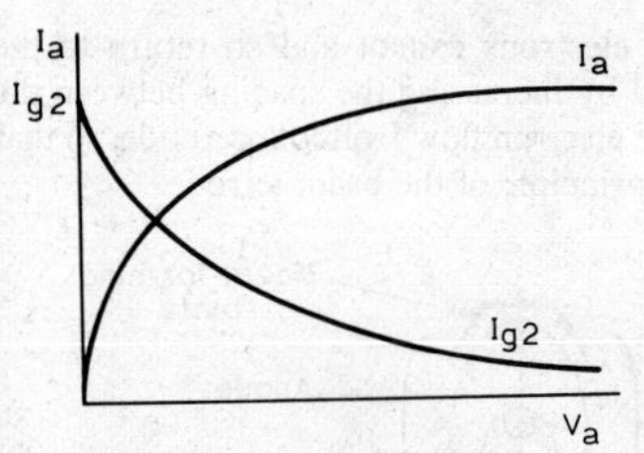

Figure 7.31 Ideal anode current and screen-grid current characteristics for tetrode

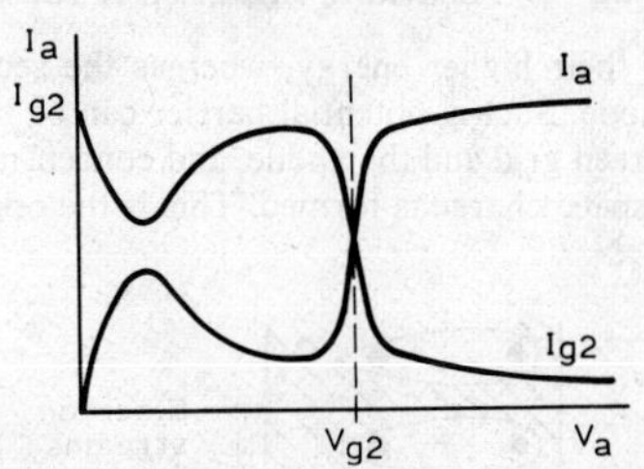

Figure 7.32 Practical tetrode characteristics

emitted will be attracted back to the anode rather than travel to the screen grid. There will be a sharp increase in the anode current, and an accompanying sharp fall in the screen-grid current. Electrons striking the screen grid can also produce secondary emission. When the anode voltage is higher than the screen-grid voltage, these secondary electrons will be attracted to the anode. The anode current will increase further, while the screen-grid current will decrease further. The practical tetrode characteristics taking secondary emission into account are shown in *Figure 7.32*.

As in the triode, the electron flow from the cathode can be controlled by the voltage on the control grid, V_{g1}. Thus the control-grid voltage will affect the anode voltage/current characteristic, as shown in *Figure 7.33*. The mutual characteristic of the tetrode, the change of anode current with control-grid voltage, is shown in *Figure 7.34*. It is similar to that of the triode, so that the value of g_m for the tetrode is similar to that of the triode (typically 1 mA/V to 7 mA/V for valves of the period). However, as the anode slope resistance r_a is the reciprocal of the slope of the anode characteristic, it can be seen from *Figures 7.32* and *7.33* that r_a is very high in the usual working region where the anode voltage is higher than the screen-grid voltage. Since the amplification factor μ is related to g_m and r_a by

$$\mu = g_m \cdot r_a$$

the value of μ is high, giving a tetrode amplifier stage a higher gain than a triode stage. Against this must be set the secondary emission which can only be prevented by operating the tetrode at a higher anode voltage than would be required by a triode stage. For this reason, methods of eliminating the 'kink' in the characteristic were investigated soon after the tetrode was introduced.

There is an energy difference between the primary electrons from the cathode and the secondary electrons produced on impact with the anode. If a 'potential barrier' is introduced between the screen grid and the anode, the primary electrons can pass through because

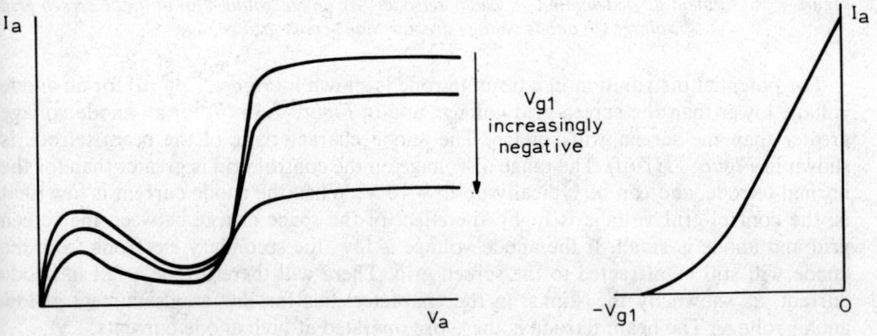

Figure 7.33 Anode characteristic of tetrode with control-grid voltage as parameter

Figure 7.34 Mutual characteristic of tetrode

of their higher energy, whereas the secondary electrons cannot and so return to the anode. Such a potential barrier can be produced by increasing the spacing between the screen grid and the anode, and concentrating the electron flow from the cathode, so that a space charge is formed. This is the operating principle of the beam tetrode.

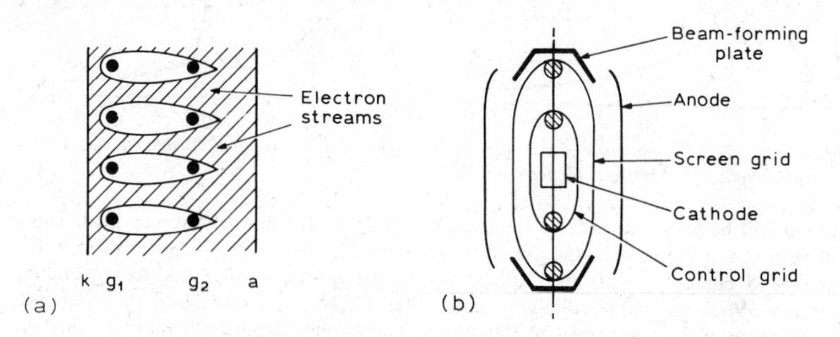

Figure 7.35 *Schematic cross-sections of beam tetrode: (a) vertical section showing electron streams (b) horizontal section showing beam-forming plates*

A schematic vertical cross-section of the beam tetrode is shown in (*a*) of *Figure 7.35*. The control grid and screen grid are wound to the same pitch, and are carefully aligned during assembly so that the screen grid is in the shadow of the control grid wires. The resulting electron streams are shown in (*a*). In addition, two 'beam-forming' plates are included in the electrode structure, as shown in the horizontal cross-section of (*b*). Because of the increased screen-grid-to-anode spacing, there is a point where the effect of the positive potentials from the anode and screen grid is weakest. Here a concentration of electron can occur, and so the required space charge is formed.

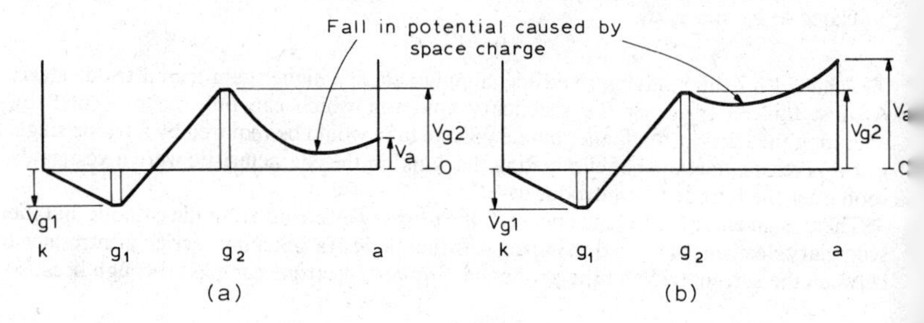

Figure 7.36 *Potential distribution in beam tetrode: (a) anode voltage lower than screen-grid voltage (b) anode voltage greater than screen-grid voltage*

The potential distribution in a beam tetrode is shown in *Figure 7.36* (*a*) for an anode voltage lower than the screen-grid voltage, and in *Figure 7.36* (*b*) for an anode voltage greater than the screen-grid voltage. The anode characteristic of the beam tetrode is shown in *Figure 7.37* (*a*). The range of voltage on the control grid is greater than for the normal tetrode, and can be typically up to −40 V. When the anode current is low (that is, the control-grid voltage is high), the effect of the space charge between the screen grid and anode is small. If the anode voltage is low, the secondary electrons from the anode will still be attracted to the screen grid. There will therefore be a fall in anode current, as shown by the 'kinks' in the characteristics for low anode current at low anode voltage. The beam tetrode is therefore operated at high anode currents.

A feature of the anode characteristic of the beam tetrode is the rapid rise of anode current. When the anode voltage is near zero, the potential minimum between the screen grid and anode is at zero. There is a virtual cathode, and the electrons passing through the screen grid towards the anode are repelled back to the screen grid. As the anode voltage increases, the potential minimum becomes positive, and accelerates the electrons towards the anode. All the electrons reach the anode, and so the anode current rapidly increases. The small increase in anode current with voltage above the 'knee' of the characteristic shows the slight effect of anode voltage on the cathode emission because of the screening effect of the grids.

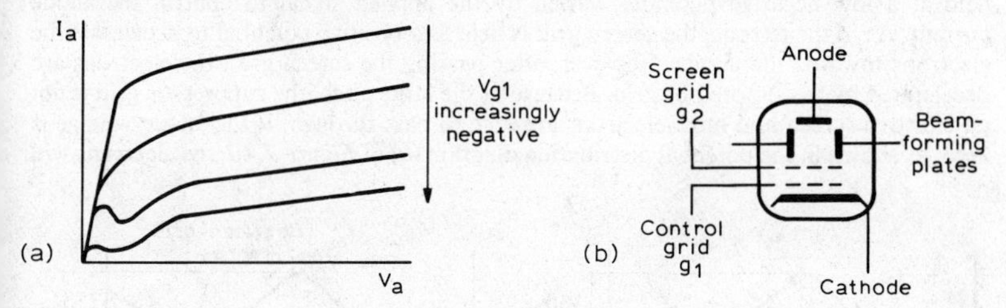

Figure 7.37 (a) Anode characteristic of beam tetrode with control-grid voltage as parameter; (b) circuit symbol for beam tetrode

As the beam tetrode has to operate at high currents, it cannot be used as a small-signal amplifier. This limitation led to the far wider use of the pentode instead of the tetrode, to be discussed in the next section. The beam tetrode did, however, find application as a power amplifier, and particularly as a small transmitting valve. The alignment required for the grids and the additional beam-forming electrodes made it a relatively expensive valve, although attempts were made to simplify the manufacture by using a screen grid with a larger pitch than the control grid to eliminate the alignment. The circuit symbol for a beam tetrode is shown in *Figure 7.37* (*b*).

Pentode

The changes to the electrode structure of the tetrode to form the beam tetrode were to encourage the formation of a space charge between the screen grid and the anode. This space charge prevented secondary electrons from the anode reaching the screen grid. The same effect can be achieved by inserting a grid held at or near the cathode potential between the screen grid and the anode. As the function of this third grid is to suppress the emission of secondary electrons, it is called the suppressor grid. It is wound to a larger pitch than the control grid and the screen grid. The circuit symbol for a pentode is shown in *Figure 7.38* (*a*); because the suppressor grid is normally operated at cathode potential, it is sometimes connected internally to the cathode as shown in the circuit symbol of *Figure 7.38* (*b*).

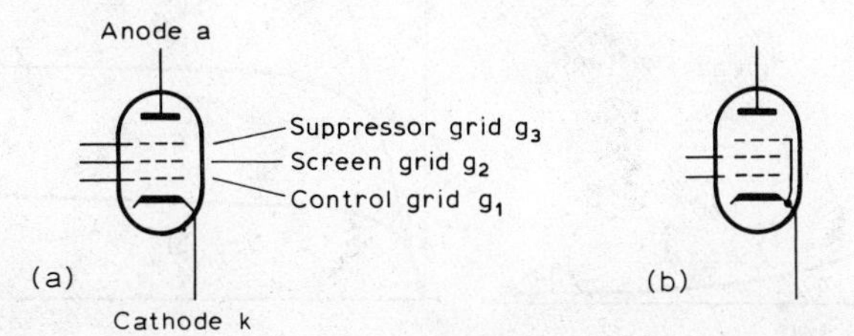

Figure 7.38 Circuit symbol for pentode: (a) separate suppressor grid (b) suppressor grid connected to cathode

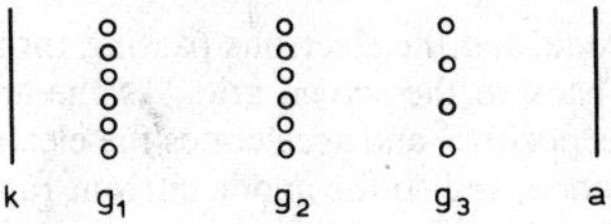

Figure 7.39 Schematic cross-section of pentode

A schematic cross-section of a pentode is shown in *Figure 7.39*. The control grid is held at a low negative potential, varied by the applied signal to control the anode current. As in the tetrode, the screen grid is held at a positive potential to accelerate the electrons towards the anode. However, after passing the screen grid, the electrons are decelerated by the suppressor grid. Because of the large pitch, the suppressor grid is not an effective screen and most electrons are able to pass through. If the anode voltage is zero, as shown in the potential distribution diagram (*a*) of *Figure 7.40*, the electrons will

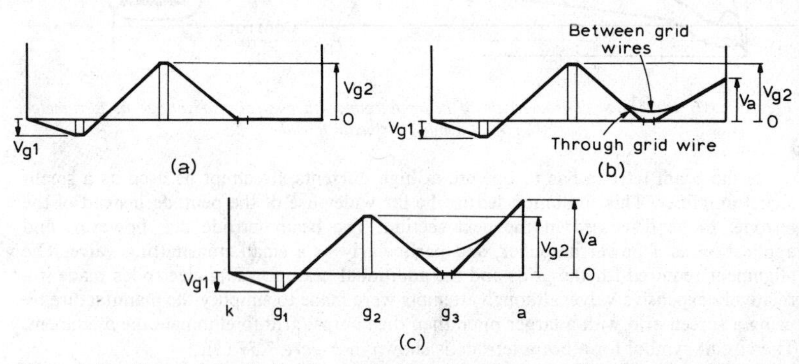

Figure 7.40 Potential distribution in pentode: (a) anode voltage zero (b) anode voltage lower than screen-grid voltage (c) anode voltage greater than screen-grid voltage

be attracted back to the screen grid (the only positive electrode), and a large screen-grid current will flow. If the anode voltage is slightly positive, (*b*), the electrons will be attracted to the anode. There will therefore be a sharp increase in anode current and a corresponding fall in screen-grid current. Secondary electrons emitted from the anode are repelled by the suppressor grid, and are attracted back to the anode. A further increase in the anode voltage, (*c*), will have only a small effect on the anode current after the initial sharp rise because of the screening effect of the three grids between the anode and the cathode.

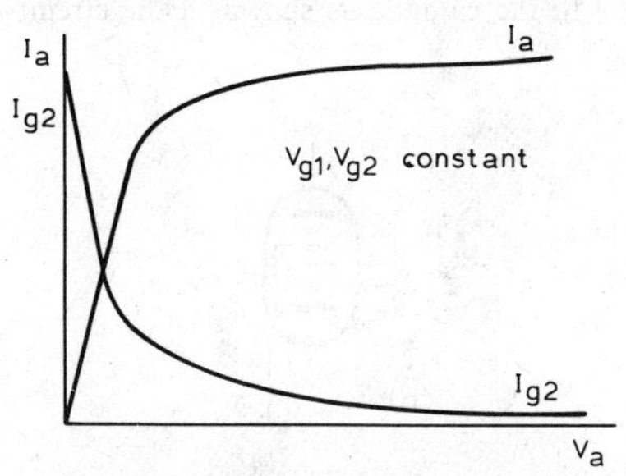

Figure 7.41 Anode current and screen-grid current characteristics for pentode

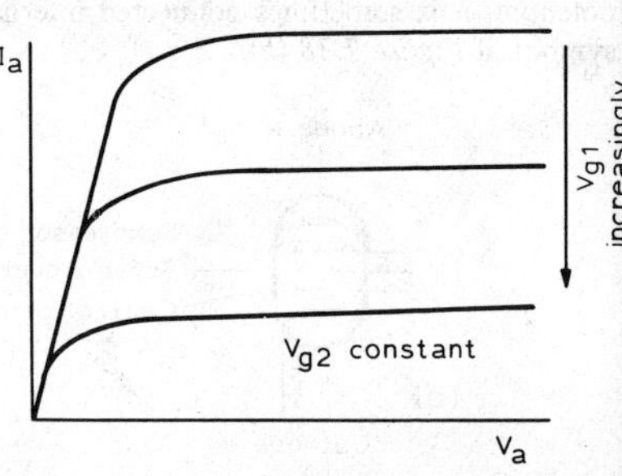

Figure 7.42 Anode characteristic of pentode with control-grid voltage as parameter

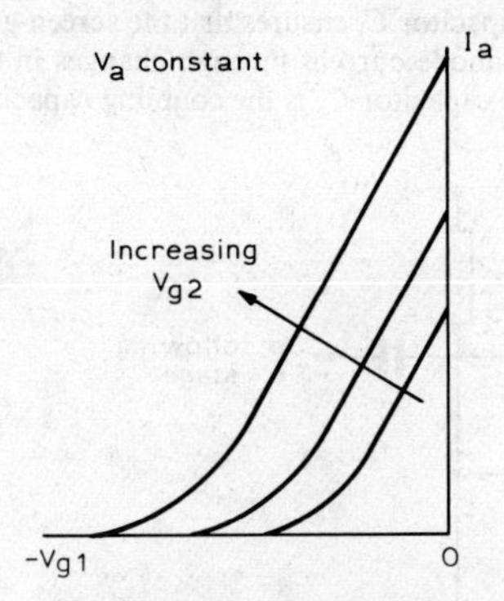

Figure 7.43 Mutual characteristic of pentode with screen-grid voltage as parameter

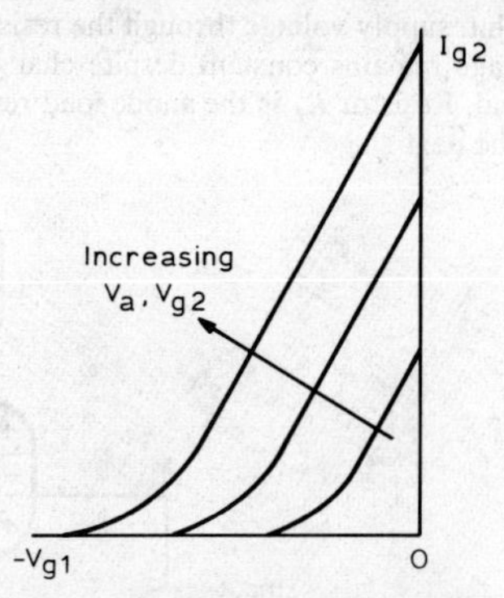

Figure 7.44 Variation of screen-grid current with control-grid voltage in pentode, with anode voltage and screen-grid voltage as parameters

The characteristic of the pentode relating the change of anode current I_a and screen-grid current I_{g2} with anode voltage V_a is shown in *Figure 7.41*. The effect of the control-grid voltage V_{g1} on the anode characteristic is shown in *Figure 7.42*. The mutual characteristic of the pentode, the change of anode current I_a with control-grid voltage V_{g1} with screen-grid voltage V_{g2} as parameter, is shown in *Figure 7.43*. The effect of the screen-grid voltage is similar to that of the anode voltage on the mutual characteristic of the triode, shown in *Figure 7.16*. The value of g_m for a pentode is similar to that of a triode. The anode slope resistance r_a, however, is much higher, as shown by the anode characteristic where the curves are nearly parallel to the voltage axis. Typical values of r_a lie between 400 kΩ and 1 MΩ. The pentode therefore has a very high value of amplification factor μ, giving a pentode stage a higher gain than a triode stage. The pentode is therefore the most commonly used valve for small-signal amplification.

The change of screen-grid current I_{g2} with control-grid voltage V_{g1}, with anode voltage V_a and screen-grid voltage V_{g2} as parameters, is shown in *Figure 7.44*.

The various classes of operation for a pentode are similar to those discussed previously for a triode. The two equivalent circuits derived for the triode and shown in *Figure 7.26* can also be used for the pentode. Because of the high value of r_a, the expression for stage gain can be simplified. The gain A is given by (from equation 3):

$$A = -\mu \frac{R_L}{R_L + r_a} = -g_m \cdot r_a \frac{R_L}{R_L + r_a}$$

where R_L is the anode load resistance. Since

$$r_a \gg R_L, \text{ then } (R_L + r_a) \simeq r_a$$

and

$$A = -g_m \cdot r_a \frac{R_L}{r_a} = -g_m R_L$$

A practical pentode amplifier stage is shown in *Figure 7.45*. Many of the components are similar to those discussed in connection with the practical triode amplifier stage shown in *Figure 7.27*. The control grid bias is provided by the cathode resistor R_k. This has the bypass capacitor C_k connected across it to provide a low-impedance path for a.c. signals so that the grid bias remains constant. The suppressor grid is connected to the cathode, while the screen grid is maintained at a positive potential by the connection to the h.t. supply voltage through the resistor R_s. Capacitor C_s ensures that the screen-grid voltage remains constant despite changes in the anode current through changes in the signal. Resistor R_L is the anode load resistor, and capacitor C_c is the coupling capacitor to the next stage.

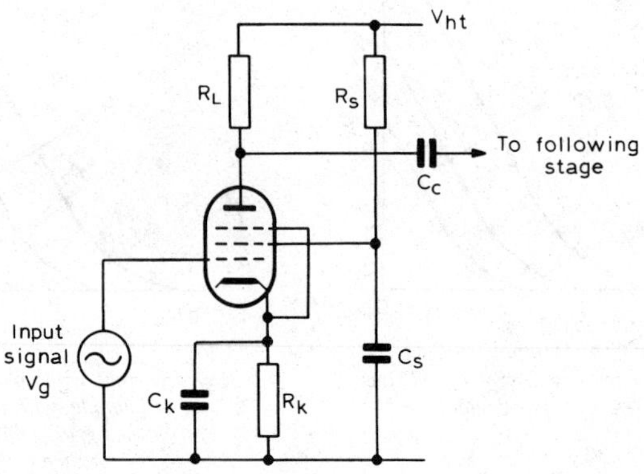

Figure 7.45 Practical pentode amplifier stage

When the pentode is used as a small-signal amplifier, changes in the control-grid voltage caused by the applied signal should produce the largest possible change in the anode current. To achieve this, the control grid is wound with a small pitch so that the grid wires are close together. The effect of the negative grid voltage on the electrons flowing between the grid wires is very marked. A small negative grid voltage will therefore prevent the flow of anode current, and the cut-off voltage is low as shown by *curve 1* in *Figure 7.46*. If the pentode is used as a power amplifier, this close control over the anode current is not required, and the control grid can be wound with a larger

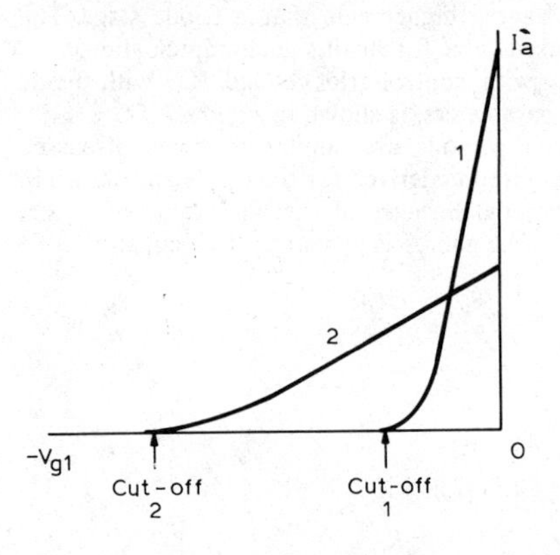

Figure 7.46 Mutual characteristic of pentode with short grid-base (curve 1), and long grid-base (curve 2)

pitch. The effect of the negative grid voltage on the electrons flowing between the wires is now not so marked, and so a large negative voltage is required for cut-off. This is shown by *curve 2* of *Figure 7.46*. Pentodes with a mutual characteristic like that of *curve 1* are called *short grid-base* pentodes, while those with a characteristic like *curve 2* are called *long grid-base* pentodes.

The two types can be combined in one to form the variable-mu pentode. The control grid of this pentode is wound with a variable pitch. At the ends of the grid, the pitch is small; in the centre, the pitch is larger. The form of the grid is shown in *Figure 7.47*. The mutual characteristic of a variable-mu pentode is shown in *Figure 7.48*. The required value of g_m (and hence of μ) can be obtained for a particular application by choosing the operating point on the characteristic. Operation near cut-off, for example, will give a low value of g_m, while operation higher up the characteristic will give a higher value. The variable-mu pentode was used extensively in radio receivers for automatic gain control (a.g.c.), also known as automatic volume control (a.v.c.).

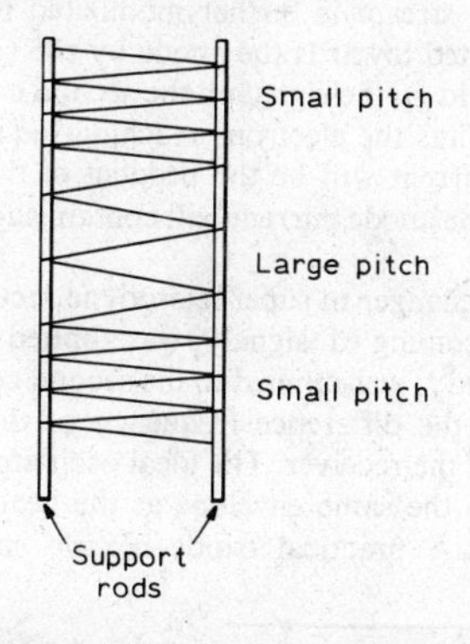

Figure 7.47 Schematic form of control grid in variable-mu pentode

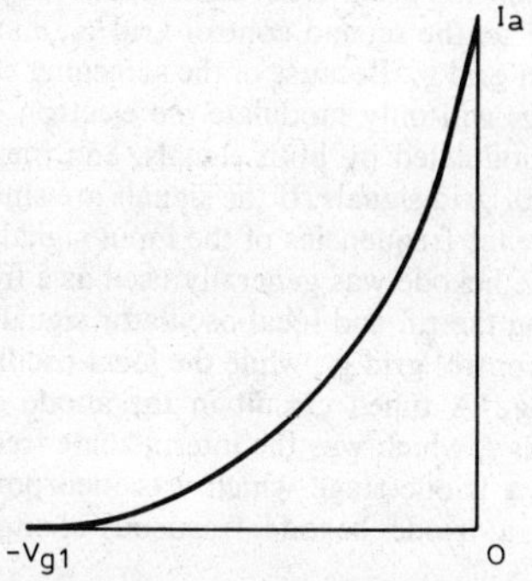

Figure 7.48 Mutual characteristic of variable-mu pentode

The published data for a pentode includes curves of the anode and mutual characteristics. Other curves that may be included are the variation of screen-grid current with control-grid voltage, with anode voltage and screen-grid voltage as parameters; and the variation of screen-grid current with anode voltage, with control-grid voltage as parameter. Characteristic values are given for particular operating conditions, the quantities listed including anode, control-grid, and suppressor-grid voltages, anode and screen-grid currents, g_m and r_a. The interelectrode capacitances are listed, typical values being 5 to 100 fF (femtofarad = 10^{-15} F) for c_{a-g1} and 2.5 to 3.0 pF for c_{g1-g2}. Limiting values of voltages and currents are also listed.

Other multi-electrode valves

After the development of the pentode, other multi-electrode valves were designed. The most commonly used types were the hexode and heptode, both intended for use as mixer valves, but octodes were also produced. Economic considerations led to two valves (electrode structures) being placed in one envelope to form a 'double valve', and

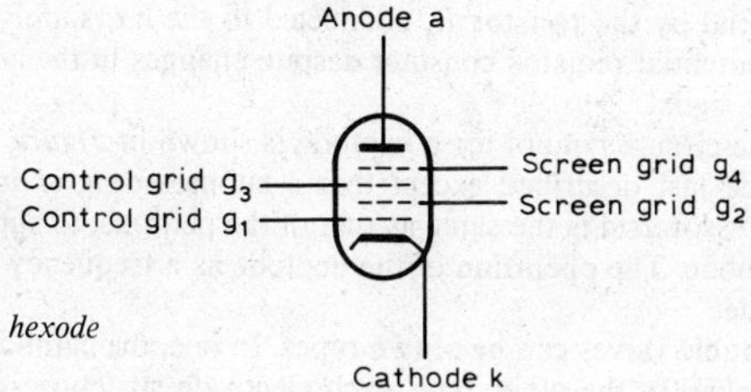

Figure 7.49 Circuit symbol for hexode

different types being placed in the same envelope to form a 'multiple valve', generally for a particular function.

The circuit symbol for a hexode is shown in *Figure 7.49*. The valve contains two control grids, g_1 and g_3, to which the signal voltages are applied; and two screen grids, g_2 and g_4, which are maintained at a constant positive potential. The electron stream from the cathode is modulated by the signal on the first control grid g_1. The first screen grid g_2 accelerates the electrons towards the second control grid g_3, and also screens the two control grids from each other. The electron stream is further modulated by the signal on the second control grid g_3, and accelerated towards the anode by the second screen grid g_4. Because of the screening effect of grid g_2, the signal on the second control grid g_3 will only modulate the electron stream. Thus the electrons reaching the anode are modulated by both signals, and the anode current will be the product of the two control-grid signals. If the signals are sinusoidal, the anode current will contain sum and difference frequencies of the input signals.

The hexode was generally used as a frequency changer in superheterodyne receivers, mixing the r.f. and local-oscillator signals. The incoming r.f. signal f_s was applied to the first control grid g_1, while the local-oscillator signal f_o was applied to the second control grid g_3. A tuned circuit in the anode extracted the difference frequency of the two signals f_i which was the intermediate frequency of the receiver. The local oscillator was often a triode stage which was incorporated into the same envelope as the hexode to form a triode–hexode frequency-changer valve. A practical triode–hexode stage is

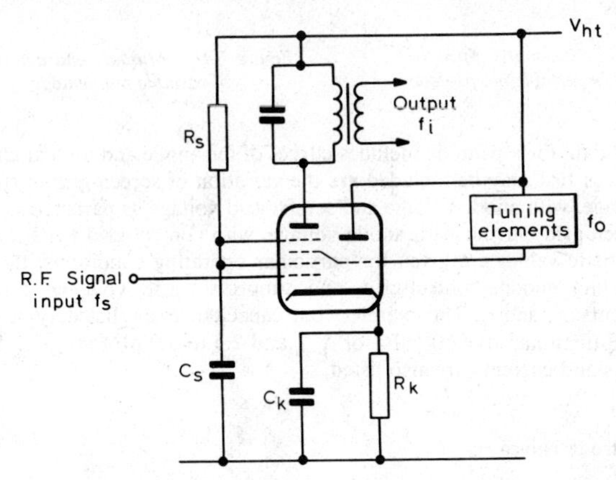

Figure 7.50 *Triode–hexode frequency-changer stage*

shown in *Figure 7.50*. A common cathode is used, and the grid of the local-oscillator triode stage is connected internally to the second control grid of the hexode stage. The r.f. signal is applied to the first control grid. The anode tuned circuit is formed by the parallel *LC* network, while the two screen grids are maintained at the required positive potential by the resistor R_s connected to the h.t. supply. The capacitor C_s ensures that this potential remains constant despite changes in the anode current caused by changes in the signal.

The circuit symbol for a heptode is shown in *Figure 7.51*. The valve is similar to the hexode just described except that a suppressor grid is included. The function of the suppressor grid is the same as that of the pentode, to suppress secondary emission from the anode. The operation of the heptode as a frequency changer is similar to that of the hexode.

Double valves can be of two types. In one, the cathode is common to both sections of the valve. In the other type, each electrode structure is separate; the valve consists of

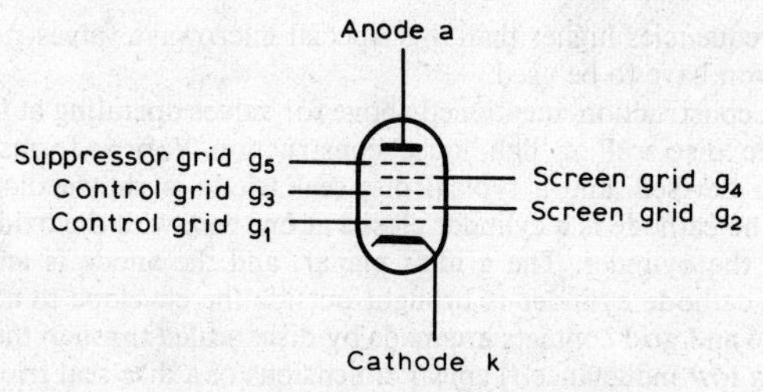

Figure 7.51 Circuit symbol for heptode

two separate valves in one envelope, often with a screen maintained at cathode potential between the two structures. In multiple valves using a common cathode, the two electrode structures were mounted one above the other, surrounding the common cathode. Examples of the double and multiple valves used in practice were double diodes, double triodes, triode–pentodes, triode–hexodes, and double pentodes.

Higher-frequency operation

The operation of a thermionic valve is limited initially by the interelectrode capacitances, particularly that between the control grid and anode. This capacitance provides a feedback path at high frequencies to produce oscillation. The grid-to-anode capacitance of a triode limits the operating frequency to a few megahertz. The reduced capacitances of a pentode enable it to be used at frequencies up to a few hundred megahertz. At frequencies above about 300 MHz, however, the inductance of the leads

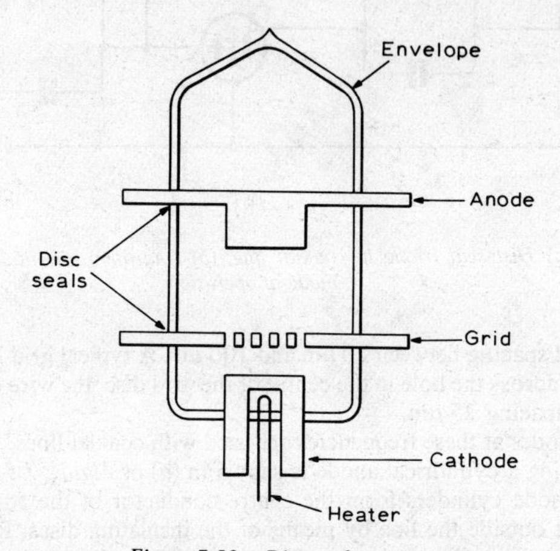

Figure 7.52 Disc-seal triode

connecting the electrodes to the valve pins becomes too high for efficient circuit operation. The electrode structure shown in *Figure 7.4* can no longer be used. A special construction for the electrodes allows the thermionic valve to be used at higher frequencies, but there is a limit to the minimum interelectrode spacing that can be achieved in a commercial valve. Hence there will be a limit to the transit time of an electron from the cathode to the anode, and this imposes an absolute limit on the operating frequency of a valve in the space-charge modulated mode. This limit is approximately 3 GHz to

4 GHz. For frequencies higher than this, special microwave valves such as the magnetron and klystron have to be used.

The 'special construction' mentioned above for valves operating at frequencies above 300 MHz is the 'disc seal' or 'lighthouse' construction. Various forms of this construction have been devised, and a typical disc-seal triode is shown diagrammatically in *Figure 7.52*. The cathode is a cylinder closed at one end, with the oxide coating formed on the end of the cylinder. The grid is planar, and the anode is another cylindrical structure. The cathode cylinder is brought outside the envelope to act as the contact, while the anode and grid contacts are made by discs sealed through the envelope. These contacts have a low inductance. Typical dimensions of a disc-seal triode operating at 3 GHz to 4 GHz are a cathode area of 20 mm^2, an anode diameter of 4.5 mm, and a

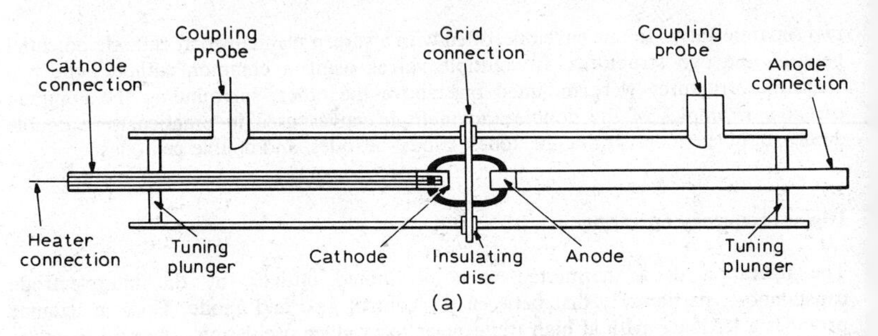

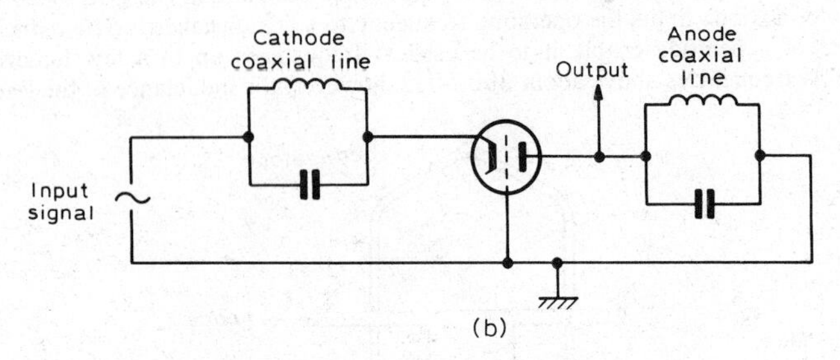

Figure 7.53 (*a*) *Disc-seal triode in coaxial line;* (*b*) *equivalent circuit showing grounded-grid mode of operation*

cathode-to-grid spacing between 20 μm and 100 μm. A typical grid is formed of parallel tungsten wires across the hole in the centre of the grid disc, the wire diameter being, say, 7 μm and the spacing 25 μm.

Disc-seal triodes at these frequencies are used with coaxial lines. An alternative form of the valve using a cylindrical anode is shown in (*a*) of *Figure 7.53*. Extensions to the anode and cathode cylinders form the centre conductor of the coaxial line, while the grid is brought outside the line by means of the insulating discs. Plungers are used to tune the anode and cathode circuits. The anode and cathode signals are coupled by probes.

Although the disc-seal triode has low-inductance connections (with consequent low circuit losses), the interelectrode capacitances are high. The effects of these capacitances are minimised by the mode of operation of the triode, the grounded-grid mode, shown in *Figure 7.53* (*b*). The grid forms an earthed screen between the anode and cathode to prevent feedback. The arrangement shown in *Figure 7.53* (*a*) can be used as an amplifier in systems operating at the lower microwave frequencies, or be adapted to form an oscillator.

FURTHER READING

OPERATION OF VALVES

Many modern textbooks contain sections on thermionic valves. A selection is listed below.

GAVIN, M. R. and HOULDIN, J. E., *Principles of Electronics*, Physical Science Texts, English Universities Press, London (1959)

HARVEY, A. F., *Microwave Engineering*, Academic Press, London (1963)

RAMEY, R. L., *Physical Electronics*, Wadsworth Publishing Company, Belmont, California (1961)

SEYMOUR, J., *Physical Electronics, An Introduction to the Physics of Electron Devices*, Electronic Engineering Series, Pitman, London (1972)

THOMSON, J. and CALLICK, E. B., *Electron Physics and Technology*, Physical Science Texts, English Universities Press, London (1959)

DEVELOPMENT OF THERMIONIC VALVE

DEKETH, J., *Fundamentals of Radio-Valve Technique*, Philips Technical Library, N.V. Philips Gloeilampenfabrieken, Eindhoven, The Netherlands (1949). Chapters 6 to 8 contain much information on the construction and manufacture of immediate post-war valves which incorporated many features dating from the 1930s

STURMEY, S. G., *The Economic Development of Radio*, Duckworth, London (1958). Chapter 2 deals with the development of the thermionic valve

Thermionic Valves 1904–1954, The First Fifty Years, Institution of Electrical Engineers, London (1955)

Textbooks of the immediate post-war period contain information on valves of the late 1930s and World War II. Two examples are given.

BOLTZ, C. L., *Wireless for Beginners*, 4th edn., Harrap, London (1949)

REYNER, J. H., *Modern Radio Communication*, Vol. 1, 8th edn., Pitman, London (1947)

HIGH-POWER TRANSMITTING VALVES

In the early years of wireless, high-power arc transmitters or high-frequency alternators were used for generating the signal and thermionic valves were only used in the receiving equipment. However the most practical way of transmitting speech and music was by using transmitters employing thermionic valves. Thus valves had to be developed capable of producing the high powers required.

The natural way to do this was by making enlarged versions of the receiving valves of that time which had the electrodes enclosed in a glass envelope but the only way these electrodes could be cooled was by radiation. To limit degassing, the glass temperature was limited to 200 °C and this put a severe restriction on the power achievable. In 1920 transmitters used banks of such valves each being about 180 mm diameter and 440 mm long and each capable of dissipating 800 watts.

The breakthrough came in 1919 when W. G. Houskeeper patented a method of sealing base metals through glass.[1] Before this the leads brought through the glass had to be wires or ribbons. With this invention it was possible to make large diameter seals using metals such as high purity copper which has high conductivity, and is easily worked into any desired shape. The expansion of copper is $165 \times 10^{-7}/°C$ and that of a common glass $52 \times 10^{-7}/°C$ so a matched seal is not possible. The technique developed by Houskeeper was to taper the metal to a feather edge and then seal the glass to the thin edge of the taper. The copper is cleaned, oxidised and then sealed to the glass. The thin copper is sufficiently flexible to equalise the difference in expansion and contraction

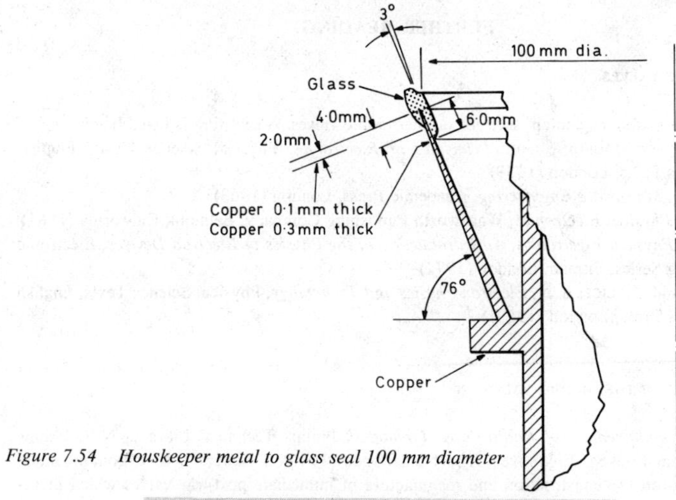

Figure 7.54 Houskeeper metal to glass seal 100 mm diameter

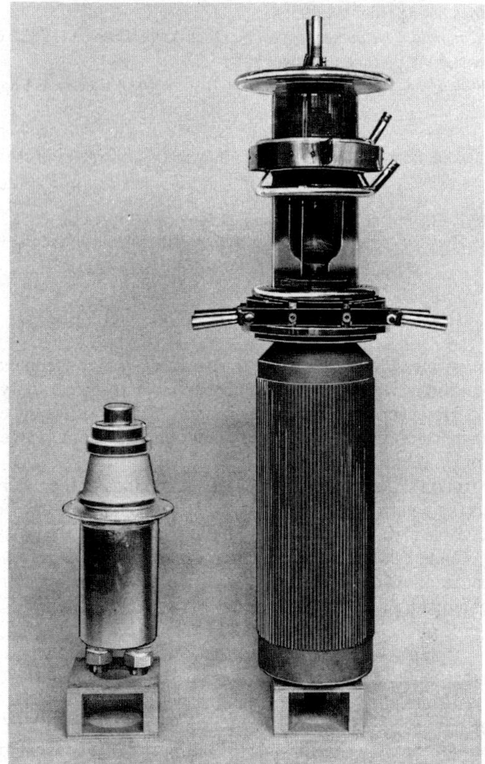

Figure 7.55 Triode CAT 17C (right) and BW1185J2 water-cooled triode (left) (Photo: English Electric Valve Company Limited)

Figure 7.56 Vapour-cooled tetrode CY1172 with 150 kW anode dissipation showing the cooling channels in the thick walled anode (Photo: English Electric Valve Company Limited)

between the glass and metal. The dimensions of a typical seal are shown in *Figure 7.54*.

With this development it became possible to make the anode part of the vacuum envelope and cool its outer surface directly. By 1925 valves with water cooled anodes were in commercial production and they mark the beginning of the era of high-power broadcast transmitters.

A high-power transmitting valve will be defined as one in which the anode forms part of the vacuum envelope which is externally cooled as described above.

A CAT 17 water-cooled transmitting valve made by the Marconi Osram Valve Co. which was in common use in 1938, is shown in *Figure 7.55*. Beside it stands a modern English Electric Valve Co. Ltd. (EEV) triode type BW1185J2 which has similar ratings and output power. These are briefly compared in *Table 7.2*.

Table 7.2

	CAT 17	*BW1185J2*
Length	1 168 mm	446 mm
Weight	35 kg	16 kg
Filament power	15 kW	4.8 kW
Anode dissipation	150 kW	120 kW
Anode voltage	15 kV	14 kV
Output power	209 kW	240 kW
Peak usable cathode current	90 A	150 A

A modern tetrode which has an anode dissipation of 150 kW and is capable of delivering an output power of 220 kW carrier with 100% anode modulation is shown in *Figure 7.56*.

Thermionic valve construction

The mechanical design and construction of high-power transmitting valves is controlled by two factors.

(*a*) The envelope must maintain a good vacuum for many thousands of hours despite repeated thermal cycling.
(*b*) The electrodes operate at temperatures above 1 700 °C and are cycled between this and room temperature. The solution to the problem of maintaining small clearances and accurate shapes over this temperature range is the secret of reliable valve design. The effects of expansion and differential expansion are of paramount importance and must always be the first consideration in the mechanical design.

ENVELOPE

Early valves all used glass as the envelope material.[2] Modern valves such as those illustrated use a high alumina ceramic. The advantages of this are:

(1) Greater strength.
(2) Higher temperature of processing leading to improved reliability and longer life.
(3) The manufacture by hand of large-diameter glass to metal seals is no longer required. This is replaced by a simple metallising followed by a brazing operation.

The ceramic commonly used contains 94% Alumina the remainder being glass-like fluxing agents. Before brazing the ceramic must be metallised. This is done by applying to the surface a thin coating of a fine powder of molybdenum or a mixture of molybdenum and manganese. The ceramic is then heated to 1 470 °C in an atmosphere of hydrogen and water vapour. The powder is bonded to the ceramic by this process in which the fluxing agents contained in the ceramic play a vital role. The metallised surface is electroplated with nickel 5 μm thick to assist in subsequent brazing.

If due consideration is given to differential expansion many metals may be brazed to the metallised ceramic. The two most important for the construction of high power transmitting valves are copper and Nilo K.

Copper has the advantage of cheapness, ease of fabrication and excellent electrical and thermal conductivity. However, its expansion coefficient is 165×10^{-7}/°C compared with 80×10^{-7}/°C for the ceramic. By taking advantage of its poor strength however satisfactory seals can be made provided that the thickness is limited to about 0.75 mm and that a ceramic is always brazed to both sides of the metal. The weakness of the thin copper necessary for the ceramic metal seal leads to other complications since it is not strong enough to withstand the forces due to atmospheric pressure. To overcome this the design shown in *Figure 7.57* has been developed.[3] In this the thrust due to atmospheric pressure is taken directly on the ceramic while expansion differences are taken up by allowing the ceramic to slide over the metal on which it rests.

Nilo K, also known as Kovar, is an alloy of nickel and iron (Fe54, Ni29, Co17) which was originally developed to match the expansion of borosilicate glasses. With a mean expansion coefficient of 95×10^{-7} over the temperature range of 30 °C–750 °C it is a reasonable match to the ceramic and it is possible to make seals to metal at least 1.5 mm thick and in many cases there is no need to have a ceramic both sides of the seal. Nilo K

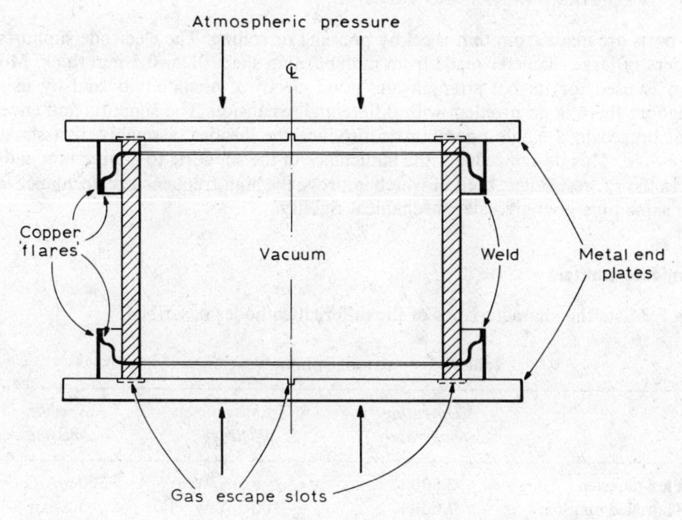

Figure 7.57 Design of vacuum envelope using ceramic metal seals

s much stronger than copper and can be a load bearing member. Its thermal and more mportant electrical conductivity however are poor compared with copper and to reduce he r.f. losses which would occur if bare Nilo K were used it is electroplated with copper.

In modern valves the envelope and electrode supports cannot be considered separ- tely. In older designs the base supports were tungsten or molybdenum wires and rods n to which additional structures were bolted or welded. Nowadays components are lesigned to be made by mass production methods. This is illustrated in *Figure 7.58* vhich shows the components for the envelope of an EEV triode type BW1185J2. All

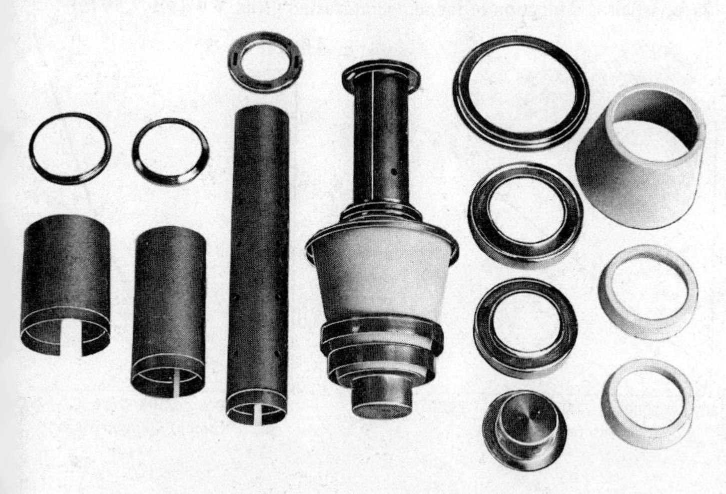

igure 7.58 Components for BW1185J2 (Photo: English Electric Valve Company Limited)

these parts are made from thin sheet by pressing or rolling. The electrode supports are cylinders of large diameter made from molybdenum sheet 0.25–0.5 mm thick. Molybdenum is used for its hot strength and good electrical conductivity and by using it throughout there is no problem with differential expansion. The supports and envelope are all brazed in a single operation to produce the 'header' assembly also shown in *Figure 7.58*. This design reduces the inductance of the supports to a minimum and also reduces the r.f. resistance, both of which improve the high-frequency performance while at the same time giving greater mechanical rigidity.

Thermionic emitters

Table 7.3 lists the characteristics of the different cathodes described.

Table 7.3 CATHODE CHARACTERISTICS

	Operating temperature	*Emission efficiency*	*Specific emission*
Pure tungsten	2 500 K	5 mA/W	300 mA/cm²
Thoriated tungsten	2 000 K	100 mA/W	5 A/cm²
Oxide coated	1 100 K	500 mA/W	10 A/cm²
Barium aluminate	1 300 K	400 mA/W	4 A/cm²

PURE TUNGSTEN

When heated to a high temperature pure tungsten emits electrons. The normal running temperature is 2 500 K when the emission efficiency is about 5 mA per watt of heating power and the total emission about 300 mA/cm².

The current to heat a tungsten wire to a given temperature can be determined from

$$I = A'd^{3/2}$$

where I = required current in amperes
d = wire diameter in cm
A' = constant is a function of the temperature and given in *Figure 7.59* (*a*).

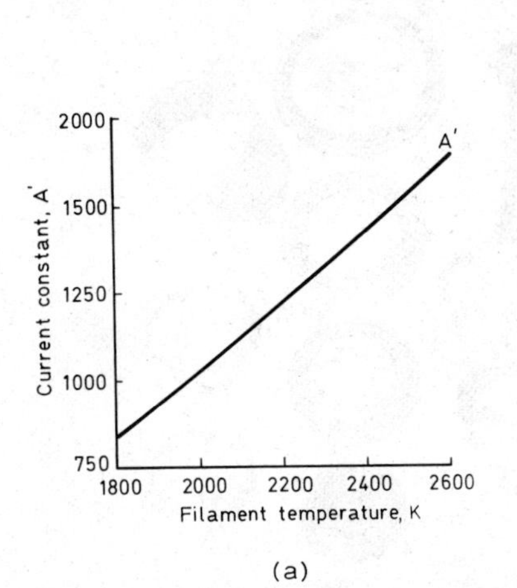

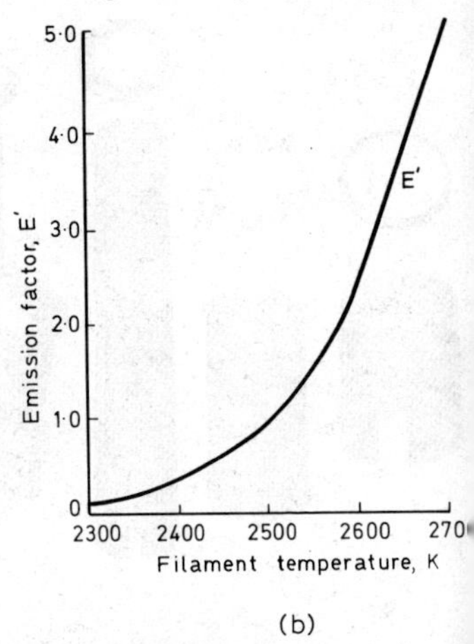

Figure 7.59 (*a*) *Graph of constant A'*; (*b*) *Graph of constant E'*

The emission E amperes is given by a similar equation.

$$E = E'd$$

E being the emission per cm length and E' a constant depending on temperature and shown in *Figure 7.59* (*b*).

Comprehensive tables on the properties of tungsten filaments were prepared by Langmuir and Jones.[4] These and the above equations refer to isolated wires and correction must be made for the mutual heating from adjacent wires and surfaces.[5]

The emission from pure tungsten is little affected by bombardment with high energy ions so that it is suitable for use in high voltage tubes with poor vacuum. It can also be operated close to current saturation and it is usual to design for a peak emission only 110% of that required.

The life of a tungsten filament can be estimated from the rate of evaporation at the operating temperature. Experience has shown that the life is the time for the filament

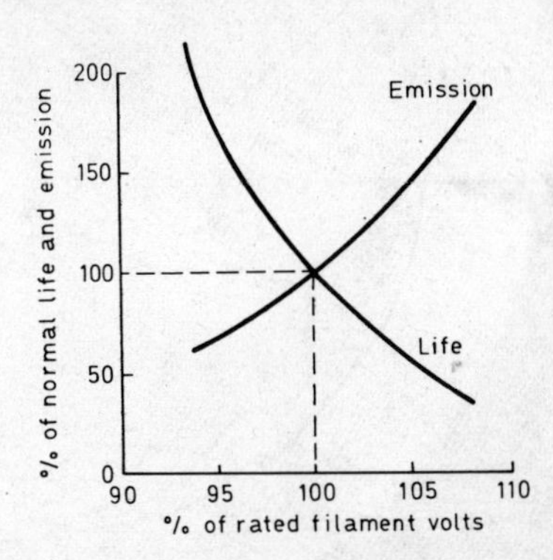

Figure 7.60 Life and emission v filament voltage for a tungsten filament

diameter to be reduced by 6%. However, the evaporation rates given by Reimann[6] should be used. The evaporation rate varies rapidly with temperature and the life is therefore critically dependent on the filament voltage. The variation is shown in *Figure 7.60* from which it can be seen that a 5% increase in filament voltage halves the life.

THORIATED TUNGSTEN

It had been known since 1914 that traces of Thoria enhanced the emission from tungsten filaments but the improvement was unstable. In 1923 it was discovered by Langmuir[7] that if the surface of the filament was carburised, i.e., converted to tungsten carbide the stability of the emission was improved and was less affected by the vacuum conditions.

Today tungsten with 1–2% Thoria is used to make the filaments of large transmitting valves almost without exception. The filament is carburised by heating to 2 200 K in a hydrocarbon atmosphere. This can be done either at a pressure of 10^{-2} to 10^{-3} Torr or the hydrocarbon vapour may be diluted in a gas such as hydrogen.[8] The hydrocarbon gas in contact with the hot filament cracks to deposit carbon on the filament which reacts with the tungsten to form tungsten carbide. If the temperature is in excess of 2 200 K other reactions can occur which are harmful to the emission. A section of a carburised filament is shown in *Figure 7.61*. It is usual to carburise between 20% and 30% of the cross sectional area. If a figure less than this is used the life is adversely effected and if it is exceeded the filament may become excessively brittle.

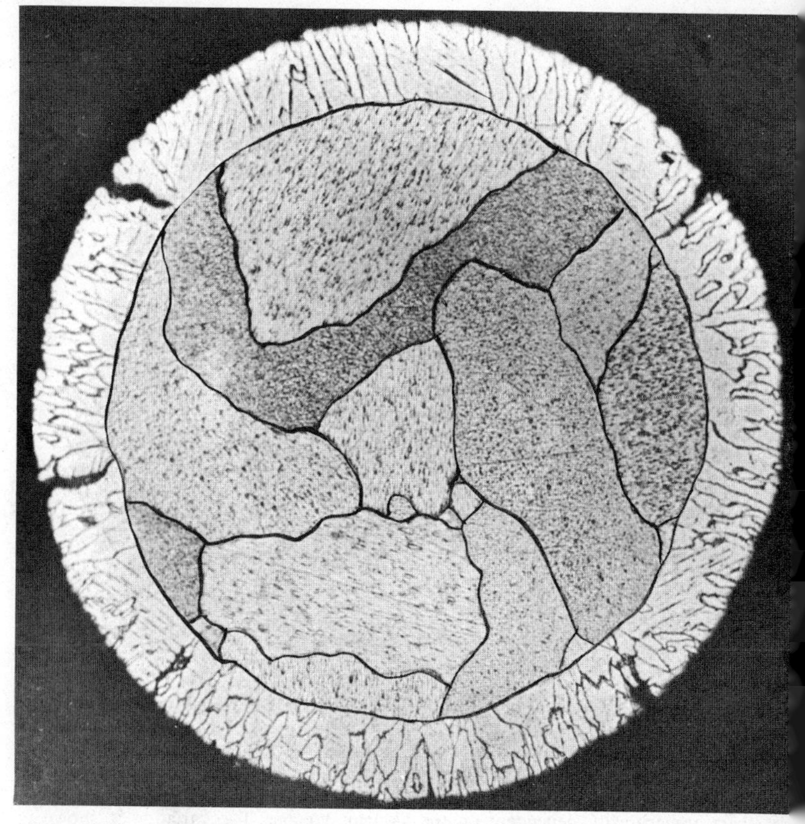

Figure 7.61 Section of a carburised thoriated tungsten filament. Magnification 400 (Photo. English Electric Valve Company Limited)

The percentage carburisation may be determined from the relationship

$$\% \text{ carburisation} = 107.5 \frac{R_2 - R_1}{R_2}$$

where R_1 = initial resistance at room temperature
R_2 = final resistance at room temperature

The filament may be designed using the equations for pure tungsten filaments since at 20% carburisation the change in resistivity cancels out the change in emissivity of the filament. However, the operating temperature is in the range 1 900 K to 2 100 K when the emission efficiency is 100 mA/W and total emission 5 A/cm^2. If the temperature is below this the replenishment of the emitting surface is too slow and the valve will fail because of low emission. If the operating temperature is too high the carbide is rapidly lost from the surface by diffusion of carbon into the core of the wire. Although greatly enhanced emission can be obtained for a short time, once the wire surface is decarburised, the emission will fail. In the former case the emission can often be recovered by running with 10% excess filament voltage for a short time. In the latter case, however, the process is irreversible.

The thoriated tungsten filament is prone to damage from high energy ion bombardment. The surface is protected by the presence of space charge and it is therefore usual to use a design safety factor of at least 2 for the saturated emission. For the permissible d.c. current which may be drawn from the cathode, a reliable figure for design is 5 milliamps d.c. anode current per watt of filament power.

OXIDE COATED CATHODES[9]

These cathodes are easily poisoned by residual gases or damaged by high energy ion bombardment. Therefore their use is restricted to the smaller tubes with anode voltages below 3 kV.

The emissive surface is a thin layer of a mixture of the oxides of Barium. Strontium and Calcium on a metal core. The usual core is made from nickel containing closely controlled amounts of activators such as silicon magnesium and manganese. The coating may be applied by spraying a mixture of the carbonates on to the base to form a layer about 0.25 mm thick. The carbonates are subsequently converted to the oxides during processing with the release of carbon dioxide.

At 1 000 K an emission efficiency of 500 mA/W and a saturated emission of 10 A/cm^2 is obtained.

BARIUM ALUMINATE CATHODE[10]

This is a recent improvement to the oxide cathode. It is made by impregnating a porous tungsten body with a mixture of Barium and Calcium Aluminate. If a film of Osmium 0.5 μm thick is evaporated on to the surface the emission and resistance to ion bombardment is still further improved.

Such cathodes run at a temperature of 1 300 K when a current of 4 A/cm^2 has been drawn for many thousands of hours.

MECHANICAL CONSTRUCTION

To obtain the required emission of several hundred amperes the filament is made from a number of separate wires. The difficulty lies in the expansion between room temperature and the operating temperature. In the case of a thoriated tungsten wire 100 mm long, operating at 1 900 K, this amounts to almost 1.0 mm. The first successful design was the free-hanging cage filament, *Figure 7.62* (*a*). Small differences in the temperature of individual wires introduce buckling forces and these are relieved by cranking the ends of the wires as shown. The lower operating temperature of thoriated tungsten permits the improvement of positional accuracy by fixing the radial position with molybdenum spiders at one or more positions.

Another solution is to tension the filament wires by means of springs. The springs cannot operate at the filament temperature so that they have to be removed to a colder place and the tension transmitted by some means which inevitably involves sliding joints, *Figure 7.62* (*b*).

The latest designs use a mesh of fine wires spot welded at each cross-over and supported at each end, *Figure 7.62* (*c*).

Grids

DESIGN

The dimensions of the grid are determined by the required characteristics while the materials used in its construction and the mechanical design are dictated by its operating environment.

Figure 7.62 *Filament designs:* (*a*) *freehanging* (*b*) *tensioned* (*c*) *mesh* (*Photo: English Electric Valve Company Limited*)

The normal electrical requirements are:

(*a*) Amplification factor μ
(*b*) Perveance or the anode current for given electrode potentials.

There are several formulae which give approximate values for μ in terms of the electrode dimensions.[11] For a cylindrical triode

$$\mu = \frac{Lg \log_n \dfrac{Sp}{Sg} - \log_n \cosh \Pi S}{\log_n \coth \Pi S}$$

where μ = amplifier factor
Sp = anode radius
Sg = grid radius
Lg = active length of grid wire per unit axial length of the valve
S = screening fraction = the area of the grid structure divided by total area of the surface containing the grid

Thus for a grid of N parallel wires diameter $2r_g$

$$Lg = N$$

$$\Pi S = \frac{Nr_g}{Sg}$$

and for a grid consisting of a square mesh of wires diameter $2r_g$ and spacing d

$$Lg = \frac{4\Pi Sg}{d}\left(1 - \frac{r_g}{d}\right)$$

$$\Pi S = Lg \frac{r_g}{Sg}$$

It should be noted that μ is independent of the grid cathode distance and this formula only holds if this distance is greater or equal to the wire spacing. Experimental corrections to μ can be made since for a small change

$$\mu \sim \frac{xr_g}{(d-2r_g)^2}$$

x = anode grid distance

The current of a cylindrical triode is given by

$$I = \frac{14.66 \times 10^{-6}\left(V_g + \dfrac{V_a}{\mu}\right)^{\frac{1}{2}}}{Sg\beta cg^2\left[1 + \dfrac{1}{\mu}\left(\dfrac{Sp\,.\,\beta cp^2}{Sg\,.\,\beta cg^2}\right)^{\frac{2}{3}}\right]^{\frac{3}{2}}} \quad \text{amps per unit length}$$

$$\beta = u - \frac{2u^2}{5} + \frac{11u^3}{120} - \frac{47u^4}{300} + \ldots \quad \text{and} \quad u = \log_n\left(\frac{S}{Sc}\right)$$

Sc being the cathode radius and S the radius of the other electrode being considered. If S/Sc is greater than 7, then β^2 is approximately 1.0.

CONSTRUCTION

The choice of materials and method of construction are determined by the requirements for:

(*a*) mechanical stability;
(*b*) low thermionic emission;
(*c*) low secondary emission.

To achieve the desired performance, the grids have to be positioned within a millimetre or so of the filament which in the case of thoriated tungsten runs at 2 000 K High-power transmitting valves are operated in the positive grid region when the grid intercepts part of the cathode current and the energy lost by the electrons hitting the grid heat it to temperatures which may be as high as 1 700 K. In a perfect valve the grid would be at a uniform temperature but in practice this is not the case. Computer studies have shown that temperature variations as little as 100 °C can produce buckling forces sufficient to cause permanent deformation of the grid. As a result of these studies designs using a mesh structure as shown in *Figure 7.63* have been produced.

The grids are wound of tantalum, molybdenum or tungsten to withstand the high operating temperatures and are spotwelded at each crossover. A large grid may have as many as 4 000 welds.

Due to the close proximity of the grid and filament, thorium evaporated from the filament is deposited on the grid and at the grid operating temperature this will cause thermionic emission from the grid. The direction of flow of this current is opposite to that of the normal current, is unstable tending to increase with life, making the circuit design for stable operating conditions difficult. Steps are therefore taken to suppress grid emission. The best-known material to do this is platinum applied either as a cladding or as an electroplated coating on the grid wire. As the temperature is raised intermetallic compounds are formed between the platinum and the wire which reduce its ability to suppress thermionic emission and finally cause dimensional changes leading to

Figure 7.63 *Square mesh grid wire diameter 0.3 mm (Photo: English Electric Valve Company Limited)*

grid distortion, *Figure 17.64*. This limits the grid dissipation to 7 W/cm^2 of grid surface.

To prevent this occurring a barrier layer may be used between the platinum and wire. One method uses a thin layer of zirconium carbide powder sintered to the wire at 1 900 °C, which is then electroplated with platinum.[12] Such grids can safely be run at 15 W/cm^2, and 25 W/cm^2 has been claimed.

Secondary emission from the grid also introduces components of negative grid current although in this case they usually get less with life. The effect is usually seen as a region of negative resistance in the grid characteristics or in severe cases the grid current may actually become negative. A bright platinum surface is the worst having a secondary emission coefficient greater than 2, while the zirconium carbide/platinum coated grid is much better.

English Electric Valve Company Limited and others[13] have recently developed grids made from pure carbon, *Figure 7.65*. Carbon has excellent properties for grid use such as high mechanical strength at high temperatures, low thermionic emission and low secondary emission. Grids made in this way can safely be run in excess of 30 W/cm^2.

Anodes

The anode collects the electrons which have passed through the grids. The energy of these electrons is the product of the instantaneous voltage and current and the anode

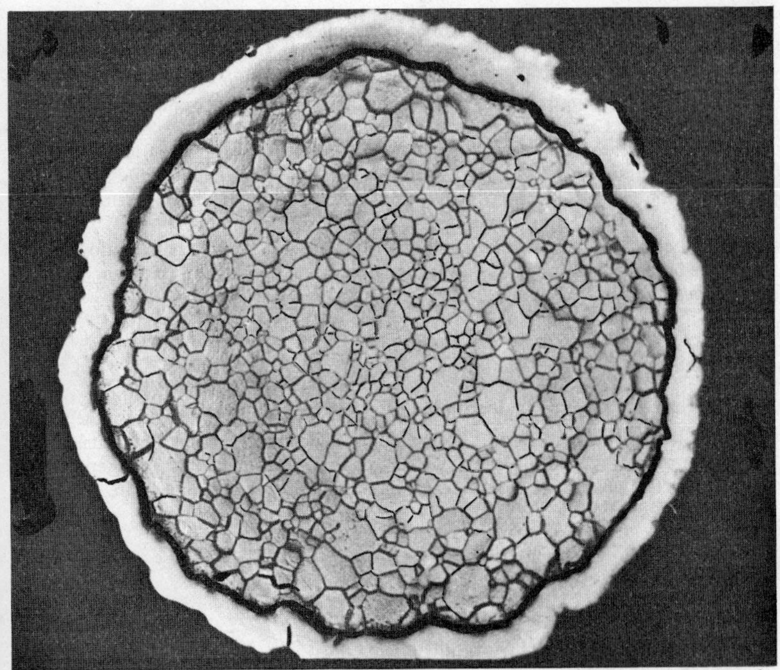

Figure 7.64 Section of a platinum clad molybdenum grid wire. The black layer between the platinum and molybdenum is an intermetallic compound formed due to overheating (Photo: English Electric Valve Company Limited)

Figure 7.65 Pyrolitic carbon shell and grid machined from it (Photo: English Electric Valve Company Limited)

dissipation is this product averaged over a time interval. Modern anodes are invariably made from copper with its good electrical and thermal conductivities. The grade of copper used is oxygen free high conductivity (OFHC) to enable brazing to be performed in hydrogen-containing protective atmospheres and the purity permits the anode to be run at 250–300 °C without problems due to degassing.

Small anodes are turned from the solid or pressed from sheet. The larger sizes are made by vacuum casting. Anodes weighing over 60 kg are made by this process. If holes are required, as shown in *Figure 7.56*, these are cast in by putting carbon tubes in the mould. After solidification the carbon tube is easily drilled out.

The anodes described dissipate anything from a few to several hundred kilowatts and are cooled by passing air or water over the surface or extracting the heat using the latent heat of evaporation.

AIR COOLING

Anodes air cooling have a number of fins brazed on to the core to increase the surface area, and are known as radiators. The maximum anode dissipation is determined by the maximum safe working temperature of the radiator core. To compute the temperature drop across the radiator it is convenient to introduce an average temperature T_{av} and the core temperature T_c where

$$\frac{Tc}{T_{av}} = \frac{\tan hQ}{Q}$$

$$Q = w\sqrt{\frac{2h}{\delta k}}$$

w = radial width of fin (cm)
h = heat transfer coefficient (cal/sec/cm^2/°C)
δ = Fin thickness cm
k = thermal conductivity of fin material (0.9 cal/sec/cm/°C for copper)

T_{av} is defined by the relationship

$$T_{av} = \frac{m}{hs}$$

m being the heat in calories generated per cm length and s the total cooling fin area in cm^2 per cm length of the radiator.

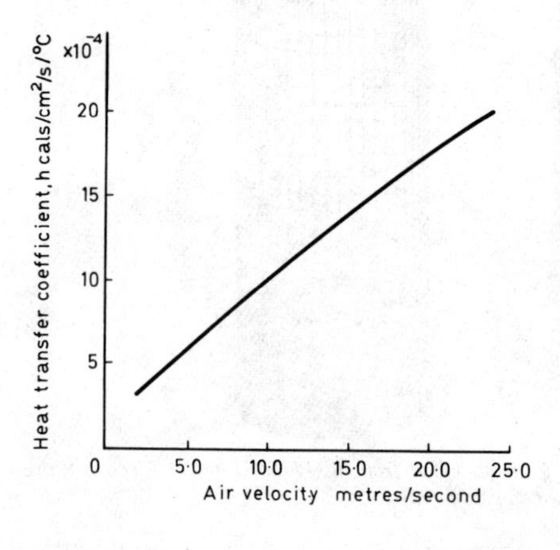

Figure 7.66 Heat transfer coefficient for a plane surface v air velocity

This analysis neglects the effect of longitudinal heat flow to the cold ends which may be corrected by a method derived by Mouromtseff.[14] The values of the heat transfer coefficient h for plane fins is given in *Figure 7.66*. If an air velocity of 20 m/s is exceeded the noise and pressure head to force the air through the radiator increase rapidly and exceed acceptable limits. By using other designs of fir the heat transfer coefficient may be increased. One design, for instance, uses louvred fins, *Figure 7.67*. While the bulk of the air is turbulent there is a thin film of laminar flow through which the heat flows by conduction. The extra turbulence introduced at the edges of the louvres reduces the thickness of the laminar flow layer and thus improves the heat transfer. The ducts between adjacent fins can behave as open ended organ pipes and to prevent this and the resultant noise a 'destabilising' tab is added, *Figure 7.67*. The louvred fin gives about double the heat transfer coefficient of a plane fin. However, this does not mean that twice the heat dissipation can be obtained, since the improved heat transfer increases the temperature gradient and reduces the effective width of the fin. To

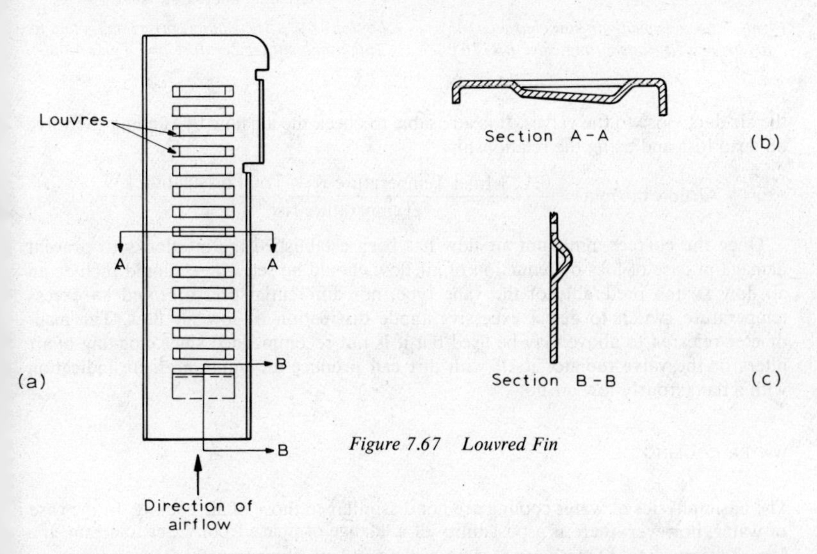

Figure 7.67 Louvred Fin

make room for the louvres the fin thickness or the number of fins may have to be reduced. The net effect is usually not to increase the dissipation but to reduce the weight. Experience has shown that above 50 kW air cooling becomes uneconomical.

The selection of a suitable air blower may be done with the aid of the curves published with the valve data. A typical example is shown in *Figure 7.68*. From these the airflow and pressure head required for a given anode dissipation are obtained. The pressure drops across any air filters, ducting or other restrictions must be added to that across the radiator to establish the fan rating necessary.[15]

At high ambient temperatures and high altitudes a correction must be made to the ground level values, *Figure 7.69*. The blower performance will also be altered and the manufacturers should be consulted.

It is usual for the direction of air flow to be from the filament connections to the anode. If the direction is reversed the hot air from the radiator will be blown into the cabinet. Not only will the valve envelope be uncooled but other components may overheat and fail.

On installation the air flow may be set by means of a water manometer connected into

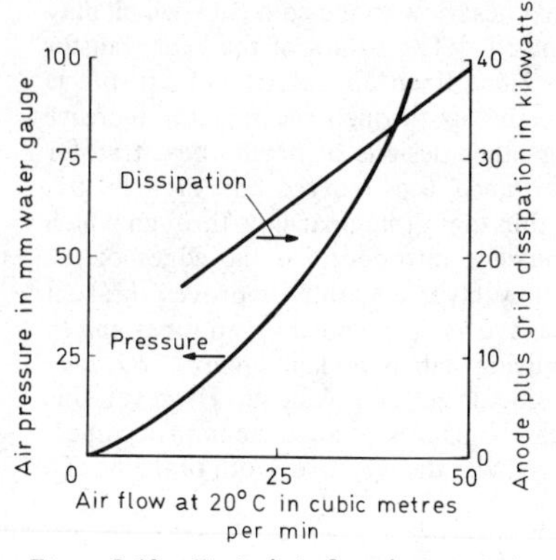

Figure 7.68 Typical air-flow characteristics for air-cooled transmitting valve BR1161

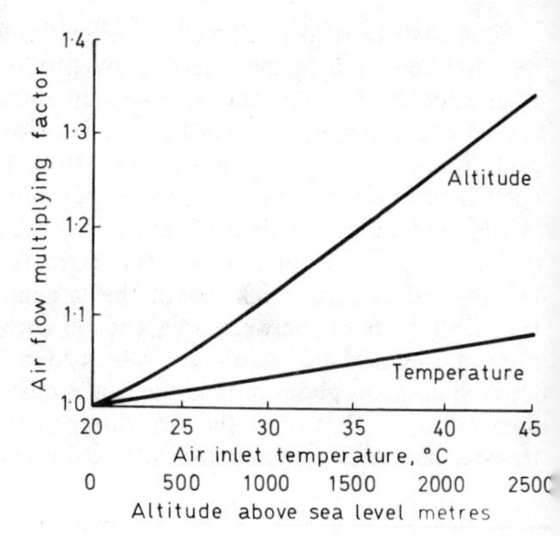

Figure 7.69 Air-cooling correction factors for high ambient temperature and high altitude

the air duct close to the valves. It is advisable to check the air flow by running the valve on dead loss and using the relationship

$$\text{Airflow m}^3/\text{min} = \frac{0.173 \text{ Inlet Temperature K} \times \text{Total dissipation kW}}{\text{Temperature rise °C}}$$

Once the correct minimum air flow has been established the interlocks to prevent damage in case of loss or reduction of air flow should be set. These should include an air-flow switch preferably of the vane type, not differential pressure, and an excess temperature switch to detect excessive anode dissipation or low air flow. The manometer referred to above may be used but it is not recommended since clogging of air filters or the valve radiator itself with dirt can produce an ample pressure indication with a dangerously low air flow.

WATER COOLING

The basic physics of water cooling are not dissimilar to those of air cooling. In the case of water, however, there is a possibility of a change of phase from water to steam. If a film of steam should form over an area of the anode the heat transfer from that spot to the water is interrupted and a hot spot is formed which can cause catastrophic failure of the valve. To prevent this happening the temperature rise is limited to 15 °C and the maximum outlet temperature to 65 °C. The water flow must be turbulent and the velocity sufficiently high to sweep away any steam bubbles as rapidly as they are formed. To do this it is necessary to work with a Reynolds number of at least 4 000

$$Re = \frac{dPv}{\eta}$$

P = density of fluid
v = velocity in cm/sec
η = viscosity
d = equivalent diameter of duct (cm)

$$= \frac{4A}{P} = \frac{4 \text{ duct cross sectional area cm}^2}{\text{wetted perimeter cm}}$$

With this design dissipations around 300 W/cm² can be achieved.

The temperature rise for a given flow rate and dissipation is given by

$$\text{Flow litres/minute} = 15 \times \frac{\text{Total Dissipation in kW}}{\text{Temperature rise °C}}$$

As in air cooling the valve should be protected against cooling water failure by a flow switch and a temperature switch.

Due to circuit considerations the anode of the valve is usual at high d.c. and r.f. potential with respect to earth. Provision must therefore be made to convey the cooling water to the anode via insulating pipes sufficiently long to reduce the leakage currents and r.f. losses to a safe level. If this is not done not only will severe corrosion occur to the water jacket and fittings but the losses in the pipes may be sufficiently high to make the water boil, blocking the pipe with steam, and causing immediate valve failure due to lack of cooling. The leakage current through the water should not exceed approximately 20 mA. A metre length of hose per kilovolt of anode potential is normally adequate.

The water used ought to have a conductivity less than 300 μmhos/cm and preferably less than 100 μmhos/cm. The dissolved solids should not be more than 30 parts per million. If poor water is used corrosion and scaling on the anode occurs. The scale partially blocks the cooling passages and is a bad conductor of heat. As a result the anode overheats leading to premature failure. Solid matter carried by the water has an equally serious result. If a supply of water of adequate quality is not available the only solution is to use a secondary circuit of good water itself cooled by a heat exchanger.

HYPERVAPOTRON-COOLING

Hypervapotron-cooling (Trade Mark of Thomson Brandt) is similar to water cooling. However, by a special design it takes advantage of the latent heat of vaporisation of water and the formation of steam bubbles to obtain greater dissipations and higher water temperatures. Typical valves operate with a 30 °C temperature rise and an outlet temperature of 100 °C. The average water temperature is therefore 85 °C and if the ambient temperature is 45 °C which is common in tropical climates the mean temperature difference is 40 °C. This is the temperature which determines the size of any heat exchanger. For conventional water cooling with inlet and outlet temperatures of 50 °C and 65 °C respectively the mean temperature is 57.5 °C and the mean temperature difference is only 12.5 °C.

It is claimed that dissipations up to 2 kW/cm^2 are achievable by this method.[16]

VAPOUR COOLING

This method of cooling takes advantage of the high latent heat of evaporation of water to remove heat from the valve anode. A valve designed for vapour cooling is fitted with a special thick-walled anode either incorporating a number of vertical cooling channels, *see Figure 7.56*, or machined to form thick ribs.[17] The anode is immersed in water in a container called a boiler. The water in contact with the anode surface boils, the steam being taken via an insulating pipe to a separate steam condenser and the condensed water is returned to the boiler by gravity, no moving parts being required, *Figure 7.70*. The whole system operates at or near atmospheric pressure and since the heat is extracted at 100 °C the mean temperature difference at 45 °C ambient is 55 °C.

The first vapour-cooled tubes were made by Beutheret[18] who used an anode with teeth approximately 10 mm square tapering to 5 mm square over 20 mm protruding from the surface. The object of the teeth was to stabilise the anode temperature and prevent a sudden catastrophic increase in anode temperature known as calefaction.

When a heat conductive wall is heated on one side and exposed to a cooling liquid at its boiling point on the other the law of heat transfer for isothermal conditions is given by a curve known as Nukiyama's curve, *Figure 7.71*. As the power is increased the temperature at the anode surface will increase following Nukiyama's curve until point

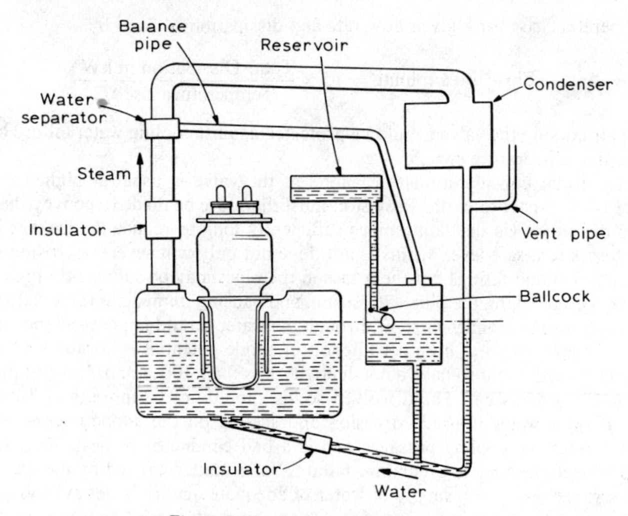

Figure 7.70 Vapour-cooling system

M is reached. Any attempt to increase the power further will force the operating point to jump to *Q* where the temperature is destructive. For an isothermal surface, region *MN* of the curve does not represent an equilibrium condition. The anodes of vapour cooled valves are designed to be non-isothermal. Part of the surface may be operating in region *MN*, part in *ML* and part in region *A*. The temperature of the surface operating in *MN* is stabilised by the temperature gradients and heat conduction to the other regions. Provided that some of the surface is operating in region *A* calefaction cannot occur, for if more power is applied the area operating in regions *MN* and *ML* increases and the temperature in region *A* increases slightly. At maximum dissipation which is about 500 W/cm² the internal temperature of the anode is 250–300 °C.

Figure 7.70 illustrates several features of the system design. The first requirement is

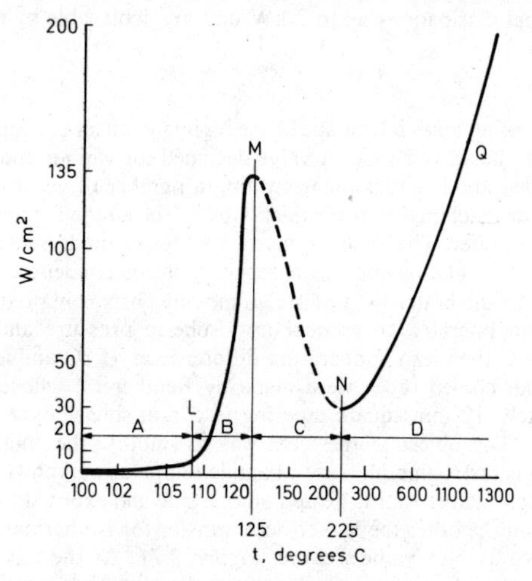

Figure 7.71 Nukiyama's curve

that the boiler should always contain adequate water and in practice it is not difficult to keep the fall in water level in the boiler to less than 25 mm. To help achieve this a balance pipe is used which balances out the pressure drop in the steam main. This pipe should be short, be connected as close to the valve as possible, tilted to prevent blocking by condensed water and at least 12 mm diameter. The other source of level drop is the excess pressure required to make the water flow back into the boiler. Although the flow rate is small, 27 cm^3/kW/min, the pressure must also be small, i.e., less than 25 mm water gauge. The return pipe must therefore be of ample diameter. It is good practice to include a thermal switch in the atmospheric vent pipe which will give warning if steam is being blown off due to a cooling failure or excess dissipation. It is a feature of the system that as long as water remains in the reservoir it may be run like this for long periods quite safely.

The system must only be filled with distilled or deionised water. The distillation process during running concentrates all the impurities in the original water in the boiler. Hence while the system water may have a very low conductivity the water actually in

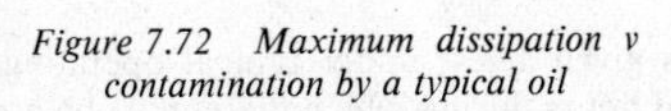
Figure 7.72 Maximum dissipation v contamination by a typical oil

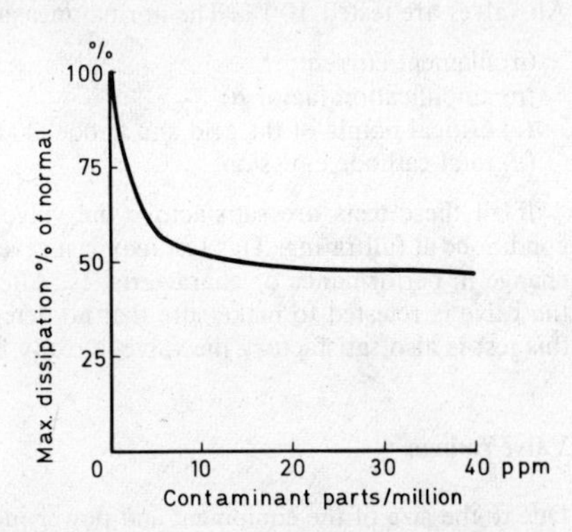

the boiler may be very bad, which can lead to severe corrosion of the valve anode. During off periods the impurities diffuse back down the inlet pipe so that on switch-on electrical breakdown in the inlet pipe can occur, and it is therefore to be recommended that the conductivity of the water in the boiler should be regularly checked. If the conductivity is worse than 100 μmho/cm the water in the boiler only should be replaced.

Traces of oil and grease can result in a severe reduction in the permissible anode dissipation due to the formation of a hydrophobic film on the anode, as shown in *Figure 7.72*.

Getters [27]

Getters are materials incorporated in the valve structure to absorb gas during the life of the valve. In particular Zirconium, Tantalum and Titanium when heated to a high temperature absorb many of the residual gases left in a pumped valve, see *Table 7.4*. The point to note with Zirconium is that hydrogen is absorbed at temperatures below 400 °C and desorbed at higher temperatures. Therefore if Zirconium is used two getters are needed, one at 800 °C or higher and the other at 400 °C.

With the high operating temperature of the electrodes in a high-power transmitting valve it is usually easy to find mounting places for the getters where they will operate at their correct temperatures.

Exhaust

After assembly the valve is evacuated, which may take several days. The pumps used commonly are oil diffusion pumps which have an ultimate pressure about 10^{-7} torr followed by a mechanical or rotary pump to maintain a pressure of one torr at the outlet of the diffusion pump necessary for their correct working. Once the pressure has reached 10^{-5} torr the temperature of the whole valve is raised slowly to 500 °C in an oven and kept at this temperature until the pressure has fallen between 10^{-5} and 10^{-6} torr, when the temperature is reduced to room temperature. Filament volts are next applied and slowly increased and held at the operating value until once again a low and constant pressure is reached. At this stage power is supplied to the grid and slowly increased until the full rated dissipation is reached. The filament and grid power are by now enough to heat the anode red hot about 750 °C. The valve is left like this until the pressure ceases to fall when the power supplies are switched off, the valve is allowed to cool and then sealed off from the pumps, the pressure being 10^{-7} torr or better.

Testing

All valves are tested 100%. The normal measurements include:

(*a*) filament current;
(*b*) amplification factor μ;
(*c*) critical points of the grid and anode characteristics;
(*d*) total cathode emission.

If all these tests are satisfactory the valve is given a test under typical operating conditions at full rating. This test may last several hours during which there must be no change in performance or characteristics. After being stored for a minimum of a week the valve is retested to make sure that no deterioration of the vacuum has occurred. If this test is also satisfactory the valve is ready for the customer.

Valve ratings

Due to the size of the equipment and power involved it is impracticable to perform life tests for all operating conditions. It is usual therefore to life test under class C oscillator conditions and to extrapolate the results to determine the ratings for other operating conditions. The theoretical and practical basis for this was developed by Spitzer.[19]

Table 7.4

	Activation temp. °C	*Operating temp.* °C	*Gettering capacity (ref)*[27] *uL/cm²*				
			CO	CO_2	Hz	Nz	O_2
Zirconium	1 300	800	14.6	12.2	—	5.8	
		350	—	—	53	—	—
Tantalum	1 200	900	9				
Titanium	1 000	700	7				

Operating conditions

CLASS B

By definition a class B amplifier is one biased to cut off so that conduction only occurs for half the electrical cycle, i.e., 180°. This operating condition is suitable for linear r.f.

amplifiers or for push–pull audio power amplifiers. The efficiency that can be achieved is about 60%.

CLASS C

In this condition the valve is biased beyond cut off so that the conduction angle is less than 180° and current only flows for part of each alternate cycle. The angle used is normally between 140° and 120° and the efficiency about 80%.

CLASS D

This is a modified class C amplifier by which efficiencies better than 90% can be obtained. *Figure 7.73* illustrates the current and voltage waveforms for class C and

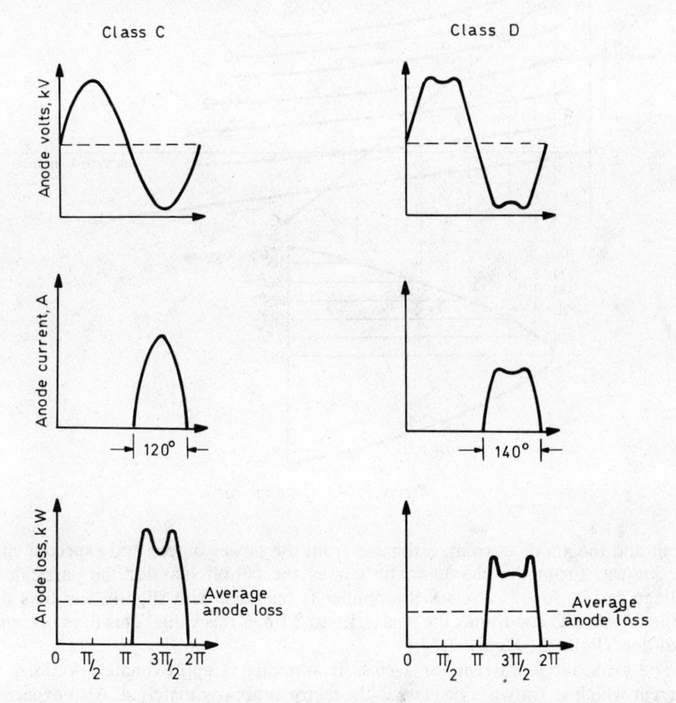

Figure 7.73 Anode voltage, anode current and instantaneous anode dissipation for class C and class D operation

class D. The instantaneous anode loss which is the product of the instantaneous anode current and voltage is also plotted for the two cases, together with the average valve. Because the anode voltage remains close to its minimum throughout the period of conduction the instantaneous and average power loss is less and the efficiency is higher. The circuit to provide the necessary waveforms and give the improvement in efficiency was devised by Tyler.[20]

Calculation of performance

For simplicity the load is considered to be a pure resistance. The load line, i.e., the locus of the points representing coincident instantaneous values of grid and anode voltage, when plotted on the valve constant-current characteristics is then a straight line. If the load is not a pure resistance the load line becomes an ellipse. The full length analysis described below must then be used rather than the quick method.

If the two end points of the load line can be determined then the load line for a resistive load can be drawn on the characteristics and the instantaneous values of the electrode voltages and currents read from the curves. The currents may be plotted against time to give the current waveforms and these may be analysed to give the d.c. fundamental and harmonic components by any of the usual methods.[21]

This process may be simplified by a method due to Sarbacher[22] and Thomas.[23] The end points of the load line are first determined as follows. The anode voltage, power

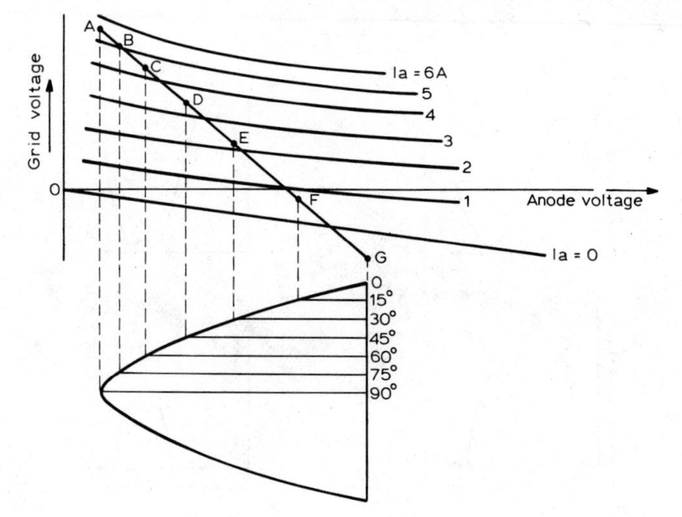

Figure 7.74 Loadline curve

input and the anode current, estimated from the power output and expected efficiency, are known. From the characteristic curves the cut-off bias for the particular anode voltage can be found. A class B amplifier is operated with slightly less bias than this while for class C conditions the bias is 1.6 to 2 times this value. This fixes one end of the load line. (Point *G*, *Figure 7.74*.)

The peak anode current for a class B amplifier is approximately π times the d.c. current which is known. For class C the factor is approximately 4. Also experience has shown that the best results are obtained when the minimum anode voltage is between 10% and 15% of the d.c. voltage. This fixes the other end of the load line (point *A*, *Figure 7.74*), which may now be drawn on the characteristics. A transparent calculator constructed as shown in *Figure 7.75* is laid over the curves between points *A* and *G* with the guide lines parallel to the load line. The instantaneous currents at points *A B C D E F G* which are the currents at 15° intervals through the electrical cycle are read off. These values are entered in the formula to give

$$\text{D.C. Current} = \tfrac{1}{12}(0.5A + B + C + D + E + F)$$

$$\text{Peak Fundamental Current} = \tfrac{1}{12}(A + 1.93B + 1.73C + 1.41D + E + 0.52F)$$

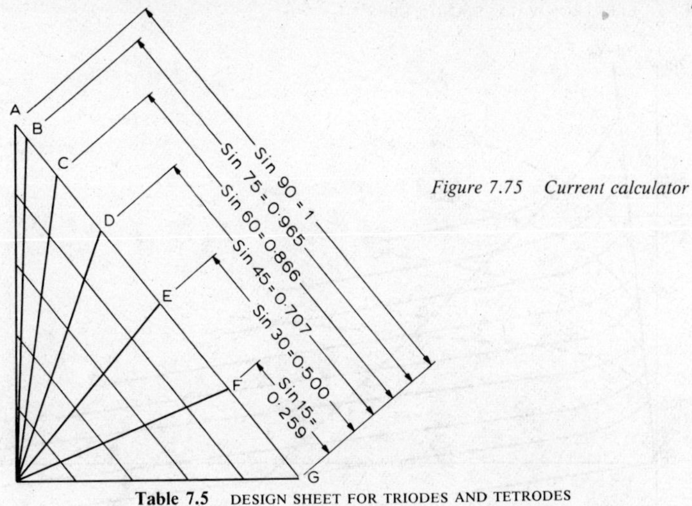

Figure 7.75 Current calculator

Table 7.5 DESIGN SHEET FOR TRIODES AND TETRODES

$V_a = 10$ kV $v_{a\,min} = 1.5$ kV	$V_{g1} = -300$ $v_d = +480$	**Valve type** BR1161
$V_a - v_{a\,min} = 8.5$ kV	$v_s = V_{g1} + v_d =$ 780	$V_{g2} =$ —

Anode Currents		Screen Currents	Grid Currents	
D.C.	**Peak Fund.**	**D.C.**	**D.C.**	**Peak Fund.**
½A = 17.5	A = 35	½A = —	½A = 5	A = 10
B = 34	**1.93**B = 65.5	B = —	B = 8	**1.93** B = 5.5
C = 30	**1.73** C = 51.9	C = —	C = 5	**1.73** C = 8.65
D = 22	**1.41** D = 31	D = —	D = 2.5	**1.41** D = 3.52
E = 9.5	E = 9.5	E = —	E = —	E = —
F = 0.1	**0.52** F = 0.05	F = —	F = —	**0.52** F = —
Totals 113.1	192.95		20.5	37.67

Divide by 12 to give:

$I_a = 9.43$ $\quad I_1 = 16.1$ $\quad I_{g2} = $ — $\quad I_g = 1.71$ $\quad I_{g1} = 3.14$

Input Power $= V_a I_a = 10.0 \times 9.43 = 94.3$ kW

Output Power $= \frac{1}{2}(V_a - v_{a\,min}) \times I_1 = \frac{1}{2} \times 8.5 \times 16.1 = 68.4$ kW

$$\textbf{Efficiency} = \frac{\textbf{Output Power}}{\textbf{Input Power}} \times \textbf{100\%} = 72.5\%$$

Anode Dissipation = Input Power − Output Power = 25.9 kW

Grid Driving Power $= v_s I_g = 780 \times 1.71 = 1.33$ kW

Loss in Bias Source $= V_{g1} I_g = 513$ W

Grid Dissipation = Grid Driving Power − Loss in Bias Source = 817 W

$$\textbf{Grid Bias Resistor} = \frac{V_g}{I_g} = \frac{300}{1.71} = 176 \text{ ohms}$$

$$\textbf{Load Resistance} = \frac{V_a - v_{a\,min}}{I_1} = \frac{8\,500}{16.1} = 529 \text{ ohms}$$

(*Courtesy: English Electric Valve Company Limited*)

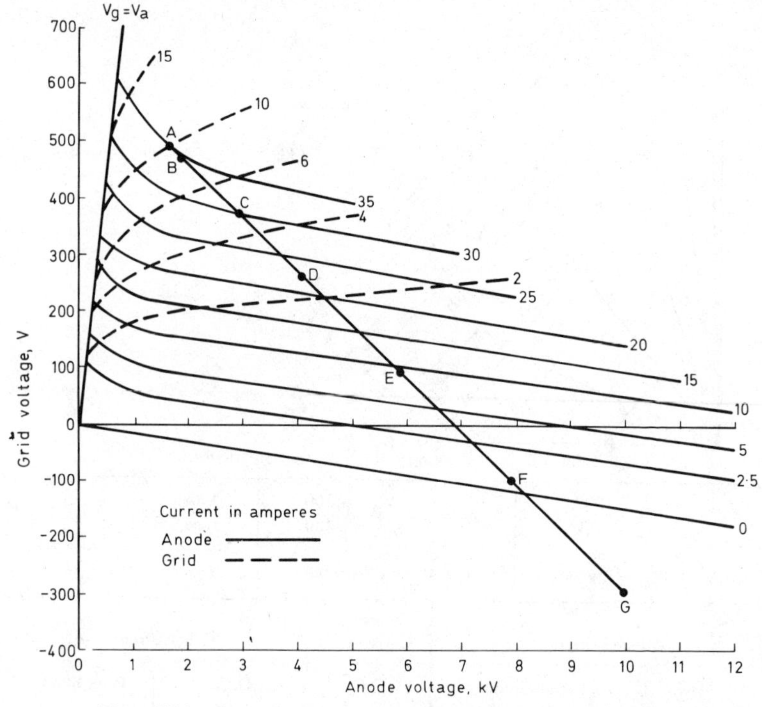

Figure 7.76 Constant current curves for air-cooled triode type BR1161

The use of this method is demonstrated by calculating (*Table 7.5*) the performance of an air cooled triode type BR1161 for class C operation from the curves given in *Figure 7.76*.

High-frequency effects

SEAL HEATING

The anode/grid capacitance of a high-power transmitting valve is typically 100 pF across which is developed the r.f. anode voltage. For example the voltage might be 12 kV and the frequency 30 MHz when the reactive current will be 160 amperes r.m.s. This current flows through the metal/envelope seal whose r.f. resistance is high due to the skin effect and the subsequent heating may cause the seal to fail. The skin depth is inversely proportional to the square root of the frequency and the heating to the square of the current. Hence to keep the seal heating constant the anode voltage must be reduced as the $(\text{frequency})^{5/4}$.

REACTANCE EFFECTS

As the frequency is raised the inductance of the electrode structures becomes important. The effect of cathode inductance is to introduce a conductive component of input admittance G_{in} given by Spangenberg.[11]

$G_{in} = \omega L_c C_{cg} G_m$
L_c = cathode inductance in henries
C_{cg} = grid cathode capacitance in farads
G_m = valve slope amperes per volt
$\omega = 2\pi \times$ frequency in Hz.

For high-power valve

$L_c \doteqdot 10^{-8}$ henry
$C_{cg} = 250$ pF
$G_m = 0.1$ amp/volt
$f = 30$ MHz

when $G_{in} = 8.9 \times 10^{-3}$ mho or $R_{in} = 112$ ohms

This conductance absorbs power and reduces the gain and increases the drive required above that calculated from the characteristic curves alone. This power is not lost but transmitted to the anode.

The interelectrode capacitance also introduces limitations. The equivalent shunt resistance of a parallel tuned circuit is given by

$$R_{sh} = \frac{1}{\omega^2 RC}$$

R being the equivalent series resistance and C the total tuning capacitance. For high-power output R_{sh} must be large and once C is reduced to the interelectrode capacitance only, R_{sh} and the power output fall as the square of the frequency.

Transit time effects

The first effect as the transit time becomes a large fraction of a period is to introduce a conductive component of the input admittance which increases as the square of the frequency. This component is not to be confused with the conductance due to cathode inductance already discussed. Efforts have been made to develop large signal theories and equivalent networks to describe these effects.[24]

As the frequency is still further increased the anode current pulse becomes distorted reducing the efficiency still further.[11]

Pulse operation

Large transmitting valves are sometimes used as high-power pulse tubes supplying currents and power for short periods far in excess of their normal ratings. It is often thought that as long as the average dissipations are within the valve ratings then operation will be satisfactory. This is not necessarily so. Although the anode has an average temperature which is easily measured, during the pulse the inner surface of the anode is being intensely heated and will rise in temperature. Experiments have demonstrated that temperature rises as little as 100 °C can lead to pitting and cracking of the anode surface. Curves which enable the surface temperature of the electrodes to be calculated for various pulse lengths and power densities are given by Singer[25] and Guernsey.[26]

REFERENCES

1. HOUSKEEPER, W. G., 'The art of sealing base metals through glass', *Jour. Am. Inst. E.E.*, **42**, 870 (1923)
U.S. Patents 1, 294, 466, 18 Feb. (1919) and 1, 560, 690, 10 Nov. (1925)

2. METALLISING MONACK A. J., 'Theory and Practise of Glass Meta Seals', *Glass Industry*, **27**, 389–582 (1946)
3. WILLIAMS, P. D., U.S. Patent 2, 879, 428, 24 March (1959)
4. LANGMUIR and JONES, *G.E. Review*, **30**, No. 6, 310 and No. 7, 354 and No. 8, 408 (1927)
5. POHL, W. J., 'Mutual heating in Transmitting Valve Filament Structures', *Proc. I.E.E.E.*, Part C, No. 3, 224 (1956)
6. REIMANN, A. L., 'The Evaporation of Atoms, Ions and Electrons from Tungsten', *Phil Mag.*, **25**, 834 (1938)
7. LANGMUIR, I., 'Thoriated Tungsten Filaments', *J. Franklin Inst.*, **217**, 543 (1934)
8. SCHNEIDER, P., 'Thermionic Emission of Thoriated Tungsten', *J. Chem. Phys.*, **28**, 675 (1958)
9. HERMANN, G. and WAGENER, S., *The Oxide Coated Cathode* (2 Vols) Chapman & Hall Ltd., London (1951)
10. ZALM, P. and VAN STRATUM, A. J. A., 'Osmium dispenser Cathodes', *Philips Tech. Rev.*, **27**, 69 (1966)
11. SPANGENBERG, K. R., *Vacuum Tubes*, McGraw-Hill (1948)
12. Australian Patent 220414, 26 Feb. (1959)
13. CFTH, British Patent 1,206,049
14. MOUROMTSEFF, I. E., 'Temperature Distribution in Vacuum Tube Coolers with Forced Air Cooling', *J. App. Physics*, **12**, 491 (1941)
15. PANUET, W. E., *Radio Installations*, Chapman & Hall Ltd., London (1951)
16. CFTH, British Patent 1,194,249
17. CFTH, British Patent 940,984
18. BENTHERET, C., 'The Vaportron Technique', *Rev. Tech. Thomson-CSF*, **24** (1956)
19. SPITZER, E. E., 'Principles of Electrical Ratings of High Vacuum Power Tubes', *Proc. I.R.E.*, **39**, 60 (1951)
20. TYLER, V. J., 'A New High Efficiency High Power Amplifier', *Marconi Review*, **21**, 96, 3rd Quarter (1958)
21. WHITTAKER and ROBINSON, *The Calculus of Observations,* Blackie & Son Ltd.
22. SARBACHER, R. I., 'Graphical Determination of Power Amplifier Performance', *Electronics*, **15**, 52, Dec. (1942)
23. THOMAS, H. P., 'Determination of Grid Driving Power in Radio Frequency Power Amplifiers', *Proc I.R.E. 21*, 1134 (1933)
24. LLEWELLYN, F. B. and PETERSON, L. C., 'Vacuum Tube Networks', *Proc. I.R.E. 32*, 144 (1944)
25. SINGER, B., 'Pulse Heating of Grids & Anodes in High Power Applications', *Cathode Press*, **23**, 24 (1966)
26. GUERNSEY, G. L., 'High Power Tube Techniques', *Advances in Electron Tube Techniques*, Pergamon Press (1963)
27. WAGENER, J. S., '4th National Conference on Tube Techniques, 1958', New York University Press

FURTHER READING

YINGST, T. E., CARTER, D. R., ESHLEMAN, J. A. and PAWLIKOWSKI, J. M., 'High-Power Gridded Tubes—1972', *Proc. I.E.E.E.*, March (1973)

KLYSTRONS

Immediately prior to World War II the development of conventional electron tubes at microwave frequencies appeared to have come up against a fundamental limitation imposed as a result of the finite transit times involved at these frequencies. One answer to this problem came in the form of the klystron—a new concept of electron tube which actually exploited transit time effects by making use of the then novel principle of velocity modulation.

Since then many important families of tubes, e.g., travelling wave tube and cross-field amplifiers, have been invented which also make use of velocity modulation but all of these involve a continuous interaction between the travelling wave front and the electron beam.

What characterises the klystron is that the velocity modulation occurs at an interaction gap, short relative to a wavelength and that this modulation is transformed into density modulation by the drifting action of electrons in an r.f. field free space known as the *drift tube*.

The klystron was first described in 1939 by the Varian brothers[1] and by Hahn & Metcalf.[2] The rapid development of radar during World War II was made possible by the advances made in producing microwave power sources. At the higher power level the klystron, limited by the technology of that time to output powers measured in tens of watts, was no match for the magnetron. However, the major shortcomings of the latter, frequency instability, emphasised the need for an electronically tunable local oscillator in the receiver. This requirement was satisfied by the invention in 1940 of the reflex klystron. After the war there was a renewed interest in producing a stable high power microwave amplifier and by the early 1950s a team at Stanford University had produced the first successful multi-cavity klystron. Since then the advances in computing techniques have made it possible to get a much better understanding of the basic operating principles of the klystron. This has led to improved design which when coupled with the advances made in technology has made it possible to build klystrons delivering hundreds of kilowatts of c.w. power and tens of megawatts of pulsed power.

Types and applications

Representative data for commercially available klystrons is shown in *Table 7.6*. Among the applications are:

High-power radars. The first amplifier klystrons were designed for this application where the high output power and the excellent frequency stability, associated with a crystal controlled drive, gave the klystron an advantage over the conventional magnetron. Modern radar systems are much more complex and demand high gain, broad bandwidth and linear phase characteristics. Klystrons are available ranging from 20 MW at L-band (1.0 GHz) to 5 MW at C-band (7.0 GHz). Gains vary from 30 to 50 dB. Although in general bandwidths are relatively narrow at approximately 2% of operating frequency, tubes have recently been announced with bandwidths as high as 10%.

Linear accelerators. Klystrons for this application are very similar to those used for high-power radars except that in general there is not a requirement for broad bandwidths.

Communications. Klystrons are widely used as the power amplifiers in satellite ground stations and tropospheric scatter systems. A feature of this application is that the equipment is often designed to be mobile and this places special emphasis on weight and efficiency. In order to satisfy this requirement special attention has been paid to focusing systems. Permanent magnet focusing is widely used and more recently electrostatically focused klystrons have been specifically developed. C.W. amplifier klystrons are available at frequencies from 200 MHz up to 30 GHz and powers range from tens of watts up to 150 kW.

U.H.F. t.v. amplifiers. Klystrons are favoured as the final amplifiers in u.h.f. t.v. systems. Their high gain 40–50 dB, results in a transmitter which is solid state except for the power amplifier stage. The resulting reliability makes it possible to adopt a system of unattended operation. T.V. systems require an 8.0 MHz bandwidth, repeatable and stable linearity and phase characteristics. Klystrons, covering 470–890 MHz, are available in power levels from 5 kW up to 55 kW of c.w. output power.

Table 7.6 TYPICAL HIGH-POWER AMPLIFIER KLYSTRONS

	Type	*Frequency* MHz	*Peak power* MW	*Average power* kW	*Beam voltage* kV	*Beam current* A	*Number of cavities*	*Bandwidth* MHz	*Gain* dB	*Focusing*	*Weight* lbs
HIGH-POWER RADAR AND LINEAR ACCELERATOR											
Varian	VA812E	400–450	20	300	230	280	5	20/30	36	Solenoid	1 500
Varian	VKL 7755	1 235–1 305	10.5	50	155	159	5	70	43	Solenoid	400
Thomson-CSF	TV2011	2 700–3 100	20	40	270	260	5		50	Solenoid	145
EEV	K 390	2 998	8	12	205	105	4		42	Solenoid	75
Valvo	YK 1110	2 998	6	9	210	100	3		30	Solenoid	240
Varian	VKC 7762	5 395–5 655	1.3	7.6	90	55	7	260	50	Solenoid	135
			C.W. power kW								
COMMUNICATIONS											
Varian	3KM50,000PA	225–400	23.0		23.0	2.6	3	1.0	36	Solenoid	2 103
Varian	3KM300LA	345–455	10.5		30.0	9.0	3	2.0	13	Solenoid	1 202
EEV	4KM50,000LR	755–985	11.5		17.0	1.7	4	6.0	30	Solenoid	404
EMI	VKP 7758	755–985	10.2		17.5	1.63	5	6.0	33	ESF	250
Varian	4K3SJ	1 700–2 400	1.0		6.0	0.54	4	6.0	36	Permanent magnet	85
Varian	VA 936G	5 900–6 400	3.0		8.0	1.05	5	50.0	40	Permanent magnet	70
Varian	VA 934	12 000–18 000	10–20		23.0	3.5	5	60.0	50	Solenoid	92
TELEVISION											
EEV	K 383/4/5	470–860	5.5		9.5	1.9	4	8.0	38	Solenoid	950
EEV	K 370/1/2	470–860	12.0		12.5	2.8	4	8.0	42	Solenoid	950
	K 3082/3/4	470–860	45		22.0	6.2	4	8.0	44	Solenoid	2 200
Varian	VA 953/4/5	470–890	55		24	7.5	5	8.0	48	Solenoid	1 040
Thomson-CSF	TH 2050/1/2	470–860	25		16.5	3.3	5	8.0	46	Solenoid	0.950

Table 7.7 TYPICAL OSCILLATOR KLYSTRONS

Type		Frequency GHz	Output power mW	Beam voltage V	Beam current mA	Description
TWO CAVITY TYPES						
EEV	K3071 series	fixed between 8.0 to 9.5	1 000–5 000	725–1 400	45–125	Fixed frequency waveguide output
Varian	VA533 series	fixed between 12.4 to 18.0	1 000–10 000	800–1 300	45–95	Fixed frequency waveguide output
REFLEX TYPES						
Raytheon	2K28	1 800–4 000	100	300	30	External cavity
EEV	K366 series	6 125–7 900	1 200	750	75	Mechanically tuned waveguide output
Mullard	KS9-20A	8.5–9.66	35	300	25	Mechanically tuned coaxial probe output
EEV	K3111	8.5–9.6	35	300	30	Mechanically tuned waveguide output
EEV	K3080	16.5–17.5	80	330	20	Mechanically tuned waveguide output
EEV	K3038	34.00–36.50	350	2 500	30	Mechanically tuned waveguide output

Oscillators and f.m. sources. The very first klystrons were used as the local oscillators in microwave receivers. Invariably the reflex klystron was used in this application as it could be easily tuned both mechanically and electronically. The electronic tuning property has also resulted in wide use of the reflex klystron as the source in microwave links. Modern reflex klystrons are of rugged construction and are used in airborne radar and missile applications.

Two-cavity klystrons are not easily tuned as both cavities have to be accurately coupled, hence they are normally supplied pre-set to the required frequency. Two-cavity klystrons are more efficient and give higher power than reflex types and are used in airborne doppler radar and other applications when higher power is required.

Typical data for commercially available oscillator klystrons is shown in *Table 7.7*.

Advantages and limitations of klystrons

GAIN

A major advantage of amplifier klystrons is their very high gain. This derives from the fact that the resonant cavities are spaced well apart by the drift tubes and the only coupling is provided by the electron beam. The resulting isolation means that the gain can be safely increased by adding on extra cavities. In this manner gains of 100 dB have been achieved. The limiting factor is either feedback in the form of returning electrons or, more usually, the difficulty in maintaining adequate isolation between the output circuit and the klystron drive.

POWER OUTPUT

The klystron is intrinsically capable of delivering much higher powers than other electron tubes. Indeed the only limitation, assuming a big enough collector, is the d.c. power in the beam.

In practice the klystron dimensions—cavity size, drift tube diameter and beam diameter—all reduce as the operating frequency is increased. As the beam diameter becomes smaller so the beam current density rises and it becomes more and more difficult to produce and constrain the resulting beam. Furthermore, as the physical size of the structure decreases so the problems of thermal stability and power dissipation become greater. The foregoing considerations limit the beam power and hence the output power at the higher frequencies. At the lower end of the frequency spectrum the available power is limited by the high voltages involved and by the problem of designing an adequate collector.

BANDWIDTH

Because klystrons employ high Q cavity resonators they are basically narrow band devices. The bandwidth can, however, be increased at the expense of gain, by stagger tuning and damping (loading) the cavities. Although unable to match the bandwidth of helix travelling wave tubes, recently developed very high power klystrons (20 MW) have been reported to give 10%, which compares very favourably with high power t.w.t.s. At lower powers it becomes more difficult to achieve good bandwidth while maintaining reasonable gain and efficiency and figures of the order of 1–2% are to be expected.

PHYSICAL SIZE

It has previously been stated that the dimensions of a klystron are largely dependent on the operating frequency. In addition the gap and drift tube lengths are related to the operating voltage. As a result a high voltage low frequency klystron can be very big. A further factor at lower frequencies is that the beam focusing system, invariably a solenoid, is also massive. An example of this is the 3KM50PA which, including the solenoid, weighs over 2 000 lb. Conversely at the higher frequencies the tube body becomes very small and the size is determined mainly by the gun and collector. It also becomes possible to use permanent magnet focusing, which results in significant weight saving. In general, any size comparison between a klystron and an alternative tube, e.g., tetrode, magnetron, of comparable power, is unfavourable but despite this the very size of the klystron conveys advantages, particularly where stability, long life and resistance to overload are considered to be important.

Principle of operation

The principle of operation is best explained by considering first the basic two-cavity klystron.

TWO-CAVITY KLYSTRON

The essential structure of a two-cavity klystron is shown in *Figure 7.77*. An electron beam (1) is formed by the electron gun (2). The electrons in this beam are accelerated to a constant velocity determined by the relation:

$$U_o = \sqrt{\frac{2e}{m} V_o} \tag{1}$$

U_o = electronic velocity
e = electronic charge 1.6×10^{-19} coulomb
m = electronic mass 9.1×10^{-31} Kg
V_o = beam voltage

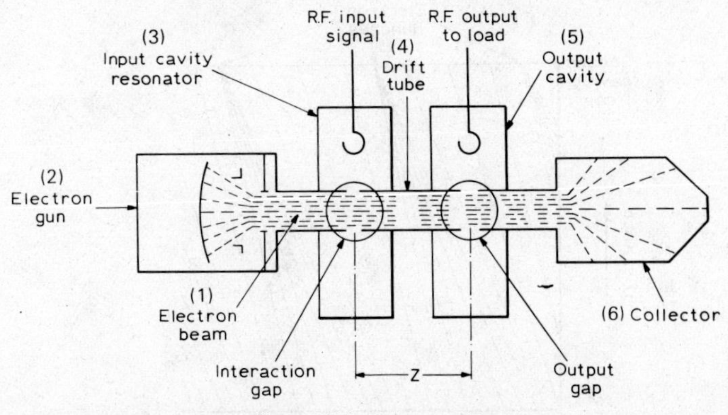

Figure 7.77 Two-cavity klystron

The electrons travelling at velocity U_o enter the input interaction gap which forms part of the input cavity resonator (3). When this cavity is driven by an r.f. signal an oscillating field is developed across this gap parallel to the beam axis. In each successive half cycle the electrons crossing the gap are either accelerated or decelerated depending on the phase of the gap voltage. If the instantaneous gap voltage is $V_1 \sin wt$, where $\omega = 2\pi f$, then the velocity of the emerging electrons will be given by

$$U = U_o\sqrt{1 + M\frac{V_1}{V_o}\sin wt} \simeq U_o\left(1 + M\frac{V_1}{2V_o}\sin wt\right) \qquad (2)$$

where M is the gap coupling co-efficient dependent on the gap geometry and beam diameter.

The velocity modulated beam enters a field free region (4) called the drift tube and here the fast electrons overtake those slowed down during the preceding half cycle. In this way electron bunches are formed and the beam becomes density modulated. This mechanism is illustrated by the time/distance (Applegate) diagram shown in *Figure 7.78*.

The first theoretical work on klystron behaviour was published by Webster.[3] In this it is shown that the value of r.f. current on the beam as a function of drift tube length is given by the relation:

$$i_{RF} = 2I_o J_1(K) \qquad (3)$$

where K the bunching parameter is given by

$$K = \frac{wz}{U_o}\frac{V_1}{2V_o}$$

I_o = beam current
J_1 = a Bessel function of the first kind
z = distance measured from the input gap
V_1 = input gap voltage

The maximum value of $J_1(K)$ is 0.58 and occurs when $K = 1.84$. It follows that for any given value of V_1 the value of $i_{r.f.}$ can be made to be a maximum by making z the appropriate length. In theory, therefore, it is possible to achieve very high gain by making z very long, but in practice space charge de-bunching[4] limits the useful value of z.

The output gap is positioned at a distance z which corresponds to the maximum value

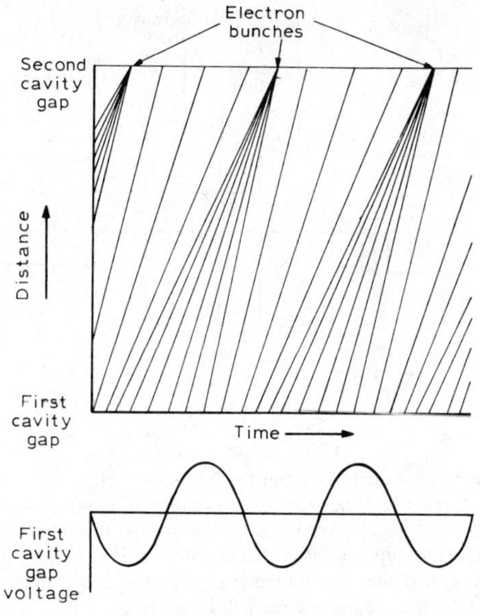

Figure 7.78 Applegate diagram for two-cavity tube

of $i_{r.f.}$. On crossing this gap the bunched beam induces currents in the output cavity (5) and delivers power to the output circuit. The voltage developed across the gap is in anti-phase with the electron bunches. Consequently, more electrons cross the gap during the retarding voltage half cycle than do during the positive cycle. There is, therefore, a nett transfer of energy during each cycle. The electrons are slowed down as they cross the gap and the maximum useful peak voltage is that which just stops the slowest electrons and is approximately the beam voltage V_o.

The spent beam, i.e., the beam that emerges out of the output gap travels into the collector (6) where the remaining energy is dissipated as heat. It is possible to predict a theoretical maximum efficiency for this idealised klystron. This will occur when the peak of $i_{r.f.}$ reaches the maximum value of $1.16 I_o$ and the peak r.f. voltage across the output gap is V_o. The output power is then

$$P_{r.f.} = \frac{1.16 I_o V_o}{2} = 0.58 I_o V_o$$

and the efficiency is 58%.

MULTI–CAVITY AMPLIFIER KLYSTRON

A multi-cavity klystron, *Figure 7.79*, can usefully be considered as a power amplifier section (the penultimate and output cavities) being driven by a driver section (all the earlier cavities driving the penultimate cavity). The power section behaves very much as described for the two-cavity klystron under large signal conditions and largely determines the linearity of the device. The driver section behaves like one or more two-cavity klystrons in cascade under small signal conditions and largely determines the gain and bandwidth of the device.

The beam is velocity modulated at the input cavity by a small r.f. voltage. The signal

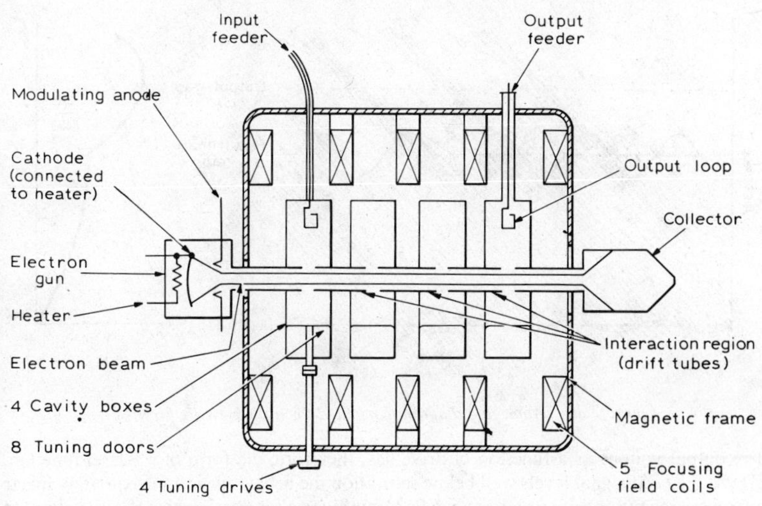

Figure 7.79 Schematic diagram of four-cavity klystron amplifier

level, and hence the electron velocities are insufficient to overcome the space charge repulsion forces between the electrons. Under these conditions the space charge wave theory of Hahn[5] and Ramo[6] is applicable and the r.f. current maximum occurs at a distance of a quarter plasma wavelength Z given by

$$Z = \frac{U_o}{2Wq} \tag{5}$$

Z = distance from the first gap
U_o = initial electron velocity
Wq = reduced plasma angular frequency

A second, lightly loaded, cavity is positioned at this distance and because of its high impedance a significantly higher voltage is developed across the second gap. This voltage now remodulates the beam and produces a higher r.f. current maximum at a third gap situated a further 0.25 plasma wavelength away. The same process is repeated at this gap resulting in a higher voltage still. Depending on the number of cavities this process continues until the penultimate gap is reached. By now the signal level at the gap will be quite large and the space charge forces will be overcome so that the basic two-cavity klystron mechanism will apply. Webber[7] has shown that under large signal conditions the optimum drift tube length is 0.1 of a reduced plasma wavelength and this figure is normally used in klystron design. A computed plot of the bunching in a three cavity klystron is shown in *Figure 7.80*.

In the driver stages the gap voltages are small relative to the beam voltage and the amplification is linear. However, in the output stage the 'drive' level (voltage at penultimate gap) is sufficiently high for the large signal conditions to apply and the r.f. current at the output gap is given as before by

$$I_{\text{r.f.}} = 2I_o J_1(Kp) \tag{6}$$

where Kp is the bunching parameter at the penultimate gap

$$Kp = \frac{wz}{U_o} \frac{V_{\text{PEN GAP}}}{2V_o}$$

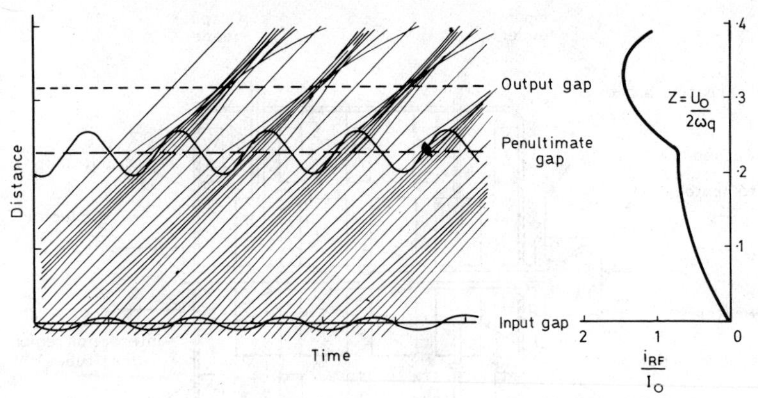

Figure 7.80 Applegate diagram computed for a three-cavity klystron

The output voltage as a function of drive has, therefore, the form of a Bessel function, *Figure 7.81*. At signal levels well below saturation the behaviour is approximately linear but on raising the power level the gain falls smoothly to a saturation value 6.0 dB below the small signal gain. The slope of this curve determines the differential gain of the klystron.

As the output power increases so kinetic energy is extracted from the beam. This results in a reduction in the velocity of the electrons which introduces a delay and hence a phase shift. This level dependent phase shift is known as a.m./p.m. conversion, *Figure 7.81*. In the particular case of television the a.m./p.m. conversion gives rise to the effect known as differential phase.[8]

Focusing. Most amplifier klystrons require a long electron beam of constant diameter. It is, therefore, necessary to provide some means of constraining the beam which would otherwise spread as a result of the electron repulsion forces. By far the most common solution is to apply a magnetic field along the axis of the beam.

For any beam there is a value of magnetic field, known as the Brillouin field, at which the centripetal forces generated by the electrons rotating in the magnetic field balance exactly the space charge repulsion forces. This value is given by the relation:

$$B_{\mathrm{BRILL}} = \frac{8.32 \times 10^{-4}(I_o)^{\frac{1}{2}}}{r_o V_o^{\frac{1}{4}}}$$

r_o = beam radius

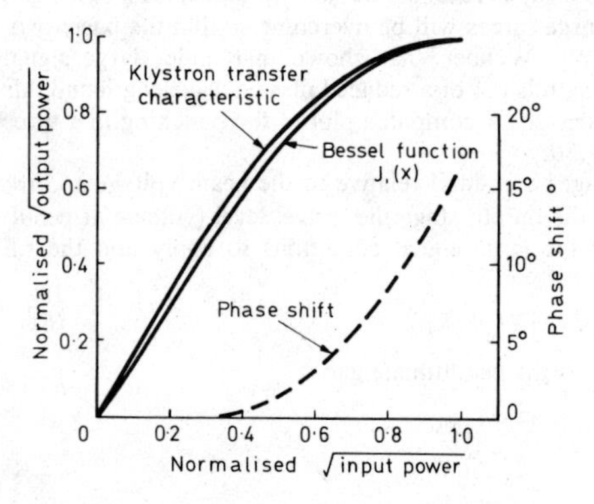

Figure 7.81 Typical klystron transfer characteristic

Unfortunately Brillouin flow requires that the current density in the beam remains constant—a condition which is not satisfied in practical klystrons. It is more usual to employ field values twice to three times Brillouin. Under these conditions the beam diameter perturbations caused by changes in current density are much reduced. This technique does, however, require that some of the magnetic flux linking the beam must also link the cathode in order to launch the beam at the correct angular velocity.

At the lower frequencies the magnetic field is generated by a solenoid but at higher frequencies it becomes practicable to use permanent magnets. In particular cases where weight is important it is sometimes possible to use p.p.m. focusing.

Other forms of focusing are used, for example that used on electrostatic focused klystrons (e.s.f.k.) where the focusing is achieved by a series of electrostatic lenses produced by fitting repellent ring electrodes, operated at cathode potential between the cavities.

On low power klystrons, particularly the oscillator types, where the beam current density is low and the drift tube length short no focusing is used. The electron gun is designed to produce a convergent beam whose focal point is midway along the drift tube section. Beyond this point the beam expands (due to electron repulsion) through a suitably sized output gap into the collector. The reflex klystron represents a further refinement in that the reflector electrode is shaped to re-focus the returning electrons back through the interaction gap.

REFLEX OSCILLATOR KLYSTRON

The structure of a typical klystron is shown in *Figure 7.82*. In this type of tube the electron beam having traversed the interaction gap enters into a region of retarding electric field. This field is generated by the reflector electrode which operates at a voltage negative with respect to the cavity, invariably earth potential. The beam is

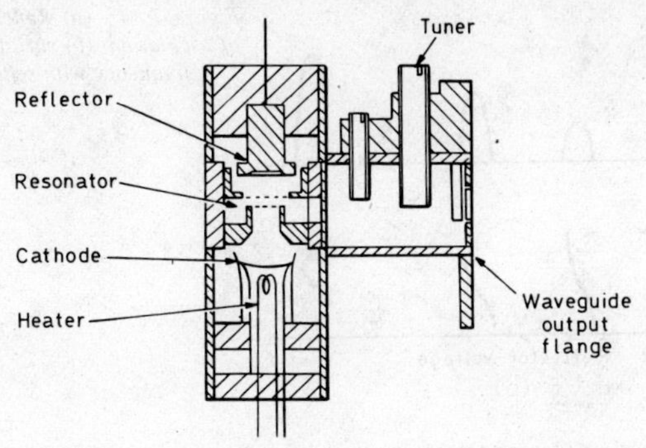

Figure 7.82 Electrode structure of a typical reflex klystron

velocity modulated on its first transit of the gap and bunching takes place. The bunches are formed in a different sense than in the two-cavity klystron because now the faster electrons travel further than the slower electrons. The time/distance diagram for a reflex klystron is shown in *Figure 7.83*. For maximum output the electron bunches must arrive back at the interaction gap so as to experience the maximum retarding voltage: this requires a drift transit time corresponding to $n + \frac{3}{4}$ cycles of r.f. where n is an integer.

In a reflex klystron the shot noise initiates the modulation which then builds up provided the drift transit time satisfies the $n + \frac{3}{4}$ relationship. Under this condition the klystron oscillates. For intermediate values the condition for oscillation is not satisfied and there is no output. The klystron, therefore, oscillates in a number of modes determined by the reflector voltage.

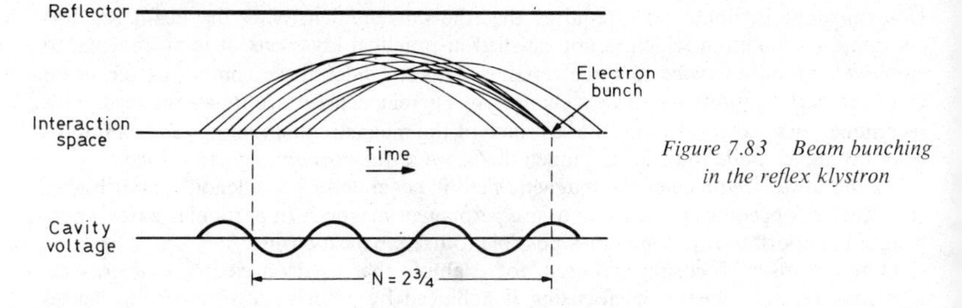

Figure 7.83 *Beam bunching in the reflex klystron*

The bunched beam presents a reactive component to the resonant cavity. The sign of this reactance is dependent on the phase of the returning bunches. In this way the frequency of the klystron can be tuned over a limited range by altering the phase of the returning bunches, i.e., by adjusting the reflector voltage. *Figure 7.84* shows the modes of oscillation and the variation of output power and frequency as a function of reflector voltage for a typical reflex klystron.

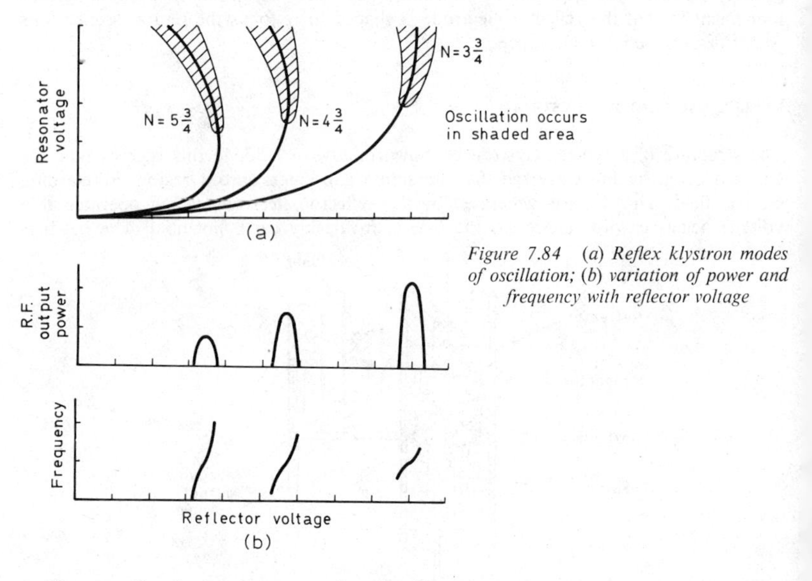

Figure 7.84 (*a*) *Reflex klystron modes of oscillation;* (*b*) *variation of power and frequency with reflector voltage*

The use of a single resonant cavity simplifies mechanical tuning arrangements and most reflex klystrons can be tuned over relatively large frequency ranges with a single control. Electronic tuning is achieved by varying the reflector voltage and as the reflector draws no current very little drive power is required for this type of tuning.

Construction

MULTI–CAVITY AMPLIFIER KLYSTRONS

The construction of a typical external cavity amplifier klystron is shown in *Figure 7.85*. The major parts are: the electron gun; the body section, made up of the interaction gaps and drift tubes; and the collector. The function of the three parts are clearly separate and can be considered individually.

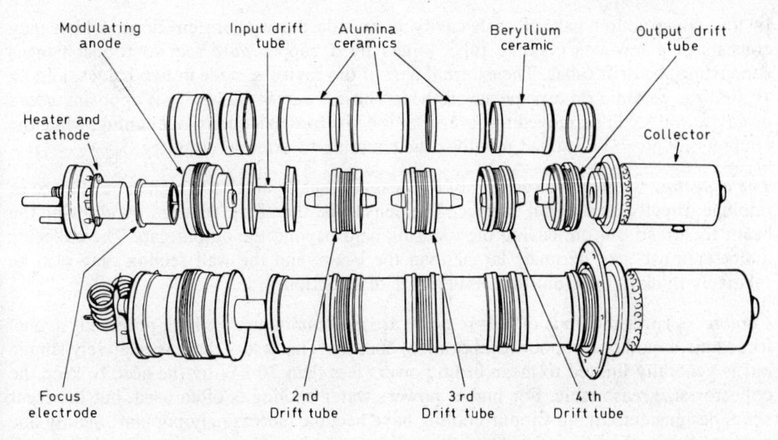

Figure 7.85 Construction of a typical external cavity amplifier klystron (*Courtesy: English Electric Valve Company Limited*)

The electron gun. The beam diameter, determined by the required drift tube diameter, is invariably such as to result in a high charge density (typically 3–10 A/cm²). In order to keep the emission density at the cathode to a reasonable level it is necessary to use a convergent electron gun. In this type of gun the cathode has a much larger surface area than the beam cross-section. The cathode and its surrounding electrodes are shaped and positioned to produce an electrostatic field which will focus the electrons through the hole in the anode. A focus electrode usually surrounds the cathode and may be run at cathode potential (by internal connection), or near it (by an external applied voltage). The electron trajectories have to be correctly aligned with the focus (magnetic) field lines and this calls for an accurate cylindrical construction and correct axial positioning of the cathode in the focusing system. The shaped anode, with its central hole may be made integral with the body or insulated from it in which case it is known as the modulating anode. Some guns employ control grids in front of the cathode but in general this is confirmed to lower power tubes.

The electron gun geometry controls the beam current for any given voltage. Under space charge limited conditions the beam current is given by the relation:

$$I_{\text{BEAM}} = KV_A^{\frac{3}{2}} \tag{7}$$

where I_{BEAM} = the beam current in amperes
V_A = the anode of modulating anode voltage in volts
K = a constant known as the microperveance (perveance)

It is important to note that the perveance is a constant determined solely by the gun geometry. A ‘perveance one’ gun produces a beam current of 1 A at 10 kV and a ‘perveance two’ gun, a beam of 2 A at 10 kV.

The body The beam formed by the gun and focused by the magnetic field passes into the body region comprising the drift tubes and the interaction gaps. At the higher microwave frequencies the body usually consists of a series of cylindrical cavities brazed together or milled from solid material, with dividing walls supporting the drift tubes. The tuner mechanism, usually an adjustable capacitance plate is fitted through a vacuum bellows. The whole body is cooled by either liquid or air blowing. In this integral configuration the coupling loops or waveguide matching sections are set during assembly and adjustments are made by varying the impedance of the output feeder or waveguide.

At lower frequencies, particularly at u.h.f. external cavity klystrons are widely used.

In this construction part of each cavity is outside the vacuum envelope, which then consists of a low-loss ceramic tube, joined by a copper *flare* seal to robust flanges supporting the drift tubes. The external part of the cavity is made in two halves, joining around the ceramic on a plane across which no r.f. current flows. Two opposing *doors* or cavity walls can be moved transverse to the klystron axis to provide tuning, while the coupling loop can be rotated to adjust the coupling to external circuitry.

The collector. The power density of the beam is usually much too high to be allowed to impinge directly on a metal surface. The density is, therefore, reduced by allowing the beam to spread out on leaving the focusing field beyond the output gap. The collector diameter must thus be made larger than the beam, and the wall section may also be relatively thick to even out the temperature distribution.

Cooling. Three methods of heat removal are in common use for high-power klystrons: forced-air, liquid and vapour (evaporative) cooling. The first of these is relatively simple but is generally limited to mean beam powers less than 30 kW by the need to keep the collector size reasonable. For higher powers water cooling is often used, but in recent years designs employing vapour cooling have become increasingly popular, mainly due to the greater reliability of the overall cooling system.

REFLEX KLYSTRON

The construction of a modern rugged reflex klystron is shown in *Figure 7.82*. The main elements are:

(*a*) *Electron Gun.* The electron gun consists of the cathode, focus electrode and anode. The purpose of the gun is to produce and focus a beam of electrons through the interaction gap.
(*b*) *Body Section.* A re-entrant cavity resonator incorporating a gap through which the electrons pass. The gap is gridded in order to provide a region of uniform axial electric field.
(*c*) *Reflector.* An electrode which is operated at a negative voltage relative to cathode, determined by the frequency mode in which the tube is being operated.

The two main types of reflex klystron are the external cavity type and the integral cavity type.

External cavity type. In this type of klystron the major part of the resonant circuit is external to the vacuum envelope. Tuning is achieved by the movement of plungers in the non-evacuated section of the resonant cavity. This configuration, much favoured on earlier designs, is seldom used on modern tubes.

Integral cavity type. The three main methods of tuning are capacitive tuning, dielectric tuning and frequency pulling using a secondary cavity.

For capacitive tuning the tube is designed such that the lower grid can be moved thus altering the gap spacing, and hence the capacity. In order to do this the wall of the cavity acts as a flexible diaphragm. A very wide tuning range is obtained by this method but for some applications it is found to be too coarse.

Dielectric tuning is achieved through the intrusion of a dielectric rod into the resonant cavity and is controlled using a micrometer screw operating through a vacuum bellows.

Frequency pulling uses a secondary cavity, external to the vacuum envelope, tightly coupled through a window to the active cavity. This secondary cavity is tuned using a capacitive probe as shown in *Figure 7.82* and its frequency pulls that of the active cavity. This is the most widely used technique on modern tubes.

Operation and characteristics

AMPLIFIER KLYSTRON

Typical power supply connections are shown in *Figure 7.86*. Where solenoid focusing is used, the klystron body is generally in direct contact with the earthed frame. Hence a body current meter in the position shown indicates also the body current for any other tube operated from the same h.t. supply.

Protection in the event of an internal arc is afforded by the modulating anode. This is an anode, isolated from the body, and connected to earth through a non-inductive 10 kW resistor. In some applications this anode is operated at an intermediate voltage,

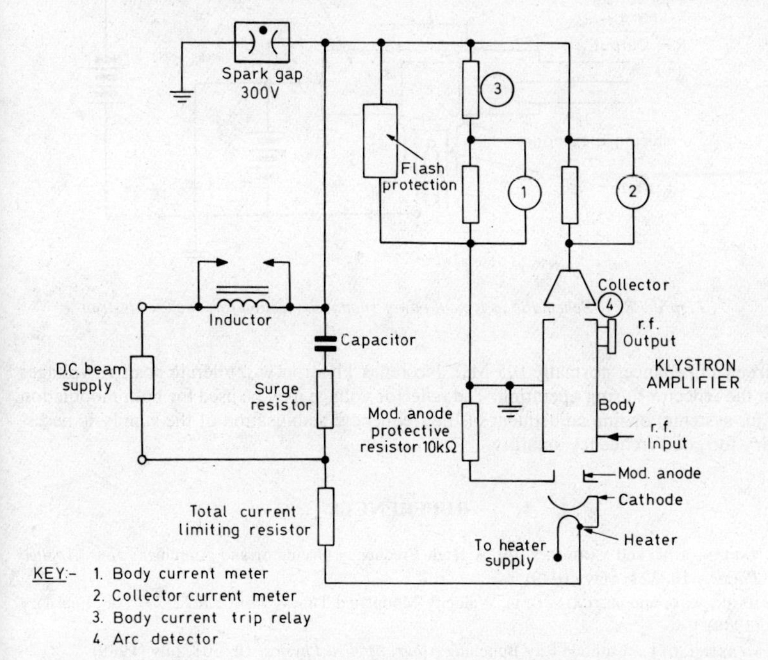

Figure 7.86 Schematic of typical power supply connection for a power amplifier klystron

derived by tapping off a potentiometer chain in order to adjust the value of the beam current.

In transmitter design it is necessary to ensure that h.t. can only be applied when gun, body and collector coolant flows are correct, and heater and focusing supplies within appropriate ranges. The h.t. supply must be interrupted rapidly if body, modulating anode or beam current rises above specified limits, or if an arc occurs in the r.f. circuitry. Additional protection may be needed in certain situations.

In setting up the r.f. circuits of a klystron amplifier, the input cavity can be tuned by observing the r.f. input power reflected from it. The penultimate cavity is usually tuned for best efficiency slightly above the higher frequency end of the required passband and an earlier cavity is correspondingly set to the low-frequency edge, to produce a flat passband response. The output cavity loading is set to ensure that the peak r.f. gap voltage approaches the d.c. beam voltage only when the maximum r.f. beam current is developed.

REFLEX KLYSTRON

The connections for a typical reflex klystron are shown in *Figure 7.87*. Most klystrons operate with the resonator at earth potential so that the cathode runs at some hundreds of volts negative to earth. This means that the heater supply is also insulated from earth.

Although the reflector current is extremely small a limit is usually placed on the

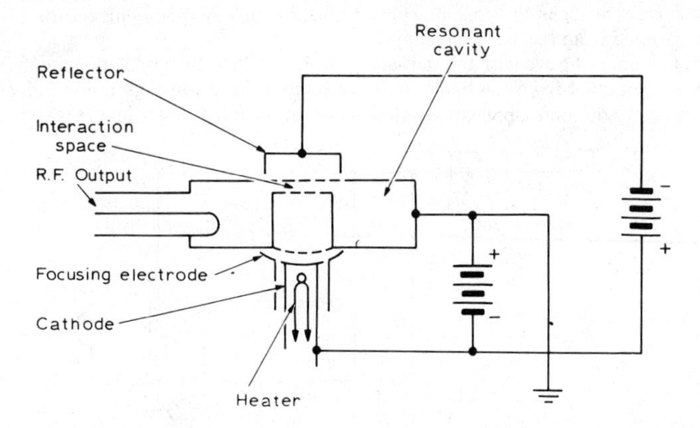

Figure 7.87 Schematic of typical power supply connection for a reflex klystron

circuit impedance, normally 0.5 MΩ. No reflex klystron will tolerate positive voltages on the reflector during operation. The reflector voltage may be used for both modulation (f.m. systems) or fine adjustment of the frequency: stabilisation of the supply is necessary for good frequency stability.

REFERENCES

1. VARIAN, R. H. and VARIAN, S. F., 'A High Frequency Oscillator and Amplifier', *Jour. Applied Physics,* **10**, 321, May (1939)
2. HAHN, W. C. and METCALF, G. F., 'Velocity Modulated Tubes', *Proc. I.R.E.*, **27**, 106, February (1939)
3. WEBSTER, D. L., 'Cathode Ray Bunching', *Jour. Applied Physics*, **10**, 501, July (1939)
4. WEBSTER, D. L., 'Theory of Klystron Oscillators', *Jour. Applied Physics*, **10**, 864, December (1939)
5. HAHN, W. C., 'Small Signal Theory of Velocity Modulated Electron Beams', *G.E.C. Rev.*, **42**, 258 (1939)
6. RAMO, S., 'Space Charge and Field Waves in an Electron Beam', *Physics Rev.*, **56**, 276 (1939)
7. WEBBER, S., 'Ballistic Analysis of a Two-Cavity Finite Beam Klystron', *Trans. I.R.E.*, ED 5, 98 (1958)
8. EDGCOMBE, C. J. and O'LOUGHLIN, C. N., 'The Television Performance of the Klystron Amplifier', *Radio and Electronic Engineer*, **41**, 405 (1971)

FURTHER READING

BECK, A. H. W., *Space-Charge Waves*, Pergamon Press (1958)
BECK, A. H. W., *Thermionic Valves*, Cambridge University Press (1953)
CHALK, G. O. and O'LOUGHLIN, C. N., *Klystron Amplifiers for Television*, English Electric Valve Co. (1965)

GITTINS, J. F., *Power Travelling Wave Tubes*, The English University Press Ltd., Chapters 4 and 5 (1965)

HAMILTON, KUPER and KNIPP, *Klystrons and Microwave Triodes*, McGraw-Hill (1948)

HARMAN, W. W., *Fundamentals of Electronic Motion*, McGraw-Hill (1953)

PIERCE, J. R., *Theory and Design of Electron Beams*, Van Nostrand (1954)

PIERCE, J. R. and SHEPHERD, W. G., *Reflex Oscillators, Bell Syst. Tech. Jour.*, **26**, 460 (1947)

SIMS, G. D. and STEPHENSON, I. M., *Microwave Tubes and Semiconductor Devices*, Blackie and Sons Ltd. (1963)

SPANGENBERG, K. R., *Vacuum Tubes*, McGraw-Hill (1948)

STAPRANS, A., MCCUNE, E. W. and REUTZ, J. A., 'High Power Linear Beam Tubes', *Proc. I.E.E.E.*, **61**, 299 (1973)

STRINGALL, R. L. and LEBACQZ, J. V., 'High Power Klystron Development at the Stanford Linear Accelerator Centre', *Proc. 8th M.O.G.A. Conf.*, Kluver-Deventer (Amsterdam), 14–13 (1970)

WARNECKE, R. R., CHODOROW, M., GUENARD, P. R. and GINZTON, E. L., *Velocity Modulated Tubes*, Advances in Electronics and Electronic Physics, 3, Academic Press, 43 (1958)

MAGNETRONS

The magnetron is a microwave oscillator with a high efficiency requiring only modest operating voltages for the generation of high-power levels. In particular it has found an enduring place in the transmitters of pulsed radar systems and it is with this type of tube that we will be mainly concerned, other types of magnetron such as the low-power voltage tunable tubes and c.w. tubes for r.f. heating described more briefly are fully listed in the bibliography.

The magnetron concept, a cylindrical diode immersed in a uniform magnetic field parallel to its axis, was first reported by A. W. Hull in 1921[1] and was later developed as a high-frequency oscillator.

The development of pulsed radar led to a pressing need for a high-power microwave source in order to improve angular and range resolution and permit the use of physically smaller antennae. The work of Prof. J. T. Randal and Dr. H. A. H. Boot[2] at Birmingham University in 1939 satisfied that need with the generation of powers at 3 GHz several orders of magnitude greater than those previously achieved.

Further war-time development in Britain and the U.S.A. provided a profusion of types at frequencies between 1 GHz and 30 GHz operating with output powers ranging from a few kilowatts to several megawatts.[2,3,4,5] Methods of mechanical tuning were devised and deployed and a considerable effort was applied, though with less success, to electronic tuning. The post-war years have seen vast improvements in operating stability and life expectancy of magnetrons, while the introduction of the long anode magnetron and the coaxial magnetron, together with the concept of rapid tuning or frequency agility, have further extended the range of application.

Principles of operation

A simplified representation of the main components of the multi-cavity magnetron in cross section is given in *Figure 7.88*. The anode, in addition to being the positive electrode and collector of electrons, also forms a structure resonant near the desired r.f. output frequency. This resonant structure is made up of a number of inter-coupled resonant cavities which provide, in the interaction space, a rotating electromagnetic wave of phase velocity below that in free space so as to give interaction with a rotating cloud of electrons emitted from the cathode.

The magnetic field, which is perpendicular both to the r.f. electric field and to the d.c. electric field (the magnetron is a 'crossed field' device), causes the electrons to rotate round the cathode in the interaction space. The strength of the magnetic field is such that, in the absence of an r.f. field, electrons would be unable to reach the anode.

With an r.f. field present, electrons in a favourable position in the rotating electron cloud will give up energy to the r.f. field and move towards the anode, thereby transforming potential energy arising from the d.c. electric field into r.f. energy. Finally they will be collected on the inner surfaces of the anode structure. Electrons in an unfavourable position in the electron cloud will take up a small amount of energy from

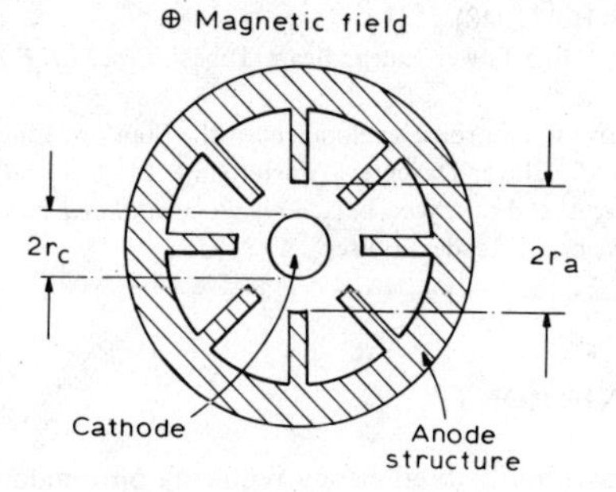

Figure 7.88 Main components of the multi-cavity magnetron

the r.f. field and will be quickly returned to the cathode surface where secondary electrons will be emitted, some of which will be useful in the generation of further r.f. power. These returning electrons expend their energy at the cathode surface, an effect known as back bombardment, thus heating the cathode and hence limiting the average power handling capacity. Electrons between the favourable and unfavourable positions described above will be accelerated or retarded into one of these groups. This process, whereby electrons are selected and phased for optimum generation of r.f. power is known as 'phase focusing' and leads to the high electronic efficiency of the magnetron.

R.F. output power is coupled from the resonant anode structure, through a vacuum window and into an external waveguide or coaxial line.

The anode voltage at which oscillation can build up is known as the threshold voltage and is given by the following expression:

$$V_{TH} = 1.01\left[\frac{r_a}{\lambda}\right]^2 \frac{T.\lambda\left(1-\dfrac{r_c^2}{r_a^2}\right)}{535N} - \frac{4\times 10^7}{N^2} \quad (1)$$

where V_{TH} = the threshold voltage (*V*)
r_a = the anode bore radius (*m*)
r_c = the cathode radius (*m*)
T = the magnetic flux density (*T*)
N = the number of resonators
λ = the free space wavelength (*m*)

The maximum electronic efficiency is given by equation 2[3]

$$\eta_{max} = 1 - \frac{1}{NX-1} \quad (2)$$

where $$X = T.\lambda.\left(1-\frac{r_c^2}{r_a^2}\right).4.67\times 10^{-5} \quad (3)$$

Magnetron construction

R.F. STRUCTURE

The r.f. structure which acts as the magnetron anode is normally made of copper and forms part of the vacuum envelope, thus providing good heat dissipation. However, temperature rise along the vane structure can be a limiting factor on power rating and in high power c.w. tubes hollow water-cooled vanes are used. In the case of magnetrons at higher frequencies the power density at the vane tips facing the cathode may be great enough to lead to quite large temperature rises during the operating pulse time.

The r.f. output system is coupled to one resonator of the r.f. structure in conventional types through an appropriate impedance matching system, either waveguide or coaxial line.

In long anode and coaxial magnetrons,[6, 7] the r.f. output is symmetrically coupled to the resonator system, in the latter case through a stabilising resonator which can materially improve the frequency pulling and pushing phenomena.

CATHODE

The cathode which operates in a mixed thermionic and secondary emitting fashion is normally equipped with non-emitting end shields or *hats* to prevent axial spread of the electron stream.

The axial magnetic field required for operation is usually provided by permanent magnets frequently with soft iron pole pieces built through the vacuum envelope in order to reduce the reluctance of the magnetic circuit. In the case of long anode magnetrons the magnetic field is normally provided by an electromagnet due to the length of uniform field required.

TYPICAL MAGNETRON STRUCTURES

A representative selection of magnetrons with their r.f. structures and cathodes are shown in *Figures 7.89* to *7.92*. *Figure 7.89* shows a high-power tunable L-band magnetron with conventional strapped r.f. structure giving a 60 MHz tuning range and a power rating of 2.3 MW peak, 3.3 kW mean, with an operating voltage of 39 kV. The tube is vapour cooled, providing excellent frequency stability and permitting the use of a very compact heat exchange system. An external permanent magnet is used.

A *Q*-band magnetron is shown in *Figure 7.90*. The r.f. structure is of the rising sun variety which permits relatively high power ratings at high frequencies, in this case 50 kW peak and 20 W mean; the operating voltage is 14 kV. The magnetic field is provided by an integral permanent magnet.

The *S*-band long anode magnetron shown in *Figure 7.91* is capable of generating a peak output of 2.5 MW with 3.75 kW mean. The operating voltage is 39 kV, the magnetic field being provided by an external solenoid, the water-cooling system of which also cools the magnetron. The length of the anode compared with the operating wavelength and the large cathode area will be noted.

Figure 7.92 shows an *X*-band coaxial magnetron with a power rating of 75 kW peak and 75 kW mean. The operating voltage is 13 kV, the magnetic field being provided by an integral permanent magnet. This design provides large dimensions for both anode and cathode for a given operating frequency. The stabilising resonator and its coupling slots surround the typical magnetron resonator system.

Figure 7.89 Vapour-cooled L-band tunable magnetron with anode and cathode assemblies (Photo: English Electric Valve Company Limited)

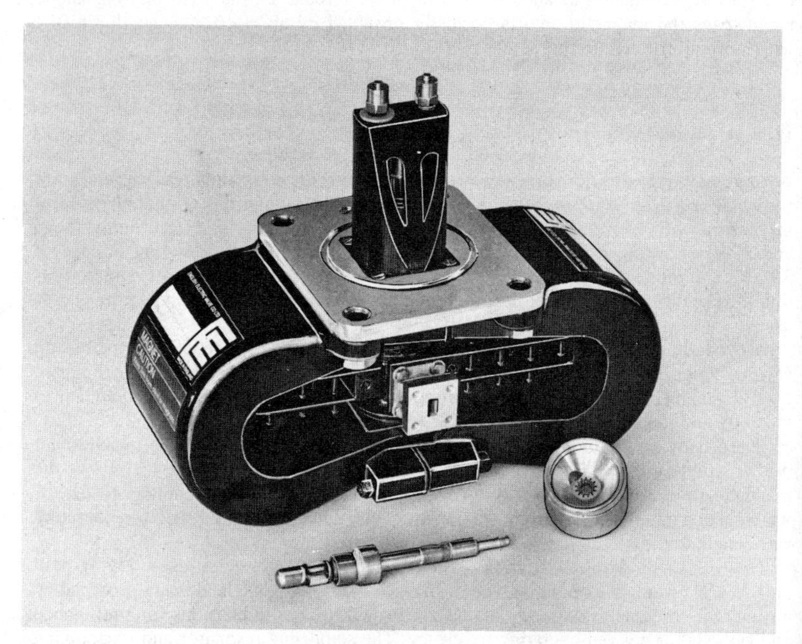

Figure 7.90 Frequency-agile Q-band magnetron anode and cathode assemblies (Photo: English Electric Valve Company Limited)

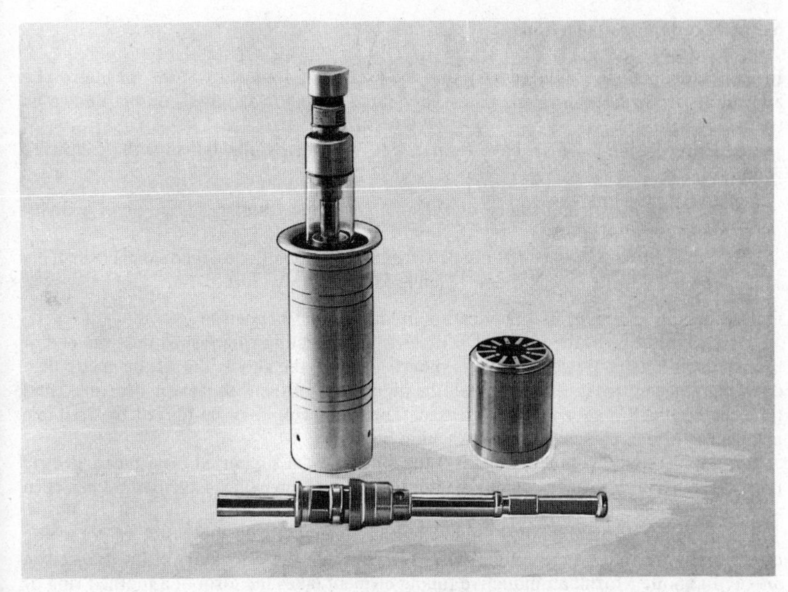

Figure 7.91 Long anode S-band magnetron with anode and cathode assemblies (Photo: English Electric Valve Company Limited)

Figure 7.92 Co-axial X-band magnetron with anode and cathode assemblies (Photo: English Electric Valve Company Limited)

Methods of tuning magnetrons

In many applications it is necessary that the oscillating frequency of the magnetron be adjustable or variable. In some cases very rapid tuning is required, this is known as frequency agility.

Methods of tuning which have been employed successfully fall into the following groups:

(*a*) Variation of the inductance or capacitance of the resonant anode or of a cavity coupled to the anode.

(*b*) Tuning by variation of the anode voltage of a specially designed magnetron. This type is known as the Voltage Tunable Magnetron (v.t.m.).[7]

Only tuners in group (*a*) find application in high-power magnetrons.

In its simplest form the magnetron tuner is a conducting ring placed near the end of the resonant anode. When the ring is moved towards the anode, the r.f. magnetic field coupling adjacent cavities is modified, the inductance of the resonator is decreased and thus the resonant frequency is increased. The tuner ring may be moved by a simple mechanism which passes through a vacuum bellows.

For rapid tuning (frequency agility) the ring may be supported on a piezo electric transducer which is placed entirely within the vacuum envelope. This method has been used successfully in magnetrons at millimetre wavelengths.[8]

The frequency range of the inductive ring tuner may be increased by attaching conducting pins to the tuner ring. These pins pass into the resonators of the magnetron anode. In another form, an inductive tuning element takes the form of a toothed ring or tube which rotates in the magnetic r.f. field providing rapid tuning. This is known as the spin-tuned magnetron.

Magnetrons have been made with capacitive tuners, i.e., tuners which act on the electric component of the r.f. field. In this case the tuning element can be a conductor or a dielectric but it must move in regions of high electric field and such tuners are prone to arcing when used directly in the anode circuit.

In some types, particularly the coaxial magnetron, the resonant anode is coupled to a second resonant cavity.[7] This cavity may be tuned as an alternative to tuning the anode. This has the advantage of separating the tuner from the anode and the tuner may even be outside the vacuum envelope. In one type of coaxial magnetron frequency agility is achieved by rapidly rotating small dielectric ‘paddles’ within the external cavity.

Operating characteristics

The major operating characteristics of the magnetron follow from two facts.

(1) There is a sharply non-linear relation between applied voltage and current through the tube.

(2) It is a self oscillator.

Input characteristics

The input voltage/current relationship results from the fact that no current flows through the device for applied voltages below the threshold value; as soon as the threshold value is exceeded oscillation starts to build up accompanied by a rapid rise in current, the operating voltage rises only slowly above the threshold point. Thus the magnetron input impedance is very high below the threshold point and very low above that point; this fact must be borne in mind when designing magnetron drive circuits. In particular the performance of drive circuits on resistive load will in general differ markedly from performance with a magnetron load.

The performance chart, *Figure 7.93*, shows this non-linear relation between applied voltage and current for several values of magnetic field, together with contours of constant efficiency. The choice of operating point in this chart may be limited by the

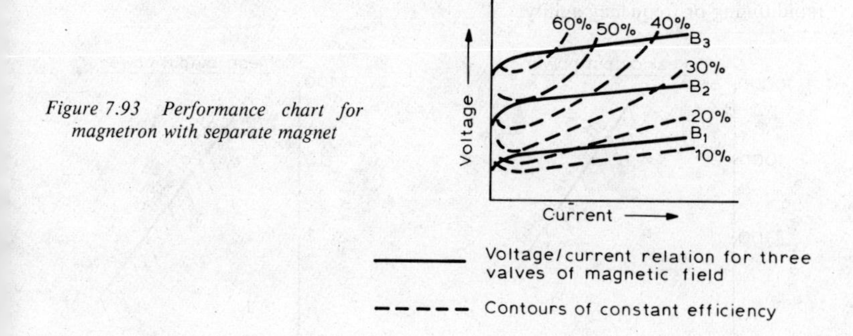

Figure 7.93 Performance chart for magnetron with separate magnet

onset of various malfunctions in certain regions. It is therefore essential that any deviation from the recommended operating point should be queried with the manufacturer.

BUILD UP PHENOMENA

In pulsed applications the oscillation must build up from noise levels at the start of each pulse. Thus there is a finite rate of build up of current and r.f. and hence the rate of build up of applied voltage must not exceed the recommended value if malfunction is to be avoided.

The front edge of the r.f. output pulse is subject to an inevitable time jitter which is of the order of 5 r.f. cycles in conventional magnetrons but may be an order of magnitude greater in the case of coaxial magnetrons.

Output characteristics

The magnetron is a self-oscillator and hence the output power and frequency of operation are affected by the r.f. load impedance, the variation of frequency is referred to as frequency pulling. Output power and frequency are also functions of the input current (the operating voltage remaining nearly constant), the variation of frequency with input current is referred to as frequency pushing.

The effects of both frequency pushing and pulling may be reduced by the use of coaxial magnetrons with their high degree of frequency stabilisation, in particular the low pushing figures obtained by these tubes may in certain circumstances be useful in ensuring minimum r.f. output bandwidth when operating with long duration pulses. The effects of frequency pulling may be reduced by the use of ferrite isolators or circulators. The use of these devices is particularly important when the susceptance of the r.f. load is a rapid function of frequency as in the case of linear accelerators or where the feeder is very long.

Thermal drift of operating frequency occurs as the magnetron warms up or if the ambient temperature changes. In this context the high-power vapour-cooled magnetrons are particularly useful as the temperature of the resonator system is stabilised at the boiling point of the cooling fluid.

Magnetron performance—practical limits

Figures 7.94 and *7.95* indicate the present limits of peak and mean power output in commercially available magnetrons as a function of r.f. frequency.

Available tunability exceeds 10% of the r.f. frequency for slow tuning and 5% for rapid tuning or frequency agility.

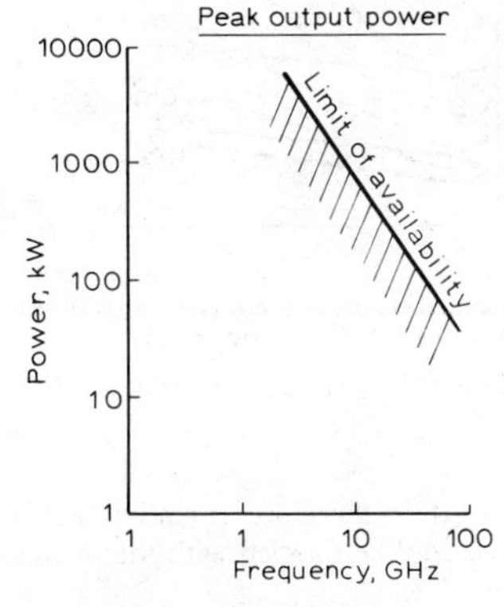

Figure 7.94 Peak output power

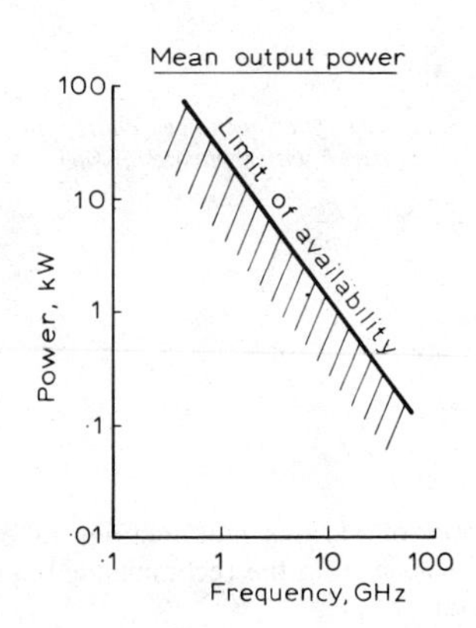

Figure 7.95 Mean output power

Both conventional and coaxial magnetrons have been used successfully in Moving Target Indication Radars. In this application there are demands on magnetron pulse-to-pulse frequency jitter and pulse-to-pulse starting jitter. Rugged magnetrons have been specially developed for airborne and missile applications where the environment can be one of high levels of shock and vibration.

REFERENCES

1. HULL, A. W., *Phys. Rev.*, 18–31 (1921)
2. RANDAL and BOOT, *Journ. Inst. Elect. Engrs.*, **93**, pt. IIIA, No. 5, 928 (1946)
3. WILLSHAW, RUSHFORTH, STAINSBY, LATHAM, BALLS and KING., *Journ. Inst. Elect. Engrs.*, **93**, pt. IIIA, No. 5, 985 (1946)
4. FISK, HAGSTRUM and HARTMAN, 'The Magnetron as a Generator of Centimetre Waves', *Bell Syst. Tech. Journ.*, 25–167 (1946)
5. COLLINS, G. B., ed., *Microwave Magnetrons*, Radiation Laboratory Series, McGraw-Hill (1948)
6. BOOT, FOSTER and SELF, *Proc. I.E.E.*, 105B, Suppl. No. 10 (1958)
7. OKRESS, E. C., ed., *Crossed Field Devices*, 2 Vols, Academic Press (1953)
8. COOPER, B. F. and PLATTS, D. C., 'Frequency Agile Magnetrons using Piezo Electric Tuning Elements', *Proc. European Microwave Conference*, Brussels (1973)

FURTHER READING

BARKER, D., 'An Experimental Investigation of the Energy Distribution of Returning Electrons in the Magnetron', *Proc. 8th M.O.G.A. Conf.*, Amsterdam (1970)

BARTON, D. K., 'Simple Procedures for Radar Detection Calculations', *I.E.E.E. Trans. on A.E.S.*, Sept. (1969)

BECK, A. H. W., *Thermionic Valves, Their Theory and Design*, Cambridge University Press (1953)

BENSON, F. A., *Millimetre and Sub-Millimetre Waves*, Iliffe Books Ltd. (1969)

BIRKENEIMIER, W. P. and WALLACE, N. D., 'Radar Tracking Accuracy Improvement by Means of Pulse to Pulse Frequency Modulation', *I.E.E. Trans. on Comm. and Electronics*, Jan. (1968)

BRODIE, I. and JENKINS, R. O., 'Secondary Electron Emission from Barium Dispenser Cathodes', *Br. Journ. of App. Phys.*, **8**, May (1957)

JEPSON, R. L. and MULLER, M. W., 'Enhanced Emission from Magnetron Cathodes', *Journ. of App. Phys.*, **22**, 9, Sept. (1951)

LATHAM, R., KING, A. H. and RUSHFORTH, L., *The Magnetron*, Chapman & Hall (1952)

LIND, G., 'Reduction of Tracking Errors with Frequency Agility', *I.E.E. Trans. on A.E.S.*, May (1968)

MALONEY, C. E. and WEAVER, F. J., 'The Effect of Gas Atmospheres on the Secondary Emission of Magnetron Cathodes', *Proc. 7th M.O.G.A. Conf.*, Hamburg (1968)

OKRESS, E. C., *Microwave Power Engineering*, 2 Vols, Academic Press (1968)

PICKERING, A. H. and COOPER, B. F., 'Some Observations of the Secondary Emission of Cathodes used in High Power Magnetrons', *Proc. 5th M.O.G.A. Conf.*, Paris (1964)

PICKERING, A. H. and LEWIS, P., 'Microwave Sources for Industrial Heating', *Proc. Bradford Conference on Microwave Heating*, Oct. (1970)

RAY, H. K., 'Improving Radar Range and Angle Detection with Frequency Agility', *Microwave Journal*, May (1966)

SIMMS, G. D. and STEPHENSON, I. M., *Microwave Tubes and Semiconductor Devices*, Blackie & Son (1963)

SLATER, J. C., *Microwave Electronics*, D. Van Nostrand (1950)

TWISLETON, J. R. G., 'Twenty-kilowatt 890 MHz Continuous-wave Magnetron', *Proc. I.E.E.*, **3**, 1, Jan. (1964)

TRAVELLING-WAVE TUBES

The travelling-wave tube (t.w.t.) is a microwave amplifier capable of amplifying over very wide frequency bands. The amplification process takes place by continuous interaction between an electron beam and an electromagnetic wave propagating along a slow-wave structure. The principle was invented by Kompfner[1] in 1943 who used a simple wire helix as a slow-wave structure. Similar tubes were then developed and first used as microwave amplifiers in microwave relay link systems. During the last 25 years travelling-wave tubes have been continuously developed using other slow-wave structures such as coupled cavities to provide c.w. output powers of tens of kilowatts and pulse powers of several megawatts with power gains of up to 60 dB. Travelling-wave tubes are usefully employed from u.h.f. to centimetric wavelengths. However, the

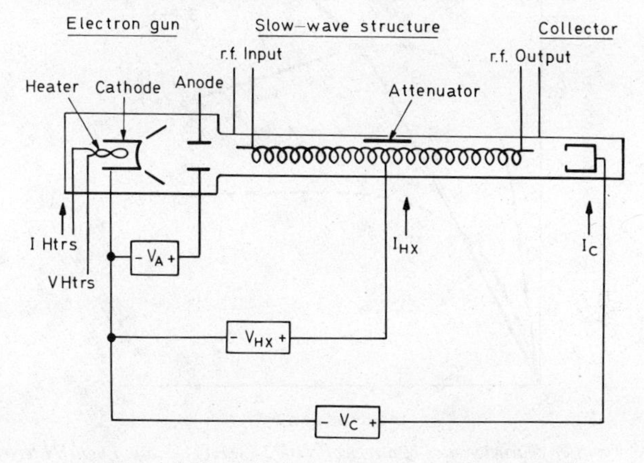

Figure 7.96 Schematic outline of a travelling-wave tube

original helix slow-wave structure is still one of the most useful due to its great bandwidth. Tubes employing helices have been made with useful amplification properties over a bandwidth greater than 2 octaves.

A travelling-wave tube consists of three main parts as illustrated in *Figure 7.96*. An electron gun [2] which is capable of producing a beam of electrons. A slow-wave structure which is in close proximity to the electron beam and which can propagate microwave signals at approximately the same speed as the electrons usually at a fraction of the speed of light. The slow-wave structure has an attenuator region approximately half way along its length to absorb r.f. energy which may propagate in the reverse direction. Without an attenuator the tube could be unstable and self oscillate. The final part is a collector which is capable of trapping the spent electron beam and dissipating this remaining energy as heat.

An axial magnetic field is required to confine the beam in the structure region and prevent it diverging under the repulsive forces between the electrons. This focusing field can be provided either by a solenoid as in all early tubes or by using permanent magnets. Most travelling-wave tubes with power outputs up to a few hundred watts use a periodic permanent magnet focusing system [3] known as p.p.m. which is lighter, more compact and has less leakage magnetic field than a uniform system.

The velocity of the electron beam in the travelling-wave tube can be adjusted by varying the voltage applied between the cathode and slow-wave structure. The magnitude of the electron beam current is generally controlled independently by varying the voltage on an additional grid or anode in the electron gun.

Amplifier mechanism

At the input and output terminals of the travelling-wave tube there are transitions which

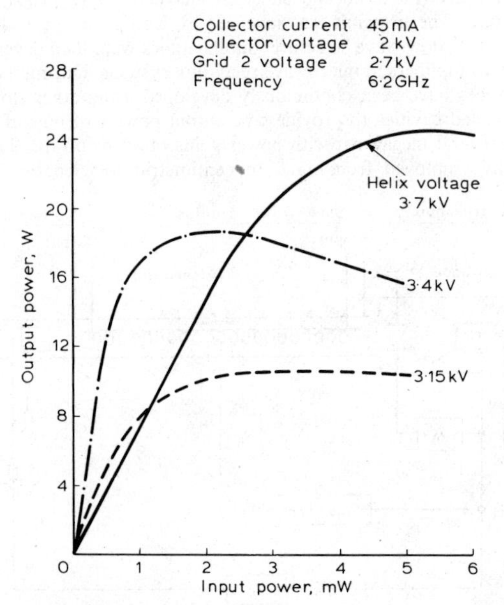

Figure 7.97 Transfer characteristics of a typical N1072 t.w.t. (Courtesy: English Electric Valve Co. Ltd.)

allow the r.f. signal to be coupled to the slow-wave structure. The axial component of the r.f. electric field at the start of the slow-wave structure accelerates or decelerates electrons as they enter. This causes a velocity modulation on the beam which gradually gives rise to a density modulation or electron bunching as the beam progresses along the slow-wave structure. As the beam and r.f. wave are in near synchronism the electron bunches induce voltages on the slow-wave structure which re-enforce those already present and thereby promote a rapid increase in the r.f. signal. To maintain this interaction the electron beam is initially travelling slightly faster than the r.f. signal wave. However, as the r.f. wave increases the average velocity of the electron bunches is reduced as more energy is extracted from the beam. Finally the electron bunches lose synchronism with the r.f. wave and no further energy is extracted from the beam. At this point the tube is said to be saturated and gives maximum power output. Any further increase of input signal level causes the output to fall. A typical travelling-wave tube input/output or transfer characteristic is shown in *Figure 7.97.*

Gain

The gain depends directly on the length of the slow-wave structure. As the helix or structure voltage is varied above or below the synchronous value the gain decreases rapidly. This characteristic is illustrated in *Figure 7.98*. The gain also increases at a rate proportional to the cube root of the beam current.

The small signal gain G in decibels is given by

$$G = BCN - A - L$$

where A, B and L are constants for a given design and they have typical values of 10, 40 and 6 respectively. C is Pierce's[4] gain parameter and N is the number of wavelengths along the slow-wave structure.

$$C = \left(\frac{KI_o}{4V_o}\right)^{\frac{1}{3}}$$

where K is called the coupling impedance and depends on the geometry of the slow-wave structure. I_o is the beam current and V_o is the beam voltage.

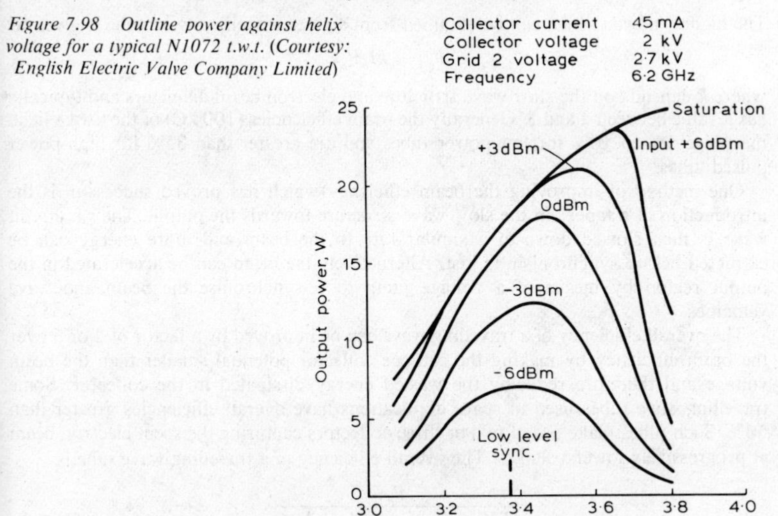

Figure 7.98 Outline power against helix voltage for a typical N1072 t.w.t. (Courtesy: English Electric Valve Company Limited)

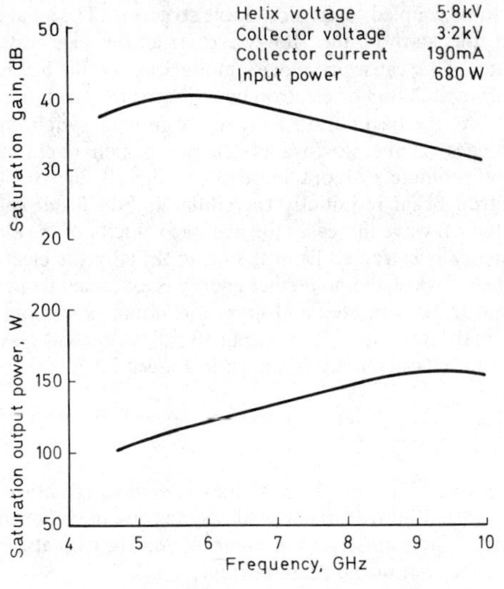

Figure 7.99 Typical performance characteristics of the N1077 t.w.t. (Courtesy: English Electric Valve Company Limited)

For a helix slow-wave structure the phase velocity of the r.f. wave is largely independent of frequency therefore the wave and beam velocity remain in synchronism and large bandwidths are possible. Typical gain and power/frequency characteristics of a broad band t.w.t. are shown in *Figure 7.99.*

Efficiency

The highest power which can be obtained from a given travelling-wave tube is given by

$$P_{r.f.} = kI_oV_oC$$

where k depends on the slow-wave structure and electron beam diameters and typically has a value between 2 and 3. Generally the beam efficiencies (100 kC) of the t.w.t.s lie in the range 5% to 10% for low-power tubes and are greater than 35% for high-power pulsed tubes.

One method of improving the beam efficiency which has proved successful is the introduction of a taper[5] in the slow wave structure towards the output. The r.f. circuit wave is then slowed down at a similar rate to the beam and more energy can be extracted before synchronism is lost. Alternatively the beam can be accelerated in the output region by means of a voltage jump to re-synchronise the beam and wave velocities.

The overall efficiency of a travelling wave can be improved by a factor of 2 or 3 over the beam efficiency by making the cathode-collector potential smaller than the beam voltage and therefore reducing the wasted energy dissipated in the collector. Some travelling-wave tubes used in space applications have overall efficiencies greater than 50%. Such tubes make use of two or three collectors capturing the spent electron beam at progressively lower voltages. The overall efficiency of a travelling-wave tube is

$$\frac{P_{r.f.}}{I_cV_c + V_{HX}I_{HX} + V_{HTRS}I_{HTRS}}$$

Noise

The main sources of noise in a travelling-wave tube arise from the shot noise or current density variations as electrons leave the cathode and secondly the random velocity fluctuations of the emitted electrons. In low noise tubes[6] the electron gun has several electrodes designed so that the potential profile between the input to the helix and the cathode can be optimised so that minimal noise in the beam is coupled to the helix. Low noise tubes have noise figures between 6 dB and 10 dB with power outputs of a few milliwatts.

By using a high magnetic focusing field over the cathode and introducing a special beam-forming electrode to produce a divergent electric field at the cathode ultra low noise tubes[7,8] are produced with noise figures between 2.5 dB and 5.0 dB.

Linearity

Since travelling-wave tube amplifiers are generally operated in the region of maximum output non-linearity of amplitude and phase distortion can occur. In some cases the saturation characteristic is put to good use and the travelling-wave tube may be employed as a limiter. However, in most other amplifier uses where linearity is important the tube is operated 2 dB to 3 dB below the saturated output level.

The conversion of amplitude modulation to phase modulation (a.m./p.m. conversion) occurs in a t.w.t. due to the reduction in the beam velocity as the input signal is increased. At saturation the a.m./p.m. conversion may rise to 5 °/dB. T.W.T. microwave link amplifiers are generally designed to have an a.m./p.m. conversion of about 1 °/dB at their operating power level.

When several signals are simultaneously amplified by a travelling-wave tube a mixing or intermodulation[9] occurs. This results in intermodulation products (i.p.) at the output with power levels dependent on the relative levels of the original signals. As in the case of a.m./p.m. conversion intermodulation distortion is significantly reduced by operating the tube in the small signal or linear region. With two equal carrier signals operating at saturation the third order i.p. is typically 10 dB below the carriers.

The construction and types of travelling-wave tube

THE ELECTRON GUN

The electron gun shown in *Figure 7.100* produces a controlled diameter beam of electrons from the cathode. The cathode has a surface with a low work function so that

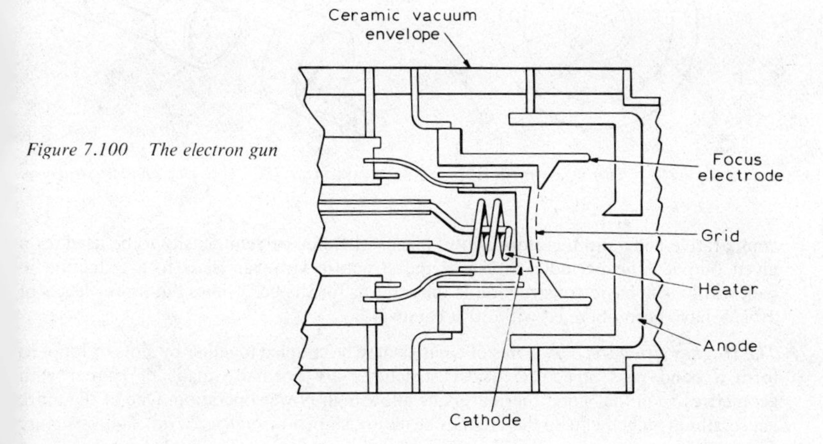

Figure 7.100 The electron gun

electrons are emitted when the cathode is heated. This surface is commonly obtained by depositing a layer of barium and strontium oxides on to a nickel base, stable emission is obtained at temperatures in the region of 750 °C. Several other types of cathode construction are in common use but one which is now finding wide application is the barium aluminate cathode. This consists of barium aluminate impregnated into a porous tungsten pellet. The temperature of operation is higher than for the oxide cathodes but higher electron emission densities are possible up to two amps/cm^2, these cathodes are also less susceptible to ion damage. The operational life of the tube is often governed by cathode life and with care in designs and production this can be in excess of 20 000 hours. The beam current may be switched on and off by modulating the voltage between the slow-wave structure and the cathode. It is also possible to introduce control grids close to the cathode that switch the beam with much smaller voltages.

THE SLOW-WAVE STRUCTURE

The helix. The helix was used in the first t.w.t.s developed and still is commonly used today. The great virtue of the circuit is its capability of producing bandwidths in excess of an octave. This capability has not been equalled by recently developed structures. The helix is however not easily cooled and this generally limits the maximum power output to below 2 kW.

The helix is usually made of tungsten and is wound with great accuracy to give a constant propagation velocity. It is supported by three or sometimes four insulating rods as shown in *Figure 7.101*. In higher-power tubes these are made of beryllia ceramic which having good thermal conductivity keep the structure cool. For power levels below 20 watts the cheaper alumina ceramics or quartz are used.

In low-power non-rugged tubes the outer vacuum envelope is made of glass but with higher power levels a metal cylinder is shrunk or brazed on to the tube rod assembly to ensure a good radial thermal path.

The ring and bar. In order to overcome the power limitations of the simple helix the contra wound helix was investigated and now finds use in the formalised version called the ring and bar structure, *Figure 7.102*. The internal beam diameter is larger than a

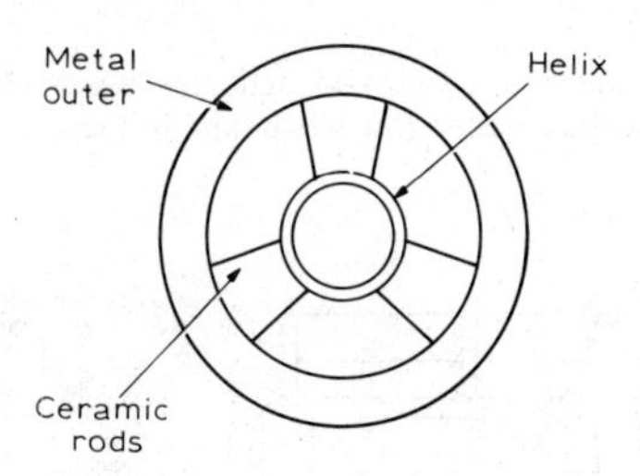

Figure 7.101 The helix structure

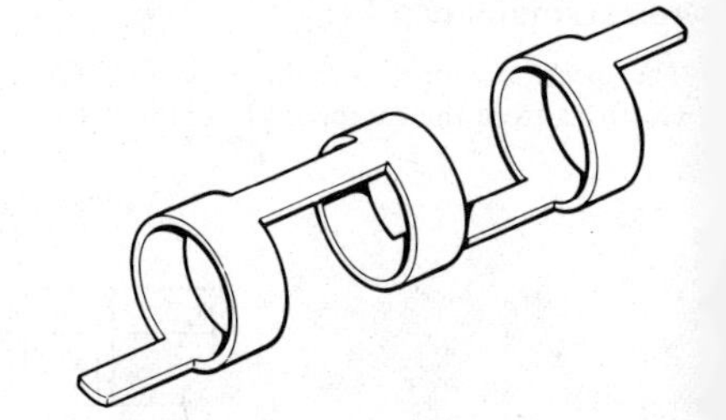

Figure 7.102 The ring and bar structure

comparative helix and therefore allows beams of lower current density to be used for a given output. The periodic nature of the structure however leads to a reduction in bandwidth. The basic construction is the same as for the helix tubes but power levels of 10 kW have been obtained with 30% bandwidth.

The Hughes structure. A series of cavities may be coupled together by slots or loops to form a band pass structure. These structures are normally made of copper with geometrics such that good thermal paths allow high-power operation. One of the more successful arrangements is the Hughes structure shown in *Figure 7.103*. This structure

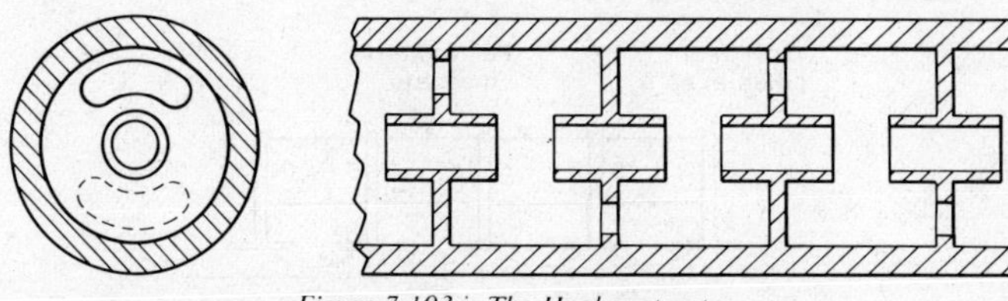

Figure 7.103 The Hughes structure

gives bandwidths of 20% when designed for use in the region of 20 kW c.w. output power. At higher powers the usable bandwidth is less.

The clover-leaf circuit. The clover-leaf circuit has four or six wedges projecting into each pill box cavity and radial slots form coupling elements into the adjacent cavities, producing a structure propagating a forward fundamental wave. The structure has found application at output power levels up to 1 megawatt peak with bandwidth of 10%. A typical tube[10] is shown in *Figure 7.104*.

THE COLLECTOR

The collector is an electrode designed to collect the spent beam of electrons after they have left the slow-wave structure. The energy of the electrons is converted into heat on

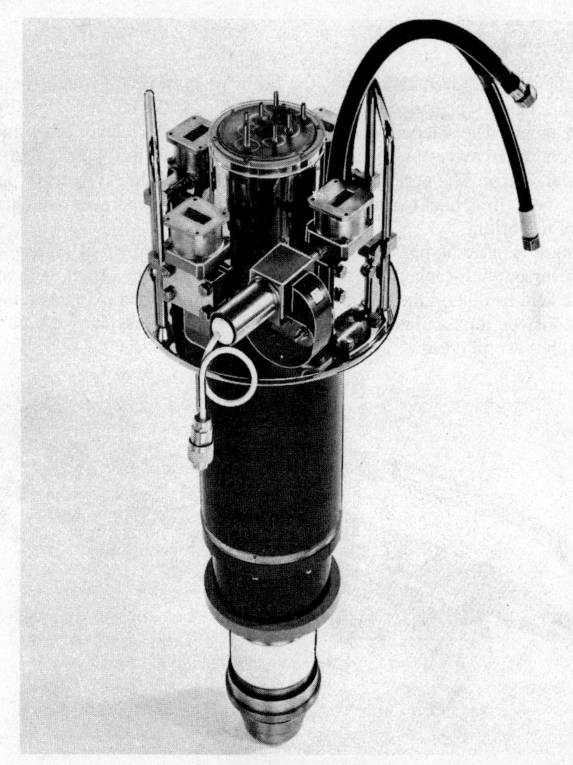

Figure 7.104 A high-power cloverleaf travelling-wave tube type N1061 (Photo: English Electric Valve Company Limited)

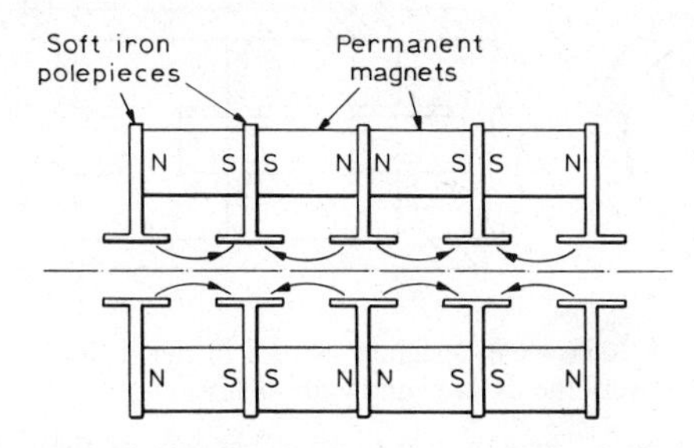

Figure 7.105 Periodic permanent magnet structure

impact with the collector surface and care is taken in the design of collectors for high-power tubes to ensure that internal surfaces of the collector are not overheated. The collector may be cooled by thermal conduction to an external heat sink, by air blown over the collector or by liquid cooling. In tubes where the overall efficiency needs to be high the collector is operated at a reduced voltage relative to the cathode. The electrons are slowed down between the slow-wave structure and the collector and thus have less energy to convert to heat on impact. Operation in this manner is called depressed collector operation.

MAGNETIC FOCUSING

A magnetic field is required to contain the beam as it travels through the slow-wave structure. This field is provided by two main methods, the solenoid and by periodic permanent magnets. The solenoid is used today on low-noise receiver tubes and on some high-power coupled cavity tubes. It is however bulky and heavy and also requires a separate power source. The periodic permanent magnet focusing systems (p.p.m.) do not suffer from these drawbacks and find wide application wherever weight, size and efficiency are important.

A diagram of a periodic permanent magnet structure is shown in *Figure 7.105*. The polarity is reversed with each cylindrical magnet so producing a stack with an alternating magnetic field profile on the beam axis. The focusing system may be an integral part of the tube construction as illustrated in *Figure 7.106* or it may form a separate mount into which tubes are plugged.

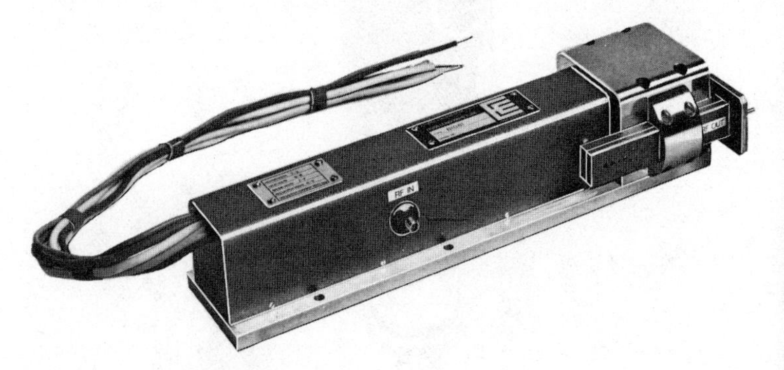

Figure 7.106 A broad-band helix travelling wave tube type N1081 (Photo: English Electric Valve Company Limited)

Application of travelling-wave tubes

Travelling-wave tubes are widely used as amplifiers in all types of pulse and c.w. radars. Where larger bandwidths are required, with greater frequency agility, travelling-wave tubes are used as the final output stage of the radar transmitter as well as in driver stages. Low-noise travelling-wave tubes are used as the first stage in radar receivers. For use in airborne radars where high gain small size and rugged construction is important the travelling-wave tube finds application. These features also make the travelling-wave tube attractive for space use.

Other applications in the military field include uses in guidance and blind landing systems and control in missile weapon systems. The broad bandwidth of the helix t.w.t. makes it particularly attractive in electronic counter measures (e.c.m.) and jamming applications.

In the civil communications field the t.w.t. is used as a transponder for television broadcasting both on the ground and in satellites. One of the widest uses is in microwave link communication systems where there may be several hundred travelling-wave tubes in one system.

A summary of examples of some typical travelling-wave tubes is shown in *Table 7.8*.

Table 7.8

Tube code No.	*Power* (W)	*Instantaneous frequency range* (GHz)	*Mode of operation*	*Slow-wave structure*	*Typical use*
N1061	1×10^6	5% bandwidth in *X*-band	Pulse	Clover leaf	Surface radar
N1096	5×10^5	6% bandwidth in *C*-band	Pulse	Clover leaf	Surface radar
N1080	200	7.0–11.0	C.W.	Helix	Airborne radar
N1081	100	9.0–16.0	C.W.	Helix	E.C.M.
N1082	2	8.0–16.0	C.W.	Helix	E.C.M.
N1078	2	4.8–9.6	C.W.	Helix	E.C.M.
N1073	20	3.6–5.0	C.W.	Helix	Radio link
TWS36	20	1.5–2.5	C.W.	Helix	Radio link
838H	1×10^5	16.0–16.5	Pulse	Coupled cavity	Surface radar
751H	5×10^4	8.8–9.7	Pulse	Coupled cavity	Surface radar
570H	1 500	2–4	Pulse	Helix	E.C.M.
830H	3 000	16.2–16.7	Pulse	Coupled cavity	Airborne radar
WJ2500–3	1	1.0–2.0	C.W.	Helix	Airborne radar
WJ340	1 000	1.0–2.0	Pulse	Helix	E.C.M.
WJ228	1.2×10^4	5.4–5.9	Pulse	Coupled cavity	Surface radar
VA134B	5 000	0.5–0.6	Pulse		Airborne radar
PT1020	4×10^6	5.35–5.85	Pulse	Clover leaf	Surface radar
214H	8	2.2–2.4	C.W.	Helix	Space communications
N1072	10	5.8–7.2	C.W.	Helix	Radio link
N1071	10	7.0–8.5	C.W.	Helix	Radio link

REFERENCES

1. KOMPFNER, R., *The Invention of the Travelling-Wave Tube*, San Francisco Press (1964)
2. PIERCE, J. R., *Theory and Design of Electron Beams*, Van Nostrand (1954)
3. STERRETT, J. E. and HEFFNER, H., 'The Design of Periodic Magnetic Focussing Structures', *Trans. I.R.E. Electron Devices*, ED 5, 1, 35 (1958)

4. PIERCE, J. R., *Travelling-Wave Tubes*, Van Nostrand (1950)
5. SAUSENG, O., 'Efficiency Enhancement of Travelling-Wave Tubes by Velocity Re-Synchronisation', *7th International Conference on Microwave and Optical Generation and Amplification Hamburg*, 16 (1968)
6. PETER, R. W., 'Low Noise Travelling-Wave Amplifier', *RCA Review*, XIII, 3, 344 (1952)
7. CURRIE, M. R. and FORSTER, D. C., 'New Mechanism of Noise Reduction in Electron Beams', *Journal of Applied Physics*, **30**, 1, 94 (1959)
8. CHALK, G. O. and JAMES, B. F., 'A Wide Dynamic Range Ultra Low Noise TWT for S-Band', *5th International Conference on Microwave and Optical Generation and Amplification Paris*, 14 (1964)
9. KUNZ, W. E., LAZZARINI, R. F. and FOSTER, J. H., 'TWT Amplifier Characteristics for Communications', *Microwave Journal*, **10**, 3, 41 (1967)
10. CHALK, G. O. and CHALMERS, P. M., 'A 500 kW Travelling-Wave Tube for X-Band', *6th International Conference on Microwave and Optical Generation and Amplification Cambridge*, 54 (1964)

CATHODE-RAY TUBES

The electrons from a heated cathode can be focused into a thin beam to travel towards a positively-charged anode. This beam can be deflected by an applied electric or magnetic field. If the beam impinges on a luminescent screen, a visible indication of the deflection of the beam by the applied field can be obtained. This is the operating principle of the cathode-ray tube.

Basic cathode-ray tube

CONSTRUCTION

The schematic cross-section of a basic cathode-ray tube (c.r.t.) is shown in *Figure 7.107*. The elongated glass envelope has a relatively narrow neck, and opens out to a larger-diameter section which contains the screen. The cathode is heated by the filament, and the emitted electrons are focused by the anode system on to the luminescent screen. A grid between the cathode and anode system enables the intensity of the electron beam to be varied, and hence the intensity of the luminescence of the screen. The cathode, grid, and anode structure is called the *electron gun*. The c.r.t. is constructed so that with no deflecting force applied to the electron beam, the beam impinges on the centre of the

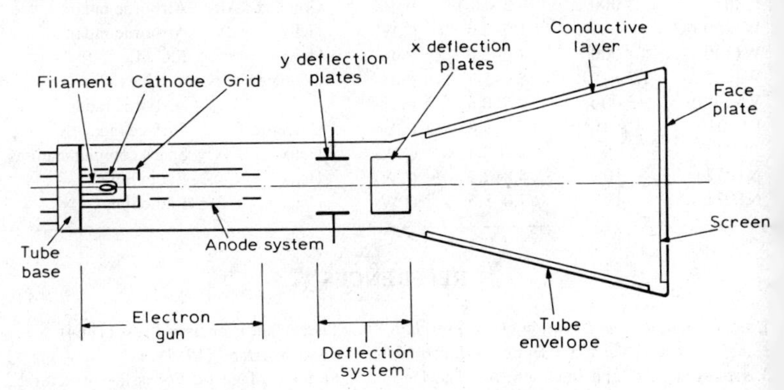

Figure 7.107 Schematic cross-section of cathode-ray tube

screen. The deflection system shown in *Figure 7.107* is an electrostatic one. Two sets of deflection plates at right angles are incorporated in the tube. The magnitude and polarity of the voltages applied to each set of plates determine the angle by which the beam is deflected in the vertical and horizontal directions. The inside of the envelope towards the screen is coated with a conductive layer, usually of graphite. This collects secondary electrons liberated from the screen by the impact of the primary electron beam, and it can also act as a post-deflection accelerator anode. The screen itself consists of a thin layer of a phosphor deposited on the inside of the faceplate of the c.r.t.

There are many variants of the basic structure described above. For example, a magnetic deflection system can be used in place of the electrostatic one. For this, coils placed around the neck of the c.r.t. provide the deflecting force on the electron beam. Two electron guns can be used to provide two electron beams, and two deflection systems can be provided, to form a *double-gun* tube. Alternatively, one gun can be used, and the electron beam split into two; this forms a *split-beam* tube. Such variants are discussed later after the components of the basic c.r.t. have been described in more detail.

ELECTRON GUN

The function of the electron gun is to collect the electrons emitted from the cathode, form the electrons into a beam, and focus the beam into a spot on the screen. The cross-section of a typical electron gun is shown in *Figure 7.108*.

The cathode k is in the form of a cylinder closed at one end. An oxide coating is applied to the closed end so that electrons are emitted from this surface when the

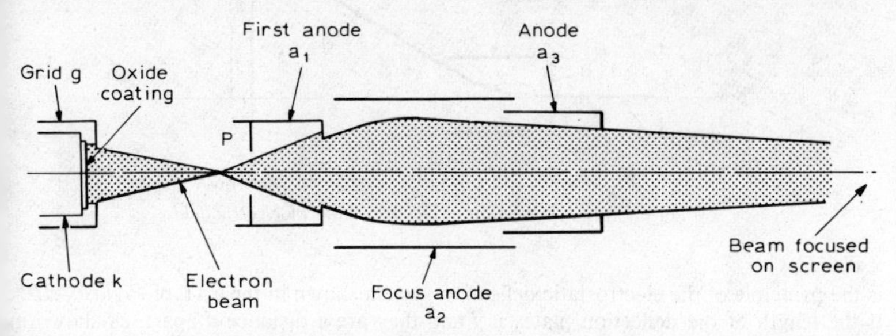

Figure 7.108 Cross-section of electron gun

cathode is heated. The heater itself is coated with aluminium oxide (alumina) to insulate it electrically from the cathode. A second cylinder surrounds the cathode, with an aperture in the end surface through which the electrons can pass. This second cylinder is the grid g of the c.r.t., sometimes known as the Wehnelt electrode. The grid is maintained at a potential between 0 V and −50 V typically, with respect to the cathode. When the grid is strongly negative, electrons cannot pass through and there is no luminescence of the screen. If the grid is less strongly negative, electrons can pass through to the anodes, the electron flow (beam current) being inversely proportional to the magnitude of the negative grid voltage. Thus the control grid can be used to cut off the beam (this is called 'blanking'), or to vary the intensity of the display on the screen.

The first anode a_1 is either a disc with an aperture through the centre for the electron beam, or a small closed cylinder with apertures in both ends. This electrode is held at a potential typically between 1 kV and 2 kV with respect to the cathode. The electric field produced by anode a_1 causes the electron beam to converge to a focus at point P. From this point, the beam diverges again but it is brought to a focus by anodes a_2 and a_3 to form a spot on the screen. Point P can be regarded as a virtual cathode, and the spot on the screen as an image of the beam at P formed by the 'lens' a_1, a_2, and a_3. Anode a_2 acts as a focus electrode, the potential on a_2 being varied about a mean value of 250 V typically. The third anode a_3 accelerates the electrons towards the screen, and is in the form of a disc or cylinder with an aperture to limit the diameter of the emerging electron beam, so that with the maximum deflection, the beam does not strike the deflection plates. Anode a_3 is maintained at the same constant potential as anode a_1.

Because the spot on the screen is the image of the electron beam at point P, the spot diameter is determined by the diameter of the beam at P rather than the diameter of the cathode or any of the apertures in the electrode system. Variations of the grid potential can affect the axial position of point P, and so can cause defocusing of the spot on the screen, but this can be corrected by adjusting the voltage on the focus electrode a_2. Because of the screening effect of anode a_1, the initial convergence of the beam at P is not affected by the focus electrode which operates only on the diverging beam. Typical spot diameters lie between 0.2 mm and 0.4 mm, while a typical cathode diameter is 2 mm. The spot diameter is also known as the line width.

DEFLECTION

An electron beam flowing between two parallel conducting plates across which a voltage is applied, is subjected to a transverse electric field and so will be deflected. This

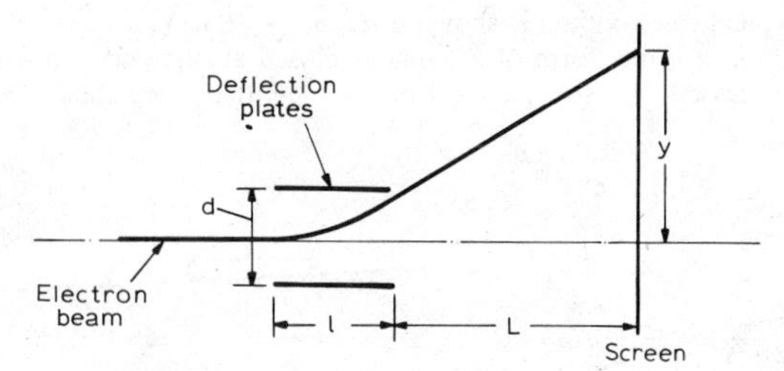

Figure 7.109 Deflection of electron beam by electric field

is the principle of the electrostatic deflection system shown in the c.r.t. of *Figure 7.107*. If the length of the deflection plates is l and they are a distance d apart, as shown in *Figure 7.109*, then the deflection y at a distance L is given by

$$y = \frac{V_d}{V_a} \frac{lL}{2d} \quad (1)$$

where V_a is the accelerating voltage (the voltage on the final anode a_3) acting on the electron beam, and V_d is the deflecting voltage between the plates with respect to the mean plate voltage (which is the same as the voltage of the final anode of the electron gun). This expression can be rearranged to give a deflection sensitivity, deflection per unit deflecting voltage:

$$\frac{y}{V_d} = \frac{lL}{2dV_a} \quad (2)$$

The deflection sensitivity is inversely proportional to the accelerating voltage since the higher this voltage, the higher is the velocity of the electron beam, and hence the shorter the time the beam is influenced by the deflecting field. Equation 2 also shows that the deflection sensitivity is proportional to the length of the deflection plates and the distance between the screen and the deflection plates, and inversely proportional to the separation of the plates.

In a practical c.r.t., limits are imposed on all the quantities in equation 2 to restrict the value of deflection sensitivity. The accelerating voltage is determined by the distance the electron beam has to travel to impinge on the screen. The distance between the deflection plates and the screen is limited by the practicable length of the c.r.t., and the increase in spot size with distance from the electron gun. The length of the deflection plates is limited by the need to prevent the deflected beam hitting the plates when leaving the deflection region. A solution to this last problem is to shape the plates to follow the pattern of the deflected beam. The minimum plate separation is then determined by the diameter of the beam entering the deflection region. It can be seen, therefore, that the design of the deflection system has to be a series of compromises between the deflection sensitivity required and limitations imposed mainly by the size of the c.r.t.

It is usual in the data for a c.r.t. to quote a deflection factor rather than the deflection sensitivity. Equation 1 can be rearranged to give:

$$\frac{V_d}{y} = \frac{2dV_a}{lL}$$

which is the inverse of equation 2. The quantity V_d/y is called the deflection factor, and generally has the units V/cm. It is a measure of the deflection voltage which must be applied to obtain a certain deflection on the screen, and typical values for practical c.r.t.s lie between 4 V/cm and 40 V/cm.

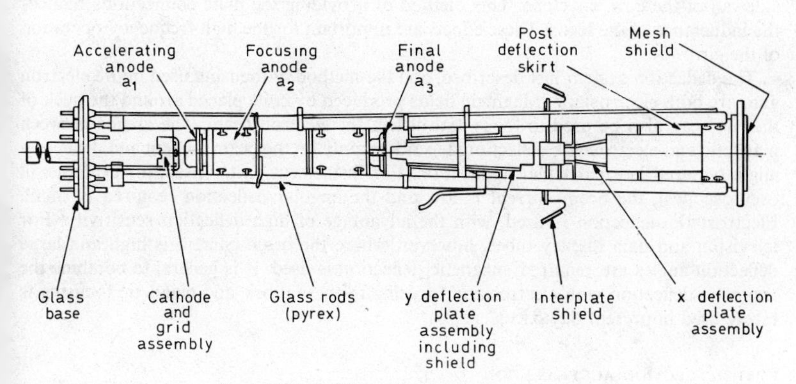

Figure 7.110 Electrode structure of typical c.r.t.

The structure of the electron gun and deflection system for a typical c.r.t. is shown in *Figure 7.110*. The cathode, grid, and anode assembly is the same as that previously described. Because the display from a c.r.t. when used in equipment such as an oscilloscope is generally required to have rectangular coordinates, two sets of deflection plates at right angles are incorporated in the electrode structure. The first set of plates provide the vertical deflection, generally called the y deflection, followed by the horizontal or x deflection. The various electrodes are manufactured from copper-nickel alloy or stainless steel, mounted on glass support rods to form a rigid structure.

If a voltage signal is applied to one set of deflection plates, say the y plates, the deflection of the beam as shown by the displacement of the spot on the screen from the centre will provide a visual indication of the magnitude of the applied voltage. If a second voltage varying at a fixed frequency is applied to the other set of deflection plates, the displacement of the spot on the screen will form a time axis. The displacement of the spot with the combined signals will show the variation with time of the signal applied to the y plates. If this signal is repetitive at a fixed frequency, and the 'time' signal is synchronised with it, a stationary waveform of the y-plate signal will be obtained.

To reduce the interaction between the *x*- and *y*-deflection plates, a screen is placed between them. This is in the interplate shown in *Figure 7.110.* A slot is incorporated in the shield corresponding to the minimum clearance required for the fully-deflected beam leaving the *y* plates. This shield is generally operated at the mean deflection plate potential, although provision can be made to vary the potential to correct distortion in the display.

A shield is also placed around the *y*-deflection plates. This is called the deflection plate shield, and its main purpose is to screen the plates from external fields. It is usually held at the mean deflection plate potential.

Because the deflection plates are not completely parallel but are shaped to allow for the deflection of the beam, the deflecting field is not always at right angles to the beam. Over the flared part of the plates, there will be a component of the deflecting field that accelerates the beam in the axial direction. Because the beam has a significant cross-sectional area in the deflection region, the result is a reduced deflecting force on electrons near the beam edge compared with that on the electrons nearer the beam axis. The displayed circular spot at the centre of the screen tends to become elliptical at the extremes of the scan on the edge of the screen. This effect is called deflection defocusing.

Because of the structure of the deflection plates, the load presented to the deflection drive circuits is mainly capacitive. The connections to the plates are often not made by pins in the tube base (with long connecting leads inside the envelope), but through the sidewall of the c.r.t. envelope. This method of providing the plate connections reduces the inductance of the leads. These effects are important for the high-frequency operation of the c.r.t.

The deflection system just described, and the method of focusing used in the electron gun, are both electrostatic. Magnetic fields produced by coils placed around the neck of the tube can also be used to focus and deflect the electron beam. The choice between magnetic and electrostatic deflection depends mainly on the beam current and deflection angle required in a particular application. For instrument c.r.t.s (used, for example, in oscilloscopes), the beam current is low and the angular deflection required is small. Electrostatic deflection is used, with the advantage of high deflection sensitivity. For television and data display tubes, however, where the beam current is high and large deflection angles are required, magnetic deflection is used. It is general to combine the magnetic deflection with electrostatic focusing in these tubes, and magnetic focusing is rarely used in present-day c.r.t.s.

POST-DEFLECTION ACCELERATION

A c.r.t. operating with high-frequency deflection signals is required to provide a display of adequate brightness for a rapidly-changing waveform. The capability of the c.r.t. to do this is called the *writing speed.* Two factors determine the writing speed of a c.r.t.: the beam current density and the final anode voltage. Current density is limited by the cathode material, while the final anode voltage is limited by the deflection sensitivity required (as shown by equation 2). The higher the potential of anode a_3 in the electron gun, the higher the energy of the electron beam passing between the deflection plates, and therefore the higher the deflection voltage required. On the other hand, the higher the accelerating voltage and energy of the electron beam, the brighter the display on the screen. This problem can be overcome by accelerating the electron beam after deflection by a fourth anode—post-deflection acceleration. A brighter display is obtained, but because the electron beam passes through the deflection plates without an increase in velocity, the deflection sensitivity is not decreased.

The graphite conductive layer inside the c.r.t. towards the screen (shown in *Figure 7.107*) can be used as the fourth anode, connected to a higher voltage than that of the final anode of the electron gun, anode a_3. The ratio of the voltage applied to the post-deflection acceleration anode and the voltage applied to the final anode of the electron gun is known as the post-deflection acceleration ratio (p.d.a. ratio).

In the first c.r.t.s incorporating p.d.a., the internal graphite layer was divided into two sections. The section nearer the deflection region was maintained at the potential of the final anode of the electron gun, while the section nearer the screen was maintained at a higher potential. Although a brighter display was obtained, the distortion at the extremes of the scan increased as the p.d.a. voltage was increased. The p.d.a. ratio was therefore limited to approximately 2. A higher ratio was obtained by replacing the two-section graphite layer by a helical layer extending from the deflection region to the screen. The end of the helix nearer the deflection region was connected to the final anode of the electron gun, while the other end was connected to the p.d.a. voltage. An increasing acceleration voltage was obtained along the length of the tube. This method, generally known as *spiral p.d.a.*, enabled a p.d.a. ratio of 6 to be obtained. It forms the basis of the generally-used method of providing p.d.a. in present-day c.r.t.s.

The electric field produced by the p.d.a. helix can penetrate the deflection region of the c.r.t., as shown in *Figure 7.111* (*a*). This penetration reduces the deflection sensitivity. If a shield is placed between the deflection plates and the helix, the field is modified to form a low-magnification diverging lens, as shown in *Figure 7.111* (*b*). The

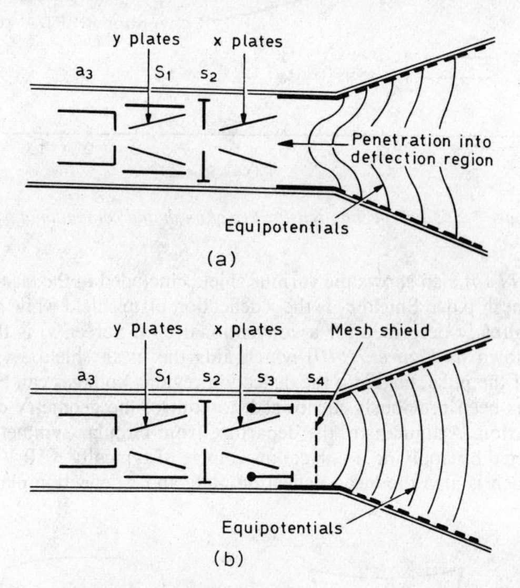

Figure 7.111 *Electric field produced by p.d.a. voltage:* (*a*) *penetration into deflection region* (*b*) *mesh shield preventing penetration and forming electron lens*

p.d.a. voltage is applied as before to the screen end of the helix, and a p.d.a. ratio as high as 10 can be obtained. Thus the voltage applied to anode a_4 of the c.r.t. can be 10 kV to 15 kV.

The shield between the deflection plates and the p.d.a. helix is usually in the form of a fine mesh, the mesh shield shown in the electrode structure of *Figure 7.110*. The position of the mesh shield along the axis of the c.r.t. can affect the size of the display. If the mesh shield is close to the deflection plates, the deflection decreases as the p.d.a. voltage increases. If the mesh shield is further away from the plates, expansion occurs with an increasing p.d.a. voltage. It is general practice to position the mesh shield so that the p.d.a. voltage has no effect on the *x* deflection, or only a deflection shrinkage of 10% at the most, but as the *y* plates are further away there will be a deflection expansion which will increase the *y*-deflection sensitivity.

The deflection sensitivities of c.r.t.s with conventional (spiral) and mesh (incorporating a mesh shield) p.d.a. are shown in *Figure 7.112*. The relative sensitivity for both x and y deflection is shown plotted against p.d.a. ratio. The more-constant sensitivity of the mesh p.d.a. c.r.t.s is clearly shown.

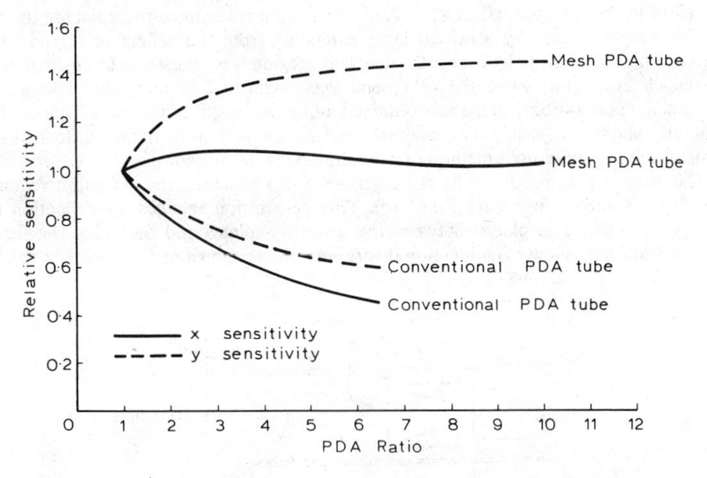

Figure 7.112 Deflection sensitivities of mesh and conventional p.d.a. c.r.t.s

Figure 7.111 (*b*) also shows the various shields included in the electrode structure of a c.r.t. with mesh p.d.a. Shield s_1 is the y deflection plate shield while s_2 is the interplate shield, both already described for a conventional c.r.t. Screen s_3 is the post-deflection skirt (also shown in *Figure 7.110*) which aids the mesh shield s_4 in preventing the penetration of the p.d.a. field into the deflection region. Voltages can be applied to these shields, as has been previously mentioned, to control the geometry of the display and correct distortion. Astigmatism, the departure from circular symmetry in the display, can be corrected by applying a correction voltage of typically ± 50 V to the voltage on anode a_3 (which is also the mean deflection plate and y-deflection plate shield voltage).

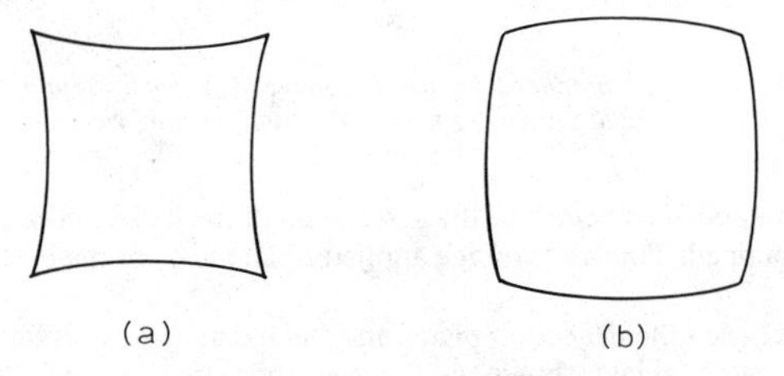

Figure 7.113 Distortion of display in a c.r.t.: (a) pin-cushion (b) barrel

A correction voltage of typically ± 15 V to the interplate shield, also held at the same potential as anode a_3, alters the geometry of the display. The two principal forms of distortion in the display are 'pin-cushion' and 'barrel' distortion, illustrated in *Figure 7.113* and caused by such factors as tolerances in the components and assembly of the electrode structure, or non-uniformity in the deflection fields. In some c.r.t.s a geometry control electrode is incorporated, between the deflection plates and the post-deflection skirt. The potential on this electrode can be varied by typically ±100 V about the mean deflection plate voltage, and used in conjunction with the interplate shield to correct pin-cushion and barrel distortion.

SCREEN

The function of the screen of a c.r.t. is to provide a visible indication of the position of the deflected electron beam at any instant. This is achieved by forming the screen from a thin layer of phosphor deposited on the inside of the faceplate of the c.r.t. A phosphor is a material which will emit light when bombarded with electrons, and by choosing particular phosphors the colour and characteristics of the screen can be matched to the requirements of an application.

The phenomenon of the emission of light when a material is bombarded with electrons is called *luminescence*. Various crystalline materials exhibit the effect, but practical phosphors are chemically-complex crystalline compounds consisting essentially of a *host* material into which impurity *activators* are introduced. Typical host materials include beryllium, cadmium, calcium, aluminium, silicon, manganese, and zinc, in the form of oxides, sulphides, silicates, selenides, halides, and tungstates. Controlled amounts of activators such as copper, magnesium or silver are introduced into the host material. Imperfections occur in the crystal structure of the host material, some being occupied by the activator and others forming electron traps. When the phosphor is bombarded by electrons, some of the electrons of the phosphor atoms are raised to a higher energy state. On falling back to their original lower energy state, light is emitted. The electrons may oscillate between the higher and lower energy states, and may be held temporarily in one of the electron traps. These factors prolong the time for which the light is emitted. Other impurity materials can be introduced into the phosphor to ensure a rapid ending to the emission of light. Such impurities are called *killers*, nickel being the most commonly-used material.

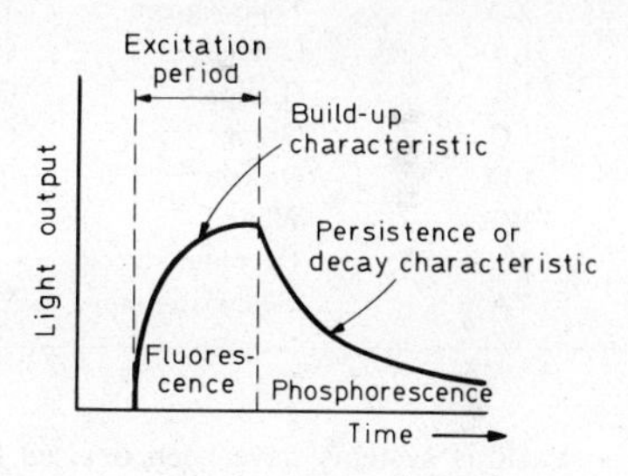

Figure 7.114 Light output characteristic of phosphor

When a point on the phosphor screen is bombarded by the electron beam, light is not mitted immediately. There is a build-up time during which the light output from the hosphor rises to a constant level. This level is proportional to the energy of the incident lectron beam, determined by the square of the velocity of the electrons on impact with he screen, which in turn is determined by the accelerating voltage. The light level will emain constant during the excitation period. The emission of light during this period is *luorescence*. When the excitation ceases (the electron beam has moved from that point), he light output does not cease immediately but decays over a finite time. The emission f light is then *phosphorescence*. A typical light output characteristic is shown in *Figure .114*. The build-up and decay times are determined by the composition of the phoshor. For very high writing speeds a short build-up time is required, but in general uild-up time is not too important a characteristic of a phosphor. The decay time, however, is of major importance. For some high-speed applications, a long decay time may be necessary to make the display visible. On the other hand, too long a decay time can make the display confused as new information is superimposed on that previously displayed. The decay of the light output from the screen of a c.r.t. is known as the *persistence* (sometimes called the afterglow). It is general to classify screens as being of short, medium, or long persistence.

Table 7.9 DESIGNATION OF PHOSPHORS

Pro-Electron designation	*Fluorescent colour*	*Phosphorescent colour*	*Persistence*	*Equivalent JEDEC designation*
BA	Purplish-blue	—	Very short	—
BC	Purplish-blue	—	Killed	—
BD	Blue	—	Very short	—
BE	Blue	Blue	Medium short	P11
BF	Blue	—	Medium short	—
GB	Purplish-blue	Yellowish-green	Long	P32
GE	Green	Green	Short	P24
GH	Green	Green	Medium short	P31
GJ	Yellowish-green	Yellowish-green	Medium	P1
GK	Yellowish-green	Yellowish-green	Medium	—
GL	Yellowish-green	Yellowish-green	Medium short	P2
GM	Purplish-blue	Yellowish-green	Long	P7
GN	Blue	Green (Infra-red excited)	Medium short (fluorescence)	—
GP	Bluish-green	Green	Medium short	P2
GR	Green	Green	Long	P39
GU	White	White	Very short	—
KA	Yellow-green	Yellow-green	Medium	P20
LA	Orange	Orange	Medium	—
LB	Orange	Orange	Long	—
LC	Orange	Orange	Very long	—
LD	Orange	Orange	Very long	P33
W	White	—	—	P4
X	Tricolour screen	—	—	P22
YA	Yellowish-orange	Yellowish-orange	Medium	—

Various systems have been devised for classifying the most commonly-used phosphors. *Table 7.9* shows one system. The phosphors are classified on the fluorescent colour, phosphorescent colour, and the persistence. The designations allocated by Pro-Electron (International Association of West European Electronic Component Manufacturers) and JEDEC (Joint Electronic Defence Executive Committee U.K.) are also included. The data for c.r.t.s will include the spectral responses and persistence characteristics for the types of phosphor used. The various terms used for persistence such as *short*, *medium-short*, and so on, are related to the time taken for the light output to fall to 10% of the initial value, as shown in *Table 7.10*.

The light output from the screen can be increased by applying a thin film of aluminium to the screen after the phosphor has been deposited. This backing film reflects forward the light from the phosphor that would otherwise shine back down the c.r.t. and be lost.

The aluminium backing also helps to prevent an effect called *sticking* of the screen

As the screen is not a good conductor of electricity, when it is bombarded with electrons there is a tendency for negative charges to accumulate on its surface. If this occurs, there is a negative potential on the screen which effectively reduces the accelerating voltage of the tube. The velocity of the electrons at impact is reduced, thereby reducing the energy imparted to the phosphor and so reducing the light output.

Table 7.10 PERSISTENCE OF PHOSPHORS

less than 1 μs	very short
1 μs to 10 μs	short
10 μs to 1 ms	medium short
1 ms to 100 ms	medium
100 ms to 1 s	long
greater than 1 s	very long

When an electron impinges on the screen, it may cause the emission of one or more secondary electrons. The ratio of the secondary to primary electrons is called the secondary emission ratio. Between certain limits of the accelerating voltage, the ratio exceeds unity, as shown in *Figure 7.115*. No negative charge can then accumulate as more electrons are leaving the screen than are arriving at it. Outside these limits the ratio falls below unity, the screen charges negatively, and sticking occurs.

The critical high voltage above which sticking occurs may lie within the operating voltage range of the c.r.t. One method of avoiding sticking if this should occur is to vary

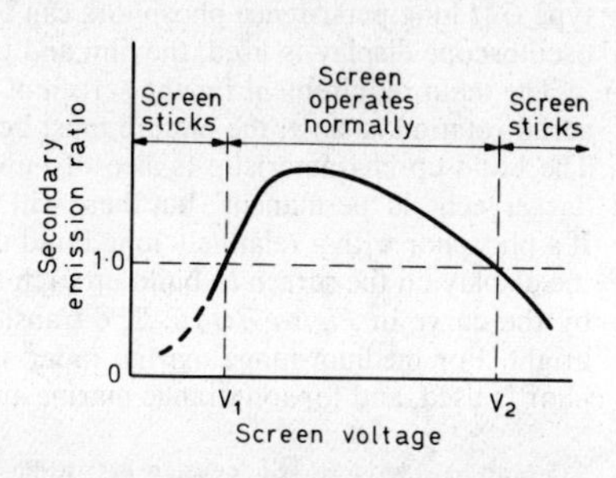

Figure 7.115 Secondary emission ratio as function of screen voltage

the composition of the phosphor to give better secondary emission. Magnesium oxide and silicon oxide can be added to the phosphor to achieve this. The additional material may, however, modify the operation of the phosphor. A better method of preventing sticking is by the use of the aluminium backing film. Any excess electrons on the screen will be conducted away, and the secondary emission properties of the phosphor are no longer important for the maintenance of the correct accelerating voltage.

The choice of a phosphor for a particular application depends on many factors. These factors will be discussed in relation to the five main application groups for c.r.t.s into which it is possible to classify them by the application and method of operation.

(*a*) Raster applications where the deflected electron beam is used to build up a raster on the screen. The writing speed of the c.r.t. will be constant, but the beam current is modulated to form bright and dark areas of the raster.

(*b*) Oscilloscope applications in which the beam current is constant during a trace but the writing speed can vary over a wide range, particularly in a wideband oscilloscope.

(*c*) Radar applications.
(*d*) Flying-spot scanners.
(*e*) Storage applications.

The chief raster applications of c.r.t.s are for television and data display. The phosphor used in c.r.t.s for television monitors and camera viewfinders have the same requirements as those for domestic receivers. The persistence of the phosphor (combined with the persistence of vision of the viewer) must obviate flicker in the picture, but the persistence must not be too great or moving objects in the picture will be smeared. Similar requirements apply to closed-circuit television. For monochrome transmission, a phosphor with a white fluorescence, type *W*, is used. For colour transmission, a tricolour screen is used, type *X*, in which dots of red, green, and blue phosphors are deposited carefully aligned with the holes in the shadow-mask of the tube. For some camera viewfinders, c.r.t.s with a green screen are used as this is least tiring on the eye. For this, type *GH* medium-short-persistence phosphor would be used. The phosphor in c.r.t.s for data display will depend on the method of transmission, particularly on the field frequency used. Flicker-free displays at low repetition rates can be obtained by using long-persistence phosphors. Where black-and-white displays are required, type *W* phosphor has to be used; where the colour of the display is not important (as in character displays), type *GH* medium-short-persistence or *GR* long-persistence phosphors can be used.

A wide range of frequencies has to be handled by an oscilloscope, and this makes conflicting demands on the screen of the c.r.t. The requirements for light output and persistence at high writing speeds conflict with those for low writing speeds, and so a compromise is necessary. As long as the phosphor used has a good luminous efficiency, a satisfactory compromise can be attained. For general-purpose oscilloscopes, type *GH* medium-short-persistence phosphor is generally used. If longer persistence is required, type *GM* long-persistence phosphors can be used. Where photographic recording of the oscilloscope display is used, the film and phosphor can be matched.

The main requirement for the screen of a radar c.r.t. is a long persistence. Where the aerial rotation is slow, the picture must be held on the screen for a relatively long time. The build-up characteristic is also of importance in a phosphor for a radar c.r.t. The 'target' echo is 'permanent', but there will be other echoes and noise such as sea clutter. If a phosphor with a relatively long build-up time is used, the permanent echo will allow the display on the screen to build up each sweep to the maximum light output, as shown by the curve in *Figure 7.116*. The transitory echoes are not additive, and so are less bright. For medium-range marine radar systems, type *LD* very-long-persistence phosphor is used, and for long-range marine and air systems, type *LC* very-long-persistence

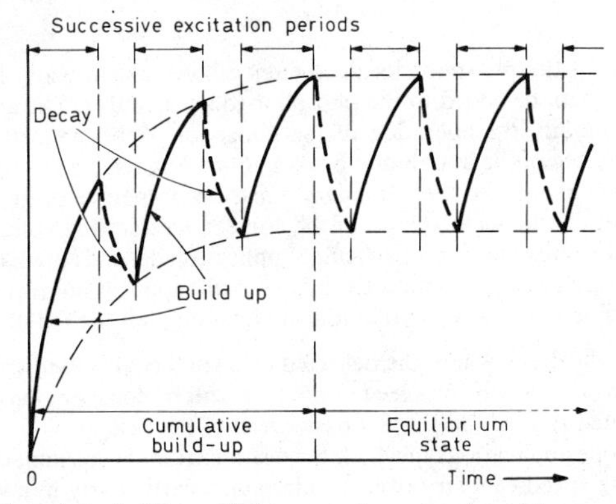

Figure 7.116 Cumulative build-up of light output from phosphor with successive excitations

phosphor is used. For river systems where target speeds are higher and the range shorter, type *LB* long-persistence phosphor is to be preferred.

In flying-spot scanners, the energy conversion efficiency of the screen in the spectral range corresponding to the response of the detecting device must be as high as possible. Very-short persistence is essential when high definition is required. For a 625-line television system with 5 MHz definition, there must be no effective light output after 0.3 μs. In a slower-speed facsimile system with a line scanning speed of 1 s, however, the persistence can be longer, up to 2 ms. For monochrome flying-spot scanner systems, type *BA* very-short-persistence phosphor is used. The peak response of this phosphor at approximately 400 nm corresponds to the response of photomultipliers with caesium-antimony *S*4 photocathodes. For colour television, type *GU* very-short-persistence phosphor is used.

The operation of the storage tube will be described later. The requirements for the screen are similar to those for an oscilloscope screen, and type *GH* medium-short-persistence phosphor is used.

The screen of a c.r.t. must consist of a uniform unblemished fine-textured layer of phosphor deposited on the inside of the faceplate. The layer must be of uniform thickness over the whole area of the faceplate, and must adhere firmly to it. The glass surface itself must be free from blemishes.

The first stage in the manufacture of the screen is the preparation of the phosphor. Because of the small amounts of the activator materials added to the host crystals, all materials used must be extremely pure, and great care is taken during manufacture to avoid the inclusion of unwanted impurities that would affect the operation. The impurity level required is less than 1 part in 10^6 otherwise the photoemission properties will be affected. The various chemical constituents are finely ground, the quantities carefully weighed, and thoroughly mixed. Uneven mixing of the constituents would result in screens with different properties in different areas. The homogenous mixture is fired under controlled conditions to form the required crystal structure, the furnace temperature being controlled to within ±10 °C at a temperature of 1 500 °C. The crystalline material is washed repeatedly to remove the chloride fluxes used in firing and other unwanted materials. The concentration of such materials must be reduced to below 0.01%. The phosphor is then dried, finely ground, and passed through a stainless steel sieve with a typical mesh size of 40 μm.

The inside surfaces of the c.r.t. envelope and faceplate are thoroughly cleaned, and the envelope is stood faceplate downwards. The exact quantity of phosphor required to form the screen is mixed with distilled water, and is poured into the envelope and allowed to settle. A small quantity of an electrolyte is added to the water to prevent uneven settling of the phosphor caused by the build-up of charges. This settling process is called screen laying, and takes several hours to be completed.

When the phosphor has settled, the distilled water is decanted by carefully tilting the envelope so that the settled layer is not disturbed. The phosphor layer is then dried by a stream of warm air.

In c.r.t.s with metal-backed screens, a thin layer of aluminium is formed over the deposited phosphor layer. A layer of lacquer is applied over the phosphor layer first, to prevent the aluminium penetrating the interstices between the phosphor particles, and the aluminium is then deposited by evaporation in a vacuum. The thickness of the aluminium is monitored during deposition to ensure the correct thickness is obtained. Too thick a film would reduce the luminous efficiency of the screen; too thin a film would be ineffective. After the correct aluminium thickness has been obtained, the lacquer is removed by baking the envelope in air.

The final stage in the manufacture of the screen and its associated components is the deposition of the graphite layer on the inside of the conical section of the envelope. The graphite can be in the form of a single layer, as in mono-accelerator tubes. It can be in more complex forms such as the helix required by post-deflection acceleration tubes.

C.R.T.s are available with circular or rectangular screens. The choice of screen shape depends on the application. With television picture tubes and tubes for data display where a rectangular image is presented, it will obviously be more economical to use a rectangular screen. On the other hand, c.r.t.s for use in radar p.p.i. systems will require a circular screen. For oscilloscope and indicator c.r.t.s, a rectangular screen may enable space to be saved on the front panel of the equipment. Typical sizes for such instrument tubes are circular screens with diameters from 3 cm (1 in) up to 13 cm (5 in), and rectangular screens with diagonals of 14 cm ($5\frac{1}{2}$ in) or 18 cm (7 in). Domestic television picture tubes have screen diagonals up to 66 cm (26 in), although monitor tubes are somewhat smaller, typically up to 38 cm (15 in).

Because oscilloscope tubes are generally used with a graticule for measurement, some c.r.t.s are fitted with an internal graticule. This is formed on the inside of the faceplate from white or grey sintered glass. The phosphor screen is then deposited on the faceplate over the graticule in the normal way.

The axes of the deflected electron beam can be aligned with those of the graticule by a magnetic field produced by a coil mounted around the c.r.t. near the deflection plates. The current through this coil can be adjusted to produce a rotation of the deflected beam of typically up to 5°. Provision is also made in some c.r.t.s to provide external illumination of the graticule.

Multi-display c.r.t.s

It is often useful in an oscilloscope to display two traces simultaneously of related waveforms with a common time scale. This is, for example, useful for comparing input and output waveforms for a particular part of a circuit, or for the alignment of a circuit. Two electron beams are required in the c.r.t. with separate *y*-deflection systems to enable the two traces to be displayed, but with a common *x*-deflection system to provide the common time scale. Separate *x*-deflections can be provided also, to form two completely independent displays.

Two methods can be used to provide the two electron beams, each with its own advantages and disadvantages. One method uses two electron guns to provide the beams, the *double-gun* c.r.t. The other method uses one electron gun but splits the beam into two, the *split-beam* c.r.t. Of the two methods, the split beam is probably the more commonly used.

The electron beam from the gun can be split in two by an electrode called the splitter plate placed between the final anode of the gun a_3 and the *y*-deflection plates. The splitter plate is horizontal, and is held at the a_3 potential. It is continued to form a screen between the two *y*-deflection plates. A deflection voltage applied to the upper deflection plate will therefore operate only on the upper electron beam, while a voltage applied to the lower *y* plate operates only on the lower beam. The two beams can then be deflected simultaneously in the horizontal direction by the *x*-deflection plates.

Because the beam is split into two, the brightness of the displays on the screen is half that of a single trace. This loss in brightness can be a disadvantage when the c.r.t. is operating at high frequencies. An alternative method of splitting the beam has been devised which overcomes this problem. The beam is split in the electron gun, the final anode having two apertures instead of one so that two beams emerge.

Symmetrical *x* and *y* deflections can be applied to split-beam c.r.t.s to give high deflection sensitivities. Post-deflection acceleration can be applied to both beams, and so the performance of a split-beam c.r.t. can be comparable to that of a single-beam tube. For the user, the most noticeable disadvantage is that only one brightness control can be fitted. If the duty cycle of the two displays differs considerably, there will be a difference in the brightness of the two traces. This can be a problem, for example, when photographing the screen. If a splitter plate is used, the position of the plate in relation to the electron gun and *y*-deflection plates has to be carefully aligned, although correction voltages can be applied to the tube to reduce the effects of any misalignment.

The double-gun c.r.t. can either have a common *x* deflection so that the two displays have a common time relationship, or have separate *x* deflection to form two separate display systems in one envelope. Both types have the disadvantage that tolerances in the electrode structure during manufacture and assembly can be additive and cause relative errors between the two displays. The need to accommodate two guns in the c.r.t. makes the envelope larger than that of a split-beam c.r.t., but each display has its own focus and brightness controls so that each can be set for optimum visibility. Double-gun c.r.t.s need careful alignment during manufacture but new manufacturing techniques have been devised in which the two guns are assembled together so that both the bulk of the c.r.t. and errors between displays can be reduced.

Higher-frequency operation of c.r.t.

In the previous discussion on the deflection of the electron beam in a c.r.t., it has been assumed that the deflecting field, and hence the signal voltage applied to the deflection plates, is constant while a particular part of the electron beam passes between the plates. Thus that part of the beam experiences a constant deflecting force to give a specific deflection from the centre of the screen. Each part of the beam will take a finite time to pass the deflection plates, the beam transit time, and so there will be a frequency limit above which the deflecting field will vary appreciably during the passage of the beam. In addition, there will be a frequency limit when the reactance of the deflection plates to the applied signal becomes significant.

When the connections to the deflection plates are made by pins in the c.r.t. base, the inductance of the connecting leads will limit the operating frequency of the deflection signal to approximately 10 MHz. If the plate connections are made by pins through the sidewall of the c.r.t. envelope, the inductance of the connecting leads is considerably reduced. Deflection signals with frequencies up to, say, 50 MHz to 80 MHz can be handled. Above these frequencies, however, both the transit-time effects of the electron beam and the reactance of the deflection plates, become significant.

If the frequency of the deflection signal is high so that the deflecting field varies considerably during the transit time, the deflection of the beam will be less than that expected. In other words, the deflection sensitivity will decrease with increasing frequency. If the deflection signal frequency is high enough for one cycle to be completed during the transit time, then that part of the electron beam will be subjected to equal and opposite deflecting forces and the net deflection is zero. If the deflection frequency is increased further, a finite deflection occurs again, reaching a maximum value when 1.5 cycles are completed during the transit time of the beam. The magnitude of the deflection, however, is only about one-fifth the d.c. value. As the deflection frequency is increased again, the deflection falls again to zero when two cycles correspond to the transit time. Although equation 2 shows that the deflection sensitivity can be increased by increasing the length of the deflection plates or decreasing the accelerating voltage of the c.r.t., neither of these methods can be used since they also increase the transit time of the electron beam. Higher-frequency operation of a c.r.t. can only be achieved by decreasing the effective transit time of the beam.

One method of achieving this is by using a series of short deflection plates connected as a transmission line. This is shown in *Figure 7.117*. Each short deflection plate has a

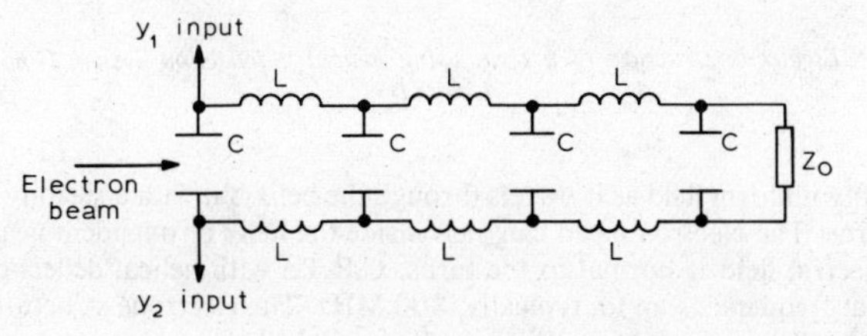

Figure 7.117 Transmission-line deflection system for higher-frequency operation of a c.r.t.

capacitance C, and is connected to the next plate by a low-value inductance L. The line is terminated by its characteristic impedance Z_0. The deflection signal is propagated along this line with a velocity equal to that of the electron beam (the propagation velocity is determined by the capacitance and inductance values). Thus the electron beam passing through the structure receives a deflecting force at each plate, the

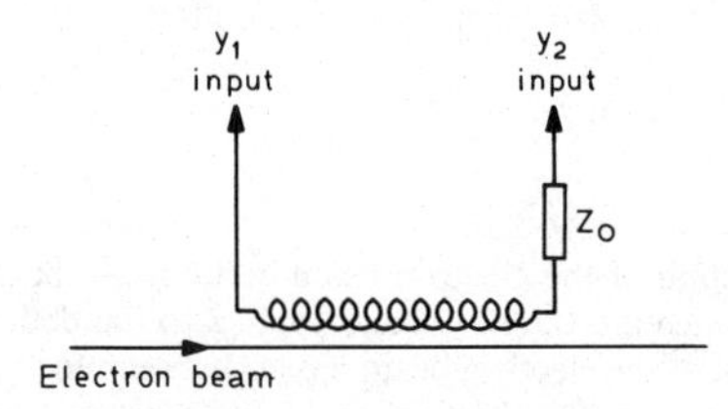

Figure 7.118 Helical deflection system for higher-frequency operation of a c.r.t.

individual deflections adding to give the final deflection value. The operating frequency of such a deflection system is limited by the resonant frequency of an individual LC element, and such transmission-line deflection systems can be used up to, typically, 500 MHz.

Higher operating frequencies can be obtained by using a helical deflection system, as shown in *Figure 7.118*. The axial propagation velocity of the deflection signal along the helix is the same as that of the electron beam. Each part of the beam, therefore, remains

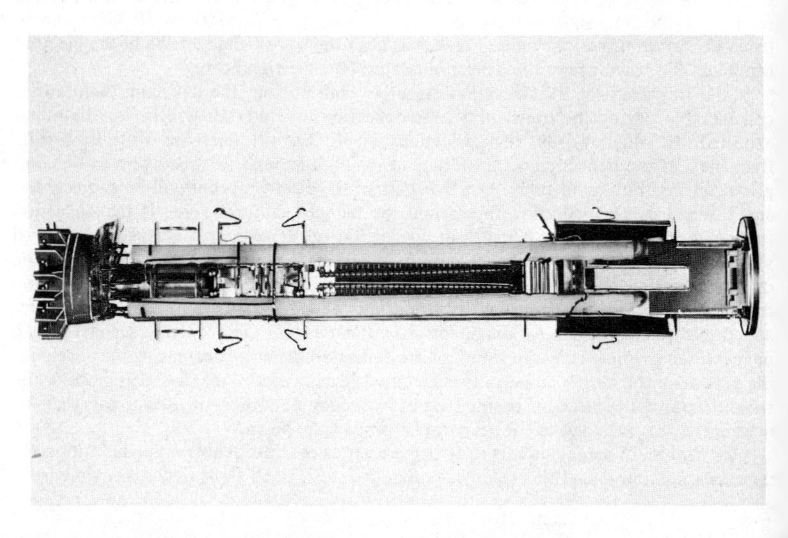

Figure 7.119 Electrode structure of a c.r.t. using helical y-deflection system (Photo: Mullar Limited)

in a reasonably uniform field as it travels through the helix but with a steadily increasin deflecting force. The electron beam can pass inside the helix or outside it near the turn where the electric field is normal to the turns. C.R.T.s with helical deflection system can operate at frequencies up to, typically, 800 MHz. The electrode structure of a c.r. with a helical deflection system is shown in *Figure 7.119*.

Data for c.r.t.s

The published data for a c.r.t. contains information on the electrical, optical, and mechanical characteristics. The electrical operating conditions specify the voltages required for beam forming (the voltages on the final anode of the electron gun a_3, on the final anode a_4 if the c.r.t. incorporates p.d.a., on the focus electrode a_2, and on the control grid for visual cut-off); the line width under specified operating conditions; the mean deflection-plate voltages and the deflection factors; and the correction voltage ranges for compensating distortion in the display. The interelectrode capacitances are given, and ratings specifying the maximum and minimum voltages that can be applied to the various electrodes are included. The optical characteristics specify the operation of the screen, the type of phosphor used and its colours, and the persistence. The mechanical information besides specifying the size and shape of the c.r.t., also includes the useful screen areas which can be used for the display.

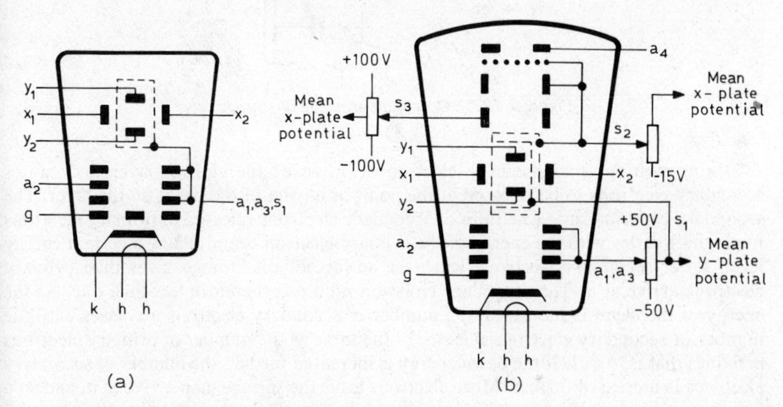

Figure 7.120 *Circuit symbols for a c.r.t.: (a) mono-accelerator c.r.t. (b) p.d.a. c.r.t. showing applied correction voltages*

The circuit symbols for two types of c.r.t. are shown in *Figure 7.120*. The symbol in (*a*) is for a simple mono-accelerator c.r.t. (not using p.d.a.). An interplate shield between the *x*- and the *y*-deflection plates is incorporated. The symbol in (*b*) is for a p.d.a. c.r.t. incorporating a post-deflection skirt and mesh shield, and a geometry control electrode. Besides the interplate shield, a *y*-deflection plate shield is also used. The correction voltages used with this type of c.r.t. are also indicated.

Storage tube

The c.r.t.s described so far produce an instantaneous visual representation of an electrical signal. If the signal changes, the visual display will also change. In some applications, it is useful to be able to store information as an electrical signal, and when required either release this information as an electrical signal or display it on a luminescent screen. A c.r.t. that can perform this function is called a storage tube.

The operating principle of the storage tube is the building up and retention of charges on an insulating surface (dielectric storage), and the removal of these charges as an electrical signal which can also be used to excite phosphors to form a visual display. The build-up of the charges is called *writing*, and the removal of the charges is called *reading*.

This operating principle is illustrated diagrammatically in *Figure 7.121*. The storage element of the tube is shown as a thin layer of insulating material with a conducting backing layer. When the electron beam of the tube impinges on a part of the insulating layer, a capacitor is formed. The electron beam and the conducting backing layer form the two plates of the capacitor, while the insulating layer forms the capacitor dielectric. If the surface of the insulating layer is initially stabilised at the potential of the cathode of the storage tube, then the charge induced in each elementary capacitor of the storage layer by the electron beam is determined by the secondary emission.

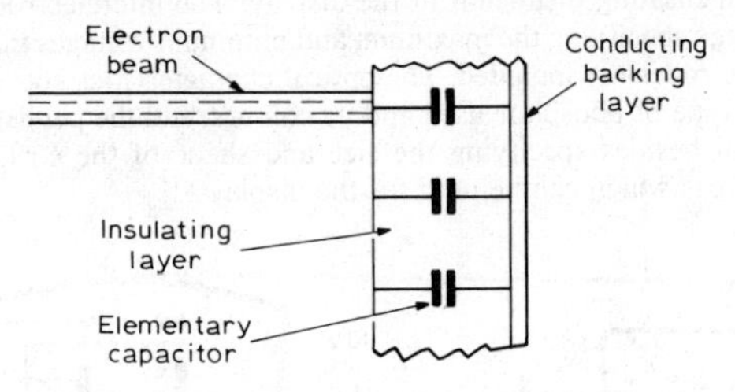

Figure 7.121 Operating principle of storage tube

During writing, a high-energy electron beam scans the storage layer and causes secondary electrons to be released at the point of impact of the beam on the layer. The secondary emission ratio (the ratio of secondary electrons released to primary electrons impinging) varies with the energy of the primary electron beam. Below a certain energy level, fewer secondary electrons leave the surface of the storage layer than primary electrons arrive at it. The secondary emission ratio δ is therefore less than one. As the energy of the beam is increased, the number of secondary electrons increases until the number of secondary electrons released is the same as the number of primary electrons arriving; that is, $\delta = 1$. If the beam energy is increased further, the number of secondary electrons is increased further. More electrons leave the surface than arrive at it, and so δ is greater than one. However, if the beam energy is increased still further, the primary electrons will penetrate below the escape depth for the insulating material, and fewer secondary electrons are released. The ratio will therefore fall towards one, and eventually fall below one. The points at which $\delta = 1$ are called the crossover points, and a typical curve of δ plotted against beam energy is shown in *Figure 7.122*. If the energy of the writing beam is sufficient to make the ratio δ greater than one, the areas of the

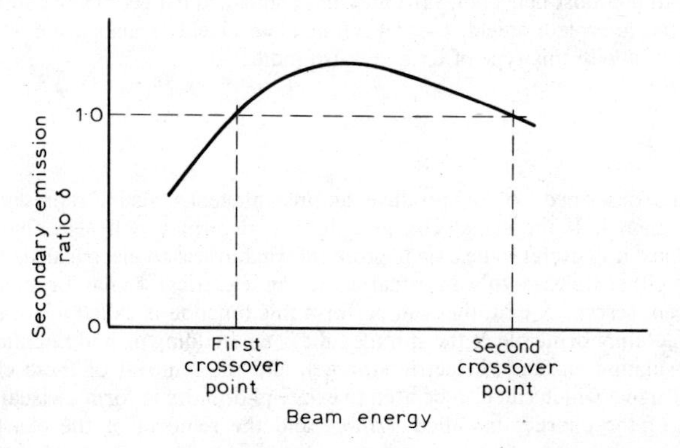

Figure 7.122 Secondary emission ratio δ plotted against energy of electron beam

storage layer on which the beam impinges will be charged positively as more electrons leave the surface than arrive at it. The required beam energy is attained by choosing the accelerating voltage and beam current to give operation above the first crossover point, and the magnitude of the induced positive charge will then be determined by the instantaneous beam current.

The cross-section of a typical practical storage tube operating in this way is shown in *Figure 7.123*. The storage layer is called the target, and the conducting backing layer forms the output electrode. A fine wire mesh is placed on the gun side of the target, separated from it by a short distance only, to collect the secondary electrons released during writing, and to ensure that the electron beam approaches the target normally. The deflection system has been omitted from the figure for clarity.

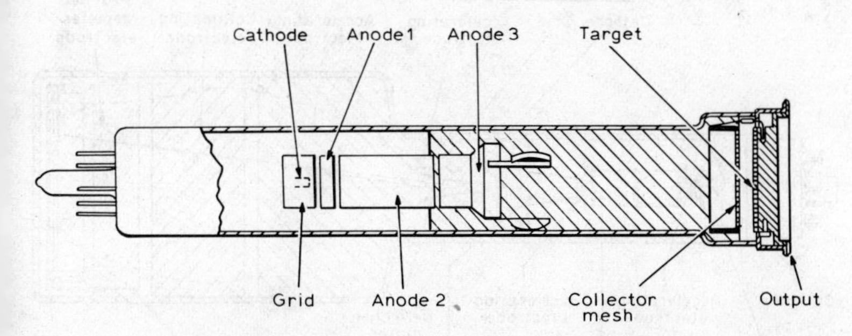

Figure 7.123 Cross-section of practical storage tube

The initial stabilisation of the target surface is achieved by scanning the target with a low-energy electron beam. The ratio δ is less than one, and so electrons will be gained by the surface until the potential becomes less positive and falls to that of the cathode. A typical cathode potential is -15 V.

To start the writing operation, the cathode potential is lowered by approximately 100 V. The grid voltage is reduced to keep the beam cut off until the writing starts, and the second and third anode voltages are also lowered to maintain the beam in focus despite the change in cathode potential. The information to be written on the target is applied to the grid as a positive signal which modulates the electron beam as it scans the target. The beam energy is now sufficient to make δ greater than one, so that the areas on which the beam impinges are charged positively by an amount determined by the instantaneous beam current. The maximum change in the potential of a small area of the target is inherently limited to only a few volts.

For reading, the cathode potential is returned to its former value near earth potential, and the grid and anode voltages changed to give a well-focused beam with the changed cathode potential. An unmodulated electron beam scans the target surface, the beam energy being low enough to produce negligible secondary emission. The areas of the target that have been charged during writing are now discharged by the beam. Because of the capacitive coupling between the target surface and the conducting backing layer, an output signal is produced in this layer. This signal will be proportional to that used to modulate the electron beam during writing. Provided a sufficiently large beam current is used for reading, the discharge of the target surface will be almost complete, leaving the surface potential equal to the cathode potential, and the target ready for the next writing operation.

Because the output signal is proportional to the input signal, this type of storage tube is called a half-tone storage tube.

When the information written into the storage tube is required to be displayed on a luminescent screen, a different construction is used for the tube. The target is not in the form of a continuous layer but the insulating material is deposited on one side of the wires of a fine conducting mesh. This storage mesh is placed a few millimetres away from a luminescent screen with the insulating material facing the electron gun. The collector mesh is placed close to the storage mesh, as in the previous type of storage tube. Two electron guns are used, one for writing and the other for display. The cross-section of a display storage tube with writing and viewing guns is shown in *Figure 7.124*. The writing beam is focused to a small-diameter spot as in a conventional c.r.t., whereas the viewing gun is required to flood the storage mesh uniformly with electrons.

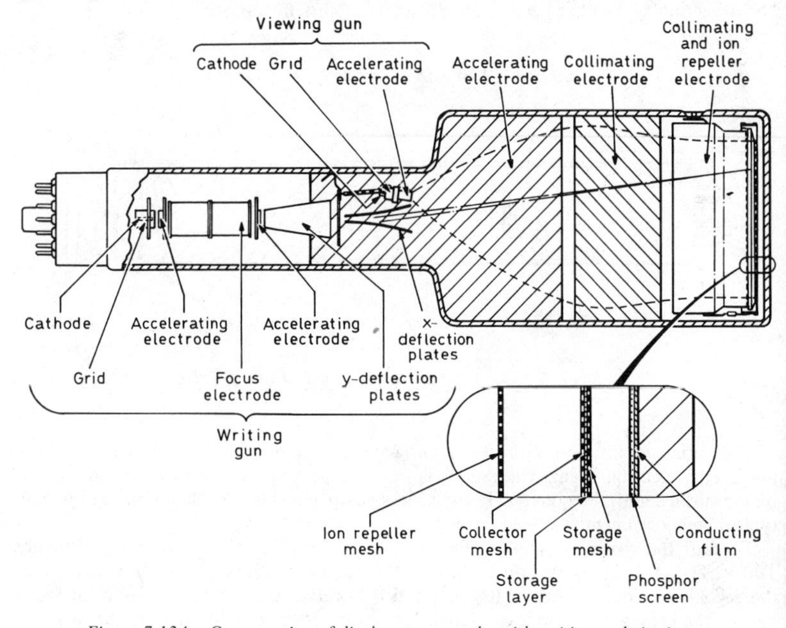

Figure 7.124 Cross-section of display storage tube with writing and viewing guns

In some tubes more than one gun is used, and these guns are then called 'flood' guns. The schematic cross-section of a display storage tube with two flood guns is shown in *Figure 7.125*. Because the electrons from the flood guns must approach the storage mesh normally, a collimation electrode is incorporated in the tube. The storage mesh has between 250 and 500 wires to the inch, and is generally operated near earth potential (typically +2 V). The cathodes of the flood guns are held at earth potential, while that of the writing gun is held at a high negative potential, typically −1.5 kV. The collector mesh is held at a positive potential of typically +100 V to ensure efficient collection of the secondary electrons, and the screen is held at a high positive potential (+6 kV) so that a high accelerating field is maintained between the storage mesh and the screen.

The writing operation of the display storage tube is similar to that of the conventional c.r.t. The beam from the writing gun is deflected by the signal applied to the deflection plates, and traces an image of the signal on the storage mesh. Although the writing beam is focused to a small-diameter spot, because of the fineness of the mesh the spot is larger than the apertures. Some electrons from the beam pass through the mesh to enter the accelerating field, and so produce a display of the signal on the screen. Other electrons strike the insulating material of the storage mesh. The beam energy is sufficient to make the secondary emission ratio δ greater than one, and so the areas of the mesh where the beam strikes will be charged positively. Thus an image of the applied signal will be formed of positive charges on the storage mesh as well as the visible image on the screen.

The flood guns operate continuously, flooding the storage mesh with electrons which have a low velocity because the potentials of the flood-gun cathodes and storage mesh are nearly the same. The beam energy is low, so that δ is less than one and the storage mesh will be charged negatively except for the areas where the writing beam produces positive charges. Where the storage mesh has a negative charge on the surface, the flood

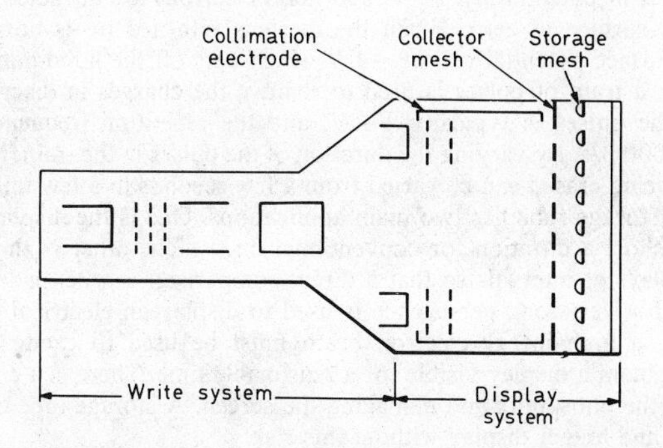

Figure 7.125 Schematic cross-section of display storage tube with writing gun and two flood guns

electrons will be repelled and are collected by the collector mesh. No display therefore appears on the screen. Where there is a positive charge, the flood electrons will be attracted and pass through the mesh into the accelerating field to reinforce the display on the screen. The higher the positive charge on the particular area of the storage mesh, the more flood electrons will pass through and the brighter the display. The intensity of the display is therefore proportional to the written signal, and so half-tone storage is obtained. The action of the storage mesh on the flood electrons for display is similar to that of the grid of a triode valve.

Once the trace has been written on to the storage mesh it will be continuously displayed on the screen, reinforced by the flood electrons passing through the mesh. The display will remain visible for a time determined by the ion current of the tube. As in all thermionic devices, there will be a small number of residual gas atoms present in the tube which will be ionised by collision with the electron beams. Positive ions produced between the screen and the collector mesh will be attracted to the areas of the storage mesh which have a negative charge. (Positive ions produced at the gun end of the tube will be repelled by the positive collector mesh.) The ions will reduce the negative charges to zero, allowing flood electrons to pass through the mesh to the screen. The negatively-charged areas of the mesh correspond to the 'black' areas of the display, and so the neutralising of the negative charge by the ions will result in a gradual increase in the background illumination. Eventually the display will be lost in the background. The time for which the display is visible is called the storage time, or more appropriately the viewing time, generally defined as the time for the background of the display to rise from zero to 10% of the saturated brightness. Typical values of the viewing time lie between one and two minutes, but this time can be considerably extended. The positive ions are produced almost entirely from collisions with the flood-gun electron beams. If these beams are pulsed rather than operated continuously, the number of ions produced is reduced considerably. With a pulse frequency between 200 p/s and 1 000 p/s, a flicker-free display of reduced brightness can be obtained for up to one hour.

If a trace is written on the storage mesh and the tube then switched off, the image can be retained for later display for up to about 7 days.

The written information on the storage mesh can be erased by raising the potential of the mesh to a higher positive value with respect to the flood-gun cathodes. Because of the capacitive coupling between the conducting backing mesh and the insulating material surface, this surface also becomes positive and attracts electrons from the flood guns. The charge pattern on the mesh is therefore destroyed, and the mesh is ready for the next writing operation.

Two modes of erasure can be used: static and dynamic. In the static mode, a single pulse of typical amplitude +4 V and duration 1 s is applied to the mesh. The storage surface increases in potential by +4 V, and flood electrons are attracted to reduce the charges on the surface to zero. When the mesh is returned to its normal operating potential, the surface potential will be −4 V which cuts off the flood-gun beam. In the dynamic mode, a train of pulses is used to remove the charges in discrete steps. The amplitude of the pulses is typically +4 V, and the repetition frequency is between 200 p/s and 1 000 p/s. By varying the duration of the pulses in the train, the persistence of the display being erased can be varied from a few seconds to a few minutes.

The display storage tube has two main applications. One is the display of electrical signals of too short a duration for conventional c.r.t.s. The other is the recording of successive displays at intervals so that a direct comparison can be made. If a conventional c.r.t. with a very-long persistence is used to display an electrical signal of very short duration, a high-energy electron beam must be used to excite the phosphor sufficiently to obtain a display visible for a reasonable time. There is a consequent risk of overloading the phosphors and damaging the screen. A storage tube can be used to give a brighter and longer display without this risk.

Television picture tube

A television picture tube can be regarded as a c.r.t. adapted for the special requirements of displaying a picture. Because a large rectangular picture is to be displayed, a rectangular rather than a circular screen is required to avoid unused screen area. The large display area requires a large deflection angle if the tube is not to be excessively long. This large deflection angle, together with the high beam current when white areas are being displayed on the screen, means that magnetic deflection has to be used instead of electrostatic deflection. It is through considerations such as these that the present-day black-and-white television picture tube has evolved.

DEVELOPMENT OF BLACK-AND-WHITE PICTURE TUBE

The picture tube in the typical pre-war television receiver had a circular screen with a diameter of 7 in, and used magnetic deflection with a deflection angle of 70°. Commercial post-war receivers when the U.K. television service was restarted in 1946 used 9-in picture tubes, but still with a deflection angle of 70°. Almost immediately there was a demand for larger pictures, and by 1950 the tube manufacturers had introduced an all glass tube with a screen diameter of 12 in, and a similar-sized tube in which a metal cone section was bonded to the glass faceplate. This glass-and-metal construction enabled a 16-in-diameter tube to be introduced, although the 70° deflection angle made the tube long. Improvements in glass technology in the next few years enabled larger all glass picture tubes to be developed, and so the metal-cone tube with its problems of metal-to-glass seals was made unnecessary.

It was realised that large-diameter circular screens for the display of a rectangular picture contained a large unused area. Thus by the early 1950s, 'rectangular' picture tubes had been introduced. The first tube to become available had a screen diagonal of

14 in; it was, in effect, a 12-in-circular screen with the corners 'pushed out'. This 14-in tube was soon followed by a 17-in tube. In both, the deflection angle had been increased to 90°, so that the increased screen area was not accompanied by a large increase in overall tube length. By the end of the 1950s a further increase in deflection angle had been made to 110°. The usual picture tubes used in receivers had screen diagonals of 17 in and 19 in.

One of the risks associated with the larger picture tubes was the increased danger of the tube imploding. An implosion, the violent disintegration of an evacuated vessel under atmospheric pressure, would result in shattered fragments of the picture tube envelope being forcibly projected. To protect the viewer from such danger, a transparent safety shield was placed in the front panel of the receiver to cover the faceplate of the tube. This shield reduced the light transmitted from the screen, caused reflections, and acted as a dust trap; effects which dimmed the viewed picture. By 1964, picture tubes with integral implosion protection had been introduced to allow direct viewing of the screen and so give a brighter picture.

Implosion of an unprotected television picture tube starts when an area under tensile or compressive stress is damaged. From the damaged area, a rapidly-growing network of cracks is propagated, until the internal and atmospheric pressures are equalised by the collapse of the tube. The area most at risk is the periphery of the faceplate, the area under the greatest tensile stress. A mild-steel shell or reinforcing band bonded to the periphery of the faceplate not only provides protection for this area, but also opposes deformation or expansion of the tube and so prevents any crack from widening and spreading.

Picture tubes with such reinforced envelopes could be mounted in receivers with the screen projecting through the front panel. The depth of the receiver was therefore decreased, giving the so-called 'slim-line' receivers of the period. Mounting lugs were also fitted to the metal reinforcing band to make the fixing of the tube in the receiver easier. This construction is used in the present-day black-and-white picture tubes. The usual sizes of picture tubes are screen diagonals of 31 cm (12 in) and 44 cm (17 in) for portable receivers, and 50 cm (20 in) and 61 cm (24 in) for standard receivers.

Besides the increase in the size of the picture tube, improvements were made over the period to the screen. More efficient phosphors were developed to give clearer and brighter pictures. The larger deflection angle combined with the shorter tube length required more efficient deflection coils. The increase in the size of the tubes also required an increase in the final anode voltage. Present-day tubes for portable receivers operate with a final anode voltage of 12 kV to 15 kV, while 50-cm and 61-cm tubes use 20 kV.

CONSTRUCTION OF PICTURE TUBE

The schematic cross-section of a present-day television picture tube is shown in *Figure 7.126*. Electrostatic focusing is used, combined with magnetic deflection. Magnetic focusing was used on some post-war picture tubes, but is not used on present-day tubes.

Electrons are emitted from the heated oxide-coated cathode k, and are brought to a focus between the grid g and the first anode a_1 by the effect of the potentials applied to these two electrodes. It is normal practice in present-day television receivers to drive the picture tube by modulating the cathode rather than the grid. The d.c. potential of the cathode is determined by the conditions of the video output stage, and this potential is modulated by the video signal. The grid is held at a potential which can be varied to adjust the brightness of the picture on the screen. The d.c. voltage applied to the grid controls the electron flow (beam current) as in any thermionic tube, and so controls the electron bombardment of the screen. When the brightness of the picture has been adjusted, the potential on the grid remains constant. If the video signal were applied to the grid, the grid potential would vary and so alter the point at which the beam was focused. Cathode drive avoids this change of focus, while the varying cathode-to-grid voltage in response to the video signal modulates the electron beam.

The potential of the first anode a_1 is typically +250 V for a 31-cm tube and +500 V for a 61-cm tube. The second anode a_2 and the fourth anode a_4 are held at the same potential, typically +12 kV for a 31-cm tube and +20 kV for a 61-cm tube. Anode a_3 is the focus electrode, with a typical control voltage range of 0 to +350 V for a 31-cm tube and 0 to +400 V for a 61-cm tube. The voltage on anode a_3 is adjusted to focus the electron beam to a small spot on the screen. The form of the anode structure is a series of cylinders.

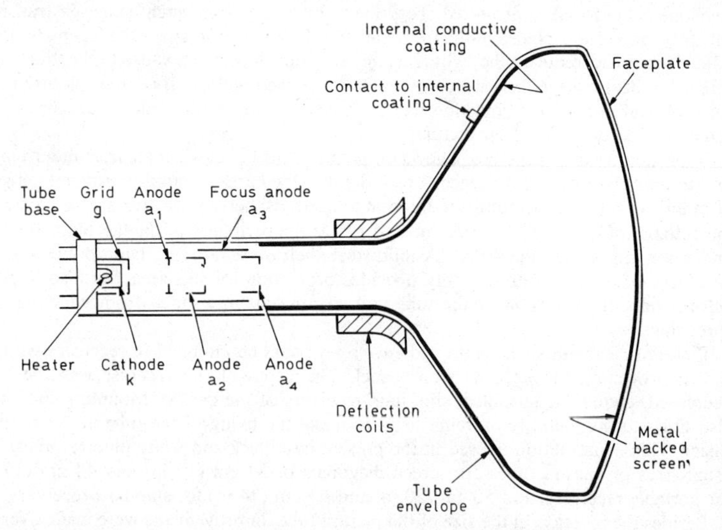

Figure 7.126 Schematic cross-section of black-and-white television picture tube (neck exaggerated for clarity)

An electron beam passing through a magnetic field is deflected along a circular path, as shown in *Figure 7.127*. If the magnetic flux density is B_d tesla (Wb/m²), and the field is uniform over a distance l, then the deflection y at a distance L is given by

$$y = B_d l L \sqrt{\left(\frac{e}{m} \cdot \frac{1}{2V_a}\right)}$$

where e/m is the electron charge-to-mass ratio, and V_a is the accelerating voltage on the beam (voltage on anode a_4). The magnetic field can be conveniently provided by a coil placed around the neck of the picture tube, the magnitude of the current flowing through the coil being chosen to give the required flux density, and hence the required beam deflection. Two sets of coils are required: one for the horizontal deflection of the beam, and the other for the vertical deflection.

The form of the coil for the horizontal deflection is shown in *Figure 7.128 (a)*. The saddle-shaped coil is placed above and below the neck of the tube between the anodes of the electron gun and the conical section of the tube. The form of the coil for the vertical deflection is shown in *Figure 7.128 (b)*. This coil is placed on the sides of the neck. In a practical deflection system, multi-turn coils are required, and the coils are wound on a former containing a ceramic ring magnet to increase the magnetic field. A practical deflection coil assembly (or yoke) is shown in *Figure 7.129*. In addition to the line and field deflection coils, the assembly also carries (at the rear) raster centring magnets, and two rod magnets mounted on 'stalks' at the sides to correct pin-cushion distortion at the

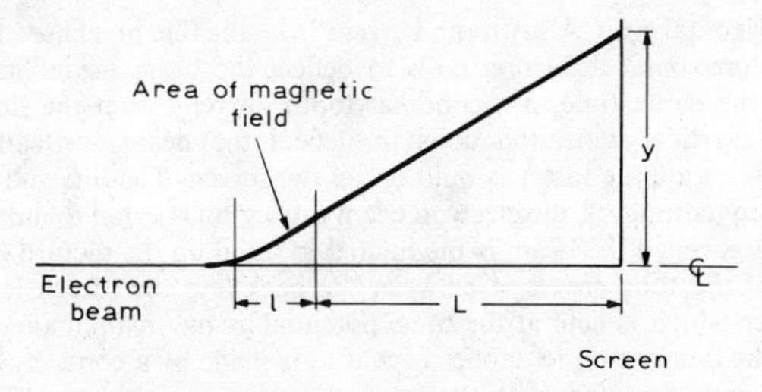

Figure 7.127 Deflection of electron beam by magnetic field

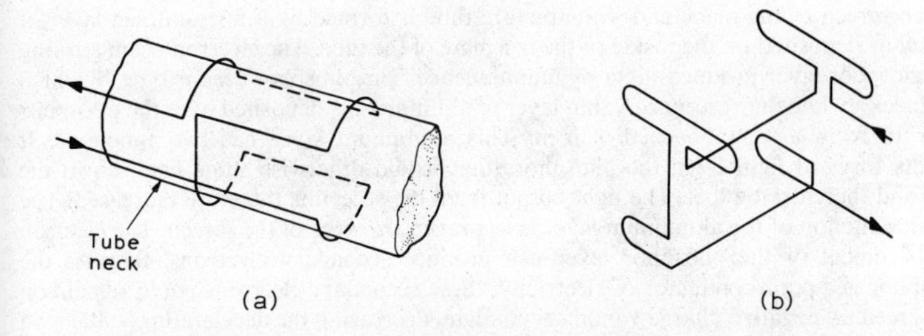

Figure 7.128 Form of picture tube deflection coils: (a) horizontal (b) vertical

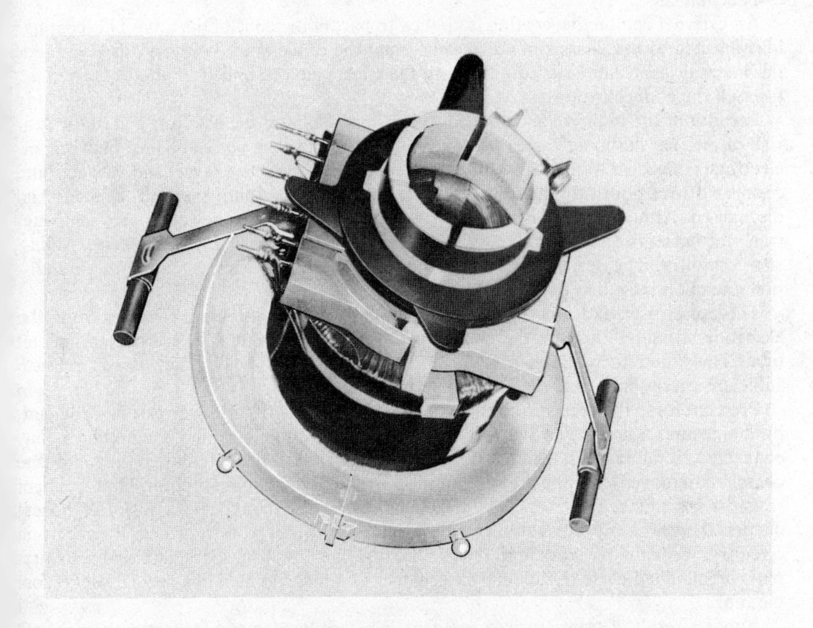

Figure 7.129 Practical deflection coils for black-and-white picture tube (Photo: Mullard Limited)

extremes of horizontal scan. A sawtooth current from the line timebase of the receiver is applied to the horizontal deflection coils to deflect the beam horizontally at the line frequency. At the same time, a second sawtooth current from the field timebase is applied to the vertical deflection coils to deflect the beam vertically at the field frequency. In this way, the raster is built up on the screen. The line and field 'flybacks' are suppressed by cutting off the electron beam during this period (blanking), and while the line is being scanned the beam is modulated to build up the picture elements.

The inside of the tube envelope between the neck and the screen is coated with a conductive layer which is held at the same potential as the final anode a_4. In fact, in a practical tube the connection to anodes a_2 and a_4 is made by a connector in the cone of the tube which makes contact with the internal coating. The coating forms a shield to protect the deflected and modulated beam from external electric fields, and connects with the metal backing of the screen to maintain the screen at a high positive potential.

The screen of the black-and-white picture tube is formed by a fine uniform layer of phosphor deposited on the inside of the faceplate of the tube. The electron beam striking the phosphor layer produces light by luminescence. The phosphor used is type *W* which produces a white fluorescence. A thin layer of aluminium is deposited over the phosphor layer to form a 'metal-backed' screen. This aluminium layer has two functions. It reflects forward light from the phosphors that would otherwise shine back down the tube and therefore be lost. The light output from the screen is therefore increased. The second function of the aluminium layer is to prevent *sticking* of the screen. The electron bombardment of the phosphor layer can produce secondary electrons. Because the phosphor is a poor conductor of electricity, these secondary electrons could remain on the screen. A negative charge would accumulate, decreasing the accelerating voltage on the electron beam, and so reducing the light output from the screen. The aluminium layer is held at a high positive potential through its connection with the internal coating and anode a_4, and so secondary electrons are conducted away and no negative charge can accumulate.

An external conductive coating is applied to part of the cone of the tube. This coating is connected to the chassis of the receiver, and the capacitance between this layer and the internal layer can be used to smooth the e.h.t. voltage applied to anodes a_2 and a_4 through the connector in the cone.

Because of the high voltage applied to the final anode of the electron gun, there is a risk of internal flashover inside the picture tube. An arc can be established between an electrode connected to the e.h.t. (through the internal conductive layer) and an electrode connected to a pin in the tube base. The arc will continue until the e.h.t. capacitor is discharged. Although the flashover may not be noticed by the viewer, it may generate transient currents and voltages sufficiently large to damage parts of the receiver solid-state circuitry. The recharging of the e.h.t. capacitor after flashover imposes an additional load on the e.h.t. generator.

The cause of breakdown is thought to be the detaching of minute particles from the electrode structure through the effects of the large electrostatic forces present in the tube. These minute particles are transported within the tube, and on impact release sufficient energy to enable the arc to be established.

Protection for the receiver circuitry against the effects of flashover can be provided by connecting a spark gap between each pin on the tube base and a common point connected to the chassis. The normal connection between the external coating and the chassis is removed, and the connection made through the common point. High-voltage resistors are connected in series with each supply lead to the tube pins. When flashover occurs, the spark gap breaks down and provides a bypass path to protect the receiver circuitry, while the resistors limit the current that can flow. The spark gaps and resistors must be mounted close to the tube base, and can be incorporated in a special socket for the tube.

Another result of transistorised receivers has been the demand for faster warm-up of the picture tube to match the 'instant on' property of solid-state circuits. Attempts to meet this demand were made in the mid-1960s by the so-called 'instant-on' circuits. In these, a low voltage was continuously applied to the heater so that almost immediately after the rest of the receiver was switched on, the cathode emission was sufficient to provide a visible picture. A more-recent development to provide a faster picture has been a redesigned cathode structure.

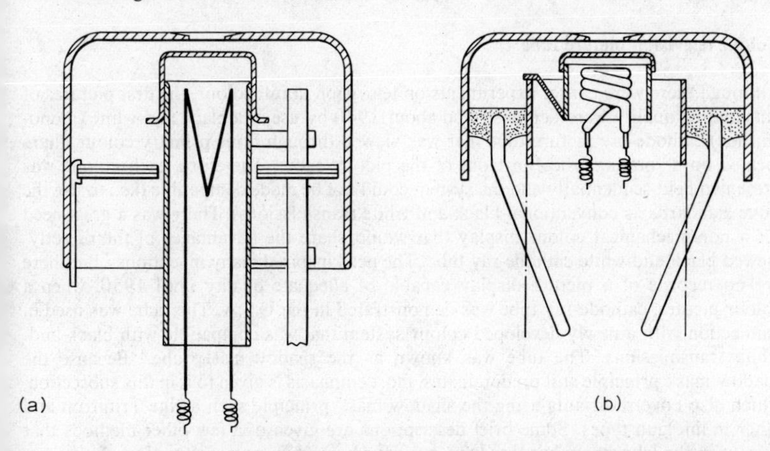

Figure 7.130 *Picture-tube cathode structures: (a) conventional structure (b) structure with fast warm-up time*

The conventional picture-tube cathode is shown in *Figure 7.130* (*a*). The heater is in the form of an M-shaped filament. The newer construction is shown in *Figure 7.130* (*b*), and here the heater is a compact bifilar winding. The efficiency of the new heater is greater than that of the old, and the new cathode structure is more compact. As a result, the warm-up time of a picture tube using the new cathode structure (that is, the time by which a reasonable picture is visible) is approximately one-third that of a conventional tube, typically less than 5 s for a 31-cm tube.

DATA FOR PICTURE TUBES

The published data for black-and-white television picture tubes specifies the electrical operating conditions for the tube, and the maximum and minimum ratings. The inter-electrode capacitances are also listed. Curves of the variation of final anode current with grid voltage (for grid modulation) and cathode-to-grid voltage (for cathode modulation) are given, together with an operating area for the tube defined by the limits of cathode-

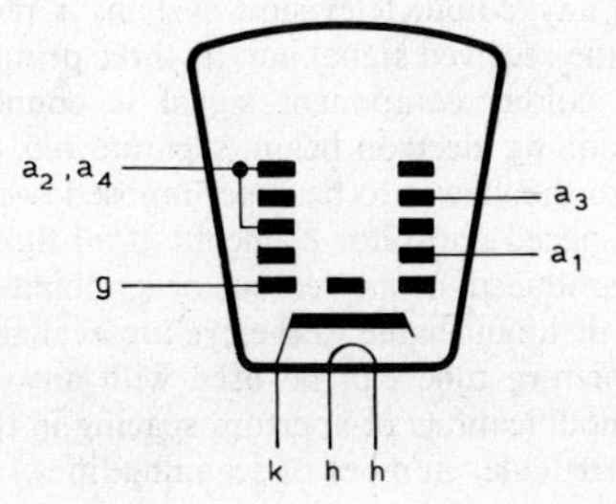

Figure 7.131 *Circuit symbol for black-and-white picture tube*

to-grid cut-off voltage (or grid cut-off voltage) plotted against the first-anode-to-grid (or first-anode) voltage. Details of the mechanical outline of the tube and the useful screen area are given.

The circuit symbol for a black-and-white picture tube is shown in *Figure 7.131*.

Colour television picture tube

Although there were earlier experiments on television in full colour, the first pictures of outstanding quality were demonstrated about 1940 by use of a black-and-white (monochrome) cathode-ray picture tube that was viewed through three primary colour filters located on a rotating wheel in front of the picture tube.[1] Therefore, each colour was presented field-sequentially and the system could not be made compatible (i.e., to use the same standards as conventional black-and-white transmissions). There was a great need for a non-mechanical colour display that would share the advantages of the directly-viewed black-and-white cathode-ray tube. The need inspired many inventions,[2] but there are no records of a picture display capable of adequate quality until 1950, when a colour-picture, cathode-ray tube was demonstrated in the U.S.A. This tube was used in connection with a newly-developed colour system that was compatible with black-and-white transmissions. The tube was known as the shadow-mask tube.[3] Because the shadow-mask principle still predominates, most emphasis is given to it in this subsection, which also covers variants using the shadow-mask principle such as the Trinitron and other in-line gun types. Some brief descriptions are given of a few other methods that survive in the laboratory but that have not yet emerged in mass production.

A complete description of the history, basic principles, theory and practice, of *all* colour-picture-tube displays, is contained in the list of the references[4] which includes a long bibliography that serves to credit many individuals and their contributions during the 1950–73 period.

In a black-and-white cathode-ray picture tube, a single, pencil-like beam of electrons is accelerated, deflected by a magnetic field into a rectangular scanning pattern and then strikes a phosphor material deposited on the inside of the front glass surface. The high-velocity electrons cause the phosphor to emit white light. There are many inorganic materials that have this light-emitting characteristic, known as cathodo-luminescence. The so-called 'white' phosphor is actually a balanced mixture of colour-emitting phosphors. A colour picture tube uses three colour phosphors, red, green and blue emitting which, if mixed in the right proportion, would also produce white. However, in a colour tube they are not mixed but are deposited on the inside of the front glass surface in many small triplet or triad groups of dots or lines. Instead of a single pencil-like electron beam, the shadow-mask tube used three such beams, closely spaced. They are accelerated and deflected just as in the black-and-white tube. The three beams converge on each other to meet at the phosphor screen. However, by use of a perforated metal mask close to the phosphor screen, with one perforation for each triplet phosphor group, each beam is cut off or 'shadowed' from two of the phosphor colours and can strike only one of them. Thus, there is one beam that produces only red, one only green and one only blue.

With any colour-television system, a receiver can be designed to demodulate or decode the received signal into its three primary component signals, red, green and blue. If each colour component signal is connected to the picture tube to control the corresponding electron beam, separate red, green and blue pictures are produced that appear to the viewer to be superimposed because his eye cannot discern the very small, closely-spaced phosphor elements. If all three beams are simultaneously excited, white can be produced, or any colour or combination of colours. Most of the entire gamut of colours distinguishable to the eye are available over a wide range of brightness. Such a colour-picture tube can be used with any system of television. (However, there are minor modifications of aperture spacing in the shadow mask that minimise moiré effect with a particular number of scanning lines.)

Figure 7.132 shows schematically a small section of the two most common forms of shadow mask. One form uses small round holes and round phosphor dots, *Figure 7.132* (*a*). The three electron beams originate from electron guns in a triangular, or 'delta' arrangement. In the other form, *Figure 7.132* (*b*), the mask openings are narrow vertical slits, and the phosphor triplets are narrow vertical lines. In the latter case, the electron guns are on a horizontal line. In both forms, mask and screen are at the same potential and electrons travel in this region in straight lines, as shown. The angle between the beams is exaggerated in *Figure 7.132*, in reality it is only about 1°. The figure does not show the deflecting system or the overall shape of the tube, both of which closely resemble those of a black-and-white tube. However, the figure does show how

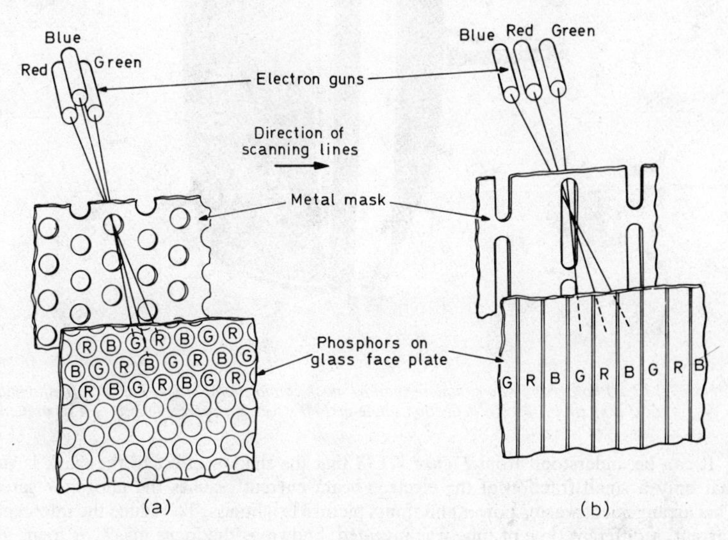

Figure 7.132 Schematic of two shadow-mask systems: (a) with round holes, and guns in a delta configuration (b) with slit openings and in-line guns. In actual tubes, mask and face-plate are slightly curved

the shadow effect is used to prevent each beam from striking more than its own colour phosphor. To be noted is that each mask aperture must be exactly in register with a trio of phosphor elements; this is a major problem that makes a colour picture tube so much more difficult to fabricate than its black-and-white counterpart.

A photograph of a complete colour picture tube is shown in *Figure 7.133* but such a picture does not disclose the type of shadow mask, or the gun configuration. The tube shown[5] produces a rectangular picture of aspect ratio 4 : 3 (horizontal size to vertical size), with a diagonal of 0.63 m, picture area of 0.20 m^2. The shadow mask is designed for best operation with either a 525- or 625-line TV system and has 564 000 round apertures spaced 0.66 mm apart. These are accurately registered with 1.7 million round phosphor dots, one third of which are in each colour. Using a power supply of 25 kV, with up to 1 mA average current, a bright, high-contrast picture is produced.

Other shadow-mask variations are possible and have been used, such as in-line guns with round shadow-mask apertures,[6] and three in-line beams from a single gun with a grill-type shadow mask[7] (the Trinitron). Another common variant, not visible in either *Figures 7.132* or *7.133*, is the use of black areas between the phosphor dots; such a screen is known as a matrix screen. These black areas can be used either to improve contrast or, more commonly, to increase brightness for a given contrast.

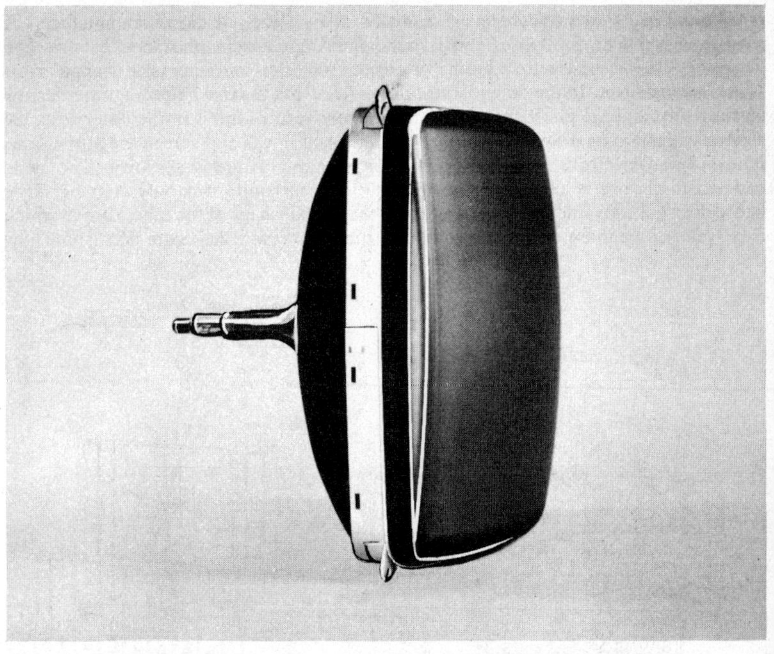

Figure 7.133 Photograph of a complete shadow-mask colour picture tube. The picture diagonal is 630 mm, and it uses a deflection angle of 110°, with an overall length of 430 mm

It can be understood from *Figure 7.132* that the shadow effect of the mask is such that only a small fraction of the electron-beam current reaches the phosphor screen. This arrangement wastes power and limits picture brightness. To reduce the intercepted current, a different type of tube was invented, known as the focus-mask, or focus-grill tube,[8] depending on whether holes or slits are used. With either design, the mask is at a much lower voltage than the phosphor screen, and each hole or slit acts as a small focusing lens. The result is that the openings can be much larger and current interception much smaller, but picture contrast is impaired. For the latter reason and also because of fabrication difficulties such tubes have remained as laboratory developments.

In a rather different approach to a colour-picture tube, there has been much laboratory effort on a single-beam type that uses no mask or grill at all.[9] With this approach, known as the beam-index tube, the phosphors are arranged in narrow vertical stripes as in *Figure 7.132* (*b*). The single electron beam must have a spot size small enough to excite only one colour stripe at a time; the control for this beam is switched from one colour signal component to another as it traverses horizontally across the stripes. To assure correct colour, a separate, fourth set of stripes is placed on the back of the colour-phosphor stripes and in registry therewith. As the beam passes this fourth, internal set of stripes, index signals are produced, either by ultraviolet radiation or by secondary electron emission. The index signals are used to control the switching phase. Since there is no shadow mask, there is no interception of beam current; however, the single beam must now be time-shared among the three colours. Because adequate guard regions are needed between phosphor stripes, and index signal difficulties are often found, the advantages are not as great as might be supposed. The beam-index tube also remains as a laboratory development and has not been produced commercially.

CHARACTERISTICS AND LIMITATIONS

A colour picture tube is characterised by colour fidelity, brightness (luminance), contrast, picture size, resolution and sharpness. Each characteristic usually involves compromise with the ideal because of various limitations, both theoretical and practical. These aspects are next described; many of the remarks are equally applicable to any of the cathode-ray types of colour tube, but all apply to the shadow-mask type.

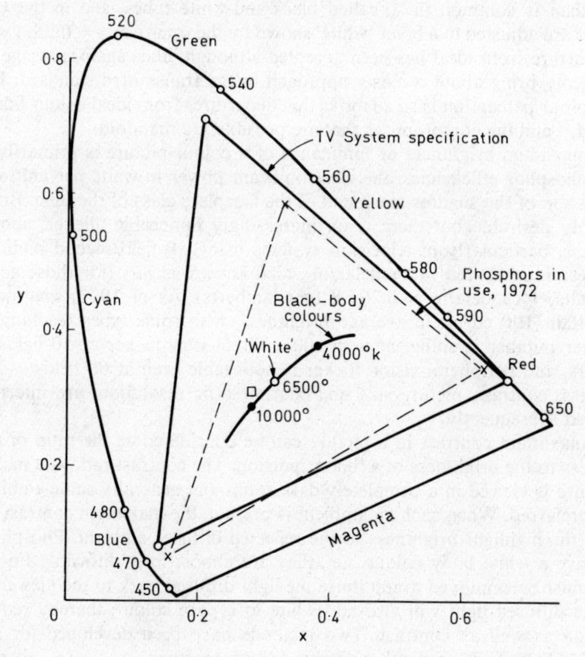

Figure 7.134 The C.I.E. colour diagram. The specified transmitted colour system primaries and white point are shown by open circles. Typical phosphors in use (1972) are shown by the crosses. In the U.S.A., the receiver 'white' point is often set at the point shown by the square

Colour fidelity is best understood in terms of the C.I.E. colour diagram[10] shown in *Figure 7.134* (see also Section 15) and derived for an average observer. All visible colours are found by x and y coordinates that lie within the horseshoe-shaped figure. One can think of the x coordinates as one of three colour stimuli, the y coordinates as another and z, not plotted, as the third, the sum being unity. Thus, 'white' is in the neighbourhood of $x=\frac{1}{3}$, $y=\frac{1}{3}$, as shown by the central open circle. Extending a radial line out from the white point, the rotational angle determines the hue, and the distance out from white is a measure of the saturation or purity. Colours and spectral wavelengths are marked on the figure. A three-colour system employs primaries that determine a triangle, and only colours within that triangle are reproduced.

In the colour-television systems used in most of the world, the colour cameras are arranged to produce the three chosen primaries that are approximated by the open circles. A 'white' point is set up, approximately as shown by the central open circle, and actually at $x=0.31$, $y=0.33$. The transmitted colour fidelity is better than that achieved by printing or photography and can be described as excellent. Ideally, the receiver picture tube should have phosphors that match these circles and, in fact, such phosphors are possible but they are not the most efficient. Research on phosphors[11] between 1951 and 1973 has more than doubled their luminous efficiency and has led to the use of the phosphors that are shown by the crosses on *Figure 7.134*. Fortunately, the compromise is hardly noticed by most viewers. Nevertheless, as phosphor research continues, one may expect closer approaches to the ideal, i.e., to the open circles.

A receiver using a colour shadow-mask picture tube, for example with the phosphors indicated by the crosses, can be adjusted to match the transmitted 'white' (open circle), by balancing the three individual beam currents. The open circle represents a *warmer* 'white' than is common in so-called black-and-white tubes, and in the U.S.A. many receivers are adjusted to a bluer 'white' shown by the square at $x = 0.28, y = 0.31$. Even this departure from ideal has been accepted although, once again, passage of time will undoubtedly bring about a closer approach to the transmitted standard. In summary, man's colour perception is so adaptive that departures from ideal colour fidelity are well tolerated,[12] and the compromises that are possible are manifold.

The maximum brightness or luminance of a colour picture is primarily determined by the phosphor efficiencies, the electron-beam power in watts per unit area, and the transmission of the shadow-mask and of the faceplate glass of the tube. Bright pictures are highly desirable but there is an increasingly noticeable 'flicker' above a certain brightness, particularly in television systems using 50 field/second picture sequence. Brightness is measured in candelas/m^2, also known as *nits* (for those accustomed to older units, 1 candela/m^2 = 0.291 8 foot lamberts). As of 1973, shadow-mask tubes easily attain 100 cd/m^2 in average brightness, with some types reaching 350 cd/m^2. The latter number is sufficient to produce direct-view flicker at 50 fields/s for many observers, and peripheral-vision flicker is noticeable even at 60 fields/s. The limits to brightness, contrast, picture size and sharpness or resolution, are interrelated, as is developed subsequently.

The maximum contrast in a picture can be considered as the ratio of the highlight brightness to the brightness of a 'black' portion. The contrast ratio is a maximum when the picture is viewed in a completely dark room, but generally some ambient illumination is preferred. When such an ambient is present, the maximum contrast becomes the ratio of the highlight brightness to the reflected diffused ambient. Phosphors in use in 1973 have a white body colour, i.e., they are almost fully diffusing. For this reason, means must be employed to minimise the light diffused back to the viewer. In a colour tube, the diffused light will also add white to a pure colour, thereby reducing colour saturation as well as contrast. Two methods have been developed for reducing the diffused light and are used either singly, or both together.

If light-absorbing material of a neutral tint is incorporated in the faceplate glass to reduce its transmission to a fraction, T, then the highlight brightness is also reduced by T. The ambient light diffused back, however, goes through the 'grey' glass once, is diffused by the phosphor and again goes through the 'grey' glass on its return to the viewer. Thus, the diffused light is reduced by the factor T^2. The contrast is then improved by the amount, $1/T$, at the expense of a brightness loss. Values of T in the range of 0.5 have been employed when this is the only method used.

A second method incorporates an apertured black coating (called a matrix) that is deposited on the rear of the faceplate glass before the phosphor is deposited.[13] The apertures in the black coating exactly correspond to the central portion of the phosphor dots. With this second method, the phosphor area that is visible from the front remains white but it now covers only a fraction, F, of the screen area, thereby cutting the diffused ambient by the same fraction. In practical designs, it is not possible to excite the entire screen area with the electron beam without danger of electron spot overlap on to undesired colours. In a conservative design (i.e., tolerance of the order of ± 0.1 aperture diameters in spot position) the useful phosphor portion is of the order of 0.5 of the entire area, i.e., F is also about 0.5, the remainder being black. This design improves contrast by just as much as grey glass with $T = 0.5$, but without the loss of highlight brightness by absorption. In practice, some matrix-type shadow-mask tubes have used

$F = 0.53$ and $T = 0.85$ in combination. Compared with grey glass only, of $T = 0.5$, such matrix tubes have a brightness improvement of $0.85/0.50 = 1.7$ and a contrast improvement of $0.5/(0.85 \times 0.53) = 1.1$. More commonly, matrix types are designed for greater brightness by increasing the useful phosphor portion to $F = 0.7$, at the expense of tighter tolerance and lower contrast. In a dark room, contrast of practical shadow-mask tubes approaches 50 : 1, which is excellent, and in a reasonably well-lit room (250 lux illumination) the contrast is still of the order of 10 : 1 or more, which is acceptable. A black matrix increases tube manufacturing cost and is used where maximum brightness is desired; because of flicker, 50 field/s systems are less likely to need matrix screens.

There are limits to brightness and, consequently, to contrast that are imposed by the electron-beam power per unit area and the necessary sharpness of the picture. To keep X-ray radiation below safe limits, even with heavy elements added to the faceplate glass, the upper voltage limit is of the order of 25–35 kV. A focused electron beam produces a spot on the screen that gets larger as the current is increased, but the effect is less in shorter tubes with large deflection angles. Thus, a 110° deflection-angle tube will be brighter than a 90° one, at the same beam spot size, because the beam current can be larger. In a given tube, if an attempt is made to increase the beam current excessively for highlights, the spot gets so much larger that picture sharpness and resolution are lost and the picture appears fuzzy. The beam spot is described as *blooming*. In general, it is desirable to have a small-enough spot to show the scanning lines, although a more important characteristic for a sharp picture is the edge appearance of large-area objects.[14, 15] With the largest shadow-mask picture sizes in use in 1973, 0.63 m diagonal, 0.20 m^2 area, and with 25 kV, 110° deflection angle, and an average beam current (the sum of all three beams) of 1 mA, it is possible to achieve an average brightness of 300 cd/m^2, contrast between 10 : 1 and 50 : 1, depending on the ambient illumination, and yet produce a sharp enough picture for satisfactory viewing. Under highlight conditions in small areas, the peak beam currents and brightness values may be several times higher without excessive deterioration of sharpness. For smaller screen sizes, it is not necessary to use as high a beam power to achieve similar results.

If all else is equal, large picture sizes are much more pleasing than small ones. The picture shape is fixed by the transmission standards, i.e., rectangular with 4 : 3 ratio of horizontal width to vertical height. The faceplate is ordinarily made in approximately this shape to conserve space. A practical limit to the size of the picture tube is imposed by cost, and by the awkwardness of length. The shortest tubes are those with the largest deflection angle; the angle was once 70°, but by 1973 only 90°, 110° and 114° were being used. The larger of these angles are close to the limit imposed by deflection difficulties as well as by practical fabrication techniques.

Both black-and-white and colour television systems compress the signal amplitudes at the transmitter by a fractional power law; usually the 'gamma' or fractional exponent is 0.46. This compression considerably improves the signal-to-noise ratio. In a colour receiver, it is particularly important that the transmitted exponent be corrected by its inverse, i.e., the light output for each colour should be approximately proportional to the 2.2-power of the signal voltage. Fortunately, it is a characteristic of cathode-ray guns that they do have about this power law and no special receiver circuits are necessary.

THE ROUND-HOLE SHADOW-MASK TUBE

The shadow mask with round holes, as in *Figure 7.132* (*a*), can be thought of as the traditional type and was the only type in use prior to 1968. The principles have already been outlined, but many details need amplification.[4, 16] *Figure 7.135* shows a cut-away view of such a shadow-mask tube with a 90° deflection angle and a 630-mm picture diagonal. The figure applies equally well to many of the designs of the 1965–73 period; the mask is spherically curved and rests behind a similarly curved glass faceplate at a spacing of about 14 mm. There are three separate electron guns in a delta arrangement, tilted slightly to form beams at an angle of about 1° from each other; when undeflected, the beams point at the centre of the faceplate. In one of the 1973 designs, the mask has about 564 000 holes in a hexagonal array, with 0.66 mm centre-to-centre spacing. Corresponding to each mask hole is a triad of red, green and blue

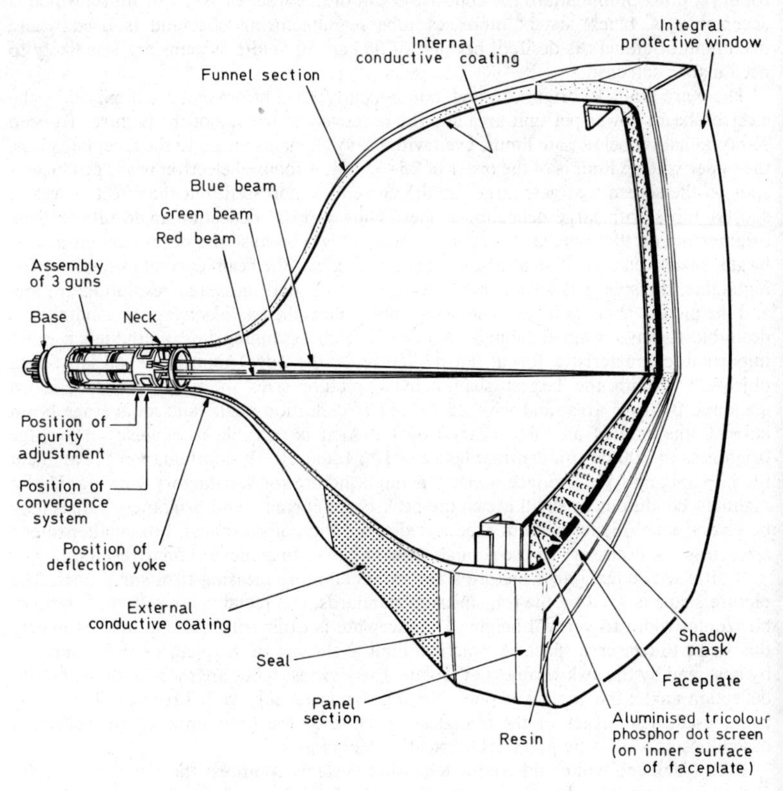

Figure 7.135 Cutaway view of a 90° deflection angle, delta-gun, shadow-mask tube

light-emitting phosphor dots superposed on a black matrix. The method of making the mask and depositing the phosphor dots on the faceplate is described later under 'Manufacturing technology'.

The figure suggests how such a tube is put together. The funnel section represents one part and the faceplate or panel section, with an integrally mounted shadow mask, represents a second part. The two are sealed together with a vacuum-tight low-melting-point glass frit and the electron gun is then sealed into the neck. The figure also shows an integral protective window on the faceplate to prevent flying glass in the event of an implosion. This window is bonded to the faceplate with a resin. Other types of protection are also in use, including a tight metal band around the rim of the faceplate.[17]

Inside the faceplate, the tri-colour arrays of phosphor dots are covered with a thin, electron-permeable aluminum layer to reflect the phosphor light in a forward direction, as with black-and-white tubes.

In operation, the deflection yoke is placed over the neck, up against the funnel section. In the tube shown, an external magnetic shield is placed around the funnel section, but some types have an internal shield. In either case, an automatic demagnetising field coil (degaussing coil) is placed around the whole, and is actuated for a short time with alternating current each time the receiver is turned on.[18]

Operating voltages are connected to the gun, and anode high voltage to the mask, phosphor screen and internal conductive coating. In most types, focusing is done electrostatically using a bi-potential design that requires an adjustable voltage for focusing of the order of 15% to 20% of the anode voltage; the beam focus electrodes for all three guns are internally connected so that only a single focus voltage is required. In some smaller-screen types, with somewhat less critical performance requirements, a so-called 'einzel-lens' focus is used; this is an electrostatic system with one element at approximately cathode potential and requires little or no adjustment.

In any three-gun colour tube it is important that the three beams coincide (converge) at the phosphor screen over the entire deflection raster, or else the colour picture would not be properly superposed. In the central region of the screen convergence is achievable by aiming the guns to one point, making minor correcting adjustments with four external magnets, combined with internal pole pieces built into the gun. Such central-region coincidence is called static convergence. When the three beams of a delta-gun-type tube are simultaneously deflected, for example to the corners of the picture, convergence errors occur. These are corrected by a number of special current wave shapes that are applied to a dynamic magnetic-field convergence system that uses internal pole pieces. These current wave shapes are derived from the deflecting system so as to be in correct time synchronism with scanning. In a typical receiver with a

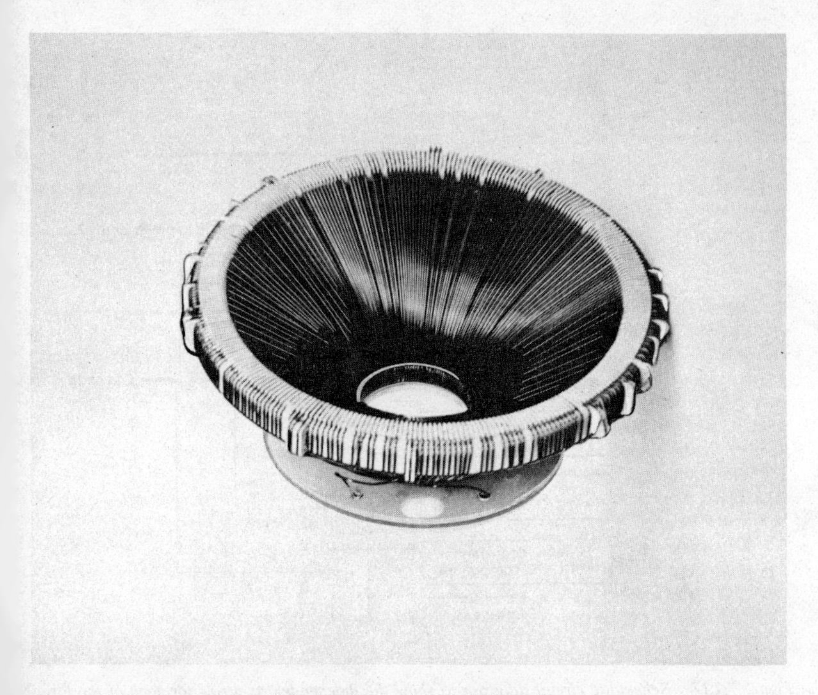

Figure 7.136 Precision toroidal yoke. Such yokes are sufficiently precise to simplify convergence of the three beams in a shadow-mask tube

delta-gun shadow-mask tube, each tube must be separately adjusted for static and dynamic convergence; this requires four static and 12 dynamic adjustments. One other adjustment is required, known as 'purity'; this is also done by an external magnet system, and serves to shift the centre-of-deflection of the three beams to the position that leads to maximum colour uniformity over the entire screen. The convergence adjustments bring the three beams together over the entire picture area, and the purity assures that each beam strikes only its correct colour phosphor dot over the entire picture area. In spite of the apparent complexity, the use of appropriate dot and cross-hatch test patterns permits these adjustments to be made by an experienced person in a matter of minutes.

The magnetic deflection yoke, as of 1973, is often of the so-called 'saddle' type, wherein there are two opposite coils for horizontal deflection, and two for vertical deflection. They are bent around to hug the neck/cone region of the tube like a saddle and are coupled with a ferrite core. The convergence magnets and coils are usually mounted on the back end of the yoke. An alternative recent development is a toroidal winding on a ferrite core, which, in its most advanced form, is lower in cost, can be considerably more precise and has major advantages particularly for 110° deflection angles (see also later this section on the precision in-line tube). A photograph of a toroidal yoke for 110° deflection is shown in *Figure 7.136*.

In most delta-gun shadow-mask tubes there are separate cathode (*K*), control-grid (*G*1) and screen grid (*G*2) electrodes for each of the three guns. Such designs can be connected to the receiver circuits in many ways. Some receivers use a luminance signal (equivalent to a black-and-white signal) on the cathodes, and three colour-difference signals on the control grids. In such circuits, the picture tube itself is part of the colour matrixing system. Other receivers fully separate the red (R), green (G) and blue (B)

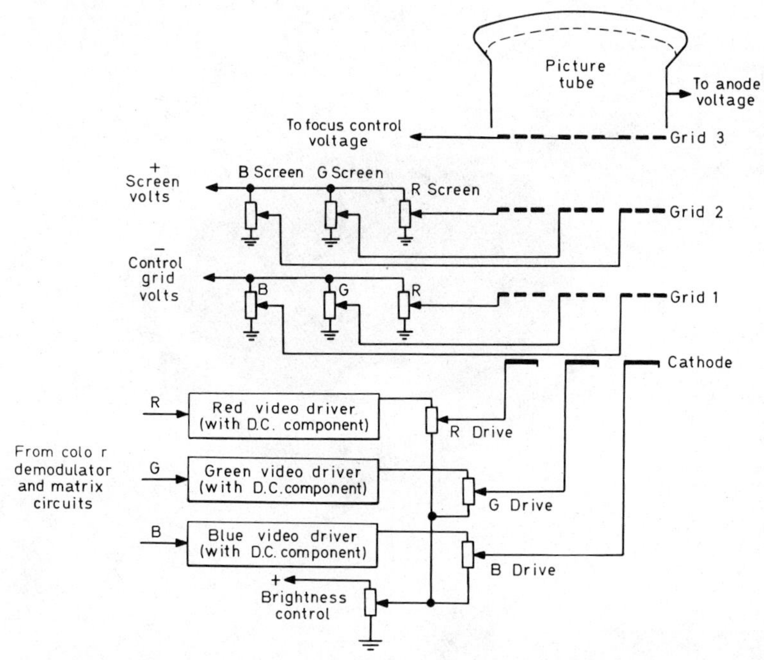

Figure 7.137 Schematic circuit diagram to show the independent controls for each of the three beams in a shadow-mask tube. Deflection and convergence circuits are omitted, and the colour saturation, tint and contrast controls precede the red, green and blue inputs

video signals, and these are applied to either the three cathodes or the three control grids. A partial circuit diagram, in schematic form, is shown in *Figure 7.137*, wherein R, G and B signals are applied to the cathodes. The deflection and convergence system is not shown and the contrast, colour and tint controls precede the parts in the figure. Because the three phosphor efficiencies are not usually equal, the three guns are required to supply different currents when 'white' is produced. An adjustable white is possible with the diagram of *Figure 7.137* because each of the three guns can be independently biased for both the control and screen grids. In addition, the R, G and B video drive amplitudes are adjustable. These adjustments are required only once for a particular tube to set up a white colour temperature that is invariant with picture brightness. When the gun characteristics are known in advance and control is exercised by the picture tube manufacturer on phosphor efficiencies, simplification is possible. It is common, for example, to omit the individual control-bias adjustment, the drive adjustments are sometimes reduced to two, since it is often known which gun requires maximum video signal.

When in-line guns are substituted for the delta guns, dynamic convergence is simplified, but the guns take up more space in the neck. A large neck is a disadvantage because greater deflection power is required. This can be overcome by using miniature in-line guns, at the expense of larger spot size because of greater lens aberration. More sophisticated in-line-gun designs use a slotted or grill-type mask, and are described separately.

In the design of the round-hole shadow-mask tube, consideration must be given to the moiré pattern that results from the scanning lines interacting with the periodic shadow-mask openings.[4] There is a minimum moiré effect for any given number of scanning lines; a 625-line t.v. system will sometimes use a different shadow-mask periodicity than a 525-line system in order to stay at the minimum.

THE TRINITRON

The Trinitron[7, 19] was the first major variation from the round-hole shadow-mask tube. In the Trinitron, the mask consists of a grill of vertical strips from top to bottom. Such a mask is not self-supporting and also cannot be spherically curved. For this reason, it is mounted under tension on a relatively massive frame and the curvature is cylindrical. The mask is used with a similarly curved faceplate and the phosphors are deposited in triads of R, G and B vertical lines (similar to *Figure 7.132* (*b*)). To prevent vibration, very small horizontal wires are placed on the convex surface of the mask, but these are so fine as not to be noticeable.

The three electron beams lie on a horizontal line, as with in-line guns sometimes used with the round-hole shadow mask, or as with the slotted mask of *Figure 7.132* (*b*). Trinitron picture-tube sizes were originally limited to diagonals of 500 mm and under, but larger sizes were announced in 1973; the Trinitron tube is considered relatively costly to manufacture. However, an important advantage of the Trinitron grill mask is that it has no structure in the vertical direction.[15] Thus, the sharpness of brightness changes in the vertical direction is comparable to that of a black-and-white tube, and is limited only by the scanning lines and the beam spot size and shape. This feature, together with the larger diameter lens that is employed in the gun, may explain the oft-expressed viewer reaction that Trinitron pictures are exceptionally sharp.

The Trinitron gun is unusual because it is not an assembly of three separate in-line guns but, instead, uses three electron beams in a single-gun system. The advantage of this lies in the larger lens diameter (with lower spherical aberration) and smaller electron-beam spot than is possible when three individual guns must be placed in the same size neck. *Figure 7.138* shows the gun configuration of a 114°-deflection Trinitron.[19] Only one electrode is used for the first grid and another one for the second grid, in distinction to the three each used in the traditional designs. To enable the gun to produce three beams, three separate cathodes are used, one for each colour signal. In the gun, the three beams are tilted to cross each other at the centre of the focus electrode; this focus system then acts in common on all three. Beyond this convergence point, the beams separate again but are electrostatically converged. When undeflected, the beams strike a common point at the centre of the screen, as indicated in the figure.

By proper yoke design, it is possible to simplify dynamic convergence in in-line tubes compared with delta-gun types. A delta-gun tube uses 12 dynamic convergence adjustments. An in-line gun reduces this to four adjustments and the 114° Trinitron gun and yoke uses two preset corrections to reduce the adjustable number to two. The precision

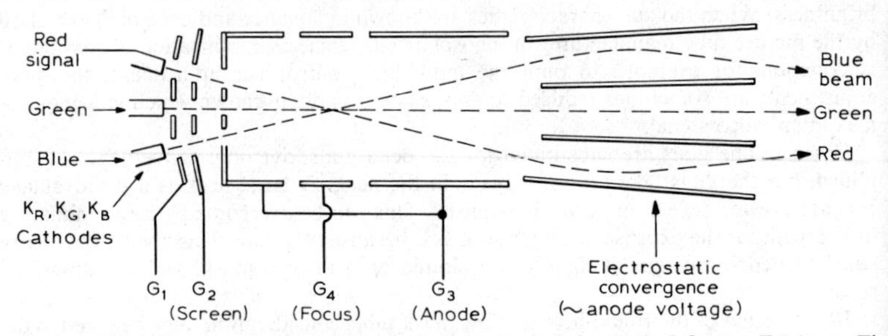

Figure 7.138 The single-gun, three-beam configuration of the 114° deflection Trinitron. The three beams are in line and are used with a vertical line screen. By crossing the beams at the focus electrode, a single large-diameter lens can be used and this reduces the spherical aberration

in-line system, to be described in the next section, eliminates *all* dynamic convergence adjustment. As of 1973, the in-line systems have been used primarily with the smaller screen sizes, i.e., below 500 mm diagonal.

THE PRECISION IN-LINE COLOUR TUBE

The precision in-line system[20] was announced in 1972 and included a group of 90° deflection-angle tubes with screen diagonals from 380 mm to 480 mm. This system is the most sophisticated variant of the shadow-mask designs. It uses a slotted mask and vertical phosphor lines, as shown in *Figure 7.132* (*b*). Since this mask is self supporting, it can be spherically curved and supported on the faceplate in the same way as with the round-hole shadow-mask. A combination of a new gun design with a new highly precise line-focus toroidal yoke made it possible to eliminate all dynamic convergence adjustments. If the yoke, and the static-convergence and purity magnets are properly positioned and adjusted, the assembly may be permanently affixed to the picture tube at the point of manufacture. This combination often requires no subsequent adjustment and may be as easily inserted in a receiver as a black-and-white picture tube; it is also relatively inexpensive. Compared with the delta-gun, round-hole shadow-mask tube with its four static, its twelve dynamic and its purity adjustments, all of which must be made for each tube in the field, the precision in-line system is a major advance.

One of the keys to elimination of adjustments lies in the magnetic deflection-yoke design. The saddle yokes in common use prior to 1972 were so non-uniform that adjustment of the dynamic convergence current wave shapes was required to compensate for the wide variability in the yoke field configuration. The toroidal yoke, used with the precision in-line system, is wound on an accurately moulded ferrite core with slotted plastic end pieces. The yoke winding is machine controlled so that each wire goes into a designated slot in the end piece. When finished, the deflection field is so closely identical from one yoke to another that the effect on beam convergence can be predetermined

With the precision in-line system, the yoke field is purposely designed to be astigmatic, i.e., under deflection it would convert a hypothetical circular bundle of electron beams into a vertical line on the screen. The three actual horizontal in-line beams from the gun are consequently converged by the yoke to the mid-point of the vertical line. *Figure 7.139* shows the convergence characteristic of the line-focus yoke; the yoke itself has an appearance similar to that in the photograph, *Figure 7.136*. As an incidental advantage, the precision toroidal yokes use relatively few turns, i.e., they are low impedance. The cost is low because so little copper is used, and such yokes achieve a better match to the solid-state deflecting systems that have largely replaced electron tube deflection.

The electron gun in the precision in-line colour tube uses a unitised construction with single-piece triple-aperture grids. In this gun, there are three cathode leads which are connected to the colour component signals and to the bias adjustments for white balance. The control and screen grids, since they are on a single piece of metal, have

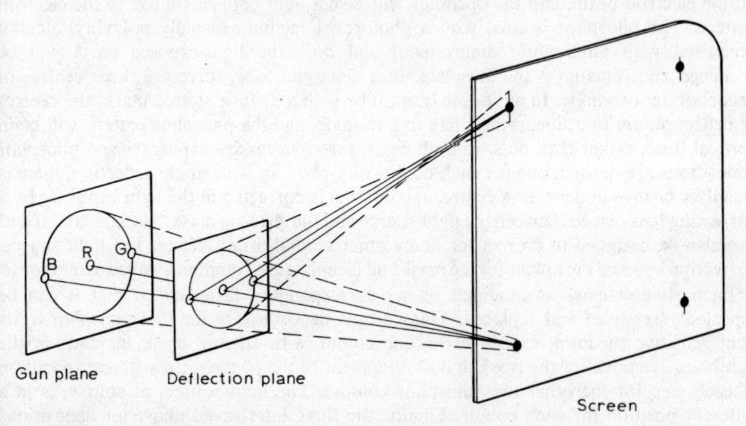

Figure 7.139 The precision in-line system uses a line-focus astigmatic yoke to produce convergence at any point of the screen

only one lead each. In addition to other advantages, the gun permits the overall tube length to be substantially shorter than that of a similar tube with conventional delta guns. The three electron beams remain parallel up to a bipotential focus lens and thereafter, by an offset of the anode apertures, are brought to approximate convergence at the screen. Final adjustment of the static convergence is done with external magnets. The purity adjustment is simpler than with the round-hole shadow-mask because, under vertical displacement, any beam remains wholly on the intended phosphor stripe.

The precision in-line system can be considered the forerunner of future advances in shadow-mask tube design. By precise design of the tube, the yoke and the convergence means as a single system, adjusted and assembled by the tube manufacturer instead of independently in the field, the system becomes simpler, shorter, more easily replaced and less costly, while retaining excellent performance.

MANUFACTURING TECHNOLOGY

In the history of electronics, the technology that made possible mass production of the colour shadow-mask tube will rank with that of the integrated circuit as among the most remarkable achievements of the century. It is interesting that both of these unrelated fields depended heavily on the art of photolithography. In the colour picture tube, the shadow-mask is made by etching a thin metal member (usually of steel) that has been coated with a photoresist and then exposed to ultraviolet radiation through a suitable master pattern. The phosphor dots or lines are deposited on each faceplate by use of other photosensitive resists, one for each colour. The faceplate resists are successively applied and exposed to ultraviolet light through a particular mask that is then identified with and used with that particular faceplate.

The key that made the shadow-mask tube possible was the recognition that electrons in a field-free region, and light, both travel in straight lines. Thus, it was possible to create, with light, exactly the same shadow pattern that the electrons would experience. Now, in fact, the electrons in a cathode-ray tube are deflected by the yoke over a wide area on the screen. However, once they have passed the deflection region, they are in a field-free environment and, if their straight line path is extended back, it is found that they appear to come from a point, known as the centre of deflection. If a point of light is placed at this centre, the shadows of the shadow-mask will correspond to those imposed on the electron beam, and the openings will give a light pattern similar to the electron pattern. The phosphor is used with a photoresist medium (usually polyvinyl alcohol sensitised with ammonium dichromate) and only the light-exposed parts will be hardened and remain on the faceplate. In a delta-gun tube, there are three centres of deflection in a triangle. In an in-line beam tube with a grill or slotted mask, the centres of deflection are in a line, rather than in a triangle, and the phosphor pattern will be in vertical lines, rather than dots. In both cases, three successive exposures and phosphor depositions are needed, one for each colour phosphor. In wide-angle deflection, there is usually a change in deflection centre, axially; this is corrected in the light exposure by a correcting lens placed between the light source and the shadow-mask. The correction lens can also be designed to correct for many other second-order effects. The light source, correction lens and the holder for the mask and faceplate are commonly called a *lighthouse*.

Each shadow-mask is mounted on its corresponding faceplate so that it can be repeatedly removed and replaced. This permits deposition of the first phosphor in its light sensitive medium, exposure for one colour with shadow-mask in place in the lighthouse, removal of the mask and development of the phosphor resist, and repetition of each step for the other two phosphor colours. The light source, of course, is in a different position for each colour. Finally, the three interleaved phosphor depositions are aluminised on the back surface to increase the light output, and the mask is replaced. This assembly, or cap, is then sealed with low-melting-point glass frit to the cone and neck and the electron guns properly aligned on the tube axis. Evacuation and test resemble those of all electron tubes. If a black matrix is used, it must be deposited using the same lighthouse arrangement, but with a reverse printing process to produce open areas for the phosphors. This involves several additional steps prior to those for phosphor deposition.[4]

Key materials in the colour tube are the phosphors, and the deposition technology is critical. These aspects are well covered in more comprehensive reviews.[4, 11] The many details of mask design, mask to screen assembly (such as means to compensate for thermal expansion), are also available.[4] It is sufficient here to point out that the accurate alignment in mass production of a tube with a half-million apertures, with three times that many colour phosphor elements, and with three electron beams, required years of painstaking development. The improvement in quality, and the cost reductions that have been achieved, have put the shadow-mask colour tube in a position of dominance that will be hard to displace.

OTHER COLOUR-TUBE TYPES

The focus-mask, or focus-grill, type of colour tube,[4, 8] to which reference has already been made, is so closely similar to the shadow-mask tube that it requires little elaboration except on one point. In the focus mask, or grill, the openings are much larger than in the shadow-mask, because each aperture acts as a small focusing lens. The electron beam striking such a mask is broken up into tiny spots or lines that are much smaller than the openings. As a result, much less current is intercepted, and very much brighter pictures can be produced for the same beam current.

This feature also explains why this system has not succeeded in the marketplace. Because the electron lenses formed by the mask or grill openings have no counterpart for light rays, the lighthouse method of phosphor deposition previously described cannot be used. Experimental tubes are made by evacuation and use of electron beams to expose the phosphor resist, or by the preparation of intermediate optical masters. The procedure is not cost competitive with shadow-mask screening. A second disadvantage is minor in comparison. Since the lens-like focus mask is at low voltage compared with the screen, secondary electrons can be attracted to the screen and reduce contrast greatly. This can be eliminated by an intermediate 'suppressor' grid or grill, but this is

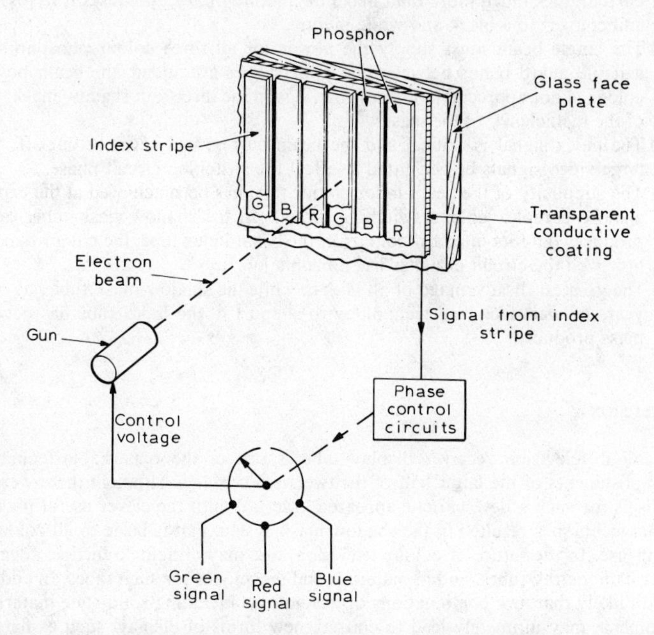

Figure 7.140 The principle of the beam-index tube. A three-colour vertical line screen is used, and index signals from a 4th-line control the switching of the single beam sequentially to the three colour signals. The deflection system is not shown

an additional costly item. Until some new invention or new technology for this system has been successfully applied in mass production, the focus-mask or focus-grill tube appears destined to remain as non-competitive.

Much more attractive in principle is the beam-index colour tube[4,9] using narrow vertical phosphor stripes. There is no mask of any kind, and there is only one electron beam so there is no convergence problem. The only accurate registration required is that the index signal stripes be accurately aligned with the phosphors. Since they are both deposited on the same surface, this is relatively simple. *Figure 7.140* shows the principle of the beam index tube. When the beam is deflected, each time it passes an index stripe, secondary electrons are released in excess of those from the phosphor, and an index signal is produced (the means of collecting the secondaries is not shown). Alternatively, the index stripes can emit ultra-violet radiation and produce a signal in a photomultiplier in back of the faceplate. In either case, the index signal pulse is an exact indication of where the beam is at a given time with respect to the phosphor triplets. By use of this index signal to switch the control electrode of the gun to the proper colour signal, a colour picture results. In *Figure 7.140*, the switch is shown as if it were a mechanically rotating one, but this representation is symbolic only.

Much effort has gone into the beam-index tube. Many kinds of index signals have been explored, and it has been found advantageous to use a 3 : 2 ratio of index stripe periodicity to phosphor stripe periodicity. Nevertheless, the superficial advantages have been outweighed by disadvantages, as follows.

(*a*) The horizontal dimensions of the beam spot must not exceed the width of a phosphor stripe plus any guard space. Any 'blooming' or overdrive of the beam current does much more than make the picture 'fuzzy'; it causes it to lose colour and convert to a black-and-white picture.

(*b*) The single beam must supply the power for all three colour phosphors. After suitable guard bands between phosphor stripes are added, the beam power per colour is not appreciably greater than it is in the three-gun shadow-mask, in spite of the inefficiency of the mask.

(*c*) The index signal is critical and the beam current can never be cut off, nor can large video signals be permitted to affect the switching circuit phase.

(*d*) The simplicity of the beam-index colour tube has been achieved at the expense of a more complex colour switching circuit. In the shadow-mask tube, each gun produces only its intended colour; in the beam-index tube, the colour depends on how well the circuit designer has done his job.

(*e*) The greatest disadvantage of all is that, while the shadow-mask tube has over 20 years of manufacturing technology to support it, the index tube has never been mass produced.

CONCLUSION

The colour television receiver display tube is one of the remarkable technological accomplishments of the latter half of the twentieth century. Although there were many proposals for such a device, none appeared practical until the clever use of photolithographic techniques resulted in the shadow-mask system that is basic to all colour tubes now in use. In the future of colour television, one may anticipate further advances in colour cathode-ray tubes, and in materials and techniques for such tubes. In addition, it appears likely that new combinations of technology, including solid-state materials and phenomena, may ultimately lead to entirely new forms of display, such as flat panels. However, until such new forms offer major improvements in size or performance, the shadow-mask tube will not easily be displaced.

Other forms of cathode-ray tube

Mention has previously been made of some more-specialised forms of c.r.t.; these are considered briefly here.

Flying-spot scanner tubes are used to produce an intense spot of light which scans a raster on the screen. A light detector, such as a photomultiplier, is used to produce an electrical signal in response to the amount of light transmitted from the spot through a photographic slide or film. A system such as this provides an accurate electrical signal corresponding to, for example, a television test card.

The requirements of a flying-spot scanner tube are high resolution and a screen with a very short persistence. Magnetic focusing is often used with this type of tube, as well as magnetic deflection. To achieve the high light output required, a high final anode voltage is necessary, typically 25 kV. As in a television picture tube, the connection to the anode is made by a connector in the cone of the tube through an internal conductive layer. An external conductive layer is also provided, and the capacitance between these two layers can provide smoothing for the e.h.t. voltage. Because of the high voltage, a shield must be placed round the tube to protect personnel from X-radiation. The phosphor used with a flying-spot scanner tube for black-and-white television is type *BA*, with very-short persistence and a purplish-white fluorescence. For colour television, type *GU* very-short persistence phosphor with a white fluorescence is used.

Projection tubes, for the large-screen projection of television pictures, are small-diameter television picture tubes producing an intense picture suitable for optical projection. A very high final anode voltage is required, typically 50 kV. Tubes with a white phosphor, type *W*, can be used for black-and-white television, or three tubes with red, green, and blue phosphors respectively (types *YA*, *GK*, and *BF*) can be used to form a colour picture.

The tubes for data display where messages are displayed on the screen are similar to television picture tubes. The electron beam scans a raster in the normal way, but with the beam blanked so that no trace is visible unless a 'bright-up' pulse is applied to the grid or cathode of the tube. Thus as a line is scanned, spots of light are produced on the screen corresponding to the bright-up pulses. As the next line is scanned, further spots of light are produced. In this way, characters are formed on the screen to make the required message. The type of phosphor used for the screen will depend on the transmission system, in particular, on the field repetition frequency of the display.

REFERENCES

1. GOLDMARK, P. C., *et al.*, 'Color Television', *Proc. I.R.E.*, **30**, 162–182 (1942) and **31**, 465–478 (1943)
2. HEROLD, E. W., 'Methods Suitable for Television Color Kinescopes', *Proc. I.R.E.*, **39**, 11, 1177–1185 (1951). See also HEROLD, E. W., *Proc. Soc. Inform. Display*, **15**, 141–149 (1974)
3. LAW, H. B., 'A Three-Gun Shadow-Mask Color Kinescope', *Proc. I.R.E.*, **39**, 11, 1186–1194 (1951)
4. MORRELL, A. M., *et al.*, *Color Television Picture Tubes*, Academic Press, New York (1974)
5. THIERFELDER, C. W., 'Large Screen Narrow-Neck 110° Color Television System', *I.E.E.E. Trans. on Broadcast & TV Receivers*, BTR-17, 141–147 (1971)
6. NARUSE, Y., 'An Improved Shadow-Mask Design for In-Line, Three-Beam Color Picture Tubes', *I.E.E.E. Trans. on Electron Devices*, ED-18, 697–702 (1971)
7. YOSHIDA, S. and OHKOSHI, A., 'The Trinitron—A New Color Tube', *I.E.E.E. Trans. on Broadcast & TV Receivers*, BTR-14, 19–27 (1968)
8. RAMBERG, E. G., 'Focusing-Grill Color Kinescopes', *I.R.E. Convention Record*, Vol. 4, Part 3, 128–134 (1956)
9. DE HAAN, E. F. and WEIMER, K. R. U., 'The Beam-Indexing Colour Display Tube', *R. Television Soc. Journal*, **11**, 278–282 (1967)
10. WRIGHT, W. D., *The Measurement of Color*, 4th edn., Van Nostrand/Reinhold (1969)
11. LARACH, S. and HARDY, A. E., 'Cathode-Ray-Tube Phosphors: Principles and Applications', *Proc. I.E.E.E.*, Vol. **61**, 915–926 (1973)
12. BARTLESON, C. J., 'Color Perception and Color Television', Journ. SMPTE, **77**, 1–12 (1968)
13. FIORE, J. P. and KAPLAN, S. H., 'The Second-Generation Color Tube Providing More Than Twice the Brightness and Improved Contrast', *I.E.E.E. Trans on Broadcast & TV Receivers*, BTR-15, 267–275 (1969)
14. HIGGINS, G. C. and PERRIN, F. H., 'The Evaluation of Optical Images', *Photog. Science and Eng.*, 2, 66–76 (1958)
15. MACHIDA, H. and FUSE, Y., 'Gain in Definition of Color CRT Image Displays by the Aperture Grill', *I.E.E.E. Conf. Record on Display Devices*, 101–108, Oct. 11–12 (1972)
16. MORRELL, A. M., 'Development of the RCA Family of 90-Degree Rectangular Color Picture Tubes', *I.E.E.E. Trans. on Broadcast & TV Receivers*, BTR-11, 90–95 (1965)

17. SPEAR, B. W. and POWELL, D. E., 'KIMCODE, A Method for Controlling Devacuations of TV Tubes', *I.E.E.E. Trans. on Broadcast & TV Receivers*, BTR-9, **1**, 25–31 (1963)
18. BLAHA, R. F., 'Degaussing Circuits for Color TV Receivers', *I.E.E.E. Trans. on Broadcast & TV Receivers*, BTR-18, 7–10 (1972)
19. YOSHIDA, S., 'A Wide-Deflection Angle (114°) Trinitron Color Picture Tube', *I.E.E.E. Trans. on Broadcast & TV Receivers*, BTR-19, 231–238 (1973)
20. BARBIN, R. L. and HUGHES, R. H., 'New Color Picture Tube System for Portable TV Receivers', *I.E.E.E. Trans. on Broadcast & TV Receivers*, BTR-18, 193–200 (1972)

FURTHER READING

BECK, A. H. W., *Thermionic Tubes*, Cambridge University Press, London (1953)
FONDA, G. R. and SEITZ, F., *Preparation and Characteristics of Solid Luminescent Materials*, Wiley/Chapman and Hall, London (1948)
GARLICK, G. F. J., *Luminescent Materials*, O.U.P., Oxford (1949)
GOLDING, J. F. (editor), *Measuring Oscilloscopes*, Iliffe, London (1971)
HARMAN, W. W., *Fundamentals of Electronic Motion*, McGraw-Hill, New York (1953)
KNOLL, M. and KAZAN, B., *Storage Tubes and their Basic Principles*, Wiley, New York (1952)
LEVERENZ, H. W., *An Introduction to the Luminescence of Solids*, Wiley/Chapman and Hall, London (1950)
PARR, G. and DAVIE, O. H., *The Cathode Ray Tube and Its Applications*, Chapman and Hall, London (1959)
SHAW, D. F., *An Introduction to Electronics*, 2nd edn., Longman, London (1970)
SOLLERT, T., STARR, M. A. and VALLEY, G. E., *Cathode Ray Tube Displays*, M.I.T. Series No. 22, McGraw-Hill, New York (1948)
SOWAN, F. A. and REID, I. A. (editors), *Screen Phosphors and Industrial Cathode Ray Tubes*, Mullard Limited, London (1964)
SPANGENBERG, K. R., *Vacuum Tubes*, McGraw-Hill, New York (1948)
THOMSON, J. and CALLICK, E. B., *Electron Physics and Technology*, Physical Science Texts, English Universities Press, London (1959)
ZWORYKIN, V. K. and MORTON, G. A., *Television*, 2nd edn., Wiley/Chapman and Hall, London (1954)

TELEVISION CAMERA TUBES

The main component of a television camera is the image sensor in which the transmitted video signal originates. The following describes practical devices in historical sequence as far as is possible. Collectively they form a family of electronic vacuum tubes to which the names 'camera-tubes' or 'pick-up tubes' have been applied.

For operation, tubes are provided with electrical power supplies, the necessary coils to generate magnetic fields, signal amplifying and processing circuits and a lens system, these being part of the television camera.

Modern television camera tubes incorporate three essential features:

(*a*) A photosensitive surface located in the image plane of the television camera object lens.
(*b*) A storage surface on which electronic charge is accumulated.
(*c*) A scanning electron beam which discharges the target to produce the amplitude modulated video signal.

The mechanism of signal production varies according to tube type and details of tube construction vary among the different manufacturers.

Earliest television systems attempted to 'dissect' or scan an optical image mechanically using a single photocell to generate the electrical signal. One of the first and best known of these was the Nipkow disc,[1] *Figure 7.141*. Other workers employed rotating mirror systems.[2]

Such schemes, apart from the severe mechanical limitations, suffered from the fact that a signal was only generated from each picture element during the instant of exposure so that, for any reasonably high resolution system, the sensitivity was very low. The problems of rotational inertia led to the use of electron scanning,[3] and the sensitivity question to the use of the storage principle.

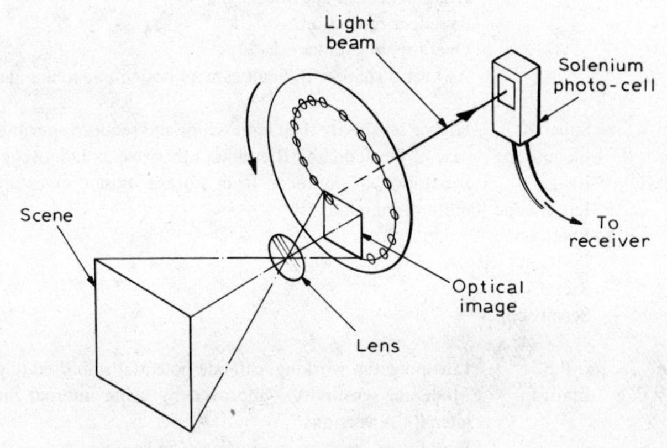

Figure 7.141 Diagram of Nipkow disc equipment

In the Farnsworth Image Dissector[4] a magnetically focused electron image from a large area photocathode was scanned over an aperture through which the photoelectrons passed to a signal electrode either directly as in early tubes or via an electron multiplier in later versions. This class of tube is still used for special purposes but has no place in ordinary television because of its inherent low sensitivity.

Approximately contemporaneously with the Farnsworth tube, Zworykin introduced the Iconoscope[5] and this can be said to be the forerunner of all modern television camera tubes. It incorporated, along with a large area photocathode, a method of storing electronic charge[6] during the non-active period, and a finely focused scanning electron beam to discharge the storage elements. The Iconoscope represents the technological breakthrough which made commercial television possible. The tube is described in detail and is used as a base to introduce the other practical derivatives eligible for inclusion in this section. These, and their various properties are summarised in *Tables 7.11*, *7.12* and *7.13*.

From the tables it can be seen that the targets of the earlier tubes were stabilised, due to secondary emission, at anode potential. This was mainly because the scanning technique had been derived from the cathode-ray tube. More recent tubes use low voltage scanning in which the secondary emission is small and targets therefore stabilise near to cathode potential. The change from 'anode potential stabilised' to 'cathode potential stabilised', was an important step in the evolution of camera tubes and was applied to both of the two main categories of tubes that were subsequently developed, i.e., photo-emissive tubes and photo-conductive tubes. Moreover c.p.s. operation gave the option of deriving the signal from that part of the low voltage beam which did not land on the target and is returned towards the gun.

Table 7.11 PHOTOEMISSIVE TUBE TYPES

Tube, first manufacturer and date	*Similar types*	*Principal generic features*
Iconoscope, RCA, 1929	Emitron	High potential working, anode potential stabilised (a.p.s.) target Very low sensitivity due to poor charge storage High content of spurious signal Excellent resolution Operational gamma < 1 Awkward shape—difficult camera design due to angular scanning
Super Emitron, EMI, 1936	Super Iconoscope Image Iconoscope Photicon Pesticon Rieseliko Scenioscope	Higher sensitivity than Iconoscope and reduced spurious signals in case of Pesticon and Rieseliko, otherwise as Iconoscope Separation of photocell from storage mosaic gives greater flexibility of manufacture
Orthicon, RCA, 1939	C.P.S. Emitron	Low potential working, cathode potential stabilised (c.p.s.) target Moderate sensitivity—improved by using internal multiplier in later RCA versions Full frame storage, minimal spurious signal but early versions unstable under high light inputs Good signal-to-noise ratio Marginal resolution Operational gamma $= 1$ Most convenient cylindrical shape—optic and scanning axes colinear
Image Orthicon, RCA, 1946	Super Orthicon	Adequate sensitivity due to image section C.P.S. target Good resolution with tendency to overemphasise information edges Some noise content in signal, especially annoying in low light areas Operational gamma varies from unity at lower light levels through to near zero at high levels
Image Isocon, RCA, 1949		High sensitivity with low noise content signal C.P.S. target Not strictly suitable for broadcast use but most useful in medium to low light industrial applications Somewhat more complicated set-up procedure than Image Orthicon from which it was derived Unity gamma

Table 7.12 PHOTO-CONDUCTIVE TUBE TYPES

Tube, first manufacturer	*Similar types*	*Principal generic features*
Vidicon, RCA, 1950	Staticon Resistron	Small size and simplicity of operation Moderate sensitivity with high signal-to-noise ratio C.P.S. target of antimony sulphide Dark current high and temperature dependent Slow response to information movement Good resolution Operational gamma varies from 0.65 to about 0.4 Susceptible to damage by high light overloads
Plumbicon[R] Philips, 1964	Leddicon Vistacon Oxycon Sensicon	C.P.S. target of lead oxide Dark current negligible and independent of temperature High sensitivity and signal-to-noise ratio Good resolution but slightly less than Vidicon Gamma is unity Some forms of lead oxide have spectral response limitations Maximum freedom from spurious signals Small size and simple operation Slight susceptibility to damage by light overload—but recovery possible
Silicon vidicon, Bell Labs., 1966	Sidicon Epicon	Similar in format to Vidicon but with c.p.s. target comprised of a silicon diode array High sensitivity with wide spectral response into infra-red region Dark current low but temperature dependent Resolution good but limited by physical dimensions of individual diodes of the target Virtually proof against light and scanning beam overload Speed of response between Vidicon and Plumbicon
Chalnicon, Toshiba, 1972		C.P.S. target of cadmium selenide Sensitivity intermediate between silicon diode array target and lead oxide—otherwise very similar to Vidicon

Table 7.13 INTENSIFIER TARGET TUBE TYPES

Tube, first manufacturer	*Similar types*	*Principal generic features*
SEC Vidicon, Westinghouse, 1966 Note: SEC is derived from Secondary Electron Conduction	Nocticon	C.P.S. target of KCl operated with high energy photoelectrons from continuous photocathode giving an electron gain of between 50 and 100 times High blemish susceptibility leading to high cost for 'clean' tubes Extremely high sensitivity, simple operation but somewhat liable to permanent high-light damage Good resolution
SIT Tube, RCA, 1970	Esicon Ebsicon	Basically similar to SEC Vidicon but using a silicon diode array target eliminating susceptibility to high-light overload damage Resolution limited by physical dimensions of diode array
NOTE: SIT indicates Silicon Intersifier Target.		Note: Ebsicon is a generic title for Electron Bombarded Silicon target tubes

Table 7.14 TUBES USED FOR TELEVISION BROADCASTING

		Photo-emissive				*Photo-conductive*				
						Vidicon				
Parameter	*Units*	*Iconoscope*	*Image Iconoscope*	*Orthicon (CPS Emitron)*	*Image Orthicon*	*Antimony trisulphide*	*Selenium*	*Lead oxide*	*Silicon*	*Cadmium selenide*
Typical operating level (note 1)	lux	25	5.0	2.5	0.1	5–100	1.0	4.0	0.5	0.10
Signal-to-noise ratio (note 2)	dB	32	40	36	38	Limited by frequency response of camera amplifier. For a broadcast bandwidth of 5 MHz a value of between 45 dB and 50 dB is usual				
Resolution (note 3)	%	100	100	40	60	50	50	45	35	50
Signal non-uniformity (note 4)	% max	60	50	20	20	15	20	10	15	10
Speed of response (note 5)	Subjective	Excellent	Excellent	Fair	Good	Poor	Poor	Very good	Good	Fair
Spectral response	Range nm Peak nm	to 750 520	Note 6	Note 6	Note 6	to 800 450	to 600 450	to 700 560	to 1 100 800	to 750 700

Comparison of tube types in terms of their operating parameters

The various tube types are compared in *Tables 7.14* and *7.15*. The figures given are approximate in all cases since no single authority has made the necessary comparative measurements and the accurate modern techniques were not available to early workers. The following notes apply to both tables.

NOTES

(1) The 'typical operating level' of tubes used for television broadcasting can vary considerably according to their detailed design. The figures given apply to the most frequently used tube types and refer to the input illumination to the photocathode to give a 'white' signal output. Photocathode and scene illumination are related by the formula

$$I_S = \frac{I_{pc} 4F^2(M + 1)^2}{TR}$$

where
I_S = Scene Illumination
I_{pc} = Photocathode Illumination
F = Lens Aperture
M = Magnification from scene to photocathode
T = Lens Transmission
R = Scene Reflectance

(2) Signal/noise ratio is calculated using the Schade formula.[7]
(3) Resolution is specified as the relative depths of modulation of 400 TV Line and 100 TV Line patterns, displayed as oscillograph line waveforms, under fully exposed input conditions.
(4) Signal non-uniformity is also dependent on the characteristics of the focus and deflection solenoids.
(5) Speed of response is given subjectively since modern accurate measuring methods cannot be applied retrospectively to early tubes.
(6) In the case of tubes with a transparent photocathode, spectral response can be varied according to choice and detail processing.
(7) At low light levels, operational emphasis concentrates on 'detectability' and 'identification' so that the broadcasting connotations of signal/noise ratio and resolution do not apply. Hence 'minimum operating level' is given as that at which a limiting resolution of 100 TV Lines is obtained—this being arbitrarily taken as necessary for a 'useful' picture.

Table 7.15 LOW LIGHT TUBES

Parameter	*Units*	*SEC Vidicon*	*Ebsicon*	*Image Isocon*
Minimum operating level (note 7)	lux	8×10^{-5}	10^{-5}	3×10^{-5}
Signal non-uniformity (note 4)	% max	40	20	20
Speed of response (note 5)	Subjective	Good	Good	Good
Spectral response	nm	According to choice of photocathode		

Iconoscope

The tube is shown diagrammatically in *Figure 7.142* and the scanning process in *Figure 7.143*.

The optical image is projected on to the photo sensitised mosaic of silver globules so that by photo-emission each element becomes charged proportionately to the elemental

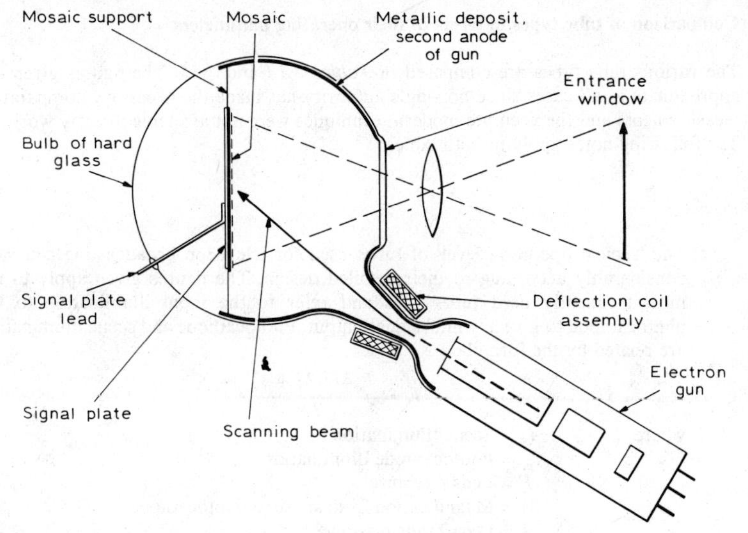

Figure 7.142 Outline schematic of the iconoscope-type tube

image brightness. Scanning by the electron beam discharges each element and produces, by capacitive coupling a series of impulses in the signal plate. These constitute the video signal.

Inherent in this theory is the assumption that each picture point is evaluated once in a complete frame period and hence, can and will store charge over that same frame time.

In practice, charge storage for a frame time does not occur—only storage for a fraction of a line time. This was explained as follows.[8]

The high potentials necessary for focusing the scanning electron beam caused the mosaic to assume a stable value of potential near to that of the second focusing anode of the electron gun. Under these conditions no field to collect the photoelectrons existed. However at the beam's point of contact with the mosaic a temporary higher potential

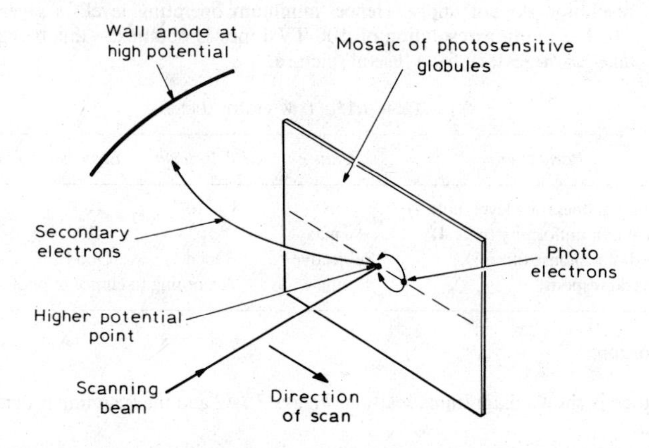

Figure 7.143 Schematic diagram of scanning process in a high velocity scanned tube

was produced by secondary emission and this attracted photoelectrons from surrounding areas and in particular from points about to be scanned. Charge storage was hence possible only during the beam's time of transit to that signal point.

This short storage period or short exposure time gave the Iconoscope a sensitivity well below the expected value.

Some derivatives of the Iconoscope are the Emitron,[8] the Super Emitron,[9] Image Iconoscope,[10] Rieseliko,[11] Photicon,[12] PES Photicon[13] (Pesticon) and Scenioscope.[14]

Of these, the Emitron at the time of its inception was somewhat more efficient than the contemporary Iconoscope. The Super Emitron and Image Iconoscope included a stage of image magnification as proposed by Lubszynski & Rodda.[9] From *Figure 7.144* it can be seen that the mosaic of this tube receives an electron picture from the simple photocell on to which is imaged the optical scene. Hence, the functions of photon

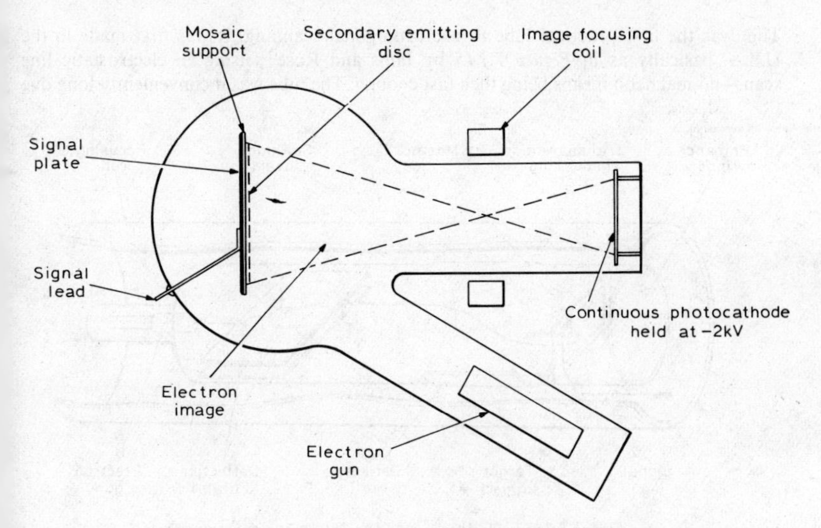

Figure 7.144 Outline schematic of the image iconoscope-type tube

detection and charge storage have been separated. The mosaic is not photo-sensitive but processed to have high secondary emission such that the charge storage process be as efficient as possible. The secondary electrons, having higher emission energy, are more easily removed from the target than the photoelectrons of simple Iconoscope-type tubes. Even so collection by the wall anode of the tube was not uniform over the whole target. Redistribution of secondary electrons on to parts of the target other than on edges of scan termination caused the latter to have high 'tilt' signals. These required complicated electronic corrective circuits. The Photoelectrically Stabilised Photicon and Riesel Iconoscope (Rieseliko) were designed to overcome this problem by providing in strategic positions two auxiliary photocathodes. These, under stimulation from externally mounted light bulbs, produced extra photoelectrons to neutralise the highly charged mosaic edges. An excellent practical tube resulted but the necessary angular separation of optic and scanning axes involved fundamental problems. These included scanned raster geometry, spurious and uncontrollable secondary emission from the high velocity scanning beam, beam focus variation, unstable target working potentials and indeterminate black level. As has been mentioned earlier, these drawbacks resulted from operation at high potentials.

It was realised[15] that an improvement could be obtained by operating the target at the stable potential of the thermionic cathode of the electron gun and in 1936 Lubszynski filed patents[16] which not only described methods by which this could be achieved but also allowed a construction in which the optical and scanning axes could be co-linear. Orthogonal beam incidence at the target was realised by using as a scanning magnetic field the resultant of two coils—the one of solenoidal construction immersing the whole tube and two auxiliary fields produced by pairs of saddle coils internal to the main solenoid. Such a system created the scanning conditions required such that the target would operate at virtually cathode potential—the only acceptably stable condition to avoid problems of non-uniform secondary emission by the scanning beam since with near zero energy none could occur.

The Orthicon

This was the first practical tube to use orthogonal scanning. It was first made in the U.S.A. basically as in *Figure 7.145* by Iams and Rose[17] using an electrostatic line scan—no magnetic means being then fast enough. The tube was inconveniently long due

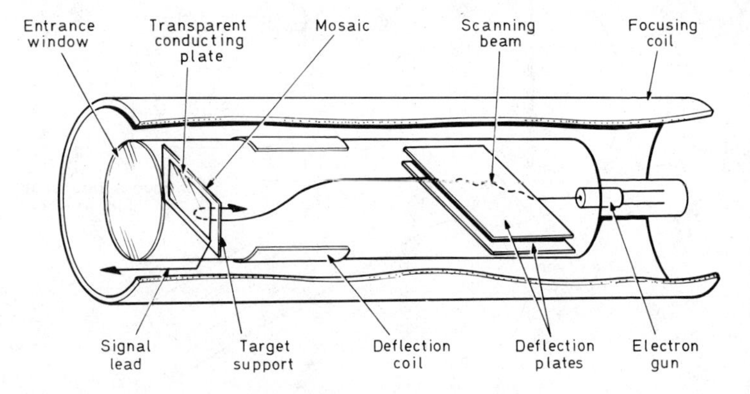

Figure 7.145 Outline schematic of the orthicon-type tube

to line and frame scan sections operating in series and had inadequate sensitivity. A British version was produced in the late 1940s—the c.p.s. (cathode potential stabilised) Emitron.[18] This tube had magnetic scan for both frame and line, a simple triode electron gun and incorporated a highly sensitive mechanically-formed mosaic. It was used operationally by the BBC for many years. The first notable broadcast was the Olympic Games in Wembley in 1948.

THE IMAGE ORTHICON

The sensitivity of the orthicon was still inadequate for many purposes because of basic limitations in the storage target. Its high capacitance produced only a small operational voltage swing causing beam discharge lag which resulted in unacceptable handling of moving objects. On the other hand higher potentials due to picture highlights tended to build up into values above the control of the scanning beam.

Both these deficiencies were overcome and combined with an image section similar to that of the Image Iconoscope to produce the Image Orthicon—announced in 1946[19] by

RCA. With this tube the majority of television services throughout the world were established and it is still the preferred camera tube for monochrome television.

Seen in *Figure 7.146* the tube incorporates three major sections:

(*a*) an image intensifier,
(*b*) a scanned storage target; and
(*c*) an electron multiplier.

The optical image projected on to the continuous photocathode generates an electron image which is focused by the surrounding solenoid and the electrostatic field in the image section upon the micro-thin glass target. In modern tubes this glass has very slight ohmic conductivity[20] compared with the ionic process of early examples. At the

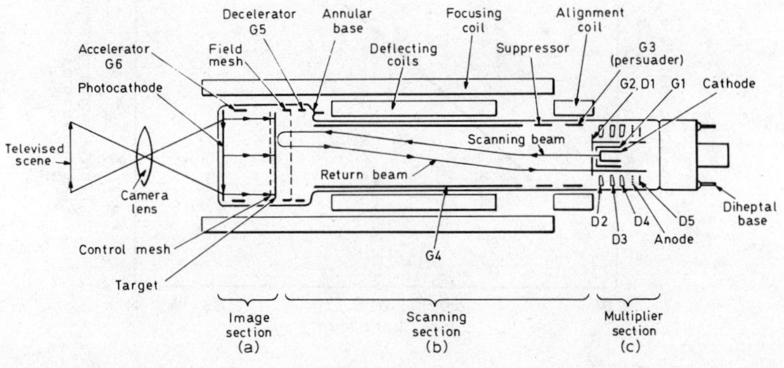

Figure 7.146 Schematic of 3″ image orthicon tube. The 4½″ version uses the same diameter optical image but incorporates a magnifying electron lens in the image section in order to completely fill the larger target

target surface secondary electron emission takes place—the secondary electrons being collected by the slightly positive and highly transparent mesh. The loss of secondaries produces a nominal electron gain and a positive charge pattern on the glass of a density varying in proportion to the original pattern of scene brightness. Charging continues in the picture highlight areas until the glass surface slightly exceeds the mesh in potential. The charge pattern is transferred to the opposite target surface for evaluation by the scanning beam. As the target is scanned some beam electrons drift back randomly to the electron gun. Other beam electrons then neutralise the target charges and the then zero potential elements resistively reduce the input face of the target to near zero potential so that a new value of charge may be assumed. The remaining beam electrons are reflected by the zero potential elements roughly along their outgoing path to the limiting aperture of the electron gun. Due to the electron-optical conditions at the target, the returning electrons impinge on the dynode surface rather than disappear into the aperture. At the dynode, secondary electrons are produced and are guided into the electron multiplier by the negative potential of the persuader. In the 5-stage electron multiplier with about 300 volts per stage a total gain of about 1 000 is achieved to give a final output signal of several microamps—well above the noise level of the head amplifier. The signal amplitude is measured from maximum output corresponding to full beam returned from picture black, to lower values created by the various shades of grey, with a minimum generated by peak white in the original scene. This negative polarity signal is one grave disadvantage of the Image Orthicon since the maximum return beam from picture black has highest noise content and this is most objectionable in those dark areas of the received picture.

Signal-to-noise ratio is optimised by making the stored target charge per picture element as high as is practicable consistent with maximum read out efficiency to give minimum lag. This is achieved by using targets with high storage capacitance—i.e., with mesh and target close together and/or large area. In this respect the series of tubes of 4½″ diameter[21] is superior to the original 3″ series. The latter is retained in high sensitivity applications where the lower capacitance target can become fully charged at lower light levels. As mentioned above, charge accumulates on the target until the potential of its input face has risen to a value just above that applied to the

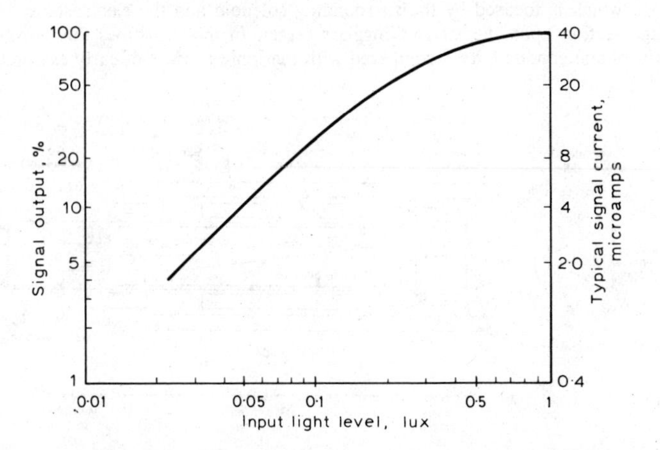

Figure 7.147 Typical transfer characteristic of an image orthicon tube. It may be noted that, according to the light level and contrast range chosen for the input values, a gamma of 1 or 0.5 or a combination may be obtained. This is of value to camera operation

mesh such that the net secondary emission is zero. Clearly, this saturation value is reached at lower light levels for wider spaced, smaller diameter systems which have lower capacity.

Graphically the Signal (S)/Illumination (I) relationship is shown in *Figure 7.147*, but variations occur according to tube type and mode of operation. With logarithmic coordinates this curve can be represented by the formula $S = \gamma I + \text{constant}$, γ or gamma, the index of proportionality is unity at lower levels of illumination and reduces at higher levels—an operational attribute of the Image Orthicon.[22, 23, 24, 25]

The Image Isocon

As mentioned earlier, the scanning of the target in an Image Orthicon gives rise to scattered electrons, in quantity proportional to the scattering charge. By using a selective electrode system, diagrammatically shown in *Figure 7.148*, these can be guided into the electron multiplier and the reflected electrons discarded to form a video signal of correct polarity and with none of the disadvantages of the image orthicon signal. By the end of 1973 such tubes began to assume great importance in low light level television systems due to the possibility of using a lower capacity storage target than is practically feasible in the image orthicon, to produce higher sensitivity without the attendant noise problems.

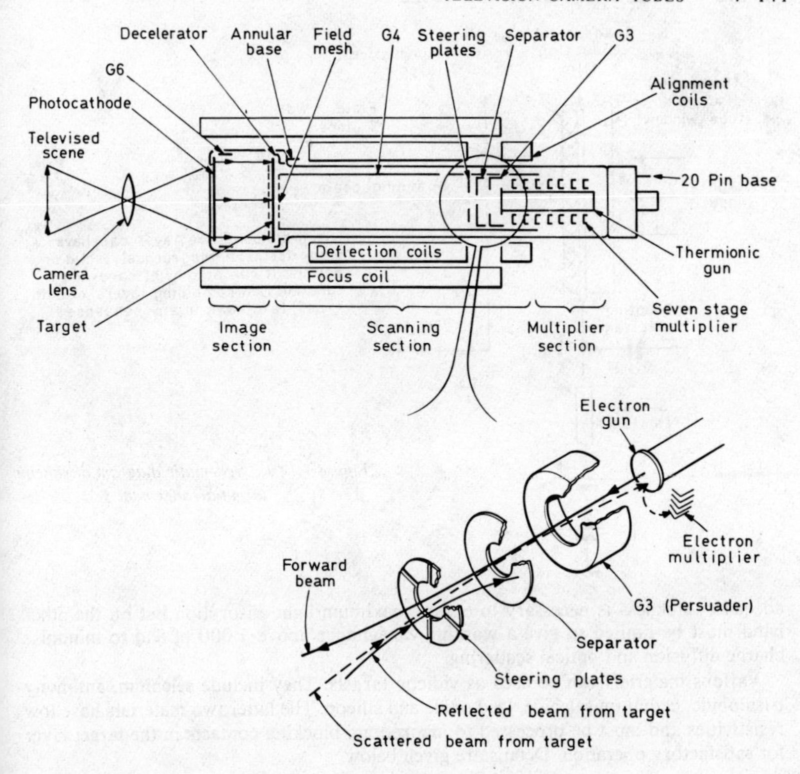

Figure 7.148 Schematic diagram of the image isocon

The Vidicon

This name, originally coined by RCA[26] is now often used as a generic term for all orthogonally scanned camera tubes using a photoconductive target—*Figure 7.149*.

In operation the laminar target suffers a proportionate reduction in resistance according to the local brightness of the incoming light image. The applied steady potential to the transparent conducting backing layer then gives rise to a potential pattern on the vacuum side of the target. When scanned to cathode potential, capacitive coupling across the target layer produces a video output current of varying amplitude proportional to the potential of the element from which it was derived and hence proportional to the elemental brightness of the original scene.

Materials for use as photoconductive targets must satisfy fairly well defined requirements.[27] There must be a maximum conversion of light into available charge carriers and to achieve this, absorption of the incoming radiation must be high and the effective band-gap of the material should correspond in energy to the longest wavelength in its spectrum. For the visible range a band-gap of less than 2 eV is required. Freedom from trapping centres is also necessary in order that the constituent molecules can return rapidly to the unexcited state.

For satisfactory operation the completed target must have a satisfactory time constant, so imposing upon the basic material a lower-limit of resistivity of 10^{12} ohm–m and certain restrictions upon thickness when used as a continuous layer. In this regard

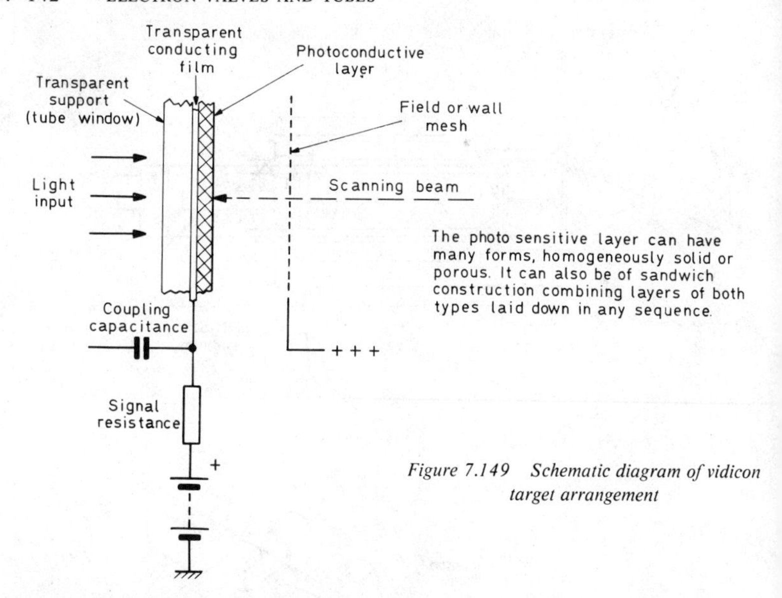

Figure 7.149 Schematic diagram of vidicon target arrangement

adequate thickness is necessary to ensure maximum light absorption but on the other hand must be limited to give a working capacitance above 1 000 pf and to minimise charge diffusion and optical scattering.

Various materials can be used as vidicon targets. They include selenium, antimony trisulphide, cadmium selenide, lead oxide and silicon. The latter two materials have low resistivities and must be processed to incorporate blocking contacts in the target layer for satisfactory operation. Details are given below.

SELENIUM

Selenium was used in first examples[26] of the vidicon-type tube. It was unstable and had a tendency to revert to an electrically conducting form after some hours of use. Resurrected briefly in the form of a Se/Te alloy in the U.S.A. in the mid-1960s it was eventually abandoned because no lasting remedy could be discovered for its instability.

ANTIMONY TRISULPHIDE

Basically of the formula Sb_2S_3, antimony trisulphide has seen extensive use in vidicons.[28] In 1973 it was the least costly material to process but has limited sensitivity and suffers from movement lag.

As a target layer, antimony sulphide can be vacuum evaporated to be thin and dense or, in an atmosphere of inert gas such as argon, to a porous film of greater thickness. The operating capacity of the latter is desirably lower as also is its sensitivity. In practise, combinations of hard and porous coatings about 1 μm to 3 μm thick are used. Controlled exposure to air is permissible to allow convenient assembly methods. Antimony trisulphide vidicons have a standing output current or dark current which is temperature dependent when not illuminated, a severe disadvantage in colour cameras. Operational gamma varies from 0.7 at low illumination to 0.4 at higher levels.

Resolving power of early antimony sulphide vidicons was inadequate and was shown

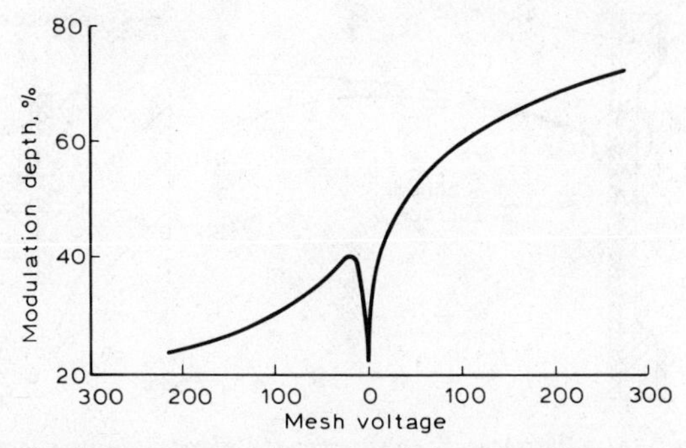

Figure 7.150 *Graph showing relation between resolution and mesh potential of a separate mesh vidicon type tube*

to result from the presence of heavy positive ions in the end region of the beam focus electrode near to its terminating field mesh.

A considerable improvement is obtained when the latter is electrically separated from the beam focus electrode and the two operated at slightly different potentials, see *Figure 7.150*.[29, 30]

The separate mesh tube allows a correction to be made for the non-linear magnetic field at the target end of the focusing solenoid. Furthermore, due to the higher field gradient between the mesh and target, higher beam currents can be used for better control of picture highlights.

In spite of several operational drawbacks such as smearing of moving objects, high temperature dependent dark current, etc. the antimony sulphide vidicon is widely used in a variety of closed-circuit television applications.

SILICON

Pure silicon exhibits very high photo-conductivity. However, due to its low ohmic resistance, a mosaic of discrete diodes must be formed[31, 32, 33] for it to serve as a camera tube target.

One technique is to first heavily oxidise the eventual mosaic surface of an optically worked lamina of single crystal silicon. Then, by photo-lithographic techniques the continuous silicon dioxide layer is converted into a fine grid a few microns in pitch—so exposing the base silicon material as a conglomerate of holes. Next a *p*-type material such as boron glass is fused over the whole surface so that each silicon hole now becomes a *pn* junction, surface insulated from its neighbours. To minimise the self-biasing action of the insulating grid when scanned, a metal, or reasonably conducting material is applied to each diode surface to reduce the effective bar width of the grid but still maintaining mutual insulation between the diode surfaces. *Figure 7.151* indicates this arrangement.

The final manufacturing operation is to chemically etch the back of the intended picture raster area to its operational thickness—about 25 μM. Sometimes this surface is given dichroic treatment to enhance a particular spectral sensitivity. The completed target is mechanically mounted in conventional vidicon position with minimum clearance between it and the inside of the tube faceplate. The inevitable small gap is of no optical consequence but does preclude the use of fibre-optic techniques (see later).

In operation, the continuous silicon backing plate is biased some 10–20 volts positive with respect to the scanning beam cathode. Scanning the mosaic stabilises it at cathode potential so that each diode assumes a reverse bias condition—the *p*-type islands just below cathode potential. Behind each one, and probably forming a continuous layer, is a region depleted of current carriers to insulate it from its neighbours.

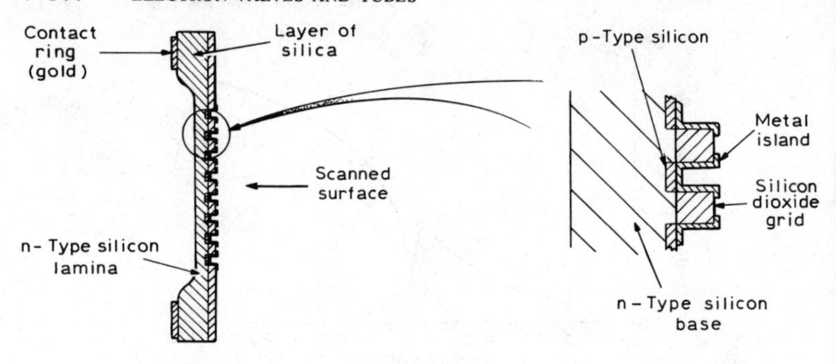

Figure 7.151 Diagrammatic cross-section of a silicon diode array target

When illuminated, current carriers, electrons and holes, are photoelectrically produced in the silicon layer. The electrons are attracted by the positive signal plate potential and the holes migrate to charge the *p*-type islands for evaluation in normal manner by the scanning beam.

The silicon vidicon has very high sensitivity and extends into the near infra-red. It is of use in this spectral band for simple surveillance work. Its large scale use was intended to be for the television telephone where its ability to survive electrical and light overloads and its anticipated long life is of extreme advantage compared with other vidicon-type tubes. The tube is used to a limited extent in the red channel of some vidicon colour cameras.

LEAD OXIDE

In recent years lead oxide of the form PbO, has become the most important photoconductor used in tubes of the vidicon format.[34, 35, 36, 37, 38] Lead oxide is a true semiconductor and can exist in *p*, *n* or intrinsic forms. In the lead oxide vidicon a porous target, structured as shown diagramatically in *Figure 7.152* exists, having been produced by evaporation under carefully controlled conditions.

In operation, the positively biased signal plate and the negative scanned surface stabilised slightly below cathode potential effectively remove all current carriers from the body of the layer to give the tube a zero output current when un-illuminated, a most valuable asset for colour signal processing. When illuminated, strict proportionality

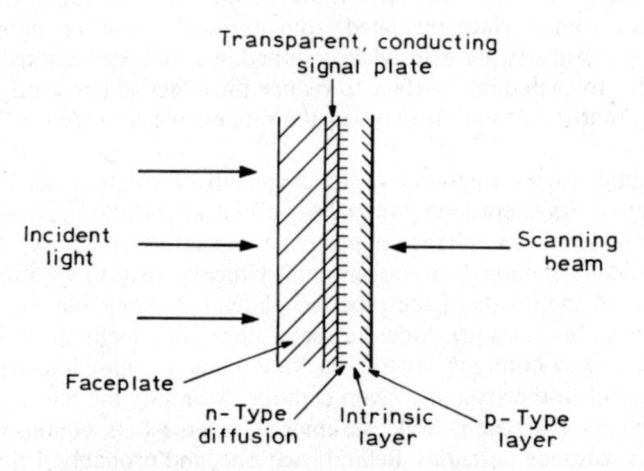

Figure 7.152 Schematic section of lead-oxide target

prevails over the operational range of scene brightness between incident illumination and signal current.

Tubes incorporating a principle known as light bias, became available[39] in the early 1970s. 'Light bias'—a small amount of target illumination—can be generated either in the optical system, in the tube socket or in an internal appendix of the tube itself. Its function is to raise the potential of the scanned surface at very low light levels so that the efficiency of the scanning beam is improved at the resulting low signal currents. Without light bias this surface stabilises slightly below cathode potential at a value at which only the higher energy beam electrons can reach the target so that discharge efficiency is seriously impaired.

Tube light bias effectively enables differential lag between individual camera tubes to be minimised.

Other relatively undesirable properties of early lead oxide vidicons have led to improved variants each of which has particular application in the television field.

Tetragonal lead oxide (litharge) has a band gap of 2 eV which gives it a wavelength threshold sensitivity at approximately 640 nanometres. This prevents it from 'seeing' long wavelength red colours. The deficiency has been overcome, where required, by a partial doping of the lead oxide, with sulphur for example, to produce a material with a band gap significantly below 2 eV. The sensitivity increase in red is roughly doubled. In general, electron trapping in the modified material tends to be higher so that image retention can be slightly greater than in the standard tube. However, the higher sensitivity and better colourimetry are advantages to be offset against this.

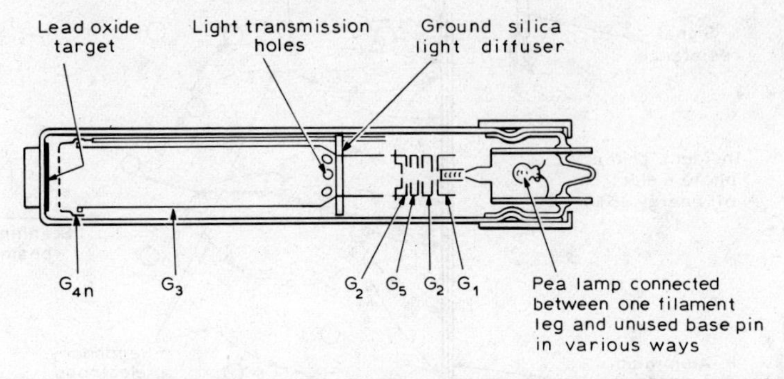

Figure 7.153 Schematic diagram of English Electric HOP/Light Bias Leddicon tube. Gun electrode G2 is, in effect, divided and an additional one, G5, interposed

The light transfer characteristic of the lead oxide target is essentially a linear function ideally suited to the mathematical signal processing required in colour television systems. However, unlike simple vidicon and Image Orthicon tubes, there is no high-light saturation—a situation leading to unwanted 'ballooning' of picture high-lights. Some alleviation can be achieved by careful camera and lighting application but up to five lens stops improvement results from special tube designs designated ACT (anti-comet-tail)[40] or HOP (high-light overload protection).[39] These types use a modified electron gun which incorporates an additional electrode, see *Figure 7.153*. A voltage pulse applied to this electrode during the line scan retrace time causes a high density beam to reach the target and remove the high surface charges resulting from high input illuminations. This technique also removes the 'comet-tail' produced when a high-light in a televised scene has movement relative to the camera.

Lead oxide vidicon tubes under various trade names and of several diameters were available in 1975 but only the 30-mm tube was available in both integral and separated field mesh construction. The improvement in resolution typical of separate mesh antimony trisulphide vidicons is not shown by lead oxide targets due to the latter's much greater thickness. However, the uniformity of resolution and highlight handling of a separate mesh tube are said to be superior.

OTHER TARGET MATERIALS

Vidicon-type tubes have been announced using other target materials. These include cadmium selenide[41] in the chalnicon by Toshiba; antimony trisulphide plus a layered structure of selenium, arsenic and tellurium in the Saticon by Hitachi and a similar structure but using zinc selenide and zinc cadmium telluride in the Newvicon by Matsushita.

The new Japanese tubes report much higher sensitivities than hitherto customary but in mid-1975 insufficient operational data was available for inclusion.

Low light level camera tubes

These are of two basic forms—either a combination of a conventional type as described above with an image intensifier[25] or a tube of Image Orthicon concept but incorporating

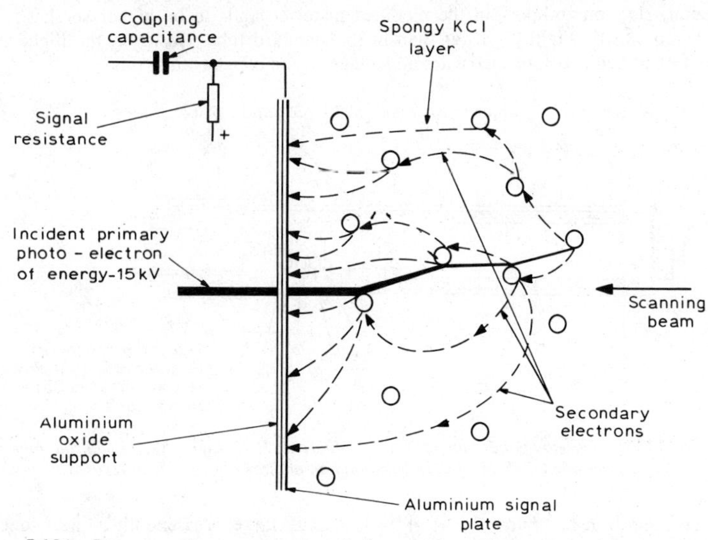

Figure 7.154 Diagrammatic representation of the secondary electron cascade in a target of spongy potassium chloride

an intensifier target,[42] *Figure 7.154*. Potassium chloride has been successfully used[43] as an intensifier target. Operated at approximately 15 kV, a cascade of secondary electrons is produced in the thickness of the target from the incoming photoelectrons such that a charge gain of approximately 50 times is achieved. Such tubes carry the generic name of SEC Vidicons—the prefix denoting Secondary Electron Conduction.

Arsenic trisulphide,[44] the silicon diode array[33] and zinc sulphide[45] are also used in this manner but in these cases the target conductivity is produced by high energy electron bombardment. Target gains of 100–500 times are typical of tubes of the Electron Bombarded Silicon Target—variously known as SIT (Silicon Intensifier Target) tubes or of the EBSICON type (Electron Bombarded Silicon) type.

In each of the above tubes the video signal is taken from the conducting support for the target mosaic—for the SEC vidicon a mesh supported aluminium film and, in the case of the EBSICON, the continuous silicon backing layer.

For tube combinations, coupling can be by lens or direct between the output phosphor and fibre optic window of the image intensifier to the fibre optic input window of the camera tube. The photocathode of the latter is specially processed to match the spectral output of the phosphor. The increase in sensitivity of the system is, naturally, the gain of the intensifier. Some reduction in resolution is introduced by the fibre optics.

Multi-colour tubes

Colour television requires the generation of three primary signals in the red, blue and green spectral regions and these must be originated in the television camera. For this purpose, apart from frame sequential systems, according to the coding principle adopted one, two, three or four video channels are employed requiring a corresponding number of camera tubes each conforming to a particular monochrome specification.

In this section tubes which, in themselves, generate composite 'coloured' signals are described. These require quite different signal processing to that used with combinations of monochrome tubes.

The simple vidicon was adapted to produce red, blue and green signals in 1960.[46] It used a target system which was in effect a superimposition of three. A sequence of colour filter stripes was applied to the inside face of the entrance window. On each stripe was deposited a transparent signal plate and the photoconductor applied as a continuous layer over the whole. A common contact was applied to each set of signal plate stripes so that when exposed and scanned in conventional manner signals from each of the three colours were produced to be analysed and reconstituted by appropriate circuitry on a shadow-mask kinescope. The tube did not reach commercial production due, probably, to the extreme precision necessary for the construction of the target and also to its very low sensitivity.

Modern tubes use similar principles but rely less on extreme mechanical accuracy and only brief constructional details are available.

Several arrangements[47, 48, 49] of internal stripe dichroic filter systems with a common signal plate have been announced but with little detailed information. One tube uses in addition an indexing signal plate in combination with the striped filters. An alternative possibility is to apply the filters to the outside face of a fibre-optic entrance window to the tube.

In all cases, when scanned these colour indexed targets give rise to composite video signals from which the required colour components are extracted by phase or frequency discriminating circuitry.

Two tube cameras[50] with an optical splitter use one tube to supply the luminance signal and the second, possessing a simple frequency coded filter system, to provide the red and blue signals. Again electronic circuitry enables the desired chrominance signal components to be derived.

In spite of their low relative cost, application of these colour tubes has been in fields other than broadcasting due to their marginal performance. However, in the education world, microscopy and limited single shot interview work their use is increasing significantly.

REFERENCES

1. NIPKOW, P., 'Electrische Teleskop', *Ger. Pat.* 30105 (1884)
2. CAREY, G., 'Design & Work', **8**, 569 (1880)
3. GARRAT, G. R. M. and MUMFORD, A. H., 'The History of Television', *Proc. I.E.E.*, Part IIIA, 25–42 (1952)
4. CAMPBELL SWINTON, A. A., *Nature*, **78**, 151 (1908) also 'Presidential Address', *J. of Rontgen Society*, **8**, 1–15 (1912)

5. ZWORYKIN, V. K., 'Television System', *U.S. Pat.* 1691324 (1928) also *U.S. Pat.* 2141059 (1938) (application date December 1923)
6. ZWORYKIN, V. K., 'Television with Cathode-ray Tubes', *Jour. I.E.E.*, **73**, 438 (1933)
7. SCHADE, O. H., 'Electro-Optical Characteristics of Television Systems', *R.C.A. Review.*, **9**, 34 (1948)
8. MCGEE, J. D. and LUBSZYNSKI, H. G., 'EMI Cathode-ray Television Transmission Tubes', *Jour. I.E.E.*, **84**, 468–482 (1939)
9. (*a*) LUBSZYNSKI, H. G. and RODDA, S., 'Improvements in or relating to Television', *Brit. Pat.* 442666 (1934)
   (*b*) CAIRNS, J. E. I., 'A small high-velocity scanning television pick-up tube', *Proc. I.E.E.*, Part IIIA, 89–94 (1952)
10. IAMS, H., MORTON, G. A. and ZWORYKIN, V. K., 'The Image Iconoscope', *Proc. I.R.E.*, 541–547 (1939)
11. (*a*) STRUBIG, H., 'Die Superikonoscop-Kamerarohre der Fernseh G.m.b.H.', Kurzmitteilungen, Fernseh G.m.b.H., Darmstadt, *Sonderheft*, 1 (1955)
   (*b*) Editorial, 'The mode of operation of the Rieseliko', *Fernseh—Communications*, **13/14**, 229–240 (1957)
12. Manufactured by Cathodeon Ltd., Cambridge.
13. COPE, J. E., GERMANY, L. W. and THEILE, R., 'Improvements in design and operation of image iconoscope type camera tubes', *Jour. Brit. I.R.E.*, **12**, 139–149 (1952)
14. SCHAGEN, P., BOERMAN, J. R., MAARTENS, J. H. J. and VAN RIJSSEL, T. W., 'The Scenioscope, a new Television Camera Tube', *Philips Tech. Rev.*, **17**, 189–198 (1956)
15. BLUMLEIN, A. D. and MCGEE, J. D., 'Mosaic stabilisation to cathode potential by scanning', *Brit. Pat.* 446661 (1934)
16. LUBSZYNSKI, H. G., 'Improvements in or relating to television and like systems', *Brit. Pat.* 468965 (1936) also *Brit. Pat.* 522458 (1958)
17. ROSE, A. and IAMS, H., 'Television Pick-up Tubes using low velocity scanning', *Proc. Inst. Radio Eng.*, **27**, 547–555 (1939)
18. MCGEE, J. D., 'A Review of some Television Pick-up Tubes', *Proc. I.E.E.*, Pt. III, **97**, 377–392 (1950)
19. ROSE, A., WEIMER, P. K. and LAW, H. B., 'The Image Orthicon—a Sensitive Television Pick-up Tube', *Proc. Inst. Radio Eng.*, **34**, 424–432 (1946)
20. BANKS, P. B., Improvements in or relating to Television Camera Cathode-ray Tubes'. *Brit. Pat.* 1048390 (1964)
21. HENDRY, E. D. and TURK, W. E., 'An Improved Image Orthicon', *J. Soc. Motion Picture and Television Engineers*, **69**, 88–91 (1960)
22. WEIMER, P. K., 'The Image Isocon—an Experimental Television Pick-up Tube based on the Scattering of Low Velocity Electrons', *RCA Review*, **10**, 366–386 (1949)
23. KLEM, A. and KINGMA, R. V., 'Low Light Level Systems developed by N.V. Optische Industrie de Oude Delft', *Proc. Electro-Optics Int.*, **71**, 304–312 (1971)
24. KLEM, A., 'Delcalix with Isocon', *Odelca Mirror*, **9**, 1–4.
25. NIXON, R. D. and TURK, W. E., 'The Image Isocon for Low Light Level Operation', *J. Soc. Motion Picture Televis. Eng.*, **81**, 454–458 (1972)
26. WEIMER, P. K., FORGUE, S. V. and GOODRICH, R. R., 'The Vidicon Photoconductive Camera Tube', *Electronics*, **23**, 70–74 (1950)
27. TURK, W. E., 'Photoconductive TV Camera Tubes—a Survey', *Journal of Science and Technology* (General Electric Company Ltd.), **37**, No 4, 163–170 (1970)
28. FORGUE, S. V., GOODRICH, R. R. and COPE, A. D., 'Properties of some Photoconductors, principally Antimony Trisulphide', *RCA Review*, **12**, No 3, 335–349 (1951)
29. LUBSZYNSKI, H. G., 'Improvements in and relating to Television and like systems', *Brit. Pat.* 468965 (1936)
   JEPSON, H. B., 'Improvements in or relating to Photoconductive Devices', *Brit. Pat.* 1030173 (1961)
30. DAWE, A. C., 'Special Types of Vidicon Camera Tubes', *Industrial Electronics*, November (1963)

31. CROWELL, M. H. and LABUDA, E. F., 'The Silicon Diode Array Camera Tube', *Bell System Tech. J.*, **48**, 1481–1528 (1969)
32. WOOLGAR, A. J. and BENNETT, E. F., 'Silicon Diode Array Tube and Targets', *J.R. Television Society*, **13**, 53–58 (1970)
33. SANTILLI, V. J. and CONGER III, G. B., 'TV Camera Tubes with Large Silicon Diode Array Targets Operating in the Electron Bombarded Mode', *Advances in Electronics and Electron Physics*, **33A**, 219–228 (1972)
34. De HAAN, E. F., 'The Plumbicon, a New Television Camera Tube', *Philips Tech. Rev.*, **24**, 57–58 (1962–63)
35. De HAAN, E. F., VAN DER DRIFT, A. and SCHAMPERS, P. P. M., 'The Plumbicon, a New Television Camera Tube', *Philips Tech. Rev.*, **25**, 133–151 (1964)
36. VAN DE POLDER, L. J., 'Target Stabilisation Effects in Television Pick-up Tubes', *Philips Res. Rep.*, **22**, 178–207 (1967)
37. DOLLEKAMP, J., SCHUT, TH. G. and WEIJLAND, W. P., 'Advances in Plumbicon Camera Tube Design', *Electron Appl.*, **30**, 18–32 (1971)
38. VAN DOORN, A. G., 'The Plumbicon compared with other Television camera tubes', *Philips Tech. Rev.*, **27**, 1–4 (1966)
39. BAILEY, P. C., 'New Lead Oxide Tubes', *Sound and Vision Broadcasting*, **11**, 19–21 (1970)
40. DOLLEKAMP, J., 'One-inch Diameter Plumbicon Camera Tube Type 19XQ', *Mullard Technical Communications*, **109**, 196–200 (1971)
41. YOSHIDA, O., 'Chalnicon—A New Camera Tube for Colour TV Use', *Japan Electronic Engineering*, 40–44 (1972)
42. GOETZE, G. W., 'Transmission Secondary Emission from Low Density Deposits of Insulators', *Advances in Electronics and Electron Physics*, **XVI**, 145–153 (1962)
43. GOETZE, G. W. and BOERIO, A. H., 'SEC camera-tube Performance Characteristics and Applications', *Advances in Electronics and Electron Physics*, **28A**, 159–171 (1969)
44. SCHNEEBERGER, R. J., SKORINKO, G., DOUGHTY, D. D. and FEIBELMAN, W. A., 'Electron Bombardment Induced Conductivity including its application to Ultra-violet Imaging in the Schumann Region', *Advances in Electronics and Electron Physics*, **XVI**, 235–245 (1962)
45. LODGE, J. A., 'A review of Television Pick-up Tubes in the United Kingdom', *Proceedings of Electro-Optic Conference, Brighton, England*, 253–264 (1971)
46. WEIMER, P. K., 'A development Tri-color Vidicon having a Multi-element Target', *I.R.E. Trans.* **Ed-7**, 147–153 (1960)
47. BRIEL, L., 'A Single-vidicon Television Camera System', *J.S.M.P.T.E.*, **79**, 326–330 (1970)
48. For example the SONY 'Trinicon' camera. I.E.E.E. Show (1973)
49. ATTEW, J. E., 'A Single-tube Colour Camera', *J. Royal Televis. Soc.*, **14**, No. 5, 123–125 (1972)
50. (*a*) JESTY, L. C., 'Recent Developments in Colour Television', *J. Royal Televis. Soc.*, **7**, 488–508 (1955)
(*b*) For example Sony colour camera DXC-5000CE.

COLD-CATHODE GAS-FILLED TUBES

The current in a cold-cathode gas-filled tube is conducted by the ionised gas filling. Practical cold-cathode tubes are essentially diodes, although many types of tube contain several cathodes. The gas filling is ionised by applying a sufficiently high voltage between the anode and cathode. Cold-cathode tubes are therefore 'on/off' devices, and as the ionised gas filling produces a glow discharge, they are often used as indicating devices.

Principle of glow discharge

Various types of discharge can occur between the anode and the cathode of a tube containing a gas filling at low pressure, depending on such factors as the gas and the

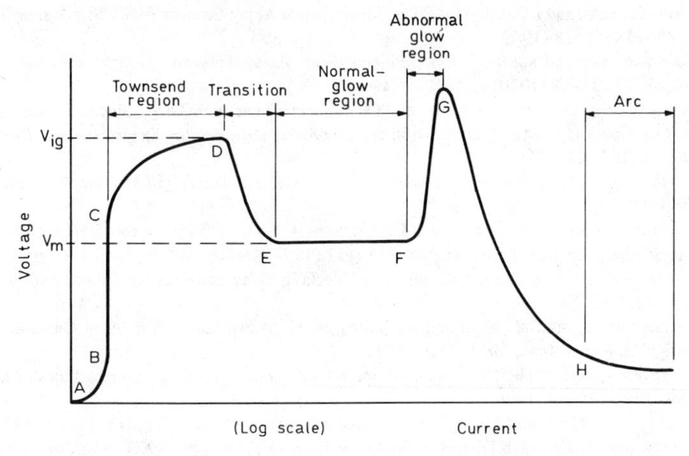

Figure 7.155 Voltage/current characteristic for cold-cathode gas-filled tube

pressure, the cathode material, the anode-to-cathode spacing, the voltage applied between the anode and the cathode, and the current through the discharge. The voltage/current characteristic for a typical gas-filled tube is shown in *Figure 7.155*. The current is plotted on a logarithmic scale, while the voltage is plotted on a linear scale.

The gas filling of the tube, even with no voltage applied between the anode and the cathode, contains some free electrons and positive ions produced by natural radiation such as cosmic rays. These ions move in a random way and recombine, so there is no net current. When a small voltage is applied, say below 10 V, a small current of the order of 10^{-14} A flows. The random movement of the electrons and positive ions is converted into a drift towards the electrode of the opposite polarity. Because the number of free electrons and ions is small, and because recombination takes place, the current is small. If the voltage is removed, the current ceases as the random movement returns. Increasing the voltage increases the current, as the electrons and positive ions move more quickly towards the electrodes and so the number of recombinations is decreased. This corresponds to the region *AB* of the characteristic, the current flowing at approximately 10 V being of the order of 10^{-12} A.

At point *B*, virtually all the free electrons and positive ions present reach the anode and cathode, and so the current does not increase as the applied voltage is increased, region *BC* of the characteristic.

A further increase in the applied voltage increases the velocity of the free electrons and positive ions. Because the mass of an electron is much less than that of an atom of the gas filling, the electrons lose little energy on collision. An electron can therefore be accelerated by the electric field to acquire eventually, despite these collisions, sufficient energy to free an electron from a gas atom on collision. The amount of energy required is called the ionisation energy, and is a property of the particular gas used in the tube. Although the incident and freed electrons have little energy after the ionisation, they both can soon acquire the ionisation energy required to free further electrons. This increase in the number of electrons is called gas amplification. The positive ions do not take a significant part in gas amplification because their mass is similar to that of the gas atoms, and so a collision results in the ion losing most of its acquired energy. Over the region *CD* of the characteristic, the current increases with an increase of applied voltage because of the increase in electrons through gas amplification. The discharge cannot be self-sustaining, however, because all the electrons formed reach the anode. Region *CD* is called the Townsend discharge region.

As the applied voltage approaches point D, the positive ions acquire sufficient energy to reach the cathode and free secondary electrons on impact. A point will be reached where, on average, each electron creates a further electron on collision with a gas atom, and the resulting positive ion travels to the cathode and creates a further electron by secondary emission. The discharge is now self-sustaining. The voltage at which this occurs (point D) is called the *ignition voltage* or *striking voltage*, V_{ig}.

After ignition, the voltage across the tube falls to a value called the *maintaining voltage*, V_m. The fall occurs because a positive space charge forms near the cathode. After ionisation, the positive ions move more slowly towards the cathode than the electrons do towards the anode. There is therefore an excess of positive ions a short distance from the cathode. This positive space charge acts as a virtual anode. The previous linear potential distribution between anode and cathode is altered, as shown in

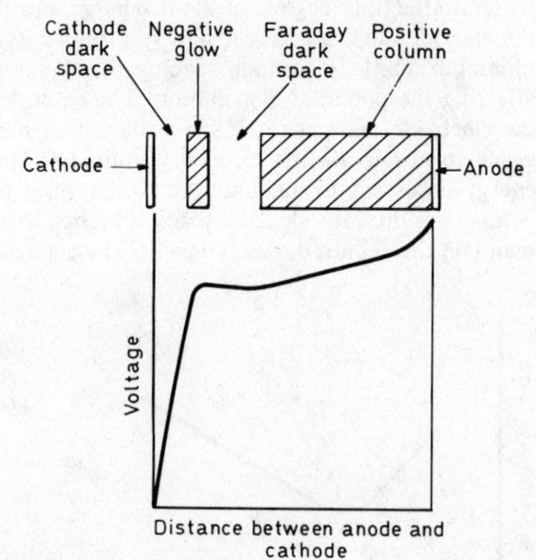

Figure 7.156 Appearance of discharge in normal-glow region, and corresponding potential distribution

Figure 7.156, to give a steep potential gradient between the cathode and virtual anode. Electrons emitted from the cathode by bombardment of the positive ions enter a strong electric field, and are given sufficient energy to cause ionisation despite the lower voltage across the tube. For practical devices, typical values of ignition voltage lie between 80 V and 200 V, and of maintaining voltage between 60 V and 180 V. The current value at ignition is of the order of 1 μA.

The discharge current increases after ignition (since it is not inherently self-limiting and in practice a limiting resistor is required), while the maintaining voltage remains reasonably constant. The characteristic is now determined by the current flow. The region EF is called the *normal-glow region* as the discharge is now visible. Initially the glow discharge covers only a small part of the cathode area. The current density in the discharge remains constant, being a function of the gas and cathode material, and as the current is increased the glow covers more and more of the cathode area. The cross-sectional area of the discharge adjusts itself so that the potential distribution in the cathode region leads to the maximum ionisation efficiency. When the glow covers the whole of the cathode area, any further increase in the current takes the tube into the *abnormal-glow region*, FG. The maximum ionisation efficiency can no longer be maintained, and so the voltage across the tube starts to rise.

Any further increase in the current causes the voltage across the tube to rise to a value greater than the ignition voltage, eventually reaching point G at which an arc is struck between the electrodes, point H. The voltage across the tube is typically 20 V, and the current density is very high.

The appearance of the discharge in the normal-glow region, and the corresponding potential distribution, is shown in *Figure 7.156*. The discharge consists of bright and dark areas, the bright areas being crosshatched in the figure. Near the cathode, is a dark region called the *cathode dark space* or *Crookes dark space*. There is then a short bright region called the *negative glow*, which is followed by a second dark space, the *Faraday dark space*. The Faraday dark space is longer than the Crookes dark space. A long bright area called the *positive column* extends to the anode. In addition, a glow on the cathode may be visible, caused by the excitation of gas atoms by the positive ions bombarding the cathode.

The potential distribution across the tube shows that most of the applied voltage is dropped across the cathode dark space, the *cathode fall*. If the length of the tube is decreased by moving the anode towards the cathode, this potential distribution initially is not greatly affected. Both the positive column and Faraday dark space can be eliminated (as is done by suitable electrode design in some practical cold-cathode tubes) without affecting the rest of the discharge and the operation of the tube. The length of the cathode dark space is inversely proportional to the gas pressure, and represents a minimum anode-to-cathode spacing that cannot be encroached by the anode without affecting the potential distribution. The cathode fall is a function of the gas filling and the electrode material; it is the voltage that must be applied across the cathode dark space to accelerate the electrons emitted from the cathode to acquire the ionisation energy necessary to maintain the discharge. If the distance between anode and cathode is less than the cathode dark space, a higher voltage has to be applied across the tube to maintain the required energy level of the electrons.

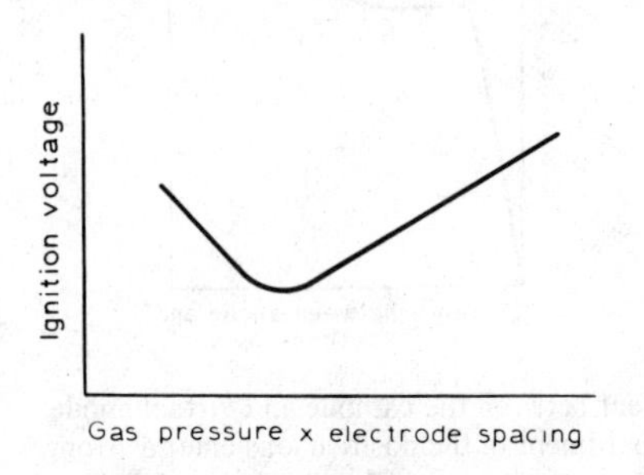

Figure 7.157 Typical Paschen curve showing relationship between ignition voltage and pressure-spacing product in cold-cathode gas-filled tube

For a given gas filling, the ignition voltage V_{ig} is a function of the gas pressure and the distance between the electrodes. This relationship is known as Paschen's Law, and strictly speaking, applies only for plane parallel electrodes. A typical Paschen curve showing the form of the relationship is shown in *Figure 7.157*. At a relatively high pressure, a positive ion loses considerable energy in frequent collisions with the gas atoms which are present in relatively large numbers. On the other hand, at relatively low pressures where there are fewer gas atoms, the number of ionising collisions are small. With a large interelectrode spacing, a large number of collisions between a positive ion and gas atoms occur, causing energy losses; with a small spacing the number of ionising collisions will again be small. For these reasons, the pressure-distance product can be chosen so that the ignition voltage required is near the minimum for that gas.

Although the discussion of gas discharges just given is greatly oversimplified, it does give an outline of the complex processes that occur in the various types of cold-cathode gas-filled tube.

Application to practical cold-cathode tubes

The previous section has indicated how the characteristics of a cold-cathode tube are determined by the gas filling and the electrode material, the gas pressure, and the interelectrode spacing. The usual electrode materials for practical cold-cathode tubes are nickel, molybdenum, and stainless steel, and the usual gas fillings are neon with a small quantity of argon added and helium with a small quantity of neon. The additional materials lower the ignition voltage of the main gas of the mixture, the Penning Effect. In the discharge, the atoms of the main gas are raised to metastable energy levels but they do not immediately lose their extra energy as light. If the added gas has an ionisation energy less than the metastable level of the main gas, the metastable atoms give up their energy to ionise the added gas atoms. The ignition voltage is therefore less than it would be for any of the gases alone. Such mixtures of gases are called *Penning mixtures*. Paschen curves can be derived for these mixtures, so that an ignition voltage can be determined on the basis of the gas pressure and interelectrode spacing. The typical gas pressure for a cold-cathode tube is 40 torr.

If a voltage greater than the ignition voltage is placed across the cold-cathode tube, the discharge does not generally appear immediately. No current flows until at least one ion has been formed by natural radiation in such a position between the electrodes that gas amplification can occur. The formation of such an ion is a random occurrence, and so no absolute time between the application of the voltage and the ignition of the tube can be stated. Instead, the published data for the tube often gives a curve of the probability that a tube will ignite in less than the time shown—the longer the time, the greater the probability that the tube will ignite in a shorter time. This time is called the *statistical delay time*, the *ignition delay time*, or the *ionisation time*. If a voltage greater than the ignition voltage is placed across the tube, the ignition delay time will be decreased.

To reduce statistical delay and to ensure reliable ignition of the tube, priming is often used. This usually takes the form of an auxilliary discharge between the cathode and a priming anode in the tube. This discharge is always present when the tube is in circuit, but the current through the discharge is limited to some tens of microamperes by a large series resistor. The priming discharge is not sufficient to cause the formation of the main discharge unless the required anode-to-cathode ignition voltage is applied across the tube. Electrons and positive ions can escape from the priming discharge, and so are present between the anode and cathode to initiate the main discharge when required. In another method of priming, a small quantity of radioactive material is incorporated in the tube to supplement the ionisation caused by natural radiation. It is also possible to coat the electrodes with a photoemissive material so that secondary electrons can be present for priming. This method is not so frequently used as the other two.

When a tube has been operating for some time and the maintaining voltage is removed, current ceases almost immediately. If a voltage greater than the maintaining voltage but lower than the ignition voltage is reapplied across the tube, the discharge may reform because of the residual ionisation. A minimum time must therefore elapse between the removal of the maintaining voltage and the reapplication of the voltage if the tube is not to be reignited. This time is called the *deionisation time* or *recovery time*. The value depends on the current that has previously been flowing through the discharge since a high current produces a larger number of ions than a low current. The time also depends on the magnitude and rise time of the reapplied voltage. Typical values of deionisation time are of the order of 1 ms or less.

A term that is sometimes encountered is *extinction voltage*. This is not a precisely-defined characteristic of a cold-cathode tube (as ignition voltage and maintaining voltage are), but rather a function of the circuit in which the tube is operating. Extinction voltage represents the value to which the anode-to-cathode voltage of the tube is reduced in the circuit to ensure the interruption of the discharge. The value of the extinction voltage is less than the maintaining voltage of the tube.

Cold-cathode tubes operating in the normal-glow region have cathodes made either from pure molybdenum or nickel, or coated with a material of low work function. Tubes using coated cathodes have lower ignition voltages than those using pure metal cathodes because of the lower work function. Ambient light causes a considerable amount of photoemission from the cathode, and the photoelectrons act as priming ions to produce prompt ignition of the discharge. However, coated cathodes can be easily damaged if sputtering occurs. (Sputtering is the removal of material from the cathode as a result of atoms being released by the positive ion bombardment, and their subsequent deposition on nearby surfaces.) A certain amount of sputter always takes place in a cold-cathode tube, and the amount depends on the current flowing. A high current produces a greater effect than a low current. As a result, tubes using coated cathodes generally have a shorter life than those using pure metal cathodes. Sputtering should be minimised, particularly in multi-electrode tubes, because of the risks of electrical leakage or short-circuits. It is, however, used in the manufacture of some cold-cathode tubes. A heavy current is passed through the tube to remove surface material from the cathode. This leaves the operating surface of the cathode as a pure metal free from any contamination that may have occurred during assembly.

The form of the voltage/current characteristic of a cold-cathode tube indicates three applications in which practical tubes can be used. The constant ignition voltage of the tube enables it to be used as a trigger device, the ignition voltage being determined during manufacture so that a range of tubes with different trigger voltages can be produced. The relatively-constant maintaining voltage despite changes in discharge current enables the tube to be used as a stabilising device. The fact that a visible discharge is produced in the normal-glow and abnormal-glow regions enables the tube to be used as an indicating device.

Indicator diode

The simplest form of cold-cathode tube is the indicator diode or 'neon lamp'. Such a diode can be made simply, often consisting of two wire electrodes enclosed in a small glass envelope filled with neon. The diode is connected across a supply so that the voltage placed across the diode when the supply is switched on exceeds the ignition

Figure 7.158 Circuit symbol for cold-cathode gas-filled diode

voltage. A glow discharge is produced, providing a visual indication that the supply is on. Because the diode is operating in the normal-glow region, an external resistor is necessary to limit the current through the device to a safe value. Indicator diodes can operate on a.c. or d.c. supplies. On a d.c. supply, the glow forms round the wire electrode which acts as the cathode, while on an a.c. supply the glow forms round the wire which is the cathode for the particular half-cycle.

The circuit symbol for a cold-cathode indicator diode is shown in *Figure 7.158*. The black dot indicates the gas filling. An older symbol uses crosshatching inside the envelope to indicate the gas filling.

Voltage stabiliser tube

The relatively constant maintaining voltage of a cold-cathode tube enables these tubes to be used as voltage stabilisers. The operation is similar to that of the indicator diode just described, except that a more rigid electrode structure is used to obtain a reproducible maintaining voltage between tubes. The usual type of structure is a cylindrical cathode surrounding a straight wire anode. The usual gas filling is neon with a small amount of argon. The electrode geometry can be varied to enable a range of tubes with different maintaining voltages to be manufactured. Typical voltage values of practical tubes lie between 75 V and 150 V. The maintaining voltage remains reasonably constant for a range of current through the tube. Typical current ranges are from 5 mA to 15 mA, or from 2 mA to 60 mA. The published data for voltage stabiliser tubes generally includes curves of the variation of maintaining voltage with current, as well as the range of maintaining voltage for the particular tube type.

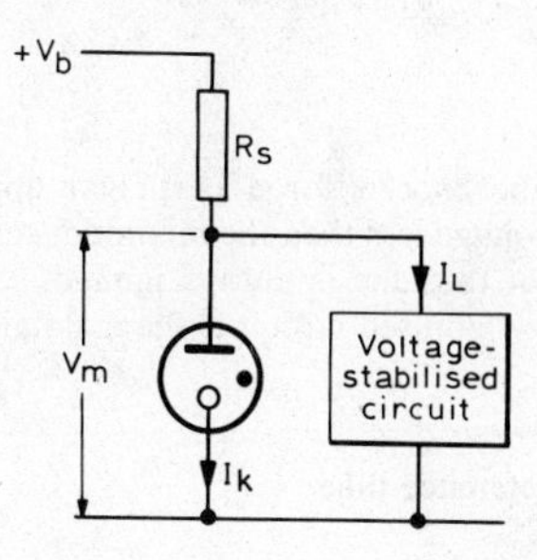

Figure 7.159 Typical voltage stabiliser circuit using cold-cathode tube

As with the indicator diode, a series resistance is required to limit the current through the stabiliser tube to a safe value. A typical stabiliser tube circuit is shown in *Figure 7.159*. The current through the tube I_k and the current through the voltage-stabilised circuit I_L are related to the value of the resistor R_s and the supply voltage V_b by the expression:

$$I_k + I_L = \frac{V_b - V_m}{R_s} \tag{1}$$

where V_m is the maintaining voltage of the tube. In the design of a practical stabiliser stage, the effects of tolerances in the resistor R_s, and the supply and maintaining voltages, must be taken into account. Thus equation 1 can be rewritten as

$$I_{k\,\max} + I_{L\,\min} = \frac{V_{b\,\max} - V_{m\,\min}}{R_{s\,\min}} \tag{2}$$

and

$$I_{k\,\min} + I_{L\,\max} = \frac{V_{b\,\min} - V_{m\,\max}}{R_{s\,\max}} \tag{3}$$

From equations 2 and 3, expressions for the maximum and minimum values of the series resistor can be derived:

$$R_{s\,\max} = \frac{V_{b\,\min} - V_{m\,\max}}{I_{k\,\min} + I_{L\,\max}}$$

and

$$R_{s\,\min} = \frac{V_{b\,\max} - V_{m\,\min}}{I_{k\,\max} + I_{L\,\min}}$$

The current through a voltage stabiliser tube varies with the supply voltage. The area of the glow discharge on the cathode therefore varies in response to the current variation. In addition, the glow also 'wanders' over the cathode surface, although the reason for this is not fully understood. If the glow reaches a discontinuity in the cathode surface, the wandering or increase in area of the glow discharge is temporarily halted. The discontinuity will be suddenly bridged, and there is a jump in the maintaining voltage characteristic as shown in *Figure 7.160*. The magnitude of the voltage jump varies with the type of tube, but typical values lie between 15 mV and 250 mV.

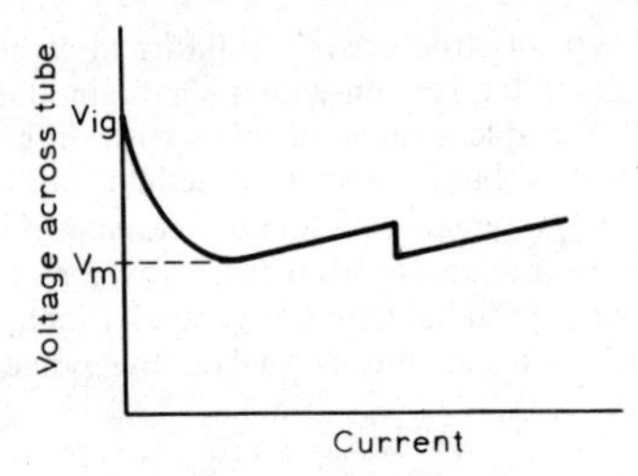

Figure 7.160 Variation of maintaining voltage with tube current showing 'jump'

The tube chosen for a particular application must have the maximum value of ignition voltage less than the minimum supply voltage applied across the tube. This is to ensure that the tube is always ignited, so that the discharge is struck and the voltage across the stabilised circuit is the maintaining voltage of the tube.

Voltage reference tube

A special form of the voltage stabiliser tube is the voltage reference tube. In this type, the current through the tube is maintained at a constant value to give a constant maintaining voltage which can be used as a voltage reference. In practice, the current can be varied over a restricted range. The corresponding change in maintaining voltage is considerably smaller than that obtained with a stabiliser tube. For example, the maximum change of maintaining voltage for a nominal 85 V reference tube operating at 2.0 mA is 3 V for a current change from 0.5 mA to 3.5 mA. In a practical circuit, the current is closely controlled so that the actual voltage change is much smaller than this.

The higher performance required from a reference tube is obtained by greater control over the construction and manufacture of the tube. In a typical voltage reference tube, the electrodes are made from pure molybdenum, and a film of molybdenum is sputtered on to the inside surface of the glass envelope during manufacture. This film decreases the outgassing from the glass, thereby maintaining the gas filling of the tube in its original pure state for a longer time.

To a large extent, both voltage stabiliser tubes and voltage reference tubes have been replaced by solid-state devices. Semiconductor voltage regulator diodes and voltage reference diodes (both also known as zener diodes) are used in solid-state equipment where the normal supply voltages are below the operating range of gas-filled devices.

Trigger tube

A cold-cathode trigger tube contains three electrodes. The additional electrode, the trigger, is placed close to the cathode so that the trigger-to-cathode gap is less than the anode-to-cathode gap. The trigger–cathode ignition voltage is therefore lower than the anode–cathode ignition voltage. Ions from the trigger-to-cathode discharge can initiate the main anode-to-cathode discharge. Thus if the anode voltage of the tube is held at a value greater than the maintaining voltage but below the ignition voltage, the main discharge can be ignited by a relatively small voltage applied to the trigger that exceeds the trigger ignition value. As the current in the main discharge can be much larger than that in the trigger-to-cathode discharge, a low-power input signal can be applied to the tube to operate higher-power circuits such as relay drive circuits.

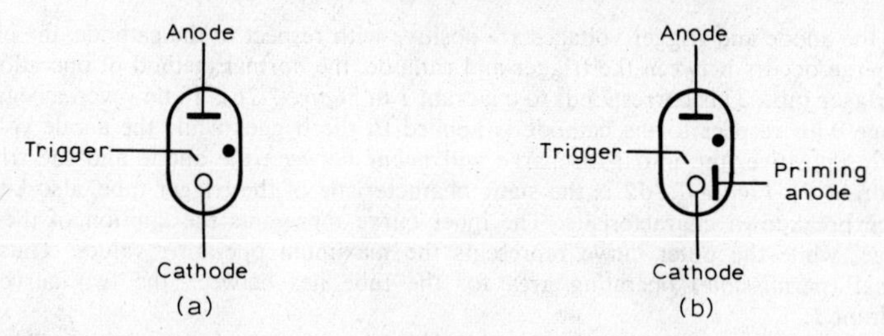

Figure 7.161 Circuit symbol for cold-cathode trigger tube: (a) basic tube (b) tube with priming anode

The circuit symbol for a cold-cathode trigger circuit is shown in *Figure 7.161* (*a*). To ensure reliable ignition of the tube, a priming electrode may be incorporated. The circuit symbol of a trigger tube with a priming anode is shown in *Figure 7.161* (*b*).

Although the usual method of operation of a trigger tube is that just described, it is possible for discharges to be ignited between the electrodes in other ways. To prevent spurious operation of a circuit, the designer must ensure that unwanted ignition and modes of operation cannot occur. The anode-to-cathode voltage is plotted against trigger-to-cathode voltage for a typical trigger tube in *Figure 7.162*. If, for example,

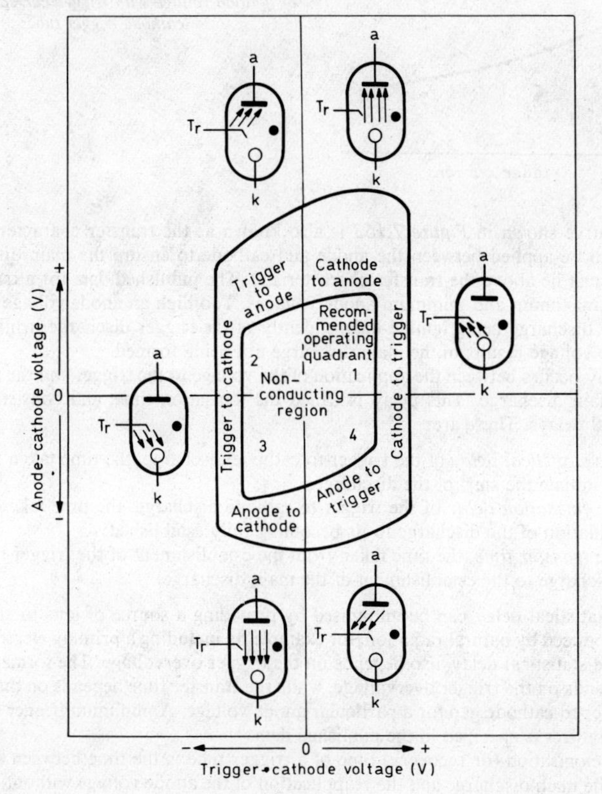

Figure 7.162 Static (or breakdown) characteristic of cold-cathode trigger tube

both the anode and trigger voltages are positive with respect to the cathode, the initial discharge occurs between the trigger and cathode, the normal method of operation of the trigger tube. This corresponds to quadrant *1* in *Figure 7.162*. If, however, a negative voltage with respect to the cathode is applied to the trigger while the anode voltage remains positive, the initial discharge will occur between the anode and the trigger (quadrant *2*). *Figure 7.162* is the static characteristic of the trigger tube, also known as the breakdown characteristic. The inner curve represents the ignition of the discharge, while the outer curve represents the maximum operating values. Thus the normal (permissible) operating area for the tube lies between the two curves in quadrant *1*.

As the discharge between the trigger and cathode provides ions to initiate the main anode-to-cathode discharge, the higher the trigger current the greater the number of ions produced. Therefore the voltage required between the anode and cathode to initiate the main discharge is reduced as the trigger current increases. The relationship between anode ignition voltage and trigger current for a typical trigger tube is shown in *Figure 7.163*. A small pre-ignition trigger current can also flow if a bias voltage is applied to the trigger, but in practical trigger tubes this is typically only 10^{-9} A.

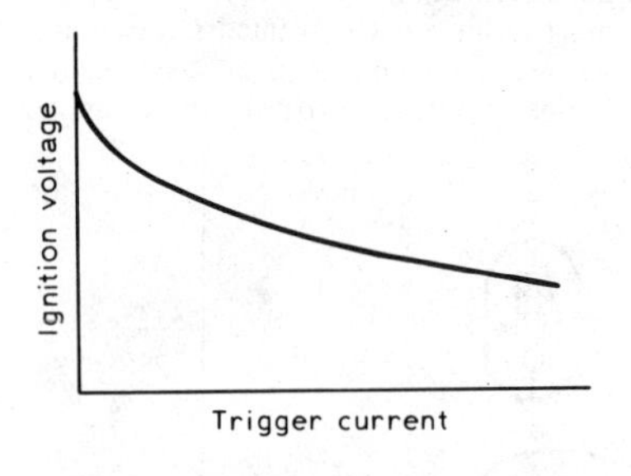

Figure 7.163 Transfer characteristic (variation of ignition voltage with trigger current) for cold-cathode trigger tube

The curve shown in *Figure 7.163* is also known as the transfer characteristic. The voltage to be applied between the anode and cathode to ensure the main discharge is ignited must lie above the transfer characteristic. The published data for a trigger tube specifies maximum and minimum anode voltages. Too high an anode voltage results in the main discharge being ignited independently of the trigger discharge, while too low an anode voltage results in the main discharge not being formed.

A delay occurs between the application of the voltage to the trigger and the formation of the main discharge. This delay is called the ionisation time, and consists of three individual delays. These are:

(*a*) the *statistical delay* of the trigger-to-cathode discharge, the time taken for an ion to initiate the start of the discharge;
(*b*) the *formation delay* of the trigger-to-cathode discharge, the time taken from the initiation of the discharge to its becoming fully established;
(*c*) the *transfer time*, the time taken from the establishment of the trigger-to-cathode discharge to the establishment of the main discharge.

The statistical delay can be minimised by providing a source of ions to supplement those produced by natural radiation, for example by including a priming electrode in the tube. The statistical delay also depends on the trigger overvoltage. The formation delay also depends on the trigger overvoltage, while the transfer time depends on the power in the trigger-to-cathode gap for a particular anode voltage. A minimum trigger current to ensure transfer is specified in the published data.

The deionisation (or recovery) time of a trigger tube is the time between the extinction of the main discharge and the reapplication of the anode voltage without reforming the discharge. It represents the time required for the space charge within the tube to

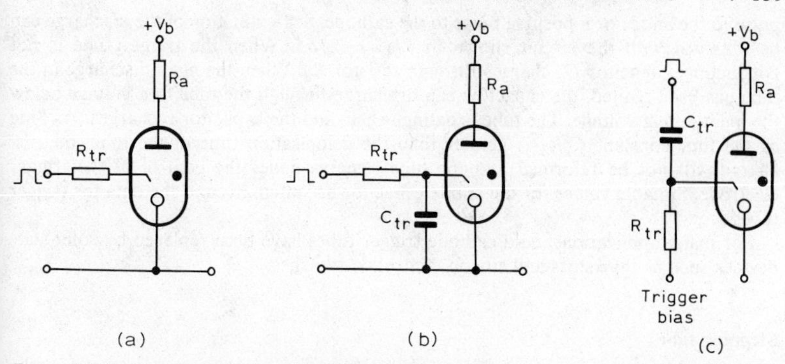

Figure 7.164 Input circuits for cold-cathode trigger tube: (a) current triggering (b) pulse triggering (c) pulse triggering with bias

disperse so that control of the tube has returned to the trigger electrode. The deionisation time therefore determines the frequency of operation of the tube.

The values of ionisation and deionisation times depend on the construction of the tube and the operating conditions. For a tube using a priming electrode, the ionisation time is typically a few milliseconds, say 2 ms. Without the priming, the ionisation time can be as long as 5 s. For tubes using coated cathodes, the ionisation time is shorter, typically 25 μs, although in darkness the time can increase to 250 μs. The value of deionisation time depends on the current the tube has been conducting. Typical values are 3.5 ms for a tube conducting 20 mA and 12 ms for 100 mA, and for tubes with coated cathodes where the current is generally smaller the deionisation time is typically 500 μs.

Typical input circuits for trigger tubes are shown in *Figure 7.164*. Resistor R_a is the current-limiting resistor for the main discharge, and resistor R_{tr} that for the trigger-to-cathode discharge. In *Figure 7.164 (a)*, the voltage pulse applied to the input $V_{tr(b)}$ exceeds the trigger ignition voltage and initiates triggering. This method is called current triggering, and the resulting current must be sufficient to cause transfer of the discharge to the anode and cathode. Pulse triggering is shown in *Figure 7.164 (b)*. Capacitor C_{tr} is charged until its voltage exceeds the trigger ignition value. The capacitor then discharges to initiate triggering. The charge stored in the capacitor must be sufficient to ensure the discharge current exceeds the minimum transfer value. Capacitor triggering with an applied trigger bias is shown in *Figure 7.164 (c)*. The bias voltage must be less than the trigger ignition voltage, but the triggering pulse can then be of much smaller amplitude than in the other methods.

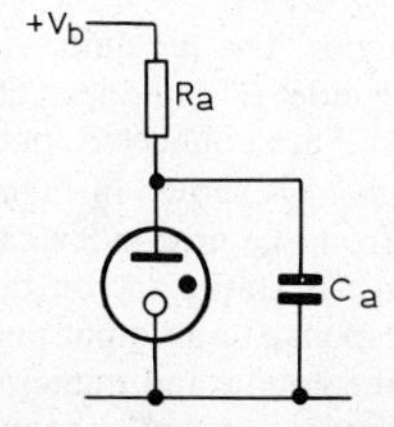

Figure 7.165 Self-extinguishing circuit for cold-cathode trigger tube

When the main discharge has been ignited, the trigger has no effect on the tube. The main discharge can only be extinguished by interrupting the current in the circuit outside the tube, or by reducing the anode-to-cathode voltage of the tube to below the maintaining value. This reduction in voltage can be achieved by applying a negative pulse to the anode or a positive pulse to the cathode. Self-extinction of the discharge can be obtained with the circuit shown in *Figure 7.165*. When the trigger tube is not conducting, capacitor C_a charges through resistor R_a. When the main discharge in the tube has been ignited, the capacitor can discharge through the tube to a voltage below the maintaining voltage. The tube is extinguished, and the capacitor recharges. As long as the time-constant C_aR_a is greater than the deionisation time of the tube, the discharge will not be reformed and the tube remains under the control of the trigger electrode. Suitable values for the anode capacitor are often given in the data for trigger tubes.

For many applications, cold-cathode trigger tubes have been replaced by solid-state devices such as thyristors and silicon controlled switches.

Stepping tube

Cold-cathode stepping tubes provide a simple method of pulse counting at moderate speeds with a visible indication of the count as it proceeds. Stepping tubes are also known as Dekatrons, although this is a proprietary name of Ericsson Telephones Limited who introduced the first stepping tube in 1949.

The electrode structure of a decade stepping tube consists of a central disc-shaped anode surrounded by thirty nickel-rod cathodes. The spacing between cathodes is approximately 2 mm, while the anode-to-cathode spacing is typically 5 mm. This structure is viewed 'end on' through the dome of the glass envelope of the tube. The cathodes are divided into three groups: index (or main) cathodes, guide *A* cathodes, and

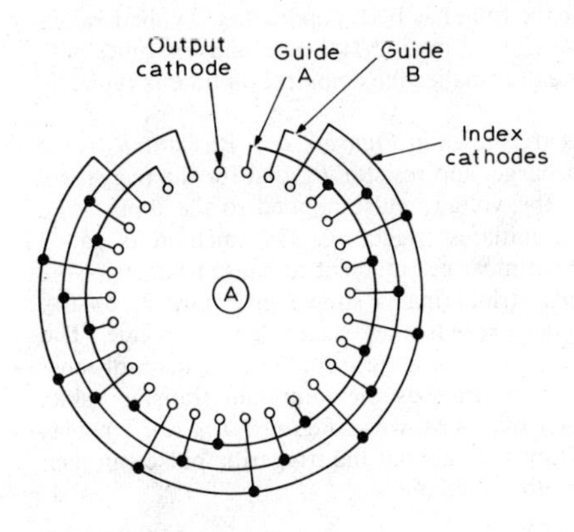

Figure 7.166 Electrode structure for decade stepping tube

guide *B* cathodes. The ten guide *A* cathodes are connected together inside the tube, as are the ten guide *B* cathodes. Of the index cathodes, those corresponding to the numerals 1 to 9 are connected together, while the '0' cathode is brought out separately. This arrangement is shown in *Figure 7.166*, and it can be seen that each index cathode is separated from the next index cathode by one guide *A* and one guide *B* cathode. The glow discharge is stationary on the index cathodes only, and 'steps' to the next index cathode in response to an input pulse. Thus the position of the glow discharge indicates the state of the count, and numerals on a panel-mounted escutcheon corresponding to the index cathodes enable the count to be read off.

A stepping tube operating in this way is called a decade counting tube. The circuit symbol for this type of tube is shown in *Figure 7.167*. Another form of the tube has the individual index cathodes brought out so that an output can be obtained from each index cathode. This form of stepping tube is called a selector tube, and the circuit symbol is shown in *Figure 7.168*. Both forms of stepping tube are called 'double-pulse' tubes from the method of operation.

A supply voltage of typically +500 V is applied to the anode of the tube, which is greater than the anode-to-cathode ignition voltage. The two groups of guide cathodes have a positive bias of typically +40 V with respect to the index cathodes. The discharge is ignited between the anode and an index cathode since the discharge seeks

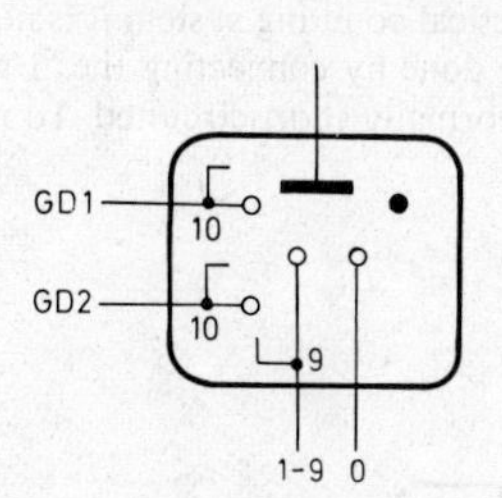

Figure 7.167 Circuit symbol for decade stepping tube

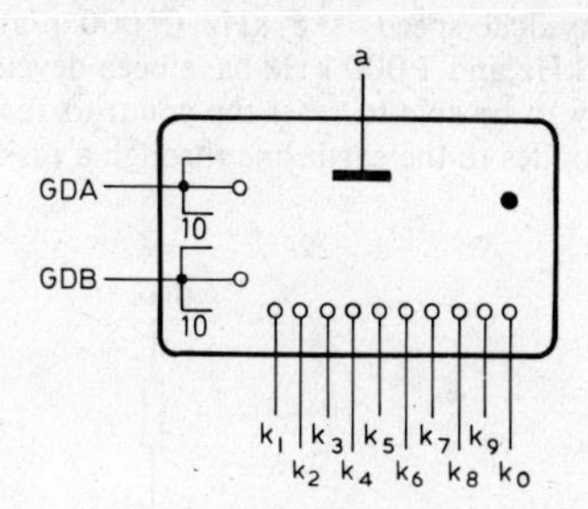

Figure 7.168 Circuit symbol for decade selector tube

the most negative electrode. The anode-to-index cathode voltage falls to the maintaining value, typically 200 V. The ions formed by the discharge prime the two adjacent cathodes to the conducting index cathode, thereby reducing the required anode-to-cathode ignition voltage. The amount of priming to cathodes which are not adjacent to the index cathode is minimised. In practical tubes, the required ignition voltage of a cathode adjacent to the conducting cathode is about 10 V above the maintaining voltage, while that of a cathode two positions away is 40 V above the maintaining voltage.

A negative-going pulse of approximately 100 V is applied to the guide *A* cathodes. When the potential on the guide *A* cathode adjacent to the conducting index cathode is about 10 V more negative than the index cathode, the discharge will transfer to that cathode. The guide *B* cathode originally adjacent to the conducting index cathode has not changed its potential, and so the discharge is not transferred to that cathode. The anode voltage falls with the negative-going pulse since the voltage between the conducting guide *A* cathode and the anode will remain constant at the maintaining value. Thus the voltage between the anode and the originally-conducting index cathode falls below the maintaining value so that the discharge is completely transferred to the guide *A* cathode. The discharge now primes the two newly adjacent cathodes. Shortly before the end of the first negative-going pulse, a second negative-going pulse of 100 V is applied to the guide *B* cathodes. The discharge transfers to the primed guide *B* cathode, the fall of anode voltage with the pulse ensuring that the discharge cannot return to the previously-conducting guide *A* cathode. Again the adjacent cathodes, including the index cathode next in the counting sequence, are primed. As the voltage on the conducting guide *B* cathode rises at the end of the pulse, the discharge transfers to the primed index cathode because of the positive bias on the guide *B* cathode. In this way, the discharge is moved 'forward' by one position in the count. It can be seen, however, that if the first negative-going pulse is applied to the guide *B* cathodes, and the second to the guide *A* cathodes, the discharge moves in the opposite direction from that just described. The count moves 'backwards' by one position. The double-pulse stepping tube can therefore operate as a reversible counter.

In practice, only one input pulse is required to step the tube one position. A typical drive circuit is shown in *Figure 7.169*. The resistor *R* and capacitor *C* form an integrating circuit which effectively delays the pulse applied to the guide *B* cathodes. Also, the charge on the capacitor maintains the guide *B* cathodes negative for a time after the pulse has been removed from guide *A* cathodes. The conditions required for stepping are therefore fulfilled.

The '0' cathode is brought out separately so that a load resistor can be incorporated in the connection to the earth line. The flow of current through the resistor when the discharge reaches the '0' cathode produces a voltage pulse which can be transferred to a second stepping tube to count 'tens'. A third stepping tube can be connected into the register to count 'hundreds', and so on. The operating speed depends on the type of tube. A typical speed is 5 kHz (5 000 p/s), but higher-frequency tubes operating up to 50 kHz and 1 000 kHz have been developed. In a practical counting system it is necessary to be able to reset the count to zero. This can be done by connecting the '1 to 9' cathodes to the earth line through a resistor which is normally short-circuited. To reset

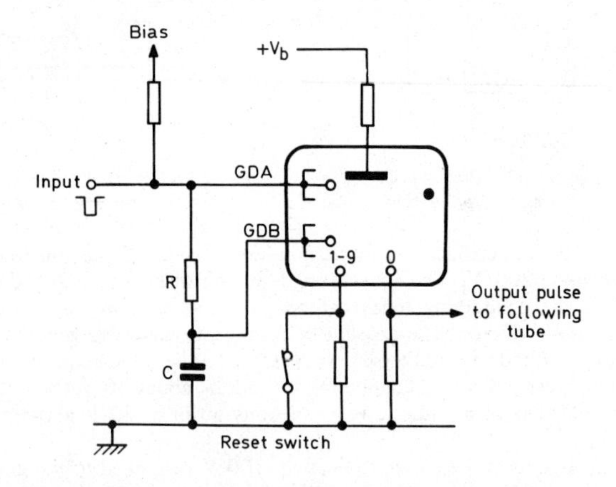

Figure 7.169 Drive circuit for decade stepping tube

the counter, the short-circuit across the resistor is removed, and the flow of current through the resistor from the discharge raises the potential of all these cathodes above earth potential. The '0' cathode is therefore more negative than the others, and the discharge transfers to this cathode.

A second type of stepping tube, called the 'single-pulse' stepping tube from its method of operation, has forty cathodes. Three guide cathodes separate each index cathode. The single negative-going input pulse is applied through a parallel *RC* network to the guide *A* cathodes, and directly to the guide *B* cathodes. In the absence of an input pulse, the guide *A* and *B* cathodes are held at a positive bias. The guide *C* cathodes are connected to earth through a resistor.

When the input pulse is applied to the guide *A* cathodes, the discharge is transferred from the index cathode to the guide *A* cathode. The current flowing through the resistor connected to the guide *A* cathode raises its potential above that of the guide *B* cathode, and so the discharge is transferred to that cathode. At the end of the input pulse, the guide *B* cathodes return to their positive bias, and so the discharge is transferred to the guide *C* cathode which is at earth potential. The flow of current through the resistor connected to this cathode raises its potential above earth, and so the discharge is transferred to the next ('forward') index cathode. The single-pulse stepping tube cannot be used for reverse counting.

Both types of stepping tube have been largely replaced by digital counters using integrated circuits.

Numerical and character indicator tubes

Numerical indicator tubes (n.i.t.s) represent the most extensively used cold-cathode tube at the present time. Although semiconductor light-emitting diodes (l.e.d.s) are coming into use as read-out devices, n.i.t.s still provide a cheap and reliable device which can be incorporated into a wide variety of read-out systems.

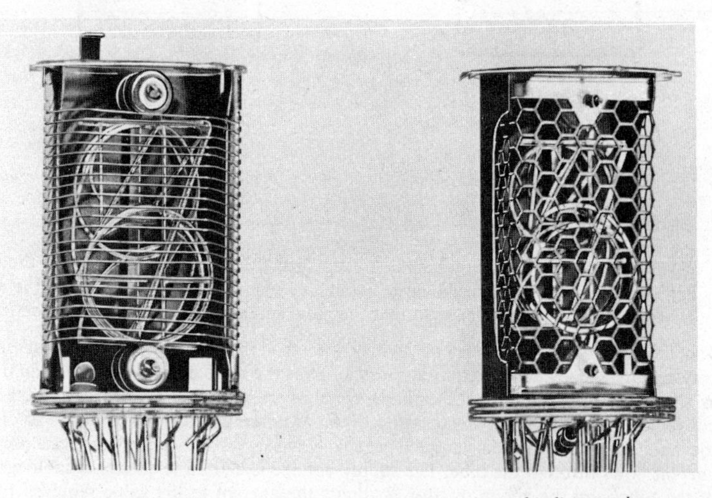

Figure 7.170 Electrode structures of side-viewing numerical indicator tubes (Photo: Mullard Limited)

In a typical n.i.t., ten thin metal cathodes (usually of nickel) are formed into the shape of the numerals 0 to 9. These cathodes are mounted on supporting rods, and separated from each other by small insulating beads. The stack is surrounded by a cylindrically-shaped anode. In side-viewing n.i.t.s, a side of the anode is formed of a wire mesh so the cathodes can be seen. In end-viewing n.i.t.s, the anode is a complete cylinder with a mesh top through which the cathodes are seen. Electrical connections are made to the anode and the individual cathodes. The gas filling used is neon with a small amount of argon, to which mercury is added for long-life tubes. Connections to the tube can be made by base pins or flying leads, and tubes with rectangular envelopes are available to allow closer spacing of the tubes in a display. The electrode structures of two typical n.i.t.s are shown in *Figure 7.170*. The height of the numerals is typically between 10 mm and 15 mm.

N.I.T.s are operated in the abnormal-glow region so that the glow discharge completely surrounds the cathode wire. The diameter of the wire is typically 0.4 mm, and this is surrounded by a glowing 'sheath' to a depth of approximately 1 mm. Thus the illuminated cathode appears as a glowing numeral 2 mm wide. Unilluminated cathodes in front of it in the stack therefore do not seriously obscure a glowing cathode.

Cathodes can be made in the shape of arithmetical symbols or any other required shape. Tubes incorporating such cathodes are called character indicator tubes. The circuit symbol for an n.i.t. is shown in *Figure 7.171*, and this symbol can be adapted for a character indicator tube simply by changing the number of cathodes.

The requirements for the operation of an n.i.t. are:

(*a*) a voltage between the anode and the cathode to be illuminated which is initially greater than the ignition voltage;

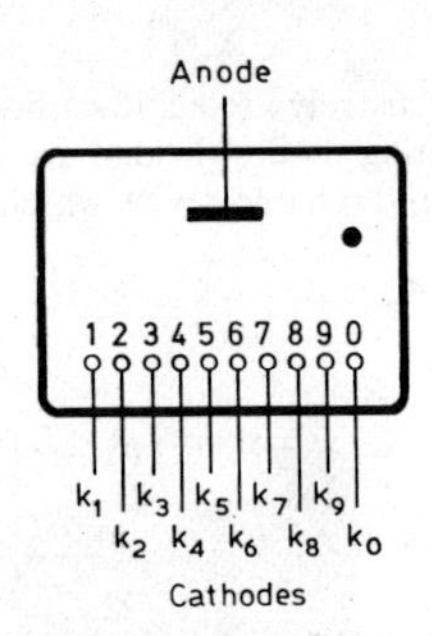

Figure 7.171 Circuit symbol for numerical indicator tube

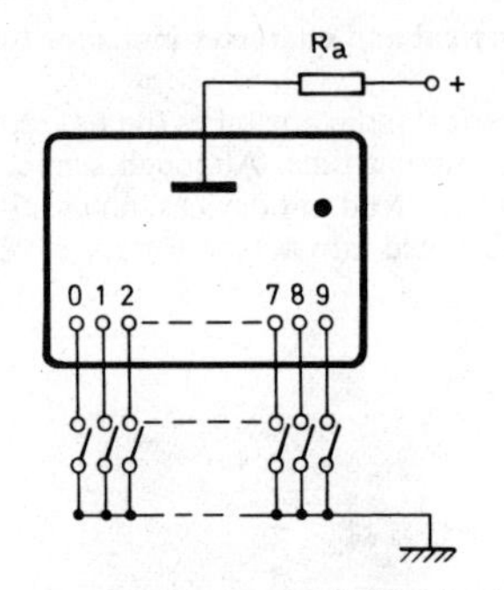

Figure 7.172 Basic display circuit for numerical indicator tube

(*b*) a voltage after ignition which is greater than the maintaining voltage of the tube;
(*c*) a current through the tube large enough to ensure that the glow covers the whole cathode, but not too large so that the glow spreads to the cathode connections.

The individual cathodes must be switched so that the igniting voltage is established between the anode and the required cathode. Thus the basic display circuit for an n.i.t. is as shown in *Figure 7.172*. The cathode switches are shown as mechanical switches for convenience, the switch corresponding to the required cathode being closed to display the numeral. The glow discharge is struck between the anode and this cathode, and current flows through the tube. The supply voltage is chosen to ensure reliable ignition of the tube, and the anode resistor R_a limits the current to the value required for the glow to cover the complete cathode at the maintaining voltage of the tube.

This simple treatment has neglected the effect of the other cathodes on the discharge between the anode and the 'on' cathode. The 'off' cathodes act as probes, taking part of the tube current. If the cathodes are isolated, as in *Figure 7.172* where mechanical switches are used, they will assume a floating potential slightly lower than that of the anode. In practice, the cathode switches are transistors or silicon controlled switches through which a leakage current can flow. The cathode potentials therefore decrease, and a greater proportion of the tube current flows through the 'off' cathodes. This results in a background haze which decreases the readability of the displayed numeral. It is necessary, therefore, to maintain the potential of the 'off' cathodes at a value high enough to ensure clear readability of the numeral. The potential must not be too high, otherwise subsidiary discharges can be formed.

The voltages and currents in a simple n.i.t. circuit like that of *Figure 7.172* are shown in *Figure 7.173*. The tube is operating on a supply voltage V_s, and a discharge has been

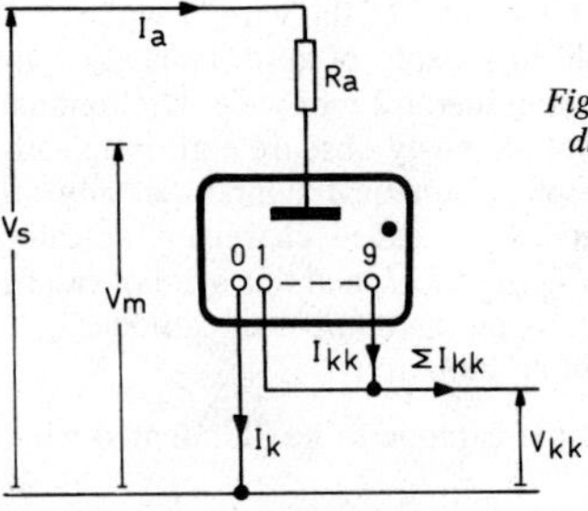

Figure 7.173 Voltages and currents in basic display circuit for numerical indicator tube

formed between the anode and the '0' cathode. A cathode current I_k flows in the '0' cathode circuit, while probe currents I_{kk} of various magnitudes flow in the other cathode circuits to form a total probe current of ΣI_{kk}. The anode current I_a is the sum of the cathode current and the total probe current:

$$I_a = I_k + \Sigma I_{kk}$$

The voltage across the tube is the maintaining voltage V_m, and the value of the anode resistor R_a is:

$$R_a = \frac{V_s - V_m}{I_a}$$

The relationship between the probe current and the voltage difference between the 'on' cathode and the 'off' cathodes, V_{kk}, is called the probe characteristic. The published data for an n.i.t. contains probe characteristics for particular 'on' cathode currents, and a typical probe characteristic is shown in *Figure 7.174*. Curves of the maximum and

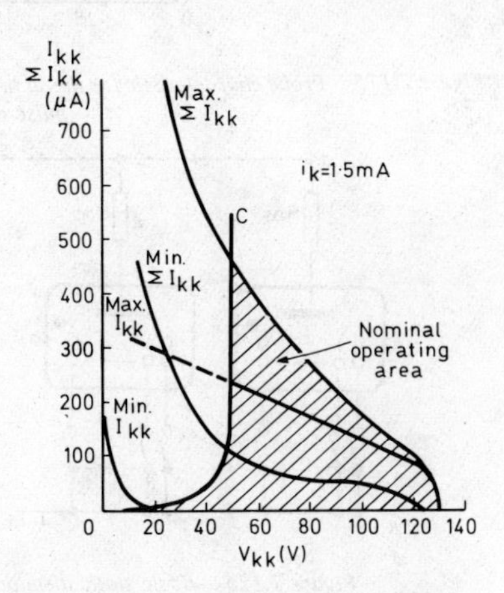

Figure 7.174 Typical probe characteristic for numerical indicator tube, d.c. operation

minimum individual probe current, max I_{kk} and min I_{kk}, are given for an 'on' cathode current of 1.5 mA. Also shown in *Figure 7.174* for comparison are the maximum and minimum curves for total probe current, max ΣI_{kk} and min ΣI_{kk}. The curve labelled *C* corresponds to a reasonable display free from background haze. This curve together with the others defines a nominal operation area. In the characteristic shown, it can be seen that a voltage difference between the 'on' and 'off' cathodes greater than 50 V is required for clear readability of the displayed numeral. Sometimes a second curve is given in the data, corresponding to less-favourable operation but still with an acceptable contrast between the displayed numeral and the background. This second curve defines a 'worst-case' operation area. As will be explained later, n.i.t.s can be operated on a d.c. supply or with pulses. The probe characteristic in *Figure 7.174* is for d.c. operation; the probe characteristic of the same n.i.t. for pulse operation is shown in *Figure 7.175*, and this also includes the curve *D* which defines the worst-case operation area. It can be seen that the voltage difference between the 'on' and 'off' cathodes for clear readability on pulse operation is greater than that for d.c. operation.

Besides the probe characteristics just described, the published data for an n.i.t. includes values of the minimum anode-to-cathode voltage for reliable ignition, the anode-to-cathode voltage below which the tube is extinguished, and the maximum and minimum values of cathode current for d.c. and pulse operation to ensure the glow covers the whole cathode. The curves given include the variation of maintaining voltage

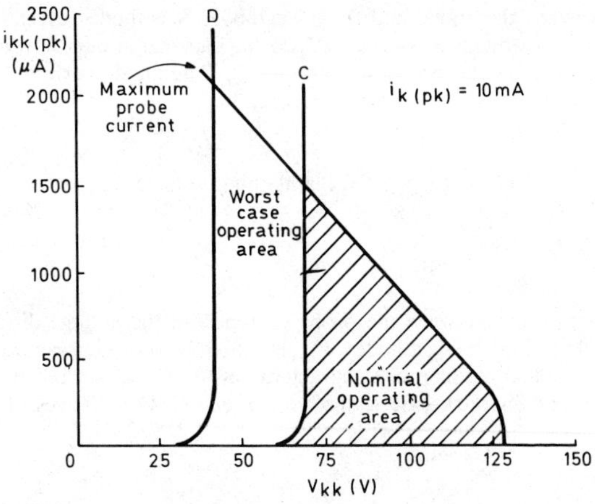

Figure 7.175 Probe characteristic for the same numerical indicator tube as in Figure 7.174 under pulse operation

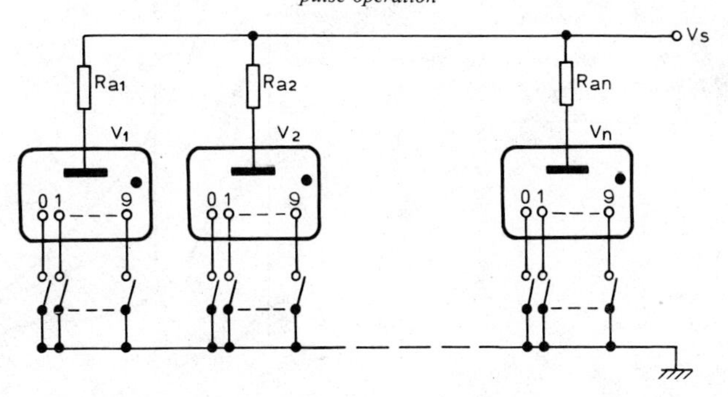

Figure 7.176 Basic static display system using numerical indicator tubes

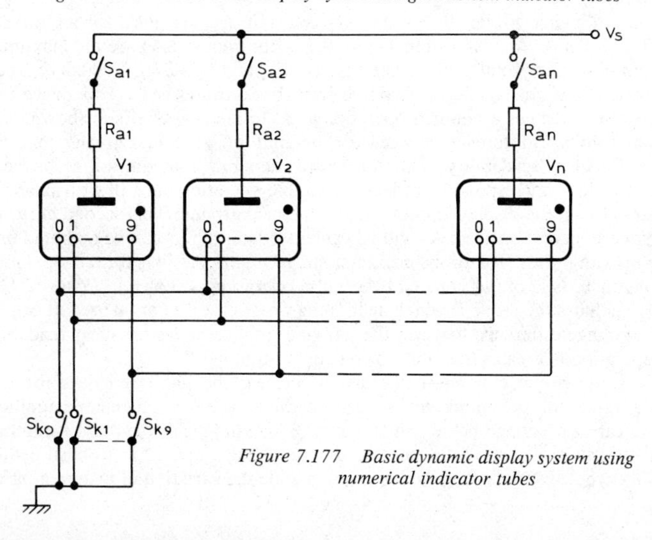

Figure 7.177 Basic dynamic display system using numerical indicator tubes

with cathode current for d.c. and pulse operation, the relationship between the supply voltage and anode load resistor for various cathode currents with d.c. and pulse operation, and the ignition delay time. The minimum duration of the operating pulses is also specified.

In a practical system, several n.i.t.s are used to form a read-out display. The basic display circuit shown in *Figure 7.172* can be converted into a practical system to operate in two ways. One way gives a static display system, in which each cathode is switched individually. The other way gives a dynamic display system, in which all the cathodes representing the required numeral are switched and the position of the numeral is selected by switching the anode of the corresponding tube. In other words, the cathode switches select *which* numeral is to be displayed, and the anode switches determine *where* it is to be displayed. The frequency at which the numerals are switched is sufficiently high for the display to appear continuous.

The basic circuit of a static display system is shown in *Figure 7.176*. Each tube has ten switches, one for each cathode. If there are n tubes in the display system, then 10 n switches are required.

The number of switches can be considerably reduced by using the dynamic system shown in *Figure 7.177*. Only ten cathode switches are used, S_{k0} to S_{k9}, but a switch has to be incorporated in the anode circuit of each tube, S_{a1} to S_{an}. To display a numeral, say 7 on V_2, switches S_{k7} and S_{a2} must both be closed. Thus the reduction in the number of cathode switches is accompanied by the complication of closing the required anode switch. The total number of switches for a display of n tubes is $(10 + n)$ switches.

Dynamic display can be achieved in two ways: by cathode scanning and by anode scanning. In cathode scanning, the cathodes of the n.i.t.s are selected sequentially in a continuous cycle, and the anode switches are closed at the appropriate time to display the required numeral. In anode scanning, it is the anodes which are selected in a continuous sequence and the cathode switches which are closed to display the required numeral. In both dynamic systems, the rate at which the anodes or cathodes are selected must be sufficiently high for the selected numeral to appear to be displayed continuously. To ensure a flicker-free display, the time taken to select all the n.i.t.s in the display must be less than 20 ms. The time taken to address each tube in a display of n tubes is therefore $(20/n)$ ms. The light output from an n.i.t. is proportional to the mean tube current. To obtain the same light output from a dynamic display of n tubes as from a static display, each tube must pass a current pulse n times the d.c. current of the static system. However, with pulse operation the glow around the illuminated cathode is confined to a narrower region than with d.c. operation. Therefore, although for a given mean current the light output from n.i.t.s under pulse and d.c. operation is the same, the displayed numeral of the pulsed n.i.t. will appear brighter. For comparable readability with a static display, the mean current of a dynamic display can be 0.8 times or less that of the static system.

Of the two methods of dynamic display, the cathode-scanning system has two disadvantages. If the same numeral is displayed by all the tubes (the zero reading is a likely occurrence), the peak anode current of all tubes flows through the one cathode switch. The cathode switches must be rated to carry this current. The second disadvantage is that the system cannot be used to display the output of passive registers such as magnetic core memory systems. For these reasons, anode-scanning systems are more generally used.

A cathode representing a decimal point is incorporated into some n.i.t.s. This cathode is operated by a separate drive circuit when required.

The development of small electronic calculators has required a display system more compact than a row of individual n.i.t.s. Several individual n.i.t. electrode structures can be combined in one cylindrical glass envelope to form a multiple indicator tube. Such a tube can contain up to, typically, 14 individual electrode structures (called 'decades') in a tube 180 mm long and 30 mm diameter. A typical numeral height is 10 mm. Such a display system has to be operated dynamically to avoid an excessive number of external

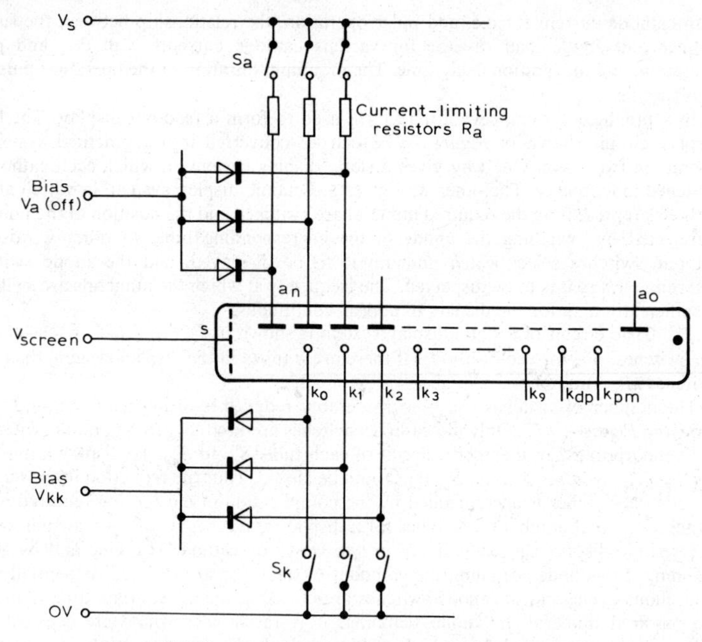

Figure 7.178 Basic drive circuit for multiple indicator tube

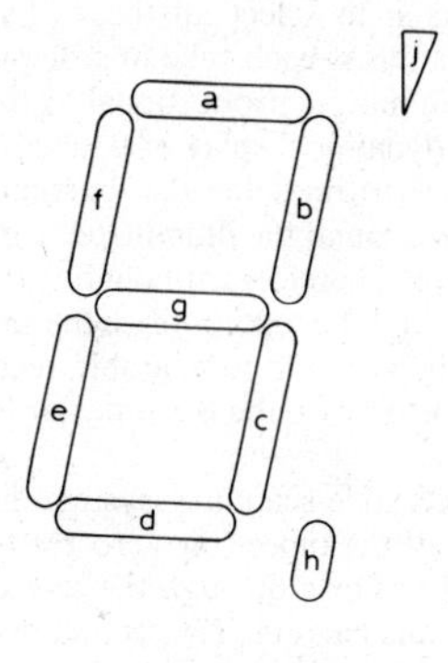

Figure 7.179 Seven-segment display incorporating decimal point and 'punctuation mark'

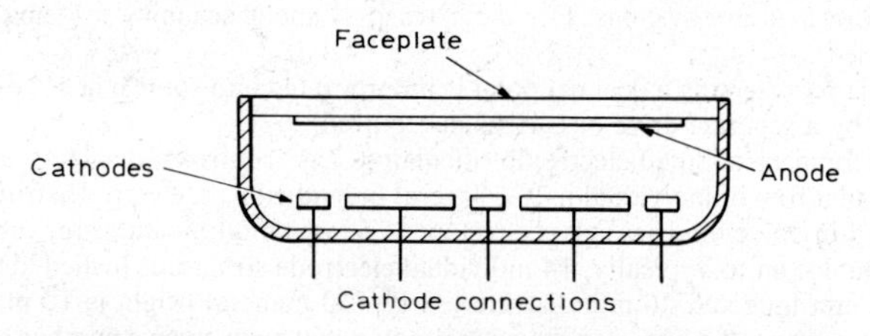

Figure 7.180 Schematic cross-section of segmented display tube

connections. The cathodes representing the same numeral in each decade are connected together so that only ten external cathode connections are required. A separate connection for the anode of each decade is required, together with connections for the decimal point cathodes, the 'punctuation mark' cathodes (which are used to indicate divisions between the 14 displayed numerals to make reading easier), and the screens separating each decade from the neighbouring ones. For a 14 decade tube, this makes a total of 27 external connections.

Anode scanning is used for the tube, with a minimum scanning frequency for a flicker-free display of typically 70 Hz and a maximum frequency determined by the ignition delay time and the time for the glow to form. This gives the minimum anode current pulse duration of typically 50 μs. It is necessary to keep the 'off' anodes as well as the 'off' cathodes biased to prevent them affecting the discharge between the 'on' anodes and cathodes. The numeral cathodes, and the decimal point and punctuation mark cathodes when required, are driven negative with respect to this bias when a numeral is to be displayed, while the required anode is driven positive with respect to the bias. The screens between the decades are maintained at a steady potential to prevent 'crosstalk' between decades. These requirements lead to the basic drive circuit shown in *Figure 7.178*.

Segmented display tube

A more recent form of indicator tube is the segmented display tube or bar matrix tube. In this type of tube the cathodes are not in the form of individual numerals or characters, but are in linear form arranged in a seven-segment display as shown by *a* to *g* in *Figure 7.179*. When a cathode is selected, a voltage greater than the ignition voltage is applied between this cathode and the anode, and a glow is produced along the cathode length. By energising the appropriate cathodes, the required numerals or characters can be formed. Additional cathodes *h* and *j* form a decimal point and a 'punctuation mark' to make the reading of long displays easier.

The schematic cross-section of a segmented display tube is shown in *Figure 7.180*. This shows the length of two cathodes and the cross-section of three others. The cathodes are contained in a shallow glass gas-filled trough. The anode is formed by a material such as tin oxide deposited on the inside of the faceplate through which the glowing cathodes can be seen. As in n.i.t.s, probe currents can be conducted by the 'off' cathodes, and it is necessary to bias the 'off' segments to prevent a background haze being formed and reducing the readability of the display. Both single and multiple tubes can be made. In multiple tubes, the similar segments of each section are connected together so that only nine cathode connections are required. Thus to form a display, anode scanning is used. The anode of one section is selected, the cathodes corresponding to the segments needed for the numeral are switched, and the required numeral for that section is displayed. The anode of the next section is selected, and the appropriate cathodes for that section switched, and so on. The time taken for all the anodes of the tube to be cycled is sufficiently short for the display to appear continuous.

Typical numerals in a segmented display tube are between 8 mm and 14 mm high, and a 12-section display tube is typically 85 mm long.

FURTHER READING

ACTON, J. R. and SWIFT, J. D., *Cold Cathode Discharge Tubes*, Heywood, London (1963)

COBINE, J. D., *Gaseous Conductors, Theory and Engineering Applications*, Dover, New York (1958)

DANCE, J. B., *Cold Cathode Tubes*, Iliffe, London (1967)

NEALE, D. M., *Cold Cathode Tube Circuit Design*, Chapman and Hall, London (1964)

HOT-CATHODE GAS-FILLED VALVES

Thermionic valves containing a low-pressure gas filling have found considerable application as rectifiers and power-control devices. Mercury-vapour and inert-gas diodes are used as power rectifiers, while mercury-vapour and inert-gas triodes (called thyratrons) are used as controlled rectifiers. Hydrogen thyratrons are used for the production of accurately-timed high-power pulses.

Gas discharge in hot-cathode tube

In a thermionic diode containing a gas filling, collisions occur between the electrons emitted from the heated cathode and the gas atoms. The result of such collisions depends on the energy possessed by the electron. If the energy of the electron is low, the collision has little effect on the gas atom while the electron itself 'bounces off' the atom. Such a collision is called an elastic collision. If the electron energy is higher, an electron in the gas atom may be temporarily displaced to a higher-energy orbit. When the gas electron returns to its original orbit, the energy absorbed during the collision is emitted as an electromagnetic radiation whose wavelength depends on the gas atom. This type of collision is known as an excitation collision. If the energy of the colliding electron is higher still, an electron can be freed from its parent gas atom to leave a positive ion. The gas is said to be 'ionised', and this type of collision is an ionising collision.

The energy at any instant of an electron emitted from the cathode is determined by its speed, which in turn is determined by the voltage acting on the electron. The voltage at which the electrons acquire sufficient energy to ionise the gas filling is called ionisation potential. The value of ionisation potential depends on the gas. In a practical tube, the voltage that has to be applied between the anode and cathode to ionise the gas filling is greater than the ionisation potential. After the tube has ignited, the voltage across the conducting valve falls to a value approximately equal to the ionisation potential.

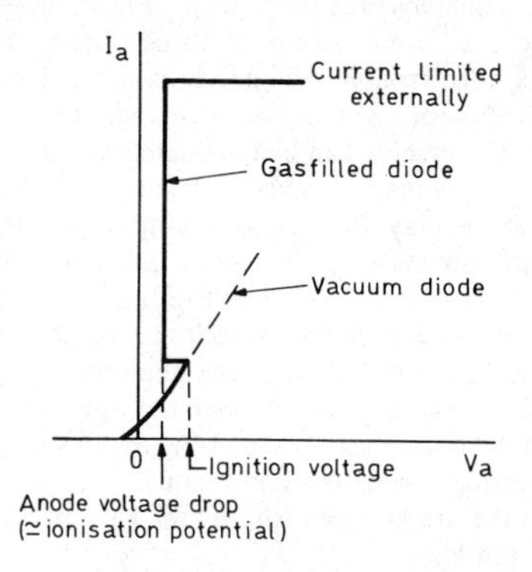

Figure 7.181 Voltage/current characteristic of hot-cathode gas-filled diode

If a positive voltage with respect to the cathode is applied to the anode of a gas-filled thermionic diode, and this voltage is increased from zero, initially only a small current flows. The voltage/current characteristic at this stage is similar to that of a vacuum thermionic diode. When the anode-to-cathode voltage reaches the ignition voltage, however, there is a sudden increase in current as the gas filling is ionised. An external current-limiting resistance is required in practical circuits to prevent the diode being damaged. The voltage across the conducting diode (after falling from the ignition value) remains reasonably constant for all permissible values of current. The diode remains conducting until the anode-to-cathode voltage falls below the ionisation value. The voltage/current characteristic for a hot-cathode gas-filled diode is shown in *Figure 7.181*. Typical values of ignition voltage for practical diodes are between 10 V and 30 V.

In the ionised gas filling, the electrons move more quickly towards the anode than the positive ions move towards the cathode. This is because of the larger mass of the positive ion. Thus a positive space charge forms at some point between the anode and the cathode. This positive space charge is in addition to the normal negative space charge that is formed near the cathode by the emitted electrons. The potential distribution between the anode and cathode is altered from the nominally linear fall from anode to cathode, giving a region with a steep potential gradient between the negative space charge near the cathode and the positive space charge. Electrons from the negative space charge move rapidly to the positive space charge, ionising further gas atoms on the way. The positive ions add to the space charge which therefore moves nearer the cathode. The potential gradient between the positive space charge and the anode is low. In this region the electrons move more slowly than between the negative and positive space charges, and there are therefore large numbers of both electrons and positive ions so that the region is highly conductive. Thus the positive space charge forms a virtual anode near the cathode. The potential distribution in a gas-filled diode, compared with that of a vacuum diode, is shown in *Figure 7.182*.

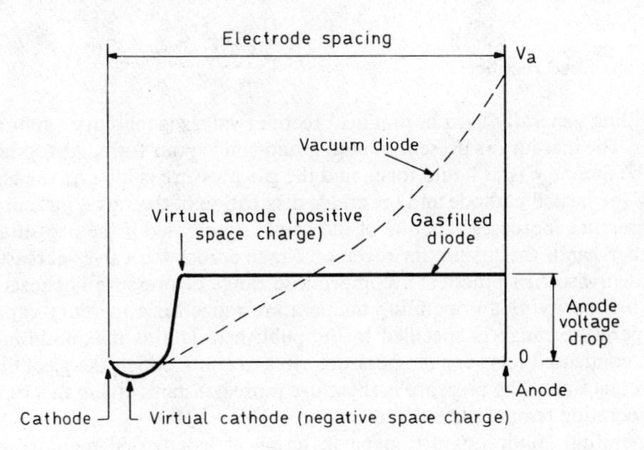

Figure 7.182 Potential distribution in conducting hot-cathode gas-filled diode compared with vacuum diode

Although the potential gradient between the positive space charge and the anode is low, the electrons still move faster than the positive ions. The larger part of the current through the diode is therefore conducted by the electrons, particularly those emitted thermally from the cathode. If no external current-limiting resistance were connected in series with the diode, the steep potential gradient between the positive and negative space charges would (as already described) cause more ions to be formed, which would result in the positive space charge moving nearer the cathode. The potential gradient would be increased, and electrons would be accelerated from the negative space charge to the positive space charge with higher energies to create more ions. This cumulative process would continue until the electron current reaches saturation and positive ions bombard the cathode, with the consequent risk of destroying the oxide coating. With a current-limiting resistance in the diode circuit, provided the saturation current for the particular cathode temperature is not reached, an equilibrium state is established. The positive space charge remains at a distance such that the potential gradient corresponds to the required electron current to be drawn from the negative space charge. If the positive space charge approaches the cathode, more electrons are drawn from the negative space charge and cancel the excess positive space charge to restore equilibrium. If the external resistance is decreased, the positive space charge moves nearer the cathode to cause extra electron current to flow. Thus the voltage across the diode is maintained reasonably constant. It can be seen that whereas in a thermionic vacuum diode the current through the valve is limited by the negative space charge near the cathode, in a thermionic gas-filled diode the current is limited by the external resistance. The voltage across a conducting gas-filled diode is called the anode voltage drop, and is approximately equal to the ionisation potential of the gas filling.

A conducting gas-filled diode is extinguished by reducing the voltage between the anode and cathode to below the ionisation potential of the gas filling. If a negative voltage with respect to the cathode is applied to the anode of the diode, there will be no thermal electron current through the valve, and the gas filling will not be ionised. With a reverse voltage applied, the valve in effect is operating as a cold-cathode gas-filled tube, and ionisation can only occur when the reverse voltage exceeds the 'cold-cathode' ignition value. With a suitable anode-to-cathode spacing, hot-cathode gas-filled diodes can be manufactured to withstand reverse voltages as high as, typically, 20 kV. Such diodes can therefore be used as rectifiers on high-voltage supplies. As the valves can be designed to conduct higher currents than those of vacuum diodes, gas-filled diodes are therefore used in high-power rectifier systems.

Practical gas-filled rectifiers

The gas filling generally used in practical rectifier valves is mercury vapour or the inert gas xenon. The mercury is present in both liquid and vapour form. At room temperature most of the mercury is in liquid form, and the gas pressure is low. As the temperature is raised, by the heated cathode and the anode dissipation of the valve, mercury evaporates and the pressure increases. The life of the valve is increased if the pressure is high, but on the other hand, the maximum reverse voltage across the valve increases as the gas pressure decreases. In practice, a compromise range of pressure is chosen, and this is achieved by specifying an operating temperature range for a mercury-vapour rectifier. This temperature range is specified in the published data as maximum and minimum values of 'condensed mercury temperature'. In a xenon rectifier, the gas filling is always in the gaseous state. The pressure is therefore more constant, giving this type of rectifier a wider operating temperature range.

The operating conditions are given in terms of 'condensed mercury temperature' because this quantity can be directly measured by the user. For practical purposes, the condensed mercury temperature is the temperature of the outside of the glass envelope of the rectifier in the area where the mercury vapour condenses into liquid, at the bottom near the base cap. The ambient conditions in which the rectifier operates should be adjusted so that the minimum condensed mercury temperature is attained when the rectifier is operating with only the heater supply applied. If cathode current is drawn before the minimum value is reached, the voltage drop across the rectifier will be too high and the cathode surface may be damaged. If the condensed mercury temperature when the rectifier is conducting current exceeds the maximum value, the reverse voltage that the valve can withstand will be less than that stated in the data, and 'arc-back' may occur.

A curve is given in the published data showing the total heating-up time required from cold plotted against ambient temperature, as in *Figure 7.183*. Heating and cooling characteristics are also given, relating the rise and fall of condensed mercury temperature above ambient temperature with time. The form of these curves is shown in *Figure 7.184*. From the known ambient temperature and the maximum and minimum values of condensed mercury temperature, the time required before current can be conducted by the rectifier can be found, and the operating conditions checked without any direct measurement of the condensed mercury temperature. If the maximum value of condensed mercury temperature is likely to be exceeded during operation, additional cooling can be applied to the rectifier. The cooling characteristic enables a check to be made that current can be conducted safely if the heater and anode voltages are re-applied when the valve has been cooling after both supplies have been removed.

In practice, it may not always be necessary for the minimum value of condensed mercury temperature to be reached. Current can be drawn when the temperature is within a few degrees of the required value. However, a minimum heating-up time is then specified in the data, and this time must elapse before current is drawn if the rectifier is not to be damaged.

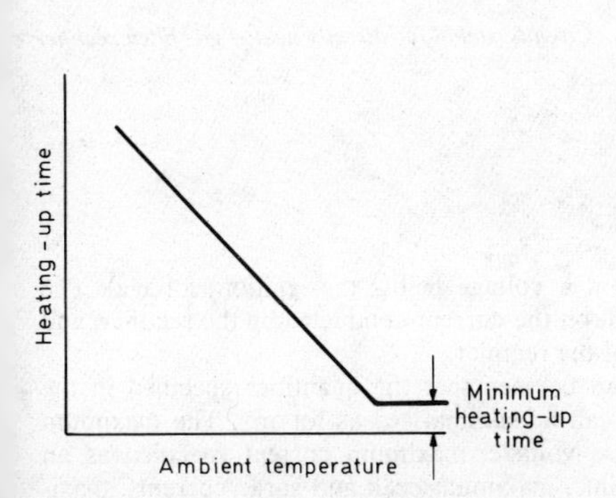

Figure 7.183 Heating-up time plotted against ambient temperature for hot-cathode gas-filled rectifier

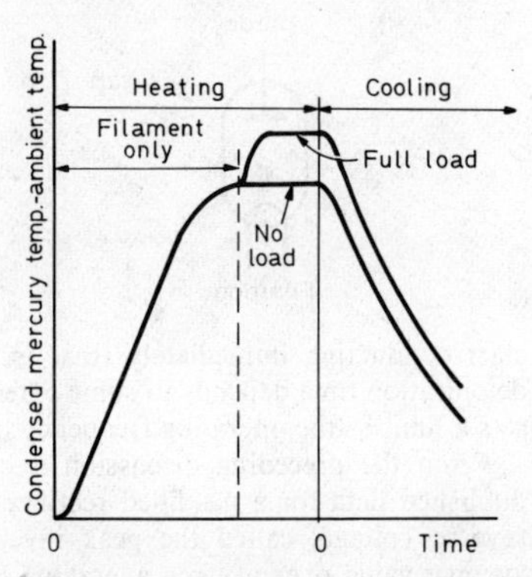

Figure 7.184 Heating and cooling characteristics for hot-cathode gas-filled rectifier

The cathode of a gas-filled rectifier can be of the directly-heated or indirectly-heated type, although the directly-heated type is more generally used. The anode connection is made by a top cap on the valve envelope, and so only two connections (for the filament) need be made at the base. These connections can be made by screw cap or large-diameter valve pins capable of carrying the heavy current. The cathode itself must be of sufficient bulk to carry the current, and for this reason requires a long heating-up time.

Because a directly-heated cathode (filament) is used, there is a voltage drop along the filament. Typical filament voltages lie between 2.5 V and 5.0 V. The voltage drop across a conducting gas-filled rectifier is low, lying between 10 V and 20 V, and so the low-potential end of the filament contributes more of the valve current than the high-potential end. If the anode and filament supplies are in phase, or 180° out of phase, the same end of the filament provides the greater part of the current, and the life of the valve is reduced. If the filament supply is 90° out of phase with the anode supply, both ends of the filament are equally used, and this ensures a longer life for the filament. In practice, the phase difference between the anode and filament supplies can generally be 90° ± 30°. The magnitude of the filament voltage must also be closely controlled to prevent variations affecting the life of the valve. Long-term variations should in general not exceed ±2.5% (measured at the base connections).

The current that can be conducted by a gas-filled rectifier must not exceed the specified value otherwise the valve may be damaged. A maximum average current is specified together with a maximum averaging time. A current higher than the maximum average rated value can be conducted for a time shorter than the averaging time provided the product of the maximum current and its duration does not exceed the product of the maximum rated average current and maximum averaging time. A peak current is specified to prevent damage to the valve which would shorten its life. The peak current limit is particularly important for capacitive loads, and affects the values of components used in smoothing filters for practical rectifier circuits. A surge current rating is also specified for the rectifier. This represents the maximum transient current that can be conducted for a specified time (typically 0.1 s) as a result of a sudden overload or short-circuit without destroying the rectifier. Several overloads of this nature, however, will appreciably shorten the life of the valve.

In common with all gas-filled tubes, gas-filled rectifiers have a deionisation, or recovery, time which represents the time which must elapse between the extinction of the discharge and the reapplication of the anode-to-cathode voltage if the valve is not to

Figure 7.185 Circuit symbol for directly-heated gas-filled rectifier

start conducting immediately (that is, at a voltage below the ignition voltage). The deionisation time depends to some extent on the current conducted by the rectifier, and sets a limit to the operating frequency of the rectifier.

From the preceding discussion it can be seen that the quantities specified in the published data for a gas-filled rectifier can be summarised as follows. The maximum reverse voltage, called the peak inverse voltage, maximum current specified as an average value over a given averaging time, maximum peak and surge currents, maximum and minimum values of condensed mercury temperature, and the maximum operating frequency. Characteristic values include the filament voltage and current, and the anode voltage drop. Curves of heating-up time and the heating and cooling characteristics are given. Typical values of peak inverse voltage can be up to approximately 20 kV, and average currents up to 40 A. The anode voltage drop is typically between 10 V and 15 V, rising towards 30 V at the end of life of the rectifier. The circuit symbol for a gas-filled rectifier is shown in *Figure 7.185*.

Gas-filled rectifier circuits

Various configurations can be used for gas-filled rectifier systems: these are similar to those used for vacuum rectifier diodes although the voltages and currents associated with gas-filled rectifiers are higher. The most commonly used configurations are shown in *Figure 7.186*. These circuit configurations show *LC* smoothing circuits connected to the rectifier system. For some applications, particularly where the three-phase full-wave configuration is used, no smoothing may be required. Where a smoothing filter is required, the component values can be obtained from the published data. The values are given for specified circuit conditions, and are often high: inductance values of, say, up to 10 H and capacitance values of 5 μF. It is also necessary to provide fuses in the output lines to protect the rectifiers from damage in the event of a short-circuit in the load circuit.

Radio frequency fields and currents can cause spurious ionisation of the gas filling of the rectifier, leading to flashover or arc-back. To prevent this occurring, the rectifiers can be screened in separate compartments, and r.f. filters connected in the anode and filament leads. Because mercury-vapour rectifiers contain liquid mercury, it is necessary to mount these rectifiers vertically and base-down. Xenon rectifiers can be mounted in any position.

Although a large number of gas-filled rectifiers are still in service, the device has been replaced to a large extent by solid-state rectifiers using avalanche diodes (see Section 8).

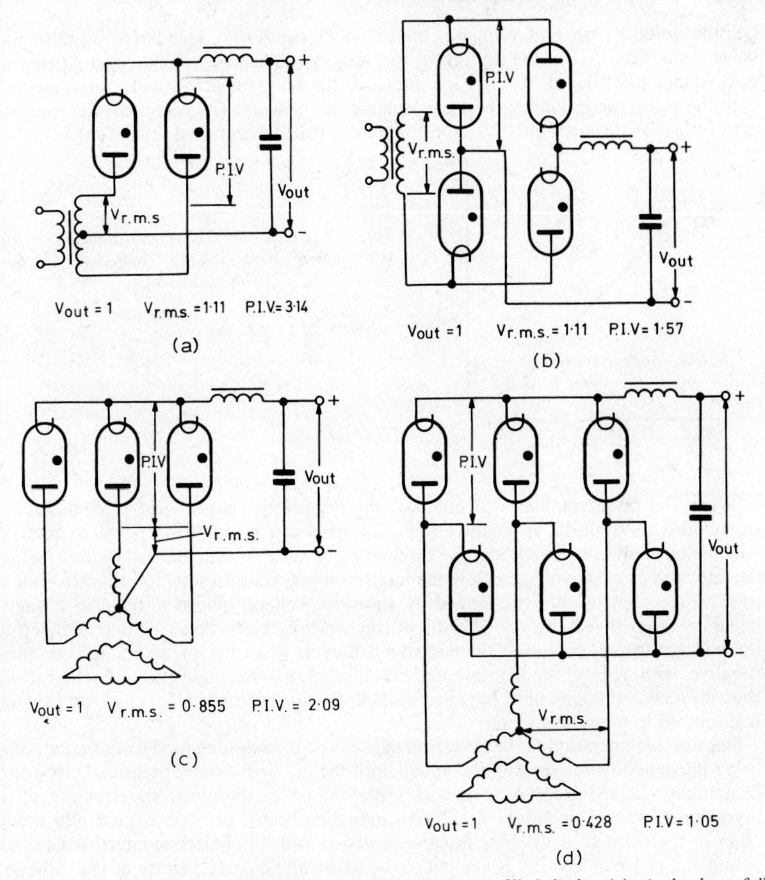

Figure 7.186 Rectifier configurations using hot-cathode gas-filled diodes: (*a*) *single-phase full wave* (*b*) *single-phase bridge* (*c*) *three-phase half wave* (*d*) *three-phase full wave*

Thyratron (hot-cathode gas-filled triode)

If a grid is placed between the anode and the cathode of a hot-cathode gas-filled diode, an additional control over the ionisation of the gas filling is obtained. Such gas-filled triodes are known as thyratrons. If a high negative voltage with respect to the cathode is applied to the grid, say −10 V to −30 V, electrons emitted by the heated cathode are prevented from travelling towards the anode and so cannot ionise the gas filling. The thyratron will not conduct even though the anode-to-cathode voltage is greatly in excess of the ignition voltage of the valve, up to values of a few kilovolts. For each value of negative grid voltage, the anode voltage must reach a corresponding breakdown value before the gas filling is ionised and the thyratron conducts. Immediately conduction starts, the voltage across the thyratron falls to the anode voltage drop, corresponding to the ionisation potential of the gas filling, and the large current flowing has to be limited to a safe value by an external resistance. The anode-to-cathode voltage required for ignition decreases as the negative grid voltage is decreased, and the variation of anode ignition voltage with grid voltage is shown in *Figure 7.187*. This curve is called the control characteristic of the thyratron. For ignition, the anode voltage for a particular grid voltage must be taken above the curve, or the grid voltage be made more positive than the value corresponding to a particular anode voltage. The instantaneous value of grid voltage that causes the thyratron to ignite is called the critical grid voltage.

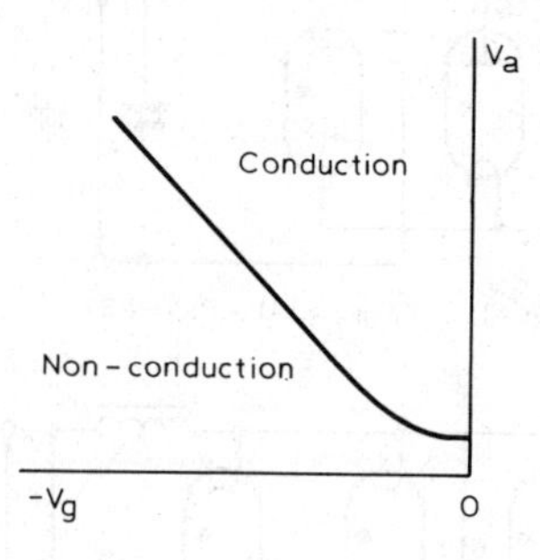

Figure 7.187 Control characteristic of thyratron (variation of anode voltage required for ignition with grid voltage)

When the thyratron starts to conduct, the negatively-charged grid is immediately surrounded by a sheath of positive ions. The grid can no longer be used to control conduction in the thyratron, and the discharge can only be extinguished by decreasing the anode-to-cathode voltage below the ionisation potential. In practice, thyratrons are used on a.c. supplies, and are ignited by a positive voltage applied to the grid at some angle in the positive half-cycle. The thyratron remains conducting for the remainder of the positive half-cycle, but in the negative half-cycle when the anode voltage becomes negative with respect to the cathode, the discharge is extinguished. The thyratron remains non-conducting until reignited in the following positive half-cycle. Grid control is discussed in more detail later.

Because the operation of the thyratron depends on electrons emitted from the cathode being prevented from reaching the anode until the device is ignited, a special electrode construction is required. A schematic cross-section of the electrode structure of a thyratron is shown in *Figure 7.188*. An indirectly-heated cathode is generally used, often in the form of a cylinder fitted with vanes, with the heater mounted inside the cylinder. The oxide coating is applied to the vanes and outside surface of the cathode. Sometimes a heat shield is fitted to increase the cathode efficiency. The grid is in the form of a disc with a hole or set of slots through which the electrons and positive ions can flow. To prevent thermally-emitted electrons from the cathode reaching the anode before the thyratron is required to be ignited, a grid shield surrounds the cathode. The anode is generally made from nickel with the surface blackened to improve the heat

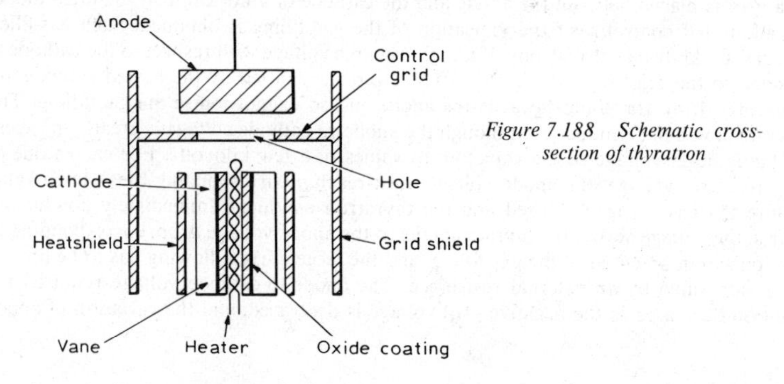

Figure 7.188 Schematic cross-section of thyratron

dissipation. The anode connection is made by a top cap, while the heater, cathode, and grid connections are made by base pins. A typical thyratron using this type of construction is shown in the photograph of *Figure 7.189*.

The construction just described is for power-control thyratrons. Small thyratrons were used as trigger devices for operation with voltage, higher than those of cold-cathode trigger tubes. As the dissipation of such 'trigger' thyratrons is small, of the same order as that of a thermionic valve, the anode, grid, and cathode can be mounted as parallel vertical plates with the required shields in a conventional valve envelope. Connections can then be made through pins in the valve base.

Figure 7.189 Typical inert-gas thyratron. The valve will conduct an average current of 2.5 A at a peak voltage of 1.25 kV; height approximately 116 mm (Photo: Mullard Limited)

The gas filling used in thyratrons is mercury vapour or an inert gas such as argon or xenon. Sometimes a mixture of mercury vapour and inert gas is used.

When the thyratron is conducting, positive ions are attracted to the most negative electrode, the grid. The grid positive-ion current must be limited to prevent the thyratron being damaged, and this is done by ensuring that the grid voltage during conduction is more positive than −10 V. The grid ion current is proportional to the anode current, and a suitable series resistor can be connected in the grid circuit to ensure the grid is maintained at the required potential. Values of this resistor may be given in the data for the thyratron, or the grid ion-current characteristic relating the grid voltage and grid current with anode current as parameter may be given, from which the resistor value can be calculated.

Much of the data for a thyratron is similar to that of a gas-filled rectifier diode. For mercury-vapour thyratrons, maximum and minimum condensed mercury temperatures are given, and for all types the peak inverse voltage, the maximum average and peak currents that can be conducted, the anode voltage drop, heating-up time, and heating and cooling characteristics are given. Quantities that are specific to thyratrons are the maximum forward anode voltage that can be applied while still maintaining grid control (equal to the peak inverse voltage); the maximum negative grid voltage that can be applied before and during conduction; the maximum positive grid current and information for selecting a suitable grid resistor; and the interelectrode capacitances. The circuit symbol for a thyratron is shown in *Figure 7.190*.

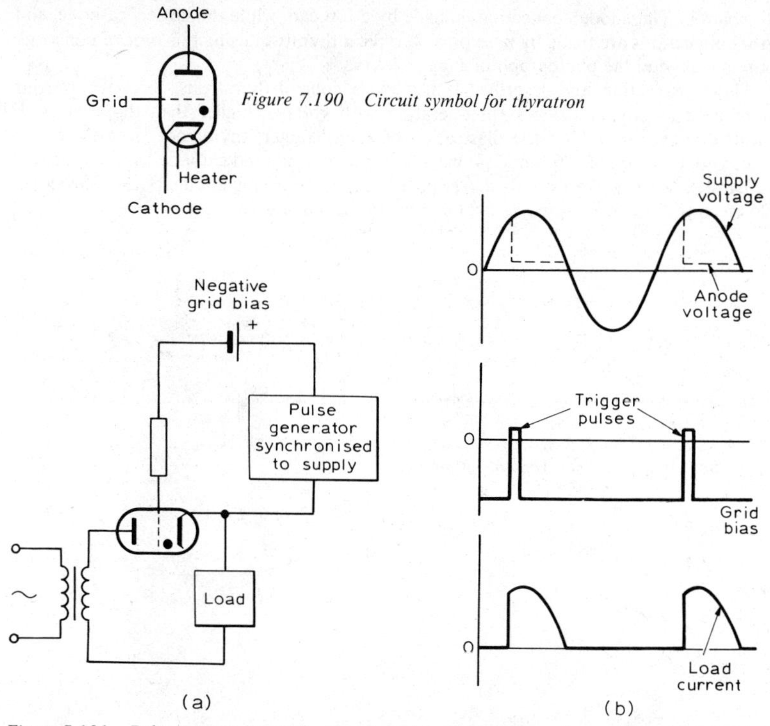

Figure 7.190 Circuit symbol for thyratron

Figure 7.191 Pulse ignition circuit for thyratron: (a) circuit (b) waveforms showing relationship of grid pulse to anode voltage

The thyratron is ignited by applying a positive pulse to the grid which momentarily overcomes a constant negative grid bias, or by a voltage which is more positive than the critical grid voltage. When the thyratron has been ignited, it conducts for the remainder of the positive half-cycle. The discharge is extinguished during the negative half-cycle, enabling grid control to be regained. If the phase position of the grid control voltage in the positive half-cycle is varied, the average load current and hence the power in the load can thereby be varied. This is the principle of power control with a thyratron. It is necessary to synchronise the grid pulse with the mains supply to the anode of the thyratron, and this can be done by deriving the grid pulse from the same supply as the anode voltage. A schematic circuit for pulse ignition is shown in *Figure 7.191* (*a*), and the related waveforms in *Figure 7.191* (*b*). The positive pulse momentarily overcomes the negative grid bias to ignite the thyratron. The phase relationship of the pulse to the anode voltage can be varied by the pulse generator. A second ignition circuit is shown in *Figure 7.192*. The resistor R and capacitor C form a phase-shift network. The magnitude of the voltage applied to the grid of the thyratron is the half-secondary voltage of transformer T_2, and the phase relationship to the secondary voltage of transformer T_1 applied between the anode and cathode of the thyratron, angle ϕ, can be varied by the variable resistor R. When the grid voltage exceeds the critical value, the thyratron will be ignited.

Although the thyratron was extensively used for power-control applications, like the gas-filled rectifier diode it has been largely replaced by a solid-state device, this time the thyristor (see Section 8).

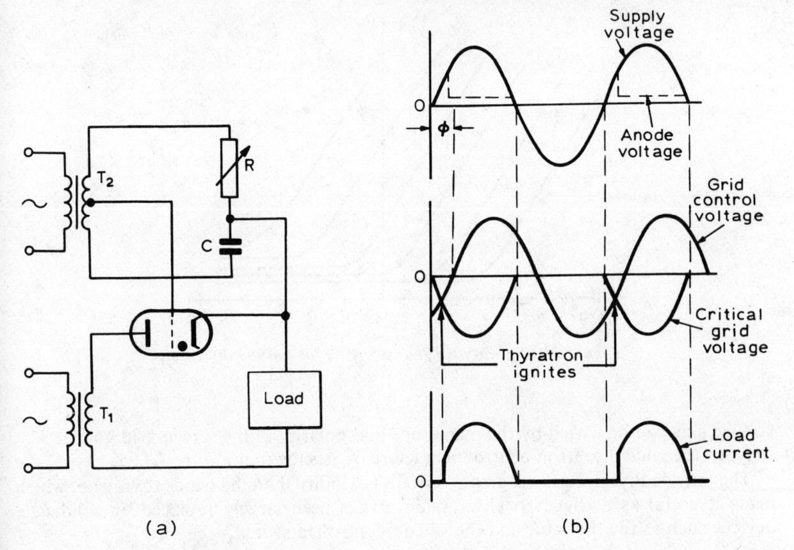

Figure 7.192 Phase-shift ignition circuit for thyratron: (a) circuit (b) waveforms showing relationship of grid control voltage to critical grid voltage and anode voltage

Tetrode thyratron

A second grid can be incorporated into a thyratron, giving the schematic electrode structure shown in *Figure 7.193*. The second grid is called the screen grid, while the original grid is known as the control grid. This type of thyratron is called a 'tetrode' thyratron, and the circuit symbol is shown in *Figure 7.194*. The thyratron described above is sometimes called the triode thyratron to distinguish it from the tetrode thyratron.

The control characteristic of a tetrode thyratron relates the anode voltage required for ignition to the control-grid voltage V_{g1}. The shape of the characteristic is similar to that of the triode thyratron, but the position of the characteristic along the control grid

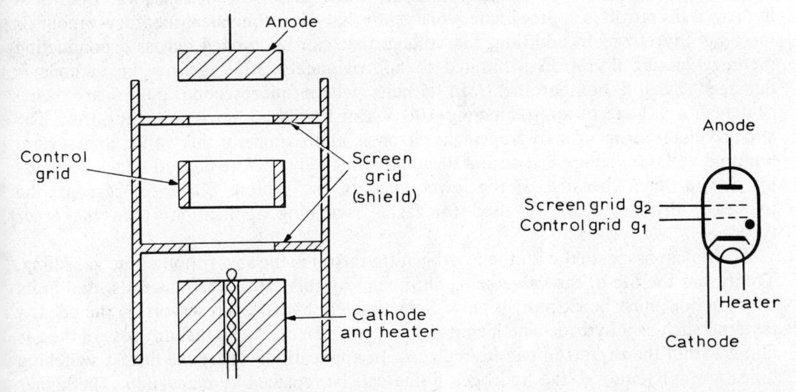

Figure 7.193 Schematic cross-section of tetrode thyratron

Figure 7.194 Circuit symbol for tetrode thyratron

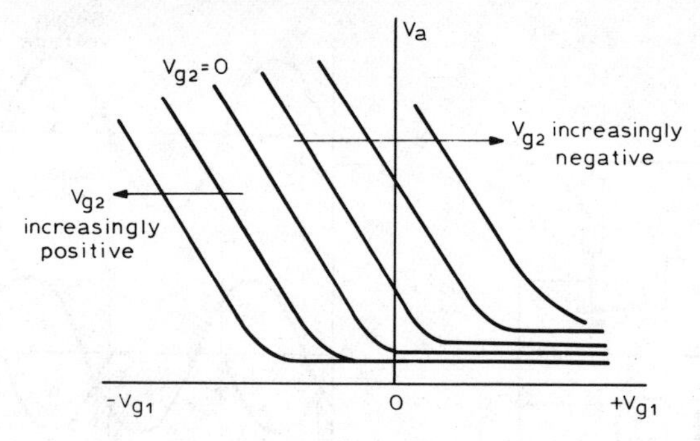

Figure 7.195 Control characteristic for tetrode thyratron

voltage axis can be varied by the magnitude and polarity of the screen-grid voltage V_{g2}. A typical tetrode thyratron control characteristic is shown in *Figure 7.195.*

The tetrode thyratron has a greater circuit flexibility than the triode thyratron, which made it useful as a trigger device. Again, it has been largely replaced by solid-state devices such as the thyristor and the silicon controlled switch.

Hydrogen thyratron

A form of thyratron that is still used in certain specialised applications is the hydrogen thyratron, for applications where accurately-timed high-power pulses are required. Originally used for switching radar pulse modulators, the hydrogen thyratron is now also used with particle accelerators and high-power energy diverter circuits.

The operation of the hydrogen thyratron is different from that of the mercury-vapour and inert-gas types already described. An initial discharge is formed between the grid and the cathode by applying a positive voltage pulse to the grid. This initial discharge then enables the main discharge between the anode and the cathode to form.

Because of the higher mobility of the proton (the hydrogen positive ion) compared with the positive ions of mercury and the inert gases, the deionisation time of a hydrogen thyratron is approximately one-tenth that of an equivalent mercury-vapour or inert-gas thyratron. In addition, the voltage that can be applied across a conducting mercury-vapour thyratron is limited to approximately 30 V before the cathode is damaged by ion bombardment. In circuits where microsecond pulses are being switched, a voltage of approximately 100 V can be placed across the thyratron. The anode voltage drop of a hydrogen thyratron is approximately this value, and several hundred volts can be applied across the conducting valve before the cathode is damaged by bombardment because of the lower mass of the proton. For these reasons, the hydrogen thyratron can be used for faster switching applications than the other types.

A problem associated with the hydrogen thyratron is the absorption of the gas filling. To prevent the life of the valve being shortened by this cause, the materials used in its construction must be extremely pure. Sometimes a 'hydrogen reservoir' is included, a material such as a hydride which emits hydrogen when heated so that any loss of the gas filling within the thyratron can be replaced. In applications where the fastest switching times are not required, the hydrogen filling can be replaced by deuterium. The higher mass of the deuterium ion reduces the absorption effects at the cost of a slightly longer deionisation time.

FURTHER READING

COBINE, J. D., *Gaseous Conductors, Theory and Engineering Applications*, Dover (1958)
CORY, B. J., ed., *High Voltage Direct Current Converters and Systems*, Trends and Developments in Engineering Series, Macdonald (1965)
LEE, R., *Electronic Transformers and Circuits*, 2nd edn., Wiley/Chapman and Hall (1955)
PARKER, P., *Electronics*, Edward Arnold (1950)
SHEPHERD, J., MORTON, A. H. and SPENCE, L. F., *Higher Electrical Engineering*, 2nd edn., Pitman (1970)
SWENNE, C. M., *Thyratrons*, Philips Technical Library, Clever-Hume (1960)

IGNITRONS

The ignitron is a special form of controlled rectifier valve used in very high current applications. The current in the valve is conducted by an arc struck in mercury vapour. For this reason it is sometimes called a mercury arc rectifier. The conduction period can be controlled by varying the angle in the positive half-cycle of the supply at which conduction starts. In this way, the power to the load circuit can be controlled. The main applications for the ignitron are in resistance welding and power rectification.

Operating principle

An ignitron consists essentially of an evacuated metal cylinder containing a mercury pool which acts as the cathode, an anode, and a rod-shaped electrode touching the mercury surface called the ignitor. A schematic cross-section of the ignitron is shown in *Figure 7.196*.

Because of the mercury pool, the space inside the ignitron contains mercury vapour. The device is therefore a special form of cold-cathode gas-filled tube. The anode-to-cathode spacing, however, is such that the gas filling cannot be ionised solely by the effect of the anode-to-cathode voltage. Before the ignitron can conduct, an initial discharge must be struck between the ignitor and the surface of the mercury, and this

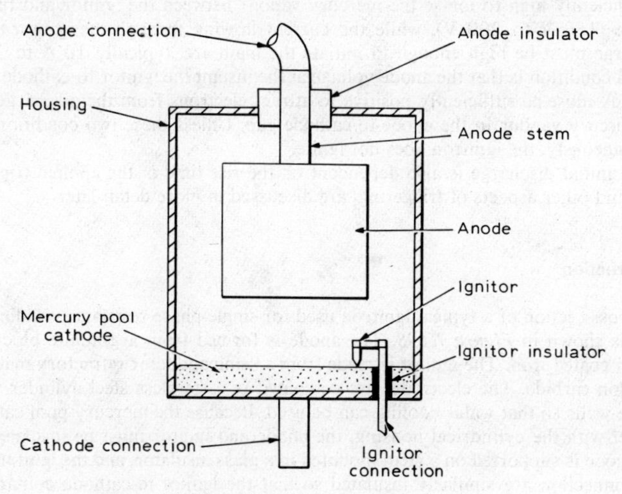

Figure 7.196 Schematic cross-section of ignitron

discharge then enables the main arc between the anode and the cathode to be struck. The arc is extinguished at the end of the positive half-cycle of the supply, and conduction can only start in the following positive half-cycle after the ignitron has been retriggered.

The ignitor electrode is in the form of a rod with a pointed end which dips into the surface of the mercury pool. Because of the effects of surface tension, a meniscus is formed as shown in *Figure 7.197*. If a positive trigger pulse is applied to the ignitor, typically 150 V to 200 V, a high electric field is developed in the space between the ignitor and the surface of the mercury in the meniscus. The mercury vapour in this space is ionised, an arc is struck and a current flows. The current causes a 'hot spot' to

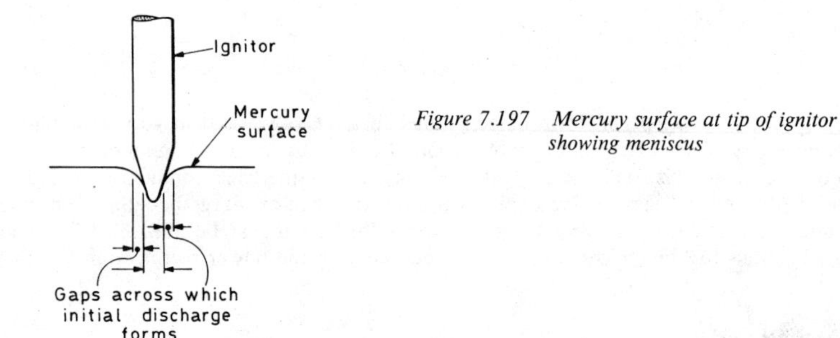

Figure 7.197 Mercury surface at tip of ignitor showing meniscus

form on the surface of the mercury, and the temperature is high enough for electrons to be emitted. If the voltage on the anode is sufficiently positive, these electrons are accelerated towards the anode and acquire sufficient energy to ionise atoms of the mercury vapour with which they collide. When sufficient ions are present, the main arc is struck between the anode and the cathode. Once the arc has been struck, the ignitor electrode has no effect on the operation of the ignitron. Control is regained during the following negative half-cycle when the arc is extinguished.

From this description of the ignition of the ignitron, it can be seen that two conditions must be met. The first condition is that the voltage applied to the ignitor must be sufficiently high to ionise the mercury vapour between the ignitor and the cathode surface (150 V to 200 V), while the current flowing through the ignitor-to-cathode discharge must be high enough to initiate the main arc, typically 10 A to 30 A. The second condition is that the anode voltage at the instant the ignitor-to-cathode discharge is struck must be sufficiently positive to attract electrons from the discharge to ionise the mercury vapour in the anode-to-cathode gap. Unless these two conditions are met simultaneously, the ignitron does not ignite.

The initial discharge is also dependent on the rise time of the applied trigger pulse. This, and other aspects of triggering, are discussed in more detail later.

Construction

The cross-section of a typical ignitron used for single-phase resistance welding applications is shown in *Figure 7.198*. The anode is formed from a graphite block or from carbon-coated iron. The ignitor is made from a semiconductor refractory material such as boron carbide. The electrodes are contained in a stainless steel cylinder which has double walls so that water cooling can be used. Because the mercury-pool cathode is in contact with the cylindrical housing, the anode and ignitor must be insulated from it. The anode is supported on a stem mounted in a glass insulator, and the ignitor electrode and connection are similarly insulated so that the ignitor-to-cathode contact is made only at the tip of the ignitor rod. The anode stem, and the anode and cathode connec-

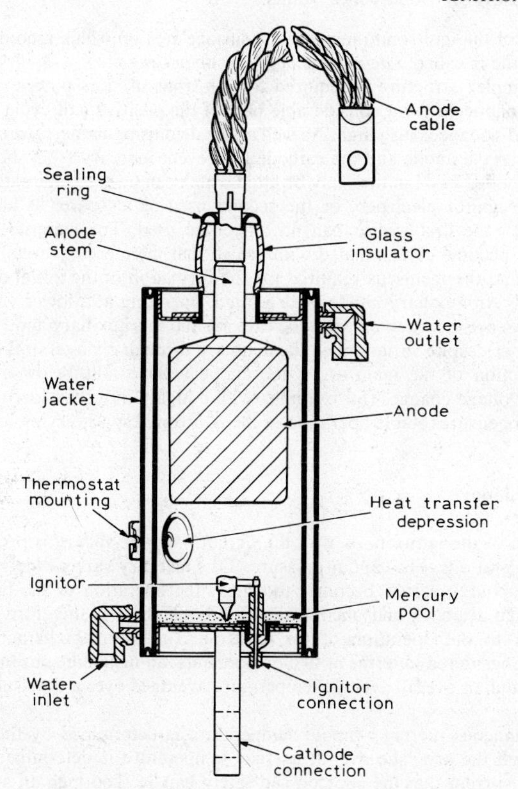

Figure 7.198 Construction of ignitron for single-phase resistance welding applications

tions, have a large cross-sectional area to carry the high currents. In some applications the cathode may not operate at earth potential, and the water inlet and outlet hoses must therefore be made from insulating material.

Water cooling of the ignitron is necessary because of the temperature rise caused by the high arc current. As the temperature of the ignitron increases, the vapour pressure of the mercury increases. Too high a pressure can lead to 'arc-back' during the negative half-cycle of the supply, thereby preventing the rectifier action of the ignitron. To ensure safe operation, the flow and temperature of the cooling water can be monitored, or the temperature of the inner wall of the ignitron can be monitored. The 'heat transfer depression' shown in the section of *Figure 7.198* provides a good thermal contact between the inner wall of the ignitron and a mounting plate on which a thermostat can be fitted. When the temperature of the inner wall, or that of the cooling water, exceeds a preset limit, a safety circuit can be operated to disconnect the ignitron from the circuit or to inhibit the triggering to prevent further operation.

The construction shown in *Figure 7.198* is used for single-phase power-control applications such as resistance welding where high peak currents of the order of 10 kA may be encountered but their duration is short. The applied voltage is also generally low. Ignitrons for three-phase power-control applications require a short deionisation time to ensure the device remains non-conducting despite the high inverse voltage which is applied after the conduction period. The required rapid recombination of the large number of ions produced by the high arc current can be achieved by fitting a baffle to the inner wall of the ignitron to increase the surface area on which recombination takes place. This baffle is called a *deionising baffle* or *deionising ring*.

A more complex structure is required for ignitrons used as power rectifiers. High currents are conducted for a considerable part of the positive half-cycle of the supply, and the applied voltage is also high. As well as the deionising baffle, an *anti-splash baffle* is fitted between the anode and the cathode to prevent mercury from the cathode pool reaching the anode. Two ignitors may be fitted which can be used alternately to lengthen the life of the ignitor electrode, or the second used as a 'spare' to be brought into operation when the first ignitor can no longer be used. The material of the ignitor electrode may become contaminated with use so that the mercury 'wets' the surface of the electrode and the meniscus required for the formation of the initial discharge is no longer formed. An auxiliary anode is also fitted, operating at a lower voltage than the main anode. An arc is struck between the cathode and this auxiliary anode to maintain a cathode hot spot despite variations in the main arc through changes in the load current, or even extinction of the main arc if the anode voltage falls below a critical value through load voltage changes (for example, with a highly inductive load). The auxiliary anode therefore ensures stable operation of the ignitron despite varying load conditions.

Practical operation

The operation of the ignitron, as with all mercury-vapour valves, is predetermined by thermal considerations. The vapour pressure of the mercury varies widely with temperature. If the vapour pressure becomes too high, the operation of the ignitron can be affected through arc-back and uncontrolled ignition. There is therefore an upper temperature limit to the operation of the ignitron. The thermal characteristic can be conveniently considered in terms of two components: an instantaneous mercury-vapour temperature, and an overall average temperature averaged over a time such as 10 or 15 minutes.

The instantaneous mercury-vapour temperature is determined by the peak current flowing through the arc. The overall average temperature is determined by the maximum average current that the ignitron can safely handle. Too high an average current decreases the heat-dissipation ability if the ignitron, leading to a gradual overheating and a deterioration in performance and a shortened life for the tube. The main effects of the gradual overheating are the liberation of gases absorbed by the anode during manufacture which decreases the quality of the vacuum, damage to the anode and ignitor glass insulators, and burning of the ignitor internal connections. Thus the peak current limits operation directly through an excessive vapour pressure and the risk of arc-back and uncontrolled ignition, while the average current limits operation through a gradual deterioration.

The electrodes of the ignitron have widely differing thermal time-constants through their different masses. The inner wall has a typical time-constant of 100 ms, while those of the cathode pool and anode are typically 1 and 5 minutes respectively. The inner wall therefore responds to rapid temperature changes produced by the peak current, while the anode temperature is determined mainly by the average current. The anode voltage therefore has an effect, the lower the anode voltage the higher the permissible value of peak current. It can be seen that the electrodes in the ignitron can have different instantaneous temperatures during operation, and this can lead to the unwanted condensation of mercury on to the anode with the consequent risk of arc-back. This is particularly so when the ignitron is brought into service from cold or when cooling down after use. To prevent this unwanted condensation, the anode stem and insulator are preheated (for convenience, using a high-wattage incandescent lamp placed close to the insulator) before load current is conducted. After operation, the cooling water flow is maintained for 15 to 30 minutes so that the inner wall of the tube is at a lower temperature than the anode as the ignitron cools.

The mercury vapour pressure, the temperatures of the inner wall and cathode pool, and the residual ionisation level in the ignitron can be combined to formulate a quantity called the *virtual temperature* T_v. The maximum permissible virtual temperature $T_{v\max}$ varies inversely with the anode voltage V_a of the ignitron, as shown in the left-hand curve of *Figure 7.199*. It can be seen that when a high peak current is to be conducted (T_v high), the value of anode voltage must be restricted. Similarly, if the anode voltage is high, the permissible value of T_v and hence of the peak current is low.

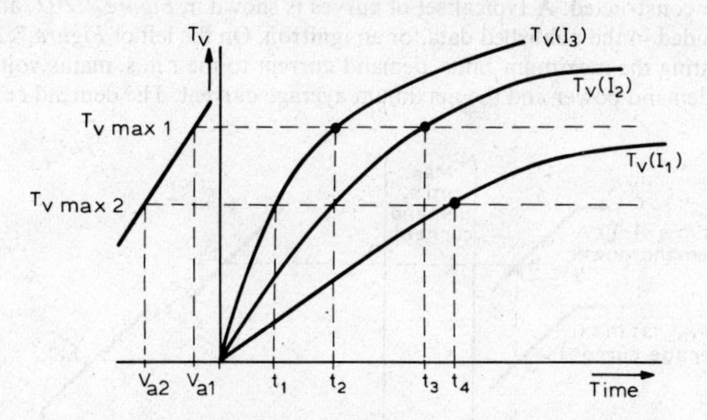

Figure 7.199 Variation of virtual temperature with anode voltage (left), and with time for different current values (right)

The variation of virtual temperature with time for different current values is shown in the right-hand curves of *Figure 7.199*. The virtual temperatures $T_{v\max1}$ and $T_{v\max2}$ corresponding to anode voltages V_{a1} and V_{a2} are marked. If the ignitron is operating at an anode voltage V_{a1}, then a current I_1 could theoretically be conducted indefinitely without exceeding the permissible limit of virtual temperature $T_{v\max1}$. On the other hand, at a higher anode voltage V_{a2}, the same current can be conducted only for a time t_4 before the permissible limit is exceeded. The corresponding permissible conduction times for higher currents I_2 and I_3 are also shown, t_1, t_2, and t_3. At the end of the

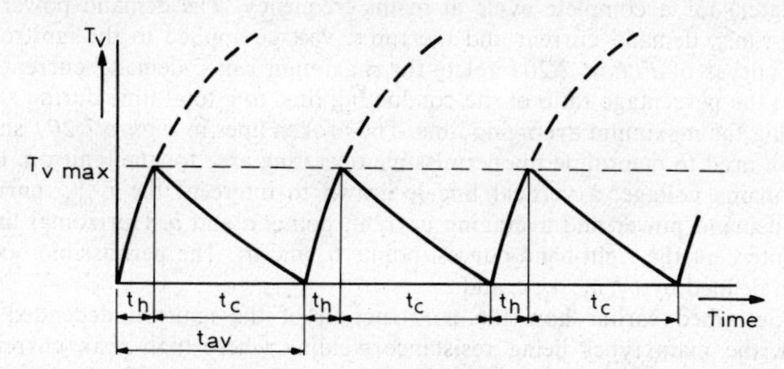

Figure 7.200 Variation of virtual temperature with time for practical operation showing heating and cooling periods

permissible period, the ignitron must cease conduction and cool down before any further conduction can take place. Thus the variation of virtual temperature with time for practical operation is as shown in *Figure 7.200*. Time t_h represents the heating time, the time for which the ignitron is conducting, while t_c represents the cooling time. The sum of t_h and t_c forms a maximum averaging time, the longest time over which the maximum average current can be calculated. The cooling time is longer than the heating time, and depends on the flow of cooling water.

From the preceding discussion it can be seen that the factors determining the operation of the ignitron are: the value of virtual temperature for a given current; the maximum permissible virtual temperature for a given anode voltage; the expected maximum average current; the expected maximum peak current; and the cooling characteristic which in turn depends on the cooling water flow. Of these factors, virtual temperature is a quantity which cannot be measured directly by the user, but from the curves of *Figures 7.199* and *7.200* practical curves showing the operating limits of an ignitron can be constructed. A typical set of curves is shown in *Figure 7.201*, and such curves are included in the published data for an ignitron. On the left of *Figure 7.201* are two curves relating the maximum r.m.s. demand current to the r.m.s. mains voltage for the maximum demand power and the maximum average current. The demand current is

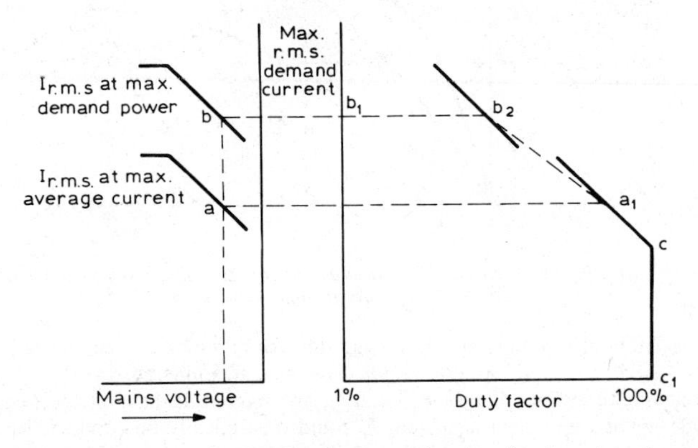

Figure 7.201 Demand current related to duty factor as a function of mains voltage for resistance welding applications

the current conducted by a pair of ignitrons in inverse-parallel (the reason for this is explained later) for a complete cycle at mains frequency. The demand power is the product of r.m.s. demand current and the r.m.s. voltage applied to the ignitron. The right-hand curves of *Figure 7.201* relate the maximum r.m.s. demand current to the duty factor, the percentage ratio of the conducting time to a total time during a period not exceeding the maximum averaging time. The broken lines in *Figure 7.201* show the construction used to determine the permissible operating area for the ignitron. For the particular mains voltage, a vertical line is drawn to intercept the $I_{r.m.s.}$ curves for maximum demand power and averaging current, points a and b. Horizontal lines are drawn to intercept the right-hand curves, points b_2 and a_1. The permissible operating area is then defined by b_1, b_2, a_1, c, and c_1.

It was described earlier how the construction of the ignitron depended on its application, the main types being resistance welding where high peak currents are conducted for short periods at a low anode voltage, and power rectification where high currents are conducted for longer periods at higher voltages. The data for ignitrons is also presented in terms of these applications. The curves in *Figure 7.201* are for resistance welding, while for rectification the peak current is plotted against average current for various peak forward and inverse voltages as shown in *Figure 7.202*. Characteristic values for welding ignitrons given in the data include the maximum demand power, maximum demand and average currents, maximum averaging time, duty factor, maximum number of cycles in the averaging time, integrated load current, and maximum surge current for a specified time, all related to the value of mains voltage. For rectifier operation, maximum peak and average currents, the maximum ratios of average to peak currents and surge to peak currents, and the maximum averaging time are related to maximum peak forward and inverse voltage. Ignitor characteristics are included for both applications, together with the required cooling water flow with maximum and minimum inlet water temperatures. Curves relating the average current to the inlet water temperature for various flow rates may also be included in the data.

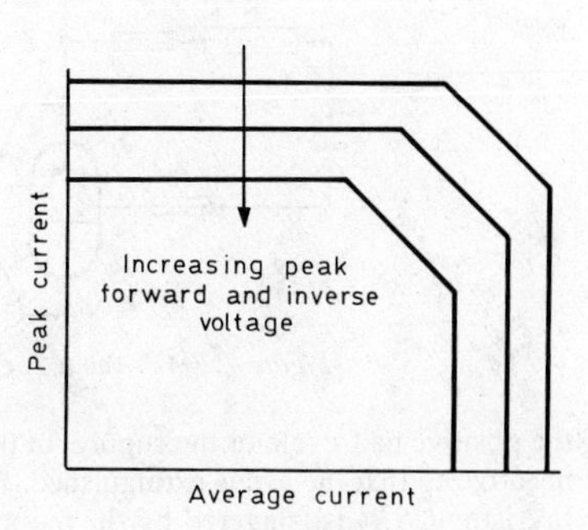

Figure 7.202 Peak current plotted against average current with peak forward and reverse voltages as parameter, rectifier operation

The voltage drop across a conducting ignitron, called the arc voltage drop, varies with the current through the tube in a similar way to that in other gas-filled valves. The value of arc voltage drop is typically 20 V to 25 V for welding ignitrons, and 25 V to 35 V for rectifier ignitrons. The higher value is a result of the baffles required by a rectifier ignitron. A curve of the variation of arc voltage drop with ignitron current is sometimes included in the data.

Triggering

Two methods of triggering ignitrons into conduction, often called *firing*, can be used according to the application. In resistance welding and other power-control applications, two ignitrons are connected in inverse-parallel as shown in *Figure 7.203*. Ignitron V_1 conducts (when triggered) for the positive half-cycle of the supply, while V_2 conducts for the negative half-cycle. The current through the load therefore consists of a series of

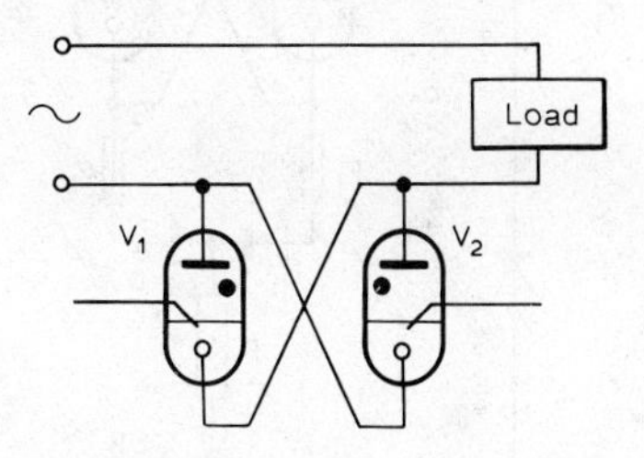

Figure 7.203 Ignitrons connected in inverse-parallel configuration

alternate pulses whose duration is determined by the angle at which the ignitron is triggered. It is for this reason the demand current for welding ignitrons is specified for two tubes in inverse-parallel. The method of triggering such an inverse-parallel pair is called *anode excitation* or *anode firing*. When ignitrons are used for power rectification, *separate excitation* has to be used.

An anode excitation circuit is shown in *Figure 7.204*. If the supply voltage is such that the anode of ignitron V_1 is positive, and the thyristor CSR_1 is triggered by an external mains-synchronised control circuit, a positive voltage is applied to the ignitor of V_1. The ignitron will be fired, and current flows through the load for the remainder of

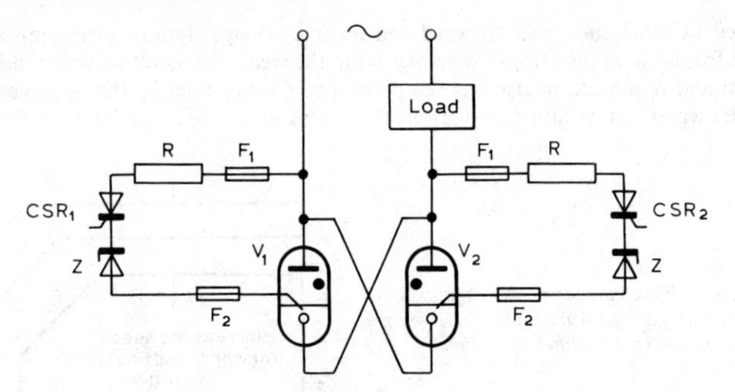

Figure 7.204 Anode excitation circuit using two thyristors

the positive half-cycle of the supply. In the following half-cycle the anode of V_1 becomes negative so that the arc is extinguished. However, the anode of V_2 is now positive, and if thyristor CSR_2 is triggered by the mains-synchronised control circuit, ignitron V_2 will be fired. Current again flows through the load. The resistor R limits the current in the ignitor circuit, values being given in the published data for various mains voltages to ensure sufficient current flows for reliable firing. The voltage regulator diode Z cuts off the thyristor when the ignitron is conducting by biasing the cathode of the thyristor at a voltage above that of the anode of the ignitron. It can be seen, however, that anode excitation depends on the value of anode voltage, and there will be a minimum value below which firing cannot take place.

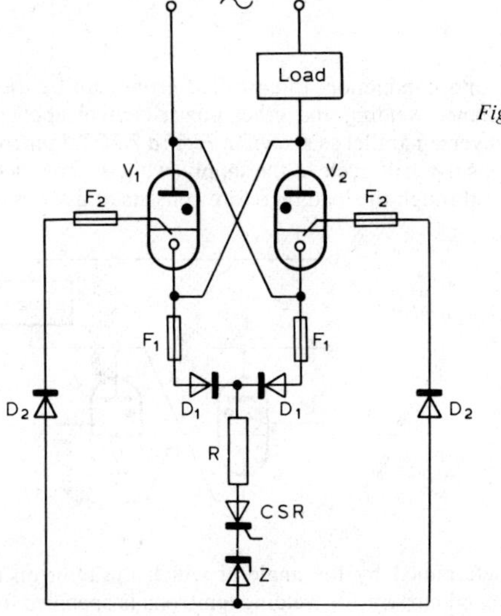

Figure 7.205 Anode excitation circuit using one thyristor

An alternative anode excitation circuit is shown in *Figure 7.205*. This circuit uses only one thyristor, but the operation is similar to that described for the previous circuit.

To ensure a short and reproducible delay between the firing of the ignitron and the anode takeover of the arc, the rate of rise of ignitor current must be sufficiently high. The variation of ignitor current for reliable firing with time is shown in *Figure 7.206*, the area enclosed by the curve being that in which ignition is ensured. Also shown in

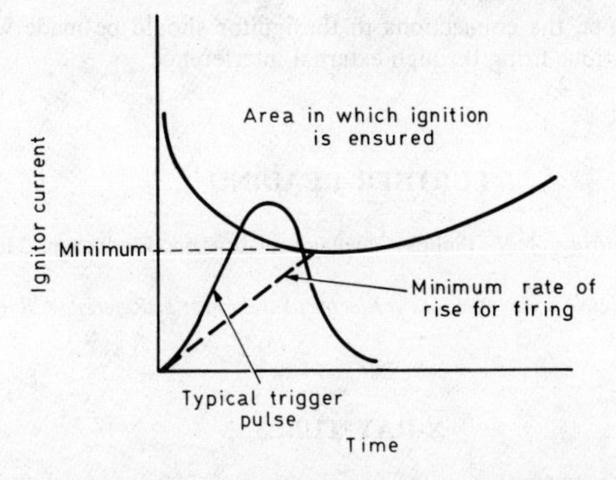

Figure 7.206 *Ignitor current characteristic for reliable firing*

Figure 7.206 is a typical trigger pulse. The pulse must enter the area in which ignition is ensured for the ignitron to be reliably fired. If the rate of rise of current is less than the slope of the line to the minimum of the curve, the ignitron will not fire. If the rate of rise of ignitor current is too high, a large current must flow to produce sufficient ionisation from the ignitor-to-cathode discharge to enable the anode-to-cathode arc to be struck. The rate of rise of ignitor current is determined mainly by the reactance of the load, and at high reactance values the rate of rise may be too low for reliable firing. Separate excitation must then be used for the inverse-parallel pair, as shown in *Figure 7.207*.

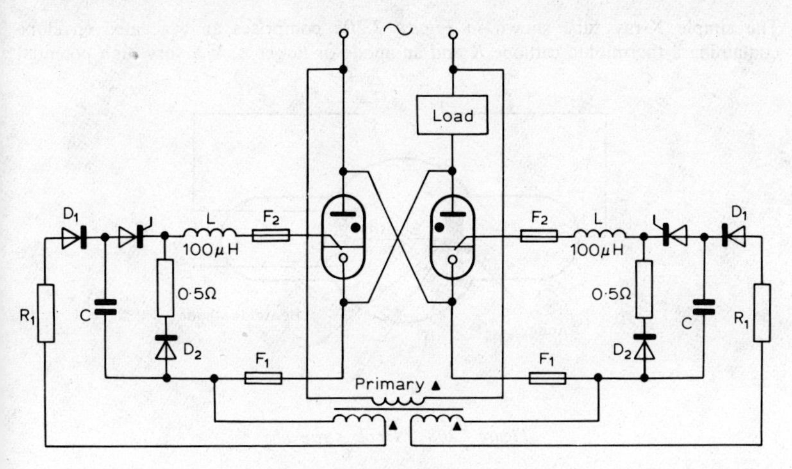

Figure 7.207 *Separate excitation circuit*

Separate excitation enables the ignitron to be fired independently of the anode circuit conditions. It is possible to control a.c. circuits of lower voltages than with anode excitation, and to control inductive loads where the low power factor would prevent satisfactory operation with anode excitation. Separate excitation is used for ignitrons operating as rectifiers, only half the ignition circuit of *Figure 7.207* then being required. The capacitor is charged through diode D_1, and when the thyristor is triggered, the capacitor discharges through the inductor into the ignitor to fire the ignitron.

In all firing circuits, the connections to the ignitor should be made with screened cable to prevent spurious firing through external interference.

FURTHER READING

BAKER, P. J., ed., *Ignitrons*, N.V. Philips Gloeilampenfabrieken, Eindhoven, The Netherlands (1968)

SAY, M. C., ed., 'Mercury Arc Rectifier', *Electrical Engineer's Reference Book*, 13th edn., Butterworths (1973)

X-RAY TUBES

It is now known that the X-ray is an electromagnetic radiation lying in the wavelength region of 10^{-10} metres (10^{12} MHz) between ultra-violet and γ radiation (Section 4).

Proof of the electromagnetic nature of X-rays came from early experiments which showed that X-rays travel in straight lines, that their intensity at a distance obeys the inverse square law, that they are not deflected by magnetic fields and that they can be transmitted through a vacuum. More recent investigations proved that X-rays can be polarised and refracted and can produce interference and diffraction.

Generation of X-rays

The simple X-ray tube shown in *Figure 7.208* comprises an evacuated envelope containing a thermionic cathode *K* and an anode or target *A*. If a very high potential

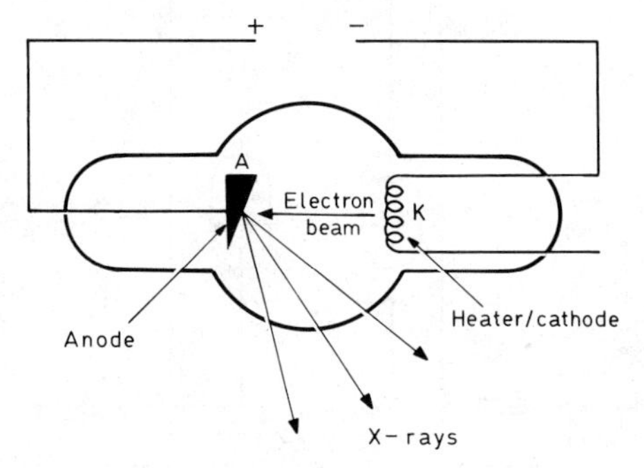

Figure 7.208 Simple X-ray tube

difference is applied between anode and cathode, with the anode positive, electrons emitted from the cathode by thermionic emission are accelerated towards the anode and strike it with considerable energy, producing X-rays, the intensity of the X-rays being a function of the current in the tube. These X-rays emanate from the surface of the anode in all directions.

In practice, of course, no modern X-ray tube is as simple as the one illustrated in *Figure 7.208*. It is sufficient to explain the principles involved and is not unlike many of the early X-ray tubes manufactured in the early part of the century.

The manner in which X-rays are produced at the anode is complex since, as will be seen, two different reactions take place.

Study of these X-rays reveals that a number of different frequencies are emitted and with the aid of a spectrometer a graph of intensity versus frequency can be built up. A typical graph is shown in *Figure 7.209*.

As will be seen, the typical spectrum consists of a continuous or 'white' range of

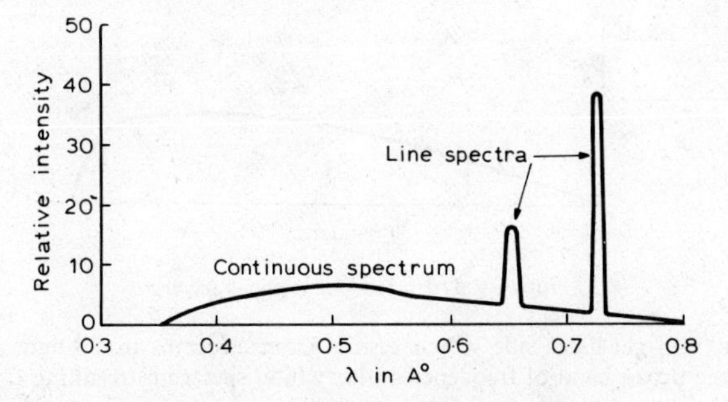

Figure 7.209 Spectrum of a typical X-ray tube

frequencies upon which are superimposed a number of sharply defined spectral lines. These two types of spectra are believed to originate in different ways and will be dealt with separately.

THE CONTINUOUS SPECTRUM

When a fast moving electron strikes an atom under the surface of the target, its energy of motion is converted into radiant energies. The electron may lose all its energy in one collision in which case a radiation of frequency given by the quantum relation $E_e = hf$ is produced, where E_e is the original energy of the electron, h is Planck's constant and f is the frequency. Thus if the original energy E_e is sufficiently high, i.e., the electron has sufficiently high velocity, then a high frequency radiation is given off as an X-ray.

Thus the highest frequency of radiation given out as a result of the electron losing all its energy in a single collision, being dependent upon the electron velocity, is a function of the accelerating force, i.e., the anode-to-cathode potential.

Thus for any given applied voltage V, a radiation of frequency f_{max} is produced where

$$V = \frac{hf_{max}}{e} \text{ (}e\text{ is the charge on the electron)}$$

or, more simply,

$$V = \frac{12.4}{\lambda_{min}}$$

where λ is the wavelength in Å and V is in kilovolts. For example, if $V = 10$ kV, $\lambda = 1.24$ Å.

However, it is possible for an electron to lose its energy not in a single collision but as a result of a number of glancing blows with a successive number of target atoms and, in this case, each collision produces a radiation, the frequency depending upon the energy lost.

This does not necessarily mean that X-rays are produced at each successive collision—some of the collisions may produce electromagnetic waves of much lower frequency, i.e., light or heat.

Thus, if an electron of energy E_e, say, strikes three target atoms before coming to rest and loses energy E_1 at the first collision, E_2 at the second and E_3 at the third, then electromagnetic waves of three different frequencies f_1, f_2 and f_3 will be produced.

The continuous spectrum illustrated in *Figure 7.210* may therefore be considered to be caused by two different effects—firstly the sharp cut off at the maximum frequency

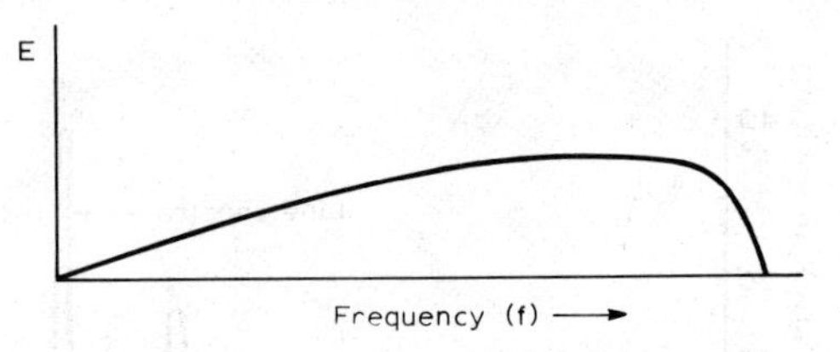

Figure 7.210 X-ray continuous spectrum

shown on the right-hand side which results from electrons in a single collision and secondly the broad band of frequencies (the 'white' spectrum) resulting from electrons losing their energy in a number of separate collisions.

LINE SPECTRA

The frequencies of the sharply defined line spectra, however, have been found experimentally to be a function of the element from which the target is manufactured and these X-rays are believed to emanate from deep within the target atoms, mostly from the electrons in the three inner shells.

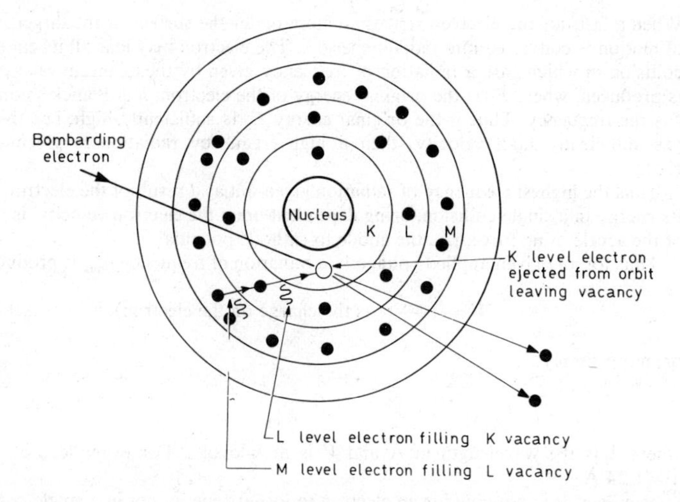

Figure 7.211 Production of line spectra

In the normal atom each of these shells contains its normal complement of electrons. It is possible, however, for a cathode-emitted electron to penetrate deep into the atom and to eject an electron which is contained in one of the inner shells, thus leaving a vacancy, *Figure 7.211*.

If this should occur, an electron from the next shell (higher energy) falls into the vacancy, giving up energy as a quantum of X radiation. As is shown in *Figure 7.211* where an electron has been ejected from the *K* shell and is filled by an *L* shell electron giving out a quantum of X radiation and leaving an *L* shell vacancy which is filled by an electron from the *M* shell producing another quantum of radiation at a different frequency.

The energy given up by an electron falling from one shell to the next is given by the energy difference between the two shells. Thus the frequency of the radiation is calculated from:

$$f = \frac{E_2 - E_1}{h}$$

where E_2 and E_1 are the energy levels of the two shells.

In the case of the *K*, *L* and *M* shells the energy difference is sufficient to produce X-ray frequencies and, since this energy difference varies from element to element, the frequency of the line spectra is a characteristic of the element. The production of line spectra requires a minimum potential difference which varies according to the atomic number of the target element. The higher the atomic number, the greater is the voltage required to give the electron sufficient energy to enter the inner shells of the element and to eject an orbital electron.

For example, the minimum potential differences to eject *K* shell electrons in copper and molybdenum are as follows:

Copper (atomic number 29)	$V_{min} = 8.9$ kV
Molybdenum (atomic number 42)	$V_{min} = 29$ kV

Of all the electrons leaving the cathode of an X-ray tube, less than 1% actually produce X-radiation. The energy of the remaining electrons is dissipated as heat and massive cooling systems must be used in practice.

Properties of X-rays

X-rays, like light rays, are electromagnetic in character and many of the properties of light are also properties of X-radiation subject to the fact that X-rays have a much higher frequency and are therefore more energetic.

Many substances which are transparent to light are also transparent to X-rays but, owing to their higher energy, X-rays can also pass through heavier or denser materials. Different elements absorb X-rays to a different degree and in fact the amount of absorption is proportional to the atomic weight and the density.

Röntgen proved that X-rays would pass more easily through flesh than bone and thus provided the medical profession with a means of investigating the human body without resorting to surgery (diagnosis).

X-rays also have the property of ionising gases. The X-ray energy produces a few high speed electrons in the gas and these, in turn, liberate a large number of low energy electrons by a collision process. From the point of impact on the material under bombardment, reflected or scattered X-rays are found to leave the material in such a manner as to obey the optical laws of reflection (angle of incidence equals angle of reflection).

Other X-rays, however, are also evident. These—the characteristic X-rays—are particular in nature to the properties of the material under bombardment. They are, however, only produced if the energy of the incident X-ray beam is greater than the minimum energy required to displace *K*, *L* or *M* electrons in the material under bombardment.

Finally, β particles or high speed electrons are also evident but can only be detected in the area of impact as a result of their limited range in a medium. As stated, these arise mainly from an inverse X-ray effect. Some of the particles, however, may result from a photo-emissive effect on the surface of the material.

X-rays themselves are affected by the medium, the attenuation as already stated being proportional to the atomic weight or density. Even in air there is a measurable attenuation.

Intense X-rays also affect living tissue by inhibiting the growth of body cells, destroying tissue and causing inflammation as a result of the production of intense heat. Although the biological action of X-rays is fundamentally injurious, it is this very property which is used in the annihilation of diseased body cells. This treatment is called radiotherapy.

SECONDARY RADIATION

A most interesting phenomenon associated with X-ray bombardment is the complex character of a secondary X-radiation which occurs. This secondary radiation comprises scattered X-rays, characteristic X-rays and β particles. The scattered X-rays may be assumed to be analogous to reflected or refracted light and the β particles, having the same velocity as the electron beam in the X-ray tube, probably result from a type of inverse X-ray effect.

The characteristic X-rays, however, arise from the ejection of electrons deep in the inner shells of the atoms of the scattering material and are produced by a process similar to the production of line spectra. The secondary radiations are therefore a characteristic of the scattering material and provide a convenient method of analysis.

Diffraction and X-ray crystallography

Two properties of X-rays which are common to all electromagnetic rays are diffraction and interference and it was in 1912 that the scientist von Laue suggested that X-rays might be diffracted if the regular arrangements of atoms in a crystal were used as a diffraction grating. In 1913 W. H. Bragg used crystal diffraction to measure the wavelengths of X-rays, calculating the crystal spacing from Avogadro's number. X-ray diffraction by crystals has led to an entirely new field of research—X-ray crystallography—three principal methods being in present day use.

LAUE METHOD

In the Laue method shown in *Figure 7.212* (*a*), a continuous spectrum of X-rays is allowed to pass through a single crystal X by way of a collimator C which narrows the

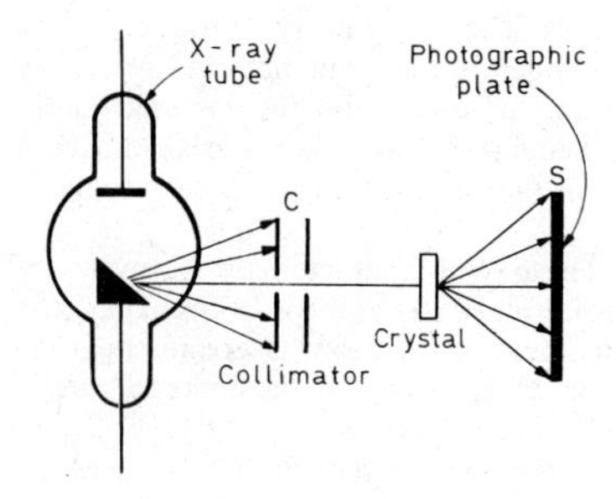

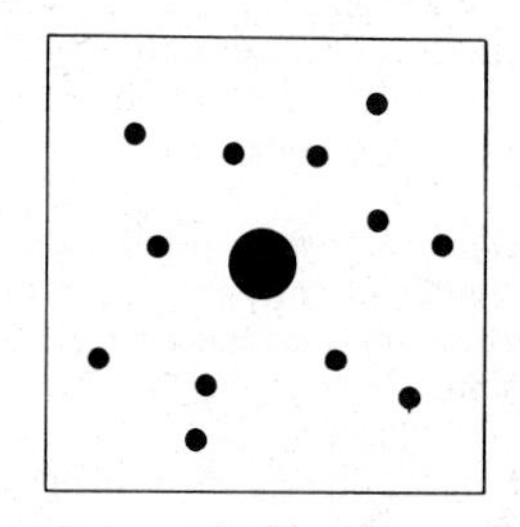

Figure 7.212 (*a*) *Laue method;* (*b*) *Laue pattern*

beam. X-ray diffraction occurs within the crystal and a number of reinforcement spots appear on a photographic screen *S* placed beyond the crystal. A typical Laue pattern is also shown in *Figure 7.212* (*b*).

BRAGG METHOD

In the Bragg method a narrow X-ray beam is reflected by a crystal. Interference occurs and the positions of the resultant spectral lines are investigated using an ionisation chamber or a photographic plate.

The principle of Bragg's apparatus, known as an X-ray spectrometer, is illustrated in *Figure 7.213* (*a*).

The incident beam of X-rays from the source *A* is limited by the collimator *B*. The reflecting crystal *C* is mounted in wax on the table which rotates, its position being determined by the vernier *V*. An arm *D*, rotating about the same axis as the table, carries the ionisation chamber *E*. In practice the ionisation chamber and the crystal are

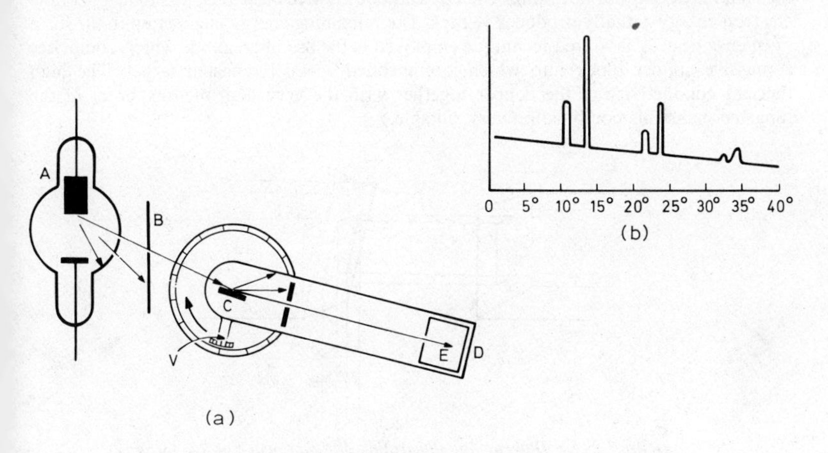

Figure 7.213 (*a*) *Bragg's apparatus;* (*b*) *Typical spectrum*

rotated slowly and the magnitude of ionisation in the chamber is noted. A graph of ionisation versus angle of rotation can then be plotted.

Figure 7.213 (*b*) shows a typical spectrum obtained in this manner. The maxima correspond to line spectra in the original beam.

POWDER METHOD

The powder method has the advantage of requiring only a very small quantity of the material under test. Crystals are ground to a very fine powder through which an X-ray beam is allowed to pass. The method is essentially that of Laue except that monochromatic X-radiation is used. The random orientation of the crystal planes in the powder produces a circular pattern which is characteristic for a particular element of compound as each circle corresponds to reflection from a given set of crystal planes. A typical powder pattern is illustrated in *Figure 7.214*.

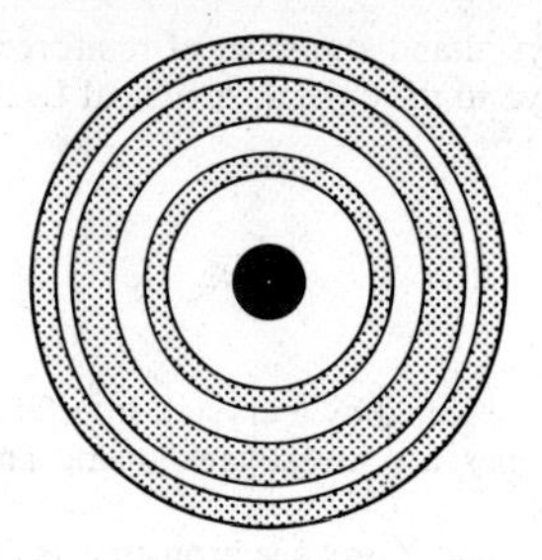

Figure 7.214 Powder method pattern

X-ray tubes

The principles of the simple X-ray tube have already been given. Here the more complete constructional details of several current types of X-ray tube are discussed.

The most difficult electrode to design is the anode or target. At the focal point, i.e., the small area of the anode bombarded by cathode-emitted electrons, less than 1% of the electron energy actually produces X-rays. The remaining energy is given up in the form of intense heat. A favoured technique employed is the use of an anode which comprises a massive copper block into which is embedded a small tungsten target. The high thermal conductivity of the copper together with the very high melting point of the tungsten make this combination very suitable.

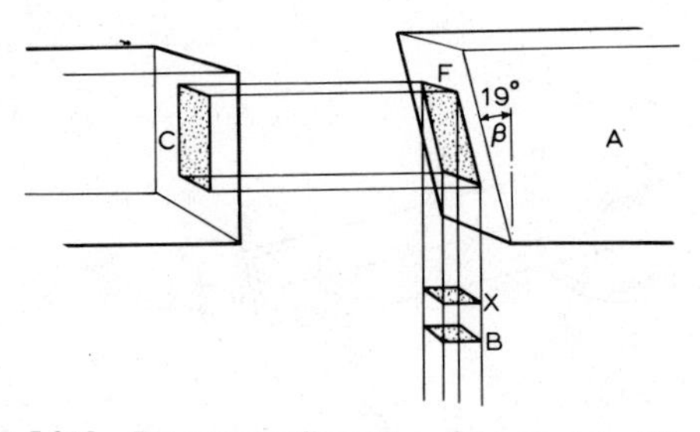

Figure 7.215 Diagramatic illustration of the principle of line focus

All X-ray pictures are shadows and so cannot be sharp unless the source is of small dimensions. In the older tubes the cathode rays were focused on a small circular area but many modern tubes have what is called a line focus, i.e., the focal area is a narrow line. This prevents excessive heating and enables the X-ray output of the tube to be increased.

Figure 7.215 illustrates the principle of line focus. If *A* is the target, *F* is the focal area and *CF* is the axis of the tube, it is seen that the electron beam travelling along the axis *CF* is rectangular in cross-section. The face of the target *A* is cut at an angle of 19° to the direction from which the X-rays are viewed, *BF*. At *B* the X-rays appear to have originated from a small square-shaped area of dimensions as shown at *X*. (In this diagram the cross-sectional areas of the electron and X-ray beams have been considerably enlarged for the purpose of clarity.)

X-ray tubes can effectively be divided into two types—those used for treatment (therapy) and those used for diagnosis.

Therapeutic tubes give rise to demands for an efficient cooling system owing to the necessity of operation over long periods of time.

Figure 7.216 shows the schematic diagram of a typical X-ray tube used for diagnosis. This comprises a substantial heater surrounded by a cup-shaped assembly *K* which is

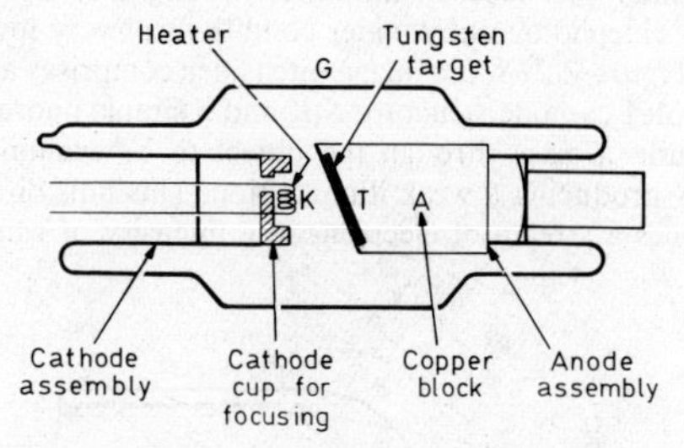

Figure 7.216 *Typical X-ray tube*

negatively charged with respect to the anode so as to form an electron beam that focuses on the target *T*. The anode comprises a massive block of copper *A*, into which a piece of tungsten *T* is embedded, this acting as the actual target for the electrons. All the electrodes are situated in an evacuated tube *G* which is itself encased with a lead shield having a small aperture *X* from which the X-rays radiate.

In addition to this type of tube, the rotating anode X-ray tube is also used for diagnosis. In the rotating anode tube, as shown in *Figure 7.217* the anode *A* is made to rotate at a speed of approximately 3 000 r.p.m. so that the area of heat is moved away

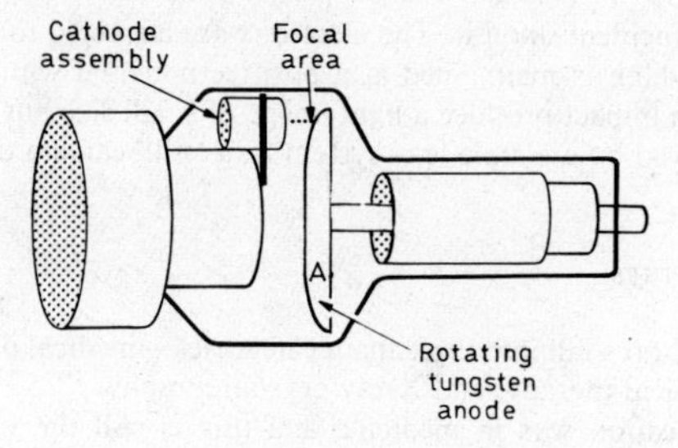

Figure 7.217 *Rotating anode X-ray tube*

from the focal point before damage to the anode can occur. In such tubes, therefore, the anode may be manufactured entirely from tungsten. Rapid rotation of the anode under vacuum conditions has given rise to lubrication difficulties. As grease or oil cannot be used, the modern rotating anode tube uses self-lubricating ballbearings. The torque is applied by electric currents induced into the anode assembly by a rotating external field, using the principle of the induction motor.

X-ray tubes may be cooled in a number of ways. Water, fan or oil may be used separately or in conjunction with each other, depending on the cooling requirements of a particular tube. Often the complete tube is immersed in an oil-filled container lined with the lead shield.

X-RAY IMAGE INTENSIFIER

Before the X-ray image intensifier was introduced, the only method of viewing X-rays directly was by means of a fluorescent screen and, unfortunately, unless the X-ray intensity is maintained at a dangerously high level, the intensity of illumination provided by the fluorescent screen is extremely low indeed.

The image intensifier provides an illumination bright enough not only for direct viewing but also for cinephotography under conditions of very low X-ray intensities.

As illustrated in *Figure 7.218*, the image intensifier comprises an evacuated envelope *E* containing a complex cathode structure *S*,*K* and a simple fluorescent screen *A*.

X-rays from a tube *X* pass through the object to be examined (*O*) and fall on a fluorescent screen *S* producing a weak illumination. This falls directly on a photocathode *K* which produces a stream of electrons, the intensity of which depends upon the

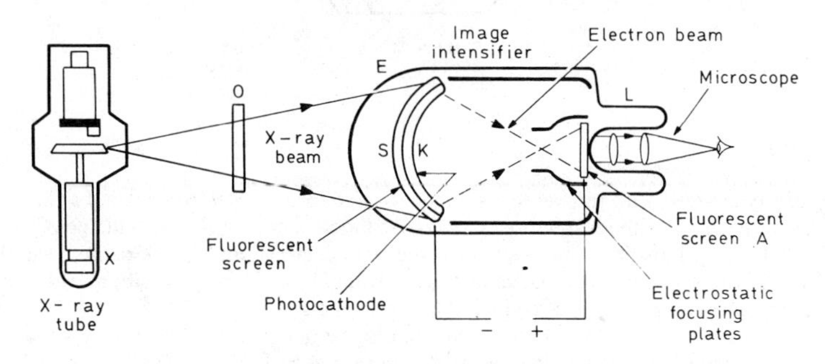

Figure 7.218 *The X-ray image intensifier*

intensity of light incident upon it. The electrons are attracted towards a second fluorescent screen *A*, which is maintained at a positive potential with respect to the photocathode, and upon impact produce a light image of small size but of high intensity. This image can be viewed by a simple lens system *L*, a 'still' camera or a cine camera.

Application of X-rays

Applications of X-rays fall into three main categories—medical diagnosis and industrial examination, medical therapy, and X-ray crystallography.

The first application was in medicine and this is still the widest use. The skilled radiologist can diagnose internal conditions with far greater certainty than can be got from external examination. The X-ray image intensifier has provided doctors with a brilliant optical picture with low intensity dose for the patient. Combined with cineradiography this makes possible permanent records on film of the functioning of internal parts of the body.

X-rays are important in industry as they offer a means of non-destructive examination and testing. Products like electrical transformers can be radiographed to show whether they have been assembled properly. X-ray photographs of steel welding can show faults in the weld that might lead to fracture. A casting may appear sound but an X-ray can reveal the presence of weakening bubbles. In aircraft examinations, parts which are heavily stressed such as a wing section are repeatedly X-rayed for fatigue defects.

Intense X-rays damage living cells and this property can be used to destroy malformed cells in the human body. For therapy treatment very high energy X-rays are used. These can be provided by a linear accelerator in which electrons are accelerated to very high velocities before they are allowed to strike the target. The X-rays thus produced are therefore of very high frequency and have, as a consequence, the power of deep penetration.

The use of X-rays to study crystal structure is an important application and much useful data concerning crystals has been obtained in this manner.

The study of X-rays has in itself led to an increase of man's knowledge in other fields of physics apart from crystallography and X-ray analysis.

The scientist Moseley has shown that if the square roots of the frequencies of the line spectra are plotted against the atomic numbers of the elements producing the lines a smooth curve is obtained, *Figure 7.219*. Moseley's results proved to be of the utmost

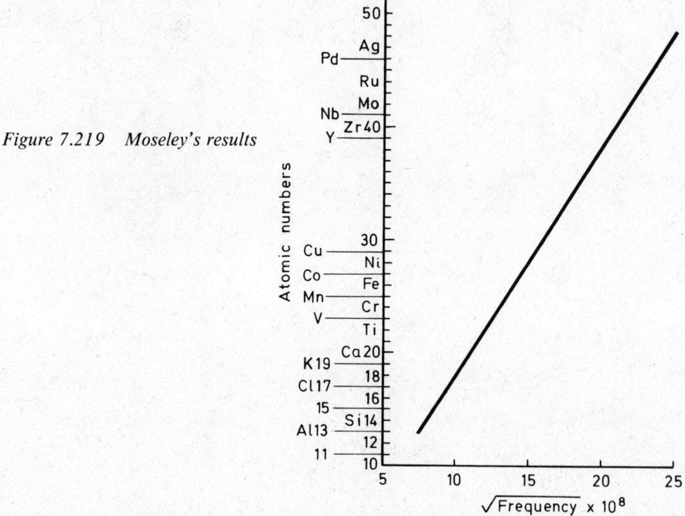

Figure 7.219 Moseley's results

importance as many of the modern theories of the atom required experimental proof at that time and Moseley's work helped considerably in predicting the existence of elements which were at that time unknown.

Additional miscellaneous applications include the location of pearls in oyster shells, examination of 'valuable' paintings for possible fraud by analysis of paints and materials, detection of contraband in luggage and examination of gems to discover imitations.

ACKNOWLEDGEMENT

The material for 'X-ray Tubes' was kindly made available by Mullard Limited.

8 SOLID-STATE DEVICES

SOLID-STATE DEVICES 8

8 SOLID-STATE DEVICES

Solid-state devices have transformed the electronics industry over the past two decades. Discrete semiconductor diodes and transistors have replaced the thermionic valve as the general-purpose electronic device, relegating valves to the most high-power and specialised applications. New types of device whose operation is based solely on the properties of semiconductor materials have been introduced. Now, the integrated circuit is beginning to replace discrete transistors and the associated passive components.

All solid-state devices depend for their operation on the controlled flow of charge carriers through semiconductor materials. A description of the properties of semiconductor materials is given in Section 6.

SEMICONDUCTOR MATERIALS

A semiconductor material is one whose conductivity lies between that of a metallic conductor and that of an insulator. This rather loose definition covers a wide range of materials including, among others, silicon and germanium, gallium arsenide, cadmium sulphide and indium antimonide, and the oxides of iron, zinc, cobalt, and nickel. Of these materials, the best known are silicon and germanium from which most present-day solid-state devices are manufactured. The other materials are used in more specialised devices. This general description of the properties of semiconductor materials will therefore be confined to silicon and germanium. The properties of the other materials are described elsewhere with the devices in which they are used.

The position of semiconductor materials between insulators and conductors can be seen from the following resistivity values. The resistivity of glass, an insulator, is approximately 2×10^{11} Ωm; that of copper, a conductor, is 17×10^{-8} Ωm. The resistivities of germanium and silicon at 27 °C are 4.7×10^{-3} Ωm and 3.0×10^{3} Ωm respectively.

Both silicon and germanium are tetravalent materials, each atom having four valence electrons in the outer electron shell. These valence electrons form covalent bonds with the valence electrons of neighbouring atoms, as shown diagrammatically in *Figure 8.1* (*a*). The larger circles represent the nucleus and inner electron shells; the smaller circles the valence electrons. Four valence electrons, one from each of the four outer atoms, combine with the four valence electrons of the central atom so that, in effect, the central atom 'shares' eight electrons. This produces a stable state giving the material a definite crystal structure. A simplified two-dimensional representation of the crystal lattice of silicon and germanium is shown in *Figure 8.1*(*b*).

The structure shown in *Figure 8.1*(*b*), strictly speaking, is true only at absolute zero. At room temperature, the atoms are oscillating about an equilibrium position in the lattice because of their thermal energy. Some of this energy is shared with the valence electrons, and occasionally an electron may acquire sufficient energy to break away from its atom. The electron is then free to drift through the lattice. The break-away electron leaves a vacancy or *hole* which can absorb a free electron, or attract another electron causing it to produce a new hole. Thus the net effect is a random movement of electrons and holes through the lattice caused by electron–hole pair generation and recombination. The number of free electrons and holes present at any instant is proportional to temperature. At room temperature the number is small, and this results in the relatively high resistivity (compared with a conductor) of intrinsic (or pure) silicon and germanium.

To provide the controlled flow of current required in a solid-state device, a large number of charge carriers must be introduced. These carriers are provided by adding precise amounts of an impurity to the semiconductor material. To allow the degree of control necessary, the intrinsic semiconductor material itself must be very pure, and the crystal structure as perfect as possible. Thus a single crystal of intrinsic silicon or germanium is grown which is cut into thin slices from which the device is manufactured. When the impurity has been added, the semiconductor material is described as extrinsic, and the addition of the impurity is called doping.

If the impurity introduced is a pentavalent material (such as arsenic, antimony, or

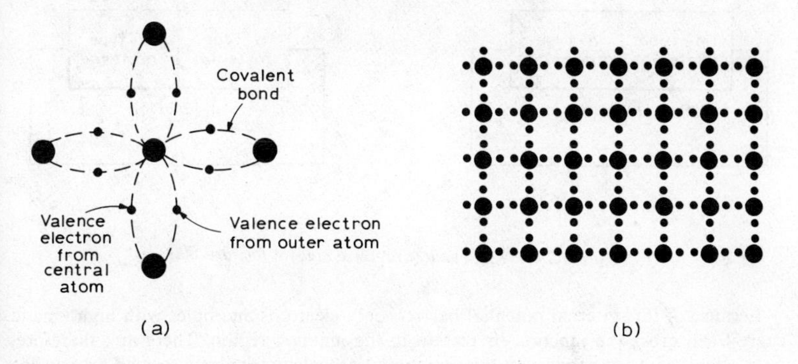

Figure 8.1 Crystal structure of silicon and germanium: (a) electron sharing between atoms in crystal (b) crystal lattice

phosphorus), it will have five valence electrons. Four of these will 'fit' into the crystal structure, leaving the fifth electron available as a charge carrier. The impurity in this case is called a donor. A semiconductor material containing a donor is called *n*-type because the available charge carriers are electrons or negative charges.

If the added impurity is a trivalent material (such as aluminium, indium, or boron), it will have three valence electrons. Three can fit into the crystal structure, leaving a hole. Because this hole can attract an electron which in turn will leave a hole, the hole forms a virtual positive charge and will act as a charge carrier. The impurity this time is called an acceptor, and a semiconductor material containing an acceptor impurity is called *p*-type because the charge carriers are holes or positive charges.

The electrons and holes formed by the addition of the donor and acceptor impurities are called majority carriers. As the name implies, these carriers are responsible for most of the current in the extrinsic semiconductor material and lower the resistivity considerably. In addition to the majority carriers, there are a small number of free electrons in *p*-type material and a small number of free holes in *n*-type material, produced by thermal agitation of the atoms in the crystal as mentioned previously. These electrons and holes are also available to carry a current, and are called minority carriers.

PN JUNCTIONS

If part of a single crystal of silicon or germanium is formed into *p*-type material, and part formed into *n*-type material, the abrupt interface between the two types of material is called a *pn* junction. As soon as the junction is formed, majority carriers will diffuse across it. Initially both types of material are electrically neutral. Holes will diffuse from the *p*-type material into the *n*-type material, and electrons from the *n*-type material into the *p*-type. Thus the *p*-type material is losing holes and gaining electrons, and so acquires a negative charge. The *n*-type material is losing electrons and gaining holes, and so acquires a positive charge. The negative charge on the *p*-type material prevents further electrons crossing the junction, and the positive charge on the *n*-type material prevents further holes crossing the junction. This space charge creates an internal potential barrier across the junction, shown diagrammatically in *Figure 8.2* by the battery in broken lines.

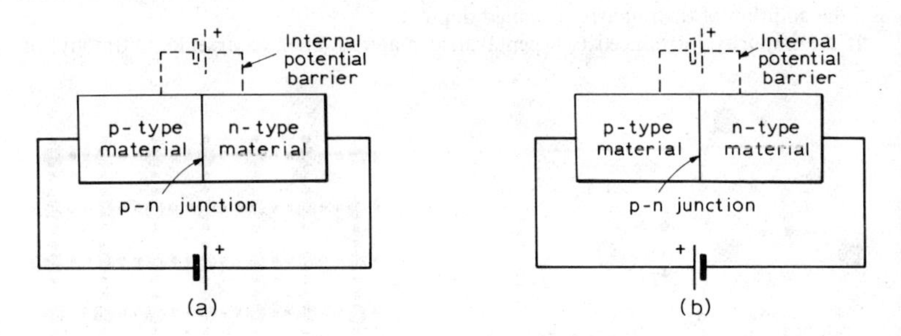

Figure 8.2 PN junction: (a) reverse bias (b) forward bias

Because of this internal potential barrier, only electrons and holes with high kinetic energy can cross the junction or remain in the junction region. There are, therefore, only a few majority carriers in the region and a depletion layer is formed around the junction. The potential barrier, however, will assist minority carriers to cross the junction.

If an external battery is connected across the *pn* junction with the positive terminal connected to the *n*-type material, *Figure 8.2(a)*, the junction is reverse-biased. The external battery adds to the internal potential barrier and prevents even the few majority carriers in the depletion layer with high energy from crossing the junction. The battery voltage helps minority carriers to cross, but because of the small number only a very small reverse current flows.

Because the depletion layer contains few charge carriers, it is effectively an insulator separating the conducting *p*-type and *n*-type regions. A parallel-plate capacitor is therefore formed. The width of the depletion layer is proportional to the reverse voltage, and so the capacitance can be varied by varying the reverse bias across the junction. Constructive use of this effect is made in variable-capacitance (varactor) diodes. On the other hand, the capacitance can limit the performance of switching diodes at high frequencies.

An equivalent circuit for a reverse-biased *pn* junction is shown in *Figure 8.3*. The depletion layer is represented by the voltage-variable capacitor C_d in parallel with the voltage-variable resistor R_j. The material outside the depletion layer has a

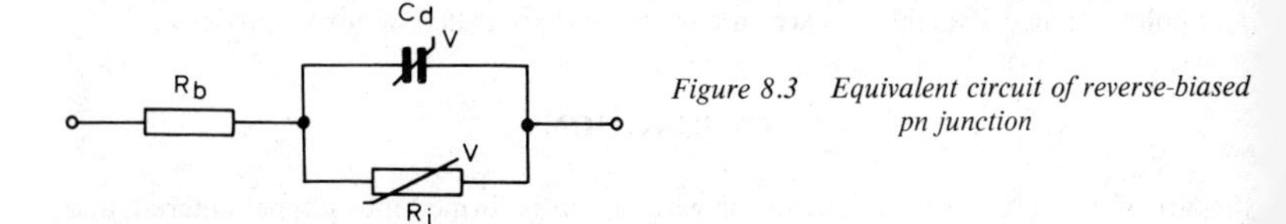

Figure 8.3 Equivalent circuit of reverse-biased pn junction

resistance, and this bulk resistance is represented by R_b in series with the junction impedance.

As the reverse voltage is increased, the reverse current remains reasonably constant until a voltage is reached at which breakdown occurs. This breakdown is caused by an increase in the number of charge carriers available so that a large, and possibly destructive, current flows.

The increase in carriers may be caused by the avalanche effect or the zener effect. Briefly, avalanche breakdown is caused by ionisation of the semiconductor material to produce the charge carriers, while in zener breakdown electrons are torn away from their atoms to become charge carriers. Constructive use is made of the two breakdown mechanisms in avalanche diodes and zener diodes, and both mechanisms will be discussed in more detail later when describing these two devices.

If the external battery is connected across the junction with the positive terminal connected to the *p*-type material, *Figure 8.2(b)* the junction is forward-biased. The battery voltage opposes the internal potential barrier, and as soon as the voltage exceeds this potential, majority carriers flow across the junction. A small increase in forward voltage therefore causes a large increase in forward current.

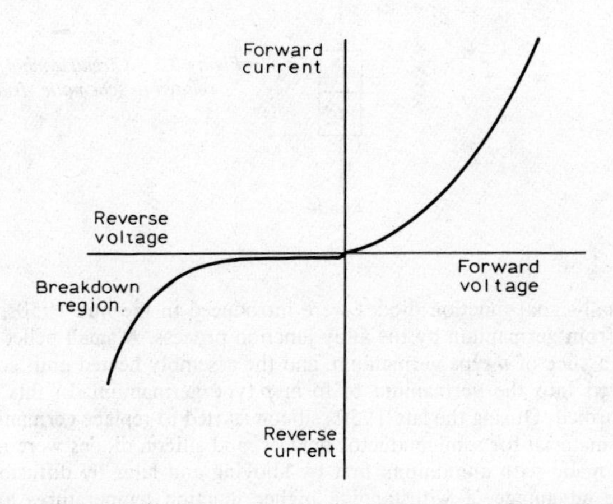

Figure 8.4 Voltage/current characteristic of junction diode

The asymmetrical current flow across a *pn* junction enables it to be used as a rectifying element. Thus a semiconductor device containing a single *pn* junction can be used as a diode. The general shape of the voltage/current characteristic of a semiconductor junction diode is shown in *Figure 8.4*.

Another cause of breakdown in a *pn* junction, besides excessive reverse voltage, is too high a junction temperature. Even with a forward-biased junction, a semiconductor diode will have a small resistance. Current flowing through the device will therefore generate heat, and will raise the temperature of the junction. Above a certain temperature the thermal agitation of the atoms in the crystal will destroy the junction. Sufficient thermal energy is transferred to the valence electrons that a large number of free electrons and holes are produced. These 'minority' carriers swamp the action of the majority carriers, and so prevent normal current flow taking place. Thus a maximum permissible junction temperature is specified for all semiconductor devices to preserve operation.

JUNCTION DIODES

All semiconductor junction diodes contain a single *pn* junction, but the properties and structure of the junction can be exploited differently to enable different types of diode to be obtained. For example, besides the normal rectifier diode, other types such as high-speed switching diodes, voltage regulator diodes and variable-capacitance diodes are available.

Small-signal diodes

Diodes which are used for demodulation and switching applications where only small amounts of power are involved can be conveniently grouped together as small-signal diodes. As such diodes are generally wired directly into circuits in the same way as resistors and capacitors, the usual construction is a glass or plastic envelopc with axial connecting wires. The circuit symbol for small-signal diodes is shown in *Figure 8.5*.

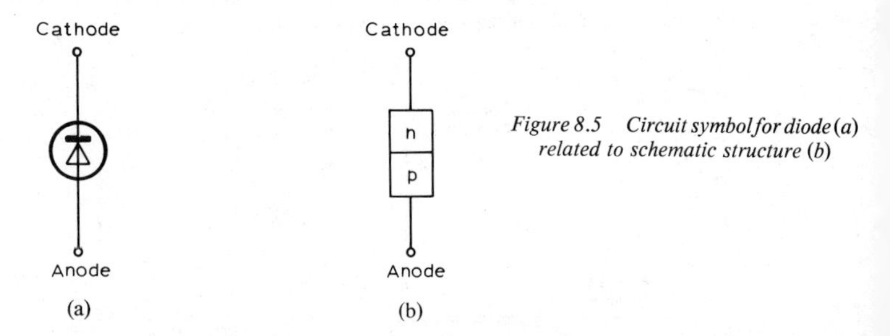

Figure 8.5 Circuit symbol for diode (a) related to schematic structure (b)

The first small-signal junction diodes were introduced in the mid-1950s, and were manufactured from germanium by the alloy-junction process. A small pellet of indium was placed on a slice of *n*-type germanium, and the assembly heated until some of the indium dissolved into the germanium to form *p*-type germanium. In this way a *pn* junction was formed. During the late 1950s silicon started to replace germanium as the generally-used material for semiconductor devices, and silicon diodes were introduced with junctions made with aluminium, first by alloying and later by diffusion. Silicon diodes had the advantage of withstanding higher junction temperatures and having lower reverse currents than germanium diodes. Typical values of the maximum junction temperature of germanium and silicon devices are 75 °C and 200 °C respectively. The reverse current for a germanium diode at a reverse voltage of 10 V at room temperature is typically 5 μA to 10 μA, while that for a silicon diode at the same voltage is typically 0·02 μA. On the other hand, the forward voltage before conduction occurs in a germanium diode is lower than that for a silicon diode, typically 200 mV instead of 600 mV. Germanium junction diodes are therefore still used today in those applications where the lower forward voltage is advantageous. The junctions, however, are made by diffusion rather than the alloying process.

The introduction of the planar technology in 1960 enabled planar and planar-epitaxial diodes to be manufactured. These methods of manufacture are described fully in the transistor section, but a brief description is given here.

A slice of *n*-type monocrystalline silicon is coated with silicon oxide. Windows are cut in this oxide, and boron diffused into the silicon through the windows to form regions of *p*-type silicon. The silicon oxide is etched away, and the slice cut up so that the individual diode elements containing a *pn* junction are obtained. Connecting leads to the *p*-type and *n*-type regions are attached, and the assembly encapsulated. This method of manufacture allows many thousands diodes to be manufactured at the same time—in fact, mass-production techniques can be applied.

Another advantage is that the diffusion through windows etched in the oxide allows precise control to be exercised over the geometry of the junction.

In the planar-epitaxial diode, the *pn* junction is formed in an epitaxial layer. This is a layer of silicon grown on the silicon slice (which now forms a substrate for the device) in which the silicon atoms take the same crystal structure as that of the underlying substrate. Thus the near-perfect crystal structure of the substrate (a slice from a single crystal) is transferred into the epitaxial layer. The layer is grown by depositing silicon from a vapour, and impurities can be added to the vapour to produce *p*-type and *n*-type epitaxial layers as required. This layer can be of high-resistivity silicon to give a high breakdown voltage for the diode. After the deposition of the epitaxial layer, the method of manufacture is the same as for planar diodes.

Two methods of construction are used for small-signal diodes. In one, the diode element is mounted on a metal backing plate with the same thermal coefficient of

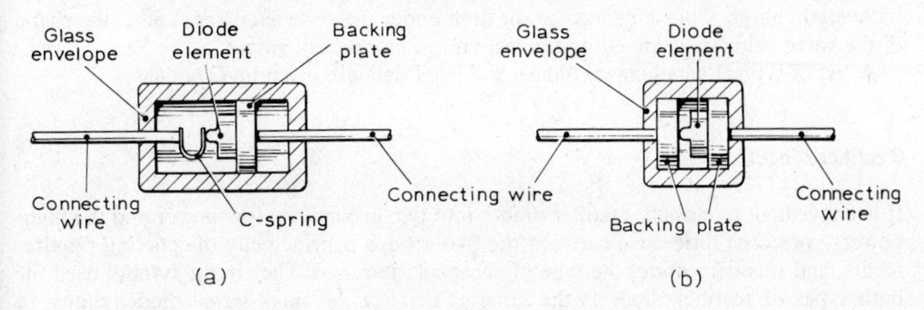

Figure 8.6 Construction of small-signal diodes: (a) spring-contact (b) whiskerless

expansion as the silicon diode element, for example, molybdenum. Contact with the other side of the diode element is made by a C-spring, as shown in *Figure 8.6(a)*. The connecting wires are attached to the backing plate and C-spring, and the whole assembly encapsulated in glass. The second method of construction produces the 'whiskerless diode'. The diode element is mounted between two metal backing plates, and the whole assembly is held rigidly by the glass envelope, as shown in *Figure 8.6(b)*. Because no spring is used in the whiskerless diode, this type has a greater reliability when used in applications where the diode is subjected to vibration.

Typical values of the forward currents that can be conducted by germanium junction diodes lie between 30 mA and 300 mA. For silicon diodes, typical forward currents lie between 100 mA and 1 A.

When diodes are used for switching, particularly at high frequencies, an important factor is the recovery time. If a diode is switched from the conducting to the non-conducting state, the current does not change instantaneously to the small reverse current but a larger reverse current spike is produced. This is caused by charge storage. When the diode is conducting, majority carriers cross the junction. Those which do not combine with majority carriers of the opposite polarity become, in effect, minority carriers. Thus a hole from the *p*-type region crossing the junction and not combining with an electron in the *n*-type region becomes a minority carrier in the *n*-type region. When the bias across the diode is reversed, these minority carriers recross the junction, and a large reverse current will flow until they have diffused or recombined to allow the depletion layer to be formed. Thus the reverse recovery time will depend on the lifetime of these minority carriers.

Another limitation on the performance of switching diodes is the capacitance associated with the depletion layer. At high frequencies, this capacitance may make the reverse impedance of the diode unacceptably low. To maintain the high-frequency performance, the capacitance can be lowered by decreasing the area of the junction (by such manufacturing techniques as mesa etching described in the transistor section). The small junction area will restrict the value of forward current that can be conducted by the diode.

Gold doping of silicon diodes reduces the lifetime of the minority carriers. By using this technique and using junctions of small area, silicon diodes capable of switching waveforms with rise times of 5 ns can be manufactured. Diodes capable of switching faster waveforms can be made by using gallium arsenide instead of silicon as the semiconductor material.

The circuit designer requires certain data on small-signal diodes to allow him to choose the right type for his needs. For the 'general-purpose' types, he must know the maximum reverse voltage the diode will withstand, the maximum forward current it will conduct continuously and for a short period, the forward voltage drop across the diode, and the reverse current. For switching diodes, additional information on the recovery time is required. This is usually given as a time under specified load and waveform conditions. An alternative method of specifying the switching characteristic is by the recovered charge. This is defined as the area under the reverse current spike, the shape of the spike being determined by the test circuit component values.

A list of typical small-signal diodes and brief data are given in *Table 8.1*.

Rectifier diodes

It is convenient to classify rectifier diodes into two groups: the low-power and the high-power types. The difference between the two groups is principally the cooling requirements, and this determines the type of encapsulation used. The circuit symbol used for both types of rectifier diode is the same as that for the small-signal diode, shown in *Figure 8.5*.

The amount of power that can be handled by a semiconductor diode is limited by the junction temperature which must not exceed a certain value if rectification is to be maintained. By 'scaling-up' small-signal diodes to provide a larger junction area, higher currents can be conducted. Such diodes can conduct average forward currents of, typically, 1 A and withstand surge currents of up to 30 A for short periods. With a reverse voltage rating of, typically, 600 V these diodes can be used for such applications as low-current power supplies from the a.c. mains, low-power inverters, and as rectifiers in the deflection circuits of television receivers. Because additional cooling is not required, the diodes can be encapsulated in plastic, and fitted with axial connecting wires for wiring into circuits.

Low-power rectifier diodes used in television deflection circuits and inverters are operating at relatively high frequencies. The diodes should therefore have short reverse recovery times, as with the switching diodes discussed previously. The information is either given as the recovered charge when switched under specified conditions or as the reverse recovery time.

A list of typical low-power rectifier diodes is given in *Table 8.2*, together with brief data.

High-power rectifier diodes were developed during the 1960s for use in rectifier systems operating from the mains supply. The aim was to replace existing systems using thermionic valves or mercury-arc rectifiers. Such diodes are required to conduct high currents and withstand large reverse voltages. Cooling, therefore, is an important requirement.

The current-carrying capacity can be increased by increasing the area of the junction and mounting the diode on a heatsink to increase the cooling and so keep the junction temperature below the limiting value. The reverse voltage can be increased by bevelling the edges of the diode element, as shown in *Figure 8.7*. When a large reverse voltage is applied across a diode, breakdown usually occurs across the junction at the edge of the

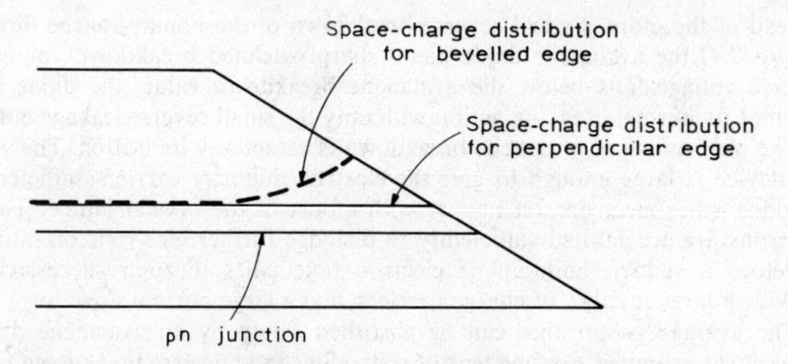

Figure 8.7 Effect of bevelled edges in high-power rectifier diode

element rather than in the centre. If the edges are bevelled, the space charge across the reverse-biased junction is deflected, as shown by the broken lines in *Figure 8.7*. The electric field produced by a large reverse voltage follows the deflection of the space charge, and so the distance over which breakdown may occur at the edge is increased. Thus the value of the breakdown voltage will be increased. Diodes can also be connected in parallel to increase the current-carrying capacity above that of the individual diodes. Similarly, diodes can be connected in series to reduce the reverse voltage across individual diodes.

Despite such improvements to the diodes and methods of connection, considerable difficulties were experienced with early semiconductor-diode rectifier systems, particularly through the effects of voltage transients. Although these transients are of very short duration, they still contained sufficient energy to destroy the diodes. These difficulties were considerably eased by the introduction of the avalanche diode.

The failure of non-avalanche diodes with large reverse voltage transients was caused by local breakdown of the junction. Small irregularities in the junction were inevitable with the alloying method of manufacture used for rectifier diodes at that time. Although these irregularities were of no consequence at lower voltages, when the diodes were subjected to the reverse transients present on the mains, the irregularities produced 'hot spots'. The high local temperature would cause that part of the junction at the 'hot spot' to break down, more current would then flow through that part of the junction raising the temperature further, until finally complete breakdown of the diode would occur.

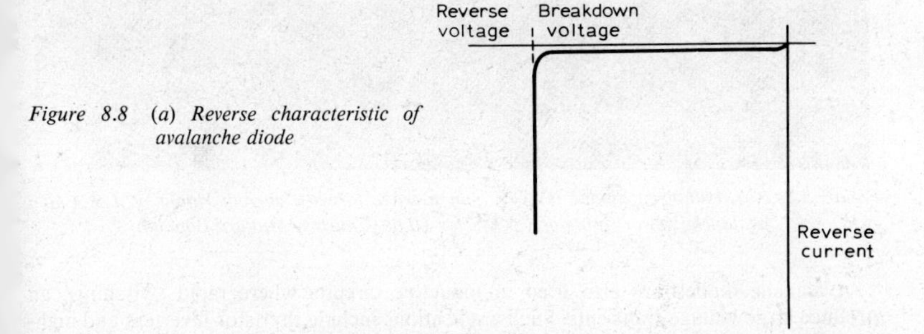

Figure 8.8 (a) Reverse characteristic of avalanche diode

The avalanche diode uses a diffused junction, and great care is taken during manufacture to ensure that the junction is uniform. When subjected to large reverse voltages, the whole of the junction breaks down at the same time so that the large reverse current is conducted evenly over the junction area. No hot spots are formed therefore. This bulk breakdown is shown by the reverse characteristic for an avalanche diode, *Figure 8.8(a)*.

Instead of the more gradual reverse breakdown of the non-avalanche diode (shown in *Figure 8.4*) the avalanche diode has a sharply-defined breakdown voltage. When the reverse voltage falls below the avalanche breakdown value, the diode recovers and resumes its normal blocking action with only the small reverse leakage current flowing.

The mechanism of avalanche breakdown is essentially ionisation. The voltage across the device is large enough to give the electron minority carriers sufficient velocity to dislodge other electrons on impact with atoms of the crystal lattice. These new free electrons are accelerated sufficiently to dislodge further electrons on impact. There is therefore a sudden build-up of electron–hole pairs through successive impacts to provide a large number of charge carriers, and a large current flows.

The average power that can be absorbed safely by an avalanche diode after the breakdown is limited to some tens of watts. Since the reverse breakdown voltage can be greater than 1 kV, the average current that can flow across the junction is only, at the most, some tens of milliamps. It should be remembered, however, that the duration of the reverse voltage transients is very short so that the pulse power the diode can withstand is considerable. Powers of up to some tens of kilowatts can be withstood for surges with durations of, typically, 10 μs.

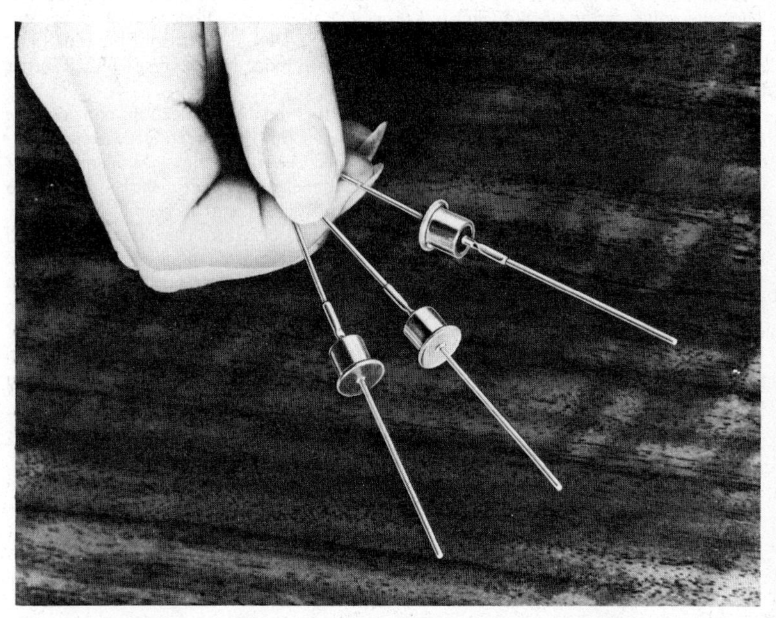

Figure 8.8 (*b*) *Avalanche diodes BYX45 with average forward current rating of 1.5 A and avalanche power rating of 1.5 kW for 10 μs* (*Courtesy Mullard Limited*)

Avalanche diodes are also used in inductive circuits where rapid switching can produce large voltage transients. Such applications include thyristor inverters and high-frequency power supplies. Avalanche diodes with an average forward current rating of 1.5 A but able to withstand an avalanche power of 5.2 kW for 10 μs are shown in *Figure 8.8*(*b*).

Stress has been laid in the discussion of rectifier diodes on the cooling requirements. Because high-power diodes need to be mounted on heatsinks, a stud-mounting construction is used. A typical high-power rectifier diode is shown in *Figure 8.9*. Normally the

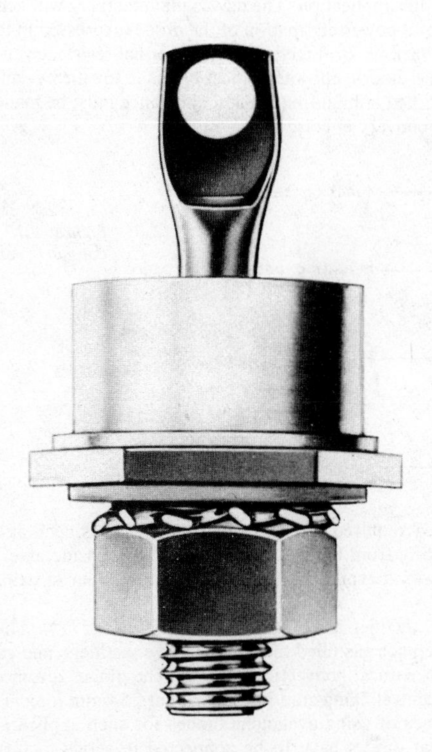

Figure 8.9 Typical high-power rectifier diode: BYX52 with average forward current rating of 40 A, height approximately 4 cm (Courtesy Mullard Limited)

cathode of the diode is connected to the stud for screwing onto the heatsink. 'Reverse-polarity' devices with the anode connected to the stud are also available.

The limiting temperature for operation, the maximum junction temperature of the diode, is quoted by the manufacturer in the published data but cannot be measured directly by the user. Additional information must therefore be given in the data, and this is in the form of thermal resistance, R_{th}. This quantity is analogous to electrical resistance, and represents the opposition to the flow of heat from the device junction. It is expressed in temperature change with power, the usual units being °C/W. Like electrical resistance, thermal resistances can be considered connected in series. Thus a chain of thermal resistance exists from the diode junction to the mounting base, $R_{th(j-mb)}$, between the mounting base and the heatsink (a contact resistance at a specified torque), $R_{th(i)}$, and between the heatsink and the ambient temperature, $R_{th(h)}$. The relationship between the thermal resistances and temperatures in the chain is shown in *Figure 8.10*.

Of the various quantities shown, the junction temperature and $R_{th(j-mb)}$ are fixed, thus fixing the mounting base temperature. The contact thermal resistance is also fixed if the specified torque is applied to the stud mounting of the diode. Therefore the heatsink temperature will be known. The ambient temperature will either be known or a limiting value known, so that the designer has to choose the appropriate thermal resistance for the heatsink to suit his application. The device manufacturer will generally publish data which relates the total power dissipation of the diode (expressed in terms of the average forward current at various form factors) to the thermal resistance and ambient temperature. With avalanche diodes, allowance must be made for the avalanche power; and for diodes that are switched at high frequencies, allowance must be made for the power loss caused by reverse recovery effects.

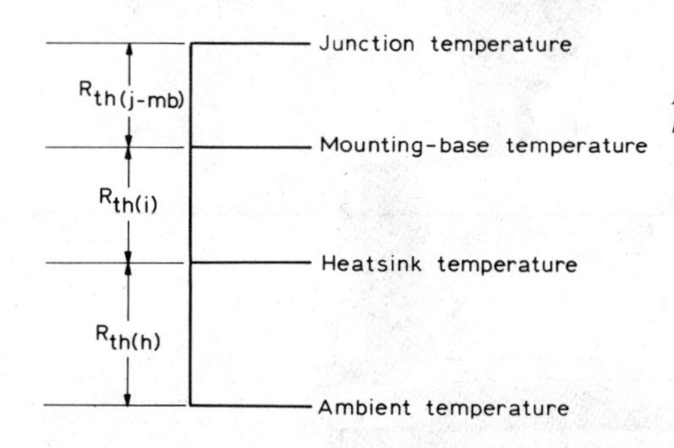

Figure 8.10 Temperatures and thermal resistances in diode mounted on heatsink

The heatsink area required for heavy-duty applications such as a high-current rectifier system operating from the mains supply may be considerable. The photograph in *Figure 8.11(a)* shows a typical three-phase bridge rectifier system mounted on three heatsinks.

Another type of rectifier stack is shown in *Figure 8.11(b)*. This is a high-voltage stack designed to replace gas-filled thermionic valve rectifiers, and capable of delivering 5 A at 12 kV with natural convection cooling. The diodes are mounted on heatsinks grouped around a central fixing stud which is about 254 mm long. This stack illustrates one of the advantages of using avalanche diodes for such applications. Because of the high voltage, several diodes have to be connected in series to form each arm of the bridge. All the diodes in an arm do not turn on at exactly the same instant, and so voltage transients are set up within the bridge. With non-avalanche diodes, equalising resistors and capacitors would be necessary to prevent breakdown of the diodes. These safety components are not required with avalanche diodes, and so the volume of the stack can be kept to a minimum.

The circuit designer, in addition to the normal data for a diode (average forward current, reverse current, and forward voltage drop), requires for high-power rectifier diodes information on the transient performance. A crest working reverse voltage V_{RWM} is given which corresponds (with a suitable safety factor) to the peak voltage of the supply on which the diode is operating. A repetitive peak reverse voltage V_{RRM} is given, corresponding to the peak value of transients occurring every cycle, and a non-repetitive peak reverse voltage V_{RSM} corresponding to transients that occur only occasionally. Similar ratings are given for forward current; a repetitive peak forward current I_{FRM}, and a non-repetitive peak forward current I_{FSM}. These current ratings specify the maximum current that the diode can conduct before breakdown, and it may be necessary to provide protective devices such as high-speed fuses to prevent damage to the diode from high-fault currents.

A list of typical high-power rectifier diodes is given in *Table 8.3*, together with brief data.

Additional information is given for avalanche rectifier diodes: the avalanche breakdown voltage and the average, repetitive, and non-repetitive avalanche power. Typical avalanche diodes are listed in *Table 8.4* with brief data.

Figure 8.11 (*a*) *Typical heavy-duty rectifier-diode stack* (*Courtesy Mullard Limited*) (*b*) *High-voltage rectifier stack using avalanche diodes: OSS9210 operating at 12 kV and delivering 5 A with natural convection cooling* (*Courtesy Mullard Limited*)

Zener diodes

The name 'zener diode' is applied to three groups of devices: voltage regulator diodes, voltage reference diodes, and surge suppressor diodes. Strictly speaking, 'zener' is not the correct description because some of the devices use avalanche breakdown rather than zener breakdown for their operation. However, the name zener seems now firmly established among electronic engineers. The circuit symbol for a zener diode is shown in *Figure 8.12*.

Figure 8.12 Circuit symbol for zener diode

The three groups of device operate on the same principle: the sudden breakdown of the junction at a specific reverse voltage, as shown by the characteristic of an avalanche diode in *Figure 8.8*. Because the characteristic after breakdown is nearly parallel to the current axis, the voltage after breakdown remains reasonably constant. The zener diode can therefore be used as a voltage stabilising device. If the voltage falls below the breakdown value, the diode will recover provided the maximum permissible junction temperature has not been exceeded.

In diodes manufactured from semiconductor material with high doping levels, the depletion layer formed when the *pn* junction is reverse-biased is very narrow. Even with a small voltage across the diode, since practically all the voltage is placed across the junction, the depletion layer is subjected to a very high electric field. Breakdown can occur through internal field emission, the covalent bonds between electrons and atoms being broken to provide a large number of charge carriers. This type of breakdown is called zener breakdown.

Reference to avalanche breakdown has already been made in the description of avalanche diodes. In diodes manufactured from materials with lower doping levels, avalanche breakdown will occur at a lower voltage than that required for zener breakdown. As an approximate guide for practical zener diodes, those with breakdown voltages below 5 V use zener breakdown, while those with breakdown voltages above 7 V use avalanche breakdown. For voltages between 5 V and 7 V, both effects are used.

It is possible, therefore, to manufacture a range of zener diodes with different breakdown voltages. The value of the breakdown voltage is determined principally by the doping level of the semiconductor material from which the diode is made. The range of voltages obtained is, typically, from 3 V to 75 V. Diodes operating at breakdown voltages below 10 V usually have the junctions manufactured by alloying; those operating at higher voltages have diffused junctions.

In practice, the reverse voltage/current characteristic may differ from that shown in *Figure 8.8*. For diodes with low breakdown voltages, the characteristic does not have such a sharply-defined knee and there is a more gradual transition into the vertical part. Also, the voltage after breakdown, the zener voltage, may not be constant but may increase slightly with the increase in current. The change in voltage is slight, but it may be significant in certain applications.

The basic characteristic of a zener diode is the dynamic characteristic of zener voltage plotted against pulsed zener current. Current pulses are used so that the junction temperature is not changed and heating effects are therefore eliminated. The slope of this characteristic is the dynamic resistance, which is specified in published data at different zener currents to allow for changes in slope. The breakdown voltage and zener voltage are affected slightly by temperature, and so a temperature coefficient is specified (also at different zener currents). This coefficient is negative in diodes with breakdown voltages below 5 V, and positive in diodes operating at higher voltages.

The static characteristic of a zener diode shows the variation of zener voltage with d.c. zener current. This characteristic takes into account the change in junction temperature through internal dissipation. The change in zener voltage with direct current depends on the dynamic resistance, the rise in junction temperature through internal dissipation and the thermal resistance between junction and ambient, and the temperature coefficient. The static characteristic can be used to determine the operating point at any zener current.

Because of tolerances during the manufacturing process, there is also a tolerance on the value of breakdown voltage and zener voltage. This tolerance is generally $\pm$ 5%, but special selections can be made to obtain zener voltages with tighter tolerances at specified currents.

When a zener diode is conducting, there is a steady-state dissipation caused by the zener current. In addition, the diode may be subjected to pulses through transients on the voltage supply rail it is stabilising. The sum of the steady-state and pulse power must not exceed the total dissipation permissible for the diode. However, the pulse power can be relatively high if the pulse duration is short.

A voltage regulator diode used in a simple stabilising circuit is shown in *Figure 8.13*.

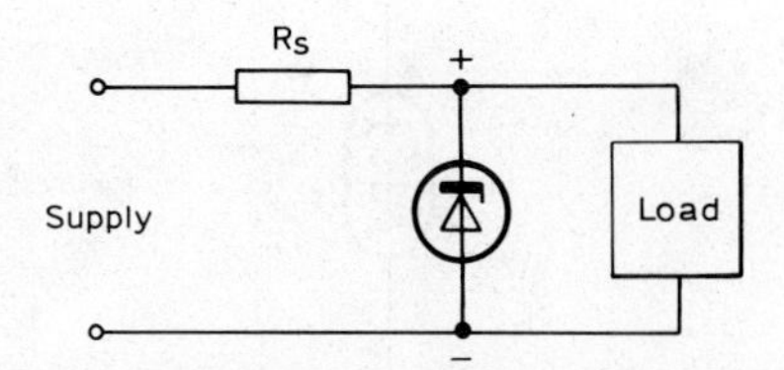

Figure 8.13 Simple stabiliser circuit using voltage regulator diode

The diode is initially biased beyond the breakdown voltage, and then clamps the supply rail at a voltage near the breakdown value. The series resistor R_s should have as high a value as possible to give good stability, but the limit values are determined by the diode. The upper limit of R_s is set by the need to maintain the voltage across the diode greater than the breakdown value when the load is drawing maximum current. The lower limit is set by the need to keep the current through the diode below the value for maximum permissible dissipation, when the load current is a minimum.

A list of typical voltage regulator diodes is given in *Table 8.5*, together with brief data.

A similar circuit is used with surge suppressor diodes to protect circuits against the effects of voltage transients. In this application, however, the diode does not conduct unless a surge is present. The diode then acts as a shunt element to absorb the energy of the surge. Provided the maximum junction temperature is not exceeded, the diode will recover when the surge energy has been dissipated.

The voltage approximately 10% less than the breakdown value is called the stand-off voltage. For a given supply, the suppressor diode is chosen so that the stand-off voltage is equal to or greater than the maximum supply voltage under normal operating conditions. Only the reverse leakage current flows through the diode, and so the power dissipated will be negligibly small. After breakdown by the voltage transient the diode reverse current rises rapidly, the maximum permissible value being the non-repetitive reverse current I_{RSM} which is specified for a particular pulse duration. The voltage also increases to a value called the clamping voltage $V_{(CL)R}$ which is generally less than twice the stand-off voltage. The dissipation at this point is the product of $V_{(CL)R}$ and I_{RSM}, and the diode must be chosen to withstand this dissipation.

The range of stand-off voltages is, typically, 5·6 V to 62 V. Pulse powers up to 13 kW can be dissipated by the diode without requiring a heatsink. The power rating, however, is a non-repetitive rating. In applications where the voltage transients are repetitive at a definite frequency (for example, switching transients occurring every cycle), a heatsink must be used. The published data gives information on the repetitive peak reverse power that can be dissipated for various surge durations and frequencies, and from this it can be verified that the diode is suitable for repetitive operation. The thermal resistance for the heatsink can then be calculated, and a suitable heatsink chosen.

A list of typical surge suppressor diodes is given in *Table 8.6*, together with brief data.

It was stated earlier that the temperature coefficient of zener diodes with breakdown voltages below 5 V was negative, while that of diodes operating at higher voltages was positive. Diodes with temperature coefficients of opposite polarity can therefore be matched to provide a combination with a very low overall coefficient. Such combinations are used to form voltage reference diodes. These provide a simple means of obtaining a voltage standard for portable measuring equipment.

In a voltage reference diode, two zener diodes are connected back-to-back as shown in *Figure 8.14*. The diodes are manufactured on separate chips but are encapsulated in

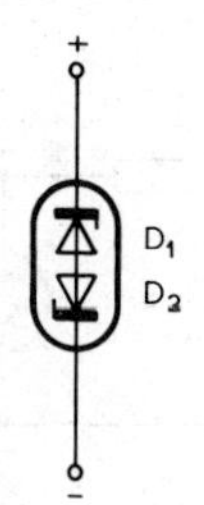

Figure 8.14 Circuit symbol for voltage reference diode

the same envelope. One diode with a zener voltage of about 5.5 V and a positive temperature coefficient operates in the reverse direction, diode D_1. The other diode, D_2, operates in the forward direction and has a negative temperature coefficient. As the voltage across the terminals is increased, D_1 breaks down. Because D_2 is conducting in the forward direction, the voltage across it is small. The main component of the reference voltage is therefore the zener voltage of D_1 which varies with current. The current through the diodes must therefore be closely controlled. To maintain a low and constant temperature gradient along the length of the reference diode, it is necessary to mount it in a temperature-controlled environment. A hole in a block of copper or aluminium is generally suitable, although for greater stability in the more accurate applications a temperature-stabilising circuit may be necessary.

This form of construction for a voltage reference diode provides a voltage standard between 6·2 V and 6·8 V. Data for voltage reference diodes are given in *Table 8.7*.

Varactor diodes

The varactor diode (variable-capacitance diode) uses the capacitance produced by a reverse-biased *pn* junction. As explained earlier, the reverse bias produces a region at the junction virtually devoid of charge carriers, the depletion layer, which forms an insulating layer between the two conducting *p* and *n* regions. A parallel-plate capacitor is therefore formed, and as the width of the depletion layer depends on the reverse bias a voltage-variable capacitance is obtained. The circuit symbol for a varactor diode is show in *Figure 8.15*.

The varactor diode has two main applications which impose slightly different constructions. One application is as a tuning element, replacing such devices as the

Figure 8.15 Circuit symbol for varactor diode

parallel-plate tuning capacitor in long, medium and short-wave radio receivers, or variable-inductance tuners in v.h.f. receivers. For this application a relatively high capacitance value is required, and so the diode is constructed with as large a junction area as possible. Typical capacitance values for this type of diode are 230 pF to 13 pF with a voltage range 1 V to 30 V, and 20 pF to 2 pF with a voltage range of 0·5 V to 25 V. Diodes intended as tuning elements are manufactured from silicon.

Varactor diodes are also used as tuning elements at microwave frequencies. These diodes are manufactured from gallium arsenide. The capacitance value is small, only a few picofarads, to give a high cut-off frequency for efficient operation. The range of capacitance is smaller than that required for the lower-frequency applications. A typical capacitance range is 5 : 1 with a voltage change from 1 V to 12 V.

The second application of varactor diodes is for frequency multiplication up to and of microwave frequencies. The variation of capacitance with voltage is non-linear, and so the application of an alternating voltage waveform across the diode produces an output waveform containing harmonics. Tuned circuits can be used to extract power at these harmonics. Generally the varactor diode is used as a doubler, tripler, or quadrupler, as at higher frequency multiplications the circuits become complex. However, a chain of varactor-diode multipliers and tuned circuits can be used to provide higher frequency multiplications, still with acceptable power levels and efficiencies.

The non-linear characteristic can also be used as a mixer to provide frequency conversion. Two input signals are applied, and the sum frequency extracted. This arrangement is called an up-converter. The extraction of the sum of an input and pump requency is analogous to the mixer action in a superhet where the difference between an input frequency and local oscillator frequency is used.

A variant of the varactor diode is the step recovery diode. A suitable doping level is chosen for the *pn* junction so that the diode can be switched from a high-capacitance forward-biased state to a low-capacitance reverse-biased state. The diode forms the capacitive element of an *LC* tuned circuit. When the diode is switched by an r.f. signal, the reverse-recovery current spike causes the inductance to produce across a resistive load a series of voltage impulses once every input cycle. The voltage train contains harmonics of the input frequency, and is used to shock-excite a resonant circuit which produces a ringing waveform of a higher frequency than the input. As sidebands are also produced, a bandpass filter is used to extract the required frequency. This type of ircuit can produce a frequency multiplication of, say, 20 times using a single diode, nd with a higher efficiency than a chain of varactor diodes and tuned circuits.

The range of capacitance required in varactor diodes for frequency multiplication is enerally less than that for tuning. In particular, at microwave frequencies the apacitance must be low to give the high cut-off frequency required for efficient peration. The junction area is therefore small to give capacitance values of a few icofarads. Varactor diodes for this application are manufactured from silicon or gallium arsenide, and those for microwave operation are available in encapsulations suitable for inclusion in waveguides.

Typical variable-capacitance and varactor diodes are listed in *Table 8.8* with brief data.

Tunnel diode, backward diode

The tunnel diode uses a *pn* junction with both *p* and *n* regions very heavily doped. The behaviour of such a junction is considerably different from that of the junction generally used in diodes. The depletion layer is extremely thin because of the very large number of charge carriers. As a result, forward current starts to flow at a voltage lower than that for a normal diode, often at zero bias. In effect, electrons *tunnel* their way through the potential barrier across the junction, behaviour that appears contrary to classical physics but which is in accordance with quantum mechanics.

The tunnel current increases with increasing forward bias until a peak is reached at approximately 0.1 V for a germanium device. As the forward voltage is increased further, the tunnel current falls to a minimum at about 0.3 V. The characteristic then becomes that of a normal forward-biased diode. The forward voltage/current characteristic of a tunnel diode is shown by the full line in *Figure 8.16* (*a*); the broken line shows the characteristic of a normal diode for comparison.

A negative resistance region exists in the tunnel current characteristic between the maximum and minimum values. This can therefore be incorporated in an oscillator circuit and frequencies up to 100 GHz have been obtained. Unfortunately, the power output is very low, typical values being 10 mW at 5 GHz and 0·2 mW at 50 GHz. These values are considerably lower than those for a Gunn-effect diode, and so the tunnel diode has not been used much in practice as an oscillator. It has found favour, however, as a negative-resistance amplifier because of its low-noise performance.

A form of tunnel diode that has been used considerably at microwave frequencies is

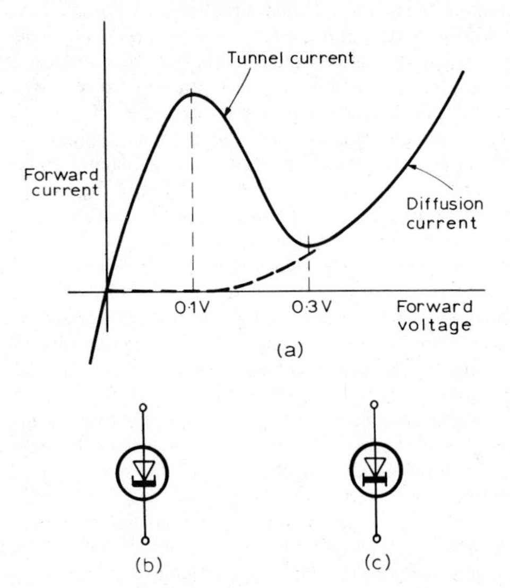

Figure 8.16 (a) Forward voltage/current characteristic of germanium tunnel diode (full line) compared with junction diode (broken line) (b) circuit symbol for tunnel diode (c) circuit symbol for backward diode

the backward diode. Because of the very high doping level (and resulting large number of current carriers), the diode breaks down at a very low reverse voltage. It is, in effect, a zener diode with a breakdown voltage of 0 V. The reverse resistance is therefore very low, lower than the forward resistance. Backward diodes are very sensitive detectors, and are used at microwave frequencies up to, typically, 40 GHz.

Tunnel diodes can be manufactured from germanium or gallium arsenide. Backward diodes are manufactured from germanium, and are encapsulated in envelopes suitable for waveguides or coaxial lines. The circuit symbol for a tunnel diode is shown in *Figure 8.16* (*b*) and that for a backward diode in *Figure 8.16* (*c*). Typical backward diodes and brief data are given in *Table 8.9*.

Schottky barrier diode

The Schottky barrier diode uses a metal-semiconductor junction as the rectifying element. In fact, the Schottky diode makes constructive use of what can be a disadvantage in many solid-state devices—the ‘contact resistance’ between the semiconductor chip and the metal leads. The semiconductor material used is highly doped silicon or gallium arsenide, on which a thin epitaxial layer of high-resistivity *n*-type material is grown. This gives the diode a low capacitance and high breakdown voltage. Planer techniques are used to deposit the metal to form a junction of very small area.

The current flow in a Schottky diode is entirely due to majority carriers. In the forward direction, electrons from the semiconductor material are injected into the metal. These electrons rapidly lose their energy, and so cannot recross the junction when the polarity is reversed. Schottky diodes do not therefore suffer from reverse recovery effects through minority carrier storage. They can therefore be used efficiently at microwave frequencies as mixer diodes and as detector diodes with low noise and high sensitivity.

Typical Schottky barrier diodes are listed in *Table 8.10*, together with brief data.

PIN diode

In the PIN diode, a layer of intrinsic silicon separates the *p* and *n* regions. Carriers leaving the doped regions have to cross the intrinsic region before reaching the opposite side of the junction. The transit time across the intrinsic region limits the operating frequency as a rectifier to approximately 100 MHz. At microwave frequencies, the PIN structure would appear simply as a linear resistance. However, if a forward bias is applied, carriers are injected into the intrinsic region and the resistance falls sharply. The difference between the resistance with forward bias and reverse bias (when no carriers are injected) can be a factor of many thousands. This property of the PIN diode enables it to be used as a modulator or switch in microwave systems. The PIN diode has also been used as a voltage-controlled attenuator. Typical PIN diodes are listed in *Table 8.11* together with brief data.

Gunn-effect diode

Strictly speaking, the Gunn-effect diode is not a diode at all as it does not contain a rectifying element. It is a *bulk effect* device, and provides one of the most convenient methods of generating power at microwave frequencies.

Gunn-effect devices can only be manufactured from certain semiconductor materials, including cadmium telluride, indium phosphide, indium arsenide, and gallium arsenide. Present-day Gunn devices are made from gallium arsenide as the manufacturing techniques with this material are well established.

In gallium arsenide (and the other semiconductor materials that can be used for Gunn devices), electrons can exist in two states. In one, the usual state, the electrons have low mass and high mobility; in the other state they have a high mass and low mobility. The electrons can be forced into the high-mass low-mobility state by subjecting them to an electric field of sufficient strength. If the distance over which the field operates is very small, such as a layer of semiconductor material some few micrometres wide, the external voltage producing the field need not be large. The critical field strength for gallium arsenide is approximately 3.5 kV/cm. For a layer 10 μm wide, this field strength is obtained with a voltage of 3.5 V across the layer.

An epitaxial layer of *n*-type gallium arsenide is grown on a gallium arsenide substrate. The substrate forms the 'anode' of the device, while an ohmic contact on the face of the *n*-type material forms the 'cathode'. A low d.c. voltage, between 3 V and 12 V, is applied across the device, the positive terminal of that supply being connected to the anode. This voltage produces in the *n*-type layer the electric field to force electrons into the high-mass low-mobility state.

Electrons reaching the cathode from the bias supply are not able to cross the *n*-type region because of their suddenly-acquired high mass and low mobility. Instead, they form 'clusters' at the cathode. These clusters are called domains, and they move across the *n*-type layer to the anode at the saturation velocity of electrons in gallium arsenide. At the anode the domain disappears, producing a current pulse, and a second domain starts to form at the cathode. This second domain rapidly crosses to the anode. Thus a series of current pulses is obtained at a frequency determined by the transit time, and therefore the width of the *n*-type layer. For a width of 10 μm, the frequency of oscillation is 10 GHz.

The Gunn device is mounted in a cavity which is shock-excited by the current pulses. If the external circuit is made to react on the Gunn device so that when the domain reaches the anode the total electric field across the device is below the critical value, the formation of the next domain is delayed. Also, if the alternating field created within the device by the external circuit is large enough to drive the net field below the critical value while the domain is crossing to the anode, then the domain is extinguished. Both these effects can be used to enable the frequency of oscillation to be controlled by the external circuit rather than the transit time of the Gunn device. The range of control possible is from about half to twice the transit-time frequency. The device for a particular application is therefore chosen to cover the operating frequency range, and the actual operating frequency is then set by tuning the cavity.

Output powers (c.w.) of up to 1 W at 8 GHz and 150 mW at 32 GHz can be obtained in production devices. Typical Gunn-effect devices are listed in *Table 8.12*, together with brief data.

Impatt diodes

The impatt diode is, at present, the most powerful solid-state device for the generation of microwave power. Its performance outstrips that of Gunn-effect devices, both in power and frequency, but it does have the disadvantage of requiring a higher voltage for operation. This voltage, above 100 V for some output powers, may prevent the use of impatt diodes in airborne or small portable systems.

The name impatt is an acronym of the method of operation: IMPact Avalanche and Transit Time. An impatt diode is a four-layer device consisting of heavily doped *p* and *n* regions separated by a less-heavily doped *n* region and an intrinsic region. The structure is shown in *Figure 8.17*. Impatt diodes can be manufactured from germanium, gallium arsenide, or gallium phosphide, but silicon is generally preferred because of thermal considerations.

A reverse bias is applied across the diode, causing avalanche breakdown to occur at the p^+n junction. Electron–hole pairs are generated, the holes flowing across the *p* region to the negative terminal of the supply, and taking no further part in the operation of the diode. The electrons enter the depletion layer which extends across the n and intrinsic layers and forms a 'drift' region.

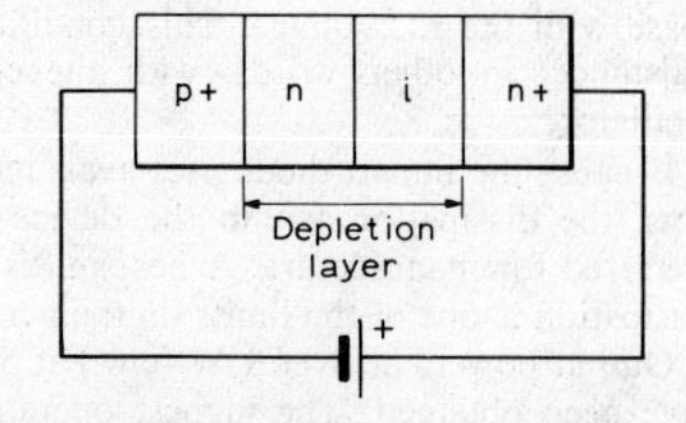

Figure 8.17 Structure of impatt diode

The ionisation rate under avalanche breakdown depends on the applied electric field. For simplicity, the curves in *Figure 8.18* are drawn for steady-state conditions when the oscillations have built up. Thus in *Figure 8.18* (*a*), the voltage across the p^+n junction, and hence the electric field producing avalanche breakdown, consists of a d.c. component and a periodic r.f. component. As the r.f. voltage increases from 0 to $\pi/2$, the ionisation rate increases; as the r.f. voltage decreases, from $\pi/2$ to π, the ionisation rate decreases. However, the carrier density does not follow the applied electric field and ionisation rate directly because the number of carriers generated depends on the number already present. After the field has passed its maximum value at $\pi/2$, the current density keeps increasing because the carrier generation rate is still above the average value. Thus the carrier density lags the ionisation rate by 90°. The result is that a sharply-defined spike of charge carriers from the avalanche breakdown appears at π and is injected into the drift region, as shown in *Figure 8.18* (*b*).

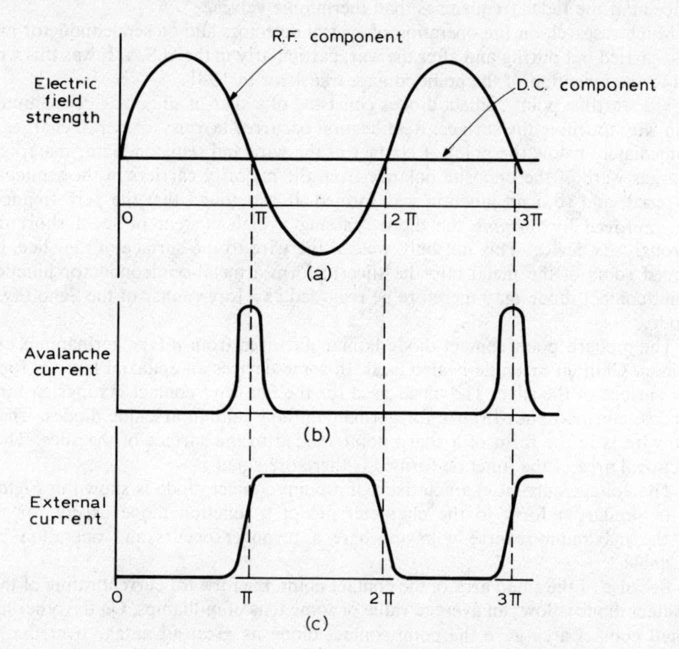

Figure 8.18 Operation of impatt diode: (a) electric field across diode (b) injected avalanche current (c) external current

The injected electrons rapidly cross the drift region at a constant velocity, and so induce a constant current in the circuit external to the diode. This current is shown in *Figure 8.18* (*c*). The width of the drift region is such that the transit time of the injected electrons is one half-cycle of the r.f. oscillation. Thus the induced current is 180° out of phase with the r.f. voltage. This condition implies that the diode presents a negative resistance; in other words, with the correct external load conditions it forms an oscillator.

Because the impatt diode uses avalanche breakdown as one of its operating mechanisms, the dissipation within the device is high. It is for this reason that silicon is preferred for manufacture. A heatsink is therefore necessary for the diode, and power dissipation is one of the limits on the performance.

Output powers of over 1 W (c.w.) at 50 GHz and pulse powers of 50 W at 10 GHz have been obtained. The highest operating frequency generated so far is about 300 GHz. Typical impatt diodes are listed in *Table 8.13*, together with brief data.

Point-contact diode

The point-contact diode is the modern form of what was probably the first commercial solid-state device—the cat's whisker of the crystal set. In the early days of radio, the detector used was a piece of a crystal of a semiconductor material onto which a fine wire was pressed. The position of the wire was adjusted until a 'sensitive spot' was found and detection occurred. The crystal detector was replaced in commercial radio receivers during the 1920s by thermionic valves, but interest remained in its use at the upper end of the radio frequency spectrum. This interest was stimulated by the development of radar just before World War II as it was found that the crystal detector was far more efficient at the radar frequencies than thermionic valves.

Much research on the operation of crystal detectors and on semiconductor materials was carried out during and after the war, particularly in the U.S.A. It was this work that led to the invention of the point-contact transistor in 1948.

The wartime point-contact diodes consisted of a slice of silicon or germanium with a thin wire touching the surface. Rectification occurred because a layer of charges formed immediately below the point of contact of the wire and semiconductor material. These charges were of the opposite polarity from the majority carriers in the semiconductor material, and so a *pn* junction was formed. It was found that the performance could be improved by *forming* the diode, passing a high current pulse of short duration through the device. This not only welded the wire to the surface of the slice, but also forced some of the metal into the slice to form a metal-semiconductor junction. The point-contact diode may therefore be regarded as a fore-runner of the Schottky barrier diode.

The modern point-contact diode is manufactured from *n*-type germanium or *p*-type silicon. Gallium arsenide is also used. In some devices an epitaxial layer is formed on the surface of the slice. The metal used for the fine wire contact is tungsten for silicon diodes, and phosphor-bronze for germanium and gallium arsenide diodes. The end of the wire is in the form of a sharp point welded to the surface of the slice. The cross sectional area of the junction formed is therefore small.

The voltage/current characteristic of a point-contact diode is shown in *Figure 8.19*. It is similar in form to the characteristic of a junction diode except for the part at the maximum reverse voltage where a turnover occurs and operation becomes unstable.

Because of the small area of the contact point, the forward current rating of the point contact diode is low: an average value of some tens of milliamps. On the other hand, the small contact area gave the point-contact diode its great advantage over the junction diode, the very low capacitance and therefore the superior performance at high frequencies. Until recently, the point-contact diode was the only detector and mixer diode that

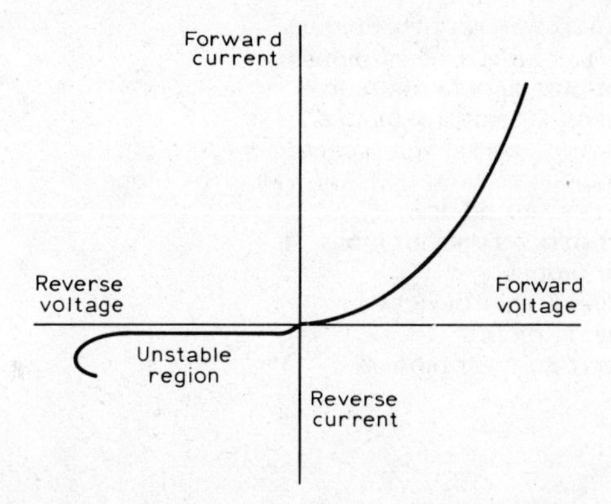

Figure 8.19 Voltage/current characteristic of point-contact diode

could be used at microwave frequencies. It is, however, now being replaced by Schottky barrier and backward diodes.

Typical point-contact diodes are listed in *Table 8.14*, together with brief data.

DIODE TABLES

Tables 8.1 to *8.14* list semiconductor diodes grouped into the various types described. They list 'typical' devices; to give the reader an idea of the range and ratings of devices available at present. It is in no way intended that the tables should be used to select devices for particular applications. The individual manufacturers should be consulted for fuller details of their ranges of devices.

The tables have been constructed from the manufacturers' 'quick-reference' data, and the help of those manufacturers who supplied information is acknowledged. Devices and values listed were correct at the time of writing. Where a figure is not given for a particular quantity, it does not imply such a figure is not available—merely that the manufacturer does not quote this quantity in his quick-reference data.

Outlines for the devices are given a reference according to JEDEC (DO and TO) or British Standard (SO) systems. Where devices do not conform to either system, a brief description is given where practicable (for example, 3 mm dia × 6.4 mm). Because microwave diodes can be made available in a variety of packages to suit waveguides, coaxial lines, or strip lines as required, no outlines have been specified.

Outline drawings and full dimensions can be obtained from the various manufacturers or from the relevant JEDEC and British Standards Institution publications.

The abbreviations used for the various manufacturers are listed below.

AEI	AEI Semiconductors Ltd.
Fa	Fairchild Semiconductors Ltd.
Fe	Ferranti Ltd.
ITT	ITT Semiconductors
M	Mullard Ltd.
MA	Microwave Associates Ltd.
P	The Plessey Company Ltd.
RCA	RCA Ltd.
TI	Texas Instruments Ltd.

List of Diode Tables

Symbols for main electrical parameters of semiconductor diodes

C_d	diode capacitance (reverse bias)
f_{co}	varactor diode cut-off frequency
I_F	continuous (d.c.) forward current
$I_{F(AV)}$	average forward current
I_{FRM}	repetitive peak forward current
I_{FSM}	non-repetitive (surge) peak forward current
I_O	average output current
I_{OSM}	non-repetitive (surge) output current
I_R	continuous reverse leakage current
I_{RRM}	repetitive peak reverse current
I_{RSM}	non-repetitive peak reverse current
I_Z	voltage regulator (zener) diode continuous (d.c.) operating current
I_{ZM}	voltage regulator (zener) diode peak current
P_{tot}	total power dissipated within device
Q_s	recovered (stored) charge
$R_{th(h)}$	thermal resistance of heatsink
$R_{th(i)}$	contact thermal resistance
$R_{th(j-amb)}$	thermal resistance, junction to ambient
$R_{th(j-mb)}$	thermal resistance, junction to mounting base
r_Z	voltage regulator (zener) diode differential (dynamic) resistance
T_{amb}	ambient temperature
$T_j$max	maximum permissible junction temperature
t_p	pulse duration
t_{rr}	reverse recovery time
$V_{(CL)R}$	surge suppressor diode clamping voltage
V_F	continuous (d.c.) forward voltage
V_{IRM}	repetitive peak input voltage
$V_{I(RMS)}$	r.m.s. input voltage
V_{ISM}	non-repetitive (surge) input voltage
V_{IWM}	crest working input voltage
V_O	average output voltage
V_R	d.c. reverse voltage
V_{RM}	peak reverse voltage
V_{RRM}	repetitive peak reverse voltage
V_{RSM}	non-repetitive (surge) peak reverse voltage
V_{RWM}	crest working reverse voltage
V_Z	voltage regulator (zenor) diode operating voltage

Type	Manufacturer	Outline	V_{RRM} (V)	I_{FRM} (mA)	$I_{F(AV)}$ (mA)	V_F (V) at	I_F (mA)	I_R (μA) at	V_R (V)	Q_s (pC)	C_d (pF)	t_{rr} (ns)
Germanium gold-bonded												
AAY30	M	DO-7	50	400	—	0.88	150	8.0	30	250	—	—
CG83H	AEI	DO-7	25	252	80	1.1	100	10	10	1 000	1.0	—
DK15	ITT	DO-7	100	120*	—	0.29	3.0	<90	100	700	—	—
DK19	ITT	DO-7	25	110*	—	0.26	1.0	<160	25	600	—	—
OA47	M	DO-7	30	150	—	0.54	30	10	30	280	—	—
Silicon junction												
BA158	ITT	Plastic 3 mm dia. × 6.4 mm	600	400	—	<1.5	1 000	<5	600	—	2.0	<300
BAV10	M	DO-35	60	600	300	1.0	200	—	—	—	2.5	6.0
BAV21	M	DO-35	250	625	200	1.25	200	—	—	—	5.0	50
BAW62	M	DO-35	75	225	100	1.0	100	—	—	—	2.0	9.0
BAX13	ITT	DO-35	50	150*	—	<1.0	20	<0.2	50	—	3.0	4.0
BAX13	M	2.4 mm dia. × 3.4 mm	50	150	75	1.0	20	—	—	—	—	4.0
BAX13	TI	DO-35	50†	75†	—	1.0	20	0.05	25	—	3.0	6.0
FDH900	Fa	DO-35	—	—	—	1.0	100	0.1	30	—	3.0	4.0
ITT600	ITT	DO-35	75	200*	—	<1.0	200	<0.1	50	—	2.5	4.0
ITT700	ITT	DO-7	30	50*	—	<1.1	50	<0.05	15	—	0.75	0.7
OA202	M	DO-7	150	250	80	1.15	30	0.01	150	—	—	—
OA202	TI	DO-35	150†	160†	—	1.15	30	0.1	150	—	—	—
1N4148	Fa	DO-35	—	—	—	1.0	10	0.025	20	—	4.0	4.0
1N4148	ITT	DO-35	75	150*	—	<1.0	10	<0.025	20	—	4.0	4.0
1N4148	M	DO-35	75	225	75	1.0	10	—	—	—	4.0	4.0
1N4376	Fa	DO-7	—	—	—	1.1	50	0.1	10	—	1.0	0.75
1S111	TI	DO-7	225†	400†	—	1.0	400	0.2	225	—	—	—
1S920	TI	DO-35	50†	200†	—	1.2	200	0.1	50	—	—	—
1S951	TI	DO-35	70†	225†	—	1.3	350	0.2	70	—	—	25

* D.C. current half-wave rectified with resistive load. † Working voltage and average rectified current.

Table 8.2 LOW-POWER RECTIFIER DIODES

Type	*Manufacturer*	*Outline*	V_{RRM} (V)	$I_{F(AV)}$ (A)	I_{FSM} (A)	V_F (V) *at*	I_F (A)	I_R (μA) *at*	V_R (V)	Q_s (nC)	t_{rr} (ns)
BY134	ITT	plastic 5 mm dia. × 9.5 mm	600	1.0	50	<1.3	2.0	< 5.0	600	—	—
BY206*	M	DO-14	350	0.4	15	1.5	2.0	—	—	0.8	—
BYX36-150, -300, -600	M	DO-15	150, 300, 600	1.0	30	1.2	1.0	1.0	V_{RRM}	—	—
MS1	AEI	2 mm dia.	60	0.2	4.0	—	—	—	—	—	—
SJ1403H	AEI	SO-101A	1 600	1.0	32	—	—	—	—	—	—
TA7895*	RCA	DO-26	800	1.0	35	1.9	4.0	—	—	—	500
ZR60	Fe	DO-1,2,3	50	0.75	58	—	—	10	50	—	—
ZS178	Fe	DO-41	800	0.75	58	—	—	5.0	800	—	—
1N3754	RCA	TO-1	100	0.125	30	1.0	0.125	—	—	—	—
1N4001 to 1N4007	Fe	DO-41	50 to 1 000	1.0	30	—	—	50	V_{RRM}	—	—
	ITT	3 mm dia. ×6.4 mm		1.0	50	<1.1	1.0	<10	V_{RRM}	—	—
	M	2.7 mm dia. ×5.9 mm		1.0	30	1.1	1.0	10	V_{RRM}	—	—
	TI	DO-41		1.0	30	—	—	10	V_{RRM}	—	—
1N4586	ITT	DO-41	1 000	1.0	50	1.0	1.0	<10	1 000	—	—
1SX173*	TI	DO-41	300	1.0	30	1.2	1.0	—	—	—	350

* Fast-recovery type.

Table 8.3 (a) HIGH-POWER RECTIFIER DIODES
($I_{F(AV)}$ 1.0 to 10 A in ascending current order)

Type	*Manufacturer*	*Outline*	$I_{F(AV)}$ (A)	I_{FSM} (A)	V_{RRM} or V_{RWM} (V)	Q_s (nC)	t_{rr} (μs)
BY127	M	SO-15	1.0	—	1 250	—	—
BYX55-350*	M	SO-15	1.2	—	300	150	—
-600					600		
BY196*	ITT	Plastic 5 mm dia. × 9.5 mm	1.2	70	100	—	<0.5
BY199*	ITT	Plastic 5 mm dia. × 9.5 mm	1.2	70	800	—	<0.5
1SO23	TI	DO-1,2,3	1.5	125	400	—	—
40266	RCA	DO-1	2.0	35	100	—	—
SJ053E	AEI	DO-4	2.0	30	60	—	—
to					to		
SJ1203K					1 400		
BYX38-300	M	DO-4	2.5	—	300	—	—
-600					600		
-900					900		
-1200					1 200		
SL-103F	AEI	DO-1,2,3	2.5	100	100	—	—
to					to		
SL-1403F					1 400		
TA7901*	RCA	—	3.0	75	800	—	0.5
1N3880*	M	DO-4	4.0	—	100	250	0.15
to					to		
1N3882R*	TI	DO-4	6.0	75	300	—	0.2
BY189	ITT	DO-4	4.0	75	850	—	<0.5
S6-103A	AEI	DO-4	6.0	130	100	—	—
to					to		
S6-1403K					1 400		
ZR200	Fe	DO-4	8.0	160	50	—	—
ZR204	Fe	DO-4	8.0	160	400	—	—
40115	RCA	DO-4	10	140	800	—	—
1S420	TI	DO-4	10	125	100	—	—
1S427	TI	DO-4	10	125	800	—	—
BYX72-150	M	Plastic †	10	—	150	—	—
-300					300		
-500					500		

* Fast-recovery type. † See manufacturer's literature for details.

Table 8.3 (b) HIGH-POWER RECTIFIER DIODES
($I_{F(AV)}$ above 10 A in ascending current order)

Type	*Manufacturer*	*Outline*	$I_{F(AV)}$ (A)	I_{FSM} (A)	V_{RRM} (V)	Q_s (nC)	t_{rr} (μs)
1N1199A to 1N1206A	RCA	DO-4	12	240	50 to 600	—	—
SL-103A to SL-1403K	AEI	DO-4	14	150	100 to 1 400	—	—
BYX46-200* to BYX46-600*	M	DO-4	15	—	200 to 600	700	0.35
ZR30C to ZR35C	Fe	SO-32A	30	800	50 to 500	—	—
SU-101A to SU-1401K	AEI	SO-32A	36	600	100 to 1 400	—	—
BYX52-300 to BYX52-1200	M	DO-5	40	—	300 to 1 200	—	—
1N1183A to 1N1190A	RCA	DO-5	40	800	50 to 600	—	—
40956* to 40960*	RCA	DO-5	40	700	50 to 600	—	0.35
S12P500E to S12P1800E	AEI	AEI †	90	2 000	500 to 1 800	—	—
S962SJU25 to S962SJU56	AEI	AEI †	335	5 400	2 500 to 5 600	—	—
DS1104SM10 to DS1104SM29	AEI	AEI †	1 500	16 000	1 000 to 2 900	—	—

* Fast-recovery type.

† See manufacturer's literature for details.

Table 8.4 AVALANCHE RECTIFIER DIODES
(Ascending current order)

Type	*Manufacturer*	*Outline*	$I_{F(AV)}$ (A)	I_{FRM} (A)	I_{FSM} (A)	V_{RRM} *or* V_{RWM} (V)	$V_{(BR)R}$ (V)
40808	RCA	DO-26	0.5	—	35	600	700 to 1 100
40809	RCA	DO-26	0.5	—	35	800	900 to 1 300
RAS310AF	ITT	DO-1	1.0	5	45	1 200	>1 250
ZAR610	Fe	DO-1,2,3	1.0	—	—	1 000	1 200 to 1 800
BYX45-600R	M	DO-1,2,3	1.5	15	40	600	—
-800R						800	
-1000R						1 000	
1AS027	TI	DO-1,2,3	1.5	—	125	800	1 000 to 1 500
1AS029	TI	DO-1,2,3	1.5	—	125	1 000	1 200 to 1 750
ZAR110	Fe	DO-5	1.5	—	—	1 000	1 200 to 1 800
RAS508AF	ITT	DO-1	5	25	100	960	>1 000
ZAR210	Fe	DO-4	8	—	—	1 000	1 200 to 1 800
RAS508CF	ITT	DO-1	10	50	200	960	>1 000
BYX40-600	M	DO-4	12	250	200	600	—
-800						800	
-1000						1 000	
BYX25-600	M	DO-4	20	440	360	600	—
-800						800	
-1000						1 000	
BYX56-600	M	DO-5	40	450	800	600	—
-800						800	
-1000						1 000	

Table 8.5 VOLTAGE REGULATOR DIODES

Type	*Manufacturer*	*Outline*	*Nominal zener voltage* (V)	*Dissipation* (W)
BZX61 series	M	DO-15	7.5 to 75	1.3
BZX70 series	M	SO-15	7.5 to 75	2.5
BZX79 series	M	DO-35	4.7 to 75	0.4
BZY88 series	M	DO-7	3.0 to 36	0.4
BZY91 series	M	DO-5	7.5 to 75	75
BZY93 series	AEI	DO-4	6.8 to 33	20
	M	DO-4	7.5 to 75	20
LR′C series	AEI	DO-35	5.1 to 39	0.4
MR′CH series	AEI	DO-7	2.7 to 7.5	0.25
VR′E series	AEI	DO-4	3.5 to 30	5.8
VR′F series	AEI	DO-1,2,3	3.5 to 30	2.25
VZ′CH series	AEI	SO-101	3.3 to 33	1.0
VZ′F series	AEI	DO-1,2,3	3.3 to 33	1.5
Z3B . . BF series	ITT	DO-1	3.3 to 100	1.5
Z5D . . BF series	ITT	DO-4	8.2 to 100	10
ZD series	ITT	DO-13	3.9 to 200	1.1
ZPY series	ITT	DO-41	3.9 to 200	1.37
1N746 to 1N759	Fa	DO-7	3.3 to 12	—
1N957 to 1N973	Fa	DO-7	6.8 to 33	—
1S2000A series	TI	DO-35	4.7 to 33	0.5
1S3000A series	TI	DO-41	7.5 to 100	1.3
1S4000 series	TI	DO-1,2,3	6.8 to 200	1.5
1S5000 series	TI	DO-4	6.8 to 200	10

Table 8.6 SURGE SUPPRESSOR DIODES
(Ascending power dissipation)

Type	*Manufacturer*	*Outline*	*Stand-off voltage* (V)	*Clamping voltage* (V)	*Dissipation* (kW)
BZW96	M	DO-1,2,3	3.9 to 7.5	6.5 to 14	0.1
BZW70	M	SO-15	5.6 to 62	9 to 104	0.4
BZW91	M	DO-5	5.6 to 62	8.5 to 86	5
BZW86	M	DO-30	7.5 to 62	12 to 92	13

Table 8.7 VOLTAGE REFERENCE DIODES

Type	*Manufacturer*	*Outline*	V_Z *at* (V)	I_Z (mA)	*Temperature coefficient* (±%/°C)	*Temperature range* (°C)
BZV10 to BZV14	M	DO-35	6.2 to 6.8	2.0	0.01 to 0.000 5	0 to +70
BZX90 to BZX94	M	DO-35	6.2 to 6.8	7.5	0.01 to 0.000 5	−55 to +100
FCT1021	Fa	TO-18	6.7 ± 5%	0.1	0.001	0 to +100
FCT1022	Fa	TO-18	6.7 ± 5%	0.1	0.002	0 to +100
FCT1025	Fa	TO-18	6.7 ± 5%	0.1	0.005	0 to +100
FCT1121	Fa	TO-18	6.7 ± 5%	0.1	0.001	−55 to +100
FCT1122	Fa	TO-18	6.7 ± 5%	0.1	0.002	−55 to +100
FCT1125	Fa	TO-18	6.7 ± 5%	0.1	0.005	−55 to +100
MR′A series	AEI	AEI*	6.6 to 11.0 ±5%	5 and 10	0.001 to 0.006	—

* See manufacturer's literature for details.

Table 8.8 VARIABLE-CAPACITANCE AND VARACTOR DIODES

Type	*Manufacturer*	V_R max (V)	C_d *at* (pF)	V_R (V)	C_d *at* (pF)	V_R (V)	*Capacitance ratio*
Variable-capacitance diodes for tuning (r.f. to u.h.f. operation)							
BA163	ITT	>14*	10	10	260	0	>26
BA182	M	35	1.0	20	—	—	—
BB105B	M	28	2.3	25	—	—	6.0
BB113	M	32	13	30	280	1.0	—
BB122	ITT	>30*	2.8	25	13	3.0	4.5 to 5.5
BBY22	Fe	100	47	4.0	—	—	8.0
DC4227C	AEI	—	68	4.0	—	—	—
DC4244C	AEI	—	350	4.0	—	—	—
ITT210	ITT	>20*	5.0	10	20	1.0	—
ZC700	Fe	30	6.8	4.0	—	—	2.8
ZC800	Fe	25	1.8	20	10	2.0	5.5
Varactor diodes for tuning (microwave operation)							
BXY53	M	>60*	1.0	4.0	—	—	4.0
BXY54	M	>60*	4.7	4.0	—	—	6.5
BXY55	M	>60*	15	4.0	—	—	7.0
DC4201B	AEI	30	2.2	0	—	—	6.0
DC4275B	AEI	30	4.7	0	—	—	6.0
MA-45064	MA	25	0.5	4.0	—	—	3.1
MA-45109	MA	45	15	4.0	—	—	5.5
MA-45149	MA	90	15	4.0	—	—	8.6
ZC751	Fe	30	3.9	4.0	—	—	2.8
Multiplier varactor diodes							
BAY96†	M	>120*	39	6.0	—	—	—
BXY56†	M	>60*	2.5	6.0	—	—	—
BXY57†	M	>60*	3.5	6.0	—	—	—
DC7201A	AEI	60	2.0	6.0	—	—	—
DC7309A	AEI	120	35	6.0	—	—	—
MA-44100†	MA	>150*	30	6.0	—	—	1.4
MA-44150†	MA	>15*	0.6	6.0	—	—	1.5
MA-44210	MA	>100*	16	6.0	—	—	1.4
ZC29	Fe	6	0.55	0	—	—	—
ZC41	Fe	20	0.5	20	—	—	—
ZC0112†	Fe	35	3.0	0	—	—	—
1N5152†	{Fe, M}	75	7.5	6.0	—	—	—

* $V_{R(BR)}$ value. † Step recovery diode.

Table 8.9 BACKWARD DIODES

Type	*Manufacturer*	*Frequency* (GHz)	*Tangential sensitivity* (dBm)
AEY29	M	12 to 18	−53
AEY31	M	1 to 18	−53
AEY32	M	18 to 40	—
DC3010	AEI	9.3	−53
DC3011	AEI	16	−53
DC3021	AEI	9.3	−53

Table 8.10 SCHOTTKY BARRIER DIODES

Type	Manufacturer	Frequency range or test frequency (GHz)	Noise figure (dB)	Tangential sensitivity (dBm)
Mixer diodes				
BAT10	M	1 to 12	7.5	—
BAV72	M	26 to 40	10.0	—
BAV96D	M	1 to 12	6.0	—
BAW95G	M	1 to 12	6.5	—
DC1304	AEI	9.3	6.0	—
DC1304A	AEI	16.0	6.5	—
DC1501F	AEI	9.3	7.0	—
DC1504E	AEI	9.3	7.5	—
MA-4861H	MA	16.0	6.5	—
MA-4882	MA	1.0	5.5	—
MA-40051H	MA	3.0	5.0	—
MA-40071I	MA	9.375	5.5	—
Detector diodes				
BAV46	M	8 to 12	10	−52
BAV75	M	8 to 14	10	−50
BAV97	M	8 to 12	10	−54
MA-40025	MA	3.0	—	−52
MA-40027	MA	10.0	—	−52
MA-40029	MA	16.0	—	−50
MA-40042	MA	3.0	—	−46
MA-40072	MA	10.0	—	−46

Table 8.11 PIN DIODES

Type	Manufacturer	$V_{R(BR)}$ (V)	C_d (pF)	P_{tot}	Carrier lifetime (ns)	Switching speed (ns)
Switching diodes						
DC2019	AEI	100	0.2	—	—	10
DC2419A	AEI	150	—	30 W c.w.	—	50
MA-47051	MA	100	0.35	250 W pk	75	5
MA-47079	MA	500	0.4	8 kW pk	1 500	100
MA-47084	MA	1 000	3.8	30 kW pk	15 000	650
ZC3202	Fe	200	0.32	3.0 W	100	—
General purpose						
DC1015	AEI	500	1.0	—	—	2 000
DC2010	AEI	250	0.5	—	—	700
DC2133A	AEI	800	0.48	—	—	4 000
MA-47100	MA	100	0.35	—	2 000	—

Table 8.12 GUNN-EFFECT DEVICES
(Ascending frequency and power order)

Type	*Manufacturer*	*Frequency range* (GHz)	*Output power* (mW)	*Operating voltage* (V)	*Operating current* (mA)
C.W. operation					
DC1203A	AEI	5 to 8	5	12	100
MA-49135	MA		25	12	150
TEO29 *	P		25	15	350
DC1253F	AEI		50	15	300
CXY17E *	M		300	10	—
MA-49139	MA		500	12	1 400
DC1201A	AEI	8 to 12	5	7	145
TEO5	P		5	8	150
CXY11C	M		15	7	—
TEO9	P		25	10	350
DC1231E	AEI		30	7	200
CXY19	M		100	12	—
DC1281G	AEI		100	10	600
TEO104 †	P		300	10	700
CXY16F	M		400	8	—
MA-49110	MA		500	9	1 900
DC1202A	AEI	12 to 28	5	6.5	140
TEO15	P		5	6.5	150
CXY14C	M		15	7.0	—
MA-49121	MA		25	8.0	200
TEO19	P		25	8.0	350
DC1232E	AEI		30	6.5	200
MA-49164	MA		250	8.0	1 100
CXY18E	M		300	6.0	—
MA-49168	MA	18 to 26	100	6	800
TEO43 †	P	26 to 36	10	3.0	350
823CXY/C	M		20	4.5	—
MA-49173 ‡	MA		50	4.0	1 300
Pulsed operation					
MA-49260	MA	4 to 8	5 W	35	3 A
TEPO24 †	P		20 W	60	10 A
MA-49265	MA	8 to 12	5 W	35	3 A
TEPO5 †	P		30 W	50	12 A
TEPO12 †	P	12 to 18	5 W	30	10 A

* 4 GHz to 8 GHz.
† Frequency selection within band.
‡ Operation up to 40 GHz.

Table 8.13 IMPATT DIODES
(Ascending frequency and power order)

Type	*Manufacturer*	*Frequency* (GHz)	*Output power* (mW)	*Operating voltage* (V)	*Operating current* (mA)
DC1183H	AEI		250	125	60
BXY60*	M	5 to 8	650	120	125
DC1183K	AEI		1 000	125	230
ML-4803	MA		2 000	135	200
BXY50	M		500	91	135
MA-46021†	MA	8 to 10	700	70	75
MA-46027†	MA		1 300	70	125
ML-4804	MA		1 500	100	225
DC1151G	AEI	8 to 12	100	84	50
DC1181K	AEI		1 000	84	200
BXY51	M		400	80	120
ML-4805	MA	10 to 12	1 200	90	200
MA-46028 ‡	MA		1 300	65	150
MA-46029 §	MA		1 300	60	150
BXY52	M	12 to 14	300	70	120
ML-4706	MA		350	65	105
ML-4707	MA	14 to 17	300	55	115
DC-1185H	AEI	18 to 26	250	45	120
DC1186H	AEI	26 to 40	250	28	180

* 6 GHz to 8 GHz. † 8 GHz to 9.5 GHz.
‡ 9.5 GHz to 11.0 GHz. § 11.0 GHz to 12.5 GHz.

Table 8.14 POINT-CONTACT DIODES
(Ascending frequency order)

Type	*Manufacturer*	*Frequency range or test frequency* (GHz)	*Noise figure* (dB)	*Tangential sensitivity* (dBm)
Mixer diodes				
CS12B	AEI	3.0	7.0	—
1N21G	MA	3.06	5.5	—
CS9B	AEI	9.375	10.0	—
MA-41202F	MA	9.375	7.0	—
MA-491E	MA	13.3	8.0	—
MA-41210F	MA	16.0	7.5	—
AAY39	M	1 to 18	6.5	—
AAY51	M	12 to 18	7.5	—
1N26C	MA	23.98	9.5	—
MA-494C	MA	34.86	9.0	—
AAY59	M	26 to 40	10.0	—
Detector diodes				
MA-4123A	MA	3.06	—	–48
MA-452A	MA	9.0	—	–51
MA-4116	MA	16.0	—	–52

TRANSISTORS

Point-contact transistor

The point-contact transistor was invented in 1948 by John Bardeen and Walter H. Brattain, members of a team at Bell Telephone Laboratories working under the direction of William Shockley. 'Invented' may not be the right word; 'discovered' may be more correct, and Shockley himself has used the words 'creative failure' in connection with the work that led to the device.

The work at Bell Telephone Laboratories immediately after World War II was directed towards producing a solid-state amplifier using semiconductor materials. Shockley had predicted from theoretical work that the resistance of a piece of semiconductor material should change when it was subjected to an electric field normal to the current flow. Experiments, however, did not confirm this. The prediction was, in fact, true but it was not until 1963 that it was verified by the invention of the insulated-gate field-effect transistor. The failure to observe the expected change in resistance was ascribed by Bardeen in 1947 to some surface states neutralising the effect of the applied electric field. Further experiments were devised to investigate the surface states of a semiconductor material.

In one of the experiments, a fine wire was used to inject a current into a slice of a germanium crystal similar to that used in a point-contact diode. A second fine wire was used to measure the potentials in the surface near the first wire. It was found that the current changes in the input wire were followed by the current in the second wire, and also that the changes had been amplified. A current gain between the input and output wires of approximately five was obtained.

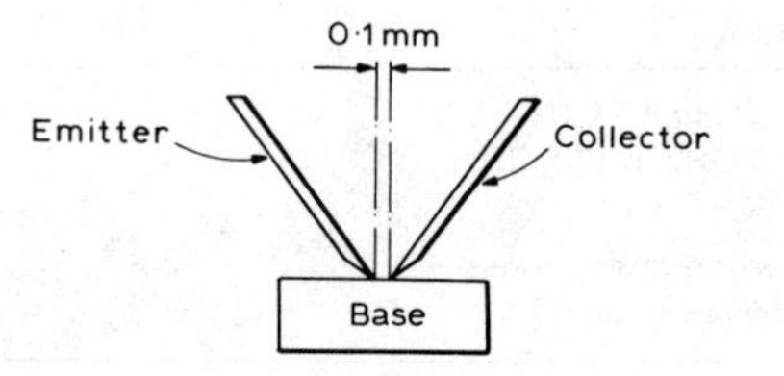

Figure 8.20 Simplified cross-section of point-contact transistor

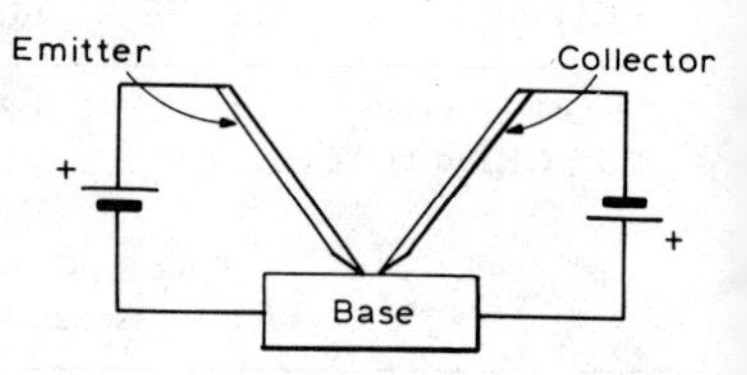

Figure 8.21 Biasing of point-contact transistor

A simplified cross-section of the point-contact transistor resulting from this discovery is shown in *Figure 8.20*. A slice of *n*-type germanium was used. Two fine wires made from tungsten with sharp points rested on the surface, the points approximately 0.1 mm apart. The method of biasing the transistor is shown in *Figure 8.21*. The emitter wire, the input wire of the experiment, was 'forward-biased' and holes were injected into the semiconductor material. These holes were 'collected' by the collector wire which was biased in the reverse direction from the emitter. The germanium slice was called the base as it was physically the base of the device. The enhancement and control of the current in the reverse-biased collector circuit by the current in the forward-biased emitter circuit was designated 'transistor action'.

Because the active area of the transistor was small, the current rating (as with the point-contact diode) was also small. The maximum collector power was limited to approximately 0.2 W. On the other hand, the small size gave the device a wide frequency range with an upper limit of approximately 10 MHz. The main disadvantage in the performance of the point-contact transistor was the very high noise figure. A typical figure for this was 60 dB with a source resistance of 1 kΩ at a frequency of 1 kHz. But perhaps the greatest disadvantage of all was the difficulty of getting the device to work. Although point-contact transistors were made available commercially, they were very much 'hand crafted' devices, requiring careful adjustment by the manufacturer.

It was thought originally that the flow of holes between the emitter and the collector was confined to the surface of the germanium slice. In 1949, however, John N. Shrive showed that the holes could flow appreciable distances through the bulk of the *n*-type material. The emitter and collector wires were placed opposite each other, one either side of a thin germanium slice, and transistor action was found still to take place. Shockley proposed an explanation of transistor action in terms of *pn* junctions. He devised an experiment to investigate the surface phenomena of the point-contact transistor, and discovered that his research structure was itself a transistor. Thus, in 1949, the junction transistor was introduced. This was a much more robust and reliable device, and its coming ended the brief life of the point-contact transistor.

Perhaps the most important function of the point-contact transistor was its demonstration that a practical amplifying device could be made using the flow of charges through a semiconductor material. It provided a verification of the theoretical work on solid-state physics that had preceded it. Much of the material technology that was to be further developed over the next two decades to enable the many new types of solid-state device to be manufactured was developed from work done for the point-contact transistor on the purification of germanium, the growing of single crystals, doping, and preparing small slices of semiconductor material. In many different ways, the point-contact transistor prepared the way for the success of the junction transistor.

Junction transistor

The junction transistor was invented by William Shockley of Bell Telephone Laboratories in 1949. The original transistor consisted of a piece of a single crystal of germanium containing two *n*-type regions separated by a *p*-type region. By analogy with the point-contact transistor, the two *n*-type regions were called the emitter and collector, and the *p*-type region was called the base. The two *pn* junctions were biased in the same way as the emitter and collector wires in the point-contact transistor had been.

TRANSISTOR ACTION

A schematic cross-section of an *npn* junction transistor is shown in *Figure 8.22*. The *n*-type emitter and collector regions are more heavily doped than the *p*-type base region.

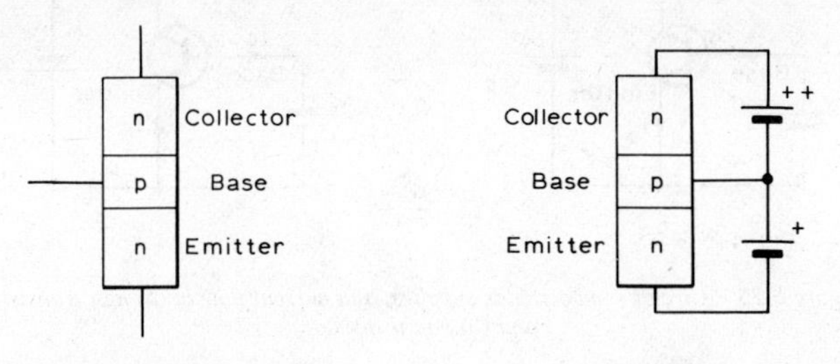

Figure 8.22 Structure of npn transistor

Figure 8.23 Biasing of npn transistor

The base region is also very narrow compared with the other two regions. The biasing of the transistor is shown in *Figure 8.23*. The base–emitter junction is forward-biased, while the collector–base junction is reverse-biased. The collector is considerably more positive than the base.

Because the base–emitter junction is forward-biased, majority carriers flow across the junction. As the base is only lightly doped, few holes cross into the emitter from the base. Most of the current is therefore provided by electrons flowing from the emitter into the base. The collector–base junction is reverse-biased, so only a small minority carrier current flows: a small number of holes from the collector into the base, and a smaller number of electrons from the base into the collector. However, because the base is narrow and the collector considerably more positive than the base, the large number of electrons crossing the base–emitter junction diffuse across the base to the edge of the collector depletion layer from which they are swept into the collector. A few of these electrons combine with holes in the base, but most cross to the collector to give rise to the collector current. The combination of electrons and holes in the base, and the holes crossing from the base to the emitter, produce the small base current.

The structure described is called an *npn* transistor from the arrangement of the *p* and *n* regions. A second arrangement is possible, the *pnp* transistor, as shown in *Figure 8.24*. This type of transistor has *p*-type emitter and collector regions separated by a

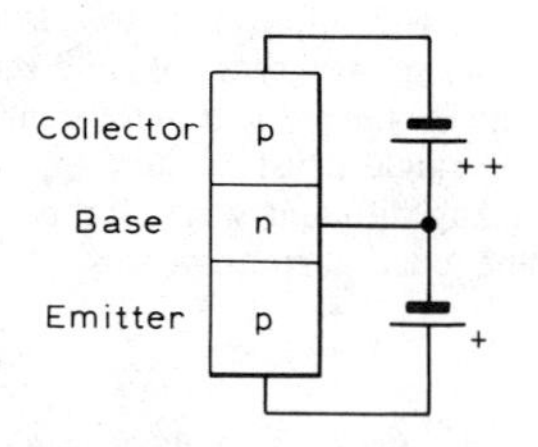

Figure 8.24 Structure and biasing of pnp transistor

narrow *n*-type base region. Again, the base–emitter junction is forward-biased, and the collector–base junction reverse-biased.

The operation of the *pnp* transistor is similar to that of the *npn* transistor except that the charge carriers are of the opposite polarity. Thus the majority carriers crossing from the emitter to the base are holes, and most of these cross the narrow base region to the more negative collector. Again, a small base current is formed of the electrons crossing from the base to the emitter, and of the combination of holes and electrons in the base region.

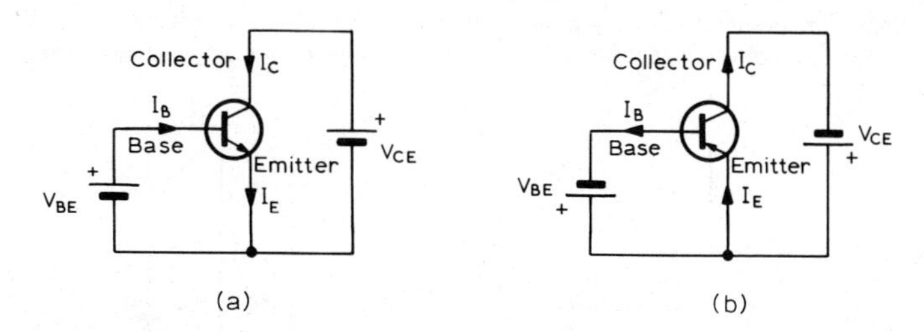

Figure 8.25 Circuit symbols, bias supplies, and current flow in (a) npn transistor and (b) pnp transistor

Circuit symbols for the *npn* and *pnp* transistors are shown in *Figure 8.25*. The direction of the arrow on the emitter shows the direction of conventional current flow. Also shown in *Figure 8.25* are the polarities of the bias supplies V_{BE} and V_{CE} for the two types of transistor, and the directions of the currents.

The transistor is a three-terminal device, but it has to be connected into circuits to give an input and an output. Thus one terminal of the device must be common to both input and output circuits. It is therefore possible to connect the transistor in three configurations, as shown in *Figure 8.26*. These configurations are called: common emitter, (*a*), because the emitter is common to both input and output circuits; common

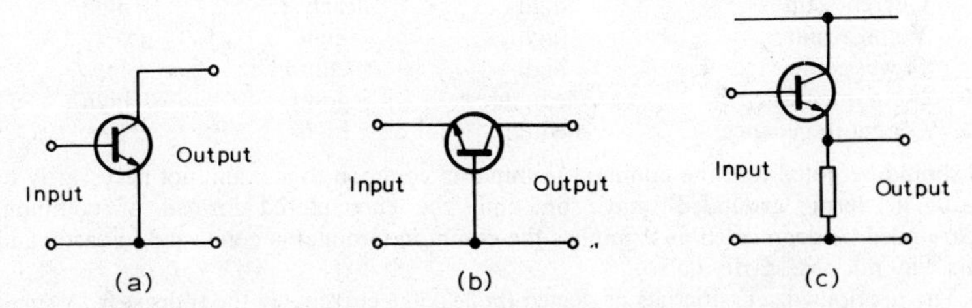

Figure 8.26 Transistor configurations: (a) common emitter (b) common base (c) common collector

base, (*b*); and common collector, (*c*). In all the configurations, the currents in the transistor are related by the equation

$$I_E = I_B + I_C$$

In the common-emitter configuration, the input is the base–emitter circuit, and the output is the collector–emitter circuit. The input current is the base current I_B, and the output current is the collector current I_C. The ratio of the two currents I_C/I_B is the current gain β, or more precisely, the static common-emitter forward current transfer ratio h_{FE}. Because the base current is small in comparison with the collector current, the value of h_{FE} is large. Thus a small change in the base current produces a large change in the collector current; in other words, the transistor forms a current amplifier.

As the base–emitter junction is forward-biased, a small change in the base-to-emitter voltage V_{BE} will produce a large change in the base current, and a larger change in the collector current. The collector–base junction is reverse-biased and so has a high impedance. A high-value load resistance can therefore be connected in the collector circuit, and the collector current will develop a high voltage across it. Thus the small change in the base voltage produces a larger change in collector voltage, and a high voltage gain is obtained. A power gain between input and output circuits is therefore obtained.

In the common-base configuration, the input is the base–emitter circuit and the output the collector–base circuit. The input current is the emitter current I_E, and the output current is the collector current I_C. The current gain α, or more precisely the static common-base forward current transfer ratio h_{FB}, is I_C/I_E and therefore a little less than unity. However, because of the low impedance of the forward-biased base–emitter junction and the high impedance of the reverse-biased collector–base junction, a high voltage gain is obtained. Thus a power gain between input and output is obtained.

The common-collector configuration is also called the emitter follower by analogy to the thermionic valve cathode follower circuit. The characteristics of this configuration differ somewhat from the other two. The input current is the base current I_B, and the output current is the emitter current I_E. The current gain, the static common-collector forward current transfer ratio h_{FC}, is high but the voltage gain is a little less than unity. The greatest difference between this configuration and the other two, and the feature that provides its main application in practical circuits, is that the emitter follower has a high input impedance. Thus this configuration will respond to voltage input signals.

The main characteristics of the three circuit configurations are summarised below.

	Common emitter	*Common base*	*Common collector*
Current gain	high	nearly 1	high
Voltage gain	high	high	nearly 1
Power gain	high	medium	low
Input impedance	medium	low	high
Output impedance	medium	high	low

It should be noted that the common terminal is common to a.c. and not necessarily to d.c. The term 'grounded' may sometimes be encountered instead of common. 'Grounded' is deprecated as it implies the common terminal is connected to earth, and this may not necessarily be so.

The previous discussion has neglected the leakage currents in the transistor. A small current flows across the reverse-biased collector–base junction caused by minority carriers. For small junction transistors at room temperature, this current is a few nanoamps in silicon devices and a few microamps in germanium devices. It is therefore considerably smaller than the normal collector current.

In the common-base configuration, the leakage current in the output circuit is I_{CBO}. (The value of I_{CBO} is quoted in published data with the emitter open-circuit.) In the common-emitter configuration, however, an increased leakage current is present through transistor action on I_{CBO} causing a current $h_{FE} \times I_{CBO}$ to flow in the emitter circuit (the base being open-circuit). Thus the total leakage current in the output circuit for the common-emitter configuration, I_{CEO}, is given by:

$$I_{CEO} = I_{CBO}\,(1 + h_{FE}),$$

and this current is considerably higher than that for the common-base configuration. The leakage current increases with temperature, and is undesirable in amplifier circuits as it limits the amount of useful power that can be derived from the transistor. One of the advantages of silicon transistors that enabled them to replace germanium transistors for many applications is the lower leakage current.

The currents I_{CBO} and I_{CEO} are described in published data as cut-off currents, being the currents flowing when the transistor is cut-off with the emitter or base open-circuit.

CHARACTERISTICS AND RATINGS

Because the transistor is a three-terminal device, there are a considerable number of characteristics relating the various currents and voltages. The data published by the transistor manufacturer contains a large number of characteristics, and the most common ones will be described. The shape of these characteristics depends to some extent on the configuration in which the transistor is operated, and the polarities of the currents and voltages depend on the type, *npn* or *pnp*. In addition, the manufacturer lists certain current and voltage values which must not be exceeded if the transistor is to remain undamaged.

The input characteristic relates the input current to the input voltage with the output voltage remaining constant. For the common-emitter configuration, the input current is the base current I_B, the input voltage is the base-to-emitter voltage V_{BE}, and the output voltage to be held constant is the collector-to-emitter voltage V_{CE}. The form of the input characteristic, as shown in *Figure 8.27*, is that of a forward-biased diode as would be expected. The reciprocal of the slope at any point gives the input impedance of the transistor when operating at that point. The voltage at which significant current starts to flow depends on the material from which the transistor is manufactured, approximately 200 mV for germanium devices and 600 mV for silicon devices.

A similar characteristic is obtained for the common-base configuration. The input current is the emitter current I_E, the input voltage is again the base-to-emitter voltage

V_{BE}, and the output voltage to be maintained constant is now the collector-to-base voltage V_{CB}. The input impedance is lower for this configuration.

The output characteristic for the common-emitter configuration relates the collector current I_C to the collector-to-emitter voltage V_{CE}. Two forms of the characteristic are usually given in published data: with base current I_B as parameter, and with base-to-emitter voltage V_{BE} as parameter. Both forms of the characteristic have a similar shape;

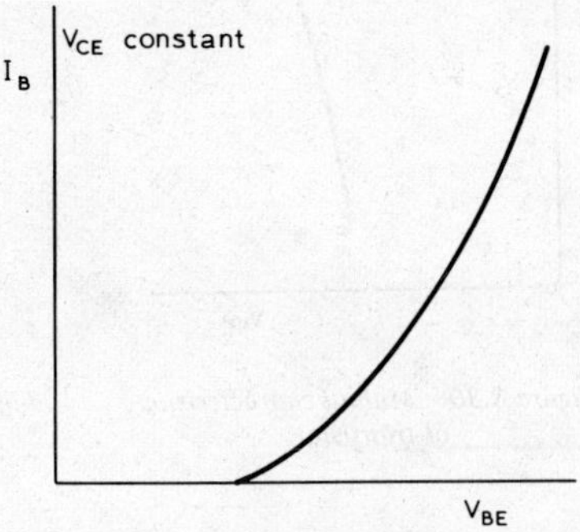

Figure 8.27 Input characteristic of transistor (common emitter)

a sharply-rising curve with a knee, followed by a straight line tending to become nearly parallel to the V_{CE} axis. For low values of V_{CE} below the knee of the characteristic, the collector is not efficient at collecting the charge carriers crossing from the emitter through the base. When V_{CE} exceeds the knee value, the number of carriers collected rises to a 'saturation' value and is nearly independent of the collector-to-emitter voltage. There is, in practice, a slight increase in collector current as V_{CE} increases which is shown by the characteristic not being truly parallel to the V_{CE} axis.

The output characteristic with base current as parameter is shown in *Figure 8.28*. The value of collector current depends on the value of base current, as would be expected through transistor action. However, a small collector current flows even with zero base current. This is the leakage current I_{CEO} already discussed. The output characteristic with base-to-emitter voltage as parameter is shown in *Figure 8.29*. In both

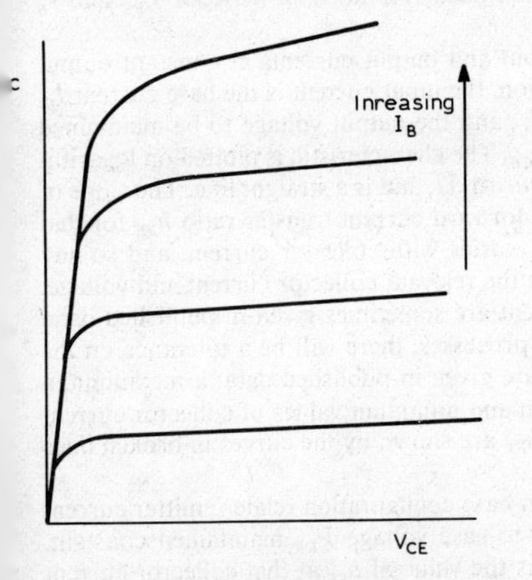

Figure 8.28 Output characteristic of transistor (common emitter) with base current as parameter

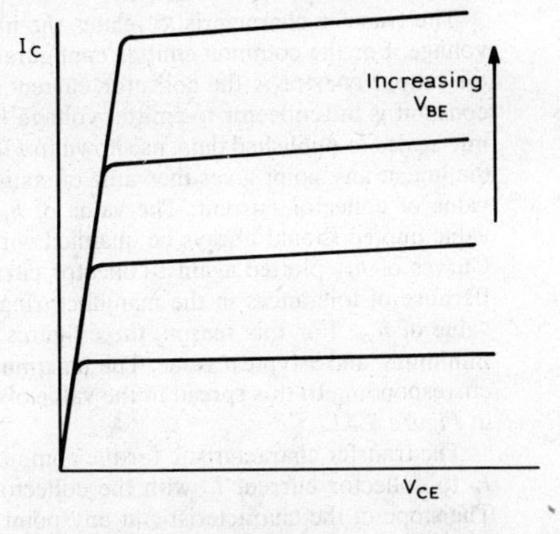

Figure 8.29 Output characteristic of transistor (common emitter) with base-to-emitter voltage as parameter

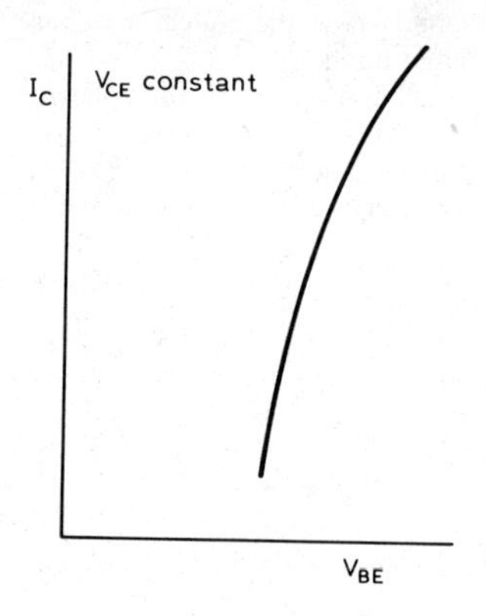

Figure 8.30 Mutual characteristic of transistor

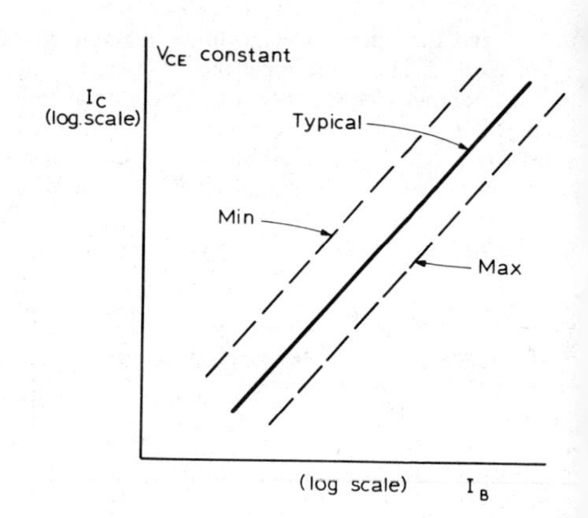

Figure 8.31 Transfer characteristic of transistor (common emitter)

forms of the characteristic, the reciprocal of the slope at any point gives the output impedance of the transistor operating at that point. This is high, as would be expected, since the collector–base junction is reverse-biased.

Similar curves are obtained for the output characteristic in the common-base configuration. The emitter current is the parameter for this configuration, and the value of output current is slightly less than that of the emitter current since the current gain is just below unity. The curves are more nearly parallel to the V_{CB} axis showing the higher output impedance obtained with this configuration.

The mutual characteristic, shown in *Figure 8.30,* relates the collector current I_C to the base-to-emitter voltage V_{BE} with the collector-to-emitter voltage constant. (The name mutual characteristic is derived by analogy with thermionic valves where the relationship between anode current and grid voltage is called the mutual conductance.) The characteristic is not linear because of the non-linear relationship between V_{BE} and I_B shown in the input characteristic.

The transfer characteristic relates the input and output currents at constant output voltage. For the common-emitter configuration, the input current is the base current I_B, the output current is the collector current I_C, and the output voltage to be maintained constant is the collector-to-emitter voltage V_{CE}. The characteristic is plotted on logarithmic scales in published data, as shown in *Figure 8.31*, and is a straight line. The slope of the line at any point gives the value of static forward current transfer ratio h_{FE} for that value of collector current. The value of h_{FE} varies with collector current, and so any value quoted should always be qualified with the relevant collector current and voltage. Curves of h_{FE} plotted against collector current are sometimes given in published data. Because of tolerances in the manufacturing processes, there will be a tolerance on the value of h_{FE}. For this reason, three figures are given in published data: a maximum, a minimum, and a typical value. The maximum and minimum values of collector current corresponding to this spread in the value of h_{FE} are shown by the curves in broken lines in *Figure 8.31*.

The transfer characteristic for the common-base configuration relates emitter current I_E to collector current I_C with the collector-to-base voltage V_{CB} maintained constant. The slope of the characteristic at any point is the value of h_{FB} at that collector current and voltage.

Transistor characteristics change with temperature, and the characteristics just discussed are therefore plotted at a constant temperature, usually 25 °C. Curves of the variation of base current and h_{FE} with temperature are sometimes given in published data. The maximum operating temperature for the transistor is set by the junction temperature at which the lightly-doped base becomes intrinsic and transistor action cannot take place.

The maximum power dissipation of a transistor P_{tot} is determined by the maximum permissible junction temperature T_j max and the ambient temperature in which the transistor operates T_{amb}. Thus:

$$P_{tot} = \frac{T_j \max - T_{amb}}{\Sigma R_{th}} \qquad (1)$$

where ΣR_{th} is the total thermal resistance.

The quantity *thermal resistance* represents the opposition to the flow of heat from the transistor. It is expressed as change of temperature with power, and the usual units are °C/W.

For low-power and medium-power transistors, the thermal resistance is specified as that between the junction and ambient, $R_{th(j-amb)}$, or that between the junction and case, $R_{th(j-case)}$. If a cooling clip is used to increase the radiating area of the transistor, a thermal resistance is given in the data for the clip, $R_{th(case-amb)}$. Thus the term ΣR_{th} in equation 1 is either $R_{th(j-amb)}$ or ($R_{th(j-case)} + R_{th(case-amb)}$). The values of T_j max and P_{tot} are given in the published data, the value of T_{amb} is known or a limit known, and ΣR_{th} is known. Thus equation 1 can be used to ensure that the cooling of the transistor (as shown by ΣR_{th}) is adequate for the total dissipation, or to calculate the maximum ambient temperature in which the transistor can operate.

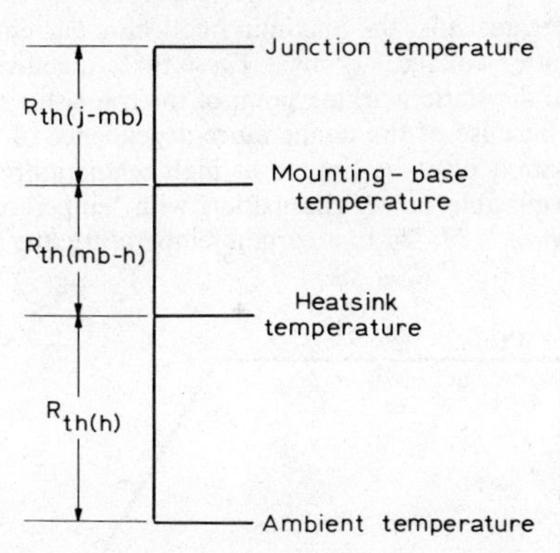

Figure 8.32 Temperatures and thermal resistances for transistor mounted on heatsink

For high-power transistors, the thermal resistance between the junction and the mounting base, $R_{th(j-mb)}$, is given. A heatsink to increase considerably the cooling area of the transistor is generally required, and a lead washer used to ensure good thermal contact between the transistor and heatsink. If the transistor case has to be insulated from the heatsink, a mica washer is used as well. For these washers, a thermal resistance $R_{th(mb-h)}$ is given. The heatsink itself has a thermal resistance $R_{th(h)}$. Thus a chain of thermal resistance exists between the junction of the transistor and ambient, as shown in *Figure 8.32*. From a given total dissipation and maximum junction temperature, equation 1 can be used to calculate the required ΣR_{th} for a particular ambient temperature, and from ΣR_{th} the thermal resistance of the heatsink calculated. A suitable heatsink can then be chosen.

The value of P_{tot} calculated from thermal considerations must be equal to the total electrical power associated with the transistor. Thus:

$$P_{tot} = V_{CE}I_C + V_{BE}I_B \qquad (2)$$

for the common-emitter configuration. Since $V_{CE} > V_{BE}$, and $I_C \gg I_B$, equation 2 can be simplified to:

$$P_{tot} = V_{CE} I_C$$

The maximum power dissipation can be plotted on the output characteristic, as shown in *Figure 8.33*. Also shown on *Figure 8.33* are the two other limits for safe operation of

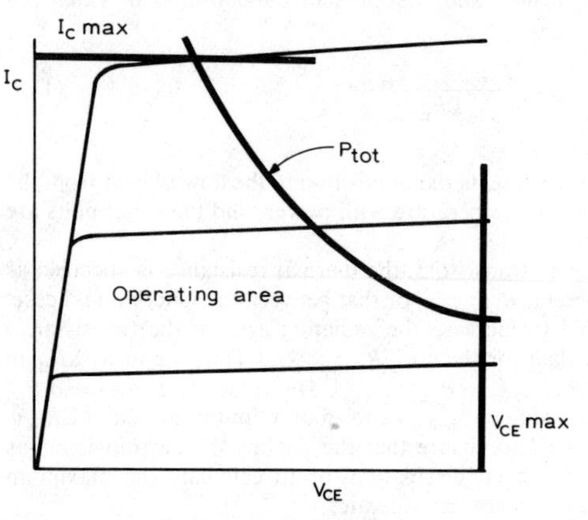

Figure 8.33 Operating area of transistor: limits of collector voltage and current, and power dissipation plotted on output characteristic (common emitter)

the transistor: the maximum collector current I_C max, and the maximum collector-to-emitter voltage V_{CE} max. These three quantities define the permissible operating area, and the static working point of the transistor must be chosen to lie within this area.

Because of the temperature-dependence of transistor characteristics, the power dissipation must be limited at high temperatures. The form of the curve for maximum permissible power dissipation with temperature given in published data is shown in *Figure 8.34*. Up to a certain temperature, the value depending on the type of transistor,

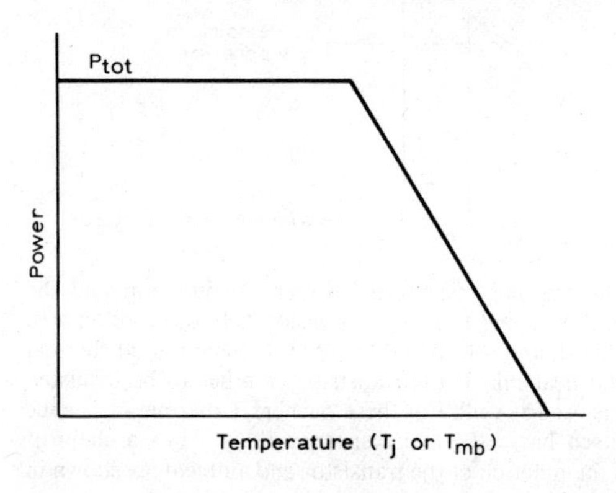

Figure 8.34 Variation of maximum permissible power dissipation with temperature

the power dissipation can be maintained at the maximum value. Above this temperature, however, the power must be limited (derated) to maintain safe operation.

The limit to collector voltage V_{CE} max shown in *Figure 8.33* is set by the reverse voltage across the collector–base junction. If this voltage is too high, avalanche breakdown occurs, as shown in *Figure 8.35*. The breakdown can occur with either forward or reverse base drive, and the effect of base drive on the breakdown is shown in *Figure 8.35*. The 'limit' curve occurs when the emitter is open-circuit. The maximum collector voltage at which the transistor can operate is therefore chosen to prevent this type of breakdown.

With power transistors, where the area of the junctions is made large to enable large currents to be conducted, a second form of breakdown can occur. Under certain voltage

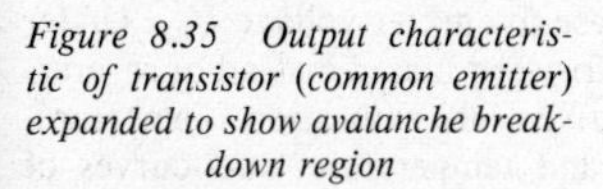

Figure 8.35 Output characteristic of transistor (common emitter) expanded to show avalanche breakdown region

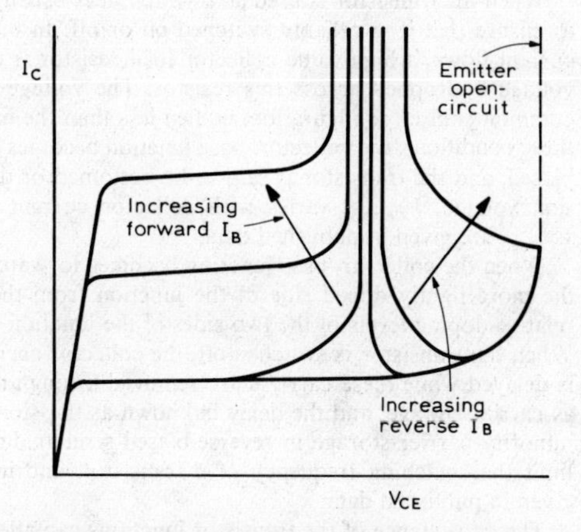

and current conditions, the emitter current tends to concentrate in one area of the emitter–base junction. This effect is called current contraction, and it can cause an increase in the local junction temperature sufficient to cause a 'hot spot', leading to local breakdown of the junction and the consequent destruction of the transistor. This form of breakdown is called *second breakdown*, and can occur with both forward and reverse base drive. With forward base drive, the breakdown usually occurs at the periphery of

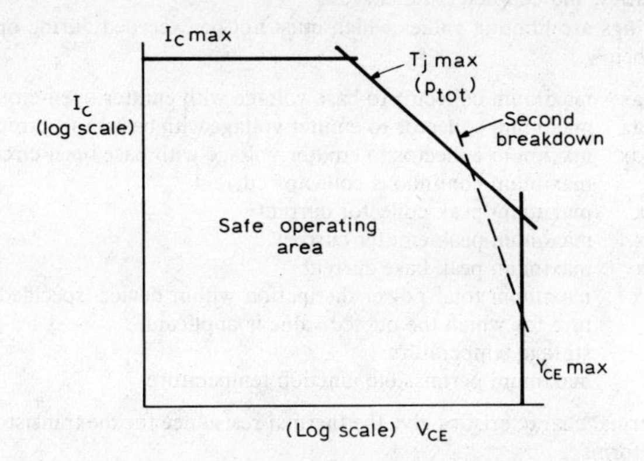

Figure 8.36 Safe operating area of power transistor showing limit imposed by second breakdown

the emitter; with reverse base drive where the transistor is being switched off after conducting, the emitter periphery is switched off first and the current is concentrated at the centre of the emitter so that breakdown occurs there. The effect of second breakdown is to add a further limit to the operating area of the transistor. The three limits shown on the output characteristic in *Figure 8.33* are shown by the full lines in *Figure 8.36*, which is plotted on logarithmic scales. The maximum collector current and voltage, I_C max and V_{CE} max, define two of the boundaries as before. The thermal limit, T_j max which is equivalent to P_{tot}, is a straight line intersecting I_C max and V_{CE} max. The fourth boundary, shown in broken lines, is the limit set by second breakdown. This safe operating area can be extended for pulse operation. Families of curves for various duty cycles and pulse durations are published in the data for power transistors.

When the transistor is used as a switch, it is usually operated with a large base drive to ensure that it is reliably switched on or off. In the on state when a large collector current flows, a high-value collector load resistor is used so that nearly all the supply voltage is dropped across this resistor. The voltage across the transistor, V_{CE} in the common-emitter configuration, is then less than the base-to-emitter voltage V_{BE}. Under these conditions, the collector–base junction becomes forward-biased instead of reverse-biased, and the transistor is said to be bottomed or driven into saturation. The saturation voltage, $V_{CE(sat)}$, varies with collector current and temperature, and curves of $V_{CE(sat)}$ are given in published data.

When the collector–base junction becomes forward-biased, carriers are injected into the more lightly doped side of the junction from the more heavily doped side. (The relative doping levels of the two sides of the junction depend on the type of transistor.) When the transistor is switched off, the collector current does not fall immediately but is delayed while these carriers are removed through the collector. This effect is known as carrier storage, and the delay is known as the storage time. (The effect is similar to minority carrier storage in reverse-biased semiconductor diodes.) Carrier storage can limit the switching frequency of a transistor, and information on switching times is given in published data.

The capacitance of the transistor junctions can affect the operation of the transistor, particularly at high frequencies. Curves showing the variation of collector–base and emitter–base capacitance with collector-to-base and emitter-to-base voltages respectively for a particular frequency are given in published data.

The preceding discussion on transistor characteristics and ratings can be conveniently summarised in the form of the information presented on a typical data sheet. The information is divided into four groups: ratings, thermal characteristics, electrical characteristics, and characteristic curves.

The ratings are limiting values which must not be exceeded during operation. They are listed below.

V_{CBO} max	maximum collector-to-base voltage with emitter open-circuit
V_{CES} max	maximum collector-to-emitter voltage with base short-circuited to emitter
V_{CEO} max	maximum collector-to-emitter voltage with base open-circuit
I_C max	maximum continuous collector current
I_{CM} max	maximum peak collector current
I_{EM} max	maximum peak emitter current
I_{BM} max	maximum peak base current
P_{tot} max	maximum total power dissipation within device, specified at a temperature for which the quoted value is applicable
T_{stg}	storage temperature
T_j max	maximum permissible junction temperature

The thermal characteristics give the thermal resistance for the transistor in one of the following forms.

$R_{th(j-amb)}$	thermal resistance, junction to ambient
$R_{th(j-case)}$	thermal resistance, junction to case
$R_{th(j-mb)}$	thermal resistance, junction to mounting base (to allow for heatsink mounting)

The electrical characteristics give information on the static performance of the transistor under specified voltages and currents at a specified ambient temperature. This information supplements that from the characteristic curves. The curves and values given for a particular transistor depend on the type and its range of applications. Typical information could include: input and output characteristics, mutual characteristic, transfer characteristic, variation of forward current transfer ratio with collector current, variation of base current with temperature, variation of collector and emitter capacitance with voltage, variation of $V_{CE(sat)}$ with collector current and temperature, and safe operating areas with respect to voltage, current, and power ratings.

In addition to the static characteristics, the published data includes *dynamic* characteristics as discussed in the following section.

A.C. CHARACTERISTICS AND EQUIVALENT CIRCUITS

To enable the behaviour of a transistor under a.c. conditions to be analysed, an equivalent circuit is required. From the earliest days of transistors, much theoretical work has been done to devise an exact equivalent circuit that will reproduce accurately the performance of a transistor in practice. Some of the circuits that have been suggested are complex. Three of the simpler circuits used extensively are described as follows. All use constant voltage or current generators in conjunction with resistors and capacitors. They are applicable to small signals only; other equivalent circuits are needed for large-signal analysis.

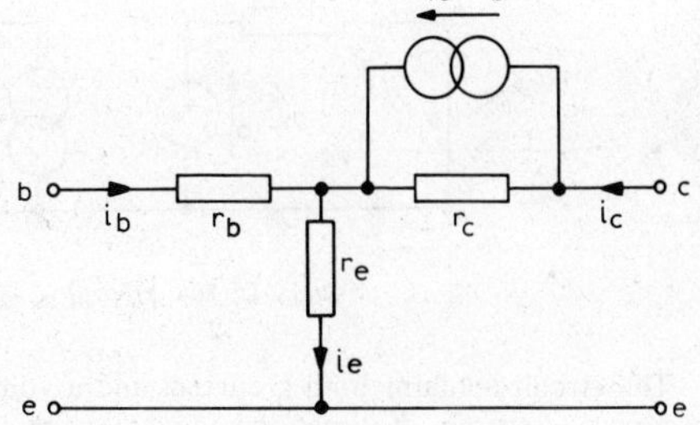

Figure 8.37 T equivalent circuit for common-emitter configuration

The T equivalent circuit is shown in *Figure 8.37* for the common-emitter configuration. Similar circuits are applicable for the other two configurations. Three resistances are connected in a T configuration: r_b representing the resistance of the base region, r_e representing the forward-biased emitter junction and r_c representing the reverse-biased collector junction. The current generator provides a current h_{fe} times the base current to represent the current gain of the transistor. Expressions relating the input and output voltages and currents can be derived in terms of the three resistances and the value of h_{fe}.

The T circuit of *Figure 8.37* applies to low and medium frequencies only. At high frequencies, the capacitances of the transistor junctions must be taken into account. Capacitances can be added to the T circuit but the expressions then become complicated. It is better to use another equivalent circuit.

The hybrid-π equivalent circuit shown in *Figure 8.38*, again for the common-emitter configuration, can be used for any frequency. The gain of the transistor is represented by the forward conductance g_m. In this circuit, a distinction is made between the base connection b and the base–emitter junction b'. Thus the resistance $r_{bb'}$ represents the bulk resistance of the base region. Resistance $r_{b'e}$ is the resistance of the forward-conducting base–emitter junction, while $c_{b'e}$ represents the emitter capacitance. Similarly, the collector–base junction is represented by resistance $r_{b'c}$ and capacitance $c_{b'c}$. Resistance r_{ce} is the output resistance of the transistor. Again, expressions relating input and output voltages and currents in terms of the resistances, capacitances, and g_m can be derived.

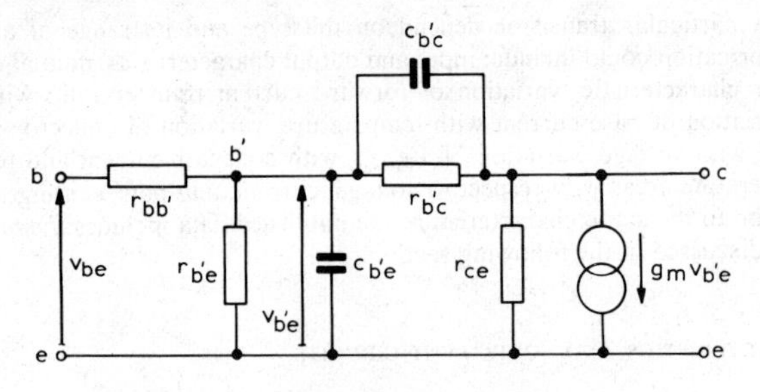

Figure 8.38 Hybrid-π equivalent circuit for common-emitter configuration

A third equivalent circuit, called the hybrid parameter network, is shown in *Figure 8.39.* This time the equivalent circuit does not represent the physical structure of the transistor, but four parameters can be derived which accurately describe the small-signal performance of the transistor at any frequency. These parameters are the h parameters quoted in published data.

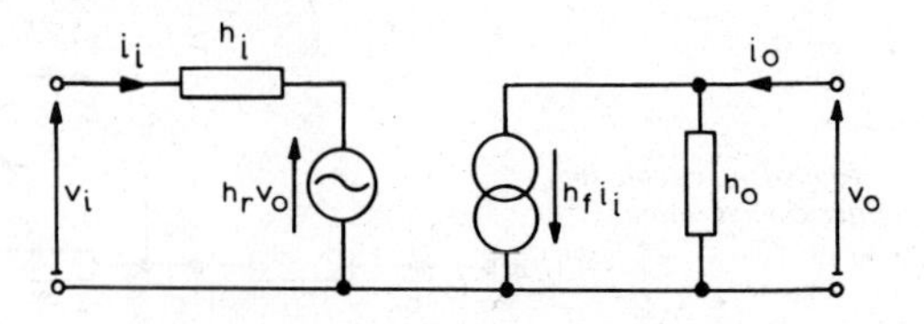

Figure 8.39 Hybrid parameter network

The circuit contains both a current and a voltage generator. The current generator produces a current h_f times the input current, representing the forward gain of the transistor. The voltage generator produces a voltage h_r times the output voltage, representing the feedback effect of the transistor. The basic equations for the network are:

$$v_i = h_i i_i + h_r v_o \tag{3}$$
$$i_o = h_f i_i + h_o v_o \tag{4}$$

If the output terminals of the network are short-circuited, $v_o = 0$.
Thus from equation 3:

$$v_i = h_i i_i \quad \text{or} \quad h_i = \frac{v_i}{i_i}$$

that is, h_i is the input impedance of the transistor. Similarly from equation 4:

$$i_o = h_f i_i \quad \text{or} \quad h_f = \frac{i_o}{i_i}$$

that is, h_f is the current gain or forward current transfer ratio of the transistor.
If the input terminals are open-circuit, $i_i = 0$. From equation 3:

$$h_r = \frac{v_i}{v_o}$$

which is the voltage feedback ratio. Similarly from equation 4:

$$h_o = \frac{i_o}{v_o}$$

which is the output admittance of the transistor.

Subscripts are added to the h parameters to indicate the transistor configuration: b for common base, c for common collector and e for common emitter. Thus h_{fe} is the forward current transfer ratio in the common-emitter configuration. It should be noted that the value of the h parameter will be different for the three configurations, for example:

$$h_{fe} \neq h_{fc} \neq h_{fb}$$

Because the h parameters apply to small signals, they are the modulus values of the ratios of small changes. For the common-emitter configuration, therefore, the parameters are:

h_{ie} input impedance $|\partial V_{BE}/\partial I_B|_{V_{CE}}$
h_{oe} output admittance $|\partial I_C/\partial V_{CE}|_{I_B}$
h_{fe} forward current transfer ratio $|\partial I_C/\partial I_B|_{V_{CE}}$
h_{re} voltage feedback ratio $|\partial V_{BE}/\partial V_{CE}|_{I_B}$

The values will change with changes in I_B and V_{CE}. Curves of the h parameters are given in published data, plotted against collector current I_C for convenience in use. The h parameters are easily measured for actual transistors, and this gives them considerable advantage over other systems of parameters that have been suggested.

The performance of the transistor at high frequencies will be affected by the capacitance of the junctions. The value of h_{fe} will fall with increasing frequency, and so will limit the operation of the transistor. Various frequency values have been suggested as 'figures of merit' for the high-frequency performance. One figure suggested was f_{hfb} or f_{hfe} (originally f_α or f_β). This frequency was defined as that at which the current gain had fallen to 0.7 times its low-frequency value. Another frequency suggested for the common-emitter configuration was f_1, the frequency at which the current gain had fallen to unity. The figure generally quoted in published data is f_T. This is the common-emitter gain-bandwidth product, the product of the frequency at which the value of h_{fe} falls off at a rate of 6 dB per octave and the value of h_{fe} at this frequency. For efficient high-frequency performance, the value of f_T should be as high as possible.

The value of f_T varies with collector current, and any value quoted should be qualified by the relevant collector current and voltage. Curves of the variation of f_T with collector current are given in published data.

ALLOYED-JUNCTION TRANSISTORS

The first transistors commercially available, in the early 1950s, were made by the germanium alloy-junction process. In this, the *pn* junctions were formed by alloying germanium with another metal. For *pnp* transistors, two pellets of indium, which would form the emitter and collector connections in the completed transistor, were placed on opposite sides of a slice of *n*-type monocrystalline germanium. The germanium slice, which formed the base, was approximately 50 μm thick. The whole assembly was heated in a jig to a temperature of about 500 °C, and some of the indium dissolved into the germanium to form on cooling *p*-type germanium. In this way, two *pn* junctions were formed, as shown in the simplified cross-section of *Figure 8.40*. Connecting wires were attached to the indium pellets for the emitter and collector leads, and a nickel tab attached to the base to provide the base connection and to mount the transistor on a header. The transistor was then encapsulated in a glass or metal envelope. *NPN* transistors were made by this process using *p*-type germanium for the base, and lead-antimony pellets instead of indium. The alloy-junction process is still used today for some audio-frequency power transistors.

The performance of an alloy-junction transistor is limited by the base width. In particular, the cut-off frequency is inversely proportional to the base width. If the assembly is heated for a longer time to allow more indium to dissolve so that the *pn* junction penetrates further into the germanium slice, there is a risk of the junctions joining and no transistor action occurring. Because of the difficulty of controlling the base width to an accuracy greater than that of the thickness of the original slice, the base width was kept to approximately 10 μm. This gave the early alloy-junction transistors a cut-off frequency of about 1 MHz, although refinements to the manufacturing process later allowed this to be raised to 5 or 10 MHz.

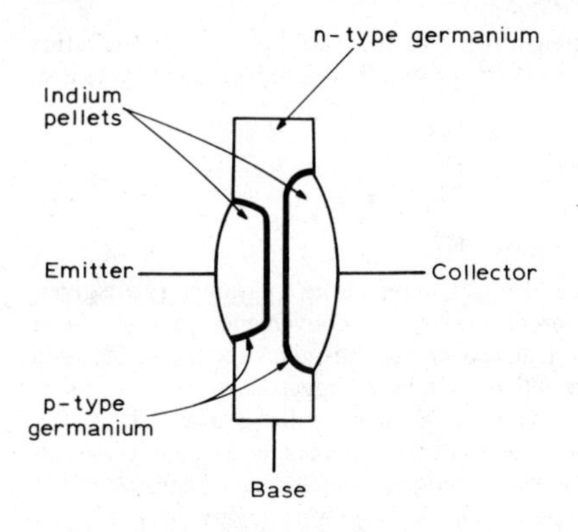

Figure 8.40 Simplified cross-section of pnp alloy-junction transistor

During the 1950s, other types of transistor were developed to overcome the limitations of the alloy-junction type. One of these was the alloy-diffused transistor. In this, two pellets which formed the emitter and base were alloyed onto the same side of a slice of *p*-type germanium which formed the collector. The base pellet contained *n*-type impurities, while the emitter pellet contained both *p*-type and *n*-type impurities. On heating the assembly to 800 °C alloying took place, but in addition diffusion of the *n*-type impurities into the *p*-type germanium ahead of the alloy front occurred. On cooling, a thin *n*-type layer linking both pellets was formed, with a *p*-type layer under the emitter pellet only. In this way, a *pnp* transistor with a narrow base layer was formed. Because of the narrow base layer, the cut-off frequency of this type of transistor was about 600 to 800 MHz. The alloy-diffused transistor was used extensively for high-frequency applications until it was superseded by the silicon planar epitaxial transistor.

Another type of transistor originally developed with germanium in 1956 was the mesa transistor. This transistor was developed to improve the switching characteristics

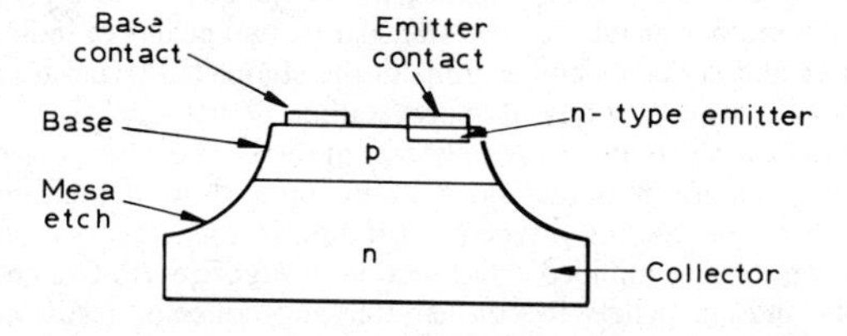

Figure 8.41 Cross-section of mesa transistor

and high-frequency performance of the then existing transistors. The principle was the etching of the edges of the transistor structure to decrease the junction areas and so reduce the capacitances. The resulting shape was a mesa or plateau, as shown in *Figure 8.41*. The original germanium mesa transistors were able to switch waveforms with rise times of 1 μs, and had a high-frequency performance comparable to that of the alloy-diffused type.

The principle of mesa etching is used extensively today in many types of transistor. It allows the edges of the transistor to be controlled, in particular the edges of the collector–base junction to be clearly defined. Mesa etching is also a prelimary process in passivation, the 'sealing' of the edges of the transistor to prevent contamination of the junction edges and the consequent change in characteristics during service.

The introduction of silicon during the late 1950s gave the transistor manufacturer a new material with considerable advantages over germanium. In particular, silicon transistors will withstand a higher junction temperature and have lower leakage currents than germanium transistors.

The alloying techniques developed for germanium were applied to silicon. Silicon *pnp* transistors were manufactured from a slice of *n*-type silicon (which formed the base) into which emitter and collector pellets were alloyed in the same way as for germanium alloy-junction transistors. The material used for the alloying pellets was aluminium.

DIFFUSED TRANSISTORS

It was soon realised that the impurity regions in silicon transistors could be diffused into the slice from material deposited on the surface from a vapour, and that this process had considerable advantages over the alloying process. In particular, the greater control possible over the process made it easier to manufacture devices with characteristics superior to those of alloyed transistors.

A silicon *npn* transistor could be made by two diffusions into an *n*-type slice which would form the collector of the completed transistor. The first diffusion, forming the base, used *p*-type impurities such as boron or gallium, and covered the whole surface of the slice. The second diffusion formed the emitter, diffusing *n*-type impurities such as phosphorus or arsenic into the already diffused base region. Electrical connections to the base were made by alloying rectifying contacts through the emitter to the base. A refinement of this manufacturing process is used today for silicon high-power transistors.

The discovery that thermally-grown silicon oxide on the surface of the slice could form a barrier to diffusion, and so could be used to define the impurity regions, led to the breakthrough in transistor manufacture—the planar process.

PLANAR TRANSISTOR

The planar process revolutionised transistor manufacture. For the first time in the manufacture of electronic devices, true mass-production techniques could be applied. The process allowed closer control over the geometry of the device, thus improving the performance. During the 1950s the transistor had first been a novelty, then regarded as an equivalent of the thermionic valve. With the introduction of the planar process in 1960, the transistor established itself as a device with a performance superior to that of the valve in 'general-purpose' electronics. Operation at frequencies up to the microwave region became possible, as did power transistors operating at radio frequencies. And it was the planar process that made possible the most important present-day semiconductor device—the integrated circuit.

The principle of the planar process is the diffusion of impurities into areas of a silicon slice defined by windows in a covering oxide layer. The various stages in the manufacture of a silicon planar *npn* transistor are shown in *Figure 8.42*.

A slice of *n*-type monocrystalline silicon (which will form the collector of the completed transistor) is heated to approximately 1 100 °C in a stream of wet oxygen. The temperature is controlled to better than ±1 deg C, and an even layer of silicon dioxide 0.5 μm thick is grown, *Figure 8.42* (*a*). The slice is spun-coated with a photoresist, an organic material which polymerises when exposed to ultraviolet light, producing a layer about 1 μm thick. The photoresist is dried by baking. A mask defining the base area is placed on the slice and exposed. On development of the photoresist, the unexposed area is removed (giving access to the oxide layer), while the exposed area remains and is hardened by further baking to resist the chemical etch. This etch removes the uncovered oxide to define the diffusion area. The remaining photoresist is dissolved away, leaving the slice ready for the base diffusion, *Figure 8.42* (*b*).

Boron is used to form the *p*-type silicon in the base region. The slice is passed through a diffusion tube in a furnace. A gas stream is passed over the slice containing a mixture of oxygen with boron tribromide BBr_3 or diboranne B_2H_6. The boron compounds decompose, and a boron-rich glass is formed on the surface of the slice. From this, boron is diffused into the silicon through the open base window. The glass is then removed by a chemical etch, and the slice heated in an oxygen stream in a second furnace. This drives in the boron further into the slice, and grows a sealing oxide layer over the surface. The depth to which the boron penetrates is determined by careful control of the furnace temperature and the time for which the slice is heated. The resulting structure is shown in *Figure 8.42* (*c*).

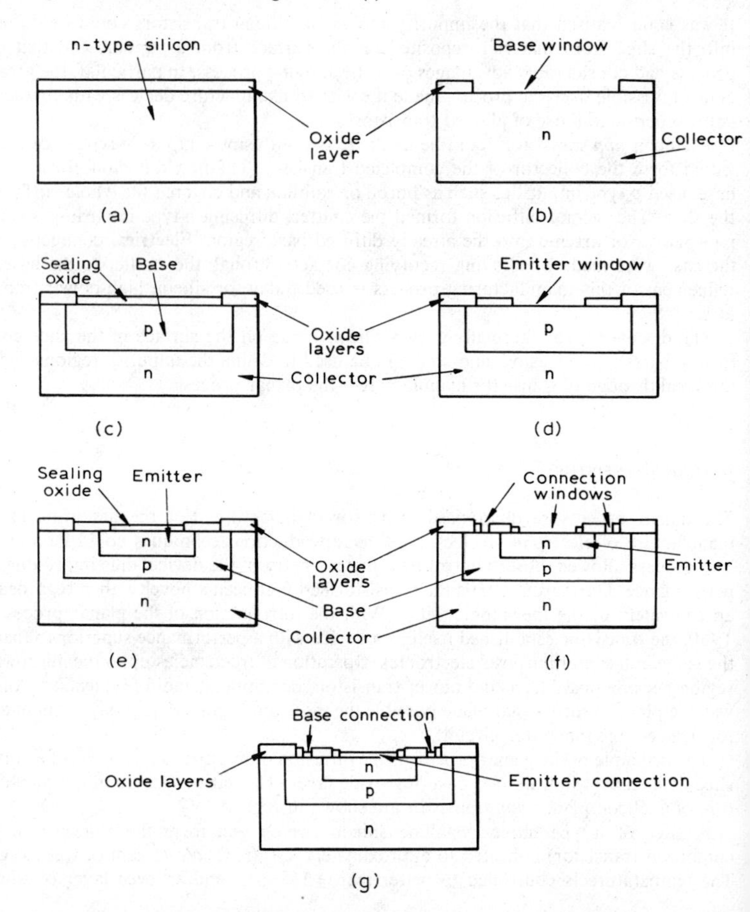

(h)

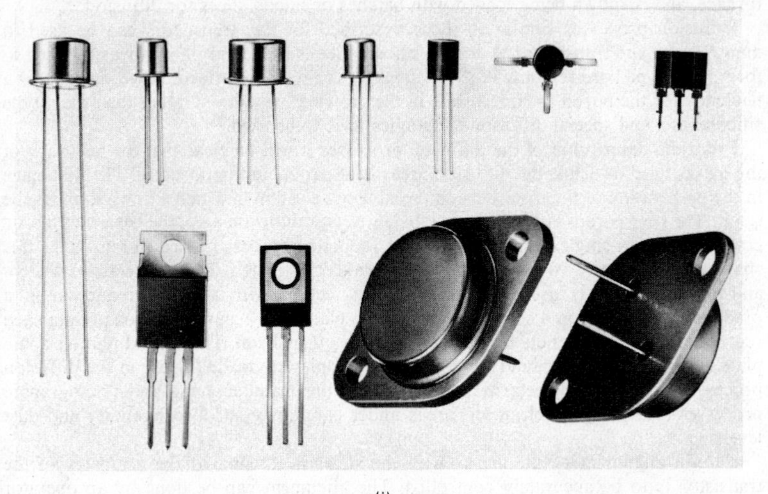

(i)

Figure 8.42 Stages in manufacture of silicon npn planar transistor (*a*) *oxide layer grown* (*b*) *base window opened* (*c*) *base diffused, sealing oxide grown* (*d*) *emitter window opened* (*e*) *emitter diffused, sealing oxide grown* (*f*) *connection windows opened* (*g*) *connection pads formed* (*h*) *typical diffusion furnace for transistor manufacture* (*Photo: Mullard Limited*) (*i*) *typical transistor encapsulations; top row, left to right: TO-5, TO-72, TO-39, TO-18, TO-92, 'T pack', 'Lockfit'; bottom row, left to right: TO-220, TO-126, TO-3 'thick base', TO-3 'thin base'* (*Courtesy Mullard Limited*)

The next stage in the planar process is the emitter diffusion. The slice is coated with photoresist as before, and a second mask used to define the diffusion area. Exposure, development, and etching are the same as for the base diffusion just described. The structure before emitter diffusion is shown in *Figure 8.42 (d)*.

Phosphorus is used to form the *n*-type emitter region. The gas stream in the diffusion furnace this time contains oxygen and phosphorus oxychloride $POCl_3$, phosphorous tribromide PBr_3, or phosphine PH_3. A phosphorus-rich glass is formed, from which phosphorus is diffused to form *n*-type silicon in the already-diffused *p*-type base region. A second furnace drives in the phosphorus to a controlled depth and forms a sealing oxide layer, as shown in *Figure 8.42 (e)*.

The final stage in this part of the manufacturing process is forming the connections to the base and emitter regions. A third mask is used to define the base and emitter contacts, the photographic and etching processes being the same as those previously described, *Figure 8.42 (f)*. The slice is coated with aluminium by evaporation onto the surface to form a layer about 1 μm thick. A fourth mask, a reversal of the third, is then used to define the areas where the aluminium is to be etched away leaving only contact pads for electrical connections, *Figure 8.42 (g)*.

The process so far has been described for clarity as though only one transistor were being manufactured. In practice, many thousand devices are manufactured at the same time. The silicon slices used at present are 5 cm in diameter, containing up to 10 000 transistors, but slices 7.5 cm in diameter are beginning to be used. More than 100 slices are processed at the same time in a diffusion furnace. Even in the early days of the planar process when silicon slices 2.5 cm in diameter containing 2 000 transistors were used, the process represented a considerable increase in production capacity over the alloyed-junction process, with a consequent drop in unit cost.

A typical diffusion furnace is shown in *Figure 8.42 (h)*. This is a dual triple-bank furnace, and is shown being loaded with batches of silicon slices for diffusion.

Diffusion processes similar to those described for *npn* transistors can be used to manufacture *pnp* transistors. A *p*-type silicon slice is used, and phosphorus diffused to form the *n*-type base regions. When the emitter regions are formed, however, there is a tendency for the boron to concentrate in the growing oxide layer rather than the *n*-type silicon base, and special diffusion techniques have to be used.

From the description of the diffusion processes it will be clear that the accuracy of the masks used to define the diffusion areas is of paramount importance. The first stage in the preparation of the masks uses a peelable opaque film in which a transparent shape is cut. The transparent area represents the particular diffusion area, the base, emitter, or contact area of a single transistor. A 20 : 1 reduction process is used to reproduce this shape on a 50 mm × 50 mm photographic plate. This plate is called a recticle. A step-and-repeat camera is used to reproduce the recticle with a 10 : 1 reduction in a predetermined array on a second photographic plate. The accuracy of positioning each reproduction of the recticle in the array is better than 1 μm. The second photographic plate is called the master, and from it working copies are made for use in the diffusion processes. To avoid defects in the masks, all the manufacturing and photographic processes take place in 'clean air' areas under carefully-controlled humidity and dust levels.

Careful alignment of the masks with the slices is essential if the geometry of the transistor is to be accurately controlled. The alignment can be done by an operator using a microscope, or in the latest equipment it is done automatically. The exposure of the mask onto the slice can be done with the mask and slice clamped together, or in some equipment by projecting an image of the mask onto the slice. The accuracy of the superimposition of the mask and the previously diffused areas on the slice is better than 1 μm. The photographic processes preceding the diffusion must take place in 'clean air' conditions.

The manufacturing processes after the transistor element has been formed are the same for both *npn* and *pnp* transistors. All the transistors in the slice are individually tested. This is done by means of probes that can move along a row of transistors on the slice testing each one, locate the edge of the slice, step to the next row, and move along this row testing these transistors. Any transistors that do not reach the required specification are marked automatically so that they can be rejected at a later stage. The slice is divided into individual transistors by scribing with a diamond stylus, and cracking the slice into individual chips. It is at this stage that the faulty transistors are rejected. The remaining transistors are prepared for encapsulation.

The oxide layer is removed from the collector side of the chip which is then bonded to a gold-plated header. The bonding action occurs through the eutectic reaction of gold and silicon at 400 °C. This contact forms the collector connection. Aluminium or gold wires 25 μm in diameter are used to connect the base and emitter contact pads to the lead-out wires in the header. Thicker wires can be used if the current rating of the transistor requires it. The final stage of assembly is encapsulation, either in a hermetically-sealed case or in moulded plastic, depending on the application of the transistor. Typical transistor encapsulations are shown in *Figure 8.42* (*i*).

The planar transistor has advantages over the alloyed-junction type besides the lower cost through mass-production manufacture. During each diffusion process, an oxide layer is grown over the edges of the junction which is not disturbed during the subsequent diffusion and assembly processes. Thus the collector-base junction, once it is formed and sealed by the oxide layer, cannot be easily contaminated by the emitter diffusion, testing and encapsulation, or during the lifetime of the transistor in service. The charge effects that occur at the bare surfaces of semiconductor devices are minimised, giving planar transistors greater reliability and stable characteristics. In addition, diffusion is a process which can be accurately controlled, and so the spacing between the three regions of the transistor can be held to a tolerance of less than 0.1 μm. Transistors with narrow base widths can be manufactured to allow operation at high frequencies, well above 1 GHz. A disadvantage of the planar transistor occurs through the resistivity of the collector. For high breakdown voltages, the resistivity should be high. On the other hand, for a high collector current the resistivity should be low. These two conflicting requirements would mean that a compromise value must be chosen for practical transistors. The conflict can be resolved, however, by the planar epitaxial process.

PLANAR EPITAXIAL TRANSISTOR

The planar epitaxial transistor, introduced in 1962, has the same structure as the planar transistor described in the previous section, but the transistor element is formed in an epitaxial layer. This layer is of high-resistivity material grown on a substrate of low resistivity, and so the conflicting requirements for the collector material can be met.

A polished slice of monocrystalline silicon, highly doped and therefore of low resistivity, forms the substrate. The slice can be *p*-type or *n*-type according to the type of transistor, and a typical resistivity figure is 1×10^{-5} Ωm. The epitaxial layer is grown on the substrate by vapour deposition in a radio-frequency heated reactor, the substrate being held at a temperature between 1 000 °C and 1 200 °C. The silicon vapour is formed by the decomposition of a silicon compound such as silicon tetrachloride $SiCl_4$ with hydrogen, and impurities can be added to the vapour to give the layer the required resistivity. For n^+ substrates, the impurities which can be used include phosphorus, arsenic, and antimony. Of these materials, arsenic and antimony are preferred because they have low diffusion constants. For p^+ substrates, the usual *p*-type impurities aluminium and gallium cannot be used because their diffusion constants are too high and the impurities would tend to migrate from the epitaxial layer into the substrate during the manufacture of the transistor. Boron is therefore used.

The silicon atoms in the epitaxial layer will take up the same relative positions as those in the substrate. Thus the near-perfect crystal lattice of the monocrystalline silicon

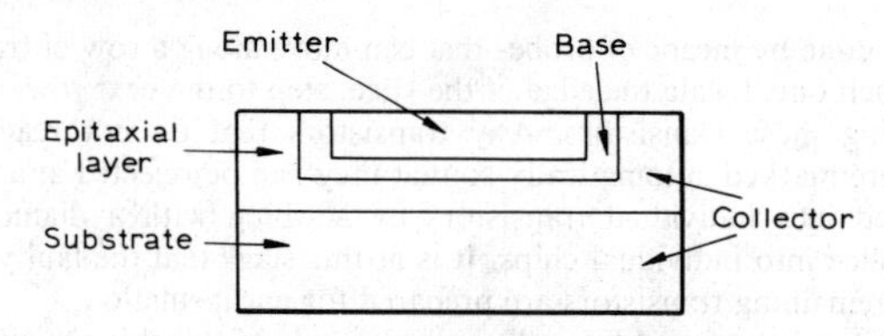

Figure 8.43 Simplified cross-section of planar epitaxial transistor

substrate is reproduced in the epitaxial layer. The thickness of the layer is between 10 μm and 12 μm, and a typical resistivity value is 1×10^{-2} Ωm. Thus the bulk of the collector of the transistor is formed by the low-resistivity substrate.

The formation of the transistor element in the epitaxial layer follows the same stages as those for the planar transistor described previously. A simplified cross-section of the completed transistor element is shown in *Figure 8.43*.

SPECIAL FORMS OF JUNCTION TRANSISTOR

The transistor manufacturer today has a variety of techniques and materials at his disposal. Special geometries for large power handling or radio-frequency operation have been developed so the operating range of the transistor has been extended. In addition, further diffusion, mesa etching, and choice of doping levels enable transistors to be manufactured with special characteristics to meet particular requirements.

Germanium power transistors were made during the early 1950s by 'scaling-up' small-signal alloy-junction transistors. The area of the junctions was increased, and the collector pellet was bonded to the case to ensure a low thermal resistance. Such transistors could dissipate 10 W but showed a rapid fall-off in gain at currents above 1 A. In the late 1950s, the indium emitter was doped with gallium to increase the emitter doping and so improve the high-current gain. Improvements to this type of transistor enable it to be used today for powers of up to 30 W.

The first silicon power transistors were introduced in the late 1950s, and used diffusion techniques. Base and emitter regions were successively diffused into one side of a slice of *n*-type silicon, and the electrical connection to the base was made by alloying rectifying contacts through the emitter. This type of transistor showed good gain up to a current of 5 A. Refinements to the manufacturing process during the 1960s led to the present-day diffused power transistor capable of handling currents of up to 30 A and powers up to 150 W. Two manufacturing processes are used for this type of power transistor, the single-diffused and the triple-diffused processes.

The single-diffused or hometaxial process uses a simultaneous diffusion on opposite sides of a homogenous base wafer, forming heavily-doped collector and emitter regions. The emitter is mesa etched to allow electrical connection to be made with the base. This type of transistor reduces the risk of hot spots through the use of a homogenous base, the wide base gives good second-breakdown properties, and the heavily-doped collector provides low electrical and thermal resistance.

Triple-diffused power transistors are manufactured by diffusing base and emitter regions on one side of a collector wafer. The third diffusion forms a heavily-doped diffused collector on the other side. This type of transistor has a high voltage rating, often able to withstand voltages of 1 kV or more.

The planar epitaxial process enables further improvements to be made to power transistors. At high current densities, current contraction can occur. (This is the cause of second breakdown, as explained under 'Characteristics and ratings'.) The edge of the emittter becomes more forward-biased than the centre, so that the current concentrates along the periphery of the emitter. It is therefore necessary to design base–emitter structures that differ from the annular or pear-drop geometries of small-signal transistors, and scaling-up can no longer be done. An emitter with a long periphery is required. Two structures that have been used successfully are the *star* and *snowflake*, the names being descriptive of the shape of the emitter. These structures could not have been produced in practical transistors without the planar technique of diffusion through a shape in the oxide layer.

More complex base–emitter structures can be produced to combine the large emitter area and long periphery required for high-power handling with the narrow spacing required for high-frequency operation. Geometries have been developed to enable power transistors to operate at radio frequencies. One such geometry is the interdigitated structure where fingers of the base interleave fingers of the emitter. Another is the overlay structure where a large number of separate emitter stripes are interconnected by metalising in a common base region. In effect, a large number of separate high-frequency transistors are connected in parallel to conduct a large current. Transistors using these structures can operate high in the radio frequencies, with typical powers of 175 W at 75 MHz and 5 W at 4 GHz.

Another structure used for power transistors is the epitaxial base or mexa structure. A lightly-doped epitaxial layer is grown on a heavily doped collector, and a single diffusion used to form the emitter in the epitaxial base layer. The resulting structure is mesa etched. Mexa transistors are rugged, and have a low collector resistance.

Power transistors are usually encapsulated in metal cases allowing mounting on a heatsink. In recent years, however, there has been a move towards plastic encapsulations. This has considerably decreased the cost of encapsulating the transistor without affecting the performance. A metal plate is incorporated in the plastic envelope to ensure good thermal contact between the transistor element and a heatsink.

A power transistor used as the output transistor in an amplifier generally requires a driver transistor to provide sufficient input power. If both transistors are mounted on heatsinks, a considerable amount of the volume of the amplifier will be occupied by these two transistors. A recent development enables space to be saved by combining the driver and output transistors on the same silicon chip in one encapsulation. This construction is the Darlington power transistor, which can have a current gain of up to 1 000 and power outputs up to 150 W.

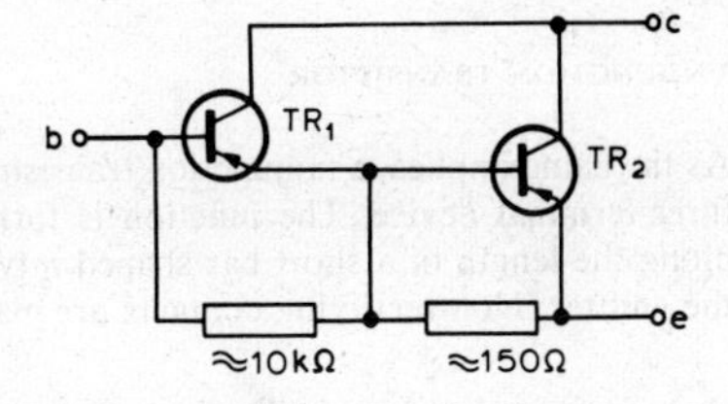

Figure 8.44 Circuit diagram of Darlington power transistor

The circuit diagram of a Darlington transistor is shown in *Figure 8.44*. The two transistors and base–emitter resistors are formed on one chip by successive diffusions using the epitaxial-base process. A diode can also be formed across the collector and emitter terminals for protection if required. The current gains of the two transistors are controlled during manufacture so that the overall gain varies linearly over a range of collector current. This linearity of gain is combined with smaller spreads than would occur with discrete transistors connected in the same circuit. These two advantages of the Darlington transistor are combined with a disadvantage, the high value of $V_{CE(sat)}$.

Transistors for high-frequency operation or fast switching must have narrow spacings between the emitter, base, and collector. Two geometries are generally used: the ring base and the stripe base. The ring-base structure is 'scaled-down' from the annular

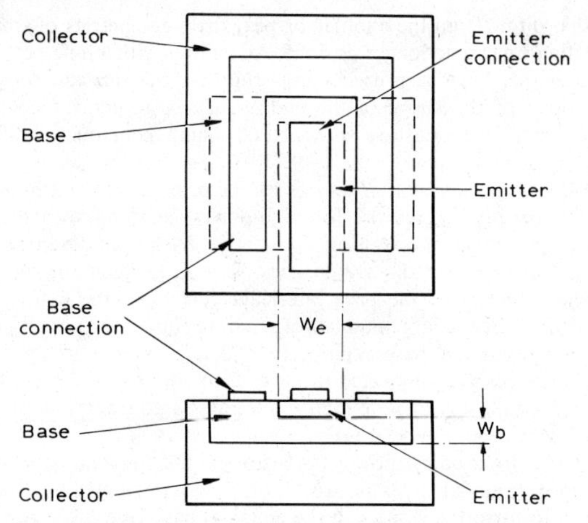

Figure 8.45 *Stripe-base structure for high-frequency transistor*

structure used for low-frequency transistors. The stripe-base structure, which is generally preferred for the higher-frequency operation, is shown in *Figure 8.45*. Many of these structures can be connected in parallel to increase the current-carrying capacity, forming the interdigitated structure already described for r.f. power transistors. The internal capacitances of the transistor, and the stray capacitances of the mounting and case, must be kept as low as possible to prevent restriction of the upper frequency limit. A planar epitaxial manufacturing process must be used to keep the collector resistance low. The doping level is chosen to suit the operating frequency and voltage.

In the stripe-base structure, two dimensions are critical for the upper frequency limit. These are the emitter stripe width (W_e in *Figure 8.45*) and the base width W_b. For present-day transistors operating up to the microwave region, the emitter width can be as low as 1 μm and the base width 0.1 μm.

UNIJUNCTION TRANSISTOR

As the name implies, a unijunction transistor contains only one junction although it is a three-terminal device. The junction is formed by alloying *p*-type impurity at a point along the length of a short bar-shaped *n*-type silicon slice. This *p*-type region is called the emitter. Non-rectifying contacts are made at the ends of the bar to form the base 1

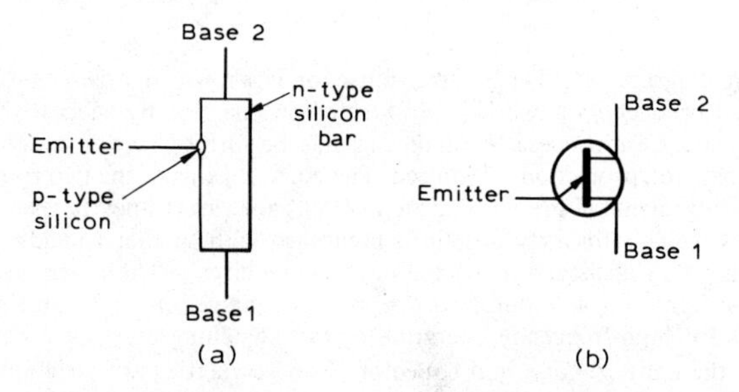

Figure 8.46 *Unijunction transistor: (a) simplified structure (b) circuit symbol*

and base 2 connections. The structure of a unijunction transistor is shown in *Figure 8.46* (*a*), and the circuit symbol in *Figure 8.46* (*b*).

The resistance between the base 1 and base 2 connections will be that of the silicon bar. This is shown on the equivalent circuit in *Figure 8.47* as R_{BB}, and has a typical value between 4 kΩ and 12 kΩ. A positive voltage is applied across the base, the base 2 contact being connected to the positive terminal. The base acts as a voltage divider, and a proportion of the positive voltage is applied to the emitter junction. The value of this voltage depends on the position of the emitter along the base, and is related to the voltage across the base, V_{BB}, by the intrinsic stand-off ratio η. The value of η is determined by the relative values of R_{B1} and R_{B2}, and is generally between 0.4 and 0.8.

The emitter *pn* junction is represented in the equivalent circuit by the diode. When the emitter voltage V_E is zero, the diode is reverse-biased by the voltage ηV_{BB}. Only the

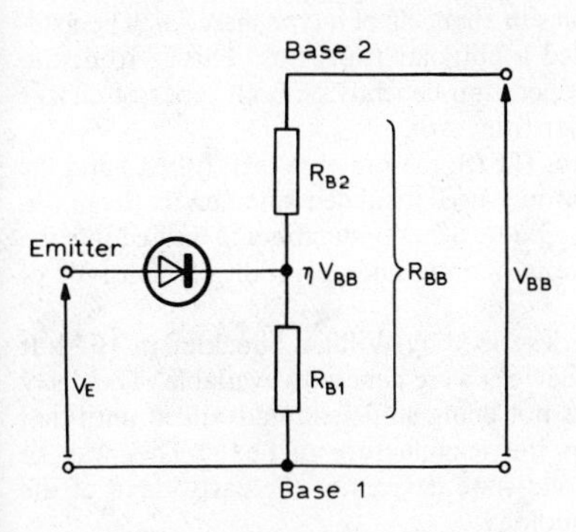

Figure 8.47 Equivalent circuit of unijunction transistor

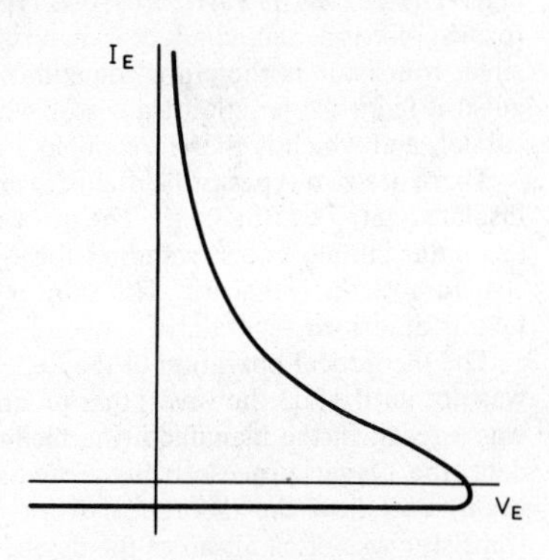

Figure 8.48 Voltage/current characteristic of unijunction transistor

small reverse current flows. If the emitter voltage is gradually increased, a value is reached where the diode becomes forward-biased and starts to conduct. Holes are injected from the emitter into the base, and are attracted to the base 1 contact. The injection of these holes reduces the value of R_{B1} so that more current flows from the emitter to base 1, reducing the value of R_{B1} further. The unijunction transistor therefore acts as a voltage-triggered switch, changing from a high 'off' resistance to a low 'on' resistance at a voltage determined by the base voltage and the value of η.

The voltage/current characteristic for a unijunction transistor is shown in *Figure 8.48*. It can be seen that after the device has been triggered, there is a negative-resistance region on the characteristic. This enables the unijunction transistor to be used in oscillator circuits as well as in simple trigger circuits.

CLASSIFICATION OF JUNCTION TRANSISTORS

Present-day junction transistors can be classified first by the semiconductor material from which they are manufactured, germanium or silicon. A further classification is then into *npn* and *pnp* types. It is then convenient to group them into low-power, medium-power and high-power types. The divisions between the groups are somewhat arbitrary, but typical values are as follows. For germanium transistors, low-power devices will have powers below 150 mW, medium-power devices up to 1 W, and high-power devices above 1 W to a typical upper limit of 30 W. For silicon transistors, low-power devices have powers below 500 mW, medium-power devices up to 10 W, and high-power devices above 10 W up to a typical limit of 150 W.

Typical germanium low-power transistors are listed in *Tables 8.15* and *8.16*; germanium medium-power transistors in *Tables 8.17* and *8.18*; and germanium high-power transistors in *Tables 8.19* and *8.20*. Silicon low-power transistors are listed in *Tables 8.21* and *8.22*; medium-power transistors in *Tables 8.23* and *8.24*; and high-power transistors in *Tables 8.25* and *8.26*. Silicon r.f. power transistors are listed in *Table 8.27*.

Field-effect transistors

A field-effect transistor consists essentially of a current-carrying channel formed of semiconductor material whose conductivity is controlled by an externally-applied voltage. The current is carried by one type of charge carrier only; electrons in channels formed of *n*-type semiconductor material, holes in channels of *p*-type material. The field-effect transistor is therefore sometimes called a unipolar transistor. This is to distinguish it from the junction transistor whose operation depends on both types of charge carrier, and which is therefore called a bipolar transistor.

There are two types of field-effect transistor (f.e.t.): the junction f.e.t. (j-f.e.t.) and the insulated-gate f.e.t. (i.g.f.e.t.). The most commonly used insulated-gate f.e.t. is the m.o.s. f.e.t., the initials m.o.s. standing for Metal Oxide Semiconductor and indicating the structure of the transistor. The shorter name is m.o.s.t. The operation of each type of f.e.t. is described separately.

The theoretical operation of the f.e.t. was described by William Shockley in 1952. It was not until 1963, however, that practical devices were generally available. The delay was a result of the manufacturing techniques not being sufficiently advanced until that date, the planar process being essential for the manufacture of f.e.t.s. This was an example of how the theoretical work on solid-state devices in the early days of the transistor was often ahead of the device technology.

OPERATION OF JUNCTION F.E.T.

The schematic structure of a junction f.e.t. is shown in *Figure 8.49*. The device shown is an *n*-channel f.e.t., formed from a bar-shaped slice of *n*-type monocrystalline silicon into which two *p*-type regions are diffused. Connections are made to the ends of the channel, the source and the drain, and to the *p*-regions, the gate.

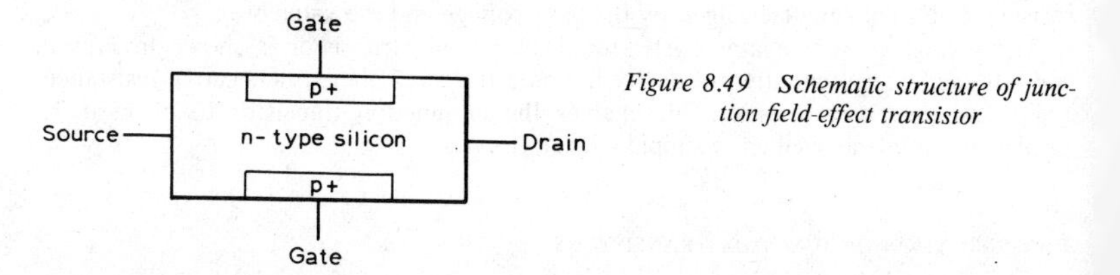

Figure 8.49 Schematic structure of junction field-effect transistor

If a voltage is applied across the channel so that the drain is positive with respect to the source, as shown in *Figure 8.50* (*a*), electrons flow through the channel from the source to the drain producing the drain current I_D. The magnitude of the drain current is determined by the conductivity of the channel and the drain-to-source voltage V_{DS}. When a negative voltage is applied to the gate, it will be reverse-biased. A depletion layer is formed about the *pn* junction between the gate and channel, as shown in *Figure 8.50* (*b*). Because the gate is more heavily doped than the channel, the depletion layer extends into the channel rather than into the gate. The depletion layer is virtually devoid of charge carriers and so has a high resistance, and therefore the channel thickness over which the current can flow is decreased. This decrease in channel thickness decreases the conductivity, and therefore decreases the magnitude of the drain current. Because the width of the depletion layer depends on the reverse bias on the gate, the drain current can be controlled by the gate voltage. This method of operation is analogous to that of the triode thermionic valve where the magnitude of the anode current is controlled by the voltage on the grid.

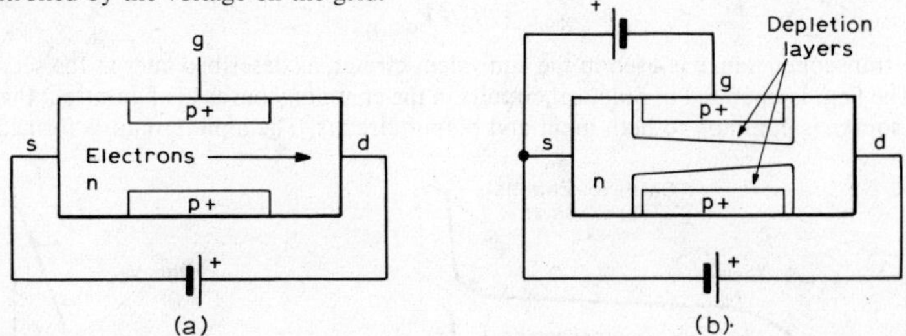

Figure 8.50 Operation of n-channel field-effect transistor: (a) bias to produce drain current (b) depletion layers produced by gate voltage and drain current voltage drop

If the gate voltage is sufficiently negative, the depletion layer will extend across the whole of the channel. The channel is then said to be 'pinched off'. The value of gate voltage at which this occurs is called the pinch-off voltage, V_P.

JUNCTION F.E.T. CHARACTERISTICS

When the gate voltage is zero, that is the gate is connected directly to the source, the flow of drain current causes a linear drop in voltage along the length of the channel. The gate-to-channel voltage will therefore form a small reverse bias at the source end of the gate, and a larger reverse bias at the drain end of the gate. The depletion layers therefore have the wedge shape shown in *Figure 8.50* (*b*). As the drain-to-source voltage is increased, the drain current will increase linearly. But the gate-to-channel voltage will also be increased so that the depletion layer will penetrate further into the channel. A point will be reached where the decrease in channel thickness has a greater effect than the increase in drain-to-source voltage. The characteristic of drain current I_D plotted against drain-to-source voltage V_{DS} will therefore form a knee. As V_{DS} is increased further, a value will be reached above which virtually no further increase in drain current will occur. This is because the depletion layers extend across the whole of the channel at the drain end of the gate and the channel is pinched off. As the value of V_{DS} is increased above the pinch-off value $V_{DS(P)}$, the point along the channel at which the depletion layers meet moves nearer the source. The drain current is maintained by electrons being swept through the depletion layer in the same way minority charge carriers are swept from the base region in junction transistors. Eventually as V_{DS} is increased further, the gate-to-channel breakdown value will be reached.

The characteristic just described is for zero gate voltage. If the gate-to-source voltage V_{GS} has a small constant negative value, the depletion layers will extend into the channel even with no drain current flowing. Thus the knee of the characteristic will occur at a lower drain current, and the voltage at which the channel is pinched-off occurs at a lower value of V_{DS}. A higher negative value of V_{GS} will decrease the knee and pinch-off value further. A family of characteristics with gate-to-source voltage V_{GS} as parameter is obtained, as shown in *Figure 8.51*. Also shown is the locus of the pinch-off value $V_{DS(P)}$, shown by the curve in broken lines.

The part of the characteristic where V_{DS} is greater than the pinch-off value is called the pinch-off or saturation region. This is the normal operating region for the f.e.t. With a constant value of source-to-drain voltage, the drain current can be varied by the gate-to-source voltage. The characteristic of drain current against gate voltage with drain-to-source voltage as parameter is shown in *Figure 8.52*. The slope of this transfer characteristic is the transconductance of the f.e.t., g_m. The value of g_m is given by:

$$g_m = |\, \partial I_D / \partial V_{GS} \,|_{V_{DS}}$$

The transconductance is used in the equivalent circuit, as described later in the section.

The f.e.t. is operated in practical circuits in the common-source configuration; that is, the source is common to both input and output circuits. The input circuit is formed by

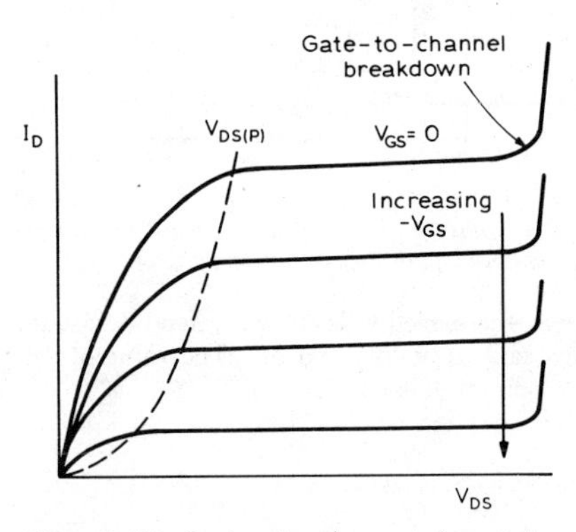

Figure 8.51 Drain voltage/current characteristics for junction f.e.t. with gate voltage as parameter

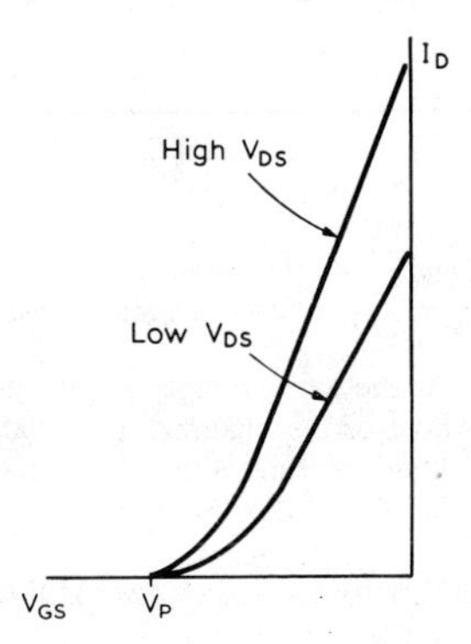

Figure 8.52 Transfer characteristic of junction f.e.t. showing effect of drain voltage

the gate and source, and the output circuit by the drain and source. Since the gate is reverse-biased, the input impedance of the f.e.t. is very high. This is in contrast with the usual configuration (common emitter) of the junction transistor, and so the f.e.t. provides the circuit designer with a useful complement to the junction transistor.

The reverse-biased gate also means that the gate current I_G flowing under normal operation is the small reverse current. The total power dissipated by the f.e.t. is:

$$P_{tot} = V_{DS}I_D + V_{GS}I_G \qquad (5)$$

Since $V_{DS} > V_{GS}$ and $I_D \gg I_G$, equation 5 can be rewritten as:

$$P_{tot} = V_{DS}I_D$$

The maximum power can be plotted on the V_{DS}/I_D characteristic to form one boundary of the operating area. Another boundary is set by the need to operate in the pinch-off region. The other two boundaries of the operating area are set by the maximum value of V_{DS} to prevent breakdown between gate and channel, and the maximum drain current I_D max that can be drawn from the device. These limits are shown in *Figure 8.53*.

Other limiting values are listed in the published data as electrical ratings. These include: the maximum drain-to-gate voltage V_{DGO} max, the maximum gate-to-source voltage V_{GSO} max, and the maximum gate current I_G max.

As with the junction transistor, the characteristics of the junction f.e.t. are temperature-dependent although to a lesser extent. Curves of the variation with temperature of drain current I_D and the gate cut-off current I_{GSS} (gate current with zero drain-to-source voltage) are given in published data.

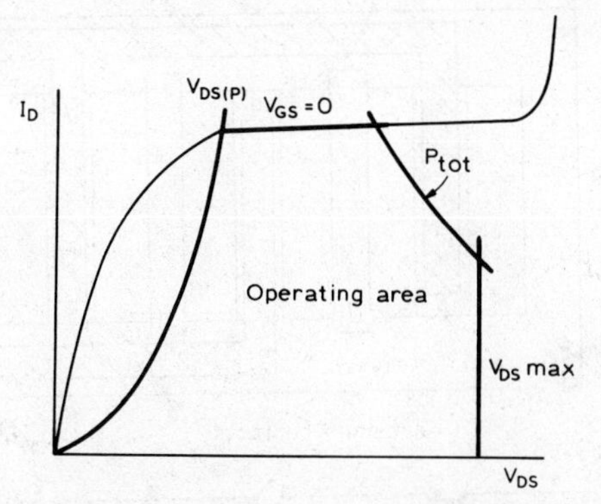

Figure 8.53 Operating area for junction f.e.t.

An equivalent circuit for the junction f.e.t. in the common-source configuration under small-signal conditions is shown in *Figure 8.54*. The gain of the f.e.t. is represented by the current generator $g_m V_{gs}$, where g_m is the transconductance of the f.e.t. and V_{gs} is the gate-to-source voltage. The input resistance and capacitance are r_{gs}, the gate-to-source resistance which is high because of the reverse-biased gate, and c_{gs}, the gate-to-channel capacitance formed by the depletion layer between the source and the pinch-off point. The value of r_{gs} is approximately $10^{11}\ \Omega$ at room temperature. Capacitance c_{gd} is the

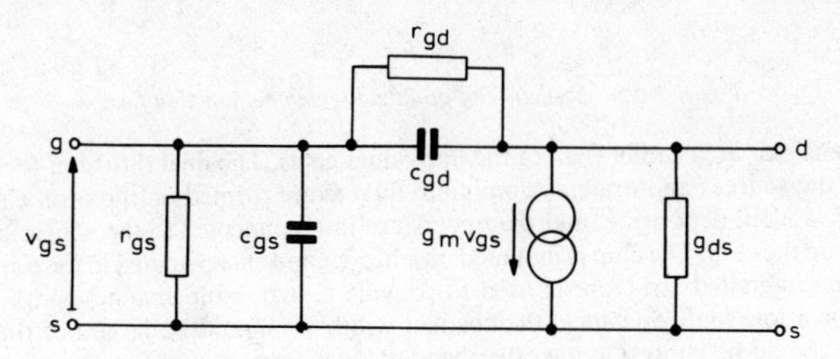

Figure 8.54 Small-signal equivalent circuit for f.e.t.

capacitance of the depletion layer beyond the pinch-off point to the drain. Both c_{gs} and c_{gd} are voltage-dependent. Resistance r_{gd} is the gate-to-drain resistance which has a similar value to r_{gs}. The output conductance g_{ds} is the drain-to-source conductance.

PRACTICAL JUNCTION F.E.T.S

The structure of a practical *n*-channel junction f.e.t. is shown in the cross-section and plan of *Figure 8.55*. An interdigitated structure is used, with interleaved fingers of drain and source between which are the channels. The gate regions are formed on top of the channels.

An *n*-type epitaxial layer is grown on a heavily-doped *p*-type substrate. Using the planar process, p^+ regions are diffused into the lightly-doped epitaxial layer to define the transistor area on the chip by isolating diffusions. A second diffusion process is used to form the p^+ upper gate regions. The individual upper gates are interconnected by the isolating diffusions, which also connect them to the substrate which forms the lower part of the gate. Connection to the gate can therefore be made by one contact at the edge

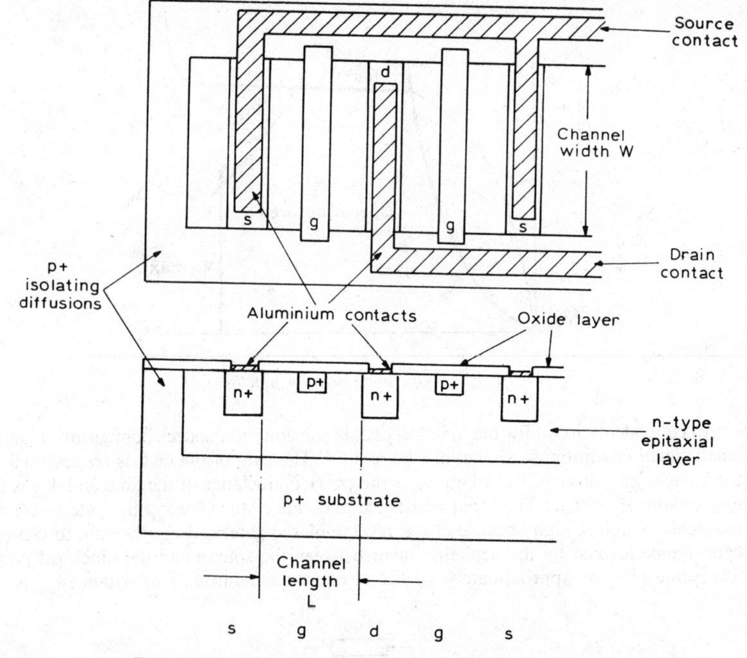

Figure 8.55 Structure of practical n-channel junction f.e.t.

of the transistor area rather than to the individual gates. The final diffusion forms the n^+ regions, the sources and drains. Aluminium fingers are formed on the source and drain areas by vacuum deposition, and the fingers are interconnected to the source and drain contacts on the chip. The chip is mounted on a header and encapsulated in the normal way.

The interdigitated structure is used to provide a favourable channel width-to-length ratio. For a low 'on' resistance, the channel width W should be large but the channel length L should be short. On the other hand, the gate capacitance is proportional to the channel area $W \times L$. Therefore the product WL should be as low as possible while the ratio W/L should be as high as possible. These requirements are met by making the channel length short, a few micrometres, but keeping the width to a practicable value by forming many channels in parallel interconnected in an interdigitated structure.

The *n*-channel junction f.e.t. has been described because this is the device most used in practice at present. *P*-channel f.e.t.s can be manufactured by growing a *p*-type epitaxial layer on an *n*-type substrate. *N*-type regions are then diffused into the epitaxial layer to form the upper gate regions, followed by a p^+ diffusion to form the sources and drains. The current in a *p*-channel junction f.e.t. is carried by holes, and the gate voltage

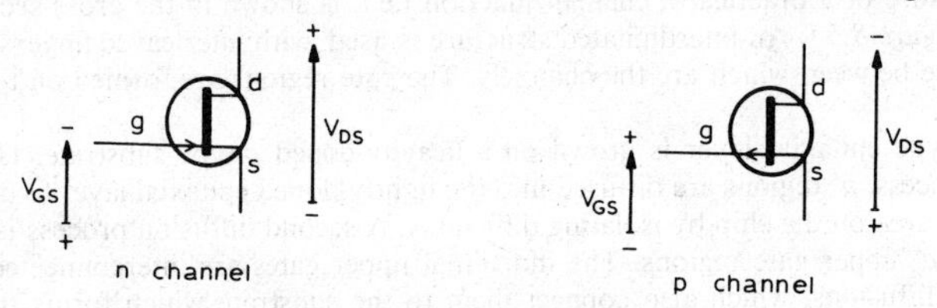

Figure 8.56 Circuit symbols for n-channel and p-channel junction f.e.t.s

and drain-to-source voltage have the opposite polarities from those used in the *n*-channel device.

The circuit symbols for *n*-channel and *p*-channel junction f.e.t.s are shown in *Figure 8.56*. The difference in the two symbols, as in the symbols for junction transistors, is the direction of the arrow, this time on the gate. The polarities of the voltages for both types of device are also shown in *Figure 8.56*. In both, the gate-to-source and drain-to-source voltages are of opposite polarity. *N*-channel and *p*-channel junction f.e.t.s conduct as soon as a drain-to-source voltage of the required polarity is applied, and the gate voltage is used to decrease the drain current. They are therefore known as depletion or normally-on devices.

OPERATION OF INSULATED-GATE F.E.T.

The structure of the insulated-gate f.e.t. differs from that of the junction f.e.t. just described in two respects: no separate current-carrying channel is built in, and the construction of the gate is different. A current-carrying channel is formed by the accumulation of charges beneath the gate electrode, as described later. The gate itself is not a diffused region but a thin layer of metal insulated from the rest of the f.e.t. by a layer usually of oxide. Thus the structure of the insulated-gate f.e.t. is successive layers of Metal, Oxide, and Semiconductor material—giving the transistor its alternative name of m.o.s. f.e.t., or the shorter form, the m.o.s. transistor or m.o.s.t. Because the semiconductor material used at present is invariably silicon, the initials m.o.s. are often interpreted as Metal Oxide Silicon.

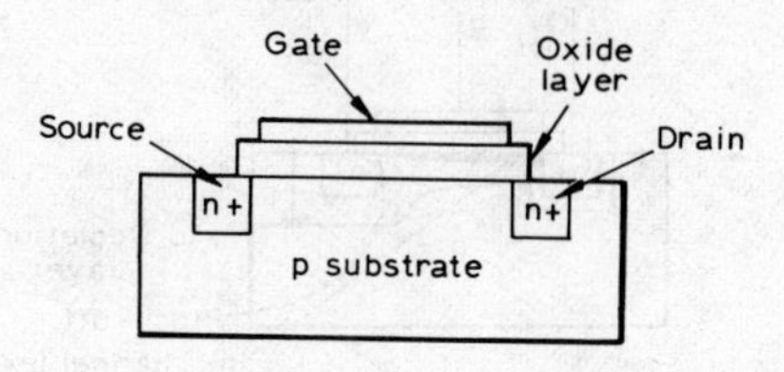

Figure 8.57 Simplified cross-section of n-channel m.o.s. f.e.t.

A simplified cross-section of an *n*-channel m.o.s. f.e.t. is shown in *Figure 8.57*. A *p*-type substrate is used, with heavily-doped n^+ regions forming the source and drain. A thin layer of silicon dioxide is grown on the surface of the substrate between the source and the drain. Aluminium is deposited on top of the oxide layer to form the gate electrode. With no voltage applied to the gate, only the depletion layers about the pn^+ junctions are present, as shown in *Figure 8.58* (*a*). Even if a large positive voltage with respect to the source is applied to the drain, no current will flow.

If a small positive voltage with respect to the source is applied to the gate, free holes in the substrate are repelled from the surface and a depletion layer is formed beneath the gate to link with the two existing layers, *Figure 8.58* (*b*). If the gate voltage is increased further, free electrons from the source will be attracted to the region under the gate oxide. In this way, a layer of electrons will be formed in the surface of the *p*-type substrate. This layer is so thin that it is treated as a sheet of charges. The layer is called an inversion layer, and is shown in *Figure 8.58* (*c*). The gate-to-source voltage at which the inversion layer is just formed is called the threshold voltage V_T. The electrons in the inversion layer can be used as charge carriers by applying a positive drain-to-source voltage. A drain current will flow whose magnitude is determined by the drain-to-source voltage and the conductivity of the channel which depends on the charge per unit area in the channel. This charge density is proportional to the voltage difference between the gate and the channel. The flow of drain current will produce a voltage drop along the channel, leading to a reduction in voltage between gate and channel towards the drain.

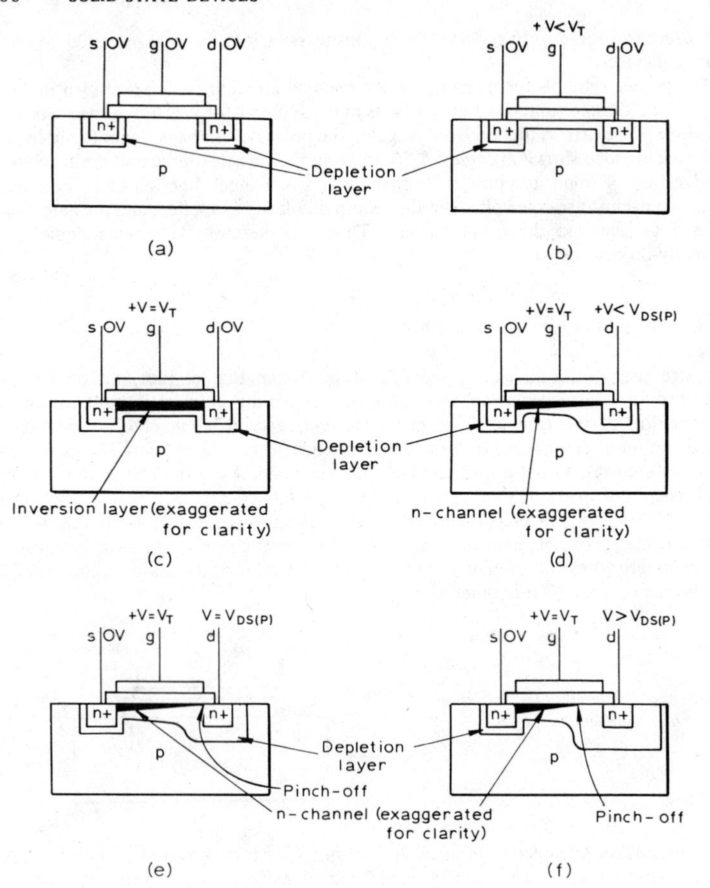

Figure 8.58 Operation of n-channel m.o.s. f.e.t.: (a) zero gate voltage (b) small positive gate voltage (c) threshold gate voltage, inversion layer formed (d) drain current flowing, drain voltage $< V_{DS(P)}$ *(e) channel pinched-off, drain voltage* $= V_{DS(P)}$ *(f) channel pinched-off, drain voltage* $> V_{DS(P)}$

The voltage distribution in the channel is then as indicated in *Figure 8.58* (*d*) in which the thickness of the depletion layer between channel and substrate is seen to increase towards the drain.

As the drain-to-source voltage is increased, the drain current increases but the voltage drop also increases, thus decreasing the channel conductivity. This decrease of conductivity amounts to a constriction of the inversion layer so that the characteristic of drain current I_D against drain-to-source voltage V_{DS} will form a knee, as in the characteristic for the junction f.e.t. When $V_{DS} = V_{GS}$, the voltage between the gate and the drain end of the channel will be zero, giving zero charge density. This is shown in *Figure 8.58* (*e*). This is the pinched-off condition, the value of V_{DS} being $V_{DS(P)}$. Any further increase in V_{DS} merely causes the drain depletion layer to expand, and the end of the channel to move towards the source, *Figure 8.58* (*f*). The drain current will be maintained by electrons being swept through the depletion layer as from the base region

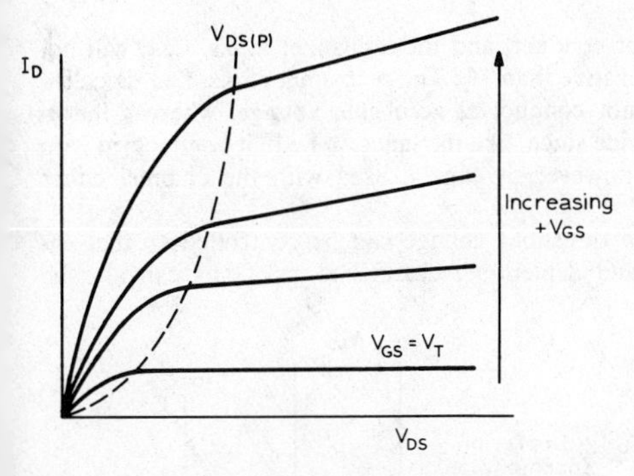

Figure 8.59 Drain voltage/current characteristics for n-channel enhancement m.o.s. f.e.t.

Figure 8.60 Transfer characteristic for n-channel enhancement m.o.s. f.e.t.

of a junction transistor, but there will be no significant increase in the drain current with increase in V_{DS}.

The channel conductivity can be varied by the gate voltage, and so a family of characteristics of drain current I_D plotted against drain-to-source voltage V_{DS} with gate-to-source voltage as parameter will be obtained, as shown in *Figure 8.59*. These characteristics are similar to those for the junction f.e.t. shown in *Figure 8.51*, and the m.o.s. f.e.t., like the junction f.e.t., is operated in the pinch-off region. The transfer characteristic of drain current plotted against gate-to-source voltage is shown in *Figure 8.60*.

The m.o.s. f.e.t. just described is called *n*-channel because the charge carriers are electrons. A *p*-channel m.o.s. f.e.t. can also be constructed. This uses an *n*-type substrate with p^+ source and drain regions. The inversion layer is formed by applying a negative voltage (with respect to the source) to the gate. The free holes from the source are attracted to the gate region to form the inversion layer. If a negative voltage with respect to the source is applied to the drain, the holes in the inversion layer will act as charge carriers and a drain current will flow. The characteristics of the *p*-channel m.o.s. f.e.t. are similar to those of the *n*-channel except that the polarities of the gate-to-source and drain-to-source voltages are negative instead of positive.

The *n*-channel and *p*-channel m.o.s. f.e.t.s described so far are ideal in the sense that a small gate voltage of the correct polarity will form the conducting inversion layer between the source and the drain. In practice, naturally-occurring positive charges in the gate oxide, and the effect of the gate material work function, will result in an *n*-type channel being formed even at zero gate voltage, *Figure 8.61*. Thus the voltage at which the device will start to conduct (V_T) will be negative, as shown in *Figure 8.62*.

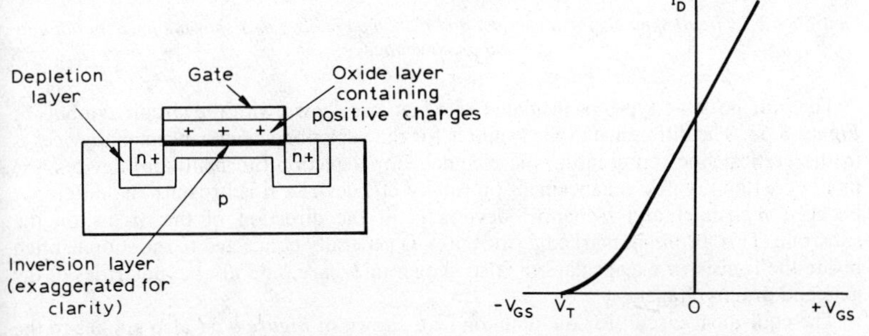

Figure 8.61 Simplified cross-section of n-channel depletion m.o.s. f.e.t.

Figure 8.62 Transfer characteristic of n-channel depletion m.o.s. f.e.t.

The *p*-channel m.o.s. f.e.t. will not conduct, and the *n*-channel m.o.s. f.e.t. will not stop conducting, until V_G is more negative than V_T. The *p*-channel m.o.s. f.e.t. is called an enhancement device as it does not conduct at zero gate voltage, whereas the *n*-channel m.o.s. f.e.t. is a depletion device since, like the junction f.e.t., it conducts at zero voltage. Unlike the junction f.e.t., however, it can be used with the channel either depleted or enhanced.

With very careful processing, the threshold voltage can be controlled so that enhancement *n*-channel m.o.s. f.e.t.s and depletion *p*-channel m.o.s. f.e.t.s can also be made.

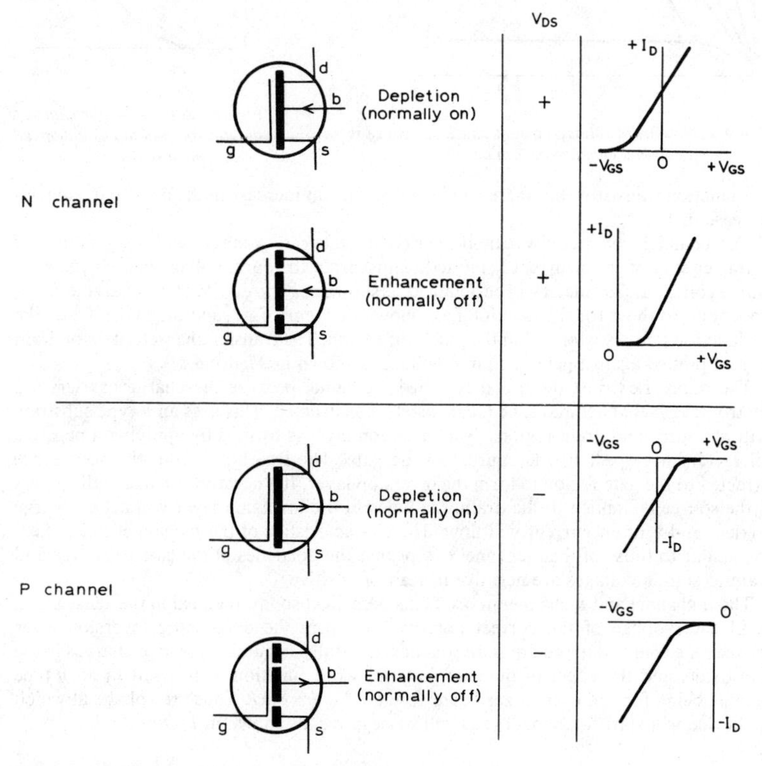

Figure 8.63 Circuit symbols for n-channel and p-channel m.o.s. f.e.t.s showing polarities of gate and drain voltages

The four possible types of insulated-gate f.e.t. are shown with the circuit symbols in *Figure 8.63*. The difference in the symbols for the depletion and enhancement devices is in the vertical line representing the channel. For depletion (normally-on) devices, the line is continuous; for enhancement (normally-off) devices, it is broken. The difference between *n*-channel and *p*-channel devices is in the direction of the arrow on the substrate. This terminal, marked *b* (for bulk), is generally connected to the source, often inside the transistor encapsulation. Also shown in *Figure 8.63* are the polarities of the gate and drain voltages.

The equivalent circuit for the junction f.e.t. shown in *Figure 8.54* also applies to the m.o.s. f.e.t. The main difference in the circuit between the two types of device is that for the m.o.s. f.e.t. the values of r_{gs} and r_{gd} at room temperature are increased to 10^{14} Ω.

PRACTICAL M.O.S. F.E.T.S

As with the junction f.e.t. discussed previously, the structure of a practical m.o.s. f.e.t. is considerably more complex than the simple cross-sections used so far. Special structures are necessary for both enhancement and depletion m.o.s. f.e.t.s.

The gate electrode in enhancement (normally-off) devices must slightly overlap the source and drain regions to ensure the inversion layer connects the two regions. This overlap can lead to relatively high parasitic capacitances unless the oxide layer underneath the overlapping parts of the gate is thicker than that above the channel. A structure like that shown in *Figure 8.64* is obtained, the thickness of the oxide under the overlapping gate being approximately 1 μm. The extensions to the source and drain regions are diffused from the doped thick oxide layer in which a window is etched for the gate oxide before the diffusion. The thin gate oxide, about 0.12 μm thick, is grown in this window and so matches exactly with the channel. The tolerances on the alignment of the gate are then much less critical.

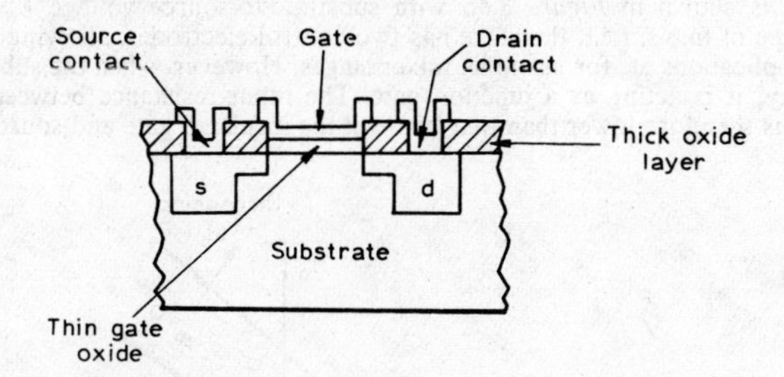

Figure 8.64 *Cross-section of practical enhancement m.o.s. f.e.t.*

As the operation of the depletion (normally-on) m.o.s. f.e.t. depends on the charges contained in the oxide layer, the structure must allow the current-carrying channel to be formed between the source and drain without allowing parasitic channels being formed. A 'closed' structure is used, the drain area being completely surrounded by the gate area, which in turn is surrounded by the source. To obtain the required high width-to-length ratio, the channel length is made short and the channel width large, but the structure is 'folded' so that it can be accommodated on the transistor chip. This structure combined with a thin oxide layer has low parasitic capacitances and a low 'on' resistance. The feedback capacitance, C_{gd} in the equivalent circuit of *Figure 8.54,* is also low, allowing operation at high frequencies.

The feedback capacitance of the simple m.o.s. f.e.t. is too high for operation at u.h.f., approximately 0.5 pF. One method of lowering the capacitance using the same manufacturing technique, to approximately 0.02 pF, is by adding a second gate to the device, to give the structure shown in *Figure 8.65* (*a*). The circuit of this dual-insulated-gate f.e.t. is shown in *Figure 8.65* (*b*). If gate 2 is earthed, the feedback between the drain and gate 1 is extremely small because of the screening effect of gate 2. In addition, the output conductance of this device is very low because the drain load resistance of the lower m.o.s. f.e.t. is the transconductance of the upper m.o.s. f.e.t. in *Figure 8.65* (*b*). Thus modulation of the drain-to-source voltage of the lower m.o.s. f.e.t. is small. The dual-gate f.e.t. can be used at lower frequencies as well. The second gate is then available for use as an a.g.c. input, for example, in the controlled stages of receivers.

Although the substrate is often connected to the source internally, in some m.o.s. f.e.t.s it is available as a fourth terminal. The drain current can be controlled by the substrate. The transfer characteristics of drain current plotted against gate-to-source

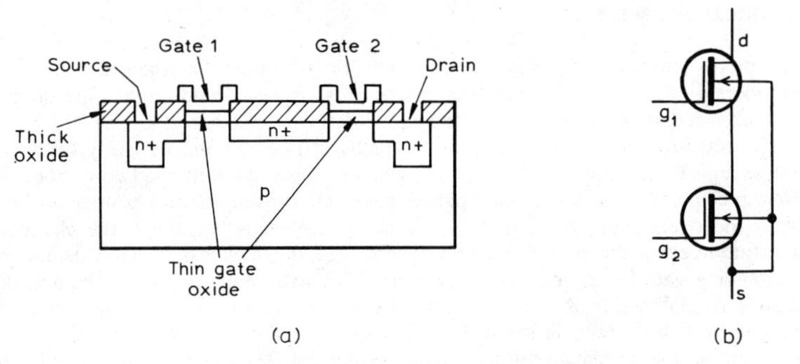

Figure 8.65 Dual-insulated-gate f.e.t.: (a) cross-section (b) circuit diagram

voltage is shown in *Figure 8.66* with substrate-to-source voltage V_{BS} as parameter. This type of m.o.s. f.e.t. therefore has two control electrodes, allowing it to be used in such applications as, for example, mixer stages. However, when the substrate is used in this way, it is acting as a junction gate. The input resistance between substrate and source is therefore lower than that between the insulated gate and source.

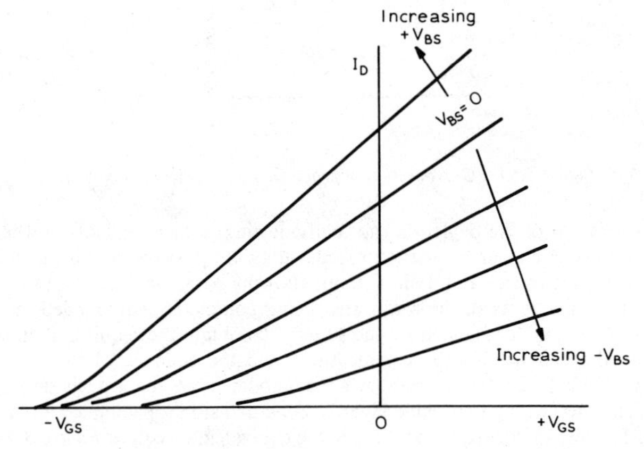

Figure 8.66 Transfer characteristics of n-channel depletion m.o.s. f.e.t. with substrate-to-source voltage as parameter

The ratings and electrical characteristics of the insulated-gate f.e.t. given in published data are similar to those of the junction f.e.t. discussed previously. When the substrate is available as a separate terminal, additional ratings are given between the substrate and the other regions of the device.

COMPARISON BETWEEN JUNCTION TRANSISTORS AND FIELD-EFFECT TRANSISTORS

The field-effect transistor can offer the circuit designer certain advantages over the junction transistor. Some of these advantages have already been mentioned. The high input resistance of the f.e.t. offers the designer a useful complement to the junction transistor whose input impedance, even in the common-collector configuration, is not as high as that previously obtainable with thermionic valves. As the f.e.t. only uses one current carrier, minority carrier storage effects do not occur to limit switching times. The f.e.t. can therefore be used for faster switching applications. The f.e.t. is also less affected by radiation since carrier lifetimes do not play an important part in its operation. The f.e.t. is also less affected by temperature than the junction transistor.

A disadvantage of the f.e.t. occurs through the thin gate oxide layer in insulated-gate devices. Electrostatic charges may accumulate on the gate lead during handling, and these can produce a large electric field in the oxide layer. This field may be large enough to puncture the layer. Once the device is connected in a circuit this danger is removed. To protect the insulated-gate f.e.t. during transport and handling, the leads are short-circuited to prevent the build-up of electrostatic charges. Another method of protection is to incorporate a diode structure from gate to substrate as a shunt protective element.

CLASSIFICATION OF FIELD-EFFECT TRANSISTORS

A simple classification of field-effect transistors is into junction and insulated-gate devices. Typical junction field-effect transistors are listed in *Table 8.28*. Typical insulated-gate field-effect transistors are listed in *Table 8.29*.

TRANSISTOR TABLES

Tables 8.15 to *8.29* list junction transistors and field-effect transistors. Junction transistors are divided into silicon and germanium devices, grouped into low-power, medium-power, and high-power types, with an indication of the intended application. Field-effect transistors are divided into junction and insulated-gate devices.

The intention is to list 'typical' devices; to give the reader an indication of the range and ratings of devices available at present. It is in no way intended that the tables should be used to select devices for particular applications. The individual manufacturers should be consulted for fuller details of their ranges of devices.

The tables have been constructed from the manufacturers' 'quick-reference' data, and the help of those manufacturers who supplied information for the tables is acknowledged. Devices and values listed were correct at the time of writing.

Outlines for devices are given a reference according to JEDEC (TO) or British Standard (SO) systems, where these apply. Outline drawings and full dimensions can be obtained from the various manufacturers or from the relevant JEDEC and British Standards Institution publications.

The abbreviations used for the various manufacturers are listed below.

AEI	AEI Semiconductors Ltd.
Fa	Fairchild Semiconductors Ltd.
Fe	Ferranti Ltd.
ITT	ITT Semiconductors
M	Mullard Ltd.
P	The Plessey Company Ltd.
RCA	RCA Ltd.
TI	Texas Instruments Ltd.

Symbols for main electrical parameters of transistors

C_{rs}	feedback capacitance in field-effect transistor
C_{Tc}	capacitance of collector depletion layer
C_{Te}	capacitance of emitter depletion layer
f_{hfb}	frequency at which the common-base forward current transfer ratio has fallen to 0.7 × low-frequency value

Symbol	Meaning
f_{hfe}	frequency at which the common-emitter forward current transfer ratio has fallen to 0.7 × low-frequency value
f_T	transition frequency (common-emitter gain-bandwidth product)
f_1	frequency at which common-emitter forward current transfer ratio has fallen to 1
G	gain
g_m	transconductance of field-effect transistor
h_{fb}, h_{fc}, h_{fe}	small-signal forward current transfer ratio for transistor configuration indicated by second subscript, output voltage held constant
h_{FB}, h_{FC}, h_{FE}	static forward current transfer ratio for transistor configuration indicated by second subscript, output voltage held constant
h_{ib}, h_{ic}, h_{ie}	small-signal input impedance for transistor configuration indicated by second subscript, output short-circuited to alternating current
h_{ob}, h_{oc}, h_{oe}	small-signal output impedance for transistor configuration indicated by second subscript, input open-circuit to alternating current
h_{rb}, h_{rc}, h_{re}	small-signal reverse voltage transfer ratio (voltage feedback ratio) for transistor configuration indicated by second subscript, output voltage held constant
i_b, i_c, i_e	instantaneous value of varying component of base, collector, emitter, current
i_B, i_C, i_E	instantaneous value of total base, collector, emitter, current
I_b, I_c, I_e	r.m.s. value of varying component of base, collector, emitter, current
I_{bm}, I_{cm}, I_{em}	peak value of varying component of base, collector, emitter, current
I_B, I_C, I_E	continuous (d.c.) base, collector, emitter, current
$I_{B(AV)}$, $I_{C(AV)}$, $I_{E(AV)}$	average value of base, collector, emitter, current
I_{BEX}, I_{CEX}	base, collector, cut-off current in specified circuit
I_{BM}, I_{CM}, I_{EM}	peak value of total base, collector, emitter, current
I_{CBO}	collector cut-off current, emitter open-circuit
I_{CBS}, I_{CES}	collector cut-off current, emitter short-circuited to base
I_{CBX}	collector current with both junctions reverse biased with respect to base
I_{CEO}	collector cut-off current, base open-circuit
I_D	drain current
I_{DM}	peak drain current
I_{EBO}	emitter cut-off current, collector open-circuit
I_G	gate current
I_{GM}	peak gate current
I_{GSS}	gate cut-off current
N	noise figure
P_{tot}	total power dissipated within device
$r_{ds(off)}$	drain-to-source off resistance
$r_{ds(on)}$	drain-to-source on resistance
$R_{th(j-amb)}$	thermal resistance, junction to ambient
$R_{th(j-case)}$	thermal resistance, junction to case
$R_{th(j-mb)}$	thermal resistance, junction to mounting base
T_{amb}	ambient temperature
T_j max	maximum permissible junction temperature
T_{mb}	mounting base temperature
T_{stg}	storage temperature

t_{off}	turn-off time
t_{on}	turn-on time
V_{BE}	base-to-emitter d.c. voltage
$V_{BE(sat)}$	base-to-emitter saturation voltage
V_{CB}	collector-to-base d.c. voltage
V_{CBO}	collector-to-base voltage, emitter open-circuit
V_{CC}	collector d.c. supply voltage
V_{CE}	collector-to-emitter d.c. voltage
V_{CEK}	collector knee voltage
V_{CEO}	collector-to-emitter voltage, base open-circuit
$V_{CE(sat)}$	collector-to-emitter saturation voltage
V_{DB}	drain-to-substrate voltage
V_{DG}	drain-to-gate voltage
V_{DGM}	peak drain-to-gate voltage
V_{DS}	drain-to-source voltage
V_{DSM}	peak drain-to-source voltage
V_{EB}	emitter-to-base d.c. voltage
V_{EBO}	emitter-to-base voltage, collector open-circuit
V_{GB}	gate-to-substrate voltage
V_{GBM}	peak gate-to-substrate voltage
V_{GS}	gate-to-source voltage
V_{GSM}	peak gate-to-source voltage
V_{GSO}	gate-to-source voltage, drain open-circuit
V_P	pinch-off voltage in field-effect transistor
$V_{GS(P)}$	gate-to-source cut-off voltage
V_{SB}	source-to-substrate voltage
V_T	threshold voltage in field-effect transistor
η	intrinsic stand-off ratio in unijunction transistor

List of Transistor Tables

Table 8.15 GERMANIUM *pnp* LOW-POWER TRANSISTORS (P_{tot} up to 150 mW)

Type	Manufacturer	Outline	Maximum ratings V_{CEO} (V)	V_{CBO} (V)	I_{CM} (mA)	P_{tot} (mW)	h_{FE}	at I_C (mA)	f_T min (MHz)	$V_{CE(sat)}$ (V)	at I_B (mA)	I_C (mA)	Features
ASY27	M	TO-5	15	25	300	150	150	20	6.0	0.2	0.4	10	Alloy-junction type for switching $t_{on} = 250$ ns, $t_{off} = 1\,000$ ns
OC71*	M	SO-2	20	20	50	140	75	3	1†	0.2	0.5	9	Alloy-junction type for general-purpose amplification
2N404	TI	TO-5	24	25	100	150	24‡	24	4.0	0.2	1.0	24	General-purpose amplification
2N1305	TI	TO-5	—	30	300	150	200	10	5.0†	0.2	0.25	10	General purpose and switching

* Obsolete type, included to show values typical of devices of the period. † f_α (f_{hfb}) value. ‡ Minimum value.

Table 8.16 GERMANIUM *npn* LOW-POWER TRANSISTORS (P_{tot} up to 150 mW)

Type	Manufacturer	Outline	Maximum ratings V_{CEO} (V)	V_{CBO} (V)	I_{CM} (mA)	P_{tot} (mW)	h_{FE}	at I_C (mA)	f_T min (MHz)	$V_{CE(sat)}$ (V)	at I_B (mA)	I_C (mA)	Features
ASY28	M	TO-5	15	30	300	150	80	20	4.0	0.2	0.33	10	Alloy-junction type for switching $t_{on} = 225$ ns, $t_{off} = 775$ ns
OC140*	M	SO-2	20	20	400	100	150	15	4.5†	0.18	0.17	7.5	Alloy-junction type for switching
2N388A	IT	TO-5	40	40	200	150	30‡	200	5.0†	—	—	—	General-purpose type
2N1308	IT	TO-5	—	25	300	150	80‡	10	15†	0.2	0.13	10	Switching

Table 8.17 GERMANIUM *pnp* MEDIUM-POWER TRANSISTORS (P_{tot} 150 mW to 1 W)

Type	Manufacturer	Outline	Maximum ratings V_{CEO} (V)	V_{CBO} (V)	I_{CM} (A)	P_{tot} (W)	h_{FE}	at I_C (mA)	f_T typ (MHz)	$V_{CE(sat)}$ (V)	at I_B (mA)	I_C (mA)	Features
AC188	M	TO-1	15	25	2.0	1.0	500	300	1.5	—	—	—	Alloy-junction type for audio-amplifier output stages
ACY17	M	TO-5	32	70	2.0	0.26	150	300	1.0	0.3	15	300	Alloy-junction type for a.f. applications

Table 8.18 GERMANIUM *npn* MEDIUM-POWER TRANSISTORS (P_{tot} 150 mW to 1 W)

Type	Manufacturer	Outline	Maximum ratings V_{CEO} (V)	V_{CBO} (V)	I_{CM} (A)	P_{tot} (W)	h_{FE}	at I_C (mA)	f_T typ (MHz)	$V_{CE(sat)}$ (V)	at I_B (mA)	I_C (mA)	Features
AC176	M	TO-1	20	32	1.0	0.7	180	500	1.0*	—	—	—	Alloy-junction type for a.f. amplifiers
AC187	M	TO-1	15	25	2.0	1.0	500	300	5.0	—	—	—	Alloy-junction type for audio-amplifier output stages

* Minimum value.

Table 8.19 GERMANIUM *pnp* HIGH-POWER TRANSISTORS
(P_{tot} above 1 W)

Type	Manufacturer	Outline	Maximum ratings							
			V_{CEO} (V)	V_{CBO} (V)	I_{CM} (A)	P_{tot} (W)	h_{FE}	*at* I_C (A)	f_T typ (MHz)	Features
AD149	M	TO-3	30	50	3.5	22.5	100	1.0	0.5	Alloy-junction type for audio-amplifier output stages
AD162	M	SO-55	20	32	3.0	6.0	320	0.5	1.5	Alloy-junction type for audio-amplifier output stages
OC28	M	TO-3	60	80	10	30	55	1.0	0.25	Alloy-junction type

Table 8.20 GERMANIUM *npn* HIGH-POWER TRANSISTORS
(P_{tot} above 1 W)

Type	Manufacturer	Outline	Maximum ratings							
			V_{CEO} (V)	V_{CBO} (V)	I_{CM} (A)	P_{tot} (W)	h_{FE}	*at* I_C (A)	f_T typ (MHz)	Features
AD161	M	SO-55	20	32	3.0	4.0	320	0.5	3.0	Alloy-junction type for audio-amplifier output stages

(P_{tot} up to 500 mW)

Type	Manufacturer	Outline	Maximum ratings V_{CEO} (V)	V_{CBO} (V)	I_{CM} (mA)	P_{tot} (mW)	h_{FE}	*at* I_C (mA)	f_T min (MHz)	$V_{CE(sat)}$ (V)	*at* I_B (mA)	I_C (mA)	Features
General-purpose types													
BC157	M	Plastic *	45	50	200	300	260	2.0	130 (typ)	0.3	5	100	Planar epitaxial
BC179	TI	TO-18	25	35	300	—	900	2.0	150	0.3	—	10	
BC252A	ITT	TO-92	25	—	200	300	170	2.0	200	0.5	5	100	Planar epitaxial
BC283	Fa	TO-18	30	—	—	400	270	50	200 (typ)	0.3	—	50	
BCY30	M	TO-5	50	64	100	250	35	20	0.25	0.55	3	20	Alloy-junction
	TI	TO-5	64	64	—	—	10 (min)	20	0.25	—	—	—	
BCY70	ITT	TO-18	40	—	200	300	50 (min)	10	275	<0.25	1	10	Planar epitaxial
	M	TO-18	40	50	200	350	50 (min)	10	250	0.25	1	10	
BFS98	Fe	SO-94	60	—	1 000	500	160	150	150	0.35	—	150	
ZTX502	Fe	SO-94	35	—	500	300	300	10	150	0.25	—	50	
ZTX538	Fe	SO-94	25	—	800	500	600	100	100	0.70	—	500	
2N3965	ITT	TO-18	60	—	200	300	600	1.0	250	<0.4	5	50	Planar epitaxial
2N4059	TI	TO-92	30	30	200	360	660	1.0	—	0.7	0.5	10	
2N4125	Fa	TO-92	30	—	—	310	150	2.0	200	0.4	—	2.0	
2N4354	Fa	TO-105	60	—	—	350	500	10	100	0.15	—	150	
Switching types													
BSV68	M	TO-18	100	110	100	250	30 (min)	25	50	0.25	2.5	25	Planar epitaxial for switching indicator tubes
BSW74	ITT	TO-18	40	75	500	400	35 (min)	10	200	<0.4	15	150	Planar epitaxial
ZTX510	Fe	SO-94	12	—	—	300	40 (min)	30	400	0.2	—	30	$t_{on} = 60$ ns, $t_{off} = 90$ ns
2N3829	TI	TO-18	20	35	500	—	120	30	350	0.18	—	10	$t_{on} = 25$ ns, $t_{off} = 65$ ns
2N4208	Fa	TO-18	12	—	—	350	120	10	700	0.15	—	10	$t_s = 20$ ns
R.F. types													
BF324	ITT	TO-92	30	—	25	250	45	1.0	350	—	—	—	N = 3 dB at 100 MHz
BF576	TI	TO-92	15	20	—	200	30 (min)	10	120 (typ)	—	—	—	For wideband amplifiers

* See manufacturer's literature for details.

Table 8.22 SILICON *npn* LOW-POWER TRANSISTORS
(P_{tot} up to 500 mW)

Type	*Manufacturer*	*Outline*	*Maximum rates* V_{CEO} (V)	V_{CBO} (V)	I_{CM} (mA)	P_{tot} (mW)	h_{FE}	*at* I_C (mA)	f_T min (MHz)	$V_{CE(sat)}$ (V)	*at* I_B (mA)	I_C (mA)	*Features*
General-purpose types													
BC107	M	TO-18	45	50	200	300	500	2.0	300 (typ)	0.25	0.5	10	Planar epitaxial
BC108A	ITT	TO-18	25	30	200	300	170	2.0	250	<0.6	5.0	100	Planar epitaxial
BC125	Fa	TO-105	30	—	—	300	30 (min)	150	40	0.5	—	150	
BC147	M	Plastic*	45	50	200	300	450	2.0	300 (typ)	0.25	0.5	10	Planar epitaxial
BC170C	ITT	TO-92	20	20	200	300	600	1.0	100	<0.4	3.0	30	Planar epitaxial
BCY43	ITT	TO-18	20	40	200	300	150	1.0	100	<0.5	2.2	50	Planar epitaxial
BFS60	Fe	SO-94	40	—	1 000	500	300	150	150	0.25	—	150	
BSW69	M	Plastic*	150	150	—	125	30 (min)	4.0	130 (typ)	4.0	1.0	20	Planar epitaxial for switching indicator tubes
ZTX107	Fe	SO-94	50	—	100	300	500	2.0	150	0.1	—	10	
ZTX384	Fe	SO-94	30	—	100	500	850	2.0	150	0.25	—	10	
2N870	Fa	TO-18	60	—	—	500	120	150	50	5.0	—	150	
2N930	ITT	TO-18	45	45	200	300	300	0.01	150	<1.0	0.5	10	Planar epitaxial
2N3706	TI	TO-92	20	40	800	360	30 (min)	50	100	1.0	5.0	100	
2N3708	TI	TO-92	30	30	200	360	660	1.0	—	1.0	0.5	10	
2N4410	Fa	TO-92	80			310	400	10	60	0.20		1.0	

* See manufacturer's literature for details.

Table 8.22—*contd.*

Type	*Manufacturer*	*Outline*	Maximum rates V_{CEO} (V)	V_{CBO} (V)	I_{CM} (mA)	P_{tot} (mW)	h_{FE}	*at* I_C (mA)	f_T min (MHz)	$V_{CE(sat)}$ (V)	*at* I_B (mA)	I_C (mA)	*Features*
Switching types													
BSW83	ITT	TO-18	25	40	500	500	70 (min)	10	200	<0.6	15.0	150	Planar epitaxial
BSX19	M	TO-18	15	40	500	360	60	10	400	0.3	0.6	10	Planar epitaxial t_{on} = 12 ns, t_{off} = 15 ns
ZTX314	Fe	SO-94	15	—	—	300	40	10	500	0.2	—	10	t_{on} = 12 ns, t_{off} = 18 ns
2N2368	TI	TO-18	15	40	500	—	60	10	400	0.25	—	10	t_{on} = 12 ns, t_{off} = 15 ns
2N4265	Fa	TO-92	12	—	—	310	400	10	300	0.22	—	10	t_s = 20 ns
R.F. types													
BF240	ITT	TO-92	40	—	25	255	—	—	—	—	—	—	
BF597	TI	Plastic*	25	40	30	360	38 (min)	7.0	550 (typ)	—	—	—	For video i.f. amplifiers
BFY90	M	TO-72	15	30	50	200	150	2.0	1 000	—	—	—	N < 3.5 dB at 200 MHz
PE5015	Fa	TO-92	20	—	—	200	—	—	300	—	—	—	N = 4.0 dB at 100 MHz
ZTX321	Fe	SO-94	15	—	—	250	—	—	600	—	—	—	Power gain = 15 dB at 200 MHz

* See manufacturer's literature for details.

Table 8.23 SILICON *pnp* MEDIUM-POWER TRANSISTORS
(P_{tot} 0.5 to 10 W)

Type	*Manufacturer*	*Outline*	Maximum ratings: V_{CEO} (V)	V_{CBO} (V)	I_{CM} (A)	P_{tot} (W)	h_{FE}	*at* I_C (mA)	f_T min (MHz)	$V_{CE(sat)}$ (V)	*at* I_B (mA)	I_C (mA)	*Features*
General-purpose types													
BC160-6	ITT	TO-39	40	—	1.0	0.75	100	100	—	<1.0	100	1 000	Planar epitaxial
BC327	M	Plastic*	45	50	1.0	0.625	600	100	100 (typ)	0.7	50	500	Planar epitaxial
	ITT	TO-92	45	—	0.8	0.625	600	0.10	—	<0.7	50	500	
BD136	M	TO-126	45	45	1.5	6.5	250	150	75 (typ)	0.5	50	500	Planar epitaxial
	TI	SOT-32	45	45	1.5	6.5	250	150	50	0.5	50	500	
BFR80	TI	TO-92	50	70	2.0	0.8	75 (min)	100	100	1.0	100	1 000	
BFT81	TI	TO-39	50	60	1.0	—	300	100	100	1.0	—	1 000	
BFX88	M	TO-5	40	40	0.6	0.6	40 (min)	10	100	0.4	15	150	Planar epitaxial
MPSA55	Fa	TO-92	60	—	—	0.625	50 (min)	100	50	0.25	—	100	
ZTX537	Fe	SO-94	45	—	0.8	0.75	600	100	100	0.7	—	500	Planar
ZTX550	Fe	SO-94	45	—	1.0	1.0	300	150	150	0.25	—	150	Planar
2N4030	Fa	TO-5	60	—	—	0.8	120	100	100	0.15	—	150	
2N4037	RCA	TO-39	40	—	—	7.0	250	150	60	1.4	15	150	
2N4235	Fa	TO-5	60	—	1.0	6.0	150	250	3	0.60	—	1 000	
2N5415	RCA	TO-5	200	—	—	10	150	50	15	2.5	5	50	
Switching types													
BSX40	ITT	TO-39	30	30	0.5	0.8	120	150	100	<0.4	15	150	Planar epitaxial
2N2905A	M	TO-5	60	60	—	0.6	300	150	200	0.4	15	150	Planar epitaxial $t_{on} = 45$ ns, $t_{off} = 100$ ns
2N5022	Fa	TO-39	50	—	—	1.0	100	500	170	0.4	—	500	$t_{off} = 90$ ns

* See manufacturer's literature for details.

Table 8.24 SILICON *npn* MEDIUM-POWER TRANSISTORS (P_{tot} 0.5 to 10 W)

Type	Manufacturer	Outline	Maximum ratings V_{CEO} (V)	V_{CBO} (V)	I_{CM} (A)	P_{tot} (W)	h_{FE}	at I_C (mA)	f_T min (MHz)	$V_{CE(sat)}$ (V)	at I_B (mA)	I_C (mA)	Features
General-purpose types													
BC337	M	Plastic*	45	50	1.0	0.625	600	100	200 (typ)	0.7	50	500	Planar epitaxial
BC338	ITT	TO-92	25	—	0.8	0.625	600	100	—	<0.7	50	500	Planar epitaxial
BD135	TI	SOT-32	45	45	1.5	6.5	250	150	50	0.5	50	500	
BD232	M	TO-126	250	500	0.5	7.0	175	50	15 (typ)	—	—	—	Diffused, line driver in TV receivers
BF259	ITT	TO-39	300	300	0.1	5.0	25 (min)	30	90	<1.0	6	30	Planar epitaxial for video output stages
BFR40	TI	TO-92	60	70	2.0	0.8	75 (min)	100	100	0.5	100	1 000	
BFY50	M	TO-5	35	80	1.0	0.8	30 (min)	150	60	0.2	15	150	Planar epitaxial
BFY52	Fa	TO-5	20	—	—	0.8	60 (min)	150	200	0.35	—	150	
BSY86	ITT	TO-39	64	120	1.0	0.9	300	150	130	<1.0	100	1 000	Planar epitaxial
MPSA06	Fa	TO-92	80	—	—	0.625	50 (min)	100	50	0.25	—	100	
TIP502	TI	TO-39	60	—	3.0	6.0	180	1 000	—	0.75	—	1 000	
ZT1480	Fe	TO-5	55	100	1.5	5.0	60	200	1.5 (typ)	—	—	—	Diffused
ZTX337	Fe	SO-94	45	—	0.8	0.75	600	100	100	0.7	—	500	Planar
ZTX450	Fe	SO-94	45	—	1.0	1.0	300	150	150	0.25	—	150	Planar
2N4239	Fa	TO-5	80	—	1.0	6.0	150	250	2	0.6	—	1 000	
2N5785	RCA	TO-5	50	—	—	10	100	1 200	1	0.75	120	1 200	Hometaxial base
40348	RCA	TO-5	65	—	—	8.75	125	300	1.4	0.75	30	300	Hometaxial base

* See manufacturer's literature for details.

Table 8.24—*continued*

Type	*Manufacturer*	*Outline*	*Maximum ratings* V_{CEO} (V)	V_{CBO} (V)	I_{CM} (A)	P_{tot} (W)	h_{FE}	*at* I_C (mA)	f_T min (MHz)	$V_{CE(sat)}$ (V)	*at* I_B (mA)	I_C (mA)	*Features*
Switching types													
BSX60	M	TO-5	30	70	—	0.8	90	500	250	0.3	15	150	Planar epitaxial $t_{on} = 40$ ns, $t_{off} = 70$ ns
2N2369	TI	TO-18	15	40	0.5	—	120	10	500	0.25	—	10	$t_{on} = 12$ ns, $t_{off} = 18$ ns
2N3724	Fa	TO-5	30	—	—	0.8	150	100	300	0.20	—	100	$t_{off} = 60$ ns
2N5262	RCA	TO-39	50	—	—	5.0	25 (min)	1 000	250	0.8	100	1 000	$t_{on} = 30$ ns, $t_{off} = 60$ ns
Darlington transistors													
BD477	TI	SOT-32	45	45	1.2	8.5	4 000 (min)	—	—	—	—	—	
BSS50	M	TO-5	45	60	1.0	5.0	1 500 (min)	500	—	1.6	1 000	1 000	Planar epitaxial $t_{on} = 400$ ns, $t_{off} = 1500$ ns
EXP4036	Fe	TO-39	60	—	1.0	7.5	10 000	500	—	1.4	10	1 000	

Table 8.25 SILICON *pnp* HIGH-POWER TRANSISTORS (P_{tot} above 10 W)

			Maximum ratings										
Type	*Manufacturer*	*Outline*	V_{CEO} (V)	V_{CBO} (V)	I_{CM} (A)	P_{tot} (W)	h_{FE} *at*	I_C (A)	f_T min (MHz)	$V_{CE(sat)}$ (V) *at*	I_B (mA)	I_C (A)	*Features*
BD234	M	TO-126	45	45	6.0	25	25 (min)	1.0	3.0	0.6	100	1.0	Epitaxial base
BD434	M	TO-126	22	22	7.0	36	50 (min)	2.0	3.0	0.5	200	2.0	Epitaxial base
BD462	TI	SOT-32	30	35	4.0	30	40 (min)	2.0	50	0.5	100	1.0	
BDX78	M	Plastic*	80	80	12	55	30 (min)	2.0	3.0	1.0	300	3.0	Epitaxial base
BUY92	Fe	TO-39	60	120	9.0	30	57 (typ)	0.5	50	1.6	—	9.0	
FRB721	Fe	TO-3	60	120	20	85	57 (typ)	0.5	50	0.7	—	20	
TA8327	RCA	TO-220	—	—	15	75	150	5.0	5.0 (typ)	—	—	—	Epitaxial base
TIP36	TI	TO-3(P)	40	40	25	125	10 (min)	15	3.0	4.0	5A	25	
2N3740	Fa	TO-66	60	—	1.0	25	100	0.25	4.0	0.6	—	1.0	
2N3790	TI	TO-3	80	—	10	150	90	1.0	—	1.0	—	4.0	
2N3792	Fa	TO-3	80	—	10	150	30 (min)	3.0	4.0	1.0	—	5.0	
2N4398	Fa	TO-3	40	—	30	200	60	15	4	2.0	—	20	
2N4399	TI	TO-3	60	—	30	200	60	15	—	1.0	—	15	
2N6126	Fa	TO-220	80	—	4.0	40	80	1.5	2.5	0.6	—	1.5	
2N6181	RCA	TO-5(P)	50	—	—	25	250	0.5	50	0.8	50	0.5	
2N6248	RCA	TO-3	100	—	—	125	100	5.0	10	1.3	500	5.0	Epitaxial base
Darlington transistors													
BDX62A	M	TO-3	80	80	12	90	1 000 (min)	3.0	2.5 (typ)	2.0	12	3.0	Epitaxial base
TA8203	RCA	TO-220	80	—	15	40	1 000 (min)	5.0	—	—	—	—	
TA8350	RCA	TO-3	80	—	20	100	1 000 (min)	5.0	—	—	—	—	
TIP127	TI	TO-66(P)	100	100	5.0	60	1 000 (min)	—	—	—	—	—	
TIP147	TI	TO-3(P)	100	100	10.0	125	500 (min)	—	—	—	—	—	

* See manufacturer's literature for details. (P) Plastic version of outline.

Table 8.26 SILICON *npn* HIGH-POWER TRANSISTORS (P_{tot} above 10 W)

Type	Manufacturer	Outline	Maximum ratings V_{CEO} (V)	V_{CBO} (V)	I_{CM} (A)	P_{tot} (W)	h_{FE}	at I_C (A)	f_T min (MHz)	$V_{CE(sat)}$ (V)	at I_B (mA)	I_C (A)	Features
BD131	M	TO-126	45	70	6.0	11	40 (min)	0.5	60	0.9	200	2.0	Planar epitaxial
BD307A	ITT	SOT-32	64	—	2.5	12.5	150	0.5	100	<1.0	250	2.5	
BD400	TI	SOT-32	100	170	1.0	15	40 (min)	0.5	90	0.7	100	1.0	
BD433	M	TO-126	22	22	7.0	36	50 (min)	2.0	3.0	0.5	200	2.0	Epitaxial base
BD461	TI	SOT-32	30	35	4.0	30	40 (min)	2.0	50	0.5	100	1.0	
BDY10	M	TO-3	40	50	4.0	150	50	2.0	1.0	0.7	400	2.0	Diffused
TA8324	RCA	TO-220	—	—	15	75	150	5.0	5 (typ)	—	—	—	Epitaxial base
TIP35	TI	TO-3(P)	40	40	25	125	10 (min)	15	3.0	4	5A	25	
ZT1483	Fe	TO-8	40	60	3.0	25	60	0.75	1.25 (typ)	—	—	—	Diffused
2N3055	Fe	TO-3	60	100	15	115	70	4.0	0.7 (typ)	—	—	—	Diffused
	M	TO-3	60	100	—	115	70	4.0	0.8	1.1	400	4.0	Diffused
2N3266	TI	TO-63	90	—	25	125	80	15	—	1.6	—	20	
2N3442	RCA	TO-3	140	—	10	117	70	3.0	0.8	1.0	300	3.0	Hometaxial base
2N3715	Fa	TO-3	60	—	10	150	150	1.0	2.5	0.8	—	5.0	
2N4912	Fa	TO-66	80	—	1.0	25	100	0.5	3.0	0.6	—	1.0	
2N5297	RCA	TO-220	60	—	4.0	36	80	1.5	0.8	1.0	150	1.5	Hometaxial base
2N5302	Fa	TO-3	60	—	30	200	60	15	2.0	2.0	—	20	
2N5579	RCA	TO-3	70	—	80	300	40	40	0.4	1.5	4A	40	Hometaxial base
Darlington transistors													
BDY63A	M	TO-3	80	100	12	90	1 000 (min)	3.0	2.5 (typ)	2.0	12	3.0	Epitaxial base
EXP4039	Fe	TO-66	60	—	5	30	10 000	0.5	—	1.75	50	5.0	
TA8201	RCA	TO-220	80	—	15	40	1 000 (min)	5.0	—	—	—	—	
TA8348	RCA	TO-3	80	—	20	100	1 000 (min)	5.0	—	—	—	—	
TIP112	TI	TO-66(P)	100	100	2.0	50	750 (min)	—	—	—	—	—	
TIP142	TI	TO-3(P)	100	100	10	125	500 (min)	—	—	—	—	—	

Table 8.27 R.F. POWER TRANSISTORS
(Ascending operating frequency)

Type	Manufacturer	Outline	R.F. performance					V_{CEO} (V)	V_{CBO} (V)	f_T min (MHz)
			Frequency (MHz)	P_{in} (W)	P_{out} (W)	Power gain (dB)	V_{CC} (V)			
BLX15	M	†	28	—	150 *	>14	50	55	110	275 (typ)
2N5709	TI	†		—	80 (PEP)	—	28	50	70	—
2N5708	Fe	†(TO-128)	30	—	40	13.0	28	50	70	—
2N6093	RCA	†(TO-217AA)		—	75 (PEP)	13.0	28	—	—	—
2N2876	RCA	TO-60	50	—	10	—	28	—	—	—
2N3950	TI	TO-60		7.95	50	—	28	35	65	—
2N5071	Fe	TO-60	76	—	24	9.0	28	30	65	—
2N3632	Fe	TO-60	175	—	13.5	5.8	28	40	65	—
	M	TO-60		—	13.5	—	28	40	65	400 (typ)
	TI	TO-60		3.5	13.5	—	28	40	65	—
2N5996	RCA	†(TO-216AA)		—	15	4.5	12.5	—	—	—
ZT5591	Fe	†(TO-128)		—	25	4.4	13.6	18	40	—
2N5643	TI	†		7.0	40	—	28	35	65	—
BLY90	M	†		—	50	>4	12.5	18	36	550 (typ)
2N5016	TI	TO-60	400	5	15	—	28	30	65	—
2N6104	RCA	†		—	30	5.0	28	—	—	—

* S.S.B. operation. † Stripline packages.

Table 8.27–*continued*

Type	*Manufacturer*	*Outline*	*R.F. performance*							
			Frequency (MHz)	P_{in} (W)	P_{out} (W)	*Power gain* (dB)	V_{CC} (V)	V_{CEO} (V)	V_{CBO} (V)	f_T min (MHz)
2N5913	Fe	TO-39		—	1·75	7.0	12	14	36	—
BLW39	Fe	†(TO-131)	470	—	2.0	7.0	12	28	50	—
*2N5915	RCA	†(TO-216AA)		—	6.0	4.8	12.5	—	—	—
BLW14	TI	†(TO-129)		2.0	7.0	—	13	18	36	—
BLX69	M	†		—	20	>4	13.5	18	36	1 000 (typ)
BLX95	M	†		—	40	>4.5	28	33	65	1 000 (typ)
BLX93	M	†		—	5.0	—	28	33	65	1 200 (typ)
2N4431	Fe	†(TO-129)	1 000	—	5.0	5.2	28	40	55	—
2N4431	TI	†(TO-129)		1.57	5.0	—	28	40	55	—
2N5765	AEI	—		—	5.0	6.0	—	—	—	1 000
TA8648	RCA	†		—	10	6.0	28	—	—	—
2N5768	AEI	†		—	5.0	8.0	—	—	—	2 000
2N6266	Fe	†	2 000	—	5.0	7.0	28	50	50	—
2N6267	RCA	†		—	10	7.0	28	—	—	—
TA8809	RCA	†	3 000	—	4.5	5.0	28	—	—	—
DC6112B	AEI	†	4 000	—	0.5	5.0	—	—	—	2 500

* S.S.B. operation. † Stripline packages.

Type	Manufacturer	Outline	Polarity	Maximum ratings V_{DGM} (V)	V_{GSM} (V)	V_{DSM} (V)	I_{GM} (mA)	$V_{GS(P)}$ (V) at	I_D (nA)	I_{GSS} max (nA)	I_{DSS} max (mA)	Features
General-purpose and r.f. amplifiers												
BF245B	M	**Plastic***	*n*-channel	**30**	30	30	10	8.0	10	5	15	**V.H.F. amplifier,** N < 1.5 dB at 100 MHz
BF246	TI	TO-92	*n*-channel	25†	—	—	—	14	—	5	—	V.H.F. amplifier, N = 1.6 dB at 100 MHz
BF256B	M	TO-126	*n*-channel	30	30	30	10	8.0	10	5	13	U.H.F. amplifier, power gain = 11 dB at 800 MHz
BF256LB	TI	TO-92	*n*-channel	30†	—	—	—	7.5	—	5	—	V.H.F. amplifier, power gain = 14 dB, N = 7.5 dB, at 800 MHz
BFW10	M	TO-72	*n*-channel	30	30	30	10	8.0	0.5	0.5	20	V.H.F. amplifier, N < 2.5 dB at 100 MHz
GAT1	P	TO-72	—	—	—	—	—	3.0 (min)	—	—	—	**Microwave amplifier**, g_m = 7 000; N = 4.0 dB, gain = 10 dB at 1 GHz
GAT2	P	L.I.D.	—	—	—	—	—	3.0 (min)	—	—	—	Microwave amplifier, g_m = 10 000; N = 4.5 dB, gain = 8.0 dB at 3 GHz

* See manufacturer's literature for details.
† V_{DGO} value.

Table 8.28–*continued*

Type	*Manufacturer*	*Outline*	*Polarity*	*Maximum ratings*								
				V_{DGM} (V)	V_{CSM} (V)	V_{DSM} (V)	I_{GM} (mA)	$V_{GS(P)}$ (V)	*at* I_D (nA)	I_{GSS} max (nA)	I_{DSS} max (mA)	*Features*
GAT3	P	L.I.D.	—	—	—	—	—	3.0 (min)	—	—	—	Microwave amplifier, $g_m = 8\ 000$; N = 6.0 dB, gain = 6.0 dB at at 8 GHz
2N2498	TI	TO-5	*p*-channel	20 †	—	—	—	—	—	10	—	N < 3.0 dB, general-purpose amplifier
2N4342	Fa	TO-106	*p*-channel	—	25 ‡	—	—	5.5	—	10	12	General-purpose amplifier
2N5485	Fa	TO-92	*n*-channel	—	25 ‡	—	—	4.0	—	1.0	10	R.F. amplifier, N = 4 dB at 400 MHz
Differential amplifiers												
BFS21A	M	TO-72	n-channel	30	30	30	10	6.0	0.5	0.5	10	Matching $V_{GS} < 10$ mV
2N5045	TI	TO-18	*n*-channel	—	50 ‡	—	—	4.6	—	0.25	—	Matching $V_{GS} = 5$ mV
Switching												
BF347	TI	TO-92	*n*-channel	—	30 ‡	—	—	1.0	—	0.5	0.5 (min)	$r_{ds(on)}$ max = 400 Ω
BSV78	M	TO-18	*n*-channel	40	40	40	50	11	1.0	0.25	50 (min)	$r_{ds(on)} < 25$ Ω
BSV80	M	TO-18	*n*-channel	40	40	40	50	5.0	1.0	0.25	10 (min)	$r_{ds(on)} < 60$ Ω
2N4860A	TI	TO-18	*n*-channel	—	30 ‡	—	—	6.0	—	0.25	20 (min)	$r_{ds(on)} = 40$ Ω

† V_{DGO} value.
‡ $V_{[illegible]}$ value.

Type	Manufacturer	Outline	Polarity	Maximum ratings V_{DS} (V)	$\pm V_{GS}$ (V)	I_{DM} (mA)	P_{tot} (mW)	$V_{GS(P)}$ (V)	I_{GSS} (nA)	$r_{ds(on)}$ (Ω)	$r_{ds(off)}$ (Ω)	Features
BFR29	M	TO-72	*n*-channel	30*	10*	50	200	—	—	—	—	A.F., i.f., and v.h.f. applications
BSV81	M	TO-72	*n*-channel	30*	10*	50	200	—	—	<50	>10^{10}	Switching and chopper applications
40467A	RCA	TO-72	*n*-channel	20	15	50	330	−8	1	—	—	General-purpose r.f. amplifier, gain = 16 dB, N = 3.5 dB at 200 MHz
3N138	RCA	TO-72	*n*-channel	35	10	50	400	−10	10 pA	180	10^{11}	Chopper applications
3N161	TI	TO-72	*p*-channel	25†	—	—	—	5	1 pA	1 000	—	Diode-protected fast-switching device
Dual-gate devices												
BFR84	M	TO-72	*n*-channel	25	6	50	300	−5	10	—	—	Protected v.h.f. and u.h.f. amplifier; gain = 14 dB, N = 4 dB at 500 MHz
3N140	RCA	TO-72	*n*-channel	20	−8 to +20	50	400	−2	1	—	—	Chopper applications
3N200	RCA	TO-72	*n*-channel	20	6	50	330	−1	50	—	—	Protected v.h.f. and u.h.f. amplifier; gain = 12.5 dB, N = 4.5 dB at 400 MHz
3N204	TI	TO-72	*n*-channel	25	—	—	—	−4	—	—	—	R.F. amplifier; gain = 14 dB at 450 MHz, N = 3.5 dB at 200 MHz

* V_{DB} and V_{GBM} values. † V_{DGO} value.

THYRISTORS

The term 'thyristor' applies to any four-layer *pnpn* semiconductor device which can operate as a switch with stable on and off states. The thyristor can have two, three or four terminals. General usage, however, has restricted 'thyristor' to the three-terminal device. The two-terminal device is called a switching diode or four-layer diode, while the four-terminal device is called the silicon controlled switch. This section will follow general usage in restricting 'thyristor' to the three-terminal device.

Switching diode

The switching diode consists of a four-layer *pnpn* structure with connections to the outer *p* and *n* regions, as shown in *Figure 8.67* (*a*). The outer *p* region forms the anode, while the outer *n* region forms the cathode.

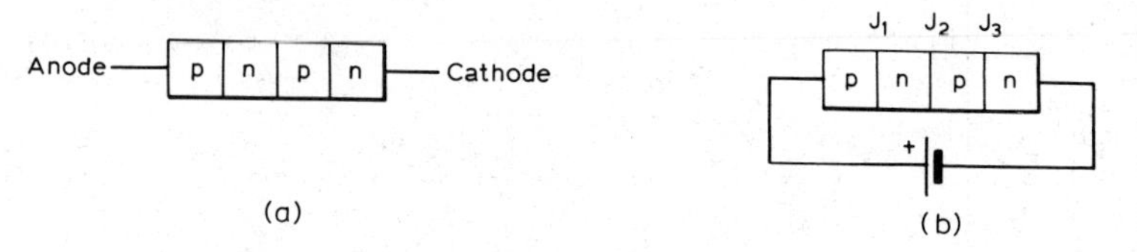

Figure 8.67 Switching diode: (a) structure (b) biasing

If a voltage is applied across the diode with the positive terminal connected to the anode, as shown in *Figure 8.67*(*b*), junctions J_1 and J_3 are forward-biased while junction J_3 is reverse-biased. Because the voltage drops across the forward-biased junctions are small, most of the applied voltage appears across the reverse-biased junction. The current through the diode will be the small reverse current of junction J_2. As the voltage across the diode is increased, the reverse current of junction J_2 will increase only slightly. Thus the forward current through the diode will be small, as shown in the forward-blocking part of the voltage/current characteristic in *Figure 8.68*. The impedance of the diode in this 'off' state is therefore high.

As the voltage is increased further, the reverse voltage across junction J_2 will reach

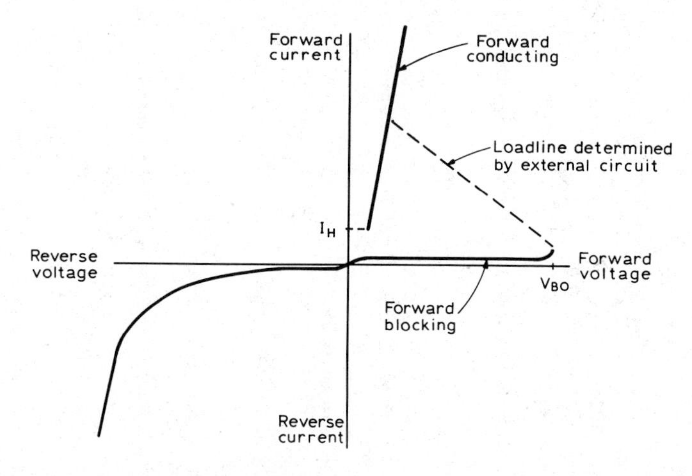

Figure 8.68 Voltage/current characteristic of switching diode

the breakdown value. Avalanche breakdown occurs, and carriers are injected into the diode. The forward current increases along a loadline determined by the external circuit, and the voltage/current characteristic changes to the forward conducting part shown in *Figure 8.68*. The switching diode now acts as a normal forward-biased semiconductor diode, with a low 'on' resistance and with the normal forward voltage/current characteristic.

The voltage at which the change from forward blocking to forward conduction occurs is called the breakover voltage V_{BO}. The switching diode will remain in the conducting state as long as the current exceeds the holding value. The diode is switched off by reducing the current to below the holding value I_H.

The reverse voltage/current characteristic of the switching diode is similar to that of a two-layer semiconductor diode.

A disadvantage of the switching diode occurs if a triggering pulse is applied to the anode or cathode to change the device to the conducting state when the voltage across it is below the breakover value. The triggering circuit is directly connected to the load circuit, which may be a high-power circuit. The switching power may also be high. These disadvantages can be overcome by switching the device with a suitable voltage applied to one of the intermediate layers in the structure. The switching diode is then changed to a thyristor.

Thyristor

The thyristor is a three-terminal *pnpn* device. As in the switching diode just described, the outer *p* and *n* regions form the anode and cathode respectively, and the third terminal, the gate, is connected to the intermediate *p* region. By analogy to the switching diode, the original name for this structure was the reverse blocking triode thyristor. As it is the most commonly-used device in the thyristor family, it has generally become known simply as the thyristor.

It is convenient to regard the thyristor as a controlled rectifier diode. Like a rectifier diode, the thyristor will conduct a load current in one direction only, but the current can flow only when the thyristor has been 'triggered'. It is this property of being rapidly switched from the non-conducting to the conducting state that enables the thyristor to be used for the control of electrical power.

The thyristor was developed during the 1960s in parallel with the silicon rectifier power diode. Both devices were intended for use in power engineering. The similarity in application led to the thyristor being called the silicon controlled rectifier SCR, or the controlled silicon rectifier which gives the device its circuit reference CSR. The circuit symbol for a thyristor is shown in *Figure 8.69*.

OPERATION OF THYRISTOR

The schematic structure of a thyristor is shown in *Figure 8.70*. If a voltage is applied so that the anode is positive with respect to the cathode, and no voltage is applied to the

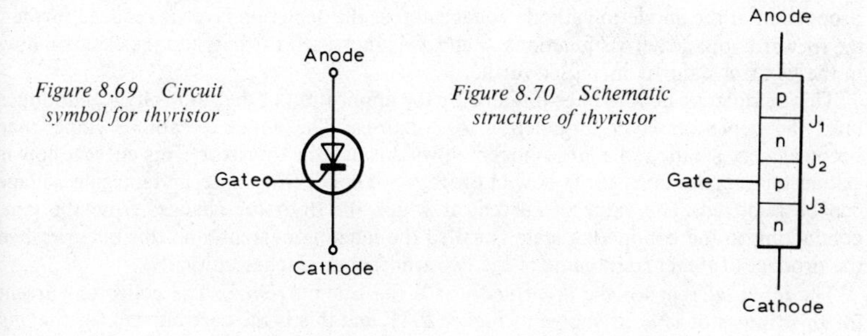

Figure 8.69 Circuit symbol for thyristor

Figure 8.70 Schematic structure of thyristor

gate, the operation will be similar to that described for the switching diode. Junctions J_1 and J_3 are forward-biased, junction J_2 is reverse-biased, and the thyristor will remain non-conducting until the voltage across it exceeds the forward breakover value and avalanche breakdown of junction J_2 occurs. The usual method of triggering the thyristor, however, is to apply a small positive voltage to the gate while the anode is positive with respect to the cathode.

The four-layer structure may be considered as forming two interconnected transistors, as shown in *Figure 8.71*. The cathode, gate and intermediate *n* region form an *npn* transistor, the cathode acting as the emitter, the gate as the base, and the *n* region as the

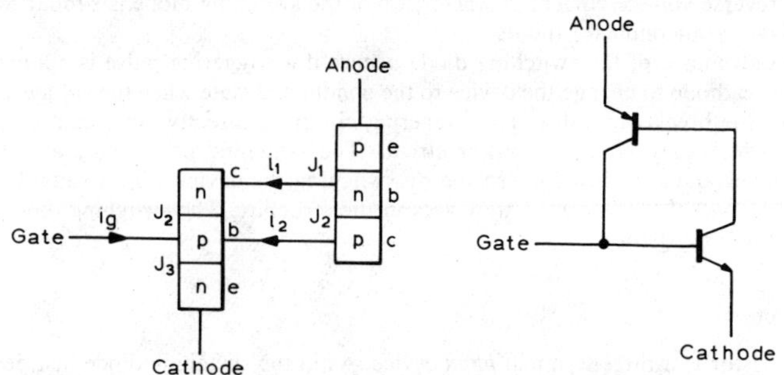

Figure 8.71 *Two-transistor analogue of thyristor*

collector. Junction J_3 is forward-biased and junction J_2 is reverse-biased as required for normal transistor operation. The anode, *n* region, and gate form a *pnp* transistor with the anode as emitter, the *n* region as base and the gate as the collector. Junction J_1 is forward-biased and junction J_2 is reverse-biased so normal transistor action can occur.

With only the anode-to-cathode voltage applied, the current through the thyristor is the small reverse current flowing across junction J_2. The thyristor is in the non-conducting forward blocking state. When the small positive voltage is applied to the gate, electrons flow from the cathode to the gate. This cathode-gate current forms the emitter–base current of the *npn* transistor, and by normal transistor action some of the electrons cross junction J_2 to enter the *n* collector region. Because the gate region is thicker than the base region of a transistor, the current gain will be less than unity.

The electron flow across junction J_2 causes the depletion layer to become narrower. The proportion of the anode-to-cathode voltage across the depletion layer is reduced, and the forward voltage across junctions J_1 and J_3 is increased. The flow of holes from the anode to the *n*-region (emitter–base current of the *pnp* transistor) is increased by this increase in forward bias across junction J_1. By normal transistor action, some of these holes cross junction J_2 to the gate (collector) region. Again, because the *n* region is thicker than the base in a transistor, the current gain is less than unity. The flow of holes across junction J_2 reduces further the width of the depletion layer. Consequently the proportion of the anode-to-cathode voltage across the depletion layer is reduced further, the forward voltage across junctions J_1 and J_3 is increased further, and the electron flow in the *npn* transistor is increased further.

This cumulative action, once initiated by the application of the gate voltage, continues until the depletion layer at junction J_2 collapses. The anode-to-cathode impedance becomes very small, and a large current flows through the thyristor. This current flow is self-sustaining, the thyristor is now in the forward conducting state, and the gate voltage can be removed. The value of current at which the thyristor changes from the non conducting to the conducting state is called the latching current, and this occurs when the product of the current gains of the two transistors reaches unity.

The input current for the *npn* transistor is the gate current i_g. The collector current for this transistor is i_1, as shown in *Figure 8.71*, and this is the base current for the *pnp* transistor. The collector current of the *pnp* transistor i_2 is fed back to the base of the *npn* transistor. If the current gain of the *npn* transistor is h_{FE1}, then:

$$i_1 = h_{FE1}(i_g - i_2) \tag{1}$$

If the gain of the *pnp* transistor is h_{FE2}, then:

$$i_2 = h_{FE2}i_1 \tag{2}$$

Substituting the expression for i_2 from equation 2 in equation 1 gives:

$$i_1 = \frac{h_{FE1}i_g}{1 - h_{FE1}h_{FE2}}$$

Substituting this expression for i_1 in equation 2 gives:

$$i_2 = \frac{h_{FE1}h_{FE2}i_g}{1 - h_{FE1}h_{FE2}}$$

The load current through the thyristor I_L is the sum of i_1 and i_2. Thus:

$$I_L = \frac{h_{FE1}i_g(1 + h_{FE2})}{1 - h_{FE1}h_{FE2}} \tag{3}$$

The load current becomes infinite, that is the value will depend on the load resistance outside the thyristor rather than on the thyristor itself, when the denominator in equation 3 is zero; that is:

$$h_{FE1}h_{FE2} = 1$$

This is the condition for the thyristor to turn on and remain in the forward conducting state.

The thyristor is turned off by reducing the current through the device to below a value called the holding current. This is often done by making the anode of the thyristor negative with respect to the cathode.

The holding current is the point of extinction of conduction in the thyristor. It is located at that part of the thyristor element which has the highest sensitivity. At this point, the transistor gains are higher than in the rest of the element, and conduction is easier. When the thyristor is triggered by the gate trigger pulse, conduction will start near the gate in a region which is not necessarily the most sensitive. The holding current in this region must be attained before the trigger pulse can be removed, and this value of holding current is the latching current for the thyristor.

The static characteristic of the thyristor is shown in *Figure 8.72*. With a reverse

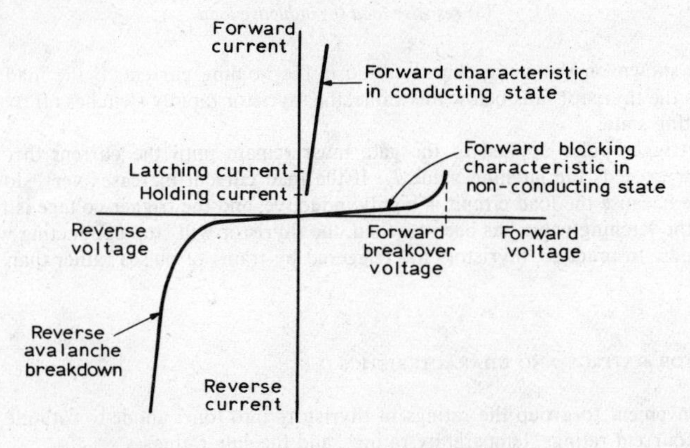

Figure 8.72 Static characteristic of thyristor

voltage applied, the voltage/current characteristic is similar to that of any reverse-biased semiconductor diode. A small reverse current flows as the reverse voltage is increased until the reverse avalanche region is reached. The reverse current then increases rapidly as the thyristor breaks down. In the forward direction if the thyristor is not triggered, a small forward leakage current flows until the forward avalanche region is reached. This is the non-conducting forward blocking state. In the avalanche region, the leakage current increases until, at the forward breakover voltage V_{BO}, the thyristor switches rapidly to the conducting state as previously described for the switching diode. If the thyristor is triggered by a positive voltage applied to the gate when the anode-to-cathode voltage is less than V_{BO}, the thyristor again switches rapidly to the conducting state. In the conducting state, the thyristor behaves as a forward-biased semiconductor diode.

When the thyristor is triggered by a gate pulse, the current increases along a loadline determined by the external circuit. The change from the non-conducting to the conducting state for resistive and inductive loads is shown in *Figure 8.73*. With a resistive load, the turn-on may go directly from a high forward voltage to a high forward current, and the holding and latching current levels will only be reached when the current through the thyristor is being reduced during turn-off. With a highly inductive load, the initial turn-on current may be below the holding and latching levels. The forward blocking voltage across the thyristor is transferred to the inductance which produces the current increase shown in *Figure 8.73* (*b*).

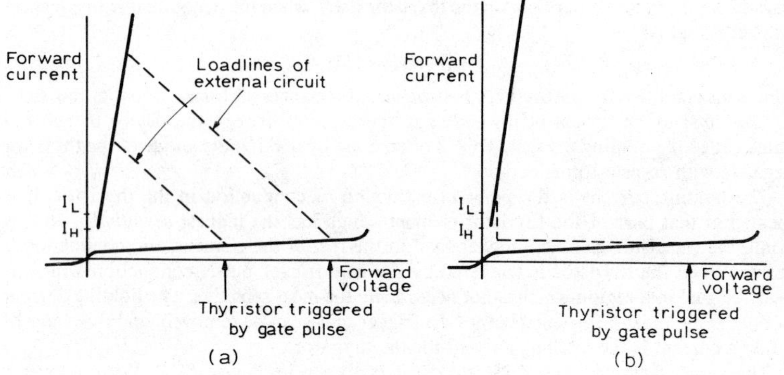

Figure 8.73 Turn-on of gate-triggered thyristor showing increase of current (broken lines): (a) resistive load (b) inductive load

Also shown on the static characteristic is the holding current. If the load current through the thyristor falls below this value, the thyristor rapidly switches off to the non-conducting state.

The trigger pulse applied to the gate must remain until the current through the thyristor exceeds the latching value I_L. If the load current increases very slowly, for example because the load circuit is highly inductive, and the trigger voltage is removed before the latching value has been reached, the thyristor will stop conducting when the pulse ends. In practice, thyristors are triggered by trains of pulses rather than a single pulse.

THYRISTOR RATINGS AND CHARACTERISTICS

It is convenient to group the ratings of thyristors into four: anode-to-cathode voltage ratings, current ratings, temperature ratings, and the gate ratings.

Because thyristors are used to control electrical power, or to control electrical machinery where safety is important, it is essential that the thyristors are always under the control of the gate signals and spurious triggering cannot occur. Thyristors often operate directly from the mains supply and are therefore subject to mains-borne voltage transients. Since the magnitude of these transients may be difficult to predict accurately, it may be necessary to provide filter circuits to ensure the thyristor is not triggered by voltages exceeding the forward breakover value. Such filter circuits will be discussed later.

As with semiconductor power diodes, three reverse voltage ratings are specified for a thyristor: continuous, repetitive, and non-repetitive values. Because thyristors may have to remain non-conducting in the forward direction, three off-state forward voltage ratings are also given. There are therefore six blocking voltage ratings specified in published data. These are listed below.

V_{RWM} crest working reverse voltage, corresponding to the peak negative value (often with a safety factor) of the sinusoidal mains supply voltage

V_{RRM} repetitive peak reverse voltage, corresponding to the peak negative value of transients occurring every cycle of the mains supply voltage

V_{RSM} non-repetitive (surge) peak reverse voltage, corresponding to the peak negative value of transients occurring irregularly in the mains supply voltage

V_{DWM} crest working off-state voltage, corresponding to the peak positive value (often with a safety factor) of the sinusoidal mains supply voltage

V_{DRM} repetitive peak off-state voltage, corresponding to the peak positive value of transients occurring every cycle of the mains supply voltage

V_{DSM} non-repetitive (surge) peak off-state voltage, corresponding to the peak positive value of transients occurring irregularly in the mains supply voltage

The repetitive and non-repetitive ratings are determined partly by the voltage limit and partly by the transient energy that the thyristor can safely absorb. The forward voltage limit is set by the need not to exceed the breakover voltage V_{BO}; the reverse voltage limit ensures that the thyristor is not driven into reverse avalanche breakdown. The transient energy that can be absorbed by the thyristor is limited (as usual) by the maximum permissible junction temperature. Therefore the duration of the repetitive and non-repetitive transients must be short compared with one half-cycle of the mains supply, and sometimes a maximum duration is specified for the transient.

The relationship between the voltage ratings and a typical mains supply waveform is shown diagrammatically in *Figure 8.74*. For clarity, the crest working reverse and off-state voltages have been shown equal to the peak sinusoidal component of the supply. It

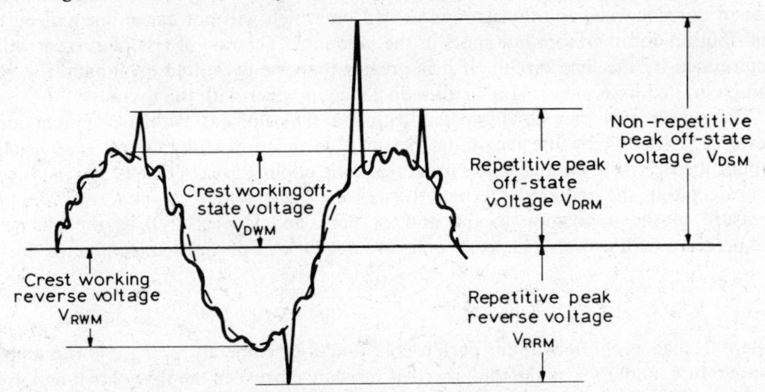

Figure 8.74 Diagrammatic relationship between thyristor voltage ratings and typical mains supply waveform with transients

is considered good practice, however, in applications where thyristors operate directly from the mains supply to choose a device in which these two ratings are twice the peak sinusoidal supply voltage.

Two other voltage ratings are usually given in published data. These are the on-state voltage V_T, the forward voltage drop across the thyristor when conducting a specified current, and the 'dV/dt' rating. This second rating specifies the maximum rate of rise of the forward off-state anode-to-cathode voltage that will not trigger the thyristor into conduction. Because of the capacitance of the depletion layer associated with the reverse-biased junction, a rapidly-changing voltage across the non-conducting thyristor will induce a current across this junction. The magnitude of this current is:

$$i = C_d \frac{dV}{dt}$$

where C_d is the capacitance of the depletion layer. If this current exceeds the latching value, the thyristor will be turned on. This spurious triggering is prevented in practice either by using thyristors with high dV/dt ratings, or by connecting suppression components across the thyristor to slow down the rate of rise of the applied voltage. These suppression circuits will be discussed later.

As with the voltage ratings, the current that can be conducted by a thyristor is specified by the continuous, repetitive peak, and non-repetitive peak values. The continuous current is also given as d.c., average, and r.m.s. values. The current ratings are listed below.

I_T	continuous on-state current at specified temperature
$I_{T(AV)}$	average on-state current at specified temperature
$I_{T(RMS)}$	r.m.s. on-state current
I_{TRM}	repetitive peak on-state current
I_{TSM}	non-repetitive peak on-state current, specified as the peak of a half-sinewave of 10 ms duration or as the amplitude of a square pulse of 5 ms duration, both at the maximum junction temperature before the surge current occurs

As well as the I_{TSM} rating, a curve of the maximum r.m.s. surge current plotted against the surge duration is given in the published data.

Two other current ratings are given in published data, the I^2t and the di/dt ratings. The I^2t rating, or surge current capability, is required for selecting fuses to protect the thyristor against excessive current being drawn from the device if a short-circuit occurs in the load circuit. The fuse has an I^2t rating, and this should be equal to or less than that of the thyristor to provide adequate protection. The di/dt rating is the maximum rate of rise of current when the thyristor is turned on which will not cause unequal current distribution and therefore hot spots in the junctions. The rate of rise of current will be determined by the load circuit. If it is greater than the permitted maximum, the value can be limited by connecting a small inductance in series with the thyristor.

The power that can be dissipated within a thyristor (as with all semiconductor devices) is limited by the maximum permissible junction temperature. It is usual to mount the thyristor on a heatsink to increase the cooling area. The heatsink is selected by calculating the required thermal resistance. The quantity *thermal resistance* is a measure of the opposition to the flow of heat, and is measured as the change in temperature with power. The usual units are deg C/W. For any semiconductor device:

$$P_{tot} = \frac{T_j \max - T_{amb}}{\Sigma R_{th}} \qquad (4)$$

where T_j max is the maximum permissible junction temperature, T_{amb} is the ambient temperature, and ΣR_{th} is the total thermal resistance between the thyristor junction and ambient. The chain of thermal resistance making up ΣR_{th} is shown in *Figure 8.75*: $R_{th(j-mb)}$, the thermal resistance between the thyristor junction and the mounting base; $R_{th(i)}$, the contact thermal resistance between the thyristor and heatsink at a specified

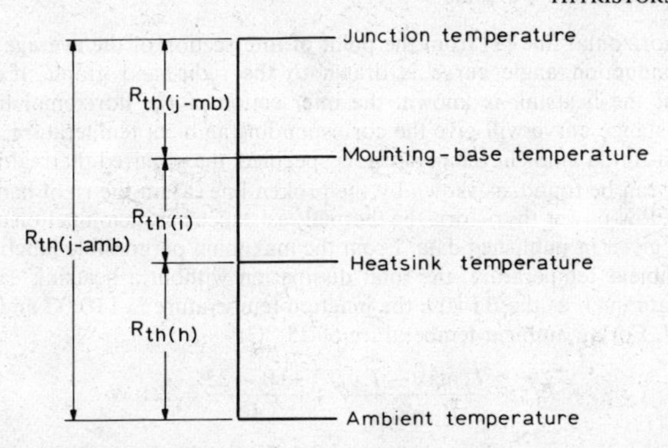

Figure 8.75 Temperatures and thermal resistances for thyristor mounted on heatsink

torque; and $R_{th(h)}$, the thermal resistance of the heatsink. Of these quantities, $R_{th(j-nb)}$ and $R_{th(i)}$ are given in published data, and T_j max is also given. The value of P_{tot} can be derived from curves in the data. Thus ΣR_{th} can be calculated from equation 4, and from this the value of $R_{th(h)}$ derived.

Curves are published in the data enabling the heatsink to be selected without the need for the calculation just described. These curves are shown in *Figure 8.76*. The left-hand graph relates the power dissipation of the thyristor P_{tot} and the average current $I_{T(AV)}$, with the conduction angle as parameter. The conduction angle is the part of the positive half-cycle of the supply for which the thyristor is conducting. As will be explained later, the thyristor is used to control electrical power by variation of the conduction angle. The right-hand graph relates the mounting-base temperature of the thyristor T_{mb} to the ambient temperature T_{amb}, with the thermal resistance of the heatsink as parameter.

For a particular application, the average current and the conduction angle are known. The vertical broken line (1) on the left-hand graph of *Figure 8.76* can therefore be

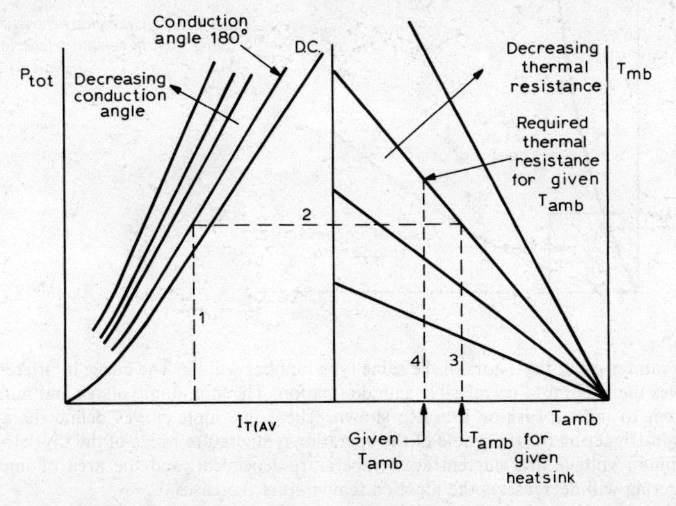

Figure 8.76 Selection of heatsink for thyristor: average current and power with conduction angle as parameter related to mounting-base temperature and ambient temperature with heatsink thermal resistance as parameter

plotted. A horizontal line (2) from the point of intersection of the average current line with the conduction angle curve is drawn to the right-hand graph. If the thermal resistance of the heatsink is known, the intersection of the horizontal line with the thermal resistance curve will give the corresponding ambient temperature, vertical line (3). If a limit to the ambient temperature is specified, the required thermal resistance of the heatsink can be found, as shown by the broken line (4) on the right-hand graph.

For some low-power thyristors, the thermal resistance between junction and ambient, $R_{th(j-amb)}$, is given in published data. From the maximum permissible junction temperature and ambient temperature, the total dissipation without a heatsink can be found. For a thyristor such as the BT109, the junction temperature is 110 °C and $R_{th(j-amb)}$ is 40 deg C/W. For an ambient temperature of 25 °C:

$$P_{tot} = \frac{T_j \max - T_{amb}}{R_{th(j-amb)}} = \frac{110 - 25}{40} = 2.1 \text{ W}$$

Therefore if the dissipation does not exceed 2.1 W, the thyristor can be used without a heatsink.

The gate ratings of the thyristor ensure that all devices of the same type number are reliably triggered over the complete operating temperature range without exceeding the permissible gate dissipation. The principal gate ratings are listed below.

$P_{G(AV)}$ average gate power, averaged over any 20 ms period
P_{GM} maximum gate power
V_{GD} maximum continuous gate voltage that will not initiate turn-on
V_{GT} minimum instantaneous gate voltage to initiate turn-on
I_{GT} minimum instantaneous gate current to initiate turn-on

A gate characteristic is often given in published data. The form of this characteristic is shown in *Figure 8.77*. The two curves in full lines define the limits between which the

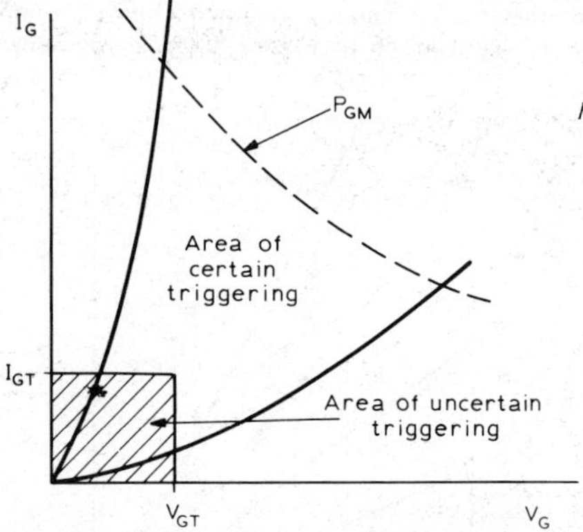

Figure 8.77 Gate characteristic showing area of certain triggering

gate ratings of all thyristors of the same type number will lie. The curve in broken lines defines the maximum permissible gate dissipation. The minimum voltage and minimum current to initiate turn-on are also shown. These five limit curves define the area of certain triggering for the whole of the operating temperature range of the thyristor. The minimum voltage and current are temperature-dependent, and the area of uncertain triggering will decrease as the junction temperature increases.

When the trigger pulse is applied to the gate of a non-conducting thyristor which has a positive anode voltage with respect to the cathode, there is a short delay, then the thyristor switches rapidly to the conducting state. The switching characteristic showing

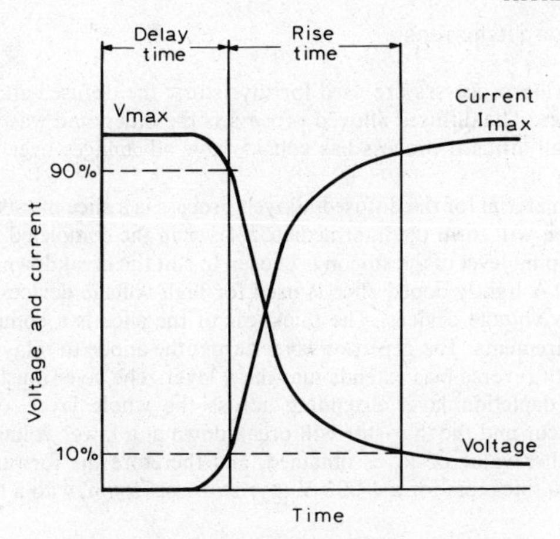

Figure 8.78 Switching characteristic of thyristor

the rise of thyristor current and fall of anode-to-cathode voltage is given in *Figure 8.78*. The rise time of the anode current is defined as the time taken for the voltage to fall from the 90% to the 10% level. The sum of the delay time t_d and the rise time t_r is the turn-on time of the thyristor t_{gt}.

The shape of the trigger pulse appreciably affects the turn-on time of the thyristor, in particular the rate of rise of the applied gate current. For most switching applications in which inverter-grade thyristors are used, a rise time for the gate current of 1 μs is satisfactory. Typical values for this type of thyristor with a trigger-pulse rise time of 1 μs are delay times between 1.0 μs and 2.0 μs, and rise times between 0.7 μs and 1.2 μs.

The thyristor is turned off by reducing the current through the device to below the holding value I_H. The value of I_H will vary with the type of thyristor, but typical values are between 10 mA and 200 mA. In practical circuits, the thyristor is often reverse-biased to ensure that it is reliably turned off. The reverse bias must be applied for a minimum time called the turn-off time t_q. For fast turn-off thyristors, the value of t_q will be between 4 μs and 25 μs. For power-control thyristors, the value of t_q will be between 20 μs and 150 μs depending on the type. The value of turn-off time increases with increase in junction temperature.

With fast-turn-off thyristors, as with semiconductor diodes, the turn-off time is limited by minority-carrier storage effects. A majority carrier from one region crossing into a region of the opposite polarity becomes a minority carrier in that region. If the bias across the junction is reversed, the minority carriers must recross the junction or combine with carriers of the opposite polarity before the depletion layer at the junction can be established. Thus the current flowing immediately after the reverse bias is applied may be high as the carriers recross the junction. This current must decay before reverse blocking of the thyristor is established. For fast switching it is necessary to reduce the lifetime of the minority carriers. The technique of gold doping the silicon material used with diodes can be applied to thyristors to give faster switching times.

The first thyristors introduced in the early 1960s were capable of handling currents up to approximately 100 A and withstanding voltages up to 300 V. By the middle of the decade, the current-handling and voltage capability had increased so that thyristors could be operated directly from the 440 V three-phase mains supply. Typical present-day thyristors are able to handle currents up to 1 000 A and withstand voltages up to 2.5 kV. Devices with dV/dt ratings up to 300 V/μs are available, enabling safe operation to be achieved with a minimum of protective components.

MANUFACTURE OF THYRISTORS

Two manufacturing processes are used for thyristors: the diffused-alloyed and the all-diffused processes. The diffused-alloyed process is the older, and was used for the first thyristors. The all-diffused process has considerable advantages over the other, and is rapidly replacing it.

The starting material for the diffused-alloyed process is a slice of *n*-type monocrystalline silicon which will form the intermediate *n* layer in the completed thyristor, *Figure 8.79* (*a*). The doping level of the silicon is chosen to suit the breakdown voltage required of the thyristor. A lightly-doped slice is used for high-voltage devices; a higher-doped slice for the low-voltage devices. The thickness of the slice is a compromise between conflicting requirements. The depletion layer across the anode-to-*n*-layer junction (J_1 in *Figure 8.70*) with reverse bias extends into the *n* layer. The layer must be thick enough to prevent the depletion layer extending across the whole layer, otherwise 'punch through' will occur and the thyristor will break down at a lower voltage. If the layer is too thick, a higher value of V_T is obtained, and therefore the forward losses will be higher. A typical thickness for a 1 000 V thyristor is 275 μm, with a typical resistivity of 0.3 Ωm.

A diffusion process is used to form *p*-type regions on both sides of the *n*-type slice, thus forming the gate and anode of the thyristor, as shown in *Figure 8.79* (*b*). The *p*-type impurity used for the diffusion is gallium or aluminium, and layers approximately 75 μm deep are diffused into the slice. The *pnp* wafer formed after the diffusion is cut into a round slice whose diameter is determined by the required current-carrying capability. For the highest current devices, the diameter may be 2.5 cm or more. The anode contact is made to one of the *p* layers using an aluminium solder to bond a molybdenum or tungsten disc to the anode, as shown in *Figure 8.79* (*c*). Because of the heavy currents that have to be conducted and the consequent large amount of heat developed, the different temperature coefficients of expansion for the silicon and the copper used for connections would lead to stresses within the device, and the possible fracture of the thyristor element. Molybdenum and tungsten have coefficients of expansion comparable with that of silicon. The edges of the *pnp* wafer may be bevelled to increase the depletion width at the surface and to ensure that any avalanche breakdown occurs in the bulk of the material rather than at the edge, *Figure 8.79* (*d*).

The final stage in the manufacture of the thyristor element is the alloying of the cathode. A foil of gold–antimony alloy and a molybdenum backing disc are alloyed in the *p*-type gate layer, *Figure 8.79* (*e*). The gold–antimony foil produces an n^+ region in the *p* layer, while the molybdenum disc matches the coefficient of expansion of the silicon to that of the copper connecting wire. The depth to which the gold–antimony is alloyed controls the base width of the *npn* transistor of the two-transistor analogue (*Figure 8.71*) and hence the gain. Too wide a base will also increase the value of the holding current. The gate connection to the thyristor element is made with aluminium wire to ensure a non-rectifying contact. The wire is welded ultrasonically to the exposed *p* layer, *Figure 8.79* (*f*). The element is etched to clean the exposed junctions J_1 and J_2 which provide the high voltage blocking characteristics of the thyristor.

The molybdenum anode contact is hard-soldered to a copper header which forms the anode connection of the device. This is usually in the form of a threaded stud to allow mounting on a heatsink, *Figure 8.79* (*g*). The cathode connection is formed by a copper lead of suitable diameter for the currents to be conducted. The lead is soldered to the cathode molybdenum disc. The device is encapsulated in a ceramic and metal top-cap which is welded to the header, as shown in *Figure 8.79* (*h*). The gate and cathode connections for the completed thyristor are made by tabs or by flying leads. Typical stud-mounting thyristors are shown in *Figure 8.80*.

The accepted convention for rectifiers is that devices with cathode studs are 'normals', and those with anode studs are 'reverses'. Under this convention, thyristors manufactured in the way just described are 'reverses'. Thyristors with cathode studs are

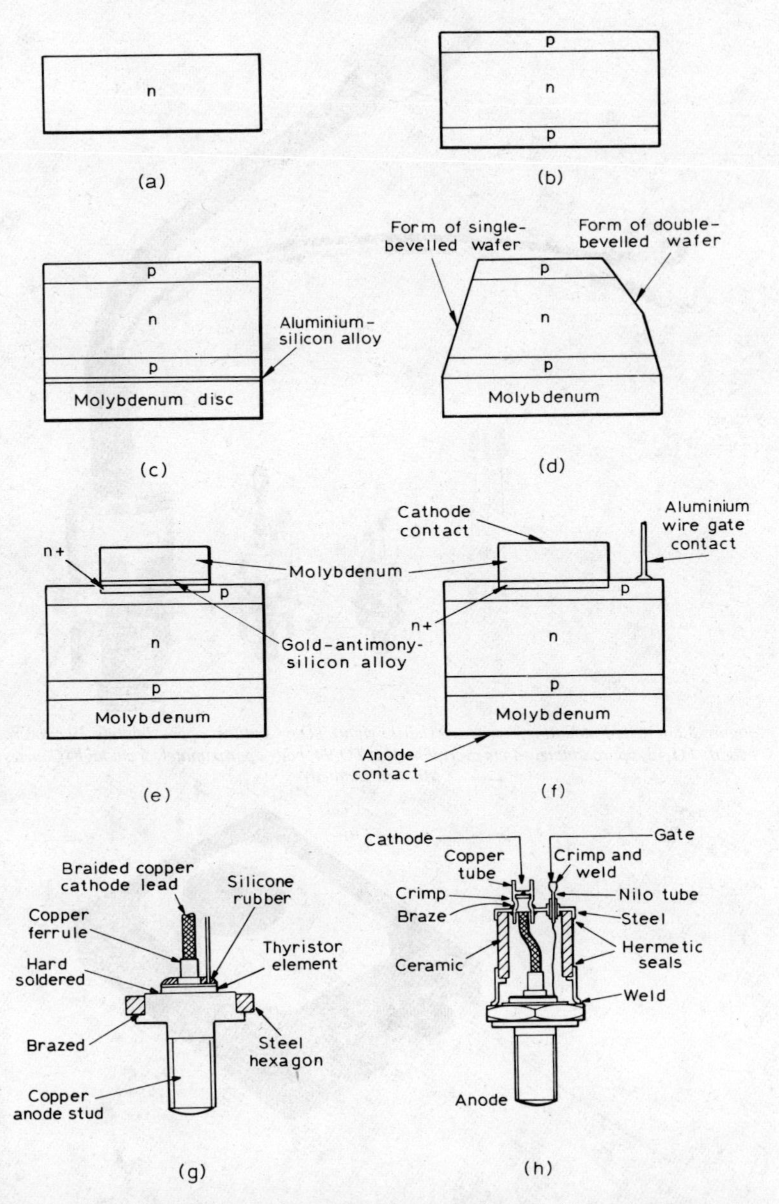

Figure 8.79 Diffused-alloyed manufacturing process for thyristor: (a) starting n-type slice (b) diffused pnp wafer (c) anode molybdenum disc attached (d) edges of wafer bevelled (showing single and double bevel) (e) cathode alloyed and molybdenum disc attached (f) gate contact fitted (g) mounted on header (h) encapsulation

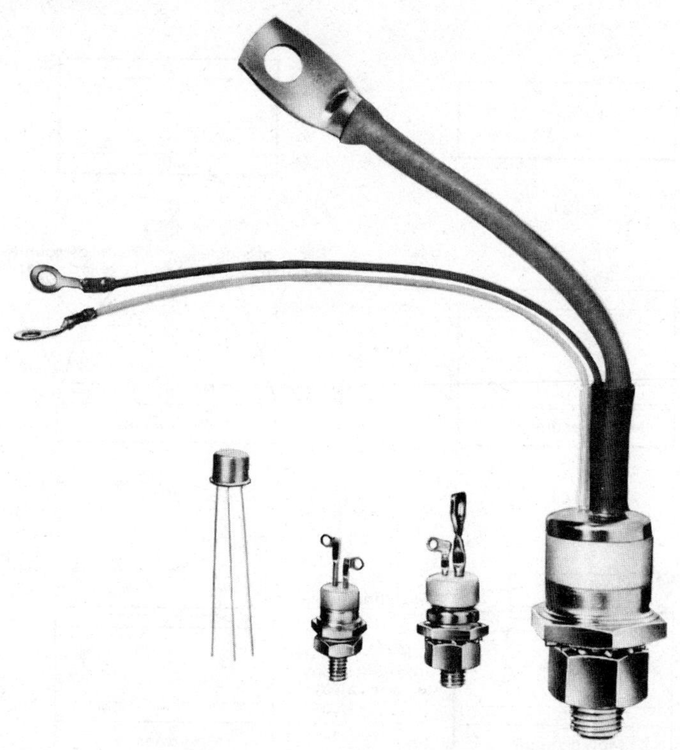

Figure 8.80 Stud mounting thyristors: (*left to right*) *TO-64 outline, approximately 3 cm overall height; TO-48, approximately 4 cm overall height; TO-94, body approximately 5 cm high* (*Courtesy Mullard Limited*)

Figure 8.81 Plastic encapsulated thyristor: *BT109 thyristor in TO-127 encapsulation showing metal plate for contact with heatsink* (*Courtesy Mullard Limited*)

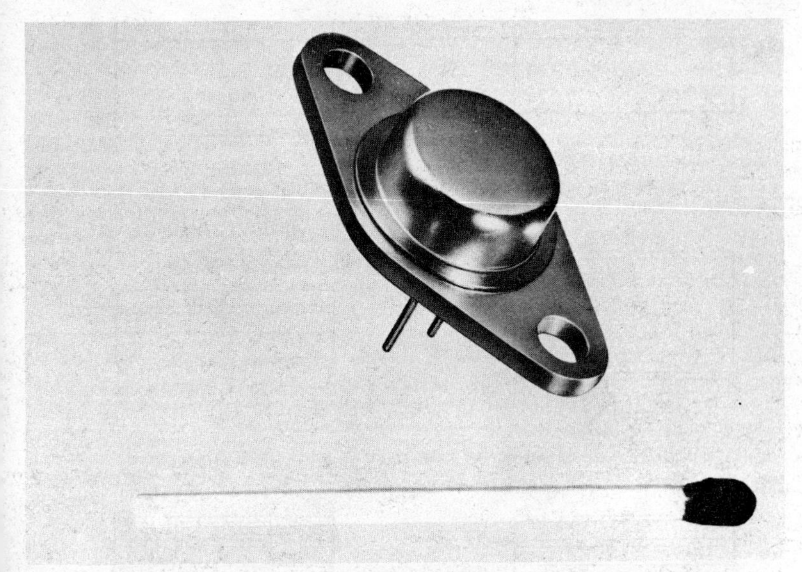

Figure 8.82 Thyristor for use in domestic equipment in TO-66 encapsulation (Courtesy Mullard Limited)

feasible, and some devices are manufactured, but they are not generally available. This is because of manufacturing problems. It can be seen from *Figure 8.79(f)* that if the cathode molybdenum disc is soldered to the header, it would be more difficult to connect the gate lead to its external terminal.

An alternative method of encapsulation can be used for low-power thyristors. This uses a moulded plastic package which is considerably cheaper than the type of encapsulation just described. The thyristor element is mounted on a metal plate which is incorporated in the plastic package so that contact can be made to a heatsink. A typical plastic-encapsulated thyristor is shown in *Figure 8.81*. Thyristors used in such applications as the line output stages of television receivers may use 'transistor' encapsulations. As an example, a thyristor in a TO-66 encapsulation is shown in *Figure 8.82*.

In the all-diffused manufacturing process, the starting material is again *n*-type silicon of similar thickness and doping level to that used in the diffused-alloyed process, *Figure 8.83* (*a*). Again, the first stage of manufacture is a *p* diffusion using aluminium or gallium as the impurity, forming a *pnp* wafer as shown in *Figure 8.83* (*b*). From this stage, however, the two processes differ. Boron is diffused into both sides of the wafer to form a thin p^+ layer, as shown in *Figure 8.83* (*c*). This layer compensates for the migration of phosphorus into the *p* layers that occurs during the contact-forming stage later in the process. The layer also imposes a control on the value of the holding current. Capsules containing wafers for this boron diffusion are shown being loaded into a furnace in *Figure 8.84* (see p. 105).

The next stage in the process is the diffusion of the cathode region. For the higher-current devices where one element is made from the *pnp* wafer, the wafer is cut to the required size. For the smaller-diameter low-current devices, many thyristor elements (say, up to 100) are made on one slice. It is therefore more economical to diffuse the cathodes for all the elements at the same time, and then separate the individual thyristor elements.

A layer of silicon dioxide is grown over the surface of the wafer by heating it in a furnace at 1 200 °C in a stream of wet oxygen. The thickness of the layer is

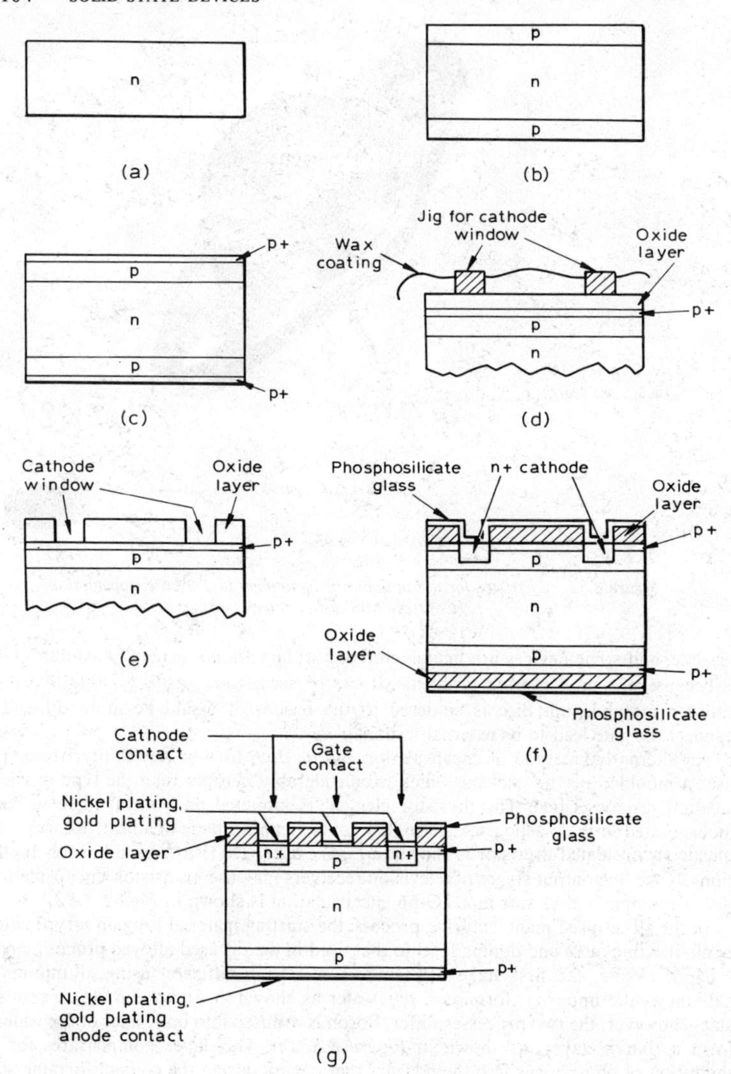

Figure 8.83 All-diffused manufacturing process for thyristor: (a) starting n-type slice (b) diffused pnp wafer (c) p^+ diffusion (d) jig and wax coating to define cathode window (e) cathode window opened (f) cathode diffused (g) contact windows opened and contact pads formed

approximately 1 μm. A window is formed in the oxide through which the cathode is diffused. Two techniques are used to form the window. The simpler one uses a jig where the window is to be, and coats the wafer with wax, *Figure 8.83* (*d*). The jig is removed, and a chemical etch used to remove the oxide which is not covered by wax. The other method uses the photographic techniques associated with the planar process. The surface of the oxide is spun-coated with a layer of photoresist about 1 μm thick. The photoresist is then dried by baking. A mask defining the cathode window is placed over

Figure 8.84 Diffusion furnace being loaded with boron capsules containing wafers during thyristor manufacture (*Courtesy Mullard Limited*)

the wafer and exposed. On development of the photoresist the unexposed area is removed, giving access to the oxide layer, while the exposed area remains and is hardened by further baking. A chemical etch is used to remove the exposed oxide. The remaining photoresist is then dissolved away, leaving the structure shown in *Figure 8.83* (*e*). The usual structure for thyristors manufactured by this process is a central gate with an annular cathode.

Phosphorus is used to form the *n*-type cathode region. The wafer is heated in a diffusion furnace to a temperature of 1 050 °C in a stream of phosphorus pentoxide vapour P_2O_5. A phosphosilicate glass is formed, about 0.5 μm thick, from which phosphorus is driven into the p^+ layer by heating in a second furnace at 1 250 °C. The phosphosilicate glass is not removed but used as the masking for the formation of the contacts in the next stage of manufacture. The resulting structure is shown in *Figure 8.83* (*f*).

The top surface of the wafer is sprayed with wax except for those areas where windows are to be cut in the glass to form the gate and cathode contacts. The exposed glass in these regions, and that on the whole bottom surface of the wafer, is etched away. The wafer is nickel plated, followed by gold plating. The nickel plating is effected by chemical displacement, and therefore nickel is only deposited on clean silicon. The contacts are formed only in the windows and on the bottom of the wafer, *Figure 8.83* (*g*). If many thyristor elements have been formed on one wafer, it is at this stage that the wafer is divided into individual elements. The edges of the element may be bevelled to increase the depletion width at the surface to ensure that breakdown takes place in the bulk of the material rather than at the edge.

The mounting and encapsulation of the element are similar to those for the diffused-alloyed types. Molybdenum backing discs are soldered to the anode and cathode to reduce thermal stress caused by the different coefficients of expansion of the silicon element and the copper connections. The element is mounted on a copper header which forms the mounting stud, and encapsulated in a ceramic and metal top cap. Alternatively, a moulded plastic package is used.

In the manufacturing process for all-diffused thyristors, especially for the higher-current types, the yield of devices is greater than with the diffused-alloyed process. In addition, the all-diffused types can have more complex cathode structures which improve the value of forward breakover voltage, the temperature stability, and the dynamic performance. It is possible, for example, to include small *p*-type regions linking with the *p* gate layer in the diffused *n*-type cathode. These small *p* regions are called *p* shorts, and the structure formed is called shorted emitter. The *p* shorts provide a means of charging the junction capacitance of the gate-to-intermediate-*n*-layer (J_2 in *Figure 8.70*) without injection of carriers from the cathode. This allows the thyristor to withstand high rates of rise of the off-stage voltage, giving the device a high d*V*/d*t* rating. This technique is not possible with diffused-alloyed thyristors, and so the all-diffused types have higher d*V*/d*t* ratings.

POWER CONTROL WITH THYRISTORS

The thyristor can be switched rapidly from the non-conducting to the conducting state. In the simple circuit shown in *Figure 8.85* (*a*), a thyristor is connected in series with the

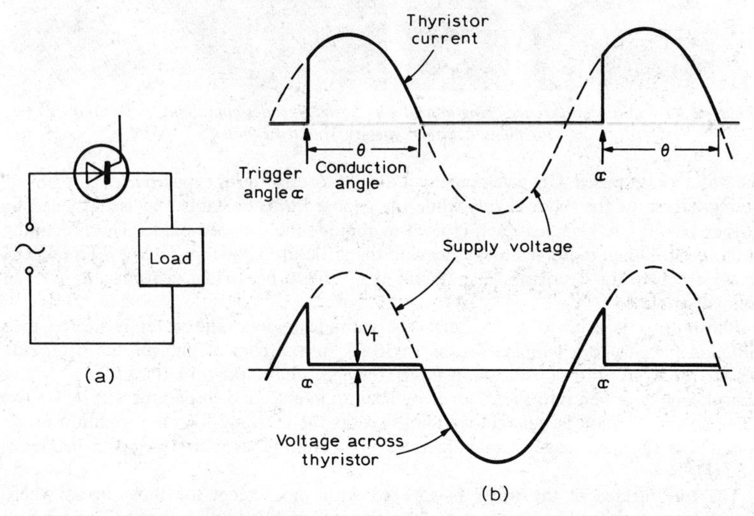

Figure 8.85 *Phase control with thyristor:* (*a*) *simple control circuit with resistive load* (*b*) *thyristor voltage and current waveforms in relation to supply voltage waveform*

supply and a resistive load. The supply voltage waveform is shown in broken lines in *Figure 8.85* (*b*). The thyristor is triggered at an angle α, called the trigger angle, in the positive half-cycle of the supply. The thyristor remains conducting for the rest of the positive half-cycle, an angle θ called the conduction angle, and is turned off when the supply voltage falls to zero. The thyristor cannot conduct during the negative half-cycle of the supply, nor can it conduct during the following positive half-cycle until it is triggered again. The load current will therefore consist of a series of positive pulses occurring once every cycle of the supply. This is shown in *Figure 8.85* (*b*), together with the voltage across the thyristor. The average value of the load current depends on the point in the positive half-cycle at which the thyristor is triggered; the average current will be greater the earlier in the half-cycle the thyristor is triggered. Thus the load current, and hence the power supplied to the load, can be varied by varying the trigger angle. This method of power control with a thyristor is called phase control.

Various configurations of thyristors are used in practice for phase control. The configurations can give an a.c. output, in which case they are called a.c. controllers, or a d.c. output, when they are called bridges. (An a.c. output consists of current pulses of

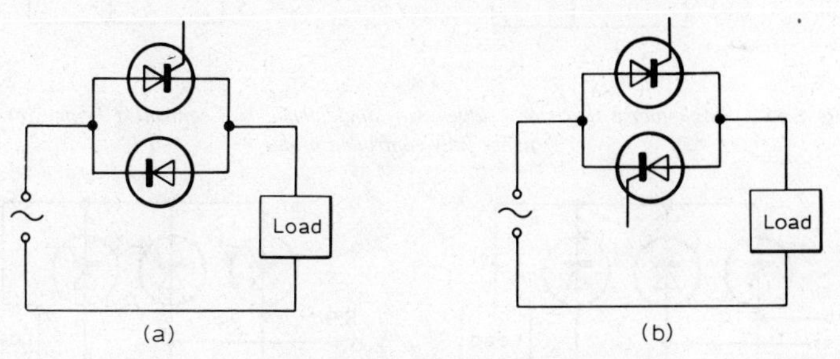

Figure 8.86 Single-phase a.c. controllers: (a) single-phase half-controlled controller (b) single-phase fully-controlled controller

alternating polarity; a d.c. output consists of current pulses of the same polarity.) The most commonly-used a.c. controllers are the half-controlled a.c. controller, *Figure 8.86 (a)*, and the fully-controlled a.c. controller, *Figure 8.86 (b)*. These are single-phase circuits; the corresponding three-phase controllers are shown in *Figure 8.87*. The two

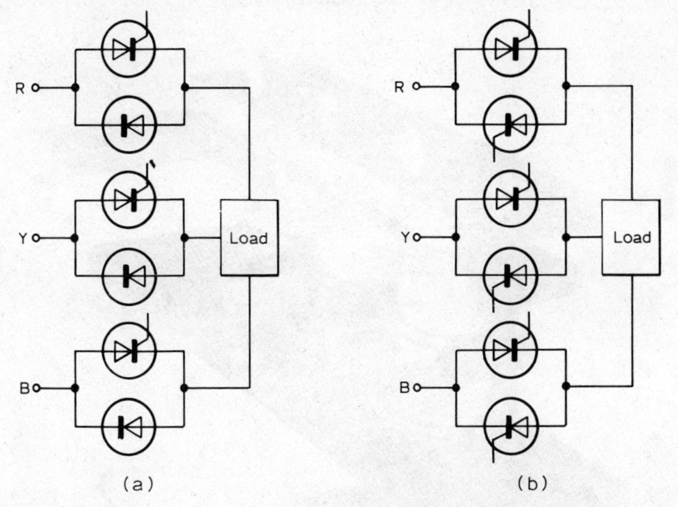

Figure 8.87 Three-phase a.c. controllers: (a) three-phase half-controlled controller (b) three-phase fully-controlled controller

common bridge circuits are the half-controlled bridge, *Figure 8.88 (a)*, and the fully-controlled bridge, *Figure 8.88 (b)*. The three-phase bridges are shown in *Figure 8.89*.

In these configurations, the current is commutated from device to device during the mains supply cycle, and no single device is conducting continuously. It is possible,

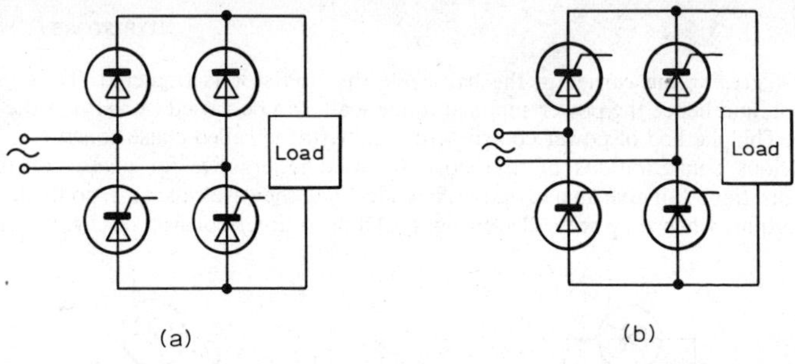

Figure 8.88 *Single-phase thyristor bridges: (a) single-phase half-controlled bridge (b) single-phase fully-controlled bridge*

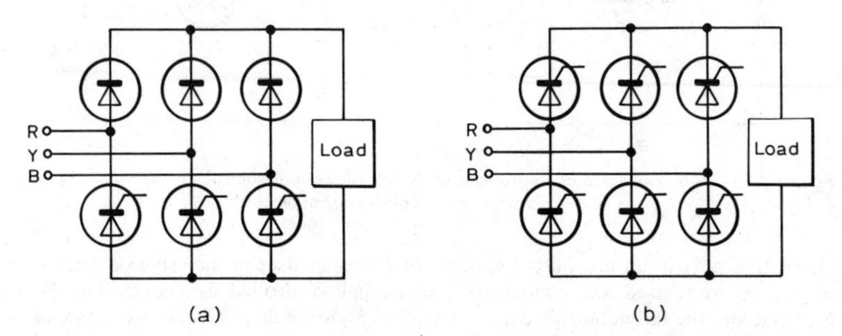

Figure 8.89 *Three-phase thyristor bridges: (a) three-phase half-controlled bridge (b) three-phase fully-controlled bridge*

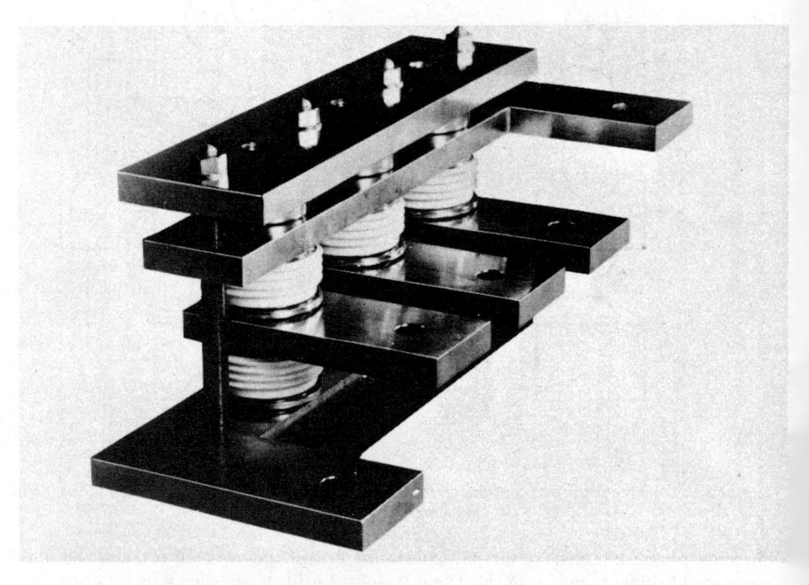

Figure 8.90 *Three-phase fully-controlled bridge capable of 2 500 A at 2.5 kV with water cooling. The stack dimensions are 228 × 228 × 152 mm (Courtesy AEI Semiconductors Ltd)*

therefore, for devices to share a heatsink. This enables a considerable saving in heatsink area to be made, particularly in high-current applications. A thyristor stack is shown in *Figure 8.90*. This example is watercooled, but generally stacks are air cooled.

A second method of power control can be used with thyristors. This is called burst triggering. Complete half-cycles of the supply are passed by the thyristor to the load for a certain time, and then the supply is blocked for a certain time. By varying the number of half-cycles conducted to those blocked, the power to the load can be controlled. Burst triggering can only be used with loads of large thermal time-constant so that the effects of the intermittent supply can be smoothed out. It is used, therefore, for such applications as electric furnaces and temperature-regulating systems. It cannot be used for applications such as motor speed control where current must be passed to the load every half-cycle. The advantages of burst triggering over phase control are that power is supplied at unity power factor, and that as the thyristor is turned on and off at the supply voltage zero, less interference is generated than with phase control.

Triac

The triac, or bidirectional triode thyristor, is a device which can be used to conduct or block current in either direction. It can therefore be regarded as two thyristors connected in the inverse-parallel configuration, but with a common gate electrode. Unlike the thyristor, however, the triac can be triggered with either a positive or a negative gate pulse.

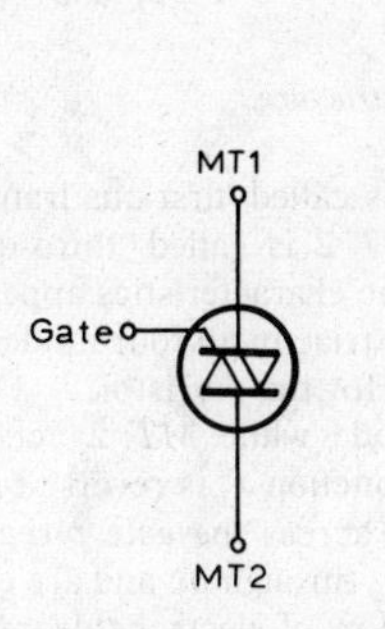

Figure 8.91 Circuit symbol for triac

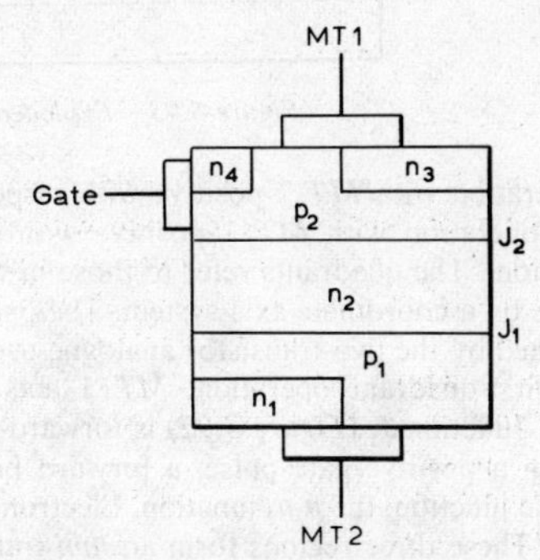

Figure 8.92 Simplified cross-section of triac structure

The circuit symbol for a triac is shown in *Figure 8.91*. The load connections of the device are called main terminals, *MT* 1 and *MT* 2. A simplified cross-section of the triac structure is shown in *Figure 8.92*. Because the current can be conducted in either direction, *MT* 1 and *MT* 2 are connected to both *p* and *n* regions. Similarly, because the triac can be triggered by both positive and negative pulses, the gate is connected to both *p* and *n* regions. The actual structure is more complex than the cross-section indicates, and is better shown by the 'exploded' view of *Figure 8.93*. However, for explaining the operation of the device, the simplified cross-section is adequate.

There are four possible modes of operation with a triac.

MT 2 positive with respect to *MT* 1, gate pulse positive
MT 2 positive with respect to *MT* 1, gate pulse negative
MT 1 positive with respect to *MT* 2, gate pulse positive
MT 1 positive with respect to *MT* 2, gate pulse negative

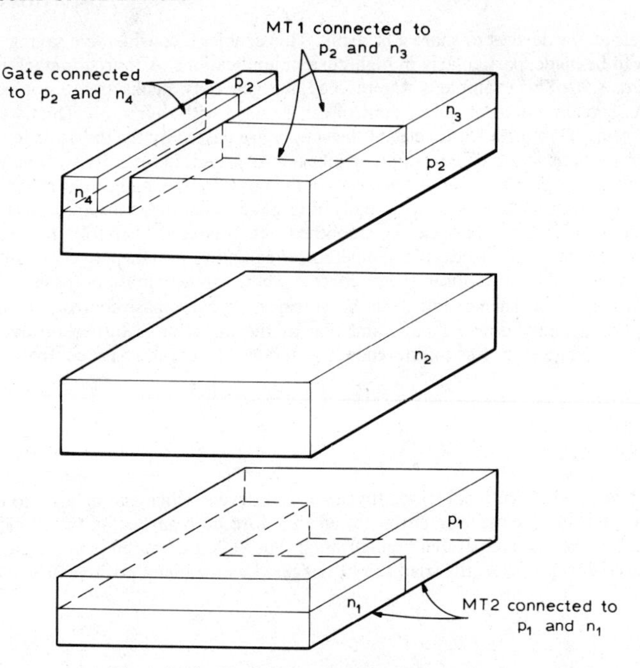

Figure 8.93 *Exploded view of triac structure*

Operation with MT 2 positive with respect to MT 1 is called 'first-quadrant operation'; operation with MT 1 positive with respect to MT 2 is called 'third-quadrant operation'. The quadrants refer to those in which the static characteristics appear when plotted on a coordinate axis system. The operation of the triac in all four modes can be explained by the two-transistor analogue used previously for the thyristor.

In first-quadrant operation, MT 1 acts as the cathode while MT 2 acts as the anode. Junction J_1 (*Figure 8.92*) is forward-biased, and junction J_2 is reverse-biased.

With a positive gate pulse, a forward bias is applied across the gate p-region and cathode junction, the p_2n_3 junction. Electrons flow from n_3, through p_2, and are collected by n_2. These three regions form an *npn* transistor. The flow of electrons decreases the reverse bias across junction J_2, thereby increasing the forward bias across junction J_1. The hole current in the *pnp* transistor formed by p_1, n_2, and p_2 increases, decreasing further the reverse bias across junction J_2 and so increasing the electron flow. A cumulative action as in the thyristor occurs, and so the triac is turned on.

With a negative gate pulse, a forward bias is applied across the p_2n_4 junction, causing electrons to flow from n_4 to n_2. The *npn* transistor for this mode is formed by n_4, p_2, and n_2; the *pnp* transistor is p_1, n_2, p_2, as with positive triggering. The same cumulative action as before occurs to turn on the triac.

In third-quadrant operation, MT 2 acts as the cathode while MT 1 acts as the anode. Junction J_1 is reverse-biased, and junction J_2 is forward-biased.

With a positive gate pulse, a forward bias is applied across the p_2n_3 junction causing electrons to flow across the forward-biased junction J_2 to n_2. The electron flow causes holes to flow from p_2, through n_2, to be collected by p_1, forming a *pnp* transistor. The hole current causes electrons to flow in the *npn* transistor n_1, p_1, n_2. Cumulative action occurs between these two transistors to turn on the triac.

With a negative gate pulse, forward bias is applied across the p_2n_4 junction causing electrons to flow to n_2. This electron flow causes holes to flow in the *pnp* transistor p_2, n_2, p_1; and electrons to flow in the *npn* transistor n_1, p_1, n_2. The same cumulative action as with positive triggering occurs to turn on the triac.

The static characteristic of the triac is shown in *Figure 8.94*. As would be expected, it consists of two thyristor forward voltage/current characteristics combined, the 'forward' current being that when the voltage at *MT* 2 is positive with respect to *MT* 1. The characteristics and ratings of the triac are similar to those of the thyristor already described, except that the triac does not have any reverse voltage ratings (a reverse voltage in one quadrant becomes the forward voltage of the opposite quadrant).

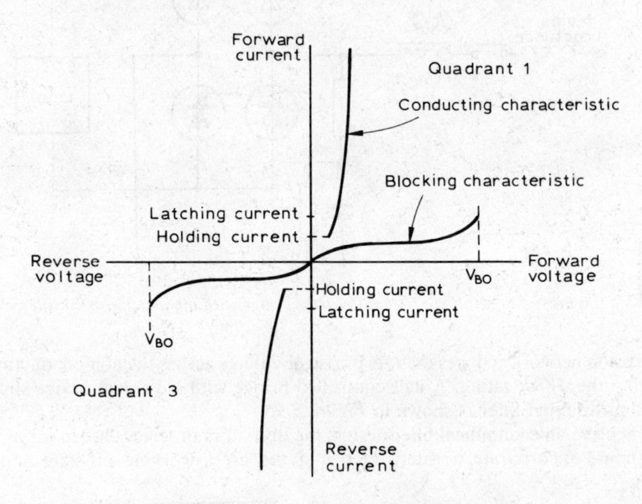

Figure 8.94 Static characteristic of triac

The triac is used in power-control applications to replace an inverse-parallel pair of thyristors, that is as a fully-controlled a.c. controller. The triac offers the advantages over the thyristor pair of easier mechanical mounting on the heatsink as only one device is used, and of a simpler triggering system.

TRANSIENT SUPPRESSION FOR THYRISTORS AND TRIACS

Thyristors and triacs in power-control applications operating directly from the mains supply are subject to voltage transients. These transients can cause the turn-on of devices, and since such turn-on is not under the control of the gate signals, it can constitute a safety hazard. There are three groups of transients: mains-borne transients, often caused by events remote from the system; transients caused by contact bounce of the system mains contactor; and transients generated within the stack by devices turning on and off. These transients can cause spurious turn-on either by exceeding the forward breakover voltage of the device, or by having a rate of rise greater than the dV/dt rating of the device. Because of the random nature of the transients, it is not possible to predict accurately their expected value. It is difficult, therefore, to choose a device economically for an application and be certain it will withstand any transient that may occur. Additional protection must be provided to suppress transients or reduce them to a safe value within the ratings of the device.

Two types of suppression circuit can provide this protection, either singly or together as required by the application. These circuits are: a shunt-capacitance input filter, and a series resistor and capacitor connected across each device. The input filter operates in conjunction with the mains supply impedance increased, if necessary, by a series line choke. The amplitude of the transient is reduced to a safe level by this circuit. A half-controlled bridge with an input filter is shown in *Figure 8.95*. The individual device

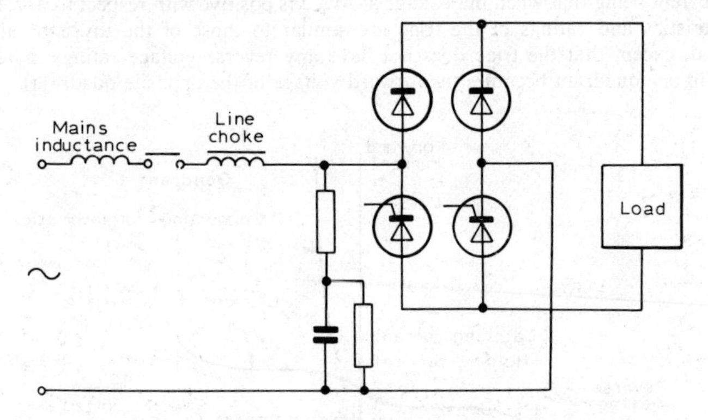

Figure 8.95 Half-controlled thyristor bridge with input suppression filter

suppression network reduces the rate of rise of voltage across the non-conducting device to within the dV/dt rating. A half-controlled bridge with individual device suppression as well as an input filter is shown in *Figure 8.96*.

To achieve an economical filter design, the thyristors or triacs chosen for an application should incorporate a safety factor in the crest working off-state and reverse

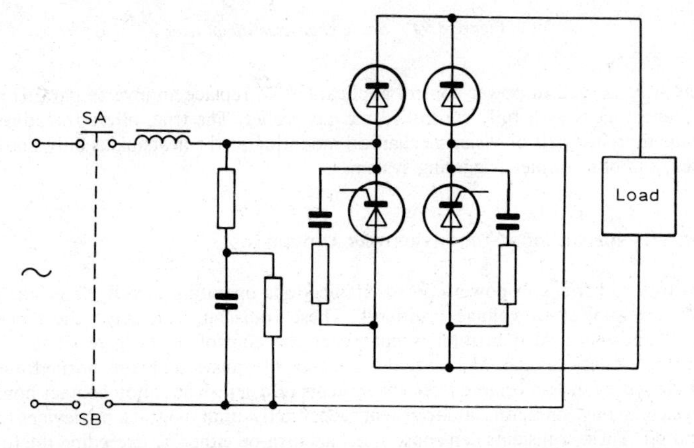

Figure 8.96 Half-controlled bridge with individual device suppression and input filter

voltages. For example, it is good practice to use 500 V, or preferably 600 V, devices for operation on the 240 V single-phase mains (peak supply voltage 360 V), and 1 000 V or 1 200 V devices for line-to-line operation on the three-phase mains. Devices with high dV/dt ratings, such as 200 V/μs, may not need individual device suppression and so can simplify the design of the stack.

TRIGGERING OF THYRISTORS AND TRIACS

Thyristors and triacs can be triggered by a single pulse, a train of pulses or a steady d.c. voltage on the gate. Pulse triggering is most frequently used; single-pulse triggering is used in specific applications and low-cost systems, and d.c. triggering only in cases of difficulty in reaching latching current. A thyristor or triac with the anode positive with respect to the cathode, and with adequate gate drive, will turn on within 10 μs. Conduction will continue irrespective of load current for as long as the gate drive is present. Conduction will continue after the gate drive ceases only if the load current has reached the latching level.

Pulse triggering is preferred to d.c. triggering for general use because the gate dissipation is lower and the trigger system is simpler. Both pulse and d.c. trigger systems must be synchronised to the mains supply, and both use a pulse generator operating through a variable delay to provide the control over trigger angle required for phase control. Whereas in a pulse trigger system the output from the pulse generator can be passed to the thyristor gate through an isolating transformer, a d.c. trigger system requires either rectification of the pulse transformer output or gating of a further supply by the pulse transformer to provide the d.c. gate drive. The more elaborate d.c. gate drive is therefore used only in applications where it is essential, for example with highly inductive loads where difficulty is experienced with latching.

The usual pulse triggering system provides a train of pulses starting at the trigger angle and continuing throughout the half-cycle of the supply. The duration of each pulse is at least 10 μs, generally 20 μs, and the repetition frequency is typically 5 kHz. With inductive loads, a flywheel diode or series RC network connected across the load can assist latching.

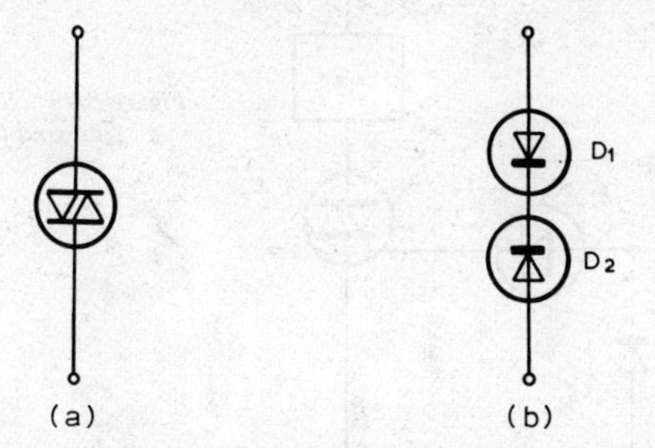

Figure 8.97 Diac: (a) circuit symbol (b) equivalent circuit

A simple trigger circuit for thyristors and triacs can be constructed with a diac. The diac is a three-layer device and therefore, strictly speaking, not a thyristor. Its function, however, is as a trigger device for thyristors, and it is generally considered with these devices.

The circuit symbol for a diac is shown in *Figure 8.97* (*a*). It can be considered as two opposed diodes connected in series, as shown in *Figure 8.97* (*b*). If a voltage is applied across the diac, say in the forward direction of diode D_1, the current through the devices will be the small reverse leakage current of diode D_2. As the voltage is increased, there will be only a slight increase in current until the breakover voltage V_{BO} for diode D_2 is reached. The current increases rapidly and the voltage falls to the working value V_W. In this condition, the diac is conducting. When the applied voltage is removed, the diac returns to its non-conducting state. If the voltage across the diac is in the forward direction of diode D_2, a similar action occurs with the current direction reversed. The voltage/current characteristic of the diac is shown in *Figure 8.98*.

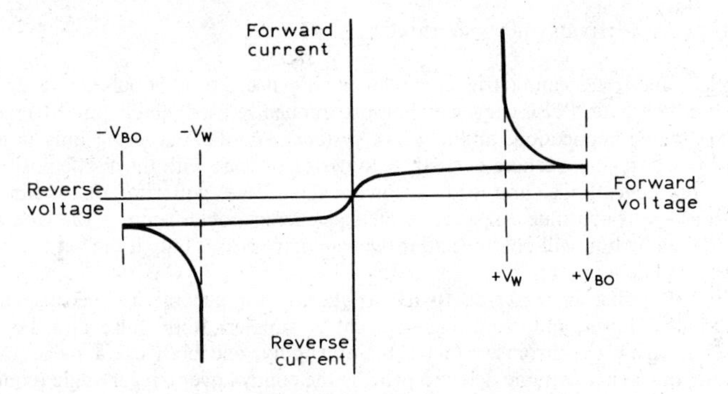

Figure 8.98 Voltage/current characteristic of diac

A simple trigger circuit for a triac using the diac is shown in *Figure 8.99*. Capacitor *C* is charged during the positive half-cycle of the supply through the variable resistor *R*. When the voltage across *C* reaches the breakover voltage of the diac, the diac changes to the conducting state and the capacitor is discharged through the triac gate to turn it on. During the following negative half-cycle capacitor *C* is recharged, the diac breaks down again, and the triac is retriggered. If a thyristor is used instead of a triac, the diode shown in broken lines must be connected across capacitor *C* to prevent negative drive

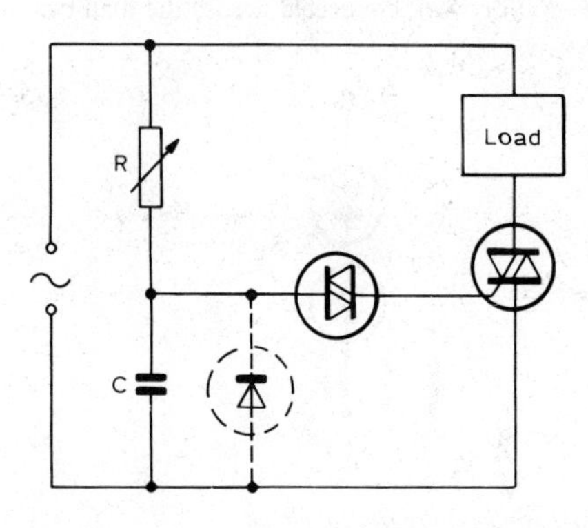

Figure 8.99 Simple trigger circuit for triac and thyristor using diac

being applied to the thyristor gate. The charging rate of the capacitor can be varied by the variable resistor *R* so that the voltage across the capacitor will reach the diac breakover voltage at different points in the half-cycle. Phase control can therefore be achieved.

The 'Ignistor'

The 'Ignistor' is a trade name for an a.c. controller specially designed to replace the pair of ignitrons used for resistance welding. Similar devices are available under other trade names, and all these devices can be grouped together under the general name of welding thyristors. Water-cooling (which would normally be available for the ignitrons being replaced) is used to enable the thyristors to deliver the high currents required for resistance welding without exceeding the maximum junction temperature. Currents up to, typically, 1 700 A r.m.s. can be obtained from an Ignistor.

The two thyristors are encapsulated in disc-shaped envelopes (sometimes called the hockey-puck outline) which allow them to be mounted between two copper bars. These bars form the internal connections between the thyristors, and flexible cables connected to the bars form the external power connections of the Ignistor. The copper bars are hollowed to form the water-cooled heat exchangers. The assembly is fitted into a metal case using a material such as glass fibre sheet to insulate the current-carrying parts from the case. A simplified cross-section of a typical Ignistor is shown in *Figure 8.100*.

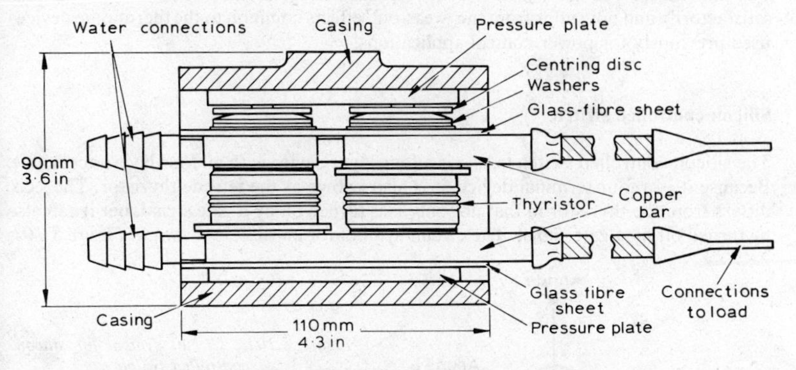

Figure 8.100 Simplified cross-section of Ignistor

Because the Ignistor is designed to replace the ignitron, it provides a clear example of the comparative merits of solid-state and thermionic devices for power-control applications. The ignitron requires an ignition voltage of approximately 150 V, and so it has to be triggered at a particular ignition angle in the half-cycle. The Ignistor requires a separate low-voltage trigger system, and can be triggered at any angle in the half-cycle. The Ignistor therefore can deliver more power in each half-cycle. The forward voltage drop of an Ignistor is smaller than that of the ignitron, typically 1 V to 2 V at a current of approximately 600 A compared with 10 V to 25 V. The operating temperature range of the Ignistor is wider than that of the ignitron. A disadvantage of the Ignistor, however, is that it requires transient suppression components to prevent spurious turn-on. Physically the Ignistor is smaller than a pair of ignitrons, and it can be mounted in any position whereas the ignitrons require vertical mounting. Thus the solid-state device is smaller, more efficient, and within comparable current ratings more power can be delivered. The main disadvantage is the need for transient suppression components.

Applications of thyristors and triacs

To complete the description of thyristors and triacs, a brief survey will be given of the applications of these devices. The range of applications is very wide, but a convenient grouping is into 'domestic' and 'industrial', corresponding roughly with the light-current and heavy-current applications.

Thyristors and triacs are used in many domestic appliances. The control of automatic washing machines where the drum has to revolve at different speeds during the various parts of the washing cycle can be conveniently performed by thyristors. Other speed-control applications include food mixers and electric drills. Thyristors are becoming increasingly used in mains power supplies and line deflection circuits of television receivers. Other light-current applications include electronic photoflashes and lamp dimmers.

The industrial applications of thyristors and triacs cover the whole field of power control. Static contactors incorporating thyristors offer considerable advantages over electromechanical types, not suffering from contact bounce or wear-out effects. A.C. and d.c. motor speed control and the control of electric furnaces can cover a wide range of powers with compact and reliable control systems. Theatre and TV studio lighting control systems use thyristors to achieve greater flexibility than is possible with other control methods. In all these applications, provided the thyristor is selected carefully so that the device ratings match the circuit requirements, the thyristor will perform satisfactorily and not suffer from the 'wear-out' effects common to the thermionic devices used previously for power control applications.

Silicon controlled switch

The silicon controlled switch (s.c.s.) is a four-layer device with all four layers accessible. Because it is a four-terminal device, it is also known as the tetrode thyristor. The s.c.s. differs from the thyristor in that not only is it turned on by a gate signal, but it can also be turned off by a gate signal. The circuit symbol for an s.c.s. is shown in *Figure 8.101*.

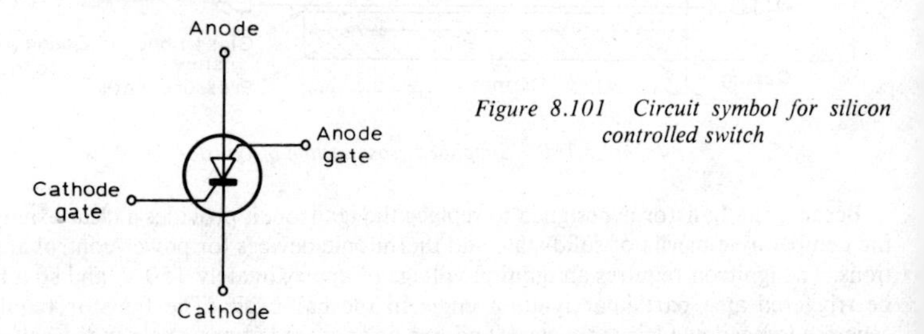

Figure 8.101 Circuit symbol for silicon controlled switch

The schematic structure of an s.c.s. is shown in *Figure 8.102(a)*. The outer *p* and *n* regions form the anode and cathode respectively, the intermediate *p* region is called the cathode gate, and the intermediate *n* region is called the anode gate. With a voltage applied so that the anode is positive with respect to the cathode, junctions J_1 and J_3 are forward-biased, and junction J_2 is reverse-biased. Because the current through the s.c.s. is the small reverse leakage current of junction J_2, the device is non-conducting.

The operation of the s.c.s. (as with that of the thyristor and triac) can be described in terms of two interconnected transistors. The two-transistor analogue for the s.c.s. is shown in *Figure 8.102 (b)*. This differs from that for the thyristor shown in *Figure 8.71*

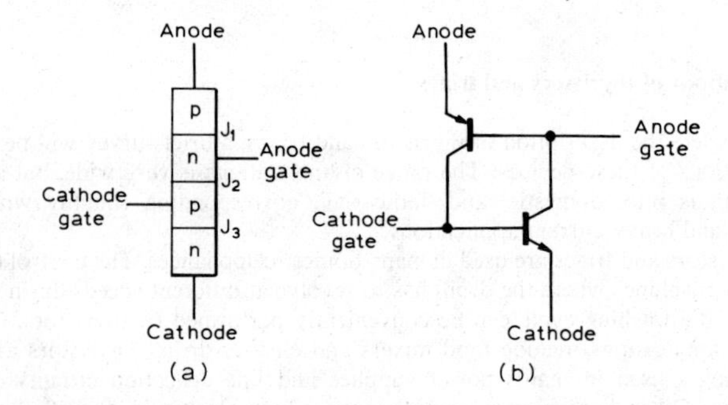

Figure 8.102 Silicon controlled switch: (a) schematic structure (b) two-transistor analogue

in that the bases of both transistors are accessible. Thus the cumulative action between the two transistors that causes turn-on can be initiated by a pulse on either base: a positive pulse on the cathode gate to start current flowing in the *npn* transistor, or a negative pulse on the anode gate to start current flowing in the *pnp* transistor. When the current through the s.c.s. exceeds the latching value (when junction J_2 changes to forward bias), the device remains in the conducting state when the gate pulse is removed.

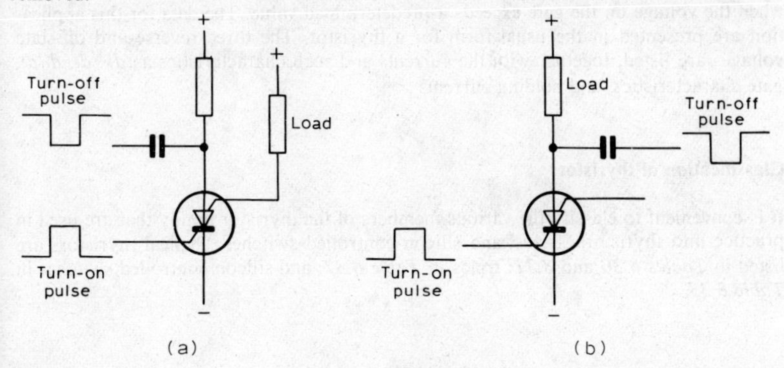

Figure 8.103 *Circuit configurations for silicon controlled switch: (a) load in anode-grate circuit (b) load in anode circuit*

The s.c.s. (like the thyristor and triac) is turned off by reducing the current through the device to below the holding value. This can be done by applying a negative pulse to the anode, a positive pulse to the anode gate (applying reverse base drive to the *pnp* transistor), or a negative pulse to the cathode gate (reverse base drive to the *npn* transistor).

In practice, when the s.c.s. is used as a fast-operating switch, the load is usually connected in the anode gate circuit, as shown in *Figure 8.103 (a)*. The device is then turned off by a negative voltage pulse to the anode which reduces the current to below the holding value for a time greater than the turn-off time. A typical value for the turn-off time t_q or t_{off} is 5 μs, although the value depends on the junction temperature and external cathode-gate resistance. To ensure that the s.c.s. is reliably turned off and to prevent spurious triggering, it is usual to maintain the anode gate and cathode gate at definite voltages rather than let them float. A positive voltage is applied to the anode gate so that the anode-to-anode-gate junction is reverse-biased. A negative voltage is applied to the cathode gate so that the cathode-to-cathode-gate junction is reverse-biased.

The positive trigger pulse to the cathode gate overcomes this reverse bias to turn on the s.c.s., the positive voltage on the anode gate assisting conduction in the *npn* transistor of the two-transistor analogue. A typical turn-on time for an s.c.s. is between 0.25 μs and 1.5 μs, depending on the external cathode-gate resistance. The turn-on time is constant over a considerable temperature range. Curves of variation of turn-on and turn-off times with temperature and cathode-gate resistance are often given in published data.

The published data for an s.c.s. in this configuration is given in the form of data for the complementary transistors of the two-transistor analogue. The forward on-state voltage and holding current are specified for the combined device. Characteristic curves showing the variation of the main voltages and currents are also given. For example, the variation of anode current with anode-to-cathode voltage and cathode-gate voltage, and the variation of anode-gate current with cathode-gate voltage.

The s.c.s. can also be used with the load connected in the anode circuit, as shown in *Figure 8.103* (*b*). The device now operates as a low-power thyristor. The anode gate is either not connected, or connected to the supply through a high-value resistor to maintain the correct operating voltage on the anode gate. If negative trigger pulses are to be used, then the cathode gate is connected to the cathode line through a high-value resistor, and the negative trigger pulses are applied to the anode gate. The s.c.s. in this configuration is used in sensing networks, being triggered to operate a relay or lamp when the voltage on the gate exceeds a predetermined value. The data for this application are presented in the usual form for a thyristor. The three reverse and off-state voltages are listed, together with the currents and such characteristics as dV/dt, di/dt, gate characteristics, and holding current.

Classification of thyristors

It is convenient to classify the various members of the thyristor family that are used in practice into thyristors, triacs, and silicon controlled switches. Typical thyristors are listed in *Tables 8.30* and *8.31*; triacs in *Table 8.32*; and silicon controlled switches in *Table 8.33*

THYRISTOR TABLES

Tables 8.30 to *8.33* list the various types of thyristor used in practice. The devices are grouped into thyristors (reverse-blocking triode thyristors), triacs, and silicon controlled switches.

The intention is to list 'typical' devices; to give the reader an indication of the range and ratings of the devices available at present. It is in no way intended that the tables should be used to select devices for particular applications. The individual manufacturers should be consulted for fuller details of their ranges of devices.

The tables have been constructed from the manufacturers' 'quick-reference' data, and the help of those manufacturers who supplied information for the tables is acknowledged. Devices and values listed were correct at the time of writing.

Outlines for devices are given a reference according to JEDEC (TO) or British Standard (SO) systems, where these apply. Outline drawings and full dimensions can be obtained from the various manufacturers or from the relevant JEDEC and British Standards Institution publications.

The abbreviations used for the various manufacturers are listed below.

AEI AEI Semiconductors Ltd.
ITT ITT Semiconductors
M Mullard Ltd.
RCA RCA Ltd.
TI Texas Instruments Ltd.

Symbols for main electrical parameters of thyristors

di/dt rate of rise of on-state current after triggering
dV/dt maximum rate of rise of off-state voltage which will not trigger any device
I_D continuous (d.c.) off-state current
I_{DM} peak off-state current
I_{FG} forward gate current
I_{FGM} peak forward gate current

I_{GaM} peak forward anode-gate current
I_{GaT} minimum anode-gate current to initiate turn-on
I_{GkM} peak forward cathode-gate current
I_{GkT} minimum cathode-gate current to initiate turn-on
I_{GT} minimum instantaneous gate current to initiate turn-on
I_{GQ} gate turn-off current
I_H holding current
I_L latching current
I_{RG} reverse gate current
I_{RGM} peak reverse gate current
I_T continuous (d.c.) on-state current
$I_{T(\mathrm{AV})}$ average on-state current
$I_{T(\mathrm{ov})}$ overload mean on-state current
$I_{T(RMS)}$ r.m.s. on-state current
I_{TRM} repetitive peak on-state current
I_{TSM} non-repetitive peak on-state current
$P_{G(\mathrm{AV})}$ average gate power
P_{GM} peak gate power
$R_{\mathrm{th}(h)}$ thermal resistance of heatsink
$R_{\mathrm{th}(i)}$ contact thermal resistance at specified torque
$R_{\mathrm{th}(j\text{-amb})}$ thermal resistance, junction to ambient
$R_{\mathrm{th}(j\text{-}h)}$ thermal resistance, junction to heatsink
$R_{\mathrm{th}(j\text{-mb})}$ thermal resistance, junction to mounting base
$R_{\mathrm{th}(\mathrm{mb}\text{-}h)}$ thermal resistance, mounting base to heatsink
T_{amb} ambient temperature
T_j max maximum permissible junction temperature
T_{mb} mounting-base temperature
T_{stg} storage temperature
t_d delay time
t_{gt} gate-controlled turn-on time
t_{off} circuit-commutated turn-off time for s.c.s.
t_{on} turn-on time for s.c.s.
t_q circuit-commutated turn-off time
t_r rise time
V_{AK} forward on-state voltage of s.c.s.
V_{BO} breakover voltage
V_D continuous off-state voltage
V_{DRM} repetitive peak off-state voltage
V_{DSM} non-repetitive peak off-state voltage
V_{DWM} crest working off-state voltage
V_{FG} forward gate voltage
V_{FGM} peak forward gate voltage
V_{GaM} peak reverse anode-gate-to-anode voltage
V_{GaT} minimum anode-gate voltage that will initiate turn-on
V_{GD} maximum continuous gate voltage which will not initiate turn-on
V_{GkM} peak reverse cathode-gate-to-cathode voltage
V_{GkT} minimum cathode-gate voltage to initiate turn-on
V_{GT} minimum instantaneous trigger voltage to initiate turn-on
V_{RG} reverse gate voltage
V_{RGM} peak reverse gate voltage
V_{RRM} repetitive peak reverse voltage
V_{RSM} non-repetitive peak reverse voltage
V_{RWM} crest working reverse voltage
V_T continuous (d.c.) on-state voltage

List of Thyristor Tables

Table 8.30 GENERAL INDUSTRIAL THYRISTORS
(Ascending current order)

Type	*Manufacturer*	*Outline*	$I_{T(AV)}$ (A)	I_{TSM} (A)	V_{RRM} (V)	V_{GT} (V)	I_{GT} (mA)	*Features*
BTX18 series	AEI, M	TO-5	1.0	10	120 to 600	2.0	5.0	
CRS1 series	ITT	TO-5	1.25	15	50 to 500*	2.5	10	$t_q = 15\ \mu s$
2N4102 series	RCA	TO-8	2.0†	60	200 to 600	2.0	15	$dV/dt = 200\ V/\mu s$
BT128	M	TO-66	3.2	50	—	4.0	40	TV line output type
CR4′01B series	AEI	TO-48	5.0	100	50 to 600	3.0	25	$dV/dt = 20\ V/\mu s$
TIC106 series	TI	TO-66(P)	5.0	30	30 to 400 ‡	1.0	200	
40888 series	RCA	TO-66	5.0†	50	700 and 750 ‡	4.0	40	Fast turn-off type for TV deflection circuits
BT109	M	TO-127	6.5	50	500	2.0	10	
BTY79A series	ITT	TO-64	10	80	50 to 500*	3.0	20	$t_q = 17\ \mu s$
BTY87 series	M	TO-48	10	140	100 to 800	3.5	65	
CR10-051B to CR10-601B	AEI	SO-28	10	120	50 to 600	3.0	80	
40740 series	RCA	—	10†	100	100 to 600	2.0	15	
TIC126 series	TI	TO-66(P)	12	100	50 to 800†	1.5	20	
BTW92 series	AEI, M	TO-48	20	320	600 to 1 600	3.5	150	$dV/dt = 200\ V/\mu s$ (AEI); $dV/dt = 300\ V/\mu s$ (M)
2N690 series	RCA	TO-48	25 ‡	150	25 to 600	3.0	50	
BTW24 series	M	§	30	600	600 to 1 600	3.5	150	
CR27′03RCA series	AEI	TO-48	32	500	800 to 1 400	3.0	60	$dV/dt = 200\ V/\mu s$
2N3899 series	RCA	—	35 †	350	100 to 800	2.0	40	$dV/dt = 100\ V/\mu s$
BTW23 series	M	TO-94	70	1 500	600 to 1 600	3.5	200	

Table 8.30–*continued*

Type	Manufac-turer	Outline	$I_{T(AV)}$ (A)	I_{TSM} (A)	V_{RRM} (V)	V_{GT} (V)	I_{GT} (mA)	Features
CR51′03AA series	AEI	TO-94	70	760	75 to 1 300	3.0	120	$dV/dt = 200$ V/μs
CR152′03A series	AEI	§	150	3 150	75 to 1 400	3.5	150	$dV/dt = 300$ V/μs
CR605SC series	AEI	§	200	3 500	75 to 1 400	3.5	150	$dV/dt = 300$ V/μs
DCR608SK series	AEI	§	270	3 000	75 to 1 400	3.5	150	$dV/dt = 300$ V/μs
CR705SC series	AEI	§	300	6 400	75 to 1 700	3.5	150	$dV/dt = 300$ V/μs
CR807SC series	AEI	§	540	9 000	75 to 2 400	3.5	150	$dV/dt = 300$ V/μs
DCR804SM series	AEI	§	870	11 000	75 to 1 500	3.5	150	$dV/dt = 300$ V/μs

* V_{RWM} value. † $I_{T(RMS)}$ value. ‡ V_{DRM} value.
§ See manufacturer's literature for details.

Table 8.31 INVERTER (FAST TURN-OFF) TRANSISTORS
(Ascending current order)

Type	Manufacturer	Outline	$I_{T(AV)}$ (A)	V_{RRM} (V)	t_q (μs)	dV/dt (V/μs)
CR5-102GB to CR5-602GB	AEI	SO-28	5.0	100 to 600	4	—
CR6-904CRB to CR6-1204CRB	AEI	SO-28	5.0	800 to 1 100	12	—
BTW30 series	M	TO-48	12	300 to 600 800 to 1 200	6 12	200
CR18-904CRB to CR18-1204CRB	AEI	SO-28	14	800 to 1 100	12	—
BTW32 series	M	†	26	800 to 1 200	25	200
2N3653 series	RCA	TO-48	35*	100 to 600	15	200
TA7395 series	RCA	TO-48	40*	200 to 600	15	200
CR70-102LA/C to CR70-1202LA/C	AEI	SO-38	60	75 to 1 100	25	—
BTW33 series	M	TO-94	65	800 to 1 200	25	200
CR120-102LC to CR120-1202LC	AEI	SO-38	105	75 to 1 100	25	—
CR707SCO101M to CR707SC1518M	AEI	†	200	75 to 1 400	30	—

* $I_{T(RMS)}$ value.
† See manufacturer's literature for details.

Table 8.32 TRIACS
(Ascending current order)

Type	Manufacturer	Outline	$I_{T(RMS)}$ (A)	I_{TSM} (A)	V_{DRM} (V)	V_{GT} (V)	I_{GT} (mA)	*dV/dt commutating* (V/μs)
TRC1′OOE series	AEI	TO-39	1.6	14	50 to 600*	4.0	40	4
2N5757 series	RCA	TO-5	2.5	25	100 to 600	—	40	8
TRC6′OOB series	AEI	TO-48	6.0	60	100 to 600*	3.0	50	2
40430 series	RCA	TO-66	6.0	100	200 to 600	—	40	10
TIC226 series	TI	TO-66(P)	8.0	70	200 and 400	2.5	50	—
TRC10′OOB series	AEI	TO-48	10	80	100 to 600*	3.0	50	2
2N5574 series	RCA	—	15	100	200 to 600	—	80	10
TIC253 series	TI	TO-3(P)	20	150	200 to 600	2.5	50	—
BTX94 series	M	TO-48	25	—	100 to 1 200	3.0	150	—
TIC263 series	TI	TO-3(P)	25	175	200 to 600	2.5	50	—
TRC25′OOB series	AEI	TO-48	25	250	100 to 600*	3.0	50	2
40672 series	RCA	—	30	300	200 to 600	—	80	20
BTW44 series BTW34 series	M	†	50	—	100 to 600 600 to 1 200	2.5	200	—
40919 series	RCA	—	80	850	200 to 600	—	150	10

* V_{RRM} value.
† See manufacturer's literature for details.

Table 8.33 SILICON CONTROLLED SWITCHES (TETRODE THYRISTORS)

Type	Manufacturer	Outline	Maximum ratings V_{GaK} (V)	V_{GaA} (V)	I_A (mA)	P_{tot} (mW)	V_A (V)	at I_A (mA)
BRY39	M	TO-72	70	−70	175	275	<1.4	100
BRY46	ITT	Plastic †	—	—	50*	330	<1.4	50
BRY56	M	Plastic †	70	−70	175	300	<1.4	100

* $I_{A(av)}$ value.
† See manufacturer's literature for details.

FURTHER READING

The descriptions of the various solid-state devices have concentrated on the physical operation of the devices. The underlying semiconductor physics have been introduced only where necessary, and mathematical treatment has been kept to a minimum. Readers wishing to study the physics of solid-state devices, or requiring more specialised descriptions of the devices, are referred to the following.

INTRODUCTORY SEMICONDUCTOR PHYSICS

AHMED, H. and SPREADBURY, P. J., *Electronics for Engineers—An Introduction*, Cambridge University Press, London (1973)
KAMPEL, I. J., *Semiconductors—Basic Theory and Devices*, Newnes-Butterworth, London (1971)
MORANT, M. J., *Introduction to Semiconductor Devices*, 2nd edition, Engineering Science Monographs, Harrap, London (1970)

ADVANCED SEMICONDUCTOR PHYSICS

ALLEY, C. L. and ATTWOOD, K. W., *Semiconductor Devices and Circuits*, Wiley, New York (1971)

GROVE, A. S., *Physics and Technology of Semiconductors*, Wiley, New York (1967)

LINDMAYER, J. and WRIGLEY, C. Y., *Fundamentals of Semiconductor Devices*, Van Nostrand, Princeton (1965)

SZE, S. M., *Physics of Semiconductor Devices*, New York (1969)

MICROWAVE DIODES AND TRANSISTORS

MATTHEWS, P. A. and STEVENSON, I. M., *Microwave Components*, Modern Electrical Science Series, Chapman and Hall, London (1968). Chapter 12 contains a concise summary of microwave semiconductor diodes.

SHURMER, H. V., *Microwave Semiconductor Devices*, Pitman, London (1971)

Proc. I.E.E.E., **59**, No. 8 (1971). This journal contains a series of articles on microwave semiconductor devices

FIELD-EFFECT TRANSISTORS

WÜSTEHUBE, J., HANSE, J. J., OVERGOOR, B. J. M., BARHAM, J. J. and GARTERS, J. A., *Field-effect transistors*, Mullard, London (1972)

THYRISTORS

ROSE, M. J. (editor), *Power Engineering Using Thyristors*, Vol. 1, 'Techniques of Thyristor Power Control', Mullard, London (1970). Chapters 1 and 2 describe the thyristor and its use in control circuits.

SCR Manual, 5th edition, General Electric Company, New York (1972)

HISTORICAL

1973, being the twenty-fifth anniversary of the invention of the transistor, saw several articles commemorating the event. Useful references are:

SHEPHERD, A. A., 'Semiconductor device developments in the 1960s', *Radio and Electronic Engineer*, **43**, 1 and 2 (1973). Other articles dealing with various aspects of solid-state devices and their effects on the electronics industry are contained in this issue

DEAN, K. J. and WHITE, G., 'The Semiconductor Story', *Wireless World*, **79**, 1447–1450 (1973)

9 PHOTO-ELECTRONIC DEVICES

9 PHOTO-ELECTRONIC DEVICES

Semiconductor materials can be used to construct many practical devices which use or produce visible or infra-red radiation. The former are described in this Section under the heading of 'Photo-electronic Devices' and the latter in Section 10 under the heading of 'Electro-optic Devices'. They are based either on electronic effects resulting from the absorption of such radiation in the material (the photoconductive, photovoltaic, and photoemissive effects), or on the emission of radiation resulting from electronic action inside the material (electroluminescence, see Section 10).

Radiation falling on a semiconductor material can produce electron-hole pairs in the material which can be used as charge carriers. Thus the conductivity of the illuminated material is increased considerably. This is the photoconductive effect, and it can be used in various solid-state devices to detect visible and infra-red radiation.

If the electron-hole pairs are generated in or near a *pn* junction, the electrons and holes will be separated by the built-in electric field of the junction. An open-circuit voltage or a short-circuit current will be generated. This photovoltaic effect can be used in photodiodes to produce both voltage and current in an external circuit. A special form of photodiode used considerably in practice is the solar cell (**9**–42). Another device using the photovoltaic effect is the phototransistor. In the phototransistor, the photovoltaic current generated at the collector-base diode is amplified by the transistor action of the emitter.

In certain materials, the energy absorbed from the radiation is sufficient not only for the creation of electron-hole pairs but gives the freed electrons enough energy to be emitted from the material. This is the photoemissive effect which is used in such devices as photoemissive tubes, photomultipliers, and image intensifiers.

Just as the absorption of radiation can generate electron-hole pairs, the recombination of electrons and holes in certain semiconductor materials can cause emission of radiation. This is the effect used in electroluminescent (light-emitting) diodes (see Section 10).

Finally, the absorption of radiation increases the temperature of the material, and various detectors have been devised using this thermal effect. One type of detector used at present is based on the pyroelectric effect, the change of electrical polarisation with temperature, observed with certain complex crystals.

PHOTOCONDUCTIVE DEVICES

Photoconductive effect

When discussing opto-electronic effects, it is convenient to regard light not as an electromagnetic radiation but rather as a stream of elementary particles. These particles are called photons. Each photon contains a certain amount of energy called a quantum. The energy of a single photon is given by the simple relationship:

$$E = h\nu, \tag{1}$$

where E is the energy of the photon, ν is the frequency of the radiation, and h is Planck's constant (6.62×10^{-34} joule-second).

When the radiation is absorbed by a material, the energy of the electrons in the material is raised by the photon energy. However, the electrons in the material can have

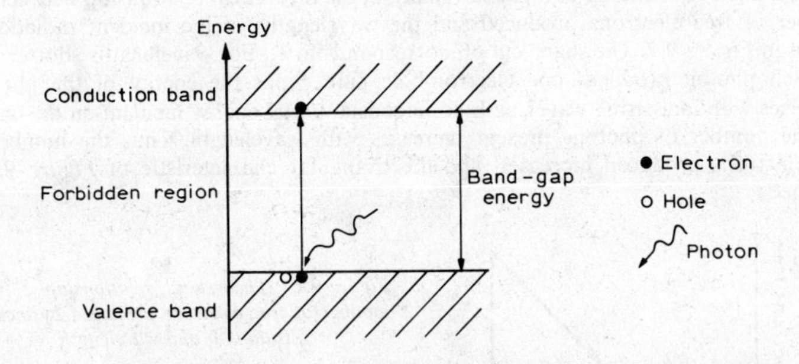

Figure 9.1 Permissible energy levels for electron in semiconductor material, generation of electron-hole pair

only certain energy levels, so the radiation is absorbed only when the photon energy can raise the electron from one permissible energy level to another. In semiconductor materials, the two energy levels or bands of interest are the valence band where the electrons are essentially bound to the parent atoms, and the conduction band where the electrons are free and so can be used as charge carriers. The difference between these two energy levels is called the band-gap energy, and is the minimum energy that will generate free carriers. Photons of energy less than the band-gap energy will not be absorbed by the material.

The permissible energy levels in a semiconductor material can be represented by an energy-band diagram, as shown in *Figure 9.1*.

If the energies of photons in the visible and infra-red regions are calculated using equation 1, it is found that the energies equal or exceed the band-gap energies of many semiconductor materials. Therefore in such materials, illumination with visible or infra-red radiation lowers considerably the resistance of the material compared to the 'dark' resistance. This is the operating principle of all solid-state photoconductive detectors. These detectors are called quantum detectors since they depend for their operation on the quantum nature of the radiation.

It is customary to describe radiation in terms of wavelength rather than frequency. Since wavelength and frequency are related by the speed of light c:

$$\nu \times \lambda = c,$$

where λ is the wavelength, equation 1 can be rewritten as

$$E = \frac{hc}{\lambda}. \qquad (2)$$

If the minimum band-gap energy for a particular semiconductor material is represented by E_g, equation 2 can be rearranged in terms of a critical wavelength λ_c corresponding to this minimum energy:

$$\lambda_c = \frac{hc}{E_g}. \qquad (3)$$

This equation can then be used to determine the wavelength of photons whose energy corresponds to the minimum band-gap energy. Since the energy of the photon is inversely proportional to wavelength, λ_c represents the longest wavelength at which photons contain sufficient energy to produce free carriers within the material. Radiation with wavelengths greater than λ_c do not produce free carriers; with a wavelength shorter than λ_c, the radiation is absorbed and free carriers are produced.

For a fixed total energy of incident radiation, the theoretical relationship between the number of free electrons produced and the wavelengths of the incident radiation is shown in *Figure 9.2*. The sharp cut-off corresponds to λ_c. For wavelengths shorter than λ_c, each photon produces one electron-hole pair. Since the energy of the photons decreases with increasing wavelength, for a constant total energy incident on the material, the number of photons present increases with wavelength. Thus the number of free electrons produced increases, and the triangular characteristic of *Figure 9.2* is obtained.

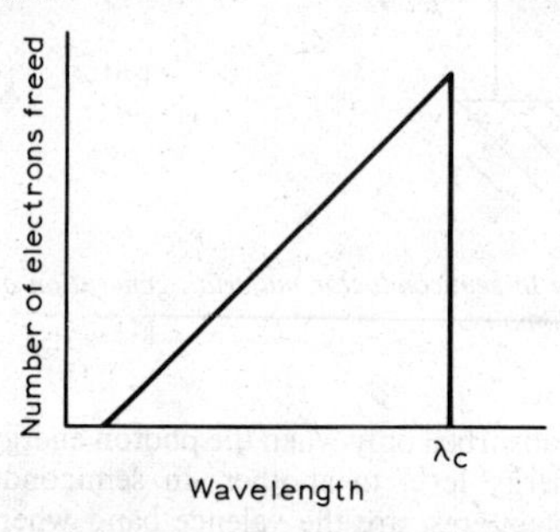

Figure 9.2 Theoretical relationship between number of free electrons produced by incident radiation and wavelength

The higher energy of the photons with shorter wavelengths may excite the electron sufficiently to not only free it from the parent atom, but also to exceed the work function of the material and so allow it to escape from the surface. Obviously such electrons cannot be used as charge carriers within the material, and so a second limiting wavelength is set for the photons λ_m, corresponding to the energy which will cause the electron to be emitted from the surface. An equation similar to equation 3 can be used to relate λ_m to this energy E_t:

$$\lambda_m = \frac{hc}{E_t}. \tag{4}$$

The wavelengths λ_c and λ_m represent the limiting wavelengths of the incident radiation for use with a particular material. The theoretical relationship between the number of usable charge carriers produced and the wavelength of the radiation is shown in *Figure 9.3*. This curve is called the spectral response of the material. The values of λ_c and λ_m can be calculated from equations 3 and 4 using the known values of E_g and E_t for the

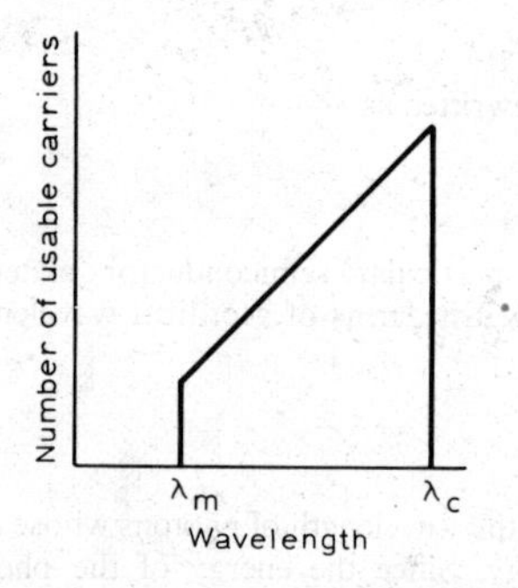

Figure 9.3 Theoretical relationship between number of usable charge carriers produced by incident radiation and wavelength (spectral response)

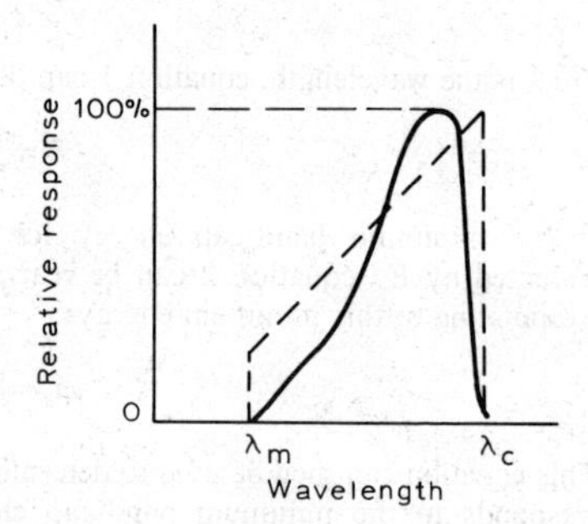

Figure 9.4 Practical spectral response (full line) compared with theoretical response (broken lines)

material. For example, the values of λ_c for silicon and germanium are calculated as 1.13 μm and 1.77 μm respectively. The range of wavelengths for visible light is from approximately 0.40 μm (violet) to 0.70 μm (red). Therefore both silicon and germanium can be used in devices operating with visible and near-infra-red radiation.

A practical spectral response is shown in *Figure 9.4* with the theoretical curve in broken lines for comparison. The cut-off wavelengths are the same for both curves since they are determined by the semiconductor material itself. The peak of the practical response, however, is shifted to a shorter wavelength. The shape of the practical curve differs from the theoretical one because the energy-band diagram from which the theoretical curve is derived is an oversimplification of what occurs in practice, and because of surface effects in the semiconductor material. These effects combine to produce the shape of the practical response curve of *Figure 9.4*.

The peak of the response curve can be controlled to some extent by doping the semiconductor material. Adding an impurity to the material creates an intermediate state for the electron between the valence and conduction states. This intermediate state is shown as the trapping level in the energy-band diagram of *Figure 9.5*. Thus an

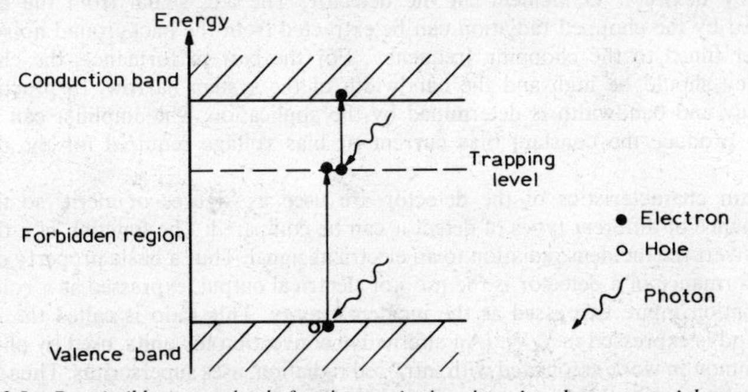

Figure 9.5 Permissible energy levels for electron in doped semiconductor material, generation of electron-hole pair in two stages

electron can be freed by a photon whose energy is less than the band-gap energy, but sufficient to raise the electron to the trapping level. A second photon can then impart further energy to the electron to transfer it from the trapping level to the conduction band. This two-stage freeing of an electron makes a hole available as a charge carrier before the electron is available. This energy effect combines with the effects mentioned previously to change further the theoretical response curve.

If a polycrystalline semiconductor material is used, the peak of the practical response curve is flattened so that a wider peak response than that shown in *Figure 9.4* is obtained. Polycrystalline material, however, contains many traps for the generated charge carriers so that a device manufactured from such material will have a slower response time than one using monocrystalline material.

It can be seen from the preceding discussion that by choosing a particular semiconductor material, and by choosing a suitable doping level, a practical device can be constructed to respond to a particular range of wavelengths.

For radiation of a particular wavelength or fixed range of wavelengths, the greater the energy the higher is the number of photons incident on the irradiated material, and hence the higher the number of charge carriers generated within the material. For visible radiation (light), the quantity of light falling on to a given surface is called the *illumination*. The unit of illumination is the lux, defined as the illumination produced when 1 lumen of luminous flux falls on an area of 1 square metre. Thus the fall in resistance of the irradiated semiconductor material is proportional to the illumination of the incident light. For infra-red radiation, it is usual to measure the radiation in terms of the power incident on an area. In this case, the fall in resistance of the irradiated material will be proportional to the radiation power per unit area.

Photoconductive detectors

The performance of a photoconductive detector is limited by noise. The noise in the detector itself is produced by the thermally-generated carriers in the semiconductor material. At very low levels of incident radiation, the thermally-generated carriers may swamp the photo-generated carriers. The effect of the thermally-generated carriers can be minimised by cooling the detector, if necessary to liquid-nitrogen or liquid-helium temperatures. There is, however, a theoretical limit to the performance of the detector set by the thermal background radiation. This radiation is noisy in character, and gives rise to a limiting noise in the detector.

Another technique to improve the performance when the signal produced by the incident radiation is near the noise level is to chop the radiation at a particular frequency before it is incident on the detector. The a.c. signal from the detector produced by the chopped radiation can be extracted from the background noise by an amplifier tuned to the chopping frequency. For the best performance, the chopping frequency should be high and the bandwidth of the system narrow. In practice, the frequency and bandwidth is determined by the application. The amplifier can also be used to produce the constant bias current or bias voltage required for the detector element.

Certain characteristics of the detector are used as 'figures of merit' so that the performance of different types of detector can be compared. The function of a detector is to convert the incident radiation to an electrical signal. Thus a basic property defining the performance of a detector is the ratio of electrical output, expressed as a voltage, to the radiation input, expressed as the incident energy. This ratio is called the responsivity, and is expressed in V/W. (An alternative convention for units, used by physicists and common in work associated with infra-red radiation, uses superscripts. Thus the unit for responsivity in this convention is VW^{-1}.) For detectors that use a constant bias voltage, a current responsivity is used, expressed as AW^{-1}.

Various noise figures can be quoted for the detector. The noise itself of the detector is usually expressed as the r.m.s. value of the electrical output measured at a bandwidth of 1 Hz under specified test conditions. Although a low noise enables smaller radiation levels to be detected, it can make the design of suitable amplifiers difficult because the amplifier noise must be low. It is therefore often important to know the ratio of detector noise to the value of Johnson noise in a resistor at room temperature equal in magnitude to the detector. This ratio may be quoted as a noise factor. A signal-to-noise ratio can be measured, but this will apply only for the test conditions under which it was obtained. A more useful quantity, which is often used as a figure of merit for a detector, is the noise equivalent poser or N.E.P. This is the amount of energy that will give a signal equal to the noise in a bandwidth of 1 Hz. It is, in general, a function of the wavelength and the frequency of the measurement. The N.E.P. is equal to the noise per unit bandwidth divided by the responsivity, and is given by the expression:

$$\text{N.E.P.} = \frac{W}{\dfrac{V_s}{V_n}[\Delta f]^{\frac{1}{2}}},$$

where W is the radiation power incident on the detector (r.m.s. value in watts), V_s is the signal voltage across the detector terminals, V_n is the noise voltage across the detector terminals, and Δf is the bandwidth of the measuring amplifier in Hz. The units of N.E.P. are $WH^{-\frac{1}{2}}$.

Both responsivity and N.E.P. vary with the size and the shape of the active element of the detector. It is found in practice, and supported by theory, that for similarly-made detectors the N.E.P. is often proportional to the square root of the area of the active element. A quantity called the area normalised detectivity or D* is commonly used, which is related to N.E.P. by the expression:

$$D^* = \frac{[A]^{\frac{1}{2}}}{\text{N.E.P.}},$$

where A is the area of the active element of the detector in cm^2. The units of D* are $cmHz^{\frac{1}{2}}W^{-1}$.

When responsivity, N.E.P., and D* are quoted in published data, certain figures are given in brackets after the quantity. Typical examples of this are responsivity (5.3 μm, 800, 1) and D* (5.3 μm, 800, 1). These figures refer to the test conditions under which the value was measured. The figure 5.3 μm refers to the wavelength of the monochromatic radiation incident on the detector, 800 to the modulation (chopping) frequency in Hz of the radiation, and 1 to the electronic bandwidth in Hz. An alternative form of the test conditions is given in terms of black-body radiation. An example of this is D* (500 K, 800, 1). The 500 K refers to the temperature of the black body from which the incident radiation was obtained. The other figures refer to the modulation frequency and bandwidth as before. Details of the other test conditions under which the quantities were measured are also given in published data. These conditions include the distance between the detector element and the source of the radiation, the aperture through which the radiation reaches the element, the operating temperature of the element, details of the chopper system producing the modulation of the radiation, and the bias conditions of the element. The electrical output signal from the detector is amplified by an amplifier tuned to the modulation frequency, which is typically 800 Hz, with a bandwidth of typically 50 Hz.

Both signal and noise vary with the bias current through the element, and so the responsivity, N.E.P., and detectivity also varies with bias current. At high bias currents the noise increases more rapidly than the signal, and therefore the signal-to-noise ratio has a peak value at some current. The form of the variation of responsivity, noise, and detectivity with bias current for a typical detector is shown in *Figure 9.6*. An optimum

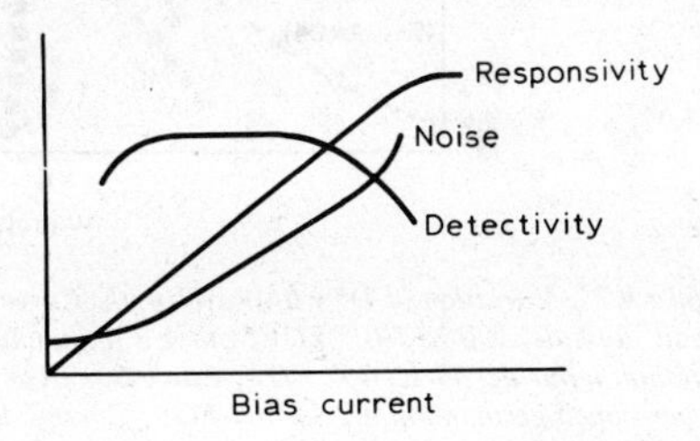

Figure 9.6 Variation of responsivity, detectivity and noise with bias current in infra-red photoconductive detector

value of bias current for the detector can be chosen from these curves. Variations between detectors of the same type may sometimes occur, and so for highly-sensitive applications a fine adjustment of the bias current may be necessary to obtain the optimum performance from the detector. Curves similar to those of *Figure 9.6*, or of the quantities plotted separately, are sometimes given in published data. Similar information may be given in a different form; for example, as the variation of signal-to-noise ratio with bias current.

The values of responsivity, N.E.P., and D* will vary with the wavelength of the incident radiation through the variations of signal and noise with wavelength. The variation of D* can be used as an indication of the spectral performance of the various types of detector, as shown in *Figure 9.7*. The range of wavelengths over which the various semiconductor materials operate can be seen, together with the relative sensitivities. Also shown in *Figure 9.7* is the theoretical limit to operation set by the thermal background radiation. This radiation, as previously explained, limits the attainable detectivity to what is called the background limited value. For wavelengths up to approximately 10 μm, the background limited value falls with increasing wavelength, and so imposes a maximum value on D*.

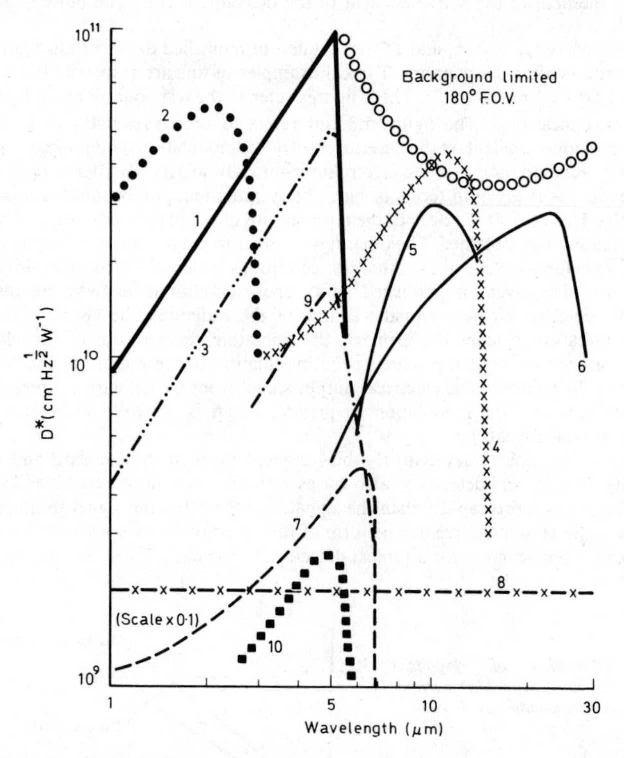

Figure 9.7 Variation of D with wavelength: Curve 1 indium antimonide, 77 K, 60° FOV; Curve 2 lead sulphide, 300 K, 180° FOV; Curve 3 indium antimonide, 77 K, 180° FOV; Curve 4 mercury cadmium telluride, 77 K, 60° FOV; Curve 5 mercury-doped germanium 35 K, 60° FOV; Curve 6 copper-doped germanium 4.2 K, 60° FOV; Curve 7 indium antimonide, 300 K (scale ×0.1); Curve 8 TGS pyroelectric detector; Curve 9 mercury cadmium telluride, 193 K; Curve 10 mercury cadmium telluride, 295 K*

One curve in *Figure 9.7* is shown operating above the theoretical limit. This is because the detector has a cooled aperture restricting the field of view and hence limiting the amount of background radiation received. Because the field of view (FOV) affects the performance, the FOV is stated as well as the temperature for the curves.

The material most suitable to detect radiation of a given wavelength is usually one with the peak spectral response equal to or slightly longer than the given wavelength. Detectors sensitive to wavelengths much longer than the given wavelength require more cooling to reduce the internal thermal noise. It has been shown previously that silicon and germanium are sensitive to radiation of wavelengths up to 1.13 μm and 1.77 μm respectively. These values lie within the infra-red region, but silicon and germanium detectors are generally used for visible radiation. These detectors, in the form of photodiodes and phototransistors, are discussed separately. The material used to detect visible radiation is cadmium sulphide CdS. The most commonly used materials for the detection of infra-red radiation are lead sulphide PbS, for radiation from the visible to that with a wavelength of 3.0 μm; indium antimonide InSb, visible to 5.6 μm when cooled, visible to 7.0 μm at room temperature; mercury-doped germanium Ge : Hg, from 2.0 μm to 13 μm; mercury cadmium telluride CdHgTe, often called MCT, from below 3 μm to 15 μm; and copper-doped germanium Ge : Cu, from 2.0 μm to 25 μm.

The choice of material for an application may often be influenced by the cooling required. Although detectors requiring cooling to liquid-nitrogen or liquid-helium temperatures are acceptable in scientific work, there are considerable disadvantages in their use for industrial applications. Thus in industrial applications, a less-sensitive detector that can operate at room temperature or with a thermoelectric cooler may be used in preference to a more-sensitive or cheaper detector requiring greater cooling.

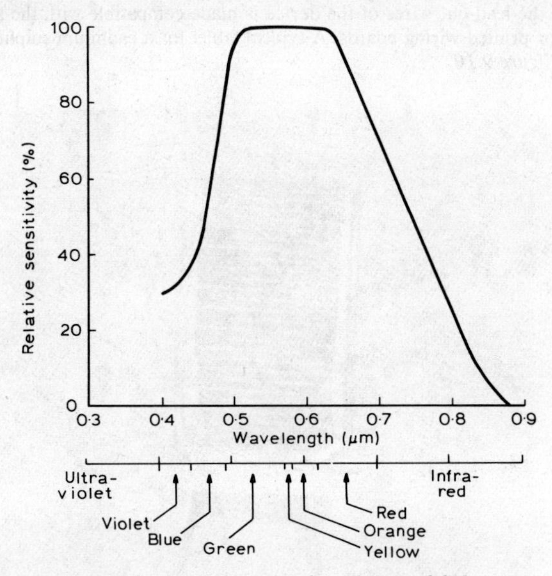

Figure 9.8 Spectral response for cadmium sulphide

PHOTOCONDUCTIVE DETECTOR FOR VISIBLE RADIATION

The semiconductor material used in photoconductive detectors for visible light is cadmium sulphide CdS. The spectral response for this material is shown in *Figure 9.8*. The detector is a two-terminal device, and is used as a light-dependent resistor. The circuit symbol for a photoconductive detector is shown in *Figure 9.9*, and consists of a resistance symbol with incident arrows representing the radiation.

The active element of the detector is usually made by sintering cadmium sulphide powder into a ceramic-like tablet. Metal electrodes are deposited on to the surface of the tablet to form non-rectifying contacts. The material used has a low resistance compared with that of the cadmium sulphide. An interdigitated structure is used for the contacts, fingers of the contacts connected to one terminal of the detector interleaving fingers connected to the other terminal. The resistance and voltage ratings of the detector are determined by the contact structure, and so detectors with different ratings can be made using the same cadmium sulphide tablet but with different electrode structures. For example, a detector with a small number of widely-spaced contact fingers has a higher resistance and voltage rating than a detector with many closely-spaced contact fingers.

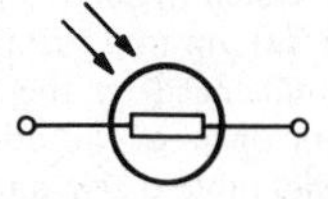

Figure 9.9 Circuit symbol for photoconductive detector (light-dependent resistor)

The tablet is encapsulated in a glass or transparent plastic envelope. The shape of the encapsulation depends on the direction of the incident light, whether an end-viewing or side-viewing device is required, and the method of termination. Some encapsulations are specially designed for mounting the detector directly on to a printed-wiring board. The spacing of the lead-out wires of the device is made compatible with the standard 2.54 mm grid for printed-wiring boards. A typical tablet for a cadmium sulphide detector is shown in *Figure 9.10*.

Figure 9.10 Typical tablet for cadmium sulphide photoconductive detector (Photo: Mullard Limited)

A second and more recent manufacturing process for the sensitive element uses individual grains of cadmium sulphide, the so-called monograin process. The grains, about 40 μm in diameter, are inserted into a thin insulating sheet of a synthetic material so that each grain projects both sides of the material. Each grain is insulated from its neighbours by the sheet, but electrical connection is made by evaporating gold contact pads on to both sides of the sheet. Terminal wires are connected to these contact pads. The complete element is then encapsulated in a transparent plastic. This method of construction enables smaller detectors to be manufactured.

The electrical characteristics of cadmium sulphide detectors depend on several factors. Some of these may be considered as 'direct' factors: the illumination, the wavelength of the incident radiation, the temperature, and the device voltage and current. Other factors, however, affect the operation of the device: the time it has been kept in darkness or the time it has been operating in a circuit, and the operation of the device in the previous 24 hours. The characteristics in the published data are therefore given for specified operating conditions, and the characteristics may vary for different operating conditions. Typical conditions specified in published data could be as follows: operation under d.c. conditions, at the start of life, in an ambient temperature of 25 °C, and with an illumination colour temperature of 2700 K. 'Colour temperature' is the temperature to which a black body must be heated to give a similar colour sensation to the source being considered. Sometimes a preconditioning is specified in the data, consisting of an illumination of the detector at a specified level for a specified time before the measurements given in the data are made.

The characteristics of most interest to the circuit designer are the fall of resistance with illumination and the device response time. Values of the 'dark' and 'illuminated' resistance are given under further conditions besides the general ones already stated. For the dark resistance, which is very high, the value is specified with a d.c. voltage applied across the device in series with a high resistance, and after the device has been in darkness for a specified time. Often two values of dark resistance are given, one after a

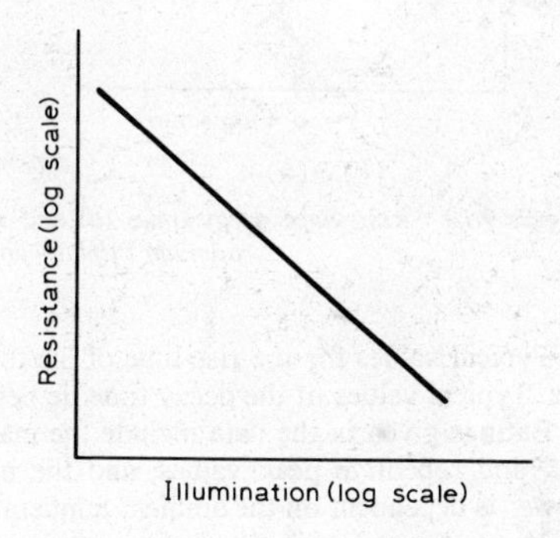

Figure 9.11 Resistance plotted against illumination for cadmium sulphide photoconductive detector

short time in the dark such as 20 s, and the other after a longer time such as 30 min. The difference between these two values of dark resistance can be considerable, the value after the longer time in the dark being at least ten times greater than the value after the shorter time. Two values are also given for the illuminated resistance. An initial value is given under a specified illumination (usually 50 lux) and an applied voltage immediately after removing the device from storage in darkness for 16 hours. (This time allows the chemical changes in the cadmium sulphide to reach equilibrium.) A second value is given after the device has been illuminated for a time such as 15 min. The difference between the two values of illuminated resistance is small. A curve of resistance plotted against illumination is generally given in published data, plotted on logarithmic scales to give a straight line as shown in *Figure 9.11*. Over an illumination range of 1 to 10 000 lux, the resistance value varies typically by three decades. A 'typical' resistance value at 50 lux is often quoted so that different types of detector can be compared.

Resistance rise and decay times are specified as the time taken to reach a given resistance value within the range of the device. The resistance decay time is specified as the time for the resistance to fall to a given value from the instant of switching on a given illumination (usually 50 lux) after the device has been in darkness for 16 hours. The resistance rise time is the time for the resistance to reach a given value from the instant of switching off an illumination of 50 lux after the device has been illuminated for a specified time. Curves of the rise and decay times with illumination as parameter are sometimes given in published data. The form of such curves is shown in *Figure 9.12*. Current rise and decay times may sometimes be specified. These are the times for the current to rise to 90% of the maximum value, or fall to 10% of the maximum value, after the switching off or on of a given illumination, with a constant voltage across the device. The conditions under which the quoted values are valid are the same as for the resistance times.

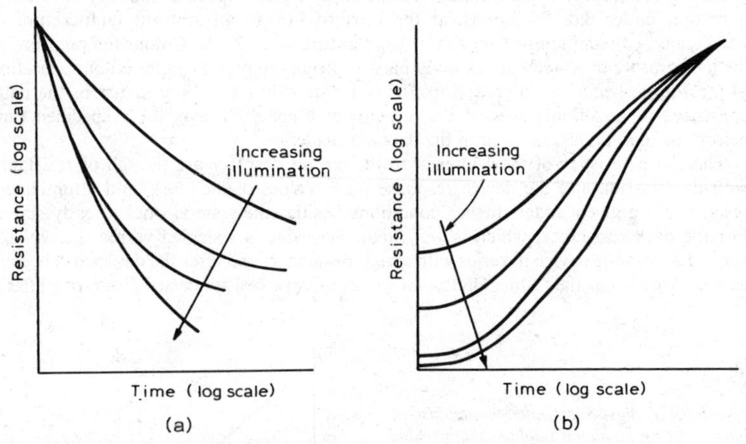

Figure 9.12 Resistance decay times (a) and rise times (b) with illumination as parameter for cadmium sulphide photoconductive detector

Typical values for the rise time of photoconductive detectors lie between 75 ms and 2 s. Typical values of the decay time lie between 25 ms and 3 s.

Ratings given in the data include the maximum device voltage and current, given as d.c. and repetitive peak values, and the maximum device power dissipation. As the power is dependent on the ambient temperature, a power derating curve may sometimes be given. A maximum illumination is also specified.

The material used to encapsulate the active element may affect the spectral response of the detector, causing it to differ from the response for cadmium sulphide shown in *Figure 9.8*. In these cases, a spectral response for the detector may be given in the published data.

When a cadmium sulphide detector is used in a practical circuit, it can be used either as an 'on/off' device or some intermediate resistance value can be used as a trigger level. In the on/off type of operation, the device is used to detect the presence or absence of a light source. The change of resistance between the illuminated and dark values can be used to operate a relay to initiate further action. Applications of this type include alarm systems operated by the interruption of a beam of light directed on to the detector, or counting systems where objects, say on a conveyor belt, interrupt a light beam to produce a series of pulses which operates a counter. In the second type of application, the detector is used to measure the light level. The resistance of the device corresponding to a predetermined light level is used as a threshold value to trigger another circuit. An example of this type of application is a twilight switching circuit. When the daylight has faded to a given level, the corresponding resistance of the detector causes another circuit to switch on the required lights.

PHOTOCONDUCTIVE DETECTORS FOR INFRA-RED RADIATION

The operating principle of photoconductive detectors for use with infra-red radiation is the same as that of detectors for visible radiation just described. Charge carriers are generated in the illuminated material, and the consequent fall in resistance provides a measure for the incident radiation. Various semiconductor materials can be used to detect infra-red radiation, enabling detectors to be chosen to suit particular wavelengths. Some detectors can be used at room temperature, while others must be cooled to low temperatures, for example with liquid nitrogen or liquid helium, to minimise the effects of thermally-generated carriers (thermal noise).

Infra-red photoconductive detectors are usually operated with a constant bias current through the active element so that the change in resistance when illuminated appears as a voltage change. For the more sensitive detectors, this voltage can be amplified to produce a larger output signal. Some detectors are operated with a constant bias voltage across the element. For these, the change in resistance produces a change in current.

The infra-red detectors that are operated at low temperatures inside some form of cooling vessel require a window through which the incident radiation reaches the active element. The transmission properties of the window may affect the spectral response, causing it to differ from that of the material of the active element. The spectral response of the detector may therefore be given in published data. A time-constant is also given. This is based on the response of the detector to a sudden application of the incident radiation. The time-constant is defined as the time taken from the application of the radiation for the output from the detector to fall to 63% of the peak value. Typical values of time-constant lie between 0.1 μs and 350 μs. Other characteristics that may be given in published data include the variations of responsivity, D*, and noise with modulation frequency and the operating temperature of the detector.

The ratings given in the data are similar to those already discussed for photoconductive detectors used with visible radiation. They include the maximum detector power, maximum bias current, and the maximum operating and storage temperatures.

The construction of infra-red photoconductive detectors depends on the material used for the active element, in particular whether or not it requires cooling. Detectors for operation at room temperature can be encapsulated in a metal envelope with a viewing window, or the element can be deposited on a flat plate. Detectors that require cooling can be deposited on a dewar vessel or arranged for mounting on to a cooling vessel. For detectors that use thermoelectric cooling, the active element is mounted on the cooler in a suitable encapsulation.

Lead sulphide detectors are of two types, depending on the method used to form the active element. The element is deposited as a film either by evaporation or by a chemical reaction. The evaporated-film type can be operated at room temperature or cooled; the chemically-deposited type is generally operated at room temperature only.

A typical construction used for the evaporated-film type is to form the active element on an inner surface of a dewar vessel, as shown in *Figure 9.13*. A metal housing can be used to protect the dewar vessel if the detector is to be used at room temperature only, while for operation below room temperature the dewar vessel can be filled with a suitable coolant. The operating temperature range of this type of lead sulphide detector is from 293 K to 193 K (+20 °C to −100 °C), while the responsivity has a peak value at 240 K and D* has a peak value at 220 K. A typical value of responsivity (500 K, 800, 1) is $2.0 \times 10^3 VW^{-1}$, and typical values of D* (2.0 μm, 800, 1) are 4.0×10^{10} $cmHz^{\frac{1}{2}}W^{-1}$ at 293 K and 2.0×10^{11} $cmHz^{\frac{1}{2}}W^{-1}$ at 230 K. The range of wavelengths over which this detector can be used is from the visible region to 3.0 μm with a peak response at 2.3 μm. The time-constant is typically 100 μs.

Because the chemically-deposited lead sulphide detector is generally operated at room temperature, the construction does not have to allow for cooling. Typical constructions used for this detector are encapsulation in a small metal envelope such as the TO-5 outline, or as a 'flat pack' in which the element is deposited on an insulating substrate. Both types of construction can incorporate a filter to modify the spectral response. This type of detector is operated with a constant bias voltage, and has a typical current responsivity (2.0 μm, 800, 1) of 200 mAW^{-1}. A typical value of D* (2.0 μm, 800, 1) is 1.0×10^{10} $cmHz^{\frac{1}{2}}W^{-1}$. The range of wavenlengths is from the visible to 2.8 μm, but this can be restricted by using a germanium window to 1.5 μm to 2.8 μm. The time-constant is typically 250 μs.

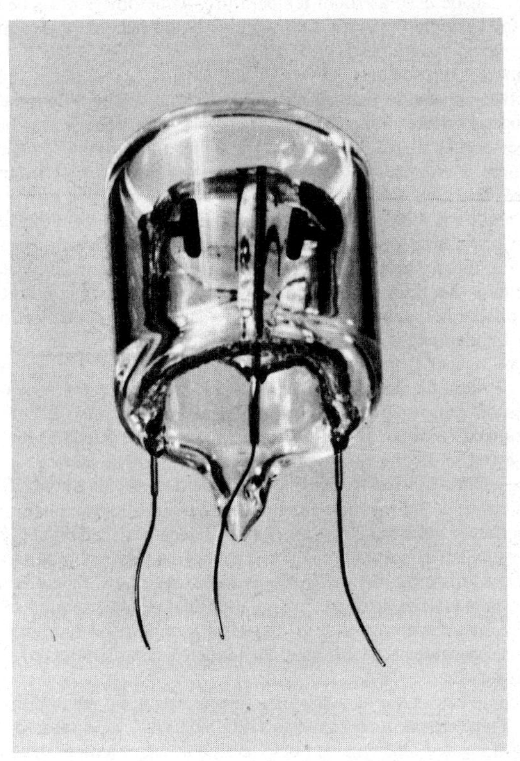

Figure 9.13 Lead sulphide photoconductive detector. The lead sulphide film is deposited on an inner surface of the dewar vessel (Photo: Mullard Limited)

Both types of lead sulphide detector have a high resistance compared to detectors using other materials. A typical resistance for the evaporated-film type is 1.5 MΩ, and for the chemically-deposited type 200 kΩ.

Indium antimonide detectors can be operated at room temperature, or cooled by liquid nitrogen to 77 K at which temperature a detectivity near the background limited value is obtained. The photoconductive material is a doped single crystal which is cut into thin slices from which the active element is cut. At room temperature the resistivity of indium antimonide is low, giving detectors with resistances of about 5Ω/square. The most useful forms of element are therefore long strips or 'labyrinths' made from a number of long strips laid in parallel but connected electrically in series. The effect of temperature on performance is considerable at room temperature, and so a good heatsink is required for the element. Typical constructions for room-temperature indium antimonide detectors mount the element either in a copper block or in a flat-pack encapsulation which can be provided with a viewing window (usually sapphire) if it is necessary to protect the element from dirty or corrosive atmospheres. A typical value of responsivity (6.0 μm, 800, 1) is between 1.0 VW^{-1} and 3.5 VW^{-1}, and a typical value of D* (6.0 μm, 800, 1) is 1.5×10^8 to 3.0×10^8 $cmHz^{\frac{1}{2}}W^{-1}$. The range of wavelengths is from the visible region to 7.0 μm, and the time-constant is typically 0.1 μs.

The elements for cooled indium antimonide detectors are made in a similar way to the room-temperature types, but are mounted on the inner surface of a glass dewar vessel. The radiation is transmitted to the element through a sapphire window. The

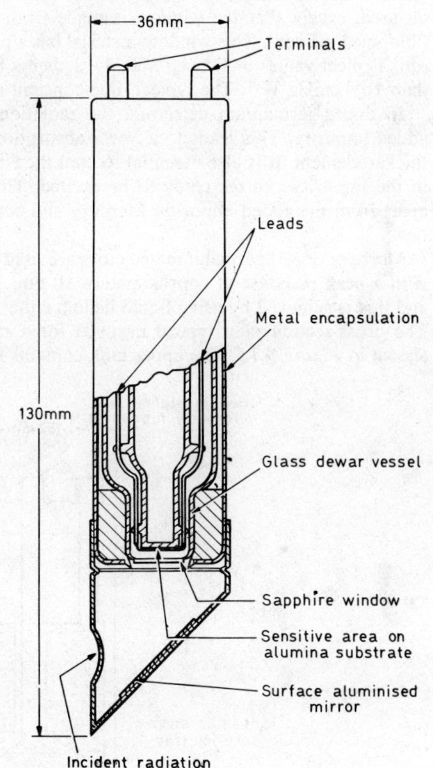

Figure 9.14 Construction of cooled indium antimonide photoconductive detector for operation at 77 K

element is cooled with liquid nitrogen either by filling the dewar vessel with the liquid or by using a miniature Joule-Thomson cooler. Cooling increases the resistance of the element to about 2 kΩ/square, and a wide range of shapes and sizes can be made for the element. Arrays as small as 0.1 mm square are possible. Because the detectors of this type are 'background limited', the use of a cooled aperture improves the detectivity. Typical values of responsivity (5.3 μm, 800, 1) and D* (5.3 μm 800, 1) are 1.2×10^4 VW^{-1} and 5.0×10^{10} $cmHz^{\frac{1}{2}}W^{-1}$ respectively. With a restricted field of view, the value of D* (5.3 μm, 800, 1) can be increased to 1×10^{11} $cmHz^{\frac{1}{2}}W^{-1}$. The range of wavelengths is from the visible to 5.6 μm, and the time-constant is typically 2μs to 5 μs.

The cross-section of a typical construction for a cooled indium antimonide detector is shown in *Figure 9.14*.

Mercury cadmium telluride is an alloy semiconductor material which forms a mixed crystal of cadmium telluride and mercury telluride. The peak spectral response of the material can be varied by the relative proportions of cadmium and mercury telluride in the crystal from 9.5 μm to 15 μm. Mercury cadmium telluride (MCT) detectors can be used to cover a range of wavelengths from approximately 3 μm to 15 μm. Detector elements are made in a similar way to those of indium antimonide just described, being cut from thin slices of a single crystal. MCT detectors can be operated at room temperature, with thermoelectric coolers at temperatures down to approximately 200 K, or with liquid nitrogen at 77 K. For operation with liquid nitrogen, a construction for the detector similar to that for the indium antimonide type shown in *Figure 9.14* can be used, except that the window is made from silicon with an anti-reflective coating ('bloomed' silicon). This window material has a peak transmission between 9 μm and 11 μm. Typical values of D* (λ_p, 800, 1), λ_p being the peak spectral response, are greater than 10^{10} cmHz$^{\frac{1}{2}}$W^{-1}. The typical time-constant is less than 1 μs.

In doped-germanium detectors, the radiation is absorbed by the electrons in the added impurity. This leads to a lower absorption coefficient, and hence the need for a thicker element. It is also essential to cool the element so that the electrons are initially in the impurity centres ready to be excited. The germanium must be extremely pure apart from the added impurity. Mercury and copper are the most widely used impurities.

Mercury-doped germanium detectors are used for wavelengths from 20 μm to 13 μm with a peak response at approximately 10 μm. The element requires cooling to 35 K, and this is achieved by using liquid helium either in bulk or in a Joule-Thomson cooler. The cross-section of a typical cryostat for a mercury-doped germanium detector is shown in *Figure 9.15*. The upper tank contains liquid nitrogen, and carries a radiation

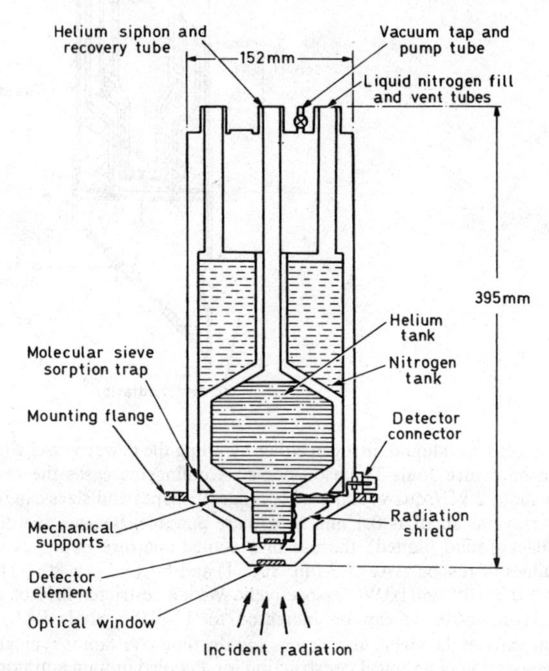

Figure 9.15 Construction of cryostat for doped-germanium photoconductive detectors operating at liquid-helium temperatures

shield which partly surrounds the lower liquid-helium tank. This minimises evaporation of the helium, and also forms a shield for the detector element. A vacuum is maintained inside the cryostat, and the quality of the vacuum is maintained by a molecular sieve trap incorporated in the radiation shield. The cooled window through which the radiation is transmitted to the element is made of bloomed germanium. The resistance of the element is typically between 10 kΩ and 60 kΩ. A typical value of D* (10 μm, 800, 1) is 1.3×10^{10} cmHz$^{\frac{1}{2}}$W^{-1}, and the time-constant is typically less than 1 μs.

Copper-doped germanium detectors are used for wavelengths between 20 μm and 25 μm with a peak response at 15 μm. This type of detector requires cooling to 4.2 K, and this is achieved with liquid helium in a cryostat similar to that shown in *Figure 9.15*. The resistance of the element is typically between 2.5 kΩ and 240 kΩ. A typical value of D* (15 μm, 800, 1) is 10×10^{10} cmHz$^{\frac{1}{2}}$W^{-1}, and the time-constant is typically 1 μs.

Photoconductive detectors have been developed for specialised applications of infra-red radiation with wavelengths extending beyond 50 μm into the sub-millimetric and microwave regions. A typical detector for these wavelengths uses the impurity level in indium antimonide, and operates at a temperature below 2 K. A complex cooling system is required for such a detector, and the cross-section of a typical cryostat is shown in *Figure 9.16*. Two glass dewar vessels are used: an inner vessel containing liquid helium which is pumped to reduce the pressure inside the vessel and hence lower

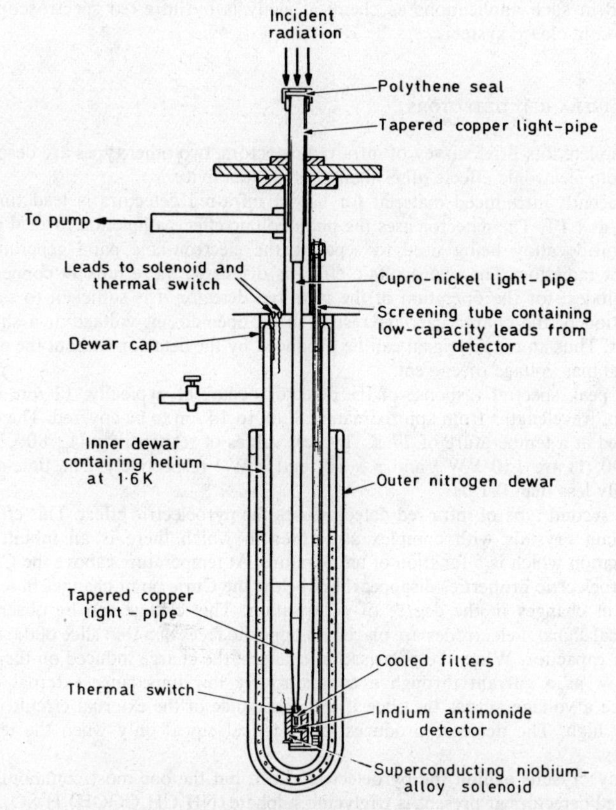

Figure 9.16 Construction of cryostat for indium antimonide photoconductive detector for sub-millimetric radiation operating at 1.6 K

the boiling point of the helium, and an outer vessel containing liquid nitrogen to minimise the evaporation of the helium. An operating temperature of 1.6 K can be obtained in this way. The detector operates in a magnetic field produced by a superconducting solenoid immersed in the liquid helium. The incident radiation is directed onto the detector element by a light pipe fitted with a polythene window. In a typical detector of this type, the diameter of the window would be approximately 2 cm, and the length of the light pipe between the window and the element approximately 60 cm. This type of detector would operate at wavelengths, say, between 0.1 mm and 10 mm with a peak response at, typically, 1 mm. A typical value of responsivity (1 mm, 800, 1) is 2×10^3 VW^{-1}, and a typical value of D* (1 mm, 800, 1) is 1×10^{12} $cmHz^{\frac{1}{2}}W^{-1}$. The time-constant is typically 1 μs.

The range of applications of infra-red photoconductive detectors is very wide, extending from simple systems using uncooled detectors to give warning of flame-failure in boilers, to complex and sophisticated systems with detectors operating at very low temperatures which are used for physical research or very precise measurements. The detection of the black-body radiation from an object can be used for such applications as detecting overheating in mechanical and electrical systems, intruder detection in security areas, temperature measurement without physical contact with the object, and thermal imaging. The varying absorption of infra-red radiation by different materials can be used in such applications as chemical analysis by infra-red spectroscopy or leak detection in closed systems.

OTHER INFRA-RED DETECTORS

To complete this brief survey of infra-red detectors, two other types are described that use photo-electronic effects other than photoconductivity.

A recently-introduced material for use in infra-red detectors is lead tin telluride, known as LTT. The detector uses the photovoltaic effect, a junction formed in the lead tin telluride alloy being used to separate the electron-hole pairs generated by the incident radiation. The photovoltaic effect is discussed more fully in connection with photodiodes; for the operation of the infra-red detector it is sufficient to say that the separation of the charge carriers results in an open-circuit voltage or a short-circuit current. Thus an output signal can be produced by the detector without the need for an external bias voltage or current.

The peak spectral response of the detector occurs at, typically, 11 μm, enabling a range of wavelengths from approximately 8 μm to 14 μm to be covered. The detector is operated at a temperature of 77 K. Typical values of responsivity (λ_p, 800, 1) and D* (λ_p, 800, 1) are 150 VW^{-1} and 8×10^9 $cmHz^{\frac{1}{2}}W^{-1}$ respectively. The time-constant is typically less than 0.1 μs.

The second type of infra-red detector uses the pyroelectric effect. This effect occurs in certain crystals with complex structures in which there is an inbuilt electrical polarisation which is a function of temperature. At temperatures above the Curie point, the pyroelectric properties disappear, but below the Curie point changes in temperature result in changes in the degree of polarisation. This change can be observed as an electrical signal if electrodes are placed on opposite faces of a thin slice of the material to form a capacitor. When the polarisation changes, the charge induced on the electrodes can flow as a current through a comparatively low impedance external circuit, or produce a voltage across the slice if the impedance of the external circuit is comparatively high. The detector produces an electrical signal only when the temperature changes.

Many crystals exhibit the pyroelectric effect, but the one most commonly used for infra-red detectors at present is triglycine sulphate $(NH_2CH_2COOH)_3H_2SO_4$, known as TGS. The incident radiation on the active element of the detector produces an increase in temperature by the absorption of energy, and hence a change in the polarisation occurs. Changes in the level of the incident radiation produce an electrical signal from the detector.

The spectral response of TGS detectors, as with other thermal detectors, is wide. It extends from 1 μm, below which incident energy is not absorbed, to the millimetric region. Filters can be used to define the range of wavelengths required for particular applications. TGS detectors generally operate at low modulation frequencies (approximately 10 Hz) but can operate at higher frequencies. This is because although the responsivity falls with increasing frequency (responsivity is inversely proportional to frequency), the noise also falls with increasing frequency. The value of N.E.P. therefore rises only slightly up to frequencies of, say, 10 kHz. The detectivity is constant over the spectral range, D* (500 K, 10, 1) being typically 2×10^9 $cmH^{\frac{1}{2}}W^{-1}$. A typical value for N.E.P. is 2×10^{-10} $WH^{-\frac{1}{2}}$.

PHOTOVOLTAIC DEVICES

Photodiode

Visible or infra-red radiation incident on a semiconductor material will generate electron-hole pairs by the normal photoconductive action. If these charge carriers are generated near a *pn* junction, the electric field of the depletion layer at the junction will separate the electrons and holes. This is the normal action of a *pn* junction which acts on charge carriers irrespective of how they are produced. However, the separation of the carriers gives rise to a short-circuit current or an open-circuit voltage, and this effect is called the photovoltaic effect. It can be used in such devices as photodiodes and phototransistors.

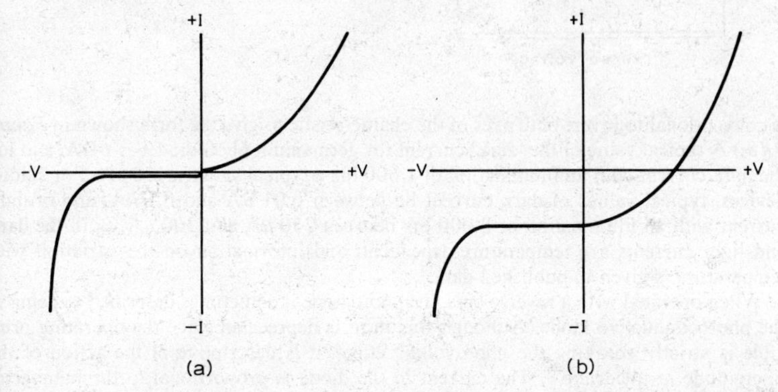

Figure 9.17 Voltage/current characteristic for photodiode: (a) not illuminated (b) illuminated

In a normal *pn* junction with no external bias applied, an equilibrium state is reached vith an internal potential barrier across the junction. This potential barrier produces a lepletion layer and prevents majority carriers crossing the junction. Minority carriers, iowever, can still cross the junction, and this gives rise to the small reverse leakage urrent of the junction diode. An external reverse voltage adds to the internal potential arrier, but the reverse current remains substantially constant because of the limited umber of minority carriers available until avalanche breakdown occurs. An external orward voltage overcomes the internal potential barrier, and causes a large majority arrier current to flow. A photodiode differs from a small-signal junction diode only in hat visible or infra-red radiation is allowed to fall on the diode element instead of being xcluded. If there is no illumination of the diode element, the photodiode acts as a ormal small-signal junction diode, and has a voltage/current characteristic like that hown in *Figure 9.17* (*a*).

With no external bias, when the n region is illuminated electron-hole pairs are generated and holes (the minority carrier in the n region) near the depletion layer are swept across the junction. This flow of holes produces a current called the photocurrent. If the p region is illuminated and electron-hole pairs generated, the electrons (the minority carrier in the p region) are swept across the junction to produce the photocurrent. In practical photodiodes, both sides of the junction are illuminated simultaneously, and the electron and hole photocurrents add together. The effect of the photocurrent on the voltage/current characteristic is to cause a displacement, as shown in *Figure 9.17* (*b*). The normally small reverse leakage current is augmented by the photocurrent.

The magnitude of the photocurrent depends on the number of charge carriers generated, and therefore on the illumination of the diode element. Thus a series of characteristics is obtained for various levels of illumination. When forward current flows through the diode, the photocurrent is swamped. Thus the part of the characteristic of interest is that where reverse current flows (quadrants 3 and 4).

With a reverse voltage across the diode (quadrant 3), the relationship between the reverse voltage and current with illumination as parameter is shown in *Figure 9.18*. (It

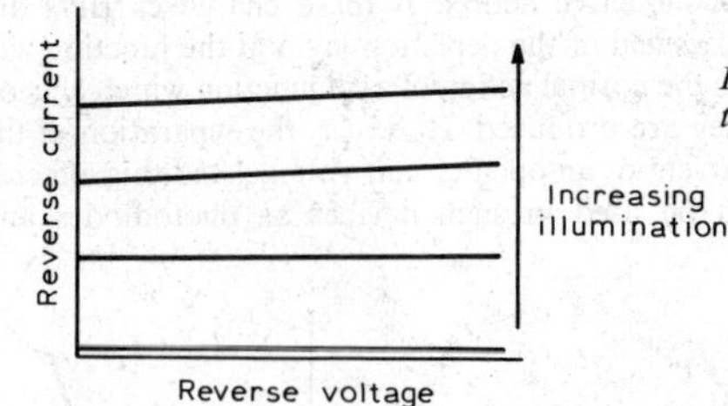

Figure 9.18 Voltage/current characteristic with illumination as parameter for photodiode operating with reverse bias

is conventional to invert both axes of the characteristic to give the form shown in *Figure 9.18*.) A typical value of the 'dark' current for germanium photodiodes is 6 μA, and for the light current with an illumination of 1 600 lux a typical value is 150 μA. For silicon devices, typical values of dark current lie between 0.01 μA and 0.1 μA, and of light current with an illumination of 2 000 lux between 250 μA and 300 μA. Both the dark and light currents are temperature-dependent, and information on the variation with temperature is given in published data.

When operated with a reverse bias, the photodiode is sometimes described as being in the photoconductive mode. Although this term is deprecated since the operating principle is strictly speaking the photovoltaic effect, it is descriptive of the action of the photodiode in quadrant 3. The current in the diode is proportional to the number of carriers generated by the incident radiation. The term also provides a convenient distinction from the operation of the diode in quadrant 4.

With no bias voltage across an illuminated photodiode, a reverse current flows, the sum of the small minority-carrier thermal current and the photocurrent. As an increasing forward bias is applied, the magnitude of the reverse current decreases as the majority-carrier forward current increases. Eventually, the magnitude of the majority-carrier current equals that of the photocurrent, and no current flows through the diode. If the forward bias is increased further, a forward current flows as in a normal junction diode. Thus the limiting values in quadrant 4 are the current at zero bias, and the forward voltage at which the current is zero. These values represent the short-circuit current and open-circuit voltage which can be obtained from the photodiode, as shown in *Figure 9.19*. (Again, it is conventional sometimes to invert this characteristic, but this time only about the voltage axis.) With a suitable load resistance connected across the photodiode, it is possible to extract power from the diode. This is the principle of the solar cell which is described later.

Because the magnitude of the reverse current depends on the illumination, the relationship between the forward voltage and the current with illumination as parameter in quadrant 4 has the form shown in *Figure 9.20*. An alternative method of presenting this information, sometimes used in published data, is a curve of the light current

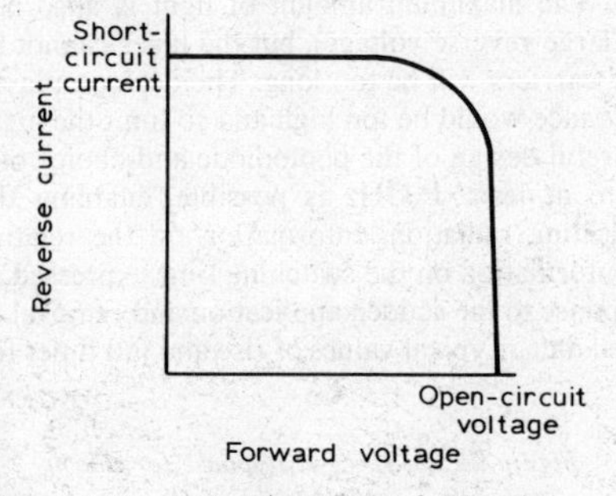

Figure 9.19 Voltage/current characteristic for photodiode with forward bias showing short-circuit current and open-circuit voltage values

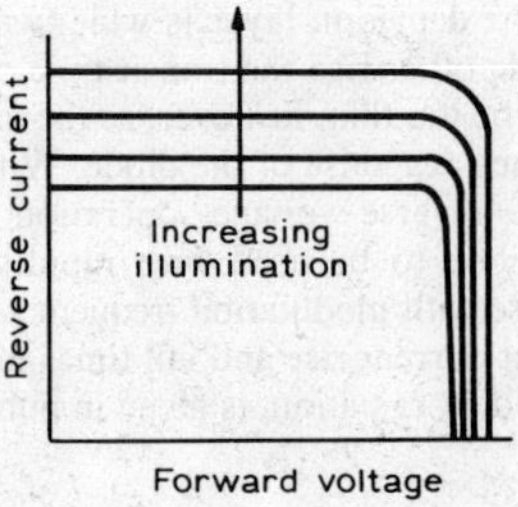

Figure 9.20 Voltage/current characteristic with illumination as parameter for photodiode with forward bias

plotted against illumination. This is plotted on logarithmic scales to give a straight line, as shown in *Figure 9.21*. A sensitivity figure may also be given, relating light current to illumination (sensitivity in μA/lux) or to the energy of the incident monochromatic radiation (sensitivity in μA/μW). Information on the variation of both light and dark currents with temperature is also given in published data.

To enable a suitable external load resistance to be chosen for the photodiode, curves of current plotted against load resistance are sometimes given, as shown in *Figure 9.22*.

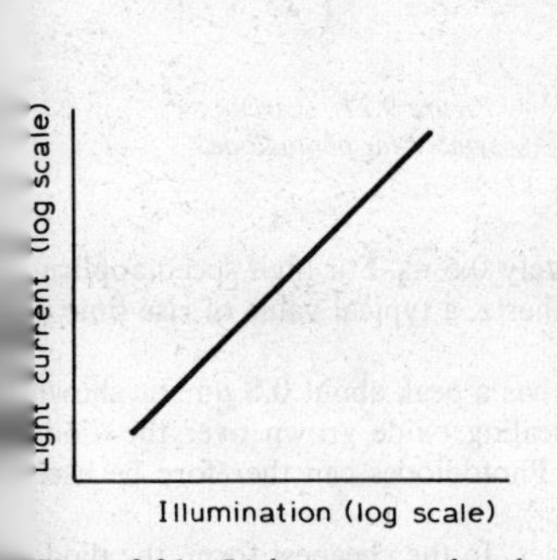

Figure 9.21 Light current plotted against illumination for photodiode with forward bias

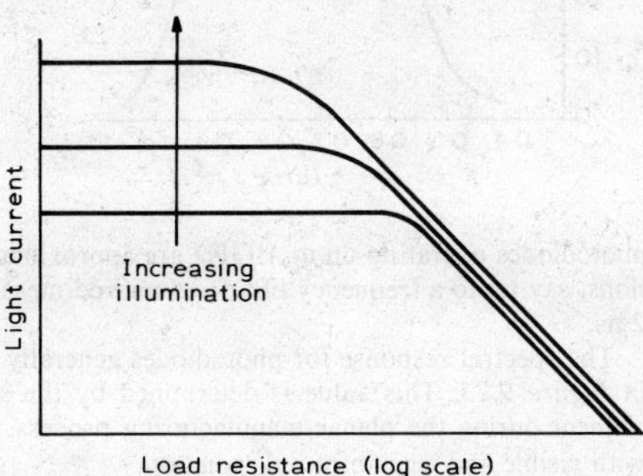

Figure 9.22 Light current plotted against load resistance with illumination as parameter for photodiode with forward bias

Ratings for the photodiode given in the data include the maximum reverse voltage, and the maximum forward and reverse currents. From these ratings and the load resistance curve, a suitable external circuit to be operated by the photodiode for the particular application can be designed.

In many optical applications, the radiation incident on the photosensitive device is modulated or is chopped by some mechanical system. The maximum frequency to which the photodiode responds is determined by three factors: diffusion of carriers, transit time in the depletion layer, and the capacitance of the depletion layer. Carriers generated in the semiconductor material away from the depletion layer must diffuse to the edge of the layer before they can be used to form the photocurrent, and this can cause considerable delay. To minimise the time of diffusion, the junction should be formed close to the illuminated surface. The maximum amount of light is absorbed when the depletion layer is wide (with a large reverse voltage), but the layer cannot be too wide otherwise the transit time of the carriers will be too long. The depletion layer cannot be too thin, however, as the capacitance would be too high and so limit the high-frequency response of the diode. With careful design of the photodiode and choice of a suitable reverse voltage, operation up to at least 1 GHz is possible, enabling the photodiode to be used with rapidly pulsating radiation. Information on the relative response with modulation frequency, or information on the switching time expressed as the light-current rise and fall times in response to the sudden application and removal of the incident radiation, is given in published data. Typical values of rise and fall times for

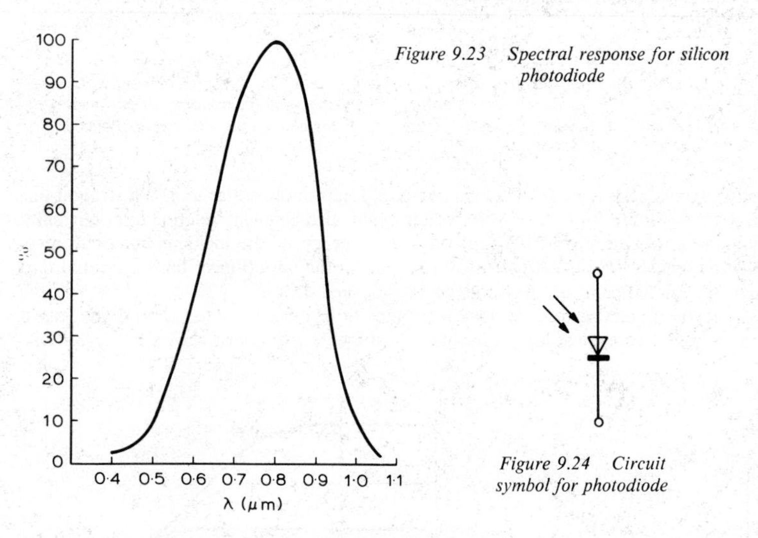

Figure 9.23 *Spectral response for silicon photodiode*

Figure 9.24 *Circuit symbol for photodiode*

photodiodes operating up to 1 GHz are approximately 0.5 ns. For high-speed applications, say up to a frequency of a few hundred megahertz, a typical value of rise time is 2 ns.

The spectral response for photodiodes generally has a peak about 0.8 μm, as shown in *Figure 9.23*. This value is determined by the sealing oxide grown over the diode element during the planar manufacturing process. Photodiodes can therefore be used with visible and near-infra-red radiation.

Various constructions can be used for photodiodes. In the cheapest form, the diode element is unencapsulated. A moulded plastic encapsulation or small metal envelope such as the TO-5 outline may also be used, with a window to allow the radiation to fall on to the diode element. A glass lens may sometimes be fitted to the device. The circuit symbol for a photodiode is shown in *Figure 9.24*.

Phototransistor

If a photodiode is formed by the collector–base diode of a transistor, the photocur rent flowing across the collector–base junction will be amplified by normal transistor action. A transistor operating in this way is called a phototransistor.

An *npn* phototransistor is shown in *Figure 9.25*. The base connection is open-circuit, that is the base is floating and the transistor is operating as a two-terminal device. The normal bias conditions are still maintained, however. With no illumination of the transistor, the current in the collector–emitter circuit is the common-emitter leakage current I_{CEO}. When the collector–base diode is illuminated, electron-hole pairs, are generated and the minority-carrier photocurrent flows across the reverse-biased collector–base junction. Electrons flow out of the base, and holes flow into the base. Thus the forward bias across the base–emitter junction is increased, which in turn increases the electron flow from the emitter, across the base, into the collector. The collector current is therefore the sum of the electron photocurrent and the electron current from the emitter which is h_{FE} times the photocurrent. In other words, the photocurrent of the

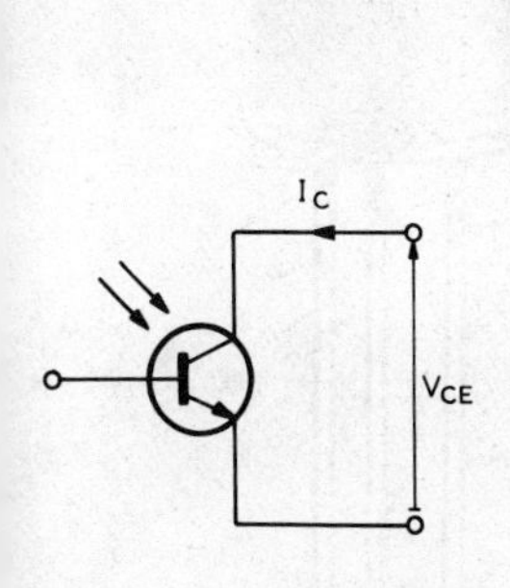

Figure 9.25 NPN phototransistor

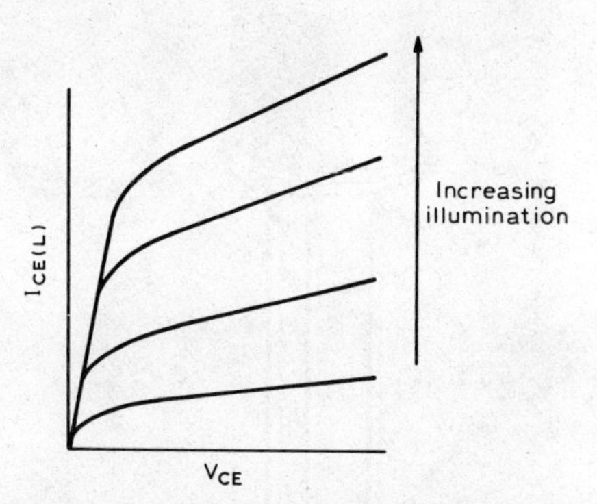

Figure 9.26 Output characteristic for phototransistor with illumination as parameter

diode has been amplified by the current gain of the phototransistor. Thus the phototransistor is used in applications where greater sensitivity is required than can be obtained with a photodiode. In some applications, the base is not left open-circuit but a relatively high value resistor is connected between the base and emitter. This does not affect the operation of the phototransistor just described, but gives improved thermal stability and improves the light-to-dark ratio.

The output characteristic of the phototransistor relates the light collector current $I_{CE(L)}$ to the collector-to-emitter voltage V_{CE} with illumination as parameter. It is similar to the output characteristic of a junction transistor in the common-emitter configuration with illumination rather than base current as parameter. The form of the characteristic is shown in *Figure 9.26*.

Other characteristics included in published data are the variation of light and dark collector currents with temperature, and the variation of collector current with illumination. Information on switching times is given, and sometimes in addition the variation of rise and fall times with emitter current. The spectral response curve is also given. As with the photodiode, the response has a peak at about 0.8 μm determined by the sealing oxide grown over the transistor element during the planar manufacturing process. The ratings for the phototransistor given in published data are similar to those for the junction transistor. They include the maximum collector-to-base and collector-to-emitter voltages, the maximum emitter-to-collector or emitter-to-base voltages, the maximum collector current, and the maximum power dissipation. A thermal resistance is given so that a suitable heatsink can be chosen if required for the application.

The usual construction for a phototransistor is a small metal envelope such as the

TO-18 outline with a plane window or a lens. Typical encapsulations for phototransistors are shown in *Figure 9.27*.

Photodiodes and phototransistors are used in applications where an electrical signal corresponding to incident visible or infra-red radiation is required. An example of this is a paper-tape reader where the photo device responds to light passing through the holes

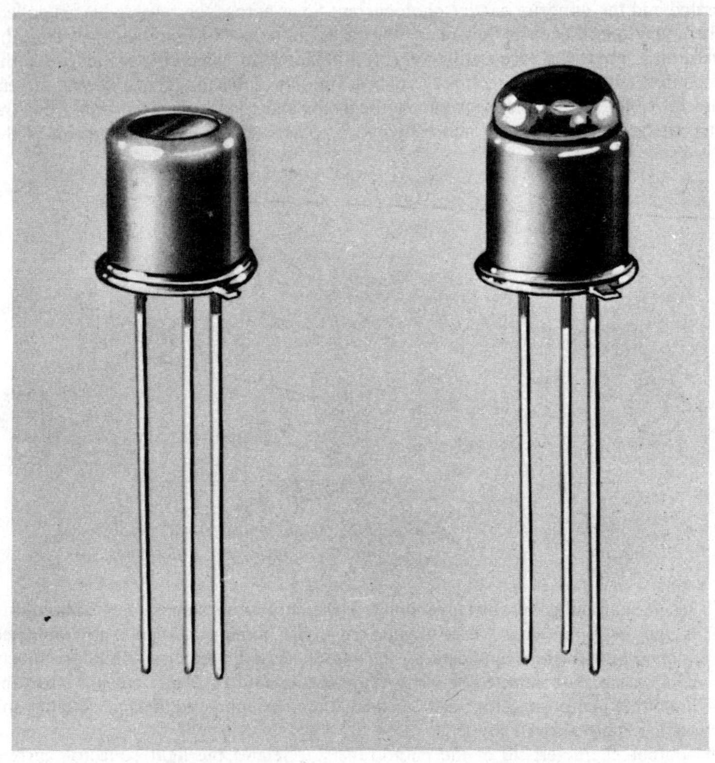

Figure 9.27 Typical encapsulations for phototransistors (*Photo: Mullard Limited*)

in the tape to produce an electrical signal based on the pattern of holes in the tape. Other applications include photographic light meters, flame-failure alarm systems for boilers, and detectors for modulated laser beams.

Light-activated silicon controlled switch

A four-layer device such as a silicon-controlled switch (s.c.s.) can be used as a photosensitive device. It can be used as a light-controlled switch, the s.c.s. being turned on when illuminated to operate another circuit. Two modes of operation are possible, as shown in *Figure 9.28*: with either the anode gate or cathode gate not connected, and a resistor connected to the other gate. Thus if the anode gate is left floating, a resistor is connected between the cathode gate and the cathode. If the cathode gate is left floating, the resistor is connected between the anode gate and the anode.

The operation of the light-activated s.c.s. can be described in terms of the two-transistor analogue which is shown, with the external resistors in broken lines, in *Figure 9.29*. When the anode gate lead is not connected, *Figure 9.29* (*a*), the base of the *pnp* transistor is left floating, and this device acts as a phototransistor. When the s.c.s. has a

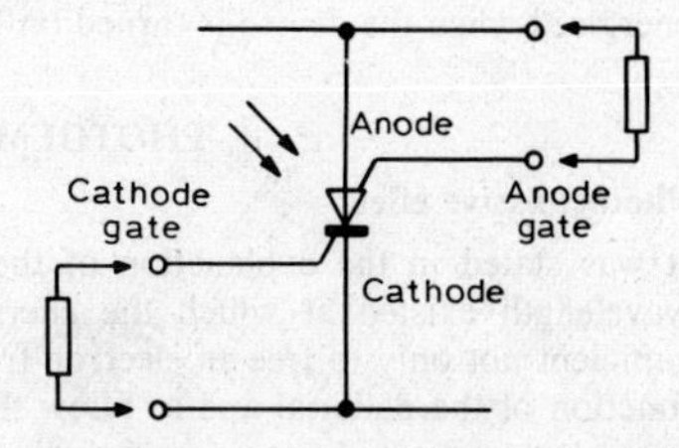

Figure 9.28 Light-activated silicon controlled switch showing alternative modes of operation

positive anode-to-cathode voltage applied, and is not illuminated, only the small leakage current flows. When the device is illuminated, the photocurrent in the *pnp* transistor flows through the resistor between thc cathode gate and cathode. A voltage is developed across this resistor so that the base of the *npn* transistor becomes positive with respect to its emitter, and so this transistor starts to conduct. (The development of the voltage drop across the resistor is equivalent to placing a small positive voltage on the cathode gate, which is one of the ways in which an s.c.s. can be triggered.) Cumulative action between the two transistors occurs, and the s.c.s. is turned on. If the cathode gate is not connected *Figure 9.29* (*b*), the base of the *npn* transistor is floating and this device acts

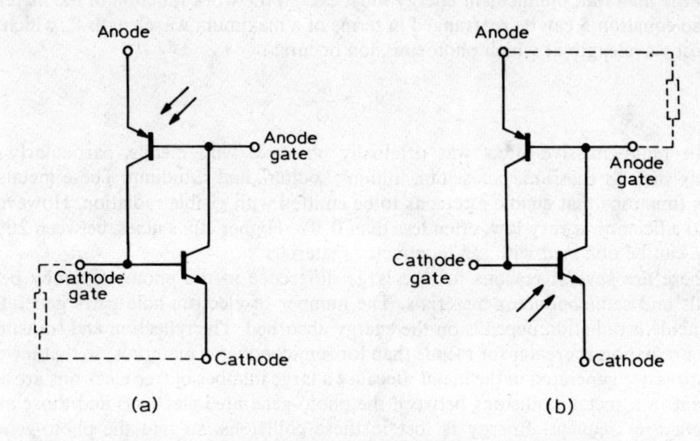

Figure 9.29 Two-transistor analogue for light-activated silicon controlled switch: (*a*) *pnp phototransistor* (*b*) *npn phototransistor*

as the phototransistor. The photocurrent develops a voltage across the resistor between the anode gate and anode, so that the anode gate becomes negative with respect to the anode, and the s.c.s. is turned on. The light-activated s.c.s. is turned off, as with the normal s.c.s., by reducing the current through the device to below the holding value.

The characteristics and ratings for a light-activated s.c.s. given in published data are similar to those for a normal s.c.s. Additional information is given on the variation of cathode-gate current with illumination, and the spectral response. Two illumination levels are specified: a minimum value $E_{\text{on min}}$ which will trigger all devices, and a maximum non-triggering value $E_{off\,max}$. These values are determined by the leakage current of the device.

The usual encapsulation for the light-activated s.c.s. is a small metal envelope such as the TO-72 outline with an end-viewing window. The device is generally used in applications as a relay driver. The relay is connected in the anode circuit, and is energised when the device is turned on by the incident radiation.

PHOTOEMISSIVE DEVICES

Photoemissive effect

It was stated in the explanation of the photoconductive effect, that a limiting short wavelength existed at which the energy absorbed from the incident radiation was sufficient not only to free an electron from its parent atom but also to exceed the work function of the material and so allow the electron to escape from the surface. Because such electrons could not be used as charge carriers within the material, this wavelength represents the shortest wavelength for operation of photoconductive devices. However, if the emitted electrons can be collected, a current will flow outside the illuminated material. The magnitude of this current is proportional to the energy absorbed, and hence to the illumination level of the incident radiation. This effect is called the photoemissive effect, and it is used in a range of practical devices.

The relationship between the energy of the incident radiation E and the wavelength λ is given by

$$E = \frac{hc}{\lambda} \tag{5}$$

where h is Planck's constant and c is the velocity of light. For electrons to be emitted from the material, the incident energy must exceed the work function of the material ϕ, and so equation 5 can be rearranged in terms of a maximum wavelength λ_m which is the longest wavelength at which photoemission occurs:

$$\lambda_m = \frac{hc}{\phi}.$$

The photoemissive effect was originally observed with metals, particularly alkali metals such as caesium, potassium, lithium, sodium, and rubidium. These metals have work functions that enable electrons to be emitted with visible radiation. However, the photo efficiency is very low, often less than 0.1%. Higher efficiencies, between 20% and 30%, can be obtained with semiconductor materials.

There are several reasons for the large difference in the photo efficiency between metals and semiconductor materials. The number of electron-hole pairs generated by the incident radiation depends on the energy absorbed. The reflection and transmission of the radiation is greater for metals than for semiconductor materials, so that fewer free electrons are generated in the metal. Because a large number of free electrons are always present in a metal, collisions between the photo-generated electrons and those already present are frequent. Energy is lost in these collisions, so that the photo-generated electrons reaching the surface may not have sufficient energy left to overcome the surface barrier and so not be emitted. Electrons near the surface of the metal suffer fewer collisions, and so lose less energy. Thus the escape depth, the distance from the surface from which electrons can reach the surface with sufficient energy to escape, is small in a metal. On the other hand, there are few free electrons in a semiconductor material. Collisions between the photo-generated electrons and the atoms of the crystal lattice (lattice scattering) results in little loss of energy for the free electron unless it has sufficient energy to generate a new electron-hole pair. In this event, energy equal to the band-gap energy is lost. For these reasons, the escape depth in a semiconductor material is greater than that in a metal, and so a larger number of electrons reach the surface to be emitted. Finally, the work functions of certain semiconductor materials are lower

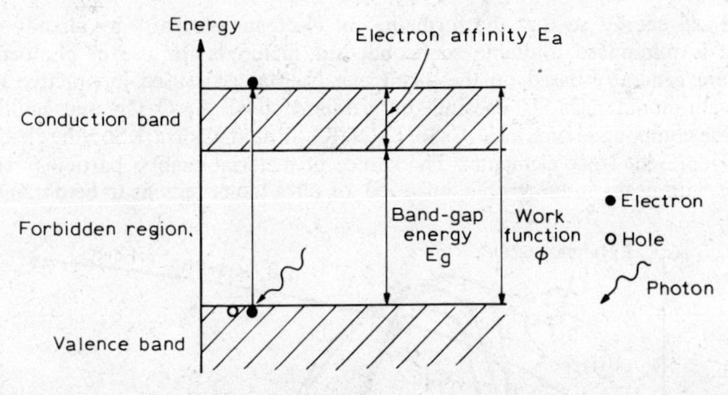

Figure 9.30 Permissible energy levels for electron in semiconductor material, generation of free electron

than those of the alkali metals, and so it is easier for electrons to be emitted. It is possible, therefore, for semiconductor materials to operate at longer wavelengths than metals.

The energy-band diagram for a semiconductor material used for photoemission is shown in *Figure 9.30*. In this diagram, E_g represents the band-gap energy, and the depth of the conduction band is represented by the electron affinity E_a. The electron affinity is the energy required for the electron to overcome the surface barrier; that is, the energy required to transform a free electron in the semiconductor material into a truly free electron outside the material. The work function ϕ is the sum of the band-gap energy and the electron affinity.

The escape depth for all semiconductor photoemissive materials is determined by lattice scattering. For the highest photoemissive efficiency, the escape depth should be as large as possible. This implies that the electron affinity should be low compared with

Table 9.1 PHOTOCATHODE MATERIALS

Spectral response designation	*Material*	*Wavelength of peak response* (nm)	*Luminous sensitivity at colour temperature of* 2854 K (μA/lm)
S1	Ag–O–Cs	800	25
S3	Ag–O–Rb	420	6.5
S4	Cs–Sb	400	40
S5	Cs–Sb	340	40
S8	Cs–Bi	365	3
S9	Cs–Sb	480	30
S10	Ag–Bi–O–Cs	450	40
S11	Cs–Sb	440	60
S13	Cs–Sb	440	60
S17	Cs–Sb	490	125
S19	Cs–Sb	330	40
S20	(Cs)–Na–K–Sb	420	150
S21	Cs–Sb	440	30
S24	Na–K–Sb	380	32
S25	(Cs)–Na–K–Sb	420	160
—	Ga–As	450	250
—	Ga–As–P	450	200

the band-gap energy so that the formation of electron-hole pairs by already-excited electrons is minimised. Suitable semiconductor materials for use in photoemissive devices are generally based on the alkali metals. Materials used in practice include caesium antimonide, Cs_3Sb, caesium on oxidised silver Ag_2O–Cs, and multi-alkali antimonide compounds such as K_2CsSb, $Cs(NaK)_3Sb$, and $(Cs)Na_2KSb$; (the elements in brackets represent trace elements). The choice of material enables particular spectral responses with peaks in the visible, infra-red, or ultra-violet regions to be obtained.

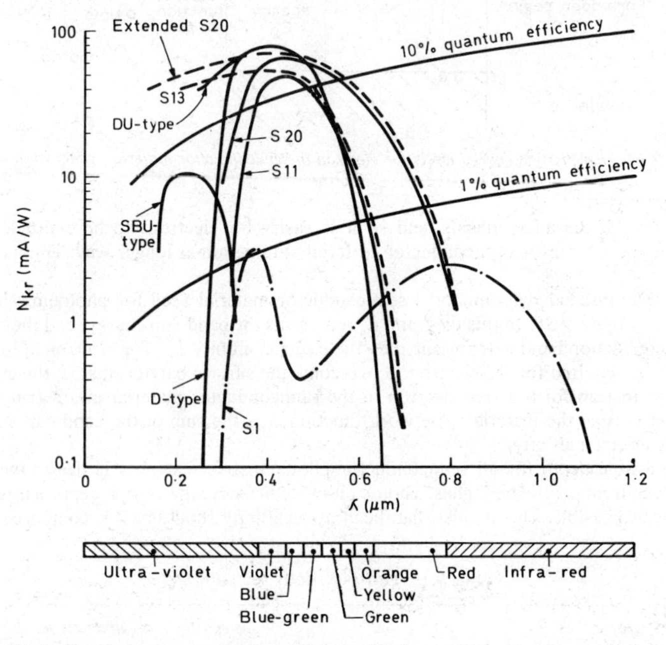

Figure 9.31 Spectral responses for typical photocathode materials. Curves not marked with S numbers are for Mullard photocathode materials. (Courtesy Mullard Limited)

The various manufacturers of photoemissive devices have their own designations for the photocathode materials used in their devices. Certain commonly-used materials have been classified by their spectral response under an S number by the Joint Electronic Defence Executive Committee (J.E.D.E.C.). A list of photocathode materials and the corresponding S number is given in *Table 9.1*, together with the wavelength of peak response and the luminous sensitivity. The spectral responses of some of these materials are shown in *Figure 9.31*.

Photoemissive tubes

Photoemissive tubes (or photoemissive cells as they were previously called) are of two types: vacuum tubes and gas-filled tubes. Both types consist of a glass envelope containing a photoemissive cathode which emits electrons when illuminated, and an anode to which the electrons flow to produce the photocurrent. The vacuum photoemissive tube is a high-vacuum device like the thermionic valve, while the gas-filled tube is filled with an inert gas at low pressure. The area of the photocathode is made as large as possible to collect the maximum amount of radiation, while the anode is a thin rod or wire so that it does not obscure much of the cathode from the incident radiation. The circuit symbol for a photoemissive tube is shown in *Figure 9.32*.

In a vacuum photoemissive tube, if the anode-to-cathode voltage exceeds a certain value called the saturation voltage, V_s, all the emitted electrons will be collected by the anode. For a fixed level of incident radiation, the photocurrent will remain substantially constant as the voltage is increased. As the level of the radiation is increased, a larger

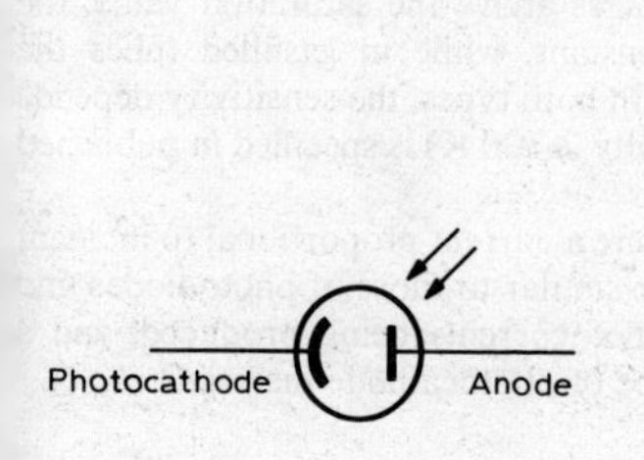

Figure 9.32 Circuit symbol for photoemissive tube

Figure 9.33 Voltage/current characteristic with luminous flux as parameter for vacuum photoemissive tube

ıumber of electrons is emitted, and so the photocurrent is increased. Thus a family of ınode voltage/current characteristics with luminous flux as parameter is obtained, as ;hown in *Figure 9.33*.

In a gas-filled photoemissive tube, the photocurrent is initially the same as in a ·acuum photoemissive tube up to the saturation voltage. As the anode voltage is ncreased, the ionisation voltage of the gas is reached. Between the saturation voltage V_s ınd the ionisation voltage V_i, the anode current remains reasonably constant. Once the onisation voltage is exceeded, however, the electrons emitted from the cathode can

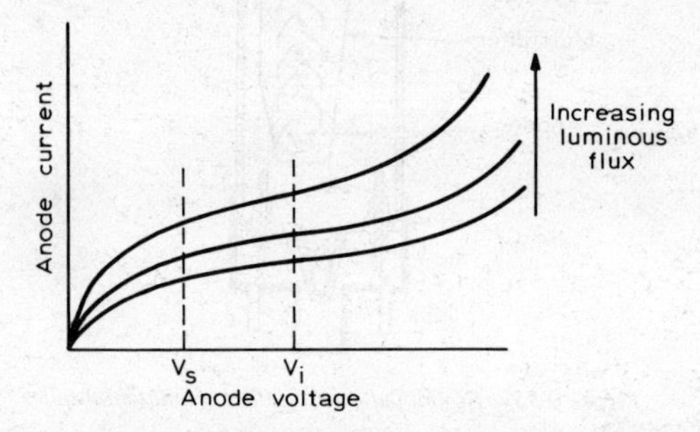

;gure 9.34 Voltage/current characteristic with luminous flux as parameter for gas-filled photoemissive tube

ɲnise the gas filling. The number of electrons is therefore increased and 'gas amplificaon' occurs. As the anode voltage is increased further, the gas amplification increases. he voltage/current characteristic for a gas-filled tube is therefore as shown in *Figure .34*.

The voltage/current characteristic is given in the published data for the photoemisve tube, together with typical loadlines so that a suitable external circuit can be designed. Limiting values of anode voltage and current are also given. The limiting voltage for a gas-filled tube is set by the need to prevent the formation of a glow discharge, and this is approximately 100 V.

A 'dark' current, the current flowing in the tube when there is no incident radiation, is specified. As would be expected, this is higher for a gas-filled tube than for a vacuum tube, typical values being less than 0.1 μA for a gas-filled tube and less than 0.05 μA for a vacuum tube. A luminous sensitivity is given, defined as the photocurrent with unit luminous flux; the generally-used unit for this sensitivity is μA/1m. From the voltage/current characteristics it can be seen that at voltages above the saturation value, the sensitivity of vacuum photoemissive tubes is constant, while in gasfilled tubes the sensitivity is higher but dependent on the voltage. In both types, the sensitivity depends on the light source, and a colour temperature (usually 2 700 K) is specified in published data.

Photoemissive tubes are used in applications where a current proportional to incident radiation is required. Many of the applications are similar to those of photodiodes and phototransistors with the advantages of larger photocurrents being produced, and a wider range of spectral response through the choice of photocathode material.

Photomultiplier

A photomultiplier is a special form of photoemissive tube in which multiplication of the electrons emitted by the photocathode occurs to produce a larger anode current. The electron multiplication is produced by secondary emission in an electrode system

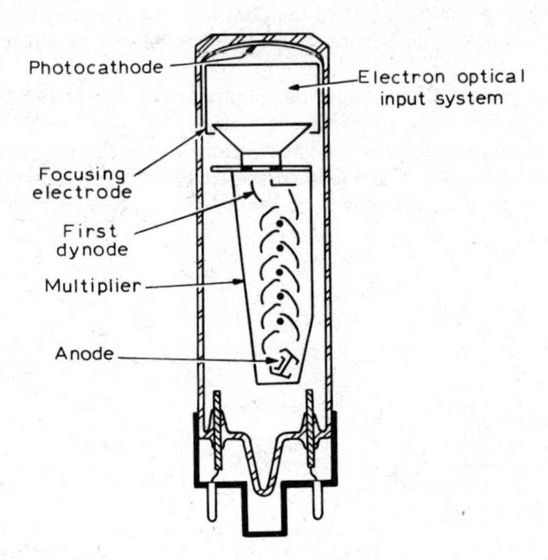

Figure 9.35 Simplified cross-section of photomultiplier

between the cathode and the anode. A simplified cross-section of a typical photomultiplier is shown in *Figure 9.35*. The incident radiation on the photocathode produce electrons which are directed by the electron-optical input system on to the first dynod Electron multiplication by secondary emission from successive dynodes occurs, so th the anode current of the photomultiplier is considerably larger than the initi photocathode current. The photomultiplier is a high-vacuum device.

Two types of photocathode can be used: the opaque and the semitransparent. A opaque photocathode consists of a thick layer of photoemissive material, and th radiation is incident on and electrons are emitted from the same side. A semitransparent photocathode is a much thinner layer deposited on the inner surface of a transparent window, and the electrons are emitted from the side away from the incident radiation. The opaque photocathode can be used for higher photocurrents than the semitransparent type, as high as several amperes. A more complex electrode structure is required, and since the photomultiplier cannot be enclosed in a glass envelope, it must be operated in a vacuum chamber. Therefore opaque cathodes are used only for specialised applications. The generally-used photomultiplier has a semitransparent photocathode.

The various semiconductor materials based on the alkali metals described earlier can be used for the photocathode. In addition, the window material can also be chosen to alter the spectral response. It is therefore possible to choose a cathode material and a window material to provide the required peak spectral response. The usual window materials are lime glass and quartz, but special glasses can also be used. Various 'standard' combinations of photocathode and window materials are tabulated, giving peak responses in the visible, infra-red, and ultra-violet regions. These combinations are listed in the published data for photomultipliers.

The electro-optic input system must ensure that as many as possible of the electrons emitted from the photocathode reach the first dynode of the electron multiplier system. In addition, the transit time between the cathode and first dynode should be independent of the point of emission of the electron. This second consideration is particularly important for photomultipliers used for very fast applications.

Photomultipliers which are not used for fast applications generally have a flat photocathode and faceplate. The focusing electrode, an aluminium ring at the same voltage as the photocathode, and the accelerating electrode, usually at the same voltage as the first dynode, direct the emitted electrons on to the first dynode. Ideally the electrons should land on the same spot (or small area) on the dynode. This is helped in fast-response photomultipliers, where variations in the transit times of electrons from different parts of the photocathode must be minimised, by making the cathode part of a sphere. The focusing electrode may also be provided with an external connection to enable fine adjustments to be made.

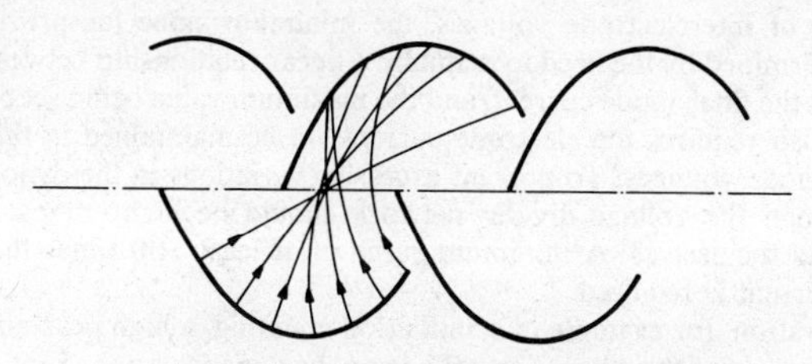

Figure 9.36 Linear cascade dynode structure

The electrons striking the first dynode produce further electrons by secondary emission. An incident electron striking a material transfers energy to electrons inside the material so that they can escape from the surface. The mechanism of this secondary emission effect is similar in many ways to the mechanism of photoemission already discussed. In a conductor, the freed secondary electrons lose much of their energy in collisions with the many free electrons in the material. Semiconductor materials with fewer free electrons enable more secondary electrons to be emitted. Because an electric field is necessary to direct the emitted secondary electrons through the multiplier, the secondary emission material must be deposited on a conducting support material. Typical dynodes are made from magnesium and caesium with an outer oxide layer on silver, although newer materials such as gallium phosphide are being introduced.

The most commonly-used structures for the electron multiplier are the linear cascade and the venetian blind. The linear cascade structure is shown in *Figure 9.36*. Each dynode acts as a reflective element, directing the secondary electrons to the succeeding dynode. The voltage between successive dynodes must be large enough to ensure a reasonable collection efficiency, and prevent the formation of a space charge which would affect the linearity of the anode current with respect to the initial photocurrent. A progressively higher voltage is applied to each dynode through the multiplier towards the anode to obtain the required voltage gradient. The venetian blind structure uses metal strips with an emissive coating placed at 45° to the tube axis, as shown in *Figure 9.37*. All the electrons even from a large photocathode strike the first dynode, but it is a relatively slow structure and can only be used when the transit time of electrons through the multiplier is not important.

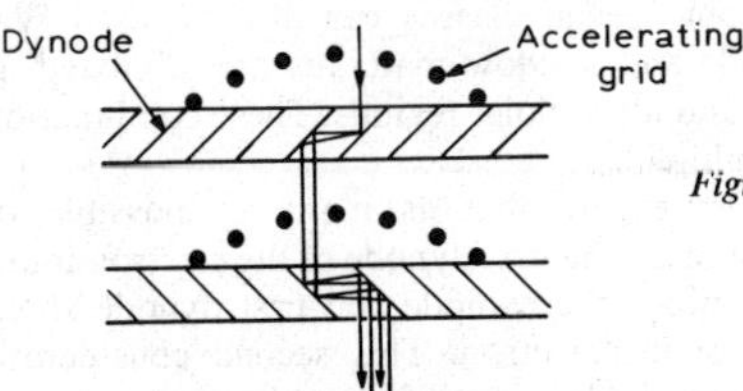

Figure 9.37 Venetian blind dynode structure

The anode of a photomultiplier is usually in the form of a grid mounted close to the final dynode. Secondary electrons from the preceeding dynode pass through the grid to strike the final dynode, and the emitted secondary electrons from the final dynode are collected by the anode.

The progressively higher voltages for the dynodes of the electron multiplier are generally obtained by a resistive voltage divider network. A stabilised high-voltage supply is required, the voltage depending on the number of stages in the multiplier. Typical values lie between 1.5 kV and 7 kV. The published data lists maximum and minimum values of interelectrode voltages, the minimum value (as previously mentioned) being determined by the need to maintain a linear relationship between the initial photocurrent and the final anode current, and the maximum value being set by flashover. Good linearity also requires the electrode currents to be maintained in the saturation region by the dynode voltages. To prevent excessive variations in the dynode voltages, the current through the voltage divider network should be high compared with the electrode currents themselves. A minimum value of at least 100 times the maximum average anode current is required.

For pulse operation, for example in scintillation counting, a high peak anode current occurs for the duration of the pulse. The voltage on the dynodes can be kept constant for the pulse duration by connecting capacitors across the resistors at the high-voltage end of the divider network. These bypass capacitors reduce the value of the current through the divider required for constant dynode voltage. However, the capacitors increase the time-constant of the divider network. At high pulse repetition rates where the intervals between pulses are short, the capacitors may not have time to discharge and considerable voltage variations may occur. The errors produced by these variations increase with the repetition frequency. For applications where a short time-constant is required at a high repetition frequency, high currents in the divider network must be accepted.

Typical voltage divider networks are shown in *Figure 9.38*. The divider in *Figure 9.38* (*a*) is designed for maximum gain with a given supply voltage; that in *Figure 9.38* (*b*) is designed for good linearity between the photocathode and anode currents, and the best time response. The progressive increase in voltage across the final stages of the photomultiplier in *Figure 9.38* (*b*) prevents premature saturation, and so improves the linearity further. The voltage on the accelerating electrode is made adjustable to allow the tube to be set up for optimum collection efficiency or optimum time response. From

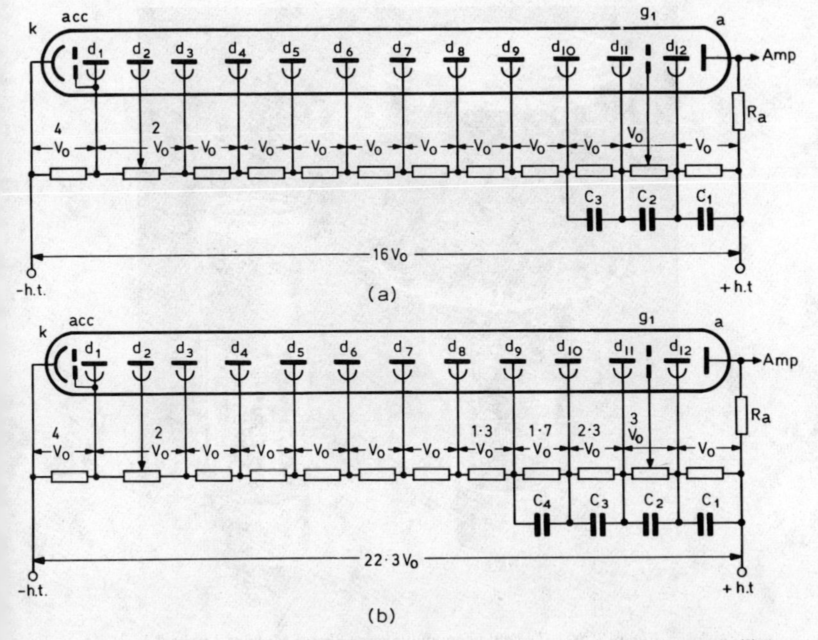

Figure 9.38 *Voltage divider network for dynodes of photomultiplier: (a) for maximum gain (b) for best linearity and time response*

the overall supply voltage, the maximum and minimum values of interelectrode voltages given in the data, the maximum anode current, and the relative values of the interelectrode voltages, suitable resistor values can be chosen for the divider network.

The published data for photomultipliers have some items in common with the photoemissive tubes already discussed. The material of the photocathode and its spectral response are given, together with the sensitivity under specified conditions. The characteristics of the tube with the two types of voltage divider network are given. For the divider network of *Figure 9.38* (*a*), the anode dark current is given, and either the gain at a specified supply voltage or the supply voltage required for a particular gain. For the other type of divider network as shown in *Figure 9.38* (*b*) the linearity, transit time, and details of anode pulse shape are given. Curves of the variation of gain, dark current, and sensitivity with supply voltage are sometimes given.

Typical applications of photomultipliers include their use in scintillation counters, laser range finders, flying spot scanners, low-level light detectors, photometry, and spectrometry. In scintillation counters, an ionising radiation such as alpha particles, beta particles, neutrons, or gamma rays reacts with the scintillation material to release energy in the form of individual flashes of light. These light flashes can be registered by a photomultiplier. Similarly, Cerenkov radiation can be detected by a photomultiplier. In both applications, the small flashes of light produced are amplified by the photomultiplier and the pulse output used in a counting or coincidence detector circuit. A fast-response photomultiplier has to be used in a laser range finder to enable the time between the transmitted and reflected light pulses to be accurately determined. In a flying spot scanner, the photomultiplier provides a convenient means of producing an electrical signal proportional to the light transmitted from the spot on the cathode ray tube through the slide or film being scanned. The low-level light detection, photometric, and spectroscopic applications all make use of the high luminous sensitivity of the

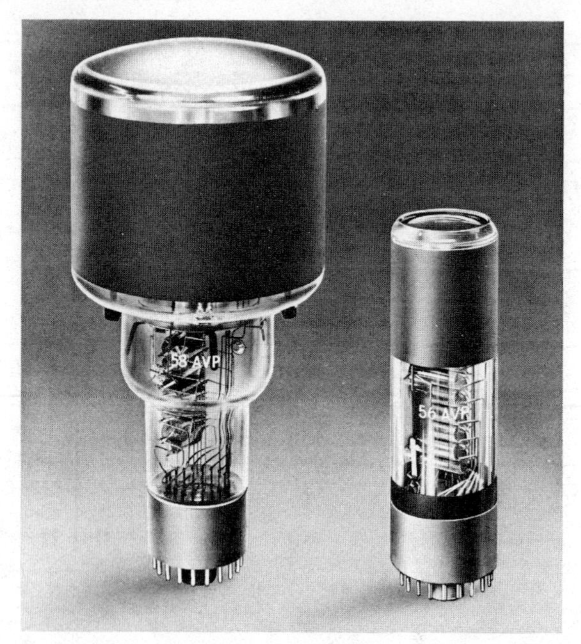

Figure 9.39 Typical photomultipliers (Photo: Mullard Limited)

photomultiplier to various wavelengths of radiation. Typical photomultipliers are shown in *Figure 9.39.*

Channel electron multiplier

The channel electron multiplier is not a photosensitive device; it responds to electrons and other ionised particles rather than photons. But as a form of this device is incorporated in one type of image intensifier described later, a brief description is given here.

A channel electron multiplier is a small curved tube made of glass, the inside wall of which is coated with a high-resistance material. If a high voltage is applied between the ends of the tube, the resistive surface becomes a continuous dynode. It is, in effect, equivalent to the combination of the separate dynodes of a photomultiplier and the resistive voltage divider network used to provide the individual dynode voltages. As the tube is open (at one end at least), the channel electron multiplier has to be operated in a vacuum. For space research, the environmental vacuum is sufficient, but for laboratory work the multiplier has to be housed in a vacuum chamber.

An electron or other particle entering the low-potential end of the multiplier generates secondary electrons on impact with the wall of the tube. The emitted secondary electrons are accelerated by the electric field produced by the applied voltage along the tube until they acquire sufficient energy to generate further secondary electrons when they strike the wall of the tube. This cumulative process produces a large number of electrons at the high positive potential end of the tube. The effect is shown in *Figure 9.40.* A channel electron multiplier therefore responds to an input of one electron by producing an output pulse of charge which can contain up to 10^8 electrons. The duration of the charge pulse (at the half-height points) is typically about 10 ns. The amplitude of the voltage pulse produced depends on the resistance and capacitance of the circuit to which the multiplier is connected. The gain is an exponential and steep function of the applied voltage up to, typically, a value of 10^7 when saturation effects begin to occur.

The multiplier responds to ions, beta particles, X-rays, and other radiation of sufficient energy. The detector efficiency is different for the different forms of radiation, but any particle or quantum which can excite an electron from the wall of the tube on the initial impact can be detected.

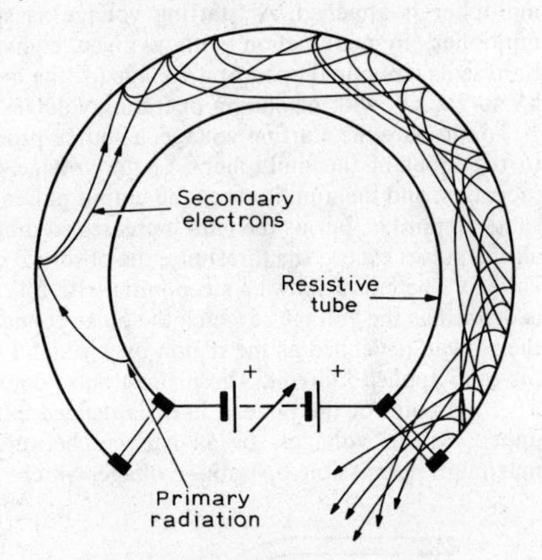

Figure 9.40 Operation of channel electron multiplier showing cumulative generation of secondary electrons

The tube forming the multiplier is curved because the gain of a straight tube would be sensitive to changes in the ambient pressure. When the first cloud of electrons reaches the output end of the tube, the cloud is sufficiently dense to ionise many of the residual gas atoms in the tube. The positive ions drift under the influence of the electric field towards the more negative input end of the tube. If the tube is straight, the ions may acquire sufficient energy before they collide with the wall of the tube to generate secondary electrons on impact. These secondary electrons produce further electrons, and so a spurious output pulse is produced. The process is then repeated, and a train of 'after pulses' occurs, lasting for about a microsecond. This effect is called ionic feedback. The spurious pulse train lasts until the field inside the tube is so distorted by the wall charging that the multiplication process cannot sustain the feedback. A 'dead time' then occurs during which the field is restored. An output pulse can then be produced. If the tube is curved, however, the positive ions strike the wall of the tube before they have sufficient energy to generate secondary electrons. Electron multiplication is not affected because the electrons need to acquire less energy before impact to generate secondary electrons. The output from a curved channel electron multiplier is therefore independent of pressure so long as the ambient pressure does not exceed 5×10^{-4} torr. Above this pressure, spurious pulses occur as with the straight channel.

The gain of the multiplier is limited by the space charge. When the total charge of the electron cloud in the channel reaches approximately 3×10^8 electrons, the gain cannot increase further. The space charge repels the emitted secondary electrons so that they strike the tube wall before acquiring sufficient energy from the field to generate further electrons. In this saturated state, the gain of the multiplier is limited to about 10^7. Increasing the applied voltage increases the amplitude of those pulses which would not otherwise have reached 3×10^8 electrons, but as the maximum charge output cannot exceed this value, the amplitude of all pulses tends to the same value. Thus the multiplier is unable to give information about the input particle. When the multiplier is operated below saturation, there is some proportionality between input and output, and information can be obtained about the energy level of the input radiation.

The materials used to form the inside coating of the channel wall can be metallic lead, which has a relatively low resistance value of about $10^9\ \Omega$, or vanadium phosphate glass which has a higher resistance value between $10^{10}\ \Omega$ and $10^{11}\ \Omega$. The inside diameter of the tube is typically 1.0 mm to 2.5 mm. An input cone of a larger diameter can be fitted. The operating voltage applied across the ends of the tube is chosen so that all the pulses produced by the multiplier are above the threshold level of the circuit to which the multiplier is attached. A 'starting voltage' is specified in the published data for the multiplier, in conjunction with a given equivalent threshold. The starting voltage represents a nominal operating voltage for the multiplier. Typical values lie between 2.0 kV to 2.5 kV, with maximum operating voltages of 3.0 kV to 4.0 kV.

To measure the starting voltage, a source producing a fixed counting rate is attached to the input of the multiplier. As the voltage across the tube is increased, the gain increases, and the amplitude of the output pulses increases. The pulses are not all of the same amplitude, but as the gain increases so more of them exceed the threshold. When all the pulses exceed the threshold, the observed count rate is plotted against the applied voltage. The curve shows a steep initial rise followed by a plateau. The starting voltage is defined as the voltage at which the pulse count rate is 90% of the plateau value, where the plateau is defined as the region over which the count rate changes by less than 10% for each applied kilovolt. The plateau pulse count rate is taken as the mid-point value. The mid-point of the plateau itself is defined either as halfway between the lower and upper plateau voltages, or as halfway between the lower plateau voltage and the maximum permissible operating voltage, whichever is the lesser.

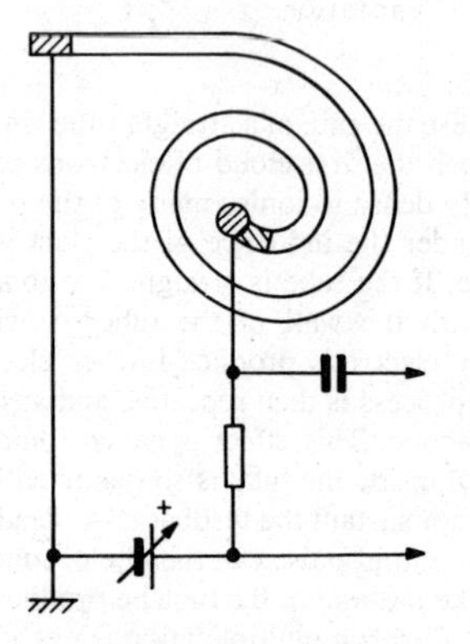

Figure 9.41 Channel electron multiplier used as pulse counter

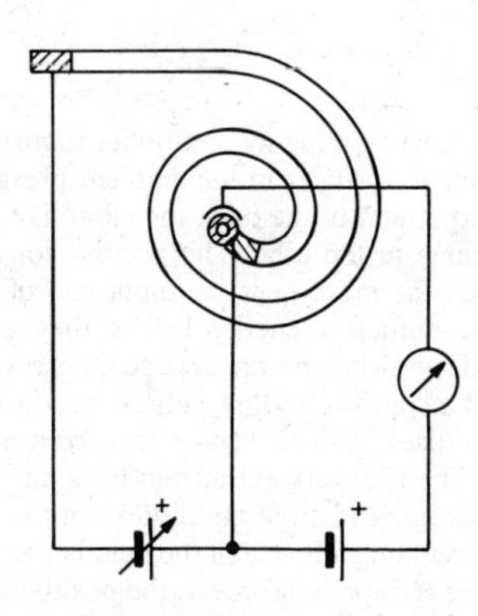

Figure 9.42 Channel electron multiplier used as current amplifier

If the equivalent threshold of the circuit to which the multiplier is connected in a particular application is higher than the value quoted in the data, then the starting voltage must be measured for the higher threshold. If the equivalent threshold of the circuit is lower, then an applied voltage lower than the starting voltage can be used.

The most common application of the channel electron multiplier is for pulse counting. A typical circuit arrangement is shown in *Figure 9.41*. The output end of the tube is generally closed to prevent any ionised gas atoms drifting round the outside of the tube to re-enter the input and so produce spurious pulses. The output charge pulse is fed to a pulse amplifier or a charge-sensitive amplifier which is connected to a counting circuit. A second application is as a current amplifier. For this, an open-ended tube is used and the charge pulse collected on an electrode, as shown in *Figure 9.42*. The current transfer characteristic departs from linearity if the output current is greater than 1% of the standing current in the tube wall. This places an upper limit on the range of the multiplier as an amplifier. For currents below 1.6×10^{-14} A (10^5 electrons per second), measurements are best made using the channel electron multiplier as a pulse counter. Currents of 10^{-11} A or more can be measured directly with commercially-available meters. For currents between these two limits, the channel electron multiplier operating as a current amplifier with a gain of 10^3 forms a useful measuring instrument.

Image intensifier and image converter

The image intensifier and image converter tubes operate on the same principle. An image of the scene to be viewed is focused on a semitransparent photocathode deposited on the window of a vacuum tube; the photocathode in response to the incident photons emits electrons corresponding to the illumination of the various parts of the scene; the photo-generated electrons are accelerated on to a luminescent screen to produce a visual image of the original scene. Because each incident photon on the photocathode gives rise to some tens of photons at the screen, a gain in the intensity of the image is obtained.

The visual image is determined by the spectral response of the photocathode. If the response lies within the visible region of the spectrum, and the sensitivity of the photocathode is greater than that of the eye, an intensified image of a dimly-lit scene will be obtained. The device is therefore an image intensifer. If the response is outside the visible region, an 'invisible' image is converted to a visible one. The device is then an image converter. Any area of the scene that cannot produce sufficient illumination of the photocathode to generate electrons, cannot be reproduced in the final image, however great the gain of the tube. Therefore the optical system producing the image of the scene on the photocathode must be as efficient as possible. For example, in a night-viewing system incorporating an image intensifier tube, the objective lens should have a large aperture to collect the maximum number of photons from the scene.

The image intensifier tube was invented during the 1930s. It consisted essentially of a photocathode placed very close to and parallel with a luminescent screen. The spacing in the original device was approximately 2 mm. The screen was held at a positive potential with respect to the photocathode, so that the electrons emitted travelled in straight lines directly to the screen. Such a system produces a distortion-free image. Although this type of image intensifier was replaced by more complex types with electrode systems between the cathode and screen, it is coming into favour again. Modern manufacturing techniques enable closer spacing of the photocathode and screen to be obtained, approximately 0.6 mm. The voltage between screen and cathode is 5 kV to 6 kV. This type of image intensifier is known as the proximity focused type, and its main advantage over the other types is the direct and distortion-free image.

The form of an image intensifier developed during the 1950s and in use today is shown in *Figure 9.43*. A spherical photocathode is used. The emitted electrons are

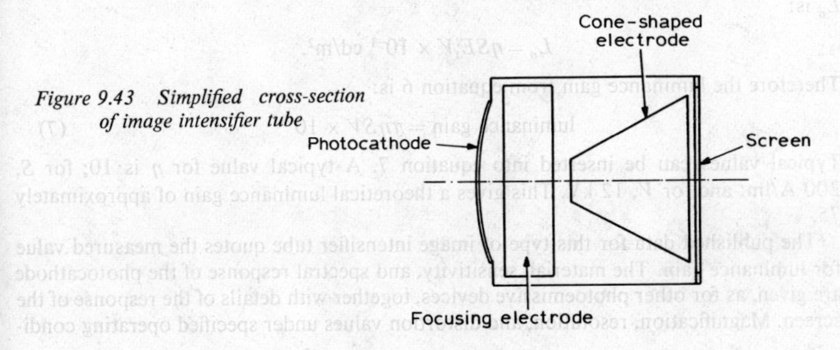

Figure 9.43 Simplified cross-section of image intensifier tube

focused by the cylindrical focusing electrode, which is held at the same voltage as the photocathode, at the apex of the cone-shaped electrode. The electrons are accelerated by the cone-shaped electrode on to the luminescent screen. Ideally the screen should also be spherical with the same radius of curvature as the photocathode, but in practice a flat screen is used. Slight distortion at the edges of the image will therefore occur. The cone-shaped electrode is held at the same voltage as the screen, and the voltage between screen and cathode is 12 kV to 15 kV. Typical dimensions for this type of image intensifier are a length of 50 mm, and a useful photocathode and screen diameter of 20 mm to 25 mm.

The phosphor used for the luminescent screen is chosen to suit the response of the eye for direct-viewing systems, or to suit films in systems where the final image is to be photographed. The back of the phosphor screen (away from the viewing side) is metallised to increase the brightness of the image, as in television picture tubes. It should be noted that the image with this type of image intensifier is inverted with respect to the photocathode. The input optical system, however, can produce an inverted image of the scene on the photocathode so the final image on the screen is correctly orientated.

The gain of an image intensifier is the ratio of the output luminance from the screen to the input illuminance of the photocathode. The exact definition is:

$$\text{luminance gain} = \frac{\pi L_o}{E_i} \qquad (6)$$

where L_o is the luminance of the screen in cadela per square metre (cd/m^2) in a direction normal to the screen, measured with an eye-corrected photometer having an acceptance angle less than 2°, and E_i is the illuminance in lux incident on a specified area of the photocathode produced by a tungsten lamp at a colour temperature of 2 856 K.

The luminance of the screen can be related to the illumination of the photocathode E_i. If the area of the illuminated part of the photocathode is A m^2, then the luminous power (called the luminous flux Φ) incident on the photocathode is:

$$\Phi = E_i \times \text{A lm}$$

If the sensitivity of the photocathode is S μA/lm, the photocurrent I_{ph} is:

$$I_{ph} = S \times E_i A \ \mu\text{A}$$

With a voltage V kV between the screen and cathode, the energy incident on the screen VI_{ph} is:

$$\text{energy} = SE_i A \times 10^{-6} \times V \times 10^3 \text{ W}$$

With a screen efficiency of η cd/W, the luminous intensity of the screen I_o is:

$$I_o = \eta \times SE_i AV \times 10^{-3} \text{ cd}$$

Assuming that the magnification between the image on the photocathode and that on the screen is 1, that is the illuminated area on the screen is A m^2, the luminance of the screen L_o is:

$$L_o = \eta SE_i V \times 10^{-3} \text{ cd/m}^2.$$

Therefore the luminance gain from equation 6 is:

$$\text{luminance gain} = \pi \eta SV \times 10^{-3} \qquad (7)$$

Typical values can be inserted into equation 7. A typical value for η is 10; for S, 200 A/lm; and for V, 12 kV. This gives a theoretical luminance gain of approximately 75.

The published data for this type of image intensifier tube quotes the measured value for luminance gain. The material, sensitivity, and spectral response of the photocathode are given, as for other photoemissive devices, together with details of the response of the screen. Magnification, resolution, and distortion values under specified operating conditions are given for the tube. Another quantity specified is the background equivalent illuminance. With the supply voltage applied but no input illumination on the photocathode, the screen will have a background luminance which may be caused by thermionic emission or field emission from the photocathode, electron or ion scintillations, or long-term phosphorescence of the screen from previous operation. This background luminance is equivalent to the 'dark' current already mentioned, for photoemissive tubes and photomultipliers, and may be regarded as the 'noise' of the system determining the smallest signal level. The quantity quoted in the data is measured as an increase in input illumination required to give an increase in screen luminance equivalent to the background luminance. The ratings include the maximum operating voltage that can be applied, and the maximum illumination of the photocathode that will not cause damage through excessive heating. The data for image converter tubes are similar.

If a night scene illuminated only by star light is to be viewed, with an image intensifier, the input illumination on the photocathode is approximately 10^{-3} lux. For comfortable viewing by the human eye, the output characteristic of the screen should have the peak response at a wavelength near the peak response of the eye (0.50 μm to 0.55 μm), and at an output luminance of 10 to 40 cd/m^2. Thus the gain required of the image intensifier is approximately 30×10^3. This gain can be achieved by connecting three image intensifiers of the type previously described in cascade. The output from the screen of the first stage forms the input for the photocathode of the second stage, and so

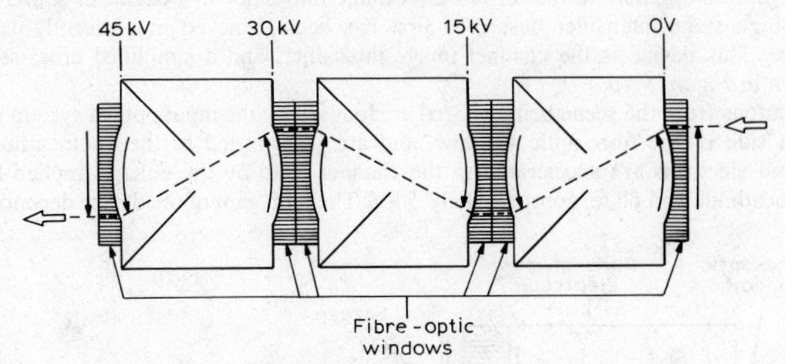

Figure 9.44 Schematic cross-section of three-stage image intensifier using fibre-optic coupling between stages

on. The stages are best coupled with plano-concave fibre-optics lenses, to give the assembly shown schematically in *Figure 9.44*. The radius of curvature of the concave side of the lens on which the photocathode or screen is deposited is chosen to optimise the electron focusing between the photocathode and the screen. The plane surface of the lens enables simple coupling of one tube to another in adjoining lenses to be achieved.

The operating voltages of the three stages are shown in *Figure 9.44*. These voltages can be obtained from a supply voltage of 2.7 kV at 1.5 kHz using a Cockcroft-Walton ladder voltage multiplier to obtain the higher voltages.

The circuit for such a voltage multiplier is shown in *Figure 9.45*. The connections to the second and third stages are made through high-value resistors. These are included to limit the brightness of the final image and to prevent discomfort to the observer in the event of a sudden and unexpected increase in the illumination of the scene. The components of the voltage multiplier circuit can be encapsulated with the image intensifier tube to form a portable assembly requiring only the 2.7 kV supply voltage.

Typical values can be inserted into equation 7 to calculate the theoretical luminance gain for the second and third stages of the assembly. If η is 10 as before; S, the green (0.50 μm to 0.55 μm) sensitivity, is 50 μA/lm; and V is 15 kV, then the gain for each of these stages is approximately 23. This gives an overall theoretical gain for the three-stage image intensifier of approximately 40×10^3.

The gain of practical image intensifiers of this type can be as high as 10^5. The typical dimensions of a practical device are a length of 200 mm and a diameter of 70 mm. The useful diameter of photocathode and screen is typically 25 mm.

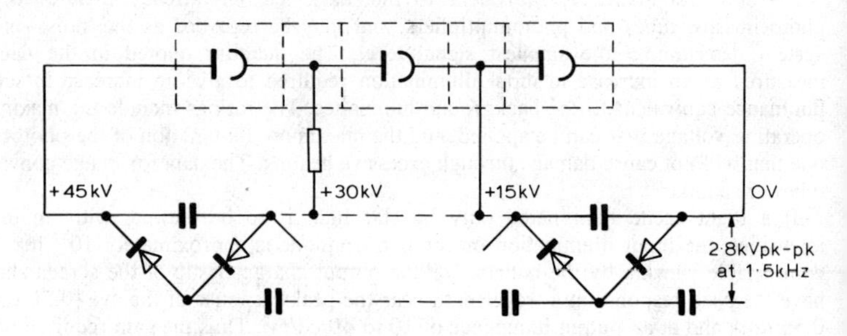

Figure 9.45 Voltage multiplier for use with three-stage image intensifier

A gain comparable to that of the three-stage intensifier in a device of similar size to the single-stage intensifier described first has been achieved in a recently-developed device. This device is the channel image intensifier, and a simplified cross-section is shown in *Figure 9.46*.

Photons from the scene being viewed are focused by the input optical system onto the plane side of the fibre-optic window, and are transmitted to the photocathode. The emitted electrons are accelerated to the channel plate by the voltage applied between photocathode and plate, approximately 5 kV. The high gain of the device depends on the

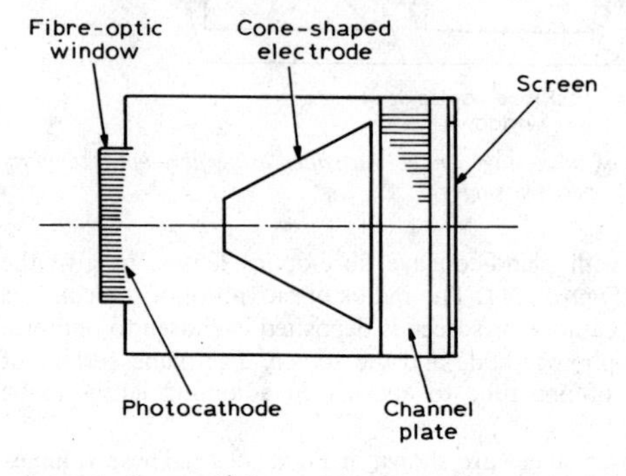

Figure 9.46 Simplified cross-section of channel image intensifier tube

channel plate, which may be regarded as a very large number of channel electron multipliers packed closely together. The plate is approximately 1 mm thick, and a voltage of 1 kV is maintained across it. The electrons from the photocathode produce further electrons by secondary emission in the channels (Section **9**–34 Channel Electron Multiplier), so that a large number of electrons are ejected from the channel plate. These electrons are accelerated to the screen by the voltage applied between the channel plate and screen, approximately 5 kV. Because of the close spacing of the channel plate and screen, between 0.5 mm and 0.8 mm, the electrons travel directly to the screen with little spreading. The resolution of the device is therefore high.

Electron gains in the channel plate of 10^3 to 10^5 can be achieved with the applied voltage of 1 kV if the channel length is 40 to 60 times the diameter, and if the secondary-emission coating on the inside wall of the channel has a resistance of approximately 10^9 Ω. The individual channels must be closely spaced in a regular array, and maintain their geometry and electrical properties, to achieve good resolution in the final image. The gain from channel to channel should be constant to prevent the generation of spatial noise. The diameter of the channel should be constant to within ±10%.

The manufacturing method for the channel plate is based on fibre-optic techniques. The material used is an electrically-conducting glass incorporating lead compounds. A glass core rod is used to draw a solid glass fibre from the conducting glass, the core rod and glass fusing together during the drawing operation. Bundles of these fibres are fused together in a regular array. Further drawing operations reduce the diameter of the fibres to the size required in the channel plate, 20 μm. The array is then cut to the required thickness of the channel plate, 1 mm, and the faces polished. The solid glass cores are dissolved away in an acid etch, and the plate heated in a hydrogen atmosphere to reduce the lead oxides in the glass to metallic lead. This metallic lead forms a thin coating, approximately 10 nm deep, on the inside wall of each channel. A nichrome electrode is

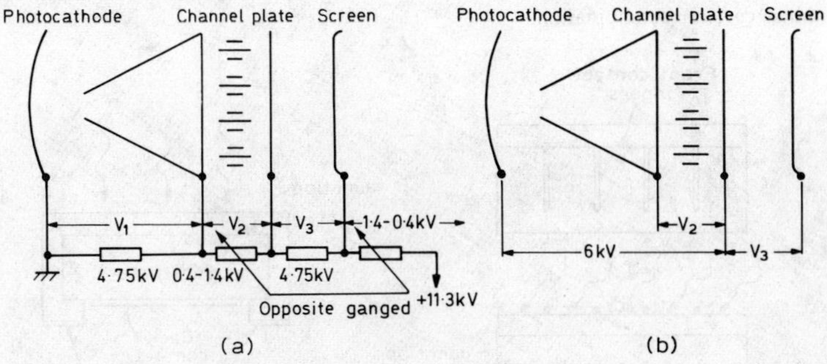

Figure 9.47 *Voltage supply for channel image intensifier tube: (a) using voltage divider network (b) using fixed channel-plate-to-screen voltage*

deposited on each face of the plate to enable the voltage to be applied across each channel. A typical channel plate has a diameter of 18 mm and contains approximately 2×10^6 channels.

The gain of the channel image intensifier can be varied by varying the gain of the channel plate. This is achieved by varying the voltage across the plate. The voltages in the channel image intensifier are shown in *Figure 9.47*. The voltage between the photocathode and the input of the channel plate is V_1, and that between the screen and the output of the channel plate is V_3. The voltage across the plate itself is V_2. Ideally the change in V_2 should not affect the other two voltages, so that the electrostatic focusing of the electrons from the photocathode, and the proximity focusing of the electrons reaching the screen, remain unchanged. A resistive voltage divider network to provide the required voltages is shown in *Figure 9.47 (a)*. Typical voltage values are indicated. An alternative supply is shown in *Figure 9.47 (b)*. In this, voltage V_3 remains constant at 4.75 kV while V_2 is varied. The voltage between the output of the channel plate and the photocathode, $V_1 + V_2$, is always equal to 6 kV. In this mode of operation, there may be a slight loss of performance but in practice this is not significant. Because of the ease with which the gain can be varied, this type of image intensifier can be incorporated in to systems to give constant gain control. The gain of the channel image intensifier plotted against channel plate voltage is sometimes given in published data.

Image converters are used to produce a visible image of 'invisible' scenes, for example to convert an infra-red image into a visible one. The applications are generally in the infra-red region, and the photocathode material is usually caesium on silver oxide. The military and security applications of image intensifiers are obvious. Other applications include the viewing of fluorescent screens when illuminated at very low levels, for example in electron microscopes, and in improving the performance of astronomical instruments. The photocathodes for these applications are made from combinations of antimony, potassium, sodium, and caesium.

SOLAR CELL

The solar cell is a form of photodiode which is optimised for operation from the sun's radiation. The operating principle is the same as that of the photodiode, the generation of electron-hole pairs by the incident radiation, and the separation of these charge carriers by the electric field of the depletion layer at a *pn* junction. The flow of minority carriers across the junction produces a short-circuit current or open-circuit voltage so that power can be extracted from the device by a suitable load resistance. The surface area of the solar cell is made as large as possible so that the maximum amount of radiation is incident on the device. The *pn* junction is formed near the surface to minimise carrier recombination.

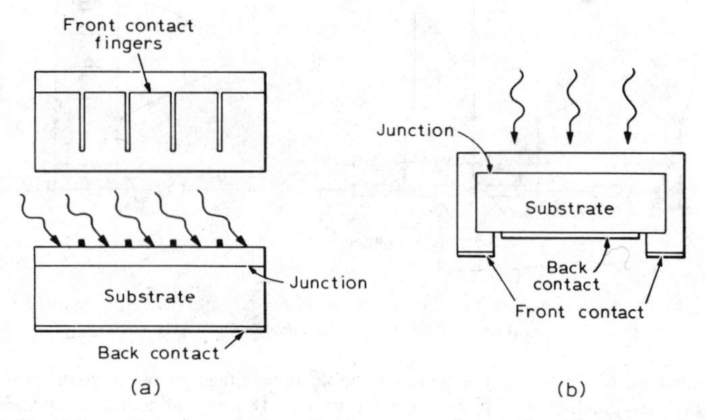

Figure 9.48 Schematic structure of solar cell: (a) contact on front surface (b) contact on back of cell

The most commonly used materials from which practical solar cells are manufactured are silicon and gallium arsenide. The cell can be constructed as a thin *p*-type layer on an *n*-type substrate, or as a thin *n* layer on a *p* substrate. Both regions of the cell are heavily doped. Non-rectifying contacts must be made to both regions. The contact of the substrate can be made on the back of the device, but the contact to the front layer must be made in such a way that the minimum surface area is obscured. Narrow contact fingers can be deposited on the front surface, as shown in the schematic structure of *Figure 9.48* (*a*), or the material of the front layer can be taken round the sides of the cell to the back of the substrate and the contact made there, *Figure 9.48* (*b*).

The relationship between the voltage and photocurrent for a solar cell is shown in *Figure 9.49* with the maximum-power rectangle superimposed. This rectangle represents the maximum amount of power that can be extracted from the cell, the product $V_{mp} \times I_{mp}$. With a suitable load resistance, this power can be 80% of the product $V_{oc} \times I_{sc}$, where V_{oc} is the open-circuit voltage and I_{sc} is the short-circuit current. Typical values of the open-circuit voltage and short-circuit current are 0.5 V and 0.1 A respectively.

Solar cells can be connected in series to provide a higher voltage than can be obtained from a single cell, and connected in parallel to provide a higher current. The main application of solar cells is for the power supplies of space vehicles and communication satellites. Large arrays of cells are used, connected in series/parallel, to provide the required values of voltage and current to drive the electronic circuitry in these vehicles.

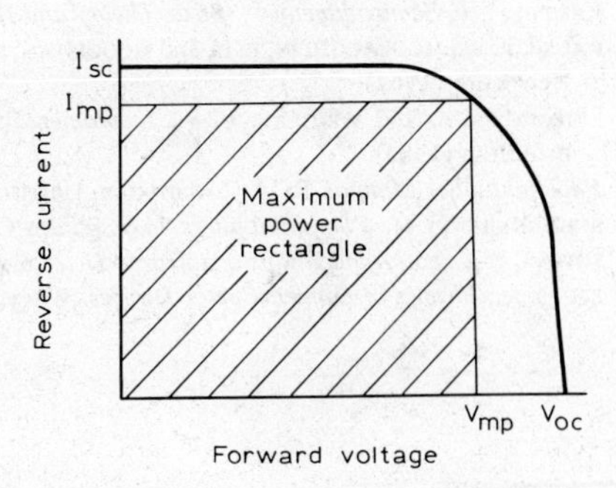

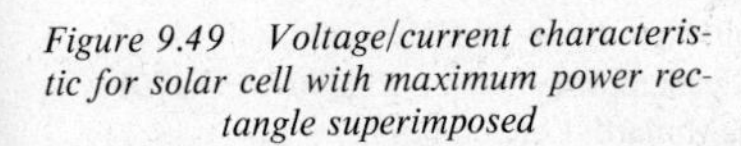
Figure 9.49 Voltage/current characteristic for solar cell with maximum power rectangle superimposed

The conversion efficiency of such batteries is between 10% and 15%. Another application of solar cells is for terrestial power supplies. In areas where extended periods of sunlight can be relied on, arrays of solar cells can be used as a power source.

SYMBOLS FOR PRINCIPAL PARAMETERS OF PHOTO-ELECTRONIC DEVICES

Symbols for the main electrical and optical parameters of photo-electronic devices, other than those common to small-signal diodes and transistors, are listed below.

Symbol	Meaning
D*	area-normalised detectivity of photoconductive detector
E	illumination
E_i	input illumination on photocathode of image intensifier tube
$I_{CE(D)}$	dark collector current of phototransistor
$I_{CE(L)}$	light collector current of phototransistor
I_{cell}	current through active element of photoconductive detector
I_{CS}	short-circuit current of solar cell
I_{RD}	dark current of photodiode
$I_{R(SC)}$	short-circuit current of photodiode
L_o	output luminence from screen of image intensifier tube
M_c	magnification of image intensifier tube at centre of screen
M_d	magnification of image intensifier tube towards outer edge of screen
N.E.P.	noise equivalent power
P_{cell}	power dissipation of active element of photoconductive detector
S, S_R	sensitivity of photodiode and phototransistor, change of current with incident radiation
T_{tablet}	operating temperature of active element of photoconductive detector
V_b	supply voltage for photomultiplier
V_{cell}	voltage across active element of photoconductive detector
V_{CO}	Open-circuit voltage of solar cell
V_d	Dynode voltage of photomultiplier
η	efficiency of screen in image intensifier tube
λ, λ_p	peak spectral response

FURTHER READING

HOUGHTON, J. T. and SMITH, S. D, *Infra-red physics*, O.U.P., Oxford (1966)

KAMPEL, I. J., *Semiconductors—Basic Theory and Devices*, Newnes-Butterworths, London (1971)

KRUSE, P. W., MCGLAUGHLIN, L. D. and MCQUISTAN, R. B., *Elements of Infrared Technology*, Wiley, New York (1962)

LINDMAYER, J. and WRIGLEY, C. Y., *Fundamentals of Semiconductor Devices*, Van Nostrand, Princeton (1965)

Photomultiplier Manual, RCA Corporation, Harrison, New Jersey (1970)

SCHONKEREN, J. M., *Photomultipliers*, N.V. Philips Gloeilampenfabrieken, Eindhoven (1970)

SOWAN, F. A., ed., *Applications of Infrared Detectors*, Mullard, London (1971)

SZE, S. M., *Physics of Semiconductor Devices*, Wiley, New York (1969)

10 ELECTRO-OPTIC DEVICES

10 ELECTRO-OPTIC DEVICES

GENERAL

The generation of an electron-hole pair in a semiconductor material by the absorption of energy from visible or infra-red radiation requires a certain minimum amount of energy to raise the electron from the valence band to the conduction band. In the recombination of an electron and hole, the electron falls from the conduction band to the valence band, and a certain amount of energy is released. Whether this energy is emitted as light or absorbed in the crystal as heat depends on the details of quantum mechanics concerning the momenta of electrons in the conduction and valence bands for the material. Silicon and germanium are generally not light-emitting; certain other semiconductor materials are. The most common of the light-emitting semiconductor materials are gallium arsenide and gallium arsenide phosphide.

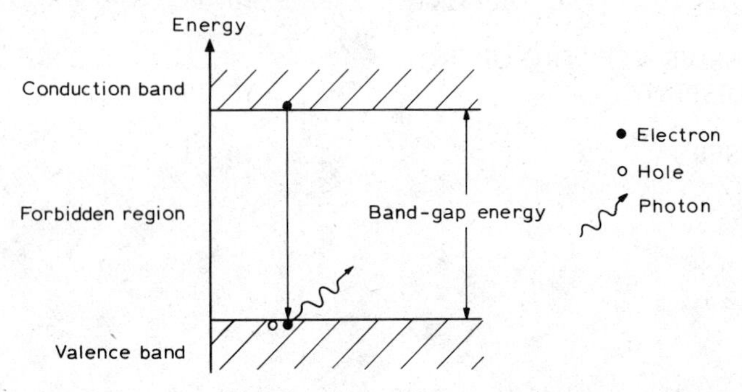

Figure 10.1 Permissible energy levels for electron in semiconductor material, electron-hole recombination

Once an electron is raised to the conduction band, it tends to fall to the lowest possible energy in the band. Thus in a semiconductor material the energy released during recombination will be the band-gap energy, as shown in *Figure 10.1*. The value of the band-gap energy E_g determines the wavelength of the emitted radiation, the wavelength λ_e being given by $\lambda_e = hc/E_g$, from the relationship between the energy of the photon and its wavelength given in equation 2 of Section 9 (page **9**–3). Gallium arsenide emits radiation with a peak response at 0.88 μm, that is in the near-infra-red region. Gallium phosphide when doped with, for example, cadmium emits either green light with a wavelength of 0.56 μm or red light with a wavelength of 0.70 μm depending on the concentration of the cadmium. Practical electroluminescent diodes (also known as light-emitting diodes or l.e.d.s) for visible radiation are manufactured from gallium arsenide phosphide. The addition of phosphorus to the gallium arsenide increases the band-gap energy, and so the wavelength of the emitted radiation is shorter, changing from infra-red to visible radiation with a wavelength of 0·65 μm. A feature of the emitted radiation, whether infra-red or visible, is the narrow bandwidth. The intensity of the radiation emitted by the diode depends on the rate of recombination of electrons and holes within the material. It is necessary therefore to ensure the greatest possible concentration of carriers is present to aid the recombination. This can easily be done by using a forward-bias diode. Electrons from the *n* region cross the junction into the *p* region where many recombine with the holes there. Since the carriers crossing the junction form the current through the diode, the number of recombinations and hence the intensity of the emitted radiation will be proportional to the forward current of the diode.

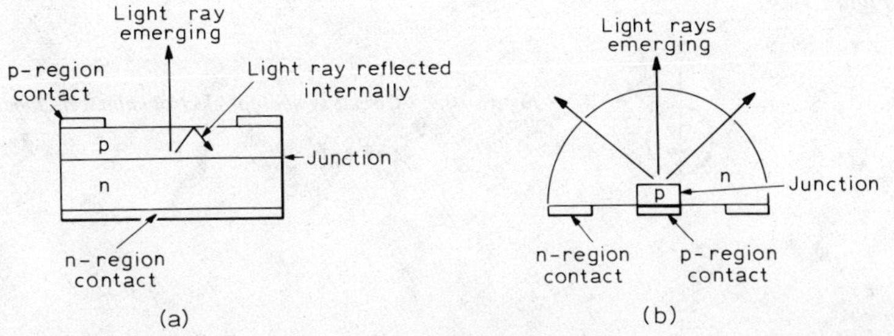

Figure 10.2 Schematic cross-section of electroluminescent diode element: (a) rectangular form (b) hemispherical form

The radiation emitted from the recombination sites will be attenuated by absorbtions in the material before the radiation emerges, and by internal reflection. The schematic cross-section of two forms of electroluminescent diode element are shown in *Figure 10.2*. In the rectangular element, *Figure 10.2 (a)*, radiation striking the front face at an angle greater than the critical angle is internally reflected. For gallium arsenide, the critical angle is 16°. Thus a narrow beam of radiation will be emitted by the rectangular structure. If the front face of the element is made hemispherical, *Figure 10.2 (b)*, the radiation is emitted as a wider beam.

The published data for electroluminescent diodes includes the characteristic of emitted radiation with forward current, and the spectral response. The relationship between the luminous intensity of the emitted light, or the radiated power of infra-red radiation, and forward current is linear over most of the current range, as shown in *Figure 10.3*. This characteristic is given for both continuous and pulsed currents.

Figure 10.3 Luminous intensity plotted against forward current for electroluminescent diode

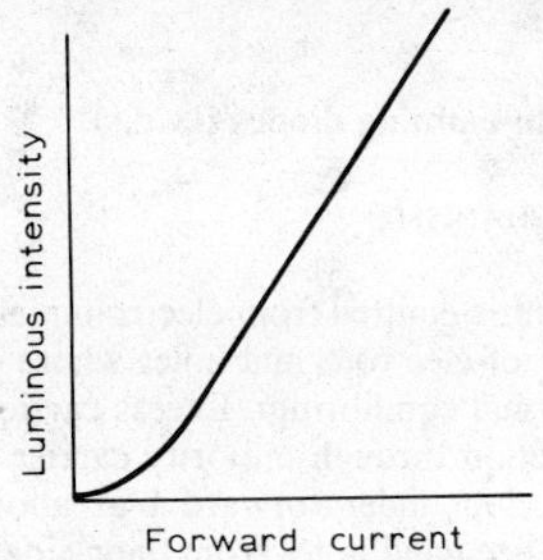

Typical values for forward current are 30–50 mA for continuous current and up to 200 mA for pulsed operation. The spectral response for both gallium arsendide and gallium arsenide phosphide diodes is narrow, with a typical bandwidth at the half-height points of 0.02–0.04 μm. The peak response for gallium arsenide occurs at 0.88 μm, in the near-infra-red region, and for gallium arsenide phosphide at 0.65 μm, in the visible red region. The variation of the emitted radiation with the junction temperature is also given in published data.

Other characteristics and ratings for electroluminescent diodes are similar to those for small-signal diodes. The forward voltage/current characteristic is given, and the ratings given include the maximum forward current, maximum reverse voltage, total power dissipation, and maximum permissible junction temperature. A power derating curve is also often given, and a thermal resistance so that a suitable heatsink can be chosen if required.

The typical encapsulation for electroluminescent diodes is a small metal envelope. A plane glass window or glass lens is fitted to the envelope to allow end-emission of the radiation. The circuit symbol for an electroluminescent diode is shown in *Figure 10.4*.

Figure 10.4 Circuit symbol for electroluminescent diode

LIGHT-EMITTING DISPLAYS

Light-emitting (active) displays differ from passive displays in that the light emitted is generated by the display itself through conversion of electrical energy. Good visibility requires that the amount of light emitted be at least comparable to any ambient light present. Therefore the amount of electrical power consumed usually scales directly with the area of the display, with ambient light, and inversely with the conversion efficiency of electrical energy to light.

Since the human eye perceives the brightness of a pulsed display as the average of the light emitted over the period (provided the repetition rate is above the perceivable flicker rate, typically ~50 Hz), an active display can have appreciable brightness even if its duty cycle ('on' to 'off' ratio) is very small. This fact is useful in multi element displays where each display element can be addressed sequentially in a time-shared mode. In passive displays, where the perceived brightness is directly related to the duty cycle ('on' to 'off' ratio), each element must be continuously addressed or exhibit storage.

Light-emitting diodes (l.e.d.s)

MECHANISM

Light is emitted from electroluminescent *pn* junctions as a result of radiative recombination of electrons and holes whose concentrations exceed those statistically permitted at thermal equilibrium. Excess carrier concentrations are obtained in a forward-biased *pn* junction through minority carrier injection: the lowering of the potential barrier of the junction under forward bias allows conduction band electrons from the *n*-side and valence band holes from the *p*-side to diffuse across the junction. These injected carriers significantly increase the minority carrier concentrations and recombine with the oppositely charged majority carriers. This recombination, which tends to restore the equilibrium carrier densities, can result in the emission of photons, i.e., light, from the junction. The recombination process, hence the amount as well as the wavelength of the light generated, is a strong function of the physical and electrical properties of the material.

Both energy and momentum must be conserved when an electron and a hole recombine to emit a photon. Since the photon has considerable energy but very small momentum ($h\nu/c$), simple recombination only occurs in direct-bandgap materials, that is, where the conduction band minimum and valence band maximum both lie at the zero momentum position. This condition was assumed to be a prerequisite for efficient electroluminescence until about 1964 when Grimmeiss and Scholz[1] demonstrated reasonably efficient electroluminescence in the indirect-bandgap material GaP.

In an indirect-bandgap material where the valence band maximum and conduction band minimum lie at different values of momentum, recombination can only occur when a third momentum-conserving particle is involved; phonons (i.e., lattice vibrations) serve this purpose. The probability of an electron-hole recombination involving both a photon and phonon is considerably smaller than the simpler process involving only a photon in a direct-bandgap material. This is clearly illustrated by the differences in recombination coefficients (R) for various materials: this coefficient (R), which relates the radiative recombination rate (r) to the excess minority (Δn) and majority (p) carrier concentrations

$$r = R(\Delta n)p$$

is of the order of 10^{-14} to 10^{-15} cm^3/s for indirect-bandgap materials (such as Si, Ge and GaP) while it is of the order of 10^{-10} to 10^{-11} cm^3/s for direct-bandgap materials (such as GaAs, GaSb, InAs or InSb).

L.E.D.s can, in principle, be made from any semiconducting compound containing a *pn* junction, and having a sufficiently wide bandgap. Despite considerable effort on other materials to achieve efficient luminescence, currently only III–V compounds are of practical interest for visible l.e.d.s.

DOPING

Donor and acceptor impurities determine the magnitude and type of conductivity and also play a major role in the radiative recombination processes. In direct-bandgap III–V compounds, donor impurities are usually chosen from group VI of the periodic table (Te, Se, S), while acceptor impurities usually belong to group II (Zn, Cd, Mg). Sometimes group IV impurities (Ge, Si, Sn) are used either as acceptors or donors depending on which of the available lattice sites (III or V) they occupy.

Optimum impurity concentrations are best determined experimentally. High doping levels, i.e., large majority carrier concentrations, are desirable (1) to lower the bulk resistivity—and consequently to minimise heating and voltage drops at high current densities—and also (2) because the recombination probability, which is directly proportional to the carrier concentration, should be as large as possible. The doping concentrations are practically limited, however, by the formation of precipitates and other crystallographic imperfections which introduce competing nonradiative recombination centres and therefore reduce the electroluminescence efficiency. Donor concentrations of 10^{17} cm^{-3} to 10^{18} cm^{-3} and acceptor concentrations of 10^{17} cm^{-3} to 10^{19} cm^{-3} are typical.

Although the luminescence efficiencies in *n*- and *p*-type material at optimum doping concentrations are essentially equal, the light emitted from the *p*-region normally dominates for several reasons: (1) electron injection into the *p*-region is favoured over hole injection into the *n*-region because of the high electron-to-hole mobility ratio in most III–V compounds: (2) the Fermi level is slightly higher than the intrinsic energy gap in *n*-type material, while it is lower in *p*-type material; the resulting heterojunction effect further favours electron injection into the *p*-region over hole injection into the *n*-region; (3) light generated in the *n*-region is usually of shorter wavelength (higher energy) than that generated in the *p*-region; therefore *n*-generated light is strongly absorbed in the *p*-region while *p*-generated light passes through the *n*-region with reduced absorption losses. This latter fact is usually taken into account in the design of efficient l.e.d.s.

In indirect-bandgap materials, e.g., GaP, recombination across the gap requires the participation of a momentum-conserving phonon and is therefore inefficient. More efficient recombination can occur when a charged carrier is first trapped at a neutral impurity centre and then used to attract the oppositely charged carrier. Momentum conservation in this case is more easily satisfied because the carrier trapped at a neutral impurity centre is highly localised in space and consequently has a wide range of crystal momentum.

Only a limited number of impurities have been found which enhance the recombination in GaP. Near-bandgap (~2.23 eV) green emission is increased by nitrogen substitution for P in GaP, but competing non-radiative recombination processes limit the internal quantum efficiency to about 1%. Red luminescence (~1.79 eV) is improved by incorporation of Zn and oxygen centres. The internal quantum efficiency in GaP : Zn,O is higher (10–20%) than for green luminescence, but the emission saturates at relatively low current densities ($\lesssim$ 10 A/cm^2) due to the limited concentration of Zn-O centres ($<10^{17}$ cm^{-3}); green luminescence in GaP : N does not readily saturate since N-concentrations of 10^{19} cm^{-3} are practical.

QUANTUM EFFICIENCY AND BRIGHTNESS

Since most of the applications of l.e.d.s involve an observer, the response of the human eye at the emitted wavelength is of primary importance. The eye sensitivity for normal (photopic) vision extends from ~4 000–7 000 Å; it peaks in the green (5 550 Å) at 680

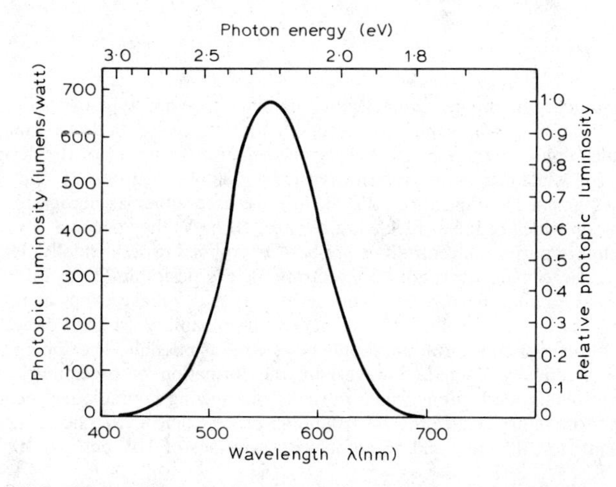

Figure 10.5 Photopic luminosity (normal vision) as a function of incident wavelength

lumens/watt and falls off towards the red and blue regions of the spectrum, *Figure 10.5*. The brightness, and to a large extent the visibility, of an l.e.d. is proportional to the product of its external quantum efficiency and the sensitivity of the eye at the emitted wavelength.

The external quantum efficiency of an l.e.d. i.e., the ratio of emitted photons to number of electrons passing through the diode; typically 0.1–7% at room temperature, is always less than the internal quantum efficiency ($\gtrsim$ 50% under optimum conditions), because all the light generated cannot exit from the diode.

The internal quantum efficiency is highly dependent of the perfection of the material near the *pn* junction. Various defects, contaminants or dislocations reduce the internal quantum efficiency by producing deep recombination centres, which lead to long wavelength radiation, or by enhancing nonradiative recombination. It is thought that the superiority of GaP:N diodes grown by liquid-phase epitaxy (l.p.e.) compared to those grown by vapour-phase epitaxy (v.p.e.) may be due to the lower density of Ga vacancies in l.p.e. material.

Poor-quality substrates are a major cause of defects. Additional imperfections can be introduced during the growth of the epitaxial layer, especially when the lattice mismatch is relatively large. These latter imperfections can be reduced by grading the composition of the film during growth from that of the substrate to the composition desired at the junction. Defects are also sometimes introduced during the diffusion process used to form the *pn* junction.

The high index of refraction of III–V compounds (typically ~3.5) leads to additional light losses. Much of the generated light suffers total internal reflection. This increases the optical path length inside the diode thereby increasing internal optical absorption which is particularly high for near-bandgap emission. Thus the external quantum efficiency can be 50–100 times smaller than the internal efficiency.

Practically, these losses can be reduced by increasing the transmissivity of the surface: anti-reflection coatings can be applied; the diode can be shaped in the form of a hemisphere, but this is very costly; or a hemispherical epoxy or acrlyic lens is used to increase the efficiency by a factor of 2–3 typically. The internal absorption is reduced at longer emission wavelengths, which are obtained by incorporating deeper acceptors, e.g., Si instead of Zn in GaAs, or in indirect-bandgap materials, where the emission is much below the energy gap. In indirect-gap materials, the reduced internal efficiency (10–20% for GaP:Zn,O as compared to ~50% for GaAs:Zn) sets an upper bound for the external efficiency of the device.

The brightness (B) of an l.e.d. in nits (candle/metre^2) is given by

$$B = \frac{3\,940\,\eta_{ext}\,LJ}{\lambda}\left(\frac{A_j}{A_s}\right)$$

where η_{ext} = external quantum efficiency
L = luminous efficiency of the eye (*Figure 10.5*) in lumens/watt
J = junction current density in A/cm^2
$\left(\frac{A_j}{A_s}\right)$ = ratio of junction area to observed emitting surface
λ = emission wavelength in μm

Brightness in excess of 3 500 nits at 10 A/cm^2 is readily achieved in commerical l.e.d.s. By way of comparison, the brightness of a t.v. picture is of the order of 300 nits; that of the surface of a frosted light bulb as much as 30 000 nits.

TERNARY COMPOUNDS

In view of the sharp peak of the eye sensitivity in the green region of the spectrum, it is very desirable to obtain yellow or green luminescence. The upper limit for the emission energy is roughly equal to the energy gap. Most direct-bandgap III–V binary compounds have energy gaps less than 1.72 eV (720 nm) and therefore can luminesce only in the infra-red. Other semiconductors, with energy gaps closer to 2.23 eV (555 nm) are therefore of great interest.

Ternary alloy systems, composed of narrow direct-gap and wider indirect-gap materials, e.g., $GaAs_{1-x}P_x$, $Al_xGa_{1-x}As$, $In_{1-x}Ga_xP$, $In_{1-x}Al_xP$, provide a monotonically increasing direct-energy gap as the relative concentration of the indirect-gap material is increased up to a critical composition x_c where the energy gap becomes indirect. This is shown in *Figure 10.6* for the $GaAs_{1-x}P_x$ system.

The radiation recombination rate decreases sharply near the critical composition because an increasing fraction of electrons is transferred to the indirect band which typically has 50–100 times as many available energy states as the direct band.

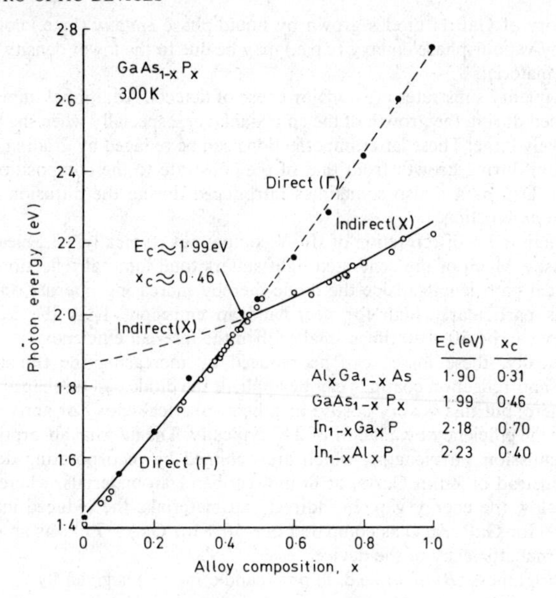

Figure 10.6 Direct (Γ) and indirect (×) conduction band minima for $GaAs_{1-x}P_x$ as a function of alloy composition, x. Closed data points are from electroreflectance measurements; open points are from electroluminescence spectra (Courtesy American Vacuum Society)

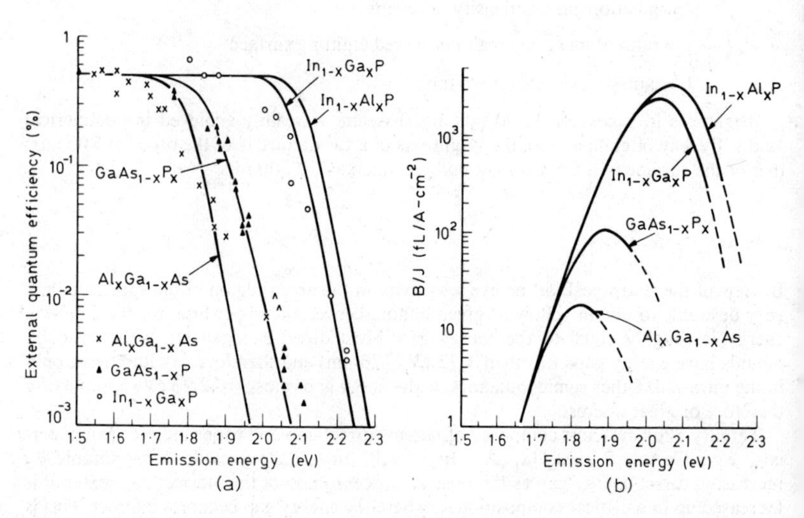

Figure 10.7 (a) Calculated external quantum efficiency for $Al_xGa_{1-x}As$, $GaAs_{1-x}P_x$, $In_{1-x}Ga_xP$ and $In_{1-x}Al_xP$ l.e.d.s as a function of emitted photon energy (b) Calculated brightness values for the same alloys (Courtesy American Vacuum Society)

From a knowledge of the energy bands, the external quantum efficiency and luminous efficiency at different wavelengths can be predicted.[2] This is illustrated in *Figure 10.7* for the ternary compounds of major importance. The alloy systems $In_{1-x}Ga_xP$ and $In_{1-x}Al_xP$ have potentially higher luminous efficiencies than either $GaAs_{1-x}P_x$ or $Al_xGa_{1-x}As$, but other aspects—such as the ease of preparation of the alloy, its defects, the extent of lattice mismatch—are also important in determining the choice of materials.

MATERIAL SYNTHESIS

L.E.D.s are generally fabricated from epitaxial material deposited on single-crystal GaAs GaP substrates. The choice of substrates depends on the lattice constant and thermal expansion coefficient of the epitaxial layer. The lattice constant of ternary alloys depends on the composition. A good match between substrate and epitaxial film can be obtained for $Al_xGa_{1-x}As$, a moderate match for $GaAs_{1-x}P_x$, and relatively poor match for $Al_xIn_{1-x}P$ and $In_{1-x}Ga_xP$.

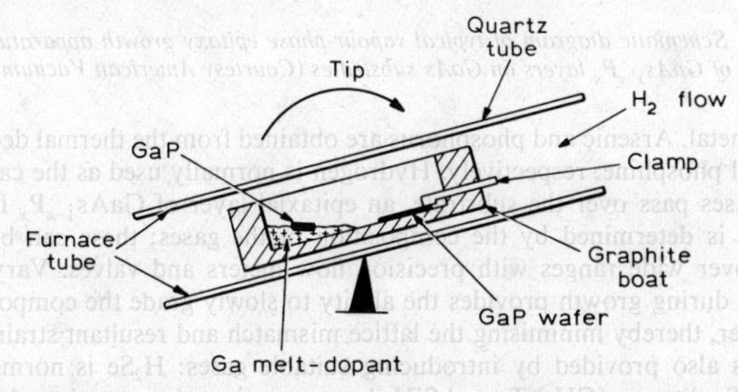

Figure 10.8 Liquid-phase epitaxy method used for growing GaP layers on a GaP seed near 1 000 °C (Courtesy American Vacuum Society)

Two techniques are widely used for the growth of the epitaxial films: liquid-phase epitaxy (l.p.e.) and vapor-phase epitaxy (v.p.e.). Each technique has advantages and disadvantages; v.p.e. is used for $GaAs_{1-x}P_x$, while l.p.e. produces the best GaP and $Al_xGa_{1-x}As$ devices at this time.

In liquid-phase epitaxy (l.p.e.), originally developed by Nelson,[3] the layer—for instance GaP—is grown from a Ga melt to which are added polycrystalline GaP and desired dopants such as Zn (*p*-type) or Te (*n*-type). The melt is positioned at one end of a graphite boat and a polished substrate at the other end, *Figure 10.8*. The boat is heated to about 1 060 °C in hydrogen and tipped so that the melt covers the substrate; upon cooling, an epitaxial GaP layer grows on the substrate. Similar techniques are used for the growth of other compounds.

Variations of the original l.p.e. technique have been developed to improve the quality of the films, ease of fabrication, or to extend the process to large-scale production. A dipping technique,[4] in which the substrate is immersed for a specified time into the melt, provides constant-temperature growth, which is advantageous for the growth of ternary or more complex compounds. A multiple-bin technique[5] was used to facilitate sequential deposition of several epitaxial layers. Also, a new technique which uses vapor-doping of the melt,[6] e.g., introduction of Zn vapour to grow a *p*-layer on the previously grown *n*-layer, has produced excellent results in the fabrication of GaP:N green l.e.d.s.

Vapour-phase epitaxy (v.p.e.) is widely used in the commercial preparation of red $GaAs_{1-x}P_x$ l.e.d.s. A typical v.p.e. system, as originally described by Tietjen and Amick,[7] is shown in *Figure 10.9*. Here the Ga is transported by flowing HCl gas over

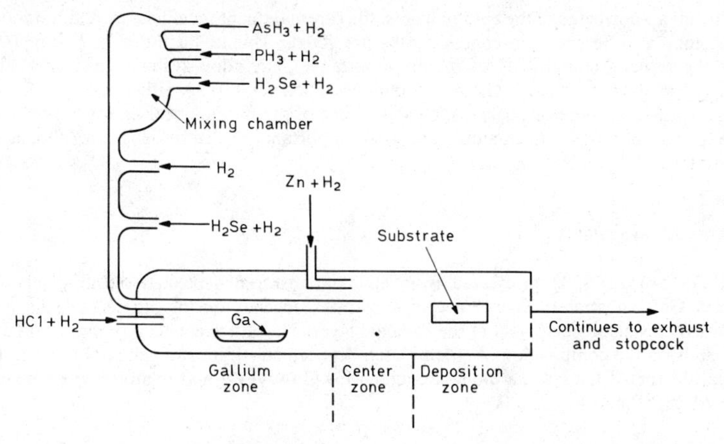

Figure 10.9 Schematic diagram of typical vapour-phase epitaxy growth apparatus used for the deposition of $GaAs_{1-x}P_x$ *layers on GaAs substrates (Courtesy American Vacuum Society)*

the molten metal. Arsenic and phosphorus are obtained from the thermal decomposition of arsine and phosphine, respectively. Hydrogen is normally used as the carrier gas.

As the gases pass over the substrate, an epitaxial layer of $GaAs_{1-x}P_x$ forms whose composition is determined by the composition of the gases; these can be controlled accurately over wide ranges with precision flow meters and valves. Varying the gas composition during growth provides the ability to slowly grade the composition of the epitaxial layer, thereby minimising the lattice mismatch and resultant strains.

Dcping is also provided by introducing suitable gases: H_2Se is normally used to incorporate Se donors; $(CH_3)_2Te$ and SiH_3 have sometimes been used to obtain Te or Si doping, respectively. Acceptor impurities can also be incorporated at the appropriate moment, e.g., by adding Zn vapour and H_2, but more often a postgrowth diffusion of Zn metal is used to form the *p*-layer; this latter approach has yielded the highest efficiencies to date for $GaAs_{1-x}P_x$ red l.e.d.s.

In addition to the original horizontal v.p.e. systems, much larger vertical reactors have been developed for mass production. These systems process many wafers simultaneously (~300 cm² of substrates at-a-time), usually rotating the substrates during growth to improve uniformity. The availability of multi-wafer-growth equipment and high quality GaAs substrates, coupled with the fact that the epitaxial layers require no

Table 10.1

Material	*Commercially available*	*Colour*	λ (nm)	η_{ext} (%)
GaP:Zn,O	Yes	Red	690	3–15*
$Al_{.3}Ga_{.7}As$	No	Red	675	1.3
$GaAs_{.6}P_{.4}$	Yes	Red	660	0.5
$In_{.42}Ga_{.58}P$	No	Amber	617	0.1
$GaAs_{.25}P_{.75}$:N	Yes	Amber	601	0.04
GaP:N	Yes	Green	550	0.05–0.7*
SiC	Yes	Yellow	590	0.003
GaAs:Si	Yes	IR	905	10–30

* Range between commercially available and best laboratory results.

postgrowth polishing, have contributed significantly to the commercialization and reductions in cost of red l.e.d.s.

V.P.E. is also used for preparation of GaN[8] and $In_{1-x}Ga_xP$[9] but is less suited for $Al_xGa_{1-x}As$ because of the reactivity of Al-containing gases.

To date, l.e.d.s made from III–V materials GaP, GaAs, $Al_xGa_{1-x}As$, $GaAs_{1-x}P_x$, $In_{1-x}Ga_xP$ are of greatest practical and commercial significance. *Table 10.1* shows some of the performance characteristics of these diodes.

$Al_xGa_{1-x}P$ diodes have also been reported,[10] with poor ($\sim 10^{-5}$) quantum efficiency. $In_{1-x}Al_xP$, which is ultimately believed to have good potential for high-brightness l.e.d.s, is difficult to synthesise and has not yet been made to luminesce.

OTHER MATERIALS

SiC diodes have been studied extensively in the past because they are, in principle, capable of emitting light throughout the whole visible spectrum. Although SiC l.e.d.s are available, the luminous efficiencies to date have been generally disappointing.

Of some interest also is the use of up-conversion phosphors. These phosphors convert two or more infra-red photons into a single higher-energy visible photon.[13] When an efficient infra-red l.e.d. (e.g., GaAs:Si with $\eta_{ext} = 10\%$ @ 930–970 nm) is coated with a thin layer of up-converting phosphor, red, green or blue light can be obtained. However, since the intensity of the emitted light varies as the second (or higher) power of the infra-red light intensity, the l.e.d. must be driven very hard in order to achieve useful overall efficiencies.

Up-converting phosphors with high efficiencies (30–50%)[14] have been prepared. However, in practice, the phosphor layers must be thin in order to minimise self-absorption; consequently, only a small portion of the emitted infra-red light is absorbed in the phosphor and the rest is wasted.

There is also renewed interest in II–VI electroluminescent diodes. Previous work was disappointing due to the inability to fabricate *pn* junctions in the wide bandgap II–VI materials of interest. More recently, luminescence in ZnSe (0.1% power conversion efficiency @ 590 nm)[11] and ZnS:Mn (0.1% power conversion efficiency @ 580 nm)[12] has been reported (see **10**–13, Electroluminescence—Destriau effect). Here the excess carriers are generated in a thin high-field region or at a metal-semiconductor barrier and consequently these materials operate at higher voltages (10–100 volts).

APPLICATIONS

Although the cost of l.e.d.s has come down considerably during the last decade, largely as a result of batch fabrication techniques, the cost of a single diode is still excessive for high resolution ($\gtrsim 10^5$ elements) displays, e.g., television. Therefore, l.e.d.s are most widely used for numeric display applications where only a few digits need to be displayed, e.g., calculators or readouts for digital equipment, and in some alarm or warning systems requiring an optical indication. The l.e.d. has the advantage over other indicators such as filament or gas-filled lamps of being operated directly by transistor or logic circuits and not requiring an additional lamp-driver circuit.

The size and geometry of the *pn* junction, and the fabrication process used depend on the intended application. Small seven-segment displays are often made in a single step on a common substrate, but the substrate cost makes this approach impractical for larger-sized displays; consequently, larger numeric l.e.d. displays are generally assembled from seven individual diodes. A seven segement display is shown in *Figure 10.10*. From the seven segments, all the numerals and certain letters such as A, H, P and U can be formed.

In monolithic displays the contact electrodes can be evaporated or photolithographically deposited on the surface parallel to the *pn* junction plane, *Figure 10.11* (*a*).

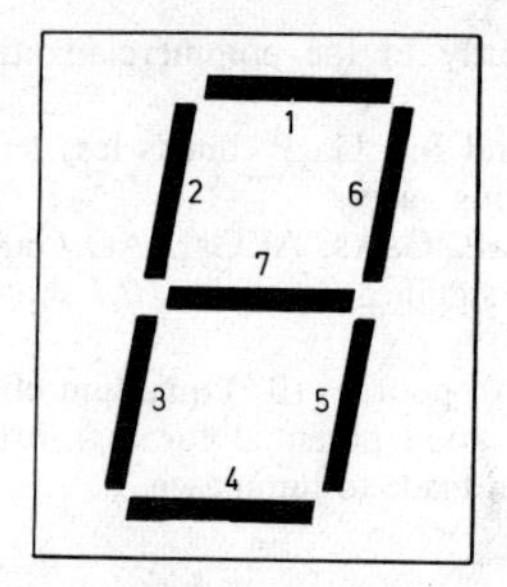

Figure 10.10 Seven-segment character display using electroluminescent diodes

Light emission is predominantly from this surface. The uppermost layer is kept thin (<5 μm) to minimise self-absorption, especially in direct-bandgap materials; it cannot be too thin, however, since current spreading to the whole junction area is desirable. In GaP displays the transparency of the substrate material leads to optical cross-coupling between segments; a special structure, *Figure 10.11* (*b*), in which the material is etched away in between the segments and coated with absorbing or reflecting films, was developed to minimise this optical coupling.

By slicing the *pn* junction into bars, *Figure 10.11* (*c*) after electroding, very high line brightnesses from the edge of the junction can be obtained. Electrical connection to the small bar segments and subsequent assembly are more complicated.

Other structures have been designed which use external means (cylindrical lenses, reflecting cavities, diffusing covers—*Figure 10.11* (*d*)) to enlarge the area from which

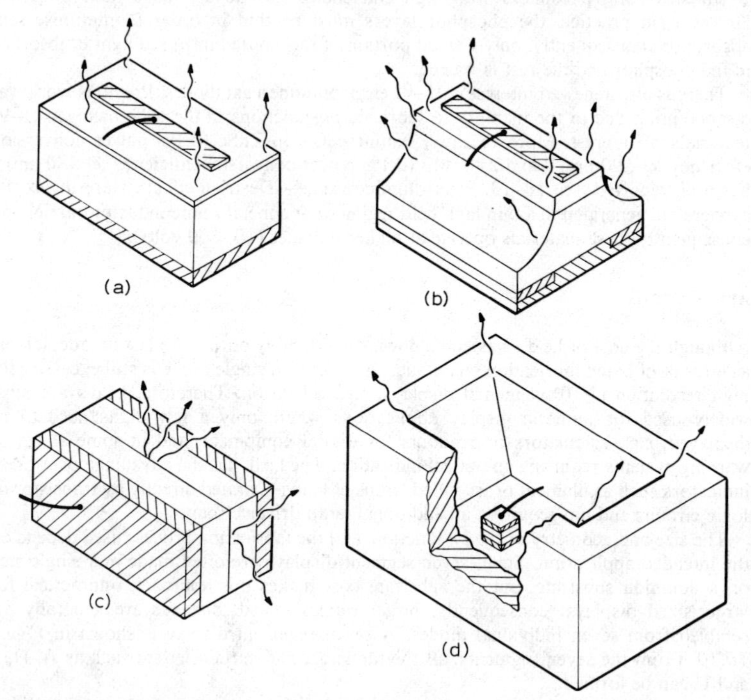

Figure 10.11 L.E.D. geometries used for the fabrication of seven-segment numeric displays: (a) and (b) surface emitters (c) edge emitter (d) cavity emitter (Courtesy American Vacuum Society)

light is emitted; these techniques are conservative of diode area but require a hybrid process for assembly and have reduced surface brightness.

The reliability, long life and compatibility with low voltage integrated driving circuitry make l.e.d.s very attractive for small display applications. However, their power consumption of typically 5–10 milliwatts/small digit is still considered high for many battery-operated products, e.g., portable calculators, electronic watches.

Additional information about materials, fabrication and applications for l.e.d.s can be found in several excellent review articles.[15, 16]

Another type of application for electroluminescent diodes is as optical coupling or isolating elements. A gallium arsenide diode emits infra-red radiation at a wavelength which corresponds to the maximum sensitivity of planar silicon photodiodes and phototransistors. If a gallium arsenide electroluminescent diode is encapsulated with a silicon phototransistor, the radiation from the diode in response to an input signal will produce an output signal from the phototransistor corresponding to the diode input signal. The two signals, however, will be isolated because there is no electrical connection between the diode and transistor. Such devices are called solid-state photorelays or photocoupled isolators, and are used in applications where isolation between input and output signals is required. The photorelay can be used as a trigger device, the transistor conducting to energise further circuitry when a light signal is received from the diode,

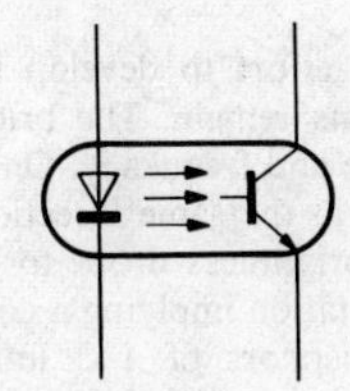

Figure 10.12 Circuit symbol for solid-state photorelay (photocoupled isolator)

or as a continuous coupling device, the phototransistor responding to modulation of the diode light output caused by changes of the diode forward current. The ratio of the output signal to the forward current of the electroluminescent diode, is called the transfer ratio. The use of a phototransistor enables a current gain to be obtained. Curves of the transistor collector current plotted against the forward current of the diode, the transfer characteristic, for various operating conditions are given in the published data for the photorelay. The output characteristic of the phototransistor with diode forward current as parameter is also given. The other characteristics and ratings for the photorelay given in the published data are those normal for phototransistors and electroluminescent diodes. The circuit symbol for a solid-state photorelay is shown in *Figure 10.12*.

Electroluminescence (EL)—Destriau effect

Some phosphors emit light when a sufficiently high electric field is applied across them. This phenomenon was first observed by Destriau[17] upon application of a changing electric field. Light is generated by recombination of electrons and holes whereby their excess energy is transferred to an emitted photon. Donors and acceptors play a major role in facilitating this radiative recombination and determining the spectral characteristics of the emitted light, i.e., the increment by which the photon energy is less than the energy gap of the phosphor material.

The mechanism by which electrons and holes are generated in the EL phosphor is not fully understood. The applied voltage is believed to be concentrated near a barrier or thin high-impedance region in the phosphor layer; the electric field strength is thereby locally increased to the point where carrier injection by field emission and possibly avalanche multiplication can occur.

A.C.-EL

Most extensively studied are a.c.-EL phosphors, usually zinc sulphides or zinc sulpho-selenides doped with copper or chlorine to produce yellow, green or blue emission; red EL phosphors have been made, but their brightness and efficiency are usually much lower.

In practice, a thin (~10–50 μm) layer of EL phosphor is sandwiched between two electrodes, at least one of which is transparent and through which the display is viewed. Both rigid (glass, ceramic) and flexible (plastic) substrates can be used. An a.c. excitation voltage, typically 50–5 000 Hz, of several hundred volts is applied to the electrodes.

An area of early interest for a.c.-EL was the combination of large-area photoconductor layers and EL phosphors for image intensification or X-Ray image storage.[18] However, saturation and trapping effects in the photoconductors and the availability of better alternatives prevented commercialisation of these devices.

Originally a.c.-EL was also believed to be of importance for large-area illumination and flat-panel television. However, in both applications luminous efficiency, average brightness and display life are of prime importance. In particular, for the television application where each display element is excited only during a small fraction of each frame time, high peak brightnesses are required in order to obtain useful average brightnesses.

Despite much effort to develop improved a.c.-EL phosphors, some fundamental materials problems remain. The brightness of EL displays increases with increasing excitation voltage and frequency. On the other hand, both life and luminous efficiency tend to decrease in the same direction. The half-life of most phosphors, i.e., the time over which the brightness drops to one-half of its initial value, varies inversely with frequency of excitation implying a constant number of cycles. Luminous efficiencies in yellow-green phosphors of 1–5 lm/W (0.1–1%) can be achieved at 35–350 nits brightness with a typical half-life of 1 000–3 000 hours[19] at 400 Hz excitation. The most important commercial applications are for low power, reliable night lights and instrument panel lighting (there is usually no catastrophic failure mechanism).

The technical feasibility of an a.c.-EL (ZnS,Se : Cu,Br) television panel has also been demonstrated;[20] however, the contrast ratio and brightness, which decrease with increasing number of scan lines due to the decreasing duty factor (~35 nits and a contrast of 10 : 1 for an 80-line panel versus ~12 nits for a 225-line panel) are not adequate for commercial television displays. Also power consumption is high due to the relatively large stray capacitance of the thin panel, and the high voltages and short pulses used to excite it.

D.C.-EL

More recently, interest in d.c.-EL materials has increased with the discovery that high brightness (several hundred nits), good life (>1 000 hrs), and modest luminous efficiency (0.1% $\approx$ 0.5 lm/W) can be achieved in ZnS doped with Mn and coated with copper sulphide.[12] The light emitted from these phosphors originates from a thin (3–5 μm) high-impedance layer near the anode; the rest of the phosphor layer is optically inactive but acts as a resistive protection layer. The high-impedance layer has to be created by a 'forming' process (high voltage and high current) and presumably results from the diffusion of Cu ions out of it.

Of particular interest has been the finding that average brightness and luminous efficiency remain high under low duty-cycle pulse excitation: for instance, an average brightness of 270 nits was achieved using 4–μs-wide 250-volt pulses at 0.5% duty cycle in a panel originally formed at 60 volts. This finding is of special significance for grey-scale displays, such as television where the duty cycle is low due to the large number of scanning lines. Thus an off-the-air 330-mm-diagonal flat d.c.-EL television display has been developed[21] with 224 × 224 elements, producing a 34-nit display with 20 : 1 contrast and a total power consumption of 150 watts. The addressing circuitry is still complex, however, and a commercial product would additionally require the development of other phosphors which can be used to obtain efficient red, green and blue emission.

Plasma displays

OPERATION

When a sufficiently high voltage is applied across two electrodes in a low pressure gas, a breakdown of the insulating properties of the gas is observed.[22] The sudden increase in conductivity is caused by impact ionisation of gas molecules by electrons that have been accelerated to sufficiently high energies by the applied electric field. Each electron leaving the cathode produces an electron avalanche travelling towards the anode and a positive ion avalanche travelling towards the cathode. When the probability of this ion stream regenerating an electron at the cathode becomes unity, then the Townsend breakdown condition is satisfied and the discharge becomes self-supporting. The flow of electrons also excites the gas molecules through collision and this energy is subsequently given off as light emission. This emitted light, which is characteristic of the ambient gas, is the basis for a variety of plasma displays.

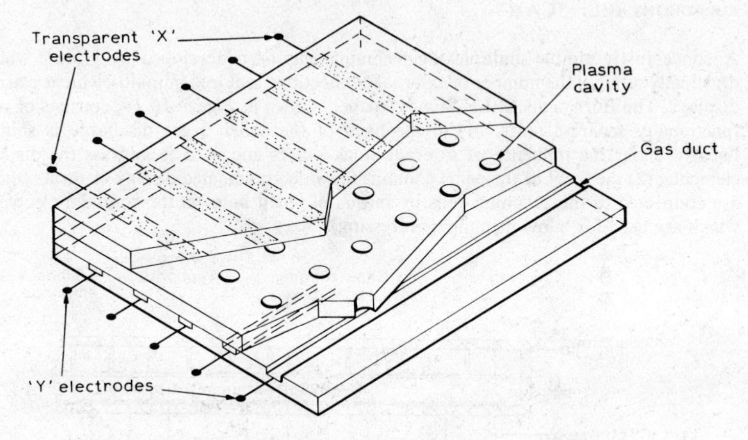

Figure 10.13 Construction of simple plasma matrix display

Plasma displays normally consist of a single glass envelope filled with a few Torr of gas, typically neon for optimum luminous efficiency (orange light at 0.1–1 lm/W); other gas additives are sometimes used, in particular Hg, to reduce sputtering of the electrodes or to obtain UV emission which can be used to excite phosphors (in order to change the colour of the display). Some internal structure is often used to isolate the display elements and to confine the discharge, *Figure 10.13*.

The extension of plasma displays to other colours can, in principle, be achieved by incorporating fluorescent phosphors which are excited by UV radiation from the discharge[23] or by using carriers from the discharge to bombard and excite a low-voltage cathodoluminescent phosphor directly;[24] suitable colour primaries, especially of the low-voltage cathodoluminescent type, are not readily available however.

The high voltage needed to fire a plasma cell (100–200 volts) is a disadvantage in many applications, especially the alphanumeric display field (calculators, digital clocks, digital meters . . .) where competition from other technologies (l.e.d., liquid crystals . . .) is strong. Nevertheless, small and moderate-size plasma displays are being used because of their inherent low fabrication cost, pleasant bright appearance and reliability.

MATRIX (MULTIELEMENT) PLASMA DISPLAYS

In larger-sized or high resolution displays, plasmas have many advantages. Since the current through a plasma, and therefore the light output is negligible until breakdown occurs, there is a built-in threshold which makes plasma displays ideal for matrix (coincidence or half-select) addressing. The display elements of a matrix display are arranged at the intersection of a set of orthogonal 'X' and 'Y' electrodes, *Figure 10.13*, so that many elements ($n \times m$) can be addressed with relatively few ($n + m$) electrode connections. Of particular interest would be a flat television display which requires about 2.5×10^5 elements (500×500), and therefore can be addressed with 1 000 (500 + 500) wire connections. This basic concept was demonstrated in a 4 000 and later 10 000-element t.v. display.[25] However, the addressing circuitry required was complex and display uniformity, as in most sampled grey-scale matrix displays, unacceptable for television.

BURROUGHS SELF-SCAN*

A conceptually simple multiplexing technique has been developed by Holz[26] which drastically reduces the number of external connections required in multi-element plasma displays. The Burroughs SELF-SCAN* panel, shown in *Figure 10.14*, consists of two functionally separate parts: (1) in the back of the panel, a gas discharge is shifted linearly under the influence of external clock pulses and used to address the display elements; (2) the front of the panel contains the video-modulated display elements; these are connected to the scanning cells by means of small holes in the common electrode which are used for 'glow priming' (addressing).

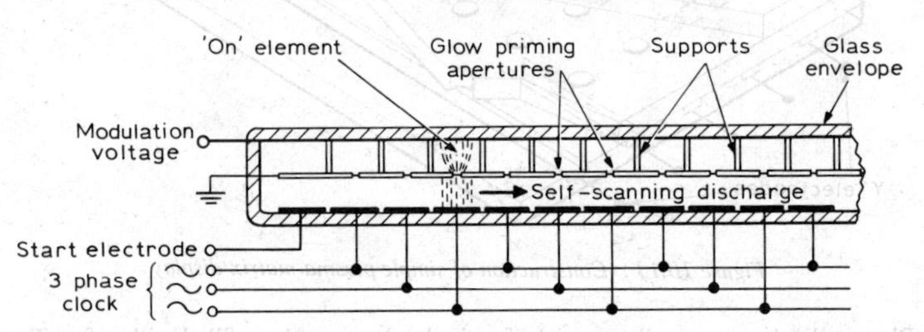

Figure 10.14 Principle of operation of Burroughs SELF-SCAN* *plasma display*

The operation of the scanning part is as follows: the cathodes of a row of plasma elements are connected alternately to one of three common electrodes, *Figure 10.14*. A three-phase sinusoidal voltage whose amplitude is just below threshold is applied to these three electrodes—consequently, none of the elements will be turned 'on'. However, if a particular element, say the first, is externally turned 'on', then metastable ions from this discharge will diffuse to the adjacent element thereby lowering its threshold for firing. This element therefore will turn 'on' when the voltage across it next reaches its maximum; in the meantime the voltage across the previous element is dropping and it

* Trademark of Burroughs Corporation.

is extinguished. In this manner the discharge is 'stepped' along by three elements for every full cycle of the sinusoidal clock voltage. In the Burroughs SELF-SCAN panel the first element of a row is used to start the discharge and the frequency of the clock voltage used to determine the rate at which it propagates down the row.

This linearly scanning discharge is then used to sequentially address ('glow-prime') the display elements which are interconnected to the scanning elements by means of a small hole in the central common electrode. The diffusion of carriers through these holes from the scanning discharge in the back is sufficient to remove the threshold in the brightness *vs.* voltage characteristic of the display element. Thus by modulating a common display anode, the brightness of the element in front of the scanning discharge can be determined: a single line of 256 elements can be addressed time-sequentially with less than 10 external leads. By scanning at a sufficiently high frequency (60 Hz) a steady, nonflickering display results.

Recently Burroughs SELF-SCAN plasma panels with 77 × 222 elements were used to produce a portion of an off-the-air television display[27, 28] which demonstrated the advantages of self-scan addressing (fewer external leads, simplified and lower-voltage circuitry, better grey-scale uniformity) for this type of analog display applications.

OWENS-ILLINOIS DIGIVUE*

A very different type of multielement plasma display was developed by Slottow and Bitzer[29] and is now marketed by Owens-Illinois under the trademark DIGIVUE*. The

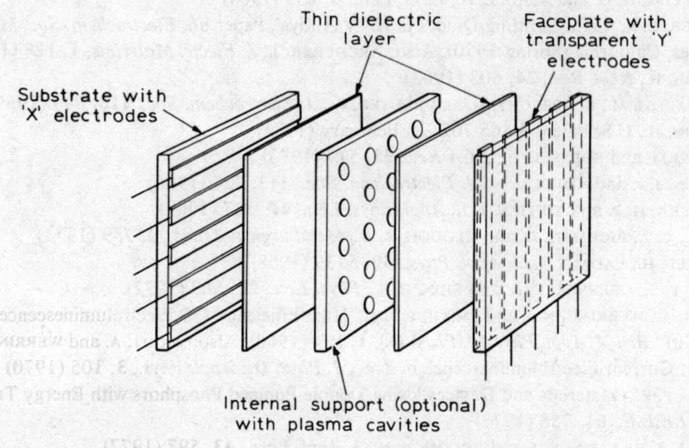

Figure 10.15 *Construction of Owens-Illinois* DIGIVUE* *plasma display*

construction of this plasma panel is illustrated in *Figure 10.15*: individual plasma elements are located at the intersection of orthogonal X-Y electrodes but these electrode strips are insulated from the gas discharge by means of a thin insulating layer—thus no d.c. current can flow through the panel.

When a sinusoidal voltage is applied to all electrodes, the voltage will divide capacitively across the insulating layers and the plasma element; if the voltage across the plasma is insufficient to cause breakdown, then the element remains 'off'. However, if any element is separately triggered 'on', then the current flow through the plasma will deposit an electric charge on the insulating layer covering the electrodes; this reduces the voltage across the plasma element causing it to extinguish itself. When the polarity of the applied voltage next is reversed, this charge on the insulator produces an

* Trademark of Owens-Illinois.

increased voltage across the display element thus allowing it to fire again and the charge is then transferred to the opposite electrode. In this manner, any element that has been triggered 'on' continues to fire at every half cycle of the applied voltage while all other elements remain 'off'. Therefore the DIGIVUE is uniquely suited for alphanumeric (on/off) display applications requiring long-term memory or slow updating.

Although the light from a single-plasma element is emitted in a very short time (10^{-7} s to 10^{-6} s), the average brightness of the panel can be quite high ($\gtrsim$170 nits) with excitation frequencies of the order of 100–500 kHz. A DIGIVUE display with more than 10^6 elements (1 024 × 1 024) has been demonstrated; simple construction techniques and optimised gas mixtures promise moderate cost and good reliability (>5 000 hrs).

Until recently the storage plasma panel was mainly considered for alphanumeric and graphics display applications where frame-storage (memory) was desirable. It has been shown possible to display grey-scale images on such a plasma panel using time modulation—that is, varying the fraction of a frame time during which each element is left 'on'.[30] Thus, in principle, off-the-air television pictures can be displayed at the expense of an analog-to-digital converter and additional storage circuitry; this approach may become economically feasible with future advances in large-scale integrated circuitry. A major advantage of this approach is that the display promises to be extremely uniform.

REFERENCES

1. GRIMMEISS, H. G. and SCHOLZ, H., *Phys. Lett.*, **8**, 233 (1964)
2. ARCHER, R. J., 'Light Emitting Diodes in III–V Alloys', Paper 66, *Electrochem. Soc. Mtg.*, Los Angeles, California (Spring 1970). Also: ARCHER, R. J., *J. Electr. Materials*, **1**, 128 (1972)
3. NELSON, H., *RCA Rev.*, **24**, 603 (1963)
4. WOODALL, J. M., RUPPRECHT, H. and REUTER, W., *J. Electrochem. Soc.*, **116**, 899 (1969)
5. NELSON, H., U.S. Patent 3 565 702, 23 February (1971)
6. LADANY, I. and KRESSEL, H., *RCA Rev.*, **33**, 517 (1972)
7. TIETJEN, J. J. and AMICK, J. A., *J. Electrochem. Soc.*, **113**, 724 (1966)
8. MARUSKA, H. P. and TIETJEN, J. J., *Appl. Phys. Lett.*, **15**, 327 (1969)
9. NUESE, C. J., RICHMAN, D. and CLOUGH, R. B., *Metallurgical Trans.*, **2**, 789 (1971)
10. KRESSEL, H., LADANY, I., *J. Appl. Phys.*, **39**, 5339 (1968)
11. PAUK, Y. S., GEESNER, C. R. and SHIN, B. K., *Phys. Lett.*, **21**, 567 (1972)
12. VECHT, A., WERRING, N. J. and SMITH, P. J. F., 'High Efficiency DC Electroluminescence in ZnS (Mn,Cu)', *Brit. J. Appl. Phys.* (*J. Phys. D.*), **1**, 134 (1968). Also: VECHT, A. and WERRING, N. J., 'Direct Current Electroluminescence in ZnS', *J. Phys. D.: Appl. Phys.*, **3**, 105 (1970)
13. AUZEL, F. E., 'Materials and Devices Using Double-Pumped Phosphors with Energy Transfer', *Proc. I.E.E.E.*, **61**, 758 (1973)
14. WITTKE, J. P., LADANY, I. and YOCOM, P. N., *J. Appl. Phys.*, **43**, 597 (1972)
15. BERGH, A. A. and DEAN, P. J., 'Light-Emitting Diodes', *Proc. I.E.E.E.*, **60**, 156 (1972)
16. NUESE, C. J., KRESSEL, H. and LADANY, I., 'Light-Emitting Diodes and Semiconductor Materials for Displays', *Proc. Symposium on Display Materials and Devices*, *J. Vac. Sci. Technol.*, **10**, 772 (1973)
17. DESTRIAU, G. J., *J. Chem. Phys.*, **33**, 620 (1936). Also DESTRIAU, G. J., *Trans. Faraday Soc.*, **35**, 227 (1939)
18. KAZAN, B. and KNOLL, M., *Electronic Image Storage*, Academic Press, New York, 418–437 (1968)
19. LARACH, S. and SHRADER, R. E., 'Electroluminescence of Polycrystallites', *RCA Rev.*, **XX**, 532 (1959)
20. ARAI, H., YOSHIZAWA, T., AWAZU, K., KURAHASHI, K. and IBUKI, S., 'EL Panel Display', *I.E.E.E. Conf. Record of the 1970 I.E.E.E. Conf. on Display Devices*, 52, New York, 2–3 December (1970). Also: ARAI, H., 'EL Panel TV', *JAEU*, **3**, 39 (1971)
21. YOSHIYAMA, M., 'Lighting the Way to Flat-Screen TV', *Electronics*, **42**, No. 6, 114 (1969).

Also: YOSHIYAMA, M., OSHIMA, N. and YAMAMOTO, R., 'A Television Display Device Utilizing DC-EL Panel', *National Technical Report*, **17**, No. 6, 670 (1971) (In Japanese)
22. VON HIPPEL, A. R., *Dielectrics and Waves*, John Wiley & Sons, New York, 234 (1954)
23. FORMAN, J., 'Phosphor Color in Gas Discharge Panel Displays', *Proc. SID*, **13**, 14 (1972)
24. KRUPKA, D.C., CHEN, Y. S. and FUKUI, H., 'On the Use of Phosphors Excited by Low-Energy Electrons in a Gas Discharge Flat-Panel', *I.E.E.E. Proc Special issue on new materials for display devices*, **61**, 1025 (1973)
25. DE BOER, TH. J., 'An Experimental 4 000-Picture-Element Gas Discharge TV Display Panel', *Proc. 9th National Symposium on Information Display*, 193, Los Angeles, May (1968)
26. HOLZ, G. E., 'The Primed Gas Discharge Cell—A Cost and Capability Improvement for Gas Discharge Matrix Displays', *1970 SID IDEA Symposium Digest of Papers*, 30, May (1970)
27. CHODIL, G. J., DE JULE, M. C. and MARKIN, J., 'Good Quality TV Pictures Using a Gas-Discharge Panel', *I.E.E.E. Conference Record of 1972 Conf. on Display Devices*, 77, October (1972)
28. CHIN, Y. S. and FUKUI, H., 'A Field-Interlaced Real-Time Gas-Discharge Flat-Panel Display with Gray-Scale', Ibid, p. 70
29. BITZER, D. L. and SLOTTOW, H. G., 'The Plasma Display Panel—A Digitally Addressable Display with Inherent Memory', *Proc. Fall Joint Computer Conf.*, San Francisco, November (1966). Also: BITZER, D. L. and SLOTTOW, H. G., 'Principles and Applications of the Plasma Display Panel', *Proc. 1968 Microelectronic Symposium, I.E.E.E.*, St. Louis (1968)
30. KURAHASHI, K., TOTTORI, H., ISOGAI, F. and TSURUTA, N., 'Plasma Display with Gray Scale', *1973 SID International Symposium Digest of Technical Papers*, **IV**, 72, New York, 15–17 May (1973)

PASSIVE ELECTRO-OPTIC DISPLAYS

Passive electro-optic devices modify light that is generated elsewhere. The parameter modulated may be the optical pathlength or the absorption, reflection or scattering parameters, or a combination of these. If the modulated quantity is wavelength dependent, then the display will be coloured. The display generally consists of a thin sheet of the material sandwiched between two planes of electrodes. The electrode configuration is such that, by applying a voltage to certain of the leads, specified areas of the display are modulated. A typical commercial display is shown in *Figure 10.16*. When only the phase of the transmitted or reflected light is modulated, then a combination of polarisers is required to convert phase changes to amplitude changes which make the display visible.

The properties of interest of an electro-optic material, in addition to the kind and amount of light modulation, are the driving voltage and power required, whether they

Table 10.2

Effect	*Liquid crystals*		*Electrophoresis*	*Electrochromism*
	Hydrodynamic	*Field effect*		
Operation in transmission	T	T		
Operation in reflection	R	R	R	R
Viewing angle	narrow	medium	no restrictions	no restrictions
Operating voltage (V)	15–40	5–10	50–100	1–2
Power consumption (mW/cm^2) (continuous operation)	0.1	0.002	0.4	
Energy consumption/cycle (mJ/cm^2)				2–100
Switching speed (s)	0.05	0.2	1	0.01

Figure 10.16 Liquid crystal display

can be excited by a.c. or d.c., the speed of response, and the life (expressed in operating life or number of addressing cycles).

To address a simple display like that shown in *Figure 10.16*, it is practical to have a separate connection leading to each segment of one electrode plane with a common connection on the other plane. However, for more complicated displays, it is necessary to reduce the number of leads by using a grid or matrix system.[1] On one plane, the segments are interconnected in horizontal rows, on the other plane, in vertical columns, so that each segment is uniquely addressed by a combination of one horizontal and one vertical lead. For this scheme to work, it is required that the electro-optic effect have a sharp threshold, fast rise time and long decay time, so that the application of two (or three) times the threshold voltage produces a large optical change, even if that voltage is only applied for a small part of the time (low duty cycle).

Presently, the most widely used passive electro-optic materials are liquid crystals. They exhibit a considerable variety of modulation effects. Other electro-optic effects that may become important are electrophoresis and electrochromism. The properties of these materials are compared in *Table 10.2*.

Liquid crystals

Liquid crystals[2] are useful in display applications because of their very large electro-optic effects.[3] The materials themselves exist in a phase which is different from both the conventional liquid and solid phases, and exhibit properties intermediate between the two. Materials having this special phase are found in the group of organic materials with moderately large rod-like molecules. In recent years, numerous compounds and mixtures have been discovered which are liquid crystalline at room temperature and over a temperature range of as much as 100 °C. This accounts for the increased practical interest in display applications.

The various material phases are distinguished by their macroscopic symmetry properties. The conventional liquid is isotropic and has no unique symmetry elements while even the lowest order crystals have complete translational symmetry. The intermediate liquid crystals retain the translational freedom of the liquid in which the molecules are free to move in varying degrees. They also exhibit long range ordering which adds certain macroscopic symmetry elements, either translational or orientational. There are presently 5 known liquid crystalline phases—nematic, cholesteric and three smectic phases. It is primarily the first two which are used in practical electro-optic devices.

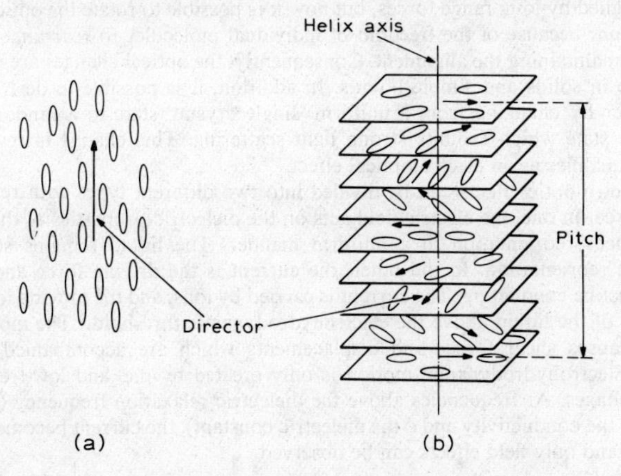

Figure 10.17 *Schematic arrangement of molecules in liquid crystals:* (*a*) *nematic* (*b*) *cholesteric. The arrow represents the director*

A nematic liquid crystal is one where all the molecules align themselves approximately parallel to a unique axis while retaining the complete translational freedom. This symmetry axis is defined locally and may vary in direction in different parts of a volume of liquid. Its direction may be described by a vector (of arbitrary sense) which is called the director, *Figure 10.17* (*a*). In the absence of external forces, the lowest energy state of a nematic is that with a uniform orientation of the director in the entire sample.

A cholesteric liquid crystal is a modification of a nematic one. In the undisturbed state, it may be described as a nematic liquid crystal which has been twisted about an axis (the helical axis) lying perpendicular to the orientation of the director at all times, *Figure 10.17* (*b*).

The optical properties of these materials are derived from their symmetry. A nematic liquid crystal is optically uniaxial, so that, as in a uniaxial crystal, two different indices of refraction apply for light propagation with the electric vector parallel and perpendicular to the axis. The same is true of the dielectric constant and the conductivity which are tensors of the same order as the index of refraction. The dielectric constant anisotropy may be positive (larger for the fields parallel to the axis than for the perpendicular direction) or negative depending on the polarisation properties of the molecules. The other anisotropies have only been observed to be positive.

The optical properties of the cholesteric fluid are more complicated. It is uniaxial with a screw-type symmetry. The axis is at right angles to the axis in the nematic (or 'untwisted' cholesteric). Light propagating parallel to the helical axis experiences optical rotation, reflection and change of polarisation, depending on the ratio of the optical wavelength to the pitch of the helix.[4] In the extreme case of very small ratio, the polarisation ellipse of the light simply rotates about the axis in the same way that the director does. Light propagating at right angles to the helical axis encounters a sinusoidal variation of index along the axis with the period equal to half the pitch. This is a phase grating which causes diffraction of the light.

The electron-optic properties of a liquid crystal are very different from those of a crystalline solid. This is what causes these materials to be of great interest in display applications. In a solid, the electric field produces small changes in location of ions and in atomic polarisation and correspondingly small changes in the refractive index. In liquids, the electric field can only polarise individual molecules and the effects are even smaller. Liquid crystals respond to an electric field like solids since the molecules remain aligned by long range forces, but now it is possible to rotate the optical axis by a large amount because of the freedom of individual molecules to rearrange themselves while still maintaining the alignment. Consequently, the optical changes are much larger than those in solids and simple liquids. In addition, it is possible to destroy the long range order by changing from a uniform 'single crystal' state to a randomised polycrystalline state which exhibits strong light scattering. This change is reversible and therefore qualifies as an electro-optical effect.

The electro-optic effects can be divided into two different types with respect to the driving force. In one, the electric field acts on the dielectric properties of the molecules and produces reorientation in a uniform manner. The liquid remains static except during the reorientation. In the other, the current is the driving force and the liquid crystal must be conducting. The current is carried by ions, and the moving ions produce movement of the liquid above the electrohydrodynamic threshold.[5] The moving liquid, in turn, causes shear forces and displacements which are accompanied by optical changes. Electrohydrodynamic motion is only created by d.c. and low-frequency a.c. driving voltages. At frequencies above the dielectric relaxation frequency ($f_d = \sigma/2\pi\epsilon$, where σ is the conductivity and ϵ the dielectric constant), the current becomes primarily capacitive and only field effects can be observed.

PRACTICAL CONSIDERATIONS

Liquid crystal electro-optic devices are generally constructed as sandwich cells with a thin layer of the liquid placed between two conducting glass plates. Because of the large electro-optic effect, the liquid layer may be very thin (3–50 μm, typically). The cells are used in two different configurations, one where the light is transmitted through the cell, the other where the incident light is reflected back at the viewer. In the latter case, the back electrode usually consists of a specularly reflecting aluminium film on a glass substrate. The front electrode is formed by a transparent conducting layer, such as SnO_2 or In_2O_3. For the transmissive cell, both electrodes must be transparent.

For practical use, the cells must be hermetically sealed, so that the liquid crystal is confined and cannot interact with the atmosphere. The major problem is water vapour which can cause decomposition and chemical reactions between the various components of the mixture. Some kinds of liquid crystal material are more susceptible to this problem than others. Chemical reactions may also take place during operation of the devices because of the electrolytic nature of the cell. This is particularly true when direct current is applied to the cell which causes transport of ions and electrochemical reactions at the electrodes. Consequently, the lifetime of the device under d.c. conditions is generally much shorter than under a.c. conditions, and most applications require the latter.

To produce any given electro-optic effect, it is necessary that the liquid crystal have a certain crystalline form or texture. The form is determined by the alignment of the molecules at the surfaces which may be perpendicular to the surface, parallel to it, or at some intermediate angle. In the parallel situation, the direction of the molecules in the plane may be required to be uniformly in a specified direction, or randomly oriented. The alignment is determined by appropriate treatments of the surface[6] in combination with chemical aligning agents.[7] In addition, the material must have a certain conductivity for the particular effect desired. The conductivity is controlled by the addition of ionic dopants.

One of the important variables in the operation of the device is its speed of response. The electro-optic effects involve a reorientation of the molecules and the response times are therefore related to the viscosity of the material, decreasing, for example, with increasing temperature. Response will be faster at higher driving fields. The equilibrium condition with no external force applied is determined by the alignment at the surface, and the speed of return to that condition increases rapidly as the cell is made thinner.

HYDRODYNAMIC EFFECTS

Domains and Dynamic Scattering. Conductivity in liquids is generally by ionic motion. At low values of conduction, the ions move through the stationary liquid but, at higher values, instabilities are produced and the liquid is set into motion.[5] It is in this region that conduction-induced electro-optic effects occur. The effects are much more important in liquid crystals than in isotropic liquids because the optical anisotropy makes the liquid motion directly visible.

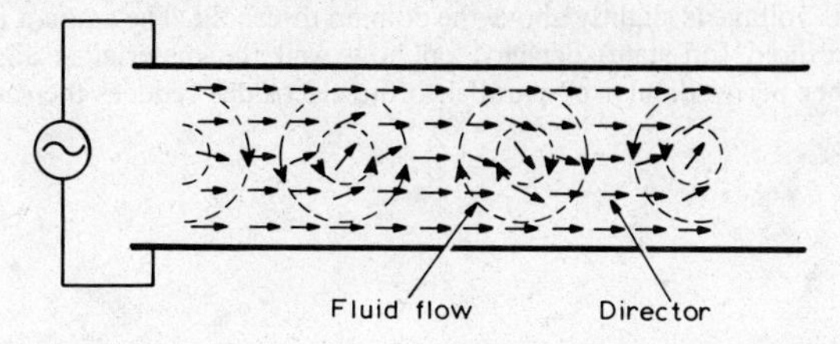

Figure 10.18 Cross section of a nematic cell exhibiting Williams' domains

The best-understood instabilities are the Williams domains.[8] They consist of rotation vortices of liquid and are observed mainly in nematic materials of negative dielectric anisotropy. A cross-section of a cell, at right angles to the vortex axes, is shown in *Figure 10.18*. The spacing between the axes is a fixed multiple of the electrode separation. The rotating fluid causes the director to tilt, as indicated. For light polarized in the plane of the figure, the optical pathlength varies depending on the location in the cell. This causes the axes and the regions separating any two vortices to become visible as an array of parallel lines. The uniform line spacing (proportional to the cell thickness) causes the cell to act as a diffraction grating.

The threshold for onset of the domain instability is determined by the interrelationship of a number of physical effects.[9] The conductivity anisotropy produces charge separation in regions where the ions are not moving parallel to the director due to the rotating liquid, *Figure 10.18*. The resulting transverse electric fields act on the charge to produce torques in the liquid. These torques reinforce the rotation. This explains why ionic conductivity is required for domain formation.

Driving voltages can be d.c. or low frequency a.c. but, as the frequency is increased, the ionic current decreases relative to the displacement current and the instability

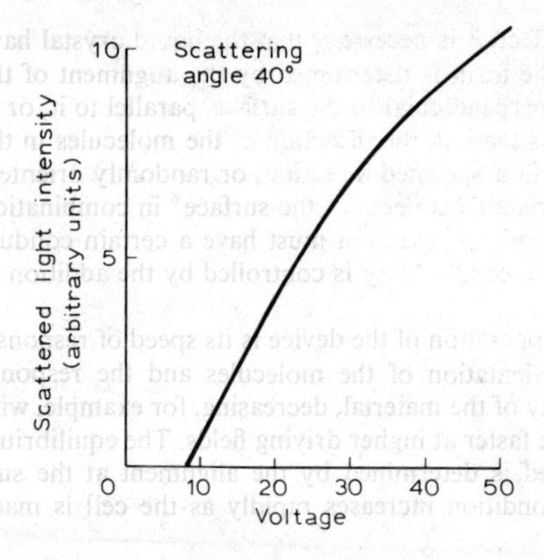

Figure 10.19 Voltage dependence of the scattered light in a dynamic scattering cell

threshold rises. Finally, near the dielectric relaxation frequency, no domains can be formed.[10] There exist other instabilities, but there are not of a hydrodynamic nature.[10]

As the driving voltage is increased beyond the threshold for domain formation, the vortices become smaller and less regular, leading finally to a turbulent state. Because the direction of the molecular axes changes rapidly from place to place, the liquid becomes highly scattering and opaque. The scattering pattern is varying continuously when observed under the microscope. This state is called *dynamic scattering*.[11] It is the most widely used electro-optic effect in liquid crystals.

The intensity of scattered light as function of the applied voltage is shown in *Figure 10.19*. It can be seen that there is a threshold voltage, below which there is little light scattering. This voltage is slightly above the domain threshold. The amount of scattering below the threshold (off-state) depends on how well the material is aligned. Good alignment, either perpendicular or parallel to the electrodes reduces the scattered light

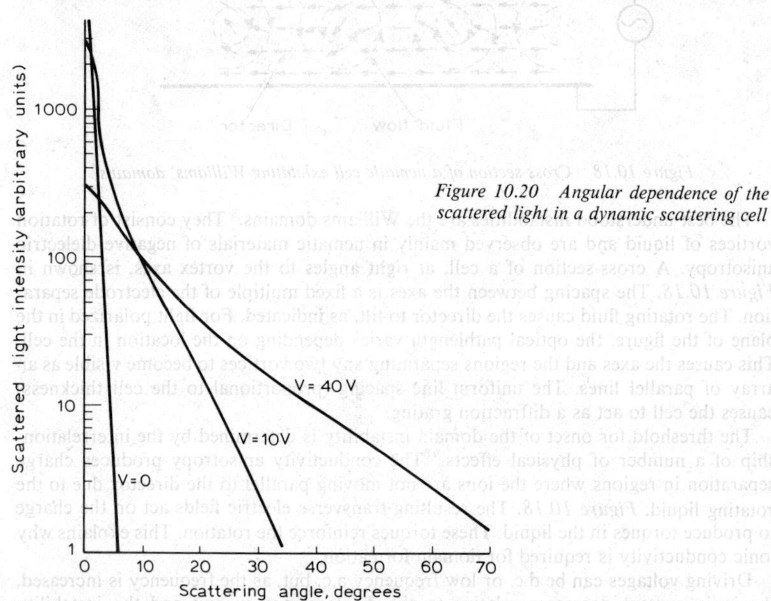

Figure 10.20 Angular dependence of the scattered light in a dynamic scattering cell

to a negligible quantity. The scattered light intensity in the on-state depends strongly on angle, as is shown in *Figure 10.20*. A determination of the contrast ratio of a display cell therefore depends on the conditions under which it is measured. The steady state scattering properties (*Figures 10.19* and *10.20*) do not depend on the thickness of the liquid crystal film over the range of 5–100 μm. The response time to application and removal of the voltages also has a strong angular dependence. It increases proportional to the square of the film thickness if the other parameters are held constant.

Figure 10.20 shows that the scattering of light takes place predominantly in the forward direction. This produces a very good display when transmitted light is used, particularly when the geometry is arranged so that the unscattered light is not viewed by the observer. The situation is not as favourable in the reflective case because the liquid crystal must be backed by a highly reflecting mirror to produce good scattering efficiency. It becomes more difficult to pevent reflected, but unscattered, ambient light from reaching the viewer. For good results, the display must be recessed behind a set of baffles which restrict the viewing angle.

Storage effect. A related hydrodynamic effect [12] is observed in cholesterics of negative dielectric anisotropy. In loosely-wound materials with a long pitch, a certain current flow will also produce the dynamic scattering instability. The cholesteric structure is completely disrupted due to the turbulence. When the field is removed, the movement stops and the cholesteric forms a 'polycrystalline' state called focal-conic texture. The individual crystallites or domains are oriented randomly with respect to each other. This state is highly scattering. To make a display, a clear state is needed and this is obtained by applying a high-frequency electric field. The frequency is well above the dielectric relaxation frequency so that there is no ionic current flowing, which would re-establish the instability. Instead, the dielectric field causes the molecules to align parallel to the electrodes which, in turn, produces a uniform alignment of the helical axis at right angles to the electrodes. This cholesteric state persists after the removal of the field (storage). Thus, the sample may be switched back and forth between the scattering and clear state by applying two different kinds of driving pulses. Optically, the two states are quite similar to the on and off state of dynamic scattering but the driving voltages are higher and the transition times longer, particularly if a long-time storage is desired.

FIELD EFFECTS

Field-Induced Birefringence. There are a number of electric field-induced electro-optic effects in liquid crystals. The simplest one, field-induced birefringence or distortion of an aligned phase,[13] is considered first, and in some detail, as the basic principles apply to all field effects. The material used is a nematic liquid with negative anisotropy of dielectric constant and negligible conductivity. The properties of the cell surfaces are controlled so that the alignment of the molecules is uniform and at right angles to the surface, *Figure 10.21* (*a*). When an electric field is applied in the perpendicular direction, it produces a torque on the individual molecules trying to turn them in a direction parallel to the electrodes because the lowest energy state of the unconstrained system is that in which the larger dielectric constant is oriented parallel to the field. However, the surface alignment is permanently fixed in the perpendicular direction and, through the elastic forces of long range order, tries to keep the bulk of the material in that same orientation. This balance between electric and elastic energy is such that below a certain threshold field E_c, the alignment remains uniformly perpendicular while, above that field, it gradually distorts towards the parallel direction as shown in *Figure 10.21* (*b*). In the figure, the rotation of the molecules from the perpendicular is shown in

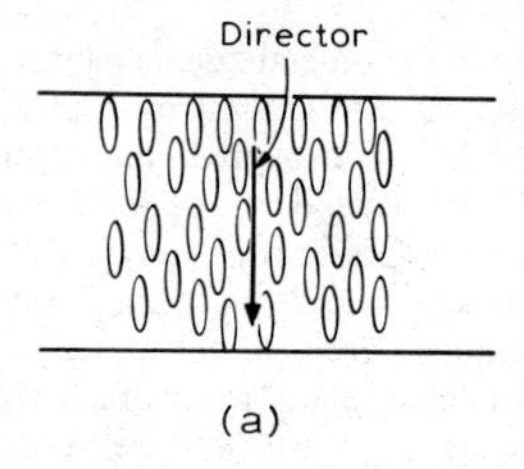

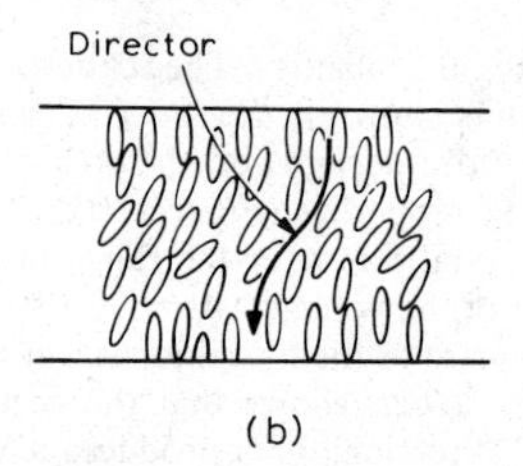

Figure 10.21 *Demonstration of the field-induced birefringence effect: (a) alignment in the absence of the field (b) when a field above the threshold field is applied between the electrodes*

one particular direction and the long range forces tend to align the entire sample in the same direction. There is, however, no reason to prefer one direction in the plane of the cell over any other and the direction may vary gradually throughout the sample.

Both states of the nematic fluid below and above the critical field are optically clear. Light propagating through the cell is subject to different indices of refraction and, therefore, phaseshifts in the two cases. These phase differences are converted to amplitude differences by positioning the cell between crossed polarisers. Then, light transmitted perpendicularly through the sample will be extinguished as long as the situation of *Figure 10.21* (*a*) applies (equivalent to light propagating along the axis of a uniaxial crystal). Above E_c, more and more light will be transmitted *Figure 10.22*,[14] until a saturation is reached when all but the surface molecules are aligned to the electrodes.

The induced birefringence effect can also be created in the opposite sense by using a material of positive dielectric anisotropy and making the surface alignment to be parallel to the electrodes. Then, the light transmission will be maximum at low fields and reduce to zero at high fields. The largest contrast is obtained if the polariser is located at 45° to the surface aligning direction.

The brightness and colour of this device depends on the difference in phase between the ight polarised along the crystal axis and that polarised at right angles to it. This means that the colour and intensity of the transmitted light observed by the viewer will not be uniform if the cell thickness is not uniform. A further disadvantage is that the transmitted light is strongly dependent on the angle of viewing.

Twisted nematic. A field effect arrangement which overcomes this disadvantage is the so-called twisted nematic cell.[15] It is similar to the second case of induced birefringence with the positive anisotropy material. The alignment is parallel to the surfaces, but the

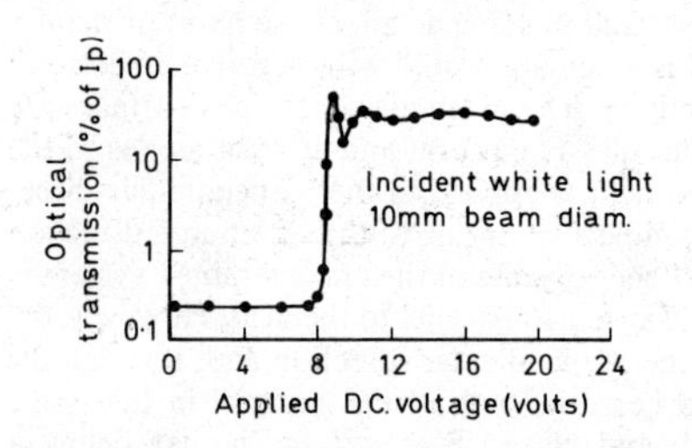

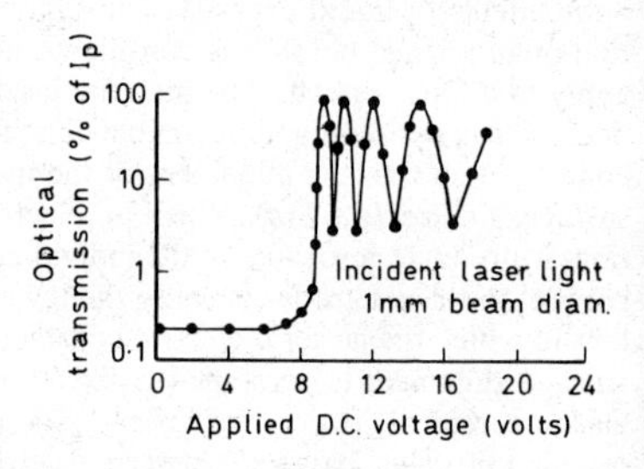

Figure 10.22 *Voltage dependence of the transmitted light for a field-induced birefringence cell at normal incidence for two different kinds of light beams* (*Courtesy American Institute of Physics*)

directions of the alignment at the two surfaces are positioned at right angles to each other. The elastic forces cause the director in the interior of the crystal to remain in the plane of the sample but to twist gradually by 90° from one electrode to the other. This produces a structure which is identical to that of a cholesteric material with rather long pitch (4 times the thickness of the sample). If plane-polarised visible light, which has a wavelength much shorter than this pitch, propagates through the cell, its plane of polarisation is rotated at the same angular rate at which the director is rotated.[4] The total angle of rotation is 90° so that, with analyser and polariser parallel to each other, no light will be transmitted. When a field above E_c is applied to the sample, the orientation will be changed to a more or less perpendicular one and the light polarisation direction no longer rotates. The light is transmitted by the analyser. If the analyser is set at right angles to the polariser, light will be transmitted by the undisturbed cell and absorbed when the field is applied.

Guest–Host Effect. Another kind of field effect in nematic liquids[16] modifies the absorption of light, rather than its phase. A pleochroic dye (in which the absorption of light varies strongly depending on whether the light is polarised parallel to the molecular axis or perpendicular to it) is added to the liquid crystal. Because such dye molecules tend to be rod-like, they orient themselves parallel to the similar-shaped liquid crystal molecules. The absorption is modulated by changing the orientation of the director and, therefore, the dye molecules. For maximum absorption by the dye, the initial alignment of a liquid crystal should be in the plane of the sample, and the incident light should be polarised parallel to the director. If the liquid crystal has a positive dielectric anisotropy, then an electric field rotates the director to a perpendicular direction where the absorption is a minimum. Note that no analyser is required. To obtain a rapid response for this effect, it is necessary that the dye molecules resemble the liquid crystal molecules as closely as possible. Also, the absorption of the dye should be as high as possible to reduce the quantity required.

Cholesteric Field Effects. A number of field effects are possible with cholesteric materials. Some produce reorientation of the existing structure as was the case with the nematics. In addition, the field may cause two kinds of changes in the cholesteric structure. These are a change in pitch of the helix and a complete disruption of the helix with a transition to the nematic phase.

A change of pitch causes optical effects if the pitch is of the order of the wavelength of light. This is because of the wavelength sensitive optical properties.[4] In particular, there is a region of total reflection for one sense of circularly polarised light propagating along the helical axis if the wavelength is close to the pitch of the helix. In such a case, the application of a field causes rapid and vivid changes of colour. However, the pitch also depends strongly on other variables, such as pressure and temperature, so that this is not a very practical electro-optic effect.

The field-induced phase change[17] is more important. In this case, use is made of the fact that, without any preferential surface alignment, the cholesteric material forms the focal-conic texture which is highly scattering and appears opaque. If the individual molecules have a positive dielectric anisotropy, then an applied field will tend to orient them parallel to the field. For a loosely wound cholesteric fluid, i.e., one with a large value of pitch, a moderate field will align the molecules into the induced nematic phase. This texture is optically clear and forms a large contrast to the zero-field opaque case. Upon removal of the field, the material returns to the scattering cholesteric phase. Cholesteric materials of large pitch are produced by mixing a short pitch cholesteric with a nematic material which, in this case, must have positive dielectric anisotropy.[18]

APPLICATIONS

Liquid crystal displays are presently used for digital watch and clock dials, for digital panel meters and other indicators, and for calculator displays, *Figure 10.16*. They are particularly attractive for small portable devices because of their low power consumption. They can be expected to find their way into numerous other indicator applications. Rise and decay times of 100 ms or less are feasible and materials that operate from 0 °C to 65 °C are being used. The information density of the displays is limited by the number of electrical connections that can be made in a practical device. Only a relatively small number of segments can be matrixed for most of the liquid crystal effects.

Whereas the first commercial displays all used dynamic scattering, most of the displays sold today use the twisted nematic field effect. They have less critical illumination requirements and a wider angle of viewability. This is particularly attractive for the watch application, where the face that the field effects require even lower drive currents is also of interest.

Other liquid crystal effects can be expected to become used in practical applications in the future.

Since electric addressing poses difficult problems in case of high information density, the only high resolution displays made so far use optical addressing schemes. In one case,[19] one electrode is replaced by a photoconducting layer with bias electrode. In the dark, the voltage drop is across the photoconductor while, under light, it appears across the liquid crystal cell and turns it on. If an optically opaque layer is placed between the liquid crystal and the photoconductor, then the same wavelength light can be used on both sides and it becomes a light amplifier.

Another optical addressing scheme[20] makes use of thermally-induced phase transitions. The effect is similar to the field-induced phase change. In the initial state, the cell is clear by having the helical axis aligned perpendicular to the plate. A focused high-intensity light beam (usually an infra-red laser beam) heats the material locally above the phase transition to the liquid phase. Upon cooling, the cholesteric forms the scattering focal-conic texture which stores the written information. The entire display can be erased by applying a field across the cell, if the cholesteric has a negative molecular dielectric anisotropy. A similar device has been constructed with smectic liquid crystal material[21] where selective thermal erasure has also been found to be practical. Writing speeds of 10^4 spots/s have been found to be feasible with these displays.

Electrophoretic displays

OPERATION

Electrophoresis—the migration of charged colloidal particles in an electric field—has been used extensively for the separation and analysis of components in biological substances.[22] It is also widely used for electrostatic coating in liquids or the development process in electrophotography.

More recently, an electrophoretic display device has been developed[23] which utilises the migration of charged pigment particles suspended in a coloured liquid to produce a reflective display. An electrophoretic suspension (25–100 μm thick), composed of pigment particles and the suspending liquid, is sandwiched between a pair of electrodes, at least one of which is transparent. This basic structure is shown in *Figure 10.23*. Under the influence of a d.c. field, the pigment particles are moved towards, and deposited on, one of the electrodes; this electrode is coloured by the pigment particles while unelectroded areas or electrodes of opposite polarity retain the colour of the suspending liquid. A high-contrast reflective display results.

The pigment particles must have the same density as the suspending liquid to avoid precipitation or separation. Organic pigment particles, e.g., hansa yellow, can be provided with the same density as that of the liquid, which consists of a mixture of

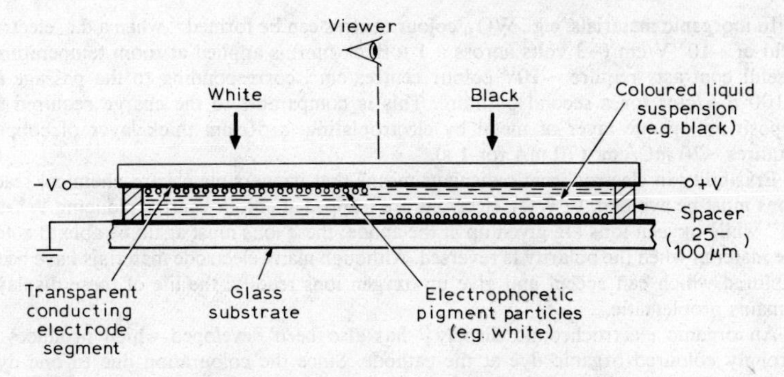

Figure 10.23 Operation of an electrophoretic display

solvents with different densities. Inorganic pigment particles, e.g., TiO_2, are denser than normal liquids, but their average density can be reduced by encapsulation with an inert plastic, e.g., polyethylene.

ELECTRICAL CHARACTERISTICS AND DEVICE PERFORMANCE

Electrophoretic displays, like liquid crystal displays, are passive and therefore attractive for viewing in high-brightness ambients. There are essentially no viewing angle restrictions, a distinct advantage over liquid crystal displays. Electrophoretic displays have memory, that is an element stays 'on' even if the power is removed, until the electric field excitation is reversed. Power dissipation (typically 0.4 mW/cm^2) is considerably higher than for liquid crystals, but low enough for moderate-sized numeric displays; the storage effect can be used to lower the average power consumption whenever the display is updated slowly.

Contrast ratios in excess of 50 : 1 are possible. A variety of colours is achievable by changing the pigment or suspending liquid. The addressing voltages (50–100 volts) however are a distinct disadvantage since high-voltage addressing circuitry is costly. Also, the high d.c. voltage can produce undesirable electrochemical effects which shorten the life of the display; operational life of 3 000 hours has been reported [23] and can be expected to improve.

ELECTROCHROMIC DISPLAYS

Electrochromism is the colouration or change in colour which occurs in certain materials upon the application of an electric field. The basis for this colouration is the formation of colour centres or an electrically induced oxidation or reduction whereby a new substance is formed which absorbs visible light. The ability to colour a material reversibly by means of a locally applied electric field can be used to produce a simple reflective display, suitable for viewing in high-brightness ambients.

In general, electrochromic displays require relatively high currents (high power) or long times to achieve useful display contrasts; thus they are best suited for small alphanumeric display applications, especially those requiring infrequent updating, e.g., digital clocks, calculators.

In inorganic materials, e.g., WO_3, colour centres can be formed[24] when a d.c. electric field of $\sim 10^4$ V/cm ($\sim$3 volts across a 1 μ thick film) is applied at room temperature. Useful contrasts require $\sim 10^{18}$ colour centres/cm^3, corresponding to the passage of $\sim$100 mA/cm^2 for a second or more. This is comparable to the charge required to deposit an opaque layer of metal by electroplating: a 50 nm thick layer of copper requires $\sim$70 mC/cm^2 (70 mA for 1 s).

Erasibility in electrochromic displays means that irreversible electro-chemical reactions must be avoided. In WO_3, electrons are trapped at colour centres reducing W^{6+} to W^{5+} while oxygen ions are given up at the anode; these ions must again be able to enter the material when the polarity is reversed. Although many electrode materials have been explored which can accept and give up oxygen ions readily, the life of these displays remains problematic.

An organic electrochromic display[25] has also been developed which produces a strongly coloured organic dye at the cathode. Since the colouration due to one dye molecule is greater than that of one metal atom or one colour centre, the electrical charge required to achieve good contrast (up to 20 : 1) in the organic system is considerably smaller (2 mC/cm^2). Life problems are claimed to be absent provided oxygen is excluded from the cell; 10^5 write-erase cycles have been achieved.[25] As with the inorganic systems, addressing voltages (1–2 volts) are compatible with integrated circuits making them of potential use for small alphanumeric applications.

REFERENCES

1. HARENG, M., ASSOULINE, G. and LEIBA, E., *Proc. I.E.E.E.*, **60**, 913 (1972)
2. GREY, G. W., *Molecular Structure and the Properties of Liquid Crystals*, Academic Press, New York (1962); a general reference of all early work
3. SUSSMAN, A., *I.E.E.E. Trans. Parts, Hybrids, and Packaging, PHP-8*, 28 (1972); a review of the current status of electro-optic devices
4. DEVRIES, H., *Acta Cryst.*, **4**, 219 (1951)
5. FELICI, N., *Rev. Gen. Elec.*, **78**, 717 (1969)
6. CHATELAIN, P., *Bull. Soc. Franc. Miner. Cryst.*, **66**, 105 (1943); JANNING, J. L., *Appl. Phys. Letters*, **21**, 173 (1972)
7. DREYER, J. F., in *Liquid Crystals and Ordered Fluids, 2nd*, JOHNSON, J. F, and PORTER, R. S., eds., Plenum Press, New York, 311 (1970)
8. WILLIAMS, R., *J. Chem. Phys.*, **39**, 384 (1963)
9. HELFRICH, W., *J. Chem. Phys.*, **51**, 4092 (1969)
10. Orsay Liquid Crystal Group, *Mol. Cryst. Liq. Cryst.*, **12**, 251 (1971)
11. HEILMEIER, G. H., ZANONI, L. A. and BARTON, L. A., *Appl. Phys. Letters*, **13**, 46 (1968)
12. HEILMEIER, G. H., and GOLDMACHER, J. E., *Proc. I.E.E.E.*, **57**, 34 (1969)
13. FREEDERICKSZ, V. and ZWETKOFF, V., *Acta Physicochimica U.R.S.S.*, **3**, 9 (1935); SCHIEKEL M. F. and FAHRENSCHON, K., *Appl. Phys. Letters*, **19**, 390 (1971)
14. SOREF, R. A. and RAFUSE, M. J., *J. Appl. Phys.*, **43**, 2029 (1972)
15. SCHADT, M. and HELFRICH, W., *Appl. Phys. Letters*, **18**, 127 (1971)
16. HEILMEIER, G. H. and ZANONI, L. A., *Appl. Phys. Letters*, **13**, 91 (1968)
17. WYSOCKI, J. J., ADAMS, J. and HAAS, W., *Phys. Rev. Letters*, **20**, 1024 (1968)
18. HEILMEIER, G. H. and GOLDMACHER, J. E., *J. Chem. Phys.*, **51**, 1258 (1969)
19. MARGERUM, J. D., NIMOY, J. and WONG, S.-Y., *Appl. Phys. Letters*, **17**, 51 (1970)
20. MELCHIOR, H., KAHN, F. J., MAYDAN, D. and FRASER, D. B., *Appl. Phys. Letters*, **21**, 392 (1972)
21. KAHN, F. J., *Appl. Phys. Letters*, **22**, 111 (1973)
22. *Electrophoresis—Theory, Methods and Applications*, Milan Bier, ed. Adacemic Press (1959)
23. OTA, I., OHNISHI, J. and YOSHIYAMA, M., 'Electrophoretic Display Device', *I.E.E.E. Conf. Record of 1972 Conf. on Display Devices*, p. 46, New York, October 11–12 (1972)

24. DEB, S. K., 'A Novel Electrophotographic System', *Appl. Opt. Suppl.*, **3**, 192 (1969)
25. SCHOOT, C. J., PONJÉE, T. J., VAN DAM, H. T., VAN DOORN, R. A. and BOLWYN, P. T., *Appl. Phys. Letters*, **23**, 64 (1973)

LASERS

The word *laser* is an acronym for Light Amplification by the Stimulated Emission of Radiation. In practice oscillators are of much greater importance than amplifiers but the acronym would then hardly be attractive. The advent of the laser has enabled visible light and infra-red radiation sources to be produced with a spectral purity and stability as good as, or better than can be achieved, in the radio spectrum, e.g., an He–Ne laser can be made with a linewidth of a few kHz or less at a carrier frequency of $\sim 5 \times 10^{14}$ Hz. Such light sources are said to have a high temporal or longitudinal coherence. The meaning of this is illustrated by the fact that the light from the source mentioned above could still interfere with another beam split off from the same source with a path difference between the two beams of up to 300 km. The corresponding figure for the best conventional light source is only 10 cm. Another property of laser beams is a high degree of transverse coherence, i.e., if one takes any section perpendicular to a laser beam, then there is a very well defined relationship between the phase of all points on the plane at all times, no matter how far from the source (this is only exactly true in free space). With normal light sources the phase relationship between different points on the source is completely random for distances greater than at most a few μm. This high transverse coherence is responsible for the very slow spreading of laser beams.

The unique properties of the laser are due to the process of stimulated emission, first postulated by Einstein in 1917 in order to explain Planck's Radiation Law. The idea of stimulated emission is that an atom in an excited state can be 'stimulated' to emit radiation by placing it in a radiation field of the same frequency as it will normally emit in going to one of the lower excited states or to the ground state. It can be seen that such a process can result in amplification, if the process of stimulated emission from atoms in the upper state is more probable than that of absorption by atoms in the lower state. Einstein showed that in general, these processes have equal probabilities for a single atom in the radiation field, thus in order to get net amplification we must have more atoms in the upper state than in the lower state. This is an inversion of the normal state of affairs and thus we speak of population inversion being necessary to achieve amplification. It can also be shown that the light produced by the stimulated emission process is in phase with the stimulating light, i.e., it is coherent with it. This process is the basis of both the laser and the maser. However at optical frequencies as distinct from microwaves spontaneous emission, i.e., the random de-excitation of upper state atoms, is a highly probable process which makes the achievement of inversion difficult. All things being equal the probability of spontaneous emission is inversely proportional to the wavelength cubed, thus inversion is much more difficult to achieve in the ultra violet than the infra-red, and extremely difficult in the X-ray region. Of course spontaneous emission is the dominant process in ordinary light sources and is responsible for the poor coherence properties mentioned above since the excited atoms emit in a quite independent and random manner, resulting in poor transverse coherence, and the emission is in the form of very short pulses, resulting in poor longitudinal coherence.

Methods of achieving population inversion

In order to achieve population inversion, we must have a means of excitation and a system of atoms or molecules having energy levels with favourable properties—e.g., the upper laser level should have a long lifetime and the lower level a much shorter one and the means of excitation should excite only the upper level. Different types of lasers are distinguished by the use of different modes of excitation. In gas lasers the primary mode of excitation is electron impact in a gaseous discharge; in solid state (crystalline and glass) and liquid lasers optical excitation is usual, whereas in semiconductor lasers the passage of a high current density through a highly doped, forward biased diode of a suitable material is the preferred mode of excitation. In most cases the excitation process is a multi-step one. For instance in many gas lasers a mixture of gases is used, one of the gases is chosen to have a metastable excited state with a high probability of excitation, whose energy coincides with that of the upper laser level, so that de-excitation by collision with a ground state atom or molecule of the other species is highly probable. Thus helium in the He–Ne laser and N_2 in the 'CO_2' laser (which usually uses a mixture of CO_2, He and N_2) both have suitable metastable levels for the excitation of Ne atoms and CO_2 molecules respectively.

In solid-state lasers, the light absorbed excites an ion to a high level which then decays very rapidly by a non-radiative process to the upper laser level which usually has a lifetime of at least several hundred microseconds, thus making inversion relatively easy to achieve on a transient basis, i.e., pulsed excitation by a flashtube. In fact solid state laser materials are rather special phosphors, chosen for the narrow linewidth of their emission. Similarly de-excitation of the lower laser level is not usually a simple radiative process. In gas lasers, more often than not, collisions with the walls of the discharge tube (He–Ne) or with another species of atom or molecule (He in the CO_2 laser) are important in de-excitation. In some solid state lasers (so-called four level materials such as neodymium doped materials) de-excitation is achieved by radiationless transfer of energy to the atoms of the lattice in which the Nd^{+++} is embedded. In the case of 'three level' materials such as ruby, the lower laser level is the ground state of the C_r^{+++} ions in an Al_2O_3 crystalline matrix. Thus 'de-excitation' is automatic, but inversion is difficult to achieve since more than half of all the C_r^{+++} ions must be excited in order to achieve inversion.

Oscillators and modes, mode selection, mode locking, etc.

If we simply have a medium in which population inversion is maintained, then it will act as a tuned amplifier resonant at a frequency corresponding to the transition concerned, with 3 dB points determined by the linewidth of spontaneous emission. If the overall amplification is very high, then amplification of spontaneous emission or super-radiance can depopulate the inversion very rapidly, with some line narrowing.

To make an oscillator, we require positive feedback; this is provided by a pair of mirrors placed at each end of the amplifier as shown in *Figure 10.24*. Usually one of these is as near 100% reflecting as possible, and the other one has a transmission typically of a few % or so. At longer infra-red wavelengths a hole in the output mirror may be used to couple power out. To obtain oscillation $G R_1 R_2$ must be greater than unity (G is the gain of the medium; R_1,R_2 are the mirror reflectivities). This arrangement is similar to a microwave resonator, except that there are no reflecting walls between the mirrors, and is therefore spoken of as an 'open' resonator.

Since the distance between the mirrors is typically more than 10^5 wavelengths, and the width of the medium is usually at least 10^3 wavelengths, then it can be seen that many more modes of oscillation are possible than in a microwave resonator. The behaviour of such modes has been described by Fox and Li and others.[1]

The modes in an open resonator of this type are what is known as transverse electromagnetic or t.e.m. waves. They are specified by three subscript mode numbers p, l, q (cylindrical symmetry) or m, n, q (rectangular). The last subscript in each case is the longitudinal mode number, i.e., a very large number. The other two refer to the transverse directions, and in practice here the low order modes (<5) are usually dominant

Transverse mode patterns corresponding to some of the low order rectangular modes

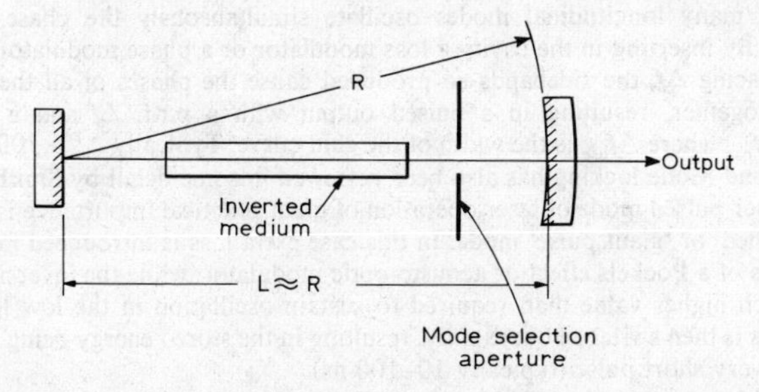

Figure 10.24 Laser with hemispherical resonator incorporating transverse mode selection

are shown in *Figure 10.25*. The 0,0q mode is of most practical importance since it has the highest degree of transverse coherence and the lowest beam divergence. The radial intensity distribution of energy in any cross section is given by a gaussian function. The propagation of such 'gaussian' beams has been treated in some detail by Kogelnik.[2]

Schemes for selecting a particular mode, usually the zero order, depend on the fact that the diffraction losses of different modes propagating in a resonator are different, and by using the hemispherical resonator configuration shown in *Figure 10.24*, the absolute magnitudes of the losses for higher order modes make it easy to select the zero order mode by the use of a suitable aperture as shown in the figure. The longitudinal mode behaviour is determined by the width of the gain curve of the inverted medium and by the frequency response of the resonator. The simplest form of resonator, using a pair of relatively broadband reflecting multilayer dielectric mirrors has very many resonances, separated by a frequency $\Delta f = C/2L$ where C is the velocity of light and L is the optical path length of the resonator.

In rare cases, such as the low pressure 'CO_2' laser, it is relatively easy to ensure single frequency operation since the linewidth of the gain curve is typically less than $C/2L$. In most types of laser there are very many resonances within the gain curve, and thus many longitudinal modes oscillate, unless a more complex resonator, having only one principle resonance within the width of the gain curve, is used. The subject of mode selection schemes has been reviewed by Smith.[3]

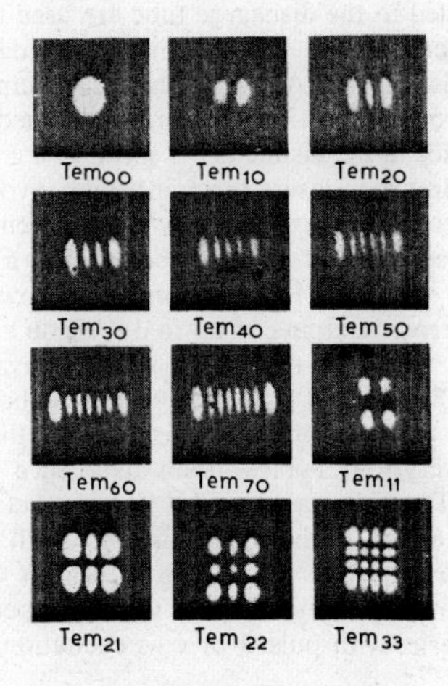

Figure 10.25 Mode patterns of a gas laser oscillator (rectangular symmetry)

When many longitudinal modes oscillate simultaneously the phases are usually random. By inserting in the cavity a loss modulator or a phase modulator, tuned to the mode spacing Δf, the sidebands so produced cause the phases of all the modes to be locked together, resulting in a pulsed output with a p.r.f. Δf and a pulse length $T = 1/\Delta f_{osc}$ where Δf_{osc} is the width of the gain curve. Typically $\Delta f \geqslant 100$ MHz, $T \leqslant 1$ nanosecond Mode locking has also been reviewed in some detail by Smith.[4]

Another pulsed mode of laser operation of great practical importance is the so called 'Q switched' or 'giant pulse' mode. In this case extra loss is introduced into the cavity, by means of a Pockels effect or acousto-optic modulator, while the inversion is built up to a much higher value than required to sustain oscillation in the low loss case. The extra loss is then switched out rapidly, resulting in the stored energy being discharged in a single very short pulse (typically 10–100 ns).

A third pulsed mode operation is 'spiking', which occurs spontaneously in pulse excited solid state lasers. The output in this case consists of a random train of sub-microsecond pulses of duration comparable with the pumping pulse. This type of operation is reviewed by Roess.[5]

Types of laser

There are a very great many types of laser, operating at wavelengths from the vacuum u.v. to the far infra-red. Thousandths of wavelengths are available directly, and even more by harmonic generation. The most important classes are as follows.

GAS LASERS

As mentioned above, these are usually excited by passing a discharge through a gas or, more often, a gas mixture. Only the most common types are described here. There are the helium-neon laser, the argon ion laser and the CO_2 laser.

The helium-neon laser usually contains a mixture of about 5 parts helium with one part Ne at a total pressure of a few torr. The discharge tube has a bore of 1–2 mm and a length of 20–100 cm, depending on the power output, usually in the range 0.5–50 mW. The current is usually ~10 mA with voltages up to a few kV. The gain of the medium is very low and losses must be kept to a minimum. Either Brewster angle windows (resulting in plane polarised output) are used with external mirrors, or integral mirrors sealed to the discharge tube are used to ensure low loss. Partly since mirror alignment is critical the sealed mirror method of construction is more difficult to engineer satisfactorily. Although the usual output wavelength is ~633 nm in the red, outputs at 1.15 μm or 3.39 μm can be obtained by using suitable mirrors. Typical beam divergence in the visible is ~1 mrad.

In argon lasers the transitions involved are between excited levels of singly ionised argon ions. Consequently high current densities (> 100 A/cm^2) are required to achieve threshold, and plasma tubes of BeO are usually used to minimise heating, erosion and gas clean-up. Brewster angle windows of fused silica are usually used with external mirrors. Outputs of up to 15 W, on a number of lines in the blue and green regions of the spectrum simultaneously, are available commercially. Principle wavelengths are ~488 nm and ~515 nm. Plasma tube-lengths of 10–100 cm, with diameters 1–2 mm are used, and more gain is available than with He–Ne. However, the efficiency, $\leqslant$0.1%, is very low. Typical beam divergence $\leqslant$1 mrad.

The 'CO_2' laser, which gives outputs at a number of wavelengths between 9.6 μm and 10.6 μm, is the most efficient of all the gas lasers (up to 30%). They are the most powerful (c.w. powers up to 30 kW or so) and the most varied in construction since there are many different versions operating at pressures from a few torr up to atmospheric, with pulsed or c.w. excitation, and even incorporating Q switching. The CO_2

laser is one of the few gas lasers in which the lifetime of the upper level is great enough to make *Q* switching worthwhile.

The most common form is still the low pressure flowing gas (He : N_2 : CO_2 :: 8 : 1 : 1 typical) system using a plasma tube ~2.5 cm diameter and up to several metres in length, depending on the power output required. Integral mirror or Brewster angle plus

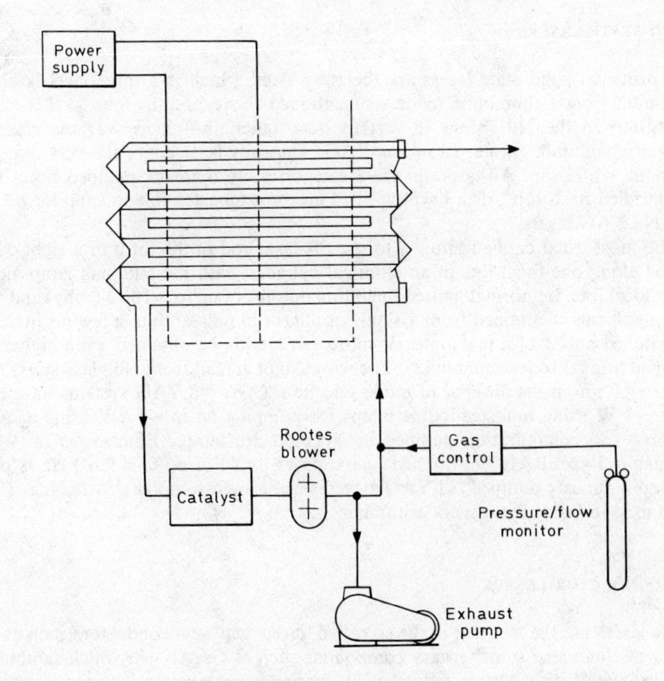

Figure 10.26 *Folded 'CO_2' laser with gas re-circulation.* (*Courtesy Marconi-Elliott Flight Automation Ltd*)

external mirror configurations are used. In the simplest form output power may be coupled out using a hole in one mirror, in conjunction with a suitable window. For longer plasma tubes, a folded configuration is used, *Figure 10.26*. A review of CO_2 laser systems has been given by De Maria.[6]

LIQUID LASERS

These use a solution of a fluorescent molecule, optically pumped either by another laser or by a flash tube in a configuration similar to that used for solid state lasers. The most common form of liquid laser is the dye laser, so called because the fluorescent molecule is a suitable dye molecule, e.g., Rhodamine 6G, in an organic solvent. Usually a substance is added to suppress the triplet excited state which would otherwise be formed, leading to unacceptable losses at the laser wavelength. Often a flowing dye solution is used to remove molecules in the triplet state from the lasing region. Dye laser outputs are usually broad band (~10 nm) unless a grating is used in the cavity. Narrow band tunable outputs can be obtained in this way over most of the visible and near i.r. spectrum, using a number of dyes. Argon lasers are usually used to pump c.w. dye lasers. Outputs up to a few watts have been obtained at some wavelengths as the excitation can be quite efficient (~20%).

SOLID-STATE LASERS

The principal solid state lasers are the ruby laser, which is a three level laser, using chromium doped aluminium oxide, as mentioned above, and the four level lasers using transitions in the Nd^{+++} ions in various host materials—chiefly various glasses and yttrium aluminium garnet, known as YAG. The ruby laser emits at ~694 nm, the Nd lasers at ~1.06 μm. All lasers in these categories are optically pumped either using a xenon filled flash tube, or a krypton filled arc or a tungsten-iodine lamp in the case of c.w. Nd-YAG lasers.

The most usual configuration is to use the laser rod in the form of a right cylinder, placed along one focal line of an elliptical cylinder, with the pumping lamp along the other focal line. In 'normal' pulsed operation outputs of up to ~100 J (ruby) and 1 000 J (Nd glass) can be attained from a single oscillator in pulses up to a few ms in duration. Q switched pulses of a few joules or more can readily be obtained, even higher values using additional rods as amplifiers. The wavelength spread from Nd glass lasers may be up to ~10 nm in the absence of mode selection. C.W. Nd-YAG systems have outputs from ~1 W using tungsten-iodine lamps for pumping up to ~1 kW using a cascaded multiple rod configuration pumped by krypton arc lamps. Efficiencies of ~1% are attainable. Repetitively Q switched operation at p.r.f.s of up to a few kHz is possible with continuously pumped Nd-YAG systems using acousto-optic Q switching. They are often used in lasers for resistor trimming.

SEMICONDUCTOR LASERS

These lasers use the fact that in the so called 'direct gap' semiconductors such as GaAs, InAs, etc. (and also some ternary compounds such as $Ga_xAl_{1-x}As$ which exhibit 'direct gap' behaviour over certain ranges of x), carrier recombination between electrons and holes occurs predominantly by the emission of radiation. By constructing a *pn* junction in which the *p* and *n* regions are very highly (degenerately) doped, population inversion between electron and hole concentration occurs in the junction region. Since the gain can be very high, up to 30 cm^{-1} or more, laser action can be achieved with cavity lengths of 1 mm merely by cleaving the crystal along faces perpendicular to the junction plane, relying on the relatively high Fresnel reflectivity at the surfaces of the high refractive index material to give sufficient positive feedback. Such cleaved surfaces act as near optically perfect mirrors.

A simple laser such as that described is only suitable for low duty cycle, short pulse operation at room temperature, since the threshold current density is of the order of 10^5 A/cm^2 and heating of the junction causes the threshold current to increase rapidly (~100 ns) to the point at which oscillation can no longer occur. At low temperatures—77 °K or lower—c.w. operation is possible and c.w. outputs of a watt or more have been obtained. More recently more complex structures involving up to five layers of different doping and composition have been devised. The inner layers usually comprise $Ga_xAl_{1-x}As$ where x is near to unity, the object being to achieve confinement of the radiation, as in a dielectric waveguide (see Section 15 optical communications) and also localisation of carrier recombination.

By these methods c.w. operation at room temperature with external efficiencies of 5% of more becomes possible. Initially the devices had very short lives, of a few hours only, but recently lives of several thousand hours have been reported. Laser action in GaAs devices is centred at 0.85–0.9 μm, depending on the temperature. $Ga_xAl_{1-x}As$ devices can work at wavelengths in the visible down to 6 500 Å or so, but no semiconductor lasers working in the green or blue have yet been made using *pn* junction methods. CdS and ZnS, which are direct gap materials emitting in the green and violet respectively, can be made to lase using optical excitation with another laser, or by injecting electrons by means of a high energy electron beam (10 kV or more) incident on the crystal surface.

REFERENCES

1. FOX, A. G. and LI, T., 'Resonant Modes in a Laser Interferometer', *Bell System Technical Journal*, **40**, 453 (1961)
2. KOGELNIK, N. and LI, T., 'Laser Beams and Resonators', *Applied Optics*, **5**, 1550 (1966)
3. SMITH, P. W., 'Mode Selection in Lasers', *Proc. I.E.E.E.*, **60**, 422 (1972)
4. SMITH, P. W., 'Mode Locking of Lasers', *Proc. I.E.E.E.*, **58**, 1342 (1970)
5. ROESS, D., *Lasers, Light Amplifiers and Oscillators*, Academic Press (1969)
6. DE MARIA, A. J., 'Review of C.W. High Power CO_2 Lasers', *Proc. I.E.E.E.*, **61**, 731 (1973)

FURTHER READING

ALLEN, L. and JONES, D. G. C., *Principles of Gas Lasers*, Butterworth (1967)
BLOOM, A. L., 'Gas Lasers', Wiley (1968)
GOOCH, C. H., ed., *Gallium Arsenide Lasers*, Interscience (1969)
GOODWIN, A. R. and SELWAY, P., 'Heterostructure Injection Lasers', *Electrical Communication*, **47**, 1, 49 (1972)
HARVEY, N. F., *Coherent Light*, Wiley (1970)
ELECCION, M., 'The Family of Lasers: Survey', *I.E.E.E. Spectrum*, **9**, 326 (1972)
RIECK, H., *Semiconductor Lasers*, MacDonald (1968)
WAGNER, W. G. and LENGYEL, B. A., 'Evolution of the Giant Pulse in a Laser', *Journal of Applied Physics*, **34**, 2040 (1963)

11 INTEGRATED CIRCUITS

INTEGRATED CIRCUITS 11

11 INTEGRATED CIRCUITS

Of the many types of solid-state device that have followed the introduction of the point-contact transistor in 1948, the most revolutionary (so far) is undoubtedly the integrated circuit. The transistor, semiconductor diode, and thyristor provided a solid-state replacement for existing thermionic devices, albeit with considerable advantages and often with an improved performance compared with the thermionic device that enabled new applications to be developed. Other solid-state devices such as opto-electronic devices created new applications for electronic devices. But the integrated circuit forced the circuit designer to look at electronic systems in a new way, and the implications of this 'new look' are still being explored today.

Popular imagination has been caught by the considerable reduction in the size of equipment possible with integrated circuits. This small size is exemplified by such photographs as that of *Figure 11.1* where an integrated circuit chip is shown in the eye of a domestic sewing needle. It is true that this reduction in the size of equipment has been important. It would not have been possible, for example, to incorporate the sophisticated electronic systems into U.S. space vehicles without the integrated circuit, and without it electronic computers would have remained impracticably large. But the improved reliability possible in equipment based on integrated circuits is the outstanding advantage.

In the manufacture of transistors, many thousand transistor elements are manufactured together, divided, and separately encapsulated. Many hundred discrete transistors are connected together with soldered joints to form, say, a logic system. Other circuit elements such as diodes and resistors also have to be wired into the system. Since all these circuit elements can be made in solid-state form, it is clear that considerable advantages result if the elements are manufactured and interconnected in one piece of silicon. The elimination of the separate encapsulations and subsequent wiring reduces the space required for the system. The replacement of soldered joints by a carefully-controlled metal film on the silicon surface improves the reliability. Instead of the many soldered connections in the system, only the supplies for the active devices and the signal inputs and outputs are required. These are the underlying principles of the integrated circuit.

TYPES OF INTEGRATED CIRCUIT

The term *microelectronics* is often used synonymously with *integrated circuits*. This is not correct, as microelectronics is a general term indicating small size. It follows such terms as miniature and microminiature coined as the size of components decreased during the development of solid-state devices. Microelectronics includes such techniques as the thin-film and thick-film processes which are described in Section 12.

Integrated circuits (i.c.s) can be of two types: monolithic and multichip. In a monolithic i.c., all the circuit elements and interconnections are formed in one piece of silicon called a chip. In a multichip i.c., more than one chip is included in the encapsulation, the chips being mounted on one header and interconnected by fine wires. The term hybrid i.c. may also be encountered. This is an i.c. using both monolithic and

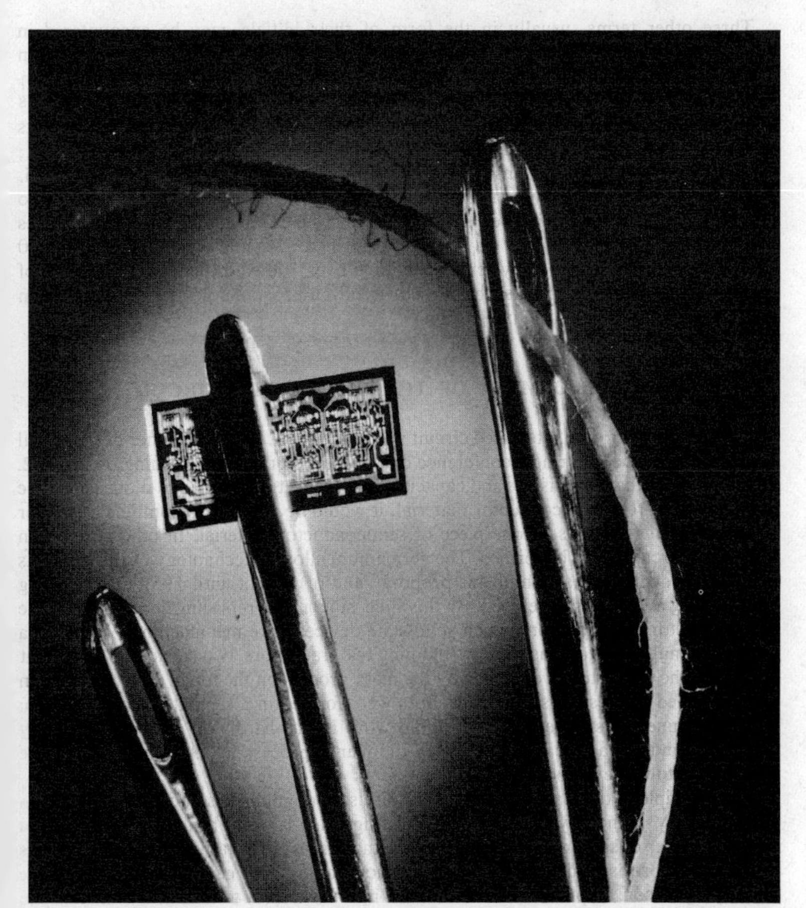

Figure 11.1 Unencapsulated i.c. in the eye of a No. 5 sewing needle. The circuit is a t.t.l. decade counter containing 120 components on a chip 1.5 mm × 3.0 mm (Mullard Limited)

thin-film techniques. The active components are formed in the chip, while the passive components and interconnections are formed by film techniques on the surface of the chip.

Integrated circuits are also classified by the type of transistor used: bipolar or m.o.s. A bipolar transistor is a junction transistor, and is so called because its operation depends on both types of charge carrier, electrons and holes. The initials m.o.s. stand (strictly speaking) for Metal Oxide Semiconductor, but are generally interpreted as Metal Oxide Silicon as the semiconductor material used at present is invariably silicon. Thus m.o.s. refers to the construction of the transistor, successive layers of silicon, silicon oxide, and a metal (aluminium or silicon). This type of transistor is a field-effect or unipolar transistor. The operation of junction and field-effect transistors is described in Section 8.

A third classification of i.c.s is by the application and mode of operation. Under this classification, i.c.s can be digital or linear. An alternative name for linear i.c. is analogue i.c.

Three other terms, usually in the form of their initials, may be encountered in connection with i.c.s. These are: small-scale integration (s.s.i.), medium-scale integration (m.s.i.), and large-scale integration (l.s.i.). The terms refer to the size of the system incorporated in the i.c., although precise definitions are not possible, particularly as manufacturing techniques are still developing. Generally, s.s.i. refers to small circuits such as flip-flops, gates, and other basic logic circuit elements incorporated into one i.c.; m.s.i. refers to small sub-systems such as counters, shift registers, and adders; while l.s.i. refers to complete logic systems and solid-state memories in one i.c. As a guide to present-day systems, s.s.i. can be taken to refer to systems containing less than 12 gates or equivalent circuit elements on one chip; m.s.i. to systems containing 12 to 100 equivalent gates; and l.s.i. to systems containing over 100 equivalent gates. Because of the manufacturing cost of i.c.s, the present trend on economic grounds is away from s.s.i. towards the larger-scale integration.

DEVELOPMENT OF THE INTEGRATED CIRCUIT

The first proposal for an integrated circuit was made by G. W. A. Dummer of the Royal Radar Establishment during a conference on the transistor held in Washington in 1952. He pointed out that as resistors and capacitors as well as transistors and diodes could be manufactured from semiconductor material, it would be possible to combine a number of these circuit elements in one piece of semiconductor material. In other words, an integrated circuit could be formed. The germanium alloying technology of the time was inadequate for the realisation of this proposal, and it was not until 1959 that anything approaching an i.c. was developed. By this time, silicon was replacing germanium as the semiconductor material from which solid-state devices were manufactured. In 1959, a solid-state multivibrator was made, containing two transistors, two capacitors, and eight resistors. The transistors were made by the mesa process on the top of a piece of silicon which formed the resistors. The capacitors were formed in a separate piece of silicon, and interconnections were made with gold wire. It was hardly an i.c. by present-day standards, and it was the introduction of the planar process in 1960 that enabled the genuine i.c. to be developed.

The principle of the planar process is the diffusion of impurities into areas of a silicon slice precisely defined by windows in a covering oxide layer. When silicon is heated in oxygen, a layer of silicon dioxide (a glass-like material) is grown over the surface. Areas of this oxide, precisely defined by photographic means, can be etched away to reveal the underlying silicon. Impurities can be diffused into these areas to form *p*-type and *n*-type regions, and because the impurities do not diffuse readily through the oxide, the *p* and *n* regions will also be precisely defined. During the diffusion a new oxide layer is grown, sealing in the diffusion and allowing new windows to be etched for further diffusions.

Bipolar transistors and diodes are made by successive diffusions of impurities to form the required structure in the silicon slice. Resistors are formed by using the resistance of an isolated diffusion, the resistance value being determined by the length and width of the diffusion. Capacitors are formed by using the capacitance of a reverse-biased diode element. In these ways, the required circuit elements can be obtained.

The first i.c.s were developed in the U.S.A. around 1960 for use in guided missiles. Initially the i.c.s were multichip types, for example with a transistor in one chip, a resistive network in a second, diodes in a third, and so on. The various chips were mounted on one header and interconnected with gold wire. The multichip technique was necessary because of the difficulty of isolating components in a single chip. By about 1962, a suitable isolation technique had been developed, and monolithic i.c.s became available.

The early i.c.s were digital circuits. This was because the control systems of missiles were largely based on digital techniques, and because it was easier to design switching circuits which have only two operating states (on and off) than linear circuits with their

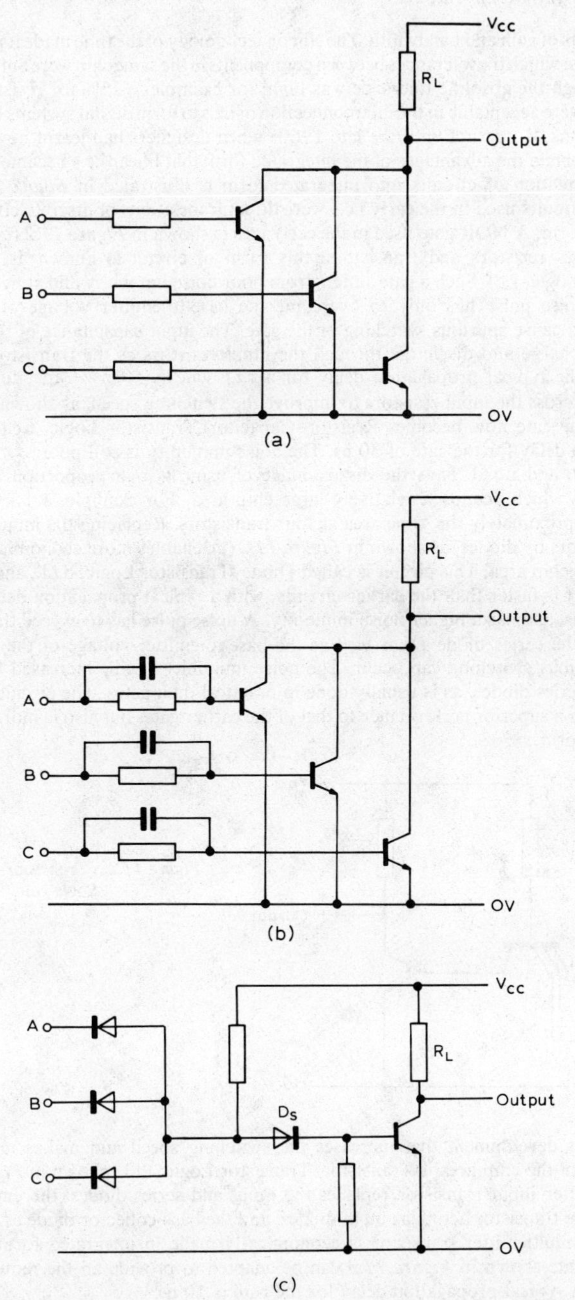

Figure 11.2 Development of logic gate circuit suitable for integrated form: (a) Resistor–Transistor Logic (b) Resistor–Capacitor–Transistor Logic (c) Diode–Transistor Logic

requirements of gain and bandwidth. The silicon technology of the time made it possible to design i.c.s in which the tolerances between components in the same chip were only a few per cent, although the absolute tolerance was high; for example, ±25% for resistors. Such tolerances were acceptable in the interconnection of i.c.s to form digital systems but not for linear systems. It was not until the late 1960s when designers had learnt how to adapt circuits to obtain the advantages of the integrated form that linear i.c.s became available.

This adaptation of circuits into integrated form is illustrated in *Figure 11.2*. The switching circuits used in the early i.c.s were direct translations of discrete circuits into integrated form. A NOR gate used in the early i.c.s is shown in *Figure 11.2* (*a*). Because the gate uses resistors and transistors, this form of circuit is known as Resistor–Transistor Logic, r.t.l. Such a gate suffers from poor noise immunity and slow operating speed. A noise pulse has only to overcome the base-to-emitter voltage of a cut-off transistor to cause spurious switching of the gate. The input capacitance of the transistor has to charge and discharge through the input resistors as the transistor changes state, and the typical propagation delay for a r.t.l. gate is 50 ns. Capacitors can be connected across the input resistors to improve the switching speed, as shown in *Figure 11.2* (*b*). The gate now becomes Resistor–Capacitor–Transistor Logic, r.c.t.l., with a propagation delay for the gate of 30 ns. The noise immunity is still poor.

Both r.t.l. and r.c.t.l. have the disadvantage of using a high proportion of passive components which occupy a relatively large chip area. For example, a 1 kΩ resistor occupies approximately the same area as four transistors. Replacing the input resistors and capacitors by diodes, as shown in *Figure 11.2* (*c*), enables more economic use to be made of the chip area. This circuit is called Diode–Transistor Logic, d.t.l., and is still in use today. It is faster than the earlier circuits, with a typical propagation delay for the gate of 25 ns, and has a higher noise immunity. A noise pulse has to exceed the forward voltage of the series diode D_s as well as the base to emitter voltage of the transistor before spurious switching can occur. The noise immunity can be increased further by using two series diodes, as is usually done in practical d.t.l. gates. The circuit therefore not only has a superior performance to that of the earlier gates but also is more suited to integrated form.

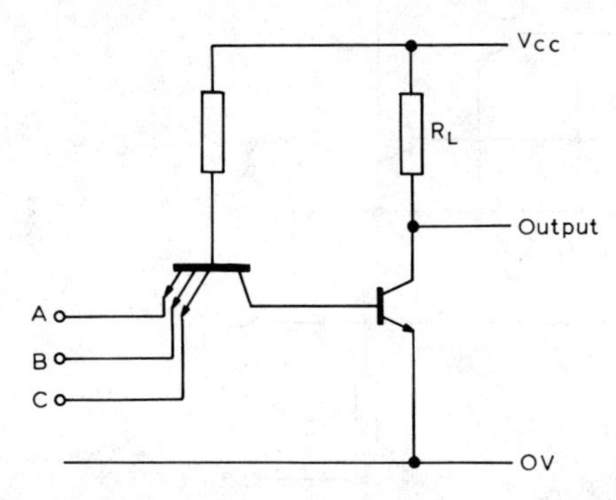

Figure 11.3 Transistor–Transistor Logic gate

A further development that increases the switching speed and makes more economical use of the chip area is Transistor–Transistor Logic, t.t.l., shown in *Figure 11.3*. A multiemitter input transistor replaces the input and series diodes, the emitter-base diodes of the transistor acting as input diodes, and the base-collector diode as the series diode. The multiemitter transistor is economically made in integrated form, and the basic t.t.l. gate shown in *Figure 11.3* can be adapted to provide all the required logic functions. A typical propagation delay for the gate is 10 ns.

Faster switching speeds can be obtained from gates using Emitter-Coupled Logic, e.c.l., and Complementary-Transistor Logic, c.t.l. An e.c.l. gate is shown in *Figure 11.4*.

The typical propagation delay for the gate is 2 ns, but the power dissipation is relatively high and a reference voltage different from the supply voltage is required. A c.t.l. gate is shown in *Figure 11.5*; the propagation delay for this gate is typically 5 ns.

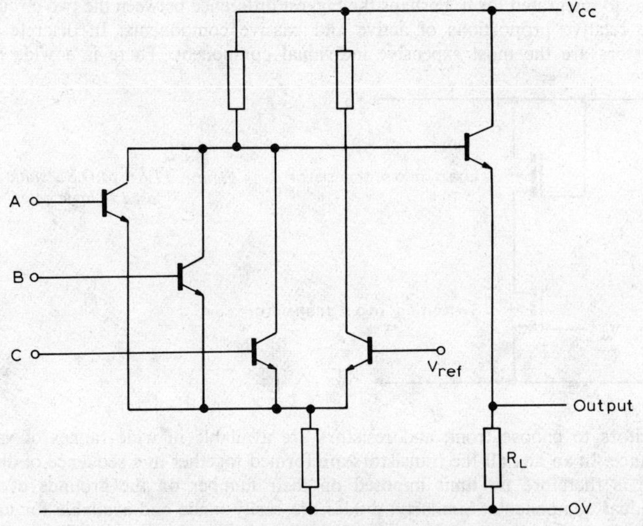

Figure 11.4 *Emitter-Coupled Logic gate*

In 1966, m.o.s. integrated circuits were introduced. The m.o.s. transistor in the integrated form offers considerable advantages over the bipolar transistor. No isolating diffusion is necessary, fewer diffusions are required to make it, and the area of the

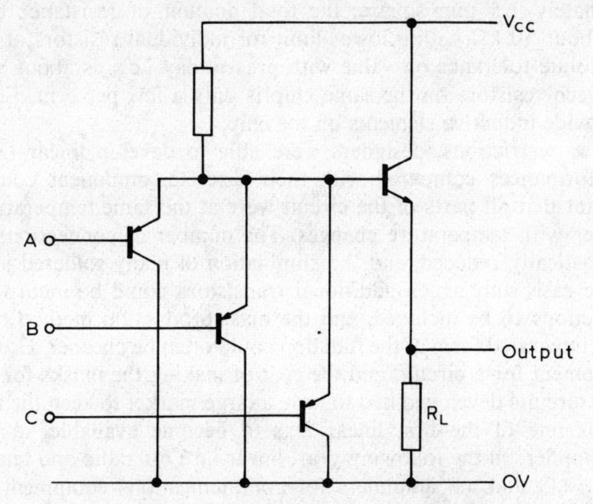

Figure 11.5 *Complementary-Transistor Logic gate*

completed transistor on the chip is smaller. An m.o.s. static inverter, the basis of all the m.o.s. logic gates, is shown in *Figure 11.6* and illustrates another advantage of the m.o.s. i.c. An m.o.s. transistor can act as the load for another transistor so that load resistors are not required. These advantages are countered by a disadvantage, the slow operating speed. The typical propagation delay for an m.o.s. gate is 250 ns.

A comparison of the gates in *Figure 11.2* shows how circuits to perform the same function can differ in their discrete and integrated forms. The experience gained with digital circuits in the early 1960s enabled designers to adapt many discrete linear circuits to integrated form. Perhaps the biggest difference between the two circuit forms is the relative proportions of active and passive components. In discrete circuits, transistors are the most expensive individual component. There is a wide range of

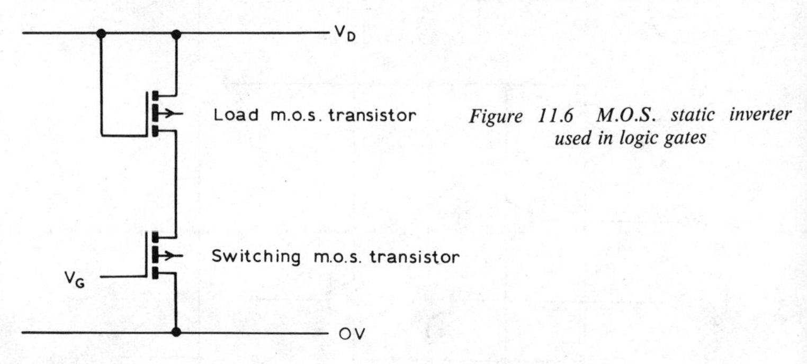

Figure 11.6 M.O.S. static inverter used in logic gates

capacitors to choose from, and resistors are available in wide ranges of value and tolerance. In an i.c., all the transistors are formed together in a sequence of diffusions. There is therefore no limit imposed on their number on the grounds of cost for individual components. Similarly, diodes are easily made and available for use in the circuit. Thus the integrated form of a linear function can be more flexible and complex than the equivalent discrete form. Against this has to be set the limitation on the values of capacitors and resistors available. The maximum value of capacitance that can be obtained is approximately 100 pF without occupying a disproportionate area of the chip. In practice, values of 20 to 30 pF are regarded as a working limit. Similarly, on a chip approximately 1.5 mm square, the total amount of resistance on the chip is restricted to about 50 kΩ with a lower limit for individual resistors of approximately 30 Ω. The absolute tolerance on value with present-day i.c.s. is about ±20%, but the tolerance between resistors on the same chip is only a few per cent. Finally, it is not possible to provide inductive elements on the chip.

Despite these restrictions, designers were able to develop linear i.c.s, often with improved performances compared with their discrete-component counterparts. The small size meant that all parts of the circuit were at the same temperature, and would change together with temperature changes. The number of connections required in a circuit was drastically reduced, and the elimination of many soldered joints improved reliability. The ease with which additional transistors could be incorporated allowed additional functions to be included, and the mass-production method of manufacture meant that the integrated form of the function could often be cheaper. However, the high cost of development for a circuit, and the cost of making the masks for the diffusions, meant that the circuits developed had to have a large market to keep the unit cost of the i.c.s low. Thus one of the first linear i.c.s to become available, in 1967, was an operational amplifier. In the following year, linear i.c.s for radio and television became available. R.F., i.f., and a.f. amplifiers for communications equipment and domestic receivers were developed, and later complete signal-processing i.c.s. By the end of the 1960s, for example, the complete sound-processing circuitry for a domestic television receiver, from the video pre-amplifier to the loudspeaker connections, could be incorporated into one linear i.c., with considerable savings in the cost of assembly for the receiver.

Throughout the 1960s, the main concern of the manufacturer was to increase the yield of satisfactory i.c.s. This implies, as will be discussed in the manufacturing section, a small chip area. The size of the chips used originally was between 1.0 and 1.5 mm square. Towards the end of the decade, the trend was towards accommodating more circuit elements on the chip, and so chip sizes had to be increased. For m.s.i., a chip 2 mm square could contain approximately 40 bipolar t.t.l. gates, while for m.o.s. i.c.s a chip 3 × 4 mm could contain up to 500 gates. As solid-state memories were developed, chip sizes of up to 5 mm square were introduced.

During the early 1970s, the range of applications for i.c.s continued to increase, often imposing more demands on the circuit manufacturer. As more elements were incorporated on a chip, techniques were devised to reduce the size of bipolar transistors so that l.s.i. bipolar i.c.s could become more competitive with m.o.s. i.c.s. Such techniques as Collector Diffusion Isolation, c.d.i., were introduced. The present applications, and the possible future development, of i.c.s will be discussed at the end of the section.

MANUFACTURE OF INTEGRATED CIRCUITS

The manufacturing process for i.c.s is an extension of the process used for planar transistors and diodes. A large number of i.c.s are manufactured at the same time on a slice of silicon by a sequence of diffusions through windows in a covering oxide layer. These windows are etched in areas precisely defined by photographic techniques by exposure through masks. The various circuit elements are interconnected by a thin metal film of the required pattern deposited on the final oxide layer. The completed i.c.s are tested on the slice, and faulty ones marked for rejection at a later stage. The slice is divided into individual i.c. chips which are encapsulated and subjected to a final test.

The manufacturing process outlined above has many stages which are common for bipolar and m.o.s. i.c.s. The following descriptions therefore apply to both types of i.c. unless it is stated otherwise.

Slice preparation

The starting material for the manufacture of i.c.s is a slice of monocrystalline silicon. Single crystals of silicon are prepared by crystal pulling techniques in which a small seed crystal is dipped into the surface of molten purified silicon and slowly withdrawn. A single crystal with a diameter of approximately 5 cm and a length of 30 cm can be obtained in this way. Impurities can be added to the molten silicon so that a *p*-type or *n*-type crystal is obtained. For bipolar i.c.s, *p*-type silicon is used; for m.o.s. i.c.s, *n*-type and *p*-type silicon is used depending on whether *n*-channel or *p*-channel devices are being manufactured.

The single-crystal ingot is cut into thin slices, about 1 mm thick, the cut being made along a crystal axis. Because the cutting damages the silicon surface, the slice is etched and polished. The final slice from which the i.c. is manufactured is about 200 μm thick, with a flat, smooth surface. Despite the great care taken in the preparation of the crystal and the slices, it is inevitable that there will be dislocations in the crystal structure and a variation in the resistivity (caused by a variation in the doping concentration) across the slice. Improvements in manufacturing techniques have enabled the number of dislocations to be reduced to about 500 per cm^2 (compared with about 30 000 per cm^2 in the early days of i.c. manufacture), and the variation of resistivity to less than 15% from the centre to the edge of the slice. An i.c. manufactured at a site on the slice containing a dislocation will almost inevitably be faulty. If the area of the individual i.c.s is small, the number of i.c.s on the slice will be large and so the proportion of faulty ones low. This was one of the main considerations dictating the small chip size of the early i.c.s.

Oxide masking

The defined diffusions that make possible the manufacture of i.c.s depend on the formation of a layer of silicon dioxide on the surface of a silicon slice, the ease with which the oxide can be etched away without etching the silicon, and the fact that the impurities used to form the *p* and *n* regions in the silicon do not diffuse readily through the oxide.

The initial oxide layer on the slice is formed by heating the slices to about 1 100 °C in a stream of wet oxygen. The temperature of the furnace is controlled to better than ±1 °C so that an even layer of silicon dioxide about 0·5 μm thick is grown, *Figure 11.7* (*a*). The surface of the slice is coated with a photoresist. A small drop of the photoresist is placed on the surface, and the slice is rotated at approximately 5 000

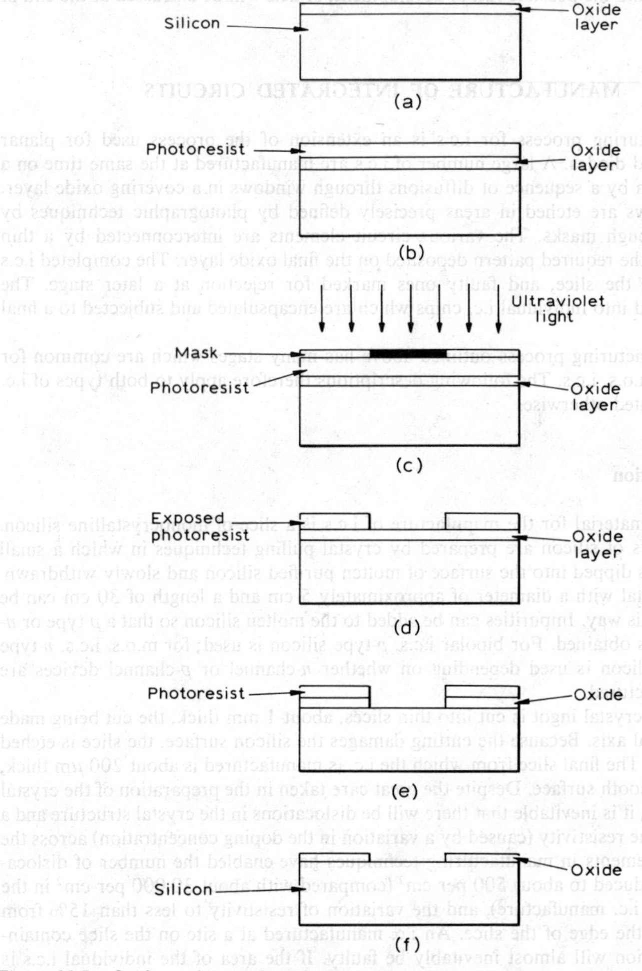

Figure 11.7 Oxide masking technique: (*a*) *silicon slice with oxide layer* (*b*) *photoresist applied* (*c*) *mask positioned and slice exposed* (*d*) *exposed photoresist etched* (*e*) *uncovered oxide etched* (*f*) *photoresist removed*

rev/min to produce an even layer about 1 μm thick, *Figure 11.7 (b)*. The photoresist is an organic material which polymerises when exposed to ultraviolet light. The polymerised material resists the action of solvents and acids, whereas the unpolymerised (unexposed) material is easily dissolved. The photoresist is dried by baking at 100 to 150 °C for a few minutes. A mask defining the areas of the oxide layer to be etched away is placed in contact with the slice, and the slice is exposed to ultraviolet light *Figure 11.7 (c)*. On development of the photoresist, the unexposed areas are removed (giving access to the oxide layer) while the exposed areas remain and are hardened by further baking, *Figure 11.7 (d)*. The slice is then etched, using a chemical such as hydrofluoric acid which dissolves the oxide but not the silicon or remaining photoresist, *Figure 11.7 (e)*. The photoresist is then dissolved away, with such chemicals as hot concentrated sulphuric or nitric acid, and the slice is ready for diffusion, *Figure 11.7 (f)*.

In all the diffusions during the manufacture of the i.c., an oxide layer forms over the diffused region. This enables the slice to be recoated with photoresist, and the exposure and etching processes to be repeated to define further diffusion windows.

Epitaxial layer

The circuit elements in a bipolar i.c. are formed in an epitaxial layer. Such a layer reproduces the near-perfect crystal structure of the *p*-type silicon slice on which it is grown, but can be of a different doping concentration (resistivity). The impurities in the layer can be of the opposite type from those in the slice, so that an *n*-type epitaxial layer can if required be grown on a *p*-type slice.

Rigorously cleaned and polished slices of the *p*-type silicon, which are now called substrates, are placed on a graphite block in a radio-frequency heated reactor. The slices are held at a temperature between 1 000 and 1 200 °C, and the epitaxial layer is grown by vapour deposition. The silicon vapour is formed by the decomposition of a silicon compound such as silicon tetrachloride $SiCl_4$ with hydrogen, and impurities can be added to the vapour to give the layer the required resistivity. For *n*-type layers the usual impurity is phosphorus, and for *p*-type layers it is boron.

The epitaxial layer is grown at a rate of approximately 1 μm/min, so the thickness of the layer can be carefully controlled. The thickness of the layer grown for bipolar i.c.s is about 20 μm. A typical value of resistivity is 5×10^{-3} Ωm.

Isolation

In an i.c., where the circuit elements are formed in silicon which is electrically conducting, it is necessary to isolate the various elements from each other. The required connections are then made by the metal interconnecting film deposited on the surface of the slice. In practice, not all the elements need to be isolated. The resistors, for example, can be grouped together in one region which is connected to the most positive voltage in the circuit to isolate them from the rest of the circuit, while the voltages of the individual resistors effectively isolates them from each other. Techniques such as this can be used to save space on the chip.

The usual method of isolation, and the only practicable one up to the end of the 1960s, is diode isolation, also known as *pn* junction isolation. The principle is to surround the components with silicon of the opposite type, and reverse-bias the resulting *pn* junction formed. The only current then flowing between adjacent components is the negligibly-small leakage current of the reverse-biased junction. The method is shown in *Figure 11.8*.

The circuit elements are formed in the *n*-type epitaxial layer grown on the *p*-type substrate, *Figure 11.8 (a)*. If p^+ regions are diffused into the epitaxial layer so that they link with the substrate, the layer is divided into islands in which circuit elements can be

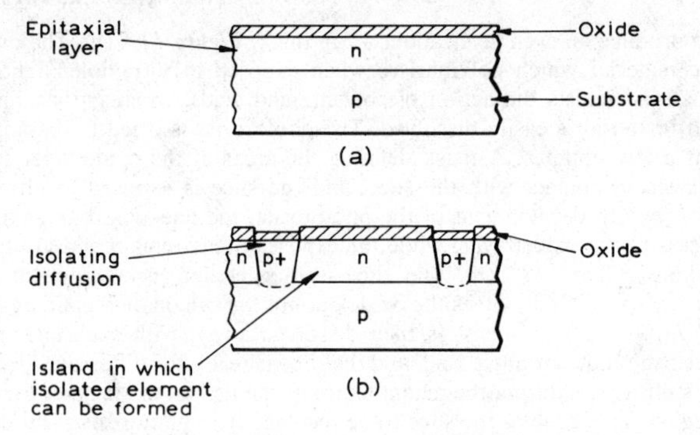

Figure 11.8 Diode (pn junction) isolation: (a) n-type epitaxial layer on p-type substrate, with masking oxide layer (b) p⁺ isolating diffusions formed

formed, *Figure 11.8 (b)*. When the completed i.c. is operated, the substrate is connected to the most negative voltage in the circuit. The *pn* junctions are then reverse-biased, and the circuit elements in the islands are isolated.

Two other isolation techniques have been introduced more recently. These are: Collector Diffusion Isolation (c.d.i.) and oxide (or dielectric) isolation. In c.d.i. a *p*-type substrate is used, into which an n^+ region is diffused. A *p*-type epitaxial layer is grown on the substrate, and n^+ regions diffused through the layer to link with the now buried n^+ region. A cross-section of the resulting structure is shown in *Figure 11.9 (a)*. If a further

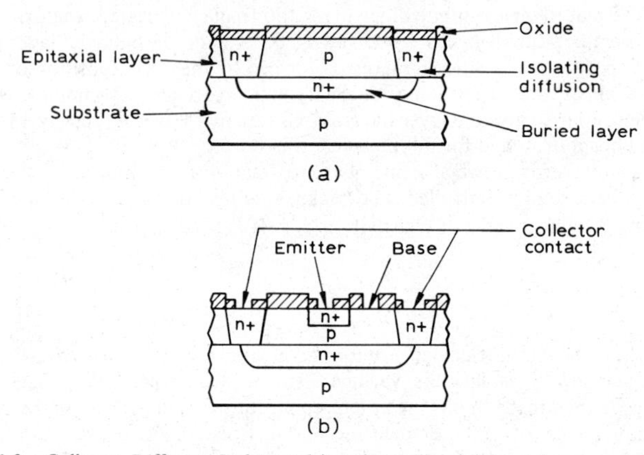

Figure 11.9 Collector Diffusion Isolation: (a) isolated island formed in p-type epitaxial layer (b) emitter diffusion to form isolated transistor

n^+ region is diffused, as shown in *Figure 11.9 (b)*, an *npn* transistor is formed in which the outer n^+ region not only acts as the collector but also provides isolation. If the substrate is taken to the most negative voltage, the biasing required for the operation of the transistor will also reverse-bias the junction between the collector and epitaxial layer.

The starting material for an early form of oxide isolation is *n*-type monocrystalline silicon. Channels are etched in the surface of the slice, *Figure 11.10 (a)*, and a layer of silicon oxide is grown over the surface, *Figure 11.10 (b)*. Polysilicon (polycrystalline silicon) is deposited on top of the oxide, as shown in *Figure 11.10 (c)*. The final stage is to invert the slice, and lap the surface until isolated regions of *n*-type monocrystalline

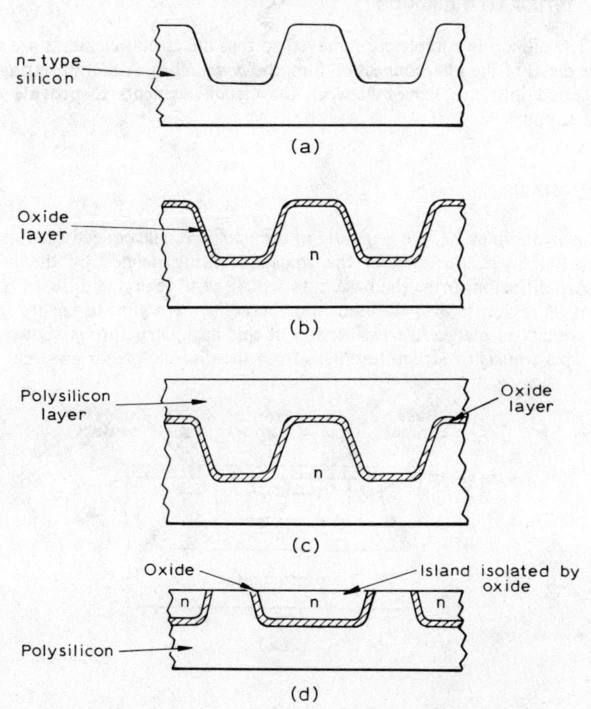

Figure 11.10 Oxide isolation: (a) n-type silicon with channels etched (b) oxide grown (c) polycrystalline silicon deposited (d) slice inverted and lapped to form isolated islands

silicon separated by oxide layers are formed, *Figure 11.10* (*d*). Circuit elements can then be diffused into these regions, and interconnected in the normal way.

A more recent oxide isolation technique is the Isoplanar process. In this process, the active p^+ isolating diffusions used for diode isolation (shown in *Figure 11.8* (*b*)) are replaced by passive oxide isolating regions. A *p*-type substrate is used, into which an n^+ region is diffused. An *n*-type epitaxial layer is grown, and the surface of the layer corresponding to the areas in which the transistor or other circuit element is to be formed, and to the collector contact of the transistor, is covered with a thin layer of silicon nitride. The exposed silicon is then etched to a depth corresponding to the thickness of the epitaxial layer, the silicon nitride masking the areas which are not to be etched. An oxide layer is then grown on the etched areas so that the original plane surface level is restored, no oxide growing on the silicon nitride. When the required oxide thickness has been grown, the silicon nitride is removed by etching with phosphoric acid. The oxide isolating regions have now been formed. A *p*-type region can be diffused into the *n*-type epitaxial layer to form the base of a transistor, and *n*-type regions to form the emitter and the collector contact area. This isolation technique is also known as Local Oxidation Of Silicon (l.o.c.o.s.).

One further isolation technique is used, but only for specialised applications, particularly at microwave frequencies. This is beam-lead isolation, also known as air isolation. The circuit elements are formed in the usual way but without isolation by any of the techniques described above. The thickness of the metal interconnection film, however, is increased considerably over that for the normal i.c. The silicon between each circuit element is then removed by lapping and etching from the back of the substrate. The silicon is completely removed so that the circuit elements are connected only by the metal of the interconnection film, the *beam leads*. A thermosetting plastic is usually inserted into the spaces between the circuit elements to provide additional mechanical support.

Circuit elements

Bipolar transistors in an i.c. are generally of the *npn* type. The collector is formed in the *n*-type epitaxial layer, the area of the transistor being defined by the p^+ isolation diffusion. A *p* diffusion forms the base, into which an n^+ region is diffused to form the emitter. An n^+ region is also diffused into the collector region to ensure that a low-resistance contact is made. A cross-section of this basic structure is shown in *Figure 11.11* (*a*). This transistor structure differs from the discrete planar epitaxial transistor

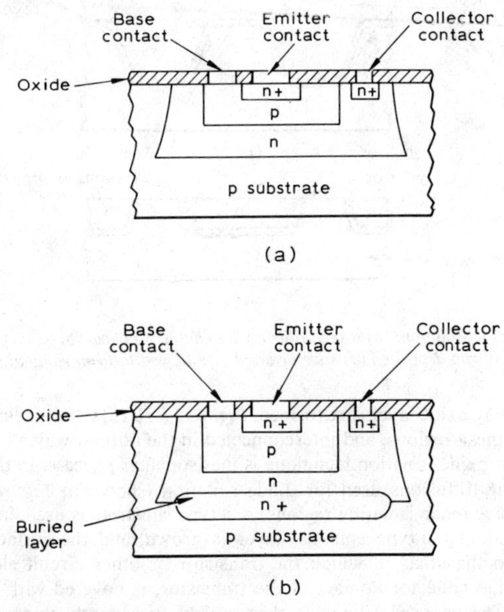

Figure 11.11 *NPN transistor in i.c.: (a) basic structure (b) buried layer added to reduce collector series resistance*

in that the collector contact has to be made on the upper surface rather than on the underside of the substrate. The collector current therefore has to flow laterally through the collector region to the contact, and so the collector series resistance is higher than in the discrete transistor. An additional n^+ region is diffused into the collector to form a low-resistance path, and reduce the length of the current flow through the higher-resistance collector region, as shown in *Figure 11.11* (*b*). This n^+ region is diffused into the substrate before the epitaxial layer is grown, and forms a *buried layer*.

It is sometimes necessary to incorporate *pnp* transistors into an i.c. There are two methods of achieving this. The *p*-type substrate can be used as the collector, and the *n*-type epitaxial layer as the base. The only additional diffusion required is a *p* diffusion to form the emitter. Such a transistor can have a reasonable current gain, but the collector is permanently connected to the substrate. An alternative method of forming a *pnp* transistor is shown in *Figure 11.12*. The buried n^+ layer and epitaxial layer now form the base, and *p*-type emitter and collector regions are diffused to form a *lateral* transistor. Because the base of such a transistor is necessarily wider than that of an *npn* transistor, the gain and cut-off frequency are low.

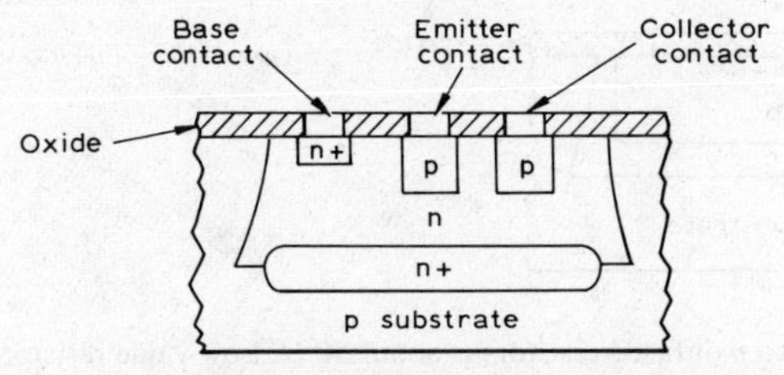

Figure 11.12 PNP transistor in i.c.

Resistors in an i.c. are made by the diffusion used to form the bases of the *npn* transistors. The *p*-type silicon has a fairly high resistivity, and the resistance value of the diffusion *R* is given by:

$$R = \frac{\rho}{t}\frac{L}{W}$$

where ρ is the resistivity of the *p*-type silicon, *t* is the thickness, *L* is the length, and *W* is the width of the diffusion. A typical resistor structure is shown in *Figure 11.13*. The

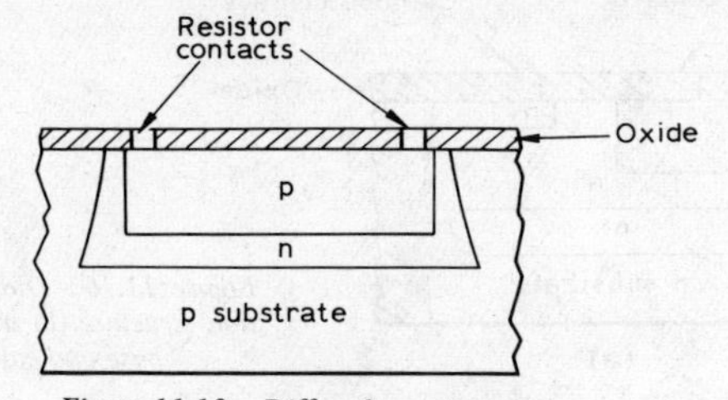

Figure 11.13 Diffused resistor in i.c.

thickness of the diffusion is set by the requirements for the bases of the transistors, and so the resistance value is determined by the relative values of width and length. The width of the diffusion may be set by the current to be conducted, and so the length of the diffusion may be the determining factor for the resistance value. High-value resistors which require a long length of diffusion are therefore *meandered*, as shown in the plan view of *Figure 11.14*. There is a limit of about 50 kΩ to a resistor of this sort, set by

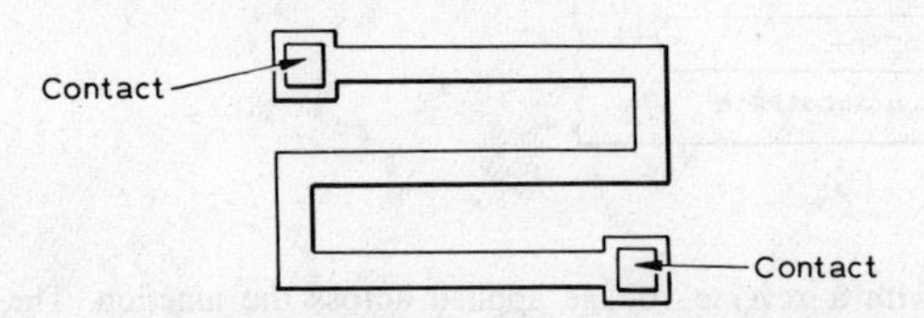

Figure 11.14 Plan view of high-value meandered diffused resistor

area available on the chip for resistors. The absolute tolerance, which is approximately ±20%, is determined by the accuracy with which the width can be defined.

Higher-value resistors can be obtained by restricting the thickness of the *p* diffusion by the n^+ diffusion used to form the emitters of the *npn* transistors, as shown in *Figure 11.15*. This structure is known as a *pinch resistor*.

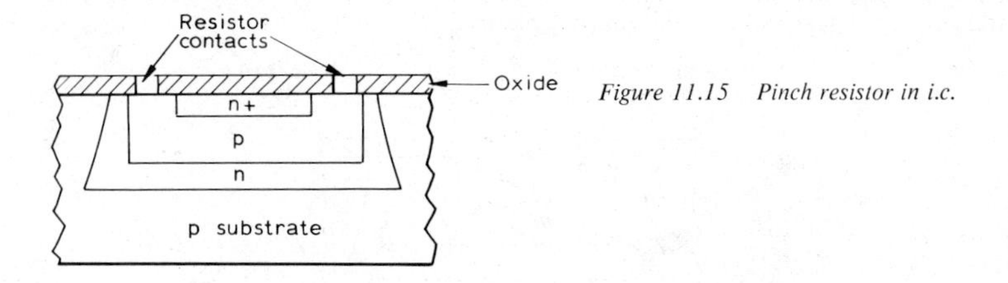

Figure 11.15 Pinch resistor in i.c.

The lower limit to *p*-diffused resistors is about 30 Ω. Low-value resistors of 1 or 2 Ω can be obtained by using the n^+ diffusion for the emitters of the transistors as the channel for the resistor.

Diodes can be formed in two ways. A general-purpose diode can be formed from a transistor structure in which the n^+ emitter has been omitted. The resulting structure is shown in *Figure 11.16* (*a*). For faster switching diodes, a normal transistor structure is used, but the base and collector regions are connected together to form the diode anode, and the emitter forms the cathode. This structure is shown in *Figure 11.16* (*b*).

Capacitors are also formed from transistor structures with regions connected

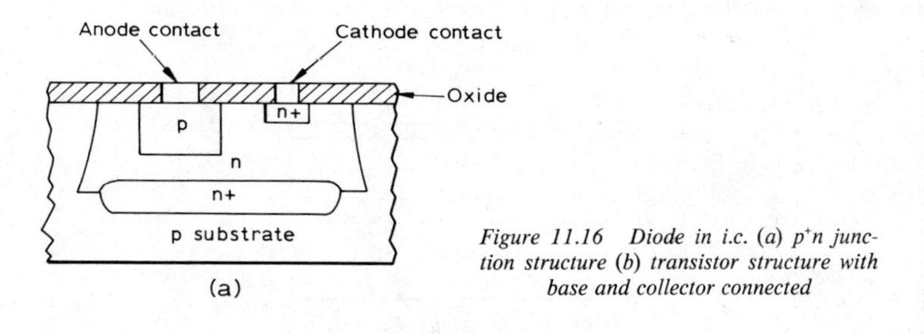

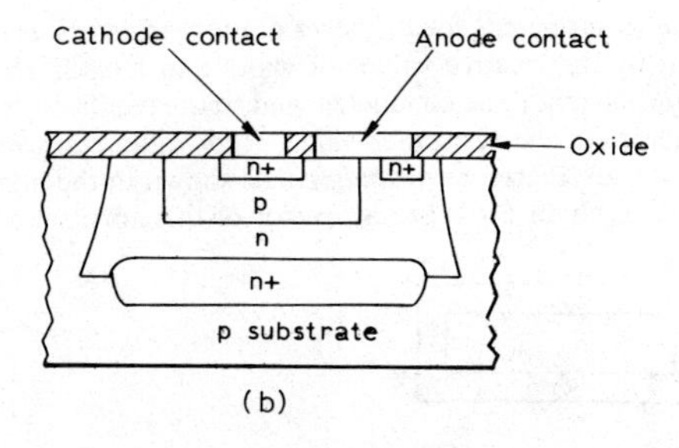

Figure 11.16 Diode in i.c. (a) p^+n junction structure (b) transistor structure with base and collector connected

together, but with a reverse voltage applied across the junction. The collector and emitter regions are connected together to form one plate of the capacitor, while the base forms the other plate, as shown in *Figure 11.17*. In this way, the capacitance of both the transistor junctions is used, but the value will vary with the reverse voltage applied. Typical capacitance values for the shorted-transistor structure are given in *Table 11.1*. It can be seen that the capacitance of the emitter-base junction is higher than that of the

Table 11.1 TYPICAL CAPACITANCE VALUES FOR SHORTED-TRANSISTOR STRUCTURE

Reverse voltage (V)	1	5	10	
Collector-base junction	180	100	70	pF/mm^2
Collector to substrate	200	90	60	pF/mm^2
Emitter-base junction	800	450	—	pF/mm^2

collector-base junction, but the reverse-bias voltage is limited. In some i.c.s, therefore, only the collector-base junction can be used.

Two other methods can be used to form capacitors. The diode structure of *Figure 11.16* (*a*), can be used with the *p* region extended to give a large junction area. This structure requires one less diffusion than the previous type, but the capacitance value is

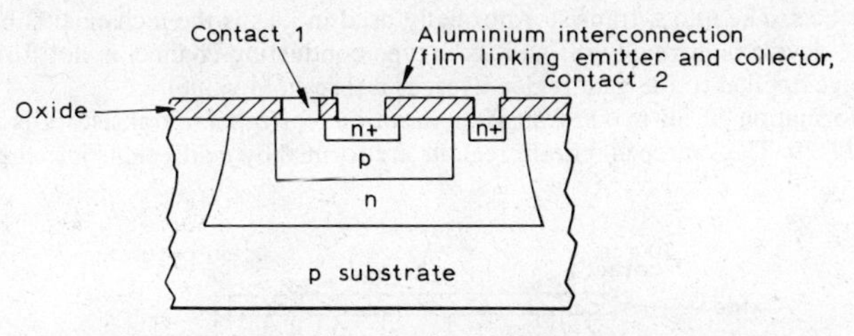

Figure 11.17 *Capacitor in i.c. formed from transistor structure*

lower. A capacitor can also be formed by using the silicon oxide as the dielectric, giving a structure like that shown in *Figure 11.18*. An extended n^+ diffusion forms one plate of the capacitor, while the other plate is a metal layer deposited at the same time as the interconnection film. The thickness of the oxide over the n^+ diffusion is controlled to provide some control over the capacitance value. This structure forms an m.o.s. capacitor. The typical capacitance value obtained is 500 pF/mm^2 which is lower than the maximum possible with the transistor structure of *Figure 11.17*, but the m.o.s. capacitor does have the advantage of a higher operating voltage and a capacitance value which is not voltage-dependent.

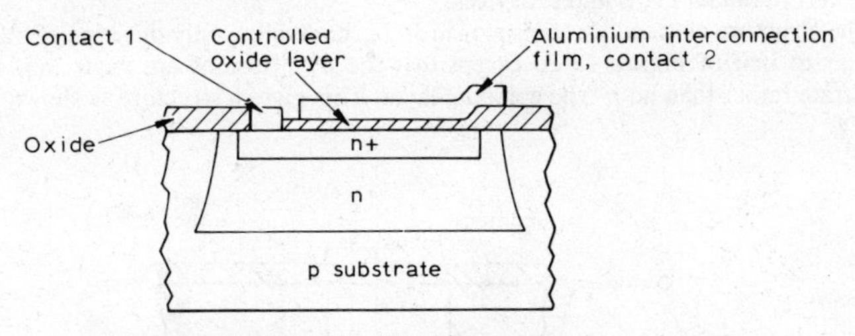

Figure 11.18 *M.O.S. capacitor in i.c.*

Although all the interconnections on an i.c. are ideally made by the metal film deposited on top of the oxide layer, there may occasionally be a need for two connections to cross. This can be done by an underpass, although this method is wasteful of chip area and should only be used when unavoidable. In the cross-section of *Figure 11.19*, connection *1* has to cross connection *2*. An n^+ region is diffused into the surface of the chip to which connection *2* is joined. Connection *1* is on top of the oxide layer, and is therefore insulated from connection *2*.

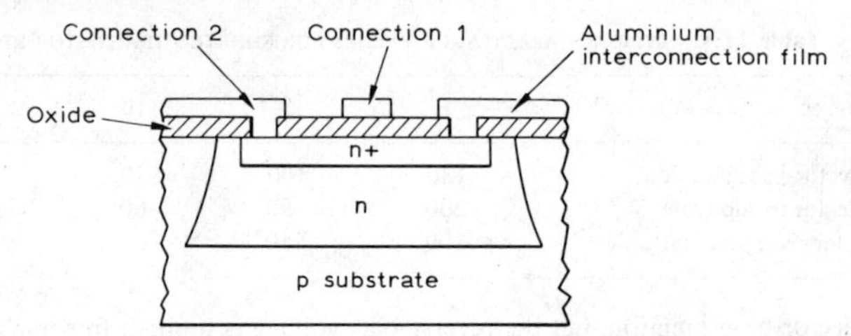

Figure 11.19 Diffused underpass in i.c.

M.O.S. transistors can be formed in i.c.s. designed for bipolar transistors if both types are required in the same circuit. It is more usual, however, to have the two types on separate i.c.s. The m.o.s. transistor normally used in i.c.s is the *p*-channel enhancement type (the device is normally off and the *p*-type conducting channel is not formed until the voltage applied to the gate region exceeds a threshold value).

The formation of an m.o.s. transistor in an i.c. for bipolar transistors is shown in *Figure 11.20*. The source and drain regions are formed by *p* diffusions into the epitaxial

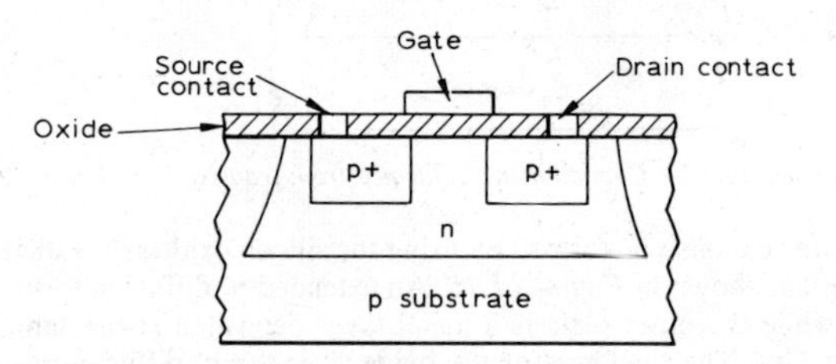

Figure 11.20 M.O.S. transistor in i.c. designed for bipolar devices

layer, while the gate is formed by part of the metal interconnection film. The insulating oxide layer between the gate and the substrate is the normal surface oxide layer of the slice. It is necessary to modify the normal bipolar manufacturing process somewhat, however, to obtain good m.o.s. devices.

The structure of an m.o.s. transistor in an i.c. containing only this type of transistor is similar to that of *Figure 11.20* except that the p^+ diffusions are made into an *n*-type substrate rather than an *n*-type epitaxial layer. This gives a structure as shown in *Figure 11.21*.

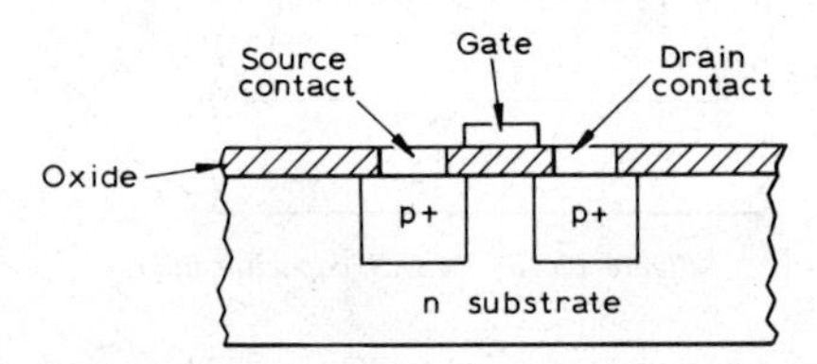

Figure 11.21 M.O.S. transistor in i.c.

The property of an m.o.s. transistor to act as a load for an m.o.s. switching transistor has already been mentioned. The region between the source and drain of an m.o.s. transistor has a resistance which depends on the gate voltage. Often the drain and gate are connected together, the bias across this structure is thus fixed, and the resistance

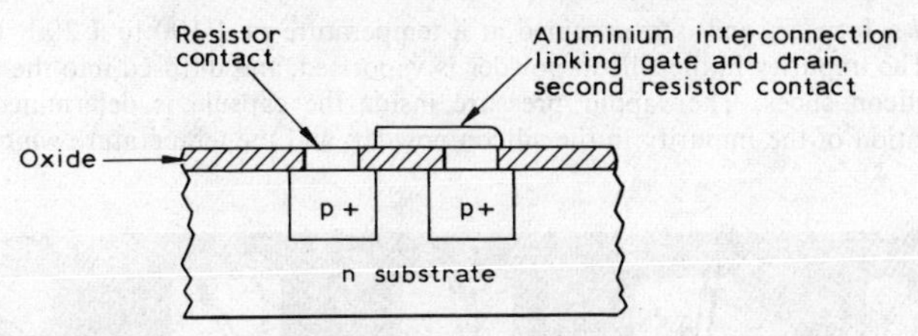

Figure 11.22 M.O.S. resistor formed from transistor structure in i.c.

value is then determined by the dimensions. The actual resistance value will still depend on the applied voltage, but this is not a significant disadvantage in practical m.o.s. i.c.s. The structure of an m.o.s. resistor is shown in *Figure 11.22*, and the ease of connecting an m.o.s. switching transistor with an m.o.s. load transistor is shown in *Figure 11.23*.

The gate capacitance of the m.o.s. transistor can be used to store charge which, although small, is sufficient for circuit operation. It can be seen, therefore, that the basic structure of the m.o.s. transistor can be used to form the required circuit elements. The

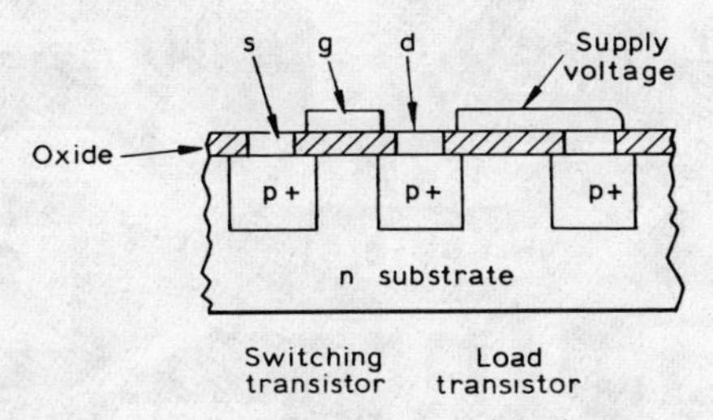

Figure 11.23 M.O.S. transistor with load transistor in i.c.

self-isolation of the m.o.s. structure makes it smaller than the bipolar structure, and in particular, an m.o.s. resistor can be 100 times smaller in area than a diffused one: m.o.s. i.c.s therefore lend themselves to large-scale integration.

A simple bipolar i.c. in which the circuit elements can be identified from the form of the aluminium interconnection pattern is shown in *Figure 11.24*. The circuit is an a.m. receiver circuit for a 1.5 W output stage.

Diffusion

Two diffusion methods can be used to form the circuit elements in integrated circuits. These are the *sealed-capsule* method, and the *open-tube* method.

In the sealed-capsule method, 300 to 500 cleaned silicon slices in which diffusion windows have been etched are placed in a quartz tube which is closed at one end. A quantity of doped silicon powder of the type required for the diffusion is also placed in the tube. The powder is produced by growing a doped single crystal of silicon of a known resistivity (and therefore a known doping level), and then crushing it into a fine powder. A close-fitting quartz stopper is placed in the open end of the tube, and this is connected to a vacuum pump to evacuate the tube. The wall of the tube around the stopper is heated so that it collapses onto the stopper to form a seal. The sealed capsule is placed in a furnace, and is maintained at a temperature of 1 100 to 1 200 °C within ±1 °C. The impurity in the silicon powder is vaporised, and diffused into the windows in the silicon slices. The vapour pressure inside the capsule is determined by the concentration of the impurity in the silicon powder, and the temperature which is held

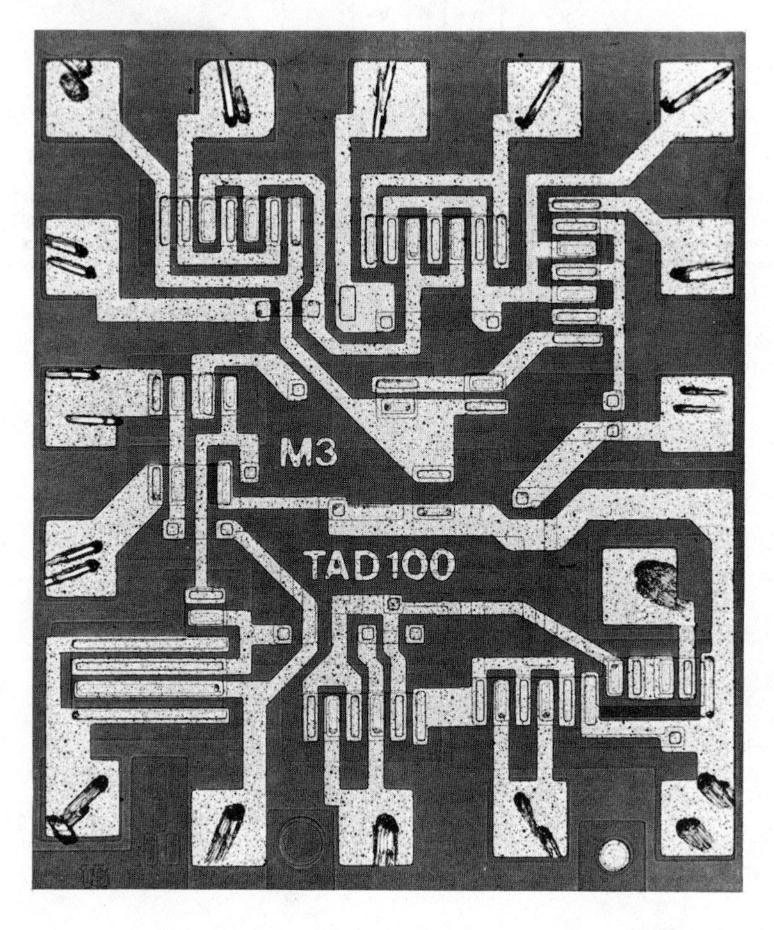

Figure 11.24 Simple bipolar i.c. in which circuit elements can be identified from form of interconnection layer (*Mullard Limited*)

constant. Therefore the depth of the diffusion will be determined by the time the slices are heated. After the required diffusion time, the capsule is removed from the furnace, allowed to cool and opened. With this method, the deposition of the impurity on the silicon surface and the drive-in to a controlled depth are combined in one furnace operation.

With the open-tube method, deposition and drive-in are performed as separate operations. In the deposition, the cleaned silicon slices are loaded into a quartz boat which is moved through a quartz diffusion tube in the furnace. A stream of gas is passed through the tube over the slices, consisting of oxygen and a vapour containing the impurity to be diffused. A glass rich in the impurity is formed on the slice with a thin diffused layer into the silicon in the windows. The furnace conditions are similar to those for the sealed-capsule method, although the time for which the slices are in the furnace is shorter. A typical deposition time is between 15 and 30 minutes, during which a glassy layer between 0.1 and 0.2 μm is formed. The number of slices processed in one passage through the furnace is 50 to 100.

The drive-in process takes place in a furnace similar to that used for the deposition. For high-resistivity diffusions, the glass is removed from the surface of the slice before heating. Only the impurity already diffused into the silicon is driven in further. For low-resistivity diffusions, the glass is left on and more of the impurity is diffused into the silicon. The glass then has to be removed to allow the sealing oxide layer to form when the slice is reheated for further drive-in. The slices are heated in the furnace first in a stream of dry oxygen and then in a stream of wet oxygen to encourage the growth of the oxide layer. The time in the furnace is determined by the depth of the diffusion required, and the required thickness of the oxide layer.

Three impurities are used in the diffusions to form an i.c. These are: arsenic for the n^+ buried layers; boron for the p regions such as isolation, bipolar transistor bases and resistors, m.o.s. transistor sources and drains; and phosphorus for the n^+ emitter regions.

The arsenic diffusion for the buried layers is generally performed by the sealed-capsule method. The arsenic is provided by the arsenic-doped silicon powder in the capsule. The diffusion time required is 16 hours at a temperature of 1 200 °C for a diffused layer 6 μm deep. If the diffusion is performed by the open-tube method, the arsenic is provided by the decomposition of a material such as arsenic trichloride $AsCl_3$. Antimony is sometimes used instead of arsenic, but it is important that the impurity chosen should have a low diffusion coefficient to reduce movement of the buried layer during subsequent high-temperature processing of the i.c.

Boron diffusion can be performed by either the sealed-capsule or the open-tube method. The boron in the open-tube method is provided by the decomposition of materials such as boron tribromide BBr_3 or boron trichloride BCl_3. The phosphorus diffusion in the open-tube method uses phosphorus oxychloride $POCl_3$.

The alignment of the masks to define the diffusion areas is critical. Apart from the first mask, the masks have to be aligned with the diffusions already made into the silicon slice. The slice and mask are held in the alignment equipment. The slice can be moved in both the x and y axes, and rotated, while the mask is held rigidly in a fixed frame approximately 100 μm above the slice. Two flat surfaces at right-angles to each other are machined on the slice, and the masks are made relative to the same two edges on the mask. An approximate alignment is obtained by positioning the flat surfaces and the corresponding edges of the mask against locating surfaces in the alignment equipment. The final alignment is carried out with a microscope, alignment marks being included in each mask so that a final accurate positioning between mask and previous diffusion is obtained. The mask in its holder is then lowered and clamped to the slice. An exposure is made using a high-intensity ultraviolet lamp, and the photographic and etching processes previously described are carried out.

Layout design and mask making

The starting point for the design of an i.c. is a performance specification. The i.c. designer then has to translate this specification into the circuit elements described previously to give the required performance. The limitations of the integrated form of the circuit (such as restricted component values and lack of inductance) have to be taken into account, as well as limitations inherent in the manufacturing process (for example,

(*a*)

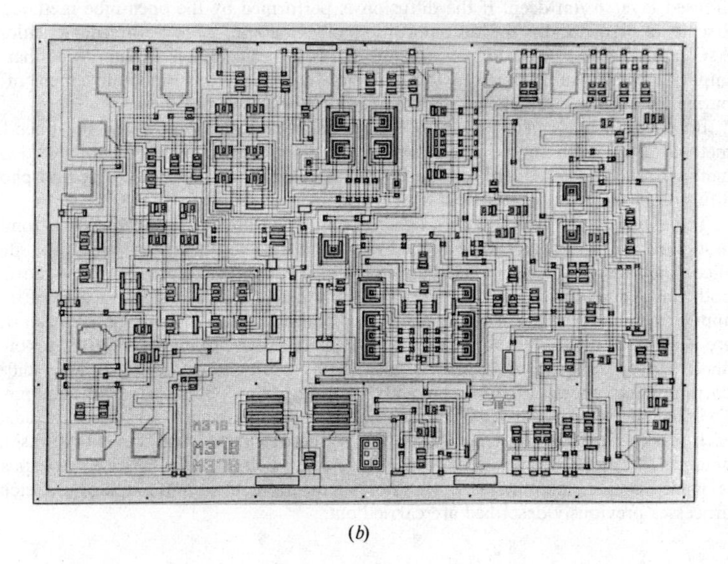

(*b*)

Figure 11.25 (*a*) *Designer using graphic display console and computer for layout of i.c.* (*Mullard Limited*) (*b*) *computer-produced composite layout drawing for i.c.* (*Mullard Limited*)

resistors are formed by the *p* diffusion which forms the transistor bases, and the depth of the diffusion is determined by the transistor performance required). Against this is set the flexibility resulting from the large number of transistors which can be used, and this flexibility can help the designer. Because of the complexity of the design, computers are used. Programs based on previous design experience have been devised to aid the design procedure, and these are particularly useful in calculating performance allowing for the stray component losses inevitable in i.c.s.

Once the circuit design has been finalised, a suitable layout for the circuit elements on the chip must be worked out. Again, computers are used for this. A rough layout is prepared by hand on which the size and approximate positions of all the circuit elements are shown. The coordinates of the elements and information about them, together with other information such as limiting values for the circuit, are fed into the computer. An accurate large-scale drawing of the layout is then produced by a computer-controlled plotter. Any alterations to the layout can be fed into the computer, and their effect evaluated. A new layout can then be produced if required. A designer using a graphic display console in devising the layout for an i.c. is shown in *Figure 11.25* (*a*) and a computer-produced composite layout drawing in *Figure 11.25* (*b*).

From the completed layout design, large-scale artwork is produced from which the individual diffusion masks are made by photographic reduction. The material used for the artwork is an opaque peelable plastic, consisting of an opaque red upper layer on a thicker transparent base layer. Parts of the upper layer can be cut and stripped away to leave transparent areas which will eventually define the diffusion areas. Because of the precision required, and to avoid human error, the artwork is made by a computer-controlled cutting table. A control tape based on the final layout design is produced by the computer, and this tape is used to guide the cutting head on the table to produce the artwork. The reduction between the artwork and the final diffusion mask can be between 200 and 1 000 times; the usual reduction is 250 : 1.

The completed artwork is mounted on a back-illuminated screen and photographed. An emulsion insensitive to red light is used, and the transparent areas in the artwork appear as dark areas on the photographic plate. A reduction of 25 : 1 is used in making this plate which is called a recticle. A step-and-repeat camera is then used to produce an array of the images of the recticle with a reduction of 10 : 1 on a second photographic plate. After each exposure by the step-and-repeat camera, the plate is stepped (moved a small amount) in one direction so that a column of images is produced. When the required column length has been reached, the plate is moved sideways the required amount, and a second column of images produced. In this way, an area corresponding to the silicon slice is covered with a regular array of images of the recticle. This second photographic plate is called a master, and from it same-size copies are made which are used as the masks in contact with the slices to define the diffusion areas.

A more recent process eliminates the artwork and first photographic reduction stages. The control tape produced by the computer is used to control a *pattern generator* instead of a cutting table. The pattern generator is similar in some respects to the step-and-repeat camera except that instead of reproducing images of the recticle on the photographic plate, a series of rectangular images of a light source is made. The plate is moved on instructions from the tape, and the size of the rectangular image is varied by the tape. In this way, a sequence of overlapping rectangular exposures is made to build up the detail of the mask on the plate. A transparency of this plate is then used in the step-and-repeat camera to produce the master.

The quality of the final diffusion masks depends on the quality and accuracy of the cameras used, and the absence of defects in any of the photographic processes. The temperature and humidity of the room in which the artwork is prepared are carefully controlled to prevent distortion of the plastic. All the processes are carried out in clean rooms, with the air supply filtered to eliminate dust particles.

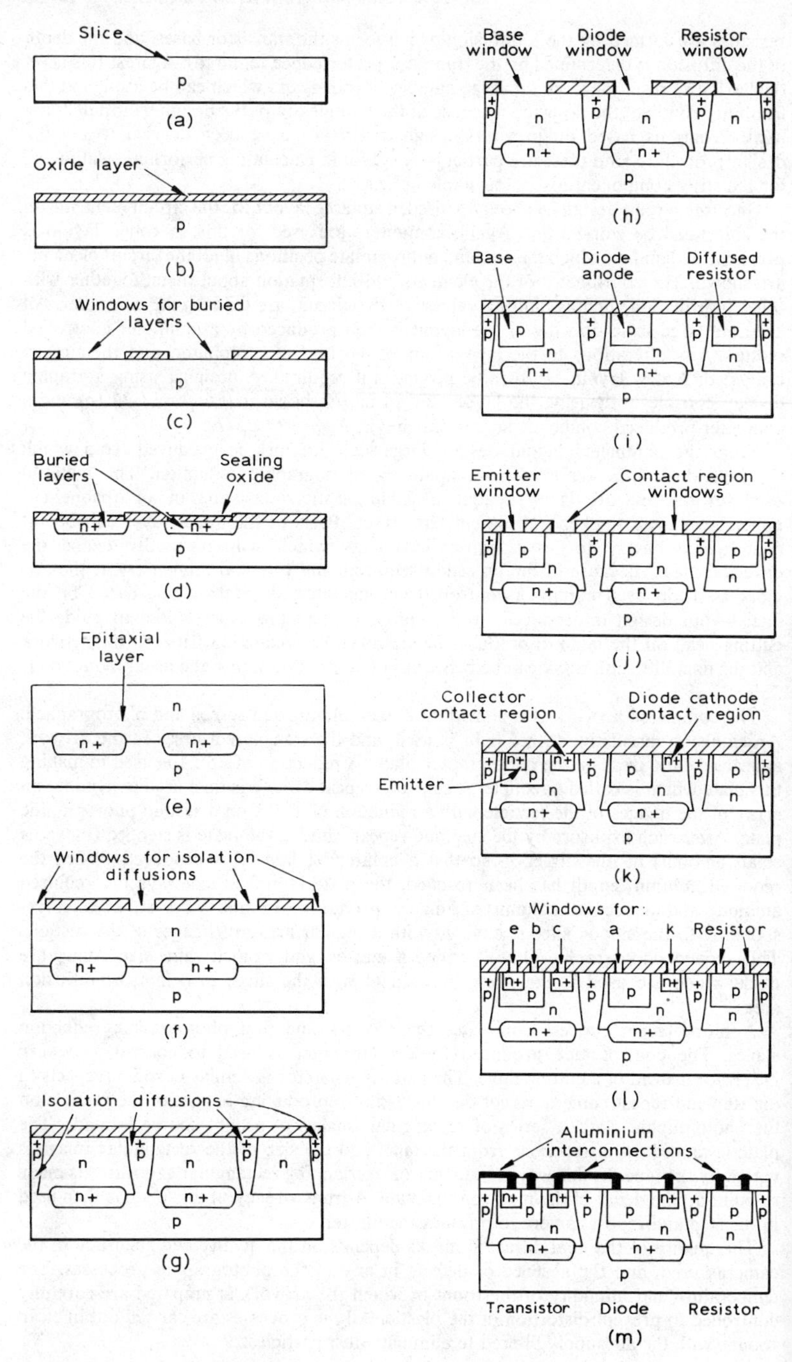
Slice
p
(a)
Oxide layer
p
(b)
Windows for buried layers
p
(c)
Buried layers
Sealing oxide
n+
n+
p
(d)
Epitaxial layer
n
n+
n+
p
(e)
Windows for isolation diffusions
n
n+
n+
p
(f)
Isolation diffusions
p+
n
n+
p
(g)
Base window
Diode window
Resistor window
n
n+
p
(h)
Base
Diode anode
Diffused resistor
p
n
n+
p
(i)
Emitter window
Contact region windows
(j)
Collector contact region
Diode cathode contact region
Emitter
(k)
Windows for:
e
b
c
a
k
Resistor
(l)
Aluminium interconnections
Transistor
Diode
Resistor
(m)

Manufacture of bipolar i.c.s

Details of various aspects of i.c. manufacture have been given, and as a guide to the manufacture of a complete bipolar i.c., the various stages resulting in the formation of part of an i.c. containing a transistor, diode, and resistor are described. Five diffusions are required, and six masks are used. These define the buried layers, isolation diffusions, base regions, emitter regions, contact windows and interconnections. The various stages are shown in *Figure 11.26*.

The starting material is a polished slice of *p*-type monocrystalline silicon (*a*). The diameter of the slice is typically 5 cm, the thickness typically 200 μm, and the resistivity 0.1 Ωm. An oxide layer is grown on the surface of the slice (*b*). The thickness of this layer is typically 1 μm. The first mask is used to define the windows for the diffusion of the buried layers, using the photographic and etching techniques already described. The resulting structure is shown in (*c*). Arsenic is used to form the n^+ buried layers (*d*).

The oxide layer is removed from the surface of the slice, and an epitaxial layer grown (*e*). The thickness of the layer is typically 20 μm. The impurity used to form the *n*-type silicon of the layer is phosphorus, and a typical value of resistivity is 5×10^{-3} Ωm. During the growth of the layer, the n^+ buried layers also grows slightly. An oxide layer is grown on the surface of the epitaxial layer, and the second mask is used to define the windows for the p^+ isolation diffusions (*f*). Boron is used for these diffusions which penetrate the epitaxial layer to link with the *p* substrate (*g*). The third mask is used to define the windows for the shallower *p* diffusions which will form the transistor base, the diode *p* region, and the resistor (*h*). Boron is used again for these diffusions (*i*). The fourth mask is used to define the windows for the n^+ diffusions (*j*). The n^+ diffusion forms the emitter and the collector contact region of the transistor, and the cathode contact region of the diode (*k*). Phosphorus is used for this diffusion. Between the deposition and drive-in stages, gold is evaporated onto the back surface of the substrate. The high diffusion coefficient of the gold allows it to diffuse throughout the slice when it is heated for the phosphorus drive-in. This gold doping is used to reduce the lifetime of minority carriers, and so increase the switching speed of the diodes and transistors.

All the circuit elements of the i.c. have now been formed and require interconnecting. A fifth mask is used to open contact windows (*l*). Aluminium is evaporated onto the whole of the top surface of the slice to form a film about 1.0 μm thick. Next, the sixth mask is used to define the interconnection pattern, and the unwanted aluminium is etched away with sodium hydroxide or phosphoric acid. The slice is heated to 530 °C to alloy the aluminium to the silicon, thus forming a good electrical contact (*m*). Aluminium can also be used to form Schottky barrier diodes. These diodes are used to clamp the collector-to-base voltage of the transistors to prevent them being driven into heavy saturation. This technique improves the switching speed, and can be used as an alternative to gold doping.

The manufacturing processes on the slice are now complete, and the top surface of a typical slice is shown in *Figure 11.27*. Five circuits different from the rest can be seen; these are the process control modules. In any diffusion production line, many different types of i.c. will be processed. It is convenient to have a *standard* circuit in each slice to enable a standard test to be carried out on all slices to monitor each process. The process control modules provide this standard circuit. They are inserted at fixed places in the masks by the step-and-repeat camera.

Figure 11.26 Manufacture of bipolar i.c.: (a) p-type substrate (b) oxide layer grown (c) windows opened for buried layers (d) n^+ buried layers diffused (e) n-type epitaxial layer grown (f) oxide layer grown and windows opened for isolation diffusions (g) isolation diffusions formed (h) windows opened for p diffusions (i) p regions diffused (j) windows opened for n^+ diffusions (k) n^+ regions diffused (l) windows opened for contacts (m) aluminium interconnection film formed

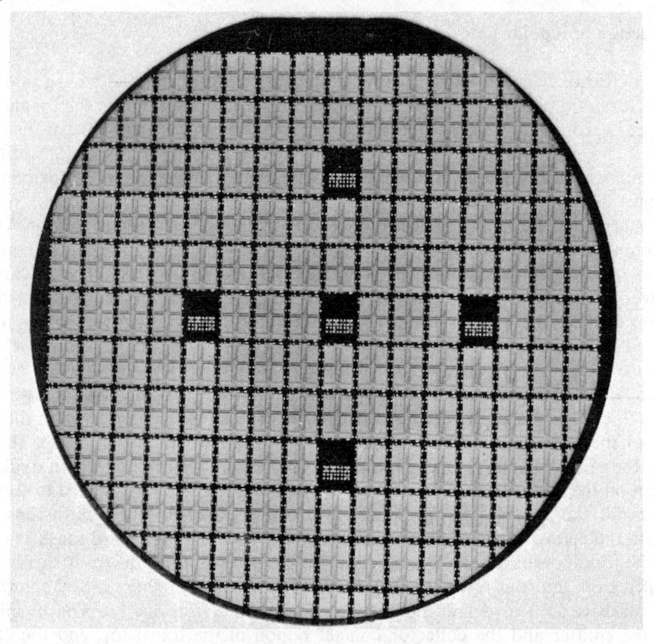

Figure 11.27 Silicon slice with completed i.c.s. The slice is 5 cm diameter, and the i.c.s are 3.0 mm × 4.0 mm (Mullard Limited)

Figure 11.28 Probe testing of i.c. on slice (Mullard Limited)

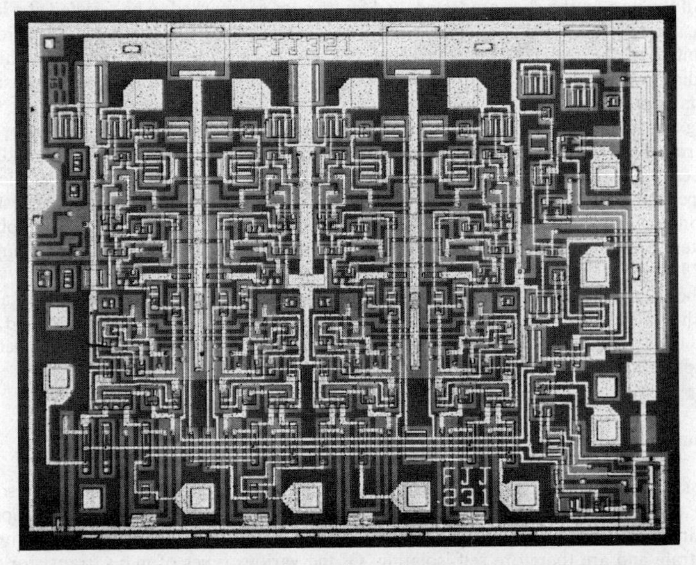

(*a*)

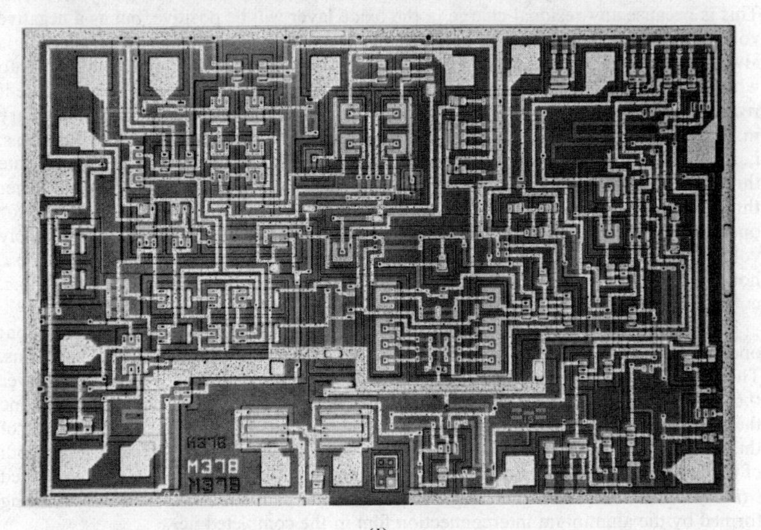

(*b*)

Figure 11.29 (*a*) *Typical bipolar m.s.i. digital i.c. chip. The circuit is the FJJ231 4-bit shift register; chip size 2.7 mm × 2.1 mm* (*Mullard Limited*) (*b*) *typical bipolar linear i.c. chip. This is the chip for which the computer-produced layout was shown* (*Figure 25*). *It is the TCA270 synchronous demodulator; chip size 3.0 mm × 2.1 mm* (*Mullard Limited*)

All the i.c.s on the slice are tested, and faulty ones are marked to be rejected at a later stage. Testing is carried out automatically, under the control of a computer. The slice is mounted in the test equipment, and probes are lowered to make contact with the bonding pads of the aluminium interconnection film. A series of 40 to 50 d.c. measurements is sufficient to test a medium-sized i.c. completely, the time taken for the test being approximately 160–200 ms. The probes are raised, the slice is moved under the control of the computer so that the next i.c. in the column is under the probes, and the probes are lowered to repeat the test. When the edge of the slice is reached, the slice is stepped so that the next column on the slice is under the probes, and the testing continues. Any i.c.s which fail the tests are marked automatically. A photograph of the test probes making contact with the bonding pads of an i.c. is shown in *Figure 11.28*.

The slice is divided into individual i.c. chips by scribing it with a diamond stylus, after which the slice can easily be cracked into the chips. Faulty i.c.s are rejected, and the good ones prepared for mounting and encapsulation. Typical bipolar m.s.i. chips before encapsulation are shown in *Figure 11.29*.

Manufacture of m.o.s. integrated circuits

One structure only is required in m.o.s. i.c.s, the transistor structure. This can also be used as a resistor, and the gate capacitance can be used as a capacitor. One diffusion is required, to form the source and drain regions which are of opposite type from the substrate and are therefore self-isolating. Of the various types of m.o.s. transistor, the easiest to make in integrated form is the *p*-channel enhancement (normally off) type. This is because any residual charge in the oxide layer will be positive, but as a negative voltage is required on the gate to form the channel, conduction in the transistor is always controlled by the gate. *P*-channel devices are made in an *n*-type substrate into which *p*-type source and drain regions are diffused. *N*-channel devices can be made if required, and both types of transistor can be combined on one chip in complementary m.o.s., c.m.o.s., integrated circuits. There are two manufacturing processes for m.o.s. i.c.s: the older aluminium-gate process which results in devices requiring a gate threshold voltage of about 2.5–4.5 V, and the newer silicon-gate process with a lower threshold voltage of about 1.5–2.0 V. Since the drain-to-source voltage required for operation of the m.o.s. transistor is four to six times the threshold voltage, a supply voltage of 30 V is required for aluminium-gate m.o.s. i.c.s. With silicon-gate i.c.s, however, supply voltages between 5 V and 20 V can be used, making the m.o.s. i.c. compatible with bipolar i.c.s.

The aluminium-gate process, *Figure 11.30*, requires four masks: for the diffusion; opening windows in the field oxide for the gate oxide; contacts; and interconnections. The starting material is a slice of *n*-type monocrystalline silicon (*a*). An oxide layer 0.6 μm thick is grown over the surface of the slice (*b*). The first mask is used to define the windows for the source and drain diffusion (*c*). Boron is deposited on the surface of the slice, and the boron glass removed before the high-temperature drive-in. The depth of the boron diffusion is about 2 μm, and an oxide layer about 1.5 μm thick is produced (*d*). This thick or *field* oxide is grown to prevent parasitic m.o.s. transistors being formed by the aluminium interconnection film in the completed i.c.

The second mask is used to open a window in the field oxide for the gate and contact areas (*e*). A second oxide layer is grown in these areas, the thin gate oxide (*f*). This oxide layer is approximately 0.12 μm thick, and is treated to reduce any absorbed charges to the lowest possible level. The third mask is used to define windows in the thin oxide for contacts (*g*). Aluminium is evaporated over the surface of the slice to form a film 1 to 2 μm thick, and the fourth mask is used to define the interconnection pattern of conductors and gates. The unwanted aluminium is etched away with sodium hydroxide or phosphoric acid (*h*). The difference in thickness of the field and gate oxide layers

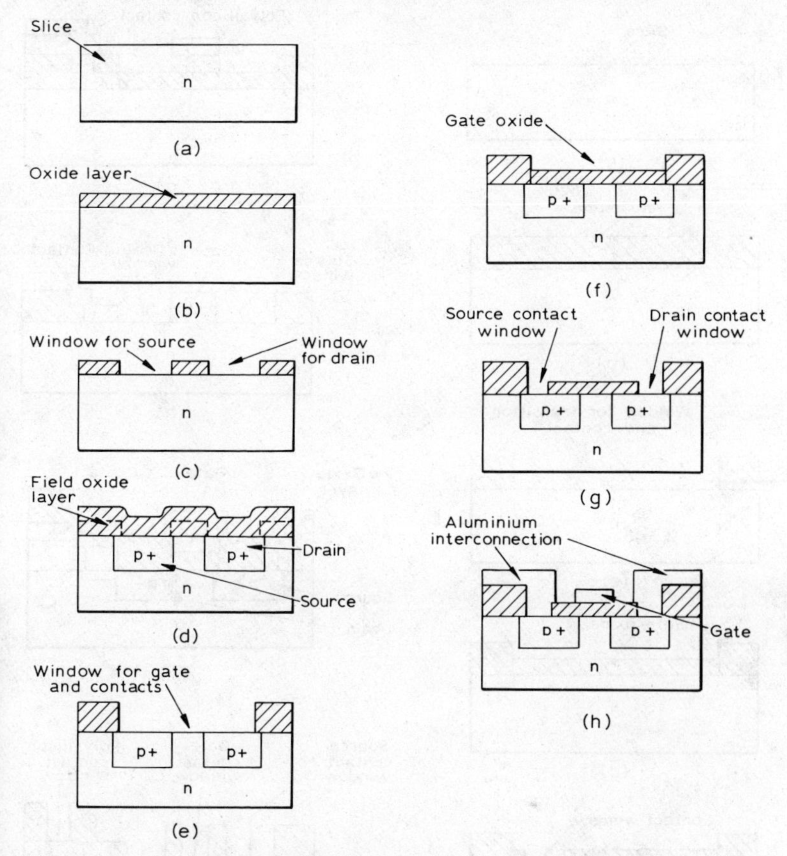

Figure 11.30 Manufacture of aluminium-gate m.o.s. i.c.: (a) n-type silicon slice (b) oxide layer grown (c) windows opened for source and drain (d) source and drain diffused, field oxide grown (e) windows opened in field oxide for gate and contacts (f) gate oxide grown (g) windows opened in gate oxide for source and drain contacts (h) aluminium gate and interconnections formed

necessitates care in the deposition of the aluminium to maintain the electrical continuity of the conductors over the various *steps* in the surface of the slice.

From this description of the aluminium-gate process, it can be seen that two masks are required to position the active gate area over the source and drain regions (the masks for the diffusions and the window in the field oxide), and a further mask is required for the etching of the aluminium gate. Careful alignment of the three masks is essential, particularly as the areas of the device are smaller than those for bipolar i.c. transistors. An additional advantage of the silicon-gate process over the aluminium-gate process (besides the lower threshold voltage) is that the gate is self-aligning.

The silicon-gate process, *Figure 11.31*, uses five masks: to open the field oxide for the transistor structure, open the gate oxide for the contacts, define the polysilicon pattern, open contact windows, and define the interconnection pattern. The starting material is again a slice of *n*-type monocrystalline silicon (*a*).

An oxide layer approximately 1.0 μm thick is grown over the surface of the slice (*b*);

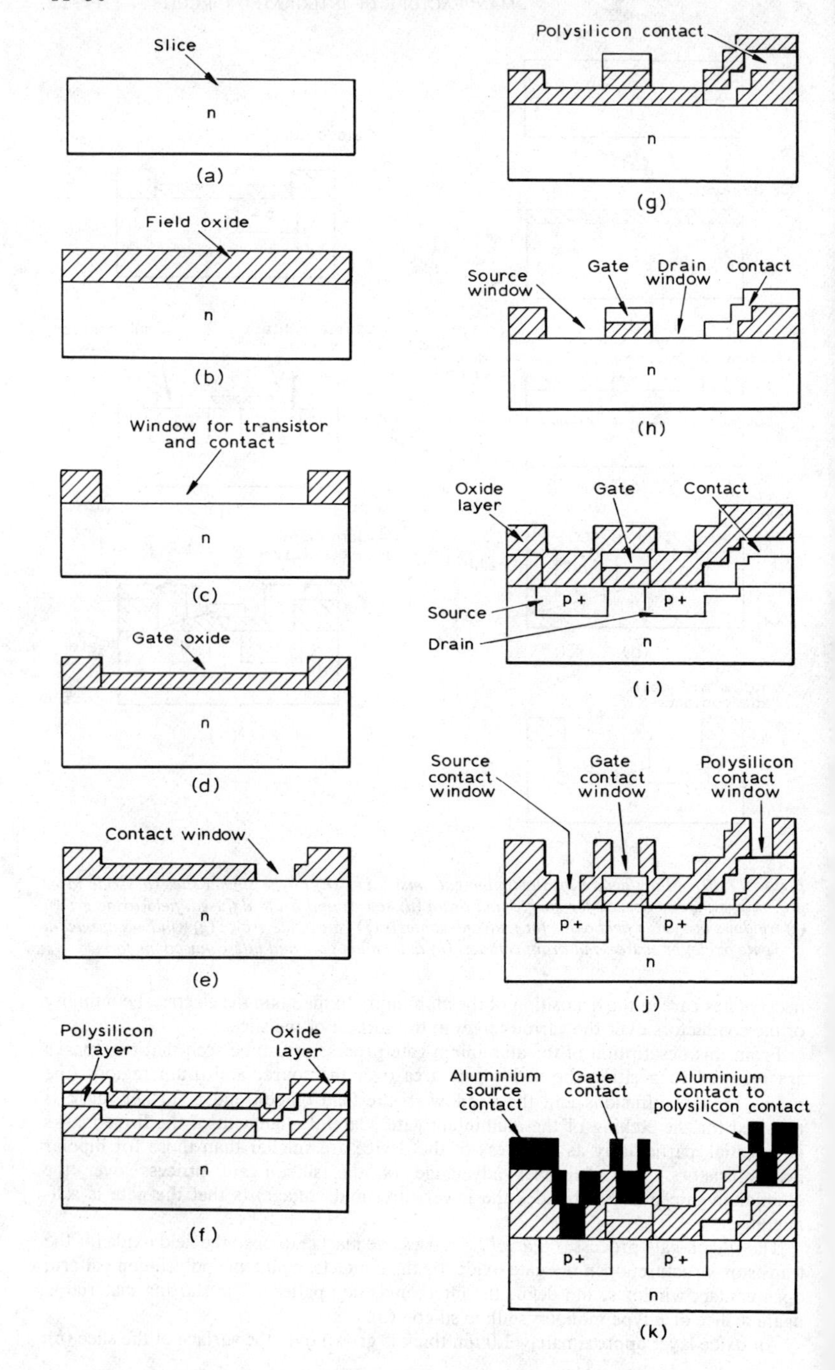
Slice
n
(a)
Field oxide
n
(b)
Window for transistor and contact
n
(c)
Gate oxide
n
(d)
Contact window
n
(e)
Polysilicon layer
Oxide layer
n
(f)
Polysilicon contact
n
(g)
Source window
Gate
Drain window
Contact
n
(h)
Oxide layer
Gate
Contact
Source
Drain
p+
p+
n
(i)
Source contact window
Gate contact window
Polysilicon contact window
p+
p+
n
(j)
Aluminium source contact
Gate contact
Aluminium contact to polysilicon contact
p+
p+
n
(k)

this is the field oxide. The first mask is used to define a window in the field oxide in which the source, gate and drain of the transistor will be formed (*c*).

The thin gate oxide layer, approximately 0.12 μm thick, is grown (*d*). The second mask is used to define a window in the gate oxide for a contact (*e*). Polycrystalline silicon (polysilicon) is deposited over the surface of the slice, and an oxide layer grown over the surface (*f*). The third mask is used to define the polysilicon pattern, the gates and silicon surface contact areas. The unwanted polysilicon is etched away, giving the structure shown in (*g*). The remaining gate oxide is etched away (*h*), and boron diffused to form the source and drain regions (*i*). The areas of the diffusions are defined by the thick oxide and the thin oxide–polysilicon region of the gate, but because there is no protective oxide over the polysilicon of the contact, some boron diffuses through it into the monocrystalline silicon of the substrate. This shallow diffusion ensures a good electrical contact between the silicon layers. There is a small sideways diffusion in the source and drain, which ensures that the gate completely covers the current-carrying channel of the transistor. The self-aligning property of the gate in this process can now be seen.

During the boron drive-in, an oxide layer about 1.0 μm is grown. This protects the m.o.s. device just formed, and provides insulation between the devices and aluminium interconnection film. The fourth mask is used to define the contact windows in this oxide layer (*j*), and aluminium is evaporated over the surface of the slice in a film about 1.5 μm thick. The fifth mask is used to define the interconnection pattern (*k*). Again it is important that the aluminium film is continuous over the steps in the surface of the slice.

A protective oxide layer may be grown over the slice after the aluminium interconnection layer has been defined. This layer is called the *glassover*, and windows have to be etched in it to expose the parts of the aluminium film forming the bonding pads by which the external connections are made.

In c.m.o.s. i.c.s, both *p*-channel and *n*-channel transistors are formed. An *n*-type substrate is used in which the *p*-channel devices are formed by the processes just described. Islands of *p*-type silicon are diffused into the substrate, and the *n*-channel devices are formed in these islands. Thus at the cost of extra diffusions, complementary m.o.s. transistors can be provided on the same chip. C.M.O.S. i.c.s can be of the aluminium-gate or silicon-gate type. The main stages in the manufacture of c.m.o.s. silicon-gate i.c.s are shown in *Figure 11.32*.

The initial stages are the growth of an oxide layer on the surface of the *n*-type monocrystalline silicon slice, and the opening of a window in this field oxide for the *p* diffusion in which the *n*-channel transistor will be formed. This deep *p* diffusion is made using boron. The resulting structure is shown in (*a*). The window for the *p*-channel transistor is opened, the gate oxide grown, and polysilicon deposited over the top surface of the slice (*b*). The gate of the *n*-channel transistor is defined by etching the polysilicon. Windows are opened in the gate oxide for the source and drain regions of the *n*-channel transistor, and the n^+ diffusions made with phosphorus (*c*). The *p*-channel transistor is made by etching the polysilicon to form the gate, opening windows in the gate oxide for the source and drain, and forming the p^+ regions with the boron diffusion (*d*). The other manufacturing stages are then the same as for the *p*-channel silicon-gate process already described.

When the manufacturing stages on the slice have been completed (whether for aluminium-gate, silicon-gate, or c.m.o.s. i.c.s), the individual i.c.s are tested on the slice by the automatic probe tester described for bipolar i.c.s. The i.c.s are separated, and the good chips are prepared for mounting and encapsulation.

Figure 11.31 Manufacture of silicon-gate m.o.s. i.c.: (a) n-type silicon slice (b) field oxide grown (c) window opened for transistor and contact (d) gate oxide grown (e) window opened for polysilicon contact (f) polysilicon deposited and oxide layer grown (g) polysilicon regions defined (h) windows opened in gate oxide for source and drain (i) source and drain diffused, oxide layer grown (j) windows opened for contacts (k) aluminium interconnections formed

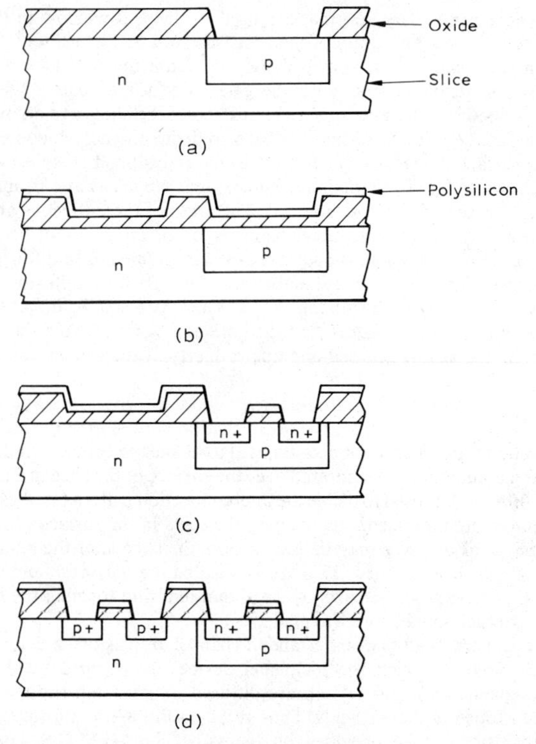

Figure 11.32 Manufacture of silicon-gate complementary m.o.s. i.c.: (a) oxide layer grown, window opened in oxide for p diffusion, p island diffused (b) window for p-channel transistor opened, gate oxide grown, polysilicon deposited (c) gate of n-channel transistor defined, windows opened for source and drain of n-channel transistor, source and drain diffused (d) p-channel transistor defined, windows opened for source and drain, source and drain of p-channel transistor diffused

Other manufacturing processes

Besides the main manufacturing processes for i.c.s outlined in the previous two sections, other processes and manufacturing techniques exist. Some of these will be outlined here.

A problem associated with the aluminium-gate m.o.s. i.c. is the high threshold voltage which can be up to 4.5 V. This means that a supply voltage of 30 V is required for the i.c. The value of threshold voltage can be lowered by using silicon nitride Si_3N_4 as the gate insulating layer. Because silicon nitride does not interface well with silicon, a thin oxide layer about 0.1 μm thick is grown on the silicon substrate, and the silicon nitride layer is formed on this. This process is called the nitride process, and results in a threshold voltage of approximately 2 V.

Another manufacturing problem associated with m.o.s. i.c.s, particularly silicon-gate types, is ensuring the continuity of the aluminium interconnection film over the many steps in the surface of the chip. A l.o.c.o.s. (local oxidation of silicon) technique can be used to reduce the number of steps. The areas on the slice to be occupied by the transistors are masked with silicon nitride, and the remaining areas etched. The field

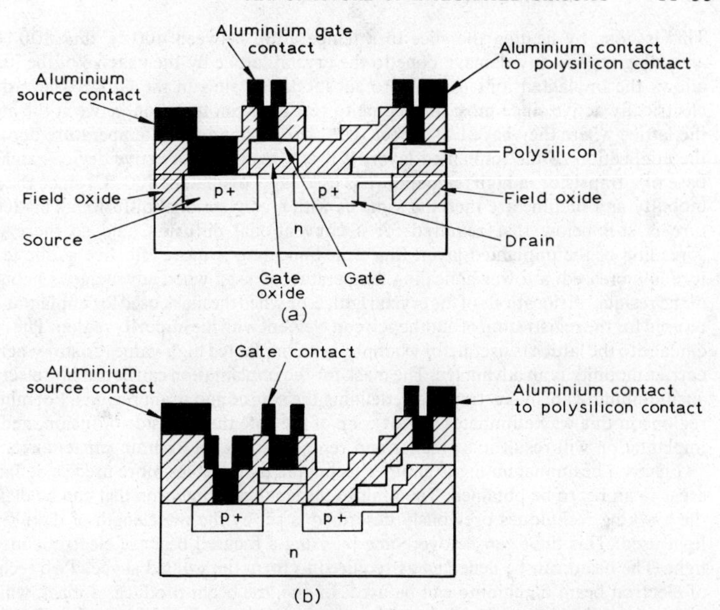

Figure 11.33 (*a*) *Structure of l.o.c.o.s. silicon-gate i.c.* (*b*) *structure of normal silicon-gate i.c. for comparison*

oxide is then grown, and the rest of the manufacturing process continues in the usual way. The structure of an i.c. using l.o.c.o.s. silicon gate is shown in *Figure 11.33* (*a*), while a similar structure for normal silicon gate is shown for comparison in *Figure 11.33* (*b*). The reduced number of steps in the surface of the slice can be seen. The l.o.c.o.s. technique is also applied to complementary m.o.s. i.c.s, forming what are known as l.o.c.m.o.s. i.c.s.

A disadvantage of diffusion processes is that the size and shape of the resulting doped region is difficult to control accurately. The impurity tends to diffuse sideways as it is driven into the substrate so that an ill-defined edge for the diffusion is produced. Also, the previously-diffused regions start to diffuse again when new diffusions are made. Although allowances are made for these effects in the design of the i.c., problems can still arise through the uncertainty of the movements. One such problem is called *base push-out*, in which the base region moves into the collector as the emitter is diffused. Thus the base width of the transistor is greater than planned and the performance of the transistor is lowered.

Some of these problems can be overcome by the technique of ion implantation. A particle accelerator is used to accelerate ions of an impurity element sufficiently to penetrate the surface of a silicon slice to a depth approaching 1 μm. The impurity ion is selected from a plasma ion source by a mass spectrometer so that only the required impurity is implanted. The concentration of the impurity is determined by the ion current and the time of the bombardment, and so it can be accurately controlled, while the depth of penetration is controlled by the accelerating voltage. The area of the impurity region is localised by a mask on the surface of the slice which is thick enough to prevent ions passing through into the slice. The mask can be made from a metal such as gold or aluminium deposited on the surface, or an insulating material such as silica; the choice will depend on what is convenient for the other processes in the manufacture of the i.c.

When the impurity region has been formed, it is necessary to anneal the silicon slice.

This is done by heating the slice to a temperature between 400 °C and 800 °C. The annealing repairs any damage done to the crystal lattice by the passage of the ions, and allows the implanted ions to move to substitutional sites in the lattice where they are electrically active since most ions come to rest after implantation between the atoms of the lattice where they have little effect. The choice of annealing temperature depends on the application of the implanted layer. If it is a region in an active device, such as the base of a transistor, a high temperature is used (approximately 800 °C) since the carrier mobility and lifetime are then the same as with a conventional diffusion. This temperature is still below that required for a conventional diffusion, and so there is little spreading of the implanted layer. Since the annealing is more effective as the temperature is increased, a lower annealing temperature is used when advantage is being taken of the residual dislocations of the crystal lattice, or when the mask used for implantation is to be used for the registration of another circuit element with the impurity region. The residual damage to the lattice is useful, for example, in an implanted high-value resistor where a low carrier mobility is an advantage. The mask for the implantation can be the gate electrode of an aluminium-gate m.o.s. transistor, defining the source and drain regions. Forming these regions in this way eliminates any overlap of the gate through side diffusion, and so ion implantation will result in a smaller and reproducible gate-to-drain capacitance.

Electron beam machining is another technique that enables more precise definition of areas in an i.c. to be obtained. The limit to the smallest dimension that can be defined by the masking techniques previously described is set by the wavelength of the ultraviolet light used. This limit can be overcome by using a focused beam of electrons instead of light. The beam can be deflected as required to form the wanted shape. Two techniques of electron beam machining can be used. In one, the beam produces a mask which can be used in the conventional way. In the second, the slice is coated with a silicon compound which can be hardened by the beam in the required areas. The rest of the compound is then removed, leaving an integral mask on the surface of the slice. With electron beam machining, line thicknesses down to 0.1 μm can be achieved.

Mounting and encapsulation

The completed i.c. chips are generally mounted on a gold-plated header which is incorporated in a suitable encapsulation. The chip is held by the edges in a vacuum chuck, and is pressed onto the gold-plated area of the header which is mounted on a heated anvil. The header is thus heated to about 430 °C, and a stream of nitrogen is used to prevent oxidation of the gold. The chip is moved with a slight scrubbing motion over the header, and a silicon-gold alloy is formed which on cooling provides a good mechanical bond.

All electrical connections are made to the top surface of the chip. Gold wire is used to connect the bonding pads of the aluminium interconnection film on the chip to the lead-out wires of the header. The wire, which is between 10 μm and 50 μm in diameter, is attached by thermocompression bonding.

The principle of thermocompression bonding is that a solid weld can be made between two metals at a temperature below their alloying temperature if pressure is applied to the junction. A sideways motion between the two contacting surfaces improves the weld. Two commonly-used adaptations of the principle for i.c. manufacture are *wedge* bonding and *nailhead* bonding.

In wedge bonding, the gold wire is placed along the surface of the heated chip in contact with the bonding pad. A blunted wedge is used to apply pressure to the wire over the pad, and the squeezing of the wire forces it along the pad to give the required sideways movement. A firm bond can be obtained with this method, but the diameter of the wire is reduced at the bond and so gives a weak point.

Nailhead bonding overcomes this disadvantage. The gold wire is contained in a heated capillary tube, and a globule of gold is formed at the end of the wire. The globule is pressed onto the bonding pad, and the compression of the globule produces the sideways movement to give a firm bond. The diameter of the bond is larger than the diameter of the wire by a factor of about three. The process is shown diagrammatically in *Figure 11.34*. The chip has been bonded to the header, and the connection between the bonding pad and the lead-out wire is about to be made (*a*). The header is mounted on a heated anvil so that the chip is at a temperature of about 300 °C. A nitrogen gas

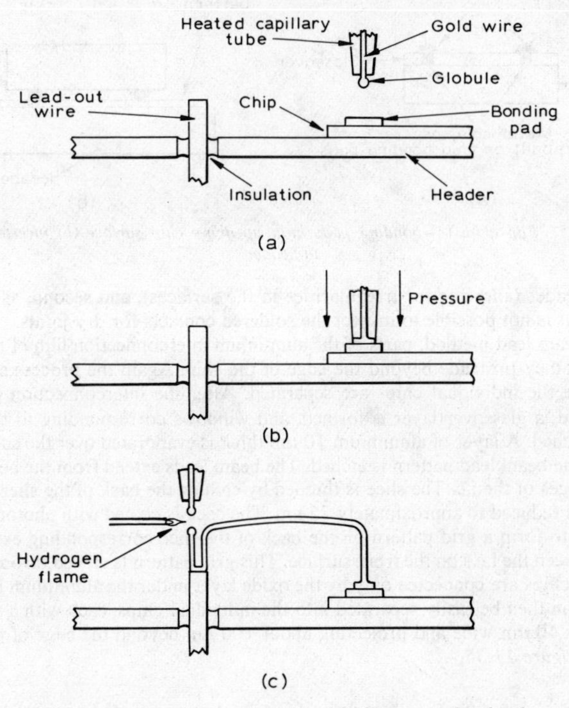

Figure 11.34 Nailhead bonding of connection wires: (*a*) *chip mounted on header, gold wire ready for bonding* (*b*) *nailhead bond between gold wire and bonding pad being made* (*c*) *wedge bond between gold wire and lead-out wire made, wire cut*

stream is used to prevent unwanted oxidation. The heated capillary tube is pressed onto the bonding pad with a force of about 30 grammes (*b*). The capillary tube is raised, leaving the wire bonded to the pad, and moved to the gold-plated lead-out wire. A wedge bond is formed here, the capillary is raised, and the wire is cut with a fine hydrogen flame. The cutting produces a globule at the end of the gold wire, ready for the next nailhead bond (*c*).

Various other methods of bonding the chip to a header have been devised. Two methods that are used in practice are the flip chip and beam lead methods.

In the flip chip method, the chip is mounted, face down, on to a header with a conductor pattern corresponding to the bonding pads of the chip. The bonding pads must be built up to make good contact with the header, and this is done on the silicon slice before the chips are separated. After the aluminium interconnection pattern has been defined, a glassover layer is formed on the slice. Windows are etched in this layer, corresponding to the bonding pads. Small metal columns are built up in these windows by plating with such sequences as aluminium, nickel, copper, finishing with tin or a tin/lead solder. A 'bump' is formed, protruding above the surface of the glassover layer, as shown in (*a*) of *Figure 11.35*. The slice is divided into individual chips in the usual way. The chip is pressed onto the heated header so that the solder bumps make contact with the conductor pattern and form soldered joints (*b*). There are two difficulties with this method. The first is the difficulty of ensuring all the solder bumps make contact

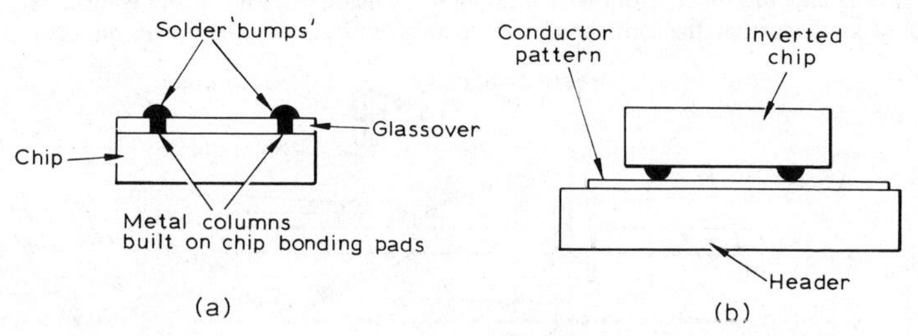

Figure 11.35 Flip chip: (a) bonding pads built up above chip surface (b) inverted chip fixed to header

with the header (allowing for irregularities in the surfaces), and second, as the chip is face down it is not possible to inspect the soldered contacts for dry joints.

In the beam lead method, parts of the aluminium interconnection film of the chip are built up so they protrude beyond the edge of the chip. Again the process starts in the slice before the individual chips are separated. After the interconnection pattern has been defined, a glassover layer is formed, and windows corresponding to the bonding pads are etched. A layer of aluminium 10 μm thick is evaporated over the surface of the slice, and the beam lead pattern is etched. The beam leads extend from the bonding pads over the edges of the i.c. The slice is thinned by etching the back of the slice so that the thickness is reduced to approximately 25 μm. The back is coated with photoresist, and a mask used to form a grid pattern on the back of the slice corresponding exactly to the spaces between the i.c.s on the front surface. This grid pattern is etched through the slice so that the chips are connected only by the oxide layer under the aluminium beam leads. The slice can then be easily separated into the individual chips, each with a set of beam leads about 40 μm wide and projecting about 100 μm beyond the edge of the chip, as shown in *Figure 11.36*.

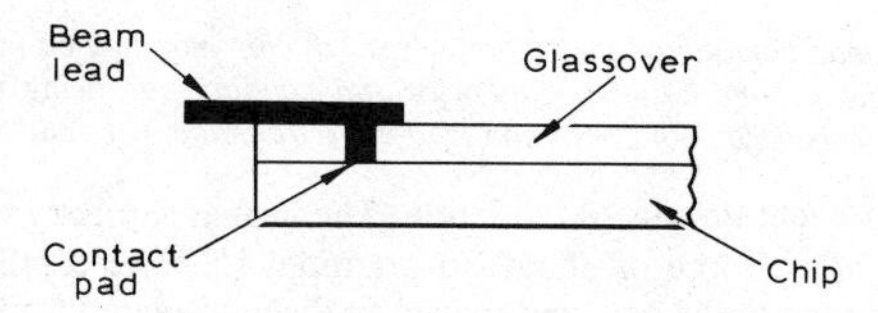

Figure 11.36 Beam lead formed on chip

Gold beam leads can also be formed and are more common. Platinum is sputtered into the bonding-pad windows, and films of titanium, platinum, and gold then sputtered over the surface of the slice. The titanium adheres to the oxide glassover layer, the platinum forms a barrier layer between the titanium and the gold, and the gold forms the low-resistance conductor. The metal layer is etched into the beam lead pattern, and the beam leads built up by electroplating with gold to 10 μm. The separation of the chips is then the same as for aluminium beam leads.

The chip is mounted face down on the header with the beam leads in contact with the conductor pattern on the header. Thermocompression bonds are made, often using a special welding tool to make all the bonds at the same time.

The mounted chips are encapsulated, either in hermetically-sealed metal envelopes or ceramic packages, or in moulded-plastic packages. Typical metal hermetically-sealed envelopes include variations of the TO-5 outline used for discrete transistors, such as the TO-74 or TO-99 outlines, or *flat packs*. Ceramic packages are of the dual-in-line or quad-in-line type, and are sealed with a glass frit. The purpose of hermetic sealing is to exclude from the area of the chip any material that might have an adverse effect on the integrated circuit, for example water vapour. The sealing is therefore carried out in an atmosphere such as dry nitrogen, and the sealed envelopes are tested for leaks. Plastic encapsulation is considerably cheaper than hermetic sealing. The mounted chip is coated with a silicon lacquer for protection, and a plastic block is moulded around the chip and header. The plastic used is generally an epoxy resin which is sufficiently water-resistant to ensure operation of the i.c. in all but the most arduous conditions. Typical encapsulations for integrated circuits are shown in the photograph of *Figure 11.37*.

Final testing

The encapsulated i.c.s are tested to ensure that all will meet the performance given in the published data. Three types of test are carried out: d.c. measurements, functional tests,

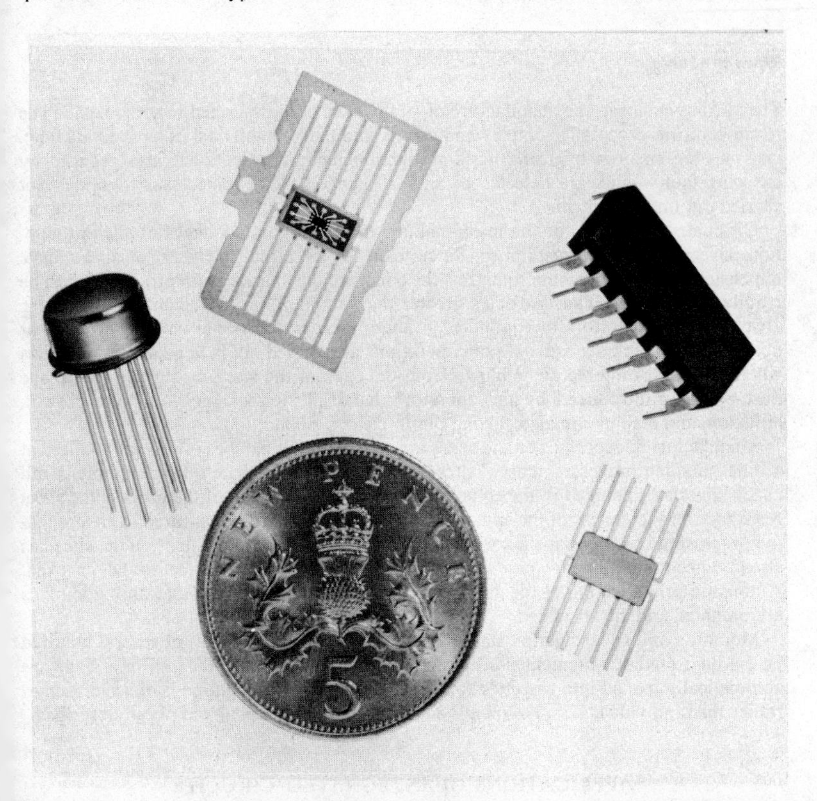

Figure 11.37 Typical encapsulations for integrated circuits: top left, 8-lead TO-99 metal envelope; top right, 14 lead dual-in-line package; bottom right, 14 lead TO-86 ceramic flat pack. The unencapsulated flat pack, top centre, shows the chip mounted on the lead-out frame which is trimmed after encapsulation (*Mullard Limited*)

and dynamic tests. The d.c. measurements are similar to those made on the completed i.c. on the slice. The tests ensure that all the leads are connected and there are no shortcircuits; that the input and output characteristics of the i.c. are correct; and that the i.c. will withstand the specified maximum voltages and currents. The functional tests ensure that the internal circuitry of the i.c. has been formed and connected correctly so that the device will perform its stated function. The dynamic tests ensure that the i.c. will operate satisfactorily at the specified speed or frequency.

As with the probe testing on the slice, these final tests are performed under the control of a computer. A particular test measurement can be compared with the test limits stored in the computer, and the i.c. passed to the next test or rejected. Test results can be stored by the computer to provide statistics on the batch of i.c.s tested; for example, the percentage of good-i.c.s, the spread in values, and so on. This information is useful for production control.

In addition to these production tests, samples of production devices are submitted to special testing for quality control and reliability purposes. The results of these tests can be used to locate weaknesses in the circuit design and production processes, as well as monitoring continuously the quality of production and providing information on the possible behaviour of the i.c. in operation.

Mass production

The various stages in the manufacture of an i.c. have been described in some detail. The manufacturing process is complex and costly, and the low unit cost of an individual i.c. can only be ensured by using mass-production methods. Costs are also reduced by ensuring faulty units are detected at an early stage so that unnecessary work is not carried out on faulty items.

Clean-air conditions for the mask-making and photographic stages of slice production have already been mentioned. Since many of the circuit elements of an i.c. have dimensions of only a few micrometres, the effects of dust particles on production can be readily appreciated. Thus particles greater than about 2 μm are excluded from these areas, and the operators of equipment in these areas wear special clothing to prevent dust particles being deposited. Many operations are carried out in laminar-flow cabinets where a 'wall' of filtered air is blown from the back of the working surface to exclude dust particles introduced by the operator's hands. Processes such as oxide growth, diffusion, and etching are also carried out in clean-air conditions.

Manufacturing processes on the slices are carried out in batches of several hundred at a time. In some processes, 'check' slices are included in the batch. For example, small check slices are included in the epitaxial reactor, and at the end of the process the sheet resistance and thickness of the epitaxial layer on these slices are measured to ensure that the processing has been satisfactory. The use of process control modules on the slice has already been mentioned; these enable each diffusion of the slice to be checked. Automatic probe testing of the completed i.c.s on the slice ensures that only good i.c.s are mounted and encapsulated.

Mounting and encapsulation stages are also performed in batches of several hundred i.c.s. Much of the equipment used is automatic, and devices at the end of a stage are automatically loaded into cassettes so that the only handling required is of large batches rather than individual i.c.s. Automatic final testing of i.c.s has already been described.

APPLICATIONS OF INTEGRATED CIRCUITS

The first i.c.s were gate circuits for logic functions. Throughout the 1960s, the trend in digital i.c.s was towards accommodating more and more circuit elements on the chip. The multiple gates and other basic building blocks for logic systems of the early 1960s (which today would be called small-scale integration, s.s.i.) were superseded about the middle of the decade by i.c.s incorporating complete sub-systems. This m.s.i. (medium-scale integration) represents a more economical use of the integrated form, and is commonly used today. A wide range of m.s.i. i.c.s is available, including shift registers and counters of various lengths, encoders, decoders, multiplexers, and arithmetic units.

The computer industry in the 1960s was developing larger, faster, and more powerful computers. This led to demands on the i.c. manufacturer to increase the number of circuit elements on the chip by reducing the size of the individual elements and increasing the chip area. The introduction of m.o.s. devices in 1966 accelerated this trend. The fact that charge could be stored on the gate electrode as long as it was periodically 'refreshed' led to the concept of 'dynamic' logic. This, and the smaller size of the m.o.s. transistor compared with the bipolar transistor, led to the higher packing density of m.o.s. i.c.s. In 1966, for example, a 100-bit shift register became available as an m.o.s. i.c.; by 1969, 2 000-bit shift registers containing some 12 000 m.o.s. transistors were available. This large-scale integration (l.s.i.) led to the development of the solid-state memory. By the early 1970s, random-access memories (r.a.m.s) with capacities of 1 kilobit became available, and read only memories (r.o.m.s) of higher capacity, say, up to 5 kilobits.

The trend towards larger-scale integration of digital i.c.s has continued up to the present time. Improved bipolar i.c. technology has enabled bipolar r.a.m.s with capacities comparable to the m.o.s. types to be manufactured. Desk calculators have provided another incentive to l.s.i. Over the past few years, the number of i.c.s required for a calculator has fallen from six to two, and single-i.c. calculators are now becoming commercially available.

The present situation in digital i.c.s is still split between the m.o.s. and bipolar technologies, and between m.s.i. and l.s.i. The axiom current at the beginning of the 1970s that m.o.s. devices were cheap but slow, while bipolar devices were fast but expensive is now being challenged. The advantages of the m.o.s. manufacturing process that make the devices cheaper than bipolar types are the fewer manufacturing stages required (only one diffusion), the self-isolating properties of the circuit elements, fewer interconnections, and the higher packing density. For the system designer, the high impedance and the lower power dissipation are useful properties. The advantages of bipolar i.c.s to the system designer are the faster switching speed, the higher current handling capability, the higher voltage gain, and the lower supply voltages required. In addition, the precautions required to protect m.o.s. devices from gate-oxide breakdown through electrostatic charges are not required by bipolar i.c.s. Approximate figures for the operating speeds of bipolar and m.o.s. i.c.s at the beginning of the 1970s are given below.

bipolar i.c.s (t.t.l.)	100 MHz
m.o.s. i.c.s, static logic	1 MHz
m.o.s. i.c.s, two-phase dynamic logic	3 MHz
m.o.s. i.c.s, four-phase dynamic logic	5 MHz

In the past few years, however, several changes have occurred. The introduction of complementary m.o.s. (c.m.o.s.) has increased the speed of m.o.s. i.c.s to near that of present-date t.t.l. bipolar i.c.s. The extra diffusions required for c.m.o.s. i.c.s increase the cost. The new bipolar manufacturing processes such as Collector Diffusion Isolation (c.d.i.) and the isoplanar process have simplified manufacture by eliminating one diffusion, and they result in smaller circuit elements and so a higher packing density. Thus bipolar r.a.m.s of 1 kilobit capacity have become available with operating speeds faster than those of m.o.s. memories or magnetic core memories. Present-day operating speeds for bipolar i.c.s using e.c.l. can be as high as 1 GHz.

The choice between l.s.i. and m.s.i. depends on the application and on the cost. Where small size is important, as in portable calculators or in aerospace applications, l.s.i. is essential. In complex logic systems where a larger number of intermediate input and

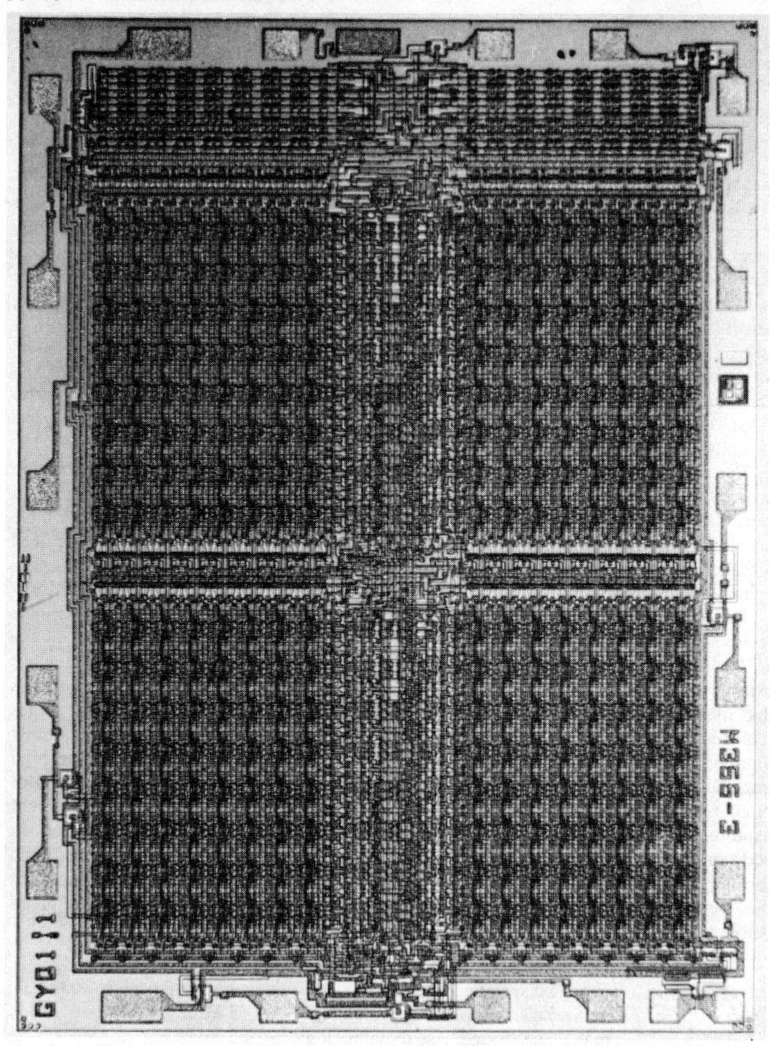

Figure 11.38 Typical present-day m.o.s. i.c. chip. The circuit is the GYQ101 1024-bit random access memory; chip size 4.0 mm × 3.0 mm (Mullard Limited)

output connections are required than can be provided by the normal i.c. packages, m.s.i. must be used. With l.s.i., the small size is obtained at the expense of system flexibility; the system designer is dependent of the function of the i.c.s he uses. With m.s.i. the designer has greater flexibility, but this is at the expense of larger size and more interconnections.

Future trends in digital i.c.s will be towards faster operating speeds and greater complexity of the circuitry included on the chip. New techniques such as injection logic seem likely to increase the packing density of bipolar i.c.s to levels comparable with m.o.s.

Development of linear i.c.s was delayed not only by the problems of adapting circuits to the integrated form, but also by the absence of a unifying application to which development could be aimed. The operational amplifier is probably the linear circuit with the widest range of applications, and so this was the first linear i.c. to become commercially available. Today i.c. operational amplifiers have found many applications in many fields. Towards the end of the 1960s, the radio and television industry with its mass-production assembly methods provided a second outlet for linear i.c.s, and the one that accounts for the largest present-day sale of linear i.c.s. As mentioned earlier, the ease with which additional transistors can be incorporated into a linear i.c. means that more complex circuits can be designed without additional cost. Thus the performance of radio and television receivers was improved over that obtained with discrete components. This improved performance combined with easier assembly of receivers offered considerable advantages to the receiver manufacturer. Today a wide range of signal-processing i.c.s for domestic receivers and sound-reproduction equipment is available.

Linear i.c.s have also found some application in professional communications equipment. Typical i.c.s for use in this field include voltage regulators, phase-locked loops, and analogue multipliers.

Future trends in linear i.c.s will be to improve the gain and bandwidth of operational amplifiers, and to increase the complexity and performance of i.c.s for domestic receivers. There will also be an increase in their use in domestic appliances. A wider range of i.c.s will become available for professional communications equipment for both radio and telephone applications.

A typical present-day m.o.s. i.c. is shown in *Figure 11.38*. This i.c. has been shown on a larger scale than the bipolar i.c.s of *Figure 11.29* because of the smaller size of the individual circuit elements. The circuit shown is a 1024-bit r.a.m.

FURTHER READING

MANUFACTURE

BEALE, J. R. A., EMMS, E. T. and HILBOURNE, R. A., *Microelectronics*, Taylor and Francis, London (1971)

BRICE, J. C. and JOYCE, B. A., 'The technology of semiconductor materials preparation', *Radio and Electronic Engineer*, **43**, No. 1/2, 21 (1973)

SMOLLETT, M., 'The technology of semiconductor manufacture', *Radio and Electronic Engineer*, **43**, No. 1/2, 29 (1973)

MANUFACTURE AND APPLICATIONS

GORE, W., ed., *Microcircuits and their applications*, Iliffe, London (1969)

GOSLING, W., 'Integrated circuits for analogue systems', *Radio and Electronic Engineer*, **43**, No. 1/2, 58 (1973)

HIBBERD, R. G., *Integrated Circuits, A Basic Course for Engineers and Technicians*, Texas Instruments Electronics Series, McGraw-Hill, New York (1969)

RODDY, D., *Introduction to Microelectronics*, Pergamon Press, Oxford (1970)

ROSE, M. J., ed., *MOS Integrated Circuits and their Applications*, Mullard, London (1973)

DEVELOPMENT OF I.C.S

BROTHERS, J. S., 'Integrated circuit development', *Radio and Electronic Engineer*, **43**, No. 1/2, 39 (1973)

DEAN, K. J., and WHITE, G., 'The Semiconductor Story, Parts 3 and 4', *Wireless World*, **79**, Nos. 1449 and 1450 (1973)

SHEPHERD, A. A., 'Semiconductor device development in the 1960s', *Radio and Electronic Engineer*, **43**, No. 1/2, 11 (1973)

12 MICROELECTRONICS

MICROELECTRONICS 12

12 MICROELECTRONICS

The term *microelectronics* is a general one indicating the small size of electronic devices. Today two techniques are classified as microelectronics: the integrated circuit and the film techniques. Sometimes the two techniques are combined to form hybrid circuits.

TECHNIQUES FOR MICROELECTRONICS

A monolithic integrated circuit is an inseparable assembly of circuit elements formed in a small piece of semiconductor material. The material used at the present time is silicon, and the piece of silicon in which the elements are formed is called a chip. Circuit elements are formed by impurity regions made by a sequence of diffusions. The diffusion regions are precisely defined by windows in a thermally-grown oxide layer over the surface of the silicon. These windows are produced by etching after the surface has been photographically exposed through a mask. This combination of photographic and etching techniques provides the precision required to form circuit elements of such small size. Completed elements are connected by an aluminium film deposited over the surface of the silicon and etched into the required interconnection pattern. The completed structure is then encapsulated. Such a device not only has a small size, but also a high reliability and, because of the mass-production techniques that can be used in the manufacturing process, a low unit cost.

The monolithic integrated circuit is the most widely used microelectronic device at the present time. Because of its importance, it is discussed in detail in Section 11.

In the film technique for microelectronics, the passive components and interconnections for a circuit are formed by metal or semiconductor-material films deposited on an insulating substrate. The active components can also be formed by films, but they require additional manufacturing stages. It is more usual, for economic reasons, to use microminiature discrete transistors and diodes which are connected into the film pattern. These active devices can either be in microminiature encapsulations, or the unencapsulated chip in which the active element has been formed is connected into the pattern and the completed film circuit encapsulated.

Two approaches to the film technique can be used: the thin-film and the thick-film processes. In the thin-film circuits, the thickness of the film is between 0.01 and 1 μm approximately; the thickness for thick-film circuits is between 10 and 50 μm approximately. There is no dimensional division between the two processes, and in practice the difference is determined by the way the film is formed. Thin films are formed by evaporating or sputtering material onto the substrate, while thick films are formed by a process similar to silk-screen printing.

THIN-FILM CIRCUITS

Substrate

The material used for the substrate of a thin-film circuit must be a good electrical insulator, and have certain physical characteristics. The coefficient of expansion should be small, and should match that of the material used for the dielectric of the capacitors. Thermal conductivity should be high to conduct away any heat generated in the film elements. The surface of the substrate should be flat and be capable of taking a smooth finish to prevent blemishes being formed in the deposited films. Materials which meet these requirements include glass and ceramics, and one which is used in practice is alumina, Al_2O_3. Substrate size will depend on the application in which the thin-film circuit is used, and the type of encapsulation required; sizes range, say, from 5 mm square up to 25 mm square.

Deposition of film

Two methods are used to deposit the film onto the substrate. These are: evaporation, in which the film material is vaporised in a vacuum chamber and the vapour condensed to form the deposit on the substrate; and sputtering, in which atoms of the film material are transferred from a cathode to the substrate which is placed on an anode.

The equipment for depositing the thin film by evaporation is shown diagrammatically in *Figure 12.1*. A substrate is placed in the vacuum chamber with the material which

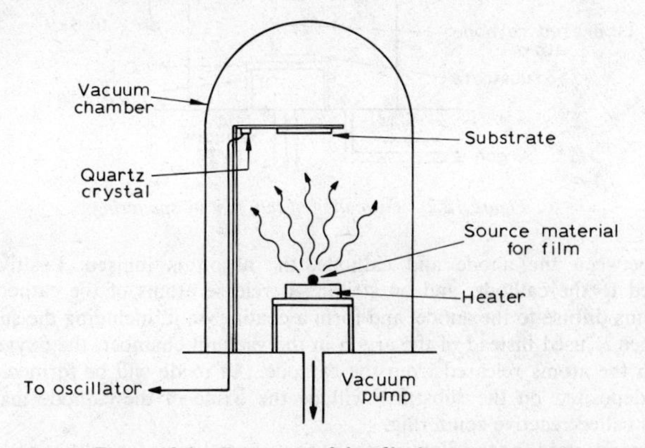

Figure 12.1 Formation of thin film by evaporation

will form the film mounted on a heater. The chamber is evacuated to a pressure of 10^{-5} torr approximately, and the film material heated until it evaporates. The vapour condenses on the inside surfaces of the vacuum chamber including the substrate. To obtain a uniform film, the substrate is placed about 10 cm away from the source of the film material. The heater used to vaporise the material depends on the evaporation temperature. Resistance heaters can be used for low-temperature materials, say with evaporation temperatures below 900 °C. For materials requiring higher temperatures, a method such as electron beam bombardment has to be used. Temperatures above 2 500 °C can be obtained with this method, and it has the advantage that impurities from the resistance material of the heater are not deposited in the film.

The thickness of the deposited film has to be monitored during the evaporation process. With resistors, where the resistance value is more important than the physical thickness of the film, the value can be monitored until the required value has been reached. For other types of film, the thickness can be measured by a quartz crystal oscillator. The quartz crystal is placed in the vacuum chamber close to the substrate and used to control the frequency of the oscillator; as material is deposited on the crystal (thereby increasing its mass), the frequency will fall. The oscillator can be calibrated directly in film thickness, although a different calibration is required for each material used.

In practical evaporation systems, the film material is deposited on many substrates at the same time in one manufacturing operation. Provision can also be made for more than one film material to be evaporated.

The equipment required for sputtering the thin film is shown in *Figure 12.2.* Two electrodes are used: the cathode is made of the material which is to be deposited on the substrate, while the anode forms a work surface on which the substrate is placed. A voltage of 2–5 kV is maintained between the anode and cathode. The pressure in the vacuum chamber is reduced to a low value, and then an inert gas such as argon is introduced to give a final pressure of approximately 0.05 torr. When the voltage is

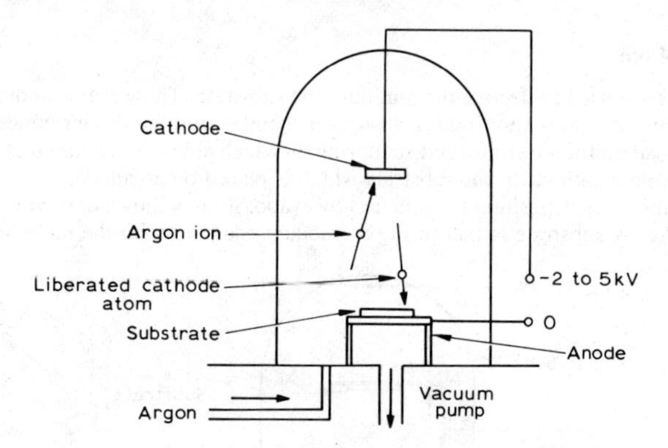

Figure 12.2 Formation of thin film by sputtering

applied between the anode and cathode, the argon is ionised. Positive ions are accelerated to the cathode, and on striking it release atoms of the cathode material. These atoms diffuse to the anode, and form a coating on it, including the substrate.

If oxygen is used instead of the argon in the vacuum chamber, the oxygen ions will react with the atoms released from the cathode. An oxide will be formed, and so the material deposited on the substrate will be the oxide of the cathode material. This process is called reactive sputtering.

The material used for the film, whether for evaporation or for sputtering, depends on the purpose of the film. For example, whether the film is to be for interconnections, resistors, or the capacitor dielectrics. The materials used in practical thin-film circuits are discussed in the next section in which the various circuit elements are described.

Thin-film circuit elements

Because it is usual to connect microminiature transistors and diodes into a thin-film circuit, the circuit elements that have to be formed by the film are resistors, capacitors, and the interconnections. The resistors and interconnections are made from single-layer films of suitable material, while the capacitors are made from successive layers of conductor–dielectric–conductor.

The resistance of a film R is given by

$$R=\rho\frac{L}{tW},$$

where ρ is the resistivity of the material forming the film, and L, W, and t are the length, width, and thickness respectively of the film. Thus by choosing a material of suitable resistivity, and by varying the relative values of L, W, and t, resistors of the required values can be obtained. Practical limits are imposed by the minimum area which can be accurately defined by a film manufacturing process, and by the maximum area of the substrate that can be allotted to resistors. Thus the range of resistance value on a thin-film circuit is typically between 1 Ω and 1 MΩ.

Typical materials used to form thin-film resistors include nichrome (nickel–chromium alloy), tin oxide, and tantalum nitride. The cross-sectional area of the film, the product tW, will be determined in some cases by the current flowing through the resistor. The resistance value will then be determined by the length of the film. High

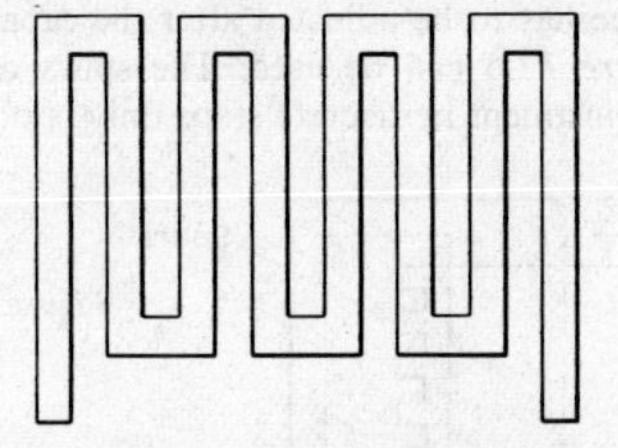

Figure 12.3 Meandered high-value thin-film resistor

value resistors can be *meandered* to make them more compact, as shown in *Figure 12.3*. Surface area of the film may also be determined by the dissipation required by the resistor. A typical value for thin-film circuits is 20–30 mW/mm^2 of resistance-film area. The tolerance on the resistance value after deposition is between 5% and 15%. The value, however, can be adjusted to within a closer tolerance as described later.

The deposited resistance films are heated for a short time to stabilise the value. Heating also produces an oxide layer over the surface which provides protection for the film. For the oxide-type resistors, this heating can be used as a fine adjustment to the deposited value. For resistors of other materials, particularly where the film area is relatively large, the value can be adjusted as shown in *Figure 12.4*. The deposited film

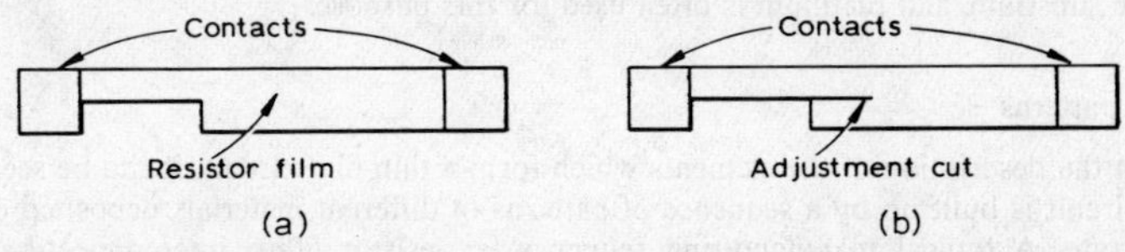

Figure 12.4 Adjustment of thin-film resistor: (a) before adjustment (b) after adjustment

has a *notch* in it, as shown in (*a*). This notch enables a diamond scribing tool to be accurately positioned to cut the film, as shown in (*b*), to provide a fine adjustment. After adjustment, the resistance value can be within $\pm 0.1\%$ of the required value.

A thin-film capacitor is shown in the plan view of *Figure 12.5*. The layers of conductor and dielectric are overlapped so that the active area is well defined. Typical materials used for the dielectric film include silicon monoxide, silicon dioxide, aluminium oxide, and tantalum pentoxide. The capacitance value is determined by the active area of the capacitor and the thickness of the dielectric film. There are practical limits imposed by the maximum area that can be occupied by a capacitor, and the minimum area of the deposited film that can be accurately defined. In addition, the minimum

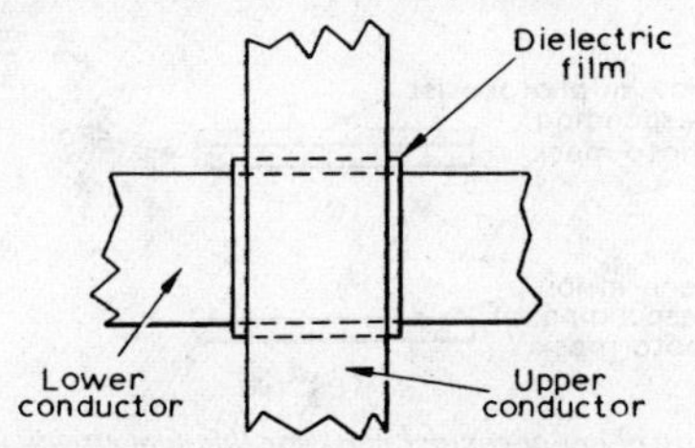

Figure 12.5 Structure of film capacitor

dielectric film thickness that can be used is about 0.02 μm. A practical lower limit to capacitance is approximately 100 pF. If the dielectric used is a tantalum oxide film 0.1 μm thick, a capacitance of approximately 2 000 pF/mm^2 is obtained. The maximum capacitance will then be determined by the area available.

If the capacitance value has to be adjusted after the capacitor has been formed, the structure shown in *Figure 12.6* can be used. The spurs can be cut to decrease the capacitance value, but adjustment in discrete steps only is possible.

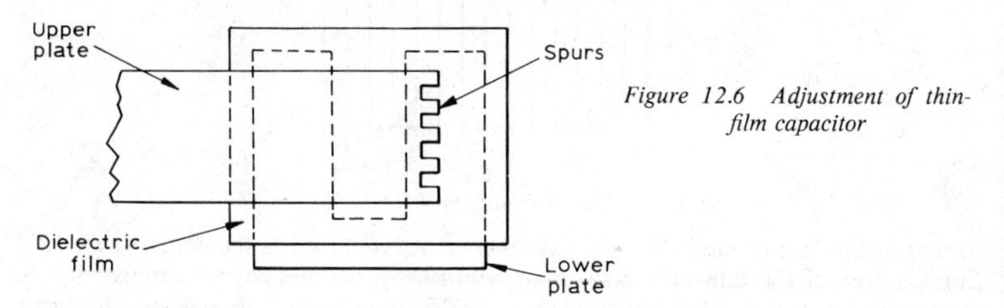

Figure 12.6 Adjustment of thin-film capacitor

As tantalum oxide is an efficient dielectric, and tantalum can be used for resistors, it is possible to combine two film-deposition stages by using this material. A pattern corresponding to the resistors and lower plates of the capacitors is formed. Parts of the film are oxidised to form the capacitor dielectric films, and parts of the final interconnection pattern are used to form the upper plates of the capacitors.

The material used for the interconnection pattern is generally aluminium or gold. If gold is used, it is necessary to use an intermediate material to provide a better adhesion to the substrate, and platinum is often used for this purpose.

Film patterns

From the description of the elements which form a thin-film circuit, it can be seen that the circuit is built up by a sequence of patterns of different materials deposited on the substrate. A typical manufacturing sequence is: resistor films; interconnections and

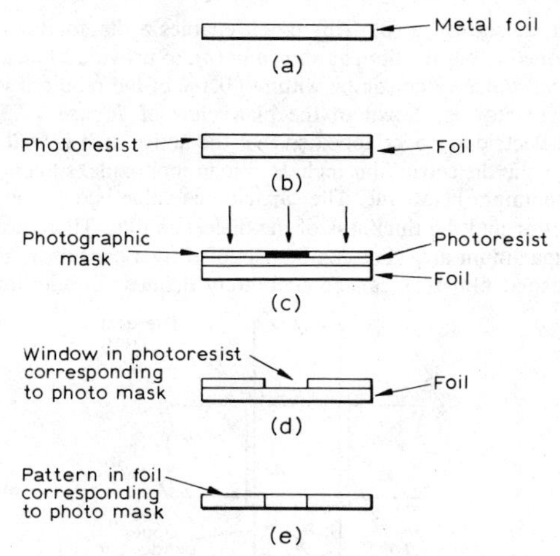

Figure 12.7 Manufacture of out-of-contact mask for thin-film circuit: (a) metal foil (b) foil coated with photoresist (c) exposure through photographic mask (d) development, unexposed photoresist removed for etching (e) completed out-of-contact mask

capacitor lower plates; dielectric films; interconnections and capacitor upper plates. There are three methods of forming the required pattern: out-of-contact masking; in-contact masking; and selective etching.

Out-of-contact masking is also known as stencil masking, a name that describes the process. A mask is placed on the substrate so that when material is deposited, the film on the substrate conforms to the pattern cut in the mask. The minimum line width that can be achieved with this method is about 50 μm. The mask itself can be made by the photographic and etching techniques used in the manufacture of integrated circuits, and the process is shown in *Figure 12.7*. The material from which the mask is made, usually a metal foil (*a*), is coated with a photoresist (*b*). The photoresist is an organic material which polymerises when exposed to ultraviolet light. The polymerised material will resist solvents and etching chemicals, while the unexposed photoresist can be easily dissolved. The coated foil is exposed through a photographic mask (*c*), and the unexposed photoresist removed to reveal the foil (*d*). The required pattern can then be etched in the foil (*e*).

A simple circuit is shown in (*a*) of *Figure 12.8*, with its translation into thin-film form in (*b*). Four out-of-contact masks are required to make the circuit. The first mask defines the resistor (*c*). The second mask forms the lower plates of the capacitors C_1 and C_2 in contact with the resistor, and terminals 1 and 3 (*d*). The third mask defines the dielectric (*e*), while the fourth mask forms the upper plates of the capacitors and terminals 2 and 4 (*f*).

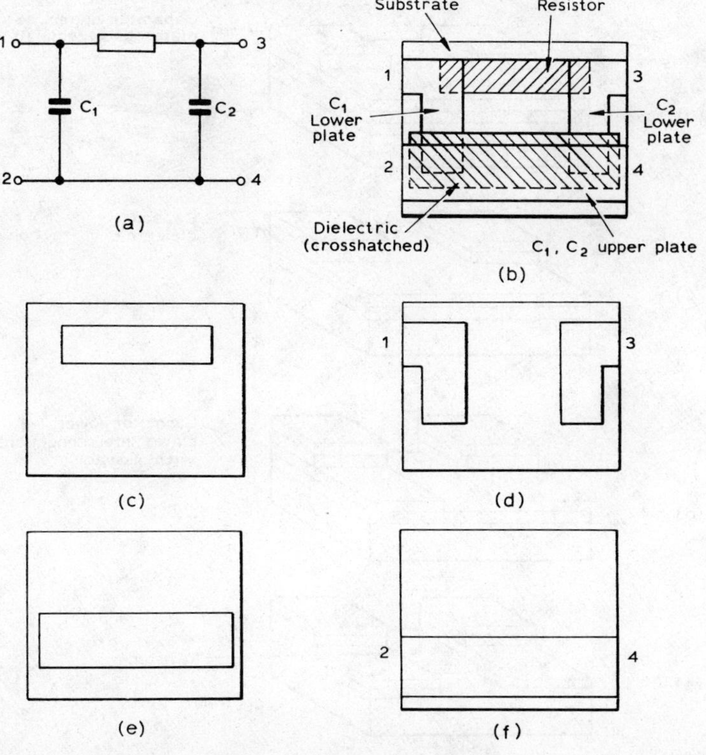

Figure 12.8 Manufacture of simple thin-film circuit using out-of-contact masks: (a) circuit (b) film pattern of completed circuit (c) mask for resistor (d) mask for capacitor lower plates (e) mask for dielectric film (f) mask for capacitor upper plates

Although the out-of-contact mask is held against the substrate, it is not in intimate contact with it (hence the name, out-of-contact). If the substrate itself is coated with photoresist which is exposed through a photographic mask and the unexposed photoresists removed, a mask is formed in-contact with the substrate. This mask can then be used to define the required pattern.

In the selective etching method, a sequence of films is deposited over the whole surface of the substrate. Each film is etched in turn through in-contact masks into the required pattern. A typical process is shown in *Figure 12.9* where the sequence of films is shown in (*a*). The first mask is used to define the capacitor upper plates and the interconnections to the terminals (*b*). The second mask defines the capacitor dielectric (*c*), and the third mask the capacitor lower plates and interconnections to the resistor

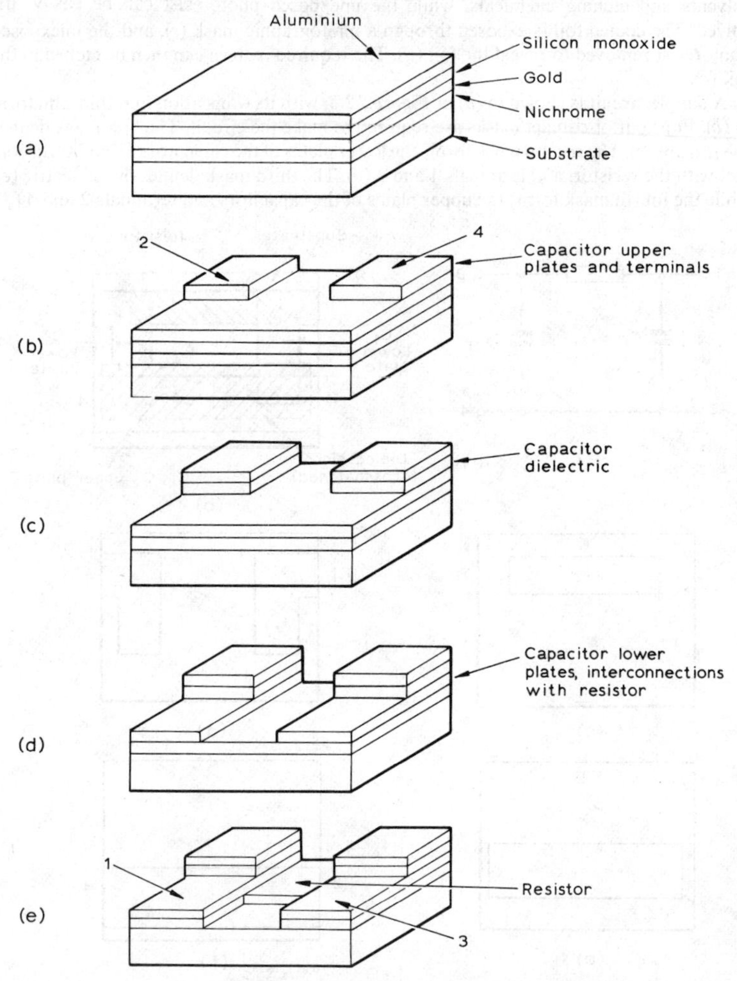

Figure 12.9 Manufacture of thin-film circuit by selective etching: (*a*) *deposited films* (*b*) *capacitor upper plates and terminals defined* (*c*) *dielectric films defined* (*d*) *capacitor lower plates and interconnections with resistor defined* (*e*) *resistor defined to complete circuit*

(*d*). The resistor itself is defined by the fourth mask, giving the completed structure shown in (*e*). The circuit formed in this way is similar to that shown in *Figure 12.8* (*a*).

The photographic masks which define the areas to be etched in this method, the in-contact masks, and the pattern in the out-of-contact masks, are made in a similar way to the diffusion masks used in integrated circuits. For thin-film circuits, however, the definition need not be so high, and so artwork can be prepared by hand rather than by computer-controlled methods. The artwork is prepared either from an opaque peelable plastic in which areas corresponding to the film pattern are cut to form transparent windows, or from tape by which the required pattern is built up. The completed artwork is photographed with a typical reduction of 20 : 1, and copies of the resulting photographic plate are used as masks.

When the film pattern has been completed, by whichever of the methods described, the active devices are connected into the circuit. The transistors and diodes are attached to the conductor films by thermocompression bonding.

THICK-FILM CIRCUITS

The principle of the thick-film circuit is that of silk-screen printing—the deposition of material through a mask on a fine screen to form the required pattern. Thick-film circuits are used in preference to thin-film circuits where the precision required is not so high, and where a higher power dissipation is required. The thick-film process is capable of greater automation than the thin-film process, and is considerably cheaper.

Circuit elements

The circuit elements formed on a thick-film circuit are the same as those on the thin-film circuit previously described. Resistors and interconnections are formed from a single layer of material of suitable resistivity. Capacitors are formed from three superimposed layers, the upper and lower plates separated by the dielectric film. The required pattern of passive components is built up by a sequence of deposited films. Transistors and diodes are added as discrete microminiature devices connected into the film pattern by thermocompression bonding.

The substrate used for a thick-film circuit is similar to that for a thin-film circuit already discussed. The size of the substrate is generally larger, say from 10 mm square up to 75 × 50 mm, and typically 1.5 mm thick. As it is necessary to heat a thick-film circuit after the films have been deposited, the substrate material should be able to withstand the heating without producing any unwanted effects. Again, alumina Al_2O_3 is a suitable material.

Thick-film deposition

The 'silk-screen' process for thick-film circuits is illustrated diagrammatically in *Figure 12.10*. The screen is made from a stainless steel fine mesh, and the screen pattern can be made in a similar way to that of normal silk-screen printing: coating the mesh with a material which prevents the passage of the ink except in the pattern area. This coating can be a photoresist which is exposed and developed in the normal way to form the screen pattern. Alternatively, an out-of-contact mask can be made and placed on the screen. The pattern is made from artwork which defines the required film, from which a photographic mask with a reduction of approximately 20 : 1 is made to produce the actual screen pattern.

The ink used for printing the thick film consists of a base to which materials are added to produce the particular film required. The base is formed by an organic filler which carries the other components of the ink through the mask pattern and the screen, a solvent which controls the viscosity of the ink, and a powdered glass frit with a low melting point which on firing produces the bond between the film and substrate. To this base can be added aluminium, gold, gold–platinum alloy, or palladium silver PdAg particles to form the conductor film. Resistors are made by adding the required proportions of palladium silver–palladium oxide PdAg–PdO to give the necessary sheet resistance. The capacitor dielectric films are made by adding barium titanate $BaTiO_3$ or titanium dioxide TiO_2.

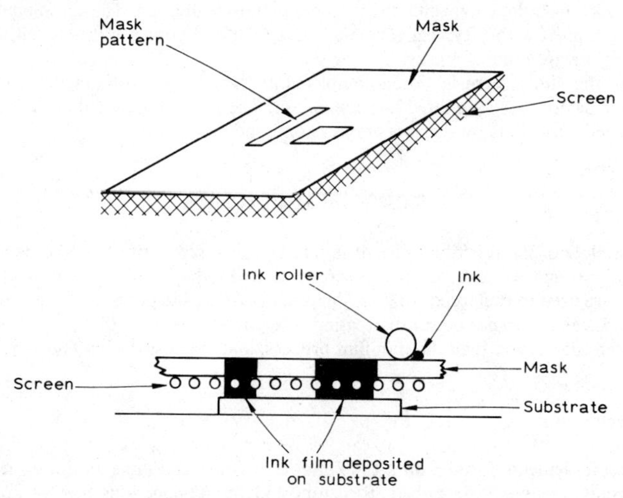

Figure 12.10 Principle of printing thick film

When a film pattern has been inked onto the substrate, the film is dried by heating to about 100 °C. This drives off the solvent from the ink. The circuit is then fired at a temperature between 700 °C and 1 000 °C. The glass frit particles sinter into the substrate and so bond the *active* particles of the film to the substrate. The firing temperature affects the properties of the film, and must be carefully controlled to within ±1 °C for reproducible films. Because a series of films is deposited, care must be taken when firing the later films not to affect those already formed.

The range of resistance values for thick-film circuits is from 0.5 Ω to 50 MΩ. The tolerance on the resistance value after firing is between ±5% and ±15%, depending on the size of the resistance film. The value can be adjusted to within a tolerance of ±1% by trimming the film area with abrasive powder. Typical dissipation values are between 60 mW/mm^2 and 1 000 mW/mm^2 of film area. This is higher than the value for thin-film circuits.

Because of the larger size of the substrate for thick-film circuits, higher capacitance values can be obtained than for thin-film circuits. In addition, discrete miniature tantalum capacitors can be attached to the substrate and connected into the film pattern to enable higher values to be obtained without occupying too large an area of the substrate.

HYBRID CIRCUITS

A hybrid circuit in the microelectronic sense is one combining both film circuits and monolithic integrated circuits. An alternative name often used is hybrid integrated circuit. The active devices, transistors and diodes, are formed by diffusions in the integrated circuit chip. The resistors and capacitors are formed by the thin-film or thick-film techniques on a substrate. An interconnection pattern is also formed on the substrate, and the integrated circuit chip connected into the pattern. The completed circuit can then be encapsulated.

The hybrid circuit just described is an extension of the normal thin-film technique where discrete transistors and diodes in microminiature chip form are connected into the film circuit. Combining the active devices into one chip enables the volume required to be reduced. The technique can also be used with monolithic integrated circuits where resistors of a higher dissipation than can be conveniently included in the chip are required.

Linear circuits operating at high frequencies can also use the hybrid technique with advantage. The capacitance associated with diffused resistors can impose limits on the gain-bandwidth product at high frequencies in some integrated circuits. Using thin-film resistors can overcome this limitation. Similarly, thin-film capacitors can be used to avoid parasitic effects that may occur with diffused capacitors. The thin-film circuit elements can be formed on the surface of the silicon chip in which the active components are diffused to form a compact device.

MICROWAVE INTEGRATED CIRCUITS

Thin-film circuits have found particular application at microwave frequencies in what have become generally known as microwave integrated circuits. Transmission lines for the electromagnetic waves are provided by microstrip lines, which are conveniently formed by thin-film techniques. A microstrip line consists of a substrate of high dielectric constant, one side of which is coated with a metal film to form a ground plane, while the other side supports a thin narrow strip. With this structure, the transmitted electromagnetic wave is contained almost entirely inside the substrate. Distributed passive components can be made simply by forming the microstrip line into the required pattern, and such components as bandpass filters, circulators, and directional couplers can be made which are considerably smaller than components using waveguides and coaxial lines. Discrete microwave semiconductor devices can be added to such components to make active components such as amplifiers, mixers, and oscillators. Lumped circuit components can also be made by thin-film techniques. Complete microwave systems can therefore be formed on the substrate which are considerably smaller than systems using conventional components.

Substrate

The substrate for a microstrip line must have a high dielectric constant (typically between 8 and 16). In addition, the material must be capable of taking a very smooth surface finish. Materials which meet these requirements and are commonly used in practice are alumina, sapphire, and ferrite. Of these three, alumina is the most widely used. Sapphire is used where very thin substrates are required, as its greater strength through the crystal structure gives it an advantage over the sintered material alumina. Ferrite is used for substrates where the magnetic properties can be used, for example in non-reciprocal isolators and circulators. Other substrate materials that are sometimes used include fused quartz and beryllium oxide.

The thickness of a typical alumina substrate for use in microstrip circuits operating at frequencies to above 10 GHz is 0.6 mm with a surface finish less than 0.2 μm. The characteristic impedance of a microstrip line is generally proportional to the ratio of strip width to substrate thickness, and so once the substrate material and thickness have been chosen, the only control over the impedance is by varying the strip width.

Accurate definition of the strip edge is therefore required. The ground plane is theoretically infinitely wide, but in practice a value three times the strip width is sufficient.

The thickness of the microstrip conductor is determined by the skin depth at the frequency of operation. The thickness generally used is four to five times the skin depth; greater thicknesses make it difficult to maintain the definition of the strip edge. For gold, the skin depth at 2 GHz is 1.6 μm, and at 100 GHz it is 0.8 μm.

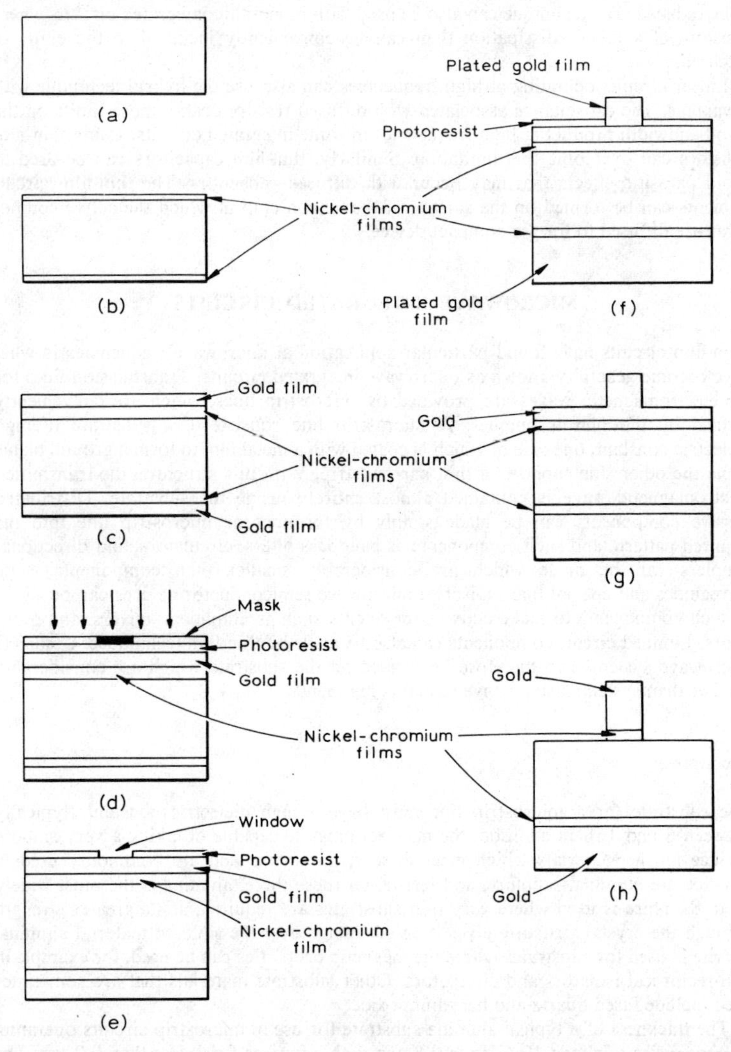

Figure 12.11 Manufacture of microstrip circuit: (a) substrate (b) nickel–chromium films deposited (c) gold films deposited (d) photoresist deposited and exposed (e) window opened for plating (f) strip and ground plane plated (g) photoresist removed (h) initial gold and nickel–chromium films removed, microstrip circuit completed

Microstrip pattern

The pattern of the microstrip line to form the required circuit component is produced on the substrate by a combination of the thin-film technique and electroplating. The main stages of the process are shown in *Figure 12.11*. The polished substrate (*a*) is cleaned, and a film of nickel–chromium about 0.005 μm thick is formed on both sides by evaporation (*b*). This layer is used to improve the adhesion between the substrate and the following gold film. Other materials are sometimes used, such as tantalum or titanium. A gold film approximately 0.05 μm thick is evaporated over the nickel–chromium film on both sides of the substrate (*c*).

The upper surface of the structure is spun-coated with photoresist, and the resulting film dried by heating to 70 °C. A mask defining the microstrip pattern is placed over the photoresist, and the substrate is exposed to ultraviolet light (*d*). On development, the unexposed photoresist is removed, revealing the underlying gold film (*e*). The remaining photoresist is cured by heating to 160 °C, and the substrate is placed in an electroplating bath. Both sides of the substrate are plated with gold until the conductor thickness is sufficient for the operating frequency (*f*). The photoresist is removed (*g*), followed by removal of the initial gold and nickel–chromium films from the top surface of the substrate. The resulting structure is shown in (*h*).

The mask defining the microstrip pattern is made in the usual way for thin-film circuits. Large-scale artwork is prepared, from which photomasks are made with a reduction of typically 20 : 1. As much of the design of microwave integrated circuits is done with the aid of computers, the program of the completed design can be used to operate a computer-controlled cutting table. Large-scale artwork for the layout can therefore be produced directly to obviate any human error.

Distributed passive components

Distributed components are made in microstrip form simply by shaping the microstrip line into the required form. For example, bandpass filters can be made by coupling a number of half-wave resonators. For wideband filters, the required tight coupling is achieved by using a series of stub resonators. The loose coupling for narrowband filters can be achieved by electromagnetic coupling between parallel resonators.

A typical pattern for a bandpass filter is shown in (*a*) of *Figure 12.12*, and a microstrip pattern for a circulator is shown in (*b*). The form generally used for the circulator is a three-port Y-junction on a ferrite substrate. A microstrip phase shifter is shown in (*c*). Other passive components can be formed in similar ways.

Active components

Semiconductor devices are available in microwave encapsulations so that they can be incorporated into microstrip components. Typical encapsulations are shown in *Figure 12.13*. The beam lead package (*a*) enables the device to be fitted directly to the microstrip conductor using thermocompression bonding. The leadless inverted device (l.i.d.) is mounted in a ceramic holder which forms a bridge between the microstrip conductors (*b*). The third encapsulation is a small capsule which can be inserted in a hole drilled in the substrate (*c*). Semiconductor devices that are available in these encapsulations include Gunn-effect devices, varactor diodes, Schottky barrier diodes, and microwave transistors. With such devices amplifiers, oscillators, multipliers, and mixers can be formed on substrates measuring 25 × 25 mm. These circuits can be combined to form receiver front-ends and Doppler radar units on substrates that may be no larger than 30 × 30 mm.

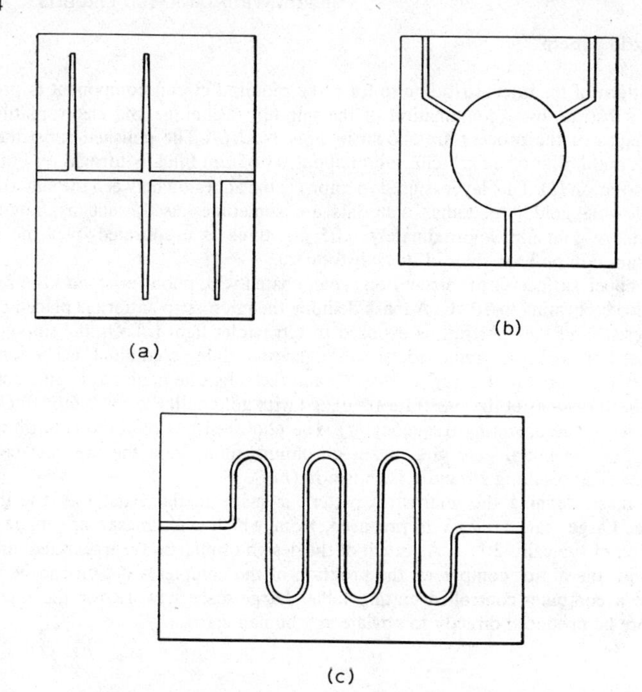

Figure 12.12 *Microstrip patterns for passive components: (a) bandpass filter pattern (b) circulator pattern (c) phase-shifter pattern*

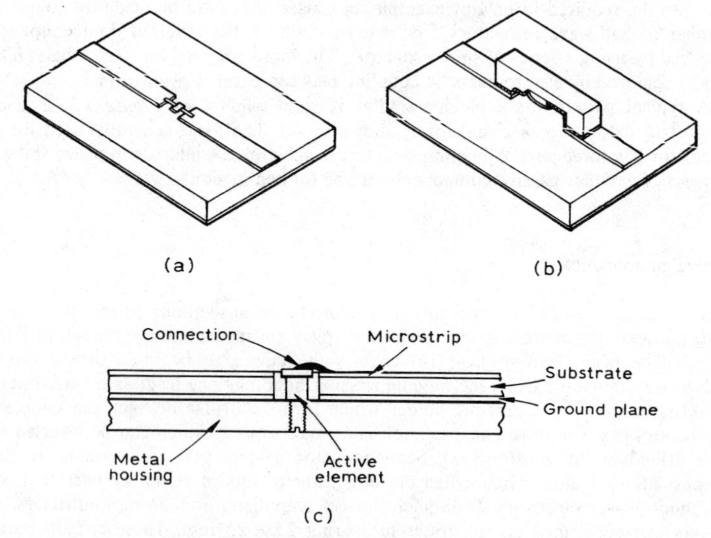

Figure 12.13 *Encapsulations of semiconductor devices for microstrip circuits: (a) beam lead package (b) leadless inverted device, l.i.d. (c) capsule for insertion into substrate*

Lumped components

Lumped components offer the advantage of being considerably smaller than the equivalent distributed components. Whereas a distributed component must have a length which is a multiple of the quarter-wavelength, the maximum size a lumped component can have before it starts to exhibit distributed effects is approximately one-tenth of a wavelength. Such dimensions can be realised by the thin-film and plating techniques used for microstrip circuits, at least for the lower microwave frequencies. Lumped components, therefore, can be used in microwave integrated circuits.

Substrates for lumped components act only as a base for the deposited component. The requirements for the substrate are therefore not so rigorous as for the microstrip substrate. Thickness can be determined solely by the required strength, and no ground plane has to be formed on the side opposite the component.

Values of inductance and capacitance required at microwave frequencies are low, only a few nanohenries and picofarads. These values can be obtained from structures like those shown in *Figure 12.14* with dimensions that enable the components to remain 'lumped'.

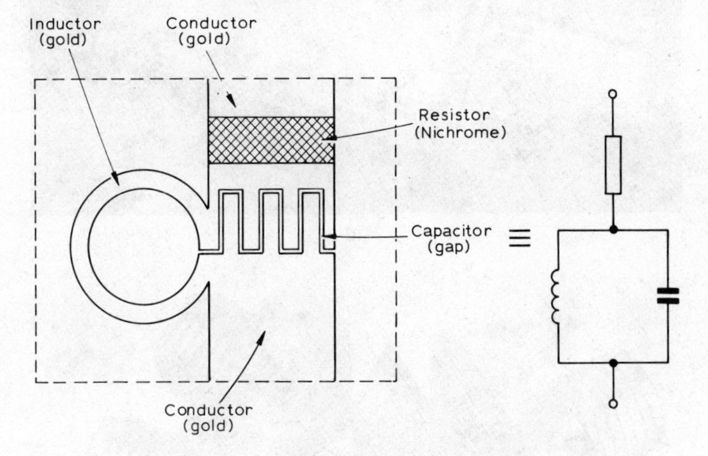

Figure 12.14 Lumped components for microwave integrated circuit

An inductor is formed by the circular structure, the inductance value being determined by the diameter and strip width. Values between 1.0 nH and 3.5 nH, typically, can be obtained. A capacitor is formed by the interdigitated structure, the capacitance being determined by the fringing field across the gap. Capacitance value is varied by changing the width of the gap and the finger pattern, and typical values lie between 0.1 pF and 1.0 pF. Larger values can be obtained by using the normal thin-film capacitor, two conductors separated by a dielectric film. This structure is used only when necessary as it involves extra manufacturing stages. Resistors are formed by the normal thin-film method, the material used being nickel–chromium which is also used as the material deposited on the substrate to improve the adhesion of the gold conductors.

Applications of microwave integrated circuits

Microwave integrated circuits enable components and subsystems to be constructed which are considerably smaller and lighter than those using conventional microwave components. Applications for these circuits are therefore those where small size and low weight are advantageous, guided missiles and airborne radar and communications systems are typical examples. Ground-based applications include portable and mobile systems.

Two typical microwave integrated circuits are shown in *Figure 12.15*. The circuit in (*a*) uses distributed components and discrete semiconductor devices mounted on an alumina substrate approximately 9 × 3 cm, with an isolator and circulator mounted on ferrite substrates 2 cm square. The circuit in (*b*) uses lumped components, and is a Doppler radar unit.

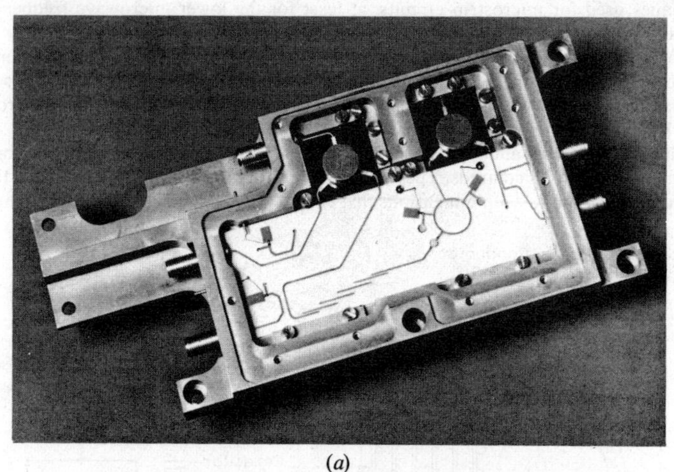

(*a*)

(*b*)

Figure 12.15 (*a*) *Microwave integrated circuit using distributed components and discrete semiconductor devices; substrate area approximately 9 × 5 cm overall (Mullard Limited)* (*b*) *Microwave integrated circuit using lumped components (Mullard Limited)*

13 BASIC ELECTRONIC CIRCUITS

BASIC ELECTRONIC CIRCUITS 13

13 BASIC ELECTRONIC CIRCUITS

Of the electronic devices described in previous sections the most used are undoubtedly the thermionic valve and the transistor, the latter as a discrete component but more recently in the form of the integrated circuit. A complete list of all the applications of these devices would be very long; the following are just a few: radio transmitters and receivers; broadcasting equipment; recording equipment; telephone equipment; public address equipment; radar; computers; medical equipment; industrial control equipment.

Most of these, even the more complex, consist of assemblies of simple circuit arrangements which generate electrical signals or perform operations upon them. For example commonly encountered circuits for generating signals are sinusoidal oscillators, blocking oscillators and multivibrators. Other well-known circuit arrangements are used for amplification, detection and frequency changing. These are all examples of analogue or linear functions, however the most rapidly expanding applications of transistors today are in digital or logic circuitry: typical circuits in this field are for gates, bistables and stores.

It is the purpose of this section to introduce the basic circuits most likely to be encountered in analogue and digital equipment. A few of the circuits will contain valves but most employ transistors. Analogue circuits will be dealt with first and the first subject will be audio-frequency amplification. A.F. equipment is required to give amplification and power output at frequencies between 30 Hz and 15 kHz and is tested with sinusoidal signals because these are typical of the waveforms the equipment has to handle.

A.F. amplifiers are a convenient subject with which to begin this section on electronic circuits because the circuits used form the basis of many other amplifying circuits. It also affords the opportunity of introducing many circuit techniques used generally in linear equipment such as those used to provide bias and d.c. stabilisation.

VALVE AMPLIFYING CIRCUITS

Single-stage voltage amplifier

The first circuit to be considered, *Figure 13.1*, is that of a single-stage *RC*-coupled voltage amplifier using a pentode. The mean cathode current is equal to the sum of the anode and screen-grid currents and, in flowing through R_k, biases the cathode positively with respect to the negative supply line. The control grid of the pentode is returned to this supply line via R_g and in this way the grid is automatically biased negatively with respect to the cathode as required for linear amplification. If the required grid bias voltage is -2 V and the mean cathode current is 1.5 mA, then the value of R_k is given by

$$R_k = \frac{\text{grid bias voltage}}{\text{mean cathode current}} = \frac{2}{1.5 \times 10^{-3}} \approx 1.3\ \text{k}\Omega$$

To avoid negative feedback (which reduces gain) R_k is decoupled by C_k. C_k should have a reactance small compared with $1/g_m$ at the lowest frequency of operation: 50 μF is a commonly-used value in an a.f. amplifier.

R_a is the anode load resistance and is often approximately 100 kΩ. R_{sg} is chosen to give the required mean value of screen-grid voltage. This is approximately equal to the mean value of the anode voltage. This gives an easy way of estimating the approximate value of R_{sg}. The screen-grid current is usually about a third of the anode current and thus R_{sg} should be three times R_a. A value of 330 kΩ is common for R_{sg}. C_{sg} is the screen-grid decoupling capacitor and its reactance at the lowest frequency of operation should be small compared with the internal screen-grid resistance of the pentode: 2 μF is frequently used in an a.f. amplifier.

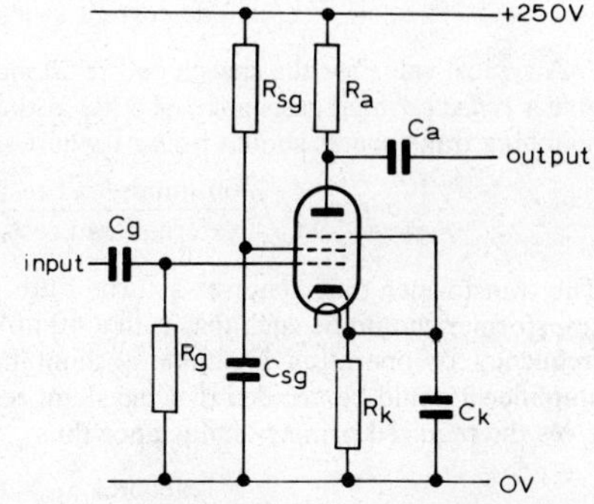

Figure 13.1 *Simple voltage amplifier using a pentode valve*

The voltage gain of an amplifier such as that shown in *Figure 13.1* is given by $g_m R_a$ approximately. Typical values for g_m and R_a are 1.5 mA/V and 100 kΩ, giving a voltage gain of 150. The input resistance of the amplifier is equal to the value of the grid resistor R_g (1 MΩ is common) and the output resistance is approximately equal to the value of R_a (100 kΩ).

Valve output amplifier

Figure 13.2 shows the circuit diagram of a pentode amplifier used to deliver appreciable power to an external load, e.g., a loudspeaker. The grid and cathode circuits are similar to those of *Figure 13.1* but the screen grid is taken directly to the positive supply rail and the external load (shown as a loudspeaker) is coupled via a matching transformer to the valve anode circuit. For a typical pentode rated for 10 W-maximum anode dissipation the mean anode current is about 40 mA. To obtain the maximum power output (4 W) the transformer must present the anode circuit with the optimum load resistance,

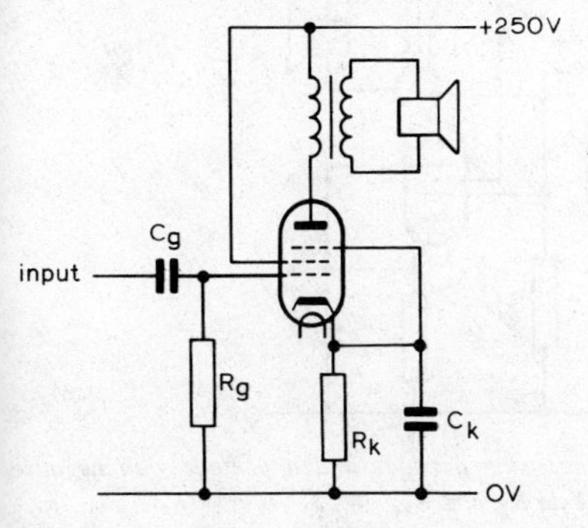

Figure 13.2 *A pentode valve used to supply power to a loudspeaker*

i.e., that value of resistance into which the valve can deliver maximum power. The I_a–V_a characteristics for $V_{sg} = 250$ V show that the anode voltage can swing to within 40 V of the cathode voltage. If we assume a grid bias voltage of 10, this is a downward voltage swing of 200 V. An equal positive-going anode voltage swing is easily possible. The anode current swing responsible for these voltage swings is probably around 35 mA (i.e., not quite equal to the mean anode current). From these values the optimum load can be calculated from the following expression

$$R = \frac{\text{anode voltage swing}}{\text{anode current swing}} = \frac{200}{35 \times 10^{-3}} \approx 6\ \text{k}\Omega$$

A typical value for the speech-coil resistance of a moving-coil loudspeaker is 3 Ω. To give a *reflected* load resistance of 6 kΩ at the valve anode circuit the turns ratio of the matching transformer should be $n : 1$ where n is given by

$$n = \sqrt{\left(\frac{\text{optimum load resistance}}{\text{external load resistance}}\right)} = \sqrt{\left(\frac{6\,000}{3}\right)} = 45$$

The transformer thus requires a turns ratio of 45 : 1. The primary inductance of the transformer should be such that with a 40 mA mean current, the reactance at the lowest frequency of operation should not shunt the optimum load unduly. For an audio amplifier it could be decided that the shunt reactance should equal 6 kΩ at 50 Hz. This gives the required primary inductance thus

$$L = \frac{\text{reactance}}{2\pi f} = \frac{6\,000}{6.284 \times 50} \approx 20\ \text{H}$$

Application of negative feedback

A pentode output stage such as that illustrated in *Figure 13.2* is capable of 4 W audio output but at a few per cent of harmonic distortion and it is usual to apply negative feedback to reduce this. The pentode is usually preceded by a voltage amplifying stage and a common technique is to apply the feedback over both stages. One possible circuit is illustrated in *Figure 13.3* in which a triode voltage amplifying stage is shown. Feedback is applied from the anode of the pentode to the cathode of the triode so that the feedback resistor R_{fb} and the cathode resistor R_{k1} form in effect a potential divider across the output circuit of the pentode and feed back a fraction $R_k/(R_{k1} + R_{fb})$ of the output voltage. If, as is usual, R_{fb} is large compared with R_{k1} the feedback fraction becomes approximately R_k/R_{fb}. This means that the voltage gain of the amplifier, from triode grid to pentode anode, is given by R_{fb}/R_{k1}. If, as an example, this is made equal to

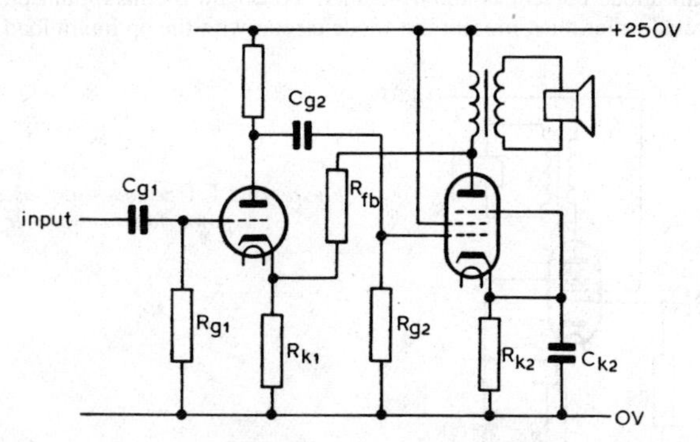

Figure 13.3 A two-stage loudspeaker amplifier using a triode and a pentode with negative feedback applied via R_{fb} and R_{k1}

200 then a 1-volt signal applied to the triode grid will give a 200-volt signal at the pentode anode: this is, of course, the condition for maximum power output. A convenient value for R_{k1} is 1 kΩ and thus R_{fb} must be 200 kΩ.

The diagram shows a direct path between the supply rails via the transformer primary winding, R_{fb} and R_{k1}. This puts a current of approximately 1.25 mA into R_{k1} in addition to the triode cathode current: both currents should be allowed for in calculating the value of R_{k1}.

Frequency-discriminating negative feedback

It is possible to use the negative feedback circuit to adjust the frequency response of the amplifier. For example for a rising bass response the feedback voltage is required to fall as frequency falls below say, 200 Hz. This can be achieved by connecting a capacitor in series with R_{fb}. A useful rise in bass response would result by making the reactance of the capacitor equal to R_{fb} at 100 Hz. If $R_{fb} = 200$ kΩ the required capacitance is given by

$$C = \frac{1}{2\pi f X} = \frac{1}{6.284 \times 100 \times 200 \times 10^3} \text{F} \approx 0.01\,\mu\text{F}$$

A falling high-frequency response can be obtained by shunting R_{fb} with a suitably-sized capacitor. This technique is often used to ensure that the amplifier does not respond at frequencies above the intended passband, e.g., it can make an a.f. amplifier unresponsive to r.f.

A rising high-frequency response can be obtained by shunting R_{k1} by a suitably-sized capacitor.

Push–pull amplification

A common circuit arrangement is to connect two similar valves in push–pull, i.e., two valves driven by antiphase input signals, the outputs being combined with the necessary

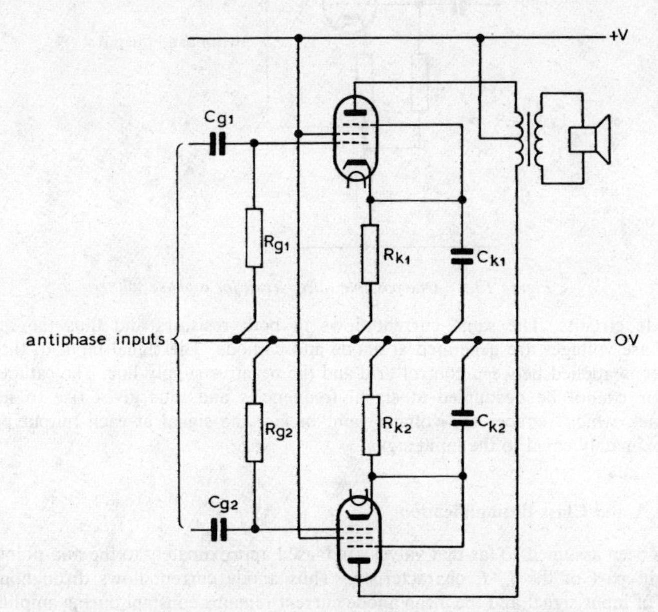

Figure 13.4 A push–pull output stage using two pentode valves

phase inversion, in a common load. The basic circuit, shown in *Figure 13.4*, has the following advantages over single-ended output stages:

(*a*) The even-harmonic distortion introduced by one valve is cancelled by that introduced by the other so that the overall distortion of a push–pull stage is less than that of the same two valves connected in parallel.
(*b*) The steady magnetic polarisation of the core of the output transformer due to the mean component of the anode current of one valve cancels that due to the other valve. Thus there is no steady magnetic polarisation of the core at all in a push–pull stage. This makes the design of the output transformer easier and also means that a smaller transformer can be used.
(*c*) By a similar process to (*b*) any ripple present on the supply terminals affects both valves equally but the effects cancel in the output transformer.

Phase splitter

To drive a push–pull stage a circuit is required which will accept the signal to be amplified and will generate from it two equal-amplitude antiphase signals. The obvious solution is to use a transformer with a centre-tapped secondary winding but a valve circuit is preferred and there are a number of possible circuits which can be used. One is shown in *Figure 13.5*: it consists of a triode valve with equal resistors in anode and

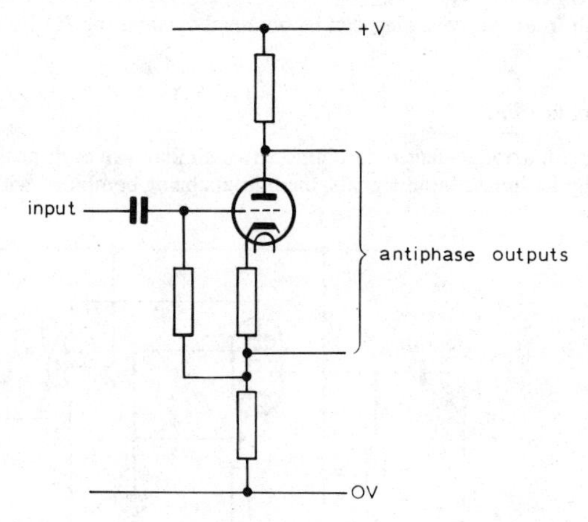

Figure 13.5 One possible valve circuit for a phase splitter

cathode circuits. The same current flows in both resistors and thus the required antiphase voltages are generated at anode and cathode. The signal input to the phase splitter is applied between control grid and the negative supply line. The cathode load resistor cannot be decoupled at signal frequencies and thus gives rise to negative feedback which reduces the voltage gain. In fact the signal at each output point is approximately equal to the input signal.

Class A and Class B amplification

It has been assumed so far that valves are biased approximately to the mid-point of the straight part of the I_a–I_g characteristic. Thus anode current flows throughout each cycle of input signal and the mean anode current remains constant during amplification at a value approximately equal to the no-signal value. This is known as *Class A amplification.* It is extravagant of h.t. power: for example the maximum undistorted output power available from a Class A stage is 25% of the power taken from the h.t. supply. This high price is often paid however to obtain the high linearity of which Class A stages are capable.

If, in a single Class A amplifier, the grid bias is increased to a point near the cut-off of anode current, the anode current waveform would no longer be a faithful representation of that of the input signal because, for the greater part of the negative half-cycle of each input sinusoid, the valve would be cut off. Positive half-cycles would, however, be reproduced without distortion if the I_a–V_g characteristic is assumed to be straight. If the valve is one of a push–pull pair, while it is cut off by negative-going signals the other valve is driven into conduction by corresponding positive-going signals. Since the outputs of the two valves are combined the overall amplification is free of distortion. An amplifier in which the valves are biased to cut off is known as a *Class B amplifier* and is often used for linear amplification provided push–pull operation is used.

A Class B amplifier has the following advantages over Class A:

(*a*) It has greater efficiency. In fact the efficiency of a Class B stage measured by (power output)/(power taken from h.t. supply) approaches 80%.

(*b*) It enables a given valve to deliver a greater output power than is possible from a Class A stage. The power output from a Class B stage is five times that available from one valve in Class A.

(*c*) The current drawn from the h.t. supply is, at any moment, only that necessary to provide the required output. The current is proportional to the output required and, in the absence on an input signal, is almost zero. This property is particularly valuable in battery-operated equipment where the cost of current is high.

The anode-to-anode optimum load of a push–pull stage is twice that for a single valve for Class A operation and four times that for a single valve for Class B operation. It is not possible to use automatic cathode bias in a Class B amplifier because the mean cathode current is not constant. The h.t. supply for a Class B stage needs good regulation because, to avoid distortion, the voltage must remain constant in spite of the wide variations in mean anode current.

BIPOLAR-TRANSISTOR AMPLIFYING CIRCUITS

In essence a bipolar transistor can amplify because its output current is controlled by its input current. This contrasts with thermionic-valve operation where the output current is controlled by the input voltage. The output current from a bipolar transistor is delivered by the collector and one other terminal: the input current is applied between the base and one other terminal. There are only three terminals and it follows that one must be common to input and output circuit: this is used as a convenient classification of the types of basic bipolar transistor amplifiers.

The input can be applied between base and emitter while the output is taken from collector and emitter: this is the *common-emitter circuit* probably the most used of all bipolar-transistor circuits.

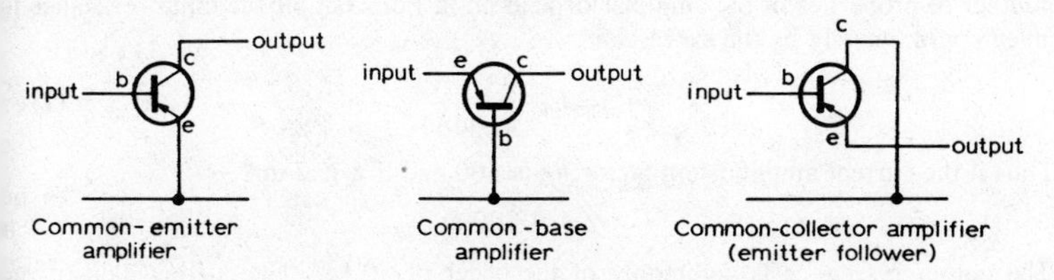

Figure 13.6 Basic forms of bipolar transistor amplifying circuits

In the *common-base circuit* the input is also applied between base and emitter but the output is taken from collector and base.

In the *common-collector circuit* the input is applied between base and collector the output being taken from collector and emitter. This circuit arrangement is more commonly known as the *emitter follower circuit.*

The basic forms of these three fundamental circuits are illustrated in *Figure 13.6.*

Common-emitter circuit

In practical versions of these circuits it is necessary to provide a source of base bias current and to include provision for stabilisation of the operating point. One way in which these requirements can be met is indicated in *Figure 13.7*. This shows a *pnp* transistor with a collector load resistor R_c, an emitter resistor R_e and base bias provided

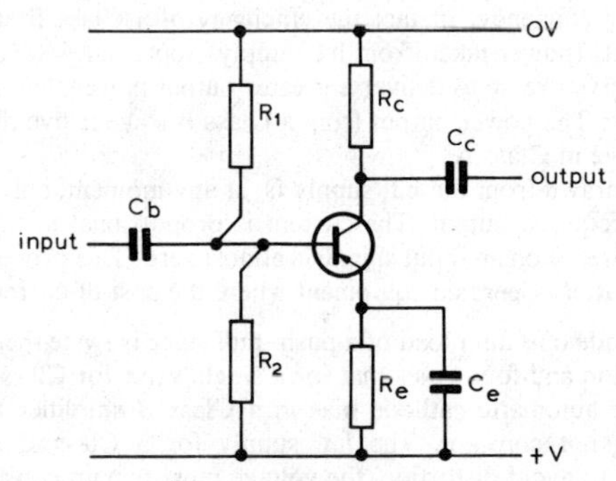

Figure 13.7 Common-emitter stage of amplification with stabilisation by potential divider and emitter resistor

by the potential divider R_1–R_2. R_e is decoupled by C_e to avoid negative feedback which would otherwise reduce the gain.

R_1 and R_2 are so chosen that there is, for example, a voltage of 5 V across R_2. If the transistor is a silicon type there is normally a voltage difference of approximately 0.7 V between base and emitter. The voltage across R_e is thus 4.3 V. The value of R_e can now be chosen to give the required mean emitter current. If this is 2 mA then R_e must be approximately 2.2 kΩ. Any change in collector current (and hence in emitter current) causes a corresponding alteration in the voltage across R_e and this produces a change in base-emitter voltage which opposes the initial collector-current change. This is the method by which this circuit achieves d.c. stability and makes the mean emitter current independent of transistor changes and temperature changes.

The mean emitter current is an important parameter of the amplifier because a number of properties of the amplifier depend on it. For example the input resistance is given approximately by the expression:

$$r_{in} = h_{fe} \cdot \frac{25\ (\text{mV})}{I_e\ (\text{mA})} \quad (1)$$

Thus if the current amplification factor h_{fe} is 100 and if I_e is 2 mA

$$r_{in} = 1\ 200\ \Omega$$

The output resistance is commonly of the order of 50 kΩ. The current gain of the amplifier shown in *Figure 13.7* approaches h_{fe} if the external load resistance connected to C_c) is small compared with R_c so that most of the output current flows through it. Spreads in h_{fe} can be very large and thus the current gain of the amplifier is not precisely known. Where it is important to have a precise value of gain it is customary to employ negative feedback. This is commonly applied over two stages and is designed also to achieve d.c. stability as explained later.

Bipolar transistors are essentially current-amplifying devices and the relationship between collector voltage and base voltage is non-linear. If, however, the base is fed from a signal source with an output resistance large compared with the input resistance of the transistor, then reasonable linearity can be obtained and the common-emitter circuit can be used as a voltage amplifier. For such a circuit the voltage gain is given by

$$\frac{V_{out}}{V_{in}} = h_{fe} \cdot \frac{R_c}{R_s} \qquad (2)$$

where R_s is the output resistance of the signal source.

Common-base circuit

Bias current and d.c. stability can be achieved by the same circuit arrangement used for the common-emitter amplifier. Thus, in *Figure 13.8*, R_1 and R_2 constitute the potential

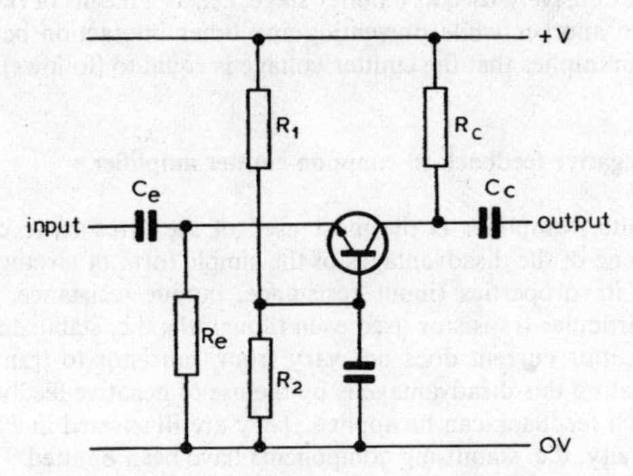

Figure 13.8 Common-base stage of amplification with stabilisation by potential divider and emitter resistor

divider, R_e is the emitter resistor and R_c the collector resistor. This circuit is shown with an *npn* transistor and a positive supply voltage is therefore required. The input resistance of such an amplifier is low, of the order of 50 Ω and the output resistance is high, commonly about 1 MΩ. The current gain is less than unity but the circuit can be used as a voltage amplifier with reasonable linearity provided the signal source has an output resistance large compared with the input resistance of the amplifier. The voltage gain is then given approximately by

$$\frac{V_{out}}{V_{in}} = \frac{R_c}{R_s} \qquad (3)$$

As an example R_s need be only 500 Ω to swamp any variations in r_{in}, and R_c can be 20 Ω giving a voltage gain of 40.

Common-base amplifiers are often used as v.h.f. and u.h.f. amplifiers: they are more stable at such frequencies than common-emitter amplifiers because of the very small capacitance linking input and output circuits (the emitter-collector capacitance).

Common-collector circuit (emitter follower)

Figure 13.9 gives the circuit diagram of a practical emitter follower using a *pnp* transistor. This circuit is characterised by low output resistance (of the order of 50 Ω), high input resistance (50 kΩ is typical), considerable current gain (nearly equal to h_{fe}) and nearly unity voltage gain.

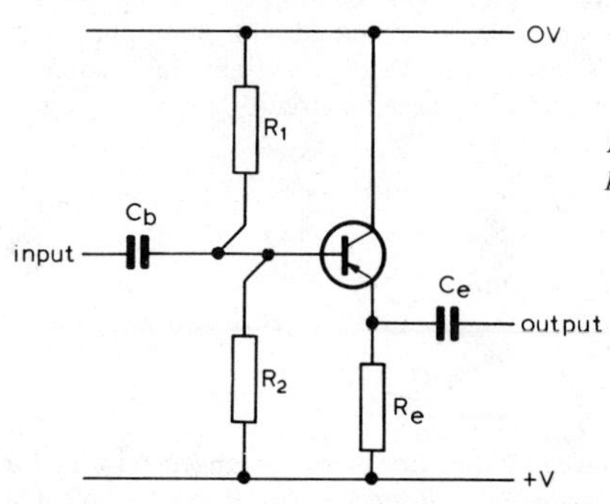

Figure 13.9 Common-collector stage of amplification (emitter follower) with stabilisation by potential divider and emitter resistor

The circuit is extensively used as a buffer stage, i.e., as a means of transferring signals from one stage to another while preventing any other interaction between the stages. Unity voltage gain implies that the emitter voltage is equal to (follows) the base voltage.

Application of negative feedback to common-emitter amplifier

The common-emitter amplifier is the most used of the three basic circuits shown in *Figure 13.6* but one of the disadvantages of the simple form of circuit shown in *Figure 13.7* is that all its properties (input resistance, output resistance, gain, distortion) depend on the particular transistor used even though the d.c. stabilising circuit ensures that the mean emitter current does not vary from transistor to transistor. The usual method of eliminating this disadvantage is by the use of negative feedback and there are two ways in which feedback can be applied. They are illustrated in *Figure 13.10* from which, for simplicity, d.c. stabilising components have been omitted.

In *Figure 13.10* (*a*) the feedback resistor R_b is connected between collector and base. It is thus in parallel with the load resistor R_c and returns a feedback current to the base

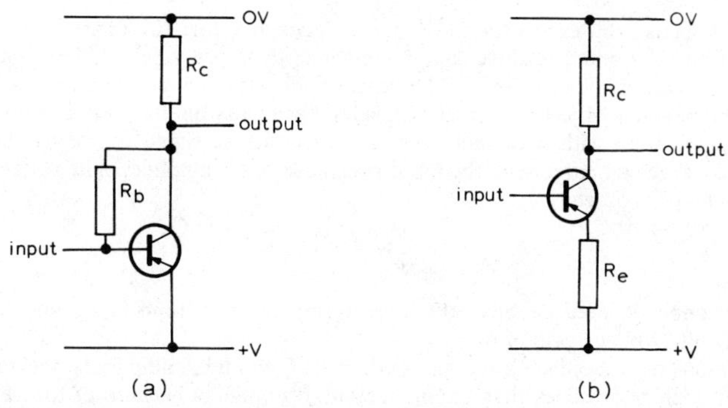

Figure 13.10 Two basic methods of applying negative feedback to a common-emitter amplifier

proportional to the output voltage. The effect of such feedback is, of course, to reduce gain and distortion: it also reduces the input resistance of the amplifier, already low, to an even smaller value and it reduces the output resistance also. The performance of an amplifier with a low input resistance is best assessed in terms of an input current because the amplifier can be connected into signal source circuits without affecting the current in them. The low output resistance implies that the output voltage of the amplifier is independent of the resistance of the external load connected to the amplifier. Thus the characteristics of amplifier (*a*) are best expressed in terms of input current i_{in} and output voltage v_{out}: the ratio of the two is in fact given by R_b and is thus independent of the characteristics of the transistor and of the value of R_c. Thus we have for the circuit of *Figure 13.10* (*a*)

$$\frac{v_{out}}{i_{in}} = R_b \tag{4}$$

In *Figure 13.10* (*b*) the feedback resistor R_e is connected in series with the load resistor R_c and returns a feedback voltage to the input circuit proportional to the output current. Such feedback reduces gain and distortion; it also increases both the input and the output resistance of the stage. The performance of an amplifier with a high input resistance is best expressed in terms of an input voltage because the amplifier can be connected across signal sources without affecting their voltage. The high output resistance implies that the output current of the amplifier is independent of the resistance of the external load. Thus the characteristics of amplifier (*b*) are best expressed in terms of input voltage and output current: the ratio of v_{in} to i_{out} is equal to R_e. Thus we have for the circuit of *Figure 13.10* (*b*):

$$\frac{i_{out}}{v_{in}} = \frac{1}{R_e} \tag{5}$$

Two-stage amplifier

By combining stages of the types shown in *Figure 13.10* it is possible to produce two-stage amplifiers of different characteristics. For example if a stage of type (*b*) is directly connected to the output of a stage of type (*a*), the input and output resistance at the inter-transistor junction are such that the output voltage of stage (*a*) is transferred without significant loss to the input of stage (*b*). The two-stage amplifier so derived has a low input resistance and a high output resistance: these are the characteristics of a current amplifier and the current gain can be derived from equations 4 and 5 by equating v_{in} with v_{out}. This gives

$$\frac{i_{out}}{i_{in}} = \frac{R_b}{R_e} \tag{6}$$

In practical examples of this current amplifier R_b is normally fed from the emitter of TR_2 instead of the collector of TR_1: this is a permissible alteration because the signal-frequency voltages are approximately equal at the two points. In fact TR_2 behaves as an emitter follower to provide the feedback voltage. The basic form of the two-stage current amplifier is shown in *Figure 13.11*. This particular circuit arrangement often results in TR_1 base voltage being too close to that of the collector. However the difficulty is easily overcome by introducing a further emitter resistor $R_{e2'}$ above R_{e2} (*Figure 13.12*) so that R_{b1} is returned to a potential divider in TR_2 emitter circuit.

One of the great advantages of transferring R_{b1} to TR_2 emitter is that it makes possible a simple but highly-effective method of ensuring stabilisation of the d.c. operating conditions in both transistors. The potential-divider method of stabilisation is given in *Figure 13.7*: this shows R_1 connected to the supply negative terminal but it could alternatively have been connected to the collector of the transistor and this would have resulted in improved d.c. stability because movements in collector voltage are in

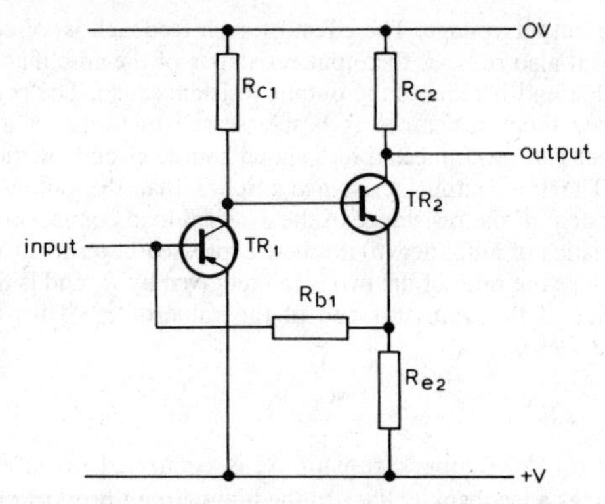

Figure 13.11 Basic form of two-stage current amplifier

antiphase with those at the base. (Such a connection would, of course, also result in signal-frequency feedback but this could be eliminated by decoupling the centre point of R_1 to the supply positive terminal.) The emitter follower action of TR_1 means that the potential divider $R_{e1}R_{e2'}$ is effectively connected to TR_2 collector and this is precisely what is required if the potential divider is to provide d.c. stability in addition to signal-frequency feedback. One component which is missing from *Figure 13.11* and which is essential in a d.c. stabilising circuit is an emitter resistor for TR_1. If we include this the circuit takes the form shown in *Figure 13.12* in which R_{e1} and $R_{e2'}$ are both decoupled to eliminate signal-frequency feedback.

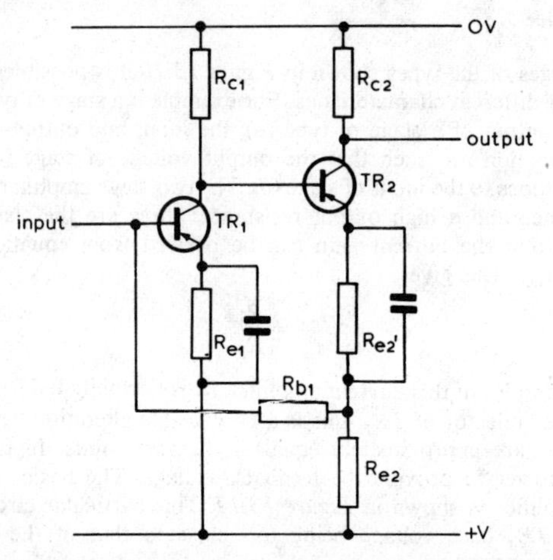

Figure 13.12 Practical circuit for a two-stage current amplifier incorporating means of ensuring d.c. stability

An alternative form of two-stage amplifier can be formed by connecting a stage of the type shown in *Figure 13.10 (a)* directly to the output of a stage of the type shown in *Figure 13.10 (b)*. The input and output resistances at the inter-transistor junction are such that the current at the output of stage (*b*) is transferred without loss to the input of stage (*a*). The amplifier so derived has the basic form shown in *Figure 13.13*: it has a high input resistance and a low output resistance. These are the properties of a voltage

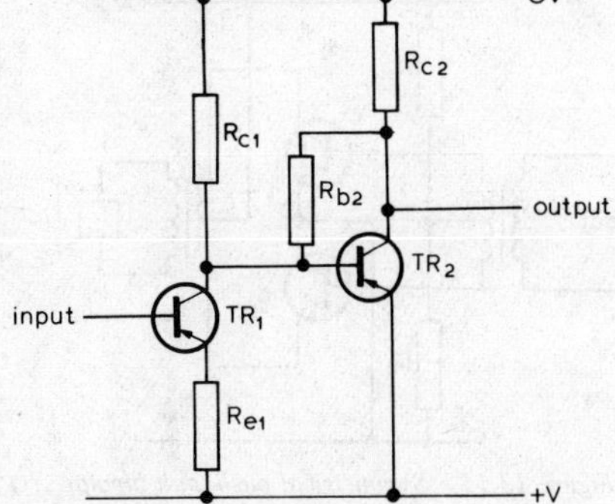

Figure 13.13 Basic form of two-stage voltage amplifier

amplifier and the voltage gain can be deduced easily from equations 4 and 5 by equating i_{in} and i_{out}. This gives

$$\frac{v_{out}}{v_{in}} = \frac{R_{b2}}{R_{e1}} \qquad (7)$$

R_{b2} has the function of returning a current to TR_2 base proportional to TR_2 collector voltage and it is more normal in a two-stage amplifier to connect R_{b2} to TR_1 emitter as shown in *Figure 13.14*. Any current put into the emitter circuit will emerge with little loss at the collector and will be transferred to TR_2 base provided R_{c1} is large compared with TR_2 input resistance.

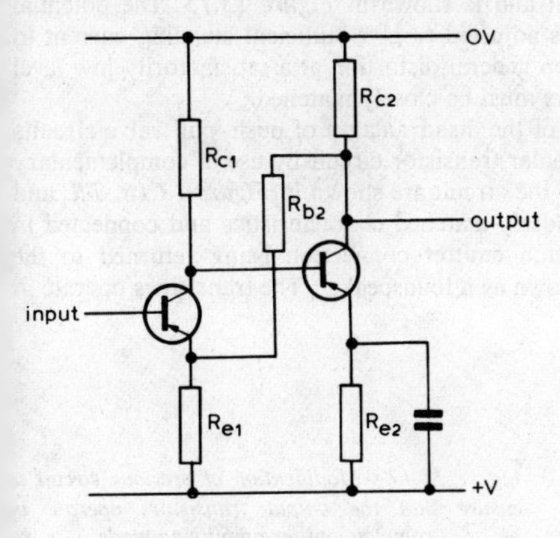

Figure 13.14 Practical circuit for a two-stage voltage amplifier incorporating means of ensuring d.c. stability

Finally, we need to ensure that the signal-frequency feedback circuit will also provide d.c. stability. We already have a potential divider $R_{b2}R_{e1}$ connected to TR_2 collector and its junction point is connected to TR_2 base via TR_1. Thus all that is necessary to ensure d.c. stability is to provide TR_2 with an emitter resistor which should, of course, be decoupled to eliminate signal-frequency feedback as shown in *Figure 13.14*.

Bipolar-transistor output stages

A common-emitter stage of the type illustrated in *Figure 13.7* can, by choice of a suitable transistor, be used to deliver several watts of power, e.g., to operate a loudspeaker. If the transistor operates in Class A there is considerable dissipation in the

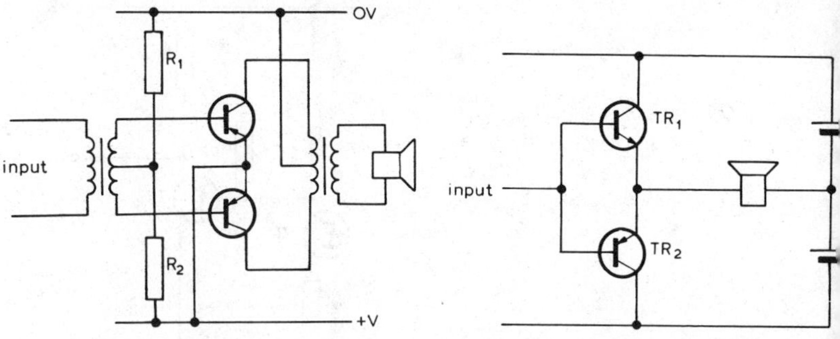

Figure 13.15 Symmetrical push–pull bipolar transistor output amplifier

Figure 13.16 Essential features of a complementary single-ended output stage

transistor itself. The resulting temperature rise must be limited otherwise the transistor can be damaged, even destroyed. The usual precaution is to mount the transistor in intimate thermal contact with a heat sink which is generally a metal structure with cooling fins. Suitable output transistors may have a mean collector current of 1 A and the optimum load is low, commonly only a few ohms, so that the speech coil can be used as a direct-coupled load and there is no need for a matching transformer.

In general, however, transistors required to deliver appreciable undistorted power are operated in push–pull. This reduces distortion by the cancellation of even harmonics and also permits the use of Class B operation which economises on power supplies. A push–pull transistor circuit can be constructed on principles similar to those of *Figure 13.4*: this is known as a symmetrical circuit and is shown in *Figure 13.15*. The potential divider R_1R_2 provides base bias and is adjusted to give sufficient standing current to eliminate cross-over distortion. To keep general distortion at a satisfactorily low level the characteristics of the two transistors must be closely matched.

The need for a phase splitter is one of the disadvantages of push–pull valve circuits but the need can be eliminated in a bipolar-transistor circuit by use of complementary transistors and the essential features of the circuit are shown in *Figure 13.16*. TR_1 and TR_2 are *npn* and *pnp* transistors of closely-matched characteristics and connected in series across the supply, their common emitter connection being returned to the mid-point of the supply via the load shown as a loudspeaker. The transistors operate in

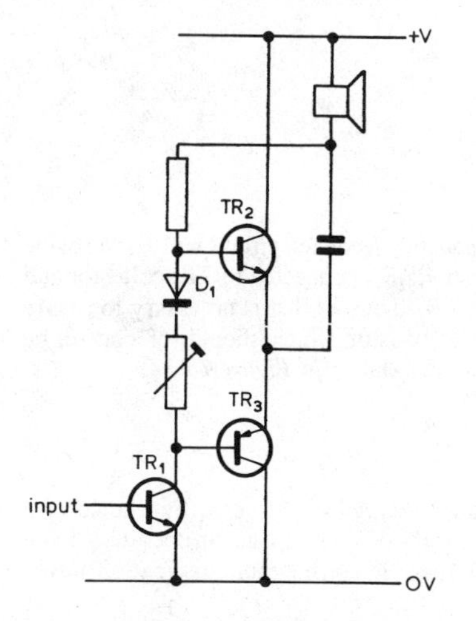

Figure 13.17 Modification of previous circuit to ensure that the output transistors operate in common-emitter amplifying mode

Class B. Positive-going signals drive TR_1 into conduction but cut TR_2 off so that the voltage of the common-emitter connection approaches supply positive value. Negative-going signals drive TR_2 into conduction but cut TR_1 off so that the voltage at the common-emitter connection approaches supply negative value. Thus the two transistors operate in push–pull without the need for a phase splitter or transformer. A difficulty of the basic circuit of *Figure 13.16* is that the loudspeaker is in the emitter circuit of the transistors which therefore operate as emitter followers and need a very large input signal to drive the loudspeaker. This disadvantage is avoided in the circuit of *Figure 13.17* which also includes means for biasing the transistors.

In this diagram the loudspeaker is fed from the common-emitter connection via a capacitor. This enables the other terminal of the loudspeaker to be connected to the positive supply line so that the junction of loudspeaker and capacitor can be used as the supply source for the driver transistor TR_1. This arrangement ensures that the signal voltage generated across TR_1 collector load resistor is applied between base and emitter of both output transistors. Expressed differently, the top end of TR_1 collector load resistor is decoupled to the common-emitter connection. Thus TR_2 and TR_3 are used as common-emitter stages and so have greater gain than the emitter follower circuit of *Figure 13.16*. The forward-biased diode $D1$ is included to compensate for the changes in base-emitter voltage of the output transistors which occur when temperature alters. The diode thus ensures greater constancy in the standing current for TR_2 and TR_3. The preset resistor enables this current to be set to the required value. This is usually known as the asymmetrical or single-ended push–pull circuit.

FIELD-EFFECT TRANSISTOR AMPLIFYING CIRCUITS

As indicated in Section 7 the properties of field-effect transistors differ considerably from those of bipolar transistors. Essentially f.e.t.s operate by virtue of an ohmic path between source and drain terminals, the conductivity of which is controlled by the voltage applied to the gate terminal. The source-drain voltage, if small, can be reversed without significant effect on the magnitude of the drain current, i.e., the internal resistance of the f.e.t. is linear which means that the f.e.t. can be used as a controllable resistance in such applications as a.g.c. circuits, remotely-operated faders, etc. Another important feature of the f.e.t. is its very high input resistance: this is of the order of 10^{11} ohms for junction-gate f.e.t.s and 10^{15} ohms for insulated-gate f.e.t.s. This makes the f.e.t. eminently suitable for use in the input stages of head amplifiers for capacitor or other types of high-impedance microphone. In general junction-gate f.e.t.s and insulation-gate f.e.t.s can be used in the same circuits but the following differences between them could affect their operation in particular circuits:

(*a*) Insulation-gate f.e.t.s can be designed to take zero drain current for zero gate voltage. A forward gate bias voltage (positive or negative depending on the nature of the conducting channel) is then necessary to give a working value of drain current. This is known as the *enhancement mode of operation* and in the graphical symbol for such transistors the channel is represented by a dashed line.

(*b*) Junction-gate f.e.t.s cannot operate in the enhancement mode. They take significant values of drain current at zero gate voltage and a reverse gate bias voltage (positive or negative depending on the nature of the conducting channel) is needed to cut off the drain current. This is known as the *depletion mode of operation* and in the graphical symbol for this type of transistor the conducting channel is represented by a solid line.

Insulation-gate f.e.t.s can also be designed to operate in the depletion mode but they will still operate in the enhancement mode also: in other words in an insulation-gate f.e.t. which takes appreciable drain current at zero gate voltage the drain current can be decreased by reverse biasing the gate and can be increased by forward biasing the gate.

Bias circuits and d.c. stability

In the depletion mode of operation the gate voltage lies outside the range of the source and drain voltages and it is therefore possible to obtain gate bias from the voltage developed across a resistor in the source circuit as shown in *Figure 13.18*. If, for example, the mean drain current is required to be 2 mA and if a reverse bias of −1 V is required to reduce the drain current to this value, then R_s should be 500 Ω. Such a resistor if not decoupled will reduce gain by negative feedback and, if this is undesirable, a decoupling capacitor must be included across R_s. It should have a reactance small compared with the internal source resistance of the f.e.t. at the lowest frequency of operation. This, of course, is similar to the automatic cathode bias circuit used with thermionic valves, one of many similarities between the properties of the f.e.t. and the valve.

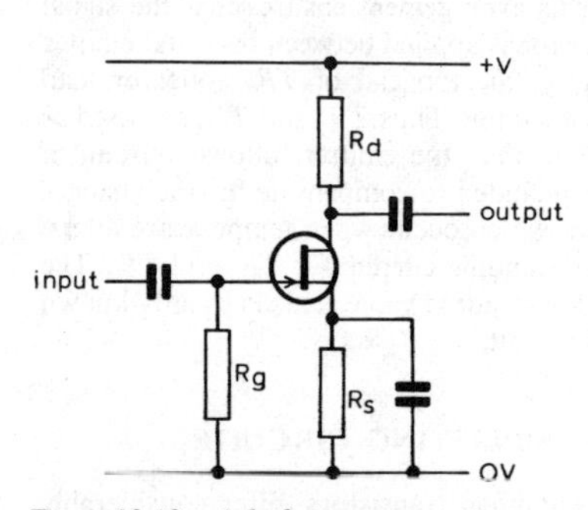

Figure 13.18 A depletion-type junction-gate f.e.t. with bias and d.c. stabilisation provided by a source resistor

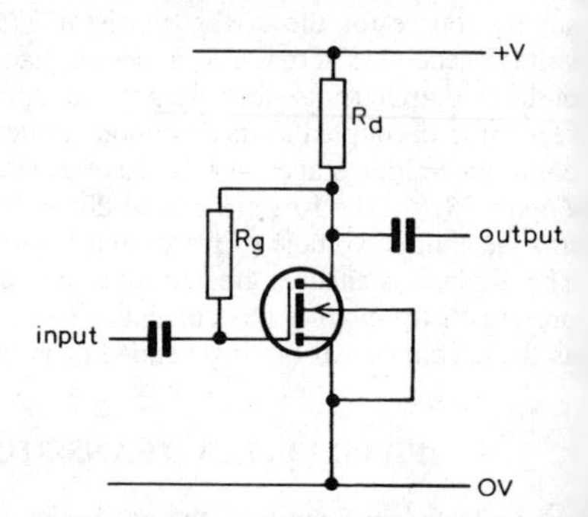

Figure 13.19 An enhancement-type insulation-gate f.e.t. with bias and d.c. stabilisation provided by a resistor in the drain circuit

This circuit provides a degree of d.c. stabilisation of the operating point because any changes in source current, whether produced by a change in temperature or by exchanging one transistor for another, produces a change in gate-source voltage which counteracts the effects of the current change.

In the enhancement mode of operation the gate voltage lies between the source and drain voltages (as in bipolar transistors) and bias can therefore be obtained from the drain circuit as shown in *Figure 13.19*. This circuit provides bias by direct-coupled negative feedback; it also gives signal-frequency feedback which, if not required, can be eliminated by decoupling the mid-point of R_g to source. R_g also gives the circuit a low input resistance (equal approximately to R_g/A where A is the signal-frequency gain of the amplifier). The d.c. stability provided by this circuit can be better than that given by the circuit of *Figure 13.18* because the external drain resistor can usually be several times the permissible value of the external source resistor.

A third bias circuit which also provides d.c. stabilisation is the potential-divider method commonly used with bipolar transistors and illustrated in *Figure 13.7*. It is suitable for enhancement and depletion-mode operation and is shown in *Figure 13.20* with a depletion type insulation-gate f.e.t.

F.E.T. applications

All three circuits (*Figures 13.18 to 13.20*) show an f.e.t. used as a simple RC-couple amplifying stage. Such stages have a high input resistance and low output resistance an can thus be used as voltage amplifiers. The voltage gain of such a stage is given by g_mR and typical values for g_m and R_d are 2 mA/V and 20 kΩ, giving a gain of 40. At audi frequencies the noise introduced by an f.e.t. is less than that of a bipolar transistor and the junction-gate f.e.t. has a lower noise figure than the insulated-gate type. Thus junction-gate f.e.t.s tend to be used in the early stages of a.f. amplifiers where it is difficult with bipolar transistors to achieve a good signal-to-noise ratio.

Because of its extremely-high input resistance the f.e.t. can also perform another useful function, that of achieving a resistance match (and thus good power transfer)

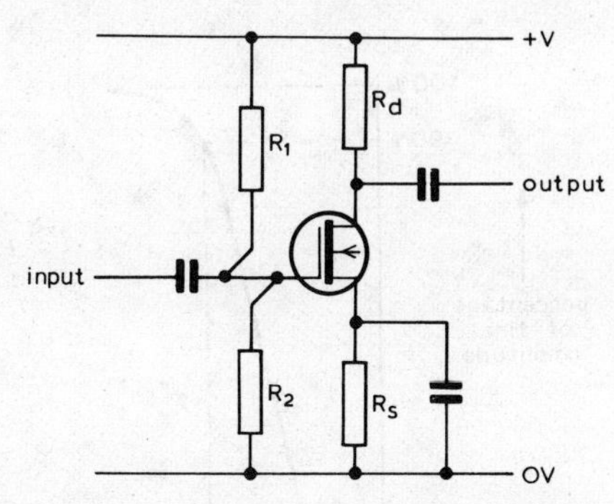

Figure 13.20 A depletion-type igfet with bias and d.c. stabilisation provided by a potential-divider circuit

between the high output resistance of certain signal sources such as capacitor microphones and ceramic gramophone pickups and the much lower input resistance of bipolar transistor amplifiers. In such applications junction-gate f.e.t.s are commonly used, connected as source followers, between the signal source and the input stage of the a.f. amplifier which is likely to be a bipolar transistor connected as an emitter follower. There is, of course, no voltage in such a circuit arrangement but there is good power transfer despite the resistance transformation from several megohms to a few hundred ohms.

WIDEBAND AMPLIFIERS

Amplifiers used, for example, in television receivers, oscilloscope *Y*-deflection circuits and radar equipment must be capable of handling pulse-type signals without distortion. The most difficult parts of such signals to amplify are vertical edges and horizontal sections: thus such amplifiers are commonly tested with rectangular pulses which are composed of such parts.

To amplify steep edges without distortion a good high-frequency response is required and a useful approximation is that the upper limit of the passband f_{max} is given by

$$f_{max} = \frac{1}{2t} \tag{8}$$

where t is the rise time of the edge, defined as indicated in *Figure 13.21*. Thus for a rise time of 0.1 μs, such as that of the edges of the line sync pulses in a 625-line television system, we have

$$f_{max} = \frac{1}{2 \times 0.1 \times 10^{-6}} \text{ Hz} = 5 \text{ MHz}$$

At the other end of the passband the low-frequency extreme f_{min} is governed by the need to reproduce horizontal parts of waveforms accurately and a useful relationship is

$$f_{min} = \frac{1}{2\pi} \times \text{percentage sag in half-cycle of 50 Hz square wave} \tag{9}$$

Thus to keep the sag to less than 2% the low-frequency limit is given by

$$f_{min} = \frac{2}{2\pi} \approx 0.3 \text{ Hz}$$

Frequently the necessary low-frequency response is achieved by using direct coupling throughout the amplifier thus extending the response down to zero frequency. It is difficult, however, to control drift in high-gain direct-coupled amplifiers. If the signal to

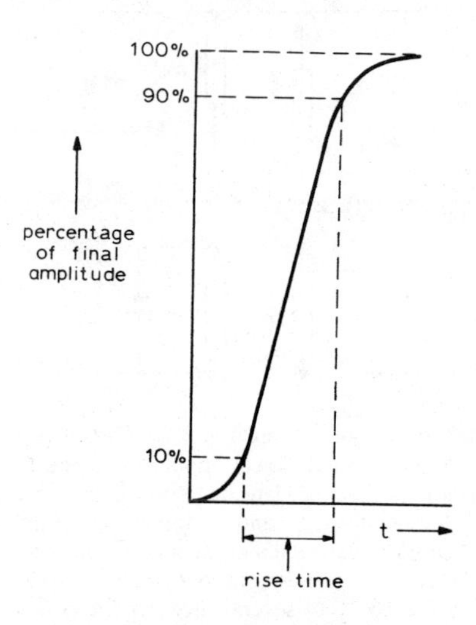

Figure 13.21 The rise time of an edge is the time taken for the signal to rise from 10% to 90% of the final amplitude

be amplified has a regular feature, such as the line-sync pulses in a video signal, it is possible, in effect, to use an amplifier with a limited low-frequency response and to restore the zero-frequency component, whenever it is required, by the techniques of d.c. restoration or d.c. clamping.

The numerical examples above show that the response of a wideband amplifier can extend from less than 1 Hz to 5 MHz, a range of more than 12 octaves and considerably wider than that required in a.f. amplifiers.

Valve amplifiers

The high-frequency response of a simple *RC*-coupled amplifier such as that illustrated in *Figure 13.1* falls off as a result of the shunting effect of input and output capacitances. One method of offsetting this effect and thus extending the high-frequency response is by the inclusion of an inductor either in shunt or in series with the valve anode circuit or by means of negative feedback. The low-frequency response can be extended by direct coupling, by exploiting the anode decoupling network or by negative feedback.

Figure 13.22 (*a*) shows the anode circuit of a stage of a wideband amplifier using shunt inductance to extend the high-frequency response. The inductance L_a must be chosen with care: if it is too low it will not be effective in extending the frequency response and if too large it will cause overshoots in the reproduced pulses. A suitable value is given by

$$L_a = 0.5\, R_a{}^2 C_t \qquad (10)$$

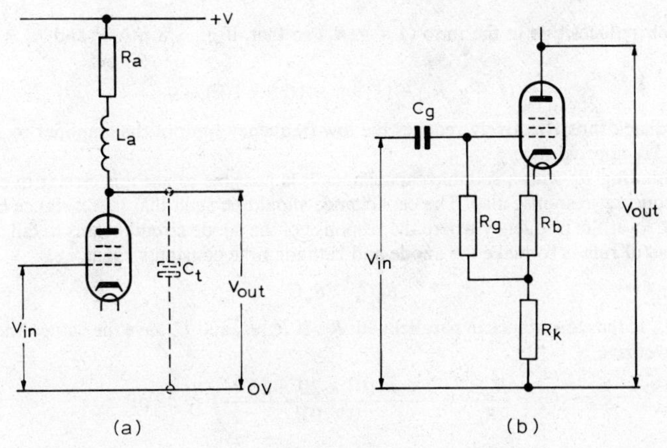

Figure 13.22 Essential features of (a) the shunt-inductance method of extending high-frequency response, and (b) one method of extending the low-frequency response of an amplifier

where R_a is the anode load resistance and C_t is the total capacitance in parallel with it (including any contribution from the next stage). The value of R_a to use can be calculated from the expression

$$R_a = \frac{1}{\omega_{max} C_t} \tag{11}$$

where ω_{max} is the maximum angular frequency required. Thus if the upper frequency limit is 5 MHz and the total shunt capacitance is 20 pF we have

$$R_a = \frac{1}{6.284 \times 5 \times 10^6 \times 20 \times 10^{-12}} \, \Omega \approx 1.6 \text{ k}\Omega$$

From this the inductance L_a is given by

$$L_a = 0.5 \times 1.6^2 \times 10^6 \times 20 \times 10^{-12} \text{ H} = 25.6 \, \mu\text{H}$$

A typical value for the mutual conductance is 8 mA/V and the voltage gain of the amplifier is given by

$$\frac{V_{out}}{V_{in}} = g_m R_a = 8 \times 10^{-3} \times 1.6 \times 10^3 \approx 13$$

Figure 13.22(*b*) gives the circuit diagram of a stage in which the low-frequency response is extended by negative feedback due to R_k which is not decoupled. This has the effect of extending the response by the factor $(1 + g_m R_k)$. Without such feedback the response would be 3 dB down at the frequency for which the reactance of C_g equals R_g. Values of C_g are unlikely in practice to exceed 0.1 μF and of R_g are unlikely to exceed 1 MΩ. For these limiting values the low-frequency limit is given by

$$\omega_{min} = \frac{1}{R_g C_g}$$

from which

$$f_{min} = \frac{1}{2\pi R_g C_g} = \frac{1}{6.284 \times 10^6 \times 0.1 \times 10^{-6}} \text{ Hz} \approx 1.6 \text{ Hz}$$

Feedback reduces this in the ratio $(1 + g_m R_k)$ so that, if $g_m = 8$ mA/V and $R_k = 1$ kΩ we have

$$(1 + g_m R_k) = (1 + 8 \times 10^{-3} \times 10^3) = 9$$

The feedback thus effectively reduces the low-frequency limit of the amplifier to 1.6/9, i.e., 0.2 Hz approximately.

By shunting R_k with a suitable capacitance it is possible to use this circuit to extend high-frequency response also. The capacitance should be such that its reactance begins to shunt R_k at the frequency where the response of the anode circuit begins to fall off: in fact a useful rule is to make the anode and cathode time constants equal, i.e.,

$$R_k C_k = R_a C_t \tag{12}$$

where C_k is the capacitance in parallel with R_k. If R_a, R_k and C_t have the values assumed earlier we have

$$C_k = \frac{R_a C_t}{R_k} = \frac{1.6 \times 10^3 \times 20 \times 10^{-12}}{1 \times 10^3} \text{ F} = 32 \text{ pF}$$

Bipolar transistor amplifier

Figure 13.23 gives a simplified form of a wideband amplifier circuit using bipolar transistors. TR_1 is an emitter-follower stage which gives the amplifier a high input resistance and TR_4 is an emitter-follower output stage which gives the amplifier a low output resistance.

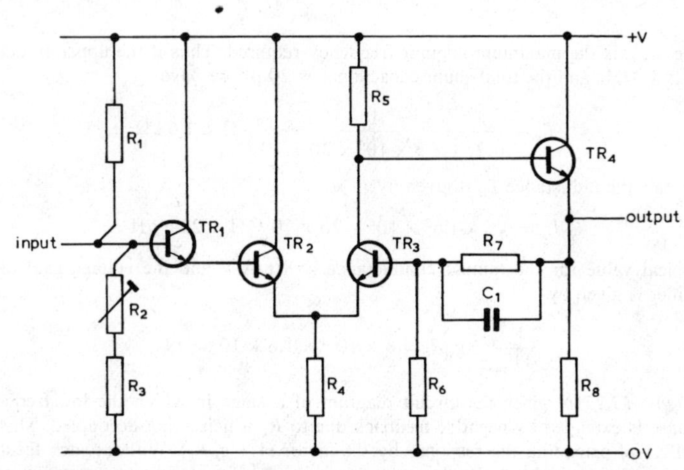

Figure 13.23 Typical circuit using bipolar transistors in a wideband amplifier

These emitter followers are direct-coupled to transistors TR_2, TR_3 which provide the voltage gain. TR_2 and TR_3 have direct emitter coupling and constitute a much-used circuit arrangement known as a *long-tailed pair.* TR_2 behaves as an emitter follower and drives TR_3 as a common-base stage. One valuable feature of the long-tailed pair is that it is possible to use a high-value resistor in the common emitter circuit to give good d.c stability without at the same time losing considerable gain as a result of signal-frequency feedback in this resistor. The resistor is shunted by the internal emitter resistance of TR_3 and feedback due to this causes a loss in signal-frequency gain of about 6 dB. In some versions of the long-tailed pair circuit the emitter resistor is replaced by the collector-emitter path of a third transistor which is stabilised by the potential divider and emitter resistor technique. All three transistors and the associated resistors are combined in one integrated circuit which ensures that all components, active and passive, share the same temperature and environment, an essential feature for good d.c. stability.

Another useful feature of the long-tailed pair is that the base of the second transistor provides a convenient point at which to inject a negative feedback signal. In *Figure 13.23* the feedback is made frequency-dependent by inclusion of the network R_6–R_7–C_1 which is designed to make the frequency response of the amplifier fall off above the required passband. The voltage gain of TR_2–TR_3 (and hence of the amplifier) is given by $g_m R_5$ where g_m is the mutual conductance of each transistor and R_5 is the value of TR_3 collector load resistance. Typical values of g_m and R_5 are 40 mA/V and 2.5 kΩ giving a gain of 100. This is reduced to the required value by the feedback network R_6–R_7. For example if a gain of 20 is required then R_7 is made equal to 20 × R_6 and if the passband limit is required to be 10 MHz then C_1 is chosen to have a reactance equal to R_7 at this frequency.

The resistor R_2 is used to set the no-signal standing output voltage to the required value.

Wideband amplifier using field-effect transistors

Because of their low noise, field-effect transistors are most likely to be used in the input stage of a wideband amplifier such as a television camera head amplifier. The high input resistance makes the f.e.t. particularly suitable for such an application because camera tubes have a high, predominantly capacitive, output impedance which needs to be loaded by a high resistance.

A single f.e.t. connected as a common-source amplifier, is not ideal because feedback via the internal drain-gate capacitance causes a fall-off in response and in signal-to-noise ratio at high frequencies. The usual solution to this problem is to feed the output of a common source amplifier into a second f.e.t. connected as a common-gate amplifier. Direct coupling is commonly used and the two transistors can be connected across the

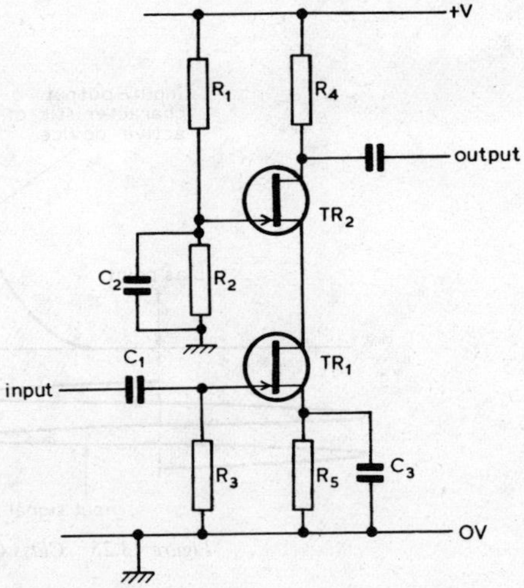

Figure 13.24 A cascode input stage for a wideband amplifier using two f.e.t.s

supply as shown in *Figure 13.24* in a circuit arrangement known as a cascode amplifier. The voltage gain of the cascode from the gate of TR_1 to the drain of TR_2 is that of TR_1 feeding the load resistor R_4 but feedback via the drain-gate capacitance is eliminated because the voltage gain of TR_1 is limited to unity, its effective load resistance being the internal source resistance of TR_2. Feedback cannot occur in TR_2 via the drain-gate capacitance because the gate is decoupled at signal frequencies by C_2. The cascode arrangement is a well-known one which has been used with thermionic valves and with bipolar transistors.

TUNED AMPLIFIERS

So far in this section we have considered circuits for amplifiers with a response which begins at zero or a very low frequency and extends for several octaves. A characteristic of such amplifiers is that they normally use *RC* coupling between the active devices.

We shall now deal with amplifiers with a passband which is small compared with the centre frequency. Typical of such amplifiers are the r.f. and i.f. amplifiers of sound and television receivers and the r.f. amplifiers used in transmitters. In such amplifiers *LC* circuits are used for inter-valve or inter-transistor coupling, the resonance frequency and the *Q* values of the inductors being chosen to give the required response. One of the basic difficulties in the design of such amplifiers is that of avoiding instability due to the positive feedback via the internal input–output capacitance of the active devices.

Amplifiers for constant-amplitude signals

The signal to be amplified may be a constant-amplitude sine wave: this applies, for example, to the carrier-wave stages in an a.m. transmitter and to many of the stages in an f.m. transmitter. In amplifying stages for such an application a linear input–output characteristic is not necessary in the amplifier because any distortion of the signal is corrected in the *LC* output circuit which offers little response to harmonics of the input signal. It is unusual, therefore, to use Class A or Class B operation in such amplifiers.

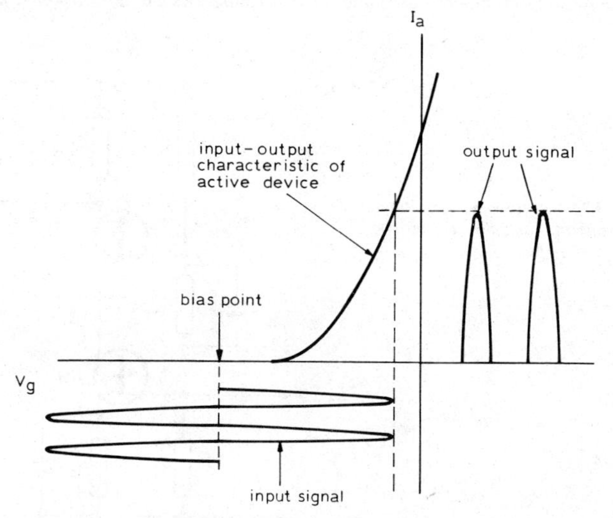

Figure 13.25 Class C operation

Higher efficiency is obtainable from Class C operation in which the active device is biased beyond the point of output-current cut off and is driven into a brief period of conduction once during each cycle of input signal. As shown in *Figure 13.25* the output signal from a Class C amplifier thus consists of a series of regular pulses which keep the output *LC* circuit in a state of steady oscillation by giving it a stimulus once during every cycle of its natural frequency.

The obvious form for a tuned amplifier is that of an active device with the input and output *LC* circuits tuned to the same frequency. The anode-grid capacitance of a triode valve is, however, high enough to cause instability even at frequencies as low as hundreds of kHz and when they are used in transmitters a neutralising circuit is necessary to offset the positive feedback by an equal amount of negative feedback. The basic form of a typical circuit is illustrated in *Figure 13.26*: the inductor of the output

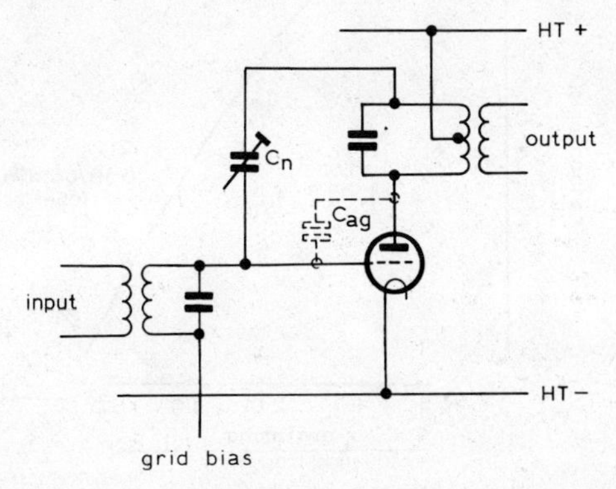

Figure 13.26 Basic form of neutralised triode r.f. amplifier used in transmitters

tuned circuit is centre-tapped and the mid-point is effectively earthed so that the signal at one end of the inductor is in antiphase with that at the anode. Thus the signal fed back to the grid via the neutralising capacitor C_n is in antiphase with that fed back via the anode-grid capacitance shown dotted. In practice C_n is made variable and is adjusted empirically to secure stable operation.

Frequency multipliers

The output circuit of a tuned amplifier can be kept in oscillation by regular stimuli applied at 1/2, 1/3 or in general $1/n$ of its resonance frequency. In other words the output circuit can be tuned to a harmonic of the input frequency. This is the basis of one form of frequency multiplier: the danger of instability in such an arrangement is not so serious because the input and output circuits are not tuned to the same frequency.

Tuned amplifiers for a.m. signals

In amplifying amplitude-modulated signals, changes in carrier amplitude which represent the modulating signal must be faithfully reproduced. The amplifiers used must therefore have a linear input–output characteristic: Class A amplifiers are used in the r.f. and i.f. amplifiers of receivers. In high-power a.m. transmitters however where economy of operation is important Class B amplifiers are used, often in push–pull to minimise radiation of even harmonics of the carrier frequency.

Use of single tuned circuits

One form of tuned amplifier which can be used to amplify a.m. signals consists of a succession of LC circuits all tuned to the same centre frequency and separated by active devices. The i.f. amplifiers of sound and television receivers are often examples of such amplifiers. The centre frequency of each tuned circuit is determined by the product LC: the rate of fall-off of the response on each side of the centre frequency is determined by

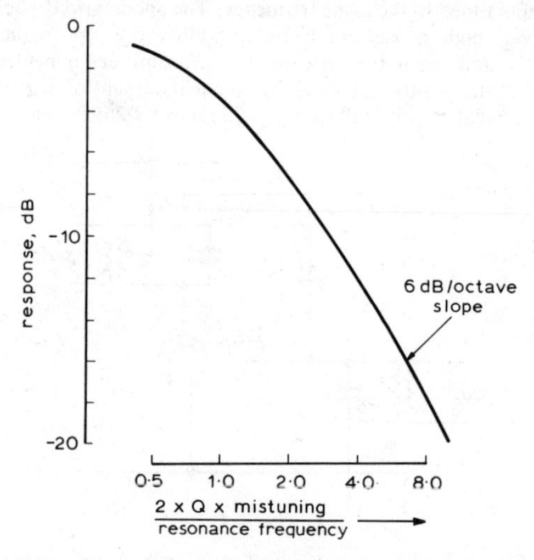

Figure 13.27 Universal curve illustrating the frequency response of a single tuned circuit

the effective Q value of the inductor. The relationship between response and mistuning Δf is illustrated in *Figure 13.17*. This shows that the loss is 3 dB when $2Q\Delta f/f_c = 1$. This gives

$$2\Delta f = \frac{f_c}{Q}$$

$2\Delta f$ is the frequency difference between the two 3-dB down points and is hence the passband. Thus we have

$$Q = \frac{\text{centre frequency}}{\text{passband}} \qquad (13)$$

This is a useful relationship which enables Q values to be assessed for each application. For example the standard centre frequency for the i.f. amplifier of a 625-line television receiver is 36.5 MHz and the bandwidth must be 6 MHz to include vision and sound signals. The required Q value in an LC circuit for such an amplifier is thus given by

$$Q = \frac{\text{centre frequency}}{\text{passband}} = \frac{36.5}{6} \approx 6.$$

It would be difficult to construct an inductor to have precisely this value of Q and the usual solution to this difficulty is to use coils of higher Q value (say 100) and to damp them by parallel resistance to give the required Q value. The input resistance of valves and transistors is usually sufficiently low at this value of centre frequency that it can conveniently be used to give the required damping. The precise damping effect can be controlled by adjusting the L to C ratio of the tuned circuit, keeping the LC product constant to maintain the value of the centre frequency.

As an example a typical value for the input resistance of a bipolar transistor is 2 kΩ. If the transistor base and emitter terminals are effectively in parallel with the *LC* circuit then the input resistance dictates the dynamic resistance R_d of the circuit. R_d is equal to $L\omega Q$ and thus we can calculate the required value of inductance from the relationship

$$L = \frac{R_d}{\omega Q} \tag{14}$$

Substituting for R_d, ω and Q we have that the inductance for the vision i.f. stage is given by

$$L = \frac{2 \times 10^3}{6.284 \times 36.5 \times 10^6 \times 6}\,\text{H} = 1.4\,\mu\text{H}$$

C can be calculated from the relationship

$$\omega = \frac{1}{\sqrt{(LC)}}$$

Substituting for ω and L we have

$$C \approx 14\,\text{pF}$$

In practice it is sometimes more convenient to use values of *L* and *C* which give a higher dynamic resistance and to reduce this to the required value by suitable choice of tapping point on the inductor for the base connection of the transistor.

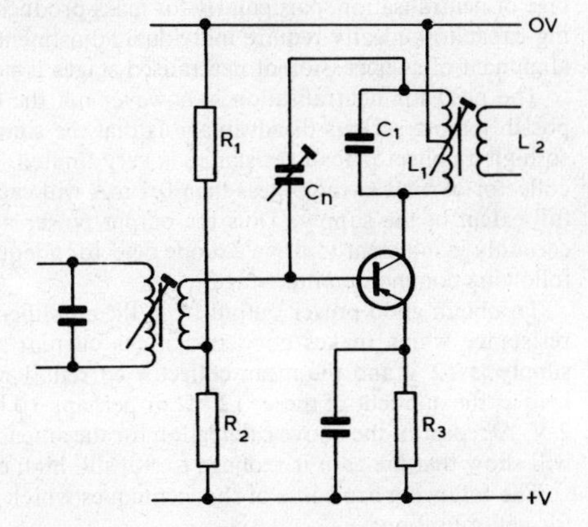

Figure 13.28 Stage of i.f. amplification using a neutralised pnp transistor

Figure 13.28 gives the circuit diagram of a typical stage of i.f. amplification using a *pnp* bipolar transistor. The single resonant circuit L_1–C_1 is connected in the collector circuit of the transistor and the output to the next stage is taken from a secondary winding L_2 tightly coupled to the primary winding. D.C. stability is ensured by the potential divider R_1–R_2 and the emitter resistor R_3.

In a receiver the next stage is likely to be a further common-emitter amplifying stage or a diode detector, both with an input resistance of the order of 2 kΩ. For maximum gain from the stage under examination the turns ratio of the r.f. transformer L_1–L_2 should be chosen to match the output resistance r_o of the transistor to the input resistance r_i of the following stage. For unity coupling between the windings the turns ratio $n:1$ is given by the usual formula

$$n = \sqrt{\frac{r_o}{r_i}} \tag{15}$$

With such a ratio the transistor is presented with a collector load resistance equal to its own output resistance. To calculate the gain available we can regard the transistor as a current generator I ($= g_m V_{in}$) feeding into its own output resistance r_o in parallel with the external load resistance (also equal to r_o). Thus the current entering the external load is equal to $\frac{1}{2}g_m V_{in}$ and the signal at the collector is equal to $\frac{1}{2}g_m V_{in} r_o$. Because of the step-down ratio of the r.f. transformer the signal at the secondary winding is given by

$$V_{out} = \frac{\frac{1}{2}g_m V_{in} r_o}{n}$$

But $n = \sqrt{(r_o/r_i)}$

Therefore
$$V_{out} = \frac{\frac{1}{2}g_m V_{in} r_o}{\sqrt{(r_o/r_i)}}$$

Therefore
$$\frac{V_{out}}{V_{in}} = \tfrac{1}{2}g_m\sqrt{(r_o r_i)} \qquad (16)$$

This can give surprisingly large gains. Typical values for a silicon planar transistor for g_m, r_i and r_o are 40 mA/V, 2 kΩ and 200 kΩ. Substitution in equation 16 shows the gain as 400. The internal collector-base capacitance of the transistor (typically 1 pF) would certainly be sufficient to cause r.f. instability and oscillation unless the stage is neutralised. One way of achieving neutralisation is illustrated in *Figure 13.28*. A disadvantage of neutralisation, particularly for mass-produced i.f. amplifiers, is that the neutralising capacitors ideally require individual adjustment for optimum results. Moreover the alignment of a succession of neutralised stages is not easy.

The need for neutralisation is however not the only disadvantage of this circuit. A possibly more serious disadvantage is that the signal-handling capacity of a stage with so high a collector load resistance is very limited. If the supply voltage is 12 V then a collector current swing of less than 0.1 mA will cause the collector voltage to swing the full extent of the supply. Thus the output power available from the stage is very low, certainly insufficient to drive a diode detector adequately and probably insufficient for a following common-emitter stage.

To obtain good power output from the amplifier it is better to present it with a load resistance which makes good use of the current and voltage swings available. If the supply is 12 V and the mean collector current 1 mA it is best to regard the optimum load as the quotient of these: 12 kΩ or perhaps 10 kΩ to allow for an emitter voltage of 2 V. A repeat of the above calculation for the amended value of collector load resistance will show that the gain is reduced to 90, still high enough to justify neutralisation.

The following are some of the techniques which may be employed to avoid the need for neutralisation:

(*a*) Use a common-base stage instead of common emitter. This helps because the internal feedback capacitance now responsible for any possible instability is that between collector and emitter and this is much smaller than the collector-base capacitance. This technique is extensively used in the r.f. stages of v.h.f. and u.h.f. receivers.

(*b*) Use a cascode stage, i.e., a common-emitter stage feeding directly into a common-base stage. A circuit diagram of a cascode stage using two f.e.t.s is given in *Figure 13.24*. This technique gives the same gain as the common-emitter stage but with stability.

(*c*) Limit the gain of the common-emitter stage to a value at which the collector-to-base capacitance cannot cause instability. It may well be that this expedient increases the number of stages necessary to give the required overall i.f. gain but there will be no neutralising capacitors to be adjusted and the alignment of the i.f. tuned circuits will be straightforward.

If a transistor has tuned circuits connected to base and collector terminals (both tuned to the same frequency) instability will result if

$$\omega c_{bc} g_m R_b R_c \text{ exceeds } 2 \qquad (17)$$

where R_b and R_c are the resistances effectively in parallel with the input and output circuits.

As an example suppose, as in the previous example, $c_{bc} = 1$ pF, $g_m = 40$ mA/V, $R_b = 2$ kΩ and $R_c = 200$ kΩ. At the standard a.m. sound intermediate frequency of 465 kHz we have

$$\omega c_{bc} g_m R_b R_c = 6.284 \times 465 \times 10^3 \times 1 \times 10^{-12} \times 40 \times 10^{-3} \times 2 \times 10^5 \times 2 \times 10^3 = 46.7$$

which is greater than 2 and hence instability, even at the low frequency of 465 kHz is inevitable unless neutralising is used.

To avoid instability, without neutralisation, the design technique used is to equate expression 17 to say 0.5 giving a protection factor of 4 : 1 against instability. Suppose we wish the stage to operate at 10.7 MHz the standard f.m. sound intermediate frequency. We have

$$R_b R_c = \frac{0.5}{\omega c_{bc} g_m} = \frac{0.5}{6.284 \times 10.7 \times 10^6 \times 1 \times 10^{-12} \times 40 \times 10^{-3}} = 1.86 \times 10^5$$

The product of R_b and R_c should now be chosen so that their product does not exceed this value. R_c should preferably be equal to the optimum load for the transistor, i.e. (collector voltage swing)/(collector current swing). Suppose, in a mains-driven receiver where current economy is not a serious consideration, the transistor has a mean collector current of 4 mA and that the collector voltage swing can be 12 V. R_c is then 3 kΩ and the maximum value of R_b is given by

$$R_b = \frac{1.86 \times 10^5}{R_c} = \frac{1.86 \times 10^5}{3 \times 10^3} = 62\ \Omega$$

The input circuit of the transistor should be designed so that the resistance between base and emitter at 10.7 MHz does not exceed 62 Ω. The base can be connected to a tapping point on the inductor so that this value of resistance is not exceeded.

Use of coupled tuned circuits

A better approximation to the ideal square-topped frequency response required for the amplification of modulated r.f. signals can be obtained by using *LC* circuits in coupled pairs. The response of two identical coupled circuits depends on the coupling between the coils. If the mutual inductance is *M*, the coupling coefficient is given by M/L and is usually represented by k. If k is less than $1/Q$ (where Q is the ratio of reactance to resistance for each coil) the response is low and has a single peak at the resonance frequency of the tuned circuits. Maximum output occurs when $k = 1/Q$, known as optimum coupling and the response has a flattened peak at the resonance frequency. If k is greater than $1/Q$ the response is still a maximum but now has two peaks at frequencies given by

$$f_1 = \frac{f_c}{\sqrt{(1-k)}} \qquad f_2 = \frac{f_c}{\sqrt{(1+k)}}$$

where f_c is the resonance frequency of both *LC* circuits. The peak separation measures the passband of the coupled coils and is given by

$$f_1 - f_2 = f_c k \qquad (18)$$

Thus the coefficient of coupling required to give a required passband at a particular centre frequency is given by

$$k = \frac{\text{passband}}{\text{centre frequency}} \qquad (19)$$

As an example consider the i.f. amplifier of an a.m. broadcast receiver. The maximum modulation frequency transmitted is often limited to 5 kHz which suggests a passband of 10 kHz but the carrier-frequency spacing on the medium waveband is 9 kHz and this figure is often taken as the bandwidth required. The centre frequency is standardised at 465 kHz and thus the coefficient of coupling required in an i.f. transformer is, from equation 19, given by

$$k = \frac{9}{465} \approx 0.02$$

Thus if the inductors have a Q value of 50 or less, this value of k corresponds to critical coupling or less and a single-peaked response is obtained. If Q exceeds 50 the response is double-humped and the two peaks become more marked as the Q value is increased.

Valve tuned amplifier

In valve tuned amplifiers, as used in receivers, instability is minimised by using valves with a very low value of anode-grid capacitance such as screened pentodes and a typical circuit diagram of a stage of i.f. amplification using such a valve is given in *Figure 13.29*. R_{sg} provides positive screen bias and is decoupled by C_{sg}. R_k provides negative

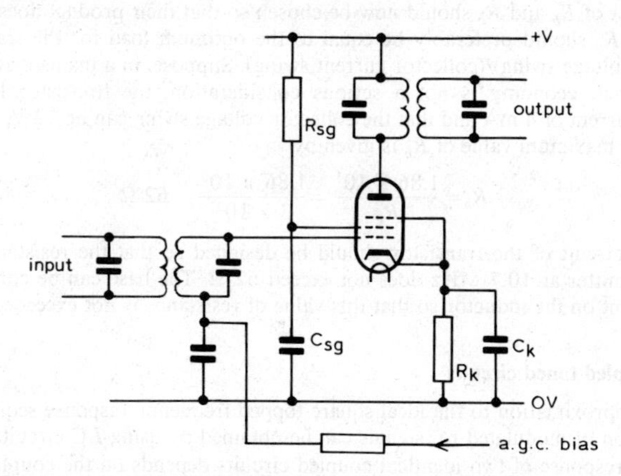

Figure 13.29 A typical stage of i.f. amplification for an a.m. valve receiver using a screened pentode

grid bias and is decoupled by C_k. Additional grid bias is injected via the secondary winding of the i.f. transformer from the detector stage to give automatic gain control, the valve being specially designed so that its gain can be controlled by adjustment of grid bias. Such valves are known as *variable-mu* types.

The input and output resistances of an r.f. pentode are so high that they impose negligible damping on the tuned windings of the i.f. transformer. Thus the transformers can be designed to give the required response (as indicated in the numerical example above) without regard to the terminating resistances.

V.H.F. and u.h.f. tuned amplifiers

At v.h.f. and u.h.f. it is common practice to use bipolar transistors as common-base amplifiers. The low input resistance is not a disadvantage and the stability of the amplifier is better than that of the common-emitter amplifier because the base acts to some extent as a screen between the emitter (input terminal) and the collector (output terminal).

A modern tendency in the r.f. stages of f.m. and television receivers is to use semiconductor diodes for tuning. A reverse-biased *pn* junction behaves as a capacitance and the value of the capacitance can be controlled by adjustment of the current through the diode. This idea is attractive to receiver designers because it enables all the r.f. circuits (including the oscillator) to be tuned remotely by adjustment of a single variable resistance carrying only d.c. It also simplifies the provision of preset tuning for only one preset potential divider is required for each station to be selected.

A simplified diagram of the r.f. circuits of a u.h.f. television tuner is given in *Figure 13.30*. The tuning inductors used in this circuit are lecher lines each consisting of a

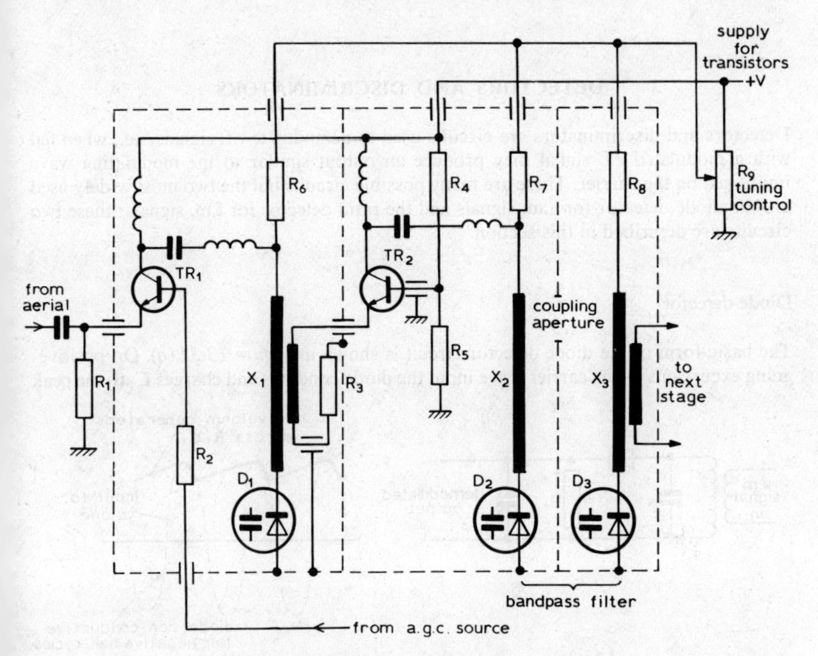

Figure 13.30 A v.h.f. trough-line tuner using varicap diodes. X_2 *and* X_3 *form a bandpass filter coupled via the aperture in the common trough-line wall*

central conductor mounted within a conducting rectangular box, an arrangement known as a trough line. One way of obtaining the coupling between inductors, which is necessary to form a bandpass filter, is to provide an aperture in the common trough wall. The capacitance between the lines then provides the required coupling.

A transmission line less than $\lambda/4$ in length and shortcircuited at one end behaves as an inductance at the other end and can be tuned by a variable capacitor (or a varicap diode) connected across the open end. Similarly transmission lines between $\lambda/4$ and $\lambda/2$ in length and open-circuited at one end behave as an inductance and can be tuned by a variable capacitance connected across the other end. Both systems are in use in v.h.f. and u.h.f. tuners but the longer lines are preferred for use with varicap diodes because the current which controls the diode capacitance can be easily introduced via the centre conductor of a trough line as shown in *Figure 13.30*.

The aerial input is applied to the emitter of the common-base stage TR_1 which is controlled from the a.g.c. line. The trough-line inductor X_1 forms the collector load for TR_1 this being tuned by the varicap diode D_1 which is controlled from the potential divider R_9 which is in fact the tuning control. X_1 is coupled to a secondary inductor in close proximity to it and this feeds the amplified signal from TR_1 to the emitter of the second common-base stage TR_2. This is stabilised by the potential divider R_4–R_5 and the emitter resistor R_3. The collector load for TR_2 is the trough line X_2, tuned by varicap diode D_2. X_2 is capacitively coupled to a similar trough line X_3 via the aperture in the common trough-line wall to form a bandpass filter. X_3 is tuned by varicap diode D_3 and all three diodes are controlled from the tuning control R_9. X_3 is coupled to a further inductor close to it and this conveys the output of the two r.f. stages to the next stage, the frequency changer, the oscillator of which is similarly tuned by a varicap diode controlled from R_9.

DETECTORS AND DISCRIMINATORS

Detectors and discriminators are circuits used to demodulate r.f. signals, i.e., when fed with a modulated r.f. signal they produce an output similar to the modulating wave impressed on the carrier. There are many possible circuits but the two most widely used are the diode detector for a.m. signals and the ratio detector for f.m. signals: these two circuits are described in this section.

Diode detector

The basic form of the diode detector circuit is shown in *Figure 13.31* (*a*). On positive-going excursions of the carrier-wave input the diode conducts and charges C_1 to the peak

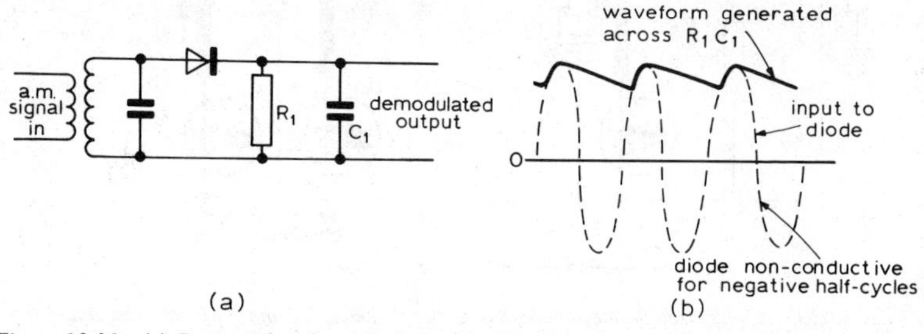

Figure 13.31 (*a*) *Basic circuit for a diode detector* (*b*) *Illustrating the action of a diode detector for a constant-amplitude input*

value of the carrier. On negative-going excursions of the carrier wave the diode is cut off and the capacitor C_1 begins to discharge through R_1. On the next positive-going signal the diode again conducts and restores the charge on C_1 as shown in *Figure 13.31* (*b*). Thus for a constant, i.e., unmodulated carrier wave input the voltage across C_1 is restored to peak value once per cycle. If the time constant R_1–C_1 is long compared with the period of one cycle of the carrier wave there is little discharge of C_1 during negative-going half cycles and to a first degree of approximation the voltage across C_1 can be regarded as steady and equal in value to the peak value of the carrier wave input to the diode.

The power dissipated in R_1 is V^2/R_1. If the diode input resistance is r_{in} we have, equating the power fed to the diode with that developed in R_1

$$\frac{v_{in}^{2}}{r_{in}} = \frac{V^2}{R_1}$$

This gives

$$r_{in} = R_1 \frac{v_{in}^{2}}{V^2}$$

But $V = \sqrt{2} v_{in}$

Therefore

$$r_{in} = \frac{R_1}{2}$$

i.e., the diode input resistance is one half the diode load resistance. The tuned circuit feeding the diode is thus effectively damped by a parallel resistance equal to R_1/R_2.

When the carrier input to the diode is amplitude modulated the voltage across C_1 is required to follow the amplitude variations of the carrier level faithfully and this implies some limitations to the value of the time constant R_1C_1. For if R_1C_1 is too large the voltage across C_1 will not be able to fall rapidly enough during negative-going half cycles to follow the slope of the steepest falls in carrier amplitude. These occur at the highest modulating frequencies and at the instant when the modulating waveform, assumed sinusoidal, crosses the datum line. The equation for the modulating waveform is $mv_{in}\sin \omega t$ where m is the modulation depth and the maximum slope of this is given by $mv_{in}\omega$. The maximum slope of an exponential waveform occurs at the initial part of the discharge and is given by v_{in}/R_1C_1. We thus have that the maximum value of time constant R_1C_1 which can be used if the highest modulation frequencies are to be reproduced without distortion is given by

$$R_1C_1 = \frac{1}{m\omega} \tag{20}$$

As an example in 625-line television the highest modulation frequency is 5.5 MHz and a modulation depth of 100% is possible at this frequency. Thus we have

$$R_1C_1 = \frac{1}{6.284 \times 5.5 \times 10^6} = 0.033\ \mu\text{s}$$

R_1 should be large compared with the diode forward resistance otherwise the diode output will be less than it need be. Nevertheless R_1 should be as small as possible to keep C_1 at a feasible value. If R_1 is made 2 kΩ we have

$$C_1 = \frac{0.033 \times 10^{-6}}{R_1} = \frac{0.033 \times 10^{-6}}{2 \times 10^3}\text{F} = 16.5\ \text{pF}$$

Complete diode detector circuit

The waveform generated across R_1C_1 for a modulated r.f. input has three components:

(*a*) a steady component proportional to the unmodulated carrier amplitude. In sound receivers this is unwanted and is removed from the a.f. output by a series capacitor (C_4 in *Figure 13.32*). The d.c. component is however used as a source of a.g.c. voltage and is returned to earlier stages in the receiver to control their gain. The a.f. signal superimposed on the d.c. is removed by the network R_3–C_3;
(*b*) the wanted modulation-frequency component;
(*c*) an unwanted r.f. ripple caused by the charging and discharging of C_1. This is removed by the network R_2–C_2 which is designed to attenuate r.f. but to leave substantially unaffected the upper frequencies of the modulation-frequency component.

All the additional components mentioned in (*a*), (*b*) and (*c*) are included in the complete circuit diagram of a diode detector shown in *Figure 13.32*. This gives a positive output

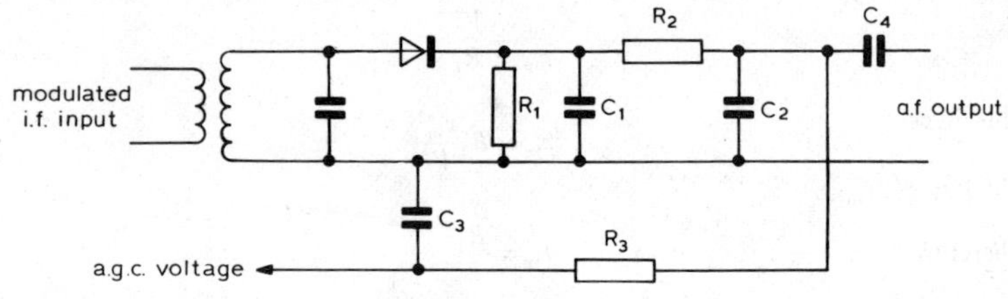

Figure 13.32 Complete diode detector circuit showing networks for r.f. filtering and a.g.c. provision

on the a.g.c. line. The polarity of output required for a.g.c. purposes depends on the type of transistor to be controlled and whether forward or reverse bias is required. Thus the a.g.c. voltage required may be positive or negative. If a negative voltage is needed this can be obtained by reversing the diode in *Figure 13.32*.

Ratio detector

This form of detector is required to be sensitive to changes in frequency of the applied carrier wave and ideally to be insensitive to any amplitude variations. Much of the noise in f.m. reception is due to amplitude variations and one of the aims in the design of f.m. receivers is to produce a degree of immunity to a.m. It is common practice to use one or more limiting stages in the i.f. amplifier before the detector but any a.m. rejection in the detector itself is clearly advantageous and the principal reason for the popularity of the ratio detector is that it can provide up to 30 dB of a.m. rejection.

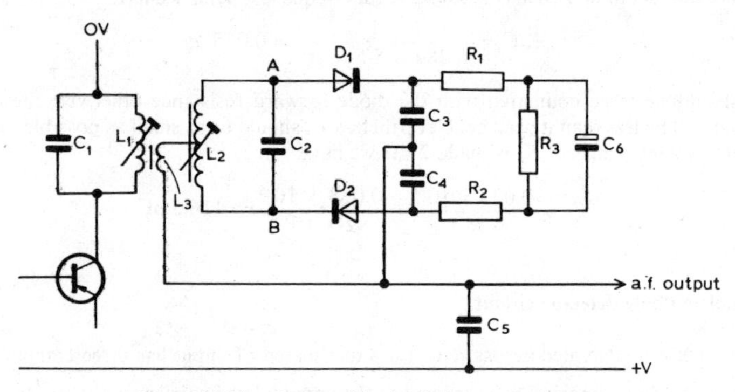

Figure 13.33 One possible circuit for a ratio detector

One form of ratio detector circuit is given in *Figure 13.33*. It depends for its action on the phase relationship between the signals developed across the tuned circuits L_1–C_1 and L_2–C_2. These are tuned to the (i.f.) carrier frequency and at this frequency the two signals are in quadrature. L_3 is tightly coupled to L_1 and the signal induced in it is in phase with that in L_1 and is injected at the mid-point of L_2. As a result the signals at the ends A and B of L_2–C_2 are equal at the carrier frequency but if the input carrier frequency is varied the signal at one end of L_2–C_2 becomes greater than that at the other. For example it could happen that end A has the greater signal for positive displacements of input frequency and B has the greater signal amplitude for negative displacements. These signals are applied to diode detectors D_1 and D_2 so that C_3 is charged to the peak value of the signal at A and C_4 is charged to the peak value of the signal at B. Because D_1 produces a positive-going output and D_2 a negative-going output the net signal at the junction of C_3 and C_4 is zero for an input signal at the carrier frequency, is positive for signals above the carrier frequency and is negative for signals below the carrier frequency. Moreover with correct design the magnitude of the net output from C_3–C_4 is proportional to the frequency displacement: thus this output is the wanted a.f. output from the detector. C_5 provides filtering of any i.f. component in this output.

The network R_1–R_2–R_3–C_6 is responsible for the a.m. rejecting properties of the detector. These components together with the diodes D_1 and D_2 constitute the circuit of a dynamic limiter connected across L_2–C_2. C_6 is charged via R_1–R_2–D_1–D_2 to the peak value of the carrier voltage across L_2–C_2 and R_3 is chosen to give heavy damping (equivalent to R_3/R_2) on L_2–C_2. Any tendency for the voltage across L_2–C_2 to increase suddenly (due for example to a noise pulse) results in even heavier damping of the tuned circuit because the charge on C_6 cannot rise instantaneously and thus the low-value resistors R_1 and R_2 are effectively connected across L_2–C_2, limiting the rise in voltage across it.

R_3–C_6 are chosen to have a long time constant. Thus if the carrier voltage across L_2–C_2 falls suddenly, C_6 holds its charge and the diodes are cut off, thus relieving L_2–C_2 of the heavy damping due to R_3. The gain of the circuit rises and largely offsets the initial tendency of the voltage across L_2–C_2 to fall. Thus upward and downward changes in carrier amplitude are limited. The values of R_1–R_2–R_3–C_6 must be carefully chosen in relation to the dynamic resistance of L_2–C_2 to give best results from this ratio detector circuit.

OSCILLATORS

Sinusoidal signals have wide application in radio and electronics. They are used to test equipment, e.g., a.f. amplifiers, to provide the carrier source in transmitters, in the frequency changers of superhet receivers, to provide bias and wiping in tape recorders: these are only a few of the many applications of sine-wave signals.

Oscillators, i.e., generators of sinusoidal waveforms contain three basic sections:

(*a*) a frequency-determining section which is normally an *LC* network but can also be an *RC* network;
(*b*) an amplifying section which can be a valve or a transistor;
(*c*) a limiting section which restricts the amplitude of the generated sine wave to a value at which it can be handled by the amplifying section: this is essential if a particularly pure waveform is required.

Hartley oscillator

In order to generate oscillations the amplifying section must contain a signal path between the output and the input circuits by virtue of which the amplifier is capable of supplying its own input: in other words the amplifier must have positive feedback. In many oscillators the frequency-determining section provides the positive feedback and if the amplifier is a single valve or transistor which inverts the input signal the correct phase relationship between input and output connections is obtained by use of a tapping point on the inductor as shown in *Figure 13.34* (*a*).

By making these three connections to the cathode, grid and anode of a triode, the oscillator circuit of *Figure 13.34* (*b*) is obtained. The input to the valve is applied between grid and cathode and the output is generated between anode and cathode. The cathode is hence the common connection and goes to the centre tap on the inductor (via the low-reactance capacitor C_2). The anode is connected directly to one end of L_1 and h.t. is

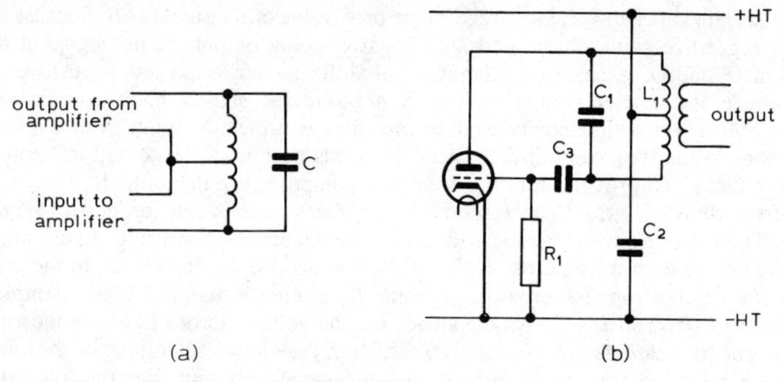

Figure 13.34 (*a*) *The three essential connections to the inductor of an LC oscillator* (*b*) *Hartley oscillator circuit using a valve*

introduced via the tapping point. The grid is connected to the other end of L_1 via the low-reactance capacitor C_3 which is necessary to isolate the grid from the h.t. supply. The grid is connected to cathode via R_1. C_3 and R_1 provide automatic grid bias. The oscillation amplitude builds up until the grid is driven positive with respect to cathode for a brief period during each cycle. During this period grid current flows and this charges C_3 (additionally to the standing charge applied by the h.t. supply) so that the grid is driven negative. During the remainder of each cycle when the grid is negative with respect to the cathode and there is no grid current, the additional charge on C_3 leaks away through R_1 but, provided R_1 is large enough, very little of it has been lost before the next cycle of oscillation restores the additional charge again. The burst of grid current momentarily applies a low-resistance across the tuned circuit and it is this which limits the amplitude of the oscillation generated. Normally the amplitude is so large that the valve is cut off for a large part of each cycle: in other words the amplifier operates in Class C. For successful results the time constant R_1C_3 should be long compared with the periodic time of the oscillation and the frequency of oscillation is given by the expression

$$f=\frac{1}{2\pi\sqrt{(L_1C_1)}}$$

This type of oscillator in which the inductor is centre-tapped is known as the Hartley. The sinusoidal oscillation is set up, of course, in L_1C_1 and the output must be taken from this circuit: a convenient method of doing this is by coupling an inductor to L_1 as shown in *Figure 13.34* (*b*).

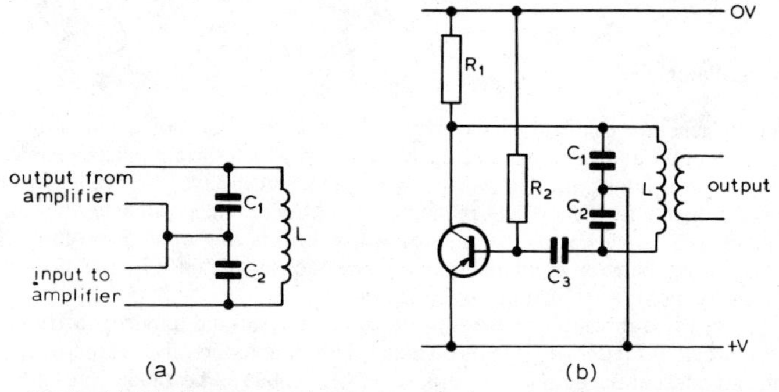

Figure 13.35 (*a*) *The three essential connections to the capacitive branch of an LC oscillator* (*b*) *Colpitts oscillator circuit using a pnp transistor*

Colpitts oscillator

It is alternatively possible to tap the capacitive branch of an *LC* circuit in order to provide the three connections essential for positive feedback: this is shown in *Figure 13.35* (*a*) and provides the basic feature of the Colpitts oscillator which thus enables oscillation to be set up in an inductor without a tapping point. If we attach these three connections to the emitter, base and collector of a *pnp* transistor we obtain the practical circuit shown in *Figure 13.35* (*b*).

The connection which is common to the input and output of the transistor is the emitter and this is connected directly to the centre point of the two capacitors. One end of the inductor is connected directly to the collector and the other to the base via the capacitor C_3 which isolates the base from the supply. The collector cannot be connected directly to the negative supply because this would result in an effective short-circuit of C_1 and so the resistor R_1 is introduced. This damps the circuit L–C_1–C_2 and thus cannot be too small. The base is biased by R_2 from the negative supply line. As in the valve circuit R_2–C_3 provide automatic bias, the transistor being driven into conduction during negative half cycles applied to the base and remaining cut off during positive half cycles, again an example of Class C operation. Oscillation occurs at the frequency

$$f = \frac{1}{2\pi\sqrt{(LC)}}$$

where

$$C = \frac{C_1C_2}{C_1 + C_2}$$

The output is again taken from L–C_1–C_2 by a coupling coil.

RC phase-shift oscillator

Generators governed by *RC* circuits can be designed to produce a wide variety of different waveforms and the production of sawtooth and rectangular waves is described later in this section. However an *RC* network can also form the frequency-determining element of a sinusoidal oscillator: two examples of such oscillators are described; the first is the *RC* phase-shift oscillator.

In the Hartley and Colpitts oscillators the *LC* circuit introduced a phase inversion between its input and output terminals. This, combined with the phase inversion of the amplifier gives the positive feedback essential to sustain oscillation. For a symmetrical wave such as a sinusoid the effect of phase inversion is the same as that of altering the phase by 180°. If therefore we can find a network which shifts the phase by 180° at a particular frequency and if the gain of the amplifier can compensate for the attenuation introduced by the network at this frequency, then we have the basis for an oscillator. A single *RC* network can at best give 90° phase shift and its attenuation is then infinite. However a network of three *RC* sections, each introducing 60° phase shift is a practical possibility: if the resistance values are equal and if the capacitance values are also equal the attenuation introduced by the 3-section network is 29 (i.e., this is the value of i_{in}/i_{out}) which can be made good by a single transistor. The circuit diagram of an oscillator operating on such principles is given in *Figure 13.36*. The frequency-determining network is R_1–C_1–R_2–C_2–R_3–C_3 in which $R_1 = R_2 = R_3 = R$ and $C_1 = C_2 = C_3 = C$. D.C. stabilisation is ensured by the potential divider R_4–R_5 and the emitter resistor R_6. The frequency of oscillation is given by $f = 1/2\pi\sqrt{6}RC$.

No means is shown in *Figure 13.36* of limiting the amplitude of oscillation and this is essential to preserve the purity of the waveform generated. It is possible, by adjustment of the value of R_6, to set the collector current at a value which only just gives sufficient current gain to make up for the attenuation in the *RC* network but this is not a good

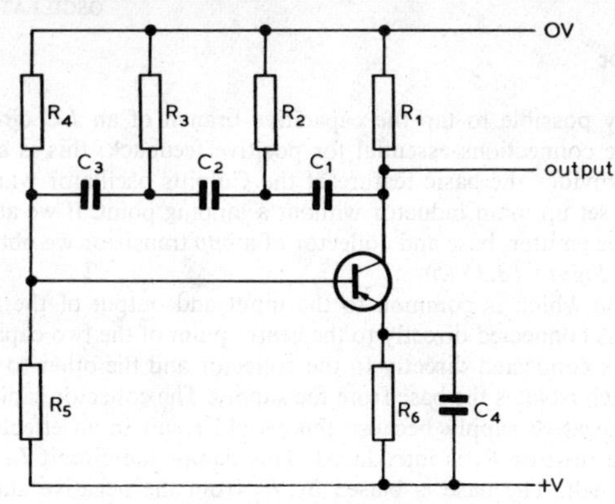

Figure 13.36 RC phase-shift oscillator

method and it would be better to use an automatic system in which the transistor is biased back when the output amplitude exceeds a predetermined value.

Wien bridge oscillator

It would be difficult to make a variable-frequency oscillator based on the circuit of *Figure 13.36*. A circuit much better suited to this purpose is the Wien bridge oscillator which is used as the basis of a number of a.f. test oscillators.

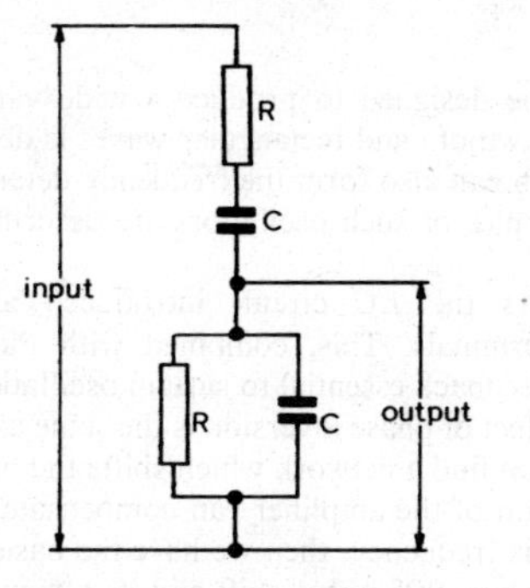

Figure 13.37 Basic Wien bridge network

The basic Wien bridge network is shown in *Figure 13.37*. It contains two equal resistors and two equal capacitors. At the frequency for which

$$f = \frac{1}{2\pi RC} \qquad (21)$$

the network has zero phase shift between input and output, and the voltage attenuation is 3, i.e., $v_{in}/v_{out} = 3$. To use such a network as the frequency-determining element in an oscillator the amplifier must also introduce zero phase shift and have a voltage gain of 3. The amplifier must, moreover, have a very high input resistance (as shown later) for it is essential to minimise any shunting effect on the parallel RC branch. Two stages at least are required in the amplifier to give the required zero phase shift and it is usual to include a third, an emitter follower, to provide a low-resistance output for the frequency-determining network and for the oscillator itself.

By using two sections of a two-gang variable capacitor for the two capacitors in *Figure 13.37* it is possible to vary the frequency of oscillation over a range of say 10 : 1. The two resistors can be varied in decade steps to produce a number of frequency ranges. Three ranges could thus cover the frequency band from 30 Hz to 30 kHz which is suitable for an a.f. test oscillator. If the variable capacitors are of 500-pF maximum capacitance then the value of R required for the lowest-frequency range can be obtained from equation 21 by rearranging it thus

$$R = \frac{1}{2\pi f C}$$

Substituting $f = 30$ and $C = 500 \times 10^{-12}$ we have

$$R = \frac{1}{6.284 \times 30 \times 500 \times 10^{-12}}\,\Omega \approx 10\ \text{M}\Omega$$

To avoid shunting this appreciably (which would increase the frequency and prevent accurate alignment of the two sections of the Wien bridge network) the input resistance of the amplifier must be very high indeed and an f.e.t. is the obvious choice for the first stage. The second and third stages can be bipolar transistors as indicated in the circuit diagram of *Figure 13.38*.

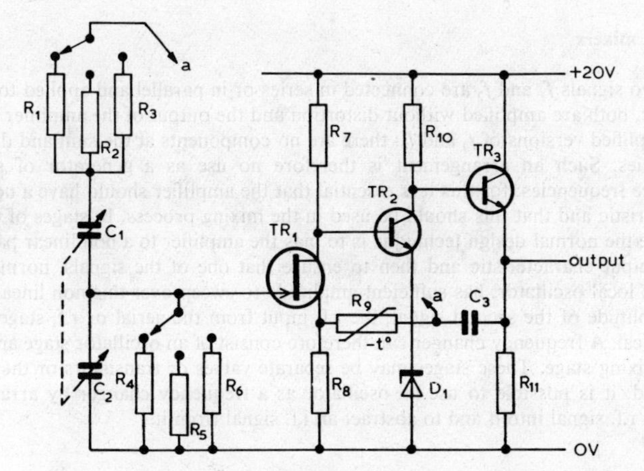

Figure 13.38 A Wien bridge oscillator with three frequency ranges

Negative feedback is applied between the emitter of TR_3 (in effect the collector of TR_2 since TR_3 is an emitter follower) and the source of TR_1 via the non-linear resistor R_9 which has a negative temperature coefficient. This has the effect of maintaining the output of the oscillator substantially constant in spite of frequency or range changes. Immediately after switching on the resistance of R_9 is high because there is no signal in it and hence no heat. The gain of the amplifier is determined by the feedback components and is given by R_9/R_8 and, at the instant of switch on, this is high, much higher than the value of 3 necessary to ensure oscillation. Oscillation therefore begins and builds up rapidly. As soon as the oscillation reaches R_9 via C_3 the resistance of R_9 falls due to the heating effect of the signal in it. The fall continues and the resistance of R_9 settles around a value approximately equal to twice R_8. The type of resistor chosen for R_9 should be such that its resistance will equal twice R_8 with the required value of oscillation amplitude across it.

The amplifier is direct-coupled throughout and an interesting design point is the inclusion of the zener diode D_1 in the emitter circuit of TR_2. The drain voltage of TR_1 is probably around 10 V, i.e., half the supply voltage which means that TR_2 emitter voltage must be approximately the same. A resistor could be included in TR_2 emitter circuit to permit such an emitter voltage to be realised but this would give negative feedback and reduce the gain of TR_2 to a very low value. This feedback could be minimised by decoupling the emitter resistor by a low-reactance capacitor but it is desirable at some point in a multistage d.c. amplifier to stabilise the voltage. By including a zener diode of a suitable voltage rating TR_2 emitter voltage can be stabilised so eliminating feedback and obtaining maximum gain from TR_2. It is true that the overall gain of the amplifier is reduced to 3 by the negative feedback due to R_8 and R_9 but the constancy of output amplitude improves as the gain of the individual transistors is increased.

FREQUENCY CHANGERS

In many examples of electronic equipment it is necessary to combine two signals of frequency f_1 and f_2 to produce new signals with frequencies of (f_1+f_2) and (f_1-f_2). Perhaps the most obvious example occurs in superhet receivers where the received r.f. signal is combined with the output of the local oscillator in the frequency-changer stage to produce the difference-frequency signal which is amplified in the i.f. amplifier.

Additive mixers

If the two signals f_1 and f_2 are connected in series or in parallel and applied to a linear amplifier, both are amplified without distortion and the output of the amplifier contains only amplified versions of f_1 and f_2: there are no components at the sum and difference frequencies. Such an arrangement is therefore no use as a generator of sum and difference frequencies: for this it is essential that the amplifier should have a non-linear characteristic and that this should be used in the mixing process. In stages of this type therefore the normal design technique is to bias the amplifier to a non-linear part of the input–output characteristic and then to ensure that one of the signals, normally that from the local oscillator, has sufficient amplitude to sweep over this non-linear region. The amplitude of the second signal, the r.f. input from the aerial or r.f. stage, is then non-critical. A frequency changer can therefore consist of an oscillator stage and a non-linear mixing stage. These stages may be separate valves or transistors or they can be combined: it is possible to use an oscillator as a frequency changer by arranging to inject an r.f. signal into it and to abstract an i.f. signal from it.

Self-oscillating mixers

As mentioned earlier many oscillators operate in Class C. The characteristic around the point of collector-current cut off is therefore used in the oscillation process and the non-linearity essential for successful mixing is present. Provided care is taken in the design of the circuit so that r.f. can be put in and i.f. can be taken out without significant detriment to the oscillating process a self-oscillating mixer can be a good frequency changer and have a conversion efficiency little short of that of a separate mixer and oscillator.

Figure 13.39 gives the circuit diagram of a self-oscillating mixer used in an f.m. receiver. The *npn* transistor is d.c. stabilised by the potential divider R_1–R_2 and the emitter resistor R_3. It operates as a Colpitts oscillator (*c.f. Figure 13.35* (*b*)) the two fundamental capacitors being C_1 and C_2 while C_3, in parallel with them, provides oscillator tuning. C_3 is ganged with other variable capacitors tuning r.f. circuits and it is desirable therefore for its moving vanes to be earthed. This necessarily earths the base of the transistor and clearly the emitter cannot also be earthed. The inductor L_2 is therefore introduced: it has high reactance at the oscillator frequency and thus permits the emitter potential to fluctuate at r.f. and also provides a convenient point at which the r.f. signal can be introduced via C_4. C_5 and C_6 are d.c. blocking capacitors permitting L_1 and C_3 to be earthed while allowing working d.c. voltages to be applied to the base and

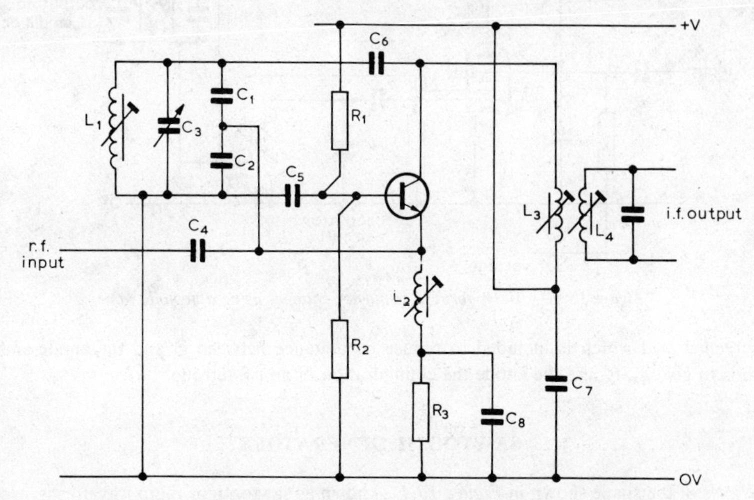

Figure 13.39 *Self-oscillating mixer circuit suitable for use in an f.m. receiver*

the collector. L_3 and L_4 constitute an i.f. transformer coupling the frequency changer to the first i.f. stage. C_7 is a decoupling capacitor and C_6 acts as the tuning capacitor for L_3, L_1 having negligible reactance at the intermediate frequency. To avoid negative feed-back at i.f. due to L_2 in the emitter circuit L_2 is made variable and is adjusted to resonate with C_8 at the intermediate frequency.

Multiplicative mixers

An alternative approach in designing a mixer stage is to adopt the multiplicative principle. If a device can be found which multiplies two inputs together instead of adding them, then the difference frequency output is obtained directly and without any need for non-linearity in the device. This can be seen directly from the trigonometrical identity

$$2 \sin \omega_1 t \,.\, \sin \omega_2 t = \cos (\omega_1 - \omega_2)t + \cos (\omega_1 + \omega_2)t$$

Figure 13.40 gives the circuit diagram of an early form of valve frequency changer using these principles: such heptode valves were commonly used in battery-operated receivers. The two control grids for the two inputs to be multiplied are g_1 and g_4. Grids g_1 and g_2 are used as the grid and anode of the local oscillator and the signal voltages generated at these electrodes are therefore in antiphase so that their effects on the electron stream are mutually destructive. For successful mixing the local oscillator frequency must be impressed on the electron stream and g_1 is therefore shaped so as to couple tightly to the stream whereas g_2 is made only just large enough to carry sufficient current to sustain oscillation. In practice g_2 often consisted simply of a number of rods. The r.f. input is applied to g_4 which is screened from g_1 and g_2 by g_3. Grid g_5 is a further

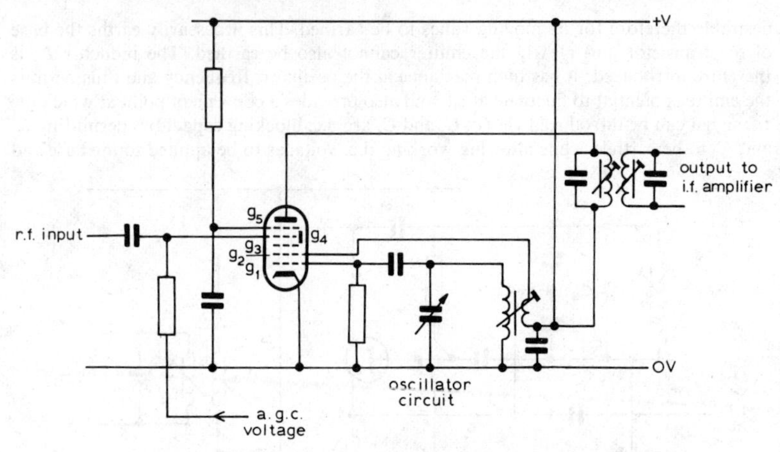

Figure 13.40 Early form of frequency changer using a heptode valve

screened grid which is included to reduce capacitance between g_4 and the anode and thus to give g_4, g_5 and the anode the configuration of an r.f. tetrode.

SAWTOOTH GENERATORS

Waves of the shape shown in *Figure 13.41*, known as sawtooth or ramp waveforms, are used for electron beam deflection in oscilloscopes, in television transmitting and receiving equipment, in digital to analogue conversion equipment and in measuring equipment. Normally the slow rise is required to be linearly related to time for this is the

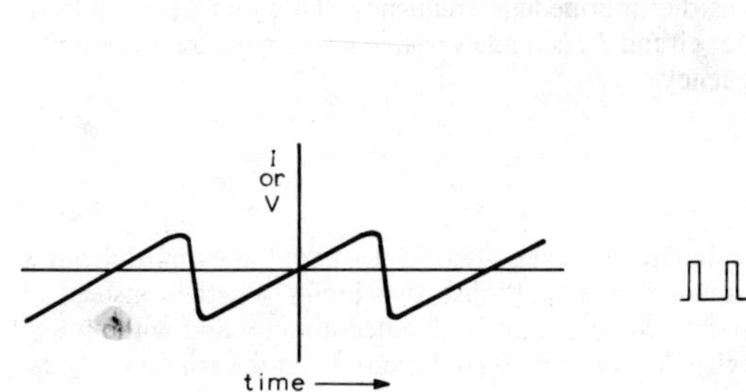

Figure 13.41 Sawtooth waveform

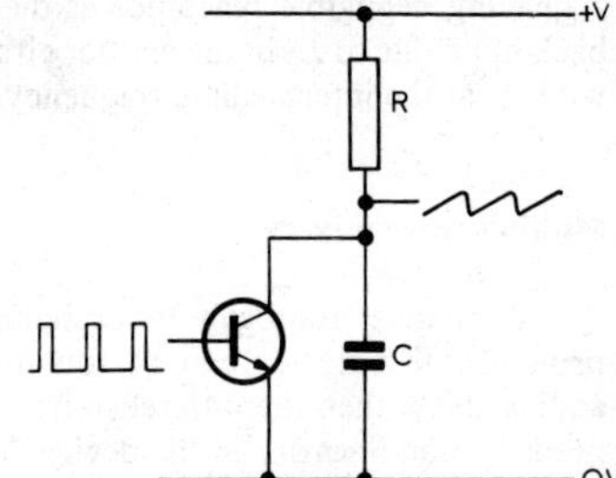

Figure 13.42 Simple discharger circuit for production of sawtooth voltages

working stroke of the waveform but the shape of the rapid fall (the return stroke) need not be linear and the shape of this transition is not usually significant.

Production of sawtooth voltages

An approximation to a sawtooth voltage waveform can be obtained from a simple circuit such as that shown in *Figure 13.42*. The required output is generated across the capacitor C as it charges from the supply via R. The flyback voltage is obtained by discharging C by the transistor which is turned on for the duration of the flyback period. Such a simple circuit has a number of serious limitations: the most important is that the rise of voltage across C is exponential not linear. The departure from linearity is not serious provided that the rise in voltage is restricted to a small fraction of the supply voltage. Normally the performance of the circuit is unsuitable and methods of improving the linearity of the working stroke are necessary. The reason for the lack of linearity is that the rise in voltage across C causes an equal fall in voltage across R and hence a fall in the charging current. Ideally, to achieve linearity, the charging current must be kept constant as charging proceeds. This is shown in the fundamental relationship:

$$V = \frac{Q}{C}$$

But $Q = It$

Therefore
$$V = \frac{I}{C} \cdot t \tag{22}$$

I must be constant to make V directly proportional to t.

Two methods of achieving a constant charging current are in common use. The first is the Bootstrap circuit illustrated in *Figure 13.43*. An emitter follower is connected across the capacitor C and delivers, at its emitter terminal, a copy of the voltage across C. The transistor therefore provides a low-resistance output terminal for the sawtooth generator. The emitter follower output is transferred via the long-time-constant circuit R_1–C_1 to the supply point for the charging resistor R. Thus as the voltage across C rises so does the voltage at the junction of R_1 and R. The voltage across R remains substantially constant throughout the charging process and thus the charging current is maintained constant.

The second method of achieving linearity is to use a source of constant current in place of the charging resistor R. A d.c.-stabilised common-emitter transistor circuit is one possible source of constant current and a circuit using this is given in *Figure 13.44*. The transistor circuit is stabilised by the potential divider R_1–R_2 connected to the base and by the emitter resistor R_3. Suppose as a numerical example it is required to generate a voltage across C which rises linearly to 10 V in 100 μs. The transistor circuit could readily be designed to give a charging current of 5 mA. We can calculate the required value for C from the simple relationship of equation 22, $C = (I/V)t$.

Substituting for I, V and t

$$C = \frac{5 \times 10^{-3} \times 100 \times 10^{-6}}{10}\ \text{F} = 0.05\ \mu\text{F}$$

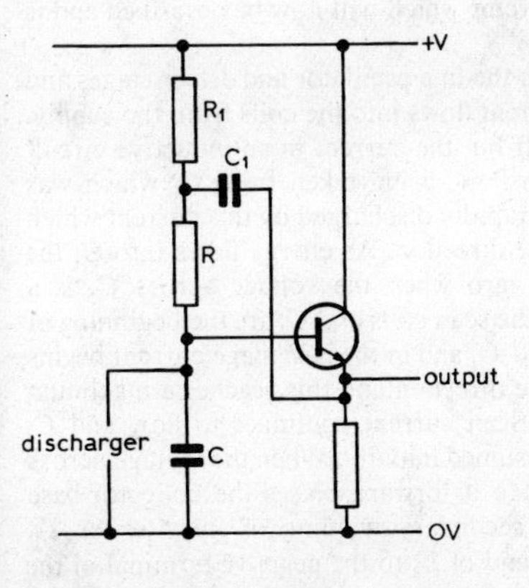

Figure 13.43 Bootstrap circuit

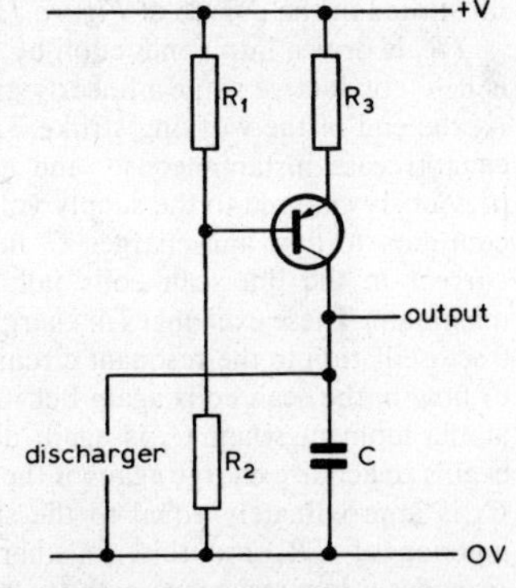

Figure 13.44 Method of achieving linearity using a transistor as a constant-current source

Production of sawtooth currents

For television camera and picture tubes a sawtooth current is required in the deflection coils to deflect the beam horizontally at line rate and vertically at field rate. At the line frequency the deflector coils are predominantly inductive and the problem thus is to generate a sawtooth current in an inductive circuit. There is a very easy solution for if an inductor is connected across a constant voltage supply the current in the inductor rises linearly with time. The voltage across an inductor is related to the rate of change of current in it according to the expression $E = -L(di/dt)$. Thus if L and E are constant then di/dt must be constant. A very simple sawtooth current generator can thus have the circuit diagram shown in *Figure 13.45*. The transistor is held in the conductive state for the required duration of the working stroke by the positive-going pulse waveform applied to the base. This gives the transistor a low-resistance path between collector and

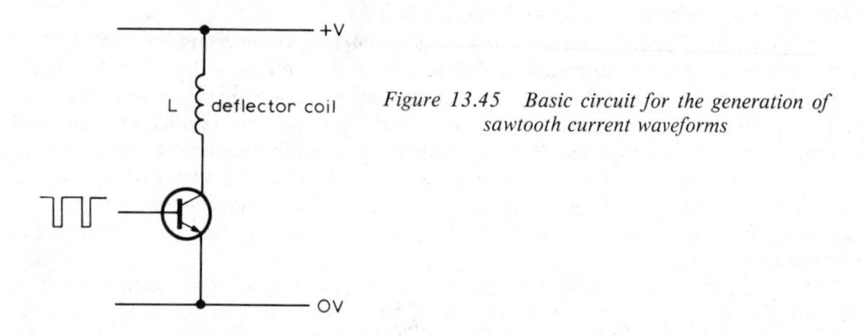

Figure 13.45 Basic circuit for the generation of sawtooth current waveforms

emitter so that, in effect, L is connected directly across the supply and current therefore grows in it linearly. At the end of the working stroke the transistor is turned off by the negative-going signal at the base and the current in L falls rapidly to zero. This is the very simple basis of the line deflection circuits used in television receivers. In practice the circuits are considerably more complex as illustrated in *Figure 13.46* (*a*).

The line scan coils L_1 and L_2 are fed from a line output transformer to ensure that no d.c. flows in the coils (this would produce an undesirable static deflection of the beam) and via the variable inductor L_3 which enables the magnitude of the line-scan current (and hence picture width) to be adjusted. The line-scan circuit is tuned by capacitor C_1: this plays a vital part in the operation of the circuit which will now be described and is illustrated in the curves of *Figure 13.46* (*b*).

TR_1 is driven into conduction by pulses from the line oscillator and driver stages and is held conductive while a linearly growing current flows into the coils from the supply. At the end of the working stroke TR_1 is cut off but the current in an inductive circuit cannot cease instantaneously and continues to flow, being taken from C_1 which was previously charged to the supply voltage. C_1 is rapidly discharged by this current which continues to flow and charges C_1 in the reverse direction. As energy flows into C_1 the current in the line-scan coils falls, reaching zero when the voltage across C_1 is a maximum. These exchanges in energy between the scan coils and C_1 are the beginning of free oscillation in the resonant circuit L_1–L_2–L_3–C_1 and in the next stage current begins to flow in the scan coils again but in the reverse direction and this reaches a maximum at the moment when C_1 is again discharged. Scan current continues to flow and C_1 begins to acquire charge again of the polarity assumed initially. When the voltage across C_1 is approximately equal to the supply voltage it forward biases the collector-base junction of TR_1 and this, together with the secondary winding of transformer T_1, provides a low-resistance path from the lower end of L_3 to the negative terminal of the supply. Thus the scanning-coil circuit is again connected directly across the supply. Now however the current in the coils is a maximum and in the reverse direction to that

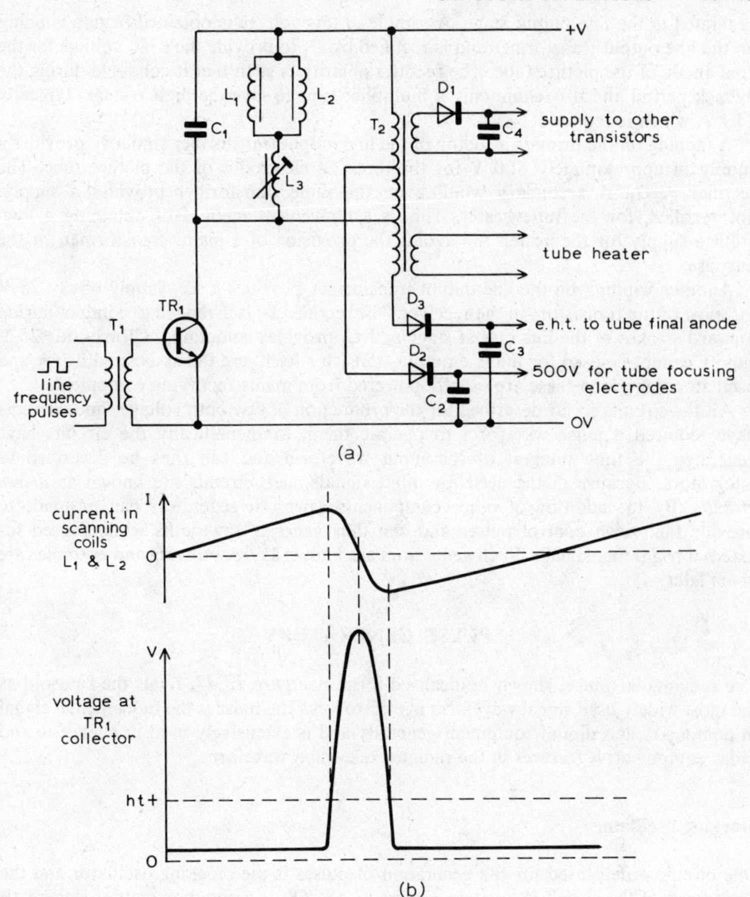

Figure 13.46 (a) More practical line output circuit for a television receiver (b) Associated waveforms

assumed initially. The current therefore falls and the constant voltage across the coils ensures that the rate of fall of current is also constant, i.e., the scan current falls linearly to zero. This current flows into the supply via TR_1 collector-base junction and the secondary winding of T_1. When the current has reached zero the circuit is in the state assumed initially and TR_1 is now turned on again to provide another period of linear growth of scan current. With proper design the linear fall and subsequent linear growth of current in the scanning coils combine to produce an uninterrupted linear change of current which constitutes the working stroke. This 'resonant return' circuit is extensively used in scanning systems and is very efficient because almost as much energy is returned by the coils to the supply during the first half of the working stroke as is taken by the coils from the supply during the second half of the working stroke. The transistor TR_1 is turned on only during the second half of the working stroke. The resonant circuit L_1–L_2–L_3–C_1 performs half a cycle of oscillation during the flyback period and this requirement enables suitable values of L_1–L_2–L_3 and C_1 to be calculated.

The rapid change of scan current during flyback causes a high voltage peak to be generated in the line output stage. A sample of this voltage is obtained from a winding on the line output transformer and is rectified by D_3 to provide the e.h.t. voltage for the final anode of the picture tube. The rectifier polarity is such that it conducts during the flyback period and it is commonly a multiplier type to give the high voltage, typically 11 kV, which is required.

A tapping on the primary winding of the line output transformer similarly provides a supply of approximately 500 V for the focusing electrodes of the picture tube. The rectifier is D_2. A secondary winding on the same transformer provides a supply, not rectified, for the tube heater. This is a convenient method of obtaining a low-voltage supply for the heater and avoids the provision of a mains transformer for the purpose.

Another winding on the line output transformer provides a d.c. supply of say 25 V for most of the transistors in the receiver. The rectifier D_1 is arranged to conduct during forward strokes of the line output stage and C_4 provides smoothing. Clearly this 25 V supply cannot be used for the line output transistor itself and the associated driver and oscillator stages and these are usually powered from mains rectifying equipment.

All the circuits so far described for the production of sawtooth voltages and currents have required a pulse waveform to operate them. Mathematically the circuits have generated the time integral of the input waveform and can thus be described as integrators. Because of the need for input signals such circuits are known as *driven circuits.* By the addition of other components sawtooth generators can be made to provide their own control pulses and can thus generate sawtooths without need for external triggering signals. Such generators are known as *free-running* and examples are given later.

PULSE GENERATORS

The rectangular pulse, shown in idealised form in *Figure 13.47*, rivals the sinusoid as the most widely used signal waveform in electronics. The pulse is the fundamental signal in computers and digital equipment generally and is extensively used in television and radar equipment: it features in the radiated television waveform.

Blocking oscillator

One circuit widely used for the generation of pulses is the blocking oscillator and the basic form of the circuit is given in *Figure 13.48*. TR_1 is a common-emitter stage with

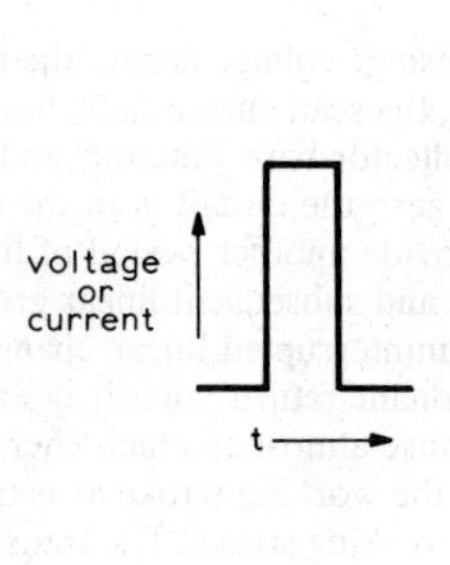

Figure 13.47 Ideal pulse waveform

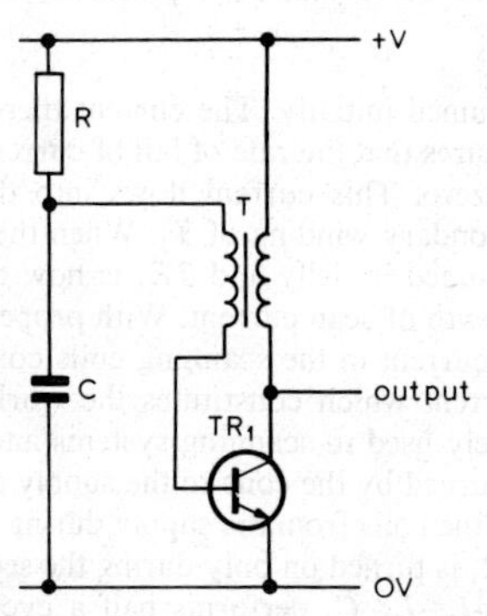

Figure 13.48 Basic circuit for a blocking oscillator

base bias provided by the resistor R. There is phase inversion between the base and collector signals and T is a transformer which also gives phase inversion between the signals delivered to the base and collector circuits. Thus a state of positive feedback exists and this results in oscillation at the resonance frequency of the collector winding, usually the larger of the two windings. One of the aims in the design of blocking oscillators is to provide considerable positive feedback so that the oscillation amplitude builds up very rapidly. As it does so TR_1 takes a burst of base current which charges up the capacitor C, the polarity being such as to drive the base negative and to cut the transistor off. This is, of course, the basis of the automatic system of biasing described in the subsection on oscillators. Ideally TR_1 should be cut off within the first half cycle of oscillation. C now discharges through R and TR_1 remains cut off until the voltage across C has fallen sufficiently for TR_1 to take base current again. This promotes another half-cycle of oscillation as a result of which TR_1 is again cut off by the negative voltage generated at the junction of R and C by the burst of base current.

Thus bursts of base current occur regularly and, of course, there are associated bursts of collector current. Negative-going voltage pulses are hence generated at the collector terminal and these can be taken as the output of the circuit. The interval between the pulses is governed by the time constant RC so that either R or C can be made variable to provide a pulse frequency control. The duration of the pulses is a function of the transformer design being primarily dependent on the inductance and self-capacitance of the primary winding.

The natural frequency of a blocking oscillator of the type illustrated in *Figure 13.48* is given by the approximate expression

$$f = \frac{n+1}{RC} \tag{23}$$

where $n : 1$ is the step-down ratio of the transformer. It is possible to control the natural frequency of a blocking oscillator by adjustment of the voltage to which R is returned

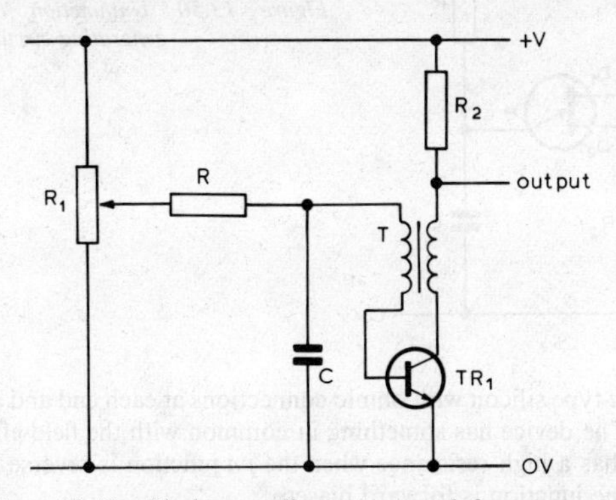

Figure 13.49 Blocking-oscillator circuit with frequency control

and this method is illustrated in the circuit diagram of *Figure 13.49*. This method of frequency control has the advantage that the potential divider R_1 carries only d.c. and can thus be situated at some distance from the blocking oscillator itself: this could be useful in a television receiver for example in which R_1 is the line-hold or field-hold control and must be readily accessible to the user. The output from the circuit of *Figure 13.49* is taken from the resistor R_2: this avoids the inclusion of any inductive effects which may be present in the output of the circuit shown in *Figure 13.48*.

Synchronisation of blocking oscillators

By contrast with the sawtooth generators described earlier the blocking oscillator is a free-running circuit, i.e., it needs no input other than the supply to make it operate. The circuit can, however, readily be synchronised at the frequency of any regularly-occurring signal applied to it. The synchronising signal is arranged to terminate the relaxation period earlier than would occur naturally. Thus to synchronise the circuits of *Figures 13.48* or *13.49* a synchronising signal could be in the form of positive-going pulses applied to the base circuit or negative-going pulses applied to the collector circuit. A blocking oscillator intended for synchronised operation should be designed to have a natural frequency slightly lower than that of the synchronising pulses.

If the natural frequency is made slightly below one half the frequency of the synchronising pulses then the blocking oscillator will be triggered into oscillation by every second synchronising signal, i.e., it will run at precisely one half of the synchronising frequency. This idea can be extended and it is possible to arrange for a blocking oscillator to run at 1/3, 1/4, 1/5, etc., of the synchronising frequency. In other words the blocking oscillator can be used as a frequency divider. Ratios much above 1/5 are a little difficult to achieve with reliability because such ratios demand close control over synchronising-pulse amplitude and natural frequency.

Unijunction transistor

The unijunction transistor (sometimes known as the double-base diode) makes possible the simple pulse-generating circuit shown in *Figure 13.50*. This semiconducting device

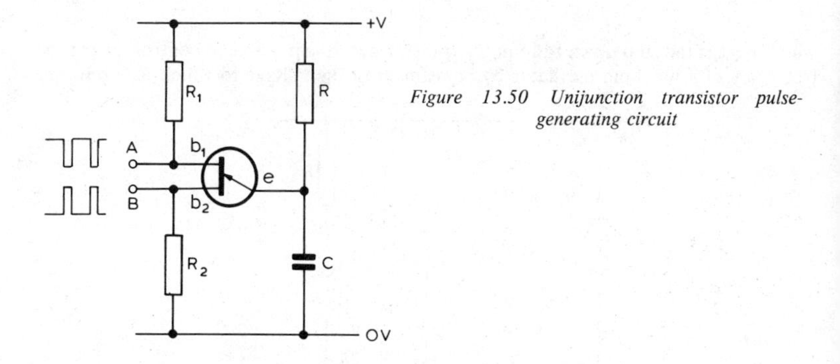

Figure 13.50 Unijunction transistor pulse-generating circuit

has a filament of *n*-type silicon with ohmic connections at each end and a *p*-type junction near the centre. The device has something in common with the field-effect transistor in that the filament has a high resistance when the *pn* junction is reverse biased but a low resistance when the junction is forward biased.

Initially C is discharged and the *pn* junction is reverse biased. The filament resistance is high and little current therefore flows through R_1 and R_2. As C charges through R the voltage across C rises. When it reaches the potential of that part of the filament which is in contact with the *p*-type emitter, the *pn* junction becomes conductive and the filament resistance abruptly drops. The enhanced conductivity is chiefly due to the injection of holes from the emitter which move towards the negative end of the filament thus increasing the forward conduction of the *pn* junction. A regenerative process is thus set up which culminates in a very rapid increase of current in the filament which discharges C. The discharge of C causes the *pn* junction to become reverse biased again with the result that the circuit just as suddenly reverts to the state originally assumed. C now begins to charge again and the cycle recommences. Each sudden burst of filament current causes a negative pulse at terminal A and a positive pulse at B. Thus the circuit produces pulses of opposite polarity at a rate determined by the time constant RC.

Multivibrators

If the transformer in the circuit of *Figure 13.48* is replaced by any other device which introduces a phase inversion between the signal received from the collector and that fed to the base, then the circuit still has positive feedback and is capable of oscillation. Such a device is a transistor connected in the common-emitter mode and the two-transistor circuit so obtained is known as a multivibrator. The behaviour of the circuit depends on the nature of the coupling circuit, i.e., whether these are direct or capacitive. There are, in fact, three basic versions of the multivibrator circuit:

(*a*) with two direct couplings;
(*b*) with one direct and one capacitive coupling;
(*c*) with two capacitive couplings.

Bistable circuit

Figure 13.51 shows the circuit diagram of a multivibrator with two direct inter-transistor couplings. The base circuit of each transistor is fed from the collector circuit

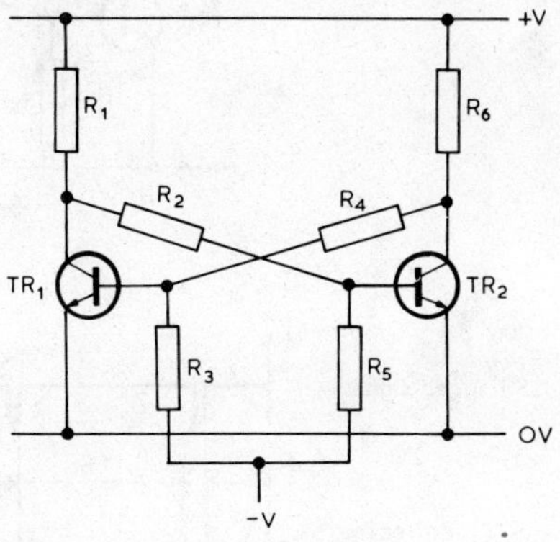

Figure 13.51 *Basic bistable multivibrator circuit*

of the other via a potential divider and the resistor values are so chosen that when one transistor is conducting, its low collector potential ensures that the other is cut off. Moreover the high collector potential of the cut-off transistor ensures that the other transistor is maintained in the conductive state. For example TR_1 may be cut off and TR_2 conducting. When the circuit is placed in this state, it will remain in it indefinitely unless compelled to change by an external triggering signal. Such a signal may consist of a positive-going pulse applied to TR_1 base to make it conduct or a negative-going pulse applied to TR_2 base to cut it off. Either form of signal will cause a change of state in both transistors. The circuit will now remain in this new state (TR_1 on and TR_2 off) indefinitely unless compelled to change it by an external signal. Persistent states such as those described are known as stable and the direct-coupled multivibrator thus has two stable states: such a circuit is known as bistable. The changes of state, once initiated by the external signal, are very fast, being accelerated by the positive feedback inherent in the circuit. The transitions from one state to the other can be made even faster by connecting capacitors in parallel with R_2 and R_4 as shown in *Figure 13.52 (a)*: C_3 and C_4 are known as speed-up capacitors.

One of the principal features of the bistable is its ability to maintain a particular state, i.e.; it has a memory. For this reason bistables are extensively used in the stores and registers of computers and similar equipment.

The bistable can also be used as a pulse generator for if it is fed with a regular stream of triggering signals it will change state with each received signal and will thus generate square waves at each collector. If the triggering signals are negative-going blips and if the first cuts TR_1 off then the next must be fed to TR_2 base to cut this off and so on, i.e., the triggers must be directed alternately to the two bases. This routing of triggers can be achieved by a diode gate circuit such as that illustrated in *Figure 13.52 (a)*. If TR_1 is cut off the diode D_1 will conduct negative triggers to TR_1 collector and thus to TR_2 base via the speed-up capacitor C_3. TR_2 is conductive and its low collector potential biases D_2 off so that the triggers cannot reach TR_1 base. Thus the triggers can only affect TR_2 which is cut off by them, causing TR_1 to be turned on. D_2 can now direct the next trigger to TR_1 base. The signals generated at the collectors are thus square waves at half the

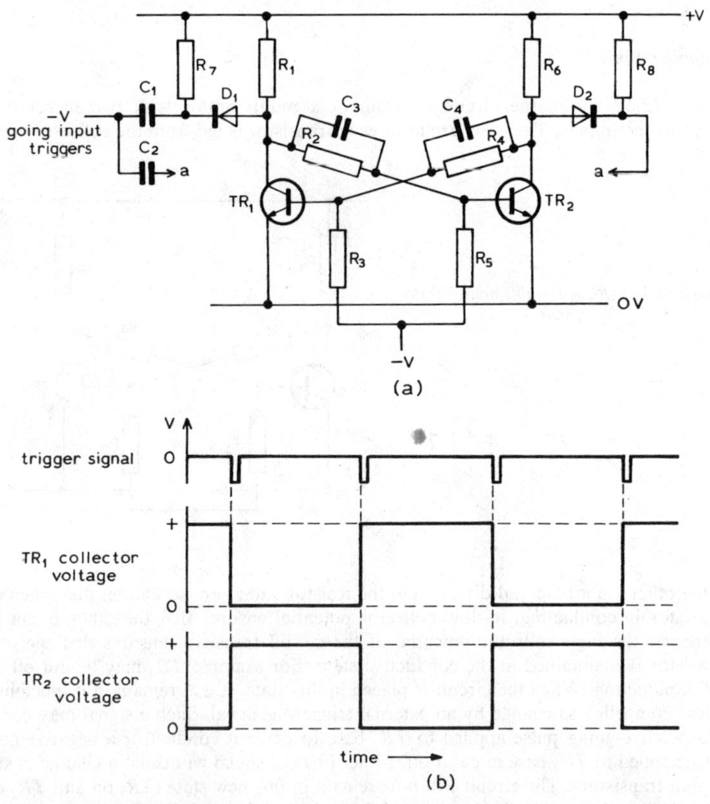

Figure 13.52 (a) Bistable circuit with speed-up capacitors and diode input gate (b) Associated waveforms

frequency of the applied triggers as shown in *Figure 13.52* (*b*). The signals at the bases are square waves also in antiphase to the collector signals and of smaller amplitude, but the speed-up capacitors can give some overshoots on the square waves which are illustrated in idealised form.

It is significant that the transistors in this circuit are either on (with a substantial collector current) or off (with zero collector current). Except for a very short period at each transition the collector currents never have any values other than these two. The transistors are used, in fact, as switches and the shape of the input–output characteristic, which is of considerable importance in analogue applications, is generally of little consequence in pulse circuits.

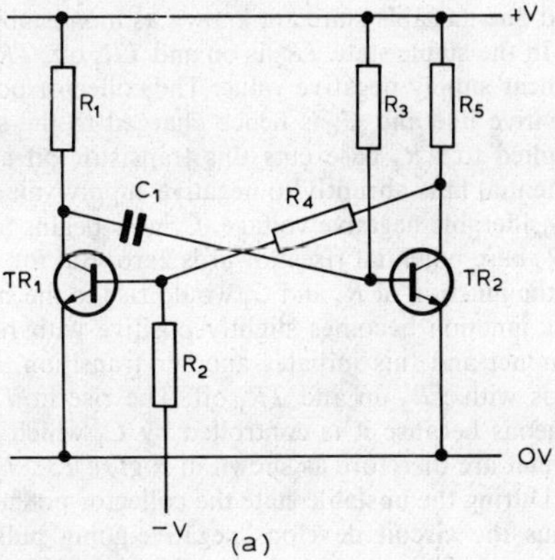

Figure 13.53 (*a*) *Basic monostable multivibrator circuit*

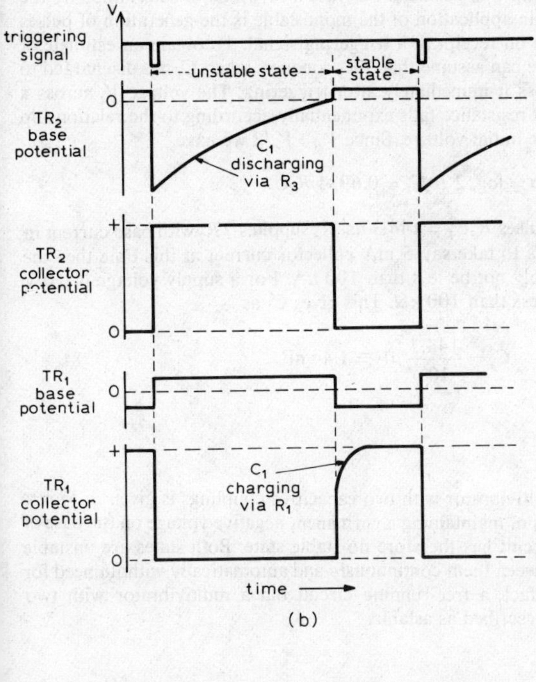

Figure 13.53 (*b*) *Associated waveforms*

Monostable multivibrator

Suppose a multivibrator has one direct coupling and one capacitive coupling as shown in *Figure 13.53* (*a*). The introduction of the coupling capacitor C_1 makes a fundamental difference to the behaviour of the circuit because there is now no means of preserving a negative voltage on the base of TR_2 to keep it non-conductive. Because the base resistor R_3 is returned to the positive supply line the circuit will always revert to the state in which TR_2 is on and TR_1 therefore off. This is therefore a stable state like those of the bistable circuit. The circuit can be triggered into the other state (TR_1 on and TR_2 off) but it cannot remain in it and the circuit will automatically return to the stable state without need of external signals to make it do so. The state in which TR_1 is on and TR_2 is off is therefore an unstable state and circuits such as this which have one stable state and one unstable state are known as monostables.

In the stable state TR_2 is on and TR_1 off. TR_2 has base current and the base potential is near supply negative value. The collector potential of TR_1 is near that of the supply positive line and C_1 is hence charged to the supply voltage. A negative-going trigger applied to TR_2 base cuts this transistor off and turns TR_1 on so that TR_1 collector potential falls abruptly to negative supply value, carrying the base potential of TR_2 to a considerable negative voltage. C_1 now begins to discharge through R_3 and as it does so TR_2 base potential rises towards zero. But for the connection to TR_2 base the potential at the junction of R_3 and C_1 would rise to the supply positive value. However as soon as this junction becomes slightly positive with respect to supply negative TR_2 begins to conduct and this initiates another transition, accelerated by positive feedback, which ends with TR_2 on and TR_1 off. The rise in TR_1 collector potential cannot be instantaneous because it is controlled by C_1 which charges via R_1. The waveforms for this circuit are therefore as shown in *Figure 13.53* (*b*).

During the unstable state the collector potential of TR_1 is low and that of TR_2 is high. Thus the circuit develops negative-going pulses at TR_1 collector and positive-going pulses at TR_2 collector during this period the duration of which is determined by the time constant R_3C_1. The main application of the monostable is the generation of pulses of a predetermined duration on receipt of a triggering signal. To obtain an estimate of the duration of the pulses we can assume that TR_2 conducts when C_1 has discharged to half the voltage it had across it immediately after triggering. The voltage V_t across a capacitor discharging into a resistance falls exponentially according to the relationship $V_t = V_o e^{-t/RC}$ where V_o is the initial voltage. Since $V_t = V_o/2$ we have

$$t = \log_e 2\, R_3C_1 = 0.6931\, R_3C_1$$

Thus if we require 100-μs pulses $R_3C_1 = 144.3\ \mu$s. R_3 supplies TR_2 with base current in the stable state and if TR_2 is to take say 5 mA collector current at this time then the base current should preferably not be less than 100 μA. For a supply voltage of 10 V therefore R_3 should not be less than 100 kΩ. This gives C_1 as

$$C_1 = \frac{144.3}{10^5}\ \mu\text{F} = 1.44\ \text{nF}$$

Astable multivibrator

The circuit diagram of a multivibrator with two capacitive couplings is given in *Figure 13.54* (*a*). There is no means of maintaining a permanent negative voltage on the base of either transistor and this circuit has therefore no stable state. Both states are unstable and the circuit oscillates between them continuously and automatically without need for triggering signals. It is, in fact, a free-running circuit but a multivibrator with two unstable states is generally described as astable.

The waveforms can be deduced from those of the monostable circuit and are given in *Figure 13.54* (*b*). The duration of one unstable state is given approximately by 0.6931 R_3C_1 and of the other 0.6931 R_2C_2 so that the natural frequency of the astable circuit is given by

$$f = \frac{1}{0.6931(R_3C_1 + R_2C_2)}$$

Although the circuit does not require external signals for its operation it can readily be synchronised at the frequency of a regular external signal. Normally synchronisation

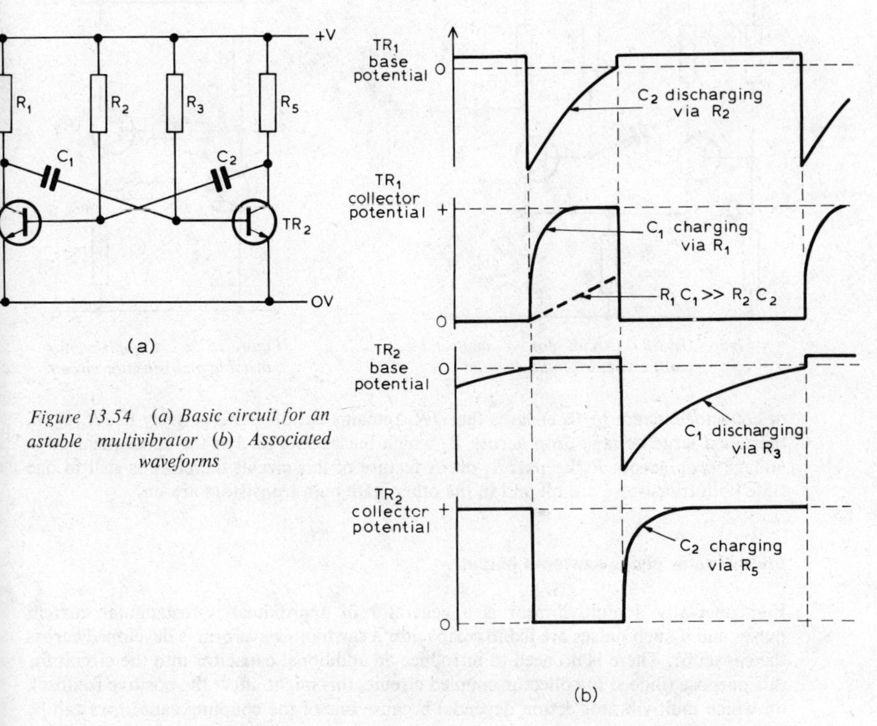

Figure 13.54 (a) Basic circuit for an astable multivibrator (b) Associated waveforms

is achieved by terminating the unstable periods earlier than would occur naturally and therefore positive-going synchronising signals are required at the bases of the *npn* transistors and the natural frequency of the astable circuit should be slightly lower than that of the synchronising signal.

Emitter-coupled multivibrator circuits

All the multivibrator circuits so far described have included two collector-to-base couplings. It is possible to replace one of these with an emitter-to-emitter coupling: in a direct coupling the emitters are simply bonded and returned to the supply via a common resistor as shown in the monostable circuit of *Figure 13.55*. Alternatively a capacitive emitter coupling can be obtained by using individual emitter resistors bridged via a capacitor. An advantage of emitter coupling is that one collector terminal is free, i.e., not involved in the provision of positive feedback and can thus be used as a convenient output point as suggested in *Figure 13.55*.

Complementary multivibrator circuits

By using a combination of *pnp* and *npn* transistors the particularly-simple bistable circuit of *Figure 13.56* is possible. When the *npn* transistor TR_1 is cut off there is no voltage drop across R_1 and therefore the *pnp* transistor TR_2 is also cut off. The absence

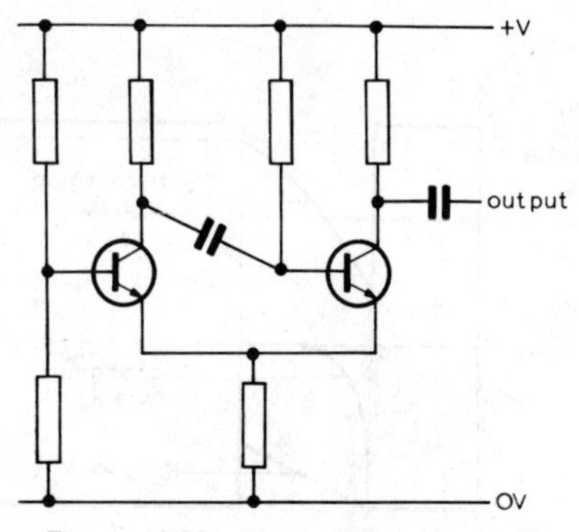

Figure 13.55 Circuit for a monostable emitter-coupled multivibrator

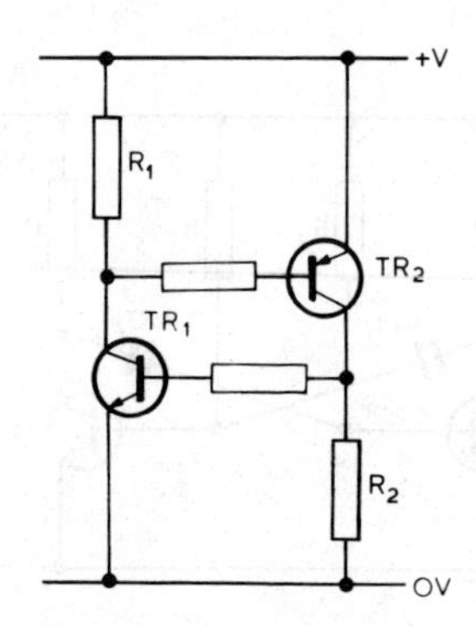

Figure 13.56 Complementary bistable multivibrator circuit

of collector current in R_2 ensures that TR_1 remains cut off. Alternatively if TR_1 is on, there is a large voltage drop across R_1 which biases TR_2 on and the consequent large voltage drop across R_2 keeps TR_1 on. A feature of this circuit therefore is that in one state both transistors are off and in the other state both transistors are on.

Multivibrator giving sawtooth output

Fundamentally a multivibrator is a generator of approximately-rectangular current pulses and if such pulses are fed to a capacitor a sawtooth waveform is developed across the capacitor. There is no need to introduce an additional capacitor into the circuit for this purpose (indeed in collector-coupled circuits this might affect the positive feedback on which multivibrator action depends) because one of the coupling capacitors can be used.

It has already been pointed out that when a transistor is cut off in a multivibrator circuit, its collector potential cannot rise instantaneously to supply voltage value because the capacitor coupling the collector to the base of the other transistor must charge during this period via the collector load resistor. Thus in *Figure 13.54* (*a*) when TR_1 is cut off C_1 charges via R_1. If the time constant R_1C_1 is made long compared with the period of non-conduction of TR_1 (determined by the time constant R_2C_2) C_1 is not fully charged during this period and the voltage generated at TR_1 collector is exponential in form. The initial part of an exponential rise is almost linear as shown in dotted lines in *Figure 13.54* (*b*). Thus the condition to be satisfied to obtain such an output is: $R_1C_1 \gg R_2C_2$. The choice of values for R_1 and C_1 is limited because the time constant R_3C_1 determines the period of non-conduction of TR_2 and hence of conduction of TR_1.

Limiting circuits

The blocking oscillator and multivibrator are two circuits capable of generating pulses. Pulses can also be obtained by shaping other waveforms such as sine waves. For example if a sinusoidal signal is applied to the circuit of *Figure 13.57* (*a*) the diode

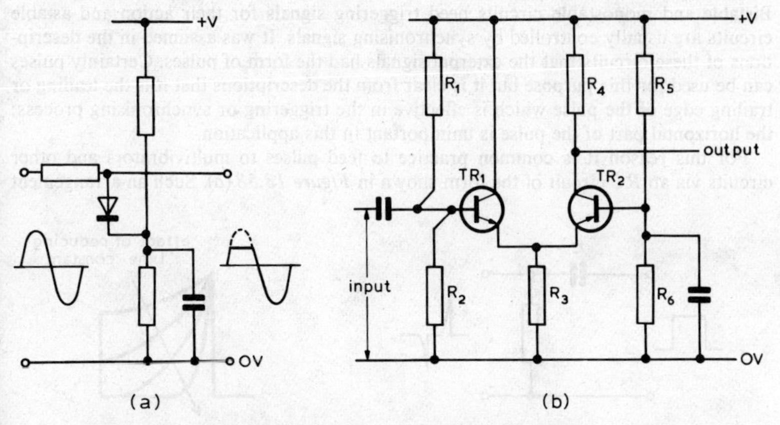

Figure 13.57 (a) Simple diode limiting circuit (b) A two-transistor limiting circuit

will conduct and hence attenuate all parts of the signal which exceed the diode cathode bias in value, giving an output of the form shown. If the negative-going excursions of this output are eliminated (for example by the use of a second diode) we are left with an approximation to a rectangular pulse: this becomes a nearer approach to the ideal form the greater the ratio of the sine-wave amplitude to the diode bias.

Better results can be obtained by using transistors in place of diodes as limiting devices and the circuit diagram of a two-transistor circuit which limits on positive and negative peaks is given in *Figure 13.57* (*b*). The two transistors are emitter-coupled by a common resistor, an arrangement known as a long-tailed pair. TR_1 is an emitter follower driving the common-base stage TR_2. Positive-going signals applied to TR_1 base appear at substantially the same amplitude at TR_2 emitter and, if large enough, cut TR_2 off, causing its collector potential to rise to supply positive value. Negative-going signals, if large enough, cut TR_1 off so that TR_2 takes a steady collector current (determined by the potential divider R_5–R_6 and the emitter resistor R_3 and the collector potential is at a steady value below that of the positive supply rail. Thus a large-amplitude signal at the input to this circuit gives an output at TR_2 collector which alternates between two steady values, i.e., it is a pulse output at the frequency of the input signal.

A simple common-emitter amplifier can, of course, be used as a limiter. By giving it a very large input signal (which, in linear amplifier operation would be described as a gross overload) the transistor is driven into cut off on one half-cycle of the input signal and into saturation on the other half cycle. Thus the collector potential is either at the positive or the negative supply value. The larger the input is made, the shorter is the time taken for the collector potential to switch from one extreme voltage to the other and the better is the shape of the output pulses. The input signal can be sinusoidal or it can be a pulse waveform which has overshoots, ripples or other undesirable features on it. These unwanted features can, with proper design, be removed in the limiting circuit which can thus be regarded as a 'cleaner stage'.

PULSE-SHAPING CIRCUITS

Differentiating and integrating circuits

Bistable and monostable circuits need triggering signals for their action and astable circuits are usually controlled by synchronising signals. It was assumed in the descriptions of these circuits that the external signals had the form of pulses. Certainly pulses can be used for this purpose but it is clear from the descriptions that it is the leading or trailing edge of the pulse which is effective in the triggering or synchronising process: the horizontal part of the pulse is unimportant in this application.

For this reason it is common practice to feed pulses to multivibrators and other circuits via an *RC* circuit of the form shown in *Figure 13.58* (*a*). Such an arrangement

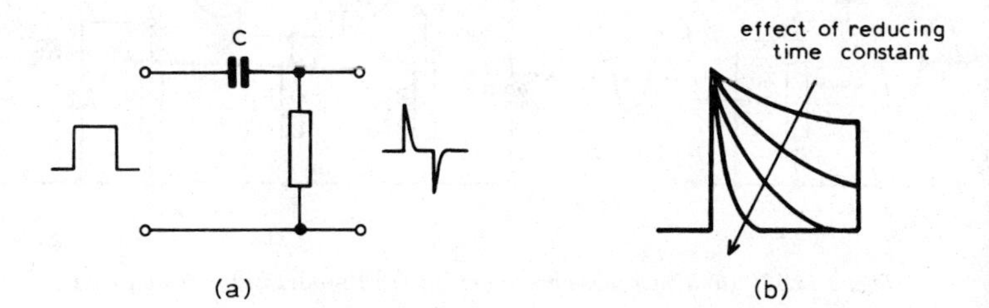

Figure 13.58 A simple differentiating circuit is shown at (*a*) *and the effect on the output waveform of reducing the time constant is shown at* (*b*)

of series capacitance and shunt resistance is also commonly encountered in the inter-transistor coupling circuits of linear amplifiers where the time constant must be so chosen that the signal is transmitted through the *RC* circuit with negligible change in waveform. For a pulse signal a very long time constant would be necessary to transmit the horizontal sections with negligible sag. However the horizontal sections are of no interest if the pulses are used for triggering or synchronising and it is therefore permissible to use a short time constant in the network used to transmit them. The effect of a time constant which is short compared with the pulse repetition period is illustrated in *Figure 13.58* (*b*). The vertical leading and trailing edges are reproduced without distortion but the horizontal section is badly distorted. If the time constant is reduced sufficiently the output becomes simply a succession of alternate positive-going and negative-going spikes. Such a waveform is, of course, quite suitable for triggering or synchronising purposes. Mathematically such a waveform is the first derivative of the input waveform and for this reason the network of *Figure 13.58* (*a*) is often called a differentiating circuit. Thus the time constants R_7C_1 and R_8C_2 in *Figure 13.52* (*a*) could both be small compared with the period of the input signal.

The circuit of *Figure 13.59*, on the other hand, gives an output waveform similar to the time integral of the input waveform, provided the time constant *RC* is long compared with the period of the input signal.

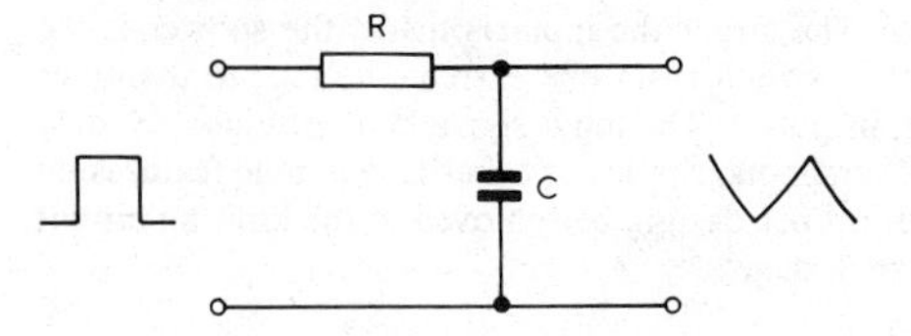

Figure 13.59 A simple integrating network. RC must be large compared with the period of the input signal

Pulse-shortening circuit

The circuit illustrated in *Figure 13.60 (a)* makes use of a differentiating network to produce short pulses. Input pulses are differentiated by RC to produce positive-going and negative-going spikes as shown. TR_1 is normally biased off by resistor R and its collector potential is therefore at supply positive value. The positive-going spike of the differentiated input signal triggers TR_1 into conduction and produces a negative-going pulse at the collector. The leading edge of this pulse is coincident with the positive-going leading edge of the pulse input to the circuit. Provided the input pulse amplitude is large

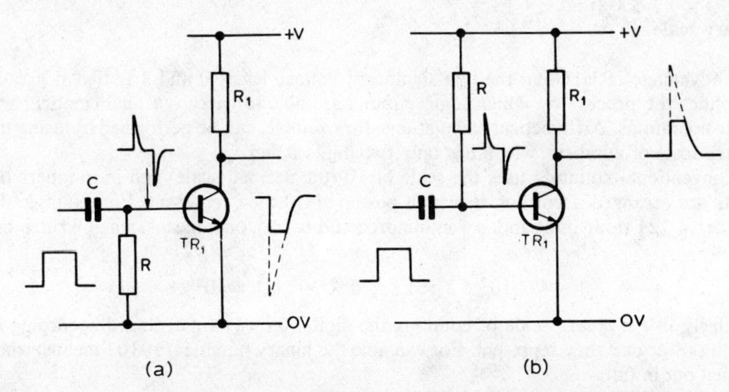

Figure 13.60 Pulse-shortening circuit giving output pulses with a leading edge coincident with (a) the leading edge and (b) the trailing edge of the input pulses

enough TR_1 will reach saturation when triggered and the width of the collector pulse waveform will depend on the time constant RC, increasing with increase in the time constant.

If the resistor R is returned to the positive supply line as shown in *Figure 13.60 (b)*, the transistor is normally biased to saturation and the collector potential is at supply negative value. The transistor now ignores the positive-going spikes of the differentiated input waveform but is cut off by the negative-going spikes so that a positive-going pulse is generated at the collector and its leading edge is coincident with the trailing edge of the input waveform. Again the width of the pulse so generated can be increased by increases in the time constant RC.

DIGITAL TECHNIQUES

Introduction

The transistors used in multivibrators and other pulse circuits are used as switches, i.e., except for the very short periods during which they are changing state they are either on (usually saturated) or off (non-conductive). It follows that the collector voltage always has one of two values, a low value (near emitter potential) when the transistor is on and a high value (near the supply potential) when the transistor is off. A vast range of circuits has been developed during the past decade or two in which diodes and transistors are used as switches and in which the signal paths always have one of two possible voltage levels. Such circuits are used to perform mathematical and logical operations on signals in computers and similar equipment, and are known as digital or logic circuits.

Logic levels

The two significant values of voltage on the signal-carrying lines are referred to as logic level 0 and logic level 1. If level 1 is more positive than level 0 the circuit is said to use the *positive logic convention* and if level 1 is more negative than level 0 the circuit is said to use the *negative logic convention.* The distinction is important because circuits can behave differently according to the logic convention chosen. This is illustrated below but it is worth stressing now that the logic convention chosen should always be stated on diagrams of logic circuits.

Binary scale

The advantage of labelling the two significant voltage levels 0 and 1 is that it greatly simplifies the process by which logic circuits are able to carry out mathematical and other operations. Arithmetical calculations, for example, can be performed by using the binary scale of numbers, which has only two digits 0 and 1.

Conventional counting uses the scale of 10 (the decimal scale) and in numbers the digits are arranged according to which power of 10 they represent. For example the number 4 721 (four thousand, seven hundred and twenty one) means, when written out in full:

$$4 \times 10^3 + 7 \times 10^2 + 2 \times 10^1 + 1 \times 10^0$$

Similarly in the binary scale of counting the digits (0 or 1) are arranged according to which power of 2 they represent. For example the binary number 110101 means, when written out in full:

$$1 \times 2^5 + 1 \times 2^4 + 0 \times 2^3 + 1 \times 2^2 + 0 \times 2^1 + 1 \times 2^0$$

i.e.,
$$32 + 16 + 0 + 4 + 0 + 1 = 53$$

The first nine numbers in the binary scale are as follows:

binary number	*decimal equivalent*
1	1
10	2
11	3
100	4
101	5
110	6
111	7
1 000	8
1 001	9

It is common practice however not to code large numbers strictly in the binary scale but instead to code each digit of the decimal number separately in the binary scale. This is known as the binary coded decimal method and in it the number 4 721 would be coded as follows:

$$100 : 111 : 10 : 1$$

i.e.,
$$4 : 7 : 2 : 1$$

This system has the advantage that after a little experience one can translate the binary coded number into decimal form on inspection and secondly it simplifies the design of equipment for coding decimal numbers into binary form and for decoding and displaying binary-coded numbers in decimal form.

Logic gates

In the simple logic circuit of *Figure 13.61*, inputs A and B may be at +10 V or 0 V, these being the standard voltage levels chosen for use in this circuit. If either input is at 0 V the associated diode conducts and, if the forward resistance of the diode is

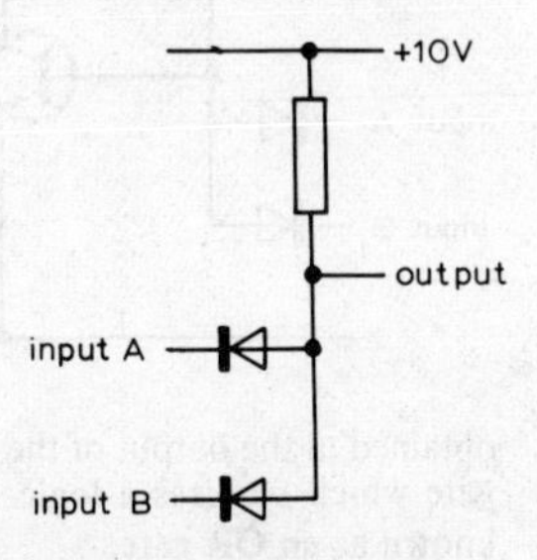

Figure 13.61 Simple diode gate circuit. It can be an AND gate or an OR gate depending on the logic convention adopted

neglected, 0 V appears at the output. For an input at +10 V the associated diode does not conduct and the circuit is isolated from this input. For two inputs there are four possible combinations of input voltage. They are shown in *Table 13.1* together with the value of the output voltage. This table shows that the output is 0 V if either or both of the inputs is at 0 V and that the only way to obtain +10 V at the output is for both of the inputs to be at +10 V.

If the positive logic convention is used +10 V is logic level 1 and 0 V is logic level 0. If we repeat *Table 13.1* in terms of logic levels 0 and 1 we obtain the result shown in *Table 13.2*. Such a table is known as a truth table and it shows that a logic 1 is obtained at the output of the circuit of *Figure 13.61* only when input A and input B have a logic-1 signal. A circuit such as this which requires a logic-1 signal at all the inputs to give a logic-1 signal at the output is known as an AND gate.

Suppose in the circuit of *Figure 13.61* we had used the negative logic convention instead. +10 V now represents logic level 0 and 0 V represents logic level 1. The truth table now has the form shown in *Table 13.3*. From this it is clear that a logic 1 is

Table 13.1 VOLTAGES IN THE CIRCUIT OF FIGURE 13.61

Input voltage		*Output voltage*
A	B	
0	0	0
0	+10	0
+10	0	0
+10	+10	+10

Table 13.2 TRUTH TABLE FOR AN AND GATE

Input		*Output*
A	B	
0	0	0
0	1	0
1	0	0
1	1	1

Table 13.3 TRUTH TABLE FOR AN OR GATE

Input		*Output*
A	B	
0	0	0
0	1	1
1	0	1
1	1	1

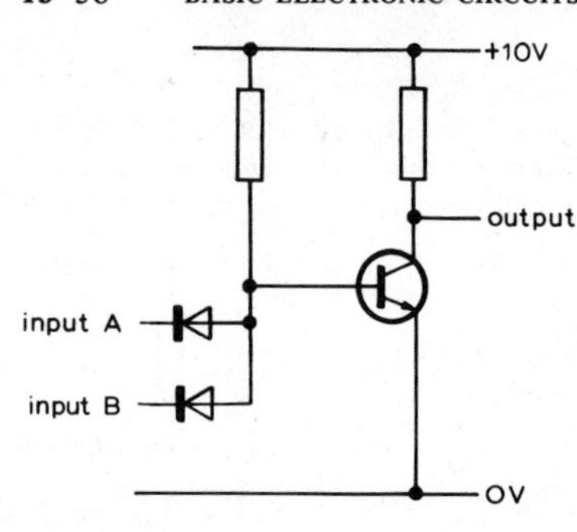

Figure 13.62 An inverter stage following the diode gate results in a NAND gate or a NOR gate depending on the logic convention adopted

obtained at the output of the gate when either input A or input B has a logic-1 signal. A gate which requires a logic-1 signal at any one input to give a logic-1 at the output is known as an OR gate.

Thus the circuit illustrated in *Figure 13.61* can behave as an AND gate or as an OR gate depending on the logic convention used. In general it is not possible to state the nature of a logic circuit until the logic convention to be used with it is known.

Suppose a common-emitter transistor amplifier stage is added after the diode gate as shown in *Figure 13.62*. The transistor circuit is so designed that the collector output is always at one or other of the two chosen voltage levels. The signal inversion introduced by the amplifier gives yet another type of behaviour. Study of the circuit shows that the relationship between the inputs and output, for positive logic, is as given in *Table 13.4*. This shows that the only way to obtain a logic 0 at the output is for all the input signals to be at logic 1. This behaviour is not surprisingly the inverse of that of the AND gate: this circuit, for positive logic, is therefore known as a NOT-AND or more simply a NAND gate.

Table 13.4 TRUTH TABLE FOR A NAND GATE

Input		*Output*
A	B	
0	0	1
0	1	1
1	0	1
1	1	0

Table 13.5 TRUTH TABLE FOR A NOR GATE

Input		*Output*
A	B	
0	0	1
0	1	0
1	0	0
1	1	0

For negative logic the behaviour of the circuit of *Figure 13.62* is illustrated in the truth table of *Table 13.5*. When any of the input signals is at logic 1 the output is at logic 0. The only way to obtain a logic 1 at the output is for all the inputs to be at logic 0. This behaviour is the inverse of that of the OR gate and this circuit, for negative logic, is known as a NOT-OR or NOR gate.

GATE SYMBOLS

For simplicity only two inputs are shown in *Figures 13.61* and *13.62* but there can, of course, be more. The block symbols for the four basic types of gate so far introduced are shown in *Figure 13.63* and in each of these three inputs are shown. The circle at the

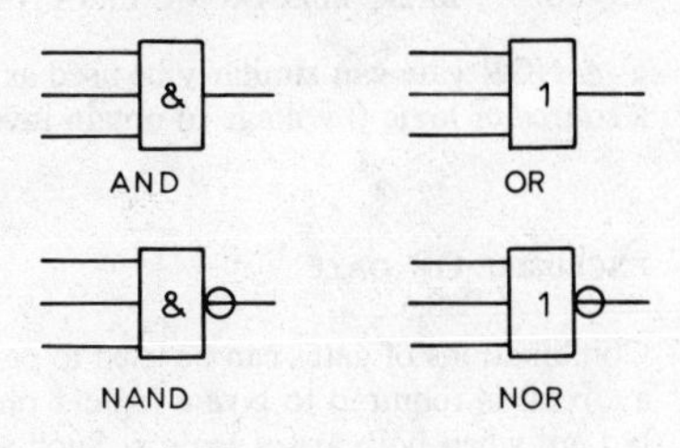

Figure 13.63 Block (or logic) symbols for the four basic types of gate

output of the NOR and NAND gates indicates the inversion of the output relative to that of the OR and AND gates.

INTEGRATED CIRCUIT GATES

It is not usual to construct gates of discrete components: normally they are in the form of integrated circuits. For example a single i.c. may contain three three-input gates. To use such an i.c. in an equipment it is not necessary to have details of the circuitry of the device. All the designer needs to know to be able to use the i.c. successfully are details of input and output signal levels, polarities, impedances and supply voltages. In preparing circuit diagrams of computers and computer-like equipments the gates and other functional units are represented by block symbols such as those for gates given in *Figure 13.63*. To help in the layout of printed wiring cards and in maintenance the inputs, outputs and supply points of the gate symbols can be labelled with the pin numbers of the i.c.s in the block diagrams which are usually termed logic diagrams.

To illustrate the versatility of logic gates a number of applications will now be considered.

GATES AS SWITCHES

Table 13.2 shows that when input A is at logic 0, the output is also 0 (irrespective of the signal on input B) whereas if input A is at logic 1 the output signal is the same as the signal on input B. Thus an AND gate can be used as a switch, a logic-1 signal on input A allowing the passage of a signal through the gate via input B. An example of such a use of an AND gate is given later.

An OR gate can be used similarly but here a logic 0 on one input allows the signal on the other input to pass through the gate.

INVERTERS

Table 13.4 shows an interesting property of a NAND gate. A logic-1 signal applied to input A gives an output from the gate which is the inverse of that on input B. A NAND gate is frequently so used, input A being connected permanently to a source of logic-1 voltage as shown in *Figure 13.64*. For some types of NAND gate i.c.s it is sufficient simply to leave input A unconnected: this has the same effect as connecting it to a logic-1 source.

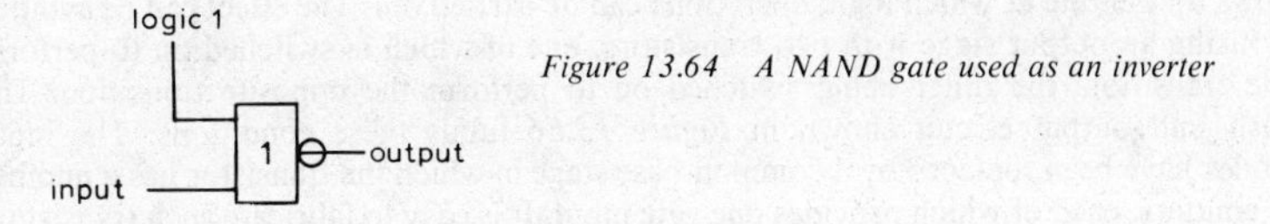

Figure 13.64 A NAND gate used as an inverter

A NOR gate can similarly be used as an inverter but one input must be connected to a source of logic-0 voltage to obtain inversion of the signal on the other input.

EXCLUSIVE-OR GATE

Combinations of gates can be used to perform desired operations. For example suppose a circuit is required to give a logic-1 output when either of the two inputs is at logic 1 but not when both are at logic 1. Such a circuit is known as an exclusive-OR gate.

There are a number of possible circuits for this and one is given in *Figure 13.65* (*a*). That this circuit gives the required performance can be checked by means of a truth

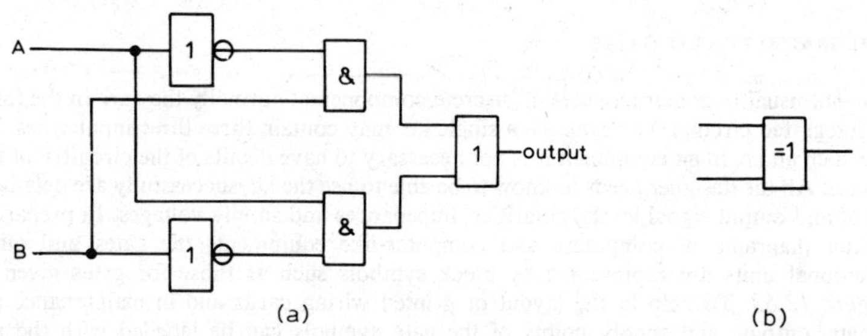

Figure 13.65 (*a*) *One possible circuit for an exclusive-OR gate, and* (*b*) *the logic symbol*

table (by listing the inputs to A, B, the inverters, the AND gates and the OR gate) or by the use of Boolean algebra and this second method is the one normally used in designing logic gate circuits. The exclusive-OR gate is usually represented by the logic symbol of *Figure 13.65* (*b*).

EQUIVALENCE ELEMENT

It is sometimes necessary to compare the logic states of two signals and to give an indication when they are the same. A circuit used for this purpose is known as an equivalence element or a comparator and is required to give a logic-1 output when the two inputs are both at logic 1 or both at logic 0. The required output is the inverse of that of the exclusive-OR gate and can be obtained from a combination of gates similar to that shown in *Figure 13.65* (*a*) but with the OR gate replaced by a NOR gate.

PRACTICAL NAND GATE CIRCUIT

The basic circuit shown in *Figure 13.62* has a number of disadvantages. One is that the speed with which the output changes from the 0 state to the 1 state (positive logic assumed) is low compared with the speed of the opposite transition. This is because a transistor can be switched on quickly but the time of switch off is appreciable. This slows up the rate at which logic operations can be carried out. The effect can be avoided by using an output stage with two transistors, one of which is switched on to perform one transition, the other being switched on to perform the opposite transition. The push–pull output circuit shown in *Figure 13.66* fulfils these conditions. The input diodes have been replaced by a common-base stage in which the transistor has a number of emitters, each of which provides one gate input. It is easy to fabricate such transistors using i.c. techniques. The second transistor is a phase splitter driving the push–pull output stage. Diode D provides base bias for TR_3.

For positive logic this circuit is a NAND gate. To confirm this assume the output is at logic 0: this requires TR_3 to be off and TR_4 on which, in turn, requires TR_2 to be on. TR_2 must therefore have a high base potential which requires TR_1 to be off. The only

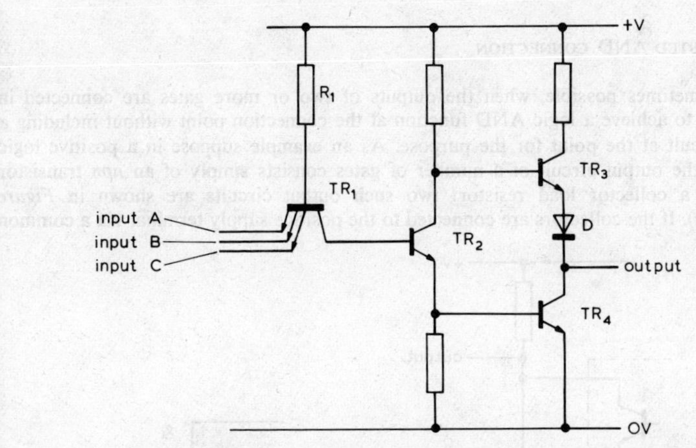

Figure 13.66 Circuit diagram of an i.c. NAND gate

way for TR_1 to be off is for *all* the emitters to be at positive supply potential, i.e., at logic 1. By definition a gate which requires all inputs to be at logic 1 to give a logic 0 output is a NAND gate.

FAN IN

In circuits containing numerous gates it may happen that several gates feed one particular gate. Several output circuits are then connected in parallel across one input circuit and it is essential that this loading should not affect the voltage level at the junction, no matter whether this is the logic-0 or logic-1 voltage. There is normally a tolerance on the voltage levels but if too many circuits are connected together the junction voltage may be outside the tolerance and normal circuit behaviour becomes impossible. The greatest number of inputs which may be connected to a gate while still permitting normal behaviour is known as the fan in. This may in practice be as high as 10 but in the gate circuits and symbols given earlier the number of inputs was shown for simplicity as 2 or 3.

FAN OUT

A junction between input and output circuits also occurs when one gate is required to feed a number of others: one output circuit is thus connected to several input circuits. Current is required to operate an input circuit. For example in *Figure 13.66* to put a logic-0 signal on an emitter of TR_1, the emitter must be brought to supply negative level (positive logic is assumed). An emitter current (determined by R_1 and the supply voltage) of, say, 1.5 mA then flows in the input circuit and the output circuit feeding it. The conductive transistor in the output circuit must be capable of passing this current without undue rise in voltage across it. Any significant rise may cause this voltage to be outside the range recognised as a logic-0 signal. The maximum collector current through the output transistor may be, for example, 15 mA which enables up to 10 gate circuits to be fed satisfactorily. This maximum number is known as the fan out of the circuit.

DISTRIBUTED AND CONNECTION

It is sometimes possible, when the outputs of two or more gates are connected in parallel, to achieve a logic AND function at the connection point without including a gate circuit at the point for the purpose. As an example suppose in a positive logic system the output circuit of a number of gates consists simply of an *npn* transistor without a collector load resistor: two such output circuits are shown in *Figure 13.67* (*a*). If the collectors are connected to the positive supply terminal via a common

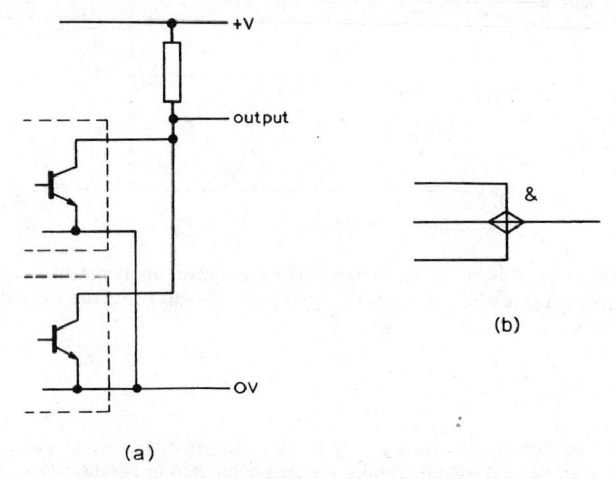

Figure 13.67 (*a*) *Distributed AND connection, and* (*b*) *the logic symbol*

load resistor then the only way in which a logic-1 signal can be obtained at the common connection point is by cutting off all the output transistors. Thus all the transistors must be in a logic-1 state to give a logic 1 at the common output point. An AND function is thus obtained at the interconnection point irrespective of the nature of the individual gates shown by dashed lines in the diagram. It is, of course, the use of free collectors in the gates and of the external load resistor which makes this possible.

An AND function so obtained is known as a distributed or wired AND: it is represented in logic diagrams by the symbol given in *Figure 13.67* (*b*).

DISTRIBUTED OR CONNECTION

It is also possible to achieve an OR function at the point where a number of gate outputs are connected in parallel. *Figure 13.68* shows an example in which the output stages are emitter followers and the emitter connections are brought out to an external load resistor. We now obtain a logic-1 output signal (positive logic assumed) when any one output stage is at logic 1; one conductive emitter follower over-rides all the non-conductive ones. This is a logic OR function and is achieved without introducing an OR

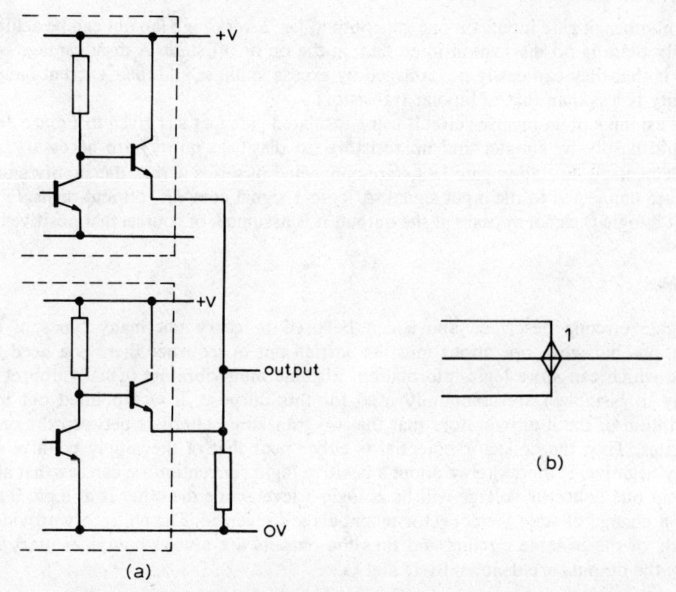

Figure 13.68 (a) Distributed OR connection and (b) the logic symbol

gate for the purpose. This is known as a distributed or wired OR connection and is represented by the logic symbol shown in *Figure 13.68 (b)*.

INSULATED-GATE F.E.T. I.C.S AS LOGIC GATES

I.C.s using the bipolar-transistor techniques so far discussed in this section are extensively used. They have the advantages that they operate satisfactorily with low collector voltages, switch-on times are low and they have a high current-handling capacity. Dissipation in the collector circuits is low in the off state because the current is nearly zero and in the on state because the collector voltage is nearly zero. However appreciable power is required in the base circuit to keep a bipolar transistor in the on state.

Insulated-gate f.e.t.s have a number of advantages over junction transistors in logic-gate i.c.s and are now being used. One advantage is that f.e.t.s can be manufactured much smaller than junction transistors and thus greater miniaturisation is possible. Secondly because of the very high input resistance of f.e.t.s it is possible to parallel a

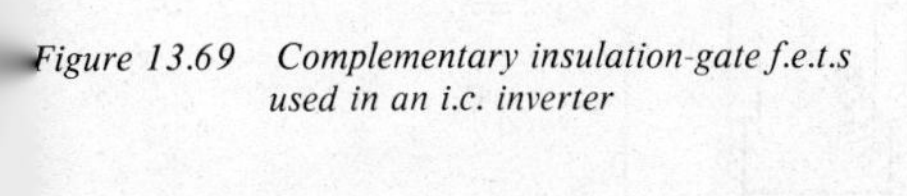
Figure 13.69 Complementary insulation-gate f.e.t.s used in an i.c. inverter

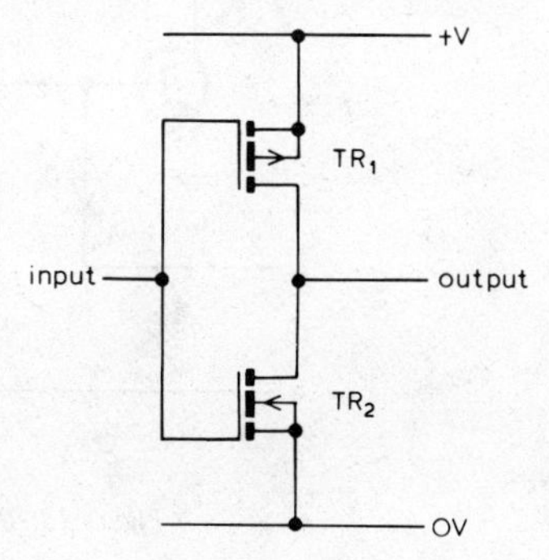

large number of gate inputs on one gate output, i.e., a very high fan out can be achieved. Thirdly there is no dissipation in an f.e.t. in the on or off state. A disadvantage of the f.e.t.s is that they can easily be damaged by excess voltages and their current-handling capacity is less than that of bipolar transistors.

An example of an inverter circuit using insulated-gate f.e.t.s is given in *Figure 13.69*. It could hardly be simpler and no resistors (to dissipate power) are necessary. Two complementary insulation-gate f.e.t.s are connected in series across the supply and the gates are connected to the input signal. A logic-1 signal cuts TR_1 off and turns TR_2 on so that a logic-0 signal appears at the output. It is assumed, of course, that positive logic is used.

Bistables

The gate circuits described above can be used to carry out many types of logic operations but when operations must be carried out in sequence there is a need for a device which can storc logic information. Bistable multivibrators (usually abbreviated simply to bistables) are commonly used for this purpose. It was pointed out in the description of the multivibrators that the two transistors alternate between the on and off states. Thus the collector potential is either near that of the supply positive or of supply negative. If therefore we adopt a positive logic convention we can say that at any moment one collector voltage will be at logic-1 level while the other is at logic-0 level. After a change of state the collector potentials are reversed. The collectors provide the outputs of the bistable circuits and thus the outputs are always complementary logic states: the outputs are denoted by Q and $\bar{Q}$.

RS BISTABLE

A simple form of RS bistable circuit is illustrated in *Figure 13.70*. Outputs are taken from the collectors of the transistors and two inputs known as S (set) and R (reset) are connected to the bases via diodes arranged to conduct negative-going signals. A

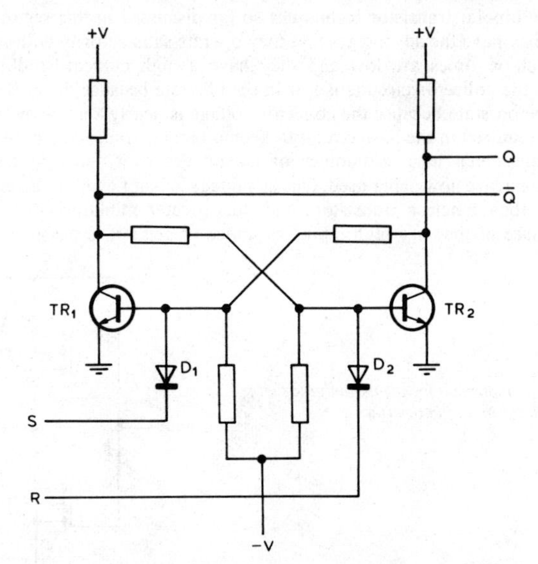

Figure 13.70 Basic RS bistable circuit

negative-going signal applied to the S input cuts off TR_1 (and hence turns TR_2 on). Similarly a negative-going signal applied to the R input cuts off TR_2 (and hence turns TR_1 on), so restoring the circuit to the state it had originally. If inputs are applied to the R and S terminals simultaneously there is no way of knowing what the resultant state of the circuit will be because it is not natural, in a bistable, to have negative signals on both bases at the same instant. RS bistables are therefore never used in circumstances where simultaneous R and S inputs are possible. If there is no input to the R or S terminals the bistable remains in its previous state.

Thus for three of the four possible combinations of R and S inputs, the resultant state of the bistable is predictable: for the fourth combination the output is indeterminate. This behaviour is summarised in the following truth table.

R *input*	S *input*	Q *output*
0	0	indeterminate
0	1	1
1	0	0
1	1	no change

The RS bistable can also be operated in a clocked mode. For this purpose a diode gate, such as that illustrated in *Figure 13.52* (D_1 and D_2) is included and the pulses fed to the diodes are known as clock pulses. These control the changes of state of the

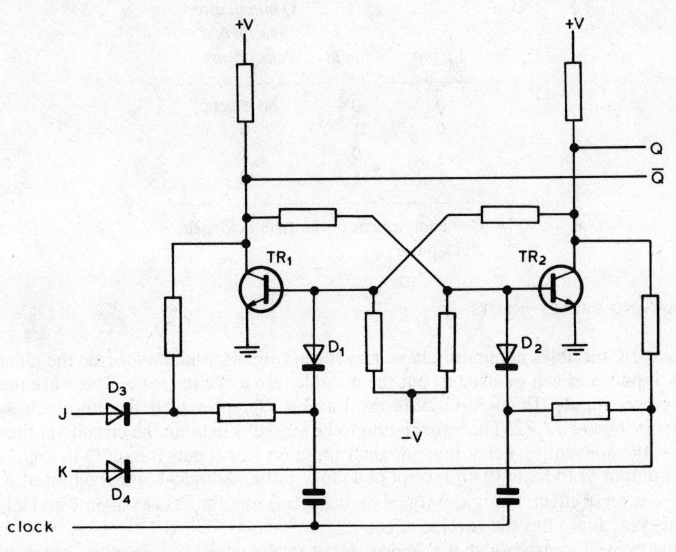

Figure 13.71 Circuit of a typical clocked JK bistable

bistable: thus when an input is applied to the R or S input, the bistable does not respond to it until a clock pulse is received. The behaviour of the circuit is still as given in the above table but the third column should be interpreted as giving the logic level of the Q output after receipt of a clock pulse.

JK BISTABLE

The JK bistable may be regarded as an improved form of the RS bistable in which there is no indeterminate state. The improvement is achieved by arranging for the signal inputs to operate on the gating diodes, D_1 and D_2, to which the clock pulses are applied. A typical circuit diagram is given in *Figure 13.71*. D_1 and D_2 are biased by the difference between the collector and base potentials of the associated transistor. If the transistor is on, there is little potential difference and the diode can conduct negative-going clock signals to the base to cut the transistor off. When the transistor is off, the considerable difference between collector and base potentials reverse-biases D_1 or D_2 so that clock signals cannot reach the base. A positive signal on the J input forward-biases D_3 and reverse-biases the clock pulse diode D_1 and thus prevents TR_1 being cut off by the clock pulses. Similarly a positive signal on the K input forward biases D_4 and reverse-biases D_2 so preventing TR_2 being cut off by the clock pulses. If J and K inputs are made positive simultaneously clock pulses are prevented from reaching either base and have therefore no effect on the bistable which remains in its former state. Negative going signals on the J and K inputs are blocked by D_3 and D_4 so that clock pulses operate on the bistable and the outputs alternate between the supply positive and negative at the clock frequency. The behaviour of the bistable is thus predictable for all four combinations of J and K inputs and is summarised in the following table in which the logic-1 signal is regarded as a change to the more positive level and a logic-0 signal as a change to the less positive level.

TRUTH TABLE FOR A JK BISTABLE

J *input*	K *input*	Q *output after receipt of a clock pulse*
0	0	no effect
0	1	0
1	0	1
1	1	C

C = complement of the state before the clock pulse

CLEAR AND PRESET INPUTS

RS and JK bistables commonly have two further inputs which over-ride the signal and clock inputs and can be used to put the bistable into a desired state. These are the clear and preset inputs. In logic circuits the bistable is represented by the block symbol shown in *Figure 13.72*. The information to be stored is fed into the circuit via the signal inputs, the convention being that a logic-1 input on J or S puts output Q to logic 1 (and hence output $\bar{Q}$ to logic 0) on receipt of a clock pulse. A logic-1 signal on input K or R has the reverse effect. A logic-0 signal on the preset input puts Q to logic 1 and a logic 0 on the clear input has the reverse effect.

This convention means that a logic-1 input to the upper square of the block symbol causes the output from that square also to go to logic 1 on receipt of a clock pulse. Clearly therefore this square cannot represent one transistor of the bistable circuit: it represents in fact the input circuit of one transistor and the output circuit of the other.

Sometimes the JK bistable has two parts known as the master and slave. On the positive edge of the clock pulse the information on inputs J and K is transferred to the

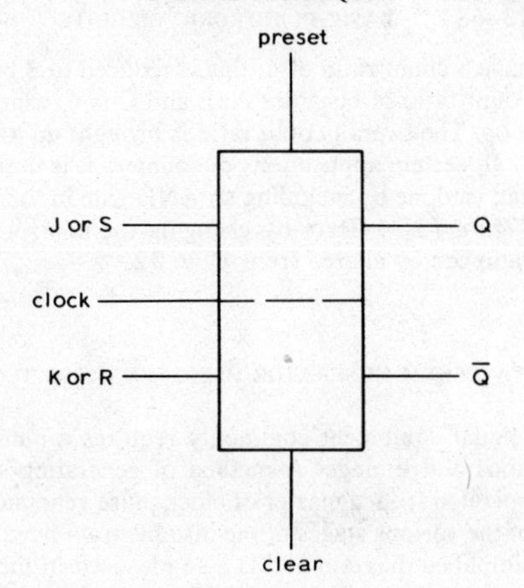

13.72 *Logic symbol for a bistable*

master section and on the negative-going edge the information is transferred to the outputs Q and $\bar{Q}$.

BINARY DIVIDERS AND COUNTERS

Bistable circuits can be used as binary dividers, giving one complete alternation in output signal (at Q and $\bar{Q}$) for every two input signals. If, in a cascade of such stages, each Q output is connected to the clock input of the following stage as shown in *Figure 13.73*, the pattern of the Q outputs follows the binary scale of numbers. For if all the Q

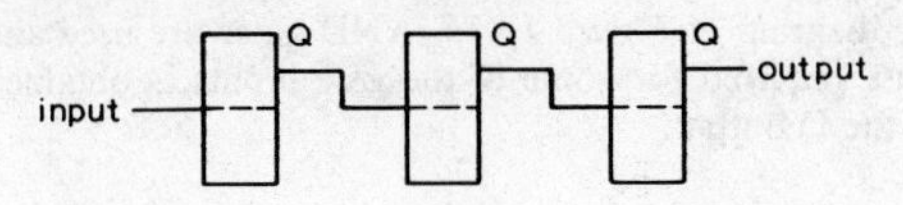

Figure 13.73 Logic diagram of a simple up-counter

outputs are initially at logic 0 then after seven input signals, all three Q outputs will be at logic 1, and 111 is the binary equivalent of 7. The next input signal causes all Q outputs to go to logic 0 and the counting process begins again. The circuit gives one output signal for every 8 input signals and so has a division or count ratio of 8. A binary counter of this type is known as an up-counter.

If the $\bar{Q}$ output of each stage is connected to the succeeding stage, the behaviour of the circuit is somewhat different. If all the $\bar{Q}$ outputs are initially at logic 1, after seven input signals they will all be at logic 0. This is an example of a down-counter.

By cascading n bistables in this way it is possible to obtain an up- or down-counter with a ratio of 2^n. Such a device is of limited application because the ratio is a power of 2 but by use of feedback it is possible to obtain any desired count ratio. The principle of the feedback or knock-back counter is illustrated in the following example.

There is inevitably a slight delay, known as propagation delay, in a bistable: in other words there is a significant time lag between the leading edge of the input signal and the leading edge of the corresponding output signal. In a cascade of bistables this delay can be appreciable and it enables the output signal to be returned to the input of the cascade and accepted as an input signal. The effect of such feedback is to reduce the number of input signals required to deliver one output signal from 2^n to $(2^n - 1)$. *Figure 13.74* gives the logic diagram of a counter employing feedback. Bistables A and B together have a count ratio of 4: this is reduced to 3 by the feedback loop embracing them. The count ratio of bistables A, B and C is 6, which is reduced to 5 by the second feedback loop. The overall count ratio is brought up to 10 by the final bistable D.

In certain applications of counters it is useful to be able to alter the count ratio. This can be done by including an AND gate in the feedback loop as shown in dashed lines in *Figure 13.74*. Here by changing the control signal from logic 1 to logic 0 the count ratio can be altered from 10 to 12.

WAVEFORM GENERATOR USING COUNTING TECHNIQUES

Digital equipment commonly requires a number of control pulses of particular durations and timings. A method of generating such pulses is to use a train of bistables operated from a master or clock pulse generator. The waveforms obtained at the outputs of the various stages of the bistable train have the form shown in *Figure 13.75*. This is a simplified diagram and is also idealised in that propagation delay is ignored but it will serve to illustrate the principle of this type of waveform generator.

Suppose a particular pulse is required to start at the time t_1. At this instant all the pulse trains shown have a negative-going edge, i.e., for the positive logic convention, they are all changing from logic 1 to logic 0. If, therefore, a gate is connected to these outputs and is designed to respond only when all its inputs are at logic 0, the output of the gate can be used to time the leading edge of the pulse to be generated. Suppose the trailing edge of the wanted pulse is required to occur at time t_2. At this instant the clock signal is at logic 0, the output of bistable A at logic 1, the output of bistable B at logic 1 and the output of bistable C at logic 0. A gate designed to detect these particular logic states will give an output signal which can initiate the trailing edge of the pulse to be generated. The pulse itself can be generated in a bistable which is turned on by the output from the first gate and turned off by the output from the second gate. This is shown in the logic diagram of *Figure 13.75*. AND gates are used and the inversion of the bistable outputs required for some of the gate inputs is obtained by using the $\bar{Q}$ output in place of the Q output.

SHIFT REGISTERS

An assembly of JK bistables is often used as a temporary store for binary information. For example suppose the binary signal 11001 (known as a binary word) is to be stored. Five bistables are needed, one for each binary digit, and the information could be fed simultaneously into the bistables by appropriate use of the preset and clear inputs. The information can now be stored for the required period and then simultaneously read out by operation of the clock input. This is the parallel method of feeding in and reading out the information.

Alternatively the bistables may be connected as a counter as shown in *Figure 13.76* and the required information can be fed into the input of the first bistable one digit at a time. The clock inputs of all the bistables are commoned and the clock rate must be the same as the rate of receiving the digits in the incoming information (the bit rate). On the first clock pulse the first binary digit (1) is transferred to the first bistable, its Q output going to logic 1. On the second clock pulse the first digit is transferred to the second bistable and the second digit (also 1) is transferred to the first bistable. Thus the process continues, the stored information moving to the right in the train of bistables until, after the fifth clock pulse, the whole binary word has entered the store. The Q outputs, in order, now read 11001. At this point the clock pulses can be stopped and the word stored for as long as necessary. When the word is required to be read out, the clock pulses are restarted and the stored information again moves to the right at clock rate

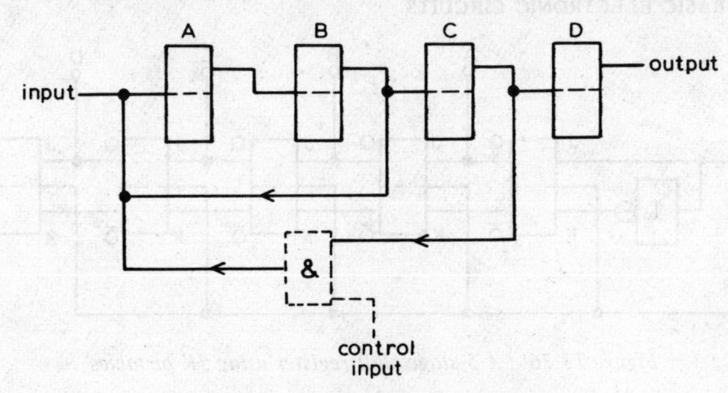

Figure 13.74 Binary divider with feedback

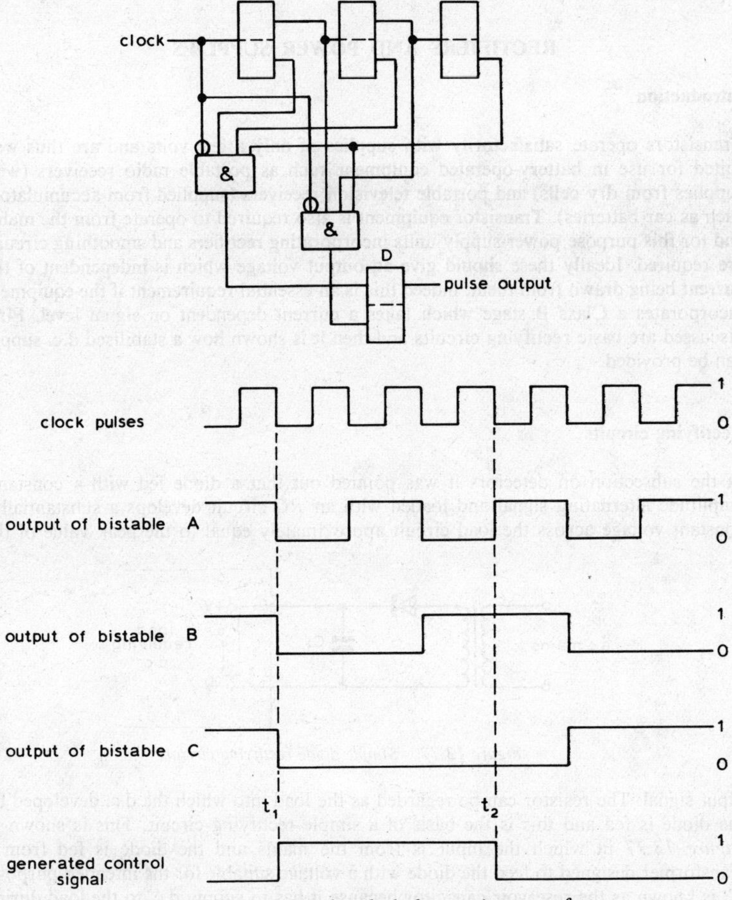

Figure 13.75 A digital method of generating a waveform

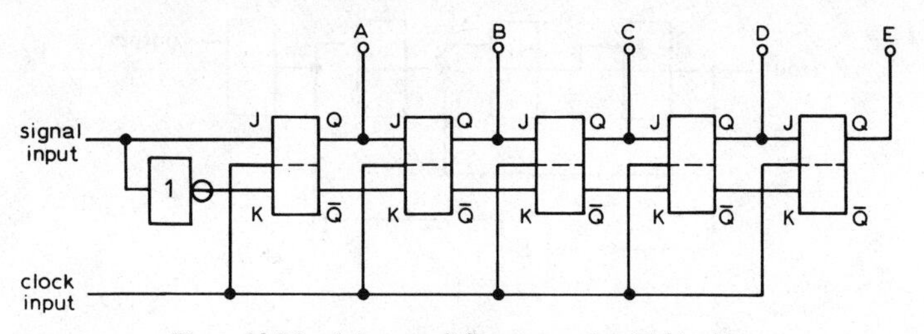

Figure 13.76 A 5-stage shift register using JK bistables

and can be read off at the Q output of the fifth bistable. This is the serial method of feeding in and reading out the information.

It is possible to enter the information into the store by parallel methods and to read it out serially: the converse is also possible.

RECTIFIERS AND POWER SUPPLIES

Introduction

Transistors operate satisfactorily with supplies of only a few volts and are thus well suited for use in battery-operated equipment such as portable radio receivers (with supplies from dry cells) and portable television receivers (supplied from accumulators such as car batteries). Transistor equipment is also required to operate from the mains and for this purpose power-supply units incorporating rectifiers and smoothing circuits are required. Ideally these should give an output voltage which is independent of the current being drawn from them: indeed this is an essential requirement if the equipment incorporates a Class B stage which takes a current dependent on signal level. First discussed are basic rectifying circuits and then it is shown how a stabilised d.c. supply can be provided.

Rectifying circuits

In the subsection on detectors it was pointed out that a diode fed with a constant-amplitude alternating signal and loaded with an *RC* circuit develops a substantially-constant voltage across the load circuit approximately equal to the peak value of the

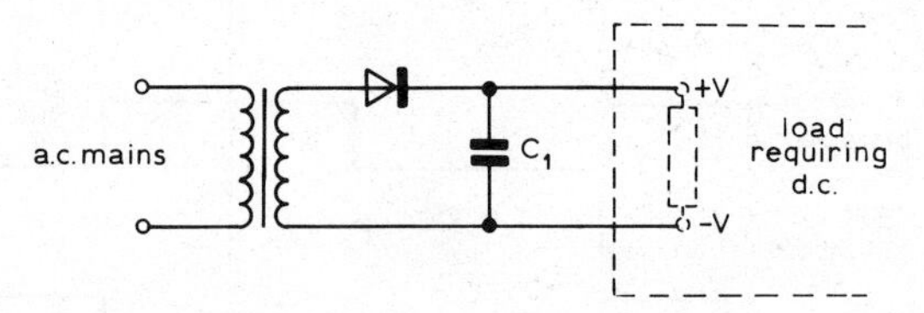

Figure 13.77 Simple diode rectifying circuit

input signal. The resistor can be regarded as the load into which the d.c. developed by the diode is fed and this is the basis of a simple rectifying circuit. This is shown in *Figure 13.77* in which the input is from the mains and the diode is fed from a transformer designed to feed the diode with a voltage suitable for the intended purpose. C_1 is known as the reservoir capacitor because it has to supply d.c. to the load during negative half cycles when the diode is non-conductive. The capacitance of C_1 depends on the input frequency and the output current and for the 50-Hz mains frequency is usually tens or hundreds of microfards. C_1 is alternately charged and discharged and for many purposes the ripple voltage across it is excessive. To reduce this a smoothing circuit of series inductance and shunt capacitance is added as shown in *Figure 13.78*.

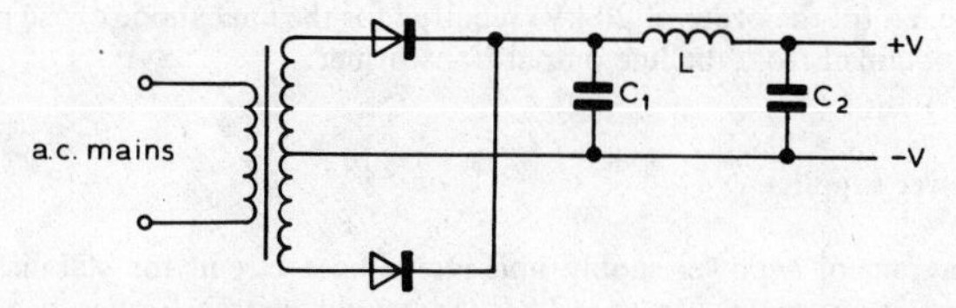

Figure 13.78 A full-wave rectifying circuit with LC smoothing

This circuit diagram also illustrates a further improvement namely the adoption of full-wave rectification: the diodes conduct on alternate half cycles of the input but feed a common smoothing circuit. The open-circuit output voltage from such a circuit is approximately $\sqrt{2}v_{in}$ where v_{in} is the r.m.s. input to each diode. There are however losses in the forward resistance of the diodes and the resistance of the transformer windings and the inductor as a result of which the output voltage for a normal load current is in practice less than the peak value of the diode input voltage.

An alternative full-wave rectifying circuit is shown in *Figure 13.79* which does not

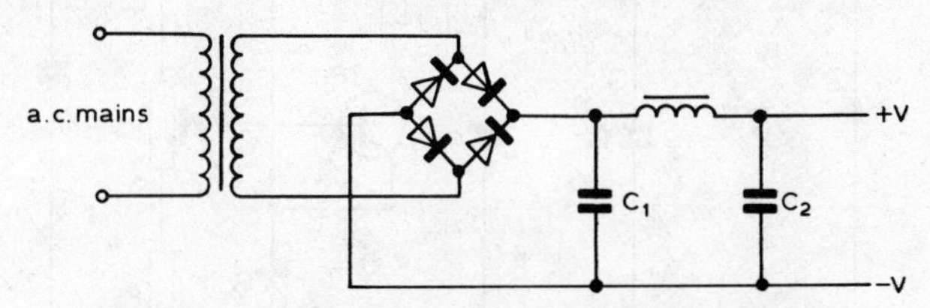

Figure 13.79 Rectifier using four diodes in bridge formation

require a centre-tapped transformer. Four diodes in bridge formation are used and two conduct during each half cycle. The open-circuit voltage from this circuit is $\sqrt{2}v_{in}$ where v_{in} is the voltage across the transformer secondary winding.

It is possible from diode circuits to obtain a d.c. output with a voltage many times the r.m.s. input. The simplest of these, the voltage-doubling circuit, is shown in *Figure 13.80* (*a*). On one half cycle of the input D_1 conducts and charges C_1 to the peak value

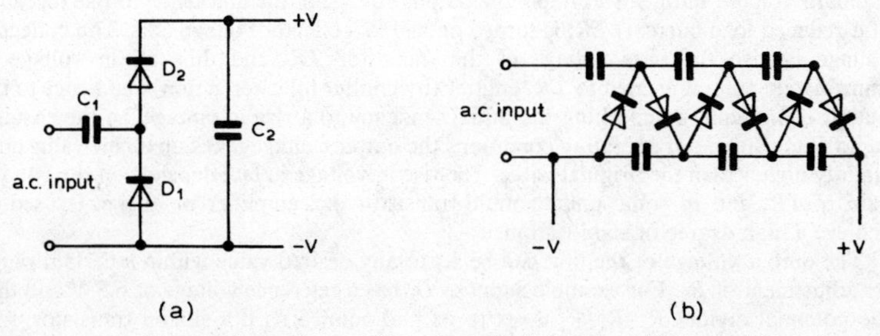

Figure 13.80 (a) Voltage-doubler rectifying circuit (b) Voltage multiplying rectifying circuit

of the alternating input. On the next half cycle the input signal together with the voltage across C_1 are applied to D_2 which conducts to give a voltage of approximately $2\sqrt{2}v_{in}$ across C_2. This form of circuit can be extended to the ladder formation shown in *Figure 13.80* (*b*) which can produce very large voltages (at small currents, however). Such circuits, known as voltage multipliers, are extensively used in television receivers to generate the e.h.t. (of the order of 20 kV) required for the final anode of the picture tube. The input is obtained from the line output transformer.

Stabilised power supplier

A circuit diagram of a power-supply unit including a circuit for stabilisation of the output voltage is given in *Figure 13.81*. A step-down transformer provides a low alternating voltage for the full-wave bridge rectifier and the network R_1–C_1–C_2 provides

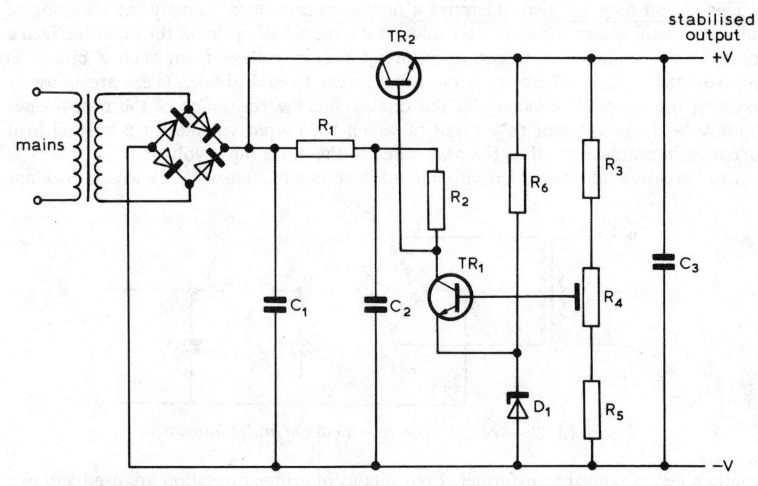

Figure 13.81 Circuit of a stabilised mains unit for transistor equipment

a measure of smoothing. The transistors TR_1 and TR_2 stabilise the output voltage of the input in the following way. The emitter voltage of TR_1 is stabilised by the voltage-reference diode D_1: the base is fed with a sample of the output voltage of the unit via the potential divider R_3–R_4–R_5. Thus TR_1 compares the output voltage of the unit with a standard voltage and if for example the output voltage of the unit tends to rise (because of a reduced load current) TR_1 is turned on and its collector voltage falls. The collector voltage is also the base voltage of the transistor TR_2 and this fall in voltage is immediately communicated to TR_2 emitter (by emitter follower action) and hence to the output of the unit, so checking the initial tendency to a rise in voltage. In the ensuing rapid readjustment of operating conditions the output voltage takes up a new value only slightly higher than the original value. The rise in voltage in fact depends on the voltage gain of TR_1 and in some units a multi-transistor d.c. amplifier or an i.c. is used to achieve a high degree of stabilisation.

The output voltage of the unit can be set to any desired value within a certain range by adjustment of R_4. For example suppose D_1 has a reference voltage of 6.3 V and that the potential divider R_3–R_4–R_5 is set to its mid-point. TR_1 if a silicon transistor, will begin to conduct at a base-emitter voltage of 0.7 V and a voltage of say 1 V is necessary to ensure a reasonable collector current which can increase or decrease to achieve stabilisation. Thus the voltage across the lower half of the potential divider is 7.3 V and the stabilised output voltage from the unit is 14.6 V. If R_3, R_4 and R_5 are equal-value resistors, by adjustment of R_4 it is possible to set the stabilised output voltage at any value within the range 11 V to 22 V.

14 ELECTRONIC INSTRUMENTATION AND MEASUREMENTS

ELECTRONIC INSTRUMENTATION AND MEASUREMENTS 14

14 ELECTRONIC INSTRUMENTATION AND MEASUREMENTS

Electronic instrumentation is being increasingly used in all branches of engineering. The wide range of electronic devices now available with sophisticated circuitry and the development of solid-state technology have contributed to the expansion of electronic measurement into every branch of science. One of the most versatile electronic instruments is the cathode-ray oscilloscope. This and other electronic instruments and measurement techniques are described in this Section, but they are examples only of a vast range.

CATHODE-RAY OSCILLOSCOPES

Introduction

A cathode-ray oscilloscope is an electronic equipment designed to display two-dimensional information on the fluorescent screen of a cathode-ray tube in a non-pictorial form. If brightness variations are employed, then to a limited extent, it can also present three-dimensional information. The horizontal axis usually represents time and the vertical axis voltage, but the latter may represent one of many quantities such as loudness or magnetic field strength.

The cost of this type of equipment varies considerably and it follows that the range of functions, facilities, performances and techniques employed is also very wide. The high degree of flexibility of the cathode-ray tube and the many versions of equipment currently available make it difficult to decide where the boundary line lies between oscilloscopes and various other specialised instruments employing cathode-ray tubes as visual display devices. In this section the emphasis is on the medium-priced conventional instrument. Some information is however provided on the more expensive and specialised instruments. The cathode-ray tube itself is described in detail in Section 7.

An oscilloscope is normally housed in a single metal case, which should be connected to the earth terminal of the mains supply when the equipment is powered from this source. The mains unit may be separate and auxiliary items such as probes and specialised graticules are sometimes provided. A probe or graticule designed for use with one type of instrument cannot usually be used with another type. Provision is sometimes made for the fitting of still and/or film cameras.

In some cases the instrument is indivisible, while in others a mainframe is provided into which various sub-units may be plugged. These sub-units are most commonly vertical or *Y* deflection amplifiers, but horizontal or *X* deflection sub-units may also be available. These provisions increase the range of facilities that may be handled by one oscilloscope. They do however increase the price and size of the instrument.

Consideration is first given to the broad outlines of a simple oscilloscope, then various sections are described in more detail. The most common use of the instrument for the display of a voltage/time curve will be considered initially. A block schematic of a typical instrument is given in *Figure 14.1*.

The voltage waveform to be examined is applied to the input. A capacitor, which may be shortcircuited by means of a switch, is located in the high-potential input connection. If the waveform is to be examined in its entirety, then the switch should be set to the d.c.

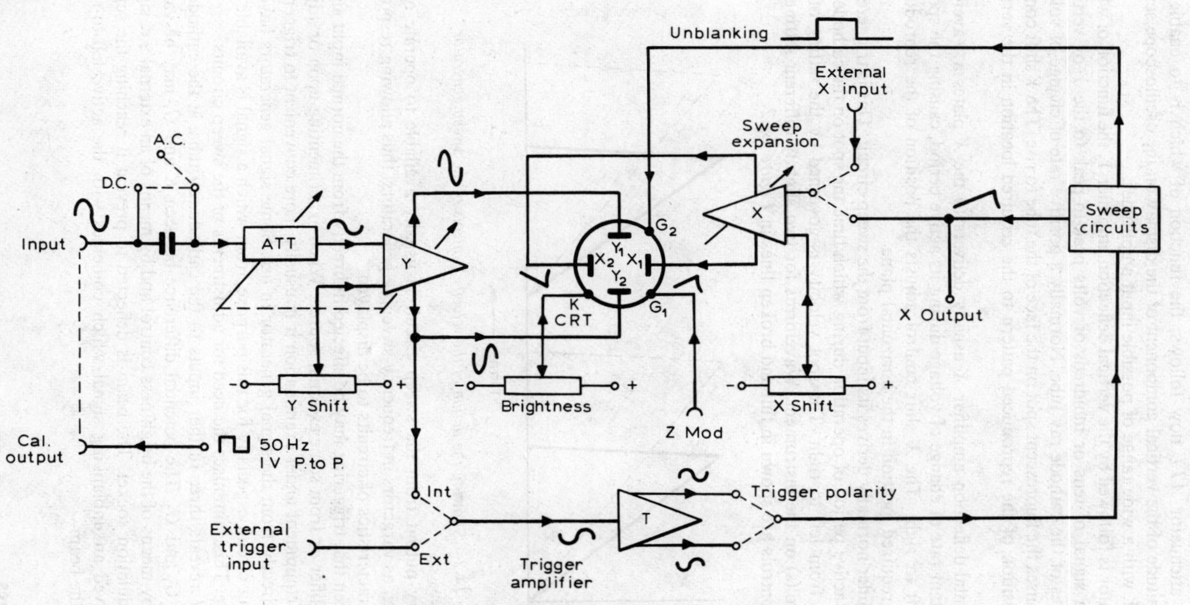

Figure 14.1 Schematic for a typical oscilloscope

position. If the signal contains d.c. and a.c. components, but only the latter is to be examined, then the switch should be set to the a.c. position.

A variable attenuator *ATT* now follows, the function of which is to enable the required magnitude of the vertical component of the display on the oscilloscope screen to be achieved, with a wide range of possible input amplitudes.

The attenuator is followed by the vertical deflection amplifier *Y*, the function of which is to deliver a signal of tens or hundreds of volts peak-to-peak to the *Y* or vertical-deflection plates of the cathode-ray tube. Normally a positive rate of change of voltage at the input causes the fluorescent spot on the face of the tube to rise. The *Y* shift control adjusts the position of the reproduced pattern to the required location in the vertical plane.

The horizontal-deflection amplifier *X* usually delivers to the *X* plates a waveform having a constant rate of change of voltage during its active period, causing the spot to move from left to right. The *X* shift control adjusts the position of the reproduced pattern to the required location in the horizontal plane.

The *X* amplifier normally derives its input from the sweep circuit. This is triggered to commence its active period of operation during which time the spot on the cathode-ray screen travels from left to right. The spot velocity is governed by the setting of the velocity control(s) on the instrument. Waveforms for two slightly different settings of the velocity controls are shown in full and broken lines in *Figure 14.2*.

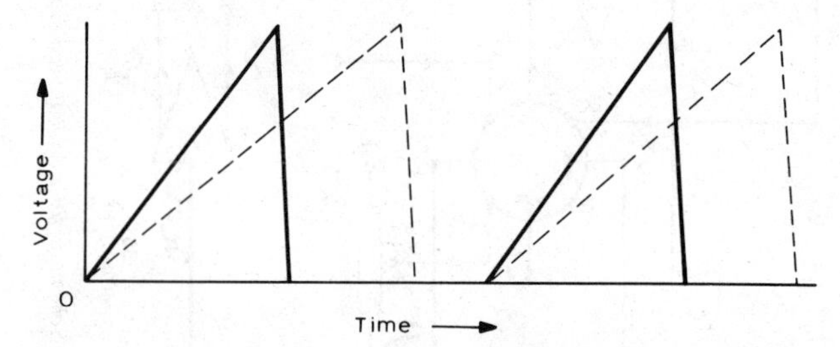

Figure 14.2 Showing the action of the velocity controls of the sweep generator

An auxiliary output from the sweep circuit is usually available to operate other equipment such as varactors in frequency-sweeping generators, thus allowing response/frequency characteristics of circuits to be displayed.

It will be seen that triggering may be effected indirectly from the normal input signal via the *Y* amplifier or from some external source. When experimenting upon or adjusting a piece of equipment under examination it is usually more convenient to trigger the oscilloscope directly from the signal generator in use. Some signal generators have an auxiliary output circuit especially for this purpose, from which a signal is available at a fixed amplitude. This minimises the need for adjustments of the sweep circuits.

Figure 14.1 reveals three further inputs to the cathode-ray tube, at the cathode *K*, and the grids G_1 and G_2. The potential difference between *K* and G_1 may be varied continuously by means of the brightness control and by means of an external a.c. signal via the *Z* modulation socket. The beam is deflected to prevent it reaching the screen until G_2 received an unblanking signal which coincides with the active left-to-right movement of the beam.

Lissajous figures

A less familiar use of the oscilloscope is for the formation of Lissajous figures by a process illustrated in *Figure 14.3*. In this case the *Y* plates receive their signals in the normal manner but the X plates receive theirs via the external X socket. The internal horizontal sweeping generator and the unblanking operation are inoperative. In the example given, it is shown that the Y plates are receiving a sinusoid while the X plates are receiving a cosinusoid of the same frequency. The two graphs in *Figure 14.3* employ similar time scales. A succession of times 1, 2, 3, etc., are similarly selected on both graphs and construction lines drawn to indicate the resulting positions of the spot. The result is a pattern which consists of an ellipse having its axes horizontal and vertical, the spot travelling in an anti-clockwise direction. Careful consideration shows that two *similar* waveforms would produce a straight

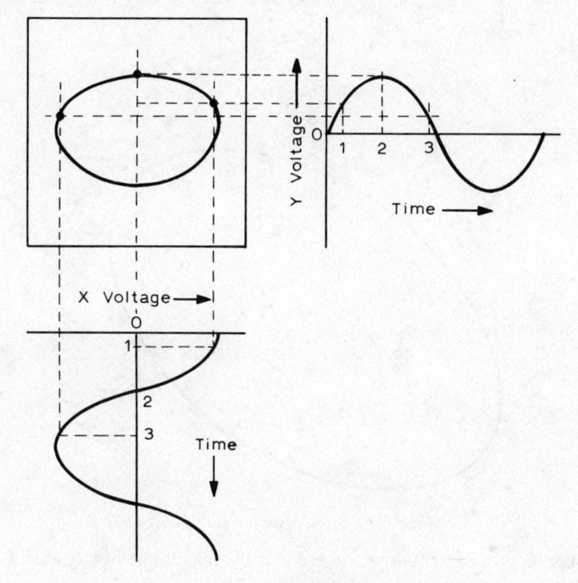

Figure 14.3 *Formation of a simple Lissajous figure*

ine with a positive slope. If one waveform be inverted, the straight line has a negative lope. By employing this principle of construction, the pattern which results may be redicted for any two waveforms of any two repetition frequencies, but in practice the echnique is normally restricted to simple waveforms having some simple frequency elationship.

The phase relationship between two sinusoidally shaped waveforms of the same requency may be determined as follows: First one of the waveforms is applied simultaneously to both X and Y inputs. At least one of these circuits should have a gain-ontrol which is continuously variable. The gain-controls are now adjusted to produce n the screen a straight line with a slope of either +45° or −45°. (The 45° slope ndicates that the deflection sensitivities are identical. The straightness of the line ndicates that the deflection circuits have identical phase shift at this frequency.) The reliminary adjustment having been made, the two signals are now fed *separately* to the orizontal and vertical deflection input sockets. The result is likely to be an ellipse, the ajor and minor axes Z_{maj} and Z_{min} being shown in *Figure 14.4*. The phase difference now given by:

$$\phi = 2 \tan^{-1} \frac{Z_{min}}{Z_{maj}}$$

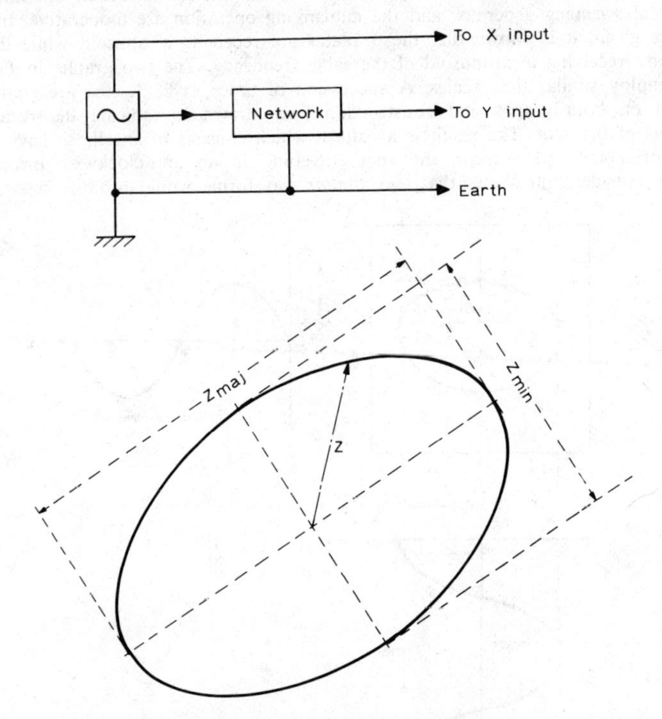

Figure 14.4 Measurement of phase shift

which may be proved as follows, where Z is the magnitude of any axis. Let the input waveforms be

$$x = \sin\theta \quad \text{and} \quad y = \sin(\theta + \phi)$$

then

$$Z^2 = \sin^2\theta + \sin^2(\theta + \phi)$$

To find Z_{maj} and Z_{min}

$$2Z\frac{dZ}{d\theta} = 2\sin\theta\cos\theta + 2\sin(\theta + \phi)\cos(\theta + \phi)$$

and

$$\frac{dZ}{d\theta} = 0$$

$$\begin{aligned}
\sin 2\theta &= -\sin 2(\theta + \phi) \\
&= -(\sin 2\theta\cos 2\phi + \cos 2\theta\sin 2\phi) \\
1 &= -\left(\cos 2\phi + \frac{\sin 2\phi}{\tan 2\theta}\right) \\
\tan 2\theta &= -\frac{\sin 2\phi}{1 + \cos 2\phi} \\
&= -\tan\phi \\
2\theta &= -\phi \quad \text{or} \quad \pi - \phi \\
\theta &= -\frac{\phi}{2} \quad \text{or} \quad \frac{\pi - \phi}{2}
\end{aligned}$$

$$Z_{\text{min}}^2 = \sin^2\left(\frac{\phi}{2}\right) + \sin^2\left(\frac{-\phi}{2}\right)$$

$$= 2\sin^2\frac{\phi}{2}$$

$$Z_{\text{maj}}^2 = \sin^2\left(\frac{\pi-\phi}{2}\right) + \sin^2\left(\frac{\pi+\phi}{2}\right)$$

$$= \cos^2\left(\frac{-\phi}{2}\right) + \cos^2\left(\frac{\phi}{2}\right)$$

$$= 2\cos^2\frac{\phi}{2}$$

$$Z_{\text{min}} = \frac{\sin\frac{\phi}{2}}{\cos\frac{\phi}{2}}$$

$$= \tan\frac{\phi}{2}$$

$$\therefore \phi = 2\tan^{-1}\frac{Z_{\text{min}}}{Z_{\text{maj}}}$$

Lissajous figures are also used for frequency comparison purposes, e.g., for checking the operation of frequency multipliers and dividers and for calibrating multifrequency oscillators from single-frequency sources. (It should be noted that the technique can be used only when the relative frequencies are stable to within about 1 Hz.) Examples where $f_X = 2f_Y$, and $f_X = f_Y/3$, are shown in *Figures 14.5* (*a*) and (*b*) respectively. For

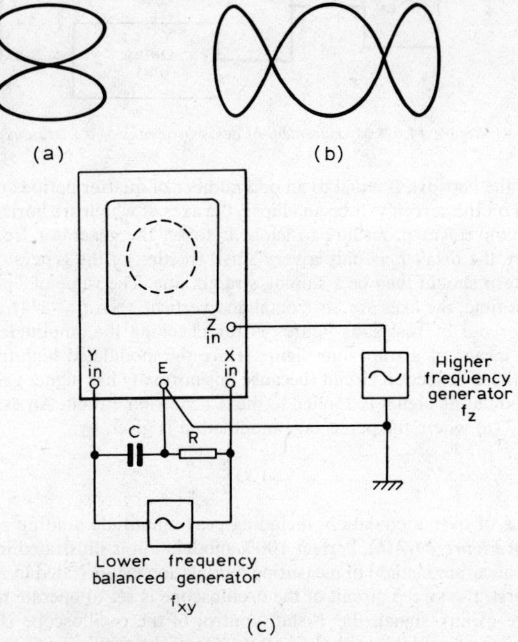

Figure 14.5 Frequency comparison using Lissajous figures
(a) $f_x = 2f_y$ (b) $f_x = f_y/3$ (c) $f_x = 15f_{xy}$

any frequency relationship however, an infinite number of patterns is available depending upon *phase* relationships. Where the frequency relationship is high, e.g., greater than six, it is more convenient to employ a technique which includes Z modulation, the necessary circuit arrangements and resulting screen pattern are shown in *Figure 14.5* (*c*). Here the lower frequency signal is applied from a *balanced* source to a phase-splitting circuit consisting of C and R in series. Assuming that the horizontal and vertical deflection systems have similar sensitivities, then the magnitude of the reactance of C should have approximately the same value as the resistance of R. This will produce an ellipse, the axes of which are horizontal and vertical. Adjustment of X or Y sensitivity or C or R will enable a circle to be produced. The higher frequency signal, having an amplitude of several volts, will cause the pattern to be broken up as shown, the number of breaks equalling f_Z/f_{XY}.

Lissajous figures may be employed for the measurement of time delay, using a simple arrangement such as is shown in *Figure 14.6*. It will be realised that whenever the time

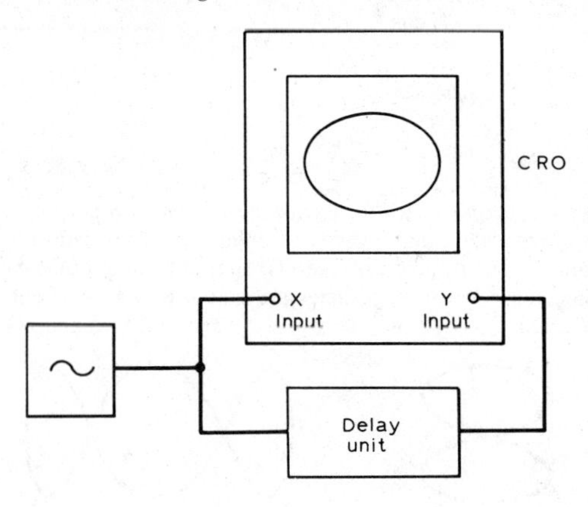

Figure 14.6 Measurement of delay by means of a Lissajous figure

delay t_d of the network is equal to an odd number of quarter periods of the source, then the pattern on the screen will be an ellipse, the axes of which are horizontal and vertical. The most convenient procedure to adopt is to set the generator frequency f to a low value where the delay t_d is only a very small fraction of the generator period $1/f$. The screen pattern should then be a sloping straight line. The value of f is now raised until, for the *first* time, the axes are horizontal and vertical, then $t_d = 1/4f$.

A further use of Lissajous figures is for checking the amplitude modulation of a carrier by means of a trapezium figure. Here the modulated high-frequency signal is applied to the Y deflection circuit (because this normally has higher gain and bandwidth) and the modulating signal is applied to the X deflection circuit. An example is shown in *Figure 14.7* (*a*) where the percentage modulation is given by

$$100\frac{a-b}{a+b}$$

An example of over modulation, including peak-amplitude limiting and carrier-cutting is shown in *Figure 14.7* (*b*). Perfect 100% modulation is illustrated in *Figure 14.7* (*c*).

A more accurate method of measuring modulation is illustrated in *Figure 14.7* (*d*), (*e*) and (*f*). First, the sweep circuit of the oscilloscope is set to operate normally. Then, in the absence of any signal, the Y shift control of the oscilloscope is set to produce a horizontal line to coincide with the bottom line of the oscilloscope graticule. The signal is now applied to the Y deflection circuit in the normal manner, and while it is

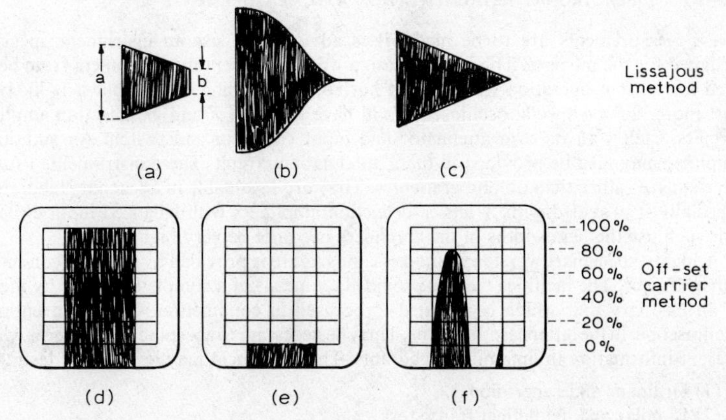

Figure 14.7 Modulation measurements

*un*modulated, the Y deflection sensitivity (or signal amplitude), is adjusted to produce a pattern which rises by 5 or 10 divisions of the graticule as shown at (*d*). This action calibrates the vertical scale. The Y shift control is now operated to lower the pattern so that its upper extremity coincides with the bottom line of the graticule as shown at (*e*). When modulation is applied, the upper extremity of the pattern indicates the percentage of the modulation, as shown in the example (*f*). It is convenient, but not essential, to synchronise the operation of the sweep circuit with the modulating signal.

A further example of a Lissajous figure, this time for displaying a B/H hysteresis loop for a specimen of iron is illustrated in *Figure 14.8*. Here a high-amplitude balanced

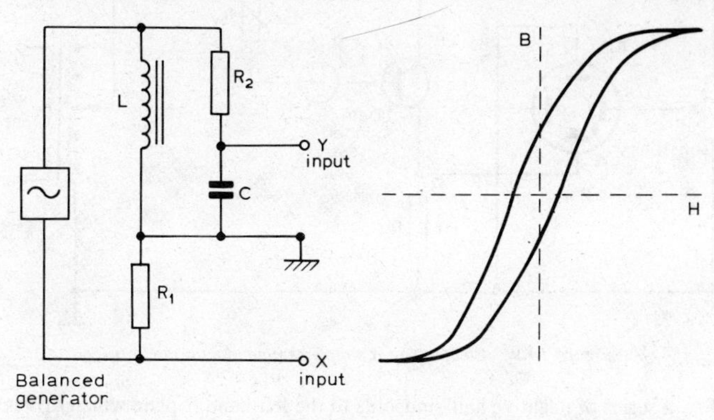

Figure 14.8 Production of a B/H curve by means of a Lissajous figure

source of power drives a current through L and R_1 in series. The resistance of R_1 should be large in comparison with the magnitude of the reactance of L. The result is a horizontal deflection, the magnitude of which is proportional to the magnitude of the magnetising force H. The potential across L, however (neglecting its own resistance), is proportional to the rate-of-change of the flux-density B, so that an integrating circuit CR_2 must now be interposed between L and the vertical deflection circuit of the oscilloscope. The product CR_2 should not be less than five times the period of the source, e.g., for 50 Hz, suitable values are $C = 1\ \mu F$ and $R_2 = 100\ k\Omega$.

The normal oscilloscope may perform many Lissajous functions satisfactorily, but where these functions are to be employed frequently and especially where precision phase measurements are to be made, it is advisable to use an instrument specially designed for the purpose. This special design will not prevent the instrument from being used for normal operation with its own horizontal sweeping circuit, but it is likely to cost more. Such a special oscilloscope will have *identical* X and Y deflection amplifier circuits, each with its own attenuator and input capacitor and switch. An additional amplifier may also be provided in the Z modulation circuit. These instruments usually have square rather than oblong graticules. They are invaluable in a teaching laboratory especially if provided with X and Y deflection amplifiers with balanced input circuits. For such use the bandwidths of the amplifiers need not be very high.

A highly specialised version of the Lissajous figure is provided by the Semiconductor Curve Tracer. The facilities may be provided by means of a complete unit, or by means of an auxiliary unit which is intended to be used in conjunction with a conventional oscilloscope. In the latter case, however, it may be necessary to accept screen patterns which present information in unfamiliar quadrants. The equipment may be used for testing

(1) Ordinary and zener diodes.
(2) Bi-polar and field-effect transistors.
(3) Voltage dependent resistors.

A very simple circuit is shown in *Figure 14.9*. (The normal X and Y amplifiers have been omitted for the sake of simplicity.) Here the mains transformer delivers, via a

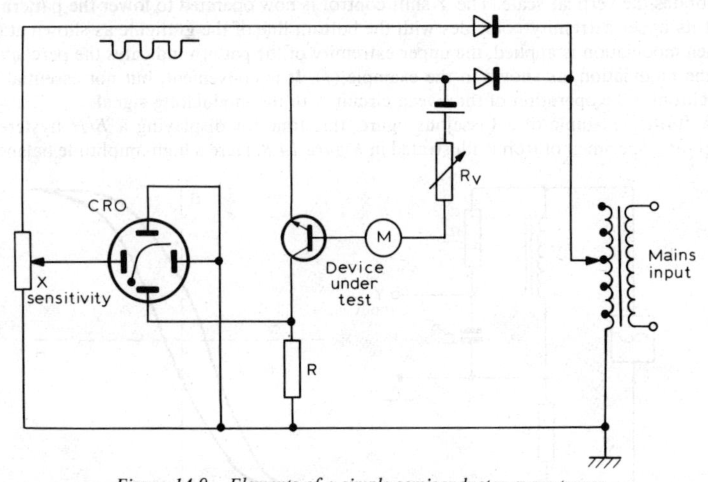

Figure 14.9 Elements of a simple semiconductor curve tracer

rectifier, a series of negative half-sinusoids to the left hand X plate which drives the spo
horizontally across the screen from its initial location which is shown here by a dot
Simultaneously there occurs a vertical deflection which is proportional to the instantan
eous value of the collector current of the transistor, passing through the transistor R
Base current is adjusted by means of R_v and monitored on the meter M.

In practice, the circuit is normally extended to provide the following facilities:

(1) Testing of *pnp* as well as *npn* transistors.
(2) Testing of field-effect transistors, both *n*-channel and *p*-channel, depletion mod and enhancement types.
(3) Testing base-emitter junction characteristics.
(4) Exploring operation with different collector or drain loads.
(5) Displays of families of curves.
(6) Matching of transistor pairs.

The following are points of interest in the design of such equipment.

(1) The transformer secondary winding is preferably earthed at one end, otherwise capacitance between windings introduces a spurious component into the circuit.
(2) Components for the production of bias for bases and/or gates should be kept at a low impedance to earth point in the circuit, so that their capacitance to earth does not produce any significant shunting effect across the resistor *R*.
(3) The use of a long-afterglow tube is preferred, especially when *families* of curves are to be displayed.
(4) Families of curves may be produced using reed-relays for switching. One technique uses a ring of reed-relays operated by a permanent magnet which is rotated by a synchronous motor.

Television oscilloscopes

Oscilloscopes used for television can be divided into three categories. The first is a relatively conventional oscilloscope with some television facilities added. The second and third are special versions, known as television wave form monitors. The latter are not suitable for more general use.

The first version is based on the conventional oscilloscope with the addition of a picture/synchronising signal separator plus a line/field frequency synchronising signal separator. The complete video-frequency signal usually covers a frequency band of 0–5.5 MHz. For the simpler work a nominal bandwidth of this order is adequate, but for critical work such as development, where a linear phase-frequency characteristic is required over the band, and where the cut-off frequencies of amplifiers, filters, etc., must be examined, the nominal bandwidth should be at least twice the nominal bandwidth of the television system. The vertical deflectional sensitivity required is not usually very high because the signal normally has an amplitude of more than one volt. The picture synchronising signal separator is used to derive from the video signal a synchronising component which is to be used to trigger the sweep circuits. It must be designed to cope with signals of both polarities, i.e., positive-going and negative-going synchronising signals, the picture signals always going in the opposite direction. An example of a circuit which will meet these requirements is given in *Figure 14.10*.

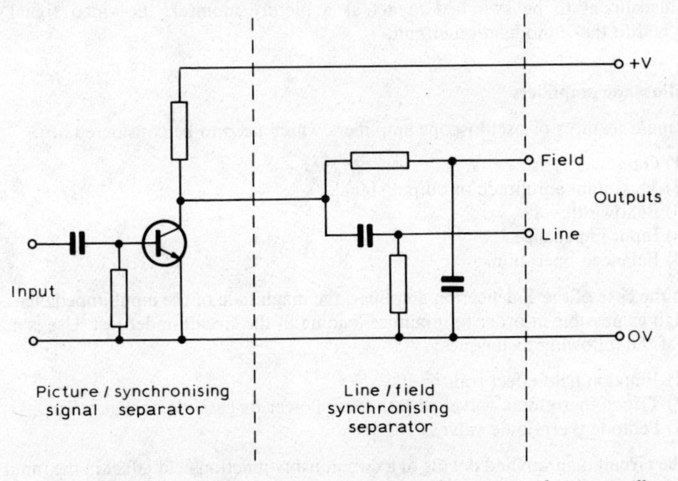

Figure 14.10 Circuit for a synchronising signal separator for use in a television oscilloscope

The oscilloscope in the second category is normally intended for permanent installation in operating positions. This also contains separators but the nature of the input signal is more rigidly defined. The signal is picture signal positive-going, synchronising signal negative-going. The amplitude is normally 1 V peak-to-peak. The signal is fed from a source having an impedance of 75 Ω and the circuit is terminated finally, either at the oscilloscope or elsewhere with another 75 Ω. The sweep circuits can normally be switched only to line frequency (15 625 Hz for most systems), or to field frequency (50 Hz for most systems). A useful feature in some instruments is that when set to line frequency the time-scale is on the horizontal axis and the signal voltage scale on the vertical axis in the conventional way, but when the instrument is set to field frequency the axes are rotated clockwise by 90 degrees. This enables the information displayed on the oscilloscope to be related more closely to that displayed on the picture monitor or to the actual scene. An example of this is shown in *Figure 14.11*.

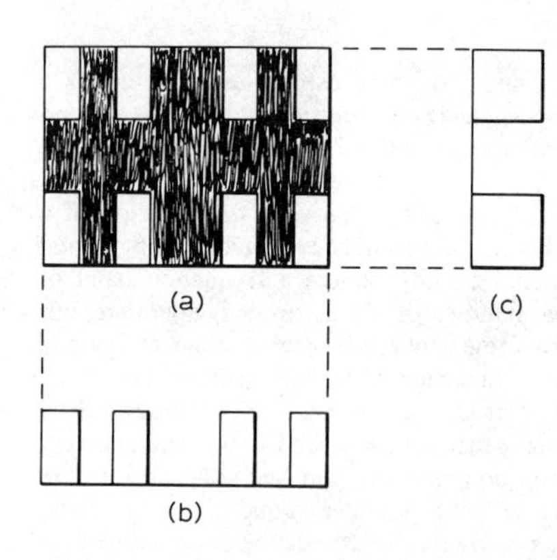

Figure 14.11 *Illustrating the use of one type of television waveform monitor. A sample picture is shown at* (*a*). *When the sweep circuit is set to operate at line frequency the screen pattern, for this picture, appears as at* (*b*). *When the sweep circuit is set to operate at field frequency, the screen pattern appears as at* (*c*).

The third version is not an oscilloscope alone but combines the features already described for the second category plus two sweep circuits, one operating at line frequency and one at field frequency, plus a wide-band *Z* modulation amplifier enabling the instrument to be switched to act as a picture monitor, the video signal being switched to the *Z* modulation circuit.

Oscilloscope amplifiers

The main features of oscilloscope amplifiers which need to be considered are:

(1) Gain.
(2) Maximum amplitude of output signal.
(3) Bandwidth.
(4) Input impedance.
(5) Balanced operation.

In the case of the *Y* deflection amplifier, the magnitude of the input impedance is kept as high as possible in order to minimise loading of the circuit under test. Use is made of one of the following techniques:

(1) Junction field-effect transistors.
(2) Triode thermionic valves in cathode-follower or cascode configuration.
(3) Pentode thermionic valves.

The circuit diagram and details of a circuit using junction field-effect in the input stage transistors is shown in *Figure 14.12*.

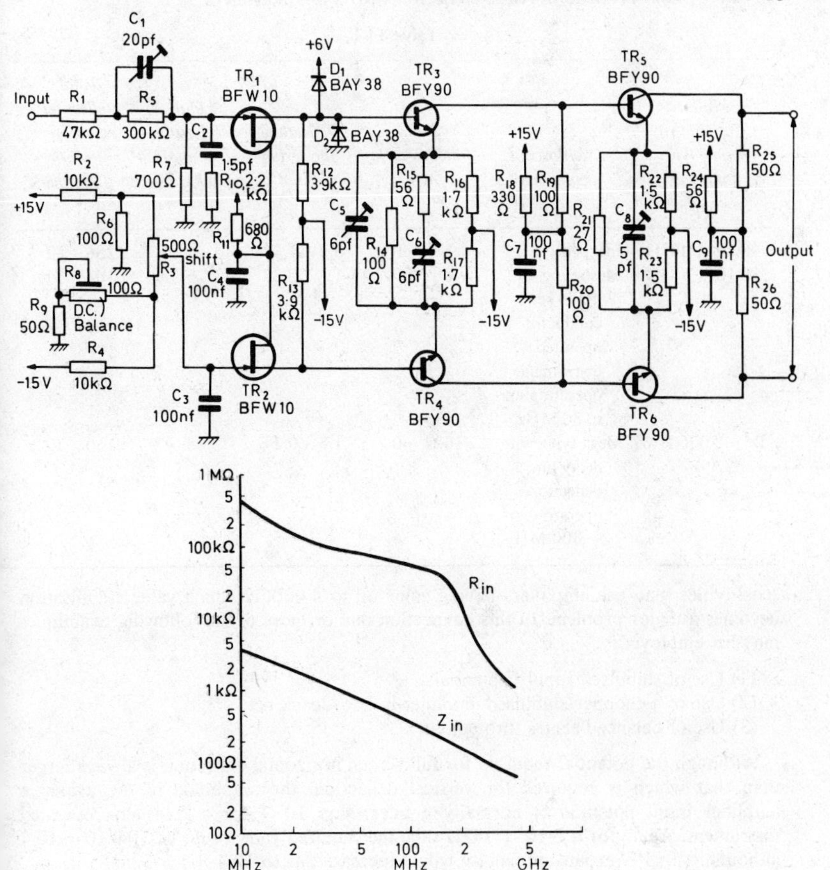

Figure 14.12 A Mullard Y deflection pre-amplifier circuit employing f.e.t. input devices. The input circuit is protected by a diode/resistor network; the variation of input impedance against frequency is shown in the graph. Gain is constant from zero frequency to 300 MHz; rise time is 1 ns; noise voltage is 0.2 mV peak-to-peak; and the voltage drift is 0.5 mV/C. (Courtesy Mullard Ltd)

The required magnitudes of the output signals from the deflection amplifiers vary from tens of volts to hundreds of volts peak-to-peak, typical values for three Mullard tubes being shown in the last two columns of *Table 14.1*.

The difference between the two deflection potentials required for a given cathode-ray tube is due to the following causes. The *X* deflection plates are nearer to the screen than the *Y* deflection plates and therefore must bend the beam through a wider angle. In many oscilloscopes provision is made for a longer deflection in the horizontal direction than in the vertical direction, typically in the ratio 4 : 3. Due to this, the two *X* deflection plates must be further apart than the two *Y* deflection plates which further reduces their deflection sensitivity proportionally.

A moderately priced oscilloscope employing a D10-170GH tube might have its sensitivity quoted as 50 mV/cm. (More strictly, this is its *inverse* sensitivity.) For a 6 cm deflection the maximum input would be 6×0.05 V leading to a gain of $21/0.3 = 63$ approximately. Instruments having high sensitivity and using tubes of lower deflection

Table 14.1

Mullard Cathode-ray tube type number	*Oscilloscope category*	*Scan dimensions in mm* X	Y	*Deflection factor in V/mm* X	Y	*Final accelerator potential in kV*	*Total deflection potentials in V peak-to-peak* X	Y
D10–160GH	Inexpensive	80	60	3.2	1.37	1.5	256	82
D10–170GH	Mesh type for use in semi-conductor operated instruments operating up to 30 MHz	80	60	1.3	0.35	6	104	21
D13–500GH/01	Mesh type, with delay line *Y* deflection operating up to 800 MHz	100	60	1.5	0.2	15	150	12

sensitivities may use amplifiers having gains up to 5 000, at which value stabilisation becomes a major problem. In this connection one or more of the following techniques may be employed:

(1) Use of stabilised supply potentials.
(2) Use of a chopper/stabilised amplifier in a feedback network.
(3) Use of balanced stages throughout.

Although the potential required for full screen horizontal deflection is always larger than that which is required for vertical deflection, the amplitude of the available amplifier input potential is normally greater, say, 10 V for a transistor operated instrument. Again for a D10–170GH tube the required gain would be $104/10 = 10.4$ although a 5 : 1 *X*-expansion facility would increase this to $10.4 \times 5 = 52$.

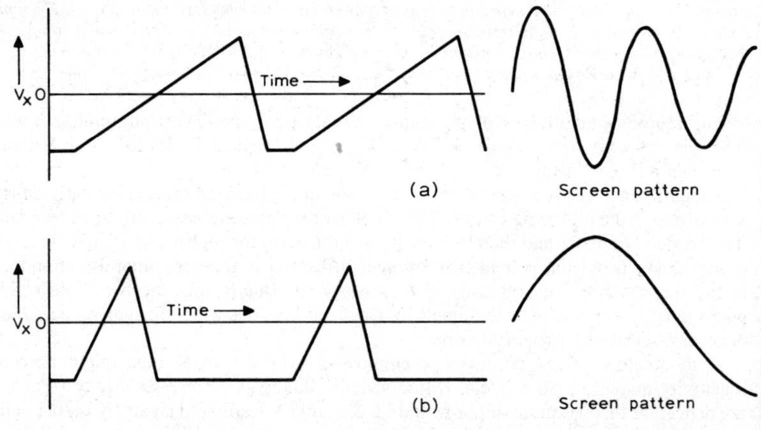

Figure 14.13 Examination of a waveform (a) without X expansion (b) with 3 : 1 expansion

The use of *X* sweep expansion may be considered, taking as an example the examination of a damped oscillation. Without *X* expansion the waveform might appear as in *Figure 14.13* (*a*). A 3 : 1 expansion would produce a pattern as in (*b*).

In the case of oscilloscopes which are specifically designed for the production of Lissajous figures, the *X* amplifier gain should be higher than the *Y* amplifier gain by a factor which is inversely proportional to the relative deflection sensitivities of the cathode-ray tube. The number of inverting stages and the manner of the connection to the *X* plates should be such that, in this particular type of oscilloscope, the application of a positive potential to the *X* input causes the spot to move from left to right.

An economically designed amplifier will have its maximum amplitude of output signal and its bandwidth determined almost entirely in the final stage, with earlier stages much more conservatively rated, i.e., these earlier stages will have greater overload capacity and greater bandwidth.

The relative bandwidths of the *X* and *Y* deflection amplifiers need consideration. If the instrument is to be used mainly for Lissajous figures and for the examination of *un*modulated waveforms then there is little logic in making the bandwidths different. Most manufacturers, however, take the view that when the *Y* deflection circuit is operating near to its maximum frequency, then the instrument is being used either as a peak-to-peak voltmeter, or is being used for viewing only *modulated* waveforms, so that some economy may be effected by providing an *X* amplifier with a lower bandwidth.

The stray capacitances in the circuit however, particularly those of the active devices, effectively shunt the load resistors and thus limit the operation of the amplifiers at the high frequency end of the spectrum. The effective bandwidth of the amplifiers may be extended by the use of one or more of the following techniques:

(1) Stages may have circuit forms which alternate between voltage-amplifier and current-amplifier, e.g., between common-emitter and common-collector types as is shown in *Figure 14.14*.

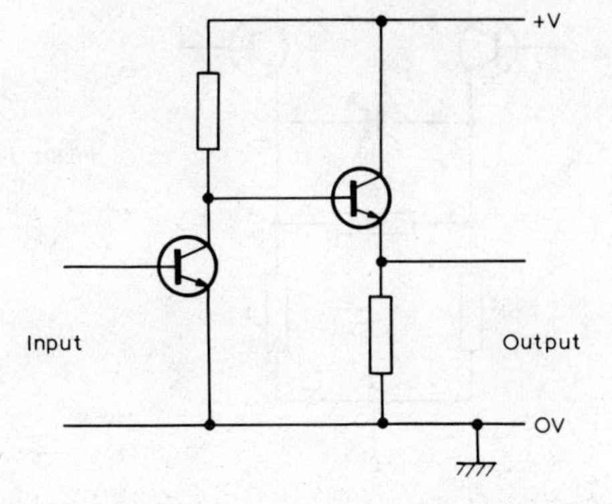

Figure 14.14 A voltage-amplifier current-amplifier combination

(2) A correction inductor may be added to the circuit, its effects offsetting partially the shunting effects of the stray capacitancies. An inductor may be connected in series with the load, this being known as shunt correction. Alternatively it may be connected between the output electrode of one active device and the input electrode of the succeeding active device, this is known as series correction.[1, 2] Typical circuits are illustrated in *Figure 14.15*.

(3) Emitter (or source), correction circuits may be employed.

(4) Distributed amplifiers may be employed. (These are normally used at frequencies exceeding 20 MHz.)

The last two forms of h.f. correction will now be discussed.

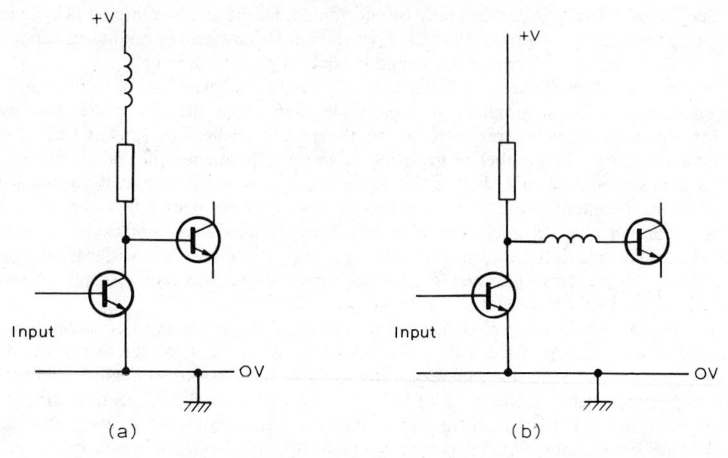

Figure 14.15 (*a*) *Shunt correction* (*b*) *Series correction*

An example of an emitter or source correction circuit is shown, with typical component values in *Figure 14.16*. At low frequencies the two emitters are partially 'held apart' by the resistor *R* so that the current gain of the stage is less than its maximum possible value due to negative feedback. At high frequencies however, due to the presence of *C*, the emitters are no longer held apart, full current gain is thus obtained, compensating thereby for the reduced values of impedances in the collector circuits at these frequencies. The technique is not entirely suitable however for use in final stages where the

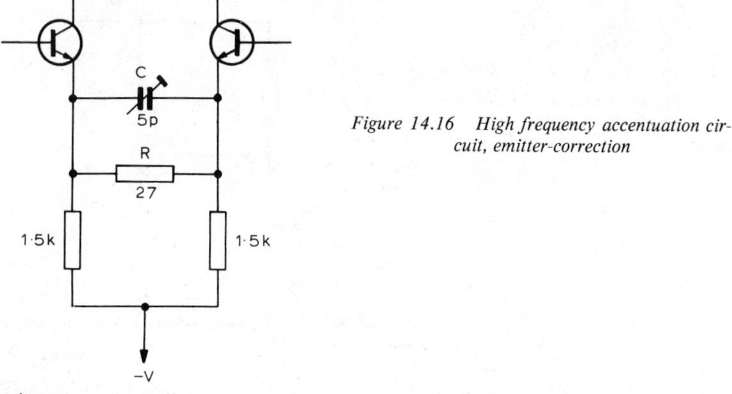

Figure 14.16 *High frequency accentuation circuit, emitter-correction*

transistors are normally driven very close to their current limits at all frequencies, so that there is no reserve of current left to call upon at the high frequency end of the spectrum.

It can be shown that the ultimate gain × bandwidth product per stage is $2.53\, g_m/\pi C$, where g_m is the mutual conductance of the active device and *C* is the stray capacitance. This sets an absolute limit to the bandwidth which may be employed with an active device using the techniques already discussed. Whereas the straightforward paralleling of active devices increases the overall value of g_m, it increases *C* proportionately and thus provides very little improvement. However, substantial improvements can be effected if the capacitancies of the devices be incorporated into artificial transmission lines, forming a distributed amplifier as shown in *Figure 14.17* (*a*).

In this case the input signal is applied to an artificial delay line from which tappings are taken at equal intervals to feed in sequence the gates of the field-effect transistors TR_1, TR_2 and TR_3 (and others if necessary). Since the gate of TR_1 is fed first, its drain

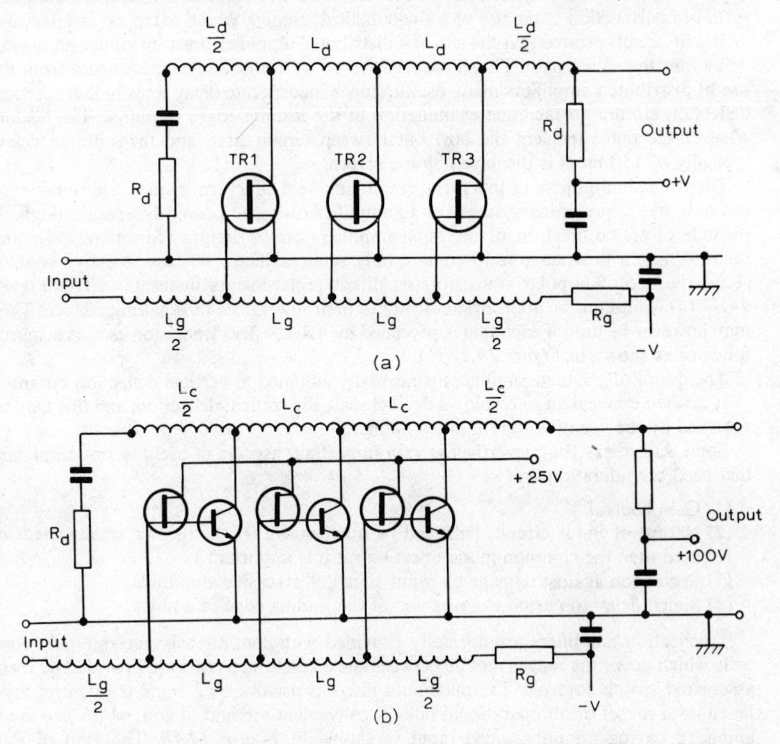

Figure 14.17 A distributed amplifier using (a) f.e.t.s (b) f.e.t.s and bipolar transistors

current is the first one to be modulated by the signal, to be followed in turn by similar components of current from TR_2 and TR_3. Each of these drain currents divides into similar components travelling to left and right to be absorbed by the two resistors R_d. The signal output is now taken from the right-hand R_d resistor. For the separate signal components that were time-split at the gates to be recombined in phase at the output, it is necessary that the time delays in the drain delay line should equal those in the gate delay line. The main characteristics of the line are given by:

$$\text{Delay time} = \sqrt{CL}$$

$$\text{Cut-frequency} = 1/\pi\sqrt{CL}$$

$$\text{Characteristic impedence} = \sqrt{\frac{L}{C}}$$

$$\text{Voltage gain} = ng_m R_d/2$$

$$\text{Output voltage} = nI_d R_d/2$$

where n is the number of transistors and I_d is the signal drain current per transistor. Note that the impedances of the active devices fall with an increase of frequency. A simple guide is that for operation up to 100 MHz a suitable characteristic impedance for a delay line is 100 Ω. For other values of maximum frequency, the impedance should be inversely proportional to that frequency.

In the circuit of *Figure 14.17* (*a*) the artificial delay lines are of the *T*-section type, but π section types also are used. The performance of the delay line may be improved by the introduction of mutual inductance between the inductors, and to this end, a common form of construction is the use of a single helical winding which is tapped at intervals.

It will be appreciated that the use of a distributed amplifier must introduce an overall delay into the signal path. This however is to be welcomed, since quite apart from the use of distributed amplifiers many oscilloscopes incorporate delay lines in their vertical deflection circuits to facilitate examination of the leading edges of pulses. The leading edge of the pulse triggers the horizontal sweep circuit first, and then after a delay, typically of 150 ns, it is displayed on the screen.

Distributed amplifiers of this form, employing field-effect transistors, are suitable for use only in the preliminary stages of an amplifier owing to power dissipation limits. If pentode valves be substituted, then such amplifiers can be designed for operation in any stage. Where a final stage is to contain only semiconductor devices as active components, the use of bi-polar transistors as direct replacements in the circuit of *Figure 14.17* (*a*) would not be appropriate owing to their low values of input impedance. They may however be used if each one is preceded by a field-effect transistor used as a source follower as shown in *Figure 14.17* (*b*).

The use of distributed amplifiers is normally confined to vertical deflection circuits.

It is often convenient to employ a decibel scale for vertical deflection, and this may be achieved by the use of a logarithmic amplifier.

Some subsidiary (but nevertheless very important) aspects of oscilloscope amplifiers that need consideration are:

(1) Gain control.
(2) Nature of input circuit, balanced or unbalanced. (If the former arrangement is used then the common-mode rejection ratio is important.)
(3) Protection against damage by input signal of excessive amplitude.
(4) Signal delay to permit examination of the leading edge of a pulse.

Y deflection amplifiers are normally provided with continuously variable gain controls which cover the ranges that lie between the discrete steps that are provided by their associated switch controls. The maximum range is usually 1 : 2.5 and the control may be either a preset or an operational one. A convenient method of control for use in an amplifier having an unbalanced input is shown in *Figure 14.18*. The gain of this amplifier is at its maximum when the gain control resistance is at its minimum. The circuits of TR_1 and TR_2 are the unbalance-to-balance amplifiers while the 'tail' of the long-tail pair is provided by the circuit of TR_3 which presents between its collector and the negative rail a dynamic resistance which is of the order of a megohm. The thermistor *TH* serves to stabilise the gain of the amplifier with change of temperature.

The shift-balance control is preset to minimise vertical movements of the trace when the gain control is operated. Vertical shift facilities are incorporated. The signal current in the drain circuit of TR_2 is less than that in the circuit of TR_1, the difference changing with the setting of the gain control. However any further long-tail pair amplifiers will serve to reduce the final inequality of the two signals to a very small proportion.

The form of gain control just described will not function when the input signal to the amplifier is balanced. Large changes of gain are best achieved by change of drain–drain (or collector–collector) load as is shown in *Figure 14.19* (*a*). Small changes of gain may be achieved by variation of the magnitude of the static value of the current which is passed by the two transistors TR_1 and TR_2. This can be done by varying the d.c. resistance of the tail transistor TR_3. In this circuit the balance control is adjusted to achieve the maximum value of common-mode rejection ratio (c.m.r.r.). When adjusting the amplifier to maximise the c.m.r.r., the oscilloscope is first driven in the common-mode, the same signal V_c of a few volts being applied in parallel to both inputs as is

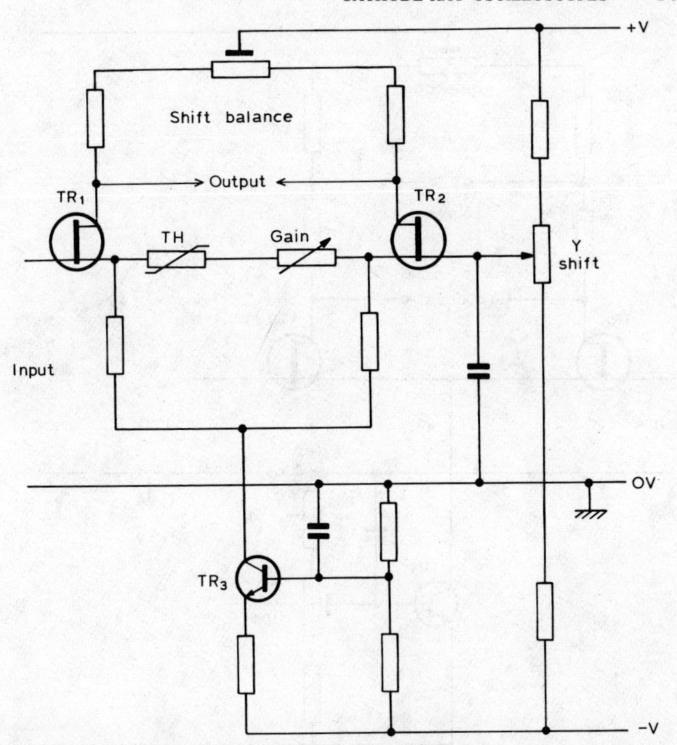

Figure 14.18 *An amplifier gain-control circuit with shift-balance and a transistor 'tail'*

shown in *Figure 14.19*(*b*). The gain-balance control is now set to *minimise* the amplitude of the vertical deflection. Secondly a signal of small amplitude V_d is applied in the differential mode from a *balanced* source as is shown in *Figure 14.19*(*c*). Its magnitude is then adjusted to produce an output signal of the same magnitude as that which existed previously. The c.m.r.r. is now V_c/V_d. At low frequencies, for relatively low values of V_c, the value of c.m.r.r. might be 1 000. Due however to limitation of supply potential, to differences between TR_1 and TR_2 and to the stray capacitance across TR_3, lower (and thus less satisfactory) values of c.m.r.r. will be experienced under other conditions.

Not only do the gains of the two sides of the circuit need to be balanced, but so also do their static potentials. This can be done by means of the coarse and fine balance controls shown in *Figure 14.20*. The resistors *R*, having values in the megohms range, allow the potentials to be varied without there being significant variations of load resistances.

The vertical deflection amplifier is usually provided with some form of protection to avoid damage occurring when it receives an input signal of significantly higher amplitude than that which would produce full-scale deflection of the spot. In this respect thermionic valves are more robust then their semiconductor counterparts, so that some oscilloscopes employ valves in this position even when the other active devices are semiconductors. Some typical protective arrangements are shown in *Figure 14.21*. In circuit (*a*) the presence of *C* and *R* in series with the grid of the valve protects the latter against the application of excessive positive potentials, the flow of grid current causing most of the excess potential to be dropped across *R*. The combination of *R* and the neon

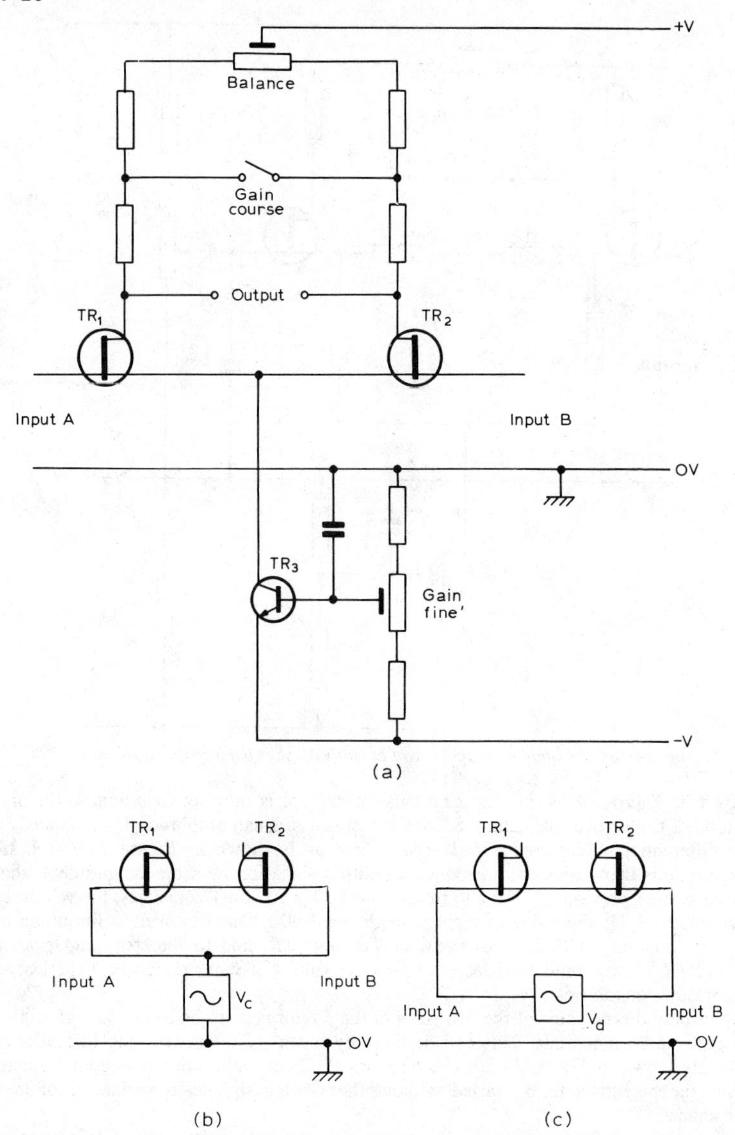

Figure 14.19 (*a*) *Gain and gain-balance controls for a balanced input amplifier.* (*b*) *and* (*c*) *Measurement of common mode rejection ratio* (*c.m.r.r.*)

Figure 14.20 *Balancing the static potentials of the Y-deflection amplifier*

Figure 14.21 Circuits for protecting the input of the Y-deflection amplifier against signals of excessive amplitude

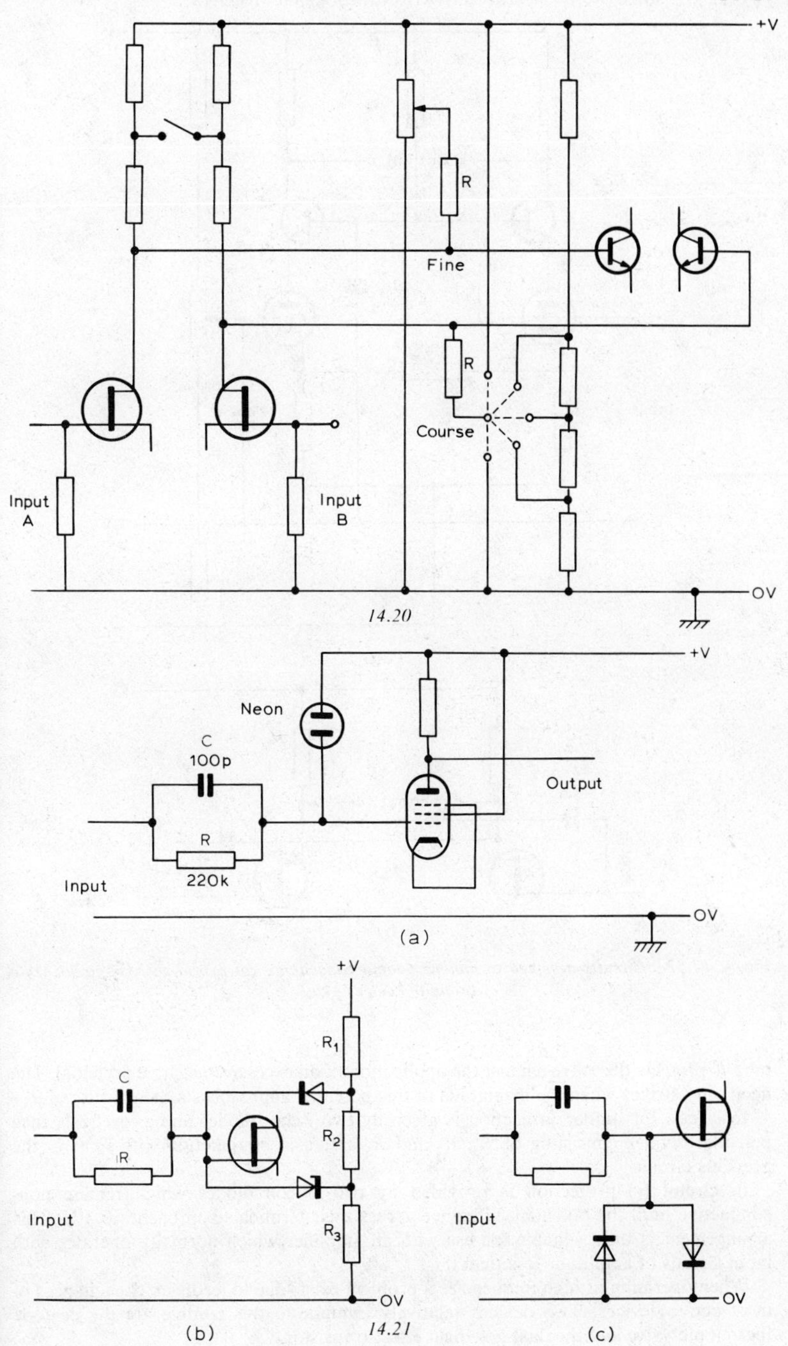

14.20

(a)

(b) *14.21* (c)

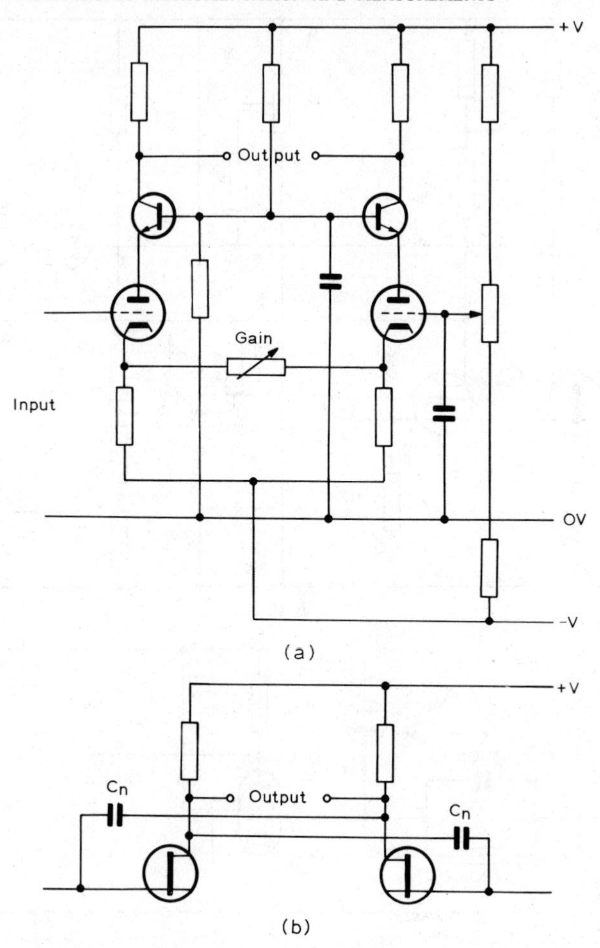

Figure 14.22 Circuits designed to minimise input capacitance (a) hybrid cascode circuit (b) a Franklin bridge circuit

tube N protects the valve against the application of an excessive negative potential. The neon only strikes when the magnitude of this potential approaches a dangerous value.

In circuit (*b*) similar protection is given by two Zener diodes and a low resistance potential divider consisting of R_1, R_2 and R_3 acting in conjunction with R as in the previous circuit.

In circuit (*c*) protection is provided by two silicon diodes which remain non-conductive until the potential difference across their terminals approaches 0.5 V. This arrangement is only suitable for use with an amplifier which normally operates with input signals of amplitude less than this.

When operating at high frequencies a problem exists due to feedback capacitances of most active devices. Two devices relatively immune to this trouble are the pentode thermionic valve and the dual-gate field-effect transistor.

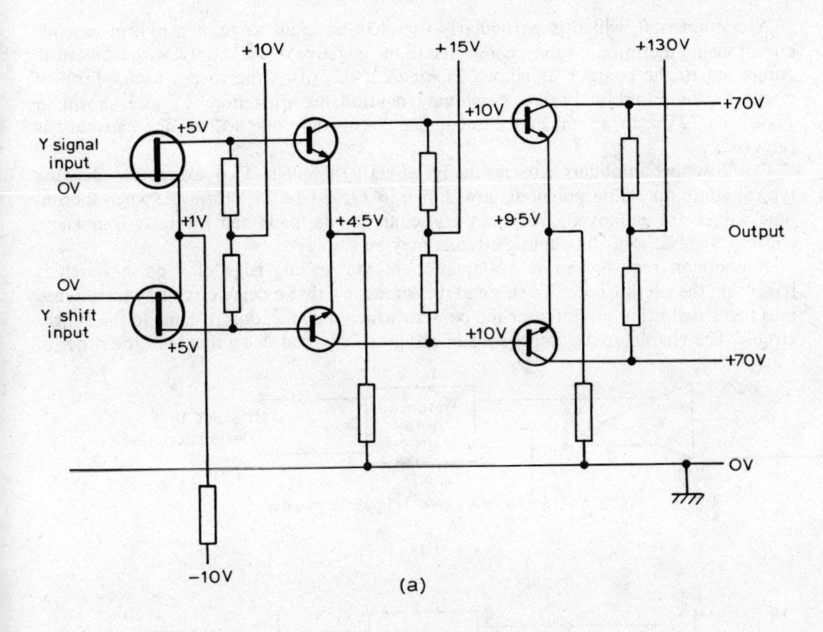

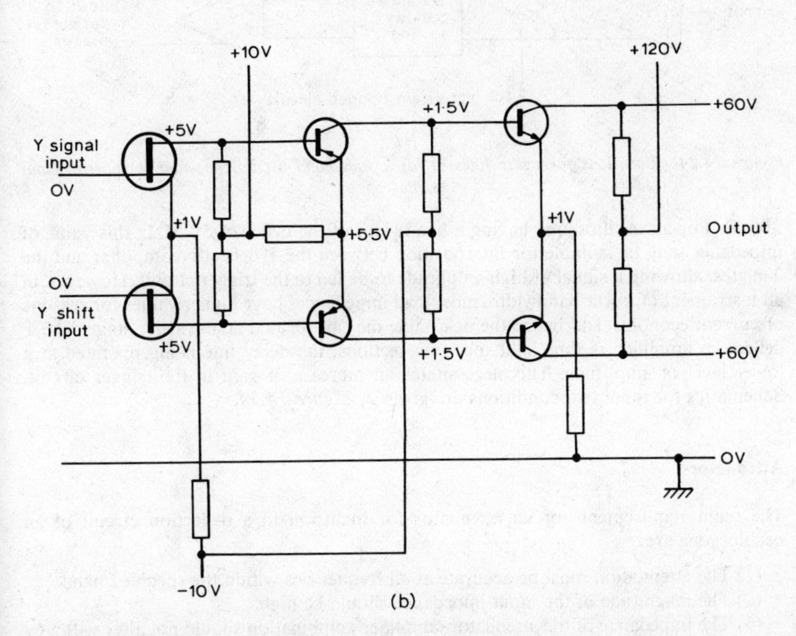

Figure 14.23 Basic circuits for typical Y-deflection amplifiers using n-channel f.e.t. input stages (a) with two npn transistor pairs and (b) with one pnp pair and one npn pair

A combination, which is particularly useful in an input stage, is a hybrid cascode circuit, using thermionic valves connected to the negative of the supply, with transistors connected to the positive as shown in *Figure 14.22* (*a*). Otherwise a useful form of circuit is the Franklin bridge, employing neutralising capacitors C_n and shown in *Figure 14.22* (*b*). In an unbalance-to-balance circuit only one neutralising capacitor is required.

Oscilloscope amplifiers are normally directly coupled. Two examples, showing typical static operating potentials are shown in *Figure 14.23*. However, when thermionic valves are employed, frequency-conscious anode loads and inversely-frequency-conscious anode-to-grid coupling circuits may be employed.[1]

A common requirement is examination of the leading edge of a pulse which is triggering the oscilloscope. To this end it is necessary that a delay circuit be introduced into the *Y* deflection circuit after the point at which a connection is made to the trigger circuit. The characteristic impedance of a typical balanced delay line is of the order of

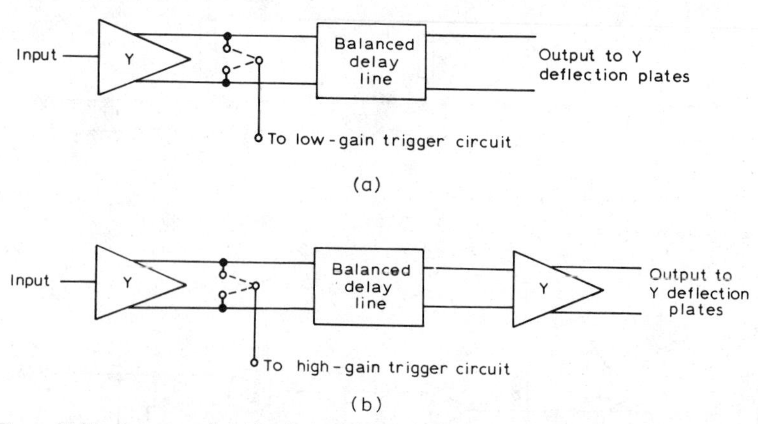

Figure 14.24 *Locations for delay lines in* (*a*) *a wide-band oscilloscope and* (*b*) *narrow-band oscilloscope*

200 Ω. For an oscilloscope having a bandwidth of the order of 50 MHz this value of impedance may be suitable for interposition between the *Y* deflection amplifier and the *Y* plates, allowing a signal of high amplitude to be fed to the trigger circuit. However, in an instrument of lower bandwidth, most load impedances have higher values for reasons of current economy and hence the delay line may be located at an earlier stage. The *Y* deflection amplifier is thus split into two sections, the delay line being operated at a lower level of amplitude. This necessitates an increase of gain in the trigger circuit. Schematics for these two conditions are given in *Figure 14.24*.

Attenuators

The main requirements of an attenuator for inclusion in a deflection circuit of an oscilloscope are:

(1) The attenuation must be accurate at all frequencies within the specified band.
(2) The magnitude of the input impedance should be high.
(3) The impedance of the attenuator–amplifier combination should not alter with any change of overall sensitivity. (This is to facilitate the design of probe units and also to maintain a constant loading upon the circuit under test.)

The attenuations, expressed as voltage ratios, are normally in the sequence 1, 2, 5, 10, 20, 50, 100, 200, 500, etc., as far as is necessary. The nine values quoted are adequate for a simple oscilloscope, but further values are necessary for sensitive instruments. In the latter case, variation of overall sensitivity is sometimes achieved by a combination of attenuation adjustment and gain adjustment of the following amplifier, the latter being used at maximum gain only when all attenuation has been switched out. There may be a reduction of bandwidth at maximum sensitivity.

The overall attenuation required is normally provided by the cascade connection of a number of attenuating sections in separately screened compartments. The nine values of attenuation are usually provided by a selection of 1, 2 or 5 in one screened compartment, cascaded with a selection of 1, 10 or 100 in another compartment. An example of a circuit giving unity or 10 : 1 attenuation is illustrated in *Figure 14.25* (*a*).

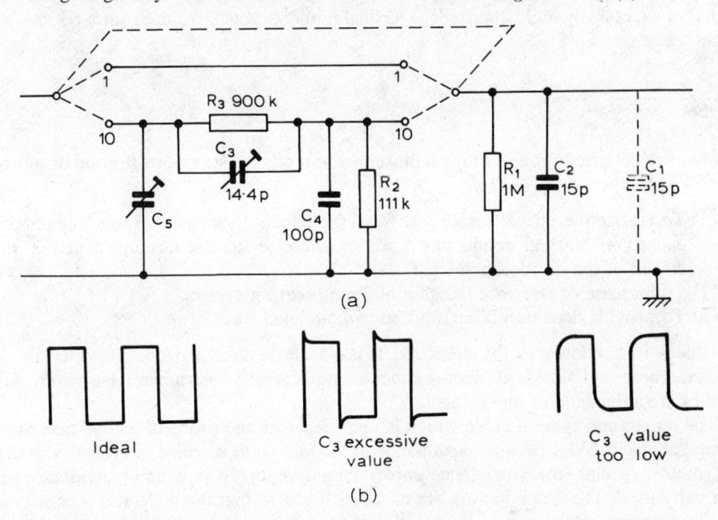

Figure 14.25 (*a*) *A basic attenuator circuit and* (*b*) *adjustment of high frequency response*

The input impedance of the amplifier is represented by the dotted C_1. Since its value is likely to be somewhat unstable it may be shunted by a capacitor C_2. The input impedance of the oscilloscope with all attenuators switched out of circuit is thus 1 MΩ shunted by 30 pF, and all these values must be maintained when the attenuators are switched into circuit. The required attenuation at low frequencies is given by

$$\frac{(R_1 \| R_2) + R_3}{R_1 \| R_2} = 10 \quad (\| \text{ signifies parallel connection})$$

While $(R_1 \| R_2) + R_3 = 1$ MΩ as required to maintain constant input resistance. To ensure that this attenuation is maintained over the specified frequency band it is necessary that the time-constants of the series and parallel combinations should be identical. Since $(C_1 + C_2)/9$ would produce the inconveniently low value of 3.3 pF for the capacitor C_3, a capacitor $C_4 = 100$ pF is added, so now the equality of time-constants can be realised by $C_3R_3 = (C_1 + C_2 + C_4)(R_1 \| R_2)$. This produces $C_3 = 14.4$ pF. The value of input capacitance which has been considered so far is 14.4 pF in series with 130 pF, i.e., 13.5 pF. To maintain a constant value of input capacitance it is therefore necessary to add $C_5 = 30 - 13.5 = 16.5$ pF. Other attenuating circuits are designed on a similar basis.

In the course of production, and possibly at long-term intervals thereafter, the values of C_3 and C_5 (and similar capacitors in other attenuator circuits), should be checked using a square-wave input to the oscilloscope. C_3 will be adjusted with only this attenuator circuit operative. An excessively high value of C_3 will produce overshoot spikes on the displayed waveform, while too low a value will produce prolonged rise-times, examples being shown in *Figure 14.25* (*b*).

Attenuators are expensive items, so that in the case of oscilloscopes having input circuits that are balanced, and hence requiring the more expensive balanced attenuators, it is usual for the manufacturers to put greater reliance on the use of amplifiers of variable gain and thus to economise on the number of circuits in the balanced attenuators.

An example of a complete attenuator circuit is shown in *Figure 14.26*. The components C_{16} and R_{10} are to safeguard the (valve) input circuit against damage from an input signal of excessively high amplitude. The dual input sockets facilitate comparison of two input signals.

Probes

Probes are designed for use with oscilloscopes and other instruments for one or more of the following purposes:

(1) To isolate the circuit under test from the effects (principally shunt capacitance) caused by normal connecting leads or cable to the oscilloscope and the input circuit of the oscilloscope itself.
(2) To increase or decrease the gain of the measuring system.
(3) To provide detection of an amplitude-modulated waveform.

Unless it is known to the contrary, it is unsafe to take a probe provided by one manufacturer and to use it with another manufacturer's instrument. However, some probes are adjustable in this respect.

The length and type of cable which is provided with the probe is critical and should not be changed. Whereas the capacitance of normal co-axial cable is typically 100 pF per metre, special low-capacitance cables employing very thin inner conductors are normally used. These conductors are usually kinked so that the dielectric is mostly air. The inductance of the cable, combined with its capacitance tends to cause ringing on signals having abrupt transitions, as components are reflected backwards and forwards along the cable. Since such a simple circuit cannot be terminated in its characteristic impedance while maintaining the other properties required, a compromise is reached by using resistance wire for the inner conductor, the resistance value being several hundreds of ohms. Careful design of the cable is necessary if microphony is to be avoided.

Some probes contain only passive components, while others contain active components which must be supplied with power along the interconnecting cable from the main instrument. Probes containing only passive components, and some containing active components also, normally attenuate the signal. In the simpler systems the attenuation is a simple round number such as 2 or 10. Since this may lead to operational errors, more complex systems are available which operate when the probe plug is inserted into the socket on the oscilloscope. This action may:

(1) Raise the gain of the vertical deflection amplifier; or
(2) change the circuit of an optical display system using light-emitting diodes or fibre-optic devices, thus indicating a change of sensitivity.

The simplest probe, containing only two components, is shown in *Figure 14.27*. Here the time-constant of the probe $5.5 \times 10^{-12} \times 9 \times 10^{6} = 50 \times 10^{-6}$ is made to equal the time-constant of the cable-plus-oscilloscope input circuit, i.e., $1 \times 10^{6} \times (25 + 25) \times 10^{-12} = 50 \times 10^{-6}$.

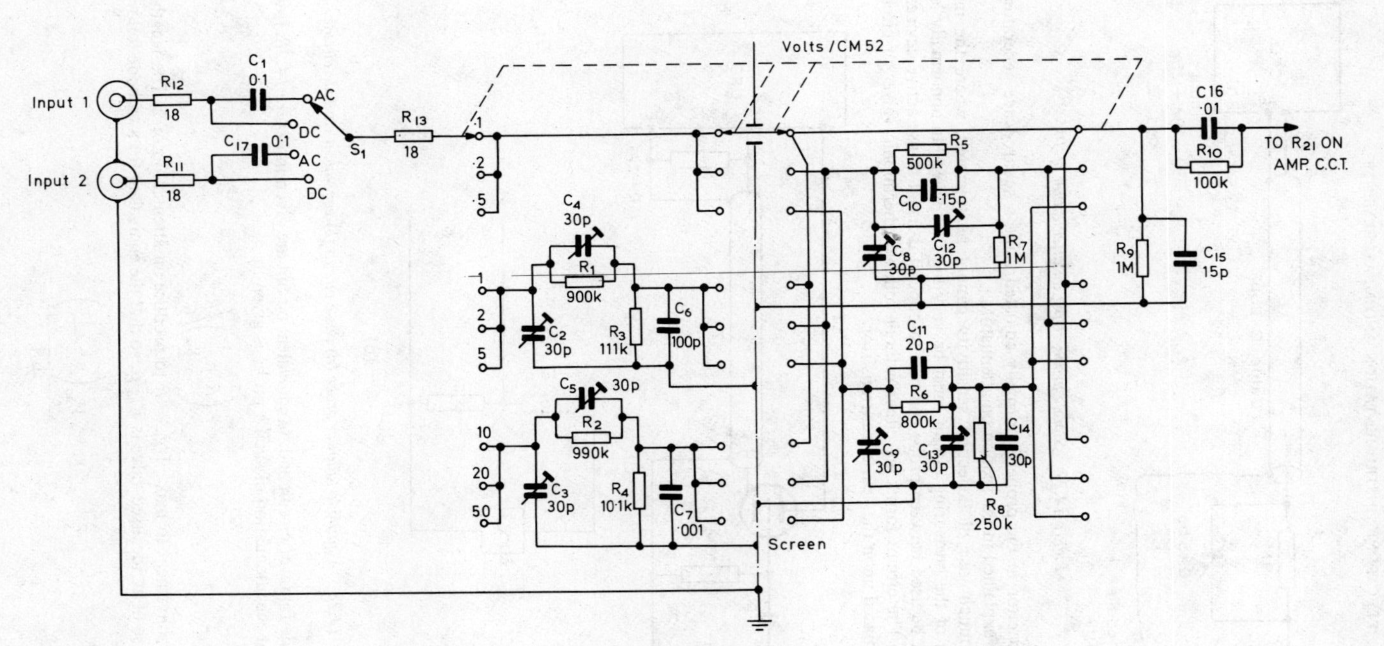

Figure 14.26 Attenuator circuit type H (Courtesy Telequipment)

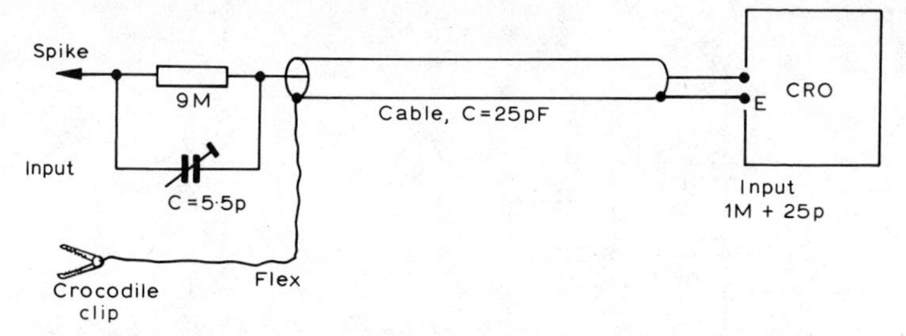

Figure 14.27 Simple probe giving an attenuation of 10

The capacitor in the probe is normally adjusted using a square-wave generator as previously explained in the section on attenuators.

Where much loss of deflection sensitivity cannot be tolerated, where the input impedance of the measuring circuit must be very high, or where an abnormally long cable must be used between probe and oscilloscope, it is usual to use active devices in the probe. The circuit forms are either cathode-follower thermionic valves or circuits which are based upon the field-effect transistor.

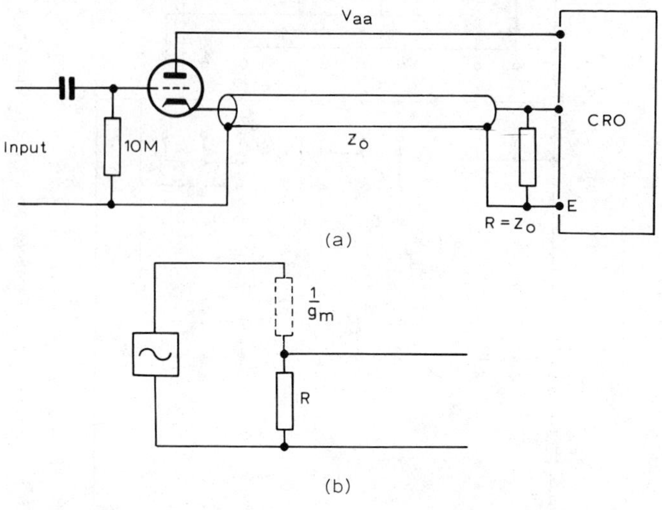

Figure 14.28 A cathode-follower probe (a) basic circuit (b) equivalent cathode circuit

A cathode-follower circuit and its equivalent circuit are given in *Figure 14.28*. It will be seen that the attenuation of such a probe is given by

$$\left(R + \frac{1}{g_m}\right) \Big/ R$$

It is often convenient to make $1/g_m = R$ to produce an attenuation of 2. The length of the grid/base of an idealised valve is V_{aa}/μ, so that the equivalent for a cathode-follower circuit is

$$\frac{V_{aa}\left(R + \dfrac{1}{g_m}\right)}{\mu R}$$

In practice the length of the grid-base is, for low-distortion operation, only about half this quantity.

It should be noted that while the value of input impedance is raised to about 10 MΩ shunt by 5 pF, operation at frequencies below about 10 Hz is sacrificed.

An alternative scheme using semiconductors is illustrated in *Figure 14.29*. Due to the lower mutual conductance of the field-effect transistor, it is necessary to employ two further semiconductor devices in similar configurations before the signal may be fed to the cable. The voltage-handling capacity of this probe is less than that using a cathode-follower but it will operate up to a higher frequency.

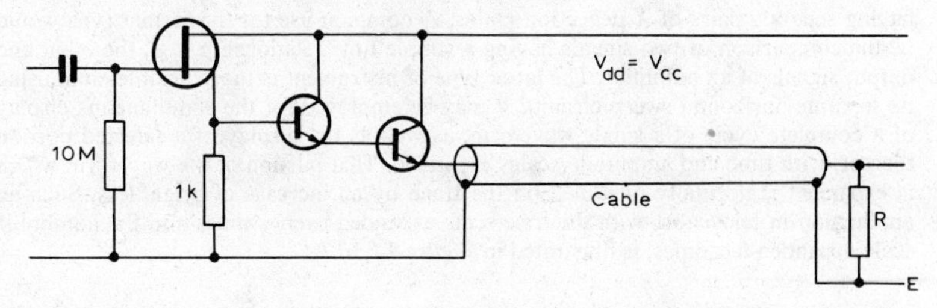

Figure 14.29 *An all semiconductor probe* (*from* Measuring Oscilloscopes *edited by J. F. Golding, Newnes-Butterworths* (*1971*))

Current-measuring probes are also available. These are basically current transformers. The current-carrying wire acts as a primary, a split ferrite core has jaws which can be opened to clip over the wire and then closed. A winding on the core acts as the secondary winding. Electrostatic and electromagnetic shielding are necessary. Such probes operate satisfactorily up to about 50 MHz, but performance is less satisfactory below 1 kHz, although frequency-compensated amplifiers may be supplied with the probe.

Multiple display facilities

Most multiple display facilities are provided by oscilloscopes specially designed for the purpose, alternatively they are provided by means of additional units used in conjunction with conventional oscilloscopes. In either case the screen size should be large and the aspect ratio preferably 1 : 1 rather than the 4 : 3 ratio commonly encountered.

Each channel must be provided with its own sensitivity control(s), its own *Y* shift control, and its own input switch. Some systems have separate horizontal sweep and shift, brightness and focus controls for the two channels, while in some cases they are common.

Multiple displays usually provide dual display facilities, sometimes with the additional facility of displaying either the sum or the difference of two quantities along a single trace. Quadruple display instruments are also available. These are particularly useful for teaching and demonstration purposes, but they are not usually precision instruments. They normally employ very large cathode-ray tubes of types designed for television or radar display purposes, and as such they are magnetically deflected, so that the bandwidth of the *Y* deflection circuits is commonly limited to about 8 kHz. Instruments employing long after-glow cathode-ray tubes are available having sweep circuits which operate with repetition frequencies as low as 2 cycles/minute. Some 4-channel oscilloscopes can be switched to provide 2-channel or 1-channel operation when required.

The multiplexing facility may be provided either by the use of a special type of cathode-ray tube or alternatively by time-division multiplex electronic circuitry. Several sub-divisions of these two techniques are available.

The earliest version of the multiple (twin) channel oscilloscope employed *X* deflection plates which were nearer to the gun assembly than the *Y* deflection plates. A splitting plate between the two *Y* deflection plates split the beam in two, leaving each half to be deflected separately by a signal applied to a single *Y* plate. The *Y* deflection plates were unbalanced and similar polarities of input signal produced opposite directions of vertical deflection.

This was followed by a type of tube having two beams produced by two holes in the final anode. Symmetrical *Y* deflection now became possible. One variation of this type of tube is available having a common pair of *X* deflection plates, another variation having separate pairs of *X* deflection plates. A common use for the former type would be the comparison of two signals having a simple time relationship, e.g., the input and output signals of an amplifier. The latter type of instrument is more flexible, and, using its separate horizontal sweep circuits, it may be employed for the simultaneous display of a complete cycle of a single waveform, as well as the display of a selected portion thereof, with time and amplitude scales expanded. That portion of the waveform which is expanded is normally identified on the trace by an increase of brightness. Such an application in television, with the time scale expanded twenty times and the amplitude scale expanded ten times, is illustrated in *Figure 14.30*.

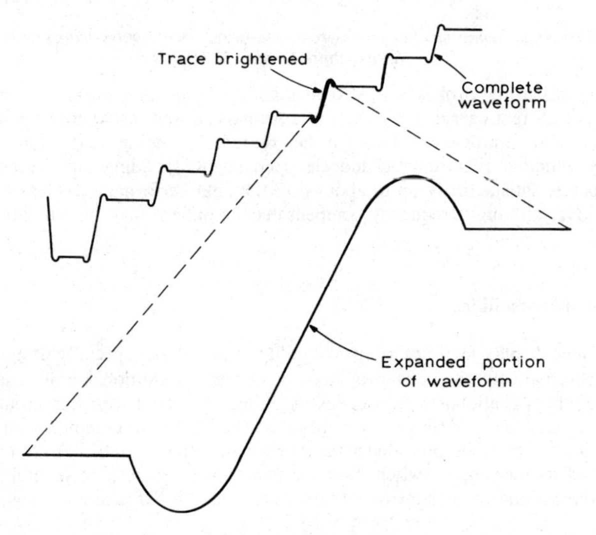

Figure 14.30 Use of twin-beam oscilloscope with double horizontal sweep circuits for simultaneous displays of one complete cycle of a waveform and an expanded version of a portion thereof

Where a single-beam cathode-ray tube is used for multiple display purposes, alternative time-division-multiplex sampling systems are available. Some oscilloscopes use one technique, some the other. Some instruments provide facilities for the use of both techniques, the choice being in certain cases left to the user and in other cases being governed by the sweep adjustment control.

In one case the switching action in the signal circuits is synchronised with the horizontal sweep, so that each sweep of the spot is devoted entirely to one signal, the various signals being written on to the screen in sequence. This is known as synchronous-mode operation, or where only two channels are provided, as alternate-mode operation. In the other case the switching action is unsynchronised, occurring usually at about 100 kHz. A typical schematic is shown in *Figure 14.31*. In this circuit the gates open alternately under the control of the bistable circuit, so that the signals are displayed in turn. The bistable circuit may be triggered either from the internal sweep circuit or from the 100 kHz internal generator.

Consider as an example two waveforms which are to be examined, a sinusoid being applied to input *A* and a triangular waveform to input *B*. To the former, a positive *Y* shift potential is added, producing V_A, and to the latter a negative *Y* shift potential,

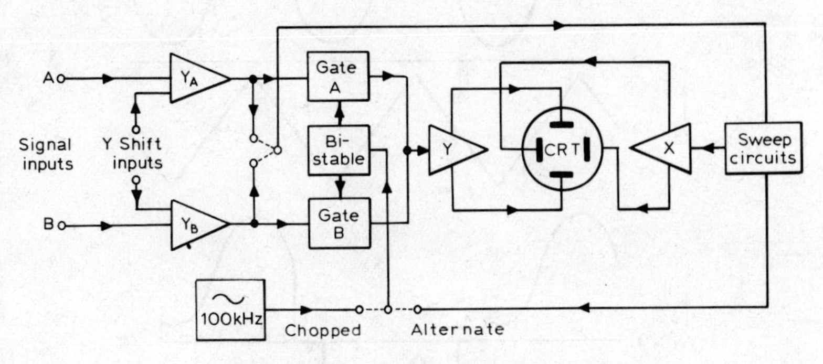

Figure 14.31 A dual-trace oscilloscope

producing V_B, both being illustrated in *Figure 14.32*. The horizontal sweep waveform is shown as V_X. If alternate mode operation is employed then the input to the common *Y* amplifier will appear as V_{Y_1}. If however chopped operation is employed, then using the chopping waveform V_{CH} the input to the common *Y* amplifier will be as shown at V_{Y_2}. (The vertical transitions of this waveform are so rapid that there is a negligible brightening of the screen in between the two waveform patterns.) The gaps which appear in both waveforms in any one horizontal sweep are filled by successive sweeps, so that irrespective of the switching employed (alternate or chopped), the final result appears as shown at the bottom of *Figure 14.32*.

Triple display oscilloscopes have been produced, the separate traces being in red, green and blue on the screen of a shadow-mask cathode-ray tube. This has the advantages that the traces can be overlapped without confusion and a very clear distinction made between the separate signals. Since the deflection is by magnetic means however, the available bandwidth is only of the order of 8 kHz.

Sampling

For the examination of waveforms having repetition frequencies between about 100 MHz and 1 GHz, or having rise-times of the order of nanoseconds it is usual to employ a sampling technique, using either a complete sampling oscilloscope or else a sampling adaptor in conjunction with a conventional oscilloscope. A simplified schematic appears in *Figure 14.33* with waveforms in *Figure 14.34*.

The input signal is applied to a trigger generator which delivers a series of pulses, normally at a frequency which is of the order of 30 kHz but at an integral submultiple of the input frequency. (This presupposes that the signal frequency exceeds 30 kHz. If however the signal frequency is less than this, then the sampling frequency is made to equal the signal frequency.)

Each trigger pulse initiates one step in the operation of a staircase generator, the output of which has its steps counted to determine the number of dots per cycle of the waveform which is to appear on the oscilloscope screen in any one horizontal sweep.

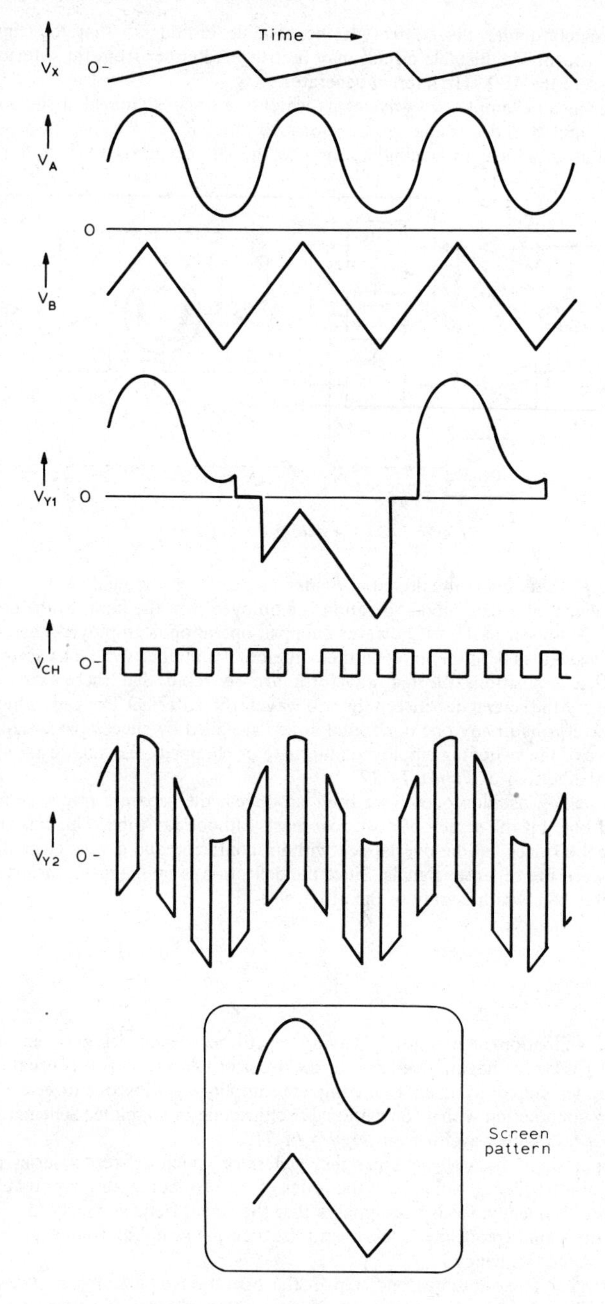

Figure 14.32 Typical waveforms and screen pattern for a single beam dual-channel oscilloscope using either alternate-mode or chopped-mode operation

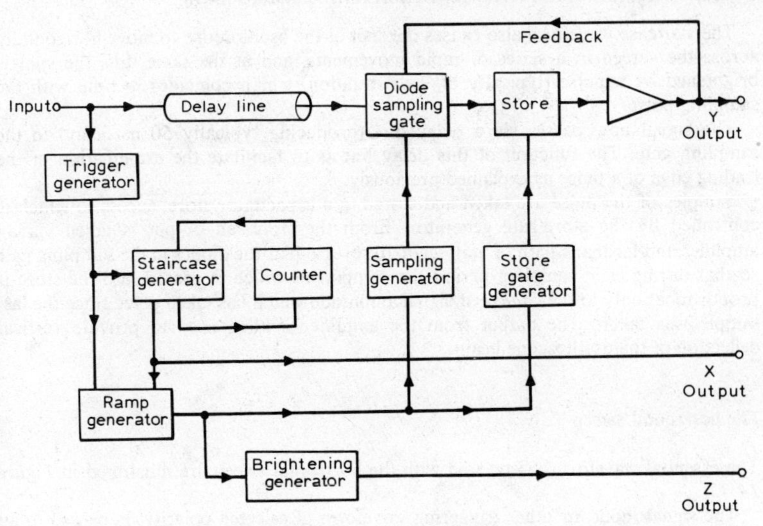

Figure 14.33 A sampling unit

after the required count, say between 200 and 1 000, the counter resets the staircase generator, which then commences the generation of a new staircase.

Each trigger pulse also causes the ramp generator to commence the generation of a very fast ramp of fixed velocity. The duration of each ramp is, however, caused to be proportional to the instantaneous magnitude of the staircase, so that during the staircase cycle the ramps are of progressively increased durations. The *termination* of each ramp now causes the sampling generator to produce a sampling pulse, typically of 350 ps, these sampling pulses occurring progressively later and later in individual periods of the test waveform.

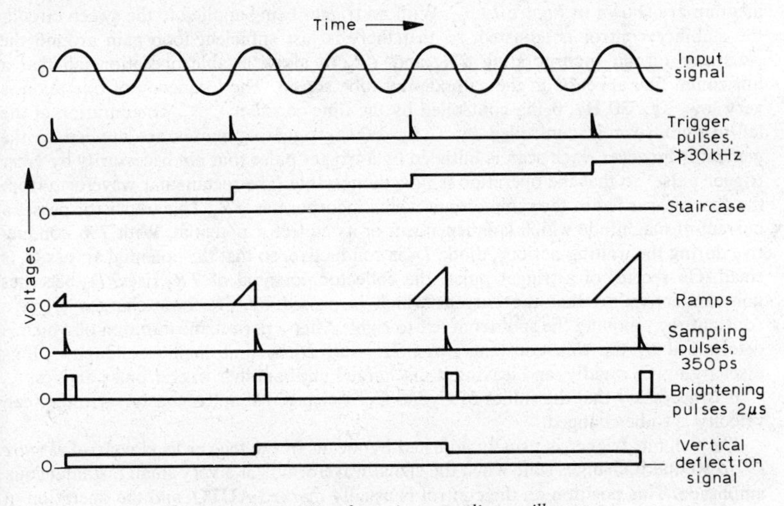

Figure 14.34 Waveforms in a sampling oscilloscope

The staircase waveform also causes the spot of the oscilloscope to move horizontally across the screen in a series of rapid movements, and at the same time the spot is brightened by a pulse (typically of 2 μs duration) which coincides in time with the sampling pulse.

The signal now passes via a delay line (producing typically 50 ns delay) to the sampling gate. The function of this delay line is to facilitate the examination of the leading edge of a pulse as explained previously.

Samples of the pulse are taken and stored in a capacitator store, access to which is controlled by the store-gate generator. From the store, an output is taken via an amplifier and feedback path of unity gain, to reverse-bias the diodes in the sampling gate so that during each sampling period the component which is passed into the store is proportional only to the *change* of signal amplitude which has taken place since the last sample was taken. The output from the amplifier is then used to provide vertical deflection of the oscilloscope beam.

The horizontal sweep

The essential waveforms associated with the horizontal sweep are illustrated in *Figure 14.35*.

The signal input, or other triggering waveform of selected polarity, is passed to an amplitude-limiting circuit, the output amplitude of which remains constant, provided that the input amplitude exceeds a certain minimum value. For internal triggering, i.e., triggering from the waveform under examination, this value usually corresponds to a vertical deflection of the spot which is not more than about 4 mm. Typical signal and limiter waveforms are shown in *Figures 14.35* (*a*) and (*b*) respectively. In order that the limiter output waveform should have fast rise-times the circuit normally takes the form of a Schmitt trigger as is shown in *Figure 14.36* (*a*).

The output of the Schmitt trigger circuit is applied to a differentiating circuit to which a diode is added to attenuate one series of spikes. The circuit is shown in *Figure 14.36* (*b*) and the waveform in *Figure 14.35* (*c*).

The spikes are applied to the input circuit of an astable/monostable generator, the output waveform of which is shown in *Figure 14.35* (*d*). A block schematic and a circuit diagram are shown in *Figure 14.37*. With no trigger being applied to the sweep circuit, the stability control is adjusted, so that there is just sufficient loop gain around the generator circuit incorporating TR_1 and TR_2 to allow astable operation, so that a horizontal line appears on the cathode-ray tube screen. The frequency of operation is very low, e.g., 20 Hz, being controlled by the time-constant C_1R_1. The duration of the left-to-right scan is controlled by C_2R_2. When triggering pulses are applied to the generator however, each scan is initiated by a trigger pulse (but not necessarily by *every* trigger pulse) so that the operation is now monostable. The rectangular waveform from the generator is fed to the ramp circuit which incorporates TR_3. This transistor passes a current of magnitude which is independent of its collector potential. With TR_2 conductive during the waiting periods, diode D_3 is conductive, so that the potential across C_3 is small. On receipt of a trigger pulse, the collector potential of TR_2 rises, D_3 becomes non-conductive, so that the current which is passed by TR_3 now charges C_3 at a constant rate, moving the spot from left to right. After a period, the duration of which is determined by the time-constant C_2R_2, TR_2 and D_3 become highly conductive, thus discharging C_3 rapidly, and leaving it discharged until another trigger pulse arrives.

It is necessary that the values of C_2 and C_3 be changed simultaneously when the scan velocity is to be changed.

The Schmitt trigger is usually adjusted by means of the trigger level control, *Figure 14.36* so that it changes state when the signal waveform is at a very small instantaneous amplitude. This position on the control is usually marked AUTO, and the operation of the sweep circuit remains unchanged when the peak-to-peak amplitude of the signal

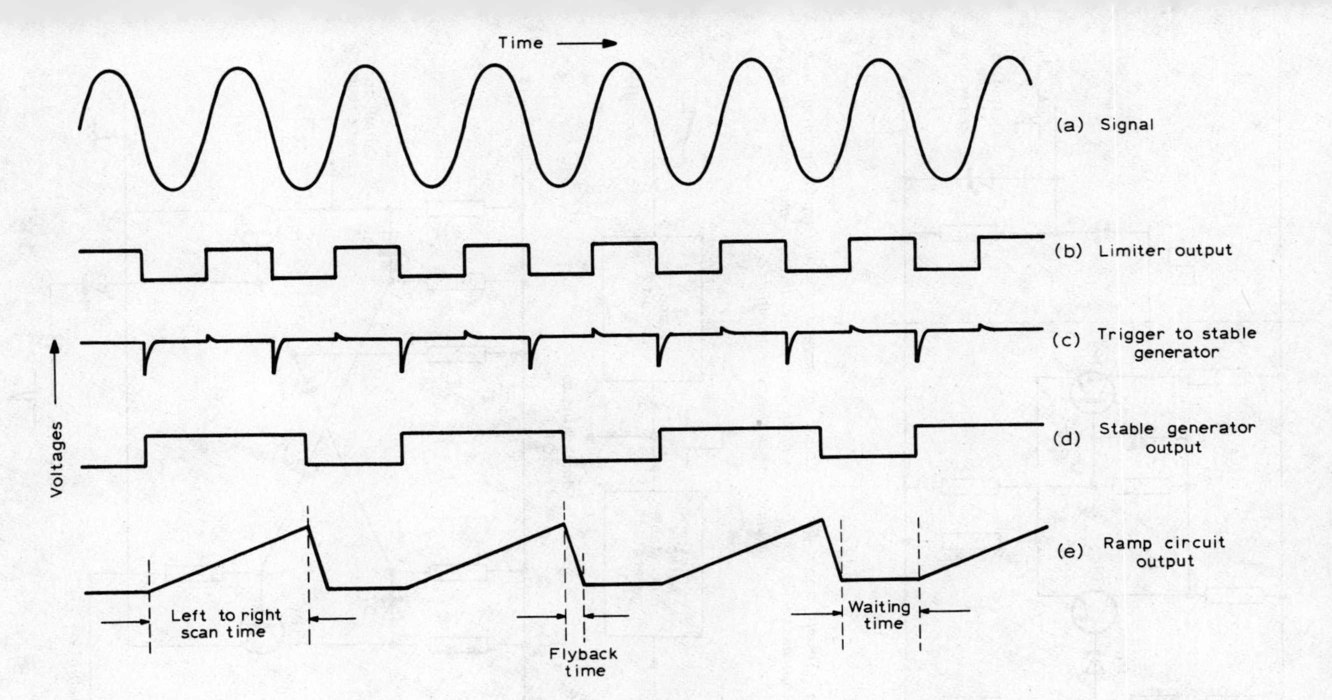

Figure 14.35 Waveforms in a horizontal sweeping circuit. In this example one and a quarter cycles of the waveform under investigation, commencing with a negative half-cycle, would be displayed on the screen. Waveform (*d*) would also be applied to the grid of the cathode-ray tube for unblanking, or alternatively its inverse would be applied to the cathode

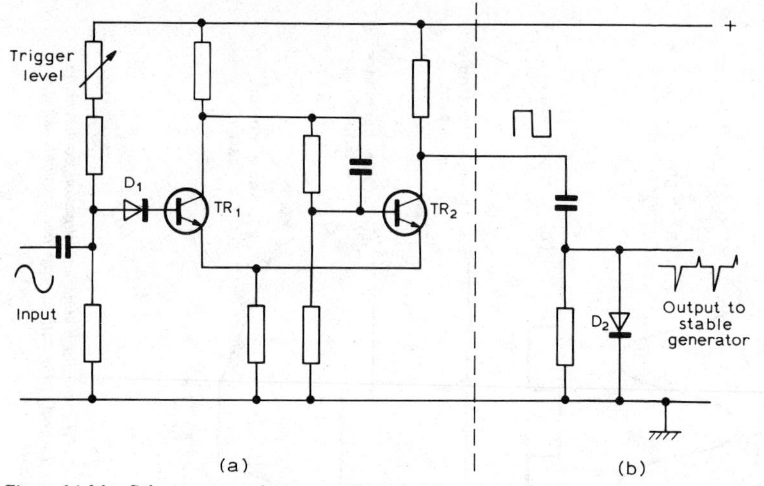

Figure 14.36 Schmitt trigger limiting circuit (a) followed by a differentiating circuit with a positive-spike attenuator (b). The function of D_1 is to protect the base–emitter junction of TR_1 against excessive negative potentials

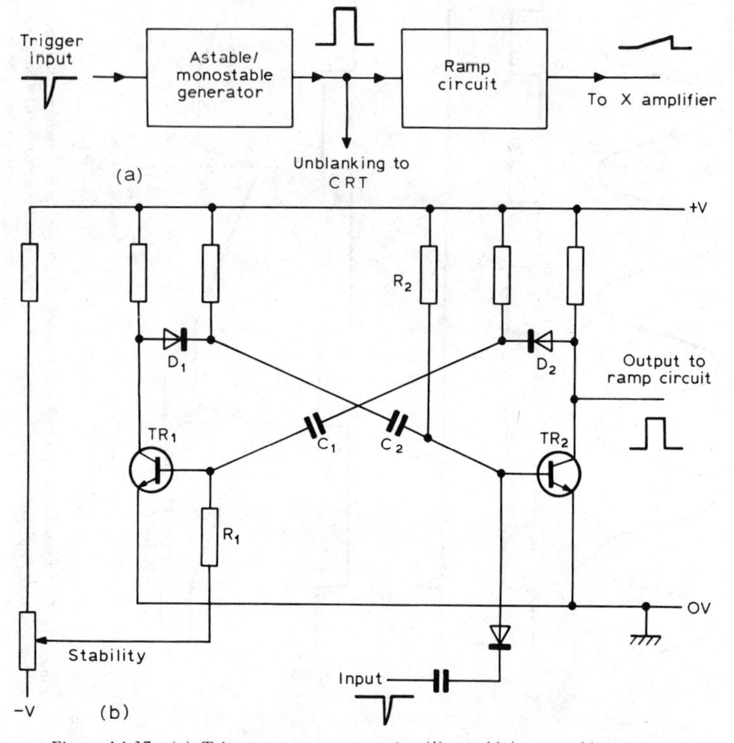

Figure 14.37 (a) Trigger to ramp conversion (b) astable/monostable generator

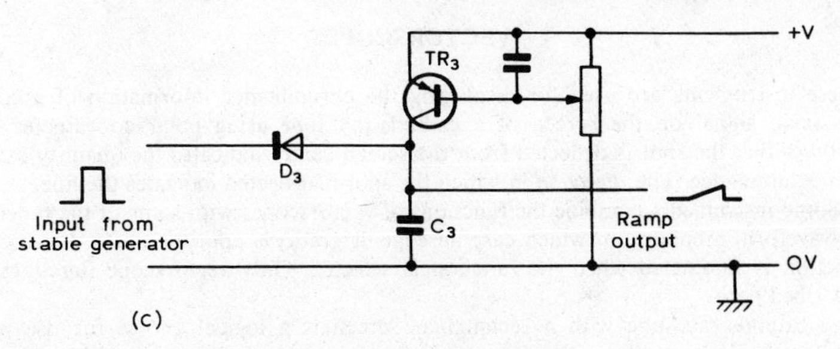

Figure 14.37 (c) ramp circuit

changes within wide limits· Exceptionally, it may be required that the display starts when the signal waveform is passing through some other instantaneous value. This may be achieved by an adjustment of the trigger level control, the Schmitt trigger circuit now delivering a waveform which is no longer symmetrical. AUTO and non-AUTO conditions are illustrated respectively in *Figures 14.38* (*a*) and (*b*). There is also a choice regarding the polarity of the first cycle of the waveform to be displayed on the screen. If the trigger polarity switch be operated, *Figure 14.1*, then the displayed waveform will be inverted as is shown in *Figure 14.38* (*a*).

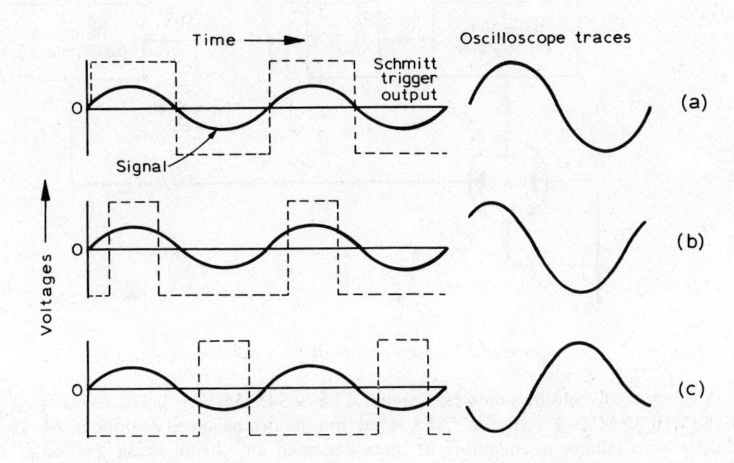

Figure 14.38 Illustrating operation of the trigger level and polarity controls

FURTHER READING

'FET Vertical Deflection pre-Amplifier for a Wide-Band Oscilloscope', *Mullard Technical Publications*, TP 1067, May (1968)

GILBERT, B., 'Timebases for sampling Oscilloscopes', *Mullard Technical Communications*, Vol. 7, No. 67, Dec. (1963)

GOLDING, J. F., ed., *Measuring Oscilloscopes*, Newnes Butterworths (1971)

HART, E. L., 'Ramp Generator', *Wireless World*, July (1965)

'Signal Gating Circuits for Sampling Oscilloscopes', *Mullard Technical Communications*, No. 56, April (1962)

TOWERS, T. D., 'Transistor pulse circuits', *Wireless World*, Jan. to Dec. (1964)

VECTORSCOPES

These instruments are used for displaying the chrominance information of a colour television signal on the screen of a cathode-ray tube using polar coordinates. The *distance* that the spot is deflected from the screen centre indicates the quantity saturation × luminance. The *direction* in which the spot is deflected indicates the hue.

Some instruments combine the functions of vectorscopes with some of the functions of waveform monitors, in which case an edge-lit graticule appropriate to the required function is luminated when the function is selected. Only vectorscope functions are described here.

A cathode-ray tube with a rectangular screen is a logical choice for use in an oscilloscope; one with a circular screen is a logical choice for use in a vectorscope.

A basic block schematic is shown in *Figure 14.39*. Line terminations, amplifiers and attenuators have been omitted for simplification. The designs for such items follow the general forms described in the sub-section on oscilloscopes.

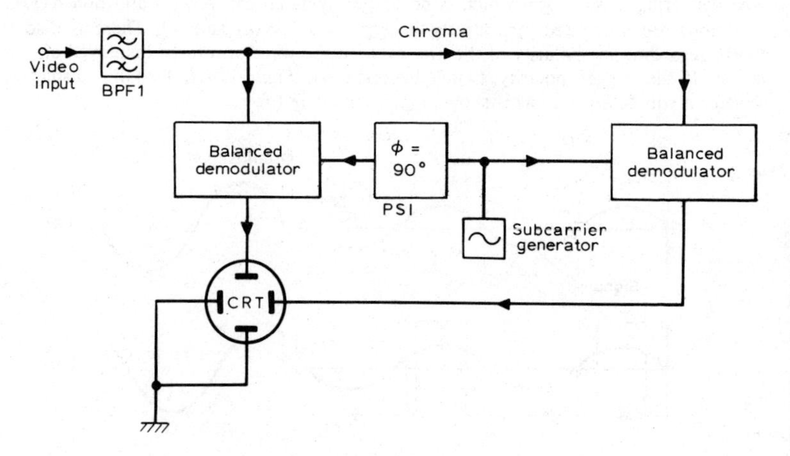

Figure 14.39 Basic schematic for a vectorscope

An input of colour subcarrier signal (3.579 545 MHz ± 1 Hz for NTSC or 4.433 618 75 MHz ± 1 Hz for PAL), is fed into the two balanced demodulators. These ideally have infinite attenuations in the absence of any signal being applied at the chroma input. If a signal is introduced at this point, the attenuations are reduced and a circular pattern, drawn anti-clockwise, will appear on the screen. (See under cathode-ray oscilloscopes; Lissajous figures.)

In operation, the chroma input is at the same repetition frequency as the sub-carrier, but at any instant it may have any phase relationship to it. The *phase* relationship of the chroma to the sub-carrier determines the *direction* in which the spot is deflected from the screen centre. For instance, if the chroma is in phase with the sub-carrier component on the right-hand plate of the cathode-ray tube, then the spot will be deflected into the first quadrant. The *magnitude* of the chroma component determines the *distance* that the spot is deflected from the screen centre.

Because the chroma component is modulated in both amplitude and phase, it carries sidebands having frequencies up to about 1.5 MHz, which are passed by the band-pass filter *BPF 1*.

As it is not always convenient to supply to the instrument a sub-carrier which is continuous, the instrument normally contains an oscillator which may be phase-locked either to:

(1) An external supply of continuous sub-carrier.
(2) The burst component of an external supply of burst plus synchronising signals.
(3) The burst component of the colour signal which is being examined.

An extended schematic illustrating the following points is given in *Figure 14.40* (*a*) with waveforms in (*b*). Video signals are fed to the synchronising signal separator, from which the leading edge of the line-synchronising waveform triggers a monostable delay generator, the trailing edge of the waveform from which operates the monostable gate generator. The latter opens the gate to pass a sub-carrier component to the band-pass filter *BPF 2* during the burst period. The component at the output of *BPF 2* may have a random phase relationship with the burst component of the input video signal, due to the difference of the lengths of cable over which these two components have travelled. It is therefore necessary to pass the component from *BPF 2* through a phase-shifter *PS 2* which covers 0°–360° and is uncalibrated. The output of this phase-shifter is applied to the phase-comparator *PC* which is part of the phase-locked loop *PLL*. The output of the voltage-controlled oscillator *VCO* in this loop is passed through a calibrated precision phase-shifter *PS 3* to the balanced demodulators. This phase-shifter is used to enable phase comparison to be made between two video signals, while using a common source for locking the internal sub-carrier generator. Some vectorscopes are provided with dual input sockets and an electronic switch which gates these two signals rapidly so that the indications appear simultaneously on the screen for comparison purposes.

Vectorscopes may be used for the examination of test signals, such as colour bars, generated while the channel is *not* being used for programme purposes or, while the channel *is* being used for programme purposes, test signals which are generated during the vertical interval periods (VITS) or programme material.

To facilitate measurements, a graticule is provided. Examples of NTSC and PAL graticules, used in connection with the display of colour-bar signals are shown in *Figures 14.41* (*a*) and (*b*). The photograph of the NTSC graticule shows the *Q* and *I* axes leading by 33° on the horizontal and vertical axes respectively. The burst signal is displayed directly to the left of the origin. The colours are represented, anti-clockwise from the 3 o'clock position by the spots in the order *MG* Magenta, *R* Red, *YL* Yellow, *G* Green, *CY* Cyan and *B* Blue. The photograph of the PAL graticule shows the swinging burst at 135° and 225°. It shows six spots located similarly to those for the NTSC case. These represent the signal during those line periods when the vertical component is *not* inverted. In addition six further spots, identified by means of lower case letters, represent the signal during those line periods in which the signal *is* inverted.

Each graticule is provided with:

(1) An outer circle with 2° and 10° markings;
(2) sets of 'boxes' within which colour bars should produce their respective spots. Small and large boxes are provided, representing respectively close-limit and wide-limit tolerances;
(3) precise points at which colour bursts and VITS should produce spots.

Colour bars are sometimes transmitted with 100% saturation and sometimes with 75% saturation. It would be inconvenient to change graticules, and it would clutter the graticule to provide duplicate sets of boxes, so provision is made for two settings of gain to enable the operator to set the burst spot(s) in accordance with the specified saturation, then the colour-bar spots should appear within their boxes.

To facilitate accurate measurement of component values, some vectorscopes are provided with means for switching off separately the vertical and horizontal deflections. For example, when the vertical deflection is switched off, all spots appear along the

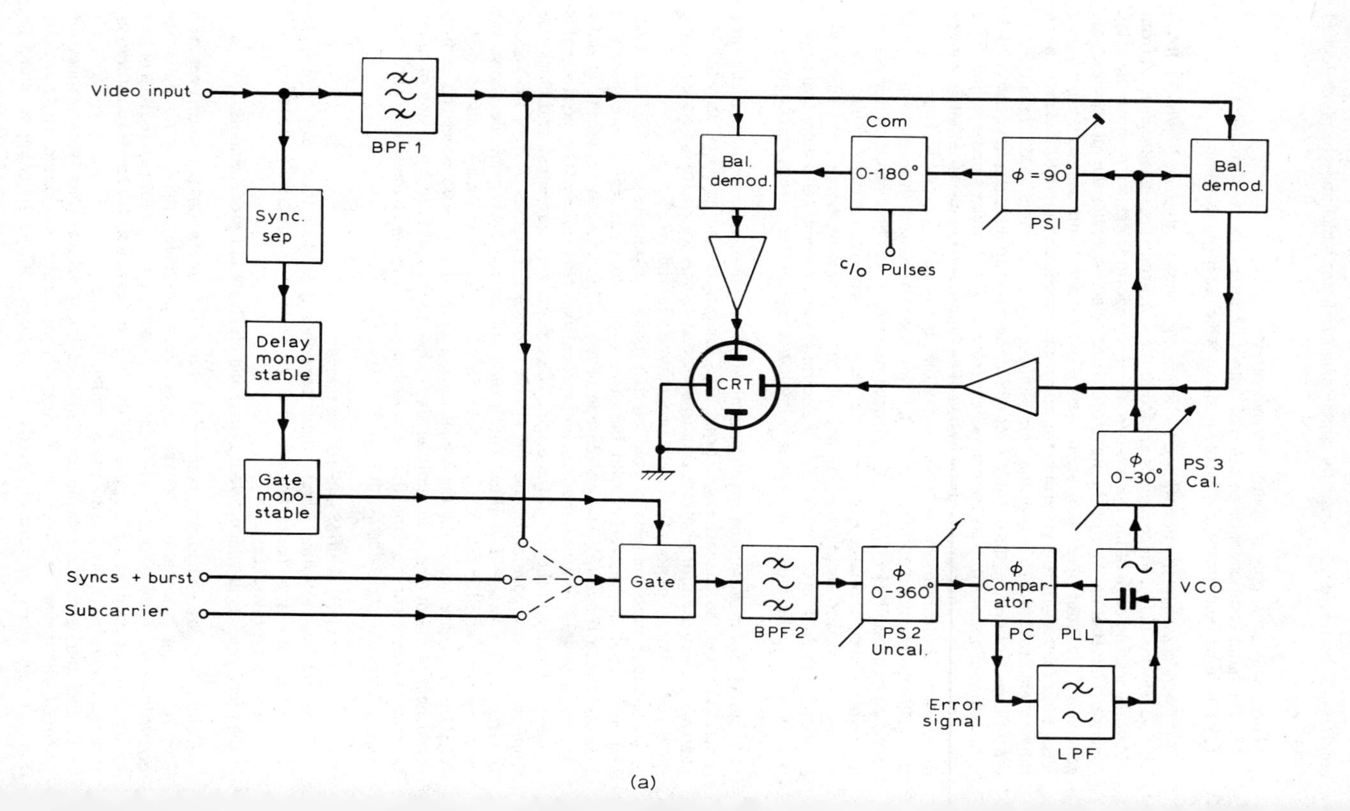

Video input
BPF 1
Sync. sep
Delay mono-stable
Gate mono-stable
Com
Bal. demod.
0-180°
ϕ = 90°
PSI
Bal. demod.
c/o Pulses
CRT
ϕ 0-30°
PS 3 Cal.
Syncs + burst
Subcarrier
Gate
BPF 2
ϕ 0-360°
PS 2 Uncal.
ϕ Compar-ator
PC
PLL
VCO
Error signal
LPF

(a)

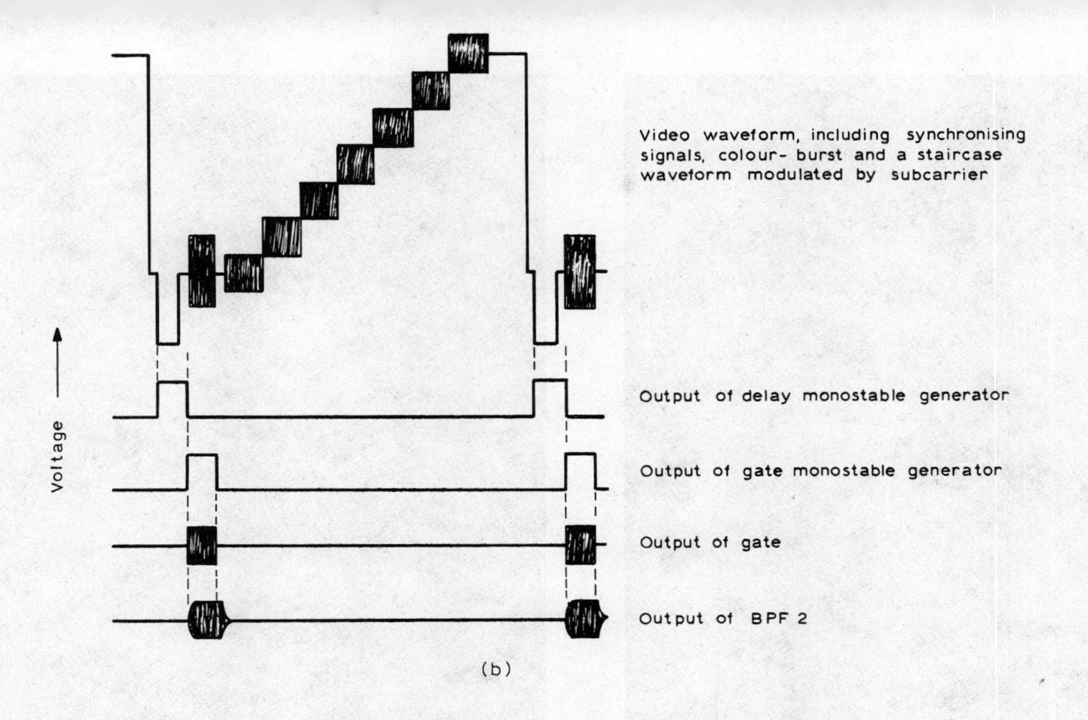

Figure 14.40 (*a*) *An expanded schematic for a vectorscope* (*b*) *production of colour burst to lock the voltage controlled oscillator* (*v.c.o.*)

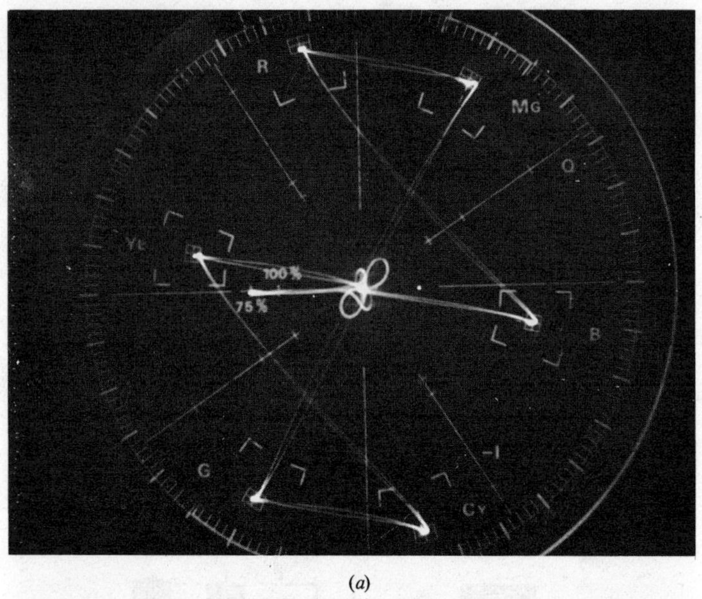

(a)

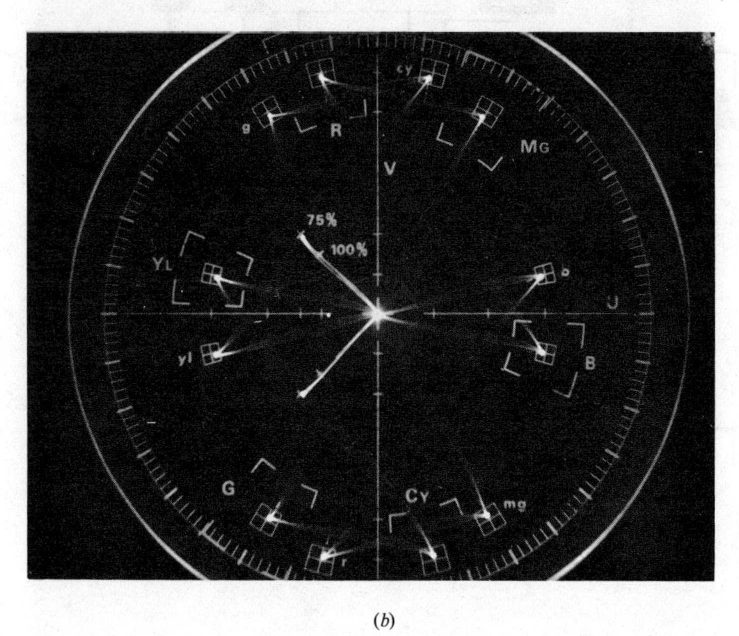

(b)

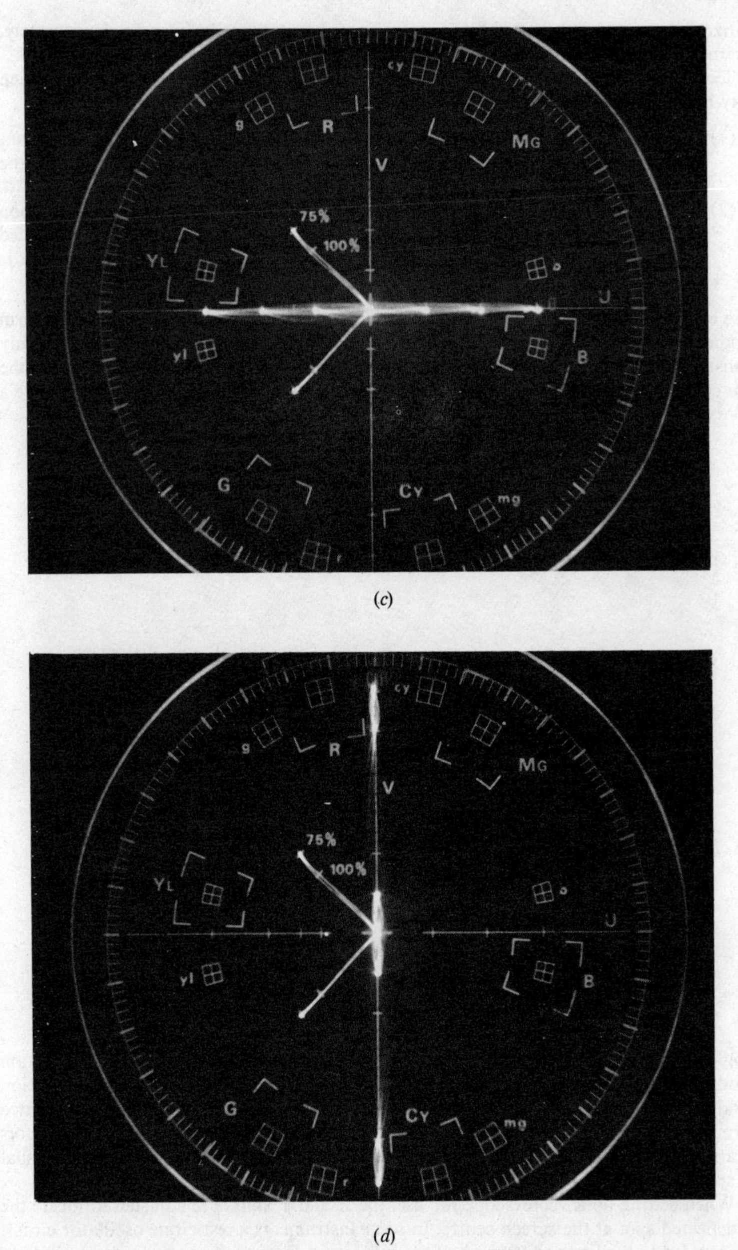

Figure 14.41 Vectorscope graticules with displays of colour bars and colour bursts (*Courtesy Tektronix*)

horizontal axis and thus their horizontal components may be measured accurately. Examples are shown in *Figures 14.41* (*c*) and (*d*).

Two forms of distortion to which colour television channels are prone and which may be examined by using a vectorscope are:

(1) *Differential gain.* The *amplitude* of the chroma component becomes dependant upon the amplitude of the luminance components, so that the saturation of the reproduced colour signal changes with change of luminance.

(2) *Differential phase.* The *phase* of the chroma component becomes dependant upon the amplitude of the luminance component, so that the hue of the reproduced colour signal changes with change of luminance. In the PAL system this is changed into a small reduction of saturation.

The effects of these distortions can be examined with the aid of a staircase waveform modulated by sub-carrier. Such a waveform, along with other components, is commonly transmitted as a VIT and this portion is illustrated in *Figure 14.40* (*b*). The phase of the sub-carrier component is normally along the $-U$ axis and ideally the result should be a

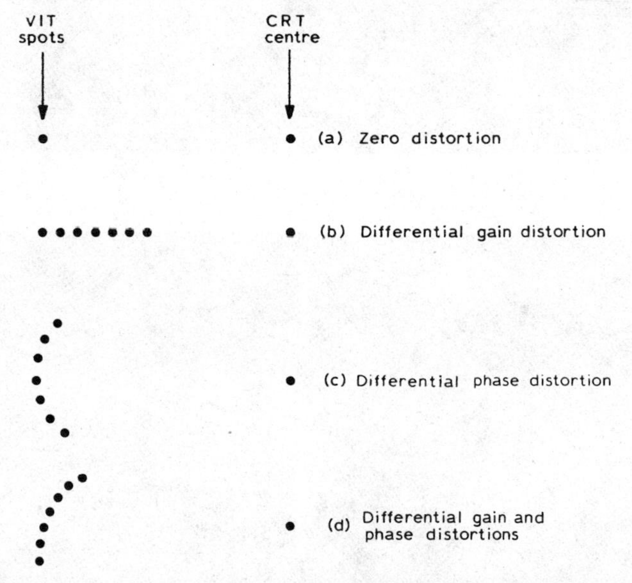

Figure 14.42 Use of the oscilloscope for the display of differential distortions

single spot along that line as is shown in *Figure 14.42* (*a*). Differential gain distortion produces a series of spots along the line as shown in (*b*). Differential phase distortion produces a series of spots along an arc as shown in (*c*), while a combination of the two forms of distortion, the condition most commonly experienced, produces a line of spots in a skewed arc as shown in (*d*). As these VITS have only a small duty cycle it is usual to apply Z modulation to them to improve visibility.

When setting-up a vectorscope for use, the X and Y shifts are adjusted to locate the undeflected spot at the screen centre. In some instruments a test-circle oscillator circuit is provided to facilitate adjustment of the sub-carrier frequency components precisely to the quadrature condition. The output of this oscillator, which has a frequency close to but not equal to sub-carrier frequency, is switched into the video input circuit. Without any further addition to the circuit, this might cause an ellipse to be drawn on the screen.

An adjustment to the phase-shift circuit *PS1*, *Figure 14.40* (*a*), would now be made to change the ellipse into a circle. However, to improve the precision with which this adjustment can be made, the input to *one* of the balanced demodulators is rapidly commutated between 0° and 180° by means of the commutator circuit *COM*. This causes the elliptical error to be reversed periodically, a typical result is shown in *Figure 14.43*. Precise adjustment is then made to *PS1* to produce two circles which overlap

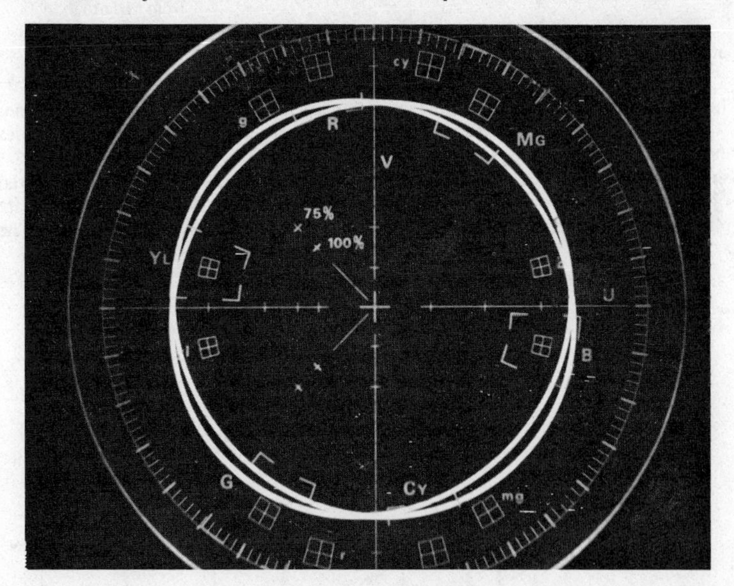

Figure 14.43 *Ellipses produced while setting to quadrature the two sub-carrier inputs to the demodulators* (*Courtesy Tektronix*)

and appear as one. At first sight it might appear to be logical to apply d.c. to the inputs of the balanced demodulators in order to produce ellipses. This, however, would produce components at the output of the demodulators at a frequency equal to that of the sub-carrier. These would not be amplified by the two deflection amplifiers shown in *Figure 14.40* (*a*) since these amplifiers are designed to handle a maximum frequency of about 1.5 MHz. As designed, the inputs to the demodulators are only at a frequency which is the *difference* between the frequencies of the test and sub-carrier oscillators.

When a signal is applied to the vectorscope, the uncalibrated phase-shifter is adjusted to cause the spot which represents the burst to appear on the horizontal axis to the left of the origin. The signal amplitude is then adjusted so that the spot for NTSC or spots for PAL are located correctly according to the specified saturation, as already described.

ANALOGUE VOLTMETERS

These instruments may be designed for the measurement of d.c. and/or a.c. In addition, resistance measuring facilities may be added. In *Table 14.2*, comparisons are made between the performances of typical modestly priced non-electronic and electronic meters.

Table 14.2 PERFORMANCE COMPARISON BETWEEN NON-ELECTRONIC AND ELECTRONIC METERS

	Non-electronic	*Electronic*	*Units*
Maximum sensitivities for full-scale deflection	100	0.1	mV
Maximum operating frequencies	10 kHz	Highly variable from MHz to GHz	
Input resistances for d.c. operation	50 000	1 MΩ	Ω/V

The values of input resistances of a.c. operation are not amenable to simple tabulation, but the values for electronic meters normally exceed by very large margins the values for non-electronic meters.

Since these instruments take less power from the circuit under test than is required to operate the moving-coil meter used for the final display, they must incorporate amplifiers. Power sources for the amplifiers may be mains units and/or batteries. The

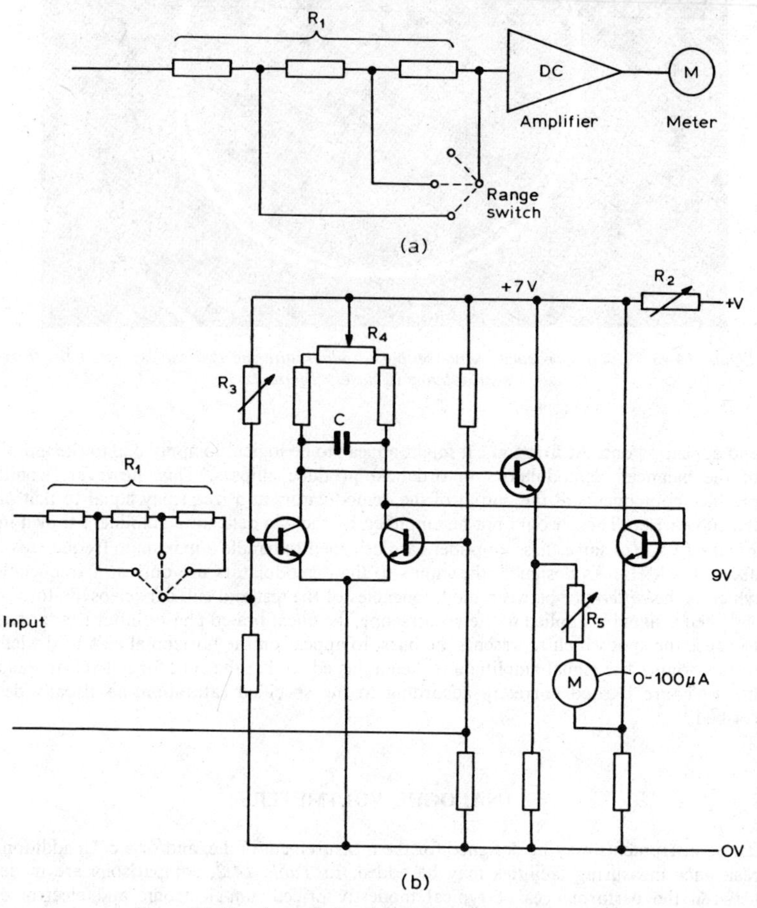

Figure 14.44 A simple d.c. electronic analogue voltmeter

batteries may be of the primary or secondary type, and if they are the latter then charging facilities are usually provided. Provision is made for checking battery voltages using the same moving-coil meters that are used for the display of signal amplitude.

Consider first operation at d.c., for which a basic schematic and simplified circuit are shown in *Figure 14.44*. R_1 is a switch-controlled resistor chain used for range changing. R_2 is first adjusted to standardise the effective supply voltage to, say, 7 V. R_3 is adjusted so that the meter reading does not change when the amplifier input circuit is alternately open and short circuited. R_4 is then adjusted to set the meter reading to zero when the input amplitude is zero. Using a dry cell as a standard, R_5 is set to standardise the meter sensitivity. Capacitor C is used to reduce meter needle fluctuations due to random noise.

The lowest voltage range may also be used as the lowest current range, typically 1 μA full scale deflection (f.s.d.).

Arrangements are sometimes made for the meter to be switchable between left hand zero and centre zero. A sensitivity of about 10 mV f.s.d. is obtainable with such a simple circuit, the limiting design factor being the stability of the d.c. amplifier. A schematic for

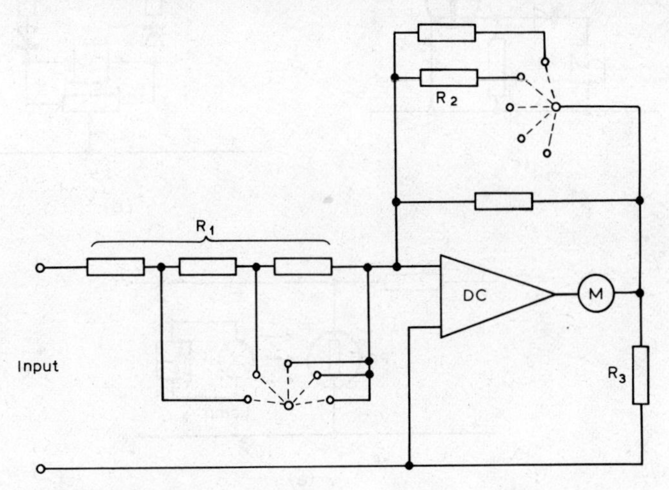

Figure 14.45 An improved voltmeter employing an operational amplifier

a more sensitive instrument is given in *Figure 14.45*. Here the voltage gain of the d.c. amplifier, as modified by negative-feedback, is given by R_2–R_1, and the current through the meter M is further controlled by R_3. Typically the value of R_2 is fixed for all except the high voltage ranges at a value of 100 MΩ, while the value of R_1 is switched in accordance with the range in use, so that the overall sensitivity of the complete circuit is 1 μA/V. For ranges above 100 V however this would involve the use of an inconveniently high value of R_1, so for these higher voltage ranges the value of R_1 may be kept fixed at 100 MΩ and reduced values of R_2 be switched into circuit, giving thereby a relatively reduced value of input resistance in terms of Ω/V.

For high values of sensitivity, amplifier stability must be correspondingly high and this is commonly achieved by the use of chopper-stabilised amplifiers. In these, the d.c. signal is modulated to produce a low-frequency signal which is amplified and then rectified. Some of the choppers available are illustrated in *Figure 14.46*.

A low-amplitude low-frequency source of power is required. This may be derived from the mains supply or a small transistor operated oscillator. Care must be taken that negligible energy from this source appears at the output of the amplifier. In addition to screening and filtering, it is helpful if the source has a negligible second harmonic

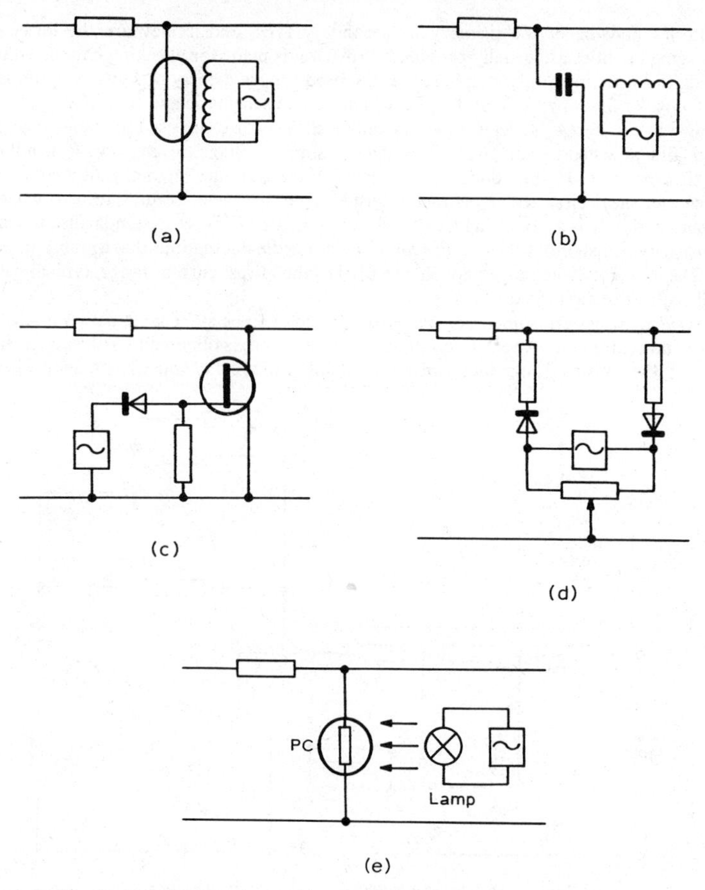

Figure 14.46 Typical chopper circuits (a) vibrating reed relay (b) vibrating variable capacitor (c) f.e.t. (d) diode bridge (e) photo-electric

content, then if the chopper operates in an unpolarised square-law manner the amplifier may be tuned to the second harmonic of the chopper-energising supply. The amplifier should have a narrow bandwidth. For operation at the lowest levels it is advisable to use synchronous detection to make the indication phase sensitive. The signal will produce deflection in one direction only, but noise will produce rectified components in both directions, causing the meter to dither about its correct indication. If a long time-constant circuit be added to the rectifier output, this noise component can be reduced, but at the expense of sluggish operation. A basic circuit is shown in *Figure 14.47*.

When measuring d.c. it is immaterial whether the positive or negative or both terminals are at high potential, but this is normally an important point which must be considered when making a.c. measurements, particularly at high frequencies and particularly when the voltmeter is being operated from a mains supply. Normally one input terminal is to be operated at high potential and has only a low value of capacitance (2–20 pF) to all other bodies. The other terminal is at low potential and has a significant

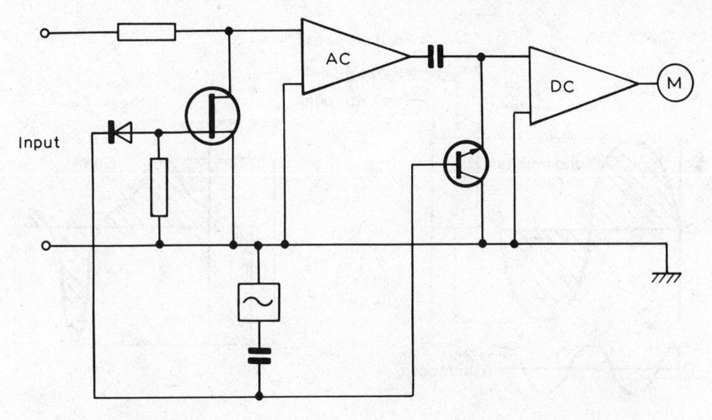

Figure 14.47 *Chopped operation with synchronous detection. The a.c. amplifier has an input signal of low level, the d.c. amplifier an input signal of high level*

value of capacitance to earth and to other bodies, particularly if the instrument is mains operated. When the instrument case is metallic, this should, for safety reasons, be connected to the earth of the mains supply when the latter is being employed. The low-potential terminal is connected to the metal case in some instruments, but not in others, although significant capacitance to the case normally exists. For use where *both* input terminals are at high potential, a special instrument with a balanced input circuit should be used.

Considering further the measurement of alternating voltages, attention must first be given to the quantity which is to be measured. Although peak-to-peak values may be required for special purposes (especially when the meter is used in conjunction with an oscilloscope), normally these voltmeters are used for the measurement of the root-mean-square values of the fundamental components of waveforms which are nominally sinusoidal, but which may contain small proportions of harmonics, noise, hum and/or other components. The influence of these additional components upon the accuracy of measurement depends upon the measurement circuit employed and sometimes upon the amplitude of the signal. The meter deflection may be proportional to:

(1) The arithmetic mean value of the waveform.
(2) The peak value.
(3) The peak-to-peak value.
(4) The root-mean-square (r.m.s.) value.

Due notice must be taken of the fact that the relationships between the r.m.s. value of the waveform and the quantities to which the meter readings are proportional are respectively $\pi/2$, $1/\sqrt{2}$, $1/2\sqrt{2}$ and 1. (Manufacturers take account of these factors when calibrating the meter scales.)

Examples of the results of adding an even order (second) harmonic, and odd order (third) harmonics to sinusoids are illustrated in *Figure 14.48*. At (*a*) a second harmonic has been added. During any half-period of the fundamental, the arithmetic mean value of the second harmonic is zero. The total arithmetic value of the waveform is thus unchanged when an even order harmonic of small amplitude is added to the fundamental. This is illustrated by the two shaded areas in (*a*), which as shown are equal for the fundamental and the sum. Variation of the phase of the harmonic relative to the fundamental will not change this.

The additions of third harmonics to the fundamentals, but with different phase relationships, are shown in (*b*) and (*c*). In (*b*) the sum waveform encloses a smaller area

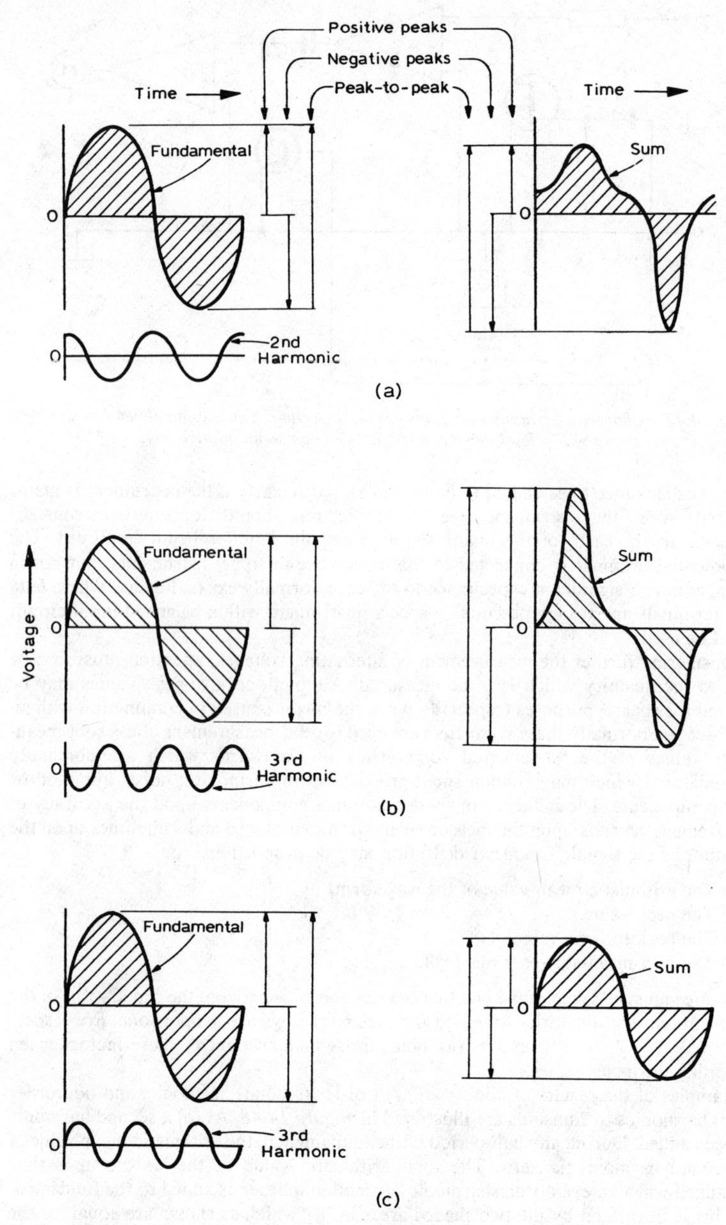

Figure 14.48 Illustrating the influence of added harmonics to a sinusoid

than the fundamental, while in (*c*) the reverse is true. Thus the circuit which responds to the arithmetic mean value of the waveform performs satisfactorily in the presence of even order harmonics, but less so in the presence of odd order harmonics, the sign of the error being unknown unless the phase relationship is known.

Consider now the measurement of peak values. Examination of (*a*) shows that for even order harmonics the sign of the error is dependent upon which peak is measured. Examination of (*b*) and (*c*) shows that while the sign of the error is independent of which peak is measured, the magnitude of the error is dependent upon the phase relationship.

Examination of the peak-to-peak values shows that these are not affected by the addition of small proportions of even order harmonics, but that positive or negative errors may occur when odd order harmonics are added.

When root-mean-square operation is employed, the fundamental and harmonic and noise amplitudes are added in quadrature. For instance a 10% harmonic will produce an error of $100\,(\sqrt{1^2 + 0.1^2} - 1)\% = 0.5\%$. The error is always positive.

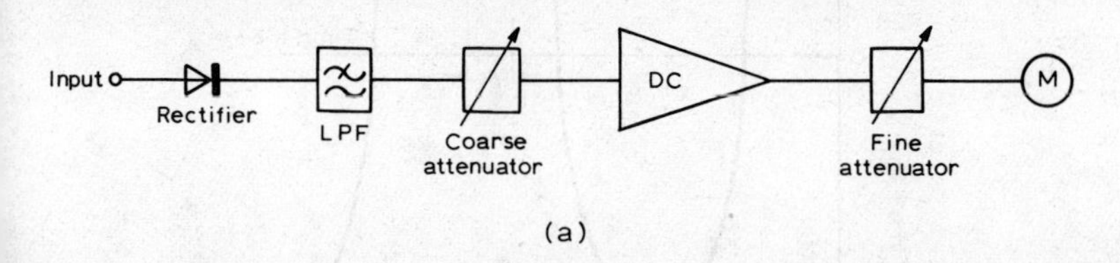

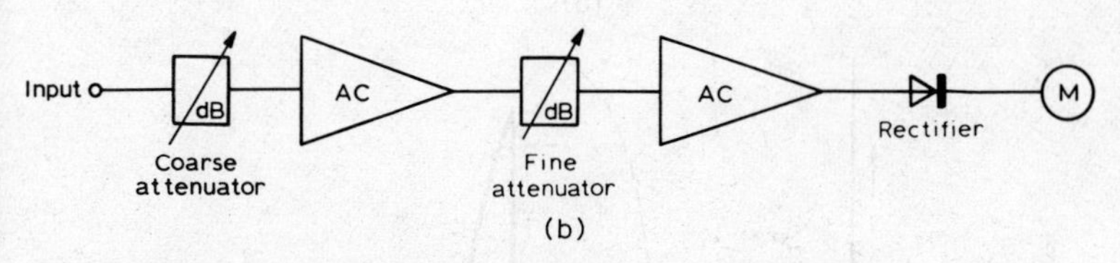

Figure 14.49 *Schematics for a.c. voltmeters* (*a*) *rectifier-amplifier type* (*b*) *amplifier-rectifier type*

Two typical basic forms of a.c. voltmeters are shown in *Figure 14.49.* In (*a*) the signal is first rectified, then, when necessary, attenuated to keep the amplitude within the operating range of the d.c. amplifier-meter combination. In (*b*), the signal is, when necessary, attenuated, amplified by the a.c. amplifiers, rectified, and a value displayed on the meter. These alternative schemes are now considered in greater detail.

A basic rectifier circuit is shown in *Figure 14.50* (*a*). Initially the following assumptions are made:

(1) Source resistance $R_s \ll$ the load resistance R.
(2) Diode resistance r when highly conductive $\ll$ the load resistance R.
(3) Resistance R_f in low-pass filter $\gg$ the load resistance R.
(4) Amplitude of the source is high, e.g., >10 V so that the diode D acts as an on-off switch.
(5) Time-constants CR and $C_fR_f \gg$ the time-period of the source t.

If the r.m.s. value of the input voltage is V_i then the capacitor voltage V_c reaches a maximum value of $\sqrt{2}V_i$, falling due to the discharge through R during the non-conductive period of the diode to $\sqrt{2}V_i - \delta V_c$ where δV_c is the change in capacitor voltage as is shown in (*b*). When $CR \gg t$ then the conductive period of the diode is

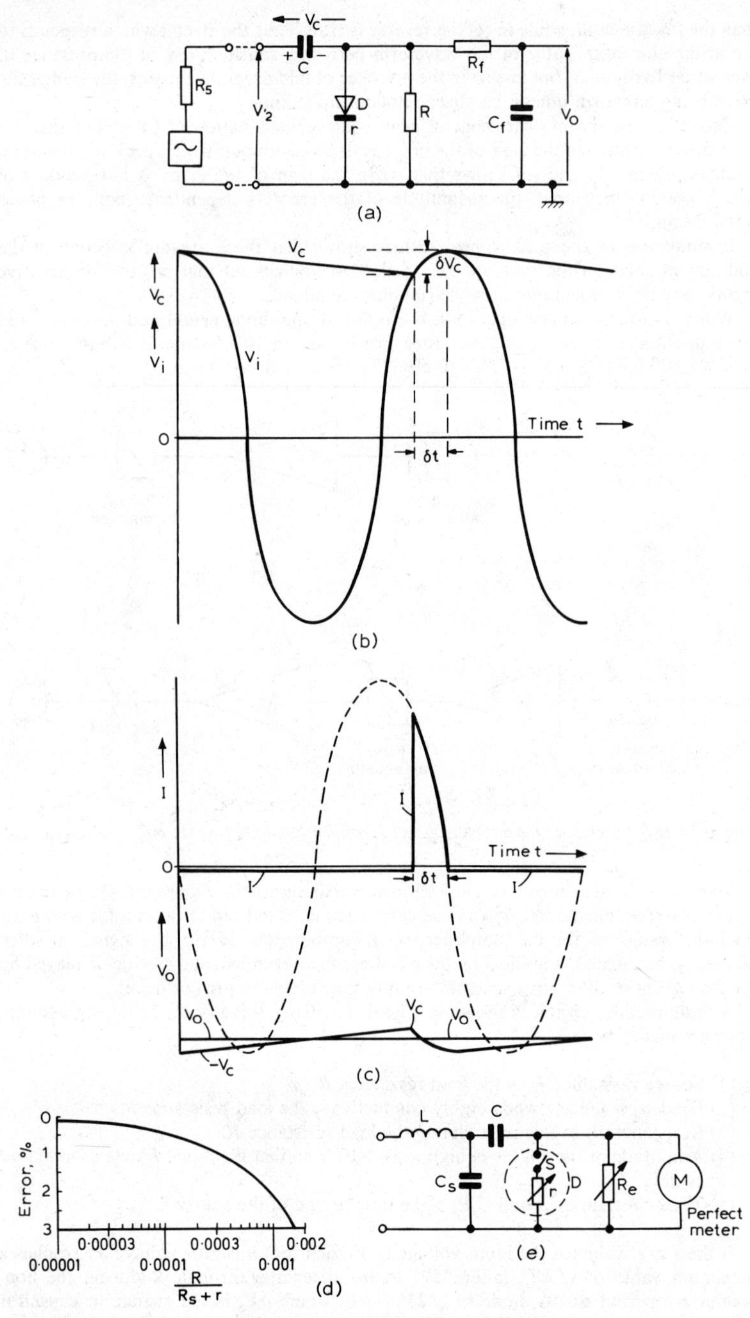

Figure 14.50 The diode rectifier circuit

$\delta t \ll t$, thus the *average* value of V_c is approximately $\sqrt{2}V_i-(\delta V_c/2)$. The circuit is nominally peak indicating so that the fractional error

$$\Delta = \frac{\text{average value of } V_c}{\text{maximum value of } V_c}$$

$$\simeq \frac{-\delta V_c}{2V_{c\max}}$$

$$= \frac{-\delta Q}{2V_{c\max}C} \quad \text{where } Q \text{ is the change in capacitor charge}$$

and if the capacitor discharge period approximates to the period t and the capacitor current is I as in (*c*) then

$$\Delta \simeq \frac{-It}{2V_{c\max}C}$$

$$= -\frac{t}{2CR}$$

or for design purposes

$$CR = \frac{1}{-\Delta 2f}$$

The required time-constant is very much longer than that which is necessary for a similar value of Δ in an amplifier coupling circuit.

When the diode is a thermionic valve, there is hardly any practical limit to the value of capacitance that may be used and hence the minimum frequency of operation. When a semiconductor diode is used, a limit occurs due to the magnitude of the charge that must be passed by the diode when a connection is made to a point where a high value steady potential exists, e.g., at the collector of a transistor or anode of a valve. The output voltage V_o is shown in *Figure 14.50* (*c*) to be the conjugate of the average value of the capacitor voltage V_c. This is due entirely to the polarity definition which has been adopted for V_c. The capacitor current I is also shown in (*c*). During the charge it takes on a shape which is a small portion of a sine wave, and the areas enclosed below and above the axis must be equal. The infinitely sharp wave-front of the current waveform during the charging period implies that the source and diode resistances have zero value. With finite values, the rise-time will of course be finite also, and the peak value of V_c will be less than that of the source e.m.f.

Because the capacitor charging current flows for only a very small fraction of the period t, its peak magnitude must be extremely high in comparison with its average value, and this results in significant errors unless the sum of the source resistance plus diode resistance is extremely small in comparison with the value of the load resistance R. The percentage error, as given by Scroggie [1] is as shown in *Figure 14.50* (*d*). In addition, a combination of high-value source resistance with intermittancy of current flow causes distortion of the waveform under investigation. Because of this short duty cycle of the diode, the value of input resistance cannot be specified by any simple figure. Arbitrarily, it could be defined as say 50 times the value of source resistance in connection with which there was a −2% error, but the value would change with the tolerance. However, when the circuit is operated from a resonant circuit of high magnification factor Q, then the input resistance approximates to $R/2$. This is because R is the only component in which there is significant power dissipation at low and medium frequencies, and current is taken from the source only when the instantaneous amplitude is close to $\sqrt{2}V_i$, so that the dissipation in R is $(\sqrt{2}V_i)^2/R = 2V_i^2/R$ so that the apparent value of R is $R/2$.

When measuring signals of high amplitude, the output voltage V_o is proportional to the input voltage V_i, so that for this condition the moving-coil meter used for display may carry, say, two scales only and multiplication factors may be used as for a simple multi-meter. For lower amplitudes the signal waveform is accommodated on the curved portion of the I_a/V_a characteristic of the diode so that the response changes from peak, through arithmetic mean, to square law, and the period of diode conduction increases. This has the advantage that errors due to harmonics are reduced and the connection of the meter introduces less distortion of the signal under test. One disadvantage, however, is due to the fact that the diode is not operating as an on-off switch, but is conducting moderately for a large proportion of the operating cycle, the dissipation within the diode increases so that the effective value of the input resistance falls. Another disadvantage is that due to the gradual change from linear to square law operation additional meter scales are required such that simple multiplying factors can no longer be applied. The proportional reading accuracy of square-law scales is less than that on linear scales. Some improvement may however be achieved by switching into circuit non-linear devices to improve the scale law, but such devices tend to be temperature sensitive. The problem of amplifier stability becomes all important at low levels of input signal.

An approximate equivalent circuit for a diode rectifier circuit is given in *Figure 14.50* (*e*). Here S and r represent the diode. At high amplitude, and when driven from low impedance or high Q circuits, the duty cycle of S is low as also is the value of r. If the amplitude is low and/or the source resistance is high, then the reverse is true. At low frequencies the actual and effective values of R_e are high, and the CR_e product limits the performance at low frequencies. At high frequencies, e.g., above 10 MHz, dielectric losses cause the value of R_e to be reduced drastically. Stray capacitance C_s and stray inductance L commonly cause a rise to occur in the response of the circuit at very high frequencies, followed by a progressive fall of response.

Extension of the linear law down to about 100 mV can be achieved by the use of the sharper changes of conductivity of the Zener diode using a circuit employing two such

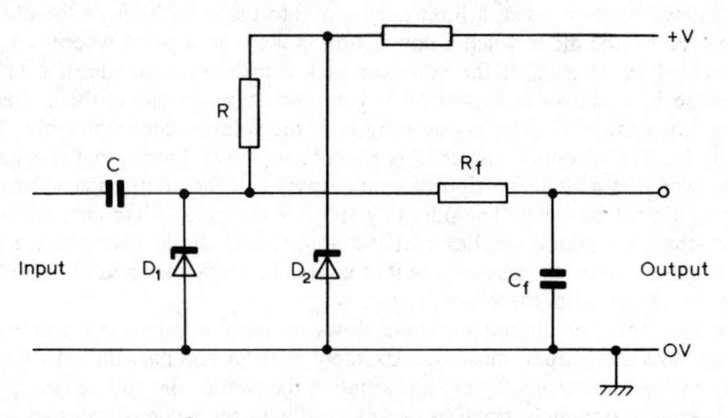

Figure 14.51 A rectifier circuit using zener diodes for operation at low amplitude

diodes of similar voltage rating, one D_1 for rectification, and another D_2 for biasing *Figure 14.51*. The output will of course be superimposed upon a steady potential. Due to the absence of hole storage, the circuit is operable up to hundreds of megahertz.

A voltage-doubler circuit, responding to peak-to-peak values has been shown to have an advantage over the single diode circuit, an example of the former being shown in *Figure 14.52* (*a*). An alternative version particularly suitable for use at levels below 1 mV is the balanced circuit of *Figure 14.52* (*b*). In order to reduce the noise generated in high value resistors, low value components may be used instead, their effective values being increased by use of positive feedback bootstrapping circuits.

Reference to *Figure 14.49* (*a*) shows that d.c. amplifiers are needed. For moderate degrees of sensitivity it is necessary either to employ frequent electrical zero setting or else a chopper-stabilised circuit in order to compensate for drift.

In order to minimise capacitance loading of the circuit under test, the diode and other components closely associated with it, are commonly housed in a probe extension from the main unit. In one type a thermostatically controlled heater prevents the operating temperature of the probe from falling below 33 °C, because rectification efficiency falls at low temperatures.

The maximum sensitivities which may be achieved using the circuit of *Figure 14.49* (*a*) are less than those for circuit (*b*). The former type of circuit is thus only popular for use at very high frequencies for which it would be difficult to design an a.c. amplifier for use in circuit (*b*).

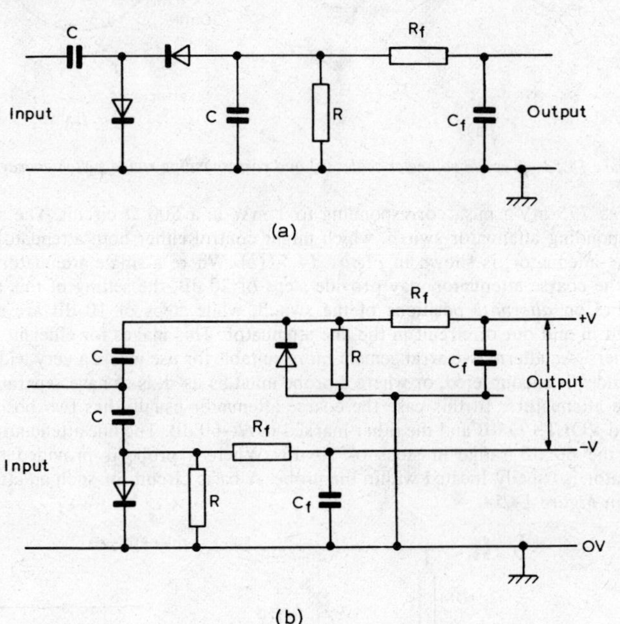

Figure 14.52 *Voltage doubler circuits* (*a*) *unbalanced* (*b*) *balanced*

For operation below a few tens of megahertz, amplification first and rectification afterwards is to be preferred, as exemplified in the circuit of *Figure 14.49* (*b*). The main advantages which this circuit has over the rectifier-amplifier combination are:

(1) An output connection may be taken off just ahead of the rectifier circuit so that the instrument may be used as a wide-band amplifier of adjustable and known value of gain.
(2) The value of the input resistance can be expressed as a simple quantity which does not change with change of amplitude of the signal.
(3) The rectifier circuit always operates at high amplitude, thereby producing high and consistent rectification efficiency.
(4) There is no need for zero setting or the use of chopper-stabilised amplifiers.
(5) Deflection is proportional to the arithmetic mean value of the signal so that the scale law is linear. This minimises the number of scales which appears on the meter face and thus permits scale multiplication factors to be used. Without overcrowding the meter face, a decibel scale may be added.

The detailed design of any instrument in this category centres largely upon the attenuator arrangement. In this connection, steps of 10 decibels are almost standard, giving a meter face as shown in *Figure 14.53* (*a*). The zero decibel value is normally

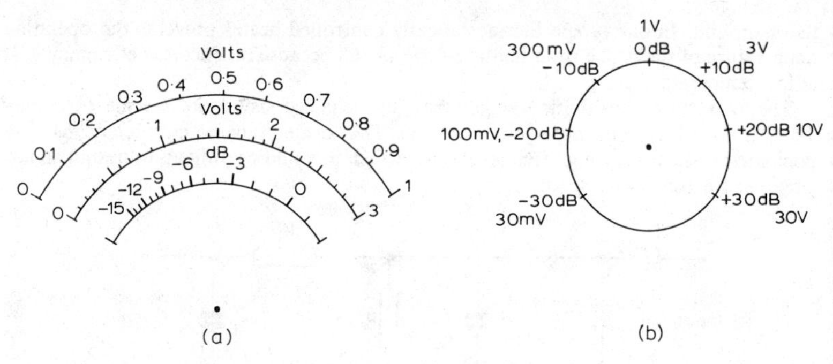

Figure 14.53 Typical voltmeter scales (a) and corresponding range switch engravings (b)

taken as 775 mV r.m.s., corresponding to 1 mW in a 600 Ω circuit. The scale for a corresponding attenuator switch, which might control either both attenuators or only the fine attenuator, is shown in *Figure 14.53 (b)*. Where a single attenuator switch is used, the coarse attenuator may provide steps of 20 dB, the setting of this attenuator changing on *alternate* positions of the switch, while steps of 10 dB are alternately brought in and out of circuit in the fine attenuator. This makes for efficient use of the amplifiers. An alternative arrangement more suitable for use where a very wide range of amplitudes is encountered, or where a probe must be used, is to have separate controls for the attenuators. In this case the coarse attenuator usually has two positions, one marked VOLTS/O dB and the other marked mV/–60 dB. The fine attenuator will then cover the 60 dB range in steps of 10 dB. Where a probe is provided the coarse attenuator is usually located within the probe. A basic circuit for such an attenuator is given in *Figure 14.54*.

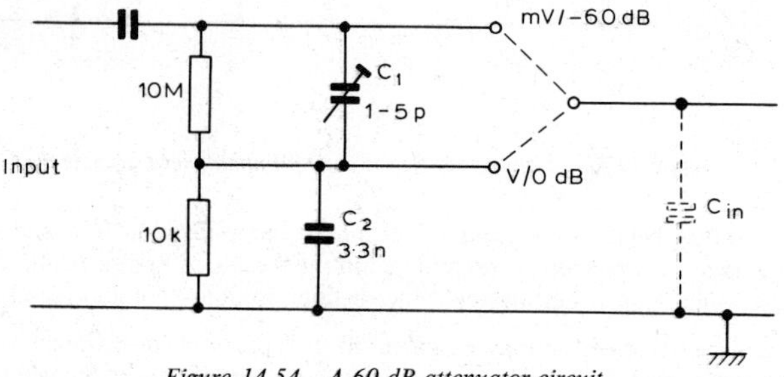

Figure 14.54 A 60 dB attenuator circuit

If the two amplifiers be designed with stages having very high values of input impedance (using say f.e.t. source-followers or compound-emitter followers), then the attenuators may take the form of frequency-compensated potential dividers, rather than constant-impedance networks.

Where the required gain × bandwidth product is small, the first amplifier may be a simple f.e.t. source-follower, feeding a potential-divider type fine attenuator. Otherwise an f.e.t. input stage and an emitter-follower output are appropriate, preferably with overall feedback. Since the forward resistances of the rectifier diodes change both with changes of current and changes of temperature, they should be fed from a high impedance source. This can be achieved in two ways. First the metering circuit is incorporated into a current negative-feedback loop which thus raises the effective value of the output impedance of the amplifier. Secondly, the output stage of the amplifier may

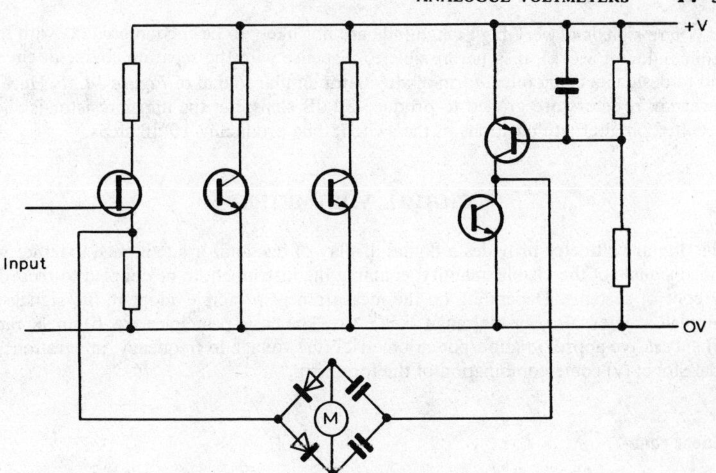

Figure 14.55 A technique for using the rectifier circuit from a constant current source

be supplied with a constant current load. The elements of a circuit incorporating both of these features is given in *Figure 14.55*.

A limiting factor with this type of voltmeter is the value of the gain × bandwidth product which may be achieved at a given cost. Thus really sensitive instruments can be produced at a modest cost for bandwidths up to about 5 MHz, beyond which frequency either the sensitivity must be reduced or the cost increased.

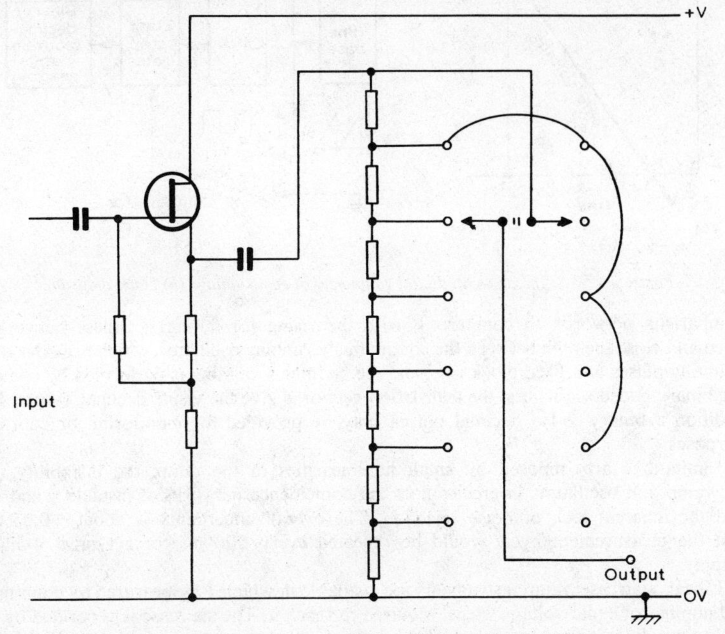

Figure 14.56 An economical circuit for use where high amplitude signals are not likely to be encountered

Where signals of very high magnitude are not likely to be encountered (as with most semiconductor work), it is permissible to dispense with the separate coarse attenuator and to design an economical form of attenuator similar to that of *Figure 14.56*. Here the lower six resistors are graded to produce 20 dB steps, but the upper resistor is short-circuited on alternate positions of the switch, thus producing 10 dB steps.

DIGITAL VOLTMETERS

The digital voltmeter provides a digital display of d.c. and/or a.c. inputs, together with coded signals of the visible quantity, enabling the instrument to be coupled to recording or control systems. Depending on the measurement principle adopted, the signals are sampled at intervals over the range 2–500 ms. The basic principles are: (i) linear ramp; (ii) successive-approximation/potentiometric; (iii) voltage to frequency, integration; (iv) dual-slope; (v) some combination of the foregoing.

Linear ramp

This is a voltage/time conversion in which a linear time base is used to determine the time taken for the internally generated voltage v_s to change by the value of the unknown voltage V. The block diagram, *Figure 14.57* (*b*), shows the use of

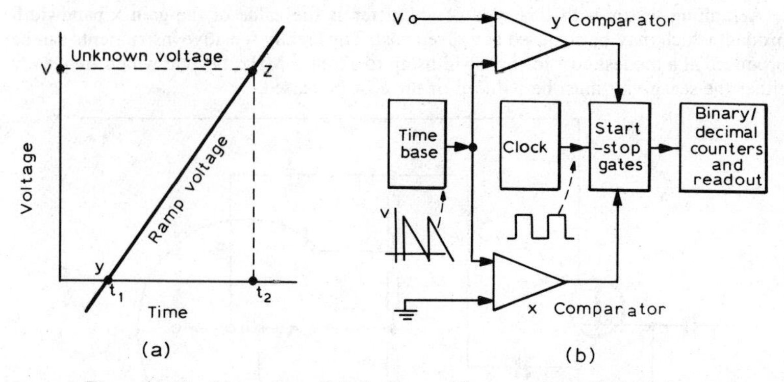

Figure 14.57 Linear-ramp digital voltmeter (*a*) *ramp voltage* (*b*) *block diagram*

comparison networks to compare V with the rising (or falling) v_s; these networks open and close the path between the continuously running oscillator, which provides the counting pulses at a fixed clock rate, and the counter. Counting is performed by one of the binary-coded sequences, the translation networks give the visual decimal output. In addition a binary-coded decimal output may be provided for monitoring or control purposes.

Limitations are imposed by small nonlinearities in the ramp, the instability of the ramp and oscillator, imprecision of the coincidence networks at instants y and z, and the inherent lack of noise rejection. The overall uncertainty is about $\pm 0.05\%$, and the measurement cycle would be repeated every 200 ms for a typical 4-digit display.

Linear 'staircase' ramp instruments are available in which V is measured by counting the number of equal voltage 'steps' required to reach it. The staircase is generated by a solid-state diode pump network, and linearities and accuracies achievable are similar to those with the linear ramp.

Successive approximation

As it is based on the potentiometer principle, this class produces very high accuracy. The arrows in the block diagram of *Figure 14.58* show the signal-flow path for one version; the resistors are selected in sequence so that, with a constant-current supply, the test voltage is created within the voltmeter.

Each decade of the unknown voltage is assessed in terms of a sequence of accurate stable voltages, graded in descending magnitudes in accordance with a binary (or similar) counting scale. After each voltage approximation of the final result has been made and stored, the residual voltage is then automatically re-assessed against smaller

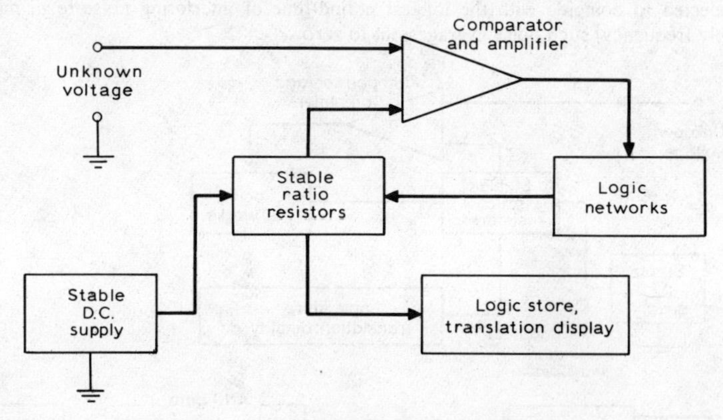

Figure 14.58 Successive-approximation digital voltmeter

standard voltages, and so on to the smallest voltage discrimination required in the result. Probably four logic decisions are needed to select the major decade value of the unknown voltage, and this process will be repeated for each lower decade in decimal sequence until, after a few milliseconds, the required voltage is stored in a coded form. This voltage is then translated for decimal display. A binary-coded sequence could be as follows, where the numerals in *italics* represent a logical *rejection* of that number and a progress to the next lower value:

Unknown analogue voltage	3	9	.	2	0	6	
Logic decisions in	*8*	8		*8*	*8*	*8*	Voltages obtained from
vertical binary	*4*	*4*		*4*	*4*	4	residual (difference)
sequences and in	2	*2*		2	*2*	2	currents across high-
descending 'order'.	1	1		*1*	*1*	*1*	stability resistors.
Decoded decimal display	3	9	.	2	0	6	

The actual sequence of logical decisions is more complicated than is suggested by the example. It is necessary to sense the initial polarity of the unknown signal, and to select the range and decimal marker for the read-out; the time for the logic networks to settle must be longer for the earlier (higher-voltage) choices than for the later ones because they must be of the highest possible accuracy; off-set voltages may be added to the earlier logic choices, to be withdrawn later in the sequence; and so forth.

The total measurement and display takes about 5 ms. When noise is present in the input, the necessary insertion of filters may extend the time to about 1 s. As noise is more troublesome for the smaller residuals in the process, it is sometimes convenient to use some different techniques for the latter part. One such is the voltage-frequency principle (see below) which has a high noise-rejection ratio. The reduced accuracy of the technique can be tolerated as it applies only to the least significant figures.

Voltage-frequency

The converter, *Figure 14.59*, provides output pulses at a frequency proportional to the instantaneous unknown input voltage, and the non-uniform pulse spacing represents the variable-frequency output. The decade counter accumulates the pulses for a predetermined time T and in effect measures the average frequency during this period. When T is selected to coincide with the longest period-time of interfering noise (e.g., mains supply frequency) such noise averages out to zero.

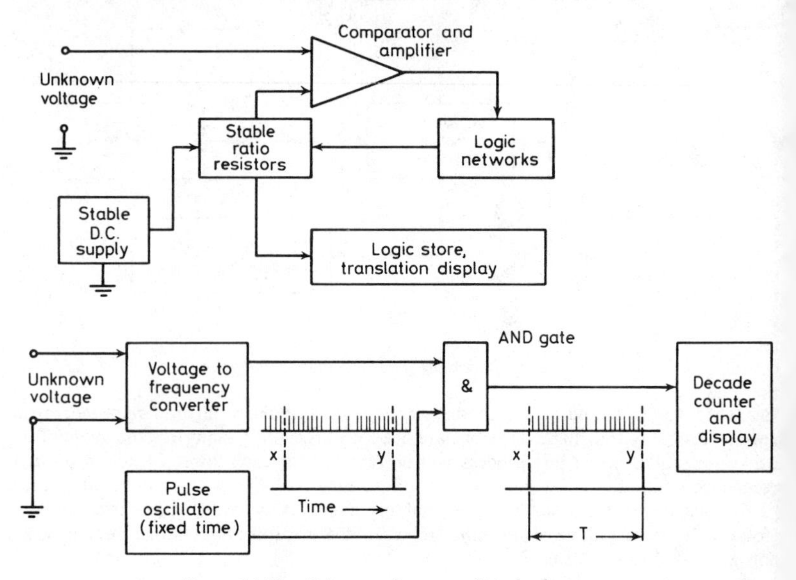

Figure 14.59 Voltage-to-frequency digital voltmeter

Instruments must operate at high conversion frequencies if adequate discrimination is required in the final result. If a 6-digit display were required within 5 ms for a range from zero to 1.000 00 V to $\pm 0.01\%$, then 10^5 counts during 5 ms are called for, i.e., a 20 MHz change from zero frequency with a 0.01% voltage/frequency linearity. To reduce the frequency range, the measuring time is increased to 200 ms or higher. Even at the more practical frequency of 0.5 MHz that results, the inaccuracy of the instrument is still determined largely by the nonlinearity of the voltage-frequency conversion process.

In many instruments the input network consists of an integrating operational amplifier in which the average input voltage is 'accumulated' in terms of charge on a capacitor in a given time.

Dual-slope

This instrument uses a composite technique consisting of an integration (as mentioned above) and an accurate measuring network based on the ramp technique. During the

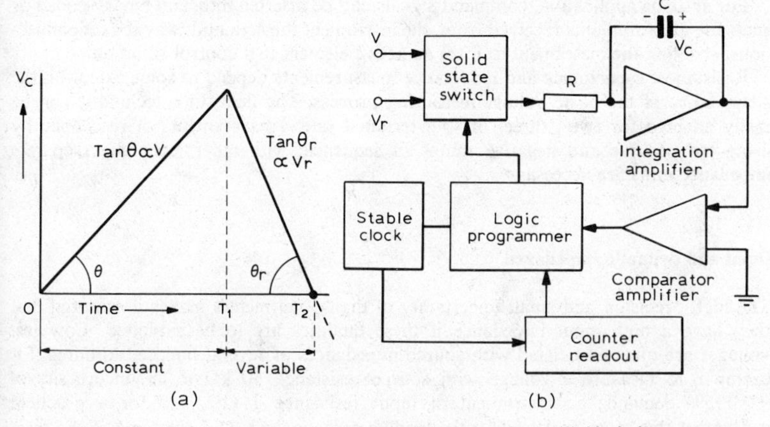

Figure 14.60 *Dual-slope digital voltmeter (a) ramp voltages (b) block diagram*

integration, *Figure 14.60*, the unknown voltage V is switched at time zero to the integration amplifier, and the initially uncharged capacitor C is charged to V in a known time T_1, which may be chosen so as to reduce noise interference. The ramp part of the process consists in replacing V by a reversed-biased direct reference voltage V_r, which produces a constant-current discharge of C; hence a known linear voltage/time change occurs across C. The total voltage change to zero can be measured by the method used in the linear-ramp instrument, except that the slope is negative and that counting begins at maximum voltage. From the diagram in *Figure 14.60* (*a*) it follows that $(\tan\theta/\tan\theta_r) = (T_2 - T_1)/T_1 = V/V_r$, so that V is directly proportional to $(T_2 - T_1)$. The dual-slope method is seen to depend ultimately on time-base linearity and on the measurement of time-difference, and is subject to the same limitations as the linear-ramp method, but with the important and fundamental quality of inherent noise-rejection capability.

Mixed techniques

Several techniques can be combined in the one instrument in order to exploit to advantage the best features of each. One accurate, precise, modern digital voltmeter is based upon precision inductive potentiometers, successive approximation and the dual-slope technique for the least significant figures. An uncertainty of 10 parts in 10^6 for a three-months period is claimed, with short-term stability and a *precision* of about 2 parts in 10^6.

Digital multimeters

Any digital voltmeter can be scaled to read d.c. or a.c. voltage, current, immittance or any other physical property provided that an appropriate transducer is inserted. The trend with instruments of modest accuracy (0.1%) is to provide a basic digital voltmeter with separate plug-in converter units as required for each parameter. Instruments scaled for alternating voltage and current normally incorporate one of the a.c./d.c. converter units mentioned in a previous paragraph, and the quality of the result is limited by the characteristics inherent in such converters. The digital part of the measurement is more accurate and precise than the analogue counterpart, but is more expensive.

For systems application, command signals can be inserted into, and binary-coded or analogue measurements received from, the instrument through multiway socket connections, enabling the instrument to form an active element in a control situation.

Resistance, capacitance and inductance measurements depend to some extent on the adaptability of the basic voltage-measuring process. The dual-slope technique can be easily adapted for two-, three- or four-terminal ratio measurements of resistance by using the positive and negative ramps in sequence; with other techniques separate impedance units are necessary.

Input and dynamic impedance

The high precision and small uncertainty of digital voltmeters makes it essential that they have a high input impedance if these qualities are to be exploited. Low test voltages are often associated with source impedances of several hundred kilohms: for example, to measure a voltage with source resistance 50 kΩ to an uncertainty of ±0.005% demands an instrument of input resistance 1 GΩ, and for a practical instrument this must be 10 GΩ if the loading error is limited to *one-tenth* of the *total* uncertainty.

The dynamic impedance will vary considerably during the measuring period, and it will always be lower than the quoted null, passive, input impedance. These changes in dynamic impedance are coincident with voltage 'spikes' which appear at the terminals due to normal logic functions; this noise can adversely affect components connected to the terminals, e.g., Western standard cells.

Input resistances of the order of 1–10 GΩ represent the conventional range of good-quality insulators. To these must be added the stray parallel reactance paths through unwanted capacitive coupling to various conducting and earth planes, frames, chassis, common rails, etc.

Noise limitation

The information signal exists as the p.d. between the two input leads; but each can have unique voltage and impedance conditions superimposed on it with respect to the basic *reference* or *ground* potential of the system, as well as another and different set of values with respect to a local *earth* reference plane.

An elementary electronic instrumentation system will have at least one ground potential and several earth connections—possibly through the (earthed) neutral of the main supply, the signal source, a read-out recorder or a cathode-ray oscilloscope. Most true earth connections are at different electrical potentials with respect to each other due to circulation of currents (d.c. to u.h.f.) from other apparatus, through a finite earth resistance path. When multiple earth connections are made to various parts of a high-gain amplifier system, it is possible that a significant frequency spectrum of these signals will be introduced as electrical noise. It is this interference which has to be rejected by the input networks of the instrumentation, quite apart from the concomitant removal of any electrostatic/electromagnetic noise introduced by direct coupling into the signal paths. The total contamination voltage can be many times larger (say 100) than the useful information voltage level.

Electrostatic interference in input cables can be greatly reduced by 'screened' cables (which may be 80% effective as screens), and electromagnetic effects minimised by transposition of the input wires and reduction in the 'aerial loop' area of the various conductors. Any residual effects, together with the introduction of 'ground and earth-loop' currents into the system are collectively referred to as *series* and/or *common-mode* signals.

Series and common-mode signals

Series-mode (normal) interference signals V_{sm} occur in series with the required information signal. Common-mode interference signals V_{cm} are present in both input leads with respect to the reference potential plane: the required information signal is the difference voltage between these leads. The results are expressed as *rejection ratio* (in dB) with respect to the input error signal V_e that the interference signals produce, i.e.,

$$K_{sm} = 20 \log (V_{sm}/V_e) \quad \text{and} \quad K_{cm} = 20 \log (V_{cm}/V_e)$$

where K is the rejection ratio.

Consider the elementary case in *Figure 14.61*, where the input-lead resistances are

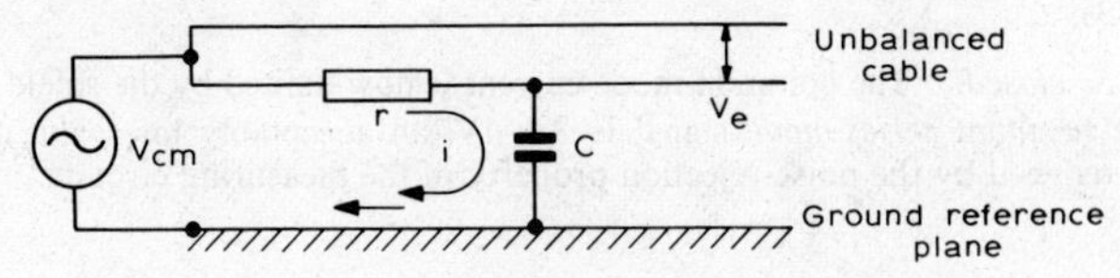

Figure 14.61 Common-mode effect in unbalanced network

unequal, as would occur with a transducer input. Let r be the difference resistance, C the cable capacitance, with common-mode signal and error voltages V_{cm} and V_e respectively. Then the common-mode numerical ratio is

$$V_{cm}/V_e = V_{cm}/ri = 1/2\pi fCr$$

assuming the cable insulation to be ideal, and $X_C \gg r$. Clearly for a common-mode statement to be complete it must have a stated frequency range and include the resistive unbalance of the source. (It is often assumed in c.m.r. statements that $r = 1\ \text{k}\Omega$.)

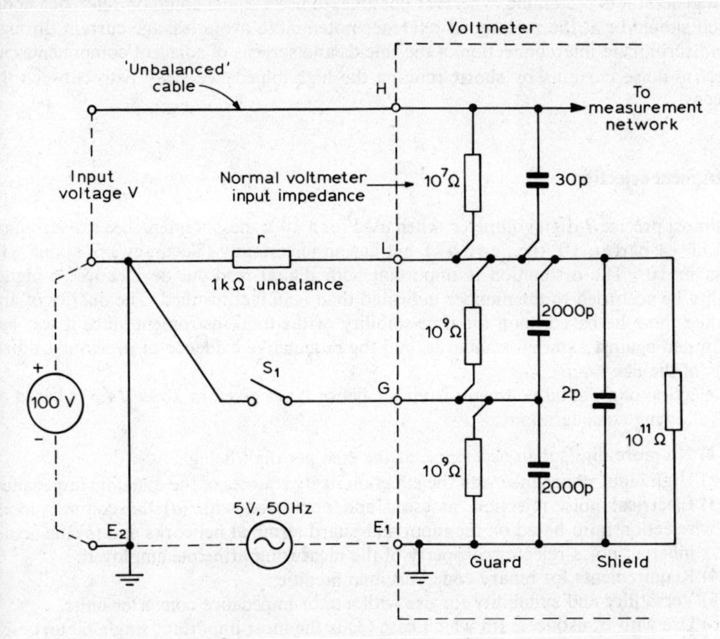

Figure 14.62 Typical guard and shield network for digital voltmeter

The c.m.r. for a digital voltmeter could be typically 140 dB (corresponding to a ratio of $10^7/1$) at 50 Hz with a 1 kΩ line unbalance, and leading consequently to $C = 0.3$ pF. As the normal input cable capacitance is of the order of 100 pF/m, the situation is not feasible. The solution is to inhibit the return path of the current i by the introduction of a guard network. Typical guard and shield parameters are shown in *Figure 14.62* for a six-figure digital display on a voltmeter with ±0.005% uncertainty. Consider the magnitude of the common-mode error signal due to a 5 V 50 Hz common-mode voltage between the shield earth E_1 and the signal earth E_2:

Switch S_1 open. The a.c. common-mode voltage drives current through the guard network and causes a change of 1.5 mV to appear across r as a *series-mode* signal; for $V = 1$ V this represents an error V_e of 0.15% for an instrument whose quality is ±0.005%.

Switch S_1 closed. The common-mode current is now limited by the shield impedance, and the resultant *series-mode* signal is 3.1 μV, an acceptably low value that will be further reduced by the noise-rejection property of thc measuring circuits.

Floating-voltage measurement

If the voltage difference V to be measured has a p.d. to E_2 of 100 V, as shown, then with S_1 open the change in p.d. across r will be 50 μV, as a series-mode error of 0.005% for a 1 V measurement. With S_1 closed, the change will be 1 μV, which is negligible.

The interconnection of electronic apparatus must be carefully made to avoid systematic measurement errors (and shortcircuits) arising from incorrect screen, ground or earth potentials.

In general it is preferable, wherever possible, to use a signal common reference node, which should be at the zero signal reference potential to avoid leakage current through r. Indiscriminate interconnection of the shields and screens of adjacent components can *increase* noise currents by shortcircuiting the high-impedance stray path between the screens.

Instrument selection

The most precise 7-digit voltmeter, when used for a 10 V measurement, has a discrimination of ±1 part in 10^6 (i.e., ±10 μV), but has an uncertainty ('accuracy') of about ±10 parts in 10^6. The distinction is important with digital read-out devices lest a higher quality be accorded to the number indicated than is in fact justified. The quality of any reading must be based upon the time-stability of the total instrument since it was last calibrated against external standards, and the cumulative evidence of previous calibrations of the like kind.

Selection of a digital voltmeter from the list of types given in *Table 14.3* is based on the following considerations:

(1) No more digits than necessary, as the cost per digit is high.
(2) High input impedance, and the effect on likely sources of the dynamic impedance.
(3) Electrical noise-rejection, assessed and compared with (*a*) the common-mode rejection ratio based on the input and guard terminal networks and (*b*) the actual inherent noise-rejection property of the measuring principle employed.
(4) Requirements for binary-coded decimal facilities.
(5) Versatility and suitability for use with a.c. or impedance converter units.
(6) Use with transducers (in which case (3) is the most important single factor).

Table 14.3 TYPICAL CHARACTERISTICS OF DIGITAL VOLTMETERS AND MULTIMETERS

1: Operating principle—
DS, dual slope;
DDS, inductive divider + dual slope;
I, integration;
ISA, mixed I and SA;
SA, potentiometer/successive approx. R, ramp.

2: Number of display digits.
3: Operating time.
4: Instrument ranges.
5: Uncertainty, smallest digits to be added in ± parts per million of maximum reading.
6: Maximum discrimination, in ± p.p.m. of maximum reading.

7: Parallel input resistance.
8: Parallel input capacitance.
9: Common (or series) mode rejection.
10: Common (or series) mode rejection, 50 Hz.
11: Recalibration period.

1	2	3 ms	4	5 p.p.m.	6 p.p.m.	7 MΩ	8 pF	9 dB	10 dB	11 months
Single-purpose Voltmeters										
R	3	500	100 mV–1 kV	5 000	1 000	10	—	90–10	40	3
I	4	2–200	100 mV–1 kV	300	10	10^3, 10	—	150	150	12
I	6	2–200	20 mV–1 kV	40	10	10^3, 10	—	—	160	6
ISA	7	1 000	1 V–1 kV	60	1	10^4, 10	40	—	160	3
DD	7	—	1 V–1 kV	10	1	10^5, 10	—	120	150	3
Small Multimeter or Panel-meter										
DS	4	200	0.2 V–1 kV 0–20 kHz 200μA–2 A 0.2–2 000 kΩ	1 000– 4 000	100	10	110	—	90	12
Modular Meters										
SA	5	2–250	0.1 mV–1 kV (d.v.)	200	10	10^4, 10	—	—	100	3
	5	2–250	1 V–1 kV 50 Hz–100 kHz	1 000– 5 000	10	1	100	—	60	3
SA	5	2–250	1 kΩ–10 MΩ	500	10	—	—	—	—	3

Calibration

It will be seen from *Table 14.3* that digital voltmeters should be recalibrated at intervals between three and twelve months. Built-in self-checking facilities are normally provided to confirm satisfactory operational behaviour, but the 'accuracy' check cannot be better than that of the included standard cell or Zener diode reference voltage and will apply only to one range. If the user has available some low-noise stable voltage supplies (preferably batteries) and some high-resistance helical voltage-dividers, it is easy to check logic sequences, resolution and the approximate ratio between ranges. The accuracy of the 1 V range can be tested with an external Weston standard cell provided that the cell voltage is known to within $\pm 3\ \mu V$ by recent NPL calibration.

THERMOCOUPLES AND THERMO-ELECTRIC EFFECTS

Thermocouples as components and the *uses* of these, and of thermistors, are described here: the thermistor as a device and the principles of thermionic emission are described in detail in Sections 6 and 7 respectively.

Thermo-electric effects are commonly used for the measurement, and frequently the subsequent control, of such quantities as temperature, power—electrical or otherwise; the level, pressure and velocity of liquids; and the pressure and velocity of gases. They may also be used to delay an operation, or in compensation for thermo-electric effects in other mechanisms. Thermo-electric effects most commonly employed can be grouped as thermo-mechanical, thermo-resistive, or thermo-voltaic.

Thermo-mechanical effect

The thermo-mechanical effect constitutes a change of dimension of a material (itself not necessarily an electrical conductor), caused by a change of temperature. Normally an

Table 14.4 COEFFICIENTS OF EXPANSION OF SOME COMMONLY USED MATERIALS

Material	*Symbol*	*Coefficient of thermal expansion* $\times 10^{-6}$	*Resistivity in ohm–centimetres* $\times 10^{-6}$	*Resistance/temperature coefficient* 'a' $\times 10^{-3}$	*Melting point* °C
Aluminium	Al	28.7	2.82	3.9	660
Advance, Constantan, Eureka (55 Cu, 45 Ni)	—	14.8	49.1	0.2	1 210
Brass (66 Cu, 34 Zn)	—	20.2	6.72	2.0	920
Copper, annealed	Cu	16.1	1.72	3.93	1 083
Duralumin	—	—	5.76	2.0	500 to 637
Iron, pure	Fe	12.1	9.65	5.2 to 6.2	1 535
Manganin (84 Cu, 12 Mn, 4 Ni)	—	—	44.8	0.02	910
Mercury	Hg	—	96.0	0.89	−38.87
Molybdenum	Mo	6.0	5.7	4.5	2 630
Nichrome (65 Ni, 12 Cr, 23 Fe)	—	—	112.0	0.17	1 350
Nickel	Ni	15.5	9.7	4.7	1 455
Phosphor-bronze (95.5 Cu, 0.5 P, 4 Sn)	—	16.8	9.4	3.0	1 050
Platinum	Pt	9.0	10.6	3.0	1 774
Tungsten	W	4.6	5.6	4.5	3 370

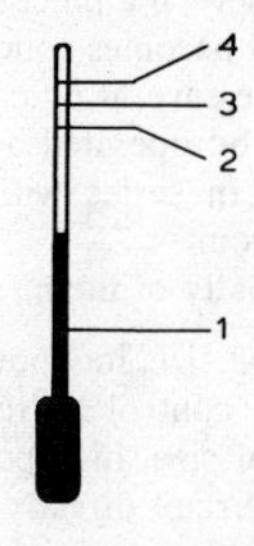

Figure 14.63 A contact thermometer

increase of temperature causes an increase of physical dimensions. The coefficients of expansion of some commonly used materials are given in *Table 14.4*.

The simplest examples are the well-known bimetallic strip thermostat and the mercury contact thermometer, *Figure 14.63*. This is used as a sensing element in thermostatically controlled ovens and refrigerated chambers. A large bulb, full of mercury, is connected to a thin stem, thus producing a large change in the length of the mercury column for a small change of temperature. Four thin platinum wires, numbered 1–4 in the diagram, are sealed into the stem. In normal operation the upper limit of the mercury column oscillates slowly between contact 3 and a point which is just below this contact. When the column fails to reach contact 3, the breaking of an electric circuit between contacts 1 and 3 causes either a heater to be switched on or a refrigerator to be switched off, as may be appropriate to the application under consideration. In the event of the temperature passing outside the permissible limit, then completion of a circuit between contacts 1 and 4, or the breaking of a circuit between contacts 1 and 2, causes an alarm to be operated. Slight changes of temperature, e.g., ± 0.1 °C are to be expected with such an on-off form of temperature control, best results being achieved when the power of the heater or refrigerator is such as to cause the on and off periods to be approximately equal. Since the mercury column is allowed to carry currents of small magnitude only, the thermometer circuit should operate to control the currents of high impedance devices such as field-effect transistors.

The thermo-resistive effect

The thermo-resistive effect constitutes a change of electrical resistance produced by a change of temperature. The value of resistance R_t at temperature t is related to its resistance value R_o at 0 °C by $R_t = R_o(1 + at + bt^2)$. For most materials (nickel excepted) the values of the coefficient b may be taken as zero, i.e., the resistance–temperature curve is linear. Examples of the property of some commonly used materials are given in *Table 14.4*.

Thermistors are described in Section 6. The following points regarding their usages are relative to the present heading. The negative-temperature coefficient (n.t.c.) thermistor is a highly developed component which is available in a very wide variety of constructions and resistance values. The positive-temperature coefficient (p.t.c.) thermistor, on the other hand, has many complex facets and is therefore not used to the same extent. N.T.C. thermistors may be connected in series but not in parallel, while p.t.c. thermistors may be connected in parallel but not in series. There is a limit to the maximum value of current for which an n.t.c. thermistor may be designed, and a limit to the value of maximum voltage for which a p.t.c. thermistor may be designed. If an n.t.c. thermistor has an excessive cross-sectional area, then by cumulative action current becomes concentrated over a small portion of that area, possibly leading to thermal overload, i.e., the large cross-sectional area n.t.c. thermistor acts like several thermistors in parallel. Conversely, if a p.t.c. thermistor has an excessive length, then by cumulative action, the potential becomes concentrated over a small proportion of the length, i.e., the component acts like several p.t.c. thermistors in series. Where appropriate a tungsten-filament lamp can be operated as a p.t.c. device. A resistor in parallel with an n.t.c. thermistor or one in series with a p.t.c. thermistor serves to reduce the speed of operation of the circuit.

Thermistors, usually of the n.t.c. type, can be classified under four broad headings:

(1) Small physical size, low-power rapidly operating types which are commonly used to indicate or control a circuit via an amplifying system;
(2) large physical size, high-power slowly operating types which give indication or control of a circuit directly without the use of amplifying systems;
(3) block thermistors, the resistance values of which are very little affected by the currents which they carry. A block thermistor may be used in one arm of a bridge to provide ambient temperature compensation for another thermistor used for sensing. Block thermistors may be bolted to surfaces, either as temperature sensors, or the surfaces may be uses as heat sinks;
(4) indirectly-heated thermistors. Here the semiconductor material is heated by means of a heater which has a negligible value of resistance-temperature coefficient. The heater resistance is usually 100 Ω.

The engineer is faced frequently with the problem of measuring the characteristics, such as the current/voltage characteristics of devices, for example, diodes and thermistors, which have non-linear relationships. Where the device operates at low power, the

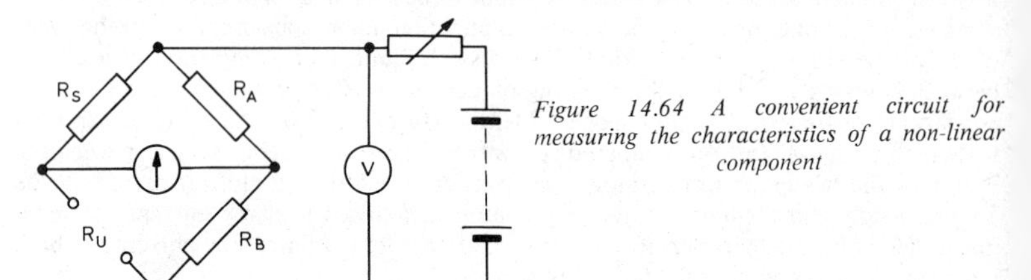

Figure 14.64 A convenient circuit for measuring the characteristics of a non-linear component

power which is required to operate conventional moving-coil meters becomes comparable with that which is required to operate the device. A solution is given in *Figure 14.64*. By making the value of $R_A = R_B$ then, *at balance*, the voltage across the unknown component R_U will equal half of that which is indicated by the voltmeter. Then $R_U = R_S$, from which the value of current can be calculated easily.

Thermo-sensitive elements having significant values of inertia may be used as delay devices. The simplest example is in the switching of large tungsten filament lamps. The resistance of the filament of such a lamp when cold is less than one-tenth of its value when hot, so that the suddenness of switching such a lamp straight on to the supply shortens the lamp life and may cause a fuse to blow. This may be avoided by fitting an n.t.c. thermistor in series with the lamp. Very little protection will be given if the lamp is switched on again soon after being switched off.

For the measurement of temperature, a Wheatstone bridge circuit is normally employed, the temperature sensor being one of the four elements. The power supply may be d.c., with a moving-coil meter used as a detector. If the supply is stabilised, then the meter may be calibrated directly in degrees Celsius or Fahrenheit. The resistance elements are selected to produce a balance either at the centre of the expected temperature range, or else at that particular temperature at which greatest accuracy is required. If the supply is not stabilised, then an uncalibrated galvanometer is used, and one of the resistance elements of the bridge contains a rheostat which is directly calibrated with a temperature scale for the balanced condition of the bridge. A high degree of sensitivity may be achieved by the use of a sensitive electronic analogue voltmeter as a detector, and, unless a diode is being used as a temperature sensor, it may be more convenient to use an a.c. supply for the bridge. A circuit in which the meter is phase-sensitive is to be preferred. Alternatively a self-balancing bridge may be employed. In this and any similar circuit it is important to note that the magnitude of the electrical power which is fed to the thermosensitive device should not be sufficient to have any significant effect upon its temperature.

For the measurement of temperature in excess of the ambient temperature a circuit is shown in *Figure 14.65*. The out-of-balance voltage across the horizontal diagonal of the

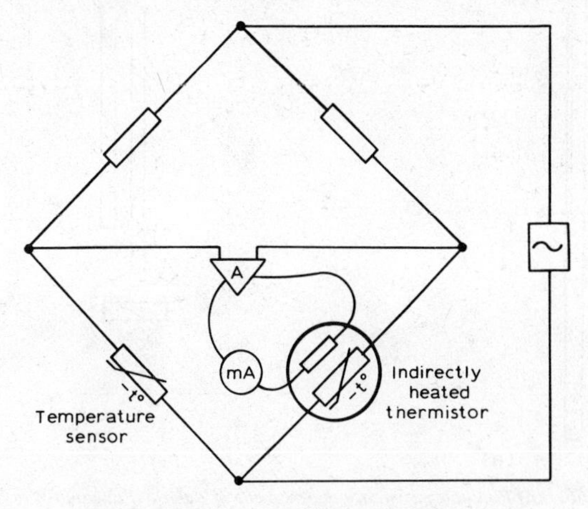

Figure 14.65 A self-balancing bridge

bridge is amplified by the amplifier *A*, the output of which is applied to the heater of the indirectly heated thermistor, producing thereby a quasi-balance of the bridge. A meter in the output circuit of the amplifier is calibrated directly in temperature. For the measurement of temperatures below the ambient temperature, either the two n.t.c. thermistors are located in the same arm of the bridge, or one thermistor is of the n.t.c. type and one of the p.t.c. type.

Commercial temperature sensors are available, usually employing platinum, some coping with temperature ranges as great as −200 °C to +1 200 °C. *Figure 14.66* shows a typical platinum resistance thermometer (*a*), and a suitable circuit (*b*). The measurement of temperature at the extreme end of the probe should be independent of the temperature along the probe stem, variations of the latter temperature being due to ambient temperature variations and wind. For this reason a loop wire, running the length of the probe, is connected in an adjacent arm of the bridge to effect compensation. These instruments take about 7 s to achieve stable operation after a change of temperature. Where these thermometers are being used to measure temperature *change*, the sensing elements are connected in the adjacent arms of the bridge, their compensating loops being cross-connected.

When measuring the temperatures of materials in the range above 1 100 °C an opto-electrical method may be employed, *Figure 14.67*. Light from the hot material and from the filament of an intervening lamp are viewed by the observer. The voltage across the lamp is adjusted to equalise the colours of the filament and the source. (The relationship between lamp temperature and voltage is almost linear.) The optical wedge or iris is adjusted to equalise the brightness. When the image of the filament can no longer be distinguished from the background, the voltage across the filament is taken as a measure

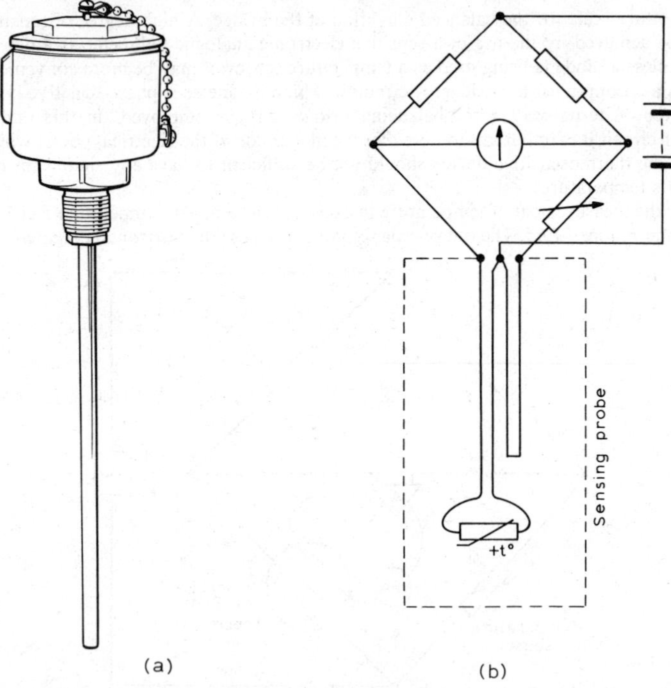

Figure 14.66 (*a*) *Platinum resistance thermometer* (*Courtesy Sangamo-Weston Ltd*) (*b*) *a suitable circuit*

of the temperature of the material. Since tungsten melts at 3 400 °C, the basic arrangement cannot be used for the measurement of very high temperatures. However, with a more complex optical arrangement, such temperatures can be measured if the colour-temperature of the lamp be raised by the use of blue filters.

Temperature sensing may in its turn be used for many other purposes, such as the measurement and/or control of liquid levels or gas or liquid velocities. These operations depend upon the fact that a heated thermo-sensitive element will be subjected to cooling if placed in a liquid, the degree of cooling being increased if the gas or liquid is flowing. One example is the maintenance of the liquid level in a tank, which is being replenished by means of an electrically driven pump, *Figure 14.68*. Currents are passed through n.t.c. elements TH_1, TH_2 and TH_3, which are located at different levels in the tank. Normally the liquid level coincides with the height of TH_2 the latter being kept cool by

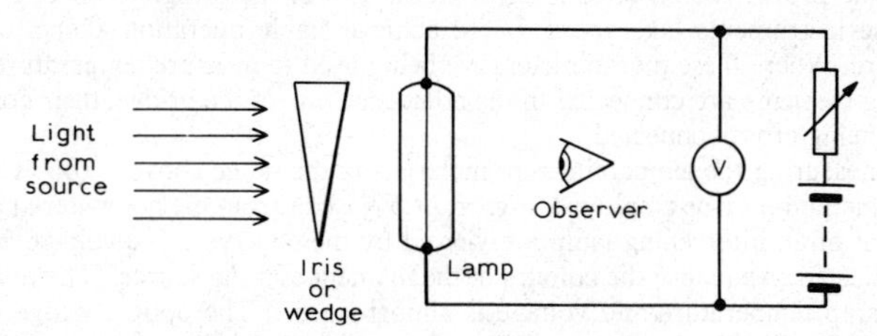

Figure 14.67 *An opto-electrical thermometer for the measurement of high temperature*

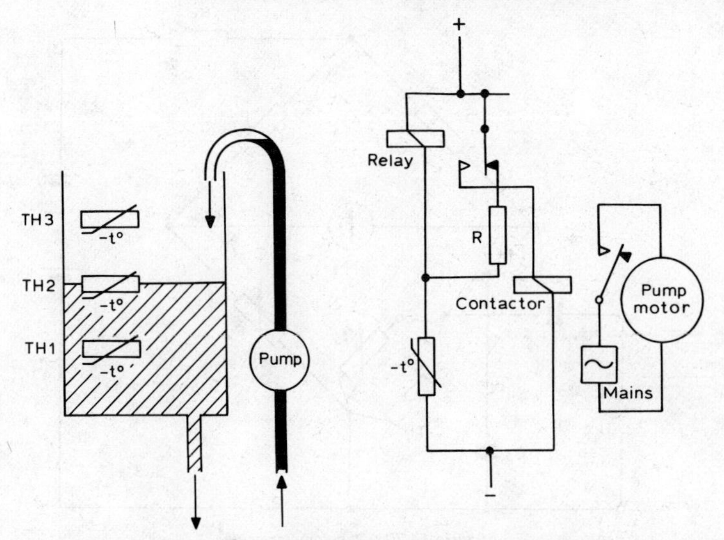

Figure 14.68 An arrangement for maintaining the liquid level in a tank

the liquid. When liquid is drawn from the tank, the temperature of TH_2 rises, its resistance falls, and via the relay and contactor, the pump motor is switched on to replenish the tank. If during the course of operation the liquid level reaches TH_3 or fails to reach TH_1, an alarm circuit is operated.

An example of gas or liquid velocity measurements (and possible control) is given in *Figure 14.69*. Here two temperature sensitive elements are used in adjacent arms of a bridge. TH_1 is in the moving stream and is cooled by it, and thereby it gauges the velocity, while TH_2 is in a pocket of stationary gas or liquid and compensates for variations of gas or liquid temperature.

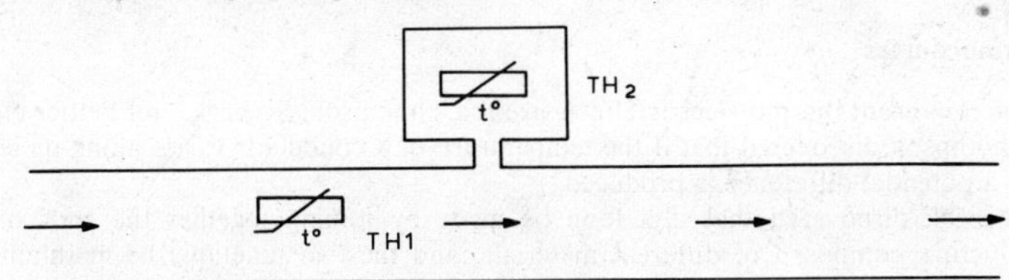

Figure 14.69 Measurement and/or control of gas or liquid velocity

A thermo-sensitive resistor *TH* may be used for the measurement of power (particularly at very high frequencies), using the bridge circuit of *Figure 14.70*. The operating range of *TH* must include the intended load resistance *R* for the a.c. source. The combination of C_1–C_2–L enables a.c. and d.c. to be mixed in *TH* without the sources feeding power to each other. With no a.c. input initially, the switch *S* is first set to position 1, R_3 is set to the value of the intended load resistance and R_4 is adjusted to produce a balance as indicated by the galvanometer *G*. *S* is then set to position 2 and R_5 is adjusted to produce a balance. *TH* now has the value *R* of the intended load resistance. The value I_1 of the current in the meter *M* is noted. The a.c. input is now applied and this changes the resistance value of *TH* thereby unbalancing the bridge. The value of R_5 is then increased to reduce the power fed in from the d.c. source and thus rebalance the bridge. The new value I_2 of the current in *M* is noted. The *reduction* of power from the d.c. source will equal the power which has been supplied by the a.c.

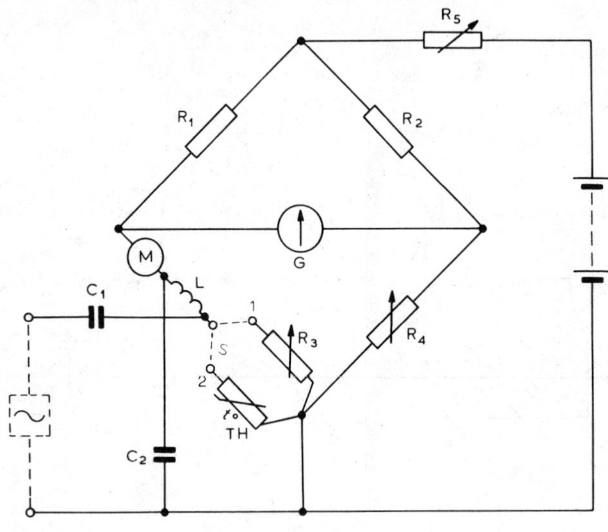

Figure 14.70 A bolometer bridge

source. Let P_t be the total power which is required to drive *TH* to the value *R*, and P_{ac} be the a.c. power which is to be measured, then,

$$P_t = I_1^2 R$$
$$= I_2^2 R + P_{ac}$$

By subtraction
$$O = I_1^2 R - I_2^2 R - P_{ac}$$

Therefore
$$P_{ac} = R(I_1 + I_2)(I_1 + I_2)$$

Thermocouples

Three important thermo-electric effects are the Thompson, Seebeck, and Peltier effects.

Thompson discovered that if the temperature of a conductor varies along its length, then a potential difference is produced.

Seebeck discovered that if a loop be made by joining together the ends of two conductors composed of different materials, and the two junctions be maintained at different temperatures, then a current will flow around the loop.

Peltier discovered that, conversely, if a current source be connected by conductors of similar material to the extremities of a third conductor of a different material, then the temperature at one end of the latter material will rise, while the temperature at the other end will fall.

An operating thermocouple circuit consists of a conductive loop of two different conductive materials, the two junctions being maintained at different temperatures. The introduction of materials beyond the minimum of two, simply adds self-cancelling e.m.f.s to the loop. An operating thermocouple loop will include a pair of self-cancelling Thompson effects.

It should be noted that whereas resistance-thermometers of the types which have been described measure temperature on an *absolute* basis, thermocouples measure temperatures on a *relative* basis, i.e., upon the basis of the *difference* of temperatures of the two junctions. When absolute temperature measurement is required, one junction may have its temperature stabilised by means of a thermostat.

When selecting materials for use in thermocouples, the main points which need consideration are the magnitude of the e.m.f. produced per degree of temperature difference, the resistivity of the materials, their tolerances of the operating environment and their operating ranges of temperatures. The relationship between the difference of temperatures of the two junctions and the e.m.f.s produced, is, within a few per cent, a linear one. The characteristics of some couples are given in *Table 14.5.*

Table 14.5 CHARACTERISTICS OF THERMOCOUPLES

Materials	*Approximate* e.m.f. *in* μV/°C *of temperature difference*	*Operating temperature range* °C
Chromel/Eureka	41	0 to 1 000
Iron/Eureka	59	−200 to +1 382
Chromel/Alumel	40	−200 to +1 200
Platinum/Platinum–Rhodium	6.5	0 to +1 450
Copper/Constantan	42	−200 to +300
Carbon/Silicon–Carbide	292 in the range 1 210 °C to 1 450 °C	0 to +2 000

Sensitivity up to 1 000 μV/°C can be achieved when one element is a thermistor.

A typical value of source resistance for a thermocouple is 10 Ω. To achieve maximum sensitivity the resistance of the thermocouple load should equal the source resistance. Since the power available from a source S equals $E_s^2/4R_s$, the power available from a combination of couples is proportional to the number of couples.

The two connections between a thermocouple and its load (typically a sensitive moving-coil meter) are usually made with the aid of special conductors having thermovoltaic characteristics similar to the materials used in the couples. This is to ensure that the effective hot and cold junctions are well separated thermally. It is important to connect each individual wire to its appropriate conductor in the couple, i.e., the two wires are *not* interchangeable. A standard colour code for identifying the type of cable for use with a given type of couple and for identifying the individual wires is given in British Standard BS 1843/1952.

The use of heavy gauge wires is recommended for use at very high temperatures, but light gauge wires are appropriate for high sensitivity and/or high speed of operation. A butt weld rather than a parallel weld produces a quicker response.

While thermocouples are sometimes used for the measurement of temperature, they are less sensitive and less convenient than resistance thermometers. Their main use is for the measurement of electric power at low amplitudes and particularly at high frequencies. The electrical power supplied to such an instrument is first converted to thermal power, then a proportion is reconverted to electric power in a form which is more convenient for measurement than the power originally supplied. The magnitude of the current which must be supplied to such an instrument normally exceeds 1 mA, a value which would be considered excessive in many electronic voltmeter applications. Such instruments are commonly employed as ammeters and milliammeters and being thermal in operation they have scale laws which are approximately square. The instruments are commonly calibrated using d.c., for subsequent use on a.c. Where a couple is not separated *electrically* from its heater, an error is about 1% may be expected on reversal of d.c.

Six different circuit arrangements for (milli)ammeters are shown in *Figure 14.71.* Example (*a*) shows a directly-connected directly-heated thermocouple, the couple acting as its own heater. The shunting of the couple by the meter, is a disadvantage. Example (*b*) shows a balanced version where this disadvantage is eliminated. Example (*c*) shows a contact type with separate heater. Example (*d*) shows an indirectly heated type where

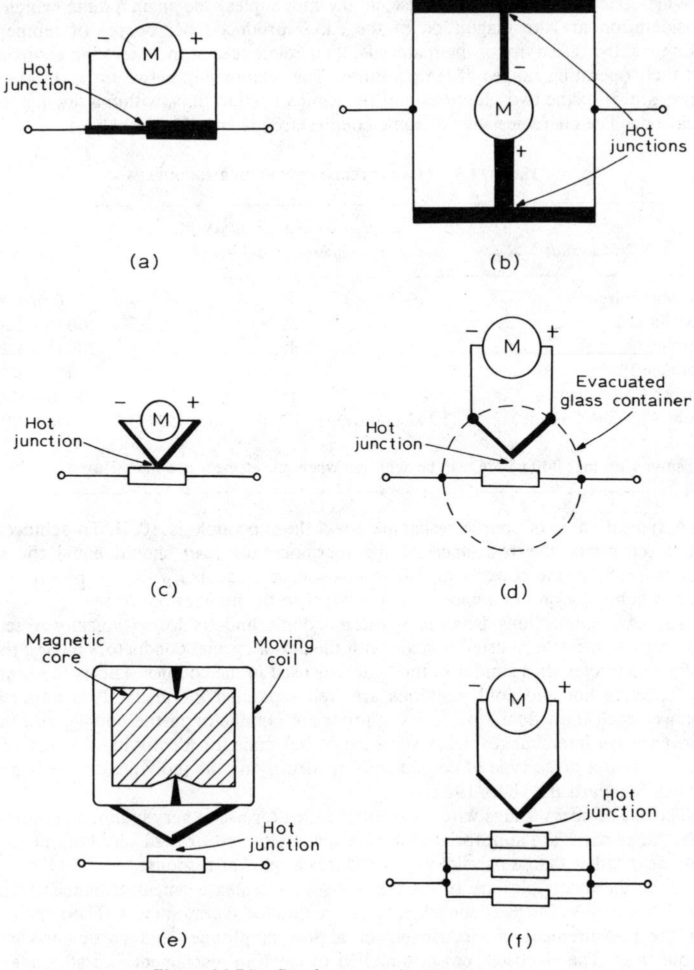

Figure 14.71 Six thermocouple arrangements

the heater and couple are embedded in a bead which is a good thermal conductor but an electrical insulator. This is the most popular arrangement. To conserve power, the elements are usually housed in a glass container evacuated down to a pressure which corresponds to about 10^{-5} metres of mercury. Example (*e*) shows a version where a couple is located above its heater and is insulated electrically and mechanically from it. The couple is mounted on the lower horizontal arm of the coil of a moving-coil meter having internal pivots, and the couple rotates with the coil. The springs of this instrument do not carry any current. Example (*f*) is used for the measurement of currents of high amplitude at high frequency. The use of several thin wires in parallel rather than one thick wire reduces errors due to skin effect. In general, due to this effect, as the operating frequency is raised (typically into the tens of megahertz), the indications have errors of a positive sign and increasing magnitude.

Meters which have heaters of relatively thick wire for the measurement of currents of high amplitude are moderately robust in operation, but those which are intended for the measurement of low values of currents are very delicate electrically, so that an overload of only 50% maintained for a fraction of a second may be sufficient to burn out the heater. Such an overload may be due to the initial charging current required for a capacitor or the current surge which occurs when switching on a tungsten filament lamp.

Due to the resistance/temperature coefficient of the heater and the relatively high temperature at which it is operated, it is inadvisable to employ shunts across the heater to give multirange operation. Instead, manufacturers commonly supply these instruments as plug-in units, which may be exchanged for use with a common moving-coil meter. Alternatively, thermocouple units frequently share a common case with a moving-coil meter. This is a common practice in the case of components which can carry several amperes safely and thus are less likely to suffer damage due to overload.

Thermocouple instruments such as that shown in *Figure 14.71* (*d*) commonly require 1–15 mW input power for full scale deflection and deliver e.m.f.s of about 6 mV from source resistances of about 10 Ω.

A circuit employing two thermocouples may be used as a wattmeter as shown in *Figure 14.72*. The resistor R_s has a relatively low value of resistance and it corresponds

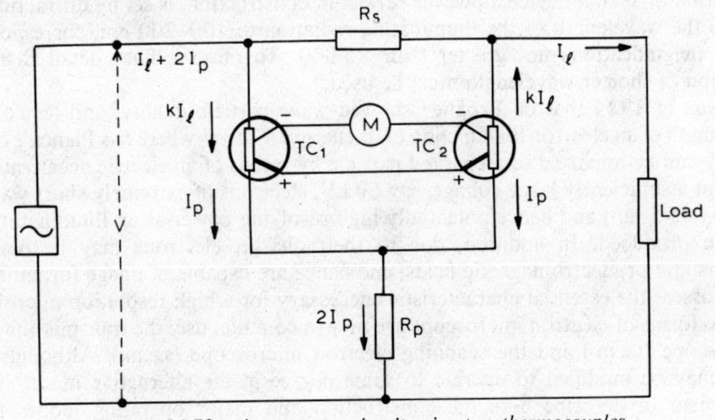

Figure 14.72 A wattmeter circuit using two thermocouples

to the shunt in a conventional ammeter. The resistor R_p has a relatively high value of resistance and it corresponds to the series resistor in a conventional voltmeter. Other symbols have the following significance:

V = supply voltage
I_l = load current
kI_l = a fixed proportion of the load current
I_p = half of the current in R_p
$kI_l + I_p$ = current in the heater of thermocouple 1
$kI_l - I_p$ = current in the heater of thermocouple 2
E_1 = e.m.f. produced by thermocouple 1
E_2 = e.m.f. produced by thermocouple 2
θ = angle of deflection of moving coil meter M

The thermocouples being connected in opposition

$$\theta \propto E_1 - E_2,$$

therefore

$$\theta \propto (kI_l + I_p)^2 - (kI_l - I_p)^2$$
$$\propto 4kI_lI_p$$

and since $I_p \propto$ supply voltage V

$$\theta \propto I_lV \quad \text{i.e., to the power dissipated in the load.}$$

FURTHER READING

HERZFELD, C. M., editor-in-chief, *Temperature Measurement and Control*, in 3 parts, Reinhold Publishing Corpn. New York, Chapman & Hall (1963)
HYDE, F. J., *Thermistors*, Newnes-Butterworths (1971)
KINZIE, P. A., *Thermocouple Temperature Measurement*, John Wiley & Sons
SCARR, R. W. A. and SETTERINGTON, R. A., 'Thermistors, their Theory, Manufacture and Applications', *IEE paper*, 3176M Han. (1960)
'Thermistors', *Mullard Technical Publication*, TP 455, Aug. (1961)

THE ELECTRON MICROSCOPE

Introduction

The principle function of any microscope is to make visible to the human eye that which is too small to be seen unaided. The oldest and most commonly used form is the light microscope, illuminating the object with visible or ultra-violet light. The limit of resolution of such a device, however perfect in construction, is set by diffraction effects due to the wavelength of the illuminating radiation to 100–200 nm, corresponding to useful magnifications not greater than ×5 000. To observe finer detail than this, a radiation of shorter wavelength must be used.

It was in 1924 that de Broglie[1] showed wave particle duality, and hence that the wavelength of an electron is a function of its energy, $E \propto \lambda/h$ where h is Planck's constant. Energy can be imparted to a charged particle by means of an electric accelerating field. Thus, at a sufficiently large voltage, say 50 kV, electrons of extremely short wavelength ($\lambda = 0.005\,5$ nm) and hence potentially high resolving power as an illuminating source can be produced. In addition, due to their charge, electrons may be focused by electrostatic or electromagnetic fields, and hence are capable of image formation. They thus possess the essential characteristics necessary for a high resolution microscope.

Two forms of electron microscope are now in common use: the transmission electron microscope (t.e.m.) and the scanning electron microscope (s.e.m.). Although each of these may be modified to operate to some degree in the alternative mode, it will be convenient to describe here the construction and major operating modes for each system considered separately.

The transmission electron microscope is applied to ultra-thin or sectioned material, through which the electron beam is projected to form an image. The whole field of view is illuminated simultaneously, and the enlarged image observed on a fluorescent screen incorporated in the microscope.

The scanning electron microscope images solid surfaces. The electron beam is focused to a single point and is then scanned over the surface. The resultant electron signal is collected and displayed as a brightness modulated image on a cathode ray tube.

The transmission electron microscope

It was in the early 1930s that the first transmission electron microscopes were built. At that time resolution was little better than that obtained with the light microscope. Present day t.e.m.s are capable of a resolving power of 0.2 to 1.0 nm.

INSTRUMENT CONSTRUCTION

Figure 14.73 shows a diagrammatic representation of a t.e.m. The electron gun contains a directly heated cathode and a Wehnelt cylinder acting as a bias shield, mounted on an

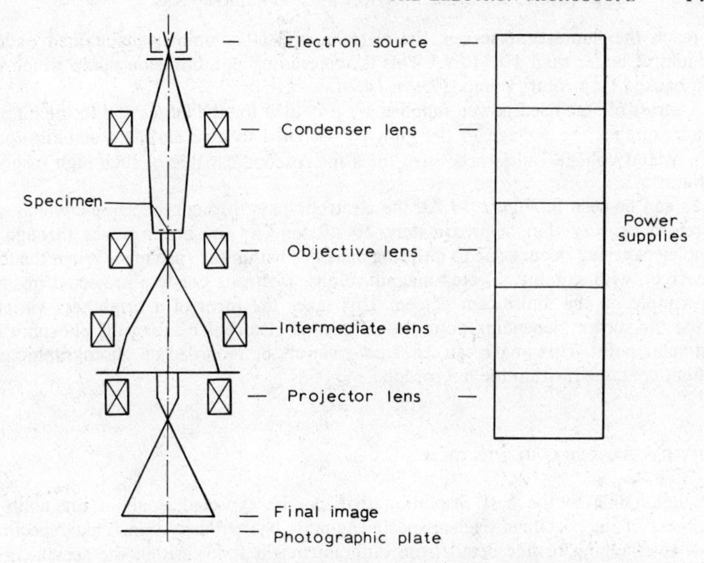

Figure 14.73 Schematic drawing of a transmission electron microscope

insulator. This, in conjunction with the anode, forms the electron gun. The accelerating potential on the gun can be varied between say 40 kV and 100 kV. A condenser lens system makes it possible to reduce the cross-section of the beam emitted from the gun and is used to illuminate the area of interest on the sample. This illumination can be varied by adjustment of the condenser lens current for differing working conditions. The objective lens current controls the focus of the image on the fluorescent viewing screen. The projector lens current is varied to provide a series of magnifications of typically ×1 000 up to ×250 000. To allow the electrons accelerated by the electron gun

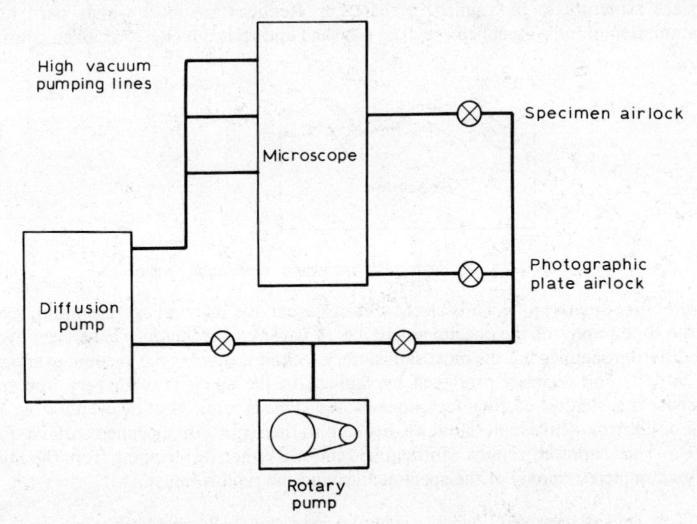

Figure 14.74 Schematic drawing of microscope vacuum system

to reach the fluorescent screen, the electron optical column is maintained under a vacuum of better than 10^{-4} torr.* This is achieved using a diffusion pump which is in turn backed by a rotary pump, *Figure 14.74*.

A series of stabilised power supplies are provided for the lenses and to the h.t. tank which supplies the voltage to the gun, all designed to give excellent stabilisation of current and voltage values necessary for a microscope capable of such high resolving power.

As can be seen in *Figure 14.73*, the electrons pass through the sample which must therefore be very thin, approximately 20–40 nm. As the beam passes through the sample scattering occurs due to differing densities within the specimen. When the focus is correct, with suitably selected magnifications, electrons create a projected image of the sample on the fluorescent screen. This takes the form of a brightness variation across the screen, depending upon the number of electrons exciting the phosphor at a particular point. This image can, after observation, be recorded on photographic plate or film, contained within the microscope.

SPECIMEN HANDLING IN THE T.E.M.

A rough guide to the best resolution that can be expected is about one-tenth the thickness of the specimen. Hence specimens must be extremely thin. Thick specimens show overlapping of fine detail from different height levels within the specimen and reduced resolution due to chromatic effects. The contrast in the final image is due to scattering of electrons within the specimen. Scattered radiation is either prevented by the objective aperture from reaching the image plane or merely contributes to the background intensity.

REPLICATION

The preparation of replicas of specimens which are opaque to electrons enables their surface structure to be studied in the t.e.m. Replicas are most widely used for the examination of bulk specimens such as polished and etched metals. A replica consists of

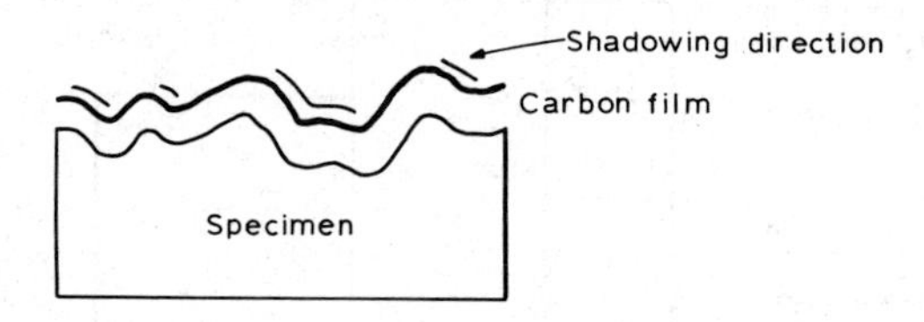

Figure 14.75 Schematic drawing of replication process

a thin film of material which is electron-transparent, the material corresponding exactly to the topography of the specimen surface. A variety of techniques have been evolved for differing samples but the most satisfactory technique used is the vacuum evaporation of carbon. The contrast produced by replicas in the t.e.m. is often very low and to increase it a 'shadow casting' technique is used. This is carried out by evaporating a thin film of electron-dense material at an angle on to the replica or specimen surface, *Figure 14.75*. The shadowed replica can then be removed either by stripping from the sample, or by complete removal of the specimen leaving the replica intact.

* Torr units of pressure (1 mm of mercury) is most universally used for measurement of high vacuum.

PREPARATION OF MATERIALS

There are two basic methods for producing specimens thin enough for the t.e.m. Building up the specimen by a deposition process, or reduction of the sample thickness by controlled removal, to a sufficient thinness for examination. The most generally useful technique is electro-polishing.

EMBEDDING AND SECTIONING

This technique is used mainly for the preparation of biological tissue, which is embedded in a resin that will remain stable under the electron beam. This sections of the embedding medium and tissue are obtained using a thin-sectioning microtome which cuts the sections by repeatedly moving the sample past a sharp cutting edge, a very small advance being made towards the knife between each cut, the amount of the advance determining the thickness of the section. Specimens about 40 nm thick can be sectioned.

MOUNTING OF SPECIMENS

Specimens are usually mounted on to 3.0 or 2.5 mm diameter 200 mesh/inch copper grids. A support film is used to cover the grid and must be able to resist heat generated by the beam and also have good mechanical strength. These films can be of carbon, collodian or formvar. The thin sample is then mounted on the support film.

Other techniques (too numerous to mention) for the preparation of samples for the t.e.m. are well documented, being mainly for the crystallographer, metallurgist, biologist, etc.

The scanning electron microscope

CONSTRUCTION AND PERFORMANCE CHARACTERISTICS

The scanning electron microscope is a recent addition to the tools available to the research worker. Postulated in the 1930s by Knoll [2] and von Ardenne,[3] serious design study started in 1948 under Oatley in Cambridge, resulting in commercial production in 1965. Such is the universal applicability of the technique that at the end of 1973 there are probably 2 000 units engaged in active research worldwide. The chief property of the s.e.m. is its ability to image rough surfaces non-destructively with a minimum of sample preparation.

In the s.e.m., *Figure 14.76* the illuminating electron beam is focused to a spot at the sample, and is then scanned over the surface in a manner analagous to television. The imaging signal results from the interaction of this incident beam with the sample surface. The electron-optical design of the microscope is therefore aimed at producing a small, high intensity beam of electrons. Most current instruments use a heated tungsten hairpin as the electron source, operating over a range of accelerating voltages of 1–50 kV; for special purposes alternative sources are already in operation, the most common being lanthanum hexaboride, giving higher brightness by an order of magnitude and improved stability; and field emission sources, which can give yet ×100 greater brightness, of particular value in high resolution microscopes. This source is then imaged by two or three electromagnetic lenses, to produce an electron beam size incident on the specimen of approximately 5–10 nm. Double deflection electromagnetic scanning coils, positioned inside the final lens, are fed from a scan generator, and cause the beam to scan over the surface of the sample, sequentially illuminating areas approximately 10 nm in diameter.

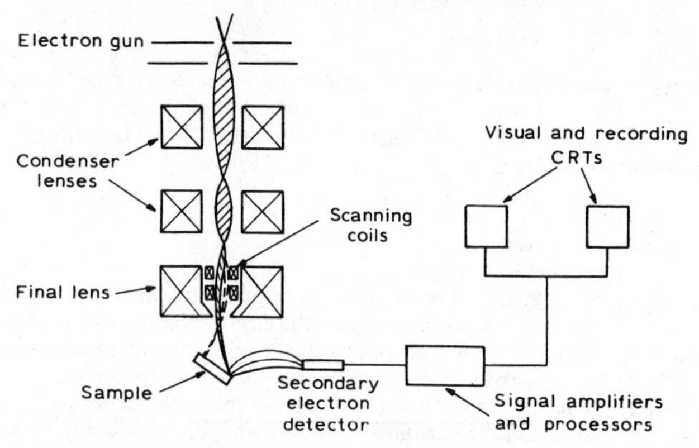

Figure 14.76 Schematic drawing of a scanning electron microscope

The high energy electrons incident on the surface interact with the atomic structure of the surface layers, generating radiations from the sample which are characteristic of its composition; a proportion of this radiation is emitted from the surface, and can be used to characterise it. Further, as the beam is scanned, the signal level of each characteristic radiation will vary with the surface composition and topography; by applying these signals as brightness modulation to a cathode ray tube scanned in synchronism with the electron beam, images of the surface characteristics can be derived.

The principal imaging mode of the s.e.m. uses secondary electrons. These are low energy electrons generated by electron-atom interactions, having a mean free path of 2–20 nm. Thus only those generated close to the surface can be re-emitted and even these are very vulnerable to absorption by the slightest surface topography. The signal produced has a spatial resolution of 10–20 nm and varies with surface topography. The detector is commonly a positively biased scintillator, accelerating the electrons on to the active area, and thence transmitting a signal via a light guide and photomultiplier to signal amplifiers and processors and finally to the cathode-ray tube. Particular properties of the scanning electron image are the large depth of focus available—some 500 times that of the light microscope—due to the very small aperture used in the final lens, and the flexible magnification range. Magnification derives simply from the ratio of area scanned on the sample to that of the cathode-ray tube and hence can be varied over a range of typically $\times 10 - \times 100\,000$ simply by changing the attenuating of the scanning coil currents while maintaining scanned area on the cathode-ray tube—a scanned area on the sample of 10 μm displayed on a 10 cm square cathode-ray tube gives a magnification of $\times 10\,000$.

Magnification is therefore independent of lens focus and can be zoomed rapidly through its range, centering on a fixed point in the surface. Although using a two-dimensional display, the images produced are characterised by their three-dimensional appearance and relative ease of interpretation by non-specialist staff. Image recording uses a second, high resolution cathode ray tube and conventional camera system.

Other imaging signals available in the s.e.m. include the following.

Backscattered electrons. High energy electrons originating from the primary beam, scattered through a large angle and re-emitted. These show less fine detail of the surface topography, but are dependent upon the mean atomic number of the material, hence showing compositional contrast.

Absorbed electrons. If the sample is not earthed, but connected to a current amplifier, an absorbed current image is produced complementary to backscattered current. However, certain materials exhibit beam-induced conductivity effects also, particularly applicable in semiconductor research.

X-rays. These are emitted under electron bombardment and can be detected by the conventional crystal spectrometer or by lithium drifted silicon solid state devices. This is the basis of the technique of X-ray microprobe analysis. The s.e.m. and microprobe analyser are alternative design optimisations of the same basic instrument.

Cathodoluminescence. Some materials fluoresce in the visible range under electron bombardment, as illustrated by the phosphors of cathode ray tubes. The radiation is in many cases a function of impurity levels within materials and is used both in semiconductor materials research and in many mineralogical investigations.

SPECIMEN HANDLING IN THE S.E.M.

Using electrons, a vacuum environment is essential in the s.e.m., thus for the most part gases and liquids cannot be examined, and special precautions must be taken with solid materials containing high proportions of liquid or gas. In addition, bombardment by negatively charged electrons ultimately builds up a surface charge, and surfaces which are not normally electrically conducting must be rendered so. Within limits imposed by the vacuum requirement, and by mounting and handling techniques, sample size is not a limiting factor and most instruments can accommodate samples up to an inch or more in diameter with ease.

Metals are mounted directly or after ultrasonic cleaning to remove loose surface debris.

Minerals, ceramics, etc., are coated by evaporation with a thin metal layer, typically gold or gold–palladium to a thickness of 10–20 nm.

Organic materials are treated similarly, but are generally examined at lower beam voltage to prevent bombardment damage.

Biological material shows drying artefact is inserted directly and is customarily subjected to freeze drying or critical point drying before metal coating. In some cases quench freezing is used and material is then examined in the frozen state on a cold stage.

Because of the large specimen chamber available, the s.e.m. is particularly well suited to *in-situ* sample processing. Behaviour of materials at high and low temperatures during tensile testing, etc., are studied using video tape to record the dynamic sample changes as they occur.

The s.e.m. is applied in both fundamental and applied research to many problems involving the physical structure of solid surfaces, and particularly those requiring greater depth of focus or higher magnification than is obtainable with the light microscope. In fractography and for all forms of failure mechanics; to study powders and compacted materials; in tribology and corrosion science. In biology and geology and increasingly in pathology, the s.e.m. is exploited both as a research tool and as a means of presentation of micrographical information to the non-specialist and student. It has also gained wide acceptance throughout the electronics industry; its range of applications in this field are detailed below.

Future advances. As a surface imaging tool the s.e.m. tends towards a basic limitation in resolution due to the mean free path of the secondary electron. As a technique for surface investigation, the combination of information derived from alternative beam-induced radiations may be more fully exploited, developing a quantitative approach to scanning microscopy and the relationship of surface structure to composition. Improved signal detectors and signal processors increase dramatically the information available from the surface.

Scanning transmission microscopy (s.t.e.m.)

This derivative of the s.e.m. uses thin, sectioned material as in the t.e.m. but retains the scanning image collection and processing, using the transmitted electron beam. Resolutions down to 0.5 nm are reported using special s.t.e.m.-optimised electron optics.

Electron microscope applications in electronics

The interaction of an s.e.m. beam with a semiconductor sample can generate many phenomena worthy of study, including some that can be observed by no other instrument. Electron beams are easy to control; to deflect in scan patterns, switch on and off, and to adjust in energy with which they bombard the specimen. Secondly, the alternative operating modes of the s.e.m. quickly answer many of the questions which arise during semiconductor research and production.

The use of electron beams in the microelectronics field has grown rapidly in the past ten years. The finely focused electron beams of the s.e.m. were first used for visual observation of transistors and integrated circuits, where their great depth of field gave a new dimension to micrographs, *Figure 14.77*. As circuits become more complex, with circuit elements reaching the limits of photolithography, the superior resolving power

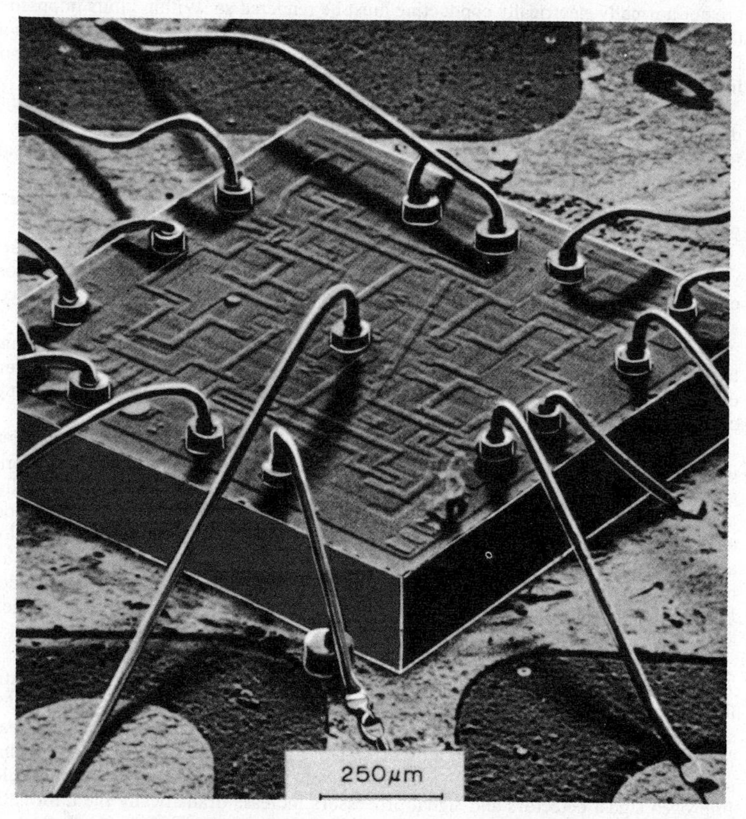

Figure 14.77 Scanning micrograph of integrated circuit

and depth of focus of the s.e.m. become a necessary part of the evaluation of new circuits and construction techniques and later became an integral part of quality control during manufacture.

As the needs of the semiconductor industry grow, so new techniques have been developed in the s.e.m., both in the handling and processing of samples and in the type of information which can be derived from them. One of the latest developments in the use of electron beams is to reduce the size of circuit elements below that which is possible by photolithography. Using the new electron beam microfabrication system, the resolution and hence minimum linear dimension of a circuit element can be reduced by at least an order of magnitude (from 1 μm to 100 nm). Therefore the packing density of a circuit can be increased one hundred fold. Applications requiring this increased resolution and packing density include magnetic bubble memories, large scale integrated circuitry, surface acoustic wave devices and very high speed m.o.s. transistors (see Section 8).

SEMICONDUCTOR INVESTIGATIONS IN THE S.E.M.

Sample preparation. This is minimal. The sample accommodation of many s.e.m.s is in excess of 50 mm square × 25 mm thick and is readily adaptable to take even larger samples. For the purposes of s.e.m. imaging, a semiconductor device is conductive, even silicon oxide and passivation layers applied to circuits exhibit induced conductivity at the beam energies normally used. Any transistor or integrated circuit can be mounted directly into an s.e.m., the only two requirements being that the package be opened to allow the circuit to be seen and that at least one termination of the circuit be grounded. S.E.M. investigations are non-destructive and thus a semiconductor can be inspected at any stage of manufacture.

The following operating modes find application in semiconductor research.

Secondary electron imaging. Used to show surface details such as etch pits, oxide steps formed during multiple diffusions, the integrity with which metal layers follow oxide steps and the nature of the bonding between contact pad and leadout. The sample handling flexibility and depth of focus of the s.e.m. allows the sample to be examined in any orientation.

The trajectory of secondary electrons leaving a surface will be affected by potential variations in that surface, leading to the phenomenon of voltage contrast. In this mode the voltage of a particular circuit element can be 'seen' and electrical failure readily observed. The potential of an element will show by its brightness on the s.e.m. display. The study of integrated circuits using voltage contrast is often done dynamically, with the circuit under simulated working conditions, e.g., an m.o.s. shift register can be run at high frequency and a beam switching system used to blank the electron beam at some sub-harmonic of the fundamental logic frequency applied to the device, effectively freezing the device in a particular logic state. The results of this may be recorded on video tape. For dynamic studies, devices can be used in their finished (but unpotted) state. For static studies and evaluation under d.c. conditions all glass passivations must be removed, *Figure 14.78*.

Absorbed electron and conductive imaging. The conduction mode depends on the conductivity increase of the target region due to the generation of extra carriers. The primary electrons entering the semiconductor create electron-hole pairs, available for conduction processes until removed by recombination.

By using this conduction mode, depletion region boundaries can be studied, both with and without bias applied to the sample, including such phenomena as depletion layer spreading, junction breakdowns and oxide pinholes. Since electron beams easily penetrate semiconductor materials, information gained in the specimen current mode is

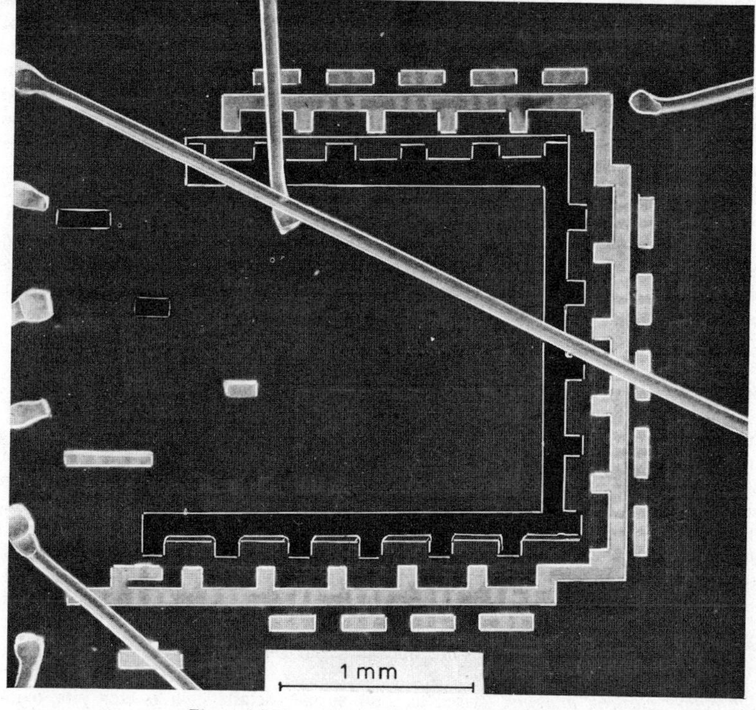

Figure 14.78 Voltage contrast on m.o.s. device

greatly influenced by beam energy. A 10 keV beam penetrates about 1 μm of silicon or silicon dioxide, a 30 keV beam penetrates in excess of 5 μm. Sub-surface phenomena are therefore visible in this mode.

The recombination process mentioned above can also lead to the emission of light photons from the sample—*cathodoluminescence*. The s.e.m. can study the light emission characteristics of phosphors, light emitting diodes and other light emitting materials. Luminescent efficiency of many materials is highly temperature dependent; special-purpose specimen stages available for the s.e.m. allow controlled temperature variation, above and below ambient during examination.

X-ray analysis. Microprobe analysis offers unique facilities for semiconductor investigation. Concentrations of elements as low as 0.01% from areas as small as 1 μm diameter can be analysed on a complete wafer or a finished device without damaging the device. Alternatively large area scans can be performed and the analysis displayed in the form of an elemental map showing the distribution of any chosen element in the sample. It provides a particularly powerful technique for the identification and location of many forms of surface contamination.

Electron channelling.[4] Using variation in angle of beam incidence on to the specimen surface, electron beam channelling occurs, giving rise to contrast related to crystallographic structure. This provides an effective, nondestructive technique for determination of grain size and orientation in crystalline materials, and the surface quality of single crystals such as silicon, germanium and gallium arsenide. Minimum areas from which such patterns are obtainable may be 2 μm in diameter.

REFERENCES

1. DE BROGLIE, L., *Phil. Mag.*, **47**, 466 (1924)
2. KNOLL, M., 'Aufladepotential und Sekundär-emission elektronbestrahlter Oberflächen', *Z. Techn. Phys.*, **2**, 467 (1935)
3. VON ARDENNE, M., 'Das Elektronen raster mikroskop', *Z. Techn. Phys.*, **19**, 407–416 (1938)
4. VAN ESSEN C. G. and SCHULSON, E. M., 'Selected Area Channelling Patterns in the Scanning Electron Microscope', *J. Sci Insts.*, **2**, 361 (1969)

FURTHER READING

KAY, DESMOND., ed., *Techniques for Electron Microscopy*, Blackwell Scientific Publication (1965)
THORNTON, P. R., *Scanning Electron Microscopy*, Chapman and Hall (1968)

NOISE AND SOUND MEASUREMENT

The words *sound* and *noise* will be treated here as synonymous from the viewpoint of measurement. Two quantities are directly measurable, the sound pressure and the particle velocity, other quantities such as sound power being derived from one or both of these two.

It is assumed, unless otherwise stated that all sound measurements are carried out in air, so that it is necessary for only one of the basic quantities, normally the pressure, to be measured, since pressure and particle velocity in a parallel wave motion are related by the equation

$$p = \rho c v \tag{1}$$

where ρ is the density of the air, c is the velocity of sound and v is the particle velocity.

Sound pressures are for practical purposes, converted into sound pressure levels, defined as

$$L = 20 \log_{10}(p/p_{ref}), \text{ in decibels (dB)} \tag{2}$$

where L is the sound pressure level and p_{ref} is a reference pressure of 2×10^{-5} N/m².

Alternatively, intensity level L_i which is the decibel equivalent of intensity, the rate of flow of sound energy across unit cross-section at a point in the sound-field, is defined as

$$L_i = 10 \log_{10}(I/I_{ref}) \text{ dB} \tag{3}$$

where I_{ref} is 10^{-12} W/m²

and

$$I = p^2/\rho c \tag{4}$$

These two reference levels are, to all practical purposes, equal for air at normal temperature and pressure.

Subjective measures of sound sensations are derived from measurements of pressure level or spectrum pressure level (pressure per unit bandwidth) throughout the spectrum.

Measurement of particle velocity

Particle velocity may be measured absolutely by means of a Rayleigh Disc,[1] which is a thin disc suspended from a torsion fibre in the sound field. The torque on the disc when held at an angle to the direction of propagation is related to the particle velocity. This method of measurement is unsuitable for use outside a specialised laboratory, and it is therefore better to measure pressure.

Measurement of sound pressure

For the measurement of sound pressure, a calibrated microphone is needed, together with an amplifier of known gain with rectifier and meter. *Figure 14.79* shows a block diagram of measuring equipment. The microphone can be calibrated by the method of reciprocity.[2] More usually, microphones for sound measurement are calibrated by comparison with a standard instrument. The two are placed side by side, or alternately in the same position, in a free-field room, i.e., one in which all the surfaces are covered with efficient sound-absorbing material to prevent disturbance of the direct sound propagation by reflections.

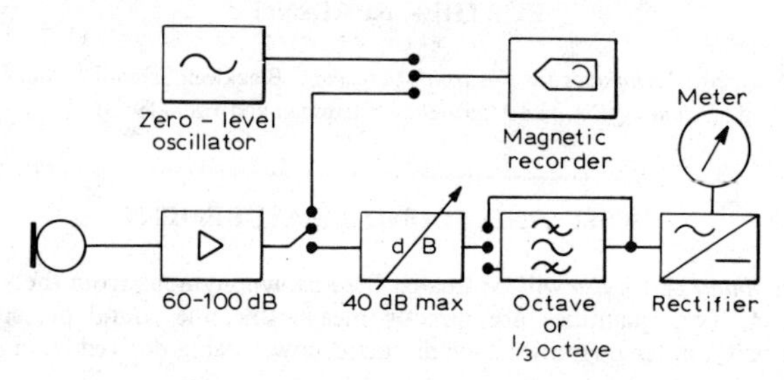

Figure 14.79 Versatile noise measuring chain

For calibration of the amplifier chain, the microphone is replaced by an inert resistor of the same resistance, and an alternating voltage derived from an oscillator is applied across the resistor, through a variable attenuator. The output voltage is compared with the unattenuated oscillator voltage, and the attenuation to the input is adjusted until the two are equal. For general use, a maximum gain of about 130 dB should be available, and the internal noise of the amplifier should be close to the minimum possible due to the thermal noise at the input resistance. The signal for the amplifier output is rectified and applied to a d.c. meter. The rectification characteristic should give an accurate indication of the root-mean-square (r.m.s.) value of the signal, which represents its power irrespective of the waveform. British Standard 3489:1962 makes the use of a r.m.s. meter obligatory for all noise measurements. To achieve a true r.m.s. reading, it is usually necessary to resort to thermal meters with an accordingly slow response. For most practical purposes, however, simple square-law rectification gives sufficient accuracy even with complex waveforms. For some purposes, peak rectification is used or an approximation to it. The ratio of the amplitude of such a waveform to that of a sinusoid with equal r.m.s. value is known as the crest factor.

As the rectifier is necessarily limited in dynamic range, it is usual to insert switched attenuators into the amplifier, by which the rectifier input, and consequently the meter indication, can be brought within a range of 10–20 dB. For the rapid measurement of steady or slowly-varying sounds, a calibrated attenuator with 1 dB steps may be used to bring the meter to a fixed reading.

High-speed level recorders

Similar in principle to the method of measurement by insertion of variable attenuation just described, is the chart level recorder by which permanent records of sound level may be made over a range of 75 dB or more. This instrument incorporates a potentiometer of which the wiper is controlled by a servo system actuated by a difference signal corresponding to the difference of the rectified output of the amplifier from an arbitrary fixed value. The wiper is driven along a linear track until it reaches a position such that there is no difference signal. The recording pen is attached to the wiper and records its position, which is directly related to the input voltage.

The chart recorder is perhaps the most versatile instrument for sound measurements. It enables noise to be recorded over a range of input levels limited only by the capabilities of the amplifier. Normally the potentiometer is arranged to plot a decibel scale. Since it is in effect a null-reading device, the accuracy of the scale depends only on the accuracy of construction of the potentiometer. For most purposes, the recorder will be used to plot a signal against time. By gearing the chart to a tone-generator dial movement, the instrument may be used for the direct drawing of response/frequency graphs.

Measurement of spectra

The spectrum of a complex sound may be measured by inserting bandpass filters into the amplifier of a sound-pressure measuring circuit, so as to obtain a series of band-pressure levels. The audible frequency range is usually divided into octave or one-third octave bands, and filters for the purpose are specified in British Standard 2475:1964. The one-third octave bandwidths, except at very low frequencies, correspond closely to the *critical bandwidths* of hearing. Thus, measurements in one-third octave bandwidths are used for loudness evaluation by the method of Zwicker,[3] and also for the great majority of development work on the silencing of machines.

Octave bandwidths are used for Stevens'[4] method of loudness evaluation and for many applications in connection with industrial and community noise. Noise containing discrete frequency components can only be analysed satisfactorily by means of narrow-band heterodyne filters, typically 10% of the centre frequency in width.

Weighting networks

A meter reading of sound pressure level by the simple measurement chain described above, i.e., one giving a total pressure for all frequencies within the range of measurement, is not well correlated with the subjective loudness of the sound because the human ear is relatively insensitive to sounds of very high and, more particularly, low frequencies.[5] Weighting networks have therefore been devised for insertion into noise measurement chains by which the low and very high frequency components are progressively attenuated. *Figure 14.80* shows three commonly used networks as specified in

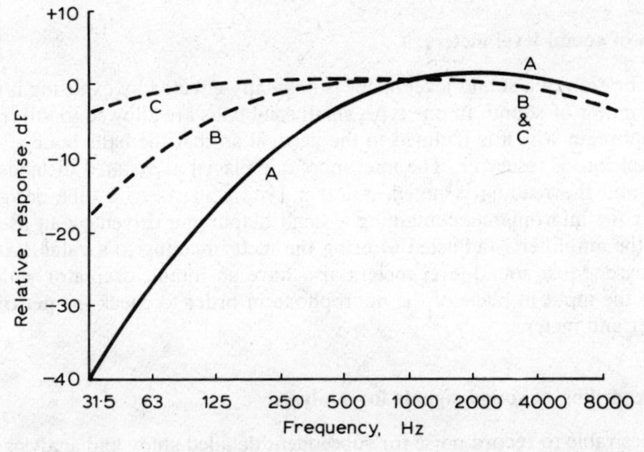

Figure 14.80 Weighting networks A, B, and C (From BS 4197:1967)

British Standard 4197:1967. The *A* network is the most generally applied; it is used to obtain single-figure measurements of traffic noise in connection with compensation procedures and it is also the basis of measurements of factory noise as required for hearing risk assessment.

Sound-level meters

A sound-level meter consists of a simple battery-operated noise-measuring circuit built into a portable case and incorporating weighting networks and a meter directly calibrated in decibels. A typical block diagram is shown in *Figure 14.81*. The requirements for such an instrument are specified in British Standard 4197:1967, (precision grade sound level meters) or British Standard 3489:1962 (for industrial grade meters). The differences are mainly in the closeness of the tolerances.

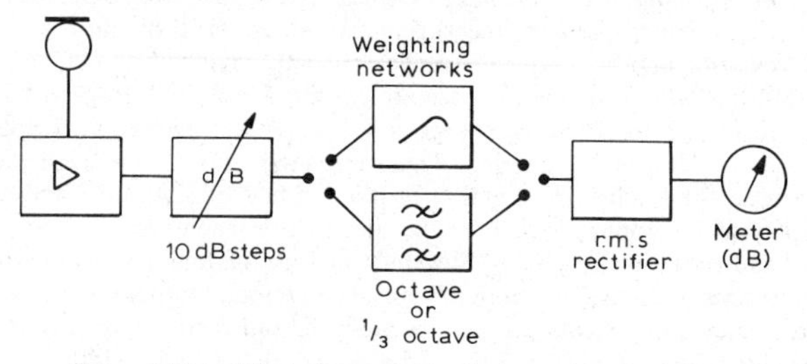

Figure 14.81 Block diagram of typical sound-level meter

The microphone, which is usually of a capacitor, moving-coil or piezoelectric type, has an omnidirectional polar diagram with controlled maximum off-axis deviations from the axial sensitivity. The main amplifier is followed by an attenuator with switched 10 dB steps and by weighting networks. (In place of the weighting networks it may be possible to connect a set of octave or one-third octave bandpass filters in series by means of input and output jacks.) The filters or weighting networks are followed by a buffer stage and a meter with r.m.s. rectification. The meter is usually calibrated from −6 to +10 dB. In use, the switched attenuators are adjusted to bring the meter reading within the scale. The reading is then added to the number shown on the attenuator switch, which represents the sound level corresponding to zero meter reading.

Calibration of sound-level meters

Overall calibration of a sound-level meter is normally effected by exposing it to a stable calibrated emitter of sound. In one type, small steel balls are allowed to fall freely on to a metal diaphragm which is inclined to the vertical so that the balls bounce off after a single impact into a reservoir. The microphone is placed at a stated distance from the diaphragm and the reading is noted. Another type consists of a tube designed to fit closely over the microphone, containing a small diaphragm driven by an oscillator. In each case, the amplifier is adjusted to bring the meter reading to a value stated on the calibrator case. Most sound-level meters also have an inbuilt oscillator which can be switched to the input in place of the microphone in order to check the performance of the amplifier and meter.

Analogue recording of sound signals for analysis

It is often desirable to record noise for subsequent detailed study and analysis. To do so may greatly speed up work on the site, where time is often severely limited. It also allows short-lived events which cannot be repeated exactly, such as the flyover of an individual aircraft, to be submitted to frequency analysis by replaying through bandpass filters.

The output of the amplifier stages of a sound-level meter or other sound-measuring equipment may be recorded on a portable reel-to-reel tape recorder. The signal should be taken from a point after any switched attenuators, filters or weighting networks, so that the system will be able to handle the whole range of levels appropriate to the equipment; these may include a calibrating signal of over 120 dB, for example, followed by background noise of, say, 40 dB (*A*). It may be necessary to insert a fixed attenuator before the recorder input, to align the maximum amplifier output (corresponding to full-scale reading) with the maximum input voltage of the recorder. The tape recorder should have a good linear range of about 40 dB within its published signal to noise ratio and adequate frequency range. Battery driven recorders with these characteristics are usually expensive, but the use of inferior equipment may give very misleading results.

Instantaneous displays of noise spectra

Equipment is now available from several manufacturers, consisting of a noise measurement chain and a set of narrow bandpass filters which are automatically scanned at regular short intervals, the levels being displayed continuously on an oscilloscope. The filters are either one-third octave or narrower, up to 400 channels being included in one instance. These instruments are designed as complete systems with a variety of facilities such as the *freezing* of individual scans for examination, the brightening of the display of one channel with simultaneous indication of its level on a numerical read-out. In one instrument using one-third octave channels, an automatic loudness summation is provided using the method of Zwicker.[3] Time-compression and digital processing may be used to speed up the analysis and increase the possible number of channels.

The principle has been extended also to record transient events in digital form at rates of up to 10^5 samples/s and replay them in digital or analogue form at very much slower speeds.

Digital and computer systems for noise measurement

Sound signals, being continuous functions of time, lend themselves readily to digital recording and analysis, either in real time or from analogue recordings. The output voltage of a microphone amplifier is converted by means of an analogue-to-digital converter into digital characters which are stored on punched paper tape or magnetic tape, with associated information such as time or filter channel. In equipment which has been recently introduced on to the market, the digital information is recorded directly on 6-mm magnetic tape in a domestic-type cassette which can be kept for subsequent direct analysis by computer or programmable desk calculator. This equipment, which is primarily intended for noise monitoring over long periods, is battery operated and accepts samples of noise voltage at half second or longer intervals, and is able to record for several days on a single cassette.

A description of rapid punched-tape system for recording sound signals was given by Moffat.[6] His equipment was the basis of a flexible system for acoustic measurements, which will be further mentioned in the following section on general acoustic measurements.

Use of equipment for noise measurements

MACHINERY NOISE

The noise output from a machine may be measured in free-field conditions by means of a sound-level meter placed in succession at a number of points distributed over a hemispherical solid angle at a constant distance from the surface of the machine. British Standard 4196:1967 describes the method. A choice of four distances from 0.3 m up to 10 m is given. Alternatively, diffuse field measurements of noise power may be made in a highly reverberant enclosure. For measurements on a machine *in situ*, where the conditions are intermediate between free-field and diffuse field, the free-field method of measurement is used and correction is made for the effect of the reverberant sound. For the case of a substantially diffuse field, the sound power level, H, is given by

$$H = (L) - 10 \log_{10} T + 10 \log_{10} V - 14 \text{ dB} \qquad (5)$$

where (L) is the mean sound pressure level, T is the reverberation time of the enclosure and V is its volume. (SI units)

ROAD TRAFFIC NOISE

Any of the methods of noise measurement described above may be used. A-weighting is required for all measurements, since the A-weighted sound pressure level is highly correlated with subjective disturbance.[7]

A number of indices have been developed for assessing traffic noise disturbance to a community. These depend upon the time distribution, rather than the maximum level, expressed as L_n, the sound level (A-weighted), exceeded for n% of the time. One such index is the Traffic Noise Index[8] (TNI) which is given by the expression

$$\text{TNI} = L_{90} + 4(L_{10} - L_{90}) - 30 \qquad (6)$$

where L_{90} is the sound level (A) exceeded for 90% of the time. TNI achieves a correlation of 0.88 with social dissatisfaction, while L_{10} measured over the 18 hours from 0600 hours to 2400 hours gives a correlation coefficient of 0.60. Eighteen-hour L_{10} has been adopted as the criterion in connection with compensation for traffic noise disturbance, in spite of its comparatively low correlation, as TNI is said to be less predictable for future situations. L_{10} and L_{90} may be computed from chart recordings or in real time by means of a level distribution analyser which accepts samples of the noise voltage at regular intervals and counts the number of samples falling within a series of 5 dB level ranges. A histogram prepared from these counts is used to estimate the required indices to within about 1 db.

AIRCRAFT NOISE

Assessment of community disturbance by aircraft noise requires a knowledge of the number of flyovers as well as the mean level of the peak sound level during individual flyovers. The two factors are combined in an index known as the Noise and Number Index, (NNI).

$$\text{NNI} = \text{Average Peak Level} + 15 \log_{10} N - 80 \qquad (7)$$

where N is the number of flyovers per day. The peak level is expressed in PN dB, which is a weighted sum of the spectral components having a high correlation with disturbance.[9]

HEARING DAMAGE RISK

Modern practice in hearing conservation for those exposed to industrial noise in work is summarised in the Department of Employment's 'Code of Practice for Reducing the Exposure of Employed Persons to Noise'.[10] It is recommended that exposure to steady sound exceeding 90 dB (A) should not be continued for more than 8 hours in any day.

For varying or intermittent exposure, the A-weighted energy level integrated for the whole working period in a day should not exceed the same integral of 90 dB(A) over 8 hours.

It is usual to make measurements in terms of the equivalent energy level for the working period where

$$L_{eq} = 10 \log (1/T) \int_o^T (p^2/p_o^2) dt \quad (8)$$

where p is the instantaneous sound pressure at time t, p_o is the usual reference pressure of 2×10^{-5} N/m^2 and T is the time of averaging. The equivalent value of L_{eq} for an exposure period may be computed from a record of sound level (A) against time, but instruments giving a total integrated energy level for a period of exposure, or even a value of L_{eq} are now in general use. These are known as noise dose meters.

ACKNOWLEDGEMENT

Figures 14.79 to *14.81* inclusive are from *Acoustics for Radio and Television Studios*, by C. L. S. Gilford, Peter Peregrinus (1972).

REFERENCES

1. KING, L. V., *Proc. Roy. Soc. A*, **153**, 17 (1935)
2. COOK, R. K., *J. Research Nat. Bur. Standards*, **25**, 489 (1940)
3. ZWICKER, E., *Acustica*, **10**, 304 (1960)
4. STEVENS, S. S., *J. Acoust. Soc. Amer.*, **28**, 807 (1956)
5. ROBINSON, D. W. and DADSON, R. S., *Brit. J. Appl. Phys.*, **7**, 166 (1956)
6. MOFFAT, M. E. B., *B.B.C. Research Dept., Report PH8* (1967)
7. MILLS, C. H. G. and ROBINSON, D. W., *Engineer*, **211**, 1070 (1961)
8. LANGDON, F. J. and SCHOLES, W. E., *Building Res. Station Current Papers*, 38/68 (1968)
9. KRYTER, K. D., *Noise Control*, **6**, 12 (1960)
10. DEPARTMENT OF EMPLOYMENT, 'Code of Practice for reducing the exposure of employed persons to noise', H.M.S.O. (1972)

ACOUSTIC MEASUREMENTS

Sound insulation

DEFINITIONS

Sound is transmitted from one point in a building to another by various paths through the air or the structure. The point of origin is described as the *source* and the other is usually in a room known as the *receiving room.* Airborne sound transmission is that in which the sound is generated in the air at the first point and received in the air at the second. The shortest path of transmission, usually through one or more intervening partitions is called the *direct path*; sound that travels by indirect paths, often including substantial distances through solid structures, is said to travel by *flanking transmission.* Sound travelling through the solid material of a building, particularly from a vibrating source in contact with it, is known as *structure borne sound. Figure 14.82* shows examples of these forms of transmission.

The sound insulating characteristic of a partition is its *transmission coefficient.* This is the ratio of the sound power radiated into the receiving room to that falling on the source room side. It is usually denoted by τ. The difference in sound power levels is

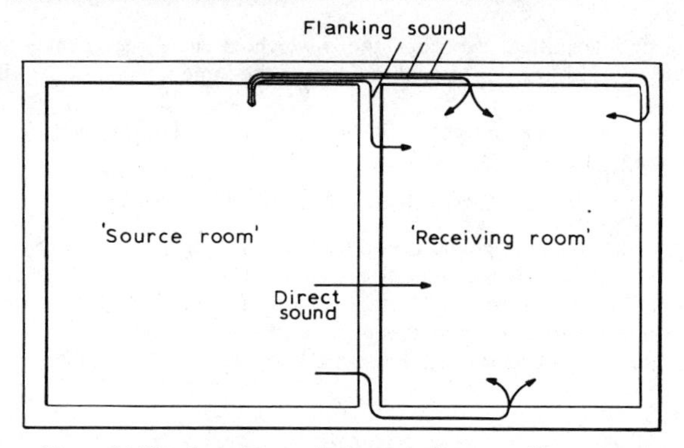

Figure 14.82 Sound transmission paths between adjacent rooms

known as the *sound reduction index* (SRI) and is the figure usually quoted. The two quantities are related by the equation

$$\text{SRI} = 10 \log_{10} 1/\tau \qquad (1)$$

The reduction of sound pressure level between two rooms is known as the *sound level reduction.*

MEASUREMENT OF AIRBORNE SOUND TRANSMISSION

To measure the sound level reduction between two rooms, it is simply necessary to place a sound source in one of them and to measure the difference between the equivalent sound pressure levels in the two rooms. To derive the SRI of the partition, a correction must be applied to the sound level reduction to take into account the absorption of sound in the receiving room. Even if the sound transmission between the two rooms is entirely by the direct path through the intervening wall, the sound level reduction will depend not only on the SRI of the partition but also on its area and the increase of sound pressure level due to reverberation in the receiving room. The relationship between the SRI and the sound level difference is:

$$\text{SRI} = \text{Sound level reduction} + 10 \log_{10} (\tfrac{1}{4} + A/S\alpha) \qquad (2)$$

where A is the area of the partition, S the total area of the interior surfaces of the receiving room and α is their average absorption coefficient. (See Measurement of Acoustic Absorption Coefficients.)

Recommendations for the sound level reduction and the derivation of the SRI are given in British Standard 2750:1956. In this document, the $\frac{1}{4}$ within the correction term is ignored, so that it becomes $10 \log_{10} (A/S\alpha)$. This simplification causes little error except when the receiving side is in the open air or a room with very great absorption. A normalising correction is recommended when assessing walls between dwellings or rooms within a dwelling where there may be multiple flanking paths. The correction term becomes $10 \log_{10} (10/S\alpha)$ or $10 \log_{10} (T/0.5)$ where T is the reverberation time of the room (see Measurement of reverberation time, page **14**–95).

To make a full assessment of the SRI of a partition, narrow bands of noise at one-third octave intervals are radiated by one or more loudspeakers into the source room. At each frequency, the sound-pressure level is measured at five microphone positions in each room by one of the methods described above in Noise and Sound Measurement, and the mean energy level computed for each room. The microphones should be in positions more than half a wavelength from any wall. The reverberation time also is

Table 14.6 POWER RATIOS FOR SOUND PRESSURE LEVELS (SPL)

SPL	*Power ratio*	*SPL*	*Power ratio*
0	1.0	7	5.0
1	1.3	8	6.3
2	1.6	9	7.9
3	2.0	10	10
4	2.5	11	13
5	3.2	12	16
6	4.0	etc.	

measured (see Measurement of reverberation time, page **14**–95), and the total absorption calculated for substitution in equation 2.

The mean energy level is computed for each room by averaging the power ratios derived from the pressure levels in the five microphone positions and reducing the average to decibels. The measured levels are expressed as decibels relative to any convenient reference, e.g., the multiple of 10 next below the lowest, and the power ratios read from a table such as *Table 14.6*.

If the spread of the five SPLs does not exceed 5 dB, they may be averaged directly, giving results less than 1 dB in error.

Table 14.7 gives the central frequencies recommended for bands of noise used for test of sound insulation and other acoustic quantities:

Table 14.7

Octave Bands	*One-third Octave Bands*		
Hz		Hz	
63	50	63	80
125	100	125	160
250	200	250	315
500	400	500	630
1 000	800	1 000	1 250
2 000	1 600	2 000	2 500
4 000	3 150	4 000	5 000
8 000	6 300	8 000	10 000

The mean SRI of the partition is defined as its average value in one-third octave bands from 100 Hz to 3 150 Hz.

PRACTICAL DETAILS OF MEASUREMENT OF SOUND INSULATION IN FIELD AND LABORATORY BY THE METHOD OF BRITISH STANDARD 2750:1956

Any method may be used for the measurement of the band pressure levels in the two rooms. Readings for all frequency bands and microphone positions may be made in one room before passing to the other. A chart recorder may be used for giving a permanent record of the levels, the source and receiver microphones being switched in alternatively. If the dial of the oscillator providing the test signal is driven by gearing from the recorder and the filters are switched by the same mechanism, each filter position gives a pair of levels corresponding to the two rooms.

As five microphone positions are normally necessary in each room to give satisfactory accuracy, one must either go to the expense of a number of matched microphones of high quality or interrupt the measurements to change microphone positions in both rooms. An alternative method is to mount each microphone on a driven rotating arm so that it traces out a circular path along which the pressure level can be sampled. In a system built by the author for measurements on scale-model samples of partitions, the microphone is mounted in this way. The microphone voltage is amplified and fed to a squaring circuit, the output of which is proportional to the power. This is integrated over a circular path by voltage-to-frequency conversion and counting.

An important point is that, whichever method is used, checks must be made frequently in measuring the receiving room level to confirm that the wanted signal is sufficiently above the noise level, whether due to electrical or ambient noise, to give reliable measurements. For this purpose the loudspeakers are momentarily switched off and the measured level should fall by at least 6 dB.

For extensive routine measurements in the BBC, digital equipment developed by Moffat [2] has been used, the test signals being played from magnetic tapes carrying also trigger signals for actuating digital processing equipment. The outputs of microphones in the two rooms are then recorded on a second tape which is then processed to yield the sound level reduction.

For laboratory measurement of sound reduction index of partitions and walls, the sample must be fixed in an opening in a heavy wall separating two rooms constituting a so-called transmission suite in which every precaution has been taken to avoid transmission of sound by any path than that through the sample partition. Each room must be of adequate size to allow the establishment of a diffuse field and the edge constraints of the sample should be similar to those in its normal situation. It should be rectangular in shape with an area of 10 m^2.

MEASUREMENT OF IMPACT SOUND TRANSMISSION

The transmission of sound from impacts within a building cannot be measured in terms of a level reduction. Instead, a standard series of impacts are delivered to a point on the structure and the mean band energy at a distant point is computed from measurements of band pressure level. Most commonly, the blows are delivered to a floor immediately above the receiving room to test the effects of footsteps in producing sound in the room below. The impacts are produced by an *Impact Machine* or *footsteps machine* specified in British Standard 2750:1956, consisting of a number of hammer heads each of 500 g weight and mechanism for allowing the hammers to fall freely in sequence on to the floor at a rate of between five and ten per second. The band pressure levels in the receiving room are plotted against frequency as a spectrum which is compared with a standard form of spectrum or converted to a single loudness figure by one of the known methods. [3,4] A disadvantage of the tapping machine described above as a representation of footsteps is that the hammers are very much lighter than legs and consequently may give misleading comparisons between structures of widely differing mechanical form, e.g., between a carpet and a floating floor as means of reducing sound transmission. Cremer and Gilg [5] have found that an electromagnetic shaker gives more reliable results.

SEPARATION OF AIRBORNE FROM STRUCTURE BORNE TRANSMISSION

The measurement of airborne sound transmission by the British Standards method described above gives no information about the path or paths of transmission. These may include structural paths and parallel transmission paths through the partition between the rooms. Part of the energy may be transmitted directly through holes, cracks, thin doors, etc., which will be greatly inferior to the rest of the partition with respect to transmission loss. *Figure 14.82* shows some of the paths by which sound may travel from a source room to an adjacent one. There are several ways in which the indirect paths may be separated from the principal path for diagnostic purposes or to determine the true sound reduction index of a partition in the presence of other paths even, if the energy transmitted through them is the greater.

A non-permeable partition cannot transmit sound except by the bodily movement of its mass and the inherent sound reduction index of a partition can therefore be determined by measuring the velocity amplitude of the surface on the receiving room side with the pressure on the source room surface. The velocity amplitude may be derived from the acceleration amplitude of the surface measured by means of accelerometers attached to it.[6] The velocity amplitude is related to the acceleration amplitude by the equation $v = q/f$, where q is the acceleration and f is the frequency of the sound. Small accelerometers are available in which the sensitive element consists of a steel disc between two discs of piezoelectric material. The acceleration of the whole causes a force difference between the steel and the two piezoelectric discs and the resulting voltage is a measure of the acceleration. The equivalent near-field sound pressure level corresponding to an accelerometer voltage level V relative to 1 volt is given by

$$L = V + 207 - 20 \log_{10} kf \tag{2}$$

where f is the frequency and k is the sensitivity in millivolts/g acceleration.

This equation breaks down below the frequency at which the speed of bending waves in the surface is equal to that of sound in air.

The accelerometer is in other ways also a powerful tool for the tracing of sound and vibration through the structure of a building, and for predicting the near-field sound pressures to be expected in a building subject to vibration.

A second general method of distinguishing a number of paths of transmission is to make use of the different times of transmission. Raes[7] used short pulses of sound as test signals and displayed the output of a microphone in the receiving room on an oscilloscope. Several separate pulses could be seen on the display at successively increasing delay times, representing the several transmission times, and he was able to assign transmission loss figures to the most important paths.

The other important embodiment of this principle is to derive the correlation function between the outputs of microphones in the source and receiver rooms using a comparatively broad band (about an octave) of noise as the signal. A variable time delay is inserted into the source-room microphone circuit and as this is slowly increased the correlation function reached a maximum value when the delay is equal to the time taken by sound to travel from the source room microphone to the receiving room microphone by any one of the paths. There may be a succession of such maxima, their amplitudes being proportional to the sound pressure levels at the receiving room attributable to the corresponding paths. The transmission losses can therefore be calculated. Details of this method have been given by Goff[8] and Burd.[9]

Measurement of reverberation time

When a steady sound is radiated into an enclosure and then suddenly cut off, the sound pressure decays according to Franklin's equation[10]

$$P = P_o e^{-kt}, \tag{3}$$

where P_o is the initial pressure amplitude and k is a constant.

The reverberation time of the enclosure is defined as the time required for the sound pressure to fall to 1/1 000 of its initial value, i.e., for the level to fall by 60 dB. Thus equation 3 yields

$$L = L_o - 60t/T \qquad (4)$$

where L, L_o are the instantaneous and initial levels and T is the reverberation time with a value of $6.9/k$. This is a linear function of time and the term *decay curve* is usually taken to refer to this form, i.e., the curve of sound pressure level against time.

A straight decay curve will be obtained only in small rooms at the frequency of a strong isolated room-mode; in all other cases it will be modified by beats, rapid fluctuations or changes in general slope as the decay proceeds. Measurement of reverberation time consists of recording or displaying a decay curve and measuring the slope of the best-fit straight line.

METHODS OF MEASUREMENT

The block diagram of *Figure 14.83* shows the commonest method of measuring reverberation time. A test signal from an oscillator or noise generator is radiated by a

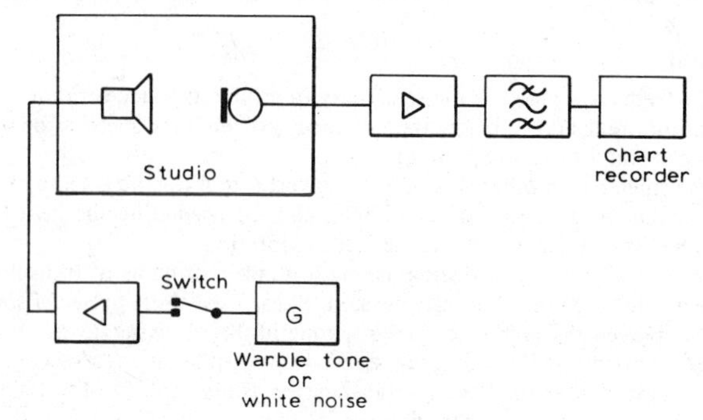

Figure 14.83 Block diagram of equipment for measurement of reverberation time

loudspeaker into the enclosure under test and received, together with the resulting reverberant sound, by a microphone. The output voltage of the microphone is amplified and fed to a logarithmic chart recorder through bandpass filters which remove noise and harmonics. The chart is started and the sound is cut off so that a decay curve is recorded. This process is repeated at a series of test frequencies, the standard frequencies being those listed in *Table 14.7* above. The series is repeated with the microphone in at least four other positions, and greater numbers in the case of large halls.

The slopes of the decay curves are measured or directly converted to reverberation times by means of a calibrated protractor, and averaged for all microphone positions at each frequency.

A more rapid method has been adopted by the BBC to meet the need for a fast direct-reading system for routine measurements in studios.[11] As shown in *Figure 14.84*, the simple switch in the loudspeaker circuit is replaced by an electronic switch which releases short bursts of sound at regular intervals. A triggering pulse is generated at the end of each burst and used to start the time base of the oscillograph. The amplified voltage from the microphone is converted to a voltage proportional to its logarithm which is displayed on the oscilloscope as the Y deflection. The reverberation time is read directly from a scale attached to a rotatable graticule, carrying a number of parallel lines which are aligned with the mean slope of the decay. A screen giving a certain degree of persistence is necessary.

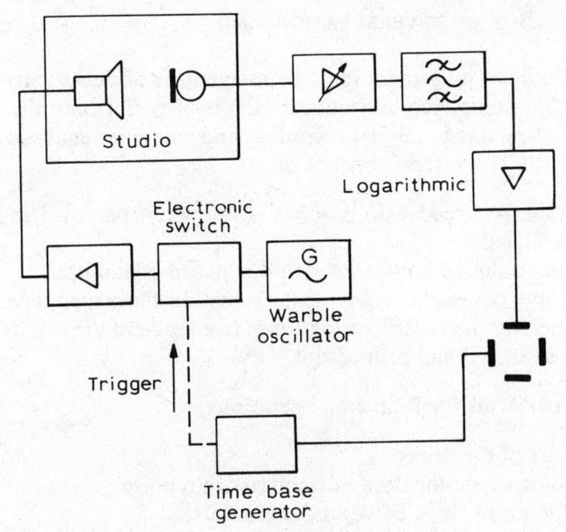

Figure 14.84 Block diagram of reverberation measuring equipment using oscilloscope

This equipment was also modified by the addition of a camera with a slow film drive to photograph the traces. If this was set into motion while the frequency of a sinusoidal test tone was slowly increased, a closely spaced succession of decay curves was produced showing characteristic patterns associated with isolated room modes and other features affecting speech quality.

SIGNALS FOR REVERBERATION MEASUREMENT

Owing to the presence of strong standing-wave systems in a room, the shapes and mean slopes of decay curves change rapidly with small variations of frequency. It is therefore essential to use a signal of finite bandwidth so that a number of room modes are simultaneously excited and a decay pattern representative of the frequency region is produced. Random noise may be used, bands of one octave or one-third of an octave being selected in the microphone circuit. A disadvantage is that full use is not made of the power handling capacity of the loadspeaker since only a small fraction of the total radiated power is selected for measurement. Frequency modulated tone may be used instead in which the oscillator frequency is varied at a rate of about 5–10 Hz and the depth of modulation is $\pm 5\%$.

Pistol shots or chords from an orchestra may be used to obtain tape recordings of wide-band decay curves for subsequent analysis through bandpass filters to obtain a graph of reverberation time against frequency. This method is frequently employed to measure the reverberation time of a concert hall in the presence of an audience; pistol shots should not be used if other methods are available as they are found by the author to give unreliable results in comparison with steady test signals.

It has been noted that the decay curves of an enclosure are seldom straight unmodulated lines and therefore some human judgement is required in assessing the best-fit slope. Moreover, background noise and the decreasing gradient often occurring at the end of a decay may influence judgement severely. By general agreement, it is now usual to quote the slope between levels 5 and 35 dB below the steady-state level reached before cut-off of the sound.

With either warble tone or bands of noise, successive repetitions of a decay vary noticeably, and a great advantage of the oscilloscope method is that the graticule can be set to mean slope of a large number of decays which follow one another on to the screen.

COMPUTER DERIVATION OF REVERBERATION TIME

The intrusion of human judgement into the assignment of decay curve slopes made it desirable to develop completely objective methods of performing the task. Equipment and methods for doing this by digital recording and computer analysis are described by Moffat and Spring.[12] The system consists of:

(1) Magnetic tape recorded with a series of bursts of narrow-band noise, together with trigger signals.
(2) An analogue-to-digital converter with a punched-tape output.
(3) Computer and peripheral equipment, accepting the paper tape data and a programme whereby the reverberation times from each of the twenty seven bands are calculated, averaged and printed out.

The programme performs the following operations:

(1) Find the start of the decay.
(2) Find the point where the decay disappears into noise.
(3) Compute the mean slope between these points.
(4) Print out a table of the reverberation times for the five microphone positions at every frequency, the average for each frequency and the standard deviation, and the level difference between the start and finish of the measured part of the decay.

SCHROEDER'S METHOD OF PROCESSING DECAY CURVES

It has been mentioned that a series of decays obtained under identical conditions when using finite-band noise such as bands of noise or warble tone, will all differ from each other because, although the successive test pulses possess a common spectrum, they represent different functions of time and hence excite the room differently. It is possible to produce identical time-functions having a predetermined effective bandwidth by the use of short trains of waves started and stopped at fixed points in their cycles.[13] With such decays, however, it was found that the variability of interpretation was greater than with more usual types of test signals with which a number of successive traces could be superimposed and the average slope estimated.

Schroeder[14] showed that by using a short tone-burst and integrating the energy in the reverberant signal from present time to infinity, a decay time was derived, representing the ensemble average of all possible decays resulting from a test signal with that particular spectrum. Curves thus obtained are free from the adventitious fluctuations which characterise ordinary decay curves.

In mathematical terms, one plots the function $\int_t^\infty p^2 dt$ where p is the instantaneous sound pressure and t the time. To obtain this integral, the decay curve can be recorded on magnetic tape which is then reversed and replayed into a squaring and integrating circuit. The output is converted to its logarithm and recorded by a chart recorder. The reverberation time is measured in the usual manner from the slope and divided by two since it represents integrated energy level instead of pressure level.

As an alternative to reversing a tape recording to obtain the integral, the first of two identical decays can be squared and integrated over its whole course and the second continuously integrated and subtracted from the whole integral before recording on the chart.

Thus,
$$\int_t^\infty p^2 dt = \int_o^\infty p^2 dt - \int_o^t p^2 dt$$

Instrumentation for this method of deriving reverberation time is available from at least one manufacturer.

REVERBERATION TIME FROM PHASE-FREQUENCY CHARACTERISTICS

Schroeder[15] has also shown that reverberation/frequency curves may be plotted by a steady-state method not requiring fast chart recorders or logarithmic amplifiers. A loudspeaker and microphone are set up at some distance from one another in the enclosure and pure tone is radiated from the loudspeaker. The microphone must be far enough away to be substantially in the reverberant field. The frequency of the tone is slowly glided upwards and the reversals of phase between the microphone signal and a similar microphone very close to the loudspeaker are logged against the frequency of the tone. This may be done by forming Lissajous figures on an oscilloscope with the outputs from the two microphones on the two sets of plates. From this the rate of change of phase with frequency, $d\phi/df$ is calculated, and the reverberation time may be shown to be

$$T = 2.2 \times d\phi/df \qquad (6)$$

Measurements of acoustic absorption coefficients

DEFINITIONS

The *absorption coefficient* of a material at a stated frequency is the proportion of sound energy incident on it which is absorbed or lost by transmission. This may be quoted for *normal* or *random incidence*. The *absorption* or *absorbing cross-section* of a finite area of a material is the product of its area and its mean effective absorption coefficient.

REVERBERATION METHOD OF MEASUREMENT

The principle of the reverberation method of measuring absorption coefficient, which yields the random-incidence coefficient, is to measure the reverberation time of a room at a series of frequencies as listed in *Table 14.7*, and then to repeat the measurements after fixing a suitable area of the sample material on to one or more surfaces of the room. The total absorption in the room is calculated from the reverberation time by the formula of Eyring[16] with and without the specimen in place. If $\bar{\alpha}$ is the mean absorption coefficient of all the room surfaces and S their area:

$$-\log_e(1-\bar{\alpha}) = 0.162\,V/TS \quad \text{(SI)} \qquad (7)$$

where V is the volume of the room, and Absorption $= S\alpha$

According to British Standard 3638:1963, the volume of the room should be between 180 and 250 m^3 and it should be surfaced with hard sound-reflecting surface finishes. Its reverberation time when empty should be at least 5 s up to 500 Hz, the lower limit diminishing by 1 s per octave above this frequency. The walls and ceiling should be heavy and rigid to avoid absorption at low frequencies by structural vibrations.

The sound should be radiated from a loudspeaker either near one corner of the room or one-third the way along a diagonal of the room, as these positions alone ensure a satisfactorily uniform excitation of all room modes.

The state of diffusion in the sound field should be enhanced by the use of sheets of sound-reflecting material hung from the ceiling. The sizes and orientations of the sheets should be distributed to obtain directional uniformity.

A single sample of 10 m^2 area is recommended for general tests of commercial materials, since this arrangement was found by Kosten[17] to result in the greatest measure of agreement between different testing laboratories. Divided samples, distributed on three or four surfaces of the room are to be recommended for tests in small areas to promote good diffusion. Subdivision increases the absorption coefficient of most materials especially at frequencies around 500 Hz. It should also be noted that subdivided distributed samples improve the diffusion of a reverberation room to such an extent that hanging sheets are unnecessary.[18] At least five measurements, from different positions of the microphone should be averaged, and it is an advantage to have two loudspeakers which are used alternately for each microphone position to increase the number of replications at each frequency. Bands of noise or warble tone are suitable as test signals.

It is usual not to make any correction for the loss of absorption of the test room surface when covered by the sample. This effect is kept small by the use of hard materials for the room surfaces, and attempts to make corrections may actually result in larger errors.

STANDING-WAVE TUBE MEASUREMENTS

Measurements of absorption coefficient by the reverberation method described above requires a large specially treated room, large test samples and facilities for the accurate measurement of reverberation time. A few establishments with a major interest in the development or use of sound absorbers are able to maintain these facilities, but most measurements of coefficients are necessarily carried out by two or three specialist consultant firms. The cost of this service makes it more suitable for checking production prototypes than as an aid for experimental product development. Useful information can, however, be obtained for the latter purpose by measurements of the normal-incidence absorption coefficient in a standing-wave tube.[19]

Figure 14.85 (*a*) shows the principle of this apparatus. The sample, a small disc of 30 to 100 mm diameter is cut from the material and held in a cap fitting tightly over one

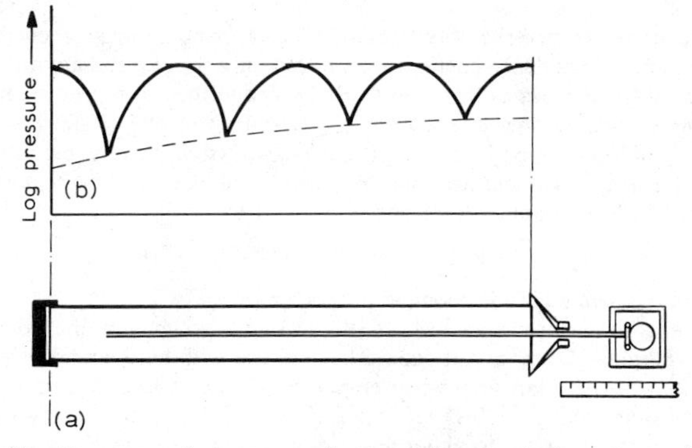

Figure 14.85 Standing wave tube apparatus for measuring sound absorption coefficient at normal incidence (a) diagram of equipment (b) variation of pressure along tube

end of a tube about 1 m long. The other end of the tube is closed by a loudspeaker diaphragm and the magnet of the loudspeaker is bored centrally to permit the insertion of a probe tube attached to a microphone. The probe tube is long enough to reach the surface of the sample and the microphone is mounted on a sliding or wheeled carriage so that the sound pressure may be measured at any point along the tube.

With the sample in place and pure tone radiated by the loudspeaker, the sound pressure along the tube shows a series of maxima and minima at intervals of a half wavelength as shown in *Figure 14.85* (*b*). The ratio of maximum to minimum pressures diminishes as distance from the sample increases owing to losses in the tube; if its value, extrapolated to the face of the sample, is n, the normal-incidence absorption coefficient is

$$\alpha_N = 4/(n + 1/n + 2) \tag{8}$$

The theory of this apparatus is given by Beranek.[20]

If the exact distance of the first minimum from the surface of the sample is measured, the real and imaginary parts of the acoustic impedance at the face of the sample can be calculated, and the random-incidence absorption coefficient derived from these parameters for many types of material.[21, 22] The calculation depends on the assumption that the impedance at the surface is independent of the angle of incidence. Although this is approximately true for a large proportion of absorbers, there is an element of doubt in the accuracy of the random-incidence coefficients arrived at in this way, and it is essential that final tests on a material to be used in acoustic treatment should be made by the reverberation method.

ACKNOWLEDGEMENT

Figures 14.82 to *14.85* inclusive are from *Acoustics for Radio and Television Studios*, by C. L. S. Gilford, Peter Peregrinus (1972).

REFERENCES

1. GILFORD, C. L. S. (for example), *Acoustics for Radio and Television Studies*, p. 52, Peter Peregrinus, London (1972)
2. MOFFAT, M. E. B., *B.B.C. Research Dept. Report PH 8* (1967)
3. ZWICKER, E., Acustica, **10**, 304 (1960)
4. STEVENS, S. S., *J. Acoust. Soc. Amer.*, **28**, 807 (1956)
5. CREMER, L. and GILG, J., *Acustica*, **23**, 54 (1970)
6. WARD, F. L., *Proc. 4th Internat. Congress on Acoustics*, Copenhagen, Paper L 11
7. RAES, A. C., *J. Acoust. Soc. Amer.*, **27**, 98 (1954)
8. GOFF, K. W., *J. Acoust. Soc. Amer.*, **27**, 233 (1955)
9. BURD, A. N., *J. Sound and Vib.*, **7**, 13 (1968)
10. FRANKLIN, W. S., *Phys Rev.*, **16**, 372 (1903)
11. SOMERVILLE, T. and GILFORD, C. L. S., *B.B.C. Q.*, **8**, 41 (1952)
12. MOFFAT, M. E. B. and SPRING, N. F., *B.B.C. Eng.*, 80 (1969)
13. GEDDES, W. K. E. and GILFORD, C. L. S., *B.B.C. Research Dept.* B 065 (1957)
14. SCHROEDER, M. R., *J. Acoust. Soc. Amer.*, **37**, 409 (1965)
15. SCHROEDER, M. R., *Proc. 3rd Internat. Congress on Acoustics*, Stuttgart, **2**, 897 (1959)
16. EYRING, C. F., *J. Acoust. Soc. Amer.*, **1**, 217 (1930)
17. KOSTEN, C. W., *Proc. 3rd Internat. Congress on Acoustics*, Stuttgart, **2**, 815 (1959)
18. GILFORD, C. L. S., *Acoustics for Radio and Television Studies*, p. 178, Peter Peregrinus, London (1972)
19. SCOTT, R. A., *Proc. Phys. Soc. London*, **58**, 253 (1946)
20. BERANEK, L. L., *J. Acoust. Soc. Amer.*, **12**, 3 (1940)
21. ATAL, B. S., *Acustica*, **9**, 27 (1959)
22. DUBOUT, P. and DAVERN, E., *Acustica*, **19**, 15 (1959)

15 TELECOMMUNICATIONS

TELECOMMUNICATIONS 15

15 TELECOMMUNICATIONS

NOISE AND COMMUNICATION THEORY

Interference and noise in communication systems

Information transmission accuracy can be seriously impaired by interference from other transmission systems and by noise. Interference from other transmission channels can usually be reduced to negligible proportions by proper channel allocation, by operating transmitters in adjacent or overlapping channels geographically far apart, and by the use of directive transmitting and receiving aerials. Noise may be impulsive or random. Impulsive noise may be man-made from electrical machinery or natural from electrical storms; the former is controllable and can be reduced to a low level by special precautions taken at the noise source, but the latter has to be accepted when it occurs. Random (or white) noise arises from the random movement of electrons due to temperature and other effects in current-carrying components in, or associated with the receiving system.

Man-made noise

Man-made electrical noise is caused by switching surges, electrical motor and thermostat operation, insulator flash-overs on power lines, etc. It is generally transmitted by the mains power lines and its effect can be reduced by:

(i) Suitable r.f. filtering at the noise source;
(ii) Siting the receiver aerial well away from mains lines and in a position giving maximum signal pick-up; and
(iii) Connecting the aerial to the receiver by a shielded lead.

The noise causes a crackle in phones or loudspeaker, or white or black spots on a monochrome television picture screen, and its spectral components decrease with frequency so that its effect is greatest at the lowest received frequencies.

Car ignition is another source of impulsive noise but it gives maximum interference in the v.h.f. and u.h.f. bands; a high degree of suppression is achieved by resistances in distributor and spark plug leads.

Natural sources of noise

Impulsive noise can also be caused by lightning discharges, and like man-made noise its effect decreases with increase of received frequency. Over the v.h.f. band such noise is only evident when the storm is within a mile or two of the receiving aerial.

Cosmic noise from outer space is quite different in character and generally occurs over relatively narrow bands of the frequency spectrum from about 20 MHz upwards. It is a valuable asset to the radio astronomer and does not at present pose a serious problem for the communications engineer.

Random noise

This type of noise is caused by random movement of electrons in passive elements such as resistors, conductors and inductors, and in active elements such as electronic valves and transistors.

Thermal noise

Random noise in passive elements is referred to as thermal noise since it is entirely associated with temperature, being directly proportional to absolute temperature. Unlike impulsive noise its energy is distributed evenly through the r.f. spectrum and it must be taken into account when planning any communication system. Thermal noise ultimately limits the maximum amplification that can usefully be employed, and so determines the minimum acceptable value of received signal. It produces a steady hiss in a loudspeaker and a shimmering background to a television picture.

Nyquist has shown that thermal noise in a conductor is equivalent to a r.m.s. voltage Vn in series with the conductor resistance R, where

$$Vn = (4kTR\Delta f)^{\frac{1}{2}} \tag{1}$$

k = Boltzmann's constant, $1.372 \times 10^{-23} J/K$
T = absolute temperature of conductor
Δf = pass band (Hz) of the circuits after R.

If the frequency response were rectangular the pass band would be the difference between the frequencies defining the sides of the rectangle. In practice the sides are sloping and bandwidth is

$$\Delta f = \frac{1}{E_o^2}\int_0^\infty [E(f)]^2 df \tag{2}$$

where E_o = midband or maximum value of the voltage ordinate and $E(f)$ = the voltage expression for the frequency response.

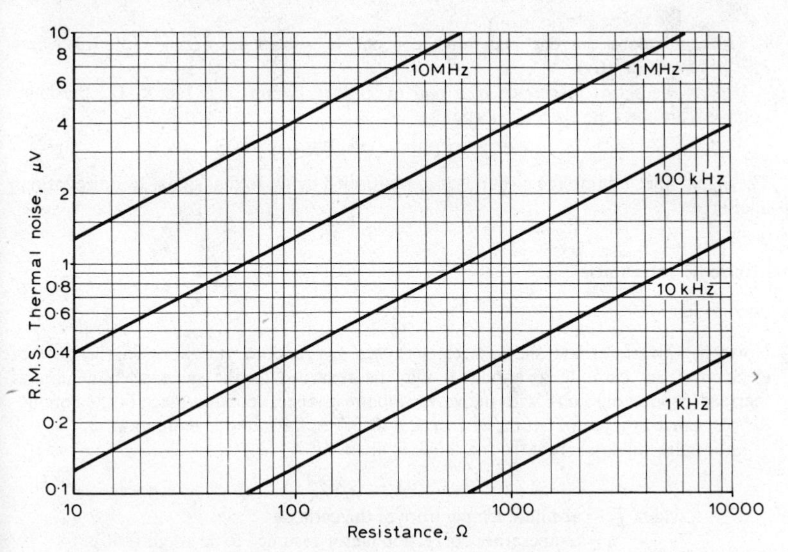

Figure 15.1 R.M.S. thermal noise (μV)—resistance at different bandwidths $T = 290$ K = 17 °C

A sufficient degree of accuracy is normally achieved by taking the standard definition of bandwidth, i.e., the frequency difference between points where the response has fallen by 3 dB.

Figure 15.1 allows the r.m.s. noise voltage for a given resistance and bandwidth to be determined. Thus for

$R = 10\,\text{k}\Omega$, $T = T_o = 17\,°\text{C}$ or 290 K, and $\Delta f = 10\,\text{kHz}$, $V \simeq 1.26\,\mu\text{V}$.

When two resistances in series are at different temperatures

$$V_n = [4k\Delta f(R_1T_1 + R_2T_2)]^{\frac{1}{2}}.$$

Two resistances in parallel at the same temperature *Figure 15.2* (*a*) are equivalent to a noise voltage

$$V_n = [4kT\Delta f \,.\, R_1R_2/(R_1 + R_2)]^{\frac{1}{2}}$$

in series with two resistances in parallel, *Figure 15.2* (*b*).

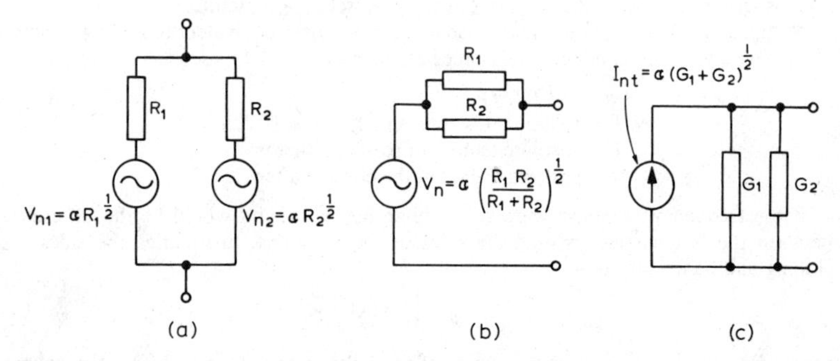

Figure 15.2 (a) Noise voltages of two resistances in parallel (b) an equivalent circuit and (c) a current noise generator equivalent $\alpha = [4kT\Delta f]^{\frac{1}{2}}$

The equivalent current generator concept is shown in *Figure 15.2* (*c*) where $I_n = [4kT\Delta f(G_1 + G_2)]^{\frac{1}{2}}$.

If R is the series resistance of a coil in a tuned circuit of Q factor, Q_o, the noise voltage from the tuned circuit becomes

$$V_{no} = V_nQ_o = Q_o(4kTR\Delta f)^{\frac{1}{2}}$$

The signal injected into the circuit is also multiplied by Q_o so that signal-to-noise ratio is unaffected.

Electronic valve noise

SHOT NOISE

Noise in valves, termed shot noise, is caused by random variations in the flow of electrons from cathode to anode. It may be regarded as the same phenomenon as thermal (conductor) noise with the valve slope resistance, acting in place of the conductor resistance at a temperature between 0.5 to 0.7 of the cathode temperature.

Shot noise r.m.s. current from a diode is given by

$$I_n = (4kaT_kg_d\Delta f)^{\frac{1}{2}} \tag{3}$$

where T_k = absolute temperature of the cathode

a = temperature correction factor assumed to be about 0.66

$g_d = \dfrac{dI_d}{dV_a}$, slope conductance of the diode.

Experiment[1] has shown noise in a triode is obtained by replacing g_d in equation 3 by g_m/β where β has a value between 0.5 and 1 with a typical value of 0.85, thus

$$I_{na} = (4kaT_k g_m \Delta f/\beta)^{\frac{1}{2}} \tag{4}$$

Since $I_a = g_m V_g$, the noise current can be converted to a noise voltage at the grid of the valve of

$$V_{ng} = I_{na}/g_m = (4kaT_k \Delta f/\beta g_m)^{\frac{1}{2}}$$
$$= [4kT_o \Delta f . aT_k/\beta g_m T_o]^{\frac{1}{2}} \tag{5}$$

Where T_o is the normal ambient (room) temperature,
The part $aT_k/\beta g_m T_o$ of expression 5 above is equivalent to a resistance, which approximates to

$$R_{ng} = 2.5/g_m \tag{6}$$

and this is the equivalent noise resistance in the grid of the triode at room temperature. The factor 2.5 in R_{ng} may have a range from 2 to 3 in particular cases. The equivalent noise circuit for a triode having a grid leak R_g and fed from a generator of internal resistance R_1 is as in *Figure 15.3*.

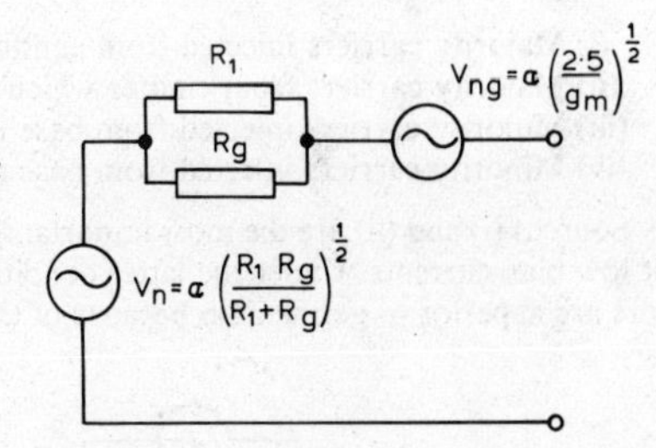

Figure 15.3 Noise voltage input circuit for a valve $a = |4kT\Delta f|^{\frac{1}{2}}$

PARTITION NOISE

A multielectrode valve such as a tetrode produces greater noise than a triode due to the division of electron current between screen and anode; for this reason the additional noise is known as partition noise. The equivalent noise resistance in the grid circuit becomes.

$$R_{ng}(tet) = (I_a/I_k)(20I_s/g_m^2 + 2.5/g_m) \tag{7}$$

Where I_a, I_k and I_s are the d.c. anode, cathode and screen currents respectively. I_s should be small and g_m large for low noise in tetrode or multielectrode valves. The factor $20I_s/g_m^2$ is normally between 3 to 6 times $2.5/g_m$ so that a tetrode valve is much noisier than a triode.

At frequencies greater than about 30 MHz, the transit time of the electron from cathode to anode becomes significant and this reduces gain and increases noise. Signal-to-noise ratio therefore deteriorates. Partition noise in multielectrode valves also increases and the neutralised triode, or triodes in cascode give much better signal-to-noise ratios at high frequencies.

At much higher frequencies (above 1 GHz) the velocity-modulated electron tube, such as the klystron and travelling wave tube, replace the normal electron valve. In the klystron, shot noise is present but there is also chromatic noise due to random variations in the velocities of the individual electrons.

FLICKER NOISE

At very low frequencies valve noise is greater than would be expected from thermal considerations. Schottky suggested that this is due to random variations in the state of the cathode surface and termed it *flicker*. Flicker noise tends to be inversely proportional to frequency below about 1 kHz so that the equivalent noise resistance at 10 Hz might be 100 times greater than the shot noise at 1 kHz. Ageing of the valve tends to increase flicker noise and this appears to be due to formation of a high resistance barium silicate layer between nickel cathode and oxide coating.

Transistor noise

Transistor noise exhibits characteristics very similar to those of valves, with noise increasing at both ends of the frequency scale. Resistance noise is also present due to the extrinsic resistance of the material and the major contributor is the base extrinsic resistance r_b'. Its value is given by expression 1, T being the absolute temperature of the transistor under working conditions.

Shot and partition noise arise from random fluctuations in the movement of minority and majority carriers, and there are four sources, viz.

(i) Majority carriers injected from emitter to base and thence to collector.
(ii) Majority carriers from emitter which recombine in the base.
(iii) Minority carriers injected from base into emitter.
(iv) Minority carriers injected from base into the collector.

Sources (i) and (ii) are the most important, sources (iii) and (iv) being significant only at low bias currents. Under the latter condition which gives least noise, silicon transistors are superior to germanium because of their much lower values of I_{co}.

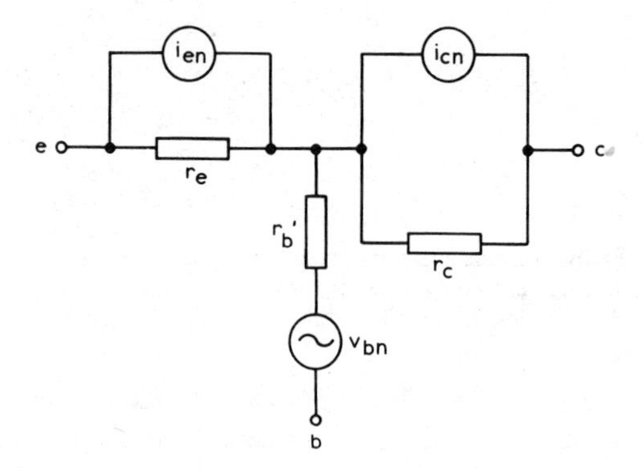

Figure 15.4 Noise circuit equivalent for a transistor

A simplified equivalent circuit for the noise currents and voltages in a transistor is that of *Figure 15.4* where

$i_{en} = (2eI_e\Delta f)^{\frac{1}{2}}$ the shot noise current in the emitter
$i_{cn} = [2e(I_{co} + I_c(1 - \alpha_o))\Delta f]^{\frac{1}{2}}$, the shot and partition noise current in the collector
$v_{bn} = (4kT \,.\, r_b'\Delta f)^{\frac{1}{2}}$, the thermal noise due to the base extrinsic resistance
e = electronic charge = 1.602×10^{-19} coulomb.

Since transistors are power amplifying devices the equivalent noise resistance concept is less useful and noise quality is defined in terms of noise figure.

Flicker noise, which is important at low frequencies (less than about 1 kHz), is believed to be due to carrier generation and recombination at the base–emitter surface. Above 1 kHz noise remains constant until a frequency of about $f_\alpha(1-\alpha_o)^{\frac{1}{2}}$ is reached, where f_α is the frequency at which the collector–emitter current gain has fallen to $0.7\alpha_o$. Above this frequency, which is about $0.15 f_\alpha$, partition noise increases rapidly.

Noise figure

Noise figure (F) is defined as the ratio of the input signal-to-noise available power ratio to the output signal-to-noise available power ratio, where available power is the maximum power which can be developed from a power source of voltage V and internal resistance R_s. This occurs for matched conditions and is $V^2/4R_s$.

$$F = \frac{P_{si}}{P_{ni}} \bigg/ \frac{P_{so}}{P_{no}}$$

$$= P_{no}/G_a P_{ni} \qquad (8)$$

where G_a = available power gain of the amplifier.

Since noise available output power is the sum of $G_a P_{ni}$ and that contributed by the amplifier P_{na}

$$F = 1 + \frac{P_{na}}{G_a P_{ni}} \qquad (9)$$

The available thermal input power is $V^2/4R_s$ or $kT\Delta f$, which is independent of R_s, hence

$$F = 1 + P_{na}/G_a kT\Delta f \qquad (10\text{ a})$$

and

$$F(\text{dB}) = 10 \log_{10}(1 + P_{na}/G_a kT\Delta f) \qquad (10\text{ b})$$

The noise figure for an amplifier whose only source of noise is its input resistance R_1 is

$$F = 1 + R_s/R_1$$

because the available output noise is reduced by $R_1/(R_s + R_1)$ but the available signal gain is reduced by $|R_1/(R_s + R_1)|^2$. For matched conditions $F = 2$ or 3 dB and maximum signal-to-noise ratio occurs when $R_1 = \infty$. Signal-to-noise ratio is unchanged if R_1 is noiseless because available noise power is then reduced by the same amount as the available gain.

If the above amplifier has a valve, whose equivalent input noise resistance is R_{ng},

$$F = 1 + \frac{R_s}{R_1} + \frac{R_{ng}}{R_s}\left[1 + \frac{R_s}{R_1}\right]^2 \qquad (11)$$

Noise figure for a transistor over the range of frequencies for which it is constant is

$$F = 1 + \frac{r'_b + 0.5r_e}{R_s} + \frac{(r'_b + r_e + R_s)^2(1-\alpha_o)}{2r_e R_s \alpha_o} \qquad (12)$$

At frequencies greater than $f_\alpha(1-\alpha_o)^{\frac{1}{2}}$, the last term is multiplied by $[1 + (f/f_\alpha)^2/(1-\alpha_o)]$. The frequency f_T at which collector–base current gain is unity is generally given by the transistor manufacturer and it may be noted that $f_T \simeq f_\alpha$, the frequency at which collector–emitter current gain is $0.7\alpha_o$.

Expression 12 shows that transistor noise figure is dependent on R_s but it is also affected by I_c through r_e and α_o. As a general rule the lower the value of I_c the lower is noise figure and the greater is the optimum value of R_s. This is shown in *Figure 15.5* which is typical of a r.f. silicon transistor. Flicker noise causes the increase below 1 kHz, and decrease of gain and increase of partition noise causes increased noise factor at the high frequency end. The high frequency at which F begins to increase is about $0.15 f_\alpha$; at low values of collector current f_α falls, being approximately proportional to

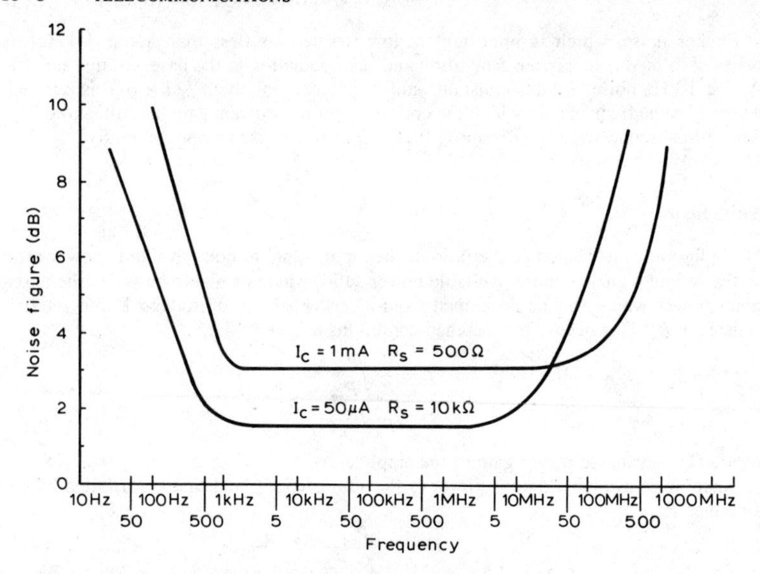

Figure 15.5 Typical noise figure—frequency curves for a r.f. transistor

I_c^{-1}. The type of configuration, common emitter, base or collector has little effect on noise figure.

Transistors do not provide satisfactory noise figures above about 1.5 GHz, but the travelling-wave tube and tunnel diode can achieve noise figures of 3 to 6 dB over the range 1 to 10 GHz.

Sometimes noise temperature is quoted in preference to noise figure and the relationship is

$$F = (1 + T/T_o) \tag{13}$$

T is the temperature to which the noise source resistance would have to be raised to produce the same available noise output power as the amplifier. Thus if $T = T_o = 290$ K, $F = 2$ or 3 dB.

The overall noise figure of cascaded amplifiers can easily be calculated and is

$$F_t = F_1 + \frac{F_2 - 1}{G_1} + \frac{F_3 - 1}{G_1 G_2} + \ldots \frac{F_n - 1}{G_1 \ldots G_{n-1}} \tag{14}$$

where F_1, $F_2 \ldots F_n$ and G_1, $G_2 \ldots G_n$ are respectively the noise figures and available gains of the separate stages from input to output. From equation 14 it can be seen that the first stage of an amplifier system largely determines the overall signal-to-noise ratio, and that when a choice has to be made between two first-stage amplifiers having the same noise figure, the amplifier having the highest gain should be selected because increase of G_1 reduces the noise effect of subsequent stages.

Measurement of noise

Noise measurement requires a calibrated noise generator to provide a controllable noise input to an amplifier or receiver, and a r.m.s. meter to measure the noise output of the amplifier or receiver. The noise generator generally consists of a temperature-limited (tungsten filament) diode, terminated by a resistance R as shown in *Figure 15.6*. The diode has sufficient anode voltage to ensure that it operates under saturation conditions and anode saturation current is varied by control of the diode filament current. A milliammeter reads the anode current I_d and the shot noise current component of this is given by $(2I_d e\Delta f)^{\frac{1}{2}}$ where e is electronic charge, 1.602×10^{-19} coulomb. The shot noise has the same flat spectrum as the thermal noise in R, and the meter is calibrated in dB with reference to noise power in R and so provides a direct reading of noise factor. R is generally selected to be 75 Ω, the normal input impedance of a receiver.

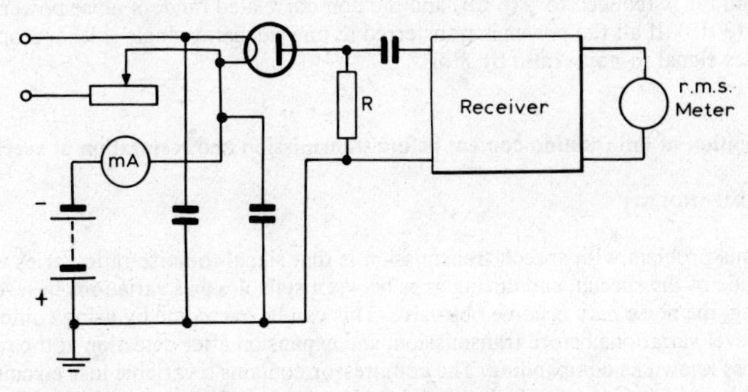

Figure 15.6 Noise figure measurement

When measuring, the diode filament current is first switched off, and the reading of the r.m.s. meter in the receiver output noted. The diode filament is switched on and adjusted to increase the r.m.s. output reading 1.414 times (double noise power). The dB reading on the diode anode current meter is the noise figure, since

$$\text{Noise output power diode off} = GP_{nR} + P_{na}$$
$$\text{Noise output power diode on} = G(P_{nR} + P_{nd}) + P_{na} = 2GP_{nd}$$
$$\text{Noise figure} = 10\log_{10}[(GP_{nR} + P_{na})/GP_{nR}] = 10\log_{10} P_{nd}/P_{nR}$$

The diode is satisfactory up to about 600 MHz but above this value transit time of electrons begins to cause error. For measurements above 1 GHz a gas discharge tube has to be used as a noise source.

Methods of improving signal-to-noise ratio

There are five methods of improving signal-to-noise ratio, viz.,

(i) Increase the transmitted power of the signal.
(ii) Redistribute the transmitted power.
(iii) Modify the information content before transmission and return it to normal at the receiving point.
(iv) Reduce the effectiveness of the noise interference with signal.
(v) Reduce the noise power.

INCREASE OF TRANSMITTED POWER

An overall increase in transmitted power is costly and could lead to greater interference for users of adjacent channels.

REDISTRIBUTION OF TRANSMITTED POWER

With amplitude modulation it is possible to redistribute the power among the transmitted components so as to increase the effective signal power. As described under 'Modulation Theory and Systems' suppression of the carrier in a double sideband amplitude modulation signal and a commensurate increase in sideband power increases the effective signal power, and therefore signal-to-noise ratio by 4.75 dB (3 times) for the same average power or by 12 dB for the same peak envelope power. Single sideband operation by removal of one sideband reduces signal-to-noise ratio by 3 dB because signal power is reduced to $\frac{1}{4}$ (6 dB) and the non-correlated random noise power is only halved (3 dB). If all the power is transferred to one sideband, single sideband operation increases signal-to-noise ratio by 3 dB.

Modification of information content before transmission and restoration at receiver

THE COMPANDER

A serious problem with speech transmission is that signal-to-noise ratio varies with the amplitude of the speech, and during gaps between syllables and variations in level when speaking, the noise may become obtrusive. This can be overcome by using compression of the level variations before transmission, and expansion after detection at the receiver, a process known as companding. The compressor contains a variable loss circuit which reduces amplification as speech amplitude increases and the expander performs the reverse operation.

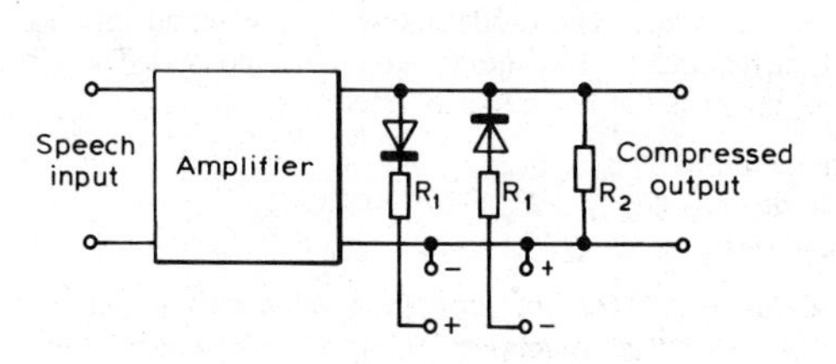

Figure 15.7 A compressor circuit

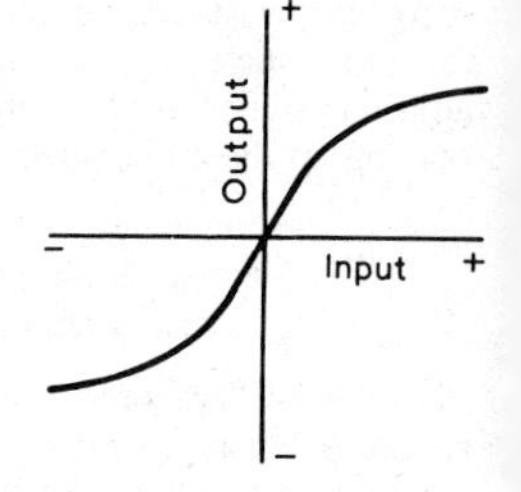

Figure 15.8 Compressor input–output characteristic

A typical block schematic for a compander circuit is shown in *Figure 15.7*. The input speech signal is passed to an amplifier across whose output are shunted two reverse-biased diodes; one becoming conductive and reducing the amplification for positive-going signals and the other doing the same for negative-going signals. The input–output characteristic is *S* shaped as shown in *Figure 15.8*; the diodes should be selected for near-identical shunting characteristics. Series resistances R_1 are included to control the turn-over, and shunt resistance R_2 determines the maximum slope near zero.

A similar circuit is used in the expander after detection but as shown, *Figure 15.9*, the diodes form a series arm of a potential divider and the expanded output appears across R_3. The expander characteristic, *Figure 15.10*, has low amplification in the gaps between speech, and amplification increases with increase in speech amplitude.

The diodes have a logarithmic compression characteristic, and with large compression the dB input against dB output tends to a line of low slope, e.g., an input variation of 20 dB being compressed to an output variation of 5 dB. If greater compression is required two compressors are used in tandem.

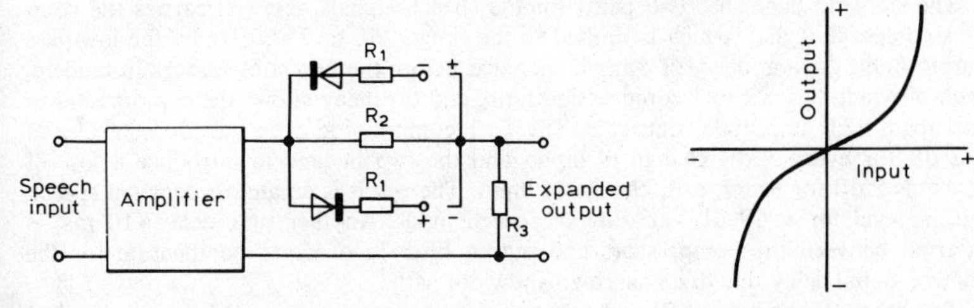

Figure 15.9 An expander circuit

Figure 15.10 Expander input–output characteristic

The collector–emitter resistance of a transistor may be used in place of the diode resistance as the variable gain device. The collector–emitter resistance is varied by base-emitter bias current, which is derived by rectification of the speech signal from a separate auxiliary amplifier. A time delay is inserted in the main controlled channel so that high-amplitude speech transients can be anticipated.

LINCOMPEX

The compander system described above proves quite satisfactory provided the propagation loss is constant as it is with a line or coaxial cable. It is quite unsuitable for a short-wave point-to-point communication system via the ionosphere. A method known as Lincompex[2] (linked compression expansion) has been successfully developed by the British Post Office. *Figure 15.11* is a block diagram of the transmit–receive paths. The simple form of diode compressor and expander cannot be used and must be replaced by the transistor type, controlled by a current derived from rectification of the speech signal. The current controls the compression directly at the transmitting end and this information must be sent to the receiver by a channel unaffected by any propagational variations. This is done by confining it in a narrow channel (approximately 180 Hz wide) and using it to frequency-modulate a sub-carrier at 2.9 kHz. A limiter at the receiver removes all amplitude variations introduced by the r.f. propagation path, and a frequency discriminator extracts the original control information.

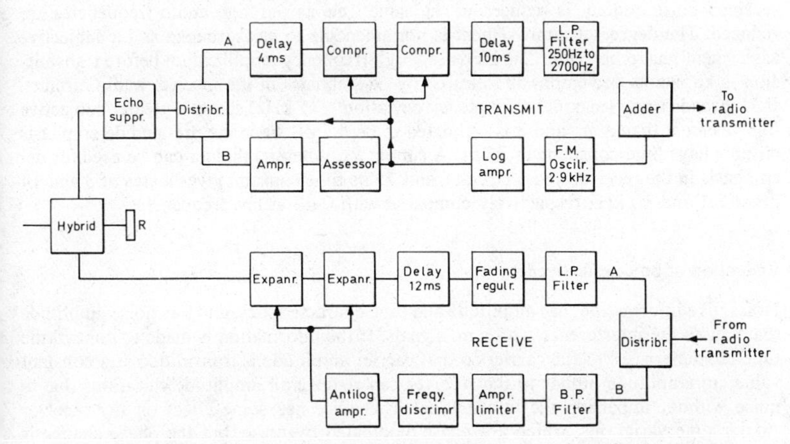

Figure 15.11 Block schematic of Lincompex Compander system for radio transmissions

The transmit chain has two paths for the speech signals, one (A) carries the compressed speech signal, which is limited to the range 250 to 2 700 Hz by the low-pass output filter. A time delay of 4 ms is included before the two compressors in tandem, each of which has a 2 to 1 compression ratio, and the delay allows the compressors to anticipate high amplitude transients. The 2 : 1 compression ratio introduces a loss of $x/2$ dB for every x dB change in input, and the two in tandem introduce a loss of $2(x/2) = x$ dB for every x dB change of input. The result is an almost constant speech output level for a 60 dB variation of speech input. Another time delay (10 ms) is inserted between the compressors and output filter in order to compensate for the control signal delay due to its narrow bandwidth path.

The other transmit path (B) contains an amplitude-assessor circuit having a rectified d.c. output current proportional to the speech level. This d.c. current controls the compressors, and after passing through a logarithmic amplifier is used to frequency modulate the sub-carrier to produce the control signal having a frequency deviation of 2 Hz/dB speech level change. The time constant of the d.c. control voltage is 19 ms permitting compressor loss to be varied at almost syllabic rate, and the bandwidth of the frequency-modulated sub-carrier to be kept within ±90 Hz. The control signal is added to the compressed speech and the combined signal modulates the transmitter.

The receive chain also has two paths; path (A) filters the compressed speech from the control signal and passes the speech to the expanders via a fading regulator, which removes any speech fading not eliminated by the receiver a.g.c., and a time delay, which compensates for the increased delay due to the narrow-band control path (B). The latter has a band-pass filter to remove the compressed speech from the control signal and an amplitude limiter to remove propagational amplitude variations. The control signal passes to a frequency discriminator and thence to an antilog amplifier, the output from which controls the gain of the expansion circuits. The time constant of the expansion control is between 18 ms and 20 ms.

PRE-EMPHASIS AND DE-EMPHASIS

Audio energy in speech and music broadcasting tends to be greatest at the low frequencies. A more level distribution of energy is achieved if the higher audio frequencies are given greater amplification than the lower before transmission. The receiver circuits must be given a reverse amplification-frequency response to restore the original energy distribution, and this can lead to an improved signal-to-noise ratio since the received noise content is reduced at the same time as the high audio frequencies are reduced. The degree of improvement is not amenable to measurement and a subjective assessment has to be made. The increased high-frequency amplification before transmission is known as pre-emphasis followed by de-emphasis in the receiver audio circuits. F.M. broadcasting (maximum frequency deviation ±75 kHz) shows a greater subjective improvement than a.m., and it is estimated to be 4.5 dB when the pre- and de-emphasis circuits have time-constants of 75 μs. A simple *RC* potential divider can be used for de-emphasis in the receiver audio circuits, and 75 μs time constant gives losses of 3 and 14 dB at 2.1 and 10 kHz respectively compared with 0 dB at low frequencies.

Reduction of noise effectiveness

Noise, like information, has amplitude and time characteristics, and it is noise amplitude that causes the interference with a.m. signals. If the information is made to control the time characteristics of the carrier so that carrier amplitude is transmitted at a constant value, an amplitude limiter in the receiver can remove all amplitude variations due to noise without impairing the information. The noise has some effect on the receiver carrier time variations, which are phase-modulated by noise, but the phase change is very much less than the amplitude change so that signal-to-noise ratio is increased.

FREQUENCY MODULATION

If the information amplitude is used to modulate the carrier frequency, and an amplitude limiter is employed at the receiver, the detected message-to-noise ratio is greatly improved. As shown under 'Modulation Theory and Systems', f.m. produces many pairs of sidebands per modulating frequency especially at low frequencies, and this 'bass boost' is corrected at the receiver detector to cause a 'bass cut' of the low frequency noise components. This triangulation of noise leads to 4.75 dB signal-to-noise betterment. Phase modulation does not give this improvement because the pairs of sidebands are independent of modulating frequency. The standard deviation of 75 kHz raises signal-to-noise ratio by another 14 dB, and pre-emphasis and de-emphasis by 4.5 dB, bringing the total improvement to 23.25 dB over a.m.

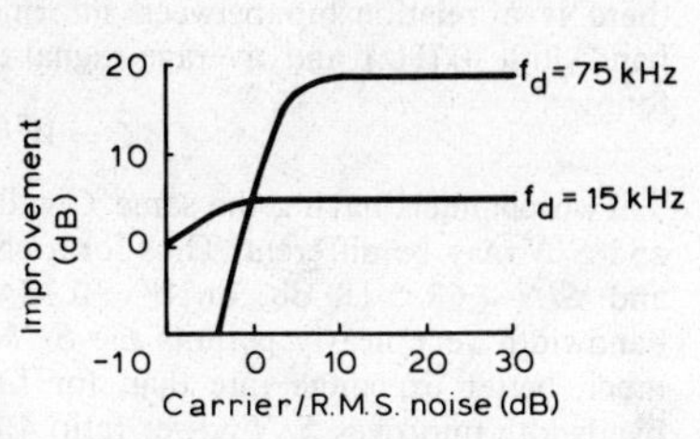

Figure 15.12 Threshold noise effect with f.m. compared with a.m.

The increased signal-to-noise performance of the f.m. receiver is dependent on having sufficient input signal to operate the amplitude limiter satisfactorily. Below a given input signal-to-noise ratio output information-to-noise ratio is worse than for a.m. The threshold value increases with increase of frequency deviation because the increased receiver bandwidth brings in more noise as indicated in *Figure 15.12*.

PULSE MODULATION

Pulse modulated systems using change of pulse position (p.p.m.) and change of duration (p.d.m.) can also increase signal-to-noise ratio but pulse amplitude modulation (p.a.m.) is no better than normal a.m. because an amplitude limiter cannot be used.

IMPULSE NOISE AND BANDWIDTH

When an impulse noise occurs at the input of a narrow bandwidth receiver the result is a damped oscillation at the mid-frequency of the pass band as shown in *Figure 15.13 (a)*.

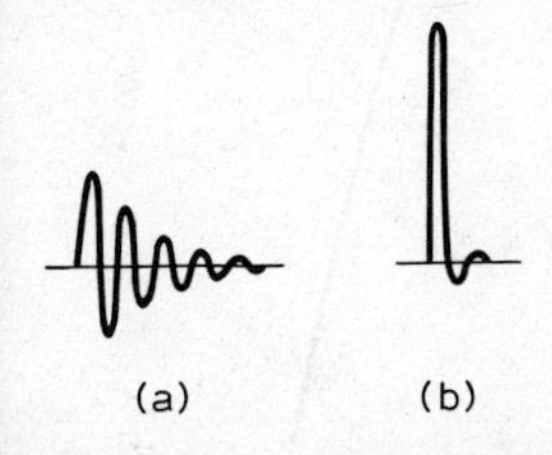

Figure 15.13 Output wave-shape due to an impulse in a (a) narrow band amplifier (b) wideband amplifier

When a wide bandwidth is employed the result is a large initial amplitude with a very rapid decay, *Figure 15.13 (b)*. An amplitude limiter is much more effective in suppressing the large amplitude near-single pulse than the long train of lower amplitude oscillations. Increasing reception bandwidth can therefore appreciably reduce interference due to impulsive noise provided that an amplitude limiter can be used.

PULSE CODE MODULATION

A very considerable improvement in information-to-noise ratio can be achieved by employing pulse code modulation[3] (p.c.m.). P.C.M. converts the information amplitude into a digital form by sampling and employing constant amplitude pulses, whose presence or absence in a given time order represents the amplitude level as a binary number. Over long cable or microwave links it is possible to amplify the digital pulses when signal-to-noise ratio is very low, to regenerate and pass on a freshly constituted signal almost free of noise to the next link. With analogue or direct non-coded modulation such as a.m. and f.m., noise tends to be cumulative from link to link. The high signal-to-noise ratio of p.c.m. is obtained at the expense of much increased bandwidth, and Shannon has shown that with an ideal system of coding giving zero detection error there is a relationship between information capacity C (binary digits or bits/sec), bandwidth W(Hz) and average signal-to-noise thermal noise power ratio (S/N) as follows:

$$C = W \log_2 (1 + S/N) \tag{15}$$

Two channels having the same C will transmit information equally well though W, and S/N may be different. Thus for a channel capacity of 10^6 bits/s, $W = 0.167$ MHz and $S/N = 63 \equiv 18$ db, or $W = 0.334$ MHz and $S/N = 7 \equiv 8.5$ dB. Doubling of bandwidth very nearly permits the S/N dB value to be halved, and this is normally a much better exchange rate than for f.m. analogue modulation, for which doubling bandwidth improves S/N power ratio 4 times or by 6 dB.

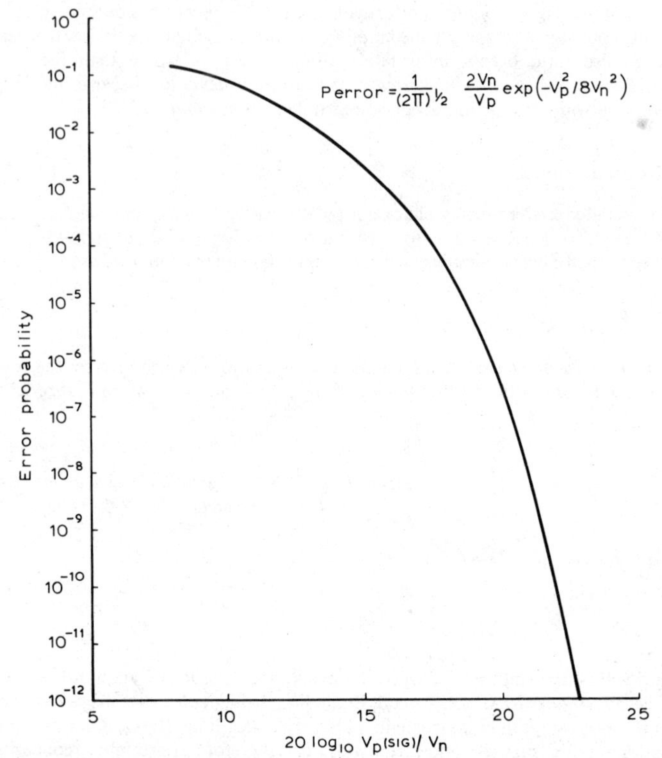

Figure 15.14 Probability of error at different V_p(sig)/r.m.s. noise ratios

In any practical system the probability of error is finite, and a probability of 10^{-6} (1 error in 10^6 bits) causes negligible impairment of information. Assuming that the detector registers a pulse when the incoming amplitude exceeds one half the normal pulse amplitude, an error will occur when the noise amplitude exceeds this value. The probability of an error occurring due to this is

$$Pe = \frac{1}{(2\pi)^{\frac{1}{2}}} \frac{2Vn}{V_p} \exp\left(-V_p^2/8Vn^2\right) \tag{16}$$

where V_p = peak voltage of the pulse
and V_n = r.m.s. voltage of the noise.
The curve is plotted in *Figure 15.14*.

An error probability of 10^{-6} requires a V_p/V_n of approximately 20 dB, or since $V_{av} = V_p/2$ a V_{av}/V_n of 17 dB. In a binary system 2 pulses can be transmitted per cycle of bandwidth so that by Shannon's ideal system

$$\frac{C}{W} = 2 = \log_2(1 + S/N), \quad \text{and} \quad S/N = 3 \triangleq 5 \text{ dB}$$

Hence the practical system requires 12 dB greater S/N ratio than the ideal, but the output message-to-noise ratio is infinite, i.e., noise introduced in the transmission path and the receiver is completely removed. There will, however, be a form of noise present with the output message due to the necessary sampling process at the transmit end. Conversion of amplitude level to a digital number must be carried out at a constant level, and the reconstructed decoded signal at the receiver is not a smooth wave but a series of steps as indicated under 'Modulation Theory and Systems'. These quantum level steps superimpose on the original signal a disturbance having a uniform frequency spectrum similar to thermal noise. It is this quantising noise which determines the output message-to-noise ratio and it is made small by decreasing the quantum level steps. The maximum error is half the quantum step, l, and the r.m.s. error introduced is $l/2(3)^{\frac{1}{2}}$. The number of levels present in the p.a.m. wave after sampling are 2^n where n is the number of binary digits. The message peak-to-peak amplitude is $2^n l$, so that the

$$\text{Message (pk-to-pk)/r.m.s. noise} = \frac{2^n l}{l/2(3)^{\frac{1}{2}}} = 2(3)^{\frac{1}{2}} 2^n.$$

$$\begin{aligned} M/N\,(\text{dB}) &= 20 \log_{10} 2(3)^{\frac{1}{2}} 2^n \\ &= 20n \log_{10} 2 + 20 \log_{10} 2(3)^{\frac{1}{2}} \\ &= (6n + 10.8)\ \text{dB}. \end{aligned} \tag{17}$$

Increase of digits (n) means an increased message-to-noise ratio but also increased bandwidth and therefore increased transmission path and receiver noise; care must be exercised to ensure that quantising noise remains the limiting factor.

Expression 17 represents the message-to-noise ratio for maximum information amplitude, and smaller amplitudes will give an inferior noise result. A companding system should therefore be provided before sampling of the information takes place.

Reduction of noise

Since thermal noise power is proportional to bandwidth, the latter should be restricted to that necessary for the objective in view. Thus the bandwidth of an a.m. receiver should not be greater than twice the maximum modulating frequency for d.s.b. signals, or half this value for s.s.b. operation.

In a f.m. system information power content is proportional to (bandwidth)^2 so that increased bandwidth improves signal-to-noise ratio even though r.m.s. noise is increased. When, however, carrier and noise voltages approach in value, signal-to-noise ratio is worse with the wider band f.m. transmission (threshold effect).

Noise is reduced by appropriate coupling between signal source and receiver input and by adjusting the operating conditions of the first stage transistor for minimum noise figure.

Noise is also reduced by refrigerating the input stage of a receiver with liquid Helium and this method is used for satellite communication in earth station receivers using masers. The maser amplifies by virtue of a negative resistance characteristic and its noise contribution is equivalent to the thermal noise generated in a resistance of equal value. The noise temperature of the maser itself may be as low as 2 K to 10 K and that of the other parts of the input equipment 15 K to 30 K.

Parametric amplification, by which gain is achieved by periodic variation of a tuning parameter (usually capacitance), can provide the relatively low noise figures 1.5 to 6 dB over the range 5 to 25 GHz. Energy at the 'pump' frequency (f_p) operating the variable reactance, usually a varactor diode, is transferred to the signal frequency (f_s) in the parametric amplifier or to an idler frequency ($f_p \pm f_s$) in the parametric converter. It is the resistance component of the varactor diode that mainly determines the noise figure of the system. Refrigeration is also of value with parametric amplification.

REFERENCES

1. NORTH, D. O., 'Fluctuations in Space-Charge-Limited Currents at Moderately High Frequencies', *RCA. Rev.*, **4**, 441 (1940), **5**, 106, 244 (1940)
2. WATT-CARTER, D. E. and WHEELER, L. K., 'The Lincompex System for the Protection of HF Radio Telephone Circuits', *P.O. Elect. Engrs. Jour.*, **59**, 163 (1966)
3. BELL SYSTEM LABORATORIES, *Transmission Systems for Communications* (1964)

MODULATION THEORY AND SYSTEMS

The transmission of information, whether speech, music, vision or data, over long distances requires the use of a carrier channel at least equal to that of the frequency spectrum of the information components. The carrier frequency must have one of its characteristics varied (modulated) by the information and the receiver must contain an information extractor (detector) designed to react to the carrier characteristic that is varied and to produce an output, which is as close a copy of the original information as possible.

If the carrier frequency is continuous the information may be employed to modulate either the amplitude (a.m.) or the time characteristic, frequency (f.m.) or phase (p.m.). The information components generally have two properties, amplitude or intensity, and frequency. *Table 15.1* shows how each is conveyed.

Table 15.1

Information property	*Amplitude modulation*	*Time modulation*: *Frequency*	*Time modulation*: *Phase*
Amplitude	change of carrier amplitude	change of carrier frequency	change of carrier phase
Frequency	rate of change of carrier amplitude	rate of change of carrier frequency	rate of change of carrier phase

The carrier may consist of a pulsed carrier, whose amplitudes (p.a.m.), positions (p.m.m.) or durations (p.d.m.) are varied by the information as shown in *Figure 15.15*. The pulsed form of carrier operation allows a number of information channels to be transmitted on the same carrier frequency by regularly allocating a pulse to a given

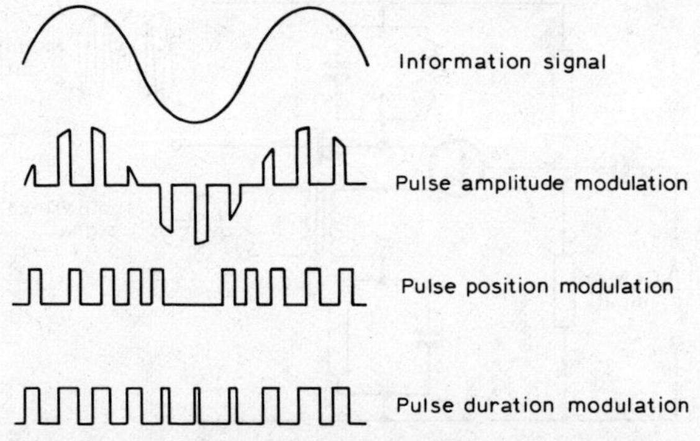

Figure 15.15 Examples of the 3 types of pulse modulation

channel. For example 8 channels can be accommodated using pulses 1, 10, 19, etc. for channel 1, pulses 2, 11, 20, etc. for channel 2. Pulses 9, 18, 27, etc. are required for synchronising a distribution gate at the receiver. The gate separates the information channels and directs them to their required destinations. Such a system is known as time division multiplex (t.d.m.).

Frequency division multiplex (f.d.m.) is used with continuous carrier in telephony, each speech channel being shifted in frequency before being added to the others occupying different frequency bands. The combined signals modulate the carrier; at the receiver the channels are filtered from each other and their frequency spectra returned to normal.

All the above systems are known as analogue modulation because the information controls the carrier directly. Considerable message-to-noise ratio advantage is gained by converting the amplitude characteristic of the information into a digital form before modulation. This is known as pulse code modulation (p.c.m.). The digital code has to be converted back to its analogue form when it is desired to interpret the information. Time division multiplex is used with p.c.m. when a number of different information channels have to be accommodated.

Continuous carrier modulation

AMPLITUDE MODULATION

The carrier amplitude may be represented by $E_c \cos \omega t$ and the modulation by $E_m \cos pt$, so that the a.m. carrier expression is

$$E_c(1 + M \cos pt)\cos \omega t \tag{18 a}$$

where M, the modulation ratio, $\leqslant 1 \propto E_m$
This is analysed into three components:

$$E_c \cos \omega t + \tfrac{1}{2}E_c M \cos(\omega \pm p)t \tag{18 b}$$

i.e., the unmodulated carrier and a pair of sidebands. Information normally consists of several frequencies, and each frequency has its own M proportional to its amplitude (the sum of all M values must not exceed unity) and produces a sideband pair.

Double sideband (d.s.b.) transmission is the simplest form of a.m. requiring only a very simple unidirectional detector at the receiver and it is therefore used for broadcasting. It requires a transmission channel bandwidth twice the highest frequency of the information components.

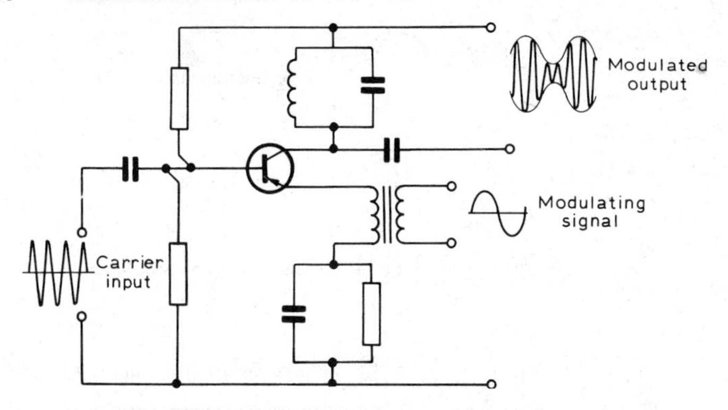

Figure 15.16 A modulated amplifier circuit

To obtain d.s.b. amplitude modulation the information voltage is used to control the gain of a carrier amplifier, and an example with a transistor amplifier is shown in *Figure 15.16*. The information could have been inserted in series with the carrier in the base, and the variation of gain is due to the non-linear characteristic of the transistor. A similar circuit can be used for an electronic tube amplifier. In a triode tube modulation can be achieved by inserting the modulating signal in the grid, cathode or anode circuit. The advantage of grid or cathode modulation is that a relatively low voltage or power is required from the modulator, but modulation envelope distortion occurs at high values of M. Anode modulation requires relatively large power from the modulator but distortion is low up to modulation percentages of 90% ($M = 0.9$). Thus broadcasting transmitters using triodes in the modulated amplifier employ anode modulation. Semiconductors are, at present, not capable of handling the high power required. A typical high-power modulator stage is shown in *Figure 15.17* (*a*); speech or music is amplified in a high-power Class B push–pull stage, which must be capable of supplying half the carrier power. Its transformer-coupled a.f. varies the anode voltage of the carrier amplifier tube from almost zero to twice the d.c. supply voltage (about 14 kV). The transformer secondary may be a.f. choke-coupled to the r.f. stage so as to divert the d.c. current of the latter from the a.f. transformer. The carrier amplifier tube in the modulated-amplifier stage operates at high efficiency in Class C, and envelope distortion is low up to $M = 0.90$; r.f. currents are bypassed from the a.f. side by the capacitor C. Envelope distortion is often reduced still further by rectifying the modulated output and applying the resulting a.f. wave as negative feedback to an earlier a.f. stage.

A single Class A a.f. amplifier may be used in series with the r.f. carrier tube anode but it suffers from low d.c. efficiency and complications due to the high d.c. voltage.

Since the modulated-amplifier r.f. stage in *Figure 15.17* (*a*) uses a triode, there will be r.f. feedback from the anode into the grid, and self oscillation is likely to occur. The stage must be neutralised by feeding into the grid circuit an inverse voltage from the end of the anode tuning coil opposite from the anode via the neutralising capacitance C_n. If the tuning coil is centre-tapped $C_n \simeq C_{ag}$, the interelectrode capacitance from anode to grid. If the r.f. output stage is in push–pull, neutralisation is secured by neutralising capacitors from the anode of one of the push–pull pair to the grid of the other.

The problem of neutralisation is much reduced if the grid of the modulated amplifier is earthed and the r.f. drive inserted in the cathode. The power output of the r.f. driver into the modulated-amplifier stage adds to the total power output, and if this is unmodulated it reduces the output modulation percentage below that of the modulated-amplifier. Thus the driver stage itself must be modulated.

Tetrode and pentode tubes can be used as modulated amplifiers with the modulating signal applied to the screen-grid or for the pentode to the suppressor-grid. Suppressor-

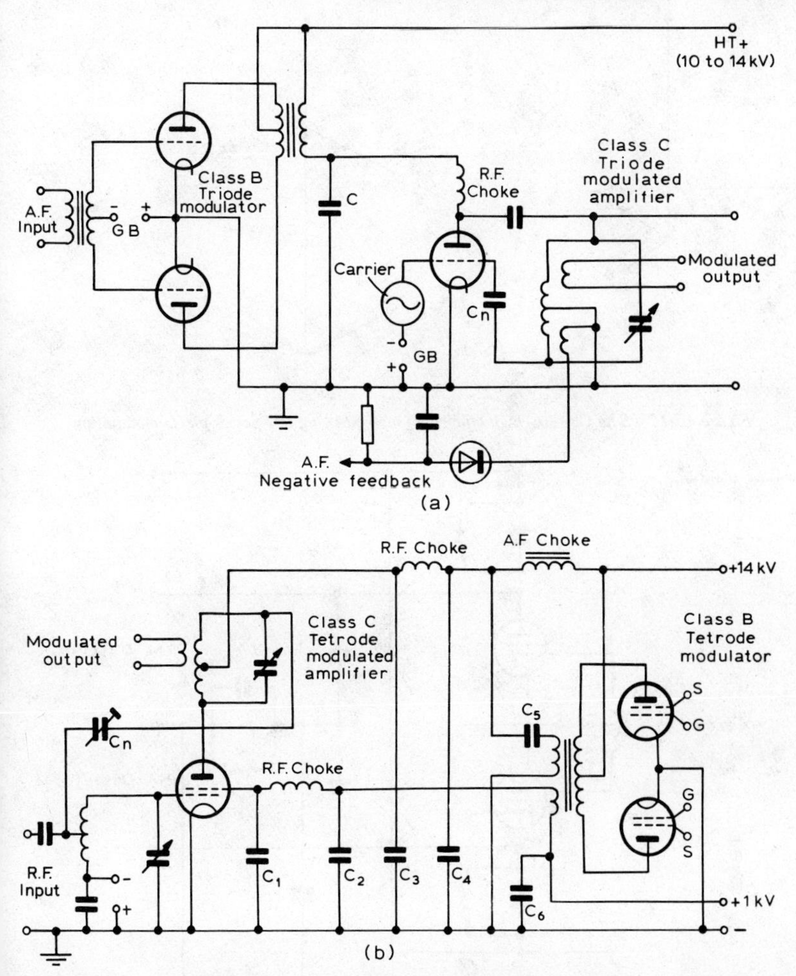

Figure 15.17 (*a*) *A high-power modulator stage using triodes* (*b*) *A high-power modulator stage using tetrodes*

grid modulation suffers from the same disadvantage as grid modulation, viz., modulation envelope distortion with high modulation.

In modern practice tetrodes using combined screen and anode modulation are tending to replace triodes in modulated-amplifier stages, because they have higher r.f. gain, often higher a.c./d.c. efficiency and, due to the r.f. earthing of the screen, reduced neutralisation problems. Their higher r.f. gain reduces the number of r.f. stages required between carrier master oscillator and modulated amplifier. Screen modulation alone is unsatisfactory because of envelope distortion but combined screen and anode modulation gives results comparable to that of anode modulation of a triode. The modulation transformer requires two secondary taps having a.f. voltage ratios approximately equal to the ratio of d.c. screen and anode voltages on the tetrode. A typical circuit is given in *Figure 15.17* (*b*), where C_1–C_4 are r.f. capacitors and C_5–C_6 are a.f. capacitors.

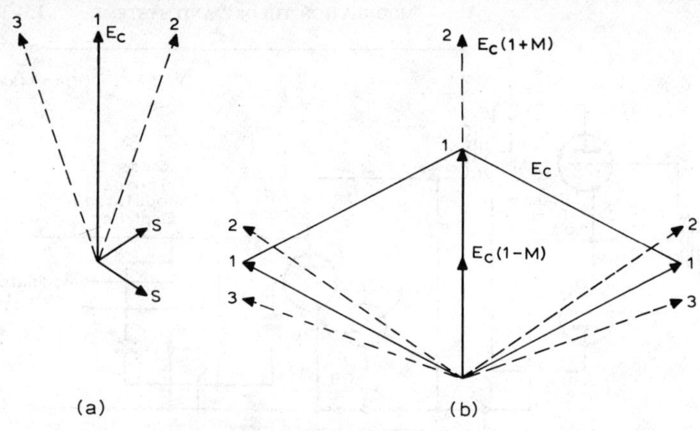

Figure 15.18 The Chireix (high-efficiency) method using low power phase modulation

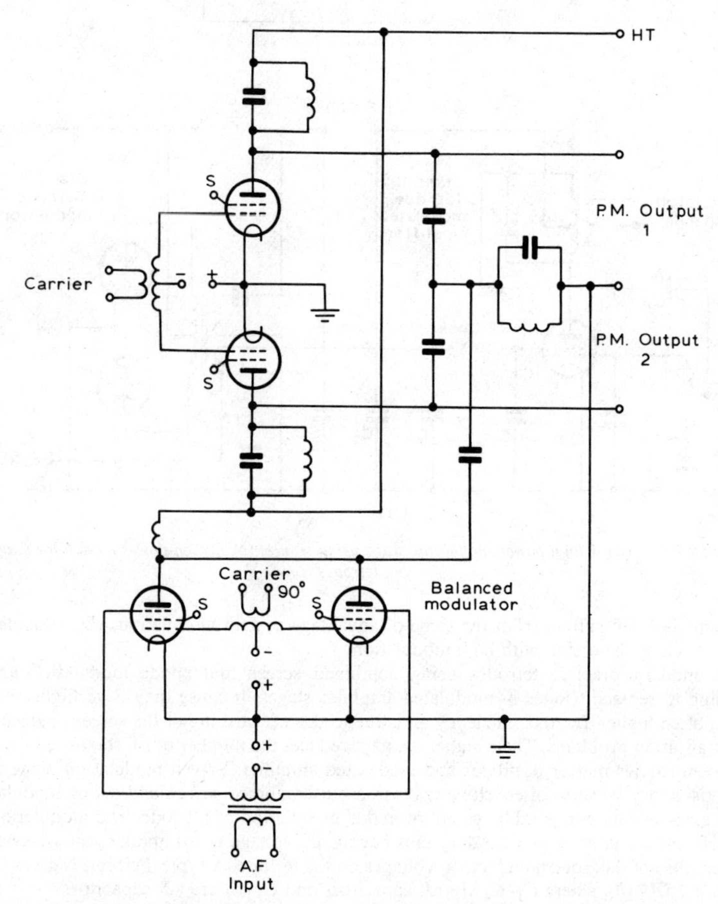

Figure 15.19 Phase modulation system for Ampliphase high-efficiency circuit

When the modulated-amplifier supplies its output direct to the aerial, the system is called high power modulation; if modulation is carried out before the final r.f. power stage, and envelope distortion is to be avoided, the modulated signal must thereafter be amplified in Class B amplifier stages, which have an appreciably lower a.c./d.c. efficiency than the high power Class C modulated amplifier.

There have been three successful attempts, known as Chireix, Ampliphase and Doherty, to use low-power modulation and approach the efficiency of the high-power modulator.

The Chireix method employs two r.f. Class C amplifying chains each carrying oppositely phase-modulated carriers of almost constant amplitude. The phase modulation is obtained by adding sidebands in phase quadrature to the unmodulated carrier as shown by the vector diagram of *Figure 15.18* (*a*). A balanced modulator permits the sidebands to be developed separately from the carrier. The two oppositely phase-modulated carriers after amplification are added at the output to generate an a.m. wave as shown in *Figure 15.18* (*b*). The unmodulated carriers must be given a phase-shift $<180°$ in order that the required unmodulated carrier amplitude (E_c) appears at the output; this is achieved by a phase-shifting network in one of the amplifying chains.

The Ampliphase system operates on the Chireix principle of combining at the output of two amplifying chains phase-modulated signals in such a way as to produce a.m. One system uses $\lambda/4$ transformers to couple the power from the anodes of the final power output stages into the common load, which is the aerial system. The two power valves operate in Class C to develop constant r.f. voltages, and these are converted by the $\lambda/4$ transformer into constant currents in the load. Phase variations of the currents cause an a.m. output in the load as shown in *Figure 15.18* (*b*). The method of combining the 90° phase-shifted sidebands with 180° carriers is illustrated in *Figure 15.19*. The carrier input to the balanced modulator is given a 90° phase shift with reference to the push–pull stage supplying the 180° carriers. One of the amplifying chains is given the requisite phase shift for obtaining carrier output in the unmodulated condition.

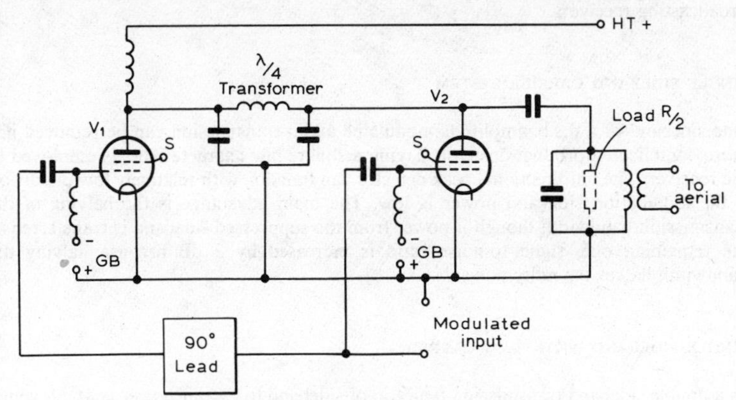

Figure 15.20 The Doherty high efficiency amplifier

The Doherty high efficiency output stage uses two valves, whose grid circuits are driven by modulated inputs in phase quadrature and whose anodes are coupled by a $\lambda/4$ network of characteristic impedance R. Both supply power into a load of $R/2$ as shown in *Figure 15.20*. The $\lambda/4$ network introduces a phase delay of 90° so that the modulated input to V_1 must be given a compensating 90° phase lead. When the carrier is equal to or less than its unmodulated value V_1 only is operative, V_2 being cut off. During positive modulation cycles, anode voltage limitation of V_1 occurs and this is transferred by the $\lambda/4$ network as a constant current into the load; V_2 begins to supply power and this increases the effective terminating impedance $R/2$. The $\lambda/4$ network, whose characteristic impedance is R, transfers this as a decreasing load seen by V_1 at the input to the network, and this changes from $2R$ when V_2 is cut-off, to R for 100% positive peak of modulation when V_2 is fully operative and supplying half the total power. Due to the change from $2R$ to R and the constant output voltage the power delivered from V_1 at 100% positive peak modulation is double the unmodulated carrier power and V_2 provides an equal power into the load, effectively increased to R by the action of V_1.

Television transmitters generally employ grid modulation since it requires much less power than anode modulation—an important consideration with wide band signals—and non-linear distortion of video signals is less serious than with speech or music. The synchronising pulse part of the video signal can be operated over the most non-linear part of the transmitter characteristic and can be predistorted to compensate for the non-linearity.

VESTIGIAL SIDEBAND TRANSMISSION

Channel bandwidth should be as small as possible to ensure economic use of the transmission frequency spectrum, and vestigial sideband transmission can represent an appreciable saving of spectrum with television signals having information components up to about 5.5 MHz. Only a part of one of the sidebands, that within about 1 MHz of the carrier is transmitted; if a simple detector circuit is employed distortion of the modulation content of all information frequencies exceeding about 1 MHz results. Fortunately their amplitudes are small and the degree of distortion is quite acceptable on vision signals. The receiver has to be detuned to half carrier amplitude so as to reduce the low-frequency sideband energy to the original level in relation to the high-frequency sidebands. Vestigial sideband operation is unsuitable for speech or music because of the distortion produced by the simple unidirectional detector used in broadcasting receivers.

SINGLE SIDEBAND CARRIER SYSTEM

One sideband of a d.s.b. amplitude-modulated audio transmission can be removed if a more complicated product detector, having a square law characteristic, is employed in the receiver. The unidirectional peak detector can function with relatively low distortion if the modulation sideband power is low. The main advantage is the halving of the transmission bandwidth though if power from the suppressed sideband is transferred to the remaining one, signal-to-noise ratio is increased by 3 dB because halving the bandwidth halves the noise power.

SINGLE SIDEBAND WITH PILOT CARRIER

In a double sideband transmission the ratio of sideband to carrier power is $M^2/2$, which is 50% at 100% modulation. The same is true of s.s.b. when the power from the suppressed sideband is transferred to the other. The carrier power therefore accounts for a very large proportion of the total transmitted power and a worth-while saving in running costs in a point-to-point communication system is realised by reducing the carrier to a low value. The carrier cannot be entirely eliminated because detection cannot be achieved in its absence. The carrier is restored at the receiver by having an oscillator locked in frequency and phase by the transmitted residual pilot carrier. If almost all the carrier power is transferred to the sideband the transmitted information content can be increased 3 times relative to the 100% modulation sideband power so that s.s.b. with pilot carrier gives a signal-to-noise ratio approaching 8 dB better than d.s.b. with full carrier (5 dB increase in signal and 3 dB decrease in noise due to halving bandwidth).

A balanced modulator permits the sidebands to be generated and the carrier to be eliminated; it may take the form shown in *Figure 15.19*, or it may be a push–pull type r.f. modulator stage with carrier frequency in the same phase to the input base circuits and the information applied in opposite phase as in *Figure 15.21*. If the characteristics

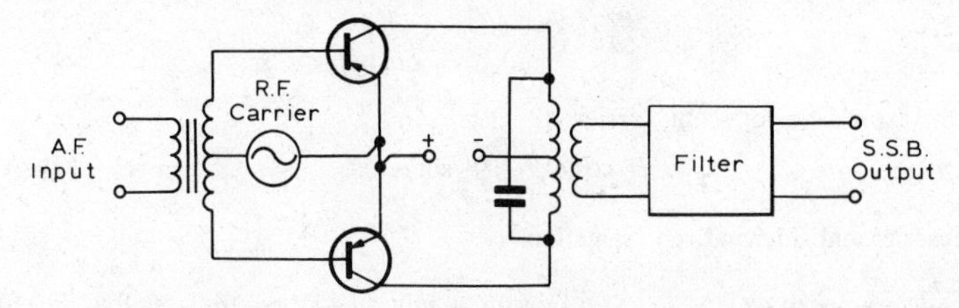

Figure 15.21 A push–pull balanced modulator for carrier elimination

of each half stage are identical the carrier component is cancelled at the output but the two sidebands are unaffected. The wanted sideband is selected by passing through a filter rejecting the unwanted.

The second sideband channel may be used to carry separate information on the same pilot carrier by means of another balanced modulator and filter; the two independent sidebands and the correct proportion of carrier are combined before power amplification to the information amplitude. The mathematical representation is
pilot carrier.

PHASE MODULATION

In phase modulation the phase angle of the carrier is advanced and retarded in proportion to the information amplitude. The mathematical representation is

$$E_c \cos(\omega_c t + \phi \cos pt) \qquad (19\text{ a})$$

where $\phi \cos pt \propto E_m \cos pt$.

Like a.m. it can be resolved into carrier and sideband components but there is more than one pair of sidebands per information frequency component, and the general expression is

$$E_c\left[J_o(\phi)\cos\omega_c t + \sum_{n=1}^{\infty} J_n(\phi)\cos\left[\omega_c + n\left(\frac{\pi}{2} \pm p\right)\right]t\right] \qquad (19\text{ b})$$

where J_o, etc. are Bessel Functions.

Theoretically there are an infinite number of sidebands spaced $\pm f_m$, $\pm 2f_m$, etc. from the carrier, but in practice the amplitudes of the higher order sidebands fall off rapidly if ϕ is not large. Thus when $\phi = 45°$ (0.787 rad), only the first two pairs ($\pm f_m$, $\pm 2f_m$) have significant amplitudes.

Phase modulation is produced by variation of the inductive or capacitive element of a buffer amplifier stage following a master oscillator. The variable element may be a varactor diode biased by the information voltage. Phase change must not exceed about $\pm 25°$ if linear modulation is to be achieved, but it can be increased by applying it as an input to a multiplier stage; thus a tripler stage multiplies the phase change by 3. A frequency changer can restore the carrier to its original frequency leaving the multiplied phase untouched.

FREQUENCY MODULATION

Frequency modulation requires the carrier frequency to be varied in accordance with the information so that carrier instantaneous frequency is

$$f = f_c + f_d \cos pt$$

where $f_d \propto E_m$ and is the frequency deviation of the carrier.

Since
$$\phi = 2\pi \int f \, . \, dt$$
$$= \omega_c t + \frac{f_d}{f_m} \sin pt$$

and the expression for a f.m. carrier is

$$E_c \cos(\omega_c t + \frac{f_d}{f_m} \sin pt) \quad (20\text{ a})$$

The carrier and sideband representation is

$$E_c \left(J_o \frac{f_d}{f_m} \cos \omega_c t + \sum_{n=1}^{\infty} J_n \left(\frac{f_d}{f_m} \right) [\cos (\omega_c + np)t + (-1)^n \cos (\omega_c - np)t] \right) \quad (20\text{ b})$$

The ratio f_d/f_m is an angle inversely proportional to f_m, so that for a given information amplitude f_d/f_m is much larger when the information frequency component is low. It is similar to phase modulation with low-frequency boost. Thus if $f_d = 10$ kHz, $f_d/f_m = 0.785$ rad., when $f_m = 12.75$ kHz and only two sidebands are significant, whereas when $f_m = 500$ Hz, $f_d/f_m = 20$ rad. and about 23 pairs of sidebands have significant amplitudes. This gives frequency modulation an advantage over noise because the latter phase-modulates a carrier whose amplitude is maintained constant by a limiter in the receiver. Hence the low frequency components of thermal noise have much less effect than the high frequency components at the output of a f.m. receiver. The normally flat frequency spectrum of thermal noise is 'triangulated' with minimum noise at low frequencies.

The carrier and sideband component amplitudes vary from maximum to minimum through zero as f_d/f_m varies, but the root-mean-square value of the amplitudes is always constant and equal to the unmodulated carrier amplitude.

A feature of f.m. is that an increase in f_d increases the information modulation content of the signal in direct proportion, so that an increase of f_d from 15 kHz (the maximum broadcast audio frequency) to 75 kHz increases information amplitude content by 5 and is equivalent to 14.25 dB increase in signal strength. Thus unlike a.m. information content is determined by the product of bandwidth and power, and there are advantages in increasing transmission bandwidth beyond $2f_m$. Channel spacing of v.h.f. transmissions does not permit $f_d > 75$ kHz and the bandwidth required is found to be 180 kHz or $2\,[f_d(\text{max}) + f_m(\text{max})]$.

Frequency modulation is obtained by varying the inductive or capacitive element of the tuned circuit of a master oscillator. The average carrier frequency must be maintained constant otherwise adjacent transmission channels will suffer interference. One way of reducing this problem is to use balanced reactance modulators, which have adding reactance changes but whose resistance components change in opposite directions so maintaining a constant total resistance component. A further improvement is gained by comparing average frequency against a crystal oscillator and converting any error into a d.c. voltage controlling the reactance modulators, or into a driving force to operate a motor controlling a tuning capacitance across the f.m. oscillator. A block schematic of a motorised correction circuit is shown in *Figure 15.22*. The f.m. oscillator frequency is changed to about 7.5 MHz with the aid of a crystal oscillator; this frequency is divided down to 15 kHz and applied to a pair of balanced modulators together with the output from a 15 kHz crystal oscillator phase-shifted by 90° to produce a 2 phase field from the balanced modulators. This field drives a synchronous motor mechanically coupled to a trimmer capacitor across the f.m. oscillator. The field

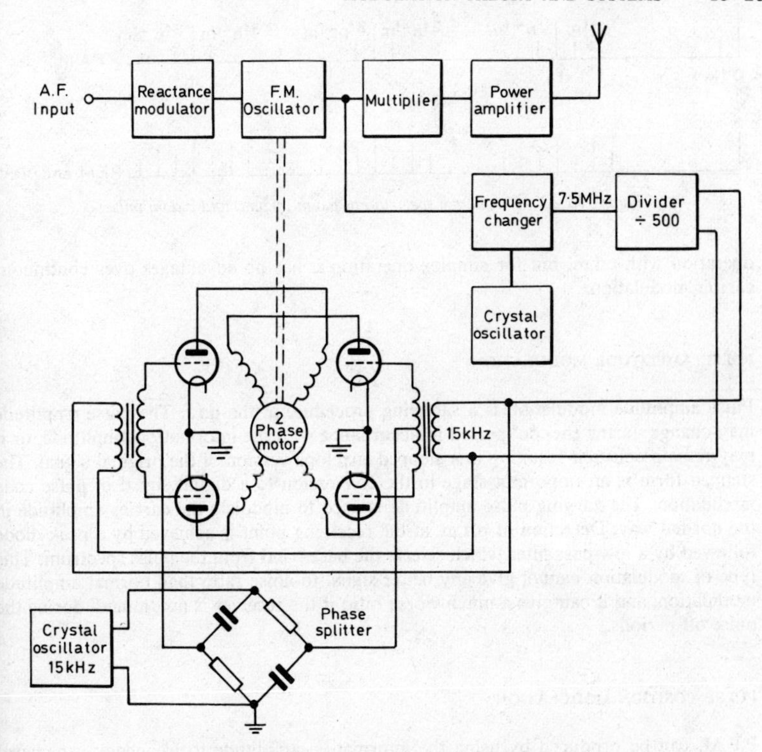

Figure 15.22 A motorised-corrected f.m. oscillator

is only present when there is an error in the divider frequency. A crystal oscillator [1] can be frequency-modulated and a highly stable average frequency maintained.

Pulsed carrier modulation

Pulsed carrier modulation is a form of double modulation in which the information is used to modulate a pulse, which is in turn used to modulate the final carrier. It requires a pulse repetition frequency (p.r.f.) at least about 2.3 times the maximum information frequency component. A series of unmodulated pulses can be resolved into a d.c. component and a theoretically infinite number of integer multiples of the p.r.f. When the pulses are amplitude-modulated the frequency spectrum can be resolved into a d.c. component and the full range of base-band information frequencies, and sideband pairs of the base band components associated with each integer multiple of the p.r.f. as indicated in *Figure 15.23*. The sideband ranges must not overlap if the signal is to be detected without distortion at the receiving point. This cannot occur with a.m. if the p.r.f. exceeds $2 f_m$(max).

When the pulse is position- or duration-modulated, harmonic sidebands of the pulse repetition integer frequencies ($nf_p \pm mf_m$) are produced but there are no harmonic frequencies of the base band, *Figure 15.23*. Detection at the receiver requires that no overlap shall occur, and if this is probable either the p.r.f. must be increased or the degree of modulation reduced. Pulsed carrier modulation is valuable for multiplex

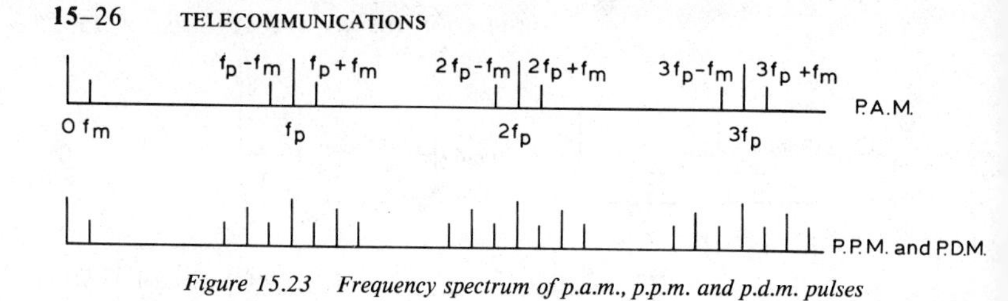

Figure 15.23 Frequency spectrum of p.a.m., p.p.m. and p.d.m. pulses

operation with t.d.m. but for simplex operation it has no advantages over continuous carrier modulation.

PULSE AMPLITUDE MODULATION

Pulse amplitude modulation is a sampling procedure at the p.r.f. The pulse amplitude may change during the 'on' period in accordance with the information amplitude, or it may remain constant resulting in a stepped envelope version of the original signal. The stepped form is an important stage in the conversion to a digital signal in pulse code modulation. The varying pulse amplitude is used to modulate the carrier amplitude in the normal way. Detection of p.a.m. at the receiving point is achieved by a peak diode followed by a low-pass filter which selects the base-band from the pulse spectrum. This type of modulation cannot give any better signal-to-noise ratio than normal amplitude modulation, and it can give a much worse ratio if the receiver is not 'muted' during the pulse off periods.

PULSE POSITION MODULATION

P.P.M. can be produced by using the information amplitude to frequency- or phase-modulate the p.r.f. Frequency modulation can be realised by a variable reactance, controlled by the information amplitude, across the p.r.f. oscillator. A limiter in the receiver removes the amplitude noise and allows the signal-to-noise improvement to be achieved. Detection by peak diode followed by a low-pass filter allows the base-band frequencies to be recovered. An alternate method is to use a frequency discriminator tuned to the p.r.f.

PULSE DURATION MODULATION

Modulation of pulse duration is obtained by varying the level at which a sawtooth voltage derived from the p.r.f. oscillator is sliced. Detection may be by peak diode followed by a low-pass filter, or the pulse duration may be converted to an amplitude variation by using the pulses to control the duration of a ramp voltage.

TELEGRAPHIC COMMUNICATION

Telegraph signals themselves are in pulse form and when used to modulate a carrier generate pulsed carrier modulation. Originally the carrier was switched on for mark and off for space, but this is unsatisfactory because the receiver reverts to full gain on space, and noise may cause interference and spurious signals. Frequency shift keying (f.s.k.) is the method now adopted and the carrier is shifted by about 100–200 Hz from mark to space.

The Morse code is satisfactory for manual operation and visual observation but is less useful when teleprinter or automated punched paper tape or magnetic tape operation is employed. A 5-unit code, which is a binary on–off system, allows 2^5 or 32 characters to be defined but gives no error indication. The 7 unit code, made up of 3 mark and 4 space elements permits 35 characters ($7C_3 = 7.6.5./1.2.3.$) to be defined, and at the same time provides a degree of error detection. The marks are counted and unless 3 are registered in each character the message must be in error and a request for a repeat can be sent from the receiver to the transmitter. The process can be automated and is then called automatic error correction (a.r.q.). A.R.Q. equipment generally records groups of 3 characters, checks for the correct number of marks in each character and sends back to the transmitter a repeat request or go-ahead signal. After checking and finding no error, the recorded signals are passed to the teleprinter. The transmitter cuts off at every 3 characters and remains inoperative until the repeat or go-ahead signal is received.

Only a single error is detected and a double error which drops a mark and inserts interference or noise simulating a mark into a space passes uncorrected.

If more than 35 characters have to be accommodated a higher unit code is required, thus an 8 unit code of 3 marks and 4 spaces could cover 56 characters or with 4 marks and 4 spaces 70 characters are possible.

The term baud is used in telegraph communication, 1 baud signifying 1 on–off period/second; a 50 baud signal has 50 pulses/second, the duration of each on–off period and of the pulse being 0.02 second and 0.01 second respectively, and the required transmission bandwidth being 100 Hz. A single telephonic speech channel can accommodate 24 such telegraph signals.

Pulse code modulation

With pulse code modulation the information amplitude is divided into a number of levels, each of which is designated by a number; it is regularly sampled and given the number of the specified level nearest the given sample. Normally a binary system of numbering is employed because this is the simplest to achieve electronically, and the digital output consists of a number of pulses per unit time corresponding to the binary number given to the sampled amplitude level. In a binary system the number of levels is 2^n where n is the maximum number of pulses per unit time; thus 8 levels require 3 pulses with zero regarded as a level, whereas 1 024 levels require 10 pulses. Since the coded signal represents a number of discrete levels, the reproduced information at the receiving point after decoding will be a stepped copy of the original information, *Figure 15.24*. The steps introduce a spurious signal, and when there are many steps of small

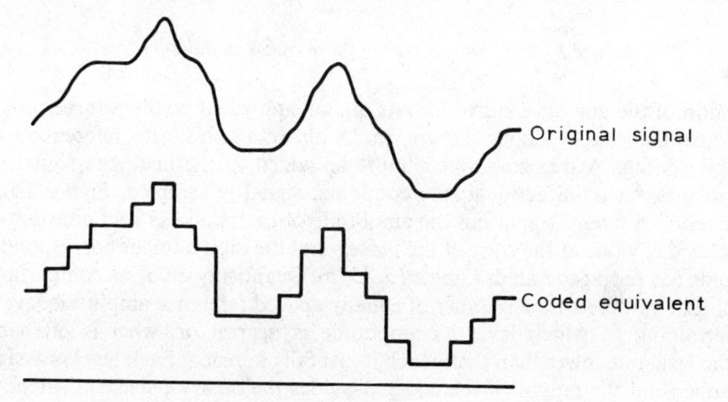

Figure 15.24 Pulse coded equivalent of the original signal

amplitude the interference appears as a hiss like random noise superimposed on the original information amplitude, and it is worst when the information amplitude lies midway between two levels. The hiss is termed quantising noise since it is caused by the discrete level quanta; its effect can clearly be reduced by increasing the number of levels, but this requires more digital pulses in a given time interval and therefore a greater transmission bandwidth.

When the number of level steps is small, i.e., when the information amplitude is very small, the step interference changes from a hiss to a seriously distorted signal and this condition is known as granular distortion. In telephonic speech, amplitude variations can be smoothed out by compression, and satisfactory communication can be achieved by a 7 digit binary code (128 levels). The same degree of compression is not permissible for high quality sound broadcasting, and subjective tests have shown that a 13 digit[2] coding (8 192 levels) is required. The 13 digit coding provides a peak signal to r.m.s. quantising noise ratio of 83 dB and a 7 digit 47 dB. It is possible to reduce the audibility of granular distortion by adding a 'dither' voltage to the information before coding. The dither voltage has two components, one a square wave of peak-to-peak amplitude equal to half a level step and at half the sampling frequency, and the other white (thermal) noise at a much lower level (about 4 dB below quantising noise). At the receiver the half sampling frequency square wave is removed by the filter selecting the information frequencies. A slightly improved signal-to-noise ratio is obtained by substituting a pseudo-random signal for the white noise and at the receiver cancelling this by subtracting an identical sequence obtained from another pseudo-random generator synchronised by the transmitted pseudo-random signal.

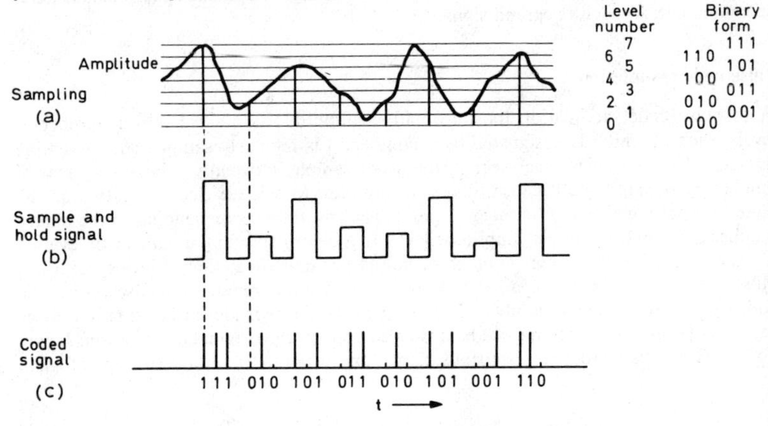

Figure 15.25 The process of pulse coded modulation

Conversion of the analogue signal into its digital equivalent involves three processes, sampling, quantising and coding. *Figure 15.25* illustrates this with reference to an 8 level 3 digit coding. A d.c. component must be added to the analogue signal before sampling to make it unidirectional; the combined signal is sampled, *Figure 15.25 (a)* and converted to a p.a.m. signal but the amplitude of each pulse is held constant at the analogue plus d.c. value at the start of the pulse, until the digit number corresponding to the amplitude has been generated, *Figure 15.25 (b)*. Quantising involves comparing each pulse amplitude sample with a number of equally-spaced reference amplitude levels and either determining to which level its amplitude is nearest or, what is often easier, selecting the level one lower than that which it just fails to reach. Each level is assigned a binary number and the output of the coder provides the binary pulse equivalent of the level determined from the unidirectional p.a.m. analogue signal, *Figure 15.25 (c)*.

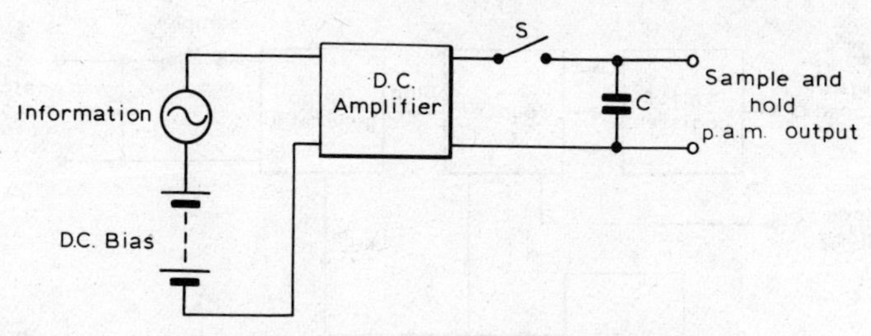

Figure 15.26 Schematic of sample and hold circuit

A simplified block diagram of the method of sample and hold, producing constant amplitude pulses, is shown in *Figure 15.26*. The d.c. and analogue combined signal is amplified and sampled by closing of the switch S, which charges the capacitor C to the sample voltage. The switch, generally a diode bridge, is operated from the sampling oscillator, whose p.r.f. is about 2.3 f_m (max). The closure time for the switch is very short compared with the period of the analogue signal and after opening, the capacitor voltage remains substantially constant at the sample amplitude until the level assessment has been made. Shortly afterwards the capacitor is fully discharged and is then ready to accept the next amplitude sample.

There are a number of systems for converting the p.a.m. signal into a digital equivalent; one uses a counter coder and another employs successive approximations. A simplified block diagram of the counter coder is illustrated in *Figure 15.27*. The held sample of information amplitude is compared against a ramp voltage of constant slope provided by charging a capacitor at a constant current. When an assessment is to be made a short-circuit S across the ramp generator capacitor C is removed and simultaneously clock pulses are fed to a counter. When the voltage across capacitor C equals

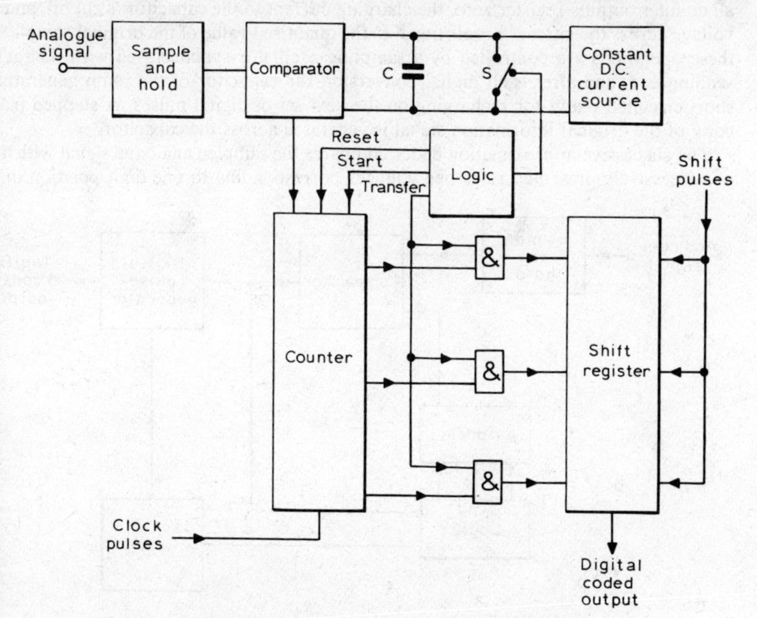

Figure 15.27 Block diagram of counter encoder

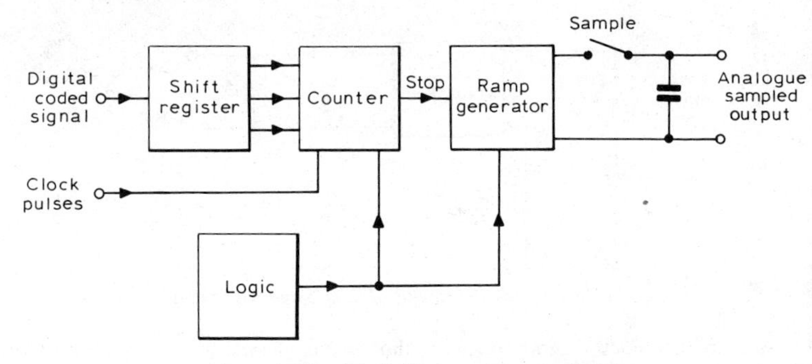

Figure 15.28 Block diagram of counter decoder

that of the sample, the comparator sends a pulse which stops the counter. Since the voltage across C increases linearly with time until equality is reached with the sample the counter registers the binary equivalent of the sample amplitude. The state of the counter is transferred to the shift register via AND gates. The counter is cleared and the ramp capacitor short-circuited in readiness for coding the next sample. Shift pulses are fed into the register to release the digital information in time sequence; the digital representation may be transmitted direct over coaxial cable or may be used to modulate a u.h.f. or s.h.f. carrier for transmission over a microwave link.

At the receiving point the reverse process may be carried out to convert the digital signal back to its analogue original. A block diagram of the counter decoder is shown in *Figure 15.28*. The coded digital signal is stored in a shift register from which it is transferred to the counter. A ramp generator—a capacitor charged from a constant current—is started simultaneously with the counter which counts back to zero. When all counter outputs register zero, the charging current to the capacitor is cut off, and the voltage across the latter is proportional to the quantised value of the original sample. All these operations are controlled by a sampling oscillator synchronised with that at the sending end, and after each digital conversion the capacitor of the ramp generator is short-circuited ready for recharging on the next set of digital pulses. A stepped p.a.m. copy of the original information signal is generated across the capacitor.

The successive-approximation coder compares the sampled analogue signal with a set of successively-presented reference voltages corresponding to one digit position in the

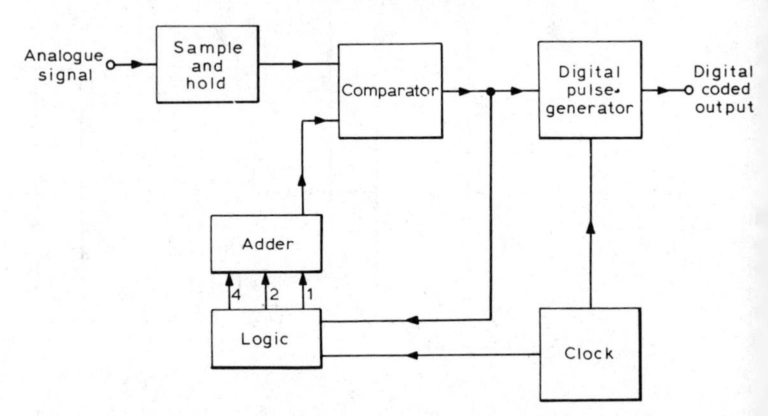

Figure 15.29 Block diagram of succession approximation encoder

code. If the maximum reference voltage is less than the sample the comparator gives a logic 1 output; the sample voltage minus the reference voltage is passed to another comparator connected to the next digit voltage reference and if the reference is more than the former the comparator gives a logic 0 output. If the maximum reference voltage is greater than the sample, the comparator registers a logic 0 output and the sample is passed on unchanged to the next reference voltage. This process is repeated until all the digit positions have been defined.

Figure 15.29 shows a modified practical version of this technique applied to a 3 digit p.c.m. Assuming a sample amplitude of 5.4 units is to be digitalised, the clock pulses are fed to the logic circuits and the maximum reference voltage of 4 units is applied to the comparator. Since the sample voltage exceeds 4 units the comparator triggers the digital pulse generator to produce a logic 1 pulse output, subtracts 4 from the 5.4 units and sets the logic circuit 2 unit reference voltage functioning. The 2 unit exceeds the sample and the digital pulse-generator transmits a logic 0 output. The 1 unit voltage is offered and as this is less than the sample a logic 1 pulse-output is produced. The binary digit coding output is therefore 101 which corresponds to a quantised voltage of 5 units. This coder is known as a series coder because the reference voltages are presented in time sequence. There is also a parallel form in which all the reference voltages are presented together and a simultaneous comparison is made. The disadvantage of the successive approximation method is that it requires accurate adjustment and maintenance of the reference voltages.

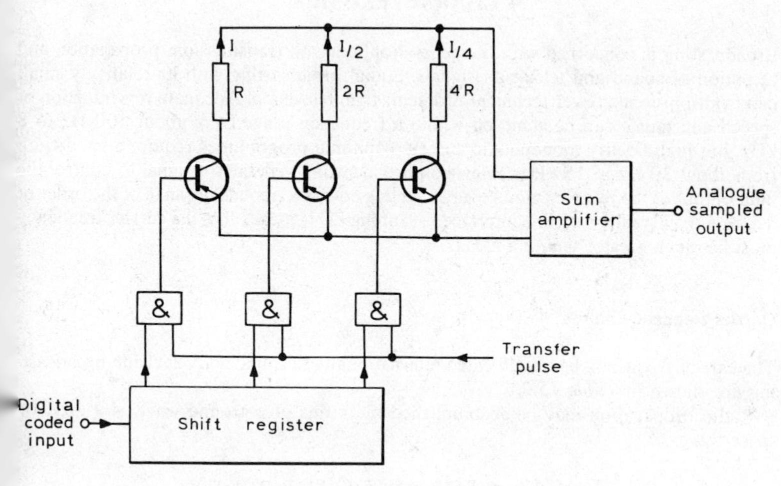

Figure 15.30 Block diagram of parallel decoder

Figure 15.30 gives a block schematic of a parallel type decoder for 3 digits. The ligital input is stored in a shift register, and each occupied digital position on command enders operative through an AND gate, a transistor providing an output current roportional to the digital position. Thus the first digit position if occupied switches on transistor supplying 4 units of current, the second 2 units of current and so on. All he currents are added in a summing amplifier, which produces a quantised output orresponding to the digital input.

Time division multiplex can readily be applied to allow several p.c.m. signals to share he same transmission channel. With the aid of electronic switches the binary coded amples of each information signal are connected in turn to the transmission system as escribed in the Introduction. All information channels are sampled once per cycle of the multiplex operation and extra binary digits called parity bits are added for synchronisation and error detection to complete the whole cycle or frame. The frame synchronising signal maintains the multiplexing and demultiplexing operations in synchronism. The demultiplexer is prevented from responding to a frame synchronising pattern occurring accidentally during the information sequence.

Errors in p.c.m. signals may be detected by transmitting an extra parity bit—with the information. The parity bit is inserted at the sending end to bring the total binary digits to an even or an odd number in each transmitted information sample. Failure at the receiver to register an odd or even number (whichever is chosen) of bits in each sample indicates an error.

REFERENCES

1. MORTLEY, W. S., 'Frequency-Modulated Quartz Oscillators for Broadcasting Equipment', *Proc. I.E.E.*, Part B, **104**, 239 (1957)
2. SHORTER, D. E. L. and CHEW, J. R., 'Application of P.C.M. to Sound Signal Distribution in a Broadcasting Network', *Proc. I.E.E.*, **119**, 1442 (1972)

BROADCASTING FREQUENCY BANDS AND PROPAGATION CHARACTERISTICS

Broadcasting is concerned with the generation, control, transmission, propagation and reception of sound and television signals. Sound broadcasting with its relatively small bandwidth presents fewer technical problems than television. Adequate reproduction of speech and music can be achieved within a frequency range from about 100 Hz to 8 kHz, but high fidelity monophonic and stereophonic programmes require a bandwidth from about 30 Hz to 15 kHz. Television broadcasting generates a signal to control the light output of the receiver picture tube and this covers a frequency range of the order of d.c. to 5.5. MHz for 625 line interlaced scanning. This means that the carrier frequency must be much greater than 5.5 MHz.

Carrier frequency bands

The carrier frequency bands allocated internationally to sound and television broadcasting are shown in *Table 15.2*.

Radio propagation may be accomplished by means of a ground wave, sky wave or space wave.

Table 15.2 CARRIER FREQUENCY BAND ALLOCATION

Frequency	*Wavelength*	*Frequency range*	*Purpose*
Low	long wave	150–285 kHz	Sound
Medium	medium wave	525–1 605 kHz	Sound
High	short wave	Bands approx. 250 kHz wide located near 4, 6, 7, 9, 11, 15, 17, 21, 26 MHz	Sound
Very high	Band I	47–68 MHz	Television
	Band II	87.5–100 MHz	F.M. Sound
	Band III	174–216 MHz	Television
Ultra-high	Bands IV and V	470–960 MHz	Television

Low frequency propagation

The low frequency range is propagated by the ground wave which follows the contour of the earth; signal strength tends to be inversely proportional to distance, the losses in the ground itself being small, and reception is possible at considerable distances though marred by noise interference. Any sky waves are absorbed in the ionospheric layers (Section 5).

Medium frequency propagation

The medium frequencies are propagated by ground wave during the day, any sky wave being absorbed by the ionospheric *D*-layer. The ground wave generates currents in the ground, and the energy loss is greatest over poor conductivity ground such as granite rocks, and is least over sea water, where propagation approaches the inverse square law. The greater the frequency the greater is the loss at a given distance. Curves of ground wave attenuation with distance for sea water and ground of good, moderate and poor conductivity, a 1 kW transmitter and a zero-gain aerial are published by the CCIR[1]. These permit the field strength (referred to 1 μV/m as zero level) of low and medium frequencies to be calculated for distances from 1 to 1 000 km and are reproduced in *Figures 15.31* to *15.34*. When using the curves shown in *Figures 15.31* to *15.34*, the following points should be especially noted.

(1) They refer to a smooth homogeneous earth.
(2) No account is taken of tropospheric effects at these frequencies.
(3) The transmitter and receiver are both assumed to be on the ground. Height-gain effects can be of considerable importance in connection with navigational aids for high-flying aircraft, but these have not been included.
(4) The curves refer to the following conditions:
  (*a*) They are calculated for the vertical component of electric field from the rigorous analysis of van der Pol and Bremmer.
  (*b*) The transmitter is an ideal Hertzian vertical electric dipole to which a vertical antenna shorter than one quarter wavelength is nearly equivalent.
  (*c*) The dipole moment is chosen so that the dipole would radiate 1 kW if the Earth were a perfectly conducting infinite plane, under which conditions the radiation field at a distance of 1 km would be 3×10^5 μV/m.
  (*d*) The curves are drawn for distances measured around the curved surface of the Earth.
  (*e*) The inverse-distance curve *A* shown in the figures, to which the curves are asymptotic at short distances, passes through the field value of 3×10^5 μV/m at a distance of 1 km.
(5) The curves should, in general, be used to determine field strength only when it is known that ionospheric reflections at the frequency under consideration will be negligible in amplitude—for example, propagation in daylight between 150 kHz and 2 MHz and for distances of less than about 2 000 km. However, under conditions where the sky-wave is comparable with, or even greater than, the ground-wave, the curves are still applicable when the effect of the ground-wave can be separated from that of the sky-wave, by the use of pulse transmissions, as in some forms of direction-finding systems and navigational aids.

As an example of the use of the curves, consider a 100 kW transmitter feeding into a 0.4 λ medium wave aerial at 1 MHz. The field strength at a point 50 km away over ground of medium conductivity from a 1 kW transmitter feeding a very short aerial is from *Figure 15.33*, 53 dB (ref. 1 μV/m). The 0.4 λ aerial has a gain of about 1 dB and the transmitter is +20 dB with reference to 1 kW, so that the field strength is 74 dB or 5.5 mV/m. When the signal path conductivity varies, field strength can be calculated by

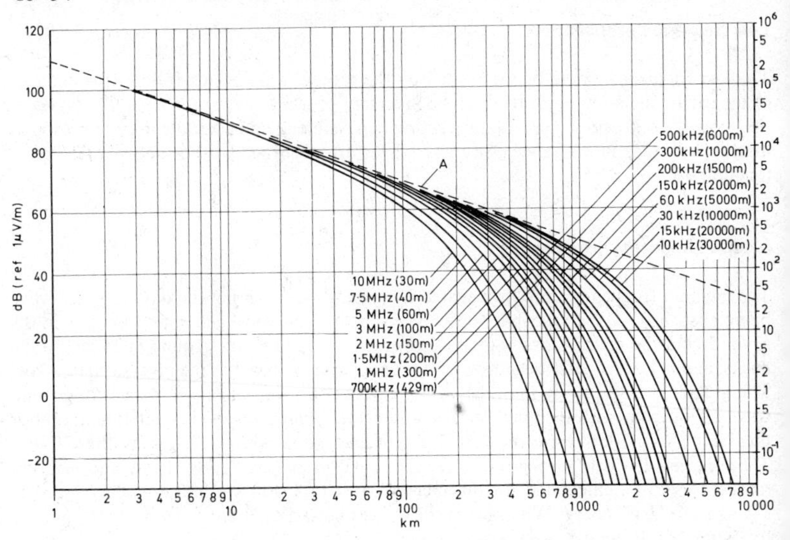

Figure 15.31 Ground-wave propagation curves over sea, σ = 4 mho/m, ϵ = 80 (Courtesy C.C.I.R.)

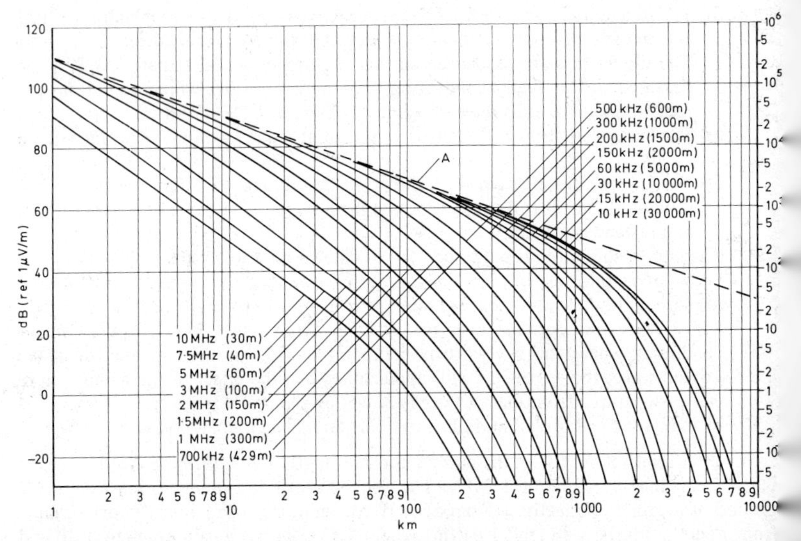

Figure 15.32 Ground-wave propagation curves over land of good conductivity, σ = 10^{-2} mho/m, ϵ = 4 (Courtesy C.C.I.R.)

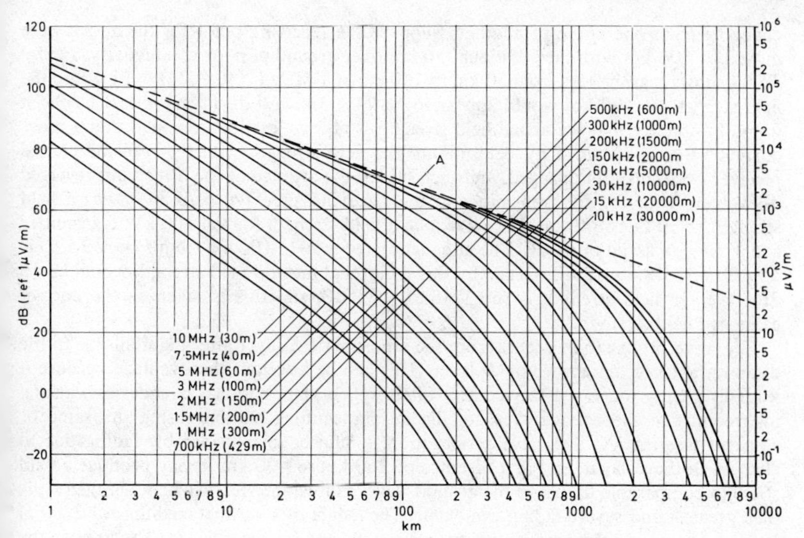

Figure 15.33 *Ground-wave propagation curves over land of moderate conductivity, $\sigma = 3 \times 10^{-3}$ mho/m, $\epsilon = 4$ (Courtesy C.C.I.R.)*

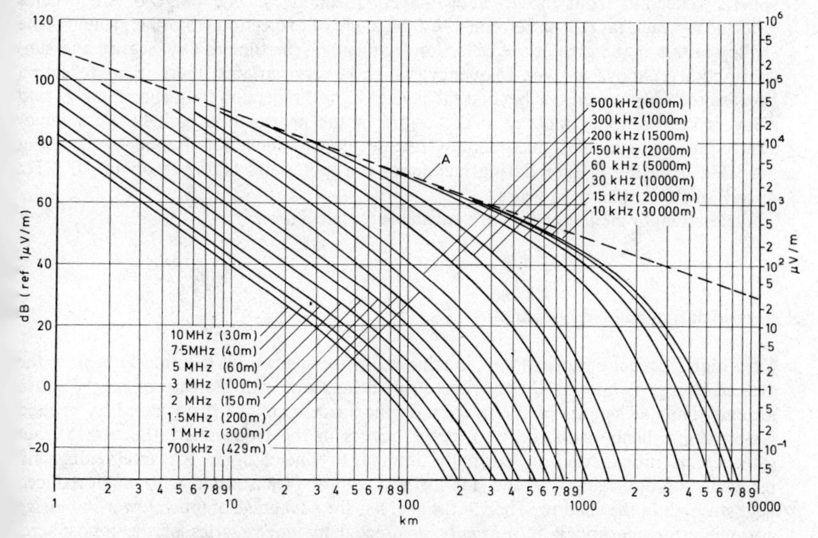

Figure 15.34 *Ground-wave propagation curves over land of poor conductivity, $\sigma = 10^{-3}$ mho/m, $\epsilon = 4$ (Courtesy C.C.I.R.)*

changing from one curve to another. Suppose the receiving point for the above were moved to 100 km with the additional 50 km over ground of poor conductivity. *Figure 15.34* shows that the loss from 50 km to 100 km at 1 MHz is $(40 - 26) = 14$ dB, so the field strength at 100 km would appear to be $74 - 14 = 60$ dB (1 mV/m). It has been found in practice that the actual field strength is always less than this and that a more correct value is obtained by determining the field strength with the transmitter and receiver positions interchanged, and then taking the average of the direct and reversed dB values. Thus for the transmitter at the receiving site *Figure 15.34* shows a field strength at 50 km of 40 dB for 1 kW, i.e., a field strength for the 100 kW transmitter and 1 dB gain aerial of 61 dB. The attenuation from 50 to 100 km from *Figure 15.33* is 53 dB to 38 dB, or 15 dB. Hence by this reciprocal method field strength would be 46 dB. Average field strength is $(60 + 46)/2 = 53$ dB, and this is taken as the correct estimated field strength.

Sky-wave propagation occurs from the ionospheric *E*-layer after nightfall due to the disappearance of the absorbing *D*-layer. This can be a serious disadvantage because it leads to greatly increased service area often well beyond national boundaries, causing interference in the service areas of distant transmitting stations using the same or adjacent frequencies. Sky-wave propagation is subject to considerable fading due to changes in the *E*-layer, and at a distance of 300 km to 400 km it may produce a field strength comparable to that of the ground wave. Mutual interference between the two is then greatest and severe fading can occur. The height of a vertical medium-wave aerial largely determines the amount of sky wave reflected by the *E*-layer. The greater the height (up to about $0.6\ \lambda$) the less the sky-wave reflection.

High frequency propagation

As the ground wave from a high-frequency transmitter is rapidly attenuated by ground absorption which increases with increase of frequency, h.f. propagation is by ionospheric reflection from the *E*- and *F*-layers (Section 5). The range of frequencies reflected by each layer is determined by the angle of projection from the ground—the shallower this is the greater the reflection frequency—the time of day, season and sun-spot activity. Above a given frequency, which is very variable, there is no reflection because the wave penetrates beyond the layer. Highest reflection frequency from *E*- and *F*-layers occurs about midday at the centre of the propagation path in the summer season and at sun-spot maximum. Greatest useable reflection frequency is about 30 MHz from the *F*-layer during maximum sun-spot years falling to about 20 MHz during sun-spot minimum years.

Field strength median values are amenable to calculation.[2]

Very-high frequency propagation

Very-high frequency propagation[1] is achieved by the space wave (Section 5); near to the ground energy is quickly lost but the effect disappears with increase of height above ground. Thus at heights of $2\ \lambda$ or greater, the space wave is little affected by ground losses. Some bending of the v.h.f. wave occurs in the troposphere (Section 5), but satisfactory propagation is normally limited to line-of-sight. An intervening hill produces a radio shadow on the side furthest from the transmitter, with much reduced field strength in the shadow. The v.h.f. signal has the advantage of much lower impulsive noise interference and it is normally unaffected by the vagaries of the ionosphere. However, fixed or moving objects comparable or greater in size than the v.h.f. wavelength can act as reflectors to cause interfering wave patterns. A fixed object may increase or decrease the field strength at a given receiving point, depending on the phase relationship between the direct and reflected wave. A moving reflecting object, such as an airliner, causes the phase relationship of the reflected wave to vary, and considerable fading and distortion of the received signal occurs.

With a.m. sound signals the change in field strength due to interference from the reflected wave of a fixed object such as a tall building or water tower would not be serious, but when f.m. is used the phase relationship of the reflected varying carrier frequency is constantly changing and this causes serious distortion (known as multipath) of the sound content. Multipath a.m. television signals produce multiple images (ghosting) on the receiver picture tube due to time delay differences between the received signals. The smaller size of v.h.f. receiving aerials (due to the shorter wavelength) permits the design of highly directional aerials, which can be angled to reduce the undesirable reflections from fixed objects.

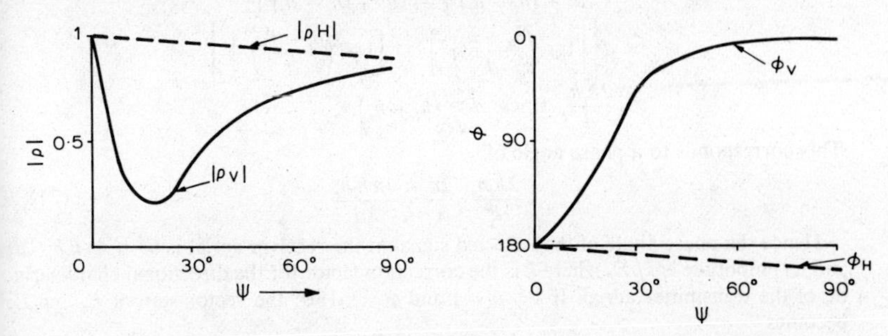

Figure 15.35 Reflection coefficient relative magnitude ρ and phase ϕ at different angles of incidence ψ to the surface (V—vertical polarisation, H—horizontal polarisation)

The energy from a v.h.f. transmitting aerial is projected at other angles as well as horizontal; some is projected upwards to penetrate the ionosphere and be lost, but some is projected downwards to the ground. Part of the energy is absorbed in the ground at the point of incidence but a good deal may be reflected to cause interference at the receiving aerial in the same manner as reflections from a fixed object. The extent of the interference depends on the amplitude and phase of the reflection coefficient and this is determined by the conductivity and permittivity of the ground at the point of reflection, the angle of incidence to the surface and the polarisation of the wave. Typical examples of the variation of the reflection coefficient of relative magnitude (ρ, the ratio of the reflected-to-incident wave amplitude) and phase (ϕ) with respect to angle to the surface (ψ) are given in *Figure 15.35*. The reflection coefficient magnitude is maximum at low values of ψ, i.e., at glancing incidence, and for the horizontally-polarised wave decreases slowly as ψ is increased. For the vertically-polarised wave the magnitude decreases relatively rapidly to a minimum and then rises again towards the horizontally-polarised value. With a perfectly conducting ground the magnitude would fall to zero at an angle ψ known as the Brewster angle. At the low angles of surface incidence and for both polarisations the reflected wave suffers a phase reversal, which changes only to a small extent at higher angles of incidence for horizontal polarisation, but decreases rapidly for vertical polarisation.

High values of reflection magnitude are only obtained when the reflecting surface is smooth, for when this is not so the reflected energy is scattered in all directions. A surface may be considered as rough if the variations in height of the surface multiplied by sin ψ are greater than $\lambda/8$, so that a land or sea surface appears rougher as the transmission frequency is increased.

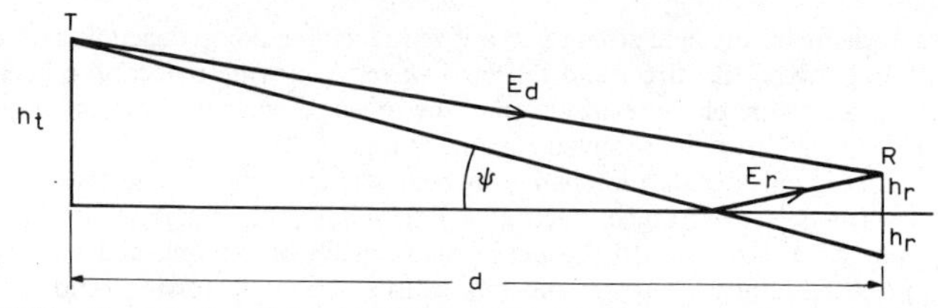

Figure 15.36 Combination of direct and reflected ground ray with v.h.f. propagation (E_d—direct field, E_r—reflected field)

The total field (E_t) at the receiving aerial is the vector sum of direct (E_d) and reflected (E_r) waves with the phase of the latter suitably modified by the extra path length it has had to travel. The extra path length is seen from *Figure 15.36* to be

$$[d^2 + (h_t + h_r)^2]^{\frac{1}{2}} - [d^2 + (h_t - h_r)^2]^{\frac{1}{2}}$$
$$= d\left[\left[1 + \left(\frac{h_t + h_r}{d}\right)^2\right]^{\frac{1}{2}} - \left[1 + \left(\frac{h_t - h_r}{d}\right)^2\right]^{\frac{1}{2}}\right]$$
$$\simeq \frac{2h_t h_r}{d} \quad \text{since } d \gg (h_t + h_r)$$

This corresponds to a phase angle of

$$\frac{2h_t h_r}{d} \cdot \frac{2\pi}{\lambda} = \frac{4\pi h_t h_r}{\lambda d} \tag{21}$$

Hence the phase angle of the reflected signal at the receiver aerial is ($\phi = 4\pi h_t h_r/\lambda d$) and its amplitude is $k\rho E_d$ where k is the correction factor for the directional characteristic of the transmitter aerial. If $k = \rho = 1$ and $\phi = -180°$ the vector sum of E_d and E_r becomes

$$E_t = 2E_d \sin(2\pi h_t h_r/\lambda d) \tag{22}$$

so that the field at the receiver aerial varies from zero at the ground ($h_r = 0$) through a series of maxima when

$$2\pi h_t h_r/\lambda d = (2n - 1) \,.\, \pi/2 \tag{23 a}$$
$$\text{or} \quad h_r = (2n - 1)\lambda d/4h_t$$

and minima when

$$h_r = 2n\lambda d/4h_t \tag{23 b}$$

Under the above conditions a transmitting aerial at a height of 300 m would give maximum field strength at a distance of 30 km when the receiving aerial is at a height of 75 m. For low height receiving aerials $2\pi h_t h_r/\lambda d$ is small and $\sin(2\pi h_t h_r/\lambda d) \rightarrow 2\pi h_t h_r/\lambda d$; since the direct wave $\propto 1/d$, the received field strength will tend to be $\propto 1/d^2$.

In practice due to tropospheric refraction it is found that propagation is always greater than line-of-sight by an amount which would be obtained with a ground profile radius of about 1.33 earth radius, and it is usual to use this value in calculations. Temperature inversions with lowest temperatures near the ground can occur from time to time to produce a ducting effect permitting propagation over considerable distances. The effect is more noticeable over sea than over land.

Diffraction of the transmitted wave can occur due to a hill between the transmitting and receiving aerials, and at grazing incidence a loss of about 6 dB in signal strength occurs. The loss increases rapidly when the hill blocks the line of sight. In order to reduce the loss to a low value, the direct path between the two aerials should clear the obstacle by

$$H = \left[\frac{\lambda d_1(d - d_1)}{d}\right]^{\frac{1}{2}} \tag{24}$$

where d_1 = distance from the receiving aerial to the obstacle.

Thus the summit of a hill halfway between transmitting and receiving aerials spaced apart 30 km should be at least

$$H = \left[\frac{3 \times 30 \times 10^3}{2}\right]^{\frac{1}{2}} = 210 \text{ m}$$

below the direct path if $f = 100$ MHz.

Ultra-high frequency propagation

The propagation of ultra-high frequencies follows the same principles as with very-high frequencies (Section 5). The reduction in wavelength means that smaller objects can produce reflections and the absorption from, for example, trees in leaf is greater; on the other hand aerials are smaller and their directivity and gain can be increased.

REFERENCES

1. *C.C.I.R., Documents of the XIIth Plenary Assembly*, Vol. II, 'Propagation', 217 to 223 (1970)
2. PIGGOTT, W. R., 'The calculation of Median Sky-Wave Field Strength in Tropical Regions', *U.K. Department of Scientific and Industrial Research*, No. 27 (1959)

BROADCASTING TRANSMITTERS

A sound or television broadcasting transmitter has two main parts, viz., the programme circuits and the r.f. amplification circuits. The programme circuits receive the signal from Central Control and amplify to the degree necessary for modulation, the last amplifier acting as the modulator. Most a.m. sound systems use high-power modulation so that the modulator has to provide a peak output power at high modulation percentages approaching that of the carrier unmodulated power, and the stage consumes a large amount of power. With f.m. sound transmitters modulation takes place at an early stage and power consumption is very low. Television transmitters employ amplitude modulation and owing to the very wide bandwidth involved low power modulation is used and appreciable r.f. amplification after modulation has to be carried out. A d.c. component has to be transmitted because black level must be clearly defined at the output of the modulator. In most television transmissions, including the U.K. 625-line transmissions, negative modulation is employed with vision peak-white at minimum carrier amplitude and sync pulses at maximum carrier amplitude. This has the advantage that interference has least effect on synchronising, and impulsive interference produces black spots on the receiver picture tube screen instead of white spots, which would be much more obtrusive.

The first unit of the r.f. stages is a master oscillator which generates a highly stable carrier frequency or submultiple thereof—a frequency stability of 1 in 10^9 or better can be achieved at low and medium frequencies. Efficient usage of the frequency channels allotted to broadcasting requires the carrier frequency to be maintained within close tolerances. Internationally agreed maximum limits are ± 10 Hz over the medium-wave range but most broadcasting organisations maintain frequency within ± 0.05 Hz. Over the long-, medium- and short-wave ranges the master oscillator frequency is amplified in a number of r.f. stages culminating in the modulated amplifier controlled by the modulator. Unmodulated carrier output powers range from 10 kW to 1 MW at medium frequencies; 500 kW is about the maximum power at present available in the short-wave band. In the v.h.f. range f.m. sound transmitter powers do not normally exceed 10 kW and in the u.h.f. range a.m. television transmitters are not greater than about 50 kW. The v.h.f. and u.h.f. aerial system can be made highly directional in the vertical plane

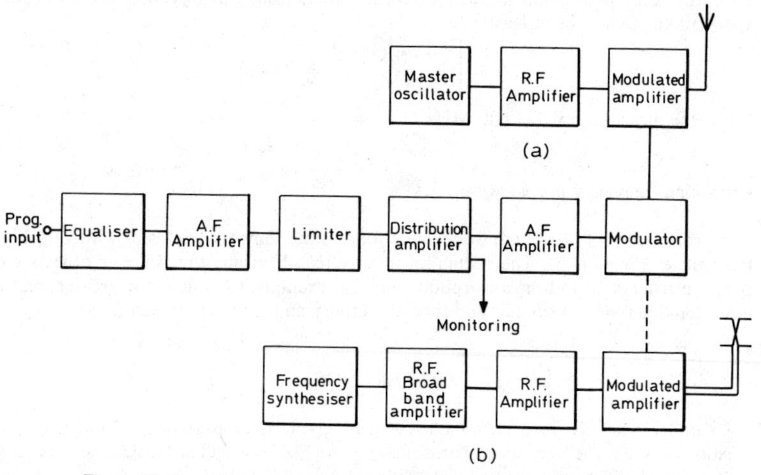

Figure 15.37 *Block schematic of (a) medium-wave sound transmitter and (b) short-wave transmitter*

and this can appreciably increase the horizontal effective radiated power (e.r.p.) beyond the transmitter output power, a ratio of 10 to 1 or greater being possible.

Figures 15.37 (*a*) and (*b*), *15.38* and *15.39* show block schematic diagrams of a medium- and short-wave a.m. transmitter, a f.m. sound transmitter operating at v.h.f., and an a.m. television transmitter at u.h.f. For long and medium frequencies the master oscillator is crystal-controlled and amplification is carried out at the master frequency throughout the r.f. chain. Broadcasting on short waves generally calls for regular changes of frequency to match ionospheric conditions and, either a number of crystals must be available to be switched as desired, or a frequency synthesiser must be used. The advantage of the latter is that any desired frequency can be selected in decade steps over a range from about 4 to 30 MHz, but its frequency stability is about an order lower than the crystal oscillator. The earliest stages following the master oscillator are transistorised and broadly-tuned to cover the short-wave range or selected parts of it. High gain tetrode valves reduce the number of stages required to one driver before the final modulated amplifier supplying 250 kW carrier power. The tuning of such a transmitter is accomplished by 8 tuning positions under the control of a memory system and a preselector switch. A single press button makes an automatic tuning change to the preselected frequency.

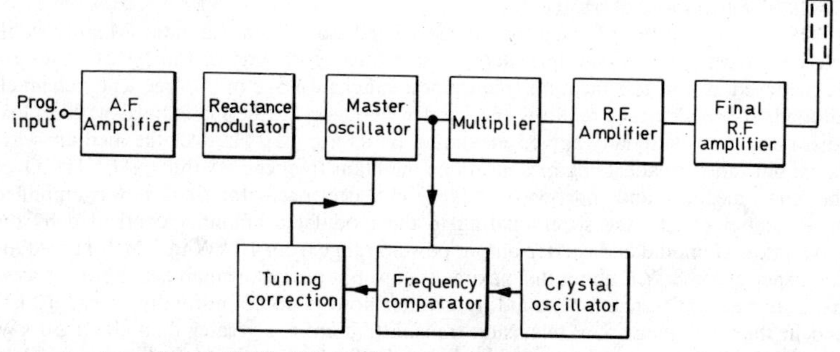

Figure 15.38 *Block schematic of a f.m. sound transmitter*

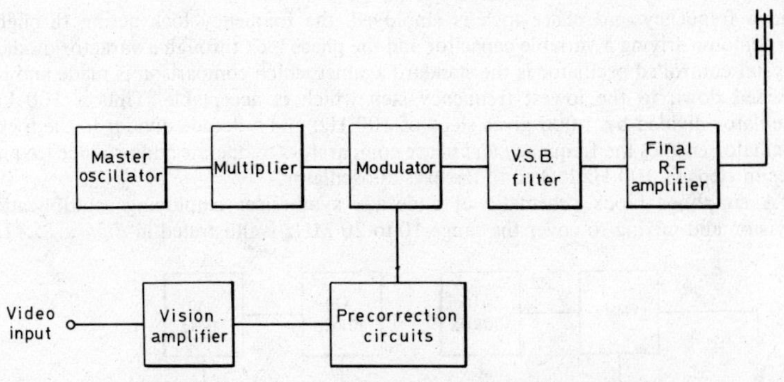

Figure 15.39 Block schematic of an a.m. television transmitter

A f.m. sound transmitter has its master oscillator at a sub-multiple of the final carrier and reactance-modulated by a very low power a.f. output stage. If it is not crystal-controlled its mean frequency will be corrected by comparing it against a separate crystal oscillator, *Figure 15.38*. The multiplier stages following the master oscillator increase the frequency deviation (and also the modulation content) by the same ratio as the carrier multiplication.

In a vision a.m. transmitter, *Figure 15.39*, the low power modulator is followed by a vestigial sideband filter giving a cut off from about 1 250 kHz below the vision carrier frequency. Waveform correction circuits are required in the vision stages before the modulator to correct for deficiencies occurring during modulation and in the klystron output stage.

The master oscillator

A schematic of a typical crystal oscillator is shown in *Figure 15.40*; the oscillation-maintaining device is a transistor as loosely coupled as possible and there is a limiter gain control to prevent excessive oscillation amplitude damaging the crystal or generating spurious frequencies. To reduce drift due to ambient temperature changes the crystal and transistor are placed in a temperature-controlled container or environment such as is found about 10 m below ground level.

Frequency synthesis can be achieved by either of two methods, direct synthesis or locked oscillator. With direct synthesis the frequency is built up from the outputs of stable frequency sources through multiplication, division and mixing (producing sum and difference frequencies). The locked oscillator method is generally used in receivers,

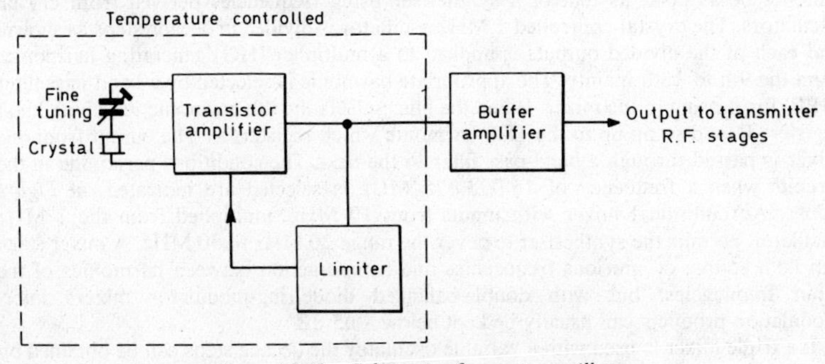

Figure 15.40 Schematic of a master oscillator

and a frequency and phase lock is employed, the frequency lock acting through a servomotor driving a variable capacitor and the phase lock through a varactor diode. A crystal-controlled oscillator is the standard against which comparison is made and it is divided down to the lowest frequency step which is acceptable. Thus a 100 kHz oscillator divided by 1 000 gives steps of 100 Hz, and a decade divider in the locked oscillator enables the frequency and phase comparators to operate and produce frequencies in steps of 100 Hz locked to the crystal oscillator.

A simplified block schematic of a decade synthesiser employing multiplication, division and mixing to cover the range 10 to 20 MHz is illustrated in *Figure 15.41*. It

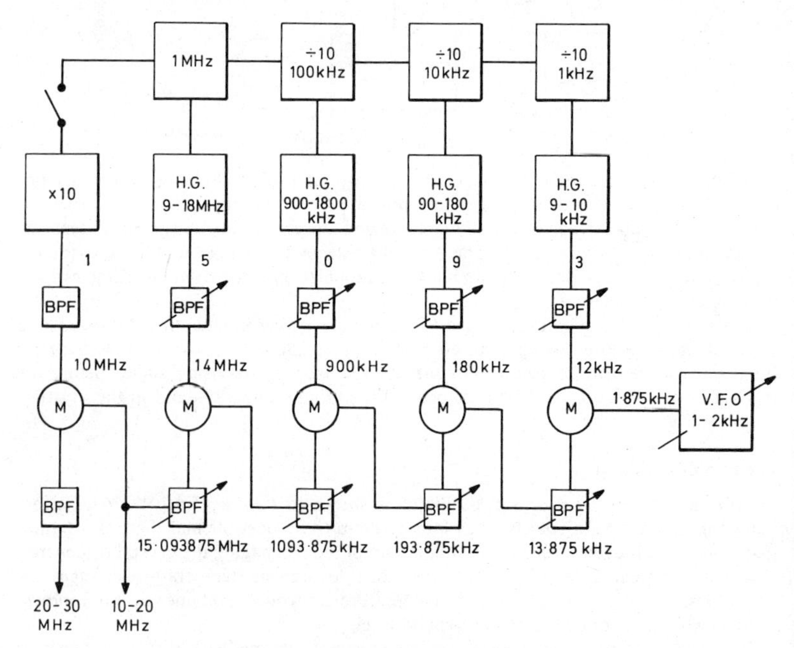

Figure 15.41 Schematic of a frequency synthesiser

uses a crystal-controlled oscillator at 1 MHz, and a variable oscillator for the range below 1 kHz. Overall frequency stability is determined by the variable oscillator and it will not be as good as that of a synthesiser using frequencies derived from crystal oscillators. The crystal-controlled 1 MHz oscillator is divided in decade steps as shown, and each of the divided outputs is applied to a multiplier (HG) generating harmonics from the 9th to 18th mainly. The appropriate harmonic is selected by a band-pass filter (BPF) for passing to the mixer. When the filter selects the 9th harmonic its tuning knob registers 0, and so on up to the 18th harmonic which registers 9. The output from one mixer is passed through a band-pass filter to the next. The conditions pertaining in the circuits when a frequency of 15.093 875 MHz is selected are indicated on *Figure 15.41*. An additional mixer with inputs from 10 MHz, multiplied from the 1 MHz oscillator, permits the synthesiser to cover the range 20 MHz to 30 MHz. A mixer stage can be a source of spurious frequencies due to interaction between harmonics of the input frequencies, but with double-balanced diode-ring-modulator mixers intermodulation products can usually be kept below –65 dB.

If a triple mixer is used with a variable oscillator the decade steps can be obtained by change of the variable oscillator with a fixed band-pass filter. The first mixer produces an output frequency of $f_o - nf_s$ the second adds in the frequency from the previous decade to give $f_o - nf_s - mfs/10$ and the third gives

$$f_o - \left(f_o - nf_s - \frac{mfs}{10}\right) = nf_s + \frac{mfs}{10}$$

which is independent of f_o, the variable oscillator frequency. Change of f_o to $f_o - f_s$ selects the $(n-1)$ harmonic of f_s because the band-pass filter following the first mixer only passes $f_o - nf_s$.

In the v.h.f. range a f.m. sound transmitter master oscillator may be crystal-controlled and reactance-modulated or it may be a variable-frequency oscillator, reactance-modulated with a correction circuit maintaining its mean frequency by comparing it against a crystal oscillator.

Crystal-controlled master oscillators are used in u.h.f. television transmitters at a sub-multiple of the final carrier frequency, the master oscillator frequency being multiplied in harmonic generators having transistors biased into cut-off and producing an output rich in harmonics. Tuned circuits in the output select the required harmonic.

The programme stages of a transmitter

An example of the programme chain for a sound transmitter is given in *Figure 15.37 (a)*. The line by which the sound signal is conveyed from the central control room may require correction for loss of high and low frequency response and this is the function of the equaliser. Equalisation always reduces signal voltage, and amplitude is restored in the amplifier; after the latter is a limiter to prevent the sound programme overloading the transmitter and perhaps causing a flash-over and damage to equipment. The limiter is a variable-gain amplifier, whose amplification is reduced when the input reaches a predetermined level. It is a form of compressor which only comes into operation at high input levels. The limiter output goes to a distribution amplifier having 2 or more separate outputs, one for the amplifiers in the modulation circuit and one for monitoring.

The programme chain for a television transmitter is normally transistorised and the main problem is that of obtaining adequate bandwidth; a.c. coupling is employed and a clamp circuit before the modulator stage restores the d.c. content and fixes black level. The precorrection circuit in *Figure 15.39* introduces clamping, linearity and group delay correction to compensate for deficiencies in the modulator and klystron final amplifier. Modulation may be performed by an absorption method at the final carrier frequency, or by a ring modulator at a much lower frequency and this is known as i.f. modulation.

The absorption modulator requires a carrier power input of the order of 10 watts, which is supplied to a 3 dB line coupler as shown in *Figure 15.42*. The tuned circuit

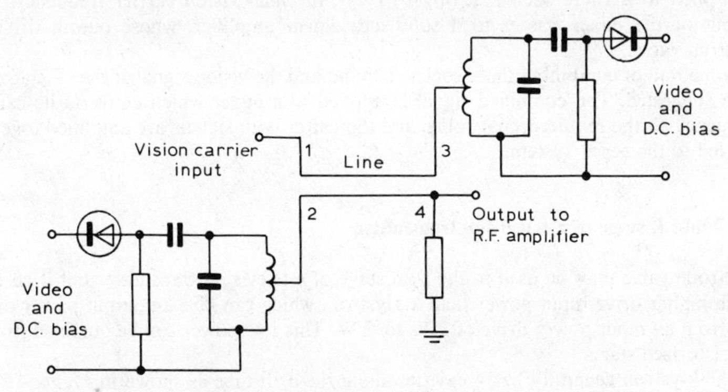

Figure 15.42 Television absorption modulator using 3 dB line coupler

diode terminations at *2* and *3* are adjusted to equal the characteristic impedance of the line so that no reflection occurs and there is zero output into *4*, the input power being divided equally between the terminations *2* and *3*. Application of the video signal to the diodes, which are biased off by about 40 V, changes the damping on the tuned circuits and causes a mismatch at *2* and *3*. The reflected wave dissipates its power in the load at *4* so that carrier power controlled by the vision signal is developed in *4*. If semiconductor diodes are used, they introduce a susceptance component which causes phase-modulation of the carrier. This phase modulation must be cancelled by an equal and opposite phase modulation of the carrier.

The i.f. modulation method is tending to replace the absorption modulator because precorrection can be carried out in the i.f. stage and this gives a better overall performance than video correction, as well as being much easier to set up. A block schematic of the i.f. modulator is shown in *Figure 15.43*. A crystal oscillator followed by a ×2 multiplier provides the i.f. carrier at a frequency (f_i) of about 37 MHz. The ring

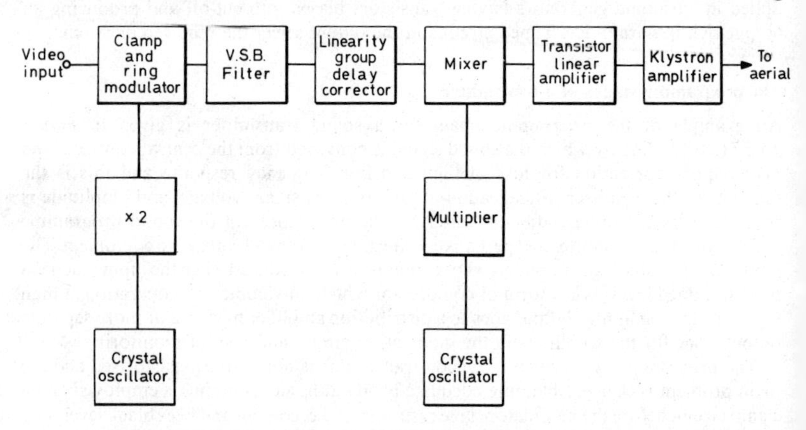

Figure 15.43 Block schematic of intermediate frequency vision modulator

modulator input has a black level clamp circuit and in its output a vestigial sideband filter. The v.s.b. modulated i.f. carrier passes through linearity and group delay precorrection circuits to a mixer where the carrier frequency is increased to its final value. The frequency of the crystal oscillator for the mixer is generally about 15 MHz and this is multiplied to a value such that $(nf_o + f_i) = f_v$, the final vision carrier frequency. The output of the mixer passes to a solid-state linear amplifier, whose output drives a klystron valve.

A method of combining the associated sound and the vision signal at the i.f. stage has been suggested. The combined signal is applied to a mixer which converts its carrier frequency to the required final value, and thereafter both signals are amplified together and fed to the aerial system.

The final r.f. stage of a television transmitter

A tetrode valve may be used in the final stage of a television transmitter but it needs a much higher drive input power than a klystron, which can give an output power of 20 kW from an input power drive of 2 W to 5 W. This low power can be obtained from a transistorised stage.

The klystron generally has 4 cavities along the drift tube as shown in *Figure 15.44*. At the initial setting up all cavities are tuned to the vision carrier f_v, then with the aid of sweep oscillator equipment they are stagger-tuned and suitably loaded to give the required vision bandwidth. Typical off-tune frequencies from f_v for input, second and third cavities are +2 MHz, −2 MHz and +6 MHz respectively. The final cavity is left tuned to f_v. As the electron beam travels down the drift tube it will tend to diverge due to repulsion between adjacent electrons, and magnetic focusing is used to correct this at the cathode, along the drift tube and at the collector. The action of thc klystron amplifier is detailed in Section 7.

The output power of the klystron depends on input signal, and beam voltage and current. If output power is reduced by a decrease in beam voltage the a.c./d.c. efficiency is reduced (its maximum theoretical value is 58%), but if beam current is reduced the efficiency remains unchanged. A modulating electrode (*M.A.* in *Figure 15.44*) is included between cathode and drift tube for this purpose.

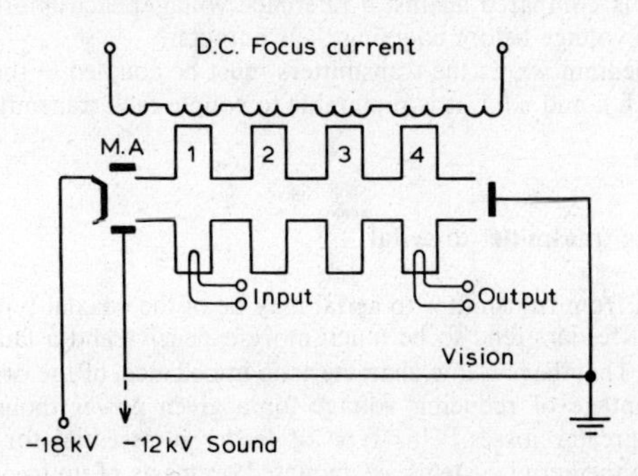

Figure 15.44 Schematic of klystron vision output stage

Sound and vision klystrons are usually identical, the lower power output needed from the sound transmitter being obtained by decreasing the voltage difference between cathode and modulating electrode. In the vision transmitter it is at the same potential as the collector.

Broadcasting transmitter measurements

With sound broadcasting transmitters the usual monitoring (aural and visual) procedure for noise, distortion and frequency response is required at certain points in the chain and also at the output to the aerial, where the original modulation is recovered with the aid of a diode detector. A modulation meter is needed for observing modulation percentage and this is achieved by the comparison of d.c. voltages from the carrier and the modulation after rectification by a diode peak detector. A cathode-ray oscilloscope is an alternative with the modulated r.f. signal applied to one pair of plates and the programme signal to the other. The dimensions of the resulting trapezium on the CRO screen register modulation percentage and curved sides indicate the presence of distortion. A dummy load to replace the aerial is needed for testing the transmitter output.

On short waves and at v.h.f. reflectometers should be provided to give warning of reflections and mismatch between aerial and feeder systems. With high-power transmitters a flashover can occur if there is a high standing wave ratio.

The television transmitter requires similar monitoring and test facilities to the television studio, and the transmitted vision envelope must be checked. A dummy load and reflectometers in the feeder system are also requirements. Mismatch between aerial and feeders can be more serious in television because it can lead to ghosting. Sweep oscillator equipment is necessary for aligning transmitter circuits.

Paralleling broadcasting transmitters

An important aim in broadcasting is continuity of service, and the use of transmitters operating in parallel from the same drive unit with the same programme input can increase reliability of service because failure of one transmitter reduces the radiated power but does not shut down the service. Two conditions must be fulfilled if parallel operation is to be successful. The r.f. voltages and the modulating voltages must be in phase. The second condition presents few difficulties since it is only necessary to ensure that the two modulators and any other paired a.f. amplifiers are as nearly as possible identical. The r.f. outputs must however be adjustable in phase from phasing units containing variable reactances included in the separated r.f. stages. When a phase comparison has to be made the output from each transmitter is switched in turn to a dummy load and is compared against a reference voltage; each is brought into phase with the reference voltage before coupling their outputs.

On long and medium waves the transmitters must be coupled to the same aerial but on short waves, v.h.f. and u.h.f. it is preferable to couple each transmitter to one half of the aerial system.

Feeders connecting transmitter to aerial

The feeder system from transmitter to aerial may be of the coaxial type or of the open-wire type. Coaxial feeders tend to be much more expensive and a fault is less easy to locate and repair. They have a low characteristic impedance, of the order of 50 Ω, and this has the advantage of reducing voltage for a given power though the increased current leads to greater losses. This type of feeder is essential for v.h.f. and u.h.f. transmission since the aerial systems are mounted on masts of up to 300 m height.

The open-wire feeder is preferred for long-, medium- and short-wave transmissions; its characteristic impedance is generally about 300 Ω though it can be designed to be as low as 150 Ω. With short-wave transmissions the frequency and therefore the aerial system must be capable of being rapidly switched. This is done by pneumatically- or electrically-operated switches in the field but remotely-controlled from the transmitter.

The main problem is to match the aerial to the feeder and so prevent high standing wave current and voltage ratios on the feeder. In order to prevent possible flashover and generally inefficient operation a standing wave ratio of about 1.2 is aimed at for the feeder.

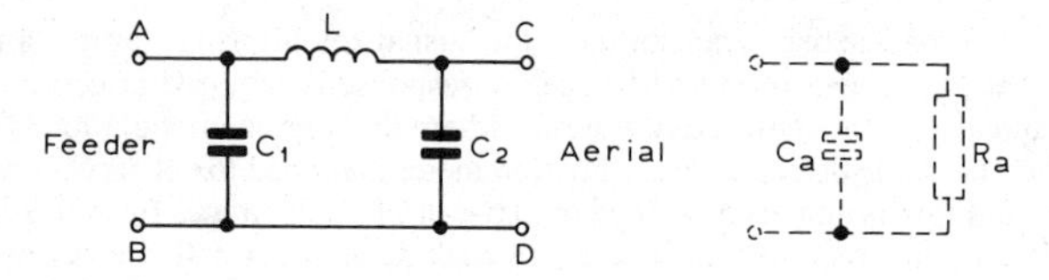

Figure 15.45 The reactance transformer for matching aerial to feeder

An aerial system has two components, resistive and reactive. The reactive component must be cancelled by an equal and opposite reactance and the resistive component must be made equal to the characteristic impedance of the feeder by a transformer or equivalent device. At low and medium frequencies matching is achieved by a reactance transformer using coils and capacitors. The reactance transformer is a π network of C and L as shown in *Figure 15.45*; the values of LC_1 and C_2 are adjusted to give a resistance equal to that of the feeder at the terminals AB. If the aerial admittance is a capacitance, C_a, in parallel with a resistance $R_a \gg R_o$ the feeder characteristic impedance, C_1 becomes zero and $L(C_2 + C_a)$ is a resonant circuit at the carrier frequency.

The circuit looking back from terminals CD is a parallel resonant circuit with R_o as the resistance in series with L, and $R_a = L/(C_2 + C_a)R_o$.

At short waves matching is done on the feeder itself. Whatever the value of the aerial terminal impedance there are always a number of points on the feeder at which the feeder looking towards the aerial connection appears as a resistance of R_o in parallel with a reactance, X. At the point nearest the aerial where this occurs a shortcircuited section of the feeder line is attached, its length being such that it gives the conjugate reactance to cancel X and leave the feeder line back to the transmitter correctly terminated. The unmatched part of the feeder line will, of course, have a standing wave. The same principle can be applied at v.h.f. and u.h.f. using shortcircuited stubs on coaxial feeders.

Transmitting aerials

Low- and medium-frequency propagation is by vertically-polarised wave from a vertical mast about 0.55 λ high, this height giving near-maximum ground-wave propagation. The gain of such an aerial over a short vertical aerial is about 3 dB, and propagation is omnidirectional. A directional pattern can be achieved by feeding two or more suitably spaced masts or by using undriven (parasitic) masts. Thus two vertical masts spaced $\lambda/2$ apart and driven in antiphase produce a figure-of-eight directional diagram. The radiation efficiency of a shorter mast may be increased by a 'capacity' top consisting of an umbrella of wires from the top of the mast. This increases the current along the vertical section; a horizontal section forming an inverted L or T aerial achieves the same object.

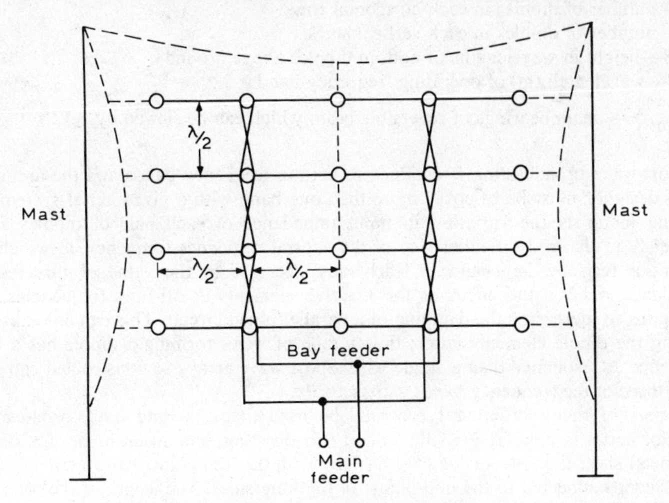

Figure 15.46 A short-wave aerial array

Short-wave broadcasting is normally intended for a particular area at a considerable distance from the transmitting site, and the propagation must be beamed in the desired direction. The beaming must be given the correct vertical angle as well, in order that it may be reflected to the area concerned. Comparatively narrow beams will cover a large area at a distance of 2 000 km; thus a beam width of 5° can cover an area approaching 25 000 sq km at 2 000 km distance. Horizontal dipoles having $\lambda/2$ elements and spaced by $\lambda/2$ in several vertical stacks and all fed in phase are the type most used, and such an array is illustrated in *Figure 15.46*. The $\lambda/2$ spacing between dipoles in a vertical stack introduces 180° phase shift so that the feeder connections must be crossed to cancel and bring the currents into phase. The greater the number of radiators in a vertical stack the narrower is the beam in a vertical plane, and the greater the number of radiators in a horizontal row the narrower is the beam in the horizontal direction.

The angle of the beam to ground is governed by the height of the bottom radiators above ground, the greater the height the smaller the angle. Long distance operation requires an angle to ground of about 5°, and the bottom radiators need to be about 1 λ above ground, whereas a relatively short hop may require an angle of about 25° with the bottom row of dipoles about 0.3 λ above ground. The horizontal direction of the beam can be slewed about 12° on either side of the normal by moving the main feeder connection on the bay feeder off-centre so that the current in one stack leads the other in time phase. The slew is to the side with the longest section of bay feeder. The aerial system of *Figure 15.46* radiates to front and back, and the back radiation can be suppressed and the front radiation increased by using a reflecting curtain of wires or a similar dipole array. With the latter either array can propagate or reflect so that propagation is reversible by switching the main feeder connections. The gain of an array of four vertical stacks each containing four dipole aerials with a reflecting curtain can exceed the gain of an isotropic radiator by at least 20 dB.

Such an aerial array is identified by a coding system. Thus HRRS 4/2/0.8/25,97°,*109°*,121°,*289°*, means

H—horizontal dipoles,
RR—reversible reflector
S—can be slewed
4—number of dipoles in each horizontal row
2—number of dipoles in each vertical stack
0.8—height in wavelengths of bottom dipoles above ground
25—wavelength (m) of operating frequency band
109°, 289° —main bearings of reversible beam which can be slewed by $\pm 12°$

Short-wave broadcasting is confined to certain frequency bands and the aim of the aerial designer must be to cover more than one band with a given aerial system. The limiting factor is the variation in input impedance of each pair of dipoles as the frequency is changed on either side of the natural resonance frequency, at which they present a resistive impedance. Clearly any decrease in the value of this resistive impedance reduces the effect of the reactive elements at off-tune frequencies. It is analogous to increasing the damping of a parallel-tuned circuit. This can be realised by making the dipole elements fatter; thus a tube of wires forming a dipole has a lower resistance at resonance than a single wire. Short-wave arrays so constructed can cover up to three of the frequency bands satisfactorily.

Aerials of many different types may be used for v.h.f. and u.h.f. broadcasting. The slot aerial is popular for v.h.f. sound broadcasting, and an example of a slot cut in a metal sheet is shown in *Figure 15.47* (*a*), with the feed points from a balanced two-wire feeder connected to the mid-points of the long sides. The length of the long sides should be about $\lambda/2$ though if the slot is cut in a cylinder the length should be greater than $\lambda/2$ to allow for the inductive loading of the end of the slot by the rest of the cylinder.

The connections for an unbalanced coaxial feeder are given in *Figure 15.47* (*b*). The electric field is horizontal across the slot so that a horizontally-polarised wave is propagated. A slot aerial system is made up of horizontal groups of four slots cut in the circumference of a metal cylinder forming the top of the masts, and all the slots are fed in phase. An omnidirectional pattern is produced in the horizontal plane, and the radiation in this plane can be increased by constructing groups of cophased four slots, one above the other in a vertical metal cylinder. Eight sets of four-slot groups so

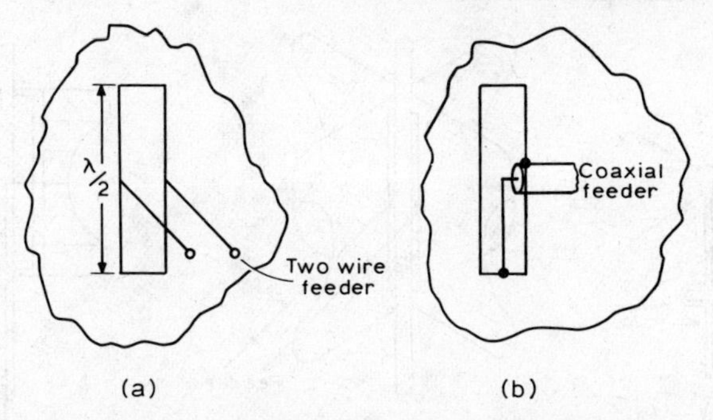

Figure 15.47 *A slot aerial with (a) balanced two wire feeder (b) unbalanced coaxial feeder*

mounted increase the horizontal gain of the aerial system by about 9 dB over the radiation by one set. The vertical spacing of the slot groups is usually 1 λ.

Vertical stacks of vertical or horizontal dipoles made up of $\lambda/4$ elements are often used for u.h.f. television transmission. Horizontal dipoles are preferable since the vertical mast has much less effect on the directional response with horizontally-polarised waves. If a large number of dipoles are used the beam may be narrowed to such an extent that it has to be tilted towards the ground to avoid skip and unsatisfactory propagation in regions near the aerial. The tilting is obtained by a phase-shift and possible amplitude variation of the vertical feeds between each horizontal row of dipoles. A wideband form of dipole such as the folded dipole of *Figure 15.48* (*a*), is required. The wider acceptance band of the folded dipole is due to a reduction in reactance variation to frequencies on either side of resonance. A further widening of the band can be obtained by using metal strip instead of wire, *Figure 15.48* (*b*).

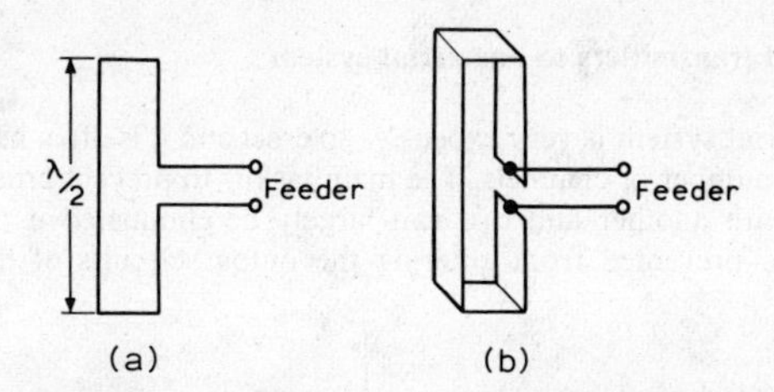

Figure 15.48 (*a*) *Folded dipole aerial* (*b*) *Wideband folded dipole*

Other possible forms of aerial are the turnstile, batwing, biconical, discone, helical and zig-zag; examples of the first two are shown in *Figure 15.49*.

The turnstile, *Figure 15.49* (*a*) consists of two $\lambda/2$ dipoles (made up of $\lambda/4$ elements) crossed at right angles and each fed 90° out of phase to the adjacent ones. Two balanced feeders are necessary and the 90° phase shift between the two feeders is achieved by including an extra $\lambda/4$ length in one of the feeders. Since the power is divided equally between the two sets of dipoles, the radiation is 3 dB less than the maximum radiation from a single dipole but the directional characteristic has the advantage of being omnidirectional instead of the figure-of-eight of the dipole.

The batwing aerial (sometimes called the superturnstile) shown in *Figure 15.49* (*b*) is an improvement on the turnstile, having wide-band characteristics and a greater radiation in the horizontal plane. The omnidirectional radiation has a value almost equal to the maximum radiation of a single dipole. The batwing is not normally constructed as a

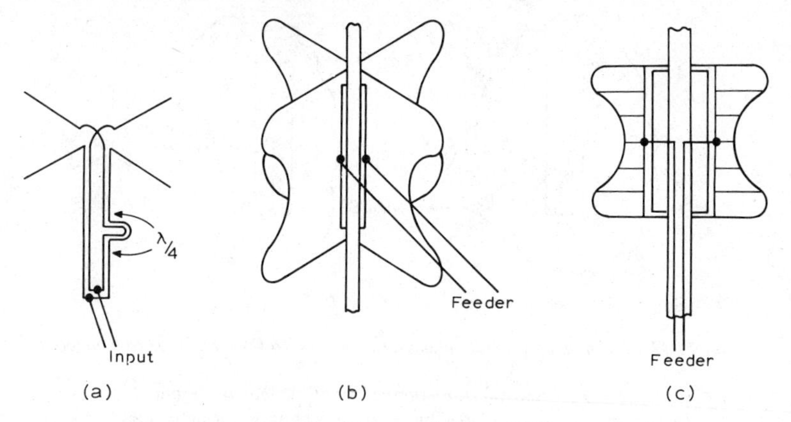

Figure 15.49 Examples of (a) turnstile (b) and (c) batwing aerials

sheet but is made up of horizontal rods connected together at each end, *Figure 15.49* (*c*). Opposite ends of each batwing are connected to the metal mast to form a slot rather greater than $\lambda/2$ long, and the feeder is joined to the centre point of the batwing across the slot. Like the turnstile each batwing is given a 90° phase shift.

Both turnstile and batwing can be used in vertical stacks to narrow the radiated beam horizontally and increase aerial gain. The turnstile vertical spacing can be $\lambda/2$ but the greater vertical dimension of the batwing demands a spacing of 1 λ. The elements are cophased vertically so that the 1 λ spacing of the batwing does not require the feeder connections to be crossed.

Connecting several transmitters to one aerial system

A broadcasting aerial system is very expensive to erect and it is often necessary to use it for transmitting a number of channels. The main task is to prevent cross modulation of one programme with another and this can largely be eliminated if output from one transmitter can be prevented from entering the output circuits of the other. In the

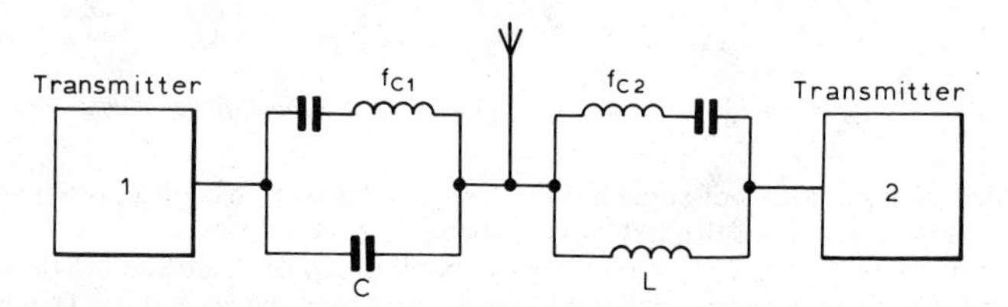

Figure 15.50 Two transmitters at two different medium frequencies feeding into a common aerial

medium-wave range this can be achieved by including in the feeder line circuits tuned to reject the band of frequencies from the other. At the same time these circuits must pass with little attenuation the band of frequencies from the first transmitter. A circuit giving this facility is shown in *Figure 15.50*. Transmitter *2* has a higher carrier frequency (f_{c2}) than transmitter *1*. The series circuits *1* and *2* are resonant at f_{c1} and f_{c2} respectively, and act as a near shortcircuit to these frequencies. Circuit *1* at f_{c2} appears as an inductive reactance and this is tuned to parallel resonance at f_{c2} by C to function as a near opencircuit to f_{c2}. Similarly circuit *2* appears as a capacitive reactance at f_{c1} and this is tuned to parallel resonance by *L*.

The same principle could be used for v.h.f. and u.h.f. with sections of feeder line placed at appropriate points on the feeders to act as a short circuit in the feeder line of transmitter *2* to all frequencies from transmitter *1*, but to approach an open circuit to frequencies from transmitter *2*. The bridge circuit known as a diplexer shown in *Figure 15.51* (*a*) is however preferred. Impedances Z_1 and Z_2, which are lengths of coaxial feeder, are opposite arms of the bridge, the other two being formed by the aerial and a balance resistor equal in value to the aerial resistance. Z_1 passes f_{c1} and stops f_{c2}, and Z_2 vice versa. Transmitter *1* is connected between points *2* and *4* and it sees the part of the

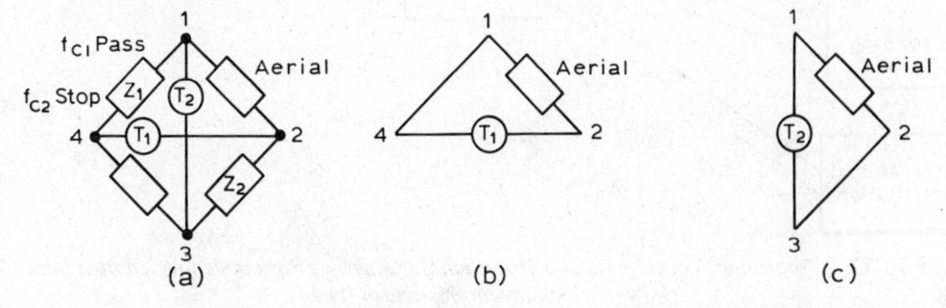

Figure 15.51 A bridge type combining circuit for two transmitters

bridge circuit shown in *Figure 15.51* (*b*), the other two arms having an opencircuit at Z_2. Similarly transmitter *2* sees the circuit of *Figure 15.51* (*c*). Thus both transmitters feed into the common aerial system without any interaction. The circuits shown in *Figure 15.50* are the lumped-constants equivalents of Z_1 and Z_2.

If more than two transmitter outputs are to be combined, another bridge circuit must be brought into service with the combined output of *1* and *2* in one of the diagonal arms and transmitter *3* in the other. Z_{12} must be a broadband short circuit to f_{c1} and f_{c2} and Z_3 a broadband opencircuit to f_{c1} and f_{c2}. Signals from the vision transmitter and its associated sound transmitter are combined in this way, with broadband open and short circuits to the vision frequency.

There are many variants of the bridge diplexer and information on these is readily available.

SOUND BROADCASTING

The equipment required for sound broadcasting

The basic equipment required for sound broadcasting would seem relatively simple, viz., a microphone, a.f. amplifier with volume control, programme line to transmitter and a feeder from transmitter to aerial. In practice the technique is much more complicated and involves a smooth change from one programme source to another, the insertion of special effects, e.g., crowd noises, motor car starting, etc., the reduction of the dynamic range of programme volume without destroying artistic values, the recording and reproduction of programme, the monitoring of output, and the provision of a communication system to coordinate and supervise the distribution to and from studios and to transmitters. A block schematic diagram illustrating a possible grouping of programme requirements is shown in *Figure 15.52*. The programme connection from each source to the control room, from which the composite programme is distributed to the transmitters, is paralleled by a link enabling communication between each source

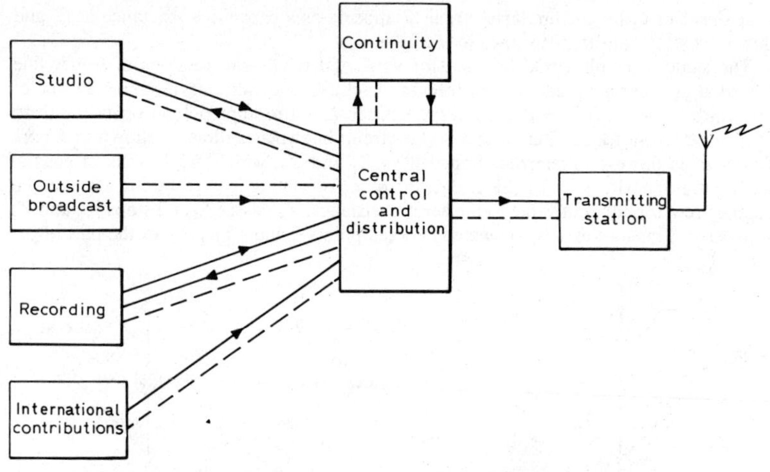

Figure 15.52 Programme collection and distribution (full lines—programme links, dotted lines—two-way communication control links)

and the distribution centre to be established, and to allow a cue feed from the preceding programme to be given when necessary.

THE STUDIO

The studio will need to be treated acoustically to give the right 'atmosphere' for performers and listeners, and this will depend on whether it is required for speech, drama or music. Reverberation time (time taken for sound to die away to 10^{-6} of its original energy intensity after it has been cut off) is important and should be reasonably constant over the a.f. range. For speech it should be of the order of 0.2 s and for large concert halls a value of about 1.5 s would be normal. Drama may call for considerable variations from a very low value, less than 0.1 s simulating an open air scene, to a high value, about 3 s simulating a large room with many reflections from relatively bare walls. Artificial reverberation is often used to cover the latter situation.

The range of audio frequencies encompasses about eight octaves and separate studio treatment has to be applied to control the low, middle and high frequencies. Acoustic design is too specialised for inclusion here.

THE MICROPHONE

A microphone consists of a diaphragm moving in a magnetic or electric field under the influence of the sound waves. The diaphragm may be metallised for operation in an electric field or attached to a coil operating in a magnetic field or attached to a piezo-electric crystal. The main problem is to secure a satisfactory response at all audio frequencies with a desired directional characteristic from an object size, which does not distort the sound field in which it is placed. This condition is most difficult to fulfil at the highest audio frequency; for example at 10 kHz, $\lambda \simeq 3$ cm, and the microphone body would need to be about 3 mm to have negligible effect on the sound field. This would be impracticable because the diaphragm would be unable to pick up enough acoustic energy to overcome the noise in the system. The main effect of size is to introduce an additional directional characteristic. The diaphragm has mass (equivalent to L) and suspension compliance (equivalent to C) so that there is a possibility of a mechanical resonance to modify the electrical output. The designer aims at introducing adequate damping to smooth out the resonance or at removing the resonance outside the a.f. band.

Microphones for broadcasting can be divided into two classes, viz., pressure-operated and differential pressure-operated. A microphone is said to be pressure-operated when the incident sound wave impinges on one side of the diaphragm only. At all frequencies having wavelengths greater than about 10 times the microphone body, the directional response is omnidirectional but at high frequencies a sound shadow is produced on the side away from the incident sound, and the incident sound pressure at the diaphragm is increased when it is facing the sound source and is decreased when it is moved through 180°. The resulting directional characteristic tends to a unidirectional response at high frequencies as shown in *Figure 15.53*. Hence reverberant sound reflected from the walls of the studio will tend to be bass 'heavy'.

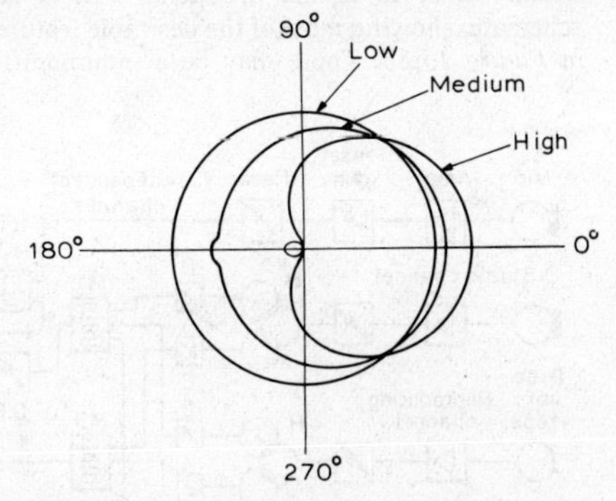

Figure 15.53 Directional response of pressure-operated microphone at low, medium and high frequencies

The differential pressure-operated microphone has both sides of its diaphragm subjected to sound pressure, and movement occurs by virtue of a time difference between the arrival of the sound wave at the front and back. An example is the ribbon microphone, which consists of a flat metallic ribbon suspended between two magnetic pole pieces. Its directional response is bi-directional or figure-of-eight as shown in *Figure 15.54* (*a*) with no pick up to sounds in the plane of the ribbon. The reverberant sound pick-up is much less than that of the pressure-operated type, and it can be used at a much greater distance from the sound source, a considerable advantage with a large orchestra having a large sound producing area.

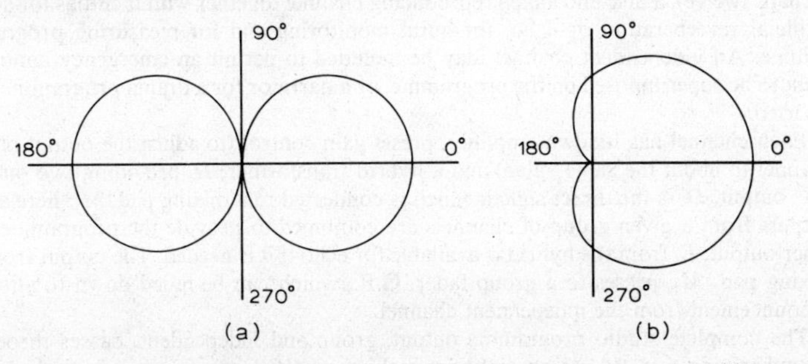

Figure 15.54 (a) Directional response of differential pressure operated microphone (b) Directional response of combined pressure and differential pressure operated microphone

The two types of operation can be combined in one microphone to produce a cardioid or heart-shaped directional response, *Figure 15.54* (*b*) and this can be employed in a theatre production to reduce audience noise with the dead side to the audience. Good quality broadcasting microphones can give a signal-to-thermal noise ratio of 70 dB or better, though their signal output is low and a high-gain a.f. amplifier is required before the volume control.

THE STUDIO APPARATUS

The requirements of one particular broadcasting organisation can differ appreciably from another; one may need a relatively simple operation of recorded disc or tape programmes, interspersed with news and advertisements, whereas another may have to cater for live programmes using many performers in large orchestral, chorus and soloist ensembles or in drama productions. It is nevertheless possible to develop a block schematic showing most of the desirable features of a studio installation and this is given in *Figure 15.55*. There may be a minimum of three microphone channels (rising to

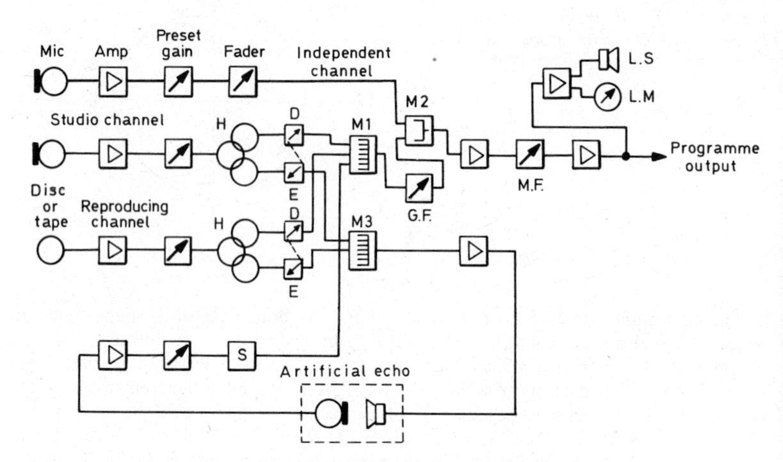

Figure 15.55 Block schematic of typical studio equipment

perhaps twelve), a disc and a tape reproducing channel together with facilities for adding artificial reverberation or echo, for aural monitoring and for measuring programme volume. An independent channel may be included to permit an emergency announcement to be superimposed on the programme, or a narrator for a drama programme to be inserted.

Each channel has its own amplifier, preset gain control (to adjust the output of each channel to about the same value) and a hybrid transformer, *H*, providing two outputs. One output, *D*, is the direct signal, which is connected to a mixing pad, M_1; here all the outputs from a given group of channels are combined to provide the programme. The other output, *E*, from the hybrid is available for echo if it is needed. The output from the mixing pad, M_1, passes to a group fader, G.F., which can be faded down to allow an announcement from the independent channel.

The complete studio programme output, group and independent, passes through a second mixing pad, M_2, to an amplifier making up the losses in the mixing pads, and thence to the main fader control *M.F.*, and another amplifier, which raises programme average level to 0 dB (reference 1 mW) for sending to the control room. A monitoring loudspeaker *L.S.*, and level meter *L.M.* (peak programme or volume) in the studio control room are fed from this point. All connections from the first amplifier in the channel are normally terminated in 600 Ω.

The microphones, being of high quality, have relatively low outputs, and a high-gain (50 dB to 60 dB), low-noise amplifier with preset gain control must be interposed before the channel fader to bring the level to between −20 dB and −10 dB relative to 1 mW. The 'echo' outputs from the hybrids enter a similar mixing pad, M_3, to the programme one, and its output after suitable amplification is connected to a loudspeaker in an echo chamber, or to an artificial reverberation plate or springs giving the desired multiple echoes. The output of the echo chamber or apparatus is amplified and preset volume controlled before being returned to the programme mixing pad (M_1) through a switch (*S*) which cuts it off when not required.

Outside sources are fed at zero level from the central control room to the studio cubicle, where they can be heard on headphones by using *prefade* keys. They are attenuated by about 70 dB to bring to about the same level as a microphone output before being plugged into the source amplifier associated with one of the microphone channels.

Talk-back may be provided from a microphone in the studio control cubicle to a loudspeaker in the studio so that the producer can issue instructions during rehearsal. During transmission the studio loudspeaker is automatically cut off, but talk-back can be provided to artists via headphones. The studio loudspeaker can be used for acoustic effects, which reproduce, for example, crowd noises so that studio background atmosphere is preserved during the insert. A change of background by fading out the studio to insert the effect may destroy an illusion.

Special distortion effects such as the simulation of telephonic speech quality may sometimes be required, and this is achieved by inserting variable frequency-response networks in a channel.

When a discussion is to be broadcast between two or more speakers in geographically distant locations, each must be able to hear the others; unless special precautions are taken 'howl-back' is a possibility. This is prevented by feeding back the output of one studio to the other from a point prior to the insertion of the other contributor. The 'clean' feed is supplied on headphones to the latter. If echo is not needed, the clean feed could be taken from the echo channel, and the second contributor treated as an external source and combined with the first contributor at the mixer before the group fader. The group fader output contains both contributions and creates the impression that both speakers are in the same studio.

OUTSIDE-BROADCAST EQUIPMENT

Outside broadcast (O.B.) equipment for important events may have to provide microphone facilities similar to those for a studio but it must be in easily-assembled transportable units with sufficient spares to deal with the normal faults. An amplifier and power supply from batteries may be carried as standby, and there may be up to four microphone inputs with amplifiers, mixing pad and gain controls, an amplifier for feeding 0 dB to line, a level meter and monitoring loudspeaker. Generally two lines will be available back to the control centre, one for programme and the other of lower quality response for control communication purposes. The latter could be used in an emergency in the event of programme line failure.

A short interview, which is not to be broadcast live, does not warrant outside-broadcast apparatus but only a good-quality tape recorder and microphone. If chance interviews with passers-by in the street are involved, the recording apparatus must be lightweight and battery-operated. A radio-car may be employed for feeding back these programmes by radio to the control centre.

CONTINUITY CONTROL

When a programme is normally broadcast nationwide, special steps may be required to maintain it at all normal broadcasting times, and at some point in the main chain before distribution to the transmitters a continuity control, *Figure 15.52*, is inserted. The apparatus is a modified version of the studio equipment with a microphone which can be switched into the circuit by an announcer. The latter can step in and take control with a special announcement or explanation for an overrun or underrun of programme or a technical fault. He will also have facilities for inserting a recorded item as a fill-in for an underrunning programme.

THE MAIN CONTROL CENTRE

The chief purpose of the main control centre is to accept programmes from many sources and route them to their destinations, either the studios or the transmitters. A degree of aural and visual monitoring is carried out but no gain control is normally performed except when a channel is being set up with test signals. Communication circuits will be available to all sources originating programme, and there will be facilities for carrying out engineering tests such as those for noise, distortion, frequency response, etc. If the majority of programmes are recorded the distribution network can be made to operate automatically using a memory storage system.

Distribution to the transmitters may be by means of telephone-type lines, coaxial lines or microwave links. The frequency response of the telephone-type lines is not usually as good as that of the coaxial or microwave link, and it may have to be corrected by using equalisers. The cost of the line links is dependent on the frequency pass range and a maximum frequency of 8 kHz to 10 kHz is regarded as adequate except for high-fidelity stereo programmes, which require a maximum frequency of 13 kHz to 15 kHz. Pulse code modulation using 13 digits with time-division multiplex is tending to replace frequency-division multiplex links because it can provide a high signal-to-noise ratio. (70 dB to 80 dB as compared with 50 dB to 60 dB for f.d.m. links.)

Stereophonic broadcasting

Monophonic broadcasting suffers from the disadvantage that no spatial sense can be given because the sound source is effectively a single ear. The aim of stereo is to make the sounds appear to come from the position they actually occupy. To do this completely would require many microphone channels with as many loudspeakers. In practice it is not feasible to use more than two channels to convey the stereo sound. The listener normally locates sound by the time difference of its arrival at the ears, with some help at high frequencies from the sound shadow of the head. This principle cannot be applied in stereo broadcasting and the illusion of space has to be created by a difference in sound amplitude from two loudspeakers, which should have identical characteristics and operate in phase. Out-of-phase sources give a diffuse sound image appearing to come from behind the head. An amplitude unbalance of about 20 dB shifts the sound image from the centre line between the loudspeakers to the extreme edge.

The stereo pair of microphones, should have identical characteristics and be located at the same spot. Spaced microphones introduce a time as well as amplitude difference and this tends to make the sound source appear to recede from the listener when the source moves from the centre to the side. Programme control also becomes more difficult since it is only possible to change amplitude. For this reason the coincident technique is employed with two directional microphones (coincident omnidirectional ones cannot create stereo) mounted at the same position with their axes at right angles.

If a pair of microphones (A and B) having figure-of-eight directional characteristics are used with axes at right angles, *Figure 15.56*, the dead axis of each is along the live axis of the other. The useful area is that bounded by the null lines of the 90° axes, and the aim should be to keep the sound sources within this area. Sounds from the back of the microphones are picked up but give a left–right inversion. There is an out-of-phase

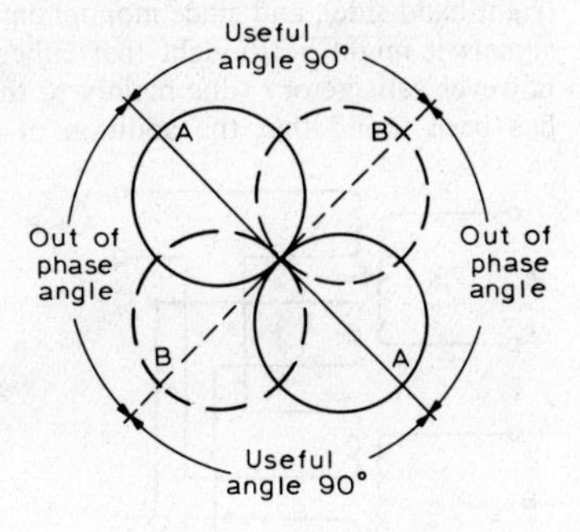

Figure 15.56 Directional response for stereo pair of figure of eight microphones

90° area outside the useful area as shown in *Figure 15.56*, where sound is picked up on the front of one microphone and the back of the other, and this is another reason for confining the sound sources to the useful area.

An improved performance can be gained by using a pair of cardioid coincident microphones with axes at 90°. As shown in *Figure 15.57* they have a much greater useful area of about 180°. Because their dead axes are at 270°, sound sources within the 45° on either side of the useful area can be accepted though their stereophonic image

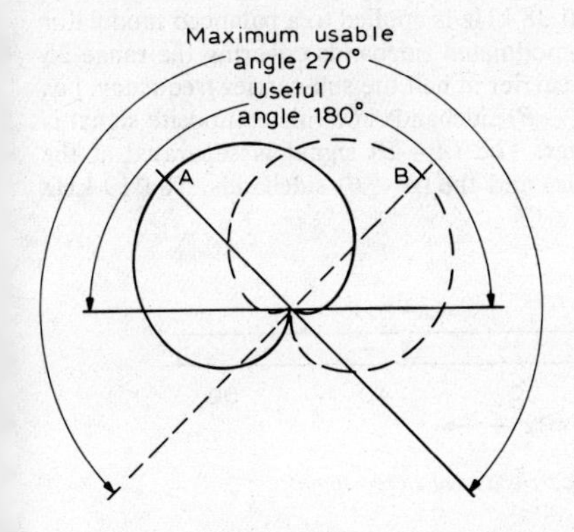

Figure 15.57 Directional response for stereo-pair of cardioid microphones

will be at the extreme left and right-hand edges of the sound space. An additional feature is that there is no out-of-phase region.

Cardioid microphones may be used in the back-to-back position but this requires careful placing of the sound sources if positional defects are to be avoided.

To prevent out-of-phase effects with natural reverberation, stereo microphones have to be used relatively close to their sound sources. When a large orchestra is involved, cardioid microphones can be used closer to the orchestra than figure-of-eight microphones because of their wider useful angle. To compensate for the low ratio of indirect (reverberant) to direct sound, artificial echo will normally be required.

Sometimes a 'spotting' microphone will be necessary for a soloist, and this may entail offsetting the mono feed into the stereo channels to one side or the other in order to maintain the correct spatial position of the soloist. This is done electronically by varying the relative amplitude of the mono signal fed into the two stereo channels by means of a panning or steering gain control (see *panpot* in *Figure 15.60*).

The two stereo microphone output channels are designated *A* (left-hand side) and *B* (right-hand side), and since monophonic receivers must be able to use the stereophonic signals it might be thought that either *A* or *B* output would be suitable. This is not however satisfactory (due mainly to the microphone directional characteristics) and it has been found that the addition of *A* and *B* gives the best monophonic version.

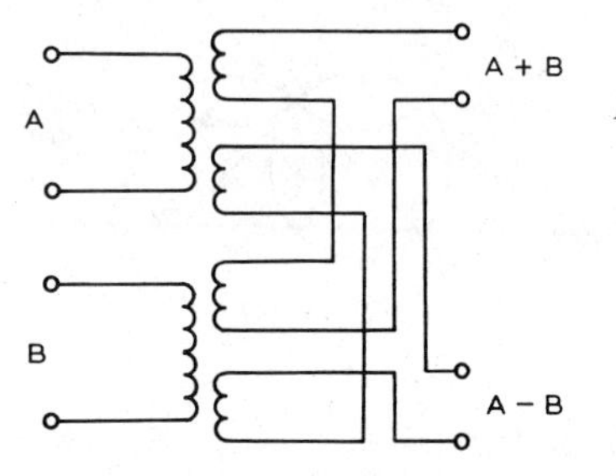

Figure 15.58 Combination of A and B signals to provide (A + B) and (A − B) outputs

Accordingly the *A* and *B* signals are converted at the transmitter into (*A* + *B*) and (*A* − *B*) signals by the transformer shown in *Figure 15.58*. A special multiplex system (the G.E. Zenith system used in many countries including the U.K. and known as the Pilot Tone system) is employed to permit both signals to be transmitted on the same carrier and to be separated at the receiver, *Figure 15.59*.

The (*A* + *B*) signal is left unchanged covering the a.f. range 30 Hz to 15 kHz but the (*A* − *B*) signal together with a sub-carrier at 38 kHz is applied to a balanced modulator whose output contains only the amplitude-modulated sidebands covering the range 23 kHz to 53 kHz. A low-amplitude pilot sub-carrier at half the sub-carrier frequency, i.e., 19 kHz is added to the (*A* + *B*) and the (*A* − *B*) sidebands and this composite signal is used to frequency modulate the transmitter. The (*A* + *B*) signal is separated at the receiver from the pilot sub-carrier (19 kHz) and the (*A* − *B*) sidebands. The 19 kHz

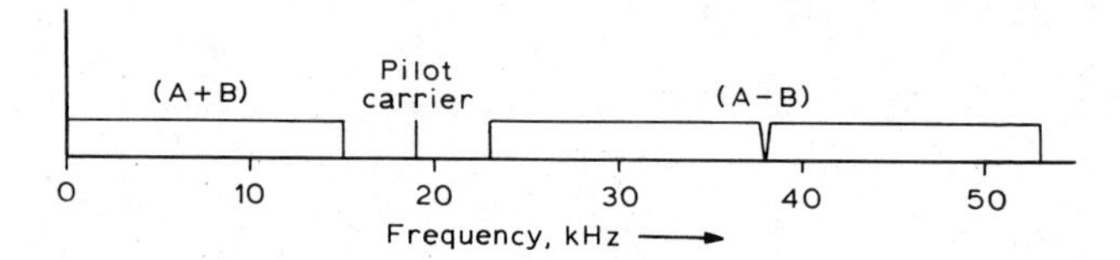

Figure 15.59 Pilot sub-carrier coded stereo signal

sub-carrier is filtered from the other signals and is amplified and multiplied by two before being applied with the (*A* − *B*) to a detector which extracts the (*A* − *B*) signal. Addition and subtraction of the recovered (*A* + *B*) and (*A* − *B*) signals recovers the original left (*A*) and right (*B*) signals. The monophonic receiver rejects the pilot carrier and (*A* − *B*) sidebands and only reproduces the (*A* + *B*) signal.

The studio equipment for stereo operation is very similar to that for mono operation, except that all channels have to be duplicated to carry the *A* and *B* signals. The electrical characteristics of the duplicated channels must be identical. A simplified schematic diagram of a stereophonic channel with facilities for inserting a soloist is shown in *Figure 15.60*. The stereo channels have six controls; immediately following the microphone amplifiers are preset ganged gain controls working in opposite directions to compensate for any difference in microphone sensitivity and place a central sound source centrally. Next is a ganged pair of faders controlling the volume of *A* and *B* equally. A reduction of apparent sound width can be obtained by cross-mixing *A* into *B* and vice versa as indicated by the width control. The channel offset position control allows the sound to be shifted to right or left by reverse ganged gain controls. Similar preset controls permit the proportion of artificial echo to *A* and *B* to be set for any value

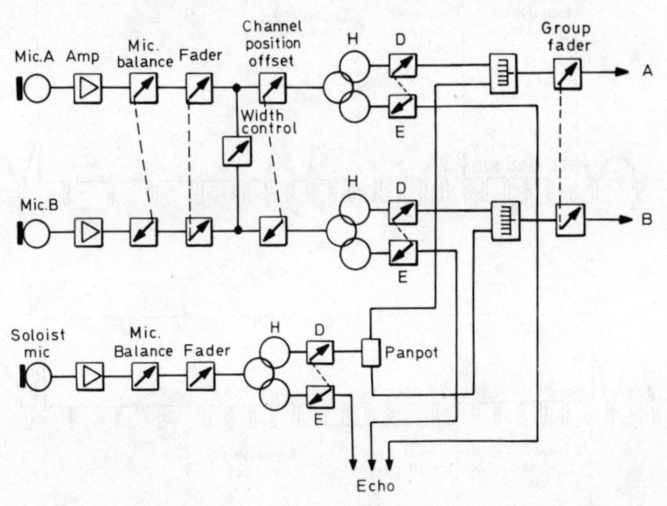

Figure 15.60 Simplified block schematic of stereo and soloist microphone studio equipment

from zero upwards in suitable steps. The group faders in both channels are ganged together. The monophonic channel is conventional except for the 'panpot' controlling the proportion of soloist added to the *A* and *B* channels and is the soloist-positioning device mentioned above.

BLACK AND WHITE TELEVISION BROADCASTING

The black and white television signal has to convey two sets of information, one is the picture signal controlling the light output of the receiver picture tube and the other is the synchronising signal for locking the line and field generators at the receiver so that the picture is reconstituted correctly. The combination of these two signals is known as the *video* signal, and modulation of the vision carrier by the video signal produces the *vision* signal. Average values of light intensity (whether a dark or light scene), i.e., a d.c. value, has to be transmitted and this permits the separation of the sync and picture signals on an amplitude basis.

The first 30% of amplitude is allotted to sync signals and the remainder (30% is black and 100% peak white) to vision. The C.C.I.R.[1] have proposed an international standard of 625 lines 50 fields interlaced, and *Figure 15.61* (*a*) shows the waveform. There are however many variants of the 625-line system in Europe; in North America, Japan and most of Central and South America a 525-line 60-field interlaced system is used.

A field sync pulse for the C.C.I.R. system, consisting of 5 broad half-line pulses, has to be inserted after lines $312\frac{1}{2}$ and 625. The broad pulses are preceded by 5 equalising half-line pulses at black level to ensure that the picture signal has no influence on the field

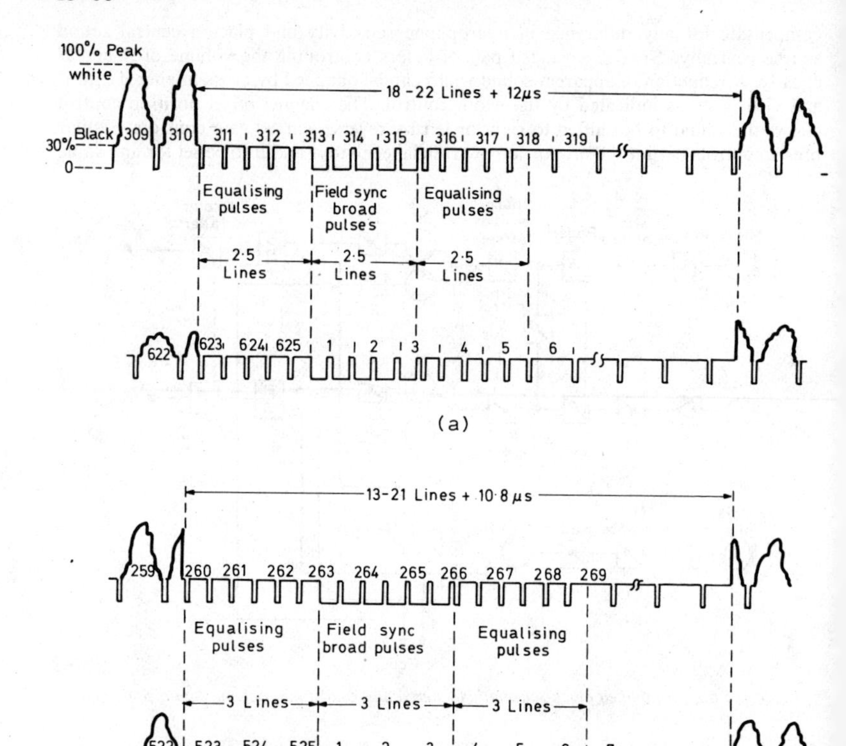

Figure 15.61 (a) C.C.I.R. 625-line/50-field interlaced black/white television waveform (b) American 525-line/60-field interlaced black/white television waveform

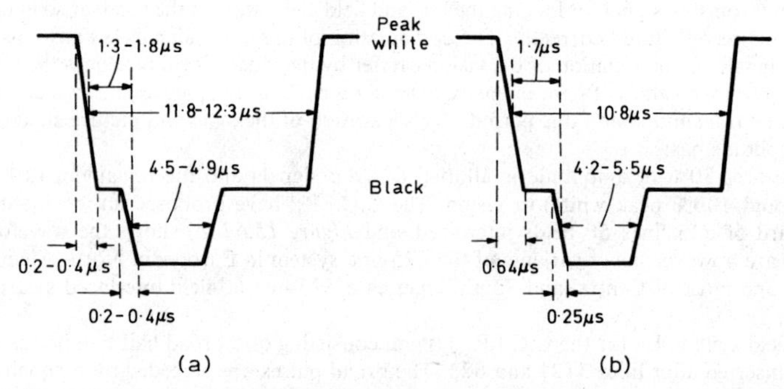

Figure 15.62 (a) C.C.I.R. line sync pulse shape (b) American line sync pulse shape

sync pulse at the receiver. Equalising pulses are continued for $2\frac{1}{2}$ lines after the broad-pulses, and black level is maintained for 18 to 22 lines plus line blanking time (12 μs); this cuts off the picture tube electron beam while the flyback of the receiver field generator returns the beam to the starting point for the next field. In the 625-line system used in the U.K. 25 lines are kept at black level. The American 525-line 60-field waveform is similar, as shown in *Figure 15.61* (*b*), except that it uses 6 equalising and 6 broad pulses, and black level is maintained over 13 to 21 lines plus line blanking time (10.8 μs). Black level starts after lines $259\frac{1}{2}$ and 522, and the field sync broad pulses after lines $262\frac{1}{2}$ and 525.

The duration of the C.C.I.R. line is 64 μs and details of the line sync pulse are given in *Figure 15.62* (*a*). Some latitude in timings is permitted as shown, and as the frequency response of the associated circuits is finite, the pulse edges are not vertical, a transition time of 0.2 to 0.4 μs being allowed. There is a front porch of 1.3 μs to 1.8 μs at black level before line sync to prevent any change in picture level at the end of the previous line causing premature triggering of the receiver line generator. A back porch of 5.1 μs to 6.5 μs ensures that a sluggish line flyback in the receiver does not cause a fold in the received picture. The width of the equalising pulse is 2.2 μs to 2.4 μs and that of the broad pulse 27.1 μs to 27.5 μs. *Figure 15.62* (*b*) shows the line sync pulse timings for the American 525-line 60-field waveform. The average timings and other details for this and the U.K. 625-line system are given in *Table 15.3*.

Table 15.3

	625 line	*American 525 line*
Line time	64 μs	63.5 μs
Line blanking	12 μs	10.8 μs
Line sync	4.7 μs	4.2 μs to 5.5 μs
Front porch	1.5 μs	1.7 μs
Back porch	5.8 μs	4.6 μs to 4.9 μs
Equalising pulse	2.3 μs	2.5 μs
Broad pulse	27.3 μs	27.1 μs
Video bandwidth	5.5 MHz	4.2 MHz
Channel bandwidth	8 MHz	6 MHz
Sound carrier frequency relative to vision	+6 MHz	+4.5 MHz
Vestigial sideband width	1.25 MHz	0.75 MHz
Vision modulation polarity	Negative	Negative
Sound modulation	F.M. ± 50 kHz 50 μs pre-emphasis	F.M. ± 25 kHz 75 μs pre-emphasis
Relative power vision/sound transmitters	5/1	10/1 to 5/1

The television programme chain

A television programme chain is shown in block schematic form in *Figure 15.63*. The studio may contain a number of cameras, whose outputs may have to be faded, cut or mixed, an operation carried out in the production control room. The time taken for the camera signals to reach the fader control point may be an appreciable proportion of line time, and either the camera cables must be of equal length or delay circuits must be introduced to ensure that all travel times are equal. Any control of the camera electronic settings and the lighting is dealt with in a vision/lighting annex at one side of programme control; at the other side is the sound control which contains all the microphone channels and disc and tape reproducing equipment for music inserts and sound effects.

The studio may require film or videotape inserts or the programme may have to be recorded for subsequent transmission; these facilities are supplied from separate telecine (film) and telerecording (videotape) areas. Outside broadcasts of a temporary nature are usually transmitted by microwave to a links terminal, which also receives programme by coaxial cable from semi-permanent sites such as large concert halls. Central control is the focus for all these activities and from here sound and video are distributed to the transmitters. The dotted lines on *Figure 15.63* represent communication circuits between the contributing and coordinating areas. The sound may be distributed on

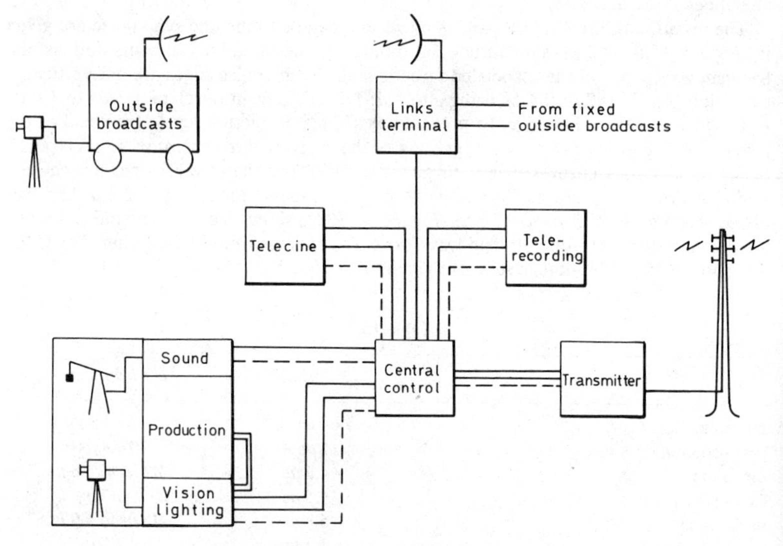

Figure 15.63 Block schematic of vision chain

separate circuits to the transmitters or it may be incorporated as a digital signal during line synchronising periods (see this Section—'High quality sound distribution systems for television' and 'Sound broadcasting systems'); the advantage of the latter is that maloperation causing the wrong sound programme to be associated with the video signal is unlikely, but equipment is required at central control and transmitters for the purpose of extracting the sound signal from the video signal before being used to modulate the respective transmitters. The outputs of the sound and vision transmitters are combined in a diplexer before feeding to a common aerial system.

Other possible facilities for the video programme are inlay and overlay. Inlay involves the replacement of a chosen area in one picture by an area from another. An electronic pulse switches the output from one camera to the other and it is derived from an opaque mask, from which the chosen area has been cut. Overlay is similar, but the switching pulse is obtained from an object in one of the pictures. The object must contrast sharply with its background in order that the silhouette pulse can be generated.

The essential elements of the vision equipment are the camera control unit and the synchronising pulse generator.

The camera control unit

The camera control unit provides power supplies, timing pulses, camera operating potentials, and the picture signal for the vision mixer and preview in production control, for the view finder at the camera and for other picture and waveform monitors. It also performs a series of operations such as blanking off the camera to black level, black level clamping and sync pulse insertion to produce the standard video (picture plus syncs) output. A simplified block schematic is given in *Figure 15.64*.

Line and field drive pulses lock the scanning generators for the camera tube electron beam and also blank off the beam during line and field flyback. The camera output is amplified before being passed to the system blanking inserter, which injects a blanking

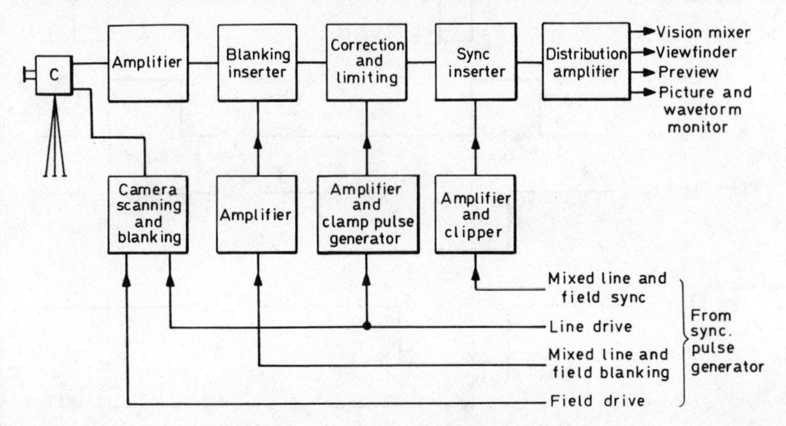

Figure 15.64 Block schematic of camera channel

period of about 12 μs at the end of each line period to accommodate the front and back porches, and the line sync pulse. It also inserts the field blanking period of 1 164 μs to 1 420 μs (18 to 22 lines plus line blanking 12 μs)—1 612 μs for the BBC 625 lines—accommodating equalising pulses, field sync broad pulses and line sync during field flyback. The blanking inserter is often preceded by an aperture corrector which compensates for the loss of the high frequencies due to the finite width of the scanning beam.

Following the blanking inserter is a limiter and black level clamp to establish peak white and black picture limits. The line and field sync pulses are next added to produce the standard video waveform which, after further amplification in a distribution amplifier is fed to the viewfinder, and picture and waveform monitors in studio production and central control. Connectors carrying picture and sync signals between apparatus are normally terminated in 75 Ω.

Talk-back via headphones is used from the producer to each camera man in the studio so that the camera man can receive directions as required.

The synchronising pulse generator

Since the field sync pulses have to start halfway through the 313th line in the 625-line system, half line pulses must be available and the master oscillator for the sync pulse generator, from which all required pulses are derived, has to operate at $625 \times 25 \times 2 = 31.25$ kHz. The pulse waveforms required from the generator are:

(i) Line drive, consisting of rectangular pulses of about 6.5 μs duration at line frequency (15.625 kHz) for locking and blanking the camera line scan.
(ii) Field drive, consisting of rectangular pulses of 480 μs ($7\frac{1}{2}$ lines) duration at field frequency (about 50 Hz) for locking and blanking the field scan.

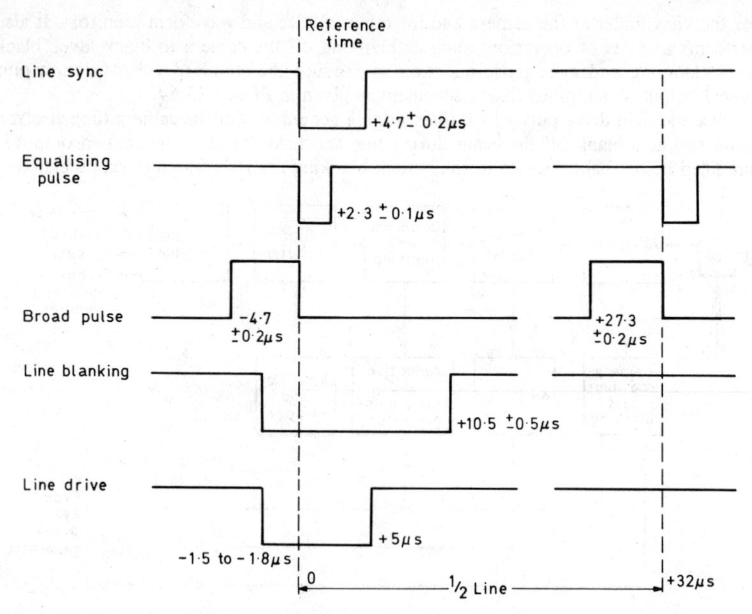

Figure 15.65 Time relationships for half-line and line waveforms for C.C.I.R. 625-line system

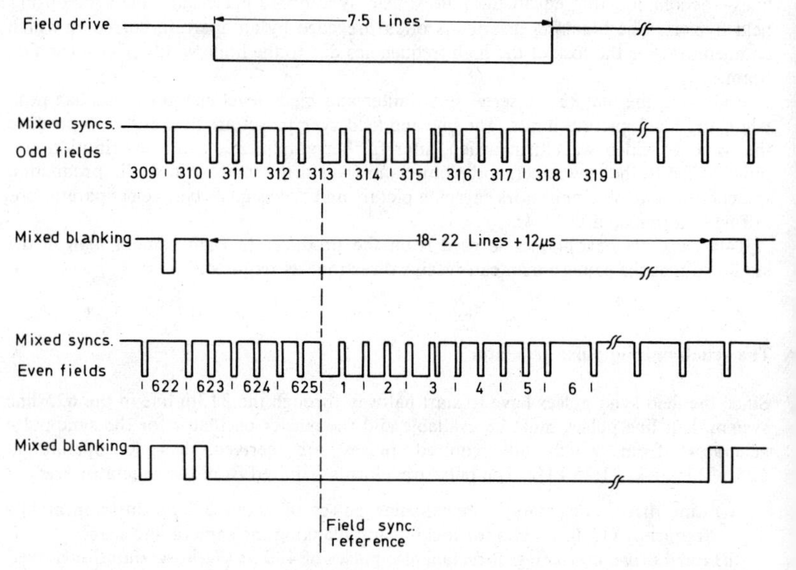

Figure 15.66 Synchronising pulse generator output waveforms for odd and even fields

(iii) Mixed line and field blanking for cutting off the camera output to allow the insertion of line and field sync pulses.
(iv) Mixed syncs of line and modified field sync pulses and the field equalising pulses.

It is best to deal with the pulse waveforms and their relative timings before considering how the waveforms are generated, and *Figure 15.65* shows the relationships of all waveforms associated with line or twice line frequency. The reference time is the leading edge of line sync pulse, and the timings given are for the 625-line 50-field system recommended by C.C.I.R. The leading edges of alternate equalising and broad pulses act as the leading edge of a line pulse to maintain the receiver line generator in sync during the time when they are gated into the waveform. Line blanking start coincides with that of line drive at 1.5 μs before line sync and it ends 10.5 μs after line sync in order to generate front and back porches. Line drive ends earlier than line blanking at 5 μs after line sync.

Field drive, mixed blanking and mixed syncs are illustrated in *Figure 15.66* for odd and even fields. The equalising pulses start at line sync at the end of odd fields and at half line at the end of even fields.

If the master oscillator at twice line frequency produces a sinusoidal output it is converted to square waves by application to a diode or similar limiting circuit. The master frequency may be obtained from an external source, an internal crystal oscillator or an oscillator locked to the 50 Hz mains supply using a divider and a phase comparator controlling a variable reactance in the master oscillator circuit as illustrated in *Figure 15.67*.

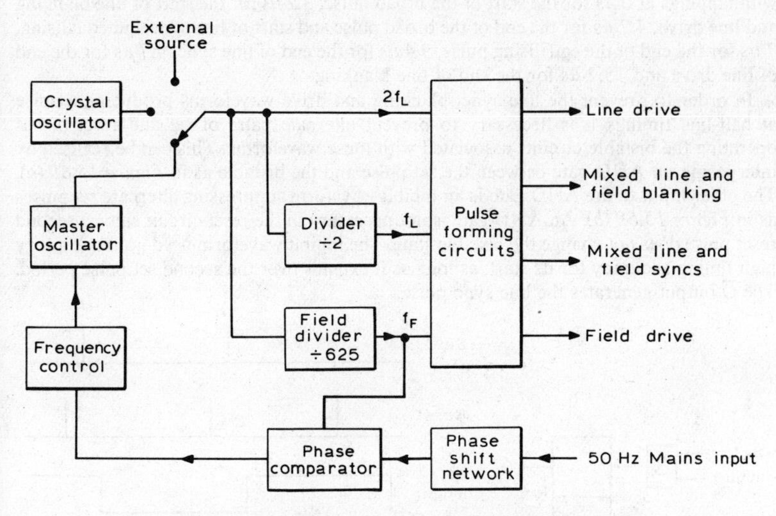

Figure 15.67 Block schematic of sync pulse generator

The pulse forming circuits require accurately-timed edge pulses, which fix the leading and trailing edges, and the bistable circuit, *Figure 15.68 (a)* is the basic element for achieving this. It has two stable states, generally designated 0 and 1. There are two outputs known as Q and $\overline{Q}$, the latter being the inverse of Q giving level 0 when Q gives level 1. Two trigger signals are needed, one sets Q to 1 and $\overline{Q}$ to 0, and the other resets Q to 0 and $\overline{Q}$ to 1. If a series of setting pulses is applied, only the first changes the state and the bistable Q output remains at 1, until a reset pulse is applied. If two pulses are applied at different times to the set and reset terminals, rectangular waveforms appear at

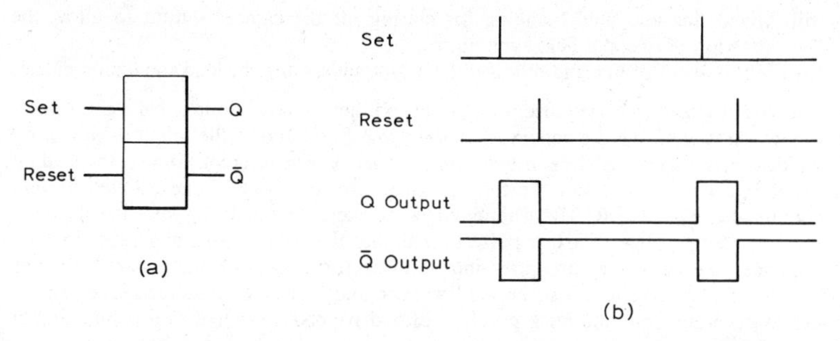

Figure 15.68 (*a*) *Bistable circuit with set and reset pulses* (*b*) *Input and output waveforms*

the Q and $\bar{Q}$ outputs as shown in *Figure 15.68* (*b*). The two pulses in *Figure 15.68* (*b*) could be derived from the leading edge of a waveform, e.g., the squared $2f_L$ output, and the leading edge delayed in time by application to a delay line. Thus the equalising pulse waveform could be generated from the $\bar{Q}$ output if the delay line introduces a 2.3 μs time delay between reset and set pulses. Similarly a broad pulse waveform can be obtained from the $\bar{Q}$ output if the delay line introduces a delay of 27.3 μs; a simpler method would be to advance the set pulse by 4.7 μs as in *Figure 15.65* and take the Q output. Thus generation of the waveforms in *Figure 15.65* would require a delay line with tappings at 0 μs for the start of the broad pulse, 3.2 μs for the start of line blanking and line drive, 4.7 μs for the end of the broad pulse and start of line sync and equalising, 7 μs for the end of the equalising pulse, 9.4 μs for the end of line sync, 9.7 μs for the end of line drive and 15.2 μs for the end of line blanking.

In order to prevent the line sync, blanking and drive waveforms producing a pulse at half-line timings it is necessary to prevent alternate pairs of set and reset pulses operating the bistable circuits associated with these waveforms. This can be realised by interposing an AND gate between the set pulse and the bistable as in *Figure 15.69* (*a*). The other input to the AND gate is an inhibit waveform suppressing alternate set pulses as in *Figure 15.69* (*b*). An AND gate is not necessary in the reset circuit since a second reset pulse does not change the bistable state. The inhibit waveform need not have a very high timing accuracy for its start, as long as it extends over the second set pulse period. The $\bar{Q}$ output generates the line sync pulse.

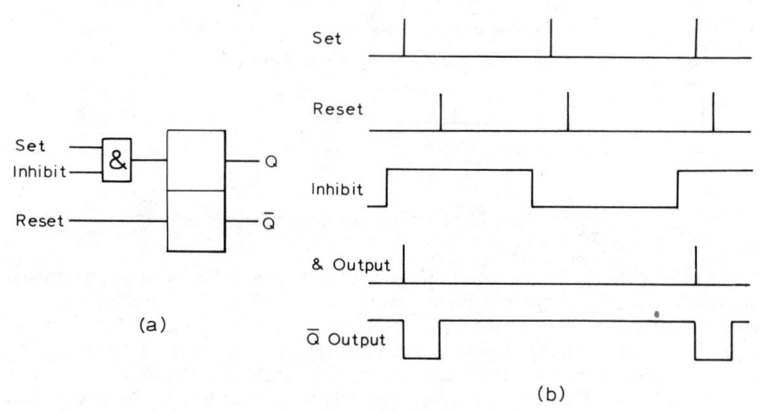

Figure 15.69 (*a*) *Bistable circuit with alternate set pulses suppressed by inhibit waveform* (*b*) *Input and output waveforms of bistable circuit with inhibit waveform*

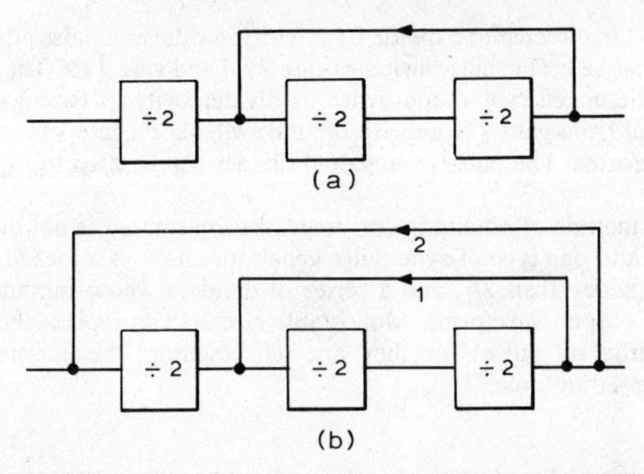

Figure 15.70 Binary divider: (a) division ratio 6 (b) division ratio 5

The field part of the waveform is obtained by dividing the $2f_L$ waveform by 625 in a binary-type divider using the bistable with appropriate 'knock-back' pulses. For example, a binary divider having a division ratio of 4 can be converted to a division ratio of 3 by feedback from the output into the input, because the output pulse itself resets the first bistable so that only 3 pulse inputs are counted. Any desired division ratio can be produced by suitable choice of 'knock-back' pulses. Thus if division by 5 is required two 'knock-back' pulses are required over a binary system normally dividing by 8 as in *Figure 15.70* (*b*). The 'knock-back' line 1 changes the division ratio from 8 to 6, *Figure 15.70* (*a*) and the knock-back line 2 converts from 6 to 5. Four such circuits in cascade divide $2f_L$ down to the field frequency f_F. The field drive waveform is derived from the divided-down field frequency; the duration of the drive pulse can be fixed through suitable addition or subtraction of waveforms generated by the dividers or it could be determined by a pulse from the mixed line and field sync system.

One method of obtaining mixed line and field sync waveforms is illustrated in *Figure 15.71*. The start of the field pulse is used to switch off gate 1 controlling the line sync waveform, and to switch on gate 2 giving free passage to the equalising pulses. An

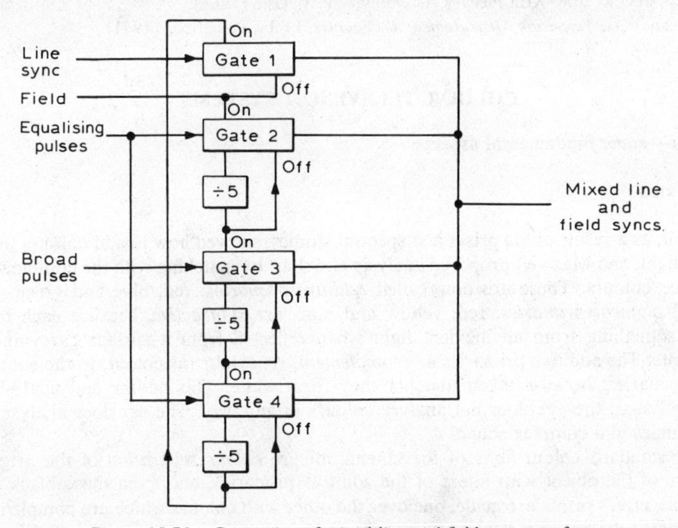

Figure 15.71 Generation of mixed line and field sync waveform

output from gate 2 is directed to a divide-by-5 which produces a pulse after 5 equalising pulses have been passed. This pulse switches gate 2 off and gate 3 on. The latter passes 5 broad pulses to the mixed sync output when its divider switches from gate 3 to gate 4. The divided output from gate 4 switches it off and switches on gate 1 to restart the cycle of line sync waveforms. The pulse from gate 4 divider can be used to fix the end of the field drive pulse.

The delay line method of obtaining correct waveform timings is not the only method that can be used, and one type of sync pulse generator employs a locked oscillator at a much higher frequency than $2f_L$, and a series of dividers whose outputs are gated to give the required output waveforms. Monostable circuits can replace the bistables and these need only the set pulses for they are self resetting, the internal components determining the resetting time.

Measurements on television circuits [2]

Measurements of level, random noise, interference, non linearity and waveform distortion of various types have to be carried out on television equipment. The cathode ray oscilloscope is the measuring tool and many types of test waveforms have been developed. Thus the pulse and bar line waveform indicates circuit performance at the bottom and top end of the frequency spectrum, and a stepped waveform can indicate variations in amplification at different amplitude levels. The two test signals (or modified forms of them) may be included on one line during the field blanking period so that a regular check, usually during a changover from one programme to another, can be made on overall circuit performance. These signals are known as insertion test signals.

Test cards enable a quick estimate of inadequate performance to be made visually on picture monitors. It is not possible here to detail the very extensive range of tests that are available but these are well documented elsewhere.[2]

REFERENCES

1. C.C.I.R. REPORT 308, XIth Plenary Assembly, Vol. V, Oslo (1966)
2. WEAVER, L. E., *Television Measurement Techniques*, I.E.E., Peregrinus (1971)

COLOUR TELEVISION SYSTEMS

Colour—some fundamental aspects

PRIMARIES

Newton, as a result of his prism and spectral studies, showed how to add colours to get white light, and Maxwell proposed analysis of colour by matching with the combination of three colours. These are now called *additive primaries*: red, blue and green. The artists' *pigment primaries*, red, yellow and blue, are *subtractive*, because each takes away something from an incident light when reflected from a surface carrying the pigments. The additive primaries are *complementary*, colour for colour, to the subtractive primaries, because taken together they form white. This adding and analysis is purely visual; the eye does not analyse colours in any way. The ear does analyse the frequencies in a complex sound.

The standard colour-films of the cinema all use *colour separation* of the original colours of the object with filters of the additive primaries, and from these black and white negatives prints are made, one over the other with colours which are complementaries of the original filter colours, i.e., subtractive primaries; in the final print each such colour takes something away from the white projection light and we are left with the required image on the screen.

In *pointillisme*, the artist uses juxtaposed dots of primary colours to add up to a visual effect; originally this was with pigments (subtractive primaries), which was erroneous.[1]

ANALYSIS OF COLOUR

This implies matching with the minimum number of adjusted colours, in practice three, so that the least amount of negative colour is required. A pure spectral colour is termed *saturated*, it has a *hue* independent of its *lightness* (*brightness*), the latter being matched for measurement against a *grey scale*, which alone can be determined objectively. Matching colours by eye for brightness is purely subjective, in exact analogy with the comparison of intensities of sounds of different frequencies by the ear.

We *desaturate* a colour by adding grey, which is merely white of a relatively low brightness, and so get pastel shades, with a *tinge* of colour before the latter fades into *off-white*. White, however, is not a definite mixture of colours, but is defined as a surface, such as of titanium oxide, white lead or barium carbonate, which reflects all incident light, regardless of frequency or wavelength over a given range, e.g., the visible range. For illumination, white is defined in terms of a colour temperature.

STANDARDISATION

For technical and scientific purposes the International Commission on Illumination (CIE, Commission Internationale l'Eclairage, ICI in U.S.), has deduced the psycho-physical system of colour specification, which is now widely used. By mixing three lights (not pigments or inks) any colour, even pure spectral lines, can be matched. A *negative* contribution of a primary additive colour merely means that it has to be added to the sample colour to achieve a match. The CIE system therefore specifies three primary colours of high saturation, of energies X, Y, Z, of primaries P_x, P_y, P_z, for mixing. By matching specific wavelengths from the spectrum, the *standard observer* is obtained from the averaged results. With this basis and a spectro-photometric energy curve, the colour specification of a sample can be computed with respect to a specified light source without any visual colour matching.[2, 3, 4]

CHROMATICITY DIAGRAM

Hue, saturation and lightness are human sensations in the brain, and hence cannot be measured by any physical *trichromatic* matching. Analogous magnitudes can, however, be found for colour matching data, e.g., *dominant wavelength*, *purity* and *luminous factor*, which form the basis of the CIE system, as expressed in the chromaticity diagram. In this the x and y coordinates are derived from

$$x = X(X + Y + Z); \qquad y = Y(X + Y + Z); \qquad \text{hence } z = 1 - (X + Y).$$

In this way the spectral colours, having the highest purity, form a *spectral locus* as the horseshoe part of the diagram; the lower rim is arbitrarily completed by a straight line between the red and the blue, representing the magentas and purples, not to be found in the spectrum. The various whites (daylight, sunlight, tungsten, specified fluorescent lights or corrected mercury) are grouped just below the centre, so that points on a radial line to a specific wavelength represent gradations of saturation, i.e., purity. Thus a colour is specified by the x and y of this chart, together with a luminous factor (lightness), given in CIE tables, and based on P_x, P_y, P_z, needed to match unit energy of each wavelength in the visible spectrum.

To standardise white sources of illumination, the CIE laid down two sources A and C. Standard Illuminant A is for artificial light, and is given by a tungsten lamp at the colour

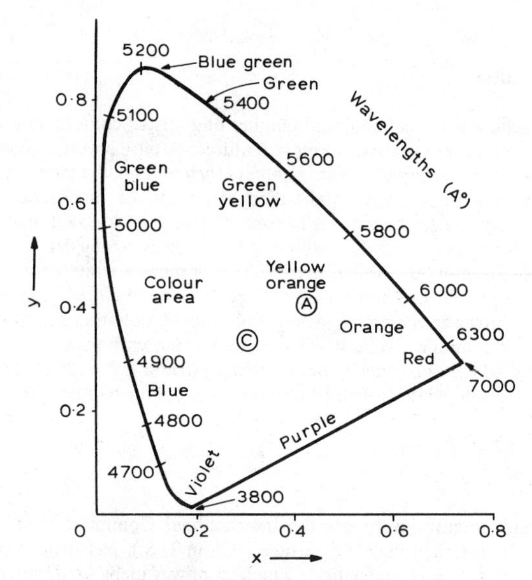

Figure 15.72 *CIE Chromaticity diagram showing Standard Illuminants A and C (wavelengths are in angstrom units)*

temperature of 2 854 K, while Standard Illuminant C is for daylight (mixture of sunlight and daylight), and is taken to have a colour temperature of 6 500 K. These are indicated on the chromaticity diagram, *Figure 15.72*.

FILTERS

These are required to change the character of a light or its colour, by rejecting what is not wanted. In an optical system, e.g., source, lenses and detector (photocell or photographic emulsion), a filter has the same effect wherever interposed. Filters are made from dyes and chemicals dispersed in gelatin or glass, or gelatine cemented between glass plates, and of various grades, either for use in scientific and technical work, or for stage or photographic usage. They are comparable with lenses and should be treated as such.

Thus a suitable filter can give tungsten light a close resemblance to daylight or sunlight, it can isolate bands from light sources, such as tungsten or mercury light, especially yellow, green, blue-violet, deep violet or ultra-violet from the latter. Filters are standardised by their makers according to numbers in a catalogue,[5] and can be relied on to be extremely uniform and repeatable. A *monochromatic* filter substantially removes all colour from an object, and gives an accurate idea of the relative luminosities of its parts.

Filters can be used to study the relation between primary colours and their complementaries as indicated in *Table 15.4*.

Table 15.4 RELATION BETWEEN COLOURS AND FILTERS

	Filter	*Absorbs*
Primary colour (lights)	Red	Green and blue
	Green	Red and blue
	Blue	Red and green
Complementary colours	Red + green = yellow	Blue
	Red + blue = magenta	Green
	Green + blue = cyan	Red

Gelatin filters require considerable care in handling, but have the advantage that they can be cut to shape (between inscribed sheets of paper) by ordinary scissors from sheet. Such filters should never be touched by hand, but held in a frame and kept in a dry flat book when not in use. Filters cemented between flat pieces of glass should be treated as lenses and kept in special boxes when not in use. They should be cleaned only with special fluid, such as can be obtained from Kodak Ltd, and no fluid should get on the rim, because it can enter the seal and possibly damage the gelatin filter proper.

Polaroid or Pola can be considered as a filter in which polarised light is controlled by a chemical dispersed in gelatin or plastic. Apart from the deliberate polarisation of light from a source, polarised light arises from the sky normal to sun's rays, and in reflections from any shiny object. Relative rotation of two pieces of such sheet material completely absorbs light passing through both. A genuine system of stereoscopic reproduction of objects through film projection uses such sheet material, with polar axes at plus and minus 45°, to separate images thrown on to a screen with the requisite polarisation; the images presented in a stereoscopic system must always be kept separate for the two eyes if the system is to depend on the parallax of the two eyes.

DENSITY

While the transmission of a filter is clearly the ratio of the intensity of the light transmitted to the intensity of that incident, and the absorption is unity less this figure, it is found more convenient when dealing with filters in photographic and photocell work to deal with *density*, i.e.

$$D = \log_{10}(1/\text{transmission})$$

Using this measurement, the densities of several filters can be added when light passes through them in series. A correction may be necessary to allow for the specular reflection between such filters, which can become serious. Lens components are cemented together with Canada balsam, of average refractive index, especially to avoid this specular reflection, and the same can be done with filters during manufacture, if required. *Blooming*, i.e., the vacuum deposition of a thin layer of magnesium fluoride on lenses, performs the same function for a glass-to-air surface.

Surfaces are shiny because they exhibit specular reflection, as desired when varnish is added to matt paint. Even the best of matt surfaces has some specular reflection at grazing incidence, as on a wet road at night. The colour of any surface is altered by the specular reflection of the incidence light, as is easily demonstrated

COLOUR TEMPERATURE

This concept arises from the black-body radiation idea, absorbing all incident radiation. For self-radiation, the black-body represents the limit of possible radiation, depending only on temperature and considered by Planck in the early days of this century. Thus all black-bodies, properly so-called, have the same colour at the same temperature, and they radiate continuously throughout the spectrum, with a dominance depending on the temperature.

A radiating body can have, by comparison with a black-body, a colour temperature which may be very different from its actual temperature; thus a tungsten lamp has a colour temperature (expressed in degrees Kelvin) some 50° above its actual temperature. Sources can have their colour temperatures altered by suitable filters. Thus it was found that a good sunshine effect could be obtained for a studio for colour-film by over-running a tungsten lamp and using a straw-coloured filter.

CHARACTERISTICS OF COLOUR VISION

Colour vision has three more or less independent characteristics:

(i) *Brightness*—by means of which colours may be located on a grey scale. Brightness sensation is determined by the sum of the responses of the eye perceptors.
(ii) *Hue*—by means of which colours may be placed in categories such as red, green, yellow, blue, etc. For spectral colours, hue is determined by the dominant wavelength.
(iii) *Saturation*—the degree by which a colour departs from a grey or neutral of the same brightness. Pale or pastel colours are much less saturated than deep or vivid colours. Saturation is determined by radiant purity, or the extent to which the light energy is confined to a narrow band of wavelengths. A saturated colour does not contain any white light.

Colours may be matched by mixtures of no more than three primary colours. These should be as widely spaced on the spectrum as practicable, and no combination of any two must be capable of matching the third. The most useful set for colour television consists of highly saturated red, green and blue. The choice of primaries for colour television is strongly dependent on the colours which are achievable with practical phosphors of cathode-ray tubes, used for displaying the colour pictures.

Hue and saturation are determined by the ratios of the stimulation of the eye perceptors, and depend on the relative distribution of light energy. In matching saturated colours, hue is controlled by the ratio of *two* of the primary colours. Hue and saturation are controlled by the proportions of the *three* primaries.

Brightness (or luminance), Y, represents the energy level by which the eye is stimulated and is controlled by the summation of the primaries. Colours of different hues and saturations may have the same or different brightness levels. In the absence of hue, colours of different brightness would appear as different shades of grey (as in black and white television). These features of controlling brightness, and matching hue and saturation, indicate the means by which signals representing the three characteristics of colour vision may be coded in a transmission system.

Basic features of colour television systems

The first practical colour television broadcasting system was developed in the United States after a variety of methods had been proposed, and in some cases tried. The evolution of this system was guided by the recommendations of the National Television System Committee and it has consequently been termed the NTSC system. Other variants have since been developed, incorporating many of the basic features of NTSC. These are PAL (Phase Alternating Line) and SECAM (Sequential Colour with Memory) systems.

The NTSC system is used in the country of its origin, Canada, Mexico, Japan and a few other countries. The PAL variant, which started in West Germany, is used in the United Kingdom and most West European countries as well as Australia, New Zealand, Brazil and South Africa.

SECAM, developed and used by the French, has been taken up by some countries with French connections as well as by the U.S.S.R. and some East European countries including East Germany.

The recommendations of the NTSC which have been adopted by all three systems were broadly as follows:

(i) *Picture quality*—the colour television system should produce high quality images with good colour fidelity, and should have a performance with respect to noise, flicker, brightness, contrast, resolution and picture texture substantially equal to black and white standards.

(ii) *System compatibility*—the system should be compatible, that is it should produce high quality images on black and white receivers without receiver modification. Conversely, colour receivers designed for the colour system must produce black and white pictures from black and white transmissions.

(iii) *Operation within existing frequency channels*—the system should operate within existing frequency channels and should make the maximum use of the frequency spectrum by assigning the available space to the various components of the colour signal in proportion to the eye's demand for the information conveyed by those components.

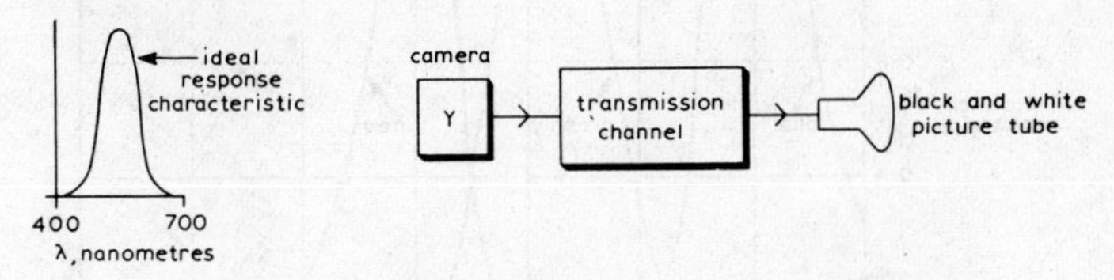

Figure 15.73 Simplified black and white television system

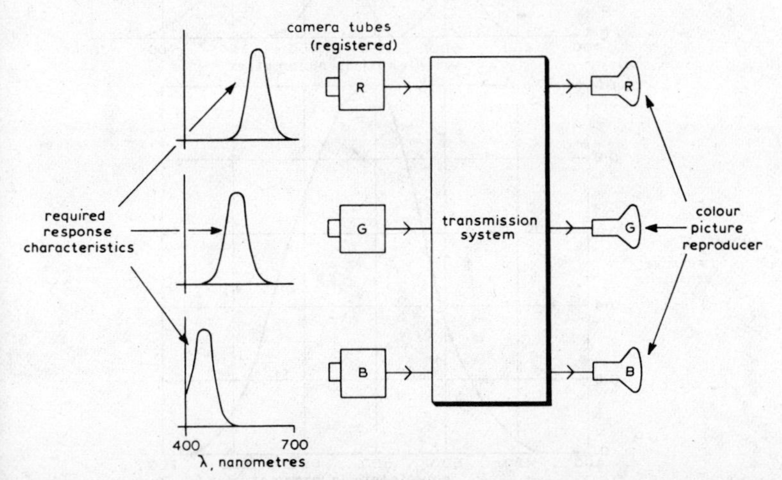

Figure 15.74 Block diagram showing the basic requirements of a colour television system. The camera scans are registered so that identical picture elements are examined simultaneously on all three camera tubes. Similar registration exists in the colour picture reproducer

In black and white television, all the necessary information about the varying brightness of different picture elements is conveyed on one signal channel. For colour, three separate signal channels are necessary, to provide information about hue, saturation and brightness.

For black and white television, the camera tube should have a spectral sensitivity characteristic which matches that of the eye, so that the relative brightness of different colours is correct (*Figure 15.73*). The simplest type of colour camera from an analytical point of view uses three pick-up tubes, one for each primary colour (*Figure 15.74*). The relative sensitivities of the three camera tubes are adjusted so that their output voltages are equal when white or neutral is being scanned. Instead of providing a separate pick-up tube for the brightness signal, a similar spectral response may be obtained by mixing together the three primaries in the right proportions.

If the spectral response curves of the R, G and B channels are added, wavelength by wavelength, in the ratio 30% R, 59% G and 11% B, the resulting characteristic has approximately the same shape as the response of an ideal black and white channel (*Figure 15.75*). This can be accomplished electrically by combining the signals from the R, G and B cameras in a simple resistive network (*Figure 15.76 (a)*). The three

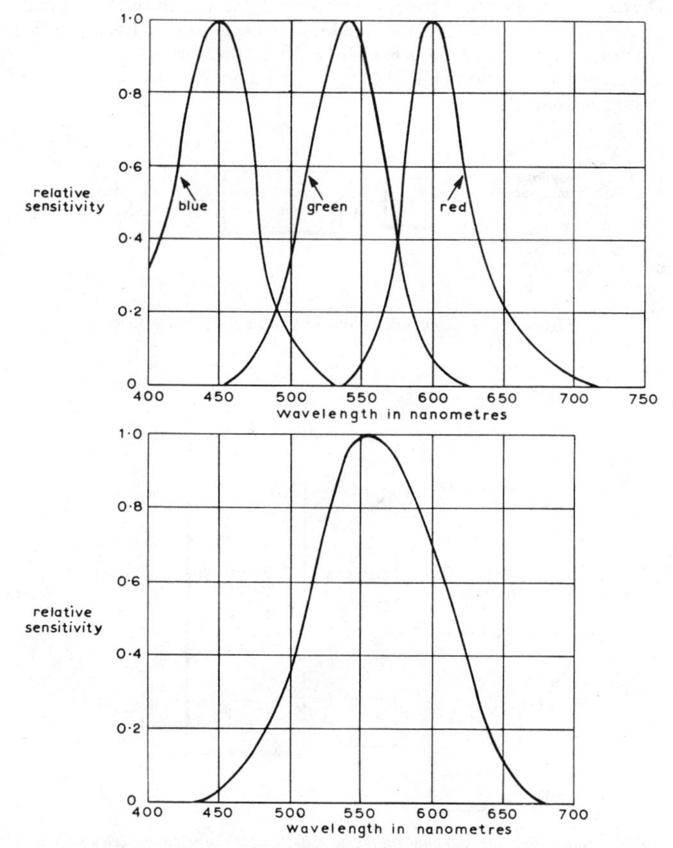

Figure 15.75 *Spectral response of red, green and blue colour channel and (below) ideal response of a black and white channel*

weighting factors have been adjusted so that their sum is unity; thus when peak white is being scanned, R = G = B = 100%, and Y also equals 100%.

If the system were linear, the Y signal would be identical to that produced by a linear black and white camera with an optimum spectral response. In practice, gamma correction is used to compensate for non-linear picture tube characteristics, but the Y signal is still a good approximation of the output of a black and white camera.

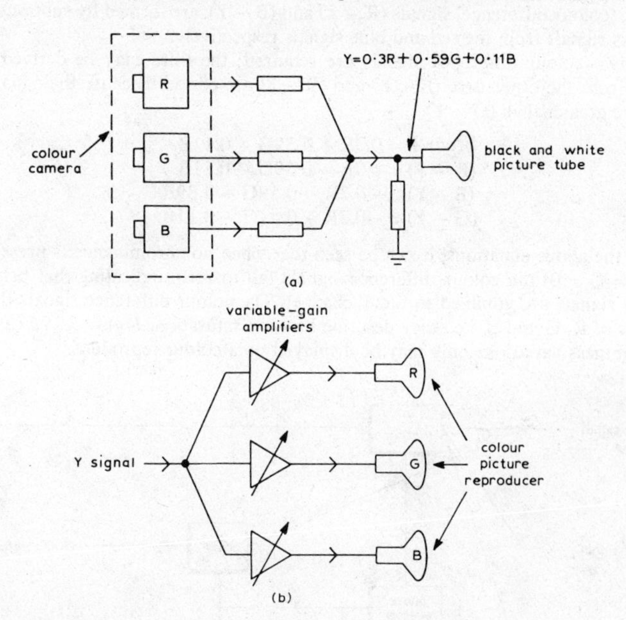

Figure 15.76 Diagrams showing (a) how a colour camera may be connected to provide a black and white signal, and (b) how a black and white picture may be displayed on a colour picture reproducer

At the receiving end of a colour system, a full colour image is produced by adding the light output of three registered images in R, G and B.

There are two basic methods for combining the primary images:

(i) Superimposition by dichroic mirrors or projection.
(ii) Intermingled colour dots, excited simultaneously by three separate cathode ray beams, and too small to be resolved separately, as in the shadow mask tube.

The system shown in *Figure 15.74* would require three times the bandwidth of a black and white transmission, and it would not be compatible. Black and white receivers could pick up the green channel signal but the pictures would not be panchromatic; greens would be too bright, reds and blues would be too dark on the black and white pictures.

If a fourth *channel* were provided for a Y signal (derived from the R, G and B cameras as shown in *Figure 15.76* (*a*), the bandwidth would be increased further and redundancy would be introduced because it is only necessary for three characteristics to be transmitted.

Ideally, the Y channel should carry all the luminance information and although three *camera tubes* are necessary, only two additional *channels* are required for hue and saturation.

The Y signal is a summation of the R, G and B signals in their appropriate proportions, whereas hue and saturation depend on their ratios; thus if two chrominance signals only are transmitted (chrominance refers to hue and saturation of colour, as distinct from brightness), the third may be derived in the receiver by subtraction from the Y signal.

To ensure that no brightness information is transmitted in the two chrominance channels, 'colour difference' signals $(R - Y)$ and $(B - Y)$, are formed by subtracting the brightness signals from the red and blue signals respectively.

Only two colour difference signals are required; the third may be derived in the receiver from the other two. $(R - Y)$ and $(B - Y)$ are chosen because their maximum values are greater than $(G - Y)$.

$$\text{From } Y = 0.3R + 0.59G + 0.11B$$
$$(R - Y) = 0.7R - 0.59G - 0.11B$$
$$(B - Y) = -0.3R - 0.59G + 0.89B$$
$$(G - Y) = -0.3R + 0.41G - 0.11B$$

From the above equations, it can be seen that when no chrominance is present (i.e., when $R = G = B$) the colour difference signals fall to zero, indicating that brightness variation signals are confined to the Y channel. The colour difference signals describe the ratios of R, G and B, i.e., they describe hue and saturation. *Figure 15.76* (*b*) shows how brightness variations only may be displayed on a colour reproducer.

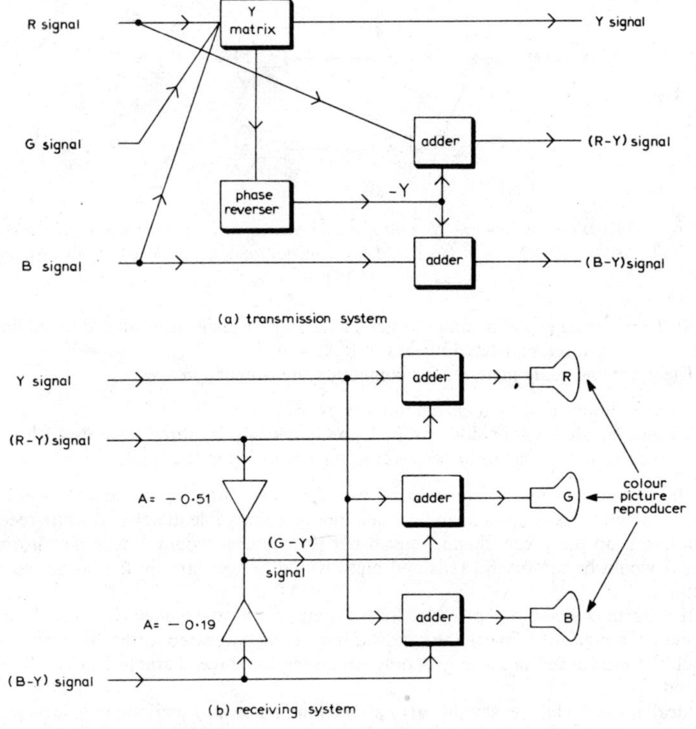

Figure 15.77 Simplified block diagram showing coding and decoding of luminance and colour difference signals

Visual acuity is very much less for hue and saturation differences than for brightness variations, which means that very fine detail is only appreciated in black and white. Consequently, the bandwidth of the chrominance channels carrying (R – *Y*) and (B – *Y*) signals may be reduced drastically in comparison with the bandwidth for the *Y* channel. Furthermore, the (R – *Y*) and (B – *Y*) signals may be used to modulate a suitably chosen sub-carrier within the normal video spectrum, which is then combined with the *Y* signal to form a composite waveform. This waveform amplitude modulates a radio-frequency carrier in the usual manner, and thus the system operates within existing radio-frequency channels.

Although the composite signal is received by the black and white receiver, use is made of the *Y* signal only. The colour receiver decodes the composite signal into its original R, G and B components. *Figure 15.77* (*a*) and (*b*) show a simple block diagram of the coding and decoding of luminance and colour difference signals, in which the (G – *Y*) signal is obtained from the (R – *Y*) and (B – *Y*) signals. [Equation 3 (see later), gives the derivation of (G – *Y*) in terms of (R – *Y*) and (B – *Y*)]. The diagram shows the connections to the adders which produce R, G and B inputs to the colour picture reproducers as follows:

(R – *Y*) added to *Y* = R
(G – *Y*) added to *Y* = G
(B – *Y*) added to *Y* = B

When brightness signals only are being received, (R – *Y*) and (B – *Y*), and consequently (G – *Y*), will be zero and the *Y* signal is applied to the three inputs to produce a black and white picture. In practice, the cathode and grid electrodes of each of the three guns of the shadow mask tube may be conveniently connected as adders; negative *Y* signals are applied to the cathodes and the appropriate colour difference signals are applied to the grids.

Having dealt with some of the basic principles, it will be useful to summarise and expand a number of points.

In a colour television camera, the scene to be televised is viewed simultaneously by three camera tubes through a lens system and an arrangement of colour filters (dichroic mirrors). One tube is provided with a red, one with a green and the other with a blue image of the scene which is thereby analysed in terms of the three chosen primaries. Exact registration of scans must be maintained so that each corresponding picture element is scanned at the same instant.

In general, the three camera channels may be identical in all respects following the light sensitive surface of each camera tube.

Practical responses for the three channels are shown in *Figure 15.75*, although these are not necessarily the only ones which are achieved or even required.

Using the maximum incident light with which the scene is to be lit, the camera is focused on to a white surface of high reflectance; this will produce maximum amplitude signals from each of the three channels. With a suitable lens aperture setting in the light path common to all three tubes (i.e., before the light is split into its three primary components), the gains of the three channels are adjusted individually so that at an early stage in the signal processing, the three signals are equal in amplitude. Whatever the actual voltage values may be, it is convenient to regard the signals as normalised to a maximum value of unity.

The arrangement is fundamentally that shown in *Figure 15.74* and although a complex coding and decoding system will be used, the essential requirement is that the effective proportions of the three primaries reflected from the scene into the camera are reproduced at the viewed output of the colour picture reproducer. To this end, the system as a whole must be linear—the problems of non-linearity and gamma correction are considered later.

For the peak white signal described, which provides equal amplitude inputs to the linear transmission system, the signals applied to the three inputs of the colour picture reproducer will also be equal. In this condition, the gains of the R, G and B displays must be adjusted until peak white of the appropriate intensity is viewed. Correct black level adjustment is also necessary. This system enables a setting up facility to be employed, in that wherever R, G and B signals appear separately (irrespective of the signals actually being transmitted), they may at least in principle, be shorted together temporarily, thus giving them the same amplitudes. This allows grey scale balance to be achieved in the subsequent parts of the chain.

Having balanced the complete system for the grey scale, including black level and peak white, the proportions of any two of the primaries at the input, i.e., R and G, G and B, or R and B, will be reproduced at the output to describe any hue within the possible range; similarly, the proportions of the third primary will describe variations of saturation.

The three signals are coded in the transmission system to achieve compatibility and bandwidth compression.

To enable black and white receivers to display a reasonably panchromatic picture, a luminance signal is transmitted which is the summation of the brightness (or luminance) of the three primary components. For transmission purposes, the electronic signals due to these primaries are amplified to equal values when white is viewed by the camera, so that with the particular primary wavelengths employed, it is necessary to use weighting factors of 30%, 59% and 11% respectively, for the red, green and blue signals when adding them together to provide a luminance signal. For instance, if a saturated blue of maximum amplitude is being transmitted, the normalised output of the blue channel is unity, but the contribution to the luminance signal is only 0.11; this is because the sensitivity of the eye to the wavelength of blue is rather low—saturated blues appear to be rather dark. Similarly, saturated green of maximum amplitude produces an output from the channel of unity, but its contribution to the luminance signal is 0.59; while saturated red of full amplitude produces an output of unity and the contribution to the luminance signal is 0.3.

Table 15.5 VALUES OF *Y*, R, G AND B AND THE COLOUR DIFFERENCE SIGNALS FOR SATURATED COLOURS AT FULL AMPLITUDE

	R	G	B	*Y*	(R – *Y*)	(G – *Y*)	(B – *Y*)
Yellow	1.0	1.0	0	0.89	0.11	0.11	–0.89
Cyan	0	1.0	1.0	0.7	–0.7	0.3	0.3
Green	0	1.0	0	0.59	–0.59	0.41	–0.59
Magenta	1.0	0	1.0	0.41	0.59	–0.41	0.59
Red	1.0	0	0	0.3	0.7	–0.3	–0.3
Blue	0	0	1.0	0.11	–0.11	–0.11	0.89
Mean arithmetical values of (R – *Y*), (G – *Y*) and (B – *Y*)					0.47	0.27	0.59

Saturated colours of other than the primary hues will produce outputs from two primary channels. For example, when a saturated yellow of maximum amplitude is being viewed, unity outputs will be obtained from the red and green channels and the contributions to the luminance signal will be 0.3 + 0.59, i.e., a total of 0.89. A less saturated yellow would produce an output from the blue channel with a further (small) contribution to the luminance signal.

The next step is to produce colour difference signals, and these will now be dealt with more fully.

The values of *Y*, (R – *Y*), (G – *Y*) and (B – *Y*) for six saturated colours at full amplitude, arranged in descending order of luminance, are given in *Table 15.5*.

In the receiver, (G – Y) is formed from proportions of –(R – Y) and –(B – Y). These may be derived as follows:
From

$$Y = 0.3\mathrm{R} + 0.59\mathrm{G} + 0.11\mathrm{B} \quad (1)$$

and letting

$$Y = 0.3Y + 0.59Y + 0.11Y \quad (2)$$

$$0 = 0.3\mathrm{R} - 0.3Y + 0.59\mathrm{G} - 0.59Y + 0.11\mathrm{B} - 0.11Y$$

$$0.59(\mathrm{G} - Y) = -0.3(\mathrm{R} - Y) - 0.11(\mathrm{B} - Y)$$

$$(\mathrm{G} - Y) = -\frac{0.3}{0.59}(\mathrm{R} - Y) - \frac{0.11}{0.59}(\mathrm{B} - Y)$$

$$(\mathrm{G} - Y) = -0.51(\mathrm{R} - Y) - 0.19(\mathrm{B} - Y) \quad (3)$$

From equation 3, the following relationships may be derived:

$$(\mathrm{R} - Y) = -1.96(\mathrm{G} - Y) - 0.373(\mathrm{B} - Y) \quad (4)$$

$$(\mathrm{B} - Y) = -2.68(\mathrm{R} - Y) - 5.26(\mathrm{G} - Y) \quad (5)$$

It is thus apparent that by choosing the combination of (R – Y) and (B – Y) for transmission instead of either (R – Y) and (G – Y), or (G – Y) and (B – Y), a simplification is achieved in the receiver; the alternatives would involve amplification instead of attenuation in deriving the third colour difference signal. This result is not unexpected when it is remembered that the response characteristic of the Y channel more closely resembles that of the green channel (near the middle of the spectrum) than those of the red and blue channels. As shown in *Table 15.5* both the maximum and mean arithmetical values of the (G – Y) signals are smaller than those of the (R – Y) or (B – Y) signals. Whereas R, G, B and Y signals are always polarised in one direction, colour difference signals may vary in either positive or negative directions and must have a definite datum of zero. Whatever method of modulating the sub-carrier is used, this datum must be retained and must be identified in the decoder.

The NTSC system

The arrangements described so far are common to all colour television systems, but the methods of modulating a sub-carrier with chrominance information differ, and these will be described separately for the NTSC, PAL and SECAM systems.

In the NTSC system, the sub-carrier is modulated in amplitude and phase by signals derived in principle from the colour difference signals (R – Y) and (B – Y). Although the modulating signals ultimately used in transmission differ from the (R – Y) and (B – Y) signals, the system will be described as if the latter were used, and corrections will be made as necessary, at the appropriate stage.

SUPPRESSED CARRIER MODULATION AND SYNCHRONOUS DETECTION

Suppressed carrier modulation, with sidebands, is used for two sub-carriers of the same frequency but differing in phase by 90°. The sub-carriers are modulated separately by the (R – Y) and (B – Y) signals respectively, and then combined to provide a resultant which varies in amplitude and phase. In the receiver, a local oscillator synchronised by a transmitted 'burst' provides separate appropriately phased sub-carriers which are combined with this resultant in two synchronous detectors, to produce completely separate outputs of the original (R – Y) and (B – Y) signals.

Figure 15.78 shows the envelopes resulting from normal (*Figure 15.78* (*a*)) and suppressed (*Figure 15.78* (*b*)) carrier modulation. It can be seen that the removal of the carrier (and the retention of the sidebands) results in a modulated envelope which varies

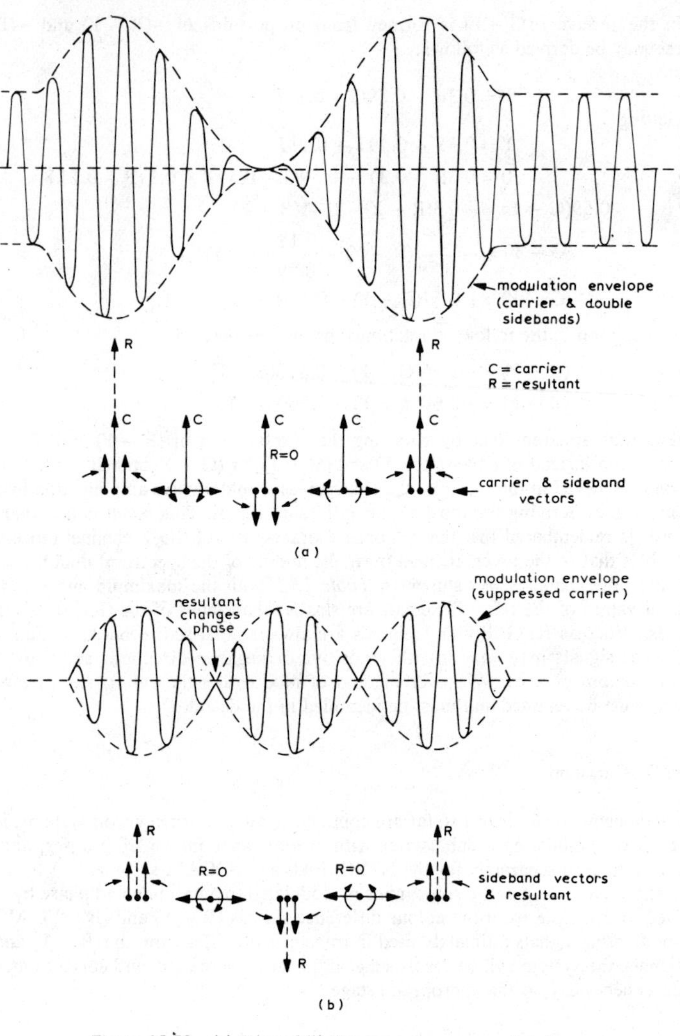

Figure 15.78 (a) Normal (b) Suppressed carrier modulation

in amplitude at twice the original modulating frequency. With normal amplitude modulation, the carrier falls to zero during the troughs of 100% modulation; with suppressed carrier modulation, however, the resultant of the sidebands falls to zero as the modulating waveform changes from positive to negative. The phase of the resultant is either 0° or 180° with respect to the original suppressed carrier when both sets of sidebands are present with their original symmetry.

The balanced ring modulator. One of the most satisfactory ways of achieving suppressed carrier modulation is by means of a balanced ring modulator, a circuit of which is shown in *Figure 15.79*. The input sub-carrier causes the diode pairs to conduct alternately, and thereby to act as switches connecting each outer end of the primary of the output transformer in turn to the centre tap of the input transformer secondary. When, for instance, the sub-carrier phase is such that point A is positive and point B is negative, D_1 and D_2 will conduct taking point D to the potential of point C by balanced bridge action. D_3 and D_4 will not conduct during this half cycle. During the next half cycle, D_3 and D_4 will conduct taking point E to the potential of point C, D_1 and D_2 now being non-conductive. If no difference of potential exists between points C and F (i.e., if there is no video input) no voltage will be impressed across DF and EF in successive

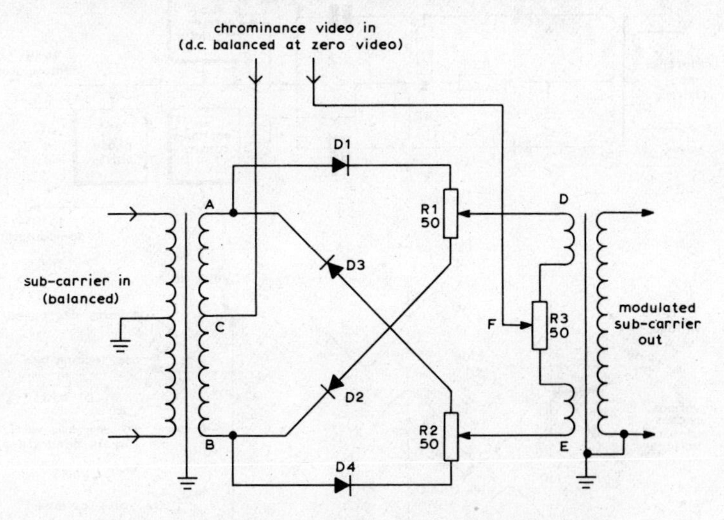

Figure 15.79 Suppressed carrier modulator for colour television

NOTE
The forward resistances of D_1 and D_2 are equalised as far as possible by R_1; similarly D_3 and D_4 are equalised by R_2. R_3 is provided to adjust circuit balance. All adjustments are made for minimum sub-carrier out with zero chrominance video in

half cycles and therefore no output will be obtained. If, however, a video input is present, causing points C and F to differ in potential, this potential will be switched across DF and EF during each alternate half cycle of the sub-carrier, creating in the transformer secondary a video modulated sub-carrier frequency output. The bandwidth of the system is such that response to sub-carrier harmonics is negligible, with the result that switching transients are largely removed and the waveform at sub-carrier frequency is substantially sinusoidal.

The balanced ring modulator has other uses besides suppressed carrier modulation. With the appropriate rectangular waveform applied to the video input terminals, it becomes a very effective sub-carrier phase reversing switch; the unit may also be connected as a synchronous demodulator.

Regeneration of the suppressed carrier. Suppressed carrier modulation must be detected synchronously in the receiver by the regeneration of the original carrier in its exactly correct phase.

To enable the sub-carrier to be regenerated, a colour synchronising burst of approximately ten cycles of the sub-carrier is inserted in each back porch of the complete television waveform (not during field sync or after equalising pulses). In the receiver, the burst is gated out and is compared with the local sub-carrier oscillator for frequency and phase. Any difference produces a correction voltage which brings the local sub-carrier automatically into its required relationship. The transmitted reference burst is 180° out of phase with the sub-carrier when it is being modulated with +(B – Y), but the action of the phase comparator produces an oscillator output which lags by 90° and is thus directly suitable for (R – Y) detection.

The manner in which the local sub-carrier oscillator in a colour television receiver may be phase controlled by the reference burst is illustrated in *Figure 15.80* (*a*), (*b*) and (*c*).

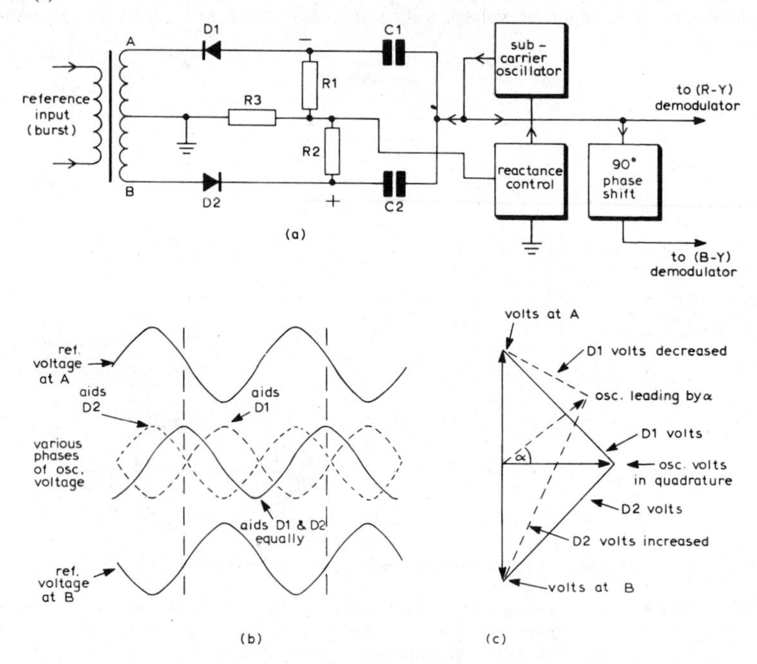

Figure 15.80 The action of a phase comparator and automatic control of oscillator phase. The controlling voltage across R_3 is only zero when the sub-carrier oscillator output voltage is in quadrature with the reference voltage, assuming both to be present

The reference burst is fixed in phase and produces equal antiphased voltages as shown at A and B in *Figure 15.80* (*b*). When rectified by D_1 and D_2, these voltages will produce equal but opposite d.c. voltages across R_1 and R_2, with no resultant d.c. voltage across R_3 to be fed to the reactance control.

Any phase of oscillator output voltage fed through C_1 and C_2 (equal low reactances) other than in quadrature with the voltages at A and B, will increase the rectified output of one diode in comparison with the other, and will thus provide a positive or negative d.c. control voltage across R_3, as shown in the waveform and vector diagrams *Figures 15.80* (*b*) and *15.80* (*c*) respectively.

Applied to the reactance control, this control voltage will shift the oscillator phase until a quadrature relationship is obtained, as indicated by the controlling voltage sinking to zero.

It should be noted that the alternative quadrature condition, 180° out of phase with the one mentioned, is unstable because *any* resultant voltage applied to the reactance control will either advance or retard the sub-carrier oscillator phase towards the required relationship.

With D_1 and D_2 and the reactance control appropriately connected, the controlled oscillator phase may be arranged to settle down, lagging the burst phase by 90°.

Synchronous detection by sampling. One method of detection is to use the regenerated carrier as a sampler as shown in *Figure 15.81* (*a*) and (*b*). In this circuit, the (R – *Y*) and (B – *Y*) detector valves V_1 and V_2 (*Figure 15.81* (*b*)) conduct on the extreme positive peaks of the locally regenerated sub-carrier applied to their grids, sampling the waveform applied to their anodes.

Taking as an example the (R – *Y*) waveform, when this is sampled at its peaks the output from the (R – *Y*) detector will be the representative sinusoidal modulating signal

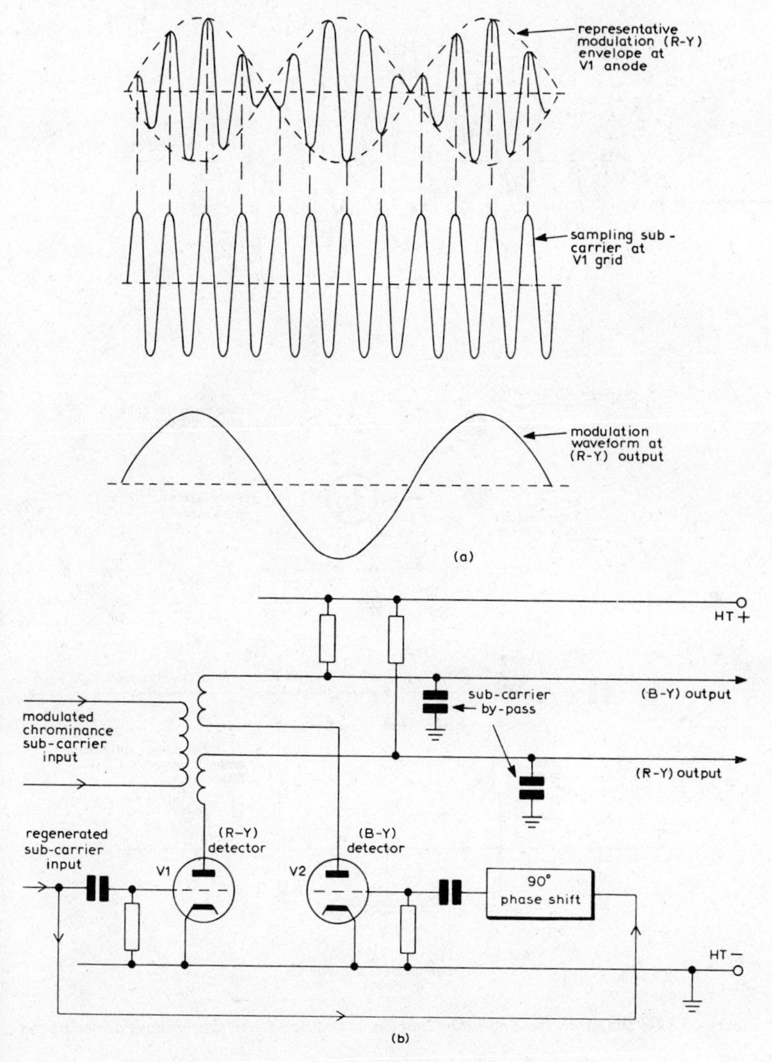

Figure 15.81 *Synchronous detection by sampling*

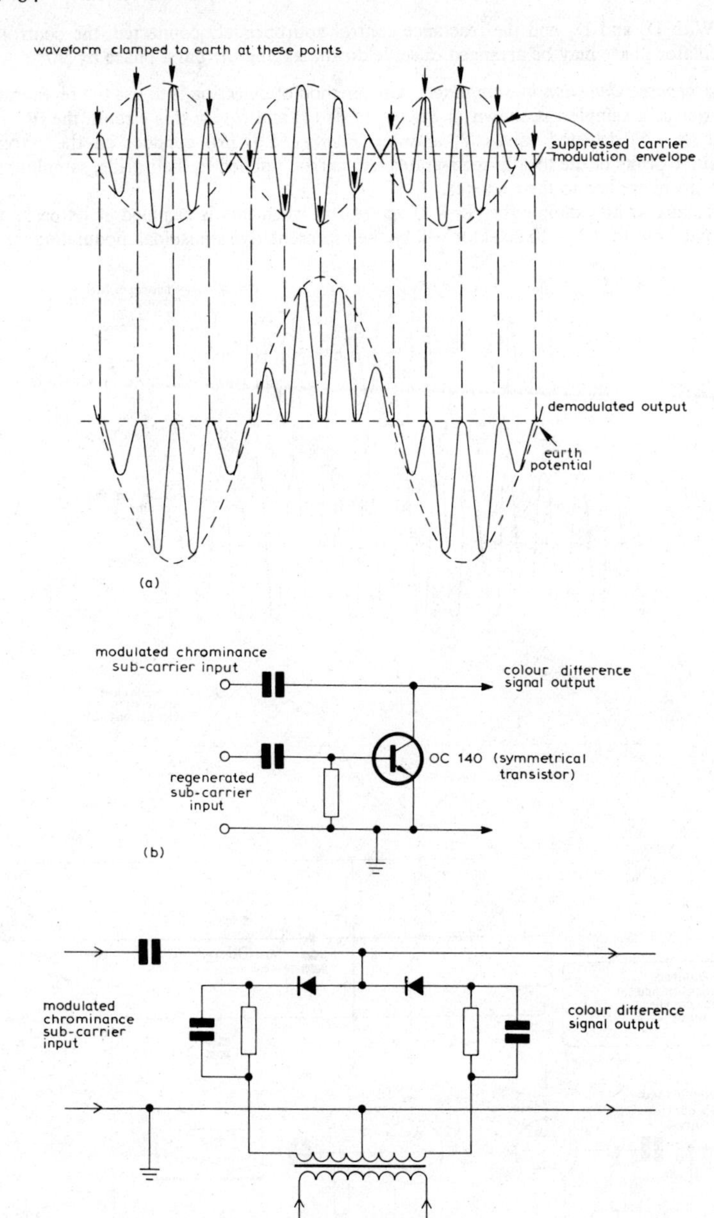

Figure 15.82 Diagrams showing action and circuits of clamp-type synchronous demodulators

as shown in *Figure 15.81* (*a*); the sampling carried out by the (B – *Y*) detector V_2, however, is in quadrature, due to the 90° phase shifting network in its grid circuit. This produces no (B – *Y*) output because V_2 conducts only when its anode signal is passing through its a.c. zero.

When the chrominance signal is applied to both V_1 and V_2 anodes, the (R – *Y*) component is detected by V_1, and the (B – *Y*) component is detected by V_2, with no mutual interference between them in the outputs.

The regenerated sub-carrier inputs are d.c.-restored at the grids of V_1 and V_2, thus biasing the two valves well into cut-off to ensure that anode conduction only takes place for brief periods during the positive peaks at the grids. This requires a regenerated sub-carrier of large amplitude.

Synchronous detection by clamping. Two forms of clamp-type synchronous demodulators are illustrated in *Figure 15.82*, one with a symmetrical transistor in an unbalanced circuit (*Figure 15.82* (*b*)), and the other using two diodes in a balanced bridge (*Figure 15.82* (*c*)). The action of both circuits is similar to that of simple clamps used commonly in television,[4] the reference sub-carrier taking the place of the usual clamp pulses. As shown in *Figure 15.82* (*a*), the sampled points are taken to earth potential by the clamps, resulting in a demodulated output from which the sub-carrier component may be filtered, leaving the original modulating waveform.

Synchronous detection with the balanced ring circuit. As mentioned previously, the balanced ring modulator circuit of *Figure 15.79* may be used for detection. If the modulated sub-carrier is applied to the output transformer terminals, a demodulated output will appear across the video terminals as provided by the sampling action of D_1 and D_2 and also D_3 and D_4 when driven into conduction at the appropriate times by the local sub-carrier applied to the input terminals. Sampling will occur at each half cycle peak of the modulated sub-carrier; this action results from the provision of the centre tap F on the output transformer.

It should be realised that although sinusoidal modulating waveforms have been drawn for convenience, in practice the modulating waveforms will vary in accordance with the (R – *Y*) and (B – *Y*) signals, which may be very complex. *Figure 15.83* shows

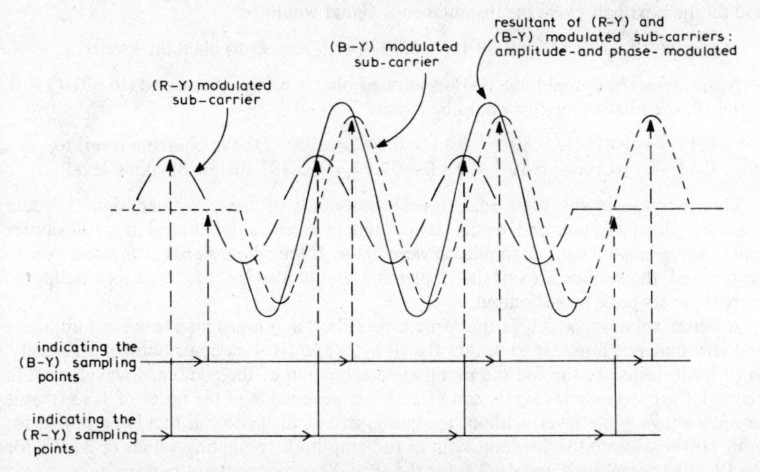

Figure 15.83 (R – *Y*) *and* (B – *Y*) *modulated sub-carriers and their resultant. The sampling points always occur at the peaks of the appropriate sub-carrier cycles, i.e., when the sub-carrier cycles of the other component are passing through zero. In this way the two contributions to the amplitude and phase modulated resultant may be separated by synchronous detection*

contributions of (R – *Y*) and (B – *Y*) modulated sub-carriers and the amplitude and phase modulated sub-carrier resultant.

When two suppressed carrier modulated signals of equal bandwidths are combined to form an amplitude and phase modulated resultant, the sidebands of each must be preserved symmetrically to avoid crosstalk between the signals (see previous).

Derivation of the (G – *Y*) *signal.* Having detected separately the (R – *Y*) and (B – *Y*) components, the appropriate negative values, as given in Equation 3, may be combined to provide a (G – *Y*) output. Alternatively, a third synchronous detector may be used, with its input common to the other two detectors as far as the modulated sub-carrier is concerned. This detector must be supplied with a local sub-carrier in phase with the (G – *Y*) component.

THE CHROMINANCE SIGNAL

Amplitude of the combined picture waveform. The (R – *Y*) and (B – *Y*) sub-carriers, separately modulated, are combined and then added to the luminance signal. The resultant amplitude of two signals combined in quadrature is the square root of the sum of their squares and therefore:

the peak amplitude of the combined chrominance sub-carriers

$$= \sqrt{[(R - Y)^2 + (B - Y)^2]}$$

When the chrominance sub-carrier is added to the luminance signal, the latter becomes the datum for the a.c. zero of each sub-carrier cycle. If blanking level is taken as the reference level (disregarding syncs), the total excursion of the waveform is from:

$$Y + \sqrt{[(R - Y)^2 + (B - Y)^2]} \quad \text{to} \quad Y - \sqrt{[(R - Y)^2 + (B - Y)^2]}$$

For example, when full amplitude 100% saturated yellow is being transmitted (R = 1.0, G = 1.0, B = 0) on one half cycle of the sub-carrier the peak amplitude would be

$$0.89 + \sqrt{[0.11^2 + 0.89^2]} = 0.89 + 0.897 = 1.79 \text{ (above blanking level)}$$

and on the next half cycle the instantaneous signal would be

$$0.89 - \sqrt{[0.11^2 + 0.89^2]} = -0.007 \text{ (with respect to blanking level).}$$

Again, when full amplitude 100% saturated blue is being transmitted (R = 0, G = 0, B = 1.0), the total excursion would be from:

$$0.11 + \sqrt{[0.11^2 + 0.89^2]} = 0.11 + 0.897 = 1.007 \text{ (above blanking level) to}$$
$$0.11 - \sqrt{[0.11^2 + 0.89^2]} = 0.11 - 0.897 = -0.787 \text{ (below blanking level).}$$

This increase in the total peak-to-peak amplitude of the combined signals would overload video circuits and would also result in over-modulation at the transmitter unless some adjustments of amplitude were made. If the whole signal as formed were to be reduced, the service area of the transmitter would also be reduced, necessitating an increase in its peak power output.

A better solution, which at the same time effects a compromise between luminance and chrominance noise, is to weight the (R – *Y*) and (B – *Y*) contributions separately, in order to utilise to the full the permissible excursion of the combined waveforms. In general, the chrominance signal can extend to a peak value of the order of 1.33 (that is to 33% above white level) without incurring serious distortion; if this is permitted for both 100% saturated yellow and cyan of full amplitude, weighting values of 87.7% for the (R – *Y*) component and 49.3% for the (B – *Y*) component are required.

Colour bars. A common test signal for colour television circuits and devices takes the form of a stepped wedge ranging from white to black through the six hues listed in *Table 15.5*, arranged in descending order of luminance.

Figure 15.84 shows unweighted (*a*) and weighted (*b*) (R – *Y*) components, and unweighted (*c*) and weighted (*d*) (B – *Y*) components of full amplitude 100% saturated colour bar signals. In these diagrams the video waveforms (to which weighting is applied) are shown in full lines and the modulated sub-carrier envelopes are shown enclosed by full and dashed lines. Normalised values are marked on the waveforms but transmission values may be read from the voltage scales (white level = 0.7 volts, blanking level = 0 volts).

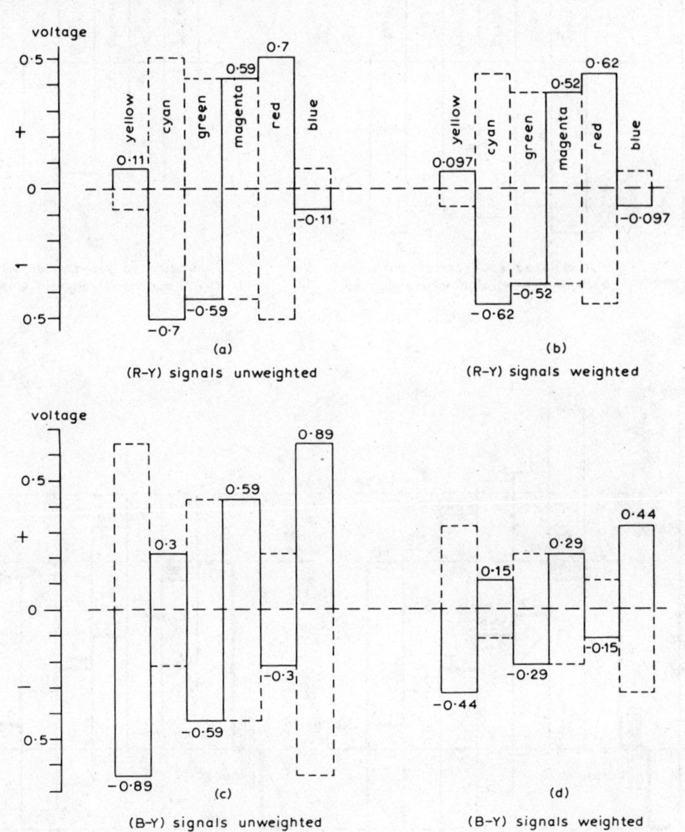

Figure 15.84 Weighted and unweighted (R – *Y*) *and* (B – *Y*) *components of full amplitude 100% saturated colour bar signals. Full lines indicate video waveforms; dashed and full lines indicate sub-carrier envelopes. Waveforms are marked with normalised values—for voltage values, refer to scale*

Figure 15.85 (*a*) shows the combination of (R – *Y*) and (B – *Y*) signals unweighted and *Figure 15.85* (*b*) shows the effects of weighting. (R – *Y*) and (B – *Y*) signals are, of course, only combined in their modulated sub-carrier conditions.

Figure 15.85 (*c*) shows the addition of the luminance signal to the waveform of *Figure 15.85* (*a*), and *Figure 15.85* (*d*) shows the addition of the luminance signal to the waveform of *Figure 15.85* (*b*). In practice, only the weighted signals are actually produced but a comparison between *Figure 15.85* (*c*) and *15.85* (*d*) gives the reduction in peak-to-peak amplitude which has been achieved.

The use of these weighting factors not only brings the sub-carrier peaks to the same

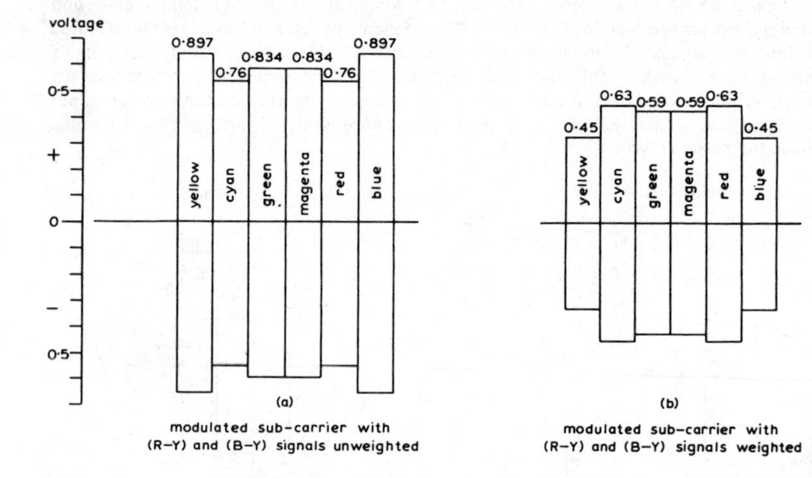

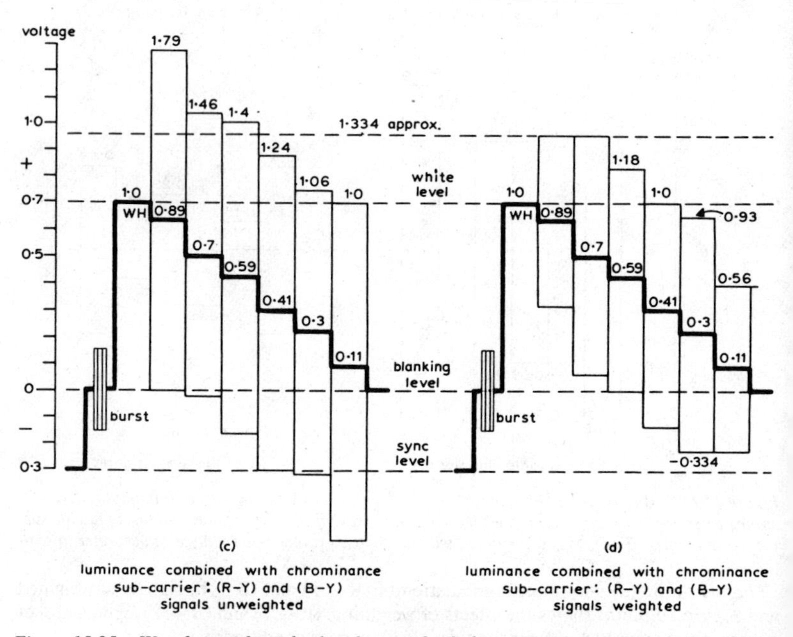

Figure 15.85 Waveforms of weighted and unweighted chrominance components both with and without the luminance signal (full amplitude 100% saturated colour bars). Waveforms are marked with normalised values—for voltage values, refer to the scale

level of 1.334 for yellow and cyan, but also brings the sub-carrier peaks in the opposite direction, for red and blue, both to a level of –0.334. These features of coincidence are useful for setting the exact quadrature relationship of the (R – Y) and (B – Y) sub-carriers in the coder. Having first adjusted the two modulated sub-carriers to the correct individual amplitudes, as shown by the dashed and full lines in *Figure 15.84* (*b*) and (*d*), then, after combination with the luminance signal, the correct phase adjustment is indicated clearly by the coincidence of the amplitudes mentioned. In this way, the need for direct phase measurement may be avoided.

Receiver chrominance values. Having modified the (R – Y) and (B – Y) signals by factors of 0.877 and 0.493 respectively on transmission, the detected (R – Y) and (B – Y) components in the receiver must be restored to their original ratios by the use of the reciprocal factors of 1.14 and 2.03. Before synchronous detection, the amplitude of the chrominance sub-carrier may be adjusted by a 'saturation control' to produce pale or vivid colour as required; after detection, however, it is essential that the (R – Y) and (B – Y) components are available in their correct proportions, otherwise errors in hue will result. Equation 3, derived on p. **15**–79 for the value of (G – Y), is in terms of the unweighted quantities of (R – Y) and (B – Y), and is used for obtaining (G – Y) in the receiver from the restored values of (R – Y) and (B – Y).

Weighted colour difference values for saturated colours are shown in *Table 15.6*, which has been compiled from *Table 15.5* by providing the weighted values of (R – Y) and (B – Y), and the amplitude and phase angle of the resultant.

Table 15.6 WEIGHTED COLOUR DIFFERENCE VALUES FOR SATURATED COLOURS AT FULL AMPLITUDE

	R	G	B	Y	0.877 (R – Y)	0.493 (B – Y)	*Resultant amplitude*	*Angle from* +(B – Y)
Yellow	1.0	1.0	0	0.89	0.097	–0.44	0.45	167°
Cyan	0	1.0	1.0	0.7	–0.62	0.15	0.63	283.5°
Green	0	1.0	0	0.59	–0.52	–0.29	0.59	240.7°
Magenta	1.0	0	1.0	0.41	0.52	0.29	0.59	60.5°
Red	1.0	0	0	0.3	0.62	–0.15	0.63	103.5°
Blue	0	0	1.0	0.11	–0.097	0.44	0.45	347°
Mean arithmetical values					0.412	0.29	0.56	

Figure 15.86 is a composite vector diagram showing the phase and amplitude of the resultant sub-carrier when describing all three primaries, and their one to one electrical mixtures as given in *Table 15.6*. There is a direct relationship between the phase of the resultant and the hue of the colour being transmitted, and an indirect relationship between the amplitude of the resultant and the degree of saturation. If the amplitude of *Y* and sub-carrier phase remain constant, a decrease in amplitude of the sub-carrier indicates a decrease in saturation.

Human vision and the I and Q axes. Studies of human vision have shown that the acuity of the normal eye is not the same for all colour combinations. Finer chrominance detail may be resolved in (approximately) orange and cyan hues than in green and magenta. To take advantage of this fact, the two chrominance sub-carriers are advanced by 33° from their (R – Y) and (B – Y) axes, when they are termed '*I*' and '*Q*' axes respectively, see *Figure 15.86*. For the *Q* signal, a comparatively narrow bandwidth is adequate and double sideband working is preserved. For the *I* signal, about double the *Q* bandwidth is used but only the lower sidebands of modulation components outside the bandwidth allotted to *Q* are transmitted. Because *Q* is double sideband, no crosstalk

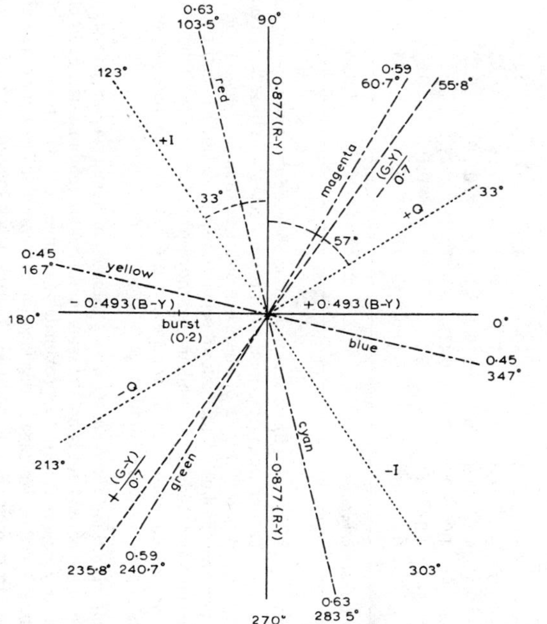

Figure 15.86 Sub-carrier phase and amplitude relationships for fully saturated colours

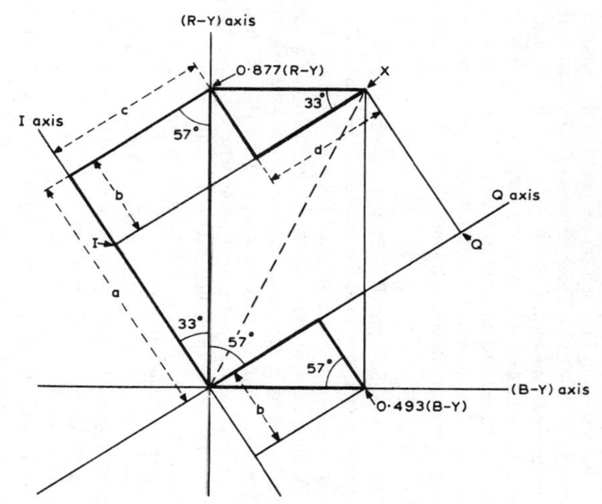

Figure 15.87 Derivation of I and Q from (R – *Y*) *and* (B – *Y*). *X shows the amplitude and phase of the signal to be described in terms of the rectangular co-ordinates* (R – *Y*) *and* (B – *Y*) *or I and Q*

$a = 0.877(\mathrm{R} - Y) \cos 33° = 0.74(\mathrm{R} - Y)$
$b = 0.493(\mathrm{B} - Y) \cos 57° = 0.27(\mathrm{B} - Y)$
$c = 0.877(\mathrm{R} - Y) \cos 57° = 0.48(\mathrm{R} - Y)$
$d = 0.493(\mathrm{B} - Y) \cos 33° = 0.41(\mathrm{B} - Y)$
$I = a - b = 0.74(\mathrm{R} - Y) - 0.27(\mathrm{B} - Y)$
$Q = C + d = 0.48(\mathrm{R} - Y) + 0.41(\mathrm{B} - Y)$

Substituting values for Y in terms of R, G *and* B

$I = 0.6\mathrm{R} - 0.28\mathrm{G} - 0.32\mathrm{B}$
$Q = 0.21\mathrm{R} - 0.52\mathrm{G} + 0.31\mathrm{B}$

takes place into the I signal. The crosstalk from I into Q only occurs for modulation frequencies outside the bandwidth of Q, and is removed by a band-pass filter. There is a change of amplitude when the mode of transmission for the I signal changes from double to single sideband and this properly requires an equalising filter in the receiver.

Determination of I and Q signals. The specifications for I and Q may be determined by projecting the (R – Y) and (B – Y) signals—at the previously determined levels—into the I and Q directions. The resultant signal produced by adding I and Q in phase quadrature is the same as that produced by adding 0.877(R – Y) and 0.493(B – Y) in phase quadrature, when:

$$I = 0.74(\mathrm{R} - Y) - 0.27(\mathrm{B} - Y)$$

and $Q = 0.48(\mathrm{R} - Y) + 0.41(\mathrm{B} - Y)$ (for derivation see *Figure 15.87*)

Expressions for Y in terms of R, G and B may be substituted in the above equations to show I and Q as functions of R, G and B, as follows:

$$I = 0.6\mathrm{R} - 0.28\mathrm{G} - 0.32\mathrm{B}$$
$$Q = 0.21\mathrm{R} - 0.52\mathrm{G} + 0.31\mathrm{B}$$

In the colour receiver:

$$(\mathrm{R} - Y) = 0.96I + 0.62Q$$
$$(\mathrm{B} - Y) = -1.1I + 1.7Q$$

and $(\mathrm{G} - Y) = -0.28I - 0.64Q$

By describing the resultant amplitude and phase modulated signal in terms of I and Q signals of unequal bandwidths, the total bandwidth necessary for chrominance is reduced but the chrominance resolution provided is fully adequate. This is advantageous, because the reduction of bandwidth enables a higher frequency to be used for the sub-carrier than would otherwise be possible.

THE CHROMINANCE SUB-CARRIER

Choice of sub-carrier frequency. Two important requirements govern the choice of the sub-carrier frequency:

(i) The presence of the sub-carrier must not degrade unduly the display of the luminance signal on colour or on black and white receivers.

(ii) The luminance signal must not interfere unduly with the chrominance signal, otherwise spurious coloured effects will be observed on the colour receiver, due to detection of luminance components by chrominance detectors.

Although a uniform response is required for the luminance signal over the video spectrum, the distribution of energy is by no means uniform. For average picture content the energy is greatest at the lower frequencies, and reduces markedly as the frequency is increased. Due to the scanning process, line, field and picture frequencies and their harmonics predominate, with the result that the spectrum is not continuously or continually occupied. The main components of the energy occur at line frequency intervals and consist of the multiples of line frequency, accompanied by little groups above and below in frequency, at picture and field frequency multiples, as if the line frequency and its multiples were accompanied by little clusters of sidebands (*Figure 15.88*). These conditions apply particularly to stationary scenes and although movement produces changes according to the degree of movement, the general nature of the pattern is retained.

A sub-carrier modulated by chrominance signals derived from the same scanning process will produce sidebands which are higher and lower in frequency and which are distributed in a kindred manner. Thus, if the sub-carrier is chosen to be an odd multiple of half the line frequency, the chrominance and luminance components will interleave with minimum interaction (*Figure 15.88*).

Effect of picture tube non-linearity. In the receiver, the chrominance sub-carrier

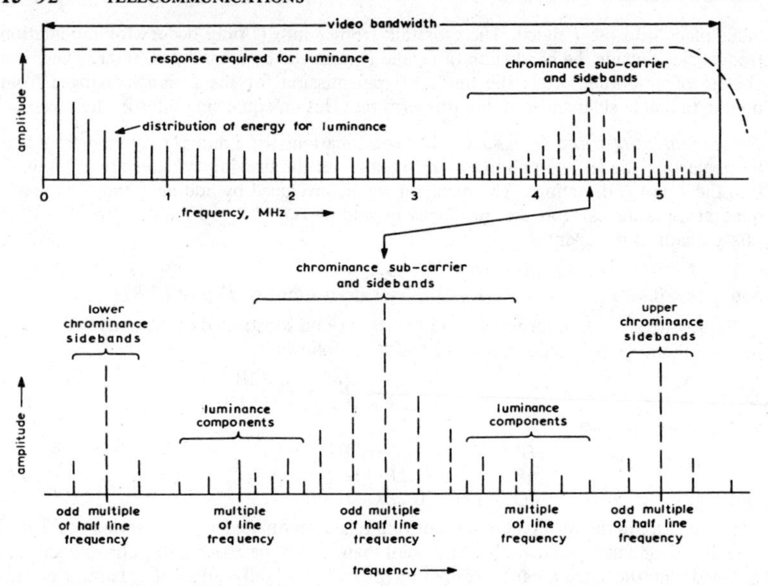

Figure 15.88 The interleaving of chrominance with luminance in colour television (*representative values*)

superimposed on the luminance signal produces a brightening and darkening of the scanning spot—a dot pattern—as it traverses the screen of the picture tube. Due to the non-linear transfer characteristics of picture tubes, an overall brightening of coloured areas takes place. (The sub-carrier is not present in neutral areas.) For a black and white receiver this effect is beneficial inasmuch as it helps to produce pictures that are more nearly panchromatic; (this aspect is discussed later). For a colour receiver, however, an unacceptable de-saturation of colours would occur, and this would necessitate the inclusion of a 'notch' filter at sub-carrier frequency in its luminance channel. This notch filter unavoidably removes luminance components in the vicinity of the sub-carrier frequency; but in general, the loss of resolution is not unduly serious.

Exact relationship of sub-carrier to line and field frequencies. In the interests of compatibility, the visibility of the dot pattern on the black and white receiver should be as low as possible. This condition is attained by selecting the sub-carrier frequency in accordance with the principles already discussed; nevertheless, it is useful to consider the nature of the patterns produced.

If the exact frequency of the sub-carrier is an odd multiple of half the line frequency, any one line will contain an integral number of cycles plus half a cycle of the sub-carrier frequency. Thus, any two complete consecutive lines will contain a complete (odd) number of sub-carrier cycles. The adjacent lines in any field (not picture) will contain an out of phase dot structure because the positive peaks of the sub-carrier on one line will lie directly above the negative peaks on the next. Odd and even lines will, therefore, contain out of phase sub-carrier cycles. As each complete *picture* contains an odd number of lines, the dot structure on consecutive pictures will be out of phase, since each complete picture contains an integral number of sub-carrier cycles plus half a cycle.

In other words, the sub-carrier frequency should also be an odd multiple of half the picture frequency. With an odd number of lines per picture this is the case. The arrangement of numbers given opposite illustrates the resulting dot pattern.

The numbers indicate the brightening half-cycles of the sub-carrier only and it can be seen that the pattern is completed in four fields. Due to stroboscopic action, the dots appear to crawl directly upwards in the example shown, and this is true for 405, 525 and 625-line systems. (It is interesting to note that with possible hypothetical systems having 403, 523 or 623 lines, for instance, the dots would appear to crawl directly downwards.) There are also less strident apparent movements of dots diagonally downwards. The non-linearity of the picture tube characteristic produces incomplete cancellation of the brightening by the darkening half-cycles of the sub-carrier, resulting in a dot pattern along each line which is effectively at double the sub-carrier frequency.

```
1 3 1 3 1 3 1 3 1 3 1 3 odd line field 1 and even line field 3
4 2 4 2 4 2 4 2 4 2 4 2 even line field 2 and odd line field 4
3 1 3 1 3 1 3 1 3 1 3 1 even line field 1 and odd line field 3
2 4 2 4 2 4 2 4 2 4 2 4 odd line field 2 and even line field 4
1 3 1 3 1 3 1 3 1 3 1 3 odd line field 1 and even line field 3
4 2 4 2 4 2 4 2 4 2 4 2 even line field 2 and odd line field 4
```

It is apparent that, as well as being an exact odd multiple of half the line frequency, the sub-carrier frequency must be sufficiently high to make the dot structure of the interference pattern as fine as possible (i.e., the number of half cycles per line should be high), and yet it must also be low enough to accommodate the upper sidebands within the upper limit of the monochrome spectrum (see *Figure 15.88*).

NTSC parameters on different line standards. The original NTSC system has been adapted for experimental use on 405 and 625-line standards, but current interest is confined to its operation on the 525-line standard. For comparison purposes, *Table 15.7* shows the relevant characteristics of the three different line standards operating on the NTSC colour television system.

Table 15.7 CHARACTERISTICS OF THE THREE DIFFERENT LINE STANDARDS OPERATING ON THE NTSC SYSTEM

Line standard	*Fields per sec.*	*Line frequency Hz*	*Sub-carrier frequency MHz*	*Sub-carrier frequency in terms of line frequency*	*Luminance bandwidth MHz*	*I bandwidth MHz*	*Q bandwidth MHz*
405	50	10 125	2.657 812 5	525/2	3.0	0.5	0.3
525	59.94	15 734.264	3.579 545	455/2	4.2	1.3	0.5
625	50	15 625	4.429 687 5	567/2	5.5	1.6	0.8

The number of half cycles of sub-carrier per line is a direct indication of the visibility of the dot pattern to be expected on black and white receivers. It may be noted that the 525-line system produces the most coarse (and therefore most noticeable) dot pattern.

Sub-carrier stability requirements. To ensure the correct relationship between line and sub-carrier frequencies, a highly stable sub-carrier oscillator is used; its output frequency is divided by a factor (see *Figure 15.89*) to provide a pulse train at twice line frequency in place of the master oscillator of the sync pulse generator.

It is not practicable to lock the system to the mains frequency (as is usual with the black and white 405-line standard) because the possible variations in mains frequency are too great to be accommodated easily when they are translated in terms of equivalent changes in sub-carrier frequency. For example, in a mains locked system, if the mains frequency varied by ±0.6% the sub-carrier would change by approximately ±21.5 kHz (525-line system).

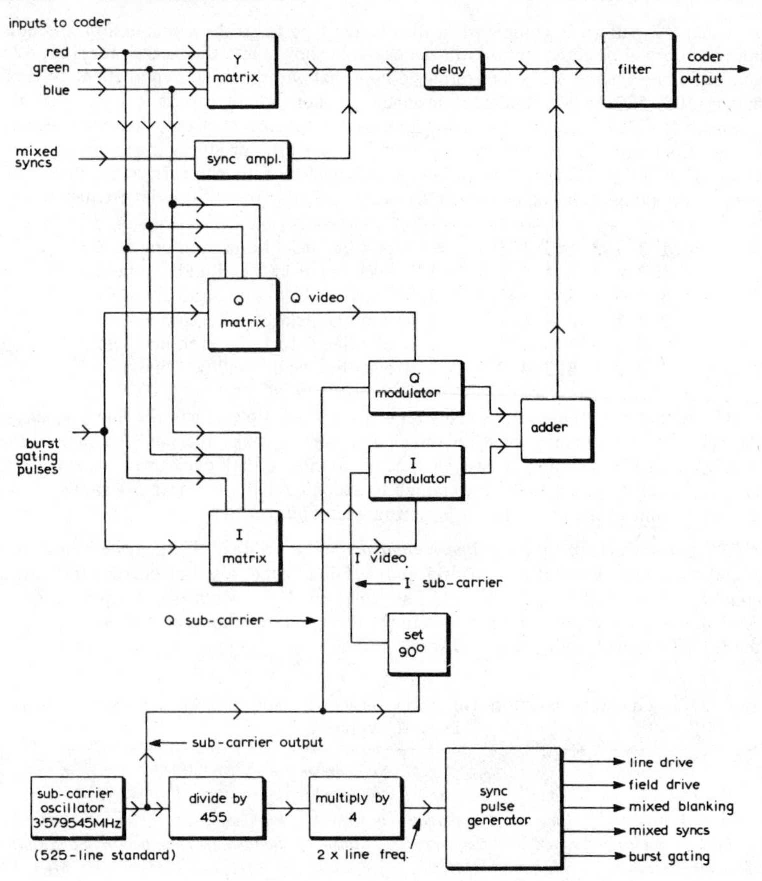

Figure 15.89 Simplified block diagram of NTSC coder and (below) diagram showing derivation of sub-carrier/line frequency relationship

The provision of a sub-carrier burst in each back porch of the line waveform results in a frequency spectrum in which the sidebands are situated at line frequency intervals above and below the fundamental sub-carrier frequency. Those sidebands near the sub-carrier are comparable in amplitude to the sub-carrier and if the latter changes in frequency (when all the sidebands would change in frequency as well) to anything like the extent given in the example, simple receiver circuits would be unable to select the fundamental sub-carrier component necessary for correct phase control. In practice, the sub-carrier frequency is transmitted with an accuracy of about $\pm 0.000\,3\%$, which permits the effective acceptance bandwidth for control purposes of the receiver to be appropriately narrow. This provides a suitable degree of noise immunity in terms of the accuracy of sub-carrier locking.

Relationship to sound carrier. In addition to selecting a chrominance sub-carrier which is an exact odd multiple of half line frequency, it is also worthwhile arranging for the spacing between the sub-carrier and the average sound carrier to be an odd multiple of half line frequency when sound and vision radio frequency carriers are finally transmitted. The beat pattern which may be apparent due to interference between the sub-carrier and the sound carrier is reduced in visibility by this relationship even when the sound carrier is frequency modulated. (This beat pattern occurs in receivers with insufficient attenuation of the sound carrier in the video frequency circuits.) With the 525-line system, to avoid altering the sound and vision carrier spacing of 4.5 MHz, line and field frequencies have been modified slightly to the values given in *Table 15.7*, producing a separation of 58.5 times line frequency between the sub-carrier and the sound carrier. With the 625-line system, the 6 MHz sound and vision carrier spacing is 100.5 times line frequency and no alteration of line and field frequencies is necessary.

THE NTSC CODER

A simplified block diagram of the NTSC coder is given in *Figure 15.89*, together with the arrangements for deriving the exact sub-carrier/line frequency relationship. *Figure 15.90* is a more detailed diagram, in which the blocks are numbered to coincide with the following list of operations and functions:

(1) *Y* matrix with switching to picture and bars (and off positions) derived from R, G and B and also to external luminance where this might be derived from a 4th camera tube.
(2) An amplifier with set luminance control.
(3) Fixed delay of about 0.25 μs added in the luminance channel to match the delays in the chrominance channels.
(4) Luminance output amplifier before the addition of the chrominance signal.
(5) Linear phase filter. Coder output.
(6) Sync amplifier, with set gain control, designed to insert mixed syncs immediately prior to the fixed delay (3).
(7) G signal amplifier providing negative G for the *Q* matrix (8) with set gain to adjust white balance at peak white.
(8) *Q* matrix forming a positive *Q* signal from –G, +R and +B inputs; a burst gating pulse is added to the waveform so that the modulator introduces the *Q* component of the burst.
(9) *Q* video amplifier, incorporating a low-pass filter and a clamp, providing an input for the *Q* modulator. A gain control sets the 33° angle between –*Q* and the burst by controlling the *Q* contribution.
See NOTE (*a*).
(10) *Q* carrier balance is obtained by the d.c. control of the current through a complementary pair of transistors identical to those in the output of (9). This ensures that the modulation terminals of the balanced modulator are presented with balanced impedances, and that the diodes of the ring modulator are working under correct conditions, although the video modulating signal is obtained from the single-ended output of (9).
See NOTE (*b*).
(11) *Q* sub-carrier tuned amplifier providing carrier input to the modulator.
(12) Ring-type suppressed carrier modulator.
(13) Simple amplifier having a collector load common with (20) forming a chrominance adder in (21).
(14) R signal amplifier providing negative R for the *I* matrix (15) with set gain to adjust white balance, at peak white.
(15) *I* matrix forming a basic negative *I* signal from –R, +G and +B inputs, thus requiring only one phase reverser. The appropriate polarity as required finally is provided by the particular way in which the modulator is connected. A burst gating pulse is added to the waveform so that the modulator introduces the *I* component of the burst.

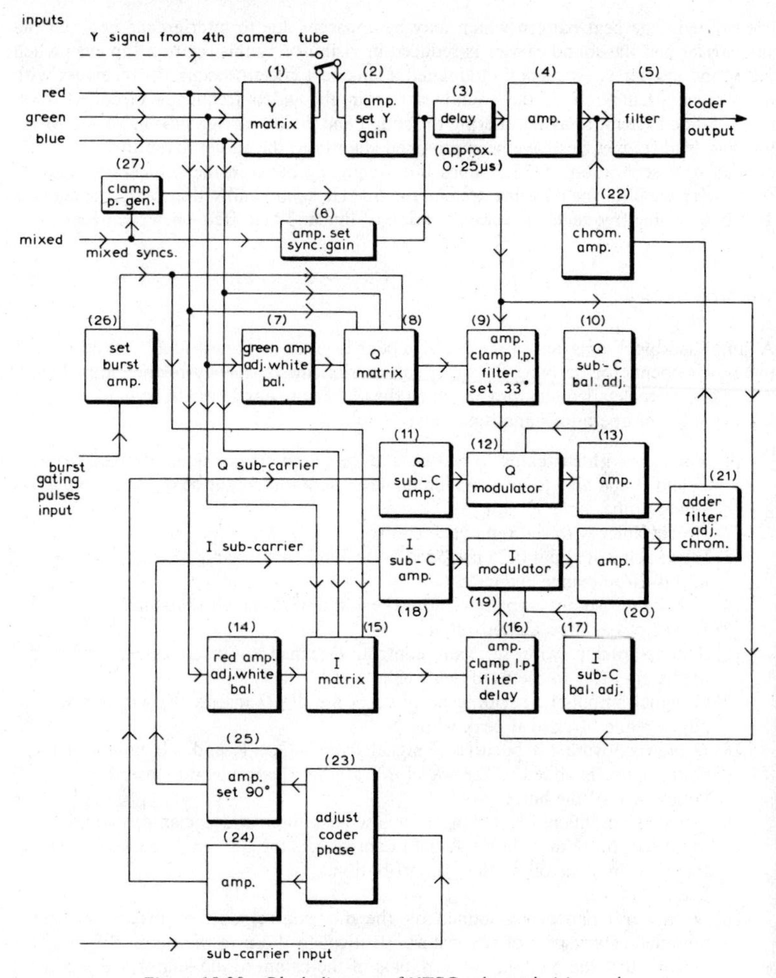

Figure 15.90 Block diagram of NTSC colour television coder:

$$-I = -0.6R + 0.32B + 0.28G$$
$$+Q = +0.21R + 0.31B - 0.52G$$

(16) *I* video amplifier, incorporating a low-pass filter and a clamp, providing an input for the *I* modulator. This is the *I* equivalent of (9).
(17) *I* equivalent of (10).
See NOTE (*b*).
(18) *I* equivalent of (11).
(19) *I* equivalent of (12).
(20) *I* counterpart of (13).
(21) Chrominance amplifier with *I* and *Q* components added at its input, incorporating a second harmonic filter and chrominance amplitude control to adjust the correct ratio of chrominance to luminance.
(22) Final chrominance amplifier forming an adder with the output emitter follower

of (4). The chrominance amplifier has a high output impedance so that the 75 Ω output impedance of the luminance amplifier (4) is seen at the coder output.

(23) Sub-carrier amplifier with phase adjustment to provide the phase matching of this coder with other coders.

(24) *Q* sub-carrier amplifier having the same input as (25).

(25) *I* sub-carrier amplifier with set 90° phase adjustment.

(26) Burst gate amplifier feeding the *I* and *Q* matrices (8) and (15) to form the burst. The amplitude of the burst gating pulse is adjusted to set burst amplitude.

(27) Clamp pulse generator deriving clamp pulses from mixed syncs for the clamps in (9) and (16).

NOTES

(*a*) The fixed proportions of the *I* and *Q* contributions to the burst are determined by the *I* and *Q* matrices, (15) and (8). Having adjusted the quadrature conditions of *I*

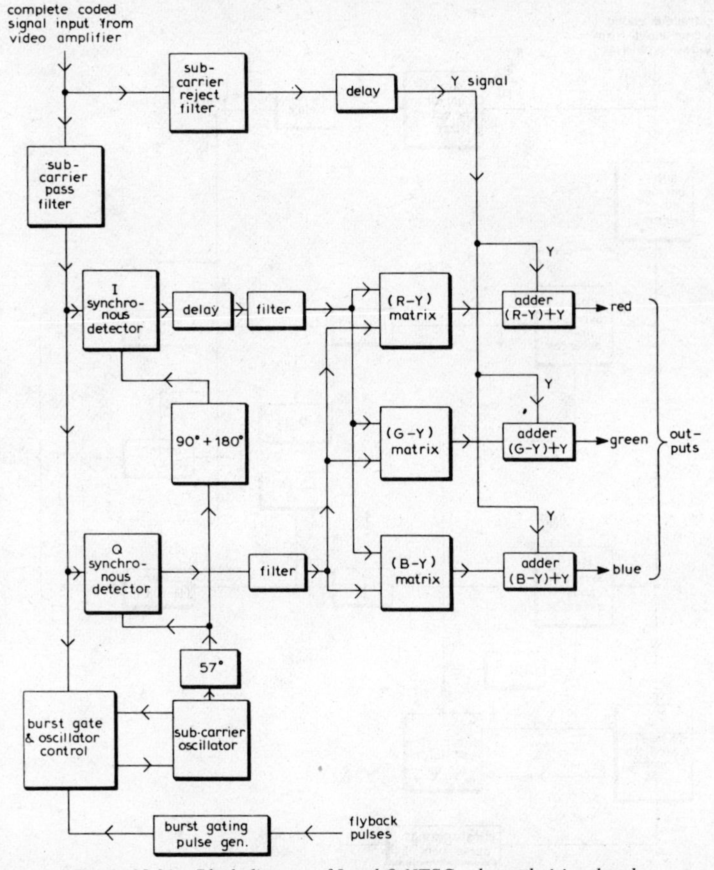

Figure 15.91 Block diagram of I and Q NTSC colour television decoder

NOTES

1. If Y is negative, the grids and cathodes of the shadow mask tube can replace the adders if desired

2. $(R - Y) = 0.96I + 0.62Q$; $(G - Y) = -0.28I - 0.64Q$; $(B - Y) = -1.1I + 1.7Q$

and Q by 'set 90°' in (25), thereafter the amplitude adjustment of Q by 'set 33°' in (9) ensures that the phase of the burst and the relative amplitudes of I and Q are correct. The amplitude of the burst is set in (26).

(*b*) I and Q sub-carriers balanced separately for zero at blanking level.

I AND *Q* RECEIVERS

In transmission, the I and Q signal are modulated on their respective sub-carriers which are then combined to form the phase and amplitude modulated resultant. For coarse chrominance detail, the maximum possible range of hues is available, but the description of fine chrominance detail (requiring frequencies outside the bandwidth of Q) is restricted to the hues of approximately orange and cyan. Full use may be made of this transmission in the receiver if detection is carried out along the I and Q axes, with the output of the Q detector filtered appropriately to the lower bandwidth to prevent

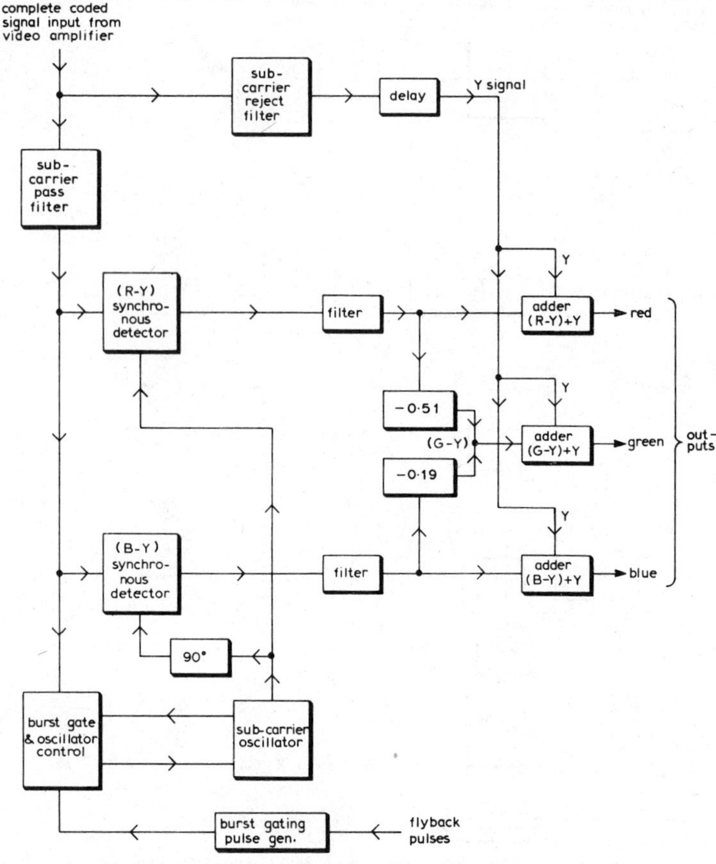

Figure 15.92 Block diagram of (R – Y) *and* (B – Y) *NTSC colour television decoder*

NOTES

1. If Y is negative the grids and cathodes of the shadow mask tube can replace the adders if desired

2. $(G - Y) = -0.51(R - Y) - 0.19(B - Y)$

crosstalk from the I signal. Maximum signal delay takes place in the narrowest bandwidth channels and thus different extra delays must be added to the luminance and I channels. A simple block diagram of the decoder section of an I and Q receiver is shown in *Figure 15.91*.

(R – Y) AND (B – Y) RECEIVERS

In the majority of receivers produced so far, no advantage is taken of the extra bandwidth of the I signal, two chrominance channels being used with bandwidths equal to that of the Q transmission. For instance, detection may take place along the (R – Y) and (B – Y) axes to provide (R – Y) and (B – Y) video signals directly, from which the (G – Y) video signal may then be derived. Only the luminance channel will require added delay and the provision of the correct phases of the reference sub-carrier is simplified. With these receivers the full range of hue is obtained in the comparatively coarse detail dictated by the bandwidth of Q; beyond this, luminance differences only are displayed. At the present state of development of the system generally, the results obtained with the simple receivers are acceptable. *Figure 15.92* shows a simple block diagram of the decoder section of a receiver using (R – Y) and (B – Y) detection.

The PAL system

One particular difficulty in television transmission and reception arises because signals which differ in frequency or amplitude take varying times to pass through a stage, network or channel; this results in what may be termed 'phase distortion', with undesirable effects on the chrominance sub-carrier.

The phase relationship between the sub-carrier transmitted during picture time and the burst transmitted within back porch periods determines the hue. If any spurious change of this relationship occurs, wrong hues will be displayed on the receiver. Phase change which results from changes in amplitude, either of the luminance signal or of the sub-carrier itself, is termed differential phase distortion.

There are a number of ways in which differential phase distortion may be produced in transmission circuits and with complex transmission chains (e.g., video tape recorders, line systems, transmitters, receiver i.f. stages); and there is some difficulty in keeping the overall distortion to within acceptable limits.

The PAL (phase alternating line) colour television system, a variant of the NTSC system, was developed to reduce the effects of differential phase errors which may be suffered by the chrominance sub-carrier in transmission and reception. With PAL, colour difference signals (R – Y) and (B – Y) of equal bandwidth are used (as opposed to the I and Q chrominance signals of NTSC), and these modulate a sub-carrier in amplitude and phase in a way similar to NTSC, except that the (R – Y) contribution to the resultant is reversed in polarity on successive lines. This is accomplished by electronic switching in the PAL coder, while the normal state is restored by a synchronous reversing switch in the receiver; differential phase errors may thus be corrected over a period of two lines.

The same value weighting factors are used for the (R – Y) and (B – Y) signals as with the NTSC system, and the weighted values are termed 'U' and 'V' as follows:

$$(\mathrm{B} - Y)\,0.493 = U$$
$$(\mathrm{R} - Y)\,0.877 = V$$

There are basically two variations of a suitable receiver, which differ in effectiveness and complexity.

SIMPLE PAL

A simple PAL receiver employs normal synchronous demodulation techniques with an electronic reversing switch operating at half line frequency, corrected if in error by the transmitted burst which is also used to synchronise the demodulators. As may be seen from *Figure 15.93*, which shows the effects of a phase error of 25°, the results over two lines may be a substantial correction of hue, but with some reduction in saturation.

With the reversal and ultimate restoration of the *V* contribution on adjacent lines, differential phase error occurring in the chain between these operations results in the hue errors on adjacent lines being in opposite directions—and this is appreciated by the

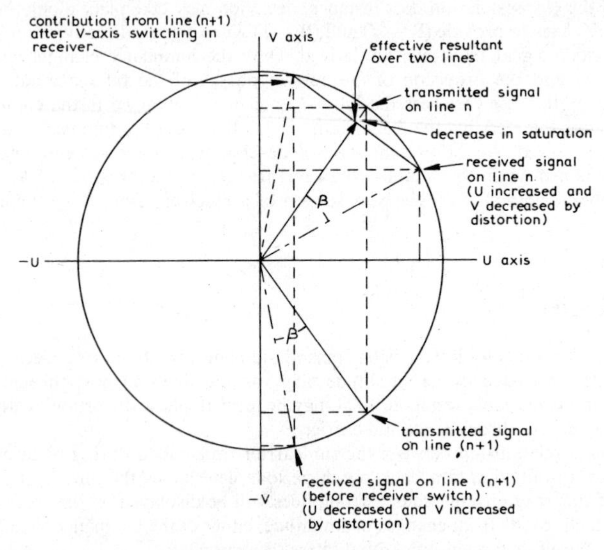

Figure 15.93 Correction of differential phase error with the PAL colour television system. Error phase angle β = 25° lagging

eye as a mixture which approximates to the original hue. It is apparent from *Figure 15.93* that, because the effective vectorial sum of the signals on adjacent lines decreases as the phase error increases, saturation is progressively reduced. When the phase error is 90° the saturation will be zero.

NOTE: It is important to realise that 'adjacent lines' refers to lines laid down in time sequence and not as they appear on the picture when the fields are interlaced. If the lines are numbered in time sequence, odd and even lines will always convey the respective polarities of the basic *V* signal. For instance, if line number one conveys a basic *V* component, then line number six hundred and twenty-six—which is superimposed on it after two fields—will convey a basic −*V* component. Line number six hundred and twenty-six may also be described as line number one of the third field and due allowance must be made for the start of a new sequence.

Hanover bars. Although the eye may accept incorrect hues on adjacent lines as a mixture which simulates the correct hues, there are, because of inherent non-linearity in the system, brightness variations associated with the display of wrong hues and these are not so acceptable; horizontal patterning becomes apparent, especially on saturated colours. The eye has greater visual acuity for brightness variations and can distinguish them although the mixture of hues may be satisfactory. The horizontal patterns appear to move, due to the strobing effects of interlace, and are often referred to as 'Hanover bars'. For small phase errors these effects may not be very significant, but they become increasingly objectionable as the errors increase.

Under good conditions the results obtained with the simple PAL and the NTSC systems are very similar. As the differential phase errors increase, however, with the NTSC system the hue errors increase, while with simple PAL saturation decreases, Hanover bars appear and vertical chrominance resolution is reduced.

DELAY LINE PAL

Instead of relying on the eye to combine the hues on adjacent lines (i.e., the hues misplaced in opposite directions by any differential phase error), the appropriate chrominance signals may be combined electronically by means of a delay line and additional circuits in the receiver.

Figure 15.94 shows a block diagram and a circuit incorporating a 64 μs delay line (a delay of one line period in the 625-line system) which separates the V and U signals effectively before they are applied to their appropriate synchronous demodulators. Each signal is a combination of the contributions from two lines—the immediate past and the present line. This results in an inherent reduction in vertical chrominance resolution, but Hanover bars are not displayed if the characteristics of the circuit are maintained accurately. As with simple PAL, an accurately timed electronic reversing switch is necessary in the V signal path and saturation decreases with increase of differential phase error unless a greater complexity of circuits is employed. *Figure 15.93* is appropriate for both simple and delay line PAL of the types described.

Methods of introducing delay. There are several ways in which a delay line may be connected to achieve the separation of V and U signals before demodulation, but the general principles are illustrated in *Figure 15.94*. The action takes place at the sub-carrier frequency and the required delay of approximately a line period is accomplished by transmitting the amplitude and phase modulated sub-carrier in an ultrasonic mode through a glass element fitted with suitable transducers. The input and output impedances may be of the order of 150 Ω, the bandwidth approximately 2 MHz and the insertion loss some 10 dB or more.

In the circuit shown, in order to obtain the addition and subtraction of the U and V signals in the appropriate adder, the delay time must coincide with an exact number of sub-carrier periods; this timing is normally achieved with an accuracy of about plus or minus 3 ns.

Complete cancellation (i.e., subtraction) in an adder will only occur if its inputs are supplied with signals of equal amplitudes and exactly opposite polarities. The signals referred to are the appropriate unwanted resultant sub-carriers modulated with either U or V components. To each adder is applied the chrominance signals of past and present lines; addition of wanted signals, and subtraction of unwanted signals, take place appropriately but the two separate outputs of U and V signals are the results of combination over two lines. Only in so far as two adjacent lines in the same field are identical, will conditions be obtained as described; and any vertical transition in chrominance will take at least two lines in the same field to be resolved properly. This means that vertical chrominance resolution is reduced to rather less than half by the use of a delay line, because it combines the signals of two lines; there is also a vertical misregistration of about one picture line pitch between the chrominance and luminance displays. As horizontal resolution is already reduced (see earlier) by bandwidth restriction, this misregistration and limitation of vertical resolution is acceptable, although at close viewing distances some unwanted effects may be observed.

Absence of Hanover bars. In the presence of differential phase distortion, the changes of hue are balanced electronically to provide lines of the correct hue. Thus, Hanover

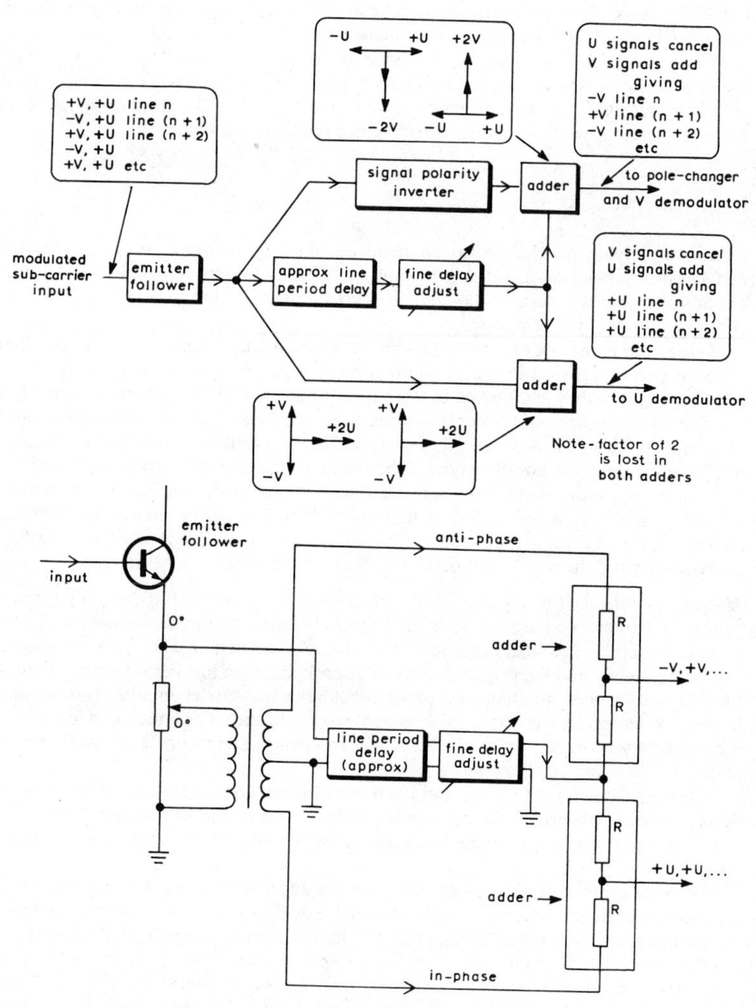

Figure 15.94 Block diagram and circuit showing delay line method of separating V and U chrominance signals in the PAL colour television system

NOTES

1. The factor of 2 is lost in both adders, i.e., 2V becomes V and 2U becomes U, because the resistive adder attenuates (this is clearly seen in the circuit)

2. The delay, which must be an exact number of cycles at sub-carrier frequency, is approximately 64 μs. In some circuits the delay is an odd number of half cycles at sub-carrier frequency with an additional phase reversal included

3. In order to compensate for the attenuation in the delay line, the undelayed inputs to the adders are reduced by the potentiometer in the output of the emitter follower

bars are not produced as they are with simple PAL, which relies on the eye to perform the integration between lines of incorrect hue.

Electronic correction. Referring to *Figure 15.93*, let the amplitude of the transmitted chrominance signal on line 'n' equal 0.5 with a phase angle of 53.1° with respect to the $+U$ axis. The amplitudes of the V and U components are 0.4 and 0.3 respectively. On line 'n + 1' the mirror image of the resultant is transmitted as shown by the reversal of the V component.

Let the error phase angle $\beta = 25°$ lagging—this applies to the received resultants on both lines.

After the V component is switched back through 180° on the appropriate lines in the receiver, the V and U components for each line will be:

On line 'n' $V = 0.5 \sin(53.1° - 25°) = 0.235$
On line 'n + 1' $V = 0.5 \sin(53.1° + 25°) = 0.49$

Average value of V over two lines = 0.36

On line 'n' $U = 0.5 \cos(53.1° - 25°) = 0.44$
On line 'n + 1' $U = 0.5 \cos(53.1° + 25°) = 0.103$

Average value of U over two lines = 0.27

The ratio of $V/U = 0.36/0.27 = 1.33$, which means that the effective angle is unchanged at 53.1°. The effective amplitude of the resultant over two lines is $\sqrt{[0.36^2 + 0.27^2]} = 0.45$ (instead of 0.5 as transmitted), or, as can be seen from the diagram, the final resultant for each line is the transmitted resultant multiplied by cos β. Thus, delay line PAL will provide electronically the average chrominance resultant of two lines (past and present) for each line displayed; in so far as succeeding lines in the same field are identical, the hue changes due to differential phase distortion will be corrected, but the effect of this distortion will be a reduction in saturation.

Chrominance errors on horizontal edges. If the scene contains a rectangular, horizontal patch of colour of even brightness and hue on a neutral (black, grey or white) background, the chrominance of the first line laid down on the display to describe the colour patch will be in error because the preceding line of the picture contained no chrominance. There will, therefore, be no contribution to the adder from the delay line. The second line laid down will be the same as the succeeding lines, but the first neutral line following the bottom of the patch will contain unwanted chrominance information supplied by the delayed previous line. These conditions obtain whether or not there is any differential phase distortion.

EFFECTS OF ASYMMETRICAL CHROMINANCE SIDEBANDS IN SIMPLE AND DELAY LINE PAL

If the response of a circuit carrying a fully coded signal is asymmetrical in the region about the sub-carrier, then the upper and lower sidebands will be unequal in amplitude and also unequally delayed. With suppressed carrier modulation, the resultant consists of sidebands only; and although there will be no change of phase of the resultant with modulation depth (i.e., no differential phase distortion) due to asymmetrical response, there will be changes of amplitude and delay with modulation frequency and the phase of the resultant will differ from that of the original unmodulated carrier before suppression; when the signal is synchronously demodulated in the receiver with a regenerated sub-carrier which is related correctly to the burst, there will be a loss of amplitude because the projection along the required axis is reduced.

Furthermore, due to this phase distortion, there will be an incorrect projection into the quadrature axis which in the NTSC system would result in crosstalk and consequent hue distortion.

In delay line PAL, *U* and *V* signals are separated before demodulation—the unwanted components being cancelled over two lines—and crosstalk is avoided. The loss in amplitude causes desaturation as referred to earlier.

With simple PAL, correction of hue errors due to asymmetrical sidebands is similar to correction of differential phase distortion, but Hanover bars are in evidence at all chrominance transitions producing frequencies outside the bandwidth within which the sidebands are symmetrical.

To illustrate how unwanted components are cancelled, the particular case may be considered of chrominance described by a resultant along the *V* axis with no *U* component. If this is subjected to phase distortion, there will be an unintended projection on to the *U* axis which would be demodulated and would produce a spurious *U* output. When on the next line the phase of *V* is reversed, the direction of this projection will also be reversed and cancellation will take place, electronically if a delay line is used, or as a function of sight with simple PAL.

If chrominance is described only by a resultant along the *U* axis, phase distortion would cause crosstalk into the *V* output. With the phase switching of *V* on alternate lines, cancellation of the unwanted component in the *V* axis would take place, which would again be removed either electronically from the *V* output, or effectively, by the observer's eye.

These examples describe the effects separately, but the principles are the same for the more general case when *U* and *V* components are present simultaneously.

CHOICE OF SUB-CARRIER FREQUENCY

Because the *V* sub-carrier contribution is reversed in phase on alternate lines, it is necessary to adopt a slightly different sub-carrier frequency from that used with the NTSC system, which is an odd multiple of half line frequency.

If, for instance, no *U* component were present in the signal, the alternately reversed phase of the *V* signal would produce a very objectionable dot pattern on a monochrome receiver because the pattern would be the same as if a sub-carrier frequency equal to a multiple of line frequency were used. If, on the other hand, the *U* component only was present, the compatibility would be the same as with NTSC.

The basic sub-carrier frequency adopted with PAL is the NTSC value plus a quarter of line frequency. This produces an effective three-quarter line offset for the *U* signal and an effective quarter line offset for the *V* signal. Thus, the dot pattern wavers diagonally downwards to the left or to the right depending on the chrominance content of the picture. By adding another half cycle per field (i.e., by adding 25 Hz) to the sub-carrier frequency, dot pattern interlacing is achieved which renders the pattern less visible.

The final precise sub-carrier frequency for the 625-line system therefore becomes:

$$\left(\frac{567}{2} + \frac{1}{4}\right) \times 15\,625 + 25\ (\text{Hz}) = 4.433\,618\,75\ \text{MHz}$$

This frequency creates a sequence which contains a whole number of cycles every eight fields; that is, the dot arrangement for a still scene on a monochrome receiver takes four pictures to complete.

Compatibility and dot crawl. Because the dot pattern appears to move in various directions, it is not practicable to describe it. It is better to observe the effects on a monochrome receiver, preferably when colour bars are being transmitted, first at close range and secondly at a distance of about six or seven times picture height. It is necessary to ensure that the receiver raster is interlaced correctly, otherwise a less favourable impression will be obtained.

It is thought by many that the compatibility of NTSC is slightly better than that of

PAL when both are on the same line standard but any differences are subjectively marginal at a distance of six or seven times picture height from the screen.

The appearance of the dot pattern varies considerably with picture content and with movement in the scene. The movement of coloured objects on a neutral background for instance, may be observed as moving areas of dots, which, although it looks quite peculiar, is not necessarily disturbing.

CHROMINANCE BANDWIDTH

The video bandwidth on transmission is about 1 MHz for both U and V signals. With systems using a luminance bandwidth of 5.5 MHz, the full double sidebands of the chrominance sub-carrier may be accommodated. When the total video bandwidth is restricted to 5.0 MHz (as it is with some Continental standards) vestigial sideband working obtains and the use of delay line colour receivers is necessary if Hanover bars are to be avoided completely.

When the characteristic changes from double to single sideband, the chrominance amplitude is halved and thus for fine chrominance detail, saturation is reduced.

From the foregoing discussion it may be concluded that, in spite of some minor disadvantages, PAL is more satisfactory than NTSC when a restriction of bandwidth is unavoidable.

THE ALTERNATING BURST

Reference signals must be transmitted to synchronise the chrominance demodulators and to identify the lines on which the basic $+V$ and $-V$ are transmitted. These requirements are provided by a burst of ten cycles of sub-carrier in each back porch of the waveform (except for omissions in and near the field syncs), the burst alternating in phase on successive lines by 45° about the phase of the $-U$ axis.

Preceding the lines during which V is basically positive, the burst will lag $-U$ by 45°; preceding the lines during which V is basically negative, it will lead $-U$ by 45°.

This alternating burst will be used in the receiver in a manner dependent on the latter's complexity. *Figure 15.95* shows sub-carrier phase and amplitude relationships for fully saturated colours.

The effective amplitude of the PAL alternating burst along its mean phase is only 0.707, in comparison with the NTSC type burst of constant phase for the same peak burst voltage. When properly designed receivers are used, however, this system can be accommodated. Circuits similar to the one described earlier may be used. Comparison of the alternating burst with the steady sub-carrier oscillator shows that the d.c. voltage across R_3 in *Figure 15.80* will vary at half line frequency, and this component is used for line identification; the half line frequency variations are smoothed out in the reactance control, thus effectively utilising the mean phase of the burst for control purposes.

Bruch blanking. In general, the sub-carrier oscillator in the receiver would be locked in quadrature (see page **15**–81) with the mean phase of the burst which is the phase of $-U$. Due to the odd number of lines per picture, if NTSC burst blanking were employed, the phase of the burst at the end of one field would only be the same as the phase of the burst at the beginning of the next for two fields out of four. During early development of the PAL system, because of these variations, certain receivers required careful setting up in order to avoid colour flicker at the top of the picture. With the most recent standards of PAL, the burst is blanked for nine lines in the vicinity of the field sync, and to avoid the burst phase variations mentioned, the timing of burst blanking is staggered progressively by one line each field for four fields and then restored to normal

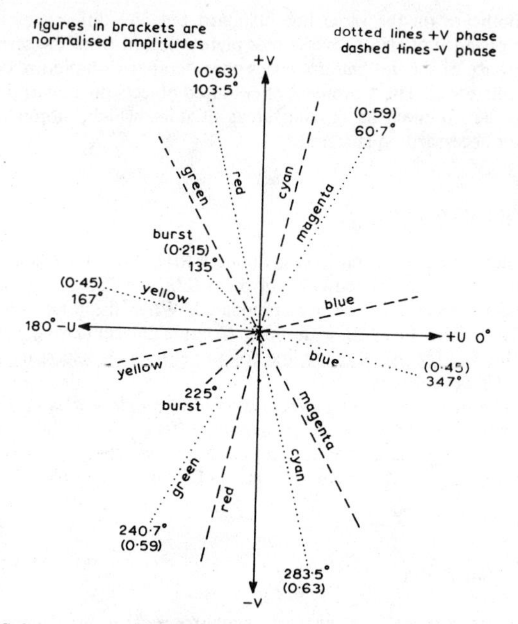

Figure 15.95 *Sub-carrier phase and amplitude relationships for fully saturated colours with the PAL colour television system (at full amplitude)*

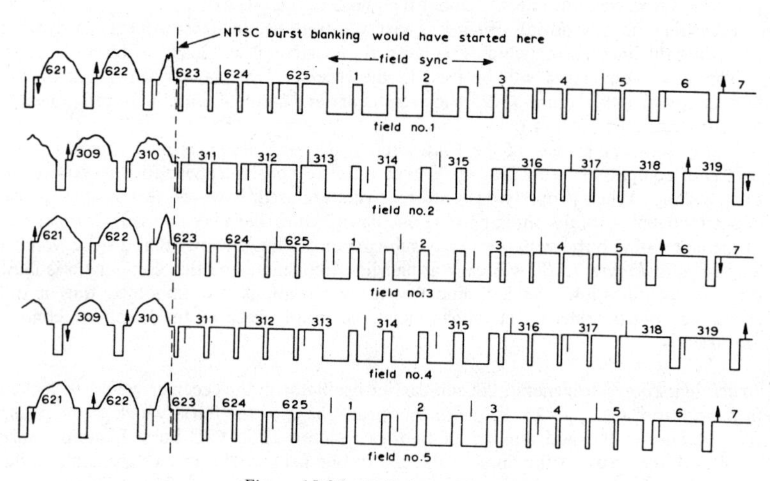

Figure 15.96 *PAL burst blanking*

NOTES

1. The arrows with heads indicate the polarity of the alternating burst, i.e., upwards = 135°, downwards = 225°

2. Lines without heads indicate the polarity in the absence of burst blanking

3. There are nine lines free of burst during each field in staggered sequence returning to starting conditions after four fields in a continuous sequence. This feature of the waveform is termed 'Bruch blanking' and a special burst gating pulse is provided for the coder from the sync pulse generator.

Bruch blanking identifies each field in a sequence of four but results in certain lines having picture information without burst information. This does not affect colour receivers, but may create difficulties in equipment which requires the correct re-insertion of synchronising bursts and pulses. Bruch (or PAL) burst blanking is illustrated in *Figure 15.96*.

THE PAL CODER

The PAL coder resembles in many respects the NTSC coder already described. The main differences are the formation of (R – *Y*) and (B – *Y*) signals instead of *I* and *Q* signals, and the addition of the phase reversing switch in the sub-carrier input to the (R – *Y*) modulator. A simplified block diagram of a PAL coder is given in *Figure 15.97*.

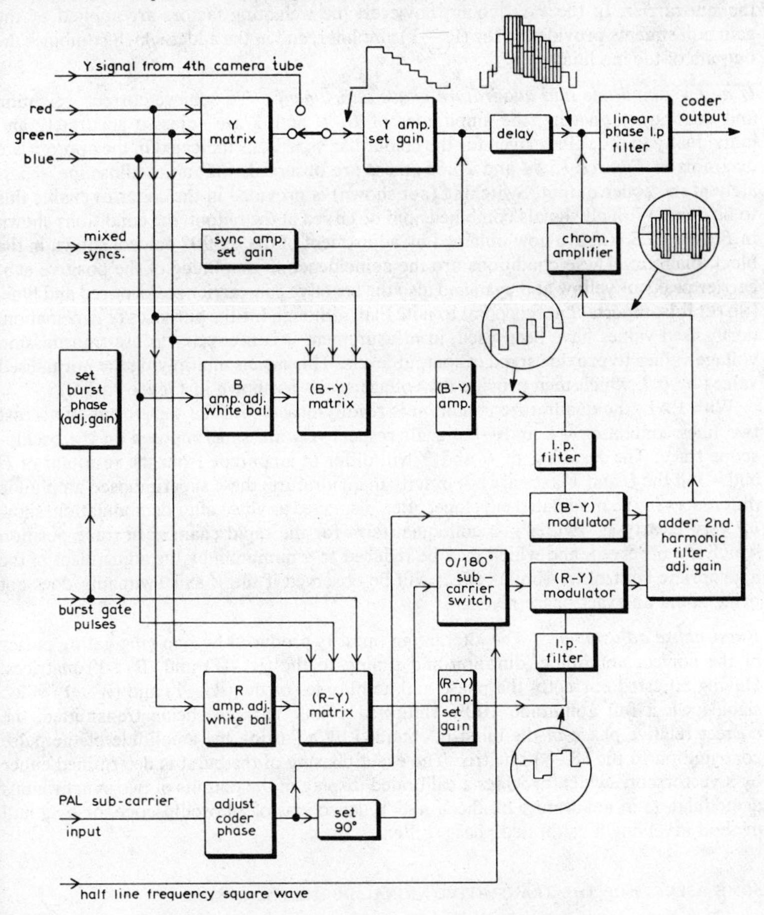

Figure 15.97 *Simplified block diagram of a PAL coder. Colour sequence in the waveforms is white, yellow, cyan, green, magenta, red, blue, black*

In the B amplifier, the blue input is phase reversed and with positive proportions of red and green in the (B – *Y*) matrix, a negative (B – *Y*) video signal is formed, the polarity of which is appropriately altered in the succeeding chain.

Similarly, in the R amplifier, the red input is phase reversed and with positive proportions of green and blue in the (R – *Y*) matrix, a negative (R – *Y*) video signal is formed, which is appropriately changed in polarity in the succeeding stages.

In each case an adjustment of the negative contribution is made to effect white balance, i.e., zero sub-carrier is obtained at the coder output when equal maximum values of R, G and B are applied to the input.

The proportions given on page **15**–76 are used, namely:

$$(\mathrm{R} - Y) = 0.7\mathrm{R} - 0.59\mathrm{G} - 0.11\mathrm{B}$$
$$(\mathrm{B} - Y) = -0.3\mathrm{R} - 0.59\mathrm{G} + 0.89\mathrm{B}$$

(NOTE: In the NTSC coder, the *I* and *Q* video signals are formed in effect from the weighted values of (R – *Y*) and (B – *Y*) (as shown in *Figure 15.87*), before modulating the sub-carrier. In the PAL coder, however, the weighting factors are applied by the gain adjustments provided in the (R – *Y*) amplifier, and in the adder which combines the outputs of the modulators.)

U and V amplitude and quadrature phase adjustment. To achieve correct weighting and quadrature phasing, the amplitudes of *U*, *V* and *Y* are adjusted accurately and individually to the values given for the colour bar signals, as indicated in the appropriate diagrams of *Figures 15.84* and *15.85*; these are observed, with an oscilloscope, separately at the coder output. Switching (not shown) is provided in the coder to enable this to be done. With all signals combined, and observed at the output, the conditions shown in *Figure 15.85* (*d*) are now obtained by adjustment of the set 90° control shown in the block diagram. These conditions are the coincidence in amplitude of the positive sub-carrier peaks of yellow and cyan and also the negative sub-carrier peaks of red and blue. (NOTE: It is important at this point to note that, although for the purposes of explanation, normalised values have been used, in measurement it is necessary to use transmission voltage values to provide standard output levels. This means multiplying the normalised values by 0.7, which then provide the voltage scales as shown in *Figure 15.85*.)

With PAL, the quadrature condition is readily observed if the waveforms of at least two lines containing $+V$ and $-V$ signals respectively are superimposed on the oscilloscope trace. The resultant of *U* and *V* will differ in amplitude from the resultant of *U* and $-V$ if the *U* and *V* axes are not exactly in quadrature; these superimposed amplitude differences in the modulated envelope, often displayed as video after demodulation, show up very clearly as 'twitter'—a colloquial term for the rapid changes in trace position which are observed, and which may be reduced to a minimum by the adjustment of the appropriate control. A similar effect will be observed if the *V* axis switching does not bring about an exact phase reversal.

Burst phase adjustment. The alternating burst is produced by applying gating pulses of the correct amplitude, duration and timing to the (R – *Y*) and (B – *Y*) matrices. Having adjusted correctly the phase and amplitudes of the (R – *Y*) and (B – *Y*) video signals when full amplitude 100% saturated colour bars are being transmitted, the correct relative phase of the burst is obtained by adjusting the amplitude of the pulse contribution to the (B – *Y*) matrix. The exact phasing of the burst is determined either by a vectorscope which provides a calibrated display of the outputs of two synchronous demodulators in quadrature on the X and Y deflections of an oscilloscope, or by a null method involving a calibrated phase shifter.

SOME ASPECTS OF THE TRANSMITTED SIGNAL IN THE PAL SYSTEM

Frequency spectrum of a colour transmission. The frequency spectrum and response characteristics of a colour transmission are shown in *Figure 15.98*. In transmission, it is usual to include a notch filter at the sub-carrier frequency in the luminance channel to reduce the interference which occurs between luminance and chrominance, although more recently the practice of including this filter has been discontinued in some countries. The high frequency luminance components are detected by the chrominance demodulators and are translated into low frequency chrominance noise.

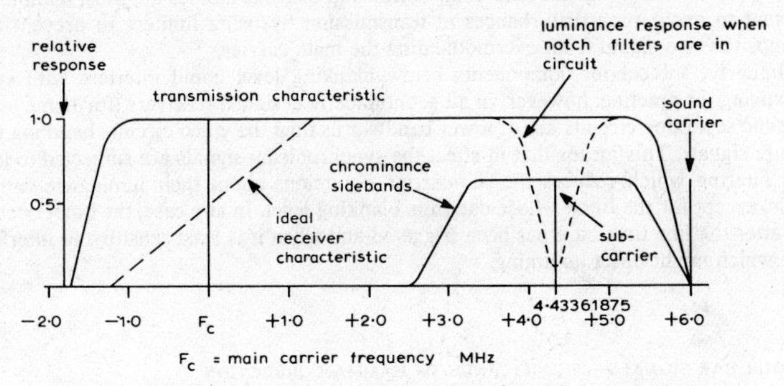

Figure 15.98 The frequency spectrum of a British colour transmission

In the luminance channel of the colour receiver there is also a sub-carrier frequency notch filter to remove the dot pattern. No attempt has been made in *Figure 15.98* to show the effects of transmission and receiver filters accurately; it is intended merely to indicate that luminance response is lacking in the region of the sub-carrier.

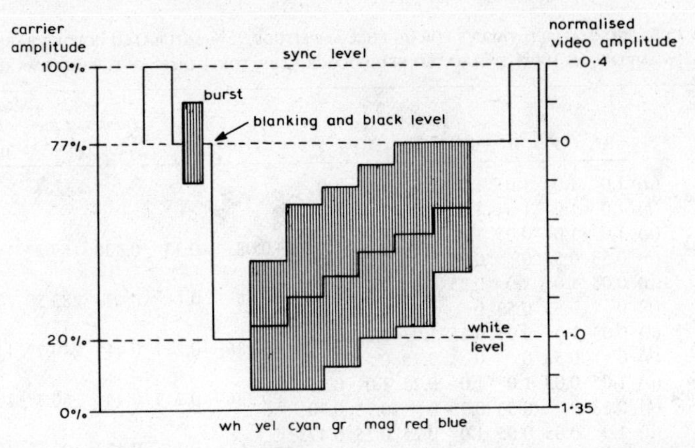

Figure 15.99 Modulation depths of the main carrier for a line waveform of full amplitude 95% saturated colour bars

Modulation of main carrier. A diagram showing modulation depths of the main carrier for a line waveform of full amplitude 95% saturated colour bars (for derivations see page **15**–111) is given in *Figure 15.99*. The sub-carrier peaks for yellow and cyan are seen to reduce the main carrier amplitude to approximately 5.5%.

If full amplitude 100% saturated bars were to be transmitted the main carrier amplitude would be reduced to about 1%, which would be a very stringent test for the amplitude level stability of the system. Providing the peaks of sub-carrier do not reduce the amplitude of the main carrier to zero or drive any amplifiers which are handling the main carrier into non-linear operation, the sub-carrier component may be removed by filtering and the performance of inter-carrier sound reception will be unaffected. Otherwise, any regular overmodulation will cause disturbances at line frequency and also at field frequency, due to its cessation during field blanking. This gives rise to 'buzz on sound', so called because the field components and their harmonics are most audible. It is usual to avoid such disturbances in transmission by using limiters to prevent the composite video signal from overmodulating the main carrier.

Similarly, sub-carrier components below blanking level could interfere with synchronising. In practice, however, in all economically designed receivers (for home use), the sync separator circuits are of lower bandwidths than the video circuits handling the picture signals. This means that in effect the synchronising signals are subjected to low pass filtering which reduces the sub-carrier excursions about their luminance datum levels, except for the burst whose datum is blanking level. In any case, the burst occurs just after the line time-base has been triggered and when it is least sensitive to interference which might upset its timing.

COLOUR BAR SIGNALS—SIGNIFICANCE OF GAMMA CORRECTION

Previously, in conjunction with *Figure 15.85* (*d*), a colour bar test signal was described in which full amplitude 100% saturated colours are represented. This signal is most useful for setting up coders but is not usually transmitted to viewers because it is an unnecessarily stringent test for transmitters and receivers. The transmitted version of the colour bar signal represents full amplitude 95% saturated colours.

Table 15.8 NORMALISED VALUES FOR (*a*) FULL AMPLITUDE 95% SATURATED COLOUR BARS, AND (*b*) 53% AMPLITUDE 100% SATURATED EBU COLOUR BARS, FOR A DISPLAY TUBE GAMMA OF 2.2

		R	G	B	R′	G′	B′	*Y*	*V*	*U*	*Amplitude*	*Phase* *Line* 'n'	*Phase* *Line* 'n + 1'
White	(*a*)	1.0	1.0	1.0	1.0	1.0	1.0	1.0					
	(*b*)	1.0	1.0	1.0	1.0	1.0	1.0	1.0					
Yellow	(*a*)	1.0	1.0	0.05	1.0	1.0	0.25	0.915	±0.08	−0.33	0.336	167.2°	192.8°
	(*b*)	0.53	0.53	0	0.75	0.75	0	0.66					
Cyan	(*a*)	0.05	1.0	1.0	0.25	1.0	1.0	0.775	±0.46	0.11	0.473	283.5°	76.5°
	(*b*)	0	0.53	0.53	0	0.75	0.75	0.53					
Green	(*a*)	0.05	1.0	0.05	0.25	1.0	0.25	0.69	±0.386	−0.217	0.44	240.7°	119.3°
	(*b*)	0	0.53	0	0	0.75	0	0.45					
Magenta	(*a*)	1.0	0.05	1.0	1.0	0.25	1.0	0.56	±0.386	0.217	0.44	60.7°	299.3°
	(*b*)	0.53	0	0.53	0.75	0	0.75	0.30					
Red	(*a*)	1.0	0.05	0.05	1.0	0.25	0.25	0.473	±0.46	−0.11	0.473	103.5°	256.5°
	(*b*)	0.53	0	0	0.75	0	0	0.23					
Blue	(*a*)	0.05	0.05	1.0	0.25	0·25	1.0	0.336	+0.08	0.33	0.336	347.2°	12.8°
	(*b*)	0	0	0.53	0	0	0.75	0.09					
Burst (*a*) and (*b*)									±0.152	−0.152	0.215	135°	225°

NOTE: $Y = 0.3R' + 0.59G' + 0.11B'$.
$R' = R^{0.45}, G' = G^{0.45}, B' = B^{0.45}$.
$V = 0.877(R - Y)$.
$U = 0.493(B - Y)$.

It should be understood that reference to saturation values in this connection is not strictly in accordance with colorimetric specification but is a convenient guide to the capability of the system. From the C.I.E. Chromaticity Diagram [6, 9] used in the specification of colour, it will be seen that fully saturated colours are described by a spectrum locus which lies outside two of the colour television primaries, green and blue. This means that no hues other than red of the same wavelength as the primary can be displayed in colour television with full saturation. Nevertheless, in practice, the maximum degree of saturation of which the system is capable is regarded as being 100%, which is the condition when there is an output from only one or two of the three channels.

British system and EBU (European Broadcasting Union) colour bar amplitudes. In order to simulate colours from a gamma corrected picture source it is necessary to apply red, green and blue waveforms of the appropriate values to the input of the coder. For full amplitude 100% saturation, there is no difference due to gamma correction because unity to the power of gamma is still unity, but values other than unity are modified.

The method of specifying saturation has already been mentioned. Taking full amplitude 95% saturated red as an example, before gamma correction this would require the normalised values of $R = 1.0$, $G = 0.05$ and $B = 0.05$ from the three sources. After gamma correction ($1/\gamma = 0.45$) these values would become $R = 1.0^{0.45} = 1.0$, $G = 0.05^{0.45} = 0.25$ and $B = 0.05^{0.45} = 0.25$.

In addition to the colour bars mentioned there is a colour bar signal used in Europe which has full amplitude white but 53% amplitude 100% saturated colours.

Table 15.8 gives the values, *Figure 15.100* shows the coded waveforms and *Figure 15.101* gives the R, G and B video waveforms which would be required at the input of a coder, for both varieties of colour bar signals.

Limitations of EBU colour bars. For transmission purposes, the EBU colour bars do not represent a sufficiently stringent test for transmitters and receivers, or for links and

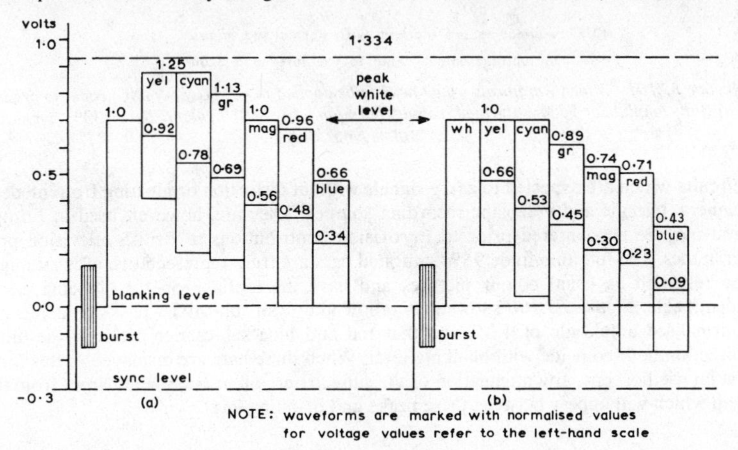

Figure 15.100 The waveforms of (a) full amplitude 95% saturated colour bars and (b) EBU colour bars

NOTE

EBU colour bars are 100% saturated with white at full amplitude, but with the luminance and chrominance of the colours calculated from fundamental R, G *and* B *amplitudes of 53%—this means that allowing for a gamma of 2.2 for the display tube, the amplitudes of the waveforms of* R, G *and* B *from the colour bar generator should be 75%*

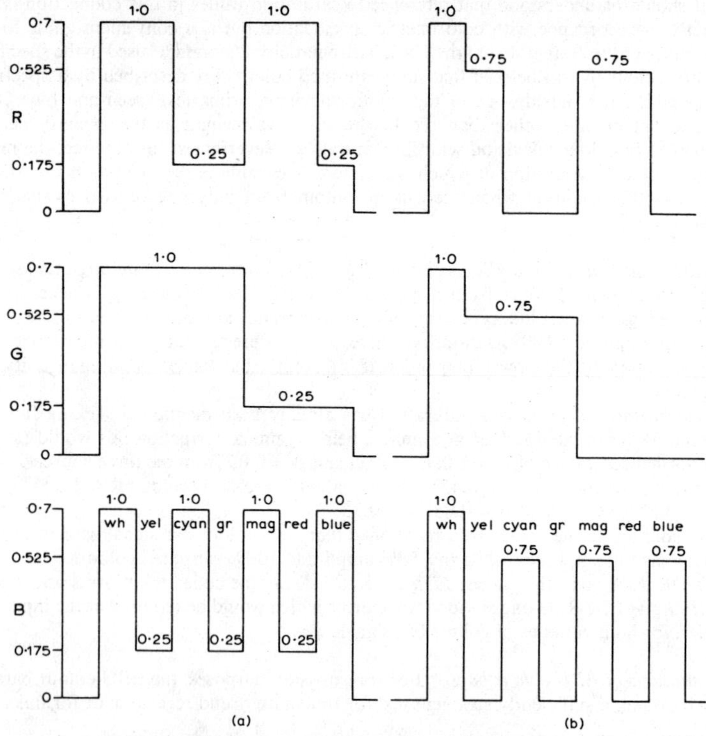

Figure 15.101 Video waveforms which need to be applied to a PAL (or NTSC) coder to produce (a) full amplitude 95% saturated colour bars and (b) EBU colour bars (53% amplitude 100% saturated)

circuits which are expected to carry signals without distortion originating from modern camera, telecine and videotape recording sources. They are, however, used in Europe and may be encountered prior to Eurovision contributions to British television programmes. The full amplitude 95% saturated bars are truly representative of what might be regarded as usual colour pictures and have the useful property of being easily identifiable. *Figure 15.100* shows that yellow and cyan sub-carrier peaks coincide at a normalised amplitude of 1.25, and that red and blue sub-carrier peaks in the other direction both coincide with blanking level. When these bars are displayed in this form on an oscilloscope, any attenuation of the sub-carrier will at once be apparent from the gap which will appear between these peaks and blanking level.

SIMPLE PAL DECODER

A simplified block diagram of a simple PAL decoder is shown in *Figure 15.102*. When line flyback pulses are available, these may be used as shown to drive the V axis 0°/180° switch; identification is then provided by the half line frequency component which results from the phase comparison between the alternating burst and the locally regenerated sub-carrier (see p. 15–105). This half line frequency component suitably changes the divide-by-two count of line flyback pulses if the switch polarity is incorrect. When line flyback pulses are not available (for example, when a decoder is not associated closely with the display tube) the switch may be operated by the half line

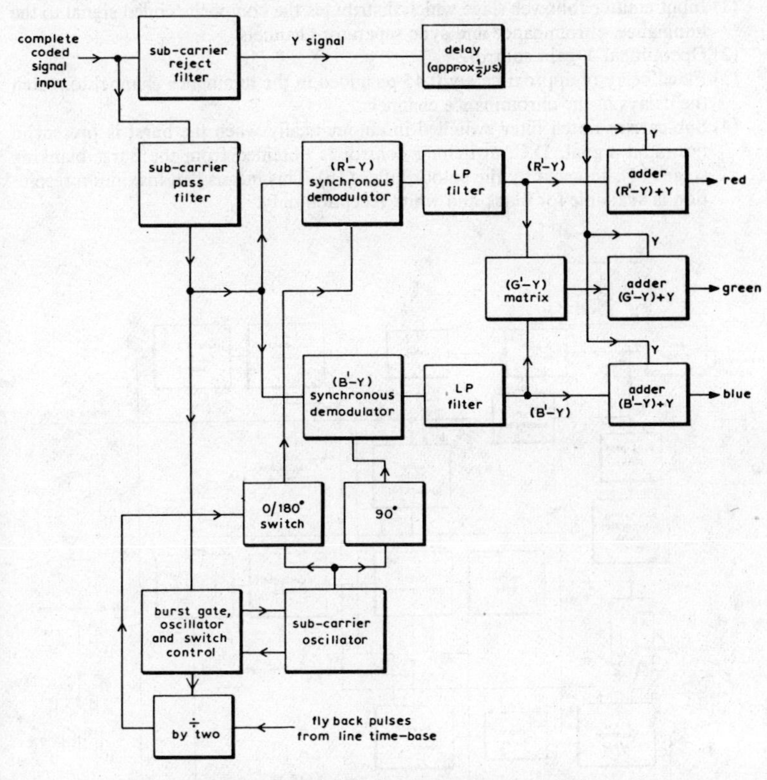

Figure 15.102 Simplified block diagram of a simple PAL decoder
NOTE: $(G' - Y) = -0.51(R' - Y) - 0.19(B' - Y)$

frequency component from the phase comparator; precise timing is provided by the addition of the trailing edges of separated line syncs. Reference to *Figure 15.96* will show that the alternating burst is blanked out preceding the last half line, one whole line and one and a half lines respectively of three fields out of four. V axis switching is maintained for these periods by a tuned circuit of appropriately high Q, associated with amplification of the half line frequency output from the phase comparator.

In this circuit the V axis switch operates on the demodulating sub-carrier input to the $(R' - Y)$ synchronous demodulator.

It should be noted that the demodulators are termed $(R' - Y)$ and $(B' - Y)$ and not V and U. The amplitude adjustments which produce the unweighted values $(R' - Y)$ and $(B' - Y)$ may be made at any part of the circuit where these two signals are handled as separate components. It is the ratio of the two signals which is of importance; their absolute values are adjusted simultaneously by means of the saturation control.

SWITCHED SIMPLE/DELAY LINE PAL DECODER

The block diagram of a PAL decoder which may be switched to the simple (S) or delay line (D) mode is shown in *Figure 15.103*. The blocks are numbered to coincide with the following list of operations and functions:

(1) Input emitter follower stage which distributes the composite coded signal to the luminance, chrominance and sync separator channels.
(2) Operational *Y* gain control.
(3) Fixed delay of approximately 0.45 μs added in the luminance channel to match the delays in the chrominance channels.
(4) Sub-carrier notch filter switched in automatically when the burst is present in the input signal. D.C. switching control is obtained from the burst blanking stage (15), actuated by the colour killer (16). This means that maximum resolution is available for black and white reception only.

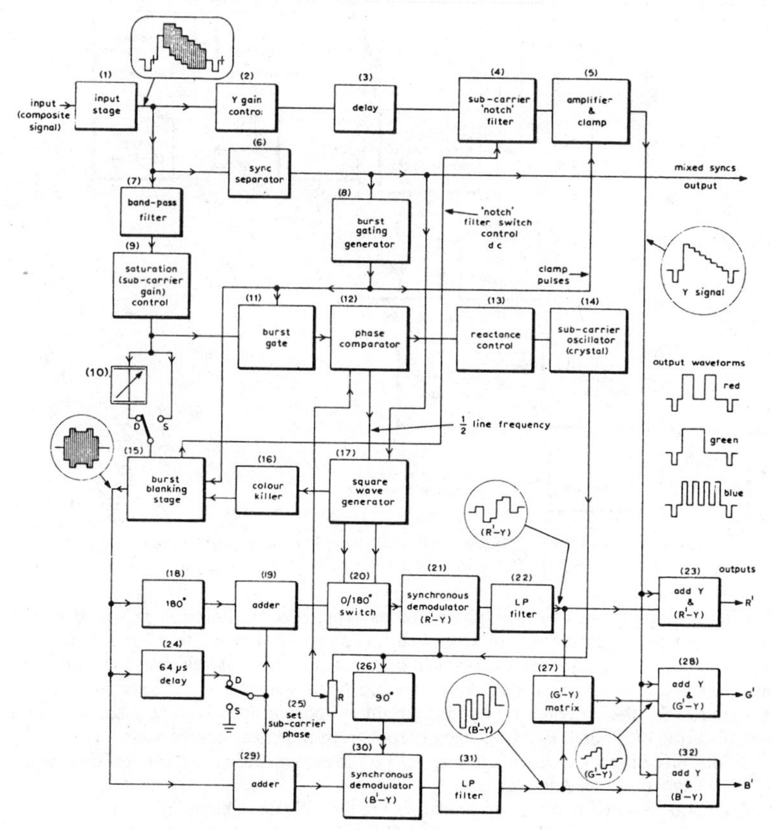

Figure 15.103 Block diagram of a PAL decoder, which can be switched to simple or delay line operation

NOTES

1. Decoder is shown switched to delay line mode (D)
2. Colour sequence in waveforms: white, yellow, cyan, green, magenta, red, blue, black

(5) Luminance amplifier and clamp. Clamp pulses are 4 μs burst gating pulses supplied by (8).
(6) Sync separator supplying external mixed syncs, timing for the burst gating pulse generator (8), and precise timing for the square wave generator (17) which actuates the 0/180° switch (20).
(7) Filter with a passband of approximately 3 MHz centred on the sub-carrier frequency.
(8) Burst gating pulse generator deriving pulse timing from mixed syncs and supplying 4 μs pulses for clamping (5), gating (11) and blanking (15). (NOTE: These pulses are all of the same duration and timing.)
(9) Saturation control—operational chrominance sub-carrier gain control.
(10) Preset variable attenuation switched into circuit by the simple delay switch in its D (delay) position to match the loss of one contribution to each adder, (19) and (29), in the S (simple) mode.
(11) Burst gate actuated by pulses from (8) allowing burst (without the chrominance signals) to be applied to the phase comparator (12).
(12) Phase comparator comparing the locally generated sub-carrier from (25) with the alternating burst from (11). The resultant of the comparison supplies a varying d.c. to the reactance control (13) and half line frequency waveform to the square wave generator (17). The mean d.c. to (13) is adjustable by means of a pre-set variable resistor to obtain the best operating range of control.
(13) Reactance control for (14) by varactor diode supplied with d.c. control voltage from (12).
(14) Quartz crystal sub-carrier oscillator controlled in frequency over ±200 or 300 Hz by (13).
(15) Burst blanking stage in which the burst is removed from the chrominance signal. This stage is inhibited by the colour killer (16) when the burst is absent from the input signal, and also provides d.c. control for the filter switch (4).
(16) Colour killer stage which provides a change in d.c. voltage for (15) depending on the presence or absence of the square wave from (17).
(17) Square wave generator (at half line frequency) which derived its precise timing from the trailing edge of line syncs in conjunction with the half line frequency output of the phase comparator (12).
(18) Phase reverser—part of delay line decoding circuit.
(19) Adder—part of delay line decoding circuit.
(20) Phase reversing switch actuated by square waves from (17). This operates on the chrominance modulated sub-carrier and not on the demodulating sub-carrier as with the decoders shown in *Figure 15.101*.
(21) (R′ – *Y*) demodulator—fed with correctly phased regenerated sub-carrier through (26). See pp. **15**–81 and **15**–85.
(22) Low-pass filter to remove the sub-carrier component from the (R′ – *Y*) demodulator (21) output.
(23) Adder to combine *Y* with (R′ – *Y*) to produce the R′ output.
(24) A glass delay line providing a delay of 63.943 μs, which is exactly 283.5 periods of the sub-carrier. The extra 180° necessary for delay line decoding is obtained by the correct polarity connection of the delay line.
(25) A variable resistance phasing control (from which the sub-carrier is taken for comparison with the burst) to ensure correct burst/regenerated sub-carrier phase relationship.
(26) Phase shifting network to ensure quadrature relationship between the demodulating sub-carriers used in (21) and (30). (NOTE: Phase shifts in both chrominance sub-carrier channels and in the regenerated sub-carrier chain are dependent on tuning (transformer inductances) and circuit constants. Phase relationships are the resultant of all these factors and must be correct at relevant points in the circuit.)

(27) Matrix which combines −0.51(R′ − Y) and −0.19(B′ − Y) to form (G′ − Y).
(28) Adder to combine Y with (G′ − Y) to produce the G′ output.
(29) Adder—part of delay line decoding circuit.
(30) (B′ − Y) demodulator fed with correctly phased regenerated sub-carrier from (14). See p. 15–85 and *Figure 15.79*.
(31) Low-pass filter to remove the sub-carrier component from the (B′ − Y) demodulator (30) output.
(32) Adder to combine Y with (B′ − Y) to produce the B′ output. (NOTE: The phase reversing switch in (20) operates on the chrominance modulated sub-carrier instead of on the demodulating sub-carrier. This avoids the changes in d.c. output which may occur on switching a high level demodulating sub-carrier when it suffers from second harmonic distortion. The waveforms shown in *Figure 15.103* are those for full amplitude 100% saturated colour bars.)

The SECAM system

GENERAL PRINCIPLES

The SECAM (Sequential Colour with Memory) system was devised from the original NTSC system by Henri de France in about 1956. It went through many phases of development, with models variously termed SECAM I, II and III, but has now been stabilised in a final form, resulting from an optimisation of SECAM III, called simply 'SECAM'.

With SECAM, the colour difference signals (R − Y) and (B − Y) are sent separately on alternate lines, and frequency modulation of the sub-carrier is used instead of amplitude and phase modulation.

In transmission during one line, (R − Y) signals are transmitted and the (B − Y) information is discarded, and on the next line (B − Y) signals are transmitted when the (R − Y) information is discarded, in a continuous sequence. In the receiver, by means of a delay line and electronic switch, a combination of past and present signals is achieved whereby the sequential information is displayed simultaneously. This results in a reduction of the vertical chrominance resolution similar to PAL, but crosstalk between the colour difference signals is avoided, as is the necessity for synchronous detection. Line identification is necessary and is provided by signals sent prior to each field.

SECAM provides good protection against differential phase distortion and against differential gain (variation of sub-carrier amplitude with changes of luminance level) because of its use of frequency modulation; there are difficulties, however, in achieving good compatibility in terms of an acceptably low degree of dot patterning on black and white receivers.

Compatibility. Because frequency modulation is used, the half line sub-carrier offset (NTSC) or three-quarter line sub-carrier offset (PAL) is not effective in reducing the visibility of the dot pattern, as this relationship would vary with the deviation of the sub-carrier. The centre (undeviated) frequency of the sub-carrier is stabilised at an even multiple of line frequency at the beginning of every line. On consecutive fields the sub-carrier is reversed in phase and this cancels the strident dot pattern which would otherwise be apparent. Due to interlace strobing, this cancellation would not be fully effective with vertical eye movement and a further phase reversal of the sub-carrier is therefore inserted on one line out of three.

As the sub-carrier is not suppressed, the dot pattern will appear in black and white as well as coloured areas of the picture; to reduce this effect, which is to the detriment of compatibility, the amplitude of the sub-carrier is made suitably low in its undeviated condition, and is increased by what is termed 'r.f. pre-emphasis' when the chrominance signal is being transmitted, to improve the chrominance signal-to-noise ratio. In the receiver, there is a filter with inverse characteristics (r.f. de-emphasis).

Video pre-emphasis. As usual for frequency modulation systems, pre-emphasis (with de-emphasis in the receiver) is used to improve the signal-to-noise ratio for the higher modulating frequencies. This, in conjunction with r.f. pre-emphasis, produces the characteristic increase in sub-carrier amplitude (as seen on the displayed waveform) at all chrominance transitions in the horizontal direction (see *Figure 15.104*).

R.F. pre-emphasis increases the amplitude of the sub-carrier in proportion to its deviation, while video pre-emphasis increases the deviation for the higher modulating frequencies.

Weighting factors. With SECAM, the weighting factors for the colour difference signals are based on the amplitude excursions required for the EBU colour bars. Reference to *Table 15.8* will show that red and cyan produce the maximum values of V, with blue and yellow producing the maximum values for U. By unweighting these values we have:

For red and cyan

$$(R' - Y) = \pm 0.525$$

For blue and yellow

$$(B' - Y) = \pm 0.67$$

These values are normalised to units by the SECAM weighting factors when, instead of V and U, the weighted colour difference signals are termed D'_R and D'_B, thus

$$D'_R = \frac{(R' - Y)}{0.525} = -1.9(R' - Y)\ (= -1.0 \text{ normalised value})$$

$$D'_B = \frac{(B' - Y)}{0.67} = 1.5(B' - Y)\ (= 1.0 \text{ normalised value})$$

(NOTE: The weighting factor for $(R' - Y)$ is made negative for reasons which will be apparent later (see below—this page).)

In the frequency modulation system, the standard deviation of the sub-carrier for D'_R lines is ± 280 kHz for $D'_R = \pm 1.0$ with maximum deviations limited to $+350$ kHz when $D'_R = 1.25$ and -506 kHz when $D'_R = -1.79$ and for D'_B lines it is ± 230 kHz for $D'_B = \pm 1.0$ with maximum deviations limited to $+506$ kHz when $D'_B = 2.18$ and -350 kHz when $D'_B = -1.52$.

Asymmetric deviation. These maximum deviations include the effects of video pre-emphasis. The possibility of making them asymmetrical for each colour difference signal, D'_R and D'_B, arises because the accurate reproduction of sharp chrominance transitions is not so important for colours of high luminance as it is for colours of low luminance. Due to the failure of constant luminance, there is a contribution of luminance through the chrominance channels which is greater for low luminance colours than for high luminance colours.

Table 15.5 shows that the low luminance colours are described by comparatively high positive values of $(R - Y)$ and $(B - Y)$ signals. Thus, asymmetric deviation of the sub-carrier is permissible for both colour difference signals.

Sub-carrier frequencies. Two sub-carrier frequencies are used, differing by 156.25 kHz.

On D'_R lines the sub-carrier frequency is 282 times line frequency, i.e., 4.406 5 MHz.

On D'_B lines the sub-carrier frequency is 272 times line frequency, i.e., 4.250 MHz.

Because of the negative sign of the weighting factor D'_R, the maximum deviation due to this signal is towards a lower frequency, whilst the maximum deviation due to the D'_B signal is towards a higher frequency. Thus, by employing two sub-carriers, the total bandwidth is not increased even though the two modulating signals are of opposite polarity—a feature which is utilised for line identification.

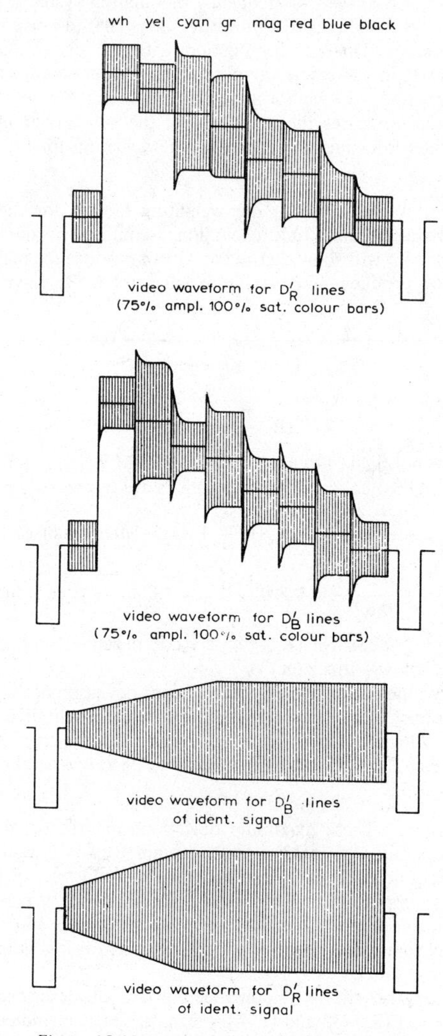

Figure 15.104 Video waveforms with SECAM

NOTES

1. These waveforms show only amplitude variations—the chrominance sub-carrier is frequency modulated. White level for the colour bars is at 75% amplitude (not 100% as with EBU bars)

2. Sub-carrier blanking leaves a few μs of undeviated sub-carrier in the back porch to prevent excessive noise at the left hand edge of the picture

3. The ident signal commences on the fourteenth line before picture information in field blanking. It consists of a negative D'_B line waveform followed by a positive D'_R line waveform, alternating for nine lines, i.e., there are five negative D'_B lines and four positive D'_R lines

Line identification (ident) signals. Line identification consists of nine lines of frequency modulated sub-carrier in the field blanking period, starting fourteen lines in advance of picture information. The modulation takes the form of a truncated sawtooth for each line; there are five D'_B lines and four D'_R lines. Because the consecutive lines contain video waveforms of opposite polarity, their order is of particular significance and may be sensed in the receiver to provide correction for its line switch should this be in error. The modulated envelopes for D'_R and D'_B lines are shown in *Figure 15.104*.

It should be realised that all the waveforms in *Figure 15.104* show only the amplitude of the sub-carrier, with no indication of its frequency deviation as a consequence of r.f. pre-emphasis. With ident signals, there is some similarity between the sub-carrier envelope and the original truncated sawtooth video modulating waveforms.

On D'_R lines of the ident signal, the signal is positive and the sub-carrier is deviated from the centre frequency of 4.406 5 MHz by 350 kHz to 4.756 5 MHz, while on D'_B lines the signal is negative and the deviation is from the centre frequency of 4.250 MHz by −350 kHz to 3.900 MHz. These conditions are shown in *Figure 15.106*.

By making the D'_B frequency modulation negative, and the D'_R frequency modulation positive, the ident signal establishes the maximum frequency deviation values of the system. This, in conjunction with the sawtooth portion of the waveshape (in terms of frequency modulation—not of amplitude variation), is useful in receiver alignment.

The normal transmitted polarity during picture time is negative D'_R and positive D'_B and therefore in the receiver chain, a phase reversal of the D'_R signal in relation to the D'_B signal must take place.

By transmitting positive D'_R and negative D'_B for identification, with the phase reversal of positive D'_R both these signals will be negative in the receiver, which, through the normal $(G' - Y)$ matrix, will provide a negative $(G' - Y)$ signal. Thus, if the switching polarity in the receiver is correct, during the period of the ident signal a negative $(G' - Y)$ pulse will be obtained which can be arranged to have no effect on the action of the switch. If, however, the switching polarity is incorrect then a positive $(G' - Y)$ pulse will be obtained which may be arranged to correct the switch action.

(NOTE: Particular polarities have been quoted for purposes of explanation—the actual polarities encountered in the receiver will depend on circuit arrangements—only the relative polarities are of significance.)

THE SECAM CODER

A simplified block diagram of the SECAM coder is given in *Figure 15.105*. The blocks are numbered to coincide with the following list of operations and functions:

(1) Matrices providing $(0.3R' + 0.59G' + 0.11B')$ for the luminance signal and $-1.9(R' - Y)$ and $+1.5(B' - Y)$ for the D'_R and D'_B chrominance signals respectively.
(2) Fixed delay in the luminance channel to ensure that its delay is the same as delays in the chrominance channels.
(3) Luminance/chrominance and sync mixer.
(4) Coder output consisting of the fully coded output with syncs.
(5) Video pre-emphasis providing about 9 dB accentuation at 800 kHz in comparison with low frequencies and video limiting for D'_R signals.
(6) Same as (5) but for D'_B signals.
(7) Frequency modulator for D'_R lines provided with the appropriate centre (undeviated) sub-carrier frequency reference at the beginning of the line.
(8) Frequency modulator for D'_B lines provided with the appropriate centre (undeviated) sub-carrier frequency reference at the beginning of the line.
(9) Electronic switch at half line frequency alternately producing D'_R and D'_B signals (on sub-carrier) as an input to (10).

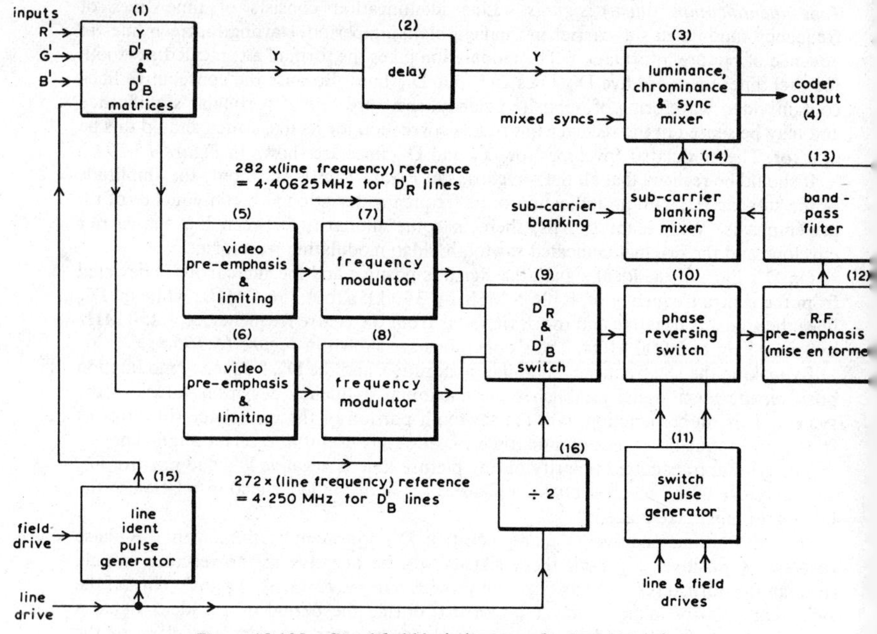

Figure 15.105 Simplified block diagram of a SECAM coder
NOTE: $D'_R = 1.9(R' - Y)$
$D'_B = 1.5(B' - Y)$

(10) Switched phase reverser which reverses the phase of the sub-carrier on succeeding fields and also on every third line.
(11) Switch pulse generator supplying the actuating pulses for (10).
(12) R.F. pre-emphasis. This takes the form of a filter which reduces the undeviated sub-carrier amplitude (peak-to-peak) to about 23% of peak white level. As the sub-carrier is deviated by the video modulating signal, its amplitude is increased. This improves chrominance signal-to-noise ratio but reduces the visibility of the dot pattern on black and white receivers in those parts of the picture which have no chrominance. Also termed 'mise en forme'.
(13) Band-pass filter of approximately ±1.5 MHz bandwidth, centred about the sub-carrier frequency.
(14) Mixer which inserts sub-carrier blanking. The latter avoids blanking out the line ident waveforms after field syncs and it also leaves a few microseconds of unmodulated sub-carrier in the back porch before the commencement of each line of picture waveform to provide the receiver limiters with a signal and so prevent excessive noise at the left hand edge of the picture.
(15) Line ident pulse generator supplying the appropriate truncated sawtooth waveforms to (1).
(16) Binary divider which derives half line frequency pulses from line drive for actuating (9).

R.F. pre-emphasis characteristics. *Figure 15.106* shows sub-carrier frequencies, deviations and amplitudes for EBU colour bars. The centre frequency of the filter, providing r.f. pre-emphasis, does not coincide with the mid-band frequency of the two sub-carriers. A centre frequency of 4.286 MHz was chosen to provide a more favour-

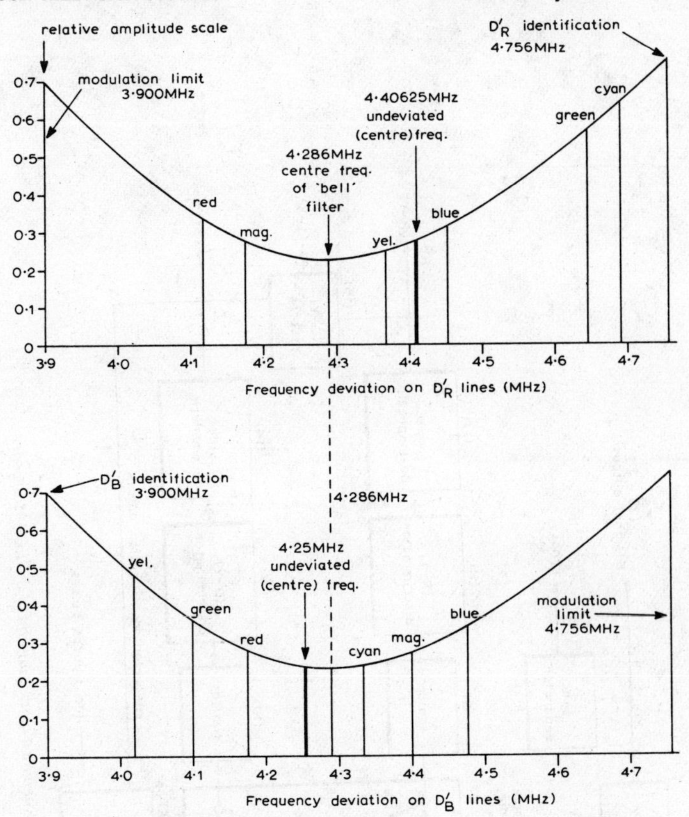

*Figure 15.106 Diagram showing sub-carrier frequencies, deviations and amplitudes for EBU colour bars, with the SECAM system**

Hue	*Frequency on D'_R lines* MHz	kHz	*Frequency on D'_B lines* MHz	kHz
Yellow	4.361 25	−45	4.020	−230
Cyan	4.686 25	+280	4.328	+78
Green	4.641 25	+235	4.100	−150
Magenta	4.171 25	−235	4.402	+152
Red	4.126 25	−280	4.172	−78
Blue	4.451 25	+45	4.480	+230

* *Information supplied by Compagnie Française de Télévision, Paris*

able distribution of noise—low frequency noise is reduced and high frequency noise is increased—also, compatibility is improved by a measure of equalisation of sub-carrier amplitudes on succeeding lines (i.e., on D'_R and D'_B lines) for colours of low luminance. The bell filter providing r.f. de-emphasis in the receiver has the same centre frequency and inverse response characteristics.

Fading a SECAM signal. Coded SECAM signals cannot be faded or mixed directly, because of the inclusion of the frequency modulated sub-carrier. It is necessary to

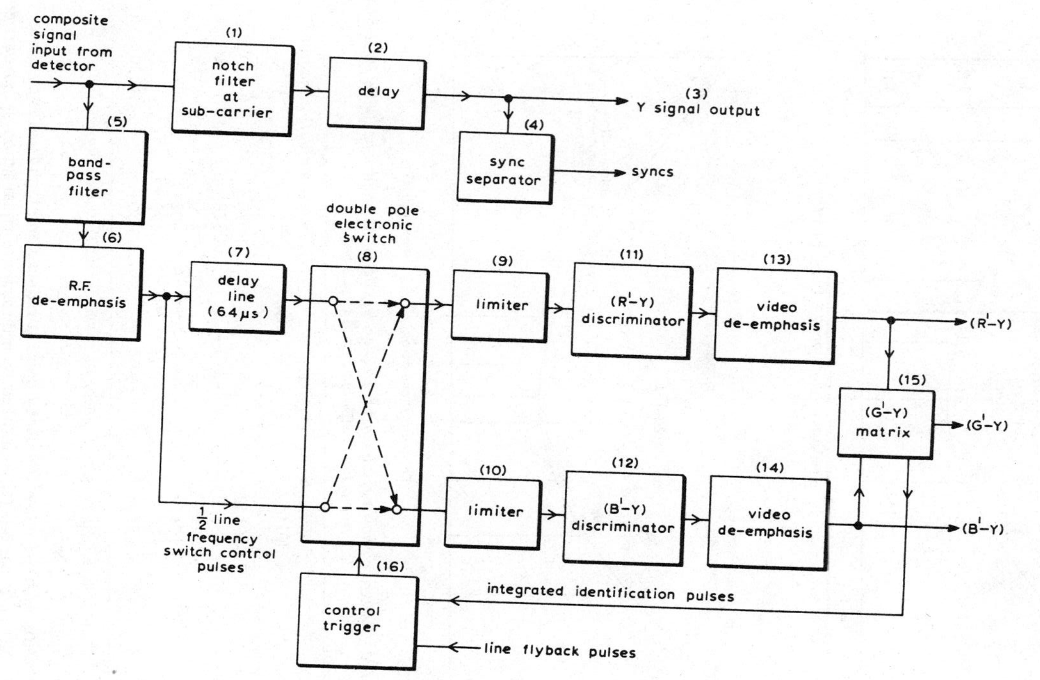

Figure 15.107 Simplified block diagram of a SECAM decoder

demodulate the sub-carriers and then carry out the operation of fading and mixing on chrominance and luminance signals separately, with re-modulation of chrominance on a new sub-carrier and a re-combination with luminance. After the operation has been completed, the system reverts to a single channel with the demodulators and re-modulators switched out of circuit.

THE SECAM DECODER

A simplified block diagram of the SECAM decoder is shown in *Figure 15.107*, in which the blocks are numbered to coincide with the following list of operations and functions:

(1) Notch filter in the luminance channel to remove the sub-carrier which would otherwise cause de-saturation by increasing the brightness, and also dot patterning.
(2) Fixed delay of about 0.8 μs added in the luminance channel to match the delays in the reduced bandwidth chrominance channels.
(3) *Y* signal output.
(4) Usual sync separator to provide line and field timing.
(5) Band-pass filter which passes only the frequency modulated chrominance sub-carrier.
(6) R.F. de-emphasis filter. Formerly termed 'remise-en-forme', currently termed 'bell' filter.
(7) SECAM delay line of 64 μs with a tolerance of ±170 ns. Compare this with the accuracy of ±3 ns required for PAL. With SECAM only the correct line by line chrominance registration is of significance; there is no requirement involving an exact number of sub-carrier half cycles.
(8) Double pole electronic switch for providing the discriminators (11) and (12), through the limiters (9) and (10), with their appropriate past and present signals.
(9) Amplitude limiter preceding (11).
(10) Amplitude limiter preceding (12).
(11) (R′ – *Y*) discriminator.
(12) (B′ – *Y*) discriminator.
(13) Video de-emphasis for (R′ – *Y*) signals.
(14) Video de-emphasis for (B′ – *Y*) signals.
(15) (G′ – *Y*) matrix. (G′ – *Y*) is formed from –0.51(R′ – *Y*) –0.19(B′ – *Y*)
(16) Control trigger for (8), operated by binary division of line flyback pulses corrected for polarity as described on p. **15**–119.

NOTE: In the decoder, the D'_R and D'_B signals must be unweighted:

$$(R' - Y) = \frac{1}{1.9} \times D'_R = 0.53\,D'_R$$

In deriving (B′ – *Y*), allowance must also be made for the reduced deviation of the D'_B sub-carrier in comparison with that made for the D'_R sub-carrier (see p. **15**–117).

$$\text{Thus } (B' - Y) = \frac{280}{230 \times 1.5} \times D'_B = 0.81\,D'_B$$

Some inherent deficiencies

Having discussed the principles of the NTSC, PAL and SECAM colour television systems it is now necessary to consider some common inherent deficiencies, some of which arise from the provision of gamma correction.

GAMMA CORRECTION

Gamma correction in television is necessary mainly to compensate for the non-linear characteristic of the picture tube. Although in monochrome television gamma correction may be desirable, in colour television it is essential, for the overall characteristic from the input to the camera to the output of the picture tube must be linear. This may be exemplified as follows:

> Let the light values at the inputs to the three camera tubes be in ratio of red light = 3 units, green light = 2 units and blue light = 1 unit. In a linear system the displayed light output would contain the same proportions. If, however, the system were non-linear to the extent of being square law (a gamma of 2.0) the proportions at the output for red, green and blue would be 9, 4 and 1 respectively. This represents a serious colour error.

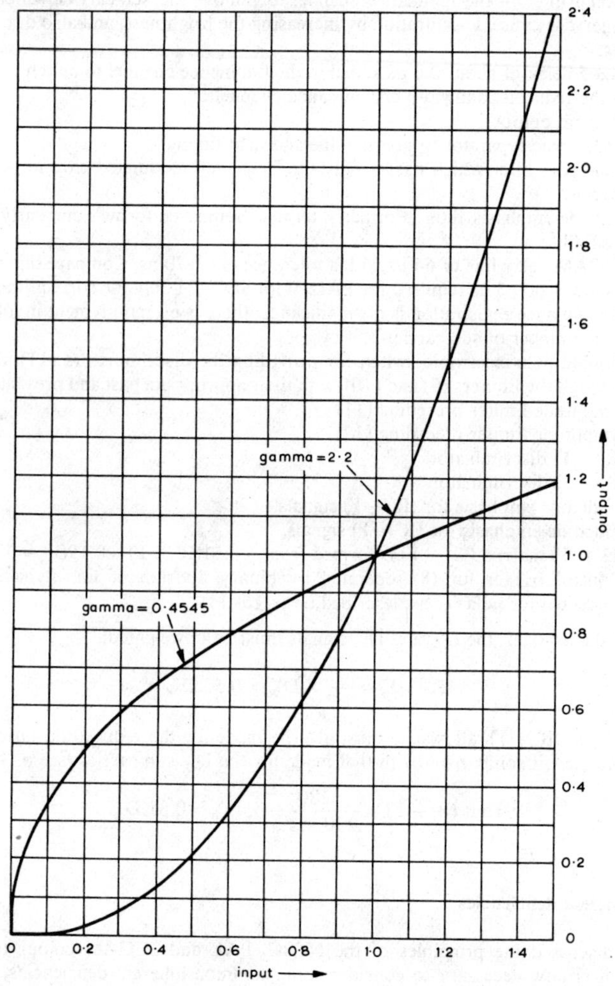

Figure 15.108 Gamma curves for use in colour television calculations

The gamma of shadow mask picture tubes has been standardised at a value of 2.2 for all three guns. (It is the triode characteristic of a gun which is responsible for non-linearity—*not* the phosphor.) This calls for a reciprocal gamma value of 0.45 in each of the three channels of the system to provide for full correction. In *Figure 15.108* curves with linear co-ordinates for gammas of 2.2 and 0.454 5 are shown to assist with calculations.

At this stage it is necessary to consider the performance of a television system in terms of contrast, the way in which the eye regards contrast and the point in the chain at which it is practicable to insert gamma correction.

Camera tube contrast ratio. The transfer characteristics of modern camera tubes are substantially linear and the maximum light contrast which may be handled satisfactorily is of the order of 50 : 1 or less. If a very low level of reflected light from any part of the scene is to be transmitted, this will be marred by noise because the signal is of such a low value; the noise itself would brighten the dark areas spuriously. The only solution is to increase the incident light so that even the black parts of the scene reflect an appreciable value of light, and thus an acceptable signal-to-noise ratio may be obtained. On the other hand, too much reflected light from the highlights will overload the camera tube. Hence, the maximum useful contrast ratio is dictated. Internal reflections in the camera lens system will also reduce the contrast at the input.

Picture tube contrast ratio. The maximum contrast available at the output of a picture tube might be of the order of 600 : 1 if conditions were ideal, but internal reflections in the glass and the reflection of ambient light from the face of the tube (the light output from the tube itself provides some ambient light) will in most cases reduce this appreciably. Although the contrast (measured by a photometer) at the tube face might suggest a high figure, many consider that subjectively, under normal viewing conditions, a maximum ratio of about 50 : 1 is usual. When it is realised that the maximum contrast presented by a glossy photographic print is about 30 : 1, the comparison indicates that television reproduction is likely to be quite acceptable.

The effects of ambient light. The way in which reflected ambient light reduces contrast is shown in the following example:

In the absence of ambient light:
Let the highlight brightness in the reproduced scene = 100 units.
Let the lowlight brightness in the reproduced scene = 1 unit.
This gives a contrast ratio of 100 : 1.
If the reflected ambient light from all parts of the scene amounts to 2% of the highlight brightness, the contrast ratio will be reduced to:

$$\frac{100+2}{1+2} = \frac{102}{3} = 34 : 1$$

Again, in another part of the scene let the original contrast equal:

$$\frac{100}{50} = 2 : 1$$

Then, with the same amount of ambient light the contrast will be:

$$\frac{100+2}{52} = 1.96 : 1 \text{ (or approximately } 2 : 1\text{)}$$

Further examples would show that reflected ambient light causes very considerable reduction of contrasts in the lowlight areas, with reduced effect in the highlights.

A non-reflective neutral density filter placed over the picture tube screen will improve contrasts which would otherwise be reduced by ambient light; this is because the ambient light reflected from the face of the tube will traverse the filter twice, whereas the direct output from the tube only passes through the filter once.

If the neutral filter has a transmission factor of 0.5 then, in the first example, the contrast ratio would be:

Highlight brightness reduced to $100 \times 0.5 = 50$ units.
Lowlight brightness reduced to $1 \times 0.5 = 0.5$ units.
Reflected ambient light reduced to $2 \times 0.5 \times 0.5 = 0.5$ units.

Resultant contrast ratio $= \dfrac{50 + 0.5}{0.5 + 0.5} = 50.5 : 1$ (instead of 34 : 1 without the filter).

Thus the filter reduces the brightness of the reproduced picture but improves contrast. It is now common practice to incorporate a neutral filter in the face of the picture tube by using glass which has a suitably low transmission factor.

Picture tube gamma and the human eye. The eye is capable of appreciating not only very small changes in light values, but also very large contrast ratios. In *Figure 15.109*, a stepped wedge diagram (*b*) of brightness values is shown which the eye would regard approximately as having steps of equal contrast. It is apparent that quite small changes in brightness in the dark areas are just as significant as very large changes in bright areas. In this example, the contrast ratio between each step is 2 : 1 (although in fact under very favourable conditions the eye is capable of detecting changes in brightness of the order of 1.01 : 1), with a maximum range of 64 : 1. To obtain this condition at the output of a picture tube with a gamma of 2.2, the waveform required at its input is that shown in *Figure 15.109* (*a*), in which the contrasts are all reduced (with a maximum range of 6.6 : 1)—in fact, all have been taken to the power of 0.45, which is the reciprocal of 2.2.

Gamma correction and noise. The question arises as to where in the chain gamma correction should take place.

If the system were linear in all its parts then the smallest signal to be handled would be

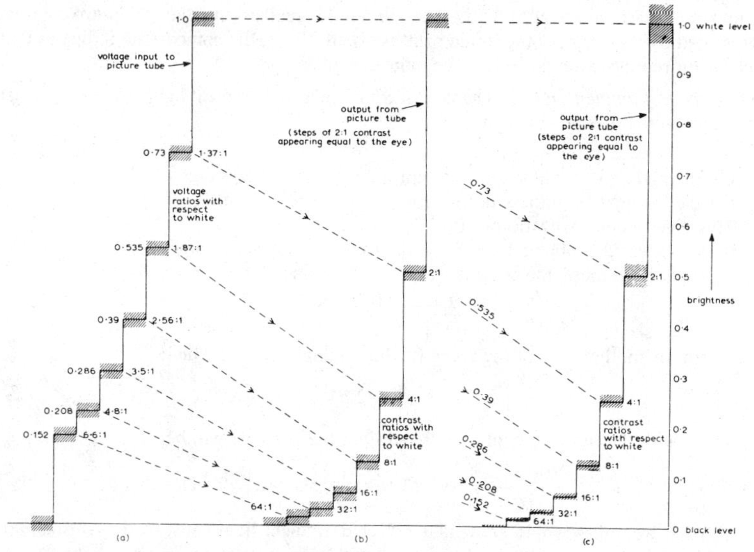

Figure 15.109 The effects of gamma correction and added noise: (a) the normalised voltage steps which produce the brightness steps of (b), assuming a tube gamma of 2.2; (b) steps of 2 : 1 change in brightness which the eye regards as having approximately equal contrast. If the tube gamma were unity, input voltage steps would be in the same ratios to white; (c) the effective output of the picture tube when the voltage steps and noise shown in (a) are applied to the tube input. The contrasts are increased but the noise in the darker areas is decreased by the non-linearity of the picture tube transfer characteristic

that pertaining to the largest contrast ratio, which in our example (chosen for easy arithmetic) is 64 : 1. To ensure an acceptable signal-to-noise ratio, any added noise must be low in comparison with one sixty-fourth of peak white. The fact that the gamma of the picture tube is greater than unity relieves the transmission situation considerably if gamma correction is provided at the beginning of the chain, for added noise (as far as high contrasts are concerned) after gamma correction must now be low only in comparison with approximately one seventh of the peak white signal. This is illustrated in *Figure 15.109* (*a*) and (*b*). Noise added to the signal before gamma correction is redistributed over the grey scale by correction, but is restored to its original distribution by the non-linear characteristics of the picture tube. Noise added after gamma correction is reduced in the darker areas by the characteristics of the picture tube in the same way that the dark grey signals are reduced (*Figure 15.109* (*a*) and (*c*)).

Thus, by introducing gamma correction at the beginning of the chain to correct for the characteristics of the picture tube at the end of the chain a compression of contrast occurs which is advantageous, as far as added noise is concerned, over the greater part of the chain. Noise is not reduced or increased appreciably, but is redistributed to make it less noticeable in the darker areas and more noticeable in the brighter areas; this is permissible to approximately the extent exemplified, because of the eye's particular characteristic.

Any non-linear device causes intermodulation of high frequency noise components and so produces low frequency noise which, in the case of television, is more noticeable; in general this is not serious unless the high frequency noise is of large amplitude.

The salient points of the foregoing discussion may be summarised and extended as follows:

(*a*) The overall characteristics of the red, green and blue channels from the light input to the camera, to the light output from the picture tube, should be linear, otherwise there will be errors in hue.
(NOTE: It may be difficult to ensure that the combination of transmission and reception characteristics provides an overall gamma of exactly unity, but it is most important that the gammas of the three channels should be matched. To this end the gamma correctors must be adjusted so that they are identical; in the receiver, the d.c. conditions and signal drive to the three guns of the shadow mask tube are also adjusted separately to provide an untinted grey scale.)

(*b*) The range of brightness contrast which is obtainable in television is severely limited in comparison with that available in nature, but nevertheless acceptable pictures are reproduced. Ambient light causes brightness contrast distortion (and de-saturation) and this may be reduced by the use of a non-reflective neutral density filter in front of or as part of the picture tube. Brightness contrasts in monochrome television are usually greater than those in colour television, but the latter has the additional feature of chrominance contrast.

(*c*) The high gamma of the picture tube means that a compression of the signal range is necessary at its input; this provides an advantage in terms of signal-to-noise ratio if the gamma correctors are placed as early as practicable in the chain. Gamma correction is carried out immediately after the signal has been suitably processed, that is, when black level has been established and blanking has been inserted. Apart from the question of signal-to-noise ratio, from an economic point of view it is obviously better to place gamma correctors in the transmission equipment than to fit them in all receivers.

(*d*) In the case of colour transmission there are inherent difficulties due to the ways in which gamma correction is applied and in this respect there are conflicts between the requirements which lead to good signal-to-noise ratio and those which provide good colour fidelity. The deficiencies of the latter in this respect are often referred to as being due to the failure of the principle of constant luminance.

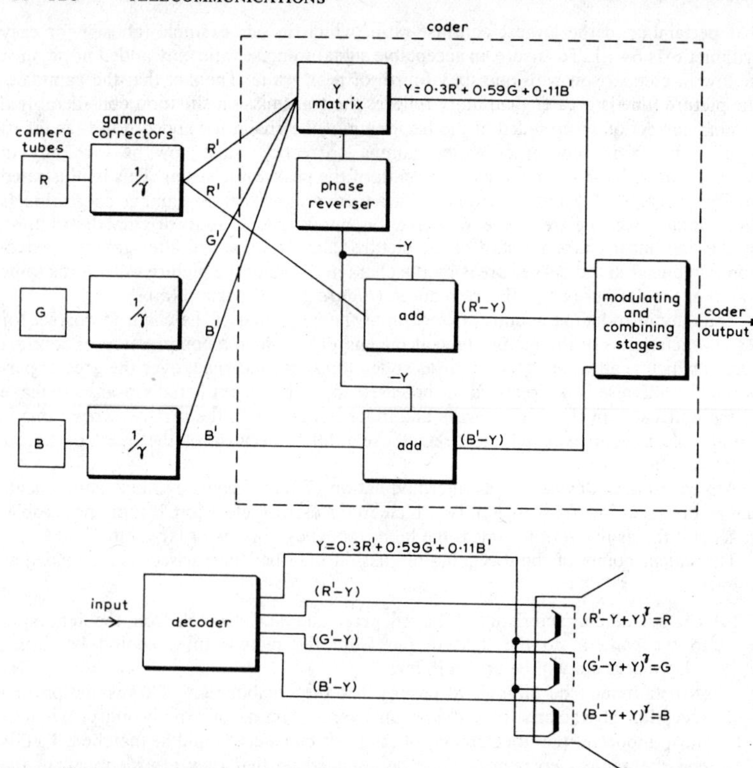

Figure 15.110 Schematic diagram of a colour television system using a three tube camera—R, G *and* B

THE PRINCIPLE OF CONSTANT LUMINANCE

In the arrangements described so far it has been assumed that the colour difference signals carry no information concerning luminance. This is an ideal state, which is said to conform to 'the principle of constant luminance'. In practice, it is not generally achieved because of the way in which gamma correction has to be applied.

Displayed brightness in uncoloured areas. The following examples will show the significance of the failure of constant luminance. The first case to be considered is that shown in *Figure 15.110* in which three camera tubes are used in the colour camera, with gamma correction applied to each signal before the formation of the *Y* signal. (R′, G′ and B′ indicate gamma-corrected R, G and B signals.)

For grey and white parts of the picture:

Let $\quad R = G = B = x \quad \text{and} \quad R' = G' = B' = x'$

Then
$$Y = 0.3R' + 0.59G' + 0.11B'$$
$$= 0.3x' + 0.59x' + 0.11x'$$
$$= (0.3 + 0.59 + 0.11)x' = x' = x^{1/\gamma}$$

The colour difference signals (R′ − *Y*) and (B′ − *Y*) are zero and in the monochrome receiver the transfer characteristic of the picture tube transforms *Y* to give the correct output,

i.e., $$(x') = x$$

This result shows that the transmitted brightness contrasts are displayed correctly on a monochrome receiver for black and white scenes.

In the colour receiver, the signals applied to the picture tube on black and white scenes are effectively:

$$(R' - Y) + x' = 0 + x' = x'$$
$$(G' - Y) + x' = 0 + x' = x'$$
$$(B' - Y) + x' = 0 + x' = x'$$

The transfer characteristics of the three guns of the shadow mask tube transform these values to give the correct outputs.

(NOTE: It must be realised that the practical signal voltage values actually applied to the shadow mask tube are dependent on its gun and phosphor characteristics. Typically, phosphor efficiencies differ and this requires unequal cathode currents and drive voltages to produce similar outputs from the three phosphors. The above example refers to normalised values which are assumed to result when adjustments of gains and operating voltages have been made to obtain a correct grey scale. It is also assumed that with the white so obtained, the brightness contributions from the red, green and blue phosphors are in the proportions of 0.3, 0.59 and 0.11 respectively.)

Presentation of saturated red. For coloured parts of the scene, however, the transmitted signals will be in error and this may be shown by taking the example of a saturated red signal at full amplitude:

When

$$R = 1.0$$
$$G = 0$$
$$B = 0$$

After gamma correction

$$R' = 1.0$$
$$G' = 0$$
$$B' = 0$$
$$Y = 0.3$$
$$(R' - Y) = 0.7$$
$$(B' - Y) = -0.3$$

In the monochrome receiver, Y is transformed to:

$$0.3^{\gamma} = 0.3^{2.2} = 0.07$$

This shows that the brightness of full amplitude, saturated red would apparently be displayed at approximately a quarter of the correct value.

In the colour receiver:

$$Y = 0.3$$
$$(R' - Y) + Y = 0.7 + 0.3 = 1.0$$
$$(B' - Y) + Y = -0.3 + 0.3 = 0$$
$$(G' - Y) + Y = -0.51(R' - Y) - 0.19(B' - Y) + 0.3$$
$$= -0.357 + 0.057 + 0.3 = 0$$

The brightness and hue displayed would apparently be correct.

Monochrome presentation of other saturated colours. The calculated results of displayed brightness on a monochrome picture tube (gamma = 2.2) for full amplitude 100% saturated colours are given in *Table 15.9*.

From the foregoing examples it is apparent that, with the arrangement used in *Figure 15.110*, the neutral areas of the transmitted picture may be received on monochrome and colour receivers without brightness distortion, but the coloured areas may be displayed with incorrect brightness on monochrome receivers.

With colour receivers the brightness and chrominance of coarse colour detail will be correct, in spite of the fact that the Y signal is incorrect for coloured areas, because the final R, G and B values are the resultant of the Y, $(R - Y)$ and $(B - Y)$ signals. This means that the brightness information for coloured parts of the scene is not confined to

Table 15.9 TRANSMITTED Y SIGNAL AND BRIGHTNESS VALUES DISPLAYED ON A MONOCHROME PICTURE TUBE ($\gamma = 2.2$) FOR FULL AMPLITUDE 100% SATURATED COLOURS

	Transmitted Y signal	*Displayed brightness*
White	1.0	1.0
Yellow	0.89	0.78
Cyan	0.7	0.46
Green	0.59	0.3
Magenta	0.41	0.14
Red	0.3	0.074
Blue	0.114	0.0086
Black	0	0

the Y channel but is conveyed partly by the chrominance channels. The latter are bandwidth-restricted and provide no contribution concerning the variations in brightness of finely coloured detail, which means that there will be a reduction in the resolution of such brightness detail.

Effect of the sub-carrier in monochrome receivers. Monochrome receivers cannot make proper use of the brightness information conveyed by chrominance signals and thus coloured areas would be deficient in brightness if no other factors were involved. There is some correction of this deficiency, however, with NTSC and PAL because of the effect of the sub-carrier and the curved characteristic of the picture tube. The sub-carrier is, in effect, 'rectified' by the non-linear characteristic to produce an increase in brightness when the sub-carrier is present, i.e., in coloured areas. This effect is mentioned on p. **15**–92. The condition in respect of SECAM is explained.

Receiver video response requirements. To achieve the increases in brightness in the black and white receiver described above, the video bandwidth must extend at least as far as the sub-carrier frequency, so that the sub-carrier appears without attenuation at the input to the picture tube. If a notch filter at sub-carrier frequency is included in the drive circuit to remove the dot pattern, then the picture will be less panchromatic. On the other hand, in the colour receiver, such a notch filter is necessary to avoid the de-saturation of colour which would occur if the dot pattern were present. Inevitably, the notch filter will remove luminance detail corresponding to frequencies in the region of the sub-carrier but in general this is not serious.

Alternative configurations. Alternative circuit arrangements have been designed to reduce the effects of the failure of constant luminance, but advantages are not generally achieved without some sacrifices; often, luminance errors are exchanged for chrominance errors and for reductions in signal-to-noise ratio.

Comparison of the systems

The original NTSC colour television system and two of its variants, PAL and SECAM, have been described. Under good conditions, all three systems are capable of providing excellent pictures and, at a viewing distance of about seven times picture height, it is often difficult to distinguish which system is in use.

The NTSC system provides good compatibility, good vertical and horizontal resolution, is rugged in the presence of noise in various degrees and forms, and is reasonably simple to transmit and receive. The main drawbacks are its susceptibility to differential phase distortion and its bandwidth requirement if crosstalk between the colour difference signals (with resultant hue distortion) is to be avoided.

The main feature of the PAL system is that it provides good protection against differential phase distortion. Transmission arrangements tend to be more complicated than those required for NTSC, and receivers also are more complicated and therefore more expensive. The advantages in connection with video tape recording outweigh the disadvantages, and transmission over long links is not as difficult as with NTSC; the effects of multi-path reception are also generally considered to be less objectionable. The main advantage of PAL over NTSC is reduced hue distortion; but this advantage is exchanged for the slightly reduced compatibility and lower vertical chrominance resolution. The latter failing is inherent in delay line receivers.

The SECAM system is relatively immune from differential phase and also differential gain distortion (both the other systems being susceptible to the latter). The receivers are comparatively simple in their decoding arrangements—the complication in this respect is a feature of the coders. The complexities involved in fading and mixing are a distinct disadvantage, but video tape recording and transmission over long links present little difficulty. The vertical chrominance resolution is inherently reduced by discarding half the chrominance information before transmission.

Before optimisation, comparative tests with NTSC and PAL revealed that SECAM gave a slightly poorer compatibility performance, a somewhat poorer performance in the presence of heavy noise and a greater susceptibility to some of the effects of multi-path reception. With the optimisation of the system, these marginal differences are reduced or removed.

The inherent stability of the sub-carrier frequencies used in SECAM need not be of a high order, although the relationships with line frequency must be maintained. In transcoding SECAM signals to PAL or NTSC, a high timing stability of synchronising signals is necessary; this demands a higher degree of stability for the SECAM sub-carriers than would otherwise be required.

REFERENCES

1. KLEIN, A. B., *Colour Cinematography*, Chapman and Hall
2. WRIGHT, W. D., *Measurement of Colour*, Hilger
3. HARDY, A. C., *Handbook of Colorimetry*, M.I.T. Mass. U.S.A.
4. LUCKIESH and MOSS, *Science of Seeing*, Macmillan
5. WRATTEN, *Colour Filters*, Kodak
6. BENSON, J. E., 'A survey of the methods and colorimetric principles of colour TV', *J. Brit. I.R.E.*, Jan. (1953). [Reprinted from *Proc. I.R.E.* (Australia), July, Aug. (1951)]
7. BRUCH, W., 'PAL—a variant of the NTSC colour television system', *Telefunken Zeitung*, selected papers (1964 and 1966)
8. CARNT, P. S. and TOWNSEND, G. B., *Colour Television*, Iliffe Books (1961)
9. HENDERSON, H., *Colorimetry*, BBC Engineering Training Supplement No. 14
10. LIVINGSTON, D. C., 'Colorimetric analysis of the NTSC colour television system, *Proc. I.R.E.*, **42**, Jan. (1954)
11. LIVINGSTON, D. C., 'Reproduction of luminance detail by NTSC colour television systems', *Proc. I.R.E.*, **42**, Jan. (1954)
12. NEUHAUSER, R. G., ROTOW, A. A. and VEITH, F. S., 'Image orthicons for colour cameras', *Proc. I.R.E.*, **42**, Jan. (1954)
13. *Proc. I.R.E.*, numerous papers, **42**, Jan. (1954)
14. SIMS, H. V., 'Black level in television', *Wireless World*, **68**, No. 1 (1962)
15. SIMS, H. V., *Principles of PAL Colour Television and Related Systems*, Newnes-Butterworths (1969)

TELEVISION STANDARDS CONVERSION

Standardisation occurs naturally in a monopoly situation. In other circumstances the problems of standardisation are less easily resolved; either because the known facts are less decisive or other pressures are at work. In early television receivers the picture suffered from 'hum bars', caused by the inadvertent introduction of power supply frequencies. These were less visible if they appeared stationary, that is, when the television system is locked to the supply frequencies. Some countries used different power supply frequencies, e.g., 50 Hz in the U.K. and 60 Hz in the U.S.A. This and other factors led to the use of different television standards and the 405-, 819-, 525- and 625-line systems were introduced. These different television standards remain and the introduction of different colour systems has further complicated the situation. Because of this standards conversion is now an accepted part of a television system.

In the early days of television, photographic film was an essential ingredient. The problem of international exchange of television programmes on different standards was largely solved by the use of film but the invention of magnetic tape recorders gave new impetus to the conversion problem. The situation became further aggravated by the international exchange of programmes via satellite. This together with the need to replace obsolescent television standards has given rise to the development of a series of new and improved standards converters.

The earliest converters employed an optical cathode ray tube image and camera. These were improved by the use of more complex techniques and new camera tubes.

All-electronic conversion was first used to convert 625-line/50-field pictures to the old 405-line/50-field standard and this was the basis of further developments in line and field rate converters operating on the colour signal. At this point the optical converter made a return employing new techniques for the conversion of colour signals and PAL and SECAM transcoders were also introduced.

In more recent years converters have been built using digital techniques. These are particularly suitable in that the signal process can be prescribed exactly, and signal storage without loss is both technically possible and economic to construct.

Principles of standards conversion

A television picture portrays a scene in terms of a scanning raster. If the same scene is scanned by two cameras on different standards the images will be very similar. However, close examination will reveal that the scanning lines cover different parts of the picture and only coincide, or cross, occasionally. Furthermore, the electrical waveform will reveal that the same, or similar parts of the picture occur at different moments in time and in extreme examples, both the line rate and the field rate may be different.[1]

When there is only one camera the image on the second standard must be replaced by a standards converted version of the original picture. This means that the signal must be modified so that it carries the correct information relative to the precise position of the second scanning raster and we must store or delay the information until the new raster requires the modified video signal. These two processes are fundamental components of all standards conversion systems.

INTERPOLATION

If we consider a very narrow vertical strip of picture, the scene is portrayed by a series of points, one from each line of the scanning raster. In general, the converter requires information for intermediate picture points and the choice of the most likely value is called interpolation. One simple form of interpolation would be to add together the magnitude of the signal on adjacent lines; weighting this magnitude according to its distance from these lines. The law must be such that, in an area of even illumination, the reconstructed picture contains no evidence of the original line structure. Nevertheless, there is an infinite variety of aperture shapes which meet this requirement. The interpolation of a sampled television system is illustrated graphically in *Figure 15.111*. The component AB is the contribution from line 1 and the part BC is the contribution from line 2. In this case there is a simple linear interpolation with a triangular aperture. More complex forms of interpolation such as the 'raised cosine' (*c*) of *Figure 15.112* gives a

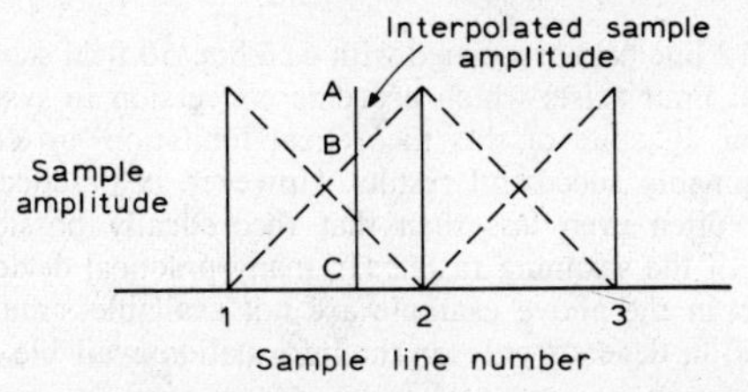

Figure 15.111 Graphical illustration of interpolation of a sampled television system

more sharply defined pass band and stop band (see *Figure 15.112* (*e*)). The quality of the interpolation is related to the number of samples employed in each calculation. In the limit an interpolating 'aperture' with a (sin *x*)/*x* shape requires an indefinitely large number of samples to be stored for calculation purposes (aperture (*d*)). These different interpolation apertures can also be illustrated in the frequency plane and a choice made which is an optimum compromise between complexity, resolution and freedom of spurious signals.

As indicated in *Figure 15.112* (*e*) it is sometimes valuable to consider the operation

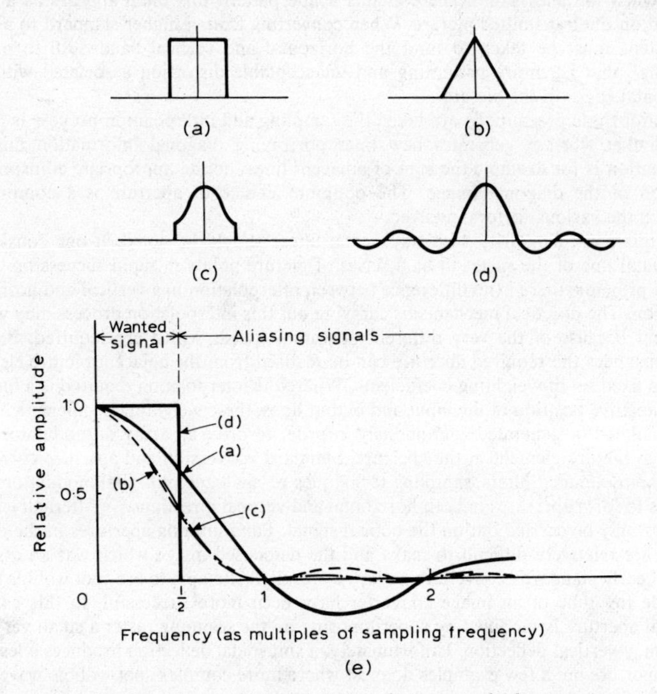

Figure 15.112 Forms of interpolation in television standards conversion

in the frequency plane. In this case a series of samples at a given frequency completely defines a signal with a bandwidth of half that frequency. If the vertical strip of picture in one field contains 262 samples it accurately portrays half a band limited 525-line/60-field, picture. Standards conversion to 202-line field of the 405-lines standard would be possible with negligible loss, but, insufficient information would be available to allow conversion to the 312-line field associated with 625-line/50-field standards. In a similar manner a theoretical limit exists which prevents conversion to systems with a higher horizontal resolution. In spite of this theoretical limitation 'up conversion' is often employed with reasonably successful results. However, in practice the quality of the picture obtained is often even less than that theoretically possible because of the interlaced structure of the scanning raster. [In many practical devices the 525-picture elements referred to in the above example are not available simultaneously and the process of interpolation depends only on the information available in one field, that is 262 lines.]

A further limit arises because of the imperfect representation of movement. Each field is different and interpolation between different fields of information gives a blurred answer. Conversion between systems of different field rate brings with it further complications, for it makes the fundamental processes of storage and interpolation more difficult and, in theory, the need for movement interpolation creates even greater problems.

The sampling process referred to earlier works satisfactorily only on a band limited picture. In practice the scene present to the camera has no well defined bandwidth in either the vertical or horizontal directions. The presence of this additional information above the frequency f_0 produces spurious signals and cross products known as *aliasing*. Thus when the television scene contains a fine pattern this often appears as a *moiré* pattern on the transmitted picture. When converting from a higher standard to a lower one, steps must be taken to limit the horizontal and vertical bandwidth to prevent 'aliasing', that is, moiré patterning and unacceptable distortion associated with near horizontal edges in the picture.

Even if these precautions are taken the sampling and interpolation process is imperfect in that when it generates new lines portraying diagonal information this new information is for example the sum of adjacent lines; not an appropriate adjustment in position of the diagonal image. The optimum choice of aperture is a compromise between the various factors involved.

At this point similarity with transversal filters should be noted. If one considers a horizontal line of the raster to be a series of picture points in rapid succession we see that in principle there is no difference between interpolation in a vertical and horizontal direction. The practical mechanisms carrying out this interpolation process may well be different because of the very different operating speeds which are required. In some circumstances the required aperture can be realised from the adjacent picture elements using a fixed set of weighting coefficients. Where the interpolation required is a function of the relative position of the input and output lines, these weighting coefficients have to be calculated or generated continuously in order to drive an array of modulators; one for each picture element in the aperture. Standard conversion and aperture correction can employ lumped filters, sampling techniques or analogue or digital modulators/multipliers to form apertures in both horizontal and vertical directions.[2, 3] Alternatively, the process may be carried out on the optical signal. Band limiting apertures in the optical world are relatively difficult to make and the defocused image which was an essential part of early standards converters gave poor results. Attempts to use spot wobble on the cathode ray tube of an image converter have been more successful. In this case the vertical aperture is modified by superimposing on the scanning raster a small very high frequency vertical deflection. Unfortunately, a sinusoidal deflection produces a less than ideal aperture but a few examples do exist where more complex spot wobble waveforms were used to give a better vertical aperture.

STORAGE

Storage of picture information is an essential part of all standards converters. In some converters it may be only necessary to store it for a very short while, typically the duration of one or two television lines. In other cases a very long storage (or delay) of many tens of milliseconds delay is essential. In each case the video signal should be stored without loss or degradation. In practical examples where this cannot be done the decay of the signal must at least be known and capable of correction.

In practice these two properties of standards converters, interpolation and storage, can be achieved using many different techniques in the optical and electrical parts of the systems. It is of course, not necessary to carry out interpolation and storage in that order. They may be carried out in any order or, in some cases, they are inextricably mixed.

Review of existing converters

OPTICAL CONVERTERS

Early converters relied upon the optical process to perform both the storage and interpolation operations.

Line rate converters were constructed employing the CPS and Image Orthicon and Vidicon camera tubes. (These converters had storage of type 1 and 2 in *Table 15.10*; interpolation type 6 in *Table 15.11.*) The resulting picture was degraded by flare (or veiling glare), optical losses, together with application signals and other spurious signals generated in the camera. It also suffered from noise added in the camera tube and consistent good performance was very dependent on operator skill. Nevertheless this type of converter was used extensively when converting between 819-, 625- and 405-line standards.[1]

Field rate converters were also built using the cathode ray tube display and camera tube. As with earlier converters the signal decayed with time and it had to be modulated in a relatively complicated manner to compensate for this loss. Both pre- and post-equalisation of amplitude was employed; the amount of correction was fixed at a calculated or predetermined gain, or alternatively, a reference stripe of known amplitude provided the basis for a form of a.g.c. which maintained the signal at constant amplitude. When converting between standards with 50 and 60 Hertz field rates, the 10 Hertz flicker signal is particularly objectionable. No practical process for completely removing this was known and the best that could be done was to reduce it to an acceptable level.

None of these converters were capable of dealing with a colour signal.

The introduction of the lead oxide photoconductive camera tube offered the possibility of a better picture and caused renewed interest in optical converters. Its value in later converters was dependent on the invention of new techniques for converting PAL (and NTSC) colour signals.

Two displays were necessary; one showing the luminance signal and the other a coded chrominance signal. All existing colour systems code the signal so that each line in the television picture is different, even when a uniform colour is displayed. However, in order to recover this successfully from a cathode ray tube, it is necessary to rescan the display, tracking it line by line. No optical interpolation would be possible. However, if the chrominance signal is put on a carrier of frequency which is a multiple of the line rate the carrier will be the same from line to line. Line tracking is no longer necessary and optical interpolation of both the luminance and separated chrominance signals is possible. Naturally the linearity of the two displays must be similar in order to avoid misregistration but the relatively low bandwidth of the chrominance signal and acceptable tolerances of misregistration make this a workable solution.[10]

ANALOGUE ELECTRONIC CONVERTERS

The problem of changing standards in the U.K. from 605-line/50-field signals to 405-line/50-field signals involved continuous use of standards conversion. All signals are now only generated on 625-line standards and the 405-line transmitters are fed with a standards converted signal. There was no optical image involved and the process of conversion can be classified as signal storage 6, *Table 15.10*; interpolation 4, *Table 15.11* or alternatively signal storage 6, *Table 15.10*; interpolation 2, *Table 15.11*. Both

Table 15.10 METHODS OF SIGNAL STORAGE

Storage mechanism	*Maximum storage period*	*Notes*
1 Phosphor afterglow	Up to several seconds	Approximately exponential signal decay throughout storage period
2 Camera tube target capacity	Milliseconds	Destructive readout process. Exponential decay of signal amplitude. Most integrating camera tubes have an output proportional to exposure time
3 Photoconductive lag in the camera tube	Milliseconds	Exponential decay of signal during the storage period
4 Magnetic tape [4]	Unlimited	Special disc and tape recorders developed to provide the necessary short-term storage
5 Propagation time of a pressure shear wave	Milliseconds	'Acoustic' or ultrasonic delay employ quartz. Silica and steel delay lines
6 Charge storage in an array of discrete capacitors	Milliseconds	Switching problems due to the large number of capacitors required
7 Charge storage in integrated semi-conductor arrays	Milliseconds	Storage of analogue signal. Variations in capacity can produce imperfections
8 Linear array of relatively large discrete capacitative stores accessed by a scanning electron beam	Microseconds	Large size giving isolation between elements
9 Shift register storage using integrated semi-conductors memories	Indefinite	Digital signal storage. Dynamic stores employing a capacitive memory, charge coupled devices or static memories employing bistable devices
10 Random access memories	Indefinite	Digital signal storage. As above but permitting more complex signal processing

converters sampled the signal and stored it in an array of 500 capacitors, but both worked well. One employed a vertical aperture which could be synthesised with a relatively simple lumped network the other used modulators and delay lines to construct the interpolated signal.[2] The signal quality was good but suffered from a background of vertical lines or striations due to minor differences in the storage capacitor and an associated switching mechanism.[5,6]

In the meanwhile pressure was growing for an improved standards converter capable of operating between standards with different field rates. In most of these earlier converters the problems of field rate conversion was overcome separately from the problems involved in the change of line rate.

The invention of a successful magnetic tape recorder offered a new alternative. Conventional transverse recorders employing 4 heads and special machines with 7 heads and double head wheels were employed to achieve field rate conversion; line rate conversion being achieved by conventional optical means.[4]

However, the growth of colour television rendered all these devices obsolete and a renewed effort was required to obtain a suitable converter. It was now possible to make large ultrasonic delays—and an array of these could be used to provide the necessary storage. This would give the opportunity for determining the optimum interpolation

Table 15.11 METHODS OF SIGNAL INTERPOLATION

	Principle	Direction of operation	Notes
1	Lumped network	Horizontal	Symmetrical aperture possible but delay correction can be difficult
2	Lumped network	Vertical	Unsymmetrical aperture; delay correction usually impractical
3	Transversal equalisers	Horizontal	Symmetrical aperture; constructed from individual picture elements appropriately modulated in amplitude
4	Transversal equalisers	Vertical	As above but individual picture elements stored until required by an array of line delays
5	Spot wobble	Normally vertical only	Symmetrical. Difficult to obtain the optimum shape. Only positive contributions to the signal amplitude
6	Defocusing	Horizontal and vertical	Uncontrolled aperture
7	Digital calculation	Vertical and horizontal	Linear and non-linear signal processing to any degree of complication is possible

between fields and, although necessarily imperfect, the best representation of movement.[7]

A simple converter was built which repeated, or omitted one field in 5 or 6 as necessary to create the required number of fields. However, this simple approach produces jerky movement. Interpolation between fields can be provided by the simple addition of signals from adjacent fields (see *Table 15.11*, 3 and 4). Theoretically interpolation involves the creation of a new intermediate image but this involves a real-time computing process which is too difficult at the moment. As with other forms of interpolation an acceptable answer can be obtained by simply adding the weighted values from adjacent fields.

So far, only passing reference has been made to the difficulties introduced by the interlaced structure of the television picture. Not only is the resolution of the converted picture often limited because of working in a field by field basis but even if both fields of information are used, it severely handicaps movement interpolation. Combining the two fields of information often causes an objectionable double image to appear. Thus, with modern cameras which have negligible lag, it would appear that there is no easy solution and alternative modes of operation may be provided which give optimum answers for near stationary or alternatively fast moving pictures.

The duration of one television line is nearly the same in the 625-line/50-field system and the 525-line/60-field system. A more advanced form of converter took advantage of this fact to carry out line rate conversion. If we add sufficient delays of different size that their combination makes any smaller delay possible, we can have the appropriate line (or lines) of the 525-line signal available for interpolation at the required time (see *Table 15.11*, 4). The error arising from this process is fortuitously within acceptable tolerances and the delays required are limited to multiples of 2 μs. Ultrasonic delays can also be used to correct the number of fields in the signal.[7]

DIGITAL STANDARDS CONVERTERS

Although the electronic line converter was a great step forward, some imperfections remained. The growth of digital integrated circuits made possible the construction of a line converter free from these defects at a competitive price.[11]

As before the signal is sampled but its amplitude is then converted into binary form ready for storage and interpolation (see for example *Table 15.10*, 9 and 10; *Table 15.11*, 7). The sampling frequency is in the region of 11–13 MHz and the calculation of the required apertures in say the 80 ns period between samples involves high speed real-time calculation. The precise frequency employed may either be a multiple of sub-carrier (3 × sub-carrier for example) or, in effect, a multiple of line rate. The system tolerances are easier if the sampling frequency is locked to sub-carrier and in some circumstances it may be advantageous to have a precise phase relationship as well. On the other hand, signal processing is often easier if the sampling frequency is locked to a multiple of line frequency.

Because of the speed limitations in some integrated digital circuits the various processes are carried out in independent sequential steps, so that the whole operation does not have to be completed in one interval between samples. In a similar manner the semiconductor storage elements forming the shift register storage may be too slow for the task. This can be overcome by dividing the signal into, say, three parts, putting each part into a separate shift register and multiplexing the signal back together when the storage process is complete.

In an attempt to minimise the complexity of the machine the bit rate requirement must be reduced to a minimum. In that the signal has to be recoded it may be convenient to separate the signal into, say, luminance and colour difference signals and then sample each signal at the minimum rate and time division multiplex the result.

Field rate conversion using digital techniques is equally possible and it is also possible to perform many of the colour systems coding and decoding operations using digital techniques.[7,8,9] In addition, digital filters using recursive techniques make it possible to construct vertical and horizontal apertures and comb filters in a more nearly ideal manner.

As with other converters it is convenient to separate the signal into luminance and chrominance (I and Q) signals. The different bandwidths of the two parts enables the adoption of lower sampling rates and digitally code the signal less accurately. Thus the cost of producing a colour capable converter, in digital form, is not excessive.

The converter operating on the 525/60 signal samples the signal at 10.7 MHz and carries out the signal process entirely in digital form. The signal is decoded to Y, I and Q signals with 10.8 and 3.6 MHz samples. Movement interpolation between fields and position interpolation between lines is carried out in the digital mode before decoding to form the standard PAL signal. The spacial aperture employed uses information from five picture points on each of three successive lines.

Two semiconductor field stores are necessary and as storage of this type does not degrade the signal an excellent result is possible.

SYNCHRONISERS

Conversion between two systems with nominally the same standard is an important problem. It is frequently necessary to bring two signals of different phase, and perhaps, only marginally different frequency into synchronism. This may be desirable to correct short term imperfections in the signal from a tape recorder or correct the long-term problems arising from, say, a non-synchronous remote contribution to a programme.[8]

In both cases a variable delay is needed. Only occasionally will it be necessary to add or delete a field of information and there is no significant problem in the interpolation involved.

Many of the examples quoted involve operating on a colour coded signal and the degradation involved in decoding and re-encoding the signal is not insignificant. The ideal synchroniser operates on the complete coded signal and does not sub-divide it into convenient parts. There seems no doubt that the digital timing corrector will be an important part of many video tape recorders.

TRANSCODERS

Unfortunately, even on a given line and field rate there is not necessarily an agreed colour system and it is necessary to convert or transcode from one system to the other. A major problem is the conversion between SECAM and PAL signals.

This can only be achieved by a process of decoding and re-encoding. 'In band colour systems' successfully conceal the coded colour information within the luminance bandwidth, but it is extremely difficult to extract it without loss for repeated coding operations. [Decoding the PAL signal successfully requires carefully designed low pass/band pass or comb filters. Successfully multiple decoding of the SECAM signal presents similar problems.]

One major problem is in one sense, incidental to the colour system. The SECAM signal is less dependent on an accurately controlled line and field rate and the signal may be generated on a less rigorous standard. Thus, unless there is prior agreement, transcoding from SECAM to PAL could result in a signal whose line and field rate are marginally in error and a full standards conversion of the colour system and line and field rates is necessary.

REFERENCES

1. LORD, A. V., 'Conversion of television standards', *BBC Quarterly*, **8**, No. 2 (1953)
2. LORD, A. V. and ROUT, E. R., 'An outline of synchronous standards conversion using a delay-line interpolator', *Television Engineering* (*I.E.E. Conf. Rep. Ser. 5*), 167 (1962)
3. MONTEATH, G. D., 'Vertical resolution and line broadening', *BBC Engineering Monograph 45* (1962)
4. N.H.K., 'Television Standards Conversion', *Asian Broadcasting Union Second General Assembly Tokyo*, **2**, E/12 (1965)
5. RAINGER, P., 'A new system of standards conversion', *Television Engineering* (*I.E.E. Conf. Rep. Ser. 5*), 54 (1962)
6. RAINGER, P. and ROUT, E. R., 'Television standards converters using a line store', *Proc. I.E.E.*, **133**, No. 9 (1966)
7. WHARTON, W. and DAVIES, R. E., 'Field-store standards conversion', *Proc. I.E.E.*, **113**, No. 6 (1966)
8. BALDWIN, J. L. E., STALLEY, A. D. and KITCHIN, H. D., 'A standards converter using digital techniques', *Royal Television Society Journal*, **14**, No. 1 (1972)
9. 'Digital Intercontinental Conversion Equipment', *IBA Technical Review*, **31** (1973)
10. WENDT, H., 'An electro-optical converter for colour television signals', *N.T.Z.*, **5**, 281 (1969)
11. EDWARDSON, S. M. and JONES, A. H., 'Digital field-store TV synchronisers and standards converters', *Wireless World*, **77**, No. 1432 (1971)

HIGH QUALITY SOUND DISTRIBUTION FOR TELEVISION AND SOUND BROADCASTING

Sound-in-syncs

The BBC distributes its television programmes between the studio centres and the transmitters by means of a network of cable and microwave links leased from the British Post Office. The audio and video components of the programme have in the past been sent separately and have quite often followed different routes and used different transmission media. The sound-in-syncs (s.i.s.) system developed by the BBC is a method of combining the audio and video signals in such a way as to enable the complete programme to be routed via the same video link. At the transmitter site the signal is split back into separate audio and video signals and transmitted in the usual way to the viewer. The composite sound-in-syncs signal offers advantages financially due to economies in audio links, and also operationally since the audio cannot become separated from the video between terminal sites.

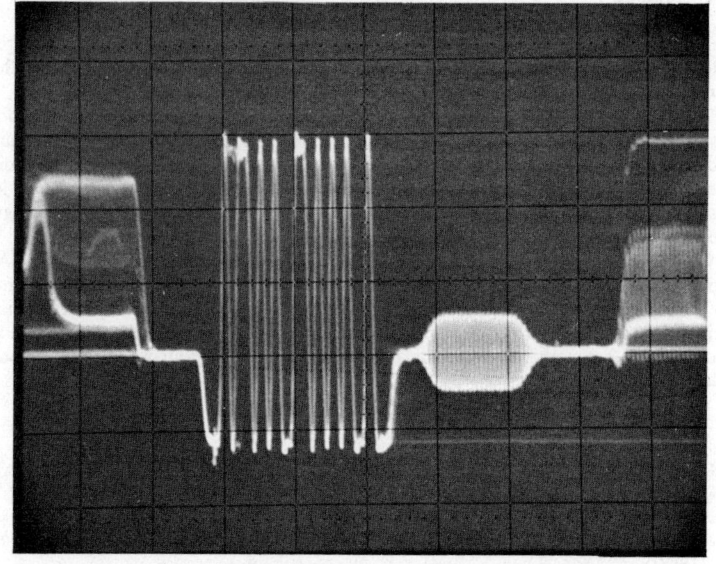

Figure 15.113 Sound-in-syncs waveform showing sound pulses inserted into the line synchronising period

The sound-in-syncs system is essentially a time-division multiplex system in which the audio signal is coded in pulse-code modulated form and the pulses sent during the line-synchronising period of the video waveform (*Figure 15.113*).

The video may be any 625-line colour signal using a field waveform comprising of 5 equalising pulses, 5 broad pulses and then a further 5 equalising pulses. The equipment, with modification, may be used with 525-line System M video signals.

The audio channel obtained has the following performance:

frequency response	50 Hz–13.5 kHz ± 0.7 dB reference 1 kHz 25 Hz–14 kHz – 3 dB
signal-to-noise ratio	65 dB (peak signal/peak weighted noise) (CCITT Rec. P53 (1972) weighting network)*
non-linear distortion	0.25% (1 kHz at full modulation)

The quality of the audio channel is independent of the noise and distortion on the bearer circuit up to a threshold point above which the audio channel rapidly deteriorates. The threshold point for s.i.s. is to a large extent determined by the ability to separate the sync pulses from the s.i.s. waveform. This is because the leading edge of the line-sync pulse is used as the timing reference for the insertion of the sound pulses.

Typical distortions permissible on the bearer circuit are:

White noise	23 dB (peak signal/r.m.s. unweighted noise)
Pulse K-rating	8%
High-frequency loss	−8 dB at 4.3 MHz

EQUIPMENT DESCRIPTION

The simplified block diagram of the complete equipment is shown in *Figure 15.114*. The audio input is first compressed in dynamic range by means of a circuit that is basically a

* Measured on a quasi peak meter with an integration time of 10 ms and bandwidth 22 kHz.

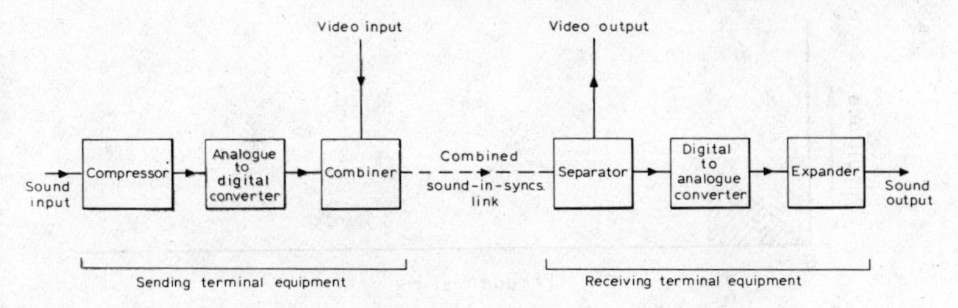

Figure 15.114 Simplified block diagram of the sound-in-syncs system

fast acting audio limiter. The signal is then sent to an analogue-to-digital converter (A.D.C.) where it is sampled at a frequency equal to twice the video line frequency (31.25 kHz) and each sample converted into a 10-bit binary code. This 10-bit code is then temporarily stored while the next sample is being converted. The two codes are combined to form a 20-bit word, a marker bit added making 21 bits total, and the complete message inserted into the following line sync pulse.

At the receiving terminal the video signal is separated from the combined sound-in-syncs signal and restored to a standard form. The 21-bit message is sent to a digital-to-analogue converter (D.A.C.) where it is split into the two 10-bit codes, each code is then separately decoded. The analogue output of the converter is then fed to the expander where the dynamic range of the audio signal is restored.

SAMPLING

It has been shown [1] that in order to transmit a complex signal containing frequencies up to f Hz, it is not necessary to send the signal continuously, it is sufficient to send a series of amplitude-modulated pulses obtained by sampling the instantaneous amplitude of the signal at a regular repetition frequency of at least $2f$ Hz. The frequency spectrums of the sampling signal and the pulse-amplitude-modulated (p.a.m.) signal are shown in *Figure 15.115*. It can be seen that the original signal may be recovered by means of a low-pass filter having a cut-off frequency of f Hz. However if the upper frequency limit of the signal extends above the half sampling frequency, the sidebands overlap producing aliasing frequencies, and the signal cannot be recovered without distortion. This means that the signal must be band-limited by means of an efficient low-pass filter prior to being sampled. In sound-in-syncs the sampling frequency employed is 31.25 kHz which when allowing for practical low-pass filters enables an audio-frequency limit of 14 kHz.

QUANTISING

The amplitude of the p.a.m. signal can follow all variations in amplitude of the input signal. In a p.c.m. system all possible amplitudes of the samples cannot be transmitted. The amplitude range is split up into a finite number of possible levels. Each of these levels can then be represented by a code. Thus after the input signal is sampled to produce a p.a.m. signal each sample is rounded off to the nearest possible level. This process of only allowing certain levels is known as quantising. The quantised signal obviously is distorted, the greater the number of quantising levels the lower the distortion. The distortion is known as quantising noise or sometimes as granular distortion. A property of quantising noise is that it only occurs in the presence of modulation. At low levels of modulation its effect on the audio signal is to produce a programme-modulated noise,[2] whereas at high levels of modulation the impairment is similar to that caused by white noise.

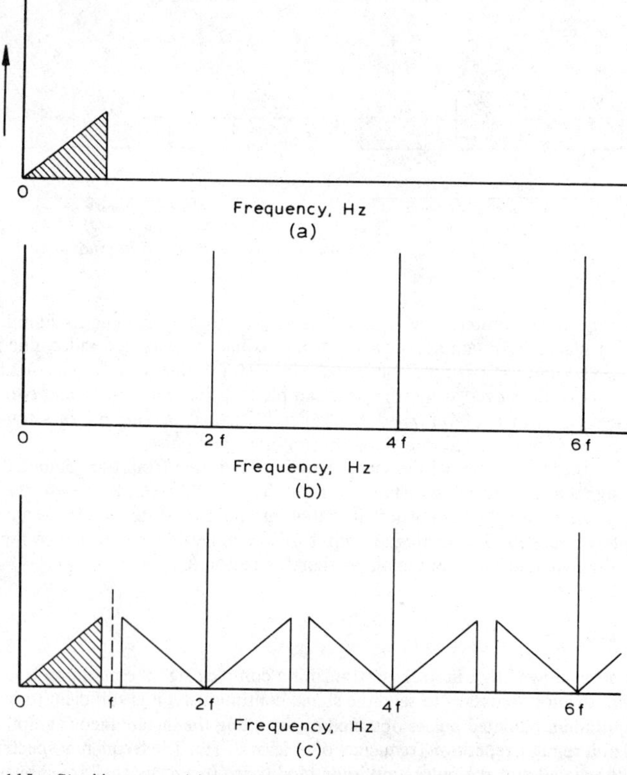

Figure 15.115 *Signal spectra: (a) spectrum of audio signal (b) spectrum of sampling signal (c) spectrum of pulse amplitude modulated signal*

The sound-in-syncs system employs a linear coding law (equal magnitude quantising levels) and uses a 10-bit binary code. The number of possible quantising levels is thus 1 024, i.e., 2^{10}. The following *Table 15.12* shows the calculated signal/quantising noise ratio[3] for systems employing 9–14 bits per sample.

Subjective tests[4] have shown that for high quality sound reproduction an overall signal-to-noise ratio of at least 60 dB (peak signal/peak weighted noise) is necessary. The distribution system alone should be better than this. This would suggest that a 12-bit code is

Table 15.12

Number of bits	*Peak signal* / *Peak q-noise*	*Peak signal* / *Peak weighted q-noise*
9	54	47
10	60	53
11	66	59
12	72	65
13	78	71
14	84	77

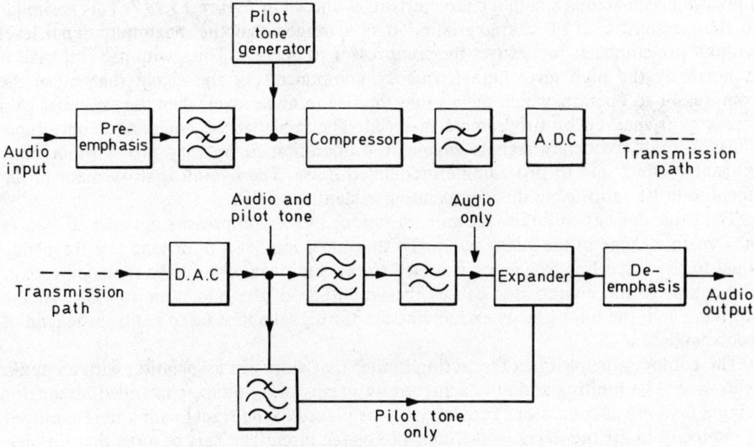

Figure 15.116 Simplified block diagram of compandor

necessary; however by using an audio compandor it is possible to improve the effective signal-to-noise ratio by 13 dB and so the p.c.m. code may be relaxed to 10-bits per sample.

COMPANDING

The compandor (compressor and expander combination) ensures that the mean-signal level in the A.D.C. and D.A.C. is as high as possible without exceeding the overload point. The block diagram is shown in *Figure 15.116*. The audio input is passed through

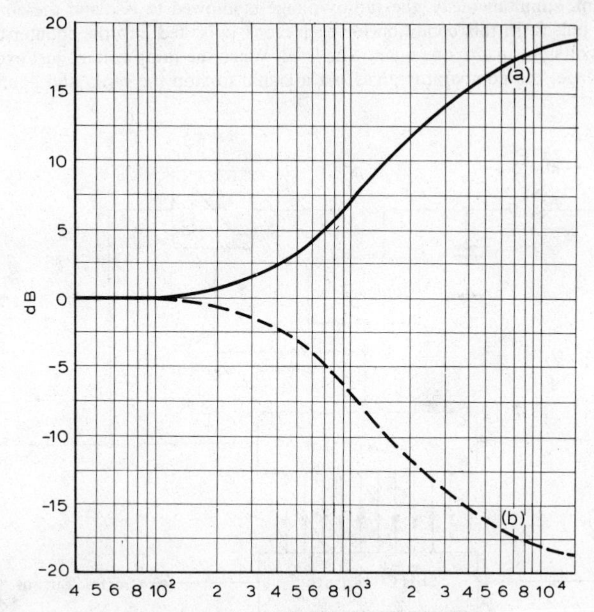

Figure 15.117 Pre-emphasis and de-emphasis characteristics used in s.i.s. system

a pre-emphasis network with a characteristic as shown in *Figure 15.117*. This is similar to the standard CCITT characteristic.[5] It is arranged that the maximum signal level without pre-emphasis just causes the compressor to operate. Thus, with pre-emphasis it is normally the high level high frequency components of the signal that cause the compressor to operate, which means that the rise in noise level when the expander gain increases is masked by programme material. The fact that the compandor only functions on high frequency signals enables the operation to be very fast without poor subjective effect due to programme modulated noise. The overall improvement in the signal-to-noise ratio using this companding system is 13 dB.

The expander has to follow the gain variations of the compressor in order to restore the dynamic range of the audio signal. To this end a low-level pilot tone at a frequency equal to the video-line frequency is added to the audio signal prior to the compressor. The action of the compressor causes the amplitude of the pilot tone to vary. At the receiving end, the pilot tone is extracted from the signal and used to control the gain of the expander.

The compressor itself is a fast acting limiter that works in association with an audio delay line. The limiting action is achieved by means of a voltage-controlled attenuator using a field-effect transistor. The delay line is placed in the signal path and is arranged to compensate for the delay in the control voltage circuitry. This ensures that limiting coincides exactly with the signal transients.

CODING AND DECODING

The analogue-to-digital converter block diagram is shown in *Figure 15.118*. This converter uses the counter-ramp principle. The compressed audio and pilot tone is sampled and the resultant voltage stored on a capacitor. This voltage is then applied to one input of a fast acting voltage comparator, while to the other input the output of a linear ramp generator is applied. In association with the ramp generator is a 10-bit up-counter. Initially the ramp voltage is in the reset condition and the counter at zero. At some instant, simultaneously, the ramp voltage is allowed to rise and the counter fed with clock pulses. In this condition for each clock pulse fed into the counter the ramp rises by a voltage equal to one quantising level. When the ramp voltage just exceeds the sample voltage, the comparator gives out a signal to stop the ramp and counter. The

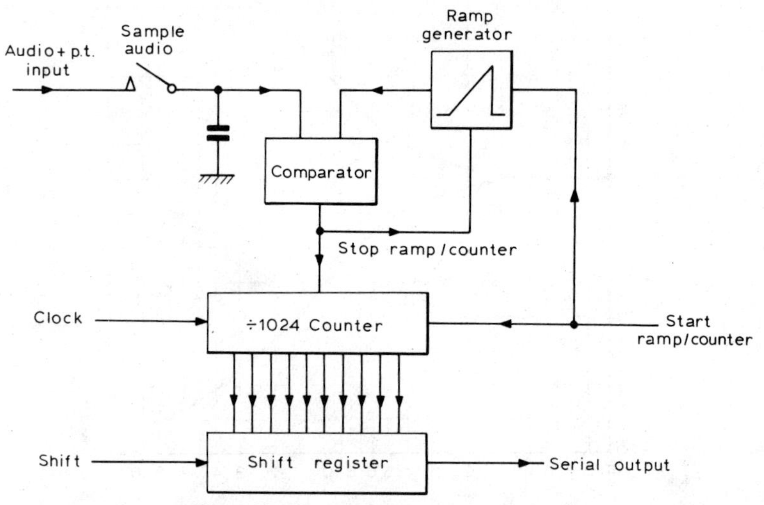

Figure 15.118 *Analogue-to-digital converter*

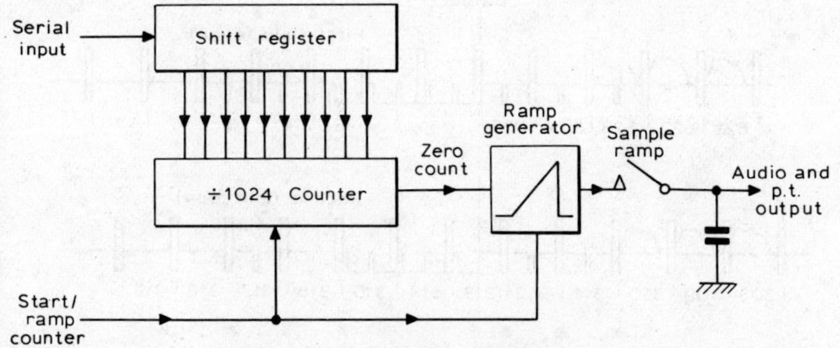

Figure 15.119 Digital-to-analogue converter

counter then contains the binary number corresponding to the sample. The 10-bit number is then transferred to a shift register where it is stored prior to serial readout.

The digital-to-analogue converter, *Figure 15.119* works in a similar manner. The 10-bit number is fed in serial form into a shift register from whence it is transferred to a 10-bit down-counter. The counter is started at the same time as the ramp generator. When the counter reaches a count of zero the ramp generator is stopped. The ramp voltage is then sampled, this voltage being the recovered audio sample.

The clock frequency used in these converters is 50 MHz, and as it may be required to provide 1 024 clock pulses in order to convert a peak positive voltage, the conversion time is approximately 20 μs.

MULTIPLEXING

Two audio samples are taken during the video line interval. These samples in 10-bit code form are combined and along with a marker pulse inserted into a single line-sync pulse at a peak-to-peak amplitude corresponding to an excursion between the bottom of a sync pulse and peak white, *Figure 15.120*. The individual pulses are shaped into sine-squared form with a half-amplitude duration of 182 ns, and a spacing of 182 ms between the middle of the pulses.

These pulses have negligible energy above a frequency of 5.5 MHz and so are quite suitable for transmission on a video link. During the field-blanking period alternate

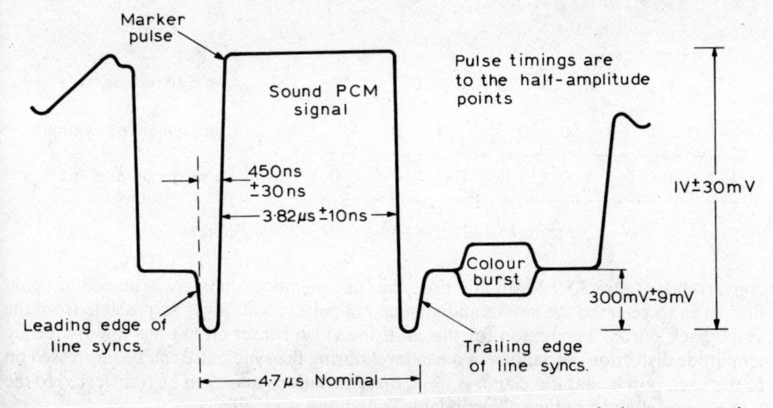

Figure 15.120 Idealised waveform showing combined pulse-group in the line-sync period

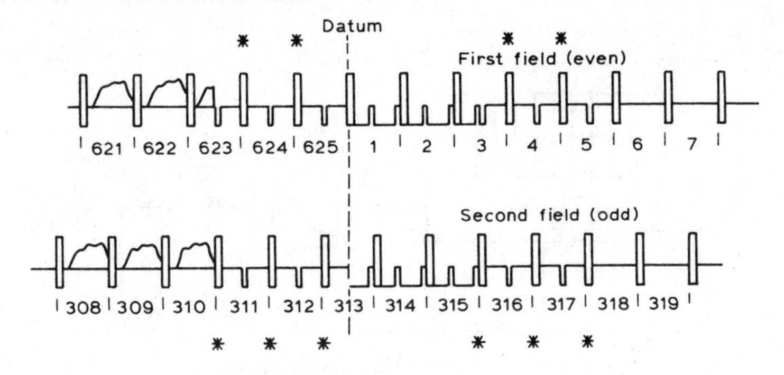

Figure 15.121 Modified field blanking waveform

equalising pulses are widened so as to accommodate the pulse groups, *Figure 15.121*. At the decoder the sync pulses are blanked out and a regenerated sync waveform, which is standard in all respects, gated in.

In order to minimise variations in the mean level of the pulse group, the 10-bit words are interleaved and one word complemented (logic 1's exchanged for a logic 0 and

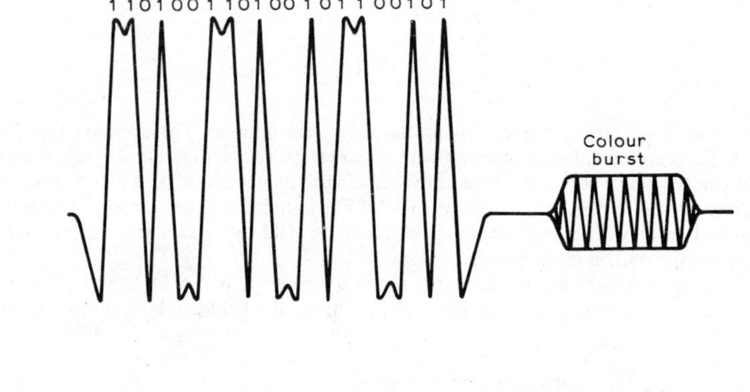

1 1 0 1 1 0 0 1 0 0	Unmodified sample
0 0 1 0 0 1 1 0 1 1	Complemented sample
1 1 0 1 0 0 1 1 0 1 0 0 1 0 1 1 0 0 1 0 1	Pulse group plus marker pulse

Figure 15.122 Details of pulse group waveform

conversely), *Figure 15.122*. In addition, the least significant bits are arranged to come first so as to keep the the most rapidly changing pulses as far away as possible from the video back porch. The reason for this is, if the video bearer circuit has low-frequency amplitude distortion, variations in mean level during the sync pulse can be impressed on to the back porch, and if a clamp is used, these variations may then be transferred to the picture period and produce objectionable sound-on-vision effects.

ERROR DETECTION

Distortion and noise on the bearer circuit can cause errors in the separation of the sync pulses and the detection of the sound pulses. To minimise the effect of these errors on the decoded audio it is necessary to monitor disturbances and take action accordingly.

There are two courses of action, either a 'hold' or an audio mute. The hold is intended for transitory errors and is a temporary process in which a faulty code is discarded and the code prior to the fault decoded instead. The audio mute is intended for sustained or serious errors. Mutes are applied in a manner that minimises the disturbance to the audio. Immediately a mute condition is found to be necessary, the audio output is shorted down to the mean audio level. When the fault clears, the mute is removed at a time corresponding to a mean-level voltage occurring in the audio signal.

Holds are applied for the following conditions:

(*a*) Failure to detect the marker pulse within 500 ns of the leading edge of line-sync pulses.
(*b*) A spurious pulse occurring immediately after the pulse group.
(*c*) The signal exceeding 3 dB above peak white or 3 dB below sync tips during the sync-pulse time.

Mutes are applied for the following conditions:

(*a*) Failure in detecting a sync pulse correctly.
(*b*) More than 3 holds occurring in a 10-line period.
(*c*) Timing generator out of lock.
(*d*) The signal exceeding 3 dB above peak white or 3 dB below sync tips for longer than 10 μs during a line period.

TIMING GENERATOR (*Figure 15.123*)

The sending and receiving terminal equipment both need timing pulses for their operation. The pulses are obtained from a timing generator that is synchronised to the input

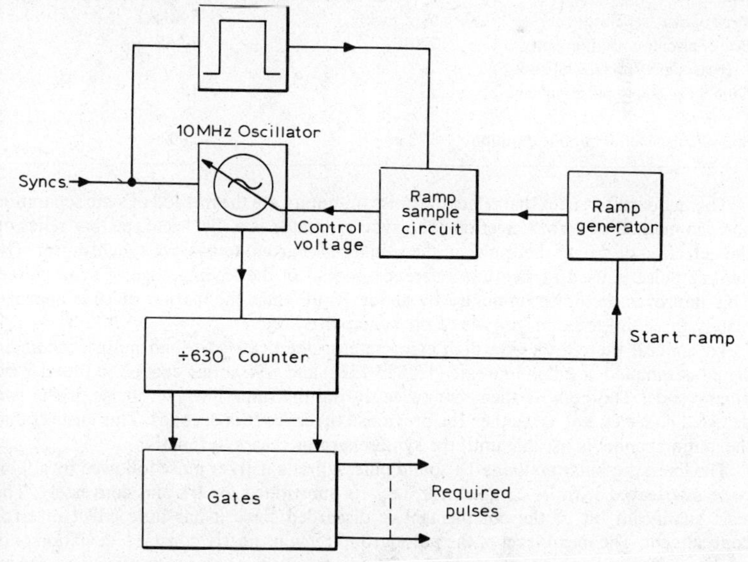

Figure 15.123 Simplified block diagram of s.i.s. timing generator

video signal. The leading edge of the separated sync pulse is used to start an oscillator whose frequency is approximately 10 MHz. This oscillator is used to clock a counter. When the counter reaches a count of 630 (i.e., after 63 μs) the counter stops and at the same instant starts a ramp generator. The ramp voltage is sampled at a time corresponding to the next leading edge of the line–sync pulse. The resultant voltage is then used to control the frequency of the oscillator so that the time between the counter stopping and the sync pulse is maintained at 1 μs. Timing pulses may then be obtained by gating various outputs of the counter.

RUGGEDISED SOUND-IN-SYNCS[6]

Television outside broadcasts sometimes have to use temporary mobile and long-distance video links of poor quality. These links are prone to multipath distortion, fading and high noise levels. Although the s.i.s. equipment will operate satisfactorily in the presence of quite high amounts of noise and distortion it can fail under these severe conditions. Ruggedised s.i.s. has been developed specifically to cope with these outside broadcast situations at the expense of a reduction in the quality of the audio signal.

The differences between the standard and ruggedised s.i.s. systems are outlined in *Table 15.13*.

Table 15.13

	s.i.s.	*Ruggedised s.i.s.*
Sampling rate	31.25 kHz	15.625 kHz
Bandwidth of audio channel	14 kHz	7 kHz
Bits/sample	10	9
Peak signal/peak weighted noise	65 dB	53 dB
Method of noise improvement	compandor (13 dB noise advantage)	CCITT pre-emphasis (6 dB noise advantage)
Type of sync separator	hard-lock	flywheel
Bearer circuit noise-immunity (peak signal/r.m.s. white noise)	23 dB	13 dB
Number of sound pulses in sync pulse	21	18
Sound-pulse half-amplitude duration	182 ns	200 ns

The major differences that effect the noise immunity are the method of sync separation and the method of error detection. The sync separator is a flywheel type that relies on the detection of the marker pulse in the sound pulse group to achieve synchronism. The marker pulse is used as the time reference instead of the leading edge of sync pulses. This improves the noise immunity by about 10 dB since the marker pulse is approximately three times the magnitude of the sync pulse.

To conceal the effects of a high error rate on the detected sound pulses, the audio signal is sampled at video line-rate (15.625 kHz) and the sample converted into a 9-bit binary code. The code is then sent twice during the same line-sync pulse, if the two detected codes do not agree then the previous correct code is decoded. This ensures that the audio channel is useable until the sync separator ceases to function.

The line-sync pulse contains 18 sound pulses, first a marker pulse followed by a 9-bit code interleaved with its complement, i.e., 1's substituted for 0's and conversely. The least significant bit of the complement is discarded since it has little effect in error concealment. The mean level of the pulse group remains nearly constant, deviation only occurring due to the least significant bit. This reduces sound-on-vision effects caused by poor low-frequency amplitude distortion on the link.

13-Channel p.c.m. sound distribution system

In order to meet the demand for high quality sound circuits capable of carrying stereo programme material, the BBC has developed a 13-channel digital time-division multiplex system. The system is intended for distribution of radio programmes between the studio centres and the main transmitters. The bearer circuit needed to convey the 13-channel p.c.m. signal is similar to that required for 625-line monochrome television.

All channels provide the same high standard of performance, and any two channels may be used as a stereo pair over long distances without loss of quality. The performance achieved on a single codec is as follows:

Frequency response	50 Hz–14.5 kHz $\pm$ 0.2 dB w.r.t. 1 kHz, typically −1.5 dB at 30 Hz and 15 kHz
Signal-to-noise ratio	69 dB, peak signal/peak weighted noise (CCITT Rec. P53 (1972) weighting network)
Non-linear distortion	0.1% (1 kHz at full modulation)

Pulse-code modulation (p.c.m.) ensures that the system is immune to all but the most severe noise and distortion on the bearer circuit. The presence or absence of the pulses may be detected up to a threshold point above which the effect on the audio channel rapidly becomes catastrophic. It can be shown [1] that the threshold for an ideal detector occurs with white noise at a peak signal/r.m.s. noise ratio of approximately 20 dB. Taking this into account a p.c.m. distribution system may be designed so that the audio performance is determined by the coding and decoding equipment, and this performance maintained throughout the complete distribution system.

FREQUENCY RESPONSE

With a p.c.m. system the highest frequency that may be conveyed is determined by the sampling frequency. The well-known rule is that the sampling frequency must be at least twice the highest frequency present in the input signal. In order to achieve an upper frequency limit of 15 kHz a sampling frequency of 32 kHz is used. This allows for the cut-off characteristics of the low-pass filters used on the input and output of the system. These filters are necessary in a sampling system in order to remove components at half-sampling frequency and above. This ensures that the audio signal may be recovered without the distortion caused by aliasing frequencies.

Since p.c.m. systems are capable of transmitting frequencies down to d.c., the lower frequency limit is determined by the analogue circuitry in the coding and decoding equipment. Likewise any variation in the frequency response throughout the band is determined by the analogue circuitry.

SIGNAL-TO-NOISE RATIO

The number of bits per sample determines the signal-to-noise ratio the system can achieve. For high quality sound distribution at least 63 dB (peak signal/peak weighted noise) is required. If allowance is made for several coding-decoding processes in tandem then it is necessary to choose a performance better than 63 dB. If a figure of 69 dB is chosen then this enables four codecs (coder-decoder combinations) to be used if the need arises. This would suggest a 13-bit code is suitable giving a theoretical signal-to-noise ratio of 71 dB. However it has been shown subjectively [2] that with critical programme material a 14-bit code is necessary in order that the granular distortion is imperceptible to the majority of people. Fortunately, a 13-bit code may be used providing a dither signal is added to the audio signal prior to the coding process. The dither signal consists of two parts, a component at half sampling frequency (16 kHz) with a peak-to-peak amplitude equal to half a quantising level, and white noise at a level 4 dB below the quantising noise. The overall effect is that granular distortion due to quantising is imperceptible at the expense of worsening the signal-to-noise ratio by 1.5 dB.

PARITY

Digit errors can occur in the bit stream due to noise and distortion on the bearer circuit. If an error occurs in the most significant bit of a word a loud click will be produced, while an error in the least significant bit will produce an almost inaudible disturbance. In order to protect the system against serious disturbances, a single parity bit is added to each 13-bit word. This provides a check on the 5 most significant bits. If an error is detected then the word is discarded and the previous word substituted in its place. If the error rate becomes high enough to impair seriously the quality of the signal the audio output is muted.

BIT-RATE

The bit rate was chosen to be 6.336 Mb/s. This enables the bit-stream to be conveyed on a bearer circuit designed to carry 625-line television signals. The individual pulses are shaped into sine squared form with a half-amplitude duration of 158 ns, and the pulses are spaced 158 ns apart. These pulses therefore have negligible energy above 6.336 MHz. It is not necessary in practice for the bearer circuit to have a bandwidth up to 6.336 MHz; the bandwidth may be reduced in exchange for a worsening of the immunity to noise and distortion. With a bandwidth up to 6.336 MHz the p.c.m. system can withstand a peak signal/r.m.s. white noise ratio of 20 dB. If the bandwidth is restricted to about 4.5 MHz then the noise immunity becomes approximately 22 dB.

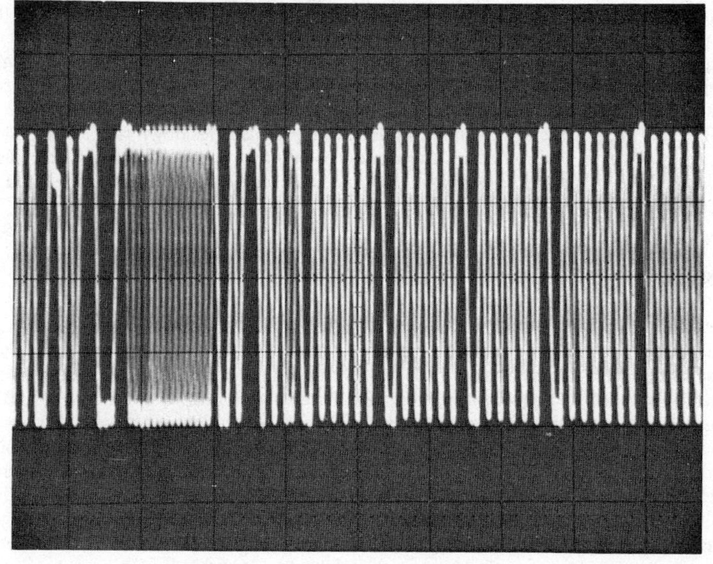

Figure 15.124 Part of p.c.m. bit stream

FRAME STRUCTURE

A bit rate of 6.336 Mb/s along with a sampling rate of 32 kHz gives a total of 198 bits per frame. The 13 audio channels each use 14-bits including the parity bit, the remaining 16 bits are used as a framing pattern (11 bits) and a data control channel (5 bits). The bit stream is shown in *Figure 15.124.*

DATA

The data channel provides the facility for remote switching of equipment at decoder sites. The data signal comprises of two parts, an address and a switching message. It is possible to switch either equipment at all the decoder sites simultaneously or to send a message to an individual site. The basic capacity of the system is 128 messages but this may be extended if the time to send a message is increased. Typical uses of the data messages are for mono/stereo switching and transmitter remote control.

CODING EQUIPMENT

The simplified block diagram of the coding equipment is shown in *Figure 15.125.*

Analogue circuitry. Each audio input is fed to a delay-line type limiter via a 15 kHz low-pass filter. The limiter is adjusted so as to operate at a level 2 dB above the nominal peak programme level. This prevents the occasional high programme peak over-modulating the p.c.m. system. If two channels are used as a stereo pair, then the limiters are interconnected so that if either limiter operates both change gain by the same amount, thus preventing shift of the stereo image. A 50 μs pre-emphasis network may be inserted prior to the limiter. This enables the channel to be used to feed a f.m. transmitter without an additional limiter at the transmitter site. The p.c.m. limiter will then prevent over-deviation of the transmitter. This is only possible due to the excellent gain stability of the audio channels.

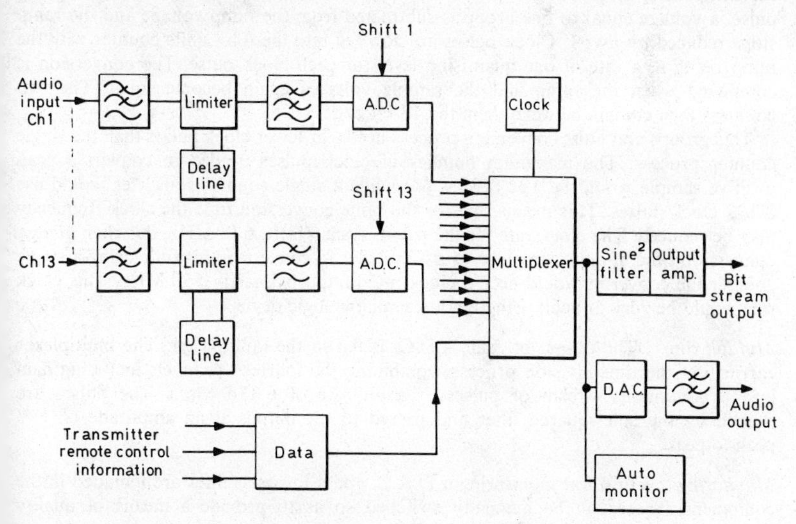

Figure 15.125 Simplified block diagram of p.c.m. coder

The audio signal passes via a second 15 kHz low-pass filter to the A.D.C. The low-pass filters are designed to pass 15 kHz with little attenuation while frequencies at 16 kHz and above are greatly attenuated.

Coding. The analogue-to-digital converter (*Figure 15.126*) uses an extension of the counter-ramp technique. The dynamic range of the converter is split up into 8 192 (2^{13}) equal magnitude quantising levels. These levels are considered to be in 128 (2^7) groups each group containing 64 (2^6) levels. Conversion is achieved by first counting the groups and then 'changing gear' to count the remaining individual quantising levels.

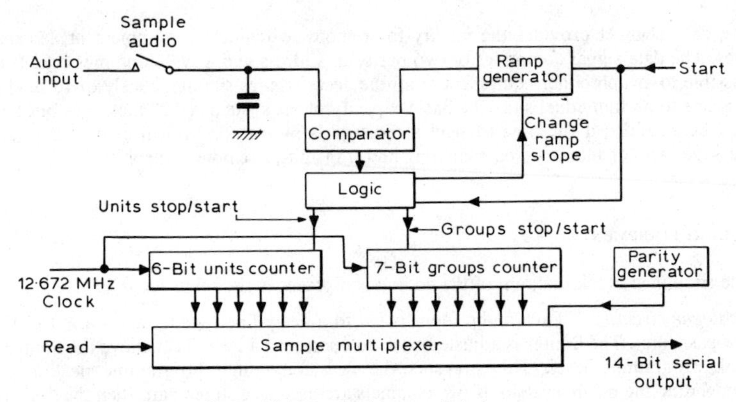

Figure 15.126 Analogue-to-digital converter

Initially clock pulses are fed into the 7-bit groups counter and the ramp generator set to a high slope. In this condition the ramp rises by a voltage equal to one group for each clock pulse entering the counter. When the ramp voltage just exceeds the sample voltage the comparator stops the groups counter. At a time corresponding to the next clock pulse, a voltage equal to one group is subtracted from the ramp voltage and the ramp slope reduced by 64 : 1. Clock pulses are now fed into the 6-bit units counter with the ramp rising at a rate of one quantising level for each clock pulse. The conversion is completed when the ramp and the sample voltages again become equal. The two counters then contain between them the 13-bit code.

This groups and units conversion process needs far fewer clock pulses than the single counter process. The maximum number of clock pulses needed to convert a peak positive sample would be 192 (128 + 64) while a single-counter converter would use 8 192 clock pulses. This means that for the same conversion time the clock frequency may be reduced. The clock rate for the p.c.m. system is 12.672 MHz, this then gives a conversion time of approximately 15 μs. In order to convert in this time using a single counter the converter would need a clock rate of approximately 550 MHz. This clock rate would be very difficult using readily available logic devices.

Multiplexing. The output of each A.D.C. is fed to the multiplexer. The multiplexer carries out the time-division process, combining the fourteen channels including data into a continuous stream of pulses at a bit rate of 6.336 Mb/s. The pulses are shaped by the sine-squared filter and passed to the output at an amplitude of 1 V peak-to-peak.

Monitoring. For aural monitoring a D.A.C. and a low pass filter are included in the equipment. These can be manually switched so as to provide a means of quality monitoring any audio channel in the bit stream.

Automatic monitoring is used to give warning of any faults occurring in the coder. The monitor is sequential and dwells on each audio channel in turn for 256 ms. During this time 256 checks are made by sampling the audio input to the channel and coding by means of a 5-bit A.D.C. The output of this A.D.C. is then compared with the 5 most significant bits of the appropriate word appearing in the bit stream at the output of the multiplexer. If more than 50% of the samples are in error then the monitor will register a fault on that particular channel.

DECODING EQUIPMENT

The simplified block diagram is shown in *Figure 15.127*.

Bit-stream processing. The bit stream in the form of sine squared pulses is fed to the input unit. This unit converts the pulses, which may be noisy and distorted, into definite logic 0 or 1 levels.

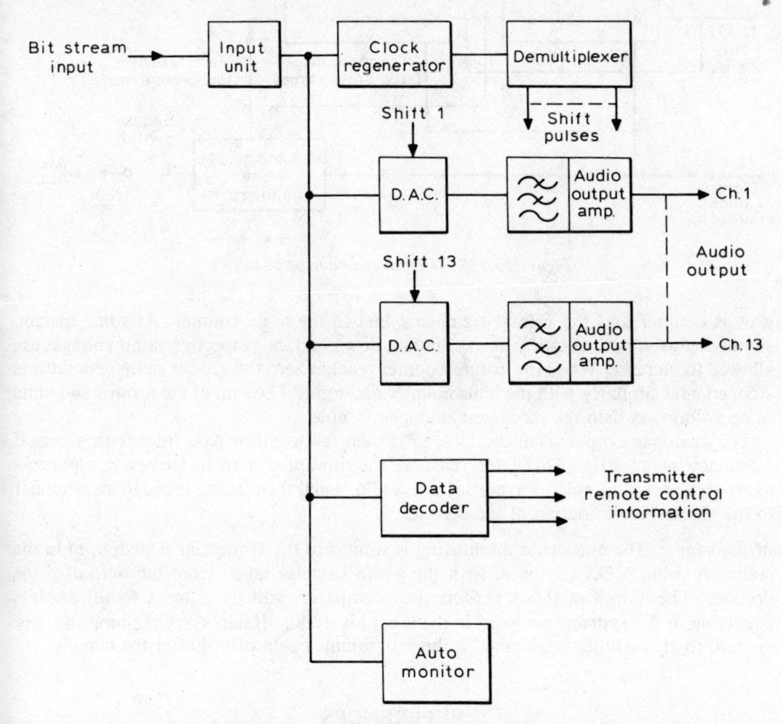

Figure 15.127 Simplified block-diagram of p.c.m. decoder

The clock regenerator produces clock pulses at 6.336 MHz locked to the incoming bit stream. The synchronisation is achieved by comparing the timing of the framing pattern occurring in the bit stream with a locally generated framing pattern, any timing error produces a change in a voltage used to control the frequency of the local timing reference.

The regenerated clock pulses are fed to a demultiplexer where groups of bit-rate shift pulses are produced. The shift pulses are used to enter the individual 14-bit pulse groups, present in the continuous bit stream, into the correct D.A.C.

Decoding. The D.A.C. (*Figure 15.128*) makes use of the groups–units technique as used in the A.D.C. with the groups and units ramps arranged to have slopes in a ratio of 64 : 1.

Initially the parity bit is checked and if correct the conversion is allowed to proceed. The 13-bit code is then transferred to the counters, the 7 most significant bits to the

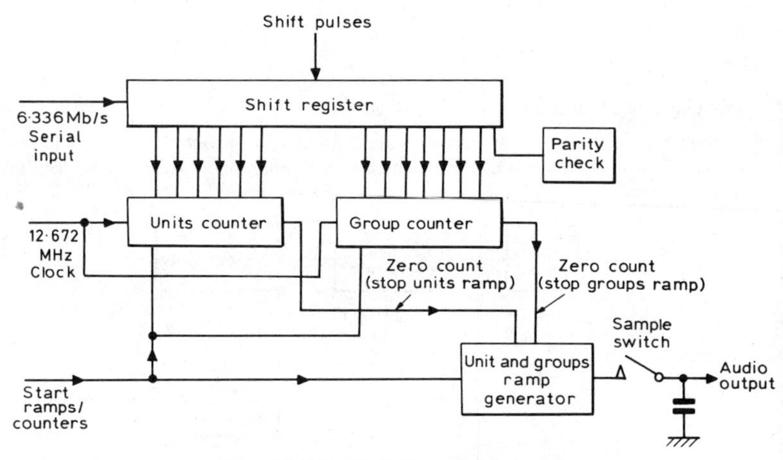

Figure 15.128 Digital-to-analogue converter

groups counter and the 6 least significant bits to the units counter. At some instant, simultaneously, both counters start counting down and the respective ramp voltages are allowed to increase. When the groups counter reaches zero the groups ramp generator is stopped, and similarly with the units counter and ramp. The sum of the groups and units ramp voltages is then the recovered analogue sample.

The analogue output from the D.A.C. is then fed to a low-pass filter with a cut-off frequency of 15 kHz. This filter removes the unwanted high frequency components associated with a sampled waveform. The audio signal then passes through an amplifier to the channel audio output of the decoder.

Monitoring. The automatic monitoring is similar to the sequential system used in the coder. A 5-bit A.D.C. is used, with the audio samples taken from the output of the decoder. The output of the A.D.C. is then compared with the 5 most significant bits occurring in the appropriate word in the input bit stream. If necessary the monitor may be used to check other equipment in the programme chain placed after the decoder.

REFERENCES

1. OLIVER, B. M., PIERCE, J. R. and SHANNON, C. E., 'The Philosophy of P.C.M.', *Proc. I.R.E.*, **36** (1948)
2. SHORTER, D. E. L. and CHEW, J. R., 'Applications of pulse-code modulation to sound-signal distribution in a broadcasting network', *Proc. I.E.E.*, **119**, No. 10 (1972)
3. HOWARTH, D. and SHORTER, D. E. L., 'Pulse-code modulation for high-quality sound signal distribution: appraisal of system requirements', *BBC Research Department report, EL10*, 12 (1967)

4. *CMTT*, Doc. 26 (June 1968)
5. *CCITT Blue book*, Vol. 3, 363 (1965)
6. DALTON, C. J., 'A P.C.M. Sound-in-Syncs system for outside broadcasts', *BBC Engineering*, **86**, 18 (1971)

CABLE TELEVISION

Historical

The use of cables to convey entertainment to the home goes back to the earliest days of the telephone: Gilbert installed it in 1882 and advised Sullivan to do the same to listen to the stage performance of their masterpieces.[1]

In Budapest in 1892 a network carried news services to homes and in the same period the Electrophone Company in London enabled subscribers to listen to theatres, balls and churches.[1]

These rather specialised services did not, however, achieve a wide acceptance and it was not until the advent of the popular demand for broadcasting in the 1920s that wired distribution set off on the path which now seems certain to lead ultimately to the fully integrated communication network carrying any kind of signals—voice, data or vision—to and from each member of the community.

In April 1924 the first commercial radio relay service was started by A. L. Bauling in Koog aan de Zaan in Holland.[1] He was followed independently by Mr. A. W. Maton[2] of Hythe, a village in Hampshire, who also started experimenting in 1924 and opened his commercial service in January 1925.

It was upon this foundation that cable television systems grew in Europe, although their recent rapid growth in North America had a different origin.

Present systems

Wired distribution systems vary considerably in their size and complexity. Perhaps the simplest possible concept is that of a single receiving aerial serving a pair of semi-detached houses or a small block of three or four flats. In a very strong signal area such a system may not need any active equipment, the signal picked up being shared between the users by means of passive dividing networks. In weaker signal areas, or where a somewhat higher number of outlets is required, it may be necessary to incorporate an amplifier before the signal from the aerial is divided up between outlets. In still more complex systems, in weak signal areas or to serve large blocks of flats, it may be necessary to change the frequency of the signal derived from the aerial before it is distributed to individual users. This is to ensure that at a television receiver interference does not occur between the signal carried over the wired system and the signal available directly from the broadcasting station.

The most sophisticated wired systems may comprise an elaborate directional aerial system, sited at a high vantage point to gather in weak signals from a distant transmitter; a device to change the frequencies of the incoming signals to other channels more suitable for distribution by wire; together with an extensive network of underground and/or overhead cables carrying the signals to many thousands of households in a given locality.

Some of the transmission systems currently in use for this type of system are described in the following sections. In most cases the input television signals are picked up at suitably sited stations receiving normal off-air broadcast signals; however, there are some cases in which signals are received by direct cable connection from the broadcasting authority. There would be no technical problem in feeding signals from other sources, such as local studios or video tape recorders, directly into the wired distribution system.

There are two main types of distribution system representing a difference of approach which arose for historical reasons. In the early 1930s a substantial business developed in the wired relay of sound radio programmes at audio frequency and the total number of subscribers reached about 1 million in 1950. This business was confined to Europe, mainly Holland and the U.K. and developed in large provincial cities, e.g., Hull, Leeds, Newcastle, Nottingham, Bristol. The cost of the cable connection, programme switch and loudspeaker was competitive with the radio receivers of the day and this, with the convenience and reliability of the system, provided the economic base for its growth. With the advent of television in the early 1950s it became necessary for the relay operators, if they were to stay in business, to develop television distribution

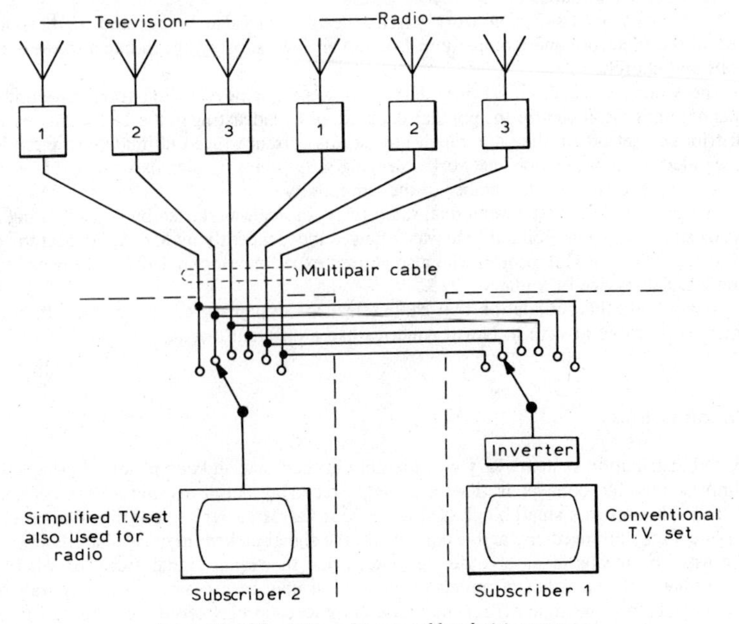

Figure 15.129 H.F. multi-pair cable television systems

methods which would, as with sound radio, be competitive with direct reception off-air in the generally good reception conditions of the large provincial cities. The result was the h.f. multipair system, *Figure 15.129*, which can serve simplified receivers directly, or conventional off-air receivers through an 'inverter'. Since the programmes are carried on physically separate channels these systems may be described as space division multiplex.

The other type of system was developed, principally in the U.S.A. for different reasons. There is no obligation on the United States broadcasting industry to serve the smaller towns where the available advertising revenue may be insufficient to support a plurality of broadcasting stations. This led to the development of systems which were purely concerned with the improvement of reception of weak signals from distant stations. In these circumstances the system does not have to compete as an alternative means of reception of local transmissions. The most suitable system for this purpose uses a single co-axial cable to which standard television receivers can be directly

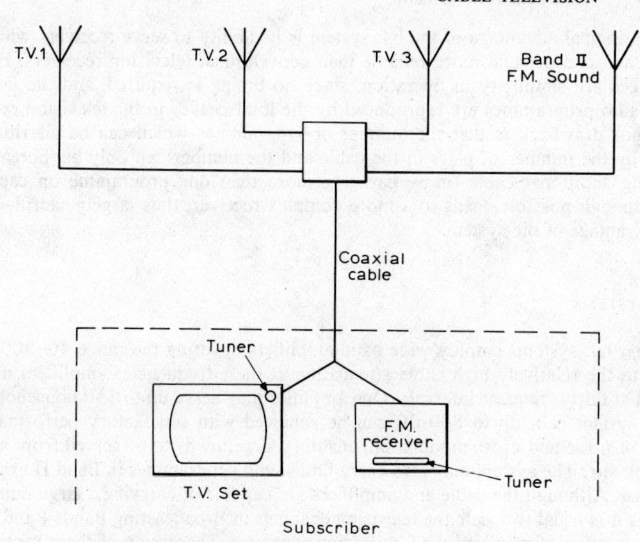

Figure 15.130 V.H.F. coaxial cable television system

connected. The different programmes are carried on different frequency channels in a manner exactly analogous to over the air broadcasting *Figure 15.130*, and may be referred to generally as frequency division multiplex.

H.F. SYSTEM

An h.f. system uses a single channel in the h.f. band for all programmes and each programme is carried on a separate pair of wires in a multipair cable. The vision signals lie in the band between 4 and 10 MHz where cable attenuation is relatively low but the frequency is high enough to avoid difficulties in the demodulation process in the receivers. The result is that a very large number of subscribers can be fed from one amplifier installation; the number ranges from several hundred to as many as 2 500. The sound accompanying the vision signal is carried at audio frequency and operates the loudspeaker directly. The sound radio programmes are also carried at audio frequency on separate pairs. Modern installations of this type employ a 12-pair cable of special construction which has capacity for six television programmes with a further six sound radio programmes. In an alternative system a 4-pair cable is used and the sound accompanying the vision signal is carried on an amplitude modulated carrier at approximately 2 MHz, enabling the same pairs to carry a further four sound programmes at audio frequency.

The distribution of sound direct to loudspeakers has represented the most economic solution for many years but with an increase in the number of programmes to be carried and the reduction in the cost of integrated circuits for f.m. demodulation and audio amplification, the position is now changing and h.f. systems in future are likely to abandon it. This will leave the audio and supersonic bands up to about 2 MHz on each pair free for two-way data or voice services.

Many occasions arise when it is desirable to serve a conventional television receiver from an h.f. system and this is done by means of a frequency changing unit which takes the vision signal from the cable and converts it into one of the standard broadcast channels in the v.h.f. or u.h.f. bands; the sound signal is fed directly to the loudspeaker.

The principal advantage of the h.f. system is its ability to serve receivers which are simpler and cheaper and more reliable than conventional television receivers. Further advantages are simplicity in operation, since no tuning is required, and the fact that sound radio programmes are reproduced by the loudspeaker in the television receiver. The major drawback is that the number of programmes which can be distributed is limited by the number of pairs in the cable and the number can only be increased by providing additional cable or by carrying more than one programme on each pair which, though possible, leads to a more complex receiver, thus largely sacrificing the main advantage of the system.

V.H.F. SYSTEMS

Modern v.h.f. systems employ wide band amplifiers covering the range 40–300 MHz. Owing to the relatively high cable attenuation at these frequencies amplifiers must be inserted at fairly frequent intervals. One amplifier may serve up to 150 households and a total system gain up to 800 dB can be achieved with satisfactory performance in respect of noise and cross-modulation, enabling large towns to be served from a single reception site. The system may also carry f.m. sound programmes in Band II exactly as broadcast. Although the cable and amplifiers are capable of carrying a large number of channels it is usual to locate the television channels in Broadcasting Bands I and III for direct operation of standard v.h.f. television receivers. The design of these receivers is such as to place a practical limit on the number of channels that may be used in these Bands. The main factors which set a limit to the number of programmes are:

(*a*) The receivers are not manufactured with adequate adjacent channel selectivity, hence adjacent channels cannot be used on the same system.
(*b*) Receivers are not well enough screened to avoid direct pick-up of strong signals from a television transmitter; hence the channels used for broadcasting in an area cannot be used in a cable system in the same area.
(*c*) Local oscillator interference from a receiver connected to the cable to a nearby receiver working from an aerial precludes the use of certain channels.

As a result of these limitations it is rare for systems to be able to distribute more than six television programmes with high technical quality in areas with strong ambient signals.

The programme required is selected by means of the selector switch or tuner fitted on the television receiver. Since the distribution system is required to serve standard domestic television receivers, some of which are only capable of receiving 405-line signals, it is usually necessary to distribute the 405-line programmes as well as those of 625-lines. A problem which has arisen in this country since u.h.f. became the primary service is that new receivers capable of operating on v.h.f. are almost unprocurable and the connection of a modern receiver to a 625-line v.h.f. system now requires the use of a frequency changer or 'up-converter' to convert the v.h.f. signal received to a u.h.f. channel to be fed to the receiver.

This problem is temporary and peculiar to the U.K. but it illustrates the difficulties which are met in devising a cable distribution system to serve receivers which were designed for a different purpose, namely, reception from an aerial.

MIXED SYSTEMS

The need to serve 405-line v.h.f. and 625-line u.h.f. receivers on the same system has led to the development of v.h.f./u.h.f. systems in which 405-line programmes are fed to the subscriber at v.h.f., 625-line programmes at u.h.f., and f.m. sound programmes on Band II. Individual 200-subscriber areas are served by a primary distribution system which may be operated at h.f. (in space division) or v.h.f. (in frequency division), the appropriate signals being converted to u.h.f. at distribution points.

U.H.F. SYSTEMS

Modern u.h.f. systems employ wideband amplifiers covering frequencies up to 860 MHz. Owing to the high cable attenuation, such systems are at present limited to areas containing about 200 subscribers. Single standard u.h.f. receivers are used and the programme required is selected by means of the selector switch or tuner fitted on the receiver. As with the v.h.f. system f.m. sound programmes are usually provided in Band II exactly as broadcast.

U.H.F. frequencies are unsuited by their nature to large cable systems and are unlikely to find more than limited use in such conditions as now exist in the U.K.

SWITCHED SYSTEMS

Higher numbers of programmes can be handled by h.f. systems with a different arrangement of the distribution network in which pairs in the cable are allocated to individual subscribers to connect them to a switching centre or programme exchange.[3, 4] Selection of the desired programme is then carried out by a switch in the programme exchange by remote control over a separate pair from the subscriber's premises, in a manner somewhat similar to that used in an automatic telephone exchange. The higher frequencies to be transmitted place a strict economic limit on the length of the subscriber's connection to the exchange so that a larger number of exchanges, each

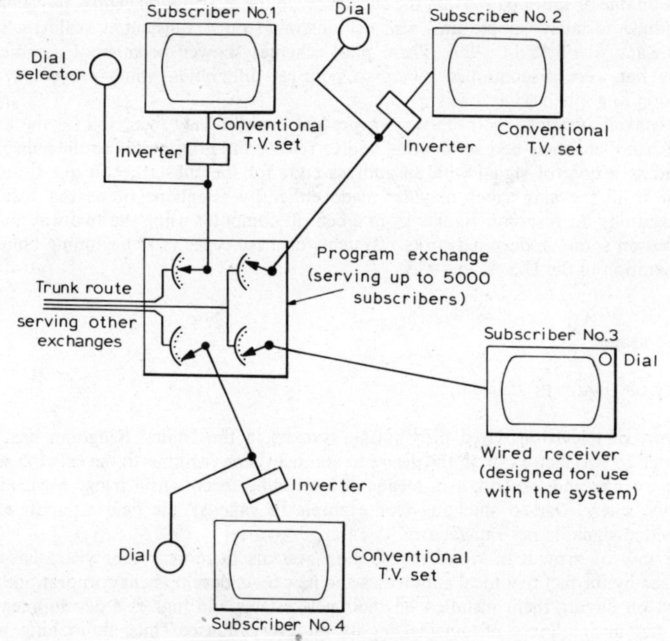

Figure 15.131 Switched systems for cable television

covering a smaller area, is required than in the case of the telephone system. A switched system can be extended to two-way working, the return path to the exchange being provided over the same subscriber's pair of wires at a higher frequency than the outgoing signal. The system is illustrated in *Figure 15.131*.

PAY-TV

Pay-TV in which the recipient pays directly for the programmes which he takes, is technically possible. The simplest form is payment on a per-channel basis; a number of channels each carrying programmes on a particular subject or group of subjects might be offered and subscribers would elect to pay a monthly subscription for the channels in which they are interested in the same way as people subscribe to magazines. This is very simply arranged on h.f. systems since the circuit carrying the chosen programme can be connected or not to each subscriber in the junction box or programme exchange. In coaxial systems the simplest scheme is to use channels to which ordinary receivers will not tune and to supply simple frequency converters to subscribers requiring the pay channels. Security against illicit use of converters is a problem and an improvement can be obtained at extra cost by scrambling the signal as transmitted and descrambling for each Pay-TV subscriber.

Payment on a per channel basis is satisfactory in many circumstances but it is necessary to achieve payment on a per programme basis if the full potential of Pay-TV is to be realised. This is because the owners of the most suitable programmes such as first-run films and important sporting events will not make them available to television unless they can be sure of receiving their share of the money paid by the public to see their particular programme, as they would from a cinema or sports arena.

Because h.f. systems lend themselves readily to the backward transmission of control signals on one or more pairs from the subscriber to the distribution centre, payment per programme is fairly simple and was demonstrated on a substantial scale in South London and Sheffield in 1966. These pilot schemes showed promise of commercial success but were discontinued owing to political difficulties which prevented their expansion to a much wider basis.

On coaxial systems it is necessary to prevent unauthorised reception by the use of 'out of band' channels and scrambling and to release the programme to the subscriber by sending a control signal with an address code for the subscriber in question. The request from the subscriber may be made either by telephone or as the result of interrogation and response signals from a central computer using the two-way facility provided on some modern networks.[6] Systems of these types were beginning commercial operation in the U.S.A. in 1973.

Current usage

EXTENT OF USAGE IN THE U.K.

The start of television wired distribution systems in the United Kingdom has been attributed to two factors: first, the desire to add television facilities to the existing sound relay networks and, second, as a means of improving reception in fringe areas and at localities where, due to shielding (for example in valleys), the field strength of the transmitted signal is not satisfactory.

The rate of growth of wired distribution systems in more recent years has been increased by the fact that local authorities and new town development corporations have insisted on having them installed on their new estates and high-rise developments to avoid the unsightliness of outside aerials on every house. Thus, flourishing wired distribution systems do exist in many areas where off-air reception is perfectly satisfactory.

The first television wired distribution system became operational in this country in 1951 and served about 300 subscribers. At the end of 1971 there were 2 778 television systems licensed, covered by 846 licences and serving 1 839 438 subscribers, which represented about 11% of the television broadcast receiving licences in the U.K. at that time. The growth rate of relay subscribers during the last decade has been approximately 12% per annum. More than 60% of the subscribers were connected to h.f. systems and in general there is a tendency for the larger systems to be h.f. using multi-pair cables, and the smaller ones to be v.h.f. using coaxial cables. The largest group of subscribers served from one receiving aerial installation is about 41 000 on an h.f. system in Hull, while the largest v.h.f. system serves about 15 000 subscribers in Swindon.

Very wide band v.h.f./u.h.f. relay systems are beginning to be used in the U.K. where only small numbers of subscribers, of the order of 100 or 200, require service within a compact area; as yet no data is available on the number of such systems.

THE REGULATORY POSITION

Before 1 October 1969, licences authorising the operation of wired distribution systems were issued by the Postmaster General under the Wireless Telegraphy Act 1949.

The Post Office Act 1969, which came into force on 1 October, re-defined the telecommunications monopoly and vested it in the Post Office Corporation, while the Government's powers under the WT Acts were assumed by the Minister of Posts and Telecommunications. In addition new provisions were made for the licensing of 'programme distribution systems' by the Minister. Consequently, activities which were previously carried out under a single licence now require separate licences from the Minister and the Post Office.

The present period of currency for both Post Office and Ministry licences is up to 31 July 1981 which coincides with the termination of the BBC Charter and the contracts of the programme companies.

Prior to January 1972 only television programmes broadcast by the BBC and ITA (now IBA) were permitted to be distributed on wired systems except for certain public service announcements which required Ministerial approval beforehand. Television programmes originating from transmissions outside the United Kingdom are prohibited, except with the consent of the Minister, from being distributed on systems within the United Kingdom except when being relayed from either a BBC or IBA television transmitter. However, five relay system companies have now been given permission to originate programmes of a local interest nature, for an experimental period up to July 1976 but the operators are prohibited from deriving any revenue from these programmes whether from the subscribers, from those using the channel to communicate with the public or from advertising.

Technical conditions are laid down by both the Post Office and the Ministry in association with the licences that are issued. The Post Office conditions are designed to minimise the risk of interference to Post Office plant in the telegraph and telephone distribution network. The Ministry conditions are intended to prevent undue interference from the system to off-air reception in the broadcasting bands and to other authorised users of the spectrum; they are also designed to prevent lethal voltages from being introduced on to the network from either distribution system equipment or subscribers' equipment. Associated with the mandatory technical conditions are recommendations for the performance standards of the relay system.

THE SITUATION IN OTHER COUNTRIES

In Canada approximately 42% of television homes in urban areas are served by cable television.[5]

In the U.S.A. as mentioned previously, the main reason for the existence of wired distribution has hitherto been the need to provide for towns too small to support a full television service. There were at the end of 1972 over 2 600 wired distribution systems serving about 6.6 million subscribers or about 9.5% of the total viewing public. Recent changes in FCC regulations, however, will allow cable operators to import programmes from distant areas into areas already well served by local stations; will require systems with more than 3 500 subscribers to originate programmes; and will require a basic minimum 20-channel capability in the most important market areas.

The importation of distant signals into the top 100 markets had previously been disallowed. Some operators believe that this relaxation will open these markets to cable but others think that the offer of 'more of the same' will not succeed and that special programmes, particularly films and sport, on a Pay-TV basis will be essential to success.

In Europe, particularly in Belgium, there has been considerable development of cable television due to the interest in the programmes from the neighbouring countries and the ability of many Belgians to speak several languages. In 1972 there were approximately 275 000 subscribers to cable systems in Belgium, representing 8% of total households. In Switzerland also there has been some development for similar reasons. In 1972 there were approximately 130 000 subscribers representing 7% of households.

In the major countries of Continental Europe, France, Germany and Italy, the development of cable systems has been prevented because the broadcasting authorities have a monopoly in the origination of programmes and the telephone administrations have a monopoly in all forms of telecommunication and have been unwilling to licence others or to spare the necessary resources themselves from their heavy programmes of telephone development. There has, however, been an interesting development in Spain where the national telephone company is now wiring extensive systems in Madrid and Barcelona. These systems will carry the existing broadcast programmes from the broadcasting authority who will, in addition, generate two further programmes solely for distribution by the cable.

In the Netherlands an experimental system has been in operation for some years which was installed by the PTT. Fairly recent legislation has permitted private operators to enter the field and a small number of experiments in local origination, similar to those in the U.K., have been authorised. Demand for cable television in the Netherlands is expected to match that in Belgium and Switzerland and to grow rapidly.

Mention must also be made of the Republic of Ireland where, although there has been a substantial demand for the British television programmes which are not easily receivable, cable television has been discouraged. However, in the last few years development has been permitted of systems limited in size to 500 subscribers provided either by the broadcasting authority or by private companies. This limit of 500 subscribers has very recently been removed and rapid development is also expected there.

The present regulatory position in the United Kingdom militates against cable systems because off-air reception in most populated areas is good and no additional programmes can be offered by the system operators, apart from the few experimental exceptions mentioned in the previous section.

The future

The future of wired distribution systems must be set against a wide range of possibilities. At one extreme lies a universal communications network, wideband, two-way and switched, capable of conveying any kind of information, visual, voice or data from anywhere to anywhere. Besides television and sound for entertainment purposes the range of services which might be provided by such a universal network include:

Telephone and viewphone facilities.
Electricity, gas and water meter reading.

Facsimile reproduction of newspapers, documents, etc.
Electronic mail delivery.
Business concern links to branch offices.
Access to computers.
Information retrieval (library and other reference material).
Computer-to-computer communications.
Special communications to particular neighbourhoods or ethnic groups.
Surveillance of public areas for protection against crime.
Traffic control.
Fire detection.
Educational and training television and sound programmes.

At the other extreme one can envisage some small advance on the present situation with a minor extension in the number of programmes distributed on wired broadcasting systems operating side by side with entirely separate switched networks for telephone and data as at present. There is no theoretical limit to the number of television programmes that could be delivered to a subscriber's premises by cable but the developments in the U.K. during the next ten years will depend upon the customer demand for additional television and other services, on the cost of providing them, and on political and legal decisions regarding licensing, copyright and other complicated issues. In the U.S.A., the Federal Communications Commission has recently laid down certain rules for the licensing of new wired distribution systems which call for at least 20 television-channel distribution capability, together with some capacity for return communication.

The technical possibilities for future wired communication systems, incorporating the distribution of sound and television programmes, are very wide. Initial studies of the various forms of these networks, now being conducted in this country and abroad, are based on the use of cable as the only transmission medium but, in the more distant future, the transmission medium might include millimetric radio links and optical fibre guides. One must also consider as a future possibility the carriage of full-bandwidth television to the home in digital form. The ultimate pattern of network connections will probably be the result of an evolutionary process whose timing will be influenced by demand considerations and by the extent and type of past investment.

For h.f. systems which employ a separate pair of wires for each programme, the practical limit to the number of programmes is about 12 owing to the large number of pairs that must be handled in junction boxes and selector switches.

For greater numbers of programmes a switched h.f. system may be used. Because this is a four-wire system (one programme pair and one control pair) it has the advantage that control of the use of certain channels (e.g., pay-television channels) is easily arranged. On the other hand only one programme at a time is available on the subscriber's premises unless a second exchange line is provided or arrangements are made to transmit a second programme over the same line by frequency division multiplex. Switched h.f. systems employing four wires can be extended to provide for all the services listed at the beginning of this chapter.

For systems operating at v.h.f. or higher frequencies a single coaxial cable should be capable of delivering up to 24 programmes. For example, the channels might be spaced at 10.5 MHz as in the Post Office educational television system used by the Inner London Education Authority. With this spacing the 24 channels might be accommodated in the band 40–300 MHz but, as the spacing is different from that of the broadcast channels, the system would be unable to serve directly ordinary receivers in use for off-air reception and either special receivers or tunable converters would be required by every subscriber. Also with 24 channels the linearity of the long chains of repeaters required in these systems needs to be of the highest order if the distortion products, particularly those of the third order, are not to degrade the picture quality to an unacceptable degree.

In order to provide many of the services mentioned at the beginning of this section systems must be capable of passing messages in both directions. In coaxial systems messages in the return direction can use the frequency space below the lowest television channel transmitted, i.e., below about 50 MHz. Initially there may be sufficient capacity on the normal simple tree topology at present used with the systems but looking farther into the future, schemes are in consideration for services other than full bandwidth television in which digital transmission would be used in conjunction with coaxial cable laid in a ring formation beginning and ending at communications exchanges. Transmission over the cable would be in one direction only and subscribers would be connected to points on the ring. The subscriber's equipment would pick out from the total digit stream that information appropriate to his needs and would re-inject into the stream appropriate return communication in digital form. A limited number of entertainment and other distributed television programmes in f.d.m. analogue form could occupy the bandwidth in the ring above the highest frequency (possibly 80 MHz) required for the both-way digital services. With a ring topology it would be possible to provide a two-way television service to a limited number of subscribers by using the remainder of the ring for the return direction of transmission.[7]

Acknowledgement

Much of the material in this section, 'Cable Television', is taken from Chapter 4 of the *Papers of the Technical Sub-committee of the Television Advisory Committee 1972*, with the permission of Her Majesty's Stationery Office.

REFERENCES

1. EXWOOD, M., 'Cable Television', *Paper to Royal Television Society Convention*, 7 September 1973
2. COASE, R. H., 'Wire Broadcasting in Great Britain', *Economica*, Vol. XV, No. 59, August 1948
3. GABRIEL, R. P., 'Cable Television and the Wired City', *I.E.E. Electronics and Power*, April 1972
4. GARGINI, E. J., 'The Total Communication Concept of the Future', *Royal Television Society Journal*, March/April 1973
5. *Canadian Radio-Television Commission*, 'The Integration of Cable Television in the Canadian Broadcasting System', 26.2.1971
6. JURGEN, R. K., 'Two-way Applications for Cable Television Systems in the 1970s', *I.E.E.E. Spectrum*, November 1971
7. HARE, A. G., 'Telecommunication 20 Years On—A Look Into Local Distribution', *Post Office Research Dept.*, May 1971

FURTHER READING

'Communication Technology for Urban Improvement', June (1971), *Proc. I.E.E.E. Special Issue on Cable Television*, July (1970)

MASON, W. F., 'Urban Cable Systems', *The Mitre Corporation* (May 1972), National Academy of Engineering, Washington D.C.

REPORT OF THE SLOAN COMMISSION, *On the Cable—The Television of Abundance*, McGraw-Hill

TELEVISION ADVISORY COMMITTEE 1972, papers of the Technical Sub-Committee, Chapter 4 H.M.S.O.

WARD, JOHN E., 'Present and Probable CATV/Broadband Communication Technology', *Report of the Sloan Commission*

SOUND AND TELEVISION RECEIVERS

All broadcast sound and television receivers work on the superheterodyne principle, i.e., they contain a local oscillator which keeps in step with the incoming signal to provide a constant heterodyne frequency, the so-called intermediate frequency, on which band the main amplification takes place prior to demodulation and final audio or video amplification. On m.f. bands and television bands oscillators work above the incoming signal, but in most British Band II (v.h.f. f.m.) receivers oscillators operate below the r.f. signal. I.F. frequencies are standardised; in Britain they are 470 kHz for m.f. bands, 10.7 MHz for

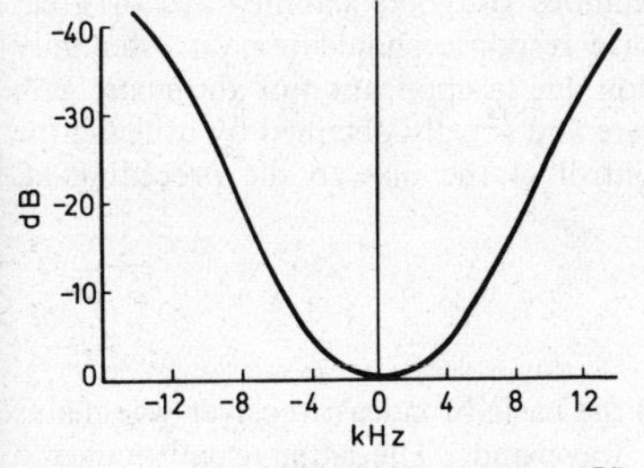

Figure 15.132 Typical m.f. receiver, 470 kHz i.f. amplifier response using 5 tuned circuits

Figure 15.133 Typical v.h.f./f.m. receiver, 10.7 MHz i.f. amplifier response using 5 tuned circuits

Band II f.m. sound receivers and 39.5 MHz (vision i.f.) for television receivers. While the bulk of gain and selectivity is obtained in the fixed frequency i.f. amplifier, some front stage gain and selectivity is needed to achieve a good signal-to-noise ratio and some amount of preselection to reduce the dangers from cross-talk, image frequency breakthrough and spurious signals due to mixing products and combinations of harmonics of the oscillator and the incoming signal. A receiver specification consists of a number of precisely defined characteristics pertaining to the rejection of all such unwanted signals.

Figures 15.132, *15.133* and *15.134* indicate the required selectivity performance in

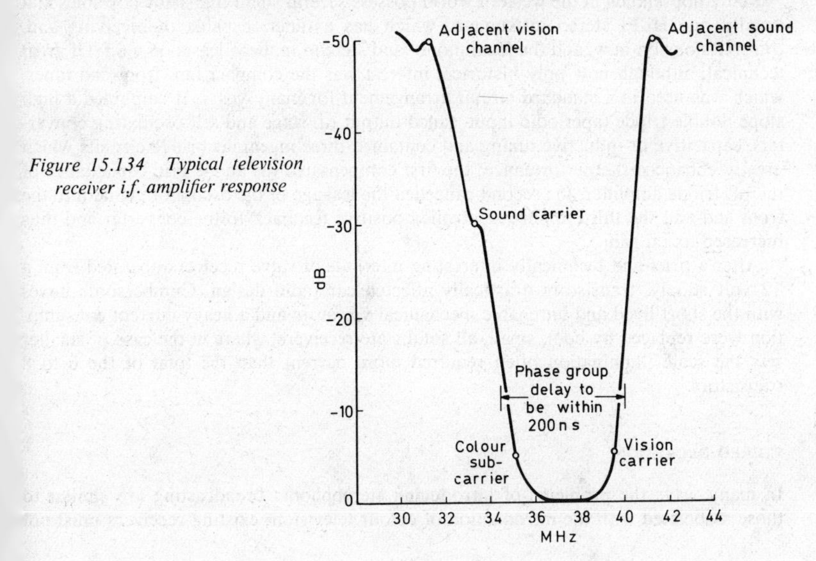

Figure 15.134 Typical television receiver i.f. amplifier response

typical broadcast- m.f., v.h.f./f.m. and television i.f. amplifiers. The response of the latter is complicated by the need for rejector circuits, to attenuate the adjacent channel vision and sound carriers and to place the wanted sound carrier on the 'intercarrier shelf' about 30 dB below the point of maximum gain. This is in order to obtain the best sound performance with the least amount of interference from intermodulation by low frequency video components (causing intercarrier 'buzz') and, in the case of colour receivers, freedom from cross-talk with the chrominance sub-carrier. The phase response in the wide band television i.f. amplifier is of crucial importance; poor phase response causes 'plastic' picture effects, overshoots and smearing of outlines; the phase response in the vicinity of the vision carrier changes extremely rapidly with slight incorrect alignment, and automatic tuning facilities or good stability are of great importance. Moreover, the amplitude and phase response should remain essentially optimised and unchanged with the gain reduction due to application of automatic gain control (a.g.c.), an integral feature of all receivers and usually obtained by utilising the rectified demodulator potential to provide control of the bias to the preceding i.f. amplifier stages.

Sound receivers

For many years (mid-1930s to the late 1950s) the basic broadcast receiver was mains operated with long, medium and, often, short wavebands. The latter usually covered from about 52 m to 16 m (5.7 to 19 MHz). This one shortwave range covered over 1 300 10 kHz channels! To improve receiver stability and ease of tuning, the so-called band spread systems were introduced, where the oscillator tuning section was so padded up by means of fixed parallel and series capacitors as to reduce the actual tuning range to a fraction of a Megahertz; in this way the narrow shortwave broadcast bands (16 m, 19 m, . . . up to the 50 m band) were separately covered. Receiver fashions began to change in the mid-fifties with the simultaneous advent of transistors—which revolutionised the portable receiver—and of high quality, interference-free f.m. broadcasts on v.h.f. (Band II). With the rapid development of television the use of sound radio declined and in the case of the great majority of listeners was used mainly for news, sport and a background of light entertainment.

Many households in the western world possess several small transistor portables and usually one Hi-Fi stereo instrument, which has a tuner, capable of receiving a.m. transmissions on m.w. and the local mono- and stereo-f.m. broadcasts on v.h.f. Of great technical, although now only historical interest was the compact f.m. front end tuner, which was used in a standard circuit arrangement for many years. It employed a high slope double-triode (aperiodic input, tuned output r.f. stage and self-oscillating converter), capacitive or inductive tuning and contained three ingenious bridge circuits which greatly enhanced the performance; the first compensated the anode-grid capacitance of the r.f. triode amplifier, the second cancelled the leakage of the oscillator signal into the front end and the third applied controlled positive feedback to the converter and thus increased its i.f. gain.

After a brief and technically interesting interlude of valve receivers operated from a 12 volt supply, transistors drastically affected car radio design. Cumbersome boxes with the short-lived and unreliable mechanical vibrators and a heavy current consumption were replaced by cool, small, all solid state receivers, where in the case of smaller sets the scale illumination often required more current than the total of the 6 to 8 transistors.

STEREO RECEPTION

In many ways the problems of introducing stereophonic broadcasting are similar to those associated with the introduction of colour television; existing receivers must not

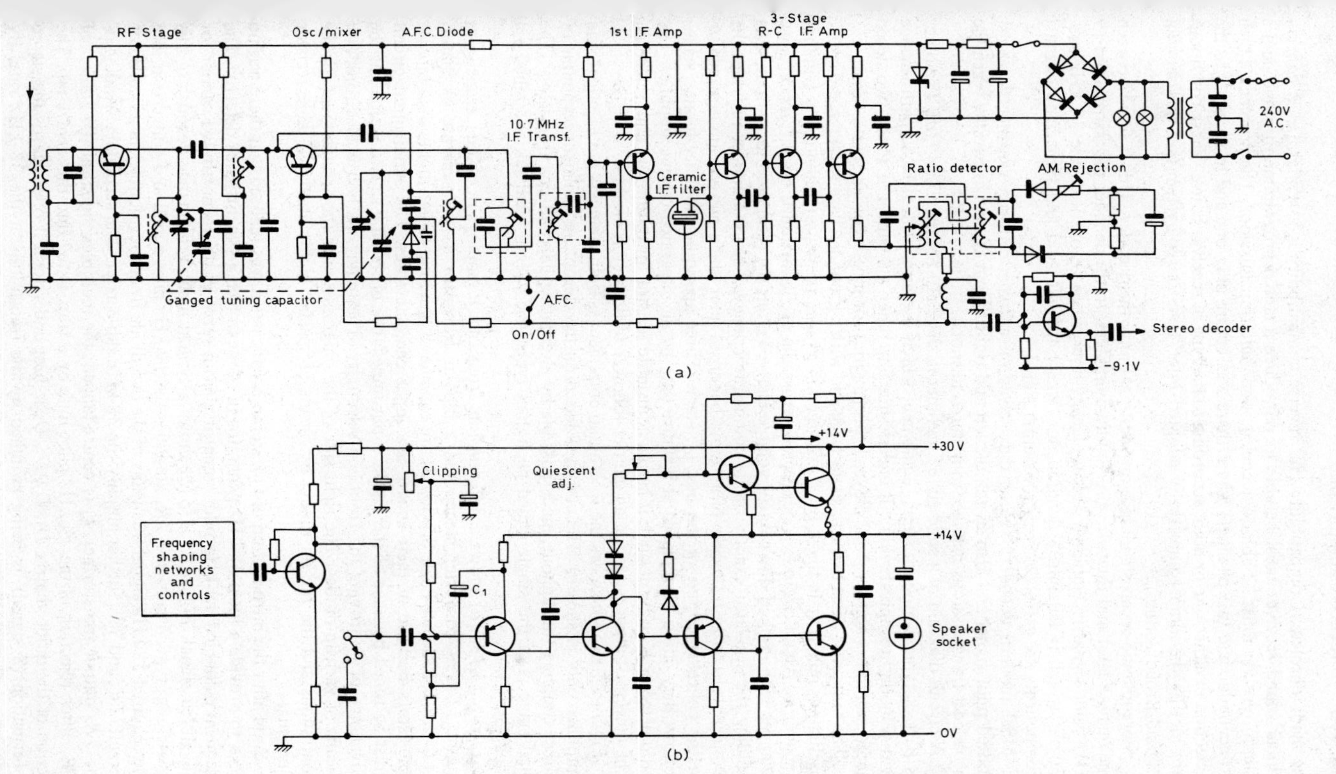

Figure 15.135 (*a*) *Decca f.m. tuner* (*b*) *Decca 6 W amplifier* (*Courtesy Decca Radio and Television Ltd.*)

be made obsolete. Compatibility and reverse compatibility is essential, i.e., a conventional sound receiver must be able to reproduce a stereo programme in mono, and a stereo receiver must be able to accept monophonic transmissions. The problem is then to accommodate two channels where there was before only one and to arrange for the selection of either one or both, according to the type of receiver in use. These conditions are met by the almost universally adopted Pilot Tone stereo system (see this section, 'Sound Broadcasting').

Stereophonic receivers include an additional unit, the decoder. This is fitted between the demodulator and the two stereo audio amplifiers. In the decoder the transmitted 19 kHz pilot signal is amplified and frequency-doubled to reconstitute the 38 kHz subcarrier which is fed into a demodulator, where the left-hand and right-hand channel components are extracted from the output signal.

Most modern decoders switch themselves on automatically on sensing the transmitted pilot signal and also switch the left and right output signals in parallel when the broadcast transmissions change from stereo to mono.

A circuit diagram of a solid state f.m. tuner is shown in *Figure 15.135* (*a*). The tuner is part of typical modern stereo equipment which has an excellent basic performance with a modest technical outlay. It includes the f.m. tuner, a 3 band a.m. tuner, two stereo audio amplifiers and a record player, with facilities for tape recorder connection.

The f.m. tuner has a two-transistor front end with a grounded base r.f. stage and a self-oscillating mixer. It is tuned by a twin-section ganged condenser, shunted by trimmer capacitors. An a.f.c. diode is connected across the oscillator tuned circuit; its capacity changes with applied d.c. control voltage, derived from the ratio detector output, thus holding the oscillator exactly on its nominal frequency. The i.f. amplifier provides the required selectivity by means of an input bandpass filter and a piezoceramic resonator in the collector circuit of the first i.f. stage. Three aperiodic amplifier-limiter stages follow; they could be replaced at a future date by an integrated circuit amplifier. The ratio detector is conventional; it has an adjustable resistor for optimising a.m. rejection. The detector output is fed via an emitter-follower to the stereo decoder.

Figure 15.135 (*b*) shows the 6 W audio amplifier (as both channels are identical, one amplifier only is shown). The circuit is based on the elegant and efficient technique of direct coupling throughout, made possible by alternating *pnp* and *npn* transistors. The final stage is of the single ended push–pull type. The overall feedback is applied via capacitor C_1.

Figure 15.136 (*a*) shows a different f.m. front end. It employs m.o.s.f.e.t. transistors in the r.f. and converter stages and a bipolar transistor as the separate oscillator. M.O.S.F.E.T. devices (see Section 8) are capable of handling larger inputs than bipolar transistors because of their square law input characteristics. The aerial input circuit must be tuned because of the higher input impedance of the m.o.s.f.e.t., which has the advantage of improving the front end selectivity. The tuning method is electronic, by means of varicap diodes. The right side of the diagram shows the arrangement for the continuous and the pre-set tuning by means of special, highly stable variable resistor elements.

Soon after the introduction of f.m. stereo broadcasts it was realised that there was no need to provide a local 38 kHz generator in order to feed a synchronous detector. The frequency-doubled 19 kHz pilot signal itself can serve as the reconstituted carrier for the stereo signal. *Figure 15.136* (*b*) shows a typical high performance decoder of this 'switching' type. Transistor TR_1 amplifies the 19 kHz pilot signal and converts it into the required 38 kHz signal by means of the two-diode frequency doubler in its collector circuit. TR_2 and TR_3 further amplify the 38 kHz sub-carrier. The secondary winding of the TR_3 output transformer T_1 is centre-tapped; the multiplex signal at the emitter of TR_1, after rejection of the 19 kHz pilot tone by a trap, is fed into this centre-tap. The action of the diode matrix D_1, D_2, D_3, D_4 is best understood if one regards the stereo information (S signal) as being switched by the transmitter at the 38 kHz rate. By connecting D_1 and D_3 to one end of the transformer secondary and D_2 and D_4 to the

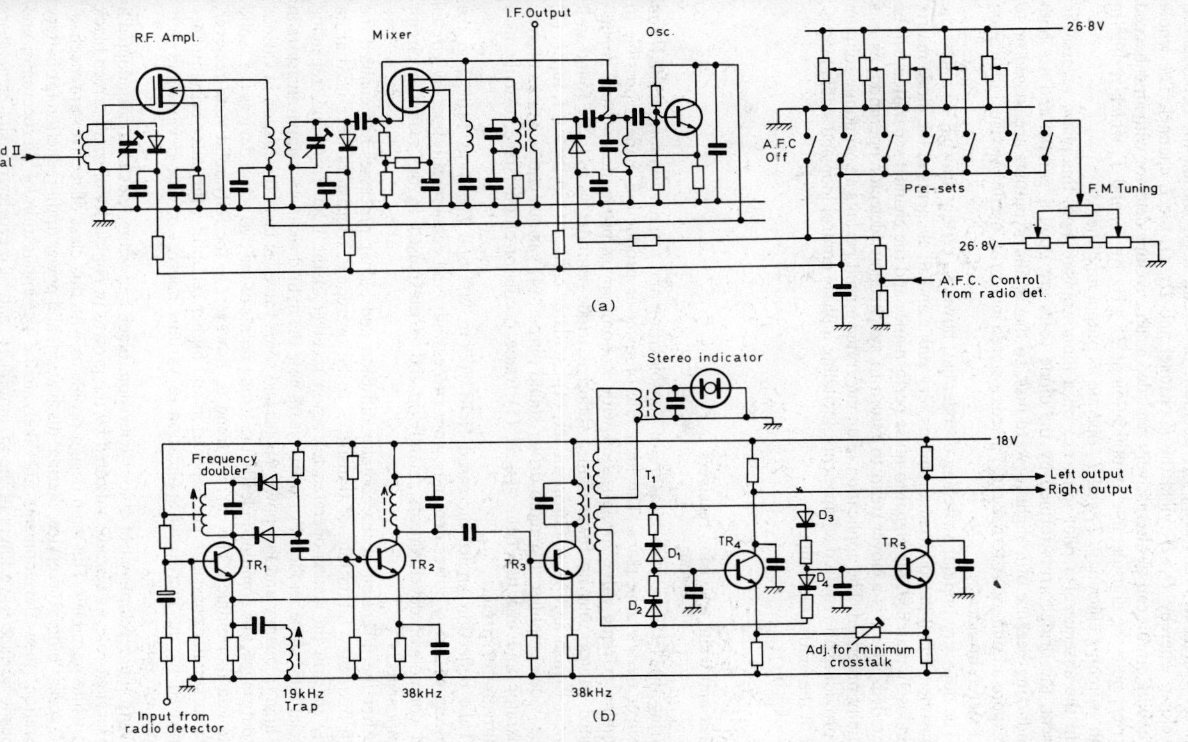

Figure 15.136 (a) Bush 'Arena' Band II varicap front end (b) Bush 'Arena' stereo decoder (Courtesy Rank Radio International)

other end, the necessary phase switching of this 38 kHz rate takes place. The effect of this reversal is to interchange the positions of L and R signals on the demodulated envelope, whereby D_1, D_2 yield the R channel and D_3, D_4 the L channel. The whole switching circuit can be alternatively regarded as an envelope detector with an enhanced carrier; note the considerable amplification at 38 kHz by T_2 and T_3 which results in high switching efficiency. T_4 and T_5 operate as output amplifiers.

In the absence of a pilot tone, i.e., on a mono-transmission, this phase reversal is absent, the bridge arms formed by the diode pairs are now in parallel and are both conducting because of a slight forward bias, so that L and R outputs are passed on to the audio amplifiers. A tertiary of T_1 feeds a stereo indicator, useful for visual indication of a stereo broadcast.

The now very popular switching decoder has, however, certain drawbacks: unless the adjustment of all tuned circuits remains very stable and all elements of the diode matrix are closely matched, the signal-to-noise performance and the channel separation will be degraded. There is a novel trend to return to synchronous detector operation, as was initially practised. The application of an integrated circuit permits an economic solution by providing all required components, including a phase control loop and stable matching of the matrix elements.

Black and white television receivers

During the 1950s a wide measure of consolidation and standardisation on television receiver design took place. The high rate of receiver production made it possible to combine extreme valve specialisation with economic mass production of very sophisticated valves; thus no fewer than 5 different triode-pentodes were to be found in a typical British and West European set: (1) oscillator-converter in the v.h.f. tuner; (2) a.g.c.-video amplifier; (3) audio driver-sound output; (4) various combinations of sync separator and line oscillator stages; (5) frame oscillator-output stage, usually in a multivibrator circuit.

Tube deflection angles increased from 70° to 90° and finally to 110°, and the growing scanning power demands were successfully and economically met by circuit and component developments, mainly in the ferrites, used in the scanning circuit components.

After the invention of the complex diffusion and planar processes had enabled the high frequency limitations of transistors to be overcome, transistors were used at an early date in the mains-operated television receiver at the u.h.f. front end, replacing special high slope triodes. For many years now the hybrid receiver, using transistors in the high frequency and low power stages and some valves of large dissipation in various output stages, has combined ideally the strong points of the two active devices (i.e., solid state and thermionic). The current trend is, however, to the fully solid state receiver, with its smaller chassis, cooler operation and higher reliability, the latter being further improved by the increasing application of various microelectronic devices and integrated circuits.

With the development of colour television, black and white television has adopted a subservient role, as already indicated by the growth in popularity of small screen black and white portables. The main design problems with this class of receiver are mechanical (accessibility for service, thermal dissipation) and power drain reduction for battery operation. New developments are likely to be the application of piezoceramic and acoustic surface wave filters in the i.f. amplifiers, and of piezoelectric e.h.t. generators for the c.r.t., particularly for smaller tubes. The predicted all solid-state flat screens ('picture on the wall') are unlikely to be a serious challenge in the foreseeable future to the well established cathode ray tube picture display.

A full circuit diagram of a modern all solid-state portable black and white receiver is shown in *Figures 15.137* (*a*) and (*b*). The u.h.f. tuner has four resonant circuits: an input circuit to the u.h.f. amplifier TR_1, a bandpass filter in its collector circuit and an oscillator circuit connected to the collector of the self-oscillating mixer TR_2. These circuits are tuned by the four-section ganged variable condenser VC_1–VC_4 and trimmed by associated four pre-set trimmer capacitors. The inductive elements are $\lambda/4$ Lecher lines in four compartments, forming high-Q cavities, the i.f. output coil and capacitors being placed in the fifth screened compartment. The primary and secondary Lecher lines of the bandpass filter are coupled through a 'window' in the separating screen. The input signal is fed via isolating and static discharge components on a separate panel to a tap on the first $\lambda/4$ line. Delayed a.g.c. is applied to the base of TR_1.

The tuner i.f. output is fed via a separate panel with four tuned circuits, which reject certain frequencies, e.g., adjacent vision and sound channels, to the three stage i.f. amplifier. The latter has wideband characteristics in the first two stages and a tuned bandpass filter in the output stage. A separate intercarrier demodulator with a 6 MHz bandpass filter leads to an integrated circuit containing a multistage amplifier-limiter and a synchronous quadrature detector, with an externally connected 6 MHz tuned circuit. A second integrated circuit with an audio amplifier-driver and an output stage follows. A number of decoupling components, associated with these two integrated circuits, are not shown in the diagram.

The video demodulator D_2 supplies via a filter network the emitter-follower TR_6 which is directly coupled to the video output stage TR_7. TR_6 also provides a signal to the first transistor of the a.g.c. unit. Tips of sync pulses cause D_3 to conduct and charge C_1. When the voltage at the base of TR_8 exceeds its emitter voltage, held constant by the Zener diode D_4, a.g.c. voltage is supplied via point A to the first i.f. amplifier TR_3, causing gain reduction due to decreased collector voltage. The second transistor in the a.g.c. unit provides a delayed control to the u.h.f. amplifier in the tuner, so that the preselection gain is not reduced before a sufficiently strong input signal assures a satisfactory signal-to-noise performance.

The collector load of TR_7 contains response shaping networks and is directly connected to the cathode of the picture tube. The tube bias is adjusted by the brightness control, and the gain of TR_7 by the contrast control. Its emitter path contains two further wavetraps, one tuned to the sound carrier (6 MHz), the other to the colour sub-carrier (4.43 MHz); sync signals from the TR_7 emitter are fed to the sync separator TR_{10}, which, together with D_5, acts as a limiter and video information remover.

Line sync pulses are fed to the phase splitter TR_{11}. A reference pulse, generated in a special winding of the line output transformer T_2, is fed to the phase detector D_6, D_7. The voltage at the point P is determined by the relative phase of the sync pulses across the collector and emitter resistors of TR_{11} and the reference pulses rectified by D_6 and D_7 and is used to initiate the precise start of line scan. The described circuit is commonly known as a *flywheel*, because it smoothes out the unavoidable slight irregularities in time base operations due to inherent internal and external noise; such irregularities show up in ragged vertical outlines of the picture content.

The line oscillator TR_{13} is tuned by L_1 and kept in step by the reactance stage TR_{12}, which is d.c. controlled by the voltage at point P. The series circuit of resistor R_1 and capacitor C_1 prevents *hunting* action of the phase control loop. TR_{13} emitter is directly coupled to the driver TR_{14}. The drive transformer T_1 is connected to the base of the line output transistor TR_{15}. The line output transformer T_2, associated with TR_{15}, feeds the line deflection coils L_2 via the adjustable linearity inductor L_3 and also generates various auxiliary supply voltages, including the e.h.t. of 12 kV for the final anode of the 14″ cathode ray tube. TR_{15} operates as an efficient switch; between input pulses it is kept in saturation, so that practically all available energy is fed into the transformer for the provision of the forward stroke current. During pulses the fast retrace takes place and the transistor is cut off, so that in both modes its dissipation is kept within modest bounds. C_2 is a 'S' corrector, which compensates for the scanning distortion occurring with all modern flat faced tubes because of the longer electron beam travel to the edges of the screen.

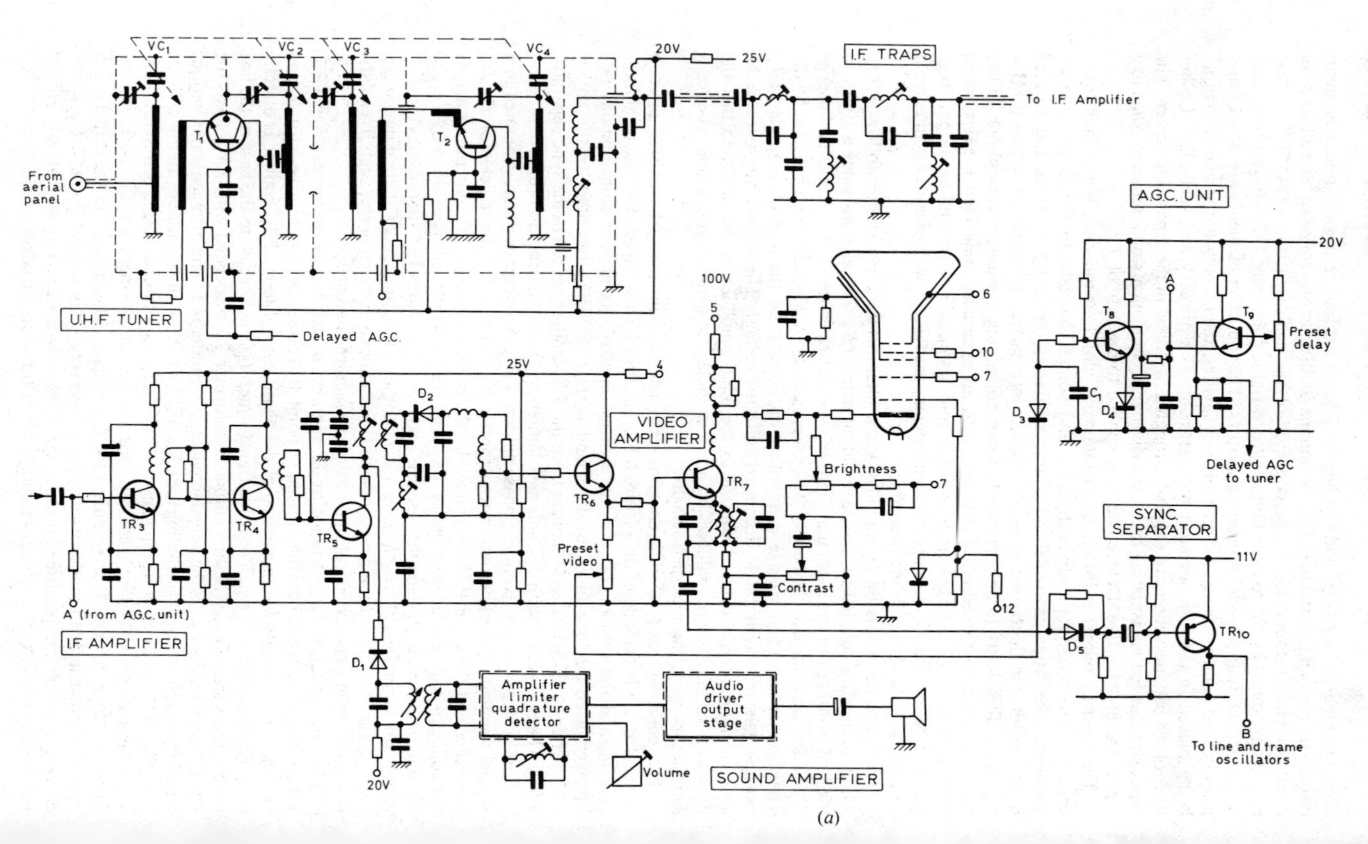
VC1
VC2
VC3
VC4
20V
25V
I.F. TRAPS
To I.F. Amplifier
T1
T2
From aerial panel
U.H.F TUNER
Delayed A.G.C.
A.G.C. UNIT
20V
A
T8
T9
Preset delay
C1
D3
D4
Delayed AGC to tuner
100V
5
6
10
7
25V
4
D2
VIDEO AMPLIFIER
Brightness
7
TR3
TR4
TR5
TR6
TR7
Preset video
Contrast
12
SYNC. SEPARATOR
11V
D5
TR10
A (from A.G.C. unit)
I.F. AMPLIFIER
D1
Amplifier limiter quadrature detector
Audio driver output stage
B
To line and frame oscillators
20V
Volume
SOUND AMPLIFIER

(a)

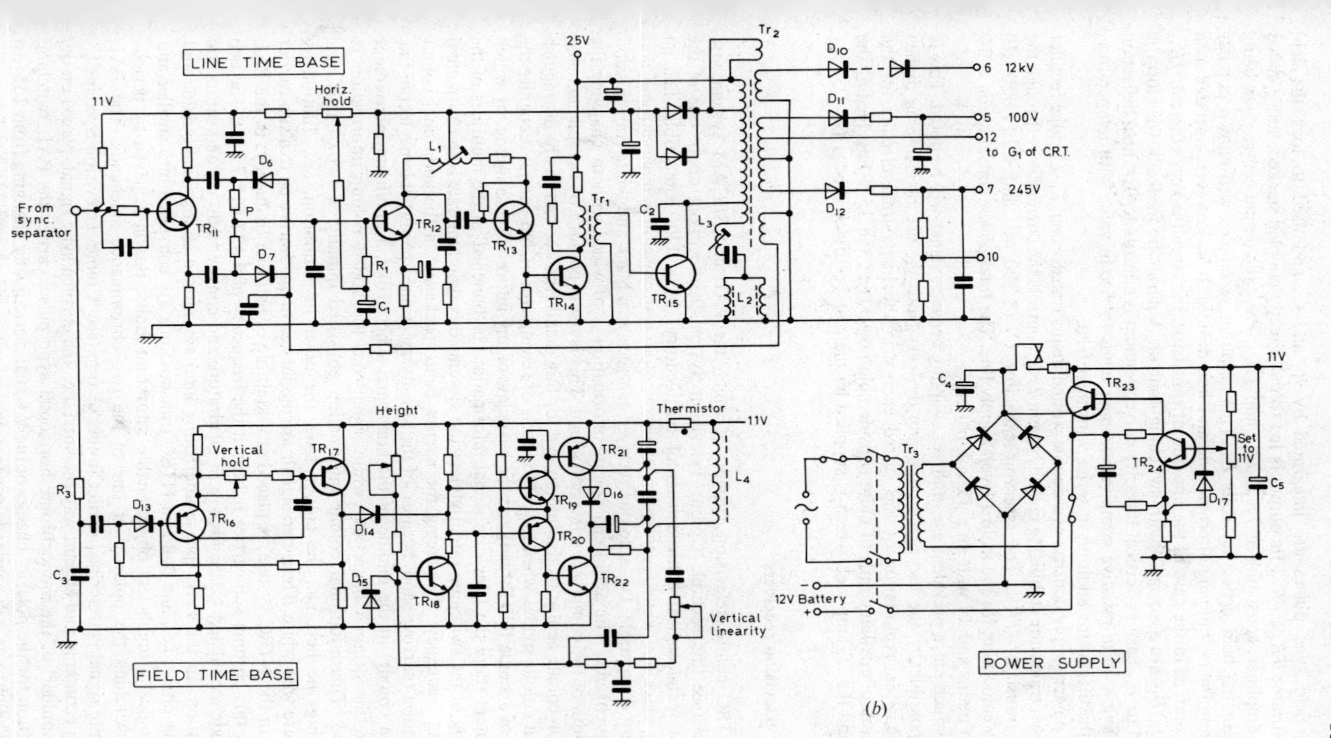

Figure 15.137 (*a*) *Circuit diagram of black and white television GEC Model 2114 (Courtesy GEC Radio and Television Ltd.)* (*b*) *Circuit diagram of black and white television GEC Model 2114 (Courtesy GEC Radio and Television Ltd.)*

The field sync pulses are integrated by R_3 and C_3 and used to lock the field multivibrator TR_{16}, TR_{17}. Its output is fed to the directly coupled stacked, single ended power amplifier, consisting of TR_{19}, TR_{21} and TR_{20}, TR_{22}, the output centre point being joined to the field deflection coils L_4. This amplifier resembles well-known popular audio amplifiers, utilising alternate *pnp* and *npn* devices. It is, however, operated as a switch and, as in the case of the line output transistor, the circuit is very efficient. TR_{18} and D_{14}, D_{15} are associated with switching control. A parabolic feedback waveform is applied to T_{18} base as vertical linearity correction and is adjustable. The field deflection coils have a large resistive component which tends to rise in value with temperature, hence the thermistor addition in series with their feed.

The power supply consists of a double-wound mains transformer T_3, a bridge rectifier and a two transistor stabilising circuit, where TR_{24} controls the current flow through the power transistor TR_{23}. The reservoir capacitor C_4 (3 300 μF) and C_5 (1 000 μF) provide smoothing for the main 11 volt supply line. The total power consumption is 36 watt on mains and 17 watt on a 12 volt battery.

The much more prevalent use of direct coupling between stages will be noted. This is made possible by the use of *npn* and *pnp* devices. Note the use of clamping diodes and the wide exploitation of the stored energy in the line output circuits for the derivation of auxiliary rails of quite considerable power, e.g., the 25 V line, feeding the tuner and the i.f. amplifier, which is supplied by the scan-rectifying diodes D_8, D_9.

Colour television receivers

The NTSC colour television system, of which the PAL and SECAM systems are variants (see this section, 'Colour Television Systems'), is without doubt one of the greatest technological achievements of our century. It is fully compatible, i.e., an existing black and white receiver can receive a colour transmission in monochrome, and reversely compatible, i.e., a colour receiver can receive a black and white transmission in monochrome. The additional colour information is provided without changing the bandwidth of the composite television signal. This is made possible by two phenomena: the poor colour acuity of the human eye for fine detail (the eye can easily distinguish different colour shades when looking at reels of sewing cotton, but cannot recognise the colour of a single thin thread) and the basic law of any information sampling, by which the content (here the television signal information) is 'bunched' at the multiples of the sampling or the switching rate. In television systems this switching rate is the horizontal scanning frequency. By placing the narrow-band 'packages' of the additional colour information (narrow-band, because of the already mentioned permissible neglect of fine detail in colour) in the middle of the horizontal scan stroke, a minimum interference between the original black and white signal and the added colour information is obtained. This additional colour information consisted initially of two signals, one describing the hue, i.e., the colour tone, the other its saturation, i.e., dilution by admixture with white. These two signals are modulated in quadrature on to a sub-carrier which, in the NTSC system, is placed at a multiple of half-line frequency; in the PAL system, this chroma sub-carrier is kept at the frequency of 4 433 618.75 Hz. The two amplitude-modulated sidebands are in quadrature and convey both hue and saturation information. In the PAL (phase alternating line) system the almost unavoidable and irritating phase, i.e., hue faults in NTSC transmissions are converted into smaller and physiologically much less objectionable saturation faults; this is achieved by periodically switching the polarity of one of the two chrominance sidebands. The PAL composite signal is described earlier in this section (see 'Colour Television Systems').

In an economic and elegant way the burst, i.e., the chroma sync signal, placed on the 'rear shoulder' of the monochrome horizontal sync pulse, carries the PAL switching information for the polarity change of the V signal, periodically changing from 135° to 225°. The chroma sub-carrier is suppressed, to further reduce possible interference in the picture; thus only saturated or near saturated colour signals are associated with a prominent chroma signal, a small signal represents pastel shades and no colour information is transmitted for black, white or any shade of grey. The receiver provides a local sub-carrier oscillator, which is locked in frequency and phase by the burst signal and used to recover the chroma modulation (the U sideband contains the red information, actually (R – *Y*); the switched V sideband the blue information as (B – *Y*); the green information G comes out after 'dematrixing' the above in combination with the luminance *Y*) in two synchronous detectors.

In order to deal with the more complex signal a colour receiver must be very well designed, manufactured and adjusted. It must also have a component and circuit stability of a whole order better than that required for black and white receivers. The following description of working principles of a modern PAL colour television receiver will illustrate this, see *Figure 15.138*. The tuner must be very stable; any mechanical or electronic channel selecting mechanisms must have a satisfactory setting-up and reset accuracy to obtain precise tuning of the vestigial single sideband transmissions; otherwise poor vision and sound quality will result and there will also be interference between the three carriers involved (vision, chroma, sound) which manifests itself in visual patterning and in spurious hum and sound distortion. Satisfactory oscillator stability is achieved by the now universal application of a.f.c. (automatic frequency control). At present all colour television receivers produced in the U.K. operate on v.h.f. bands IV and V, 625 lines only. Later, when the duplicated black and white 405-line transmission on v.h.f. cease and bands I and III are freed, it is planned to re-engineer these bands for new 625-line channels. Combined v.h.f./u.h.f. tuners, as currently in use in Western Europe, will then be employed in British receivers.

The vision i.f. amplifier is in principle the same as for black and white receivers, only again the demands on stability and phase response are higher. Much more care must be taken with demodulators. The economic usage of a common vision- and sound-i.f. detector and of the common amplification of the demodulated video signals and the intercarrier sound (this latter can be regarded as the 6 MHz sideband of the video wideband signal) in the video output stage are no longer permissible. Instead, to achieve a minimum of distortion and cross-modulation, separate vision and sound demodulators are employed. In most modern sets the video demodulator is a synchronous detector which can handle faithfully the varying modulation depth of the composite television waveform much better than the conventional envelope detector. The intercarrier sound output is fed to a multistage amplifier-limiter and then to a demodulator, usually no longer a ratio detector but again a synchronous (the so-called quadrature) detector, then to an audio driver and power output stage.

The demodulated video signal is separated into the luminance signal and the chroma signal, which consists of both sideband components and of the colour burst signal. The chrominance signal is amplified in a tuned amplifier which is protected from overload by automatic chroma gain control action (a.c.c.), obtained by a.m. rectification of the burst signal and usually applied as control bias to the first stage of the amplifier, and then fed to the two synchronous chroma demodulators, where the (R – *Y*) and (B – *Y*) signals are recovered.

The colour burst signal is also amplified and fed to the crystal-controlled reference oscillator at the sub-carrier frequency of 4 433 618.75 Hz. Note the extreme accuracy required, down to the fraction of one cycle! The reference is doubly controlled by the received burst, first in a relatively wide frequency control loop, then by a high gain narrow band phase control loop, and its output is applied as the steering vector to the synchronous (R – *Y*) and (B – *Y*) demodulators.

A separate oscillator, the so-called PAL switch, produces square pulses at half-line frequency, kept in step by suitable pulses from the line time base and pulled into correct phase by the PAL identification circuit, which analyses the phase of the sub-carrier burst. The PAL switch operates the addition and subtraction of the V signals which are obtained directly at the output of the chroma amplifier and also at the output of the PAL

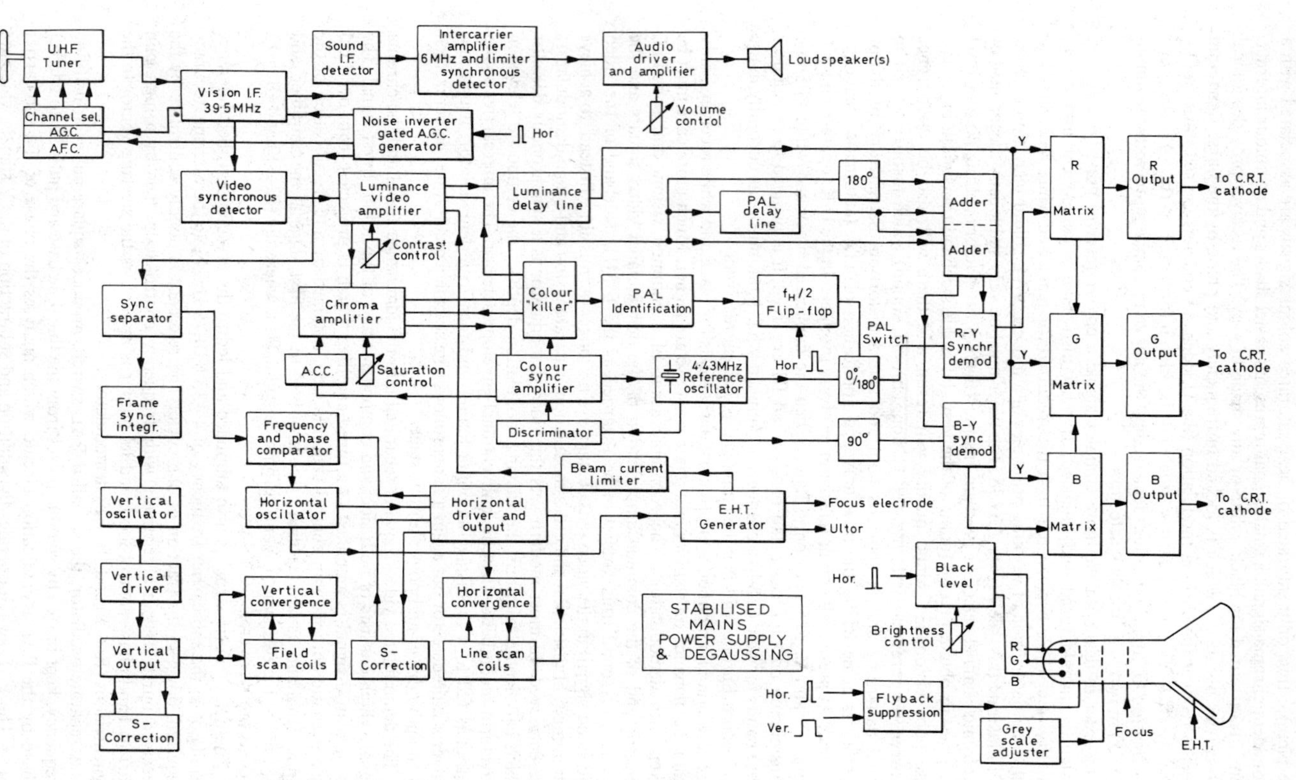

Figure 15.138 Block diagram of a PAL television receiver

delay line, the phase alternation principle relying on imperceptible information change between two successive lines in time.

The luminance signal from the video stage passes a delay line of about 1 μs which is needed to achieve simultaneous arrival of the wideband low Q luminance processing circuit signals and the narrower band chrominance signals at the output of the chroma synchronous demodulators. In an accurately dimensioned and stable resistive matrix the luminance signal Y is added to the (R – Y) and (B – Y) signals, the green signal G is recovered and the three chroma signals R, B and G are amplified in separate stages and fed to the three cathodes of the shadowmask display tube. Before the application of transistors and integrated circuits it was usual, for economic reasons, to feed the colour difference signals immediately to the cathodes of the display tube, using the grid-cathode system itself in order to recover the three colour components. Nowadays a more satisfactory chroma processing is obtained in the specially provided output stages.

In black and white transmissions the chroma burst signal is absent; this is recognised by a special circuit which initiates the *colour killer*. The colour killer turns off the chroma amplifier and 'fills out' the special sub-carrier notch filter, which is included in the luminance signal path to further reduce the visibility of the sub-carrier. Cancelling the effect of this notch filter in black and white transmissions improves the video response.

The main sync signal is fed via limiters, which remove the video content, to a conventional sync separator stage. The line sync pulses are applied to the flywheel circuit of the horizontal oscillator, where the arriving sync pulses are compared with the noise-immune gated pulses from the line time base and lock the line oscillator. Frame sync pulses, after passing an integrator stage, lock the vertical oscillator.

The time base driver and output stages are similar in principle to those in black and white receivers except for the prodigious power demands needed to scan a 110° deflection colour tube at 25 kV, and for much higher linearity demands. Modern circuit arrangements are distinguished by a large variety of designs. The vertical output stage is frequently a push–pull one, and often directly matched to the frame deflection coils. The line output stage employs one or more transistors, often in a single ended 'bean stalk' arrangement.

The stored energy during the line flyback is fully utilised for energy (current) recovery; several auxiliary windings on the output transformer provide additional supply rails for various sections of the receiver and also positive and negative going line frequency pulses for various control and protection functions. E.H.T. is either generated off the flyback as in black and white sets or obtained from a separate power stage, pulsed at line frequency rate and electronically stabilised. The timebase power stages have to be designed for fail-safe condition and be amply protected against electrical and thermal overloads. Many receiver sections rely on the control pulses and supplies from the line timebase power circuits, and special starting-up circuits are often required to prevent a 'lock-out' situation.

Once their response lag to fast switching modes was overcome solid-state electronic relays—thyristors—have been successfully applied in line output stages in place of power transistors. Their current capability is such that it is possible to provide *all* power supply requirements of the rest of the receiver from the stored-up energy during the flyback (often two thyristors, working in counter stroke, are provided).

To achieve an acceptable picture, no fewer than five classes of correction are necessary. These are: 'pin-cushion' correction, S-correction, colour purity adjustment, static convergence and dynamic convergence. 'Pin-cushion' correction is required because of the flat face of the wide angle tube, no compromise between the focusing performance and the distortionless rectangular reproduction as in black and white units being feasible. This correction is obtained electronically by modulating the scanning currents with suitable additional waveforms which are processed in a transductor and have the effect of 'drawing' the 4 tube corners in. Unfortunately, this pin-cushion distortion in colour sets is not symmetrical as for instance in high grade black and white monitors, but has a superimposed trapezoidal distortion. This explains the considerable complexity of the correction circuits and components.

S-correction in both scan directions is required, as in black and white receivers, in order to correct for the longer beam travel to the edges of the spherically curved tube face, it is undertaken by broadly similar circuit means, not requiring active components.

Colour purity adjustment aims at directing the red gun beam at the red phosphor dots, the blue gun beam at the blue phosphor dots and the green gun beam at the green phosphor dots on the tube face. It is undertaken by means of low strength magnetic rings around the tube neck, which can be jointly and separately rotated.

Static convergence deals with the correct registration of the three colour beams in the centre of the tube face, and dynamic convergence with the correct registration of the beams in the outer parts of the tube face. Correction components are assembled on a three-pronged yoke mounted on the tube neck immediately behind the scanning coils assembly, with twin coils for each colour mounted 120° apart and with permanent magnets on their outside for the static convergence adjustment. The coil currents have a precisely processed shape, built-up from sawtooth- and parabolic waveforms and provided by auxiliary circuits associated with the time base output stages.

Colour television receivers have further circuits for alignment of the three gun systems of the tube and horizontal and vertical picture shift adjustments by variable d.c. currents.

It is customary to bring out all convergence and picture-shift controls on a small control panel, so that all necessary adjustments, which have to be done in a strict and sometimes repeated sequence, can be conveniently undertaken while observing a test signal.

The receiver power supply is stabilised and equipped with fast and delayed action fuses, thermal cutouts and electronic protection devices. Where wide mains fluctuations are expected, as for instance in some export markets, more efficient supplies operate on an automatically controlled chopper type system, a technique first developed in professional equipments.

Each time a modern colour receiver is switched on, its tube is automatically *degaussed*, i.e., deleterious effects of spurious external magnetic fields, e.g., the terrestrial field, are removed by a brief application of a heavy alternating current through special coils surrounding the tube.

It is impossible to give a full description of the complete circuit diagram of a modern colour television receiver within the framework of this section, and the interested reader is advised to consult the available technical literature on the subject (see 'Further Reading'). However, some selected sections of the modern Thorn 4000 receiver are briefly explained. This receiver is distinguished by its wide application of integrated circuits and thick film techniques and by modular design for ease of servicing, as clearly shown by photograph of *Figure 15.139*. The set is intended for global use, wherever the PAL transmission systems exist; it has therefore a combined v.h.f./u.h.f. tuner and is capable of operating from fluctuating mains voltages.

The tuner's programme selection is by finger touch; 8 touch pads are connected to a high impedance input m.o.s. integrated circuit and associated active and passive networks, which switch to appropriate bands and channels, provide neon lamp indication, disable a.f.c. and mute the speaker during the tuning process. The tuner has no moving mechanical parts or switches; tuning is by voltage dependent varicaps, band switching from v.h.f. to u.h.f. is by transistor logic, band switching within the v.h.f. band is by d.c. actuated PIN diodes. A modern television front end is a mass produced sub-assembly of optimised performance and requires very elaborate test equipment for alignment and checks of noise performance and stability.

Figure 15.140 is a photograph of a typical piece of specialised test equipment, a large

Figure 15.139 Rear view of Thorn type 4000 colour receiver (Courtesy Thorn Consumer Electronics Ltd.)

Figure 15.140 Test equipment for the alignment and checking of a television receiver tuner (Courtesy Thorn Consumer Electronics Ltd.)

display wobbulator for the alignment and checking of such a tuner. Crystal-controlled markers are provided for the exact setting of various tuning circuits and the large and controlled dynamic range of the *Y*-amplifier permits precise checking of various rejection and attenuation characteristics.

The three-stage i.f. amplifier is conventional; it is followed by a complex integrated circuit which contains a synchronous video detector, a series of a.g.c. sensing and delayed circuits for the gain control of the first and second i.f. stages and a separate control of the tuner r.f. stage, the a.f.c. discriminator and amplifier, gated noise suppression circuits and two video output stages, one for the luminance and sound signals, the other for chrominance and sync signals.

The sync separator and horizontal oscillator with its associated frequency and phase control circuits are part of another special integrated circuit. The frame time base uses a *npn* and a *pnp* push–pull pair of power transistors, directly coupled to the scanning coils. The line timebase is split at the driver transformer level; one secondary feeds the line output transistor which has a collector-emitter rating of 1 500 V and can pass a 5 A peak current. It supplies the deflection current for the line scan coils and various auxiliary waveforms and voltages with the aid of a ferrite-cored output transformer. The other secondary is linked to the slightly less prodigious transistor associated with a Cockcroft–Walton tripler for the generation of 25 kV for the c.r.t. final anode; the stage also supplies high voltages for the focus electrode and the first anode of the c.r.t. Both timebases make extended use of high-dissipation, high-stability thick film components. The e.h.t. is electronically stabilised. Both line and field pin-cushion correction systems employ active components for the shaping of the required parabolic currents which are fed to the bases of the respective output transistors.

The schematic diagram of the static and dynamic systems for the radial and lateral convergence is shown in *Figure 15.141.*

The three-cornered yoke unit around the c.r.t. neck is situated immediately behind the deflection toroid and carries on each of its pole pieces a pair of convergence coils: three large windings for static convergence and three small coils for the dynamic convergence. The sub-assembly also carries the blue 'lateral' coil and purity adjusting magnets.

Figure 15.142 indicates circuit techniques adopted for parts of these convergence circuits. The unit shown is based on two thick film assemblies, with their specific

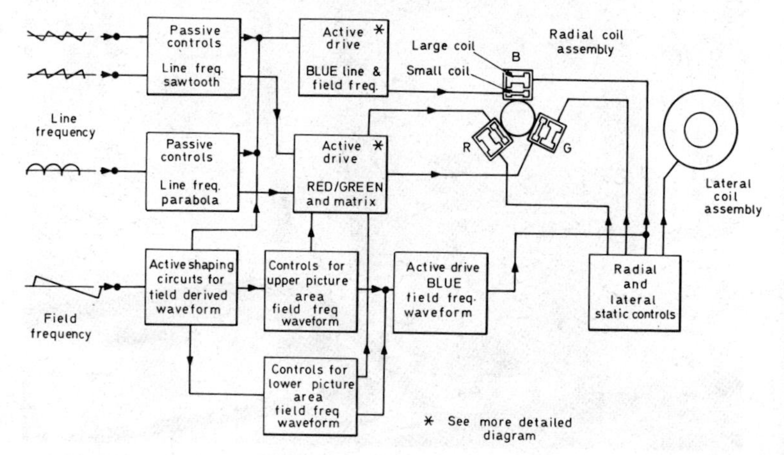

Figure 15.141 Thorn type 4000 colour receiver general schematic of the convergence system (Courtesy Thorn Consumer Electronics Ltd.)

advantages of excellent thermal conductivity, precision-trimmed and very stable resistor elements and of thermal balance due to the common substrate. While the left (green) and the right (red) amplifier pair associated with the first thick film assembly look very similar, their bias and feedback arrangements are different and so chosen that an input signal at point H generates identical current changes (increasing or decreasing) in the small red and green convergence coils; thus picture elements of the green and red rasters shift horizontally with respect to each other.

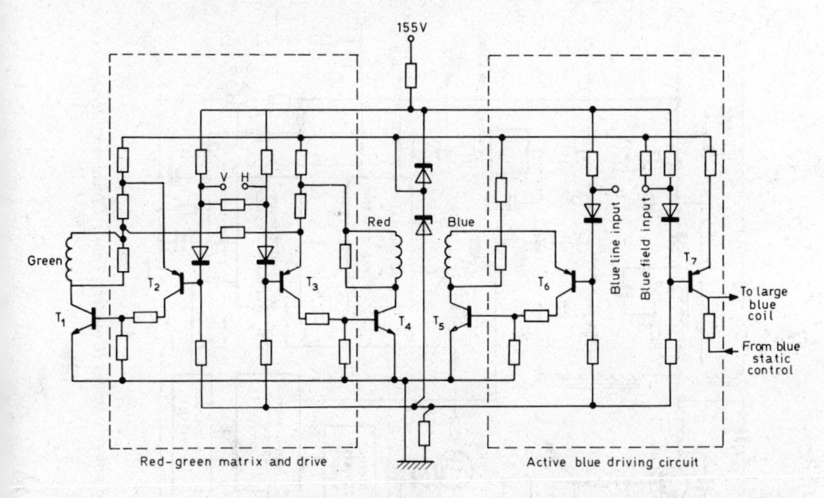

Figure 15.142 Convergence circuits of the Thorn type 4000 colour receiver (Courtesy Thorn Consumer Electronics Ltd.)

An input signal at point V causes similar currents to flow, but in mutually opposite directions; thus horizontal green and red raster lines shift mutually in a vertical direction.

The second thick film assembly contains the drive for the small blue convergence coil and also the generator for the large blue coil. Note the different arrangements for blue controls. 'Blue' is the odd man out in colour receivers, both in the chroma signal processing (decoding and matrix circuits) and in the attainment and adjustment of static and dynamic convergence.

Figure 15.143 shows the complex power supply system, capable of working the receiver off 180 V to 265 V a.c. Direct mains voltage is fed via the auto-transformer-acting primary of the small transformer T_1 to the bridge rectifier B_1. The potential across the reservoir capacitor C_1 can vary between −230 V and −350 V. The chopper-mode controlled supply system is based on the power transistor TR_1, driven through T_2 by the driver TR_2 from circuits in an integrated circuit. This integrated circuit contains the line oscillator, phase detector and the elements of the flywheel for feeding the line timebase drive, a line-pulse-timed and receiver-load regulated Schmitt trigger mark-space generator, various safety devices such as an electronic switch responding to overloads and fault conditions, a slow warm-up current limiter, which is controlled from the direct mains supply via the bridge rectifier B_2, an internal 13 V stabilised supply and even the sync separator (not shown).

The basic principle of the chopper mode stabilisation is to control the current flow into the transformer T_3 and hence its power output by pulsing the current source TR_1. This pulse rate is varied by the mark-space ratio drive supplied from the integrated circuit after sensing various input parameters. The stabilisation process is highly

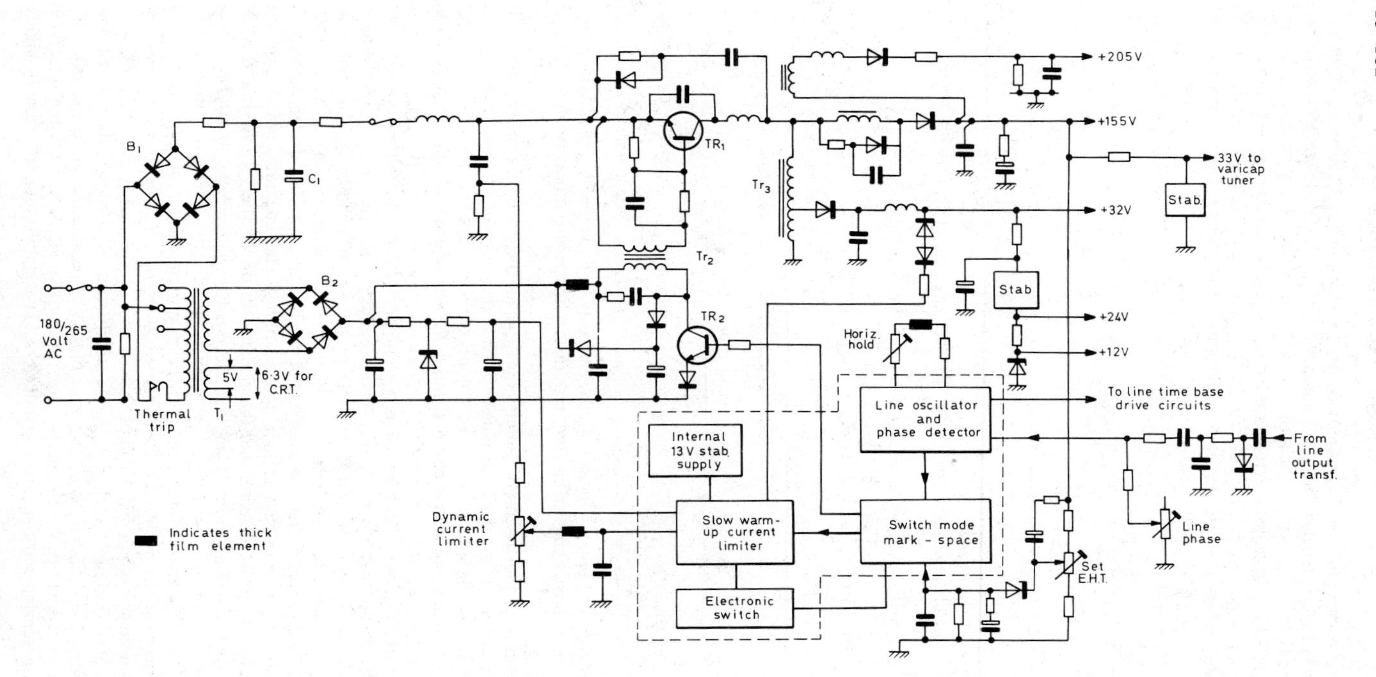

Figure 15.143 Thorn type 4000 colour receiver—schematic of power supplies (Courtesy Thorn Consumer Electronics Ltd.)

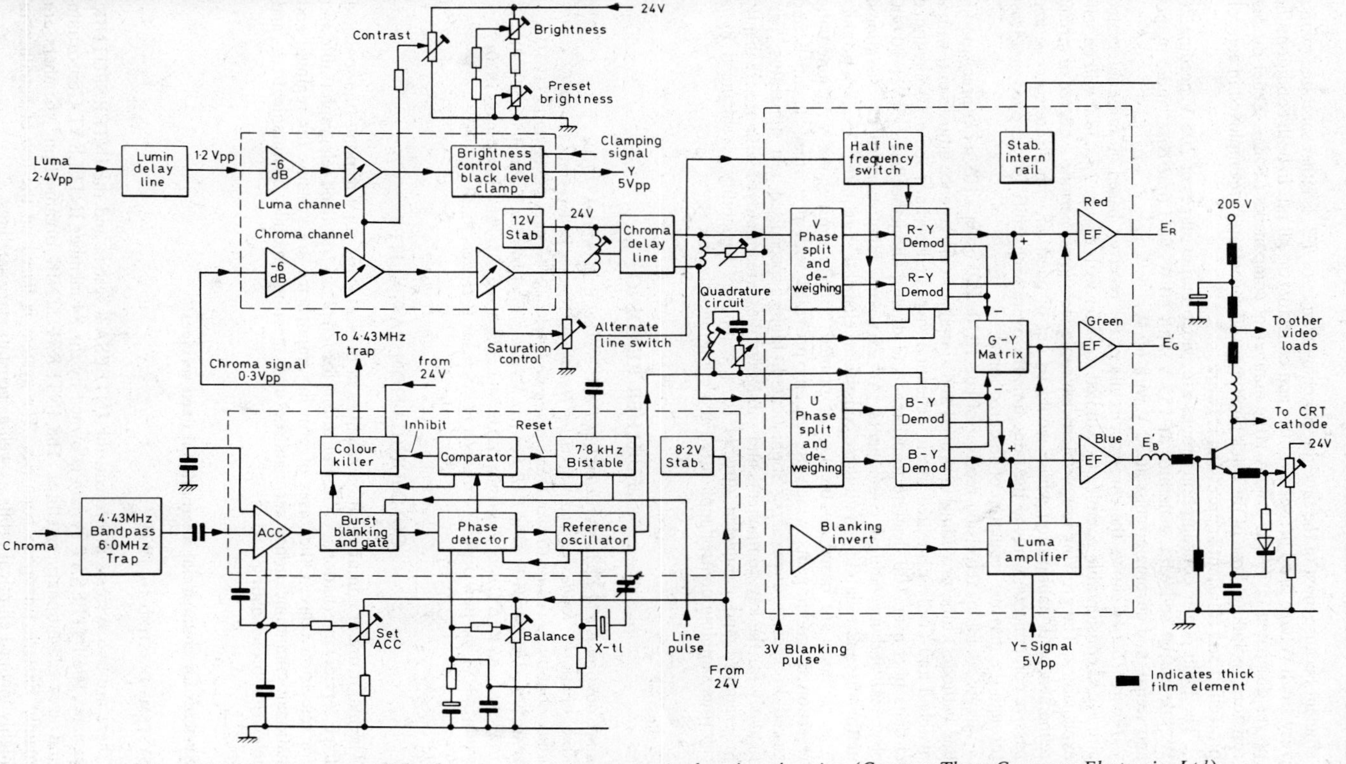

Figure 15.144 Thorn type 4000 colour receiver—chroma circuits, based on three i.c.s (*Courtesy Thorn Consumer Electronics Ltd.*)

efficient because most of the time the switching transistor TR_1 is either saturated (with a very small voltage drop between emitter and collector) or cut off. In both conditions the power dissipated in TR_1 is low. A large number of components (diodes, Zener diodes, resistors, including those of the thick film type, and capacitors) are employed in control, protection and transient-suppression networks.

The main power consumption groups are: video output stages at 205 V needing 20 W, line output and e.h.t. generator at 155 V with 80 W, field time base, audio output and signal- and chroma-circuits at 32 V with 60 W.

Figure 15.144 shows the chroma circuits of the receiver, based on three integrated circuits of considerable complexity. Studied together with *Figure 15.138* it will be self-explanatory. The growing use of microelectronic techniques permits a degree of circuit refinement and stabilisation which would be out of question with discrete active and passive components, while at the same time the overall reliability is improved, not least through the massive reduction of soldered joints.

A typical colour television receiver of late 1970s will use a 110° p.i.l. (precision-in-line cathodes) tube providing improved and simplified convergence adjustment, with a matched and firmly cemented toroid deflection yoke and perhaps some 10 integrated circuits plus over 100 discrete transistors and diodes, altogether having over 10 000 electrical and mechanical components. An extremely high standard of operational uniformity and reliability in all these constituent elements is required. Modern production methods are based on the use of printed circuits throughout, with automatic insertion of components which are usually provided with preformed terminals and stored in special dispensers. Sophisticated, highly specialised test apparatus, particularly for the key components and sub-assemblies, are employed with are frequently programmed and connected directly to a computer. Reliability is enhanced by a rigorous sampling control application and comprehensive life tests.

FURTHER READING

KING, GORDON J., *FM Radio Servicing Handbook*, Newnes-Butterworths (1970)
KING, GORDON J., *Radio and Audio Servicing Handbook*, Newnes-Butterworths (1970)
KING, GORDON J., *Newnes Colour Television Servicing Manual*, Newnes-Butterworths (1973)
SEAL, D. J., *The Mazda Book of PAL Receiver Servicing*, Foulsham Technical Books (1972)
TOWNSEND, B., *PAL Colour Television*, I.E.E. Monograph Series 3, Peter Peregrinus (1970)
WHARTON, W. and HOWARTH, D., *Principles of Television Reception*, Pitman (1967)

COMMUNICATION SATELLITES

The major use of communication satellites at present is for point-to-point (the so-called fixed service) international communications for telephony, data and telegraph traffic and for television relay. Other uses of growing importance are systems for regional and national communications, aeronautical and maritime mobile communications and television broadcasting.

Systems for commercial international fixed services[1]

SYSTEMS EXISTING IN 1975

There are two systems at present, the INTELSAT system and the INTERSPUTNIK system. The INTELSAT system is owned by the 84-member INTELSAT Consortium and uses geostationary satellites. The INTERSPUTNIK organisation[2] members are mainly Eastern European countries. The system is based on the use of satellites in highly inclined elliptical orbits as well as geostationary satellites.

FREQUENCY BANDS[3]

Of the bands allocated to communication satellites the following, or parts of them, are the ones used or likely to be used for commercial systems:

Up paths (GHz)	*Down paths* (GHz)
5.925–6.425	3.7–4.2
14.0–14.5	10.7–10.95 and 11.2–11.45
12.5–12.75 (use depends on region)	
17.7–21.2	27.5–31.0

Most of these bands are shared with terrestrial services in which cases there are limits on transmitted powers and the power flux density a satellite may set up at the Earth's surface.[3] In the 4 GHz band the limit is -152 dBW/m^2/4 kHz below 5° arrival angle, rising to -142 dBW/m^2/4 kHz at 25° arrival angle and above. In the 11 GHz bands it is 2 dB greater than in the 4 GHz band, and in the shared part of the 20 GHz band approximately 11 dB greater still (but specified for a 1 MHz rather than 4 kHz bandwidth).

ORBITAL CONSIDERATIONS

By far the most useful orbit for communication satellites is the geostationary satellite orbit, a circular equatorial orbit at approximate height 36 000 km for which the period is 23 hours 56 mins, the length of the sidereal day. A satellite in this orbit remains approximately stationary relative to points on the Earth's surface. There are two main perturbations to the orbits of such satellites.

(1) A drift in orbit inclination out of the equatorial plane due to effects of the Sun and Moon. This is at the approximately linear rate of 0.8° a year for small inclinations. If not corrected it causes the satellite to move around a progressively increasing daily 'figure of eight' path as viewed from the Earth.

(2) Acceleration of the satellite in longitude towards one of the two stable points at 75° E and 105° W longitude. This is caused by non-uniformity of the Earth's gravitational field.

LAUNCHING OF SATELLITES

The launcher generally places the satellite first into a low altitude circular inclined orbit and then into an elliptical transfer orbit with apogee at the altitude of the geostationary orbit. At an appropriate apogee the apogee boost motor, generally solid fuelled and forming part of the satellite itself, is fired to circularise the orbit and remove most of the remaining orbit inclination. The satellite's own control system is used to obtain the final desired orbit and maintain it.

Some U.S. launchers suitable for communications satellites, together with their approximate geostationary orbit payload capabilities, are the Delta 2914 (340 kg), Atlas Centaur (790 kg) and Titan IIIc (2 300 kg). A Delta 3914 with 400 kg capability is expected to become available before long.

SATELLITE STABILISATION AND CONTROL

Stabilisation of a satellite's attitude is generally necessary since, for high communications efficiency, directional aerials must be used and pointed at the Earth. Most geostationary communication satellites are spin-stabilised but stabilisation in three axes (body-stabilisation) is expected to be used for some missions after the mid-1970s.

In a spinning satellite the body of the spacecraft spins at typically 50–100 r.p.m. about an axis perpendicular to the orbit plane, but the aerial system is generally de-spun. A reference for the control system is usually obtained primarily by infrared earth sensors supplemented by sun sensors. Antenna pointing accuracy of typically $\pm 0.3°$ or better is obtained through the antenna despin control electronics and by occasional adjustments to the direction of the satellite's spin axis.

Body-stabilised designs generally employ an internal momentum wheel with axis perpendicular to the orbit plane. Control about the pitch axis is through the wheel's drive motor electronics, while control about the yaw and roll axes (necessary because of perturbations to the wheel axis direction) may be by gimballing the wheel or by use of hydrazine monopropellant thrusters to correct the axis direction. In any case thrusters must be used for occasional dumping of momentum.

The orbit of the satellite must also be controlled and this is achieved by ground command of hydrazine thrusters. The mass of hydrazine required for a 7-year life, expressed as a percentage of total satellite mass, is approximately 12% for N–S station-keeping (inclination control) and less than 1% for E–W station-keeping. In a spinning satellite axial thrusters are used for inclination corrections and radial thrusters, operated in a pulsed mode, for longitude corrections. In the future electric thrusters are likely to be used for N–S station-keeping.[4]

SATELLITE POWER SUPPLIES

Silicon solar cells are the accepted source of primary power, except during eclipse when power is maintained by Nickel–Cadmium cell batteries. Eclipse occurs about eighty nights a year with maximum duration about seventy minutes.

Spinning satellites have body-mounted solar cells producing about 25 W per square metre of cell array at end of life (seven years assumed). Body-stabilised satellites can more readily use extendible arrays, rotated so as to always face the Sun and therefore giving approximately three times the power output for the same array area. Outputs of up to 50 W per kg mass have been claimed for extendible arrays using special lightweight cells.

TELEMETRY AND COMMAND (T & C)

T & C signals generally occupy a narrow channel within the communication band but outside the communication channels. T & C facilities are essential for control and monitoring of many functions within the satellite. For example in the INTELSAT IV satellite provision is made for 223 command channels and for 64 8-bit telemetry words.

SATELLITE AERIALS

A circular radiation pattern of 17.5° beamwidth is just adequate to cover the area of the Earth visible from a geostationary satellite. This corresponds to a beam-edge gain (relative to isotropic) of approximately 16 dB and is generally provided by means of a horn antenna. Spot-beam aerials, which may be steerable, are used for increasing the gain and hence the effective isotropically radiated power (e.i.r.p.). On more advanced satellites spot-beam aerials are used for both reception and transmission to permit re-use of the frequency band by relying on the high degree of isolation possible if the beams have adequate angular separation. Spot-beam aerials are generally front-fed paraboloids. The beam-edge gain of such aerials is approximately $41-20\log_{10}$ (beam-width in degrees) dB, regardless of frequency.

The aerial types used in INTELSAT satellites to date have been as follows. The INTELSAT I (Early Bird) and INTELSAT II satellites used linear arrays of dipoles (not despun) producing toroidal radiation patterns, the INTELSAT III satellite has a despun-global coverage horn while INTELSAT IV has global horns and two 4.5° spot-beam paraboloids all mounted on a despun-platform which also carries the repeaters.

A particular aerial implementation combining large area coverage with frequency re-use between east and west is to be used on the INTELSAT IV-A satellite. The coverage areas in this case are obtained by composite spot beams formed with multiple feeds illuminating large paraboloids, selected feeds being fed in quadrature with others.

SATELLITE REPEATERS

The essential purpose of the repeater is to translate the received signals to a new frequency band and amplify them for transmission back to the Earth. In early satellites this process was carried out broadband but modern satellites generally separate the signals into a number of channels at least for part of the amplification. The INTELSAT IV repeater[5] will be described in order to illustrate typical techniques. *Figure 15.145* shows the basic elements. Signals in the band 5.932 to 6.418 GHz from the receive aerial are first amplified in a 6 GHz tunnel diode amplifier (t.d.a.) and then frequency translated by 2 225 MHz for further broadband amplification in the 4 GHz t.d.a. and low-level travelling-wave tube (t.w.t.). This part of the repeater sub-system has fourfold redundancy to achieve high reliability. The signals are then split by a circulator-filter dividing network into 12 channels of 36 MHz bandwidth and 40 MHz spacing between centre frequencies. The channels have gain-control attenuators with eight 3.5 dB steps, except channels 9 to 12 where there are only 4 steps. These are followed by the high-level (6 watt) t.w.t.s (and standby) and then the switches selecting the transmit aerials. Channels 1, 3, 5 and 7 can be connected to global-beam aerial or spot-beam aerial no. 1, channels 2, 4, 6 and 8 can be connected to the other global-beam aerial or to spot-beam aerial no. 2 and channels 9, 10, 11 and 12 are permanently connected to the global-beam aerials. No two adjacent channels are ever connected to the same aerial which simplifies the output multiplexer filter design.

The principle characteristics of the repeater and aerial sub-system are as follows.

Receive system gain-to-noise temperature ratio (G/T) −17.6 dB/K.
Flux density for t.w.t. saturation −73.7 to −55.7 dBW/m^2.
Transmit e.i.r.p.—global beam 22 dBW per channel.
—spot beam (4.5°) 33.7 dBW per channel.
Receive aerials are left-hand circularly polarised.
Transmit aerials are right-hand circularly polarised.

Other specifications relate to gain stability, gain slope, group delay response, amplitude linearity and amplitude to phase modulation transfer (a.m. to p.m.).

The above relates to INTELSAT IV but alternative arrangements are possible. In the European Orbital Test Satellite (OTS) for example,[6] besides operating in the 11/14 GHz rather than 4/6 GHz frequency bands the receiver employs a parametric amplifier instead of a t.d.a. and a double conversion repeater is used, the bulk of the amplification taking place in a broad-band intermediate frequency amplifier operating at u.h.f. Also the channels connected to a 2.5° spot-beam aerial have 108 MHz bandwidth instead of the 36 MHz bandwidth used for channels connected to the *Eurobeam* aerial. Finally, the repeater syb-system is completely duplicated since the satellite employs frequency re-use through polarisation discrimination.

In future satellites the repeater arrangements are likely to be further complicated by frequency re-use between spot beams as well as, or instead of, re-use on separate polarisations. The use of more than one set of frequency bands in the same satellite is also likely—e.g., 4/6 GHz and 11/14 GHz—possibly interconnected with each other.

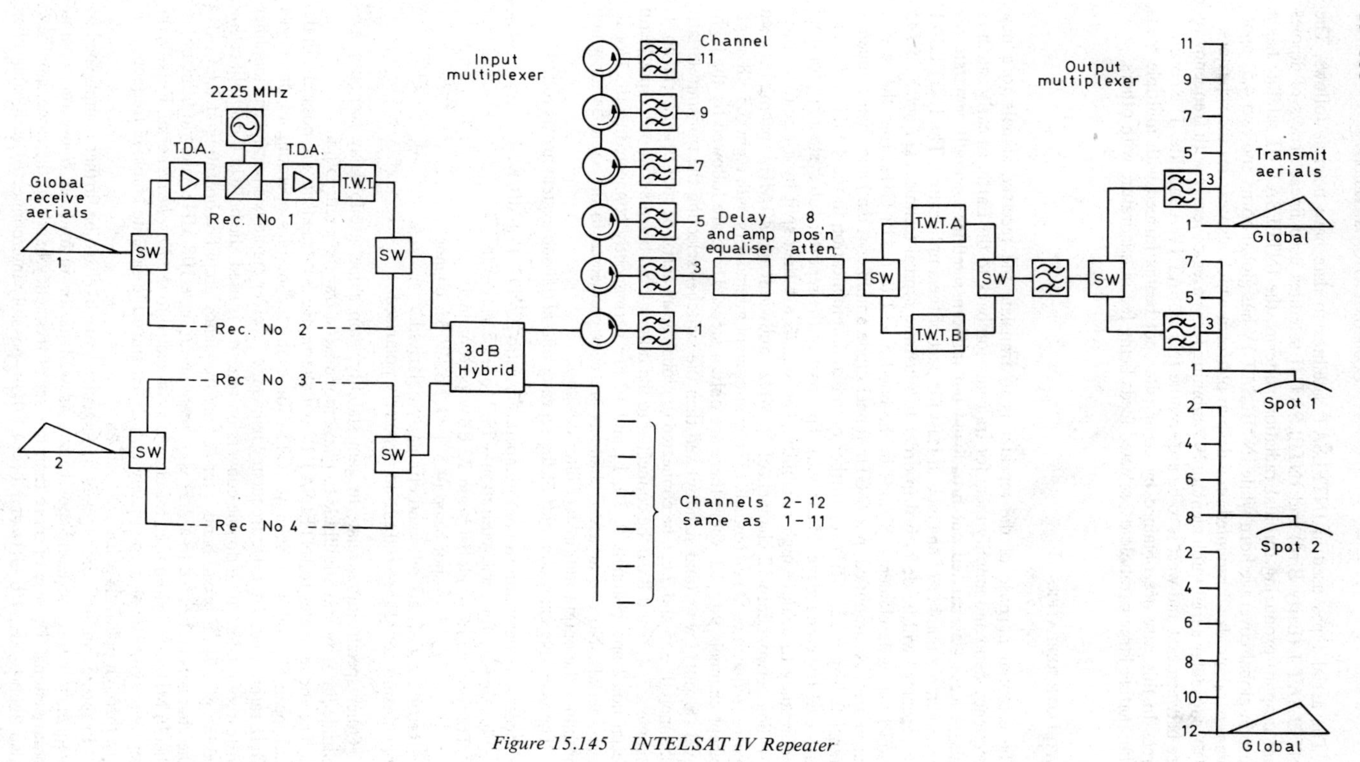

Figure 15.145 INTELSAT IV Repeater

In all these cases the main complication to the repeaters arises through the need to achieve an adequate degree of connectivity between receive and transmit beams, together with flexibility to meet changing requirements.

In the even longer term satellite repeaters may be completely revolutionised by the use in the satellite of time-division switching between beams.

MULTIPLE ACCESS METHODS

In a communication satellite system many earth stations require access to the same repeater sub-system and usually also to the same r.f. channel in a channellised satellite. Currently frequency division multiple access (f.d.m.a.) is used in which the r.f. carriers from the various stations are allocated frequencies according to an agreed frequency plan. When several carriers are passed through the same t.w.t. the non-linearity and a.m. to p.m. conversion cause intermodulation which adds considerably to the noise level.[7, 8] It becomes necessary to reduce the t.w.t. drive level well below the saturation point—so reducing the output power. At the optimum operating level the intermodulation noise is usually of the same order as the down-path thermal noise. Typical output power back-off for INTELSAT IV is 4–5 dB for global-beam channels and 8–9 dB for spot-beam channels.

Alternatively, for digital signals, time division multiple access (t.d.m.a.) can be used,[9] each station transmitting its digital traffic in a short burst during the portion of the overall time frame allocated to that station. The frame format of a typical t.d.m.a.

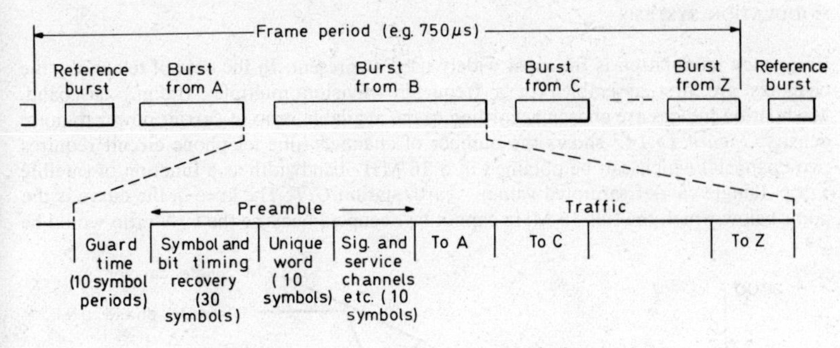

Figure 15.146 Frame and burst format of t.d.m.a. system

system is shown in *Figure 15.146*. A station synchronises its bursts by monitoring the position of the burst received back from the satellite relative to the position of the reference burst. One particular station is nominated as the reference station.

Most multiple access systems have fixed assignment of telephone channels, with enough channels allocated to meet the peak demand on each route. For low capacity routes it may be more economic to use 'demand assignment' systems in which channels from a pool are allotted to pairs of earth stations as the demand arises. The INTELSAT SPADE system[10] is of this type.

TRANSMISSION EQUATION

The carrier-to-noise power ratio, C/N, in bandwidth B Hz, for either the up-path or down-path is given by:

$$C/N = \text{e.i.r.p.} - L - LA + G/T - 10 \log_{10} k - 10 \log_{10} B \text{ dB}$$

where

e.i.r.p. = transmitter power × aerial gain, dBW.

L = free-space path loss, $20 \log_{10} (4\pi d/\lambda)$, dB.
LA = atmospheric loss (0.3 dB at 5° elevation in clear weather at 4 GHz).
G/T = receive system aerial gain-to-noise temperature ratio, dB/K.
k = Boltzmann's constant, 1.37×10^{-23} W/Hz/K.

The C/N ratio available in a satellite link is used to determine the channel capacity and performance of the link.

PERFORMANCE OBJECTIVES

The *C.C.I.R.* recommends performance objectives for international telephone and television connections via satellite.[11] For a telephone channel the objective most important for the system designer is that the psophometrically weighted noise power must not exceed 10 000 pW for more than 20% of any month. For television the corresponding objective is for the weighted luminance channel signal-to-noise ratio not to exceed 56 dB (525-line system M), or 52 dB (625-line systems B, C, G, H and I) for more than 1% of any month.

Corresponding objectives for digital systems have not yet been recommended by *C.C.I.R.* For p.c.m. telephony system designers frequently adopt bit-error rates of 10^{-7} or better for 20% of any month and 10^{-4} for 0.1% or 0.3% of any month.

MODULATION SYSTEMS

Frequency modulation is the most widely used at present. In the case of telephony the channels are first assembled in a frequency division multiplex (f.d.m.) baseband. Modulation indices are chosen according to the available ratio of carrier power to noise density. *Figure 15.147* shows the number of channels (one telephone circuit requires two channels) which can be obtained in a 36 MHz bandwidth as a function of satellite e.i.r.p. (single carrier saturated value) + earth station G/T. The knee in the curve is the point below which the full 36 MHz cannot be occupied because the C/N ratio would be

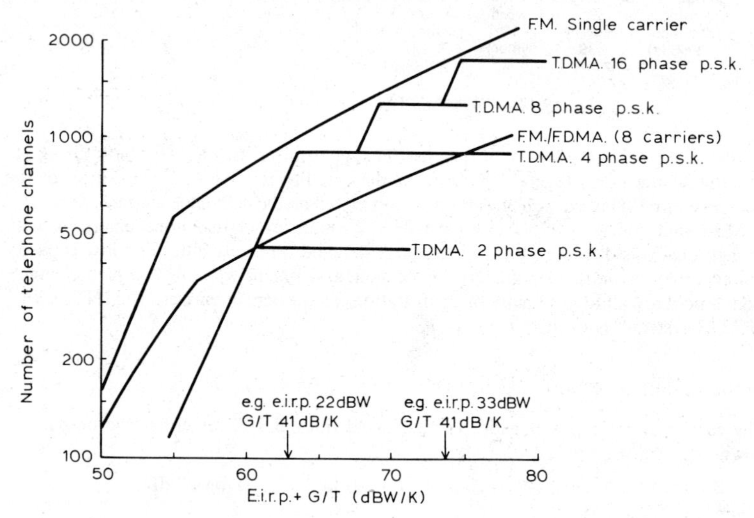

Figure 15.147 Telephone channel capacity in 36 MHz channel

below threshold, even assuming the use of threshold-extension demodulators. Methods of calculating performance, etc., are given in references 12 and 13.

A typical allocation of the 10 000 pW of noise allowed in the telephone channel would be as follows:

Up-path thermal	1 000 pW	Down-path thermal	3 000 pW
Earth station-intermod.	500 pW	Satellite intermod.	2 500 pW
Earth station—other	1 000 pW	Terrestrial interference	1 000 pW
Other satellite system interference	1 000 pW		

In the case of television frequency modulation is again the universally adopted method at present. Reference 13 gives a method for calculating performance.

Table 15.14 gives the parameters used in INTELSAT IV for a typical f.m. carrier (132 channels). Other telephony carriers used range from 24 channels (in 2.5 MHz global) to 1 872 channels (in 36 MHz spot-beam). Television is normally transmitted in 30 MHz bandwidth giving 54 dB *S/N* ratio. Alternative TV parameters for 17.5 MHz bandwidth are also used allowing two 49 dB *S/N* ratio video signals to be passed through a single global-beam transponder.

Table 15.14 EXAMPLE OF MODULATION PARAMETERS USED WITH INTELSAT IV

Carrier type	*Bandwidth allocated*	*Bandwidth occupied*	*Baseband limits*	*Deviation* (rms)	*Overall C/N ratio*
132 ch. global	10 MHz	7.5 MHz	12–552 kHz	1 020 kHz	12.7 dB
132 ch. spot	5 MHz	4.4 MHz	12–552 kHz	529 kHz	20.7 dB

The modulation method for digital signals is generally phase-shift keying (p.s.k.). The theoretical minimum *C/N* ratio for 4 phase p.s.k., the most commonly preferred system, is 11.4 dB in a bandwidth equal to the symbol rate (half the bit rate) for a bit-error rate of 10^{-4}. Practical bandwidth requirements are greater than the symbol rate by typically 20% and the practical *C/N* ratio is greater than the theoretical minimum value by typically 2–3 dB. *Figure 15.147* shows also the capacity obtainable from 36 MHz using p.s.k. modulation for p.c.m. telephony.

The parameters of a t.d.m.a. system [9] of which INTELSAT proposes to hold a field trial are:

—Channel encoding at 64 K bit/s (8 bits per sample) with or without digital speech interpolation (d.s.i.). Alternatively multichannel f.d.m. direct encoding or direct digital interface with terrestrial network.
—4 phase p.s.k. modulation at 60 M bit/s.
—900 channels approx. per 36 MHz (1 800 with d.s.i.).
—750 μs frame period.

The main parameters of the SPADE system [10]—a single channel digital system using f.d.m.a.—are as follows:

—56 k bit/s p.c.m. (7 bits per sample, 8 kHz sampling rate).
—64 k bit/s transmission rate (including synchronising bits).
—4 phase p.s.k. modulation, 45 kHz channel spacing.
—800 channels in 36 MHz transponder (using voice switching of the carriers).

SUMMARY OF INTELSAT SATELLITES (*Table 15.15*)

Table 15.15 MAIN FEATURES OF INTELSAT SATELLITES

				R.F. channel characteristics			*Total telephone channel capacity (Note 1)*
Series	*Number successfully launched*	*Date of first launch*	*Mass* (kg)	*No.*	*Bandwidth* (MHz)	*Saturation output e.i.r.p.* (dBW)	
INTELSAT I	1	April 1965	39	2	25 MHz	10	480
INTELSAT II	3	Jan. 1967	82	1	125 MHz	15.5	480
INTELSAT III	5	Dec. 1968	127	2	225 MHz	22	3 000
INTELSAT IV		Jan. 1971	700	12	40 MHz	22, 33.7	7 000 (typ)
INTELSAT IVA		(1975)	750	20	40 MHz	22, 26, 29	12 000 (typ)

EARTH STATIONS FOR INTERNATIONAL TELEPHONY AND TELEVISION[1] (*Figure 15.148*)

Aerial system.[14] Economic trade-off studies of Earth station receive system performance against satellite power show an optimum value of G/T that depends on the number of channels required by the Earth station. The value of 40.7 dB/K chosen as the G/T for standard INTELSAT Earth stations is close to the economic optimum with current satellites for all except stations with very small numbers of channels.

To achieve a G/T of 40.7 dB/K at 4 GHz requires an aerial diameter of at least 26 m, but modern earth station aerials are more commonly about 30 m diameter to give more flexibility in the positioning of the low-noise receivers by allowing longer waveguide runs (with the associated extra loss and noise). Modern large earth stations also adopt the Cassegrain construction in which the use of a sub-reflector permits the feed to be located at the centre of the main reflector where it is readily accessible and from where the waveguide runs to the transmitters and receivers are relatively short.

Aerial steering—generally in azimuth and elevation axes—is necessary because the satellites are not perfectly stationary when considered in terms of the very narrow beamwidth (approximately 0.2° at 6 GHz) of a 30 m aerial. Small beam-pointing adjustments can also be obtained by moving the feed. The satellite beacon is generally used to derive the control signal for the auto-track steering system, a common method being by the detection in the feed system of higher order modes which undergo rapid rate of change as the received signal direction varies near the axis.

Low-noise receiving system.[15] The wide bandwidth and low noise temperature of the parametric amplifier makes it the universal choice as the first stage amplifier. The full satellite bandwidth of 500 MHz can be readily covered. Cooling to 20 K or below using closed-cycle cryogenic systems with gaseous helium produce amplifier noise temperatures of the order of 15 K. The total receiving system noise temperature however is more typically 70 K at 5° elevation angle, the extra being made up of about 25 K due to atmospheric absorption, 15 K due to sidelobe pick-up and a similar amount due to losses in the feed and waveguide.

High-power amplifiers. Transmitter powers are relatively modest but the transmitter arrangements are frequently complicated by the need to transmit more than one carrier. In a small or medium size station transmitting only one or two carriers a klystron for each carrier may be the best arrangement. For a large station the klystron's bandwidth of 50 MHz or so leads to a preference for t.w.t.s covering the whole 500 MHz band. A t.w.t. may be used with multiple carriers but a large station can still need several t.w.t.s, connected through a high power combining network to the transmit waveguide.

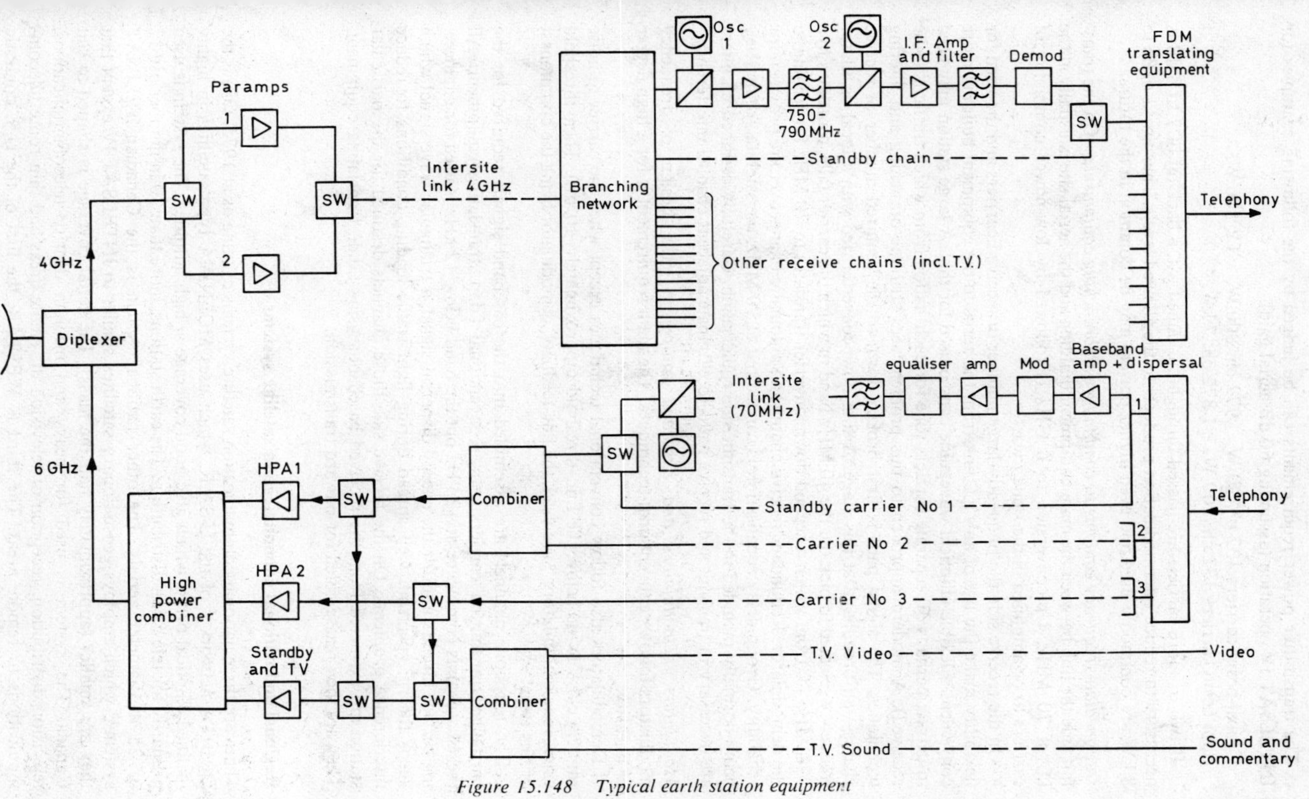

Figure 15.148 Typical earth station equipment

The transmitter power requirements can be judged by the following examples for INTELSAT IV operation (assuming 60 dB aerial gain).

Global-beam carriers 132 ch. 90 W, 972 ch. 800 W, TV 500 W.
Spot-beam carriers 132 ch. 200 W, 1 872 ch. 7 kW.

If two or more carriers are passed through the same t.w.t. a back-off of 7 to 8 dB is necessary to reduce intermodulation products to the allowed levels. Thus t.w.t.s of up to 8 kW are commonly used and even higher powers may be common in the future.

Intermediate frequency equipment, combining and branching equipment. On the transmit side the link between the main equipment building and the aerial site is usually at the i.f. of 70 MHz. Up-conversion to 6 GHz is followed by low-power combining of carriers to be amplified in the same t.w.t.

On the receive side the frequently large number of received carriers and the need for flexibility make the use of an s.h.f. link from the aerial to the equipment building more convenient. Flexible elliptical waveguide can be used for this. A large station may need to extract perhaps 30 separate carriers (one for each station from which traffic is to be received). A modern approach to this problem uses stripline branching and filtering techniques. The signal path is first broken down to the required number of receive chains—all broadband at this stage. A first downconverter, fed with a local oscillator frequency selected to pick out the 40 MHz band in which the carrier falls, translates to a 770 MHz i.f. from which a second downconverter translates to 70 MHz for feeding to the demodulator. Redundant receive chains and automatic changeover are provided for reliability. Group-delay equalisation is carried out at 70 MHz, the satellite group delay being normally included in the transmit side equalisation. Carefully designed i.f. filters are necessary in transmit and receive paths to eliminate adjacent channel interference.

Modulators, demodulators and baseband equipment. The design of frequency modulators follows conventional techniques. The main requirement is for a high degree of linearity.

Demodulators also follow conventional techniques except when the carrier-to-noise ratio is close to or below the f.m. threshold (approximately 10 dB). Then threshold extension demodulators [16] are used, the f.m. feedback demodulator being the commonest of this type.

The telephone channels are assembled into the baseband signals required for the particular transmitted carriers using conventional f.d.m. translating equipment. All these basebands commence at 12 kHz but the band below 12 kHz is used for engineering service circuits and for the energy dispersal signal, a triangular wave signal which keeps the r.f. spectrum well spread during light traffic loading conditions to reduce interference problems. On the receive side those channels destined for the particular station are extracted from the received basebands using f.d.m. translating equipment. They are then reassembled for onward transmission.

Regional and national communication satellite systems

There are three national communication satellite systems existing as at 1975. One is the ORBITA network [17] of the U.S.S.R. which uses MOLNIYA type satellites in highly inclined elliptical orbits which give good coverage at high latitudes. The system is used primarily for television distribution. The earth stations, more than 30 in number use 12 or 25 m diameter aerials. The other national system is the Canadian TELESAT system [18] which employs geostationary satellites similar to INTELSAT IV except that they are smaller (approximately half the mass). The aerial coverage is shaped to suit Canada. The system is used for heavy and medium density telecommunications, television distribution, reception of television in remote areas and thin route telecommunications to remote areas. The third is WESTAR, the first of the U.S. domestic systems. The WESTAR satellites are similar to those in the Canadian system.

Some general principles concerning regional and national systems can be stated. Such systems are likely to be justified

(i) where the terrain is very difficult for terrestrial communications,
(ii) where there are no existing facilities and distances are large or
(iii) where terrestrial facilities exist but large volumes of traffic have to be carried over large distances.

Regional systems may differ from long distance international systems in the types of traffic to be carried and also in the economic factors affecting earth station design. In particular, the use of spot beams can give high e.i.r.p. so that smaller earth station aerials can be used, especially if the traffic is modest or the number of stations very large. Also, if a regional system is to prove economic for shorter distance circuits than international systems then it is important to locate the earth stations as close as possible to the traffic centres. This may give rise to problems in the sharing of frequency bands with terrestrial services and encourage the use of higher frequency bands such as 11/14 GHz or 20/30 GHz.

Satellite systems for aeronautical and maritime mobile communications [19]

There are no satellite systems existing in 1975 for commercial mobile communications although some experiments have been conducted.

In the case of communication with aircraft the requirement is likely to be for a fairly small number of voice channels and perhaps some channels for automatic data transmission as a complement to a radio-determination system which would allow closer spacing between aircraft on heavily loaded oceanic routes. Voice channels would probably use narrow-band f.m. or Delta modulation with p.s.k. The largest single factor influencing system performance is the aircraft antenna which necessarily has very wide beamwidth and hence low gain since it cannot be steerable, except perhaps for rudimentary steering with a phased array or switching between elements. This constraint means that only a handful of channels can be obtained from a satellite which would give thousands of channels to standard INTELSAT Earth stations.

Communication with ships poses very similar problems although in this case modest aerial gain can be achieved if the aerial is mounted on some form of stabilised platform. Accurate stabilisation is expensive and the final choice of aerial gain is likely to be a compromise between costs and complexity on the one hand, and channel capacity obtainable with given satellite power on the other. The choice is likely to result in a G/T in the range -10 dB/K to -6 dB/K. At a G/T of -10 dB/K a satellite transmitter power of up to 5 W may be required per voice channel of adequate quality. This assumes global coverage aerials on the satellite since there is little scope for the use of spot beams.

The frequency bands of prime interest for maritime and aeronautical communication satellites are:

Maritime-mobile satellites 1 535 to 1 543.5 MHz and 1 636.5 to 1 645 MHz
Aeronautical-mobile satellites 1 542.5 to 1 558.5 MHz and 1 644 to 1 660 MHz

For the links between the satellites and the land-based stations frequencies in the 4/6 GHz or 11/14 GHz bands are likely to be used although use of the above mentioned u.h.f. bands is permitted.

Television broadcasting satellite systems [20]

These are also at the study and experimental phases. Owing to the very large number of receiving terminals the economic balance favours a very high satellite e.i.r.p. and relatively low-gain, high noise temperature receiving systems. The requirements on the satellite can be eased considerably by the sharing of a larger earth receiving aerial among many television sets through the use of wired distribution community antenna schemes.

Typical parameters for television broadcasting direct to the home or to community receivers are given in *Table 15.16*.

Table 15.16 EXAMPLES OF PARAMETERS FOR F.M. TELEVISION BROADCAST SATELLITE AT 12 GHZ

Service type	*Receive aerial diameter*	*System noise figure*	*S/N weighted*	*Bandwidth per channel*	*Satellite e.i.r.p.*	*Satellite r.f. power for 2° beam*
Home	1 m	6 dB	45 dB	20 MHz	64 dBW	1 kW
Community	3 m	4 dB	47 dB	22 MHz	57 dBW	200 w

The first application of television broadcasting satellites is likely to be for large developing countries[21] where the cost of setting up a terrestrial network would be prohibitive. Later applications may arise in countries which have a u.h.f. terrestrial broadcast network but which require more channels than can be obtained from the u.h.f. bands.

The bands allocated to satellite broadcasting are 2.5 to 2.6 GHz (community reception only) and 11.7 to 12.2 GHz (11.7 to 12.5 GHz in Europe and Africa).

Other satellite communication topics

Information on the following further topics can be found in the references given.

—Propagation factors in satellite communication.[22]
—Interference with terrestrial or other satellite systems.[23, 24]
—Efficiency of use of the geostationary orbit.[24, 25]
—Effects of propagation delay and echo.[26, 27]

REFERENCES

1. BACK, R. E. G., WILKINSON, D. and WITHERS, D. J., 'Commercial Satellite Communication', *Proc. I.E.E., I.E.E. Reviews*, **119**, No. 8R, 929 (1972)
2. PETROV, L., 'Intersputnik, International Space Communication System and Organisation', *Telecommunications Journal* (ITU, Geneva), **39**, No. XI, 679 (1972)
3. *Radio Regulations* (1968 edition revised after the Space Conference of 1971) General Secretariat of the ITU, Geneva (1971)
4. FREE, B. A., 'Chemical and Electric Propulsion Tradeoffs for Communications Satellites', *Comsat Technical Review*, **2**, No. 1, 123 (1972)
5. JILG, E. T., 'The INTELSAT IV Spacecraft', *Comsat Technical Review*, **2**, No. 2, 271 (1972)
6. BARTHOLOME, P., 'OTS—A Forerunner of a European Communication Satellite System', *Presented at the European Conference on Electronics*, Amsterdam, April (1974)
7. WESTCOTT, R. J., 'Investigation of Multiple FM/FDM Carriers Through a Satellite TWT Operating Near to Saturation', *Proc. I.E.E.*, **114**, No. 6, 726 (1972)
8. CHITRE, N. K. M. and FUENZALIDA, J. C., 'Baseband Distortion Caused by Intermodulation in Multicarrier FM Systems', *Comsat Technical Review*, **2**, No. 1, 147 (1972)
9. SCHMIDT, W. G., 'The Application of TDMA to the INTELSAT IV Satellite Series', ibid., **3**, No. 2, 257 (1973)

10. EDELSON, B. I. and WERTH, A. W., 'SPADE System Progress and Application', ibid., **2**, No. 1, 221 (1972)
11. *C.C.I.R. Recommendations*, 353–2, 421–3 and 451–2 (ITU, Geneva, 1974)
12. HILLS, M. T. and EVANS, B. G., *Telecommunications System Design, Vol. I Transmission Systems*, George Allen and Unwin (1973)
13. BARGELLINI, P. L., 'The Intelsat IV Communications System', *Comsat Technical Review*, **2**, No. 2, 437 (1972)
14. 'Design and Construction of large steerable aerials', *I.E.E. Conference Publication*, **21** (1966)
15. DAGLISH, H. N., ARMSTRONG, J. G., WALLING, J. C. and FOXELL, C. A. P., *Low-noise microwave amplifiers*, Cambridge University Press (1968)
16. 'Earth Station Technology', *I.E.E. Conference Publication*, **72**, 321, 333 and 340 (1970)
17. *C.C.I.R. Report*, 207–3, 'Active Communication Satellite Systems. Characteristics of Experimental and Operational Systems' (ITU, Geneva, 1973)
18. 'Canadian Domestic Communication Satellite', *Telecommunication Journal* (ITU, Geneva), **40**, No. 1, 12 (1973)
19. *C.C.I.R. Report*, 515–1, 'Technical Characteristics of Systems Providing Communication and/or Radiodetermination using Satellite Techniques for Aircraft and Ships' (ITU, Geneva, 1974)
20. *C.C.I.R. Report*, 215–3, 'Feasibility of Sound and Television Broadcasting from Satellites' (ITU, Geneva, 1974)
21. NERURKAR, B. Y., 'Educational Applications of Satellite Television and the Indian Project', *Telecommunications Journal* (ITU), **38**, No. V, 325 (1971)
22. 'Propagation of Radio Waves at frequencies above 10 GHz', *I.E.E. Conference Publication*, No. 98 (1973)
23. JOHNS, P. B. and ROWBOTTOM, T. R., *Communication Systems Analysis*, Butterworth, London (1972)
24. *C.C.I.R. Report*, 388–2, 'Techniques for Calculating Interference Noise in Radio-Relay Systems and Communication Satellite Systems Carrying Multichannel Telephony' (ITU, Geneva, 1974)
25. *C.C.I.R. Report*, 453-1, 'Technical Factors Influencing the Efficiency of Use of the Geostationary Satellite Orbit by Communication Satellites Sharing the Same Frequency Bands' (ITU, Geneva, 1974)
26. HUTTER, J., 'Customer Response to Telephone Circuits Routed via a Synchronous-orbit Satellite', *Post Office Elec. Engrs J*, **60**, Part 3, 181 (1967)
27. CAMPANELLA, S. J., SUYDERHOUD, H. G. and ONUFRY, M., 'Analysis of an Adoptive Impulse Response Echo Canceller', *Comsat Technical Review*, **2**, No. 1, 1 (1972)

OPTICAL COMMUNICATION USING LASERS

Much of the initial interest in the use of lasers (see Section 10) for communication was aroused by the thought of the very high carrier frequencies involved and the enormous possibilities for the transmission of information if even a very small proportion of the total available spectrum, say 1%, could be used for communication. Even on a single carrier of frequency 3×10^{14} Hz, 1% bandwidth would amount to 3 000 GHz, i.e., much more than the whole radio spectrum up to the mm wave region. In fact progress has been very slow towards realising even a fraction of this largely owing to the difficulties involved in very wide band modulation, detection and amplification. Work on laser communications has been concentrated in three main areas; firstly, short distance communication links in the atmosphere; secondly spacecraft to spacecraft and spacecraft to ground links; and lastly but probably the most important, towards guided-wave systems for terrestrial telecommunications; the work in this field is now concentrated in the area of fibre optic links using gallium arsenide laser transmitters.

Modulation

GENERAL

Modulation of light is a relatively difficult process at bandwidths much greater than the audio region of the spectrum, where mechanical means can be used (e.g., choppers, variable area slits, etc.). One other simple form of modulation is modulation of the current driving the light source; this can be used to modulate He–Ne gas lasers up to perhaps 100 kHz and semiconductor lasers up to 1 GHz or more. For other forms of laser more indirect methods must be used.

ELECTRO-OPTIC MODULATORS

Most modulators in this category depend on the change in refractive index induced by an applied electric field in suitable (usually single crystalline) media. There are two other possible kinds of modulator, of relatively small importance at present, depending on either electric field induced scattering of light, e.g., in liquid crystals or ferroelectric ceramics, or electric field induced absorption (Franz Keldysh effect) in semiconductors near the band edge. The most common types are amplitude modulators using the Pockel effect, in which the induced change in refractive index is directly proportional to the electric field, as distinct from the Kerr effect in which the refractive index change is proportional to the square of the electric field. Kerr effect modulators, formerly much used, are of little practical importance at the present time.

Pockels effect modulators are of two principal types, i.e., 'longitudinal' modulators in which the applied electric field is parallel to the direction of propagation of the light, and transverse modulators in which the electric field is perpendicular to the direction of light propagation. In both cases the action of the modulator is based on the use of light which is polarised in such a direction that the two orthogonally polarised components of light propagating in the crystal are of equal amplitude and undergo a relative phase shift when the electric field is applied, resulting, in the general case, in initially plane polarised light being converted into elliptical polarised light. This is passed through an analyser, crossed with respect to the incident polarisation. Thus for no applied volts, zero light passes. If a phase shift of π radians is introduced, corresponding to a path length difference of half wavelength and an applied voltage V_π, then the light coming out is plane polarised in the direction of the analyser and the transmission is 100%. For intermediate voltages V a fraction $\sin^2 V/V_\pi \, \pi/2$ is transmitted. Thus linear modulation can only be achieved for small percentage modulation unless special drive circuits are used to correct for the inherent nonlinearity.

In longitudinal modulators, either transparent conducting electrodes or 'ring' electrodes must be used. In the latter case the electric field on axis is somewhat reduced. Even in the case of transparent electrodes the voltage required for 100% modulation/V_π, is usually several kV irrespective of the dimensions of the cell. In longitudinal modulators, the voltage requirement can be reduced by a factor of one half of the length/height ratio of the crystal, where the height is measured in the direction of the applied field. However there is a disadvantage since in this configuration the crystal is usually naturally birefringent along the direction of light propagation, i.e., there is a relatively large path difference for the two orthogonally polarised components even in the absence of an applied field, which is not the case for longitudinal modulators. Since this birefringence is temperature dependent, it is usually necessary to maintain the temperature of the modulator crystal constant to 0.1 °C or better if a single crystal is used. It is possible to obtain a large measure of compensation by using a second crystal so arranged that birefringence in the second cancels that in the first although the field induced changes add.

Table 15.17 gives some of the properties of the most commonly used modulator

Table 15.17 PROPERTIES OF ELECTRO-OPTIC MODULATOR MATERIALS

Formula	KH_2PO_4	KD_2PO_4	$NH_4H_2PO_4$	$LiNbO_3$	$LiTaO_3$
Abbreviation	KDP	KD*P	ADP		
Growth method	Aqueous solution	Aqueous solution	Aqueous solution	Czochralski melt	
V_π (transverse) kV	17.6 l/h	7.2 l/h	7.74 l/h	5.88 l/h	5.68 l/h
V_π (longitudinal) kV	8.8	3.6	10.6	2.94	2.84
Cost increasing in	2	3	1	4	5
Notes	hygroscopic	hygroscopic	hygroscopic	These two materials are subject to refractive index damage	

l = Length of crystal. h = Distance between electrodes.
Values for half-wave voltages are for most favourable crystal cuts in each case at a wavelength of 540 nm.

crystals together with an indication of their relative advantages. The 'damage' referred to for $LiNbO_3$ and $LiTaO_3$ is a localised change in refractive index produced by exposure to short wavelength (500 mm or less) laser light, resulting in trapped photoelectrons. It is thought that these effects are due to impurities and will eventually be eliminated.

ACOUSTO-OPTIC MODULATORS

These depend on changes in refractive index induced by the passage of ultrasonic waves through a liquid or solid medium. There are two possible modes of operation, the Debye–Sears or Raman–Nath type of modulation, in which the incident light beam is perpendicular to the direction of the ultrasonic waves, and the so-called 'Bragg' modulators in which some of the incident light is diffracted by the ultrasonic wave fronts in a similar manner to the diffraction of X-rays by crystal planes. This only occurs for certain angles of incidence given by $n\lambda/2\Lambda = \sin\theta$ where λ is wavelength of light, Λ is the wavelength of the sound waves and n is an integer.

In the Debye–Sears modulator some of the incident light is also diffracted, but instead of all the light being diffracted into a single beam, it is diffracted into two or more directions through angles $\theta \approx \pm n\lambda/\Lambda$. Thus in this case it is usual to select the undiffracted light which undergoes 'negative' modulation. In Bragg modulators either the diffracted wave or the undiffracted wave can be selected, depending on whether positive or negative modulation is desired. In either case amplitude modulation is produced by amplitude modulating an r.f. carrier of suitable frequency with the desired modulation and then applying this to a transducer coupled to the modulating medium. For relatively low carrier frequencies (<20 MHz) and relatively short interaction lengths (<1 cm or so) between sound and light, the Debye–Sears effect predominates. For higher frequencies (>50 MHz) and longer interaction lengths the Bragg effect predominates. In between there is an intermediate region which cannot be simply described.

Most practical modulators use the Bragg effect, since this gives a larger bandwidth capability. The modulation in this case is proportional to $\sin^2 P/P_\pi\, \pi/2$ where P_π is the power required to produce a maximum path length of change of $\lambda/2$. The law for Debye–Sears modulators is more complex.

The materials for practical acousto optic modulators are usually solids. These are chosen to give a large acousto-optic interaction and a low acoustic loss so that the ultrasonic wave is not appreciably attenuated in traversing the optical beam. A large acousto-optic interaction requires a material with a high refractive index n (the effect is proportional to n^6 or n^7), low sound velocity and a high value of the interaction parameter p. Most modulators marketed so far either use a high refractive index glass or single crystal lead molybdate, although a number of other promising materials are being developed. Modulation bandwidths up to about 10 MHz with efficiencies up to ~70% have been achieved. In an optimum design acoustic and optical angular beamwidths should be approximately equal and chosen to fulfil the Bragg condition over the required bandwidths so as to achieve a reasonably flat frequency characteristic. The piezoelectric transducer which is used to couple acoustic power into the medium should be as efficient as possible consistent with sufficiently broad band operation; $LiNbO_3$ is commonly used. Drive powers are usually a few watts or less at carrier frequencies of 50 MHz or more.

Receivers

GENERAL

Laser communication systems comprise a transmitter, transmission medium and a receiver. Two types of receiver system have been used, depending on the characteristics of the laser and the transmission medium. These are *direct* detection and *coherent* detection respectively.

DIRECT DETECTION RECEIVERS

The majority of laser communications systems constructed so far use direct detection, i.e., they comprise a lens or mirror system used to collect the transmitted light, a filter arrangement to reject unwanted light from other sources, photoelectric detector and an amplifier. Such an arrangement corresponds approximately to a TRF radio frequency receiver, whereas the coherent detection receiver corresponds to a superheterodyne or synchrodyne receiver. The sensitivity attainable with an ideal direct detection receiver is only a factor of two lower than that of an ideal coherent detection receiver and it is in principle much simpler to engineer.

However, the ideal direct detection receiver is not easy to achieve, since it requires a detector with a unit quantum efficiency (one photo electron per incident photon) and zero dark current, followed by a noise-free amplifier. In addition filtering of unwanted light must be perfect. The nearest approach to this situation is obtained using a photomultiplier chosen for high quantum efficiency and low dark current, together with a narrow band optical filter and a focal plane stop which acts as a *spatial filter* to eliminate unwanted light. There is difficulty in finding efficient photo cathodes at laser wavelengths longer than 700 nm and in any case photomultipliers and associated power supplies are too large and insufficiently rugged for some applications. In this case avalanche photodiodes, which are to some extent a solid state analogue of photomultiplier, can be used. However in this case the internal amplification M is far from noiseless and the noise current increases as $M^{2.3}$ for silicon diodes which are the most common type. Thus in practice the useful value of M is no larger than 100, and the sensitivity is ultimately decided by the noise of the following wideband amplifier. However, it does give a sensitivity up to 100 times larger than a receiver using a conventional photodiode and it seems destined to be very widely used in fibre optic systems.

COHERENT DETECTION

The use of coherent detection implies that the transverse and longitudinal (i.e., temporal) coherence of the transmitting laser must be adequate initially, and that the transverse coherence across the receiver aperture must not be appreciably disturbed by the transmission medium. This condition has effectively limited the use of coherent detection, in atmospheric transmission systems, to the CO_2 laser wavelength of 10.6 μm. It is relatively easy to obtain stable single frequency operation with low pressure CO_2 lasers, and atmospheric turbulence does not affect the transverse coherence so severely as at shorter wavelengths. Furthermore the sensitivity of direct detection schemes is relatively much worse than in the visible since no avalanche multiplication is available with the narrow band gap semiconductors such as Hg–Cd Te (no photoemissive devices exist at 10.6 μm), and filtering of background radiation is more difficult. An additional factor favouring the use of coherent detection at 10.6 μm is the relative ease of alignment of the signal and local oscillator wavefronts, to ensure a constant phase difference across the wavefront and hence a high conversion efficiency.

Transmission media

DEEP SPACE

This is the ideal transmission medium for utilising the very small beamwidths which can be achieved with reasonably small 'dishes'. Thus a laser system would be very useful for high data rate transmissions from planetary probes to an orbiting earth satellite using conventional means for transmission to earth through the atmosphere. Two or three laser systems for use in satellites are at present under development in the U.S.A. These are, however, intended to investigate the usefulness of high data rate (up to 1 GHz) satellite–Earth links.

THE TERRESTRIAL ATMOSPHERE

The atmosphere is a very imperfect transmission medium, in that it can give rise to very high transmission losses (up to 80–100 dB/km in dense fogs) over the visible spectrum, and indeed, for many types of droplet fog, in the near infrared out to 10 μm or more. Even in the relatively clear atmosphere absorption by water vapour can be quite severe in high humidity conditions (up to 20 dB/km) at 10.6 μm, and molecular scattering losses due to normal atmospheric constituents are far from negligible in the visible. In addition to these 'microscopic' effects, there are quite severe effects due to macroscopic turbulence, particularly in the daylight hours around noon. These give rise to large fluctuations in signal strength with a log-normal distribution, and, as mentioned above, break up the coherence of the received wavefronts severely.

Anomalous propagation effects are caused by ray bending due to refractive index gradients occurring in calm, non-turbulent conditions near dawn and sunset when an inversion layer is present.

GUIDED PROPAGATION

Open waveguides for guided propagation of laser radiations, of the type first suggested by workers at Bell Laboratories, comprise an equi-spaced array of lenses, the purpose of which is to counteract the spreading of the beam due to diffraction. In practice a lens spacing of about 100 m would appear to be sufficient, enabling losses of a few dB/km to be achieved. However, the required mechanical stability of the lenses appeared to be unattainable in realistic underground environments. (It was envisaged that the guides would be contained in buried metal pipes and for this and other reasons this type of guide has been abandoned.) Later a great deal of work was carried out in the U.S.A., U.K. and Japan on the production of sufficiently low loss glass fibre waveguides, and losses down to less than 10 dB/km at 0.8 μm and 1.06 μm have been attained using very pure starting materials and improved methods of fibre production. Losses of these magnitudes require that impurity levels of copper and iron should be reduced to 1 part per million or less, and also that the OH content of fused silica, used in most cases for the core material, should be kept very low.

Glass fibre waveguides usually consist of a cylindrical core, and a concentric cladding of a glass with a slightly lower refractive index. Propagation of light in fairly large core diameter waveguides (>10 μm) can be thought of in terms of rays undergoing periodic total internal reflections at the core-cladding interface.

In waveguide terms, such a waveguide is capable of sustaining many modes, hence it is spoken of as a *multimode* fibre. Since each mode has a characteristic velocity, then it is obvious that considerable spreading of a light pulse travelling through the fibre will occur, unless some scheme for eliminating at least the higher order (i.e., slower) modes can be devised. Such spreading would limit the information rate to less than 100 M bit/s for 1 km repeater spacing: one means of reducing this dispersion is the graded index fibre of the 'Selfoc' type, invented in Japan. Here an uncladded fibre with a parabolic radial index variation, produced by leaching ions from the outer layers of thc fibre, acts like a lens and largely equalises the transit time of rays travelling with different inclinations to the axis. It is thought that information rates of several hundred M bit/s may be possible using such fibres, although losses are presently 60 dB/km, and losses of 20 dB/km or less will be necessary for practical applications.

The other method of reducing pulse spreading is the use of glass fibres capable of transmitting only a single low order mode. Such fibres can be made with a core diameter of a few microns, i.e., many wavelengths, providing the cladding has a refractive index only about 1% smaller than the core. This is large enough to make a fibre which can be easily handled, but makes it difficult to align the laser sources, e.g., GaAs laser diodes sufficiently accurately to give good lower coupling efficiency. Multimode fibres commonly have a core diameter of $\sim$50 μm, the cladding in each case is a few μm thick.

Laser communication systems

SHORT DISTANCE ATMOSPHERIC LINKS

Semiconductor laser systems. The simplest type of laser communication system, which has been demonstrated in the U.S.A., U.K. and Japan uses a low duty cycle (0.1%) pulsed GaAs laser in the transmitter and a silicon avalanche diode or ordinary silicon photodiode in the receiver. In some versions the transceiver is incorporated into a pair of binoculars. The maximum range of such systems is from 2 to 8 km with 50.8 mm optics, the range depending on the weather conditions. Usually a pulse position modulation scheme is used with very short (10–20 ns) pulses, to obtain the maximum data rate ($\sim$200 k bit/s) with the low available duty cycle.

Gas laser systems. Experimental communications links operating over distances up to 5 km, at data rates of 100 M bit/s, have been developed in the U.S.A. and Japan using a mode locked helium neon laser operating at 6 328 A° to generate a steady stream of pulses which are then selected for transmission by an external modulator of $LiTaO_3$. The receivers use a mirror, narrow band filter and a photomultiplier.

A 10.6 μm laser communication system, working at 5 M bit/s, has been operated over distances up to 20 miles in the U.S.A. It uses optical heterodyne detection with a mercury–cadmium telluride detector operating at a temperature of 77 K and an i.f. frequency of 30 MHz. The transmitter laser and the receiver local oscillator lasers are both frequency stabilised to one of the rotational transitions of the CO_2 molecule. When tested over a 20-mile path, variations in signal to noise ratio of 40 dB or so were observed at times due to atmospheric scintillation, although at other times the mean value was less than 1 dB. The transmitter power was 1 watt with a beamwidth of $\frac{1}{2}$ minute of arc. The receiver aperture had a diameter of 10 cm. A similar system is reported to have been developed by Siemens in Germany.

SPACECRAFT TO SPACECRAFT OR SPACECRAFT TO EARTH LINKS

No space qualified laser systems for this purpose have yet been made, however two laboratory versions of 1 G bit/s systems designed for such use have been made in the U.S.A. Both are designed to use c.w. frequency doubled Nd YAG lasers operating at 530 nm. One uses direct wideband modulation, the other uses external modulation of a 500 M bit/s pulse stream produced by mode locking. Lithium niobate modulators were used in both systems.

In the c.w. wideband case, a phase shift keyed modulated microwave carrier (1.5 GHz) was applied to the optical modulator. In the mode locked case a 1 G bit/s stream was produced by splitting the beam and delaying one portion by half a bit interval with respect to the other. These were then separately modulated and recombined. Both receiving systems use dynamic crossed field photomultipliers as a means of obtaining adequate bandwidth and signal to noise ratio. The mode locked systems were tested using four digital television signals and a 500 M bit/s synchronous data source.

FIBRE OPTIC COMMUNICATION SYSTEMS

These are still very much at the research and development stage; a number of experimental systems have been demonstrated in the U.S.A., Japan, Germany and the U.K. These range from a 6.3 M bit/s system, including repeaters, demonstrated at Bell Laboratories, to a 200 M bit/s system employing double heterojunction, GaAlAs diodes, using a relatively short length of multimode fibre, demonstrated in Japan. In the U.K. a 100 M bit/s system has been demonstrated by S.T.L. using 100 m of single mode fibre, and an analogue colour television link transmitting over ~1 km of liquid cored multimode fibre has been developed by Gambling and others at Southampton University.

FURTHER READING

GORDON, E. I., 'A review of Acousto-Optical Deflection and Modulation Devices', *Applied Optics*, **5**, 1629 (1966)

KAMINOW, I. P. and TURNER, E. H., 'Electro-Optic Light Modulators', *Applied Optics*, **5**, 1612 (1966)

KOMPFNER, R., 'Optics at Bell Laboratories—Optical Communications', *Applied Optics*, **11**, 11 (1972)

MASLOWSKI, S., 'Activities in Fibre Optical Communications in Germany', *Opto-Electronics*, **5**, 275 (1973)

PRATT, W. K., *Laser Communications Systems*, Wiley (1969)

RAMSEY, M. M., 'Fibre Optical Communications within the United Kingdom', *Opto-Electronics*, **5**, 261 (1973)

ROSS, M., *Laser Receivers, Devices, Techniques, Systems*, Wiley (1966)

Digest of 1973 I.E.E.E./O.S.A. Conference on Laser Engineering and Applications

DATA SYSTEMS

Data communications

DATA COMMUNICATIONS SYSTEMS

In less than two decades data communications has developed from an off-shoot of telegraphy to a point where its influence is bringing together two major industries—telecommunications and computing. The transmission of data cannot now be seriously studied in isolation from the total systems of which data transmission forms a part.

The term data communications has a much wider meaning than the transmission of encoded information over some kind of electrical transmission system. It embraces the whole area of remote computing and a data communications system includes many elements not least of which is the people who use it. The design of such systems therefore demands the close cooperation and mutual understanding of people from a wide variety of different disciplines ranging from the communications engineer to the behavioural scientist.

DATA COMMUNICATIONS CODES

The number of characters which can be derived from a binary code is given by 2^n where n is the number of units in the code: 128 different characters can therefore be derived from a code with seven binary digits per character. The five unit CCITT (International Telegraph and Telephone Consultative Committee) No. 2 code used on the Telex network employs two shift characters 'letters' and 'figures' which when transmitted determine the meaning ascribed to the characters which follow. In this way the normal capacity of a five unit code is extended from 32 to 60 characters. The efficiency of codes employing shifts will depend on the application in which they are used. Although CCITT No. 2 is an efficient code when used for ordinary message traffic it may be inefficient for some data communications messages where letters and figures may be frequently contiguous. Although some data communications codes now used do include a shift character (DLE or Data Link Escape in CCITT No. 5) this is intended to give some flexibility to the code by changing the meaning of a small number of control characters.

DATA COMMUNICATIONS CODING PRINCIPLES

Using binary notation, decimal symbols can be converted as follows:

Decimal equivalent	2^3	2^2	2^1	2^0
0 =	0	0	0	0
1 =	0	0	0	1
2 =	0	0	1	0
3 =	0	0	1	1
4 =	0	1	0	0
5 =	0	1	0	1
6 =	0	1	1	0
7 =	0	1	1	1
8 =	1	0	0	0
9 =	1	0	0	1

These are known as binary coded decimals, and the four binary digits (bits) needed to contain the information are described as a *basic quartet*. By splitting the alphabet into three groups, numbers and letters can be obtained within six binary digits thus:

	Control duet XY	Basic quartet
Group 1—A to I (01)	01	0001 = A
Group 2—J to R (10)	10	0101 = N
Group 3—S to Z (11)	11	0011 = U

It will be seen that the 'X' and 'Y' components in the code indicate which group of letters is involved, and the basic quartet gives the position of the letter within the group. Thus the letter 'A' is indicated by binary 0 in the 'X' column, and binary 1 in the 'Y' column followed by a basic quartet the decimal equivalent of which is 1: therefore the letter is the first letter in the first group—'A'. A number can be shown by having binary 0 in the 'X' and 'Y' columns followed by the basic quartet.

Although most data communications codes in use today are based on the principles described in practice group 3 is usually adjusted slightly. There are only eight letters in this group (S to Z) and the first letter in the group S may be represented by 11 0010, T by 11 0011 and so on. In addition to alpha-numeric information there is a need to have other characters for transmission control, to change format and to separate information. This has resulted in the use of seven and eight unit codes and the seven unit CCITT No. 5 has been developed as a step towards code standardisation. In this and other codes a *control trio* is commonly used instead of a *control duet* in order to derive the additional characters required. EBCDIC (extended binary coded decimal interchange code) uses four bits for control in addition to the basic quartet.

ABBREVIATION OF CONTROLS IN CCITT ALPHABET NO. 5

Transmission controls

1. ACK Acknowledge
2. BEL Bell
3. DLE Data link escape
4. ENQ Enquiry
5. EOT End of transmission
6. ETB End of transmission block
7. ETX End of text
8. NAK Negative acknowledge
9. SOH Start of heading
10. STX Start of text
11. SYN Synchronous idle

Information separators numbered in order of heirarchy

4. FS File separator
3. GS Group separator
2. RS Record separator
1. US Unit separator

Format effectors

1. BS Back-space
2. CR Carriage return
3. FF Form feed
4. HT Horizontal tabulation
5. LF Line feed or carriage return line feed
6. VT Vertical tabulation

Other abbreviations

1. CAN Cancel
2. DEL Delete
3. EM End of medium
4. ESC Escape
5. SI Shift in
6. SO Shift out
7. SUB Substitute
8. NUL Blank tape

Device controls

1–4 DC On/off instruction to a device

DATA TERMINALS

The progressive reduction in costs of logic circuitry and the need to optimise the distribution of intelligence in data communications networks is encouraging the production and use of 'intelligent' terminals. Many of these might be regarded as computers in their own right and it becomes increasingly difficult to define the term 'data terminal'. Again, almost any measuring device such as a temperature gauge or an electrocardiograph machine can be regarded as a potential data terminal if there are benefits to be gained in reading the information it provides remotely. Although definition is impossible categorisation is merely difficult as there are broad classes of terminals which have sufficient distinguishing features for an attempt to be made. There are grey areas, however, and the reader should be prepared to encounter data terminals which do not fall clearly into any of the broad categories described below. Descriptions here are confined to the main types now in use in the United Kingdom.

DATA PRINTERS

These are typewriter terminals using data communications codes such as CCITT No. 5. They can produce hard copy of data passing in either direction. Start/stop synchronisation is usually employed and a typical character format for a 10 character per second machine is shown in *Figure 15.149.*

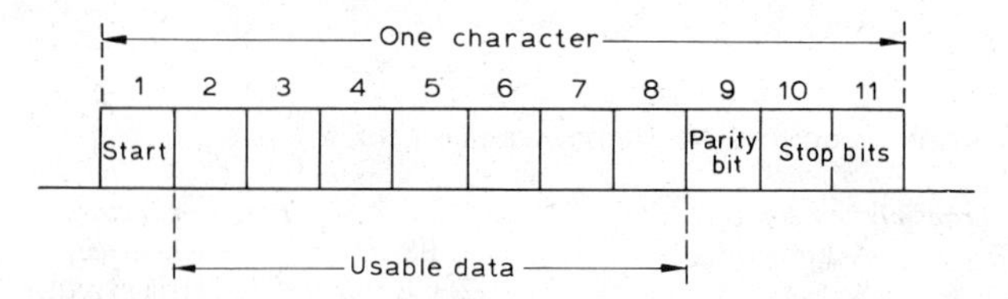

Figure 15.149 Typical character format for a 10 character per second data printer machine

Of the 11 bits transmitted for each character only 7 bits contain usable data and the maximum information transfer rate is therefore 70 bits per second. As 11 unit signal elements are required to transmit each character, a 10 character machine requires a circuit with a modulation rate of 110 baud. The parity bit is rarely used for error control on 10 character per second machines and therefore 36% of the signals transmitted are redundant from the user's point of view. Although start/stop synchronisation is simple and therefore relatively cheap in terms of machine costs this is gained at the expense of increased redundancy in transmission. However as these machines are used mainly for 'conversational' computing there are long natural gaps between manual key depressions. The transmission redundancy becomes a significant cost factor only in cases where long messages are sent using automatic tape reading facilities.

Machines range in speed from 6 to over 100 characters per second.

Communications requirements. Leased telegraph circuits are available in the U.K. which have a modulation rate of 110 baud and therefore allow for 10 characters per second operation without the need for modems. Alternatively, the Datel 200 service can be used on the public switched telephone network; this normally provides for a modulation rate of 200 baud. Above this speed the character format of machines changes and only one stop element is used giving only 10 unit signal elements per character. A 30 character machine therefore requires a modulation rate up to 300 baud and the Datel 200 service has been extended so that transmission of these rates can often (but not always) be achieved.

Applications. Data printers are used extensively in the timesharing bureaux field in an interactive or enquiry/response mode. They are also used for many other purposes such as file updating or data retrieval. The work they do is essentially low volume requiring hard copy. Faster machines have been developed not so much to speed data entry, which is usually geared to the keying speed of the operator (less than 5 characters per second) but to enable speedier replies to be obtained from the computer system. Thirty characters per second is generally regarded as a comfortable speed for the human eye and brain to follow a line of type.

Ancillary facilities. The basic facility offered by a typewriter terminal is Keyboard Send and Receive (K.S.R.). This enables the operator to transmit data manually (with local copy) and to receive hard copy.

Two additional facilities which may be provided are:

(*a*) Receive Only (R.O.)—no keyboard is provided; but hard copy messages can be received. A typical application for this device is where a large network exists and 'incoming only' lines increase the efficient use of a multiline route.
(*b*) Automatic Send and Receive (A.S.R.). The operator has the facility to compile a message on the terminal, editing it as required. The data is stored on paper or magnetic tape and this tape is then connected to an automatic transmitter for transmission to line at maximum speed.

Buffered terminals. Strictly speaking, any machine which has a tape cutting facility could be termed *buffered*, however in a buffered terminal the buffer is an integral part of the hardware of the machine; data is held in store, rather than on a removable tape or card. Such a terminal offers automatic send and receive facilities but additionally the buffer can hold data and transmit when called upon to do so by a polling computer at higher speeds than the terminal itself.

Tape readers and printers. Paper tape is commonly used as a storage medium with automatic transmission. Data can be stored for repeat messages or used to provide local hard copy. Received messages can be printed on paper tape as hard copy is being produced on the machine.

Generally the paper tape reader, which is used to transmit the message, operates at the maximum speed of the machine and offers the following advantages:

(*a*) Operator verification of the text before transmission.
(*b*) Maximum utilisation of line time.

An alternative to paper tape is magnetic tape cassettes, these being used increasingly for they offer similar facilities to paper tape but are easier to handle, and store.

Punched cards may also be used with the appropriate reader and printer but their usage is less common than paper tape.

VIDEO DISPLAY UNITS

Video display units fall into two main categories, alpha-numeric display terminals and graphic display terminals.

Alpha-numeric displays. An alpha-numeric display terminal consists basically of a cathode ray tube on which data in the form of letters or figures is displayed and a keyboard for data entry. Data is entered from the keyboard and displayed on the screen providing a visual error check for the operator before the data is passed to the computer. Replies from the computer are displayed on the screen and a wide variety of editing and formatting facilities are usually available.

Figure 15.150 shows schematically a basic alpha-numeric display terminal. Timing is required to synchronise the data flow within the system and the information on the screen must be refreshed by the character regenerator at 50 times per second, typically.

Screen sizes vary and can be measured either by the dimensions of the tube or more usefully by the number of characters a screen can display. The number of lines per screen vary between 12 and 40 and characters per line between 30 and 80, giving a capacity range of between 360 to 3 200 characters per screen. The factors affecting capacity are the dimensions of the screen, the size of characters, the distance between characters and the distance between lines. Although *Figure 15.150* shows a 'stand alone' terminal economies can sometimes be achieved by grouping a number of terminals in one location into clusters, *Figure 15.151*.

Graphic displays. Graphic displays are used mainly for design work, the designer entering the information using a device such as a light pen or rolling ball. Drawings are displayed on the screen and in some cases can be adjusted to alter perspective. The designer works interactively with the computer by amending the design with the input device, the amendments being instantly displayed on the screen. Graphic terminals usually have a minicomputer associated with them when working to a remote central computer.

Communication requirement. The information transfer rate requirements vary with the amount of data to be transmitted and the *response time* needed. This is the time between the data being entered by the operator pressing the transmit key and the first character of the reply being displayed on the screen. Response time is a critical factor in the design of the on line data communications system in which alpha-numeric displays are often used and is determined by both human and economic factors. The requirement will vary with each application but systems with a mean response time of 1.5 seconds are not uncommon. Most systems in use today use transmission facilities which give information transfer rates of between 600 and 4 800 bits per second.

Applications. Alpha-numeric displays are used in a wide number of applications. In *off-line* systems, where a communications link may or may not be involved alpha-numeric displays are often used for data preparation. A typical example might be a 'key to magnetic tape' facility where the input data is displayed and checked visually before being committed to magnetic tape.

For *on-line systems*, data entry as a means of 'overnight update' is used for many business activities such as banking and mail order. Orders or transactions are entered via an alpha-numeric display terminal, validated and transmitted to the central processor. Here the data is stored until end of business and the updating of the main file is carried out overnight.

On line–real time systems may employ alpha-numeric displays for enquiry/response. A central data base is accessed by the terminal to obtain up-to-date information; the Reuters Stockmaster service being a typical example. On line file update systems, often employ alpha-numeric displays. Input data is checked on the screen before transmission to line and then used to update the main file. Interactive display terminals are widely used in the seat and hotel reservation field.

REMOTE BATCH TERMINALS

Normally in batch processing relatively large quantities of data are collected to be processed in one or more computer *runs*. Batch processing requires high speed peripheral equipment such as magnetic tape decks, line printers, card readers, etc.

Many companies cannot justify obtaining their own central processor for batch processing. However, remote batch processing terminals consisting of remote computer

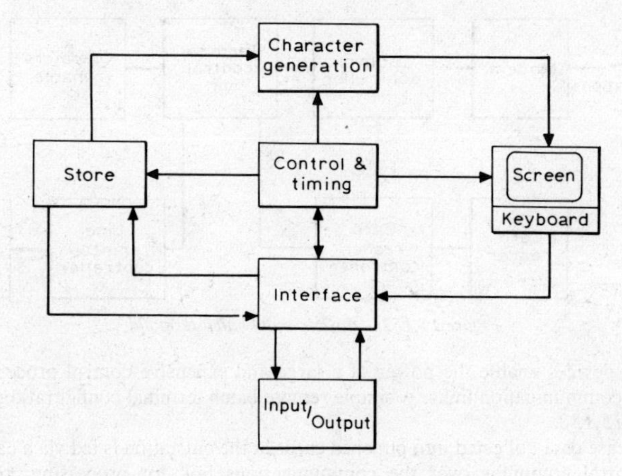

Figure 15.150 Basic alpha-numeric display terminal

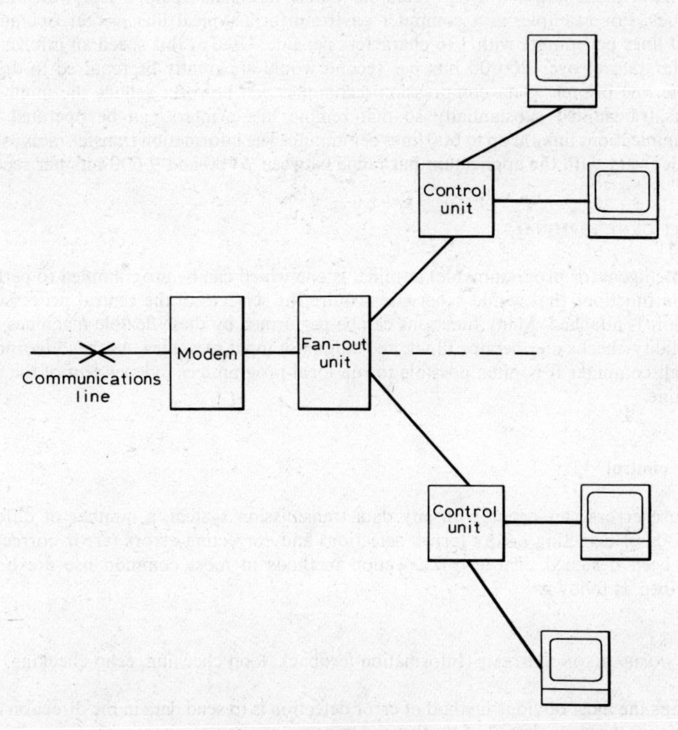

Figure 15.151 Clustered displays

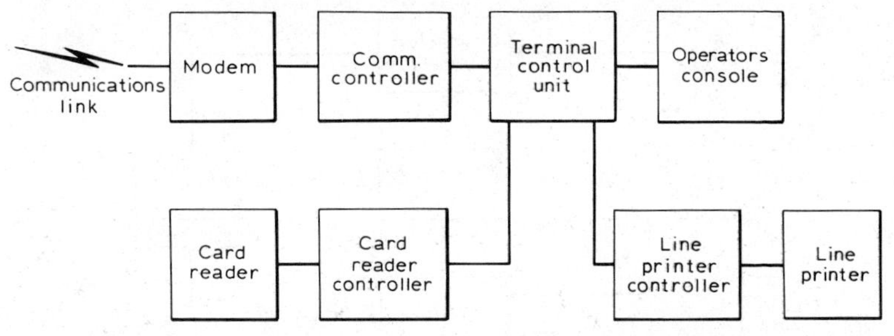

Figure 15.152 Basic remote batch terminal

peripheral devices enable the power of a large and expensive control processor to be used over communication links. A simple remote batch terminal configuration is shown in *Figure 15.152.*

In this case data collected into punched cards at the outstation is fed via a card reader to the central computer over the communications link for processing; results are returned from the central computer and received on the line printer. Some remote batch terminals include a small computer so that small jobs can be handled completely by the terminal and others requiring more computing power are processed centrally by the main machine.

Communications requirements. Remote batch terminals require fast transmission facilities. For example, in a computer environment a typical line printer operates at 1 200 lines per minute with 136 characters per line. Used at this speed an information transfer rate of over 20 000 bits per second would apparently be required to drive a remote line printer. Data compression techniques can however reduce the number of blanks transmitted substantially so that remote line printers can be operated over communications links at up to 600 lines per minute. The information transfer rates used in practice vary with the application but range between 2 400 and 9 600 bits per second.

INTELLIGENT TERMINALS

An intelligent (or programmable) terminal is one which can be programmed to perform certain functions that would otherwise require the services of the central processor to which it is attached. Many functions can be performed by these flexible machines such as validity checks on operator input, and editing of input messages. As they incorporate a small computer it is often possible to run local programmes independent of the main machine.

Error control

Because errors can occur with any data transmission system, a number of different methods of detecting errors (error detection) and correcting errors (error correction) have been designed. The error detection methods in most common use are briefly described as follows.

FULL COMPARISON SYSTEMS (Information feedback, loop checking, echo checking)

Perhaps the most obvious method of error detection is to send data in the direction A–B and return the same data B–A so that the transmit end can compare what was originally sent with what is returned after transmission. This method may be used with a block system where the data is sent a number of characters at a time. This avoids the retransmission of the whole message in the event of an error; the optimum block length being decided by the number of errors likely to occur in the particular transmission system. Block markers are generally used with this method to indicate the beginning, or end of a block or both. Duplex transmission is employed and transmission continues until an error is detected; when this happens the block in error is retransmitted. A weakness of full comparison systems is that an error in the forward direction can be compensated by an error in the reverse direction and so remain undetected.

TOTAL SYSTEMS

If data is in the form of figures, a total can be sent at the end of each block; by checking this total against the figures received the receiver can detect a proportion of errors. This method is particularly useful in banking when cash details are to be sent. There are variations of this method, but none of them can give full protection against compensating errors and it is therefore usually employed in combination with other error detection techniques.

PARITY SYSTEMS

Vertical (character) parity. Both odd and even parity systems are in use. If for example a system employs odd parity an additional binary digit is added to each character so that the total number of 'one' bits is always odd; all characters should therefore contain an odd number of binary ones when they arrive at the receiver. If a character is found to contain an even number of binary ones this is recognised by the receiver as an error.

A single parity system of this kind has a number of weaknesses; for example it cannot detect an even number of compensating errors within a character. If a character is sent from a transmitter thus: 1001111 and arrives at the receiver so: 0101111, an error will not be detected. The effectiveness of this kind of system depends largely on the nature of the error bursts. If characteristically a line is prone to single bit errors a system of this kind will be extremely effective. This however is highly unlikely; transmission errors on telegraph and speech circuits occur in bursts and therefore a high percentage of errors may escape detection if a simple character parity check is used.

Two co-ordinate parity systems. An additional character may be added to each block viz:

1 010010	1 101101	
1 100010	1 010010	
0 001110	1 101011	
1 000110	1 101011	
1 100010	1 010001	
1 010010	1 100010	
1 010001	1 000110	
0 110010	1 101101	
0 000010		
1 0xx011		
0 1xx000	0 100110	Horizontal parity
0 101100	x = error	

In the above example a two co-ordinate system is employed, parity bits being added both horizontally and vertically. By using these combined parity methods the chances of an error escaping detection are substantially reduced, but some errors patterns will not be detected as will be seen from the above example.

'n' OUT OF 'M' CODING

A group of error detection codes known as 'n' out of 'M' codes have been used in data transmission. An example of such a code is given below:

'4 out of 8' code. The character set is constructed so that each character always contains four '1' bits and four '0' bits. The receiver checks the number for 1 bits it receives and if more or less than four are received in any character a negative acknowledgement will be returned at the end of the transmission block, the block is then re-transmitted.

As with all n out of M codes there is a high proportion of redundancy which is not immediately obvious. Each character has 8 bit positions and an 8 unit information code could provide a character set of 256 characters (2^8). However, the character set in the 4 out of 8 code in the example is limited by the structure of the code. The number of characters which can be derived is given by the formulae:

$$\frac{M!}{n!(M-n)!} \quad \text{or} \quad \frac{8!}{4!(4)!} = 70 \text{ characters}$$

CYCLIC CODING (CYCLIC REDUNDANCY CHECKING)

This method is so called because of the employment of cyclic shift registers to produce the checking information. The arrangement of the check bits can be optimised for differing circumstances by the use of polynomial equations and cyclic codes are particularly efficient in data transmission where errors usually occur in bursts. Cyclic coding methods are being used increasingly in data transmission systems as the costs of logic circuitry is progressively being reduced.

ERROR CORRECTION

Most of the data transmission error control systems in use today rely on *decision feedback* to correct errors which are detected. Typically an acknowledge character (symbol—ACK in CCITT No. 5 code) is returned by the receiver to the transmitter on receipt of an error-free block of data; the next block is then sent. If an error is detected the receiver returns a negative acknowledge character (NAK) to the transmitter and the complete block of information is sent again.

A number of cyclic error correction codes have been designed by Hamming, Hagelburger and others which can enable errors to be corrected as well as detected by the receiver. The complexity and therefore the costs of methods using these codes tends to rise with the number of error bits in an error burst which can be corrected. The use of forward error correction codes has so far been restricted either to simplex transmission systems where decision feedback cannot be employed (e.g., radio) or in situations where error bursts are short in duration (e.g., in transmission between a control processor and its peripherals equipment).

The datel services

Datel is a generic title covering P.O. data transmission services comprising the provision of a suitable line and where necessary a modulator/demodulator (modem) together with maintenance support. An extensive range of Datel services have been developed together with other supporting services to meet customers needs.

DATEL 100

This is the slowest speed service operating at up to 50 bits per second over the telex network and up to 110 bits per second over private telegraph circuits. Standard Post

Office teleprinters are available for speeds up to 75 bits per second and facilities are available for error control, switching in private transmitting equipment if necessary and to facilitate the use of non CCITT No. 2 codes. This service is used for reservations, stock control, data collection and, with the advent of automatic calling and answering equipments, for time sharing computer bureau operation. Although the speed is relatively slow it is often quite adequate where the amount of data transfer is small. The terminal equipment costs are comparatively inexpensive.

Terminal equipment. Datel 100 terminal equipment can include the following:

(*a*) A *teleprinter* fitted with a tape-punching attachment; it can be used for preparation and/or reception of 5-bit data in punched paper tape form and may also be used in the normal way for sending messages manually from the keyboard. Incoming signals received on this type of teleprinter can produce simultaneous page copy and punched paper tape. A modification can be arranged which permits transmission and reception of any 5-bit alphabet in punched paper tape form. A teleprinter can be provided for use on 75 baud telegraph circuits.

(*b*) A *tape reperforator*, for receiving data transmitted in any 5-bit code as fully punched perforations in paper tape (not available on telex). A similar machine, the printing reperforator, will receive data coded in International Telegraph Alphabet No. 2 as partially punched perforations and printed characters in the same tape.

(*c*) An *automatic transmitter*, which is a punched paper tape reader device. Data stored in punched paper tape can be transmitted at a maximum rate of 50 bits per second.

(*d*) An *error detection unit* working on the information feedback principle' which reduces undetected errors on 50 baud circuits to between 1 in 10^5 and 2 in 10^6 characters.

(*e*) A *switching device* which allows connection to customers' privately-owned equipment to telex and telegraph private circuits for transmitting data in codes other than 5-bit alphabets.

Notes

(i) Transmission of data in both directions at once (duplex working) is possible on telegraph private circuits.

(ii) Provided that data is transmitted in International Telegraph Alphabet No. 2, output from Post Office teleprinters can be in punched paper tape form and, simultaneously, printed page copy.

(iii) All other 5-bit alphabets can be reproduced faithfully at the receiving end in punched paper tape form but the page copy will be unintelligible.

DATEL 200

This service uses the Modem No. 2 operating at speeds up to 300 bits per second bothways simultaneously and is used mainly to service the time sharing computer bureau market, *Figure 15.153*. Most customers use teletypes operating at 110 bits per second. Although the Modem No. 2 was designed for a maximum operating speed of 200 bits per second it has been found as a result of tests that speeds of 300 bits per second should be possible on the majority of calls made.

Operational characteristics of the Datel Modem No. 2. The modem is designed to accept from the input equipment, d.c. signals of specific characteristics and use them to modulate a voice-frequency carrier signal generated within the modem. The modulated signal is fed to line and reconverted to d.c. pulses by the demodulator in the modem at the distant terminal. The modem provides 2 transmission channels, one to transmit data at any rate up to 300 bits per second, and the other to receive data at any rate up to 300

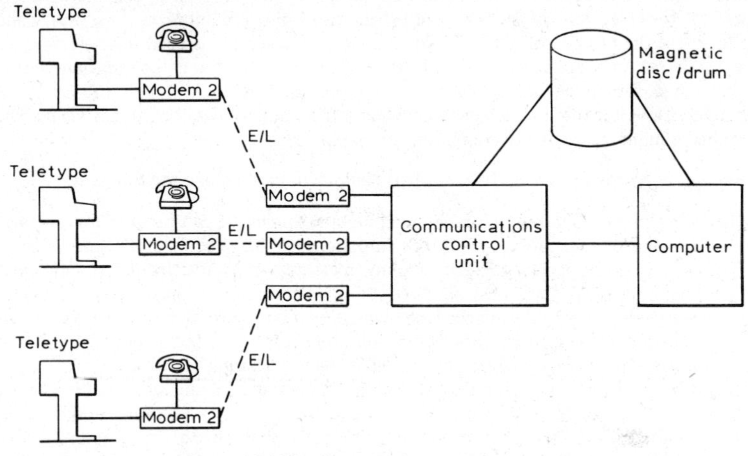

Figure 15.153 Computer bureau arrangements

bits per second. The signals produced by the modem to represent the binary 0 (space) and binary 1 (mark) conditions of the data have the following frequencies:

Channel 1	binary 0	1180 Hz
Channel 1	binary 1	980 Hz
Channel 2	binary 0	1850 Hz
Channel 2	binary 1	1650 Hz

Normally the modem is conditioned to transmit on Channel 1 and receive on Channel 2. For data to be exchanged between 2 terminals, however, one modem must be conditioned to receive on Channel 1 and transmit on Channel 2, and by international agreement this must be the modem at the called terminal, *Figure 15.154*. The Modem No. 2 switches automatically on the reception of ringing current from an automatic telephone exchange.

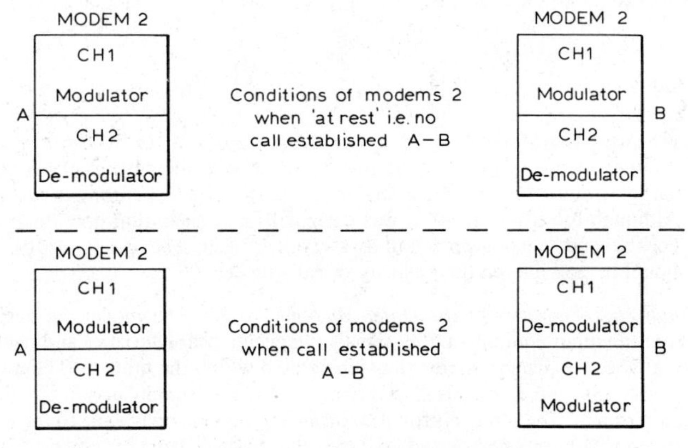

Figure 15.154 Modem 2 channel switching

DATEL 600

This offers transmission speeds up to 1 200 bits per second and uses the Modem No. 1. Where the public telephone network is used, speeds up to 600 bits per second should always be possible in one direction (on some calls it is possible to transmit up to 1 200 bits per second). Where private circuits are used the full 1 200 bits per second rate is available and data can be transmitted in both directions simultaneously. The service is used widely throughout industry and commerce for both off-line and on-line applications.

Operational characteristics of the Modem No. 1. Frequency shift keying (f.s.k.) is used. The signals produced by this process to represent the binary 0 and binary 1 conditions of the data have the frequencies shown below. The selection from the two ranges available, viz. up to 600 bits per second and up to 1 200 bits per second, is controlled from a switch on the customer's terminal equipment via a connection on the interface cable.

Data signalling rate	*Frequency used for data channel*
Up to 600 bits per second	binary 1 1300 Hz
	binary 0 1700 Hz
Up to 1 200 bits per second	binary 1 1300 Hz
	binary 0 2100 Hz

The optional slow speed return channel, operating at up to 75 bits per second enables supervisory and control signals to be returned to the sending station. The frequencies employed for this channel are:

Binary 1	390 Hz
Binary 0	450 Hz

These frequencies conform to internationally agreed standards and make optimum use of the available bandwidths on telephone circuits, in particular those of the switched public network. During data transmission the operation of the modem is remotely controlled from the private terminal equipment.

Control signals are provided by the modem to the terminal equipment enabling line disconnections, etc., to be brought to notice quickly.

Transmission facilities

(*a*) Data in one direction and simultaneous return path supervisory channel.
(*b*) Data in either direction, but not both ways simultaneously.
(*c*) As (*b*) with supervisories.
(*d*) Data in both directions simultaneously.
(*e*) As (*d*) with supervisories.

Note: Facilities (*d*) and (*e*) cannot be offered over the public telephone network.

DATEL 2400

This service uses the Modem No. 7 and provides a transmission speed of 2 400 bits per second in both directions simultaneously over high grade speech circuits. Alternatively operation at 600 and 1 200 bits per second over the public Telephone Network can be provided. The service is used for batch data transfer, information retrieval, for specialist time sharing applications and also, together with multiplexors, to enable a number of low speed data channels to be derived from a single circuit. The circuits necessary for Datel 2400 are special quality 4 wire private telephone circuits provided by the Post Office. Alternative working can be provided over the public telephone network.

Operational characteristics of the Datel Modem No. 7. The Datel Modem No. 7 employs a 4 phase differential modulation technique for synchronous transmission at 2 400 bits per second. Each pair of bits (dibits) in the originating bit stream are compared at the modulator with the preceding pair of bits using coding logic. Changes in the data result in 180° changes in the phase angle of one or both of two 1 800 Hz carrier sources, one being lagged 90° behind the other. After filtering, the combined signal is transmitted to line. As each pair of bits requires only one unit signal element the modulation rate is 1 200 baud.

At the de-modulator a reverse process detects and interprets the phase changes in the incoming signal, decoding logic reconstructing the original bit pattern. Frequency shift keying techniques are used synchronously to give alternative working over the public switched telephone network at 600/1 200 bits per second. Frequency shift keying is also used to provide a 75 bits per second supervisory channel.

Transmission facilities

(*a*) 2 400 bits per second both ways simultaneously only.

(*b*) 2 400 bits per second both ways simultaneously plus alternative working 600/1 200 bits per second in either direction but not simultaneously. Unattended answering is available on the public telephone network.

(*c*) 2 400 bits per second and 75 bits per second in both directions simultaneously only.

(*d*) 2 400 bits per second and 75 bits per second in both directions simultaneously, plus alternative working 600/1 200 bits per second and 75 bits per second in both directions simultaneously using two public telephone circuits. Unattended answering is available on the public telephone network.

(*e*) 2 400 bits per second forward channel and 75 bits per second on the backward channel simultaneously only.

(*f*) 2 400 bits per second forward channel and 75 bits per second on the backward channel simultaneously, plus alternative working 600/1 200 bits per second forward channel and 75 bits per second backward channel simultaneously. Unattended answering is available on the public telephone network.

A Post Office Datel Modem No. 7 will be provided at each end of the line. To meet the facilities offered in (*c*) and (*d*) two Datel Modems No. 7 will be required at each end of the line. Each Datel Modem No. 7 must have access to an exchange line, except in (*c*) where single exchange line access will serve two modems.

DATEL 2400 DIAL UP

The Post Office offers a Datel 2400 Dial Up Service within the United Kingdom. This service is separate from the Datel 2400 Service and is limited to operation over the inland public switched telephone network. It offers synchronous communication facilities at 2 400 bits per second or at a reduced rate of 1 200 bits per second between any two points on the telephone network in either direction alternately, but not in both directions simultaneously. The customer can switch to an alternative mode of operation which provides for speeds of 600 or 1 200 bits per second, and is identical with the fall back mode of the Datel 2400 Service. Communication between both services is therefore possible in this mode.

No. 75 bits per second supervisory channel is available.

DATEL 48K

General. Datel 48K provides facilities for the simultaneous bothway transmission of serial binary data at fixed rates of 40.8 and 48 kilobits per second over 48 kHz wideband circuits. The service is available nationally on a point to point basis only.

To optimise on line plant arrangements two modulator/demodulator equipments have been developed—the Datel Modem No. 8 which is sited at the customer's terminal premises and the Datel Modem No. 9 which is normally sited at the nearest suitable PO HF repeater station. A complete Datel 48K link generally comprises two Datel Modems No. 9 inter-connected by a 48 kHz wideband circuit and two Datel Modems No. 8 which are connected to the respective Datel Modems No. 9 by a suitable cable: a 4 wire circuit will be provided at each local end to facilitate simultaneous bothway data transmission. The system configuration for a Datel 48K link is shown diagrammatically in *Figure 15.155*.

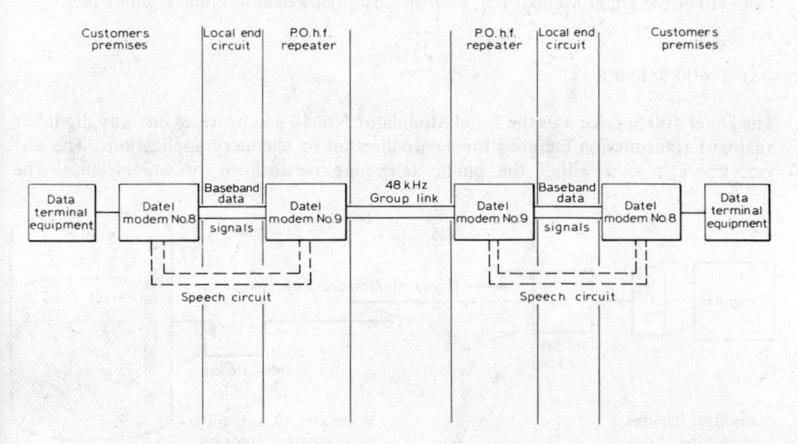

Figure 15.155 System configuration for a Datel 48k link

The service is intended for computer users who require computer to computer connection, magnetic tape to magnetic tape transfer or where there is a requirement for computer load shedding. In addition, applications can be foreseen in the computer graphics field where it is often necessary for a very high speed link between the computer and the display terminal. A further application is in conjunction with multiplexors to enable a number of separate high speed channels to be derived.

Operational characteristics of the Datel Modems Nos. 8 and 9. The Datel Modem No. 8 is designed to accept from the customer's data terminal equipment binary signals at a nominal isochronous rate of 40.8 and 48 kilobits per second. In the modem the signals are encoded to produce a pseudorandom data pattern which spreads the signal energy over a frequency spectrum independent of the input data, this technique is known as suppressed d.c. binary (s.d.c.b.). The signal from the Modem No. 8 may be transmitted via local lines to a Modem No. 9 where it is re-shaped before amplitude modulating a 100 kHz carrier. The modulated signal is then filtered to produce a signal occupying the frequency band 60–104 kHz, the 100 kHz carrier being transmitted at a reduced level. Transmission between two Modems No. 9 will be via group transmitting equipment (the signal occupying one group band circuit). On the receive side the Modem No. 9 demodulates the signal using a carrier which is frequency and phase locked to the carrier component present in the received signal and reforms a s.d.c.b. signal for transmission to a Modem No. 8. In the Modem No. 8 the signal is re-shaped, the d.c. component restored and it is decoded before being passed to the customer's terminal equipment. Full duplex operation is possible.

As stated above the data channel uses the band 60 to 104 kHz. This allows a telephone circuit to be provided as well over the same group link. The Datel Modem

No. 9 includes the necessary modulation and demodulation equipment allowing the use of a telephone channel in the 104–108 kHz band. A speech quality circuit additional to the data circuit is necessary to convey the speech channel between Datel Modems No. 8 and 9. In-band signalling facilities using 2 280 Hz are included in the modems, together with power supplies for the associated telephone.

During data transmission the operation of the Datel Modem No. 8 is controlled from the customer's data terminal equipment. Control signals are provided by the modem to the terminal equipment giving notification of conditions such as 'Ready for Sending' and 'Receive Line Signal Fail'. The timing source for isochronous operation may originate from either the Datel Modem No. 8 or the customer's data terminal equipment.

DATEL 400 SERVICE

The Datel 400 Service uses the Datel Modulator No. 10 and provides one way digital or analogue transmission facilities for data collection or telemetry applications. The service operates over either the public telephone network or private circuits. The

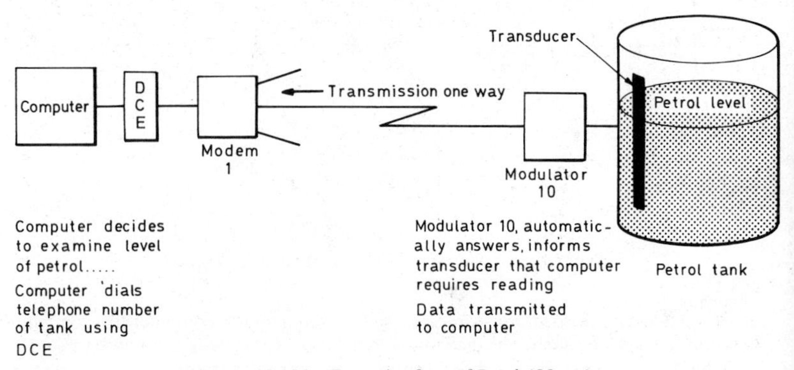

Figure 15.156 Example of use of Datel 400 series

Modulator No. 10 can be provided in a weatherproof case, if required, for siting in exposed locations. *Figure 15.156* illustrates a potential use of the Datel 400 service.

Facilities

(*a*) Digital data transmission at speeds up to 600 bits per second in one direction only, i.e., between the outstation and the central computer station.
(*b*) The Modulator No. 10 can also accept analogue data signals in the range ± 0.5 volts.
(*c*) Automatic answering facilities are provided by the Modulator No. 10.
(*d*) Speech facilities at the outstation terminal are optional.
(*e*) The Modulator No. 10 is powered from telephone line current and no mains power supply is required.

Operation. The Modulator No. 10 operating in a digital mode is fully compatible with the Datel 600 Modem (No. 1). Transmission frequencies of 1 300 Hz (Binary 1) and 1 700 Hz (Binary 0) are used. In the analogue mode it is necessary for a modified type of Modem No. 1 to be fitted at the central station which has the capability of receiving analogue signals. The analogue mode has a 300 Hz frequency spread. The modified Modem No. 1 will also accept digital signals and is therefore suitable where a system comprises both digital and analogue outstation terminals.

Environmental considerations. The Modulator No. 10 can be accommodated in either a plastic case, where normal office conditions apply, or in a weatherproof case for external siting.

The Modulator No. 10 will operate satisfactorily between −20 °C and 50 °C and up to 95% relative humidity. An associated telephone will normally be provided where office type accommodation is provided but only exceptionally where external siting is required.

Power supplies. For most locations telephone line current will be used to power the modem, but where this is not possible a Post Office provided mains power unit is used. In circumstances where no mains power is available, i.e., on a mountain, then dry batteries may be provided.

ANCILLARY FACILITIES AVAILABLE WITH POST OFFICE DATEL SERVICES

Automatic calling and answering—data control equipment (DCE)

DCE 1 Provides automatic calling facilities over the public telephone network in conjunction with the Datel 200, 600 and 2400 dial up services.

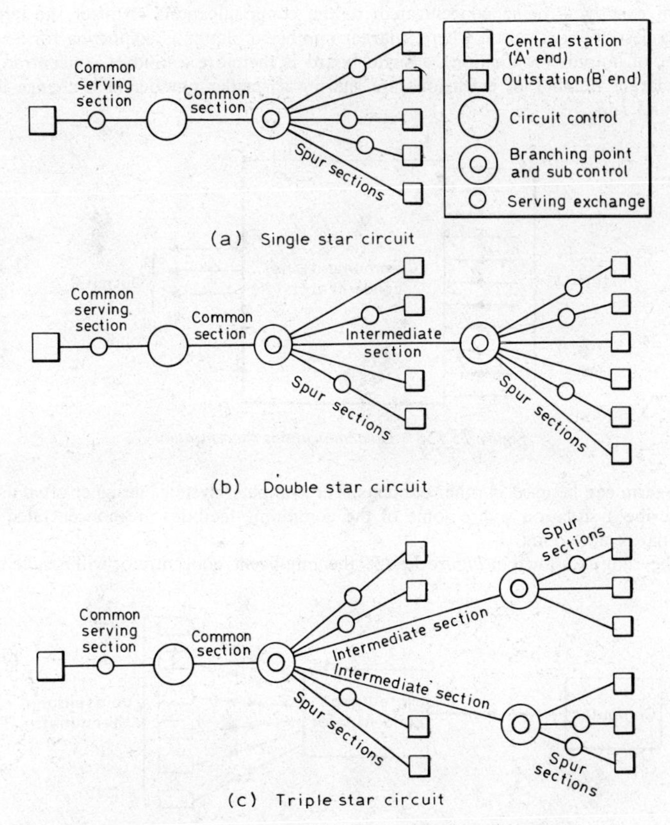

Figure 15.157 Typical Multipoint circuit configurations

DCE 2 — Provides automatic international answering facilities with Datel 200 and 600 services.

DCE 3 — Provides automatic answering at a Datel 100 termination and subsequent switching to customer's own equipment; this unit obviates the need for a teleprinter.

DCE 3 (With dialling unit) — Provides automatic calling and answering facilities for the Datel 100 service over the Telex Network.

Midnight line service. This service enables the customer to make unlimited use of the inland STD network between the hours of midnight and 6.00 a.m. daily for a fixed annual rental.

Multipoint system. The Post Office offers a systems package whereby up to 12 terminations on a system can be concentrated through a common point (branching panel) on to a high capacity circuit to a nominated centre. Maintenance is facilitated by locating the branching points in continuously manned Post Office premises. Typical arrangements are shown in *Figure 15.157*.

Concentration. Much confusion arises from the use of this term. The problem lies in what exactly is being concentrated: to the communications engineer the term is used to describe a situation where a larger number of inputs is competing for a small number of outputs (contention), a switchboard is therefore a kind of concentrator in that a larger number of extensions are sharing a smaller number of exchange lines, *Figure 15.158*.

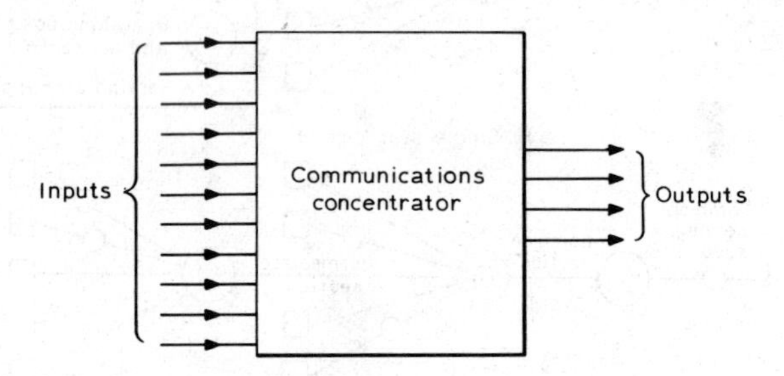

Figure 15.158 Communications concentrator

The term can be used in other contexts. The computer systems designer often uses it to describe a situation where some of the computing facilities are concentrated at a particular remote point.

In the example shown in *Figure 15.159* the 'intelligent' concentrator will handle some

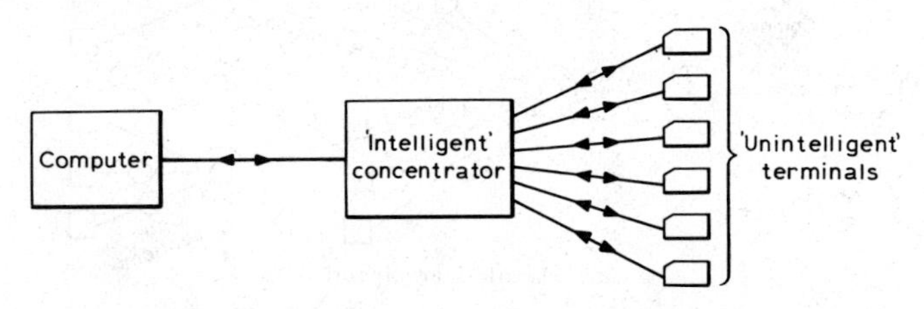

Figure 15.159 'Intelligent' concentrator

of the jobs which otherwise might be done by the main processor. Examples of some of the functions which might be performed are as follows:

(i) Assembly of complete message blocks for transmission over the main link.
(ii) Stripping and adding of start/stop signals—necessary for the terminals but not the main processor.
(iii) Error checking.
(iv) Data compression—reducing the amount of data transmitted over the main link by eliminating blank spaces, etc.

These devices are often referred to as *hold and forward*, or *store and forward* concentrators.

Multiplexing. Multiplexing is the division of a common path into a number of separate circuits. There are two types of multiplexing in use today. These are frequency division multiplexing (f.d.m.) and time division multiplexing (t.d.m.).

With *frequency division multiplexing* a relatively wide bandwidth (range of frequencies available for signalling) is divided into a number of smaller bandwidths to provide more channels of communication. For example the bandwidth required for the transmission of 50 baud telegraph signals is only 120 Hz, while the bandwidth of a good quality speech circuit is about 3 000 Hz—using f.d.m. 24 telegraph circuit can be derived from one speech circuit.

Time division multiplexing is a process whereby a circuit which is capable of a relatively high information transfer rate is divided into a number of time slots to provide a number of lower speed circuits. For example a line which is capable of carrying 2 400 bits per second could, by the use of t.d.m. theoretically be divided into four, 600 bits per second circuits, or a combination of different speed circuits up to a maximum of 2 400 bits per second. It will be noticed that in both forms of multiplexing there is no *contention*, i.e., the number of inputs is the same as the number of outputs. This is the major difference between *concentrating* and *multiplexing*.

Comparison of t.d.m. and f.d.m. systems used in data communications. Each type of multiplexor has its advantages and disadvantages. The main advantage of f.d.m. is its low cost and simplicity. However, where more than about 12 low speed circuits are required t.d.m. systems begin to be more attractive on a cost basis, the effective cost per circuit reducing as the number increases.

An increasing advantage of t.d.m. over f.d.m. is the facility to multiple medium speed asynchronous data, i.e., 30, 60 characters per second as well as the usual 10 characters per second speeds of the teletype. This is particularly important with the advent of higher speed printing terminals and other such terminal devices. Some t.d.m. systems also have the ability to intermix synchronous as well as asychronous data. This could be useful to service high speed line printers and video terminals, etc., on the same link as a number of lower speed terminals.

Some advantages of t.d.m. over f.d.m. are:

(*a*) Stripping of start–stop bits for increased transmission efficiency;
(*b*) Improved monitoring and diagnostic facilities;
(*c*) Constant performance on all low speed circuits;
(*d*) Flexibility to permit system and circuit configuration changes.

THE POST OFFICE DATAPLEX SERVICES

Dataplex System 1. Dataplex System 1 employs frequency division multiplexing techniques and enables up to 12 separate circuits, each operating up to 110 bits per second to be derived from a good quality speech circuit. See *Figure 15.160*.

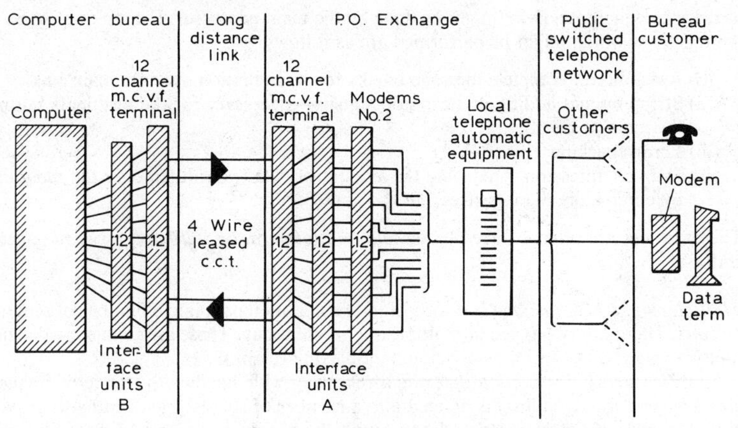

Figure 15.160 Dataplex System 1

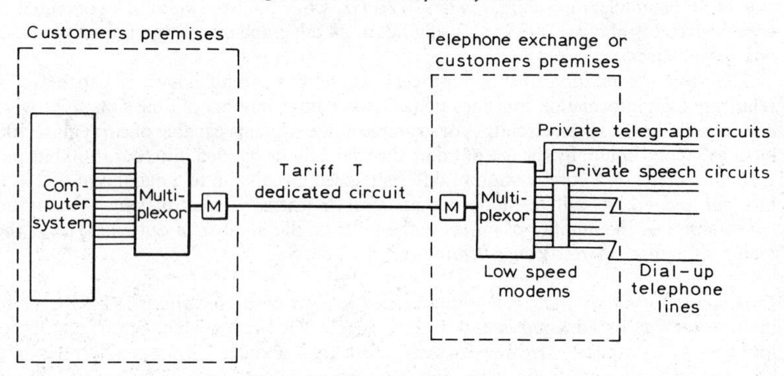

Figure 15.161 Dataplex System 2

Dataplex System 2. Dataplex System 2 employs time division multiplexing and enables up to 51 circuits operating at up to 110 bits per second to be derived from a Tariff T circuit operating at a bearer rate of up to 4.8 kilobits, *Figure 15.161*. Varying speeds of up to 1 200 bits per second can be multiplexed simultaneously.

FURTHER READING

ADAMS, G. N., 'A Modem for the Datel 2400 Service', *P.O.E.E. Journal*, **62**, 156 (1969)

BIGG, R. W., 'Modems for 48 K bit/s Data Transmission', *P.O.E.E. Journal*, **64**, 116 (1971)

BOTT, A. J. and CASTLE, W. C., 'Automatic Calling and Automatic Answering for Datel Services', *P.O.E.E. Journal*, **63**, 253 (1971)

BOTT, A. J. and GARRETT, W. F., 'Automatic Calling and Answering of Datel Calls on the Telex Network', *P.O.E.E. Journal*, **64**, 105 (1971)

CCITT Green Book Vol. VIII

CHESTERMAN, D. A., 'International Telegraph Alphabet No. 5', *P.O.E.E. Journal*, **62**, 89 (1969)

CULLEN, B. C., *The Post Office Handbook of Data Communications* (19)

HART, J. J., 'The Maintenance of Datel 600 Multipoint Networks', *P.O.E.E. Journal*, **65**, 115 (1972)

HUFFMAN, A., 'A method for the Construction of Minimum-Redundancy Codes', *Proc. I.R.E.*, **4**D, **9**, 1098 (1952)
MARTIN, J., *Systems Analysis for Data Transmissions*, Prentice Hall (1972)
PRITCHARD, *An Introduction to On-line Systems*, NCC Publications (1973)
ROBERTS, L. W. and SMITH, W. G., 'A Modem for the Datel 600 service—Datel Modem IA', *P.O.E.E. Journal*, **59**, 75 (1966)
RUSSELL, J. D., 'Data Transmission in the Terminal Orientated Environment', *Data and Communications Design*, August–September (1972)
RENTON, R. N., *Data Telecommunications*, Pitman (1973)
SPANTON, J. C. and CONNELLAN, P. L., 'Modems for the Datel 200 Service—Datel Modems 2A and 2B', *P.O.E.E. Journal*, **62**, 1 (1969)
YOURDON, E., *Design of On-line Computer Systems*, Prentice Hall (1972)

ELECTRONIC TELEPHONE EXCHANGES

Since the introduction of automatic switching early this century the majority of telephone exchanges throughout the world have used electro-mechanical equipment; Strowger step-by-step and crossbar being the most popular systems. Rapid developments of Radar and other electronic techniques during World War II made available components and circuitry which telecommunications engineers subsequently utilised in experimental switching systems. Some of these experiments led to the successful introduction of electronic elements within electro-mechanical exchanges, while others paved the way for the electronic telephone exchanges now being installed in many parts of the world.

Although many technical problems had to be solved, the electronic exchange development engineers encountered two other problem areas which influenced their work. Firstly it was found that existing specifications of electro-mechanical equipment described each unit of equipment in detail but, taken collectively, they did not define the basic system requirements: initially much effort had to go into the identification and definition of these requirements, with special attention being given to the problems of interworking with a variety of subscribers apparatus and conventional exchange equipment, using many different signalling and line transmission systems. The second, and perhaps more difficult problem was an economic one in which it was necessary to satisfy the operating companies and administrations that the relatively expensive electronic exchange equipment offered a viable commercial proposition. Although it was clear that a reduction in moving parts would reduce maintenance costs it was difficult to place a financial value on this, and even more difficult to forecast the cost benefit of a system which offered enhanced performance and potential for improved facilities. The threat of continually rising labour costs and the potential for cost reduction of electronic equipment as new devices became available, slowly swayed the arguments in favour of electronically controlled exchanges.

At this point in time several countries have electronic exchanges in production. Most of these systems use electronic control of space division analogue switching networks, employing either reed relays or crossbar contacts. More advanced systems, some using time division and digital switching, are under development.

Basic types of electronic exchange

The functions of all types of telephone exchange can be divided between those that control the setting up and release of calls and those that supervise a switched connection. In the Strowger system the switching and control functions are largely combined in the two motion selector with a poor utilisation of equipment elements. In attempting to improve the utilisation, designers have made progressively more use of time sharing techniques so that appropriate equipment elements are allocated only when needed during a call and are otherwise available to service other calls. The ultimate in this approach is reached when computer-like equipment is used to control a switching network which may also be time shared.

Switching networks can be divided into two types. One type provides capacity for switching analogue transmission circuits and generally utilises space division techniques, although there have been experimental systems using time-division-multiplexing (t.d.m.). The other type provides digital switching to interconnect circuits carrying coded samples of speech using pulse code modulation (p.c.m.).

Analogue switching systems

The switching network of a space-division exchange is usually built up of a number of coordinate arrays of switching elements, each array being referred to as a switch. *Figure 15.162* (*a*) shows diagramatically a switch in which any one of n inputs may be connected to any one of m outputs by operating a switching device at the appropriate intersection or crosspoint. *Figure 15.162* (*b*) shows a typical trunking for interconnecting switches, with *A*-, *B*- and *C*-switches being arranged to provide three stages of switching. Exchange lines and junctions could be connected to the n inputs of the several *A*-switches, and a connection between two lines would be switched via six crosspoints connected in the order *A-B-C-C-B-A*. The *A*-switches may be regarded as a concentration or access stage, each line having crosspoints to provide access to common *B*-switches. Trunks between *B*- and *C*-switches may be so arranged that every n input to a *B*-switch may obtain access to every m output of *C*-switches. Connections from the latter are turned back to other *C*-switches via a link circuit and this link provides a convenient place for equipment to supervise the connection. The trunking is similar to conventional crossbar in many respects but the ability to vary the dimensions n and m enables smaller switches to be made and consequently trunking configurations requiring fewer crosspoints may be devised. Considerable effort has been expended on the development of economic trunkings, and in providing efficient service even in the event of individual switches becoming unserviceable. Additional variations are possible by arranging the individual switches so that they provide limited availability, i.e., each of the n inputs is given access to a limited number of m outputs and vice versa. This economises in crosspoints per switch but construction and control become more complicated.

A principal feature of conventional space-division switching systems is a two wire electrical connection through the switching network, as provided by selector wiper-to-bank contacts in the Strowger system and relay spring contacts in crossbar switches. Since these contacts offer very high resistance when open and almost zero resistance when closed, it has been difficult to find suitable electronic alternatives. Gas diodes, gas triodes, semi-conductor diodes, transistors and *pnpn* devices have all been tried but generally the systems have been uneconomic because of the cost of essential per-line equipment for impedance matching and protection against high line voltages. The use of these electronic devices also leads to crosstalk problems.

The search for a genuine electronic crosspoint continues and a promising recent development is a family of solid state semiconductor devices known as light emitting diodes (l.e.d.s). These devices can be packaged alongside photo-sensitive detectors to provide opto-isolation of very high resistance but useful coupling when energised: it remains to be seen whether they can be of use in analogue switching matrices.

A switching device being used in many of the new space division analogue exchanges is the reed relay in which an electromagnetic coil surrounds a number of glass-encapsulated reed inserts. Energisation of the coil causes the contact blades within each insert to be magnetically attracted and the resultant contact between gold plated surfaces provides a noise-free connection of less than 200 milliohms resistance. The construction of the reed relay makes it suitable for electronic equipment production

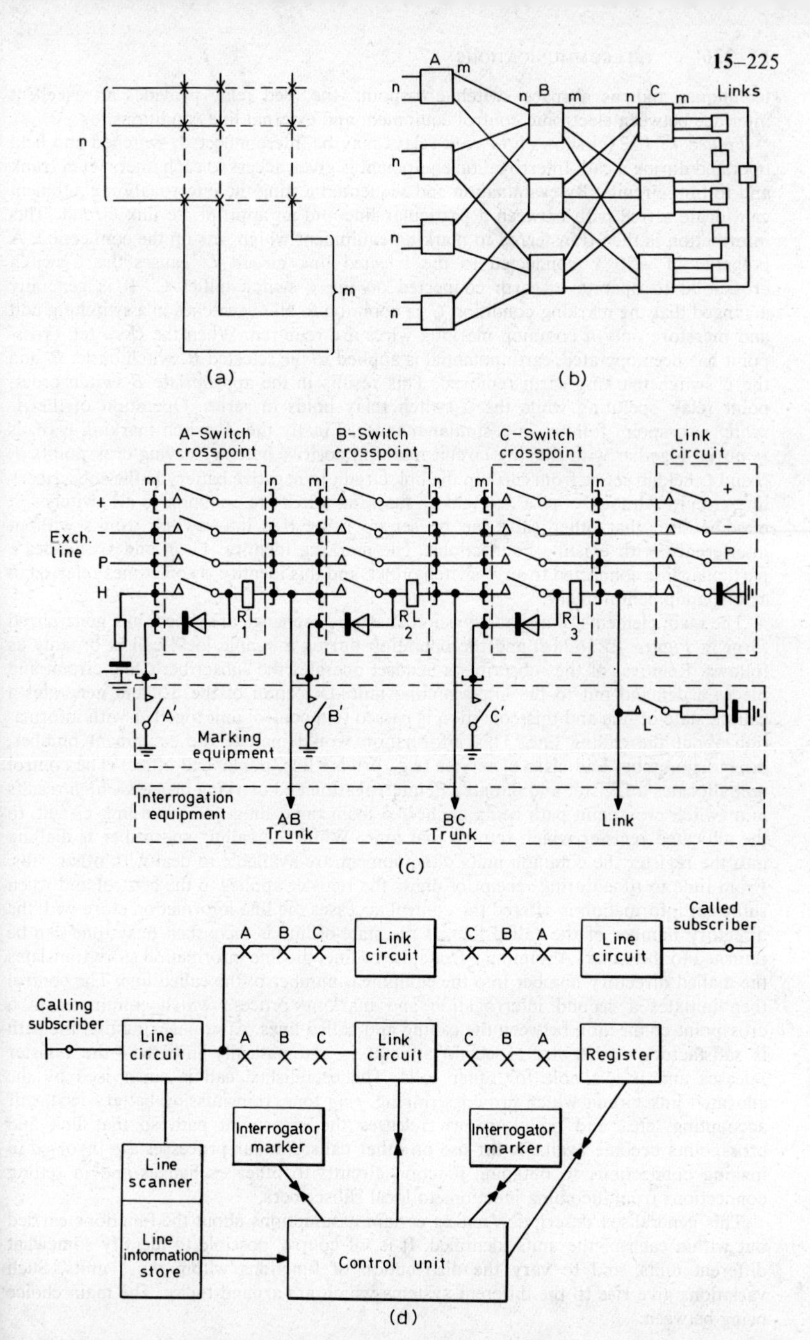

Figure 15.162 (a) Basic space-division switch for connecting any one of n inputs to any one of m outputs (b) Trunking plan using space-division switches (c) Method of marking, switching and holding reed-relay crosspoints in a space-divisions exchange (d) Main elements of a space-division analogue-switching telephone exchange.

techniques and, as a space switch crosspoint, the reed relay provides an excellent interface between electronic control equipment and external line conditions.

Figure 15.162 (*c*) shows how reed relays may be interconnected, switched and held operated during a call. Interrogation equipment is given access to each interswitch trunk and to link circuits. By examination and sequential testing the interrogation equipment can locate a free path between a particular line and an appropriate link circuit. This information is then transferred to marking equipment which sets up the connection. A potential of +50 V connected to the selected link circuit L' causes the C-switch crosspoint to operate to earth connected on the C-switch outlet C'. It is generally arranged that the marking condition C' is common to all C-switches in a switching unit and therefore only n common marking wires are required. When the C-switch crosspoint has been operated, earth potential is applied to the selected B-switch outlet B' and the C-switch marking earth removed. This results in the appropriate B-switch crosspoint relay operating while the C-switch relay holds in series. Operation of the A-switch crosspoint follows in a similar manner. Finally the A-switch marking earth is removed together with the link circuit marking positive battery, leaving crosspoints A, B and C held in series from earth in the link circuit to negative battery in the subscribers line circuit. Although the A', B' and C' marking wires are common to all switches, it may be seen that other calls can be set up using free inter-switch trunks without interference with existing connections. The marking identity A', B' and C' defines a particular line connected to an A-switch outlet, and this identity is sometimes referred to as the equipment number.

The main elements of a space-division analogue exchange are shown in a generalized form in *Figure 15.162* (*d*) and the operation during a simple local call is broadly as follows. Removal of the subscriber's handset operates the subscriber's line circuit and places a demand out to the line scanning unit. Detection of the demand generates a calling state-of-line and this condition is passed to a control unit together with information about the calling line. This information would include the equipment number, directory number and class of service (e.g., private line, public call office). The control unit allocates a register and initiates an interrogation and marking process which results in a switch crosspoint path being connected from the calling line, via a link circuit, to the allocated register which returns dial tone. While the calling subscriber is dialling into the register the common units of equipment are available to deal with other calls. From time to time during receipt of digits the register applies to the control and when sufficient information is offered the control accesses the line information store with the directory number of the called line. If the state-of-line is *busy* then busy-tone can be returned to the caller. Assuming a free state-of-line, the line information store translates the dialled directory number into the equipment number of the called line. The control then initiates a second interrogation and marking process which culminates in a crosspoint connection between the calling and called lines. After checking that the path is satisfactory—otherwise a second attempt is automatically initiated—the register releases and is available for other calls. The established call is supervised by the allocated link circuit which provides ringing, ring tone, transmission battery feed, call accounting, etc., and on cleardown releases the crosspoint path so that link and crosspoints become available for use on other calls. Similar processes are involved in making connections to outgoing junction circuits to other exchanges and in setting connections from incoming junctions to local subscribers.

This generalised description makes certain assumptions about the functions carried out within each of the units identified. It is, of course, possible to identify somewhat different units, and to vary the distribution of functions within those units. Such variations give rise to the different systems which are around today. The main choice being between.

(*a*) Progressive concentration of functions into powerful central control units which are efficient but insecure and very complicated.

(*b*) Distribution of functions into smaller control units which are simple and secure, but less efficient.

Digital switching systems

In recent years there has been a strong upsurge of interest in transmission systems using p.c.m. transmission techniques. Many telecommunications administrations are introducing 24 or 32 time-slot systems which allow a substantial increase in the number of circuits carried by valuable cable pairs. In order to use p.c.m. transmission, a multiplex terminal is introduced in which the waveforms of several analogue speech circuits are sampled and coded on to a binary bit-stream in such a way that up to 30 circuits are interleaved. The latest European system uses 8 kHz sampling, 256 levels, 8-bit coding and 125 microsecond repetition rate with 30 speech and 2 auxiliary channels byte interleaved on a 32 time-slot bearer operating at 2 048 k bit/s. Higher order systems operating at 8 448 k bit/s and 120 M bit/s are under development.

At present p.c.m. transmission systems are normally used on point-to-point links with analogue/digital conversion at each end of the link. Interconnection of channels between systems is then performed by analogue space-division switching. There is increasing activity on the development of digital switching systems in which the 2 048 k bit/s bit streams would be connected directly to a time-shared digital switch, permitting interconnection of channels without the digital-analogue-digital conversion process.

In its simplest form as shown in *Figure 15.163* (*a*), a digital switching system would contain a single time-switch store having areas allocated for each incoming time-slot of all p.c.m. line systems on the exchange. The time-switch store would be scanned cyclically so that each 8-bit byte appearing in the incoming bit stream could be written into the appropriate store area. A connection through this time-switch would be set up by inserting the address of the incoming store position against the outgoing time-slot in the outgoing address store. The addresses in this outgoing store are used to direct the acyclic read process so that the appropriate 8-bit byte can be read out into the outgoing bit stream. For example, the 8-bit byte labelled *D* in time-slot 4 of the incoming bit stream would be written into the time-switch store position labelled *S4*, alongside cyclic address position 04. When the cyclic scan of the outgoing address store reaches outgoing time-slot 8, the address *S4* is used to give acyclic access to the 8-bit byte labelled *D* in the time-switch store. The byte *D* is then read out into the outgoing bit stream in time-slot 8. Thus a time switching action has taken place between time-slot 4 of one system and time-slot 8 of another. The reverse direction of transmission would be similarly switched. By providing multiplexers and operating the highways to the time-switch store at a relatively high speed, it would be possible to connect several 2 048 k bit/s line systems on to one time-switch store. However the 2 048 k bit/s stream allows only 488 nanoseconds for each bit and, since any multiplexing reduces this period, the capacity of a single time-switch is severely limited by the basic store technology. One technique for increasing the capacity is to convert from the 8-bit serial stream as on the line, into single bit parallel on each of 8 parallel highways into the time-switch store. Even greater capacity can be obtained by using a number of time-switches in association with time-divided space switches to provide full availability between all slots in all systems. If the space switches are inserted between the line systems and central time-switches a space–time–space (STS) configuration emerges as shown in *Figure 15.163* (*b*). Alternatively, 2 time-switch stages can be connected to a central space switch to give the time–space–time (TST) arrangement in *Figure 15.163* (*c*). The relative costs of components for time and space switching has changed over the last few years so that early enthusiasm for STS has been replaced now by a general acceptance of the TST configuration.

Figure 15.163 (*d*) shows a typical TST system in which 8 systems are terminated on a unit which provides frame alignment, signal extraction, time-switch storage, etc. The

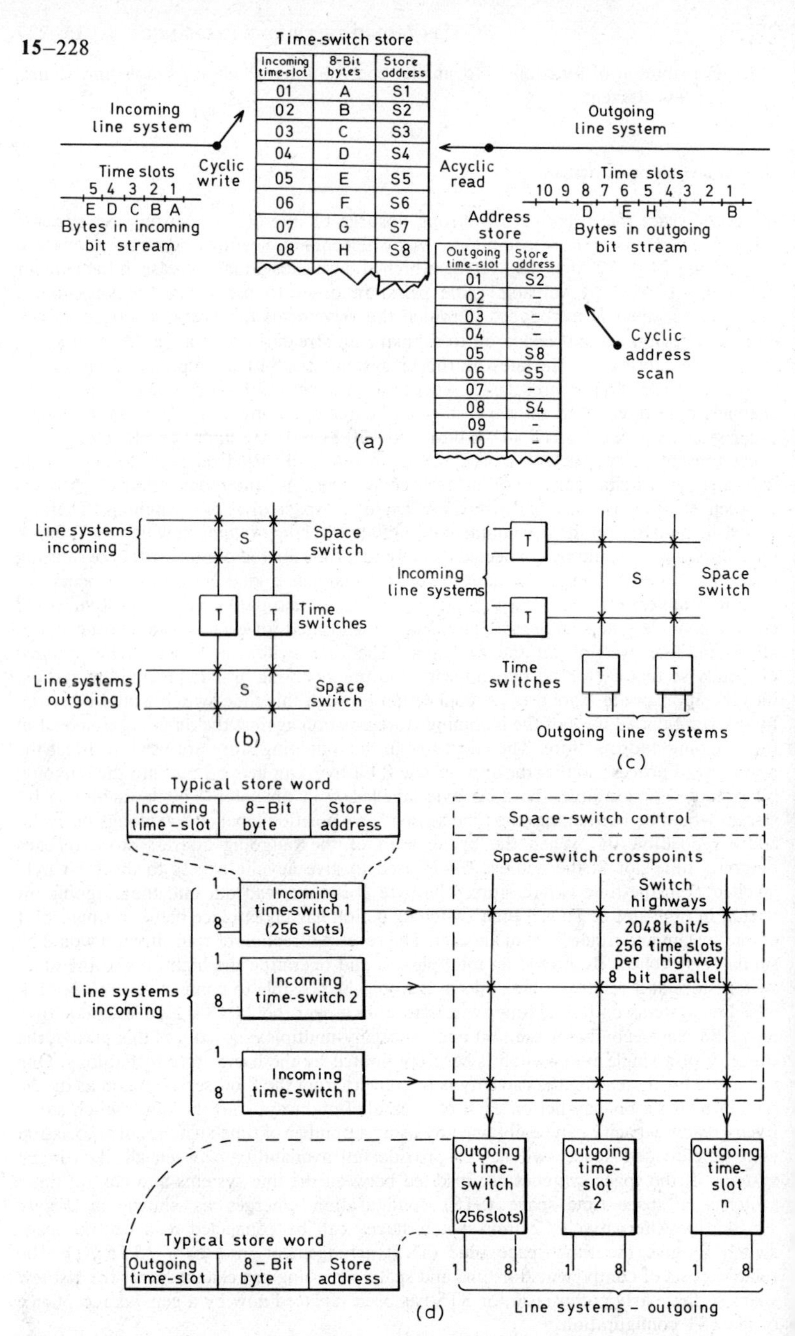

Figure 15.163 (*a*) *A single time-switch providing flexible interconnection of time-slots between two digital line systems* (*b*) *Space–time–space configuration* (*c*) *Time–space–time configurations* (*d*) *Digital switching system: typical time–space–time configuration*

8-bit bytes from each incoming line system would be written into the input time switch stores in a way similar to that described above. However, a connection would be made by inserting the incoming time-switch address and an outgoing time-switch address into the space-switch control store. The switching action is then for the space-switch control to operate the appropriate space-switch crosspoints so that an 8-bit byte can be read out of the incoming time-switch store and written into the outgoing time-switch store, from where it is subsequently read out into the bit stream of an outgoing line system. The operation of the space-switch crosspoints and the transfer of bytes takes place rhythmically, once every 125 microseconds. The actual time taken to transfer the bytes through the space switch will vary with the order of multiplexing but the typical figures shown in *Figure 15.163* (*d*) allow only about 400 nanoseconds for this operation. Again the reverse direction of transmission is switched in a similar way although some designers favour the operation of one crosspoint path for a simultaneous both-way transfer of bytes, while others treat the two directions separately and provide independent transfers of bytes.

Other exchange configurations are feasible using various combinations of time and space switching with serial or parallel highways having bit rates ranging from a conservative 2 M bit/s to an ambitious 100 M bit/s. Although very high speed space-switching is feasible, such speeds are not compatible with currently available mass storage systems, and the high order of multiplexing produces large capacity units which would be a serious security risk.

Network synchronisation

The use of integrated digital transmission and switching equipment in an extensively meshed network introduces a new concept of network synchronisation. If the clocks at digital exchanges are not closely controlled it is possible for phase changes to develop in which digitally coded samples would be either lost or repeated. While such 'slips' are not likely to have a noticeable affect on speech transmission quality, the possible use of digital connections for data transmission purposes has prompted some administrations to propose a network synchronisation system in which the frequency of the clock at each exchange is controlled to a mean level at which slips do not occur. Network synchronisation would be maintained by equipment at each digital exchange which would monitor the incoming 2 048 k bit/s bit stream on nominated links and compare the phase with that of the local clock: phase differences would be recognised and clock frequency adjustments made following an interchange of control signals between the relevant exchanges.

Control equipment

Electro-mechanical space-division systems generally use distributed control in which the functions associated with the setting-up, supervision and release of a call are performed by relatively slow elements scattered throughout the exchange. The utilisation of these elements is low and simple replacement by higher speed electronic elements does not improve matters. Hence the introduction of electronic devices in telephone exchange equipment has led the system designers to progressively greater concentration of functions into time shared control equipment. One stage of this progression is illustrated in *Figure 15.164* (*a*) where a number of relatively simple registers are being controlled by a common serial processor. Each register is connected in turn to the processor via gate circuits actuated by pulses P_1 to P_n in a regular sequence; during each pulse period the register offers information to the processor which carries out the logic functions, stores data and returns further instructions to the register. A similar

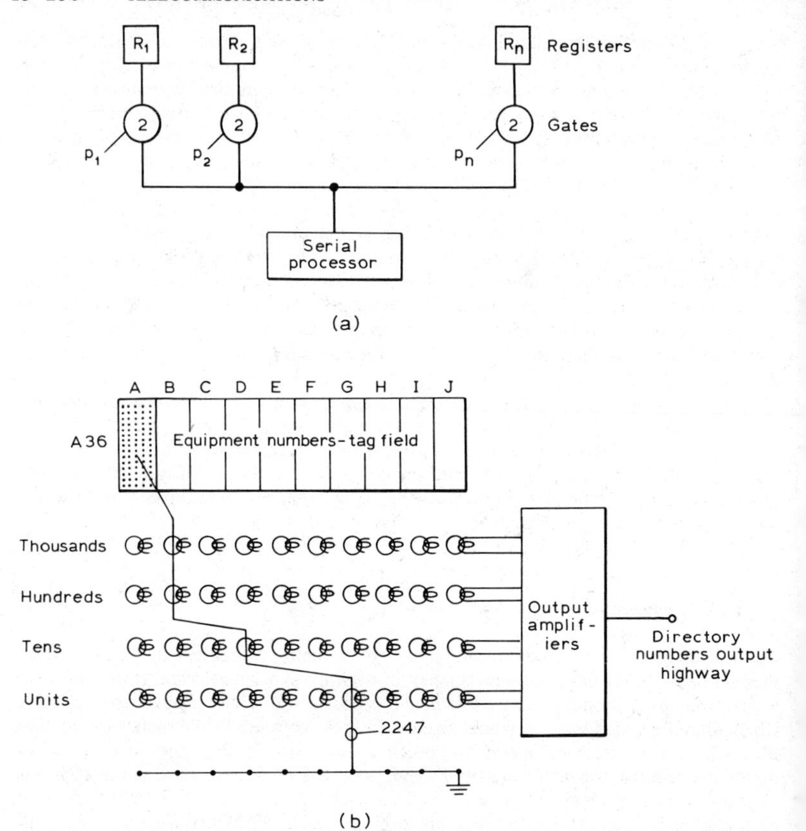

Figure 15.164 (*a*) *Basic arrangement for serial processing many individual equipments by common equipment* (*b*) *Equipment number to directory number translation using a threaded core store*

arrangement can be used to control a number of simplified supervisory relay sets, performing such functions as the connection and disconnection of ringing and tones, line signalling, call accounting and various time-outs.

In its relatively simple form a serial processor would use wired logic and although this offers a considerable improvement over distributed control, any changes to introduce additional circuits or new facilities still requires wiring modifications. An improved arrangement has been introduced on several of the electronically controlled reed relay exchange systems in which individual wires are threaded through rows of transformer cores in a Dimond ring store, in such a way that a current pulse in a single wire generates a discrete set of output pulses. *Figure 15.164* (*b*) shows the principle in use for translating a subscriber's equipment number—as identified by the tag wiring—into the directory number—as generated in the output circuits. The ability to change a translation by simple rethreading of the wire gives considerable flexibility for the allocation of exchange equipment. The technique can also be used to provide several thousand words of program in a wired-program-store to control the function of other units of exchange equipment such as registers and interrogator/makers.

While the threaded core technique offers clear advantages over wired logic, telephone exchange designers have for some time, been aware of the further advantages in speed and flexibility available in computer-like control systems. In a typical stored-program-control (s.p.c.) system the switching-system hardware would be controlled by a number of central processing units (CPUs), using ferrite core main storage for programs and fast-access data, supported by magnetic drum or disc stores for slow-access data and additional programs. Extensive use of software and electrically-alterable storage provides a system which is easily updated with changes in exchange data and which can readily accommodate new facilities.

The progressive concentration of functions into control units of ever increasing complexity presents a very real security problem when applied to public telephone exchanges, and considerable effort is going into the development of processing system configurations capable of giving long periods of uninterrupted service. A typical design objective would be 100 years mean-time-between-failures of the complete processing system.

Summary

The evolution of automatic telephone exchanges has so far moved from electro-mechanical space-division analogue systems using distributed control, through to electronically controlled reed relay systems having progressively more sophisticated processing systems and exploiting stored program control techniques. Fully electronic digital switching systems, suitable for telephone and perhaps other services such as data communication, are now being developed but there are already signs of a reduction in the amount of centralised control. Apart from the security problems mentioned earlier, the effort required to develop software for a complex central processing system runs into hundreds of man-years, and having completed the programs it is not unusual to find the exchange size restricted by the power of the processing system. To ease the load on a central processing system, smaller pre-processors are being introduced to carry out more of the simple logic functions so that ample central processing system power is available for the really difficult tasks. This swing back towards distributed control is being encouraged by the availability of medium and large scale integrated circuits, programmable logic elements and other novel techniques.

For electronic telephone exchanges the development scene is set for the next few years with the present trend being towards stored program control of digital switching systems. However, the increasing rate of change of technology, together with the insatiable demands of modern society for more services and new facilities, will no doubt stimulate further developments. The extent to which these demands can be met will depend largely on the inertia of the existing networks and the need to interwork, nationally and internationally, with a plethora of ancient and modern systems.

FURTHER READING

ATKINSON, J., *Telephony*, Pitman (1950)

HARTLEY, G. C., *et al.*, *Techniques of Pulse Code Modulation in Communication Networks*, C.U.P./I.E.E. (1967)

16 SOUND AND VIDEO RECORDING

SOUND AND VIDEO RECORDING 16

16 SOUND AND VIDEO RECORDING

SOUND RECORDING ON MAGNETIC TAPE

The recording of sound started in 1876 when Thomas Alva Edison invented his tin-foil phonograph. It had little or no commercial value but out of this small beginning sprang the great recording industry of today. In the years that followed improvements were made in acoustic recording and reproduction, but it was not until the early 1920s when amplification was introduced into electrical communication, that any great advancements were made possible in the field of sound recording and reproduction.

The principle of magnetic recording was developed in Germany in 1924, and has been used in public service broadcasting since 1930.

Early systems of magnetic recording used a steel wire but this proved both bulky and difficult to join. Between the years 1939–45 experiments were conducted using a plastic tape impregnated with iron dust. Improvements in plastic tape have continued to produce the high quality tape available today.

Basic principles of magnetic recording

In the basic process of sound recording sound waves are converted by a microphone into varying e.m.f.s which correspond in frequency and amplitude to the original sound waves. These e.m.f.s are amplified and applied to the coil of an electromagnet which forms the recording head, the magnetic field of which varies in direct relationship to the sound waves. A plastic tape coated with a ferrous oxide material is moved at a constant speed past the head over a gap in the pole pieces where the magnetic field is concentrated. The time taken for a particle of tape to move past the head is small and each particle of tape is left with a remanent flux density the value of which bears a relationship to the original sound waves, providing the following conditions are fulfilled.

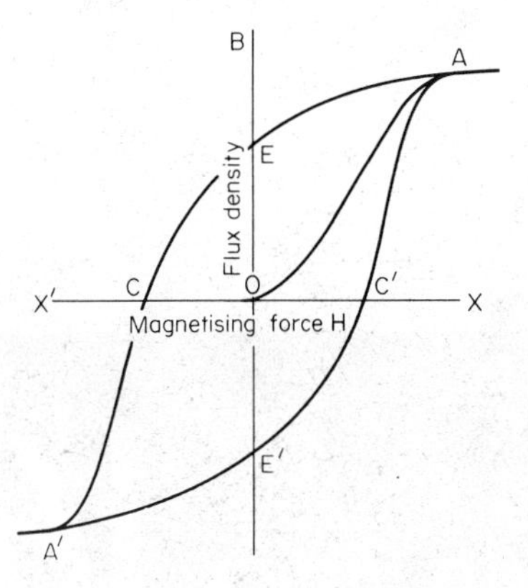

Figure 16.1 Magnetic hysteresis loop

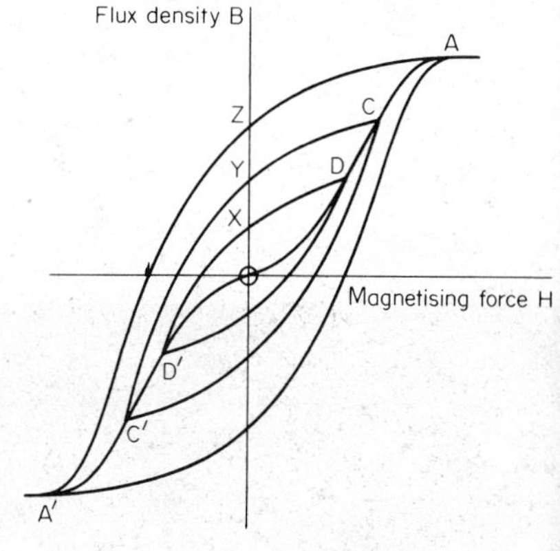

Figure 16.2 Family of hysteresis loops

(*a*) All the elements in the tape must be in the same magnetic state when they reach the recording head.
(*b*) The remanent flux density on the tape must be proportional to the magnetising force which produces it.

The first condition is achieved by completely demagnetising the tape prior to the recording head. The second, by ensuring that the linear part of the B/H curve for the material is used, *Figures 16.1* and *16.2*.

REPRODUCTION

If the magnetised tape is passed across the gap of an electromagnet similar to the recording head the remanent flux densities on the tape will induce corresponding e.m.f.s in the coil of the electromagnet. These are amplified and fed to a loudspeaker which converts them into sound waves which are a replica of the original.

HYSTERESIS LOOP

If a demagnetised material is subjected to a magnetising force H the flux density B in the material increases as H is increased. This is shown by the curve OA, *Figure 16.1*. At point A any further increase in H does not produce a corresponding increase in B. In this condition the material is said to be saturated. OA is known as the B/H curve of the material and A as the saturation point. If the magnetising force H is now reduced to zero the flux density falls to E and OE is the remanence for the material used. Should H now be increased in the opposite direction to C the flux density reduces to zero. OC is known as the coercitivity of the material. A further increase of H towards X' causes the flux density B to rise but in the opposite direction until it reaches saturation point A'. Decreasing H to zero now causes B to fall to E'. Increasing H to C' causes B to fall to zero. If H is increased still further B continues to rise until A is reached.

This cycle of magnetisation produces the hysteresis loop $AECA'E'C'A$. If the hysteresis loops for a material are determined for differing maximum values of magnetising force they are found to lie within one another as shown in *Figure 16.2*. It will be seen that the values of remanent flux density depend on the maximum magnetising force. With loop A remanent flux density is Oz whereas with loop D remanent flux density is OX.

If an electromagnet is bent in the form of an annular ring with a small gap between the pole pieces, the flux density is concentrated within the gap and varies as shown depending on the value of the magnetising force.

This is the basic principle on which a sound recording head operates.

TAPE TRANSPORT SYSTEM

The fundamental requirement of the tape transport system is to take the tape past the head assembly at a contant speed, i.e., with minimum *wow* and *flutter*. Wow is a long term speed variation caused by any uneveness in the driving system. This can be caused by wear or lack of cleanliness. Flutter is a short term variation which can be caused by excessive pressure between tape and heads, or uneveness in tension on the take-up spool. Wow and flutter are expressed as a percentage of the tape speed and a typical figure for a professional machine is 0.06%

Figure 16.3 shows a tape transport system comprising feed spool with tape, strobe wheel to determine if the machine is running at the correct speed, pressure or pinch wheel, idler and take-up spool. The pressure wheel is made of rubber and holds the tape against the drive capstan to provide drive and prevent tape slip. When the machine is at rest the pressure wheel is held clear of the drive capstan. Were the two to remain in contact when the machine is stationary flats would develop on the pressure wheel resulting in uneven drive.

A spring loaded idler is fitted on some machines, this prevents variations in tension on the take-up spool being reflected back to the drive system. It is common practice to fit a microswitch to the idler, so that in the event of the tape running out the drive mechanism is released and the spool motors switched off.

Various methods of operating the tape transport mechanism are in use. These include

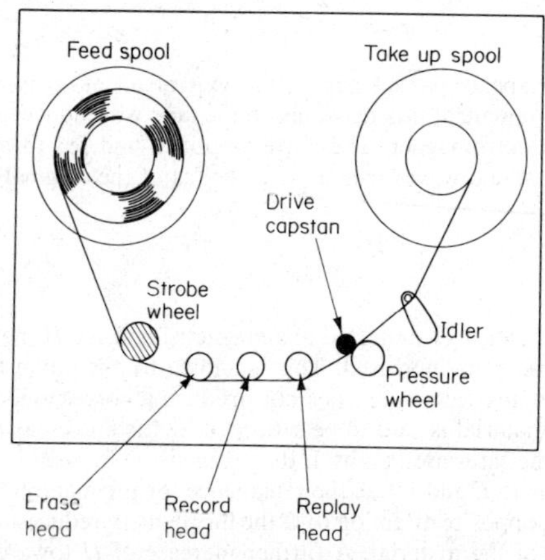

Figure 16.3 Tape transport system

belts, pulleys, flywheels and separate motors. The particular method employed plays a large part in determining the price of the machine. All professional machines employ separate motors for the feed spool, take-up spool and drive capstan. Most modern machines are capable of operating at more than one speed, some form of speed change mechanism is therefore necessary. In the fast forward or fast rewind, the tape is held clear of the head assembly in order to prevent undue head wear. An additional requirement on modern machines is the *edit* facility. This partly releases the braking system and allows the tape to remain in contact with the head while leaving the main drive inoperative.

HEAD ASSEMBLY

The head assembly comprises an erase head, record head and replay head. These may be fitted as separate items or as a complete head block assembly. The advantage of the latter is in the ease of replacement. All professional machines are fitted with separate record and replay heads enabling continuous monitoring of the recording as it is being made. There is of course a small delay between the actual sound and the recorded sound, the duration of which depends upon the recording speed and the distance between the record and replay heads. On domestic type tape machines it is usual to have a common head which serves for both record and replay. This appreciably reduces the cost of the equipment for not only is there a saving of a head but also a saving of one amplifier. This arrangement has the disadvantage that there are separate requirements for the record and replay heads. For recording purposes the impedance of the recording head coil should be low to ensure a constant current independent of frequency, whereas for replay the impedance of the coil should be high to obtain maximum output voltage from the tape. When a common head is employed a compromise has to be adopted both in terms of impedance and width of head gap (normally smaller for replay).

The magnetic head is constructed from laminations of high permeability material, such as mu-metal, to ensure minimum losses to the magnetising force and complete demagnetisation on the removal of the magnetising current. The pole pieces are formed by the two ends of the laminated core and the space between them is filled with a low permeability material offering a high magnetic resistance.

THE EFFECT OF HEAD GAP SIZE AND TAPE SPEED

Both the width of the gap and the tape speed affect the frequency response of the system. When a magnetic tape recording is being reproduced the output depends upon the rate of change of magnetic flux across the head gap. The rate of change is proportional to the frequency of the recorded sound, and for a given tape speed and head gap size the output increases with frequency at the rate of 6 dB per octave (i.e., the voltage doubles as the frequency is doubled). This is in fact the case only within certain limits proscribed by the tape speed and head gap size. Head gap size is concerned with the wavelength (λ) of the recorded signal which is related to the recorded signal frequency (f) in Hz by the expression λ (cm) $= v/f$, where v is the tape speed in cm/s. For example, at a tape speed of 19.05 cm/s the recorded wavelength of a 1 000 Hz tone is $\frac{19.05}{1000}$ or 0.019 05 cm.

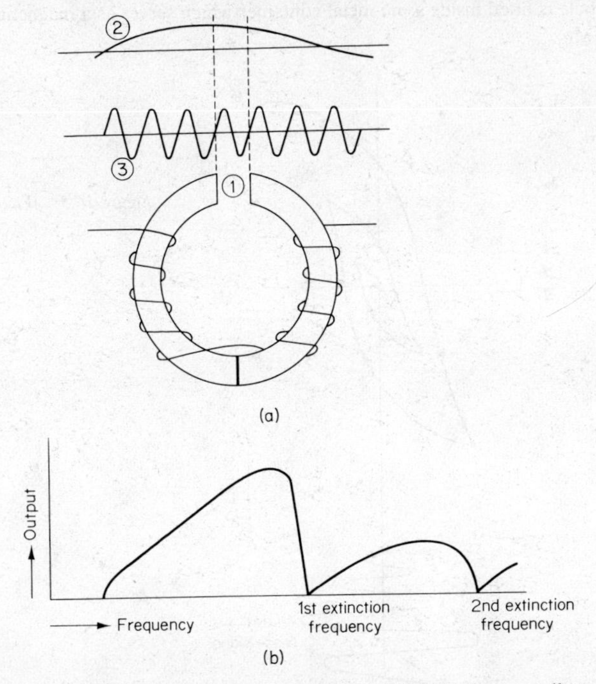

Figure 16.4 (a) 1. head gap 2. tape passing head gap with low frequency (long wavelength) recorded, showing small rate of change across the gap 3. tape passing head gap with 'extinction' frequency recorded, a complete wavelength is the same size as the gap. (b) curve relating output level to frequency for a recording made with constant current input and tape speed

The relationship between head gap size and wavelength is illustrated in *Figure 16.4* (*a*). At very low frequencies the wavelength is long compared with the gap and the rate of change of flux across it is small. The low-frequency response is therefore limited by the signal/noise ratio of the system. If the tape speed is held constant and the frequency is increased the wavelength on the tape of the recorded signal approaches the physical size of the head gap. When they are equal complete cancellation occurs and the output drops to zero. The frequency of the recorded signal when this condition is reached is known as *extinction frequency*. *Figure 16.4* (*b*) shows a typical frequency response characteristic for a recording made with a constant current input and tape speed. It will be noted that until the first extinction frequency is reached, the output increases over most of the frequency range at approx 6 dB per octave. This is modified to some extent as the frequency is increased by the effect of tape thickness and the impedance of the head. Further increase in frequency produces a series of rather similar response characteristics with progressively smaller amplitudes, eventually limited by the head impedance and tape characteristics. Normally only the range up to the first extinction frequency is used.

The upper frequency limit can be increased by increasing the tape speed or narrowing the head gap. In both cases this will be at the expense of the low frequency limit but a normal margin of about 10 octaves is available which can easily cover the audio range of 30 Hz to 15 000 Hz. The trend has been towards lowering tape speeds to increase playing times resulting in a requirement to reduce head gap size. Head materials have been improved and the gap in present-day magnetic heads is measured in microns. There is also a second air gap at the rear of the laminations which improves the B/H characteristic. The head is assembled with a separate coil on each stack of laminations and the whole is fitted inside a mu-metal container which serves as a magnetic shield.

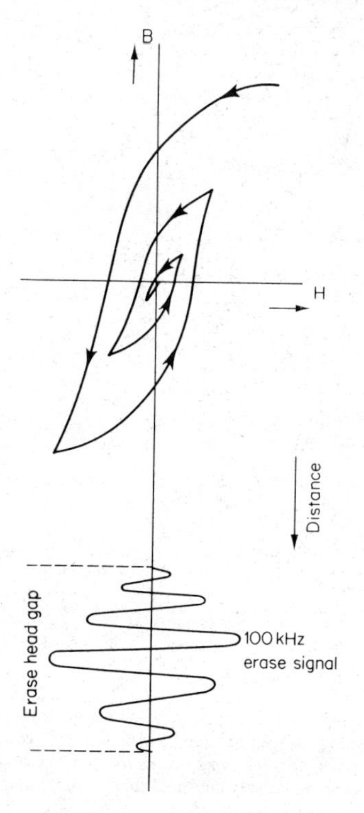

Figure 16.5 Wiping Process

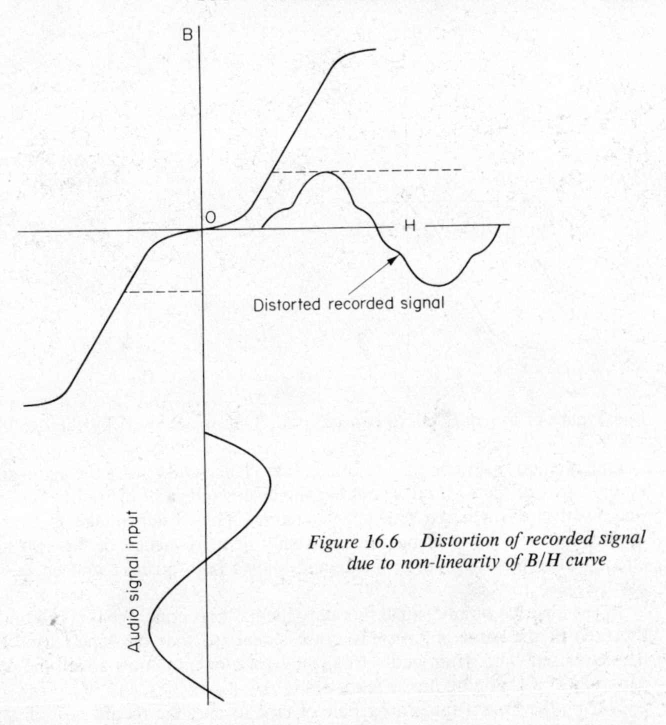

Figure 16.6 Distortion of recorded signal due to non-linearity of B/H curve

ERASING PROCESS

As previously stated one of the conditions that must be fulfilled for a satisfactory sound recording to be made is the complete demagnetisation of the tape prior to it passing the recording head. This process of demagnetisation is known as wiping or erasing. The output of an oscillator operating in the region of 100 kHz is fed to the coil of the erase head developing a high flux density, which alternates in polarity. At any given point in time the flux density is a maximum at the centre of the gap, falling away quickly on either side. Referring to *Figure 16.2* a particle of tape passing the erase head is subjected to a constantly reversing magnetisation due to the alternating nature of the magnetising force, *Figure 16.5*. This steadily increases until the particle of tape reaches the centre of the erase head gap when it is subjected to maximum flux density. At this point it becomes saturated and the effects of any previous magnetisation are removed. Further movement of the tape across the erase head subjects it to a decreasing magnetisation and hence it leaves the erase head completely demagnetised. This is illustrated by the decreasing hysteresis curves, *Figure 16.5*. The increasing field is not shown for the sake of clarity.

The 100 kHz signal for erasing also provides bias for the record head.

RECORDING PROCESS

The tape having been wiped now approaches the recording head in a completely demagnetised state. If an audio signal is now applied to the recording head distortion will occur because of the non-linearity of the B/H curve near the point of origin, O, *Figure 16.6*. As previously stated the audio signal will be recorded faithfully only if the

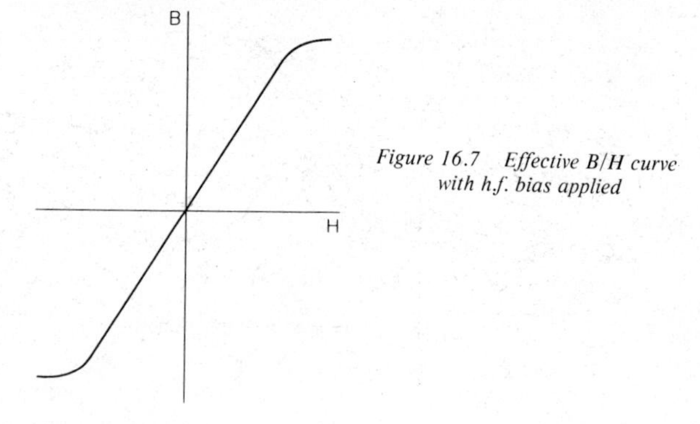

Figure 16.7 Effective B/H curve with h.f. bias applied

linear part of the B/H curve can be used. This is achieved by the use of bias, *Figure 16.7*.

One method would be the use of d.c. bias. This would move the audio signal on to the straight part of the B/H curve but has the disadvantage of restricting the signal to either the positive or negative part of the curve. This disadvantage is overcome by the application of a h.f. biasing signal, usually at a frequency of 80–100 kHz, which in effect makes the B/H curve linear and allows both positive and negative parts to be used, *Figure 16.7*.

In the absence of any signal flux only the h.f. bias component is present. This acts on the tape in the same way as the erase signal, so that the tape leaves the head in a demagnetised state. If an audio frequency component is now added the particle of tape experiences a changing flux, *Figure 16.8 (a)*.

During the time it takes a particle of tape to pass the record head it is assumed that the audio frequency component is constant. The a.f. field across the gap is shown by the dotted line. The h.f. field across the gap is as shown in the lower half of *Figure 16.5* and the combined h.f. and a.f. field is shown by the solid waveform in *Figure 16.8 (a)*. This is in effect the h.f. signal displaced by using the a.f. field as a datum. The particle of tape experiences the hysteresis excursion shown in *Figure 16.8 (b)* (*oabc–pqrst*) and due to

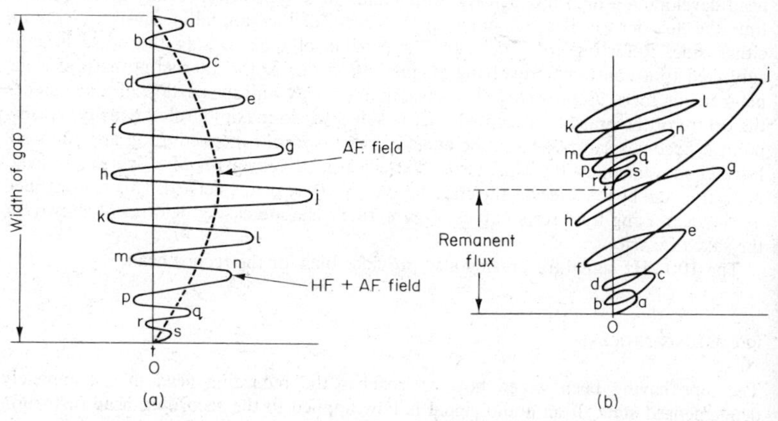

Figure 16.8 Effect of combining h.f. bias and recording fields (a) showing field displaced to one side of zero axis (b) resulting hysteresis curve

the asymmetrical nature of the applied signal leaves the record head with a remanent flux density equal to *ot*. This value is closely proportional to the instantaneous value of the audio frequency signal over a wide range.

In common with all magnetic materials there is a tendency towards self demagnetisation in the tape as poles of opposite polarity approach each other. This self demagnetisation increases with frequency.

TAPE EDITING

It is universally accepted practice that all recordings including those destined for commercial discs are originally made on magnetic tape. The advantage of this is that it provides ease of editing and consequently a more polished end product is achieved. The process of editing includes the removal of unwanted material, i.e., hesitations and mistakes. It also enables items which are recorded out of sequence to be placed in the correct order.

Editing can either be done electronically, by dubbing from one machine to another, or it can be accomplished by physically cutting the tape. In the latter method a pair of non-magnetic scissors or a single-sided razor blade should be used. When using a razor

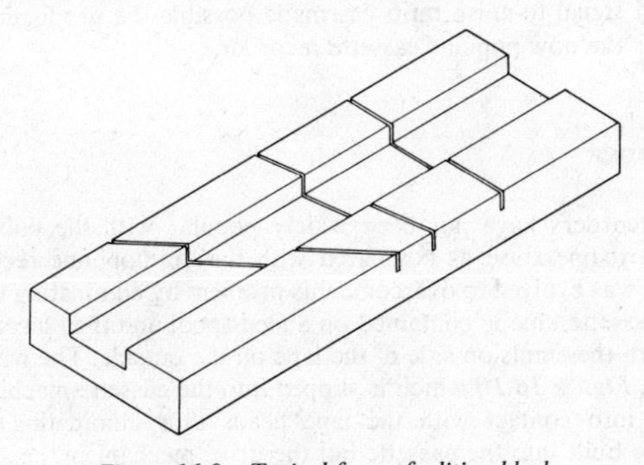

Figure 16.9 Typical form of editing block

blade it will be found advantageous to employ an editing block, *Figure 16.9*. This consists of a block of metal with a groove milled along the centre to provide a snug fit for the the tape. The block shown provides cutting guides at 90°, 60° and 45° to the tape. The 90° cut is little used as there is a tendency to introduce clicks on to the recording due to the possibility of a sudden change in level of any material recorded at the join. It is more usual to use one of the oblique cuts thereby providing a smooth transition from one section of the tape to the next. The oblique angle virtually amounts to a fade over from one tape to the other.

When editing stereophonic recordings it is advisable to use the 60° cut in order to minimise the risk of introducing small phasing errors between the two tracks.

The method of joining the tape is to place the two ends of the tape to be joined in the editing block and to stick them together with a short length (about 3 cm) of special jointing tape. Ordinary adhesive tape should not be used as the adhesive will be squeezed out when the tape is rewound, resulting in some of the turns becoming stuck together. For the same reason care must be taken to remove any jointing tape which may protrude over the edge of the magnetic tape. Cut editing may of course only be carried out on tapes containing more than one track if all the tracks are in the same time sequence, e.g., a stereophonic recording. All other multitrack recordings must be edited electronically if the material on the other tracks is to be preserved.

Developments in tape recorders

Two of the major factors governing the performance of a tape recorder are:

(*a*) The frequency response.
(*b*) Ability to maintain a constant speed, i.e., low percentage of wow and flutter.

If it is accepted that the design of amplifiers with good audio frequency response presents no problems, then the frequency response of the tape recorder will depend on its extinction frequency. As previously stated this is determined by the width of the recording head gap and the tape speed.

Improvements in head design during the past few years have made it possible to manufacture heads with gaps having widths of 10 μm or less and manufacturers claim a frequency response of 30 Hz to 14 kHz at a tape speed of 4.76 cm/s.

Advances have also been made in the design of synchronous drive motors and it is now possible to produce relatively small motors capable of a constant speed such that wow and flutter figures of 0.1% and 0.2% even at 4.76 cm/s are claimed.

These factors coupled with the use of transistors and integrated circuits and the introduction of the Dolby noise reduction system which allows the use of 6.25 mm wide tape with a good signal-to-noise ratio has made possible the production of small tape recorders such as the now popular cassette recorder.

CASSETTE RECORDER

Hitherto tape recorders have not been widely popular with the public due to their relatively awkward operation as compared with the gramophone record player. The cassette recorder was evolved to overcome this problem by eliminating the need for tape threading. The cassette tape is contained on a feed spool and then laced through to the take-up spool with the emulsion side of the tape on the outside. The whole is fitted in a plastic container, *Figure 16.10* which is slipped into the cassette machine such that the tape is brought into contact with the tape heads thus eliminating tape threading. Pressure pads are built into the cassette but the drive mechanism, i.e., the capstan and pressure roller are an integral part of the recorder.

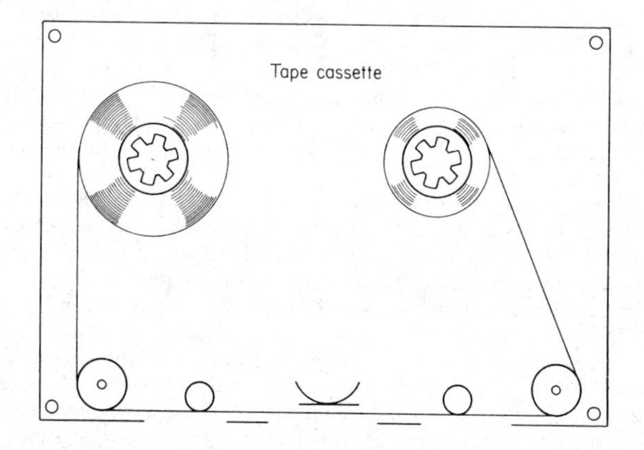

Figure 16.10 Arrangement of tape in a typical cassette recorder

Cassette playing times

Type C60 holds 90 m (300 ft) which gives 30 min playing time per track.
Type C90 holds 135 m (450 ft) which gives 45 min playing time per track.
Type C120 holds 180 m (600 ft) which gives 60 min playing time per track.

NOISE REDUCING SYSTEMS

A widely used system for obtaining an improved signal-to-noise ratio in recordings is the Dolby Noise Reduction System. This system has two classifications – Dolby A and Dolby B. In Dolby A the signal-to-noise ratio of a signal recording system is improved by 10 dB in the range 20 Hz to 5 kHz rising to 15 dB at 20 kHz. The simplified Dolby B system gives up to 10 dB improvement at the upper audio frequencies. Either system reduces tape noise without changing the characteristics of the signal.

The Dolby 'A' system. The Dolby A system is most commonly used in multiple tape recording where one signal may pass through the record/replay process several times or where a large number of tracks are mixed to make one or two tracks. Most professional recordings are made using the Dolby A system. It is applied only to the intermediary process, e.g., dubbing, not to the final output. Basically the system takes advantage of the following facts.

(1) It is pointless to apply volume compression over the whole dynamic range of the system to combat noise when this only concerns a small percentage of the volume range at the bottom of the scale.
(2) The ear has a tendency to mask quiet sounds while, and for a brief time after, being exposed to much louder sounds of similar frequency.
(3) Most sound sources, including music, have a very unequal and varying distribution of spectral energy.
(4) Any undesirable effects associated with compression are likely to be much more obvious at high volume than at low.

It could therefore be advantageous to divide the volume range into bands and arrange for the action to affect only the low end of the scale (below about −40 dB) where noise is a problem. It could also be an advantage to divide up the frequency spectrum and arrange that this level—selective process is applied to each division in proportion to its individual need, taking advantage of the ears' masking properties whenever possible.

Ideally the audio spectrum should be divided into an infinite number of separately controlled bands, but in practice four are found to provide a reasonable compromise between cost and efficiency for most applications. Clearly more could be used to meet particular circumstances possibly some being finely tuned to deal with specific narrow band noise.

The Dolby A301 system uses four bands.
Band 1. 80 Hz low pass (to deal with hum and rumble).
Band 2. 80–3 000 Hz band-pass (to deal with mid-band noise and print through).
Band 3. 3 000 Hz high-pass (to suppress hiss and modulation noise).
Band 4. 9 000 Hz high-pass (to suppress hiss and modulation noise).

Unlike ordinary compressors the Dolby system provides two parallel paths for the signal. One via a linear amplifier and the other via a differential network, the output of which is added to the output of the linear amplifier in the record mode and subtracted from it in the replay mode, *Figure 16.11*. The differential network consists of four band-splitting filters which divide the frequency spectrum as stated above. These are followed by linear compressors and finally by peak clippers before being recombined and either added or subtracted from the straight through output, *Figure 16.12*. The units are adjusted so that the low-level output from the processor is uniformly 10 dB higher than the input up to about 5 kHz above which the increase in level rises smoothly to 15 dB at

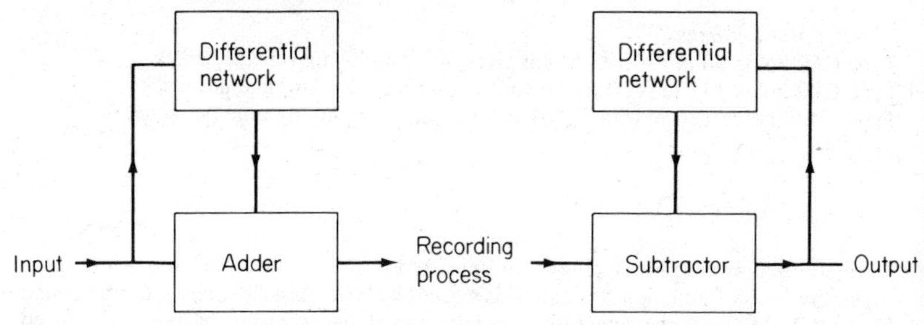

Figure 16.11 Arrangement of Dolby A system. The output of the differential network is added for recording and subtracted for replay

15 kHz. The clippers are there to prevent overshoot due to the relatively long attack time of the compressors. The transient distortion they introduce is masked by the output of the linear amplifier.

There are problems concerning the choice of attack and recovery time of compressors. Short time constants can result in modulation distortion particularly at low frequencies. Long attack times can result in overshoot on sudden peaks and long recovery times give rise to 'breathing' effects (the noise level appears to rise after peaks and thus becomes more intrusive). The Dolby system uses relatively long attack and decay time constants for moderate signal level changes so that unwanted modulation products are low. At high levels the attack times are decreased by means of non-linear detection circuits. The differential unit for adding is identical to that used for subtracting.

The net effect is to enable the frequency bands containing only weak signals to be raised in level with respect to peak level by about 10 dB before recording and then reduced to their proper relationship afterwards. The recording noise is correspondingly reduced in the process. The noise introduced by the recording process is therefore made less significant in relation to the wanted signal. Signals above −40 dB reference peak level are progressively less affected because the side chain is very heavily compressed above this value, so that their additive (and later subtractive) contribution to the combined output is correspondingly reduced.

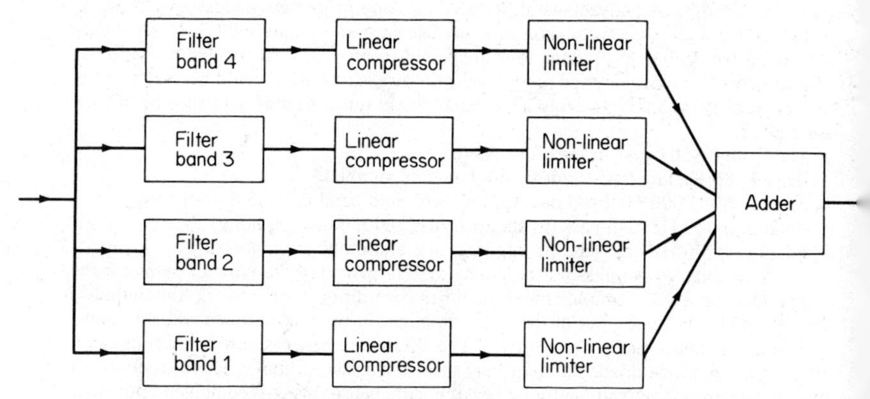

Figure 16.12 Dolby A system. Block diagram of differential network, consisting of four splitting filters and low level compressors. The low level signal is added to the straight through signal in record (send) mode and subtracted from it in the replay (receive) mode

Consideration of the average spectral energy distribution of music will make it evident that the system can make a very worthwhile improvement in music recording and transmission. Whereas musical sound energy tends to be centred on the lower middle frequency band the most common noises such as tape hiss tends to afflict the ear in a sensitive region of its response too far removed in frequency from the louder sounds for their masking effect to mitigate.

Dolby B system. For normal domestic tape recorders the cheaper and simpler Dolby B system is used. This functions in the same way as Dolby A in that the low-level signals are increased in level in the recording process and reduced during the playback process. The main difference between Dolby A and B systems is that the latter operates only in one frequency band and provides reduced improvement below 5 kHz. The signal passes first through a 19 kHz low-pass filter then divides, one path going direct to the

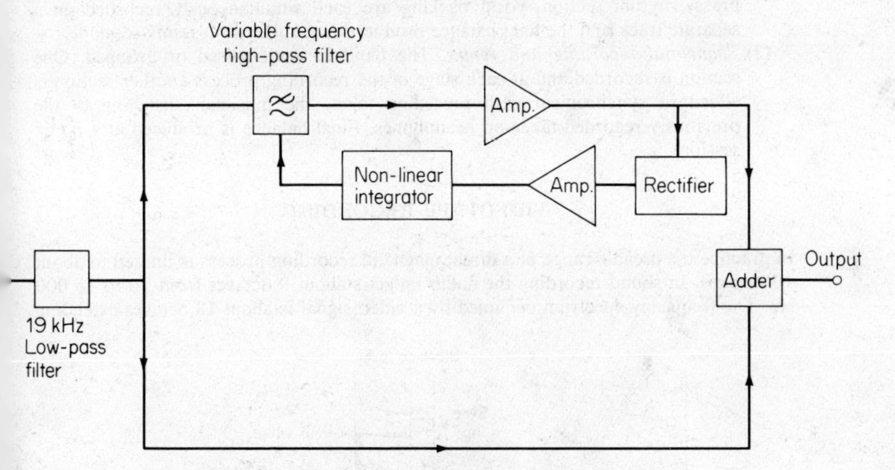

Figure 16.13 Arrangement of Dolby B noise reduction system

adder while the differential signal is fed via a variable frequency high-pass filter and amplifier, *Figure 16.13*.

The side-chain rectifier, amplifier and integrator provides a d.c. signal to alter the cut-off frequency of the variable filter. When a low-level signal is fed to the Dolby B system the cut-off frequency of the high-pass filter is reduced and a large part of the input signal from about 2 kHz upwards is processed. When the input level is high the cut-off frequency is high and the Dolby effect is small. Thus the Dolby B uses a variable bandwidth system which has the advantages of a multiple system without the complexities of filters and multiple circuit paths.

Multitrack recording in professional recording practice

Voice and orchestra are usually recorded on separate tracks and mixed afterwards. In this way the vocalist can do several takes of his performance without having to extend to an expensive band session. It also gives the sound mixer the opportunity of experimenting with mixes and electronic treatments. This technique has been extended and it is quite common in recording studios to record on 4, 8 or 16 tracks and manufacturers are now producing 24 track machines. The soloists and each section of the orchestra (or each musician) are recorded on separate tracks which are later mixed to produce a final reduction copy of 1, 2 or 4 tracks.

This technique is also used in multitracking, a method building up an orchestration by using a particular recording several times. A musician or orchestral section playing one part, e.g., rhythm, is recorded and further parts are added and re-recorded on subsequent playbacks. This is possible with a twin track tape recorder but leads to a noisy final result because of the number of times the first sound is reproduced and recorded. Moreover mixing is required at each re-recording and it is extremely difficult to control the balance of the final result.

The usefulness of multitracking is illustrated in the following examples which assume the use of an eight track tape recorder.

(1) *Record and remix.* The band is split up into any number of groups up to eight, or sixteen, according to the recording system in use. Vocal, strings, woodwind, brass, rhythm section, vocal backing are each simultaneously recorded on a separate track and the final balance produced in a subsequent remix session.
(2) *Sequential recording and remix.* The band is sectionalised or grouped. One section is recorded and at each stage of the recording process another section is added in synchronism with preceding takes, the musicians listening to the previously recorded takes on headphones. Final balance is produced at a remix session.

VIDEOTAPE RECORDING

In practice the useable range of a direct magnetic recording process is limited to about 10 octaves. In sound recording the audio range is about 9 octaves from 30 to 15 000 Hz. The frequency spectrum occupied by a video signal is about 18 octaves extending

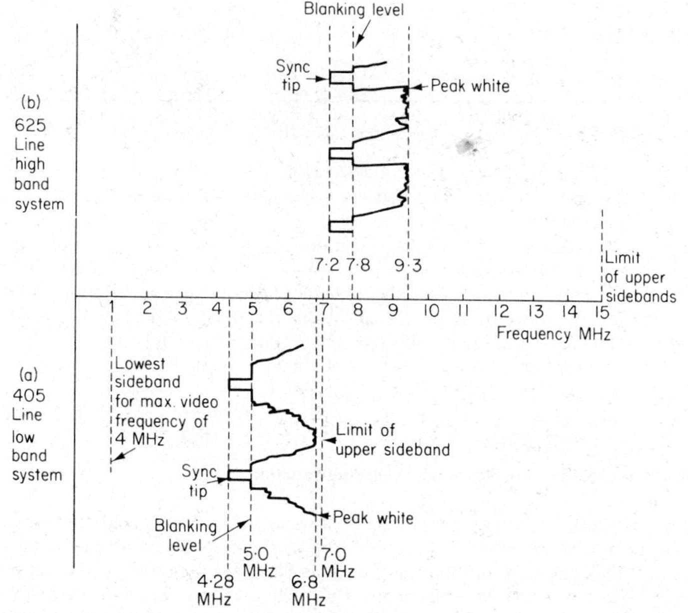

Figure 16.14 Frequencies used in (a) low band and (b) high band videotape recording systems

from practically zero to about 5.5 MHz for a high-definition colour signal on European 625-line television standards. To make possible the satisfactory recording of the video signal within the ranges available to the magnetic recording/reproduction process, modulation techniques are used which move the spectrum to a higher range where the ratio of frequency variation can be contained within fewer octaves. In practice less than four.

Frequency modulation is chosen for this purpose because the recovered signal does not suffer distortion due to amplitude variations such as could be caused by slight variations in tracking. Amplitude limiting can also be used to improve further the general stability of the signal, because with f.m. it does not matter if the amplitude shape of the reproduced waveform differs significantly from that recorded. Provided that the waveform crossovers coincide the signal can be satisfactorily reproduced. The output is not affected by the non-linearity of the B/H characteristic provided that it is symmetrical about the H axis so there is no need for bias to be used as in sound recording.

Low-band modulation

The original (low-band) modulation system used for videotape recording of 405-line black-and-white pictures, had black level based on 5.0 MHz with peak white extending to 6.8 MHz and sync tips at 4.28 MHz, i.e., a maximum deviation of 2.52 MHz. With such a low-modulation index and comparatively little energy at high frequency the sidebands are virtually limited to a single pair. For a bandwidth of 4 MHz the lowest sideband is 1 MHz (5 – 4) and the upper sideband can be limited to just above the upper deviation frequency, i.e., 7 MHz so that the whole video signal can be contained within the range 1 to 7 MHz which is less than 3 octaves, *Figure 16.14* (*a*).

High-band modulation

The low-band system is not suitable for 625-line colour television recording because of the extended bandwidth (a flat response up to 5.5 MHz is required) and the large amount of h.f. energy in the colour sub-carrier signal at 4.43 MHz. If a low carrier frequency is used for the recording system spurious noise pattern effects can be produced by the colour sub-carrier because the lower sidebands, which contain substantial energy extending to below zero frequency (relative to the carrier), can become 'folded back' in the modulation process to beat with the carrier. For this reason the 'high band' system is used for 625-line colour picture recording. The system is based on blanking level clamped at 7.8 MHz with sync tips at 7.16 MHz and peak white 9.3 MHz.

In practice it is necessary for the video replay head to reproduce frequencies up to about 15 MHz in order to recover the necessary upper sidebands, *Figure 16.14* (*b*).

The recording standards which were laid down in the early days of videotape recording used heads with an effective gap of about 0.1 mil and aimed at an extinction frequency of 16 MHz. Better design has since enabled heads to be produced with gaps of half this size enabling a better h.f. response without appreciable reduction in overall sensitivity.

Tape transport

The required tape velocity = extinction frequency × effective head gap = $16 \times 10^6 \times 0.000\,254$ cm/s = 4 064 cm/s. A tape transport system with this speed of traverse is unpracticable. One hour of recording would require a reel of tape over 146 000 m long.

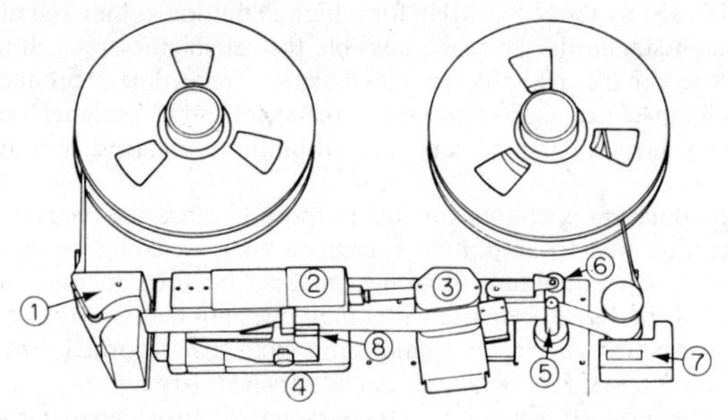

Figure 16.15 Videotape transport system 1. master erase head 2. head-drum, incorporating motor, head-drum, tachometer and pick up transformer or sliprings 3. sound and cue recording heads 4. vacuum tape guide 5. capstan 6. pressure roller 7. tape timer 8. control track head

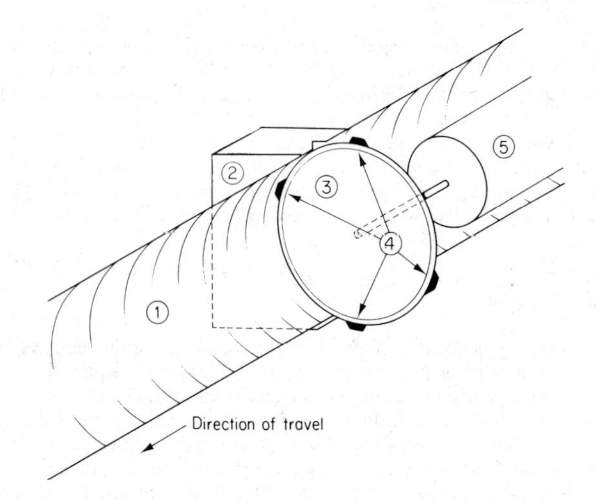

Figure 16.16 Head-drum arrangement for transverse-scan recorder 1. tape, formed into a curve around head-drum by vacuum guide 2. vacuum guide 3. head-drum (or head wheel) 4. recording heads projecting from head-drum 5. head-drum motor

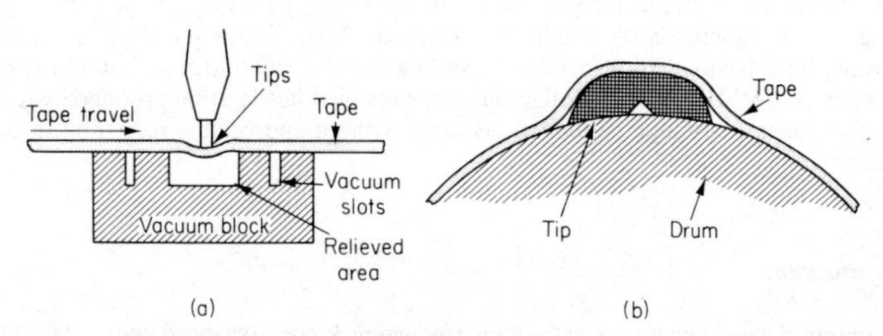

Figure 16.17 Arrangement of tape and head-drum (a) cross section of tape guide vacuum block, showing penetration of recording head tips below normal level of tape (b) enlarged side view of head-drum showing tip projection

To overcome this problem videotape recorders employ heads mounted on drums which rotate transversely across the tape which moves through the machine at normal tape recorder speeds, e.g., 38.1 cm/s. In this way the required head-to-tape velocity of around 4 000 cm/s is achieved.

Quadruplex machines

Early V.T.R. machines and those mainly in use for professional purposes are of the quadruplex type, i.e., with four heads, using a tape 50.8 mm wide running at 19.05 cm/s and 38.1 cm/s, the latter being the normal standard. The tape is transported from reel to reel by means of a capstan and pinch roller in the manner of a professional magnetic tape sound recorder. For video recording the four heads are mounted in quadrature on the periphery of a head drum (sometimes called a headwheel) approximately 50 mm in diameter which rotates at right angles to the length of the tape, *Figure 16.15*. The tape is made to curve around the head-drum by a guide of similar diameter into which the tape is sucked by a vacuum, *Figure 16.16*. The guide is slotted to enable the tip of the head-drum to penetrate the normal line of the tape, stretching it slightly at the point of contact. This stretching action ensures consistent head-to-tape contact and makes it possible to compensate for head wear, which would otherwise cause timing errors, *Figure 16.17*. The head drum rotates at five times the television field frequency, i.e., 250

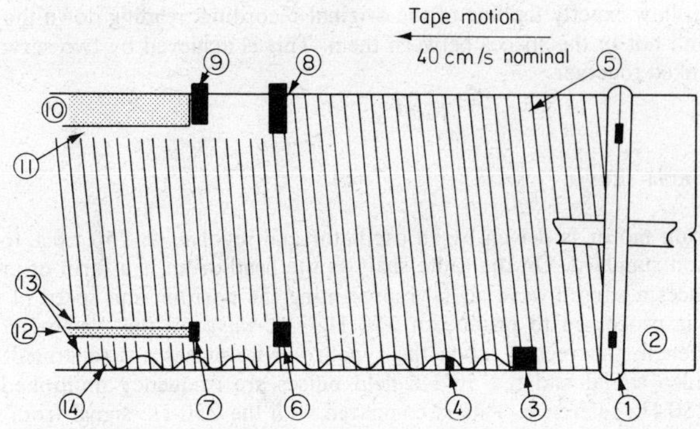

Figure 16.18 Quadruplex videotape recorder-arrangement of tracks. 1. head-drum 2. tape 3. control track head 4. edit pulse 5. video tracks 0.25 mm wide with 15.9 mm guard band 6. cue track erase head 7. cue track record/reproduce head 8. audio track erase head 9. audio track record/replay head 10. audio track, 1.78 mm wide 11. guard band, 0.5 mm wide 12. cue track 0.5 mm wide 13. cue track bands 0.25 mm wide 14. control track 1.27 mm wide

r.p.s. for the 625-line standard or four times field frequency, i.e., 240 r.p.s. for the 525-line system.

For the 625-line television standard the heads sweep across the tape in turn covering an arc of 114° at a head-to-tape speed of 4 013.2 cm/s laying down 4 × 250 tracks per second. The tracks are 0.25 mm wide, and due to the movement of the tape at 38.1 cm/s are spaced apart 0.14 mm, i.e., the centre-to-centre spacing is 0.4 mm, *Figure 16.18*.

In the recording mode all four heads are fed with the signal continuously so that recording takes place over almost the whole width of the tape, the information recorded at the bottom of one track being duplicated at the top of the next. On reproduction each head must be switched in sequence otherwise the noise produced by the non-productive heads would be excessive. Although only 90° of rotation is required of each head in turn representing a head-to-tape sweep of 4.01 cm in fact a sweep of 4.62 is used so as to allow an overlap at the beginning and end of each track in which the switching can take place. This is done in the line blanking periods of the video wave-form to prevent flashing due to transients. With the 625-line standard each 90° of the head drum corresponds to 15.625 lines.

SOUND TRACK

The sound track is recorded and read longitudinally along the length of the tape in the conventional manner of a sound recorder. The relatively high tape speed and the width of the sound track helps in some measure to compensate for the fact that the magnetic particles are orientated transversely to favour the video recording. Similarly, near the bottom of the tape there is a cue track 0.5 mm wide with 0.25 mm guard bands on either side. This is a low-quality track used mainly for editing purposes, e.g., address codes can be recorded to assist in the rapid location of sequences or even to control thc process. Sometimes sound effects of sporting events are recorded on the cue track to provide a source for subsequent mixing with commentaries in various languages. Below this, along the bottom edge of the tape is the control track. This acts rather like the sprocket holes of a film to control the lateral movement of the tape regardless of possible stretch or slip. It is necessary for the movement of the tape to be perfectly synchronised with the rotation of the head-drum so that on reproduction the heads can be made to follow exactly the line of the original recording, reading down the centre of the tracks and not in the spaces between them. This is achieved by two servo systems which are linked together.

THE HEAD-DRUM SERVO

The head-drum motor is driven by an oscillator and revolves at 250 r.p.s. for the 50-field television standard. On the same shaft as the head-drum is a form of tachometer which produces a square-wave signal representing the position and speed of the head-drum. This is processed to produce a 250 Hz sine-wave. In the Ampex V.T.R., for example, when in the record mode the synchronising pulses are separated from the incoming video signal and the 50 Hz field pulses are frequency multiplied by 5 to produce a 250 Hz reference. This is compared with the 250 Hz signal from the head-drum tachometer by a comparator which adjusts the frequency of the drive oscillator, thereby altering the speed of the synchronous drum motor to bring the two signals into phase. The 250 Hz tachometer signal is also frequency divided by 4 electronically to produce 62.5 Hz. This is amplified and used to power the synchronous motor driving the tape transport capstan. In this way the movement of the tape is locked to the rotation of the head-drum.

CONTROL TRACK

During the recording process the 250 Hz sine-wave signal derived from the head-drum tachometer is recorded as a control track along the bottom of the tape. Edit pulses of 12.5 Hz, i.e., half picture frequency with the 50-field standard, are also recorded on the control track. These are necessary to locate the precise points where edits can be made without disturbing the video waveform. In the case of a colour signal it is necessary to preserve the colour-burst signal-phase relationship. The pulses are also used to assist in achieving approximate synchronism during the locking-up sequence in the first few seconds of replay.

REPRODUCTION

In the replay mode the head-drum oscillator is locked to the studio synchronous signals or to some other reference such as mains frequency on the incoming video signal. The capstan is driven by its drive oscillator, the frequency of which is compared with the reproduced signal from the control track so that the same relationship exists between tape and head-drum motion as in the original recording. When the movement of the head-drum and capstan are synchronised there is still the possibility of the heads reading between, rather than down the middle of, the tracks. This is corrected by a manual control (called 'Tracking').

The signals from the four heads are picked up by slip rings (or by a rotary transformer) and amplified in separate head amplifiers. They are then fed to an electronic gating circuit which switches each one in sequence during the 'front porch' of the line blanking period following each period of 15 or 16 lines which represents the 90° rotation of a head. The f.m. signal from the gating circuit is limited to remove any amplitude variations before being demodulated and processed. In the processor the synchronous pulses are stripped from the video signal, and cleaned up. New blanking pulses are generated and the whole recombined to provide the output.

COLOUR SYNCHRONISING PROBLEMS

The reproduction of colour signals requires a stable sub-carrier with timing errors within about 4 ns. This is achieved partly by careful mechanical design, e.g., the head-drum runs in air bearings to reduce friction thus increasing sensitivity to correction and by frequent measurement of positional errors and correction by the servo systems. The head-drum motor servo is capable of synchronising the video output to a reference signal to within about ± 100 ns. Further adjustment to within the permitted tolerance is made electronically using variable delay times.

These methods can only correct the timing at the beginning of each line and hold it for the duration of the line.

VELOCITY ERRORS

Timing errors, such as may be caused by head-drum speed jitter or head/tape misalignment can accumulate over the length of each line depending upon the rate of change of timing (i.e., velocity) error. Velocity errors due to head-drum jitter tend to be small, as the rate of change is small, and random in position so that they tend to cancel out.

Errors due to head/tape guide geometry are related to head-drum rotation which is locked to the field frequency so that they appear in the same position repeatedly and are therefore visible. If the tape and head-drum are not properly concentric, due, for example, to incorrect adjustment of the tape guide height, the head/tape velocity, and hence the frequency, will vary along the length of each track consisting of some 16 lines. Although correction is applied at the start of each line there can still be a considerable rate of change of phase resulting in a change of hue across each line of picture. The hue error will vary according to the position of each line in relation to the sweep of the head with the result that the picture is broken up into bands of approximately 16 lines with a marked change of hue between them. With the delay-line PAL colour system the hue errors are translated into saturation errors which, due to the effect of the banding, are equally objectionable.

Velocity errors become particularly significant when dubbing from one machine to another is required, e.g., for editing purposes, because the velocity errors build up and, not being accompanied by corresponding synchronous and colour burst signal errors are not easily corrected.

When tapes recorded on one machine are played on another it is unlikely that the recording and reproducing heads will match exactly. If there is any difference in the length of the individual head tips this is equivalent to incorrect head tip insertion with consequent velocity error and banding. To make possible complete interchangeability of tapes as well as dubbing between machines it is necessary to employ velocity compensation.

VELOCITY COMPENSATION

The velocity compensator applies timing correction by electronically adjusting a variable delay line continuously throughout the length of each line. It works on the principle that detected timing errors are regularly repeated for each revolution of the drum and are therefore predictable. Timing errors are measured by comparing the reproduced line sync pulses with a reference signal. The amount of correction required at each correction step represents the velocity error along each line. An error signal is produced with every line-synchronous pulse so that correction is applied in a series of steps. The error is, however, continuously changing during each line and the difference between the error waveform and the stepped correction waveform represents the velocity component. This can be simulated by a waveform consisting of a series of ramps, starting from zero at the beginning of each line and changing in slope and direction to the value of the next correction step. Such a waveform, representing changes of timing error, can be created by storing the value of the correction signals for each line during a full rotation of the head-drum. Sixty-four such stores are required, representing the maximum number of lines that are reproduced by one rotation of the drum. As the errors tend to be repeated each revolution of the head-drum the correction waveform built up in the stores can be applied to the appropriate lines during successive rotations of the drum thereby effecting continuous timing correction. The stores are capacitors with a comparatively slow charging time (about 0.5 s) so that the correction signal builds up over a number of revolutions of the drum and is not susceptible to spurious time errors.

TAPE EDITING

Tape editing with quadruplex videotape machines can be achieved either by physically cutting the tape or by electronic means.

Physical editing. Two-inch video tape can be physically cut and joined by sticking adhesive tape on the back as in a sound tape recorder but the tape must be cut and joined very accurately during the vertical blanking interval between picture frames. With the tape travelling at approximately 19.05 cm/s. These occur every 0.762 cm at 25 pictures per second (European standard).

To enable the correct point at which to cut the tape to be located edit pulses are recorded on the control track. These occur at a frequency of 25 Hz for monochrome (European standard) or 12.5 Hz for PAL colour pictures. In the latter case it is necessary, when making to join, to preserve the colour burst phase relationship which, in the PAL system is a four-field sequence. A cut can therefore only be made at intervals of two complete interlaced pictures. Thus editing cannot be quite so fine with colour as with monochrome pictures.

The cutting points are located by marking the tape at the precise moment of the programme cue and when the machine is stopped painting over the area of the control track with a liquid containing powered iron in suspension. The edit pulse can then be seen with the aid of a microscope and by means of a special splicer the tape is cut and joined to the next sequence which has also been cut at an edit pulse.

The operation is complicated by the fact that the sound recording head is approximately 23 cm in advance of the vision recording so that if a cut is made on a vision cue approximately 0.6 s of the sound from the cut material will remain on the wanted tape. similarly, if the cut is made at the start of a vision cue 0.6 s of the new sound will be lost. Whenever possible, therefore, for cut editing the cutting point should be chosen to coincide with a gap in sound of at least 0.6 s. If this is not possible it will be necessary to 'lift off' that part of the sound by recording it on to 6.25-mm tape and re-recording it over the join.

Electronic editing.[1] Editing of videotape by physical cutting of tape is being superseded by electronic methods. Involving the use of an electronic editor which controls the switching logic of a videotape machine to enable it to switch instantly from replay mode to record or vice versa. This entails precise control of the erase current and video record signal to make allowance for the distance between the erase head and the record head-drum so that picture continuity is maintained. The electronic editing system can be operated in sound or vision separately so that it is possible to insert new material on the sound track while retaining the original vision recording or vice-versa.

Dubbing. The main use for electronic editing is in dubbing. Programmes are often recorded in a series of sequences, not necessarily in the proper order. These can be assembled in the correct sequence by dubbing on to a master tape on another machine, the edits being made electronically so that the master tape is complete and without physical joins. This will, of course, be a second-generation recording with the possibility of consequential degradation but provided care is taken at every stage the impairment should not be serious. In order to save the expense of operating quadrature machines for rehearsal and selection purposes use is often made of relatively inexpensive helical-scan machines using 2.54 cm wide tape (see below). The original material is copied on to one of these machines and all the time-consuming operations of relating material and establishing editing points is accomplished. The cutting cues can be recorded on the 2.54-cm tape and these can eventually be used to programme an electronic editor controlling the quadruplex machines for the final assembly.

Helical scan machines

Professional quadruplex transverse-scan machines described above, although differing to some extent in detail and design philosophy, all work to agreed standards so that recorded tapes are interchangeable. This has not been the case with the less costly helical-scan machines, nevertheless they have many useful applications and the technology is advancing rapidly such that helical scan machines have reached a high standard of performance and are being increasingly used.

In general helical scan recorders employ rotating head-drums to increase the head-to-tape speed but instead of rotating transversely across the tape they rotate at an oblique angle (typically 3 or 4 degrees) to the longitudinal axis of the tape. This is achieved by wrapping the tape around the head-drum in the form of a helix. Various configurations are used with different machines but the object is to produce long oblique head sweeps each containing a complete picture field, with the change-over occurring during field blanking or at the bottom of picture frame (which in this case would be omitted from the reproduced picture). Most helical scan recorders use a low tape speed typically between 15.24 and 25.4 cm/s so that a large head-drum is required to achieve the necessary head-to-tape velocity. The head-drum can consist of two fixed drums with a slit between through which the rotating record/reproducing head (or heads) protrude. In some types the heads are fixed to the upper half of the drum which rotates in the same direction as the tape motion and thereby reduces friction.

JITTER

Tape friction plays an important part in the performance of helical scan recorders. Variation in friction or tension can give rise to rapid fluctuations in tape speed which causes a form of instability known as jitter. This was a characteristic of the early helical-scan recorders and was often manifest as an annoying effect known as 'flagging', where the vertical elements in the top of the picture bend over. Some helical-scan machines employ sensitive servo systems to control tape tension and sophisticated electronic time correction circuits to mitigate this problem.

SLOW SPEED AND STOP ACTION

Helical-scan machines that have their synchronous pulses aligned so that they coincide in adjacent tracks are capable of being used for reproduction in slow or stop motion.

Video cassette recorders (VCR)

Since 1970 VCR machines have been developed and have proved their usefulness in a number of applications ranging from industrial and educational to domestic use. Many problems had to be solved in order to meet the requirements of low-cost, light-weight, simple operation, good quality sound and vision including full colour capability, suitability for use with a domestic type television receiver and able to provide up to one hour of playing time from one cassette. Machines meeting these requirements are now available and there is an increasing demand for pre-recorded television programmes in cassette form for both educational and domestic use.

Pre-recorded video discs

While tape cassettes are eminently suitable for pre-recordings and have the advantage that the user can make his own recordings on them they are expensive to manufacture and duplicate. Where large numbers of each recording are required, by far the cheapest type of 'software' is the disc, which, once the initial master has been made, can be stamped out in quantity.

Several different systems of video disc have been developed, some of which are discussed below.

THE PHILIPS VLP SYSTEM

This system uses discs of similar size and material to an ordinary L.P. record but instead of the usual groove there is a spiral track of minute indentations, *Figure 16.19* (*a*). These are all of equal depth and width but their length and the distance between them along the track varies such that a type of pulse modulation system containing all the necessary information to reproduce a colour television picture with synchronous sound is provided. The disc revolves at 25 revolutions per second so that each revolution contains one complete image. Playing time is 30–45 minutes. The surface of the disc is coated with a thin reflective metallic layer after pressing. A very small spot of light from a helium–neon laser source is focused on to the track and is reflected back, modulated in accordance with the pattern of indentations, into a photodiode. The resulting signal after amplification and suitable processing can be fed directly to the input of a television receiver.

The light beam is centred on to the track by means of an opto-electronic control system. Thus there is no need for a mechanical groove and the track pitch can be made extremely small, *Figure 16.19* (*b*). The opto-electronic mechanism does not have to

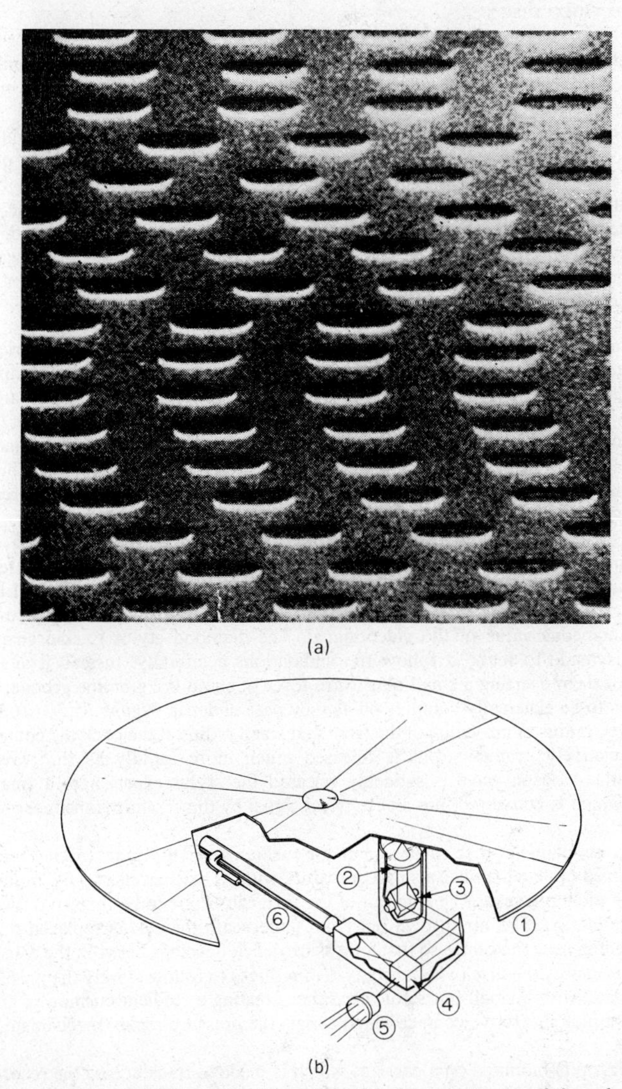

Figure 16.19 Philips VLP disc system (*a*) *surface of the long-playing record through a scanning electron microscope* (*b*) *schematic representation 1. video long-playing record 2. spring suspended lens with automatic focusing of the light beam 3. hinged mirror for following the track 4. beam-splitting prism 5. photodiode* (*detector*) *6. light source* (*Courtesy Philips Electrical Limited*)

remain synchronised to the track pitch and it is possible to speed up or slow down or reverse the motion or to hold a still frame. As there is no mechanical contact between the pick-up and the disc this can be accomplished, as can normal replay, without any wear of the record or the pick-up.

THE TED VIDEO DISC[2]

This system, developed by the Teldec research organisation of AEG Telefunken and Decca is intended to produce video recordings on discs, made of conventional materials, and inexpensive to duplicate in quantity.

The discs are 21 cm in diameter with a maximum playing time of 10 minutes. Two models of the player are in production, one for playing single discs, the other for magazines to play 12 discs automatically in sequence with no discernible break in programme continuity. The system uses what might at first appear to be the conventional method of stylus tracking in a groove as with a gramophone record, but the frequency range requirements are enormously increased. In fact a video bandwidth of 3 MHz and an audio bandwidth of 30 Hz to 15 kHz is provided, using a special modulation and colour coding system.

Modulation system. Picture and sound frequency modulate separate carriers. The sound carrier frequency is about 1 MHz with a deviation of ±25 kHz at an amplitude about 30 dB below that used for the vision carrier. The picture carrier has an amplitude of 0.5 μ and a frequency deviation range from 2.75 MHz (sync tips) to 3.75 MHz (peak white). The minimum recorded wavelength is about 2 μ and the maximum signal velocity is 11.5 m/s, about 100 times greater than for audio records. This would be impossible to read by normal stylus displacement methods used for sound reproduction. A novel method known as 'pressure scanning' has been adopted.

The signal is recorded as vertical undulations in a V-shaped spiral groove 3.5 μm wide with an included angle of 135°. There is no land between the grooves which are spaced approximately 140 per radial mm compared with about 12–13 for a stereo LP sound recording. The undulations represent the carrier whose frequency varies according to the instantaneous value of the video signal. The diamond stylus is skid shaped. No attempt is made to make it follow the undulations accurately. Instead it sits over a number of them exerting a small downward force of about 0.2 g on the groove, causing the waves to be elastically compressed as they pass under it, *Figure 16.20* (*a*). Because of the large radius of curvature at the front and small radius at the back the compression occurs relatively gradually but is released much more rapidly as the waves pass underneath. As each wave is suddenly released the stylus experiences a small force impulse which is converted into an electrical signal by the ceramic transducer to which it is attached.

The 21-cm-diameter disc rotates on an air cushion at 1 500 r.p.m. (50 field standard) above a fixed table which has a small curvature with its axis parallel to the radius along which the pickup tracks. The rotation of the disc causes air to be thrown outwards by centrifugal action. The air thrown out from underneath the disc is replaced through a small opening near the centre so that a pressure deficit develops between the disc and the plate. This causes the disc (which is only 1 mm thick) to follow closely the profile of the 'saddle plate' with a small but stable clearance creating a resilient cushion of air under the area of pickup/groove contact which assists the pressure replay mechanism, *Figure 16.20* (*b*).

The pickup is mounted on a carriage which is made to track across the record at the mean groove pitch by means of a thin wire belt driven through a reduction drive from the disc rotating spindle. The diamond stylus and its associated ceramic transducer are connected to the carriage by means of a highly compliant Duralumin tube, set at a tangent to the groove. This defines the static vertical stylus force of about 0.2 g and

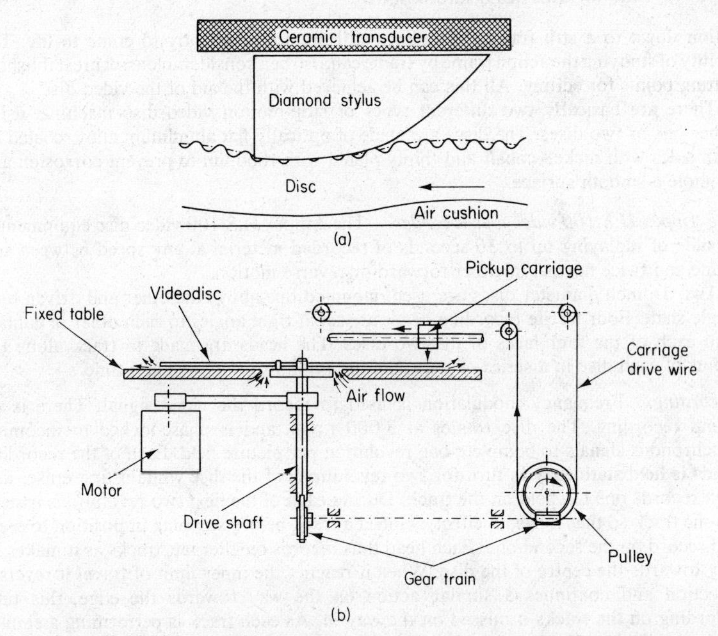

Figure 16.20 (a) Elastic deformation of videoscan groove by replay stylus (b) mechanical layout of video disc player (Courtesy Decca Record Company Limited)

allows for some lateral movement to accommodate slight departures from the mean advance rate such as might be caused by disc eccentricity.

Still Frame. If the carriage drive is held stationary the stylus will repeat a groove and as each revolution of the disc contains one complete frame of video signal (two fields) reasonable continuity of synchronising signals is maintained and a still frame picture obtained. Although this action involves the stylus jumping the groove each revolution the groove pitch is so shallow and the mass of the replay element so small that the wear is insignificant unless it is prolonged unreasonably.

Colour Coding System. Because of the limited bandwidth (less than 3 MHz) and the fact that the timing stability of the disc system is not good enough, normal methods of colour coding are not practicable.

A special coding system called 3 PAL is used which is very tolerant of timing errors. The colour information is placed at the low end of the frequency spectrum, the red, green and blue components being recorded sequentially in bands up to 500 kHz and the luminance information from 500 kHz upwards to the limit of the system. By the use of two one-line-period delay elements and a 3 PAL switch continuous RGB signals can be derived. These can be encoded in the player to produce a composite u.h.f. PAL signal which can be reproduced on a standard PAL colour TV receiver.

SLOW-MOTION, STOP-ACTION DISC RECORDERS

In televising all types of sporting events, instructional programmes and sometimes even drama there can be a requirement for slow-motion replay or the ability to slow the action down to a still frame, or cause a still picture apparently to come to life. The facility of studying the action frame by frame can also be a considerable asset in establishing cutting points for editing. All this can be achieved with the aid of the video disc.

There are basically two different types of stop-motion video disc machine, using either one or two discs. The discs are made of optically-flat aluminium alloy, coated on both sides with nickel–cobalt and thinly plated with rhodium to prevent corrosion and promote a smooth surface.

The Ampex H.S.100 video disc recorder. The Ampex H.S.100 video disc equipment is capable of replaying up to 36 seconds of recorded material at any speed between still frame and twice normal in either forward or reverse motion.

Two 16-inch diameter discs are used, mounted one above the other and driven by a single shaft. Four ferrite recording heads are set at right angles to each other in contact with each of the four faces of the two discs. The heads are made to track along the radius of each disc in a series of steps, ten-thousandths of an inch at a time.

Recording. Frequency modulation is used to record the video signal. There is no sound recording. The disc rotates at 3 000 r.p.m. and is phase-locked to incoming synchronous signals to complete one revolution per picture field. Each of the recording heads is held stationary in turn for two revolutions of the disc while it first erases and then records one t.v. field on the track. During each of the next two revolutions it steps on one track so that it has, in effect, skipped a track before stopping in position to erase and record on the second one. Each head thus records on alternate tracks as it makes its way towards the centre of the disc. When it reaches the inner limit of travel it reverses direction and continues a similar action on the way towards the edge, this time recording on the tracks it missed on the way in. As each track is performing a similar sequence displaced by one function the recording proccss is continuous.

It takes 36 seconds to complete the cycle starting from any point on the disc to the time when the respective head is retracing the track recording new material, so that from the moment that recording is stopped the previous 36 seconds of material is available for reproduction.

Reproduction. For reproduction the rotation of the discs can be related either to incoming synchronousing or local reference pulses. As each revolution of the disc contains one picture field a still frame can be achieved by simply holding the heads stationary on the required track and selecting the output of the right head so that it is repeated continuously. Slowing down or speeding up the motion is achieved by altering the stepping rate of the heads, e.g., half-speed can be obtained by causing the heads to remain on each track for two revolutions instead of one.

Reverse motion can be obtained by reversing the direction of traverse of the heads and switching their output in reverse order. The machine controls provide three fixed speeds, normal, half and one-fifth speed as well as continuously variable control, stop motion and reverse. Means are provided for rapid search for the required material such that any part of the recording can be reached within 4.5 seconds.

A disadvantage of obtaining slow motion in this manner is that, unlike filmed slow motion which is achieved by increasing the taking speed and reproducing at normal picture rate, the slow-motion disc records at normal rate and can reproduce at a different rate. At slow speed this can result in poor definition and blurred, jerky motion.

TELEVISION FILM RECORDING

The earliest method of recording television pictures consisted of photographing the picture display of a cathode-ray tube with a moving picture camera. Although film has been largely superseded by magnetic videotape recording, as a method of recording video signals, it still has many applications in broadcasting and research, etc.

The principal uses for television film recording, or telerecording as it is known in the U.K. (kinescope recording in America), nowadays stem from the fact that many of the world's small broadcasting organisations, who buy or exchange recorded programmes, are not equipped with the expensive quadruplex videotape machines that provide the only current international recording standard. Or they may have differing television standards from the originating area and are not equipped with the elaborate equipment required for standards conversion. On the other hand equipment for scanning optical film (telecine) is relatively cheap and universal.

There are two main problems associated with filming broadcast television signals: firstly, due to the use of interlaced scanning it is necessary to expose each film picture frame for two video fields, and secondly, the extremely short interval available between fields for the film to be pulled down and realigned for the next exposure which, in the case of the 625-line television standard amounts to 1.2 ms, corresponding to only about 10.8 degrees of shutter rotation (normally a film camera shutter is closed for 180° of its 360° rotation). The necessary acceleration would be impossible to achieve with 35-mm film, which would have to be accelerated at 4 000 G for 0.7 ms and then decelerated at 4 000 G for 0.7 ms. The more modest requirement of 1 600 G for 16-mm film, which is lighter and does not have to move so far, is just about attainable. Nevertheless this would place a considerable strain on the film and transport mechanism.

Various continuous-motion systems were tried in the early days and discarded due to problems of mechanical alignment (in relation to the stringent requirement to lay down interlaced scans so that they interleave accurately) and the light and resolution losses caused by halation at the various glass-to-air interfaces.

Suppressed field system

In an early system the pull-down problem was overcome by simply missing out alternate fields and using the blanked-out field time to move the film. This could be achieved using a normal 180° camera shutter, but the picture only contained half the number of lines ($188\frac{1}{2}$ instead of the 377 picture lines available in the 405 system in use at the time).

Suppressed field telerecording gave better results than might have been expected but there was inevitably some loss of definition, movement tended to be jerky and diagonal lines had a stepped appearance. Special flat-faced tubes are used for telerecording to minimise optical distortion and maintain maximum resolution right to the edges. They usually have blue (P11) or Ultra Violet (P16) phosphors. 'Spot wobble' is applied to fill in the gaps in the line structure. This is especially necessary to avoid stroboscopic action (Moiré patterning) with the scanning process when the film is reproduced by telecine. When standard movie film is reproduced or recorded on television the difference in frame rate between the two systems must be taken into account. This presents no problem with the U.K. television system which has a frame rate of 25 per second, standard film rate being 24 frames per second (f.p.s.). The film can be run at the slightly faster speed with little noticeable effect, except that the programme timing will be different and some people who possess perfect pitch have been known to complain that music from the sound track is sharp. Films made specifically for television and film telerecordings are shot at 25 f.p.s. so the problem does not arise. It is a different matter, with the American 525-line, 60-field television standard, where the 30 f.p.s. rate must be made compatible with the film rate of 24 f.p.s. In this case there is no point in making a change in the camera rate because there is a useful mathematical relationship between the two speeds. Five television fields are scanned in 5/60 second (1/12 second). Two film frames occupy 2/24 second (1/12 second) which should coincide with 4 television fields. By dropping one field in every five, i.e., converting every five television fields into two frames, compatibility is established. This can be accomplished by a process known as the 'Midfield Splice', *Figure 16.21*. The sequence is as follows: one television frame

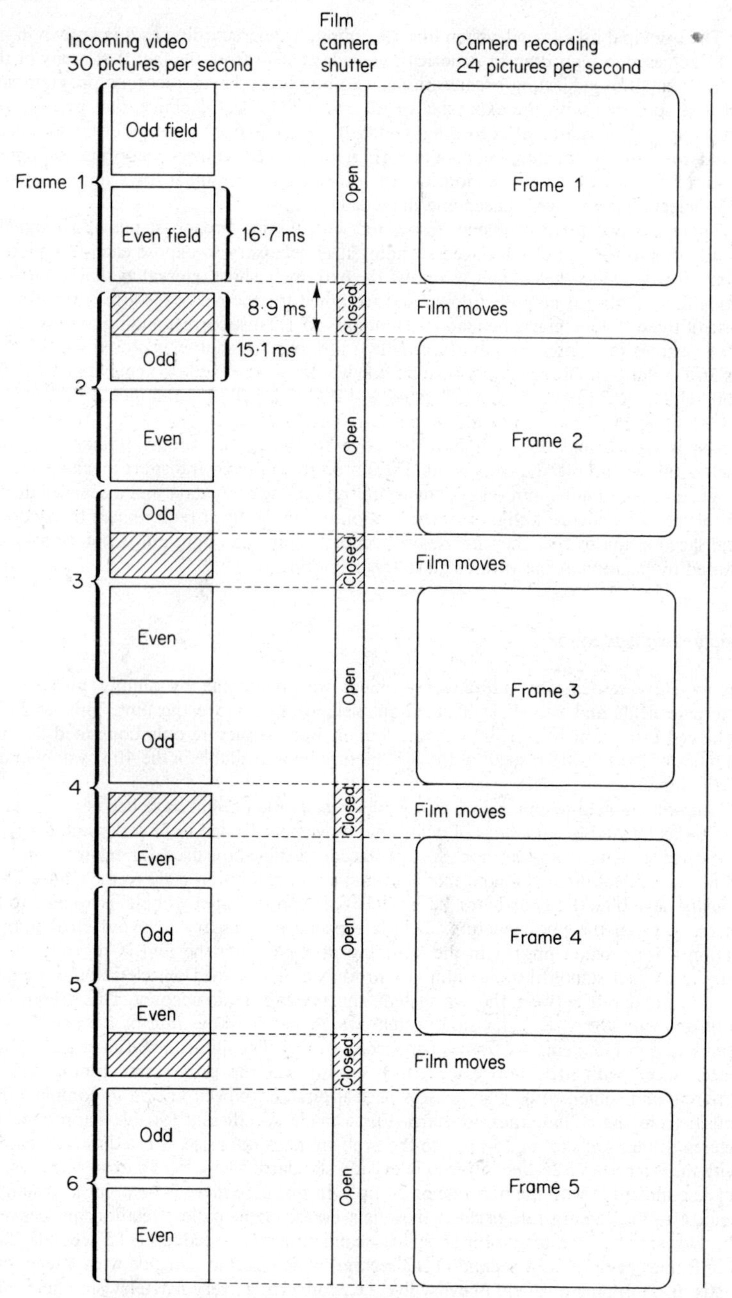

Figure 16.21 Midfield splice method of converting from 30 p.p.s. video frame rate to 24 p.p.s. film picture rate

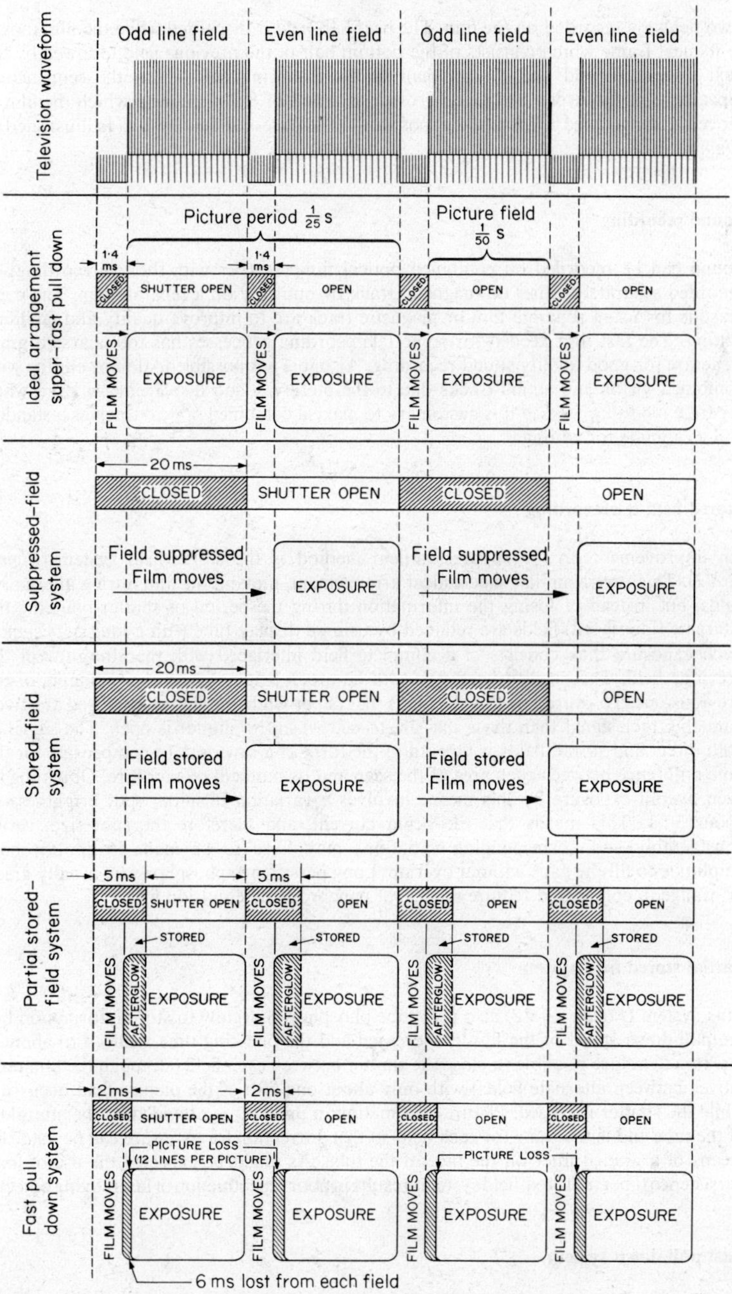

Figure 16.22 Comparison of timing between the video signal, blanking and camera shuttering for various telerecording systems

(two fields) is recorded on the film. The next half field is discarded while the film moves to its next frame which consists of the bottom half of the previous field followed by the next interlaced field and the top half of the following field. Then the sequence is repeated. The discarded half fields provide a period of 8.9 ms during which the film is moved and realigned for the next exposure. The suppressed-field system is illustrated in *Figure 16.22*.

Sound recording

Sound can be recorded on combined optical film together with the telerecording, or recorded separately either on magnetic track or optically on a separate film. The main reasons for using separate film or magnetic track are to improve quality and facilitate editing. The fast film needed for some telerecording processes has too coarse a grain structure for good quality sound recording. Also it is not possible to do fine editing with combined vision and sound tracks due to the picture/sound displacement. Even when separate recording is used it is customary to make a combined optical track as a standby and as a guide for editing.

Stored field telerecording

An improvement on the suppressed-field method is the stored-field system (*Figure 16.22*). This uses a similar mechanical arrangement, moving the film during alternative fields, but instead of losing the information during the period of shutter blanking the alternate (interlaced) fields are retained by using a display tube with a long persistence. Each exposure thus consists of a complete field interlaced with the afterglow of the previous field. The newly traced field will be much brighter than the remnants of the previous one. To overcome this the fields that occur while the shutter is closed are given a much larger signal than those that are traced when the shutter is open. The signal to both direct and stored fields is tapered in the form of a sawtooth to compensate for the time difference between each area of the scan and its moment of exposure. Obtaining an even overall exposure by this means involves a variation in initial spot brightness of about 3 : 1. This means that the beam current, and therefore the spot size, varies considerably and corresponding variations may have to be made in spot wobble amplitude to fill the gaps without overlap. Long persistence phosphors are usually green or orange in colour and require the use of panchromatic emulsion film.

Partial stored field system

This system (*Figure 16.22*) also uses tube phosphor afterglow to store information but the pull-down speed of the film is increased and the blanking time reduced to about 5 ms. It is therefore possible to close the shutter between each field (although the film only moves between alternate fields) with only about one fifth of the picture field occurring while the shutter is closed. Picture information in this area is provided by the afterglow of the tube and is the same for each field so that correction for exposure can be made by means of a graded filter on the face of the tube. As might be expected the use of long persistence tubes in stored-field systems results in poor reproduction of fast moving objects.

Fast pull down system

By restricting telerecording to 16-mm film, which is much lighter and requires a smaller movement than 35 mm, it is possible to devise mechanical drives that can pull down the film between frames in about 2 ms. This movement can be accommodated in an extended field blanking interval. Normal field blanking time is 1.2 ms for 625-line 50-field systems. A low-persistence tube is used and although a few lines are lost from the picture this is not very significant. Arrangements can be made in the picture composition to allow for the cut-off. New camera traction systems are becoming available which move the film entirely by means of blasts of air. These are capable of pulling the film down within the normal field blanking interval without losing any picture information. The system is illustrated in *Figure 16.22*.

Electron beam recording

Recording off the face of a cathode-ray tube requires a fast film and this produces a grainy picture. Also the variations in spot size involved with systems using tube afterglow result in poor definition. To overcome this a system of electron beam recording has been devised in which the electron beam is projected straight on to the film emulsion. It is necessary for the scanned portion of the film to be run in a vacuum (about 10^{-3} mm) when the exposure is made. To reduce the pressure on the air seals the whole magazine is evacuated to about 15 mm. This also helps to prevent the build-up of electrostatic charges which can deflect the scanning beam. The use of a laser beam as a scanning medium can produce extremely fine resolution without the necessity to run the film in a vacuum.

Colour

Colour telerecordings can be made on negative film stock or reversal film by adjusting the polarity and characteristics of the signal fed to the display device to match.

The display device for colour can be a standard shadow-mask colour television tube, but this tends to give poor definition due to the dot structure. Also it is difficult to achieve good colour response due to the problem of matching the film colour sensitivities and dyes to the tube phosphors.

Alternatively a trinoscope can be used. This is a system which combines three monochrome tubes in register. The tubes are reflected on to the film by a system of plain and dichroic mirrors so that they produce red green and blue images respectively which combine to produce a single coloured image.

A method of telerecording has been developed called Vidtronics which is akin to the three strip technicolor process. First a colour videotape recording is made of the original material. This recording is then replayed three times in succession to make three monochrome telerecordings, one corresponding to the red components in the picture, one to the green and one to the blue. These three black and white negative films are used as colour separation negatives which are superimposed by printing in register to produce the final colour print.

REFERENCES

1. HIGGS, C. R., 'From Manual Splicing to Time Code Editing', *BBC Engineering*, Number 95, September (1973)
2. THORNE, K. G., and BAYLIFF, R. W., *Royal Television Society Journal*, **Vol. 14**, Number 6 (1972)

FURTHER READING

ALKIN, E. G. M., *Sound with Vision: Sound Techniques for Television and Film*, Newnes-Butterworths (1973)

ENNES, HAROLD E., *Television Broadcasting, Equipment Systems and Operating Fundamentals*, Foulsham Sams (1971)

JONES, A. H., 'Digital Television Recording: A Review of Current Developments', *BBC Engineering*, Number 8, May (1974)

KEMP, W. D., 'Video Recording and Reproduction Techniques', *Electronics and Power*, **Vol. 18**, October (1972)

PALMER, A. B., 'The Technical Problems of Television Film Recording', *Journal of the SMPTE* December (1965)

ROBINSON, J. F., *Videotape Recording*, Focal Press (1975)

WHITE, G., *Video Recording: Record and Replay Systems*, Newnes-Butterworths (1972)

WOOD, C. B. B., ROUT, E. R., LORD, A. V. and VIGURS, R. F., *The Suppressed Frame System of recording*, BBC Monograph, June (1955)

17 ELECTRONIC MUSIC

ELECTRONIC MUSIC 17

17 ELECTRONIC MUSIC

ELECTRONIC ORGANS

History

Through the centuries man has evolved a great variety of music-producing instruments. Large numbers of performers assemble in orchestras to play some of these instruments. Or one performer may sit at a console of keys and pedals arranged to operate perhaps thousands of windblown pipes and reeds which sound like many of these instruments. This is the heritage and tradition behind the new field of electronic organs.

Beginning with the 1900s there were many attempts to offer electric or electronic substitutes for organs, but none enjoyed any degree of commercial success. They were based on photo-optics, magnetic or electrostatic prerecordings, vacuum tube or neon lamp oscillators, or amplified blown reeds, etc. One of these, called the Telharmonium, was invented by Thadius Cahill[1] and demonstrated in 1908. The size of a small power-generating station, it consisted of almost a hundred alternator generators for all the frequencies of the scale. Then through a console of switches, synthesised musical signals were transmitted over telephone lines without benefit of amplifiers.

In 1935 Mr. Laurens Hammond,[2] based on his synchronous electric clock, invented the first commercially successful mass-produced electric organ that started an industry.

Since then, many manufacturers all over the world have joined this industry, offering instruments in a variety of sizes and prices that have transformed modern music.

Organ Configurations

Electronic organs are classified as to size. Keyboards have the traditional 5 octaves (61 keys) or are shortened to 49, 44 or 37, etc. By the numbers of upper keys, lower keys and bass pedals, typically we have

Consoles: 61-61-25 or 61-61-32
Spinets: 44-44-13 or 37-37-13

Consoles are the traditional-size organ, whereas spinets, by their smaller size and cost, have been most widely accepted. Practically all spinets and most consoles contain their own speakers. Originally all console-size electronic organs required external tone cabinets.

Usually the upper manual is for playing melody or solo, the lower is for accompaniment, and the bass pedals supply the deep rhythmic accenting. Occasionally organs will have 3 or 4 manuals for greater player versatility.

Bass pedalboards traditionally had 32 pedals and were radiating and concave. Now the radiating but flat 25-pedal configuration is the more popular console arrangement. Spinets usually have only 12 or 13 pedals which are part of the organ and not detachable as they are on consoles.

The control tabs turn on the many voices for each division as well as control most of the other features of the organ. These controls, sometimes called stops, are usually finger tabs, sometimes rocker tabs and occasionally drawknobs. Tabs are usually in a straight row above the manuals, except where the cinema organ tradition forms them in a wrap-around horseshoe. Some institutional organs have tabs or drawknobs on side jambs as well.

Another very successful form of electronic organ has been the chord organ, which uses chord buttons to play accompaniment chords, instead of a second keyboard.

Recently, portable organs, usually having one or two spinet-size manuals, have become popular. They usually require separate tone cabinets.

Musical Requirements

The objective in electronic organ design is to maximise the variety of tonal effects; that is, to approach or exceed the variety of an orchestra or a multi-rank pipe organ, with preferably a single tone-generating system. To this end a fascinating variety of techniques have evolved.

In organ design the useful musical range of frequencies (fundamentals and harmonics) is generally between 32 Hz and 8 kHz. This is divided into 8 octaves and each octave is divided into 12 equal intervals called semitones. To deal with pitch accuracies, each semitone may be divided into 100 parts called cents.

An *octave* is any interval of 1 : 2 ratio. A *semitone* is any interval of $\sqrt[12]{2}$ ratio or 5.95% ($\sqrt[12]{2} = 1.0594631$). One *cent* is a 0.0595% interval or 1 part in 1 682.

Occasionally pedal tone fundamentals include the octave of 16 Hz to 32 Hz and harmonics of some keyboard tones are higher than 8 kHz.

Musical Tones

Helmholtz's law states that a musical tone is made up of a pure fundamental (sine wave) tone and various amplitudes of harmonics (also sine waves) whose frequencies are integers times the fundamental. Except at very low frequencies, the ear does not detect the phase relationships of these harmonics.

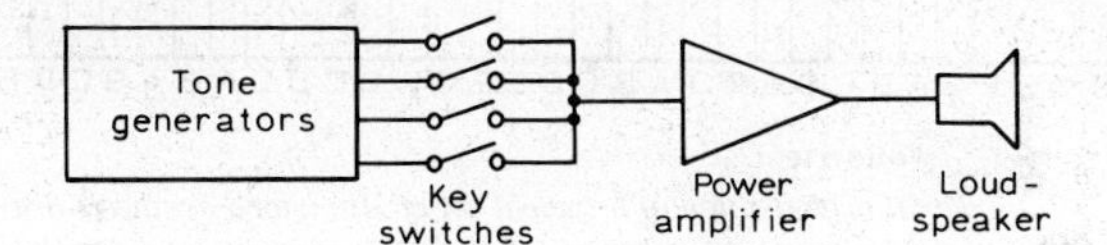

Figure 17.1 A simple electronic organ system

Musical tones can be created by either starting with all sine waves and mixing their amplitudes as desired (this is called *synthesis*); or starting with a complex tone or 'mother' tone which contains many harmonics and alter the harmonic content by subtraction, or filter rejection (this is called the *formant* system). A third method (called *digital synthesis*) described later, is different from, but has attributes of, both above systems.

A simple electronic organ would consist of a series of tone generators (oscillators) connected through switches under the playing keys to a power amplifier and loudspeaker as shown in *Figure 17.1*. This system would produce but a single type of tone. To produce a variety of tones, voicing circuits must be added. These can be divided into two major systems just described.

Synthesis Organs

The generator in this system contains all the sine waves covering the useful range described—approximately 96 from 32 Hz to about 8 kHz. They will be heard as individual fundamentals as well as harmonics of other fundamentals. A set of nine drawbars regulate their loudnesses in nine steps including silence. They control sub-fundamental, sub-third, fundamental and the 2nd, 3rd, 4th, 5th, 6th and 8th harmonics. Note that the odd harmonics used are not exact multiples of the fundamental, but are

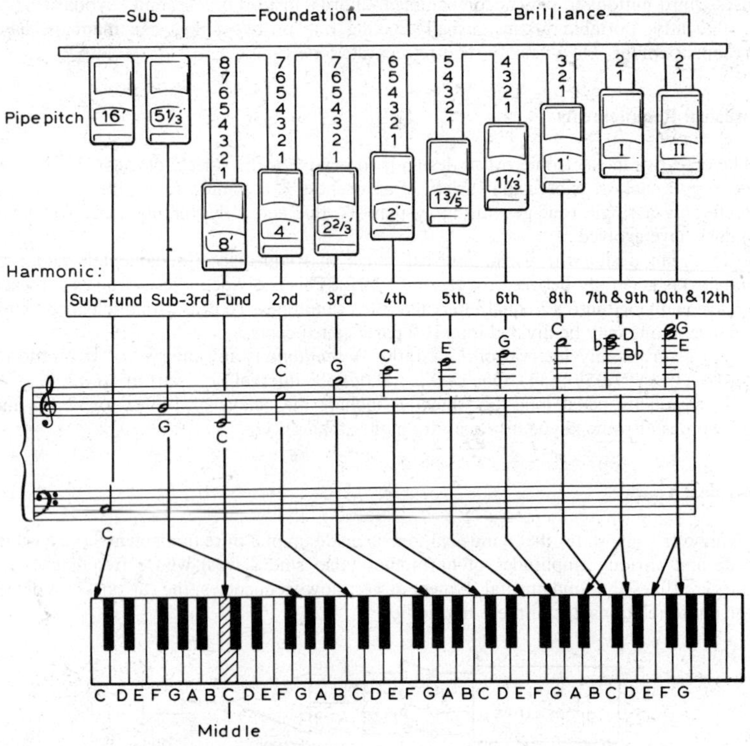

Figure 17.2 Tonal organisation of a typical set of Hammond synthesis drawbars

very close frequencies already used for other notes on the organ. Thus there are no interferences or 'temperament beat' patterns produced between natural harmonics and other notes on the organ, since these harmonics come from the same source.

Illustrated in *Figure 17.2* is a typical set of nine drawbars (plus two extra drawbars on deluxe models), the harmonics they control, and the tone sources borrowed for each harmonic (assume fundamental is middle C).

A simple synthesis system is shown in *Figure 17.3*.

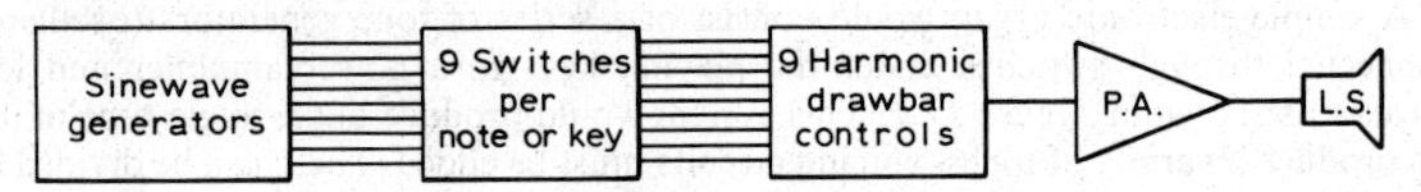

Figure 17.3 A simple synthesis organ system

The essential waveform for synthesis is the pure sine wave. At the low end of the range some harmonic development is helpful. But in the mid and high range, any harmonic content weakens the illusion of synthesis and the mixtures tend not to 'cohere'. Modern low-cost divider chains yielding square waves require complex group band pass filters to yield suitable sines.

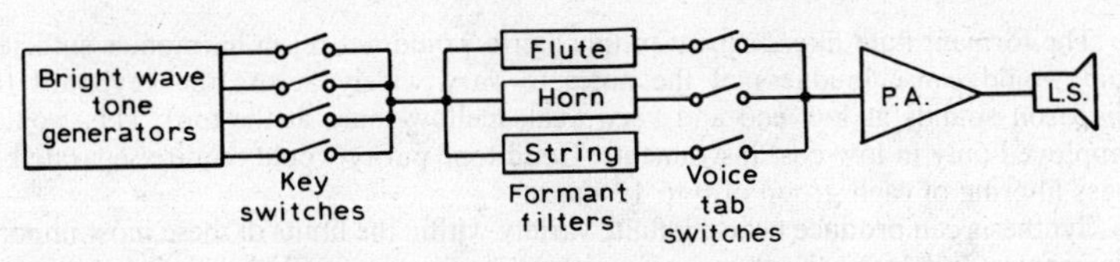

Figure 17.4 A simple formant organ system

Formant Organs

This voicing method starts with a tone of rich harmonic content and creates variety by subtraction or selective filtering. Low-pass filters suggest mellow flute and diapason tones, all-pass string tones, and band-pass resonators create reeds and horns. These filters are connected to the amplifier with the voicing switches as shown in *Figure 17.4*.

The preferred single waveshape for a formant organ system is the sawtooth, which contains all harmonics in descending amplitudes. The modern trend in electronics has brought in the lower cost flip-flop dividers which produce the less desired square wave containing only the odd harmonics. A technique called 'stairstepping' can restore the missing evens to the square. When the square is desired in a generator producing only saws, the 'outphasing' technique can cancel out the even harmonics.

Other useful generic waveforms include various pulse shapes, triangle or pyramid, etc. A most versatile organ would contain generators of two or more waveforms, appropriated to their most suitable voice usages.

Tone Families

Both synthesis and formant systems can produce the typical tone families shown in *Figure 17.5*.

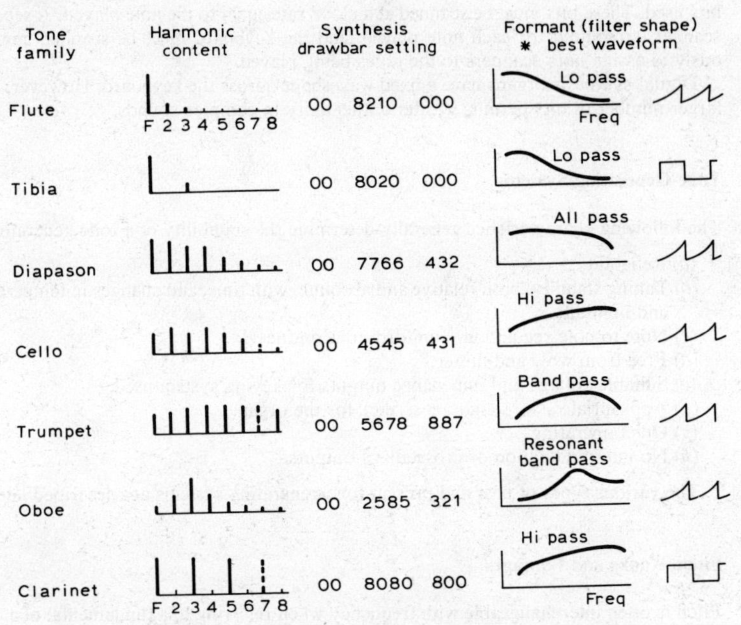

Figure 17.5 Synthesis and formant organ tone family harmonic character

The formant flute filters shown in this figure would not reject harmonics sufficiently and would cause loudness of the notes to vary widely across the keyboard (loud diapason sounds at low end and very weak mellow tones at the top). This would be employed only in low-cost instruments. Good tone purity would require separate band-pass filtering of each group of 6 to 12 notes.

Synthesis can produce nearly infinite variety within the limits of these most important harmonics, but lacks the 'buzz' or real high brilliance possible with formanting and characteristic of some string and reed tones.

Synthesis reproduces good flutes, mellow horns and unusual harmonic combinations (used by pop musicians).

Formanting is a lower-cost system but variety is limited to the complexity of filtering. Unless the keyboard is filtered by groups of notes, the note loudness (or regulation) varies across the keyboard. Synthesis tends to maintain the same harmonic content and loudness across the keyboard (sometimes called a 'travelling formant', such as a small flute at the high end and a big one at the low end).

Formant systems contain temperament beats inherent in natural polyphonic instruments. But synthesis tones lack these beats and are more suited to rotating speaker-type animations described later.

Both synthesis and formant systems continue in the marketplace with their separate advantages and proponents.

Digital Synthesis Organs

Digital synthesis organs are now making their appearance.[3] This very complex technique has become feasible only through micro-miniature integrated circuitry. They construct a waveform by bits in the time domain rather than by harmonics in the frequency domain. The quality of the reproduced tone is determined by the number of bits used. These bits must be scanned at a clock rate equal to the note played. A separate scanner is required for each note played, and the keyboard must be scanned continuously to assign note scanners to the notes being played.

Digital synthesis organs have a fixed waveshape across the keyboard. However, their larger number of bits permits greater authenticity in complex sounds.

Tone-Generating Systems

The following characteristics generally determine the suitability of a tone generator:

(*a*) Reliability.
(*b*) Tuning stability, both relative and absolute, with time, and changes in temperature and humidity.
(*c*) Note-to-note regulation—uniform tone and level.
(*d*) Free from wow and flutter.
(*e*) Suitable voltage and impedance outputs for keying system used.
(*f*) Appropriate size, weight, cost, etc., for the organ.
(*g*) Quiet operating.
(*h*) No intermodulation or crosstalk of outputs.

The various types of past and present tone-generating systems are described later.

Pitch, Ranks and Footages

Pitch is often interchangeable with frequency when referring to a fundamental of a tone. When, as on a piano, the middle C key sounds middle C note (and all notes above and

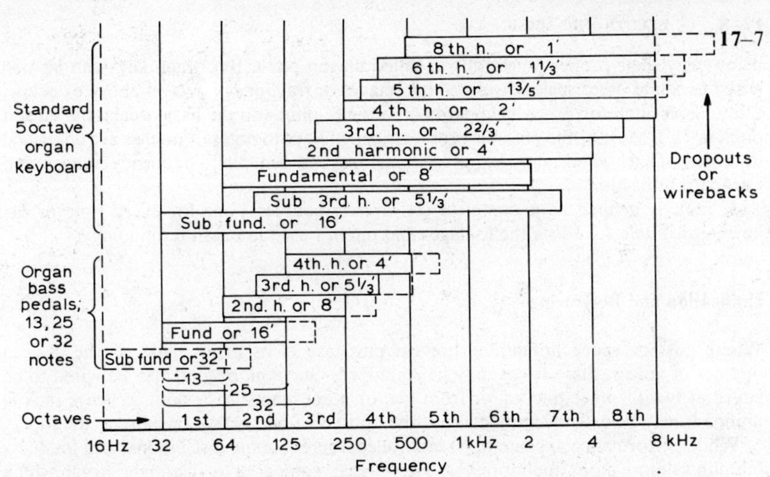

Figure 17.6 Frequency ranges of pitches or harmonics on keyboard and pedalboards

Table 17.1 RELATING HARMONICS, FOOTAGE, PITCH, NAMES

Harmonic name:	*Pitch footage on manuals* (note 1)	*Equivalent scale letter (assume fund.* = C)	*Scale interval name*	*Pitch names in pipe organ tech.*	*Exact freq. of each harmonic (assume fund.* = C_M)	*Hammond drawbars (handle colours)*
Sub. fund.	16′	C	—	Sub octave	130.8125	Brown
Sub. 3rd	$5\frac{1}{3}'$	G	5th	Quint	392.4375	Brown
Fund.	8′	C	1st	Unison	261.625	Ivory
$1\frac{1}{4}$ (note 2)	$6\frac{2}{5}'$	E	3rd	—	327.00	(note 2)
2nd	4′	C	8th	Octave	523.25	Ivory
3rd	$2\frac{2}{3}'$	G	12th	Nazard	784.875	Black
4th	2′	C	15th	Super octave	1 046.50	Ivory
5th	$1\frac{3}{5}'$	E	17th	Tierce	1 308.125	Black
6th	$1\frac{1}{3}'$	G	19th	Larigot	1 569.75	Black
7th	$1\frac{1}{7}'$	between A & A♯	♭21st	Septiéme	1 831.375	(notes 3 and 4)
8th	1′	C	22nd	Super super octave	2 093.00	Ivory
9th	$\frac{8}{9}'$	D	23rd	—	2 354.625	(note 3)
10th	$\frac{4}{5}'$	E	24th	24th	2 616.25	(note 3)
11th	$\frac{8}{11}'$	between F & F♯	♭25th	—	2 877.875	(notes 3 and 4)
12th	$\frac{2}{3}'$	G	26th	26th	3 139.50	(note 3)

Notes

1 Pitch footage (on manuals) $= \dfrac{8'}{\text{harmonic } \#}$. On pedals $= \dfrac{16'}{\text{harmonic } \#}$.

2 This $1\frac{1}{4}$ harmonic, not on a drawbar, is in some organs to synthesise a *chime* tone.

3 These harmonics are on some Hammond organs.

4 On the Hammond Grand-100, these harmonics were accurately produced by an extra tone generator.

below play their proper notes) this is called unison pitch. But organ keys can be also wired to sound other scales displaced by octaves or fractions of octaves above or below. The unison pitch for organ keyboards is called 8′ pitch (on the bass pedals, the unison pitch is 16′). Most of the possible cross wirings of keys to notes or pitches are described in *Figure 17.6*, which shows the relationships between pitch frequency ranges and keyboard utilisation.

A rank is defined as a complete set of tone sources—one for every note of the keyboard. *Table 17.1* lists the footages and pitches used in organ terminology.

Unification and Borrowing

Where cost or space limitations prevent purchase of as many ranks as the desired number of voices, these tricks may be employed. One rank of tone may be wired to be keyed at two or more pitches, or from two or more manuals, or both. When a rank is unified it may be extended an octave at the bottom or one octave or more at the top.

While performance appears to be multiplied, there is some loss in tonal and loudness buildup as more stops and more keys are pressed, compared to a 'straight' organ which has a separate rank for every stop.

Couplers

Couplers are used on organs to increase the capability of the organ player by making multiple use of the number of keys and pedals that he can depress at one time. Thus, a 'lower to pedal' coupler connects the voices called for on the lower manual to the pedal keys in addition to the pedal voices already called for. The disadvantage of this is that once the note has been played via a pedal, playing it again on the lower manual results in no more output. This causes 'lost' notes as more keys and couplers are used, similar to unification. Some examples of couplers are:

(*a*) *Lower to pedal.* Pedals play lower manual voices.
(*b*) *Upper to lower*. Lower manual plays upper manual voices.
(*c*) *Octave*. The keys play tones an octave higher or lower than normal or unison.
(*d*) *Unison off*. A form of coupler that disconnects the normal or unison pitch connections so that only coupled pitches speak.

Keying

A.C. keying is the commonest form of keying. In this system, a contact under the key connects the proper frequency to the output wire or 'bus'. This requires a separate contact and bus for each pitch (or 'footage') to allow for subsequent filters and controls. Since the key is rarely closed at the instant the voltage crosses the zero axis, a transient 'key click' is generated which is suppressed by limiting the high frequency response of the output system.

D.C. keying uses a tube, a diode or a transistor to connect the required frequency to the output in response to d.c. voltage received from the key contact closure. Thus, only one contact is required under a key. Instead, additional keyers are added as the number of pitches is increased. Most keyer systems result in some d.c. shift in the signal axis as the keyer is turned on, so the turn-on (attack) is slowed down to prevent an objectionable 'thump' in the output.

The advent of d.c. electronic keyers made it possible to cause a note to linger on or 'sustain' for a short period of time after the key was released by charging a capacitor when the key was played and allowing it to slowly discharge after release for a pleasing musical effect.

Percussion

Percussion tones are generated by adding an exponential envelope to the proper voice. Electronic keyers made this possible.

An *independent* percussion system uses an envelope-shaper or keyer for each note and thus each note plays and decays independently.

Where product cost prohibits making each note independently percussive, a single *unitary* gate can apply a percussive envelope to some or all of the voice outputs of a manual or the pedals. This gate may be triggered by only the first note played, or the first and all subsequent notes played, to approximate the independent percussion keyers. In the latter mode all the previously played notes are re-keyed along with the new note, which is the chief limitation of this system.

Reiteration or Repeat

This is often used on organs which include percussion (either unitary or independent polyphonic). This function causes the notes to repeat at a rate between 4 and 15 repeats per second, resembling banjo strumming. Sometimes harmonics are alternated to produce a twin mallet marimba or xylophone effect.

When steady state tones are given a percussive envelope they take on a different character. For example, a diapason becomes a piano, a flute becomes a harp, etc.

Percussion or sustain envelopes are sometimes applicable to individual harmonics or drawbars offering synthesis as well as bright wave variety.

Arpeggiator *

This feature, when coupled with an independent percussion or sustain tone system, enables the amateur to play single and multiple arpeggios that may exceed a professional's capability.[4] When passing a fingertip over its narrow playing surface, arpeggios up to four octaves long are heard based on the chord held on the lower with the left hand. With Arpeggiomatic,* a chromatic glissando and either whole tone glissandos may also be played.

High-Select and Low-Select Systems

It is sometimes expedient or musically desirable to have a polyphonic or monophonic tone system speak only one note at a time. The human voice and many orchestral instruments are examples. The tone source(s) may be playable from a keyboard as either high-note or low-note select. Examples are bass pedal systems on most spinets and a few consoles. They are monophonic to reduce the number of tone generators and to eliminate the chance of a novice playing two adjacent pedal notes which would be discordant and also overload the output system.

On simple instruments like the Piper Autochord,* there is no pedal keyboard, so the lowest note played by the left hand is selected by a series logic circuit and divided down two octaves to produce the pedal tone.[5] In the Autochord* mode, also, the highest note played by the left hand is likewise selected and divided down to produce a complimentary pedal tone on alternate beats in the measure. Then percussive gating of these two bass notes, as well as the manual tone, is triggered by the automatic rhythm clock.

The Solovox and Clavioline are examples of three-octave keyboard products, usually attached to pianos, that imitated many orchestral voices one note at a time.

* Trademark of Hammond Corp.

Currently popular synthesisers are products based on one-note-at-a-time, which maximise modern electronics with v.c.o.s (voltage-controlled oscillators), v.c.f.s (voltage-controlled filters) and v.c.a.s. (amplifiers) towards creating usual and unusual sounds.

Animation

Vibrato is the most common form of animation. An example is the violinist's finger rapidly changing the length of the string while bowing. It is a smooth sinusoidal modulation of all the musical frequencies above and below normal by up to 2%. The rate is usually between 5 and 7 Hz. If this modulation is applied to the master oscillator of a cascaded divider generator system, all the outputs will have the same vibrato width. Some organs may use an 'after vibrato' device effective after the tones are generated and mixed. An example is the rapid scanning back and forth of an electrical delay line (of about 1 ms delay). Other methods employ one or more cascaded stages of phase shift circuits.

Delayed vibrato is a feature whereby the player can control the slow starting up of the vibrato, as does a vocalist or violinist.

Tremolo is a less popular and less desirable effect and is an amplitude modulation, of the same rates as vibrato, sinusoidal and of depths of up to ± 6 db. This is usually achieved by varying the gain of a preamp.

Rotating loudspeakers or baffles produce a combination of amplitude, frequency and formant modulation in conjunction with a sense of motion, which is very popular.[6] These may be rotated at a vibrato rate of about 6 Hz or at a *Celeste* or *Chorale* rate of 0.6 Hz which simulates the slow beat patterns of multiple ranks of pipes in large pipe organs. These devices may be inside the organ or in the tone cabinet.

Pistons-Presets

Since the ability to change the voices is one of the most charming aspects of an organ, it is desirable to be able to set up the desired combination of voices before playing a composition and instantly switch this combination in at the appropriate time in the piece. These devices which turn off one combination of voices and turn on another are called presets or pistons. They may appear as reverse colour keys, buttons between the keyboards or large knobs above the pedal keyboard.

Expression Pedals

The precedent in reed and pipe organ for the player to regulate or 'express' the music loudness while playing was a set of moving shutters in front of the tone sources. In electronics variable gain circuits are employed which usually include compensation for listener hearing characteristics (less than normal dynamic change at the low and high ends of the frequency spectrum).

Reverberation

Listeners are conditioned to associate impressively large pipe organs and orchestras with large reverberant halls and cathedrals. Electronics offers several methods of simulating this echo effect in small absorbent rooms. Reverberation may be simulated by electromechanical devices using coil springs[7] or sheets of metal, or re-entrant record/playback devices using magnetic or electrostatic recording.

Rhythm

Electronic circuits can generate rhythm voices such as cymbal, snare, blocks, drums, etc. These rhythm voices may be keyed by the lower manual or pedal keys in tempo with the music. Or they can be sounded individually and momentarily by separate pushbuttons.

Digital logic circuits have also made it possible to key these rhythm voices in various popular rhythm patterns (waltz, fox-trot, Latin, etc.). A master clock the rate of which is adjusted by the player is divided down to eighth, quarter, half and whole note periods from which AND gates trigger the rhythm voices in the pattern selected by the player.

Amplifiers

It is necessary that all amplifiers used in an organ system have very low distortion (less than 0.1% harmonic distortion). When a player holds down two notes close together, the intermodulation difference beat note is a lower frequency, not necessarily harmonically related to the generating notes. The 'sum and difference' frequencies produced by distortion are unpleasant and very obvious to the listener.

Loudspeakers

The choice of speakers, their number and size, their characteristics and their placement within the organ or tone cabinet enhance the listening qualities of the music.

Since very low frequencies are contained in the pedal tones, the loudspeaker used must be able to handle this low frequency power without exceeding the limits of its linear excursion. Separate channels are often used to minimise the intermodulation between different sections of the organ.

Accessories

A recent popular accessory has been a built-in cassette recorder/player. It permits the player to record his own playing as well as play prerecorded tapes including self-teaching instructions.

Occasionally, player-roll-type player mechanisms have been combined into electronic organs.

An inbuilt headphone jack on most organs permits the player to practice without being heard.

Organ Technology

Technology has played a major role with regard to both performance and cost of electronic organs. Few, if any, consumer products have been so greatly influenced by technical change, while at the same time maintaining traditional or characteristic performance established by earlier instruments. The major electrical functions within the organ are tone generation, keying means, voicing, circuits, animation and amplification. However, the foundation of all systems is the generation and keying means.

Electromechanical Systems

The tone wheel generator discussed earlier involves a variable reluctance magnetic pickup which senses the teeth on a gear-like driven wheel. Each tone used in the organ is generated by a separate wheel with an appropriate number of teeth. However, all the tone wheels (typically 91) are combined into one assembly and driven by a common synchronous motor. This concept provides two unique tuning benefits. Since all of the tone wheels are locked together within the mechanical drive train, the relative tuning (accuracy) between notes is fixed. Also, since the drive motor is synchronous, the absolute tuning accuracy is determined by the frequency of the power line system which is quite stable with less than 0.1% error. Therefore, 'never needs tuning' has been a major advantage of the tone wheel system.

This type of generation scheme for tones and appropriate individual filters for each note to obtain very pure sine waves is the foundation of the synthesis-type organ. In order to combine the appropriate tones for synthesis, multiple contact a.c. key switches and drawbars switches are required. The vitality of this technical concept has been outstanding as evidenced by the fact that organs are still being made this way.

Other electromechanical schemes have also been tried but none have developed the wide acceptance of the tone wheel/synthesis approach. Several electrostatic approaches based on wave shape generation by means of mechanical variable capacitors have been developed.[8]

Optical sensing wave generation schemes have had continuing interest as tone generators.[9] Usually the approach has been to make optical wheels containing equivalent wave shape recordings of an instrument, such as a pipe organ flute, for each note. An optical transducer is then used to reproduce the wave, hopefully, to the same quality of the original instrument on command from the keyboard. Generally, these systems have not had practical success because of the large number of wave shapes and the extremely tight tolerances required to minimise undesirable wow and flutter. Also, tone colour and recognition characteristics of traditional instruments, be they organ or orchestral, are closely associated with keying envelope transients and noise on attack and decay, which are difficult to reproduce. The extreme skill and manufacturing cost associated with overcoming these problems have prevented optical generator organs from becoming a significant force in the electronic organ marketplace.

However, on the other extreme from a cost standpoint, a rather interesting application of optical generation in a musical product has had some success.[10] Although it is not an organ in the usual sense, it has a keyboard for the melody part and buttons to play accompaniment and rhythm. In this device the melody keyboard sounds, say for a given instrument like a piano, optically recorded similar to a movie film sound track on a phonograph type transparent disc. Similar tracks are recorded on the disc for the rhythm and accompaniment buttons, except that normally an ensemble of instruments will be included on the tracks. Playback is achieved by means of a light source, photocell and keying means associated with each track. In order to change melody and accompaniment voicing, the player must change records. Due to the relatively slow operating speed of the disc, 'once-around' problems are minimised. However, care must be exercised to maintain disc alignment and avoid dirt accumulation within the optics system to avoid cross talk and noise problems.

Transistor/Diode Systems

The advent of low-cost silicon transistors and diodes in the 1960s dramatically changed the approach to designing formant-type electronic organs. As noted in previous discussions, formant instrument systems have been based on generation of twelve notes in the top octave by means of independent master oscillators. The octavely related notes for the lower tones are then generated by dividing the master oscillator frequency by means of flip-flop divider circuits. Transistors are readily adaptable to both the oscillator and divider applications and quickly replaced relatively expensive and less reliable vacuum tubes in these applications.

Formant instruments require complex or generic wave forms for many of their voices, and transistor generation techniques to distort or combine octavely related square waves to obtain the desired enrichment were developed. Also, early formant instruments were primarily a.c. keyed systems. However, the availability of low-cost diodes with high back-to-forward resistance ratios permitted the wide usage of sustain-type circuits using d.c. type keyers. The relative ease with which pitches can be added to formant instruments by either a.c. or d.c. keying means resulted in a proliferation of various size organs.

Discrete transistors also have found wide application in other organ circuits such as preamplifiers, power amplifiers, active filters, phase shift and reverberation circuits, and keying-type amplifiers.

The technological impact of solid state discrete transistors and diodes on performance and cost of formant instruments as compared to synthesis types was very significant. This change permitted the formant organ to achieve a dominant position in the marketplace over the full gamut of organs.

Integrated Circuits/L.S.I. Systems

The extremely high redundant circuit requirements of electronic organs in the generation and keying areas have made them responsive to technology changes which permit cost and/or handling improvements with circuits of this character. The development of first integrated circuits in the late 1960s and then large-scale integrated circuits (l.s.i.) in the early 1970s appeared to be the answer to the redundancy problems. Integrated circuit dividers readily found application in many organs, and component manufacturers made them available on an 'off-the-shelf' basis. However, beyond that point very little across-the-board application of standard i.c. packages has been made. This is primarily due to significant different musical philosophies between organ designers with regard to wave shape generation, d.c. keyers, and voice filter organisation. Also, when l.s.i.s appeared on the scene, some additional hurdles had to be overcome to take advantage of their potential. In order to achieve low function cost with l.s.i., it was necessary to incorporate multiple functions on a given chip. To achieve this need meant development of some new circuit and system concepts to fit the characteristics of the l.s.i./m.o.s.f.e.t. technology. Since these circuits are generally applicable to only one organ type and manufacturer, a custom l.s.i. package is usually required. This approach is practical only if the product volume is sufficient to cover custom tooling costs and create the need for a large production volume of l.s.i. devices. As a result of these factors, only a few organ manufacturers have been able to apply l.s.i. in its most efficient mode which is custom applications.

One exception to this situation is an l.s.i. device which most organ manufacturers have elected to use. This is a top octave generator which generates all twelve notes of the highest octave from a single very high frequency master oscillator. Several ways of doing this are possible, but the scheme that has been most acceptable is to use straight division for each note.[11] Typically the master oscillator will run at a frequency around 2 MHz and produce a top octave in the 4 kHz to 8 kHz range. With this scheme relative accuracies of better than $\pm 1\%$ are achieved. Absolute accuracies are, of course, dependent on the master oscillator stability. Some organs have used crystal oscillators for this purpose, which places them in the 'never needs tuning' category.

In examining ways to apply custom l.s.i. devices to organ systems, some new concepts have evolved. As discussed earlier, one approach has been to apply digital concepts throughout the system complete to digitalising the wave form of the musical voice. This approach to making organ music has found application primarily in the institutional organ area due to the relatively large number of l.s.i. devices required. Another major effort to utilisation of custom l.s.i.s has found broad application from the simplest spinet organ to the most deluxe professional console instruments. This concept is based on concentrating the redundancy functions within the l.s.i. device.[12] In this approach tone generation, keying, wave shaping and extensive bus wiring for as many as three notes and all their octavely related counter parts needed to scale a complete spinet organ are included on one l.s.i. chip.[13] Typically one of these custom l.s.i. devices involves 600 transistors on a single chip mounted in a dual-in-line package measuring 15 mm × 51 mm with 40 leads.

On the other end of the product spectrum, this approach to custom l.s.i. has made it practical to make a synthesis organ without tone wheels while at the same time expanding its performance to incorporate sustain and/or percussion sustain keying. A deluxe console organ in this category, which is being manufactured, has 30 000 transistors contained within 47 l.s.i. packages.

One very appealing promise of l.s.i. devices is improved reliability. Process methods imply potential failure rates of l.s.i. should approach that of individual transistors. If this goal is achieved, orders of magnitude improvement in product field reliability should result. Early reports[14] from custom l.s.i. users indicate this potential is becoming a reality.

Over the next few years, it is anticipated that custom l.s.i. will revolutionise the way in which electronic organs are designed and manufactured. In addition, the lower cost to performance ratio resulting from volume use of these devices should greatly broaden the scope of the organ market.

Organ System

Figure 17.7 illustrates a typical modern electronic organ. New custom l.s.i. packages generate and key the desired waveshapes and pitches. All packages receive the top 12 frequencies from a multiple derivative divider system controlled by a 2 MHz master oscillator. Vibrato may modulate this oscillator. The outputs representing upper and lower and pedal drawbars may be heard through the main channel or the tremolo animated channel. The percussions, rhythms, reverberation and cassette are heard only through the main channel. The white noise type rhythm sounds and cassette output are kept out of the reverberation device. The three audio channels are simultaneously loudness controlled by the expression pedal through variable transmission devices. Accessory switches on the sides of the expression pedal may silence the rhythm, or cancel vibrato or animation, or add keyboard sustain. The low frequencies in the tremolo channel are diverted to the woofer. The headphone jack when used, silences the speakers. An optional remote tone cabinet cable receptacle is provided. The main amplifier output is divided to tweeter and woofer speakers.

Future

From a musical standpoint, it is anticipated that the electronic organ is rapidly reaching maturity in both synthesis and formant areas. Instrument families have all been explored and produced electronically. Further effort will be aimed at improving authenticity of instrument voices and extent of performance within a product. Extensive performance capability will require more consideration for human engineering needs so the player can take advantage of the available musical potential.

Automatic accompaniment concepts such as Autochord* will be improved to make them easier for the player to use and applied more broadly to attract the novice player.

Rhythm unit accessories have reached a high level of musical sophistication and future effort will be aimed at reducing their cost via the l.s.i. route.

Synthesisers built into the organs will become a major feature in the more deluxe units.

* Trademark of Hammond Corp.

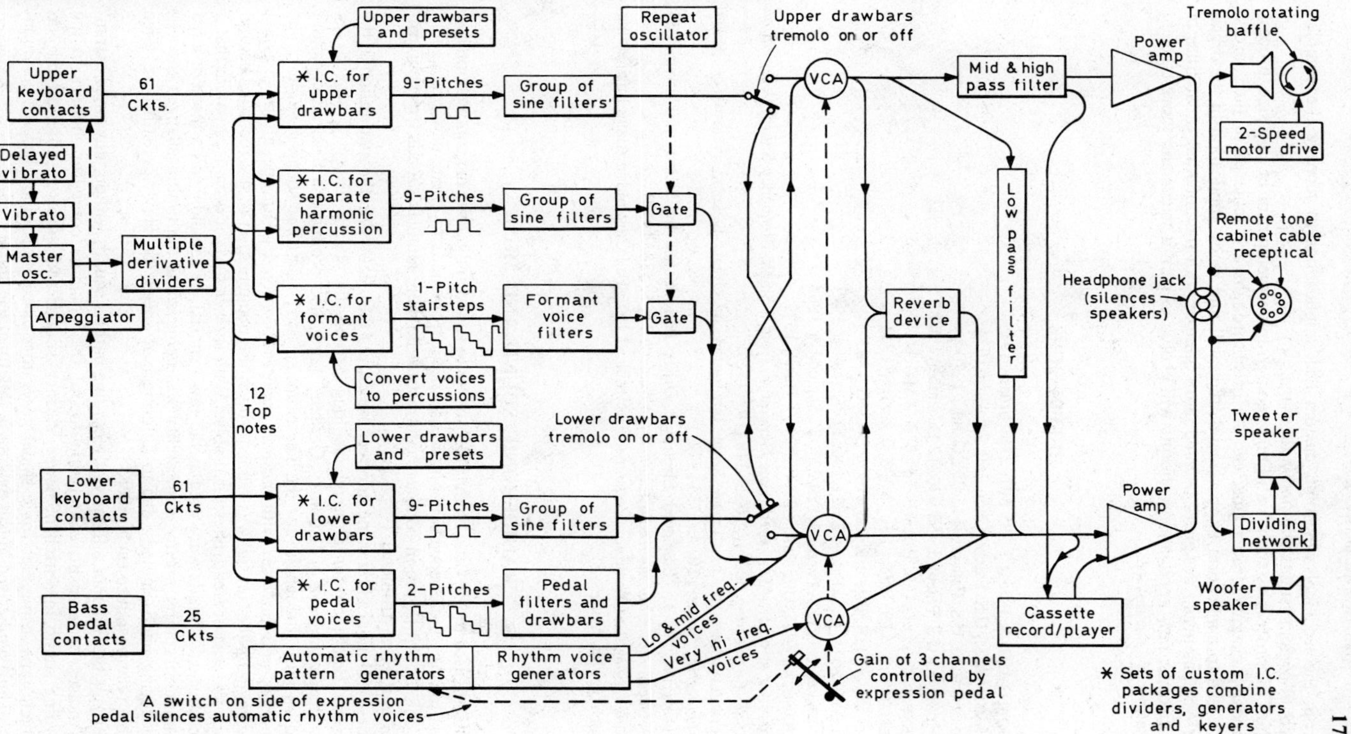

Figure 17.7 Block diagram—typical modern electronic organ

From the technical viewpoint, custom l.s.i. applications to manage the complex aspects of organ systems will find wide usage.

The l.s.i. process as such is still undergoing rapid development in chip size, process methods, packaging, lead frame, etc. As breakthroughs are achieved in these areas, which lead to application improvements, they will find ready application in the organ field. Also, as organ designers become more familiar with the potential capabilities of l.s.i. circuits, new application concepts will evolve to permit wider and more efficient utilisation of l.s.i. devices.

REFERENCES

1. CAHILL, T., U.S. Patent 1 295 691, 25 February 1919
2. HAMMOND, L., U.S. Patent 1 956 350, 24 April 1934
3. DEUTSCH, R., U.S. Patent 3 515 792, 2 June 1970
4. YOUNG, A., U.S. Patent 3 358 070, 12 December 1967
5. TENNES, C., *et al.*, U.S. Patent 3 567 838, 2 March 1971
6. LESLIE, D. J., U.S. Patent RE 23 323, 9 January 1951
7. YOUNG, A., U.S. Patent 3 106 610, 8 October 1963
8. IRASTORZO, J., U.S. Patent 3 621 106, 11 November 1971
9. ZIEHLKE, R. J., U.S. Patent 3 150 227, 22 September 1964
10. CHANG, R., *et al.*, U.S. Patent 3 647 927, 7 March 1972
11. COTTON, R. B., Jr., 'Tempered Scale Generation from a Single Frequency Source,' *Journal of Audio Engineering Society*, **20**, 5, 376 (1972)
12. SCHRECONGOST, R. B., 'Large Scale Integration in Organ Design,' *Journal of Audio Engineering Society*, **20**, 4, 275 (1972)
13. BRUNSTING, W. J., 'Enhanced Electronic Organ Performance by Means of Custom LSI Devices,' *Journal of the Audio Engineering Society*, **21**, 5, 373 (1973)
14. JONES, D. A., 'LSI Reality in Consumer Electronics,' 1973 Fall Convention Audio Engineering Society, New York

RADIOPHONIC SOUND AND MUSIC

This term, used in the U.K. by the BBC, refers to sound and music, the composition and realisation of which depends upon the use of electronic aids to create new sounds and tone colours which are in complete contrast to the more familiar sounds from conventional musical instruments.

These techniques first came into real use during the 1950s with the maturation of the magnetic tape recorder although as long ago as the 1930s Percy Grainger, the Australian composer of 'Country Gardens' fame, had produced a brief composition based on pure frequencies for the Theremin, an early electronic sound generator. The beginnings were however with *musique concrete* pioneered in Europe, although as the name suggests the 'music' was made through the manipulation of pre-recorded natural sounds and was in fact orchestrated noise. With further study it became apparent that if more 'musical' sounds were used as the raw material, i.e., sounds with a more ordered harmonic structure, greater malleability was achieved as the timbre changes encountered during pitch changes, due to differing tape speeds on playback, still bore some audible relationship to each other.

Perhaps the inherent difficulty of recording natural sounds without unwanted extraneous noises or acoustic colouration forced the would-be composer indoors to the recording studio in search of newer 'cleaner' sounds and soon recording directly on magnetic tape, using electronic signal generators as sound sources, became commonplace. Thus synthesised music-making in this form was born.

It was soon found that pure tones were unrewarding. Being of a fundamental frequency they contained no harmonics and offered nothing to interest the ear. So other waveforms were investigated. Square and triangular waveforms were used but filtration networks were needed to modify or temper the abundance of harmonics that these provided. A band of random frequencies termed *white noise*, analogous to white light containing all colours, and sounding like escaping steam also became a stock sound source. This when subjected to filtration produced bands of sounds often designated as *pink noise*, *brown noise* and so on.

Once sound, whatever the source, is recorded on magnetic tape it can be subjected to many treatments. Whole chromatic sequences can be constructed by replaying the original recording at different speeds and re-recording. Mention has already been made of filtration as a means of changing the quality of the sound and it is of interest that highly sophisticated filters are not necessarily the best for this purpose. More often a coarse filter which will 'chop the sound apart' is needed. Artificial reverberation, often referred to as *echo*, can also be applied. This can be done by various means such as an echo room, echo plate, echo springs, revolving magnetically coated drum or by another magnetic tape recorder with a predetermined delay. But by far the most creative effect that can be imposed on pre-recorded sound is achieved by manipulation of the tape itself.

The shape of the sound can be changed considerably. Attack and decay of the sound can be modified. Transients can be physically modified or even removed by editing, i.e., by cutting off pieces of the tape, and any instrumental sound can be effectively disguised by removing the tell-tale identification marks at each note's outset. Sounds can be reversed by the simple process of playing the tape in the reverse direction. This process can also be used to change the shape of the sound. A sound with a normally fast attack and slow decay can be modified by reversal to have a slow attack and a fast decay, or two identical sound recordings can be edited together 'back to back' to give a long attack and delay.

Musical sequences can be built up using only one original sound, each note being recorded from the master note via a variable speed tape machine and then edited into the required order with its duration and characteristic adjusted to suit the tempo and tone colouring of the composition. Several musical lines can be constructed separately (using a common time scale) and then replayed simultaneously and re-mixed on to a final recording. Alternatively, with the advent of the multi-track recorder, each line can be stored on a separate track for re-mixing at a later time.

During the 1960s several attempts were made to integrate electronic sounds, pre-recorded on magnetic tape, into orchestral works.[1] Once the initial technical difficulties

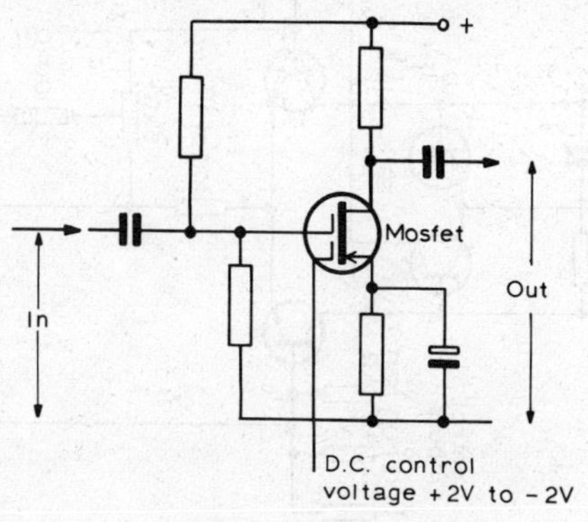

17.8 Simple voltage control

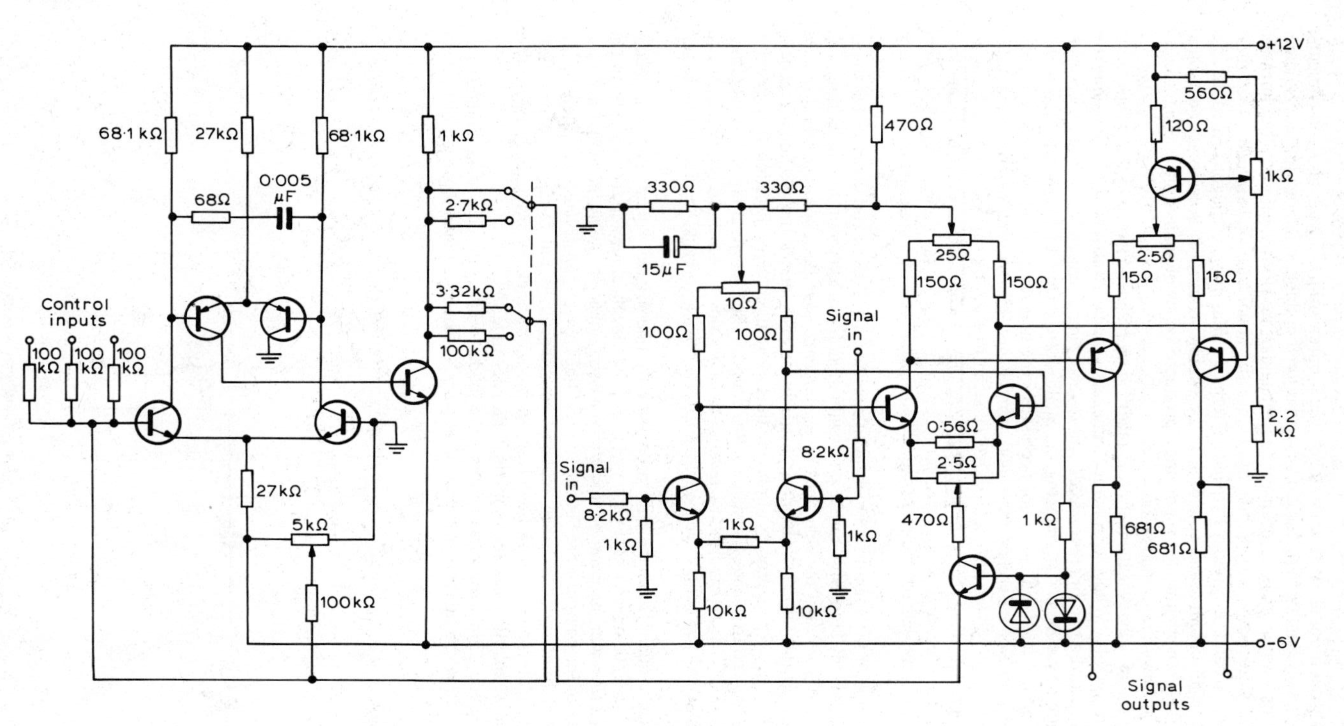

17.9 Voltage controlled amplifier

were overcome the results were quite successful. There is however the problem that a pre-recorded tape has a fixed time scale and cannot, as in the case of a human performer, be made to make changes in tempo which may be required by the conductor during a 'live' performance. There is also the problem of audible reproduction to be considered in that the tape must hold its own against the output of a full symphony orchestra and still be capable of giving the subtle nuances suggested in the score. Prior to these experiments 'electronic-sounding' music could be produced by means of the electronic organ and this was the only electronic musical instrument at the composers and performers disposal. Electronic music composition generally needed the medium of the tape recorder to act as a catalyst.

During the late 1960s various electronic *synthesisers* began to appear which were specifically designed for the electronic music composer. Some included a conventional keyboard which some schools of thought saw as a confinement to conventional composition rather than the hoped-for freedom from orthodox methods. Other types of synthesisers included retrieval or memory systems which 'remembered' the musical sequence of events but allowed the composer to change the tonal quality of the 'instruments' or even the 'instruments' themselves without disturbing the 'tune'. In these synthesisers, made practicable by the development of transistors and integrated circuitry, the major key to success is a type of circuit known as *voltage control* which provides greater flexibility of operation. A simple voltage control circuit is shown in *Figure 17.8*. The simple circuit diagram represents a voltage-controlled amplifier stage, with the controlling d.c. voltage being applied to the second trigger of the field effect transistor. Varying this voltage correspondingly alters the gain of the amplification stage. The device, need not be an amplifier—oscillators and filters can have their frequency responses controlled in this fashion.

A single voltage can control several devices simultaneously, or conversely (by means of voltage adding networks) many different voltages can control one device.

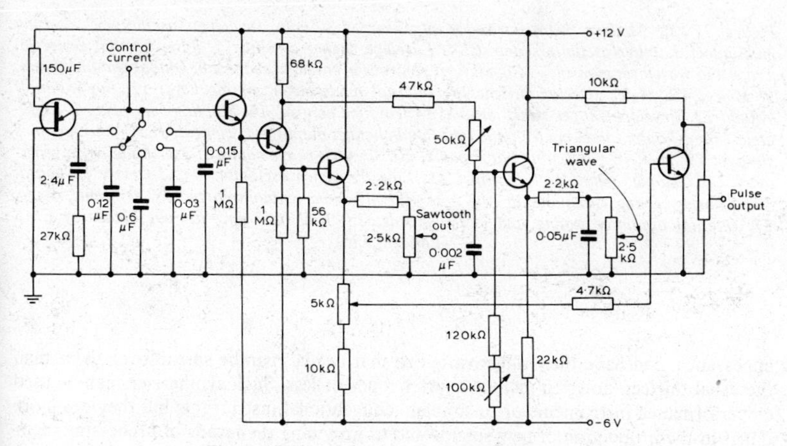

Figure 17.10 Voltage controlled oscillator

With this facility outputs from the electronic signal generators are not only used as audible material but can also be used as controlling voltages to vary other parameters such as level control, filter response settings or even their own frequencies. A circuit of a voltage controlled amplifier is shown in *Figure 17.9* and a voltage controlled oscillator in *Figure 17.10*.

The means of 'shaping' sound is provided by an envelope shaper which provides control of 'attack', 'on time' and 'decay'. Keyboards, which may have a conventional

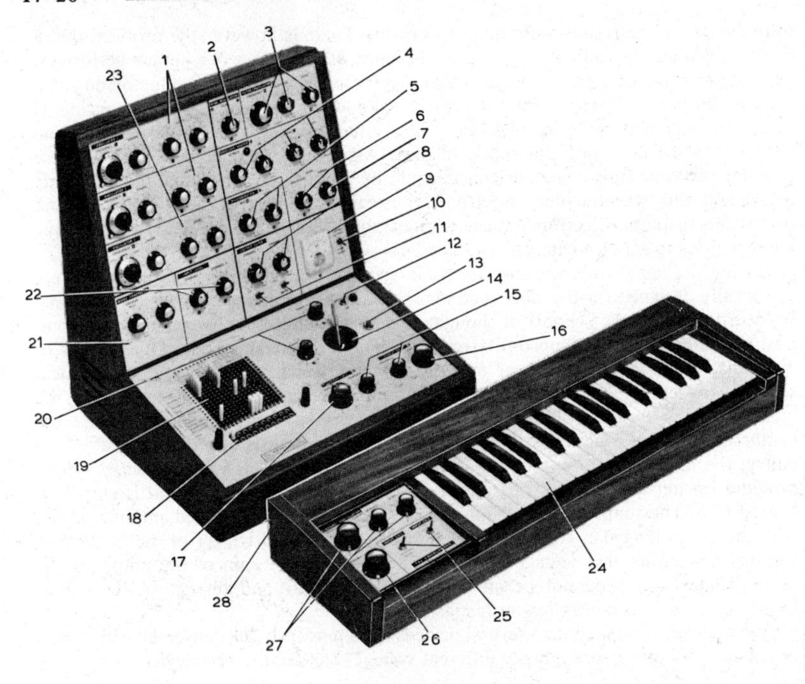

Figure 17.11 Modern transistorised synthesiser: 1. Audio frequency oscillators; 2. Ring modulator; 3. Filter/oscillator controls; 4. Envelope shaper controls; 5. Reverberation device; 6. Trapezoid waveform control; 7. Signal level control for envelope shaper; 8. Output channel filters; 9. Meter; 10. Meter selector switch; 11. Internal loudspeaker muting keys; 12. Power on/off switch; 13. Envelope shaper 'start' switch; 14. Joystick control; 15. Panning controls; 16. Right-hand channel output level control; 17. Left-hand channel output level control; 18. Connecting pins storage; 19. 16 × 16 connecting patch panel; 20. Joystick range controls; 21. White noise generator; 22. Channel input level controls; 23. Low frequency oscillator; 24. 3-Octave keyboard; 25. Switches giving two notes or one note with dynamic output voltage; 26. Dynamic range; 27. Internal oscillator tuning and level controls; 28. Umbilical plug at rear (Courtesy EMS (London) Ltd)

appearance, can have their range varied so that octaves can be spread over more than the usual thirteen notes to include quarter tones or less. Such synthesisers can be used as performance instruments or to simulate conventional instruments but they are more suited to the production of new sounds and to providing the means of processing them electronically. As can be seen from *Figures 17.11* and *17.12* the modern transistorised synthesiser embodies a comprehensive range of sound sources, filters and amplifiers sufficient to form the basis of a *sound studio*, a term normally applied to the whole studio premises. It is not long since a whole room would have been required to accommodate its valved counterpart. The synthesiser illustrated in *Figure 17.11* measures only 438 mm × 444 mm × 419 mm and weighs 10.2 kg.

At the time of the emergence of *musique concrête* there were several studios on the Continent devoting time to this new artform. In 1958 the BBC inaugurated the first studio in the U.K. to be totally committed to the creation of special sounds and electronic music for the support of radio and television programmes. Since then 'Radiophonic Sound and Music' has become a standard part of broadcasting. The output of the BBC's Radiophonic Workshop is, due to its nature and conception, an applied artform but as the creative demands have grown from special sound sequences to complete electronic incidental music scores, so more and more of the commitments can be taken out of their original programme contexts and take their place as compositions in their own right.[2]

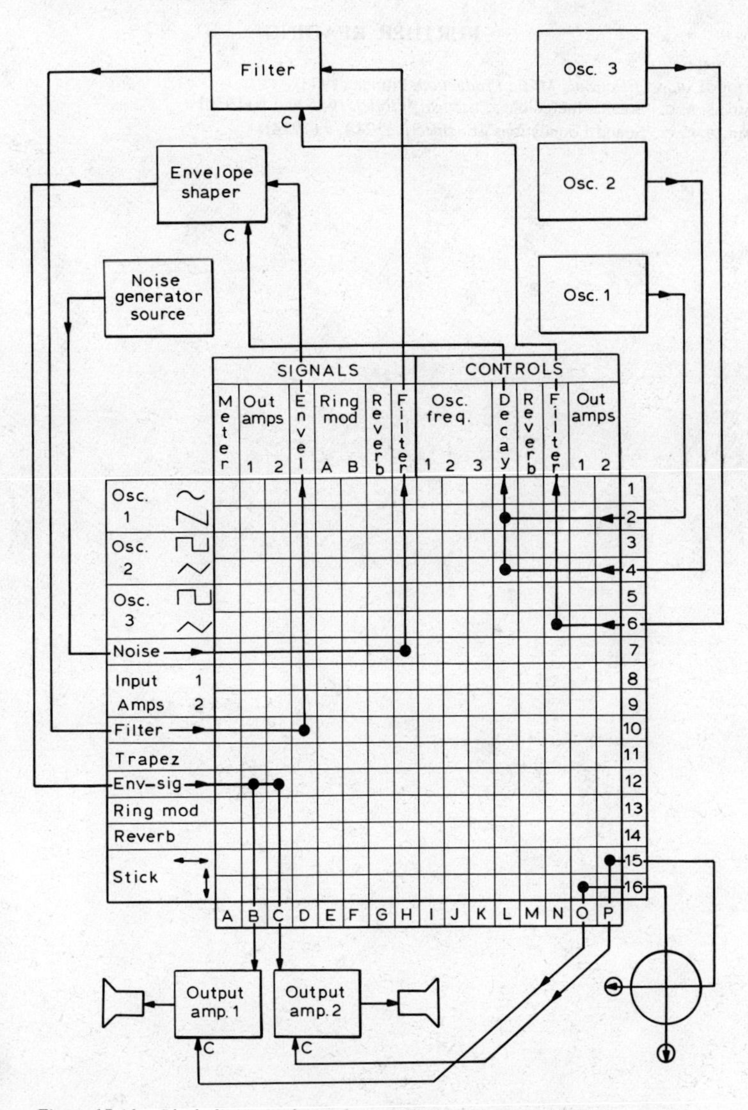

Figure 17.12 Block diagram of a modern synthesiser (*Courtesy EMS* (*London*) *Ltd*)

REFERENCES

1. Symphony 3 (Collages), Roberto Gerhard. Royal Festival Hall 1966, BBC Symphony Orchestra, Conductor Rudolf Schwartz. Also at Royal Albert Hall, Conductor Frederik Prausnitz. Available on HMV record ASD 2427
2. 'BBC Radiophonic Music' record No. REC 25M. 'Fourth Dimension' record No. RED 93S. 'The Radiophonic Workshop' record No. REC 196. BBC Radio Enterprises.

FURTHER READING

DOUGLAS, A., *Electronic Music Production*, Pitman (1973)
MILLS, R. C., 'Sounds Incredible', *Practical Wireless*, **6**, 5 and 6 (1970)
MILLS, R. C., 'Sound Foundations', *Engineering*, **214**, 6 (1974)

18 RADAR SYSTEMS

RADAR SYSTEMS 18

18 RADAR SYSTEMS

PRIMARY RADAR

Definitions and scope

Radar techniques quickly spread into a number of branches different from their original application in early pre-World War II days. The word 'RADAR' is taken generally to be an acronym of U.S.A. origin of the phrase *RA*dio *D*etection *A*nd *R*anging. The word 'detection' appears to have been replaced, at the time of its its being first coined, by the term 'direction-finding' in the cause of security. The primary radar technique, from which others have grown, is based upon two phenomena.

(*a*) That radio energy impinging upon a discontinuity in the atmosphere is reflected by the discontinuity.
(*b*) That the velocity of propagation of radio waves is constant.

Exploitation of these phenomena led to transmission of regular pulses of radio energy, range being obtained from measurement of the round-trip time of transmitted pulses. Further pulse transmission and reception engineering led to systems which although radar-based, or inspired are not true radar, e.g., Oboe, Gee, Omega, etc.—Since these are pure distance measuring systems they will not be treated here. A close relative of primary radar which does not use the reflection of transmitted energy is, however, so important that it merits separate treatment. This is the Secondary Surveillance Radar (SSR) system wherein pulse transmissions from the ground are received in aircraft, detected and decoded in the aircraft's transponder. The aircraft's transponder then transmits coded pulses back to the ground after a short fixed delay. Thus the operational benefit of primary radar is maintained and many bonuses accrue which are explained later.

History

Radar's history is currently incompletely written. The most prolific historian to date is the generally acknowledged 'father of radar' Sir Robert Watson-Watt.[1, 2, 3] He rightly observes the radar principle as being 'often discovered and always rejected'. Early workers, even Heinrich Hertz himself, made observation of the reflection of radio waves from metallic surfaces. In 1922 Marconi introduced publicly the notion of this effect being put to use in the detection and location of the direction of ships. If there is any one point in time at which we can say radar was invented it comes from Sir Robert's Memorandum of 27 February 1935. This laid down the basic principle we know as Primary Radar and also pointed to the need and possibility of SSR by realising how necessary it would be to distinguish 'Friend from Foe'. Emphasis was placed also on the need for reliable ground–air communications. This Memorandum was stimulated by a previous U.K. Air Ministry request to investigate the possibility of destroying the attacking power of aircraft by radiation. It was concluded that the aim could not be achieved at that time—but in any case the aircraft had first to be located. This would be entirely practicable and Sir Robert elaborated upon this in the second Memorandum.

In the few months following February 1935, remarkable progress was made, culminating in detection and ranging of aircraft to half mile accuracy out to nearly 60 miles. Height measurement was also achieved out to 15 miles using range and elevation angle. All this early work was conducted on wavelengths between 50 and 25 metres. No measurement of bearing was attempted at this time, it being thought adequate to rely upon independent range measurements taken from a chain of stations of known location. However, around January 1936, the application of receiver DF techniques, using goniometer principles provided a crude facility which was immediately successful. Similar work was carried out on the European continent, and in the U.S.A. Various reports ascribe the lower level of progress, in Germany for instance, to the German High Command view that the bomber was unstoppable and bombing, a very quick way to victory. In the U.S.A., the defence needs were held to be much less urgent than in the U.K. This inhibited progress in the early days of radar history.

By 1939 in the U.K. a radar chain (CH) was established, operating east-wards and ranging from the Orkney Isles to Portsmouth. By September 1941 the chain encircled the whole of England, Scotland and Wales. By this time, the needs and possibilities of Radar led to the production of equipments for airborne use, rotatable aerials, and the Plan Position Indicator (PPI). In 1940, stimulated by the need for small equipment Randall and Boot successfully operated their resonant cavity magnetron and thus revolutionised the radar technique. Their device,[4] producing at that time 50 kW peak power at 10-cm wavelength, is now the most widely used generator of microwave power. The use of very short wavelengths made possible the development of small aerials with high discrimination, and equipments with a wide range of power weight and size. The study of radar applications in defence and aircraft navigation, continued in peacetime, the latter blossoming into the field of civil aviation electronics, air traffic control, satellite communications. By the end of World War II, radar engineering had produced many different equipments with differing attributes. Almost immediately the concept of systems engineering took concrete shape and became virtually a separate discipline. The possibility of mixing computer technology into signal and radar processing has led to the present point of the next revolution in radar.

Fundamentals

The radar system can be represented in universal form as the functional diagram of *Figure 18.1*.

In any radar system the generated power directed into free space by an aerial will be intercepted by discontinuities in the atmosphere. These are typified as follows:

(*a*) The ground–air interface.
(*b*) Hills, mountains, buildings, etc.
(*c*) Clouds (precipitating and non-precipitating).
(*d*) Rain, snow, hail, etc.
(*e*) Aircraft, ships and vehicles.
(*f*) Birds, insects, dust clouds, atmospheric discontinuities (i.e., sharp changes of refractive index).

Although some power will be absorbed on impact, most is reflected and some will travel back to the power gatherer, the receiving aerial. The received power is amplified in a receiver, the signal competing with random noise gathered by the aerial and added to that generated at the receiver input (galactic noise, interference signals from electric devices and receiver noise). After frequency selective amplification, the signal plus noise is passed to a signal processor. Here various characteristics of the numerous types of received signal are exploited to separate the wanted from unwanted, e.g., moving targets from stationary targets, large from small, long from short, etc.

The filtered signals are then passed to a display equipped to register the signals in such a manner as to allow their position to be measured by means of range, bearing and sometimes height, scales. The signal processor is being increasingly used to serve a data processor which also accepts external data on targets, situations and various other criteria. Its output can furnish modified display presentations to augment the radar display, and categorise and clarify the total radar-sensed situation.

MONOSTATIC, BISTATIC AND ADAPTIVE RADARS

Referring to *Figure 18.1* the modulator and power generator constitute the transmitter in all of the above categories. Monostatic systems differ from the bistatic in that a single aerial is used for power radiation and power gathering, *Figure 18.2*. A Duplexer is used to separate the transmit and receive functions, both in the time and power amplitude domains. In the bistatic system, *Figure 18.3*, two separate aerials are used, one for the transmit and the other for the receive function. These aerials may sometimes be many miles apart.

Normally the operating parameters of a radar system are fixed or there can be a few selectable changes during operation, e.g., type of modulation, data gathering rate, power output, receiver sensitivity and selectivity. In the so-called adaptive radar system, parameters of operation (within limits) are varied as a function of the radar's performance on a number of targets. For example, feedback is generated by the processor or data processor to increase information on specific targets by, for instance directing the aerial to targets in a specific order and governing the dwell-time on each to improve data quality; varying the output pulse rate or spectrum. These systems are not yet in general use.

The most commonly used radar system is the monostatic and this section therefore concentrates upon this. Further information on bistatic technique and adaptive radar will be found in references 5 and 6.

THE RADAR EQUATION

There are many forms of the radar equation to be be found in the literature. It is considered therefore helpful to develop it from fundamental ideas in order that these various forms, each with their own subtleties and idiosyncrasies, can be better understood. Taking the simple case of free space performance, the maximum range of the radar is a function of:

P_t Peak transmitted power
G Gain of the aerial in the direction of the target
σ Effective reflecting area of the target
λ Wavelength of radiation
S_{min} Minimum received signal power required to be detectable above system noise
A_r Effective area of receiving aerial

Consider a target at range R. The power density at R will be equal to

$$\frac{P_t G}{4\pi R^2} \tag{1}$$

The target will intercept this power over its equivalent area of σ and reflect it back over the same distance, R, to the aerial which now becomes the receiver in the monostatic system. Thus the power gathered by the aerial equals

$$\frac{P_t G \sigma A_r}{4\pi R^2 4\pi R^2} \tag{2}$$

Now A is related to the aerial gain in the following way:

$$A = \frac{G\lambda^2}{4\pi} \tag{3}$$

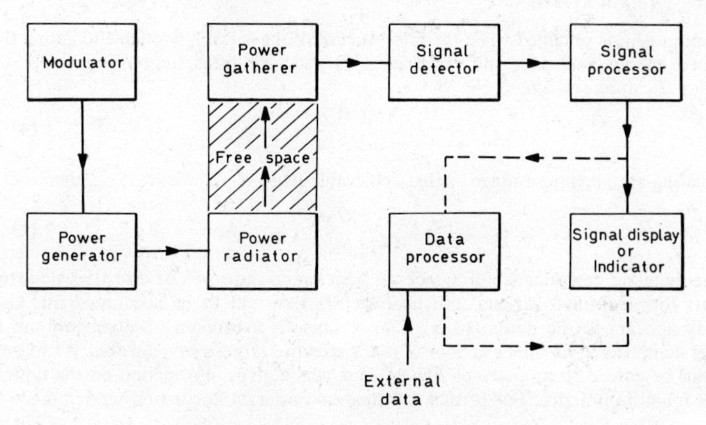

Figure 18.1 Generalised radar system

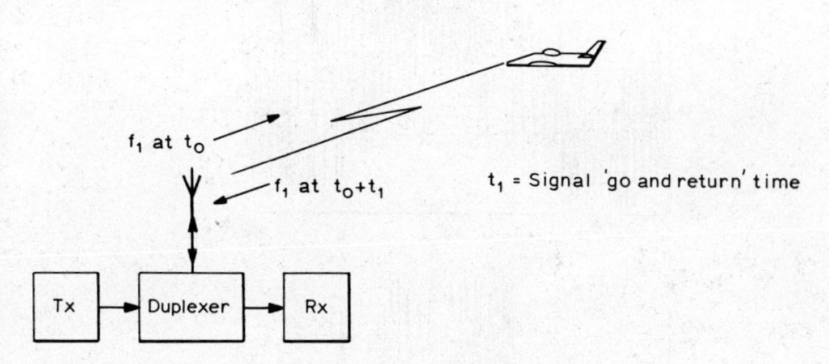

Figure 18.2 Monostatic radar

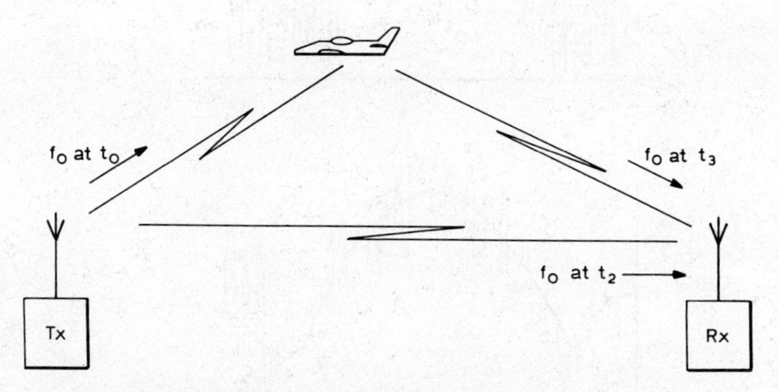

Figure 18.3 Bistatic radar

Maximum gain is obtained when the full aperture of the aerial is available to gather the returned energy and since the same aerial is used for transmission and reception, equation 2 can be rewritten:

$$P_r = \frac{P_t G^2 \lambda^2 \sigma}{(4\pi)^3 R^4} \tag{4}$$

Postulating the maximum range as that achievable when P_r reduces to S_{min}, then

$$R_{max}{}^4 \; \frac{P_t G^2 \sigma \psi^2}{(4\pi)^3 S_{min}} \tag{5}$$

The dynamic performance of a radar system cannot however be directly calculated by this formula since various statistical factors have yet to be accounted for. One already appears in the notion of S_{min}, since noise is a random phenomenon and is further complicated by the behaviour of σ, the effective target's echoing area. An idea of this can be gained from study of *Figure 18.4* which gives information on the typical values found in practice. For further information on target fluctuation see reference 6.

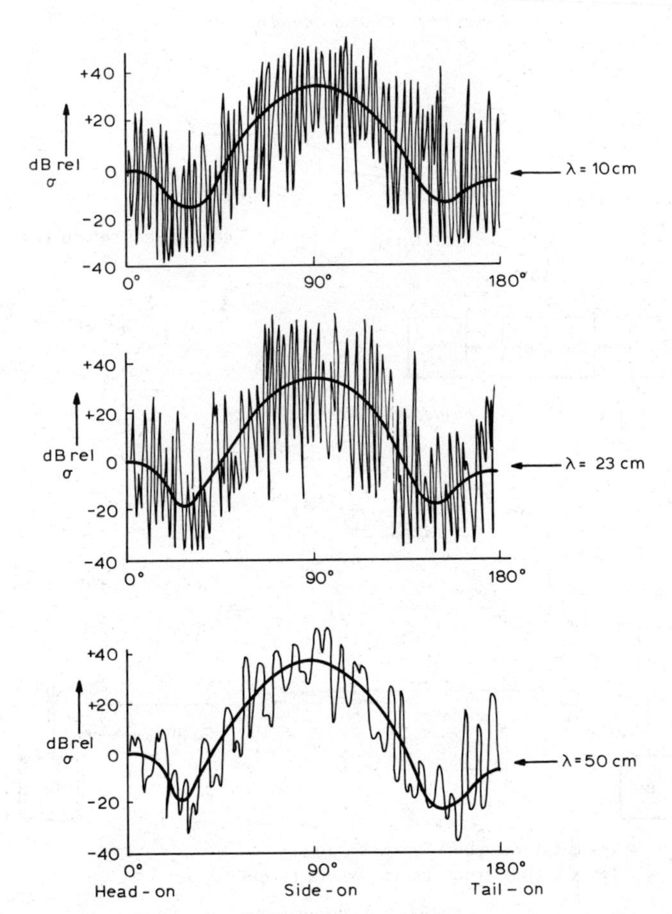

Figure 18.4 Showing typical variations in effective echoing area (σ) of aircraft targets

The value of σ can change violently over very small azimuth increments at microwave frequencies. As the target attitude to the radar can change even during interpulse periods, there is a range of probabilities that the echoing area will produce a returned signal of a given strength. The distribution of these probabilities is taken variously as Gaussian, Raleigh or Exponential. Radar engineers are generally concerned with probability of target detection of the order 80% and so for all practical purposes the difference in these distributions is negligible. These probabalistic factors, together with others associated with display, operator, atmospheric loss and system loss factors are all combined to modify the final calculation. Two commonly used formulations are those due to L. V. Blake [7] and W. M. Hall. [8]

PROPAGATION FACTOR

In all radar systems, theory begins with free space propagation conditions. These almost always do not pertain since radiations, from and to the aerial via the target and its environment, are seldom via a single path. The most common effect is that of the ground above which the radar aerial is mounted. Consider the situation in *Figure 18.5* (*a*) where the phase centre of the aerial is at height h above the ground. Energy reaches the distant target by the direct ray R and that reflected at G. Over the elevation range of which ϕ is one example, the direct and reflected rays will be alternatively in and out of phase with each other. The free-space pattern of *Figure 18.5* (*b*) will thus be modulated, a series of

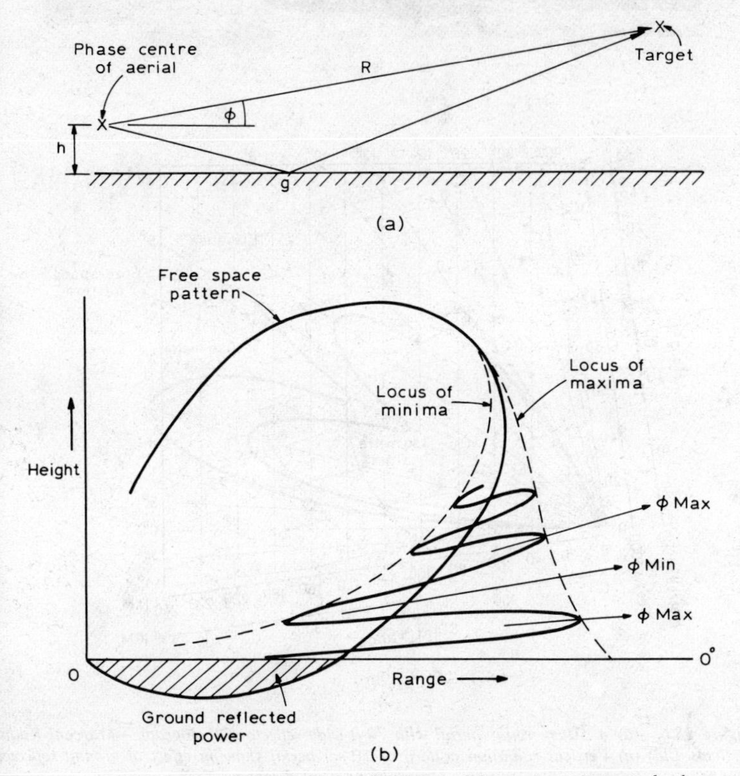

Figure 18.5 (*a*) *Ground reflection mechanism* (*b*) *Showing lobes and gaps in vertical polar pattern*

(a)

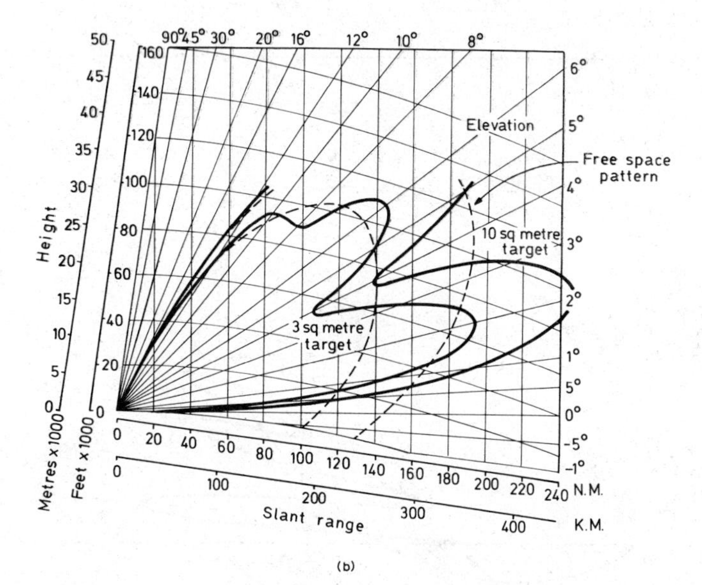

(b)

Figure 18.6 (a) A 50-cm radar aerial with 70-ft-wide reflector (Photograph—Marconi Radar Systems Ltd) (b) Vertical radiation pattern of 50-cm aerial showing effect of ground reflected energy in providing extended range

lobes being formed. The position of these is governed by h and the wavelength in the following way:

$$\phi_{max}=\frac{\lambda n_1}{4h} \quad \phi_{min}=\frac{\lambda n_2}{42h} \tag{6}$$

Where λ = wavelength, h = height and n_1 = odd integers, n_2 = even integers. Max and min indicate peaks and troughs of lobes respectively.

This model and theory holds for very flat ground or water but becomes a complicated calculation as ground roughness and departures from flat plane increase. Vincent and Lynn have made the problem tractable.[9]

Obviously these lobes are useful providers of extra range but the gaps are less welcome. The modulations can be reduced by several means.

(*a*) Putting less power into the ground.
(*b*) Increasing the aerial tilt in elevation.
(*c*) Intercepting the reflected power and dissipating or scattering it in incoherent fashion.

An interesting alternative is to be found in 50-cm radars such as are used by the Civil Aviation Authority for almost all of the U.K. airways surveillance system. Here the wavelength and mean aerial height combine to give a long low lobe embracing most of the required long range airspace, *Figure 18.6* (*a*) and (*b*). The first null is at an altitude and of such a depth as is entirely tolerable.

Dependence is placed upon a very flat site, the tolerable roughness being great because of the longer (in the microwave sense) wavelength.

A useful comparison of ground reflection effects across the microwave band is to be found in reference 10. The pattern propagation factor which thus modulates the free space pattern is also embodied in range calculation.

Types of radar

Figure 18.1 represents many types of radar; some of the more important of these, starting with the most simple, are now considered.

C.W. RADAR

Here there is no modulation provided in the transmitter. The movement of the target provides this in its generation of a doppler frequency. The system is illustrated in *Figure 18.7*. The presence of the target is indicated by the presence of the doppler frequency of value

$$f_d=\frac{89.4V_r}{\lambda} \tag{7}$$

where V_r = radial speed, λ = wavelength and f_d at 10 cm = approx. 9 Hz per m.p.h.

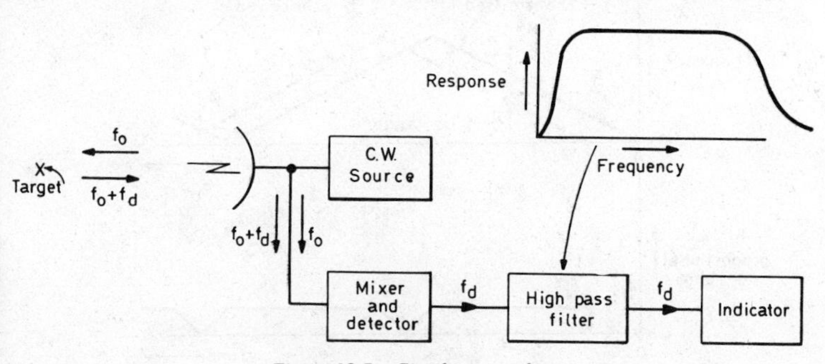

Figure 18.7 Simple c.w. radar

Range is not measurable with one c.w. element; but certain modern equipments based on this principle do not need to indicate range, e.g., police speed traps using radar techniques. The technique is widely used for velocity measurements and provided no great range is required, very modest power output can be used, e.g., 10 to 15 watts at 10 cm with a small dish aerial of some 3-ft diameter can effectively operate up to 10 to 15 miles on aircraft.

A.M./C.W. RADAR

This variant of the c.w. technique employs two slightly differing c.w. transmissions. With two frequencies f_1 and f_2 a single target at given range will return $f_1 \pm f_d$ and $f_2 \pm f_d$. The doppler frequency difference can be made small if $f_1 \sim f_2$ is small. However the range domain from the radar is characterised by a phase difference between f_1 and f_2 which is linearly proportional to distance and unambiguous up to phase differences of 2π. By detecting the phase difference of the two almost identical doppler components of the returned signals the range can be measured, being:

$$R = \frac{1C\phi}{4\pi(f_1 - f_2)} \tag{8}$$

where R = range to target, C = velocity of propagation, $f_1 - f_2$ = frequency difference of the two transmitted c.w.s and ϕ = phase difference between the two doppler frequencies.

The complication of dealing with a number of targets simultaneously present in the system is resolved by erecting a number of doppler filters, each with its own phase measuring element.

F.M./C.W. RADAR

A simple form of this is illustrated in *Figure 18.8*. Here the range continuum has its analogue in frequency deviation from a starting point. Comparison of the frequency of

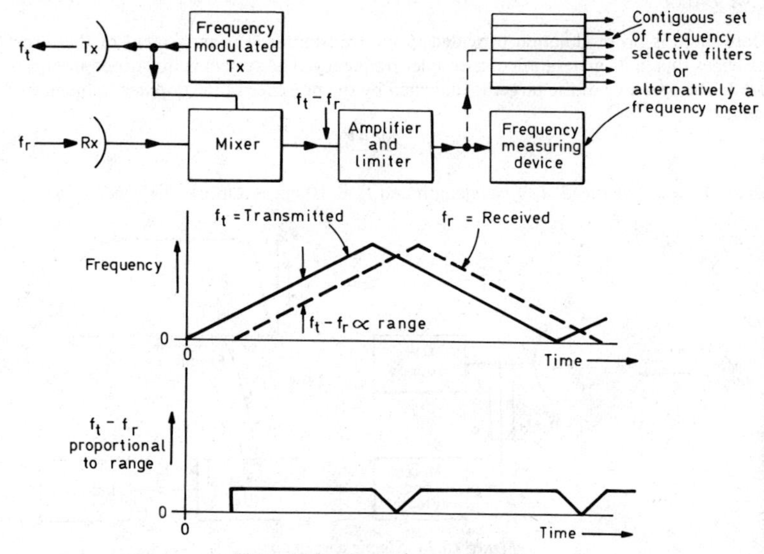

Figure 18.8 An f.m./c.w. radar system

the returned signal and the transmission frequency gives a difference which is directly proportional to range. Range accuracy is determined largely by the bandwidth of the frequency measuring cells into which the range scale is broken and the stability of frequency deviations. The technique has been used for a radio altimeter, the earth being the radar's 'target', and the range being the height of the aircraft above the reflecting surface.[11]

In all the systems described above there was a practical limitation of performance due to the limited c.w. power that could be generated at very high frequencies so that small aerials of high directivity could be used. This limitation of power was eventually

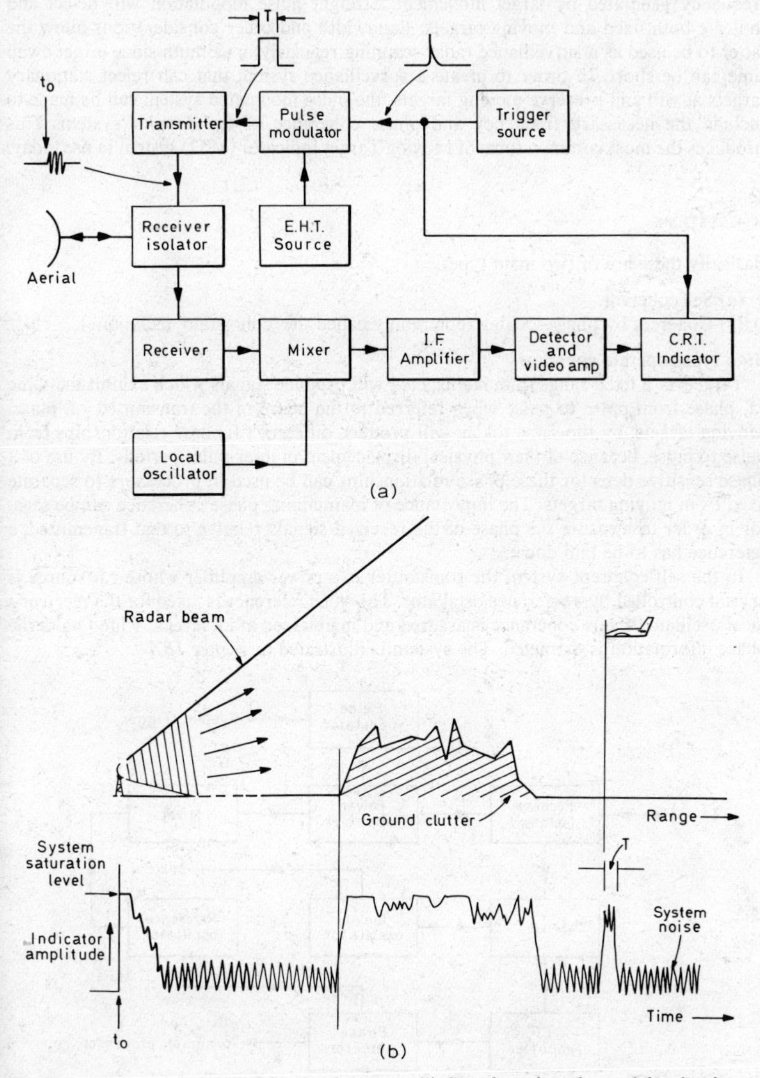

Figure 18.9 (a) Pulse modulated radar system (b) Signals in the pulse modulated radar

overcome by using pulse modulation techniques. In this way very high peak power could be produced within the device's mean power capability. When pulse modulation is used the transit time of the pulses to and from targets is the direct measure of range. A pulse modulated system is illustrated in *Figure 18.9* (*a*) and (*b*).

PULSE MODULATED DOPPLER SYSTEM

In the normal doppler system, target detection is based upon sensing the doppler frequency generated by target movement. Straight pulse modulation will detect and indicate both fixed and moving targets. Bandwidth and other considerations allow the latter to be used as a surveillance radar scanning regularly in azimuth since target dwell time can be short. In order to create a surveillance system that can reject stationary targets at will and preserve moving targets, the pulse modulated system can be made to include the necessary frequency and phase coherence of the doppler system. This produces the most common form of Moving Target Indicator (MTI) system in use today.

MTI SYSTEMS

Basically these are of two main types.

(*a*) Self coherent.
(*b*) Coherent by phase-locking (commonly called the 'coho-stalo' technique).

Both types operate as follows.

Targets at a fixed range (mountains, etc.) will produce signals which exhibit the same r.f. phase from pulse to pulse when referred to the phase of the transmitted r.f. pulse. Moving targets, by the same token, will produce different r.f. phase relationships from pulse to pulse, because of their physical displacement in inter-pulse periods. By use of a phase sensitive detector these phase relationships can be used in processors to separate fixed from moving targets. The importance of maintaining phase coherence can be seen, for in order to measure the phase of the received signals relative to that transmitted, a reference has to be laid down.

In the self-coherent system the transmitter is a power amplifier whose r.f. source is crystal controlled by a reference oscillator. The same reference is taken for the receiver's local oscillator. Thus coherence is assured and maintained at i.f. level at which point the phase information is extracted. The system is illustrated in *Figure 18.10*.

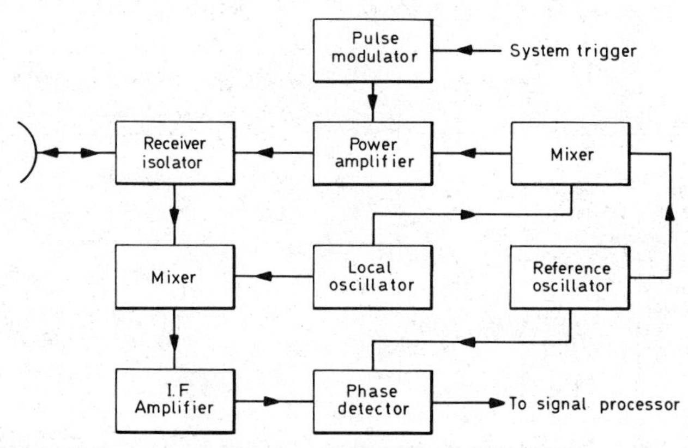

Figure 18.10 Self-coherent pulse radar

In the 'coho-stalo' system, the transmitter is a self-oscillating magnetron whose phase at the on-set of oscillating is varying from pulse to pulse. Coherence is achieved as follows.

The receiver has an extremely stable local oscillator (stalo). At each transmission, a sample of the r.f. output power is taken at low level. This, after mixing with the stalo, forms an i.f. locking pulse which is used to prime an oscillator which becomes the coherent reference (the coho). This oscillator is switched off during the time the lock pulse is injected. Upon switching on, the stored energy of the lock pulse starts the oscillation in a controlled manner. Thus the coho preserves the phase reference for a

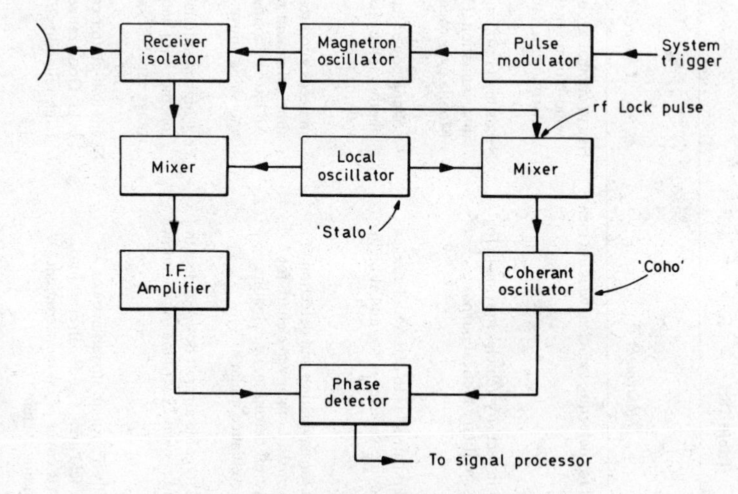

Figure 18.11 Coho-stalo system

pulse period. The coho is then switched off just prior to the arrival of the next lock pulse and the process is repeated. The system is illustrated in *Figure 18.11*.

CHIRP

Radar performance is dependent upon the mean power of the system. If this is increased by lengthening the output pulse duration, immediately the range discrimination is reduced (12.36 μs is equivalent to 1 nautical mile in primary radar terms). The *chirp* or pulse compression technique is a means of obtaining high mean power by use of longer pulse output while retaining high range discrimination. It operates as follows.

The transmitter output pulse is frequency modulated, usually linearly, throughout its duration. Received signals which bear this modulation are passed through a pulse compression filter which has a delay characteristic which is frequency dependent. This characteristic is made to be of opposite sense to that used in transmission so that the frequencies occurring later in the transmitted pulse are delayed least. Those occurring early in the output pulse are delayed most. By this means the received energy is compressed in time. If the original pulse duration is τ the output pulse duration from the compression filter is $1/B$ where $B = f_{max} - f_{min}$ used in the frequency modulation. The peak power using this technique is effectively increased by a factor of $B\tau$ which is called the pulse compression ratio or dispersion factor.[6]

Table 18.1

Operational role	*Typical wave-length*	*Typical peak power and pulse length*	*Deployment*	*Characteristics*
Infantry manpack	8 mm	F.M.–C.W.	Used for detecting vehicles, walking men	Has to combat ground clutter and wind-blown vegetation etc. Hostile environment. Battlefield conditions
Mobile detection and surveillance	3 cm	20 kW at 0.1 μs to 0.5 μs	Used in security vehicles moving in fog also military tactical purposes	Needs high resolving power for target discrimination. Hostile environment. Mobile over rough ground
Airfield surface movement indicator	8 mm	12 kW at 20 ~ 50 ns	Used on airfields to detect and guide moving vehicles in fog and at night	Resolving power has to produce almost photographic picture. Can detect walking men
Marine radar (civil and military)	3 cm	50 kW at 0.2 μs	Used for navigation and detection of hazards. Longer wavelength for aircraft detection, i.e., 'floating surveillance' radars	3 cm needs good resolution and very short minimum range performance. All have to contend with sea clutter and ships' movement
	10 cm and 23 cm			
Precision approach radar (PAR)	3 cm	20 kW at 0.2 μs	Used to guide aircraft down glidepath and runway centreline to touchdown	Very high positional accuracy required together with very short minimum range capability and high reliability
Airfield control radar (ACR)	3 cm	75 kW at $\frac{1}{2}$ μs	Usually 15 r.p.m. surveillance of airfield area. Guides aircraft into PAR cover or on to instrument landing system	Needs anti-clutter capability. 10 cm has MTI. Good range accuracy. High reliability required
	10 cm	0.4 MW at 1 μs		

Airborne radar	3 cm	40 kW at 1 μs	Used for weather detection and storm avoidance	Forward-looking sector scanning. Storm intensity measurement system usually incorporated
Met. radar	3 cm 6 cm 10 cm	75 kW to $\frac{1}{2}$ MW $\frac{1}{2} \sim 1\ \mu$s	Surveillance and height scanning gives range/bearing/height data on weather. Also balloon-following function	Good discrimination and accuracy. Rain intensity measuring capability
Air traffic control—terminal area surveillance (TMA)	10 cm 23 cm 50 cm	$\frac{1}{2}$MW to 1 MW 1 μs to 3 μs	Surveillance of control terminal areas. Detection and guidance of aircraft to run-ways and navigational aids. 60n miles range	Needs good discrimination and accuracy MTI system necessary. Display system incorporates electronic map as reference
Air traffic control—long range	10 cm 23 cm 50 cm	$\frac{1}{2}$ MW to 2 MW 2 μs to 4 μs	Surveillance of air routes to 200n miles range. Monitoring traffic in relation to flight plans	Has to combat all forms of clutter. MTI essential. Circular polarisation necessary except at 50 cm
Defence—tactical	10 cm 23 cm	$\frac{1}{2}$ MW to 1 MW 1 μs to 3 μs	Mobile or transportable. Used for air support and recovery to base	As for TMA radar plus ability to combat jamming signals
Defence—search	10 cm 23 cm	2 MW to 10 MW 2 μs to 10 μs	Used for detection of attacking aircraft monitoring of defending craft including direction for interception. Usually a height finder operated together with search	As for ATC long range plus ability to combat all forms of jamming. Forward stations report data to defense centre

Operational roles

Radar's ability to 'see without eyes' has placed the technique at the service of a wide variety of users. A resumé of operational roles with pertinent data is given in *Table 18.1*. It will be seen that a wide range of engineering is required to furnish the requisite hardware.

Transmitters

Across the band of microwaves from 8 mm to 50 cm transmitters of peak power from watts up to 10 MW are found. For the higher powers required, magnetrons are almost exclusively used for wavelengths up to 23 cm. Klystrons and other forms of power amplifiers have been employed at wavelengths of 10 and 23 cm. For 50-cm wavelengths high power klystrons are universally employed. The advantage of klystrons and power amplifiers is their ability to be driven with crystal-controlled sources. This produces automatic frequency and phase coherence in the system which is called for in all MTI systems. Magnetrons have to be excited by a very high voltage pulse.

This is produced from a source the energy of which can be transferred as a pulse of controlled rate of rise and fall, at a known time, for a known duration. This is usually effected by a pulse-forming network the stored energy of which is released to the magnetron by a trigger device via a pulse transformer (usually a thyratron, although hard valve modulators are still used). The released energy, in pulses of some 30–60 kV of up to 10 μs duration, is restored by a charging circuit which can be resonant, for greater efficiency, at the repetition frequency of the transmitted pulses. Care has to be exercised in pulse shaping to avoid magnetron moding and unwanted frequency modulations. Purity of frequency spectrum is difficult to achieve unless great care in modulator design is exercised.[5] Klystrons are virtually power amplifiers and still need high voltage pulses to provide the amplification during the required pulse output duration.

Other methods of providing high peak power at microwaves are possible but very rarely found in practice, examples are as follows.

High power klystrons. These are available for operation at B (250–500 MHz) L (0.5–2.0 GHz), S (1.5–5.0 GHz) and C (4.0–6.0 GHz) band with output power up to 20 MW at L band. Power gain typically 40 dB.

High power travelling wave tube (TWT). These can be operated at L, S and C band giving output of up to 2 MW peak with gains typically 45 dB.

Twystron. So called because it uses a klystron-type cathode gun assembly and the TWT technique of slow wave restriction to provide gain of up to 40 dB at S band. It has the very desirable property of some 200 MHz instantaneous bandwidth, allowing frequency agility to be used without transmitter complication.

Crossed-field amplifiers (CFA). These are found for use across the X to C (4.0–11.0 GHz) band and provide typically 10 dB gain. They are a form of amplitron and use cold cathodes.[6]

Receivers

Across the microwave band, receivers almost always use superheterodyne technique. This is to simplify signal handling in processors, etc. The design aim is to preserve dynamic range to prevent amplitude and phase distortion outside the designer's control. Parametric amplifier techniques are now common which produce noise figures of some 3 dB at 50-cm and 6 dB at 3-cm wavelength.

These commonly employ variable capacitance diodes which 'noiselessly' extract power from a separate microwave c.w. oscillator source called the *pump* and convert it to the signal frequency, thus achieving amplification of signal without introduction of extra noise. The gain of a parametric amplifier of this kind is proportional to the ratio of pump to signal frequency. When integrated into a high power radar system care has to be used to protect the sensitive diodes from damage due to the leakage of power during transmission.

Very nearly theoretical limits of sensitivity have been reached in some receiver designs (i.e., noise figures approaching 0 dB indicating the receiver operating at the limit of KTB*). This is achieved by super-cooling of the receiver input elements incidentally producing a novel engineering problem embracing physics, electronics, mechanics, chemistry and refrigeration disciplines. These types of parametric amplifiers are to be found in radio telescopes the powers of which are determined to a large degree by receiver performance.[5]

I.F. AMPLIFIERS

It is common in modern transistorised equipment to find 100 MHz bandwidth easily achieved and bandwidths subsequently restricted to optimum by design of a filter at input or output stages. Dynamic characteristic-forming is usually done at i.f. and can produce linear, logarithmic, limiting or compression characteristics fairly readily over 80 dB of dynamic range above noise.

Transmitter–Receiver devices

Until very recent times high power radars used gas discharge tubes as a means of producing isolation between transmitter and receiver. A simple form is illustrated in

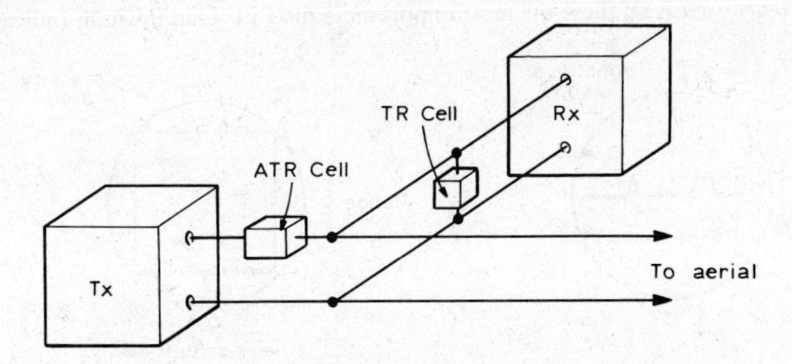

Figure 18.12 Simplified diagram showing receiver protection devices

Figure 18.12. When the transmitter fires, the gap in the ATR cell breaks down and allows power to progress to aerial and receiver. Power across the TR cell will cause the cell's tube to strike and impose a short circuit across the receiver input terminals. This short circuit is so placed that matching to the aerial is not disturbed. Upon cessation of the transmission the ATR device will restore to quiescence and thus disconnect the transmitter from the system. The TR cell's short circuit is also removed and thus returned signals can reach the receiver. Disadvantages of this technique are:

(*a*) Leakage past the TR device into the receiver.
(*b*) Long recovery time after firing which provides attenuation to received signals. This limits minimum range performance.

* K = Boltzmann's constant, T = temperature in degrees absolute, B = bandwidth.

These devices are being replaced in modern equipment by high-power-handling diodes using the same technique of open and shortcircuit lines. They have better recovery times and less insertion loss but produce problems associated with the need to switch bias voltages in synchronism with transmission in order to prevent destruction of the diodes.

Aerials

The gain of an aerial is related to its area. But the maximum gain is not the only, nor even the prime, consideration. The radar designer attempts to achieve gain in desired directions and thus beam shape in both horizontal and vertical planes becomes extremely important. The beam may be formed in a variety of ways as follows:

(*a*) Use of groups of radiators (dipoles, unipoles, radiant rods, helices, etc.).
(*b*) Reflector illuminated by various elements, e.g., dipole, horn, poly-rod, linear feed, etc.
(*c*) Slotted waveguides.
(*d*) Microwave lens system.

By far the most common is group (*b*). There are here many choices of illumination, distribution and reflector shape to permit practically any radiation pattern to be produced.[12]

Radar displays

Almost exclusively these use cathode ray tubes to represent various related dimensions on its screen, e.g., range and azimuth, range and height, range and relative amplitude, etc. Two types of modulation are used: amplitude and intensity. Typical examples are given in *Figure 18.13*. Of these the most important is the PPI (Plan Position Indicator).

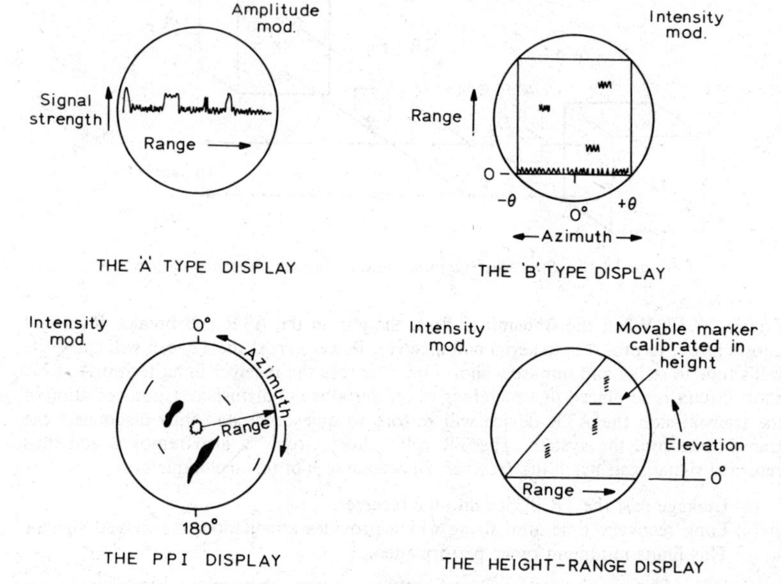

Figure 18.13 Various radar displays

THE PPI

Although *Figure 18.14* shows magnetic deflection, electrostatic deflection can also be used. The deflection coils, nowadays of sophisticated design with correction for many types of error and abberation due to tube geometry and manufacture, are mounted around the neck of the tube and produce orthogonal fields across the path of the electron beam. Variation of these fields deflect the beam across the face of the tube which it strikes and excites the phosfor on its surface producing an intense glow at the point of contact. This glow is visible through the glass of the screen.

Assume the tube centre is to represent the position of the radar and its edge, the radar's maximum range. The display system generates a *bright-up* waveform synchronised to the radar transmission time t_o, and with its amplitude and level set to give a threshold producing very small excitation of the phosfor in the absence of modulating signals. This waveform is also used to govern the period over which the time-base

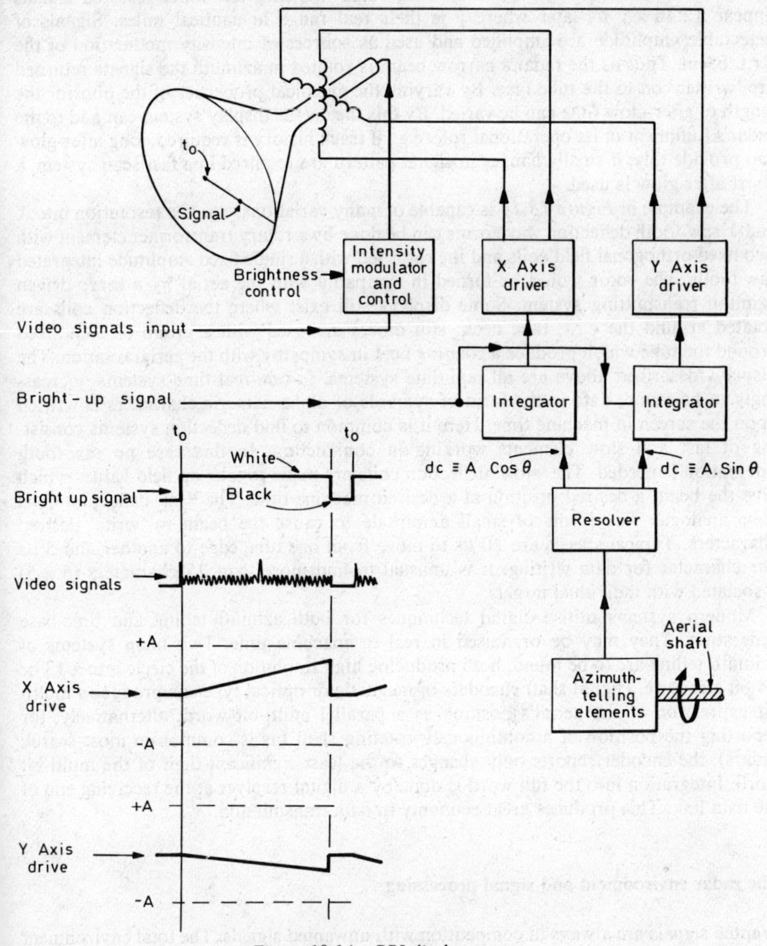

Figure 18.14 PPI display system

integrators operate. There are identical chains for the X and Y axes and both perform integration of a d.c. voltage expressive of the sine and cosine of the bearing to be indicated. These d.c. voltages can be generated in a number of ways, a common method being a 'sin/cos' potentiometer. The resolver is made to rotate in synchronism with azimuth tell-back elements mechanically geared to the aerial turning system. This may be done for example with torque-transmitting rotary transformers ('selsyns') or a servo-driven synchro system, again using rotary transformer technique. As the aerial turns, the values of the resolver's d.c. outputs will change, going through one cycle of maxima and minima per aerial revolution. The action of the integrators will therefore produce X and Y saw-tooth waveforms whose amplitude and polarity change in sympathy with the aerial's rotation thus automatically reproducing the fields necessary to cause the electron beam to move across the tube at the azimuth of the aerial. By making the X and Y drive waveforms highly linear and governing the overall gain of the deflection system, the tube face can be made to represent areas to calculable scales in a very linear fashion. Since the time-base is synchronised with radar transmission time, received signals appear $12.36 \times y$ μs later where y is their real range in nautical miles. Signals of detectable amplitude are amplified and used as sources of intensity modulation of the c.r.t. beam. Thus as the radar's narrow beam is rotated in azimuth the signals returned are 'written' on to the tube face. By varying the chemical properties of the phosfor the length of after-glow time can be varied. By this means the display system can add to the radar's fulfilment of its operational role, e.g., if track history is required, long after-glow can provide this; if small changes in signal pattern are required in a fast-scan system, a short after-glow is used.

The example in *Figure 18.14* is capable of many variations, e.g., the resolution into X and Y saw-tooth deflection waveforms can be done by a rotary transformer element with two fixed orthogonal field coils and the rotor fed with a single fixed amplitude integrated saw-tooth. The rotor would be turned in sympathy with the aerial by a servo-driven azimuth transmitting system. Some displays still exist where the deflection coils are rotated around the c.r.t. tube neck; still others are used with a group of coils fixed around the tube which produce a rotating field in sympathy with the aerial rotation. The displays described above are all real-time systems. In non-real time systems, increasingly to be found, data in the form of symbols or alpha-numeric characters is written upon the screen in machine time. Here it is common to find deflection systems consisting of fast and slow elements working in conjunction. In this case no saw-tooth integration is needed. The 'slow' deflection coils are made to take up field values which give the beam a desired position at a desired machine time. The 'fast' deflection coils have analogue waveforms of small amplitude to cause the beam to 'write' desired characters. Typical speeds are 20 μs to move from one tube edge to another and 5 μs per character for data writing. It is unusual to find more than 25 characters (5×5) associated with individual targets.

Modern systems utilise digital techniques for both azimuth-telling and time-base generation. They may be organised in real or machine time. Two main systems of azimuth-telling are to be found, both producing high resolution of the circle into a 13 or 14 bit structure. Digital shaft encoders of magnetic or optical type can produce a multi-bit expression of the aerial's position as a parallel multi-bit word; alternatively, for reporting the position of a continuously rotating shaft (as is common in most search radars), the encoder reports only changes to the least significant digit of the multi-bit word. Integration into the full word is done by a digital resolver at the receiving end of the data link. This produces great economy in data transmission.

The radar environment and signal processing

Wanted signals are always in competition with unwanted signals. The total environment can be appreciated by reference to *Figure 18.15*. This is typical of the general case and illustrated need to separate the wanted aircraft signals from those simultaneously arriving from, in this radar's role, unwanted or clutter sources. The radar system also suffers total blindness due to shadowing by solid objects and attenuation due to gases, dust, fog and similar particular media, e.g., rain and clouds. The radar aerial attempts to concentrate its radiation in a well-defined beam. But there are always sidelobes present which produce radiation simultaneously with the main beam in unwanted directions. These also gather unwanted signals.

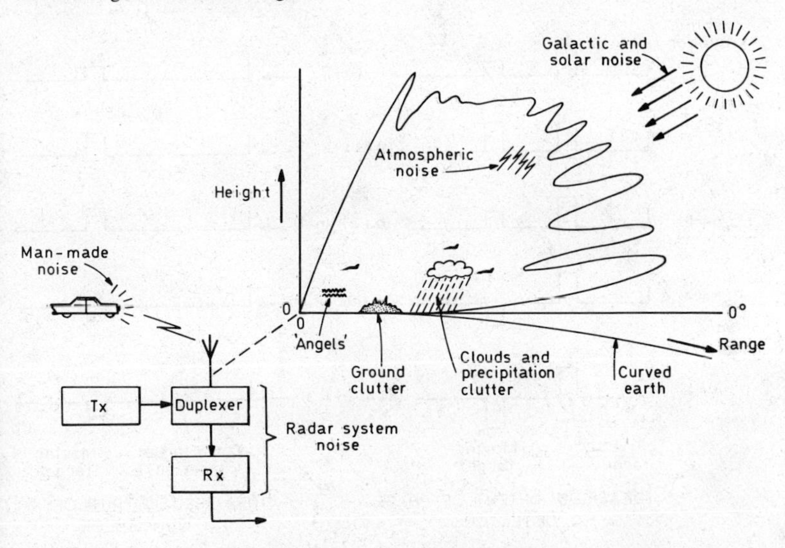

Figure 18.15 The radar environment

Signal processing can be used to discriminate in favour of wanted signals. The basic radar system is also designed to provide some bias in favour of the wanted signals, e.g., by using as narrow a radar beam as possible, by the use of the longest wavelength possible to minimise signals from cloud and rain, by the use of the shortest pulse possible at the highest repetition rate, etc.

DISCRIMINATION AGAINST GROUND CLUTTER

A number of techniques are available ranging from the very simple to the highly sophisticated.

Map blanking. An electronic video map of the clutter pattern from a fixed site is used to produce suppression signals which inhibit the radar display, thus producing a 'clean' picture containing only moving targets. This method is not effective with moving targets in regions of clutter and for this reason is seldom used.

Pulse length discrimination (*PLD*). Here use is made of the difference in the range continuum in the duration between aircraft signals and the general mass of ground clutter. All clutter greater in duration than twice the transmitted pulse length is totally rejected, those up to this limit are displayed thus preserving aircraft targets and rejecting almost all ground clutter. This is a relatively simple video signal process which has the advantage of providing supra-clutter visibility. Thus if the wanted signal is of sufficient strength to show above the clutter, it is displayed in clutter regions. A logarithmic receiver is usually employed ahead of the process to increase the dynamic range before signal amplitude limiting takes place.

Moving target indicators (MTI). In this case the processing is closely linked with modulator and transmitter design and is therefore more of a radar system than pure signal processing. The purpose of an MTI system is to discriminate between fixed and moving targets.

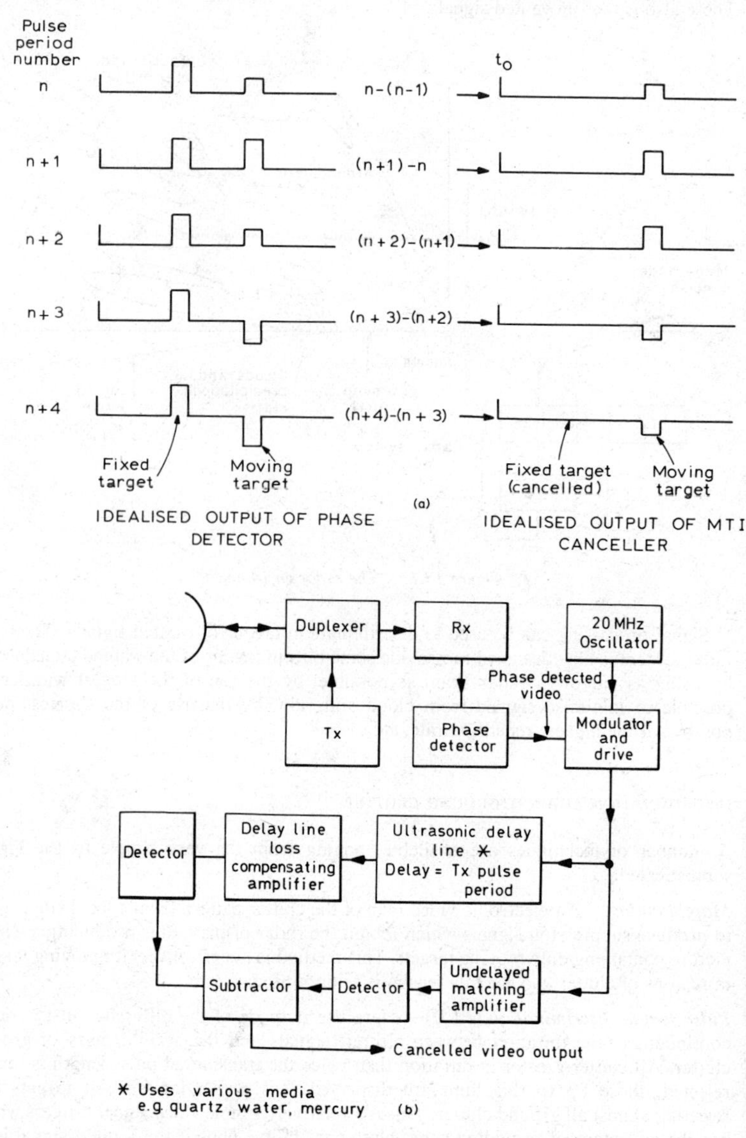

Figure 18.16 *(a) The MTI system's phase detector-canceller principle (b) A simple MTI radar*

The discrimination is made in the following manner. Consider the radar aerial stationary and illuminating a fixed target, e.g., the face of a mountain. Successive signals received from regular pulse transmissions will be at fixed range R. In the continuum this range can be expressed as $R = n\lambda + d\lambda$ where λ = wavelength of transmission, n = the number of whole wavelengths to target and d = fraction.

By establishing a phase reference at each transmission and preserving it for a whole reception period it is possible to give coherent meaning to the above expression using a phase sensitive detector the output of which for any target is characterised by three factors·

(i) its occurrence in time relative to transmission (due to the value of $n\lambda$) to the nearest whole wavelength.
(ii) its peak to peak amplitude due to the usual radar parameters, e.g., target echoing area, etc.
(iii) its amplitude and polarity within the peak to peak range due to the value of $d\lambda$.

In the case cited the output would be as illustrated in *Figure 18.16* (*a*). A moving target would produce different values of $d\lambda$ for relatively slow speeds and different values of $n\lambda$ as speed increased. Thus the phase detected output might behave as illustrated. By storage and comparison technique, successive received signals can be subtracted from each other. Both analogue and digital techniques are employed.

In the example given it will be seen that the signals from fixed targets can be reduced to zero and moving targets will produce non-cancelling outputs. The rate at which the moving targets change their amplitude from pulse to pulse is analogous, and equal in value, to the Doppler frequency referred to in equation 7. A simple pulse doppler MTI is illustrated in *Figure 18.16* (*b*). It has a number of limitations which tend to off-set its major advantage of providing sub-clutter visibility (SCV) (the ability to see moving targets superimposed upon fixed targets where the moving target is smaller than the fixed). Values of 20 dB for SCV are common in many surveillance radars using this type of MTI. Some of the limitations referred to are as follows:

(*a*) Blind speeds—where a moving target behaves as though it was stationary, i.e., $n\lambda$ changes by an integer in a pulse period. In this condition the doppler frequency is a multiple of the pulse repetition frequency of the transmitter.
(*b*) Blind phases—where a moving target produces the same output from the phase detector whose characteristic is symmetrical in the phase axis.
(*c*) Phase difference masked by noise or swamped by phase distortion in the receiver.

The system disadvantages can be overcome in varying degrees, e.g., blind speeds by staggered p.r.f. systems; blind phases by multiple phase detectors.

Other disadvantages are systematic and less tractable. For example 'fixed' targets are seldom found. The 'mountain' referred to above commonly has trees growing on it and wind causes fluctuations which spoil complete cancellation. Also the aerial is rotating in practice and its horizontal polar pattern modulates the target's amplitude again spoiling cancellation.[6]

DISCRIMINATION AGAINST INTERFERENCE

P.R.F. discrimination. Use is made of the fact that radar transmission is at regular rates, i.e., a known repetition frequency. Using a storage technique similar to that of the MTI system, successive signals are compared. Those due to the station's own transmission will correlate in range, those due to interference will do so only if harmonically or randomly related to the station's p.r.f. Instead of a subtraction process, coincidence gating is used and so wanted signals are released and unwanted interference rejected. In systems not using a fixed p.r.f., use is made of the known time of transmission. Range correlation of wanted targets is used as detailed above, de-correlation again provides rejection of unwanted interference.

DISCRIMINATION AGAINST CLOUD AND PRECIPITATION

To a certain extent the PLD technique can be used, but only when clutter is very severe. It is more general to use the Circular Polarisation (CP) technique. Here the radiation from the radar is changed from linear to circular by either waveguide elements or quarter wave plates.[13] Rain drops and cloud droplets return nearly equal components in vertical or horizontal planes. When put into the aerial system these components of the signals from rain are of opposite phase and if of equal amplitude, will cancel. Signals from irregular targets have inequality of vertical and horizontal components and thus do not cancel. However, there will be some loss of wanted target strength (typically 2–3 dB). Rejection of rain signals can be as high as 20 dB thus discrimination in favour of wanted targets is of the order 17–18 dB.

The choice of wavelength is critical in achieving discrimination against rain, the longer wavelengths increasingly provide protection due to an inverse fourth power law relating echoing area and wavelength.[5, 14]

DISCRIMINATION AGAINST ANGELS

Again, wavelength is an important factor, angels being more apparent at the shortest wavelengths. Angels are now generally recognised as being due to birds, in isolation and in great flocks.[15] Use is made of the target signal amplitude difference between angels and aircraft, angels being smaller in general. Range dependent signal attenuation or receiver desensitisation is used to reduce angel signals to noise level; the coincident aircraft signals are also reduced at the same time but being stronger are retained above noise.

Plot extraction

Radar signals from given targets exist, in real time, for a few microseconds in periods of milliseconds. From this data the operator reads the signal's centre of gravity as displayed upon his p.p.i. The data on a given target is repeated each time the radar beam illuminates the aircraft—typically, every 4 to 15 seconds. Radar data commonly has to be sent over long distances, e.g., from the aerial's high vantage point to a control centre sometimes many miles away. To transfer this data in real time, inordinately large bandwidth must be wastefully employed.

In order to avoid this, and simultaneously to provide positional data in a form more suitable to computer handling, plot extraction equipment is fast emerging and being operationally used. The plot extractor accepts radar signals in real time from the processor, together with the digital expression of the aerial azimuth. The extractor performs logical checks such as range correlation, azimuth contiguity, etc., after signals cross a pre-set threshold and become digital. The extractor derives the range and azimuth centre of all plots which meet the logic criteria and stores these in azimuth and range order ready for release as range and azimuth words in digital form; typically 12 bits for range of 200 miles and 13 bits for 360° of azimuth.

The assembled data is read from store at rates suitable for transmission over telephone bandwidths. By this means 100 to 200 plots can be reported in a few seconds. Computer programmes have been constructed which convert plot data into track data, deriving speed and heading as by-products.[16] Track data may be fed into p.p.i. type displays at a high refresh rate (approximately 40 times per second) in machine time producing a picture which may be viewed in high ambient lighting.

SECONDARY SURVEILLANCE RADAR (SSR)

General

The need for automatic means of distinguishing friend from foe was foreseen by Sir Robert Watson-Watt in his original radar memorandum (see p. **18**–2). The SSR system which gives automatic aircraft identity started its life as the wartime IFF system (Identification Friend or Foe); the term IFF is used universally for the military application of the technique. The civil version is known variously as SSR and, more commonly in the U.S.A., as the Radar Beacon System (RBS). Both IFF and SSR use the same system parameters these days and no distinction is made in the following text unless necessary.

Basic principles

Regular transmissions of pairs of pulses are made from the ground station at a frequency of 1 030 MHz via a rotating aerial with a beam shape narrow in azimuth and wider in elevation. A pulse pair constitutes an interrogation and modes of interrogation

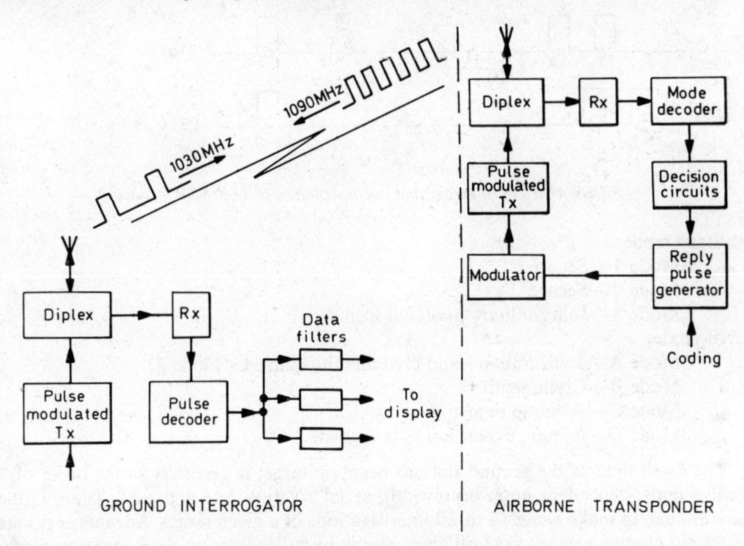

Figure 18.17 The SSR principle

are characterised by coding of the separation of the pair, internationally designated P_1 and P_3. Aircraft carry transponders which detect and decode interrogation. When certain criteria are met the transponders transmit a pulse position coded train of pulses at 1 090 MHz via an omni-directional aerial. The ground station receives these replies and decodes them to extract the data contained. This process of interrogation and reply is illustrated in *Figure 18.17*.

MODES OF INTERROGATION

These are shown in *Figure 18.18*. Through the agency of the International Civil Aviation Organisation (ICAO) there is internationally agreed spacing and connotation for the civil modes. Various military agencies treat the military modes in the same fashion:

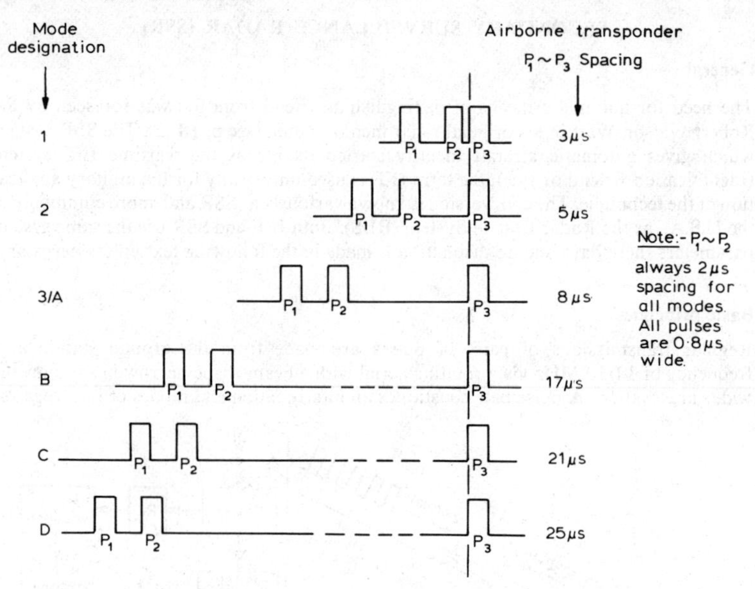

Figure 18.18 Interrogation pulse structure in IFF/SSR systems

Military modes:
- Mode 1—Secure
- Mode 2—Secure
- Mode 3—Joint military and civil identity

Civil modes:
- Mode A—Joint military and civil identity (same as Mode 3)
- Mode B—Civil identity
- Mode C—Altitude reporting
- Mode D—System expansion (unassigned)

The dwell-time of the ground stations beam on target is generally of the order of 30 milliseconds, dependent upon beamwidth, aerial rotation rate and p.r.f. There is thus time enough to make some 15 to 20 interrogations of a given target. Advantage is taken of this to execute repetition of different modes by mode interlacing. Repetition confers the necessary redundancy of data in the system so that data accuracy is brought to a high level. By this means, for example, the identity of an aircraft and its altitude can be accurately known during one passage of the beam across the target. Hence in one aerial revolution this data is gathered on all targets in the radar cover.

REPLIES

Aircraft fitted with transponders will receive interrogation during the dwell-time of the interrogating beam. The transponder carries out logic checks on the interrogations received. If relative pulse amplitudes, signal strength, pulse width and spacing criteria are met, as specified in Reference 18 replies are made within 3 μs of the receipt of the second pulse (P_3) of the interrogation pair. Replies are made at a frequency of 1 090 MHz. This is 60 MHz higher than the interrogation frequency. In air traffic control terms this provides an important advantage with SSR vis-à-vis primary radar as can be seen from *Figure 18.19* illustrating SSR's clutter-freedom.

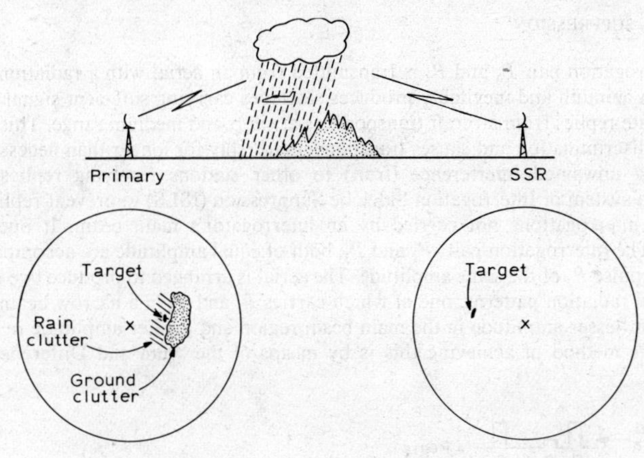

Figure 18.19 The clutter-freedom of SSR

The pulse train of a reply is constructed, again according to internationally agreed standards, and allows codification of any or all of 12 information pulses contained between two always-present 'bracket' or 'framing' pulses designated F_1 and F_2. The pulse structures possible may be deduced from *Figure 18.20*. The code is octal based; the four groups of three digits are designated in order of significance as groups A, B, C and D. Of the 4 096 possible codes, ICAO specifies that code 7700 is reserved for signalling emergencies, code 7600 to signal communication failure and code 3100 to signal 'Skyjack'.

The codification which signals altitude is formulated by a Gilham code pattern giving increments of 100 ft. The tabulation can be found in the ICAO document Annex 10[18] and covers altitudes from −1 000 ft to +100 000 ft.

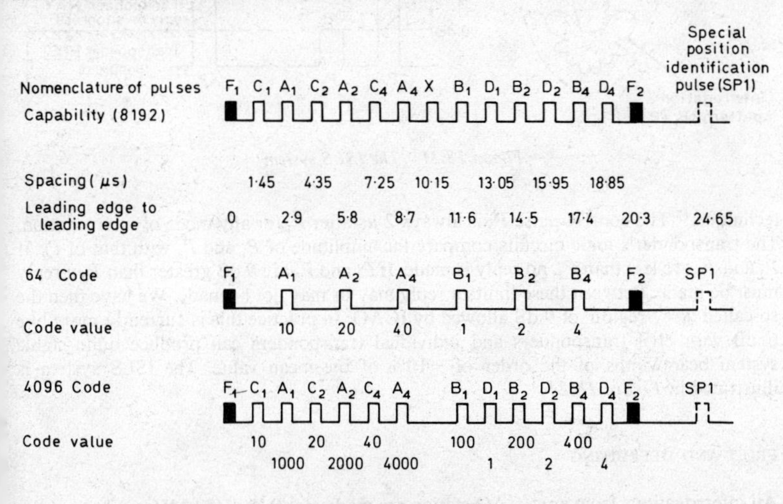

Figure 18.20 The ICAO reply format in SSR

SIDELOBE SUPPRESSION

The interrogation pair P_1 and P_3 is transmitted from an aerial with a radiation pattern narrow in azimuth and inevitably produces sidelobes carrying sufficient signal strength to stimulate replies from aircraft transponders at short and medium range. This impairs azimuth discrimination and causes transponders to reply for longer than necessary thus generating unwanted interference (fruit) to other stations receiving replies. ICAO specifies a system of Interrogation Sidelobe Suppression (ISLS) to prevent replies being made to interrogations not carried by an interrogator's main beam. It operates as follows. The interrogation pair, P_1 and P_3, both of equal amplitude are accompanied by a control pulse P_2 of the same amplitude. The aerial is arranged to produce two different horizontal radiation patterns, one of which carries P_1 and P_3 in a narrow beam and the other P_2 at lesser amplitude in the main beam region and greater amplitude in sidelobe regions. A method of achieving this is by means of the 'Sum and Difference' aerial

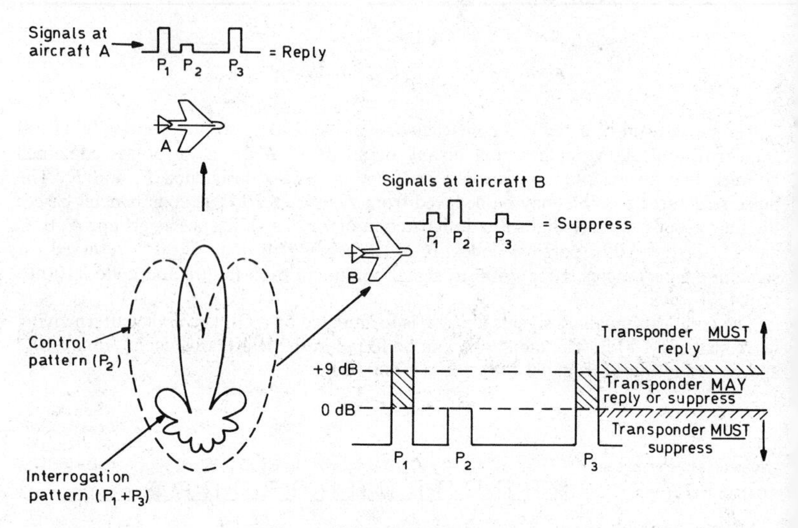

Figure 18.21 The ISLS system

technique.[17] The control pulse P_2 is always 2 μs after P_1 for any mode of interrogation. The transponder's logic circuits compare the amplitude of P_1 and P_3 with that of P_2. If P_1 and P_3 are less than P_2, no reply is made. If P_1 and P_3 are 9 dB greater than P_2 a reply must be made. Between these limits a reply may or may not be made. We have then the so-called 'grey region' of 9 dB allowed by ICAO. In practice this is currently more like 6 dB with civil transponders and individual transponders can produce quite stable system beamwidths of the order of $\pm 10\%$ of the mean value. The ISLS system is illustrated in *Figure 18.21*.

FRUIT AND DEFRUITING

All interrogations from any ICAO station are made at 1 030 $\pm$ 0.2 MHz and all replies from all ICAO standard transponders are made at 1 090 $\pm$ 3 MHz.

Interrogations are made via beams narrow in azimuth, replies are made omnidirectionally. Thus replies to one interrogator station can be received by other stations via their main beam or receiver sidelobes (unless receiver sidelobe suppression is used). To prevent ambiguous range and bearing information being gathered by these means, each interrogator is assigned a specific p.r.f. or interrogation rate. A synchronous detector at each station can then separate replies due to its own interrogations from those due to other interrogators. Asynchronous replies are known as *fruit* and the process of filtering them out is known as *defruiting*. There are various defruiting techniques but range oriented digital comparators are mostly used.

GARBLING AND DEGARBLING

If two aircraft on one bearing are close enough in slant range, their replies can overlap because of their simultaneity. This condition is known as synchronous garbling. It is common to find SSR decoders with fast dual registers operated by a commutating clock. These enable all but the most severe pulse masking to be tolerated. In view of the dubiety of data under garbled conditions it is common to find data labelled in such a way that garble conditions are indicated to the operator.

REAL TIME DECODING

Replies are standardised as shown in *Figure 18.20.* All pulses are 0.45 ± 0.1 μs in duration separated by 1.45 μs and contained within bracket pulses spaced 20.3 μs apart. Thus reply detectors of various sorts can be constructed which sense pulse width, position and spacing. Decoders operating in this manner pass all valid code data to operators who have control units upon which wanted codes may be set by mode/code switches. Every reply contains F_1/F_2 bracket pulses. Decoders are generally arranged to seek coincidence of F_1 and F_2 by delaying F_1 by 20.3 μs. When coincidence is found the decoder produces an output pulse similar to that of a primary radar signal. Thus p.p.i. display responses from SSR appear similarly to the operator, giving range and azimuth data.

REAL-TIME ACTIVE DECODING

The signals indicating aircraft 'presence' and position can be used by the operator to call out the specific code data of any aircraft reply. By use of a light-pen or real-time gate associated with a steerable display marker, any individual target can be isolated and code data of its replies examined by the decoder. The code structure and the mode of interrogation are indicated to the operator on separate small alpha-numeric displays as four octal digits representing the four groups of information pulses referred to above. A similar indicator is used to show automatically reported altitude of targets in decimal form in units of 100 ft (flight levels).

REAL-TIME PASSIVE DECODING

Aircraft identity codes can be discovered by use of active decoding as described above. Commonly, identities will be known by either flight planning or radio telephone request. Operators are usually given sets of controls upon which they can set any mode and code combinations, again using the four digits octal structure. When the control is activated the decoder looks for correspondence between the set values and all replies. When correspondence occurs a further signal is generated by the decoder to augment that due to detection of F_1/F_2. Thus unique identity of a number of targets can be established.

AUTOMATIC DECODING

The processes described above operate in real time and require the operator to filter wanted data from that given, by use of separate controls and indicators. Automatic decoding carries out the process of active decoding on all targets continually, storing in digital form all the mode and code data, together with positional data. Using modern fast p.p.i. display techniques this data can be converted into alpha-numeric form and other symbols for direct display upon the p.p.i. screen, the position of the data being associated with the position symbol of the relevant target. By this means a fully integrated primary and secondary radar system can be produced. This concept is widely being implemented for air traffic control use.

Important distinctions between primary and secondary radar

The following is a brief resumé of the advantages and disadvantages of SSR relative to primary radar.

Positional data. Both systems give this, although quality of data in terms of resolution is better with primary radar. SSR reply time (20.3 μs) and wide horizontal beamwidth precludes high range and azimuth resolution. The SSR pulse width (0.45 μs) however, provides high range accuracy potential.

Height data. In primary radar systems either a separate height-finding equipment or a multibeam surveillance system must be deployed. In SSR the height-finding facility is in-built but dependent upon aircraft transponders being fully equipped with height reporting elements (this is not yet mandatory but may become so).

Identity of targets. This is inherent in SSR giving 4 096 individual identities in a service area. In primary radar, identity has to be established by request of manoeuvre or unreliably inferred from position reports.

Dependence upon aircraft size. In SSR all transponders have ICAO standard performance hence the signals from small craft are as strong as those from large aircraft. In primary radar the equivalent echoing area is dependent upon target size and shape. Although both systems suffer from target 'glint', SSR suffers least and can be improved by use of multiple aerials.

Constraints of ground clutter. Primary radar has to include clutter rejecting systems with inclutter target detection capability. This always has a limit which can inhibit target detection in regions of high level clutter. SSR is completely free from this constraint since the ground station is not tuned to receive on its transmitting frequency; reflected energy is thus rejected.

Constraints of weather clutter. The same is true here as for ground clutter, i.e., SSR will not receive energy reflected from weather clutter. Two-way attenuation of microwaves is a factor for consideration in both primary and secondary radar but is less significant in the SSR system which, by ICAO specification, is organised with power in hand and thus more able to cope with losses.

Constraints of angel clutter. SSR has complete freedom from angel clutter. This freedom can be achieved in primary radar only by sacrifice of detection of very small targets, or complication of signal processing.

Dependence upon target cooperation. SSR is entirely dependent upon target cooperation and is totally impotent unless aircraft carry properly working transponders. Primary radar is dependent only upon the presence of a target within its detection range.

Range capability. The range performance of SSR is amenable to calculation as with primary radar. A major difference is that SSR range is the compound of two virtually independent inverse square law functions and primary radar range, when a single aerial is used, is an inverse fourth power law function. With SSR it is required to calculate

(*a*) Maximum interrogation range ($R_i$max).
(*b*) Maximum reply range ($R_r$max).

$$R^2_{i\max} = \frac{P_i G_i G_r \lambda_i^2}{(4\pi)^2 S_{a/c} L_i L_{a/c}} \tag{9}$$

and

$$R^2_{r\max} = \frac{P_r G_r G_i \lambda_r^2}{(4\pi)^2 S_r L_i L_{a/c}} \tag{10}$$

where P_i = Peak of interrogator power output
G_i = Gain of the interrogation pattern
G_r = Gain of the aircraft transponder aerial
$S_{a/c}$ = Signal necessary at aircraft aerial to produce satisfactory interrogation
L_i = Losses between interrogator and its aerial terminals
$L_{a/c}$ = Losses between aircraft aerial and transponder receiver terminals
λ_i = Wavelength of interrogation (29 cm)
P_r = Transponder peak output power
G_r = Gain of aircraft aerial
S_r = Reply signal level necessary at responser aerial for satisfactory decoder operation
λ_r = Wavelength of reply (27.5 cm)

The following points should be noted relating to the above:

(1) *Interrogator aerial gain G_i.* It is usual to find aerials of very small vertical dimension, of the order of 46 cm to 61 cm ($1\frac{1}{2}$ to 2 wavelengths approximately). This produces beams which are very wide in the elevation plane and consequently a great deal of power is directed into the ground. The resultant vertical polar diagram is therefore subject to very large modulation about the free-space pattern by the mechanism described earlier.

(2) *Aircraft aerial gain G_r.* The aircraft aerial is intended to be omnidirectional. Due to aerodynamic restrictions and varying shape of the airframe, the polar pattern of the aerial is modulated away from the desired shape and, more importantly, subject to shadowing and sometimes total obscuration during aircraft manoeuvring. Average reported gain performance is ±7 dB about a mean of zero. There is evidence of wide variations over small solid angles of −40 dB and +20 dB. For these reasons it is usual to find figures of unity gain assumed in calculations.

(3) *Range performance.* If an interrogator is operated at the maximum permitted power ($52\frac{1}{2}$ dBW ERP) and a transponder of nominal performance is taken, maximum interrogation free space ranges of 300 to 400 nautical miles can be achieved. Allowing the system to have some beamwidth (say equivalent to that at −3 dB relative to the horizontal polar pattern peak), the range is still of the order 200 to 280 nautical miles.

Taking practical values for system loss and ground receiver (responser) sensitivity, reply ranges in excess of 300 nautical miles are obtained. Thus the system has gain in hand on the reply path, ensuring high signal strength of received data.

Future extensions of SSR

The SSR system, as with any other, suffers limitation and has its own special systematic problems, e.g., over-interrogation of transponders; over-suppression of transponders;

corruption of data by fruit; and poor azimuth discrimination. These have been variously reported and analysed.[19, 20, 21, 22]

In an attempt to lessen the effects of these shortcomings a number of variations on the beacon theme are currently under intensive investigation with a view to future implementation on the same internationally agreed basis as current SSR.

These variants (notably DABS and ADSEL; acronyms for Direct Addressed Beacon System and the Addressed Selected System) postulate the ability of a ground station to send an interrogating train of pulses to selected aircraft of known position. The interrogation would contain the selected aircraft's code of identity. If the aircraft found correspondence between the addressed code and its own identity it would issue its reply data which, it is claimed, can be more intense than at present and contain for instance speed, heading, altitude, fuel state, etc. 23, 24, 25.

All the currently proposed variants are organised to be compatible with existing SSR specifications (which doubtless will be protected by ICAO for many years to come) and use normal SSR to obtain the necessary positional information. In automated radar systems this form of SSR is ideally suited to operational needs since, under computer control, aircraft data may be organised to be obtained when the system needs it and in the form required. All SSR systems are eminently suited for integration into data handling systems because of the digital nature of its signals.

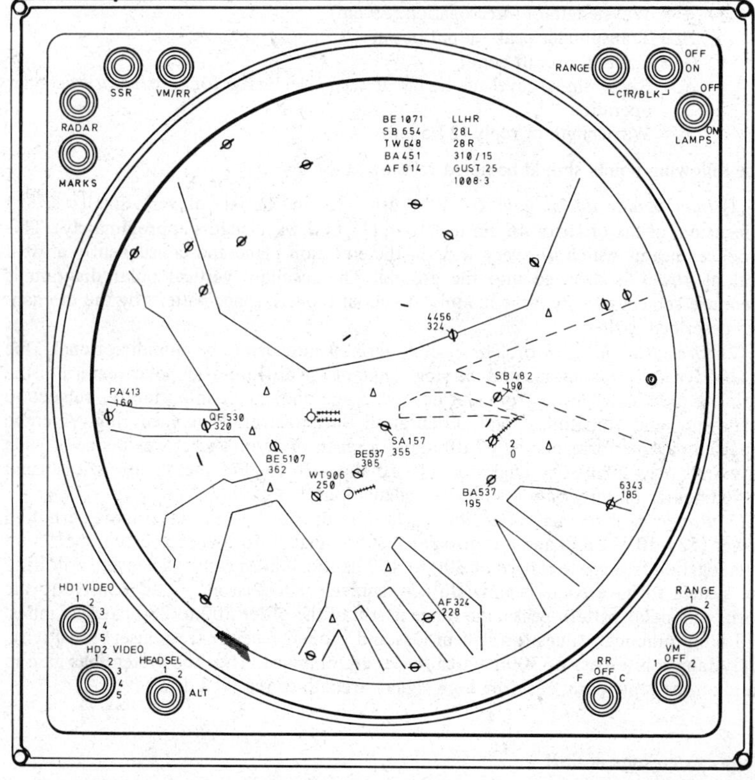

Figure 18.22 Air Traffic Control system radar display. Example of modern PPI display capable of presenting real time signals (arcs) together with machine time data (circles, alpha-numerics, video map and symbols). (Marconi Radar Systems Ltd)

Modern operational methods

The radar art and technology have reached a point where practically any civil operational requirement can be met. In the military sphere electronic countermeasures are developed to great sophistication. As an example of the former, *Figure 18.22* illustrates the way in which primary and secondary radar, display and data handling systems can be combined to give all the information an air traffic control system needs for the safe and expeditious handling of air traffic. Data on selected aircraft is available to operators by simple controls organised in such a way that displays are not cluttered with masses of available data but which is on call at the operator's discretion.

REFERENCES

1. WOOD, D. and DEMPSTER, D., *The Narrow Margin*, Arrow Books, Revised Illustrated Edition (1969)
2. WATSON-WATT, SIR ROBERT, *Three Steps to Victory*, Odhams Press (1957)
3. WATSON-WATT, SIR ROBERT, 'Radar in War and Peace', *Nature*, **156**, 319 (1945)
4. RANDALL, J. T. and BOOT, H. A. H., 'Early Work on Cavity Magnetrons', J.I.E.E., **93**, 997 (1946)
5. SKOLNIK, M. I., *Introduction to Radar Systems*, McGraw-Hill, International Student Edition (1962)
6. SKOLNIK, M. I., *The Radar Handbook*, McGraw-Hill (1970)
7. BLAKE, L. V., 'Recent Advances in Basic Radar Range Calculation Technique', *I.R.E. Trans.*, **MIL.5**, 154 (1961)
8. HALL, W. M., 'Prediction of Pulse Radar Performance', I.R.E., **44**, 224 (1956)
9. VINCENT, N. and LYNN, P., 'The Assessment of Site Effects of Radar Polar Diagrams', *The Marconi Review*, **28**, No. 157 (1965)
10. HANSFORD, R. F., *Radio Aids to Civil Aviation*, Heywood & Co., 112 (1960)
11. RIDENOUR, L. N., *Radar System Engineering*, McGraw-Hill, 143 (1947)
12. SILVER, S., *Microwave Antenna Theory & Design*, McGraw-Hill (1949)
13. RIDENOUR, L. N., *Radar Systems Engineering*, McGraw-Hill, 81–86 (1947)
14. BARTON, D. K., *Radar Systems Analysis*, Prentice Hall, 105 (1964)
15. EASTWOOD, SIR ERIC, *Radar Ornithology*, Methuen (1967)
16. HOWICK, R. E., 'A Primary Radar Automatic Track Extractor', *I.E.E. Conference Publication*, No. 105, 339 (1973)
17. SKOLNIK, M. I., *The Radar Handbook*, McGraw-Hill, 38.10 (1970)
18. Annex 10 to the Convention on International Civil Aviation (International Standards & Recommended Practices—Aeronautical Telecommunications) H.M.S.O., **2**, April (1968)
19. COLE, H. W., 'S.S.R.—Some Operational Implications in Technical Aspects', *World Aerospace Systems*, **1**, No. 10, 468, Oct. (1965)
20. ULLYATT, C., 'Secondary Surveillance Radar in ATC', *I.E.E. Conference Publication*, No. 105 (1973)
21. ARCHER, D. A. H., 'Reply Probabilities in SSR', *World Aviation Electronics*, Dec. (1971)
22. HERRMANN, J. E., 'Problems of Broken Targets and What to do about them', *Report on the 1972 Seminar on Operational Problems of the ATC Radar Beacon System*, National Aviation Facilities Experimental Centre, Atlantic City, N.J., 31 (1973)
23. BOWES, R. C., GRIFFITHS, H. N. and NICHOLS, T. B., 'The Design and Performance of an Experimental Selectivity Addressed (ADSEL) SSR System', *I.E.E. Conference Publication*, No. 105, 32 (1973)
24. STEVENS, M. C., 'New Developments in SSR', *I.E.E. Conference Publication*, No. 105, 26 (1973)
25. AMLIE, T. S., 'A Synchronised Discrete-Address Beacon System', *I.E.E.E. Transactions on Communications*, **COM-21**, No. 5, 421 (1973)

Modern operational methods

The computer and technology have reached a point where practically any type of operational [illegible] can be built [illegible] military [illegible] developed [illegible] example of the [illegible] Figure 18.22 illustrates the way [illegible] can be [illegible] system [illegible] for the safe and [illegible] of available data [illegible] at the operator's discretion.

REFERENCES

1. [illegible] *The Radar* [illegible] Arrow Books [illegible] Edition (1960)
2. [illegible]
3. [illegible] 156, 319 (1945)
4. [illegible] McGraw-Hill [illegible] (1968)
5. [illegible] *Introduction to Radar Systems*, McGraw-Hill International Student Edition (1962)
6. [illegible] *The Radar Handbook*, McGraw-Hill (1970)
7. [illegible] Radar Range Calculation [illegible] *Trans.* MIL-8, 145 (1964)
8. [illegible] 44, [illegible] (1956)
9. [illegible] The Assessment [illegible] Radar Polar Diagrams', *Marconi Review* 28, [illegible] (1965)
10. [illegible] 113 (1966)
11. [illegible] McGraw-Hill (1962)
12. [illegible] *Antenna Theory* [illegible] (1949)
13. [illegible] Engineering, McGraw-Hill [illegible] (1947)
14. [illegible] 105 (1963)
15. [illegible]
16. [illegible] Radar [illegible] (1975)
17. [illegible] Handbook, McGraw-Hill [illegible] (1970)
18. [illegible]
19. [illegible]
20. [illegible]
21. [illegible] (1971)
22. [illegible]
23. [illegible] DABS [illegible] (1975)
24. [illegible] (1973)
25. [illegible] *Transactions on Communications*, COM-21, No. 5 [illegible] (1973)

19 ELECTRONICS IN WEATHER FORECASTING

ELECTRONICS IN WEATHER FORECASTING 19

19 ELECTRONICS IN WEATHER FORECASTING

The science of meteorology in general, and weather forecasting in particular, is at an extremely exciting and important stage in its development. This is not identifiable with any one specific cause, but the impact of our high technology society is making a major contribution both in the demands it is making upon weather forecasting and in the tools and techniques that it is providing. Perhaps the greatest single contribution of all has been made by the electronic computer. Its introduction is allowing the replacement of subjective inferences, which rely heavily on past experience and empirical relationships, by objective numerical forecasting. This certainly does not imply that the subjective methods have no part to play in solving the overall problem but rather that the emphasis is changing. The meteorologist is no longer totally reliant upon such techniques but uses the numerical forecast to predict the large and even medium scale atmospheric properties, in combination with his local knowledge and experience, to satisfy the particular needs of the various sections of the community.

The process of numerical weather forecasting, which has been used in the U.S.A. for example since 1955, and operationally in the U.K. since 1965, consists of observing and recording the physical properties of the atmosphere at a particular time and of using the laws of physics to predict the state of the atmosphere at some future time. Atmospheric motion is very complicated and extensive, occurring on many scales from planetary disturbances of the order of 5 000 km across to turbulence on the centimetric scale and below. One of the main difficulties of developing mathematical methods of forecasting is to decide how to simplify the problem sufficiently to make it tractable while retaining the essential controlling factors. There are also problems in the time available to complete the forecasting process. A prediction for the following day is clearly of little value if it requires 24 hours to collect the data, make the computations and provide the results in a suitable form for the user.

The following describes the manner in which electronic equipment is having an impact on the problem of providing the basic data set, the communication of these to a central processor, the solution of the computational problem, its interpretation and the dissemination of the results.

THE BASIC DATA SET—RADIOSONDES

The basic data set for quantitative definition of the 'present' state of the atmosphere for computational purposes is provided by a network of radiosondes. These sondes are expendable balloon borne packages conveying transducers to measure the local atmospheric pressure, temperature and humidity. Each sonde also contains an r.f. transmitter for telemetering the measured data to the base station; normally at the launch site. By monitoring the sonde position the variation of wind with height can also be obtained. The problems of specifying the optimum spacing of this network, the altitude range over which measurements are required and the necessary accuracy and resolution of the meteorological data have been discussed by Gardin *et al.*[1] and Hawson.[2] Hawson, for example, points out that for a typical station separation of 300 km it is unnecessary to measure temperature to better than ± 0.15 °C in the troposphere (lower atmosphere) and to ± 0.3 °C in the stratosphere. He also suggests that measurements with errors as large as 1 or 2 °C, depending upon the latitude, are of very little value.

Some of the basic limitations of current sondes arise from a lack of uniformity of response in the array of sondes which make up the necessary network, particularly between one type of sonde and another as used by different nations. This is of major importance as atmospheric dynamics on the scale of the network are largely governed by horizontal gradients in the meteorological parameters. This problem has led to the introduction of relatively high cost reference sondes in an attempt to provide recognised standards and means of comparison. Some of the difficulties of this approach have been pointed out by Rider [3] for example and in the U.K. the problem is being tackled by the introduction of a new improved radiosonde designated the Mk 3. This sonde has been designed to improve the quality of the received data at a cost commensurate with an operational rather than research or reference sonde.

The Mk 3 sonde is described in some detail by Hooper [4] but basically the earlier bi-metallic strip temperature transducer has been replaced by a fine tungsten wire whose essentially linear resistance-temperature characteristics are used in an *RC* phase shift oscillator to generate an a.f. modulation whose period is linearly related to temperature. The transducer has an extremely low thermal time constant and a predictable response to solar radiation and self heating. The pressure transducer consists of an aneroid capsule whose deflection during an ascent moves a ferrite rod along the axis of a ferrite inductor. Time multiplexing is used in the sonde so that this inductor forms part of an *LC* circuit which is switched into the same phase shift oscillator to generate an a.f. modulation which is a function of pressure. The expansion and contraction of a strip of goldbeater's skin with changes in humidity again modifies the inductance in a further *LC* circuit which is switched in turn into the oscillator. Reference signals also modulate the transmitter during a commutation cycle. The transmitter is crystal controlled in the 27.5 to 28.0 MHz band. The whole circuit is built from discrete components and uses six low grade transistors instead of the three valves of the earlier Mk 2 sonde.

Other types of transducer are in use in other countries. For example, humidity data are obtained in the U.S.A. by using the change in resistance produced when a carbon hygristor experiences a varying humidity, see for example Brousaides,[5] and in many European countries rolled hair replaces goldbeater's skin.

Advances are also being made in the treatment and analysis of data received from sondes in that attempts are being made to achieve quality control and to extract meteorologically significant data automatically. The majority of meteorological services in Europe and North America at least are seeking to make progress in this direction. Hooper [6] has pointed out the problems of such an approach, encroaching as it does into the difficult field of pattern recognition. The U.K. system is built around a dedicated computer programmed to maintain the sonde data precision through the necessary calculations and to reduce the arithmetic skills required of staff. Some complicated judgements as to data validity are made on an interactive basis, by the operator, but the majority of decisions and all computations are made automatically.

Windfinding in the U.K. and in several other countries is achieved by tracking a passive reflector, incorporated into the sonde train, by radar. This is not the only approach. For example, the approximately 25 MHz transmissions from the Finnish radiosonde are received by an array of aerials spaced several wavelengths apart. The direction of propagation of the essentially plane wave from a distant sonde may then be obtained from phase comparison between signals received at the various aerials. This method provides azimuth and elevation only, and a distance, usually sonde height, must be obtained by some other method. The measured temperature and pressure profile may be used for this purpose. The early U.S.A. Rawin system operated in a similar direction finding manner, except that the equipment operated at a high frequency of 1 680 MHz so that a steerable, narrow beam, paraboloid aerial was used incorporating a conical scan generated by an off-axis rotating hemisphere. A modulated signal proportional to the amount the aerial system was 'off target' was therefore generated as an error signal for repositioning. Later versions of the AN/GMD system, as it is known, have incorporated a range finding technique in which the sonde receives a modulated signal at 403 MHz from a yagi aerial mounted coaxially with the paraboloid. The modulation signal is detected and used to re-modulate the 1 680 MHz carrier. Phase comparison between the ground transmitted and received signals allows the determination of range as described by Todd and Peterson.[7] Phase shifts produced in the sonde are relatively constant and may be 'zeroed' out.

The system is now basically a radar (see Section 18), as the ranging capability exists, with the sonde acting as a transponder. Such an arrangement is known as secondary radar to distinguish it from primary radar where a passive reflector is used. The main advantage of secondary radar lies in its capability of operating over long ranges with a much lower power expenditure than is practicable with primary radar. Other similar types of equipment are described in Handbook of Meteorological Instruments.[8] Current improvements to the U.K. Meteorological Office system are largely concerned with the provision of automatic data print out and with interfacing the radar with the computer needed for the automatic analysis of sonde data.

All of the above tracking methods necessitate the measurement of one or more angles with respect to some reference plane or planes. This implies firstly that a stable reference exists and secondly that position errors contain a range dependent term irrespective of radar power. It is invariably the angular error or resolution limit which dictates the useful range of such systems. As the demand for data from greater heights increases so too does the need for greater ranges from the tracking system. There are also instances, for example over the oceans, where stable reference platforms are difficult and expensive to provide. Similarly, although dropsondes ejected from aircraft offer the opportunity for achieving an extensive observational array, particularly for research purposes, the difficulty of making angular measurements to obtain position has in the past necessitated either doing without wind data or restricting operations to small areas where ground based tracking radars, for example, were available.

Angular measurements become unnecessary as the number of reference points increases, and this of course has been appreciated for some time by the designers and users of the long range navigation aids, Decca, Loran C and Omega. The principle of operation of these systems is described in Section 25 and it is only necessary here to recall that positional information is obtained from the difference in time of arrival (or phase) of radio waves from coherent transmitters. A minimum of three transmissions is required when the system is used in its normal mode. At first sight this would appear to be an unrealistic solution as the necessary decoding and analysis electronics, as used on ships and aircraft for example, is certainly too expensive to be incorporated into each operational sonde. Fortunately it proves to be unnecessary to perform the analysis on the sonde. To achieve long ranges, all the navigation aids operate at frequencies in the l.f. and v.l.f. bands (approximately 100 kHz for Decca and Loran C and 10 to 13 kHz for Omega) so that sufficient bandwidth can be obtained on a u.h.f. link at say 400 MHz to re-transmit the navigational aid signals, while preserving the necessary phase relationships, to a ship, aircraft or ground station where the analysis equipment is located.[9] The sonde merely time-freezes the navigational aid data appropriate to its position for re-transmission and provided only that the u.h.f. link is adequate to achieve the necessary signal to noise ratio at the receiver the attitude or movement of the latter and the addition of phase shifts are unimportant so long as time differences are retained. For windfinding purposes where the rate of change of time or phase difference is required, rather than an absolute value, coherent transmissions are not essential and it is now being appreciated that any stable frequency transmission in the v.l.f. band may be used for this purpose. For example NAA Cutler, GBR Rugby and NBA Balboa are capable of being used for windfinding in the North Atlantic as well as the Omega transmissions from North Dakota, Norway and Trinidad.

The techniques developed for the use of Loran C, etc., for navigation purposes have found application in windfinding systems. For example phase locked loops of one form or another are used to generate stable internal oscillations from which time differences are measured. It is, however, important to realise that as the first time differential is required rather than the absolute value used in navigation, the extension is certainly non-trivial. The quite well documented stability data available for Loran C as a navigational aid cannot be used directly to assess the accuracy of Loran C as a windfinding aid. However it is clear that of the options available Loran C does offer the greatest potential accuracy. This arises because the transmissions are pulsed rather than continuous wave and in principle at least, by working at the front of the pulse, skywave contamination may be reduced. Tests by Acheson[10] in the U.S.A. and by the U.K. Meteorological Office have demonstrated that winds averaged over one minute may be obtained during daytime, in favourable geometrical areas, with r.m.s. errors as low as 0.2 metres/second. The situation at night time is less good however as there is evidence that even towards the leading edge of a Loran C pulse, interactions between mutually shifting ground and sky wave contributions can produce significantly increased errors. Nevertheless to the north west of Scotland, where the transmitters of Sandur, Ejde and Sylt provide good coverage, an area of 600 000 km^2 exist where one minute day time winds may be expected to be determined to $\pm 0{\cdot}5$ metres/second. Omega/v.l.f. offers the potential of world wide cover, unlike Loran C, but at reduced accuracy. It is important to appreciate that windfinding errors are again a function of geometry and time of day. The receiving location is probably also of importance because atmospheric activity is quite high at these low frequencies in the tropics.

Omega windfinding was used for the first time operationally in the Global atmospheric research programme Atlantic Tropical Experiment (GATE) during the northern hemisphere summer of 1974. This was an international programme in which 37 ships and 14 aircraft from 9 nations collected meteorological and oceanographic data over the tropical Atlantic extending from 5° S to 20° N. Of these only the 12 U.S.S.R. ships used radar tracking and the remainder of the synoptic wind data from ship and aircraft sondes were obtained from Omega/v.l.f. tracking. The accuracy of determined winds vary within the area and with time of day but typically 2 minute winds to $\pm$ 3 metres/second are expected.

This experiment indicates the importance which is attached to improving our understanding of meteorological processes which occur in the tropics and of their significance for weather forecasting both at low and other latitudes. The importance of this experiment arises from a realisation that long term forecasts must include observations made over large, perhaps global, areas and particularly over the oceans. At middle and high latitude it is possible to obtain an estimate of the wind field and hence mass flow from a knowledge of the pressure field by assuming that the pressure gradient force is balanced by the apparent coreolis force produced when movement occurs on the rotating earth. This is known as the geostrophic approximation. The assumption is reasonable where the coreolis and pressure differential forces are much larger than friction or acceleration forces for example. The assumption is not valid at low latitudes where the coreolis force becomes very small. It is therefore particularly important to make measurements of wind speed and direction directly in the tropics.

Efforts have been made to obtain soundings from commercial ships; so called 'ships of opportunity', in addition to those available from the standard but limited and expensive ocean weather ship network. Wind measurements are particularly difficult in this case as commercial shipping in general does not normally carry tracking equipment. The introduction of navigation aid tracking is an important advance in this respect. However large gaps remain in the data as ships tend to follow well defined routes. Alternative methods such as the U.S. Global Horizontal Sounding Technique (GHOST) have been suggested. In its latest configuration this is designed around a 'mother' balloon floating in equatorial regions at about 40 mb (22 km) carrying up to 100 small dropsondes. The expected life of such a balloon is about two months. Upon command from a satellite, a dropsonde will be released. This will take some 40 minutes to reach the sea surface during which time Omega and meteorological data will be transmitted to the 'mother' package for onward transmission via a geostationary satellite to the ground, Frykmann and Lally.[11] As an example of the impact of present technology, the production of a low cost reliable lithium primary cell with twice the capacity and low temperature performance orders of magnitude better than any other cell, has had a large effect on the viability of this concept, Lally.[12]

THE BASIC DATA SET—ARTIFICIAL EARTH SATELLITES

The major advance in obtaining world wide observations for weather forecasting and atmospheric research has occurred with the advent of artificial earth satellites. All the methods described so far have involved measurements being made in the immediate vicinity of the sounding instrument. Satellite measurements are made by remote sounding techniques which involve the measurement of atmospheric parameters from a distance. An excellent review by Houghton and Taylor[13] is available on this subject.

All remote sounding techniques are based upon the detection and measurement of electromagnetic energy after it has been emitted from, scattered by or partially absorbed by some part of the atmosphere. The simplest form of remote sensing is the provision of images of the underlying surface, be it ground or clouds, in a spectral region in which the atmosphere is essentially transparent. These images do not form part of the numerical weather forecasting data set but provide valuable information for the preparation of surface analyses, together with the other surface observations available from meteorological stations all over the world. In particular they provide valuable confirmation of the basic data set. For example it is entirely possible for the present observing grid to miss or provide insufficient data of some weather system until it arrives unheralded in the eastern North Atlantic. The earliest meteorological satellites were the TIROS series (Television and Infrared Observation Satellite) which carried 500-line vidicon television cameras to provide images of the cloud patterns. The first satellites were spin stabilised with their axis direction fixed in space rather than relative to the earth. This meant that the earth was only visible to the camera for part of each orbit and the rather poor resolution images were subject to considerable distortion. The pictures were tape recorded and replayed to a central ground station in the U.S.A.

Since those early days vast improvements have been made. Three axis stabilisation of the spacecraft, relative to local earth axes, has provided the imaging device with a continuously favourable view of the surface, and greatly simplified the task of geographically locating the image data. By placing the satellite in a circular orbit with a high inclination to the equatorial plane at a height of about 1 500 km, the entire earth's surface may be viewed at least twice per day. By carefully matching the inclination and the satellite's height, the orbital plane can be arranged to precess, under the influence of the non-sphericity of the earth, by 360° per year. In this type of orbit (known as 'sun synchronous') the satellite always crosses the sun-lit equator at the same local solar time, which can be chosen to provide ideal conditions of illumination for visible imagery. Night-time images can be generated by an infra-red radiometer sensitive in the spectral region around 11 μm, where the atmosphere is almost transparent. The intensity of this radiation is a function of the temperature of the ground surface or cloud top from which it is emitted. It therefore provides a direct means of distinguishing between clouds at different temperatures and hence heights, which is equally useful by day and by night.

Finally, steps have been taken to enable potential users to receive image data directly from meteorological satellites whenever they are within range. This APT (Automatic Picture Transmission) facility was initially implemented by providing an additional internal long persistence storage layer within each vidicon, which could be scanned at a low rate (4 lines/second) requiring a signal bandwidth of only 2.4 kHz. This signal could be received and displayed as an image with relatively simple and inexpensive ground equipment. Several hundred such APT stations have been established around the world. Although the current ITOS satellites (Improved TIROS Operational Satellites) carry radiometers which view the earth through a rotating mirror, which scans a strip at right angles to the ground track of the satellite, they still provide direct transmissions which are receivable at up-graded APT stations. The ITOS satellites also carry a more sophisticated radiometer (VHRR) which generates visible and infra-red images with a resolution of about 1 km. However, reception of data from this instrument, transmitted at 1 697 MHz as an analogue signal with a bandwidth of 35 kHz, requires a much more expensive ground station. A very readable account of the details of the earlier satellites has been given by Widger [14] and the ITOS satellite has been described by Schwalb. [15]

The view of the earth obtained from a geostationary satellite, in orbit at a height of 35 000 km above a fixed point on the equator, also has meteorological applications. The chief virtue of this orbit derives from the fact that it allows an essentially continuous view of one part of the earth. Two satellites in the ATS series (Advanced Technology Satellite) carried experimental radiometers which provided visible imagery for the central Pacific and North and South America for about six years. The radiometers used the satellite rotation, coupled with a north–south stepping mechanism, to build up a raster image of the disc with about 2 000 line resolution over a period of about 20 minutes. A second and more sophisticated generation of these satellites, which will produce both infra-red and visible imagery, is under preparation. One such satellite, known as Meteosat, is being developed by the European Space Research Organisation, and is scheduled to be launched at the end of 1976.

This type of satellite allows the observation of the development and movement of the clouds over a very wide area, extending to about 60° of geocentric arc from the satellite sub-point. Apart from the use of this data in short term forecasting, for example in providing warnings of tornadoes in the U.S.A., the images may be used to obtain a measure of wind speed and direction by considering the clouds as tracers of air motion. There are difficulties in assigning a height to these clouds, and hence winds, and in separating the effects of development and net movement. It is, however, possible to identify cloud types from the images and therefore to assign crude heights. For example, cirrus cloud which occurs in the upper troposphere has a very different appearance to that of low cloud. An alternative method is available when the cloud structure is also observed in the infra-red to provide data on cloud top temperature. From this information an approximate height may be inferred.

A second and equally powerful means of remote sensing is to observe emitted radiation in the infra-red or microwave regions of the spectrum at wavelengths where the atmosphere partially absorbs. It is beyond the scope of this section to enter into the details of this method and the interested reader may refer to Houghton and Taylor. [13] It is however clear that if the upwelling radiation observed at the satellite is at a wavelength which is strongly absorbed then this must largely originate high in the atmosphere as any from below is attenuated. By observing at wavelengths which are only weakly absorbed we probe deeper into the atmosphere. Thus, recalling again that infra-red and microwave emissions are temperature dependent, a mean temperature appropriate to various altitudes may be obtained by observing the upwelling radiation at a range of differently absorbed wavelengths. The absorbing/emitting constituent must of course be uniformly mixed to remove any dependence on its concentration. Carbon dioxide and molecular oxygen fulfil this role in the lower atmosphere at least.

This is the principle of operation of the atmospheric temperature soundings obtained on Nimbus 3 and 4 for example using the SIRS (Satellite Infra-red Spectrometer), on Nimbus 4 using the SCR (Selective Chopper Radiometer), and on the ITOS series using the VTPR (Vertical Temperature Profile Radiometer), by observing the radiation received at several wavelengths in the vicinity of the 15 μm absorption of carbon dioxide. There are quite severe practical problems of retrieving a temperature profile from such measurements because the observed intensity at any wavelength is dependent on the emissions from a wide range of heights, and the presence or absence of cloud has a profound effect on the interpretation of the measurements. In this respect, molecular oxygen, which absorbs at wavelengths close to 5 mm, provides some improvement as clouds not containing large liquid drops or ice cyrstals are substantially transparent at microwave frequencies. However it is more difficult to obtain good horizontal resolution at these longer wavelengths. Infra-red data are now beginning to be used operationally and are providing a useful addition to the basic data set.

TELECOMMUNICATIONS

Although the process of obtaining the data set for numerical forecasting has been stressed so far this forms a relatively small part of the observations required by a meteorological service. Data on surface weather conditions are obtained from a very much closer spaced network to allow short term checking of numerical forecasts and to assist in their interpretation. For global studies radiosonde ascents are made throughout the world at 00 and 12 GMT every day at about 700 stations. Surface observations are made at regular intervals throughout the 24 hours at some 4 000 stations. For regional and national purposes a denser network of surface observations is required and observations are made more frequently, at least every three hours in general but at hourly or half hourly intervals on some airfields for example. This provides a rather unusual constraint on the telecommunications network as the system must accept messages for transmission largely at the same time, at all centres in the system, but dissemination patterns vary according to the global, national or regional responsibilities of receiving centres. As previously explained a very tight transmission schedule is necessary to produce worthwhile forecasts. For example on a global basis the target time for completion of collection and dissemination of synoptic information is two hours from the standard time of observation.

Normally in a country there is a National Meteorological Centre (NMC) acting as a focal point of its meteorological services. Internationally there is a complementary overlay of centralised services. These are provided by World Meteorological Centres (WMC) and Regional Meteorological Centres (RMC). The WMCs are at Washington, Moscow and Melbourne and their task is to process meteorological information on a global scale while the RMCs are required to process the information on a scale sufficient to meet the demands of their respective regions. The U.K. Meteorological Office headquarters in Bracknell is a RMC on the main trunk route from Washington to Moscow serving part of northwest Europe, the U.K., the Irish Republic, Iceland and Greenland.

A forecast is useless until it reaches the user and in this respect the telecommunications system must also handle processed information consisting largely of charts of the current and forecast weather, pressure patterns, etc. These are transmitted as analogue facsimile data at present, in which a source chart is scanned at between 1 and 4 lines per second for reproduction at a similar speed at the receiver. A single chart may thus occupy some 20 minutes of transmission time. Observing stations produce alpha numeric data, however, and the mixture of alpha numeric and analogue facsimile data also places an important constraint on the telecommunications system.

In the U.K. the majority of data is transmitted via 50 baud teleprinter and facsimile land lines, owned by the GPO but leased for meteorological purposes. The Ocean Weather Ship data is received by hand speed morse and a limited amount of information is provided on telephone links. International traffic on the main Washington–Moscow trunk is transmitted by the GPO telecommunications system but again land lines are used for the majority of the RMC requirements. In addition radio-telegraph and radio-facsimile broadcasts are provided, and APT satellite transmissions are received at an outstation close to Bracknell. An automatic complex based upon two dedicated computers has been constructed to interface with the processing computer discussed below, to automate message routing, preparation and storage and to perform error detection and correction. Transmission speed conversions are also made as the main trunk circuit carries data at 2 400 bits/second while a transmission speed of 1 200 bits/second is used on some regional circuits terminating at Bracknell.

COMPUTATION

The input data set received from the telecommunications system is not in a suitable form for use as the specification of the atmosphere for computational purposes because the observations are scattered non-uniformly through the volume of interest. The first task therefore is to interpolate the data to a set of grid points, and to test its meteorological validity. This process is known as objective analysis. Where data are scarce, previously forecast data may be used to maintain a level of continuity. In the U.K. approximately 3 000 grid points are currently used in two different models. One of these covers practically the whole of the northern hemisphere with a horizontal grid spacing of about 300 km at 50 °N and a vertical separation of 100 mbs from 1 000 mb to 100 mb. The other is on a finer mesh, in the horizontal, of 100 km spacing and covers much of the North Atlantic and Europe, while retaining the same resolution in the vertical.

Given the initial meteorological parameters the mathematical laws of hydrodynamics and thermodynamics are then used to predict these parameters some short time ahead. These new data are then used for the next time step and so on. This process is necessary because analytical solutions to the equations are not possible and approximations have to be made which, without careful control, can lead to a build up of mathematical errors. The hemisphere model currently uses a time step of about 6 minutes while the fine mesh model uses approximately 2.5 minutes. A new mathematical integration scheme is however allowing expansion of this to 12 minutes with a commensurate saving in computer time as indicated below. Approximately ten thousand million numerical operations are necessary to generate a 72-hour forecast on the hemispheric model.

The U.K. Met. Office computer comprises a central processing unit (CPU) with a basic cycle time of 54 nanoseconds. The main core store consists of 1 megabyte (each byte comprising 8 binary digits) with a cycle time of 756 nanoseconds. The CPU also has a 32 kilobyte buffer store again with a 54 nanosecond cycle time. Additional storage of 11.2 megabytes is provided on a fixed head disc and 1 200 megabytes on 12 exchangeable magnetic disc drives. Data can be transferred at a maximum rate of 1.5 megabytes per second to and from the fixed head disc and at half this rate for the others. A range of peripheral equipments are available from magnetic tape units to on or off line plotters. A more detailed description may be found in the Meteorological Office Annual Report for 1972.[16]

Two operational runs start each day at 03 and 15 GMT using data obtained at 00 and 12 GMT to produce forecasts for up to 72 hours ahead. Additional runs are made at 1245 and 0045 to utilise the greater amount of data for 00 and 12 GMT which is received after the 03 and 15 GMT operational runs. The 72-hour hemisphere forecast requires approximately 40 minutes of computer time while the fine mesh model takes 50 minutes to produce a 36-hour forecast although this has been reduced to 12 minutes with the latest integration scheme. These times can be compared with the four hours required to produce a 24-hour forecast over a smaller area using an earlier research computer.

INTERPRETATION

As stated above the minimum horizontal resolution in use at the present time is about 100 kms. In order to interpret the forecast on a finer scale than this the meteorologist must be aware of the manner in which these relatively large scale features control and are affected by atmospheric behaviour on the much smaller scale of the individual user.

Much of research meteorology is directed at this general problem. Although many techniques are used at present, including past observational experience and persistence of current weather detected by the surface observation network, the research meteorologist's aim is to seek a physical understanding of atmospheric behaviour on this sub-synoptic scale. The multitude of research tools and topics being studied precludes an exhaustive review but the use of weather radar is characteristic of the general approach.

It is well known that radar performance at wavelengths of 10 cm or less is greatly affected by meteorological conditions (see Section 18). In particular areas of rain produce significant echoes by scattering electromagnetic radiation back to the transmitting aerial. It can be shown that the power reflected from raindrops is proportional to the fourth power of the wavelength used (see for example Mason)[17] and therefore a 3 cm radar has greater sensitivity than one operating at 10 cm, all other factors remaining the same. However absorption increases with decreasing wavelength and this can lead to attenuation of the beam and obscuration of more distant rainfall. The 10 cm radar of course requires a larger aerial to maintain a given beamwidth. The reflected power is also proportional to the sixth power of the mean droplet radius so that the relatively small cloud droplets produce a very small return compared with that from raindrops. Empirical relationships have been developed between rainfall rate and reflectivity and with a typical transmitted peak power of 600 kW at 10 cm, with a beam width of 2° and pulse length of 2 μs, a relatively small rainfall rate of the order of 0.1 mm/hr is detectable at ranges up to 50 km. A weather radar PPI display indicates the distribution of precipitation over a large area, and quantitive estimates of rainfall at the surface have been produced by, Harrold *et al.*,[18] for example, by quantising and integrating the video signal to obtain average echo intensities from volume elements covering an area of 10 000 km^2.

Properties of the windfield may also be inferred by detecting the Doppler frequency shift produced when precipitation has a component of motion along the radar beam. In this mode the radar scans a conical surface. Precipitation is moving towards the radar on the upwind side of the cone and away from it on the downwind side so that a plot of Doppler velocity against azimuth produces a sine wave whose amplitude is a function of wind speed and whose phase can be used to obtain the wind direction. If the sine wave is not symmetrical about the zero velocity axis then there is either a net motion into or out of the circle defined by the cone at a particular height, i.e., convergence or divergence is occurring. Doppler and precipitation radars have been used together with an array of aircraft windfinding dropsondes to study the sub-synoptic properties of weather systems, Browning *et al.*[19]

REFERENCES

1. GARDIN, L. S., ALAKA, M. A., MASHKOVICH, S. A. and LEWIS, F., 'Design of Optimum Networks for Aerological Observing Stations', *World Weather Watch*, Planning Report, No. 21, WMO (1967)
2. HAWSON, C. L., 'Performance Requirements of Aerological Instruments', WMO, No. 267. TP 151. Technical Note, No. 112 (1970)
3. RIDER, N. E., 'Upper Air Instruments and Observations', *Proceedings of the WMO Technical Conference, Paris, September 1969*, 329 (1970)
4. HOOPER, A. H., 'Upper Air Instruments and Observations', *Proceedings of the WMO Technical Conference, Paris, September 1969*, 364 (1970)
5. BROUSAIDES, F. J., 'An assessment of the Carbon Humidity Element in Radiosonde Systems', *AFCRL-TR-73-0423*, Instrumentation Paper, No. 197, U.S.A.F. (1973)
6. HOOPER, A. H., 'Upper Air Instruments and Observations', *Proceedings of the WMO Technical Conference, Paris, September 1969*, 347 (1970)
7. TODD, W. and PETERSON, A., *Proceedings of the Fifth Weather Radar Conference*, Fort Monmouth, New Jersey, 207 (1955)

8. *Handbook of Meteorological Instruments*, Part II, Her Majesty's Stationery Office (1961)
9. BEUKERS, J. M., *I.E.E.E. Transactions on geoscience electronics*, GE-6, No. 3, 143 (1968)
10. ACHESON, D. T., *ESSA Technical memorandum WBM EDL 11*, U.S. Dept of Commerce (1970)
11. FRYKMANN, R. W. and LALLY, V. E., *A Carrier Balloon for Tropical Soundings*, NCAR TN/EDD 63, National Centre for Atmospheric Research (1971)
12. LALLY, V. E., *Proceedings of the WMO Technical Conference, Tokyo 1972*, Reports on Marine Science Affairs No. 7, 564 (1973)
13. HOUGHTON, J. T. and TAYLOR, F. W., *Reports on Progress in Physics*, **36**, 827 (1973)
14. WIDGER, W. K., *Meteorological Satellites*, Holt, Rinehart and Winston (1966)
15. SCHWALB, A., *NOAA Technical Memorandum NESS 35*, U.S. Dept of Commerce (1972)
16. *Annual Report on the Meteorological Office, 1972*, Her Majesty's Stationery Office (1973)
17. MASON, B. J., *The Physics of Clouds*, 2nd edn., Clarendon Press (1971)
18. HARROLD, T. W., ENGLISH, E. J. and NICHOLASS, C. A., *Quarterly Journal of the Royal Meteorological Society*, **100** (1974)
19. BROWNING, K. A., HARDMAN, M. E., HARROLD, T. W. and PARDOE, C. W., *Quarterley Journal of the Royal Meteorological Society*, **99**, 215 (1973)

20 RADIO ASTRONOMY

RADIO ASTRONOMY 20

20 RADIO ASTRONOMY

EARLY HISTORY

Until the late 1940s almost all astronomy was carried out in the narrow atmospheric window in the electromagnetic spectrum which occurs over about a 2 : 1 frequency range at optical wavelengths. The first indication that there existed another atmospheric window which could be used for astronomy came in 1932 when K. G. Jansky of the U.S.A. carried out observations which showed that background noise affecting short-wave communications at around 15-m wavelength varied with a period equivalent to 1 sidereal day, indicating that the maximum radiation came from a fixed point in space which turned out to correspond to the densest part of the Milky Way. (A sidereal day is equal to 23 hrs 56 mins solar time.)

However it was 1940 before anyone attempted to actually make use of radio-wavelengths for astronomy, when another American, G. Reber, built the first large steerable parabolic radio-telescope, 9 m in diameter and operating at a wavelength of about 6.5 m. Reber confirmed Jansky's observations and used the steerability of his antenna to make the first radio maps of the galaxy.

Reber's work was not directly followed up due to the outbreak of World War II. However during the course of the war two important radio-astronomical discoveries were made as a consequence of investigations into problems associated with the development of radar.

The first of these occurred when J. S. Hey in England and G. C. Southworth in the U.S.A. were, like Jansky 10 years previously, searching for the cause of intense radio interference, though at much shorter wavelengths than Jansky. In 1942 Hey realised that intense variations in noise at metre wavelengths were coming from the sun. He had made what we now know to be the first identification of a major solar outburst, although retrospectively it was realised that solar radio noise had been observed and commented on back in the 1930s, but was not identified as being of solar origin.

The second important discovery was connected with the unexpectedly high attenuation of radar signals at cm wavelengths. This was subsequently identified with molecular line emissions associated with the rotation of H_2O and O_2 molecules in the earth's atmosphere, though the direct importance of the occurrence of molecular emission lines in the radio region to astronomy was not demonstrated until many years after the solar observations developed into a major branch of astronomy.

Since these early observations radio astronomy has expanded to cover virtually the whole of the radio window, from about 30 m to 1 mm wavelength, a range of about 3×10^4 to 1 as opposed to the 2 to 1 range in the optical window. The radio window is limited by ionospheric reflection at the long-wave end and by the molecular attentuation at the short-wave end, where it merges with the far infra-red region of the electromagnetic spectrum.

Radio-astronomy is now accepted as a major branch of astronomy rivalling optical astronomy in importance. Indeed many of the more recent astronomical discoveries have originated in radio rather than optical observations. These are described later in the text.

RADIO-TELESCOPES

Single antenna telescopes

Most large radio-telescopes today are basically long-wavelength versions of the parabolic optical reflecting telescopes, although at metre wavelengths sometimes large arrays of dipoles are used. In some the receivers, which incorporate a small feed antenna for collecting the focused energy, are located at the focus of the paraboloid. Others, following the optical Cassegrain systems, use a hyperbolic sub-reflector placed near the focus to reflect the energy to receivers located close to the vertex of the main reflector.

The largest single element radio-telescope in the world is the 305 m diameter one at Arecibo in Puerto Rico. This antenna has been built in a modified natural circular valley and the main reflector is not steerable, though the overall antenna beam can be directed through a small angular range by adjustment of the feed system at the focus. In order to simplify its construction the main reflector is actually spherical rather than parabolic. This results in the presence of spherical abberation, which can however be corrected to a large extent by use of a more complex feed system based on slotted waveguides.

For many years the largest steerable reflector antenna was the 76-metre diameter Mark 1 telescope at Jodrell Bank in England. Originally this was built for operation at wavelengths down to 21 cm, but recently its surface has been improved to enable it to operate at somewhat shorter wavelengths. More recently this honour has gone to the 100-m diameter antenna at Effelsberg in the Federal Republic of Germany. Not only is this antenna considérably larger than the Jodrell Bank one, but it has a much more accurate surface. It can be used down to 5-cm wavelength over the whole of its surface, and currently down to 1.7 cm over the inner 80 m. Eventually it is hoped that it will be usable in part down to 0.8 cm.

Its configuration is of the Gregorian type rather than the more conventional Cassegrain system. In this system an ellipsoid sub-reflector is used instead of a hyperboloid. This has the advantage that the sub-reflector does not obscure the focal point of the main reflector and thus the receiver can either be located at the focus of the main reflector (prime focus) or near the vertex of the main reflector (secondary focus). This antenna is the first large one to be built using the principle of homology. This means that it has been designed so that as the elevation angle is altered the antenna distorts under its own weight but always keeps a parabolic shape, although the form of the parabola is continually changing. This results in a continual but calculable change in focal length, which can be countered by automatically moving the position of the receiver.

Some other notable large steerable radio-telescopes are the 64 m ones at Parkes, Australia and Goldstone, California, plus several others in the 30–40 m range in Canada and the U.S.A.

The three most precise radio-telescopes in the world are the 100-m Bonn telescope, the 22-m one at Lebedev, U.S.S.R. and the 11-m one at Kitt Peak, Arizona. Although differing widely in size they all have r.m.s. surface error to diameter ratios of between 1 and 2×10^{-5}.

The gain G of an antenna of diameter D and r.m.s. surface error ϵ is given approximately by

$$G = 0.65 \frac{\pi D^2}{\lambda} \exp - \left(\frac{4\pi\epsilon}{\lambda}\right)^2 \qquad (1)$$

where λ is the operating wavelength.

The spatial resolution θ (width of the main antenna beam at its half-power points) is given approximately by

$$\theta = \frac{\lambda}{D} \text{ radians} \qquad (2)$$

i.e., the resolving power is proportional only to the diameter of the paraboloid measured in wavelengths.

Interferometers

The resolving power of any antenna system depends only on the dimensions of the instrument relative to the wavelength at which it is used. As we have seen, in the case of a parabolic antenna it is proportional to the diameter D. However, the same angular resolution could be obtained by suitably combining the outputs of a number of smaller antennae distributed over an area corresponding to a circle of diameter D. The gain of such a system would not be equivalent to the single antenna but would depend on the total area of all the smaller antennae, and would be less than that of the single large antenna.

In its simplest form a radio interferometer consists of two small antennae located a distance D apart, and it is a counterpart of Michelson's optical interferometer. Such a system has a resolving power equal to a single antenna of diameter D, though it has a much lower gain. Its antenna pattern consists of a series of interference fringes or lobes resulting from changes in the relative path length from a distant source to the two antennae as the source position varies relative to the antenna. The number and width of these lobes can be varied by moving one antenna relative to the other, and the separation θ between adjacent lobes is given by

$$\theta = \frac{\lambda}{D} \text{ radians} \qquad (3)$$

(compare this with equation 2).

Simple 2 element interferometers give very good resolution along the line joining them, but at right angles to it the resolution is only that of one of the component antennae. More intricate versions use a long line of small antennae so that a range of different spacings can be obtained electronically, without the need to move one antenna relative to the other, and the most elaborate type involves two such lines at right angles intersecting in the centre of each line (a cross interferometer, such as the one at Fleurs, Australia). Such an instrument has a basic cross-shaped beam, but by combining the signals from the two arms in phase and in anti-phase, only those sources contained in the common centre of the cross are observed, thus avoiding confusion between multiple sources which would possibly be contained in the full cross-pattern. As with the simpler interferometers this system has good spatial resolution but low sensitivity.

Very long baseline interferometry (VLBI)

Originally radio interferometers consisted of either 2 small or 1 large and 1 small antennae working together on one site with perhaps only a few hundred metres separation, and with the antenna signals carried to the central receiver by cable. By using microwave radio links to transmit local oscillator and intermediate frequency signals the technique was extended to distances of over 100 km, giving values of D/λ of up to 2×10^6. The problem is essentially one of maintaining very precise phase relationships and the technique became limited by problems in significantly extending the length of the microwave links. However, the development of extremely stable rubidium, caesium and hydrogen-maser clocks has enabled signals to be recorded separately at the two antennae on magnetic tape and then subsequently combined to produce the interference pattern. The advantage of this has been enormous since it now allows pairs of very large single antennae on different continents to be combined as interferometers, thus increasing not only the spatial resolution but also the sensitivity of interferometry. It has enabled values of D/λ of up to 2×10^8 to be obtained, with angular resolutions down to less than one thousandth of an arc second, comparable to that of the best optical measurements.

Apart from radio-astronomy the precision of VLBI techniques has been used for accurate measurement of the distances between the component antennae and for measuring the position of earth satellites.

Aperture synthesis

Aperture synthesis is a technique designed to obtain the sensitivity and spatial resolution of an extremely large antenna by using a combination of physically small antennae. It is in fact used to obtain collecting areas far larger than could conceivably be physically built with a single large steerable antenna.

Basically it involves only two small steerable antennae, one of which is fixed in location and the other of which can be moved relative to it. A sequence of observations is made with the antenna in different locations over a wide area, and after suitably combining with results of all the observations (the data processing requires the use of a large computer) the performance of a much larger single antenna can be simulated.

The effective spatial resolution obtained in this technique corresponds to that of a single antenna of diameter equal to that of the furthest separation of the two antennae, and the gain corresponds to that of a single antenna whose area is the geometrical mean of that of one of the small antennae and of the total area covered by the movable antenna. Since observations on a single region can take many weeks by this technique it can only be used on sources whose output remains constant over that period.

RADIO ASTRONOMY RECEIVERS

Radio astronomy is basically concerned with the measurement of very weak thermal-type signals, that is random noise signals similar in nature to the noise produced within the receiving system itself. Full advantage is taken of the availability of low-noise amplifiers such as masers, parametric amplifiers and travelling wave tubes, for reducing the receiver noise itself. However one is still left with the problem of detecting noise signals which are very much smaller than the receiver noise.

The basic solution to this problem was devised in 1946 by R. H. Dicke who developed what is now commonly referred to as the Dicke radiometer. Dicke's original radiometer operated at 1.25 cm, for which wavelength r.f. amplification was not then possible. Basically it consisted of a horn antenna followed by a modulator (in this case a rotating disc half of which was transparent and half coated with a resistive material) which chopped the incoming signal before it became added to the combined receiver noise produced by the local oscillator, diode mixer and i.f. amplifier. This means that the signal noise remained identifiable even after being added to the large set noise.

After detection and audio amplification the combined signal passed through a phase sensitive detector which operated synchronously with the modulator and which recovered the envelope of the modulated signal. Finally a simple RC circuit enabled the output to be integrated.

This basic technique is still in very wide use even today. It is really a comparison technique since the output is proportional to the difference between the noise temperature of the signal and that of a matched load (the resistive half of the disc) at ambient temperature. At radio wavelengths the noise power P is related to the noise temperature T by:

$$P = kTB \tag{4}$$

where k is Boltzmann's constant and B is the i.f. bandwidth. The sensitivity of such a receiver (the r.m.s. minimum detectable ΔT in noise temperature difference that can be observed) is given by

$$\Delta T = K\frac{(T_R + T_S)}{(B\tau)^{\frac{1}{2}}} \tag{5}$$

where T_R is the receiver noise, T_S is the source noise at the output of the antenna and τ is the output time constant, and K is a constant which depends on the exact form of the receiver. T_S is normally much less than T_R, but is sometimes greater than it (e.g., for a low noise system pointed at the sun).

The comparison technique means that the receiver is being calibrated at the modulation rate, so reducing considerably the effects of gain changes with periods greater than that of the modulation system.

Another type of radiometer is the correlation receiver in which the signal is passed through two separate i.f. systems and the outputs are correlated by a multiplier, the idea being that the signals in the two channels are correlated but the receiver noises are uncorrelated.

For studies of atomic or molecular line emissions (see later this section) modified versions of the Dicke receiver can be used. For example the switching can be done by actually changing the local oscillator frequency so that the comparison is between the signal with the receiver centred on the line and the signal with the receiver tuned off the line. The output section of line receivers usually incorporates a spectrometer, which may be either analogue (a bank of up to several hundred filters) or digital. In the latter case the phase sensitive detection and integration can also be carried out digitally.

The most sensitive radio astronomy receivers can detect temperature changes of the order of 10^{-3} K.

RADAR ASTRONOMY

Most radio astronomy is concerned with the passive reception of radiation generated by the astronomical sources themselves. However within the solar system considerable use has been made of radar techniques for mapping the moon and the planets and also for studying the solar atmosphere and for detecting meteorites entering the earth's atmosphere.

Radar echoes were first obtained from the moon in 1946, but it was not until 1961 that echoes were obtained from Venus, followed by Mercury in 1962 and Mars in 1963. Since then no other planets have been detected though attempts have been made to observe Jupiter. The difficulties involved are illustrated by the fact that even at its closest approach to the earth there would be a delay of over one hour between the transmission of the pulse and the reception of the reflected signal back on earth.

The mapping technique makes use of the facts that signal time delay measurements enable distances of points in a plane perpendicular to the line of sight from the earth to be measured, while Doppler frequency shift effects due to different line-of-sight velocities over the planet resulting from the planet's rotation enable points to be located on the circumference of a circle whose plane is parallel to the line of sight direction.

Doppler-delay maps, as these are called, have been successfully made of the surfaces of the moon and Venus, and to some extent of Mercury. The surface of Venus cannot be observed at optical wavelengths due to the thick clouds surrounding the planet. The clouds, however, are nearly transparent to radio wavelengths. Mars has a very rapid rotation rate, making it unsuitable for this type of mapping, although time delay measurements have been made.

Radar observations have also enabled the rates of rotation of the inner planets to be measured and in particular have shown that Venus has a retrograde motion of about 250 days period and that Mercury rotates in the normal sense but in 57 days and not 88 days as was previously thought.

SOLAR SYSTEM RADIO ASTRONOMY

Sun

Radio emission from the sun consists of three basic types, namely (1) quiet sun emission, (2) the slowly varying component and (3) bursts.

The quiet sun emission comes from the undisturbed chromosphere, a few thousand km above the visible solar disc (mm and cm wavelengths) and from the corona out to several solar radii (dm and m wavelengths) and is the basic thermal radiation of the sun.

The slowly varying component is associated with localised optical active regions on the sun, particularly sun-spot groups. As these regions tend to persist for more than one solar rotation period of 27 days and are hotter than the general background their effects appear as a 27 day modulation of solar radio-emission integrated over the whole solar disc. Their radio spectra peak at a wavelength between about 6 and 10 cm and as a consequence the intensity of the sun at 10.7 cm wavelength is now widely used as an indicator of solar activity, and correlates very closely with the optical sunspot number.

Solar radio bursts are closely associated with optical solar flares. They occur over a very wide range of wavelengths from hectometres to millimetres and exhibit considerable variety in intensity variation, duration and spectra. Individual events can last from less than a second (simple impulsive burst at centimetre wavelengths) to many hours or days (noise storms at metre wavelengths), and their form can change rapidly with both time and wavelength. They can also be extremely intense. For example, although the optical sun is never observed to even double in brightness, single localised bursts can result in the average radio intensity of the sun increasing by up to a million times at metre wavelengths. The bursts result from bremsstrahlung, synchrotron and possibly Cerenkov radiation, together with plasma oscillations, and any one event can involve more than one of these mechanisms.

The bursts are often associated with other solar emissions, especially X-rays which cause Sudden Ionospheric Disturbances (SIDS) at the earth, and some types are closely related to the emission of protons, which result in Polar Cap Absorption (PCA) of high frequency radio communications on earth. Special instruments, such as the giant radioheliograph at Culgoora in Australia, which consists of 96 antennae each nearly 14 metres in diameter and arranged in a circle of 3 km circumference, are used to take radio snapshots of the bursts as they propagate out through the solar atmosphere.

Moons and planets

With the exception of Jupiter, all radio emission from the moon and planets is thermal in origin. Whereas infra-red radiation from the moon tends to come from the surface, radio radiation comes from different depths within the moon, with the optical thickness of the lunar rocks increasing with wavelength. Thus radio observations can be used to study different layers with the moon.

The angular size of the planets is much less than that of the moon and hence they are much more difficult to detect. Nevertheless thermal radio emission from all the planets out to Saturn has been observed. In addition non-thermal bursts of radiation have been observed from the planet Jupiter on metre wavelengths. This radiation is associated with the intense ionosphere surrounding Jupiter, but also appears to be strongly affected by tidal effects of one of Jupiter's moons, Io, which moves through the Jovian radiation belts.

GALACTIC RADIO ASTRONOMY

Continuum radiation

This is the radiation originally discovered by Jansky in the 1930s, and is now known to be non-thermal synchrotron emission from high energy electrons in the presence of a very weak magnetic field. It is most intense at metre wavelengths and maps of the galaxy made in the metre band show considerable detail up to high galactic latitudes.

A particularly useful feature of this radio radiation is that it enables the centre of the galaxy to be studied. At optical wavelengths starlight from the centre of the galaxy is almost completely obscured by clouds of inter-stellar dust. Although there are a few very small gaps in these clouds the picture that is obtained of the centre is very incomplete. However this inter-stellar matter is almost completely transparent to radio waves and this has enabled a great deal of information to be obtained which suggests that our galaxy has spiral arms and a concentrated nucleus similar to that of the Andromeda galaxy.

Discrete sources

For the most part localised regions of high intensity radio emission, or 'radio stars', are not related to the optical stars seen with the naked eye, but rather with somewhat peculiar astronomical sources, some of which are described below. In fact all the discrete radio 'stars' found in the early years of radio-astronomy are now known to be clouds of radiating gas (nebulae), and not stars at all.

FLARE STARS

It has already been mentioned that at radio wavelengths the sun can increase many times in brightness during the course of flares. Even so, stars whose characteristics are similar to the sun are far too faint to be detected at radio wavelengths. However a certain type of star, which unlike the sun exhibits sudden increases in optical brightness, has also been found to emit very intense bursts of radio waves, indicating events up to a million times more powerful than those occurring on the sun.

SUPERNOVAE

Some of the most intense radio sources outside the solar system are now known to be the remains of supernovae, that is stars which become unstable and explode, resulting in a great increase in brightness and producing an envelope of gas which recedes from the explosion at high speed.

The most intense of these is the radio source Cassiopeia A, which results from a supernovae which occurred in about AD 1700. Although there are no records of this event being observed optically at that time, other intense radio sources have been clearly identified with historically recorded optical events. The Crab nebula, for example, corresponds to a supernova which was recorded by Chinese astronomers in AD 1054 as a 'guest star'. Other old stellar explosions whose remains have been detected at radio wavelengths include Tycho's Supernova of AD 1572 and Kepler's supernova of AD 1604. Another well-known localised radio source, the Cygnus loop, is probably the remains of a much more ancient explosion which occurred about 50 000 years ago.

In all these cases the radio emission is due to synchrotron radiation from the still expanding gas shells centred on the point where the star originally existed.

PULSARS

Pulsars are one of the most astounding discoveries to have come from radio astronomy. In 1967 radio-astronomers at Cambridge found that a galactic source they were investigating emitted a very regular train of pulses with a period of about one second. Following this discovery over 80 other pulsating stars, or pulsars, have been found.

The pulse is typically about 5% of the total cycle, and is highly polarised, and although the pulse amplitude and shape can vary considerably the pulse duration, which ranges between 33 millisecond and 3.8 seconds, is surprisingly stable. The period of all pulsars so far discovered is however progressively slowly increasing (the rate of increase is typically a few nanoseconds per day), indicating a gradual slowing down of the pulse mechanism.

It was also found that pulsars existed at the centres of some of the known supernova remnants, such as the Crab nebula already mentioned. It has further been discovered that the centre of the Crab nebula also pulses at optical and X-ray wavelengths.

It is now thought that pulsars are in fact rotating neutron stars, that is stars which have collapsed under gravity into a state where the atomic nuclei are packed tightly together instead of being relatively widely separated as in a normal star. Such stars can have masses similar to that of the sun and yet have diameters of only about 10 km. They also have extremely intense magnetic fields of up to about 10^{12} G. These stars are believed to be formed in supernova explosions, although most pulsars are not associated with detectable supernova remnants. However this can be attributed to the outward moving gas clouds merging with the general galactic background in a time much less than the lifetime of the central pulsar.

THE 21-CM HYDROGEN LINE

Apart from emitting and absorbing radiation at optical wavelengths corresponding to changes in electronic energy levels, the neutral hydrogen atom also emits and absorbs at the radio wavelength of 21 cm, corresponding to a change in energy from when the electron spin is in the same sense as that of the nucleus to when it is opposed to that of the nucleus (a hyperfine transition). Since hydrogen atoms are very widely distributed throughout the galaxy observations of emission or of absorption of radiation from background continuum sources at 21 cm constitute a powerful tool for investigating the structure of the galaxy, and in particular for measuring the velocity of the hydrogen clouds.

The wavelength of the line is known very accurately and any movement of the hydrogen cloud relative to the observer results in an apparent shift of the line wavelength (Doppler shift). The difference in wavelength is a measure of the velocity of the cloud and also indicates whether it is receding or approaching the observer. Thus observations of the neutral hydrogen line provide the best means for studying the structure of our galaxy, and have played a major part in showing that it has a spiral form.

Although many other interstellar radio lines have now been observed, the 21-cm line was the first to be discovered (in 1951) and is by far the most widely studied, and hydrogen is the only atom whose radio emission has been detected in interstellar space.

MOLECULAR LINES AND MILLIMETRE ASTRONOMY

The second radio line, and the first from a molecule, to be found in the galaxy was the 18-cm line of hydroxyl (OH) in 1963, some 12 years after the first detection of the hydrogen line. The only other molecular lines known in interstellar gas clouds were discovered in 1937 to 1940. These however were ultra-violet lines, not radio ones, and came from the diatomic molecules CH, CH^+ and CN (methylidyne and cyanogen). For five years OH was the only molecule observed at radio wavelengths. Then in 1968/69 ammonia (NH_3), water vapour (H_2O) and formaldehyde (H_2CO) were discovered at wavelengths down to about 12 mm. However, since then about 20 further molecules have been detected, many of them in more than one transition and/or in isotopic forms, bringing the number of known lines to nearly 100.

Although many of these lines occur at cm wavelengths a large number, including some of the most important, occur at mm wavelengths, giving a big impetus to the development of millimetre-wavelength astronomy and bridging the gap between radio astronomy and infra-red astronomy. Already molecular line studies have revealed a wide range of interesting features:

(*a*) They are predominantly organic and similar to products obtained in the synthesis of amino-acids. The most complex found so far have 7 atoms (methy-acetylene, CH_3C_2H and acetaldehyde, CH_3CHO), although previous to 1968 it was believed impossible for molecules of more than 2 atoms to be able to exist in interstellar space.
(*b*) Emission from the H_2O molecule is both extremely intense and also time-dependent, suggesting a maser-type amplification process.
(*c*) Formaldehyde (H_2CO) has been observed in absorption against the 3 K microwave background of the universe, suggesting some sort of inverse-maser action, i.e., refrigeration.
(*d*) Some, at least, of the molecules (e.g., carbon monoxide, CO) are very widely distributed and could be used to map the galaxy at different distances to those observed with the 21-cm H line.
(*e*) Some molecules which are extremely unstable on earth appear to be stable in the interstellar environment.
(*f*) Some lines have been detected which have not yet been identified with any known terrestrial molecules.
(*g*) The molecules appear to be closely related to the occurrence of interstellar dust particles and it is probable that some at least of the molecules are synthesised on the surface of these dust grains.

The general nature and number of molecules so far discovered has resulted in the opening up of a field in which there is a direct overlap of interests between radio astronomers, optical astronomers, ultra-violet astronomers, physicists, chemists and biologists, involving both the study of the chemistry of the molecules and the use of the molecules as probes of the astrophysical conditions such as ambient densities, temperatures and cloud velocities. It is expected to be a field in which there will be a considerable growth of activity in the next few years.

EXTRA-GALACTIC RADIO ASTRONOMY

The 3 K cosmic background

During the course of measuring atmospheric radiation at centimetre wavelengths and comparing the results with theoretical estimates, a residual noise temperature of 3 K was found which could not be accounted for, and which further measurements showed to be isotropic. One particular theory of the origin of the universe suggests that it began as an enormous fireball, the remnants of which would be expected to result in a universal background temperature of about just this amount. Further measurements at other wavelengths showed that the radiation corresponds to a universal black-body at 3 K, and it is now generally accepted that these radio observations have made a major contribution to studies of the origin of the universe, by strongly supporting the 'big-bang' theory of the Universe (as opposed to the continuous creation theory).

Galaxies

Of the several thousand discrete radio sources that have now been discovered, several hundred have been positively identified with external galaxies, and it is generally accepted that in fact most of these sources are external to our own galaxy. Those sources which are extended, i.e., subtend angles of about 1 to 2 arc seconds are probably galaxies, that is compact groups of stars sometimes similar in optical appearance to our own galaxy, but sometimes very different, for example having an elliptical slope, or having a double nucleus. Galaxies are in general larger at radio than at optical wavelengths indicating the presence of a halo of gas.

Those galaxies which resemble our own are known as 'normal' galaxies. There are however many other galaxies, some of them similar optically to ours, whose radio emission is very different, and these are generally called 'radio' galaxies. The most notable of these is the double source Cygnus A which, although a very intense radio source, is optically very faint and is at an extremely large distance, which as measured by the Doppler red-shift of its optical spectra is about 6×10^8 light years.

The most remote radio source so far discovered is nearly 100×10^8 light years distant and is barely detectable optically. It is thus likely that some of the radio galaxies are too far away to be detected optically and therefore are the most remote objects so far observed.

Quasars

Quasar is the shortened title given to some extremely unusual and apparently extremely distant optical sources, or quasi-stars, which are associated with discrete radio sources. At first they were thought to be ordinary stars, but spectral studies showed that they have very unusual spectra including very large redshifts, suggesting very high recession velocities and correspondingly great distances from the earth.

Very long base-line radio interferometer measurements however have shown that the sources have extremely small angular diameters, and are often not resolvable even with inter-continental baselines, and those that are resolvable often seem to have more than one centre of emission. Some also show changes of optical and radio intensity on time scales of months or years, again suggesting a very small size, particularly in relation to normal galaxies.

The origin of these very mysterious objects is still unknown, and further confusion has been caused by the discovery of two apparently associated quasars which have however widely differing red-shifts. Some astronomers are beginning to doubt if the widely accepted relation between red shift and distance holds for these objects and suggest that they are in fact relatively close, possibly within our own galaxy, and that the large red-shift is caused by some physical process which is as yet completely unknown to us.

FURTHER READING

CHRISTIANSEN, W. N. and HÖGBOM, J. A., *Radiotelescopes*, Cambridge (1969)
EVANS, J. V. and HAGFORS, T., *Radar Astronomy*, McGraw-Hill (1968)
HEY, J. S., *The Evolution of Radio Astronomy*, Elek Science (1973)
KRAUS, J. D., *Radio Astronomy*, McGraw-Hill (1966)
KUNDU, M. R., *Solar Radio Astronomy*, Interscience (1965)
PACHOLCZYK, A. G., *Radio Astrophysics*, Freeman (1970)
STEINBERG, J. L. and LEQUEUX, J., *Radio Astronomy*, McGraw-Hill (1963)
Various, *Proc. I.E.E.E.*, **61**, complete issue (September 1973)

21 ELECTRONICS IN SPACE EXPLORATION

ELECTRONICS IN SPACE EXPLORATION 21

21 ELECTRONICS IN SPACE EXPLORATION

Often all the electronics found in space exploration vehicles are categorised under the term instrumentation yet there are few applications which utilise so many diverse branches to provide for a single result, retrieval of decipherable research data. Spacecraft electronics consist of the following systems and circuit design disciplines: control; communications; high frequency (microwaves); antennas; telemetry and tracking; data handling, processing and storage; signal and power conditioning; all working to particularly rigorous reliability, environmental and packaging specifications.

Three principal types of space exploration vehicle are the sounding rocket, the earth orbiting or geostationary spacecraft and the deep space probe for interplanetary excursions. The electronics in each vehicle perform the same function but the techniques differ as do the individual operational time durations, the communication power budgets

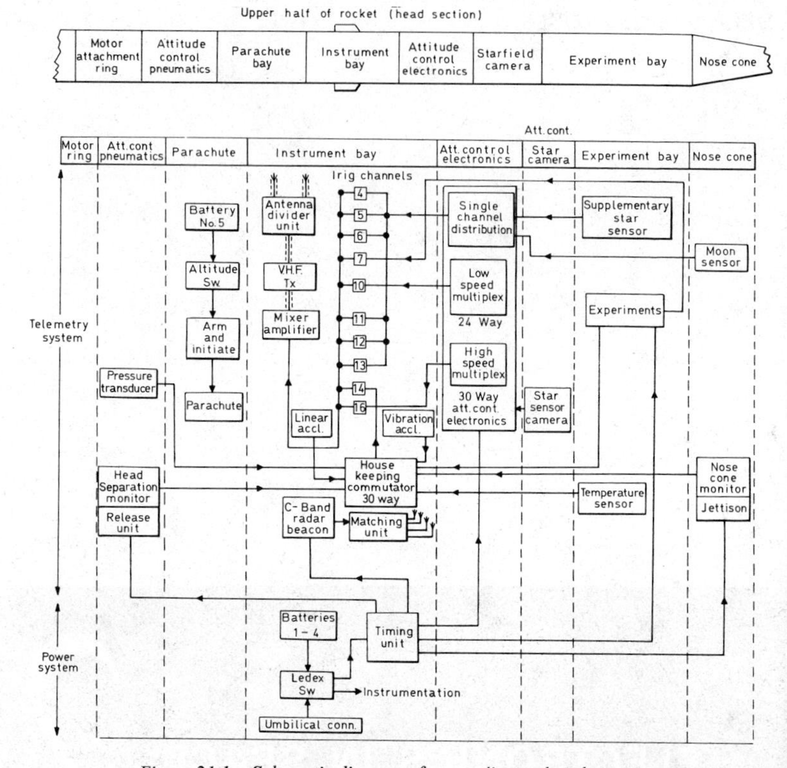

Figure 21.1 *Schematic diagram of a sounding rocket electronics*

and the periods of isolation between the far range spacecraft and their associated ground facilities. A typical electronic fit for each of the three vehicles is described below with the aid of schematic diagrams. One electronic function is selected for each and indicated more fully; attitude control for the sounding rocket, data handling for the relatively near-Earth research satellite, and communications for the interplanetary probe.

SOUNDING ROCKET ELECTRONICS

This type of rocket is used for vertical exploration of the upper atmosphere and to provide platforms for observation of extra-terrestrial phenomena. Typical experiments carried are for solar spectroscopy, X-ray photography, spectro heliography and stellar ultra violet photography. Maximum altitudes are in excess of 200 km and total flight times are up to 10 minutes. *Figure 21.1* is a schematic diagram of the complete electronics as housed in the upper (head) section of the rocket and where the more expensive sections such as the attitude control system and certain experiments are generally recoverable by parachute. The photograph, *Figure 21.2* shows the upper half under test and gives some idea of physical dimensions. On the launcher primary power supplies are connected through an umbilical cable to serve any functions requiring gyro spinning, warm-up or last minute calibration.

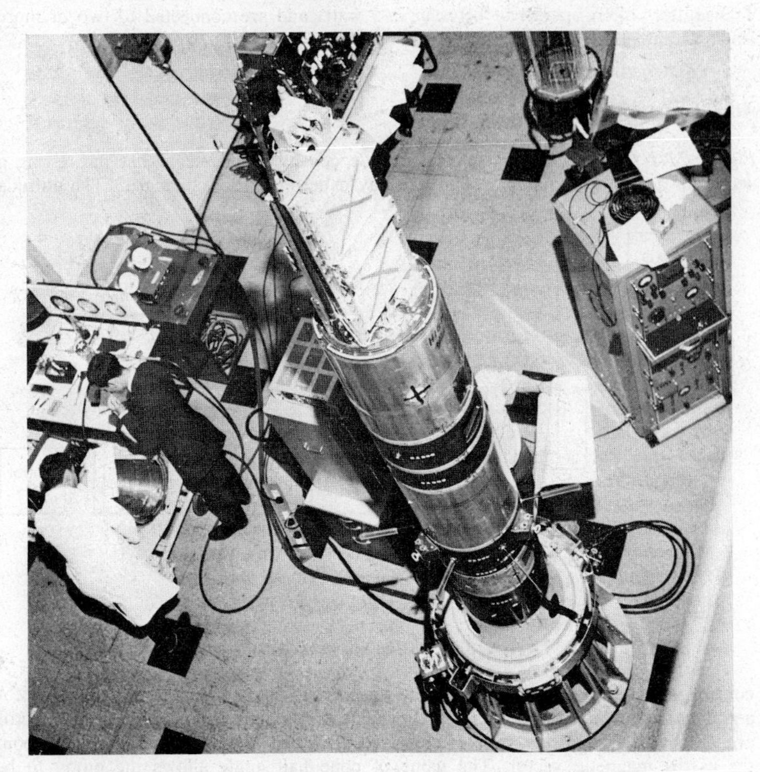

Figure 21.2 Skylark rocket under test (Courtesy British Aircraft Corporation, Bristol)

At launch the umbilical is broken and internal batteries, typically silver-zinc or nickel cadmium cells, are brought into circuit. A timing unit, generally of the electro-mechanical type continues to allocate supplies to the post launch electronic units and the experiments as a sequential switch. Many of the blocks within the schematic diagram are self-explanatory. The timer also triggers the pyrotechnics which separate the rocket head from the boost motor and which jettison the nose cone covers to permit visibility of experiment and some attitude control sensors. Invariably a radar beacon is carried and operates at a wavelength, nominally in C-band (4–6 GHz) to permit tracking and identification by a ground radar.

Generally rocket systems operate with a single communications link. The link provides real time telemetry; there is no on-board data storage and, unlike satellites, there is no up link for guidance or functional commands. As with satellites, however, the telemetry system accepts inputs from two main sources, the payload and the housekeeping. The former is the essential relaying of experimental data together with monitoring of operational parameters within the experimental and attitude control bays. The latter is monitoring of temperatures, vibration or acceleration values, power levels, transmitter oscillator stabilities and so on. Most European rockets use analogue telemetry and adopt a degree of standardisation to characteristics stipulated by IRIG (Inter-Range Instrumentation Group) a committee with representatives from all the leading U.S.A. telemetry ranges and having extensive working group support. The current frequency band in European usage is 215–260 MHz and a popular modulation format is a.m./f.m./p.m. with the sub-carriers in particular confirming to IRIG standard bands. Transmitters often operate at between 2–5 watts and are connected to two or more isotropic antennas.

Attitude control

Figure 21.3 shows a fairly complex attitude control arrangement but one giving a degree of accuracy required by modern day experimentation standards. The attitude

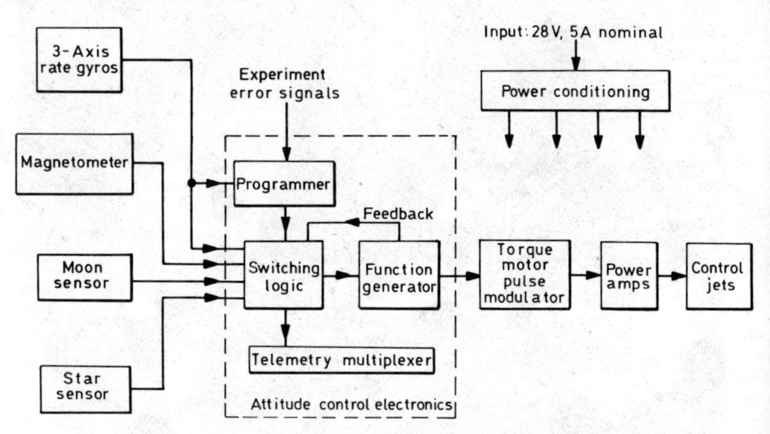

Figure 21.3 Attitude control system

control procedure commences when the rocket reaches a sufficiently high altitude for aerodynamic torques to be negligible at which point the nose cone and the boost motor are jettisoned. The gas jets are turned on and the rockets set for a coning motion about the earths magnetic vector. The value of cone half angle allows the moon to be encountered and the lunar sensor assumes control of the torques in the pitch and yaw axes until the vehicle is aligned with the moon. The diminishing error signals from the lunar sensor provide resetting of the rate gyros and at zero error they cause the vehicle to roll until the target star lies in one of the lateral control planes. A scan pattern is initiated and the position of the brightest (target star) is stored. At the end of the scan pattern the vehicle is aligned to the target star. The electronic functions performed in the vehicle alignment manoeuvre therefore comprise d.c. and power amplification, gyro control, extensive relay or logic switching, gas jet pulsing, telemetry multiplexing and power conditioning. The switching and timing circuitry is sufficiently flexible to allow re-alignment after a prescribed period on to an alternative star or for the full acquisition procedure to be repeated.

A moon camera operates by focusing the lunar image on to a quad of silicon photocells. The difference signals (A + B) – (C + D) and (B + C) – (A + D) provide for torques about the y (pitch-lateral) and z (yaw-lateral) axes, respectively, as shown in *Figure 21.4*.

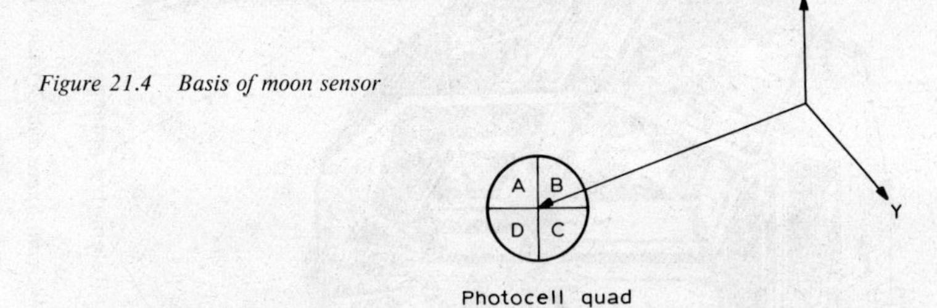

Figure 21.4 Basis of moon sensor

A sufficiently accurate star sensor consists of a 3 in cassegrain telescope with a $\frac{1}{2}$ in field of view. The telescope forms a defocused star image in the plane of a D-shaped chopper blade which is fixed at right angles to a rotor aligned with the telescope axis. Unless the star image is at the field-of-view cross-wires the light output flickers, as seen from behind the rotating blade. This modulation rate when compared in phase sensitive detectors with a reference signal coupled from the blade rotor is the basis of an error signal resolved into two components. A photomultiplier provides optical filtering and light to electrical conversion prior to high performance amplification. *Figure 21.5* shows the basic components of the system.

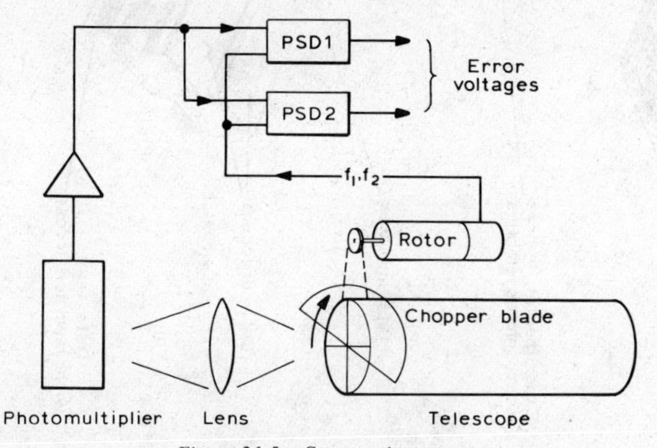

Figure 21.5 Star sensing system

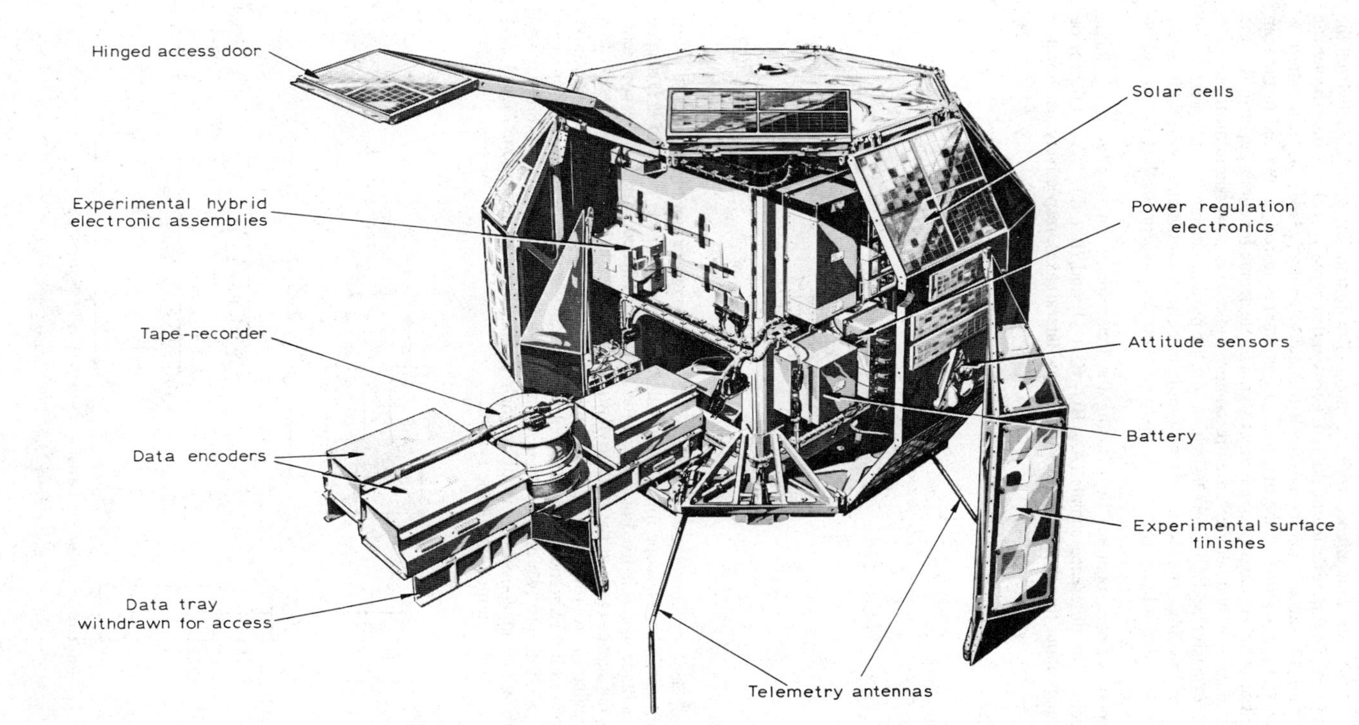

Figure 21.6 U.K. Prospero Satellite (Courtesy Marconi Space and Defense Systems Ltd.)

NEAR EARTH SPACECRAFT

This group contains research satellites placed into elliptical orbits or into circular orbits at heights of 250 km to 35 000 km, the geostationary altitude at which spacecraft circle the earth once every 24 hours and can therefore be arranged to be stationary above a given equatorial point. Operating time durations are between several months and five years and examples of experiments carried are for measurement of upper atmosphere meteorological concentrations, electron density, X-ray astronomy and low frequency magnetic intensity.

Figure 21.6 is an exploded diagram of the U.K. Prospero satellite, launched in October 1971 and *Figure 21.7* shows an early design of the European geostationary research satellite, Geos, currently under development and due for launch in August 1976. The principal electronic units are annotated in both cases.

The total electronics complement in a research satellite provides for all the services shown in the rocket schematic *Figure 21.1*, but where the attitude control system is generally less elaborate and where there are two additional major functions provided. The first is a command receiver and encoder which, with the telemetry transmission, enables a closed loop to be formed for continuous satellite operation by ground control. The second is an on board data storage facility to enable experimentation to continue throughout the periods when the telemetry downlink is out of range with its ground

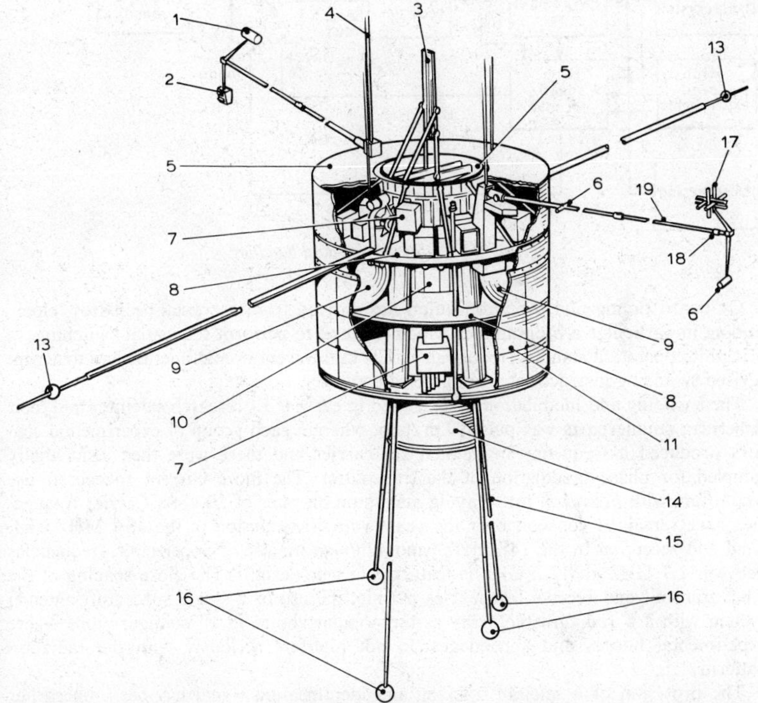

Figure 21.7 Early design of the European GEOS Satellite. 1 magnetic sensor, 2 electrostatic analysers, 3 u.h.f. antenna, 4 v.h.f. antennas, 5 thermal shield, 6 four electron guns, 7 spectrometer sensor, 8 equipment platforms, 9 hydrazine tanks, 10 thrust tube, 11 solar array, 12 electric dipole, 13 spherical probe, 14 apogee motor nozzle, 15 axial booms, 16 electric sensors, 17 magnetic sensor, 18 magnetic sensor preamplifier, 19 short radial booms (Courtesy STAR Space Engineering Consortium)

receiver. Both these items are contained within *Figure 21.8* which is derived from Reference 1, but which is also a representation of several recent data handling systems within U.K. satellites.

The high speed encoder handles the real time telemetry inputs and during the launch phase accepts inputs from the various stages of the launch rocket, together with the immediate post launch housekeeping parameters of the satellite. Once in the operational orbit command signals from the ground determine which encoder output is transmitted. The low speed encoder handles the slow experiment and housekeeping inputs and feeds a data store which, for example, is a specialised tape recorder having a record to playback ratio of up to 1 : 50.

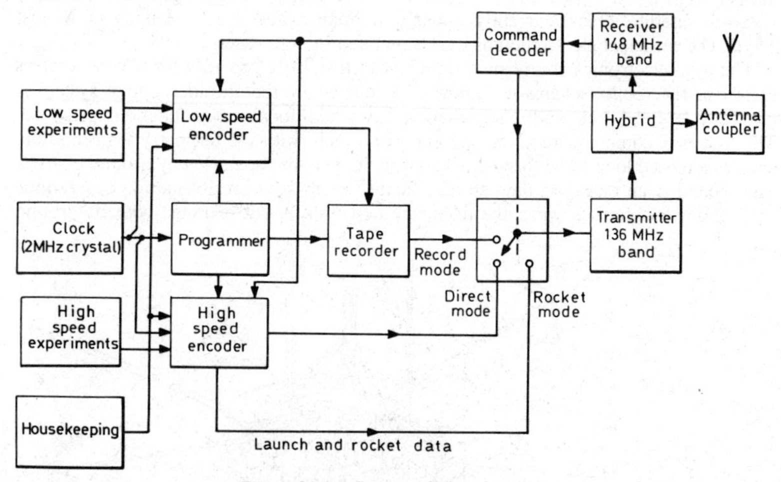

Figure 21.8 Spacecraft data handling

On-board timing and synchronisation are derived from a crystal-oscillator 'clock' feeding in particular a programmer which is preset to perform sequential switching, to bring into operation standby (redundant) units in the event of malfunction and to accept revised switching instructions from ground control.

The encoding and modulation system used in early U.K. research satellites and their American counterparts was pulsed f.m./p.m. wherein each group of experimental sensors produced a frequency modulated sub-carrier and these were then sequentially sampled for phase modulation of the transmitter. The more current spacecraft use p.c.m./f.m. with provision for relaying maximum bit rates of 50 kb/s. Carrier frequencies have remained constant over the years with transmission in the 136 MHz v.h.f. band and reception in the 148 MHz band although the use of upper u.h.f. frequencies between 1.7 GHz and 2.2 GHz is featuring in new designs. The close spacing of the v.h.f. transmit and receive frequencies permits the use of a single spacecraft antenna system with a 4 rod turnstile array as the popular choice in all configurations where experimental booms and appendages do not interfere seriously with the radiation patterns.

The provision of a satellite tracking and identification signal has been, on earlier spacecraft, a secondary function of the telemetry downlink and has often restricted the number of sub-carrier frequencies so that a constant power distribution was maintained in a narrow band centred on the carrier. As space technology has improved the amount of primary power available from solar cell systems has increased and separate tracking links are now possible.

Spacecraft computers

The ever-increasing quantity, complexity and speed requirements of spacecraft electronic functions together with the possible exclusive use of p.c.m. and digital techniques for command and telemetry are leading towards the implementation of a computer to perform all the services listed below within a single programmable unit.

Payload control
—to command the sequence of measurements
—for adaptive control of experiments
—to implement calibration and check routines
Scientific data processing
—for data collection
—for data reduction and correlation
—for data compression
—for data formatting
Station keeping (orbital positioning)
—to aid attitude control
—to aid thermal control
—to aid power control
Spacecraft monitoring
—for housekeeping data compression
—for housekeeping data formatting
Command
—for interpretation and distribution of ground commands
—for storage of ground commands for delayed execution

and, as in most satellites the computer will probably be the most complex electronic unit on-board and therefore the one most prone to failure it must provide for an element of self repair (see Reference 2). *Table 21.1* is a list of characteristics representative of a machine capable of performing all the above services but where the relatively slow execution time of 2.8 μs is compromised for lower power consumption.

Table 21.1 ON-BOARD COMPUTER CHARACTERISTICS

Type	General purpose, stored program	
Operation	Binary, parallel	
Word length	16 bits	
Arithmetic	Twos complement, fixed point	
Control	Microprogrammed	
Memory	Expandable up to 64 K	
Input/output	Program controlled and direct memory access channels	
Program Interrupt	16 expandable request lines	
Introduction set	Full arithmetic and logic capability	
	Shift/rotate provision	
	Inter-register transfers	
	Conditional/unconditional branching	
Addressing	Single-address instructions, fixed page, index, indirect, immediate	
Speed	1.4 μs cycle time	
Typical Execution Times	Load/store	2.8 μs
	Add/subtract	2.8 μs
	Multiply	28.0 μs
	Divide	51.8 μs

Data storage systems

A consideration of the design aims for future storage system gives a good indication of the problems peculiar to spacecraft usage and also of the permissible on-board technologies. The requirements in the past have been met by tape recorders or by core stores where the amount of data retained has been in the megabit region. Several other technologies could now offer a more attractive solution, the candidates being semiconductors, electron beam recording and magnetic bubbles.

Semiconductors of principle interest are m.o.s., c.m.o.s., m.n.o.s. (see Section 8) and charge coupled devices in that they offer a high bit density per chip with a low power consumption, at least in the standby or data retention mode. Electron beam storage tubes appear to offer the capability of storing the equivalent of 10^7 bits with the information in image form. This is the order of storage required by a small scientific satellite and if the desired capacity could not be met by a single tube a number of them could be used. This would have the advantage of providing a certain amount of redundancy, i.e., duplication for reliability. The nature of the data is generally one or more channels of analogue and/or digitised information which is essentially uncorrelated with the storage mechanism therefore the use of a conventional tube would require accurate beam registration in both the read and write modes.

Magnetic bubble memories appear to be ideal for those space applications requiring storage capacities well in excess of 10^7 bits. Points in favour are the high packing density in bits per slab and the potentially high packaging density in bits per unit volume, the low power versus read/write speed product, the almost zero standby power requirement and the non-volatility characteristic.

Systems design, component selection and reliability

The desired operational lifetime of the electronics within a space vehicle can be realised by attention to several factors throughout a development programme. Reliability is as much dependent upon such factors as data processing and modulation methods as it is on careful selection of electronic components. As illustration of this point consider the following two paragraphs.

P.C.M. COMMAND STANDARDS

As satellites become more complex the need arises to transmit more commands to control them, with greater security against spurious commands. Increasing complexity also means that commands have to be generated by computer which yields output in digital form. All of these requirements are met by a p.c.m. system and since p.c.m. is already used on telemetry downlinks the principles are well established but additional safeguards are necessary to ensure that only authentic and correct commands are accepted.

A p.c.m. telecommand system has been drawn up by the European Space Research Organisation (ESRO) for use in future satellites where the basic command element is an 8-bit binary word which can assume one of 256 values, each of which can be decoded as an unique address to a control line performing an on-off function in the satellite. Alternatively the bit pattern can be transferred as a number to a storage location in the data system. To ensure the rejection of mutilated commands each 8-bit word is augmented by 4 bits which provide a parity check. Using this system errors of up to 3 bits can occur in a command without generating a pattern which could be decoded as another command, such errors will therefore cause the command to be rejected. It can be expected that an error rate greater than 3 bits in 12 will cause the whole transmission to be rejected during other checks.

Command words, now 12 bits long, are assembled into a frame which is a block of 10 words and constitutes the basic command sequence. The first word, of 16 bits, is the address and is a pattern unique to the satellite being commanded. Next come two 4-bit mode words which are identical and which inform the decoder that the command is to be executed immediately, stored for execution later or transferred as a number to experiments or to an on-board computer. Of the 16 combinations available from a 4-bit word only 6 are used. This ensures that loss of 1 bit and some combinations of 2 bits will invalidate the frame.

The main body of the frame is the 6 command words, comprising 3 commands each sent twice. One word of each pair must be recognised as valid or the frame is rejected. Since the checks previously noted are not fully effective against bit slippage (i.e., the loss or insertion of a bit period during a noise burst) the address word is re-transmitted and must be detected as having the correct pattern and timing.

On receipt of the second address word the decoder initiates an execute sequence which, according to the code word sent, may consist of (*a*) three single (on-off) commands executed in succession, (*b*) three 8-bit words transferred to a time-tagged store for later use, (*c*) an 8-bit word decoded in parallel as an experiment address with the other 2 words transferred to the selected location as a 16-bit serial word, or (*d*) all 3 words sent as a 24-bit serial word to an on-board computer.

Failure of any of the error checks results in the decoder being stopped so that the execute sequence will not be reached. This gives a high degree of security against false commands. The address word at the end of the frame may be the end of the command sequence or the first word of another frame. Thus many frames may be sent in a single transmission, permitting a very high rate of command transfer.

MULTI-CHIP INTEGRATED (M.C.I.) CIRCUITS

M.C.I. is the term given to a packaging method using thick film circuits (see Section 11) on ceramic substrates to which are mounted active and passive components. The active devices include linear and digital bipolar integrated circuits, complementary and single metal oxide silicon (m.o.s.) transistors and bipolar transistors.

M.C.I. microcircuits are potentially attractive for spacecraft electronic systems for three major reasons:

(1) They offer a major contribution to the solution of the problem of making reliable systems. Any system is only as good as its most poorly manufactured component and where systems are complex and use a large number of components from a variety of different sources, the final quality is very difficult to control. Semiconductor integrated circuits for example are often procured from assembly lines which are geared to high volume production at competitive prices and reasonable reliability. The majority of spacecraft requirements are for relatively small quantities of items having an extremely reliable performance. To overcome this problem special lines are set up to make these highly reliable items. This, however, creates difficulties where quantities are small and results in high costs and long lead times for procurement. Experience has shown that a large proportion of faulty components have failed not through complex mechanisms but due to poor quality control brought about by the commercial pressures on the production lines. M.C.I. microcircuit technology can help to overcome some of these problems since by purchasing devices at the pre-encapsulation stage and then subjecting them to more detailed quality control the electronics designer can optimise the quality of the overall product.

(2) Hybrid microcircuits offer substantial reductions in weight over the more common form of electronic packaging. Weight saved in this way can result in smaller launch vehicles or a greater electronic capability for a given payload.

(3) With m.c.i. circuit technology the circuit designer has greater freedom to choose the range of devices most suited to his need. This is not always the case with standard large and medium scale integrated circuits (l.s.i. and m.s.i.). Designers are able to lay out circuits with a particular performance in mind, combining the best attributes of different circuit technologies. They can use the components with which they are most familiar, and at the same time use a compatible packaging technique. Compared with monolithic technology the cost of tooling is minimal, allowing relatively low cost changes to be made if required. This is particularly useful where only small quantities are required, as in satellite applications.

DEEP SPACE PROBES

Programmes of planetary exploration have been undertaken by the U.S.A. and U.S.S.R. for over 12 years. In 1961 the Russian Venus-1 was launched into an orbit which passed within 100 000 km of Venus, making measurements of radiation belts, ionised gases and magnetic fields and also measuring cosmic rays, solar corpuscular radiation and interplanetary dust while en route. These experiments were repeated at closer range (24 000 km) in 1965 by Venus-2. The first successful soft landing took place in 1967, when a capsule parachuted from Venus-4 transmitted data for 94 minutes during its descent to the surface. Two similar capsules in 1969 radioed information to earth on the pressure, density, temperature and composition of the Venusian atmosphere, and the Venus-7 capsule which soft-landed in December 1970 established that the surface temperature was about 475 °C, the atmospheric pressure was 90 times that of the earth's and that Venus rotates in the opposite direction to the earth (from east to west).

The first spacecraft launch in the direction of Mars was the Russian Mars-1 in 1962 the object of the flight was rather to make a long-distance exploration of interplanetary space than to examine Mars itself. The first U.S.A. success with a Mars probe was when it passed within 9 000 km of Mars, taking 22 photographs and making measurements of radiation, magnetic fields and meteoroids.

More recently the second U.S.A. Pioneer Jupiter probe was launched in April 1973. The spacecraft is hexagon-shaped, $2\frac{3}{4}$ metres in diameter and 1 metre long, with a mass (or weight as measured on earth) of 250 kg. Plutonium-fuelled radioisotope generators provide 155 watts of electric power at the time of launch, falling to 140 watts at the time of rendezvous with Jupiter and 100 watts after five years due to the radioactive decay of the plutonium. Solar cells cannot be used to provide power as in earth-orbiting spacecraft since Jupiter is five times as far from the sun as the earth and solar energy is correspondingly only one twenty-fifth at that great distance.

To transmit scientific data and t.v. signals to earth, antennas are fitted giving high, medium and low gains. The high gain parabolic antenna is $2\frac{3}{4}$ metres in diameter and half a metre deep. The Pioneers are directionally stabilised by spinning around the axis of this high gain antenna at 4.8 revolutions per minute so that the axis can be kept pointing earthwards.

The temperature within the scientific experiment compartment is controlled to between −25 °C and +40 °C; any excess heat generated by the electronics is dissipated through thermostatically-controlled louvres in the base of the spacecraft. Individual control of temperature for certain experiments is achieved by thermal insulation and electric and radioisotope heaters.

As expected the longer the operating range the more the antennas tend to dominate the entire spacecraft configuration. The extraordinarily high free-space path losses, see *Table 21.2*, make every programme an exacting communications exercise for both telemetry and command with the down link as the most critical in power budget of the two. The satellite data storage capabilities have often to accept both experimental data and television pictures where the video is stored in real time and played back into the down link transmission, for reconstitution on the ground, at a rate commensurate with the t.v. bandwidth allocation.

Table 21.2 FREE SPACE SINGLE PATH LOSSES AT 2.3 GHZ

	dB
Moon	210
Sun	270
Venus	255–272
Mars	252–270
Jupiter	275–278
Pluto	295

Most probe system designs commence with rigorous trade-off examinations which allocate the available mass, volume, fuel and power to the payload and satellite subsystems. Two prime restrictions with this type of vehicle are the necessary use of the deep space 'window' frequencies around 2.2 GHz and the shroud limits of the launch rocket. Launch costs account for the major part of total programme expense so the choice of boost rocket and therefore its payload limits is generally the very first design constraint. An objective that immediately follows is to provide the greatest possible antenna directivity so that the transmitter power necessary to achieve a given radiation level (e.r.p.) can be less. Directivity is dependent upon surface area and in satellites which derive prime power from solar radiation there is obviously a conflict in finding maximum space for both antennas and cells.

For active devices by far the most popular is the travelling wave tube (t.w.t.) (see Section 7) for the transmitter power stage and the tunnel diode (see Section 8) amplifier (t.d.a.) as the first stage of the receiver. However, innovations in both power and low noise transistors are constantly researched and the superior reliability and attendant electrical performance of t.w.t. and t.d.a. circuits will eventually be equalled.

The t.w.t. in particular is far from ideal as a Space device. Its main attribute lies in the predictability of its ageing characteristic but the disadvantages are a relatively fragile construction, its need for several operating potentials, some of them at e.h.t., its localised thermal generation and its low efficiency which is worsened when combined with the conversion efficiency of the tube power supply and made less still in single tube multicarrier systems where non-optimum biassing (back-off) must be used in order to reduce intermodulation products. There have been satellite programmes where, in providing for an operational lifetime of only a few months, the ampliton (travelling wave magneton) having three times the efficiency of the more normal t.w.t. but far less reliability, has been used, and in addition where the uncooled parametric amplifier has been seriously considered as a receiver pre-amplifier. The greatest current development effort is devoted to solid state circuitry for the critical transmit and receive stages where much attention is presently being afforded to perfecting power transistor coupling methods such that the failure of one transistor does not seriously reduce the total power output of a multi-device transmitter stage.

REFERENCES

1. SMITH, L. *et al.*, 'U.K. Satellite Review', *Journal of British Interplanetary Society*, **26**, No. 3 (1973)
2. Editor, 'STAR—Self Test and Repair in Satellites', *Electronics*, 30 March (1970)

FURTHER READING

BALAKRISHNAN, A. V., *Space Communications*, McGraw-Hill (1963)
FILIPOWSKY and MUEHLDORF, *Space Communication Systems*, Prentice-Hall (1965)
GRUENBERG, E., *Handbook of Telemetry and Remote Control*, Sec. 14, McGraw-Hill (1967)
LINDEN, A. E., *Printed Circuits in Space Technology*, Prentice-Hall (1962)

22 ELECTRONIC AIDS IN EDUCATION

ELECTRONIC AIDS IN EDUCATION 22

22 ELECTRONIC AIDS IN EDUCATION

GENERAL

Extensive practical work is an essential feature of any useful course of study involving electronics. For this purpose there is available a wide range of specially designed educational electronic equipment to assist the student in the study of the experimental aspects of the subject. This equipment covers most aspects of electronic devices and their basic applications, from the simplest of concepts to quite complex systems.

For schools the needs for educational equipment are very well defined. For the Nuffield Schools Science courses for example, fairly rigid equipment specifications have been suggested. Outside secondary school education however, the requirements are less well defined. In a technical college for example individual lecturers will often decide what equipment is suitable for particular courses. Consequently many equipment manufacturers cater mainly for schools but this does not preclude the use of their equipment elsewhere and, in general, it is equally suitable for technical education, industrial training and for home-study purposes.

Specialised equipment for use outside the secondary school area (which does not of course necessarily preclude its use in schools) appears to be used mainly outside the United Kingdom, particularly in the developing countries. There are also available a number of home-construction kits. The following are some typical examples of educational electronic equipment available. They include only equipment which is strictly peculiar to the teaching and learning of *electronics* as apart from standard laboratory equipment (oscilloscopes, oscillators, etc. see Section 14) which could be used for purposes other than educational or training. More detailed information can be obtained from manufacturers specialising in this field, a list of which can be supplied by the Engineering Teaching Equipment Manufacturers Association (ETEMA) Leicester House, 8 Leicester Street, London WC2.

TYPES OF EQUIPMENT

Most of the equipment available may be broadly classified under three main headings:

(1) Complete equipment;
(2) Modular equipment;
(3) Constructional kits.

Complete Equipment

In this category are completely assembled pieces of equipment designed to meet particular educational or training needs. Often such equipment is designed to demonstrate a particular technique or system (e.g., an educational digital computer). In some cases provision is made (on the front panel) for changing the interconnections between sub-systems. However access to individual components is not generally provided for and such equipment is therefore suitable in the main for training at systems level. The range of this type of equipment, because of its specialised nature, is not extensive. Examples are illustrated in *Figures 22.1* and *22.2*.

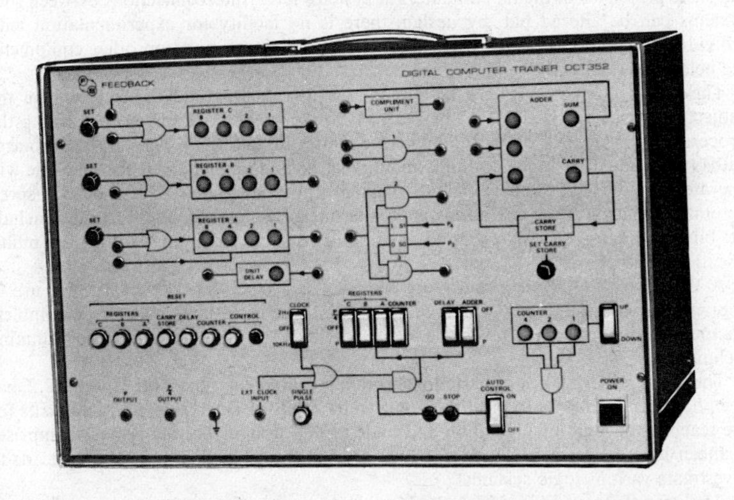

Figure 22.1 *Digital computer trainer.* (*Photograph—Feedback Instruments Limited*)

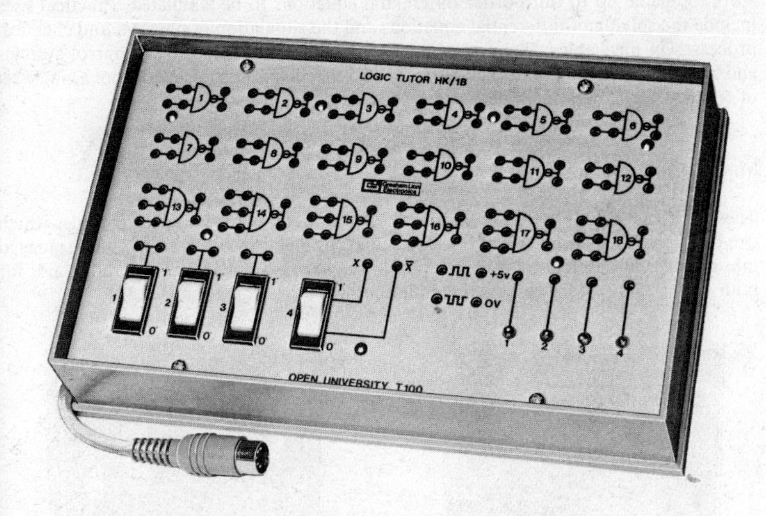

Figure 22.2 *Logic tutor* (*Photograph—A. M. Lock & Co.*)

Digital computer trainers of the type shown in *Figure 22.1* are designed for teaching the basic principles of digital computers at systems level. Interconnections between sub-systems can be altered but, by design, there is no facility for experimentation with individual components. (Such experimentation is best performed on other equipment, see below.)

The digital computer trainer illustrated is a useful visual aid for lecturers on the subject. It forms the basis of a number of laboratory exercises for those learning the processes by which all digital computers operate. In setting up the data and control paths, the student is forced to think about what he is doing and how the machine will operate. Operational modes include manual single step, and low and high-speed automatic control. Demonstrations and experiments which can be performed include the binary number system, setting numbers into registers, addition, subtraction, multiplication, division, etc.

A Core Memory Trainer and Core Memory Interface unit, is available for use in conjunction with the Digital Computer trainer. These three units make up a complete computer system which extends the work possible into core store and programming techniques.

There is a wider choice of basic logic training equipment, One unit, *Figure 22.2*, has been designed primarily to meet the home study needs of Open University students for the teaching of logic manipulation and logic circuit design. The instrument comprises eighteen logic gates, NAND gates, which can be wired together with plug-in leads to investigate various logic systems.

Analogue Tutors, designed for teaching the basic principles of analogue and hybrid computing, are available. A typical unit comprises six solid state operational amplifiers, each with an input potentiometer. Three of the amplifiers can be used as integrators or summers and the remainder can be used as summers, comparators or function generators to enable up to third-order differential equations to be simulated. Practical uses include the solving of differential equations and the simulation of physical and chemical processes in mechanics, thermodynamics, reaction kinetics, feedback control systems and so on. Two or more units can be connected together and controlled from a single set of controls for simulating higher order systems.

Modular Systems

These systems include a variety of discrete components, sub-systems, or modules which can be interconnected by means of cables, links, or plugs and sockets, or can be plugged into a universal base board, rack or consul. They are extremely useful training aids for both circuit and system design, and fault finding.

Figure 22.3 Single valves for experiments and demonstrations (Photograph—Unilab Limited)

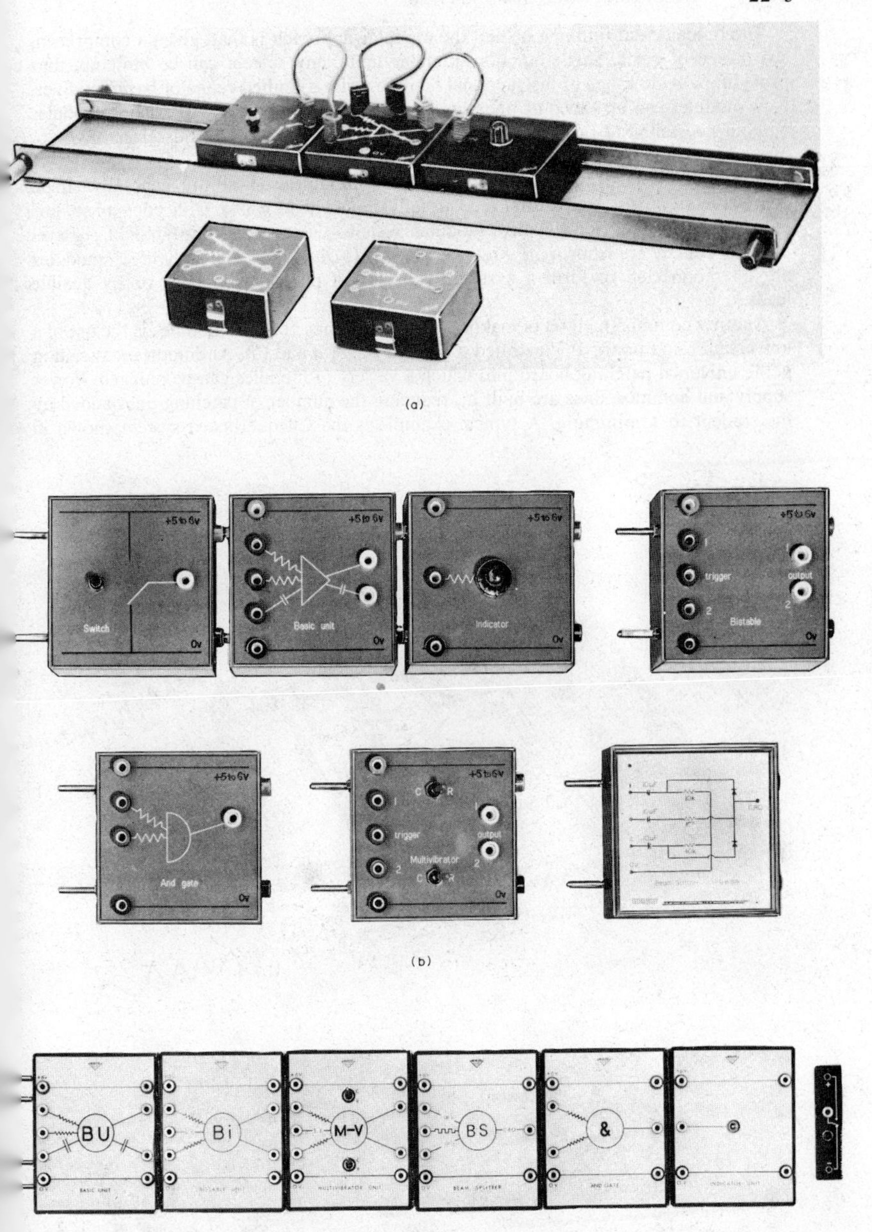

Figure 22.4 Typical examples of simple modular systems designed primarily for the Nuffield A-Level Physics syllabus (Photographs—(a) Philip Harris Ltd. (b) Unilab Ltd. (c) White Electrical Instrument Co. Ltd.)

The fundamental thinking behind the modular approach is that, given a comprehensive range of basic, single-function modular units, any system can be built up, thus providing a wide range of instructional facilities with a limited range of basic modules. New modules can be added or updating modifications can be made to existing modules without necessitating revision of the entire range, which is vast in this class of equipment. The following are typical examples.

Figure 22.3 illustrates two of the simplest modules. These are generally single components (valves, transistors, etc.) usually mounted in a box with convenient terminals brought out. Some typical modular systems designed for the Nuffield A-Level physics course are shown in *Figure 22.4* (*a*), (*b*) and (*c*). The individual modules are interconnected to form a system by means of plugs and sockets or by flexible leads.

Another common method of making interconnections between modules is the use of a universal circuit board as illustrated in *Figure 22.5* (*a*) and (*b*). An important variation is the universal patching board into which a variety of modules can be plugged. Power supply and common lines are built in, reducing the number of patching lines added by the student to a minimum. A typical example is the Chequerboard system shown in

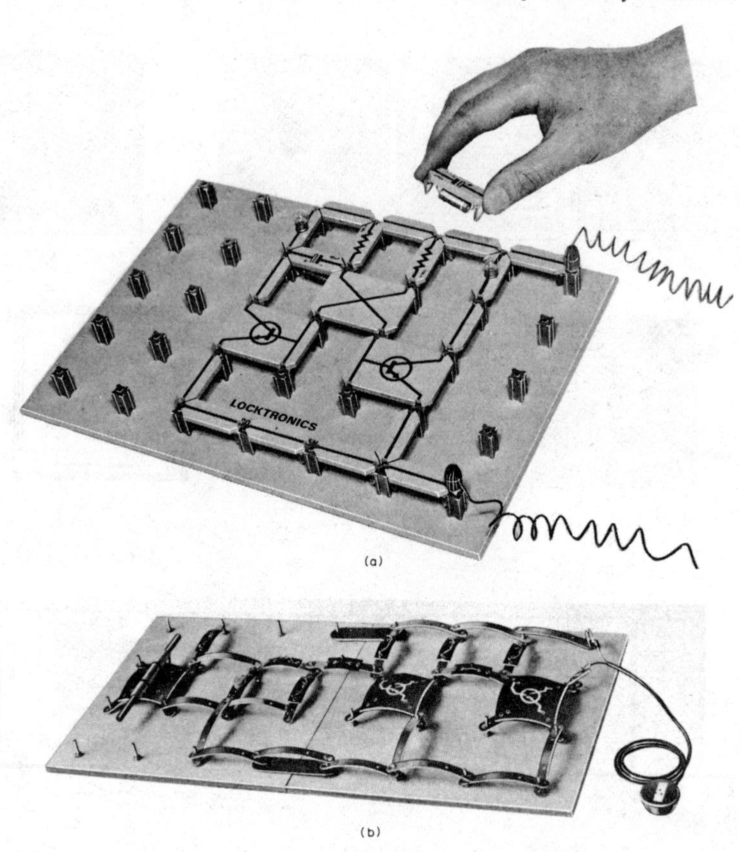

Figure 22.5 Typical universal circuit boards (Photographs—(a) A. M. Lock & Co. Ltd. (b) Philip Harris)

Figure 22.6 Example of a patching board (Photograph—Fairhurst Instruments Ltd.)

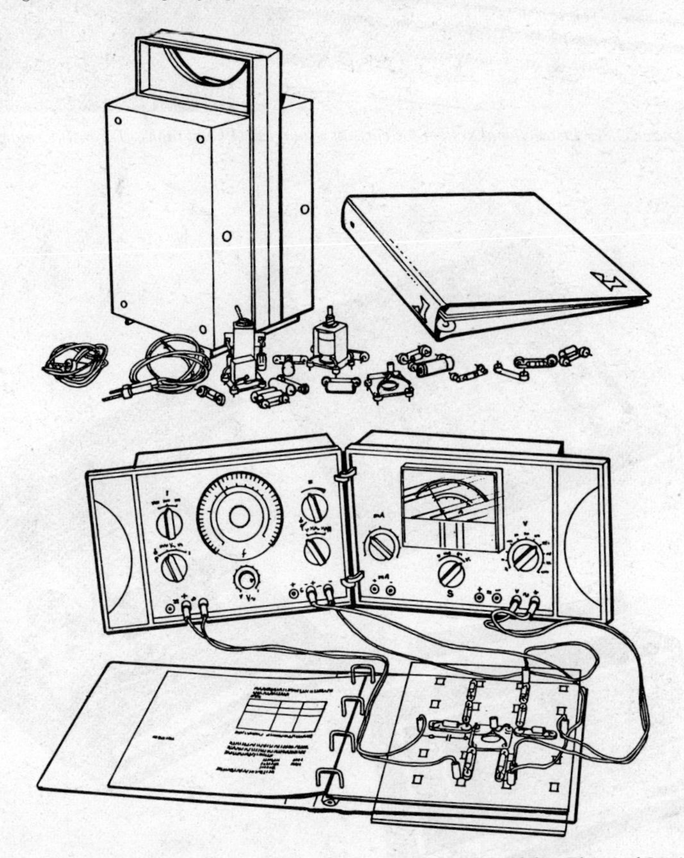

Figure 22.7 A complete course in practical electronics (Courtesy Philips Electrical Limited)

Figure 22.8 *Instructional system for class or group use (Photograph—Taran International)*

Figure 22.9 *Three completed items from a range of constructional kits (Photograph—Heath (Gloucester) Limited)*

Figure 22.6. A choice of integrated circuit modules or discrete components can be plugged in to the patchboard.

Figure 22.7 shows apparatus designed for a complete course in practical electronics. It comprises a set of instructional books which are used in conjunction with a transparent circuit matrix board, and the basic instruments necessary to perform a large number of experiments, together with a set of plug-in components. The transparent matrix is designed to overlay a circuit diagram and enables the student to build up the circuit using the components supplied. Specially designed measuring apparatus and power sources bring the circuits to life, thereby integrating theory and practice. This type of portable self-contained instructional equipment enables the student to proceed through the course with little or no supervision and at his own pace.

In this category, there are also systems designed primarily for class or group instruction purposes, rather than for individual student use. These systems are often similar in concept to the universal circuit boards (i.e. various modules or components can be plugged into a base board, rack or consul) but on a larger scale. Such a system is shown in *Figure 22.8*. The apparatus is the demonstrator phase or 'working wallchart' part of a complete training system which also includes individual student units.

Constructional Kits

These are complete kits of component parts with which to construct a useful piece of equipment such as an oscilloscope, television camera, hi-fi sound reproduction system, etc. The main objective is usually to obtain a working result without involving experimentation. Adjustments requiring technical appraisal, e.g., the alignment of an r.f. stage in a receiver are usually avoided by the use of pre-aligned sub-assemblies. Thus the constructor usually ends up with a very professional-looking piece of working apparatus, and although this gives him a sense of achievement he may learn little about

Figure 22.10 A CCTV Camera (assembled from a kit) based on a Mullard design (Photograph—Crofton Electronics)

electronics from the exercise. However there are exceptions to this with some professionally produced construction kits such as those illustrated in *Figure 22.9*, or for more elementary projects such as Morse code units, metal detectors, audio amplifiers and simple electronic organs.

A simple CCTV Camera is also available in kit form, see *Figure 22.10*. With this project a satisfactory end result is assured by an instruction manual which leads the student through the various steps of construction, testing and alignment, in such a way that an understanding of the techniques employed is essential to progress.

23 PUBLIC ADDRESS AND SOUND REINFORCEMENT SYSTEMS

PUBLIC ADDRESS AND SOUND REINFORCEMENT SYSTEMS 23

23 PUBLIC ADDRESS AND SOUND REINFORCEMENT SYSTEMS

There is a difference between public address and sound reinforcement. To some extent the terms are self-explanatory: sound reinforcement is necessary in a large hall, auditorium or room where the natural voices of the speakers or performers need electronic reinforcement to cover the areas involved in listening, and public address (P.A.) refers to any area, where a large audience, perhaps in the open air, cannot be reached with the unaided human voice. P.A. obviously covers outdoor rallies, sports meetings, railway station announcements, factory sound systems and so on. Technically, the systems overlap and the chief differences are in amplifier output powers and associated loudspeaker designs and numbers.

Good speech intelligibility is often difficult to obtain in enclosures with long reverberation times, such as churches and some concert halls, where the architectural acoustics are usually created to optimise the musical sounds rather than speech requirements. Thus speech reinforcement should be considered where effortless sustained listening is involved.

A design objective must be to make the reinforced speech so natural and unobtrusive that the listeners are not specially aware of its contribution. Ideally, the listeners should only be aware of it when the system is switched off. The criterion should be, although rarely attained, that every listener in the room should be able to hear the person speaking with the same intelligibility and loudness that would occur if the two were talking face to face. This degree of clarity and loudness should be maintained irrespective of the actual separation between the person speaking and the listener.

When talking in a medium or small-sized room, which has good acoustic guidelines, the unamplified voice should be adequate for normal communication, but this is not always the case unless special attention has been given to acoustical design.[1] The reasons are usually:

(1) Even with the best plans, acoustics perfection is an elusive quality, because most spaces reflect a number of compromises between functional requirements, aesthetics, space allocation and budget.
(2) Some speakers have difficulty in voice projection or they tire easily.
(3) Hearing impairment varies with age.
(4) Ambient noise, non-uniform acoustic absorption and poor reflective surfaces within an enclosure can degrade speech intelligibility.

As acoustical engineers will agree, sound reinforcement systems are not a universal panacea for all the defects and oversights in a listening enclosure. Nevertheless, such systems can prove very helpful when structural or furnishing changes in a hall or room are not aesthetically or economically feasible.

Poor P.A. system sound quality, namely, a loud, often booming, indistinct sound, perhaps approaching the verge of howlback, is still found in many situations where modern reinforcement techniques should be used.

With the growth of hi-fi amplifiers and transducers, some sound systems are aimed at amplifying the entire sound spectrum, rather than those selective frequencies or bands of frequencies required to enhance natural speech quality. One of the latest and most effective techniques available to the sound system designer has many names, from 'Graphic Equaliser' to 'Acousta-Voicing' units, incorporated in commercial designs with proprietary names. One model has a 24 section, one-third octave band equaliser, which offers critical tuning of the entire electro-acoustic system for optimum sound quality and prevention of acoustic feedback. Another widely used unit is known as a 'Feedback Controller', which has eight dip filters with depth individually adjustable from zero to approximately 12 dB centred at around 63, 125, 250, 500, 1 000, 2 000, 4 000 and 8 000 Hz. Switches insert 6 dB roll-offs outside the frequencies covered by the dip filters.

ACOUSTIC FEEDBACK

In a sound distribution system where the microphone is not in the same field, or its acoustic path is isolated in some way, the problem of acoustic feedback does not arise. However, if the positive or forward gain of the amplifier exceeds the loss in the path between loudspeaker and microphone, the system will feedback and oscillate or 'howl round', as this effect is sometimes called.

The frequency at which the system oscillates will depend on several parameters, mainly acoustic features of the enclosure, but is usually within the range 250 to 750 Hz. In more detail the factors influencing acoustic feedback include:

(1) Surround acoustics of the enclosure.
(2) Relative phase between microphones and loudspeakers.
(3) Extent of the resonances in the microphone(s) and loudspeaker(s).
(4) Positioning of the microphone(s) and loudspeaker(s).
(5) Directional characteristics of the microphone(s) and loudspeaker(s).
(6) Frequency characteristic of the amplifier and the bass/treble tone control settings.
(7) Overall gain of the amplifier.

It is unlikely that the sound systems engineer can modify the acoustics of the auditorium involved, but to some degree the other six points can be controlled. For example, by employing directional microphones and loudspeakers, less unwanted sound from the loudspeakers will reach the microphones. Microphones with a cardioid response pattern are the most commonly used to cope with acoustic feedback problems, as their pick-up is minimal from the rear. Smoothness of the microphone frequency response is another important factor; allied to a frequency response free from serious resonances in the loudspeaker array this can add to the effectiveness of a sound amplification system.

Speech realism will not be aided by amplifying frequencies below 300 Hz or so. Because of the persistence of low frequencies in some auditoria, it is often advisable to attenuate the lower end of the spectrum, consistent with the overall fidelity required, to maintain intelligibility. In this way, it is often possible to get a 2–3 dB increase in gain of the system before the acoustic feedback point is reached.

Some of the limiting factors in a sound reinforcement system have been described. If a further margin is still required, a frequency shifting device can be used. As already stated, when microphones and loudspeakers are in the same vicinity acoustic feedback (howl-round) occurs when the amplification exceeds a critical value. By shifting the audio spectrum fed to the loudspeakers by a few Hertz the reproduced signal differs from that fed into the input and the tendency to howling at room resonances is reduced. A gain increase of 6–8 dB is possible before the onset of feedback, a worthwhile improvement in difficult conditions.

100 VOLT LINE LOUDSPEAKER DISTRIBUTION SYSTEM

This system of sound distribution[2] is based on constant voltage rather than on constant output impedance.

From the expression watts (W) = voltage $(V)^2$/impedance (Z), the relationship between power output and output impedance for a 100 V line voltage is as follows:

Power Output (*watts*)	*Output Impedance* (*ohms*)
1	10 000
10	1 000
50	200
100	100
500	20

With the use of the 100 V line system and the most widely adopted amplifiers capable of producing 25–100 W of audio power the line impedances provide values most suited to capacity and copper loss. The system assumes that the maximum voltage will be 100 V and all calculations are made on this figure. A 70 V distribution system has also been used.

This method of distribution can be likened to a mains power supply with a constant voltage to which is connected various appliances of different wattages.

Provided that the system is not overloaded the connection of one appliance should not significantly change the mains voltage as the internal impedance of the generating source is much less than any device likely to be connected. Thus overloading will not occur unless too many devices are connected. The same applies to loudspeaker systems, in which the coupling transformers usually have tapped primaries and secondaries, the secondaries being rated by impedance (3, 8, 5, 15 ohms, etc.) and the primary tappings in watts.

The wattage consumed by a loudspeaker from a 100 V line will be the same whatever the power output rating of the amplifier, but with a small amplifier obviously fewer loudspeakers can be supplied. The impedance varies with the amplifier power rating, whereas the line voltage remains constant. The amplifier output impedances at various powers for 100 V and 70 V operation are as follows:

Power (*watts*)	100 V (*ohms*)	70 V (*ohms*)
20	500	245
25	400	196
30	333	163
40	250	122
50	200	98
75	133	65
100	100	49

CODES OF PRACTICE AND RECOMMENDATIONS FOR SOUND SYSTEMS

British Standards Institution publication CP. 327 (1964)[3] deals with the various aspects of a sound installation. The performance parameters related to three main categories of equipment are set out, the principal features being:

(1) *Frequency response* must not vary by more than 2 dB over the following frequency ranges:

Type 1	20 Hz to 15 kHz at 0.25 rated output
	40 Hz to 12 kHz at rated output
Type 2	60 Hz to 12 kHz at 0.25 rated output
	100 Hz to 10 kHz at rated output
Type 3	150 Hz to 7 kHz at 0.25 rated output
	200 Hz to 5 kHz at rated output

(2) *Distortion*, at the rated output power, shall not exceed the following:

Type 1 0.2% at 1 kHz and 0.5% at 40 Hz and 7.5 kHz
Type 2 2% at 1 kHz and 4% at 70 Hz and 5 kHz
Type 3 5% at 1 kHz and 10% at 150 Hz and 2.5 kHz

(3) *Hum and noise* levels with reference to full rated output should not exceed:

Type 1–60 dB
Type 2–50 dB
Type 3–40 dB

BSI publication, CP. 1020 (1973)[4] includes a section dealing with wired distribution systems for sound and television programmes.

It is important that the power handling capacity of the sound system is sufficient so that the input circuit is not overloaded at high signal conditions. A margin of at least 10–15 dB in output power should be allowed when considering the difference between speech reinforcement and the level required for a vocalist or instrumentalist.

The Association of Public Address Engineers[5] publish Recommendations for P.A. Amplifiers, Equipment Tests, Microphone Measurements, Amplifier Measurements and Tests, and Loudspeaker Measurements. This Association has formulated an agreed test procedure for microphones, amplifiers and loudspeakers based on BSI and IEC Standards. For example, the P.A. amplifiers section covers microphone inputs, amplifier output, phasing, output regulation, terminating connectors, and identification.

MICROPHONES

Detailed information on microphones, types and applications are given in reference 6. The main groups can be classified according to their directional characteristics: omnidirectional (all round pickup), bi-directional (figure-of-eight polar response), cardioids (heart-shaped response) sometimes referred to as unidirectional types, and highly directional microphones. The last type are substantially dead to sound from the side or rear. They comprise either a large parabolic dish concentrating sounds on to a microphone element near the focus, or, more commonly today they are the so-called gun microphone which operates by providing many different length sound paths (along tubes) to the microphone transducer. Sound waves arriving from the front have the same path length and so aid each other whereas those from other directions interfere and cancel thus producing less output.

Radio microphones which operate with small low-powered radio transmitters are used for some locations where trailing leads are undesirable. To use these in the U.K. a licence is necessary from the Ministry of Posts and Telecommunications. Among the conditions is the requirement that they shall be used with emissions within the frequency band 174.1 MHz to 175 MHz.

Among the latest audio control devices, the 'Voicegate' is of interest. This is a voice-activated microphone gain control with response-shaped voice frequency sensor. Unwanted background noise is blocked and the unit is adjustable to keep the microphone 'on' up to 30 seconds during conversational pauses.

LOUDSPEAKERS

Apart from the special application loudspeakers ranging from megaphones, siren megaphones (for fog-horns or alarms) to ceiling units, P.A. and sound reinforcement speakers can be classified, in increasing order of directivity, as follows:

(1) Ordinary moving coil cone models, in cabinets or with baffles.
(2) Horn loudspeakers.
(3) Line source or column loudspeakers.

Ordinary dynamic cone speakers are of low efficiency (typically around 1% or less), hence the big amplifiers used to drive loudspeaker arrays for 'pop' groups. With a good horn system the efficiency rises to at least 25%.

The line source or column system comprises an array of loudspeaker units mounted in line in a cabinet 1.5–2 m or more in length. The radiation of sounds, particularly at high and middle frequencies, is directed along the plane at right angles to the column, in a fan-like pattern. If the column is mounted vertically unwanted radiation of sound upwards or downwards on to the heads of the nearby listeners is minimised. This makes possible a saving in power. With this arrangement however no marked directional effect is achieved in the horizontal plane. If this is necessary, the column must be tilted or even set up horizontally.

For optimum directional effectiveness, horn loudspeakers in a cluster overhead with a 50% dispersion pattern overlap can be used. This arrangement will provide a generally uniform sound pressure over the seated listening area but the cluster must be carefully sited to avoid acoustic coupling to open microphones. With very long or wide, low ceiling, rooms or halls it is helpful to use an array of ceiling speakers, sometimes called a low level or distributed speaker system. The lower the ceiling height, the greater the number of loudspeaker units required to provide uniform sound coverage.

CONTROLLED TIME DELAY SYSTEMS

The Haas (or precedence) effect[7] is a sound delay phenomenon primarily concerned with room acoustics but it has had considerable influence on the design of speech reinforcement systems. In the past magnetic tape recording systems (using loops or discs) and acoustic-pipe devices have been employed for time-delay correction. But because of the inherent limitation of these methods the time-lag in sound systems is controlled by digital delay lines in the most advanced designs. These new digital techniques have been made practical by the use of l.s.i. (large scale integration) methods which allow a multiplicity of complex circuitry to be housed in a small space.

Design trends

The development of solid-state amplifiers with compact modular construction and operation from low voltage d.c. has made possible simple remote operation of sound systems. Solid-state power amplifiers have been designed enabling very high sound levels to be attained. It is reported that one 'pop' group has created a sound system which develops 26.4 kW of audio power producing an acceptable sound at a distance of a quarter of a mile, and a figure of 36 kW audio power has been given for a gathering at Wembley.

The continuing development in the electronics field has led to the design of sophisticated P.A. and Sound Reinforcement equipment and systems capable of meeting the requirements of any situation from a local gymkhana to a large concert hall or a 'pop' festival covering many acres.

REFERENCES

1. IRVING W. WOOD, 'Guidelines for good sound and AV system automation', *BKSTS Journal*, July (1974)
2. F. POPERWELL, 'The 100 Volt line Loudspeaker Distribution System', *APAE Technical Bulletin*, **No. 1.** (1963)
3. BSI. CP.327: Telecommunication facilities in Buildings: Part 3: 1964, Sound Distribution Systems
4. BSI. CP.1020. The reception of sound and television broadcasting: 1973

5. Association of Public Address Engineers, 47 Windsor Road, Slough, Berks, SL1 2EE
6. NISBETT, ALEC, *The Use of Microphones*, Focal Press (1974)
7. H. HAAS, English translation of original German article appeared in Journal of the Acoustical Society of America, **Vol. XX**, pp. 145–159, March (1972)

FURTHER READING

ALKIN, E. G. M., *Sound With Vision*, Butterworths (1973)
BORWICK, J., *Sound: Facts & Figures*, Focal Press (1963)
BURROUGHS, L., *Microphones: Design & Application*, Sagamore Publishing Co., Inc., New York (1974)
DAVIS, D., *Acoustical Tests and Measurements*, Foulsham (1965)
Journal of the Audio Engineering Society, 60 East 42nd Street, New York 10017
WARREN, H., *Technical Matters* feature, 'Public Address', *J. of APAE*, October (1974), March (1975)

24 ELECTRONICS IN INDUSTRY AND BUSINESS

ELECTRONICS IN INDUSTRY AND BUSINESS

24

24 ELECTRONICS IN INDUSTRY AND BUSINESS

AUTOMATION, CONTROL AND MEASUREMENT

History of automation

Automation has had an enormous economic, political and social effect during the last 20 years. Indeed, it is difficult to see how so much that we take for granted today would otherwise be possible.

Although automation is commonly regarded as a contemporary development the use of mechanical techniques to reduce or eliminate the labour required in labour-intensive processes dates back over a long period. Weaving and printing, both labour-intensive processes, had reached advanced stages of automation using mechanical methods long before the Industrial Revolution. Windmills have been used to convert energy from the wind to mechanical power to grind corn for hundreds of years with a form of closed loop feedback to steer the sails into the wind to obtain the maximum power output.

The advent of the Industrial Revolution resulted in the development of complex mechanical linkages and mechanical control systems typified by the centrifugal governor to regulate the speed of steam engines, however, the relatively bulky and cumbersome nature of mechanical control systems limited their application. Electrical machinery developed in the latter part of the last century further extended applications of automation in industry. The ability to control the speed and power of a d.c. electric motor by a relatively small control current in the field winding enabled automatic feedback systems of various kinds to be developed for application in rolling and paper mills, conveyor drives and traction.

The application of the thermionic valve in the 1920s and the effects of World War II resulted in further developments although the high cost, relatively fragile nature and the need for skilled servicing of electronic equipment limited applications to large-scale processes where the advantages were considerable. A few early applications of computers using thermionic valves were developed in the 1950s in such areas as the steel, chemical and electrical supply industries.

More recently the impact of the semiconductor followed by the integrated circuit has had an enormous effect on the growth of automation. Previously where automation may hitherto have been possible at a high cost and relatively low reliability, for the first time low-cost control systems capable of trouble-free operation in the industrial environment are now available. The full extent of this revolution has yet to be seen. More recently still the advent of low-cost computers capable of adaptive control of complete processes is likely to have a marked impact on industry in the future.

The potential availability of low-cost automation based on single chip microprocessors applicable to processes for which automation has hitherto been regarded as of too high a capital cost will be considered here as well as the large scale processes which have become synonymous with automation. The principles are common and it is only the degree of complexity that varies.

It is perhaps too early to assess the eventual impact of large scale integration of electronic computers on process control. The real possibility of low-cost mini computers, in which the sensor, peripheral equipment and software may be more costly will result in an enormous expansion in the application of computers in process control in the next decade.

Effects of automation

The advantages of automation are considerable. Some of these are illustrated schematically in *Figure 24.1*. The most obvious advantage is the reduction of the labour required. This has led to fears of redundancy which have often been unfounded and the effect of automation has often been quite the reverse of the initial fears for security of employment. The benefits of automation have resulted in bigger markets, new and better jobs together with better working conditions. The labour released from work which may have been tedious and repetitive has been used for more congenial and fulfilling work.

In many cases the introduction of automation has been accelerated by the non-availability of suitable labour or by social or environmental considerations. Examples are in jobs which may be boring and repetitive but require the skill of a craftsman, the drift of suitable labour into other more highly-paid industries and safety requirements in

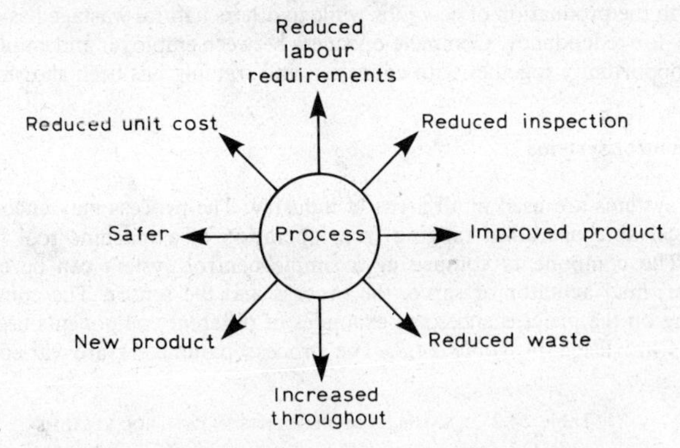

Figure 24.1 Some advantages that can be obtained from automation

hazardous processes. In other cases it has made new processes possible. The operation today of a chemical plant, power station, steel works or car factory is inconceivable without an advanced degree of automation. Automation has enabled technical advances in processes possible. The elimination of human error is another important advantage enabling greater precision, repeatability and safety to be achieved. Subjective assessment and tedious and dangerous jobs can be eliminated.

Automation may be justified on social, environmental or economic considerations. Introduction of industry in areas in which it has not previously existed and where no suitable labour exists may necessitate introduction of automated processes whereas in other parts of the country an adequate workforce may be available. Dangers from hazards in new processes may be unacceptable so that the use of automated systems is dictated as a prerequisite.

Automation may often be used to assist in labour-intensive processes to obtain improved product quality and increase throughput. These considerations can be readily assessed on economic grounds if the true process costs are available. In many cases an approximate analysis can often give an indication if an economic advantage is possible. Some of the factors that should be considered are listed in *Table 24.1*. Some of the advantages of automation may be obvious and can be readily quantified. Other factors such as an improved product enabling a premium price to be obtained, reduced waste—due to closer production specification, ability to quantify factors which was not previously possible enabling savings to be made, e.g., stock control and feedback of information during processing enabling waste to be reclaimed, are some of the benefits

Table 24.1 FACTORS IN THE SELECTION OF AUTOMATED SYSTEMS

Capital cost	Reduced waste
Depreciation	Improved quality and precision
Overheads	Increased throughput
Running costs	Elimination or reduction in number of processes
Labour costs	Reduction in inspection

that may also result. In some cases it has been subsequently shown that these have even outweighed the initial advantages that were thought to be important.

The necessity for cooperation in the introduction of automated systems between management and the labour force cannot be over emphasised. Successful examples of cooperation have shown that fears of unemployment have often been unfounded. In some cases the increased output in part of a process that has been automated has resulted in the production of new jobs while in others natural wastage has eliminated the necessity for redundancy. Complete openness between employer and employee from the earliest opportunity together with education and training has been shown to work.

Linear control systems

Control systems are used in all areas of industry. The process may encompass almost any conceivable operation ranging from operation of a machine tool to filling milk bottles. The components comprising a simple control system can be categorised as controller, final actuator or servo, the process and the sensor. The components vary depending on the process and some examples of different components used in different processes are listed in *Table 24.2*. The process parameters are varied by the final

Table 24.2 EXAMPLES OF ELEMENTS IN CONTROL SYSTEMS

Process	*Control variable*	*Controller*	*Servo*	*Sensor*
Milling	position	on-off	stepper motor	encoder
Oil flow to burner	flow	proportional	d.c. shunt motor	turbine flowmeter
Rolling mill	velocity	proportional and derivative	Ward-Leonard drive	tachometer
Batch furnace for heat treatment	temperature	proportional, derivative and integral control	triac	thermocouple

actuator. The final actuator may be controlled in turn by a secondary actuator, e.g., the speed of a hydraulic drive may be increased with a motorised valve. The control setting may be varied with the controller which may be mechanical, electromechanical, hydraulic or electronic. The setting (set point) of the controller is fixed by the operative in charge of the process.

Such a system is known as an open-loop system and is illustrated schematically in *Figure 24.2* (*a*). The sensor indicates the state of the process. Open-loop systems are used in all areas of industry where fluctuations in the process variables occur within acceptable limits or occur slowly and can be corrected by adjusting the set point by the operator. If now a feedback signal is obtained from the sensor this can be used to maintain the process within set limits. The feedback path is shown in *Figure 24.2* (*b*). This now becomes a closed-loop control system and is known as a servomechanism. The action of the controller is now to compare a reference signal corresponding to the

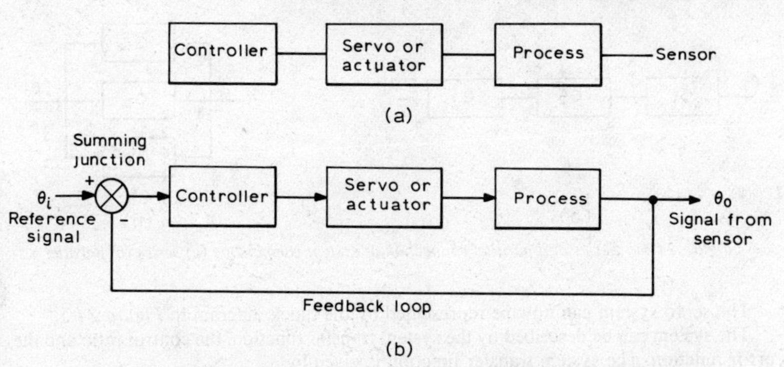

Figure 24.2 Basic components of a control system (a) open loop (b) closed loop

set point with the feedback signal, the difference being the error signal E. This is normally amplified and used to restore the process to the set point. A simple example of a closed-loop control system is the self balancing potentiometer illustrated in *Figure 24.3*. Depending on the polarity of the out-of-balance voltage the servomotor varies the tapping position on the slide wire. The error signal tends to zero as the balance position is reached so that in practice a small error exists corresponding to a region either side of the balance position over which no control action occurs known as the dead space. This is reduced by the amplifier. If the gain A of the amplifier is high and the feedback is large so that the amplification is small the sensitivity of the system is increased by a factor $1/A$.

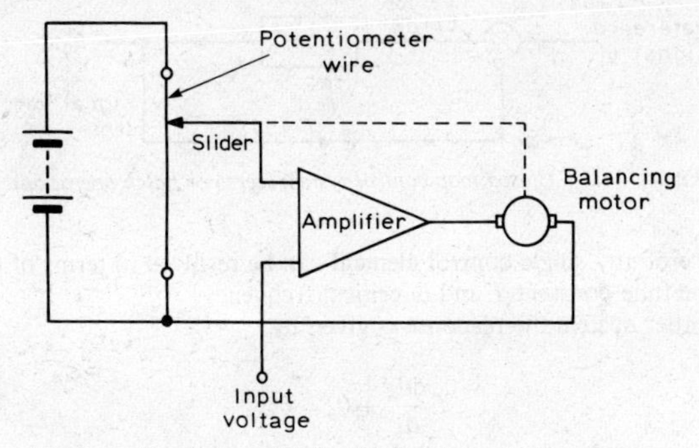

Figure 24.3 Servo-assisted self balancing potentiometer

Each element of a control system can be typified by its response in terms of the change in output over the input to which will be a time-dependent function and can be expressed in terms of first or higher order differential equations.

By taking the Laplace transform of the response the transfer function of a system can be readily determined by combining the transfer functions of the individual elements of the system in *Figure 24.4*. The transfer function of elements connected in series is equal to their product

$$G = G_1 \cdot G_2 \cdot G_3$$

and in parallel

$$G = G_1 + G_2 + G_3$$

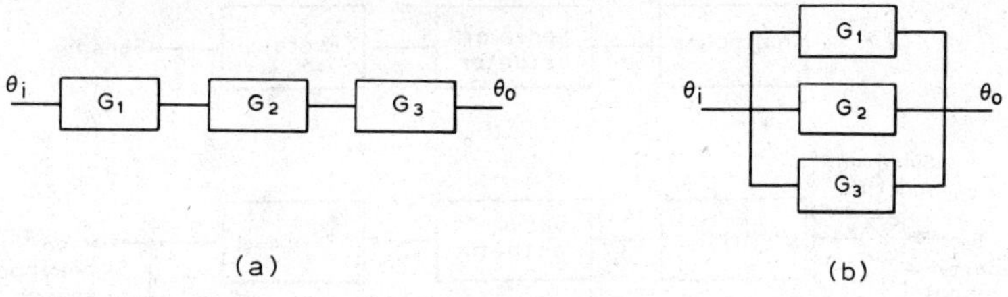

Figure 24.4 *Series and parallel connection of system components (a) series (b) parallel*

The servo system can now be represented by the block diagram in *Figure 24.5*.

The system can be described by the system transfer function, the control ratio and the error function. The system transfer function is given by

$$G = \frac{\theta_o}{E}$$

The control ratio is given by

$$\frac{\theta_o}{\theta_i} = \frac{G}{1 + G}$$

and the error ratio

$$\frac{E}{\theta_i} = \frac{1}{1 + G}$$

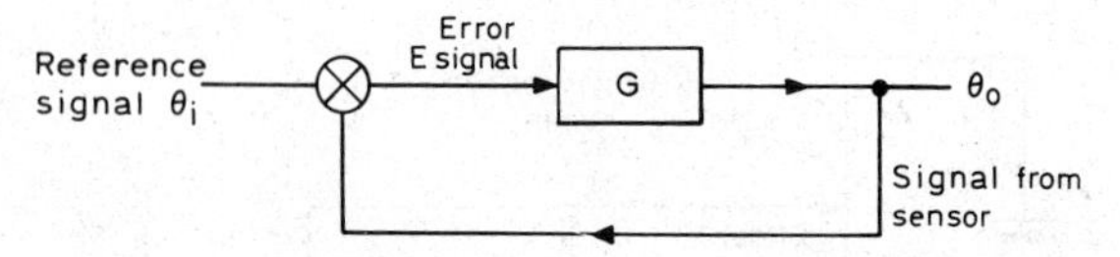

Figure 24.5 *Closed loop control system (servo or batch operation)*

The response of any single control element can be resolved in terms of the damping coefficient ξ the time constant T and ω critical frequency.

For a first-order system the response is given by

$$T\frac{d\theta_o}{dt} + \theta_o = \theta_i$$

and

$$a_2\frac{d^2\theta_o}{dt^2} + a_1\, d\theta_o + a_o\theta_o = a_o\theta_i$$

where

$$T = \frac{2a_2}{a_1} \qquad \xi = \frac{a_1}{2\sqrt{a_o a_2}}$$

and

$$\omega_n = a_o/a_2$$

from which it can be seen that

$$\frac{1}{T} = \xi\omega_n$$

Table 24.3 EXAMPLES OF CONTROL SYSTEMS AND THEIR TRANSFER FUNCTIONS

System or process	*Transfer function*
First-order	$\frac{1}{Ts + 1}$
Second-order	$\frac{1}{T^2\zeta^2 s^2 + 2T\zeta^2 s + 1}$
Integral controller	$\frac{1}{Ts}$
Proportional controller	K_c
Rate controller	Ts
Linear final control element with a first-order lag	$\frac{K_v}{Ts + 1}$

By combining the Laplace transforms of each element it is possible to determine the response of the complete system. The Laplace transforms corresponding to various common control system elements are tabulated in *Table 24.3*. The response of the system can be solved analytically for first- and second-order systems. Since the order of the system is equal to the sum of the order of each component in the system, systems with third and higher order responses are often encountered which need to be solved by iterative techniques.

Another example of a single closed-loop control system is the variable load or regulator system shown in *Figure 24.6*. This corresponds to a process with varying but controlled parameters, e.g., throughput. In this case assuming the control level θ_i is fixed the transfer function becomes

$$\frac{E}{L} = \frac{G_2}{1 + G}$$

$$\frac{\theta_o}{L} = \frac{G_2}{1 + G}$$

In any closed-loop control system a high degree of precision in terms of the error ratio is required. One factor that limits this is the system stability which decreases as the amplitude ratio is increased. If a system becomes unstable it goes out of control which may have disastrous effects.

For any control system there is a small region either side of the set point over which no control action occurs. This may be due to mechanical back-lash, thermal inertia, etc.: if the dead space is made very small instability may occur. This is known as hunting and occurs where the control element is in effect oscillating about the set point which can give rise to instability and is an important consideration in the design of precision control systems.

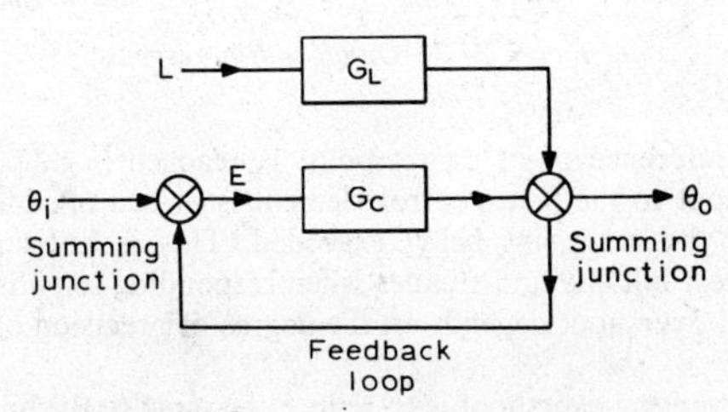

Figure 24.6 *Regulator or variable load operation*

The stability of a system can be determined if the damping coefficient time constant and critical frequency are known. By varying the relative magnitudes of these parameters the stability can be changed. The cause of instability can be considered in terms of the open-loop frequency response of a system. If delays which exist in the system result in phase lags which are in anti-phase with the input, the output signal is fed back to the input so as to increase the error rather than decreasing it in the same way as positive feedback causes instability in an amplifier.

The magnitude of the delays in a control system is often an important factor governing the stability of the system as well as the degree of precision of control that may be obtained. Delays occur with each element of the system, e.g., measurement, controller, actuator and process. Delays also occur in transmission between each element although this will often be small. Depending on the system involved the relative magnitudes of the delays may be important or not. In complex systems the response of individual elements may not be known or the overall response of the entire system may be too complex to analyse. In these cases the response may be measured by determining the response of the system or a model of it to standard input signals. These are usually the response to a continuous sinusoidal input, a ramp function and a square wave input.

The response to a sinusoidal input as a function of frequency enables the amplitude frequency response to be obtained which when plotted on a log scale of frequency is referred to as a Bode diagram. The plot of the phase shift as a function of frequency in polar coordinates is referred to as the Nyquist diagram from which the system stability can be obtained.

CONTROL MODES

The response of an on-off control system used for a closed-loop temperature control system is shown in *Figure 24.7*. Other control systems can be considered in similar

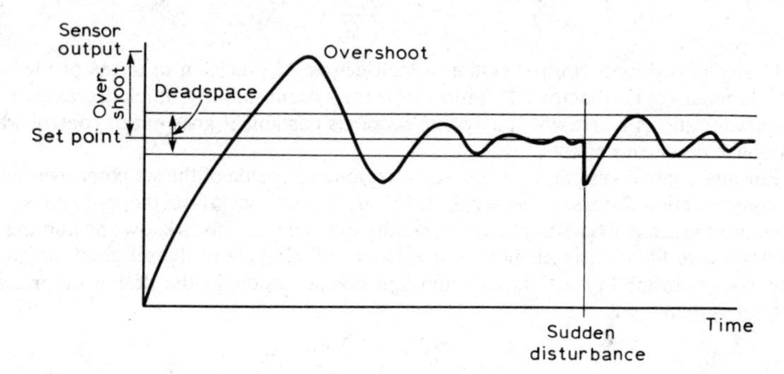

Figure 24.7 On-off control response

ways. When the reference point or set point is reached (e.g., position, temperature, speed, etc.) the input to the final control element is turned off. Mechanical or thermal inertia results in the set point being exceeded. This is known as overshoot. The overshoot is greatest initially and smaller when responding to subsequent changes. The acceptable level of overshoot depends on the degree of precision of control required by the process.

One way of preventing overshoot is to reduce the input to the final control element as the set point is approached so that it is zero when the set point is reached. This is known as proportional control, in which the input to the control element is reduced according to the relation

$$\theta_c \,\alpha\, \frac{\theta_s - \theta}{\theta_s - \theta_o} W$$

where θ_c is the control signal, θ_s is the set point, θ the intermediate position, θ_o is the initial condition and W is the maximum control action.

This suffers from the limitation that as the set point is approached the control signal is zero and it would take an infinitely long time to reach this point. The effect of any small changes about the set point would also take a long time to correct. This can be overcome by applying proportional control only over a limited range $\theta_s - \theta_p$ (the proportional band).

Implicit within the concept of proportional control is a preset value corresponding to the initial value of the input signal. If now the power changes for some reason such as increase in friction, load, increase in heat losses or other factors outside the control system the set point may never be reached since although an error signal exists no further input to the control element is available. The difference in error is known as droop. The effect of this can be reduced by arranging the proportional band so that the set point is always reached before the input to the control element is zero. This results in some overshoot but will be less than for on-off control.

The effect of fluctuations in variables outside the control system can be reduced by integral control. The function of integral control is to integrate the deviation of the control point from the set point over a long period and correct the input W. The effect of integral control compensating for changes in ambient temperature in a temperature control system is illustrated in *Figure 24.8* (*a*).

In any temperature control system since control only occurs during the proportional band and due to time lags and effects of integral control the degree of overshoot is

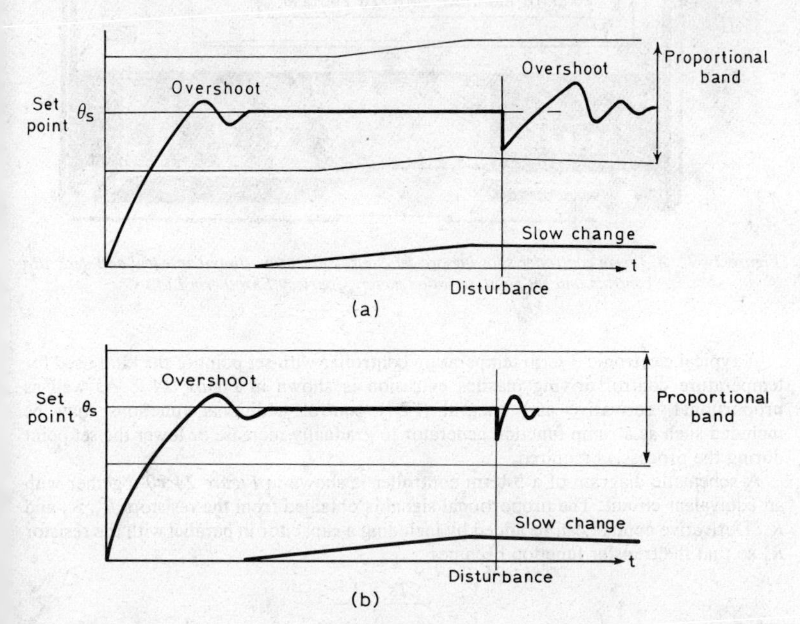

Figure 24.8 PDI control (*a*) *proportional derivative and integral* (*b*) *proportional and derivative*

considerable. The effect of integral action may be such as to move the proportional band so far up scale that the set point may be reached before the proportional band. To offset the effect of integral control during start up the proportional band may be automatically shifted to a lower temperature.

The effect of a sudden fluctuation about the set point can only be slowly corrected by proportional control. This can be overcome by differential control. This responds to fluctuations over only a short period and results in a restoring signal much greater than that due to proportional control alone. The effect of derivative control in response to a sudden change is shown in *Figure 24.8* (*b*).

Proportional control is often used alone, however where precise control is required a combination of proportional and integral or derivative or both is used. Where all three are used the controller is referred to as a 3 term controller.

Analog control techniques

THE ELECTRONIC CONTROLLER

The electronic analog controller is based on proportional control to which derivative or integral action may be added. Other functions may also be incorporated to modify the response.

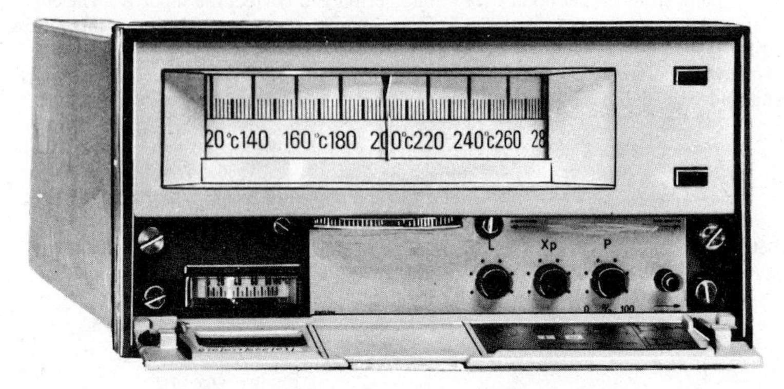

Figure 24.9 A 3 term controller showing pre-set controls for proportional, integral and derivative controls and set point deviation meter (*Courtesy Eurotherm Ltd.*)

A typical electronic 3 term temperature controller with set point of the kind used for temperature control driving plastics extrusion is shown in *Figure 24.9*. As well as proportional, derivative and integral (PDI) control additional functions may be included such as a ramp function generator to gradually increase or lower the set point during the process as required.

A schematic diagram of a 3 term controller is shown in *Figure 24.10* together with an equivalent circuit. The proportional signal is obtained from the resistors R_1, R_2 and R_3. Derivative control can be added by including a capacitor in parallel with the resistor R_1 so that the transfer function becomes

$$\frac{V_o}{V_c} = \frac{Ts + 1}{Ts + \dfrac{R_1}{R_2} + 1}$$

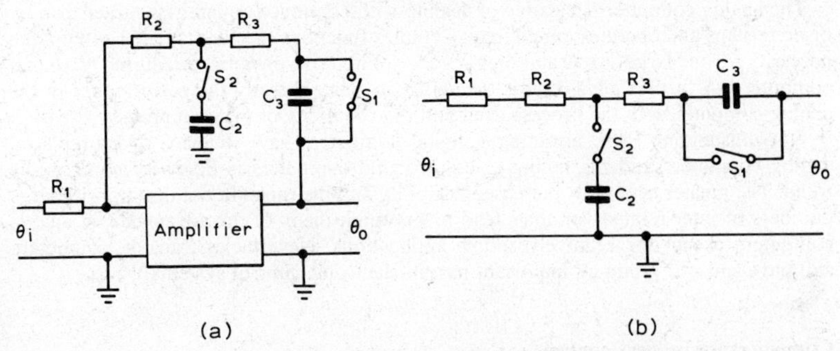

Figure 24.10 (a) *Schematic diagram of 3 term controller and* (b) *simplified equivalent circuit*

Integral action is obtained by connecting a capacitor across R_2 so that

$$\frac{V_o}{V_i} = \frac{1}{Ts + \frac{R_1}{R_2} + 1}$$

which if $\frac{R_1}{R_2} < Ts$ tends to

$$\frac{V_o}{V_i} = \frac{1}{Ts}$$

By using an amplifier the load on the reference and sensor can be reduced and the sensitivity increased. If the gain of the amplifier is large the input to the amplifier A can be considered as a virtual earth. The transfer function of the controller is

$$\frac{\theta_o}{\theta_i} = -\left| \frac{R_2 + R_3}{R_o} + \frac{R_2 C_2}{R_1 C_3} + \frac{1}{R_1 C_{3p}} + \frac{R_3}{R_1} R_2 C_{2p} \right|$$

If $R_2 + R_3 = R_1$, $R_3 = R_2$, and $C_2 \gg C_3$ then

$$\frac{\theta_o}{\theta_i} \simeq -\left(1 + \frac{1}{T_i} + T_d\right)$$

where T_i and T_d are the integral and derivative time constants

$$T_D \simeq \frac{R_2 C_2}{2} \qquad T_i \simeq R_3 C_3$$

i.e., the integral and derivative action can be adjusted by varying C_3 and C_2 independently.

ANALOG COMPUTERS IN CONTROL SYSTEMS

The analog computer is based on the use of operational amplifiers to carry out arithmetical operations. These include summation, integration, differentiation and scaling. Multiplication and division and non-linear functions can also be obtained.

Electronic analog computers may be used to simulate complex systems. Changes in the parameters may be made by varying resistance and capacitance. These may be at preset values or varied according to a preset programme. Analog computers are usually designed for one specific application although some degree of versatility may be achieved by varying the interconnection of the computer modules.

The analog computer is capable of dealing with continuous quantities rather than by discrete steps and operates in real time. Complex functions can be simulated without the necessity for the large stores and high number of iterative operations required by digital computers. A further advantage is the ability to relate directly the performance of the analog computer with the process itself and the simplicity of programming.

Notwithstanding these advantages digital computers have in many areas replaced analog computers and the falling price of digital computers is likely to increase this trend. The greater accuracy, increased flexibility and the capability of dealing with large numbers of inter-related variables tend to outweigh many of the advantages of analog computers except for relatively simple applications. Nevertheless, analog computers still have and will retain an important part in electronic control systems.

Computers for process control

Computers used for process control have stemmed from the large multi-purpose computers developed for data processing for which high operating speeds, large stores, considerable versatility and the capability of carrying out a wide range of tasks is required. The most obvious application of computers is for multi-variable process control. The relatively low cost of small computers also makes them competitive with applications where several analogue controllers are required.

A computer for process control can be considered in terms of:

(1) the store;
(2) the processor;
(3) peripheral equipment.

The specification for any given process will depend on the requirements of the process.

The outputs from the individual sensors may be sampled in sequence and fed into an analog to digital converter. The output of the converter is then fed into the process computer which has been appropriately programmed. The output of the computer is converted to a suitable analog signal for the actuator.

Most simple control processes are analog systems, however where a number of control variables exist it can be more economical to use a single digital computer rather than a number of separate analog controllers. The computer is inherently more versatile than analog controllers and can be programmed in an infinite variety of ways. Where several interconnected control loops corresponding to interrelated process variables exist the computer can also carry out control functions which are not possible with single-loop analog controllers enabling the optimum combination of process variables to be obtained.

The development of the minicomputer has greatly extended the application of computers for process control where previously the high cost of a full-scale computer could not have been justified. The minicomputer used for process control is distinguished from the general purpose digital computer by its low cost and high performance/cost ratio and efficiency when used to perform specialised tasks. Because of its narrow range of task it is also possible to respond rapidly to changes in variables. Although the overall performance of a minicomputer may be inferior to that of a large machine its performance of any one function may be comparable or even better.

The advances that have made the minicomputer possible are the introduction of medium- and large-scale integration and in particular the use of semiconductor memories. This has already resulted in significant reductions in computer prices and considerable further reductions are likely. The situation has already arisen in some cases where the use of a minicomputer can be justified on cost grounds as an alternative to analog control systems.

A minicomputer system is built around a central processor unit in which logical and arithmetic functions are performed controlled by the program, and a main memory. A

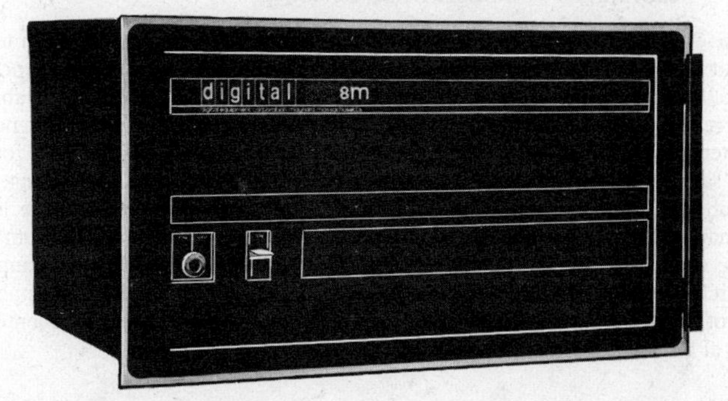

Figure 24.11 *A minicomputer for process control with 4 K store* (*Courtesy Digital Equipment Ltd.*)

typical low-cost processor with a 4 K store size is shown in *Figure 24.11*. In addition to this various other components will be required which are referred to as peripherals.

The number and choice of peripherals will be dictated by the process. These may include

(1) Input–output devices which enable the operator to control the computer and insert computer programs as well as some kind of readout. Examples are tape punches, an optical tape or card reader and a typewriter to print the output.
(2) The central processor will normally have only a small store of between 4 K and 32 K words. Bulk storage may be required to supplement the main memory such as magnetic tapes or discs. Large quantities of information can be stored in this way at low cost but has to be transferred to the main memory before it can be used by the central processing unit so that the response time is slow.
(3) Analog to digital and digital to analog converters to enable the computer to interpret analog signals received from the process and convert the control signals back to the analog form in which they are required.

As well as these, communication equipment to enable the computer to interface with other computers either within the plant or externally via GPO lines may be required.

While for a large complex process a large amount of peripheral equipment may be required for some processes a central processor, analog to digital converter tape punch and reader may be sufficient.

The principal application of minicomputers in industry excluding small scale data processing include process control, automatic testing and inspection, numerical control of machine tools, and materials handling.

Adaptive control

While it is possible to control a single variable or a process with a number of variables separately if the optimum operating conditions are known this is not always the case. In some cases external uncontrolled parameters of this process may vary outside the limits of control of the process so that optimum conditions cannot be obtained or in the worst case loss of control may occur. The effect of the process variables on the output may be interrelated and the required operating conditions of each variable may not be accurately known and may vary. A simple example of this is the operation of a plastics extruder where for example the output depends on the speed, torque, throughput and temperature during the extrusion process. In a chemical process the output may be dependent on large numbers of indirectly related process variables.

Adaptive control is used to determine the optimum operating conditions in terms of the output parameters of the process. Such a multi-variable process may be represented by a two, or more, dimensional surface corresponding to the different variables, the peak corresponding to the optimum condition. The process conditions are measured and related to a performance index and the effect of incremental changes found by repeated sampling of the process. This is sometimes referred to as hill climbing. Where more than one peak exists or the curve is discontinuous the shape of the hill and approximate location of the optimum value may be found first by a random search technique over a wide range of process variables. The subject of adaptive control is a topic which is a source of extensive research, the principal objectives being establishing a model of the system, relating actual and required performance and implementation of the control command without causing instability.

Non-linear control systems

So far only linear control systems for which the superposition principle applies have been considered. Non-linear systems may arise in the system due to static friction, hysteresis and other undesirable effects although their magnitude may be such that they can be ignored. The system components may have a non-linear response, e.g., the output voltage from a generator, however the application of feedback will tend to reduce this.

Non-linear components may be deliberately used in systems. A common type of non-linear system encountered is the on-off controller which has the advantage of low cost and the ability to obtain maximum effect when reacting to a disturbance. Alternatively non-linear characteristics may be introduced to modify for example the amplitude response of a system to stay within certain limits.

Although the transfer function completely describes the linear system the behaviour of a non-linear system can only be completely analysed by integrating the differential equation describing it. As a result it is difficult to generalise.

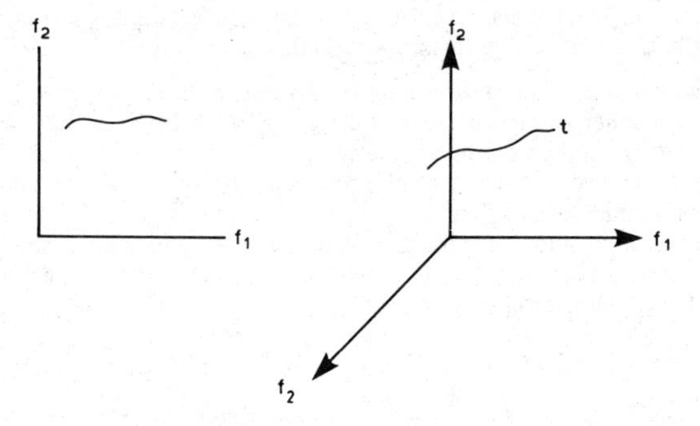

Figure 24.12 Phase plane representation with 2 and 3 states

The two principal techniques used to analyse non-linear systems are the phase plane and the describing function. The system can be described in terms of its state, which is the minimum numbers of variables or numbers required to define past and future states. In a two- or three-state system this can be described by a two- or three-dimensional geometrical model. This is illustrated in *Figure 24.12*. The time variance of any two parameters can be represented in one place such as the on-off controller.

The second technique is the describing function which is defined as the complex ratio between the output and the input amplitudes of the same frequency components. From these the stability and other factors of non-linear systems can be analysed (see under Further Reading).

The control of electric power

Variation of electrical power is a common method of controlling such processes as electrical heating and motor drives of all kinds. Methods used range from transistors at milliwatt levels to air blast circuit breakers at megawatt levels.

ON-OFF CONTROL

On-off or discontinuous control is often used in such areas as electric heating and chemical processes, where the inertia of the process is high. A thermostat used for temperature control is an example. A dead space corresponding to the cut-in (on) and cut-out (off) temperatures exists. Advantages of on-off control are its relatively low cost and simplicity, and the application of the full control action when needed which reduces errors due to friction, and effects of mechanical or thermal inertia. The large overshoot and dead space which are characteristic of on-off control can to some extent be reduced by incorporating anticipatory action. In the case of a thermostat this may be a small heater built into the thermostat housing which heats up while the thermostat is off. This results in the thermostat switching at a lower temperature than the set point which if correctly set reduces overshoot and the dead space band.

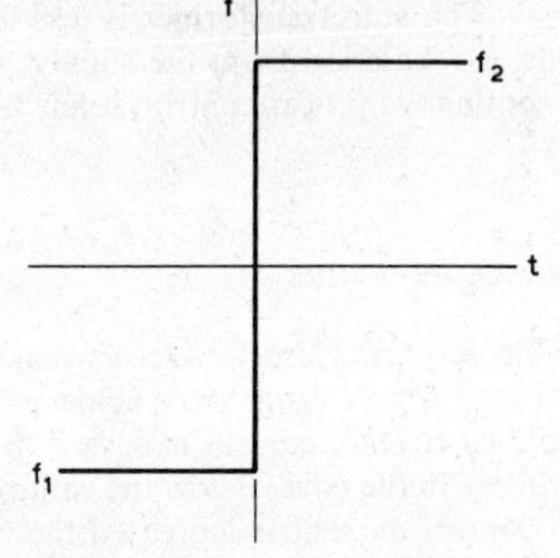

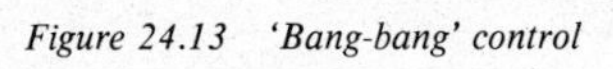
Figure 24.13 'Bang-bang' control

Discontinuous control may also be used in the so-called 'band-bang' mode. This is illustrated by the temperature control process shown in *Figure 24.13* in which below the set-point heating occurs while above the set-point cooling takes place.

On-off control is easily incorporated in automatic systems and for high power loads is often the cheapest method of control. The electromagnetically operated relay or contactor can be operated by a control signal so as to open and close as required. Switching times of a few milliseconds are easily achieved.

More recently thyristors have been used for on-off switching. Two thyristors connected back to back so as to form an a.c. switch controlled by the signal at their gates are sometimes referred to as a Triac (see Section 8). These are capable of rapid and repetitive switching that would result in an unacceptable contact life of contactors. Proportional control can also be applied by varying the mark–space ratio. This control mode is often referred to as burst firing. Although it overcomes many of the disadvantages associated with on-off control it is not suitable for processes with time constants of the order of only a few cycles such as tungsten filament heaters, lamps and precision mechanical servomechanism.

On-off control, since it is discontinuous by nature, is difficult to analyse mathematically and is the subject of extensive research.

TAPPED TRANSFORMER

The output voltage of the transformer is varied by changing the voltage tapping on the primary or secondary. This can be carried out automatically. Special precautions are taken with on-load tap changers to prevent sparking and arcing at the contacts. The voltage output of single and multiphase transformers may be varied in this way. The high cost of large numbers of tappings and multicontact tap changers usually limits the use of tap changers to relatively small ranges or coarse adjustments.

As well as tapping the primary or secondary the connections of 3-phase transformers can be varied from star to delta connection. A change of star to delta results in a change of $\sqrt{3}:1$ so that if both primary and secondary are switched a total variation of $\sqrt{3}:1/\sqrt{3}$ is possible in two steps. This is often used for starting large motors, to reduce the starting current of large motors and for induction furnaces.

AUTOTRANSFORMERS

The autotransformer, or voltage regulator, is often used in single or 3-phase form for continuous voltage regulation. The transformer winding is tapped by a moving contact which can be driven by a motor. A continuously variable voltage output is obtained in this way. The auto transformer is less flexible than the double wound transformer and provides no isolation from the supply. Formerly one of the few methods of obtaining truly continuous power control, it has been largely superseded by thyristor control.

SATURABLE REACTORS

Saturable reactors, also known as magnetic amplifiers, enable continuous control of current over a wide range to be achieved. A saturable reactor is shown schematically in *Figure 24.14*. D.C. current is passed through the control winding which increases the flux density in the core. Below the saturation level the reactor has a high impedance but as the control current is increased the reactance decreases. Very high currents can be controlled in this way with small control signals using a variable resistance or thyristors to control the current in the control winding.

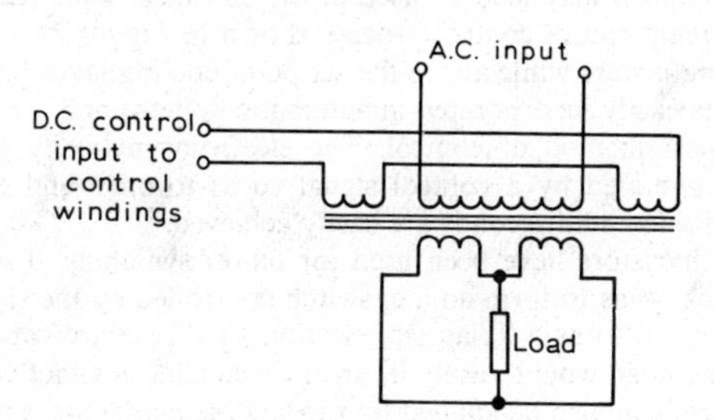

Figure 24.14 *Saturable reactor with balanced d.c. control winding*

RESISTANCE CONTROL

The use of switched or continuously variable resistors is one of the simplest and cheapest methods of current and hence power control but results in unwanted power dissipation. This is both inconvenient and costly in the case of continuous processes. Despite this resistors are extensively used to control motor field current and for starting d.c. motors. More recently thyristors have been used for controlling the field current of d.c. motors. An interesting version of the resistance starter is the liquid resistance which vaporises during the starting process preventing excessive overloading and cutting itself out after the motor has started.

THYRISTORS

Thyristors have enormously extended the ability to control current and power in a.c. circuits. The solid state thyristor (see Section 8) has to a large extent superseded the ignitron (see Section 7), thyratron and mercury arc control rectifier and except at the highest powers, variable and tapped transformers and saturable reactors.

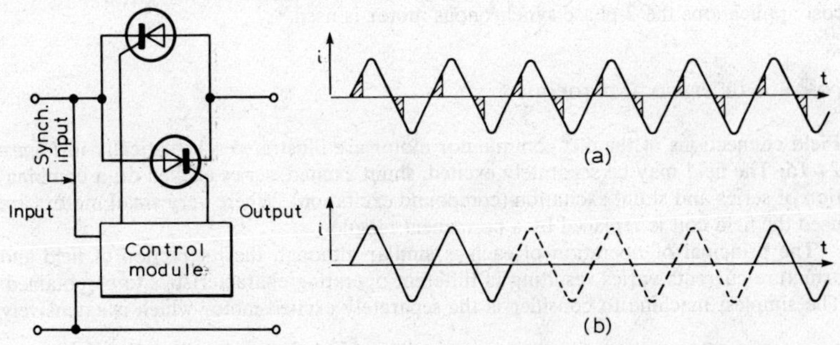

Figure 24.15 Thyristor power controller (a) phase shift control (b) burst firing or rapid sequence control

Control of a.c. power is obtained by connecting the thyristors in opposed pairs or back to back as shown in *Figure 24.15*. Such an arrangement can be produced as two separate interconnected devices or combined in one device known as a triac. It is also possible to turn off d.c. power using an oscillatory circuit to provide a reverse bias across the thyristor.

The conduction cycle is initiated by applying a forward biasing pulse on the gate. Conduction continues until the anode current reverses. The turn-on time is very short but the turn-off time is limited to about 20 μs which limits the maximum frequency at which thyristors can be used to about 10 kHz. Peak-to-peak voltage ratings of up to 1 800 V and current ratings of 1 250 A can be achieved. Series and parallel interconnection for use at higher voltages or currents is possible.

The thyristors are triggered by output pulses synchronised with the power supply shown in *Figure 24.15*. The choice of phase shift or burst firing control is often dictated by the process, for example motor speed control and the temperature control of loads of low thermal inertia normally required the use of phase shift control. The control module may also enable a slow automatic increase from zero current to the required value (soft start).

Operation is possible as an on-off switch, or for cyclic switching in the burst firing mode in which the work–space ratio can be continuously varied, or for phase shift control. Operation in the burst firing mode is often preferred as phase shift control can result in the production of undesirable harmonics on the supply.

Transistors may also be used for power control enabling continuous control through the conduction cycle and control of d.c. power to be achieved. Up to now power ratings have been limited to power of the order of 100 W but parallel connection has enabled currents of several hundred amps in welding arcs to be controlled.

Servomotors

Servomotors are used in an enormous variety of applications ranging from miniature motors used in data recorders to motors with ratings in excess of 1 MW for winding machines used in mines. Both d.c. and a.c. motors are used.

The d.c. commutator motor is the most commonly encountered servomotor drive. Capable of continuous variation in speed from zero to full speed, the torque and power output can be continuously controlled over wide ranges. Although the a.c. squirrel cage induction motor is most extensively used in industry where constant speed drives are required it is not suitable for applications where wide variations in speed are needed. For speed control of a.c. motors a wound rotor with slip ring outputs is necessary or special field coil designs such as those used in the Schrage motor. For low-power low-cost applications the 2-phase synchronous motor is used.

CONTROL OF SPEED AND TORQUE

Field connections of the d.c. commutator motor are illustrated schematically in *Figure 24.16*. The field may be separately excited, shunt excited, series excited or, a combination of series and shunt excitation (compound excitation). Where very small motors are used the field coil is replaced by a permanent magnet.

The principal of operation of each is similar although the interaction of field and armature currents varies resulting in different operating characteristics to be obtained. The simplest machine to consider is the separately excited motor which is extensively

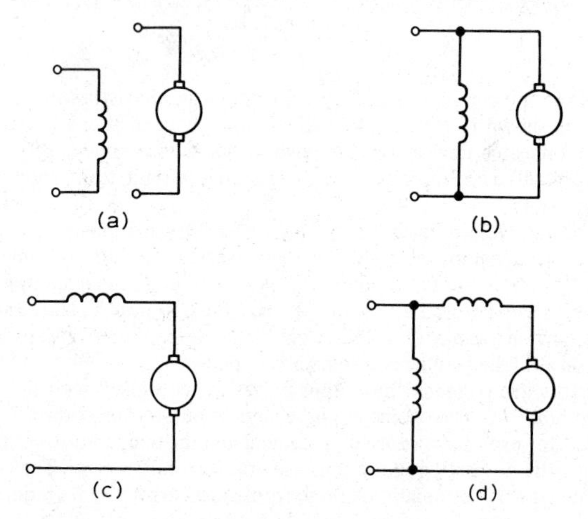

Figure 24.16 *Field connections of d.c. motors (a) separately excited (b) shunt excited (c) series excited (d) compound excited*

used in control systems. The flux from the field coil is proportional to the field current, i.e., $\phi \propto I_f$. The induced voltage (back e.m.f.) in the armature is proportional to

$$E \propto \frac{N\mathrm{d}\phi}{\mathrm{d}t}$$

which in turn is proportional to $V - I_a R_a$.

Hence

$$N \propto \frac{V - I_a R_a}{\phi}$$

and the torque $T \propto \phi I_a$.

The rotor should be capable of being rapidly accelerated where the acceleration is given by

$$\frac{\mathrm{d}^2\theta}{\mathrm{d}t^2} = \frac{Tg}{J} \text{ rad/s}^2$$

where θ is the angular displacement and J is the moment of inertia.

The rotor should be sufficiently damped so as to reduce overun where the damping coefficient is given by

$$D = \frac{\mathrm{d}T}{\mathrm{d}N} gs$$

And the time constant of the rotor should be sufficiently short where

$$t_n = \frac{J}{D} \text{ s}$$

The time required to reverse is approximately 1.7 t_n.

Speed control of motors can be derived directly from the armature current at constant field current. Speed control to an accuracy of 5–15% can be achieved in this way. If armature compensation is used accuracies of $2\frac{1}{2}$–5% are possible

Change of the direction of rotation is obtained by reversing the direction of either the armature current or field current. Split field windings are sometimes used to enable continuous variation of speed through zero. The motor is stationary when the currents in the field are such that the fluxes oppose each other. The speed and direction is a function of the magnitude and direction of the out of balance current.

D.C. motor drives for servomechanisms may be rather different from those used for other applications. Often the use is intermittent so that power dissipation may not be important but high starting torque, rapid acceleration and uniform running torque may be more critical. The control of field current can be carried out by incorporating a variable resistance in the field supply. This is wasteful of power and is normally not used for continuous control at high powers. Where rectified power frequency is used to supply the field current a variable reactor on the primary side of the rectifier may be used. A variation of this is the saturable reactor. More recently the availability of thyristors which combine rectification and current control have greatly extended the application of d.c. motor drives.

Where continuous independent variation of both torque and speed is required over large ranges Ward–Leonard drives are used. A Ward–Leonard system is shown in *Figure 24.17*. The d.c. generator is driven at constant speed by an a.c. motor or some other prime mover. The output current of the generator is connected to the armature of a d.c. motor. A d.c. supply is provided for both fields which can be separately controlled. As the field current of the motor increases from zero the motor rotates and delivers torque. The magnitude of the armature and torque and hence current is controlled by the generator field current. The direction of rotation of the motor is controlled by the direction of the generator field current.

Holding the armature current of the motor constant (by varying the field current of

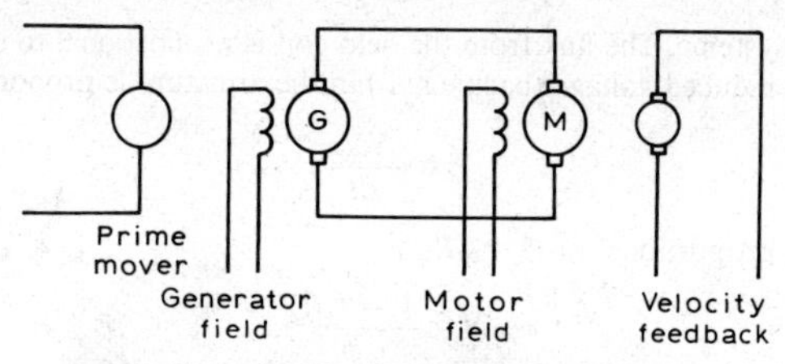

Figure 24.17 Ward–Leonard drive

the generator) while holding the field current of the motor constant results in a constant torque output independent of speed. Keeping the motor armature voltage constant and varying the field current enables a constant power output to be achieved. A further advantage, where fluctuating loads or large overloads occur, is obtained with a flywheel on the drive shaft of the prime mover. Applications of Ward–Leonard drive systems include drives of rolling mills, traction drives and winding machines.

POSITION CONTROL

The synchro or resolver normally consists of a polyphase stator with a single or polyphase rotor. A single polyphase stator with single phase rotor is shown in *Figure 24.18*. The stator is excited and the voltages in the rotor are a function of its angular position.

If the receiver (which is identical) is also excited by the same source the rotor will be

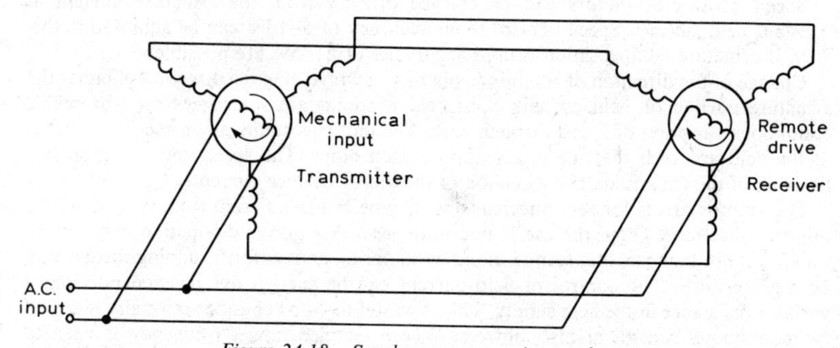

Figure 24.18 Synchronous transmitter and receiver

driven by the transmitter until the induced e.m.f. is equal but in opposition to that from the transmitter. A resolution corresponding to about 0.3° is possible, however as the receiver torque decreases as the required position is approached the error signal decreases and effects due to friction and load torque limit the accuracy obtainable. Amplification of the error signal from the rotor enables this to be minimised.

Synchronous motor drives may be used where a very high degree of precision is not necessary and only a small torque is required. The rotational speed is a function of the frequency and power or higher frequency supplies may be used. Resolution to about 1/100 of a revolution is possible. A multiple tooth permanent magnet rotor is driven by a 2-phase field winding at power or higher frequencies. The 2-phase supply can be derived from a single-phase supply as shown in *Figure 24.19* or with a Scott-connected transformer. A 100-tooth rotor operated at 50 Hz has a slew speed of 60 r.p.m.

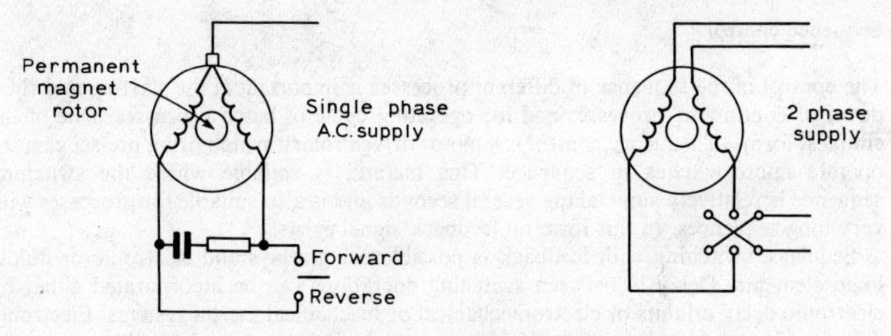

Figure 24.19 Synchronous motor drives

A stepper motor translates a digital signal into precise angular movement, each digital signal corresponding to an incremental movement of the shaft. Variable reluctance motors use a soft iron rotor in which teeth and slots are cut. A stepper motor is illustrated schematically in *Figure 24.20*. The wound stator has corresponding teeth and slots. When the stator is energised the motor aligns so as to minimise the reluctance of the air gap. Rotation is controlled and synchronised by input pulses to the stator winding. Variable reluctance motors are capable of very high stepping rates at small stepping angles making them ideally suited for control of position to a high degree of

Figure 24.20 Variable reluctance stepper motor

precision. An additional advantage of variable reluctance motor drives is the small compact dimensions that can be achieved. Higher speeds with lower rotor inertia than permanent magnet motors are obtainable, however, the efficiency and power output are lower. Typical step angles vary between 0.9° to 90°.

An alternative form of construction uses a permanent magnet rotor and wound stator. The rotor rotates until it reaches an equilibrium position opposite to an opposing pole. If the stator windings are excited in sequence the rotor rotates. An alternative construction uses a toothed soft iron rotor and a stator comprising a permanent magnet and control winding. The control winding is used to alter the distribution of the field resulting in stepwise rotation. A feature of this design is the high residual torque obtained when the control winding is not energised.

The drive signals for stepper motors are usually obtained from pulses derived from logic circuits operating in sequence. In its simplest form where the motor is operated in on-off mode within its synchronous range the drive circuit is in the form of a transistor switch for each motor coil. This is controlled by a logic circuit to switch the coils in the correct sequence. The complexity of the logic circuit will depend on the number of stator windings. The voltage and current required to drive the motor may vary from a few volts and 50 mA to more than 20 A at 100 V. Stepping speeds from zero to more than 10 000 revs/second are possible.

Sequence control

The control of the sequence of different processes is important at the start-up and shut-down of continuous processes and for operating cycle of batch processes. One of the simplest forms of sequence control is a motor driven rotary switch using pre-set cams to operate microswitches in sequence. This method is suitable where the switching sequence is relatively slow taking several seconds and is also suitable for processes with very long sequences. In this form no feedback signal exists.

Sequence switching with feedback is possible using relays and electronic or fluidic logic elements. Delay in between switching operations can be incorporated either by electronic delay cricuits or electromechanical or mechanical inertia systems. Electronic delay can be achieved by simple *RC* systems which together with an amplifier can easily enable delays of several seconds to be obtained or by electromagnetic solenoids with hydraulic delay. Motor driven timers are used to obtain longer delays.

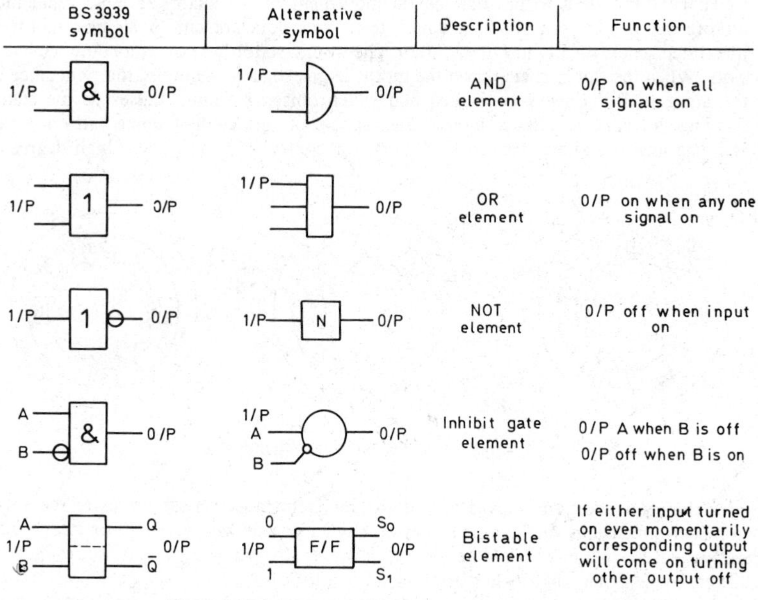

BS 3939 symbol	Alternative symbol	Description	Function
1/P & O/P	1/P O/P	AND element	O/P on when all signals on
1/P 1 O/P	1/P O/P	OR element	O/P on when any one signal on
1/P 1 O/P	1/P N O/P	NOT element	O/P off when input on
A B & O/P	1/P A B O/P	Inhibit gate element	O/P A when B is off O/P off when B is on
A 1/P B Q $\bar{Q}$ O/P	0 1/P F/F 1 S_0 O/P S_1	Bistable element	If either input turned on even momentarily corresponding output will come on turning other output off

Figure 24.21 Examples of logic circuit elements

Feedback is often necessary in the control of a complex process. This can be achieved using logic circuits using electromagnetic relays or semiconductor switching circuits. Some typical functions that can be obtained are illustrated in *Figure 24.21*. A sequence controller comprising a series of these control elements can be made to carry out a series of operations in a predetermined order only if each operation is verified by an appropriate feedback signal and if necessary corrective operation can be incorporated. An example of the application of sequence control is for resistance welding where the position, pressure and current flow during the welding process are varied at each cycle.

An alternative technique to electromagnetic or electronic switching is the use of fluid flow in logic elements known as fluidic elements. A fluidic and logic element is illustrated in *Figure 24.22*. In a fluidic element the direction of fluid flow can be altered by a control flow using the boundary or attachment effects. Most of the functions that

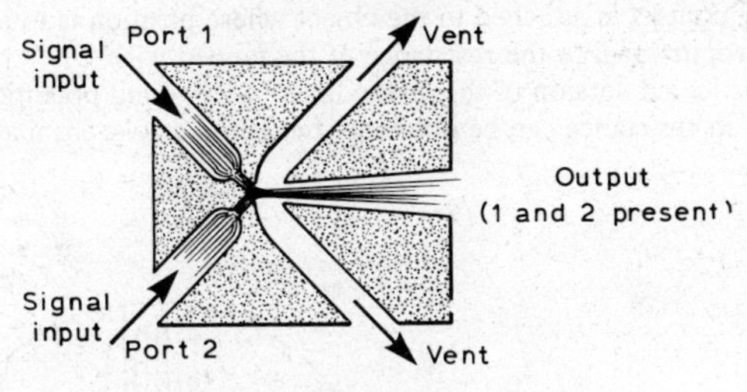

Figure 24.22 AND *fluidic logic elements*

can be carried out by electronic elements can be obtained using fluidic elements. Principal applications for fluidic devices are in hostile or noisy environments where electronic systems may not be acceptable.

Parameters of transducers

One of the most important parameters of transducers is the sensitivity. A high sensitivity is advantageous so as to reduce the need for amplification, however with the introduction of cheap reliable and high-gain amplifiers this is no longer so important. Where precision control is required a high sensitivity enabling small changes to be detected may be necessary which may be limited by the transducer design rather than the amplifier gain.

The range of operation is often important where wide fluctuations in the measured variable are encountered. A linear response is normally required and in some cases where the output is non-linear it may be necessary to incorporate some degree of linearisation. Stability is another important parameter of a transducer enabling consistent and repeatable measurements to be obtained.

SENSING ELEMENTS

The sensing element has an important function in the control circuit and the characteristics of the sensing element often limit the precision of control that can be achieved. A very large number of sensing elements exist and it is not intended to discuss the merits of all the different types here but rather to give examples of the application of representative types of sensors.

The measurement of the control parameter is often the major problem, since with semi-conductor amplifiers it is now relatively easy to amplify the signal and use it for controlling the process. A desirable feature is that a linear output should be obtained but the availability of linearising circuits has reduced the importance of this. In all the transducers described here, provided the variable can be measured to the required sensitivity it is implicit that the transducer can be used in a control system.

POSITION

The measurement of position is important in many processes. It is also often used for the indirect measurement of other parameters such as pressure, force, acceleration and strain.

One of the simplest methods of position measurement is the use of a resistance slide wire. The sliding contact is attached to the object where position is to be measured, the distance being proportional to the resistance of the tapped wire.

A more sophisticated version of this is the linear wirewound potentiometer in which larger variations in resistance can be obtained. Length of stroke from a few millimetres

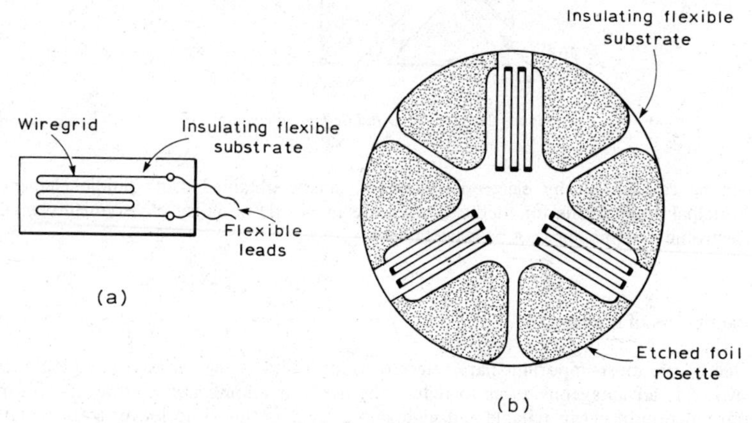

Figure 24.23 Strain gauges for measuring strain in one or more directions

to over a metre are possible with a resistance of several thousand ohms. The rotary potentiometer can be used for measuring angular or rotational position. Gearing can be used to translate linear to rotational movement.

The strain gauge transducer is capable of measuring very small deflections. The sensor is a small wire wound or etched foil resistance on a flexible flat support with a resistance of about 50 Ω–1 000 Ω which is attached to the body to be strained. Examples of wirewound and etched foil strain gauges are shown in *Figure 24.23*.

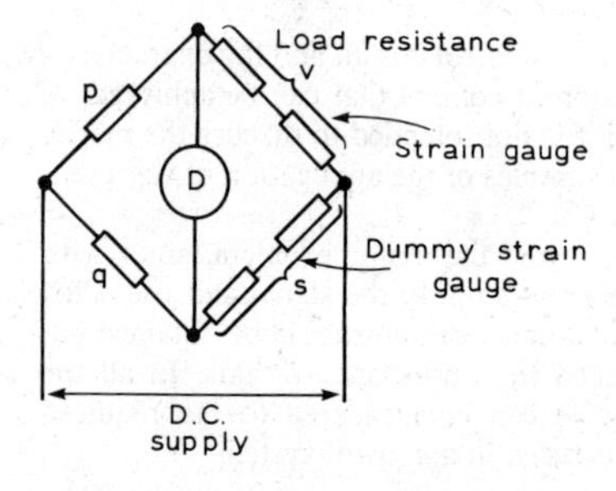

Figure 24.24 Wheatstone bridge for strain or temperature measurement

Simultaneous measurement of strain in more than one direction is possible with multiple gauges. The strain gauge is connected as one arm in a Wheatstone bridge shown in *Figure 24.24*. Compensations for changes in temperature can be achieved by connecting an identical (but unstrained) dummy gauge in the other arm of the bridge. Extension of the sensor results in a linear change in resistance which is a measure of the strain. The deflection of loaded structures and measurement of pressure and force can be measured in this way. Large deflections can be measured using strain gauges mounted on a cantilever to reduce the strain.

Semiconductor strain gauges utilise the piezoresistive effect in crystal filaments of germanium and silicon. A change in stress results in a change in resistance. The sensitivity is generally higher than foil strain gauges and the gauge can be made very small.

Capacitive strain gauges have a relatively low sensitivity but are capable of operation up to temperatures in excess of 700 °C. A typical value of unstrained capacitance is 1 pf.

The use of inductive and capacitive transducers enables small distances to be measured without contact and they are sometimes referred to as proximity effect devices. Inductively coupled transducers enable effects of contacts to be eliminated and virtually infinite resolution dependent only on the measuring circuit to be achieved. Linear resolution over distances of up to 600 mm using the eddy current effect can be obtained. Differential transformers in which one limb of the transformer core usually

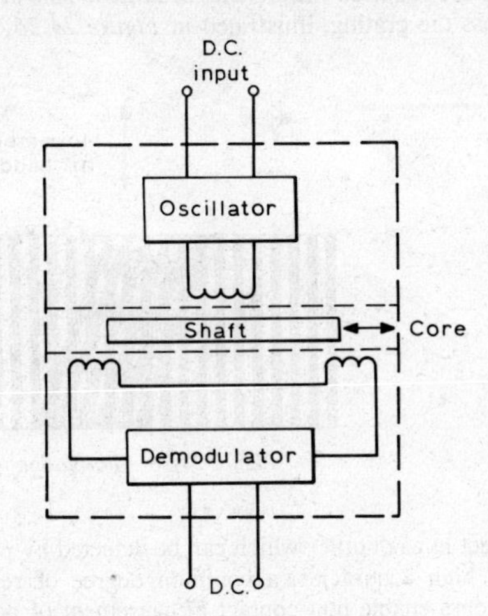

Figure 24.25 Inductive transducer

the I of an EI construction is attached to the moving object and the E section is held stationary may be used to detect movement over small distances. A large out-of-balance signal is obtained from windings connected in anti-phase on the two outer limbs. A schematic circuit is shown in *Figure 24.25*. The circuit is entirely self contained and a linear d.c. output which is a function of the displacement is obtained. Typical ranges of measurement are from 0.125 mm to 75 mm. A similar construction can be used to measure angular displacement up to 300°.

Capacitive transducers, in which the capacitance between a probe and the surface of the object is used, are capable of measuring very small distances. Sensitivity of measurement of 10^{-8} pF which corresponds to a relative movement of 2.5×10^{-7} mm between two capacitor plates each of 10 cm^2 spaced 10 mm apart is possible. Linear tubular capacitors can be used for measuring relatively large distances of up to 400 mm with a resolution of 10^{-4} mm.

The recent availability of compact semiconductor microwave transmitters and detectors utilising the Gunn effect (see Section 8) has enabled the application of radar techniques to be extended to measuring distance in industrial processes. The output at about 10.3 GHz is modulated and the time taken for the signal to be reflected back to the detector is measured.

Photoelectric methods in which a beam of light is interrupted may also be used to measure distance. A single source and sensor only enables a signal corresponding to whether the distance is within or outside given limits. Multiple detectors enable more precise location and photosensitive resistance elements are now available which enable continuous measurement of distance to be obtained. The limit of operation and accuracy of optical methods is governed by divergence of the beam and by diffraction. The low divergence of lasers has enabled the use of optical methods to be extended.

Electrical and optical encoders are used for precision measurement of linear or rotational movement enabling a binary output to be obtained. Electrical coding in which contact brushes alternatively make and break the electrical circuit are capable of a resolution better than 1/200 of a revolution over 100 revolutions with suitable gearing.

Optical coding enables non-contact reading of the encoder and very high precision to be achieved. The detector consists of a light source with a single or multiple detector. A refinement of this is the use of moiré fringes which are formed when two diffraction gratings are superimposed. These occur as constructive interference of a parallel beam of light takes place. Instead of two complete plates only one plate and a small part of a plate are required if the plates are tilted. This produces alternating light and dark bands across the grating, illustrated in *Figure 24.26*, which move as the gratings move with

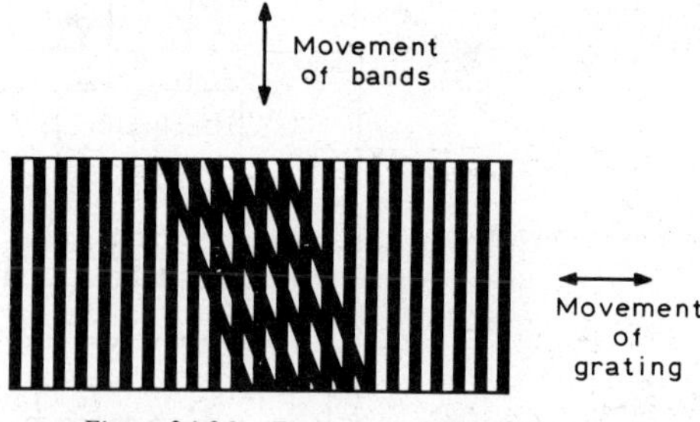

Figure 24.26 *Formation of Moiré fringes*

respect to each other which can be detected by photo cells. The gratings can be made to very high accuracies, a common degree of resolution being 254 lines/mm. Optical encoders enable non-contact measurement of position with very high resolution to be obtained. Resolution up to 1/40 000 of a revolution is possible. Absolute encoders use a second track with associated light source and detector.

Applications of encoders include measurement of distance and velocity in such applications as machine tools, weighing machines, automatic cut to length processes and remote control of position.

The ultimate (at least at present) in precision is obtainable by optical methods. The interferometer which relies on the constructive and destructive interference of light from a coherent source have been used as a laboratory measurement tool for many years. More recently the availability of lasers with depth of coherence of more than a metre has enabled interferometer methods to be taken from the laboratory and these are now in use in such applications as precision machining and calibration. Measurements to accuracies of a fraction of a wavelength are possible enabling measurements of distance over several metres to within 0.1 μm to be made.

Measurement of liquid level and volume

One of the simplest methods of measuring liquid level is by measurement of displacement such as with a ball float coupled to a potentiometer. Variation in level is translated to angular and hence rotational movement resulting in a change in resistance.

Other techniques for the measurement of liquid level using floats enable vertical displacement to be obtained which can be measured by many of the methods for distance measurement. Capacitive transducers can be used for measuring the level of polar liquids which have a high permittivity such as aircraft fuel oils. Where a difference in refractive index of transparent liquids exists, photoelectric methods may be used. Thermistors, using the self heating effect and hence the change in resistance that occurs when the thermistor is in or out of the liquid may also be used to indicate if the liquid level is above or below the set level.

Measurement of velocity

Measurement of rotational velocity is possible using small generators or tachometers (sometimes referred to as tacho generators). The generated e.m.f. is directly proportional to the rotational speed. D.C. tachometers using permanent magnets have no error at zero speed and may be made compact. To reduce ripple the armature plots are skewed and the pole faces contoured. Response to sudden changes is not possible due to the inherent inductance of the system and difficulties exist at low speeds due to the low and varying voltage that is generated. Photo-electric detectors (see Section 9) may be used to detect the light through a rotating slit or returned from a reflecting mark. By counting the output over a specified period the signal which is proportional to the rotational velocity is obtained.

One type of linear velocity transducer uses a magnet coupled to the moving object which moves along the axis of a solenoid coil. This generates an e.m.f. proportional to

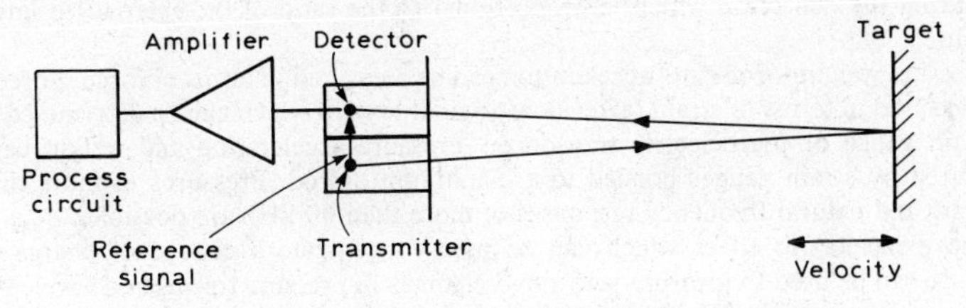

Figure 24.27 Use of Doppler effect for velocity measurement

the velocity but is limited to relatively short path lengths. An advantage of moving the magnet rather than the coil is the elimination of the need for flexible leads.

Velocity may also be measured by the Doppler effect at microwave frequencies. A Doppler velocity system is illustrated schematically in *Figure 24.27*. The Doppler velocity is given by

$$\Delta f = \frac{2Uf}{C} \cos \theta$$

where Δf is the change in frequency, U is the target velocity, λ is the wavelength of the transmitter, θ is the direction relative to the signal and C is the velocity of light. The operating frequency normally used is 10.7 GHz so that a velocity of 1 m/s corresponds to a beat frequency of 70 kHz which can be easily amplified. The reference and reflected signal are combined together at the detector. The Doppler source and receiver may be made small and compact.

Measurement of fluid flow

Fluid flow can be measured by a rotating vane transducer in the flow line coupled to a tachometer or by the pressure drop across a calibrated orifice. These have the disadvantages of interference with the liquid flow and difficulty of measurement of low velocities, or velocity of viscous or corrosive fluids. Non-linearities may also exist and external friction also limit the accuracy.

Variable area flowmeters, in which a rotating float is supported by the fluid flow are extensively used to measure low and high flow rates of liquids and gases. On-off control signals may be obtained by optical or magnetic coupling of the float in an external circuit. A scanned optical system can be used to give a continuous measurement of fluid flow.

Alternative methods of measuring fluid velocity include thermistors and hot wires. Both have an electric current passed through them and rely on the rate of cooling due to the liquid flowing past them which is a function of its velocity. Very accurate measurement of velocities of fluids over large ranges can be made using the Doppler effect at optical wavelengths. The laser Doppler velocimeter is a non-contact method and utilises the scattering caused by particles or local changes in refractive index to obtain a forward or backscattered beat frequency which is proportional to the velocity.

Measurement of pressure, weight, force, torque and acceleration

Pressure can be measured from sub atmospheric to very high pressures with Bourdon gauges. The accuracy is relatively low, the frequency response poor and an electrical output signal is not easily obtained. Aneroid capsules can be used to measure small changes in pressure, for example for use in altimeters and an output signal obtained by measuring the deflection which is proportional to the ratio of the external to internal pressure.

Pressure weight torque and acceleration can be measured in terms of force. Force can be measured in terms of strain (extension/original length) which can be determined with a strain gauge or piezoelectric transducer. Pressure acceleration and weight can be measured by strain gauges bonded to a diaphragm or rod. Pressures of more than 6 kilobars and natural frequency responses of more than 60 kHz are possible.

The piezoelectric effect which can be used to generate a change of charge with pressure can be used to measure very rapid changes in pressure torque and acceleration over very large ranges. The construction of a piezoelectric transducer is illustrated in *Figure 24.28*. Since the capacitance is small (typically 20 pF) an amplifier with a high

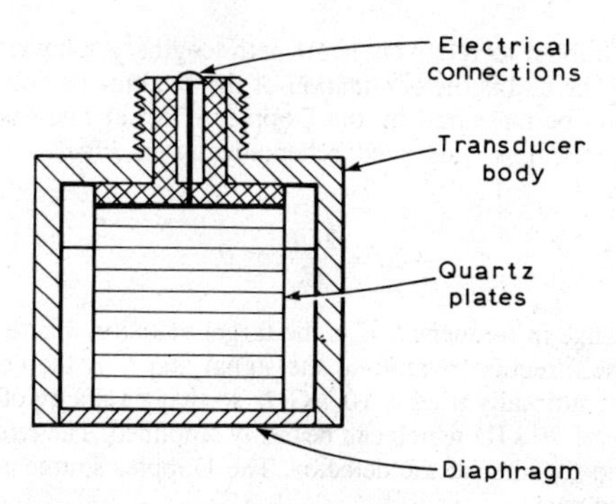

Figure 24.28 Construction of a piezoelectric transducer

input impedance is necessary. Resolution of 1 part in 10^6 is possible. Differential pressure transducers using an inductive transducer to measure the deflection of a diaphragm are suitable for use up to about 1 bar.

Measurement of very low pressure is more difficult. The McLeod gauge traps a sample of the gas in the system which is compressed by a column of mercury of known height. The pressure is related to the volume occupied by the compressed gas. Measurement down to about 10^{-4} Torr are possible but the operation is difficult to carry out automatically. The thermal gauge is suitable for operation over the range $1–1.10^{-4}$ Torr. A resistance heater is inserted in the vacuum; the power dissipation by convection is a function of the pressure and gas. Alternatively the resistance may also be used to indicate the pressure.

Ionisation gauges are capable of measuring very low pressures from 10^{-3} Torr to 10^{-10} Torr. As the gas pressure is reduced through the Paschen minimum the voltage gradient increases and can be used to measure the pressure. Thermocouple gauges can be used over the range $1–10^{-4}$ Torr.

Measurement of Humidity, Water Content and Density

The continuous measurement of humidity is relatively difficult. Electrical resistance methods in which the resistivity of a hygroscopic medium is measured are inaccurate due to the very non-linear response that is obtained. Capacitive effects which rely on the high dielectric constant of the polar water molecule ($\epsilon_r = 80$) and its high loss factor are generally preferred. The water vapour is trapped in a porous dielectric of low permittivity and loss and a few microns thickness with a capacitance of up to 0.05 μF.

The water content of various materials and in particular paper is often required to be measured during manufacture. Infra-red techniques in which the transmission of the paper at wavelengths at the infra-red absorption band of water at 1.935 μm corresponding to the hydroxyl ion are used.

Liquid density can be measured by passing the liquid through a vibrating tube. The resonant frequency of the tube is determined by the density of the liquid.

Temperature measurement

Temperature control is important in most fuel fired or electric heating processes. Examples of applications where precision temperature control is required are in heat treatment of metals and in extrusion or injection moulding of plastics.

The thermocouple is very widely used in control systems. The hot junction of the thermocouple produces an e.m.f. relative to the cold junction which is a function of the temperature difference. Different thermocouple materials and methods of construction are used depending on the temperature, environment and sensitivity. Examples of some typical thermocouple materials and conditions of usage are given in *Table 24.4*. Only a relatively small output, of the order of 10–20 μV/°C is obtained and it is necessary to amplify the output. Although over short ranges of temperature the output may be approximately proportional to the temperature, over wide ranges it is non-linear and linearising circuits are necessary for the different thermocouple materials.

The output voltage is with respect to the cold junction e.m.f. which must be held constant. This is normally carried out by a built-in temperature compensated reference voltage. The advent of stable high sensitivity solid-state amplifiers with built-in linearisation has enabled temperatures to be measured to within 0.1% often in excess of that required and has extended the application of thermocouples.

Principal sources of error in temperature measurement with thermocouples result from temperature gradients at the junction and the effect of the junction on the temperature distribution. A further source of error common to most temperature

Table 24.4 COMMONLY USED THERMO-COUPLE MATERIALS

Thermocouple	*Range*	*Typical output* (μV/°C)	*Comments*
Noble metals			
Platinum–platinum 10% Rhodium ($Pt–Pt_{90}Rh_{10}$) and ($Pt–Pt_{87}Rh_{13}$)	0–1 600 °C	6 (0–100 °C) 12 (1 600 °C)	Capable of operation at high temperatures in oxidising atmospheres. High resistance to chemical attack. Very stable. Similar couples but with I_r or R_e in place of R_h used with greater output but less stable. Used at temperatures above maximum for base thermocouples and when optical pyrometers are unsatisfactory, e.g., difficulty in sighting source.
Base metals			
Chromel–alumel ($Ni_{90}Cr_{10}–Ni_{94}$ + Al,Si,Mn)	−200–1 300 °C	35 (400 °C) 25 (150–1 300 °C)	Deteriorates rapidly in atmospheres containing H_2, S or CO at high temperatures. Wide range of uses tending to replace other types except for special applications.
Iron–constantan ($Fe–Cu_{57}Ni_{35}$)	−190–800 °C	26 (−190 °C) 63 (800 °C)	Usable in oxidising atmosphere up to 760 °C and in a reducing atmosphere up to 1 000 hours.
Copper–constantan (Cu–Adams' Constantan)	−200–350 °C	15 (−200 °C) 60 (350 °C)	Oxidation of copper occurs above 350 °C can be made very small and flexible. Useful below 0 °C.
Chromel–Constantan	0–700 °C	68 (100 °C) 81 (500 °C) 77 (900 °C)	Highest output of commonly used thermocouple. Used for differential temperature measurement and thermopiles where calibration not important.

control systems is the interpretation of a temperature measured at one point in a process such as a furnace since the temperature often varies over wide limits in the furnace.

Series and parallel connection of thermocouples are occasionally used to give mean temperatures. Thermocouples can also be connected in series to increase the output at low temperatures and large numbers may be connected to form thermopiles. Differential connection of thermocouples in which two junctions are connected so that the outputs oppose each other are sometimes used for direct indication of water temperature rises of, for example, cooling water. Another application of differential connection, for example, where the output may be used for control of combined heating and cooling systems, is by mounting thermocouples at different depths in the walls of pressure vessels or the barrels of extruders or injection-moulding machines. The temperature gradient and hence the direction of heat flow is measured in this way.

The resistance thermometer is capable of extremely high accuracies and of measuring very small changes in temperature. Essentially similar in construction to the strain gauge it comprises a sensor in one arm of a Wheatstone bridge (*Figure 24.24*). Lead-compensation in the other ratio arm is incorporated. The bridge may be made self balancing over limited ranges with reduced accuracy and the out of balance current can be amplified and used to indicate the temperature. The relatively large bulk of the resistance thermometer and narrow range of use limit it primarily to measurement of liquid or gas temperatures up to about 600 °C.

The thermistor, or temperature sensitive resistor is a semiconductor junction whose resistance is a known function of temperature. The temperature resistance characteristic

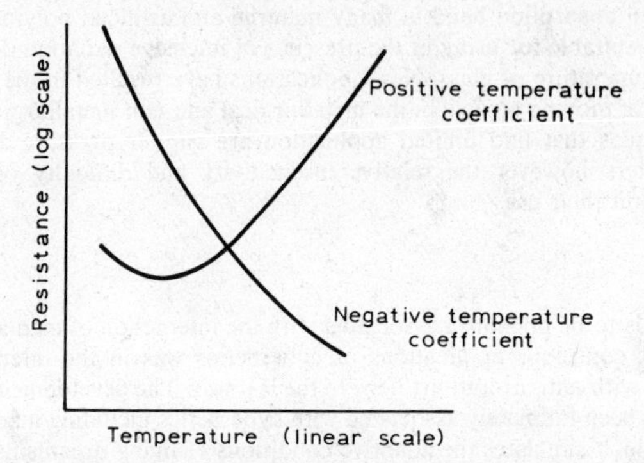

Figure 24.29 Typical thermistor characteristics with positive and negative temperature characteristics

is highly non linear, and the temperature coefficient of resistance may be positive or negative.

Typical thermistor characteristics with positive and negative temperature characteristics are shown in *Figure 24.29*. The resistance at any temperature θ_1 is given by

$$R_{\theta_1} = R_{\theta_2} e^{\frac{B}{\theta_1} - \frac{B}{\theta_2}}$$

where R_{θ_2} is the resistance at temperature θ_2 and B is a constant determined by calibration.

Thermistors are available encapsulated in a variety of forms including glass, metal and ceramic construction. Ceramic insulated thermistors are capable of use up to 1 000 °C. Unfortunately the high non linearity of the temperature coefficient and the difficulty of reproducing thermistors with identical characteristics tends to limit their application. The maximum operating temperature is generally about 350 °C although thermistors capable of operation up to 600 °C have been developed.

The non-linearity and difficulty of reproductibility can to some extent be offset by using a ratio bridge and built-in reference sources for change of thermistor are possible. Thermistors are however extensively used for control between temperature limits and applications in the self-heating mode. In this case a small current is passed through the thermistor causing the junction to heat up. Changes in heat transfer such as by immersion in a liquid or change in gas flow result in large changes in heat losses and large changes in resistance and thermistors are extensively used for fluid flow measurement and liquid level detection.

Optical pyrometers are used for the non-contact measurement of temperature over wide ranges. The subjective output of the disappearing filament pyrometer is unsuitable for control purposes but the total radiation pyrometer enables an output to be obtained which is a function of temperature. The detector is normally a black body detector such as a thermopile so that a voltage output is obtained. The output is highly non-linear and it is normally only suitable for use about about 700 °C.

Other forms of radiation pyrometer use detectors which are sensitive to only part of the radiated spectrum normally in the near infra-red over the range 0.7–5 μm. This results in increased sensitivity with regard to background illumination and noise and enables lower temperatures than those measured by total radiation detectors to be measured. A typical sensitivity band includes 2–2.6 μm which enables temperatures down to 130 °C to be measured and is suitable for general purpose applications. Alternative variations include narrow band absorption at 3.43 μm corresponding to the carbon–hydrogen absorption band in many material and artificial polymeric materials and 4.8–5.2 μm suitable for using in the presence of intensive radiation down to 30 °C including the temperature of glass. New applications have resulted in the measurement of temperatures of moving objects in the metallurgical and non-metallurgical industries.

Other techniques that find limited application are vapour pressure and bimetallic strip thermometers however the relative insensitivity and difficulty of obtaining a control signal limit their use.

Cybernetics

Cybernetics tends to be popularly associated with the interaction of man and machines. One of the first conscious applications of cybernetics was in the interaction of the human operator with anti-aircraft artillery in the last war. The development of automata (robots) has also been intimately associated with cybernetics including machines such as the haemostat which simulates the adaptive conditions of living organisms.

The term cybernetics was originally coined by Wiener from a Greek word meaning steersman to encompass an interdisciplinary field 'Centring about communication, control and statistical mechanics whether in machine or living tissue.'

Up to now the application of control theory in terms of technical processes has been considered. Most of the examples have been well defined closed-loop problems however many apparently open-loop systems can be considered as closed if we take into account other self-limiting factors. An example is the temperature control of an electric oven without a thermostat. This is controlled by the heat losses which are in turn governed by ambient conditions. The control loop is therefore completed by an external feedback loop although it may be difficult to determine the transfer functions of elements in the system. Alternatively, the loop may be closed by an operator who turns off the power source when the required temperature is reached and turns it back on at a lower temperature. If the power is left on continuously the temperature eventually reaches a maximum value. The application of the principles of control theory is however still applicable. Cybernetics is the term used to encompass the broad application of control principles in the broadest sense.

More general examples of cybernetic systems are the interaction of man and machine, a society or section of a society, the interaction between nations, defence strategies (war games), economic systems and biological behaviour. The control and automation of processes is simply a narrow (but important) facet of the broad field of cybernetics. Often the principal difficulty in cybernetic systems is the problem of describing the system performance. Unlike the complex control systems which can be broken down into a series of individual control loops which, if a sufficiently large computer were available could be precisely analysed—a cybernetic system may consist of numerous interrelated systems and feedback loops some of which may not even be known to be present.

In many cases since the transfer function of the components of the system loop may not be accurately known it is necessary to use statistical theory, probability and other information-processing techniques. The feedback loop in animals or humans can be provided by communication and the techniques of communication theory may also be involved.

Electronic control of electric motors

Electric motors are used for an almost infinite variety of applications and with a corresponding large range of sizes from, for example, drive motors for tape recorders to the provision of power for steel-rolling mills. In some cases it is required that the speed remain constant over a range from almost zero to full load, in others it is necessary that the speed can be varied while the load remains substantially constant. Until recently this control has been difficult to achieve in practice due to the limitation imposed by the fundamental theoretical characteristics of the motor. These implied that the required operational needs could only be satisfied by a reduction in efficiency or by the use of motors extremely complicated in design. Even then the range of control was limited. The use of electronic techniques has increased the control possibilities enormously and has allowed the designer of complex machine drives, in which the motor is part of an integrated system, much more freedom in the selection of techniques.

MOTOR TYPES

The direct current motor has a wound armature and a separate field winding. There is some degree of flexibility in operation depending on how the two are interconnected. In the series motor as the name implies the two windings carry the same current; the speed is inversely proportional to the current in the field winding and the output power is proportional to the product of field and armature currents, hence the speed is a direct function of load as shown in *Figure 24.30*, the speed falling as the load increases. If the field is connected in parallel with the armature as in the shunt motor the field current is independent of the armature current and therefore the load and is a function of the voltage applied to the terminals and the resistance of the field winding. In practice therefore the shunt motor is essentially a constant speed machine, *Figure 24.31*, the speed falling off

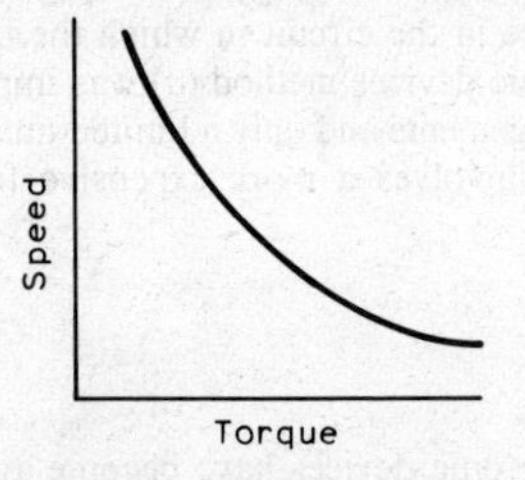

Figure 24.30 D.C. series field motor characteristics

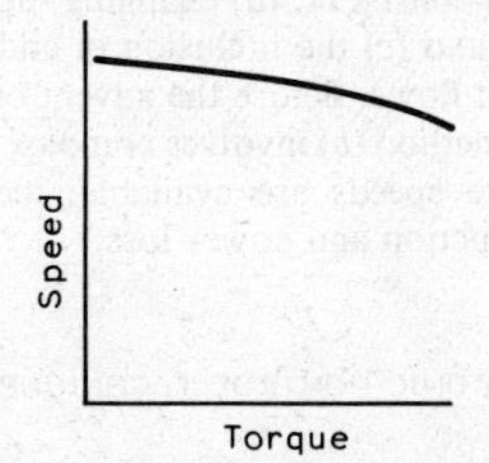

Figure 24.31 D.C. Shunt field motor characteristics

gradually as the load increases due to magnetic saturation effects. If the applications engineer wishes to take advantage of the characteristics of these machines a direct current supply must be available which usually means added complexity and expense. A further problem is that, on starting, these motors draw a very large current from the supply and special arrangements need to be made to limit this current to an acceptable value.

Alternating current machines may be connected directly to the mains supply without the need for A.C./D.C. conversion but the speed of these machines is basically related directly to the frequency of the supply and to the arrangement of windings in the motor. The synchronous motor relies for its operation on alternating current drawn from the mains and a separate direct current field winding; it is essentially a constant speed machine although the output power can be varied by control of the field current. Its main disadvantage, apart from the necessity for a separate d.c. supply, is the fact that it is not self starting. These two limitations are overcome in the induction motor, the operation of which relies on the interaction between two magnetic fields, one produced by a current drawn from the mains which also induces a current in a second winding resulting in a

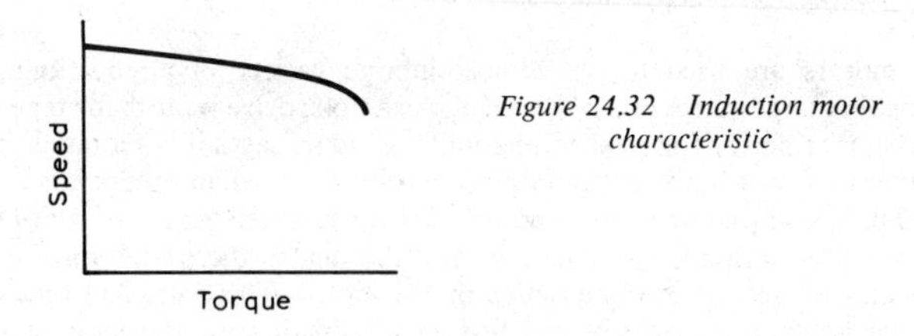

Figure 24.32 Induction motor characteristic

second field. The speed characteristic of the machine exhibits a falling off with load, *Figure 24.32*, in practice the speed reduction from zero to full load is usually quite small.

Other special machines have been developed giving characteristics differing from those described above but in all cases the lack of flexibility is a major disadvantage.

METHODS OF SPEED CONTROL

Speed control of the d.c. series motor can be achieved by changing the field current usually by diverting some of the total current through an adjustable resistor in parallel with the field; this of course means a power loss. In the shunt motor the field current variation may be obtained by connecting a resistor in series with the field circuit. The latter carries a relatively small current and the power dissipation is not large. Alternatively the energy input to the motor can be controlled. In the past this has usually meant the inclusion of a volt-dropping resistor in the supply circuit with again a large power loss.

In so far as a.c. is concerned, in particular the induction motor, the control possibilities consist of: (*a*) changing the supply frequency; (*b*) changing the winding arrangement; and (*c*) the inclusion of additional resistance in the circuit in which the induced current flows. Before the advent of power electronic devices method (*a*) was impracticable; method (*b*) involves complex switching arrangements and only a limited number of discrete speeds are available; method (*c*) again involves a more expensive form of construction and power loss.

ELECTRONIC CONTROL TECHNIQUES

As higher capacity and more reliable power electronic devices have become available the whole position with regard to control has changed. One important area is the control of d.c. machines by varying the energy input. This is achieved by using thyristors (which have largely replaced thyratrons used in earlier systems) to control the supplied energy by varying the conduction angle, thus supplying the machine with a pulse of d.c. energy which may vary in duration from a complete half cycle to zero. The thyristors act as rectifiers as well as power controllers. The field is supplied with rectified a.c. via power diodes. An alternative is to use a burst-firing technique to supply an interrupted chain of half-wave pulses to the motor, the mark-space ratio of the chain being adjusted by the controller which thus achieves a variation in power input over a period of time. A reference is required with which to compare the speed. The comparison is made between a voltage proportional to the set value, either the voltage developed by a small tacho-generator mounted on the motor shaft which produces an output proportional to speed, or by direct measurement of voltage developed across the motor armature which is also a function of speed.

A.C. machine control techniques make use of thyristor static inverters which enable induction machines to be used for variable speed drives giving a wide range of speeds and good accuracy. The basic principle is that the incoming a.c. supply is rectified using a thyristor bridge, smoothed and then inverted to variable frequency a.c. using thyristor switching. The output frequency is controlled by varying the frequency at which the thyristors are switched.

A further example of the use of electronics in motor control is the use of logic circuits in complex industrial drives where several motors are used to perform different functions. Starting operations and the control of the individual motors can be performed in response to the output from a system employing conventional logic techniques.

D.C. MOTOR CONTROL

A simple speed control system for a d.c. shunt motor is shown in *Figure 24.33*. The a.c. supply is rectified by the bridge circuit, the output of which supplies the field circuit directly and the armature of the motor via the thyristor T_1. At the beginning of each half cycle T_1 is in the *off* state and the capacitor C_1 starts to charge through the armature,

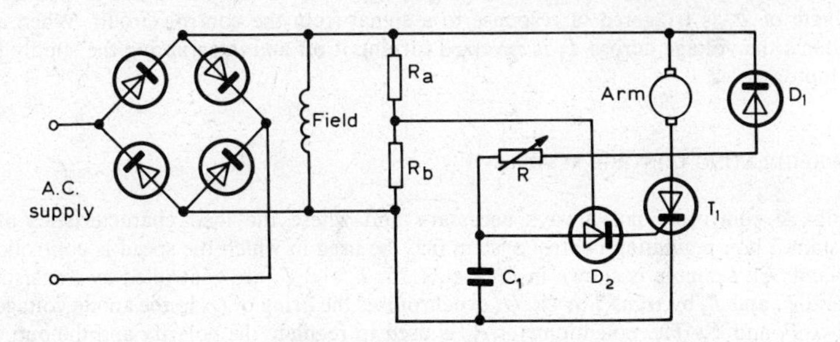

Figure 24.33 Simple speed control system for a d.c. shunt motor

rectifier D_1 and the resistor R which is used to set the control level. When the voltage across C_1 reaches the conduction voltage of the trigger diode D_2 (after a time determined by the value of R and the voltage across T_1) a pulse is applied to the gate of T_1 so applying power to the armature for the rest of the cycle. At the end of each half cycle C_1 is discharged through D_2 and the resistors. Since the voltage across T_1 is equal to the total input voltage less the voltage developed across the motor armature, which in turn is dependent on the motor speed, the charging of C_1 is also dependent on the motor speed. If the motor runs at a slower speed the armature voltage decreases and the voltage applied to the charging circuit is higher. The time required to trigger T_1 is therefore reduced, thus increasing the portion of the cycle during which T_1 conducts and consequently the power supplied to the motor so compensating for the fall-off in speed due to load increase. The rectifier D_3 provides a circulating path for the energy stored in the armature inductance when T_1 turns off. Without this the current circulating through T_1 and the bridge rectifier would prevent T_1 turning off. At lower speeds and higher armature currents D_3 remains conducting for a longer period of time at the beginning of each half cycle. This causes even faster charging of C_1 providing compensation relative both to speed and armature current.

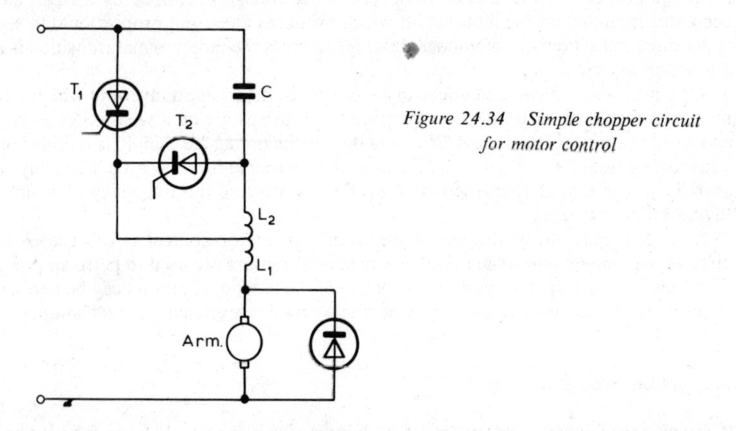

Figure 24.34 *Simple chopper circuit for motor control*

The circuit shown in *Figure 24.34* is a simple 'chopper'. T_1 is the main power thyristor and, when the gate is triggered, current flows through the tapped choke L_1 and the motor armature circuit. At the same time this current induces a voltage in the second winding of the choke L_2 so charging up the capacitor C, this charge is held until the gate of T_2 is triggered in response to a signal from the control circuit. When T_2 conducts the voltage across T_1 is reversed turning it off and interrupting the supply to the motor.

NON-REGULATING CONTROL SYSTEMS

Feedback control is not always necessary and where the load characteristics are constant a non-regulating control system may be used in which the speed is controlled by hand. An example is shown in *Figure 24.35*. T_2 and T_3 are controlled by transistor Q_1 and T_1 and T_4 by transistor Q_2, Q_3 synchronises the firing of Q_2 to the anode voltages across T_1 and T_4. The potentiometer R_1 is used to regulate the polarity and the output voltage across the armature which in turn controls the direction of rotation and speed of the motor. When R_1 is at its central position neither transistor fires and no voltage is applied to the armature. As the potentiometer arm is moved towards one end of R_1, Q_1 and its associated thyristors begin to fire, the point at which firing takes place, and therefore the voltage applied to the load, depends upon the position of the arm. If the arm is moved in the opposite direction a similar action takes place except for the fact that the polarity of the applied voltage is reversed. R_2 and R_3 together with the commutating reactor L and capacitor C are used for protection purposes, the resistors to limit fault current in case a transient should cause T_1 and T_3 (or T_2 and T_4) to be fired simultaneously and the reactance to limit the rate of rise of voltage which one pair of thyristors can impress upon the opposite pair.

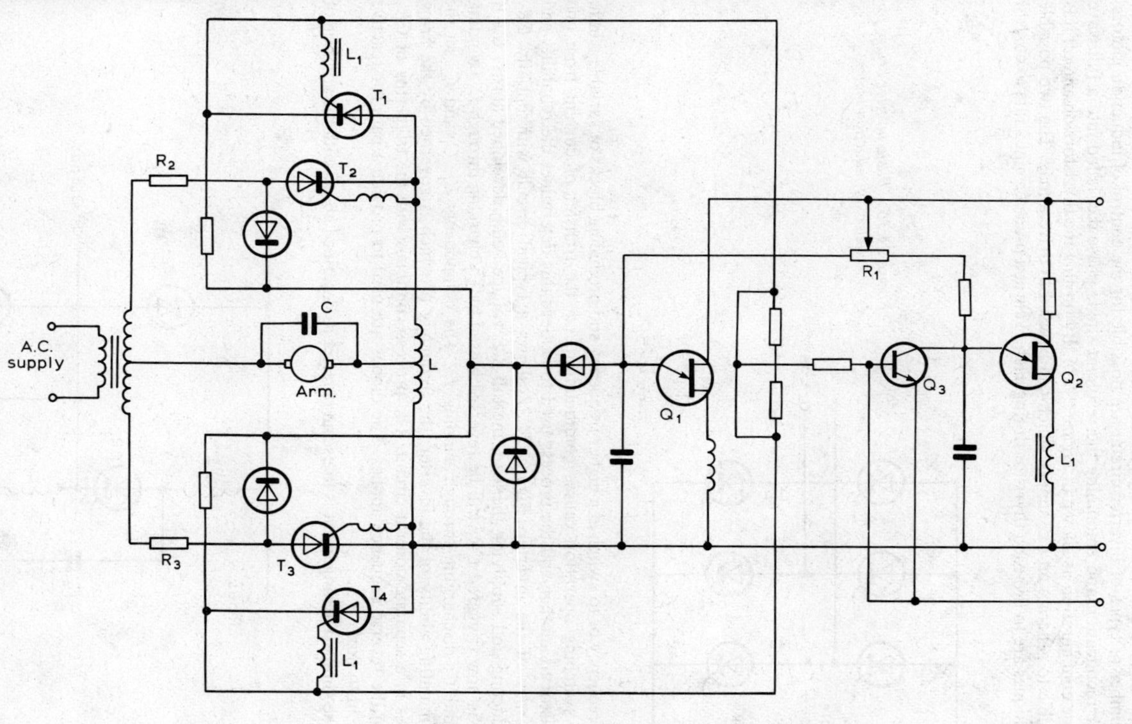

Figure 24.35 Non-regulating motor control system

INDUCTION MOTOR CONTROL

A number of types of inverter circuit are possible for the control of induction motors. After rectification and smoothing the resultant single phase d.c. is fed into a full wave bridge configuration as shown in *Figure 24.36*. By turning the thyristors on and off it is possible to generate an a.c. supply which can be used to drive a load. The two possible techniques are to fire each thyristor during each cycle at a predetermined rate to control

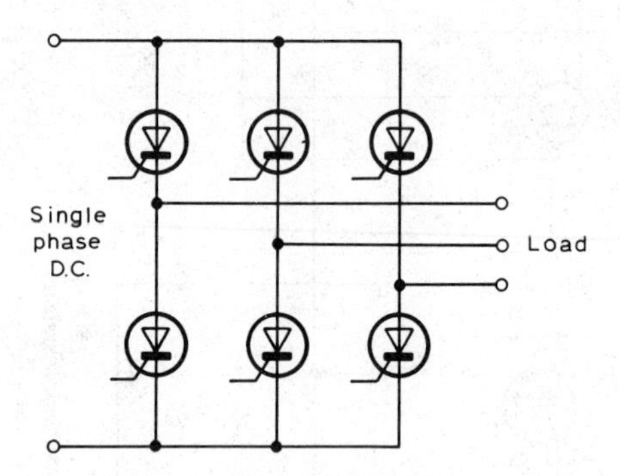

Figure 24.36 Basic inverter bridge for control of induction motors

the frequency or to switch at a higher frequency and producing blocks of variable width. The switching operation causes commutation, i.e. the transfer of current from one thyristor to another and the process used to achieve this determines the reliability and efficiency of the inverter. *Figure 24.37* shows a typical circuit with initially CR_1 conducting and supplying the load, the full d.c. voltage being developed across C_2. If CR_2 is now triggered CR_1 will be reversed biased by C_1 charging through L_1 (winding *A*) and C_2 discharging through winding L_2. The components L_1, C_1 and C_2 set the commutating ability of the inverter, the frequency of which is governed by the firing pulses in a separate control circuit. The process continues with the triggering of CR_1 when the reverse sequence applies. The diodes are used to provide a path for reactive current to flow.

The simplest arrangement is to use an inverter as described above with the addition of

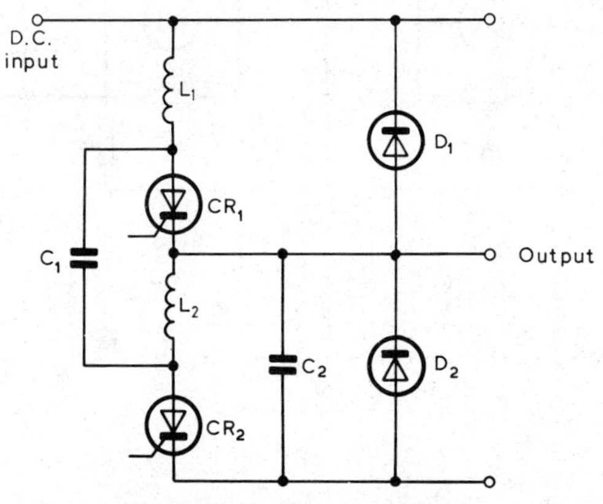

Figure 24.37 Commutator circuit

a variable voltage output transformer so that both supply voltage and frequency can be varied. An alternative is to apply triggering control to the bridge rectifier used for initial conversion to d.c. thus varying the portion of the cycle during which this conducts and so controlling the d.c. voltage which in turn changes the output voltage of the inverter. A separate d.c. chopper circuit can also be used replacing the bridge-controlled thyristors. A different approach is on the basis of a pulse-width modulated circuit which produces an output voltage consisting of constant amplitude—variable width pulses which, when applied to a motor having an appreciably internal inductance, are effectively 'integrated' to produce a wave-form which is very nearly sinusoidal. A waveform of the required output frequency and whose amplitude is proportional to that of the output voltage is generated by the control circuit and compared with a high frequency triangular waveform, the cross over points of the two being used to generate the firing pulses of the main thyristors.

Lighting control

During the past fifteen years electrical discharge lamps have been developed to take the place of incandescent lamps in many circumstances. In the industrial, commercial and street lighting sectors the takeover has been almost complete. Control problems associated with incandescent lamps merely involve some method of regulating the light output and since the latter is a function of the filament temperature this can be done by adjusting the supply voltage in order to change the filament current. Ideally this should be achieved at low capital cost and with a minimum reduction in efficiency. Discharge lamps present other problems; in addition to the control of light output some means must be provided for starting the lamp and for the maintenance of a stable discharge.

DISCHARGE LAMPS

The simplest and most common type of discharge lamp is the fluorescent tube, *Figure 24.38*. This consists of a glass tube, the inner surface coated with fluorescent powder,

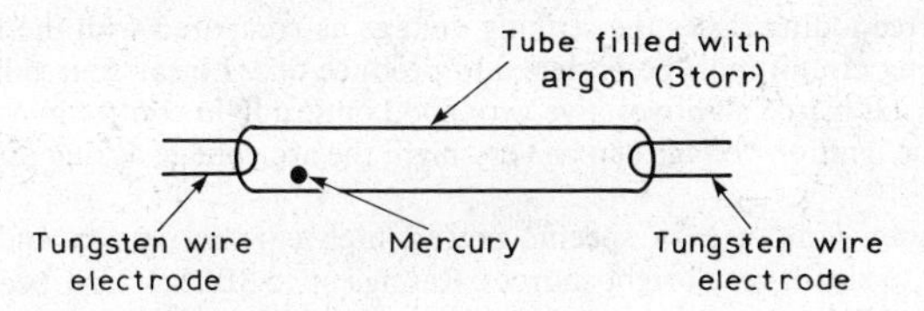

Figure 24.38 The fluorescent tube

filled with argon and containing a small amount of mercury. The discharge is maintained between two tungsten electrodes and the argon gas is necessary for starting purposes. The control gear must provide current to heat the cathodes, produce sufficient voltage to start a discharge between them and limit the lamp current during operation. The life of these lamps is reduced with an increasing number of starts. This disadvantage is overcome in the cold-cathode tubes although in these devices approximately 2 kV is needed to start the lamp, the necessary voltage reducing to half of this when the lamp is running. Sodium lamps present much the same problem to the design engineer, sodium replacing mercury in the tube, a high voltage is again required for starting. In the high-pressure type this may be up to 2 kV.

A number of types of mercury lamps are in use, they are all based on electric discharges in tubes filled with mercury vapour. As the lamp heats up the vapour pressure of the mercury increases affecting the quality of the light output. The simplest form of mercury discharge lamp is the MB type where the final pressures are between 2 atm and 10 atm. Like most discharge lamps the resistance characteristic is negative and some device is needed to control the current. The starting voltage is higher than the operating voltage and the external circuit must be designed to allow for this. One problem arises from the fact that when the lamp is switched off the pressure is high and, unless an ignitor which can supply pulses of several kilovolts is available, restarting is difficult for several minutes until the lamp has cooled. If a high-intensity beam is required, for instance in cinema or theatre work, the simple MB lamp is not adequate and must be modified either by reducing the length of the discharge (MD type) or increasing the voltage (ME type). In all types a.c. operation is simpler but d.c. may give improved discharge stability, the operating current of MD and ME lamps is usually higher than for the MB lamp and the control gear is more expensive.

Increased light output can be achieved by using lamps filled with a metal halide. Glass envelopes have been used with ratings up to 3.5 kW. In the iodide variety the

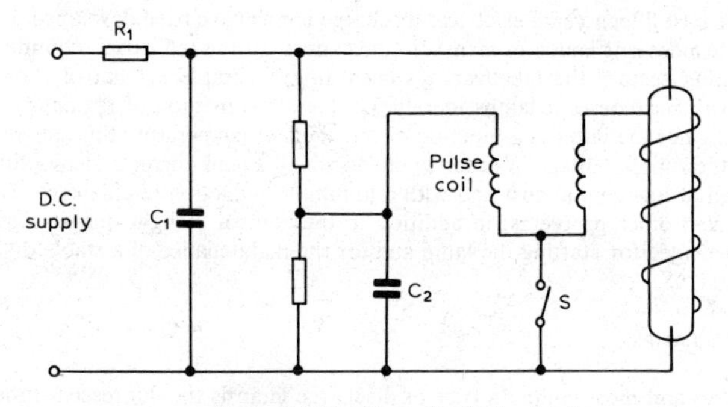

Figure 24.39 Xenon flashtube circuit

presence of free iodine raises the striking voltage as compared with the mercury lamp and the starting circuit must be designed to produce this. Linear source lamps containing similar metal halide additives give very good optical light control in ratings up to 10 kW. Again the ignition voltage can be very high; the arc voltage during operation is also high.

In the Xenon lamp a high specific power, high current density and high vapour pressure lead to small, very bright sources. Ratings up to 300 kW have been described in the literature. Althouch a.c. operation may be used d.c. is to be preferred especially at ratings above 500 kW. If successful results in terms of long life are to be obtained the d.c. must be smooth with an r.m.s. ripple of less than 5%. A high voltage, high frequency pulse starter producing pulses of 30–40 kV is used to initiate the arc. The Xenon device may be used to produce a single high intensity light pulse, i.e., as a flashtube. A circuit similar to that shown in *Figure 24.39* is used to initiate the pulse. The energy stored in C_2 promotes the starting of the tube and at the same time releases the energy in C_1 to be dissipated as light in the tube. Trigger voltages of up to 16 kV having a rise time corresponding to 40 kHz are required. Although electrolytic capacitors are more compact paper capacitors give a shorter flash duration due to their lower internal resistance. Pulse coils which are air-cored and must have low resistance to promote rapid discharge should be located near to the tube to prevent h.f. losses. If the tube is to be used in stroboscopic applications the time constants of the circuits become of greater importance. The basic circuit is shown in *Figure 24.40*. This gives a repetition rate of

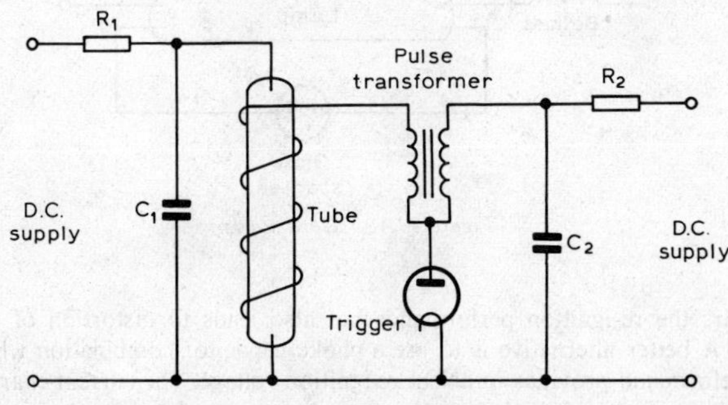

Figure 24.40 Xenon stroboscopic circuit

up to 300 flashes/second using a cold cathode trigger tube driven by a multivibrator. If the required flashing rate is higher than this a series control valve such as a thyratron is used, *Figure 24.41*, the thyratron being triggered at the appropriate frequency by an external circuit.

Pulsed Xenon lamps give high-power pulses each half-cycle from a normal a.c. supply. The lamp is started by a 10–15 kV pulse from a transformer which ionises the Xenon. Energy stored in a capacitor is then released to produce a light flash whose duration is dependent upon the inductance and capacitance of the circuit. This is

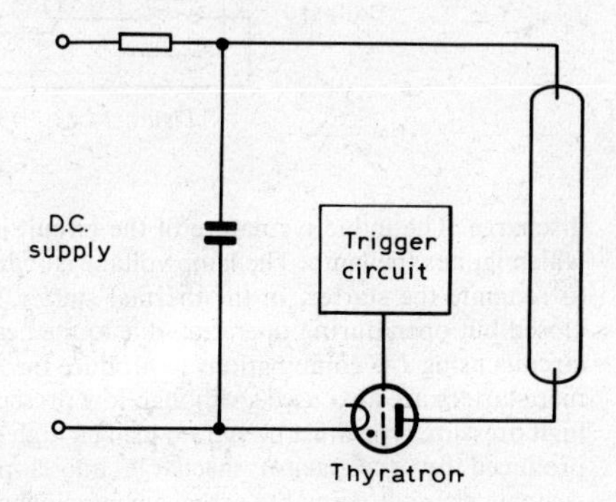

Figure 24.41 High rate Xenon stroboscopic circuit

repeated for a few seconds, the capacitor being recharged during the alternate half-cycles, after which the trigger pulses are stopped and the tube becomes self-operating. Other circuits use triacs to switch the power on and off during the cycle to control lamp power.

STARTING AND DISCHARGE CONTROL

In addition to starting devices which must produce a high voltage gradient in all or part of the lamp at the instant of switching on some form of ballast or stabiliser is required to limit the arc current. The ballast converts the inherent negative current/voltage characteristic to one having a positive slope. In addition to limiting the current the result is to sustain the electrode voltage which is necessary to maintain lamp operation through re-ignition. Simple resistors result in power loss and cause delay in re-ignition which not

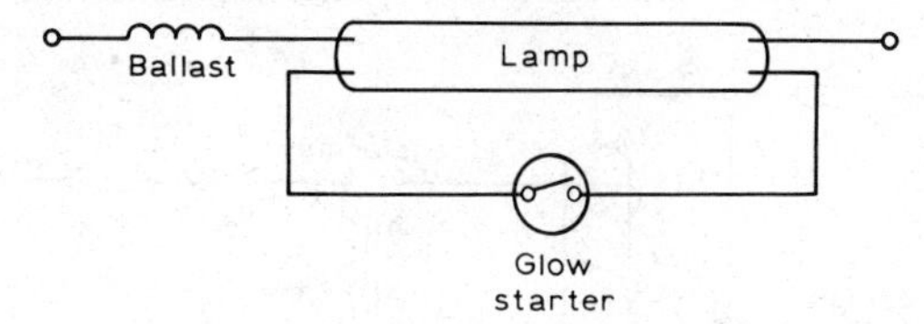

Figure 24.42 Glow starter

only impairs the re-ignition performance but also leads to distortion of the current waveform. A better alternative is to use a choke/capacitor combination which gives a better waveform and provides sufficient re-ignition voltage. The current characteristic is very nearly constant and the lamp is less sensitive to supply voltage variations. Radio frequency interference in the range 100 kHz to 10 MHz is still possible due either to direct radiation or mains interference and it may be necessary to fit suppressor capacitors and even to screen the lamp.

For fluorescent lamps a simple ignition device such as a glow starter is used, *Figure 24.42*. When the voltage is applied a small discharge is initiated in the glow tube. The resultant current heats the contacts which expand and touch, so extinguishing the

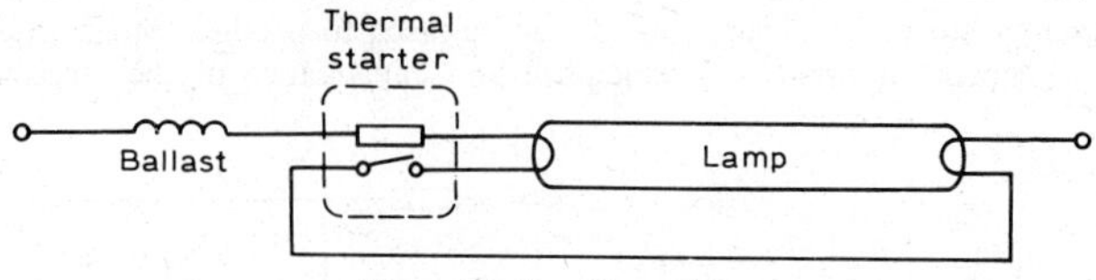

Figure 24.43 Thermal starter

discharge. The inductive nature of the circuit leads to the production of a high voltage which ignites the lamp. The lamp voltage is reduced on ignition and is not then sufficient to re-ignite the starter. In the thermal starter, *Figure 24.43* the contacts are normally closed but open during operation due to the heat from a small resistance coil. Resonant circuits using *LC* combinations to produce the required voltage conditions and transformer starters are also used. Although low pressure sodium lamps can be cold started the high pressure type must be started using a high voltage, low energy pulse. The pulses are produced by a capacitor discharge into a pulse transformer via a semiconductor switching circuit using the arrangement shown in *Figure 24.44* which includes a series

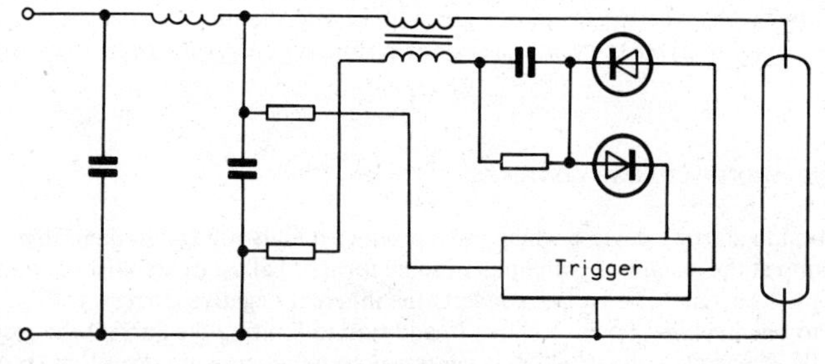

Figure 24.44 Starting circuit for high-pressure sodium lamp

pulse transformer although a shunt unit or even the main ballast choke may be used. The high voltage pulses for iodide lamps are obtained from a vibrator circuit used in conjunction with a series connected ferrite core pulse transformer.

LAMP DIMMING

A simple method of controlling the light output, i.e., lamp dimming, involves the use of a large series variable resistor but this implies a large power loss. A more efficient technique is to replace the resistor by a transductor. This is more efficient but the device is bulky and expensive if a wide range is required. Variable auto-transformers driven by an electric motor are also used.

Solid-state semiconductor switches used to vary the conducting period during each half-cycle (phase control) are more efficient and compact. The Triac, *Figure 24.45*,

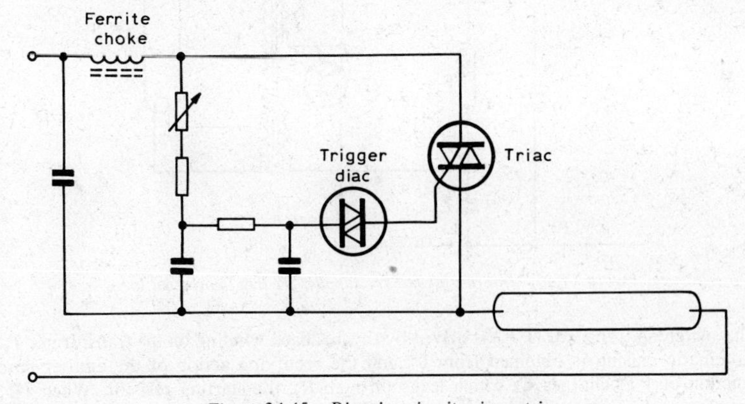

Figure 24.45 Dimming circuit using a triac

conducts when a pulse (positive or negative) is fed into the gate. The device is restored to its blocking state when the load current is reduced to zero. The triggering pulses are controlled by the Diac which only conducts when the gate potential exceeds a set value. The instant at which this occurs is regulated by the RC components of the circuit. Only a limited control range is possible and the use of two thyristors connected in reverse parallel gives a better performance. This circuit is used both for incandescent lamps and for hot cathode fluorescent tubes in which case a small heater transformer is needed. An additional starting pulse is provided during each half-cycle through a series choke, this assists re-ignition and increases the dimming range. Power factor correction capacitors must not be connected to the output side of the semiconductor dimmer.

ELECTRONIC BALLAST

Conventional ballast circuits may be replaced by a transistor chopper. The lamp current is regulated to a constant average value by the controlled high frequency switching of a single transistor. The circuit is more compact and efficient, stability is better and flicker is almost eliminated.

LAMP POWER SUPPLIES

Although mains a.c. is usually used for simplicity this is sometimes not possible particularly in the case of transport and portable lighting applications. In the latter case the main supply is a battery and semiconductor inverters produce a.c. with frequencies often higher than 50 Hz to lower the cost of circuit components and improve the efficiency. There may be an individual inverter for each lamp or a single inverter supplying a group. An individual inverter is illustrated in *Figure 24.46*. It is operated from a 12 V battery and the use of frequencies up to 20 kHz leads to high efficiency.

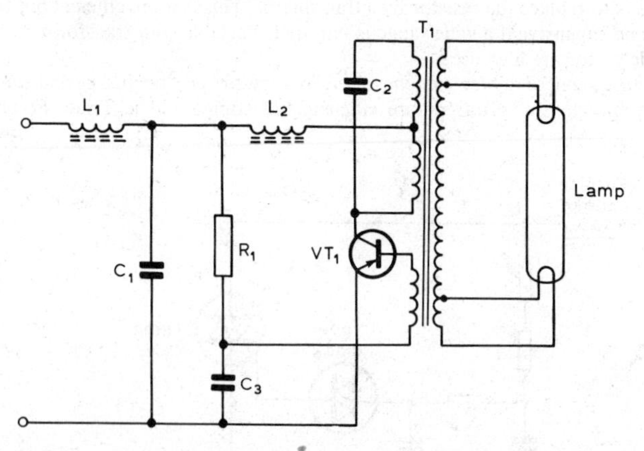

Figure 24.46 Inverter for battery supplies feeding fluorescent lamp

The switching transistor (VT_1) is driven by the feedback winding on the transformer T_1. Automatic biasing is obtained from C_3 and the rectifying action of the emitter base junction of VT_1 charges C_3 which leaks through R_1, the starting resistor. When VT_1 switches on the resonant circuit formed by C_2 and the primary winding of the transformer is excited, generating a sinusoidal high frequency. The loosely coupled secondary winding of the transformer provides the starting high voltage and supplies the two heaters. After starting, the secondary voltage falls due to the leakage reactance and stabilises the lamp current. The choke L_2 limits the inrush current to the capacitor and improves transistor performnce. The filter choke L_1 and the smoothing capacitor C_1 are used to limit interference.

FURTHER READING

Automation

LUKE, H. D., *Automation for productivity*, Wiley, New York (1972)

Control theory

ELGEROD, O. I., *Control Systems Theory*, McGraw-Hill, New York (1967)

HEALEY, M., *Principles of Automatic Control*, 2nd edn., English Universities Press, London (1970)

PIKE, C. H., *Automatic Control of Industrial Drives*, Newnes-Butterworth, London (1971)

PRIME, H. A., *Modern Concepts in Control Theory*, McGraw-Hill, London (1969)
RAVEN, F. H., *Automatic Control Engineering*, 2nd edn., McGraw-Hill, New York (1968)
TABAK, D. and KUO, B. C., *Optimal Control by Mathematical Programming*, Prentice-Hall, New Jersey (1971)
YOUNG, A. J., *An Introduction to Process Control System Design*, Longmans, London (1955)

Instrumentation

BELSTERLING, C. A., *Fluidic Systems Design*, Wiley-Interscience, New York (1971)
HARRY, J. E., *Industrial Lâsers and Their Applications*, McGraw-Hill, London (1974)
JONES, E. B., *Instrument Technology*, Vol 1, 3rd edn., Newnes-Butterworth, London (1974)
OLIVER, F. J., *Practical Instrumentation Transducers*, Pitman, London (1972)

Cybernetics

WIENER, N., *Cybernetics*, M.I.T. Press, Cambridge, Mass., 2nd edn. (1962)
YOUNG, JOHN F., *Cybernetic Engineering*, Butterworths (1973)
YOUNG, JOHN F., *Robotics*, Butterworths (1973)

Electronic motor control

'Electrical Variable Speed Drives', *I.E.E. Conference Publication*, Number 93 (1972)
GUTZWILLER, F. W., *Silicon Controlled Rectifier Manual*, General Electric, U.S.A. (1967)
Mullard Ltd., *Power engineering using thyristors*, Mullard, London (1970)
PELLEY, B. R., *Thyristor Phase Controlled Convertors and Cycloconvertors*, Wiley-Interscience (1971)
'Power Thyristors and their Application', *I.E.E. Conference Publication*, Number 53 (1969)
Proceedings of Conference, Industrial Static Power Conversion, *Institution of Electronic and Electrical Engineers*, New York (1965)

Electronic lighting control

ALLPHIN, W., *Primer of Lamps and Lighting*, Addison-Wesley Publishing Co., U.S.A. (1973)
AMICK, C. L., *Fluorescent Lighting Manual*, McGraw-Hill (1964)
BEAN, A. R., and SIMMONS, R. H., *Lighting Fittings, Performance and Design*, Pergamon Press (1968)
ELENBAAS, W., *Fluorescent Lamps*, Macmillan (1971)
HENDERSON, S. T. and MARSDEN, A. M., *Lamps and Lighting*, Edward Arnold (1972)
TUKE, R. L. C., *Thorn Lighting Technical Pocket Book*, Thorn Lighting, London (1972)

INDUSTRIAL PROCESS HEATING

The use of electricity for industrial process heating is continually growing. In most cases it is in direct competition with gas and oil both of which are usually thought of as cheaper fuels. In many instances the ease of control and other factors have led to electricity being preferred while in others the unique qualities of electroheat have made it the only real choice. The major difference between electrical heating and gas or oil is that heat can be generated within the body of the workpiece; it does not need to be conducted from the surface and therefore through-heating is possible without the necessity for long process times or excessively high surface temperatures. Other factors leading to the selection of electricity for a particular process are high heating rates and the possibility of using temperatures higher than those which can be obtained with gas or oil.

The chief mechanism for the generation of heat when the workpiece is a reasonably good electrical conductor is the Joule effect, i.e., the I^2R losses developed when a current flows through the material. The hysteresis effect in magnetic materials is also a source of heat. A somewhat similar property exists in insulators (i.e., non-conducting materials); this is the polarisation loss which takes place when such materials are placed in a high-frequency field. There is also the possibility of heating by electron bombardment and lasers (see Sections 10 and 24).

Electrical energy at all frequencies from direct current to the microwave region (2 450 MHz) is used in heating applications as shown in *Table 24.5*. At the lower end

Table 24.5 FREQUENCY BANDS USED FOR INDUSTRIAL PROCESS HEATING

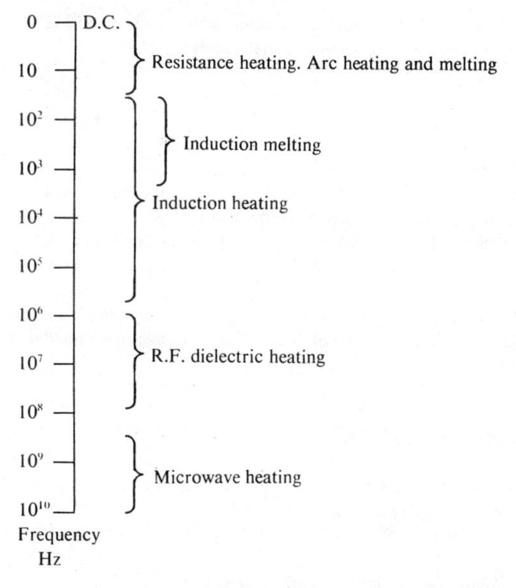

the power is derived either from the mains or from motor generator sets but as the frequency is increased electronic devices must be used. Recent developments include high-power thyristor inverters for induction heating applications at frequencies of 10 kHz and above. The use of such techniques is now being extended into the mid-frequency heating band up to 100 kHz. A major use of electronic equipment for induction heating purposes has been for surface hardening applications at around 400 kHz where class-C power amplifier circuits using power triodes are employed, the outputs ranging up to 500 kW. New devices, including the magnetically focused triode are now on the market. Dielectric heating at still higher frequencies (up to 40 MHz) employs techniques which are basically similar but in these cases the power requirements are usually lower.

When the frequency is increased still further into the microwave band the usual power source is the magnetron although klystrons are occasionally used. Electron beam and laser equipment is highly specialised and in many cases is still in the process of

Table 24.6 FREQUENCIES AND TOLERANCES ALLOCATED UNDER INTERNATIONAL REGULATIONS FOR DIELECTRIC AND MICROWAVE HEATING

	Frequency (Hz)	*Tolerance* (%)
Dielectric heating	13.65×10^6	± 0.05
	27.12×10^6	± 0.6
	40.68×10^6	± 0.05
	461.04×10^6	± 0.2
Microwave	915×10^6	± 2.5
	2.45×10^9	± 2

development. These are highly specialised techniques and are not dealt with here, the reader is referred to page **24**–62, 'Further Reading'.

At the shorter wavelengths the frequency used must be one of those agreed internationally (to avoid communications wavebands) as listed in *Table 24.6*.

Principle of induction heating

Induction heating relies on the Faraday effect. The material to be heated, which must be an electrical conductor, is placed inside a work-coil supplied with current at the appropriate frequency. The work-coil current creates a magnetic field causing a current to flow in the workpiece, the latter acting effectively as a one-turn secondary circuit. The work-piece is then heated by the resulting I^2R losses. In general the current distribution and generation of heat is not uniform. To a first approximation the whole of the current can be assumed to flow in a depth δ, given by

$$\delta = \frac{1}{2\pi}\sqrt{\frac{\rho}{\mu f}}$$

where ρ is the work-piece resistivity, $\mu = \mu_o \mu_r$, $\mu_o = 4\pi \times 10^{-7}$ and μ_r is the relative permeability, f = supply frequency (Hz) and ρ and μ_r are expressed in S.I. units.

For a given material the current penetration depth and therefore the heating depth can be controlled by varying the frequency. When induction techniques are used for metal melting and through-heating the frequency is kept low in an attempt to achieve uniformity but if surface heating only is required the frequency must be high, of the order of 300–400 kHz. The rate of heat input must be high and the workpiece must be cooled rapidly as soon as the required temperature is reached. These latter requirements are necessary for metallurgical reasons and to prevent excessive heating of the centre due to heat conduction from the surface layer.

The heating power, P, developed in the workpiece is a function of the volume, the physical parameters and the supply frequency. It also depends on the magnetic field. Thus a high value of magnetic field is necessary and, since the value of the magnetic field is equal to the product of the current in the work coil and the number of turns on the work coil, for high powers when the number of turns is limited the required current from the high-frequency generator can be very high. The coil has another influence on the generator operation in that it serves to couple the work-piece load into the generator circuit. The impedance thus introduced into the output depends on the coefficient of coupling between the work-piece and the coil and on the number of turns on the latter.

The load on the generator may vary during the heating cycle since the resistivity of metals, i.e., the value of ρ, increases as the temperature increases and the value of μ may also change. This is particularly true of ferrous materials which have a high value of μ_r until the temperature corresponding to the Curie point is reached after which the relative permeability becomes unity.

Induction heating power sources

MEDIUM FREQUENCY

The main interest to the electronic engineer is the use of thyristors in the control of rotary generators or directly as part of an inverter.

Thyristor regulators are common for machine control using a circuit similar to that in *Figure 24.47*, the main advantages being the close degree of regulation control

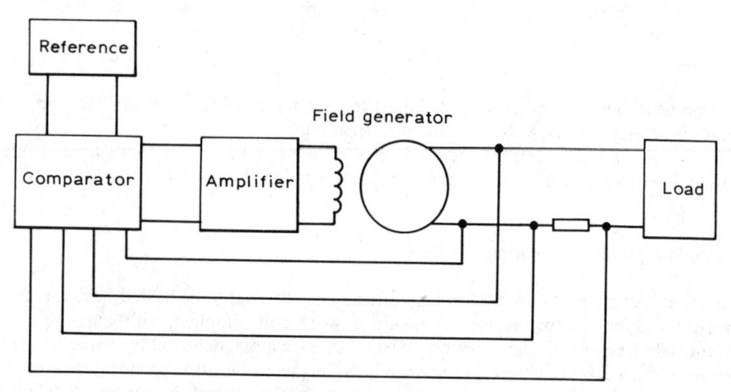

Figure 24.47 Field control for medium frequency generator

possible, the avoidance of damaging inrush currents, high efficiency, ease of protection and relatively small size. The technique has been extended to larger installations, up to 100 MW. These regulators, as do other devices of this type, generate harmonics which may damage the regulator or introduce undesirable oscillations in the supply system; filters are therefore necessary. Burst firing techniques are usually preferred as an alternative to phase angle control.

The replacement of the traditional power source by a thyristor inverter overcomes

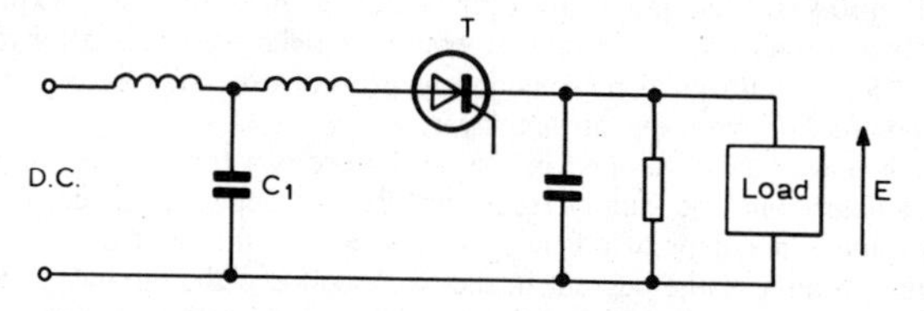

Figure 24.48 Simple one thyristor inverter

the problem of changes in load characteristics during the melting cycle. The use of thyristors enables a continuous control of all parameters, including frequency and current, to be achieved. The equipment is physically smaller and a better furnace utilisation and heating efficiency is claimed. One possible scheme is shown on a single-phase basis in *Figure 21.48*, the thyristor being controlled from the load voltage E. The current drawn from the supply compensates for the losses in the load resistance R and the input frequency is controlled by the capacitance C_1. An alternative is to use a

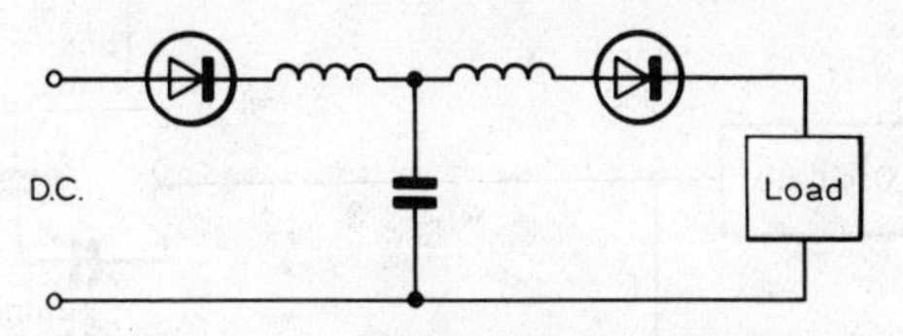

Figure 24.49 *Frequency divider inverter*

frequency divider circuit, *Figure 24.49* in which the upper frequency limit is controlled by the dynamic parameters, i.e., the turn-on and turn-off times of the thyristor. A control circuit allows the sequence firing of the transistors in a selected ratio of the input voltage, i.e., one firing in 3, 5, 7, etc., zero crossings of the input. The result is an output voltage which is a reasonable approximation to a sinewave.

RADIO FREQUENCY

The standard power source is a vacuum triode designed to operate in an industrial environment with a high degree of pollution, the possibility of excessive vibration, high ambient temperatures and a lack of skilled maintenance. The unit will often be part of a complex production process and reliability must be high. The load is variable and the generator must operate satisfactorily over a wide range of unmatched conditions. Unlike transmitting valves the properties of wide bandwidth, low distortion and high power gain are not important and a valve having relatively low power gain combined with a grid structure having low primary and secondary emission gives very nearly constant output power over a wide range of operating conditions. In a typical case the output power may only vary by 4% for a $2\frac{1}{2}$ times variation in load impedance. The power efficiency is important and values of at least 75% should be aimed for while keeping the anode voltage as low as possible. Both ceramic and glass envelopes are used with adequate cooling. Forced air is adequate for tubes of outputs below 10 kW, but for higher ratings the valve is usually cooled either by the flow of water in a jacket integral with the valve or by a vapour cooling system.

Power supplies are obtained using a suitably rectified single- or three-phase mains supply. It is common practice for this supply to feed other industrial loads in the factory and variations in input voltage must be taken into account when the generator unit is designed. The ripple should be kept low since ripple voltages can modulate the r.f. energy which may give rise to h.f. radiation from the set and affect the quality of the output product. Solid state rectifier stacks are used for anode voltage supplies. These avoid the need for high-voltage filament transformers, rectifier tubes and complex protective circuits. The diodes used in these circuits have a mean forward current of approximately 20 A and a reverse voltage rating of 1 kV. The working voltage should be kept low and the anode current high in order to reduce the number of diodes connected in series.

The variations in load during the processing cycle have an important effect on the frequency stability of the oscillator. The load circuit reactances reflected through the coupling coil should be as small as possible in comparison with tank circuit impedances in order to minimise the tank circuit frequency shift. When the coupling factor of the coil is high large circulating currents and losses may be present in the tank circuit. An alternative is to introduce frequency control by monitoring the output frequency, *Figure 24.50*, and using the resultant error signal to adjust the tank circuit tuning.

A relatively new development is the magnetically focused triode, *Figure 24.51*, in which the electron flow from cathode to anode is focused by means of the field of an external magnet. The conventional wound grid is replaced by a robust 'gate' electrode positioned so as to intercept few electrons. The power gain of the valve is approximately

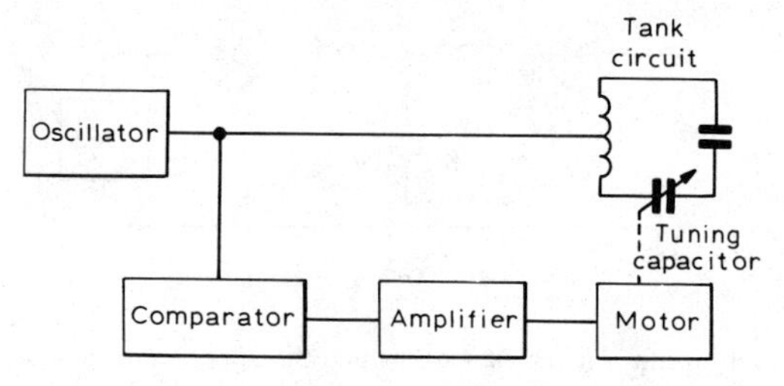

Figure 24.50 Automatic tuning for r.f. generator

30 dB with low driving power. The magnetically focused triode can withstand interruption of oscillation resulting from faults since the anode dissipation at full h.t. and zero bias is well within the valve capability. There are no heavy current surges to damage the rectifiers in such an event.

Induction heating involves frequencies up to 500 kHz and any of the well-known tank circuits and feedback arrangements may be considered with preference given to

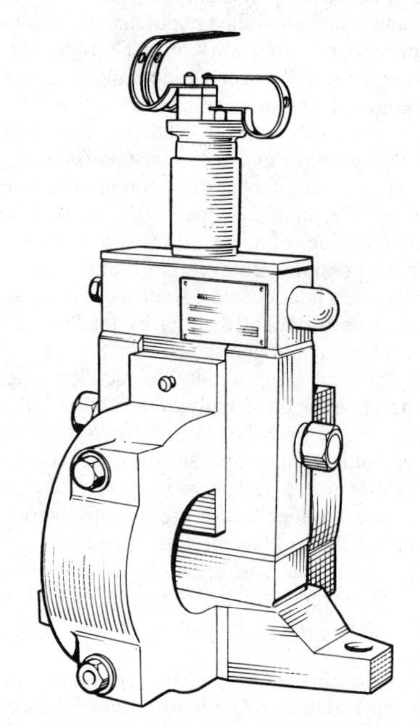

Figure 24.51 Magnetically focused triode (Courtesy ITT Ltd.)

those based on the Hartley configuration (see Section 13); the load inductance can then become an integral part of the circuit. Where high powers are required it may be necessary to resort to a multiple valve arrangement. In some cases two small valves in parallel may provide more suitable matching than a single larger one but it is necessary to prevent interaction between one valve circuit and the other. This can be achieved in practice by connecting damping resistors of a few ohms between the grids as shown in

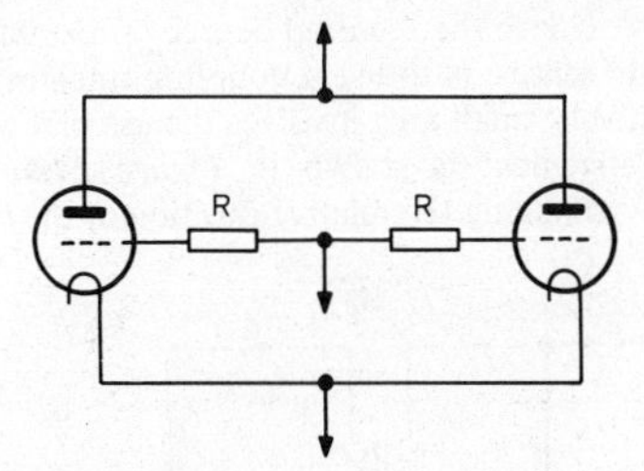

Figure 24.52 Parallel valves with damping resistor

Figure 24.52. A push–pull arrangement may also be used. For the lower frequencies appropriate to induction heating discrete coils and capacitors are normally used for the tank circuit. The capacitors, which may be water cooled, should be capable of withstanding the full r.f. voltage and should be protected against transient over-voltages which may arise in industrial situations due to sudden changes in loading conditions.

The necessary ampere-turns on the load coupling coil are found by substituting $H=nI$ in the power equation. In practice the coupling factor is often much less than unity. Its maximum value for a cylindrical work-piece surrounded by a cylindrical coil can be taken as

$$K=\frac{d^3}{D^3}$$

where $d=$ diameter of workpiece and $D=$ diameter of coil. The coil and workpiece parameters can then be calculated and transferred to an equivalent circuit which in turn may be used to calculate the valve operating conditions.

Figure 24.53 Directly coupled induction heater—a simple arrangement

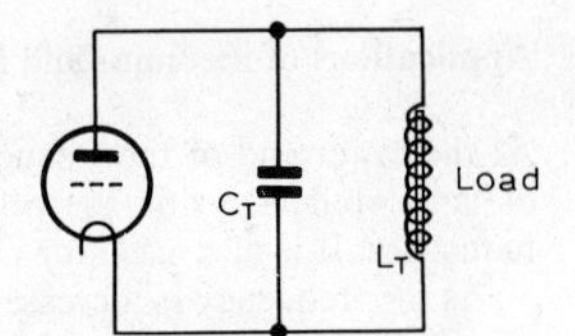

In some simple applications the tank circuit coil can act directly as the load coil, *Figure 24.53*, but this is only suited to single-purpose generators. Greater flexibility is achieved by making the load coil only a part of the tank circuit inductance thus enabling the load coil to be changed if necessary. Tappings may then be provided on the main circuit inductor to compensate for the load coil change, *Figure 24.54* (*a*). Alternatively the load coil may be placed in parallel with the tank coil, *Figure 24.54* (*b*) enabling a very high Q to be obtained in the tank circuit.

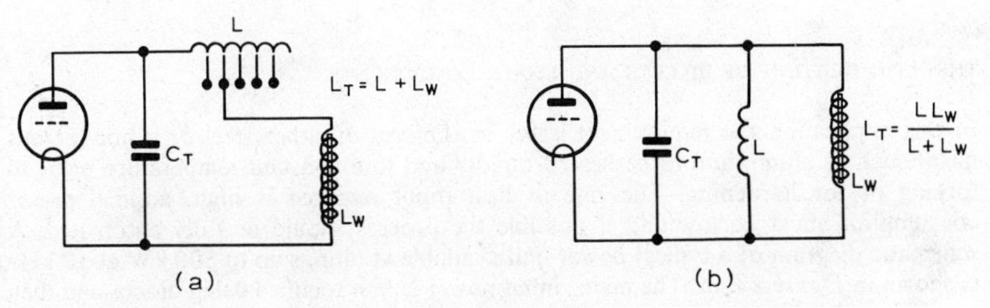

Figure 24.54 Directly coupled induction heaters, giving greater flexibility

If it is not possible to achieve the required degree of matching with the above forms of direct coupling the alternative is to use a coupling transformer, as is the case when surface heating of a relatively small area involves the use of a work coil having only one or two turns. The arrangement is shown in *Figure 24.55*. Power control in the workpiece is achieved by changing the relative position of the two coupling transformer

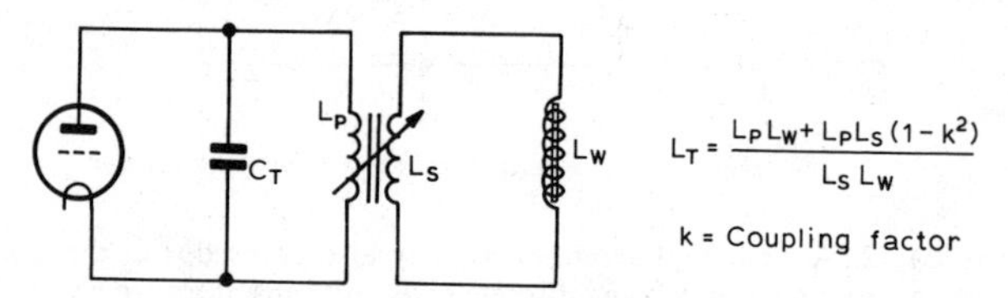

Figure 24.55 *Transformer coupled induction heater*

windings so changing the coupling coefficient. Tuned coupling transformers are also used.

The following methods are used for the control of the r.f. power:

(1) Adjustment of the h.t. supply voltage by a variable transformer or a series reactor.
(2) Physical movement of the inductors in the tank and load circuits so as to change the coupling.
(3) The use of thyristor control on the primary of the input transformer.
(4) Grid pulsing of the r.f. valve using thyristors or similar devices.

Applications of medium- and high-frequency induction heating

At the lower end of the frequency range induction heating is used for through heating of small billets, prior to rolling or forging, in the ferrous and non-ferrous metals industries. It is also used for upset forging in the general engineering industry.

As the frequency is increased the surface heating properties become of interest. Many automobile components such as gears, axles and crankshafts are hardened using induction techniques. Tempering and annealing can be carried out, as also can brazing and soldering, particularly in the more sophisticated areas such as fixing diamond and other work-tips to machine tools. Since the work piece can, if necessary, be placed in an air-tight enclosure with the coil wound around the outside, heating may be carried out in a vacuum or under an inert atmosphere. This leads to a number of applications in the electronics manufacturing industry such as the crystal growing of semiconductor materials. A relatively new application is for sintering metal powders and ceramics.

THROUGH HEATING OF BILLETS AND SMALL COMPONENTS

In this application the requirement is for workpieces of either steel or a non-ferrous metal such as aluminium to be heated rapidly and to a constant temperature prior to forging or for hardening. The rate of heat input required is high, no-load power consumption must be low and if possible the process should be fully automated. A schematic diagram of a typical power unit available at ratings up to 500 kW at 10 kHz is shown in *Figure 24.56*. The mains input power is first rectified using diodes and then supplied to the thyristor inverter which is of a standard design. The output frequency is

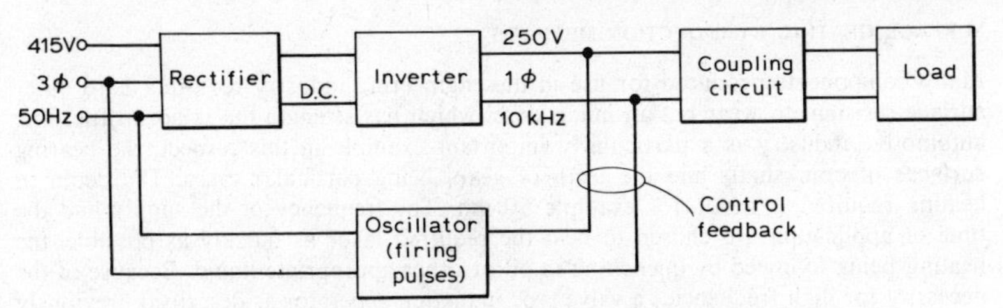

Figure 24.56 Power unit for m.f. induction heating (Courtesy Stanelco Ltd.)

determined by a small oscillator which provides firing pulses for the thyristors. Power control is dependent on the tuning of the inverter in response to the resonant characteristics of the work coil and the compensating capacitors. The nearer the inverter frequency to the resonant frequency of the load circuit, the greater the power output. In practice the system is adjusted to a required power level and subsequently it tends to remain at that level since changes in load characteristics are automatically compensated for by frequency shift. A minimum load circuit Q of 10 is normally available with the standard capacitors provided although this can be increased if required by using additional capacitors. The equipment is water cooled and an adequate supply must be available. As with all forms of induction heating the actual load is highly inductive. However, this is compensated by the capacitors in the tuned output circuit and the makers claim that the input power factor is 0.9 or better, which is important from electricity supply and cost calculations. The efficiency of the unit can be over 85% at full load.

In many applications, especially where through heating is involved, lower frequencies can be used and generators with rating up to 1 MW at a frequency of 1 kHz are available, the design being based on a principle similar to that described above. The machine illustrated in *Figure 24.57* employs a series of coils through which the billet is transported, being heated as it travels. The forward speed of the conveying mechanism, in this case a walking beam, and the total length of the coil assembly together with the power rating determine the total heat input to the billet. With the machine shown, which has a heating zone approximately 3.8 m long, a throughput of 2.25 tonnes per hour can be handled.

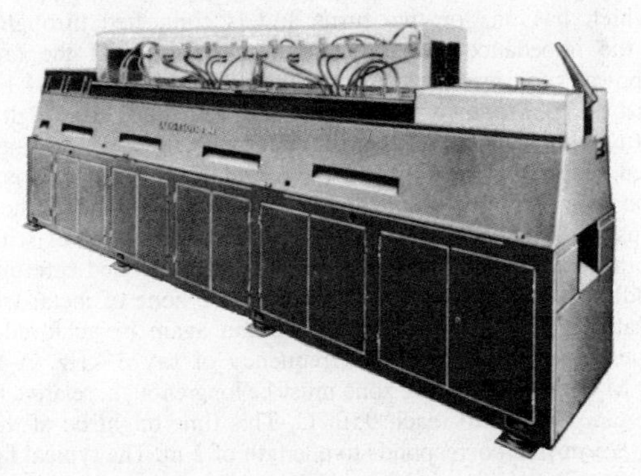

Figure 24.7 Induction billet heater with protective covers of heating coils removed (Courtesy AEG-Elotherm Ltd.)

SURFACE HEATING BY INDUCTION METHODS

Many components produced for use in the engineering industry require a hard outer surface resistant to wear but an inner 'core' which has strength but is not brittle. The automotive industry is a particularly important example in this respect, the bearing surfaces of crankshafts and the teeth of gears being particular cases. The depth of heating required is small, for example 50 μm. The frequency of the supply and the time of application are chosen to heat the required layer as quickly as possible, the heating being followed by quenching in oil or other appropriate liquid. Because of the necessity for high frequencies a valve type induction generator as described previously and operating in the 400 kHz range is used.

A typical example is the production of ring gears for motor vehicles. The process consists of the hardening of the gear teeth and then pressing the whole gear on the flywheel. A high frequency induction generator rated at 45 kW is used, the whole gear wheel being placed within the work-coil and the flywheel mounted above it. The work coil is energised and the gear is rotated within it to ensure diametral uniformity. In approximately 12 seconds the gear teeth have been raised to the required temperature of 900 °C extending to about 1 mm below the root of the teeth. During this time some heat has been conducted to the centre of the wheel producing a thermal expansion of the bore, the flywheel is then inserted in the gear and the whole assembly lowered into a quench tank. The rapid reduction in temperature completes the hardening process at the circumference and also shrinking the gear on to the flywheel. Using hand loading techniques the whole process might be completed in 30 seconds including loading and unloading. Since in the above process the generator, the most capital-intensive part of the equipment, is only used during heating, i.e., for less than 50% of the time, an obvious improvement is to use a single generator feeding two work coils alternately. One coil is energised and heating takes place while in the other station the flywheel is being fitted followed by the quenching operation.

INDUCTION WELDING OF TUBES

A common method used for the manufacture of large quantities of steel tubes is to begin with a continuous strip which is passed through rollers and bent across its width to form a circle. The two edges thus brought together are welded to each other and a continuous pipe is produced. The production requirements are a weld which is both mechanically and metallurgically sound and a high production speed. One method which is adopted is to heat the edge region using high frequency induction techniques at the same time pressing the two edges together. The strip is first bent and then passed through the inductor coil which has one or two turns and is connected through rigid copper connections to the impedance-matching output transformer of the radio frequency generator. The power requirements can be very high; units of up to 1 MW are in use and the coil must be robust since the tube is passing through it at high speed, in one design a single turn, uninsulated, is used. In order to confine the heating action to the edges as required a current concentrator, for example a ferrite core, is placed in the pipe. For a good weld the pipe speed must be at least 25 m/min. The inductor and concentrator must be correctly positioned and both the angle at which the edges approach each other and the welding pressure must be controlled carefully.

After the welding process has been carried out the zone of metal which has been disturbed by heating must be normalised. This can again be achieved by induction heating techniques, this time at a lower frequency of say 3 kHz. A typical power requirement is 2 MW and the heating zone must be long enough, relative to the speed of the tube, for the temperature to reach 950 °C. This time might be of the order of 15 seconds which at 25 m/min corresponds to a length of 2 m. The typical figures referred to above are based on data for a machine which produces pipe of 320 mm diameter and with a wall thickness of 5 mm.

CRYSTAL GROWING USING INDUCTION HEATING TECHNIQUES

A crystal-growing furnace has been designed by the R.R.E. to produce single crystals, semiconductors and optical materials usually by the Czochralski ('pulling') technique although it is possible to use the Stockbarger or 'floating zone' method. The operation is carried out under vacuum and a number of interchangeable work chambers are provided including a metal unit designed to work at pressures up to 2 000 p.s.i. and a device to facilitate the use of electromagnetic stirring techniques.

An 18 kW unit generating r.f. power at 500 kHz is used, the power being controlled by a saturable reactor to maintain constant r.f. coil current. No attempt is made to control the crucible temperature directly since it is appreciated that it is the rate of change of temperature rather than absolute temperature stability which is important in crystal growing. A small pick-up loop monitors the coil current. This is then rectified using a diode, smoothed and compared with a standard voltage, the resultant error signal being fed into the solid state control system. The input power is 33 kW with a power factor of approximately 0.9.

Theory of dielectric heating

When a non-conducting material is placed in an electric field there is a movement of the electrical charges in the material such that they tend to align themselves with the field. This effect is most pronounced in those materials made up from molecules which have distinct centres of action of positive and negative charges. Such materials are said to be dipolar. When placed in the field dipolar polarisation is said to occur. If the field is alternating this alignment takes place continuously as the direction of the field is changed and is accompanied by the generation of heat within the body of the material. The amount of heat produced per unit volume of the material is given by

$$P = E^2\, 2\pi f\, \epsilon_o \epsilon_r \tan\delta \text{ watts}$$

where E is the field strength (volts/metre), f is the frequency (Hz), ϵ_o is the permittivity of free space (8.854×10^{-12} F/m), ϵ_r is the relative permittivity of the material and tan δ is its loss angle. Tan δ varies appreciably with the type of material; it is particularly high in those materials which exhibit dipolar polarisation. Water is the most common of such materials; it has a value of tan δ of 0.022 at room temperature and a frequency of 1 MHz, the corresponding value of ϵ_r being 80. For all materials the values of both ϵ_r and tan δ vary with frequency, that of ϵ_r decreasing as frequency increases while tan δ exhibits a resonant peak at a particular value of frequency. The two quantities are also temperature dependent.

While there is a particular value of frequency at which the product (ϵ_r tan δ) is maximised, nevertheless, since the power equation also includes a frequency term, there is a theoretical advantage to be gained from using as high a frequency as possible. The available frequencies for use are 13 MHz, 27 MHz and 40 MHz in the r.f. range, while 900 MHz and 2.45 GHz may be used in the microwave band. Inspection of the power equation shows that the field strength should be as high as possible but it should not be so high as to cause electrical breakdown between the electrodes. Such breakdown is likely to damage or even destroy the material which is being processed and in certain cases, for example in the drying of wool fibres, an explosion may result. Since the equipment is designed for use in an industrial situation there is likely to be some pollution present such as dust or moisture vapour which will reduce the breakdown strength of the air.

In order to achieve uniform heating the field strength through the material should be uniform. This is not always possible particularly if the material is not homogeneous. Inclusions having a different relative permittivity from the main body of the material will cause a distortion of the field leading to regions of high stress with non-uniform heating and even local breakdowns.

Dielectric heating generators

For dielectric heating in the r.f. range the design principles used for the power source are similar to those involved in r.f. induction heating although circuit techniques must be adapted to take into account the higher frequencies used. Radiation of r.f. frequencies becomes even more of a problem at these frequencies due to regulatory requirements in respect of the allowable bandwidth. Care must be taken to achieve frequency stability and to eliminate unwanted harmonic resonances. At these higher frequencies it is now commonplace to use cavity forms of circuit for the tank inductance and capacitance these give an improved high frequency performance.

The workpiece is now capacitive in nature although due either to a possible change in permittivity as the temperature is increased (see above) or the fact that the application may be one of drying where the material changes its characteristics as water is driven off, this capacitance may be variable. It may still, however, be used as part of the oscillatory circuit in which case some form based on the Colpitts circuit (see Section 13) is usually more convenient.

Applicators for r.f. dielectric heating

The load coupling circuit is designed to provide the chosen voltage at the work piece and to give satisfactory matching between the generator and load. If a higher voltage than that provided by the oscillator is needed a step-up transformer (design of which is appropriate to the frequency used) can be used. A reduction of potential may be brought about by the use of a capacitance potential divider, lower arm of which is formed by the load material. In some applications the space between the electrodes is entirely filled with the load material. This is the case in plastic welding where solid electrodes are used to maintain the required contact pressure as well as to establish the electric field. The voltages involved are usually low (up to 1 kV) and a divider arrangement as mentioned above is used. In other cases there is an air-gap between the electrodes and the material. These air-gaps act as series capacitors and reduce the voltage available to set up the field in the material. A higher voltage is therefore needed from the source.

A directly coupled system is illustrated in *Figure 24.58*. The work material is an

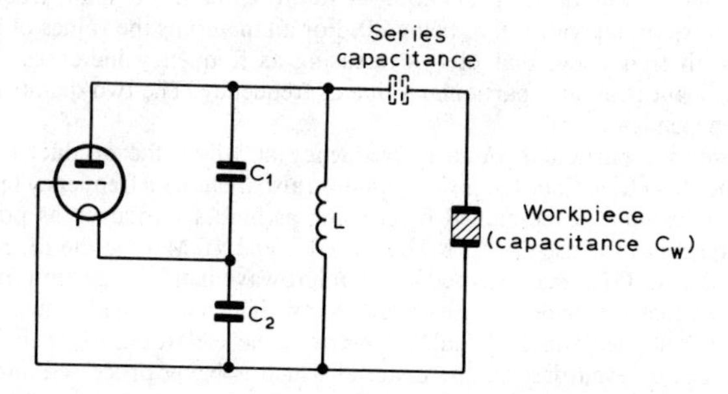

Figure 24.58 Directly coupled dielectric heating generator

integral part of the frequency controlling components and changes in the load material during the processing cycle may cause a shift of the operating frequency. This shift can be limited by making the tank circuit capacitor large in comparison to the load or by using a series capacitance which is small compared with the load. The radiation of harmonic frequencies is a problem since there is no possibility of including filtering circuits between the generator and the load. Greater flexibility can be achieved if an

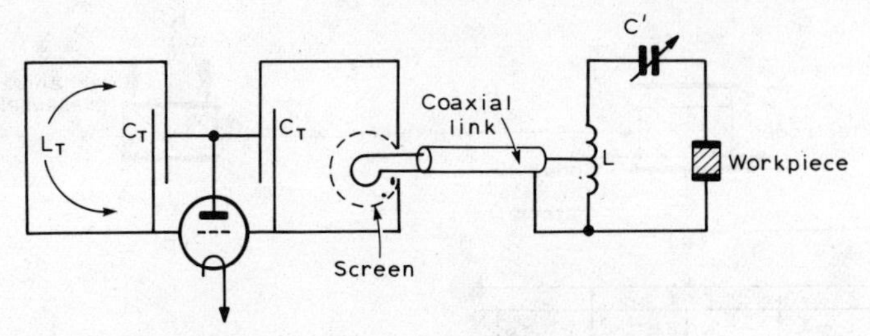

Figure 24.59 Coaxial coupling for r.f. generator

indirectly coupled system is used. Athough transformer coupling is possible the system shown in *Figure 24.59* uses a coaxial link, the frequency stability of such circuits is high and rematching during the cycle may be achieved by adjustment of C_1. This circuit uses a cavity construction and screens are inserted to minimise harmonic power transfer. Some detuning will take place as the characteristics of the load material change during the cycle and if it is desired to maintain a constant output power in such circumstances the peak output power of the generator must be increased. A completely different technique is the use of an integrated coupling system in which the generator and load components are joined to form one resonant circuit with unity coupling. *Figure 24.60*

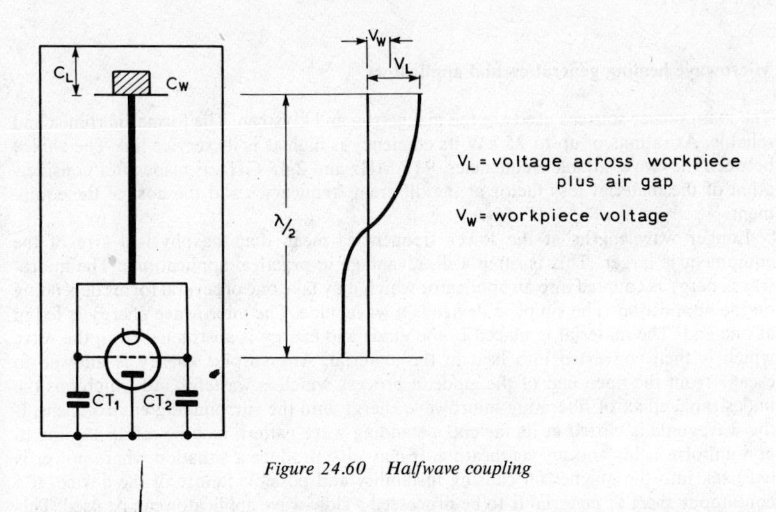

Figure 24.60 Halfwave coupling

shows a system consisting of a half wavelength line capacitively shorted at one end by the tank capacitance and at the other by the load. A transformation from generator to load can be achieved by suitable selection of the end capacitance. Frequency stability is improved and the circuit is relatively free from parasitic oscillations.

The simplest form of electrode system consists of two flat plates. The work material is placed between them and the field passes completely through the material. An alternative is the stray field system in which rods are placed on one side of the material, alternate rods being connected to opposite output terminals of the generator. This is suitable when the area involved is large and some degree of ventilation is required as in a drying process. A similar applicator is used for edge gluing of wood. The wood and

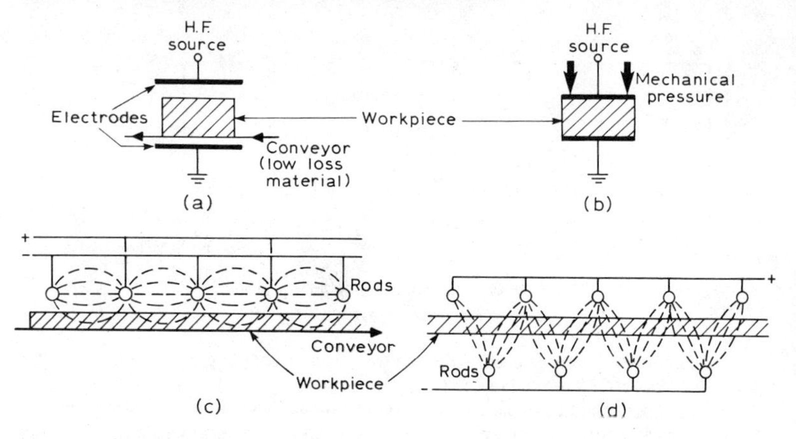

Figure 24.61 *Dielectric heating applicators (a) platers for through heating (b) platers with applied pressure (c) strayfield system (d) rid system for through heating*

glue appear as parallel capacitors and if an appropriate glue with a high loss factor is used the process is very efficient. These and other forms of applicator are illustrated in *Figure 24.61*.

Microwave heating generators and applicators

The main power sources used are the magnetron and klystron. The former is robust and reliable. At ratings of up to 25 kW its efficiency is high as is its service life. The choice between the two available frequencies, 915 MHz and 2.45 GHz, is made after consideration of the material loss factor at the different frequencies and the cost of the equipment.

Longer wavelengths at the lower frequencies mean that the physical size of the equipment is larger. This is often a disadvantage in practical applications. The microwave energy is coupled into an applicator which may take one of several forms depending on the application. The simplest device is a waveguide. The microwave energy is fed in at one end. The material is placed in the guide and energy is abstracted from the wave which is then converted into heat in the material. Any surplus energy is allowed to escape from the open end of the guide, a process which is wasteful and which has the undesirable effect of liberating microwave energy into the surrounding environment. If the waveguide is closed at its far end a standing wave pattern may be set up leading to non-uniform fields and uneven heating. It may also produce a situation where power is fed back into the magnetron causing instability and possible failure of the device. If a continuous sheet of material is to be processed a slow-wave applicator can be used. This consists of a waveguide as mentioned above but bent back upon itself repeatedly in a serpentine arrangement, *Figure 24.62*. The material is introduced in a slot at the side of the first section of guide and passes through the remaining sections in turn, absorbing power at each stage until all the power is extracted from the guide and utilised in the material. A dummy water-load is situated at the end of the last section of guide in order to absorb any surplus power. This is necessary since the generator may be switched on with no load material in the applicator or if the applicator is only partly loaded, for instance at the beginning and end of the processing cycle.

If the material to be heated is bulky rather than laminar in form, the type of applicator described above is impractical and a cavity is used. The cavity is a rect-

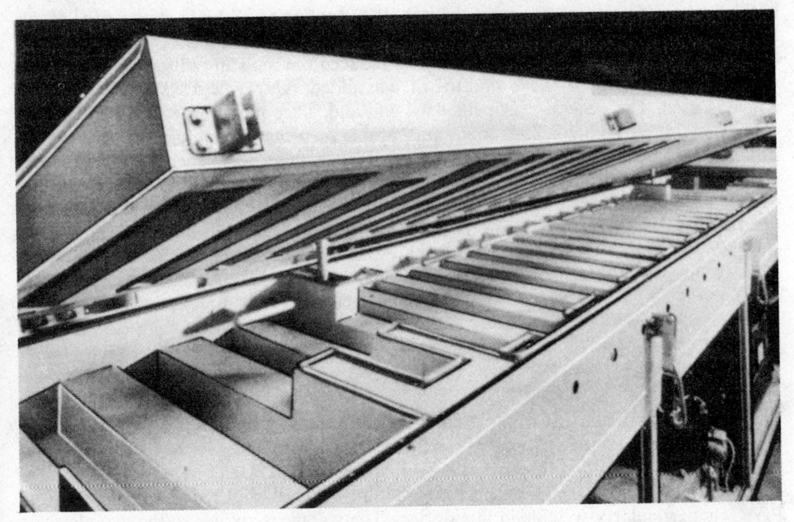

Figure 24.62 Serpentine applicator (*Courtesy Magnetronix Ltd.*)

angular box, the microwave energy being inserted by a coupling device at one point. Standing waves are set up in the cavity leading to non-uniform fields which in turn means non-uniform heating. The number of resonances occurring depends on the physical dimensions of the cavity and mathematical techniques are available to calculate the optimum dimensions. However, there is still some non-uniformity and in any case the introduction of the load will disturb the field pattern. In order to increase the uniformity of field distribution the technique of 'mode stirring' is adopted. The microwave energy is directed at a slowly rotating fan continuously changing the angle of reflection, so varying the field pattern and providing for more uniform heating over the whole of the load.

Applications of dielectric heating

R.F. dielectric heating is used widely for the welding and shaping of plastics and in the shoe industry for preforming. It is applied in the woodworking industry where it is used for glue drying and in food manufacturing for the drying and baking of cereal products. In the automobile manufacturing industry it is used for forming and welding operations on linings and seat covers.

The main application of microwave heating is in domestic and commercial food preparation. It is used for the curing of rubber and leather and in the drying of textiles after dyeing. Other applications have been shown to be possible but further development is required before the techniques can be used in industry.

R.F. DRYING OF PAPER

Paper is normally made from cellulose materials suspended in water. The bulk of this water must be removed during the manufacturing process to leave the material with a uniform moisture content of around 7%. The usual method is to squeeze the paper between large steam-heated rollers. This is effective apart from leaving non-uniformity in the final moisture content across the web. R.F. heating at 27 MHz is now used as the final stage in the drying process. In one application, processing 100 kgm of paper per hour, the addition of 100 kW of r.f. drying reduced the moisture differential from 7% without the r.f. energy to 0.5% after the r.f. was added. A stray field system as described above is used. The generator is a standard unit and, since the paper is fed through in a continuous sheet, the loading conditions are approximately uniform so easing the problem of matching. R.F. has been chosen instead of microwave energy since although the rate of heating of the complete paper-water system is higher with the latter due to the overall higher loss factor nevertheless, at the lower frequency, the differential between the loss factors of paper and water is much greater with the result that most of the heat is developed in the water in the areas where the moisture content is higher. Since the process involves the removal of water by vaporisation the equipment must be well ventilated in order that drying may proceed satisfactorily.

WELDING OF PLASTICS

This is probably the largest application of dielectric heating in the United Kingdom and although the size of the individual units is relatively small (down to 1 kW), the total number is so large that the overall load is very high. Many types of goods are manufactured, the only requirement being that the material must have a high loss factor. P.V.C. is particularly suitable in this respect. The voltage is applied to the two plates of a press which also act as dies and the material is welded together according to the die pattern. The shoe industry uses the flow forming technique with cheap electrodes manufactured from silicone rubber producing embossed materials with surfaces very similar to those of natural materials.

In the automobile industry 100 kW generators capable of pressures up to 70 tonnes are used to produce a complete set of seating for a car in less than a minute.

The size of the electrodes in this as in other applications must not be larger than a quarter wavelength at the frequencies used otherwise standing waves may be set up with the consequent risk of uneven field distribution and non-uniform heating. At 27 MHz the size is limited to 200 cm but, at the lower frequency of 13 MHz, platens of 300 cm have been used to produce large welded items such as inflatable boats.

GLUE DRYING

The furniture manufacturing and woodworking industries assemble finished components by gluing sections together. If normal glue drying is used the manufacturer must store the finished assemblies carefully for a long period until the glue sets. This requires large storage areas and ties up a considerable amount of capital. Dielectric heating is used to accelerate the process.

MICROWAVE HEATING IN THE RUBBER CURING INDUSTRY

In the manufacture of synthetic rubber products the material is first extruded into a flat sheet which is raised to the curing temperature and held there for some time until the chemical reactions involved in the curing process have taken place. A microwave curer has been developed to act on-line with the extruder to give a continuous extrusion-curing process of compact dimensions, *Figure 24.63*. Microwave energy is used to give total heat penetration of the material. In order to avoid low surface temperatures due to cooling, conventional surface heaters of the resistance type are incorporated. This combination gives a rapid increase to curing temperature which is typically 120 °C above the extruder temperature. The frequency used is 896 MHz allowing an applicator

Figure 24.63 Microwave rubber curing. The photograph shows a complete production line incorporating a 90-mm extruder, a single channel 25 kW 'Magnacure' and water cooling bath (Courtesy Magnetronix Ltd.)

to be designed with an inlet aperture of 102 mm by 355 mm. When a power unit of 25 kW rating is used the throughput capability is 450 kg/hour. The material is fed on to a conveyor belt into the heating zone which is a specially developed form of serpentine waveguide. The material crosses a series of twenty waveguides at right angles. The microwave energy flows along the adjacent channels continually crossing the extrusion and generating heat so that the temperature of the material is gradually increased along the length of the heating zone. An absorbing water load is included at the end of the final channel to absorb any surplus energy. Considerable pains are taken to ensure that the heating is as uniform as possible and the series of waveguides are effectively separate units connected together by coupling straps and slightly displaced in position by half a wavelength so that any standing wave effects are equalised. As mentioned above the surface is heated by electric resistance heaters and hot air is also circulated within the unit. A hot tunnel is fitted at the end of the device to provide for the dwell time. This tunnel is typically 5 metres long and is designed to maintain the rubber at curing temperature until the cure is complete. The material is carried on a low-loss glass fibre/p.t.f.e. belt the speed of which can be varied up to 36 m/min.

MICROWAVE COOKING

There has been a continuous increase in the use of microwave ovens for food cooking both in the domestic and commercial sectors. The main advantages are the fact that since the heat is generated within the material, unlike conventional ovens where it needs to be conducted from the surface, the final product is more uniform without over-cooking of the surface and the speed of operation is much higher. All the natural qualities of the food are preserved, the main disadvantage being that the surface finish is inferior to that normally expected from a conventional oven. This is due to the absence of surface heating, although current developments of a combined hot-air/microwave system overcome this problem without sacrificing any of the advantages of the micro-waves alone.

Magnetrons rated at up to 2 kW are used as the basic high frequency source and the major design emphasis is on reliability, simplicity and safety. The only control required is a timer due to the short and therefore critical times involved. This is arranged to switch off the power source after the appropriate period. Anode current stabilisation is usually considered to be necessary due to the possibility of major fluctuations in the mains voltage. A typical power supply unit has a separate heater transformer which is energised as soon as the oven is switched on to the mains. The h.t. transformer is energised at the instant when the cooking process is started. At the same time the heater supply voltage is reduced since, during load operation, electron bombardment contributes to the cathode temperature. A simple series resonant circuit utilising a capacitor and the leakage reactance of the anode transformer is used to provide a good regulation characteristic. This gives a form factor of about 1.15 and a high power factor.

The oven itself is a cavity usually fitted with a mode-stirrer. From the safety aspect the main requirement is a good interlock which prevents the magnetron being energised when the door is open, together with door seals which prevent the egress of microwave energy. One method is to use quarter-wave slots around the edge of the door. These are filled with a dielectric to prevent the accumulation of dirt in the slot.

FURTHER READING

CABLE, J. W., *Induction and Dielectric Heating*, Reinhold, New York (1954)
COPSON, D. A., *Microwave Heating*, AVI Publishing Co. Conn. (1962)
DITTRICH, H. F., *Tubes for R.F. Heating*, Philips Application Book, Philips Ltd. (1971)
HARRY, J. E., *Industrial Lasers and their Applications*, McGraw-Hill (1974)
J. and Trans. International Microwave Power Institute, Edmonton, Canada
LOZINSKI, M. G., *Industrial Applications of Induction Heating*, Pergamon Press (1969)
OKRESS, E., *Microwave Power Engineering*, Vols I and II, Academic Press (1968)
POUND, J., *R.F. Heating for the Wood Industry*, Heywood and Co., London (1957)
Proc. VIIth International Congress on Electroheat, Warsaw (1972), International Union of Electroheat, Paris
Proc. 1969 and 1973 I.E.E.E. Conferences on Electric Process Heating in Industry, I.E.E.E., New York
PUSCHNER, H., *Heating with Microwaves*, Philips Technical Library, Cleaver Hulme Press (1966)
SIMPSON, P. G., *Induction Heating*, McGraw-Hill (1960)
THE ELECTRICITY COUNCIL, *Induction and Dielectric Heating*, The Electricity Council, London (1962)
VON HIPPEL, A. R., *Dielectric Materials and Applications*, Chapman and Hall (1954)

INDUSTRIAL LASER APPLICATIONS

Materials processing

The use of laser materials processing is at present rather limited in scope, largely owing to the high cost and low efficiency of available lasers. This means that there are relatively few applications for which the potential advantages of the laser, e.g., non-contacting, low kerf, etc., cutting, make it also economically attractive. The use of lasers for material processing depends on the very high power densities, or energy densities (in the case of pulsed lasers) available in focused laser beams. Power densities of 10^7 W/cm^2 from c.w. lasers, and energy densities of 10^7 Joules/cm^2 or more with pulsed lasers (in pulses down to 10 nanoseconds or less) are readily attained. These figures mean that all known materials can be vaporised. Of course vaporisation is not the only process of importance in materials processing using lasers. Indeed in welding it is important to keep the power density sufficiently low that vaporisation does not take place. Even in cutting materials, ejection of liquid material by the vapour is often of importance, and sometimes the laser is used merely as a concentrated heat source in conjunction with a concentric oxygen jet which then removes the material (usually metallic) by oxidation, *Figure 24.64*.

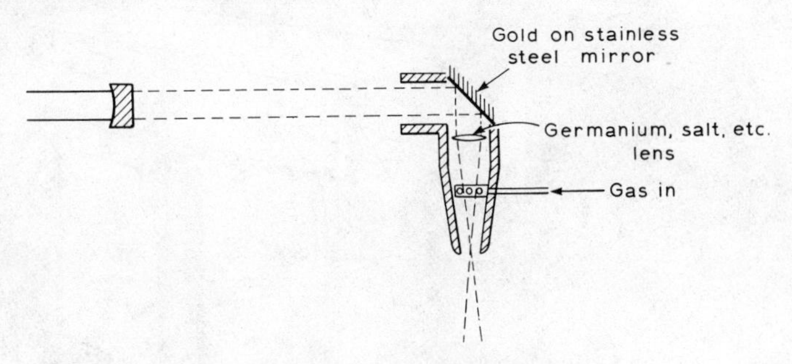

Figure 24.64 Gas jet nose piece for CO_2 laser

Only two types of laser are presently of interest for materials processing, gas lasers and solid stage lasers; of these the CO_2 laser and ruby, Nd glass, and Nd YAG systems are the only ones of any importance so far.

APPLICATIONS IN THE ELECTRONICS INDUSTRY

The most extensive use of lasers in this area is in the trimming of thick-film resistors to tight tolerances by the selective removal of material. The preferred laser for this application is the continuously pumped repetitively Q switched Nd YAG laser, which gives pulses a few hundred nanoseconds in length with peak powers of 1 kW or so with a pulse rate of up to a few kHz. The substrate is usually moved relative to a fixed laser beam in a predetermined pattern. A microscope used with a closed TV system is usually used for initial setting up, for safety reasons, and a built-in monitor stops the cut when the trimming is complete.

Another application is the use of a focused c.w. laser beam (usually CO_2) for 'scribing' semiconductor slices into dice. This uses the induced thermal stress to crack the wafer along the required lines. This is usually done by moving the slice. Pulsed lasers have found some applications in the welding of connections to printed circuit boards and even in the repair of broken connections inside vacuum tubes with transparent envelopes.

OTHER INDUSTRIES

Pulsed ruby and Nd glass or Nd YAG lasers are used in the drilling of very hard refractory materials such as ruby and diamond. Synthetic ruby bearings are extensively used in the production of watches and clocks. These require the drilling of a hole about 0.1 mm in diameter in a ruby blank ~1 mm thick and several pulsed Nd YAG systems with outputs of ~0.5 joules/pulse are in use for this purpose. Pulsed ruby and Nd glass lasers with outputs of several joules per pulse have found applications in the drilling of diamond dies for wire production and in the dynamic balancing of chronometers and small gyroscopes, by selective removal of small quantities of material from the moving

Figure 24.65 500 W CO_2 laser cutter (Courtesy Elliott Bros (London) Ltd.)

part. C.W. CO_2 lasers are in use in the U.S.A. for cutting out garments and are reported to be undergoing evaluation for the welding of car body panels. A commercial laser for cutting applications is shown in *Figure 24.65*.

Metrology and non-destructive testing

ALIGNMENT DEVICES

Low power He–Ne lasers are almost universally used for this role. They utilize the low beam divergence to provide a long weightless straight edge which cannot be accidentally destroyed. It is used in many civil engineering applications such as tunnelling, road grading, pipe laying, etc., and for these applications a rugged laser mounted on a theodolite base, using visual sighting of the beam centre, is adequate. For short distances, the unmodified laser beam can be used (divergence ~1 mrad, initial diameter ~1 mm). For larger distances, an inverted telescope is used to expand the laser beam by a factor equal to the power of the telescope, *Figures 24.66* and *24.67*. This reduces the beam divergence by the same factor, thus a ×10 telescope would give a beam diameter of ~1 cm, expanding to ~10 cm at 1 km under ideal conditions. In practice the beam would usually be appreciably larger than this due to atmospheric turbulence.

For more precise alignment work, such as the alignment of the bearings of a large turbo-attenuator, more precise methods of determining the beam centre are necessary. A quadrant photodiode is usually used for this purpose. For very precise work, where angular accuracies of better than a few seconds of arc are required, the direction of the laser beam is not sufficiently stable, and a passive prism system can be used to give a beam stability of ~0.1 second of arc.

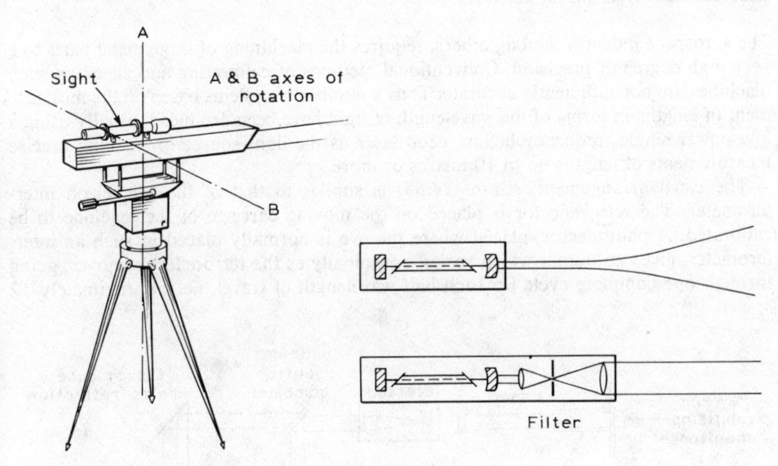

Figure 24.66 Alignment laser

Figure 24.67 Alignment laser in use (*Courtesy Elliott Bros* (*London*) *Ltd.*)

PRECISION LENGTH MEASUREMENTS

The aerospace industry, among others, requires the machining of large metal parts to a very high degree of precision. Conventional methods of calibrating and checking such machines are not sufficiently accurate. Thus a number of systems based on the measurement of lengths in terms of the wavelength of light have been produced, usually using a low power single frequency helium neon laser as the light source for making precise measurements of lengths up to 10 metres or more.

The usual arrangement, *Figure 24.68*, is similar to that of the Michelson interferometer. The retroreflector is placed on the moving carriage of the machine to be calibrated. A photodector, placed where the eye is normally placed in such an interferometer, gives an output which varies sinusoidally as the retroreflector moves, going through one complete cycle for each half wavelength of travel, i.e., approximately 12

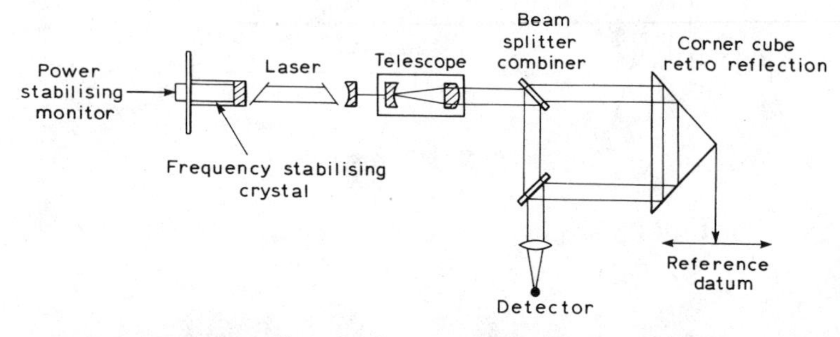

Figure 24.68 Laser interferometer

microinches. A bidirectional counting circuit is used to find the total length traversed. Since the wavelength of light depends on the refractive index of the air which it traverses, which in turn depends on the atmospheric pressure and humidity, in very precise measurements, 1 part/million or better, it is necessary to measure these quantities and to correct for them. In one machine part of the reference path can be evacuated. The resulting path change is measured and an integral computer corrects the data automatically.

NON-DESTRUCTIVE TESTING APPLICATIONS

These comprise vibration analysis, strain analysis, contour measurement and fault detection. Most of these applications require the use of holography, for which a laser is essential. The most common lasers used for these purposes are He–Ne and Argon–Ion lasers; the latter are usually only required for large specimens (area 1 m^2 or more).

A hologram consists of a recording of an interference pattern between light scattered from the object under test and a reference beam derived from the same light source. The optical path difference between the scattered light and the reference beam must be less than the coherence length of the light from the laser. For objects of large extent, this means that a single frequency laser operating in a single longitudinal mode must be used. The usual method of recording the interference pattern is a high resolution photographic plate, such as the Agfa Scientia E 75, and the laser power required must be sufficient to give an adequate exposure of the photographic plate in a reasonable time, say 1 minute or less. The laser and other optical components must be mounted on a vibration-free platform, such that optical path differences do not change by more than an eighth of a wavelength during the exposure.

Holograms can be used for investigating vibration and strain by replacing the processed holographic plate and the object exactly in the positions they occupied during the exposure and then viewing the object through the hologram, as if it were a window, when changes in position due to an applied stress or to induced vibration can be seen in the form of interference fringes superimposed on the test object.

In the case of vibration analysis stroboscopic viewing synchronised to the vibration frequency can be used to give time resolved information; otherwise time averaged information is obtained. *Live fringe* holography as this method is called, is not often used because of the difficulty of replacing the photographic plate exactly in the same place without *in situ* processing. Instead a second exposure is taken, before processing, giving a superimposed hologram of the strained or vibrating object. On development the fringes due to the movement or to the time averaged vibration are visible and the amplitude of vibration or the movement of the surface can be calculated, although it should be pointed out that for movement in more than one plane the interpretation of the fringe pattern is not simple.

More recently, a method of holography in which the reference beam interferes with a focused image of the (stationary, unstrained) object under test, at a small angle so that fringes of such low spatial frequency, as can be resolved by a TV camera, are formed. The resulting hologram is stored on videotape and the signal subtracted from a 'live' hologram of the vibrating or strained test object, to give a display of the 'live fringe' interference pattern on a TV monitor. Unfortunately rather higher laser powers are required than for photographic holography. The sensitivity of holographic methods of measuring distortions of objects due to applied stresses, temperature changes, etc., are often too high for engineering use, and methods using moiré effects using projected fringe patterns, obtained by interfering two laser beams (from the same laser) at a small angle, have been devised.

LASER VELOCIMETERS

There are many industrial situations where it is required to measure the velocity of moving objects by a non-contact method, or to measure velocities in liquids or gases without interfering with the flow. There are a number of different optical systems used in laser velocity measurements. *Figure 24.69* shows one of these in which it is seen that two beams, derived from the same laser, intersect on the moving surface, producing fringes. The light scattered from a point moving in the surface is modulated as it passes through the fringe pattern, and the receiver, whose axis bisects the angle defined by the two beams, picks up light modulated at a frequency determined by the fringe spacing and the velocity.

Other arrangements can be interpreted in terms of Doppler shifted light being heterodyned with unshifted light. In most cases a frequency tracking filter is used to improve the signal-to-noise ratio.

Miscellaneous applications

PRINTING APPLICATIONS

A considerable amount of development work is being carried out aimed at the development of improved printing (both black and white and colour printing) and copying systems using lasers. For instance, considerable improvements in the speed of scanners for three-colour printing can be obtained because of the greater light intensities available

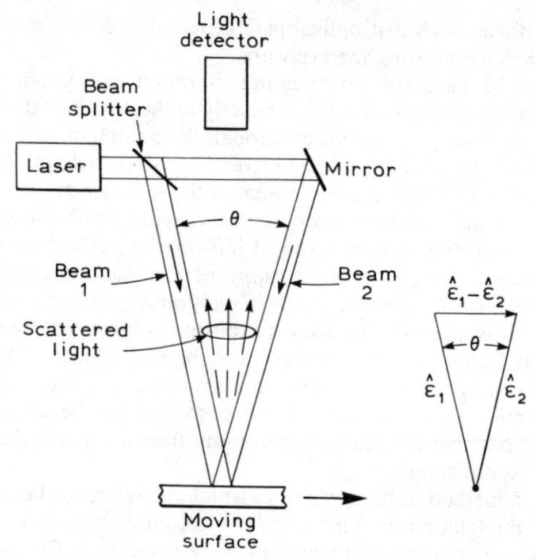

Figure 24.69 Differential Doppler velocity sensor (Conference Papers on Laser Engineering and Applications, I.E.E.E. (1969))

with laser sources. Very high speed printing (>10^5 characters/second) is possible using zinc oxide coated electrophotographic papers and an argon ion laser, together with an ultrasonic light deflector and modulator.

POWER GENERATION

One of the two main lines of research aimed at achieving thermonuclear power generation for peaceful purposes proposes the use of pulsed lasers to 'ignite' a small speck of lithium deuteride, or other suitable material, to a sufficiently high temperature. Then by means of shock wave pressure, to contain the plasma initially generated for a sufficiently long time to ensure that a substantial fraction of the available thermonuclear energy is produced. A thermonuclear reactor would comprise such means for ignition, on a repetitive basis, and for extracting the energy produced, probably in the form of heat initially, to 'fire' the boilers of conventional generating equipment. Theoretical estimates as to the minimum laser energy requirements vary from 10^4 joules to $>10^5$ joules. The pulse length must be of the order of 1 nanosecond or less and the pulse profile must be tailored to ensure proper confinement. There appears to be some difference of opinion as to the optimum in wavelength, with a preference for the ultra violet; however, various groups in the U.S.A. are building prototype systems using CO_2 and Nd glass lasers to try out the principles involved (an array of ten or more lasers is envisaged, each pointing at the fuel pellet position), so some experimental evidence may be available before very long.

FURTHER READING

BEESLEY, M. J., *Lasers and their applications*, Taylor and Francis (1971)

ELECCION, M., 'Materials Processing with Lasers', *I.E.E.E. Spectrum*, **9**, 4, 62 (1972)

ROBERTSON, E. R., ed., 'Engineering uses of Holography', *Proc. Conf. Strathclyde University, Sept, 1968*, C.U.P. (1970)

ROSS, M., ed., *Laser Applications*, Academic Press (1971)

WANG, C. P., 'A Unified Analysis on laser Doppler Velocimeters', *J. of Physics E.*, **5**, 721 (1972)

COMPUTERS

Analogue computer

Analogue and digital computers are now widely used in many fields. The two types of computer differ in fundamental concept. The analogue machine may be regarded as a model of a physical or mathematical problem. The values of the variables are represented in the machine by a physical quantity and the result is obtained by the measurement of another quantity. A slide rule is a simple example of an analogue computer, utilising the rule $\log AB = \log A + \log B$. Lengths proportional to $\log A$ and $\log B$ are added by moving the slide to give a length proportional to $\log AB$. The value of AB can then be determined by reading the scale, i.e. converting the length proportional to $\log AB$ into AB. The planimeter, for measuring area, is another example. The mechanical differential analyser and its electronic equivalent are useful for solving complex differential equations and electronic analogue computing machines are widely used in industry. Analogue computers are usually designed for one application, although some machines can be adapted to a range of problems by changing interconnections between their various units. The analogue machine, although limited by the accuracy with which the various quantities may justifiably be measured, can deal with continuous variables; in the digital machine the variables must change in discrete steps, though these may be small.

Analogue computation is applied to solving the behaviour of a system by representing its variables by analogues in some more convenient physical form, usually electrical quantities and time. Some kind of analogue computation enables the engineer to obtain approximate solutions to his problems with a speed and ease that greatly enhance the design value of mathematical equations otherwise difficult and tedious to solve. The accuracy is very considerably below that of a digital computer, but there are several compensating advantages.

The commonest form is the electronic differential analyser. The present state of development of this instrument is such that only an elementary knowledge of electronics is demanded of the operator. A full appreciation of the mathematical significance of the problem to be solved is, of course, essential.

ELECTRONIC ANALOGUE COMPUTER

All variables are represented by time-varying voltages, so that relatively simple recording equipment (oscillographs or curve-plotters) is required.

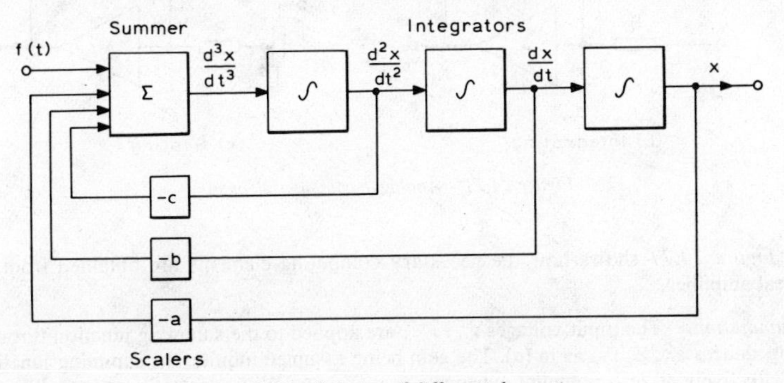

Figure 24.70 Solution of differential equation

The solution of a problem cast in the form of a differential equation involves the key process of integration. *Figure 24.70* shows the formulation of the differential equation

$$ax + b(dx/dt) + c(d^2x/dt^2) + (d^3x/dt^3) = f(t)$$

which can be restated as

$$(d^3x/dt^3) = f(t) - ax - b(dx/dt) - c(d^2x/dt^2)$$

The process can be followed by starting with the highest-order derivative of x, which is successively integrated to form the lower-order derivatives and, finally, x itself. Examination of the differential equation will now show that the highest derivative, which is to be fed to the beginning of the chain of integrators, may be obtained by adding together the variable x and its lower derivatives in appropriate proportions.

The summing and integrating units are based on high-gain electronic amplifiers with appropriate input and feedback impedances.

OPERATIONAL AMPLIFIER

An ideal amplifier for the purpose has an infinite voltage gain, zero input current (i.e., infinite input impedance) and zero drift. A typical electronic amplifier may have a gain of 10^6, an input current of 100 pA and a drift of 1 mV; it also has a sign reversal between input and output. It is usual to have one terminal common to both input and output circuits, at earth potential so that all voltages are relative to earth.

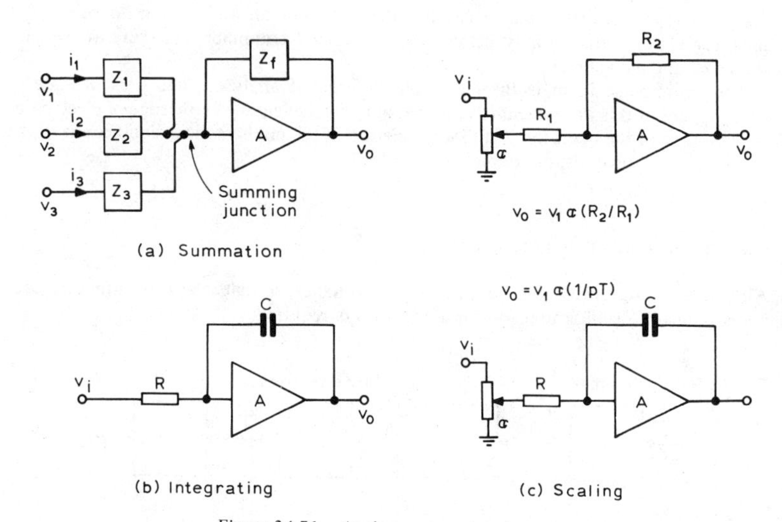

Figure 24.71 Analogue computer elements

Figure 24.71 shows how the necessary computing elements are obtained from an ideal amplifier.

Summation. The input voltages $v_1, v_2 \ldots$ are applied to the summing junction through impedances $Z_1, Z_2 \ldots$ as in (*a*). The gain being assumed infinite, the summing junction is effectively at zero potential, whence $i_1 = v_1/Z_1$, etc. With infinite input impedance to the amplifier A, the input current is zero and consequently $i_1 + i_2 + \ldots + i_f = 0$; that is, $v_1/Z_1 + v_2/Z_2 + v_o/Z_f = 0$, whence

$$v_o = -v_1(Z_f/Z_1) - v_2(Z_f/Z_2) - \ldots$$

If all impedances Z are resistive, their ratios are pure real numbers, and the output voltage v_o is the sum of the input voltages multiplied by any desired numerical coefficient.

Integration. In (*b*) the feedback impedance Z_f is a capacitor C. Then writing p for d/dt,

$$v_o = -v_i(Z_f/Z_i) = -v_i(1/RC\mathrm{p}) = -v_i(1/\mathrm{p}T)$$

which means that v_o is the time-integral of v_i divided by $T = CR$ (the time-constant) and reversed in sign.

Scaling. This is shown in (*c*), with potential-dividers giving the fraction α of the input voltage v_i. For the summer, α acts as a simple fractional multiplier, but with the integrator it changes the time constant in effect to CR/α.

PRACTICAL AMPLIFIER

The expressions above are for ideal amplifiers of infinite gain. Practical devices cannot match this condition. Further, it is clearly not feasible to have too great a range of capacitor and resistor values, nor are variable components of the appropriate values readily available; but a limited range will serve if adjustments are made by potentiometers, normally of value between 20 and 100 kΩ. Then most requirements can be satisfied by one value of capacitor (1.0 μF) and two of resistors (0.1 and 1.0 MΩ), with amplifiers having a voltage range of ±100 V and output currents up to ±10 mA.

Example

Suppose that it is required to solve the second-order differential equation

$$ax + b(\mathrm{d}x/\mathrm{d}t) + c(\mathrm{d}^2x/\mathrm{d}t^2) = \mathrm{f}(t)$$

The equation is rearranged as

$$\mathrm{d}^2x/\mathrm{d}t^2 = (1/c)\mathrm{f}(t) - (a/c)x - (b/c)(\mathrm{d}x/\mathrm{d}t)$$

The analogue computer is now set up as in *Figure 24.72*. As a sign-reversal is introduced between the input and output of each amplifier, care is necessary to ensure

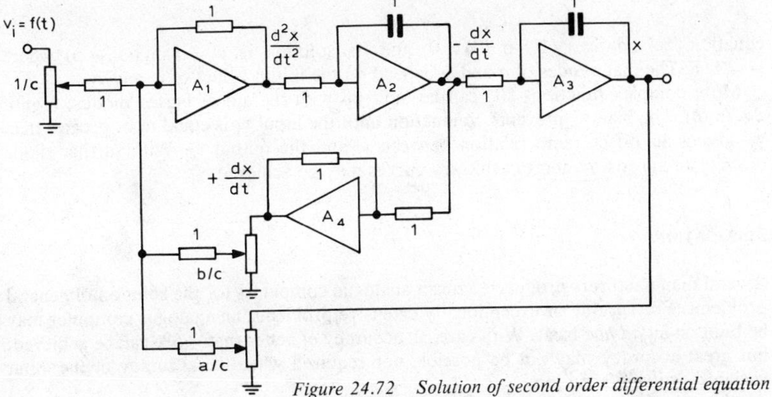

Figure 24.72 Solution of second order differential equation

that each loop is closed with the correct sign. If a change of sign is required, it is introduced by an additional amplifier having a gain of unity, as for A_4. Amplifier A_1 acts as a summer, while A_2 and A_3 are integrators. The figures marked on the resistor and capacitor components refer respectively to MΩ and μF.

Integrators require to be reset to a starting condition, either of zero or of some finite initial voltage. This is done by connecting the output and input terminals together through a resistor which has the input end connected through a second resistor to a voltage source of appropriate value.

FUNCTION GENERATORS

The time-variation of f(t) may require to take a variety of forms. There is no difficulty with a step-function. A sine function can be developed by means of operational amplifiers as in *Figure 24.73* (*a*), where A_1 is a summer and A_2 and A_3 are integrators. The

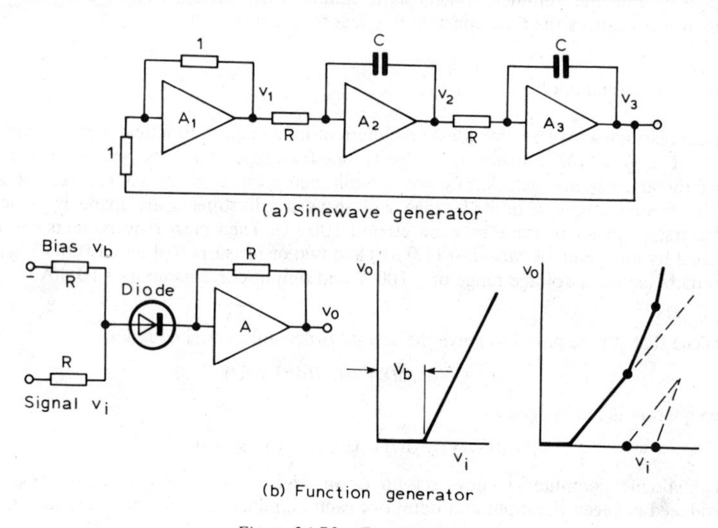

Figure 24.73 Function generation

equation 'solved' is $v_3(1 + p^2T^2) = 0$, and its solution is $v_3 = a \sin(\omega t + \phi)$ where $\omega = 1/T$. The magnitudes of a and ϕ depend on the initial values of v_2 and v_3.

More complex functions f(t) can be built up with the aid of biased diodes, *Figure 24.73* (*b*). The bias V_b prevents conduction until the input v_1 is equal to or greater than v_b, giving an offset ramp relation between v_i and the output v_o. With further diode circuits an approximation to arbitrary curves may be achieved.

APPLICATIONS

Several manufacturers produce compact analogue computers for the solution of general problems. For specific or exceptionally extensive problems, an analogue computer may be built on an *ad hoc* basis. With care, an accuracy of better than ±5% can be achieved; but great accuracy may not be possible nor required when the accuracy of the input variables is itself fairly low.

A system may be set up to solve a problem in 'real time', in which case actual components of the real prototype can be included if their analogue representation is difficult. This simulation may often be very useful, as in the design of a system in which certain components of a complicated nature are unavoidable. They can be fitted into an analogue computer in order to evaluate and optimise the system behaviour by extracting the required characteristics of those other parts of the system capable of adjustment in subsequent design.

When all parts of a prototype can be represented by analogue computer elements, the analogue can be arranged to operate with a time scale faster or slower than that of the prototype, a facility having clear advantages.

Among the very wide range of problems that have been studied by analogue computer the following are representative. Apart from cases that include simulation with actual components, any mechanical, electrical, biological or even economic system dynamics involving motion or variation in time may be studied. Provided that the equations defining the behaviour can be formulated, then the analogue may be set up on the computer and studied, even before the prototype exists physically.

EXAMPLES

The main areas of use can be summarised as (1) optimisation of parameter values in system design studies; (2) dynamic behaviour of open- and closed-loop control systems; (3) simulation; (4) stability of structures and of networks under dynamic-load conditions; (5) data reduction.

Optimisation. Study of bridge design and behaviour under wind loads. Adjustment of parameters in a feedback system for best performance. Heat transfer by natural convection.

Dynamic behaviour. Control regulator performance. Production processes in which operation is based on continuous-flow methods. Car-suspension systems. Aircraft vibration problems in flight.

Simulation. Nuclear reactor control systems. Chemical reactions. Servomechanisms with some real parts incorporated into the computer.

Stability. Electric power system studies.

Data reduction. Chemical plant analogue with a variety of input signals, and output arranged through transducers to alter the inputs.

Digital computer

The digital computer performs three major roles. As a *calculating machine* it is capable of executing computations of considerable complexity and finds widespread application in all branches of science and engineering. It is used extensively for *data processing* in commerce and industry. In this activity the calculating power of the machine is less important than its ability to store and rearrange large quantities of information. The third role is in the *monitoring and control* of industrial processes and communication systems. The on-line process control computer has been used in chemical plants, generating stations, road-traffic control and process industries.

COMPUTER STRUCTURE AND ORGANISATION

The basic digital computer has four identifiable sections: the store, arithmetic unit, control and input/output devices. The *store* contains the numerical quantities and data which are to be processed and the program or list of instructions which are to be performed. The *arithmetic unit* normally performs the operation of addition, subtraction, multiplication and division and certain other special operations. The *control unit* directs the whole machine and determines when a change from the normal sequence in execution of the list of instructions is required. The *input and output devices* provide a means by which information may be supplied and obtained from the computer. The normal input media for a digital computer are punched paper tape or punched cards while results are usually obtained from a printer.

BINARY NUMBER SYSTEM

Information within the computer is normally represented in binary form, because it is convenient to use circuits which have two clearly defined states. Thus a transistor can be fully conducting or non-conducting, and magnetic material can be saturated, in one direction or the other to represent the two binary states 0 or 1. In the binary system the successive digit positions represent the powers of 2 in increasing significance. Thus the decimal number 115 is represented in binary form as

$$1110011 = 1(2^6) + 1(2^5) + 1(2^4) + 0(2^3) + 0(2^2) + 1(2^1) + 1(2^0)$$

The conversion is performed by successive subtraction of the highest power of 2 less than the decimal number. The conversion of decimal 115 to binary proceeds as follows

$115 - 2^6 = 115 - 64 = 51$	hence number contains	$1(2^6)$	1000000
$51 - 2^5 = 51 - 32 = 19$		$1(2^5)$	100000
$19 - 2^4 = 19 - 16 = 3$		$1(2^4)$	10000
3 is less than 2^3		$0(2^3)$	0000
and less than 2^2		$0(2^2)$	000
$3 - 1^1 = 3 - 2 = 1$		$1(2^1)$	10
$1 - 2^0 = 1 - 1 = 0$		$1(2^0)$	1
		Sum	1110011

BINARY-CODED DECIMAL

This is less common. Each decimal digit is represented by four binary digits having weights 2,4,2,1. Thus the decimal digits are represented as

Decimal:	0	1	2	3	4	5	6	7	8	9
Binary:	0000	0001	0010	0011	0100	0101	0110	0111	1110	1111

BINARY OPERATIONS

Extensive use is made of the basic logical operations on binary digits within the computing and control circuits of the computer. The most elementary is the *unary* operator 'not', which is defined as follows:

if $z = \text{not } x$, then $z = 1$ for $x = 0$, and $z = 0$ for $x = 1$.

The other operations are performed with two input digits and are defined for $z = x$ 'operator' y by the following, in which the 'operator' is 'and', 'or', or 'exclusive or'

Operator	$x =$ $y =$	0 0	0 1	1 0	1 1
'and'	$z =$	0	0	0	1
'or'	$z =$	0	1	1	1
'exclusive or'	$z =$	0	1	1	0

In symbolic writing the operation 'and' is represented by the full point '.', 'or' by the plus sign '+' and 'exclusive or' by an enclosed plus sign '⊕': while 'not x' is represented by a superscript bar, '$\bar{x}$'. The latter operation may be defined in terms of the other operators

$$x \oplus y = (x + y) \,.\, (\overline{x \,.\, y})$$

RULES

The general rules for operations on binary variables have been defined in the laws of Boolean algebra. These are described by a set of logical equations or assertions which state that if the left part of the equation is true (i.e., has the logical value 1) then the right part must be true. The first four laws deal with one logical variable p and the digits 0 and 1.

$$p + 0 = p \quad p + 1 = 1 \quad p \,.\, 0 = 0 \quad p \,.\, 1 = p$$

The first of these states that the value of $p + 0$ (that is, p 'or' 0) is p; thus when $p = 0$ it has the value 0 and when $p = 1$ it has the value 1.

The *idempotent* laws are

$$p + p = p \qquad\qquad p \,.\, p = p$$

The operations 'or' and 'and' are *commutative*, i.e.,

$$p + q = q + p \qquad\qquad p \,.\, q = q \,.\, p$$

and they are also *associative*, i.e.,

$$(p + q) + r = p + (q + r) \qquad\qquad (p \,.\, q) \,.\, r = p \,.\, (q \,.\, r)$$

The *distributive* laws are

$$(p \,.\, q) + (p \,.\, r) = p \,.\, (q + r) \qquad\qquad (p + q) \,.\, (p + r) = p + (q \,.\, r)$$

The *complementary* law is

$$p + \bar{p} = 1 \qquad\qquad p\bar{p} = 0$$

and the *involution* law is $(\bar{p}) = p$. A useful pair of postulates known as de Morgan's theorem is

$$(\overline{p \,.\, q}) = \bar{p} + \bar{q} \qquad\qquad (\overline{p + q}) = \bar{p} \,.\, \bar{q}$$

All these laws may be verified by substitution of the binary values 0 and 1. Thus

when $p = 0$ and $q = 0$, then $(\overline{p \,.\, q})$ is 1 and $\bar{p} + \bar{q}$ is 1
when $p = 0$ and $q = 1$, ,, 1 ,, 1
when $p = 1$ and $q = 1$, ,, 0 ,, 0

LOGIC CIRCUITS

The logical operations 'not', 'and' and 'or' can be implemented by transistor and diode circuits. In *Figure 24.74* when a positive current I_i of sufficient magnitude flows into the

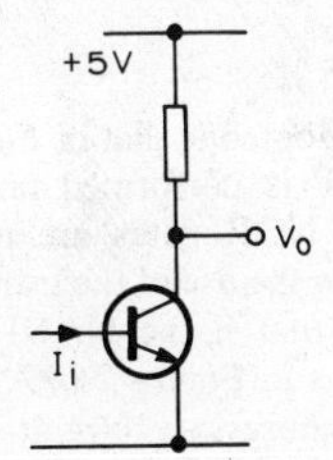

Figure 24.74 *Transistor inverter performing a 'not' operation*

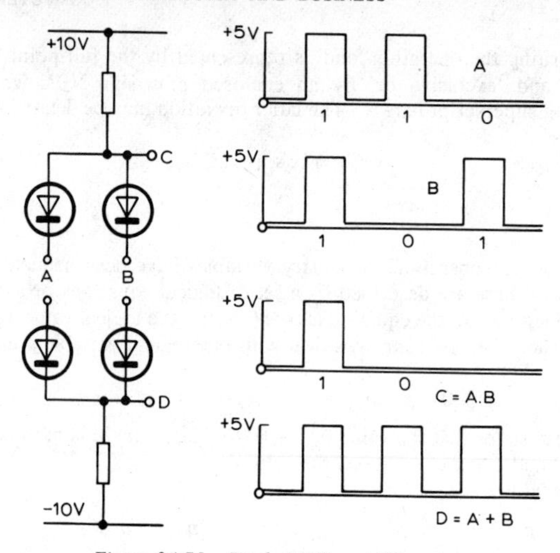

Figure 24.75 Diode AND *and* OR *gates*

base, the transistor will be fully turned on (*saturated*) and the output voltage V_o will be equal to V_{sat}, i.e., 0.1–0.2 V; but under conditions of zero input current $V_o = +5$ V. The transistor can thus be used to invert the polarity of an input signal, and with a suitable circuit a +5 V input results in an output of approximately zero, while input zero results in an output of +5 V. Representing logic 1 by +5 V and logic 0 by zero voltage it is seen that the transistor circuit performs the logical operation 'not'.

The logical operations 'and' and 'or' can be implemented by the diode circuits of *Figure 24.75*: they are normally referred to as AND and OR gates.

The logical operations are available in a number of ranges of integrated circuits. The

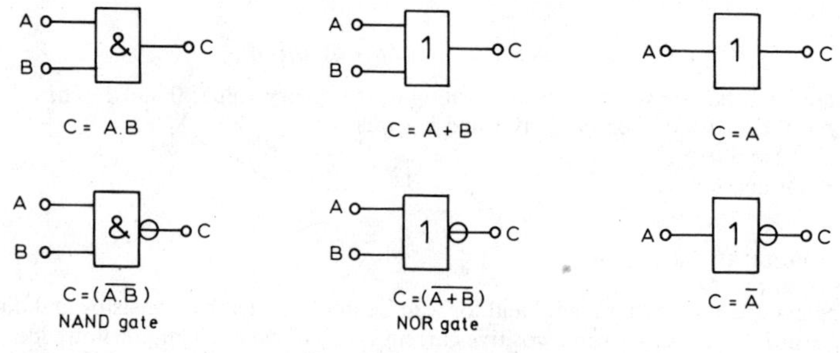

Figure 24.76 Logic gates

symbolism adopted is that of *Figure 24.76*. NAND is equivalent to NOT AND, i.e., the operand 'and' is performed on the two inputs followed by the operation 'not'. The NAND and NOR gates are most common because there is an inherent inversion resulting from the use of the transistor amplifier. By reference to de Morgan's theorem it can be seen that if the NAND operates with the inverse inputs, $\bar{A}$, $\bar{B}$ the result is C = A + B, as in *Figure 24.77*.

The other necessary circuit element is the *single-bit store*. The *clock* is a narrow

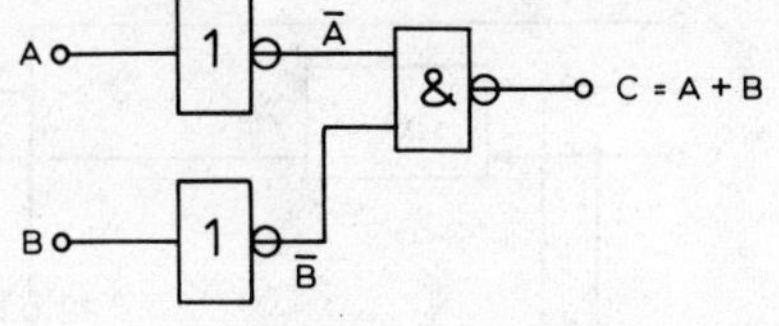

Figure 24.77 NAND *gate with inverted inputs*

positive pulse. If at the time of the occurrence of the clock pulse, the input =0, then the output becomes 0 and remains in this state until the next clock pulse. Similarly an input 1 at the time of the clock will set the output to 1. A reset signal is sometimes also provided, this resets the output to zero.

BINARY ADDITION

This can be implemented by a combination of the basic circuit elements. An example is shown in *Figure 24.78* (*a*). The operation proceeds from right to left. It will be recalled that the four columns correspond to the binary weights $2^3 = 8$, $2^2 = 4$, $2^1 = 2$ and $2^0 = 1$. The first operation is the addition 1 + 1. This results in a zero sum ($s = 0$) and a carry c which is held over to the next digit position. The next operation is performed

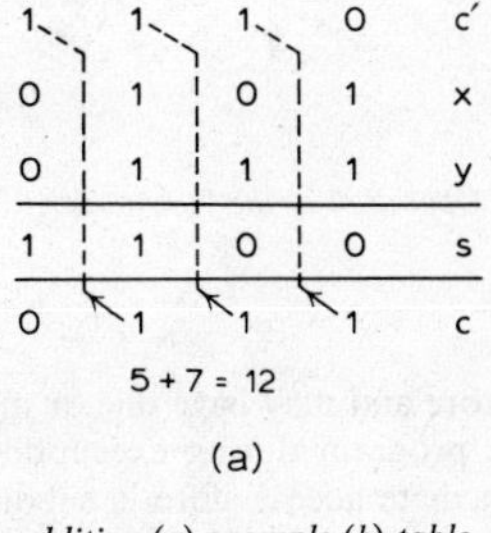

x	y	c′	s	c
0	0	0	0	0
1	0	0	1	0
0	1	0	1	0
0	0	1	1	0
1	1	0	0	1
1	0	1	0	1
0	1	1	0	1
1	1	1	1	1

(b)

Figure 24.78 *Binary addition* (*a*) *example* (*b*) *table*

with $x = 0$, $y = 1$ and the carry from the previous digit position $c' = 1$. The result is $s = 0$ and $c = 1$. The values of x, y and c' can be represented as inputs, and s and c as outputs of the adder. The complete binary addition table showing values of the outputs for all input combinations is shown in *Figure 24.78* (*b*): it gives the sum s and carry c digits formed as a result of adding all possible combinations of the digits of two numbers, x and y, and the carry c' from the previous digit place.

The addition table can be implemented from a combination of the basic circuit elements. The EXCLUSIVE OR is used within the adder as an element. This takes the form shown in *Figure 24.79*.

The full adder, with a switch for converting it to subtraction is shown in *Figure 24.80* together with a table of operations with intermediate results.

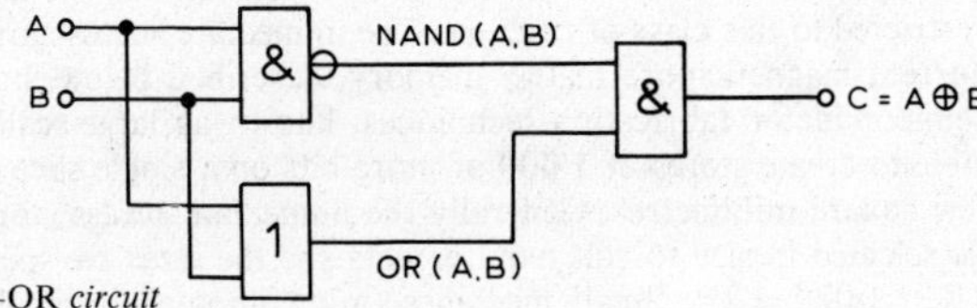

Figure 24.79 EXCLUSIVE-OR *circuit*

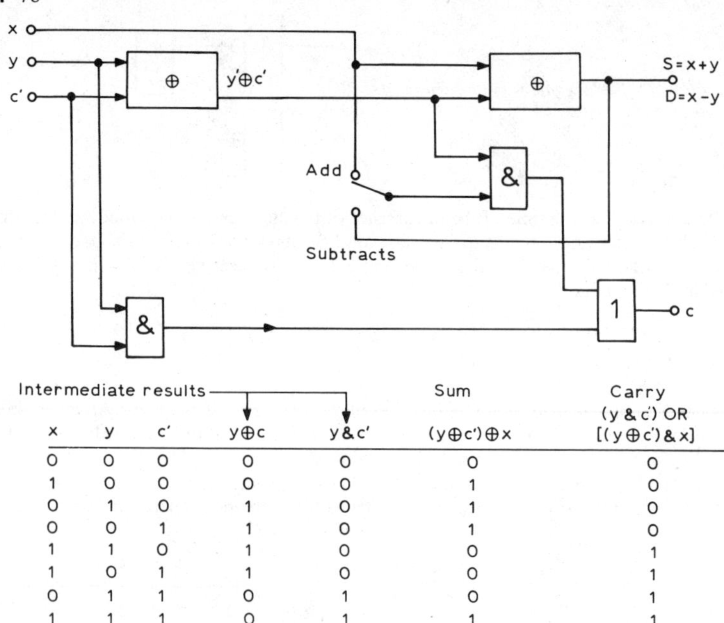

			Intermediate results		Sum	Carry
x	y	c′	y⊕c	y&c′	(y⊕c′)⊕x	(y & c′) OR [(y⊕c′)&x]
0	0	0	0	0	0	0
1	0	0	0	0	1	0
0	1	0	1	0	1	0
0	0	1	1	0	1	0
1	1	0	1	0	0	1
1	0	1	1	0	0	1
0	1	1	0	1	0	1
1	1	1	0	1	1	1

Figure 24.80 Operation of the logic adder

Storage

A computer has an immediate access store and may have one or more auxiliary stores. The *immediate access* store holds the program during execution together with data associated with the program. The immediate access store is subdivided into locations each with an identifying number or *address*. In a word-organised machine each location contains one word which is equivalent to an instruction or a number. Word lengths of 12, 16, 24 and 32 or more bits are used, the longer word lengths normally being employed in large computers for scientific applications, the greater length giving increased precision. When shorter word lengths are used, improved precision is obtained by using two or more words for each number. Certain machines primarily intended for commercial data-processing have a word length corresponding to one byte. A *byte* is between 6 and 10 bits and corresponds to a storage unit necessary to contain the binary code for the set of *alphanumeric* characters, that is, the alphabetic, the numeral and other special characters normally found, for example, in the typewriter keyboard. Computers normally employ one of a small number of international agreed standard character sets. A stored instruction or number will occupy several consecutive bytes and provision is made in byte-organised machines for operations with a variable number of bytes.

The majority of machines are word-organised and the description which follows is restricted to this class of machine. The immediate access store normally is a coincident-current magnetic-core matrix memory, described below, but current developments in semiconductor fabrication techniques, known as large-scale-integration, make it possible to create stores of 1 000 or more bits on a single slice of silicon with an area of a few square millimetres. Generally the immediate-access store of a given computer can be selected in size to suit requirements and the sizes are specified in K words in which $1K = 1\,024 = 2^{10}$. Small machines (minicomputers) are obtainable with a 4K store extendable in units of 4K up to 32K or 64K with a word length of 12 or 16 bits. Larger machines for data processing and scientific applications have immediate-access store ranging from 32K to 256K. The size of the immediate-access store must match the expected work load of the machine; this is considered in a later section.

The access time of the immediate store is of the order of 1 μs; this is the time to gain access to any one of the several thousand locations. The size of the immediate-access store is limited by cost.

MAGNETIC-CORE STORE

The magnetic-core matrix form of store employs ferrite rings of outer diameter in the range 0.2–2 mm, arranged in a square array, *Figure 24.81*. In the simplest scheme each core is threaded by three wires (horizontal, vertical, diagonal). The diagonal wire, for reading, passes through every core in the matrix. The cores are magnetised by currents

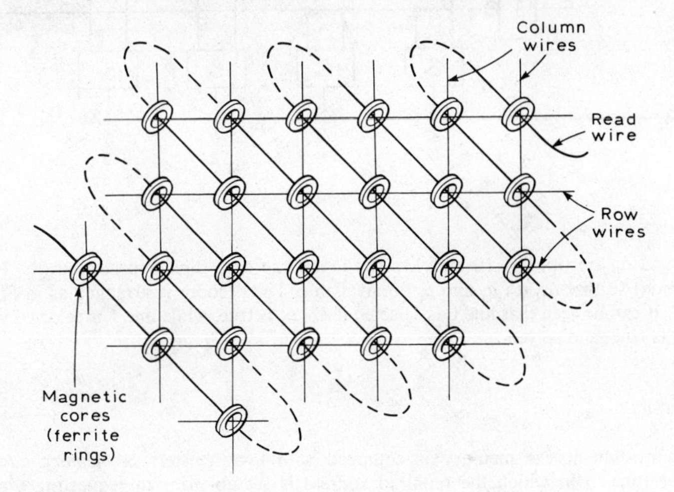

Figure 24.81 *Section of a magnetic core matrix store*

in the wires, and the two digit states are represented by the two remanent states of magnetisation. In reading, the magnetisation of a core is determined by attempting to change the state from 1 to 0: if an actual change of state occurs, then the previous state was a 1 and an e.m.f. is induced in the reading wire. If the original state is to be preserved for later reference the reading action must be followed by one that changes the state back to 1.

Selection of an individual core is based on the rectangular hysteresis loop, in that negligible change of state occurs for a current $\frac{1}{2}I$, but a reversal for a current I (in the appropriate sense), where I is of the order of 1 A. To select a given core, a current of $\frac{1}{2}I$ is passed down the column and another current of $\frac{1}{2}I$ simultaneously along the row that together locate the core. Only this core is subjected to a total current I, and if I is in the appropriate direction the core will change state.

A typical store module for a 16-bit word machine consists of 16 planes each consisting of a 64×64 array of cores. This module has thus a capacity of 4096 words (4K) and large memory sizes can be provided by using a number of such modules. The ability to select an individual word by directing a half-current pulse through one of the column wires and through one of the row wires of each of the 16 planes is an important feature of the matrix store. The addresses of the words in the store are 0,1,2 . . . 4094,

4095 and the required address is specified to the store as a 12-bit number as 000000000000 = 0 and 111111111111 = 4095. The address is divided into two groups of 6 bits, the less significant half being used to specify the row, and the more significant half the column. It is necessary to provide decoders which translate a 6-bit number into a choice of 1 out of 64. The principle of a decoder can be illustrated with a 2-bit number for simplicity: if the number is $a_1(2^1) + a_0(2^0)$ then the four possible values

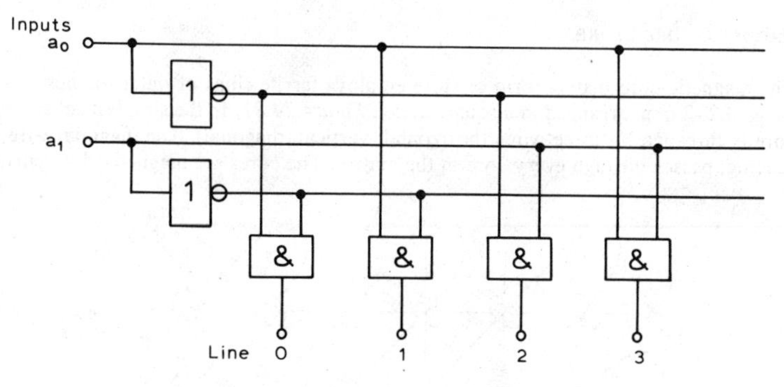

Figure 24.82 Decoder

of a_1 and a_0 are 00, 01, 10 and 11. The inverse of each digit is produced (the NOT operation) so that a_1, a_0, $\bar{a}_1$ and $\bar{a}_0$ are available. The decoder is arranged as in *Figure 24.82*. It can be seen that line 0 is selected if $\bar{a}_0 . \bar{a}_1$ is true, while line 1 is selected when $a_0 . \bar{a}_1$ is true, and so on.

REGISTERS

The immediate-access memory is equipped with two registers, a *memory address register* (m.a.r.) in which the required address is set up prior to requesting a store access, and a *memory data register* (m.d.r.). When a word is to be read from the memory, the address is set in the m.a.r. and a read request command is given, in response to which the contents of the required address is set up on the m.d.r. When it is required to write a word into a location, the word is set up in the m.d.r., the address in m.a.r. and a write command is issued. It is also possible to read out the contents of a location and write in a new word. The time to perform any one of these operations is referred to as the *cycle time* of the store and this time (which is important in deciding the overall speed of the machine) is a function of the type, size and cost of the immediate-access memory. Large core memories (256K) with cycle times of 1.5 to 2 μs are employed and smaller core memories (4K to 64K) with cycle times of 0.75 μs are available, while at the present time small integrated semiconductor memories with cycle times of 0.03 μs are being used, often as a small special memory to hold frequently used parameters and variables of a calculation. As a very rough guide the speed of a computer may be deduced from the cycle time of the immediate access store by noting that a typical instruction may require two accesses to the store, one to extract the instruction and the other to access data. The number of instructions which can be performed per second is thus the reciprocal of twice the cycle time of the store. However, there are many other factors which influence the speed of the computer and it is difficult to formulate a method by which the speed of one machine may be compared with that of another.

AUXILIARY STORES

Even quite large immediate-access stores have inadequate capacity in many applications, and a number of storage devices are available which provide very large storage capacity. Generally these stores employ a magnetic recording surface which is moved past one or more magnetic recording and reading heads. The surface may be in the form of a cylinder as in the drum store, a flat circular plate as in the disc store or a magnetic tape.

In contrast with the immediate access store there is a finite access time to information recorded on an auxiliary store. This is the elapsed time between making a request and the particular record becoming available under the reading head. With a drum or a disc the average access time is equal to half the period of revolution, but with magnetic tape the access time depends on the position of the required record on the tape relative to the record currently scanned. The three classes of auxiliary stores are intended for different applications.

MAGNETIC DRUM

The drum with many fixed pick-up heads has the shortest access time and is used as an extension of the immediate-access store. Typical drum speeds are 1 500, 3 000 or 6 000 rev/min, giving mean access times respectively of 20, 10 and 5 ms. The capacity of a drum ranges from 32K to 4 million words.

MAGNETIC DISC

In applications that require access to large files of data a disc memory is appropriate. One form uses interchangeable disc packs with 5 discs (i.e., 10 recording surfaces) mounted on the same access. The capacity of each disc pack is 7.25 million bytes (each of 8 bits) and the average access time is under 90 ms.

MAGNETIC TAPE

Magnetic tapes are used for very large files and for permanent records. As it is possible that the contents of a disc pack may be inadvertently lost by overwriting it is good practice to transfer periodically the contents of the disc to successive sections of a magnetic tape so that the information may be retrieved from the tape if necessary. The width of a standard tape is 12.7 mm (0.5 in) and the usual length is 730 m (2 400 ft). A byte (a character of 7 or 9 bits) is recorded longitudinally across the tape and a multiple head reads or records all the bits simultaneously. The density of recording depends on the tape transport: standard options are 8, 22, 32, 44 and 63 characters per linear mm (200, 556, 800, 1 100 and 1 200 per in). It is normal to record information in blocks of a few thousand characters, and an inter-record gap is left between each block to allow the tape to be started or stopped between a block. Each record is given its own identification to permit a specific block to be retrieved. However, it is time-consuming to make random accesses to blocks, and in normal usage the blocks are recorded and read serially. Magnetic tape is a low-cost storage medium and the capacity of a normal reel of tape is 5 to 20 million characters, depending on packing density.

General organisation

The block schematic diagram of a simple computer is given in *Figure 24.83*. Data paths between units are shown; the control unit also transmits control signals to all other units.

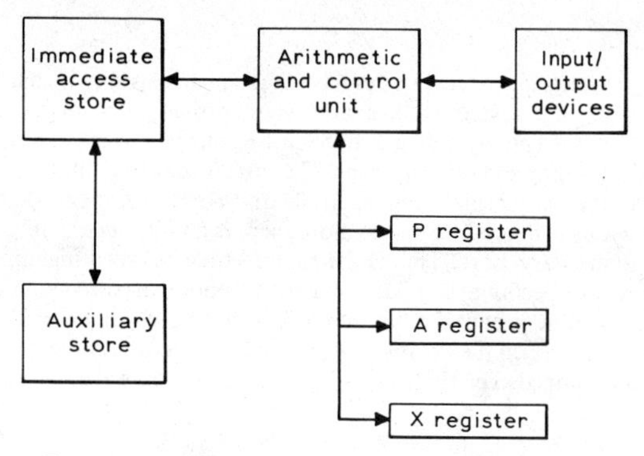

Figure 24.83 *General organisation of a digital computer*

OPERATION

The computer operates under the control of a stored program consisting of a list of instructions held in the immediate-access store (subsequently referred to as the 'store'). In a word-organised machine each instruction occupies one word and is stored in coded form as a set of binary digits. Essentially the instruction has two parts, a function and an address. The *function* specifies the type of operation to be performed and the *address* specifies one location of the store. Four simple instructions are

LDA (n) STA n ADD (n) SUB (n)

LDA is the mnemonic for 'load A' and the effect of the instruction is to access the store location with address *n*, read out the contents and write this into the single word store known as the A-register. The contents of the store location *n* are not altered in the process. The brackets enclosing *n* denote the 'contents of' location A. STA stores the contents of the A-register into the location *n* again without altering the contents of the A-register. ADD (n) and SUB (n) respectively add and subtract the contents of location *n* to or from the contents of the A-register. A 'literal' instruction LDAn loads the number *n* into the A-register. Most machines except the smallest are equipped with an automatic multiplier and divider and appropriate instructions are available for these functions.

Widespread use is made of repetition of short sequences of instructions. Consider the simple problem of forming the sum of 100 numbers contained in consecutive locations in the store. This could be achieved with the sequence of 100 instructions (conventionally arranged in a vertical sequence)

LDA (n), ADD (n + 1), ADD (n + 2), ... ADD (n + 99)

However, this is both laborious for the program and wasteful in storage space. This task is simplified by the use of the index of X-register. When the X-register is specified in an instruction, the address selected in the store is obtained by forming the sum of the contents of the X-register and the address specified in the address part of the instruction. The actual address to which the instruction refers is modified by the contents of the X-register. The instruction may be repeated, but if on each occasion the contents of the X-register is changed, the instruction will refer to a different location each time.

An instruction, TDX, is provided which tests if the X-register is zero and then decrements the contents of the register by one (i.e., subtracts one from the contents). If the X-register is zero the next instruction is skipped and the instruction in the next-but-one location after the TDX instruction is obeyed. The instruction

JMP*+n or JMP*−n

causes a jump in sequence, the next instruction being taken from the location n places beyond (or before) that location in which the JMP instruction is placed. The program using the X-register is

LDX 99, LDA 0, ADD (100) x, TDX, JMP*−2, STA 200

The set of numbers to be added is stored in locations 100 to 199. The first instruction loads the X-register with the number 99 and the next loads zero into the A-register to ensure that no result of a previous computation remains in this register. The instruction ADD(100) x is modified by the X-register and the location accessed is (100 plus the contents of the X-register) i.e., location 199.

The fourth instruction tests if the X-register contains zero and then decreases the contents to 98. The JMP instruction causes a return to the third instruction which has the effect ADD(198). The sequence repeats until the third instruction is obeyed as ADD(100), the TDX instruction then finds that the contents of the X-register are now zero and the JMP instruction is skipped over. Finally STA 200 deposits the sum in location 200. The X-register is a useful facility and some larger machines are equipped with several such registers.

HIGH-LEVEL LANGUAGES

The natural instruction set (often referred to as 'machine language') of a computer bears little resemblance to the algebraic form in which expressions are normally written. A number of special 'problem-oriented' languages have been developed by means of which a programmer is able to instruct a computer in a more natural and simpler way. A program written in a high-level language is translated by a special program known as a *Compiler*, into machine language. This translation is considerably more complex than the conversion of the mnemonic forms used above (referred to as 'Assembly language') which is performed by the 'Assembler'.

The most widely employed language for application in science and engineering is FORTRAN (FORmula TRANslator). The essential features are described; for a full account it is necessary to consult one of the many standard texts.

Fortran. Two basic types of variables, namely *real* variables and *integer* variables, are available. Integer variables may assume any integer value between prescribed limits determined by the word length of the computer. For a machine with n-bit words the range is from -2^n to $+(2^n - 1)$. Even for a short word length this is generally adequate for operations involving integers. However, the range is inadequate for many calculations and no direct provision is made for handling fractions. To overcome this difficulty, larger machines intended mainly for numerical calculation have special hardware to allow the manipulation of numbers in a *floating point* from $a(2^b)$ where a is always in the range $-\frac{1}{2}$ to $+\frac{1}{2}$ and b is a positive or negative integer.

The programmer can choose names for variables subject to the constraint that a variable with a name commencing with one of the letters, I,J,K,L,M or N is treated as an integer, while a variable with a name commencing with A to H or O to Z is treated as a *real* variable; that is, one represented in floating-point form. The standard method of input of programs employs punched cards or punched paper tape and there are certain limitations imposed by the keyboards of the machines used for the preparation of input data using the media. The arithmetic operations are represented as

+addition * multiplication
−subtraction / division ** exponentiation

so that the expression

Y = A * B / (C * D) + X ** (P + 1) represents $y = (ab/cd) + x^{p+1}$

If numerical values have been provided for the variables on the right-hand side of the expression, the r.h. side will be evaluated and assigned to the left-hand side.

There are facilities for inputting quantities from paper tape or card reader using the READ statement and for printing results using the PRINT statement.

Conditional statements will allow alternative actions within a program. An example of a conditional statement with alternative actions is

```
      IF (X – Y) 4, 5, 6
   4  Z = Y
      GO TO 7
   5  Z = O
      GO TO 7
   6  Z – X
   7  END
```

The first statement forms the value of (X – Y): if it is negative the statement labelled 4 is obeyed; if zero the statement 5; and if positive the statement 6 is obeyed. The effect is

if (X – Y) > O then Z = X; if (X – Y) = 0 then Z = O; if (X – Y) < O then Z = Y

An example of a repetitive sequence in a program forming the sum of a set of 100 numbers A(1), A(2) . . . A(100), is

```
      S = 0
      DO 5  I = 1, 1, 100
   5  S = S + A(I)
```

The first statement sets the value of the variable S to zero. The 'DO' statement causes the execution of all the following statements down to and including that labelled S (only one statement in this case) repeated for values of I starting with with 1 and increasing by steps of 1 until 100. The program has the effect of forming the sum S = a_i for all integral values of i between $i = 1$ and $i = 100$.

Print-out. FORTRAN provides facilities for printing results with a convenient layout and in a desired format. For example the statements

```
      PRINT 10, L
  10  FORMAT (7HLENGTH =, 14)
      PRINT 12, A
  12  FORMAT (5HAREA =, F3.2)
```

would result in a print-out

```
LENGTH = 25
AREA = 625.00
```

Other languages. Other languages have been developed. Those in common use are ALGOL, which is intended for the same application area as FORTRAN; COBOL (a COmmon Business Oriented Language) which as its name implies is intended for commercial data processing applications; and PL/1 which combines features of other programming languages in a single language of more generality.

The COBOL language allows statements to be expressed in a subset of the ENGLISH language, allowing programs to be read (but not written) by persons unfamiliar with the rules of the COBOL language. It also makes provision for handling the types of data structures normally required in commercial work.

OPERATING SYSTEMS

In early computer systems the programmer operated the computer from the typewriter console. This led to uneconomic utilisation of the machine. Further, when the results were printed the computer was restricted by the slow rate of printing. Considerable improvement in work throughput is achieved by *multi-programming*, in which several users share the total system resources. While one user's program is being executed by the central computing unit, that of another user is being loaded by the input device and the results for a third are being printed. Input and output devices are partly mechanical, and slow compared with the central processor: it is therefore a desirable strategy to keep these devices fully occupied. One technique, known as spooling (Simultaneous Peripheral Operation On Line) operates under the control of a master or *executive* program. Each input and output device has an associated program which organises transfers of information between the device and a disc backing store. The latter acts as a buffer store and the success of spooling depends on a suitable job mix in which peripheral and central processor operations can be overlapped.

INPUT-OUTPUT PROCESS

In a computer system with only one central processor the precise way in which the input (and output) operations are performed is as follows, with reference to the functioning of a paper-tape reader. At some stage an input operation is initiated by the executive. The input program is called into action and reserves a block of consecutive locations in the core store and then activates the paper-tape reader. The input program then hands back control of the computer to the executive which, in turn, executes another user's program. When the first character is read from the tape reader, the 8 bits are written directly into the first location of the block reserved in the core store. This operation requires one core-store cycle, typically 1 μs. The paper tape advances in the reader and, with a high-speed device operating at 1 000 characters/s, the next character becomes available 1 ms later. A further core-store cycle is utilised to enter this character into the next location in the store, an automatic counting mechanism being provided to advance to the next location and to report when the block is full. Since only 1 core-store cycle in every 1 000 is used in this way, the tape reader effectively reduces the central processor speed by 0.1% by making the store unavailable while entering characters by this mechanism, which is referred to as an *autonomous transfer*. When the allocated block has been filled an interrupt signal is generated by the mechanism. This signal causes an interruption to the execution of the current program, and the machine instead performs the interrupt service routine for the peripheral (which among other actions, allocates a further core block for the device). When sufficient data have been received from the input device, an autonomous transfer from core to disc store will be initiated. Ideally all the peripherals for which there is a queue of work and the central processor are kept occupied the whole time; in this way a maximum throughput of work is achieved.

EQUIPMENT

The basic media are punched paper tape and punched cards.

Standard paper tape is about 25 mm (1 in) wide and has 8 data channels and one sprocket hole across the width. Normally one data channel is allocated for a *parity check*. If a hole represents a 1 and the absence of a hole represents a 0, than an *odd* parity check is obtained by arranging that the total number of 1's in the seven data positions, together with the one parity position, is odd. An even total is required for *even* parity. It should be noted that a single parity bit can detect only one error (or, strictly, an odd number of errors) in the data, but no indication is available to correct the error.

Several error detecting and correcting codes have been devised, but they require greater redundancy. Tape-readers operate at up to 1 000 characters/s.

Standard punched card, measuring 185 mm × 82.5 mm ($7\frac{3}{8}$ in × $3\frac{1}{4}$ in), has 80 columns and 12 rows. The decimal digits are represented each by one hole in the appropriate row, while the alphabetic symbols are recorded by punching holes in two rows. Certain symbols require holes in three rows. Card-readers operate at 300 or 900 cards/min.

Magnetic tape encoders are available, with operation direct from a keyboard, where (as in data-processing applications) considerable amounts of information must be prepared for input to the computer.

Special input/output facilities are needed for real-time interactive applications. For example, in airline seat-reservation systems several hundred agents must have access to common files in which the booking lists for each aircraft are held. Since the agent must make a booking for a waiting client, an immediate response is required. Use is made of a video display unit which resembles a television receiver but is equipped with a keyboard and is capable of displaying alphanumeric characters. The display units are normally provided with local storage and can display a format into which the operator can insert the information relating to the transaction.

Character-recognition systems attempt to read typed and hand-written symbols directly, to save the considerable labour expended in data preparation. The MICR (Magnetic Ink Character Reader) is widely used for reading the account and branch numbers which are printed at the bottom of a cheque with magnetic ink. This is not a true recognition scheme because a special character font has been designed and the scheme can be regarded as a method of magnetic recording which can also be understood by the human reader. Similar special fonts have been devised for optical character recognition in which normal ink is usable, in this case the amount of reflected light when the character is scanned by a light beam is measured. True character recognition does not pre-suppose special fonts and the practical systems devised to date are expensive and limited in scope.

Printing devices are available ranging from teleprinters (speeds of 10–30 characters/s) to line printers, which print lines of 120 or more characters at rates up to 1 000 lines/min. Graphical plotters can be used for the production of line diagrams.

Applications

A few typical applications out of many of interest to the engineer are briefly described.

COMPUTER-AIDED DESIGN

This has been extensively developed for several sectors. Programs have been devised for the analysis of passive and active electrical networks. Means are provided for 'sketching' the network diagram with a 'light-pen' on a graphical display. Provision is made for determining the frequency and transient response of the networks and for determining the stability of feedback networks. Other classes of programs prepare component and wiring layouts of printed circuit boards and design masks for integrated circuit production.

POWER SUPPLY SYSTEMS

In electrical power engineering programs have been developed for power-flow calculations and for determining the stability of a network. Process-control computers have been installed in most modern generating stations and employed for monitoring temperatures and other variables. Further applications include alarm analysis, by means of which fault conditions in a plant are observed and the cause diagnosed. The computer is also used in certain cases to control and monitor the start-up and shut-down of the generator and auxiliaries. In nuclear stations the computer is used extensively for monitoring and detecting rate of change in parameters.

CONTROL SYSTEMS

In a control application it is generally necessary to convert analogue quantities into digital form. For example, temperature measurements are obtained using thermocouples giving an output of a few milliwatts while other transducers may have outputs up to 20 mA. Several forms of analogue/digital converter have been devised. In one type the input analogue quantity (e.g., voltage) is converted into the form of a frequency signal by means of an oscillator whose frequency is voltage-dependent. The digital output is obtained by counting the number of cycles occurring in a fixed period. By arranging that the counting period is equal to the period of the 50 Hz mains supply, an interfering component due to mains pickup in the analogue input can be eliminated. Higher speed can be obtained from the successive-approximation analogue/digital converter. In this a register is provided comprising a number of bistables. Each flip-flop represents successive powers of 2 of the result, and controls a current source. The lowest-order digit of the register controls a current of unit value, the next a current of two units, successive digits control 4,8,16 ... 512 units of current. The current sources are summed and converted to an analogue voltage which is compared with analogue input. Initially all digits in the register are set to zero and the highest-order digit is set to one; the output is then compared with the input. If the output exceeds the input the highest-order digit is reset to zero but otherwise remains set to one. Lower-order digits are tested successively and the final setting of the register is the binary digital equivalent of the analogue input.

ON-LINE COMPUTERS

The digital computer is applied to complex calculations which involve large sets of simultaneous equations, or which concern statistical data and demand logical decisions. Computers to which data and instructions are fed, yielding answers after the necessary calculations to a read-out equipment, are *off-line* machines. In contrast, *on-line* computers are integrated into a production process; they take information direct from the process and can either give instructions to the human operator or control the process directly, as shown schematically in *Figure 24.84* (*a*).

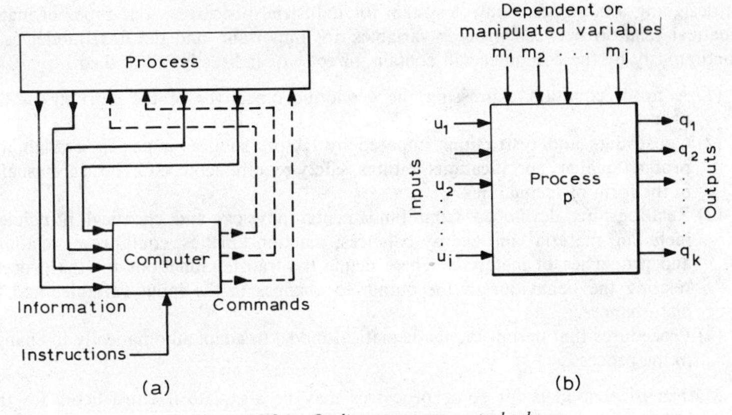

Figure 24.84 On-line computer control scheme

The installation of an on-line computer calls for considerable capital outlay. Careful investigation is therefore necessary to ensure economic viability. This involves a detailed study by a team comprising engineers from both the production company and the computer manufacturer.

If a digital computer is to exercise extensive control, it must be very reliable; this is especially important where a shut-down has results (as in hot-rolling steel) that prevent immediate restart after the fault has been cleared. In some cases (for example, nuclear generating stations) the financial penalty caused by a shutdown is so high that a standby computer is justified.

PROCESS CONTROL

An on-line computer-controlled process can be designed to give any of the following fully-automatic types of service, singly or in combination:

(1) Optimisation with respect to a specified criterion (e.g., fuel consumption, yield of product) using either (*a*) a mathematical model of the process, or (*b*) an empirical technique for searching methodically step-by-step for an optimum.
(2) Feedforward control with computed compensations for those independent variables that can be measured (such as variations in load, composition of input material, etc.).
(3) Feedback control based on performance, computed from measured variables and measurements made by continuous analyses on the output of the process.
(4) Adaptive control, in which the computer adjusts the mathematical model of the process to take into account slow changes in its behaviour.
(5) Development, or refinement, of a mathematical model of the process, based on automatic analysis of information provided by the process itself.
(6) Controlled start-up and shut-down of process plants.

In addition to these automatic control functions, the computer can carry out such data-diagnosis or data-processing operations as: monitoring, warning, recording, statistical analysis, determination of correlations, instrument calibration, detection and indication of equipment failure, anticipation of the need for maintenance or replacement of elements in the process chain.

MATHEMATICAL MODELS

In designing a computer control system for industrial processes, four types of mathematical relation between process variables are important, and the final model used continuously by the computer will contain, directly or indirectly, all of them:

(1) A profit equation expressing the economic objectives of the working of the process.
(2) Constraints and restrictions imposed by factors such as materials availability, product quality, specifications, ratings, safety requirements, etc. These are usually in the form of inequalities.
(3) Relationships developed from fundamental physical and chemical principles, including material and energy balances, reaction kinetics, equilibrium relations and properties of materials. These define the transfer functions for the process relating the behaviour of the output to changes in the input variables and to disturbances.
(4) Procedures that permit the mathematical model to adapt automatically to change in the process.

Mathematical models for some processes may be adequate if constructed for the steady state, i.e., as algebraic equations not time-dependent. Dynamic models are needed, expressed in the form of the differential equations of the process, where the operation of the process is such as to demand them.

OPTIMISATION

Suppose that *Figure 24.84* (*b*) represents a process having a performance p. Then p will be a function of a number of independent input variables u, other dependent or manipulated variables m, and the system output or computed variables q, so that

$$p = \mathrm{f}(u_i, m_j, q_k).$$

The system variables will be related by mathematical equations

$$g_n(u_i, m_j, q_k) = 0$$

where $n = 1, 2, 3 \ldots \mathrm{N}$, expressing physical laws, empirical relations or constraints. The excess of variables over the number of equations yields the number F of degrees of freedom. For an optimum solution of the process it is necessary to solve for as many functions g_n as possible, and to deal with the remainder by forming partial differentials of p with respect to the manipulated variables m_j.

The methods of obtaining optimum solutions to the equations include:

(1) Direct solution when an analytical model is available, perhaps simplified by reducing the number of variables considered.
(2) Langrangian multipliers for inequalities; trial-and-error procedures are necessary for constraints with upper and lower limits.
(3) Linear programming.
(4) Calculus of variations, with trial-and-error techniques where necessary.
(5) Dynamic programming, formulating the problem as a set of decision processes.
(6) Gradient method, in which change in performance is evaluated in relation to step changes made by the computer in manipulated variables.

COMPUTER CHARACTERISTICS

Whereas computers used for accounting and scientific purpose work under human guidance, those for on-line process control may have no human intervention (apart from maintenance) after initial setting-up. The computer derives its input directly from the process, makes the necessary calculations, draws the conclusions and sends appropriate signals to the regulating devices on the plant. From time to time it will print out records for the information of the operator. These functions impose special conditions on the design and programming of the computer, beyond those found in a scientific or business machine. The additional characteristics are: the ability to accept and produce a great number of signals; a permanent store for the many programs needed; a high degree of reliability; and simplicity of operation.

Process-control programs have more steps, more parallel routes and more safeguards. Certain forms of programs are always required, including routines for scanning and recording incoming signals, correcting measured values to 'absolute' values and comparing measurements with predetermined limits.

PERSONAL ELECTRONIC CALCULATORS

No treatment of the subject of electronics could be considered complete without specific mention of small 'personal' electronic calculators because since about 1970, these have come to touch nearly every person's life, to a degree reminiscent of the personal transistor radio of the 1950s. Accordingly, the following is a description of the design and use of typical personal calculators, of the mid-1970s.

The term, 'personal' calculator, refers to a device small and light enough to be carried on the person—in a pocket or briefcase—as opposed to the larger and heavier 'commercial and scientific' desk-top models. Many of the larger desk-top calculators have achieved the level of minicomputers (or at least 'microcomputers'), complete with programs and peripherals.

The development of personal calculators has been revolutionary. Though functionally related to larger calculators, they are in effect an entirely new breed that emerged about 1970. Being able to carry a calculator in a pocket or briefcase sets it worlds apart from a calculator as big and heavy as a large electric typewriter. Not only the concept, but also the marketing and use of personal calculators have been revolutionary. Before 1970, there were only two electronic calculators that could be called 'personal', and very few were in use. By 1974, the number of different models available was in the hundreds, at least. And as the low end of the price range declines steadily, personal calculators are being used in increasing numbers; estimated worldwide sales for 1974 (including small desk-top models) being around 25 to 30 million units.

Furthermore, and perhaps most important, personal calculators are revolutionising the way the world copes with numbers. This applies not only to the housewife checking her domestic budget, but also to the engineer designing an electronic filter or a distillation column. Give an engineer a calculator he can take with him anywhere, and you free his mind from the tyrannical drudgery of numbers, permitting him to concentrate his talents on higher concepts than mere arithmetic.

Design requirements

For the benefit of readers as yet unfamiliar with personal electronic calculators, it might be appropriate to begin with a description of the design requirements. First, the calculator must perform as many arithmetic operations as possible—at least add, subtract, multiply and divide—with as many significant figures as possible (at least six), and if possible, it should be programmable to perform a predetermined sequence of such operations. Second, it should be small and light enough to be easily carried on the person. To be personal in every sense, it should be able to operate on self-contained batteries with a lifetime of at least several hours. To be available to the 'mass market', it should be low priced.

Beyond this point, trying to describe personal calculators is rather like trying to describe a speeding train as seen by a nearby stationary observer. Calculators are month by month getting not only more economical but also more sophisticated and capable. This is primarily because of improvements in the technology and manufacture of integrated circuits, in terms of more function and less cost per chip. The task of description is made even more difficult by the fact that the category of 'personal' calculators extends from the most rudimentary four-function, six-digit machines to magnetic-card-programmable 'near-microcomputers'.

General operation

For the sake of example, let us consider a fairly representative (though hypothetical) pocket calculator, built around a Texas Instruments calculator 'chip' called TMS 0100, which is a *p*-channel MOS large-scale-integrated circuit. Custom-programmed modifications of the 'parent' chip of this series have formed the basis for many calculators from various assemblers since about 1971.

The inside of such a calculator might appear as shown in *Figure 24.85*. Beside the single highly complex integrated circuit for all calculating and code-converting functions, there would be a keyboard for data entry and control, an array of seven-segment l.e.d. (light-emitting diode) character displays for data output, a couple of simple bipolar i.c.s for driving the l.e.d. display, a few incidental discrete devices for such functions as

Figure 24.85 Internal components of a typical pocket calculator (Courtesy Texas Instruments Incorporated)

clock generation and power-supply regulation, a rechargeable battery pack and a plastic case.

Figure 24.86 shows typical circuitry outside the main chip. The main chip requires a ±7 V power supply and a 250 kHz clock signal. To keep the number of pins on the device within the standard complement of 28, all data inputs and outputs are time-multiplexed as shown, using only 23 pins. Both the keyboard matrix and the l.e.d. array have eleven lines in common. One of these lines at a time is energised for about 160 ms during a *scanning cycle* of about 1.7 ms.

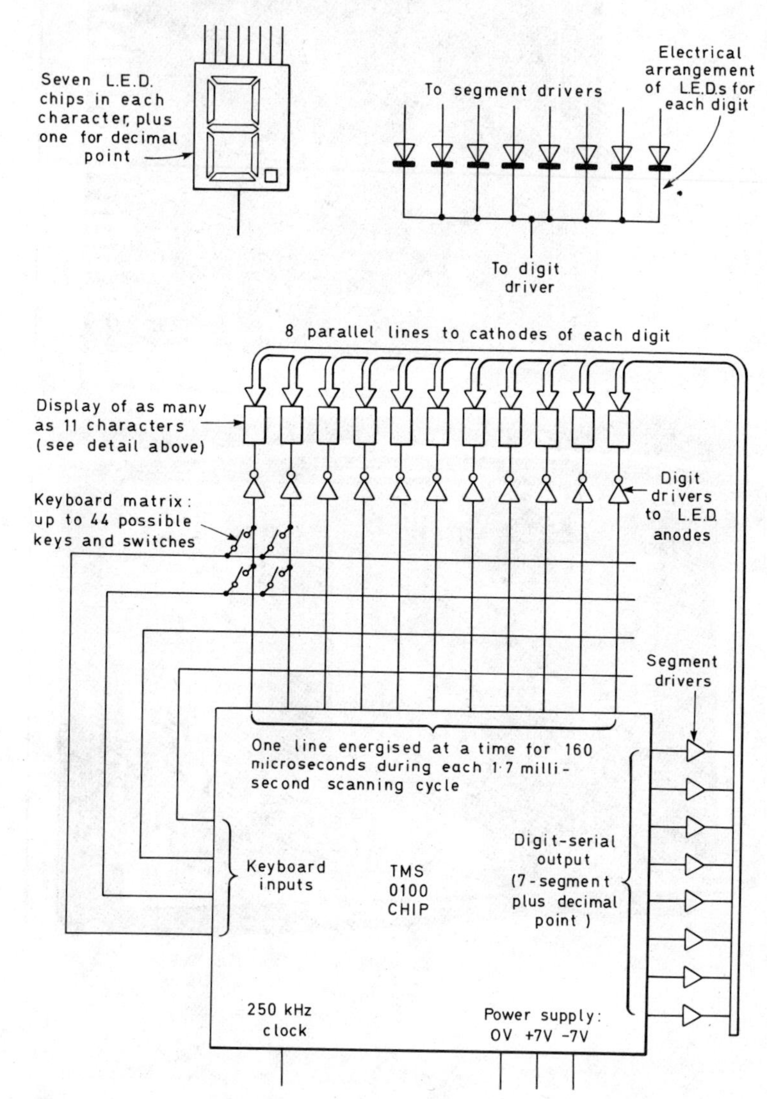

Figure 24.86 Possible circuit concept for a calculator based on the TMS 0100 chip

While a given *scan line* is energised, the corresponding character in the display may be held *on* by a *digit-driver* buffer connected to the l.e.d. anodes for that character. The definition of the character (if any) is determined by the seven-segment-coded output that goes through *segment-driver* buffers to the seven l.e.d. cathodes in each character. The eighth segment-line provides for a decimal-point l.e.d. by each digit. Thus, each displayed character flickers at a rate of about 600 Hz, being *on* for about 9% of the time.

The eleven scan lines and four keyboard-input lines form a matrix of 44 possible intersections, each one corresponding to a possible push-button or control switch on the keyboard. However, calculators using this chip typically only have about 20 push-buttons and switch positions. If a switch be held closed, then when the turn comes for its scan line to be energised, a signal is passed to its keyboard-input line, which is in turn decoded within the chip. Contact-bounce effects are eliminated by modified scan routines and internal programming that are actuated upon detection of a newly changed keyboard switch.

Organisation and programming

With this general picture in mind of typical circuitry external to the main chip (or chips), the following is a description of what *really* makes a calculator work—the architecture, or organisation, of the chip, and the algorithms, or numerical routines, that it uses to achieve the desired arithmetic and logical functions.

For this purpose the *Texas Instruments* SR-50 is taken as an example. This is a 'slide-rule' calculator with complete logarithmic, exponential, trigonometric and hyperbolic functions, plus one *accessible* memory register and *internal* memories for storing intermediate products for summing. A simplified conceptual diagram of the SR-50 architecture (contained on two separate chips) is shown in *Figure 24.87*.

Like most calculators, the SR-50 has many architectural elements that are reminiscent of full-scale computer architecture. The broad arrows represent several lines in parallel, typically a *bus*, with many points of input and output that can be selectively activated. The functional heart of the system is the arithmetic logic unit (a.l.u.), which is capable of adding or subtracting numbers selected from two registers and placing the results (shifted, if required) into one register. By means of a properly programmed sequence of adding, subtracting and shifting data in the various registers, any arithmetic or logical function desired may be performed.

Control over all sub-systems is exercised by the *instruction decoder and control logic* (or simply *decoder/controller*), according to a word placed in the instruction register from the 13-kilobit read-only memory. Each instruction word typically causes one *operation*, say, the addition of the two-digit exponent in one designated register to that in a second, with the results placed in a certain third register.

Sequences of instructions, *microinstructions*, stored in the 13-kilobit r.o.m. constitute *microprogrammed* routines similar to those in full-scale computers. According to information in the instruction currently being executed, the address of the next instruction word may come either from the keyboard register or by adding 'one' to the current address or to a segment of the current instruction. The keyboard register, in turn, is fed from the keyboard encoder or the a.l.u. Typically, the address of the first instruction of a microprogrammed routine is placed in the keyboard register. During execution of the routine then, the address register is incremented by one after each instruction is executed.

The *constant r.o.m.* contains numerical constants that are required, such as pi, the natural-log base, and coefficients of various numerical-expansion series. Selection of a given constant is achieved in much the same way as the addressing of an instruction in the larger r.o.m. The selected constant passes into one of the registers.

Any of several *flag bits* may be set and consulted by the decoder/controller. The flags

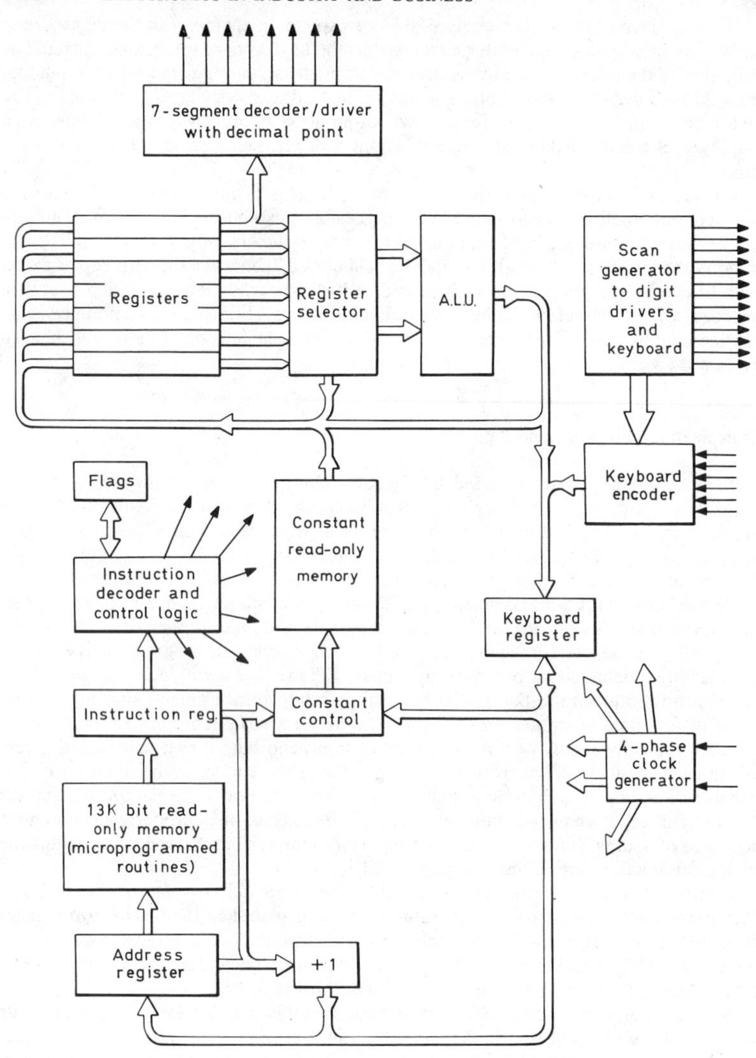

Figure 24.87 Simplified architectural concept of the SR-50 calculator, contained on two separate chips

constitute, in effect, a small random-access memory with one-bit words that is available to the decoder/controller. Typically a flag stores status information that is consulted to condition further action within a certain routine.

The numerical contents of one of the registers is communicated to a decoder-driver circuit that produces 7-segment character code for the display, plus an eighth line for the decimal point at each digit. Separate segment-driver buffers are not required as with the TMS 0100 chip shown in *Figure 24.86*.

The keyboard and display have a number of *scan lines* in common, in a fashion similar to that of *Figure 24.86*. The keyboard encoder interprets signals in the keyboard-input lines, together with information on which scan line is currently activated, and generates codes for numerical digits and control signals. These codes pass into the keyboard register for action. For example, the code for a decimal digit calls a sequence of instructions that causes the generation of that digit in binary-coded-decimal (b.c.d.) form in the appropriate register.

The algorithms (arithmetic routines) that the SR-50 uses are no different in concept from algorithms used by a computer. Each algorithm (say, the one for adding two numbers) is based on a *flow chart* that anticipates all eventualities and specifies a routine for each. Since the numbers involved are typically in either floating-point or exponential (power-of-ten) form, even the simplest arithmetic operations may require quite a few passes through the a.l.u.—which after all can only add, subtract and shift. The SR-50 routine for adding two numbers, for example, contains more than thirty *microinstructions*. And chips in the TMS 0100 series (depending on how they are programmed and on just what numbers are involved) may take from 10 000 to 15 000 clock cycles to perform a complete addition or subtraction.

The algorithms for functions other than addition, subtraction, multiplication and division, are constructed from these four basic functions by standard algebraic approximations to whatever degree of accuracy is called for. The approximations involve either series expansions (such as Taylor or Fourier) or numerical-analysis techniques, akin say, to the Newton–Raphson method.

Most other personal electronic calculators are similar in concept to the two discussed here. In fact, the only essential difference between some calculators is different microprogramming on an otherwise identical chip. Once a new set of microprograms has been developed (which is no simple job), the new instructions can be committed to the chip by changing only one photomask during manufacture.

Probable developments

The discernible trend in future development of personal electronic calculators is towards greater calculating power and lower prices, as a result of cramming more circuitry on to fewer chips. On the one hand, a very cheap basic calculator with *all* its electronics contained on just one chip could emerge. On the other hand, it is probable there will be inexpensive fully programmable pocket microcomputers complete with thousands of words of random-access memory, and optional peripheral equipment such as printers, strap-on memories and tape drives. Some manufacturers are already working on calculators small enough to wear like a wristwatch. Displays are expected to become larger and easier to read, perhaps through improvements in liquid-crystal technology, which would also pay dividends in terms of reduced power consumption and longer battery life.

ELECTRONIC CLOCKS AND CLOCK SYSTEMS

Electronic clock movements

The field of timekeeping has been revolutionised over the last two decades with the widespread use of semiconductor devices. With the introduction of the germanium transistor, horologists were able to perform switching functions hitherto requiring mechanical contacts which reduced the accuracy of pendulums and balance wheels. Further development of semiconductor techniques has led to the introduction of integrated circuits, performing the functions previously requiring many tens or even hundreds of discrete components, yet occupying minimal space. These devices have led to widespread production of electronic wrist watches, wall clocks and master clocks for controlling large clock systems.

This introduction of electronics into the horological field has produced three main types of clock movement for use as independent time pieces—balance wheel, tuning fork and quartz crystal.

BALANCE WHEEL CONTROLLED MOVEMENT

Clocks regulated in this manner are based on the well-known principle of balance-wheel oscillation. The balance wheel is coupled to an escapement and gear train driving the clock hands.

Figures 24.88 and *24.89* show the basic electronic switching circuit and a simplified

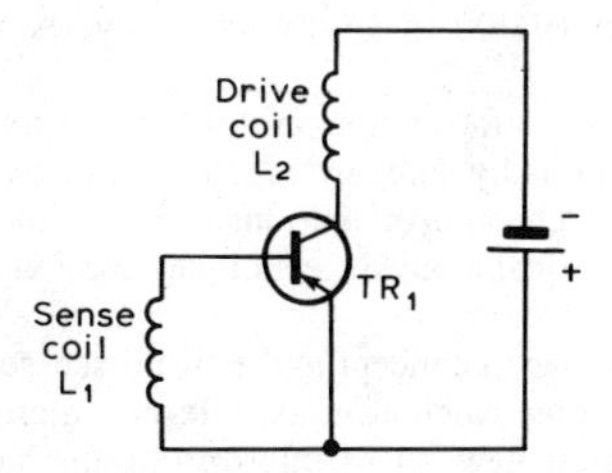

Figure 24.88 Balance wheel switching circuit

mechanical layout of the balance wheel assembly. Rotation of the balance wheel causes the permanent magnet *A* to generate a voltage in L_1 with a polarity dependent upon its position relative to L_1. As the magnet approaches L_1, transistor Q_1 is switched on and the electromagnetic field associated with L_2 attracts the permanent magnet *B*. This increases the rate of acceleration of the balance wheel as the magnet approaches the magnetic field. Once the permanent magnet *A* crosses the centre of the coil L_1, the polarity of the induced voltage changes and Q_1 is switched off, allowing the balance wheel to travel freely to the extremity of its arc.

A similar sequence of events occurs each half cycle of balance wheel rotation.

In commercially available movements, coils L_1 and L_2 are generally combined into one assembly with high flux permanent magnets above and below the coils. This method of construction gives good magnetic flux linkage which assists in maintaining the

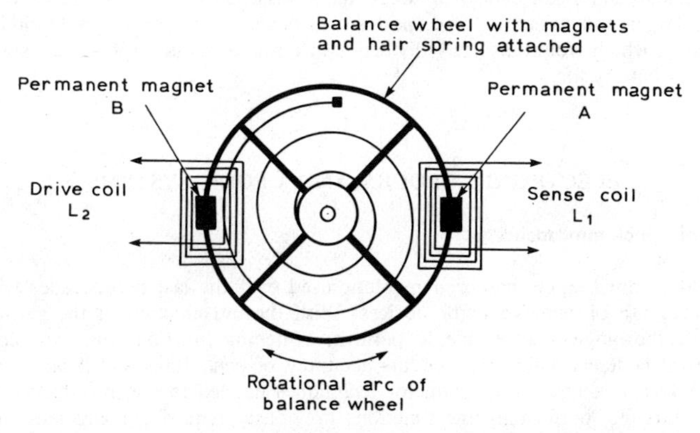

Figure 24.89 Simplified balance wheel assembly

balance wheel rotational arc at a constant amplitude. Further refinements include voltage regulation to compensate for changes in battery potential and usually some form of electronic temperature compensation to improve the time keeping versus temperature characteristic of the clock.

TUNING FORK CONTROLLED MOVEMENT

Clock movements employing a tuning fork as the time base have, for some years, been available and can be capable of time keeping accuracies of a second or so per day.

The tuning fork is caused to oscillate by a similar circuit to that of the balance wheel movement, the magnets being attached to the prongs or 'tines' of the tuning fork. Coupling of the tuning fork to the mechanical movement is generally performed magnetically. A forked arm, fitted with small, powerful permanent magnets forming a strong magnetic field within the fork, is attached to one of the tines in such a manner that a ferromagnetic wheel with a sinusoidal or other suitably shaped edge is caused to rotate by the oscillatory movement of the magnetic field about the edge of the wheel. The gear train driving the clock hands is connected to the magnetically rotated wheel.

QUARTZ CRYSTAL CONTROLLED MOVEMENT

Recently, the quartz crystal controlled oscillator has become an accepted timebase for both clocks and watches.

Battery operated clocks with an average time-keeping accuracy of better than one minute per year are readily available at moderate cost and need occupy no greater volume than more conventional clock movements.

The use of crystal-controlled oscillators become practical with the advent of the low power integrated circuits and small crystal packages. *Figure 24.90* shows a typical

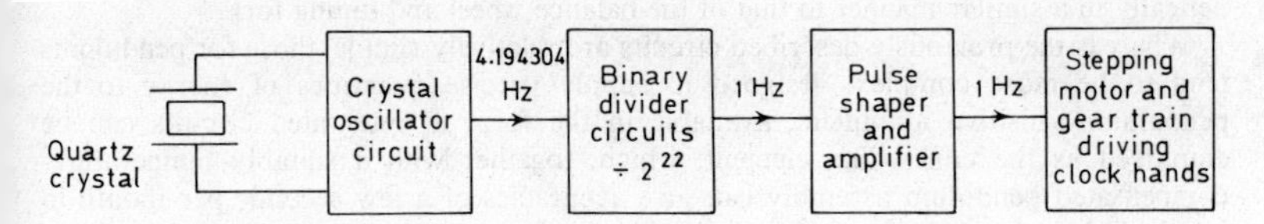

Figure 24.90 Schematic drawing of a typical crystal controlled clock movement

schematic circuit utilising a 4 194 304 Hz crystal oscillator feeding 22 binary dividers (÷2) connected in series. This results in an output frequency of 1 Hz which after shaping and amplifying by appropriate circuits is applied to a stepping motor to drive the clock hands.

The choice of the crystal frequency is determined by the temperature versus frequency characteristics required of the clock, and also the integrated circuit and type of electro-mechanical clock mechanism employed to drive the hands. With the advent of m.o.s. i.c.s, it has become possible to include many transistors in one small package requiring little power and a very low operating voltage. A typical device currently in use is mounted in a 14 pin dual in line package having a total volume, excluding pins, of approximately 500 mm^2 and contains most of the oscillator components, 22 binary dividers and the electro-mechanical mechanism driving circuit. The device operates on two 1.5 volt dry cells and requires an operating current of only a few tens of micro-amps which allows the whole clock to operate for a period in excess of a year before battery replacement becomes necessary.

A further form of this type of clock utilises a crystal oscillator and divider circuit with an output frequency of the order of 50–60 Hz. This output frequency is amplified and drives a low power synchronous motor allowing the seconds hand to 'sweep' round rather than step round. The main disadvantage is that a relatively high power is required to drive the motor due to it being permanently energised whereas a stepping motor requires only brief impulses of energy to perform each stepping operation. The increased power consumed entails the use of larger batteries or more frequent replacement which increases the operational cost.

Electronic master clocks

Although the installation of individual clocks within a building reduces the total wiring required, it is often advantageous to ensure that all clocks are synchronised to a master clock. It is easy to imagine the resulting chaos if all the clocks at an airport or large factory indicated different times. In such locations a master clock with good time-keeping properties is used to ensure all clocks are synchronised. The slave clocks always indicate the time shown by the master clock, and it is a simple matter to perform correction of a complete clock system for any slight inaccuracies by means of controls at the master clock.

Depending upon the accuracy required of the clock systems the three types of electronic master clock generally in use are the electronically maintained pendulum, the quartz crystal controlled and the atomic.

ELECTRONICALLY MAINTAINED PENDULUM MASTER CLOCK

One of the major problems associated with pendulums is the variation of friction between the pendulum and the associated escapement due, for example, to changes in lubricant viscosity. The problem is partly overcome by driving the pendulum magnetically in a similar manner to that of the balance wheel and tuning fork.

Whereas the previously described circuits are relatively simple, those for pendulums tend to be more complex, designed to supply precise quantities of energy to the pendulum. Sensitive amplifiers, available in the form of integrated circuits can be employed as the controlling element, which, together with a suitably temperature-compensated pendulum assembly can give accuracies of a few seconds per month in relatively low cost commercial master clocks.

The use of electromagnetic pendulum drive has also been applied to precision pendulum master clocks utilised for scientific purposes and relatively high degrees of accuracy have been achieved. The Fedchenko Clock[1] designed in the U.S.S.R. is claimed to have an r.m.s. error of daily operation of less than 0.2 ms to 0.3 ms which is less than that of most commercial crystal master clocks. This and similar precision pendulum clocks, however, require many refinements to achieve these high accuracies and consequently their use is limited.

QUARTZ CRYSTAL CONTROLLED MASTER CLOCK

The problems involved in maintaining accurate time keeping with pendulum master clocks, designed for general use, are overcome by utilising a quartz crystal as the time base in a similar manner to that previously described for the individual clock.

A schematic diagram of a typical crystal controlled master clock is shown in *Figure 24.91*. The crystal, and in most cases, the oscillator components, can be housed in a temperature-controlled oven to ensure that the temperature-dependent characteristics of all components governing the frequency of oscillation remain stable.

The most important item is the quartz crystal itself, an *AT* cut crystal (see Section 6) is generally employed due to its low temperature coefficient which, within certain limits, can be linear. This tends to result in self compensation for slight changes in oven temperature above and below the mean. Such crystals, by virtue of their form, have relatively high oscillation frequencies, greater than 500 kHz, and a frequency is generally chosen which can easily be divided by decades to allow the indication of fractions of a second when required. In consequence, the integrated circuits generally used are decade dividers rather than simple binary dividers. Decade division is achieved by the grouping of four binary dividers which are reset to zero, by suitable gating, after reaching the count of nine. The integrated circuits may also be reset to zero by external

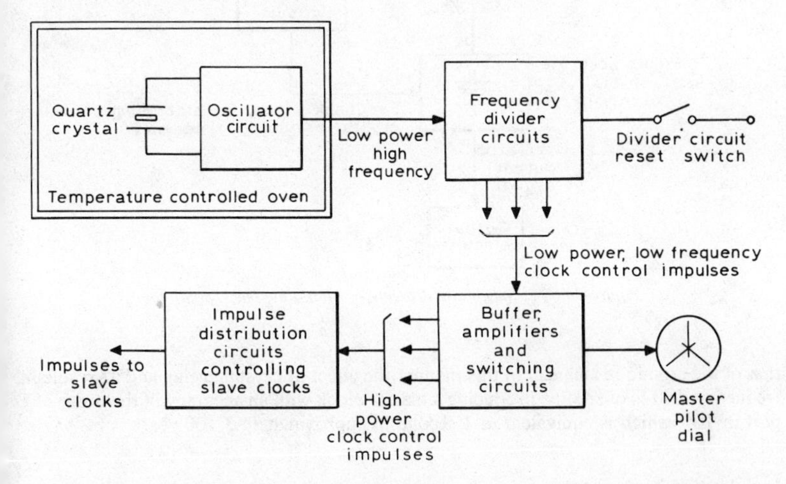

Figure 24.91 Schematic drawing of a typical crystal controlled master clock

switching to allow precise setting of the master clock to a standard time. Having reset to zero, all the divider stages must then receive input pulses generated by the oscillator before an output pulse is achieved.

Having divided the oscillator frequency to lower frequencies, the divider stages pass low power signals to buffer and switching circuits which control the impulses to the master pilot dial indicating the master clock time. The impulses distributed to the slave clock network are synchronised with those driving the master pilot dial, so ensuring all clocks within the system indicate the same time.

ATOMIC MASTER CLOCK

At the 13th General Conference of Weights and Measures, the second was defined internationally by reference to the frequency of oscillation of Caesium 133 which is 9 192 631 770 Hz. The principle of exciting and then detecting and measuring the oscillations of the atoms is that used in the atomic clock, which is illustrated in schematic form in *Figure 24.92*.

The oscillatory circuit consists of a resonator tube containing the material employed as the standard, for example Caesium 133. The material is formed into a beam which is then excited by means of the highly accurate crystal oscillator and the resultant atomic oscillation is picked up by a detector.

The accuracy of the exciting oscillator is corrected by means of the detected signal using a feedback loop technique so ensuring the accuracy of the crystal oscillator to be

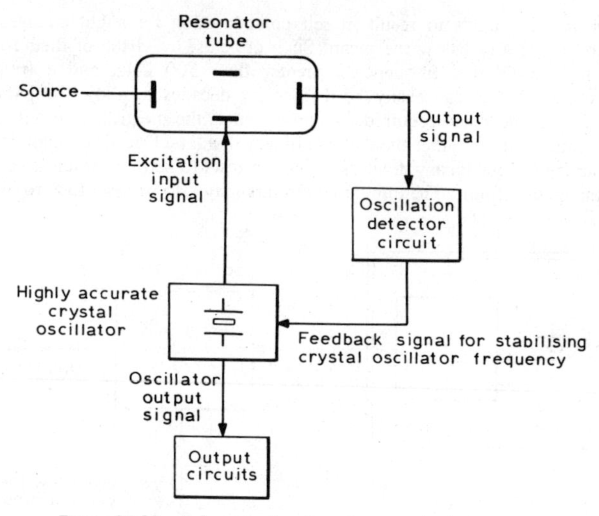

Figure 24.92 Schematic drawing of an atomic clock oscillator

that of the atomic resonance. In this manner the effect of crystal ageing and its resultant frequency drift is overcome producing a master clock with an accuracy of the order of 1 part in 10^{11} which is equivalent to 1 second in approximately 3 100 years.

Impulse distribution systems

Having generated timing impulses with the master clock, it is necessary to transmit these, in a suitable form, to the slave clocks. Each year consists of 525 600 minutes and 31 536 000 seconds; the life of an electromechanical relay is very limited when switching at these rates and significant increases in system reliability can be achieved with electronic switching circuits.[2] Dependent upon the type of mechanical slave clocks employed, two basic forms of electronic switching circuits are generally used—unidirectional impulse, and alternating polarity impulse.

UNIDIRECTIONAL IMPULSE SYSTEM

This is the simplest method of impulse distribution, controlling slave clock movements operated by the limited movement of an armature under the influence of an electromagnetic field.

A typical control circuit is shown in *Figure 24.93*. Low power impulses, generated by the master clock, are supplied to the base circuit of transistor TR_1. The resultant current impulses from the collector of TR_1 energise the slave clock electromagnets to drive the clock hands or other indicating devices.

Switches SW_1 and SW_2 are included in the circuit for stopping and manually advancing the slave clocks in the event of error occurring, for example, due to maintenance or fuse failure. The circuit also includes a transient suppression diode D_1 to limit the reverse voltage developed by the slave clocks during the collapse of their associated magnetic fields. The capacitor C_1 is included as a simple method of increasing the length of the driving pulse where highly inductive series connected slave clocks are used. In such circuits, the current flowing through the electromagnets may take several hundred

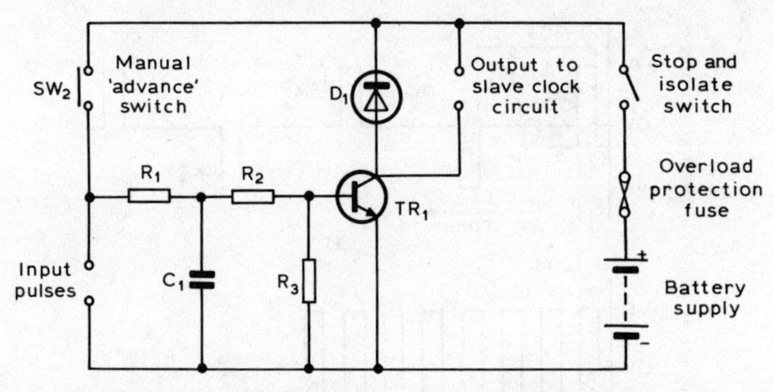

Figure 24.93 Unidirectional impulse control circuit

milliseconds to reach a steady level and the effective length of the drive pulse may be reduced below a safe level to ensure operation of all movements. A capacitor of several hundred microfarads will increase the duration of the applied impulse.

THE ALTERNATING POLARITY IMPULSE SYSTEM

Figure 24.94 shows the rotor and stator assembly of a clock movement operating on this principle. The stator consists of a four-pole electromagnet arranged by the linking of 'U' shaped pole pieces to opposite ends of a coil with a central ferrous core. In the centre of the stator is a rotor consisting of four specially shaped pole pieces polarised by means of a permanent magnet in such a manner as to give alternate north and south poles.

When the electromagnet is de-energised the rotor rests with its poles aligned with the stator poles in any one of the four stable positions. By polarising the electromagnet so

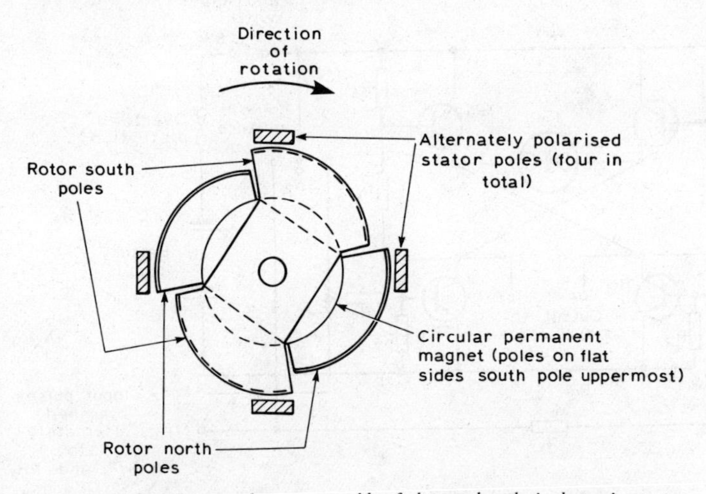

Figure 24.94 Rotor and stator assembly of alternately polarised stepping motor

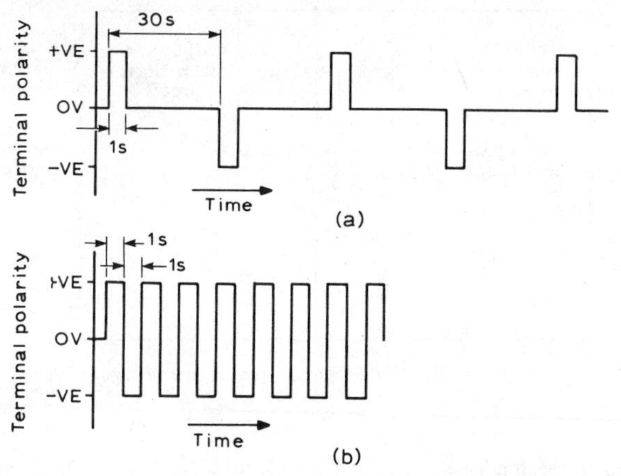

Figure 24.95 Voltage waveforms applied to stepping motor (a) ½ minute impulse system (b) seconds impulse system

that adjacent stator and rotor poles are of like polarity, the rotor turns in a clockwise direction to align itself with the closest unlike stator poles. Rotation by 90° takes place to produce the condition of smallest air gap between stator and rotor poles and hence maximum flux density within the air gap. On removal of the stator field, the rotor remains stationary until the field is reversed when a further 90° rotation occurs. *Figure 24.95* shows the voltage waveform applied to the electromagnet for both half minute and seconds impulse systems where the current through the electromagnet reverses every second without any significant zero current period.

The circuit diagram of *Figure 24.96* shows basic details of a suitable forced inverter for driving such slave clock movements.

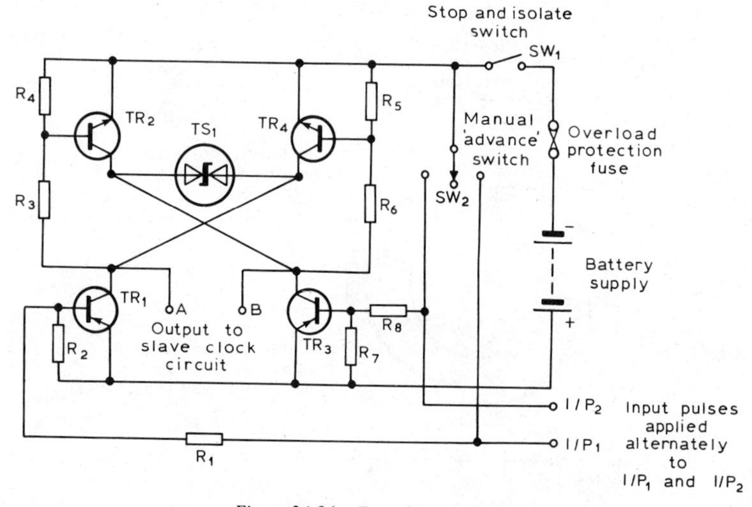

Figure 24.96 Forced inverter circuit

The four transistors TR_1 to TR_4 are connected in such a manner as to allow only two transistors to conduct at any time. By alternately applying impulses from the master clock to input circuits I/P_1 and I/P_2 transistors TR_1 and TR_2 followed by TR_3 and TR_4 conduct. This causes the output voltage at terminals A and B to change polarity with each change of input. For half minute impulse systems one of the power lines may also be switched to produce the waveform of *Figure 24.95* (*a*).

Switches SW_1 and SW_2 are included for stopping and manually advancing the slave clocks and TS_1 limits reverse voltages generated by the electromagnets within the slave clock movements during the collapse of their magnetic fields.

Electronic digital readout displays

For some applications, it has been found that the standard form of mechanical clock readout incorporating hour, minute and in some cases seconds hands is either not precise enough or is difficult to read. Electronic digital readout clocks have been introduced which allow precise sub division of the second for stop clock applications

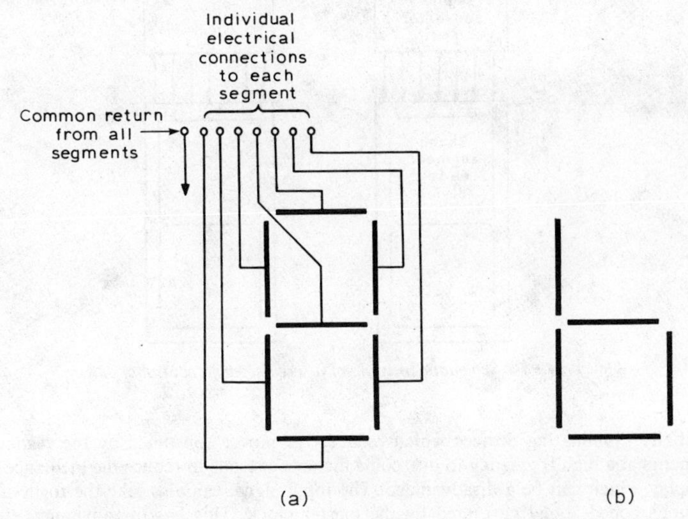

Figure 24.97 *Seven segment readout device* (*a*) *electrical connections* (*b*) *figure '6' illuminated*

and high definition displays for instrumentation panels and other exacting applications. The devices generally employed for readout are:

(1) Cold cathode discharge devices (Section 7).
(2) Light emitting diodes (Section 10).
(3) Liquid crystals (Section 10).
(4) Fibre optics (Section 10).

With the exception of Cold cathode 'Nixie' tubes, almost all other digital read out devices are of the seven-segment type illustrated in *Figure 24.97*. Using selected combinations of the three horizontal and four vertical segments any number between 0 and 9 may be displayed.

Figure 24.98 shows in schematic form the basic circuit necessary for controlling the readout device from a succession of input pulses.

The binary coded decimal (b.c.d.) counter circuit consists of four series connected binary dividers, each giving an output signal equivalent to a binary 0 or 1 at terminals *A*, *B*, *C* and *D*, for example after six input pulses the output sequence at these terminals would be 0 1 1 0. The b.c.d. to seven-segment converter, converts the binary coded outputs from *A*, *B*, *C* and *D* to a series of output voltages to illuminate the correct segments of the readout device. *Figure 24.97* (*b*) shows the number 6 displayed and it may be seen that five of the segments are utilised.

Refinements incorporated into digital readout clocks may include a circuit for

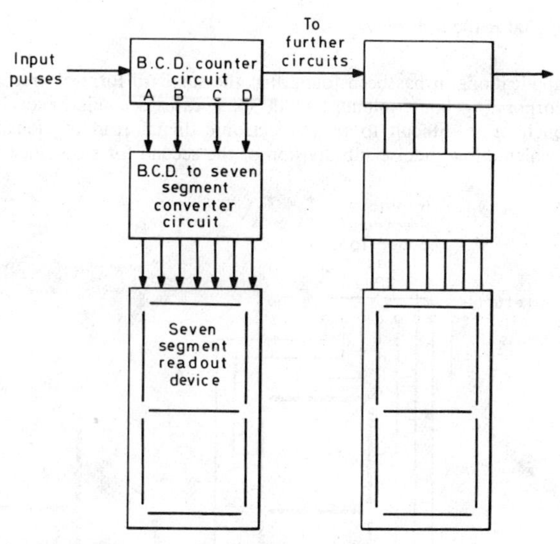

Figure 24.98 Schematic drawing of a seven segment control circuit

sequentially illuminating devices which reduces the power consumed by the readout. This occurs at a high frequency to overcome flicker but tends to reduce the brilliance of the display which can be a disadvantage. The input signal can also take the form of a continuous coded signal generated by the master clock. This has the advantage that should the input signal cease for any reason, it is not necessary to re-synchronise the readout circuit. It also overcomes problems that may be experienced due to the coupling of stray transient voltages to the input cable from external sources. Such signals may either cause the b.c.d. counters to receive spurious inputs or counteract the genuine input signal causing a loss of count.

Each of the readout devices previously mentioned has been used for digital readout clocks but in each case the device has drawbacks. The cold cathode tube, while capable of excellent definition, even in high ambient light, requires a high voltage to ionise the gas. This causes some problems as master clock systems generally operate from low voltage supplies. The light-emitting diode operates from a low voltage supply but can be rather difficult to read in conditions of high ambient light. The liquid-crystal display, while operating on low voltage and low current generates no light and must therefore either incorporate some form of illumination or rely upon ambient light. These devices can also be difficult to read under certain lighting conditions. Finally the fibre-optic readout usually relies on tungsten filament lamps for the illumination of the fibres. These have the disadvantage of a limited life particularly when being switched on and off at frequent intervals. The advantage however, is the definition of the display.

For the reasons stated, different readout devices have different applications. Cold cathode devices can be used for clocks on aircraft flight decks where legibility under all conditions is essential. For portable stop clocks, however, where power is at a premium, liquid crystal displays may be utilised.

Electronic time distribution system

Having described the circuit elements of electronic clock systems, the incorporation of some of these items into an overall system is shown in *Figure 24.99*. This is based on the Model C17 Crystal Controlled Master Clock manufactured by *Gent & Co. Ltd.*, and incorporates the main features required in comprehensive systems.

The whole clock system normally derives its power from the mains supply, suitably transformed, rectified and smoothed to a nominal 24 V d.c., by the battery charger (*22*), a standby battery (*21*) is incorporated to supply power in the event of mains failure.

The crystal controlled master clock incorporates a synchronising circuit (*1*) to enable the whole clock system to be synchronised with an external time standard, for example the Greenwich time signals. This increases the long term accuracy to that of the external time standard which is usually controlled by an atomic clock. The low frequency output impulses from the crystal master clock are linked to a switching unit (*13*) to allow a standby master clock (*23*) to take over time control in the event of failure or for servicing of the first master clock.

The operation of the switching unit is controlled automatically by a sensor (*9*) connected to the 0.5 Hz alternating polarity signal controlling the master pilot dial (*10*).

The output impulses from the switching unit are fed to the seconds and half minute slave clock distribution circuits (*15*), (*16*), (*18*) and (*19*) and also to an electromechanical comparator unit (*14*) for comparing the long-term accuracy of the 'mains' power supply frequency with that of the crystal master clock. Such a comparison is of value in generating stations and power distribution centres in order to maintain the accuracy of mains operated clocks and timing mechanisms by varying the 'mains' frequency.

Such a clock system may be utilised for controlling many circuits of slave clocks each arranged, as shown, with its own individual switching circuit for time correction and forced inverter for controlling the polarised clock movements. Systems with the capacity for driving thousands of slave clocks can be built in this manner.

Past and future development

Since the 1840s when Alexander Bain and Sir Charles Wheatstone first investigated the possibility of applying electrical energy to clock mechanisms,[3] man has been striving to improve the time-keeping properties of clocks and watches. It was not, however, until the transistor became generally available[4] in the mid-1950s that commercially available electronic clocks started to be developed. Over the past two decades the horological industry has made great advances in the field of electronics, in many cases as a result of spin-offs from the defence and computer industries. This has led to greater accuracy and reliability of watches and clocks, both the individual clock movement and the master clock system.

There can be little doubt that in the future, the power necessary to drive these devices will be further reduced and greater use will be made of electronic digital readout devices which are already being incorporated in wrist watches. Developments in the field of accuracy can be foreseen with greater use of time signal transmissions for the synchronisation of clocks, even those used in the domestic environment. This would give

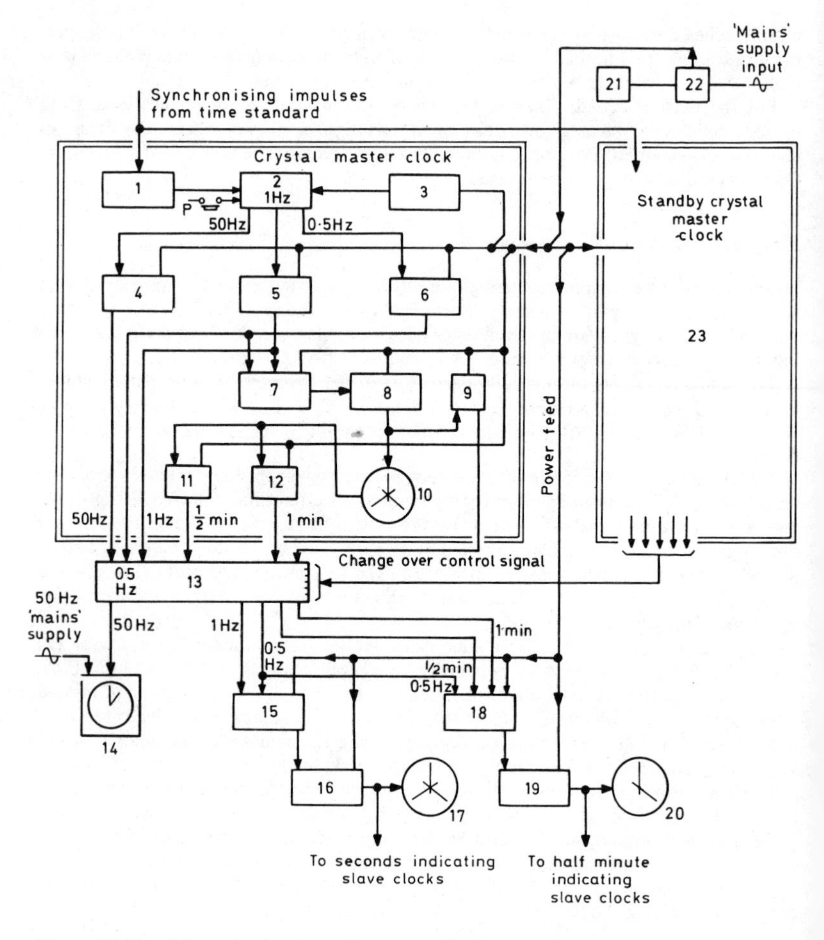

Figure 24.99 Schematic drawing of a time distribution system (Courtesy Messrs Gent and Co. Ltd.)

1 Time synchroniser
2 Crystal oscillator and divider circuits with external manual synchronising push P
3 Voltage regulators
4 50 Hz power amplifier
5 1 Hz power amplifier
6 0.5 Hz power amplifier
7 Master pilot dial stop/advance switching circuit
8 Master pilot dial forced inverter
9 Master clock monitor circuit
10 Master pilot dial incorporating ½ minute and 1 minute contacts
11 ½ minute power amplifier
12 1 minute power amplifier
13 Master clock change-over switching circuits
14 Frequency comparator
15 Seconds slave clock stop/advance switching circuit
16 Seconds slave clock forced inverter
17 Second slave clock pilot dial
18 ½ minute slave clock stop/advance switching circuit
19 ½ minute slave clock forced inverter
20 ½ minute slave clock pilot dial
21 24 volt d.c. standby battery
22 Constant potential battery charger
23 Standby master clock

the capability of atomic clock accuracy, within the home, by means of a small battery-powered clock containing a special radio receiver for synchronisation and digital readout indicating time to the nearest second.

REFERENCES

1. PLEASURE, M., 'The Fedchenko Clock', *Horological Journal*, **116**, 3, 3 (1973)
2. HOLMES, P. G., 'Semiconductors in Controlled Clock Systems', *Electrical Times*, 17 March (1966)
3. HOPE-JONES, F., 'Electric Clocks'; *N.A.G. Press Ltd* (1949)
4. SPARKES, PROF. J. J., 'The first decade of transistor development: a personal view', *The Radio and Electronic Engineer*, **43**, 1 and 2, 3–9 (1973)

ELECTRONIC FIRE DETECTION SYSTEMS

For the early detection of a fire, a device is used that senses, by changing from one electrical state to another, the products released during a burning process, and then signals by means of control equipment a warning before a conflagration, where serious damage or injury results.

The control equipment is used to monitor the detecting devices and their associated wiring as well as to convert the signal from the detecting head, to a visual and audible warning which is used to indicate the location of the fire. Further signals may be used summoning the fire brigade direct but this requirement depends on the type and complexity of the installation.

Products of a fire

The products released during the burning process that can readily be 'sensed' comprise:

(*a*) Invisible particles (combustion gas).
(*b*) Visible particles (smoke).
(*c*) Heat (rise in temperature).
(*d*) Flame (infra-red or ultra-violet light).

Figure 24.100 refers to the sequence in which these products are produced from a smouldering fire onwards. It must be noted, however, that the time for this sequence varies considerably for different types of fires.

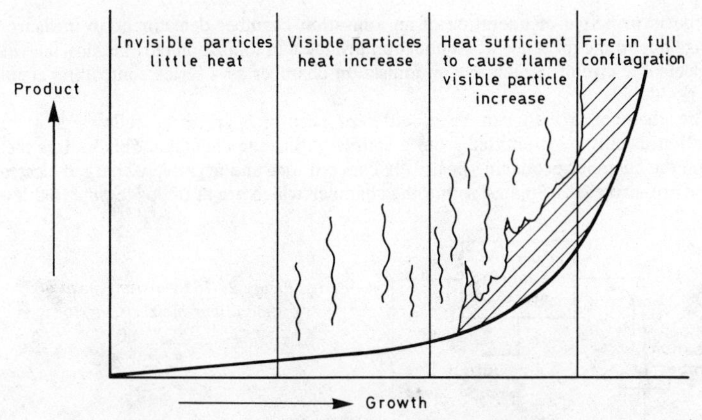

Figure 24.100 *The order in which products of combustion are produced in a typical cellulosic fire*

During the burning process combustion products are produced which consist of very small suspended particles of both solid and liquid in air (aerosols). These particles at the very early stages of combustion measure between 0.01 μm to 10.0 μm diameter. (1 μm is equivalent to 40×10^{-6} in). As the Aerosol ages the particles conglomerate, become visible in the form of smoke and eventually precipitate out of suspension.

Detectors

Four basic products of combustion can be readily detected. These are combustion gas, smoke, flame and change in temperature. The means by which these products are detected are varied but categorised by:

(*a*) Ionisation chamber detector (combustion gas and smoke).
(*b*) High impedance grid detector (combustion gas and smoke).
(*c*) Optical detectors (smoke and flame).
(*d*) Thermal detectors (changes in temperature).

These in turn are further subdivided by their different principles of operation and comprise:

(1) *Ionisation*
Single or double ionisation chamber.
(2) *High impedance grid*
Single or double high impedance grid.
(3) *Optical*
Obscuration of light.
Scattering of light.
Ultra-violet or infra-red measurement.
(4) *Thermal*
Displacement of two dis-similar metals.
Expansion of air within a chamber.
Change in magnetic flux.
Change in resistance of an element within an electronic circuit.

IONISATION CHAMBER DETECTORS

The basic principle of operation of an ionisation chamber detector is, to measure the change due to the presence of combustion products in ionisation current, flowing within an electronic circuit, which has an ionisation chamber as a series controlling element, *Figure 24.101*.

The chamber, which can be of different shapes, operates as follows: low value radiation usually α, is emitted from a source within the chamber. This ionises the air within the chamber producing positively charged ions and negatively charged electrons. These are attracted to plates within the chamber which are at opposite potential levels,

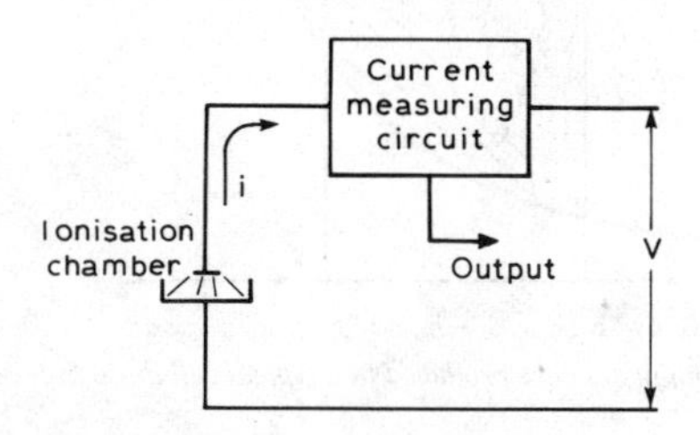

Figure 24.101 Arrangement of ionisation chamber detector

one positive, one negative. The result is that positive-charged ions drift towards the negative plate while the negative-charged electrons drift to the positive plate effecting a current flow, *Figure 24.102* (*a*). This 'ionisation current' is very small and in the order of 10^{-8} to 10^{-12} amperes.

When very small particles in the order of 0.01 μm to 1.0 μm diameter, produced during combustion, enter the chamber the ions within the chamber are attracted to them, causing the drift of charged ions to slow down, thus reducing the current, *Figure 24.102* (*b*). The reduction in current is used to trigger a current- or voltage-sensing

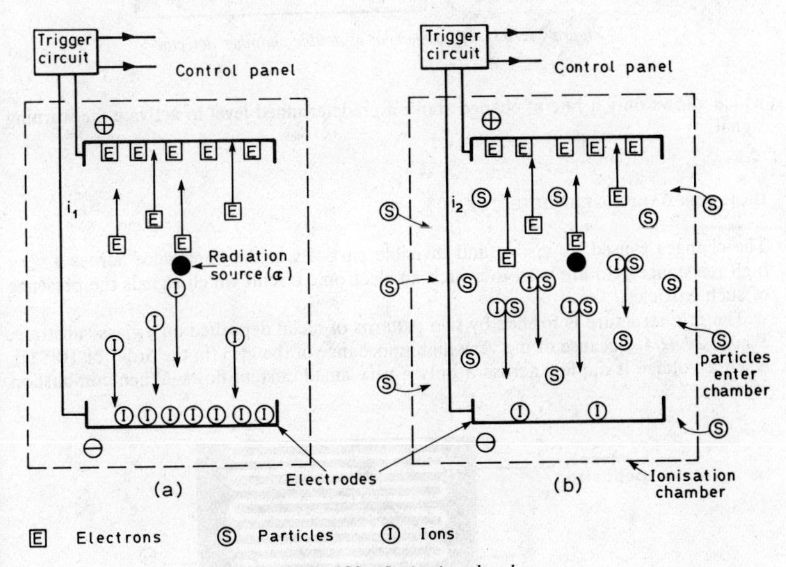

Figure 24.102 Ionisation chamber

circuit, the level of which can be varied. From this an appropriate output warning signal suitably amplified is transmitted to the control panel.

Because of the high impedance involved in the chamber it is necessary to use suitable electronic components which are compatible with these impedances. Devices commonly used are the field effect transistor (f.e.t.) and cold cathode tube (see Sections 7 and 8).

It is interesting to note that as the particle size increases, the sensitivity of such a chamber reduces, to the degree that doubling the diameter of the particle reduces the sensitivity by a factor of four.

The single chamber described has a limitation in its use since any effect that slows down the ionised particle drift sufficiently, causes the device to operate. To overcome ambient conditions that exist which have this effect, two sensors are used and it is the difference between the two that is measured. Thus with a double chamber detector, *Figure 24.103*, for very slow increases in conditions that affect the ion flow, both chambers are affected approximately by the same degree. Above a predetermined level, the exposed chamber changes at a greater rate than the unexposed chamber causing the detector to operate when the difference between the two reaches the threshold level set in the electronic sensor circuit. By using two chambers in this way compensation for non-fire conditions is achieved. It is also possible to use an 'electronic compensation'

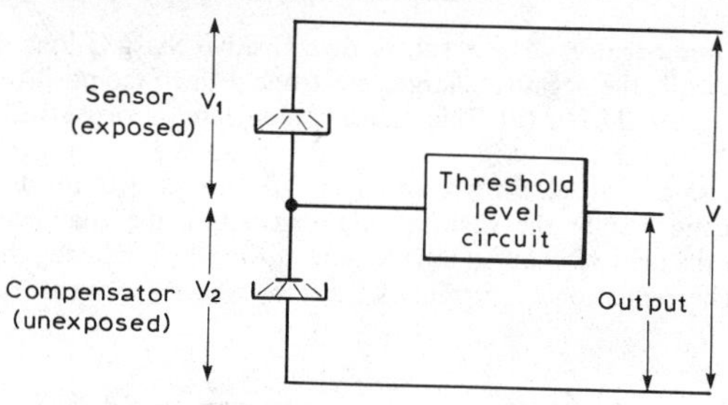

Figure 24.103 Arrangement of double chamber detector

which allows only a rate of change above a predetermined level to activate the warning signal.

HIGH IMPEDANCE GRID DETECTOR

The changes caused by visible and invisible particles in the impedance across a very high resistance grid are used to switch an electronic circuit which signals the presence of such particles.

The grid structure is formed by two patterns of metal deposited on a glass substrate, *Figure 24.104*. Because of the very high impedance of the grid (in the order of 10^{10} Ω), when a voltage is applied across it only a very small current flows. When combustion

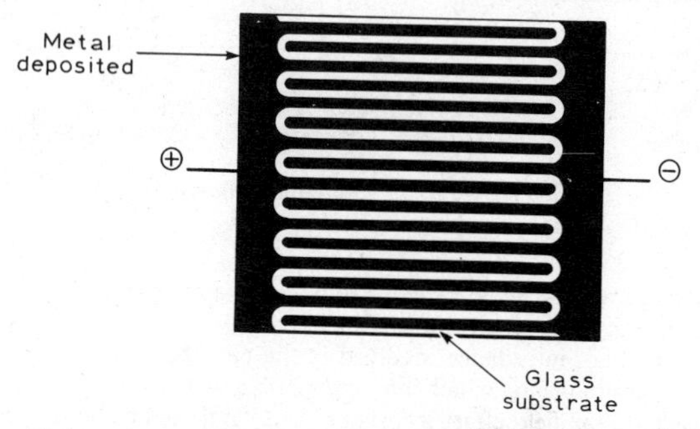

Figure 24.104 High impedance grid. The resistance between the two metal conductors is the surface resistance of the glass

products of carbonous matter and water vapour impinge on the grid its impedance falls causing the current to increase. The grid is connected in series with a second high impedance element which can be fixed in value, or with a similar grid not exposed to the same degree. This forms a potential divider, the voltage of which is used to switch a threshold level circuit, when the potential at the centre point reaches a pre-selected value. The input of the threshold level circuit must be compatible with the high impedances of the grids and for this reason f.e.t.s are used to amplify the signal to a useful level for switching. A simple circuit is shown in *Figure 24.105* (*a*). The second grid acts as a compensator for changes below required levels. With requirements becoming more complex and the need for very sensitive devices which are not prone to

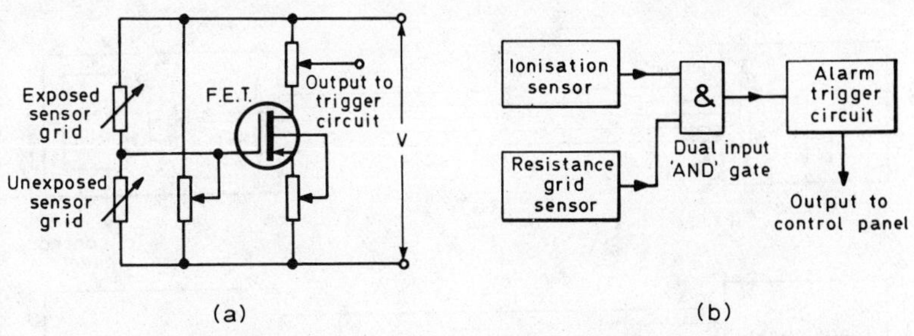

Figure 24.105 (a) Simplified circuit of high impedance grid detector (b) dual gate detector

false alarms, detectors are being produced with elements that sense two separate products of combustion and will alarm only when a signal is transmitted from both sensors simultaneously, *Figure 24.105 (b)*.

OPTICAL DETECTORS

As previously stated, when small particles conglomerate visible smoke is produced. To detect visible smoke a range of optical detectors have been developed that use a beam of light and measure changes in that light to warn of the presence of smoke. The unit used

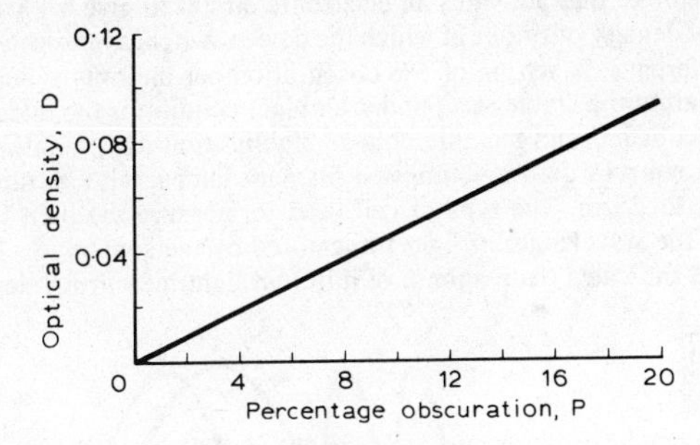

Figure 24.106 Relationship between optical density and percentage obscuration

in calibrating such detectors is the percentage in obscuration per metre of a light beam. This is a measure of the smoke density and is defined as

$$P = 100\,(1 - e^{-2.303D})$$

when P = percentage obscuration and D = the optical density of the suspended particles, $D = \text{Log}\ 10\ I/I_0$ where I_0 = Initial intensity of light and I = Intensity of light transmitted.

Figure 24.106 shows the relationship between optical density and percentage obscuration. By using optical filters of given optical density, P may be calculated.

OBSCURATION OF LIGHT

A beam of light is used where the intensity of the light is attenuated by the visible smoke particles passing between the light source and a measuring device such as a photocell, *Figure 24.107*. The light beam may be within a light chamber or across the area to be

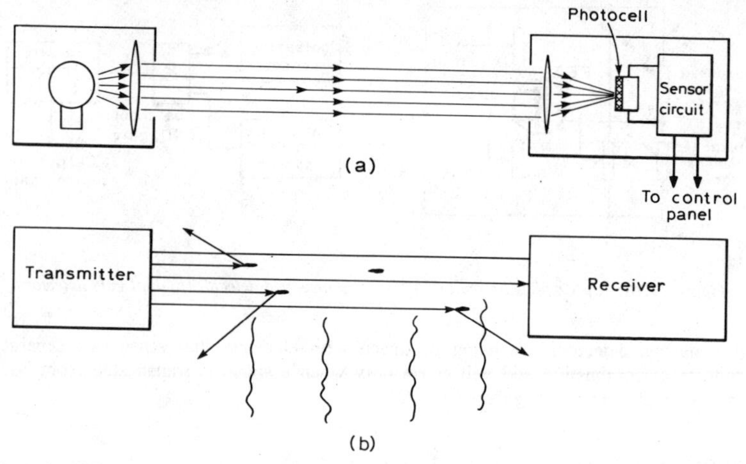

Figure 24.107 Principle of optical smoke detector (a) light beamed across protected area (b) light reflected off particles causing attenuation

monitored, in both cases, it is the change in state of the light measuring device caused by the presence of smoke that activates an electronic circuit to give a warning signal. The threshold level of density of smoke at which the device switches is adjusted within the circuit against a reference. A setting of 5% obscuration per metre is widely used.

In order to maintain a stable state under ambient conditions the intensity of the light source must be constant and suitable voltage stabilisation and monitoring circuits are used. Most light sources used are tungsten filament lamps, with an output wavelength between 0.3 μm to 2 μm. The type of cell used to measure the light is selected to be compatible with the wavelength of light transmitted by the source.

Figure 24.108 indicates the response of different light measuring elements. For long

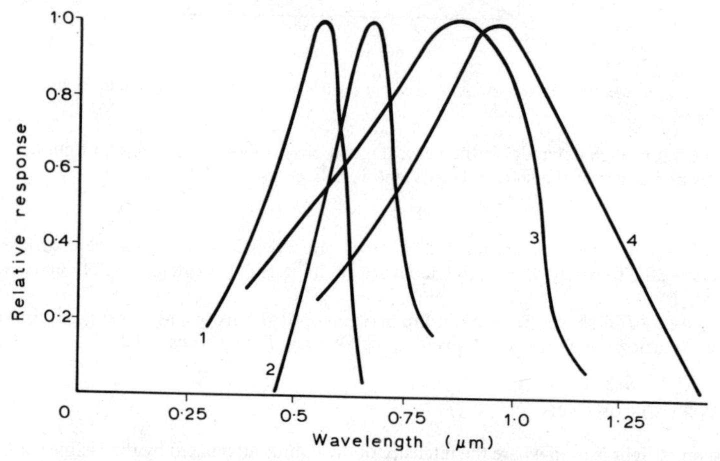

Figure 24.108 Light detecting elements. 1. Selenium photovoltaic cell 2. Cadmium sulphide photoconductor 3. Silicon photovoltaic cell 4. Silicon phototransistor, photodiode

path lengths across rooms, etc., filters may be used to eliminate ambient effects such as bright sunlight.

SCATTERING OF LIGHT

With this type of detector a beam of light is focused across a blackened chamber which allows smoke to enter but not ambient light. A screen or inner chamber is used to stop the beam of light falling directly on to a photo cell or similar device, *Figure 24.109* (*a*). When particles of sufficient size enter the chamber the light is scattered on to the

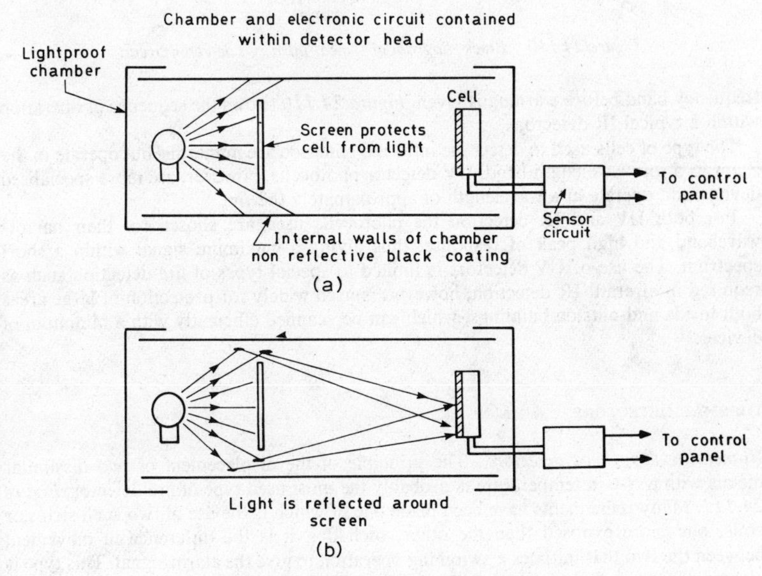

Figure 24.109 Principle of light scatter

photocell as it reflects off the particles, *Figure 24.109* (*b*). As the cell changes resistance and the threshold level is reached in the alarm trigger circuit a warning signal is transmitted to the associated control equipment. The size and colour of the smoke particles regulate the amount of light scattered but devices of this type are set at a threshold level in the order of 5% obscuration per metre.

INFRA-RED AND ULTRA-VIOLET FLAME DETECTORS

When a flame occurs during combustion the presence of that flame can be detected by measuring the infra-red or ultra-violet radiation emitted from it. The measurement can be of a quantitive nature using an electronic filter to eliminate normal ambient radiation, so that the signal operating the alarm circuit is actuated by a narrow, defined, frequency band common only to a flame.

Further refinements are to measure the flickering effect of the flame which again is filtered and analysed by the internal electronic circuit and must be within a defined

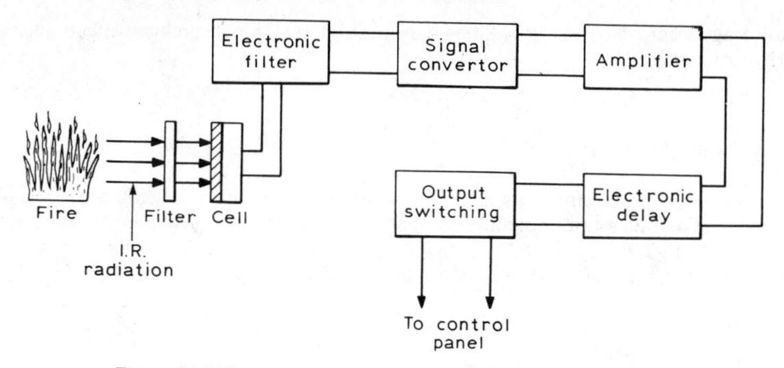

Figure 24.110 Block diagram of typical infra-red detector circuit

frequency band before warning is given. *Figure 24.110* shows the sequence of operation within a typical IR detector.

The type of cells used to detect the infra-red emission are numerous but operate in the 2 μm to 5 μm wavelength band. UV detector photocells, however, are more specialised devices and operate at a wavelength of approximately 0.2 μm.

For both UV and IR detection the photocells used are chosen for their narrow waveband and high peak of response, thus giving a maximum signal within a short spectrum. The use of UV detectors is limited to special types of fire detection such as required in aircraft. IR detection, however, is used widely for protection of large areas both inside and outside buildings, which can be scanned efficiently with a minimum of devices.

THERMAL DETECTORS

Bi-metallic strip heat detector. The principle of the displacement of two dissimilar metals with a rise in temperature is probably the most used type of fire detector, *Figure 24.111*. Many refinements have been made one of which is the use of two such strips or coils, one more exposed than the other, such that it is the difference in movement between the two that initiates a switching operation to give the alarm signal. This type is classed as a 'rate of rise' detector compared with the threshold 'fixed temperature' type. Most rate of rise detectors have a built-in fixed temperature threshold level which activates when the crictical temperature is reached even if the rate of rise level is not reached. The movement of the bi-metallic strip operates an electric contact which may be a sealed mercury tube, reed or simple precious metal contact.

Expanding air chamber heat detection. The principle of operation is similar to that of the bimetallic strip but the movement which activates the contact giving the warning

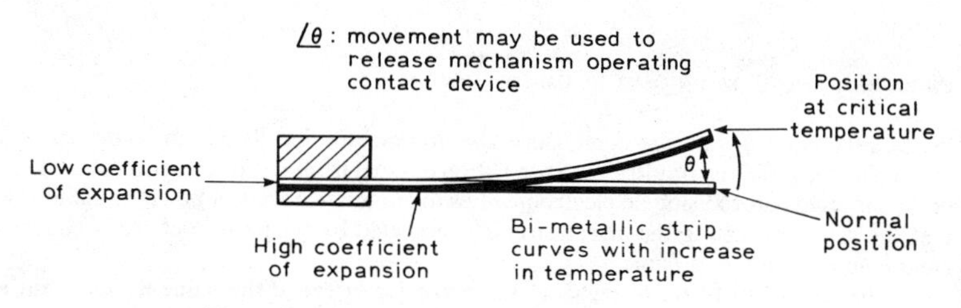

Figure 24.111 Basic bi-metallic strip heat detector

signal to control equipment is that of a diaphragm, which flexes as the air within a chamber expands with the rise in temperature, *Figure 24.112*. To allow for normal ambient rises in temperature a bleed valve can be used which determines the rate of escape of expanding air from the chamber. The point at which the device signals a warning can be adjusted to allow for different levels of the 'rate of rise of temperature' required for different locations.

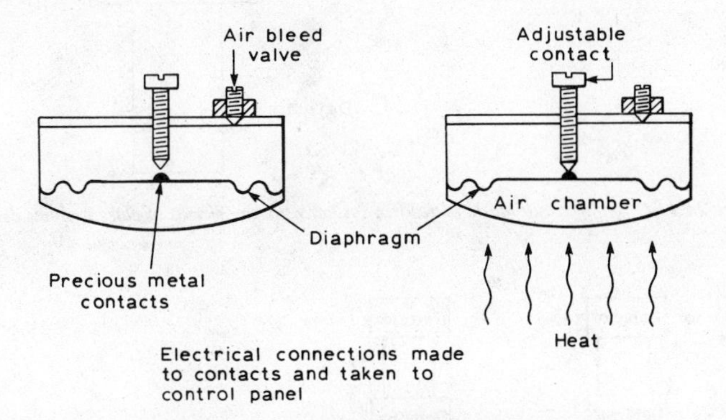

Figure 24.112 *Air chamber heat detector*

Magnetic effect heat detection. This type of detector uses the principle of the change in magnetic effect of a material with change in temperature, *Figure 24.113*. A magnetic material is used whose 'Curie point' is reached at the threshold level of temperature which is considered a fire risk. (Most heat detectors have a fixed threshold level of 57.5 °C.) The sudden change in the reluctance of the magnetic circuit is used to cause a mechanical displacement which operates an electrical contact.

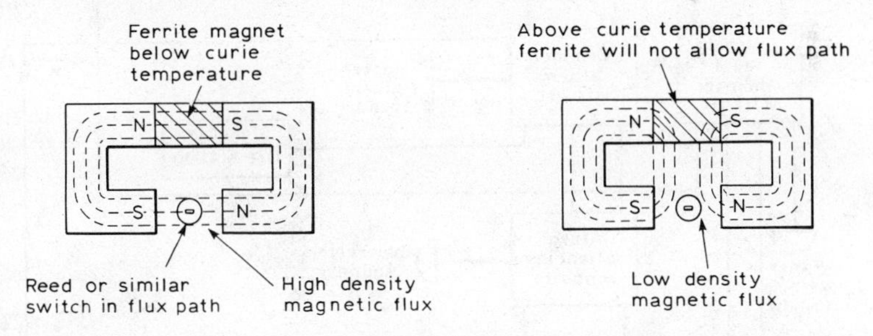

Figure 24.113 *Magnetic flux change due to temperature*

Resistance change detector. An electronic switching circuit is used which contains a thermistor, thermo-couple or similar junction device that changes resistance with temperature. When the critical level of temperature is reached the element causes an electronic circuit to operate in a similar manner to that of the optical detection circuit. The rate of change in temperature can be measured by using two devices in series as a potential divider having one more exposed than the other and using the voltage between the two to operate a voltage sensing circuit.

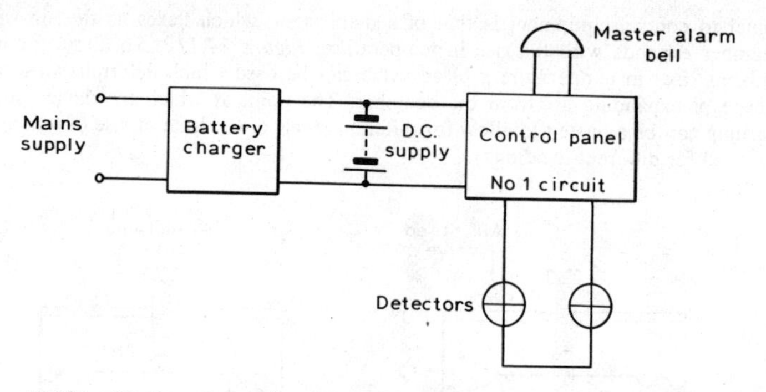

Figure 24.114 Simple automatic fire-alarm system with one circuit of closed circuit detectors

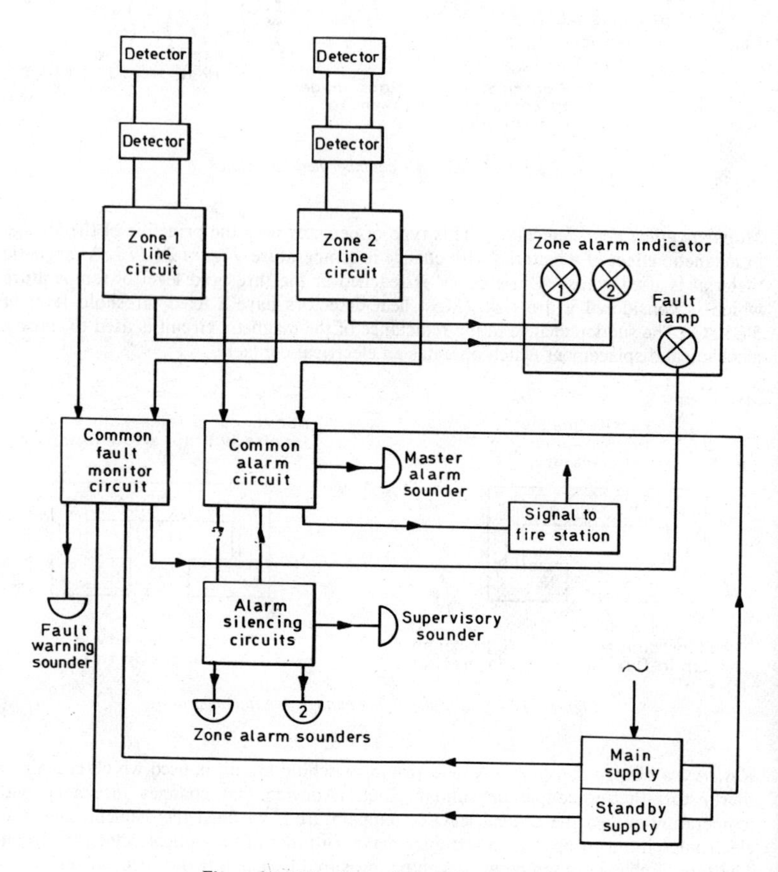

Figure 24.115 Control circuit for two zones

Fire alarm systems

The types of fire alarm systems vary considerably, from a simple single detector connected to a bell and battery, to highly sophisticated control centres monitoring large buildings and complexes.

The function of an alarm system is to warn of the presence of a fire condition. Signals from detectors or alarm points are converted to visible and audible warnings by a control panel, indicating the area from which the signal came.

To increase the reliability of the circuits, continuous monitoring for open circuit, short circuit or earth faults may be included. In the event of their occurrence, visual and audible fault signals are given within the control panel and test facilities can be provided for regular checks of the system to be made.

A fire alarm system must be ready for operation at all times. For this reason a standby power supply is automatically switched in if the mains voltage to the system fails. The standby supply usually consists of secondary cells (Nickel–Cadmium or lead acid batteries), maintained in a fully charged state by a charger fed from the mains voltage supply. The layout of a simple system is shown in *Figure 24.114*, and *Figure 24.115* shows some of the features of more sophisticated control panels.

ELECTRONICS IN SECURITY SYSTEMS

Security is defined as: the condition of being secure. The condition of being protected from or not exposed to danger. Freedom from doubt. Well-founded confidence or certainty. Freedom from care, anxiety or apprehension; a feeling of safety. Property deposited or made over, or bonds, or the like entered into, by or on behalf of a person to secure his fulfilment of an obligation.[1]

Self-preservation is mankind's strongest instinct. Security of the person, of property and of intangibles is of prime importance. Security is a loaded term often loosely used and taking on meanings not within the definition, e.g., security man. Until the end of the nineteenth century, security was mostly the province of a country's armed forces and the police. But the increasing need to guard against the effects of human error in a sophisticated civilisation, the toxic effects of civilisation itself, the need to introduce more efficient and speedy methods, and the problems of policing as the result of manpower shortages coupled with the growth of organised and technologically competent criminal elements, have all served to produce a continually growing demand for devices and systems to supplement the traditional forces of law and order. Many procedures prone to human failings can, with the help of electronic technology, be replanned on a relatively error free foundation with enhanced cost reduction; this is specially so in the realm of banking and money technology. In the field of security as a whole the further application of electronics is delayed only by the absence of suitable sensors or transducers[2] capable of interpreting certain conditions or parameters, e.g., as yet there is no really practical alternative to a dog's nose to follow a scent, and there is an urgent need for transducers that will respond directly to environmental–compositional changes reversibly and in a quasi-specific manner. And, most intractable of all obstructions, is many private individuals' and firms' reluctance to trust electro-mechanical equipment, or to treat the cost of installing or hiring it as an essential extension of their usual insurance.

There is no strict division between security and some other applications for components and equipment. Therefore, reference should also be made to, e.g., noise and sound measurement (Section 14); cardiac pacemakers and other medical devices (Section 27); fire detection (Section 24); traffic control (Section 25); etc. These all have a modicum of 'exposed to danger' avoidance.

Before the planning of any new security system, or contracting in detail for an installation, a search should be made for relevant international, national or local requirements. For example the British Standard Institution's list of specifications and codes of practice, Acts of Parliament with a possible bearing, and local bye-laws should be consulted. Few people who have suffered hours of an unattended alarm bell ringing know that Section 87 of the *London Government Act* (1963), forbids the sounding of an alarm for more than a reasonable time, or that other local authorities are proposing to enforce this bye-law. It is said that new additions to the *Factory Act* will prohibit excessive industrial noise. It is noteworthy that not all police forces allow warning devices to be wired to an annunciator in their stations. Engineers worried about the privacy of the individual and computer security legislation may find some guidelines on what to expect in *The Swedish Data Act 1973:289*.[3]

The choice of suitable security devices and equipment depends on the characteristics of the location and whether the installation is to be of a peripheral, trap, interior or specific function type, and fitted either outside, or in, a building. The likely climate and degree of normal wear-and-tear must be taken into account. Also whether conditions vary between night and day; what the available power supplies are and how well guarded; what facilities there are to change batteries or install a standby generator; the likely types of danger; the degree of automation desirable and practicable; local environmental requirements; whether devices are to be hidden or blatantly displayed as an additional deterrent; and the type of articles, etc., or person(s) to be protected.

It will be appreciated that a cardinal characteristic of efficient security is that detailed data shall not be published. Consequently only an outline of some devices is given here.

Intruder alarms

Much ingenuity has gone into making intruder alarms and surveillance systems. The primary elements designed to detect the presence of an intruder take many forms. Choice of the right one for a particular position depends on whether it is to be installed in a peripheral guard location outdoors (where sensitivity may have to withstand the rigours of climate, pollution and long periods between servicing as well as encasement in a construction strong and inaccessible enough to discourage sabotage); or in a trap position where the progress of the intruder is noted but not thwarted; or within the target area. Primary elements include devices depending on the doppler effect of a moving object in a microwave radiation field; infra-red or laser beam generators with sensors; proximity, stress, strain, inertia and normal open or closed switches; vapour, smoke, gas, air-pressure and temperature-change detectors; closed-circuit television cameras; pressure-pad mats; special locks and fastenings; etc., or any combination of these. Some elements passively survey and others await activation by the intruder.

A typical installation comprises a network of elements positioned near to external doors, gates, fences, walls, etc. A trap area network positioned near to paths, driveways, etc. of a yard, garden or grounds. And a target area network guards house or building doors, windows, airbricks (to warn of possible introduction of explosives or listening devices under floorboards) and special items of importance such as a safe, antiquities, pictures, etc. These elements may be discrete individual self-powered units with a built-in alarm, or wired into a central monitoring and control desk which may, in turn, be connected by line or radio to a central or district control and action station. The local control desk will have circuits to block off risk of spurious alarms caused by small random disturbances, birds and small animals or equipment failure.[4] British Standard Specification 4737:1971 *Intruder Alarms*; and 4166:1967 *Automatic Intruder Alarm Terminating Equipment in Police Stations* are worth consulting. The National Supervisory Council for Intruder Alarms Limited has been instrumental in establishing standards of commercial integrity.

A recent development in c.c.t.v. surveillance utilising existing cameras—if desired—is likely to supersede the practice of fitting an individual sensing element to each danger location. As long as the area surveyed by a t.v. camera (by daylight or infra-red light, etc.) is quiescent, i.e., nothing moves, the picture waveform remains constant and the monitoring screen can, in fact, be blacked out automatically to indicate a 'safe' condition. But immediately anything moves within the area surveyed by the camera, the video waveform changes, the monitoring screen flashes into life and automatic countermeasures are taken such as the sounding of an alarm, the closing off of the particular area where the intruder is, together with identification on an annunciator panel of the precise location. Spurious signals caused by insignificant movements such as the flight of a bird or insect are eliminated by the picture circuits. Continuous surveillance may be videotaped for records.

Antiques and objets d'art

The display of valuables such as pictures, ceramics, tapestries and artefacts often requires security arrangements additional to the normal. For example, provision may have to be made to facilitate change of display position, and there is nearly always the need for strict automatic control of environmental conditions.[5] Many art gallery authorities are totally opposed to the use of ultrasonic devices although recent research has shown harm is unlikely to result to ceramic, glass and picture surfaces.[6] Detection of forgeries, alterations and repairs may cause more damage to the valuable than is tolerable, so it is noteworthy that reliable results are obtained with specimens of less than 1 mg using neutron activation analysis techniques.[7]

Picture theft countermeasures range from monitoring the infra-red radiation from the picture surface under normal lighting conditions, to the attachment of a specific protection device to the back of the frame or canvas; e.g., a pressure-sensitive switch between the picture and wall which activates an alarm if the former is pulled away from the wall, or the fixing of a loop of self-adhesive metal foil which forms a closed circuit with the monitor. Picture suspension wires may be hung from a weight or vibration sensitive device hidden in the picture rail. Another system consists of battery-operated responders, or reflectors mounted unobtrusively on each picture frame, which is scanned in turn with other pictures, artefacts, etc., by a centrally placed interrogator coupled to the central monitor.

Ceramic pieces, glass, etc., may be protected by being placed on a weight or vibration sensitive support. Identification of stolen property is helped by the placing of a microdot of a specially resistant paint on the article. The paint is identifiable with a particular owner as it contains a combination of trace-elements exclusive to that particular person, and a specimen is kept by the manufacturer; comparison of the two by spectrographic analysis indicates ownership positively.

Tapestries deteriorate if fixed rigidly[8] and must be protected by devices that allow a latitude of movement. There is often the need to discourage the public from handling these very valuable fabrics. This is accomplished by attaching a vibration or strain-sensitive fine-wire netting to the back of the hanging in addition to setting up an invisible screen of infra-red beams and detectors some distance away from, and not impinging on, the front of the fabric. Ultra-violet sensitive devices should never be employed.

Crime detection and prevention

BOMB AND EXPLOSIVES DETECTION

The first aim is to find an object which may be a bomb, and the second is to determine its composition and the best way to defuse it. A search for luggage and letter bombs may be made in two stages with a view to filtering off non-suspects. The contents of luggage, letters, people's clothes, wallets, purses and handbags are first screened by a detector capable of finding all types of metallic objects including compounds used in firearms.

The detector works on the eddy-current principle and gives off an audible or other danger signal. A sophisticated development of the principle is pulse-operated and has the advantage of being far less influenced by stray fields radiated by industrial equipment, car ignition, electric motors, etc. The detector may be small, self-contained and portable, or incorporated unobtrusively into a special doorway through which suspects are required to pass, or else be a component of a luggage-conveyor belt, etc. The same type of equipment is used to find metal objects hidden in ceilings, floors, walls, the ground, etc., and gives a response depending on the size and depth of the metal object. It is claimed to have no effect on persons with heart pacemakers, etc.

Suspect packets and persons are then—or routinely—exposed to a harmless dose of short-duration X-rays (less than 0.3 mR for about 30 ns). The resulting image is intensified electronically, stored and displayed on a cathode-ray tube screen, where an observer notes the presence of any suspect object, including plastic and ceramic shapes, and has the option of storing the image on videotape for possible prosecution evidence. Smaller and simpler units for office use, but with complete protection for the operator, are used for routine examination of letters.

There is a good chance of rendering a bomb harmless by the use of recently developed techniques to identify the materials. Instruments are now extremely versatile and so reliable and sensitive that a positive identification can be obtained from the seat of a car on which a box of explosive material has rested, or from the clothing or hands of a handler who believes he has washed away all traces. Full details of these instruments is classified, but it may be said that there are two main types of sensor or 'sniffer'. One comprises a radioactive substance and is used first to confirm the presence of an explosive and then to identify it. The other depends upon a change in a micro-organism when exposed to any of a number of different vapours or pollutants, including narcotics such as heroin.

IDENTIFICATION TECHNIQUES

A speech analysis diagram called a *voiceprint* proven to be identifiable as a personalised record of the characteristic utterances of a particular person, however much he may try to change his voice, is helping to identify authors of bogus and threatening telephone calls. The equipment can produce spectrographic analysis from standard magnetic tape and, used as a comparator system, provides a means of identifying a credit card holder when a bank of recorded voices has been set up, or of validating the authenticity of applicants wishing to enter a restricted area.[9]

A fingerprints identification system semi-automated and easily learned by a policeman with little knowledge of computer usage has been developed by the U.S.A. Federal Bureau of Investigation. The policeman describes the fingerprint of a suspect in a simple and computer-readable code, and transmits his query by line or radio to the FBI's central, computerised file where a rapid search is made; if the print is matched the computer responds with a record of the suspect.[10, 11]

Considerable progress has been made in Britain in the reading of difficult fingerprints, e.g., those on cloth. The technique depends on the measurement of the electrical characteristics of microscopic skin and sweat-gland debris.[12] Laser photography is used to outline, for example, invisible footprints on a carpet trodden hours earlier; this effect is produced by the slow movement back to normal of carpet fibres.

Lie detectors are suitably treated with reserve. A typical instrument may record galvanic skin response, blood pressure, pulse and respiration rate, eyelid and finger or hand activity and vocal responses of the subject under investigation. A new device neglects these characteristics and merely analyses speech performance in terms of the stress exhibited to 'awkward' questions.

LISTENING DEVICES

Listening devices may operate at carrier frequencies up to 2 000 MHz, but more usually are one-tenth of this frequency. Unless powered by the telephone or mains supply the average working life without battery replacement is from 50 to 200 hours; the effective range may be up to 800 metres. External listening devices are more difficult to apprehend but often as effective; one type is a directional microphone and the other depends on the vibrations set up in the window glass of the room where speech is to be monitored. A laser beam reflected off the outside of the glass is modulated by the vibrations and, therefore, readable by a suitable sensor.[13]

A typical surveillance instrument comprises a portable radio receiver with an automatic frequency band sweep of about 5 minutes and manual fine adjustment. Headphone and visual presentation circuits monitor both amplitude and frequency modulation clandestine transmissions, using the receiver as a direction finder. Likely accessories include modules which locate telephone operated or powered, and also mains-carrier or mains-wiring remotely controlled transmitters; and a tone generator to modulate transmitters within audible range. Minute self-powered and long-playing tape recorders are a special hazard. One may be found in a bogus book with the microphone in the 'spine' camouflaged by the 'title'.

Aircraft protection

The 'black-box' flight recorder has proved one of the most successful of all electronic safety devices, for it produces information to prevent similar accidents.

A recent aircraft intruder prevention system comprises a portable generator of a magnetic field around the aircraft; a person moving across this field is apprehended immediately.

Vehicle protection

VEHICLE SAFETY

A car radar has been developed to reduce rear-end collision risks by giving a warning in poor visibility when a driver's speed exceeds the braking safety margin, with an obstacle 100 m to 200 m ahead. The radar takes automatic account of the road surface conditions.[14] Another car radar collision predictor provides for automatic tightening of seat-belts and concurrently inflates an air-bag buffer between passengers and the fascia.[15] American car radar development has been concentrated more on collision prevention at highway cruising speeds, and is centred on automatic braking and accelerator systems with audible and visual warning to a reckless driver, but with the option for him to over-ride the automatic system while concurrently being subjected to warning indicators.

With effect from 1974, U.S.A. Federal legislation requires all American-made cars to be fitted with an interlocking system that inhibits engine starting until anyone in the front seats weighing 47 lb (20 kg) or more has fastened their shoulder and lap belts and all doors are properly closed. Each component of the seat-belt is independently monitored by a sensor linked to an integrated circuit, which handles the complex sequential logic and malfunction indicators.[16]

Experiments are in progress in Europe (with official support) with a car function central information system that constantly monitors ambient conditions and visually presents, in hierarchical importance, through logic analysers 12 potential danger areas of the engine and body mechanism, and gives both threshold and actual danger warnings.[17]

VEHICLE THEFT AND IDENTIFICATION

Systems are numerous and of vastly varying effectiveness and so the greatest deterrent to thieves is perhaps the ability to locate and stop a stolen vehicle. The most promising of these is the automatic vehicle identification (a.v.i.), and the automatic vehicle monitoring (a.v.m.) systems. The former system comprises a transponder mounted unobtrusively on the underside of the vehicle, and interrogatable by a buried induction loop signal operating in the frequency band of 96 kHz to 200 kHz, and radiating a working signal to about 380 mm above road level. The interrogator commands the transponder to transmit the vehicle's coded road registration details; this is received by the induction loop and relayed by line or radio to a central monitoring station where a computer automatically matches the data against 'wanted' vehicles, and alerts local mobile police when a stolen vehicle is spotted.

A.V.M. provides immediate automatic reports on the location of specific vehicles direct to, e.g., a company's 'dispatch' office. The system utilises existing mobile radio equipment adapted to the transmission of information in a time-saving digital code format. Low-priced transmitters positioned in a grid format attached, perhaps, to overhead lamp standards, transmit location codes to vehicles passing by, and these are triggered to respond automatically to their base offices with a message showing grid reference position and, optionally, sending any information of current importance. In addition to thwarting hijacking this system improves the efficiency of taxi, commercial vehicle and police deployment.

Personal identification

Credit cards originated as a means of reducing the demand for, and the carrying of, cash, but have been extended to control entry and exit to security areas. Experience of use in the former application has resulted in the embodiment of electronic equipment ideas to provide greater reliability for the latter use. The card has to be relatively inexpensive to produce and yet practically impossible to forge or to misuse. Consequently most card holder systems require the user to identify himself by keying in manually and concurrently a secretly memorised code, and the card reader provides him with a means of aborting the action if he is obliged to use it under duress, i.e., he is being forced to use the card against his will. Card design may be extremely complex and include codes requiring the automatic reader to be capable of discerning information expressed individually or as a combination of materials possessing magnetic, optical, electrostatic, punched, radio-isotope, fluorescent or other properties.

Experiments are being made with a peripheral reader which automatically scans the image of the card holder and compares the image with the card photograph. Card reader peripherals are normally connected direct to a central control which identifies codes by comparing them with codes in issue, and sounds an alarm while concurrently blocking the card in the reader mechanism if it appears to be bogus, or the holder signals he is obliged to use it under duress.

In the U.S.A. a key corporation is installing in hotels a system which issues a card punch-encoded anew for each new user of a bedroom. A duplicate card is encoded by the receptionist and inserted for the period of room occupancy in a key card code comparator located in the hotel's reception desk area. Although each bedroom door has to be wired to the comparator, or multiplexed with the room's telephone and TV relay cabling, it is claimed that the security inherent in the issue of a virtually non-repeatable key code to each new guest, and the elimination of the risk of a departing guest not surrendering his key, more than offsets the cost of the system.

Property protection

Shoplifting or so-called 'shrinkage' has developed in recent years at an uncomfortably costly speed. There is little chance of reducing this trend until shop architects can be persuaded to help persuade the public to use entry and exit doors, and not to shun shops with narrow doorways. The most satisfactory method of prevention is to attach an electronically detectable identifying tag to each item, which is removed at the point of sale, or identified by a sensor located in the doorway when an article is removed illegally. The threat of detection is often a sufficient deterrent, especially if a seeming 'victim' is seen to be apprehended!

Similar techniques are being introduced in public lending libraries to stop the pilferage of books; the tag may be hidden permanently in the book spine as it is so much easier to monitor persons passing through the customary entrance/exit passageways of the typical library. In this application, the tag may be encoded to transmit data for the library records, and so facilitate the projected Authors' Lending Right or Public Lending Right system being sponsored jointly by the Writers' Action Group and various computer manufacturers, and supported by the Rowntree Trust.

A most ingenious fail-safe electronically-actuated proximity 'key' with an almost infinite number of possible combinations and the size of a large felt-pen, in the form of a hand-held battery-operated encapsulated probe, is now increasingly applied to the securing of car and van doors and windows, garage and outer domestic doors, filing cabinets and safes, cash registers and drawers. An electro-mechanical lock mechanism is unobtrusively incorporated in each item to be locked, and has an i.c. receiver which actuates the mechanism into an 'unlocked' state only when a specified train of signal pulses is received from the probe, which has a free-air range of about 1 m. If the probe is switched off, or moved more than 1 m away, the mechanism automatically locks.

Banking and money

The only true money is gold. Banknotes and coins are symbolic money and vary in value in relation to the value of gold, but have to be safeguarded with equal care. In fact, the problems of guarding cash often tends to cost as much as the amount protected. The most economical and efficient way of safeguarding money is to take it out of circulation in exchange for credit, which in a computer-centred age is a much more manageable form of spending and saving power, for it can be identified with its owner, which coins and banknotes can never be (it is illegal to deface cash with the name and/or address of an owner). Moreover, credit notes may easily be made automatically readable for computer input and processing. During the last decade increasing pressure has been exerted on the public to accept cheques in payment of wages, and to arrange payments through banks by either the so-called 'bank giro', or the National Giro operated by the Post Office.[18]

Britain was the last major country in Europe to introduce a postal giro and therefore had the advantage of benefiting from the accumulated experience of Sweden, the Netherlands and West Germany, whose credit transfer systems are the most secure and efficient of the 40-odd countries using this advanced monetary and banking method both nationally and internationally.[19]

Electrostatic storm protection

A major security problem successfully overcome by the designers of the rocket assembly buildings and tower launchers in the U.S.A., was the virtual elimination of risk from lightning strikes. The area protection system developed specially for this purpose has proved so successful that it satisfies the requirements of insurance companies' policy regulations for radio and television mast, office building, etc., protection, and is now commercially available at economic rates. The equipment harmlessly discharges thunder clouds over any pre-determined area.

The future of electronics in security systems

The application of electronics in security systems has progressed far during the last decade. There remains the need to bend the imagination and inventiveness of an ever increasing number of electronics engineers in the direction of security in its many guises, including accident prevention and safety improvement. Very few security systems are so perfect that they can be regarded as absolute. Particularly in the sector concerned with possible criminal activity there is a need to regard, and to treat, every problem as an individual case; once a criminal has broken one type of security he holds a key to similar types.

REFERENCES AND ACKNOWLEDGEMENTS

1. ONIONS, C. G., *The Shorter Oxford English Dictionary Based on Historical Principles*, 3rd edn., Oxford University Press (1970)
2. O'KEEFE, A. E., 'Needs in Electronic Instrumentation for Air-Pollution Analysis', Report of the *Joint Conference on Sensing of Environmental Pollutants, Palo Alto, Calif., USA, November 8–10 (1971)*
3. VINGE, P. G., *Swedish Data Act 1973:289*, Federation of Swedish Industries (1973)
4. Anon., 'Failure-to-danger potentialities of solid-state logic control systems', *The Radio and Electronic Engineer*, **43**, 4, 266 (1973)
5. HARVEY, G., 'Air-conditioning for museums', *Museums Journal*, **73**, 1, 11–16 (1973)
6. STEPHENS, R. W. B., 'Ultrasonic alarms in picture galleries', *Security Gazette*, December, 445 (1973)
7. GORDUS, A., 'Activation analysis, artefacts and art', *New Scientist*, 17 October, 128–31 (1968)
8. THOMSON, W. G., F. P. and E. S., *A History of Tapestry*, 3rd edn., 547, EP Publishing Ltd (1973)
9. DUMMER, G. W. A., ROBERTSON, J. M. and THOMSON, F. P., 'Voice identification equipment', *Banking Automation* II, 776–9, Pergamon Press (1971)
10. Anon., 'Detectives who fight crime in the laboratory', *Science Horizons*, **113**, April, 18–19 (1970)
11. HIPWELL, C., 'The hard labour of building a secure police system', *Computer Weekly*, 10 May, 13 (1973)
12. THOMAS, G. L., 'The physics of fingerprints', *The Criminologist*, **8**, 30, 21–38 (1973)
13. GLEESON, T., 'The electronic cancer', *Police Review*, **81**, 4192, 700 (1973)
14. Anon., 'Collision avoidance radar has all gear on one car', *Electronics*, 10 January, 3E–4E (1974)
15. Anon., 'Radar in auto on collision course tightens passengers' seat-belts', *Electronics*, 7 February, 107–11 (1974)
16. CURRAN, L., 'Seat-belt interlock deadline nears', *Electronics*, 1 March, 70–1 (1973)
17. Anon., 'Dashboard indicator alerts driver to 12 car malfunctions', *Electronics*, 7 February, 2E (1974)
18. THOMSON, F. P., *Giro Credit Transfer Systems: Popular International Facilties for Economic Efficiency*, 3, 100–2. Pergamon Press, Oxford (1964)
19. THOMSON, F. P., *Money in the Computer Age*, 62–131, Pergamon Press, Oxford (1968)

25 ELECTRONIC APPLICATIONS IN TRANSPORT

ELECTRONIC APPLICATIONS IN TRANSPORT 25

25 ELECTRONIC APPLICATIONS IN TRANSPORT

AVIATION

Historical background

Modern electronics feature widely in aircraft today, as the sciences of electronics and aviation have grown up together. At the end of the nineteenth century when aviation pioneers were struggling to beat gravity in powered flight, Marconi was engaged in experiments aimed at communicating over a distance by the use of radio waves. In 1899 a successful demonstration was carried out in which a spark transmitter was used to send messages from an army observation balloon. In November 1910 the first wireless assisted air-sea rescue took place. The airship America ran into trouble while attempting to cross the Atlantic, and had to be ditched. Distress signals sent by radio from the airship were received by nearby ships, and all the airship's crew were safely rescued. Also in 1910, wireless messages were first exchanged between an aeroplane and the ground. During World War I, great strides forward were made in the fields of both aviation and wireless. By 1918 spark transmitters in aircraft had been replaced by valve transmitters, and telephony introduced for air to air communication. Some of these advanced wireless sets had been specially developed for small aircraft and weighed as little as 10 kg.

Between 1918 and 1939, air transport expanded, and the use of wireless kept pace. All airlines used wireless to keep in touch with their aircraft and in many instances provided a radio telephone link for the use of passengers.

The use of fixed radio direction finding stations on the ground and airborne direction finding equipment using loop aerials proved of great assistance in the navigation of these airliners.

During World War II, the success of operations like the Battle of Britain depended on effective communications both between aircraft and between aircraft and the ground. New equipments were developed, the main new approach being the use of v.h.f. (100–156 MHz) which gave a high quality voice channel over line-of-sight paths. Concurrently with this, several new navigation techniques were developed to guide aircraft to their targets at night or in cloud. These were mainly based on the use of narrow radio beams or hyperbolic techniques in which the difference in time of arrival at the aircraft of pulses transmitted from several well-spaced beacons was measured and used to calculate the aircraft's position. Radar was also developed into an airborne equipment during this time, as well as transponder beacons to be carried on aircraft to enable differentiation between friendly and hostile craft.

Following 1945, civilian air fleets recommenced operations and the skies soon became crowded with airliners. As a result, there grew a demand for increased accuracy of navigation, and it is in this field that the greatest expansion of the use of electronics in aircraft has occurred in the last 20 years. A list of recently developed electronic navigation aids would include such items as fully Automatic Direction Finding equipments, V.O.R. and D.M.E. navigation systems, Doppler Navigators, Inertial Navigation systems, Instrument Landing Systems, Microwave Landing Systems, Omega, Radio Altimeters and Collision Avoidance Systems. The main advances in communication systems have been in modulation systems, use of higher frequencies, use of transistors and integrated circuits to replace valves (making equipments smaller and consuming less power) and the development of remote tuning systems. This latter development has allowed all electronic equipments to be housed in purpose built bays, with only controllers and read-out devices needing to be mounted in the cramped confines of the flight deck instrumentation panels. Concurrently with these advances, a range of electronic engine monitors and controls has evolved. The navigation and engine control units have come together to some extent in providing automatic pilots which have relieved the human pilots of much of the concentration required in flying the aircraft for long hours on steady courses.

In the military field, in addition to the above uses of electronics, a large field has opened up in the form of specialised radar equipments with target marking facilities. These feed into purpose built computers which control weapon delivery to ensure maximum effectiveness of the aircraft, whether fighter or bomber. The situation has now been reached that for a modern strike aircraft the cost can be split into three approximately equal parts, one for the airframe, one for the engines and one for the electronics.

Modern jet transport aircraft

To the present-day passenger on the world's airlines there are few sign of the presence of electronics on the aircraft. From the outside the aerials festooned round the fuselage can be seen, and inside announcements and music from the public address system will be heard. However, these outward manifestations of the art are but the tip of a huge iceberg. In the electronic bays there is a subdued 400 Hz hum from the vast array of 'black boxes' that are used by the flight deck crew for navigation and communication purposes. *Figure 25.1* shows the multiplicity of aerials on a typical Boeing 707. *Figure 25.2* illustrates a portion of the electronics bay of a Trident aircraft. This latter shows how the use of standard size boxes enables a high packing density of electronic units to be achieved. These boxes, in widths that are fractional multiples of 254 mm, have been specified by Aeronautical Radio Incorporated (ARINC). This is a body composed of representatives of aircraft manufacturers, electronic equipment manufacturers and airlines to draw up specifications to ensure interchangeability and correct function of all electronic equipment used in civil aircraft.

COMMUNICATION EQUIPMENT

For long-range communication with the ground use is made of the h.f. band 2 MHz to 30 MHz. The modern transmitter for this band generates up to 1 kW of output power, and is fully transistorised with the exception of a high-power forced air cooled valve output stage. The aerial used is not the long wire of earlier days, with its problems of drag, but normally takes the form of a notch cut into the aircraft skin in the area of the tail fin. A tuning unit fitted with variable capacitors and inductors is fitted to the notch to match its impedance to the transmitter. Once the transmitter is energised, phase and amplitude detectors built into the tuning unit monitor the mismatch presented by the aerial to the transmitter and operate servo mechanisms to provide an optimum match. When this condition is achieved, the maximum possible transmitter power is radiated from the aircraft.

The h.f. transmitters use amplitude modulation normally in the single sideband mode. This mode has two advantages. Firstly, it enables twice as many communication channels to be accommodated in a given band compared with double sideband operation, while at the same time reducing transmitter power. Secondly, the quality of the received speech is improved. For a double sideband system, selective fading and distortion is caused by differential phase changes between the upper and lower sidebands, but this cannot occur with a s.s.b. system. In addition, the absence of a continuous high-power carrier means that the receiver does not suffer from continuous audio beat

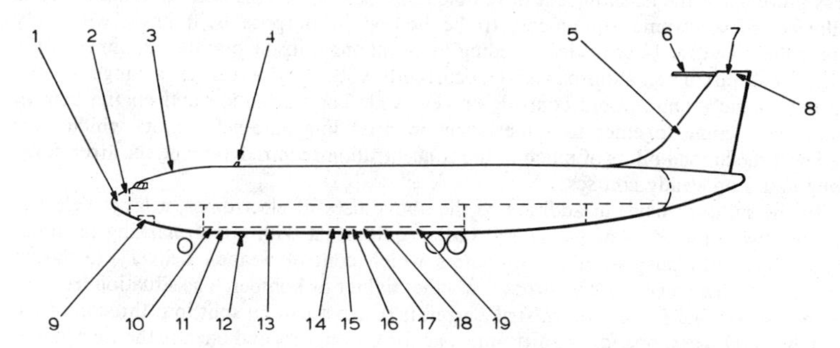

Figure 25.1 Aerial locations on Boeing 707: 1 Weather radar 2 Glide slope 3 A.T.C. transponder 4 V.H.F. comm No 1 5 Omni range 6 H.F. comm 7 H.F. receiver 8 Loran 9 Doppler 10 Radio altimeter transmitter 11 D.M.E. No 1 12 V.H.F. comm No 2 13 Radio altimeter receiver 14 A.D.F. loop No 2 15 D.M.E. No 2 16 A.D.F. loop No 1 17 Marker beacon 18 A.D.F. sense No 2 19 A.D.F. sense No 1

Figure 25.2 Part of Electronics Bay of Trident aircraft

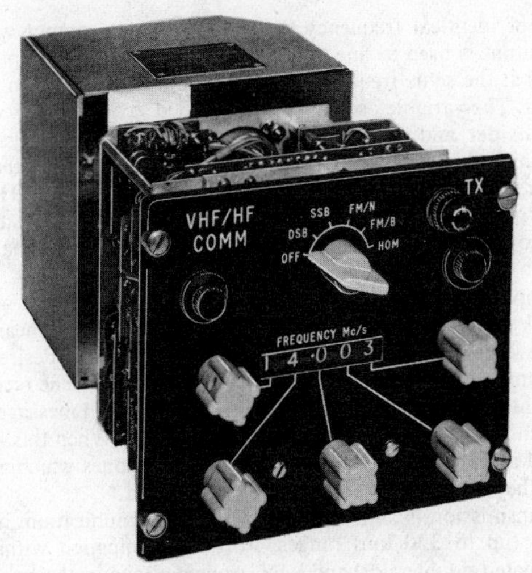

Figure 25.3 Marconi controller for AD1400 h.f./v.h.f. communications system

notes caused by heterodyne interference from other transmissions. Again, in the interests of making maximum use of the available frequency spectrum, it is important that the transmitter and receiver can be precisely tuned to the desired frequency.

This is accomplished on the flight deck by means of a small digital controller, *Figure 25.3*, which drives a frequency synthesiser in the transceiver. *Figure 25.4* shows the block diagram of a typical synthesiser. The oscillator, which is used either for the transmitter drive or for the receiver local oscillator, is tuned over the desired range by variation of the voltage applied to one or more varactor diodes (diodes whose capacitance is inversely proportioned to the square root of the reverse bias). When a frequency is selected on the controller, a discrete combination of control lines is earthed or left open circuit, and this information is decoded for two purposes. Firstly, it provides a voltage to tune the oscillator approximately to the selected frequency and secondly, it is used to set the programmable counter to the required division ratio. The low-frequency output of this programmable counter is fed to a phase detector, whose other input is fed from the divided down output of the crystal oscillator. Unless these

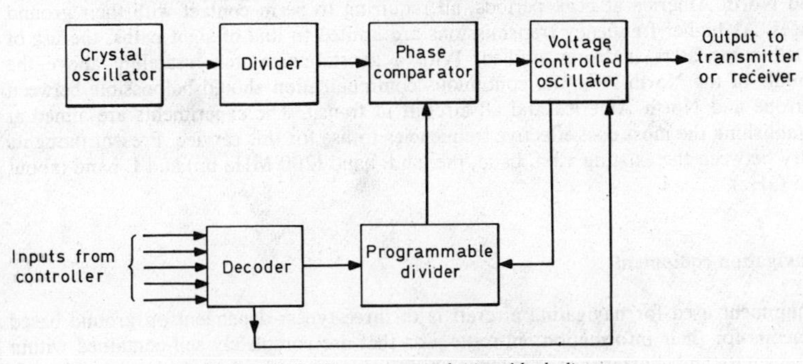

Figure 25.4 Frequency synthesiser block diagram

two inputs are of identical frequency, the phase detector output will be continually varying. This output is used to fine-tune the oscillator until the output of the programmable divider is at the same frequency and at fixed phase relative to the counted down crystal oscillator. The variable oscillator is thus tuned to a frequency determined by the programmable divider, and with the accuracy and stability of the crystal oscillator. The increments in which this oscillator can be tuned are determined by the frequency of the counted down crystal oscillator, which in some cases is as low as 1 kHz. The majority of the circuitry of such a frequency synthesiser can be readily obtained in integrated circuit packs, so enabling the complete synthesiser to form a small sub-unit of the transceiver.

The speech input to the transmitter is obtained from the headset microphone of one of the flight deck crew via the intercom system. This provides a means of communication between crew members as well as enabling selection of the desired transmitter and receiver for communication with the ground or other aircraft. The receiver output is fed to the crew via the intercom. To avoid the need to listen to all messages transmitted, the h.f. receiver often operates into a 'Selective Calling Unit'. When this is in use messages from the ground are preceded by a coded pair of audio tones which activate a warning light or bell on the flight deck of the aircraft being called.

While h.f. transmission is suited for long range communication, it provides a poor service at short (up to 320 km) ranges. For communication within terminal areas amplitude modulated double sideband v.h.f. transmissions in the band 118–136 MHz are used. *Figure 25.5* shows the block diagram of a typical aircraft v.h.f. transceiver, and *Figure 25.6* the construction of a modern system. The transmitter output stage generates 25 watts of r.f.

If the aerial presents a poor match to the transmitter (possibly due to physical damage), it is sensed by a reverse power detector which reduces the gain of the transmitter driver stage to avoid damage to the output transistors. Under receive conditions, the signal from the aerial is fed to a r.f. amplifier which is tuned by varactor diodes fed with a voltage derived from the frequency synthesiser. The signal is then mixed with the local oscillator (also provided from the synthesiser), to produce a 20 MHz i.f. signal which is amplified, filtered and detected. The audio output is then passed to the intercom system via a squelch gate operated from the receiver a.g.c. line and only opened to pass the audio signal when the r.f. signal is above the threshold level. This ensures that circuit noise does not pass into the intercom system when no usable signal is being received. The aerial used is commonly a blade, projecting about 380 mm from the airframe, which with a suitable matching network provides a load to the transmitter of within 2 : 1 of 50 ohms over the 118 MHz to 136 MHz band.

Because of the present and increasing congestion in the h.f. band, experiments are being undertaken in the use of higher frequencies for long range communication.

By 1980 there are expected to be up to 200 aircraft in transit between Europe and North America at peak periods, all requiring to be in contact with their ground bases. As higher frequency transmissions are limited to line of sight paths, the use of satellite repeaters is being studied. With a geostationary relay satellite above the middle of the North Atlantic, continuous communication should be possible between Europe and North America and all aircraft in transit. The experiments are aimed at establishing the most cost effective frequencies to use for this service. Present thoughts vary between the existing v.h.f. band, the u.h.f. band (200 MHz up) and L band (about 1.5 GHz).

Navigation equipment

Equipment used for navigating aircraft is of three types; dependent on ground based beacons for their information, en-route aids that are completely self-contained within the aircraft, and shorter range navigation aids, the more sophisticated types of which are

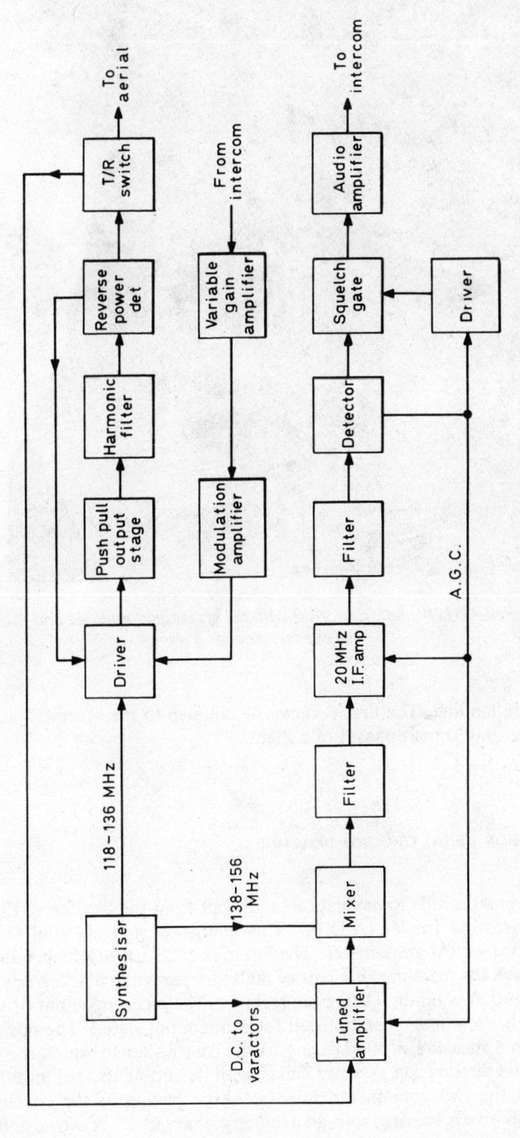

Figure 25.5 V.H.F. transceiver block diagram

Figure 25.6 Marconi AD170 v.h.f. transceiver with the synthesiser removed and with a controller in the foreground

used for automatic landing. The first two can be coupled to the aircraft's autopilot for control during the long transit phases of a flight.

MEDIUM-RANGE AIDS USING GROUND BEACONS

One of the most versatile aids to navigation in aircraft from the smallest to the largest is the Automatic Direction Finder (A.D.F.). This comprises an m.f. band (170 kHz to 1 800 kHz) receiver and a goniometer. The latter is a r.f. transformer with two fixed input windings, each fed from one of a pair of mutually perpendicular ferrite rod aerials, and a rotateable output winding feeding the receiver. The receiver output drives a servo loop to position the goniometer output coil for a minimum signal. The position of the output coil is then a measure of the bearing of the transmitter to which the receiver is tuned relative to the ferrite rods. A sense aerial, with its output also fed into the receiver resolves which of the two goniometer nulls (at 180° spacing) is the correct one. An A.D.F. system will give a bearing with an accuracy of about 2° of a transmitter within range. Simultaneous use of two A.D.F. systems tuned to different transmitters enable manual plots of aircraft position to be made rapidly to an accuracy of about 16 km at a range of 480 km.

Another system of obtaining bearing information from ground beacons is Very High Frequency Omnidirectional Range (V.O.R.). This operates in the 112 MHz to 118

MHz band with special ground beacons that radiate a carrier with two modulations at 30 Hz. The phase difference of these two modulations is proportional to the angular direction of the point of reception relative to magnetic north. The airborne equipment consists of a v.h.f. receiver fed from a slot or blade aerial which resolves the phase difference between the two modulations and gives the bearing of the transmitter and/or the deviation from a desired bearing to an accuracy of about 4°. Recent improvements in ground transmitters and airborne receivers have made possible an accuracy with a 95% probability of better than 2°.

A further navigation aid used in conjunction with V.O.R. is the Distance Measuring Equipment (D.M.E.). This is a pulse system operating at about 1 GHz. The aircraft transmits a series of coded pulses which are received at the ground beacon and re-transmitted on a new frequency 50 microseconds later. By timing the period from transmission of a pulse to the reception of the beacon reply, a measure is obtained of the distance of the aircraft from the beacon. The accuracy of a typical D.M.E. is better than half a mile. The V.O.R. and D.M.E. ground beacons are commonly co-located, and between them give range and bearing information simultaneously to the aircraft up to a range of 320 km at 12 000 m altitude.

A military system, Tactical Air Navigation (TACAN) is based on V.O.R. and D.M.E. but only uses a single ground beacon at each location. The distance is measured as in D.M.E. and the ground beacon reply pulses carry amplitude modulation in a manner analogous to V.O.R. which conveys bearing information. Using TACAN, an

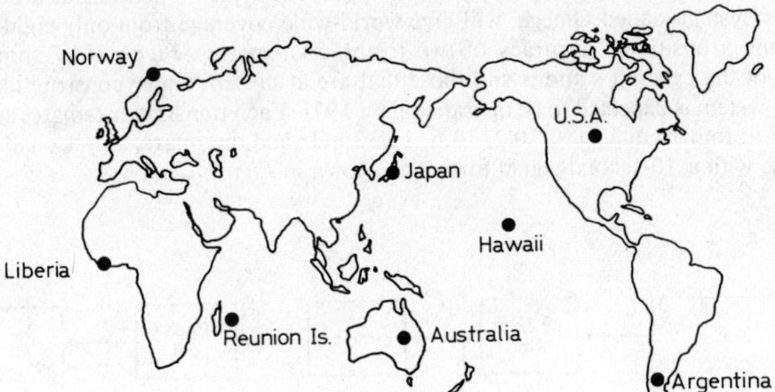

Figure 25.7 Omega ground stations—approximate location

aircraft can rapidly obtain range and bearing information from a ground beacon or from a second similarly equipped aircraft.

A further series of position fixing aids are the so-called hyperbolic ones. In these, a series of ground stations transmit coded signals and a receiver in the aircraft measures either the phase difference or difference in time of arrival of pulses from these stations. A given phase or time difference of signals from one pair of transmitters locates the receiver along a discrete hyperbola. Thus, use of 3 or 4 transmitters can give a positional fix. Two systems that are well established in this field are the Decca Navigator and LORAN (Long Range Navigation). The Decca system uses a chain of phase locked c.w. transmitters operating between 70 kHz and 130 kHz. Each chain of transmitters has a range of about 400 km, and these at present cover Western Europe, the Atlantic Seaboard of the U.S.A., Japan and a few other areas. The use of fully automatic receivers gives continuous positional information to an accuracy of better than 1.6 km within the area of coverage of a chain. LORAN on the other hand is a pulse system, operating with a carrier frequency of about 100 kHz. Chains of high power

Transmission interval	0·9 (Start)	1·0	1·1	1·2	1·1	0·9	1·2	1·0	0·9 (Start, Etc.)
Norway	10·2	13·6	11·33	f1					10·2
Liberia	f2	10·2	13·6	11·33	f2				
Hawaii		f3	10·2	13·6	11·33	f3			
U.S.A.			f4	10·2	13·6	11·33	f4		
Reunion				f5	10·2	13·6	11·33	f5	
Argentina					f6	10·2	13·6	11·33	f6
Australia	11·33					f7	10·2	13·6	11·33
Japan	13·6	11·33					f8	10·2	13·6

10 seconds; 0·2 seconds

Figure 25.8 *Omega ground stations—positions in signal format*

LORAN transmitters are sited to cover the North Atlantic area, and provide an accuracy around 40 km at a range of 1 920 km from the transmitters. This poorer accuracy is due to the interaction of ground and sky waves at long ranges, and thus is inherent in the LORAN system.

Recent work has been carried out using a hyperbolic system operating at around 10 kHz. This system, called Omega, will give world-wide coverage from only eight transmitters, and a positional accuracy of two to three kilometres. *Figure 25.7* shows the locations of the existing stations and those that are at present under construction. The complete system is expected to be operational by 1976. Each transmitter radiates a power of 10 kW at four frequencies, 10.2, 13.6, 11.33 kHz and one spare, as yet unallocated frequency, with a 10-second signal format as shown in *Figure 25.8*.

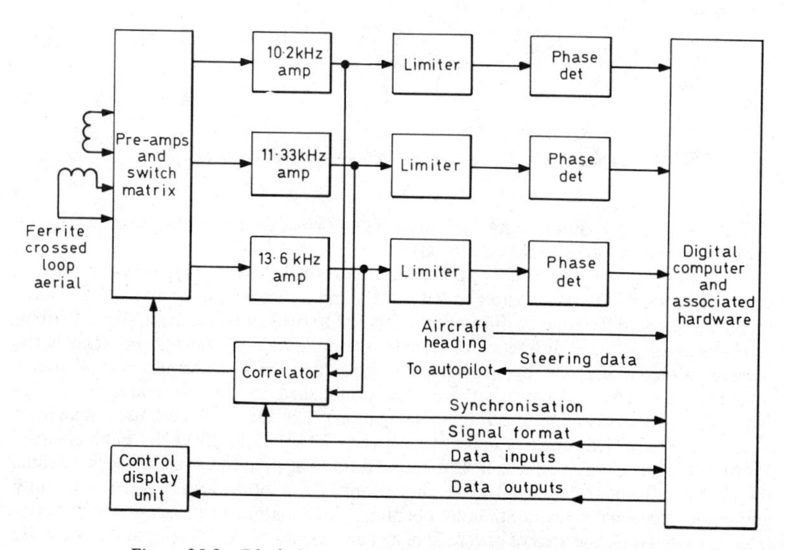

Figure 25.9 *Block diagram of airborne Omega navigation system*

These frequencies are derived at each transmitter from four caesium beam frequency standards whose phase is continuously monitored. Propagation of these signals around the globe is by a waveguide mode, the waveguide being formed by the earth's surface and the ionosphere. The velocity of propagation is therefore constant, although subject to a predictable diurnal variation as the height of the ionosphere varies throughout the 24-hour day. At the airborne receiver, phase comparison of the received signals enables a position fix to be made to an accuracy by day of 1.6 km and by night of 3.2 km. A block diagram of a typical receiver is shown in *Figure 25.9*. *Figure 25.10* shows a

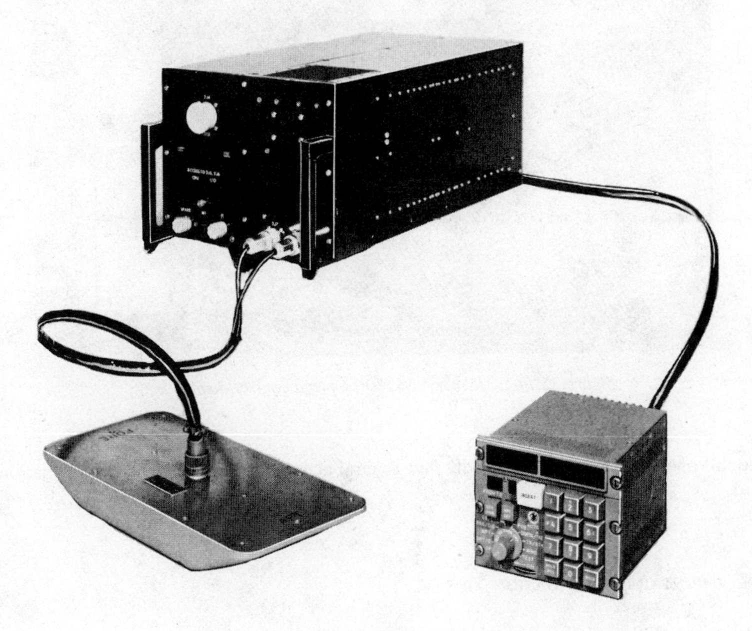

Figure 25.10 Marconi AD1800 fully automatic Omega system

typical modern airborne Omega receiver. In this receiver, there are three signal channels, one each tuned to 10.2, 11.33 and 13.6 kHz. Their outputs are first fed to a correlator circuit which operates using the signal format of *Figure 25.8* and enables each transmitter's time slot to be identified for each of the frequencies. A small special purpose computer accurately measures the phase difference between the three pairs of strongest signals. From this information a positional fix is calculated automatically. As there is a repetition of the phase pattern from any two transmitters every 115 km, the operator has to insert the aircraft's approximate position after the equipment has been first switched on. From then on, a highly accurate readout of aircraft position is continuously available.

In this particular equipment, advantage has been taken of the built-in computer to solve other navigational problems. *Figure 25.11* shows the Control/Display unit which gives position in latitude and longitude, range and bearing to any one of several preselected waypoints or destinations, speed and track angle, time to destination and distance travelled along and at right angles to a preselected track. This type of equipment is presently being fitted to some long range aircraft, and when the ground stations

Figure 25.11 Marconi AD1800 Omega control/display unit

are all operating it will undoubtedly find general acceptance as an invaluable navigation aid.

SELF CONTAINED NAVIGATION AIDS

Prior to 1950, if there was no access to ground beacons (as for example when flying over undeveloped countries or large oceans) recourse to dead reckoning navigation based on measured airspeed and compass heading had to be made. An unexpectedly strong

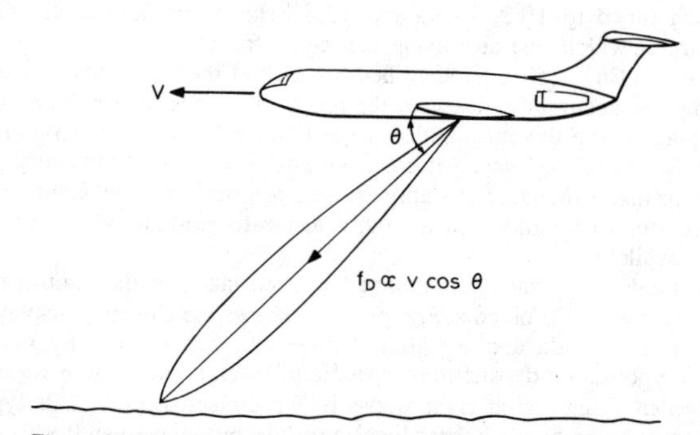

Figure 25.12 Single aerial beam for Doppler speed measurement

wind could cause an error of several hundred kilometres at the end of a long flight, and this could be disastrous. During the 1950s, a new aid was developed, operating on the Doppler principle, *Figure 25.12*. A narrow radio bean radiated from the aircraft is partly reflected back to the aircraft. Due to the relative motion of the aircraft and the

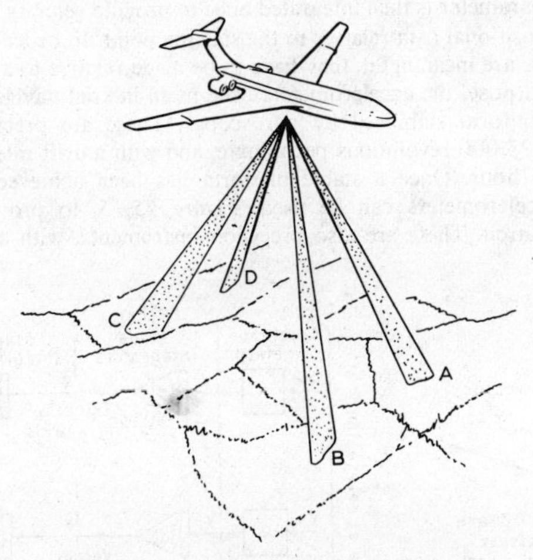

Figure 25.13 Typical doppler aerial beams

point of reflection the returned signal undergoes frequency shift proportional to the velocity vector of the aircraft resolved in the direction of the transmitted beam. For the example of *Figure 25.12*, the shift frequency f_D is given by $f_D = (2V \cos \theta)/\lambda$, where V is the aircraft velocity and λ is the wavelength of the transmitted beam. For a practical Doppler system operating at 13.3 GHz, this shift frequency is 1 000 Hz to 1 500 Hz

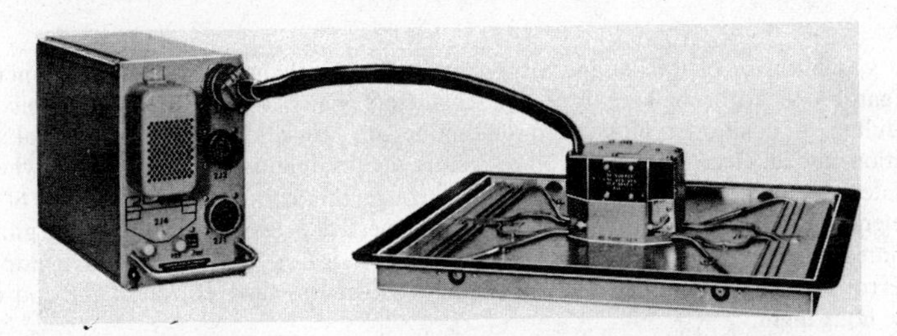

Figure 25.14 Marconi AD2510 Doppler velocity sensor

for each 100 knots of speed. By utilising a system which radiates four beams to the ground, *Figure 25.13*, the components of aircraft velocity along, across and perpendicular to heading, or total velocity vector magnitude and direction, can be resolved. The important factor is that this velocity readout is relative to the ground, and so can be used for accurate navigation where no other aids are available.

By integration of the velocity information, the present position of the aircraft relative to its starting point can be determined. *Figure 25.14* shows a typical modern Doppler navigation system, suitable for operation over the whole range of aircraft speeds, with an accuracy of positional readout of 0.15% of total distance travelled.

Another self-contained navigation aid developed in the 1950s is the Inertial Navigator. It is extremely simple in its basic concept, but the practical implementation has proved to be very complex. Basically, the system measures the acceleration of the aircraft and this parameter is then integrated once to provide velocity data and a second time to provide positional data relative to the starting point. In order that the acceleration measurements are meaningful, they have to be made relative to a fixed direction in space. For this purpose, the accelerometers used in an inertial navigator are normally mounted on a platform stabilised by gyroscopes. These are precision instruments rotating at about 25 000 revolutions per minute, and with a drift rate of no more than 0.01 degrees per hour. Once a stable platform has been achieved, three mutually perpendicular accelerometers can be used, *Figure 25.15*, to provide velocity and positional information. These are also precision instruments, with a typical range of

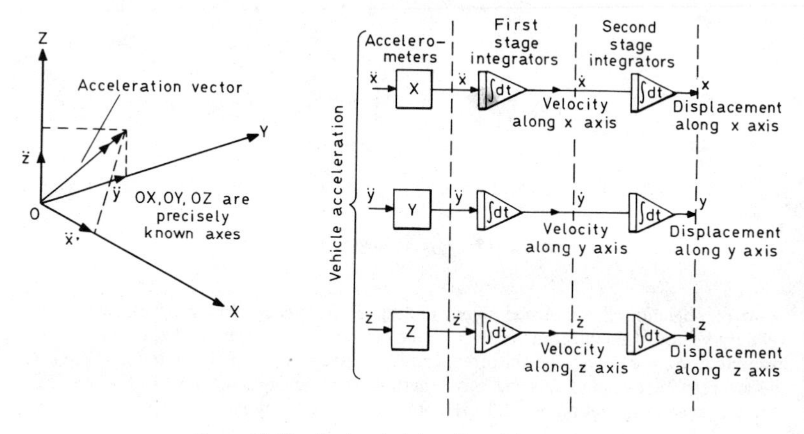

Figure 25.15 Basic principles of inertial navigation

± 10 g, a linearity of 0.01% and a threshold of better than 10^{-5} g. A typical inertial navigator uses force balance type accelerometers which consist of a freely pivoted pendulum, a positional pick-up to determine any deviation from the normal rest position, and an electromagnetic coil to restore the pendulum to its rest position when it deviates due to acceleration. The current through this coil is then a measure of the acceleration to which the instrument is being subjected. *Figure 25.16* shows the gimbal mounted gyroscopes of a typical modern inertial navigator. The Inertial Navigator has an error rate corresponding to a positional error of less than 16 km at the end of a three hour flight.

Two other items of equipment that are of use in navigation without the assistance of ground beacons are the Weather Radar and the Radar Altimeter. The Weather Radar is a pulse system operating at X-Band (around 9 500 MHz) using a dish type aerial mounted behind a radome in the nose of the aircraft. This rotates in the azimuth plane and is normally fixed in elevation to look along the line of flight. A small cathode ray tube indicator on the flight deck enables radar reflections from storm clouds to be observed at ranges up to 240 km long before the clouds are visible. The aerial dish can also be tilted down below the line of flight so that echoes from the terrain ahead appear on the c.r.t. This ground mapping mode is a useful navigational aid e.g., the identification of a coast line at the end of a long oceanic flight.

The normal aircraft altimeter operates by measuring the barometric pressure. While this is adequate at higher altitudes, it is often desirable at low altitudes to be able to measure actual ground clearance. For this purpose a radar altimeter is used which, in its simplest form, transmits a series of pulses to the ground at a frequency of about

Figure 25.16 Elliott E3 inertial platform

4.3 GHz from a suppressed horn aerial. These pulses are reflected from the ground to the receiver and the ground clearance of the aircraft is computed from the time between transmission and reception. Another type of altimeter, which does not require the high peak pulse powers of the above type uses a frequency modulated c.w. transmission.

In the receiver the phase of the modulation of the signal reflected from the ground is compared with the phase of the modulation on the transmitter. This phase difference is a direct measure of the aircraft's ground clearance. Current radar altimeters measure ground clearance up to 1 500 m to an accuracy of about 2%.

Flight control system

For a large modern aircraft, the effort required to operate the control surfaces (rudder, elevators and ailerons) is such that a pilot controlling these directly from the control column and rudder bar would quickly tire. These control surfaces are operated by servo-mechanisms which may be hydraulic, electrical or a combination of the two, which also form the basis for automatic control of the aircraft. The Flight Control System,

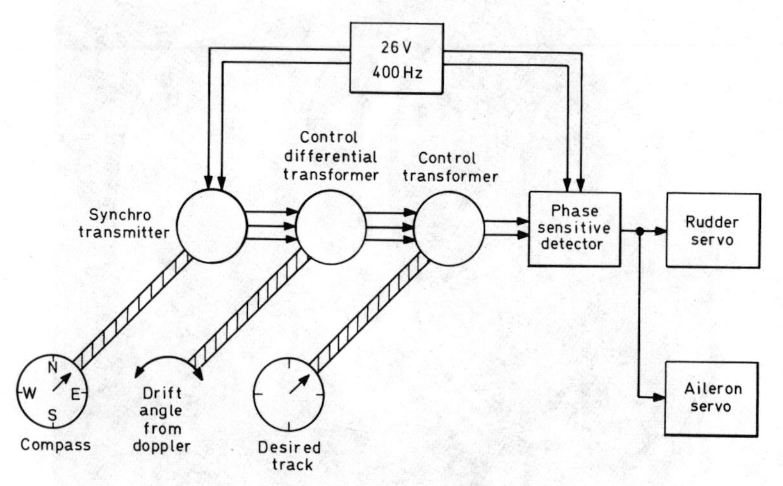

Figure 25.17 Automatic track angle flying

basically, is formed of a series of comparison circuits in which desired and actual flight conditions are compared and which produce outputs to feed the control servos. *Figure 25.17* shows, in a simplified form, one form of automatic control of the track angle being flown. The aircraft compass is provided with a 26 V transmitting synchro the 3 wire output of which is a function of the aircraft heading. This signal is fed to a control differential transformer mounted in the Doppler navigator, the shaft angle of which is an analogue of aircraft drift angle (due to wind). The resulting 3 wire output is a function of the aircraft's actual track angle (an algebraic summation of heading and drift angles), and this is fed to the stator winding of a control transformer. The shaft of this transformer is manually rotated so that its pointer indicates the desired flight track angle. The output from the control transformer rotor is a 400 Hz voltage with an amplitude which is a function of the error angle between the desired and actual tracks, and whose phase, 0° or 180° relative to the compass synchro excitation, is a measure of the sign of the error angle. Phase sensitive rectification of the signal can thus be used to provide an input to the servo mechanisms controlling the aircraft's rudder and ailerons. As a result, the aircraft will automatically fly along the track angle that has been set up. Similar channels with inputs from the airspeed indicator and altimeter can be used to fly at pre-set airspeeds and heights. Limits are placed on the rate and extent of movement of controls to ensure that the safety of the aircraft and the comfort of the passengers are not jeopardised. The whole system is designed so that any failures cause the affected channel to be disconnected from the system. The servo design is such that if there is any failure of the primary electrical or hydraulic supply system the aircraft can be flown manually.

The next stage is to control the aircraft to fly along pre-selected routes. A special purpose digital computer is normally used, forming the basis of an Area Navigation (RNAV) System which takes inputs from a variety of navigation sensors, such as V.O.R., D.M.E., Doppler, Inertial or Omega, as well as from airspeed and altitude. The Control/Display Unit of an RNAV system is similar to that of the Omega system shown in *Figure 25.11*. It has a keyboard which is used to enter the latitude and longitude of a number of waypoints in the sequence required for the flight. The computer accepts the inputs from the various sensors, and uses these continuously to calculate the position of the aircraft. The readout at the top of the unit can display present position, position of any selected waypoint and several other useful items, such as range and bearing to a waypoint, groundspeed and estimated time of arrival, local wind, or distance along and across the track between two waypoints. Signals are fed from the RNAV system to the autopilot giving deviation between the aircraft's present position and the track it should be flying between the two selected waypoints. The autopilot then corrects as necessary.

By using such a system the en-route period of a flight can be carried out completely automatically. As each waypoint is reached, the computer switches to the next leg and the aircraft turns towards the next waypoint, and so on until the destination terminal area is reached, when manual control is resumed.

Automatic landing

Automatic landing systems rely upon electronic control effected by means of signals from ground beacons. These beacons have been used for some years to give guidance to the aircrew during the landing approach, but accuracies have been improved to enable coupling to the aircraft's flight control systems. The standard Instrument Landing System (I.L.S.), shown diagrammatically in *Figure 25.18*, involves five ground transmitters. The Runway Localiser transmits two beams in the 108 MHz to 112 MHz band, one with 90 Hz modulation and one with 150 Hz modulation. These beams overlap to give an equi-signal zone which is on the extended runway centre line. On the aircraft, the two signals are received and compared. If the aircraft is correctly lined up for a landing, the signals are equal. If to the left of the correct line, the 90 Hz signal is larger, and if to the right, the 150 Hz signal predominates. These relative amplitudes are detected and used either to operate a steering indicator, or in the case of automatic landing to feed the autopilot which corrects any error in the aircraft's position. The Glide Slope transmitter at about 330 MHz, is used to provide guidance in the vertical plane to maintain an approach angle of 2.5°. Three fan marker transmitters, operating at 75 MHz, enable the distance from touchdown to be monitored. As the aircraft crosses the outer marker, about 6 km from touchdown, a 400 Hz modulated signal is received, which is fed to headsets and also used briefly to light a blue warning lamp. At one kilometre or so from touchdown, a 1 300 Hz signal is received and lights an amber light, and finally, about 75 m before touchdown a 3 kHz signal is received and lights a white light.

For fully automatic landings, inputs are also required from the airspeed indicator, aircraft attitude references and the radar altimeters. Using this system, the aircraft automatically flies down the path determined by the Glide Slope and Localiser transmitters with its altitude and speed under autopilot control. As the runway threshhold is reached, the rate of descent is checked and the aircraft touches down smoothly, whereon the throttles are closed. To ensure a high standard of reliability (no more than 1 failure in 10^8 landings) all control channels are triplicated and continuously compared. If any discrepancy appears, there is immediate warning, so that manual control can be resumed. Similarly, if at any time for any reason the landing is to be aborted, the manual opening of the throttles initiates an automatic overshoot and climb away.

While the standard I.L.S. can provide adequate guidance for automatic landing, it can suffer from errors of distortions of the radio beams due to multiple reflections from nearby objects. Calibration of a site can enable corrections to be made for reflections from fixed objects, but variable errors can still occur due to reflections from other aircraft, either in flight or on the airport. In addition, as airports and their surrounding skies become more crowded, there is becoming a need for even greater accuracy of

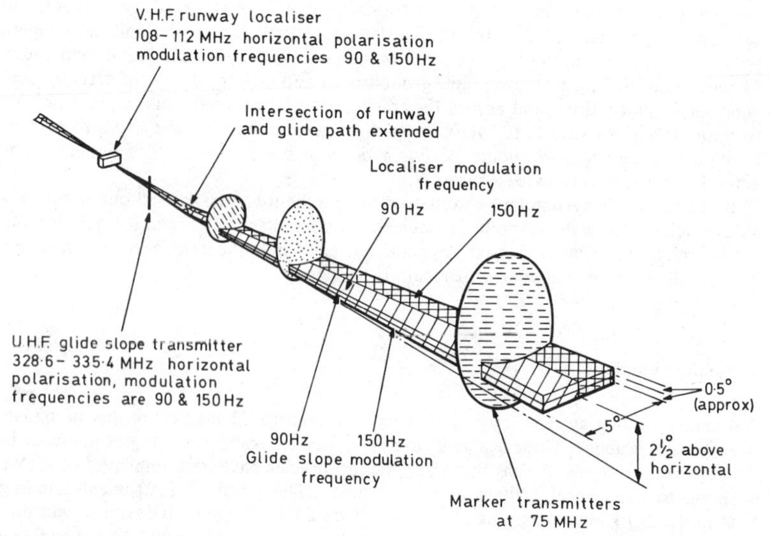

Figure 25.18 Instrument landing system (I.L.S.)

landing aids to make fuller use of the terminal area airspace. As a result, much work has been carried out recently on Microwave Landing Systems, as the use of these higher frequencies enables clearly defined aerial beams to be obtained, so obviating most of the errors due to reflections from objects outside the main beam. One type of system presently being investigated is the Commutated Doppler System. In this, a linear array of radiators are fed sequentially from a microwave power source, which is equivalent to the source moving along a linear track. An aircraft will receive the signal with a doppler shift which depends on the component of the apparent velocity of the source towards the aircraft, and is proportional to the cosine of the angle between the aircraft and the line of the transmitting array. To eliminate effects of frequency drift and doppler shift due to the movement of the aircraft, the same r.f. signal is simultaneously radiated from a fixed aerial. The measurement of doppler shift is thus merely a matter of comparing the frequencies of the signals received from the two transmitting aerials, and defines an angle. Azimuth guidance is obtained from a horizontal transmitting array, while a vertical array gives guidance in the elevation plane. A measure of distance from touchdown is obtained from a system developed from the standard D.M.E. By the use of shorter, closer spaced pulses and improved measuring techniques a distance measurement to within 6 m may be achieved. This total system, of course, requires much refinement but indications are that it will eventually be able to provide the accuracy required for fully automated blind landings, followed by guidance for taxiing in the densest fog.

FURTHER READING

BECK, G. E., *Navigation Systems*, Van Nostrand Reinhold, London (1971)

BERGER, F. B., 'The Design of Airborne Doppler Velocity Measuring Systems', *Trans. I.R.E.*, 157–75, December (1957)

HANSFORD, R. F., *Radio Aids to Civil Aviation*, Heywood & Co., London (1960)

HOWARD, R. W., 'Automatic Flight Controls in Fixed Wing Aircraft—the First 100 years', *Aeronautical Journal*, November (1973)

KAYTON, M. and FREID, W., *Avionic Navigation Systems*, John Wiley (1969)

SWANSON, E. R. and TIBBALS, L., 'The Omega Navigation System', *Navigation*, **12**, 1, 24–35 (1965)

Articles relating to aircraft navigation using electronics appear in most issues of the following two journals:

The Journal of Navigation, published quarterly by The Royal Institute of Navigation, London.
Navigation, published quarterly by The Institute of Navigation, Washington D.C., U.S.A.

Specifications relating to electronic equipment for fitting to aircraft are published by Aeronautical Radio Inc (ARINC) Annapolis, Maryland, U.S.A.

MARINE AND OCEANOGRAPHY

This sub-section describes electronic equipment and applications which are specifically marine; i.e., that which does not have an exact counterpart elsewhere. Marine radio communication equipment, for example, is not essentially different from that used on land and is not included although it is undoubtedly the principal application of electronics at sea. However, radio direction finding is covered because the problems it presents at sea are different from those encountered elsewhere. The same applies to marine radar. The engineering techniques and design principles of radar are covered in Section 18 but some notes on the design parameters and the utilisation of a typical marine radar are presented here which may be useful to radar engineers who wish to gain some idea of marine radar requirements.

The characteristic feature of the rest of the equipment described is its involvement with water—used either as a transmission medium for signals or as a source of data.

Marine direction finding

The object of direction finding is to measure the bearing of a transmitting station from the ship. The direction finding receiver is fitted with an aerial system having well-defined nulls in its polar diagram. The simplest system uses a rotation loop aerial, which ideally has the polar diagram shown in *Figure 25.19*. Normally the axis of rotation of the antenna is vertical as it is assumed that the signal is arriving in a horizontal direction.

For accurate direction finding it is essential that the signal is vertically polarised (as is the normal case with ground wave signals in the medium frequency range). Any horizontally polarised component, due, for example, to sky wave propagation, will affect the sharpness of the nulls.

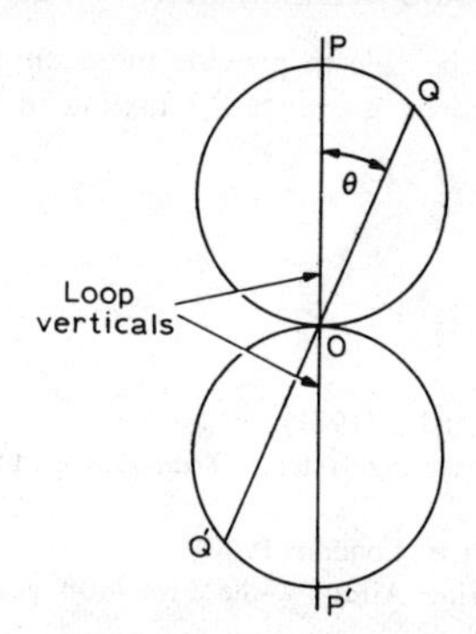

Figure 25.19 Horizontal polar diagram of loop antenna OQ = OP cos θ OQ and OP are in opposite phase

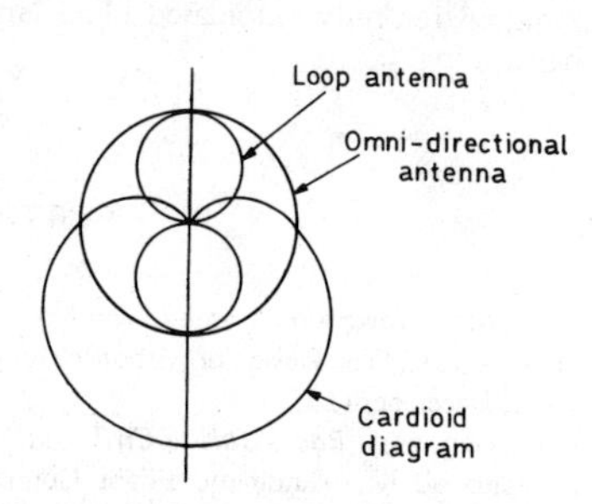

Figure 25.20 Cardioid polar diagram of combined vertical antenna and loop

This arrangement, having two nulls, gives an ambiguity in the bearing of the transmitter. In order to resolve this, an input from an ordinary vertical aerial is combined with that of the loop, the resultant polar diagram being cardioid in form (*Figure 25.20*). This is achieved by phasing the two inputs relative to each other so that they add for signals arriving in one direction along the loop; they will then be out of phase for signals arriving in the opposite direction. The inputs have to be balanced in order to achieve this. *Figure 25.21* shows a method of coupling the sense antenna.

The cardioid polar diagram has one minimum signal direction which is at right angles to the loop minimum; generally a separate scale or pointer is provided for use with the cardioid diagram, this being marked 'sense'.

Because of the necessity of balancing the signals to obtain a good null on the cardioid and because in any case the cardioid does not change so rapidly near the null position as does the figure of eight diagram for the loop, it is the invariable practice to use the loop alone for accurate direction finding and the combined aerials simply to determine which of the two possible bearings is the correct one.

If for any reason the loop or its associated circuits tend to act as an omni-directional antenna in addition to their correct mode of operation, the polar diagram becomes distorted, the zeroes moving slightly or becoming blurred. This is usually known as *antenna effect* or *vertical effect*. A common cause of this trouble is impedance unbalance, often due to stray capacity to earth. To overcome stray pick-up, the loop may be electrostatically screened, *Figure 25.22*, as it operates by magnetic pick-up. Or adjustable capacitors

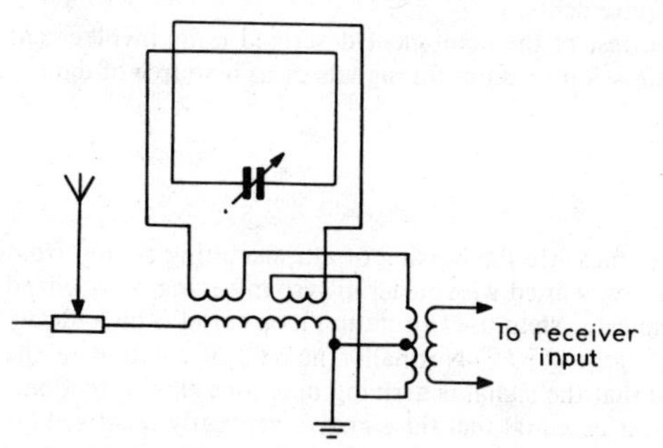

Figure 25.21 Method of coupling vertical antenna and loop

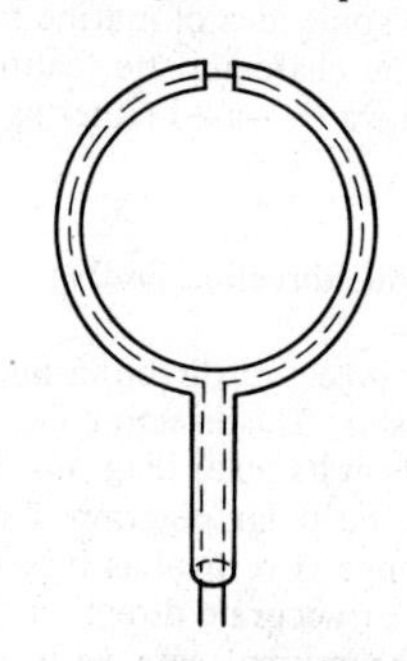

Figure 25.22 Screened direction finder loop antenna

may be provided to balance the impedances. Another method of compensation is to inject a small signal permanently from the sense antenna, this being set to give sharp nulls.

The figure of eight polar diagram is only correct for aerials whose dimensions are small compared with a quarter of the wavelength of the transmitted signal. Even so, where air cored loops are used, they tend to be somewhat bulky if they are to pick up adequate signals, since most direction finding is carried out on the medium- and long-wave bands. Instead of rotating a large loop, it is more common to employ two fixed loops at right angles to each other. These are connected through a device known as a goniometer which consists of two identical coils of wire fixed at right angles to one another. Within these two coils is a third coil which can be rotated. The signals from the two loops are combined in varying phase as the coil is rotated, the resultant output having exactly the same characteristics as that from a rotating loop. In other words, the goniometer is equivalent to a small loop aerial direction finder with the search coil acting as the loop. This system is known as the Bellini-Tosi system after its inventors. A sense input can be injected into the system from a vertical aerial in exactly the same way as for the rotating loop.

This system has the enormous advantage that the loops can be positioned where their performance is least affected by their surroundings. In fact the rotating loop system is generally only found in small vessels.

For pleasure boats and the like the rotating loop system has taken on a new form in which the loop aerial is wound on a ferrite rod of the type commonly used in small domestic receivers. Battery driven hand-held devices, sometimes incorporating a compass, are available, some of them cleverly engineered and designed, but hardly of an accuracy that would commend them to the professional navigator.

At the other end of the scale, a number of automatic systems have been produced. In the majority of these a servo-driven goniometer automatically comes to rest at the correct null point. In others, the goniometer spins continuously at around 1 000 r.p.m. and the resultant output is displayed on a cathode-ray tube or in one case by a flashing neon light spinning synchronously with the goniometer revolving behind a compass rose.

Sources of errors in direction finders

The errors present in any direction finder installation may be divided into two classes: constant and variable. It may not be practicable to remove all constant errors but where necessary they can be allowed for by producing a calibration chart for the equipment. This is the usual practice. Constant errors are often classified as: *Quadrantal*, so called because the error reaches a maximum four times in a complete revolution of the loop caused by electromagnetic linkage of the direction finder loop antenna with a loop formed by the ship's hull. *Field alignment*, caused by asymmetrical surroundings to the direction finder loop. *Semi-circular*, this error is so called because it reaches a maximum twice in a revolution of the loop or goniometer. This is usually caused by electromagnetic linkage of the loop with the re-radiated field from a nearby antenna or stay or other vertical structure.

The most serious causes of variable error are: *Night Effect*, the name usually given to the effect of sky waves on bearing accuracy which has already been described. *Interference*, which may produce undetected errors on automatic and manual equipment. The best guard against interference is to listen to the signal being used for direction finding. *Unearthed Screening*, a broken or poor earth connection on the loop or lead-in cables which produces large variable errors and very probably blurred zeros. *Variation in Impedance of Local Structures*, wood and rope have very different conductivity when wet than when dry. *Movable Parts on Superstructure*; accurate calibration and bearings can only be achieved when movable objects are swung into definite positions, preferably with their centre lines fore and aft.

It has already been mentioned that resonant structures may cause trouble. A structure 7.5 m high has a quarter wave resonance of 10 MHz and if it has a lot of top hamper the resonant frequency is reduced and it also radiates more readily—it forms a 'capacity top' antenna. Stays should be insulated at the top and bottom and also at intervals to ensure that their resonant frequency is outside the upper direction finding frequencies.

Marine radar

Situations in which marine radar proves its usefulness are:

(1) Navigation in confined spaces, for example in an estuary or approaching harbour, particularly if visibility is poor.
(2) Observance of other shipping and so ensuring safety from collision.
(3) Positioning the ship relative to landmarks and coastlines. This is particularly important for trawlers, many of whom identify their favourite fishing grounds by radar fixes.

By and large the ship's radar provides a combined positioning and early warning service, particularly the latter.

The transmitter consists of a small fixed frequency magnetron (see Section 7) operating in the 3-cm waveband. 10-cm installations are sometimes found but generally the extra bulk is too heavy a price to pay for some improvement in long-range performance. Transmitted power (peak pulse) may lie between 15 kW and 60 kW. Typical display ranges provided might be 1, 5, 10, 20 and 40 nautical miles.*

The aerial might be slotted wave guide, 'cheese' or 'orange peel' reflector fed by a horn. The horizontal beam width would typically be about 2° and the vertical beam width about 20°. The vertical beam width must be fairly broad to allow for the movement of the ship.

The local oscillator is a klystron (see Section 7) tuned mechanically and electrically to operate at a slightly higher frequency than the magnetron, the intermediate frequency being about 10 MHz. The mixer is usually of simple design, often a single crystal though balanced mixers are becoming more common.

The receiver and video amplifiers are standard. The video usually includes a differentiator which can be switched in at will to cope with 'sea clutter', i.e., reflection from the sea surface.

The display is normally a PPI (see Section 18) incorporating fixed range rings and a variable strobe marker which enables ranges to be read off on a calibrated scale. Often the display is stabilised in the North upwards position, the necessary alignment information being derived from the ship's gyrocompass. A further refinement is the 'true motion' display on which the ship's position, instead of being fixed in the middle of the display, moves across it in accordance with information derived from the ship's compass and the ship's log.

Echoes from coastlines or buoys remain stationary on such a display. This makes it very useful when bringing a ship into harbour because the landmarks remain clear instead of becoming blurred as the ship moves.

It is not generally realised that if another ship is on a collision course with the ship having a true motion display, the collision course will be revealed by plots made on the true motion display even if the 'own course' information fed to the display is not accurate, because that information affects all plots alike and their relationship to each other is unaltered. It is this useful fact which makes a true motion display such a valuable aid to navigation.

Many marine radar displays are fitted with a dichroic mirror system which forms images of plotting marks (actually made on a glass surface above the cathode-ray tube) in the plane of the phosphorescent screen of the tube so that these plots appear to be on the display itself and parallax is eliminated.

* The nautical mile is an internationally accepted unit despite metrication, being in fact the mean length of arc subtended by one minute of latitude at the earth's surface. This length decreases as the latitude increases, the mean value being subtended at about lat. 48°.

Echo sounding

INTRODUCTION AND DESIGN PARAMETERS

Excepting only the radio transmitter and receiver, the echo sounder is by far the most widely fitted marine electronic device. It is used not only for navigational and survey purposes but also for fish finding, and in this connection numerous specialised displays have been developed incorporating highly sophisticated mechanical and electronic technology.

Echo sounding depends on the transmission of short pulses of ultra-sound through the water. Sea water does not transmit electromagnetic waves to any appreciable degree, but ultrasonic waves are propagated fairly well. *Table 25.1* (taken from Beranek) gives typical attenuation figures.

Table 25.1

F (kHz)	dB/metre
10	0.000 7
20	0.002 7
30	0.005 5
40	0.009 1
60	0.016 5
80	0.022 9
100	0.320 0
200	0.073 2

These figures, it should be remembered, represent absorption of the sound power by sea water and take no account of spreading. It is clear that frequencies of around 200 kHz can only be used where the depth to be measured is not very great. In practice such frequencies are used only for small inshore sounders, while for great depths 10 kHz is often chosen. For the normal run of echo sounders, whether for fish finding or depth sounding, the most usual frequency is around 30 kHz.

Figure 25.23 based on the results of various experimenters, notably A. B. Wood, shows the effect of temperature, pressure and salinity on the velocity of sound in water.

The basic echo sounding set-up is simple. At the moment of transmission of an ultrasonic pulse from a transducer set in the bottom of the ship, a moving light or in the case of a recording instrument a moving stylus crosses the zero of a scale. After a time proportional to the depth of water below the ship, an echo of the pulse is picked up by a receiving transducer (often the same transducer that transmitted the pulse), detected and amplified in the echo-sounding receiver. The receiver output causes the light to flash or passes an electric current through the stylus which marks the recording paper. If the speed of sound in water is taken to be 1 460 m/s then for every fathom* of depth the sound pulse has to travel 3.66 m which takes it 2.5 ms. On the display, therefore, the stylus or moving light must traverse a scale length corresponding to 1 fathom in 2.5 ms also.

The ultrasonic transducer used in an echo sounder is invariably resonant to ensure the maximum transfer of power into the water. Formerly it was common practice to use a magnetostrictive transducer and discharge a capacitor across the winding, very often by means of a mechanical relay or alternatively through a thyratron or small ignitron (see Section 7). Though simple, this method was inefficient and often a source of interference with the ship's radio and it has now largely fallen into disuse. Most echo sounder transmitters now employ an oscillator which produces a pulse of controlled

* The fathom (1.83 m) will be with us for many years yet as a depth measurement, because replacement of charts marked in fathoms by metric units will take a long time.

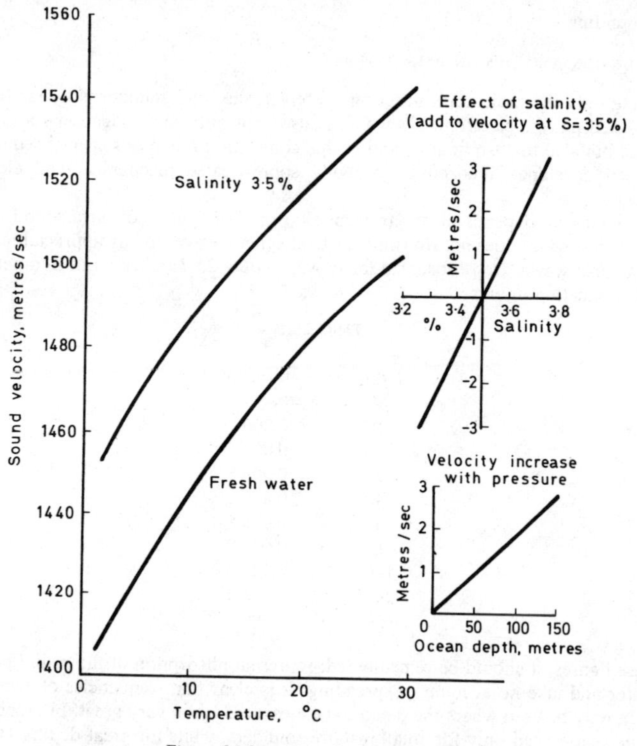

Figure 25.23 Sound velocity in water

length when triggered. The transmitter is triggered by a contact (or, very often nowadays, a magnetic pick-up which does not rely on physical contact) which is driven by or synchronised with the display. Although so far it has been assumed that triggering of the transmitter always takes place when the stylus or light is at zero on the display, this need not necessarily be the case. By triggering the transmitter at the correct time before the stylus reaches the scale, the zero can be off-set and by this means the scale can be used to indicate varying ranges; a typical choice of range might be 0–60 fm, 50–110 fm, 100–160 fm and 150–210 fm.

Where the echo sounder is required to give information about fish echoes, displays become more elaborate. There are two reasons for this. Firstly, the recorder gives very limited indication of echo strength, and secondly, many deep sea fishermen use trawls which trail along the bottom, and they are only interested in fish which are about 1.8 m or less from the bottom, this in a depth of perhaps 250 fm. One solution to these problems is to use a cathode-ray tube display, triggered, not by the transmitted pulse, but by the bottom echo itself. Before appearing on the display the received signals are passed through a delay line having a delay time of 7.5 ms so that when the time base is triggered off by the undelayed bottom echo a signal received from a fish 5.5 m above the bottom is just reaching the x plates of the tube (in this application the time–base is taken to the y plates so that the bottom echo appears at the bottom of the tube).

In one highly ingenious application of this principle the delay line is replaced by a drum carrying a magnetic tape on which the received signal is recorded. One pick-off head triggers the cathode-ray tube time–base while another transmits the signal to the x

plates. With this arrangement the recorded echo can be displayed again and again, thus overcoming one of the principal disadvantages of echo sounding—the slow sounding rate which is an inevitable consequence of the comparatively slow speed of sound in water.

It is of course not possible to run the recorder stylus fast enough to expand the recorder scale sufficiently to show two or three fathoms over the whole length of the scale, but the problem of displaying the bottom echo and signals from two fathoms or so above it on the recorder has been solved in two ways. In one equipment the incoming echoes are digitised and stored in a magnetic core memory which is addressed as the recorder stylus passes over the paper and feeds out the signals in synchronism with the movement of the stylus. Another equipment makes use of a recorder in which a 'comb' of separate styli is in contact with the paper. These styli do not themselves move, but each is taken to a separate terminal and the incoming signal is switched electrically from one stylus to the next which gives the effect of a rapidly travelling single stylus. The switching sequence is initiated by an undelayed bottom echo and the signal presented to the switching system is delayed so that the first stylus receives a signal corresponding to a point a fathom or so above the bottom, as in the case of the cathode-ray tube display employing a delay line described above.

The echo sounding receiver is a comparatively straight forward device. Its bandwidth should be matched to the length of the transmitted pulse, and the designer should take care to avoid overloading and paralysis due to breakthrough of the transmitted pulse. Ideally, of course, the receiver should be fully operative as soon as the transmitted pulse is finished, but reduced gain immediately after the end of transmission is no disadvantage and in fact means of reducing it automatically are often built in. The biggest problems arise in the circuits used to drive the stylus of a dry-paper recorder. It is very difficult to get a reasonable relationship between gradation on dry paper and signal strength. In this respect 'wet' paper has the advantage, though this is not generally felt to outweigh the inconvenience of using it. In one application its use is still essential—the multi-stylus display described above, where the high voltage and arcing associated with dry paper cannot be tolerated on the closely spaced styli.

The ultrasonic transducer of the echo sounder corresponds to the radar antenna, and by a curious coincidence the wave lengths involved are much the same. However, the movement of the ship and the infrequency of transmission combine to dictate that the beam width shall not be too narrow. A very common size and shape for an echo sounder transducer is a 3λ by 2λ rectangle, giving a beam width of about 17° fore and aft (the long axis being placed in this direction) and 27° transversely. The wavelength is, of course, the signal wavelength in water—not the wavelength in the transducer material. Note that a good approximation to the beam width to half power points of a linear transducer along an axis n wave lengths long is $51/n°$.

Typical pulse lengths encountered in high-power echo sounders (and for fish finding the highest possible power is essential) are 1 ms for short ranges and 5 ms for greater depths. It seems surprising that a pulse 1.8 m long gives information about fish shoals but in fact this is found to be the case. It is true that in most cases there is quite a distance between individual fish. The receiver bandwidth should be tailored to suit the pulse as in radar.

Another surprise is found when considering the maximum power an echo sounder is able to transmit. This is limited by cavitation, and theoretically cavitation should occur when the peak sound pressure in the water exceeds 1 atmosphere, since under this condition the water should vaporise on the minima. The sound power density corresponding to this condition is in theory 0.3 W/cm^2 or thereabouts, but in practice with the short pulses used in echo sounding it is possible to obtain power densities of almost 10 times this figure without any noticeable cavitation effect. For example, on a 3λ by 2λ transducer where λ was 6 cm the transmitted power into the transducer was about 1 200 W; the transducer was about 50% efficient giving 600 W over an area of 216 cm^2 without any noticeable cavitation effect. When the power was increased to 1 800 W the performance of the equipment fell off considerably.

Sonar

Similar equipment is used to search horizontally for mid-water fish. The transducers, instead of pointing downwards, are lowered beneath the bottom of the ship and can be tilted as well as rotated. The commonest displays are paper recorders, often identical to those used for normal sounding. The direction and tilt of the transducer must, of course, be indicated and this leads to some complexity of repeaters and servo-mechanisms. With the increased use of mid-water trawling the demand for equipments such as these is likely to increase.

Sonars perform the task universally associated with this equipment—the detection of 'targets', generally submarines, torpedoes or objects on the sea bed. Another sonar role is the monitoring of ships' speed, which may be measured by observing the doppler frequency shift produced by the relative velocity of the ship and the object causing the echo.

The doppler shift in the frequency received from a source of sound waves moving relative to the receiver in sea water is in round figures 0.3 Hz per kHz of sound frequency per knot* of approach speed. Where the transmitter is approaching a stationary echo source and carries the receiver with it, this figure is doubled. Clearly, if frequencies of the order of 100 kHz are used, very small approach speeds produce easily detectable doppler shifts, and such a system shows to best advantage when used in checking the approach of large ships to the dock-side; it can of course measure the sideways speed as well as the fore and aft speed of a ship, depending on the direction in which the transmitting transducers (on a large ship there are often four or more) point. The normal echo source is scattered reflection from the sea bed, though in deep water for navigational purposes it is possible to use sound reflected by the ocean itself (known as 'reverberation' and caused by the presence of plankton or other small organisms in the water). The difficulty with reverberation as a reference is that very often the source of reverberation may itself be moving, thus introducing inaccuracy. In processing the received signal to determine the doppler shift, allowance must be made for the fact that different parts of the sound beam from the transducer hit the bottom at different angles and so their velocity relative to the bottom is different. The received echo signal therefore contains a spectrum of frequencies and the determination of relative speed from this calls for a fair amount of sophistication in the electronics.

Ship's logs

The electromagnetic log has now virtually superseded the traditional Pitot tube as a means of measuring the speed of a ship relative to the water. The measuring device comprises a flat coil enclosed in a streamlined casing through which an electric current is passed to produce a magnetic field which extends in to the surrounding water. Water flowing past the casing acts as a conductor cutting the magnetic lines of force, and this gives rise to an e.m.f. across electrodes in contact with the water. In a typical arrangement the coil is wound to produce vertical magnetic lines, i.e., it is wound round a vertical axis, the water streams past the casing from front to back, and the electrodes are placed on an axis at right angles to both these directions, causing an e.m.f. to appear across them.

In order to avoid distortion, no iron is used in the magnetic circuit. To get a useful magnetic field, a fairly high voltage is applied across a large number of turns of fine wire. It is quite possible to get an e.m.f. of 10 μV per 0.1 knot of ship's speed. Experimental apparatus producing this figure utilised a 50 gauss magnetic field and the conduction path between the electrodes was 4 cm.

In order to avoid polarisation effects at the electrodes, the magnetising coil is energised with alternating current. This has the added advantage that the output from the electrodes is alternating and thus easier to handle.

* The knot is a measure of speed, not distance. 1 knot = 1 nautical mile (1.582 km) per hour.

A possible source of trouble with electromagnetic logs is cathodic protection when this is fitted to a ship. This produces stray electric fields in the surrounding water which may affect the readings obtained from the electrodes of the log. If the cathodic protection system is energised by current rectified from the ship's mains, the fields will carry an appreciable alternating component, and if the ship's mains are also used to energise the coil of the log there may be trouble. For this reason there is advantage in energising the coil from a separate source of known frequency and fitting suitable filters in the amplifier.

FURTHER READING

'Advances in Marine Navigational Aids', *I.E.E. Conference Publication* No. 87 (Conference held 25–27 July 1972)

BERANEK, L. L., *Acoustics*, McGraw-Hill (1954)

CAMP, L. W., *Underwater Acoustics*, Wiley Interscience (1970)

HAIGH, Brig. J. H., *Radiolocation Techniques* (*The Services Textbook of Radio, Vol. 7*), H.M. Stationery Office

HORTON, J. W., *Fundamentals of Sonar*, United States Naval Institute

Proceedings of British National Conference on the Technology of the Sea and the Sea-bed, 5–7 April (1967)

Proceedings of I.E.E. Conference on Electronic Engineering in Oceanography, September (1966) (2 Vols)

URICK, R. J., *Principles of Underwater Sound for Engineers*, McGraw-Hill (1967)

WATSON, D. W. and WRIGHT, H. E., *Radio Direction Finding*, Van Nostrand Reinhold (1971)

MOTOR VEHICLES

Until the early sixties, only mechanical, electrical and electro-mechanical techniques were employed in the various motor vehicle systems to execute monitoring, control and command functions. Improvements in engine design and performance, trends in vehicle styling providing less room in the engine compartment and changing vehicle usage patterns have meant that alternative means of performing these functions, particularly in certain vital systems, has had to be sought. The electronic approach offers potential in terms of equipment size and weight reductions, improved performance and increased reliability.

Electronic equipment intended for motor vehicle applications has to be designed for operation in a severe climatic and electrical environment—a factor which has inhibited its employment. Depending upon the site of the equipment, some or all of the following conditions will have to be withstood without deterioration in performance.

(1) Engine compartment temperatures of −30 °C to +93 °C.
(2) Thermal shock.
(3) Thermal cycling.
(4) Humid and water splash conditions.
(5) Contamination from petrol, diesel fuel, detergents, anti-freeze, mud and salt.
(6) Sand abrasion.
(7) Mycological growth.
(8) Vibration.
(9) Supply voltage 12–15 V.
(10) Possible ripple voltage of 2 V peak-to-peak on the supply.
(11) Supply voltage excursions down to 5 V during engine cranking and up to 20 V when the battery is on boost charge.

(12) Load dump surges.
(13) Voltage transients, the major ones emanating from the ignition system.

The equipment has also to survive possible maltreatment (for example reverse polarity).

Alternator output rectification and control

Alternators have now extensively replaced dynamos as the source of generated electricity on motor vehicles.[1] The most common type of alternator employed has a rotating wound field and a laminated stator with a three-phase output winding. Rectification of the stator output for battery charging is achieved by a three-phase full-wave bridge forming a rectifier pack built into the machine. Rectifier packs of many current alternator designs also contain three 'field' diodes which together with three of the output diodes, form a second bridge through which rectified current from the stator windings passes to give self-excitation of the field at charging speeds. The three additional diodes eliminate the need for relays, necessary with 6-diode separately-excited alternators, to break the field circuit when the engine is switched off and to operate a 'charge failure' warning lamp.[1]

Alternators, with their self-regulating current characteristic and integral rectifier pack, present fewer problems of control than do dynamos. No cut-out relay or current regulator is required, only a voltage regulator to control d.c. output. The most common way of achieving this has been by switching the field current.

Regulators used with early 6-diode separately-excited alternators were remotely mounted units and employed the principle of the d.c. electro-mechanical types, but differed in that they had transistor-assisted contacts to switch the field current. These were superseded in the early 1960s when transistor economics enabled an all-electronic discrete component regulator of the switching type to be a viable proposition. However, costs limited the regulator to a two transistor circuit.[2]

In 1967 a thick-film microcircuit regulator was developed which could be incorporated into the alternator.[3,4] The cost saving of having a single encapsulation for the complete circuit enabled a third transistor to be incorporated to obtain a higher gain, thus permitting the system voltage to be sensed through a permanent connection instead of as previously through the ignition switch.

More recently, the advent of integrated planar semiconductor devices (see Section 8) has permitted further development of a discrete component regulator with a reduced number of circuit elements, resulting in increased reliability.[5] In standard form the regulator is suited to use with alternators in which the system voltage is sensed within the machine, *Figure 25.24*. Operation of the regulator is as follows.

When the ignition switch is moved to 'on', transistor TR_2 receives base current through the warning light and R_4, and is switched on; this in turn switches on TR_3. With TR_3 switched on field current can flow and the alternator output builds up until full excitation is obtained. According to the state of charge of the battery and other electrical loads, the system voltage rises until the regulating voltage is reached. At this point the voltage regulator diode D_1 conducts turning on TR_1, which causes TR_2 and thereby TR_3 to turn off. As TR_3 switches off, the instantaneous field current is diverted via the field recirculating diode D_2, and commences to fall. Consequently system voltage commences to fall until TR_1 is switched off and the cycle of events is repeated.

Over the regulating band the regulator will oscillate between TR_3 on and TR_1 off and TR_3 off and TR_1 on at a frequency dependent upon the internal time constant of the circuit, and the alternator field current (and therefore the alternator voltage) is controlled by modulation of the mark/space ratio of the oscillation brought about by the variations in the base current of TR_1.

The positive feedback circuit consisting of R_5 and C_1 ensures that TR_3 is never held

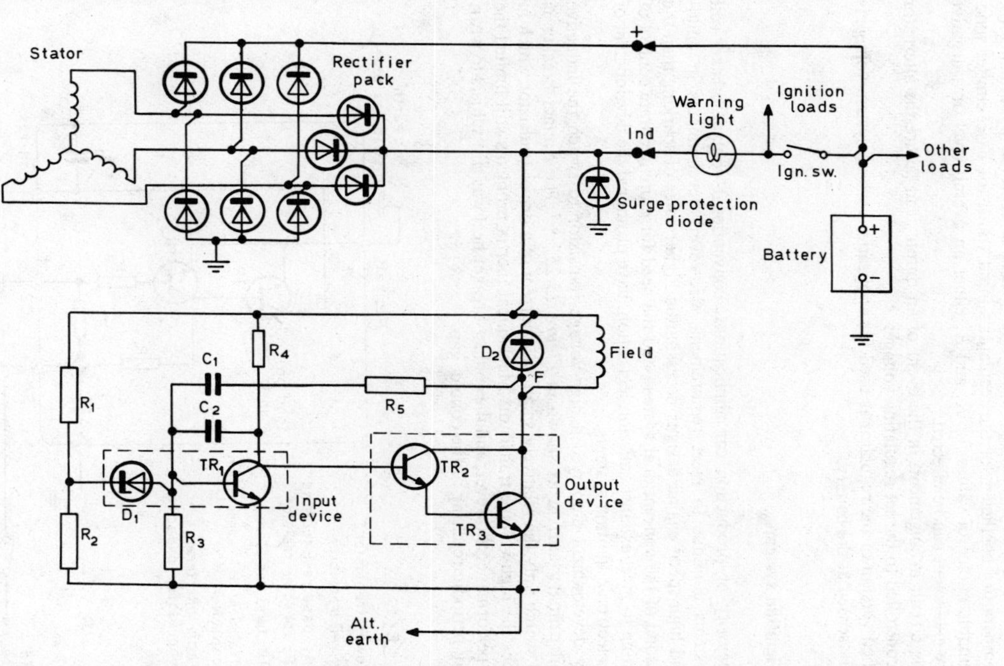

Figure 25.24 A vehicle generating system employing an alternator which incorporates a regulator that senses system voltage within the machine

in a high power dissipation state, and is switched quickly to either the fully on or the fully off condition. D_2 prevents high inductive voltages from being applied to the collector of TR_3 as this transistor switches off.

Temperature compensation of the regulating voltage is achieved by utilising the negative temperature coefficient of the regulator's input transistor base–emitter junction. By suitable choice of values for R_1 and R_3 either an overall level or a negative temperature coefficient can be effected.

With this circuit arrangement, in the event of an output lead failure the alternator voltage is controlled to the set regulating voltage.

A modified version of the regulator is used for applications where system voltage is permanently sensed at the battery.[5]

Electronic ignition systems

For the majority of production car applications, conventional contact breaker-coil ignition systems continue to meet requirements. However, developments in multi-cylinder and high-speed engine designs necessitating higher rates of sparking than the 400 per second of the conventional system, and the need for more accurate control of ignition timing to meet exhaust emission legislation, have prompted the development of alternative electronic ignition systems.

An early development employed a high-voltage transistor to switch the inductive ignition coil primary current as shown in *Figure 25.25*.[5,6,7] The contact breaker is required to switch only the relatively small base current of a driver transistor which in turn switches the high-voltage transistor. Mechanical inertia problems still limited the high-speed performance, however, and the contact-breaker heel was still subject to wear, necessitating periodic resetting of the contact gap.

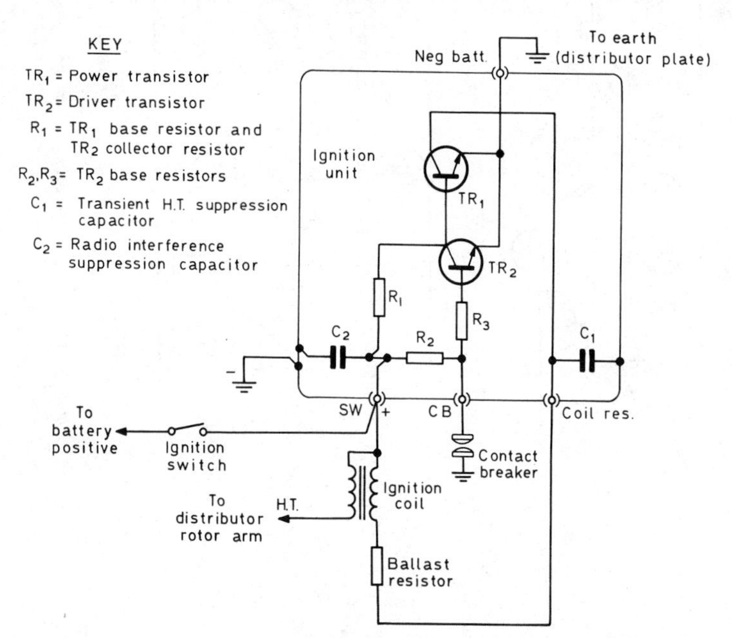

Figure 25.25 A transistor-assisted contact breaker system for negative earth vehicles

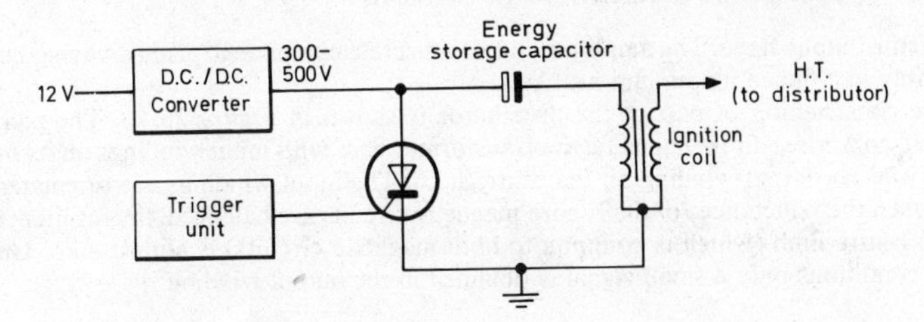

Figure 25.26 Schematic of a capacitive discharge system

In capacitive discharge systems, the energy to fire a cylinder's charge is stored on a capacitor and discharged into the primary of an ignition coil through a thyristor, *Figure 25.26*. Systems have been developed in which the timing pulse to trigger the thyristor is either obtained from a mechanical contact-breaker, or generated in a magnetic pick-up by a rotating engine member. Capacitor discharge systems enable high secondary voltages to be achieved at high spark rates, and fast times to be obtained to fire fouled sparking plugs. On the other hand, they are more costly than inductive discharge systems and have certain inherent problems concerning short spark duration and the ignition of lean mixtures.

In contact-breakerless systems, methods that have been investigated by which a suitable timing pulse could be obtained, have involved various pick-up/rotating engine member arrangements utilising electromagnetic, electrostatic or photo-electric principles operating in conjunction with amplifying/switching circuitry.[8] The circuit of one system is shown in *Figure 25.27*, and comprises an amplifier unit, a distributor, a ballast resistor unit and an ignition coil. The amplifier unit consists of a continuously operating fixed-frequency (600 kHz approx.) oscillator and trigger circuit which are transformer coupled by way of a pick-up housed in the distributor, and a high-voltage

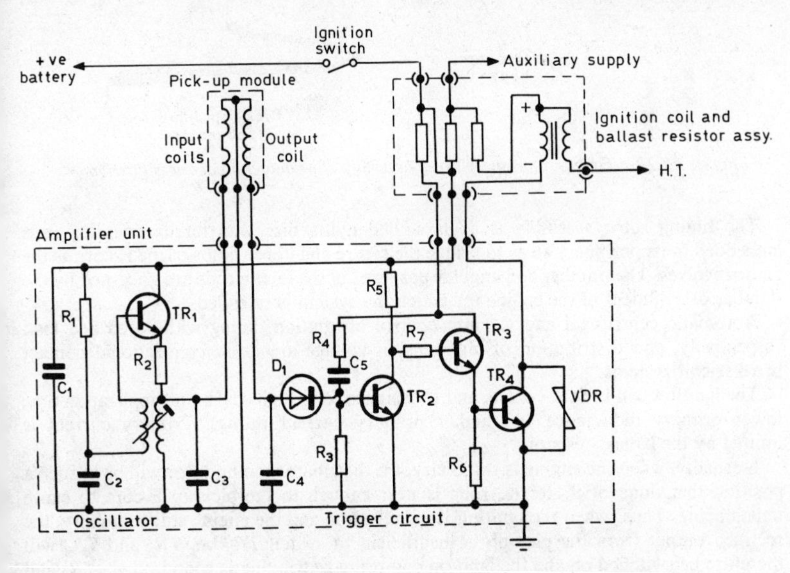

Figure 25.27 Circuit of a contact-breakerless ignition system

transistor output stage. The amplifier unit is a discrete component/printed wiring board assembly in a finned aluminium heat sink.

The construction of part of the distributor is shown in *Figure 25.28*. The pick-up module comprises an E-shaped ferrite transformer core with input windings on its outer limbs and an output winding on the centre limb. The input windings are orientated so that when the reluctances of the E-core magnetic circuits are balanced, the resultant flux in the centre limb (which is common to both magnetic circuits) is almost zero. Under these conditions only a small signal is obtained in the output winding.

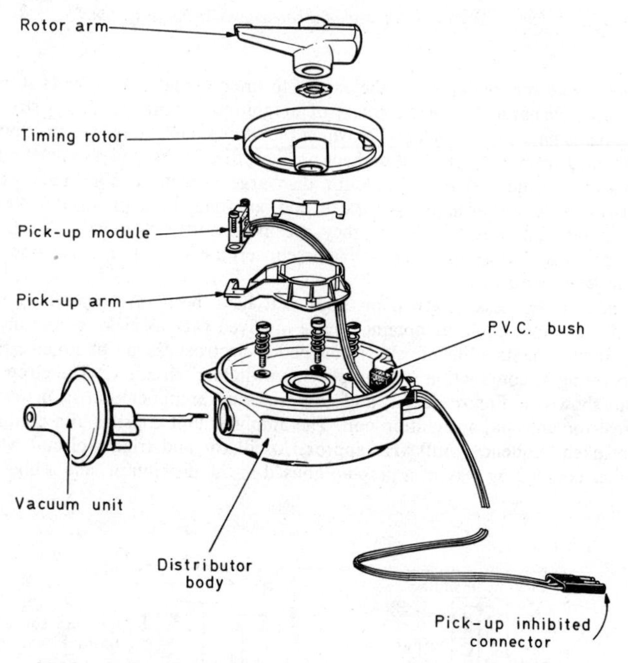

Figure 25.28 Timing-rotor and pick-up assembly of a contact-breakerless distributor

The timing rotor assembly is a glass-filled nylon disc with ferrite coupling rods embedded in its periphery so as to bridge the centre and upper limbs of the E-core as the rotor revolves. The number and angular positions of the ferrite rods are governed by the number of cylinders of the engine for which the system is intended.

Automatic centrifugal and vacuum control of ignition timing with speed and load respectively, and distribution of high-tension voltage are as in conventional contact breaker-coil systems.

The ignition coil is fluid-cooled, and is designed for a high spark rate application (i.e., lower primary inductance and higher primary current rating). Primary current is limited by the ballast resistor.

Normally, when the engine is stationary, the distributor timing rotor will be in such a position that none of its ferrite rods is near enough to the pick-up E-core to cause unbalancing. Thus, when the ignition is switched on and the engine still stationary, the rectified output from the pick-up is insufficient to switch TR_2 on. TR_3 and TR_4 will therefore be switched on and the ignition coil primary winding is energised via TR_4 and the ballast resistor.

When the engine is cranked, and a ferrite coupling rod in the timing rotor passes across the face of the E-core, the magnetic circuits are unbalanced; an increased signal voltage thus appears in the output winding of the pick-up module. This rectified signal causes TR_2 to be switched on and thereby TR_3 and TR_4 to be switched off. The ignition coil primary is thus broken, inducing a high voltage in the coil secondary and a spark at a plug in the usual manner. The positive feedback circuit comprising R_4 and C_5 ensures that TR_2 and TR_3 are rapidly switched between their 'on' and 'off' states. Although the output signal from the pick-up is an oscillating one, TR_4 remains switched off for as long as the increased signal voltage is present at the trigger circuit input. Thus, one spark is produced each time a ferrite rod passes across the face of the pick-up.

Maximum spark rate of the system is 800 sparks/s, and as there are no parts to wear routine maintenance is eliminated and spark timing maintained at its original setting.

Further development of this system has resulted in the electronic circuit of the previously separate amplifier unit being included with the timing rotor and stationary pick-up in an otherwise standard distributor body.

Engine speed limiters

Advances in multi-cylinder and high-speed engine design have resulted in greater possible acceleration rates, with the increased likelihood of damage to engines from excessive speeds. Engine speed limiters have been developed to prevent this eventuality.[9] The circuit of one such speed limiter is shown in *Figure 25.29*.

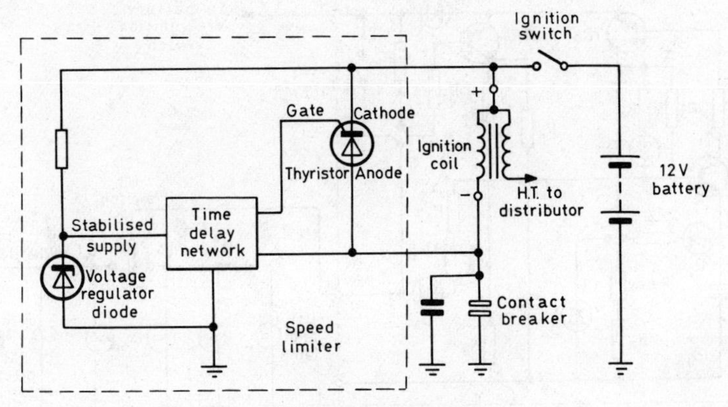

Figure 25.29 *An engine speed limiter*

At engine speeds below the set limit, and with the contact-breaker closed, the anode and gate terminals of the thyristor are both negative with respect to the cathode, and the thyristor is therefore switched off. Thus current flows through the ignition coil primary winding.

As the contact-breaker opens it induces an e.m.f. in the ignition coil and propagates in the time delay network a brief pulse which is subsequently used to unclamp the thyristor gate. The polarity of the e.m.f. induced in the ignition coil primary makes the thyristor anode positive with respect to its cathode; but because the delayed gate unclamping pulse propagated by the previous opening of the contact-breaker has passed through the time delay network and is already over, the gate is at a negative potential and the thyristor does not switch on. A h.t. voltage is thus induced in the secondary of the coil, and a spark produced at the appropriate plug. The cycle of events recommences with each opening of the contact-breaker. To prevent system voltage variations from affecting the delay of the pulse through the delay circuit, the supply voltage of the delay circuit is stabilised by a voltage regulator diode.

When engine speed rises to the set limit, the limiter operates in a similar manner—except that the opening of the contact-breaker occurs at the same time as the delayed pulse produced by the previous opening of the contact-breaker unclamps the thyristor gate. The thyristor therefore switches on, limiting the h.t. voltage induced in the ignition coil secondary winding to a value insufficient to ignite the cylinder's charge.

The speed limiter is a single static unit of discrete component/printed wiring board pattern. The desired engine speed limit is set during manufacture by selecting a resistor of appropriate value. Compared with previous mechanical and electronic speed limiters it has a fast response time, and therefore virtually no speed overshoot, and no speed hysteresis or vibration problems. It is suitable for use with either conventional ignition systems or contact-breakerless types such as the one previously described.

Electronic flasher units

In performing the dual function of direction indicator and hazard warning units, an electronically-controlled relay flasher unit has advantages over a thermal type in that it provides greater control of flashing frequency and mark/space ratio, it is easier to produce, and the relay contacts are less subject to arcing.

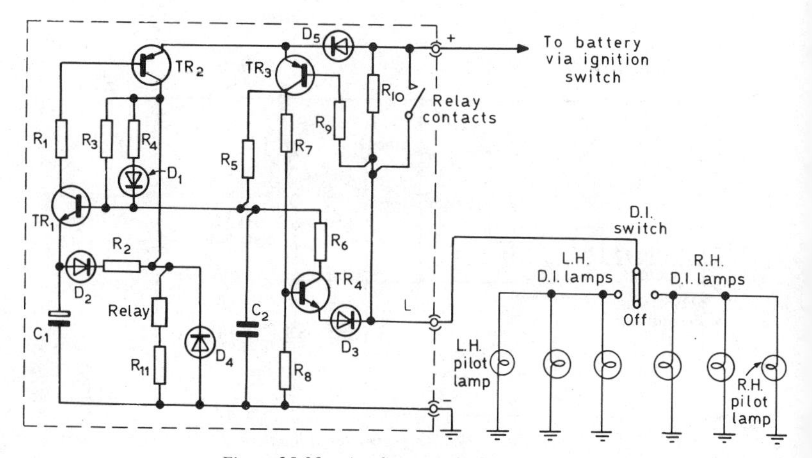

Figure 25.30 *An electronic flasher unit*

Figure 25.30 shows the internal circuit and external wiring of an electronic flasher unit in a direction indicator system with a fixed two-bulb load. The unit comprises basically a relay with normally-open contacts which flash the direction indicator bulbs, an oscillator which controls the operation of the relay, and a sensing circuit to detect bulb failure.[10]

When the direction indicator switch is operated for a left or right hand turn, transistor TR_3 is switched on followed by TR_1 and then TR_2 which energises the relay. With the closing of the relay contacts and illumination of the direction indicator bulbs, TR_3 switches off. C_1 charges up until TR_1 emitter voltage reaches a value approximating to its base voltage, determined by R_3, R_4 and R_5. (R_6 has no effect because TR_4 is switched off.) At this point, TR_1 switches off. TR_2 is consequently switched off and the relay de-energised. Its contacts thus open, allowing TR_3 to switch on again and the voltage across the bulbs to fall to a low enough value to also switch on TR_4. Whereupon the base voltage of TR_1 falls to a lower value, now determined by R_5, R_6 and R_3. In conjunction with the voltage across C_1, this results in TR_1 being held switched off until C_1 has discharged through R_2 and D_2 to a voltage slightly less than TR_1's new base voltage. As TR_1 and hence TR_2 switch on, TR_3 and TR_4 switch off and the cycle of events is repeated to give a flashing rate of 60–120 flashes per minute.

In the event of one of the two bulbs failing, the voltage across the remaining bulb during the 'off' period is increased, due to the greater effective impedance of the bulb circuit. This increase in voltage is sufficient to keep TR_4 switched off. The lower base voltage of TR_1 is now determined only R_5 and R_3, and is raised in value. The voltage range over which C_1 charges and discharges is therefore reduced, as is the time between switching of TR_1, TR_2 and the relay. Failure of a bulb is therefore indicated by an increase in bulb flashing rate to approximately 200 per minute.

When a flasher unit is required for hazard warning purposes (for example in the event of the vehicle being involved in an accident), closure of the hazard warning switch connects L.H. and R.H. direction indicator bulbs to the flasher unit which operates in the manner described for the normal complement of direction indicator bulbs.

The flasher unit has a discrete component/printed wiring board construction. It is being fitted as standard equipment (as is a 'variable load' flasher unit of similar design[11]) to large-volume production line vehicles.

Fuel-injection systems

Impending exhaust emission control legislation in the U.S.A. has created fresh interest in fuel-injection systems (for engines operating on the Otto cycle), because of their ability to provide better air/fuel matching and thereby give cleaner exhaust emissions.[12, 13, 14, 15] To mechanically process signals representing the essential engine operating parameters required to achieve the legislated emission standards involves a complex arrangement of levers, cranks, governors and three-dimensional cams. By comparison, an electronic approach yields a static control system which is compact, reliable and accurate, with facility at a later date to increase the number of monitored parameters without much difficulty and additional expense.

Figure 25.31 is a schematic diagram of one prototype-electronic fuel injection system,[12, 13, 14] although others have been developed and some fitted as standard equipment to production vehicles.[16, 17] A system of fuel control is employed which is responsive basically to engine speed and to throttle angle position (i.e., to engine load). These two engine parameters provide a more accurate assessment of engine fuel requirements over the operating range than do the combination of engine speed and manifold pressure.

An engine's fuel requirements at various speeds and under various loads are first determined on a test bed by means of a specially developed electronic Performance and Demand Analyser—PANDA. The information obtained is then programmed into a m.o.s. read-only memory of an electronic control unit for carrying on the vehicle, each seven-bit word in the 16×16 memory matrix representing a different quantity of fuel. When the vehicle is operating, the appropriate fuel quantity for given engine speed and load conditions is obtained from the memory by feeding analogue signals representing these operating parameters to the electronic control unit; here they are shaped and added to analogue interpolation data before undergoing the necessary comparison, digital conversion and counting operations for addressing the correct word in the memory. Interpolation of input data is necessary to obtain a smoothly varying fuel requirement from the memory, which stores this information for discrete engine speeds and throttle angle positions.

The fuel requirement read out of the memory is modified according to correction

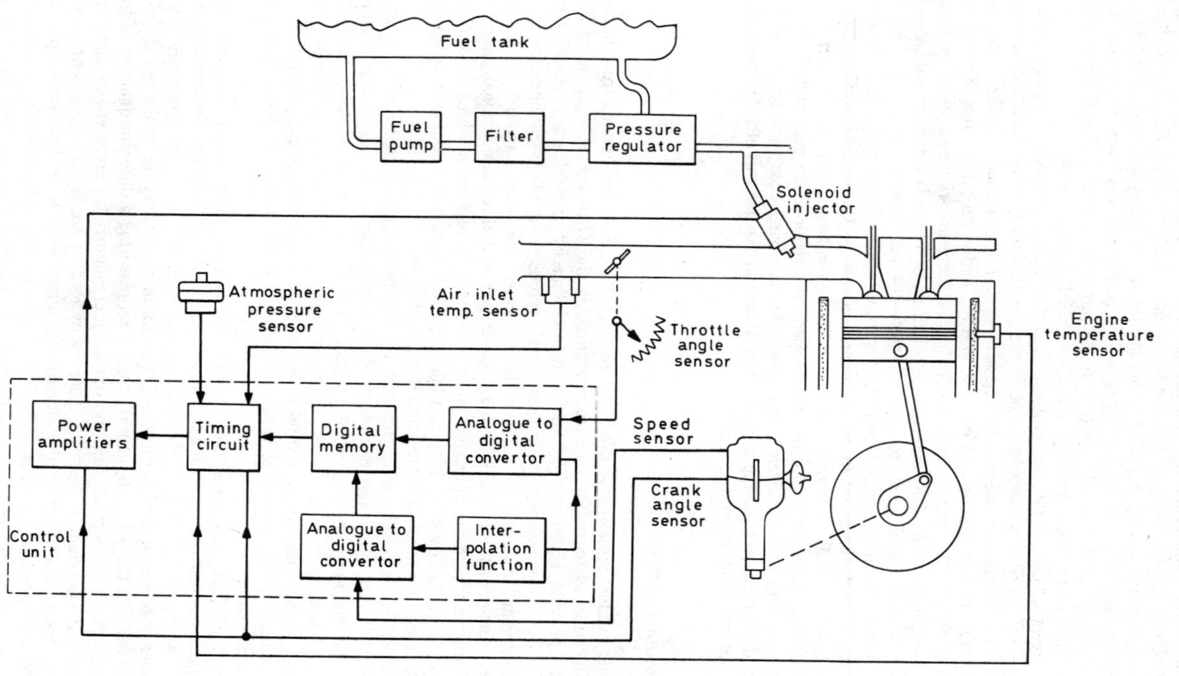

Figure 25.31 Schematic of an electronic petrol injection system

signals for engine temperature, atmospheric pressure and air inlet temperature, and converted to a pulse whose duration is proportional to the fuel required. This pulse is then used to energise electromagnetic fuel injectors. Since the fuel pressure is constant, the amount of fuel injected will depend only on the time for which an injector is held open.

Readily available discrete components, and digital and analogue integrated circuits were employed in the prototype electronic control unit.

Safety systems

Concern has increased in recent years over the safety of motor vehicles and their occupants. This has prompted the development of several safety subsystems, e.g., wheel-slide protection, vehicle condition monitoring and headway control.

WHEEL-SLIDE PROTECTION

As vehicle skidding and/or loss of steering control (arising from too fierce an application of the brakes) continues to be a factor in a large percentage of road accidents, despite improved braking system performance and tyre adhesion and improvements to vehicle handling characteristics, one important safety subsystem will undoubtedly be Wheel-Slide Protection (WSP). A completely mechanical WSP system has the disadvantages of bulk, weight and cost-factors which do not favour its application to small family cars. Electronic techniques can provide less bulky and cheaper systems, more suited to a small car, with the added advantage that faster response times can be obtained.[18, 19]

On a good road surface, deceleration of a car can rarely exceed 1.3 g. Wheel deceleration in excess of this amount means that the wheel is decelerating faster than the car, and will stop rotating before the car has stopped, with resultant skidding. Imminent wheel slide can therefore be detected by monitoring wheel deceleration. In the system illustrated in *Figure 25.32*, this is achieved in the following manner. Pulses whose frequency is proportional to wheel speed are induced in wheel sensors and fed to the electronic module, where they are converted to an equivalent d.c. signal. This d.c. signal is then differentiated to obtain wheel deceleration and compared with a set threshold value of 1.3 g. When wheel deceleration exceeds this threshold value, a power amplifier stage is switched on to energise the control valve solenoid.

Under normal braking conditions, one side of the actuator piston is permanently supplied with brake fluid at a pressure of 2.068 MN/m^2 (300 lb/in^2), while the other side is connected to the low-pressure reservoir via the control valve. Energising the control valve solenoid shuts off the low-pressure brake fluid line from the reservoir to the actuator, and at the same time connects the previously low-pressure side of the actuator to the 2.068 MN/m^2 pressurised side. The resulting movement of the actuator piston isolates the brake from the brake line and provides additional volume into which the trapped fluid in the brake line may expand. Pressure at the wheel is therefore reduced and the brake released.

Compared with a wheel on a good braking surface, one on a slippery surface decelerates faster when braked, requires the brake pressure to fall to a lower level before it starts to accelerate as the brakes are released, and takes longer to reach its rolling speed. Ideally, therefore, re-application of the brakes should be withheld for a period dictated by the nature of the road surface. This is achieved by storing on a capacitor a measure of the wheel speed lost up to the point when the brakes are released. The signal for the brakes to be re-applied is not given until the wheel has substantially regained this lost speed, the time the wheel takes to do so depending on the road surface. It is not generally possible for the wheel to recover this speed entirely, since the vehicle is slowed

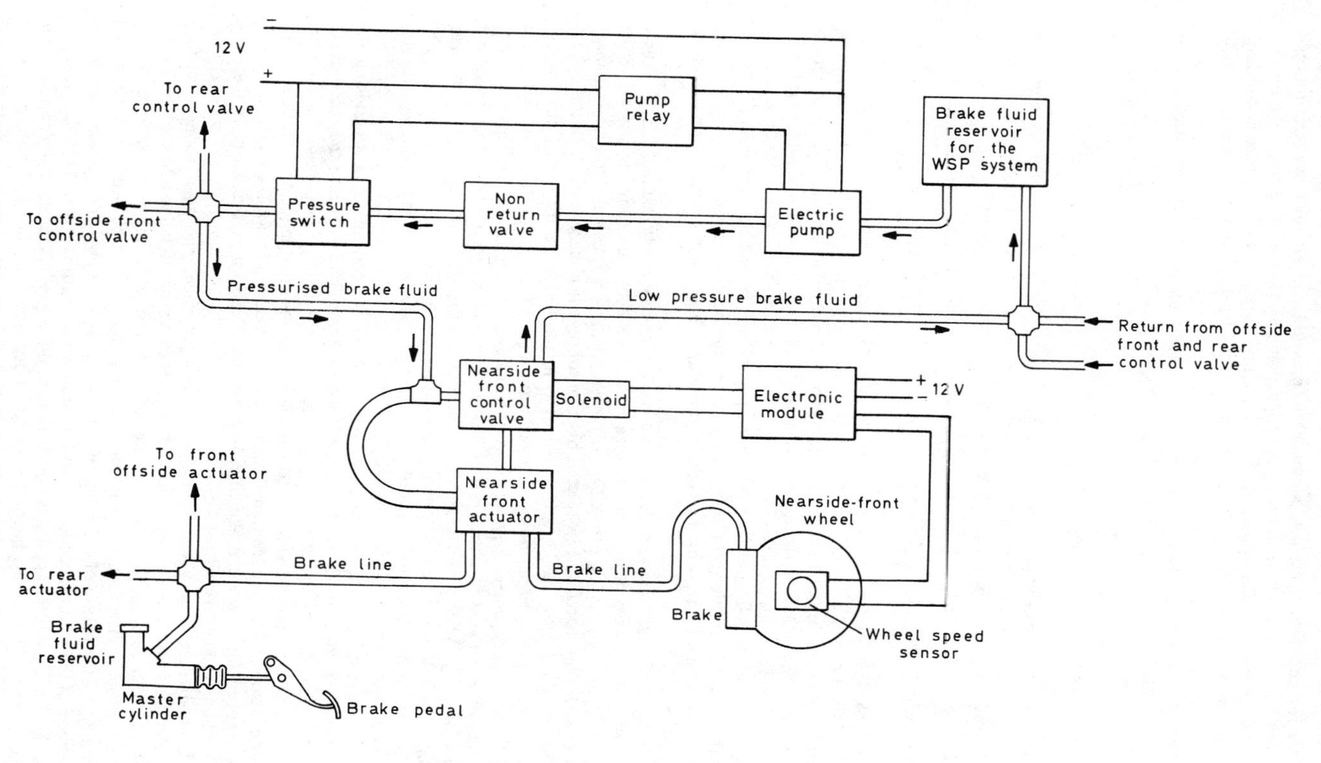

Figure 25.32 A wheel-slide protection system

down by the brakes on other wheels. For this reason the memory action of the capacitor is modified to signal re-application of the brakes at a lower speed, whereupon the electronic module de-energises the solenoid for the control valve to reconnect the low-pressure reservoir to the actuator. The actuator piston then moves to exhaust fluid in its low-pressure side back to the reservoir, and to connect the brake with the brake line—causing the brake to be re-applied. The system cycles in this way at up to 10 times a second (while any tendency towards wheelslide exists).

VEHICLE CONDITION MONITORING

Reliable, continual and comprehensive monitoring of vital vehicle systems (e.g., the braking system), and communication of their condition to the driver, makes an obvious and valuable contribution towards vehicle safety. Advantages are also to be gained from monitoring other categories of systems such as those subject to legislation (lighting and

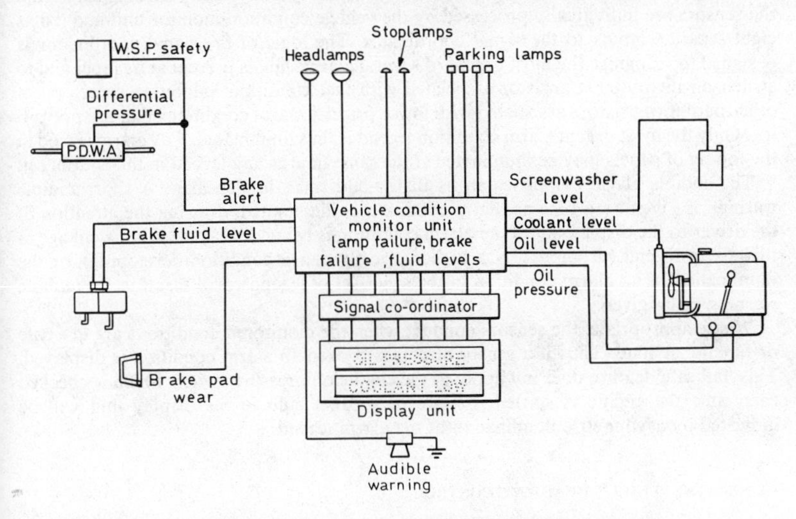

Figure 25.33 Schematic of a vehicle condition monitoring system

screen-washing equipment) and those essential to a vehicle's 'health' (lubrication and coolant systems).

A typical vehicle-condition monitoring system comprises various sensors for monitoring the different vehicle conditions, a vehicle condition monitor unit, a signal coordinator unit, a display unit and an audible alarm, *Figure 25.33*.

The types of sensors used are as follows:

(1) Brake alert: a pressure differential warning actuator (which warns of pressure failure in the brake system), and a wheel-slide protection safety circuit (which indicates any of twelve faults in the anti-lock braking system previously described).
(2) Oil pressure: a standard oil pressure switch.
(3) Lighting: a small-valued series resistor in the supply to each bulb.
(4) Brake pad wear: a contact inserted in the brake pad.
(5) Brake fluid level: a conductivity probe which fits into the cap of the reservoir.
(6) Oil level: a small ferrite magnet, carried by a float in the sump, which operates a

reed switch if the level falls. Because the level of the oil normally falls when the engine is running, the signal from this sensor is required when the engine is stationary. The signal from the sensor is therefore inhibited when the engine is running by the signal from the oil pressure switch.

(7) Coolant level: two conductivity probes, operated in parallel to avoid water surge effects, fitted in the radiator header tank.

(8) Headlamp and Screenwash fluid levels: conductivity probes are fitted to the two reservoir bottles.

The display unit consists essentially of eight transparent panels each marked with an alarm legend, and arranged one behind another in two sets of four, one set mounted above the other. Each panel is separately illuminated (when required) by two bulbs, one at each panel edge, each switched through one of eight output transistors in the signal coordinator unit.

Both discrete component/printed wiring board and integrated circuit technologies are used in the vehicle condition monitor and signal coordinator units. Alarm signals from the sensors are individually processed by the vehicle condition monitor unit and fed as eight separate inputs to the signal coordinator. The logic of the signal coordinator is designed to recognise the most urgent of several alarm signals present at its input and to switch on the output transistor associated with that condition, inhibiting switch-on of other output transistors associated with lower priority alarm conditions that are prevailing. Only the most urgent alarm condition legend is thus illuminated. Any alarm legend in the top set of panels may be illuminated at the same time as any legend in the bottom set.

The audible alarm comprises an oscillator and small loudspeaker. A short audible warning is given each time an alarm condition is illuminated, drawing the attention of the driver to the display unit. If an alarm condition is being displayed, either a change in this alarm legend (prompted by a higher priority alarm condition developing), or the illumination of an alarm condition in the other set of legends, will result in the audible warning being given.

Where appropriate, the sensors conduct when the monitored conditions are in a safe or functional state. Thus if a sensor circuit is broken an alarm condition is displayed. This 'fail safe' feature does not apply to the engine oil pressure switch which is checked each time the engine is started. Failure of a panel bulb in the display unit will be indicated by asymmetric illumination of the alarm legend.

AUTOMATIC VEHICLE HEADWAY CONTROL

The growing incidence of multiple vehicle collisions, caused by vehicles travelling too close to one another and too fast in conditions of poor visibility, has demonstrated the need for an automatic adaptive speed control system.

In the prototype system to be described,[20] one of several headway control systems that have been developed,[21, 22, 23, 24] a frequency modulated continuous wave (f.m.c.w.) radar (see Section 18) is preferred to a pulsed radar as the means of acquiring range and relative velocity information with regard to the target vehicle. Compared with pulsed radar, the large bandwidth required for the measurement of small ranges is more easily obtained within the bandwidth of the very high frequency microwave circuits of a continuous wave system.

In a f.m.c.w. radar the frequency of the continuously transmitted microwaves is varied with time. The returning wave from a relatively stationary target therefore has a slightly different frequency to the wave being transmitted. This difference frequency is proportional to the target range. Where relative velocity exists between the transmitter and target, a Doppler shift in the frequency of the return wave occurs proportional to the relative velocity. In these circumstances, the difference frequency between the received wave and that being transmitted contains both range and relative velocity information.

The system, *Figure 25.34*, developed primarily for motorway operation, comprises the following:

(1) Radar unit: a Gunn diode (see Section 8) oscillator is used as the continuous wave source, and is frequency-modulated by mechanically altering the length of the cavity.
Constraints placed upon the electrical design of the system, and the range, resolution and response times demanded, made it necessary to operate the radar in the Q-band at a frequency of about 35 GHz (wavelength 8 mm). A study of motorway geometry, lane widths and bridge heights, etc., indicated that a narrow radar beam width (approx. 5°) was required if single-target operation was to be achieved. The small dimensions of the transmitting and receiving horn antennae required by the narrow radar beam are compatible with automotive styling requirements.

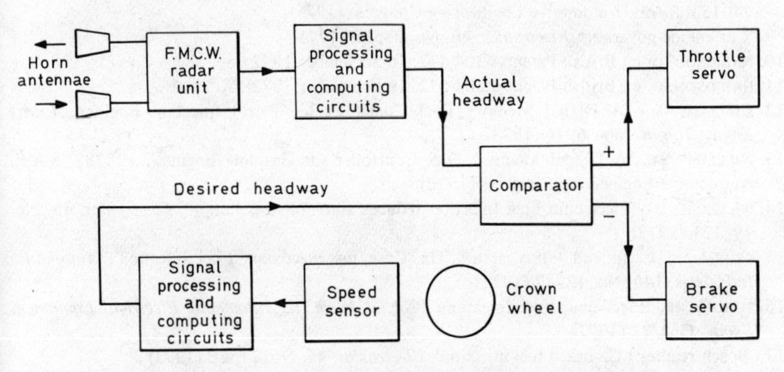

Figure 25.34 Schematic of vehicle headway control system

(2) Magneto diode speed sensor: the sensor is mounted inside the rear-axle housing, together with a permanent magnet. Rotation of the crown wheel causes a change of magnetic flux across the diode, resulting in a signal (whose frequency is proportional to true vehicle speed) being induced in the latter.
(3) Computing and controlling electronic circuits: similar digital and analogue circuits are employed to process signals from the radar unit and speed transducer to produce two analogue voltages representing actual headway and desired headway. These are compared to yield a positive or negative error signal which is used to actuate the throttle or brakes respectively, thereby providing the necessary safe vehicle spacing.
(4) Throttle and brake actuators with their associated servo controls: manual operation of the throttle over-rides the automatic system without disengaging it, while manual operation of the brakes returns the system to the 'standby' condition. A loss of radar signal automatically returns the vehicle to manual control, with an appropriate warning to the driver.

The above system has been developed and miniaturised to a degree where it is no larger than a headlamp.[25] Other features include the employment of the latest advances in microwave technology; a flat plate aerial (located centrally in the front grille of the test vehicle) which replaces the two horn antennae of the earlier version; and an advanced digital micro-computer which calculates range and relative velocity from the information contained in the reflected radar beam and relates this to the speed of the vehicle, making adjustment as necessary to the throttle or brakes to maintain a safe distance.

REFERENCES

1. EDWARDS, L. E. and WINKLEY, A. W., 'Alternators on British Passenger Cars', *Lucas Engineering Review*, **4**, 58 (1968)
2. SAY, M. G., *Electrical Engineers Reference Book*, 13th edn., Butterworths, **21**–110 (1973)
3. WINKLEY, A. W., 'An Integrated Circuit Regulator for Automotive Alternator Systems', *S.A.E. Automotive Engineering Congress* (1968)
4. NOLAN, R. W. and WINKLEY, A. W., 'An Integrated Microcircuit Voltage Regulator for Passenger Car Alternators', *Lucas Engineering Review*, **4**, 74 (1969)
5. COX, A., 'Electronics in Motor Vehicles', *Electrotechnology*, **1**, No. 3, 3 (1973)
6. SHARPE, J. W., 'Electronic Ignition Systems', *Lucas Engineering Review*, **1**, No. 2, 2 (1964)
7. SHARPE, J. W., 'Transistorised Ignition for High Speed Gasoline Engines', 650498, S.A.E. Mid-year Meeting (1965)
8. MOORE, J. H. and LONGSTAFF-TYRRELL, J., 'Engine Position Transducers for Electronic Ignition', 740153, S.A.E. Automotive Engineering Congress (1974)
9. 'Car-engine governor', *Electronics*, **46**, No. 23, 8E (1973)
10. HARRISON, D. B., British Patent 4 304 672, 16 September 1972
11. HARRISON, D. B., British Patent 4 304 572, 16 September 1972
12. WILLIAMS, M., 'A Digital Memory Fuel Controller for Petrol Injection Engines', *Lucas Engineering Review*, **6**, 16 (1973)
13. WILLIAMS, M., 'A Digital Memory Fuel Controller for Gasoline Engines', 720282, S.A.E. Automotive Engineering Congress (1972)
14. WILLIAMS, M., 'Electronic Fuel Injection reduces automotive pollution', *Electronics*, **45**, No. 19, 121 (1972)
15. ESHELMAN, R. H. and POND, J. B., 'The Case for Electronic Fuel Injection', *Automotive Industries*, **146**, No. 12, 47 (1972)
16. SCHOLL, H., 'Electronic Fuel Injection, *Proc. I. Mech. E. Automobile Electrical Equipment Symp.*, C61/72 (1972)
17. 'Bosch readies I.C.–based fuel injection', *Electronics*, **46**, No. 21, 5E (1973)
18. CURTIS, A., 'Taking the Skid out of Braking', *New Scientist and Science Journal*, **51**, 358 (1971)
19. ELLIOTT, D. R. and SLAVIN, M., 'Safety Related Electronics in the Automotive Environment', *Proc. I. Mech. E. Automobile Electrical Equipment Symp.*, C107/72 (1972)
20. IVES, A. P., *et al.*, 'Vehicle Headway Control', *Proc. I. Mech. E. Automobile Electrical Equipment Symp.*, C88/72 (1972)
21. 'Collision Avoidance Radar', *Electronics*, **47**, No. 1, 3E (1974)
22. HAROKOPUS, W. P., 'Radar hits road', *Electronics*, **45**, No. 2, 54 (1972)
23. HOLDER, F. W., 'Anti-Collision Systems for Autos', *Popular Electronics*, **2**, No. 4, 48 (1972)
24. KLENSCH, R. J. and SHEFER, J., 'Harmonic Radar Helps Autos Avoid Collisions', *I.E.E.E. spectrum*, **10**, No. 5, 38 (1973)
25. IVES, A. P., 'A Vehicle Headway Control System using Q-band Primary Radar', 740097, S.A.E. Automotive Engineering Congress (1974)

MOTOR VEHICLE TESTING

The introduction, and subsequent increase, in the use of electronic test equipment in the motor industry is due to a number of reasons. In the early days of motoring, when cars were an unusual sight, servicing was performed by enthusiastic owners and skilled mechanics, the engines and mechanical assemblies were fairly straightforward and money and time were lavished on these rare machines. Now, with a United Kingdom vehicle population in excess of twenty million, many with engines requiring accurate tuning, complicated transmissions and suspensions, relatively few highly skilled mechanics and modern economic pressures, any process or device able to simplify or shorten servicing or manufacture is of great value. The use of electronic techniques enables sophisticated, yet robust and cheap, test equipment to be produced to the advantage of those working with motor vehicles.

The application of such equipment falls into several categories; economy, pollution, safety and performance. Examples of these are:

(1) Economy; plug condition, timing, carburettion.
(2) Pollution; exhaust gas analysis, CO (carbon monoxide) measurement.
(3) Safety; brake testing, headlamp alignment, performance.
(4) Power output at road wheels using a dynamometer.

The equipment described is typically that of one manufacturer. Variations in construction and measuring range will naturally occur among other manufacturers.

Engine ignition analysis (petrol)

Engine analysers may be divided into two categories, oscilloscope units and meter units. The oscilloscope unit provides more diagnostic information but requires more skilful interpretation than the purely meter type.

The main parameters which require measurement are:

L.T. (low tension) voltage	0–2 V, 0–20 V
H.T. (high tension) voltage	0–30 kV
Contact breaker dwell angle	0–90 degrees
Ignition timing angle	0–80 degrees
Engine speed	0–6 000 rev/min.

Other measurements often made are:

Air fuel ratio	10:1–16:1
Power balance speed drop	0–300 rev/min.

These ranges are typical for a 12 volt, 4 cylinder, 4-stroke petrol engine. Variations in the ranges may be required for different engine configurations.

Connections are made to five basic points in the engine ignition system. Direct connections are made to the vehicle chassis (earth), battery live or coil SW (switch) terminal and the CB (contact breaker) terminal at the distributor or coil. An inductive probe is connected to a particular cylinder spark plug lead, usually number one, as a timing reference and a capacitive probe is connected to the distributor 'king lead' (coil H.T. output) to permit measurement of H.T. voltages.

The direct connections are used to check the L.T. voltages at the battery, coil SW terminal and CB terminal under conditions of ignition load, starter load and the charging voltage with the engine running at, or above, 1 500 rev/min.

The waveform derived from the CB terminal, when the engine is running, is filtered to produce a square wave whose edges coincide with points opening (p.o.) and points closing (p.c.) conditions, *Figure 25.35*. The square wave is fed to a moving-coil meter in parallel with a large capacitor, the electrical and mechanical combination of capacitance and meter inertia providing integration and the meter reading is proportional to the mark/space ratio. If the meter is adjusted to read 90 degrees at full scale with the points closed, then the dwell angle will be indicated, for a 4 cylinder engine, when the engine is running. The dwell angle is the angle through which the distributor turns while the contact-breaker is closed for one cylinder. The distributor turns through 360 degrees for a complete combustion cycle on a 4-stroke engine and, in the case of a 4-cylinder engine, each cylinder utilises 90 degrees of this rotation. Other scales may be marked on the meter, e.g., 0–60 degrees for a 6-cylinder engine, etc.

Either edge of the dwell waveform may be used to trigger the oscilloscope timebase to provide superimposed traces running from p.o. to p.o. or p.c. to p.c. The CB waveform is applied directly to the y plates of a 220-mm electrostatic cathode-ray tube

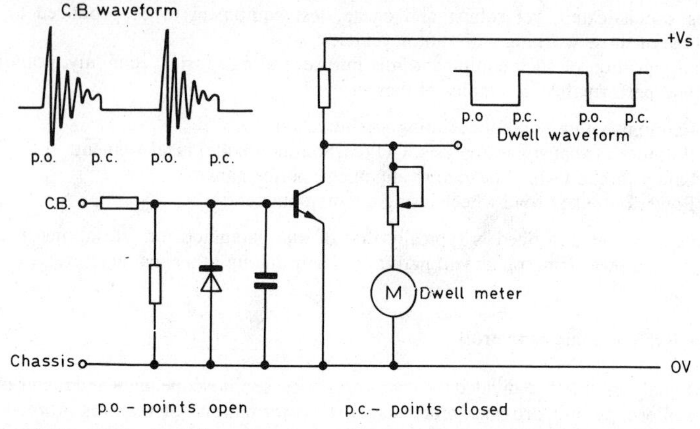

Figure 25.35 *Basic circuit for measurement of dwell angle, derived from contact breaker signal*

as the sensitivity of 35 V/cm provides an adequate trace amplitude. On larger tubes, or electromagnetically deflected types, amplifiers may be necessary in order to achieve a satisfactory trace. Inspection of the traces reveals any difference in individual traces and shows the condition of coil, condenser, contact breaker points and distributor, *Figure 25.36*. For interpretation of the various waveforms associated with ignition faults see *Corrective Servicing.*[1]

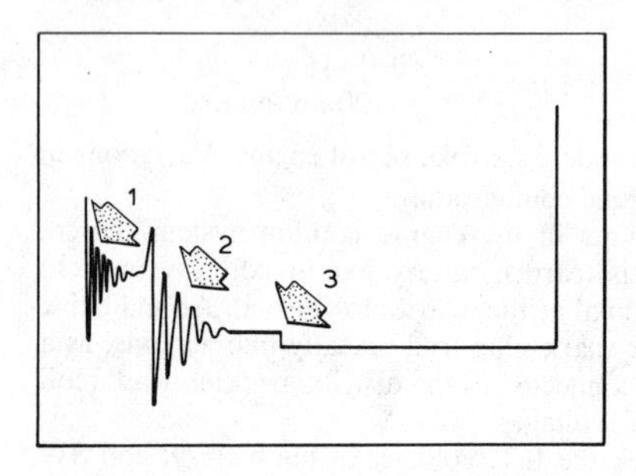

Figure 25.36 *Primary (CB) waveform from points open to points open, showing 1. coil and condenser oscillations during spark duration 2. coil and condenser oscillations after spark extinguishes 3. points closing.*

The dwell waveform is used to control the width of the timebase, *Figure 25.37*. The output of a mono-stable circuit, triggered from the dwell waveform, is integrated to produce a d.c. voltage proportional to the CB repetition frequency. The voltage is used to control a constant current circuit which charges the timebase capacitor. The capacitor is charged to the same voltage, over a wide range of engine speed, before being discharged by means of the silicon controlled rectifier (s.c.r.), which is triggered from either the p.o. or p.c. edge of the dwell waveform.

The signal from the inductive probe on the number one plug lead is filtered and used to trigger a monostable circuit. This is used for tachometer, timebase triggering and timing circuitry. One pulse is produced for every two rotations of the crankshaft on a 4-stroke engine, regardless of the number of cylinders, and a tachometer is produced by integration of the pulses and displayed on a moving coil meter. Manual switching enables the pulse height or width to be halved so that the same system can be used on a 2-stroke or Wankel engine, where one pulse is produced for one revolution of the crankshaft.

A second, variable, monostable circuit is triggered from the front edge of the first monostable. The trailing edge of the second monostable is used to fire an s.c.r. which triggers a xenon flash tube via a trigger transformer. The pulse transformer steps up the

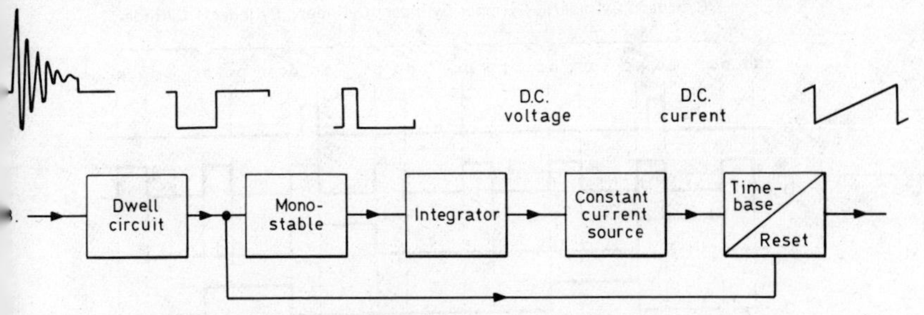

Figure 25.37 Block diagram of controlled width time base circuit

pulse from 150 V to 4 kV. A capacitor of 2 μF is charged to 600 V through 10 kΩ (to prevent permanent ionisation of the flash tube) and the discharge of the stored energy into the flash tube produces a brilliant white light lasting about 5 μs. The repetition rate must not exceed that at which the flash tube dissipates 8 W for a typical tube. The tube is mounted in a 'timing gun' and the light is focused and used to illuminate the crankshaft pulley, or wherever the timing marks are located, on the vehicle under test. The timing marks appear 'frozen' by the stroboscopic action of the timing light and the relative position of the fixed and rotating marks may be varied by a control on the 'gun' which alters the monostable period and delays the flash after the firing of the number one spark plug. Integration of the monostable pulses enable a meter reading of the amount of timing 'advance' to be obtained, the meter usually being calibrated in crankshaft degrees.

The number one cylinder pulse is also used to trigger the timebase to produce a

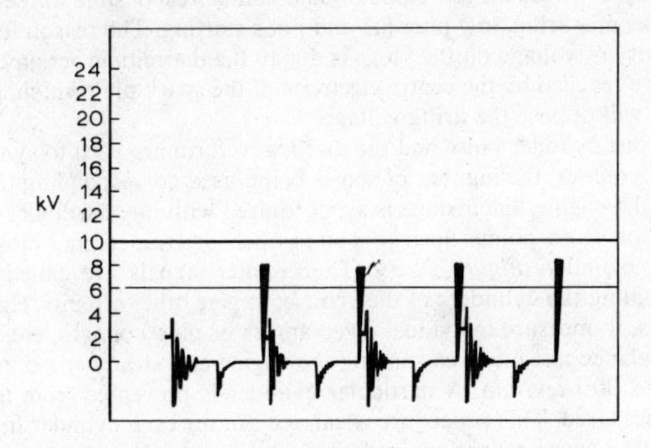

Figure 25.38 Secondary (H.T.) parade waveform showing plug firing voltages of 8 kV

'parade' pattern, *Figure 25.38*. The y deflection is obtained from the capacitive probe on the coil h.t. output and the pattern shows all the plug firing waveforms serially.

The 'kV' probe is basically a 20 pF 50 kV capacitor, the output is capacitively divided down to provide signals of the desired level for oscilloscope deflection, or, in the case of a meter analyser, to feed a peak reading circuit which indicates the plug firing voltage on a suitably scaled meter. A calibrated graticule on the oscilloscope unit enables a measurement of the plug firing voltage to be made.

Information concerning the condition of the spark plugs, plug leads, king lead, distributor rotor, coil, timing, carburettion and compression may be derived from the parade pattern by a skilled operator.[1] The important feature of coil polarity may also be

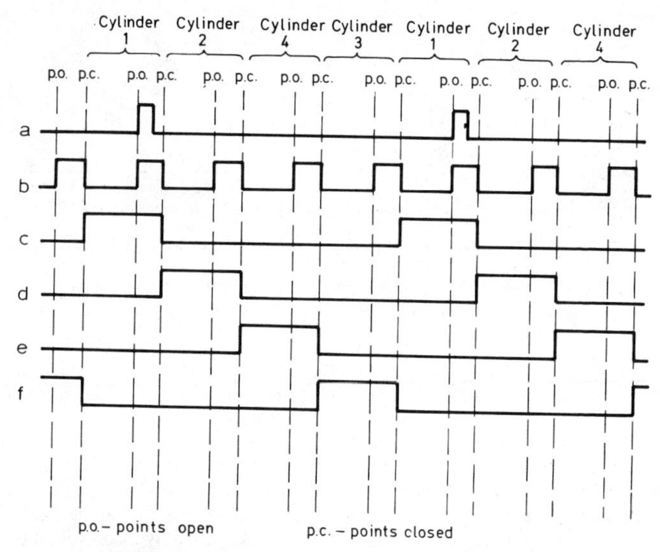

Figure 25.39 *Waveforms for 4 cylinder, 4-stroke engine, firing order 1-2-4-3, produced by* (*a*) *number one plug trigger signal* (*b*) *Dwell* (*CB*) *signal,* (*c*), (*d*), (*e*), (*f*) *counter stages 1, 2, 3 and 4 corresponding to cylinders, 1, 2, 4 and 3 respectively*

observed as the coil output should always be negative, regardless of vehicle earth polarity; reversed coils are all too common and can be responsible for several ignition problems such as mis-firing, low plug life and poor starting. The reason for requiring a particular polarity of voltage on the plugs is due to the thermionic action caused by the high temperature reached by the centre electrode of the spark plug which, in the case of a reversed coil, will oppose the firing voltage.

The number one cylinder pulse and the dwell waveform are used to synchronise and clock on a ring counter, the number of stages being used corresponding to the number of cylinders in the engine. Each stage is synchronised with one particular cylinder and may be turned on from points open to points open or from points closed to points closed for that cylinder, *Figure 25.39*. The counter signals are called up by push buttons, representing the cylinders of the vehicle, to gate other circuits, e.g., to unblank individual traces, to measure individual dwell angles or plug voltages, etc.

The power balance test involves running the engine at a steady speed, usually in the range 1 000 to 2 000 rev/min. A particular cylinder is prevented from firing and the speed drop is measured. This procedure is carried out for each cylinder in turn and the individual resulting speed reductions are compared for similarity, i.e., cylinder power balance. The blanking out of a cylinder from firing may be achieved by operating a relay or triac, connected across the contact breaker points, controlled by the appropriate counter output. The engine speed before shorting a cylinder is memorised by an electronic circuit which then detects the change in speed during the test and this is indicated on an extra tachometer scaled 0–300 rev/min, the reading increasing for a fall in true speed.

Exhaust analysis

Air fuel ratio measurement. The E.G.A. (exhaust gas analyser) may be included in the engine analyser or used as a separate instrument. The unit relies on the fact that a rich or lean mixture of air and fuel into the engine will result in an excess of hydrogen or carbon dioxide respectively in the exhaust gases and that different gases have differing thermal conductivities. Hydrogen conducts heat about seven times better than air, carbon dioxide conducts heat only half as well as air. A sample of exhaust gas, after passing through a condenser to lower the temperature and remove moisture, is passed over a heated platinum wire in a balanced Wheatstone Bridge. A similar filament, subjected only to atmosphere, compensates for variations in ambient temperature. Any change in filament temperature of the cell receiving the gas sample unbalances the bridge and a meter can be calibrated to the air/fuel ratio of the mixture entering the engine. The readings must be considered in relationship to the results of other tests, as poor combustion could lead to indication of a rich mixture.

Pollution measurement. The catalytic CO (carbon monoxide) tester also consists of a heated filament in a bridge arrangement but, in this instrument, the filament is heated to 900 degrees Celsius and oxidises the excess CO and H (hydrogen). The assumption is made that the normal ratio of CO to H exists in the exhaust gas and the CO content will contribute to about 45% of the change in temperature. The resultant rise in filament temperature causes an unbalance of the bridge which results in a meter reading. This may be calibrated directly in percentage of CO, typical scaling 0–10%.

A more sophisticated but less robust CO tester is the infra-red type. Emission from an infra-red source passes along a tube from 80 mm to 250 mm long to an infra-red detector fitted with a 4.7 μm optical filter. By comparing the transmitted signal level for atmosphere containing negligible amounts of CO with the signal obtained when exhaust gas may be indicated on a meter. This system responds purely to CO and not to other constituents of the exhaust gas.

Brake testing

Various methods of vehicle brake testing exist but many of these involve moving the vehicle for some distance, often on public highways, under varying road conditions and it is difficult to obtain consistent results. The roller brake tester provides the necessary control in that the surface on which the tyres run and the speed at which they rotate are sensibly constant for a given machine. The rollers are in parallel pairs, one pair for the

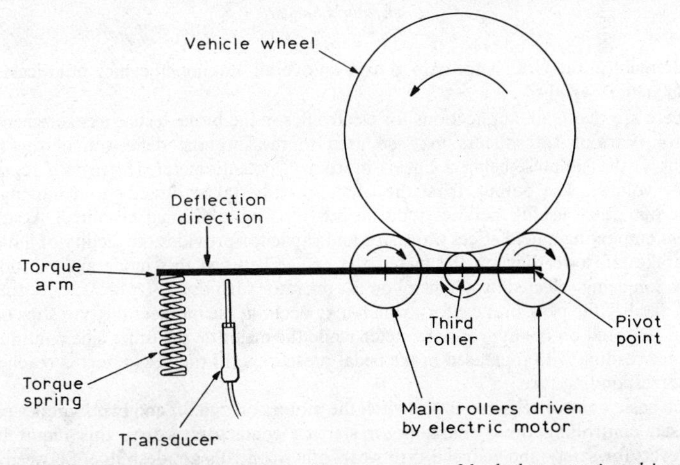

Figure 25.40 Basic brake tester system. Torque reaction of braked tyre against driven main rollers, produces a deflection of the torque arm against the coil spring

near side wheel and one pair for the off side wheel. The diameter of each roller is, typically, 200 mm and the separation 400 mm, the surface may be slotted steel, grit, concrete or suitably rough material. In the case of a low-speed roller brake tester the rollers are driven through a gearbox by a 4 kW motor and turn the tyre at a surface speed of 5 km/h. The complete roller and motor assembly is allowed to pivot about the axis of one of the rollers but is restrained from movement by means of a coil spring, *Figure 25.40*. High-speed brake testers run up to 150 km/h the basic principles being identical to the low speed type.

The front or rear wheels of the vehicle under test are located between the rollers and the nearside, offside, or both motors are started. When the brake is applied the braking force, at the surface where the tyre and roller meet, creates a torque acting on the roller assembly which causes the spring to be compressed. Provided the percentage of compression remains small compared to the total length the movement will be proportional to the torque and, hence, the brake effort. On a car tester an individual wheel brake effort up to 1 000 kg may be measured, and the sum of the efforts on each wheel may be compared with the weight of the vehicle to give the brake efficiency. An efficiency of 100% indicates a potential deceleration of 1 g. The current legislation

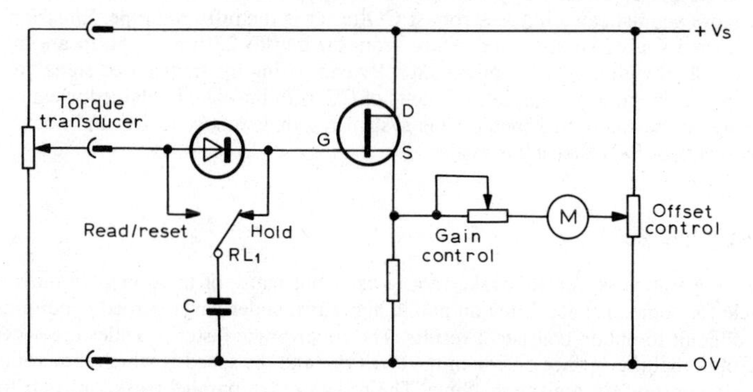

Figure 25.41 Read/reset circuit for brake tester. Relay RL is operated by a control on the operator's handset

requirement in the U.K. is for cars to have an overall braking efficiency of at least 50% of the vehicle weight.

There are two main applications for electronics in the brake tester, measurement and control. Various transducers may be used to measure the deflection of the roller assembly, the simplest being a linear, or rotary, potentiometer, fed from a regulated supply voltage. The output, from the wiper, may be taken directly to a moving coil meter but other useful facilities may be provided by additional circuitry. A storage system employing a field effect transistor and capacitor provides the facility of 'holding' the brake effort reading on the meter, the choice between this mode and the normal indication being selected by a control on the operator's handset, *Figure 25.41*. Using the 'hold' facility the peak brake effort, which may occur at the moment the tyre slips on the roller, is stored on the appropriate meter while the meter on the other side continues to increase reading with increased brake pedal pressure until the peak effort is reached on the corresponding tyre.

The basic control required is to switch the motors on and off and two switches on the handset, controlling the 3-phase motor starting contactors, satisfy this requirement. However, for safety and reduced tyre wear, other controls are desirable. Between each main pair of rollers is fitted a 'third roller', of a smaller diameter and free to rotate. When a vehicle wheel is placed on the tester this 'third roller' is depressed but maintains contact with the tyre. The downward movement of the roller closes a microswitch on each side of the tester and the motors will only start when both switches are closed, indicating that a vehicle is on the tester. When the motors are running the 'third rollers' attain the same peripheral speed as that of the tyre and main rollers. If the tyre slips, on the roller being driven by the motor, damage might occur very rapidly due to the rough surface of the roller, such surface being necessary to provide a reasonable coefficient of friction. The change in speed of the type also occurs on the 'third roller' and using a fixed opto-electronic source and detector, and a segmented rotor on the 'third roller', a speed signal of 400 Hz may be obtained when no slip is present. Using a missing pulse, or other fast frequency sensing technique, the contactors may be removed in less than 5 milli-seconds compared with some 2 seconds to 3 seconds using electromagnetic systems. The system may be wired so that slip occurring on either side will de-energise both contactors or just the one controlling the side on which the slip occurs.

Headlamp alignment

One form of headlamp aligner in widespread use is the condenser lens type in which the beam pattern is focused on a screen some 250 mm square, usually within a shielded light box to provide good contrast in conditions of high ambient light. The screen is marked with a graticule comprising a vertical and horizontal line, intersecting at the centre and also a slanting line at 15 degrees above each horizontal, starting at the same intersection point. The angled lines indicate the upper cut-off surface for an asymmetric dipped beam headlamp pattern, one for left-hand and one for right-hand drive conditions, *Figure 25.42*. Degree or mm markings above and below the horizontal indicate the amount of positional error in the aim of the headlamp beam based on the location of the cut-off surface of the dipped beam. When the headlamp aligner is correctly positioned, with respect to the vehicle under test, the screen is equivalent to a flat, vertical surface perpendicular to the vehicle at a distance of 10 m (30 m is used in most Continental countries).

An array of photo-detectors arranged above and below the dipped beam upper cut-off surface may be biased to switch t.t.l. logic gates even for poorly defined cut-off

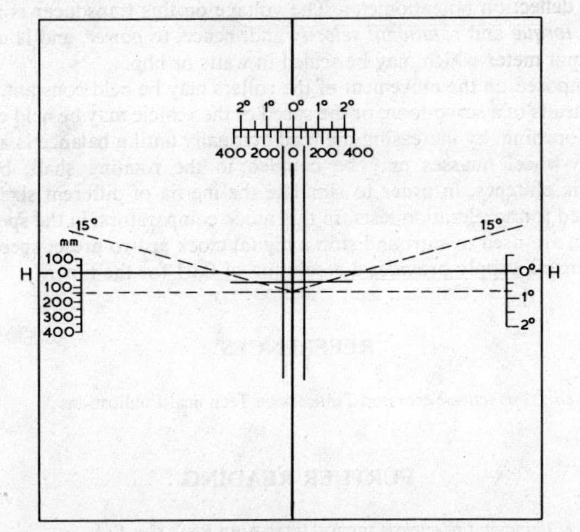

Figure 25.42 Headlamp aligner screen showing graticule for both left hand and right hand driving conditions

patterns. The gates can then be inter-connected to analyse the pattern position. Logic output signals can be arranged to feed one of three lamps on the aligner, indicating whether the headlamp is adjusted too high, correctly or too low.

A photo-detector, mounted on the centre vertical line a little above the horizontal, is used to determine the intensity of the light. A suitable device should have an optical response similar to that of the human eye. If this is not so, it is possible to obtain lower readings for a tungsten-halogen lamp than for a conventional lamp due to the variation in the wavelength distribution of the emitted light and contrary to the opposite deduction from human observation. An output of 400 mV for 100 lx may be obtained from a silicon photo-detector in the photo-voltaic mode, sufficient to drive a 25 μA meter.

Ranges of 0–6 lx, 0–50 lx and 0–100 lx are used for measurement of dipped beam, conventional main beam and tungsten-halogen main beam intensities respectively. Regulations limiting the maximum and minimum intensities of the two beams exist in some countries.

Vehicle dynamometry

The vehicle dynamometer requires a pair of rollers for each wheel to be tested, at least one roller of each pair being able to absorb a considerable amount of power by means of some form of brake. A hydraulic or electro-magnetic brake is generally used, capable of absorbing some 150 kW to 250 kW (200 b.h.p. to 300 b.h.p.). The rollers, for high speed and high power use, are 500 mm or more in diameter to reduce the hysteresis heating of the tyres caused by the deformation at the power transfer surface. The torque reaction on the roller assembly, which is pivoted, deflects the assembly against a spring and the amount of deflection is detected by a transducer. The transducer is a linear potentiometer whose supply voltage is proportional to roller speed.

One way of achieving this is to generate pulses from an electromagnetic pick-up, located close to a protuberance on the end plate of a roller. The number of such pulses is proportional to roller speed and, hence, to road speed and when integrated provides a d.c. voltage proportional to road speed.

The output is used to drive a meter scaled 0 km/h to 200 km/h in addition to supplying the deflection potentiometer. The voltage on this transducer is now proportional to both *torque* and *rotational velocity* and, hence, to *power*, and is used to drive the power output meter which may be scaled in watts or bhp.

The load imposed on the movement of the rollers may be held constant, irrespective of speed, by means of a servo-loop, or the speed of the vehicle may be held constant, at a given throttle opening, by increasing the load gradually until a balance is achieved.

Various ‘fly-wheel’ masses may be coupled to the rotating shaft, by means of electromagnetic clutches, in order to simulate the inertia of different sizes of vehicle, these being used for acceleration tests. In this mode comparators, in the speed measurement circuitry, are used to start and stop a digital clock at two preset speeds. Division of the 50 Hz mains supply provides a resolution of 0.01 for the timer.

REFERENCES

1. EAGLE, C. D., ed., *Corrective Servicing*, Transervice Technical Publications

FURTHER READING

BERRY, J. B., ed., *Engine and Electrical service*, Inter Auto Book Co. Ltd.
OSBORNE, H. C., *Automotive Technology Series, Vol. 7, Vehicle Operation and Testing*, Iliffe (1969)
YOUNG, A. P. and GRIFFITHS, L., *Automobile Electrical Equipment*, 8th Edn., Butterworth (1970)

ROAD TRAFFIC CONTROL

Electronic equipment is to be found in ordinary traffic signals, vehicle detectors and counters, car park control equipment, motorway signals and access control systems, traffic surveillance systems including closed circuit television, emergency telephone systems, identification and location systems for emergency vehicles and buses, bridge toll systems, monitoring and control of tunnels, etc. Similar equipment could be used in future for road pricing, transmission of route information to vehicles, etc.

Large motorway- or traffic-signal systems are controlled centrally, and this can involve substantial data transmission networks and on-line computer systems. The connection of traffic signals to computers is known as Area Traffic Control (ATC), and in a large city several control systems may be brought together in one integrated traffic control centre. In the U.K. such schemes are the responsibility of Local Authorities, using equipment to Government national standards, and are often operated jointly with Police Forces.

Traffic-signal controllers

Linking. A traffic-signal controller is the equipment in a roadside cabinet which controls the local lights at a particular intersection. At an isolated intersection the controller works on its own. Where signalled intersections are relatively close in urban areas, and platoons of traffic leaving one intersection are not unduly dispersed by the road system before they arrive at the next intersection, then the controllers are linked via multicore cables to a master controller to obtain a progression. Better coordination over a wide area can be achieved by a computer if more than about 30 signalled intersections are involved. Local linking then need only be provided as a standby to the computer. Controllers now use digital counters to derive the signal timings. By using the 50 Hz mains as the reference, controllers can remain in synchronism if the computer link fails without the need for local cables and master controller.

Safety. The local controller is always responsible for the basic sequence of signal aspects; for a route gaining right of way—red, red and amber ('starting amber'), green; for a route losing right of way—green, amber ('leaving amber'), red. The controller distributes right of way in a cyclic sequence, the time taken to return to the same point in the cycle is called the cycle time. The cycle progresses one *stage* at a time for each change in right of way; stages do not overlap. Each signalled traffic movement of different duration is a *phase*; a phase may last for one or more stages and overlap another phase.

The time between the finish of green on one stage and the start of green on the next is called the intergreen. The intergreen is thus composed of a leaving amber, any 'all-red' necessary to allow traffic to clear the intersection and a starting amber. The controller has a number of safety functions which must not be overridden by linking or central control:

(1) There must be interlocking to prevent conflicting greens being shown.
(2) If the mains power is interrupted, the controller must start in a safe way.
(3) Intergreen timings must be preserved.
(4) Once a green is shown, it must be maintained for a safe period known as the minimum green. The next change must not begin until the minimum has expired.
(5) Phases may have a critical sequence in the cycle to allow traffic to clear, which must not be overridden.
(6) Certain signal aspects must be correctly terminated, e.g., a filter arrow into a leaving amber.

Fixed time and vehicle actuation. To calculate fixed time settings, it is necessary to know for each phase the ratio y of actual flow divided by saturation flow, and the sum of the y values for the whole intersection Y. Saturation flow is the maximum steady flow over a stop line fed with a constant queue, and is proportional to road width, corrected if necessary for gradient and site conditions. Knowing the lost time per cycle (due to vehicle starting delays and amber periods) the optimum cycle time giving least average delay can be worked out[1] within practical limits. The minimum cycle time cannot be less than the sum of the minimum greens and intergreens, which for a complex intersection may be the limiting factor in a linked system where a short cycle is required. Delay increases sharply below the optimum value. Maximum cycle times should rarely exceed 120 seconds. The cycle time less the lost time can then be split among the phases in the proportion of y to Y. Hence, relative to other phases, the greater the flow, or the narrower the road, the greater the value of y, and the greater the share of the cycle time.

To allow for a wider range of traffic conditions, the standard controllers used in Great Britain have been vehicle actuated, using detectors to measure the traffic on each arm of the intersection. The detectors have two main functions. Firstly they demand a stage, so that if there is no traffic calling for a stage (say a side road at night) that stage may be omitted. The controller then remains where it is, or goes to the next stage in cyclic order for which there is a demand. Detectors may be specially positioned to call turning phases. Secondly the detectors may extend the green time of a stage by an increment for each vehicle that is detected while the stage is running. The next stage change can proceed when a gap is detected in the traffic. Consider the case of stage 1 just as it is gaining right of way.

The starting amber is followed by green, which must then run for the minimum period. Any other demand, say on stage 2, will be stored and remain inoperative during 1's minimum. If there is no further traffic over 1's detectors at the end of the minimum, 2's demand will initiate the change to stage 2. Or if there is traffic, each vehicle over 1's detectors will add an 'extension' to 1's green, still preventing 2's demand from acting. When extensions cease, 2's demand will proceed. To prevent continuous extensions causing unacceptable delays to other stages, there is a preset maximum green time for each stage. When this expires the next stored demand becomes operative regardless of extensions. In the event of continuous traffic on all stages, the controller will therefore work fixed time at the maximum settings. If after a minimum and a maximum there is no other demand stored, the controller will stay where it is, and will respond directly to the next demand, or it may be set to return to the arterial route. The equipment has been further sophisticated to allow the extensions to be speed timed, depending on the time taken for the vehicle to pass between two detectors, and to provide variable minimum green, variable all-red and variable maximum periods.

Vehicle actuated controllers may be linked to a master controller to form a 'flexible progressive' system, in that each intersection has some local flexibility dependent on its detectors, whereas the master ensures a traffic progression along a through route. The master has to prevent local vehicle actuation at some parts of the cycle to be sure that the controller can accept a progressive forced change on time—otherwise it could find that the controller had just started the minimum of an unwanted stage.

Equipment. The nationally standardised controller for general use in permanent signal installations in Great Britain is vehicle actuated, and is equipped for the number of phases, and special facilities required for the site. Many installed controllers employ thermionic valves in *RC* timing circuits, and PO 3 000-type relays for switching, but this type is no longer in production. Current types are solid state, using digital counting to derive timings wherever possible. Signal aspects are switched at mains voltage by relays, with electrical interlocking. Low voltage tungsten-halogen lamps are used in high intensity signals, with a transformer for each lamp in the signal head. At night, a common dimming transformer is switched in at the controller.

Electromechanical controllers used pneumatic detectors. Depression of a rubber tube in the carriageway operated a pressure sensitive switch in a footway box. The tubes were in pairs to operate uni-directional logic and to provide speed timed extensions.

These detectors are currently being replaced by the inductive loop type. A cable is buried in a slot cut into the road surface, forming a turn or two of a coil in an oscillator circuit operating up to some 150 kHz. The presence of a metallic vehicle (though not necessarily ferromagnetic) causes a phase shift which is detected as a voltage change and which operates a relay. If permanently occupied, say by a parked vehicle, the detector restores by means of a long time constant *RC* circuit to allow further vehicles to be detected. Loops are frequently placed at an angle to increase the sensitivity to 'magnetically narrow' vehicles like bicycles. Several loops are used for speed timing.

Area traffic control (ATC)

Along a progressive route, the time between the appearance of successive green signals is called the 'offset'. For a given speed the offsets can be obtained from a time-distance diagram in the case of a simple route. Where a larger network of signals is to be controlled, as in ATC, offsets, cycle times and splits are optimised using a computer method such as TRANSYT.[2] According to the environmental and planning needs of the area, these methods can give a reduction in journey time of more than 10%,[3,4] special priority to reduce bus journey times by allowing for bus stops between signals, or even a reduction in traffic capacity for special control schemes.

A set of timings for a number of junctions is called a plan. Individual junctions, or clusters of closely spaced junctions which must always by synchronised, form sub-groups. Sub-groups are formed into a group for an area which has particular traffic characteristics, e.g., requires a plan with a common cycle time. Plans are changed according to the time of week, and are provided for morning and evening peaks, off-peak, night time, special contingencies, etc. An ATC area will thus contain many groups. The basic task for the central computer is to be able to load a variety of plans into a backing store, and according to a timetable for the area, to call down the correct plan for each group, and output the timings to the signals.

The basic electronic requirements are thus (*a*) modification of controllers for central control, (*b*) a data link between each controller and the computer control centre, and (*c*) the central computer system.

Modification of controllers for central control. For a vehicle actuated controller the computer must be able to call for any stage change immediately (subject to safety requirements), or permit a selected stage or stages to respond to a local demand if there is one. To do this it is necessary to override the controller stage sequence, terminate any maximum green running, and initiate the stage change by a computer 'demand' if a forced change is required. If a vehicle actuated change is required, the appropriate stage is selected, the computer demand omitted, and one from the street awaited. The controller is also told whether to obey the computer, or to operate on local control (linking). Any failure to receive new and correct control messages must cause the controller to revert to local control. The coding of the bits of the control message from the computer varies from scheme to scheme, but an 8-bit word is sufficient for a 4-stage controller. The green indication for each stage is monitored and fed back to the computer on a matching 8-bit reply word. Second 8-bit words can be used for complex controllers. Signal lamps themselves are not monitored as in Great Britain the junction layout always incorporates repeat aspects. To allow the computer sufficient accuracy of timing, the transmission cycle must not exceed one second.

The current trend is away from fully vehicle actuated controllers, as the computer can produce a wide variety of plans. Partial vehicle actuation is provided so that fixed time stages can be skipped if there is no demand (e.g., on a side road, or pedestrian stage). Hence a much simpler controller can be used at most junctions in ATC. These controllers contain a special mains clocked register known as a group timer containing a single standby plan whose timings are an acceptable traffic compromise for normal conditions. This plan runs continuously, even though the controller is being controlled centrally, and is occasionally synchronised by the computer in relation to neighbouring controllers. If computer control is lost, the controller reverts to the group timer, and remains in synchronism during the fault. An additional option allows the standby plan to be programmed from the computer rather than set up locally.

Data transmission system. The requirement is to transmit about 16 bits to and from each junction each second. A large system may have 1 000 junctions. If it is practical and economic in a particular city to lay special purpose cable ducts, then multicore cable can be installed, and a simple parallel d.c. transmission system used. If rented telephone circuits have to be used, then multiplexing is necessary to allow one pair per junction to be used, or less with some kind of omnibus network. For short distances between the signals and the control centre, the most practical method is to use an individual radial line to each junction unless a direct link can be provided between a pair of close junctions. For distances over about 6–8 km, it is worth sharing several junctions per line, for example by using a common bearer circuit to a convenient junction and local rented branches to a group of surrounding junctions. The branching point can employ digital concentration of data (particularly if local processing of data is required), but otherwise the most practical method is to use transmission equipment to match the lines, in conjunction with f.d.m. channels for outstations. For individual data links, typical techniques are:

(1) T.D.M. 50 bit/s, half duplex, per outstation using (*a*) d.c. transmission for short lines synchronised to the mains for timing, or (*b*) using a modulated v.f. start-stop system with single bit parity.
(2) T.D.M. and f.d.m., using a slow t.d.m. scan of 8 or less timeslots per second defined by parallel f.d.m. channels, with full duplex transmission of data in each direction in each timeslot by further f.d.m. channels, coded for security. Reply bits can be timed at the computer round the loop from the outgoing control bits, as the long timeslot periods make the skew in transmission delays neglible.
(3) T.D.M. 200–600 bits/s, a single v.f. channel, timeshared between 4 or more outstations, each polled by its own address, and operating on a start–stop basis with parity.

1 (b) and 3 are best used where C.C.I.T.T. requirements have to be met. 1 and 2 are slow enough for the control word v.f. transmitters to be connected directly to the computer output channels, and driven by the software in the main computer at transmission speed. 3 requires a dedicated processor, or special hardware, as it would be difficult for the general software to be sufficiently accurate in providing the high bit rate, but less v.f. equipment is necessary at the centre.

At the outstation, the outstation transmission unit (OTU) has once a second to staticise the control words into a parallel register driving a relay interface with the signal controller. If an error is detected, the last message is retained. The OTU contains a 'watchdog', which drops the signals on to local control if new and correct messages are not received for 3 or 4 seconds.

Computer system. A small system may control typically 100 signals, a large one, 1 000. Accordingly the computer system comprises:

(1) A single central processor doing all jobs, or several dedicated processors.
(2) A standby computer system.
(3) An interface to the instation data transmission equipment with (automatic) changeover to either computer.
(4) A teleprinter and paper tape station per computer.
(5) A backing store (small fixed head discs, or large moving head discs with removeable discpacks for a large system, with switching for disc copying and maintenance).

(6) For a large system, a line printer for listing programmes, signal plans, data, etc. (The teleprinter is fully loaded with alarms, logging and input on a large system.)
(7) For a large system, a card reader to input small packages of frequently changed data too small for tape, particularly relating to individual junctions.
(8) A display system, possibly a lamp drive to a mimic diagram, or if the area is too large for a diagram, an interface to Visual display unit (v.d.u.) terminals for the control operators. V.D.U.s with inbuilt storage, full graphical capability, and facility for remote operation are most practicable. All system commands can then be executed via the v.d.u. keyboard.
(9) A job scheduling executive capable of accepting new jobs on-line, and designed to give protection to the basic traffic control programmes which must remain operational without risk during the continuous development and housekeeping associated with a large system. The same applies to programmes loading signal plan data, v.d.u. data, etc.

Surveillance systems. The two main forms are closed circuit television and detectors. In a large system where it would be impracticable to monitor say 100 TV cameras at once, detectors are used to produce alarms of congestion via the computer, to pinpoint which camera to monitor to follow up the incident manually.

Detectors are of the inductive loop type. Congestion is derived from the proportion of time that a single loop is occupied. The loop is sited to be as sensitive as possible to unusual conditions, e.g., in the weaving section of a gyratory systems over which vehicles would not normally pass unless diverted by congestion. The length of the detector pulse is measured on site and transmitted back to the centre in binary with the controller bits, or is transmitted back in its raw analog form and measured centrally, by software if desired. In either case the pulse effectively gates standard clock pulses into a counter. Thus a stopped vehicle would store a complete second's worth of clock pulses which would be proportioned to give 100% occupancy. Following suitable smoothing of the one-second values, an alarm threshold is set at a percentage corresponding to the congested condition of the particular site.

For counting traffic on a carriageway of n lanes, $n + 1$ adjacent loops are used, whose outputs are gated logically. A single vehicle straddling lanes can thus be distinguished from two parallel vehicles.

Detectors have been used to modify signal plans on-line to try to produce a more optimum solution than that given by pre-computed fixed time plans. Their widespread and effective use for this has yet to be realised in Great Britain.

At critical intersections, t.v. cameras with remote control of pan, tilt and zoom are provided on masts or buildings, avoiding the level of street lights. Control is normally by joystick at the centre; with a pan speed of 15° per second the camera can be positioned by the operator without feedback, providing no significant backlash occurs (e.g., due to sampling in the control system). Pre-set shot systems avoid this problem, but the extra cost and complexity of monitoring the pan and tilt head, and the associated servo, is not normally justified for traffic use. Despite the advances in high sensitivity camera tubes for night use, conventional vidicons are still most widespread.

For small systems, over distances of a few miles, baseband video is transmitted over suitably equalised circuits. For larger systems camera signals are connected to a convenient point (e.g., a local telephone exchange) at video frequency, and there modulate carriers for transmission to the control centre on a common wideband bearer channel. When an operator selects a camera, a switching matrix selects the required bearer, and a demodulator associated with each monitor tunes to the required camera channel frequency.

Separate speech quality pairs are provided to carry coded signals for camera control.

REFERENCES

1. WEBSTER, F. V. and COBBE, B. M., *Traffic Signals*, HMSO (1966)
2. ROBERTSON, D. L., 'TRANSYT: A Traffic Network Study Tool', *Transport and Road Research Laboratory, Crowthorne*, Report LR 253 (1969)
3. HOLROYD, J. and OWENS, D., 'Measuring the Effectiveness of Area Traffic Control Systems', *Transport and Road Research Laboratory, Crowthorne*, Report LR 420 (1971)
4. KAPLAN, J. and POWERS, L., 'Results of SIGOP–TRANSYT Comparison Studies', *Traffic Engineering*, September (1973)

FURTHER READING

SMITH, E., 'Recent Developments in Road Traffic Control', *The Journal of the Society of Electronic and Radio Technicians*, **8**, 7, August (1974)

RAILWAYS

Railways are a transportation system. The obvious aim is to move people or goods from one geographical location to another. A practical system, must meet certain specifications, including economic factors, thus defining a 'quality of service'. Factors defining this quality are, e.g., flexibility, reliability of service, transit time, safety, energy consumption, etc.

Electronics plays an increasingly important part in achieving these specifications. The essential elements of a railway system, are depicted in *Figure 25.43*, in the shape of a hierarchical control system. In practice, each of the loops can be redrawn in considerably more detail, incorporating various nonlinearities, adaptive paths, dependent variables and variable coefficients. The railways have the advantage of having control over the whole system, i.e., infrastructure, rolling stock, personnel, marketing, etc.

Controlling the movement of individual trains

BASIC INFORMATION REQUIREMENTS

The control of the movement of an individual train is represented in *Figure 25.43* primarily by the letters *T* (traction) and *B* (braking). Even if there is only one vehicle in the whole system, the safety aspect of the maximum permissible speed must be taken into consideration, thus requiring the knowledge of the position and values of speed limits associated with the infrastructure and/or the vehicle, values of the track gradient, length of the vehicle and its dynamic characteristics. When more than one train makes use of the same track, measures have to be taken to ensure that conflicts cannot result in an accident. This can be achieved by means of any of a number of control strategies, and the one adopted by all the major railways in the world is that of the Fixed (space interval) Block.

In this system, colour lights are used to indicate the distance at which the speed of the vehicle must be zero. Different combinations of colour lights are used to give advanced information and increase track capacity.

COMMUNICATIONS BETWEEN TRACK AND TRAIN

In addition to fixed lineside signals or traffic lights, there are many possibilities for conveying information between a land based control system and moving vehicles. These include:

(1) Free-space radio.
(2) Radio communication using leaky transmission lines.

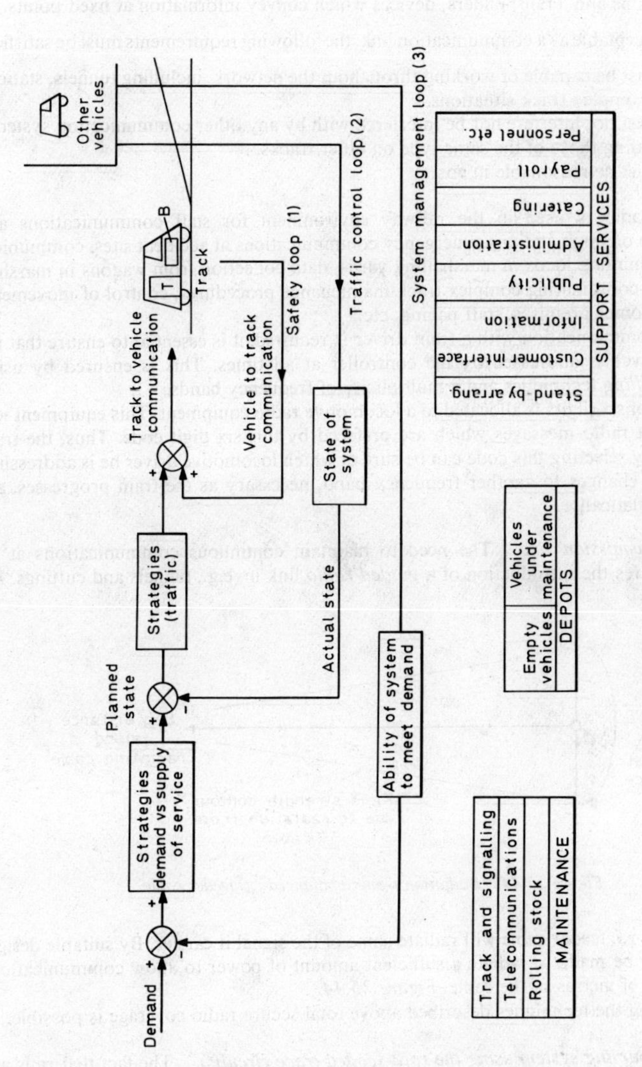

Figure 25.43 Elements of a railway system

(3) Inductive coupling systems using metal rails (coded track circuits).
(4) Inductive coupling system using track conductors.
(5) Beacons and Transponders, devices which convey information at fixed points.

To be acceptable as a communication link, the following requirements must be satisfied:

(1) It must be capable of working throughout the network, including tunnels, stations and complex track situations.
(2) It must not interfere nor be interfered with by any other communication systems, including those of the same type on other tracks.
(3) It must be reasonable in cost.

Radio. Radio is used in the railway environment for staff communications and supervision on large stations, emergency communications at accident sites, communication with shunting locos in marshalling yards, data collection from wagons in marshalling yards, coordinating complex track maintenance procedures, control of movements within customer premises, staff paging, etc.

When communication with a train driver is required, it is essential to ensure that the correct driver is addressed by the controller at all times. This is ensured by using *selective calling* techniques and a multiplicity of frequency bands.

A code of six digits is allocated to a locomotive radio equipment. This equipment will only accept radio messages which are prefaced by this six digit code. Thus, the train regulator by selecting this code can be sure of which locomotive driver he is addressing. Frequency changes to another frequency band, necessary as the train progresses, are made automatically.

Leaky transmission line. The need to maintain continuous communications at all times requires the introduction of a *guided radio* link in, e.g., tunnels and cuttings. An

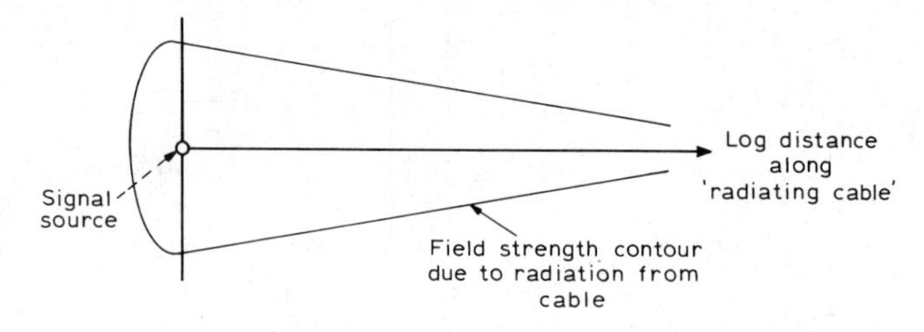

Figure 25.44 Radiation from unbalanced r.f. feeder cable

unbalanced r.f. feeder cable will radiate some of the signal it carries. By suitable design, a cable can be made to radiate a sufficient amount of power to allow communications within tens of metres of the cable, *Figure 25.44*.

By mixing the techniques described above total secure radio coverage is possible.

Inductive coupling system using the rails (*coded trace circuits*). The fact that rails are metallic implies that an electric circuit can be established and therefore, that a modulated signal can be transmitted. The electronic aspects of modern track circuits are described in more detail later in this section (**25**–62).

Coded (modulated) track circuits, the essential elements of which are shown in *Figure 25.45*, have been seen in use for several years in the U.S.A., London Transport and some European railways. This technique is presently limited by the electrical characteristics of the track and track bed.

This limitation makes itself apparent in the maximum length of track that can be

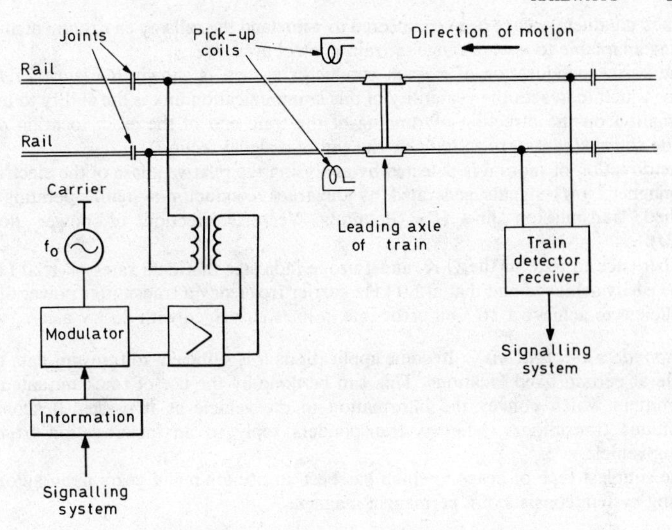

Figure 25.45 Elements of coded track circuits

energised using reasonable power (e.g., under 15 watts), which in the range up to 2 500 Hz is in the order of 1 km.

At frequencies above the mid-audio range, track attenuation becomes excessive and therefore, the information bandwidth of this communication link is limited. The problems arising from electrical interference are described later (**25**–63).

Inductive coupling system using track conductors. Most of the disadvantages of coded track circuits are removed by providing a separate information carrying path. A pair of parallel conductors is placed on the track to form a two-wire transmission line. Track conductors perform well in the frequency range 20 kHz to 150 kHz, and offer the additional advantage of allowing vehicle to track communication.

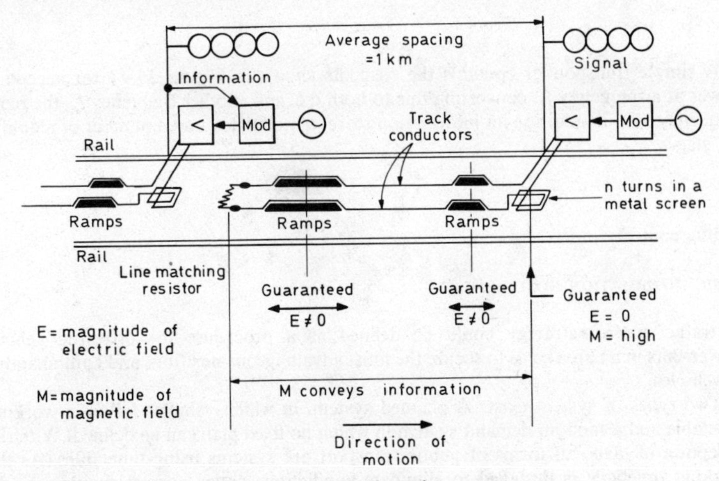

Figure 25.46 Track conductor system

Track conductors have been engineered to withstand the railway environment and are proving adaptable to a wide range of train control systems.

The basic configuration of a track conductor system is shown in *Figure 25.46*. A feature which increases the versatility of this communication link is the ability to impart information on the direction of running of the train and of the exact location of the lineside signals (necessary to indicate the end of a signal section).

The direction of motion is detected by analysing the relative phase of the electric (E) and magnetic (M) signals generated by the track conductor system, operating as a matched transmission line ($P =$ (Poynting Vector, direction of power flow) $= E \times M$).

Experience to date in the U.K. and Europe indicates that data rates of 1 200 bits/s can be easily achieved and that at 60 kHz carrier frequency a transmitter power of 2 W is sufficient to achieve a 10^{-8} bit error rate in lines electrified with 25 kV a.c.

Transponders and beacons. In some applications it is sufficient to transmit data to the vehicle at certain fixed locations. This can be done by the use of track mounted data transmitters which convey the information to the vehicle as it passes. Beacons are continuous transmitters, whereas transponders reply to an interrogation from the passing vehicle.

The simplest type of beacon, which has been in use for many years in an automatic warning system consists of a permanent magnet.

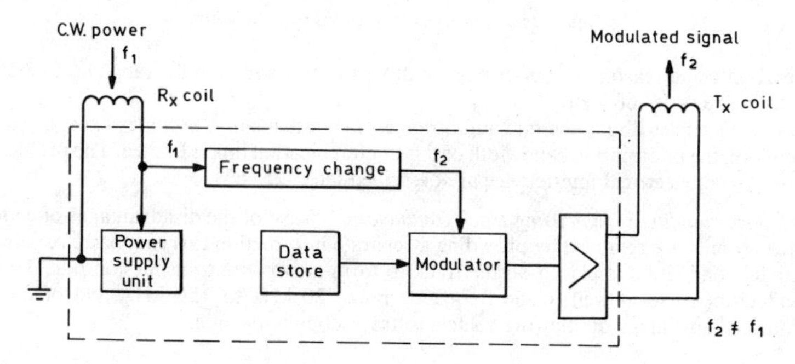

Figure 25.47 Transponder elements

A simple transponder contains the elements shown in *Figure 25.47* for accepting power at a frequency f_1, converting this to both d.c. and another frequency f_2, the reply frequency, and in addition an information store, a modulator and a number of amplifying stages.

Traffic control

BASIC INFORMATION REQUIREMENTS

A traffic control strategy could be defined as a procedure for directing vehicle movements in an area so as to secure the most advantageous positions and combinations of vehicles.

Two types of system exist: A planned system, in which vehicles follow a working timetable and a random-demand system in which no fixed plan can be defined. With the exception of taxis, all forms of public transport are systems using timetables. As the working timetable is designed to eliminate conflictions under normal running conditions, it should therefore be regarded as a control strategy, and given that the system remains undisturbed, there is no need to resort to any other strategy.

Control is effected by the traffic control loop (number *2* in *Figure 25.43*), by comparing the stage of the system (identity, position and performance of each vehicle) with the planned state, and using the 'error signal' thus obtained as an input to a 'strategy' function such as to adhere as closely as possible and to establish conditions such as the working timetable is again the only necessary control strategy. Such a fixed-plan system assumes that in reasonably normal operating conditions, the infrastructure and the vehicles will be able to meet the plan. In the case of an 'on demand' system, assumptions about the nature of the random demand will be necessary prior to system design.

NATURE OF THE PERTURBATIONS AND POSSIBILITIES FOR OPTIMISATION

Perturbations in transport systems manifest themselves as delays. An event may delay initially one vehicle, and this is called a *primary* delay. Reaction of this on other vehicles may give rise to *secondary* delays.

If delay is to be used as a basis for a strategy it must be determined how, that is where and when it is to be measured. One way of achieving this is to choose sampling points and determining the planned passing time. In general the sampling points can be chosen in the vicinity of those points in the network where it is more important to minimise delays, i.e., junctions, stations and terminals. A large number of constraints limits the optimisation possibilities, and only the most important are mentioned here:

(1) Safety: no vehicle can be allowed to operate faster than it is safe to do so.
(2) Passenger comfort: rates of acceleration and braking must be limited.
(3) Reliability of services: when working to a plan, vehicles cannot be allowed to depart earlier than their booked time nor to alter their routes too significantly.

In the case of railways, the use of 'train describers' has proved an invaluable tool for traffic control.

In the case of dense operations, the stability of the control loop is just as important as the system capacity, and from a review of the published material considerably more attention is being given to track capacity than to system stability.

System management

Although the control loops described are essential parts of any transport system, their operation in the real world depends on a higher order control loop of 'system management', the main functions of which are:

(1) To allocate vehicles and staff to specific duties and make the necessary alterations when failures or maintenance so require.
(2) To inform and advise when the system is unable to meet the demand.
(3) To monitor the location and status of all vehicles in the system and to supply the appropriate information on demand.
(4) To assist in long-term planning, by advising how many vehicles will be required to meet the estimated demand, taking into account vehicle utilisation and market response to inadequate ability to meet demand.

In any practical system, economic considerations require high vehicle utilisation as this reduces capital investment, overall energy consumption, 'parking' place when not in use, maintenance, etc.

This is an appropriate point to return to the subject of the 'on demand' services previously mentioned, since the traffic control function and the system management function are very much closer integrated than in the case of 'planned' services.

On demand transport systems are not entirely new, since all the freight consignments under 20 ton which are unsuitable for containerisation carried by the railways, are generated more or less at random at all points in the network for transport to other random destinations.

The traditional solution of collecting sufficient such consignments to form an 'economic length' train through a number of classification stages results in both poor vehicle utilisation and long transit times.

Both should be significantly improved since the commissioning in 1975 of a system already well tried in the U.S.A. called TOPS (see **25**–75). The functions of TOPS are precisely those listed earlier, and it has potential for further development, e.g., by including vehicle maintenance records to the information it contains.

Applications to railway signalling

ELECTRONIC TRACK CIRCUITS

Track circuits have been used for many years for the purpose of detecting whether or not a portion of track is occupied by a railway vehicle.

In their simplest form, they consist of a d.c. circuit, which is closed by the axles of a vehicle, *Figure 25.48*.

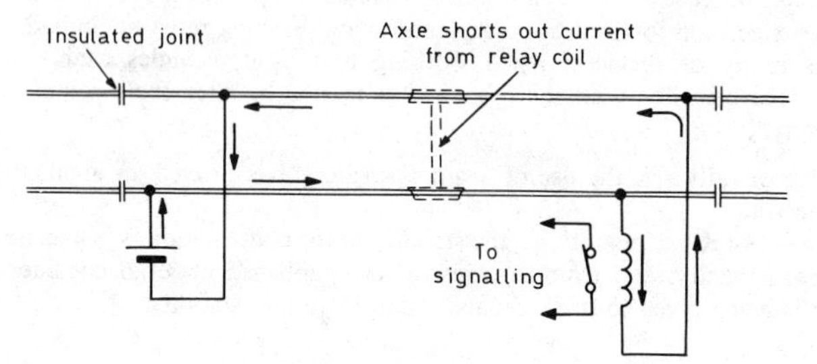

Figure 25.48 Electronic track circuit

The use of d.c. or low frequency a.c. necessitates the use of insulated joints between signal sections in order to detect the presence of a vehicle on an identifiable position of the track.

Today's high speeds, impose stringent demands upon rail joints and continuously welded rails not only overcome this problem but also present additional advantages.

The problem of train detection without joints led to two separate developments: jointless track circuits and axle counters.

Jointless track circuits. The removal of the rail joint implies that the other means of defining the 'ends' of a signal section have to be found. A number of successful techniques are in use, but for the sake of illustration, the principle of resonating a portion of the track at the signal frequency is described:

Figure 25.49 illustrates how a tuned circuit can be built by joining the two rails with coils and capacitors to achieve resonance. To differentiate between adjacent sections, different frequencies are used. These frequencies are sufficiently separated so that a circuit tuned to f_1 does not affect signals of frequency f_2 and vice-versa.

These techniques can be applied without difficulty to non-electrified railways. In the case of electric railways, where traction return currents flow partly through the running rails, interference may occur between a harmonic of the supply frequency and the track circuit frequency.

Although in principle, it could be possible to choose track circuit frequencies that

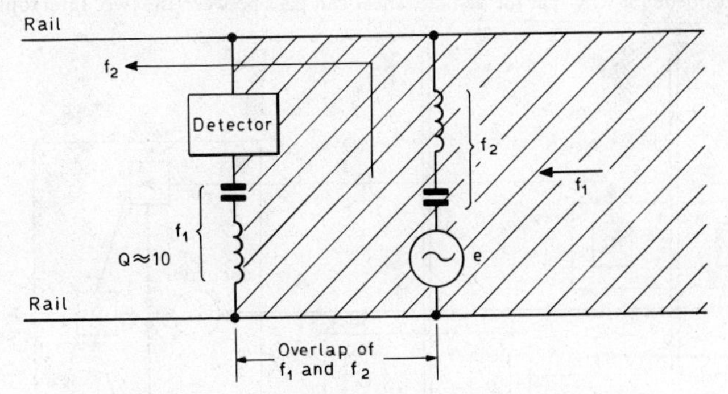

Figure 25.49 Use of tuned circuit in jointless track circuit

do not coincide with traction harmonics, in practice, fluctuations of the supply frequency render this approach useless at frequencies of approximately 1 200 Hz and above.

Coded jointless track circuits. In addition to the coded track circuits previously described a number of systems have been tried, in which the track circuit signal is frequency modulated in order to differentiate it from the traction return current harmonics.

Application of short track circuits to the detection of wheel flats. If a brake application causes a wheel to lock, it will slide on the rail and develop a flat segment generally called a *wheel flat*, which could, if left unattended cause damage to both track and train.

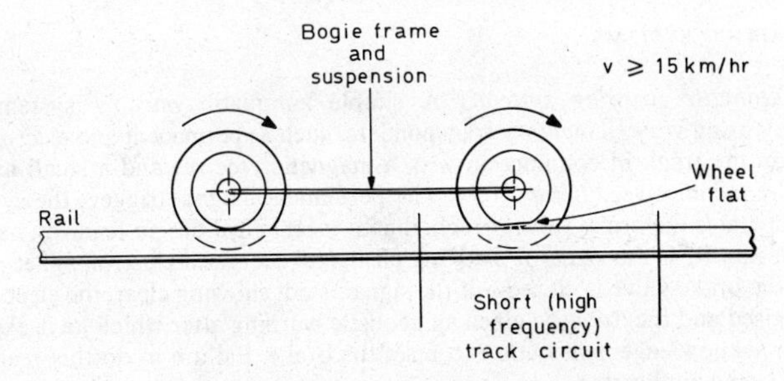

Figure 25.50 Method of detecting wheel flats

Figure 25.50 shows how in a bogie vehicle a wheel flat, at running speeds momentarily loses contact with the rail. If a short track circuit, short enough to recognise a single wheel is installed, an interruption of current will indicate the presence of a wheel flat. Furthermore, such a short track circuit will also enable the number of axles passing over it to be counted (axle counter).

AXLE COUNTERS AND HOT BOX DETECTORS

An axle counter, *Figure 25.51*, does not have to rely on direct electrical interaction with the rails but instead it electrically interacts with the metallic wheels of vehicles passing next to it.

A magnetic circuit, consisting of a transmitter with an a.c. signal and a receiving coil is placed in such a way that the vehicle wheel can pass between the two. Interruption of

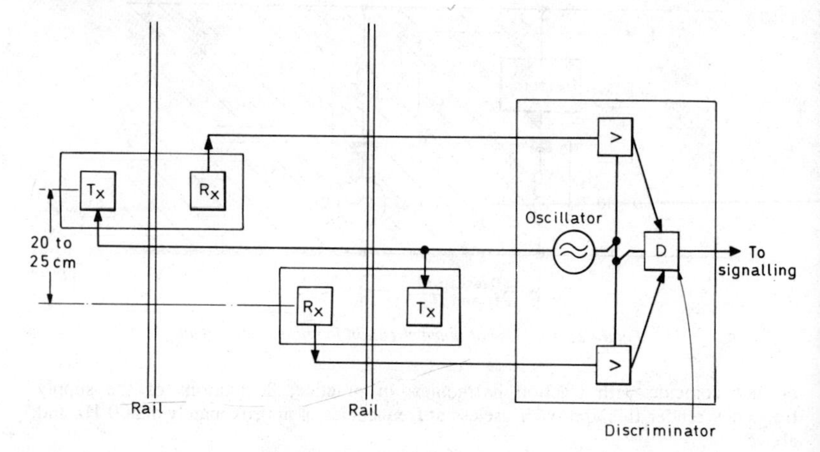

Figure 25.51 Axle counter system

the magnetic circuit indicates the presence of an axle. By using two such units staggered, it becomes possible to identify the direction of motion of the vehicle.

By adding an infra-red detector and a threshold circuit to an axle counter it becomes possible to detect whether any axle bearing exceeds the permitted temperature, which is in itself an indication of the onset of bearing failure. Axle counters cannot be used—as track circuits are—to detect broken rails.

CAB SIGNALLING SYSTEMS

Simple automatic warning system. A simple automatic warning system can be obtained by using very elementary transponders, such as permanent and electromagnets mounted on the track in conjunction with a magnetic receiver and a small amount of logic on the locomotive, *Figure 25.52*. The permanent magnet triggers the system and the logic operates the brake valve, which, due to mechanical design features, has a small delay built into it. If the signal is showing clear, the energised electromagnet resets the logic and the brake valve is closed—if the signal is not showing clear, the electromagnet is de-energised and the driver is given an acoustic warning after which he is expected to operate an acknowledgement button to reset the brake. Failure to do this results in an automatic brake application.

The system is also 'fail safe' since no failure in the equipment can result in an indication of a clear signal.

Full cab signalling system. Since colour light signals may show indications other than *clear* or *stop* it may, in certain cases be desirable to extend the simple cab signalling system described to indicate to the driver the exact aspect he is approaching. One

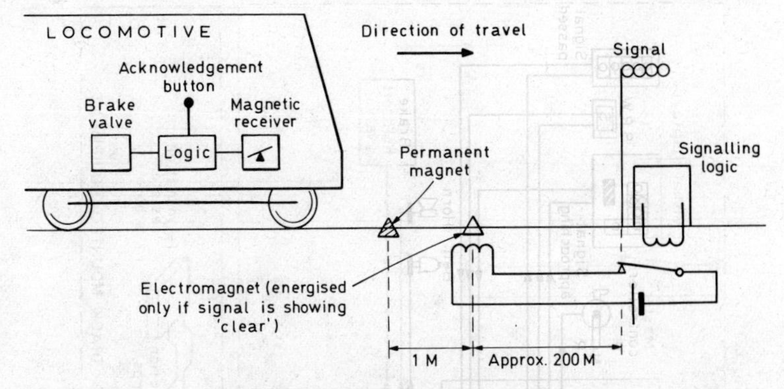

Figure 25.52 Automatic warning system

possible solution has undergone service trials in the Southern Region of British Railways. It uses as a supplement to the permanent magnet, a track conductor loop, some 250 m long (see **25**–66).

The information used to drive the lineside signals is also used to select two (out of four) audio tones generated by vibrating reeds and these tones in turn, modulate in frequency a signal of 45 kHz. A 15 W signal is applied to the track conductors and on the train electronic and relay equipment is used to decode and display the information as well as control the brake valve to safeguard against human error.

The main elements of this system are shown in *Figure 25.53*. Two attractive features of this system are firstly, its potential for expansion attributable to the high information capacity of the communication link and second that it is an overlay system, in other words, a failure in the cab signalling system does not degrade in any way the operation of the signalling system itself.

TELECOMMUNICATIONS

In order to communicate control signals from main centres (power signal boxes) to remote outstations (local interlocking situations) and to transmit monitor information from these outstations back to the main centres, use is made of telemetry equipment.

Both time division multiplex (t.d.m.) and frequency division multiplex (f.d.m.) systems are used in order to carry many channels on single wire pairs.

T.D.M. This form of multiplexing necessitates allocating each individual information source a period of time. During this time period, information relating to that information source is communicated along the signal circuit at carrier frequency. The signal has use of the full available carrier bandwidth for this time. Clearly, the amount of time allocated to an individual information channel is a function of the number of channels to be carried by the system and the required intervals between updates of the information. Typical t.d.m. systems in railway signalling use have time allocations per channel of 50 ms hence for a 10-channel system this would mean an update time of 0.5 s. Carrier frequencies are typically 5–30 kHz and are allocated in accordance with C.C.I.T.T. (International Telegraph and Telephone Consultative Committee) recommendation.

F.D.M. If a band of frequencies is divided up into smaller bands, these smaller bands may be used to convey information relating to control signals or monitor status. Such

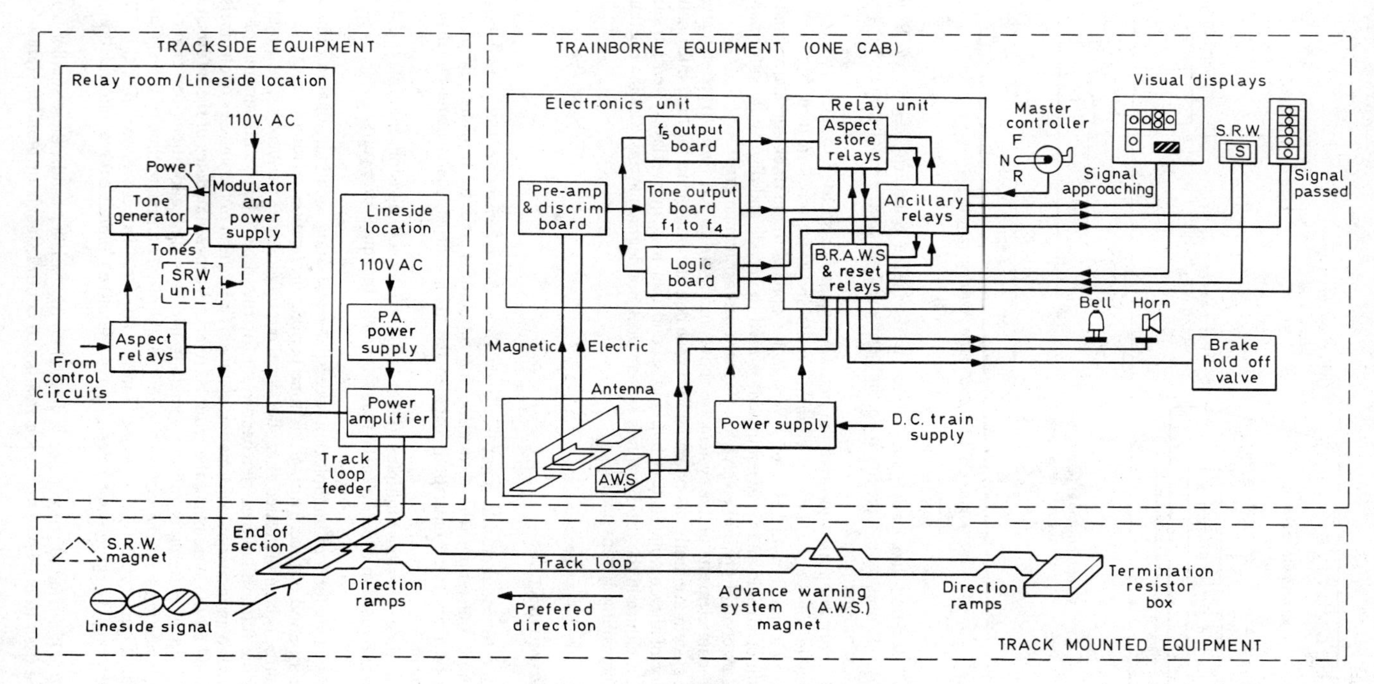

Figure 25.53 Full cab signalling system

systems are used on the railways typically occupying some 400 Hz of bandwidth to carry up to 75 information channels.

Computer based train describer. A standard component in the modern signalbox, the train describer is a system by means of which the train identity or *headcode*, entered at the fringe area of the control zone either by the train describer of the adjacent control zone, or by hand where this latter train describer has not yet been fitted, is displayed on a control panel together with an indication of the train position and its progress, supplied by the track circuits used for train detection.

Current train describer technology uses a fairly simple computer installation with back up facilities to ensure adequate system availability.

The displays most commonly used are miniature cathode-ray tubes with a screen size of 25 × 50 mm and four characters are displayed, but desk top video display units are increasingly being used to present information other than headcodes.

From the electronic engineer's point of view, the installation of computing equipment in a signalbox full of relays and often very close to electric traction supplies, demands great care in fitting all the inputs to the interface between signalling and train describers, as well as in some cases the use of Faraday cages.

Apart from presenting an instant picture of railway traffic at all times the computer based train describer is an important tool in railway operations, since it can also be used for: train reporting; train reporting by exception (those trains not running to plan); automatic route selection; automatic control of platform indicators, and pre-recorded public address.

Where sufficient storage capacity exists, e.g., by means of magnetic discs, it will be possible to include the real time analysis of operation through junctions and complex terminal stations. By programming a set of traffic priority rules for the various types of trains in use under different perturbation conditions it would be possible to optimise platform allocation and train operations through junctions.

Electronics in traction equipment

Judicious replacement of electro-mechanical components used in the control of traction equipment by electronic circuits can lead to improved equipment performance and reliability. It is of course essential that the electronic circuits thus used are designed and 'protected' with a full knowledge of the electrical and mechanical environment within which they are required to function, for this potential performance and reliability advantage to be gained.

In some areas of application, the use of electronics enables the designer to offer significant energy savings to the operators.

THYRISTOR CONTROL OF D.C. TRACTION MOTORS, THE 'CHOPPER' CIRCUIT

The thyristor chopper, *Figure 25.54*, replaces the starting resistor network used in conventional equipment with the advantages that:

(1) Energy is saved—no longer dissipated by the starting resistors.
(2) Smooth acceleration is achieved with minimal wheel-slip.

The thyristor chopper is usually switched at a fixed frequency (of the order of 200 Hz is typical) and the on–off ratio sets the average current level through the traction motor. Electronic control of this on–off ratio takes account of maximum permitted acceleration rates for passenger comfort, and can include automatic compensation for wheel-slip.

A further benefit can be obtained from the chopper when it is used in a 'regenerative braking' mode, *Figure 25.55*. Here the thyristor is arranged in the circuit such as to

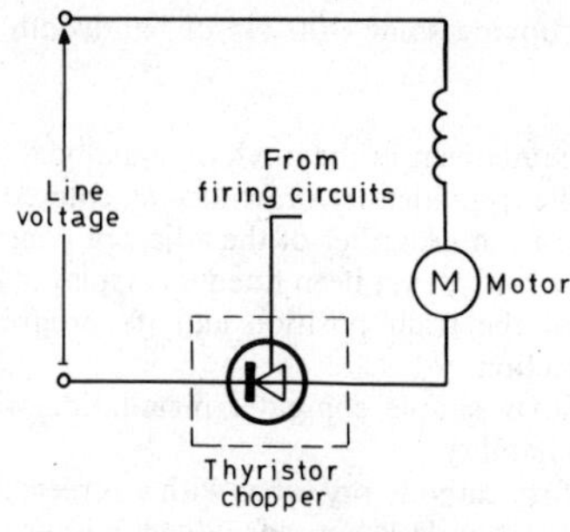

Figure 25.54 Thyristor chopper circuit controlling traction motor

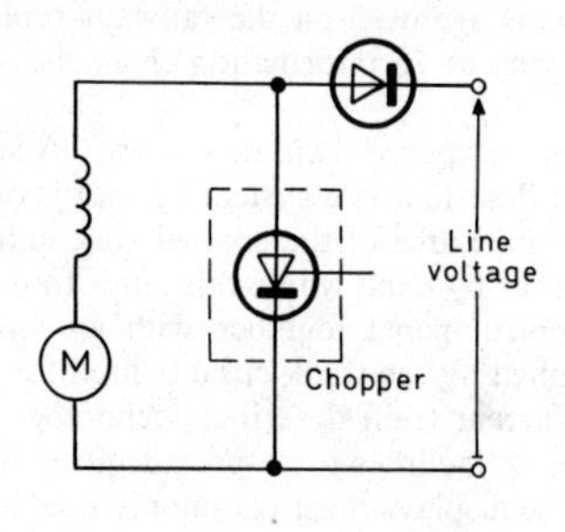

Figure 25.55 Thyristor chopper circuit in 'regenerative braking' mode

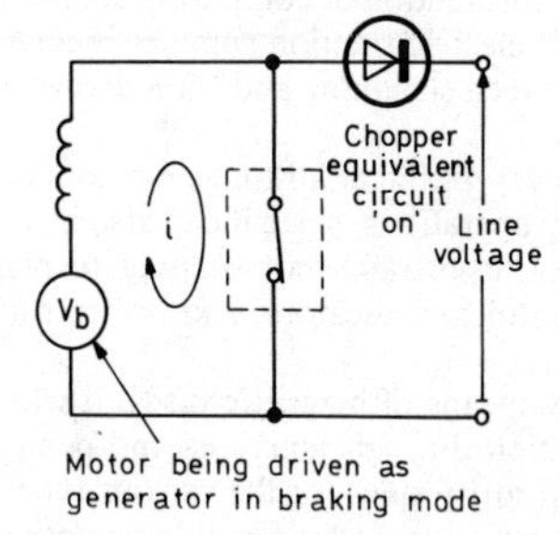

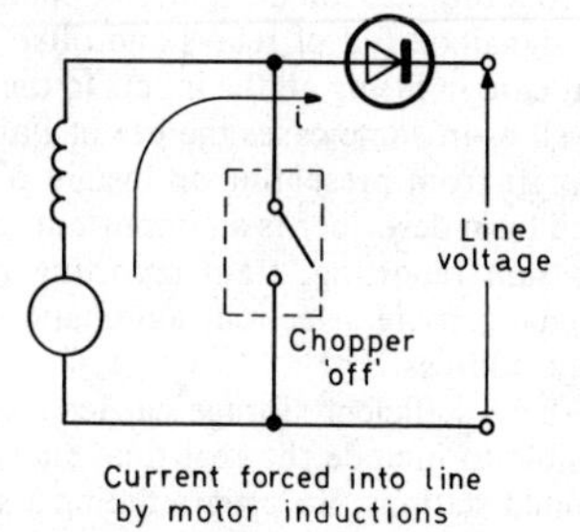

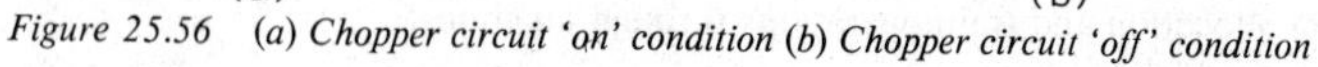

Figure 25.56 (a) Chopper circuit 'on' condition (b) Chopper circuit 'off' condition

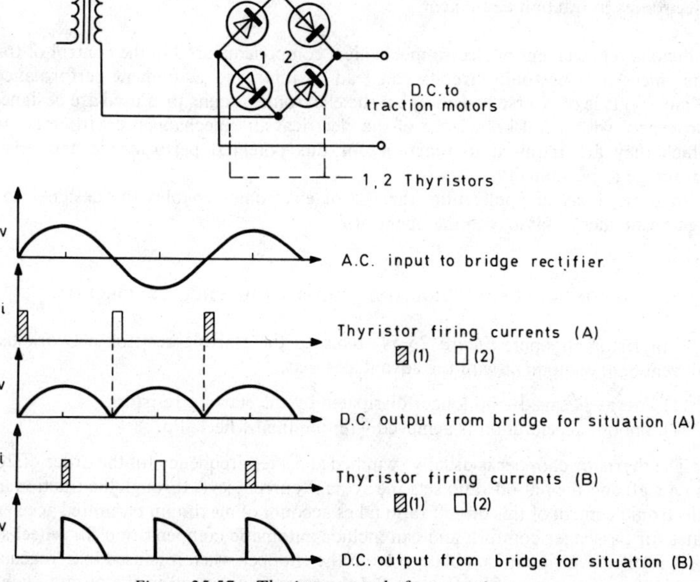

Figure 25.57 Thyristor control of a.c. traction

return electrical energy generated by the traction equipment during braking to the power lines. In the braking mode, *Figure 25.56*, the chopper, when 'on', shorts out the motor thus causing an increase in the current. Shutting the chopper 'off' causes the motor circuit inductance to drive this current back into the supply system.

Logic to control the thyristor firing circuits typically utilises integrated circuit techniques to minimise circuit complexity and to improve reliability.

THYRISTOR CONTROL OF A.C. TRACTION

Typical electromechanical control for a.c. traction involves transformer tap-changing. This however leads to poor utilisation of available adhesion between wheel and rail owing to the step changes in tractive effort applied when tap changes are made.

Thyristor control, *Figure 25.57*, enables 'stepless' control to be achieved. Clearly, the switch-on point of the thyristors can be determined by the controller—or driver—hence smooth control can be exercised over the full range of available power. Rate of application of power can also be easily controlled by logic circuitry, together with automatic control of wheel-slip—by load sensors feeding back information to the logic circuitry.

THYRISTOR INVERTERS

The advent of thyristors and the associated control electronics has led to widespread application on the railways of solid-state inverters for lighting purposes (in particular, fluorescent illumination). In addition to this, some work has been done on the application of this technique to the traction area. In particular, the use of solid state equipment for the conversion of single phase fixed frequency supply to multiphase variable frequency supply for induction motor driving, *Figure 25.58*. Some advantages are claimed for the use of induction motors in the traction application.

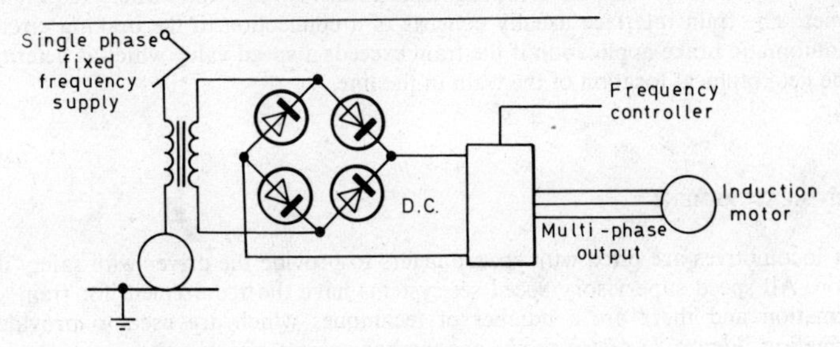

Figure 25.58 Single-phase frequency conversion to multi-phase variable frequency supply

Electronics in traction

PROCESSING EQUIPMENT FOR CAB SIGNALLING AND SPEED CONTROL/SUPERVISION SYSTEMS

It is considered necessary in some railway applications to interface the signals received from cab signalling equipment and speed supervisory equipment, with the train controls. In the case of cab signalling systems this may take the form of a simple relay interface with the brake circuit. Logic connected with this relay determines whether or not a brake application shall be made depending upon the signal aspect received and the acknowledgement made by the driver (see **25**–65).

A more sophisticated system, *Figure 25.59*, employs a binary-coded converter which accepts a binary coded demand from a logic circuit and converts it into a brake-pressure demand. Such control may be necessary when speed-set systems are employed. Here, the desired speed of running is selected by a locomotive driver and a servo control loop ensures that the train achieves and maintains that speed.

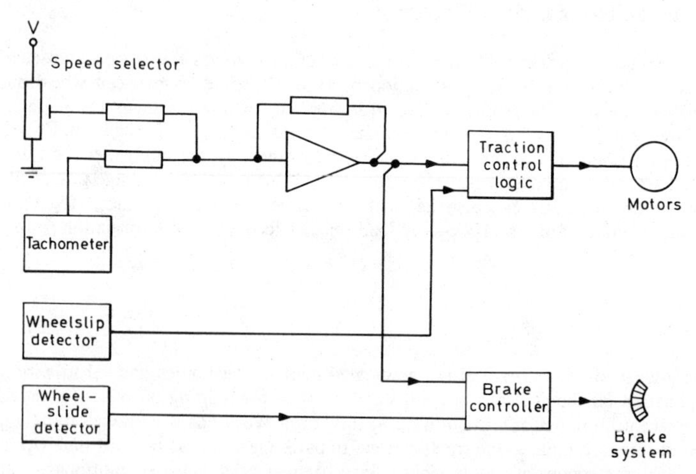

Figure 25.59 Cab signalling and speed control supervision system

Speed supervision exists on some high-speed railways to ensure safety requirements are met. The train interface usually consists of a connection to the braking circuit to give automatic brake application if the train exceeds a speed value which is determined by the geographical location of the train in the line.

SPEED MEASUREMENT

Most locomotives are fitted with speedometers to provide the driver with safety information. All speed supervisory/speed set systems have the requirement for train speed information and there are a number of techniques which are used to provide the information. *Figure 25.60* (*a*) shows one method using a simple voltage generator, and *Figure 25.60* (*b*) a more elaborate toothed wheel, variable reluctance generator technique. Both have the disadvantage that adjustments are necessary as the train wheel diameter changes through wear or replacement.

Doppler radar. Some work has been done to evaluate Doppler radar speed measurement in the railway application and equipment has been built which provides speed measurement equipment which is quite independent of wheel diameter and wheel slip/slide effects.

Noise—correlation type speed measurement. A novel technique for speed measurement involves a correlation technique, continuous comparison of noise patterns from reflected electromagnetic energy. By correlating reflections from physically displaced points, the time delay between the two points is deduced, and hence the speed of passage of the train between the points.

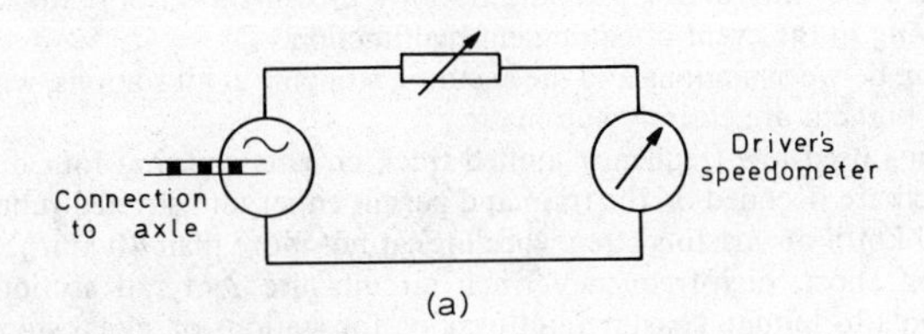

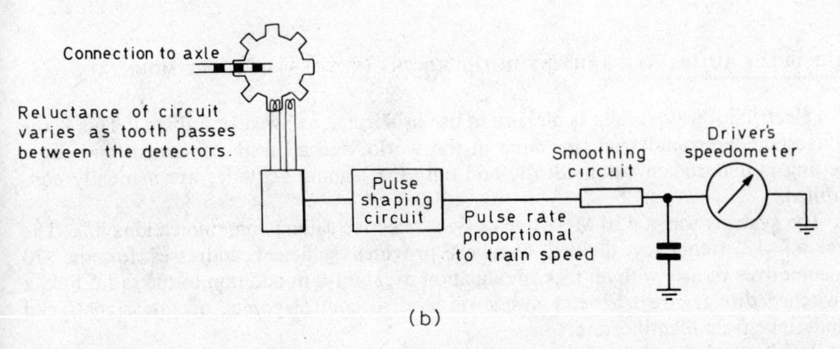

Figure 25.60 (a) Speedometer using voltage generator (b) Speedometer using toothed wheel variable reluctance generator

OTHER TRACTION APPLICATIONS

There are many other applications of electronics in railway traction, some of which are listed below

Automatic voltage regulations, and load regulators for diesel electric traction.
Driver vigilance devices.
Field weakening control circuiting.
Stored programme public address.
Contactor timers.
Moving map systems.
Slip and slide detectors.

Automatic train operation—(ATO)

ATO integrates some of the systems described in applications of electronics to signalling and to traction. Railways lend themselves by their guided nature and controllability, to the application of automation techniques. Two established systems are described to illustrate the wide range of possible approaches:

VICTORIA LINE (LONDON TRANSPORT)

In operation since 1960 the Victoria Line operates almost in a fully automated mode. The functions of the still present train attendant are to drive in and out from the depot, open and close the train doors, start the train by pressing a set of buttons and, to take over the driving in the event of equipment malfunction.

The driving between stations and the accurate stopping at all stations, with a stopping accuracy of ± 60 cm are entirely automatic.

The systems used low frequency jointed track circuits coded at four different rates, three of which are decoded on the train and permit either unrestricted running, running limited to 40 km/h or coasting (free wheeling) at not more than 40 km/h.

In addition short, high-frequency track circuits use 3-m rail sections known as command spots to initiate coasting and braking for stations or signal stops. A number of command stops placed at short distances on the approach side of a station permit a fine speed control leading to accurate stopping.

THE FULLY AUTOMATED RAILWAY IN THE KIRUNA (SWEDEN) IRON ORE MOUNTAIN

An electric-railway system is already in use in Kiruna, in Swedish Lapland. This is the longest underground iron ore mine in the world, and all train movement and route setting is undertaken automatically and only the loading activities are manually controlled.

The systems was a 420 MHz line of sight selective calling communications link. The use of 12 frequency divided channels provides sufficient addresses for the 20 locomotives in use, with an f.s.k. modulation capability. In addition to the radio link, a switched direct wire telemetry system is used to control points, operate signals and undertake train identification.

Problems of signal attenuation in the wire are overcome by using several transmitters. To avoid interference adjacent transmitters operate at slightly different transmission frequencies and automatic diversity switching units on the locomotives select the best signal at any time.

The safety of train movements is achieved by standard track circuits using relays.

Example of systems combining a number of electronic techniques

AUTOMATION IN A MARSHALLING YARD

The problem. Marshalling yards are installations where freight wagons arrive in train form from an origin A to be classified and reformed into new trains with destinations X, Y, Z, etc. *Figure 25.61* shows a typical layout for such an installation.

Incoming wagons are inspected (visually and with hot axlebox detectors), their numbers are read, the wagons are weighed (strain gauge bridges) and, after decoupling pushed over a hump and directed to the classification yard appropriate to their destination by conventional route controls (points machines).

Ideally, the wagon should roll down the hump with a speed such that it will stop next to the previous wagon on the track. There is a practical constraint to the speed with which wagons are allowed to collide, which in order to avoid damage to consignments is only 1 m/s. Since all wagons have different loads, rolling resistance and external factors such as wind speed and direction are uncontrollable, a number of systems has been developed to achieve this result.

Systems. There are three main variables determining the contact speed of a wagon with the previous wagon sent on that track: The distance to run $l_T - l_w$ (the length of free track minus the length already occupied), the wagon weight and its rolling resistance, the latter one determined by its own dynamics, wind, track curves, etc. Since the rolling resistance is the most complex one to establish the speed control problem is divided into two.

Devices called *retarders* usually consisting of friction arrangements making contact with the wheels are placed in two different places: one at the end of the hump portion and the other after the track work on the straight portions of each classification track.

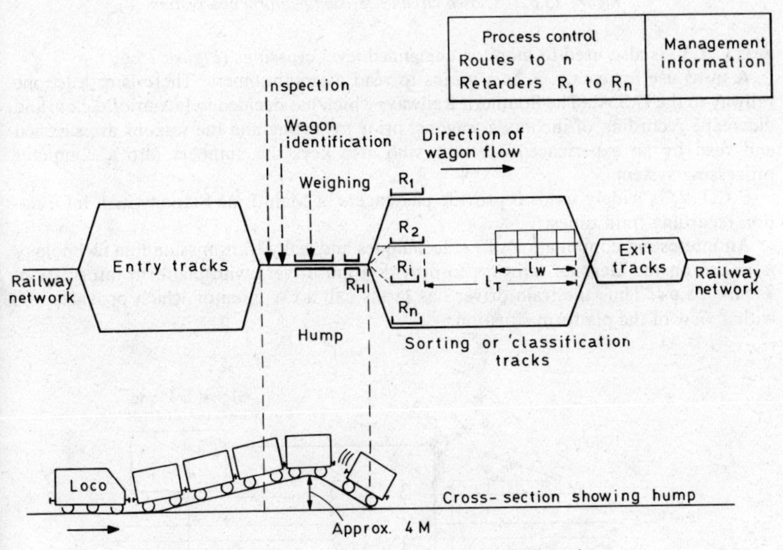

Figure 25.61 *Basic layout—marshalling yard*

In these, friction is applied until the wagon speed, measured for example by a Doppler radar system or by the transit time between two sensors reached a predetermined value. In practice, the speed at the end of the first retarder R_H is constant, e.g., 4 m/s and the second retarders are controlled by a minicomputer by using a formula in which the rolling resistance of the vehicle is estimated from its performance at the first retarder.

Systems of this nature have been in operation in the U.S.A. and Europe for a number of years with a regularity of performance superior to that of an experienced operator.

CLOSED CIRCUIT TELEVISION (C.C.T.V.)

There are a number of ways in which c.c.t.v. is used in several railway administrations. In London Transport's Victoria Line a layout, *Figure 25.62* is used to show the train attendant whether the doors can be closed. Other screens are placed in the central control room, where the controller can select, by means of a switch any of the platforms in the system for observation.

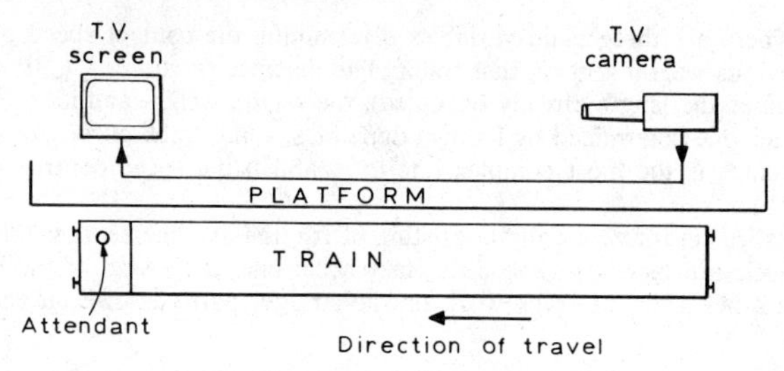

Figure 25.62 Closed circuit television platform observation

C.C.T.V. is also used to monitor unmanned level crossings (*Figure 25.63*).

A third use for c.c.t.v. is as a means to read wagon numbers. There is at least one railway in the U.S.A. (The Southern Railway) which has decided in favour of c.c.t.v. and videotape recording of incoming wagons, prior to sorting and the wagons are scanned and read by an experienced operator who also keys the numbers into a computer processing system.

C.C.T.V. is widely used to provide passengers at Main-Line Stations with information regarding train times.

An interesting combination of t.v. techniques and leaky transmission line technology is in use on the German railways to provide train drivers with platform information, *Figure 25.64*. Thus the train driver has in his cab a t.v. monitor which provides him with a view of the platform situation.

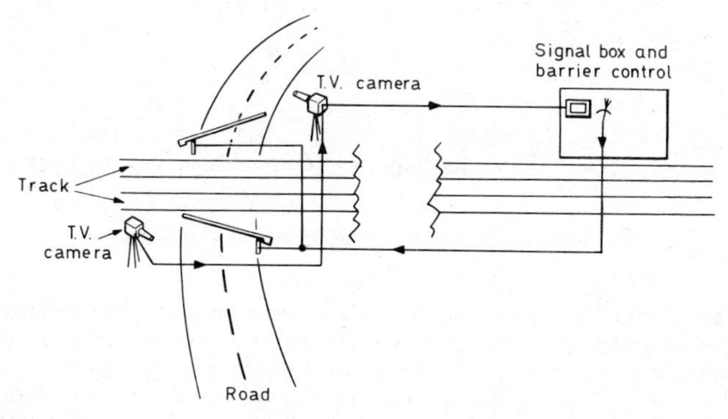

Figure 25.63 Closed circuit television—observation of unmanned level crossing

AUTOMATIC WAGON IDENTIFICATION

In order to permit sorting operations in the movement of freight, each freight wagon, and even each container is given a number—and in British Railways there are in excess of 300 000 wagons. To improve both the speed and accuracy of this reading, an automatic wagon identification system is desirable. Since on a multinetwork basis such as in the U.S.A. or Europe a much larger number of wagons needs to be handled, high information capacity, high reliability of reading, high equipment reliability and low cost per wagon are essential requirements.

There is universal agreement that any practical system must use a passive device for the numbering of wagons and a diversity of approaches is possible ranging from

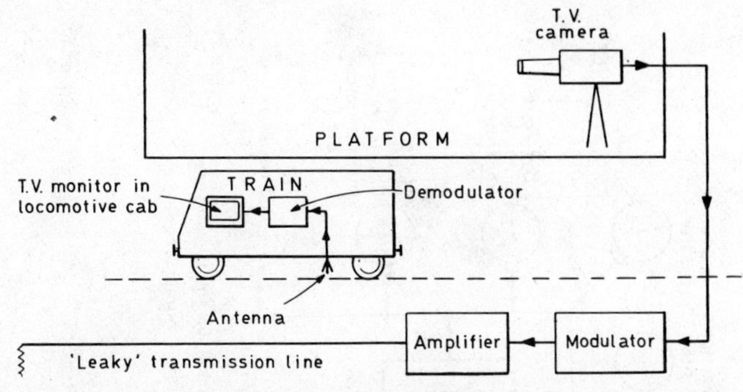

Figure 25.64 *Method of providing train drivers with platform information using t.v. techniques*

coloured labels, using white, blue and red segments in a number of permissible combinations to devices which operate by modifying and reflecting microwave energy.

The *transponder* is attached to the wagon to be interrogated and read at scanning points, usually on the approach to sorting yards, *Figure 25.65*.

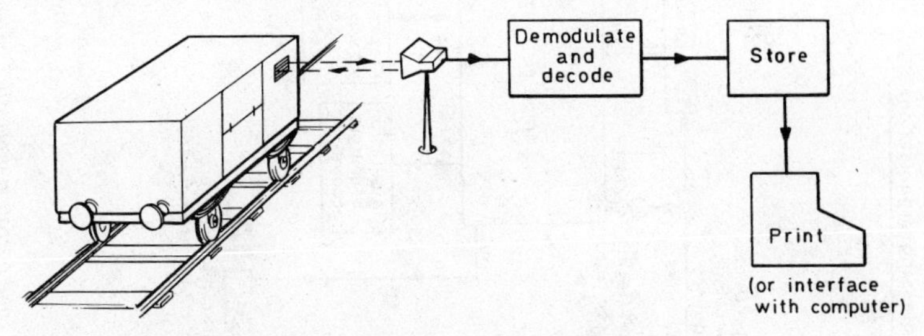

Figure 25.65 *Reading system for wagon identification*

TOTAL OPERATION PROCESSING SYSTEM—TOPS

Currently in use in the U.S.A. and in the U.K., this system provides an on-line service to managers covering the following aspects of railway operations:

(1) Information regarding location of rolling stock.
(2) Information regarding location of motive power.
(3) Maintenance record and progressing for the above.
(4) Crew rostering.
(5) Waybill information (wagon load identity + transportation details).
(6) Train consist (composition of train).
(7) Fault reporting plus likely delays.

The system, illustrated in *Figure 25.66*, has up to date knowledge of all features of the actual operations enabling the actual operation to be monitored on a real time basis by the TOPS network.

Freight managers have at their elbows a database which enables them to offer a reliable service to customers and enables the customers to receive accurate reports from the freight managers on the location of their loads.

Use of rolling stock is optimised since wagons can be directed to areas of demand. Marshalling yards can be automatically prepared for train sorting operations; maintenance areas have accurate schedules for attending to rolling stock and motive power.

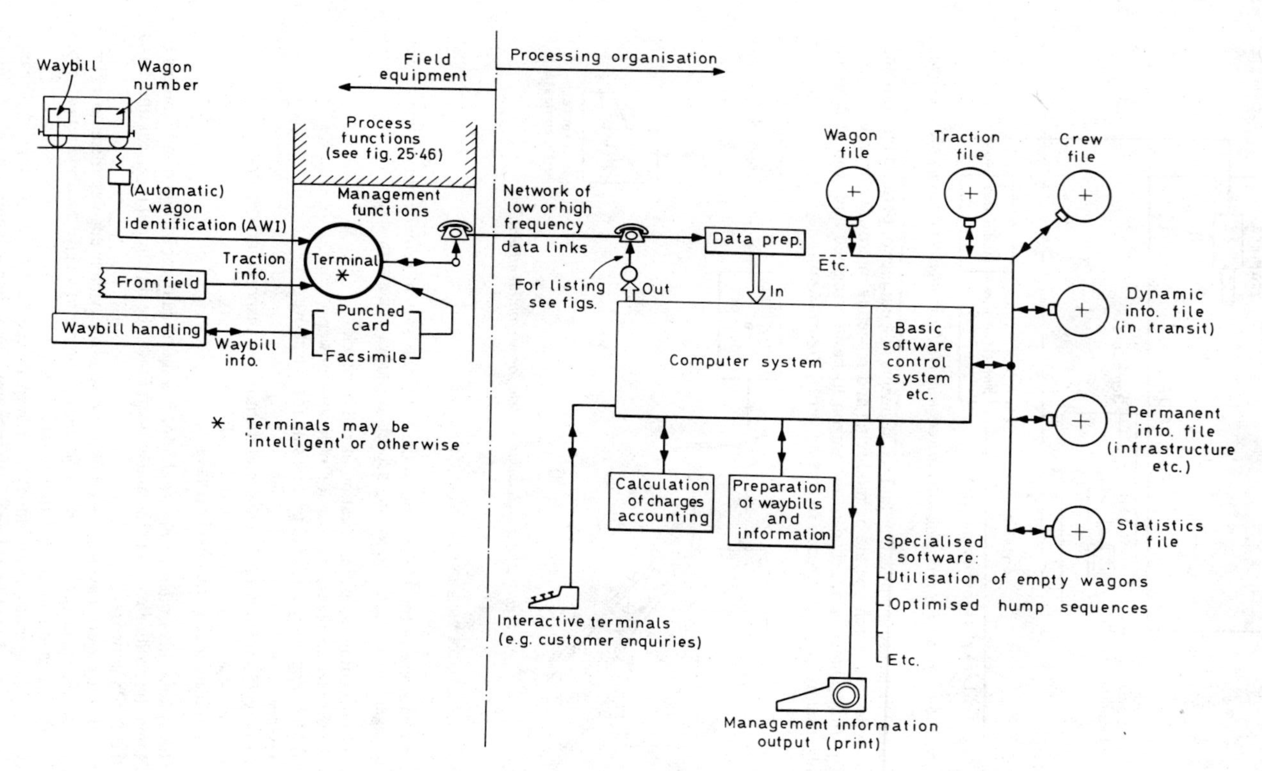

Figure 25.66 Total operation processing system (TOPS)

An embodiment of TOPS operating in the U.S.A. uses two IBM 370/165 computers —one of which is continually on-line, the other acting as a standby system. Basic data, up dated continuously from field location, is held on file at the computer centre. The reliability of the data on file is ensured by a sophisticated monitoring system.

The data can not only be used to control movements through the Railway system, but also to forecast demands for wagons and motive power thus enabling the Management to plan distribution to meet likely demands.

Financial control—invoicing customers is a facility available to the freight management also.

26 PARTICLE ACCELERATORS FOR NUCLEAR RESEARCH

PARTICLE ACCELERATORS FOR NUCLEAR RESEARCH 26

26 PARTICLE ACCELERATORS FOR NUCLEAR RESEARCH

Modern accelerators produce high-energy beams of electrons, ions, X-rays, neutrons or mesons for nuclear research, X-ray therapy, electron irradiation and industrial radiography. If a particle of charge e is accelerated between electrodes of potential difference V it acquires a kinetic energy eV electron-volts (1 MeV = 1.6×10^{-13} J). Accelerators are classified as *Direct*, in which the full accelerating voltage is applied between the two electrodes; *Indirect*, in which the particles travel in circular orbits and cyclically traverse a region of electric or magnetic field, gaining energy in each revolution; and *Linear*, in which the particles travel along a straight path, arriving at correct phase at gaps in the structure having high-frequency excitation, or move in step with a travelling electromagnetic wave.

DIRECT ACCELERATORS

The Cockcroft–Walton Multiplier circuit has two banks of series capacitors, alternately connected by rectifiers acting as change-over switches according to the output polarity of the energising transformer. The upper limit of energy, about 2 MeV, is set by insulation. A typical target current is 100 μA.

The Van de Graaff electrostatic generator is capable of generating a direct potential of up to about 8 MV of either polarity. It has an endless insulating belt on to which charge is sprayed from 'spray-set' needle-points at about 50 kV. The charge is carried upwards to the interior of the h.v. electrode, a metal sphere, to which it is transferred by means of a second spray set. H.V. insulation difficulties are overcome by operating the equipment in a tank filled with a high-pressure gas, e.g. nitrogen-freon mixture at 1 500 kN/m^2. In two-stage Van de Graaff generators for higher energies, negative hydrogen ions are accelerated from earth potential to 6 MeV, then fired into a thin beryllium foil 'stripper' which removes the electrons from the outer shells of the atom, and leaves the remanent ions moving on with little change of energy but with a positive charge. The second stage brings these ions back to earth potential and the total energy gain is 12 MeV. To bring ions on to a small target the accelerating and deflecting fields must be accurately controlled, and scattering limited by evacuating the accelerator tubes to very low pressure. The energies are sufficient for the study of nuclear reactions with the heaviest elements.

INDIRECT (ORBITAL) ACCELERATORS

These may have orbits of approximately constant radius with a changing magnetic field (betatrons and synchrotrons) or orbits consisting of a series of arcs of circles of discrete and increasing radii in a constant magnetic field (cyclotrons and microtrons).

Betatron

The betatron is unique, in that the magnetic field not only directs particles into circular orbits but also accelerates them. The magnet has an alternating field of which only one quarter period is used. Electrons are accelerated in an evacuated toroidal chamber between the poles of the magnet. They are injected at an energy corresponding to a low magnetic field, which bends them on circular orbits round the toroid. A cross-section of the poles and vacuum chamber is shown in *Figure 26.1*. As the magnetic flux through an electron orbit increases during the cycle of alternation, the electron experiences a tangential force, and its gain in energy per revolution is the voltage that would be induced in a loop of wire in the orbit. As the electron gains energy, the magnetic guide field intensity at the orbit increases at a suitable rate. To keep the electron on a constant radius from injection to peak energy requires the rate of change of intensity at the orbit to be half that of the mean flux per unit area within the orbit. At peak energy (or earlier) the electrons are caused to move away from their equilibrium orbit and strike a target inside the vacuum chamber, producing X-rays or corresponding energy. The output consists of short pulses of radiation whose repetition rate is the frequency of the magnet excitation. Energy limitations are set by the size and cost of the magnet and the radiation loss when a high-energy electron has circular motion.

Synchrotron

This uses an annular magnetic guide field which increases as the particles gain energy, as in the betatron, so that they maintain a constant orbit radius. Electrons are initially accelerated by the action of central 'betatron bars' which saturate when the main magnetic field corresponds to an energy of 2–3 MeV when electrons travel at a velocity only 1–2 per cent less than the velocity of light. Further gain of energy is produced by r.f. power at the frequency of orbital rotation (or a multiple of it), fed to resonators inside the vacuum chamber. The particles become bunched in their orbits so that they pass across the accelerating gap in the resonator at the correct phase of the r.f. field. The limitation on electron acceleration is now mainly set by radiation losses due to circular motion.

Protons are injected at about 500 keV, which produces a velocity of only 3 per cent of that of light. Further acceleration changes the frequency of orbital rotation. For a proton synchrotron the magnetic guide field strength and the r.f. power frequency have to be varied accurately over large ranges. The arrangement of a 300 MeV synchrotron is shown in *Figure 26.2* and the injection system of a 3 GeV proton synchrotron (cosmotron) in *Figure 26.3*.

Cyclotron

This early form of accelerator consists of a vacuum chamber between the poles of a fixed field magnet containing two hollow 'D' shaped electrodes which load the end of a quarter-wave resonant line so that a voltage of frequency 10–20 MHz appears across the accelerating gap between the D's. Positive ions or protons are introduced at the centre axis of the magnet, and are accelerated twice per rotation as they spiral out from the centre. The relation between particle mass m, charge e, magnetic flux density B and frequency f is $f = Be/2\pi m$. Energy limitation is set by the relativistic increase of mass, which limits the speed of high-energy particles so that their phase retards with respect to the r.f. field.

Synchro-cyclotron

The energy limitation of the cyclotron can be removed by modulating the oscillator frequency to a lower value as a bunch of particles gains energy.

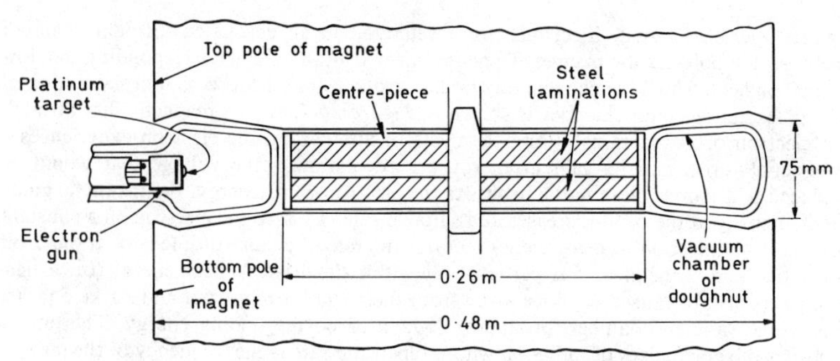

Figure 26.1 20 MeV betatron: cross section of magnet and electron gun

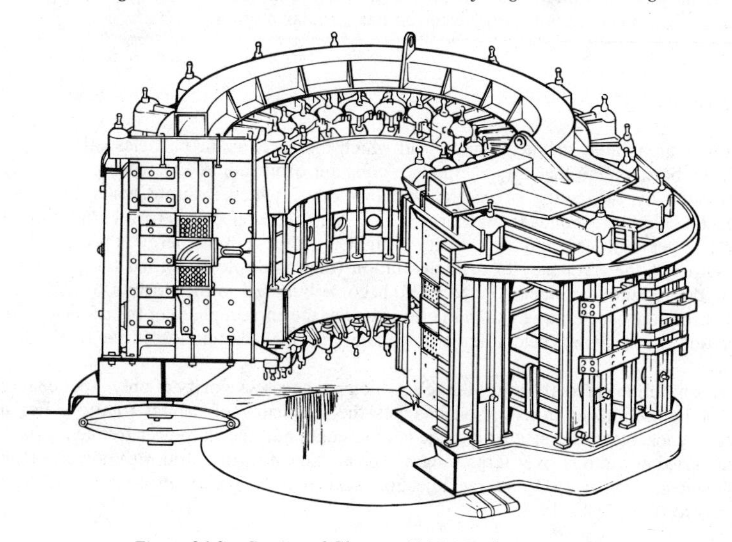

Figure 26.2 Section of Glasgow 300 MeV electron synchrotron

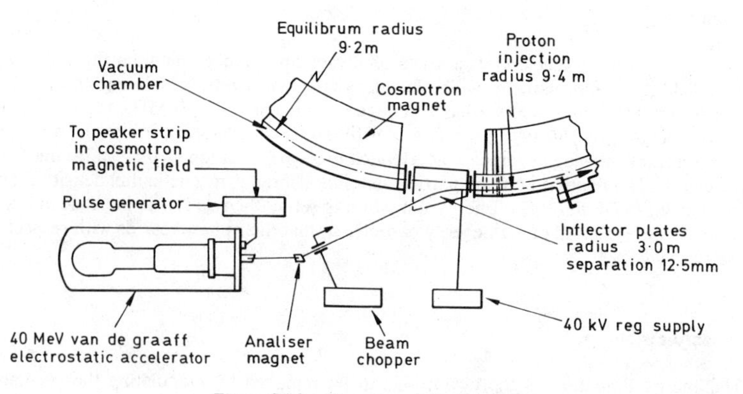

Figure 26.3 Cosmotron injection system

Microtron

In the microtron, or electron cyclotron, electrons are accelerated in a vacuum chamber between the poles of a fixed field magnet. The orbits consist of a series of discrete circular arcs which have a common tangent at a resonant cavity in which the electrons gain their successive increases of energy from an r.f. electric field. The highest energy achieved with such a machine is 6 MeV, and mean currents are less than 1 uA.

LINEAR ACCELERATORS

Indirect accelerators of protons have so far used a resonant cavity in which drift-tube electrodes are introduced that distort the fields and enable particles to be shielded from

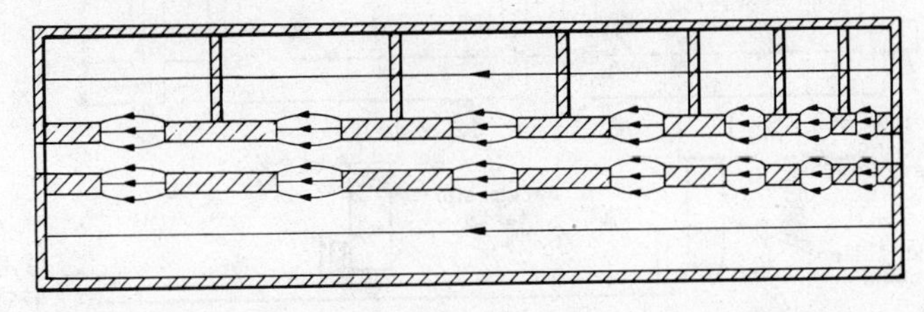

Figure 26.4 Field resonant cavity with drift tubes for proton acceleration

field reversals. Particles are accelerated between centres of successive gaps in one complete period of oscillation (*Figure 26.4*). Oscillators operating at about 200 MHz and a pulse power of 1–2 MW are used to excite the cavity for some hundreds of microseconds. Injection is by a Cockcroft–Walton or Van de Graaff device.

An important device for electron acceleration is the travelling-wave accelerator, using megawatt pulses of r.f. power at 3 000 MHz. The power is propagated along a cylindrical waveguide loaded with a series of irises. A travelling wave is set up with an axial electric-field component, and correct dimensioning of the iris hole radius a and the

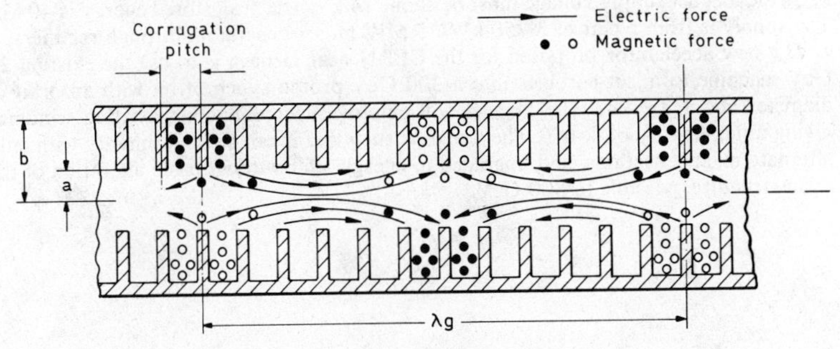

Figure 26.5 Fields in corrugated waveguide

waveguide radius b, *Figure 26.5* enables the propagation velocity and the field-intensity/power-flow relation to be varied. An electron injected along the axis with an energy of the order of 45 keV is accelerated by the axial field, and as its velocity changes it remains in correct phase with the travelling field, the propagation velocity of which is varied to match. A fixed axial field is required to provide for electron focusing.

High-energy machines with low beam-currents have been used in the U.S.A., low-energy machines with high beam-currents in the U.K. The 25 MeV Harwell accelerator has a length of 6 m divided into six sections, each fed by a 6 MW klystron amplifier to give a peak beam-current of 1 A and a mean output power of 30 kW. A simplified diagram of the arrangement is given in *Figure 26.6*.

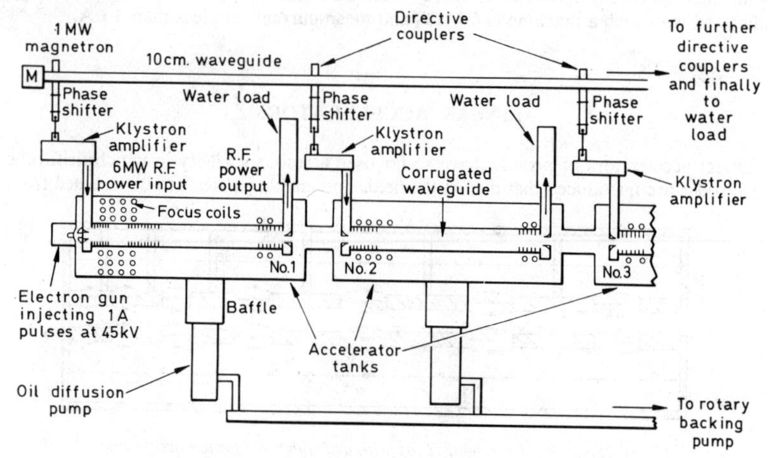

Figure 26.6 Part of 25 MeV high intensity electron accelerator

LARGE MACHINES

The Harwell proton synchrotron gives the particles an energy of 7 GeV in an orbit of radius 19 m within a 7 000 ton magnet. The magnet takes 10 kA to raise the orbit flux density to about 1.4 T in 0.75 s, hold this value for 0.25 s, and reduce it to zero in 0.75 s, with a repetition frequency of about two cycles per hour. The inductance of the magnet is about 1.1 H, and to produce a rate of change of current of $10/0.75 = 13.3$ kA/s the magnet supply voltage must be about 14 kV. The peak stored energy is 40 MJ. The supply is from a pair of 3 750 kW/75 MW motor/generators through rectifiers.

The new accelerator proposed for the CERN near Geneva is to use the existing 25 GeV machine to inject particles into a 300 GeV proton synchrotron with an orbit of diameter about 2.2 km. The magnet will employ superconducting exciting windings giving a flux density of 4–6 T. The design is such that it can be built initially with only alternate magnet sections, and upgraded in energy later without basic alteration of the main structure, possible to 800 GeV.

27 ELECTRONIC AIDS TO MEDICINE

ELECTRONIC AIDS TO MEDICINE

27

27 ELECTRONIC AIDS TO MEDICINE

The widespread use made of electronic techniques and the various medical domains in which they are utilised, is indicated in *Table 27.1*. The use of transducers, ultrasonics, etc., is well covered in current literature.[1,2,3] This section illustrates the necessary broad approach in the application of electronics to medical problems. New developments are well covered in the Current Awareness Publications.[4,5] Biomedical Engineering[6] is an introduction to the subject and includes a useful bibliography and list of journals. Reference 7 is a more advanced general reference with a good bibliography. Reference 8 is much more detailed, and includes an extensive bibliography.

There is in many cases a large choice of high-quality instruments to choose from. Those mentioned in this section are illustrative of current good practice. A publication listing U.S.A. equipment is available.[9]

The economic importance of the medical electronics market can be gauged by looking at world market figures. A recent market survey[10] shows that the main world market sales for 1972 *excluding* the U.S.A., were $1 400m, a rise of nearly $2 500m being predicted for 1976. These figures cover Medical and Analytical Instruments and X-ray equipment. Estimated expenditure in the U.K. for 1972 was the comparatively small sum of £24m.[10]

DIAGNOSIS

General

Diagnosis is the initial problem of a doctor confronted by a patient. Computers are being used to assist with the complexity of this task even to the extent of programming the computer to interrogate the patient.[11] To assist in the diagnosis an analysis of one or more of the body fluids will probably be undertaken in the Medical Laboratories. (See 27–18.)

An extension of diagnosis on the grand scale is Automated Multiphasic Health Testing (AMHT) where everything from interrogation, analysis of recordings and fluids, to final diagnosis, is computerised.[11,12]

At a less exalted level the mercury in a glass thermometer, for reasons discussed in *Health Devices*[13] has yet to be displaced extensively by electronic products, though it is being supplemented.[14]

Other techniques which may be used are as follows.

Radiology[15]

One of the first medical applications of electronics was that of X-rays and in modern hospitals the Radiology department is one of the largest and most expensively equipped. Simple X-ray sets are giving way to systems in which image intensifiers in conjunction with closed-circuit television are being used which improve image quality, increase operator convenience and safety, as well as minimise the radiation dose. Another approach to greater detail in the image is that of television subtraction whereby the difference between two films, one taken before and the other after injection of a contrast medium, is obtained by simultaneous scanning with two cameras, the difference signal being presented on a video monitor.

In one system of electron radiography silver-halide film is replaced by a xerographic

Table 27.1 CHART SHOWING MEDICAL AREAS AND ELECTRONIC TECHNIQUES EMPLOYED

		Anaesthesia	*Cardiology*	*Clinical chemistry (L)*	*Gastrology*	*Haematology (L)*	*Neurology*	*Obstetrics*	*Opthalmology*	*Otorhinolaryngology*	*Paediatrics*	*Physical medicine*	*Physiology*	*Psychiatry*	*Radiology/diagnosis*	*Surgery*	*Urology*
Electro-analytical instruments		×		×									×				
Aparatus		×	×				×	×		×	×	×		×	×	×	
C.C.T.V.									×						×		
Computers/ data handling		×	×	×		×	×								×		
Lasers						×			×						×		
Measuring instruments		×	×	×		×		×		×			×	×			×
Nucleonics			×	×								×			×		
X-ray	Radiation														×		
I.R.	Radiation			×					×			×			×		
Microwave	Radiation										×	×			×		
R.F.	Radiation										×	×				×	
Recorders			×				×	×	×	×		×	×		×		
Telemetry			×		×		×					×	×				×
Transducers		×	×		×	×	×	×	×	×			×				×
Ultrasonics			×				×	×	×	×		×			×		

technique in which the electrostatic image is formed in a new design of ionisation chamber.[16] It is claimed that the system affords greater resolution as well as being about twice as sensitive.

In a scanner developed for the diagnosis of brain disease, a computer is utilised to process the information obtained as the X-ray tube and detector are moved sequentially to and fro and step by step around the head.[17]

Radioisotopic methods[18]

These supplement X-ray methods particularly for studying the functions of internal organs, heart, lungs, kidneys, thyroid, etc. A suitable isotope is administered and its

Figure 27.1 Whole body monitor for studying the functions of internal organs (J & T Engineering (Reading) Ltd.)

uptake and distribution studied in the body, either by suitably placed detectors or by means of a scanning system. The largest and most expensive of these are whole body monitors in which two or more scintillation detectors can scan over any part of the body, *Figure 27.1*. Another instrument for *in vivo* studies is the gamma camera,[19] in which the radiation received by an array of detectors is utilised to form an image of internal organs to detect, for example, tumours.

Another vast range of bodily functions can be studied by what are known medically as *in vitro* methods, i.e., in which a body fluid is analysed in a test-tube. In this case the progress of a labelled substance administered to the patient is followed by taking samples of blood, or whatever, which are counted in a scintillation spectrometer.

Thermography[20, 21, 107]

This is a method of imaging which makes use of the infra-red radiation emitted by the body. An optical scanning system focuses the radiation on to one or more detectors to give a signal which modulates the intensity of a cathode-ray oscilloscope. Originally the interest and incentive was for the detection of breast cancer, as a non-harmful and perhaps cheaper alternative to X-rays. Difficulties have been encountered and it has not supplanted other techniques. Other uses include the study of burns and detection of vascular disease. The required spatial resolution, temperature resolution and speed of scan are not widely agreed upon or easily obtained. A fast scanning system has been

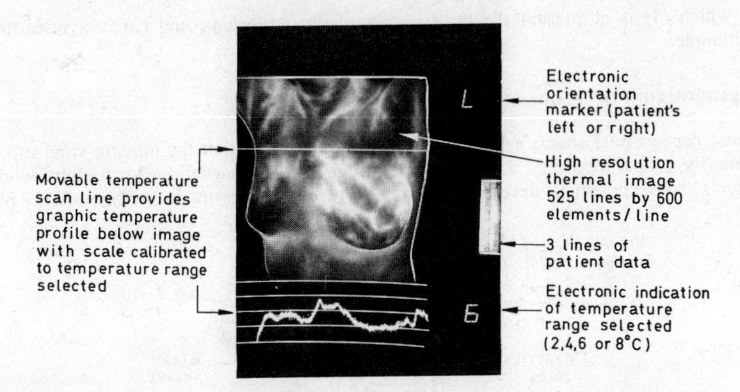

Figure 27.2 *High resolution thermogram obtained with General Electric (U.S.A.) Spectrotherm 1000 (Courtesy General Electric (U.S.A.))*

developed (Rank Precision Industries Ltd.) which provides 46 frames/second each having 120 lines of 170 elements. *Figure 27.2* indicates the resolution that has been achieved (525 lines of 600 elements in 2 seconds) in a slower system. Both have a temperature resolution of 0.2 °C.

Colour displays with isotherms are obtainable with some systems.

Ultrasonic imaging[22]

Sonar systems have advantages over X-rays in being quite safe and over thermography in showing details of deep-body organs (see **27–15**). Colour displays are a recent development.[23]

ELECTRONIC INSTRUMENTS

Cardiovascular instruments

The heart is the most vital organ in the body and much effort is devoted to studying and controlling its action. Instruments in the following categories are described. In practice they may be conveniently combined, e.g., a phonoelectrocardiograph:

(1) Blood-flow measurement;
(2) Blood-pressure measurement;
(3) Cardiac output measurement;
(4) Defibrillators;
(5) Electrocardiographs;
(6) Pacemakers;
(7) Pulse rate;
(8) Heart sounds.

BLOOD FLOW MEASUREMENT [24, 99]

Blood flow is at present detected by three direct methods:

(1) Electromagnetic;
(2) Ultrasonic;
(3) Thermal, i.e., Hot-wire;

of which (1) is at present the most commercially important, and (2) is a research technique.

ELECTROMAGNETIC FLOWMETERS [1, 8]

These depend on Faraday's Law of Electromagnetic Induction the moving conductor being the blood, *Figure 27.3*. It is not easy to get intense magnetic fields within blood vessels and the potentials developed are small. To eliminate errors due to the presence of

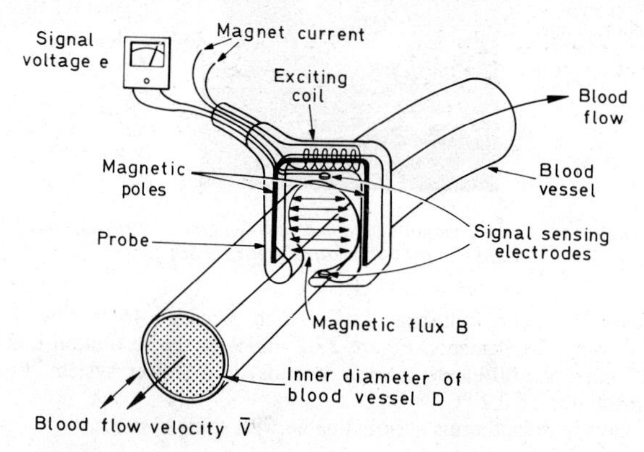

Figure 27.3 Principle of electromagnetic blood flowmeter. As blood flows through the magnetic field (in the vessel), a voltage expressed in the following equation is induced between the two sensing electrodes

$$e = BD\bar{V} \times 10^{-8} \text{ volts}$$

where B = Magnetic flux density in gauss; D = Inner diameter of vessel in centimetres; V = Blood flow velocity in centimetres per second (Reprinted from Japan External Trade Organisation Publication 'Medical Systems and Equipment')

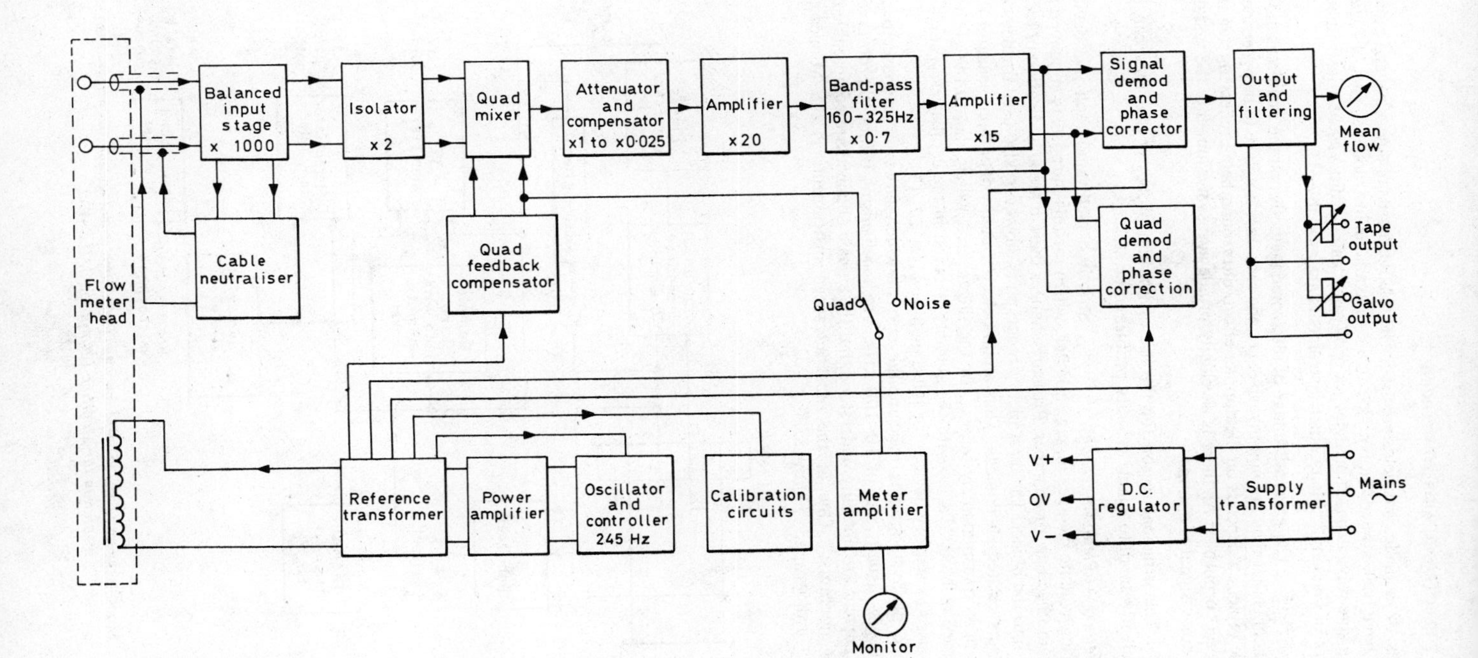

Figure 27.4 Block diagram of an electromagnetic flowmeter (*Courtesy SE Laboratories* (*EMI*) *Ltd.*)

electrolytes, oxygen, etc., a.c. fields are used. The relative merits of sine-wave and pulsed excitation have been argued but both are in use in available instruments. That of *Figure 27.4*, using sine-wave is based on the work of Wyatt[25] and Cardio-Vascular instruments on the work of Gasking.[26] Many other instruments are available.[27]

Ultrasonic.[8, 24, 100] The great advantage of ultrasonic methods and the main reason for pursuing them is that they are non-invasive, i.e., the skin does not have to be penetrated. Their mode of action depends on the doppler effect, ultrasound being reflected from the moving blood corpuscles. As there is a velocity profile across the lumen of the vessel a somewhat noisy signal is obtained.

Thermal.[24] The principal reason for the use of hot-film probes is that they can be made sufficiently small to enable velocity profiles to be studied.

Indirect methods—Plethysmography.[7, 8, 28, 101] Strictly a plethysmograph is a device for measuring volume changes in, e.g., a finger, which for this purpose, is contained in an air or fluid filled bag, or chamber. Pressure changes therein are an indication of local blood flow into the object. The term plethysmograph is, however, also used for devices which measure related phenomena. Thus a photoelectric plethysmograph measures the amount of light transmitted by the ear, or the finger, etc., which, due to the greater optical density of blood, varies with the amount of blood in the capillaries. They are commonly used as sensory devices for pulse monitors (see **27–12**).

Impedance plethysmographs measure the current which flows in response to an applied voltage—usually a.c. of 10–100 kHz. The mechanisms behind the observed changes in current are not fully understood, but the signals obtained are comparable to true plethysmograms.[29] One of the most advanced instruments is the Minnesota Impedance Cardiograph,[30] *Figure 27.5*.

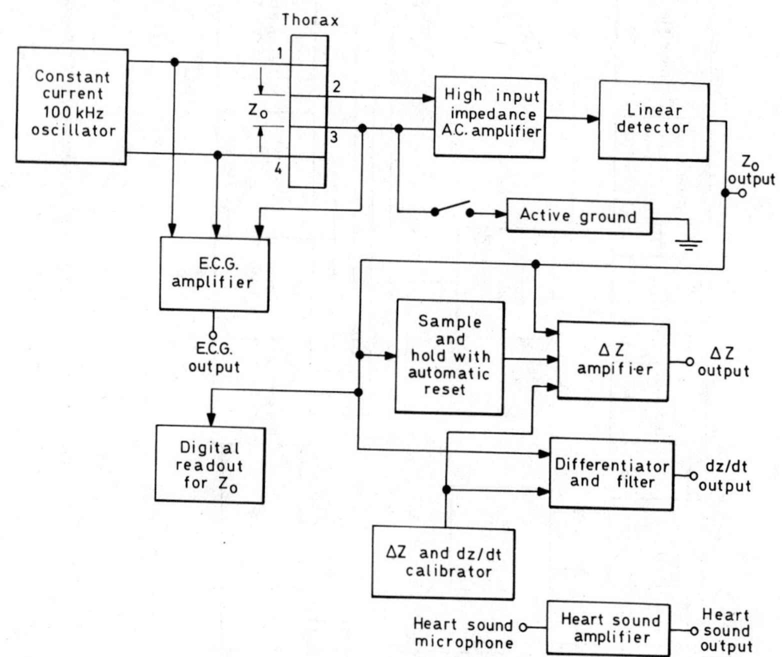

Figure 27.5 Block schematic of IFM/Minnesota impedance cardiograph (Courtesy *Instrumentation for Medicine Incorporated, U.S.A.*)

Other indirect methods of measuring mean flow are referred to under Cardiac Output (27–9 and 27–10).

BLOOD PRESSURE MEASUREMENT

Techniques of measurement again fall into 'invasive' and 'non-invasive' categories. As yet, continuous measurement can only be obtained with the former which are utilised only when knowledge of blood pressure is of essential importance, e.g., in coronary care, or diagnosis of heart disease.

Non-invasive.[31] The obvious method of applying electronics is to automate what a doctor does, i.e., listen to the sounds in the artery as it is occluded by an external cuff. These rather curious noises (Korotkoff sounds), although utilised by numerous instruments, do not produce results that are 100% reliable, but by employing sophisticated techniques for gating and filtering the sounds have been utilised.

A method[32] that is independent of noises within or without the patient is utilised by Roche in their arteriosonde in which movement of the arterial wall is detected by an ultrasonic doppler method, *Figure 27.6*.

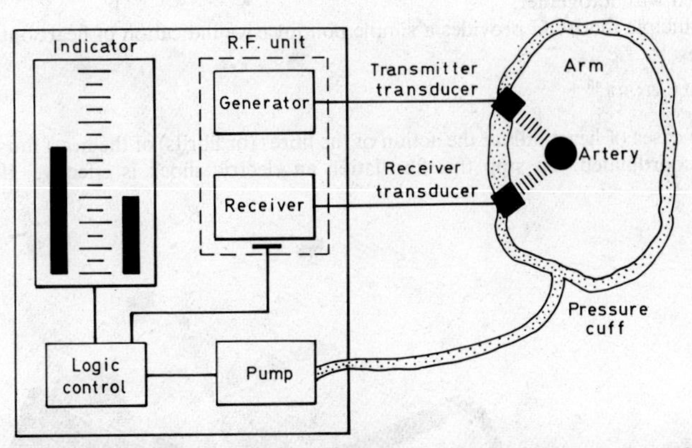

Figure 27.6 Block diagram of an ultrasonic blood pressure monitor. Transmitting and receiving crystals are placed beneath a Riva-Rocci cuff on the upper arm. Cuff and crystals are connected to a monitor which inflates and deflates the cuff, generates ultrasound, receives and interprets reflected ultrasonic energy and displays blood-pressure readings (Courtesy ASP Biological and Medical Press, Amsterdam)

Direct.[1, 8] Direct methods of continuous blood pressure measurements have relied on strain gauge or capacitance transducers. These are connected into the artery via a catheter and consequently require extremely small volume displacements (<0.1 mm^3/100 mm Hg) in order to attain the requisite frequency response (0–100 Hz *in situ*). Current developments include sub-miniature semiconductor sensors[33, 34] which can be attached to catheters and inserted within the heart.

CARDIAC OUTPUT

Dye dilution.[7, 8] This is one of the established methods which depends on the rapid injection of a quantity of dye through a catheter into the heart. Its progress is monitored photo-electrically and from the resulting dilution curve the output can be calculated either by analog or digital methods. A further development of this technique makes use of a fibre-optic catheter to permit *in vivo* measurements of oxygen saturation and dye-dilution combined. In each case the result is obtained by computing the ratio of the back-scattered light at two different wavelengths.[35] The instrument is called the In vivo Haemoreflectometer.

Isotopic dilution. This is a variant in which a radiochemical is injected in place of the dye and its progress followed with a ratemeter.[36]

Thermal dilution. Repeated injections of either dye or isotope are undesirable and are avoided by injecting cold saline and monitoring changes in the outflowing blood temperature with a thermistor.

While relatively little harm accrues from repeated injections of saline it is still discomfiting and slow. Another approach is to calculate the output from the area of the aortic pressure/time curve.[37]

Ballistocardiograph (BCG). The mechanical effect of the heart in moving the body has been studied for a long time but the requirement for a platform suspended from the ceiling in a vibration-free environment has militated against its use. In one recent development[38] a more advanced suspension has been employed. In another[39] it is dispensed with altogether.

In principle the BCG provides a simple non-invasive indication of heart output.

DEFIBRILLATORS[40]

In some cases of heart failure the action of the fibres (or fibrils) of the heart muscle is no longer coordinated. To stop this fibrillation an electric shock is effective. If applied

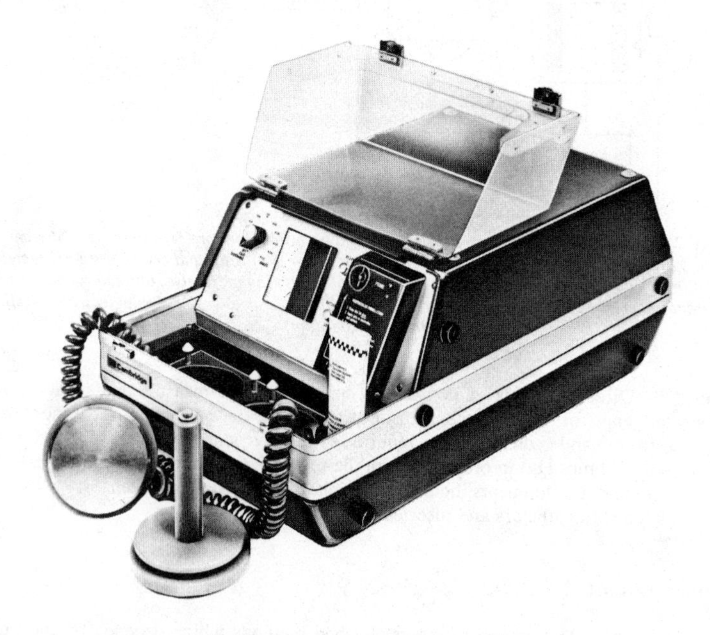

Figure 27.7 Defibrillator (Courtesy Kent Cambridge Medical Ltd.)

externally, a d.c. discharge of up to 400 joules is required at voltages up to about 5 kV d.c.

A desirable feature is that of being able to initiate the discharge at a particular point of the heart cycle and for this purpose a means of synchronisation is provided. *Figure 27.7* shows a typical instrument. The large electrodes (paddles) are provided to convey the current safely to a large area of the patient without any of it reaching the doctor.

ELECTROCARDIOGRAPH (ECG)

The ECG, which is a record of the electrical activity of the heart, is perhaps the most important clinical measurement involving electronics. It is informative, non-invasive and, with today's techniques, simple and safe.

Safety. The attachment of current-carrying wires to a patient can be dangerous. This potential hazard has been eliminated by the development of 'fully floating'[41] input circuits for ECG amplifiers. With proper design the maximum current which in any circumstance can flow to the patient is thereby limited to around 10 μA.

Simple. Current technology has overcome the difficulties of making reliable amplifiers of sufficiently high input impedance (>5 MΩ) and common mode rejection (>60 dB) largely surmounting troubles from electrodes and mains borne artifacts.

Informative. The information content of ECG records is prodigious and the ultimate step would seem to be interfacing the patient directly to a computer. There remains the

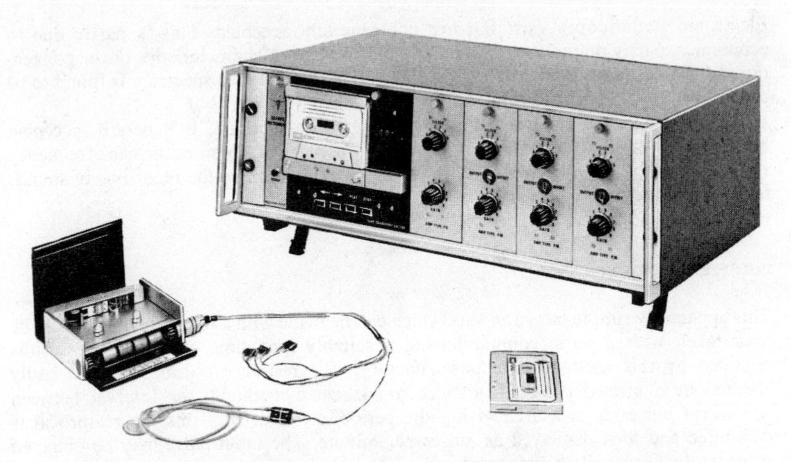

Figure 27.8 'Medilog' recorder and playback system (Courtesy Oxford Instruments)

problem of programming the computer to be as effective as a trained person in evaluating the data.[42] A more dedicated instrument is the Neilson arrythmia monitor which detects irregularities in the heart rhythm.[43] Modern miniature components have permitted the development of recorders which may be carried by the patient thus providing 24 hr records which may be subsequently analysed.[44] One of these is shown in *Figure 27.8*.

PACEMAKERS[8, 45, 46]

Under normal circumstances the natural pacemaker of the heart, located in the atrium, supplies electrical impulses which travel through the internal nerve system of the heart to the ventricles which then contract to provide the required pumping action. Failure of the nerve system (heart block) results in a slowing, or cessation, of the heart beat and inadequate circulation. Electrical impulses to stimulate ventricular contractions are required and while an external source is usable in the short-term, for long-term use an implant becomes desirable. The first of these were of fixed rate and are suitable in cases of complete heart block. Synchronous pacemakers which have in effect a built-in ECG to detect atrial activity have been developed but difficulties with these have led to the design of a 'demand' unit. In these, stimulus pulses are only produced in the absence of natural contractions. This result is achieved by incorporating circuits to detect the normal peak deflection (QRS wave) which accompanies each contraction. Each wave detected inhibits the production of a stimulus for approximately one second, consequently a stimulus is only delivered if no contraction takes place in this interval, i.e., on demand. Much greater detail on the currently available units, of which there are over 50, is given in reference 102.

The environment within the body is very aggressive. Special epoxy resins have been developed for encapsulation but their permeability to water vapour has led to hermetically sealed circuits being incorporated in one example and hermetically sealed titanium capsules in another.

Altogether some 150 000 are estimated to be in use. One problem remaining is adequate life from the power supply.[46, 47]

HEART SOUNDS[1, 8]

Electronic stethoscopes exist but are not generally accepted. This is partly due to economics, partly due to the questionably useful filtering characteristics of the conventional form with which doctors are taught, so that the extended spectrum is found to be confusing.

The recording of the heart sounds, the phonocardiogram, is generally accepted however as useful in the diagnosis of heart disease, particularly since the wide frequency response of sub-miniature pressure transducers have permitted the recording of sounds from within the heart.

PULSE RATE METERS

This apparently simple measurement, which can be made with a watch, can obviously be undertaken with a pulse counter having a suitably long time constant. The results obtained by this approach are unsatisfactory. The response is too slow and badly affected by a missed pulse. In a more sophisticated method[48] the interval between successive pulses is measured giving the periodicity and from this the reciprocal is computed and then displayed as pulses per minute. The input pulse may be obtained from the ECG or a plethysmograph.

Respiratory instrumentation[8]

The study of respiratory and pulmonary function are branches of physiology. In the hospital environment they are of particular interest to the anaesthetist whose concern it is to keep the patient breathing. Instruments exist for the following tasks:

(1) Pulmonary function studies;
(2) Respiratory gas analysis;
(3) Blood gas analysis;
(4) Ventilation.

PULMONARY FUNCTION STUDIES

The range of instruments—and their nomenclature—is very wide. Impedance pneumographs,[2] which record respiratory movements, have been developed to the stage of being respirometers[49] which indicate volume. Then there are respiratory (or lung, or pulmonary) function analysers which will measure or compute parameters such as Peak Flow, Vital Capacity (VC), Forced Expiratory Volume in one second (FEV_1) and this, as a percentage of VC, FEV%. The Lung Function Analyser by Mercury Electronics is of this type.

By addition of a gas analyser, as in the Hewlett Packard Respiratory Computer VR6100, a rather tedious procedure to measure the efficiency of the lungs, denoted the 'nitrogen washout', is computed in minutes in addition to all the previously mentioned quantities.

RESPIRATORY GAS ANALYSIS[8, 50]

Most medical gas analysers are adaptations of industrial instruments and are largely non-electronic. For analysis of individual breaths a time constant of < 0.1 s is desirable. NDIR Analysers are used for CO_2 and anaesthetic gases.

Several mass spectrometers have been specially designed for medical use. One of these makes use of a membrane in the sampling system, and can be used for blood gas as well as respiratory gas analysis. The problems[51] are not so much with electronic as sampling techniques.

BLOOD GAS ANALYSIS[8]

Recent developments have sought to make the analysis of blood for pH, pCO_2, etc., more simple to carry out and the latest instruments are semi-automatic. The prevailing trend is to present the results in digital form.

VENTILATORS

These are servo-controlled pumps to assist or take-over from the lungs and are mostly mechanical. One instrument makes use of electronics to give greater flexibility and additional facilities for recording and alarms. The desired respiratory pattern is set up on the front panel controls and the servo system ensures that this is maintained. Provision also exists for the patient to trigger the start of the respiration.

Patient monitoring in intensive treatment and coronary care units (ITU/CCU)

The function of patient monitoring equipment[28] is to relieve the nursing staff of the minute by minute task of watching over the vital functions of the critically ill. The parameters which are commonly monitored include temperature, respiration, blood pressure, pulse rate and ECG.

Single patient monitors exist but systems are commonly for 4 to 20 beds, the nurse in charge being provided with a console equipped with multi-channel monitor scopes for ECGs, alarm and facilities to monitor continuously other parameters on a selected

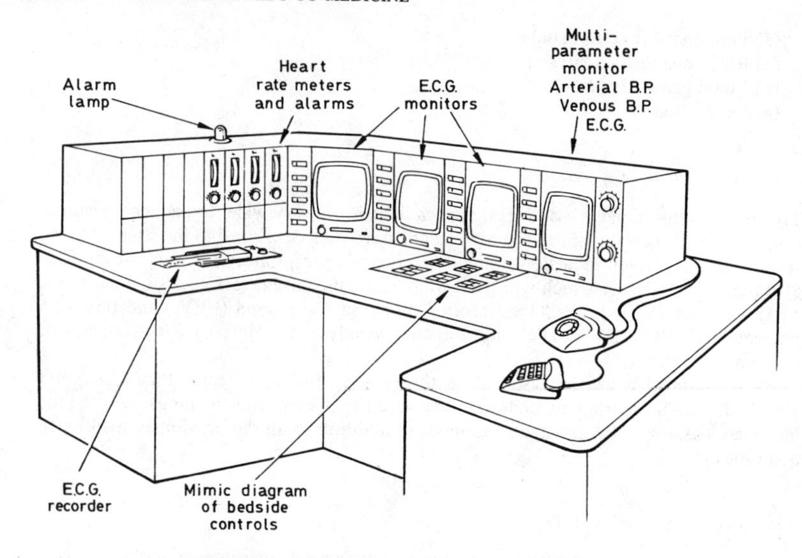

Figure 27.9 General layout of a nurse central station

patient. *Figure 27.9* shows the layout of a custom-built central station. Many systems are available in the U.K.[52] Some of the recent and useful developments are as follows

MEMORY SCOPES[53]

Monitor scopes for ECGs with repetition rates around 10 per minute, i.e., a scan duration of 3 seconds, present viewing problems. To overcome these the signal is sampled and successive values stored. Normally the memory is continually updated, but the facility to stop this allows the trace to be 'frozen'—in the event of a significant event occurring—to await the arrival of a specialist or to permit readout to a recorder.

TELEMETRY SYSTEMS

In Coronary Care Units (CCU) particularly the use of a telemetry link for ECG is advocated on the grounds that heart malfunction is more likely to occur when the patient starts to move around.

COMPUTERISED SYSTEMS

In a multi-bed unit a computer can cope with the task of sampling and recording the data, and performing routine calculations. Decisions based on these can be used to initiate automatic action to improve the patient's condition.[54]

Instruments for neurological use

An instrument for the detection and recording of the electrical action of the nerve cells within the brain—brain waves—is the electroencephalograph (EEG).[3]

A recent U.K. development is the cerebral monitor[55] which makes use of 'brain-waves' to monitor the state of critically ill patients.

Another measurement of growing importance is that of intracranial pressure. Details of the various approaches are given in reference 103 and include the use of implantable transducers complete with telemetry transmitters.

Widely used devices are stimulators for inducing nerve/muscle action.[46,47]

Ultrasound has been used for some years in determining the mid-line of the brain and instruments for this purpose are termed echoencephalographs.[56]

Instruments for Obstetric and Paediatric use

In order to minimise the possibility of radiation damage, ultrasonic methods are being used in place of X-rays for the study of the foetus. Ultrasonic doppler foetal heart monitors[57] can detect the presence of the foetal heart beat at 10 to 12 weeks. Ultrasonic scanners can be used to make measurements of foetal size and position.[58] One such is the diasonograph.

The cardiotocograph[59] simultaneously measures intra-uterine pressure and foetal heart rate and is used to assess foetal well-being.

Once born it may be necessary to ensure that breathing continues, and an interesting variety of apnoea monitors have been developed for this purpose. Most make use of the technique of impedance pneumography.[2,49] A much cheaper and simpler device which does not require electrode attachment is based on the detection by a thermistor of airflows from one part to another of a suitably divided mattress.[60] Another contactless system employs a millimetric doppler radar to detect movement.[108] A recent design employs a plastic foam mattress, impregnated with a conductive material. This is pressure sensitive and hence responsive to the baby's movements.[61]

Physical medicine—applications

The electromyograph (EMG) for recording muscle potentials is widely used. It is usually associated in use with nerve and muscle stimulators. The required frequency response exceeds that of most chart recorders. One method uses a fibre-optic recorder with response to 100 kHz.

An unusual application is the load measuring sandal, *Figure 27.10*, which serves to telemeter the load on the foot. The sole and heel of the sandal is a transducer of the condenser type. The equipment is used to indicate the vertical component of the body weight transmitted by the feet when walking. The relative values of the heel and toe peak loadings can be ascertained.

Therapeutic equipment makes use of l.f., short-wave and micro-wave radiation as well as ultrasonics.[62] Potentially of great importance to the disabled is the myo-electric control of prostheses.[3]

Instruments for ear, nose and throat

EAR

The primary tool is the audiometer the function of which is to determine the acuity, i.e., the equivalent of the frequency-response, of the ear.[63] One of the difficulties is that of knowing accurately whether the subject can really hear any given signal. A self-recording audiometer in which the subject himself adjusts the sound level to maintain audibility as the frequency is scanned, is one approach.

Uncooperative subjects can have their response measured by the evoked response

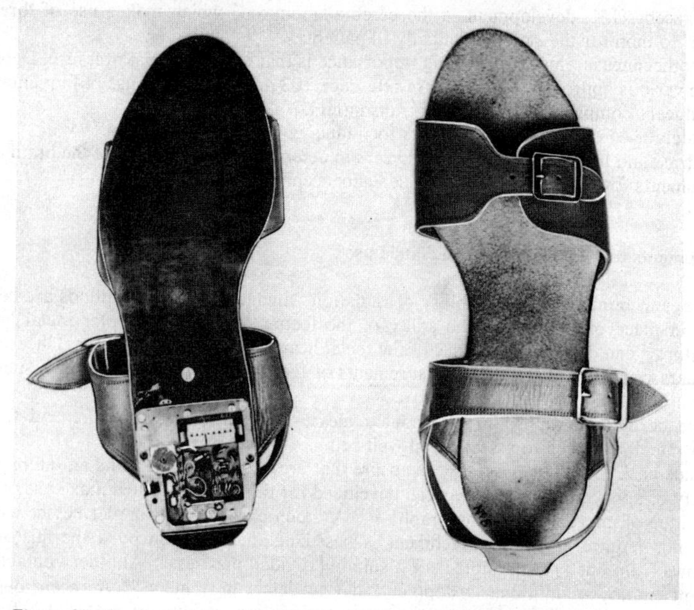

Figure 27.10 Load measuring sandals showing telemetering equipment in the heel (Courtesy Institute of Orthopaedics)

audiometer [3] in which signal averaging techniques applied to the EEG (see **27–14**), indicate when an audible signal has been generated.

While the measurement of noise is not normally of medical interest the fact that intense noise is a cause of deafness makes audio dosimeters, which measure personal exposure to noise, a relevant device. Like the SAMIs (see **27–20**) one type makes use of an E-Cell* as an integrating memory device. It also gives an indication if the wearer is exposed to levels in excess of 115 dBA (sound level in dBs weighted as per BS 4198). (Pop groups are reported to produce noise levels up to 125 dBA.)

Hearing aids.[104] A still widely used hearing aid available from the NHS is based on the work described in the MRC Special Report No. 261 'Hearing Aids and Audiometers' published in 1947. One of the latest developments in hearing aid microphones is the 'electret condenser'. This is a permanently polarised dielectric and units made from a new electret film with integral f.e.t. amplifier, approximately 2 × 5 × 8 mm in size, are available. They are claimed to be insensitive to shock and vibration.

NOSE

Electronics are currently little used in connection with the nose and its function.

THROAT[64]

The most complex instrument that has been developed is the laryngograph which provides a display of frequencies in the voice versus time. It is being used to teach deaf children to speak.

* Registered trade mark of Bissett–Berman Incorporated, U.S.A.

Pocket size amplifiers have been devised for larnyngectomised patients and have proved useful to others with weak voices. The artificial larynx,[65] which is a small tone generator applied to the throat, is another aid.

Electrosurgical instruments

Current designs of electrosurgical equipment utilise solid state electronics[8] in place of the valve and spark-gaps circuits that have long been in use. The interference from spark-gap units is intolerable. An improved valve design[66] which appears electrically and surgically acceptable has been developed. An extensive study in which contemporary designs were compared and evaluated, including their safety and clinical effectiveness has been carried out in the U.S.A.[67]

Psychiatric applications

An early application of electrical impulses was in the treatment of mental disorders, electro-convulsive therapy (ECT) is now less popular.

Of continuing interest is the galvanic skin reflex (GSR), popularly known as the lie detector. The electrical resistance of the skin is related to one's psychic state, and can be monitored with suitable equipment.

For studying perception, electronically controlled tachistoscopes are available.[7]

Eye instruments

When the eye moves, potentials are generated which can be used to provide a record. Instruments for this purpose are the electronystagmogram or electro-oculogram.[1] A non-contacting eye-movement monitor is available which makes use of a modulated infra-red illumination to sense its position. Another instrument, incorporating a special purpose closed-circuit television, can be used to measure pupil diameter—a parameter of use in behavioural studies.

The pressure within the eyeball is used to diagnose glaucoma and an electric tonometer has been developed which accomplishes this.[68] Potentials produced when the eye is illuminated can be recorded by means of the electroretinogram[69] and used in diagnosis.

Instruments for gastrology

An early application of telemetry was in the 'radio-pills' which were designed to measure pressure and temperature and pH.[70, 71, 72, 73]

The 'pill' designs are still used. More recently the smallness of semiconductor pressure sensors has led to the development of esophageal motility sensors.

Urological instruments

The measurement of pressure, force, volume and flow in the bladder and associated parts of the body is now denoted by the term cystometry.[105] One of the few systems available records two pressures, usually the bladder and intra-abdominal, the urine flow rate and the EMG signal from the sphincter muscle.

An ingenious application is a device for shattering bladder stones *in situ*. Essentially this is a battery operated e.h.t. supply with arrangements for discharging the reservoir capacitor within the bladder. Operation of the Lithotriptor is unfortunately not without risk nor is there universal agreement as to the desirability of having a large stone replaced by small sharp fragments.

Medical laboratory instruments

The following is concerned mainly with the applications of electronics to instruments which measure the properties of blood, sera or other specimens.

For blood measurements there is the following range of instruments:

(1) Blood cell counters;
(2) Blood clotting instruments;
(3) Haemoglobinometer;
(4) Analytical instruments.

BLOOD CELL COUNTERS

Blood contains an enormous number of cells. The number present in a sample is an indication of the quality, and as about 5×10^6 are present in 1 ml automatic counting is essential to cope with the numbers of samples taken. Roughly speaking, blood contains two types of cell, red and white (erythrocyte and leucocyte). The simplest instruments merely differentiate between these, generally by passing diluted blood through an orifice. The electrical resistance from one side of the orifice to the other is monitored, and cells, which partly occlude it, cause variations in resistance, which are counted.[74] The amplitude and shape of the electrical impulses are related to the volume of the cells and more sophisticated instruments use this information to derive the mean cell volume and the Haematocrit (the percentage, by volume, of red cells in blood). The optical technique of darkfield scatter is utilised in one instrument.[75] This counts platelets—a small cell that plays a part in blood clotting. There are several sorts of white cells and an instrument for differential white-cell counting makes use of a computer-based pattern recognition system which operates on the image received from a microscope slide via a closed-circuit television system.[11, 76, 77]

BLOOD CLOTTING

The speed at which blood clots is another vital measurement. Many parameters are involved in blood clotting and instruments are available for measuring one or more of these. In one type a reagent is added to the blood sample contained in a thermostatted test-tube and the time of formation of a clot is determined from the optical transmission which is monitored photo-electrically. (Platelet aggregometer.)

Another instrument, which is an automated device for prothrombin test (PT) and prothrombin thromboplastin (PTT) measurement also relies on photo-electric sensing of the clot. Other instruments use combined mechanical and optical means, e.g., a magnetic stainless steel ball is held in the optical path until the clot, forming in the vertically oscillated test tube, drags it out of the light path. In another example, particulate iron oxide is added to the sample and agitated by a rotating magnetic field rendering it opaque. At the end-point the particles are entrapped and the sample rapidly clears.

It is possible to detect clot formation by monitoring the electrical impedance.[1, 78, 79]

HAEMOGLOBINOMETERS

Haemoglobin is the constituent of red cells that by combining with oxygen or carbon dioxide effects their transport within the blood. These, and other, compounds of haemoglobin all have differing light absorption properties. A standard method of measurement involves dilution of the sample with potassium ferricyanide to form cyanmethaemoglobin. It's absorption at 540 nm is a measure of the concentration.

Another method makes use of the fact that all the compounds have the same absorption at 548.5 nm. A single measurement at this wavelength is thus a measure of total haemoglobin. Another instrument which by means of analog computation upon the measurements made simultaneously at three wavelengths, determines not only haemoglobin and oxyheamoglobin but also HbCO, which is the compound formed with the poisonous gas carbon monoxide.

ANALYTICAL INSTRUMENTS

Automatic analysers [80] fall into two classes, continuous flow and discrete.

Continuous flow. These rely upon the discovery that samples may be effectively separated in a flowing system if bubbles are introduced at regular intervals, Auto Analysers based on this are in worldwide use.[81]

Discrete. In these a discrete quantity of substance is placed in a test tube to which reagents are then added. There are many analysers of this type.[82, 109] They vary

Figure 27.11 Vickers 3000 Multichannel Analyser (Courtesy Vickers Ltd.)

enormously in size, cost and complexity, ranging from a single channel unit performing 120 anaylses/hour, to one which carries out up to 6 000 tests per hour. *Figure 27.11* shows a multichannel analyser.

These larger instruments are so complex that a digital computer is at the heart of their function to process the patient data, to control the sequence of the machine operations so as to carry out the chosen tests on each sample, and to compute the results and present them in an assimilable form.

Safety and standardisation

While early application of electronics to medical problems were applauded because of the possibilities they introduced, much more importance is now attached to the safety of electronic devices.[41, 83, 84] Not only are more devices attached to a patient at one time thus multiplying the dangers due to earth faults, etc., but with the use of intracardiac catheters the maximum current that can be withstood falls drastically from 100 mA for externally applied potentials[83] to about 60 μA for the catheterised subject.[85]

Equipment for the U.K. should conform to the recommendations of H.T.M. 8 'Safety Code for Electro-Medical Apparatus'.[86] A technical committee of the IEC[87] has prepared a standard on 'Common aspects of electrical equipment used in medical practice' and others on specific instruments may be expected. Changes taking place in the U.S.A. are reported,[88, 89] and a review of standards relating to other countries has been compiled.[106]

Other related areas where safety limits have been recommended are microwave radiation,[90] and laser radiation.[91] The situation with regard to ultrasonic radiation has been reviewed.[92] An additional advantage of having instruments that conform to a standard of performance is that measurements made anywhere can be directly compared.

Miscellaneous

Telemetry[7, 93] has been the subject of considerable expenditure by N.A.S.A. so as to make possible the monitoring of astronauts. The advanced state of patient monitoring in the U.S.A. is in part due to 'spin-off' from N.A.S.A. Another example is the Impedance Cardiograph (see 27–8).

Analytical techniques which can be expected to have a growing impact are that of the laser microprobe[94] and neutron activation analysis,[95] both of which, in their capability to deal with small samples, have applications in forensic medicine. Less serious misdemeanours are detected by the Alcolmeter[96] for breath alcohol testing which employs a special fuel cell as the sensing element.

Socially acceptable monitoring instruments (SAMI) is the generic name for a group

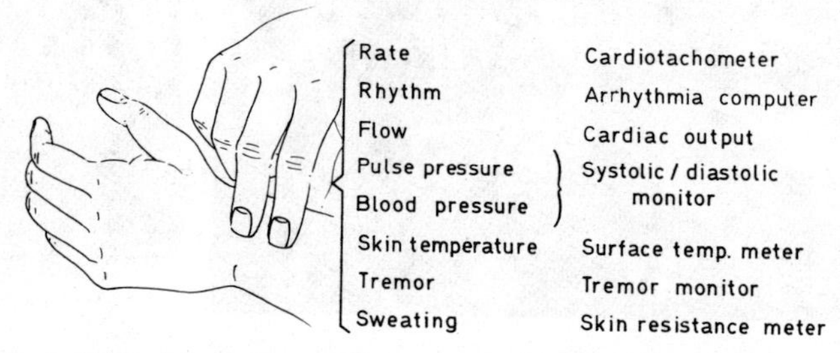

Figure 27.12 Information available to a skilled practitioner when feeling the pulse

of instruments designed to be worn continuously without discomfort.[97] Temperature and heart rate can at present be monitored. Their purpose is to gain information unavailable in a clinic.

The cost of medical instrumentation has largely confined its use in the U.K. to hospitals, and it has not to any extent penetrated the G.P.'s surgery. As has been pointed out,[98] the G.P. with his finger on the pulse can make the qualitative assessments shown in *Figure 27.12*. It is not anticipated that a 'cheap substitute' for this impressive array will suddenly appear, but that quantitative instruments will gradually supplement it.

THE LASER IN MEDICINE AND ITS SAFE USE

It does not appear that the laser will have anything like as great an impact on medicine as X-rays or electronic aids to diagnosis. Up to the present it has only found application on any scale in eye surgery, notably in the treatment of detached retinas. Experimental use of pulsed solid state lasers in treatment of skin cancer (melonomas) has had some success, as has the use of a CO_2 laser 'knife' in surgery in areas where there is a high concentration of blood vessels (highly vascular areas). In medico-biological research a finely focused ultra-violet laser beam has been used experimentally for the excision of cell nuclei. The use of the laser in eye surgery is, of course, intimately connected with the safety aspects of lasers which is therefore dealt with in this section.

General surgery

Much of the work in this field has been done by Professor L. Goldman and his colleagues at the University of Cincinnati in the U.S.A.[110, 111] They have used both solid state lasers and CO_2 lasers in their research. Solid state lasers have chiefly been investigated for use in treatment of melanomas. One example of successful treatment was of a man with a highly malignant melonoma of the cheek, about 3 cm in diameter. This was treated with about nine overlapping shots each of about 90 joules from a ruby laser. The tumour disintegrated in about ten days and complete healing of the area had taken place in about six weeks. There was no recurrence or evidence of secondary deposits. In a second example a woman with about 60 secondary deposits from a melanoma had each growth treated separately with a ruby laser and regression of the growth occurred in about a week. After thirteen months there was no recurrence of disease.

Initial work using a CO_2 laser was done on similar cases to the second example above, with equal success, using a 25 W c.w. laser. Later work was done with a 50–300 W laser fitted with a flexible beam manipulator and an endoscope and gastroscope. With this, liver surgery was carried out on experimental animals; it was claimed to be superior to the high frequency surgical knife, in control of capillary bleeding. The same laser system was used to excise two malignancies in man, primarily because the area of the cancer was highly vascularised. One of these was a 2 cm diameter basal cell epithelioma of the neck, the other a lobulated angiosarcoma of the finger. Both operations were successful and no recurrence had been observed after 9 months.

Eye surgery

The most widely used application of lasers in medicine so far has been in the treatment of eyes in which some part of the retina has become detached from the choroid, *Figure 27.13*. These detached areas produce 'blind spots'. If the retina can be flattened against the choroid, however, a re-attachment operation may be attempted, or, if this is not possible, the existing area of the detachment must be sealed off to prevent it progressing

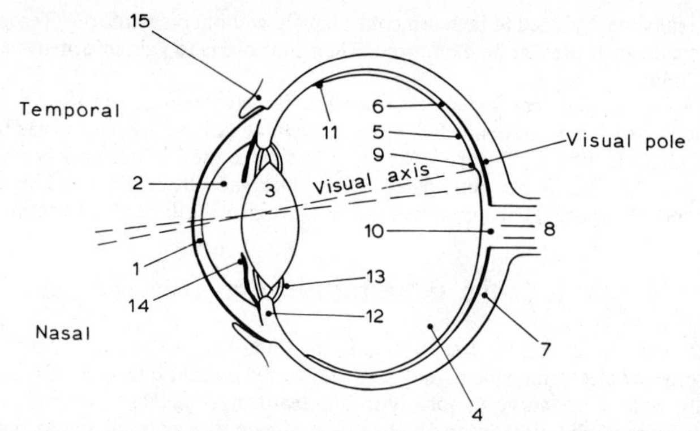

Figure 27.13 Schematic near-horizon median section of the right eye seen from above: (1) Cornea (2) Aqueous humour (3) Crystalline lens (4) Vitreous humour (5) Retina (contains photosensitive end-organs) (6) Choroid (vascular and pigmented membrane) (7) Sclera (outer coat of the eyeball) (8) Optic nerve (9) Fovea (10) Optic disc (papilla) (11) Ora serrata (front edge of the retina) (12) Ciliary body (13) Zonule (14) Iris (15) Ocular conjunctiva (B.S.4803:1972)

to total detachment. Essentially all the procedures used by opthalmologists aim at producing a localised welding of the retina and the choroid. Many techniques are available such as diathemy, cryosurgery and the xenon arc coagulator. The use of a pulsed ruby laser can be regarded as an improvement on the xenon arc coagulator. The latter requires that the patient be anaesthetised so that the eye can be immobilised, since an exposure of several seconds is required. The laser pulse only lasts ~1 ms, so there is no need to immobilise the eye, and as a miniaturised laser combined with an ophthalmoscope is usually used, outpatient treatment is possible.

The safe use of lasers

From the applications of lasers in medicine described above, it is obvious that the use of lasers in industry or otherwise may be hazardous if proper precautions are not

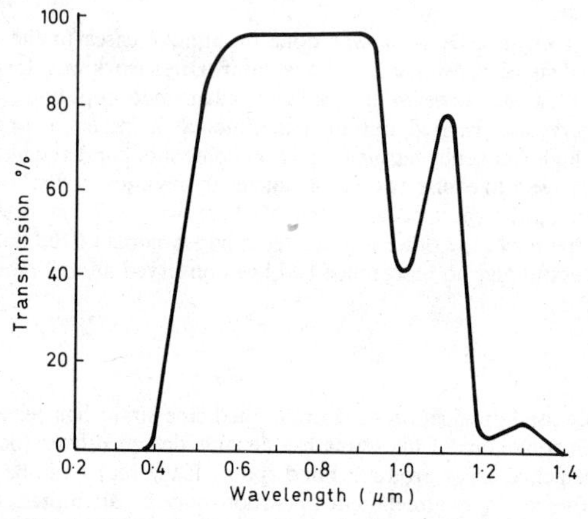

Figure 27.14 Spectral transmission in the human eye (B.S.4803: 1972)

... FOR OCULAR EXPOSURE TO LASER RADIATION (0.4–1.4 μm) FOR A 7-mm PUPIL

Organization	*Laser wavelengths*, nm	*Exposure duration*, t	*Corneal radiant exposure*, J . cm^{-2}	*Corneal irradiance*, W . cm^{-2}
	Intrabram viewing of a collimated beam			
U.S. Departments of the Army and Navy (February 1969)	400–1 400	5–50 ns	10^{-7}	
		Approx. 1 ms	10^{-6}	
		Continuous		10^{-6}
ACGIH (1971)	694.3	1 ns to 1 μs	10^{-7}	
	694.3	1 μs to 0.1 s	10^{-6}	
	400–750	>0.1 s		10^{-5}
U.S. Department of Labor 29 CFR 1518.54 (1971)	632.8	Incidental (1s)		10^{-23}
		Continuous		10^{-26}
U.S. Department of the Air Force (September 1971)	400–700	10–100 ns	1.3×10^{-6}	
		200 μs to 2 ms	10^{-5}	
		2–10 ms		5×10^{-3}
		10–500 ms		2.5×10^{-3}
	1 064	10–100 ns	6×10^{-6}	
		200 μs to 2 ms	5×10^{-5}	
		2–10 ms		2.5×10^{-2}
		10–500 ms		1.3×10^{-2}
ANSI Z-136 proposed (February 1972)	400–700	1 ns to 18 μs	5×10^{-7}	
		18 μs to 10 s	$1.8 \times 10^{-3} . t$	
		10–10^4 s	10^{-2}	
		>10^4 s		10^{-6}
	700–1 060	1 ns to 18 μs	$5C_1 \times 10^{-7}$	
		18 μs to 10 s	$1.8C_1 \times 10^{-3} . t$	
		10–100 s	$C_1 \times 10^{-2}$	
	700–800	100 – [$10^4/(\lambda - 699$ nm)] s	$C_1 \times 10^{-2}$	
		>[$10^4/(\lambda - 699$ nm)] s		$C_1 (\lambda - 699 \text{ nm}) \times 10^{-6}$
	800–1 060	>100 s		$C_1 \times 10^{-4}$
	1 060–1 400	1 ns to 100 μs	5×10^{-6}	
		100 μs to 10 s	$9 \times 10^{-3} . t$	
		10–100 s	5×10^{-2}	
		>100 s		5×10^{-4}

Note: t is in seconds, λ is wavelength in nanometers and $C_1 = \exp\{[(\lambda - 700 \text{ nm})/224]\}$.

Reprinted from I.E.E.E. Spectrum No. 8, 1973

Table 27.3 PROPOSED LASER GUIDELINES FOR SKIN PROTECTION FROM VARIOUS ORGANISATIONS

Organisation	*Continuous-wave (UV excepted)* W/cm²	*Non-Q-switched* J/cm²	*Q-switched* J/cm²	*Other classification*
Weston (1965), British Ministry of Aviation	0.1	0.1	0.1	Not specified
Electronic Engineering Association of Great Britain (1966)	0.1	0.1	0.1	Not specified
U.S. Atomic Energy Commission, Nevada Operations Office (1967)	1.0	0.1	0.1	UV, visible, IR
American Conference of Governmental Industrial Hygienists (ACGIH) (1968)	5.0	0.05	0.005	Visible
Bell (1968), Cincinnati Laser Safety Conference	1.0	0.1	0.1	Not specified
U.S. Departments of the Army and Navy (1969)	0.1	0.1	0.01	Visible, IR

Reprinted from I.E.E.E. Spectrum No. 8, 1973

observed. Although damage to the eye is the chief hazard, skin damage is also a real danger where high power c.w. lasers or high energy pulsed lasers are being used. The relative hazards in fact depend largely on the wavelength of the radiation concerned, since both the transmission of the eye and skin reflectivity vary with wavelength as is shown for the former in *Figure 27.14*.[112] At wavelengths at which the eye is transparent, damage is primarily due to retinal burns. Damage can be significant at quite low-power levels, since the power level on the retina can be 10^5 times greater than that incident on the cornea, due to the focusing action of the eye. *Tables 27.2* and *27.3* give the recommended limits for ocular and skin exposure respectively, proposed by various organisations.[91]

REFERENCES

1. GEDDES, L. A. and BAKER, L. E., *Principles of Applied Biomedical Instrumentation*, John Wiley & Sons Inc., N.Y. (1968)
2. WATSON, B. W. (ed.) *I.E.E. Medical Electronics Monographs*, 1–6, Peter Peregrinus Ltd., London (1971)
3. HILL, D. W. and WATSON, B. W. (eds.), *I.E.E. Medical Electronics Monographs*, 7–12, Peter Peregrinus Ltd., London (1974)
4. *BECAN*, Biomedical Engineering Current Awareness Notification, Project FAIR, Bioengineering Division, Clinical Research Centre, Watford Road, Harrow, Middlesex, HA1 3UJ
5. *TOPICS*, Inspec Marketing Department, Savoy Place, London, WC2R 0BL
6. WOLFF, H. S., *Biomedical Engineering*, World University Library, Weidenfeld & Nicolson, London (1970)
7. CROMWELL, L., WEIBELL, F. J., PFEIFFER, E. A. and USSELMAN, L. B., *Biomedical Instrumentation and Measurements*, Prentice-Hall Inc., N.J. (1973)
8. HILL, D. W., *Electronic Techniques in Anaesthesia and Surgery*, 2nd edn., Butterworths, London (1973)
9. Medical Electronics & Equipment News, *Dictionary & Buyers Guide*, Chilton Co., Chilton Way, Radnor Pa. 19089

10. *Global Market Survey—Biomedical Equipment*, U.S. Department of Commerce, Superintendent of Documents, U.S. Government Printing Office, Washington, DC. 20402
11. HAGA, E. (ed.), *Computer techniques in biomedicine and medicine*, Auerbach Publications Inc., Phil. (1973)
12. 'Development of Multiphasic Health Administration in Japan', *Trade and industry of Japan*, Vol XX, No. 9, 54 (1971)
13. 'Mercury vs. Electronic Thermometers', *Health Devices*, **2**, 3 (1972)
14. SOLMAN, A. J. and DALTON, J. C. P., 'New thermometer for deep tissue temperature', *Biomedical Engineering*, **8**, 432 (1973)
15. JAUNDRELL-THOMPSON, F. and ASHWORTH, W. S., *X-Ray Physics & Equipment*, 2nd edn., Blackwell, Oxford (1970)
16. 'Electron Radiography', *Laboratory Weekly 1*, 8, 1 Nov. (1973)
17. HOUNSFIELD, G. F., 'Computerised Transverse Axial Scanning (and Tomography) Pt. 1', *British Journal of Radiography*, **46**, 1016 (1973)
18. BELCHER, E. H. and VETTER, M. (eds.), *Radioisotopes in Medical Diagnosis*, Butterworths, London (1971)
19. PRINGLE, D. H., *British Journal of Radiology*, **46**, 824 (1973)
20. MAXWELL-CADE, C., 'Principles and Practice of Clinical Thermography', *Radiography XXXIV*, No. 398, 23 (1968)
21. BARNES, R. B., 'Diagnostic Thermography', *Applied Optics*, **7**, 1673 (1968)
22. WELLS, P. N. T. (ed.), *Ultrasonics in Clinical Diagnosis*, Churchill Livingstone (1972)
23. HIROMU YOKOI and KEN-ICHI ITO, *Japan Electronic Engineering*, **7**, No. 76, 38 (1973)
24. ROBERTS, C. (ed.), *Blood Flow Measurement*, Sector Publishing, London (1972)
25. WYATT, D. G., *I.E.E. Medical Electronics Monographs*, 1–6, p. 181, Peter Peregrinus Ltd., London (1971)
26. GASKING, J., *Biomedical Engineering*, **7**, 474 (1972)
27. TERRY, H. J., *Biomedical Engineering*, **7**, 466 (1972)
28. CRUL, J. F. and PAYNE, J. P., *Patient Monitoring*, p. 84, Excerpta Medica, Amsterdam (1970)
29. BAKER, L. E., *I.E.E. Medical Electronics Monographs*, 1–6, p. 1, Peter Peregrinus Ltd., London (1971)
30. KUBICEK, W. G. *et al.*, 'The Minnesota impedance cardiograph—theory and applications', *Biomedical Engineering*, **9**, 410 (1974)
31. GREATOREX, C. A., 'Indirect methods of blood-pressure measurement', *I.E.E. Medical Electronics Monographs*, 1–6, p. 43, Peter Peregrinus Ltd., London (1971)
32. POPPERS, *Patient Monitoring*, p. 82, Excerpta Medica, Amsterdam (1970)
33. GIELES, A. C. M. and SOMMERS, G. H. J., *Philips Technical Review*, **33**, 14 (1973)
34. MILLAR, H. D. and BAKER, L. E., *Medical & Biological Engineering*, **11**, 86 (1972)
35. GAMBLE, W. J., HUGENHOLTZ, P. G., MONROE, R. G. and POLANYI, M., *Circulation XXXI*, p. 328 (1965)
36. BELCHER, E. H. and VETTER, M. (eds.), *Radioisotopes in Medical Diagnosis*, p. 576, Butterworths, London (1971)
37. NICHOLS, W. W., *Biomedical Engineering*, **8**, 376 (1973)
38. KENEDI, R. M. (ed.), *Perspectives in Biomedical Engineering*, p. 255, Macmillan (1973)
39. WRIGHT, B. M., *Coop Cardiologica*, No. 33, p. 35 (1974)
40. SEIFFERT, S. V., 'Recent Developments in defibrillators', *Proc. of the 1972 I.E.E.E. Region Six* (U.S. Western Region) *Conference*, San Diego, California, 19–21 April 1972, pp. 92–95
41. 'Patient Safety', *Hewlett-Packard Application Note AN 718*
42. STACY, R. W. and WAXMAN, B. D. (eds.), *Computers in Biomedical Research*, Vol. III, Academic Press, N.Y. (1969)
43. NEILSON, J. M., 'Special Purpose Hybrid Computer for Analysis of ECG Arrhythmias', *I.E.E. Conference Publication*, No. 79, p. 151, September 1971
44. MARSON, G. B. and MCKINNON, J. B., 'A miniature tape recorder for many applications', *Controls and Instrumentation*, **4**, No. 11, 46 (1972)
45. KENNY, J., *I.E.E. Medical Electronics Monographs*, p. 82, Peter Peregrinus Ltd., London (1971)

46. CLYNES, M. and MILSUM, J. H. (eds.), *Biomedical Engineering Systems*, McGraw Hill (1970)
47. ALBERT, R., VOGT, W. and HELBIG, W. (eds.), Digest of the Xth International Conference on Medical and Biological Engineering, Pub. by Conf. Committee, Dresden (1973)
48. NEILSON, J. M., *World Med. Electronic & Inst.*, **3**, No. 7, 274–275 (1965)
49. HAMILTON, L. H. and RIEKE, R. J., *Medical Research Engineering 11*, No. 3, 20 (1973)
50. VERDIN, A., *Gas Analysis Instrumentation*, Macmillan, London (1973)
51. FOWLER, K. T., 'The Respiratory Mass Spectrometer', *Physics in Medicine & Biology*, **14**, 185 (1969)
52. HAYES, B. and HEALY, T. E. J., *British Journal of Hospital Medicine—Equipment Supplement*, **9**, 4 (1973)
53. *Hospital Equipment Information*, **45**, Section 4/73 (1973), Issued by Dept. of Health and Social Security
54. SHEPPARD, L. C. *et al.*, 'The Digital Computer in Surgical Intensive Care Automation', *Computer*, **6**, No. 7, 28 (1973)
55. MAYNARD, D., PRIOR, P. F. and SCOTT, D. F., *British Medical Journal*, **4**, 545 (1969)
56. WHITE, D. N., *Biomedical Engineering*, **8**, 242 (1973)
57. PENDLETON, H. J., 'Foetal Heart Monitoring', *British Journal of Hospital Medicine*, **3**, 509–512, 515 (1970)
58. BLACKWELL, R. J., *Biomedical Engineering*, **7**, 356 (1972)
59. HAMMACHER, K., Gerbutsh. u. Frauenhk., **22**, 1542 (1962)
60. LEWIN, J. E., *The Lancet*, **II**, 667 (1969)
61. SCOPES, J. W. and SMITH, J. E., *The Lancet*, **II**, 545 (1972)
62. WATKINS, A. L., *A Manual of Electrotherapy*, Lea and Febiger, 3rd edn., Philadelphia (1968)
63. PICKETT, J. M., 'Status of speech-analysing communication aids for the deaf', *I.E.E.E. Trans. on Audio and Electroacoustics*, **20**, 3 (1972)
64. GREENE, M. C. L. and WATSON, B. N., 'Electronic Equipment in Speech Therapy', *British Journal of Hospital Medicine*, Equipment Supplement, **9**, 68, May (1971)
65. Londesborough Scientific Instruments, Birmingham, U.K.
66. COX, A. G., LOWE, L. W. and SOLMAN, A. J., 'The Northwick Park Diathermy', *The Lancet*, **I**, 239 (1973)
67. 'Electrosurgery—A Hot Topic', *Health Devices*, 1, 183
68. MACKAY, R. S. and MARG, E., *Electronics*, **33**, No. 7, 115 (1960)
69. BEHRMAN, J., 'Electrodiagnostic methods in Opthalmology', *Design Electronics* (1966)
70. ROWLANDS, E. N. and WOLFF, H. S., *British Communications & Electronics*, **7**, 598 (1960)
71. WOLFF, H. S., MCCALL, J. and BAKER, J. A., *British Communications and Electronics*, **9**, 120 (1962)
72. JACOB, B., RIDDLE, H. and WATSON, B. W., *Biomedical Engineering*, **8**, 292 (1973)
73. WELLER, C., 'An inexpensive diversity receiving system for medical telemetry', *Biomedical Engineering*, **8**, 157 (1973)
74. COULTER, W. H., British Patent 722, 418
75. SIMMONDS, SCHWABBAUER and EARHART, *J. Lab. Clin. Med.*, **77**, 656 (1971)
76. SAGE, B. H., Laboratory Equipment Digest, **11**, No. 11, 113 (1973)
77. PRESTON, K., 'Use of the Cellscan/Glopr System in the Automatic identification of White Blood Cells', *Biomedical Engineering*, **7**, 226 (1972)
78. UR, A., *Biomedical Engineering*, **5**, 342 (1970)
79. UR, A. and BROWN, F. J., 'Rapid Detection of bacterial activity using impedance measurements', *Biomedical Engineering*, **9**, 18 (1974)
80. WHITE, W. L., ERICKSON, M. M. and STEVENS, S. C., *Practical Automation for the clinical laboratory*, C. V. Mosby, St. Louis, U.S.A. (1972)
81. SKEGGS, L. T., 'An automatic method for Colorimetric Analysis', *American Journal of Clinical Pathology*, **28**, 311 (1957)
82. MITCHELL, F. L., 'Mechanisation in Clinical Chemistry: Trends and Equipment', *Biomedical Engineering*, **5**, 534 (Pt. I), 589 (Pt. II) (1970)
83. DOBBIE, A. K., 'Electricity in Hospitals', *Biomedical Engineering*, **7**, 12 (1972)

84. POCOCK, S. N., 'Earth-Free Patient Monitoring', *Biomedical Engineering*, **7**, pp. 21 and 34 (1972)
85. GREEN, H. L., RAFTERY, E. B. and GREGORY, I. C., *Biomedical Engineering*, **7**, 408 (1972)
86. 'Safety Code for Electro-Medical Apparatus', *Hospital Technical Memorandum No. 8*, H.M.S.O., London, 1963, Revised 1969
87. International Electrotechnical Commission, Sub-Committee 62a, 62A-WG1 (Secretariat) 34
88. 'Electrical Safety: "What's coming in standards?"', *Health Devices*, **2**, 231 (1973)
89. *Instrumentation Technology*, **20**, No. 10, 5 (1973)
90. MRC 70/1314, Exposure to Microwave and Radio-Frequency Radiations: Medical Research Council Recommendations
91. B.S.4803: 1972, Guide on Protection of personnel against hazards from Laser Radiation, British Standards Institution, London
92. ULRICH, W. D., 'Ultrasound Dosage for Nontherapeutic Use on Human Beings', *I.E.E.E. Transactions on Biomedical Engineering, BME 21*, 48 (1974)
93. KIMMICH, H. P. and VOS, J. A. (eds.), *Biotelemetry—International Symposium*, Meander, N.V. Leden (1972)
94. MELA, M. J., *I.E.E.E. Transactions on Biomedical Engineering, BME 13*, 70 (1966)
95. COLEMAN, R. F. *et al.*, 'Detection of trace elements in Human Hair by Neutron Activation and the application to Forensic Medicine', London H.M.S.O. for A.W.R.E. (1967)
96. Lion Laboratories Ltd, Cardiff
97. BAKER, J. A., HUMPHREY, S. J. E. and WOLFF, H. S., *Journal of Physiology*, **188**, No. 2, 4–5 P (1967)
98. WOLFF, W. S., *Perspectives in Biomedical Engineering*, p. 306, Macmillan (1973)
99. ROBERTS, V. C., 'A review of non-invasive measurements of blood flow', *Biomedical Engineering*, **9**, 332 (1974)
100. MCCARTY, K. and WOODCOCK, J. P., 'The Ultrasonic Doppler Shift Flowmeter—a new development', *Biomedical Engineering*, **9**, 336 (1974)
101. WOODCOCK, J. P., 'Plethysmography', *Biomedical Engineering*, **9**, 406 (1974)
102. CHAMBERLAIN, D. and ENGLISH, M., 'Pacing and Pacemakers', *British Journal of Hospital Medicine—Equipment Supplement*, **10**, 4 (1973)
103. COOPER, R., *British Journal of Hospital Medicine—Equipment Supplement*, **10**, 18 (1973)
104. BERGER, K. W., *The hearing aid—its operation and development*, National Hearing Aid Society, Michigan (1970)
105. BROWN, M., *British Journal of Hospital Medicine—Equipment Supplement*, **10**, 33 (1973)
106. Survey of Standards for Electrical Equipment used in Medical Practice, FME (Federation of the Mechanical and Electrical Engineering Industries), Technical Department, 13, Nassaulan, The Hague, The Netherlands (1972)
107. RING, E. F. J., 'Equipment for Medical Thermography', *British Journal of Hospital Medicine*, equipment supplement, **5**, 24 (1971)
108. CARO, C. G., *British Journal of Hospital Medicine*, equipment supplement, **9**, 60 (1973)
109. NORTHAM, B. E., *British Journal of Hospital Medicine*, equipment supplement, **5**, 44 (1971)
110. GOLDMAN, L. and ROCKWELL, R. J. jun., 'Laser Systems and their applications in Medicine and Biology', **1**, *Advances in Biomedical Engineering and Medical Physics*, Levine, S. N. (ed.), Wiley, New York
111. GOLDMAN, L. and ROCKWELL, R. J. jun., *Lasers in Medicine*, Gordon and Breach, New York
112. ELECCION, M., 'Laser Hazards', *I.E.E.E. Spectrum*, **10**, No. 8, 32 (1973)

FURTHER READING

WOHLBART, M. L. (ed.), *Laser Applications in Medicine and Biology*, Plenum, New York (1971)

INDEX

Numerals in bold indicate section numbers.